Analog Circuit Design

Volume 2

Immersion in the Black Art of Analog Design

Analog Circuit Design
Volume 2

Immersion in the Black Art of Analog Design

Edited by

Bob Dobkin

Jim Williams

AMSTERDAM • BOSTON • HEIDELBERG • LONDON • NEW YORK • OXFORD
PARIS • SAN DIEGO • SAN FRANCISCO • SINGAPORE • SYDNEY • TOKYO

Newnes is an imprint of Elsevier

Newnes is an imprint of Elsevier
The Boulevard, Langford Lane, Kidlington, Oxford OX5 1GB, UK
225 Wyman Street, Waltham, MA 02451, USA

First edition 2013

British Library Cataloguing in Publication Data
A catalogue record for this book is available from the British Library

Library of Congress Cataloging-in-Publication Data
A catalog record for this book is availabe from the Library of Congress

ISBN: 978-0-12-397888-2

Printed and bound in the United States of America
13 14 15 10 9 8 7 6 5 4 3 2

For Jerrold R. Zacharias, who gave me the sun, the moon and the stars.
For Siu, who is the sun, the moon and the stars.

In memory of Jim Williams, a poet who wrote in electronics.

Contents

Publisher's Note

This book was compiled from Linear Technology Corporation's original *Application Notes*.

These Application Notes have been re-named as chapters for the purpose of this book. However, throughout the text there is a lot of cross referencing to different *Application Notes*, not all of which have made it into the book. For reference, this conversion table has been included; it shows the book chapter numbers and the original Application Note numbers.

CHAPTER NUMBER	APPLICATION NOTE
1	2
2	104
3	133
4	29
5	44
6	32
7	118
8	65
9	95
10	107
11	15
12	74
13	132
14	4
15	5
16	10
17	13
18	14
19	16
20	18
21	21

CHAPTER NUMBER	APPLICATION NOTE
22	27A
23	38
24	79
25	123
26	128
27	131
28	85
29	91
30	129
31	12
32	45
33	52
34	57
35	61
36	67
37	75
38	87
39	98
40	105
41	113

Trademarks

These trademarks all belong to Linear Technology Corporation. They have been listed here to avoid endless repetition within the text. Trademark acknowledgment and protection applies regardless. Please forgive us if we have missed any.

Linear Express, Linear Technology, LT, LTC, LTM, Burst Mode, Dust Networks, FilterCAD, LTspice, OPTI-LOOP, Over-The-Top, PolyPhase, SwitcherCAD, TimerBlox, μModule and the Linear logo are registered trademarks of Linear Technology Corporation. Adaptive Power, Bat-Track, BodeCAD, C-Load, Direct Flux Limit, DirectSense, Dust, Easy Drive, Eterna EZSync, Filter-View, Hot Swap, isoSPI, LDO+, LinearView, LTBiCMOS, LTCMOS, LTPoE++, LTpowerCAD, LTpowerPlanner, LTpowerPlay, MicropowerSwitcherCAD, Mote-on-Chip, Multimode Dimming, No Latency ΔΣ, No Latency Delta-Sigma, No R_{SENSE}, Operational Filter, PanelProtect, PLLWizard, PowerPath, PowerSOT, PScope, QuikEval, RH DICE Inside, RH MILDICE Inside, SafeSlot, SmartMesh, SmartMesh IP, SmartStart, SNEAK-A-BIT, SoftSpan, Stage Shedding, Super Burst, SWITCHER+, ThinSOT, Triple Mode, True Color PWM, UltraFast, Virtual Remote Sense, Virtual Remote Sensing, VLDO and VRS are trademarks of Linear Technology Corporation. All other trademarks are the property of their respective owners.

Acknowledgments

Analog Circuit Design, Volume 2 emerges a year after the first volume through the efforts of a dedicated team. For many of us, this is a labor of love, giving further legs to the timeless application notes of the late Jim Williams and many colleagues at Linear Technology. Thanks first to the authors, who do the heavy lifting—in the lab and through their insightful writing. Also to the dedicated graphics and editorial team that ensured that the application notes are clear and consistent—Gary Alexander and Susan Dale. We are indebted to the efforts of the professionals at Elsevier/Newnes, including publisher Jonathan Simpson and the production efforts of Pauline Wilkinson and Fiona Geraghty. Finally, to Bob Dobkin for his insight, time and belief in this project.

John Hamburger
Linear Technology Corporation

Introduction

Why I Write

In the early 1980s we wanted prospective customers to know our name and what we were up to. The real issue was finding a way to productively use the seeming dead time before product availability. What readers wanted was a series of credible, full length technical articles in the language of relevant, working circuits.

I moped for weeks over this problem before a possible solution became apparent. Instead of waiting for products, I'd simply go into the lab, develop the applications, and then write the articles. The key to this approach was to synthesize the expected products using available ICs and discretes to build rough equivalents on small plug-in boards. We could develop functional applications and write most of the text. We'd then shelve the manuscript and breadboards. Later, when products became available, we could put them into the breadboards and implement the attendant final changes. Once we had done these tasks, we could drop scope photos and specifications into the waiting text, tweak the manuscript, and ship it off. This approach would speed publication by perhaps a year and synchronize the article's appearance with product introduction.

Initially, the whole scheme appeared absurd and eminently unworkable, with uncountable technical and editorial sinkholes. Getting started was much more difficult than I had imagined. Synthesizing the hardware for our unborn ICs proved tricky; my methods, clumsy and stumbling. Breadboarding the applications was laborious and slow, primarily because I wasn't sure how accurately I was mimicking the forthcoming IC's performance. Writing was equally painful. Text flow was staccato and disjointed because of the gaps that occurred while I waited for results with actual products. I had to keep separate notes directing me to unfinished text when we finally dropped the products into the breadboards.

The first article took almost two months, but things slowly became easier. Tricks to move along the lab work evolved, and I found ways to write more efficiently, making the manuscripts inherently adaptable to the planned additions and changes. Soon, I was producing an almost-finished article every two weeks or so, roaring along, powered by adrenaline, solder, pencils, paper, and pizza.

During the next year, life was a dizzy seven-day-a-week blur of breadboards and manuscripts shuttling between work and my home lab. My diet was a cardiologist's nightmare. I don't recall having a meal at home. The refrigerator was devoid of food but well-provisioned with Polaroid film to feed the oscilloscope camera. All this frenetic bustle boiled off any semblance of a normal social life. At dinner in San Francisco, while nominally listening to my date describe her job intricacies, I silently calculated the optimum chapper-channel crossover frequency in a composite amplifier. The regimen of madness continued for about a year, resulting in 35 full-length feature articles between June 1983 and November 1987.

I still write, although at a significantly less frenetic pace. Now, when the kids in our lab complain to me about writing technical material, I try not to sound like the curmudgeon I am not so slowly becoming. I think that mad tear almost 25 years ago contributes to my current lack of empathy. These kids today, with a catalog full of products, they don't know what they've got.

Jim Williams
Staff Scientist
Linear Technology Corporation

Note: This essay originally appeared in EDN magazine.

Foreword

We appreciate the enthusiastic response to the original book, *Analog Circuit Design: A Tutorial Guide to Applications and Solutions*. The acceptance of this book emphasizes the need for making good circuit design applications accessible. These writings on applications fill a vacuum, since the majority of application notes and magazine articles do not have sufficient depth to teach analog design.

At the time I learned analog design, there were not yet any analog ICs. Circuits were all transistorized (maybe some tubes) and the circuit explanations in magazines and books were more complete than many are today. In those days equipment manuals included schematics for repair. I was fortunate to work at a large company with a huge calibration lab for their equipment. I spent many lunch hours perusing the calibration and repair manuals for analog systems. I thank HP, Tektronix and many other companies for their tutorials on analog circuit design that I discovered in the calibration lab. It is interesting to note that Jim Williams in his early years at MIT spent much time repairing electronic test equipment that was nonfunctional. Today's manuals are more sparse.

It is relatively hard to find completed analog designs to study. Books may include analog designs, but they are not necessarily complete circuits with test results. Likewise most application notes are very specific to the device they are showing and don't provide a general breadth of information that is useful in the analog world.

I am pleased that this book series is becoming a teaching handbook of applications. An application should be useful, have multiyear life and be easy to duplicate. Any good application note should have a description of the application and discussion of where it will be used. It should include ancillary information, such as temperature range, power systems, lifetime and other key data usually excluded in textbooks that just focus on principles. The application block diagram needs to be explained so that the approach to the solution is understandable. Unless you know where you are going, it is difficult to understand how you got there.

The circuits for the applications need to be fully developed, have part types and construction information. The solutions may include specialized components that require the reader to have knowledge of the function and special properties. The circuit should be shown and the properties and function of each item in the circuit explained with sufficient detail that readers can extract that information and use it again.

The origin of "breadboard" dates back to a time when actual bread cutting boards were used to assemble electronic circuits for testing. Components were screwed to the board and wired with Fahnestock clips, such as the one in this diagram.

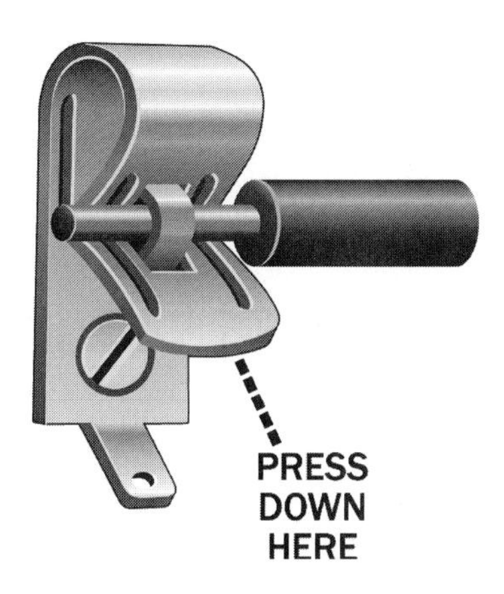

Many analog circuits are layout-sensitive and if not properly laid out, the circuits may not work. I have seen many circuits constrained by this problem.

Finally, designers need a detailed set of test results for the circuit. These show how the circuit works when it's working properly and provides a comparison guide for anyone trying to duplicate the circuit. Without all of these, the applications fail as part of a teaching system.

Analog design is challenging. There are many ways to get from input to output, and the circuitry in the middle can lead to divergent results. Analog design is like learning a language. When you first learn a language, you begin with a vocabulary book and then analyze writings in that language by looking up words one by one as you encounter them. Likewise, in analog design you learn the basics of the circuit, as well as the function of different devices. You can write node equations and determine what the circuit is doing by studying each of the individual circuits.

With analog circuit design, you end up using the basic circuit configurations you have learned—differential amplifiers, transistors, FETs, resistors, and previously studied circuits—to achieve the final circuit. As with a new language, it takes many years to learn to write poetry, and the same is true of analog circuit design.

Today's system manuals rarely have any circuit design. They contain block diagrams and hookup schematics with thousands of leads to different blocks. So where do people go for analog circuit design? If you're just starting, or if you encounter a problem that's outside your experience, appropriate references are difficult to find.

Hopefully, these books provide some answers, as well as circuit design and test and lab techniques for design and duplication of the circuits.

Today, analog design is in greater demand than ever before. Analog design is now a combination of transistors and ICs that provide high functionality in analog signal processing. This volume focuses on fundamental aspects of circuit design, layout, and testing. It is our hope that the talented writers of these application notes shed some light on the "black art" of analog design.

Bob Dobkin
Co-Founder, Vice President, Engineering,
and Chief Technical Officer
Linear Technology Corporation

PART 1

Power Management

Section 1

Power Management Tutorials

Performance enhancement techniques for three-terminal regulators (1)

This application note describes a number of enhancement circuit techniques used with existing 3-terminal regulators which extend current capability, limit power dissipation, provide high voltage output, operate from 110VAC or 220VAC without the need to switch transformer windings, and many other useful application ideas.

Load transient response testing for voltage regulators (2)

Semiconductor memory, card readers, microprocessors, disc drives, piezoelectric devices and digitally based systems furnish transient loads that a voltage regulator must service. Ideally, regulator output is invariant during a load transient. In practice, some variation is encountered and becomes problematic if allowable operating voltage tolerances are exceeded. This mandates testing the regulator and its associated support components to verify desired performance under transient loading conditions. Various methods are employable to generate transient loads, allowing observation of regulator response. This application note presents open and closed loop transient load testing circuitry with measured performance taken under various conditions. Practical considerations for a memory supply voltage regulator are reviewed. Four appended sections cover capacitor parasitics and their effects on load transient response, output capacitor selection, probing techniques and a stabilized transient load tester.

A closed-loop, wideband, 100A active load (3)

Digital systems, particularly microprocessors, furnish transient loads in the 100A range that a voltage regulator must service. Ideally, regulator output is invariant during a load transient. In practice, some variation is encountered and becomes problematic if allowable operating voltage tolerances are exceeded. 100A load steps, characteristic of microprocessors, exacerbate this issue, necessitating testing the regulator and associated components under such transient loading conditions. To meet this need, a closed-loop, 500kHz bandwidth, linearly responding, 100A capacity active load is described. Study of this approach is prefixed by a brief review of conventional test load types and noting their shortcomings.

Performance enhancement techniques for three-terminal regulators

1

Jim Williams

Three terminal regulators provide a simple, effective solution to voltage regulation requirements. In many situations the regulator can be used with no special considerations. Some applications, however, require special techniques to enhance the performance of the device.

Probably the most common modification involves extending the output current of regulators. Conceptually, the simplest way to do this is by paralleling devices. In practice, the voltage output tolerance of the regulators can cause problems. Figure 1.1 shows a way to use two regulators to achieve an output current equal to their sum. This circuit capitalizes on the 1% output tolerance of the specified regulators to achieve a simple paralleled configuration. Both regulators sense from the same divider string and the small value resistors provide ballast to account for the slightly differing output voltages. This added impedance degrades total circuit regulation to about 1%.

Figure 1.2 shows another way to extend current capability in a regulator. Although this circuit is more complex than Figure 1.1, it eliminates the ballasting resistor's effects and has a fast-acting logic-controlled shutdown feature. Additionally, the current limit may be set to any desired value. This circuit extends the 1A capacity of the LT®1005 multifunction regulator to 12A, while retaining the LT1005's enable feature and auxiliary 5V output. Q1, a booster transistor, is servo-controlled by the LT1005, while Q2 senses the current dependent voltage across the 0.05Ω shunt. When the shunt voltage is large enough, Q2 comes on, biasing Q3 and shutting down the regulator via the LT1005's enable pin. The shunt's value can be selected for the desired current limit. The 100°C thermo-switch limits dissipation in Q1 during prolonged short circuits by disabling the LT1005. It should be mounted on Q1's heat sink.

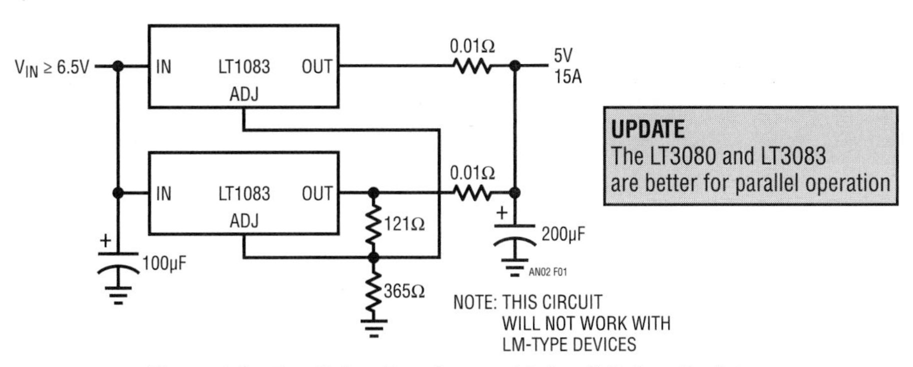

Figure 1.1 • Paralleling Regulators with Small Ballast Resistors

Analog Circuit and System Design: Immersion in the Black Art of Analog Design. http://dx.doi.org/10.1016/B978-0-12-397888-2.00001-8

Boosted regulator schemes of this type are often poorly dynamically damped. Such improper loop compensation results in large output transients for shifts in the load. In particular, because Q1's common emitter configuration has voltage gain, transients approaching the input voltage are possible when the load drops out. Here, the 100μF capacitor damps Q1's tendency to overshoot, while the 20Ω value provides turn-off bias. The 250μF unit maintains Q1's emitter at DC. Figure 1.3 shows that this "brute force" compensation works quite well. Normally the regulator sees no load. When Trace A goes high, a 12A load (regulator output current is Trace C) is placed across the output terminals. The regulator output voltage recovers quickly, with minimal aberration.

While the 100μF output capacitor aids stability, it prevents the regulator output from dropping quickly when the enable command is given. Because Q1 cannot sink current, the 100μF unit's discharge time is load limited. Q4 corrects this problem, even when there is no load. When the enable command is given (Trace A, Figure 1.4) Q3 comes on, cutting off the LT1005 and forcing Q1 off. Simultaneously, Q4 comes on, pulling down the regulator output (Trace B), and sinks the 100μF capacitor's discharge current (Trace C). If fast turn-off is not needed, Q4 may be omitted.

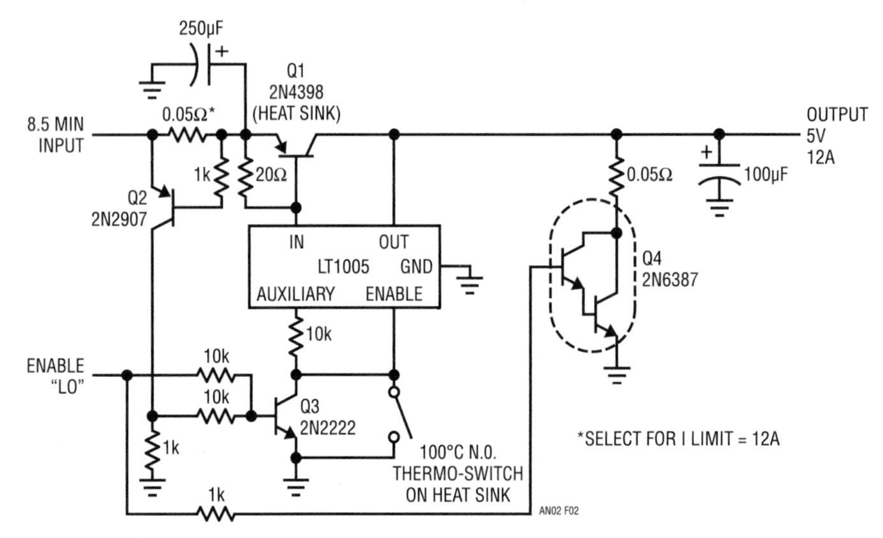

Figure 1.2 • Switched High Current Regulator with Fast Turn-off

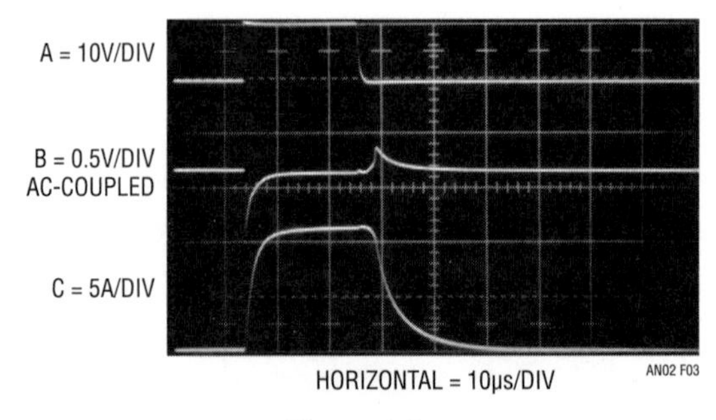

Figure 1.3 •

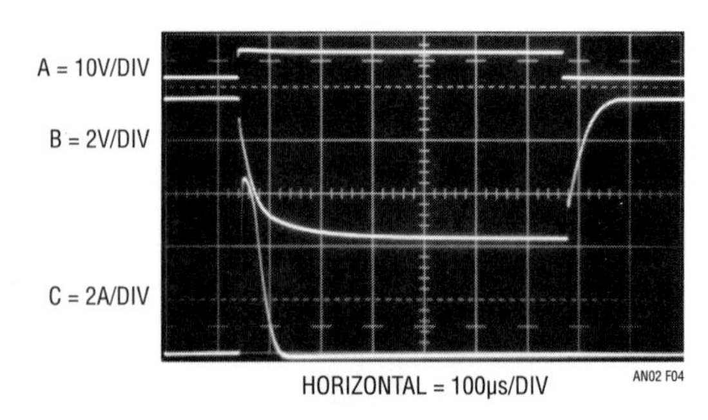

Figure 1.4 •

Power dissipation control is another area where regulators can be helped by additional circuitry. Increasing heat sink area can be used to offset dissipation problems, but is a wasteful and inefficient approach. Instead, the regulator can be placed within a switched-mode loop that servo-controls the voltage *across* the regulator. In this arrangement the regulator functions normally while the switched-mode control loop maintains the voltage across it at a minimal value, regardless of line or load changes. Although this approach is not quite as efficient as a classical switching regulator, it offers lower noise and the fast transient response of the linear regulator. Figure 1.5 details a DC driven version of the circuit. The LT350A functions in the conventional fashion, supplying a regulated output at 3A capacity. The remaining components form the switched-mode dissipation limiting control. This loop forces the potential across the LT350A to equal the 3.7V value of V_Z. When the input of the regulator (Trace A, Figure 1.6) decays far enough, the LT1018 output (Trace B) switches low, turning on Q1 (Q1 collector is Trace D). This allows current flow (Trace C) from the circuit input into the 4500μF capacitor, raising the regulator's

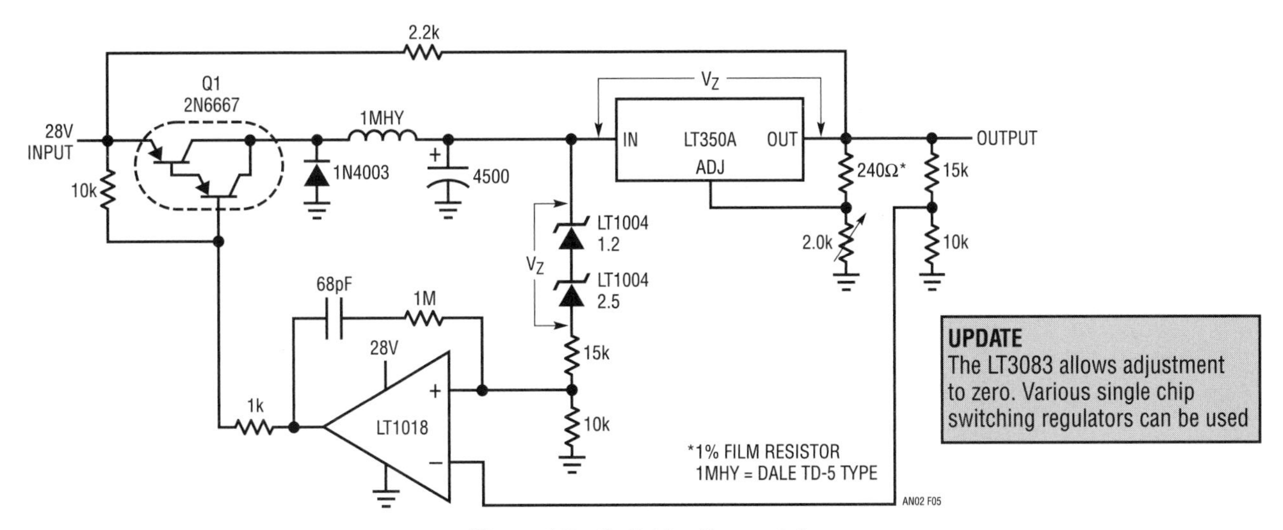

Figure 1.5 • Switching Preregulator

UPDATE
The LT3083 allows adjustment to zero. Various single chip switching regulators can be used

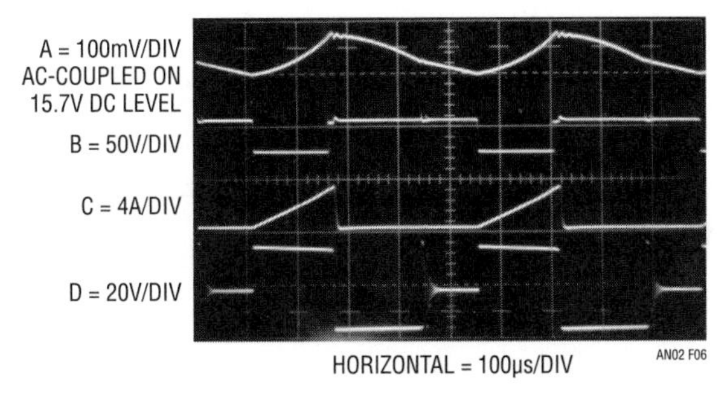

Figure 1.6 • Switching Waveforms

input voltage. When the regulator input rises far enough, the comparator goes high, Q1 cuts off and the capacitor ceases charging.

The 1N4003 damps the flyback spike of the current-limiting inductor. The 4.7kΩ unit ensures circuit start-up and the 68pF-1MΩ combination sets loop hysteresis at about $80mV_{P-P}$. This free-running oscillation control mode substantially reduces dissipation in the regulator, while preserving its performance. Despite changes in the input voltage, different regulated outputs or load shifts, the loop always ensures the minimum possible dissipation in the regulator.

Figure 1.7 shows the dissipation limiting technique applied in a more sophisticated circuit. The AC-powered version provides 0V-35V, 10A regulation under high line-low line (90VAC-140VAC) conditions with good efficiency. In this version, two SCRs and a center-tapped

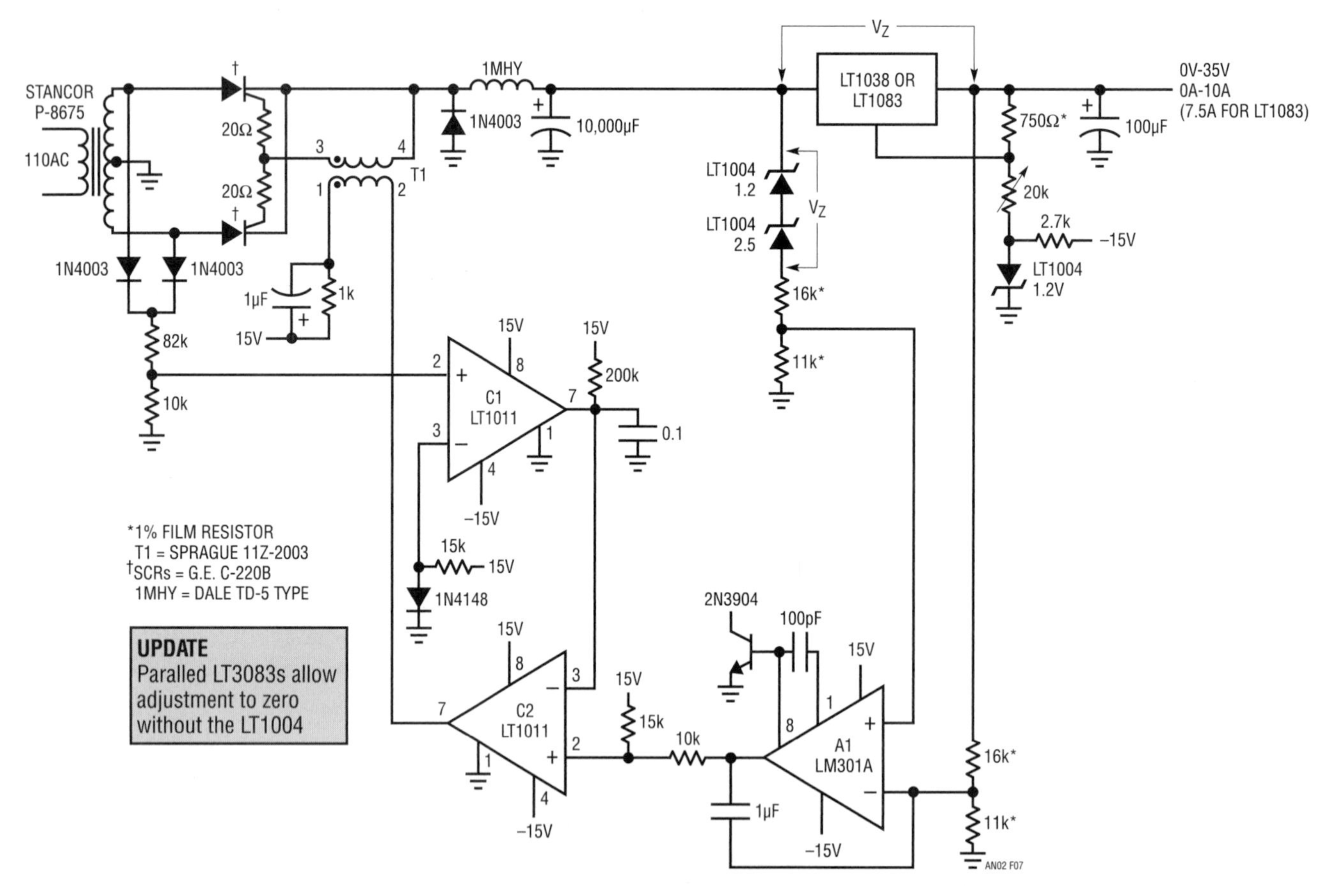

Figure 1.7 • Phase Controller Preregulator for 50 or 60 Hz Power Input

transformer source power to the inductor-capacitor combination. The transformer output is also diode rectified (Trace A, Figure 1.8), divided down, and used to reset the 0.1μF unit (Trace B) via C1. The resulting AC line synchronous ramp at C1's output is compared to A1's offset output by C2. A1's output represents the deviation from the V_Z value that the loop is trying to force across the LT1038. When the ramp output exceeds C2's "+" input value, C2 pulls low, dumping current through T1's primary (Trace C). This fires the appropriate SCR and a path from the main transformer to the LC pair occurs (Trace D). The resultant current flow (Trace E) is limited by the inductor and charges the capacitor. When the AC line cycle drops low enough, the SCR commutates and charging ceases. On the next half cycle the process repeats, except that the alternate SCR does the work. In this fashion, the loop controls the phase angle at which the SCRs fire to keep the voltage across the LT1038 at V_Z (3.7V). As a result, the circuit functions over all line, load and output voltage conditions with good efficiency. The 1.2V LT1004 at the LT1038 allows the output voltage to be set down to 0.00 and the 2N3904 clamp at A1 prevents loop "hang-up". Figure 1.7A shows a way to trigger the SCRs without using a transformer.

Although A1's output is an analog voltage, the AC-driven nature of the circuit makes it approximate a

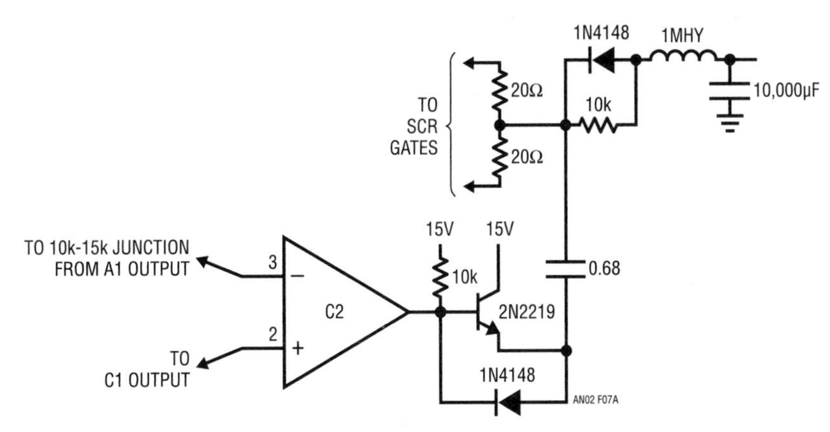

Figure 1.7A • Triggering SCR without a Transformer

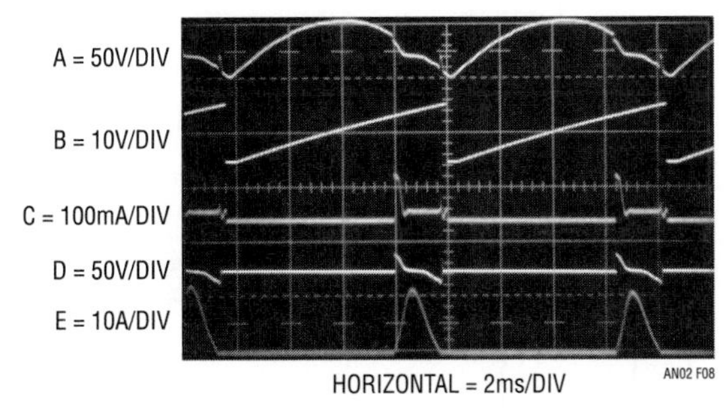

Figure 1.8 • Trigger Waveforms

smoothed, sample loop response. Conversely, the regulator constitutes a true linear system. Because these two feedback systems are interlocked, frequency compensation can be difficult.

In practice, A1's 1μF capacitor keeps dissipation loop gain at a low enough frequency for stable characteristics, without influencing the LT1038's transient response characteristic. Trace A, Figure 1.9 shows the output noise while the circuit is operating at 35V into a 10A load (350W). Note the absence of fast switching transients and harmonics. The output noise is made up of residual 120Hz ripple and regulator noise. Reflected noise into the AC power line is also negligible (Trace B) because the inductor limits current rise time to about 1ms, much slower than the normal switching supplies. Figure 1.10 shows a plot of efficiency versus output voltage for a 10A load. At low output voltages, where the static losses across the regulator and SCRs are significant, efficiency suffers, but 85% is attained at the upper extreme.

High voltage output is another area for regulator enhancement. In theory, because the regulator does not have a ground pin, it can regulate high voltages. In normal operation the regulator floats at the supply's upper level, and as long as the V_{IN}–V_{OUT} maximum differential is not exceeded there are no problems. However, if the output is shorted, the V_{IN}–V_{OUT} maximum is exceeded and device destruction will occur. The circuit of Figure 1.11 shows a complete high voltage regulator that delivers 100V at 100mA and withstands shorts to ground. Even at 100V output the LT317A functions in the normal mode, maintaining 1.2V between its output and adjustment pin.

Under these conditions the 30V Zener is off and Q1 conducts. When an output short occurs, the Zener

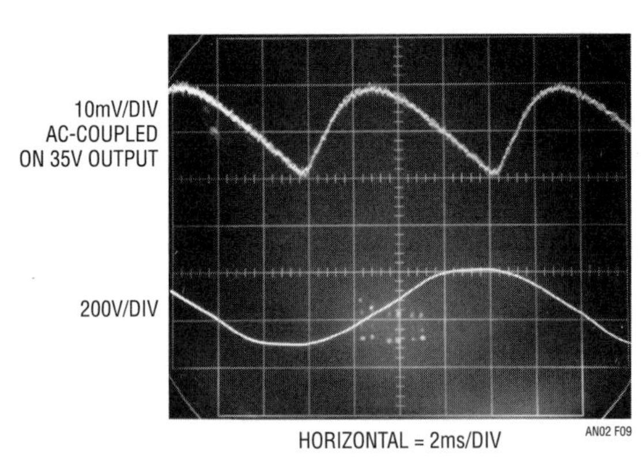

Figure 1.9 •

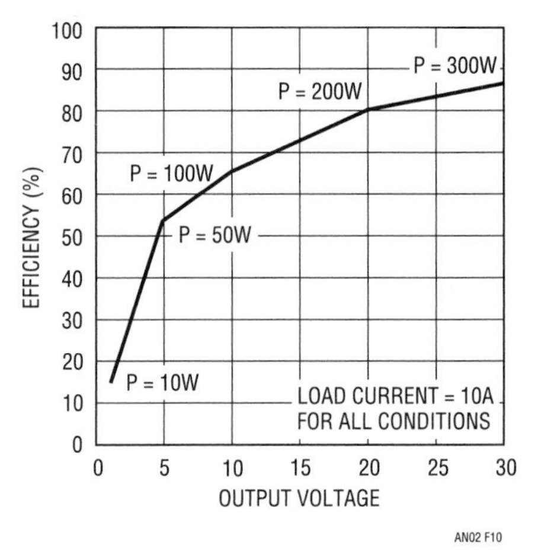

Figure 1.10 •

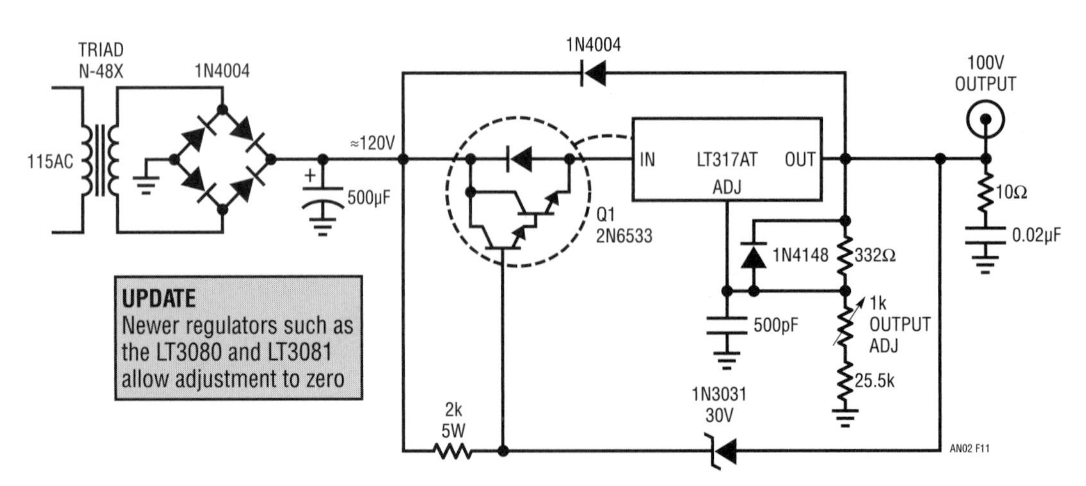

Figure 1.11 • Preregulating and Buffering the Voltage to the IC Allows Operation with High Voltage

conducts, forcing Q1's base to 30V. This causes Q1's emitter to clamp 2 V_{BE}s below V_Z, well within the V_{IN}–V_{OUT} rating of the regulator. Under these conditions, Q1, a high voltage device, sustains 90V V_{CE} at whatever current the transformer and the regulator's current limit will support. The transformer specified saturates at 130mA, keeping Q1 well within its safe area as it dissipates 12W. If Q1 and the LT317A are thermally coupled, the regulator will soon go into thermal shutdown and oscillation will commence. This action will continue, protecting the load and the regulator as long as the output remains shorted. The 500pF capacitor and the 10Ω-0.02μF damper aid transient response and the diodes provide safe discharge paths for the capacitors.

This approach to high voltage regulation is primarily limited by the power dissipation capability of the device in series with the regulator. Figure 1.11A uses a vacuum tube (remember them?) to achieve very high short-circuit dissipation capability. The tube allows high voltage operation and is extremely tolerant of overloads. This circuit allows the LT317A to control 600W at 2000V (V1's plate limit is 300mA) with full short-circuit protection.

Power is not the only area in which regulator performance can be augmented. Figure 1.12 shows a way to increase the stability of a regulator's output over time and temperature. This is particularly useful in powering strain gauge-based transducers. In this circuit the output voltage is divided down and compared to the 2.5V reference by A1, a precision amplifier. A1's output is used to force the LT317A's adjustment pin to whatever voltage is required to maintain the 10V output. A1 contributes negligible error. The resistors specified will track within 5ppm/°C and the reference contributes about 20ppm/°C. The regulator's internal circuitry protects against short circuits and thermal overload.

Figure 1.13's circuit allows a regulator to remotely sense the feedback voltage, eliminating the effects of voltage drop in the supply lines. This is a concern where high currents must be transmitted over relatively long supply rails or PC traces. Figure 1.13's circuit uses A1 to sense the voltage at the point of load. A1's output, summed with the regulator's output, modifies the adjustment pin voltage to compensate for the voltage lost across R_{DROP}. The feedback divider is returned through a separate lead from the load, completing the remote sensing scheme. The 5μF capacitor filters noise and the 1k value limits bypass capacitor discharge when power is turned off.

A final circuit allows voltage regulator-powered circuity to run from 110VAC or 220VAC without having to switch transformer windings. Regulator dissipation does not increase for 220VAC inputs. In Figure 1.14, when T1 is

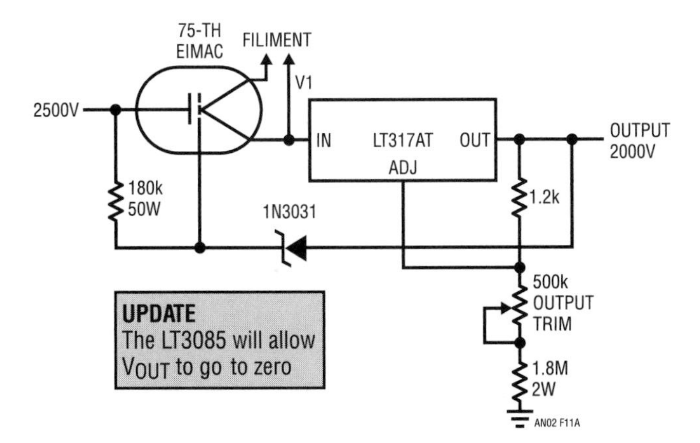

Figure 1.11A • Very High Voltage Regulator

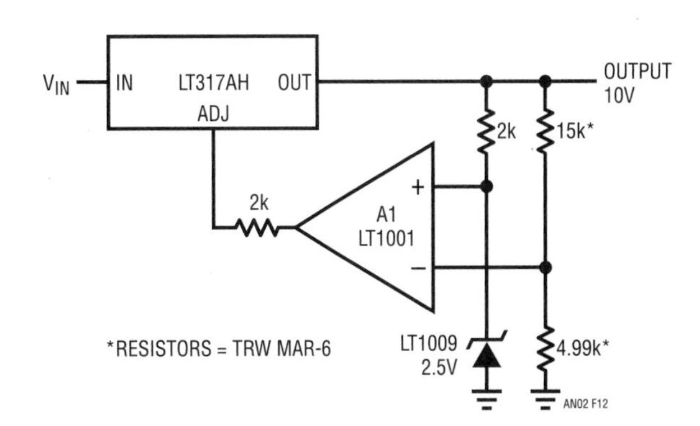

Figure 1.12 • Using a Regulator as a Protected Power Stage for a Reference

driven from 110VAC, the LT1011 output goes high, allowing the SCR to receive gate bias through the 1.2k resistor. The 1N4002 is off. T1's output is rectified by the SCR and the regulator sees about 8.5V at its input. If T1 is plugged into a 220VAC source, the negative input at the LT1011 is driven beyond 2.5V and the device's output clamps low. This steers the SCR's gate bias to ground through the LT1011's output transistor. The diodes in the LT1011 output line prevent reverse voltages from reaching the SCR or the LT1011 output. Now, the SCR goes off and the 1N4002 sources current to the regulator from T1's center tap. Although T1's input voltage has doubled, its output potential has halved and the regulator power dissipation remains the same. Figure 1.15 shows the AC line input versus regulator input voltage transfer function. The switch to center tap drive occurs midway between 110VAC and 220VAC. The hysteresis, a desirable characteristic, occurs because T1's output voltage shifts with the step change in loading.

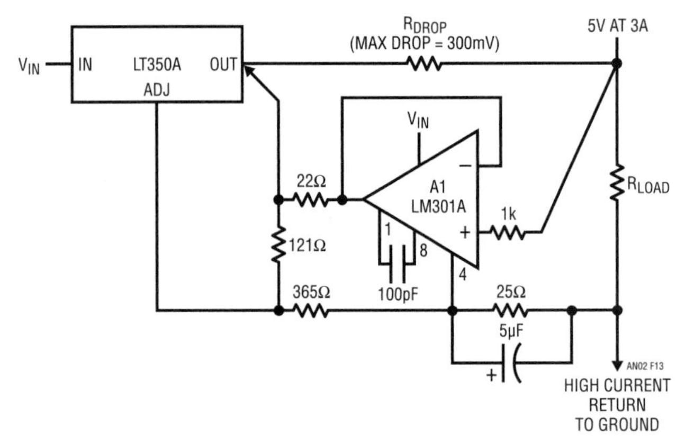

Figure 1.13 • Remote Sensing the Load Voltage for Better Regulation

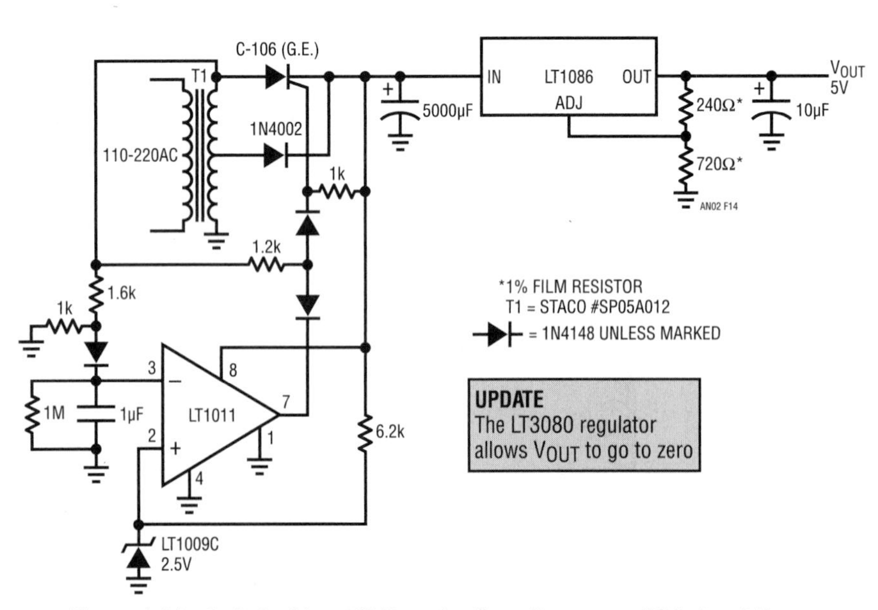

Figure 1.14 • Switched Input Voltage for Operating over a Wide Input Range

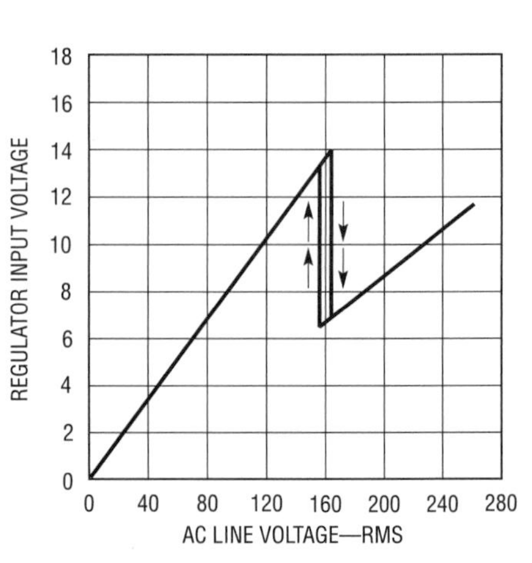

Figure 1.15 •

Load transient response testing for voltage regulators

Practical considerations for testing and evaluating results

2

Jim Williams

Introduction

Semiconductor memory, card readers, microprocessors, disc drives, piezoelectric devices and digitally based systems furnish transient loads that a voltage regulator must service. Ideally, regulator output is invariant during a load transient. In practice, some variation is encountered and becomes problematic if allowable operating voltage tolerances are exceeded. This mandates testing the regulator and its associated support components to verify desired performance under transient loading conditions. Various methods are employable to generate transient loads, allowing observation of regulator response.

Basic load transient generator

Figure 2.1 diagrams a conceptual load transient generator. The regulator under test drives DC and switched resistive loads, which may be variable. The switched current and output voltage are monitored, permitting comparison of the nominally stable output voltage versus load current under static and dynamic conditions. The switched current is either on or off; there is no controllable linear region.

Figure 2.2 is a practical implementation of the load transient generator. The voltage regulator under test is augmented by capacitors which provide an energy reservoir, similar to a mechanical flywheel, to aid transient response. The size, composition and location of these capacitors, particularly C_{OUT}, has a pronounced effect on transient response and overall regulator stability.[1] Circuit operation is straightforward. The input pulse triggers the LTC1693 FET driver to switch Q1, generating a transient load current out of the regulator. An oscilloscope monitors the instantaneous load voltage and, via a "clip-on" wideband probe, current. The circuit's load transient generating capabilities are evaluated in Figure 2.3 by substituting an extraordinarily low impedance power source for the regulator. The combination of a high capacity power supply, low impedance connections and generous bypassing main-

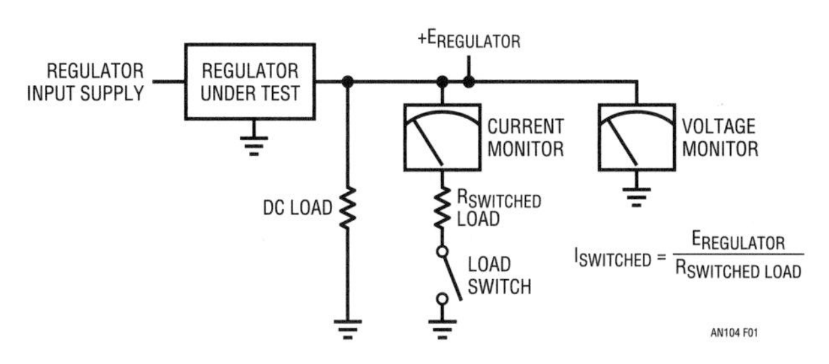

Figure 2.1 • Conceptual Regulator Load Tester Includes Switched and DC Loads and Voltage/Current Monitors. Resistor Values Set DC and Switched Load Currents. Switched Current is Either On or Off; There is No Controllable Linear Region

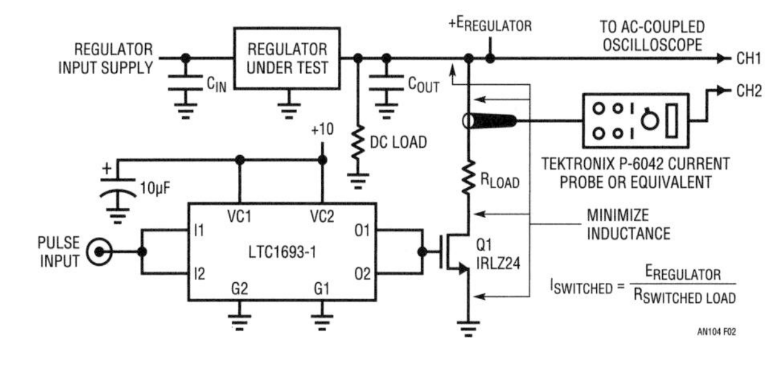

Figure 2.2 • A Practical Regulator Load Tester. FET Driver and Q1 Switch R_{LOAD}. Oscilloscope Monitors Current Probe Output and Regulator Response

Note 1: See Appendix A, "Capacitor Parasitic Effects on Load Transient Response" and Appendix B, "Output Capacitors and Stability" for extended discussion.

Analog Circuit and System Design: Immersion in the Black Art of Analog Design. http://dx.doi.org/10.1016/B978-0-12-397888-2.00002-X

tains low impedance across frequency. Figure 2.4 shows Figure 2.3 responding to the LTC1693-1 FET driver (Trace A) by cleanly switching 1A in 15ns (Trace B). Such speed is useful for simulating many loads but has restricted versatility. Although fast, the circuit cannot emulate loads between the minimum and maximum currents.

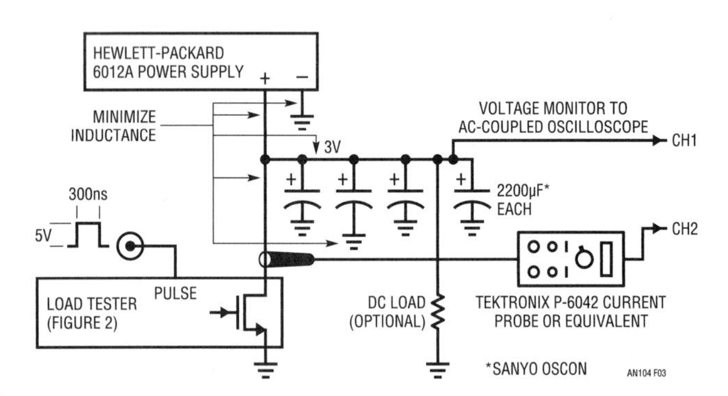

Figure 2.3 • Substituting Well Bypassed, Low Impedance Power Supply for Regulator Allows Determining Load Tester's Response Time

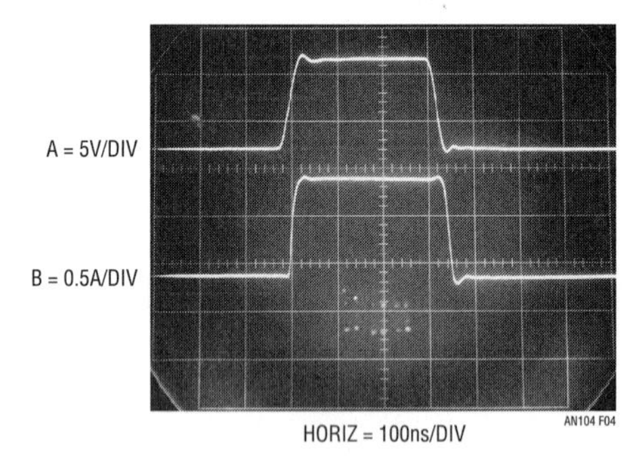

Figure 2.4 • Figure 2.2's Circuit Responds to FET Driver Output (Trace A), Switching a 1A Load (Trace B) in 15ns

Closed loop load transient generators

Figure 2.5's conceptual closed loop load transient generator linearly controls Q1's gate voltage to set instantaneous transient current at any desired point, allowing simulation

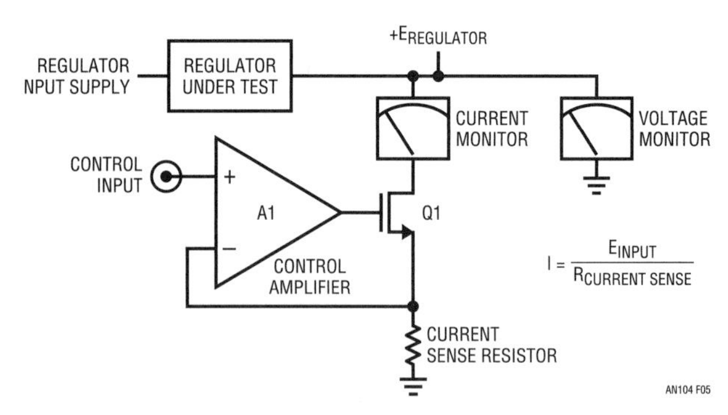

Figure 2.5 • Conceptual Closed Loop Load Tester. A1 Controls Q1's Source Voltage, Setting Regulator Output Current. Q1's Drain Current Waveshape is Identical to A1 Input, Allowing Linear Control of Load Current. Voltage and Current Monitors are as in Figure 2.1

of nearly any load profile. Feedback from Q1's source to the A1 control amplifier closes a loop around Q1, stabilizing its operating point. Q1's current assumes a value dependant on the control input voltage and the current sense resistor over a very wide bandwidth. Note that once A1 biases to Q1's conductance threshold, small variations in A1's output result in large current changes in Q1's channel. As such, large output excursions are not required from A1; its small signal bandwidth is the fundamental speed limitation. Within this restriction, Q1's current waveform is identically shaped to A1's control input voltage, allowing linear control of load current. This versatile capability permits a wide variety of simulated loads.

FET based circuit

Figure 2.6, a practical incarnation of a FET based closed loop load transient generator, includes DC bias and waveform inputs. A1 must drive Q1's high capacitance gate at high frequency, necessitating high peak A1 output currents and attention to feedback loop compensation. A1, a 60MHz current feedback amplifier, has an output current capacity exceeding 1A. Maintaining stability and waveform fidelity at high frequency while driving Q1's gate capacitance necessitates settable gate drive peaking components, a damper network, feedback trimming and loop peaking adjustments. A DC trim, also required, is made first. With no input applied, trim the "1mV adjust" for 1mV DC at Q1's source. The AC trims are made utilizing Figure 2.7's arrangement. Similar to Figure 2.3, this "brick wall" regulated source provides minimal ripple and sag when step loaded by the load transient generator. Apply the inputs shown and trim the gate drive, feedback and loop peaking adjustments for the cleanest, square cornered response on the oscilloscope's current probe equipped channel.

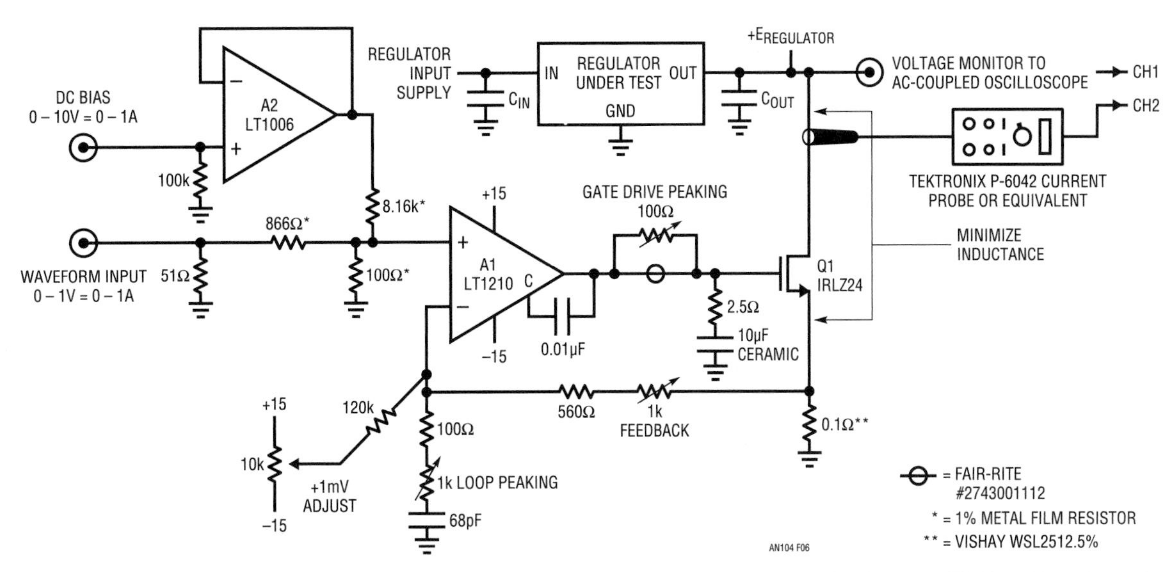

Figure 2.6 • Detailed Closed Loop Load Tester. DC Level and Pulse Inputs Feed A1 to Q1 Current Sinking Regulator Load. Q1's Gain Allows Small A1 Output Swing, Permitting Wide Bandwidth. Damper Network, Feedback and Peaking Trims Optimize Edge Response

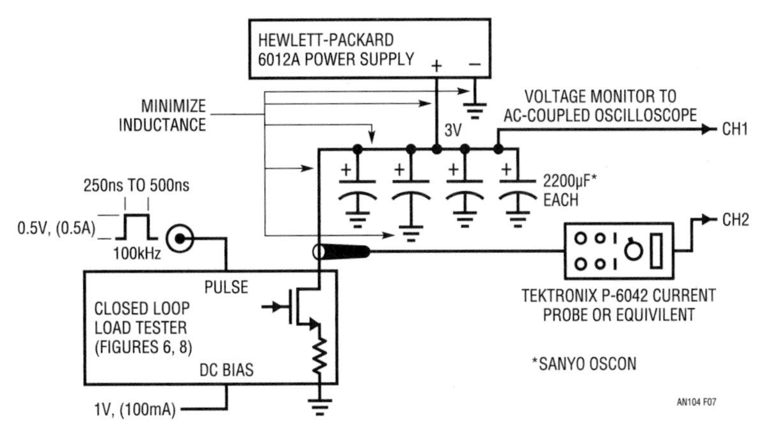

Figure 2.7 • Closed Loop Load Tester Response Time is Determined as in Figure 2.3. "Brick Wall" Input Provides Low Impedance Source

Bipolar transistor based circuit

Figure 2.8 considerably simplifies the previous circuit's loop dynamics and eliminates all AC trims. The major trade-off is a 2× speed reduction. The circuit is similar to Figure 2.6, except that Q1 is a bipolar transistor. The bipolar's greatly reduced input capacitance allows A1 to drive a more benign load. This permits a lower output current amplifier and eliminates the dynamic trims required to accommodate Figure 2.6's FET gate capacitance. The sole trim is the "1mV adjust" which is accomplished as described before[2]. Aside from the 2× speed reduction the bipolar transistor also introduces a 1% output current error due to its base current. Q2 is added to prevent excessive Q1 base current when the regulator supply is not present. The diode prevents reverse base bias under any circumstances.

Note 2: This trim may be eliminated at some sacrifice in circuit complexity. See Appendix D, "A Trimless Closed Loop Transient Load Tester."

Closed loop circuit performance

Figures 2.9 and 2.10 show the two wideband circuits' operation. The FET based circuit (Figure 2.9) only requires a 50mV A1 swing (Trace A) to enforce Trace B's flat-topped current pulse with 50ns edges through Q1. Figure 2.10 details the bipolar transistor based circuit's performance. Trace A, taken at Q1's base, rises less than 100mV causing Trace B's clean 1A current conduction through Q1. This circuit's 100ns edges, about 2× slower than the more complex FET based version, are still fast enough for most practical transient load testing.

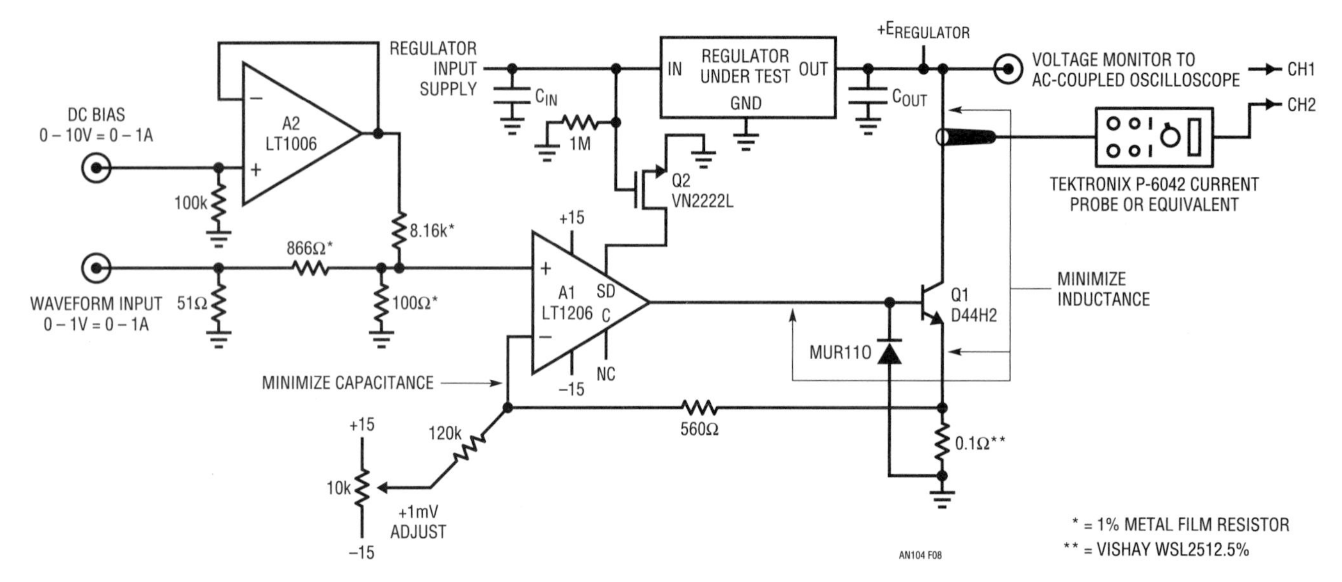

Figure 2.8 • Figure 2.6 Implemented with Bipolar Transistor. Q1's Reduced Input Capacitance Simplifies Loop Dynamics, Eliminating Compensation Components and Trims. Trade Off is 2× Speed Reduction and Base Current Induced 1% Error

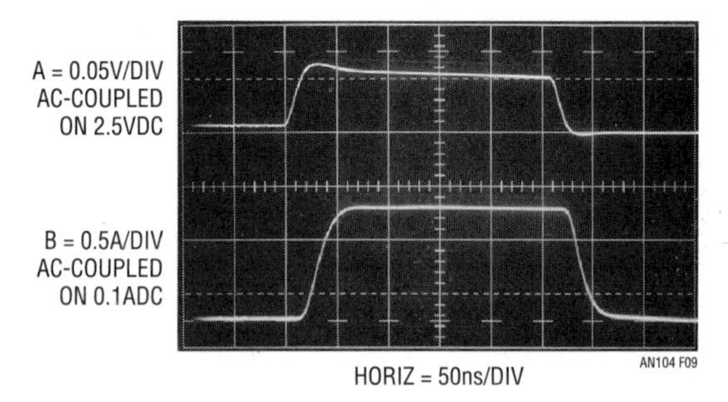

Figure 2.9 • Figure 2.6's Closed Loop Load Tester Step Response (Q1 Current is Trace B) is Quick and Clean, Showing 50ns Edges and Flat Top. A1's Output (Trace A) Swings Only 50mV, Allowing Wideband Operation. Trace B's Presentation is Slightly Delayed Due to Voltage and Current Probe Time Skew

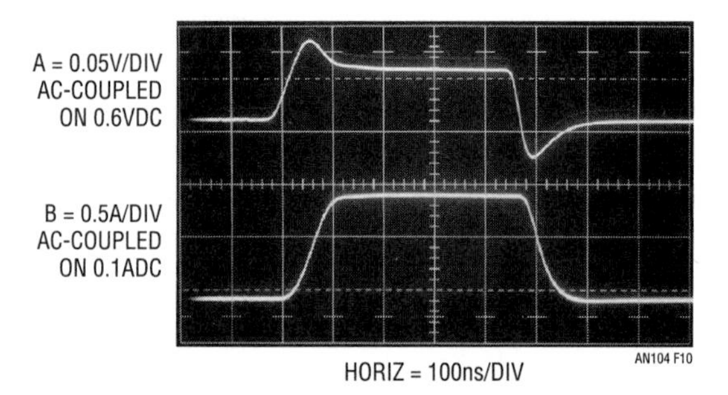

Figure 2.10 • Figure 2.8's Bipolar Output Load Tester Response is 2× Slower than FET Version, but Circuit is Less Complex and Eliminates Compensation Trims. Trace A is A1's Output, Trace B is Q1's Collector Current

Load transient testing

The previously discussed circuits permit rapid and thorough voltage regulator load transient testing. Figure 2.11 uses Figure 2.6's circuit to evaluate an LT1963A linear regulator. Figure 2.12 shows regulator response (Trace B) to Trace A's asymmetrically edged input pulse. The ramped leading edge, within the LT1963A's bandwidth, results in Trace B's smooth $10mV_{P-P}$ excursion. The fast trailing edge, well outside LT1963A passband, causes Trace B's abrupt disruption. C_{OUT} cannot supply enough current to maintain output level and a $75mV_{P-P}$ spike results before the regulator resumes control. In Figure 2.13, a 500mA peak-to-peak 500kHz noise load, emulating a multitude of incoherent loads, feeds the regulator in Trace A. This is within regulator bandwidth and only $6mV_{P-P}$ of disturbance appears in Trace B, the regulator output. Figure 2.14 maintains the same conditions, except that noise bandwidth is increased to 5MHz. Regulator bandwidth is exceeded, resulting in over $50mV_{P-P}$ error, an 8× increase.

Figure 2.15 shows what happens when a 0.2A DC biased, swept DC-5MHz, 0.35A load is presented to the regulator. The regulator's rising output impedance versus frequency results in ascending error as frequency scales.

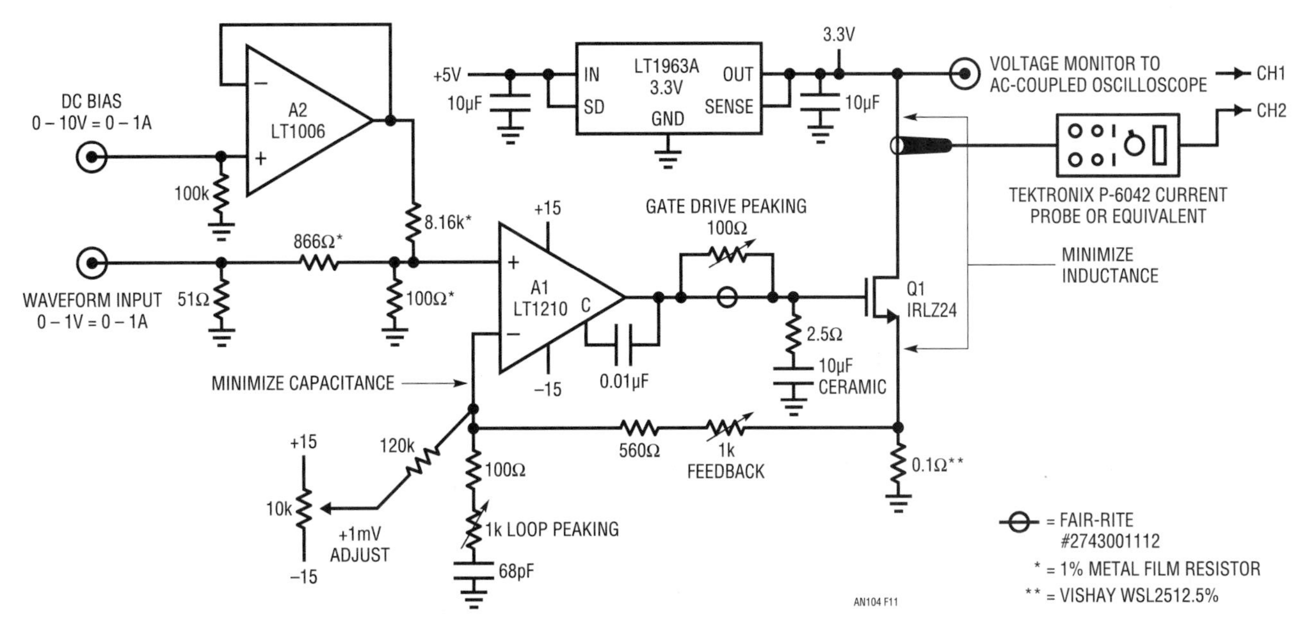

Figure 2.11 • Closed Loop Load Tester Shown with LT1963A Regulator. Load Testing for a Variety of Current Load Waveshapes is Possible

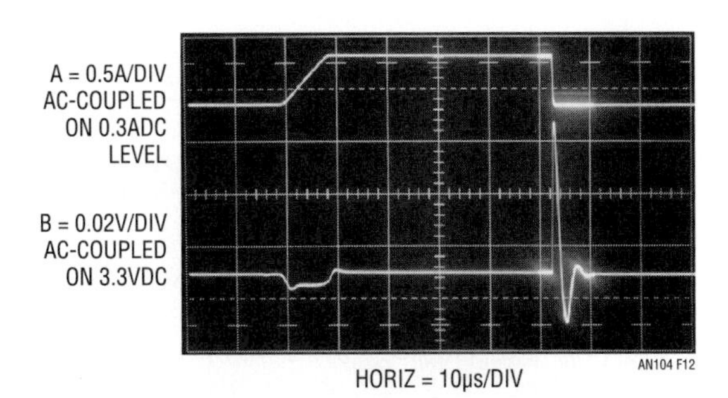

Figure 2.12 • Figure 2.11 Responds (Trace B) to Assymetrically Edged Pulse Input (Trace A). Ramped Leading Edge, Within LT1963A Bandwidth, Results in Trace B's Smooth 10mV$_{P-P}$ Excursion. Fast Trailing Edge, Outside LT1963A Bandwidth, Causes Trace B's Abrupt 75mV$_{P-P}$ Disruption. Traces Latter Portion Intensified for Photographic Clarity

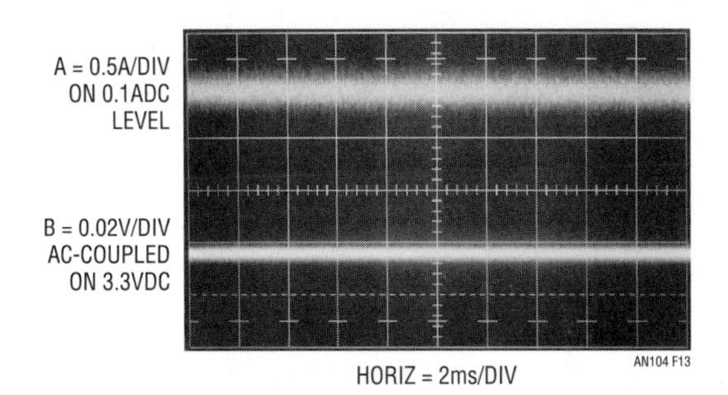

Figure 2.13 • 500mA$_{P-P}$, 500kHz Noise Load (Trace A), Within Regulator Bandpass, Produces Only 6mV Artifacts at Trace B's Regulator Output

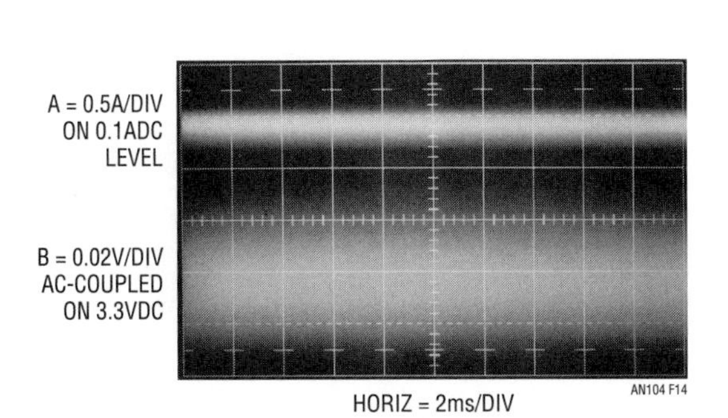

Figure 2.14 • Same Conditions as Figure 2.13, Except Noise Bandwidth Increased to 5MHz. Regulator Bandwidth is Exceeded, Resulting in 50mV$_{P-P}$ Output Error

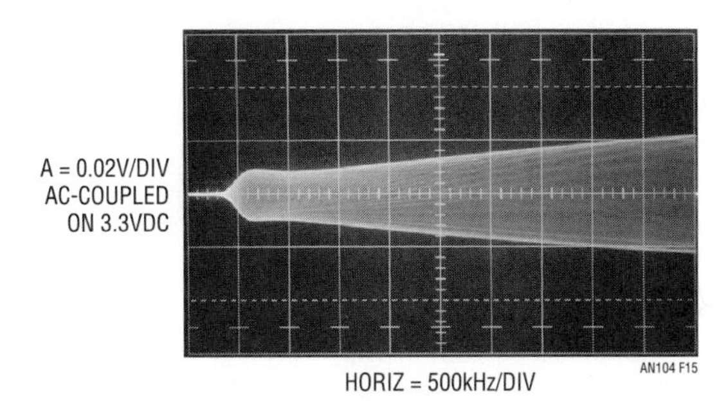

Figure 2.15 • Swept DC –5MHz, 0.35A Load (On 0.2ADC) Results in Above Regulator Response. Regulator Output Impedance Rises with Frequency, Causing Corresponding Ascending Output Error

This information allows determination of regulator output impedance versus frequency.

Capacitor's role in regulator response

The regulator employs capacitors at its input (C_{IN}) and output (C_{OUT}) to augment its high frequency response. The capacitor's dielectric, value and location greatly influence regulator characteristics and must be quite carefully considered.[3] C_{OUT} dominates the regulator's dynamic response; C_{IN} is much less critical, so long as it does not discharge below the regulator's dropout point. Figure 2.16 shows a typical regulator circuit and emphasizes C_{OUT} and its parasitics. Parasitic inductance and resistance limit capacitor effectiveness at frequency. The capacitor's dielectric and value significantly influence load step response. A "hidden" parasitic, impedance build-up in regulator output trace runs, also influences regulation characteristics, although its effects can be minimized by remote sensing (shown) and distributed capacitive bypassing.

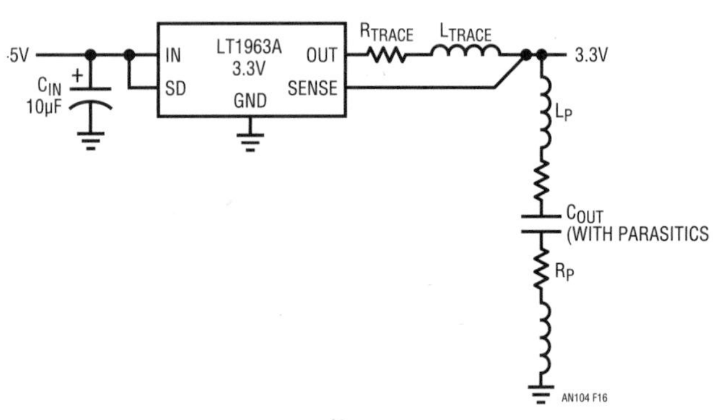

Figure 2.16 • C_{OUT} Dominates Regulator's Dynamic Response; C_{IN} is Much Less Critical. Parasitic Inductance and Resistance Limit Capacitor Effectiveness at Frequency. Capacitor Value and Dielectric Significantly Influence Load Step Response. Excessive Trace Impedance is Also a Factor

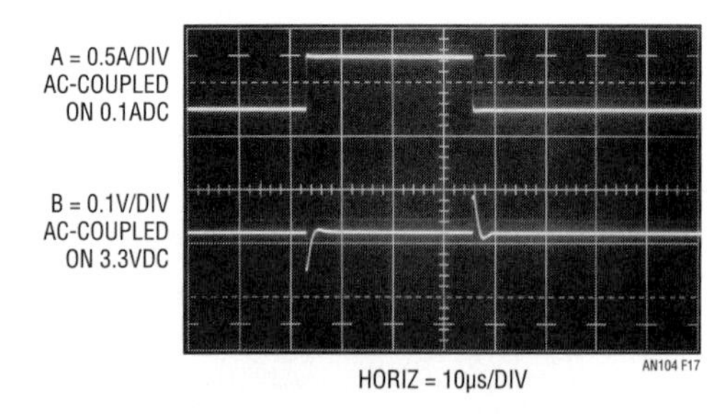

Figure 2.17 • Stepped 0.5A Load to Figure 2.16's Circuit (Trace A) with $C_{IN} = C_{OUT} = 10\mu F$ Results in Trace B's Regulator Output. Low Loss Capacitors Promote Controlled Output Excursions

Note 3: See Appendices A and B for extended discussion of these concerns.

Figure 2.17 shows Figure 2.16's circuit responding (Trace B) to a 0.5A load step biased on 0.1A DC (Trace A) with $C_{IN} = C_{OUT} = 10\mu F$. The low loss capacitors employed result in Trace B's well controlled output. Figure 2.18 greatly expands the horizontal time scale to investigate high frequency behavior. Regulator output deviation (Trace B) is smooth, with no abrupt discontinuities. Figure 2.19 runs the same test as Figure 2.17 using an output capacitor

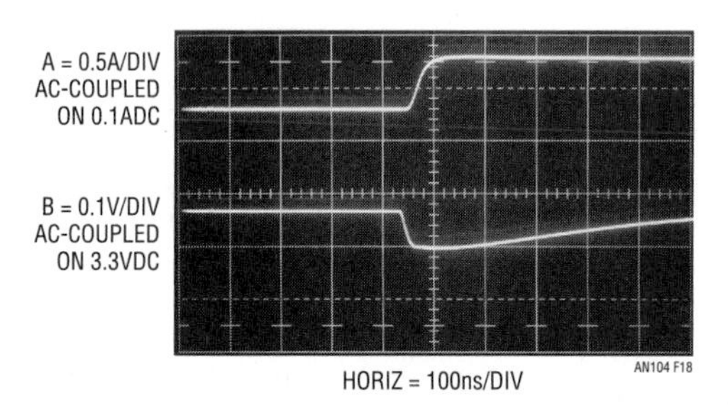

Figure 2.18 • Expanding Horizontal Scale Shows Trace B's Smooth Regulator Output Response. Mismatched Current and Voltage Probe Delays Account for Slight Time Skewing

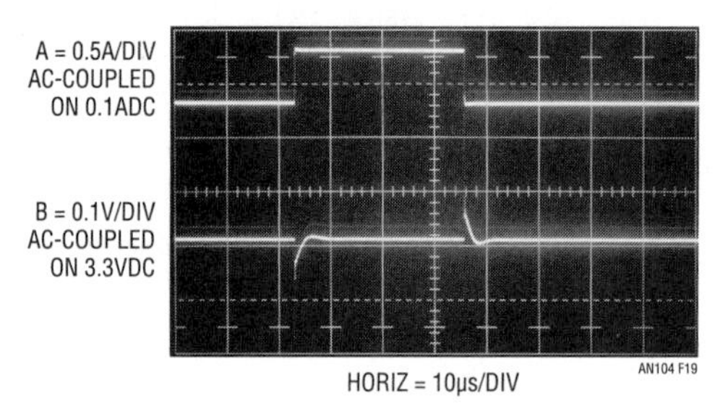

Figure 2.19 • "Equivalent" 10µF C_{OUT} Capacitor's Performance Appears Similar to Figure 2.17's Type at 10µs/DIV

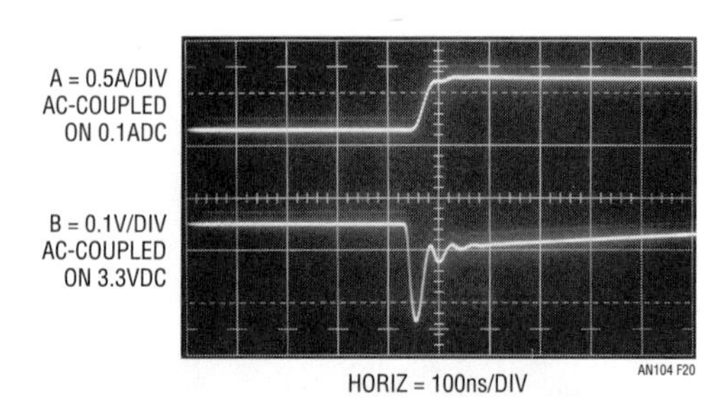

Figure 2.20 • Horizontal Scale Expansion Reveals "Equivalent" Capacitor Produces 2x Amplitude Error vs Figure 2.18. Mismatched Probe Delays Cause Time Skewing Between Traces

claimed as "equivalent" to the one employed in Figure 2.17. At 10μs/division things seem very similar, but Figure 2.20 indicates problems. This photo, taken at the same higher sweep speed as Figure 2.18, reveals the "equivalent" capacitor to have a 2× amplitude error versus Figure 2.18, higher frequency content and resonances.[4] Figure 2.21 substitutes a very lossy 10μF unit for C_{OUT}. This capacitor allows a 400mV excursion (note Trace B's vertical scale change), >4× Figure 2.18's amount. Conversely, Figure 2.22 increases C_{OUT} to a low loss 33μF type, decreasing Trace B's output response transient by 40% versus Figure 2.18. Figure 2.23's further increase, to a low loss 330μF capacitor, keeps transients inside 20mV; 4× lower than Figure 2.18's 10μF value.

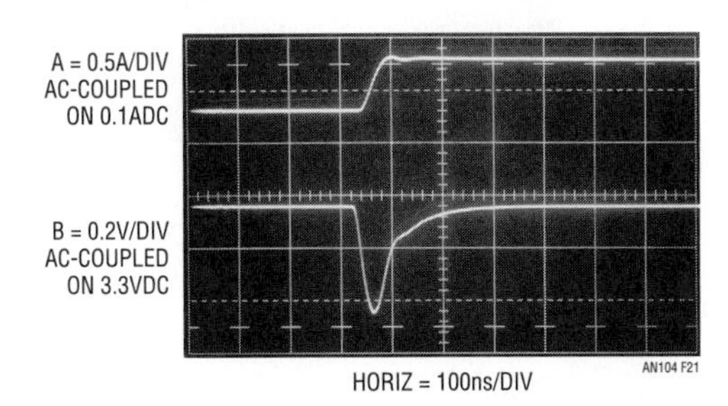

Figure 2.21 • Excessively Lossy 10μF C_{OUT} Allows 400mV Excursion –4× Figure 2.18's Amount. Time Skewing Between Traces Derives from Probe Mismatch

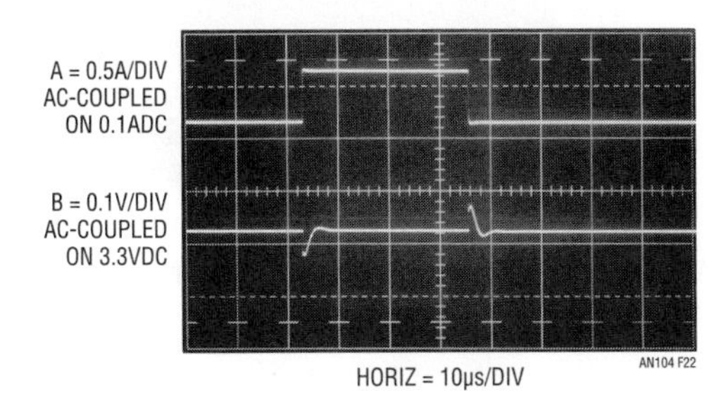

Figure 2.22 • Increasing C_{OUT} with Low Loss 33μF Unit Reduces Output Response Transient by 40% Over Figure 2.17

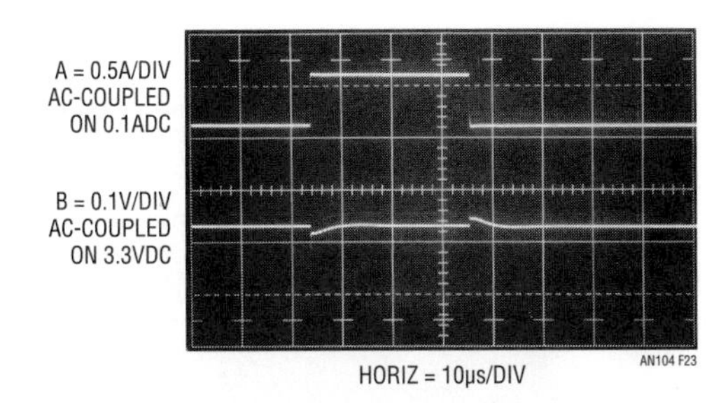

Figure 2.23 • Low Loss 330μF Capacitor Keeps Output Response Transients Inside 20mV –4× Lower than Figure 2.17's 10μF

Note 4: Always specifiy components according to observed performance, never to salesman's claims.

The lesson from the preceding study is clear. Capacitor value and dielectric quality have a pronounced effect on transient load response. Try before specifying!

Load transient risetime versus regulator response

The closed loop load transient generator also allows investigating load transient risetime on regulation at high speed. Figure 2.24 shows Figure 2.16's circuit (C_{IN} = C_{OUT} = 10μF) responding to a 0.5A, 100ns risetime step on a 0.1A DC load (Trace A). Response decay (Trace B) peaks at 75mV with some following aberrations. Decreasing Trace A's load step risetime (Figure 2.25) almost doubles Trace B's response error, with attendant enlarged following aberrations. This indicates increased regulator error at higher frequency.

All regulators present increasing error with frequency, some more so than others. A slow load transient can unfairly make a poor regulator look good. Transient load testing that does not indicate some response outside regulator bandwidth is suspect.

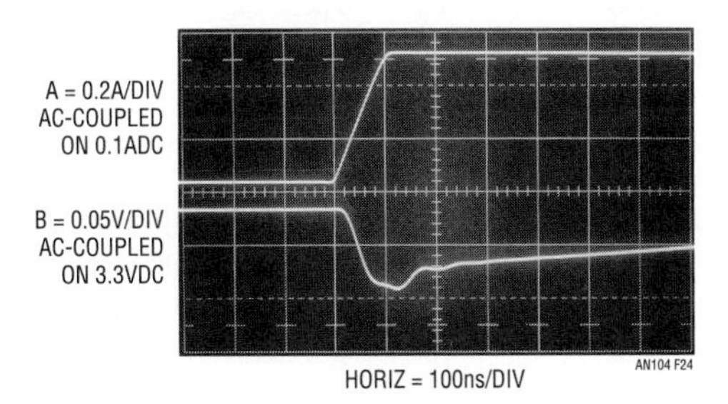

Figure 2.24 • Regulator Output Response (Trace B) to 100ns. Risetime Current Step (Trace A) for C_{OUT} = 10μF. Response Decay Peaks at 75mV

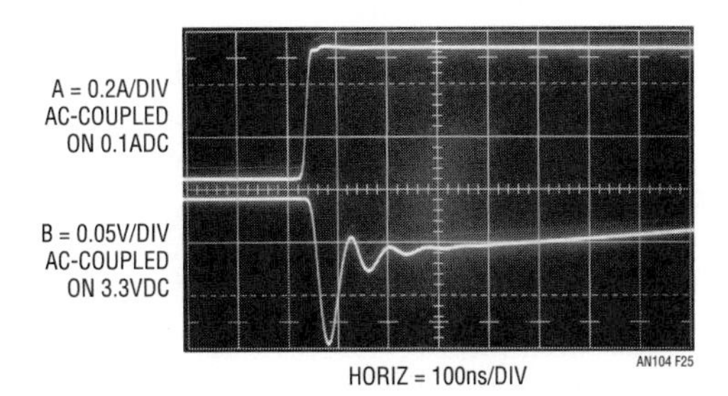

Figure 2.25 • Faster Risetime Current Step (Trace A) Increases Response Decay Peak (Trace B) to 140mV, Indicating Increased Regulation Loss vs Frequency

A practical example – Intel P30 embedded memory voltage regulator

A good example of the importance of voltage regulator load step performance is furnished by the Intel P30 embedded memory. This memory requires a 1.8V supply, typically regulated down from +3V. Although current requirements are relatively modest, supply tolerances are tight. Figure 2.26's error budget shows only 0.1V allowable excursion from 1.8V, including all DC and dynamic errors. The LTC1844-1.8 regulator has a 1.75% initial tolerance (31.5mV), leaving only a 68.5mV dynamic error allowance. Figure 2.27 is the test circuit. Memory control line movement causes 50mA load transients, necessitating attention to capacitor selection.[5] If the regulator is close to the power source C_{IN} is optional. If not, use a good grade 1μF capacitor for C_{IN}. C_{OUT} is a low loss 1μF type. In all other respects the circuit appears deceptively routine. A load transient generator provides Figure 2.28's output load test step (Trace A).[6] Trace B's regulator response shows just 30mV peaks, >2× better than required. Increasing C_{OUT} to 10μF, in Figure 2.29, reduces peak output error to 12mV, almost 6× better than specification. However, a poor grade 10μF (or 1μF, for that matter) capacitor produces Figure 2.30's unwelcome surprise. Severe peaking error on both edges occurs (Trace B's latter portion has been intensified to aid photograph clarity) with 100mV observable on the negative going edge. This is well outside the error budget and would cause unreliable memory operation.

Intel P30 Embedded Memory Voltage Regulator Error Budget

PARAMETER	LIMITS
Intel Specified Supply Limits	1.8V ± 0.1V
LTC1844 Regulator Initial Accuracy	±1.75% (±31.5mV)
Dynamic Error Allowance	±68.5mV

Figure 2.26 • Error Budget for Intel P30 Embedded Memory Voltage Regulator. 1.8V Supply Must Remain Within ±0.1V Tolerance, Including All Static and Dynamic Errors

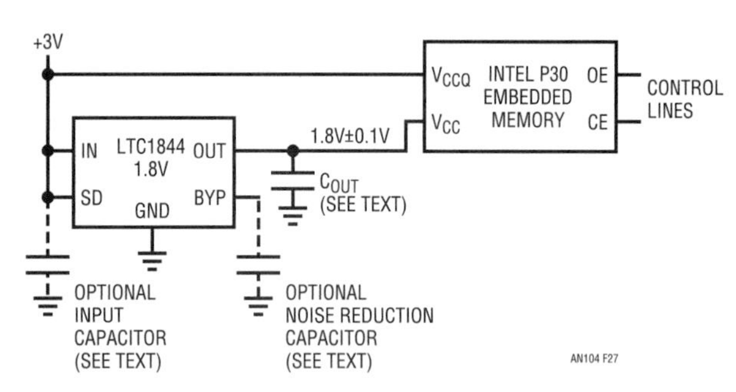

Figure 2.27 • P30 Embedded Memory V_{CC} Regulator Must Maintain ±0.1V Error Band. Control Line Movement Causes 50mA Load Steps, Necessitating Attention to C_{OUT} Selection

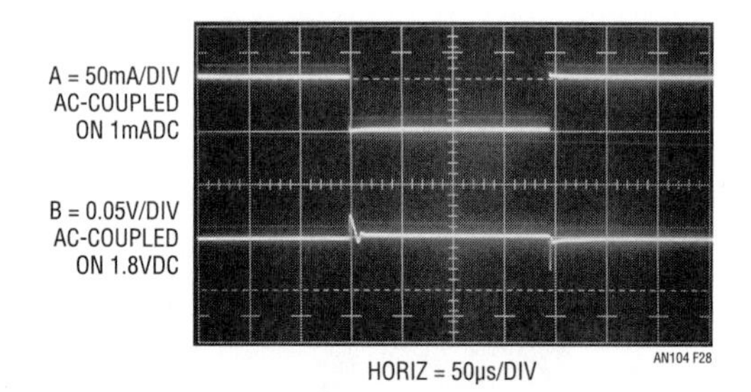

Figure 2.28 • 50mA Load Step (Trace A) Results in 30mV Regulator Response Peaks, 2× Better than Error Budget Requirements. C_{OUT} = Low Loss 1μF

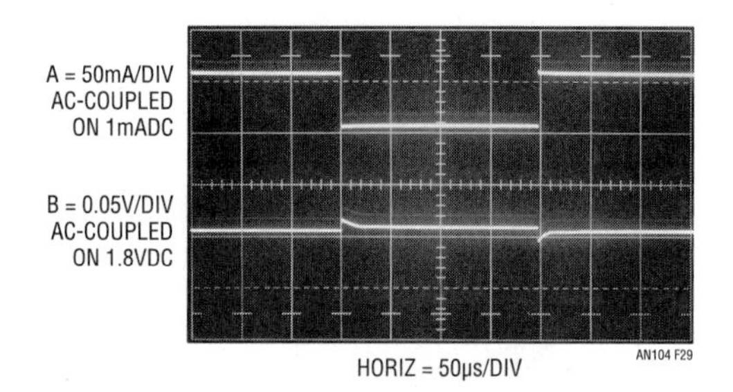

Figure 2.29 • Increasing C_{OUT} to 10μF Decreases Regulator Output Peaks to 12mV, Almost 6× Better than Required

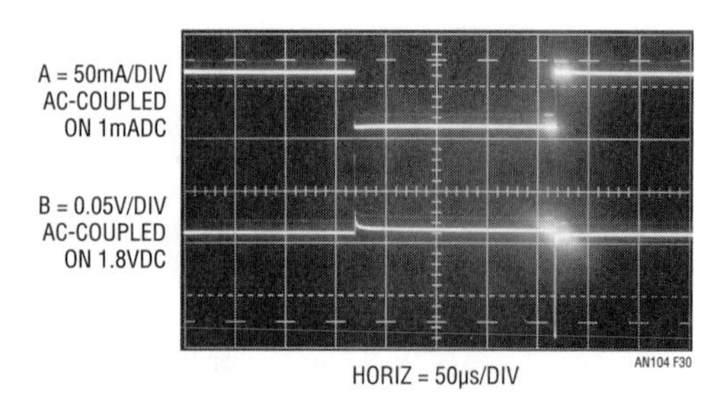

Figure 2.30 • Poor Grade 10μF C_{OUT} Causes 100mV Regulator Output Peaks (Trace B), Violating P30 Memory Limits. Traces Latter Portion Intensified for Photographic Clarity

Note 5: The LTC1844-1.8's noise bypass pin ("BYP") is used with an optional external capacitor to achieve extremely low output noise. It is not required for this application and is left unconnected.

Note 6: Figure 2.8's circuit was used for this test, with Q1's emitter current shunt changed to 1Ω.

Note. This application note was derived from a manuscript originally prepared for publication in EDN magazine.

References

1. LT1584/LT1585/LT1587 Fast Response Regulators Datasheet. Linear Technology Corporation
2. LT1963A Regulator Datasheet. Linear Technology Corporation
3. Williams, Jim, "Minimizing Switching Residue in Linear Regulator Outputs". Linear Technology Corporation, Application Note 101, July 2005
4. Shakespeare, William, "The Taming of the Shrew," 1593-94

Appendix A
Capacitor parasitic effects on load transient response

Tony Bonte

Large load current changes are typical of digital systems. The load current step contains higher order frequency components that the output decoupling network must handle until the regulator throttles to the load current level. Capacitors are not ideal elements and contain parasitic resistance and inductance. These parasitic elements dominate the change in output voltage at the beginning of a transient load step change. The ESR (equivalent series resistance) of the output capacitors produces an instantaneous step in output voltage.($\Delta V = \Delta I \bullet ESR$). The ESL (equivalent series inductance) of the output capacitors produces a droop proportional to the rate of change of output current ($V = L \bullet \Delta I/\Delta t$). The output capacitance produces a change in output voltage proportional to the time until the regulator can respond ($\Delta V = \Delta t \bullet \Delta I/C$). These transient effects are illustrated in Figure A1.

The use of capacitors with low ESR, low ESL, and good high frequency characteristics is critical in meeting the output load voltage tolerances. These requirements dictate high quality, surface mount tantalum, ceramic or organic electrolyte capacitors. The capacitor's location is critical to transient response performance. Place the capacitor as close as possible to the regulator pins and keep supply line traces and planes at low impedance, bypassing individual loads as necessary. if the regulator has remote sensing capability, consider sensing at the heaviest load point.

Strictly speaking, the above are not the only time related terms that can influence regulator settling. Figure A2 lists 7 different terms, occurring over 9 decades of time, that can potentially influence regulation. The regulator IC must be carefully designed to minimize regulator loop and thermal error contributions.

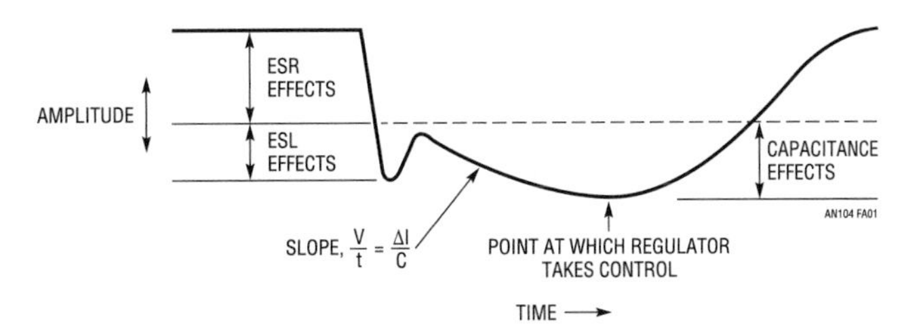

Figure A1 • Parasitic Resistance, Inductance and Finite Capacitance Combine with Regulator Gain-Bandwidth Limitations to Form Load Step Response. Capacitors Equivalent Series Resistance (ESR) and Inductance (ESL) Dominate Initial Response; Capacitor Value and Regulator Gain-Bandwidth Determine Responses Latter Profile

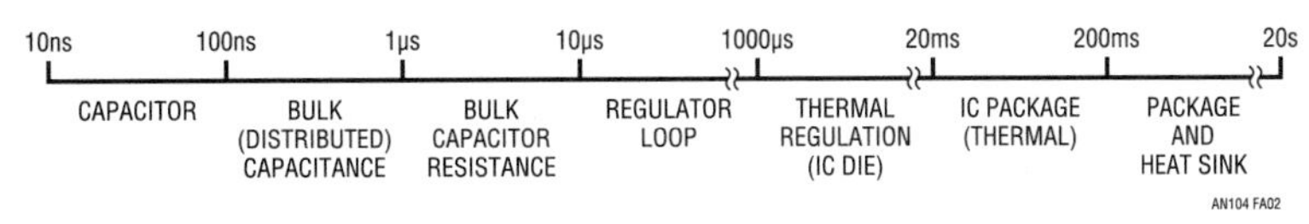

Figure A2 • Time Constants Potentially Influencing Regulator Settling Time After a Load Step are Electrical and Thermal. Effects Span Over 9 Decades

Appendix B
Output capacitors and loop stability

Dennis O'Neill

Editorial Note: The following text, excerpted from the LT1963A datasheet, concerns the output capacitor's relationship to transient response. Although originally prepared for LT1963A application, it is generalizable to most regulators and is presented here for reader convenience.

A voltage regulator is a feedback circuit. Like any feedback circuit, frequency compensation is needed to make it stable. For the LT1963A, the frequency compensation is both internal and external – the output capacitor. The size of the output capacitor, the type of the output capacitor, and the ESR of the particular output capacitor all affect the stability.

In addition to stability, the output capacitor also affects the high frequency transient response. The regulator loop has finite bandwidth. For high frequency transient loads recovery from a transient is a combination of the output capacitor and the bandwidth of the regulator. The LT1963A was designed to be easy to use and accept a wide variety of output capacitors. However, the frequency compensation is affected by the output capacitor and optimum frequency stability may require some ESR, especially with ceramic capacitors.

For ease of use, low ESR polytantalum capacitors (POSCAP) are a good choice for both the transient response and stability of the regulator. These capacitors have intrinsic ESR that improves the stability. Ceramic capacitors have extremely low ESR, and while they are a good choice in many cases, placing a small series resistance element will sometimes achieve optimum stability and minimize ringing. In all cases, a minimum of 10μF is required while the maximum ESR allowable is 3Ω.

The place where ESR is most helpful with ceramics is low output voltage. At low output voltages, below 2.5V, some ESR helps the stability when ceramic output capacitors are used. Also, some ESR allows a smaller capacitor value to be used. When small signal ringing occurs with ceramics due to insufficient ESR, adding ESR or increasing the capacitor value improves the stability and reduces the ringing. Figure B1 gives some recommended values of ESR to minimize ringing caused by fast, hard current transitions.

V_{OUT}	10μF	22μF	47μF	100μF
1.2V	20mΩ	15mΩ	10mΩ	5mΩ
1.5V	20mΩ	15mΩ	10mΩ	5mΩ
1.8V	15mΩ	10mΩ	10mΩ	5mΩ
2.5V	5mΩ	5mΩ	5mΩ	5mΩ
3.3V	0mΩ	0mΩ	0mΩ	5mΩ
≥ 5V	0mΩ	0mΩ	0mΩ	0mΩ

Figure B1 • Capacitor Minimum ESR

Figures B2 through B7 show the effect of ESR on the transient response of the regulator. These scope photos show the transient response for the LT1963A at three different output voltages with various capacitors and various values of ESR. The output load conditions are the same for all traces. In all cases there is a DC load of 500mA. The load steps up to 1A at the first transition and steps back to 500mA at the second transition.

At the worst case point of $1.2V_{OUT}$ with 10μF C_{OUT} (Figure B2), a minimum amount of ESR is required. While 20mΩ is enough to eliminate most of the ringing, a value closer to 50mΩ provides a more optimum response. At 2.5V output with 10μF C_{OUT} (Figure B3) the output rings at the transitions with 0Ω ESR but still settles to within 10mV in 20μs after the 0.5A load step. Once again a small value of ESR will provide a more optimum response.

At $5V_{OUT}$ with 10μF C_{OUT} (Figure B4) the response is well damped with 0Ω ESR.

With a C_{OUT} of 100μF at 0Ω ESR and an output of 1.2V (Figure B5), the output rings although the amplitude is only $20mV_{P-P}$. With C_{OUT} of 100μF it takes only 5mΩ to 20mΩ of ESR to provide good damping at 1.2V output. Performance at 2.5V and 5V output with 100μF C_{OUT} shows similar characteristics to the 10μF case (see Figures B6 and B7). At $2.5V_{OUT}$ 5mΩ to 20mΩ can improve transient response. At $5V_{OUT}$ the response is well damped with 0Ω ESR.

Capacitor types with inherently higher ESR can be combined with 0mΩ ESR ceramic capacitors to achieve both good high frequency bypassing and fast settling time. Figure B8 illustrates the improvement in transient response that can be seen when a parallel combination of ceramic and POSCAP capacitors are used. The

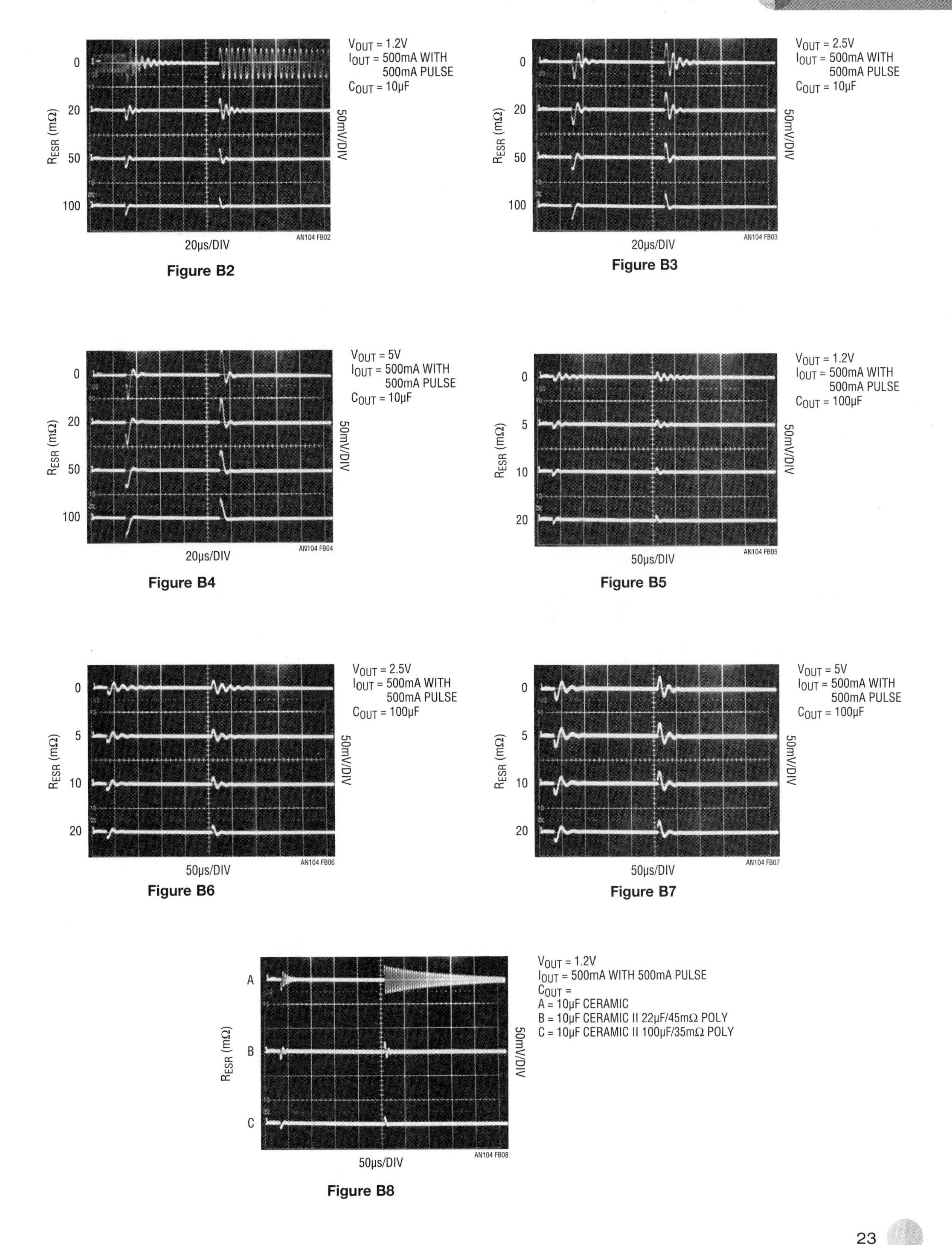

Figure B2

Figure B3

Figure B4

Figure B5

Figure B6

Figure B7

Figure B8

output voltage is at the worst case value of 1.2V. Trace A with a 10μF ceramic output capacitor, shows significant ringing with a peak amplitude of 25mV. For Trace B, a 22μF/45mΩ POSCAP is added in parallel with the 10μF ceramic. The output is well damped and settles to within 10mV in less than 20μs.

For Trace C, a 100μF/35mΩ POSCAP is connected in parallel with the 10μF ceramic capacitor. In this case the peak output deviation is less than 20mV and the output settles in about 10μs. For improved transient response the value of the bulk capacitor (tantalum or aluminum electrolytic) should be greater than twice the value of the ceramic capacitor.

Tantalum and polytantalum capacitors

There is a variety of tantalum capacitor types available, with a wide range of ESR specifications. Older types have ESR specifications in the hundreds of mΩ to several Ohms. Some newer types of polytantalum with multi-electrodes have maximum ESR specifications as low as 5mΩ. In general the lower the ESR specification, the larger the size and the higher the price. Polytantalum capacitors have better surge capability than older types and generally lower ESR. Some types such as the Sanyo TPE and TPB series have ESR specifications in the 20mΩ to 50mΩ range, which provide near optimum transient response.

Aluminum electrolytic capacitors

Aluminum electrolytic capacitors can also be used with the LT1963A. These capacitors can also be used in conjunction with ceramic capacitors. These tend to be the cheapest and lowest performance type of capacitors. Care must be used in selecting these capacitors as some types can have ESR which can easily exceed the 3Ω maximum value.

Ceramic capacitors

Extra consideration must be given to the use of ceramic capacitors. Ceramic capacitors are manufactured with a variety of dielectrics, each with different behavior over temperature and applied voltage. The most common dielectrics used are Z5U, Y5V, X5R and X7R. The Z5U and Y5V dielectrics are good for providing high capacitances in a small package, but exhibit strong voltage and temperature coefficients as shown in Figures B9 and B10. When used with a 5V regulator, a 10μF Y5V capacitor can exhibit an effective value as low as 1μF to 2μF over the operating temperature range. The X5R and X7R dielectrics result in more stable characteristics and are more suitable for use as the output capacitor. The X7R type has better stability across temperature, while the X5R is less expensive and is available in higher values.

Voltage and temperature coefficients are not the only sources of problems. Some ceramic capacitors have a piezoelectric response. A piezoelectric device generates voltage across its terminals due to mechanical stress,

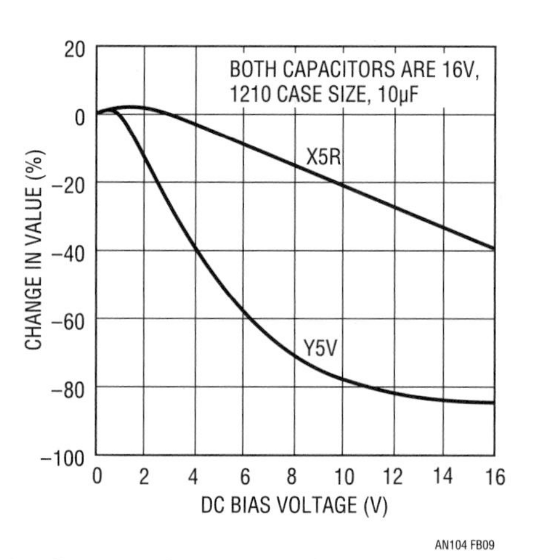

Figure B9 • Ceramic Capacitor DC Bias Characteristics

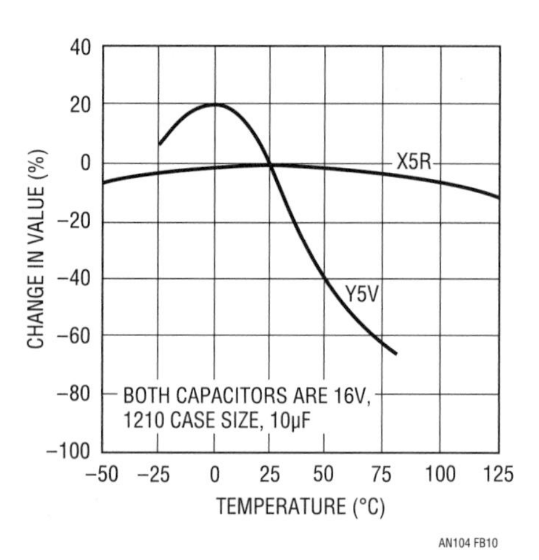

Figure B10 • Ceramic Capacitor Temperature Characteristics

similar to the way a piezoelectric accelerometer or microphone works. For a ceramic capacitor the stress can be induced by vibrations in the system or thermal transients.

"Free" resistance with PC traces

The resistance values shown in Figure B11 can easily be made using a small section of PC trace in series with the output capacitor. The wide range of non-critical ESR makes it easy to use PC trace. The trace width should be sized to handle the RMS ripple current associated with the load. The output capacitor only sources or sinks current for a few microseconds during fast output current transitions. There is no DC current in the output capacitor. Worst case ripple current will occur if the output load is a high frequency (>100kHz) square wave with a high peak value and fast edges (<1μs). Measured RMS value for this case is 0.5 times the peak-to-peak current change. Slower edges or lower frequency will significantly reduce the RMS ripple current in the capacitor.

This resistor should be made using one of the inner layers of the PC board which are well defined. The resistivity is determined primarily by the sheet resistance of the copper laminate with no additional plating steps. Figure B11 gives some sizes for 0.75A RMS current for various copper thicknesses. More detailed information regarding resistors made from PC traces can be found in Application Note 69, Appendix A.

		10mΩ	**20mΩ**	**30mΩ**
0.5oz C_U	Width	0.011" (0.28mm)	0.011" (0.28mm)	0.011" (0.28mm)
	Length	0.102" (2.6mm)	0.204" (5.2mm)	0.307" (7.8mm)
1.0oz C_U	Width	0.006" (0.15mm)	0.006" (0.15mm)	0.006" (0.15mm)
	Length	0.110" (2.8mm)	0.220" (5.6mm)	0.330" (8.4mm)
2.0oz C_U	Width	0.006" (0.15mm)	0.006" (0.15mm)	0.006" (0.15mm)
	Length	0.224" (5.7mm)	0.450" (11.4mm)	0.670" (17mm)

Figure B11 • PC Trace Resistors

Appendix C
Probing considerations for load transient response measurements

Signals of interest in load transient response studies occur within a bandwidth of about 25MHz (t_{RISE} = 14ns) This is a modest speed range but probing technique requires some care for high fidelity measurement. Load current is measured with a DC stabilized (Hall Effect based) "clip on" current probe such as the Tektronix P-6042 or AM503. The conductor loop placed in the probe jaws should encompass the smallest possible area to minimize introduced parasitic inductance, which can degrade measurement. At higher speeds, grounding the probe case may slightly decrease measurement aberrations, but this is usually a small effect.

Voltage measurement, typically AC-coupled and in the 10mV to 250mV range, is best accomplished with Figure C1's arrangement. The measured voltage is fed to a BNC fixtured 50Ω back terminated cable, which drives the oscilloscope via a DC blocking capacitor and a 50Ω termination. The back termination is strict practice, enforcing a true 50Ω signal path. Practically, if its ÷2 attenuation presents problems, it can usually be eliminated with only minor signal degradation in the 25MHz measurement passband. The termination at the oscilloscope end is not negotiable. Figure C2 shows a typical observed load transient with no back termination but 50Ω at the oscilloscope. The presentation is clean and well defined. In C3, the cable's 50Ω termination is removed, causing a distorted leading edge, ill-defined peaking and pronounced post-event ringing. Even at relatively modest frequencies the cable displays unterminated transmission line characteristics, resulting in signal distortion.

In theory, a 1× scope probe using a probe-tip coaxial connection could replace the above but such probes usually have bandwidth limitations of 10MHz to 20MHz. Conversely, a 10× probe is wideband, but oscilloscope vertical sensitivity must accommodate the introduced attenuation.

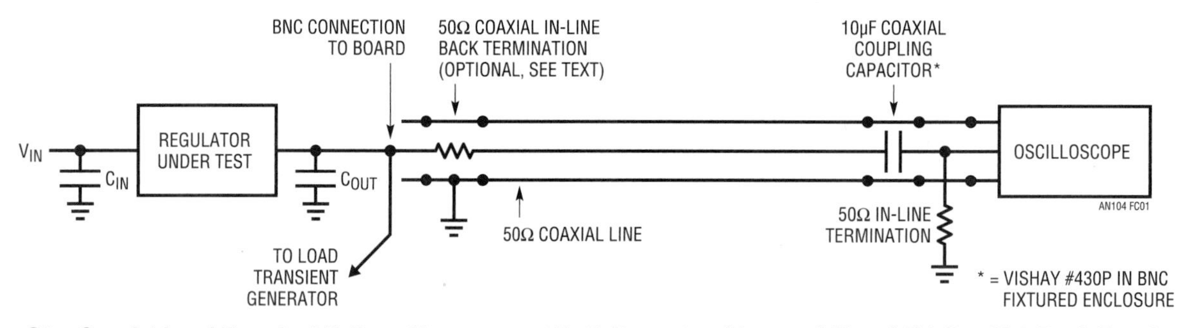

Figure C1 • Coaxial Load Transient Voltage Measurement Path Promotes Observed Signal Fidelity. 50Ω Back Termination May Be Removed with Minimal Impact on 25MHz Signal Path Integrity. 50Ω Termination at Oscilloscope Cannot Be Deleted

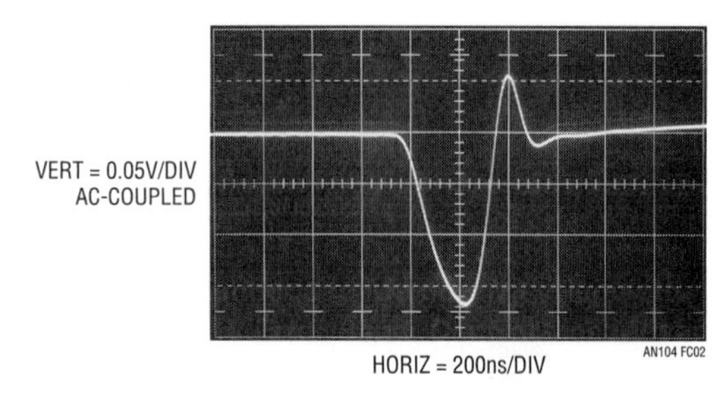

Figure C2 • Typical High Speed Transient Observed Through Figure C1's Measurement Path. Presentation is Clean and Well Defined

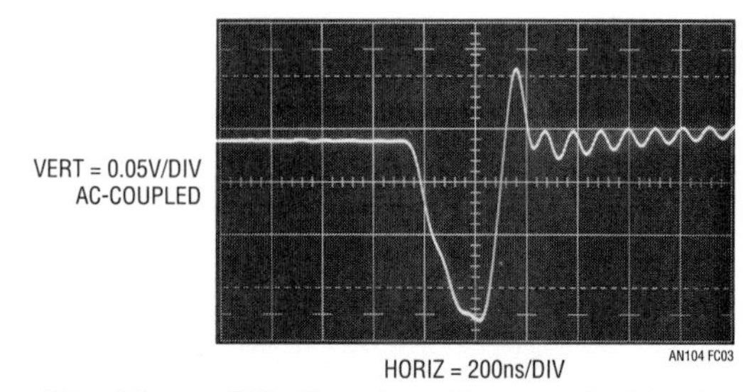

Figure C3 • Figure C2's Transient Measured with 50Ω Oscilloscope Termination Removed. Waveform Distortion and Post-Event Ringing Result

Appendix D
A trimless closed loop transient load tester

Text Figure 2.8's circuit is attractive because it eliminates the FET based design's AC trims. It does, however, retain the DC trim. Figure D1 trades circuit complexity to eliminate the DC trim. Operation is similar to text Figure 2.8's circuit except that A2 appears. This amplifier replaces the DC trim by measuring the circuits DC input, comparing it to Q1's emitter DC level and controlling A1's positive input to stabilize the circuit. High frequency signals are filtered at A1's inputs and do not corrupt A1's stabilizing action. A useful way to consider circuit operation is that A2 will balance its inputs, and hence the circuit's input and output, regardless of A1's DC input errors. DC current bias is set to any desired point by a variable reference source directed to A2's positive input. This network's resistors are arranged for a minimum load current of 10mA, avoiding loop disruption for currents near zero.

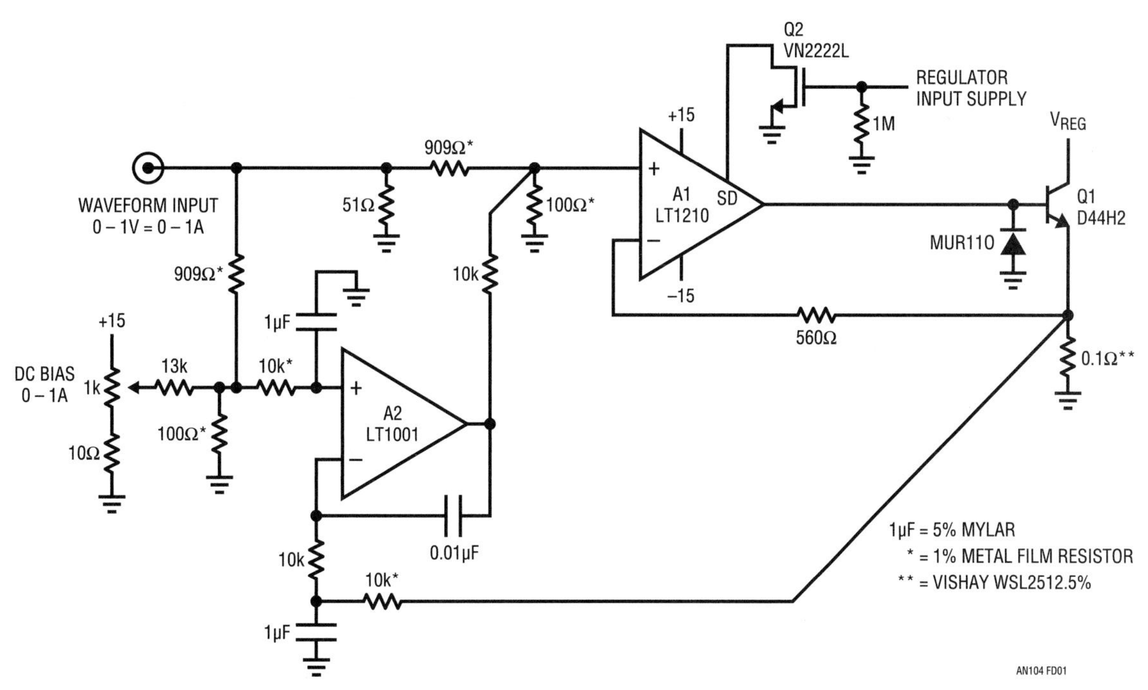

Figure D1 • A2 Feedback Controls A1's DC Errors, Eliminating Text Figure 2.8's Trim. Filtering Restricts A2's Response to DC and Low Frequency

A closed-loop, wideband, 100A active load

Brute force marries controlled speed

3

Jim Williams

Introduction

Digital systems, particularly microprocessors, furnish transient loads in the 100A range that a voltage regulator must service. Ideally, regulator output is invariant during a load transient. In practice, some variation is encountered and becomes problematic if allowable operating voltage tolerances are exceeded. 100A load steps, characteristic of microprocessors, exacerbate this issue, necessitating testing the regulator and associated components under such transient loading conditions. To meet this need, a closed-loop, 500kHz bandwidth, linearly responding, 100A capacity active load is described below.

Study of this approach is prefixed by a brief review of conventional test load types and noting their shortcomings.[1]

Basic load transient generator

Figure 3.1 diagrams a conceptual load transient generator. The regulator under test drives DC and switched resistive loads, which may be variable. The switched current and output voltage are monitored, permitting comparison of the nominally stable output voltage versus load current under static and dynamic conditions. The switched current is either on or off; there is no controllable linear region.

Figure 3.2 develops the concept by including electronic load switch control. Operation is straightforward. The input

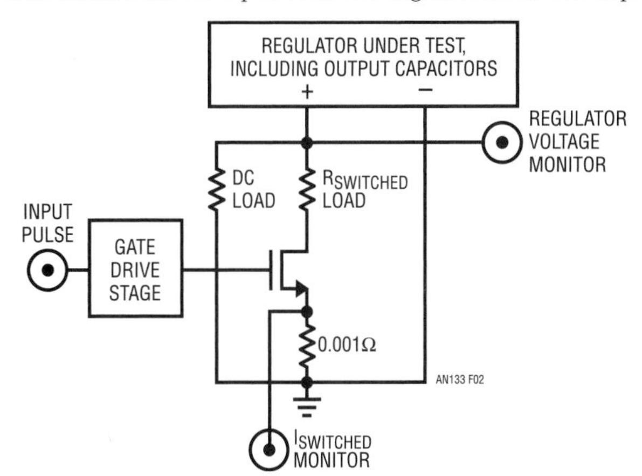

Figure 3.2 • Conceptual FET Based Load Tester Permits Input Pulse Controlled Step Loading. As Before, Switched Current Is Either On or Off; There Is No Controllable Linear Region

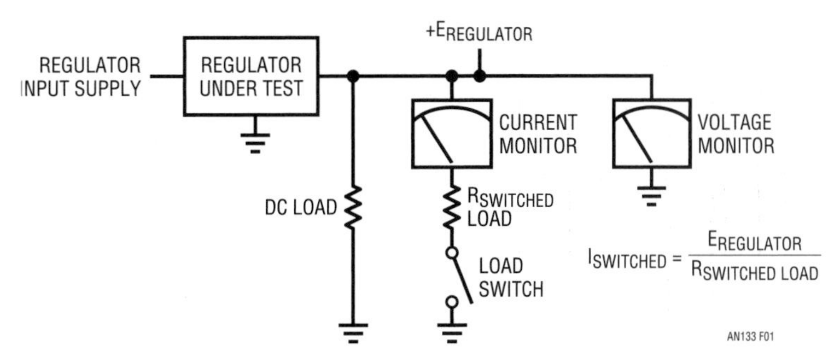

Figure 3.1 • Conceptual Regulator Load Tester Includes Switched and DC Loads and Voltage/Current Monitors. Resistor Values Set DC and Switched Load Currents. Switched Current Is Either On or Off; There Is No Controllable Linear Region

Note 1: See Reference 1, from which the immediately following material partially derives, for details and descriptions of very wideband load transient generators, albeit at much lower currents.

Analog Circuit and System Design: Immersion in the Black Art of Analog Design. http://dx.doi.org/10.1016/B978-0-12-397888-2.00003-1

pulse switches the FET via a drive stage, generating a transient load current out of the regulator and its output capacitors. The size, composition and location of these capacitors has a pronounced effect on transient response and must be quite carefully considered. Although the electronic control facilitates high speed switching, the architecture cannot emulate loads between the minimum and maximum currents. Additionally, FET switching speed is uncontrolled, introducing wideband harmonic into the measurement, potentially corrupting the oscilloscope display.

Closed-loop load transient generator

Placing Q1 within a feedback loop allows true, linear control of the load tester. Figure 3.3's conceptual closed-loop load transient generator linearly controls Q1's gate voltage to set instantaneous transient current at any desired point, allowing simulation of nearly any load profile. Feedback from Q1's source to the A1 control amplifier closes a loop around Q1, stabilizing its operating point. Q1's current assumes a value dependent on the instantaneous input control voltage and the current sense resistor over a very wide bandwidth. Note that once A1 is biased to Q1's conduction threshold (by the "DC Load Set"), small variations in A1's output result in large Q1 channel current changes. As such, large output excursions are not required from A1; its small signal bandwidth is the fundamental speed limitation. Within this restriction, Q1's current waveform is identically shaped to A1's input control voltage, allowing linear control of load current. This versatile capability permits a wide variety of simulated loads.

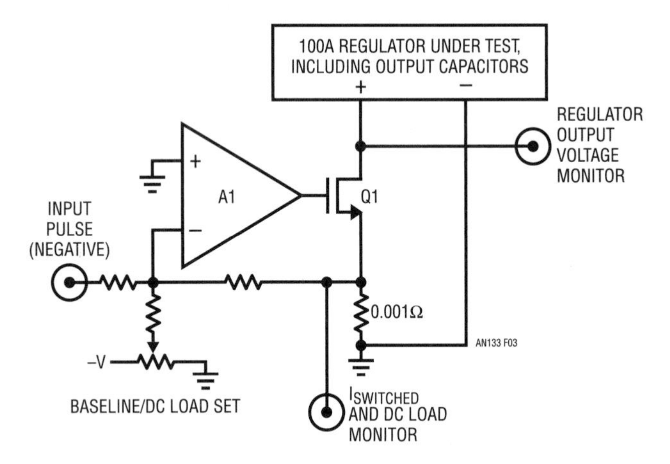

Figure 3.3 • Feedback Controlled Load Step Tester Allows Continuous FET Conductivity Control. Input Accommodates Separate DC and Pulsed Loading Instructions

Figure 3.4 further develops Figure 3.3, adding new elements. A gate drive stage isolates the control amplifier from Q1's gate capacitance, maintaining amplifier phase margin and providing low delay, linear current gain. An X10 differential amplifier provides high resolution sensing across the 1mΩ current shunt. The dissipation limiter, acting on the averaged input value and Q1 temperature, shuts down gate drive to preclude excessive FET heating and subsequent destruction. Amplifier associated capacitors tailor bandwidth and optimize loop response.

Detailed circuitry discussion

Figure 3.5's detailed schematic of the 100A capacity load tester follows Figure 3.4's outline. A1, responding to DC and pulse inputs and current indicating feedback from A3, sets Q1's conductivity via the actively biased Q4-Q5 gate drive stage. A2 determines stage bias under all conditions by comparing Q5's averaged collector voltage to a reference and controlling Q3's conduction, closing a loop. C1, instructed by the average input value, shuts down FET gate drive via

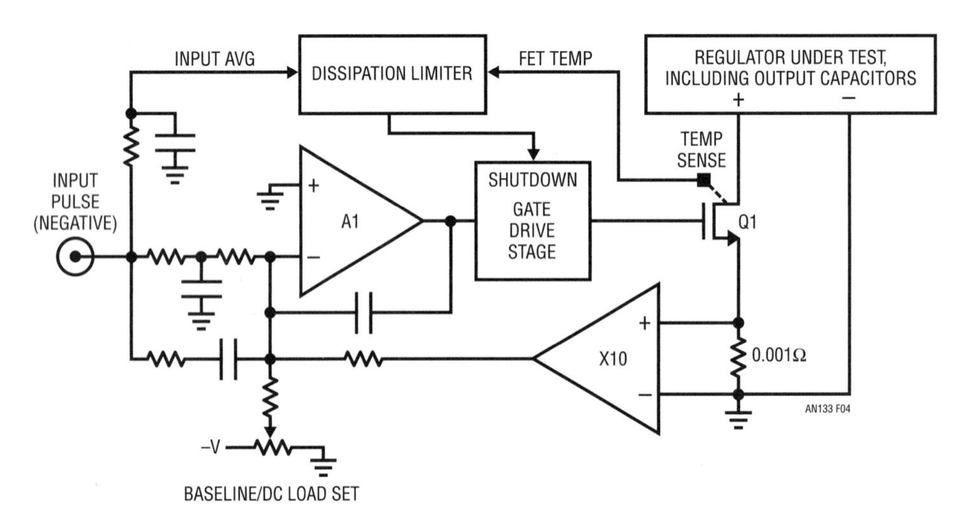

Figure 3.4 • Developed Form of Figure 3.3. Differential Amplifier Provides High Resolution Sensing Across Milliohm Shunt. Dissipation Limiter, Acting on Average Input Value and FET Temperature, Shuts Down Gate Drive, Precluding Excessive FET Heating. Amplifier Associated Capacitors Tailor Bandwidth, Optimize Loop Response

Q2 under harmful conditions.[2] SW1, sensing heat sink temperature, contributes thermally activated Q1 shutdown. Q6 and the Zener prevent Q1 turn-on if the −15V supply is not present by diverting Q4 bias. A1's 1k resistor precludes A1 damage due to 15V supply loss. Trims optimize dynamic response, determine loop DC baseline idle current, set dissipation limit and control gate drive stage bias. The DC trims are self-explanatory. The "loop compensation" and "FET response" AC trims at A1 are more subtle. They are adjusted for the best compromise between loop stability, edge rate and pulse purity. A1's loop compensation trim sets roll-off for maximum bandwidth while accommodating phase shift introduced by Q1's gate capacitance and, to a lesser extent, by A3. The "FET response" adjustment partially compensates Q1's inherent nonlinear gain characteristic, improving front and rear pulse corner fidelity.[3]

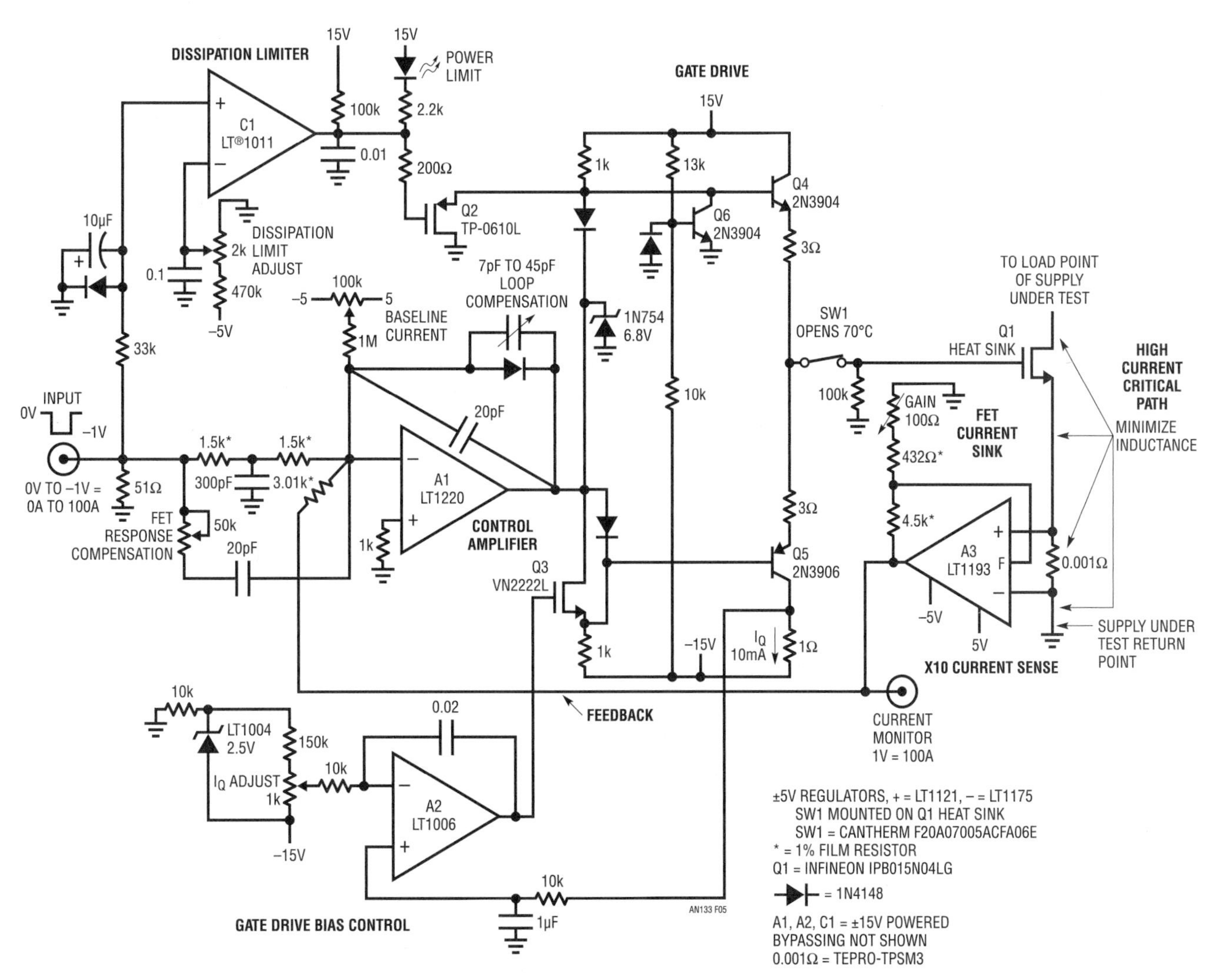

Figure 3.5 • Detailed Circuitry Follows Figure 3.4's Concept. A1, Responding to DC and Pulsed Inputs, Sets Q1 Conductivity via Actively Biased Gate Drive Stage. A3, Sensing Q1 Current, Closes A1's Feedback Loop. C1, Instructed by Average Input Value, Shuts Down Q1 Gate Drive Under Harmful Conditions. Q6 Guards Against Lost −15V Supply. SW1 Adds Thermal Limiting. Trims Optimize Dynamic Response, Determine Loop Baseline Idle Current, Set Dissipation Limit and Control Gate Drive Stage Bias

Note 2: The protection scheme is patterned after techniques utilized in high power pulse generators. See Reference 2 and 3.

Note 3: The trimming procedure is not described here in order to maintain text flow and focus. For detailed trimming information see Appendix B, "Trimming Procedure".

Circuit testing

The circuit is initially tested using a fixture equipped with massive, low loss, wideband bypassing shown in Figure 3.6. Additionally, the importance of exceptionally low inductance layout in the high current path cannot be overstated. Every attempt must be made to minimize inductance in the 100A path; such current density at high speed demands low inductance if waveform purity is desirable. Figure 3.7 shows results with the circuit properly trimmed and a minimized inductance high current path. The 100A amplitude, high speed waveform is exceptionally pure with just discernible top-front and bottom-rear corner infidelities.[4] AC trim effects on waveform presentation are studied by deliberate mis-adjustment. Figure 3.8's overdamped response is typical of excess A1 feedback capacitance. The current pulse is well controlled but edge rate is slow. Figure 3.9's inadequate A1 feedback capacitance decreases transition time but promotes instability. Further capacitance reduction will cause loop oscillation

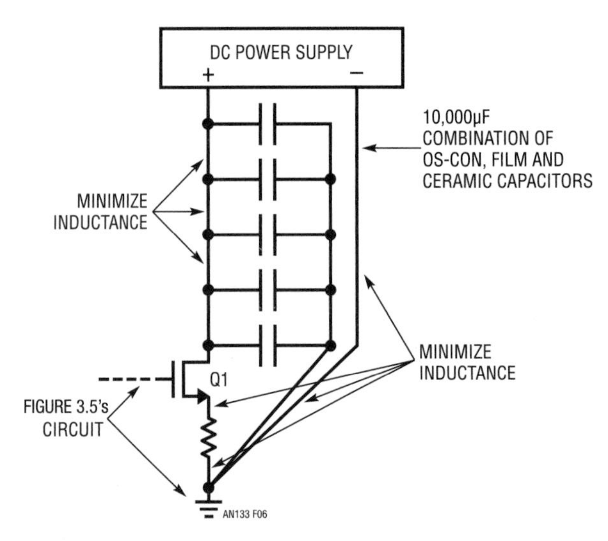

Figure 3.6 • Fixturing Tests Figure 3.5's Dynamic Response. Massive, Broadband Bypassing Combined with Low Inductance Layout Provides Low Loss, High Current Power to Q1

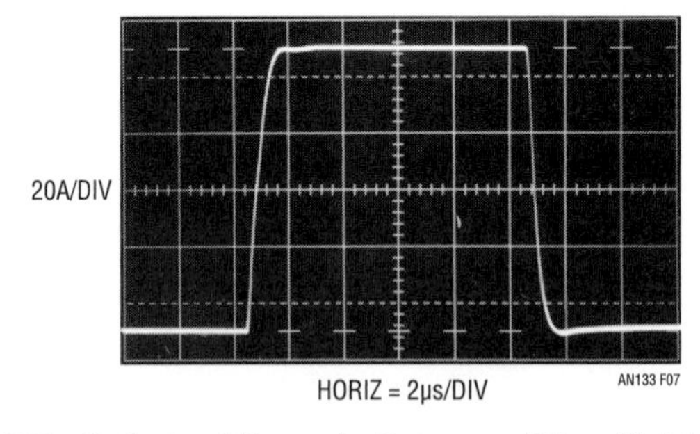

Figure 3.7 • Optimized Dynamic Response Trims Yield Exceptionally Pure, 100A Q1 Current Pulse. Residual Top-Front and Bottom-Rear Corner Infidelities Are Just Discernable

Note 4: See Appendices A, "Verifying Current Measurement" and C, "Instrumentation Considerations" for guidance on obtaining Figure 3.7's performance level.

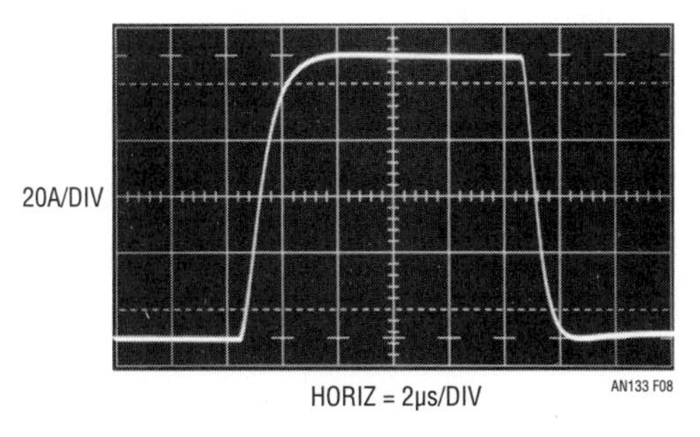

Figure 3.8 • Overdamped Response Characteristic of Excessive A1 Feedback Capacitor Value

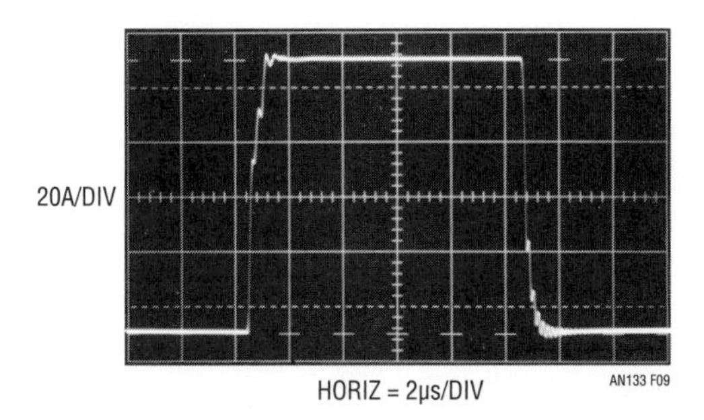

Figure 3.9 • Inadequate A1 Feedback Capacitor Decreases Transition Time But Promotes Instability. Further Capacitor Reduction Causes Oscillation

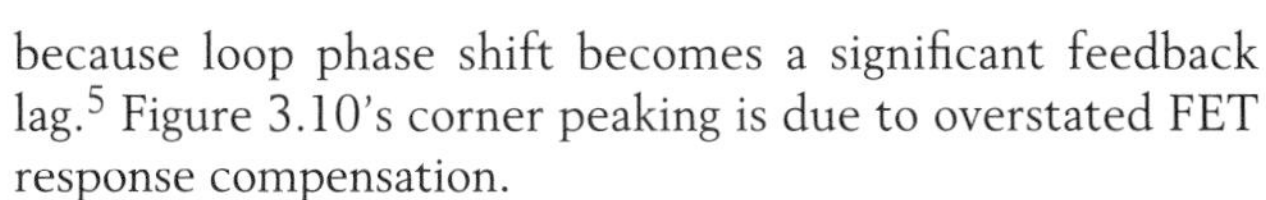

because loop phase shift becomes a significant feedback lag.[5] Figure 3.10's corner peaking is due to overstated FET response compensation.

If the AC trims are restored to nominal values, Figure 3.11's leading edge indicates 650 nanosecond rise time, equivalent to ≈540kHz bandwidth. The trailing edge (Figure 3.12), taken under the same conditions as the previous figure, is somewhat faster at 500 nanosecond fall time.

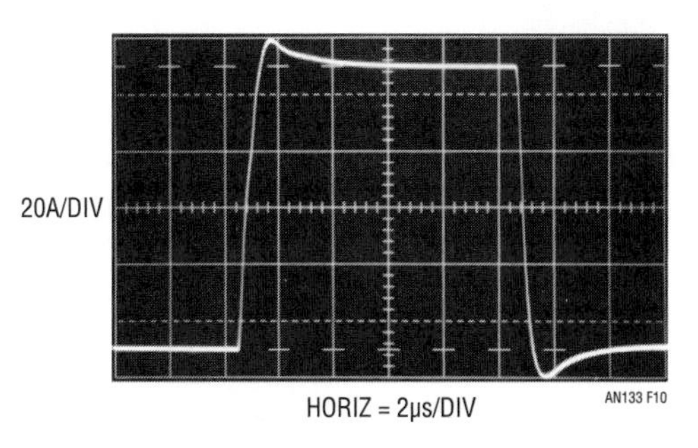

Figure 3.10 • Corner Peaking Due to Overstated FET Response Compensation

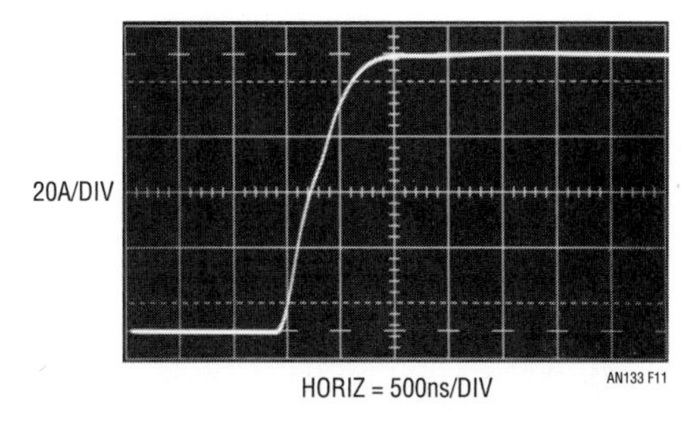

Figure 3.11 • Optimized Dynamic Trims Allow 650 Nanosecond Rise Time, Corresponding to ≈540kHz Bandwidth

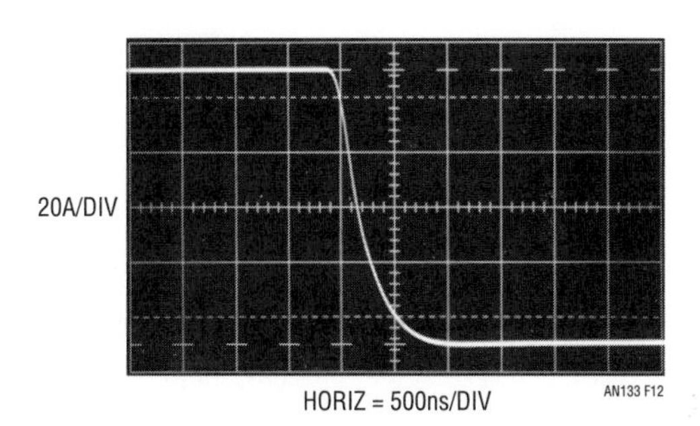

Figure 3.12 • Same Conditions as Figure 3.11 Show 500 Nanosecond Fall Time

Note 5: Sorry, no photo available. Uncontrolled 100 Ampere amplitude loop oscillation is too thrilling for documenting.

Layout effects

None of the previous responses, even the mis-compensated examples, can be remotely approached if parasitic inductance is present in the high current path. Figure 3.13 deliberately places only 20nH parasitic inductance in Q1's drain path. Figure 3.14a displays enormous waveshape degradation deriving from the inductance and the loop's subsequent response. A monstrous error dominates the leading edge before recovery occurs at the pulse middle-top. Additional aberration is evident in the falling edge turn-off. It is especially noteworthy that the photo's horizontal scale is 5× slower than Figure 3.7's optimized response, repeated here for emphasis as Figure 3.14b. The lesson is clear. High speed 100A excursions do not tolerate inductance.

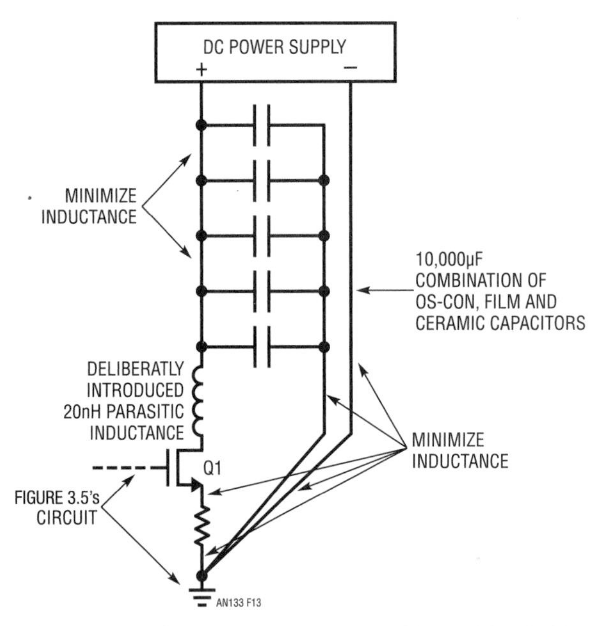

Figure 3.13 • Deliberately Introduced Parasitic 20nH Inductance Tests Figure 3.5's Layout Sensitivity

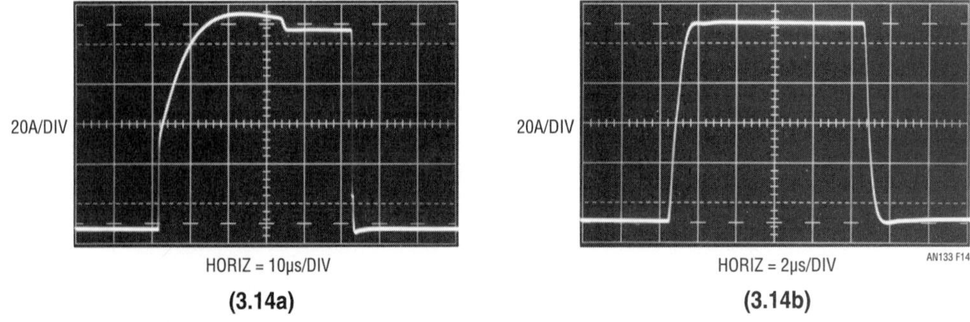

Figure 3.14 • 1.5" of 0.075" (20nH) Flat Copper Braided Wire Completely Distorts (Figure 13.4a) Text Figure 3.7's (Shown Here as Figure 3.14b) Response. Note 5× Horizontal Scale Change

Regulator testing

Regulator testing is possible after the above compensation and layout issues have been addressed. Figure 3.15 describes a test arrangement for an LTC®3829 based, 6-phase, 120A buck regulator.[6] Trace A, Figure 3.16 shows the 100A load pulse. Trace B's regulator response is well controlled on both edges.

The active loads true linear response and high bandwidth permit wide ranging load waveform characteristics.

Note 6: This regulator is conveniently incarnated as LTC demonstration board 1675A.

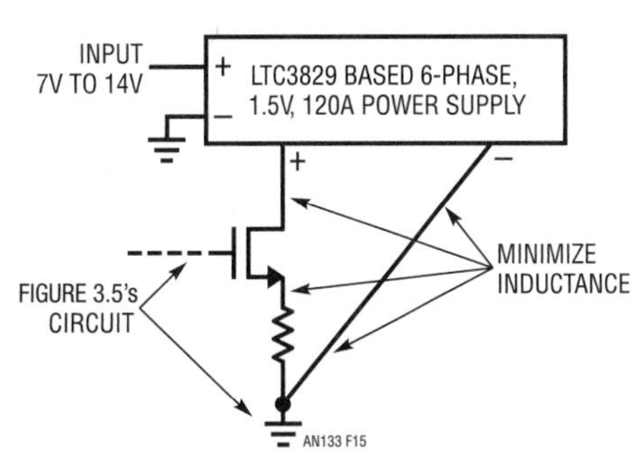

Figure 3.15 • Test Arrangement for LTC3829 Based 6-Phase, 120A Buck Regulator Emphasizes Low Impedance Connections to Figure 3.5's Circuit

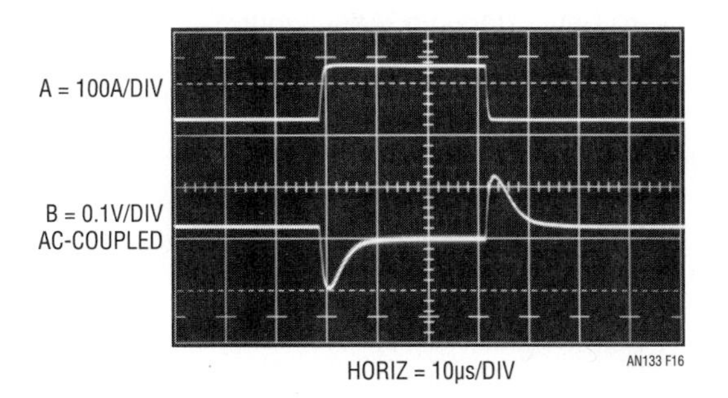

Figure 3.16 • Figure 3.5's Circuit Subjects Figure 3.15's Regulator to 100A Pulsed Load (Trace A). Response (Trace B) Is Well Controlled on Both Edges

While the step load pulse shown above is the commonly desired test, essentially any load profile is easily generated. Figure 3.17's burst of 100A, 100kHz sinewaves is an example. Response is crisp, with no untoward dynamics despite the high speed and vcurrent involved. In Figure 3.18, an 80 microsecond burst of 100A peak-peak noise forms the load. Figure 3.19 summarizes active load characteristics.

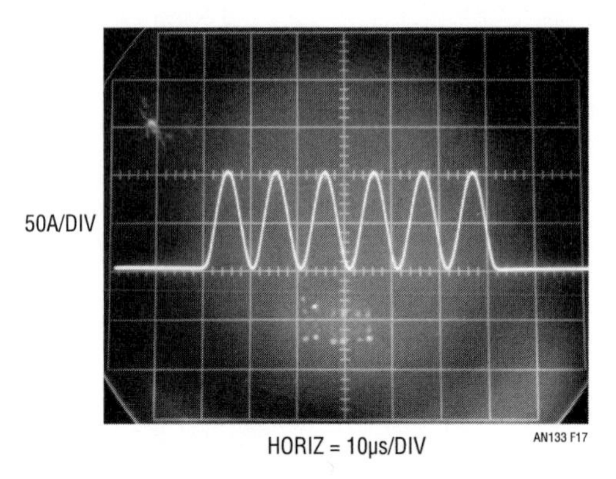

Figure 3.17 • Figure 3.5 Supplies 100kHz, 100A Sinewave Load Transient Profile. Circuit's Wide Bandwidth and Linear Operation Permit Wide Ranging Load Waveform Characteristics

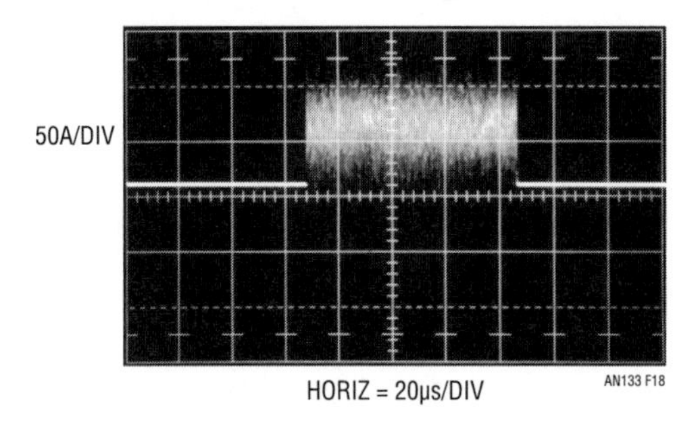

Figure 3.18 • Active Load Circuit Sinks 100A Peak-Peak in Response to Gated Random Noise Input

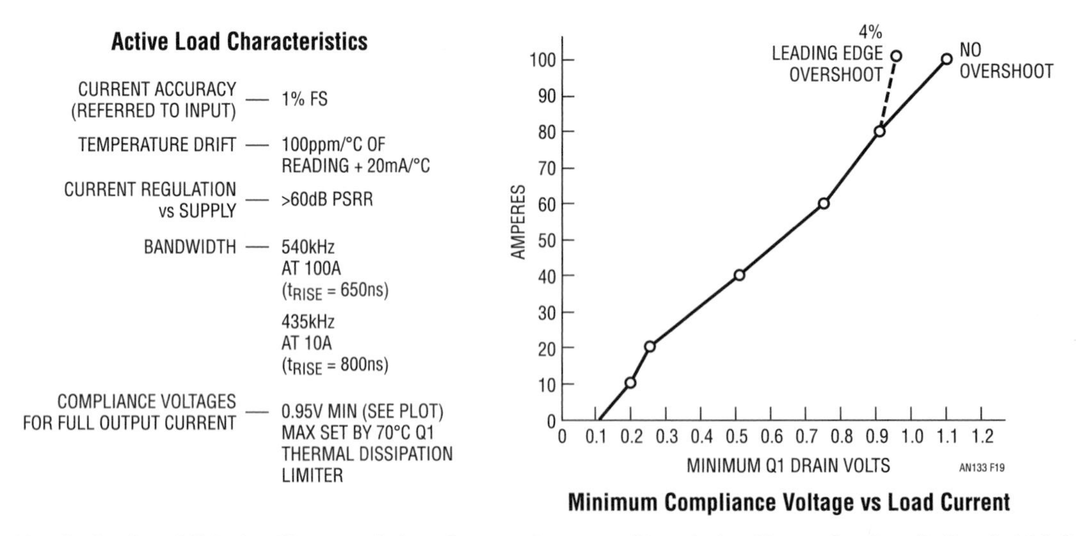

Figure 3.19 • Active Load Tabular Characteristics. Current Accuracy/Regulation Errors Are Small, Bandwidth Mildly Retards at Low Current. Compliance Voltage Is Below 1V at 100A with 4% Leading Edge Overshoot, 1.1V with No Overshoot

References

1. Williams, Jim, "Load Transient Response Testing For Voltage Regulators", Linear Technology Corporation, Application Note 104, October 2006
2. Hewlett-Packard Company, "HP-214A Pulse Generator Operating and Service Manual", "Overload Adjust", Figure 5-13. See also, "Overload Relay Adjust", pg. 5-9. Hewlett-Packard Company, 1964
3. Hewlett-Packard Company, "HP-214B Pulse Generator Operating and Service Manual", "Overload Detection/ Overload Switch…", pg. 8-29. Hewlett-Packard Company, March, 1980

Appendix A
Verifying current measurement

Theoretically, Figure 3.5's Q1 source and drain current are equal. Practically, questions center on potential effects of residual inductances and the huge (28000pF!) gate capacitance of the 1.5mΩ FET specified. If these or other terms impact drain-source current equivalence at speed A3's indicated instantaneous current could be erroneous. Verification is necessary and Figure A1 diagrams a method. An added "top-side" 1mΩ shunt and its attendant X10 differential amplifier duplicate the circuit's resident "bottom- side" current sensing section. Figure A2's results happily eliminate concern over dynamic Q1 current differences. The two 100A pulse outputs are identical in amplitude and shape, promoting confidence in circuit operation.

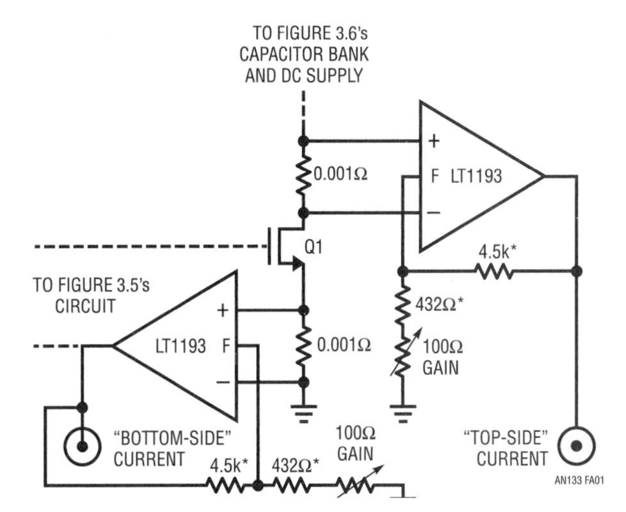

Figure A1 • Arrangement For Observing Q1's "Top" and "Bottom" Dynamic Currents

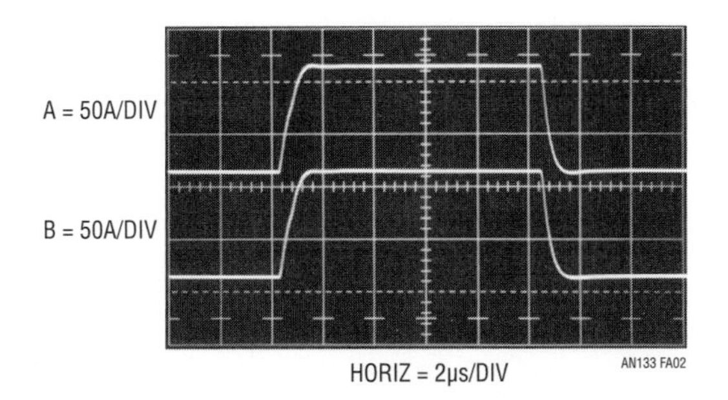

Figure A2 • Q1 "Top" (Trace A) and "Bottom" (Trace B) Currents Show Identical Characteristics Despite High Speed Operation

Appendix B
Trimming procedure

Trimming Figure 3.5's circuit is a 7-step procedure which should be performed in the numerical order listed below. Out-of-sequence adjustments are permissible assuming the dissipation limiter circuitry is working and adjusted.

1. Set all adjustments to mid-range except A1's feedback capacitor, which should be at full capacity.
2. Apply no input, bias Q1's drain from a 1V DC supply, turn on power and trim "Base Line Current" for 0.5A through Q1. Monitor this current with an ammeter in Q1's drain line.
3. Turn power off. Lift Q2's source lead and let it float. *This disables the dissipation limit circuitry, leaving Q1 vulnerable to damage from inappropriate inputs. Follow the remainder of this step in strict accordance with the instructions given.* Turn power on, bias Q1's drain from a 1V supply and apply a −0.1000V DC input while monitoring Q1's drain current with an ammeter. Trim the "Gain" adjustment for a 10.50A meter reading.[1] Make this adjustment fairly quickly as Q1 is dissipating 10 Watts. Turn power off and reconnect Q2's source lead.[2]
4. Turn power on with no input applied and Q1's drain unbiased. Trim the "I_Q" adjustment for +10mV at A2's positive input measured with respect to the −15V rail. Turn off power.
5. Bias and bypass Q1's drain in accordance with text Figure 3.6. Set the drain DC power supply for 1.5V output and turn on power. Apply a 1kHz, −1V amplitude, 5μs wide pulse. Slowly increase pulse width until C1 trips, shutting down circuit output and illuminating the "Power Limit" LED. Tripping should occur at about 12μs to15μs pulse width. If it does not, adjust the "Dissipation Limit" potentiometer to bring the trip point within these limits. This sets allowable full-amplitude (100A) duty cycle at about 1.5%.
6. Under the same operating conditions as Step 5, set input pulse width at 10μs and adjust A1's capacitive trim for the fastest positive going edge obtainable at A3's output without introducing pulse distortion. Pulse clarity should approach text Figure 3.7 with somewhat degraded top-front and bottom-rear corner rounding.
7. Adjust "FET Response Compensation" to correct the corner rounding noted in Step 6. Some interaction may occur with Step 6's adjustment. Repeat Step 6 and 7 until A3's output waveform looks like text Figure 3.7.

Appendix C
Instrumentation considerations

The pulse edge rates discussed in the text are not particularly fast, but high fidelity response requires some diligence. In particular, the input pulse must be cleanly defined and devoid of parasitics which would unfairly make circuit output pulse shape look bad. The pulse generators pre-shoot, rise time and pulse-transition aberrations are well out of band and filtered by A1's 2.1MHz (t_{RISE} = 167ns) input RC network. These terms are not a concern; almost all general purpose pulse generators should perform well. A potential offender is excessive

Note 1: The tight trim targets are mandated by trimming gain at only 10% of scale. This is certainly undesirable but less painful than trimming at 100% of scale which would force astronomical (and brief) dissipation in Q1 and the milliohm shunt.

Note 2: It is worth mention that the primary uncertainty necessitating gain trimming is the sense lines mechanical placement at the milliohm shunt.

tailing after transitions. Meaningful dynamic testing requires a rectangular pulse shape, flat on top and bottom within 1% to 2%. The circuit's input band shaping filter removes the aforementioned high speed transition related errors but will not eliminate lengthy tailing in the pulse flats. The pulse generator should be checked for this with a well compensated probe at the circuit input. The oscilloscope should register the desired flat top/bottom waveform characteristics. In making this measurement, if high speed transition related events are bothersome, the probe can be moved to the band limiting 300pF capacitor. This is defensible practice because the waveform at this point determines A1's input signal bandwidth.

Some pulse generator output stages produce low level DC offset when their output is nominally at its zero volt state. The active load circuit will process such DC potentials as legitimate signal, resulting in DC load baseline current shift.

The active loads input scale factor of 1V = 100A means that a 10mV "zero" state error produces 1A of DC baseline current shift. A simple way to check a pulse generator for this error is to place it in external trigger mode and read its output with a DVM. If offset is present it can be accounted for, nullified with the circuit's "baseline current" trim or another pulse generator can be selected.

Parasitic effects due to probe grounding and instrument interconnection should be kept in mind. At pulsed 100A levels parasitic current is easily induced into "grounds" and interconnections, distorting displayed waveforms. Probes should be coaxially grounded, particularly at A3's output current monitor and preferably anywhere else. Additionally, it is convenient and common practice to externally trigger the oscilloscope from the pulse generator's trigger output. There is nothing wrong with this, in fact it is recommended to insure a stable trigger as probes are moved between points. This practice does, however, potentially introduce ground loops due to multiple paths between the pulse generator, the circuit and the oscilloscope. This can falsely cause apparent distortion in displayed waveforms. This effect is avoidable by using a "trigger isolator" at the oscilloscope external trigger input. This simple coaxial component typically consists of isolated ground and signal paths, often coupled to a pulse transformer to provide a galvanically isolated trigger event. Commercial examples include the Deerfield Labs 185 and the Hewlett-Packard 11356A. Alternately, a trigger isolator is easily constructed in a small BNC equipped enclosure as shown in Figure C1.

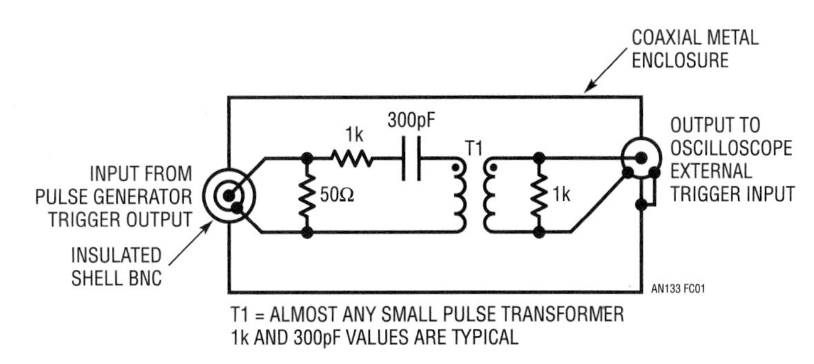

Figure C1 • Trigger Isolator Floats Input BNC Ground Using Insulated Shell BNC Connector. Capacitively Coupled Pulse Transformer Avoids Loading Input, Maintains Isolation, Delivers Trigger to Output. T1 Secondary Resistor Terminates Ringing

Section 2

Switching Regulator Design

Some thoughts on DC/DC converters (4)

This note examines a wide range of DC/DC converter applications. Single inductor, transformer, and switched-capacitor converter designs are shown. Special topics like low noise, high efficiency, low quiescent current, high voltage, and wide-input voltage range converters are covered. Appended sections explain some fundamental properties of different types of converters.

Theoretical considerations for buck mode switching regulators (5)

This note discusses the use of the LT1074 and LT1076 high efficiency switching regulators. These regulators are specifically designed for ease of use. This application note is intended to eliminate the most common errors that customers make when using switching regulators as well as offering insight into the inner workings of switching designs. There is an entirely new treatment of inductor design based upon simple mathematical formulas that yield direct results. There are extensive tutorial sections devoted to the care and feeding of the positive step-down (buck) converter, the tapped inductor buck converter, the positive-to-negative converter and the negative boost converter. Additionally, many troubleshooting hints are included as well as oscilloscope techniques, soft-start architectures, and micropower shutdown and EMI suppression methods

Some thoughts on DC/DC converters

4

Jim Williams   Brian Huffman

Introduction

Many systems require that the primary source of DC power be converted to other voltages. Battery driven circuitry is an obvious candidate. The 6V or 12V cell in a laptop computer must be converted to different potentials needed for memory, disc drives, display and operating logic. In theory, AC line powered systems should not need DC/DC converters because the implied power transformer can be equipped with multiple secondaries. In practice, economics, noise requirements, supply bus distribution problems and other constraints often make DC/DC conversion preferable. A common example is logic dominated, 5V powered systems utilizing ±15V driven analog components.

The range of applications for DC/DC converters is large, with many variations. Interest in converters is commensurately quite high. Increased use of single supply powered systems, stiffening performance requirements and battery operation have increased converter usage.

Historically, efficiency and size have received heavy emphasis. In fact, these parameters can be significant, but often are of secondary importance. A possible reason behind the continued and overwhelming attention to size and efficiency in converters proves surprising. Simply put, these parameters are (within limits) *relatively easy to achieve!* Size and efficiency advantages have their place, but other system-oriented problems also need treatment. Low quiescent current, wide ranges of allowable inputs, substantial reductions in wideband output noise and cost effectiveness are important issues. One very important converter class, the 5V to ±15V type, stresses size and efficiency with little emphasis towards parameters such as output noise. This is particularly significant because wideband output noise is a frequently encountered problem with this type of converter. In the best case, the output noise mandates careful board layout and grounding schemes. In the worst case, the noise precludes analog circuitry from achieving desired performance levels (for further discussion see Appendix A, "The 5V to ±15V Converter — A Special Case"). The 5V to ±15V DC/DC conversion requirement is ubiquitous, and presents a good starting point for a study of DC/DC converters.

5V to ±15V converter circuits

Low noise 5V to ±15V converter

Figure 4.1's design supplies a ±15V output from a 5V input. Wideband output noise measures 200 microvolts peak-to-peak, a 100× reduction over typical designs. Efficiency at 250mA output is 60%, about 5% to 10% lower than conventional types. The circuit achieves its low noise performance by minimizing high speed harmonic content in the power switching stage. This forces the efficiency trade-off noted, but the penalty is small compared to the benefit.

The 74C14 based 30kHz oscillator is divided into a 15kHz 2-phase clock by the 74C74 flip-flop. The 74C02 gates and 10k-0.001μF delays condition this 2-phase clock into non-overlapping, 2-phase drive at the emitters of Q1 and Q2 (Figure 4.2, Traces A and B, respectively). These transistors provide level shifting to drive emitter followers Q3-Q4. The Q3-Q4 emitters see 100Ω-0.003μF filters, slowing drive to output MOSFETs Q5-Q6. The filter's effects appear at the gates of Q5 and Q6 (Traces C and D, respectively). Q5 and Q6 are source followers, instead of the conventional common source connection. This limits transformer rise time to the gate terminals filtered slew rate, resulting in well controlled waveforms at the sources of Q5 and Q6 (Traces E and F, respectively). L1 sees

Analog Circuit and System Design: Immersion in the Black Art of Analog Design. http://dx.doi.org/10.1016/B978-0-12-397888-2.00004-3

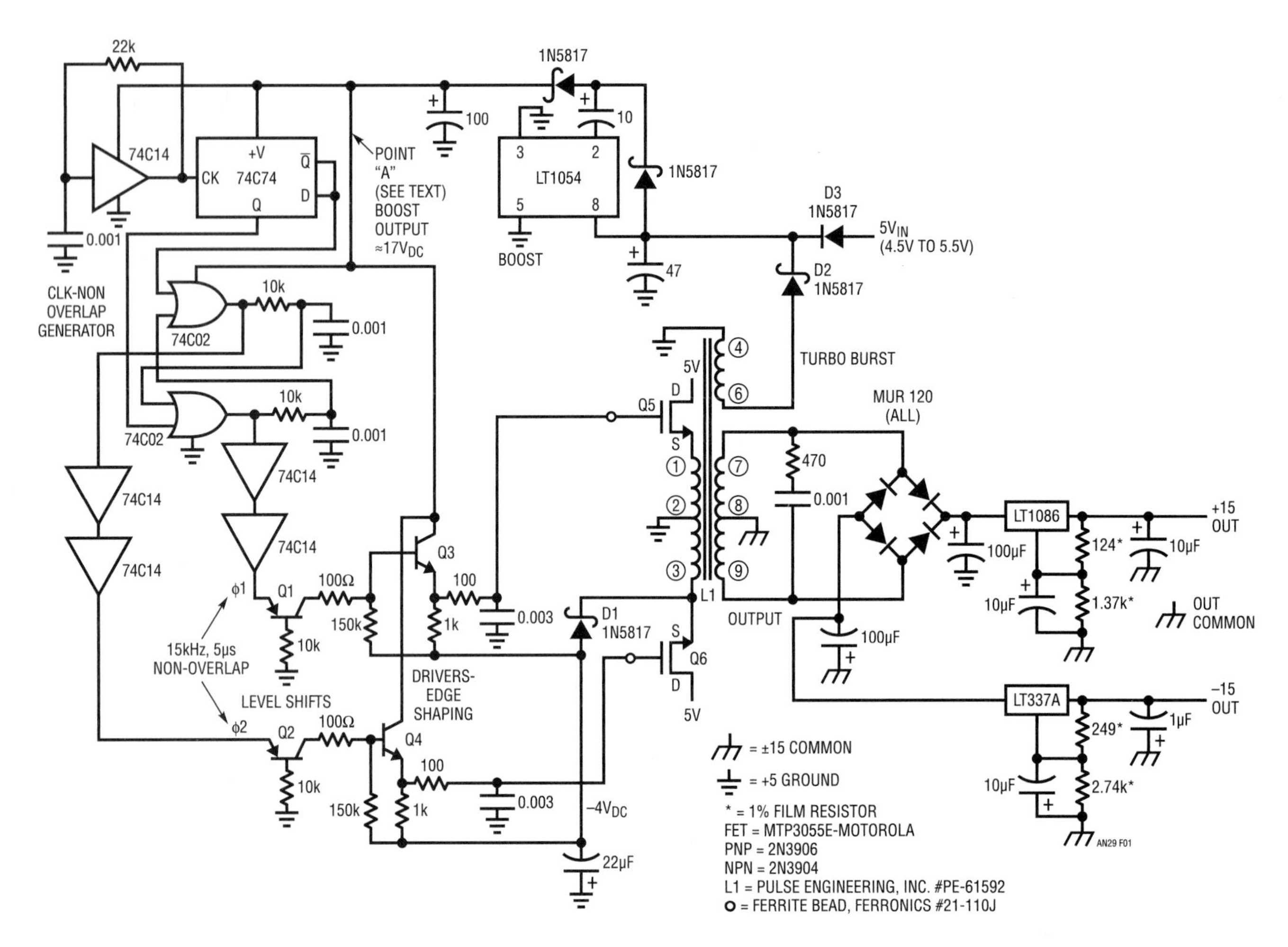

Figure 4.1 • Low Noise 5V to ±15V Converter

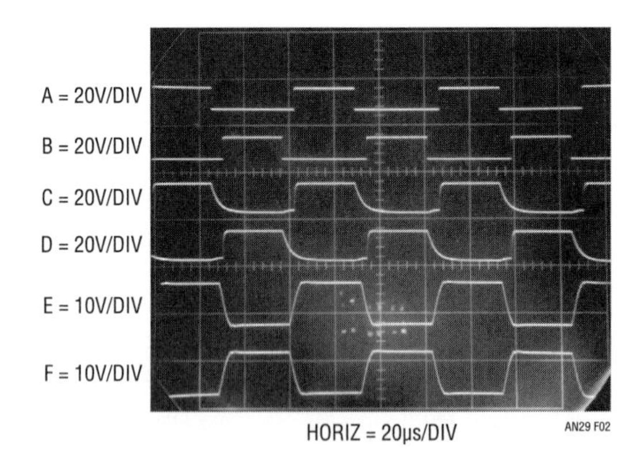

Figure 4.2 • 5V to ±15V Low Noise Converter Waveforms

complimentary, slew limited drive, eliminating the high speed harmonics normally associated with this type converter. L1's output is rectified, filtered and regulated to obtain the final output. The 470Ω-0.001µF damper in L1's output maintains loading during switching, aiding low noise performance. The ferrite beads in the gate leads eliminate parasitic RF oscillations associated with follower configurations.

The source follower configuration eases controlling L1's edge rise times, but complicates gate biasing. Special provisions are required to get the MOSFETs fully turned on and off. Source follower connected Q5 and Q6 require voltage overdrive at the gates to saturate. The 5V primary supply cannot provide the specified 10V gate—channel bias required for saturation. Similarly, the gates must be pulled well below ground to turn the MOSFETs off. This is so because L1's behavior pulls the sources negative when the devices turn off. Turn-off bias is bootstrapped from the negative side of Q6's source waveform. D1 and the 22μF capacitor produce a −4V potential for Q3 and Q4 to pull down to. Turn-on bias is generated by a 2-stage boost loop. The 5V supply is fed via D3 to the LT®1054 switched capacitor voltage converter (switched capacitor voltage converters are discussed in Appendix B, "Switched Capacitor Voltage Converters — How They Work"). The LT1054 configuration, set up as a voltage doubler, initially provides about 9V boost to point "A" at turn-on. When the converter starts running L1 produces output ("Turbo Boost" on schematic) at windings 4-6 which is rectified by D2, raising the LT1054's input voltage. This further raises point "A" to the 17V potential noted on the schematic.

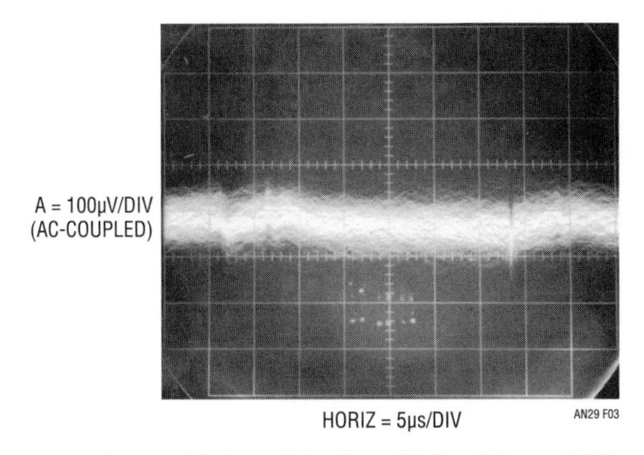

Figure 4.3 • Output Noise of the Low Noise 5V to ±15V Converter. Appendix H Shows a Modern IC Low Noise Regulator

These internally generated voltages allow Q5 and Q6 to receive proper drive, minimizing losses despite their source follower connection. Figure 4.3, an AC-coupled trace of the 15V converter output, shows 200μV_{P-P} noise at full power (250mA output). The −15V output shows nearly identical characteristics. Switching artifacts are comparable in amplitude to the linear regulators noise. Further reduction in switching based noise is possible by slowing Q5 and Q6 rise times. This, however, necessitates reducing clock rate and increasing non-overlap time to maintain available output power and efficiency. The arrangement shown represents a favorable compromise between output noise, available output power, and efficiency.

Ultralow noise 5V to ±15V converter

Residual switching components and regulator noise set Figure 4.1's performance limits. Analog circuitry operating at the very highest levels of resolution and sensitivity may require the lowest possible converter noise. Figure 4.4's converter uses sine wave transformer drive to reduce harmonics to negligible levels. The sine wave transformer drive combines with special output regulators to produce less than 30μV of output noise. This is almost 7× lower than the previous circuit and approaches a 1000× improvement over conventional designs. The trade off is efficiency and complexity.

A1 is set up as a 16kHz Wein bridge oscillator. The single power supply requires biasing to prevent A1's output from saturating at the ground rail. This bias is established by returning the undriven end of the Wein network to a DC potential derived from the LT1009 reference. A1's output is a pure sine wave (Figure 4.5, Trace A) biased off ground. A1's gain must be controlled to maintain sine wave output. A2 does this by comparing A1's rectified and filtered positive output peaks with an LT1009 derived DC reference. A2's output, biasing Q1, servo controls A1's gain. The 0.22μF capacitor frequency compensates the loop, and the thermally mated diodes minimize errors due to rectifier temperature drift. These provisions fix A1's AC and DC output terms against supply and temperature changes.

A1's output is AC coupled to A3. The 2k –820Ω divider re-biases the sine wave, centering it inside A3's input common mode range even with supply shifts. A3 drives a power stage, Q2-Q5. The stages common emitter outputs and biasing permit 1V_{RMS} (3V_{P-P}) transformer drive, even at V_{SUPPLY} = 4.5V. At full converter output loading the stage delivers 3 ampere peaks but the waveform is clean (Trace B), with low distortion (Trace C). The 330μF coupling capacitor strips DC and L3 sees pure AC. Feedback to A3 is taken at the Q4-Q5 collectors. The 0.1μF unit at this point suppresses local oscillations. L3's secondary RC network adds additional high frequency damping.

Without control of quiescent current the power stage will encounter thermal runaway and destroy itself. A4 measures DC output current across Q5's emitter resistor and servo controls Q6 to fix quiescent current. A divided portion of the LT1009 reference sets the servo point at A4's negative input and the 0.33μF feedback capacitor stabilizes the loop.

L3's rectified and filtered outputs are applied to regulators designed for low noise. A5 and A7 amplify the LT1021's filtered 10V output up to 15V. A6 and A8 provide the −15V output. The LT1021 and amplifiers give better noise performance than three terminal regulators. The Zener-resistor network clips overvoltages due to start-up transients.

L1 and L2 combine with their respective output capacitors to aid low noise characteristics. These inductors are outside the feedback loop, but their low copper resistance does not significantly degrade regulation. Trace D, the 15V

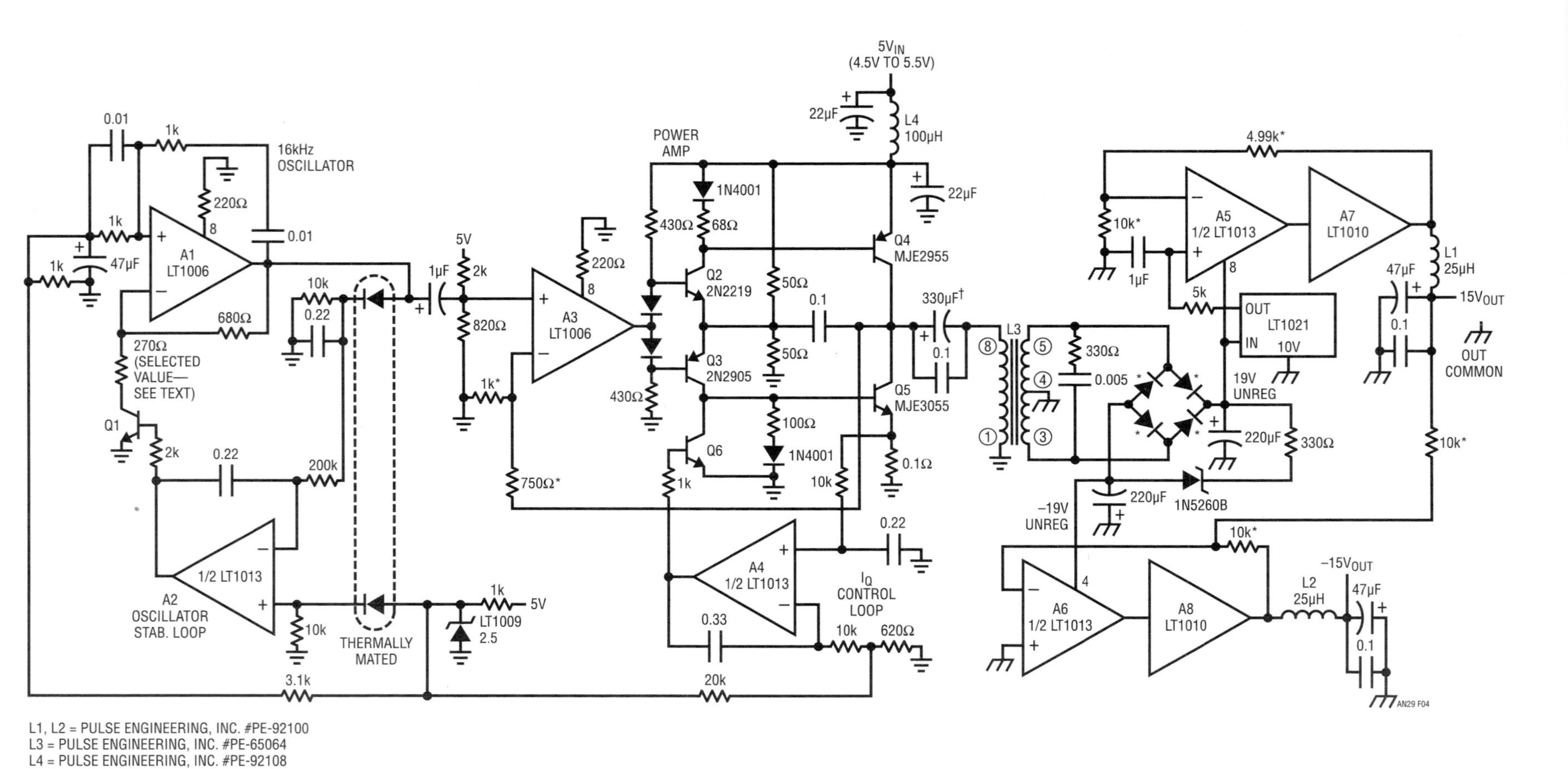

Figure 4.4 • Ultralow Noise Sine Wave Drive 5V to ±15V Converter

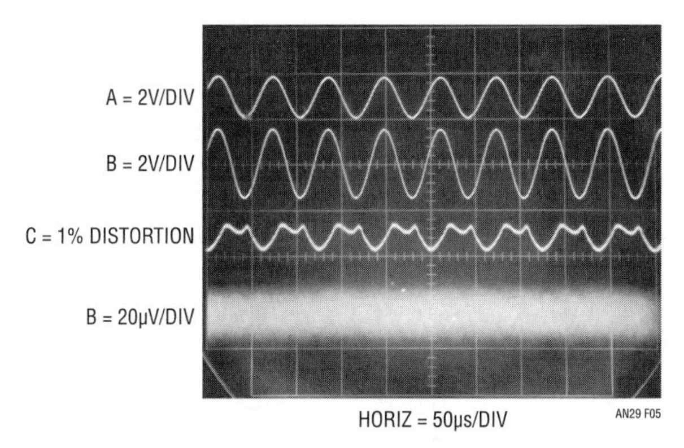

Figure 4.5 • Waveforms for the Sine Wave Driven Converter. Note that Output Noise (Trace D) is Ony 30μVP-P

output at full load, shows less than 30μV (2ppm) of noise. The most significant trade-off in this design is efficiency. The sine wave transformer drive forces substantial power loss. At full output (75mA), efficiency is only 30%.

Before use, the circuit should be trimmed for lowest distortion (typically 1%) in the sine wave delivered to L3. This trim is made by selecting the indicated value at A1's negative input. The 270Ω value shown is nominal, with a typical variance of ±25%. The sine wave's 16kHz frequency is a compromise between the op amps available gain bandwidth, magnetics size, audible noise, and minimization of wideband harmonics.

Single inductor 5V to ±15V converter

Simplicity and economy are another dimension in 5V to ±15V conversion. The transformer in these converters is usually the most expensive component. Figure 4.6's unusual drive scheme allows a single, 2-terminal inductor to replace the usual transformer at significant cost savings. Tradeoffs include loss of galvanic isolation between input and output and lower power output. Additionally, the regulation technique employed causes about 50mV of clock related output ripple.

The circuit functions by periodically and alternately allowing each end of the inductor to flyback. The resultant positive and negative peaks are rectified and filtered. Regulation is obtained by controlling the number of flyback events during the respective output's flyback interval.

The leftmost logic inverter produces a 20kHz clock (Trace A, Figure 4.7) which feeds a logic network composed of additional inverters, diodes and the 74C90 decade counter. The counter output (Trace B) combines with the logic network to present alternately phased clock bursts (Traces C and D) to the base resistors of Q1 and Q2. When ϕ1 (Trace B) is unclocked it resides in its high state, biasing Q2 and Q4 on. Q4's collector effectively grounds the "bottom" of L1 (Trace H). During this interval ϕ2 (Trace A) puts clock bursts into Q1's base resistor. If the −15V output is too low servo comparator C1A's output (Trace E) is high, and Q1's base can receive pulsed bias. If the converse is true the comparator will be low, and the bias gated away via Q1's base diode. When Q1 is able to bias, Q3 switches, resulting in negative going flyback events at the "top" of L1 (Trace G). These events are rectified and filtered to produce the −15V output. C1A regulates by controlling the number of clock pulses that switch the Q1-Q3 pair The LT1004 serves as a reference. Trace J, the AC-coupled −15V output, shows the effect of C1A's regulating action. The output stays within a small error window set by C1A's switched control loop. As input voltage and loading conditions change C1A adjusts the number of clock pulses allowed to bias Q1-Q3, maintaining loop control.

When the ϕ1 and **ϕ2** signals reverse state the operating sequence reverses. Q3's collector (Trace G) is pulled high with Q2-Q4 switching controlled by C1B's servo action. Operating waveforms are similar to the previous case. Trace F is C1B's output, Trace H is Q4's collector (L1's "bottom") and Trace I is the AC-coupled 15V output. Although the two regulating loops share the same inductor they operate independently, and asymmetrical output loading is not deleterious. The inductor sees irregularly spaced shots of current (Trace K), but is unaffected by its multiplexed operation. Clamp diodes prevent reverse biasing of Q3 and Q4 during transient conditions. The circuit provides ±25mA of regulated power at 60% efficiency.

Low quiescent current 5V to ±15V converter

A final area in 5V to ±15V converter design is reduction of quiescent current. Typical units pull 100mA to 150mA of quiescent current, unacceptable in many low power systems.

Figure 4.8's design supplies ±15V outputs at 100mA while consuming only 10mA quiescent current. The LT1070

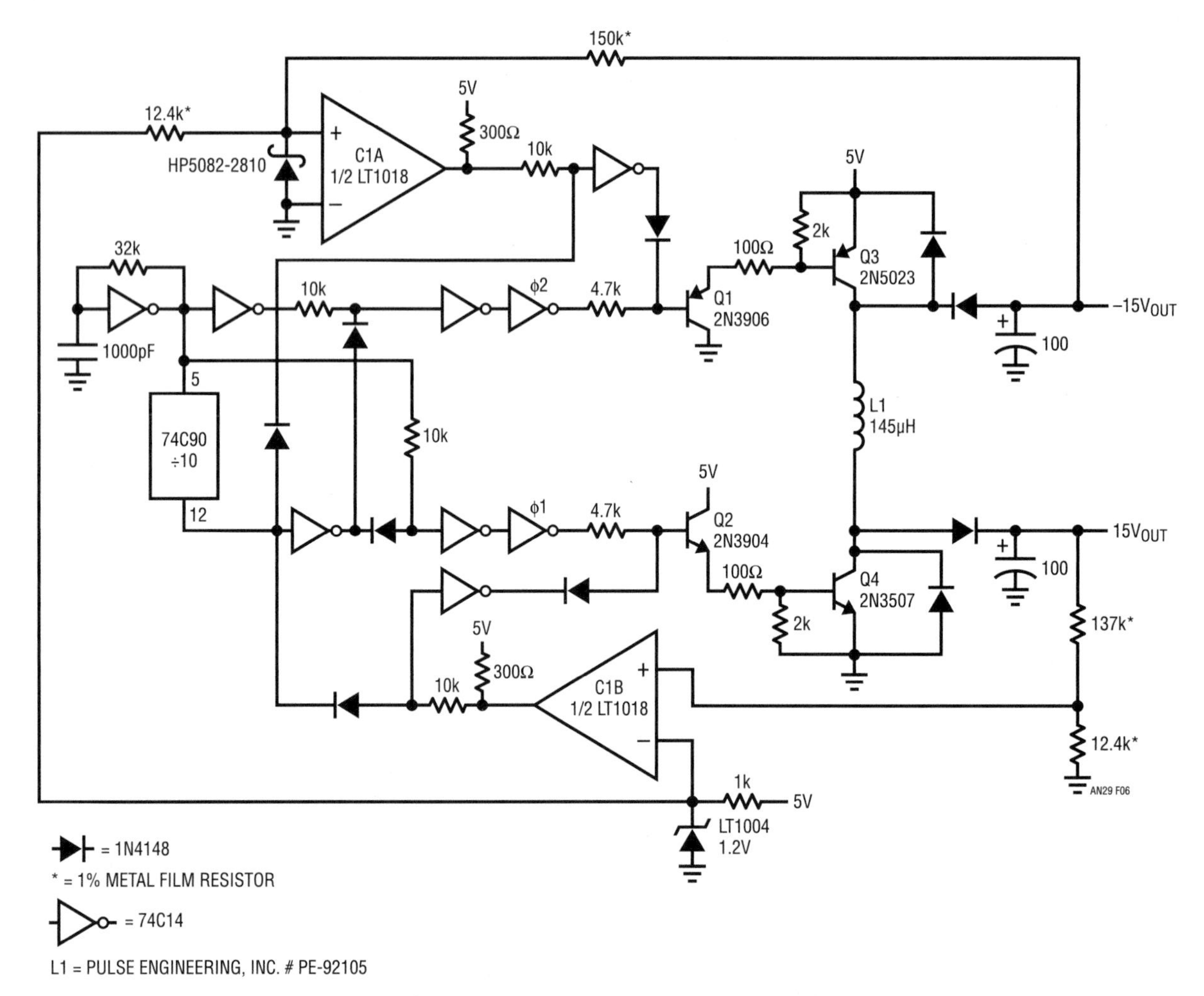

Figure 4.6 • Single Inductor 5V to ±15V Regulated Converter

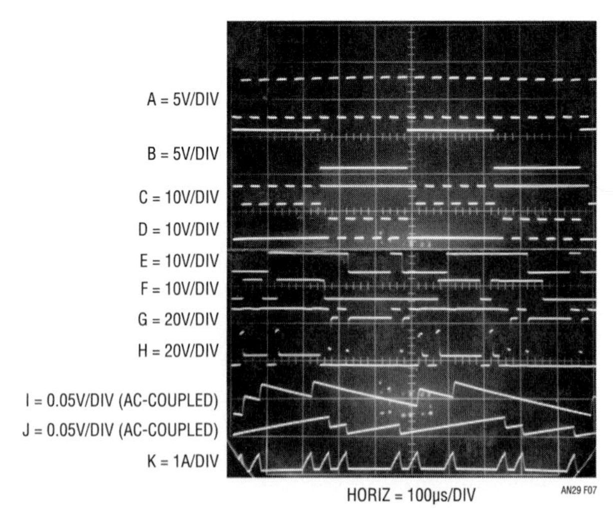

Figure 4.7 • Waveforms for the Single Inductor, Dual-Output, Regulated Converter

switching regulator (for a complete description of this device, see Appendix C, "Physiology of the LT1070") drives L1 in flyback mode. A damper network clamps excessive flyback voltages. Flyback events at L1's secondary are half-wave rectified and filtered, producing positive and negative outputs across the 47μF capacitors. The positive 16V output is regulated by a simple loop. Comparator C1A balances a sample of the positive output with a 2.5V reference obtained from the LT1020. When the 16V output (Trace A, Figure 4.9) is too low, C1A switches (Trace B) high, turning off the 4N46 opto-isolator. Q1 goes off, and the LT1070's control pin (V_C) pulls high (Trace C). This causes full duty cycle 40kHz switching at the V_{SW} pin (Trace D). The resultant energy into L1 forces the 16V output to ramp quickly positive, turning off C1A's output. The 20M value combined with the 4N46's slow response

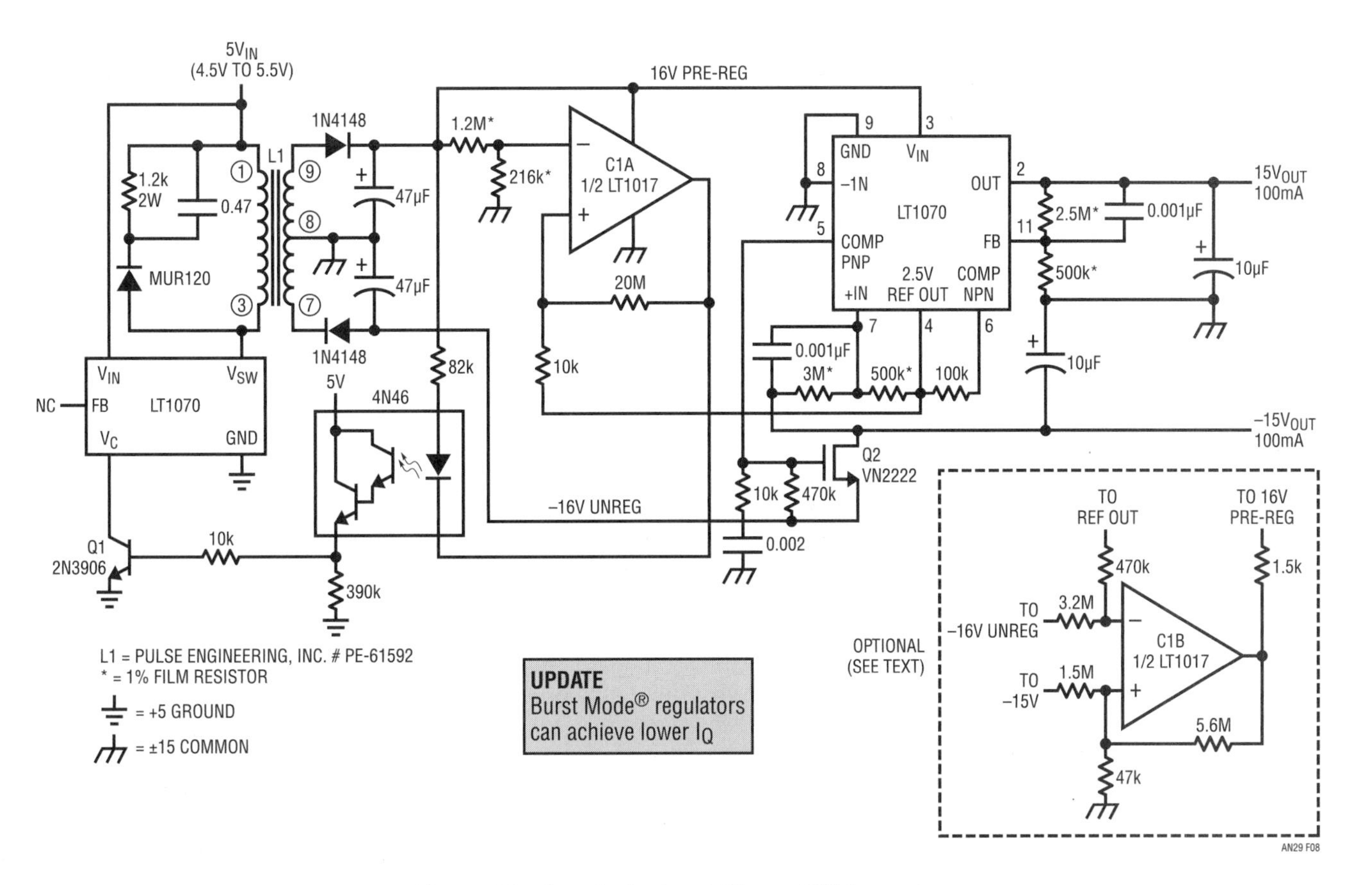

Figure 4.8 • Low I_Q, Isolated 5V to ±15V Converter

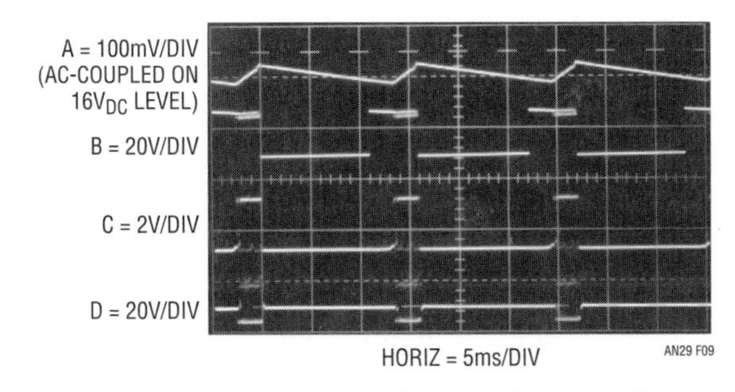

Figure 4.9 • Waveforms for the Low I_Q 5V to ±15V Converter

(note the delay between C1A going high and the V_C pin rise) gives about 40mV of hysteresis. The LT1070's on-off duty cycle is load dependent, saving significant power when the converter is lightly loaded. This characteristic is largely responsible for the 10mA quiescent current. The opto-isolator preserves the converters input-output isolation. The LT1020, a low quiescent current regulator with low dropout, further regulates the 16V line, giving the 15V output. The linear regulation eliminates the 40mV ripple and improves transient response. The −16V output tends to follow the regulated −16V line, but regulation is poor. The LT1020's auxiliary onboard comparator is compensated to function as an op amp by the RC damper at Pin 5. This amplifier linearly regulates the −16V line. MOSFET Q2 provides low dropout current boost, sourcing the −15V output. The −15V output is stabilized with the op amp by comparing it with the 2.5V reference via the 500k-3M current summing resistors. 1000pF capacitors frequency compensate each regulating loop. This converter functions well, providing ±15V outputs at 100mA with only 10mA quiescent current. Figure 4.10 plots efficiency versus a conventional design over a range of loads. For high loads results are comparable, but the low quiescent circuit is superior at lower current.

A possible problem with this circuit is related to the poor regulation of the −16V line. If the positive output is lightly loaded L1's magnetic flux is low. Heavy negative output loading under this condition results in the −16V line falling below its output regulators dropout value. Specifically, with no load on the 15V output only 20mA is available from the −15V output. The full 100mA is only

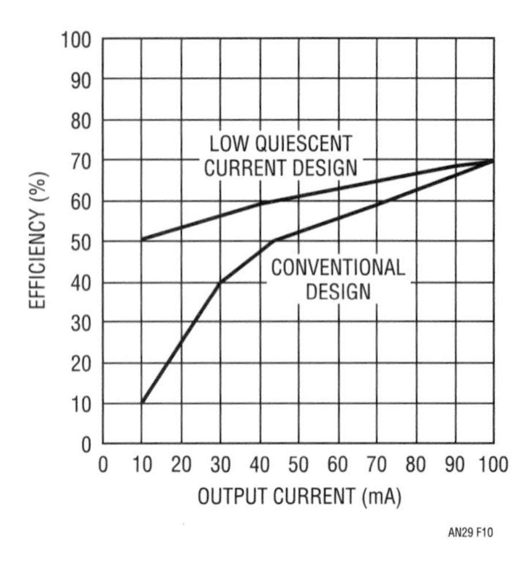

Figure 4.10 • Efficiency vs Load for the Low I_Q Converter

available from the −15V output when the 15V output is supplying more than 8mA. This restriction is often acceptable, but some situations may not tolerate it. The optional connection in Figure 4.8 (shown in dashed lines) corrects the difficulty. C1B detects the onset of −16V line decay. When this occurs its output pulls low, loading the 16V line to correct the problem. The biasing values given permit correction before the negative linear regulator drops out.

Micropower quiescent current converters

Many battery-powered applications require very wide ranges of power supply output current. Normal conditions require currents in the ampere range, while standby or "sleep" modes draw only microamperes. A typical laptop computer may draw 1 to 2 amperes running while needing only a few hundred microamps for memory when turned off. In theory, any DC/DC converter designed for loop stability under no-load conditions will work. In practice, a converter's relatively large quiescent current may cause unacceptable battery drain during low output current intervals.

Figure 4.11 shows a typical flyback based converter. In this case the 6V battery is converted to a 12V output by the inductive flyback voltage produced each time the LT1070's V_{SW} pin is internally switched to ground (for commentary on inductor selection in flyback converters see Appendix D, "Inductor Selection for Flyback Converters"). An internal 40kHz clock produces a flyback event every 25μs. The energy in this event is controlled by the IC's internal error amplifie r, which acts to force the feedback (FB) pin to a 1.23V reference. The error amplifiers high impedance output (the V_C pin) uses an RC damper for stable loop compensation.

This circuit works well but pulls 9mA of quiescent current. If battery capacity is limited by size or weight this may be too high. How can this figure be reduced while retaining high current performance?

A solution is suggested by considering an auxiliary V_C pin function. If the V_C pin is pulled within 150mV of ground the IC shuts down, pulling only 50 microamperes. Figure 4.12's special loop exploits this feature, reducing quiescent current to only 150 microamperes. The technique shown is particularly significant, with broad implication in battery powered systems. It is easily applied to a wide variety of DC/DC converters, meeting an acknowledged need across a wide spectrum of applications.

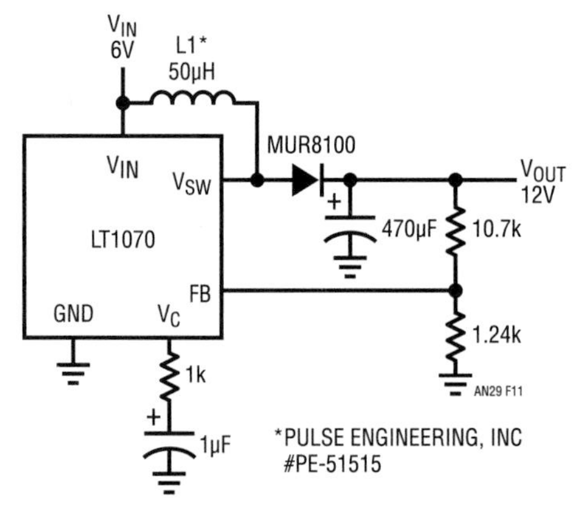

Figure 4.11 • 6V to 12V, 2 Amp Converter with 9mA Quiescent Current

Figure 4.12's signal flow is similar to Figure 4.11, but additional circuitry appears between the feedback divider and the V_C pin. The LT1070's internal feedback amplifier and reference are not used. Figure 4.13 shows operating waveforms under no-load conditions. The 12V output (Trace A) ramps down over a period of seconds. During this time comparator A1's output (Trace B) is low, as are the 74C04 paralleled inverters. This pulls the V_C pin (Trace C) low, putting the IC in its 50μA shutdown mode. The V_{SW} pin (Trace D) is high, and no inductor current flows. When the 12V output drops about 20mV, A1 triggers and the inverters go high, pulling the V_C pin up and turning on the regulator. The V_{SW} pin pulses the inductor at the 40kHz clock rate, causing the output to abruptly rise. This action trips A1 low, forcing the V_C pin back into shutdown. This "bang-bang" control loop keeps the 12V output within the 20mV ramp hysteresis window set by R3-R4. Diode clamps prevent V_C pin overdrive.

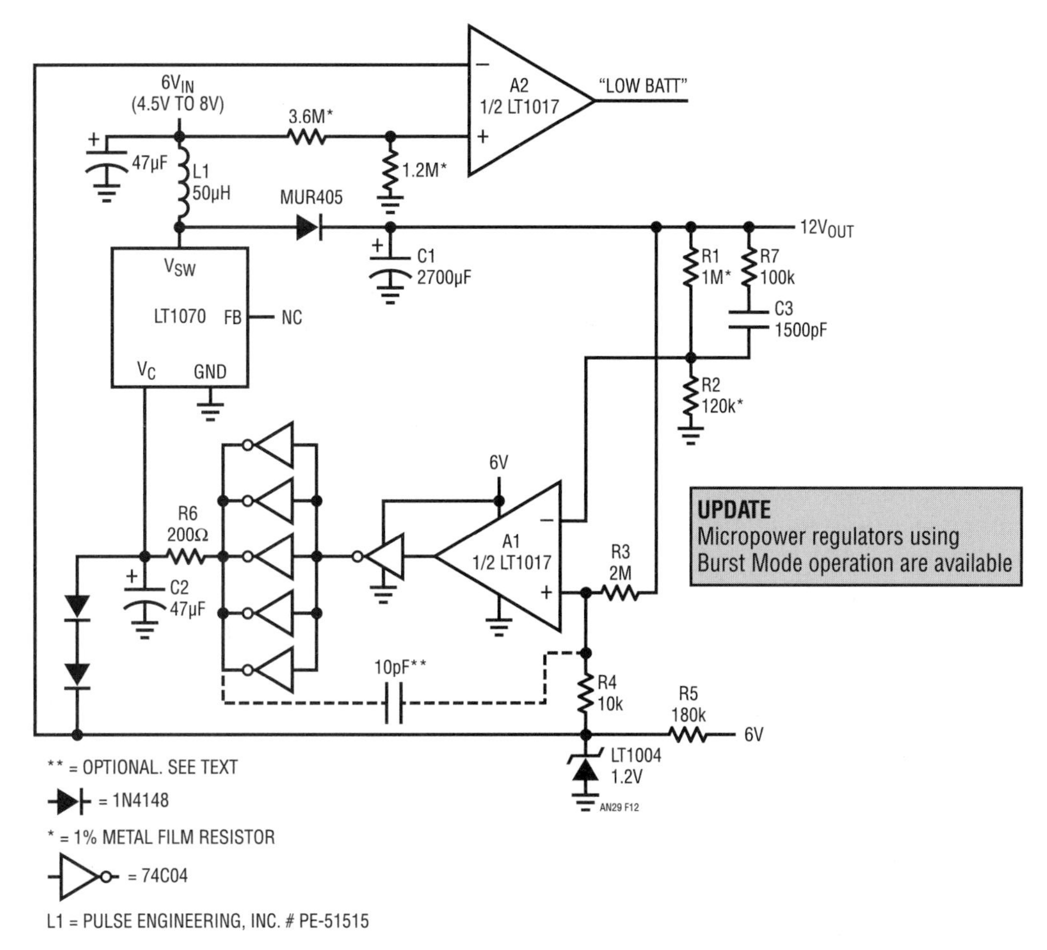

Figure 4.12 • 6V to 12V, 2 Amp Converter with 150µA Quiescent Current

Note that the loop oscillation period of 4 to 5 seconds means the R6-C2 time constant at V_C is not a significant term. Because the LT1070 spends almost all of the time in shutdown, very little quiescent current (150µA) is drawn.

Figure 4.14 shows the same waveforms with the load increased to 3mA. Loop oscillation frequency increases to keep up with the loads sink current demand. Now, the V_C pin waveform (Trace C) begins to take on a filtered appearance. This is due to R6-C2's 10ms time constant. If the load continues to increase, loop oscillation frequency will also increase. The R6-C2 time constant, however, is fixed. Beyond some frequency, R6-C2 must average loop oscillations to DC. Figure 4.15 shows the same circuit points at 1 ampere loading. Note that the V_C pin is at DC, and repetition rate has increased to the LT1070's 40kHz clock frequency. Figure 4.16 plots what is occurring, with a pleasant surprise. As output current rises, loop oscillation frequency also rises until about 500Hz. At this point the R6-C2 time

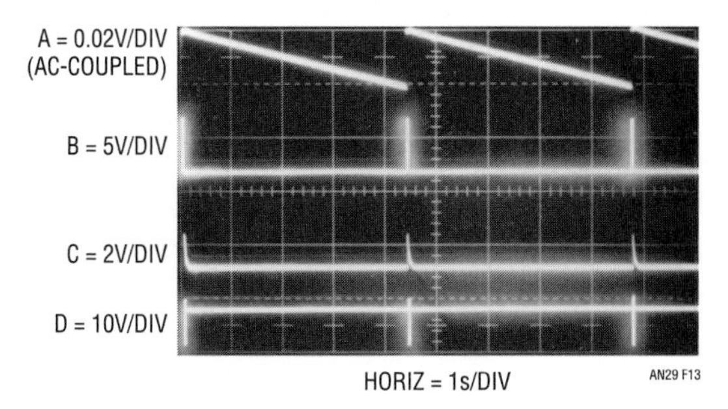

Figure 4.13 • Low I_Q Converter Waveforms with No Load (Traces B and D Retouched for Clarity)

constant filters the V_C pin to DC and the LT1070 transitions into "normal" operation. With the V_C pin at DC it is convenient to think of A1 and the inverters as a linear error amplifier with a closed-loop gain set by the R1-R2 feedback divider. In fact, A1 is still duty cycle modulating, but at a rate far above R6-C2's break frequency. The phase error contributed by C1 (which was selected for low loop frequency at low output currents) is dominated by the R6-C2 roll off and the R7-C3 lead into A1. The loop is stable and responds linearly for all loads beyond 80mA. In this high current region the LT1070 is desirably "fooled" into behaving like Figure 4.11's circuit.

A formal stability analysis for this circuit is quite complex, but some simplifications lend insight into loop operation. At 100μA loading (120kΩ) C1 and the load form a decay time constant exceeding 300 seconds. This is orders of magnitude larger than R7-C3, R6-C2, or the LT1070's 40kHz commutation rate. As a result, C1 dominates the loop. Wideband A1 sees phase shifted feedback, and very low frequency oscillations similar to Figure 4.13's occur[1]. Although C1's *decay* time constant is long, its *charge* time constant is short because the circuit has low sourc-ing impedance. This accounts for the ramp nature of the oscillations.

Increased loading reduces the C1 load decay time constant. Figure 4.16's plot reflects this. As loading increases, the loop oscillates at a higher frequency due to C1's decreased decay time. When the load impedance becomes low enough C1's decay time constant ceases to dominate the loop. This point is almost entirely determined by R6 and C2. Once R6 and C2 "take over" as the dominant time constant the loop begins to behave like a linear system. In this region (e.g. above about 75mA, per Figure 4.16) the LT1070 runs continuously at its 40kHz rate. Now, the R7-C3 time constant becomes significant, performing as a simple feedback lead[2] to smooth output response. There is a fundamental trade-off in the selection of the R7-C3 lead network values. When the converter is running in its linear region they must dominate the DC hysteresis deliberately generated by R3-R4. As such, they have been chosen for the best compromise between output ripple at high load and loop transient response.

Despite the complex dynamics transient response is quite good. Figure 4.17 shows performance for a step from no load to 1 ampere. When Trace A goes high a 1 ampere load appears across the output (Trace B). Initially, the output sags almost 150mV due to slow loop response time (the R6-C2 pair delay V_C pin response). When the LT1070 comes on (signaled by the 40kHz "fuzz" at the bottom extreme of Trace B) response is reasonably quick

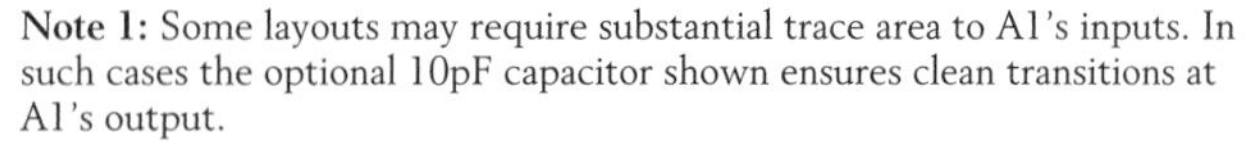

Note 1: Some layouts may require substantial trace area to A1's inputs. In such cases the optional 10pF capacitor shown ensures clean transitions at A1's output.

Note 2: "Zero Compensation" for all you technosnobs out there.

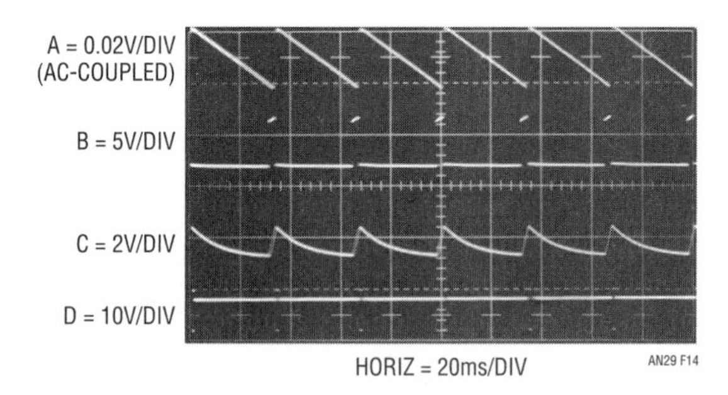

Figure 4.14 • Low I_Q Converter Waveforms at Light Loading

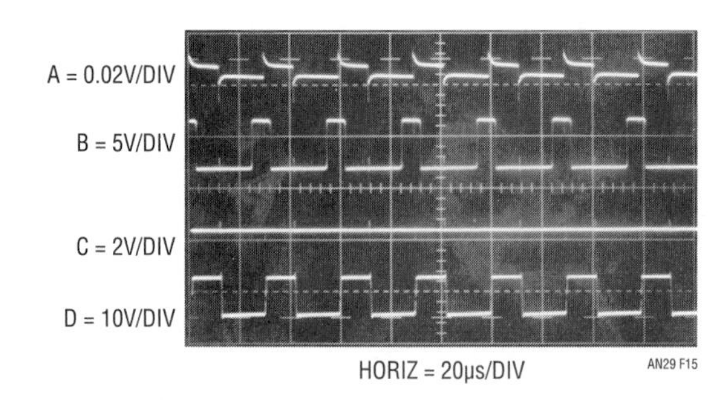

Figure 4.15 • Low I_Q Converter Waveforms at 1 Amp Loading

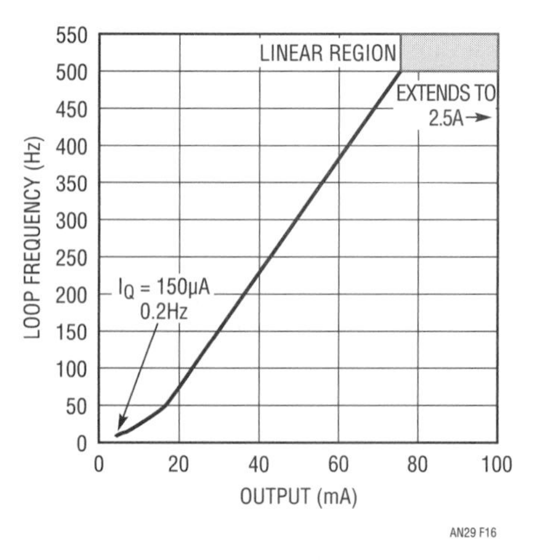

Figure 4.16 • Figure 4.12's Loop Frequency vs Output Current. Note Linear Loop Operation Above 80mA

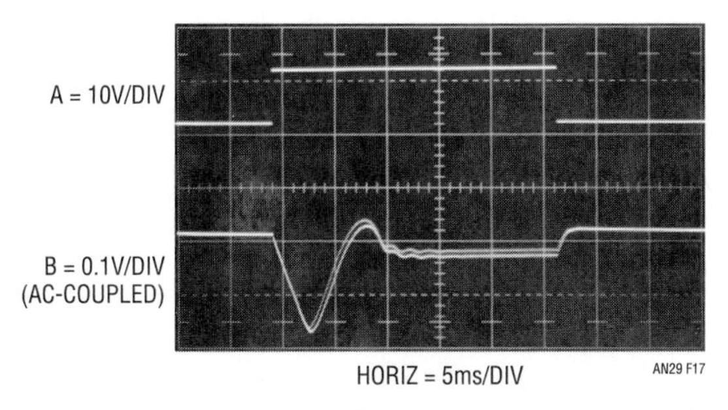

Figure 4.17 • Load Transient Response for Figure 4.12's Low I_Q Regulator

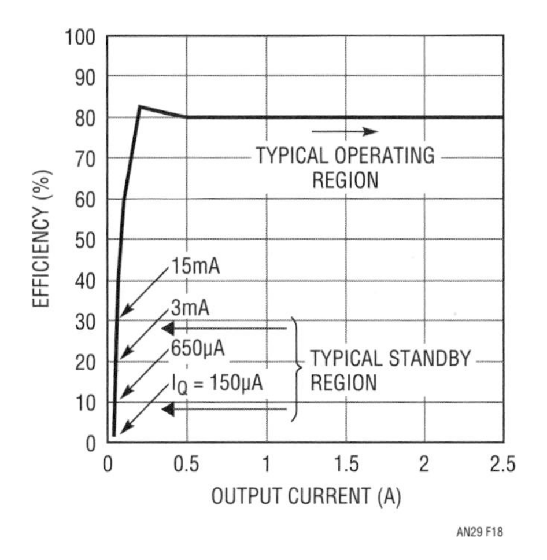

Figure 4.18 • Efficiency vs Output Current for Figure 4.12. Standby Efficiency is Poor, But Power Loss Approaches Battery Self Discharge

and surprisingly well behaved considering circuit dynamics. The multi-time constant decay[3] ("rattling" is perhaps more appropriate) is visible as Trace B approaches steady state between the 4th and 5th vertical divisions.

A2 functions as a simple low-battery detector, pulling low when V_{IN} drops below 4.8V.

Figure 4.18 plots efficiency versus output current. High power efficiency is similar to standard converters. Low power efficiency is somewhat better, although poor in the lowest ranges. This is not particularly bothersome, as *power* loss is very small.

This loop provides a controlled, conditional instability instead of the more usually desirable (and often elusive) unconditional stability. This deliberately introduced characteristic lowers converter quiescent current by a factor of 60 without sacrificing high power performance. Although demonstrated in a boost converter, it is readily exportable to other configurations. Figure 4.19a's step-down (buck mode) configuration uses the same basic loop with almost no component changes. P-channel MOSFET Q1 is driven from the LT1072 (a low power version of the LT1070) to convert 12V to a 5V output. Q2 and Q3 provide current limiting, while Q4 supplies turn off drive to Q1. the lower output voltage mandates slightly different hysteresis biasing than Figure 4.12, accounting for the 1MΩ value at the comparators positive input. In other respects the loop and its performance are identical. Figure 4.19b uses the loop in a transformer based multi-output converter. Note that the floating secondaries allow a −12V output to be obtained with a positive voltage regulator.

Low quiescent current micropower 1.5V to 5V converter

Figure 4.20 extends our study of low quiescent current converters into the low voltage, micropower domain. In some circumstances, due to space or reliability considerations, it is preferable to operate circuitry from a single 1.5V cell. This eliminates almost all ICs as design candidates. Although it is possible to design circuitry which runs directly from a single cell (see LTC® Application Note 15, "Circuitry For Single Cell Operation") a DC/DC converter permits using higher voltage ICs. Figure 4.20's design converts a single 1.5V cell to a 5V output with only 125µA quiescent current. Oscillator C1A's output is a 2kHz square wave (Trace D, Figure 4.21). The configuration is conventional, except that the biasing accommodates the narrow common mode range dictated by the 1.5V supply. To maintain low power, C1A's integrating capacitor is small, with only 50mV of swing. The parallel connected sides of C2 drive L1. When the 5V output (Trace A) coasts down far enough C1B goes low (Trace B), pulling both C2 positive inputs close to ground. C1A's clock now appears at the paralleled C2 outputs (Trace C), forcing energy into L1. The paralleled outputs minimize saturation losses. L1's flyback pulses, rectified and stored in the 47µF capacitor, form the circuits DC output. C1B on-off modulates C2 at whatever duty cycle is required to maintain the circuits 5V output. The LT1004 is the reference, with the resistor divider at C1B's positive input setting the output voltage. Schottky clamping of C2's outputs prevents negative going overdrives due to parasitic L1 behavior.

Note 3: Once again, "multi-pole settling" for those who adore jargon.

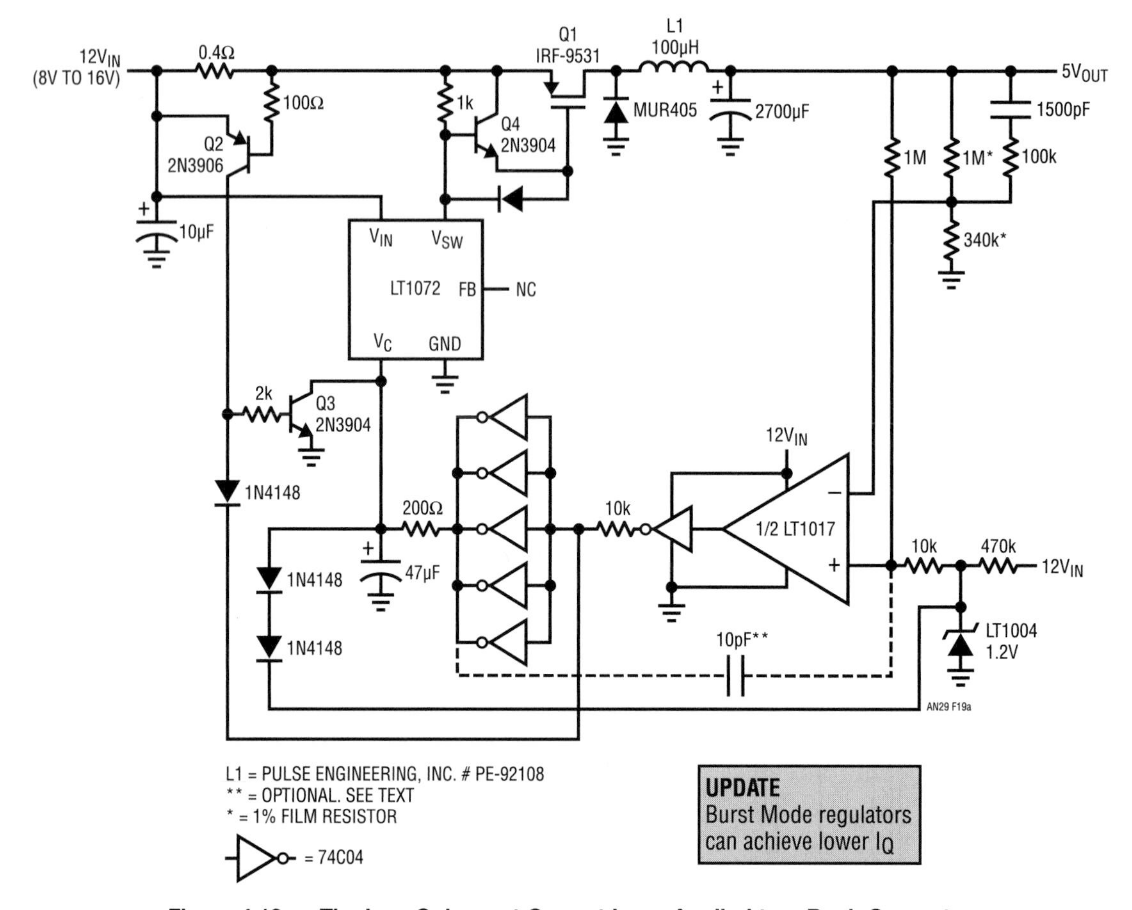

Figure 4.19a • The Low Quiescent Current Loop Applied to a Buck Converter

The 1.2V LT1004 reference biasing is bootstrapped to the 5V output, permitting circuit operation down to 1.1V. A 10M bleed to supply ensures start-up. The 1M resistors divide down the 1.2V reference, keeping C1B inside common mode limits. C1B's positive feedback RC pair sets about 100mV hysteresis and the 22pF unit suppresses high frequency oscillation.

The micropower comparators and very low duty cycles at light load minimize quiescent current. The 125μA figure noted is quite close to the LT1017's steady-state currents. As load increases the duty cycle rises to meet the demand, requiring more battery power. Decrease in battery voltage produces similar behavior. Figure 4.22 plots available output current versus battery voltage. Predictably, the highest power is available with a fresh cell (e.g., 1.5V to 1.6V), although regulation is maintained down to 1.15V for 250μA loading. The plot shows that the test circuit continued to regulate below this point, but this cannot be relied on in practice (LT1017 V_{MIN} = 1.15V). The low supply voltage makes saturation and other losses in this circuit difficult to control. As such, efficiency is about 50%.

The optional connection in Figure 4.20 (shown in dashed lines) takes advantage of the transformers floating secondary to furnish a −5V output. Drive circuitry is identical, but C1B is rearranged as a current summing comparator. The LT1004's bootstrapped positive bias is supplied by L1's primary flyback spikes.

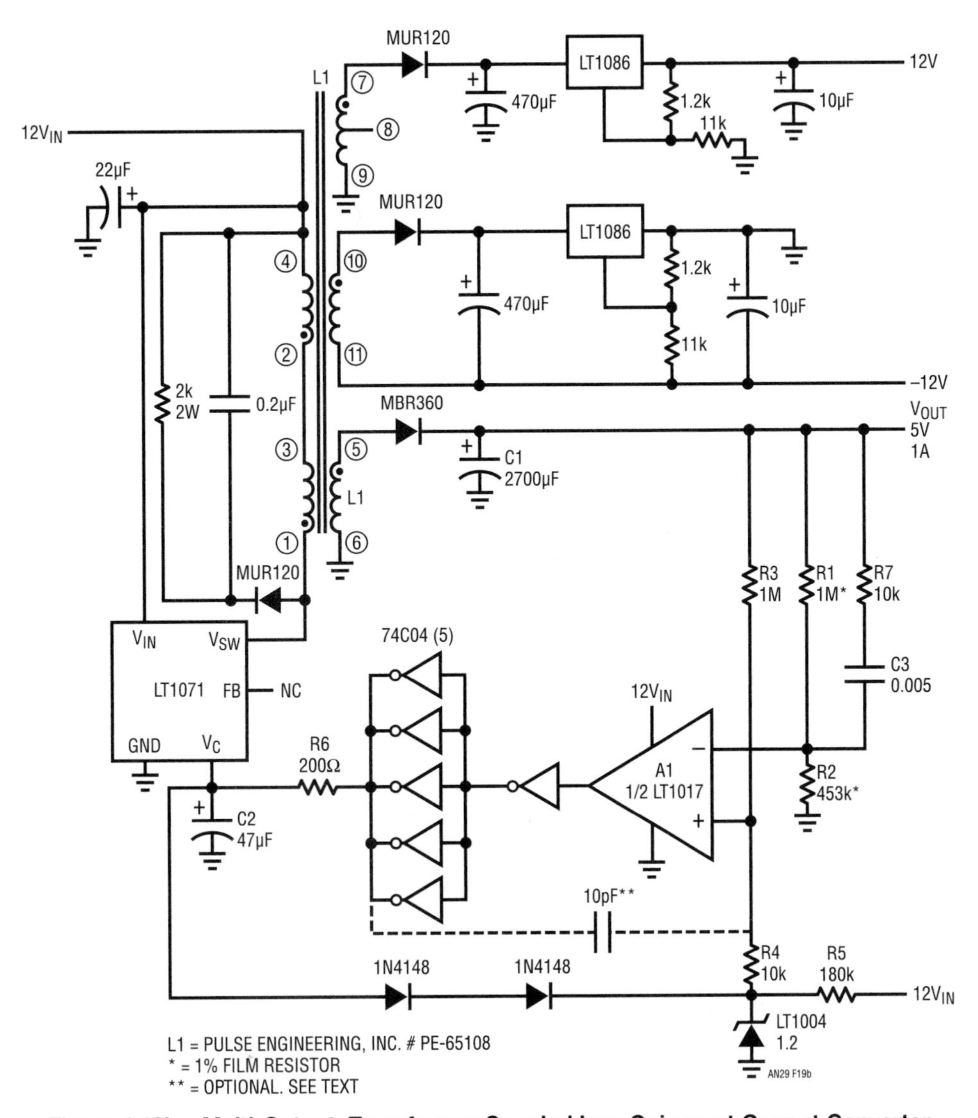

Figure 4.19b • Multi-Output, Transformer Coupled Low Quiescent Current Converter

200mA output 1.5V to 5V converter

Although useful, the preceding circuit is limited to low power operation. Some 1.5V powered systems (survival 2-way radios, remote, transducer fed data acquisition systems, etc.) require much more power. Figure 4.23's design supplies a 5V output with 200mA capacity. Some sacrifice in quiescent current is made in this circuit. This is predicated on the assumption that it operates continuously at high power. If lowest quiescent current is necessary the technique detailed back in Figure 4.12 is applicable.

The circuit is essentially a flyback regulator, similar to Figure 4.11. The LT1070's low saturation losses and ease of use permit high power operation and design simplicity. Unfortunately, this device has a 3V minimum supply requirement. Bootstrapping its supply pin from the 5V output is possible, but requires some form of start-up mechanism. Dual comparator C1 and the transistors form a start-up loop. When power is applied C1A oscillates (Trace A, Figure 4.24) at 5kHz. Q1 biases, driving Q2's base hard. Q2's collector (Trace B) pumps L1, causing voltage step-up flyback events. These events are rectified and stored in the 500µF capacitor, producing the circuit's DC output. C1B is set up so it (Trace C) goes low when circuit output crosses about 4.5V. When this occurs C1A's integration capacitor is pulled low, stopping it from oscillating. Under these conditions Q2 can no longer drive L1, but the LT1070 can. This behavior is observable at the LT1070's V_{SW} pin (the junction of L1, Q2's collector and the LT1070), Trace D. When the start-up circuit goes off, the LT1070 V_{IN} pin has adequate supply voltage and

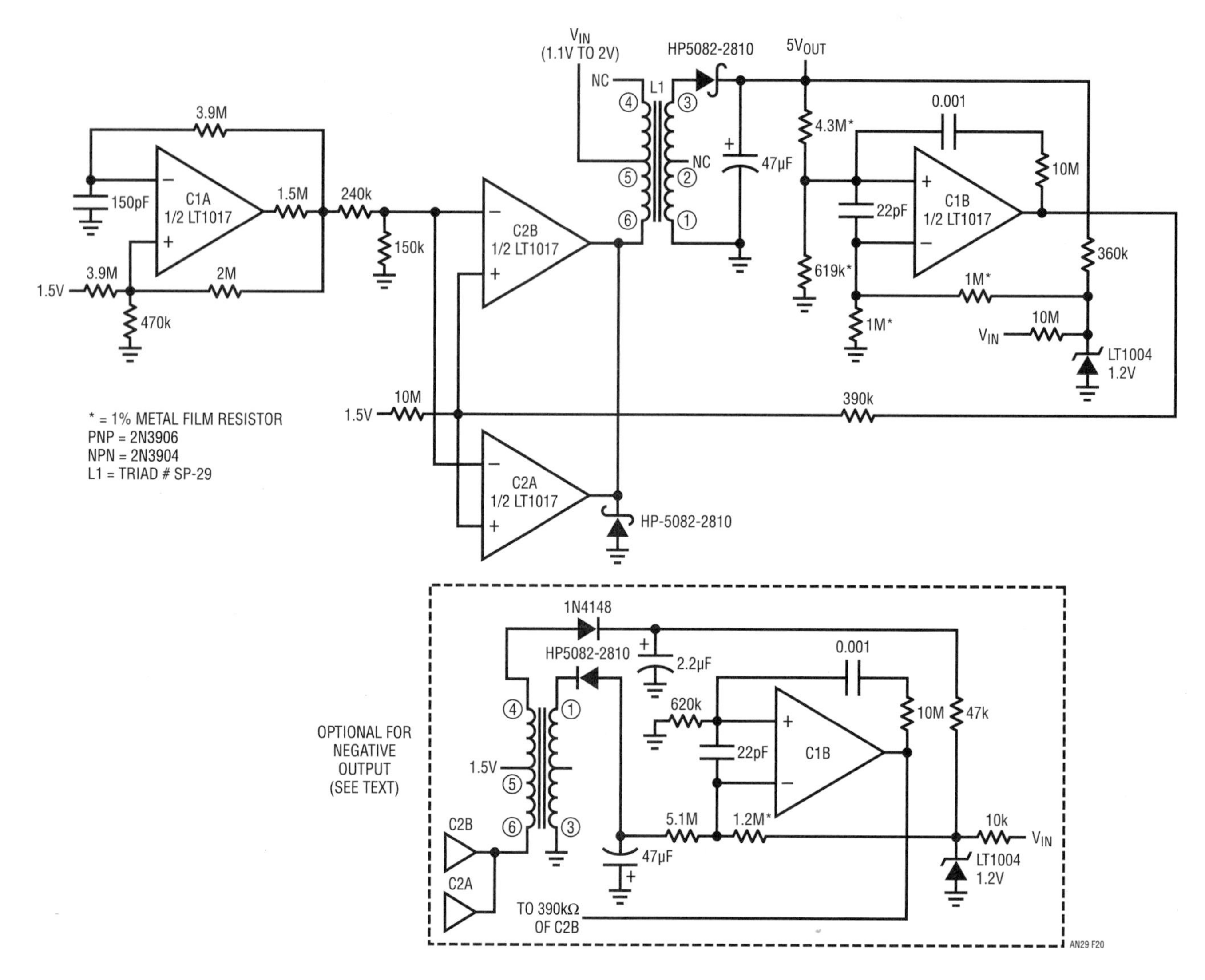

Figure 4.20 • 800μA Output 1.5V to 5V Converter

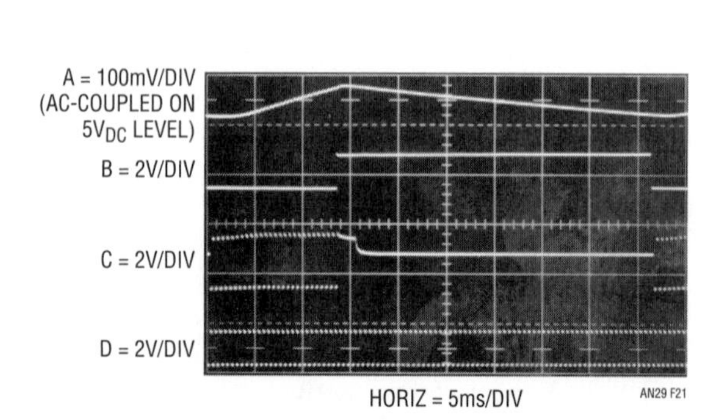

Figure 4.21 • Waveforms for Low Power 1.5V to 5V Converter

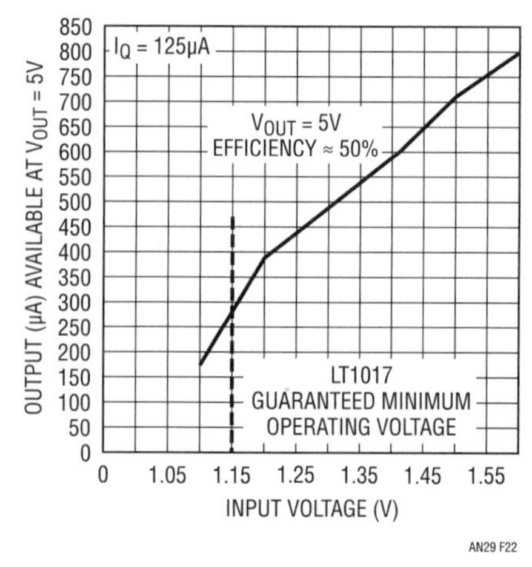

Figure 4.22 • Output Current Capability vs Input Voltage for Figure 4.20

it begins operation. This occurs at the 4th vertical division of the photograph. There is some overlap between start-up loop turn-off and LT1070 turn-on, but it has no detrimental effect. Once the circuit is running it functions similarly to Figure 4.11.

The start-up loop must be carefully designed to function over a wide range of loads and battery voltages. Start-up currents exceed 1 ampere, necessitating attention to Q2's saturation and drive characteristics. The worst case is a nearly depleted battery and heavy output loading. Figure 4.25 shows circuit output starting into a 100mA load at $V_{BATTERY} = 1.2V$. The sequence is clean, and the LT1070 takes over at the appropriate point. In Figure 4.26, loading is increased to 200mA. Start-up slope decreases, but starting still occurs. The abrupt slope increase (6th vertical division) is due to overlapping operation of the start-up loop and the LT1070.

Figure 4.27 plots input-output characteristics for the circuit. Note that the circuit will start into all loads with $V_{BATTERY} = 1.2V$. Start-up is possible down to 1.0V at reduced loads. Once the circuit has started, the plot shows it will drive full 200mA loads down to $V_{BATTERY} = 1.0V$. Reduced drive is possible down to $V_{BATTERY} = 0.6V$ (a very dead battery)! Figures 4.28 and 4.29, dynamic XY crossplot versions of Figure 4.27, are taken at 20 and 200 milliamperes, respectively. Figure 4.30 graphs efficiency at two supply voltages over a range of output currents. Performance is attractive, although at lower currents circuit quiescent power degrades efficiency. Fixed junction saturation losses are responsible for lower overall efficiency at the lower supply voltage. Figure 4.31 shows quiescent current increasing as supply decays. Longer inductor current charge intervals are necessary to compensate the decreased supply voltage.

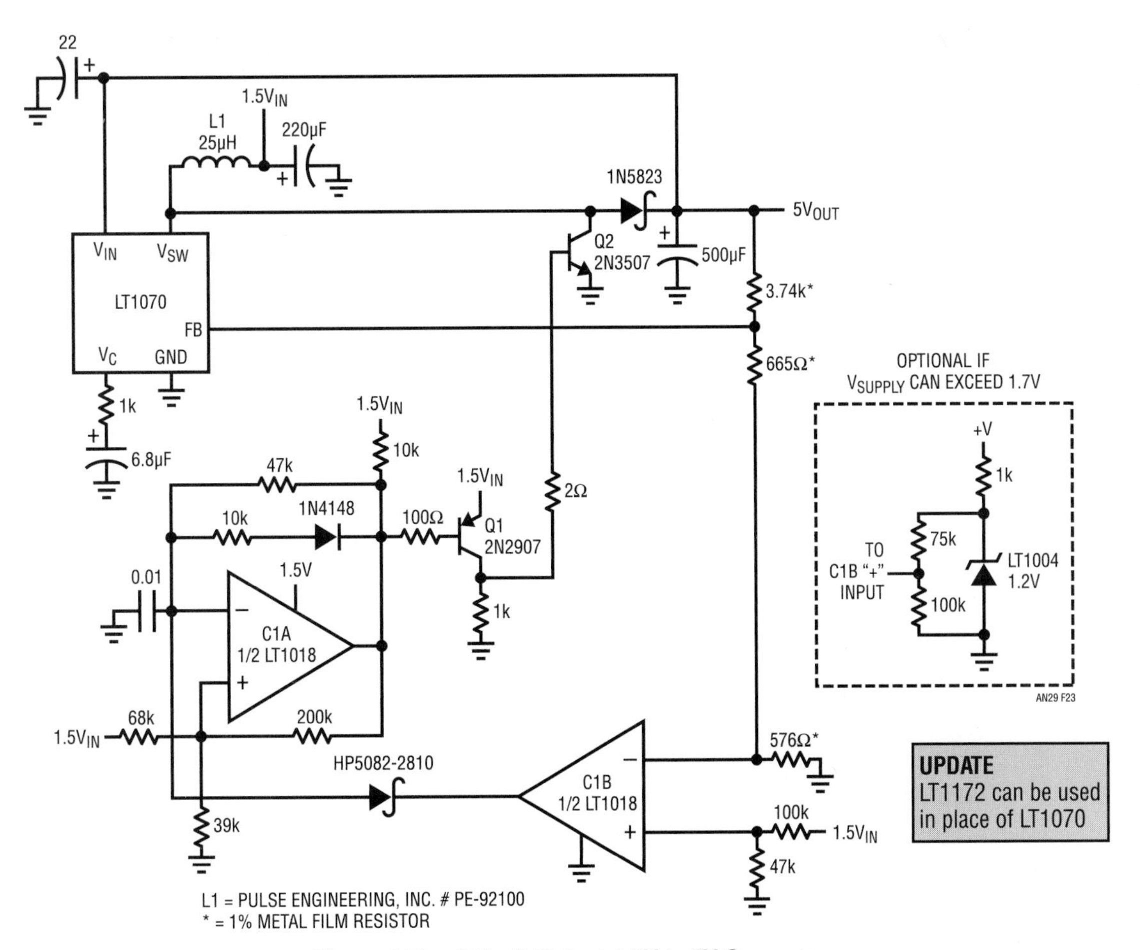

Figure 4.23 • 200mA Output 1.5V to 5V Converter

High efficiency converters

High efficiency 12V to 5V converter

Efficiency is sometimes a prime concern in DC/DC converter design (see Appendix E, "Optimizing Converters for Efficiency"). In particular, small portable computers frequently use a 12V primary supply which must be converted down to 5V. A 12V battery is attractive because it offers long life when all trade-offs and sources of loss are considered. Figure 4.32 achieves 90% efficiency. This circuit can be recognized as a positive buck converter. Transistor Q1 serves as the pass element. The catch diode is replaced with a synchronous rectifier, Q2, for improved efficiency. The input supply is nominally 12V but can vary from 9.5V to 14.5V. Power losses are minimized by utilizing low source-to-drain resistance, 0.028Ω, NMOS transistors for the catch diode and pass element. The inductor, Pulse Engineering PE-92210K, is made from a low loss core material which squeezes a little more efficiency out of the circuit. Also, keeping the current sense threshold voltage low minimizes the power lost in the current limit circuit. Figure 4.33 shows the operating waveforms. Q5 drives the synchronous rectifier, Q2, when the V_{SW} pin (Trace A) is turned "off". Q2 is turned off through D1 and D2 when the V_{SW} pin

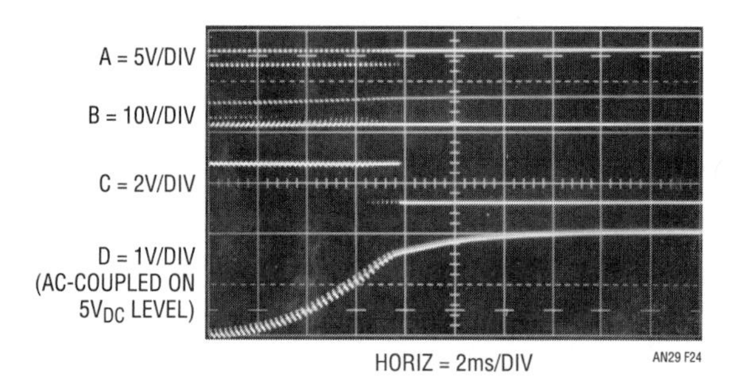

Figure 4.24 • High Power 1.5V to 5V Converter Start-Up Sequence

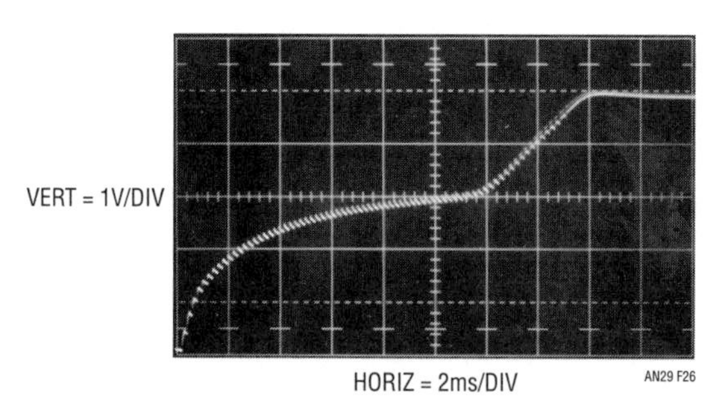

Figure 4.26 • High Power 1.5V to 5V Converter Turn-On Into a 200mA Load at V_{BATT} = 1.2V

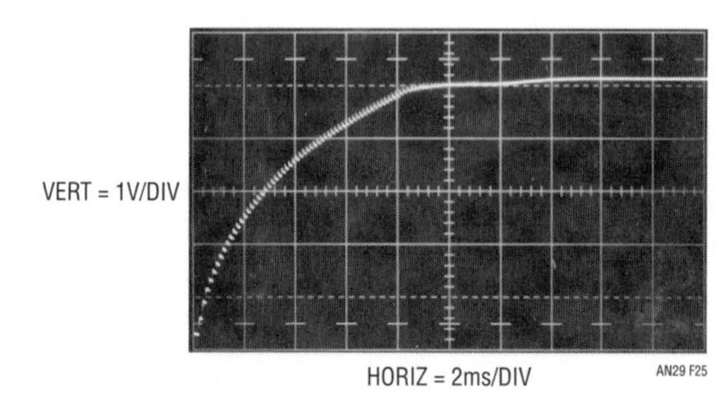

Figure 4.25 • High Power 1.5V to 5V Converter Turn-On Into a 100mA Load at V_{BATT} = 1.2V

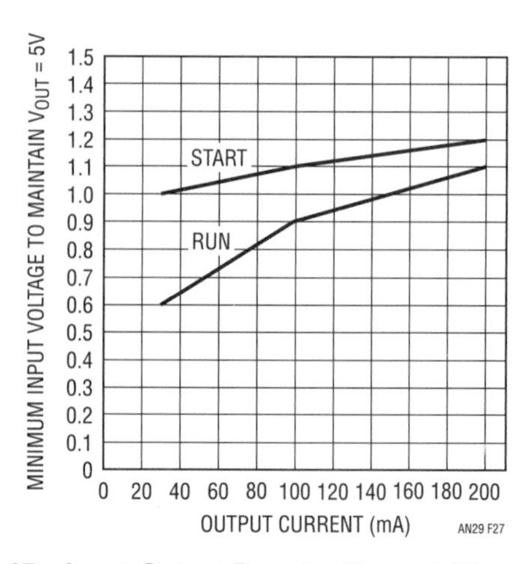

Figure 4.27 • Input-Output Data for Figure 4.23

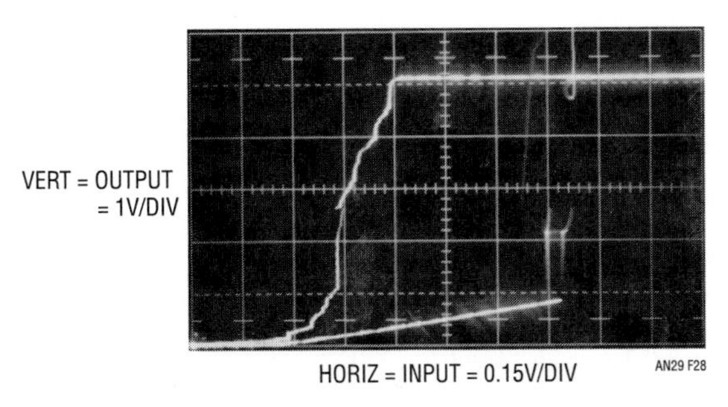

Figure 4.28 • Input-Output XY Characteristics of the 1.5V to 5V Converter at 20mA Loading

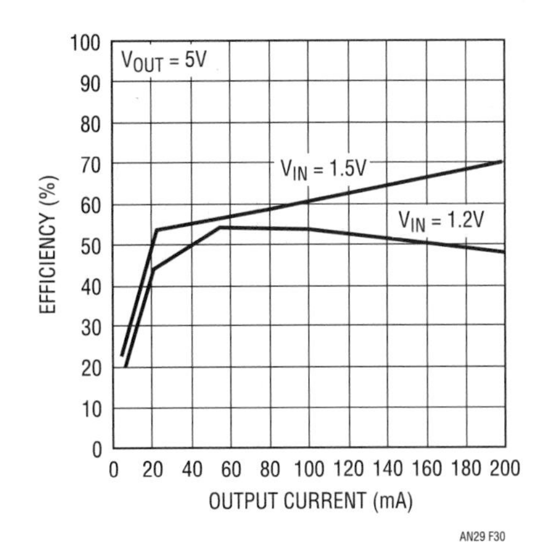

Figure 4.30 • Efficiency vs Operating Point for Figure 4.23

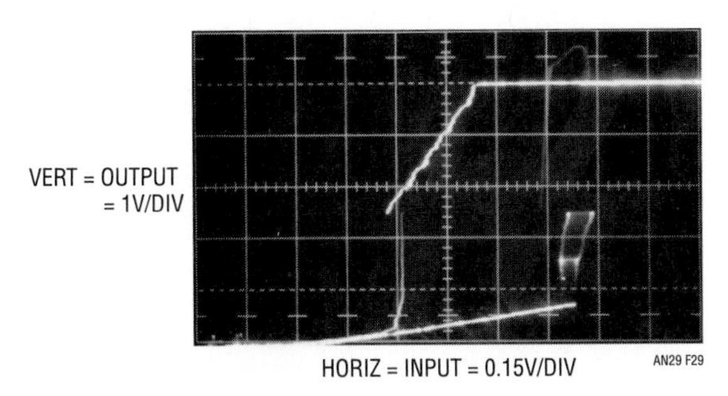

Figure 4.29 • Input-Output XY Characteristics of the 1.5V to 5V Converter at 200mA Loading

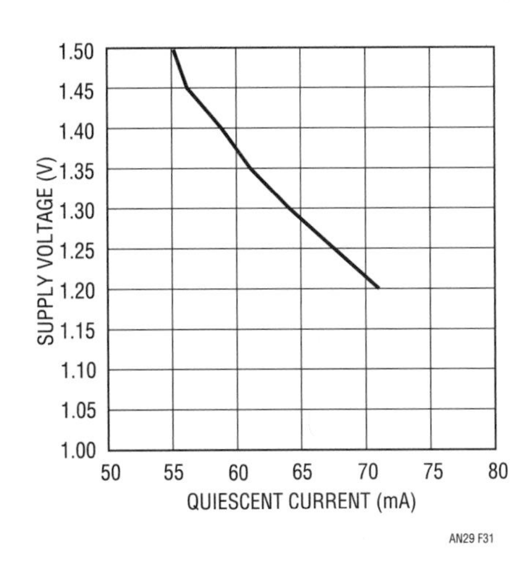

Figure 4.31 • I_Q vs Supply Voltage for Figure 4.23

is "on". To turn on Q1, the gate (Trace B) must be driven above the input voltage. This is accomplished by bootstrapping the capacitor, C1, off the drain of Q2 (Trace C). C1 charges up through D1 when Q2 is turned on. When Q2 is turned off, Q3 is able to conduct, providing a path for C1 to turn Q1 on. During this time, current flows through Q1 (Trace D) through the inductor (Trace E) and into the load. To turn Q1 off, the V_{SW} pin must be "off". Q5 is now able to turn on Q4 and the gate of Q1 is pulled low through D3 and the 50Ω resistor. This resistor is used to reduce the voltage noise generated by fast switching characteristics of Q1. When Q2 is conducting (Trace F), Q1 must be off. The efficiency will be decreased if both transistors are conducting at the same time. The 220Ω resistors and D2 are used to minimize the overlap of the switch cycles. Figure 4.34 shows the efficiency versus load plot for the circuit as shown. The other plots are for non-synchronously switched buck regulators (see indicated Figures).

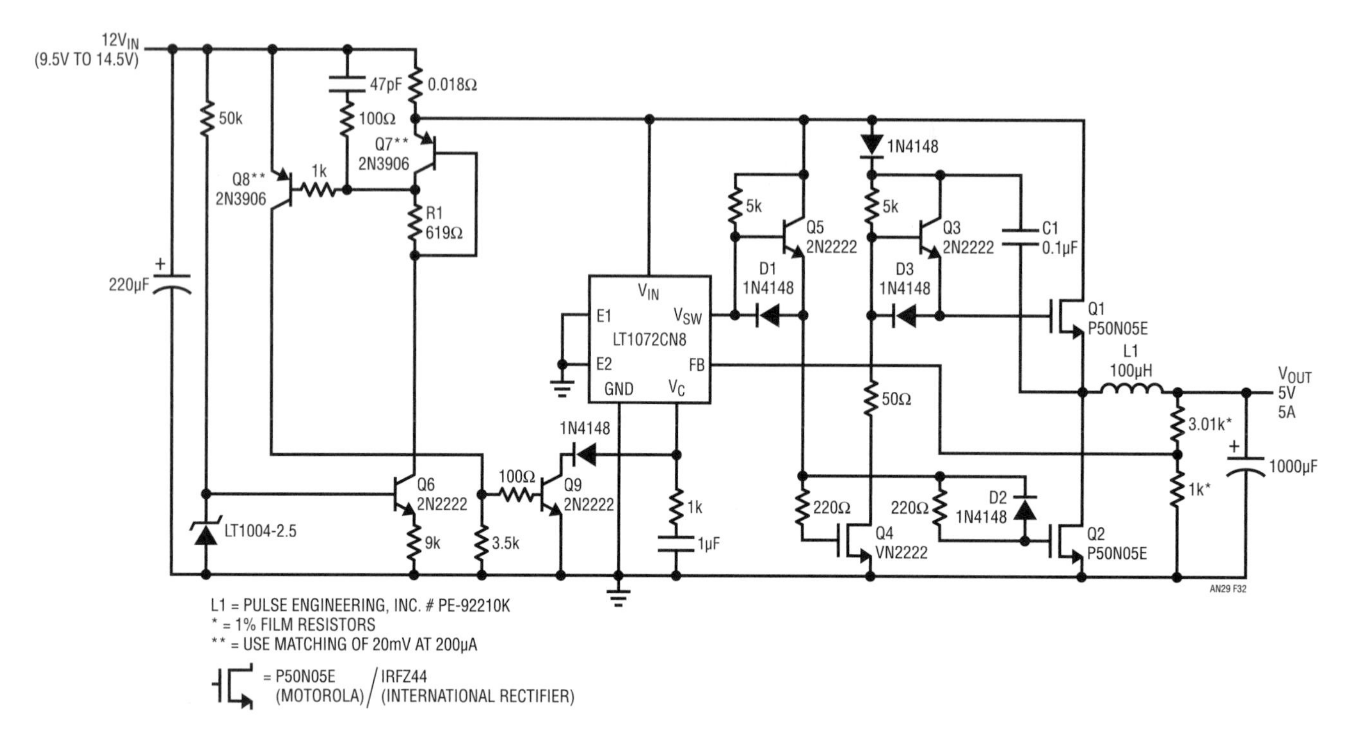

Figure 4.32 • 90% Efficiency Positive Buck Converter with Synchronous Switch

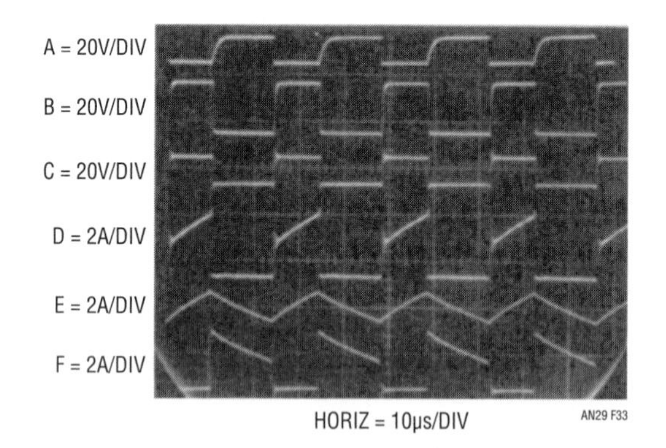

Figure 4.33 • Waveforms for 90% Efficiency Buck Converter

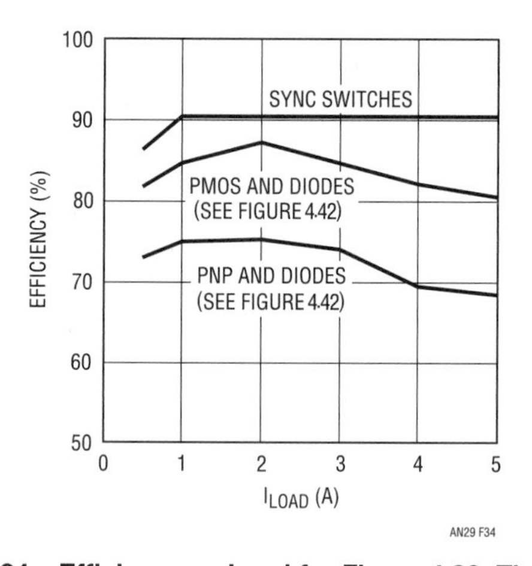

Figure 4.34 • Efficiency vs Load for Figure 4.32. The Synchronous Switches Give Higher Efficiency than Simple FET or Bipolar Transistors and Diodes

Short circuit protection is provided by Q6 through Q9. A 200μA current source is generated from an LT1004, Q6 and the 9k resistor. This current flows through R1 and generates a threshold voltage of 124mV for the comparator, Q7 and Q8. When the voltage drop across the 0.018Ω sense resistor exceeds 124mV, Q8 is turned on. The LT1072's V_{SW} pin goes off when the V_C pin is pulled below 0.9V. This occurs when Q8 forces Q9 to saturate. An RC damper suppresses line transients that might prematurely turn on Q8.

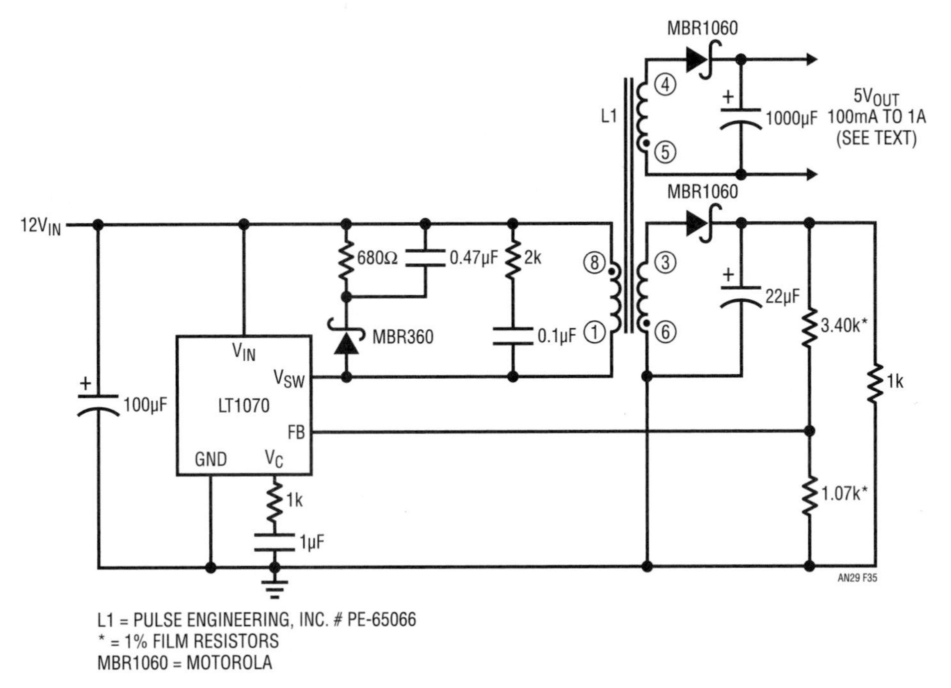

Figure 4.35 • High Efficiency Flux Sensed Isolated Converter

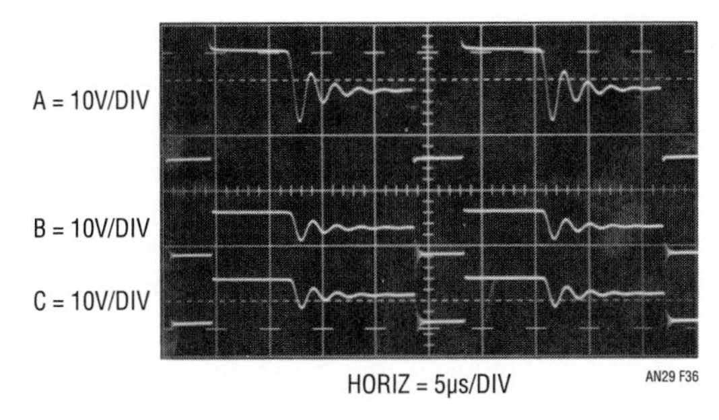

Figure 4.36 • Waveforms for Flux Sensed Converter

High efficiency, flux sensed isolated converter

Figure 4.35's 75% efficiency is not as good as the previous circuit, but it has a fully floating output. This circuit uses a bifilar wound flux sensing secondary to provide isolated voltage feedback. In operation the LT1070's V_{SW} pin (Trace A, Figure 4.36) pulses L1's primary, producing identical waveforms at the floating power and flux sensing secondaries(Traces B and C). Feedback occurs from the flux sense winding via the diode and capacitive filter. The 1k resistor provides a bleed current, while the 3.4k-1.07k divider sets output voltage. The diode partially compensates the diode in the power output winding, resulting in an overall temperature coefficient of about 100ppm/°C. The oversize diode aids efficiency, although significant improvement (e.g., 5% to 10%) is possible if synchronous rectification is employed, as in Figure 4.32. The primary damper network is unremarkable, although the 2k-0.1µF network has been added to suppress excessive ringing at low output current. This ringing is not deleterious to circuit operation, and the network is optional. Below about 10% loading non-ideal transformer behavior introduces significant regulation error. Regulation stays within ±100mV from 10% to 100% of output rating, with excursion exceeding 900mV at no load. Figure 4.37's circuit trades away isolation for tight regulation with no output loading restrictions. Efficiency is the same.

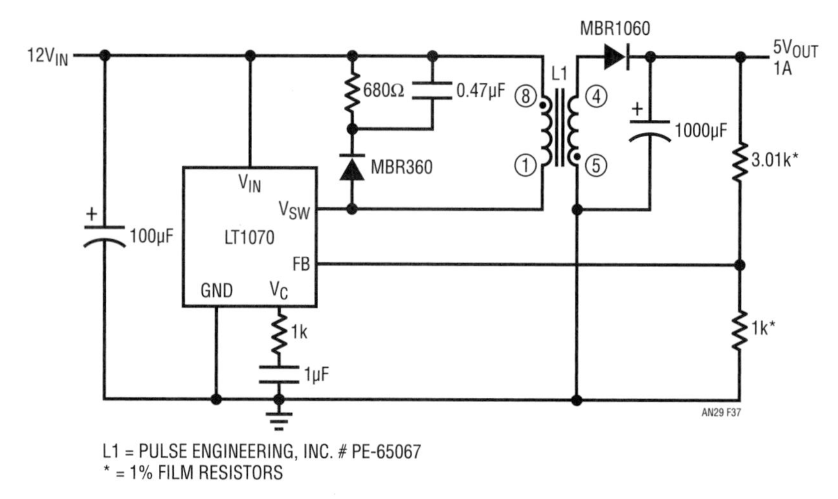

Figure 4.37 • Non-Isolated Version of Figure 4.35

Wide range input converters

Wide range input −48V to 5V converter

Often converters must accommodate a wide range of inputs. Telephone lines can vary over considerable tolerances. Figure 4.38's circuit uses an LT1072 to supply a 5V output from a telecom input. The raw telecom supply is nominally −48V but can vary from −40V to −60V. This range of voltages is acceptable to the V_{SW} pin but protection is required for the V_{IN} pin (V_{MAX} = 60V). Q1 and the 30V Zener diode serve this purpose, dropping V_{IN}'s voltage to acceptable levels under all line conditions.

Here the "top" of the inductor is at ground and the LT1072's ground pin at −V. The feedback pin senses with respect to the ground pin, so a level shift is required from the 5V output. Q2 serves this purpose, introducing only −2mV/°C drift. This is normally not objectionable in a logic supply. It can be compensated with the optional appropriately scaled diode-resistor shown in Figure 4.38.

Frequency compensation uses an RC damper at the V_C pin. The 68V Zener is a type designed to clamp and absorb excessive line transients which might otherwise damage the LT1072 (V_{SW} maximum voltage is 75V).

Figure 4.39 shows operating waveforms at the V_{SW} pin. Trace A is the voltage and Trace B the current. Switching is crisp, with well controlled waveforms. A higher current version of this circuit appears in LTC Application Note 25, "Switching Regulators For Poets."

3.5V to 35V_{IN}–5V_{OUT} converter

Figure 4.40's approach has an even wider input range. In this case it produces either a −5V or 5V output (shown in dashed lines). This circuit is an extension of Figure 4.11's basic flyback topology. The coupled inductor allows the option for buck, boost, or buck-boost converters. This circuit can operate down to 3.5V for battery applications while accepting 35V inputs.

Figure 4.41 shows the operating waveforms for this circuit. During the V_{SW} (Trace A) "on" time, current flows through the primary winding (Trace B). No current is transferred to the secondary because the catch diode, D1, is reverse biased. The energy is stored in the magnetic field. When the switch is turned "off" D1 forward biases and the energy is transferred to the secondary winding. Trace C is the voltage seen on the secondary and Trace D is the current flowing through it. This is not an ideal transformer so not all of the primary windings energy is coupled into the secondary. The energy left in the primary winding causes the overvoltage spikes seen on the V_{SW} pin (Trace E). This phenomenon is modeled by a leakage inductance term which is placed in series with the primary winding. When the switch is turned "off" current continues to flow in the inductor causing the snubber diode to conduct (Trace F).

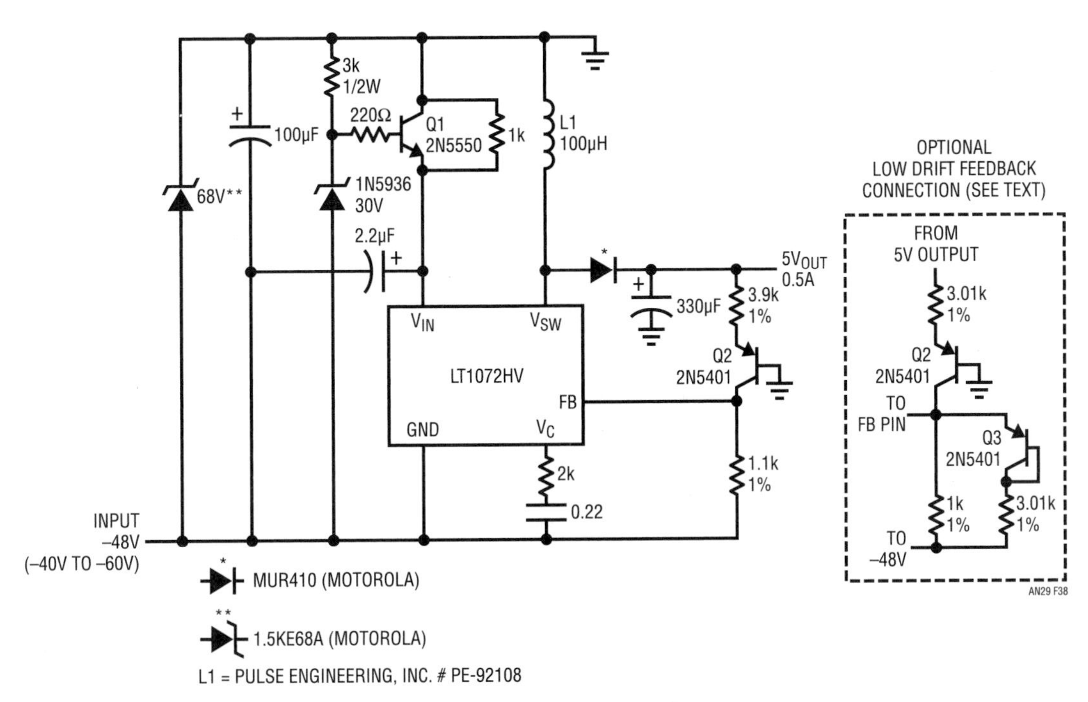

Figure 4.38 • Wide Range Input Converter

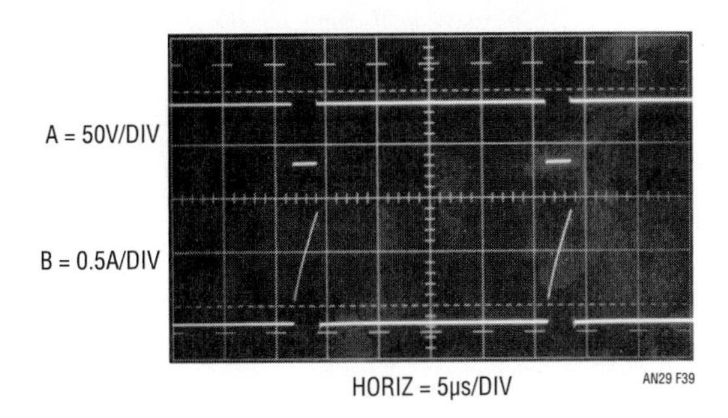

Figure 4.39 • Waveforms for Wide Range Input Converter

The snubber diode current falls to zero as the inductor loses its energy. The snubber network clamps the voltage spike. When the snubber diode current reaches zero, the V_{SW} pin voltage settles to a potential related to the turns ratio, output voltage and input voltage.[4]

The feedback pin senses with respect to ground, so Q1 through Q3 provides the level shift from the −5V output. Q1 introduces a −2mV/°C drift to the circuit. This effect can be compensated by a circuit similar to the one shown in Figure 4.38. Line regulation is degraded due to Q3's output impedance. If this is a problem, an op amp must be used to perform the level shift (see AN19, Figure 29).

Note 4: Application Note 19, "LT1070 Design Manual," page 25.

Wide range input positive buck converter

Figure 4.42 is another example of a positive buck converter. This is a simpler version compared to the synchronous switch buck, Figure 4.32. However, efficiency isn't as high (see Figure 4.34). If the PMOS transistor is replaced with a Darlington PNP transistor (shown in dashed lines) efficiency decreases further.

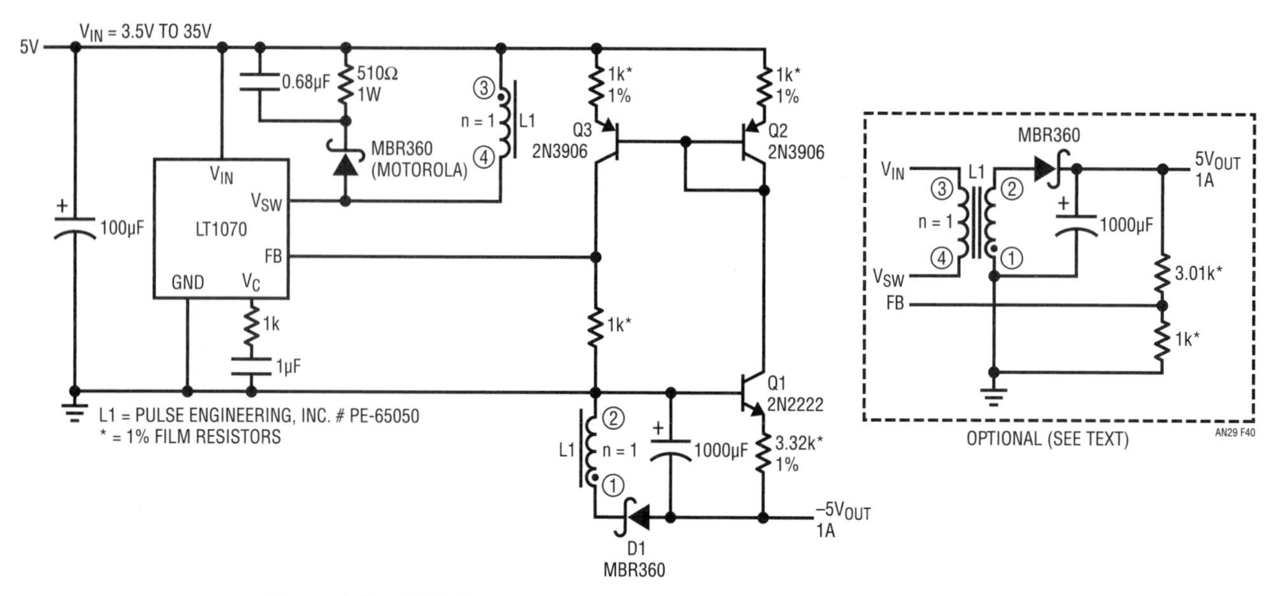

Figure 4.40 • Wide Range Input Positive-to-Negative Flyback Converter

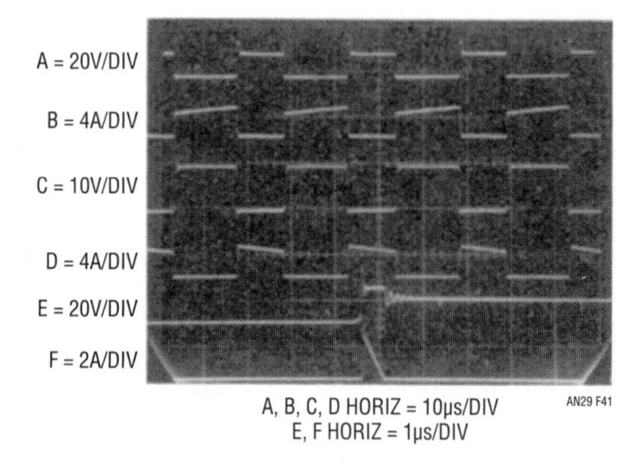

Figure 4.41 • Waveforms for Wide Range Input Positive –5V Output Flyback Converter

Figure 4.43a shows the operating waveforms for this circuit. The pass transistor's (Q1) drive scheme is similar to the one shown in Figure 4.32. During the V_{SW} (Trace A) "on" time, the gate of the pass transistor is pulled down through D1. This forces Q1 to saturate. Trace B is the voltage seen on the drain of Q1 and Trace C is the current passing through Q1. The supply current flows through the inductor (Trace D) and into the load. During this time energy is being stored in the inductor. When voltage is applied to the inductor, current does not instantly rise. As the magnetic field builds up, the current builds. This is seen in the inductor current waveform (Trace D). When the V_{SW} pin is "off," Q2 is able to conduct and turns Q1 off. Current can no longer flow through Q1, instead D2 is conducting (Trace E). During this period some of the energy stored in the inductor will be transferred to the load. Current will be generated from the inductor as long as there is any energy in it. This can be seen in Figure 4.43a. This is known as continuous mode operation. If the inductor is completely discharged, no current will be generated (see Figure 4.43b). When this happens neither switch, Q1 or D2, is conducting. The inductor looks like a short and the voltage on the cathode of D2 will settle to the output voltage. These "boingies" can be seen in Trace B of Figure 4.43b. This is known as discontinuous mode operation. Higher input voltages can be handled with the gate-source Zener clamped by D2. The 400 milliwatt Zener's current must be rescaled by adjusting the 50Ω value. Maximum gate-source voltage is 20V. The circuit will function up to $35V_{IN}$. At inputs beyond 35V all semiconductor breakdown voltages must be considered.

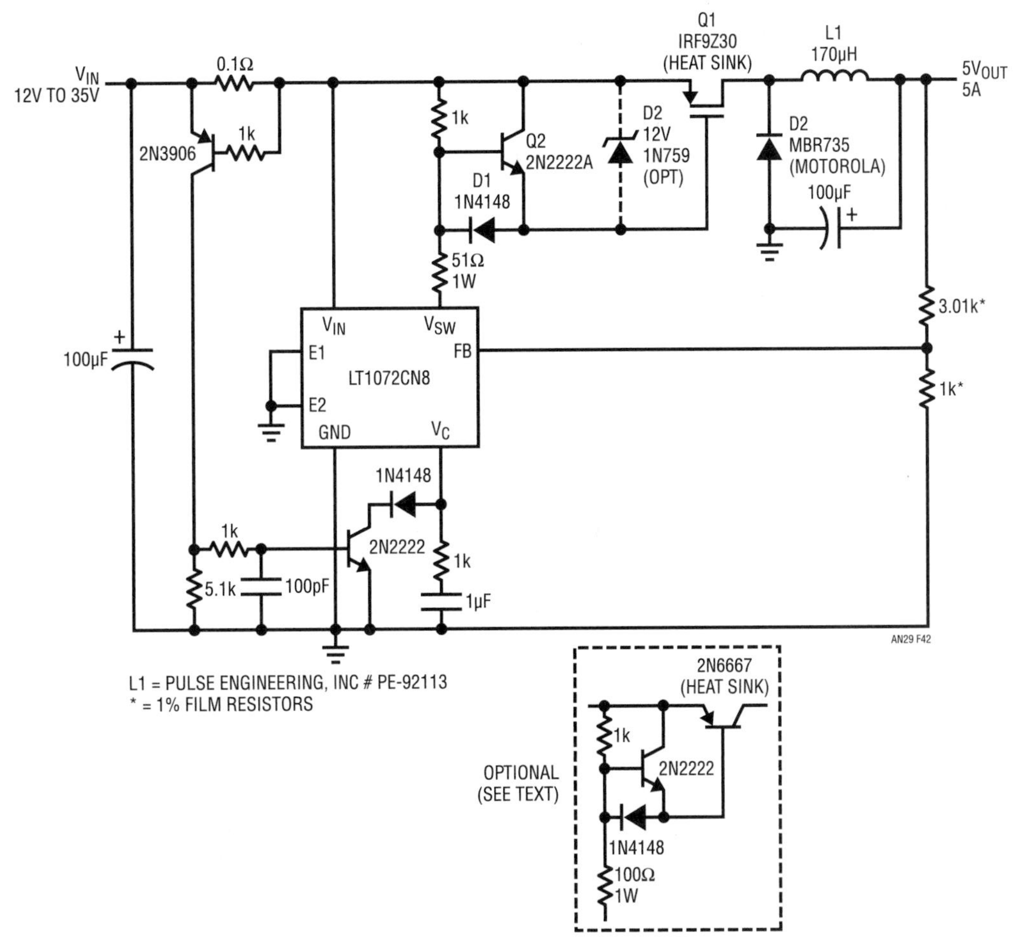

Figure 4.42 • Positive Buck Converter

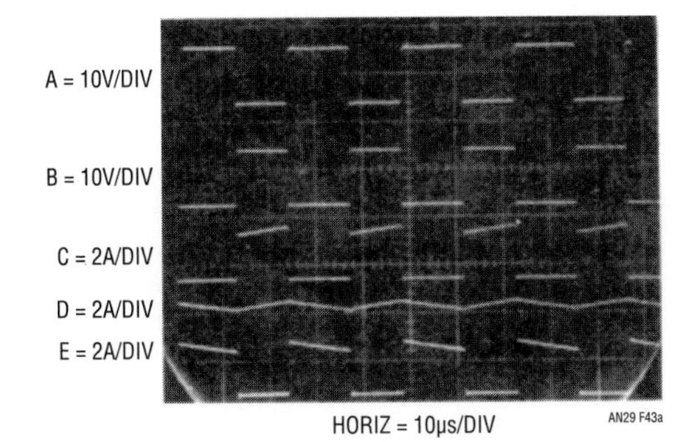

Figure 4.43a • Waveforms for Wide Range Input Positive Buck Converter (Continuous Mode)

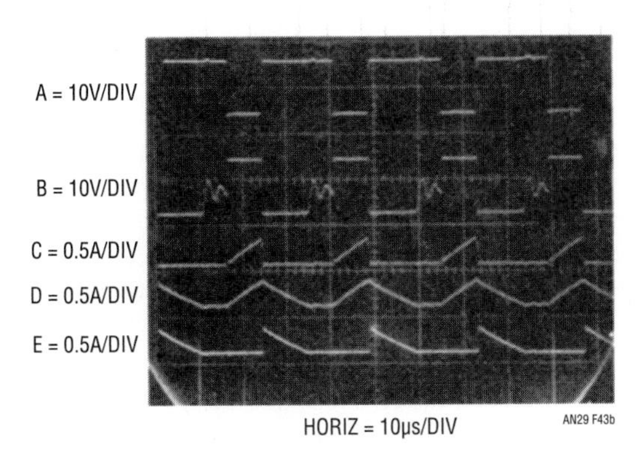

Figure 4.43b • Waveforms for Wide Range Input Positive Buck Converter (Discontinuous Mode)

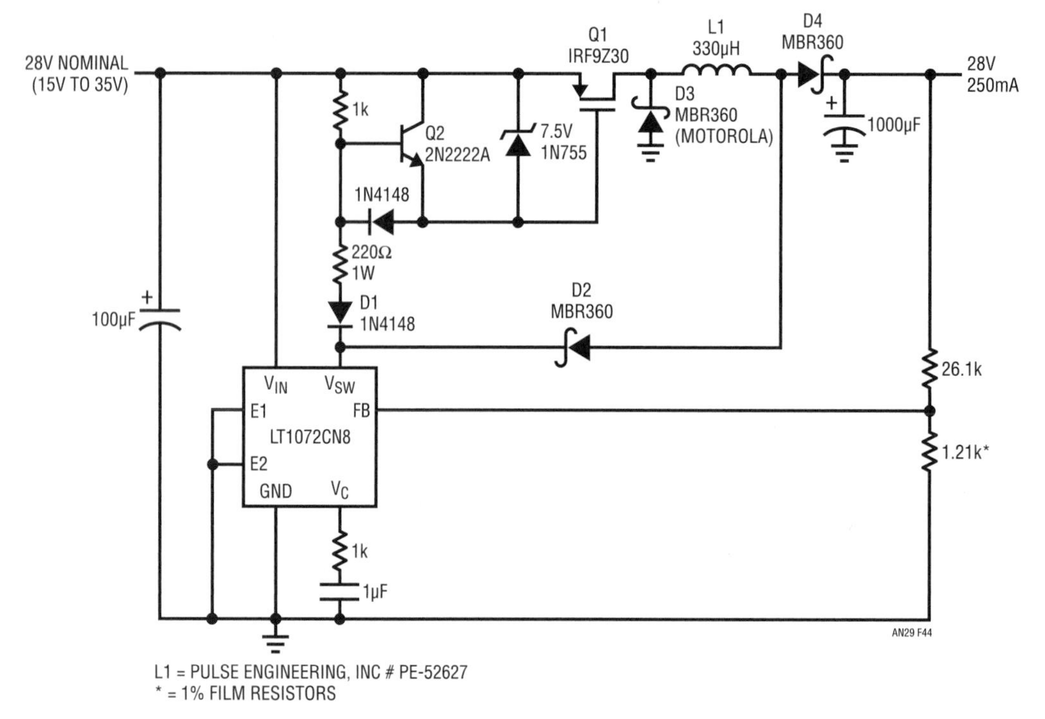

Figure 4.44 • Positive Buck-Boost Converter

Buck-boost converter

The buck boost topology is useful when the input voltage can either be higher or lower than the output. In this example, Figure 4.44, this is accomplished with a single inductor instead of a transformer, as in Figure 4.40 (optional). However, the input voltage range only extends down to 15V and can reach to 35V. If the maximum 1.25A switch current rating of the LT1072 is exceeded an LT1071 or LT1070 can be used instead. At high power levels package thermal characteristics should be considered.

The operation of the circuit is similar to the positive buck converter, Figure 4.42. The gate drive to the pass transistor is derived the same way except the gate-source voltage is clamped. Remember, the gate-source maximum voltage rating is specified at ±20V. Figure 4.45 shows the operating waveforms. When the V_{SW} pin is "on" (Trace A), the pass transistor, Q1, is saturated. The gate voltage (Trace B) is clamped by the Zener diode. Trace C is the voltage on the drain of Q1 and Trace D is the current through it. This is where the similarities between the two circuits end. Notice the inductor is pulled to within a diode drop, D2 above ground, instead of being tied to the output (see Figure 4.42). In this case, the inductor has the input voltage applied across it, except for a Vbe and saturation losses. D4 is reverse biased and blocks the output capacitor from discharging into the V_{SW} pin. When the V_{SW} pin is "off" Q1 and D2 cease to conduct. Since the current in the inductor (Trace E) continues to flow, D3 and D4 are forward biased and the energy in the inductor is transferred into the load. Trace F is the current through D3. Also, D2 keeps Q1 from staying on if the circuit is operating in buck mode. D1, on the other hand, blocks current from flowing into the gate drive circuit when operating in boost mode.

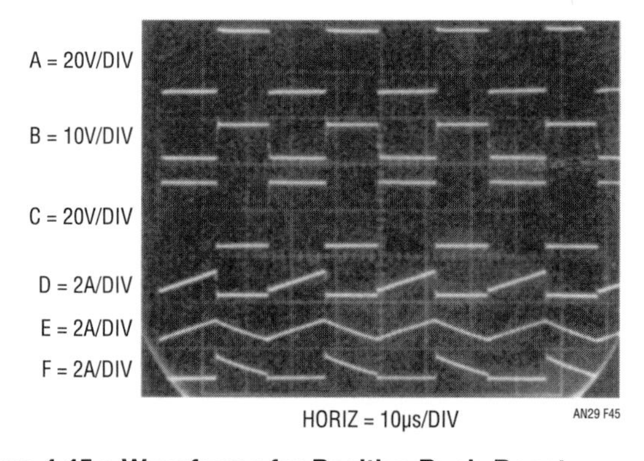

Figure 4.45 • Waveforms for Positive Buck-Boost Converter

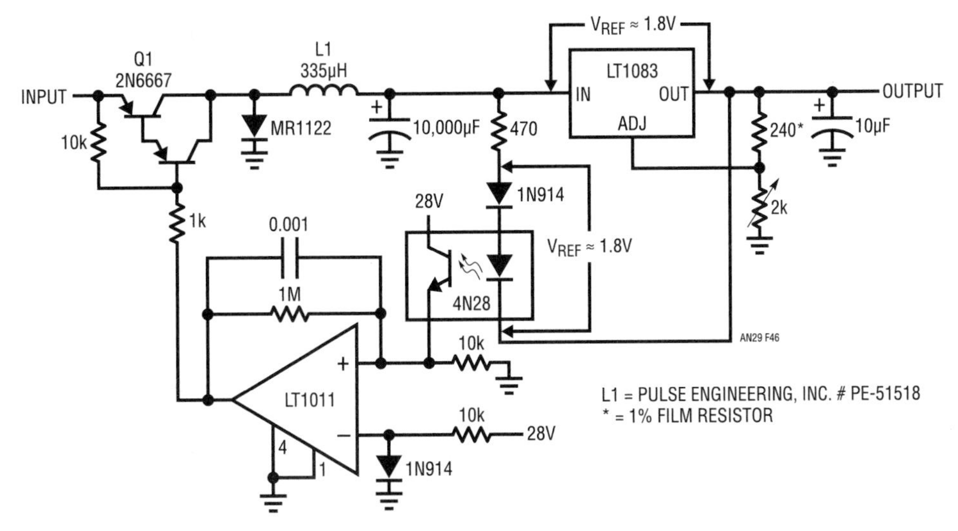

Figure 4.46 • High Power Linear Regulator with Switching Pre-Regulator

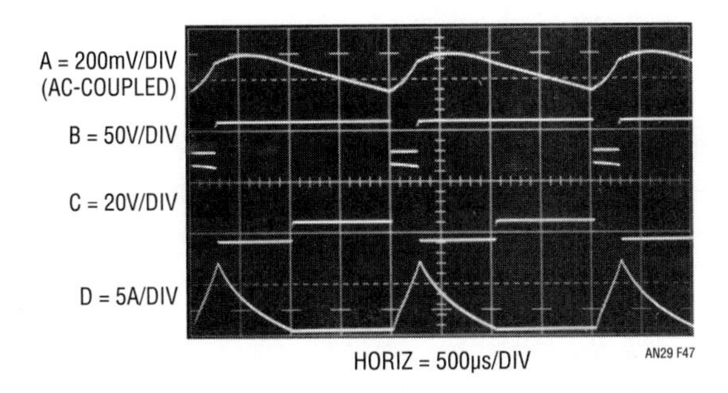

Figure 4.47 • Switching Pre-Regulated Linear Regulators Waveforms

Wide range switching pre-regulated linear regulator

In a sense, linear regulators can be considered extraordinarily wide range DC/DC converters. They do not face the dynamic problems switching regulators encounter under varying ranges of input and output. Excess energy is simply dissipated at heat. This elegantly simplistic energy management mechanism pays dearly in terms of efficiency and temperature rise. Figure 4.46 shows a way a linear regulator can more efficiently control high power under widely varying input and output conditions.

The regulator is placed within a switched-mode loop that servo-controls the voltage *across* the regulator. In this arrangement the regulator functions normally while the switched-mode control loop maintains the voltage across it at a minimal value, regardless of line, load or output setting changes. Although this approach is not quite as efficient as a classical switching regulator, it offers lower noise and the fast transient response of the linear regulator. The LT1083 functions in the conventional fashion, supplying a regulated output at 7.5A capacity. The remaining components form the switched-mode dissipation limiting control. This loop forces the potential across the LT1083 to equal the 1.8V value of V_{REF}. The opto-isolator furnishes a convenient way to single end the differentially sensed voltage across the LT1083. When the input of the regulator (Trace A, Figure 4.47) decays far enough, the LT1011 output (Trace B) switches low, turning on Q1 (Q1 collector is Trace C). This allows current flow (Trace D) from the circuit input into the 10,000μF capacitor, raising the regulator's input voltage.

When the regulator input rises far enough, the comparator goes high, Q1 cuts off and the capacitor ceases charging. The MR1122 damps the flyback spike of the current limiting inductor. The 0.001μF-1M combination sets loop hysteresis at about $100mV_{P-P}$. This free-running oscillation control mode substantially reduces dissipation in the regulator, while preserving its performance. Despite changes in the input voltage, different regulated outputs or load shifts, the loop always ensures minimum dissipation in the regulator.

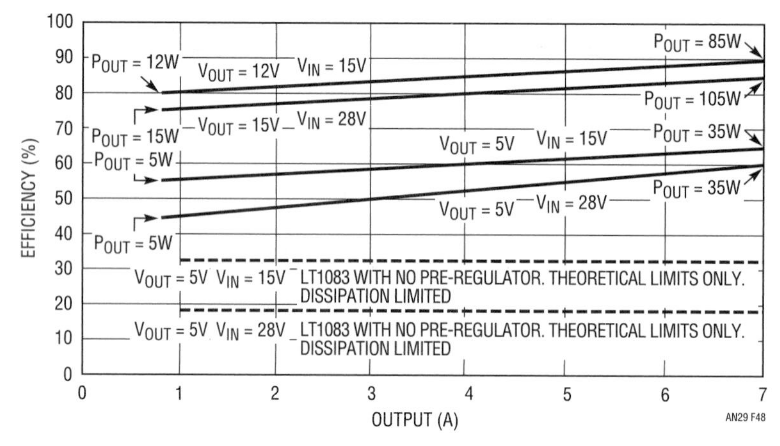

Figure 4.48 • Efficiency vs Output Current for Figure 4.46 at Various Operating Points

Figure 4.48 plots efficiency at various operating points. Junction losses and the loop enforced 1.8V across the LT1083 are relatively small at high output voltages, resulting in good efficiency. Low output voltages do not fare as well, but compare very favorably to the theoretical data for the LT1083 with no pre-regulator. At the higher theoretical dissipation levels the LT1083 will shut down, precluding practical operation.

High voltage converters

High voltage converter—1000V$_{OUT}$, nonisolated

Photomultiplier tubes, ion generators, gas-based detectors, image intensifiers and other applications need high voltages. Converters frequently supply these potentials. Generally, the limitation on high voltage is transformer insulation breakdown. A transformer is almost always used because a simple inductor forces excessive voltages on the semiconductor switch. Figure 4.49's circuit, reminiscent of Figure 4.11's basic flyback configuration, is a 15V to 1000V$_{OUT}$ converter. The LT1072 controls output by modulating the flyback energy into L1, forcing its feedback (FB) pin to 1.23V (the internal reference value). In this example loop compensation is heavily overdamped by the V_C pin capacitor. L1's damper network limits flyback spikes within the V_{SW} pin's 75V rating.

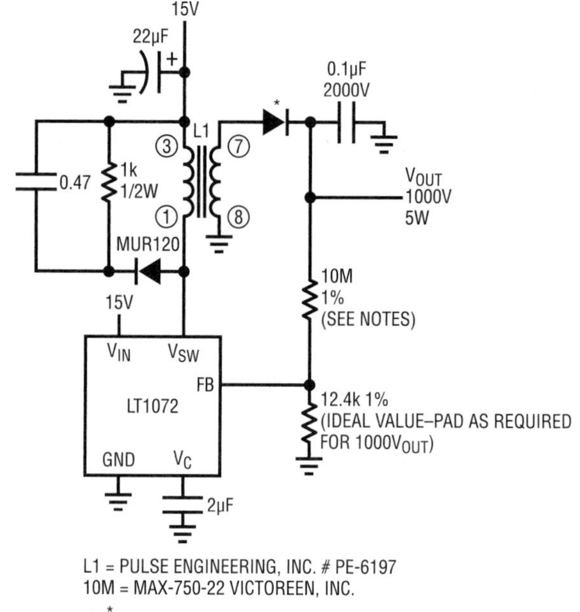

Figure 4.49 • Nonisolated 15V to 1000V Converter

Fully floating, 1000V$_{OUT}$ converter

Figure 4.50 is similar to Figure 4.49 but features a fully floating output. This provision allows the output to be referenced off system ground, often desirable for noise or biasing reasons. Basic loop action is as before, except that the LT1072's internal error amplifier and reference are replaced with galvanically isolated equivalents. Power for these components is bootstrapped from the output via source follower Q1 and its 2.2M ballast resistor. A1 and the LT1004, micropower components, minimize dissipation in Q1 and its ballast. Q1's gate bias, tapped from the output divider string, produces about 15V at its source. A1 compares the scaled divider output with the LT1004 reference. The error signal, A1's output, drives the opto-coupler. Photocurrent is kept low to save power. The opto-coupler output pulls down on the pin, closing a loop. Frequency compensation at the V_C pin and A1 stabilizes the loop.

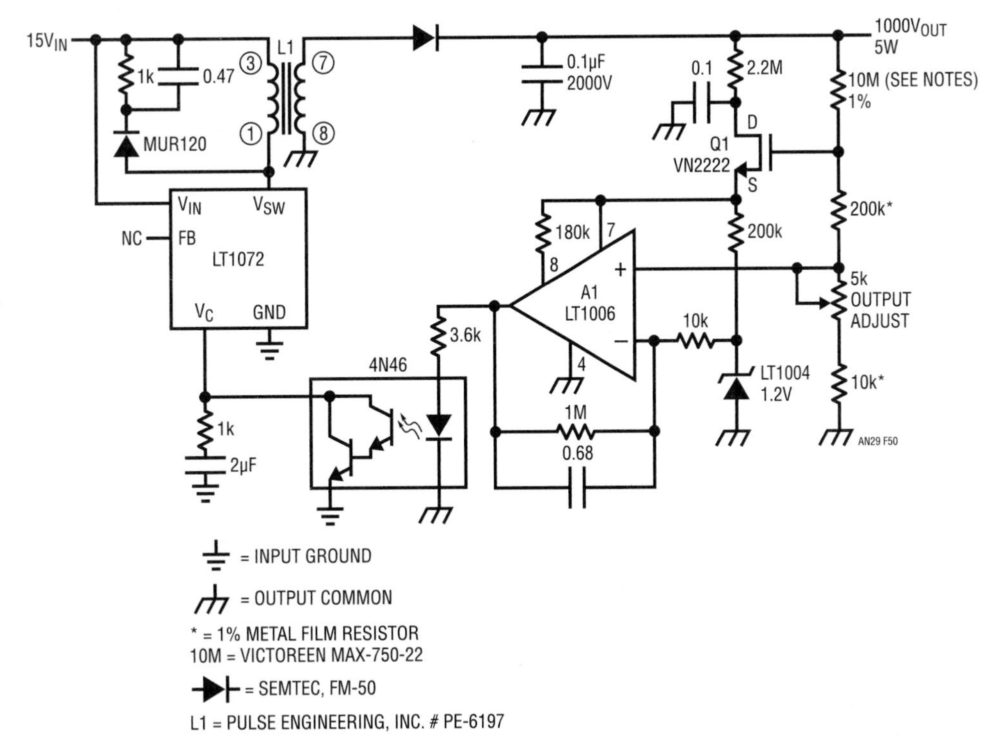

Figure 4.50 • Isolated Output 15V to 1000V Converter

The transformers isolated secondary and optical feedback produce a regulated, fully galvanically floating output. Common mode voltages of 2000V are acceptable.

20,000V$_{CMV}$ breakdown converter

Figure 4.50's common mode breakdown limits are imposed by transformer and opto-coupler restrictions. Isolation amplifiers, transducer measurement at high common mode voltages (e.g., winding temperature of a utility company transformer and ESD sensitive applications) require high breakdowns. Additionally, very precise floating measurements, such as signal conditioning for high impedance bridges, can require extremely low leakage to ground.

Achieving high common mode voltage capability with minimal leakage requires a different approach. Magnetics is usually considered the only approach for isolated transfer of appreciable amounts of electrical energy. Transformer action is, however, achievable in the acoustic domain. Some ceramic materials will transfer electrical energy with galvanic isolation. Conventional magnetic transformers work on an electrical-magnetic-electrical basis using the magnetic domain for electrical isolation. The acoustic transformer uses an acoustic path to get isolation. The high voltage breakdown and low electrical conductance associated with ceramics surpasses isolation characteristics of magnetic approaches. Additionally, the acoustic transformer is simple. A pair of leads bonded to each end of the ceramic material forms the device. Insulation resistance exceeds $10^{12}\Omega$, with primary-secondary capacitances of 1pF to 2pF. The material and its physical configuration determine its resonant frequency. The device may be considered as a high Q resonator, similar to a quartz crystal. As such, drive circuitry excites the device in the positive feedback path of a wideband gain element. Unlike a crystal, drive circuitry is arranged to pass substantial current through the ceramic, maximizing power into the transformer.

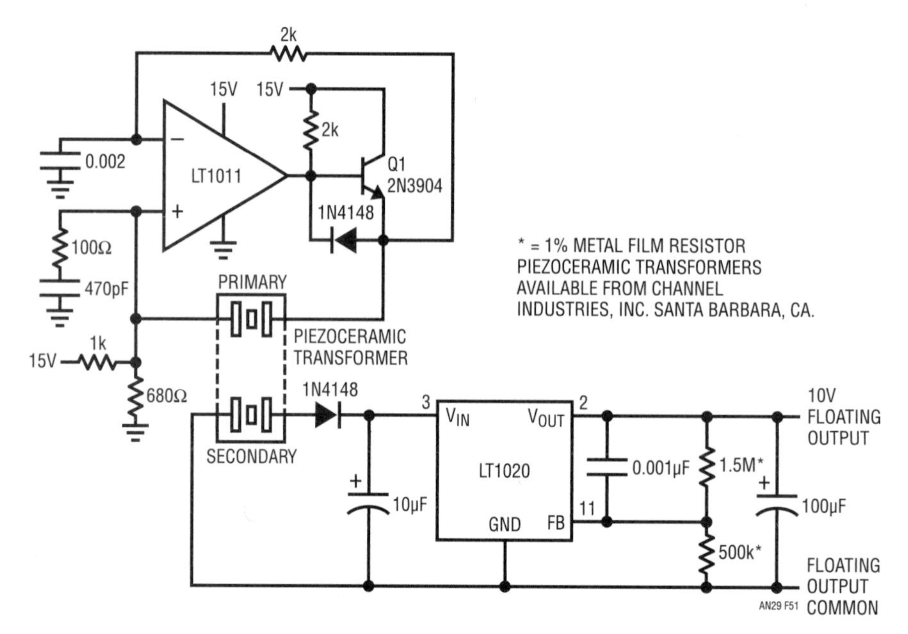

Figure 4.51 • 15V to 10V Converter with 20,000V Isolation

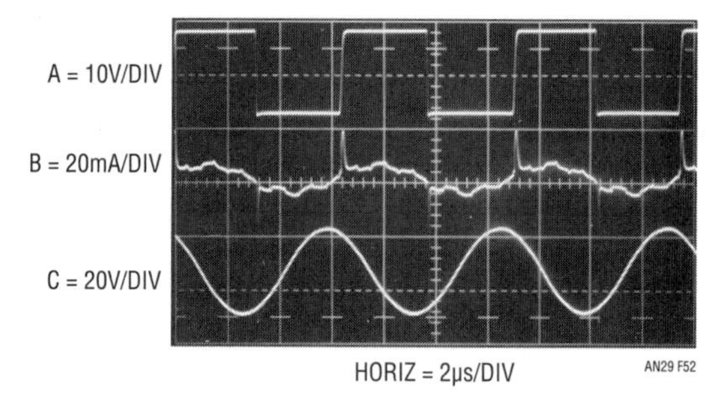

Figure 4.52 • Waveforms for the 20,000V Isolation Converter

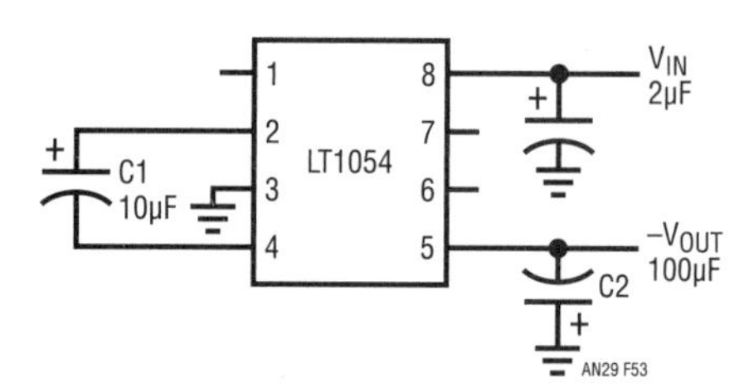

Figure 4.53 • A Basic Switched-Capacitor Converter

In Figure 4.51, the piezo-ceramic transformer is in the LT1011 comparators positive feedback loop. Q1 is an active pull-up for the LT1011, an open-collector device. The 2k-0.002μF path biases the negative input. Positive feedback occurs at the transformers resonance, and oscillation commences (Trace A, Figure 4.52 is Q1's emitter). Similar to quartz crystals, the transformer has significant harmonic and overtone modes. The 100Ω-470pF damper suppresses spurious oscillations and "mode hopping." Drive current (Trace B) approximates a sine wave, with peaking at the transitions. The transformer looks like a highly resonant filter to the resultant acoustic wave propagated in it. The secondary voltage (Trace C) is sinusoidal. Additionally, the transformer has voltage gain. The diode and 10μF capacitor convert the secondary voltage to DC. The LT1020 low quiescent current regulator gives a stabilized 10V output. Output current for the circuit is a few milliamperes. Higher currents are possible with attention to transformer design.

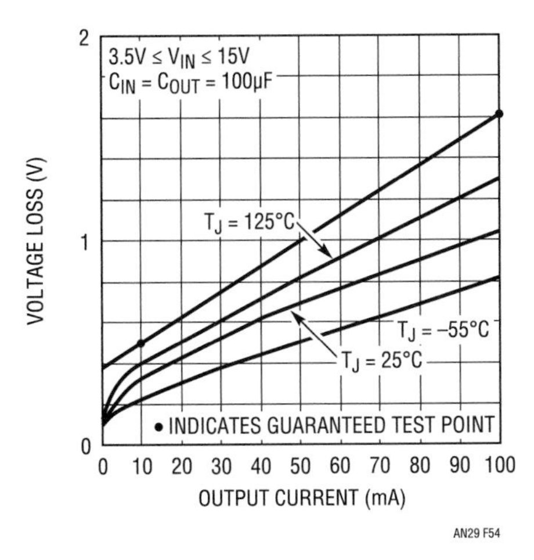

Figure 4.54 • Losses for the Basic Switched-Capacitor Converter

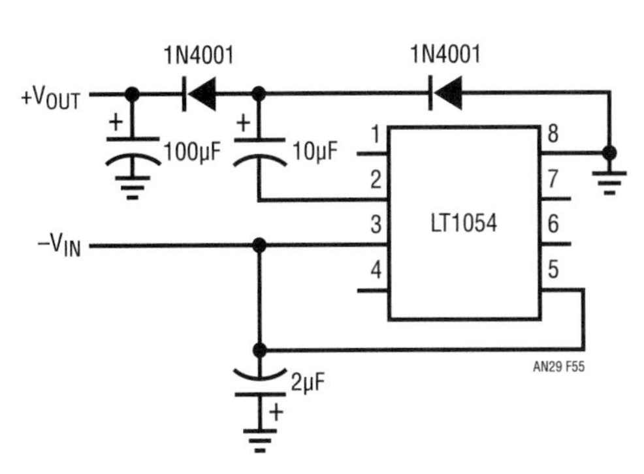

Figure 4.55 • Switched-Capacitor $-V_{IN}$ to $+V_{OUT}$ Converter

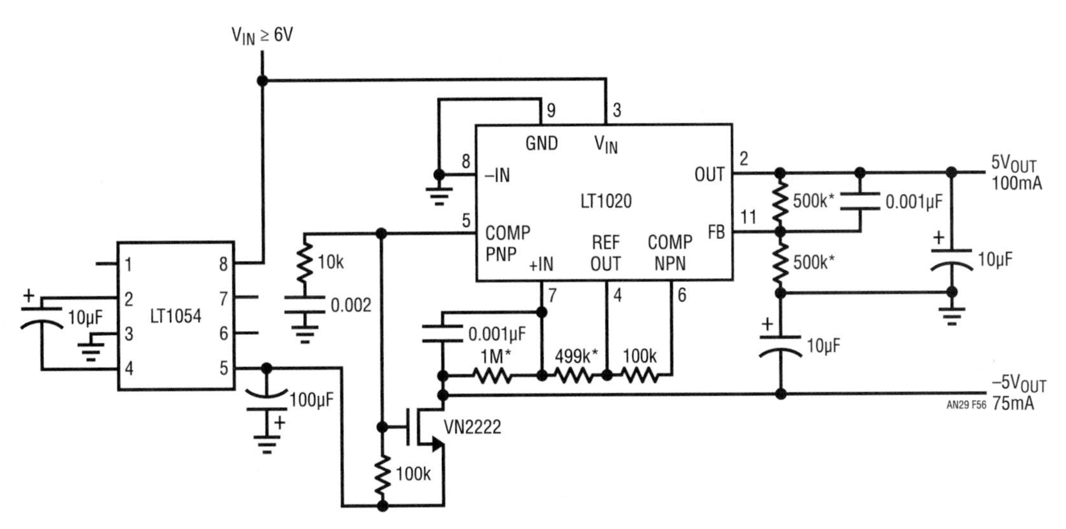

Figure 4.56 • High Current Switched-Capacitor 6V to ±5V Converter

Switched-capacitor based converters

Inductors are used in converters because they can store energy. This stored magnetic energy, released and expressed in electrical terms, is the basis of converter operation. Inductors are not the only way to store energy with efficient release expressed in electrical terms. Capacitors store charge (already an electrical quantity) and as such, can be used as the basis for DC/DC conversion. Figure 4.53 shows how simple a switched-capacitor based converter can be (the fundamentals of switched-capacitor based conversion are presented in Appendix B, "Switched-Capacitor Voltage Converters—How They Work"). The LT1054 provides clocked drive to charge C1. A second clock phase discharges C1 into C2. The internal switching is arranged so C1 is "flipped" during the discharge interval, producing a negative output at C2. Continuous clocking allows C2 to charge to the same absolute value as C1. Junction and other losses preclude ideal results, but performance is quite good. This circuit will convert V_{IN} to $-V_{OUT}$ with losses shown in Figure 4.54. Adding an external resistive divider allows regulated output (see Appendix B).

With some additional steering diodes this configuration can effectively run "backwards" (Figure 4.55), converting a negative input to a positive output. Figure 4.56's variant gives low dropout linear regulation for 5V and −5V outputs from $6V_{IN}$. The LT1020-based dual output regulation scheme is adapted from Figure 4.8. Figure 4.57 uses diode steering to get voltage boost, providing $\approx 2V_{IN}$. Bootstrapping this configuration with Figure 4.54's basic circuit leads to Figure 4.58, which converts a 5V input to 12V and −12V outputs. As might be expected output current capacity is traded for the voltage gain, although 25mA is still available. Figure 4.59, another boost converter, employs a dedicated version of Figure 4.58 (the LT1026) to get regulated ±7V from a 6V input. The LT1026 generates unregulated ±11V rails from the 6V input with the LT1020 and associated components (again, purloined from Figure 4.8) producing regulation. Current and boost capacity are reduced from Figure 4.58's levels, but the regulation and simplicity are noteworthy. Figure 4.60 combines the LT1054's clocked switched-capacitor charging with classical diode voltage multiplication, producing positive and negative outputs. At no load ±13V is available, falling to ±10V with each side supplying 10mA.

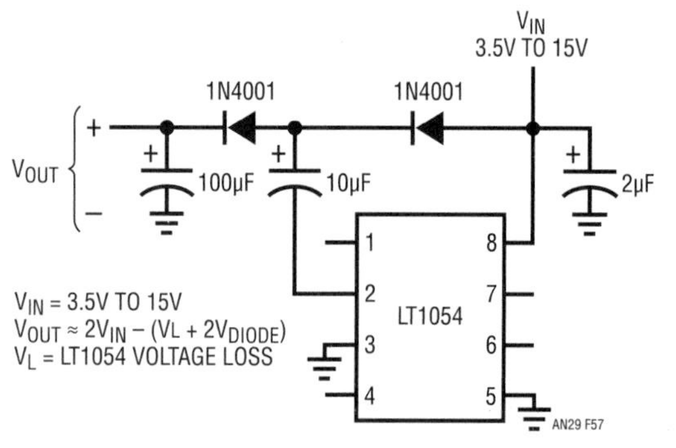

Figure 4.57 • Voltage Boost Switched-Capacitor Converter

High power switched-capacitor converter

Figure 4.61 shows a high power switched-capacitor converter with a 1A output capacity. Discrete devices permit high power operation.

The LTC1043 switched-capacitor building block provides non-overlapping complementary drive to the Q1-Q4 power MOSFETs. The MOSFETs are arranged so that C1 and C2 are alternately placed in series and then in parallel. During the series phase, the 12V supply current flows through both capacitors, charging them and furnishing load current. During the parallel phase, both capacitors deliver current to the load. Traces A and B, Figure 4.62, are the LTC1043-supplied drives to Q3 and Q4, respectively. Q1 and Q2 receive similar drive from Pins 3 and 11. The diode-resistor networks provide additional non-overlapping drive characteristics, preventing simultaneous drive to the series-parallel phase switches. Normally, the output would be one-half of the supply voltage, but C1 and its associated components close a feedback loop, forcing the output to 5V. With the circuit in the series phase, the output (Trace C) heads rapidly positive. When the output exceeds 5V, C1 trips, forcing the LTC1043 oscillator pin (Trace D) high. This truncates the LTC1043's triangle wave oscillator cycle. The circuit is forced into the parallel phase and the output coasts down slowly until the next LTC1043 clock cycle begins. C1's output diode prevents the triangle down-slope from being affected and the 100pF capacitor provides sharp transitions. The loop regulates the output to 5V by feedback controlling the turn-off point of the series phase. The circuit constitutes a large scale switched-capacitor voltage divider which is never allowed to complete a full cycle. The high transient currents are easily handled by the power MOSFETs and overall efficiency is 83%.

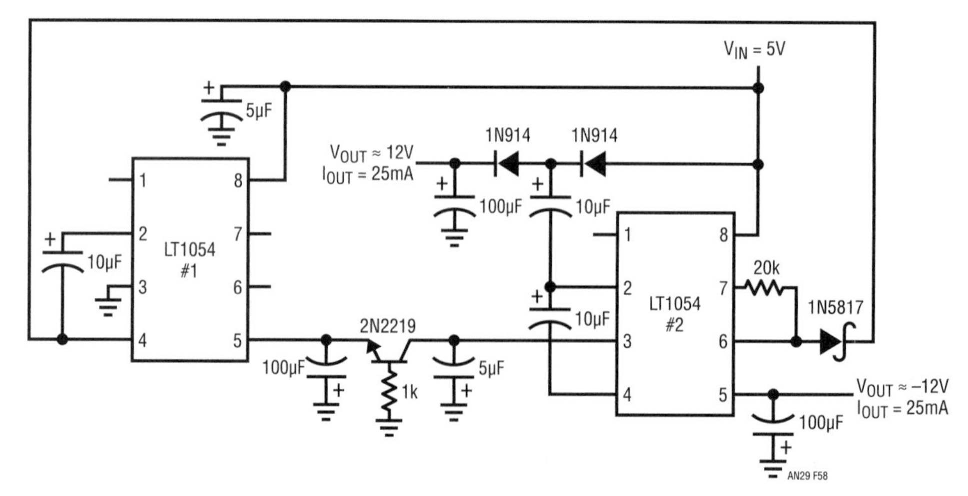

Figure 4.58 • Switched-Capacitor 5V to ±12V Converter

References

1. Williams, J., "Conversion Techniques Adopt Voltages to your Needs," EDN, November 10, 1982, p. 155
2. Williams, J., "Design DC/DC Converters to Catch Noise at the Source," Electronic Design, October 15, 1981, p. 229
3. Nelson, C., "LT1070 Design Manual," Linear Technology Corporation, Application Note 19
4. Williams, J., "Switching Regulators for Poets," Linear Technology Corporation, Application Note 25
5. Williams, J., "Power Conditioning Techniques for Batteries," Linear Technology Corporation, Application Note 8
6. Tektronix, Inc., CRT Circuit, Type 453 Operating Manual, p. 3-16
7. Pressman, A. I., "Switching and Linear Power Supply, Power Converter Design," Hayden Book Co., Hasbrouck Heights, New Jersey, 1977, ISBN 0-8104-5847-0
8. Chryssis, G., "High Frequency Switching Power Supplies, Theory and Design," McGraw Hill, New York, 1984, ISBN 0-07-010949-4
9. Sheehan, D., "Determine Noise of DC/DC Converters," Electronic Design, September 27, 1973
10. Bright, Pittman, and Royer, "Transistors as On-Off Switches in Saturable Core Circuits," Electronic Manufacturing, October, 1954

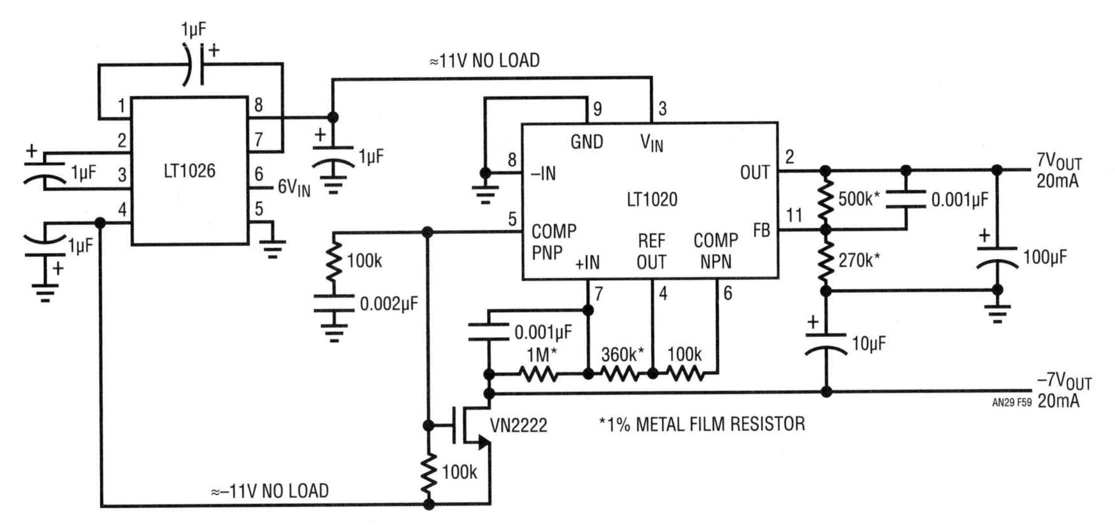

Figure 4.59 • Switched-Capacitor Based 6V to ±7V Converter

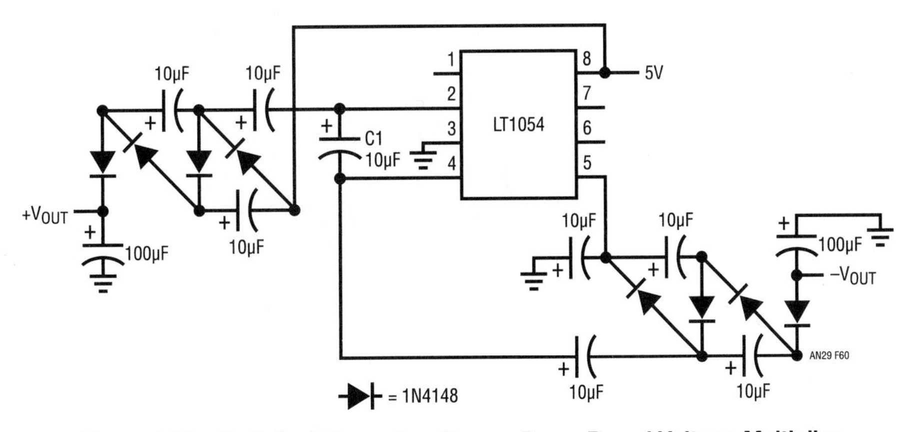

Figure 4.60 • Switched-Capacitor Charge Pump Based Voltage Multiplier

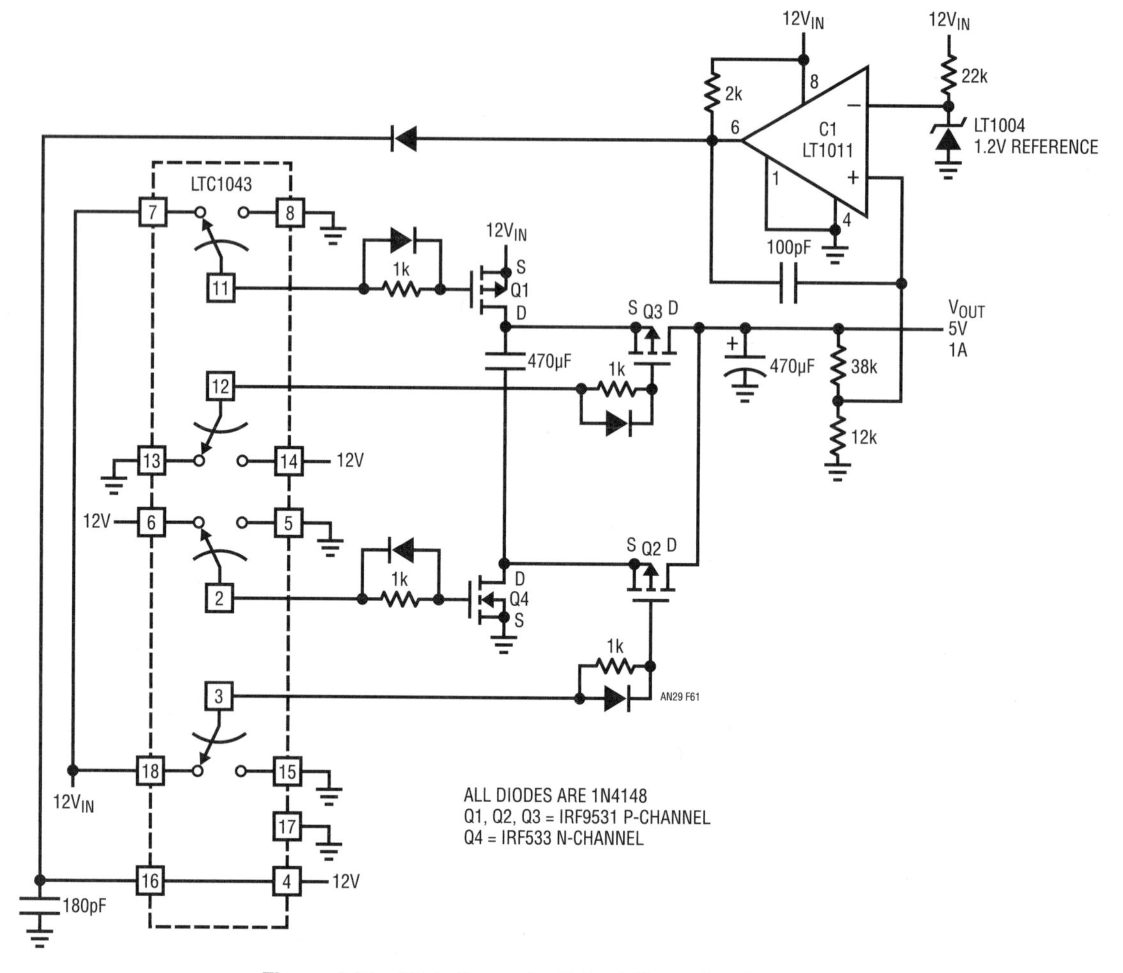

Figure 4.61 • High Power Switched-Capacitor Converter

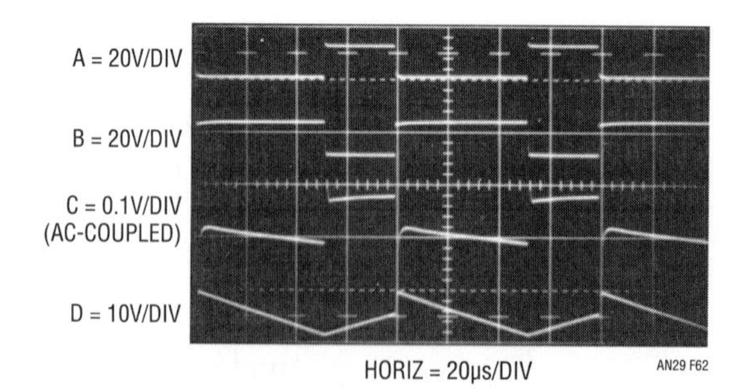

Figure 4.62 • Waveforms for Figure 4.61

Appendix A
The 5V to ±15V converter—a special case

Five volt logic supplies have been standard since the introduction of DTL logic over twenty years ago. Preceding and during DTL's infancy the modular amplifier houses standardized on ±15V rails. As such, popular early monolithic amplifiers also ran from ±15V rails (additional historical perspective on amplifier power supplies appears in AN11's appended section, "Linear Power Supplies—Past, Present and Future"). The 5V supply offered process, speed and density advantages to digital ICs. The ±15V rails provided a wide signal processing range to the analog components. These disparate needs defined power supply requirements for mixed analog-digital systems at 5V and ±15V. In systems with large analog component populations the ±15V supply was and still is usually derived from the AC line. Such line derived ±15V power becomes distinctly undesirable in predominantly digital systems. The inconvenience, difficulty and cost of distributing analog rails in heavily digital systems makes local generation attractive. 5V to ±15V DC/DC converters were developed to fill this need and have been with us for about as long as 5V logic.

Figure A1 is a conceptual schematic of a typical converter. The 5V input is applied to a self-oscillating configuration composed of transistors, a transformer and a biasing network. The transistors conduct out of phase, switching (Figure A2, Traces A and C are Q1's collector and base, while Traces B and D are Q2's collector and base) each time the transformer saturates.[1] Transformer saturation causes a quickly rising, high current to flow (Trace E). This current spike, picked up by the base drive winding, switches the transistors. Transformer current abruptly drops and then slowly rises until saturation again forces switching. This alternating operation sets transistor duty cycle at 50%. The transformers secondary is rectified, filtered and regulated to produce the output.

This configuration has a number of desirable features. The complementary high frequency (typically 20kHz) square wave drive makes efficient use of the transformer and allows relatively small filter capacitors. The self-oscillating primary drive tends to collapse under overload, providing desirable short-circuit characteristics. The transistors switch in saturated mode, aiding efficiency. This hard switching, combined with the transformer's deliberate saturation does, however, have a drawback. During the saturation interval a significant, high frequency current spike is generated (again, Trace E). This spike causes noise to appear at the converter outputs (Trace F is the AC-coupled 15V output). Additionally, it pulls significant current from the 5V supply. The converters input filter partially smooths the transient, but the 5V supply is usually so noisy the disturbance is acceptable. The spike at the output, typically 20mV high, is a more serious

Note 1: This type of converter was originally described by Royer, et al. See References.

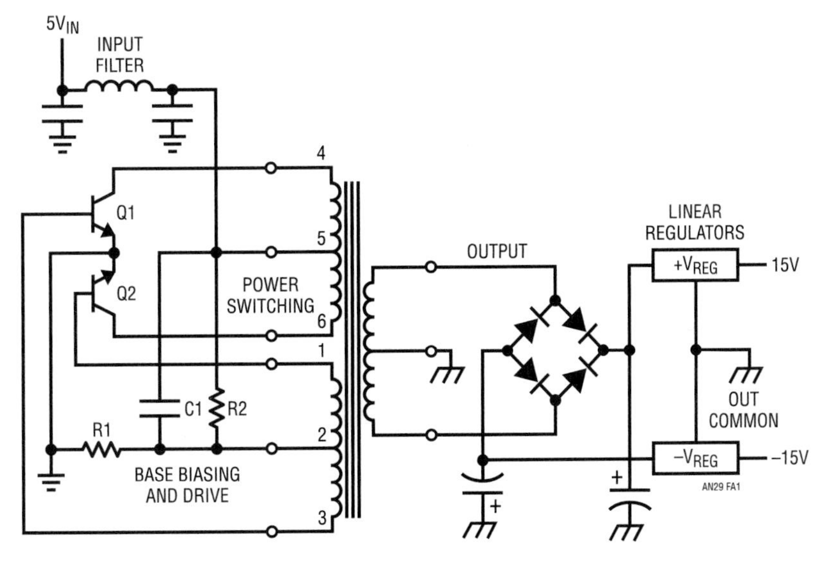

Figure A1 • Conceptual Schematic of a Typical 5V to ±15V Converter

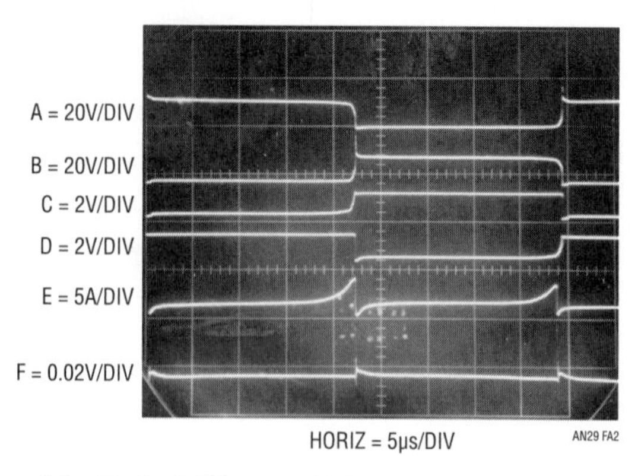

Figure A2 • Typical 5V to ±15V Saturating Converters Waveforms

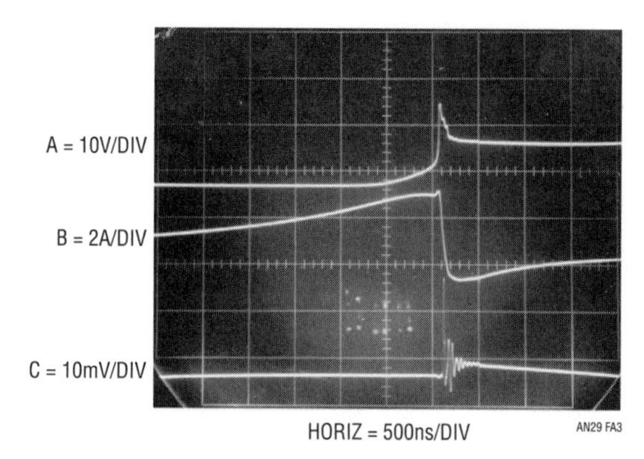

Figure A3 • Switching Details of Saturating Converter

problem. Figure A3 is a time and amplitude expansion of Figure A2's Traces B, E and F. It clearly shows the relationship between transformer current (Trace B, Figure A3), transistor collector voltage (Trace A, Figure A3) and the output spike (Trace C, Figure A3). As transformer current rises, the transistor starts coming out of saturation. When current rises high enough the circuit switches, causing the characteristic noise spike. This condition is exacerbated by the other transistors concurrent switching, causing both ends of the transformer to simultaneously conduct current to ground.

Selection of transistors, output filters and other techniques can reduce spike amplitude, but the converters inherent operation ensures noisy outputs.

This noisy operation can cause difficulties in precision analog systems. IC power supply rejection at the high harmonic spike frequency is low, and analog system errors frequently result. A 12-bit SAR A-to-D converter is a good candidate for such spike-noise caused problems. Sampled data ICs such as switched capacitor filters and chopper amplifiers often show apparent errors which are due to spike induced problems. "Simple" DC circuits can exhibit baffling "instabilities" which in reality are spike caused problems masquerading as DC shifts.

The drive scheme is also responsible for high quiescent current consumption. The base biasing always supplies full drive, ensuring transistor saturation under heavy loading but wasting power at lighter loads. Adaptive bias schemes will mitigate this problem, but increase complexity and almost never appear in converters of this type.

The noise problem is, however, the main drawback of this approach to 5V to ±15V conversion. Careful design, layout, filtering and shielding (for radiated noise) can reduce noise, but cannot eliminate it.

Some techniques can help these converters with the noise problem. Figure A4 uses a "bracket pulse" to warn the powered system when a noise pulse is about to occur. Ostensibly, noise sensitive operations are not carried out during the bracket pulse interval. The bracket pulse (Trace A, Figure A5) drives a delayed pulse generator which triggers (Trace B) the flip-flop. The flip-flop output biases the switching transistors (Q1 collector is Trace C). The output noise spike (Trace D) occurs within the bracket pulse interval. The clocked operation can also prevent transformer saturation, offering some additional noise reduction. This scheme works well, but presumes the powered system can tolerate periodic intervals where critical operations cannot take place.

In Figure A6 the electronic tables are turned. Here, the host system silences the converter when low noise is required. Traces B and C are base and collector drives for one transistor while Traces D and E show drive to the other device. The collector peaking is characteristic of saturating converter operation. Output noise appears on Trace F. Trace A's pulse gates off the converter's base bias, stopping switching. This occurs just past the 6th vertical division. With no switching, the output linear regulator sees the filter capacitor's pure DC and noise disappears.

This arrangement also works nicely but assumes the control pulse can be conveniently generated by the system. It also requires larger filter capacitors to supply power during the low noise interval.

Other methods involve clock synchronization, timing skewing and other schemes which prevent noise spikes from coinciding with sensitive operations. While useful, none of these arrangements offer the flexibility of the inherently noise free converters shown in the text.

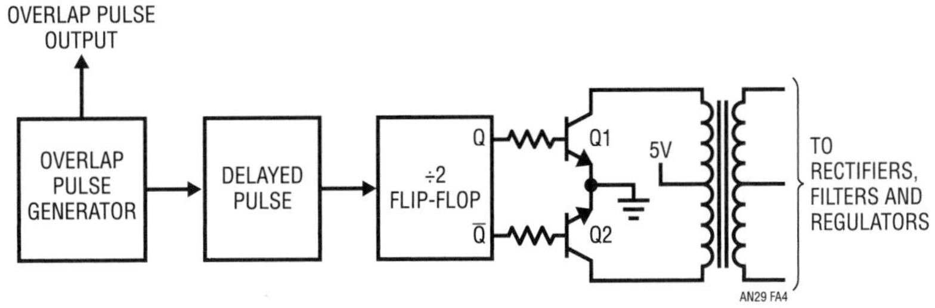

OVERLAP PULSE OUTPUT
OVERLAP PULSE GENERATOR
DELAYED PULSE
÷2 FLIP-FLOP
Q
$\overline{Q}$
Q1
Q2
5V
TO RECTIFIERS, FILTERS AND REGULATORS
AN29 FA4

Figure A4 • Overlap Generator Provides a "Bracket Pulse" Around Noise Spikes

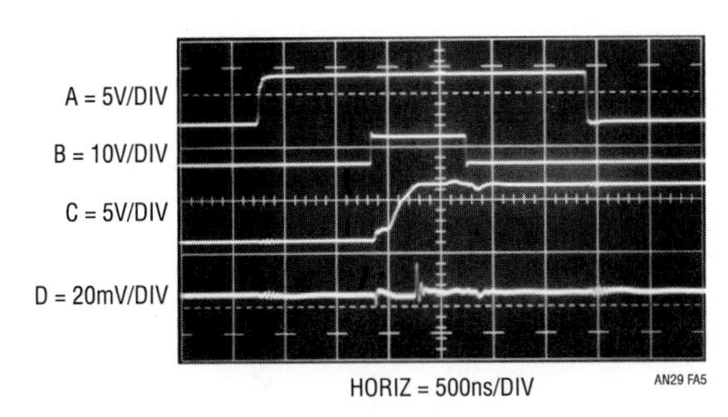

Figure A5 • Waveforms for the Bracket Pulse Based Converter

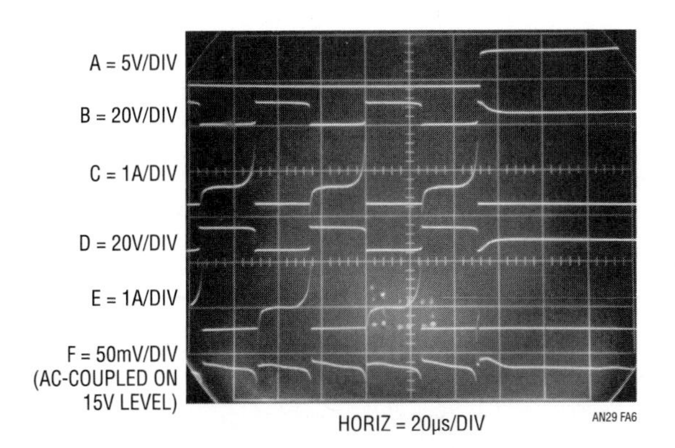

Figure A6 • Detail of the Strobed Operation Converter

Appendix B
Switched capacitor voltage converters—how they work

To understand the theory of operation of switched capacitor converters, a review of a basic switched capacitor building block is helpful.

In Figure B1, when the switch is in the left position, capacitor C1 will charge to voltage V1. The total charge on C1 will be Q1 = C1V1. The switch then moves to the right, discharging C1 to voltage V2. After this discharge time, the charge on C1 is Q2 = C1V2. Note that charge has been transferred from the source, V1, to the output, V2. The amount of charge transferred is:

$$Q = Q1 - Q2 = C1(V1 - V2)$$

If the switch is cycled f times per second, the charge transfer per unit time (i.e., current) is:

$$I = f \bullet Q = f \bullet C1(V1 - V2)$$

To obtain an equivalent resistance for the switched-capacitor network we can rewrite this equation in terms of voltage and impedance equivalence:

$$I = \frac{V1 - V2}{\frac{1}{fC1}} = \frac{V1 - V2}{R_{EQUIV}}$$

A new variable, R_{EQUIV}, is defined such that $R_{EQUIV} = 1/fC1$. Thus, the equivalent circuit for the switched-capacitor network is as shown in Figure B2. The LT1054 and other switched-capacitor converters have the same switching action as the basic switched-capacitor building block. Even though this simplification doesn't include finite switch on-resistance and output voltage ripple, it provides an intuitive feel for how the device works.

These simplified circuits explain voltage loss as a function of frequency. As frequency is decreased, the output impedance will eventually be dominated by the 1/fC1 term and voltage losses will rise.

Note that losses also rise as frequency increases. This is caused by internal switching losses which occur due to some finite charge being lost on each switching cycle. This charge loss per-unit-cycle, when multiplied by the switching frequency, becomes a current loss. At high frequency this loss becomes significant and voltage losses again rise.

The oscillators of practical converters are designed to run in the frequency band where voltage losses are at a minimum. Figure B3 shows the block diagram of the LT1054 switched-capacitor converter.

The LT1054 is a monolithic, bipolar, switched-capacitor, voltage converter, and regulator. It provides higher output current then previously available converters with significantly lower voltage losses. An adaptive switch drive scheme optimizes efficiency over a wide range of output currents. Total voltage loss at 100mA output current is typically 1.1V. This holds true over the full supply voltage range of 3.5V to 15V. Quiescent current is typically 2.5mA.

The LT1054 also provides regulation. By adding an external resistive divider, a regulated output can be obtained. This output will be regulated against changes in input voltage and output current. The LT1054 can also be shut down by grounding the feedback pin. Supply current in shutdown is less than 100μA.

The internal oscillator of the LT1054 runs at a nominal frequency of 25kHz. The oscillator pin can be used to adjust the switching frequency, or to externally synchronize the LT1054.

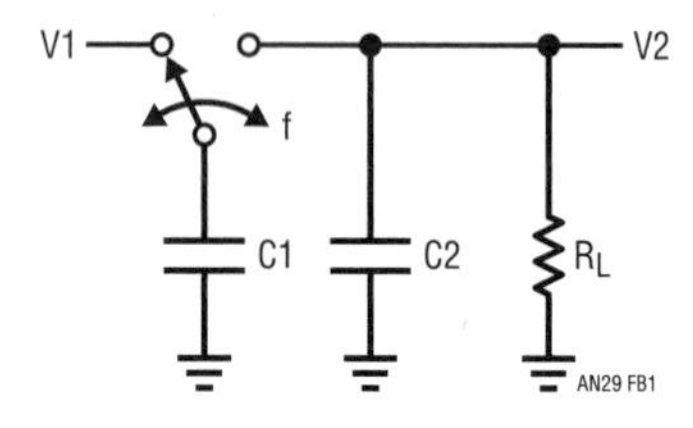

Figure B1 • Switched-Capacitor Building Block

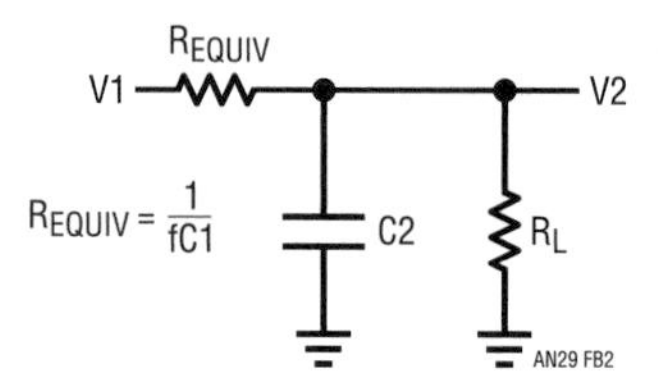

Figure B2 • Switched-Capacitor Equivalent Circuit

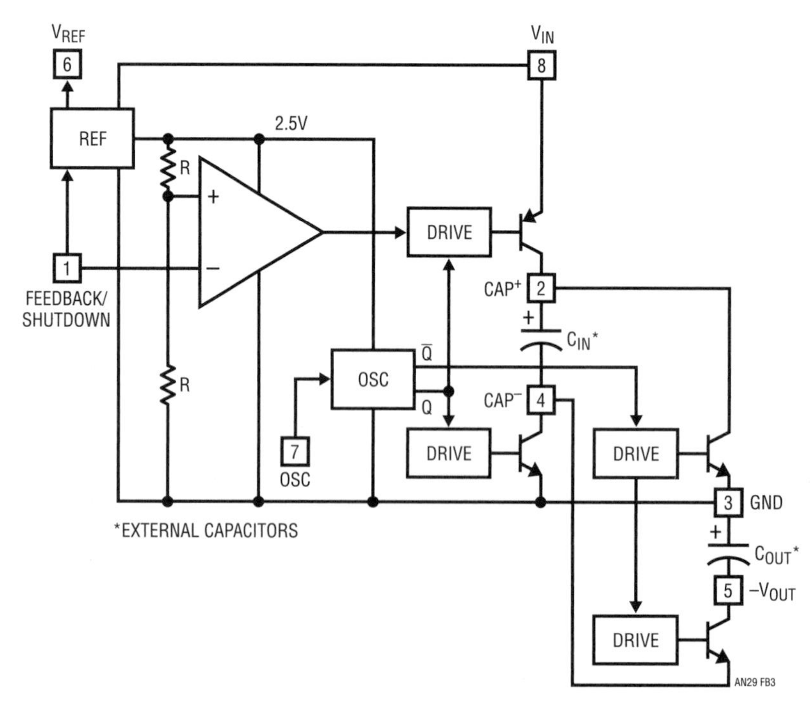

Figure B3 • LT1054 Switched-Capacitor Converter Block Diagram

Appendix C Physiology of the LT1070

The LT1070 is a current mode switcher. This means that switch duty cycle is directly controlled by switch current rather than by output voltage. Referring to Figure C1, the switch is turned on at the start of each oscillator cycle. It is turned off when switch current reaches a predetermined level. Control of output voltage is obtained by using the output of a voltage-sensing error amplifier to set current trip level. This technique has several advantages. First, it has immediate response to input voltage variations, unlike ordinary switchers which have notoriously poor line transient response. Second, it reduces the 90° phase shift at mid-frequencies in the energy storage inductor. This greatly simplifies closed-loop frequency compensation under widely varying input voltage or output load conditions. Finally, it allows simple pulse-by-pulse current limiting to provide maximum switch protection under output overload or short conditions. A low dropout internal regulator provides a 2.3V supply for all internal circuitry on the LT1070. This low dropout design allows input voltage to vary from 3V to 60V with virtually no change in device performance. A 40kHz oscillator is the basic clock for all internal timing. It turns on the output switch via the logic and driver circuitry. Special adaptive antisat circuitry detects onset of saturation in the power switch and adjusts driver current instantaneously to limit switch saturation. This minimizes driver dissipation and provides very rapid turn-off of the switch.

A 1.2V bandgap reference biases the positive input of the error amplifier. The negative input is brought out for output voltage sensing. This feedback pin has a second function; when pulled low with an external resistor, it programs the LT1070 to disconnect the main error amplifier output and connects the output of the flyback amplifier to the comparator input. The LT1070 will then regulate the value of the flyback pulse with respect to the supply voltage. This flyback pulse is directly proportional to output voltage in the traditional transformer-coupled flyback topology regulator. By regulating the amplitude of the flyback pulse the output voltage can be regulated

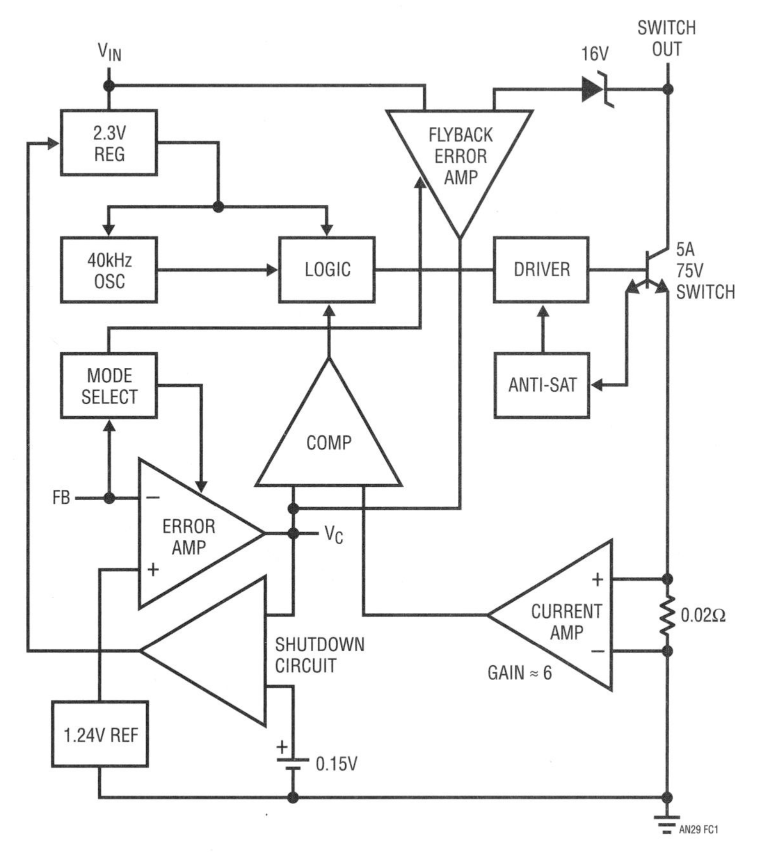

Figure C1 • LT1070 Internal Details

with no direct connection between input and output. The output is fully floating up to the breakdown voltage of the transformer windings. Multiple floating outputs are easily obtained with additional windings. A special delay network inside the LT1070 ignores the leakage inductance spike at the leading edge of the flyback pulse to improve output regulation.

The error signal developed at the comparator input is brought out externally. This pin (V_C) has four different functions. It is used for frequency compensation, current limit adjustment, soft-starting, and total regulator shutdown. During normal regulator operation this pin sits at a voltage between 0.9V (low output current) and 2.0V (high output current). The error amplifiers are current output (g_m) types, so this voltage can be externally clamped for adjusting current limit. Likewise, a capacitor-coupled external clamp will provide soft-start. Switch duty cycle goes to zero if the V_C pin is pulled to ground through a diode, placing the LT1070 in an idle mode. Pulling the V_C pin below 0.15V causes total regulator shutdown with only 50μA supply current for shutdown circuitry biasing. For more details, see Linear Technology Application Note 19, Pages 4-8.

Appendix D Inductor selection for flyback converters

A common problem area in DC/DC converter design is the inductor, and the most common difficulty is saturation. An inductor is saturated when it cannot hold any more magnetic flux. As an inductor arrives at saturation it begins to look more resistive and less inductive. Under these conditions current flow is limited only by the inductor's DC copper resistance and the source capacity. This is why saturation often results in destructive failures.

While saturation is a prime concern, cost, heating, size, availability and desired performance are also significant. Electromagnetic theory, although applicable to these issues, can be confusing, particularly to the non-specialist.

Practically speaking, an empirical approach is often a good way to address inductor selection. It permits real time analysis under actual circuit operating conditions using the ultimate simulator—a breadboard. If desired, inductor design theory can be used to augment or confirm experimental results.

Figure D1 shows a typical flyback based converter utilizing the LT1070 switching regulator. A simple approach may be employed to determine the appropriate inductor. A very useful tool is the #845 inductor kit[2] shown in Figure D2. This kit provides a broad range of inductors for evaluation in test circuits such as Figure D1.

Figure D3 was taken with a 450μH value, high core capacity inductor installed. Circuit operating conditions such as input voltage and loading are set at levels appropriate to the intended application. Trace A is the LT1070's V_{SW} pin voltage while Trace B shows its current. When V_{SW} pin voltage is low, inductor current flows. The high inductance means current rises relatively slowly, resulting in the shallow slope observed. Behavior is linear, indicating no saturation problems. In Figure D4, a lower value unit with equivalent core characteristics is tried. Current rise is steeper, but saturation is not encountered. Figure D5's selected inductance is still lower, although core characteristics are similar. Here, the current ramp is quite

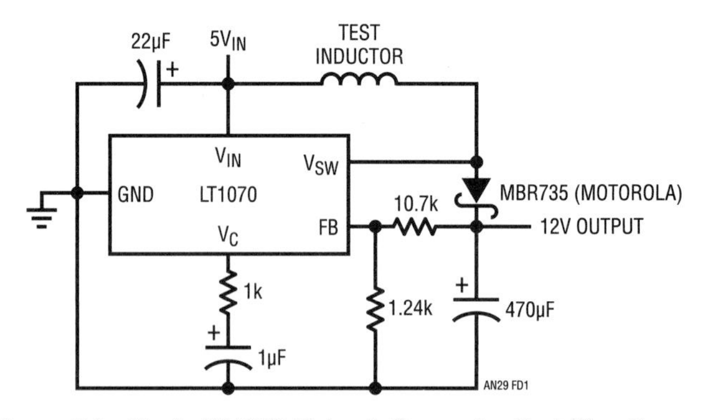

Figure D1 • Basic LT1070 Flyback Converter Test Circuit

Note 2: Available from Pulse Engineering, Inc., P.O. Box 12235, San Diego, CA 92112, 619-268-2400.

pronounced, but well controlled. Figure D6 brings some informative surprises. This high value unit, wound on a low capacity core, starts out well but heads rapidly into saturation, and is clearly unsuitable.

The described procedure narrows the inductor choice within a range of devices. Several were seen to produce acceptable electrical results, and the "best" unit can be further selected on the basis of cost, size, heating and other parameters. A standard device in the kit may suffice, or a derived version can be supplied by the manufacturer.

Using the standard products in the kit minimizes specification uncertainties, accelerating the dialogue between user and inductor vendor.

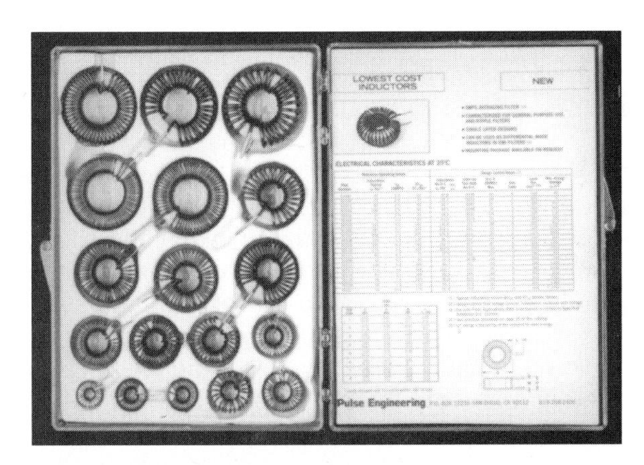

Figure D2 • Model 845 Inductor Selection Kit from Pulse Engineering, Inc. (Includes 18 Fully Specified Devices)

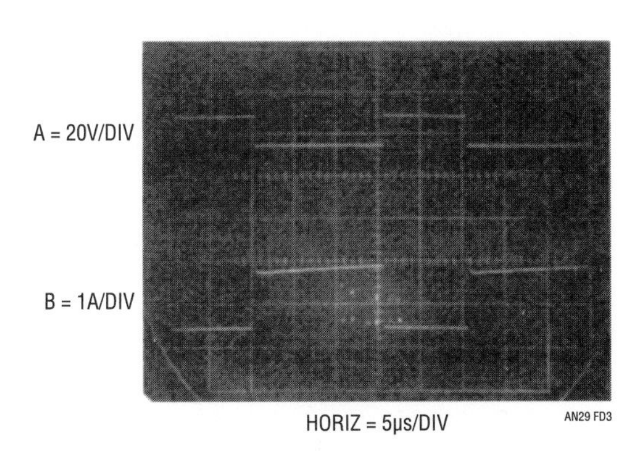

Figure D3 • Waveforms for 450μH, High Capacity Core Unit

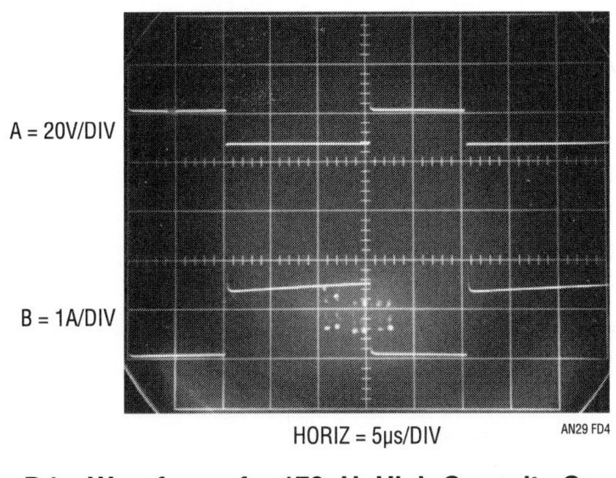

Figure D4 • Waveforms for 170μH, High Capacity Core Unit

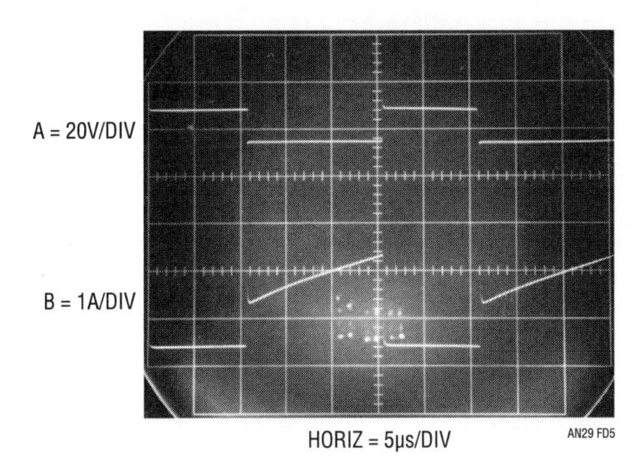

Figure D5 • Waveforms for 55μH, High Capacity Core Unit

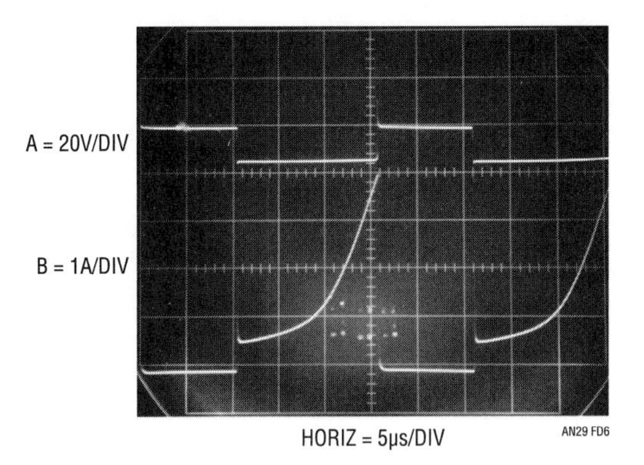

Figure D6 • Waveforms for 500μH, Low Capacity Core Inductor (Note Saturation Effects)

Appendix E
Optimizing converters for efficiency

Squeezing the utmost efficiency out of a converter is a complex, demanding design task. Efficiency exceeding 80% to 85% requires some combination of finesse, witchcraft and just plain luck. Interaction of electrical and magnetic terms produces subtle effects which influence efficiency. A detailed, generalized method for obtaining maximum converter efficiency is not readily described but some guidelines are possible.

Losses fall into several loose categories including junction, ohmic, drive, switching, and magnetic losses.

Semiconductor junctions produce losses. Diode drops increase with operating current and can be quite costly in low voltage output converters. A 700mV drop in a 5V output converter introduces more than 10% loss. Schottky devices will cut this nearly in half, but loss is still appreciable. Germanium (rarely used) is lower still, but switching losses negate the low DC drop at high speeds. In very low power converters Germanium's reverse leakage may be equally oppressive. Synchronously-switched rectification is more complex, but can sometimes simulate a more efficient diode (see text Figure 4.32). When evaluating such a scheme remember to include both AC and DC drive losses in efficiency estimates. DC losses include base or gate current in addition to DC consumption in any driver stage. AC losses might include the effects of gate (or base) capacitance, transition region dissipation (the switch spends some time in its linear region) and power lost due to timing skew between drive and actual switch action.

Transistor saturation losses are also a significant term. Channel and collector-emitter saturation losses become increasingly significant as operating voltages decrease. The most obvious way to minimize these losses is to select low saturation components. In some cases this will work, but remember to include the drive losses (usually higher) for lower saturation devices in overall loss estimates. Actual losses caused by saturation effects and diode drops is sometimes difficult to ascertain. Changing duty cycles and time variant currents make determination tricky. One simple way to make relative loss judgements is to measure device temperature rise. Appropriate tools here include thermal probes and (at low voltages) the perhaps more readily available human finger. At lower power (e.g., less dissipation, even though loss percentage may be as great) this technique is less effective. Sometimes deliberately adding a known loss to the component in question and noting efficiency change allows loss determination.

Ohmic losses in conductors are usually only significant at higher currents. "Hidden" ohmic losses include socket and connector contact resistance and equivalent series resistance (ESR) in capacitors. ESR generally drops with capacitor value and rises with operating frequency, and should be specified on the capacitor data sheet. Consider the copper resistance of inductive components. It is often necessary to evaluate trade-offs of an inductors copper resistance versus magnetic characteristics.

Drive losses were mentioned, and are important in obtaining efficiency. MOSFET gate capacitance draws substantial AC drive current per cycle, implying higher average currents as frequency goes up. Bipolar devices have lower capacitance, but DC base current eats power. Large area devices may appear attractive for low saturation, but evaluate drive losses carefully. Usually, large area devices only make sense when operating at a significant percentage of rated current. Drive stages should be thought out with respect to efficiency. Class A type drives (e.g., resistive pull-up or pull-down) are simple and fast, but wasteful. Efficient operation usually requires active source-sink combinations with minimal cross conduction and biasing losses.

Switching losses occur when devices spend significant amounts of time in their linear region relative to operating frequency. At higher repetition rates transition times can become a substantial loss source. Device selection and drive techniques can minimize these losses.

Magnetics design also influences efficiency. Design of inductive components is well beyond the scope of this appended section, but issues include core material selection, wire type, winding techniques, size, operating frequency, current levels, temperature and other issues.

Some of these topics are discussed in Linear Technology Application Note 19, but there is no substitute for access to a skilled magnetics specialist. Fortunately, the other categories mentioned usually dominate losses, allowing good efficiencies to be obtained with standard magnetics. Custom magnetics are usually only employed after circuit losses have been reduced to lowest practical levels.

Appendix F
Instrumentation for converter design

Instrumentation for DC/DC converter design should be selected on the basis of *flexibility*. Wide bandwidths, high resolution and computational sophistication are valuable features, but are usually not required for converter work. Typically, converter design requires simultaneous observation of many circuit events at relatively slow speeds. Single ended and differential voltage and current signals are of interest, with some measurements requiring fully floating inputs. Most low level measurement involves AC signals and is accommodated with a high sensitivity plug-in. Other situations call for observation of small, slowly changing (e.g., 0.1Hz to 10Hz) events on top of DC levels. This range falls outside the AC-coupled cutoff of most oscilloscopes, mandating differential DC nulling or "slide-back" plug-in capability. Other requirements include high impedance probes, filters and oscilloscopes with very versatile triggering and multi-trace capability. In our converter work we have found a number of particularly noteworthy instruments in several categories.

Probes

For many measurements standard 1× and 10× scope probes are fine. In most cases the ground strap may be used, but low level measurements, particularly in the presence of wideband converter switching noise, should be taken with the shortest possible ground return. A variety of probe tip grounding accessories are available, and are usually supplied with good quality probes (see Figure F1). In some cases, directly connecting the breadboard to the 'scope may be necessary (Figure F2).

Wideband FET probes are not normally needed, but a moderate speed, high input impedance buffer probe is quite useful. Many converter circuits, especially micropower designs, require monitoring of high impedance nodes. The 10MΩ loading of standard 10× probes usually suffices, but sensitivity is traded away. 1× probes retain sensitivity, but introduce heavier loading. Figure F3 shows an almost absurdly simple, but useful, circuit which greatly aids probe loading problems. The LT1022 high speed FET op amp drives an LT1010 buffer. The LT1010's output allows cable and probe driving and also biases the circuit's input shield. This bootstraps the input capacitance, reducing its effect. DC and AC errors of this circuit are low enough for almost all converter work, with enough bandwidth for most circuits. Built into a small enclosure with its own power supply, it can be used ahead of a 'scope or DVM with good results. Pertinent specifications appear in the diagram.

Figure F4 shows a simple probe filter which sets high and low bandwidth restrictions. This circuit, placed in series with the 'scope input, is useful for eliminating switching artifacts when observing circuit nodes.

An *isolated* probe allows fully floating measurements, even in the presence of high common mode voltages. It is often desirable to look across floating points in a circuit. The ability to directly observe an ungrounded transistor's saturation characteristics or monitor waveforms across a floating shunt makes this probe valuable. One probe, the Signal Acquisition Technologies, Inc. Model SL-10, has 10MHz bandwidth and 600V common mode capability.

Current probes are an indispensable tool in converter design. In many cases current waveforms contain more valuable information than voltage measurements. The clip-on types are quite convenient. Hall effect based

Figure F1 • Proper Probing Technique for Low Level Measurements in the Presence of High Frequency Noise

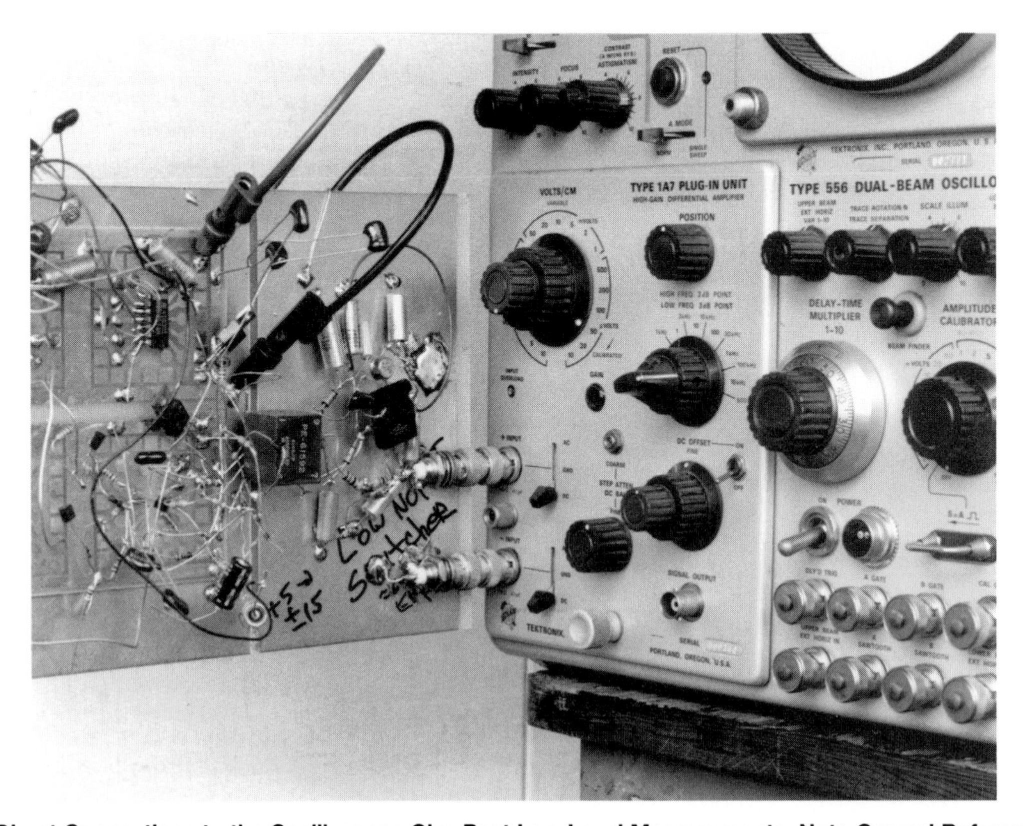

Figure F2 • Direct Connections to the Oscilloscope Give Best Low Level Measurements. Note Ground Reference Connection to the Differential Plug-In's Negative Input

versions respond down to DC, with bandwidths of 50MHz. Transformer types are faster, but roll off below several hundred cycles (Figure F5). Both types have saturation limitations which, when exceeded, cause odd results on the CRT, confusing the unwary. The Tektronix P6042 (and the more recent AM503) Hall type and P6022/134 transformer based type give excellent results. The Hewlett-Packard 428B clip-on current probe responds from DC to only 400Hz, but features 3% accuracy over a 100μA to 10A range. This instrument, useful for determining efficiency and quiescent current, eliminates shunt caused measurement errors.

Oscilloscopes and plug-ins

The oscilloscope plug-in combination is an important choice. Converter work almost demands multi-trace capability. Two channels are barely adequate, with four far preferable. The Tektronix 2445/6 offers four channels, but two have limited vertical capability. The Tektronix 547 (and the more modern 7603), equipped with a type 1A4 (2 dual trace 7A18s required for the 7603) plug-in, has four full capability input channels with flexible triggering and superb CRT trace clarity. This instrument, or its equivalent, will handle a wide variety of converter circuits with minimal restrictions. The Tektronix 556 offers an extraordinary array of features valuable in converter work. This dual beam instrument is essentially two fully independent oscilloscopes sharing a single CRT. Independent vertical, horizontal and triggering permit detailed display of almost any converters operation. Equipped with two type 1A4 plug-ins, the 556 will display eight real time inputs. The independent triggering and time bases

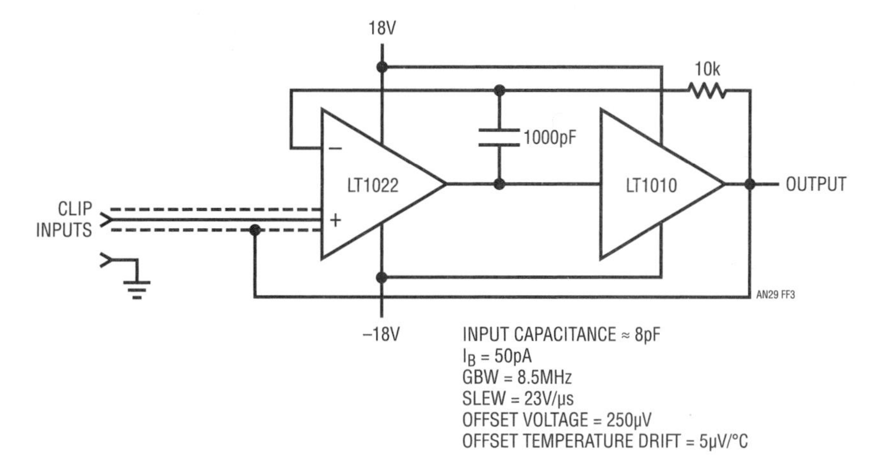

Figure F3 • A Simple High Impedance Probe

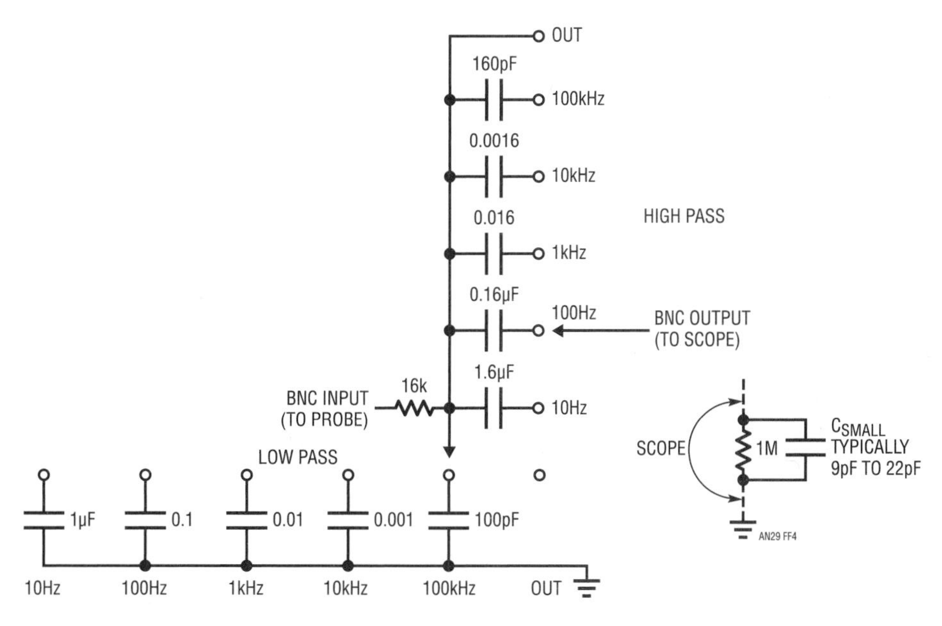

Figure F4 • Oscilloscope Filter

allow stable display of asynchronous events. Cross beam triggering is also available, and the CRT has exceptional trace clarity.

Two oscilloscope plug-in types merit special mention. At low level, a high sensitivity differential plug-in is indispensable. The Tektronix 1A7 and 7A22 feature 10μV sensitivity, although bandwidth is limited to 1MHz. The units also have selectable high and low pass filters and good high frequency common mode rejection. Tektronix types W, 1A5 and 7A13 are differential comparators. They have calibrated DC nulling ("slideback") sources, allowing observation of small, slowly moving events on top of common mode DC.

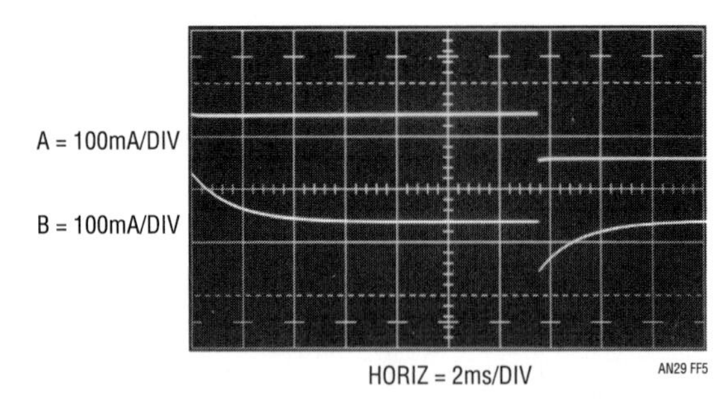

Figure F5 • Hall (Trace A) and Transformer (Trace B) Based Current Probes Responding to Low Frequency

Voltmeters

Almost any DVM will suffice for converter work. It should have current measurement ranges and provision for battery operation. The battery operation allows floating measurements and eliminates possible ground loop errors. Additionally, a *non-electronic* (VOM) voltmeter (e.g., Simpson 260, Triplett 630) is a worthwhile addition to the converter design bench. Electronic voltmeters are occasionally disturbed by converter noise, producing erratic readings. A VOM contains no active circuitry, making it less susceptible to such effects.

Appendix G
The magnetics issue

Magnetics is probably the most formidable issue in converter design. Design and construction of suitable magnetics is a difficult task, particularly for the non-specialist.

Figure G1 • Magnetics for LTC Applications Circuits are Designed and Supplied as Standard Product by Pulse Engineering, Inc.

It is our experience that the majority of converter design problems are associated with magnetics requirements. This consideration is accented by the fact that most converters are employed by non-specialists. As a purveyor of switching power ICs we incur responsibility towards addressing the magnetics issue (our publicly spirited attitude is, admittedly, capitalistically influenced). As such, it is LTC's policy to use off-the-shelf magnetics in our circuits. In some cases, available magnetics serve a particular design. In other situations the magnetics have been specially designed, assigned a part number and made available as standard product. In these endeavors our magnetics supplier and partner is;

Pulse Engineering, Inc.
P.O. Box 12235
7250 Convoy Court
San Diego, California 92112
619-268-2400

In many circumstances a standard product is suitable for production. Other cases may require modifications or changes which Pulse Engineering can provide. Hopefully, this approach serves the needs of all concerned (see Figure G1).

Appendix H
LT1533 ultralow noise switching regulator for high voltage or high current applications

The LT1533 switching regulator[1, 2] achieves 100μV output noise by using closed-loop control around its output switches to tightly control switching transition time. Slowing down switch transitions eliminates high frequency harmonics, greatly reducing conducted and radiated noise.

The part's 30V, 1A output transistors limit available power. It is possible to exceed these limits while maintaining low noise performance by using suitably designed output stages.

High voltage input regulator

The LT1533's IC process limits collector breakdown to 30V. A complicating factor is that the transformer causes the collectors to swing to twice the supply voltage. Thus, 15V represents the maximum allowable input supply. Many applications require higher voltage inputs; the circuit in Figure H1 uses a cascoded[3] output stage to achieve such high voltage capability. This 24V to 5V (V_{IN} = 20V–50V) converter is reminiscent of previous LT1533 circuits, except for the presence of Q1 and Q2.[4] These devices, interposed between the IC and the transformer, constitute a cascoded high voltage stage. They provide voltage gain while isolating the IC from their large drain voltage swings.

Normally, high voltage cascodes are designed to simply supply voltage isolation. Cascoding the LT1533 presents special considerations because the transformer's instantaneous voltage and current information must be accurately transmitted, albeit at lower amplitude, to the LT1533. If this is not done, the regulator's slew-control loops will not function, causing a dramatic output noise increase. The AC-compensated resistor dividers associated with the Q1–Q2 gate-drain biasing serve this

Note 1: Witt, Jeff. The LT1533 Heralds a New Class of Low Noise Switching Regulators. Linear Technology VII:3 (August 1997).

Note 2: Williams, Jim. LTC Application Note 70: A Monolithic Switching Regulator with 100μV Output Noise. October 1997.

Note 3: The term "cascode," derived from "cascade to cathode," is applied to a configuration that places active devices in series. The benefit may be higher breakdown voltage, decreased input capacitance, bandwidth improvement or the like. Cascoding has been employed in op amps, power supplies, oscilloscopes and other areas to obtain performance enhancement.

Note 4: This circuit derives from a design by Jeff Witt of Linear Technology Corp.

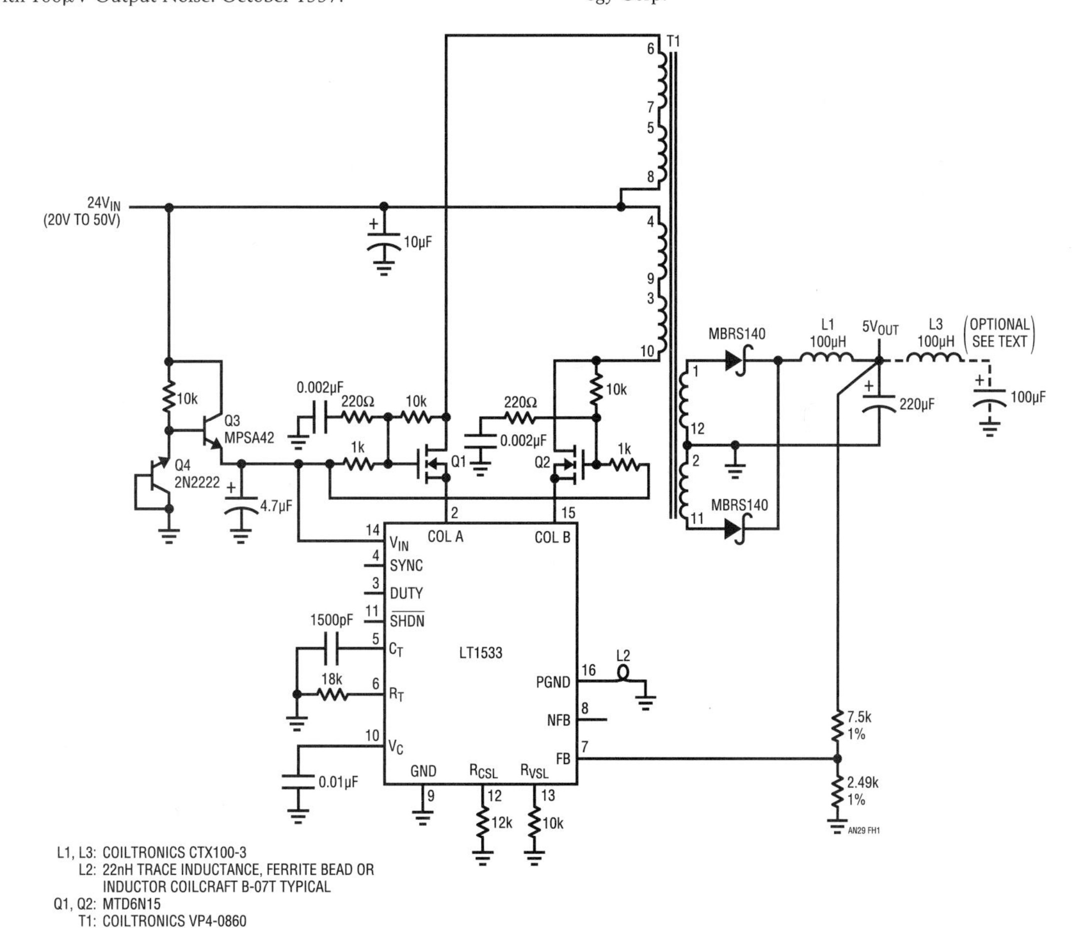

Figure H1 • A Low Noise 24V to 5V Converter (V_{IN} = 20V–50V): Cascoded MOSFETs Withstand 100V Transformer Swings, Permitting the LT1533 to Control 5V/2A Output

purpose, preventing transformer swings coupled via gate-channel capacitance from corrupting the cascode's waveform-transfer fidelity. Q3 and associated components provide a stable DC termination for the dividers while protecting the LT1533 from the high voltage input.

Figure H2 shows that the resultant cascode response is faithful, even with 100V swings. Trace A is Q1's source; traces B and C are its gate and drain, respectively. Under these conditions, at 2A output, noise is inside 400μV peak.

Current boosting

Figure H3 boosts the regulator's 1A output capability to over 5A. It does this with simple emitter followers (Q1–Q2). Theoretically, the followers preserve T1's voltage and current waveform information, permitting the LT1533's slew-control circuitry to function. In practice, the transistors must be relatively low beta types. At 3A collector current, their beta of 20 sources ≈150mA via the Q1–Q2 base paths, adequate for proper slew-loop operation.[5]

The follower loss limits efficiency to about 68%. Higher input voltages minimize follower-induced loss, permitting efficiencies in the low 70% range.

Figure H4 shows noise performance. Ripple measures 4mV (Trace A) using a single LC section, with high frequency content just discernible. Adding the optional second LC section reduces ripple to below 100μV (trace B), and high frequency content is seen to be inside 180μV (note ×50 vertical scale-factor change).

Note 5: Operating the slew loops from follower base current was suggested by Bob Dobkin of Linear Technology Corp.

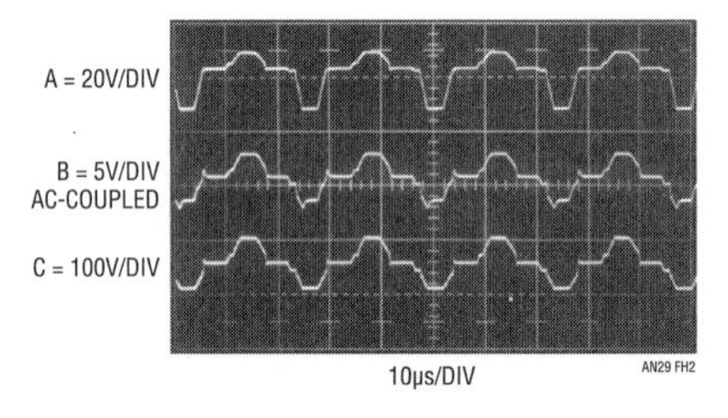

Figure H2 • MOSFET-Based Cascode Permits the Regulator to Control 100V Transformer Swings While Maintaining a Low Noise 5V Output. Trace A Is Q1's Source, Trace B Is Q1's Gate and Trace C Is the Drain. Waveform Fidelity Through Cascode Permits Proper Slew-Control Operation

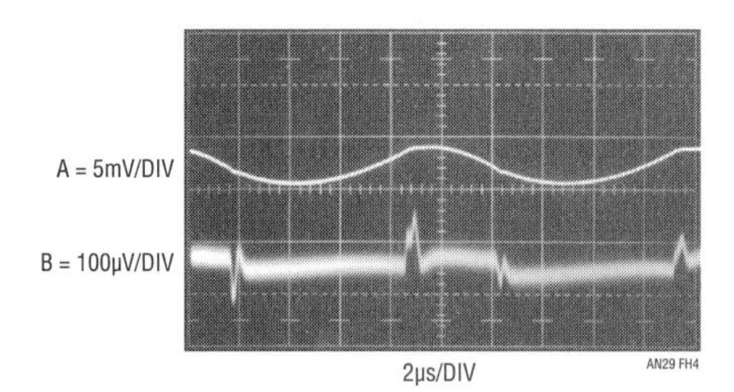

Figure H4 • Waveforms for Figure H3 at 10W Output: Trace A Shows Fundamental Ripple with Higher Frequency Residue Just Discernible. The Optional LC Section Results in Trace B's 180μV$_{P-P}$ Wideband Noise Performance

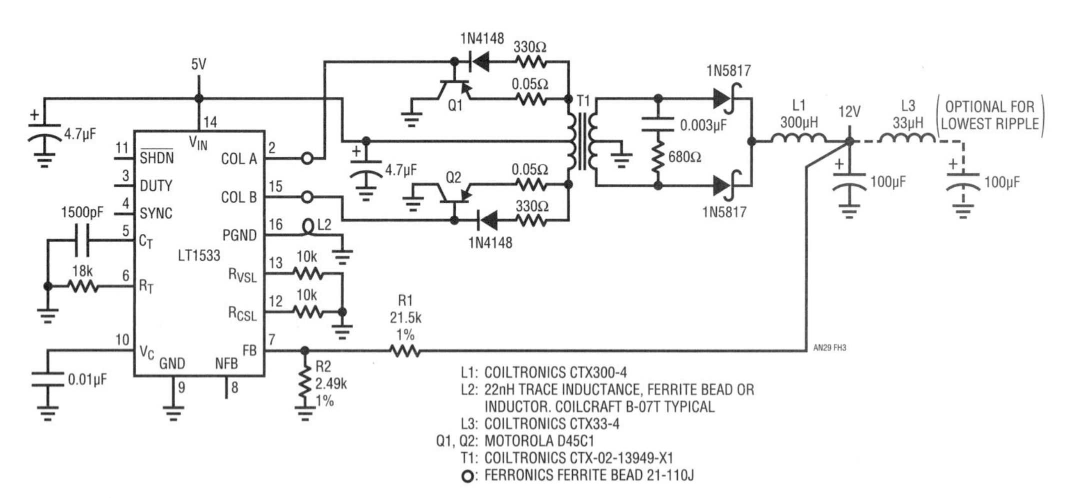

Figure H3 • A 10W Low Noise 5V to 12V Converter: Q1–Q2 Provide 5A Output Capacity While Preserving the LT1533's Voltage/Current Slew Control. Efficiency Is 68%. Higher Input Voltages Minimize Follower Loss, Boosting Efficiency Above 71%

Theoretical considerations for buck mode switching regulators

5

Carl Nelson

Introduction

The use of switching regulators increased dramatically in the 1980's and this trend remains strong going into the 90s. The reasons for this are simple; heat and efficiency. Today's systems are shrinking continuously, while simultaneously offering greater electronic "horsepower." This combination would result in unacceptably high internal temperatures if low efficiency linear supplies were used. Heat sinks do not solve the problem in general because most systems are closed, with low thermal transfer from "inside" to "outside."

Battery-powered systems need high efficiency supplies for long battery life. Topological considerations also require switching technology. For instance, a battery cannot generate an output higher than itself with linear supplies. The availability of low cost rechargeable batteries has created a spectacular rise in the number of battery-powered systems, and consequently a matching rise in the use of switching regulators.

The LT®1074 and LT1076 switching regulators are designed specifically for ease of use. They are close to the ultimate "three terminal box" concept which simply requires an input, output and ground connection to deliver power to the load. Unfortunately, switching regulators are not horseshoes, and "close" still leaves room for egregious errors in the final execution. This application note is intended to eliminate the most common errors that customers make with switching regulators as well as offering some insight into the inner workings of switching designs.

There is also an entirely new treatment of inductor design based on the mathematical models of core loss and peak current. This allows the customer to quickly see the allowable limits for inductor value and make an intelligent decision based on the need for cost, size, etc. The procedure differs greatly from previous design techniques and many experienced designers at first think it can't work. They quickly become silent after standard laborious trial-and-error techniques yield identical results.

There is an old adage in woodworking—"Measure twice, cut once." This advice holds for switching regulators, also. Read AN44 through quickly to familiarize yourself with the contents. Then reread the pertinent sections carefully to avoid "cutting" the design two, three, or four times. Some switching regulator errors, such as excessive ripple current in capacitors, are time bombs best fixed *before* they are expensive field failures.

Since this paper was originally written, Linear Technology has produced a CAD program for switching regulators called LTspice.® A spice simulator, LTspice, has been developed and optimized for switching regulator simulation. IC models for switching regulators with fast transient simulation allow regulator circuits to be simulated for transient response without resorting to linearized models.

Once the basic design concepts are understood, trial designs can be quickly checked and modified on the simulator. Start-up, dropout, regulation, ripple and transient response are available from the simulator. The output correlates well with the actual circuit on a well laid-out board.

LTspice can be downloaded free from www.linear.com.

Analog Circuit and System Design: Immersion in the Black Art of Analog Design. http://dx.doi.org/10.1016/B978-0-12-397888-2.00005-5

Absolute maximum ratings

Input Voltage
- LT1074/LT1076 .... 45V
- LT1074HV/LT1076HV 64V

Switch Voltage with Respect to Input Voltage
- LT1074/LT1076 .... 64V
- LT1074HV/LT1076HV 75V

Switch Voltage with Respect to Ground Pin (V_{SW} **Negative**)
- LT1074/LT1076 (Note 6) 35V
- LT1074HV/LT1076HV (Note 6) 45V

Feedback Pin Voltage .... −2V, +10V

Shutdown Pin Voltage (Not to Exceed V_{IN}) 40V

Status Pin Voltage 30V
(Current Must Be Limited to 5mA When Status Pin Switches On)

I_{LIM} Pin Voltage (Forced) 5.5V

Maximum Operating Ambient Temperature Range
- LT1074C/76C, LT1074HVC/76HVC .... 0°C to 70°C
- LT1074M/76M, LT1074HVM/76HVM .... −55°C to 125°C

Maximum Operating Junction Temperature Range
- LT1074C/76C, LT1074HVC/76HVC .... 0°C to 125°C
- LT1074M/76M, LT1074HVM/76HVM .... −55°C to 150°C

Maximum Storage Temperature .... −65°C to 150°C

Lead Temperature (Soldering, 10 sec.) .... 300°C

Package/order information

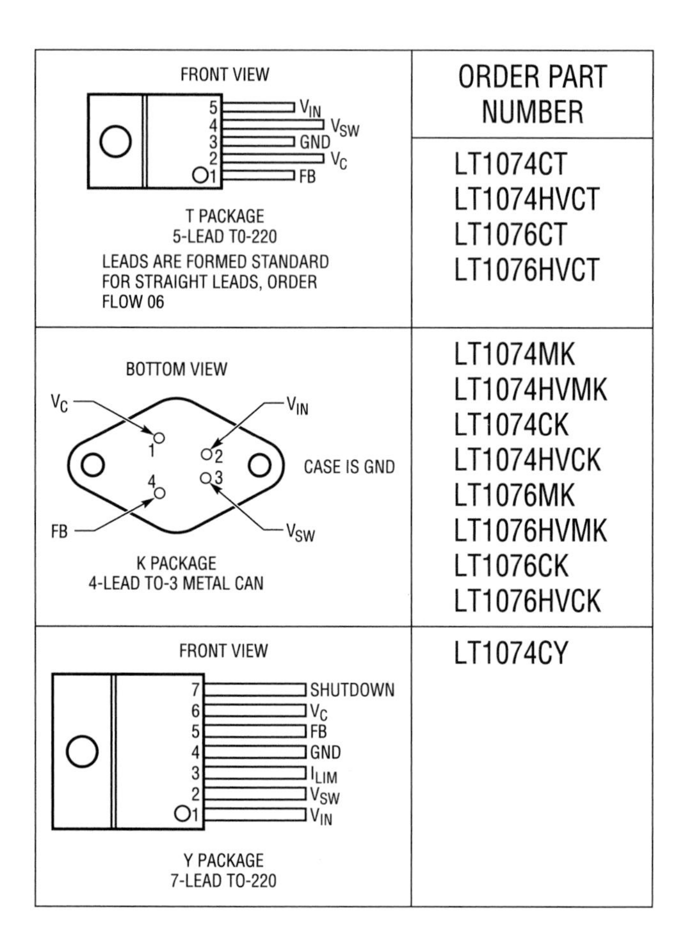

ELECTRICAL CHARACTERISTICS
T_J = 25°C, V_{IN} = 25V, UNLESS OTHERWISE NOTED.

PARAMETER	CONDITIONS			MIN	TYP	MAX	UNITS
Switch On Voltage (Note 1)	LT1074	$I_{SW} = 1A, T_J \geq 0°C$				1.85	V
		$I_{SW} = 1A, T_J < 0°C$				2.1	V
		$I_{SW} = 5A, T_J \geq 0°C$				2.3	V
		$I_{SW} = 5A, T_J < 0°C$				2.5	V
	LT1076	$I_{SW} = 0.5A$	•			1.2	V
		$I_{sw} = 2A$	•			1.7	V
Switch Off Leakage	LT1074	$V_{IN} \leq 25V, V_{sw} = 0$			5	300	μA
		$V_{IN} = V_{max}, V_{sw} = 0$ (Note 7)			10	500	μA
	LT1076	$V_{IN} \leq 25V, V_{SW} = 0$				150	μA
		$V_{IN} = V_{max}, V_{sw} = 0$ (Note 7)				250	μA
Supply Current (Note 2)	$V_{FB} = 2.5V, V_{IN} \leq 40V$		•		8.5	11	mA

PARAMETER	CONDITIONS		MIN	TYP	MAX	UNITS
	$40V < V_{IN} < 60V$	•		9	12	mA
	$V_{SHUT} = 0.1V$ (Device Shutdown) (Note 8)	•		140	300	μA
Minimum Supply Voltage	Normal Mode	•		7.3	8	V
	Start-Up Mode (Note 3)	•		3.5	4.8	V
Switch Current Limit (Note 4)	LT1074 I_{LIM} Open	•	5.5	6.5	8.5	A
	$R_{LIM} = 10k$ (Note 5)			4.5		A
	$R_{LIM} = 7k$ (Note 5)			3		A
	LT1076 I_{LIM} Open	•	2	2.6	3.2	A
	$R_{LIM} = 10k$ (Note 5)			1.8		A
	$R_{LIM} = 7k$ (Note 5)			1.2		A
Maximum Duty Cycle		•	85	90		%
Switching Frequency			90	100	110	kHz
	$T_J \leq 125°C$	•	85		120	kHz
	$T_J > 125°C$	•	85		125	kHz
	$V_{FB} = 0V$ Through $2k\Omega$ (Note 4)			20		kHz
Switching Frequency Line Regulation	$8V \leq V_{IN} \leq V_{MAX}$ (Note 7)	•		0.03	0.1	%/V
Error Amplifier Voltage Gain (Note 6)	$1V \leq V_C \leq 4V$		2000			V/V
Error Amplifier Transconductance			3700	5000	8000	μmho
Error Amplifier Source and Sink Current	Source ($V_{FB} = 2V$)		100	140	225	μA
	Sink ($V_{FB} = 2.5V$)		0.7	1	1.6	mA
Feedback Pin Bias Current	$V_{FB} = V_{REF}$	•		0.5	2	μA
Reference Voltage	$V_C = 2V$	•	2.155	2.21	2.265	V
Reference Voltage Tolerance	V_{REF} (Nominal) $= 2.21V$			±0.5	±1.5	%
	All Conditions of Input Voltage, Output Voltage, Temperature and Load Current	•		±1	±2.5	%
Reference Voltage Line Regulation	$8V \leq V_{IN} \leq V_{MAX}$ (Note 7)	•		0.005	0.02	%/V
V_C Voltage at 0% Duty Cycle				1.5		V
	Over Temperature	•		-4		mV/°C
Multiplier Reference Voltage				24		V
Shutdown Pin Current	$V_{SH} = 5V$	•	5	10	20	μA
	$V_{SH} \leq V_{THRESHOLD}$ ($\cong 2.5V$)	•			50	μA
Shutdown Thresholds	Switch Duty Cycle = 0	•	2.2	2.45	2.7	V
	Fully Shut Down	•	0.1	0.3	0.5	V
Status Window	As a Percent of Feedback Voltage		4	±5	6	%
Status High Level	$I_{STATUS} = 10μA$ Sourcing	•	3.5	4.5	5.0	V
Status Low Level	$I_{STATUS} = 1.6mA$ Sinking	•		0.25	0.4	V
Status Delay Time				9		μs
Status Minimum Width				30		μs
Thermal Resistance Junction to Case	LT1074				2.5	°C/W
	LT1076				4.0	°C/W

The • denotes the specifications which apply over the full operating temperature range.

Note 1: To calculate maximum switch on voltage at currents between low and high conditions, a linear interpolation may be used.

Note 2: A feedback pin voltage (V_{FB}) of 2.5V forces the V_C pin to its low clamp level and the switch duty cycle to zero. This approximates the zero load condition where duty cycle approaches zero.

Note 3: Total voltage from V_{IN} pin to ground pin must be $\geq$8V after startup for proper regulation.

Note 4: Switch frequency is internally scaled down when the feedback pin voltage is less than 1.3V to avoid extremely short switch on times. During testing, V_{FB} is adjusted to give a minimum switch on time of 1μs.

Note 5: $I_{LIM} \approx \frac{R_{LIM}-1k}{2k}$ (LT1047), $I_{LIM} \approx \frac{R_{LIM}-1k}{5.5k}$ (LT1076)

Note 6: Switch to input voltage limitation must also be observed.

Note 7: V_{MAX} = 40V for the LT1074/76 and 60V for the LT1074HV/76HV.

Note 8: Does not include switch leakage.

Block Diagram

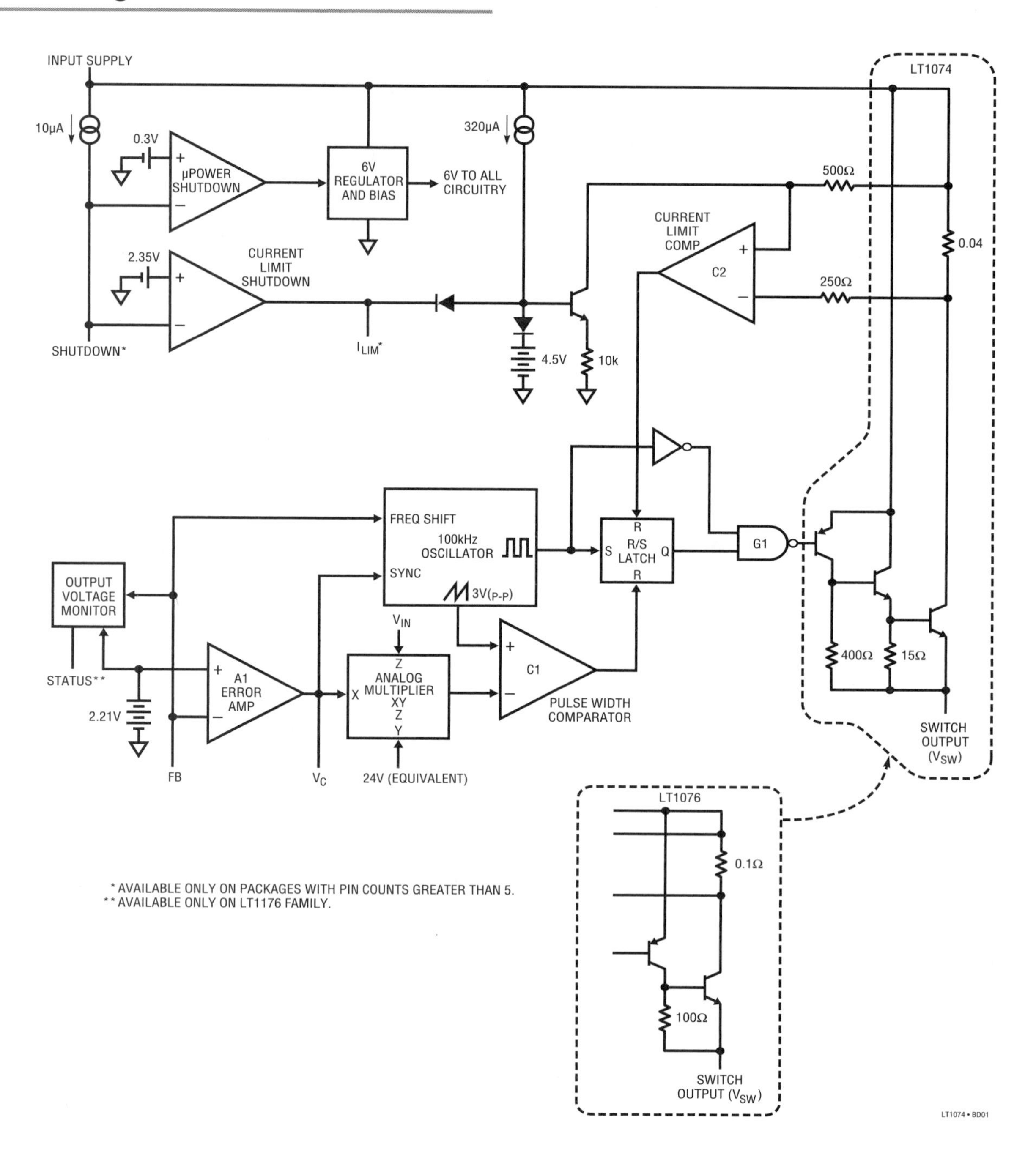

* AVAILABLE ONLY ON PACKAGES WITH PIN COUNTS GREATER THAN 5.
** AVAILABLE ONLY ON LT1176 FAMILY.

Block diagram description

A switch cycle in the LT1074 is initiated by the oscillator setting the R/S latch. The pulse that sets the latch also locks out the switch via gate G1. The effective width of this pulse is approximately 700ns, which sets the maximum switch duty cycle to approximately 93% at 100kHz switching frequency. The switch is turned off by comparator C1, which resets the latch. C1 has a sawtooth waveform as one input and the output of an analog multiplier as the other input. The multiplier output is the product of an internal reference voltage, and the output of the error amplifier, A1, divided by the regulator input voltage. In standard buck regulators, this means that the output voltage of A1 required to keep a constant regulated output is independent of regulator input voltage. This greatly improves line transient response, and makes loop gain independent of input voltage. The error amplifier is a transconductance type with a G_M at null of approximately 5000μmho. Slew current going positive is 140μA, while negative slew current is about 1.1mA. This asymmetry helps prevent overshoot on startup. Overall loop frequency compensation is accomplished with a series RC network from V_C to ground.

Switch current is continuously monitored by C2, which resets the R/S latch to turn the switch off if an overcurrent condition occurs. The time required for detection and switch turn-off is approximately 600ns. So minimum switch on time in current limit is 600ns. Under dead shorted output conditions, switch duty cycle may have to be as low as 2% to maintain control of output current. This would require switch on time of 200ns at 100kHz switching frequency, so frequency is reduced at very low output voltages by feeding the FB signal into the oscillator and creating a linear frequency downshift when the FB signal drops below 1.3V. Current trip level is set by the voltage on the I_{LIM} pin which is driven by an internal 320μA current source. When this pin is left open, it selfclamps at about 4.5V and sets current limit at 6.5A for the LT1074 and 2.6A for the LT1076. In the 7-pin package an external resistor can be connected from the I_{LIM} pin to ground to set a lower current limit. A capacitor in parallel with this resistor will soft-start the current limit. A slight offset in C2 guarantees that when the I_{LIM} pin is pulled to within 200mV of ground, C2 output will stay high and force switch duty cycle to zero.

The shutdown pin is used to force switch duty cycle to zero by pulling the I_{LIM} pin low, or to completely shut down the regulator. Threshold for the former is approximately 2.35V, and for complete shutdown, approximately 0.3V. Total supply current in shutdown is about 150μA. A 10μA pull-up current forces the shutdown pin high when left open. A capacitor can be used to generate delayed startup. A resistor divider will program "undervoltage lockout" if the divider voltage is set at 2.35V when the input is at the desired trip point.

The switch used in the LT1074 is a Darlington NPN (single NPN for LT1076) driven by a saturated PNP. Special patented circuitry is used to drive the PNP on and off very quickly even from the saturation state. This particular switch arrangement has no "isolation tubs" connected to the switch output, which can therefore swing to 40V below ground.

Typical performance characteristics

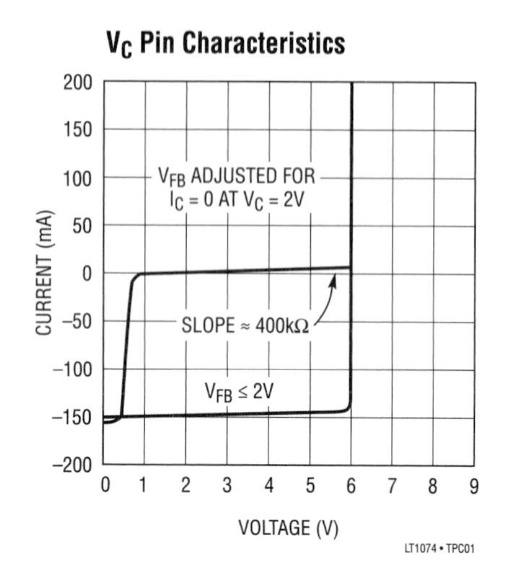

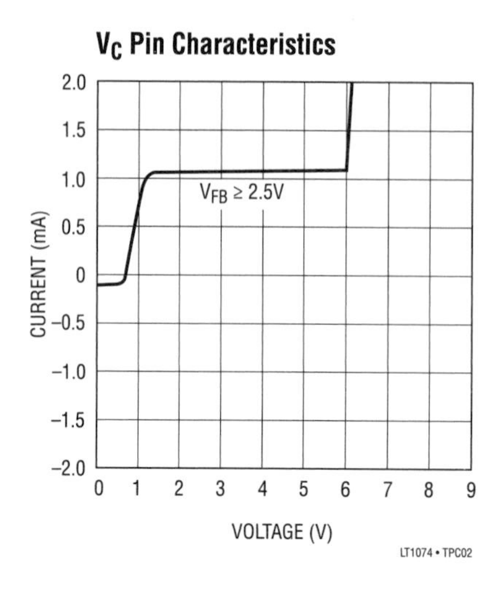

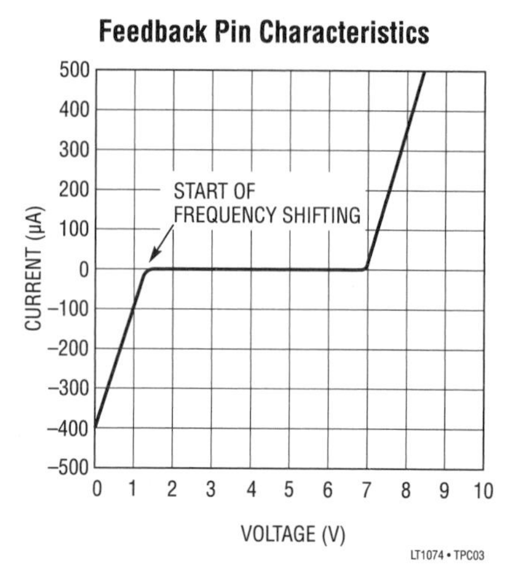

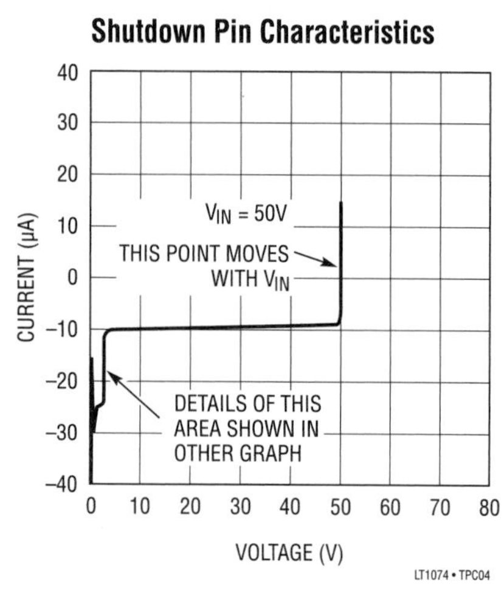

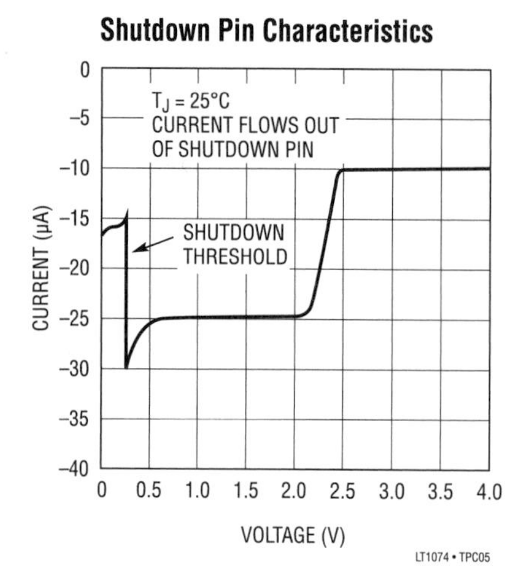

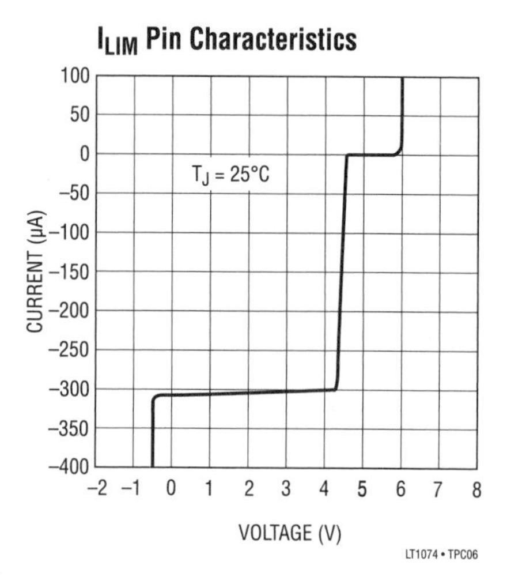

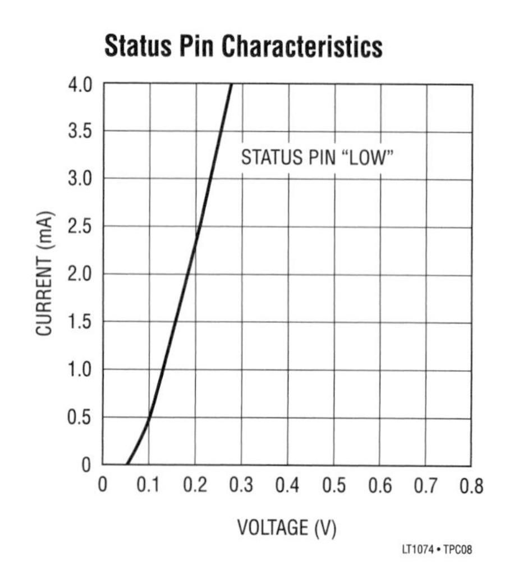

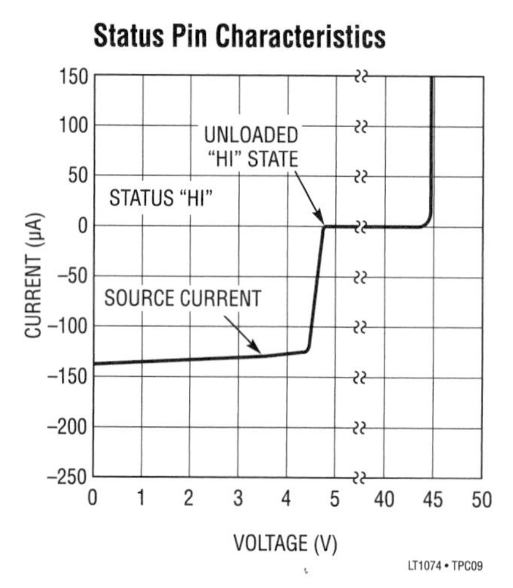

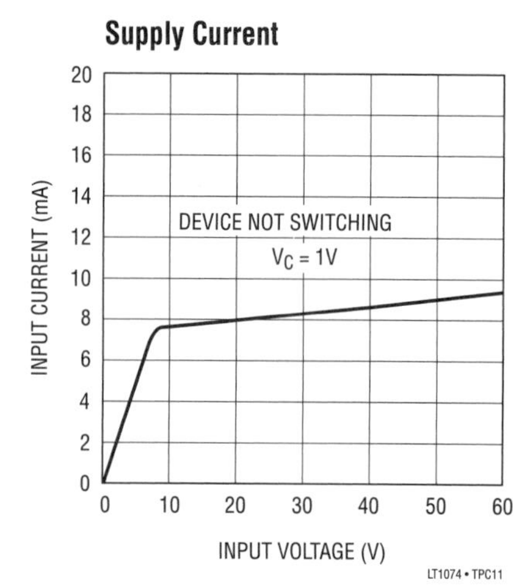

Typical performance characteristics

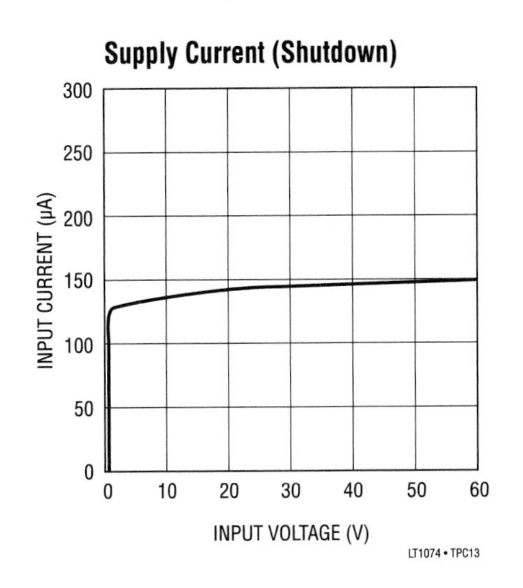

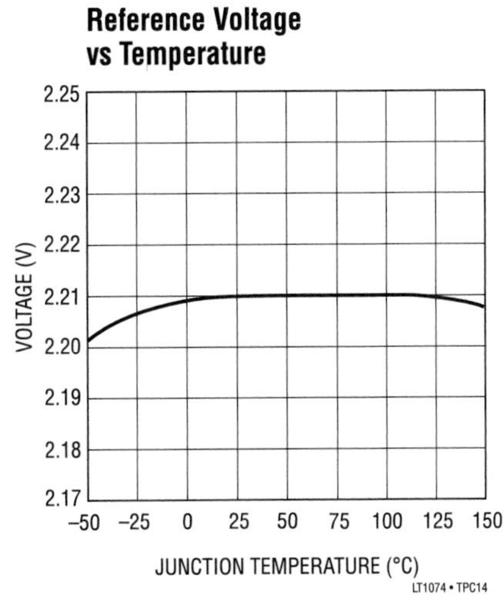

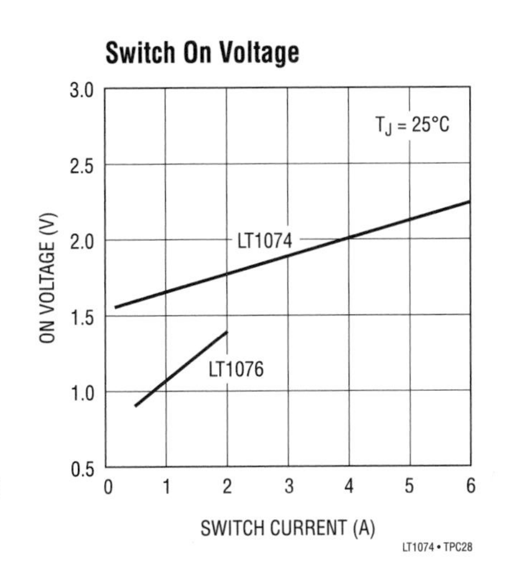

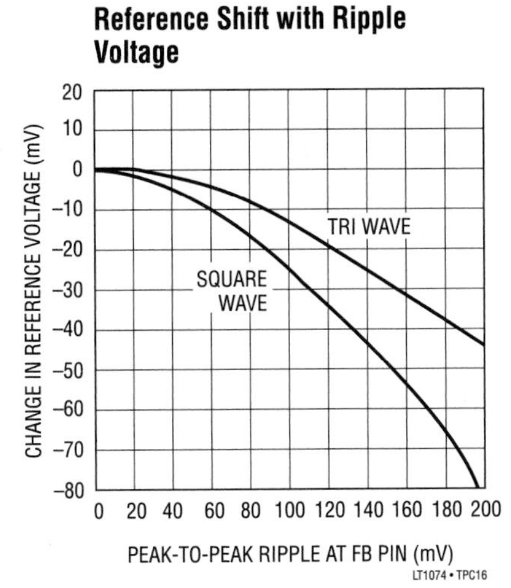

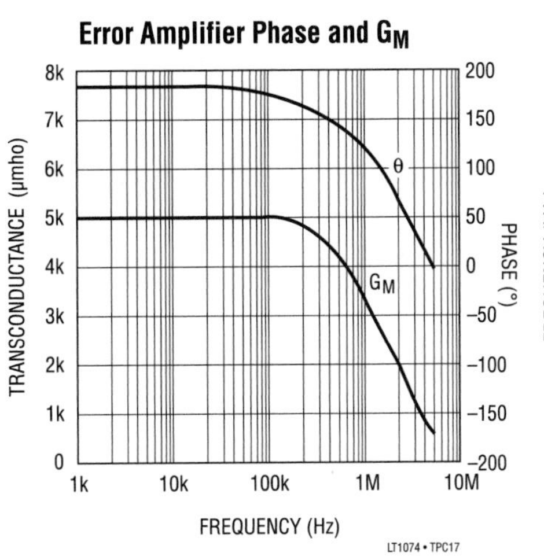

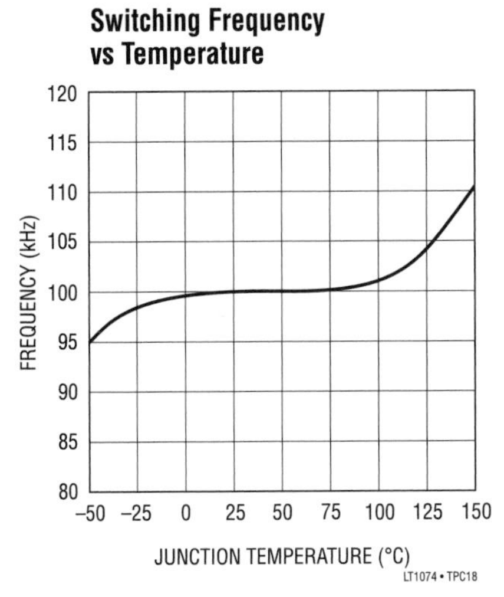

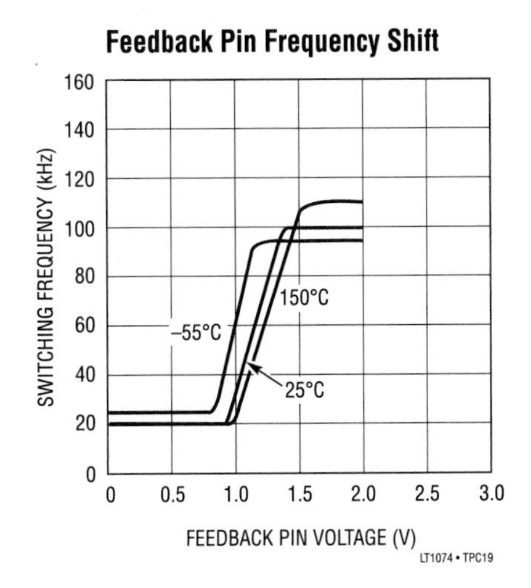

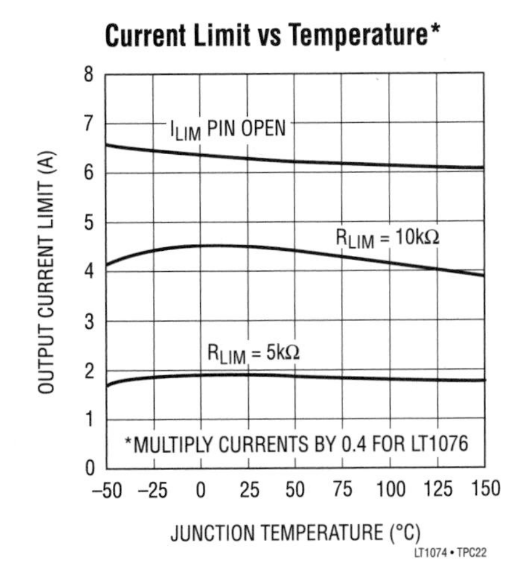

Typical performance characteristics

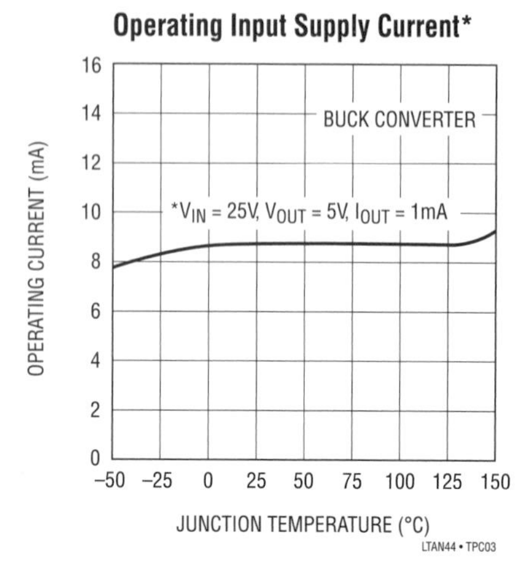

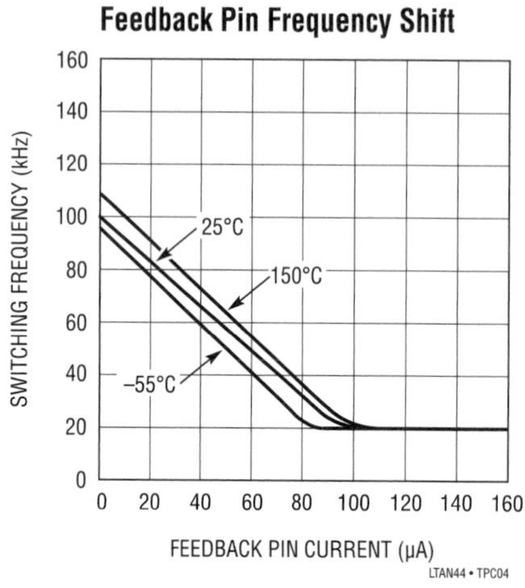

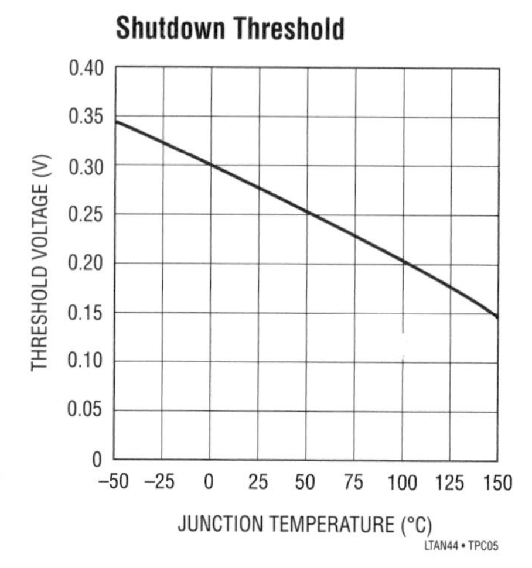

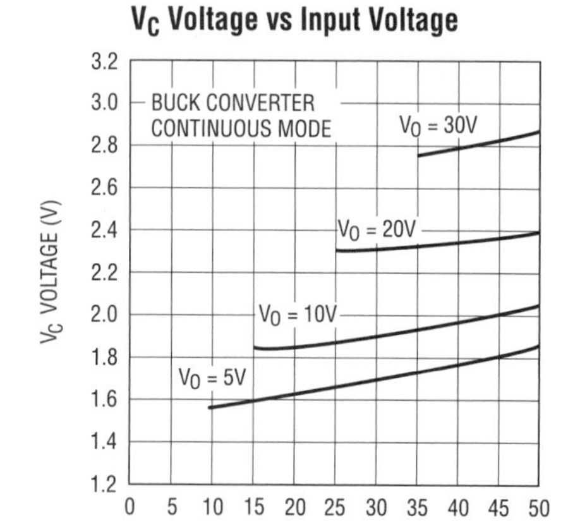

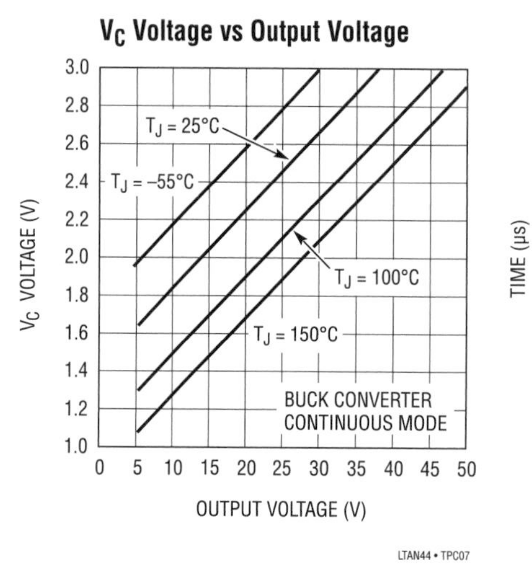

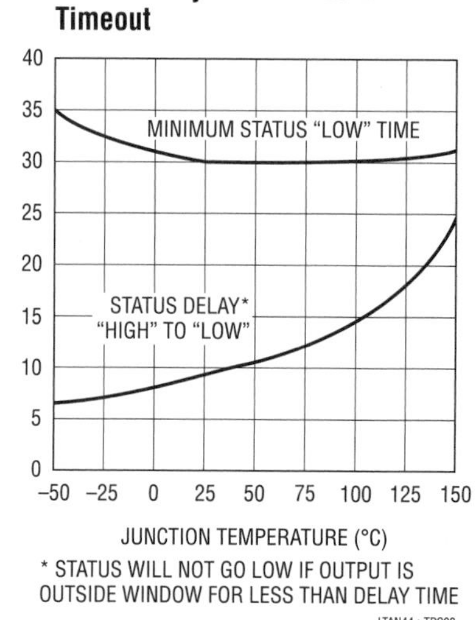

* STATUS WILL NOT GO LOW IF OUTPUT IS OUTSIDE WINDOW FOR LESS THAN DELAY TIME

LTAN44 • TPC08

Pin descriptions

V_{IN} pin

The V_{IN} pin is both the supply voltage for internal control circuitry and one end of the high current switch. It is important, *especially at low input voltages*, that this pin be bypassed with a low ESR, and low inductance capacitor to prevent transient steps or spikes from causing erratic operation. At full switch current of 5A, the switching transients at the regulator input can get very large as shown in Figure 5.1. Place the input capacitor very close to the regulator and connect it with wide traces to avoid extra inductance. Use radial lead capacitors.

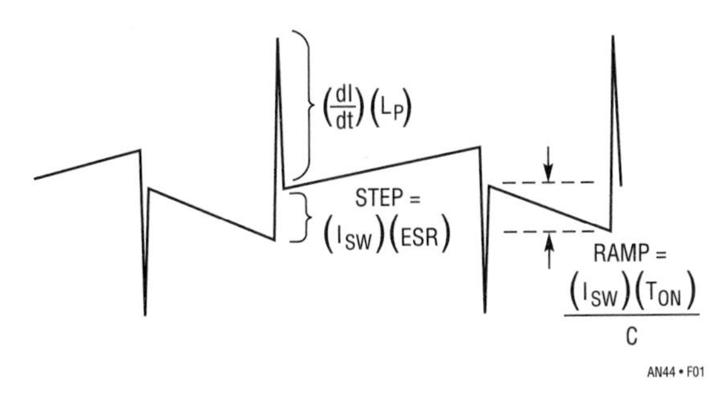

Figure 5.1 • Input Capacitor Ripple

L_P = Total inductance in input bypass connections and capacitor.

"Spike" height is $\left(\frac{dI}{dt} \bullet L_P\right)$ approximately 2V per inch of lead length.

Step = 0.25V for ESR = 0.05Ω and I_{SW} = 5A is 0.25V.

Ramp = 125mV for C = 200μF, t_{ON} = 5μs, and I_{SW} = 5A is 125mV.

Input current on the V_{IN} Pin in shutdown mode is the sum of actual supply current (≈140μA, with a maximum of 300μA) and switch leakage current. Consult factory for special testing if shutdown mode input current is critical.

Ground pin

It might seem unusual to describe a ground pin, but in the case of regulators, the ground pin must be connected properly to ensure good load regulation. The internal reference voltage is referenced to the ground pin; so any error in ground pin voltage will be multiplied at the output;

$$\Delta V_{OUT} = \frac{(\Delta V_{GND})(V_{OUT})}{2.21}$$

To ensure good load regulation, the ground pin must be connected directly to the proper output node, so that no high currents flow in this path. The output divider resistor should also be connected to this low current connection line as shown in Figure 5.2.

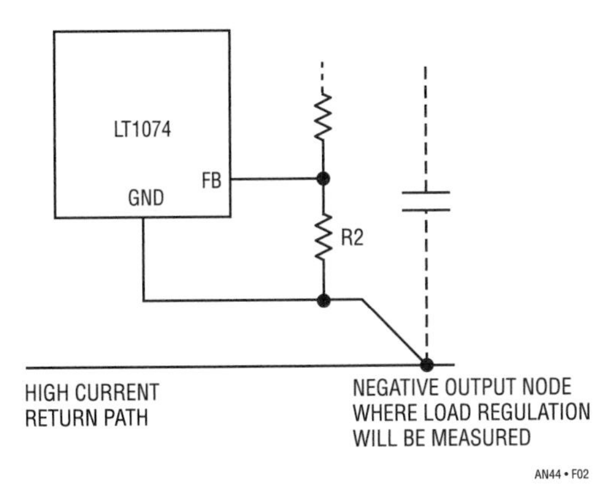

Figure 5.2 • Proper Ground Pin Connection

Feedback pin

The feedback pin is the inverting input of an error amplifier which controls the regulator output by adjusting duty cycle. The noninverting input is internally connected to a trimmed 2.21V reference. Input bias current is typically 0.5μA when the error amplifier is balanced (I_{OUT} = 0). The error amplifier has asymmetrical G_M for large input signals to reduce start-up overshoot. This makes the amplifier more sensitive to large ripple voltages at the feedback pin. 100m$V_{P\text{-}P}$ ripple at the feedback pin will create a 14mV offset in the amplifier, equivalent to a 0.7% output voltage shift. To avoid output errors, output ripple ($_{P\text{-}P}$) should be less than 4% of DC output voltage at the point where the output divider is connected.

See the Error Amplifier section for more details.

Frequency shifting at the feedback pin

The error amplifier feedback pin (FB) is used to downshift the oscillator frequency when the regulator output voltage is low. This is done to guarantee that output short-circuit current is well controlled even when switch duty cycle must be extremely low. Theoretical switch on time for a buck converter in continuous mode is;

$$t_{ON} = \frac{V_{OUT} + V_D}{V_{IN} \bullet f}$$

V_D = Catch diode forward voltage (≈0.5V)

f = Switching frequency

At f = 100kHz, t_{ON} must drop to 0.2μs when V_{IN} = 25V and the output is shorted (V_{OUT} = 0V). In current limit, the LT1074 can reduce t_{ON} to a minimum value of ≈0.6μs, much too long to control current correctly for V_{OUT} = 0. To correct this problem, switching frequency is lowered from 100kHz to 20kHz as the FB pin drops from 1.3V to 0.5V. This is accomplished by the circuitry shown in Figure 5.3.

Q1 is off when the output is regulating (V_{FB} = 2.21V). As the output is pulled down by an overload, V_{FB} will eventually reach 1.3V, turning on Q1. As the output continues to drop, Q1 current increases proportionately and lowers the frequency of the oscillator. Frequency shifting starts when the output is ≈60% of normal value, and is down to its minimum value of ≅20kHz when the output is ≅20% of normal value. The rate at which frequency is shifted is determined by both the internal 3k resistor R3 and the external divider resistors. For this reason, R2 should not be increased to more than 4k, if the LT1074 will be subjected to the simultaneous conditions of high input voltage and output short circuit.

Shutdown pin

The shutdown pin is used for undervoltage lockout, micropower shutdown, soft-start, delayed start, or as a general purpose on/off control of the regulator output. It controls switching action by pulling the I_{LIM} pin low, which forces the switch to a continuous off state. Full micropower shutdown is initiated when the shutdown pin drops below 0.3V.

The V/I characteristics of the shutdown pin are shown in Figure 5.4. For voltages between 2.5V and ≈V_{IN}, a current of 10μA flows out of the shutdown pin. This current increases to ≈25μA as the shutdown pin moves through the 2.35V threshold. The current increases further to ≈30μA at the 0.3V threshold, then drops to ≈15μA as the shutdown voltage falls below 0.3V. The 10μA current source is included to pull the shutdown pin to its high or default state when left open. It also provides a convenient pull-up for delayed start applications with a capacitor on the shutdown pin.

When activated, the typical collector current of Q1 in Figure 5.5, is ≈2mA. A soft-start capacitor on the I_{LIM} pin will delay regulator shutdown in response to C1, by ≈(5V)(C_{LIM})/2mA. Soft-start after full micropower shutdown is ensured by coupling C2 to Q1.

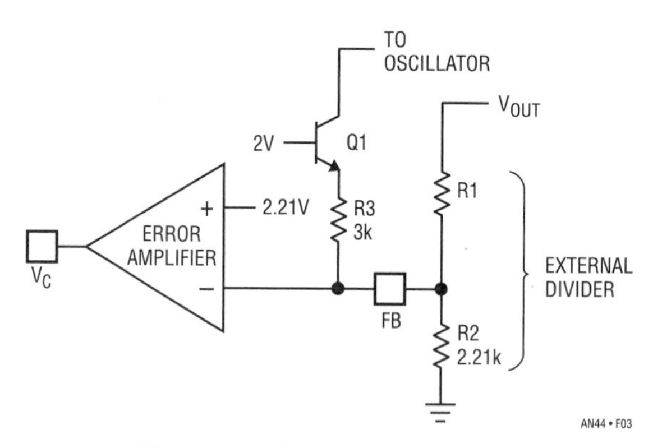

Figure 5.3 • Frequency Shifting

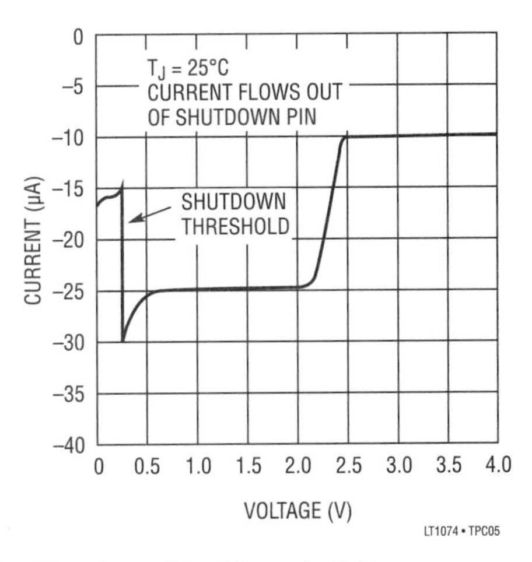

Figure 5.4 • Shutdown Pin Characteristics

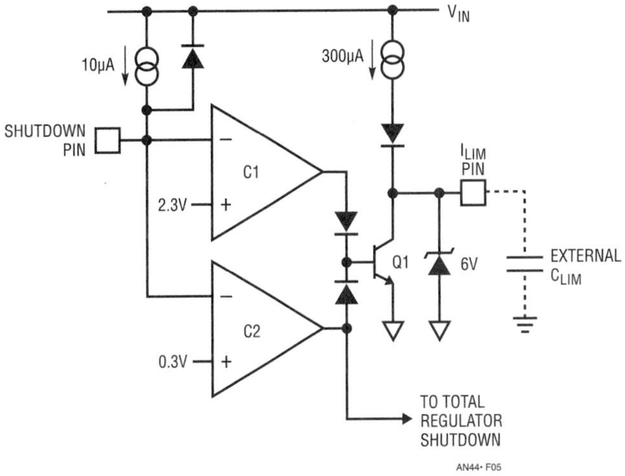

Figure 5.5 • Shutdown Circuitry

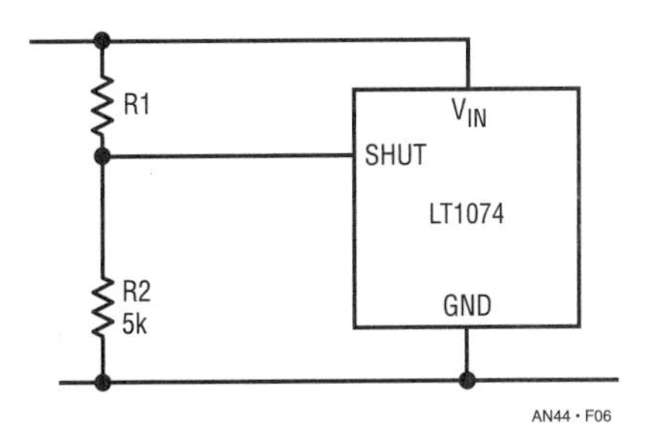

Figure 5.6 • Undervoltage Lockout

Undervoltage lockout

Undervoltage lockout point is set by R1 and R2 in Figure 5.6. To avoid errors due to the 10μA shutdown pin current, R2 is usually set at 5k, and R1 is found from:

$$R1 = R2\frac{(V_{TP} - V_{SH})}{V_{SH}}$$

V_{TP} = Desired undervoltage lockout voltage.

V_{SH} = Threshold for lockout on the shutdown pin=2.45V.

If quiescent supply current is critical, R2 may be increased up to 15k, but the denominator in the formula for R2 should replace V_{SH} with $V_{SH} - (10\mu A)(R2)$.

Hysteresis in undervoltage lockout may be accomplished by connecting a resistor (R3) from the I_{LIM} pin to the shutdown pin as shown in Figure 5.7. D1 prevents the shutdown divider from altering current limit.

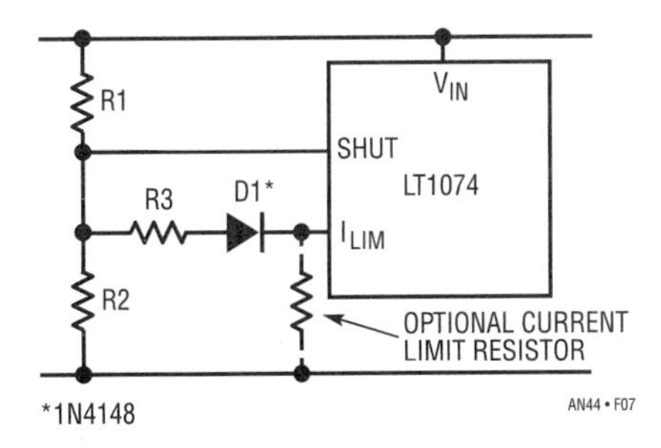

Figure 5.7 • Adding Hysteresis

$$\text{Trip Point} = V_{TP} = 2.35V\left(1 + \frac{R1}{R2}\right)$$

If R3 is added, the lower trip point (V_{IN} descending) will be the same. The upper trip point (V_{UTP}) will be:

$$V_{UTP} = V_{SH}\left(1 + \frac{R1}{R2} + \frac{R1}{R3}\right) - 0.8V\left(\frac{R1}{R3}\right)$$

If R1 and R2 are chosen, R3 is given by:

$$R3 = \frac{(V_{SH} - 0.8V)(R1)}{V_{UTP} - V_{SH}\left(1 + \frac{R1}{R2}\right)}$$

Example: An undervoltage lockout is required such that the output will not start until $V_{IN} = 20V$, but will continue to operate until V_{IN} drops to 15V. Let R2 = 2.32k.

$$R1 = (2.32k)\frac{15V - 2.35V}{2.35V} = 12.5k$$

$$R3 = \frac{(2.35 - 0.8)(12.5)}{20 - 2.35\left(1 + \frac{12.5}{2.32}\right)} = 3.9K$$

Status pin (available only on LT1176 parts)

The status pin is the output of a voltage monitor "looking" at the feedback pin. It is low for a feedback voltage which is more than 5% above or below nominal. "Nominal" in this case means the internal reference voltage, so that the ±5% window tracks the reference voltage. A time delay of ≈10μs prevents short spikes from tripping the status low. Once it does go low, a second timer forces it to stay low for a minimum of ≈30μs.

The status pin is modeled in Figure 5.8 with a 130μA pull up to a 4.5V clamp level. The sinking drive is a saturated NPN with ≈100Ω resistance and a maximum sink current of approximately 5mA. An external pull-up resistor can be added to increase output swing up to a maximum of 20V.

When the status pin is used to indicate "output OK," it becomes important to test for conditions which might create unwanted status states. These include output overshoot, large-signal transient conditions, and excessive output ripple. "False" tripping of the status pin can usually be controlled by a pulse stretcher network as shown in Figure 5.8. A single capacitor (C1) will suffice to delay an output "OK" (status high) signal to avoid false "true" signals during start-up, etc. Delay time for status high will be approximately (2.3×10^4) (C1), or 23ms/μF. Status low delay will be much shorter, ≈600μs/μF.

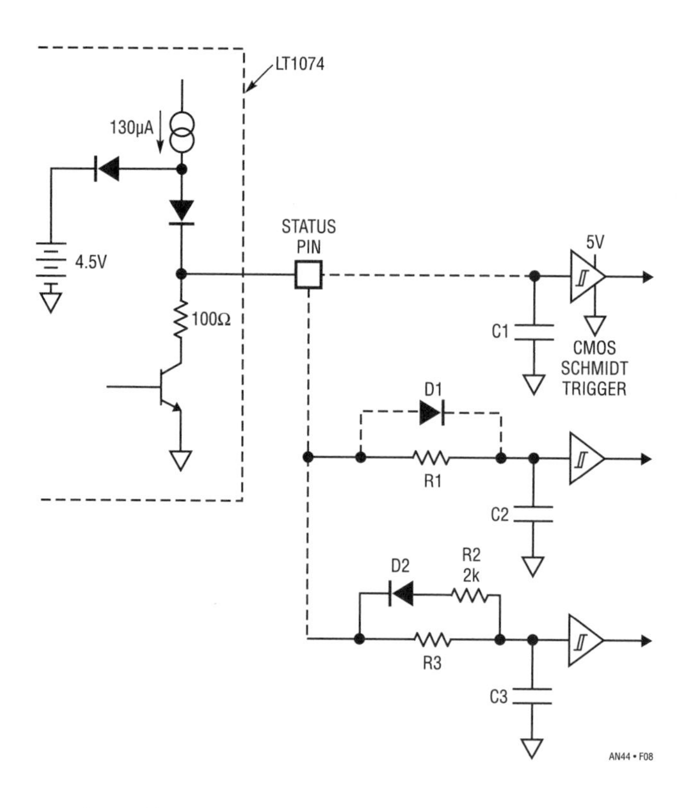

Figure 5.8 • Adding Time Delays to Status Output

If false tripping of status low could be a problem, R1 can be added. Delay of status high remains the same if R1 ≤ 10k. Status low delay is extended by R1 to approximately R1 • C2 seconds. Select C2 for high delay and R1 for low delay.

Example: Delay status high for 10ms, and status low for 3ms:

$$C2 = \frac{10\text{ms}}{23\text{ms}/\mu\text{F}} = 0.47\mu\text{F} (\text{Use } 0.47\mu\text{F})$$

$$R1 = \frac{3\text{ms}}{C2} = \frac{3\text{ms}}{0.47\mu\text{F}} = 6.4\text{k}\Omega$$

In this example D1 is not needed because R1 is small enough to not limit the charging of C2.

If very fast low tripping combined with long high delays is desired, use the D2, R2, R3, C3 configuration. C3 is chosen first to set low delay:

$$C3 \approx \frac{t_{LOW}}{2\text{k}\Omega}$$

R3 is then selected for high delay:

$$R3 \approx \frac{t_{HIGH}}{C3}$$

For $t_{LOW} = 100\mu s$ and $t_{HIGH} = 10ms$, C3 = 0.05µF and R3 = 200k.

I_{LIM} pin

The I_{LIM} pin is used to reduce current limit below the preset value of 6.5A. The equivalent circuit for this pin is shown in Figure 5.9.

When I_{LIM} is left open, the voltage at Q1 base clamps at 5V through D2. Internal current limit is determined by the current through Q1. If an external resistor is connected between I_{LIM} and ground, the voltage at Q1 base can be reduced for lower current limit. The resistor will have a voltage across it equal to (320µA) (R), limited to ≈5V when clamped by D2. Resistance required for a given current limit is:

$$R_{LIM} = I_{LIM}(2\text{k}\Omega) + 1\text{k}\Omega \text{ (LT1074)}$$

$$R_{LIM} = I_{LIM}(5.5\text{k}\Omega) + 1\text{k}\Omega \text{ (LT1076)}$$

As an example, a 3A current limit would require 3A (2k) + 1k = 7k for the LT1074. The accuracy of these formulas is ±25% for 2A ≤ I_{LIM} ≤ 5A (LT1074) and 0.7A ≤ I_{LIM} ≤ 1.8A (LT1076), so I_{LIM} should be set at least 25% above the peak switch current required.

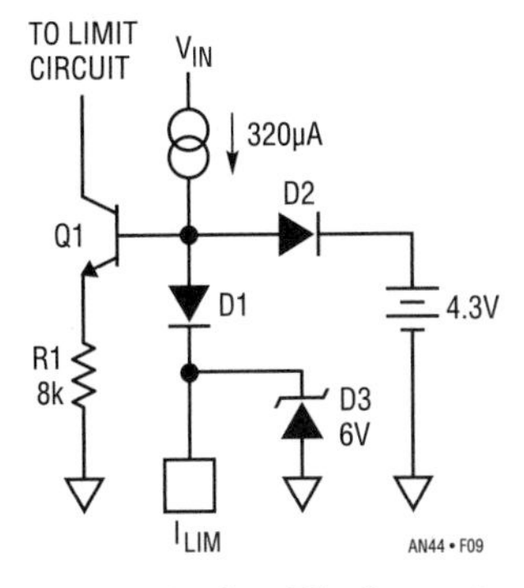

Figure 5.9 • I_{LIM} Pin Current

Foldback current limiting can be easily implemented by adding a resistor from the output to the I_{LIM} pin as shown in Figure 5.10. This allows full desired current limit (with or without R_{LIM}) when the output is regulating, but reduces current limit under short-circuit conditions. A typical value

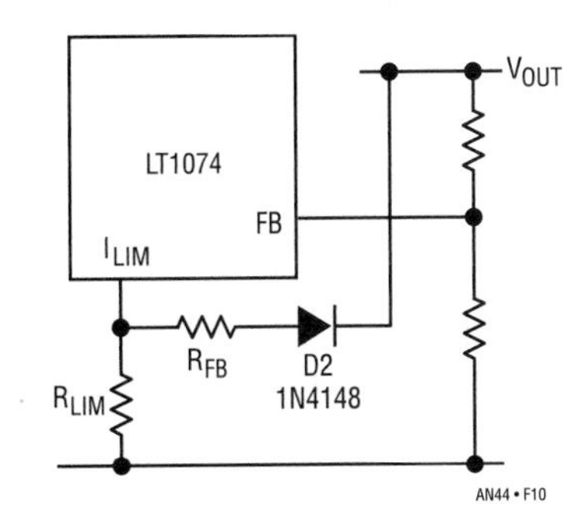

Figure 5.10 • Foldback Current Limit

for R_{FB} is 5k, but this may be adjusted up or down to set the amount of foldback. D2 prevents the output voltage from forcing current back into the I_{LIM} pin. To calculate a value for R_{FB}, first calculate R_{LIM}, then R_{FB}:

$$R_{FB} = \frac{(I_{SC} - 0.44^*)(R_L)}{0.5^*(R_L - 1k\Omega) - I_{SC}} = R_L \text{ in } k\Omega$$

*Change 0.44 to 0.16, and 0.5 to 0.18 for LT1076.

Example: $I_{LIM} = 4A$, $I_{SC} = 1.5A$, $R_{LIM} = (4)(2k) + 1k = 9k$:

$$R_{FB} = \frac{(1.5 - 0.44)(9k\Omega)}{0.5(9k - 1k) - 1.5} = 3.8k\Omega$$

Error amplifier

The error amplifier in Figure 5.11 is a single stage design with added inverters to allow the output to swing above and below the common mode input voltage. One side of the amplifier is tied to a trimmed internal reference voltage of 2.21V. The other input is brought out as the FB (feedback) pin. This amplifier has a G_M (voltage in to current out) transfer function of ≈5000µmho. Voltage gain is determined by multiplying G_M times the total equivalent output loading, consisting of the output resistance of Q4 and Q6 in parallel with the series RC external frequency compensation network. At DC, the external RC is ignored, and with a parallel output impedance for Q4 and Q6 of 400kΩ, voltage gain is ≈2000. At frequencies above a few hertz, voltage gain is determined by the external compensation, R_C and C_C.

$$A_V = \frac{G_m}{2\pi \bullet f \bullet C_C} \text{ at mid-frequencies}$$

$$A_V = G_m \bullet R_C \text{ at high frequencies}$$

Phase shift from the FB pin to the V_C pin is 90° at mid-frequencies where the external C_C is controlling gain, then drops back to 0° (actually 180° since FB is an inverting input) when the reactance of C_C is small compared to R_C. The low frequency "pole" where the reactance of C_C is equal to the output impedance of Q4 and Q6 (r_O), is:

$$f_{POLE} = \frac{1}{2\pi \bullet r_O \bullet C} \quad r_O \approx 400k\Omega$$

Although f_{POLE} varies as much as 3:1 due to r_O variations, mid-frequency gain is dependent only on G_M, which is specified much tighter on the data sheet. The higher frequency "zero" is determined solely by R_C and C_C:

$$f_{ZERO} = \frac{1}{2\pi \bullet R_C \bullet C_C}$$

The error amplifier has *asymmetrical* peak output current. Q3 and Q4 current mirrors are unity gain, but the Q6 mirror has a gain of 1.8 at output null and a gain of 8 when the FB pin is high (Q1 current = 0). This results in a maximum positive output current, of 140µA and a maximum negative (sink) output current of ≅1.1mA. The asymmetry is deliberate—it results in much less regulator output overshoot during rapid start-up or following the release of an output overload. Amplifier offset is kept low by area scaling Q1 and Q2 at 1.8:1.

Amplifier swing is limited by the internal 5.8V supply for positive outputs and by D1 and D2 when the output goes low. Low clamp voltage is approximately one diode drop (−0.7V – 2mV/°C).

Note that both the FB pin and the V_C pin have other internal connections. Refer to the frequency shifting and synchronizing discussions.

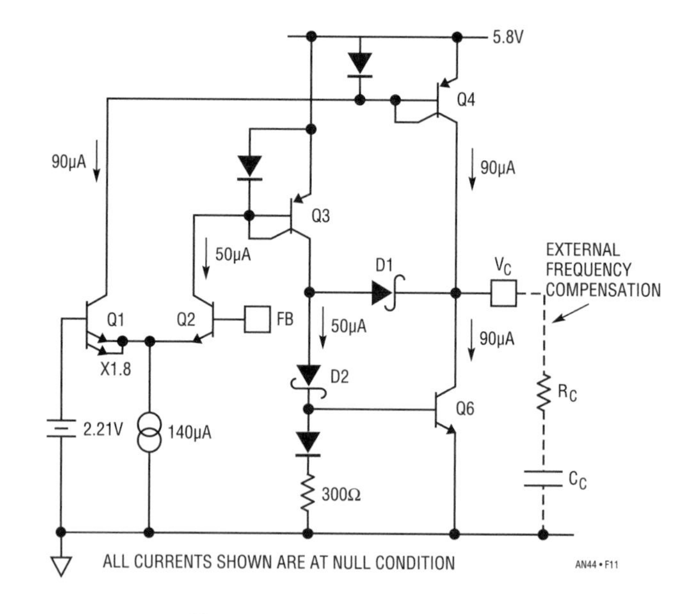

Figure 5.11 • Error Amplifier

Definition of terms

V_{IN}: DC input voltage.
V_{IN}': DC input voltage minus switch voltage loss. V_{IN}' is 1.5V to 2.3V less than V_{IN}, depending on switch current.
V_{OUT}: DC output voltage.
V_{OUT}': DC output voltage plus catch diode forward voltage. V_{OUT}' is typically 0.4V to 0.6V more than V_{OUT}.
f: Switching frequency.
I_M: Maximum specified switch current $I_M = 5.5A$ for the LT1074 and 2A for the LT1076.
I_{SW}: Switch current during switch on time. The current typically jumps to a starting value, then ramps higher. I_{SW} is the average value during this period unless otherwise stated. It is not averaged over the whole switching period, which includes switch off time.
I_{OUT}: DC output current.
I_{LIM}: DC output current limit.
I_{DP}: Catch diode forward current. This is the peak current for discontinuous operation and the average value of the current pulse during switch off time for continuous mode.
I_{DA}: Catch diode forward current averaged over one complete switching cycle. Ida is used to calculate diode heating.
ΔI: Peak-to-peak ripple current in the inductor, also equal to peak current in the discontinuous mode. ΔI is used to calculate output ripple voltage and inductor core losses.
V_{P-P}: Peak-to-peak output voltage ripple. This does not include "spikes" created by fast rising currents and capacitor parasitic inductance.
t_{SW}: This is not really an actual rise or fall time. Instead, it represents the effective overlap time of voltage and current in the switch. t_{SW} is used to calculate switch power dissipation.
L: Inductance, usually measured with low AC flux density, and zero DC current. Note that large AC flux density can increase L by up to 30%, and large DC currents can decrease L dramatically (core saturation).
B_{AC}: Peak AC flux density in the inductor core, equal to one-half peak-to-peak AC flux density. Peak value is used because nearly all core loss curves are plotted with peak flux density.
N: Tapped-inductor or transformer turns ratio. Note the exact definition of N for each application.
μ: Effective permeability of core material used in the inductor. μ is typically 25-150. Ferrite material is much higher, but is usually gapped to reduce the effective value to this range.
V_e: Effective core material volume (cm^3).
L_e: Effective core magnetic path length (cm).
A_e: Effective core cross sectional area (cm^2).
A_w: Effective core or bobbin winding area.
L_t: Average length of one turn on winding.
P_{CU}: Power dissipation caused by winding resistance. It does not include skin effect.
P_C: Power loss in the magnetic core. PC depends only on ripple current in the inductor not DC current.
E: Overall regulator efficiency. It is simply output power divided by input power.

Positive step-down (buck) converter

The circuit in Figure 5.12 is used to convert a larger positive input voltage to a lower positive output. Typical waveforms are shown in Figure 5.13, with $V_{IN} = 20V$, $V_{OUT} = 5V$, $L = 50\mu H$, for both continuous mode (inductor current never drops to zero) with $I_{OUT} = 3A$ and discontinuous mode, where inductor current drops to zero during a portion of the switching cycle ($I_{OUT} = 0.17A$). Continuous mode maximizes output power but requires larger inductors. *Maximum* output current in true discontinuous mode is only one-half of switch current rating. Note that when load current is reduced in a continuous mode design, eventually the circuit will enter discontinuous mode. The LT1074 operates equally well in either mode and there is no significant change in performance when load current reduction causes a shift to discontinuous mode.

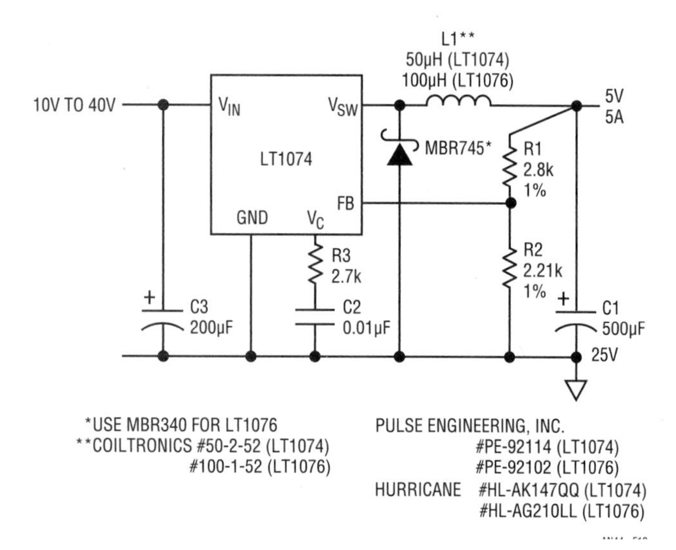

Figure 5.12 • Basic Positive Buck Converter

Duty cycle of a buck converter *in continuous* mode is:

$$DC = \frac{V_{OUT} + V_f}{V_{IN} - V_{SW}} = \frac{V_{OUT'}}{V_{IN'}} \qquad (1)$$

V_f = Forward voltage of catch diode

V_{SW} = Voltage loss across on switch

Note that duty cycle does not vary with load current except to the extent that V_f and V_{SW} change slightly.

A buck converter will change from continuous to discontinuous mode (and duty cycle will begin to drop) at a load current equal to:

$$I_{OUT(CRIT)} = \frac{(V_{OUT'})(V_{IN'} - V_{OUT'})}{2 \bullet V_{IN'} \bullet f \bullet L} \qquad (2)$$

With the possible exception of load transient response, there is no reason to increase L to ensure continuous mode operation at light load.

Using the values from Figure 5.12, with $V_{IN} = 25V$, $V_f = 0.5V$, $V_{SW} = 2V$

$$DC = \frac{5 + 0.5}{25 - 2} = 24\% \qquad (3)$$

$$I_{OUT(CRIT)} = \frac{(5.5)(23 - 5.5)}{2(23)(10^5)(50 \bullet 10^{-6})} = 0.42A$$

The "ringing" which occurs at some point in the switch off cycle in discontinuous mode is simply the resonance created by the catch diode capacitance plus switch capacitance in parallel with the inductor. This ringing does no harm and any attempt to dampen it simply wastes efficiency. Ringing frequency is given by:

$$f_{RING} = \frac{1}{2\pi\sqrt{L \bullet (C_{SW} + C_{DIODE}}} \qquad (4)$$

$C_{SW} \approx 80pF$

$C_{DIODE} = 200pF$ to $1000pF$

No off-state ringing occurs in continuous mode because the diode is always conducting during switch off time and effectively shorts the resonance.

A detailed look at the leading edge of the switch waveform may reveal a second "ringing" tendency, usually at frequencies around 20MHz to 50MHz. This is the result of the inductance in the loop which includes the input capacitor, the LT1074 leads, and the diode leads, combined with the capacitance of the catch diode. A total lead length of 4 inches will create $\approx 0.1\mu H$. This coupled with 500pF of diode capacitance will create a damped 25MHz oscillation superimposed on the fast rising switch voltage waveform. Again, no harm is created by this ringing and no attempt should be made to dampen it other than minimizing lead length. Certain board layouts combined with very

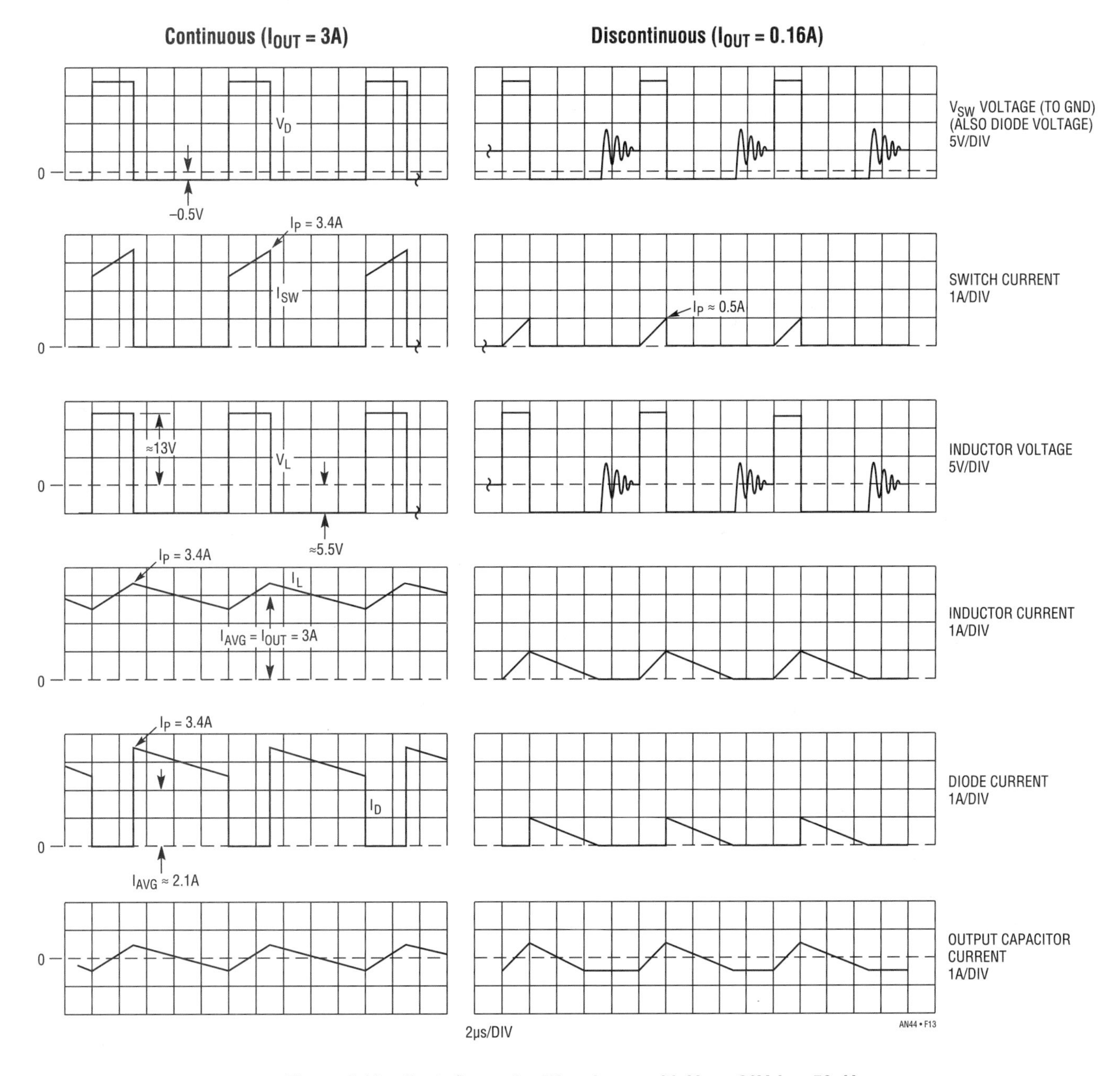

Figure 5.13 • Buck Converter Waveforms with V_{IN} = 20V, L = 50μH

short interconnects and high diode capacitance may create a tuned circuit which resonates with the switch output to cause a low amplitude oscillation at the switch output during on time. This can be eliminated with a ferrite bead slipped over either diode lead during board assembly.

It is interesting to note that standard silicon fast recovery diodes create almost no ringing because of their lower capacitance and because they are effectively damped by their slower turn-off characteristics. This slower turn-off and the larger forward voltage represent additional power loss, so Schottky diodes are normally recommended.

Maximum output current of a buck converter is given by Continuous Mode

$$I_{OUT(MAX)} = I_M - \frac{V_{OUT}(V_{IN} - V_{OUT})}{2f \bullet V_{IN} \bullet L} \quad (5)$$

I_M = Maximum switch current (5.5A for LT1074)
V_{IN} = DC input voltage (maximum)
V_{OUT} = Output voltage
f = Switching frequency

For the example shown, with L = 50μH, and V_{IN} = 25V,

$$I_{OUT(MAX)} = 5.5 - \frac{5(25-5)}{2(10^5)(25)(50 \times 10^{-6})} = 5.1A \quad (6)$$

Note that increasing inductor size to 100μH would only increase maximum output current by 4%, but decreasing it to 20μH would drop maximum current to 4.5A. Low inductance can be used for lower output currents, but core loss will increase.

Inductor

The inductor used in a buck converter acts as both an energy storage element and a smoothing filter. There is a basic trade-off between good filtering versus size and cost. Typical inductor values used with the LT1074 range from 5μH to 200μH, with the small values used for lower power, minimum size applications and the larger values used to maximize output power or minimize output ripple voltage. The inductor must be rated for currents at least equal to output current and there are restrictions on ripple current (expressed as volt • microsecond product at various frequencies) to avoid core heating. For details on selecting an inductor and calculating losses, see the Inductor Selection section.

Output catch diode

D1 is used to generate a current path for L1 current when the LT1074 switch turns off. The current through D1 in continuous mode is equal to output current with a duty cycle of $(V_{IN} - V_{OUT})/V_{IN}$. For low input voltages, D1 may operate at duty cycles of 50% or less, but one must be very careful of utilizing this fact to minimize diode heat sinking. First, an unexpected high input voltage will cause duty cycle to increase. More important however, is a shorted output condition. When V_{OUT} = 0, diode duty cycle is ≈1 for any input voltage. Also, in current limit, diode current is not load current, but is determined by LT1074 switch current limit. If continuous output shorts must be tolerated, D1 must be adequately rated and heat sunk. 7 and 11-pin versions of the LT1074 allow current limit to be reduced to limit diode dissipation. 5-pin versions can be accurately current limited using the technique shown in Figure 5.20.

Under normal conditions, D1 dissipation is given by:

$$P_{DI} = I_{OUT}\frac{(V_{IN} - V_{OUT})}{V_{IN}} \bullet V_f \quad (7)$$

V_f is the forward voltage of D1 at I_{OUT} current. Schottky diode forward voltage is typically 0.6V at the diode's full rated current, so it is normal design practice to use a diode rated at 1.5 to 2 times output current to maintain efficiency and allow margin for short-circuit conditions. This derating allows V_f to drop to approximately 0.5V.

Example: $V_{IN(MAX)}$ = 25V, I_{OUT} = 3A, V_{OUT} = 5V, assume V_f = 0.5V:

Full Load

$$P_{DI} = \frac{(3)(25-5)(0.5V)}{25} = 1.2W$$

Shorted Output

$$P_{DI} = (\approx 6A)(DC = 1)(0.6V) = 3.6W$$

The high diode dissipation under shorted output conditions may necessitate current limit adjustment if adequate heat sinking cannot be provided.

Diode switching losses have been neglected because the reverse recovery time is assumed to be short enough to ignore. If a standard silicon diode is used, switching losses cannot be ignored. They can be approximated by:

$$Pt_{rr} \approx (V_{IN})(f)(t_{rr})(I_{OUT}) \quad (9)$$

trr = Diode reverse recovery time

Example: Same circuit with t_{rr} = 100ns:

$$Pt_{rr} = (25)(10^5)(10^{-7})(3) = 0.75W \quad (10)$$

Diodes with abrupt turn-off characteristics will transfer most of this power to the LT1074 switch. Soft recovery diodes will dissipate much of the power within the diode itself.

LT1074 power dissipation

The LT1074 draws about 7.5mA quiescent current, independent of input voltage or load. It draws an additional 5mA during switch on time. The switch itself dissipates a power approximately proportional to load current. This power is due to pure conduction losses (switch on voltage times switch current) and dynamic switching losses due to finite switch current rise and fall times. Total LT1074 power dissipation can by calculated from:

$$P = V_{IN}[7mA + 5mA \bullet DC + 2 I_{OUT} \bullet t_{SW} \bullet f] + DC\left[I_{OUT}(1.8V)^* + 0.1\Omega^*(I_{OUT})^2\right] \quad (11)$$

$$DC = \text{Duty Cycle} \approx \frac{V_{OUT} + 0.5V}{V_{IN} - 2V}$$

t_{SW} = Effective overlap time of switch voltage and current

$\approx 50ns + (3ns/A)(I_{OUT})$ (LT1074)

$\approx 60ns + (10ns/A)(I_{OUT})$ (LT1076)

Example: $V_{IN} = 25V$, $V_{OUT} = 5V$, $f = 100kHz$, $I_{OUT} = 3A$:

$$DC = \frac{5 + 0.5}{25 - 2} = 0.196 \quad (12)$$

$$t_{SW} = 50ns + (3ns/A)(3A) = 59ns$$

$$P = 25\begin{bmatrix} 7mA + 5mA(0.196) + \\ (2)(3)(59ms)(10^5) \end{bmatrix} + 0.196\left[3(1.8) + 0.1(3)^2\right]$$

$$= \underbrace{0.21W}_{\text{Supply Current Loss}} + \underbrace{0.89W}_{\text{Dynamic Switching Loss}} + \underbrace{1.24W}_{\text{Switch Conduction Loss}} = 2.34W \quad (13)$$

*LT1076 = 1V, 0.3Ω

Input capacitor (buck converter)

A local input bypass capacitor is normally required for buck converters because the input current is a square wave with fast rise and fall times. This capacitor is chosen by ripple current rating—the capacitor must be large enough to avoid overheating created by its ESR and the AC RMS value of converter input current. For continuous mode:

$$I_{AC,RMS} = I_{OUT}\sqrt{\frac{V_{OUT}(V_{IN} - V_{OUT})}{(V_{IN})^2}} \quad (14)$$

Worst case is at $V_{IN} = 2V_{OUT}$.

Power loss in the input capacitor is not insignificant in high efficiency applications. It is simply RMS capacitor current squared times ESR:

$$P_{C3} = (I_{AC,RMS})^2(ESR) \quad (15)$$

Example: $V_{IN} = 20V$ to $30V$, $I_{OUT} = 3A$, $V_{OUT} = 5V$.

Worst case is at $V_{IN} = 2 \bullet V_{OUT} = 10V$, so use the closest V_{IN} value of 20V:

$$I_{AC,RMS} = 3A\sqrt{\frac{5(20 - 5)}{(20)^2}} = 1.3A \text{ RMS} \quad (16)$$

The input capacitor must be rated at a working voltage of 30V minimum and 1.3A ripple current. Ripple current ratings vary with maximum ambient temperature, so check data sheets carefully.

It is important to locate the input capacitor very close to the LT1074 and to use short leads (radial) when the DC input voltage is less than 12V. Spikes as high as 2V/inch of lead length will appear at the regulator input. If these spikes drop below ≈7V, the regulator will exhibit anomalous behavior. See V_{IN} Pin in the Pin Descriptions section.

You may be wondering why no mention has been made of capacitor value. That's because it doesn't really matter. Larger electrolytic capacitors are purely resistive (or inductive) at frequencies above 10kHz, so their bypassing impedance is resistive, and ESR is the controlling factor. For input capacitors used with the LT1074, a unit which meets ripple current ratings will provide adequate "bypassing" regardless of its capacitance value. Units with higher voltage rating will have lower capacitance for the same ripple current rating, but as a general rule, the *volume* required to meet a given ripple current/ESR is fixed over a wide range of capacitance/voltage rating. If the capacitor chosen for this application has 0.1Ω ESR, it will have a power loss of $(1.3A)^2$ $(0.1\Omega) = 0.17W$.

Output capacitor

In a buck converter, output ripple voltage is determined by both the inductor value and the output capacitor:

Continuous Mode

$$V_{P\text{-}P} = \frac{(ESR)(V_{OUT}\left(1 - \frac{V_{OUT}}{V_{IN}}\right)}{(L1)(f)} \quad (17)$$

Discontinuous Mode

$$V_{P\text{-}P} = ESR\sqrt{\frac{(2I_{OUT}(V_{OUT})(V_{IN} - V_{OUT})}{L \bullet f \bullet V_{IN}}}$$

Note that only the ESR of the output capacitor is used in the formula. It is assumed that the capacitor is purely resistive at frequencies above 10kHz. If an inductor value has been chosen, the formula can be rearranged to solve for ESR to aid in selecting a capacitor.

Continuous Mode

$$ESR(MAX) = \frac{(V_{P\text{-}P})(L1)(f)}{V_{OUT}\left(1 - \frac{V_{OUT}}{V_{IN}}\right)} \quad (18)$$

Discontinuous Mode

$$ESR(MAX) = V_{P\text{-}P}\sqrt{\frac{L \bullet f \bullet V_{IN}}{2I_{OUT}(V_{OUT})(V_{IN} - V_{OUT})}}$$

Worst-case output ripple is at highest input voltage. Ripple is independent of load for continuous mode and proportional to the square root of load current for discontinuous mode.

Example: Continuous mode with $V_{IN(MAX)} = 25V$, $V_{OUT} = 5V$, $I_{OUT} = 3A$, $L1 = 50\mu H$, $f = 100kHz$. Required maximum peak-to-peak output ripple is 25mV.

$$ESR = \frac{(0.025)(50 \bullet 10^{-6})(10^5)}{(5)\left(1 - \frac{5}{25}\right)} = 0.03\Omega \quad (19)$$

A 10V capacitor with this ESR would have to be several thousand microfarads, and therefore fairly large. Trade-offs which could be made include:

A. Paralleling several capacitors if component height is more critical than board area.

B. Increasing inductance. This can be done at no increase in size if a more expensive core (molypermalloy, etc.) is used.

C. Adding an output filter. This is often the best solution because the additional components are fairly low cost and their additional space is minimized by being able to "size down" the main L and C. See the Output Filter section.

Although ripple current is not usually a problem with buck converter output capacitors because the current is pre-filtered by the inductor, a quick check should be done before a final capacitor is chosen—especially if the capacitor has been "downsized" to take advantage of an additional output filter. RMS ripple current into the output capacitor is:

Continuous Mode

$$I_{RMS} = \frac{0.29(V_{OUT})\left(1 - \frac{V_{OUT}}{V_{IN}}\right)}{L1 \bullet f} \quad (20)$$

From the previous example:

$$I_{RMS} = \frac{0.29(5)(1 - \frac{5}{25})}{(50 \bullet 10^{-6})(10^5)} = 0.23A\ RMS \quad (21)$$

This ripple current is low enough to not be a problem, but that could change if the inductor was reduced by two or three to one and the output capacitor was minimized by adding an output filter.

The calculations for discontinuous mode RMS ripple current were considered too complicated for this discussion, but a conservative value would be 1.5 to 2 times output current.

To minimize output ripple, the output terminals of the regulator should be connected directly to the capacitor leads so that the diode (D1) and inductor currents do not circulate in output leads.

Efficiency

All the losses except those created by the inductor and the output filter are covered in this buck regulator section. The example used was a 5V, 3A output with 25V input. Calculated losses were: switch, 1.24W; diode, 1.2W; switching times, 0.89W; supply current, 0.21W; and input capacitor, 0.17W. Output capacitor losses were negligible. The sum of all these losses is 3.71W. Inductor loss is covered in a special section of this Application Note. Assume for this application that inductor copper loss is 0.3W and core loss is 0.15W. Total regulator loss is 4.16W. Efficiency is:

$$E = \frac{I_{OUT} \bullet V_{OUT}}{I_{OUT} \bullet V_{OUT} + \Sigma P_L} = \frac{(3A)(5V)}{(3A)(5V) + 4.16} = 78\% \quad (22)$$

When considering improvements or trade-offs of particular loss terms, keep in mind that a change in any one term will be attenuated by efficiency squared. For instance, if switch loss were reduced by 0.3W, this is 2% of the 15W output power, but only a $2(0.8)^2 = 1.28\%$ improvement in efficiency.

Output divider

R1 and R2 set DC output voltage. R2 is normally set at 2.21kΩ (a standard 1% value) to match the LT1074 reference voltage of 2.21V, giving a divider current of 1mA. R1 is then calculated from:

$$R1 = \frac{R2(V_{OUT} - V_{REF})}{V_{REF}} \quad (23)$$

If R2 = 2.21kΩ, R1 = $(V_{OUT} - V_{REF})$ kΩ

R2 may be scaled in either direction to suit other needs, but an upper limit of 4kΩ is suggested to ensure that the frequency shifting action created by the FB pin voltage is maintained under shorted output conditions.

Output overshoot

Switching regulators often exhibit start-up overshoot because the 2-pole LC network requires a fairly low unity-gain frequency for the feedback loop. The LT1074 has asymmetrical error amplifier slew rate to help reduce overshoot, but it can still be a problem with certain combinations of L1C1 and C2R3. Overshoot should be checked on all designs by allowing the output to slew from zero in a no-load condition with maximum input voltage. This can be done by stepping the input or by pulling the VC pin low through a diode connected to a 0V to 10V square wave.

Worst-case overshoot can occur on recovery from an output short because the V_C pin must slew from its high clamp state down to ≈1.3V. This condition is best checked with the brute force method of shorting and releasing the output.

If excessive output overshoot is found, the procedure for reducing it to a tolerable level is to first try increasing the compensation resistor. The error amplifier output must slew negative rapidly to control overshoot and its slew rate is limited by the compensation capacitor. The compensation resistor, however, allows the amplifier output to "step" downward very rapidly before slewing limitations begin. The size of this step is ≈(1.1mA)(R_C). If R_C can be increased to 3kΩ, the V_C pin can respond very quickly to control output overshoot.

If loop stability cannot be maintained with R_C = 3kΩ, there are several other solutions. Increasing the size of the output capacitor will reduce short-circuit-recovery overshoot by limiting output rise time. Reducing current limit will also help for the same reason. Reducing the compensation capacitor below 0.05µF helps because the V_C pin can then slew an appreciable amount during the allowable overshoot time.

The "final solution" to output overshoot is to clamp the V_C pin so that it does not have to slew as far to shut off the output. The VC pin voltage in normal operation is known fairly precisely because it is made independent of everything except output voltage by the internal multiplier:

$$V_C \text{ Voltage } \approx 2\phi + \frac{V_{OUT}}{24} \quad (24)$$

$\phi = V_{BE}$ of internal transistor = 0.65V – 2mV/°C

To allow for transient conditions and circuit tolerances, a slightly different expression is used to calculate clamp level for the V_C pin:

$$V_{C(CLAMP)} = 2\phi + \frac{V_{OUT}}{20} + \frac{V_{IN(MAX)}}{50} + 0.2V \quad (25)$$

For a 5V output with $V_{IN(MAX)} = 30V$:

$$V_{C(CLAMP)} = 2(0.65) + \frac{5}{20} + \frac{30}{50} + 0.2 = 2.35V \quad (26)$$

There are several ways to clamp the V_C pin as shown in Figure 5.14. The simplest way is to just add a clamp Zener (D3). The problem is finding a low voltage Zener which does not leak badly below the knee. Maximum Zener leakage over temperature should be 40μA at $V_C = 2\phi + V_{OUT}/20V$. One solution is to use an LM385-2.5V micropower reference diode where the calculated clamp level does not exceed 2.5V.

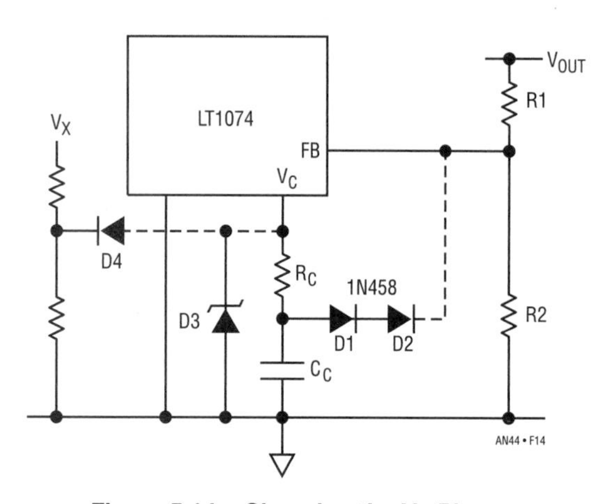

Figure 5.14 • Clamping the V_C Pin

A second clamp scheme is to use a voltage divider and diode (D4). V_X must be some quasi-regulated source which does not collapse with regulator output voltage. A third technique can be used for outputs up to 20V. It clamps the V_C pin to the feedback pin with two diodes, D1 and D2. These are small signal non-gold doped-diodes with a forward voltage that matches ϕ. The reason for this is start-up. V_C is essentially clamped to ground through the output divider when $V_{OUT} = 0$. It must be allowed to rise sufficiently to ensure start-up. The feedback pin will sit at about 0.5V with $V_{OUT} = 0$, because of the combined current from the feedback pin and V_C pin. The V_C voltage will be $2\phi + 0.5V + (0.14mA)(R_C)$. With $R_C = 1k\Omega$, $V_C = 1.94$. This is plenty to ensure start-up.

Overshoot fixes that don't work

I know that these things don't work because I tried them. The first is soft-start, created by allowing the output current or the V_C voltage to ramp up slowly. The first problem is that a slowly rising output allows more time for the V_C pin to ramp up well beyond its nominal control point so that it has to slew farther down to stop overshoot. If the V_C pin itself is ramped slowly, this can control input start-up overshoot, but it becomes very difficult to guarantee reset of the soft-start for all conditions of input sequencing. In any case, these techniques do not address the problem of overshoot following overload of the output, because they do not get "reset" by the output.

Another common practice is to parallel the upper resistor in the output divider with a capacitor. This again works fine under limited conditions, but it is easily defeated by overload conditions which pull the output slightly below its regulated point long enough for the V_C pin to hit the positive limit (≈6V). The added capacitor remains charged and the V_C pin must slew almost 5V to control overshoot when the overload is released. The resulting overshoot is impressive—and often deadly.

Tapped-inductor buck converter

Output current of a buck converter is normally limited to maximum switch current, but this restriction can be altered by tapping the inductor as shown in Figure 5.15. The ratio of "input" turns to "output" turns is "N" as shown in the schematic. The effect of the tap is to lengthen switch on time and therefore draw more power from the input without raising switch current. During switch on time, current delivered to the output through L1 is equal to switch current—5.5A maximum for the LT1074. When the switch turns off, inductor current flows only in the output section of L1, labeled "1," through

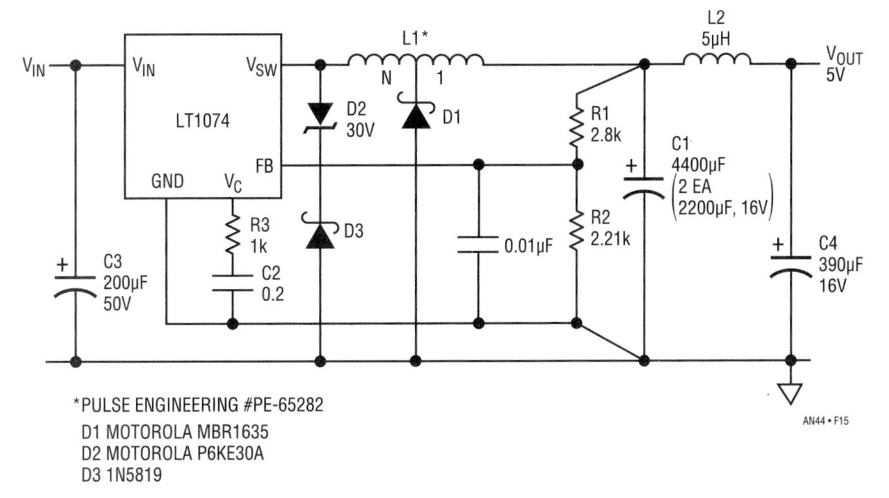

Figure 5.15 • Tapped-Inductor Buck Converter

D1 to the output. Energy conservation in the inductor requires that current increase by the ratio (N + 1):1. If N = 3, then maximum current delivered to the output during switch off time is (3 + 1)(5.5A) = 22A. Average load current is increased to the weighted average of the 5A and 22A currents. Maximum output current is given by:

$$I_{OUT(MAX)} = 0.95\left[I_{SW} - \frac{(V_{IN'} - V_{OUT'})(1+N)}{2Lf\left(N + \frac{V_{IN'}}{V_{OUT'}}\right)}\right]\left[\frac{N+1}{1+\frac{N \bullet V_{OUT'}}{V_{IN'}}}\right] \quad (27)$$

L = Total Inductance

The last term, (N + 1)/(1 + N • V_{OUT}/V_{IN}) is the basic switch current multiplier term. At high input voltages it approaches N + 1, and theoretical output current approaches 18A for N = 3. For lower input voltages the multiplier term approaches unity and no benefit is gained by tapping the inductor. Therefore, when calculating maximum load current capability, always use the worst-case low input voltage. The 0.95 multiplier is thrown-in to account for second order effects of leakage inductance, etc.

Example: $V_{IN(MIN)}$=20V, N=3, L=100µH, V_{OUT}=5V, Diode V_f = 0.55V, f =100kHz. Let I_{SW} = Maximum for LT1074 = 5.5A, V_{OUT}' = 5V + 0.55V = 5.55V, V_{IN}' = 20V – 2V = 18V:

$$I_{OUT(MAX)} = 0.95\left[5.5 - \frac{(18-5.55)(1+3)}{2(10^{-4})(10^{5})\left(3+\frac{18}{5.5}\right)}\right]\left[\frac{3+1}{1+\frac{3(5.55)}{18}}\right]$$

$$= 0.95[5.5 - 0.4][2.08] = 10.08A \quad (28)$$

Duty cycle of the tapped-inductor converter is equal to:

$$DC = \frac{1+N}{N + \frac{V_{IN'}}{V_{OUT'}}} \quad (29)$$

Average and peak diode currents are:

$$I_{D(AVG)} = \frac{I_{OUT}(V_{IN'} - V_{OUT'})}{V_{IN'}} \quad (30)$$

(Use Maximum $V_{IN'}$)

$$I_{D(PEAK)} = \frac{I_{OUT}(NV_{OUT'} + V_{IN'})}{V_{IN'}}$$

(Use Minimum $V_{IN'}$)

Average switch *current during switch on time* is:

$$I_{SW(AVG)} = \frac{I_{OUT}(N \bullet V_{OUT'} + V_{IN'})}{V_{IN'}(1+N)} \quad (31)$$

(Use Minimum $V_{IN'}$)

Diode peak reverse voltage is:

$$V_{DI(PEAK)} = \frac{V_{IN} + N \bullet V_{OUT}}{1+N} \quad (32)$$

(Use Maximum V_{IN})

Switch reverse voltage is:

$$V_{SW} = V_{IN} + V_Z + V_{SPIKE} \quad (33)$$

(Use Maximum V_{IN})

V_Z = Reverse breakdown of D2 (30V)

V_{SPIKE} = Narrow (<100ns) spike created by rapid switch turn-off and the stray wiring inductance of C3, D2, D3, and the LT1074 V_{IN} and switch pins. This voltage spike is approximately $I_{SW}/2$ *volts per inch* of total lead length.

Using parameters from the maximum output current example, with $V_{IN(MAX)} = 30V$, $I_{OUT} = 8A$:

$$\text{DC at } V_{IN} = 20V = \frac{1+3}{3+\frac{18}{5.55}} = 64\% \quad (34)$$

$$I_{D(AVG)} = \frac{(8)(28-5.55)}{28} = 6.7A$$

$$I_{D(PEAK)} \text{ at } V_{IN} = 20V = \frac{(8)(3 \bullet 5.55 + 18)}{18} = 15.4A$$

$$I_{SW(AVG)} \text{ at } V_{IN} = 20V = \frac{(8)(3 \bullet 5.55 + 18)}{18(1+3)} = 3.85A$$

Note that this is the average switch current during on time. It must be multiplied by duty cycle and switch voltage drop to obtain switch power loss. Total loss also includes switch fall time (rise time losses are minimal due to leakage inductance in L1).

$$P_{SWITCH} = (I_{SW})(DC)[1.8V + (0.1)(I_{SW})] + (V_{IN'} + V_Z)(I_{SW})(f)(t_{SW})$$

$$t_{SW} = 50ns + 3ns \bullet I_{SW}$$

$$= (3.85)(0.64)[1.8 + (0.1)(3.85)] + (20+30)(3.85)(10^5)(62ns)$$

$$= 5.3W + 1.19W = 6.5W \quad (35)$$

$$V_{DI(PEAK))} = \frac{30+3.5}{1+3} = 11.25V \quad (36)$$

$$V_{SW} = 30 + 30 + \frac{3.85}{2}(2'')^* = 64V$$

*This assumes 2" of lead length.

Snubber

The tapped-inductor converter requires a snubber (D2 and D3) to clip off negative switching spikes created by the leakage inductance of L1. This inductance (L_L) is the value measured between the tap and the switch (N) terminal with the tap shorted to the output terminal. Theoretically, the measured inductance will be zero because the shorted turns reflect "0" ohms back to any other terminals. In practice, even with bifilar winding techniques, there is ≥1% leakage inductance compared to total inductance. This is ≈1.2μH for the PE-65282. L_L is modeled as a separate inductance in series with the "N" section input, which does not couple to the rest of the inductor. This gives rise to a negative spike at the switch pin at switch turn-off. D2 and D3 clip this spike to prevent switch damage, but D2 dissipates a significant amount of power. This power is equal to the energy stored in LL at switch turnoff, ($E = (I_{SW})^2 \bullet L_L/2$) multiplied by switching frequency and a multiplier term which is dependent on the *difference* between D2 voltage and the normal reverse voltage swing at the inductor input:

$$P_{D2} = \frac{(I_{SW})^2 \bullet L_L}{2}(f)\left(\frac{V_Z}{V_Z - V_{OUT'} \bullet N}\right) \quad (37)$$

For this example:

$$P_{D2} = \frac{(3.85)^2(1.2 \bullet 10^{-6})(10^5)}{2}\left(\frac{30}{30 - 5.55 \bullet 3}\right) = 2W \quad (38)$$

Output ripple voltage

Output ripple on a tapped-inductor converter is higher than a simple buck converter because a square wave of current is superimposed on the normal triangular current fed to the output. Peak-to-peak ripple current delivered to the output is:

$$I_{P\text{-}P} = \frac{I_{OUT}(N \bullet V_{OUT} + V_{IN})(N)}{V_{IN}(1 + N)} + \frac{(1 + N)(V_{IN} - V_{OUT})}{f \bullet L\left(N + \frac{V_{IN}}{V_{OUT}}\right)} \quad (39)$$

(Use Minimum V_{IN})

A conservative approximation of RMS ripple current is one-half of peak-to-peak current.

Output ripple voltage is simply the ESR of the output capacitor multiplied times $I_{P\text{-}P}$. In this example, with ESR = 0.03Ω

$$I_{P\text{-}P} = \frac{(8)(3 \bullet 5 + 20)(3)}{20(1 + 3)} + \frac{(1 + 3)(20 - 5)}{(10^5)(10^{-4})(3 + \frac{20}{5})} = 11.4A \quad (40)$$

$$I_{RMS} = 5.7A$$

$$V_{P\text{-}P} = (0.03)(11.4) = 340mV$$

This high value of ripple current and voltage requires some thought about the output capacitor. To avoid an excessively large capacitor, several smaller units are paralleled to achieve a combined 5.7A ripple current rating. The ripple voltage is still a problem for many applications. However, to reduce ripple voltage to 50mV would require an ESR of less than 0.005W—an impractical value. Instead, an output filter is added which attenuates ripple by more than 20:1.

Input capacitor

The input bypass capacitor is selected by ripple current rating. It is assumed that all the converter input ripple current is supplied by the input capacitor. RMS input ripple current is approximately:

$$I_{IN(RMS)} \approx \frac{(I_{OUT})(V_{OUT'})}{(V_{IN'})(1 + N)}\sqrt{(1 + N)\left(\frac{V_{IN'}}{V_{OUT'}} - 1\right)} \quad (41)$$

(Use Minimum V_{IN})

$$= \frac{(8)(5.5)}{(18)(1 + 3)}\sqrt{(1 + 3)\left(\frac{18}{5.5} - 1\right)} = 1.84A\ RMS$$

The input capacitor value in microfarads is not particularly important since it is purely resistive at 100kHz; but it must be rated at the required ripple current and maximum input voltage. Radial lead types should be used to minimize lead inductance.

Positive-to-negative converter

The LT1074 can be used to convert positive voltages to negative if the sum of input and output voltage is greater than the 8V minimum supply voltage specification, and the minimum positive supply is 4.75V. Figure 5.16 shows the LT1074 used to generate negative 5V. The ground pin of the device is connected to the negative output. This allows the feedback divider, R3 and R4, to be connected in the normal fashion. If the ground pin were tied to ground, some sort of level shift and inversion would be required to generate the proper feedback signal.

Positive to negative converters have a "right half plane zero" in the transfer function which makes them particularly hard to frequency stabilize, especially with low input voltage. R1, R2, and C4 have been added to the basic design solely to guarantee loop stability at low input voltage. They may be omitted for $V_{IN} > 10V$, or $V_{IN}/V_{OUT} > 2$. R1 plus R2 is in parallel with R3 for DC output voltage calculations. Use the following guidelines for these resistors:

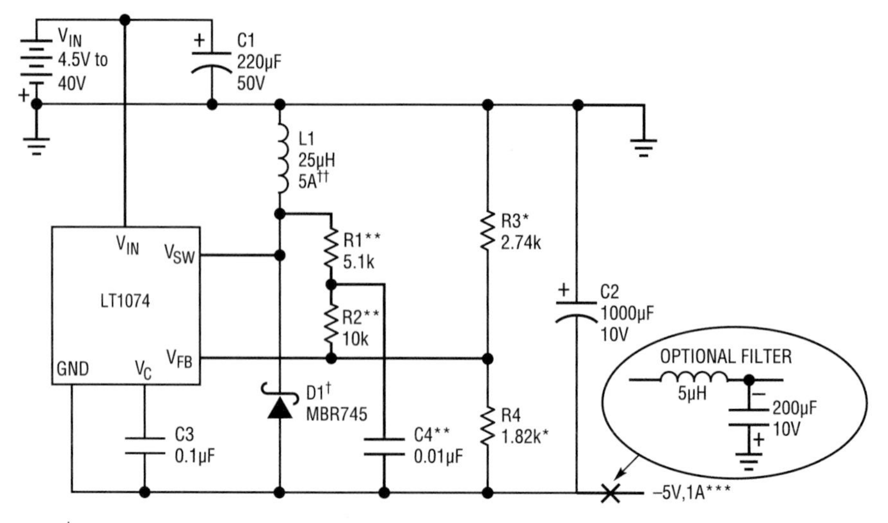

* = 1% FILM RESISTORS
D1 = MOTOROLA-MBR745
C1 = NICHICON-UPL1C221MRH6
C2 = NICHICON-UPL1A102MRH6
L1 = COILTRONICS-CTX25-5-52

† LOWER REVERSE VOLTAGE RATING MAY BE USED FOR LOWER INPUT VOLTAGES. LOWER CURRENT RATING IS ALLOWED FOR LOWER OUTPUT CURRENT.

†† LOWER CURRENT RATING MAY BE USED FOR LOWER OUTPUT CURRENT.

** R1, R2, AND C4 ARE USED FOR LOOP FREQUENCY COMPENSATION, BUT R1 AND R2 MUST BE INCLUDED IN THE CALCULATION FOR OUTPUT VOLTAGE DIVIDER VALUES. FOR HIGHER OUTPUT VOLTAGES, INCREASE R1, R2 AND R3 PROPORTIONATELY;
R3 = V_{OUT} –2.37 (kΩ)
R1 = (R3) (1.86)
R2 = (R3) (3.65)

*** MAXIMUM OUTPUT CURRENT OF 1A IS DETERMINED BY MINIMUM INPUT VOLTAGE OF 4.5V. HIGHER MINIMUM INPUT VOLTAGE WILL ALLOW MUCH HIGHER OUTPUT CURRENTS.

AN44 • F16

Figure 5.16 • Positive-to-Negative Converter

$$R4 = 1.82k$$
$$R3 = |V_{OUT}| - 2.37 \quad \text{(In k}\Omega\text{)}$$
$$R1 = R3\ (1.86)$$
$$R2 = R3\ (3.65)$$

If R1 and R2 are omitted:

$$R4 = 2.21k$$
$$R3 = |V_{OUT}| - 2.21 \quad \text{(In kQ)}$$

A +12V to −5V converter would have R4 = 2.21k and R3 = 2.74k.

Recommended compensation components would be C3 = 0.005µF in parallel with a series R_C of 0.1µF and 1kΩ.

The converter works by charging L1 through the input voltage when the LT1074 switch is on. During switch off time, the inductor current is diverted through D1 to the negative output. For continuous mode operation, duty cycle of the switch is:

$$DC = \frac{V_{OUT'}}{V_{IN'} + V_{OUT}} \tag{42}$$

(Use absolute value (for V_{OUT})
Peak switch current for continuous mode is:

$$I_{SW(PEAK)} = \frac{I_{OUT}(V_{IN'} + V_{OUT'})}{V_{IN'}} + \frac{(V_{IN'})(V_{OUT'})}{2f \bullet L(V_{IN'} + V_{OUT'})} \tag{43}$$

To calculate maximum output current for a given maximum switch current (IM) this can be rearranged as;

$$I_{OUT(MAX)} = \frac{V_{IN'} - (I_M)(R_L)}{V_{IN'} + V_{OUT'}}\left[I_M - \frac{(V_{IN'})(V_{OUT'})}{2f \bullet L(V_{IN'} + V_{OUT'})}\right] \tag{44}$$

(Use Minimum $V_{IN'}$)

Note that an extra term ($I_M \bullet R_L$) has been added. This is to account for the series resistance (R_L) of the inductor, which may become a significant loss at low input voltages.

Maximum output current is dependent upon input *and* output voltage, unlike the buck converter which will supply essentially a constant output current. The circuit shown will supply over 4A at V_{IN} = 30V, but only 1.3A at V_{IN} = 5V. The $I_{OUT(MAX)}$ equation does not include second order loss terms such as capacitor ripple current, switch rise and fall time, core loss, and output filter. These factors may reduce maximum output current by up to 10% at low input and/or output voltages. Figure 5.17 shows $I_{OUT(MAX)}$ versus input voltage for various output voltages. It assumes a 25µH inductor for V_{OUT} = -5V, 50µH for V_{OUT} = –12V, and 100µH for V_{OUT} = –25V.

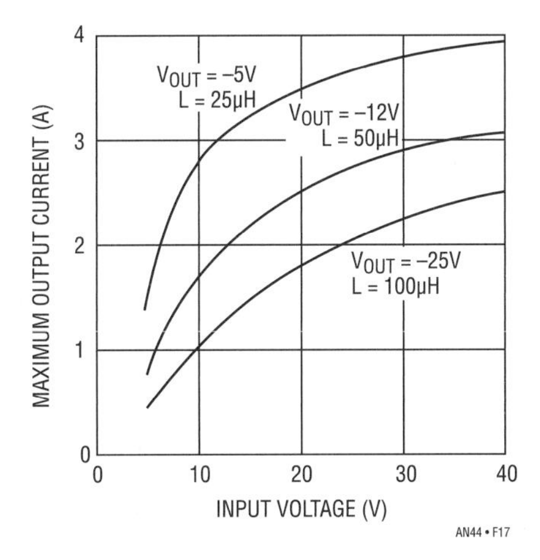

Figure 5.17 • Maximum Output Current of Positive-to-Negative Converter

If absolute minimum circuit size is required and load currents are not too high, discontinuous mode can be used. Minimum inductance required for a specified load is:

$$L_{MIN} = \frac{2I_{OUT}(V_{OUT'})}{(I_M)^2 \bullet f} \quad (45)$$

There is a maximum load current that can be supplied in discontinuous mode. Above this current, the formula for L_{MIN} is invalid. Maximum load current in discontinuous mode is:

Discontinuous Mode

$$I_{OUT(MAX)} = \left(\frac{V_{IN'}}{V_{IN'} + V_{OUT'}}\right)\left(\frac{I_M}{2}\right) \quad (46)$$

(Use Minimum V_{IN})

Example: V_{OUT} = 5V, I_M = 5A, f = 100kHz, Load Current = 0.5A. Diode Forward Voltage = 0.5V, giving $V_{OUT'}$ = 5.5V. V_{IN} = 4.7V to 5.3V. Assume $V_{IN'(MIN)}$ = 4.7V – 2.3V = 2.4V.

$$I_{OUT(MAX)} = \left(\frac{2.4}{2.4 + 5.5}\right)\left(\frac{5}{2}\right) = 0.76\text{A} \quad (47)$$

The required load current of 0.5A is less than the maximum of 0.76A, so discontinuous can be used:

$$L_{MIN} = \frac{2(0.5)(5.5)}{(5)^2(10^5)} = 2.2\mu\text{H} \quad (48)$$

To ensure full load current with production variations of frequency and inductance, 3µH should be used.

The formula for minimum inductance assumes a high peak current in the inductor (≈5A). If the minimum inductance is used, the inductor must be specified to handle the high peak current without saturating. The high ripple current will also cause relatively high core loss and output ripple voltage, so some judgment must be used in minimizing the inductor size. See the Inductor Selection section for more details.

To calculate peak inductor and switch current in discontinuous mode, use:

$$I_{PEAK} = \sqrt{\frac{2 \bullet I_{OUT} \bullet V_{OUT'}}{L \bullet f}} \quad (49)$$

Input capacitor

C3 is used to absorb the large square wave switching currents drawn by positive to negative converters. It must have low ESR to handle the RMS ripple current and to avoid input voltage "dips" during switch on time, especially with 5V inputs. Capacitance value is not particularly important if ripple current and operating voltage requirements are met. RMS ripple current in the capacitor is:

Continuous Mode

$$I_{RMS} = I_{OUT}\sqrt{\frac{V_{OUT'}}{V_{IN'}}} \quad (50)$$

(Use Minimum $V_{IN'}$)

Discontinuous Mode*

$$I_{RMS} = \frac{(I_{OUT})(V_{OUT'})}{V_{IN'}}\sqrt{\frac{1.35\left(1-\frac{m}{2}\right)^3}{m} + 0.17m^2 + 1 - m} \quad (51)$$

$$m = \frac{1}{V_{IN'}}\sqrt{2L\,fI_{OUT}V_{OUT'}}$$

*This formula is a test for calculator students

Examples: A continuous mode design with $V_{IN} = 12V$, $V_{OUT} = -5V$, $I_{OUT} = 1A$, $V_{OUT'} = 5.5V$, and $V_{IN'} = 10V$.

$$I_{RMS} = (1)\sqrt{\frac{5.5}{10}} = 0.74A\ RMS \quad (52)$$

Now change to a discontinuous design with the same conditions and L = 5μH, f = 100kHz:

$$m = \frac{1}{10}\sqrt{(2)(10 \bullet 10^{-6})(105)(1)(5.5)} = 0.33 \quad (53)$$

$$I_{RMS} = \frac{(1)(5.5)}{10}\sqrt{\frac{1.35(1-0.165)^3}{0.33} + 0.17(0.33)^2 + 1 - 0.33}$$
$$= 0.96A\ RMS$$

Notice that discontinuous mode saves on inductor size, but may require a larger input capacitor to handle the ripple current increase. The 30% increases in ripple current generates 70% more heating in the capacitor ESR.

Output capacitor

The inductor on a positive to negative converter does not operate as a filter. It simply acts as an energy storage device so that energy can be transferred from input to output. Therefore, all filtering is done by the output capacitor, and it must have adequate ripple current rating and low ESR. Output ripple voltage for continuous mode will contain three distinct components; a "spike" on switch transitions which is equal to the rate of rise/fall of switch current multiplied by the effective series inductance (ESL) of the output capacitor, a square wave proportional to load current and capacitor ESR, and a triangular component dependent on inductor value and ESR. The spikes are very narrow, typically less than 100ns, and often "disappear" in the parasitic filter created by the inductance of pc board traces between the converter and load combined with the load bypass capacitors. One must be extremely careful when looking at these spikes with an oscilloscope. The magnetic fields created by currents transitions in converter wiring will generate "spikes" on the screen even when they do not exist at the converter output. See the Oscilloscope Techniques section for details.

The peak-to-peak sum of square wave and triangular output ripple voltage is:

$$V_{P\text{-}P} = ESR\left[\frac{I_{OUT}(V_{IN'} + V_{OUT'})}{V_{IN'}} + \frac{(V_{OUT'})(V_{IN'})}{2(V_{OUT} + V_{IN})(f)(L)}\right] \quad (54)$$

(Use Minimum $V_{IN'}$)

Example: $V_{IN} = 5V$, $V_{OUT} = -5V$, L = 25μH, $I_{OUT(MAX)} = 1A$, f = 100kHz. Assume $V_{IN'} = 2.8V$, $V_{OUT'} = 5.5V$, and ESR = 0.05Ω.

$$V_{P\text{-}P} = 0.05\left[\frac{(1)(2.8+5.5)}{2.8} + \frac{(5.5)(2.8)}{2(5.5+2.8)(10^5)(25 \bullet 10^{-6})}\right]$$
$$= 172mV \quad (55)$$

For some applications this rather high ripple voltage may be acceptable, but more commonly it will be necessary to reduce ripple voltage to 50mV or less. This may be impractical to achieve simply by reducing ESR, so an output filter (L2, C4) is shown, The filter components are relatively small and low cost, both of which are additionally offset by possible reduction in the size of the main output capacitor C1. See the Output Filters section for details.

C1 must be chosen for ripple current as well as ESR. Ripple current into the output capacitor is given by:

Continuous Mode

$$I_{RMS} = I_{OUT}\sqrt{\frac{V_{OUT'}}{V_{IN'}}} \quad (56)$$

Discontinuous Mode

$$I_{RMS} = I_{OUT}\sqrt{\frac{0.67(I_P - I_{OUT})^3}{I_{OUT}(I_P)^2} + \frac{0.67(I_{OUT})^2}{(I_P)^2} + 1 - \frac{2I_{OUT}}{I_P}} \quad (57)$$

where I_P = Peak Inductor Current:

$$= \sqrt{\frac{2I_{OUT}(V_{OUT'})}{L \bullet f}}$$

For the Continious Mode example:

$$I_{RMS} = (1A)\sqrt{\frac{5.5}{2.8}} = 1.4A\ RMS \quad (58)$$

with Discontinuous Mode using a 3μA inductor, with I_{OUT} = 0.5A:

$$I_P = \sqrt{\frac{(2)(0.5)(5.5)}{(3 \bullet 10^{-6})(10^5)}} = 4.28$$

$$I_{RMS} = (0.5)\sqrt{\frac{(0.67)(4.28 - 0.5)^3}{(0.5)(4.28)^2} + \frac{(0.67)(0.5)^2}{(4.28)^2} + 1 - \frac{2(0.5)}{4.28}} = 1.09A\ RMS \quad (59)$$

Notice that output capacitor ripple current is over twice the DC output current in this discontinuous example. The smaller inductor size obtained by discontinuous mode may be somewhat offset by the larger capacitors required on input and output to meet ripple current conditions.

Efficiency

Efficiency for this positive to negative converter can be quite high for larger input and output voltages (>90%), but can be much lower for low input voltages. Losses are summarized below for a continuous mode design. Discontinuous losses are much more difficult to express analytically, but will typically be 1.2 to 1.3 times higher than in continuous mode.

Conduction loss in switch = P_{SW} (DC):

$$P_{SW}(DC) = \frac{(I_{OUT})(V_{OUT'})}{V_{IN'}}\left[1.8V + \frac{(0.1)(I_{OUT})(V_{OUT'} + V_{IN'})}{V_{IN'}}\right] \quad (60)$$

Transient switch loss = P_{SW} (AC):

$$P_{SW}(AC) = \frac{I_{OUT}(V_{OUT'} + V_{IN'})^2 2(t_{SW})(f)}{V_{IN'}} \quad (61)$$

where t_{SW} = 50ns + 3ns ($V_{OUT'}$ + $V_{IN'}$)/$V_{IN'}$. The LT1074 quiescent current generates a loss called P_{SUPPLY}:

$$P_{SUPPLY} = (V_{IN'} + V_{OUT'}\left[\frac{7mA + 5mA(V_{OUT'})}{(V_{OUT'} + V_{IN'})}\right] \quad (62)$$

Catch diode loss = P_{DI} = (I_{OUT})(V_f)

where V_f = Forward Voltage of D1 at a current equal to:

$$I_{OUT}(V_{OUT'} + V_{IN'})/V_{IN'}$$

Capacitor losses can be found by calculating RMS ripple current and multiplying by capacitor ESR. Inductor losses are the sum of copper (wire) loss and core loss:

$$P_{L1} = R_L\left[\frac{(I_{OUT})(V_{OUT'} + V_{IN'})}{V_{IN'}}\right]^2 + P_{CORE} \quad (63)$$

R_L = Inductor Copper Resistance

P_{CORE} can be calculated if the inductor core material is known. See the Inductor Selection section.

Example: V_{IN} = 12V, V_{OUT}=-12V, I_{OUT} = 1.5A, f = 100kHz. Let L1 = 50μH, with R_L = 0.04Ω. Assume ESR of input and output capacitor is 0.05Ω. $V_{IN'}$ = 12V – 2V = 10V, $V_{OUT'}$ = 12V + 0.5V = 12.5V.

$$P_{SW}(DC) = \frac{(1.5)(12.5)}{10}\left[1.8 + \frac{(0.1)(1.5)(12.5+10)}{10}\right] = 4W \quad (64)$$

$$P_{SW}(AC) = \frac{(1.5)(12.5+10.5)^2}{10}\left[2(50ns+3ns)\frac{(12.5+10)}{10}\right](10^5) = 0.86W$$

$$P_{SUPPLY} = (12+12)\left[7mA + \frac{5mA(12.5)}{12.5+10}\right] = 0.23W$$

$$P_{DI} = (1.5)(0.5) = 0.75W$$

$$I_{RMS(INPUT\ CAP)} = 1.5\sqrt{\frac{12.5}{10}} = 1.68A\ RMS$$

$$P_{C3} = (1.68)^2(0.05) = 0.14W$$

$$I_{RMS(OUTPUT\ CAP)} = I_{OUT}\sqrt{\frac{(12.5)^2+(12.5)(10)}{(10)(12.5+10)}} = 1.68A\ RMS$$

$$P_{C1} = (1.68)^2(0.05) = 0.14W$$

$$P_{L1} = 0.04\left[\frac{(1.5)(12.5+10)}{10}\right]^2 = 0.46W$$

Assume $P_{CORE} = 0.2W$

$$\text{Efficiency} = \frac{I_{OUT}V_{OUT}}{I_{OUT}V_{OUT} + \Sigma P_{LOSS}}$$

$$\Sigma P_{LOSS} = 4 + 0.86 + 0.23 + 0.75 + 0.14 + 0.14 + 0.46 + 0.2 = 6.78W$$

$$\text{Efficiency} = \frac{(1.5)(12)}{(1.5)(12) + 6.78} = 73\%$$

Negative boost converter

Note: *All equations in this section use the absolute value of V_{IN} and V_{OUT}.*

The LT1074 can be configured as a negative boost converter (Figure 5.18) by tying the ground pin to the negative output. This allows the regulator to operate from input voltages as low as 4.75V if the regulated output is at least 8V. R1 and R2 set the output voltage as in a conventional connection, with R1 selected from:

$$R1 = \frac{V_{OUT} \bullet R2}{V_{REF}} - R2 \quad (65)$$

Boost converters have a "right-half plane zero" in the forward part of the signal path and for this reason, L1 is kept to a low value to maximize the "zero" frequency. With larger values for L1, it becomes difficult to stabilize the regulator, especially at low input voltages. If $V_{IN} > 10V$, L1 can be increased to 50μH.

There are two important characteristics of boost converters to keep in mind. First, the input voltage cannot exceed the output voltage, or D1 will simply pull the output unregulated high. Second, the output cannot be pulled below the input, or D1 will drag down the input supply. For this reason, boost converters are not normally considered short-circuit protected unless some form of fusing is provided. Even with fuses, there is the possibility of damage to D1 if the input supply can deliver very large surge currents.

Boost converters require switch currents which can be much greater than output load current. Peak switch current is given by:

$$I_{SW(PEAK)} = \frac{I_{OUT} \bullet V_{OUT'}}{V_{IN'}} + \frac{V_{IN'}(V_{OUT'} - V_{IN'})}{2L \bullet f \bullet V_{OUT'}} \quad (66)$$

For the circuit in Figure 5.18, with $V_{IN} = 5V$, ($V_{IN'} \approx 3V$), $V_{OUT'} \approx 15.5V$, with an output load of 0.5A:

$$I_{SW(PEAK)} = \frac{(0.5A)(15.5)}{3} + \frac{3(15.5-3)}{2(25\mu H)(10^5)(15.5)} = 3.07A \quad (67)$$

This formula can be rearranged to yield maximum load current for a given maximum switch current (I_M)

$$I_{OUT(MAX)} = \frac{I_M \bullet V_{IN'}}{V_{OUT'}} - \left(\frac{V_{IN'}}{V_{OUT'}}\right)^2 \frac{V_{OUT'} - V_{IN'}}{2L \bullet f} \quad (68)$$

For $I_M = 5.5A$, this equation yields 0.82A with $V_{IN} = 4.5V$, 1.8A with $V_{IN} = 8V$, and 3.1A for $V_{IN} = 12V$.

The explanation for switch current which is much higher than output current is that current is delivered to the output only during switch off time. With low input voltages, the switch is on a high percentage of the total

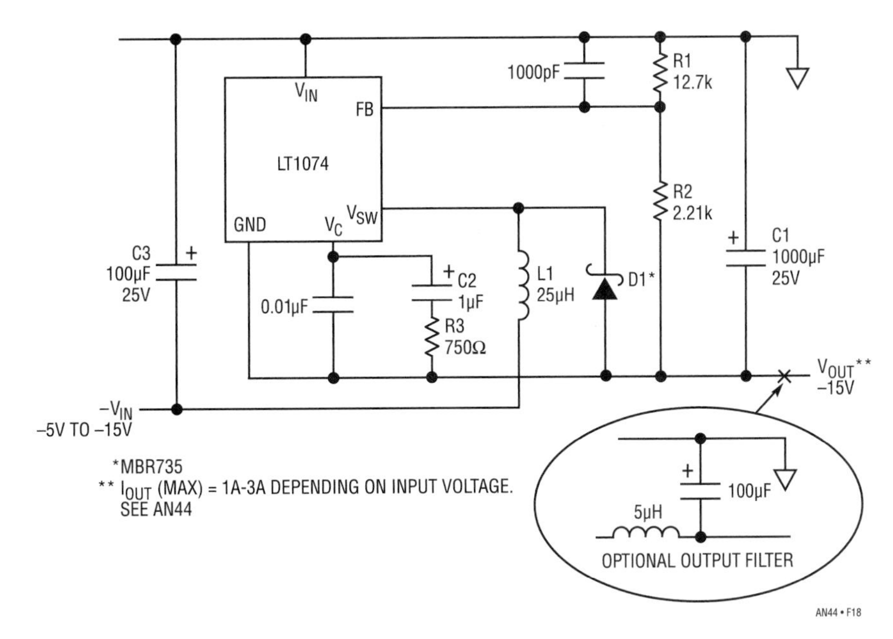

Figure 5.18 • Negative Boost Converter

switching cycle and current is delivered to the output only a small percent of the time. Switch duty cycle is given by:

$$DC = \frac{V_{OUT'} - V_{IN'}}{V_{OUT'}} \tag{69}$$

For $V_{IN} = 5V$, $V_{OUT} = 15V$, $V_{IN'}$ - 3V, $V_{OUT'} = 15.5V$ and:

$$DC = \frac{15.5 - 3}{15.5} = 81\% \tag{70}$$

Peak inductor current is equal to peak switch current. Average inductor current in continuous mode is equal to:

$$I_{L(AVG)} = \frac{I_{OUT} \bullet V_{OUT'}}{V_{IN'}} \tag{71}$$

A 0.5A load requires 2.6A inductor current for $V_{IN} =$ 5V.

Along with high switch currents, keep in mind that boost converters draw DC input currents *higher* than the output load current. Average input current to the converter is:

$$I_{IN}(DC) \approx \frac{(I_{OUT})(V_{OUT'})}{V_{IN'}} \tag{72}$$

with $I_{OUT} = 0.5A$, and $V_{IN} = 5V$ ($V_{IN'} \approx 3V$):

$$I_{IN}(DC) = \frac{(0.5)(15.5)}{3} = 2.6A \tag{73}$$

This formula does not take into account secondary loss terms such as the inductor, output capacitor, etc., so it is somewhat optimistic. Actual input current may be closer to 3A. *Be sure the input supply is capable of providing the required boost converter input current.*

Output diode

The *average* current through D1 is equal to output current, but the peak pulse current is equal to peak switch current, which can be many times output current. D1 should be conservatively rated at 2 to 3 times output current.

Output capacitor

The output capacitor of a boost converter has high RMS ripple current so this is often the deciding factor in the selection of C1. RMS ripple current is approximately:

$$I_{RMS(C1)} \approx I_{OUT}\sqrt{\frac{V_{OUT'} - V_{IN'}}{V_{IN'}}} \tag{74}$$

for $I_{OUT} = 0.5A$, $V_{IN} = 5V$:

$$I_{RMS} \approx 0.5\sqrt{\frac{15.5 - 3}{3}} = 1A\ RMS \tag{75}$$

C1 must have a ripple current rating of 1A RMS. Its actual capacitance value is not critical. ESR of the capacitor will determine output ripple voltage.

Output ripple

Boost converters tend to have high output ripple because of the high pulse currents delivered to the output capacitor:

$$V_{P\text{-}P} = ESR\left[\frac{I_{OUT} \bullet V_{OUT'}}{V_{IN'}} + \frac{V_{IN'}\,(V_{OUT'} - V_{IN'})}{2L \bullet f \bullet V_{OUT'}}\right] \quad (76)$$

This formula assumes continuous mode operation, and it ignores the inductance of C1. In actual operation, C1 inductance will allow output "spikes" which should be removed with an output filter. The filter can be as simple as several inches of output wire or trace and a small solid tantalum capacitor if only the spikes need to be removed. A filter inductor is required if significant reduction of the fundamental is needed. See the Output Filter section.

For the circuit in Figure 5.18, with $I_{OUT} = 0.5A$, $V_{IN} = 5V$; and an output capacitor ESR of 0.05Ω:

$$V_{P\text{-}P} = 0.05\left[\frac{(0.5)(15.5)}{3} + \frac{3(15.5 - 3)}{2(25 \bullet 10^{-6})(10^5)(15.5)}\right] = 153mV \quad (77)$$

Boost converters are more benign with respect to input current pulsing than buck or inverting converters. The input current is a DC level with a triangular ripple superimposed. RMS value of input current ripple is:

$$I_{RMS(C3)} \approx \frac{V_{IN'}\,(V_{OUT'} - V_{IN'})}{3L \bullet f \bullet V_{OUT'}} \quad (78)$$

Notice that ripple current is independent of load current assuming that load current is high enough to keep the converter in continuous mode. For the converter in Figure 5.18, with $V_{IN} = 5V$:

$$I_{RMS} = \frac{3(15.5 - 3)}{3(25 \bullet 10^{-6})(10^5)(15.5)} = 0.32A\ RMS \quad (79)$$

C3 may be chosen on a ripple current basis to minimize size. Larger values will allow less conducted EMI back into the input supply.

Inductor selection

There are five main criteria in selecting an inductor for switching regulators. First, and most important, is the actual inductance value. If inductance is too low, output power will be restricted. Too much inductance results in large physical size and poor transient response. Second, the inductor must be capable of handling both RMS and peak currents which may be significantly higher than load current. Peak currents are limited by core saturation, with resultant loss of inductance. RMS currents are limited by heating effects in the winding. Also important is peak-to-peak current which determines heating effects in the core itself. Third, the physical size or weight of the inductor may be important in many applications. Fourth, power losses in the inductor can significantly affect regulator efficiency, especially at higher switching frequencies. Last, the price of inductors is very dependent on particular construction techniques and core materials, which impact overall size, efficiency, mountability, EMI, and form factor. There may be a significant cost penalty, for instance, if more expensive core materials are needed in "minimum size" applications.

The issues of price and size become particularly complicated at higher frequencies. High frequencies are used to reduce component size, and indeed, the inductance values required scale inversely with frequency. The problem with a scaled-down high frequency inductor is that total core loss increases slightly with frequency for constant ripple current, and this power is now dissipated in a smaller core, so temperature rise *and* efficiency can limit size reductions. Also, the smaller core has less room for wire, so wire losses may increase. The only solution to this problem is to find a better core material. Common low cost inductors use powdered iron cores, which are very low cost. These cores exhibit modest losses at 40kHz with a typical flux density of 300 gauss. At 100kHz, core losses can become unacceptably high at these flux densities. Reducing flux density requires a larger core, canceling part of the advantage gained in reducing inductance at the higher frequency.

Molypermalloy, "high flux," Kool Mμ (Magnetics, Inc.), and ferrite cores have considerably lower core loss, and can be used at 100kHz and above with higher flux density, but these cores are expensive. The basic lesson here is that attention to inductor selection is very important to minimize costs and achieve desired goals of size and efficiency.

A special equation has been developed in the following section which shows that for a given core material, total core loss is dependent almost totally on *frequency* and *inductance* value, not physical size or shape. The formula is arranged to solve for the inductance required to achieve a given core loss. It shows that, in a typical 100kHz buck converter, inductance has to be increased by a factor of three over the minimum required, if a low cost powdered iron core is used.

"Standard" switching regulator inductors are toroids. Although this shape is hardest to wind, it offers excellent utilization of the core, and more importantly, has low EMI fringing fields. Rod or drum shaped inductors have very high fringing fields and are not recommended except possibly for secondary output filters. Inductors made with "E-E" or "E-C" split cores are easy to wind on the separate bobbin, but tend to be much taller than toroids and more expensive. "Pot" cores reverse the position of winding and core—the core surrounds the winding. These cores offer the best EMI shielding, but tend to be bulky and more expensive. Also, temperature rise is higher because of the enclosed winding. Special low profile split cores (TDK "EPC," etc.) are now offered in a wide range of sizes. Although not as efficient as EC cores in terms of watts/volume, these cores are attractive for restricted height applications.

The best way to select an inductor is to first calculate the limitations on its minimum value. These limitations are imposed by a maximum allowed switch current, maximum allowable efficiency loss, and the necessity to operate in continuous versus discontinuous mode. (See discussion elsewhere of the consequences related to these two modes.) After the minimum value has been established, calculations are done to establish the operating conditions of the inductor; i.e., RMS current, peak-to-peak ripple current, and peak current. With this information, next select an "off-the-shelf" inductor which meets all the calculated requirements, or is reasonably close, Then ascertain the physical size and price of the selected inductor. If it fits in the allowed "budget" of space, height, and cost, you can then give some consideration to increasing the inductance to gain better efficiency, lower output ripple, lower input ripple, more output power, or some combination of these. If the selected inductor is physically too large, there are several possibilities; select a different core shape, a different core material, (which will require recalculating the minimum inductance based on efficiency loss), a higher operating frequency, or consider a custom wound inductor which is optimized for the application. Keep in mind when attempting to shoehorn an inductor into the smallest possible space that output overload conditions may cause currents to increase to the point of inductor failure. The major failure mode to consider is winding insulation failure due to high winding temperature. IC failure caused by loss of inductance due to core saturation or core temperature is not usually a problem because the LT1074 has pulse-by-pulse current limiting which is effective even with drastically lowered inductance.

The following equations solve for minimum inductance based on the assumption of limited peak switch current (I_M).

Minimum inductance to achieve a required output power

Buck Mode Discontinuous, $I_{OUT} \leq \frac{I_M}{2}$, Use Maximum V_{IN} (80)

$$L_{MIN} = \frac{2 \bullet I_{OUT} \bullet V_{OUT}(V_{IN'} - V_{OUT})}{f(I_M)^2(V_{IN'})}$$

Buck Mode Continuous, $I_{OUT} \leq I_M$, Use Maximum V_{IN} (81)

$$L_{MIN} = \frac{V_{OUT}(V_{IN'} - V_{OUT})}{2 \bullet f \bullet V_{IN'}(I_M - I_{OUT})}$$

Inverting Mode Discontinuous, $I_{OUT} \leq \frac{I_M \bullet V_{IN'}}{2(V_{IN'} + V_{OUT'})}$ (82)

$$L_{MIN} = \frac{2 \bullet I_{OUT} \bullet V_{OUT'}}{(I_M)^2 \bullet f}$$

Inverting Mode Continuous, $I_{OUT} \leq \frac{I_M \bullet V_{IN'}}{(V_{IN'} + V_{OUT'})}$ (83)

$$L_{MIN} = \frac{(V_{IN'})^2 \bullet V_{OUT'}}{2 \bullet f(V_{OUT'} + V_{IN'})^2 \left(\frac{I_M \bullet V_{IN'}}{V_{IN'} + V_{OUT'}} - I_{OUT}\right)}$$

$$\text{Boost Mode Discontinuous, } I_{OUT} \leq \frac{I_M \bullet V_{IN'}}{2 \bullet V_{OUT'}} \quad (84)$$

$$L_{MIN} = \frac{2 \bullet I_{OUT}(V_{OUT'} - V_{IN'})}{(I_M)^2 \bullet f}$$

$$\text{Boost Mode Continuous, } I_{OUT} \leqslant \frac{I_M \bullet V_{IN'}}{V_{OUT'}} \quad (85)$$

$$L_{MIN} = \frac{(V_{IN'})^2 (V_{OUT'} - V_{IN'})}{2 \bullet f(V_{OUT'})^2 \left(\frac{I_M \bullet V_{IN'}}{V_{OUT'}} - I_{OUT}\right)}$$

$$\text{Tapped Inductor Continuous, } I_{OUT} \leqslant \frac{I_M(N+1)(V_{IN'})}{V_{IN'} + NV_{OUT'}} \quad (86)$$

$$L_{MIN} = \frac{V_{IN} \bullet V_{OUT}(V_{IN} - V_{OUT})(N+1)^2}{I_M \bullet 2f \bullet V_{IN}(N+1)(V_{IN} + NV_{OUT}) - I_{OUT}(V_{IN} + NV_{OUT})^2(2f)}$$

Minimum inductance required to achieve a desired core loss

Power loss in inductor core material is not intuitive at all. It is, to a first approximation, independent of the *size* of the core for a given inductance and operating frequency. Second, power loss *drops* as inductance increases, for constant frequency. Last, raising frequency with a given inductor will *decrease* core loss, even though manufacturer's curves show that core loss increases with frequency. These curves assume constant flux density, which is not true for a fixed inductance.

The general formula for core loss can be expressed as:

$$P_C = C \bullet B_{AC}^p \bullet f^d \bullet V_C \quad (87)$$

C, d, p = Constants (see Table 5.1)
BAC = Peak AC Flux Density (1/2 peak-to-peak) (gauss)
f = Frequency
V_C = Core Volume (cm^3)

The exponent "p" falls in the range of 1.8-2.4 for powdered iron cores, ≈2.1 for molypermalloy, and 2.3-2.8 for fer- rites. "d" is ≈1 for powdered iron and ≈1.3 for ferrite. A closed form expression can be generated which relates core loss to the basic requirements of a switching regulator; inductance, frequency, and input/output voltages. The general form is:

$$\text{Continuous Mode } P_C = \frac{a \bullet b^p}{f^{p-d} \bullet L^{p/2}} \quad (88)$$

$$\text{Discontinuous Mode } P_C = a \bullet f^{d-1} \bullet e \quad (89)$$

a,d, p = Core Material Constants (see Table 5.1)
b,e = Constants Determined by Input and Output Voltages and Currents
L = Inductance

These formulas show that core material, inductance, and frequency are the only degrees of freedom to alter core loss in the continuous mode case. For discontinuous mode, even inductance disappears as a variable, leaving frequency and core material. Further, the constant "d" is close to unity for many core materials, yielding a discontinuous mode core loss independent of all user variables except core material!

The following specific formulas will allow calculation of the inductance to achieve a given core loss in continuous mode and will indicate actual core loss for the discontinuous mode.

When using these formulas, assume initially that the term $V_e^{p-2/p}$ can be ignored. It is close to unity for a relatively wide range of core volumes because the exponent (p-2)/2 is less than 0.1 for commonly used powdered iron and molypermalloy cores. After an inductor is chosen and Ve is known, the term $V_e^{p-2/p}$ can be calculated to double check its effect on the value for L_{MIN}, usually less than 20%:

Continuous Mode

$$L_{MIN}^* = \frac{a \bullet \mu_e \bullet V_L^2}{(P_C)^{2/p} \bullet f^{\left(2-\frac{2d}{p}\right)} \bullet V_e^{\left(\frac{p-2}{p}\right)}} \quad (90)$$

Buck Mode Discontinuous

$$P_C = \frac{a \bullet \mu_e(0.4\pi)f^{d-1}}{10^{-8}}(V_L \bullet I_{OUT}) \quad (91)$$

*A strict derivation

a, d, p = Core loss constants. Use Table 5.1.
μ_e= Effective core permeability. For ungapped cores, use Table 5.1. For gapped cores, use manufacturer's specification, or calculate.
V_L = An equivalent "voltage," dependent on input voltage, output voltage, and topology. Use Table 5.2.
P_C = Total core loss in watts.
L = Inductance.
V_e = Effective core volume in cm^3.

Table 5.1 Core Constants

		C	a	d	p	μ	Loss at 100kHz, 500 Gauss (mW/cm^3)
Micrometals							
Powdered Iron	#8	4.30E-10	8.20E-05	1.13	2.41	35	617
	#18	6.40E-10	1.20E-04	1.18	2.27	55	670
	#26	7.00E-10	1.30E-04	1.36	2.03	75	1300
	#52	9.10E-10	4.90E-04	1.26	2.11	75	890
Magnetics							
Kool Mμ	60	2.50E-11	3.20E-06	1.5	2	60	200
	75	2.50E-11	3.20E-06	1.5	2	75	200
	90	2.50E-11	3.20E-06	1.5	2	90	200
	125	2.50E-11	3.20E-06	1.5	2	125	200
Molypermalloy	−60	7.00E-12	2.90E-05	1.41	2.24	60	87
	−125	1.80E-11	1.60E-04	1.33	2.31	125	136
	−200	3.20E-12	2.80E-05	1.58	2.29	200	390
	−300	3.70E-12	2.10E-05	1.58	2.26	300	368
	−550	4.30E-12	8.50E-05	1.59	2.36	550	890
High Flux	−14	1.10E-10	6.50E-03	1.26	2.52	14	1330
	−26	5.40E-11	4.90E-03	1.25	2.55	26	740
	−60	2.60E-11	3.10E-03	1.23	2.56	60	290
	−125	1.10E-11	2.10E-03	1.33	2.59	125	460
	−160	3.70E-12	6.70E-04	1.41	2.56	160	1280
Ferrite	F	1.80E-14	1.20E-05	1.62	2.57	3000	20
	K	2.20E-18	5.90E-06	2	3.1	1500	5
	P	2.90E-17	4.20E-07	2.06	2.7	2500	11
	R	1.10E-16	4.80E-07	1.98	2.63	2300	11
Philips							
Ferrite	3C80	6.40E-12	7.30E-05	1.3	2.32	2000	37
	3C81	6.80E-14	1.50E-05	1.6	2.5	2700	38
	3C85	2.20E-14	8.70E-08	1.8	2.2	2000	18
	3F3	1.30E-16	9.80E-08	2	2.5	1800	7
TDK							
Ferrite	PC30	2.20E-14	1.70E-06	1.7	2.4	2500	21
	PC40	4.50E-14	1.10E-05	1.55	2.5	2300	14
Fair-Rite	77	1.70E-12	1.80E-05	1.5	2.3	1500	86

Table 5.2 Equivalent Inductor Voltage

TOPOLOGY	V_L
Buck Continuous	$V_{OUT}(V_{IN} - V_{OUT})/2V_{IN}$
Buck Discontinuous	
Inverting Continuous	$V_{IN}' \bullet V_{OUT}'/[2(V_{IN}' + V_{OUT}')]$
Inverting Discontinuous	
Boost Continuous	$V_{IN}'(V_{OUT}' - V_{IN}')/2V_{OUT}'$
Boost Discontinuous	
Tapped-inductor	$(V_{IN} - V_{OUT})(V_{OUT})(1 + N)/2(V_{IN} + NV_{OUT})$

Example: Buck converter with $V_{IN} = 20V$ to $30V$, $V_{OUT} = 5V$, $I_{OUT} = 3A$, $f = 100kHz$, maximum inductor loss $= 0.8W$.

3A is more than $I_M/2$, so continuous mode must be used. Maximum input voltage is used to calculate L_{MIN} from Equation 81:

$$L_{MIN} = \frac{5(30-5)}{2(10^5)(30)(5-3)} = 10.4\mu H \qquad (92)$$

Now calculate minimum inductance to achieve desired core loss. Assume 1/2 total inductor loss in winding and 1/2 loss in the core ($P_C = 0.4W$). Try Micrometals #26 core material. V_L (from Table 5.2) = 5(30 – 5)/(2 • 30) = 2.08

$$L_{MIN} = \frac{(1.3 \bullet 10^{-4})(75)(2.08)^2}{(0.4)^{0.985} \bullet (10^5)^{2-1.34}} = 52\mu H \qquad (93)$$

The inductance must be five times the minimum to achieve desired core loss. Let's assume that 52μH is too large for our space requirements and try a better core material, #52, which is only slightly more expensive.

$$L_{MIN} = \frac{(4.9 \bullet 10^{-4})(75)(2.08)^2}{(0.4)^{\frac{2}{2.11}} \bullet (10^5)^{\frac{2-2(1.26)}{2.11}}} = 35\mu H \qquad (94)$$

To see if an off-the-shelf inductor is suitable, calculate inductor currents and V • t product using Table 5.3.

$$I_{RMS} = I_{OUT} = 3A \qquad (95)$$

$$I_P = 3 + \frac{5(30-5)}{2(35 \bullet 10^{-6})(10^5)(30)} = 3.6A$$

$$V \bullet t = \frac{5(30-5)}{(10^5)(30)} = 42V \bullet \mu s$$

Table 5.3 Inductor Operating Conditions

	I_{AVG}	I_{PEAK}	$I_{P\text{-}P}$	$V \bullet \mu s$
Buck Converter (Continious)	I_0	$I_0 + \frac{V_0(V_I-V_0)}{2 \bullet L \bullet f \bullet V_I}$	$\frac{V_0(V_I-V_0)}{L \bullet f \bullet V_I}$	$\frac{V_0(V_I-V_0) \bullet 10^6}{f \bullet V_I}$
Positive to Negative (Continuous)	$\frac{I_0(V_I+V_0)}{V_I}$	$\frac{I_0(V_0+V_I)}{V_I} + \frac{V_I \bullet V_0}{2 \bullet L \bullet f(V_I+V_0)}$	$\frac{V_I \bullet V_0}{L \bullet f(V_I+V_0)}$	$\frac{V_I \bullet V_0 \bullet 10^6}{f(V_I+V_0)}$
Negative Boost (Continuous)	$\frac{I_0 \bullet V_0}{V_I}$	$\frac{I_0 \bullet V_0}{V_I} + \frac{V_I(V_0-V_I)}{2L \bullet f \bullet V_0}$	$\frac{V_I(V_0-V_I)}{L \bullet f \bullet V_0}$	$\frac{V_I(V_0-V_I) \bullet 10^6}{f \bullet V_0}$
Tapped-Inductor*	$\frac{I_0(N \bullet V_0+V_I)}{V_I(1+N)}, \frac{I_0(N \bullet V_0+V_I)}{V_I}*$	$\frac{I_0(N \bullet V_0+V_I)}{V_I(1+N)} + \frac{(V_I-V_0)(1+N)(V_0)}{2L \bullet f(N \bullet V_0+V_I)}*$	$\frac{(V_I-V_0)(1+N)(V_0)}{L \bullet f(N \bullet V_0+V_I)}*$	$\frac{10^6(V_I-V_0)(1+N)(V_0)}{f(N \bullet V_0+V_I)}$
Buck Converter (Discontinuous)	$\frac{1}{4}\sqrt{\frac{(I_0)^3 \bullet V_0(V_I-V_0)}{f \bullet L \bullet V_I}}$	$\sqrt{\frac{2I_0 \bullet V_0(V_1-V_0)}{L \bullet f \bullet V_I}}$		$10^6\sqrt{\frac{2 \bullet L \bullet I_0 \bullet V_0(V_I-V_0)}{f \bullet V_I}}$
Positive to Negative (Discontinuous)	$\frac{1}{4}\sqrt{\frac{I_0^3 \bullet (V_I+V_0)^2}{V_I \bullet f \bullet L}}$	$\sqrt{\frac{2I_0-V_I}{f \bullet L}}$		$10^6\sqrt{\frac{2I_0 \bullet V_0 \bullet L}{f}}$
Negative Boost (Discontinuous)	$\frac{1}{4}\sqrt{\frac{I_0^3 \bullet V_0^2(V_0+V_I)}{V_I^2 \bullet L \bullet f}}$	$\sqrt{\frac{2I_0(V_0-V_I)}{L \bullet f}}$		$10^6\sqrt{\frac{2I_0 \bullet L(V_0-V_I)}{L \bullet f}}$

*Values given for tapped-inductor I_{AVG} are average current through entire inductor during switch on time (first term), and average current through output section during switch off time (second term). To calculate heating, these currents must be multiplied by the appropriate winding resistance and factored by duty cycle.

I_{PEAK} is used to ensure the core does not saturate and should be used with the entire inductance.

Peak-to-peak current is used with the entire inductance to calculate core heating losses. It is the equivalent value if the inductor is not tapped.

This inductor must be at least 35μH, rated at 3A and ≥42V • μs at 100kHz. It must not saturate at a peak current of 3.6A.

Example: Inverting mode with V_{IN} = 4.7-5.3V, V_{OUT} = -5V, I_{OUT} = 1A, f = 100kHz, maximum inductor loss = 0.3W. Let V_{IN}' = 2.7V, $V_{OUT'}$ = 5.5V. Maximum output current for discontinuous mode (Equation 82) is 0.82A, so use continuous mode:

$$L_{MIN} = \frac{(2.7)^2(5.5)}{2 \bullet 10^5(5.5+2.7)^2\left(\frac{5 \bullet 2.7}{5.5+2.7}-1\right)} = 4.6\mu H \quad (96)$$

Now calculate minimum inductance from core loss. Assume core loss is 1/2 of total inductor loss, (P_C = 0.15W):

$$V_L(\text{From Table 5.2}) = \frac{(2.7)(5.5)}{2(2.7+5.5)} = 0.905 \quad (97)$$

Assuming Micrometals type #26 material:

$$L_{MIN} = \frac{(1.3 \bullet 10^{-4})(75)(0.905)^2}{(0.15)^{\frac{2}{2.03}} \bullet (10^5)^{2-\frac{2.72}{2.03}}} = 26\mu H \quad (98)$$

This value is over five times the minimum of 4.6μH. Perhaps a higher core loss is acceptable. Here's how to do a quick check. If we assume total efficiency is ≈60% (+ to – conversion with a 5V input is inefficient due to switch loss), then input power is equal to output power divided by 0.6 = 8.33W. If we double core loss from 0.15W to 0.3W, efficiency will be 5W/(8.33 + 0.15) = 59%. This is only a 1% drop in efficiency. A core loss of 0.3W allows inductance to drop to 12μH, assuming that the 12μH inductor will tolerate the core loss plus winding loss without overheating. Inductor currents are:

$$I_{RMS}(\text{From Table 5.3}) = \frac{1A(2.7+5.5)}{2.7} = 3A \quad (99)$$

$$I_P = \frac{(1A)(2.7+5.5)}{2.7} + \frac{(2.7)(5.5)}{2(12 \bullet 10^{-6})(10^5)(2.7+5.5)} = 3.8A$$

$$V \bullet t = \frac{(2.7)(5.5)}{(10^5)(2.7+5.5)} = 18V \bullet \mu s \text{ at} 100Hz$$

Micropower shutdown

The LT1074 will go into a micropower shutdown mode, with I_{SUPPLY} ≈ 150μA, when the shutdown pin is held below 0.3V. This can be accomplished with an open-collector TTL gate, a CMOS gate, or a discrete NPN or NMOS device, as shown in Figure 5.19.

The basic requirement is that the pull down-device can sink 50μA of current at a worst-case threshold of 0.1V. This requirement is easily met with any open-collector TTL gate (not Schottky clamped), a CMOS gate, or discrete device.

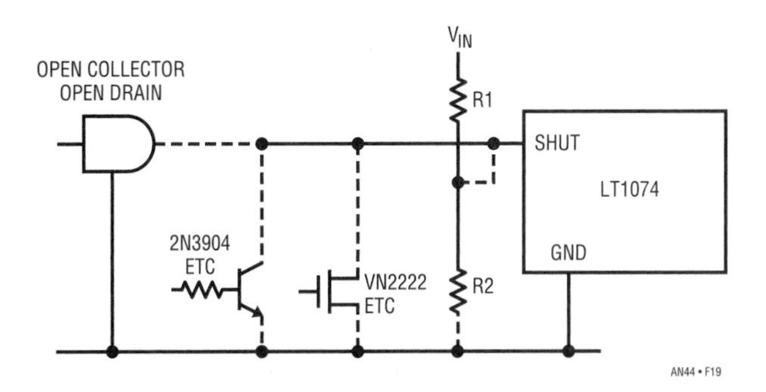

Figure 5.19 • Shutdown

The sink requirements are more stringent if R1 and R2 are added for undervoltage lockout. Sink capability must be 50μA + V_{IN}/R1 at the worst-case threshold of 0.1V. The suggested value for R2 is 5k to minimize the effect of shutdown pin bias current. This sets the current through R1 and R2 at ≈500μA at the undervoltage lockout point. At an input voltage of twice the lockout point, R1 current will be slightly over 1mA, so the pull-down device must sink this current down to 0.1V. A VN2222 or equivalent is suggested for these conditions.

Start-up time delay

Adding a capacitor to the shutdown pin will generate a delayed start-up. The internal current averages to about 25μA during the delay period, so delay time will be = (2.45V)/(C • 25μA), ±50%. If more accurate time out is required, R1 can be added to swamp out the effects of the internal current, but a larger capacitor is needed, and time out is dependent on input voltage.

Some thought must be given to reset of the timing capacitor. If a resistor to ground is used, it must be large enough to not drastically affect timing, so reset time is typically ten times longer than time delay. A diode to V_{IN} resets quickly, but if V_{IN} does not drop to near zero, time delay will be shortened when power is recycled immediately.

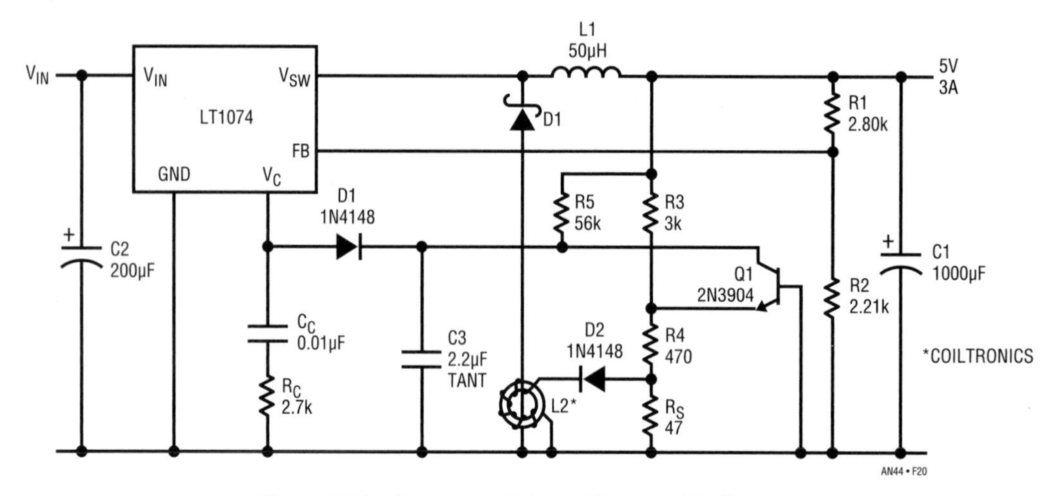

Figure 5.20 • Low Loss External Current Limit

5-pin current limit

Sometimes it may be desirable to current limit the 5-pin version of the LT1074. This is particularly helpful where maximum load current is significantly less than the 6.5A internal current limit, and the inductor and/or catch diode are minimum size to save space. Short-circuit conditions put maximum stress on these components.

The circuit in Figure 5.20 uses a small toroidal inductor slipped over one lead of the catch diode to sense diode current. Diode current during switch off time is almost directly proportional to output current, and L2 can generate an accurate limit signal without affecting regulator efficiency. Total power lost in the limit circuitry is less than 0.1W.

L2 has 100 turns. It therefore delivers 1/100 times diode current to RS when D1 conducts. The voltage across Rs required to current limit the LT1074 is equal to the voltage across R4 plus the forward biased emitter base voltage Q1 (≈600mV at 25°C). The voltage across R4 is set at 1.1V by R3, which is connected to the output. Current limit is set by selecting R_S:

$$R_S = \frac{R_4 I_X + V_{BE}}{\frac{I_{LIM}}{100} - I_X} \qquad (100)$$

$$I_X = \frac{V_{OUT} + V_{BE}}{R_3} + 0.4\text{mA}$$

V_{BE} = Forward biased emitter base voltage of Q1 at I_C = 500μA (≈600mV).

N = Turns on L2.

I_{LIM} = Desired output current limit. I_{LIM} should be set ≈1.25 times maximum load current to allow for variations in V_{BE} and component tolerances.

The circuit in Figure 5.20 is intended to supply 3A maximum load current, so I_{LIM} was set at 3.75A. Nominal V_{IN} is 25V, giving:

$$I_X = \frac{5 + 0.6}{3000} + 0.4 \bullet 10^{-3} = 2.27 \bullet 10^{-3} \qquad (101)$$

$$R_S = \frac{(470)(2.27 \bullet 10^{-3}) + 0.6}{3.75/100 - 2.27 \bullet 10^{-3}} = 47\Omega$$

This circuit has "foldback" current limit, meaning that short-circuit current is lower than the current limit at full output voltage. This is the result of using the output voltage to generate part of the current limit trip level. Short-circuit current will be approximately 45% of peak current limit, minimizing temperature rise in D1.

R5, C3, and D3 allow separate frequency compensation of the current limit loop. D3 is reversed biased during normal operation. For higher output voltages, scale R3 and R5 to provide approximately the same currents.

Soft-start

Soft-start is a means for ramping switch currents during the turn on of a switching regulator. The reasons for doing this include surge protection for the input supply, protection of switching elements, and prevention of output overshoot. Linear Technology switching regulators have built-in switch protection that eliminates concern over device failure, but some input supplies may not tolerate the inrush current of a switching regulator. The problem occurs with current limited input supplies or those with relatively high source resistance. These supplies can "latch" in a low voltage state where the current drawn by the switching regulator in much higher than the normal input current. This is shown by the general formula for switching regulator input current and input resistance:

$$I_{IN} = \frac{(V_{OUT})(I_{OUT})}{(V_{IN})(E)} = \frac{P_{OUT}}{(V_{IN})(E)} \quad (102)$$

$$R_{IN} = \frac{-(V_{IN})^2(E)}{(V_{OUT})(I_{OUT})} = \frac{-(V_{IN})^2(E)}{P_{OUT}} \text{ (Note negative sign)}$$

E = Efficiency (≈0.7-0.9)

These formulas show that input current is proportional to the reciprocal of input voltage, so that if input voltage drops by 3:1, input current increases by 3:1. An input supply which rises slowly will "see" a much heavier current load during its low voltage state. This can activate current limit in the input supply and "latch" it permanently in a low voltage condition. By instituting a soft-start in the switching regulator which is *slower* than the input supply rise time, regulator input current is held low until the input supply has a chance to reach full voltage.

The formula for regulator input resistance shows that it is negative and decreases as the square of input voltage. The maximum allowed positive source resistance to avoid latch-up is given by:

$$R_{SOURCE(MAX)} = \frac{(V_{IN})^2(E)}{4(V_{OUT})(I_{OUT})} \quad (103)$$

The formula shows that a+12V to −12V converter with 80% efficiency and 1A load must have a source resistance less than 2.4Ω. This may sound like much ado about nothing, because an input supply designed to deliver 1A would not normally have such a high source resistance, but a sudden output load surge or a dip in the source voltage might trigger a permanent overload condition. Low V_{IN} and high output load require lower source resistance.

In Figure 5.21, C2 generates a soft-start of switching current by forcing the I_{LIM} pin to ramp up slowly. Current out of the I_{LIM} pin is ≈300μA, so the time for the LT1074 to reach full switch current ($V_{LIM} \approx 5V$) is ≈$(1.6 \bullet 10^4)$

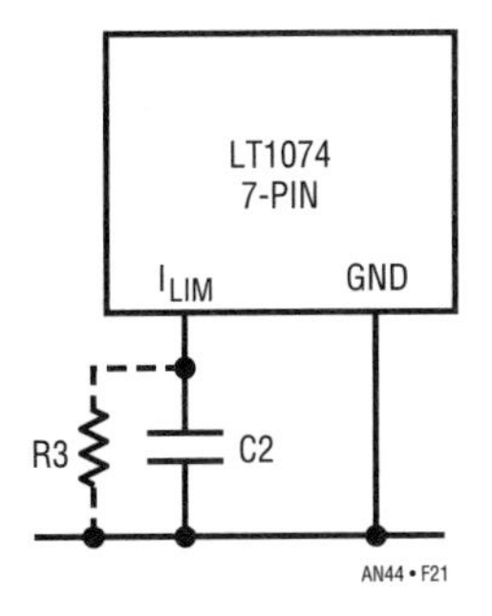

Figure 5.21 • Soft-Start Using I_{LIM} Pin

(C). To ensure low switch current until V_{IN} has reached full value, an approximate value for C2 is:

$$C2 \approx (10^{-4})(t)$$

t = Time for input voltage to rise to within 10% of final value.

C2 must be reset to zero volts whenever the input voltage goes low. An internal reset is provided when the shutdown pin is used to generate undervoltage lockout. The undervoltage state resets C2. If lockout is not used, R3 should be added to reset C2. For full current limit, R3 should be 30k. If reduced current limit is desired, R3's value is set by desired current limit. See the Current Limit section.

If the only reason for adding soft-start is to prevent input supply latchup, a better alternative may be undervoltage lockout (UVLO). This prevents the regulator from drawing input current until the input voltage reaches a preset voltage. The advantage of UVLO is that it is a true DC function and cannot be defeated by a slow rising input, short reset times, momentary output shorts, etc.

Output filters

When converter output ripple voltage must be less than ≈2% of output voltage, it is usually better to add an output filter (Figure 5.22) than to simply "brute force" the ripple by using very large output capacitors. The output filter consists of a small inductor (≈2μH to 10μH) and a second output capacitor, usually 50μF to 200μF. The inductor must be rated at full load current. Its core material is not important (core loss is negligible) except that core material will determine the size and shape of the inductor. Series resistance should be low enough to avoid unwanted efficiency loss. This can be estimated from:

$$R_L = \frac{(\Delta E)(V_{OUT})}{(I_{OUT})(E)^2} \quad (105)$$

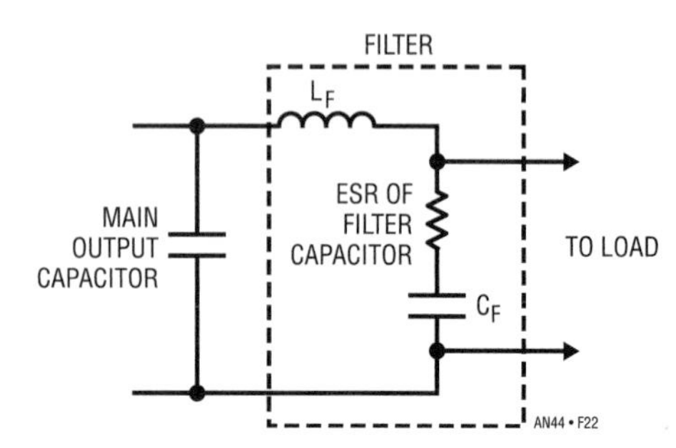

Figure 5.22 • Output Filter

"E" is overall efficiency and ΔE is the loss in efficiency allocated to the filter. Both are expressed as a ratio, i.e., 2% ΔE = 0.02, and 80% E = 0.8.

To obtain the required component values for the filter, one must assume a value for inductance or capacitor ESR, then calculate the remaining value. Actual capacitance in microfarads is of secondary importance because it is assumed that the capacitor will be basically resistive at ripple frequencies. One consideration on filter capacitor value is the load transient response of the converter. A small output filter capacitor (high ESR) will allow the output to "bounce" excessively if large amplitude load transients occur. When these load transients are expected, the size of the output filter capacitor must be increased to meet transient requirements rather than just ripple limits. In this situation, the main output capacitor can be reduced to simply meet ripple current requirements. The complete design should be checked for transient response with full expected load change.

If the capacitor is selected first, the inductor value can be found from ripple attenuation requirements.

Buck converter with triangular ripple into filter:

$$L_f = \frac{(ESR)(ATTN)}{8_f} \quad (106)$$

All other converters with essentially rectangular ripple into filter:

$$L = \frac{(ESR)(ATTN)(DC)(1 - DC)}{f} \quad (107)$$

ESR = Filter capacitor series resistance.

ATTN = Ripple attenuation required, as a ratio of peak-to-peak ripple IN to peak-to-peak ripple OUT.

DC = Duty cycle of converter. (If unknown, use worst-case of 0.5).

Example: A 100kHz buck converter with $150mV_{P\text{-}P}$ ripple which must be reduced to 20mV. ATTN = 150/20 = 7.5. Assume a filter capacitor with ESR = 0.3Ω

$$L = \frac{(0.3)(7.5)}{8(10^5)} = 2.8\mu H \quad (108)$$

Example: A 100kHz positive to negative converter with output ripple of $250mV_{P\text{-}P}$ which must be reduced to 30mV. Assume duty cycle has been calculated at 30% = 0.3, and ESR of filter capacitor is 0.2Ω

$$L = \frac{(0.2)(250/30)(0.3)(1 - 0.3)}{10^5} = 3.5\mu H \quad (109)$$

If the inductor is known, the equations can be rearranged to solve for capacitor ESR:

Buck Converter:

$$ESR = \frac{8f(L)}{ATTN} \quad (110)$$

Square Wave Ripple In:

$$ESR = \frac{f \bullet L}{(ATTN)(DC)(1 - DC)}$$

The output filter will affect load regulation if it is "outside" the regulator feedback loop. Series resistance of the filter inductor will add directly to the closed-loop output resistance of the converter. This closed-loop resistance is typically in the range of 0.002Ω to 0.01Ω, so a filter inductor resistance of 0.02Ω may represent a significant loss in load regulation. One solution is to move the filter "inside" the feedback loop by moving the sense points to the output of the filter. This should be avoided if possible because the added phase shift of the filter can cause difficulties in stabilizing the converter. Buck converters will tolerate an output filter inside the feedback loop by simply reducing the loop unity gain frequency. Positive-to-negative converters and boost converters have a "right-half plane zero" which makes them very sensitive to additional phase shift. To avoid stability problems, one should first determine if the load regulation degradation caused by a filter is really a problem. Most digital and analog "chips" in use today tolerate modest changes in supply voltage with little or no effect on performance.

When the sense resistor is tied to the output of the filter, a "fix" for stability problems is to connect a capacitor from the input of the filter to a tap on the feedback divider as shown in Figure 5.23. This acts as a "feedforward" path around the filter. The minimum size of C_X will be determined by the filter response, but should be in the range of 0.1μF to1μF.

C_X could theoretically be connected directly to the FB pin, but this should be done only if the peak-to-peak ripple on the main output capacitor is less than $75mV_{P-P}$.

A word about "measured" filter output ripple. The true ripple voltage should contain only the fundamental of the switching frequency because higher harmonics and "spikes" are very heavily attenuated. If the ripple as measured on an oscilloscope is abnormally high or contains high frequencies, the measurement technique is probably at fault. See the Oscilloscope Techniques section.

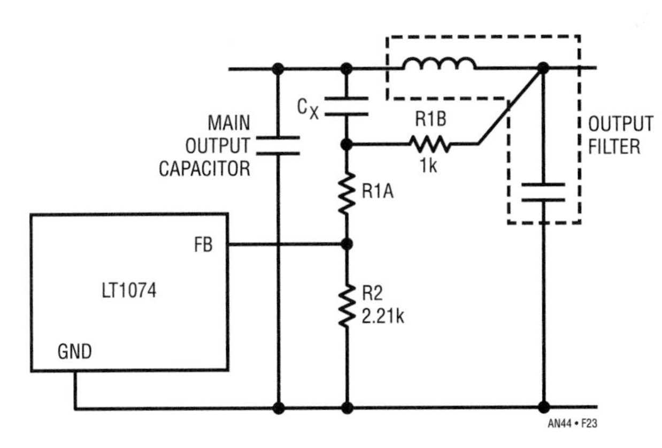

Figure 5.23 • Feedforward when Output Filter is Inside Feedback Loop

Input filters

Most switching regulators draw power from the input supply with rectangular or triangular current pulses. (The exception is a boost converter where the inductor acts as a filter for input current). These current pulses are absorbed primarily by the input bypass capacitor which is located right at the regulator input. Significant ripple current can still flow in the input lines, however, if the impedance of the source, including the inductance of supply lines, is low. This ripple current may cause unwanted ripple voltage on the input supply or may cause EMI in the form of magnetic radiation from supply lines. In these cases, an input filter may be required. The filter consists of an inductor in series with the input supply combined with the input capacitor of the converter, as shown in Figure 5.24.

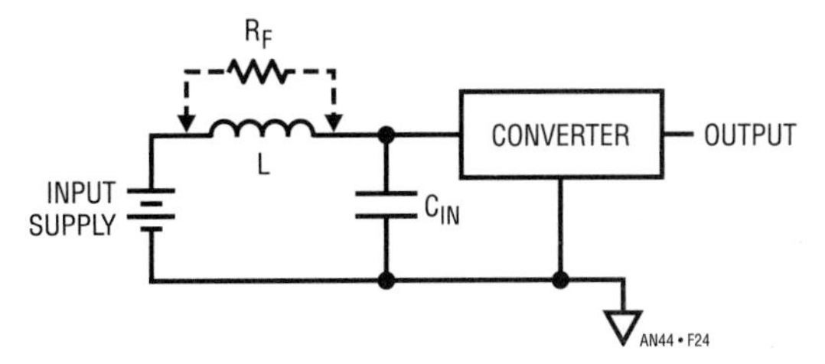

Figure 5.24 • Input Filter

To calculate a value for L requires knowledge of what ripple current is allowed in the supply line. This is normally an unknown parameter, so much hand waving may go on in search of a value. Assuming that a value has been arrived at, L is found from:

$$L = \frac{ESR(DC)(1 - DC)}{f\left(\frac{I_{SUP}}{I_{CON}} - \frac{ESR}{Rf}\right)} \quad (111)$$

ESR = Effective series resistance of input capacitor.
DC = Converter duty cycle. If unknown, use 0.5 as worst case.
I_{CON} = Peak-to-peak ripple current drawn by the converter, assuming continuous mode. For buck converters, $I_{CON} \approx I_{OUT}$. Positive-to-negative converters have $I_{CON} = I_{OUT} (V_{OUT}' + V_{IN}')/V_{IN}'$. Tapped-inductor $I_{CON} = I_{OUT} (N \bullet V_{OUT}' + V_{IN}')/[V_{IN}' (1+N)]$.
I_{SUP} = Peak-to-peak ripple allowed in supply lines.
R_f = "Damping" resistor which may be required to prevent instabilities in the converter.

Example: A 100kHz buck converter with $V_{OUT} = 5V$, $I_{OUT} = 4A$, $V_{IN} = 20V$, $(DC = 0.25)$. Input capacitor ESR is 0.05Ω. It is desired to reduce supply line ripple current to 100mA(P-P). Assume R_f is not needed $(= \infty)$

$$L = \frac{(0.05)(0.25)(1 - 0.25)}{10^5\left(\frac{0.1}{4} - 0\right)} = 3.75\mu H \qquad (112)$$

For further details on input filters, including the possible need for a damping resistor (R_f), see the Input Filters section in Application Note 19.

The current rating of the input inductor must be a minimum of:

$$I_L = \frac{(V_{OUT})(I_{OUT})}{(V_{IN})(E)} \text{Amps} \qquad (113)$$

(Use Minimum V_{IN}).
For this example;

$$I_L = \frac{(5)(4)}{(20)(E \approx 0.8)} = 1.25A$$

Efficiency or overload considerations may dictate an inductor with higher current rating to minimize copper losses. Core losses will usually be negligible.

Oscilloscope techniques

Switching regulators are a perfect test bed for poor oscilloscope techniques. A "scope" can lie in many ways and they all show up in a switching regulator because of the combination of fast and slow signals, coupled with both large and very small amplitudes. The following Rogue's Gallery will hopefully help the reader avoid many hours of frustration (and eliminate some embarrassing phone calls to the author).

Ground loops

Good safety practice requires most instruments to have their "ground" system tied to a "third" (green) wire in the power cord. This unfortunately results in current flow through oscilloscope probe ground leads (shield) when other instruments source or sink current to the device under test. Figure 5.25 details this effect.

A generator is driving a 5V signal into 50Ω on the breadboard, resulting in a 100mA current. The return path for this current divides between the ground from the signal generator (typically the shield on a BNC cable) and the secondary ground "loop" created by the oscilloscope probe ground clip (shield), and the two "third wire" connections on the signal generator and oscilloscope. In this

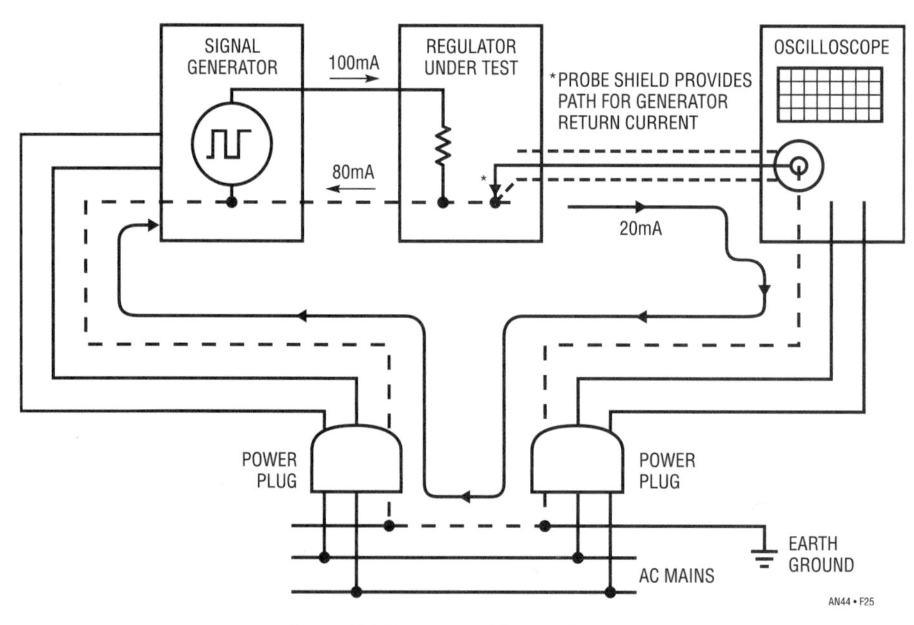

Figure 5.25 • Ground Loop Errors

case, it was assumed that 20mA flows in the parasitic ground loop. If the oscilloscope ground lead has a resistance of 0.2Ω, the screen will show a 4mV "bogus" signal. The problem gets much worse for higher currents, and fast signal edges where the inductance of the scope probe shield is important.

DC ground loops can be eliminated by disconnecting the third wire on the oscilloscope (its called a cheater plug, and my lawyers will *not* let me recommend it!) or by the use of an isolation transformer in the oscilloscope power connection.

Another source of circulating current in the probe shield wire is a second connection between a signal source and the scope. A typical example is a trigger signal connection between the generator trigger output and the scope external trigger input. This is most often a BNC cable *with its own grounded shield connection*. This forms a second path for signal ground return current, with the scope probe shield completing the path. My solution is to use a BNC cable which has had its shield intentionally broken. The trigger signal may be less than perfect, but the scope will not care. Mark the cable to prevent normal use!

Rule #1: *Before making any low level measurements, touch the scope probe "tip" to the probe ground clip with the clip connected to the desired breadboard ground.* The "scope" should indicate flatline. Any signal displayed is a ground loop lie.

Miscompensated scope probe

10X scope probes must be "compensated" to adjust AC attenuation so it precisely matches the 10:1 DC attenuation of the probe. If this is not done correctly, low frequency signals will be distorted and high frequency signals will have the wrong amplitude. In switching regulator applications, a "miscompensated" probe may show "impossible" waveforms. A typical example is the switching node of an LT1074 buck converter. This node swings positive to a level 1.5V to 2V below the input voltage, and negative to one diode drop below ground. A 10X probe with too little AC attenuation could show the node swinging above the supply, and so far negative that the diode forward voltage appears to be many volts instead of the expected 0.5V. Remember that at these frequencies (100kHz), the wave shape looks right because the probe acts purely capacitive, so the wrong amplitude may not be immediately obvious.

Rule #2: *Check 10X scope probe compensation before being embarrassed by a savvy tech.*

Ground "clip" pickup

Oscilloscope probes are most often used with a short ground "lead" with an alligator clip on the end. This ground wire is a remarkably good antenna. It picks up local magnetic fields and displays them in full color on the oscilloscope screen. Switching regulators generate lots of magnetic fields. Switch wires, diodes, capacitor and inductor leads, even "DC" supply lines can radiate significant magnetic fields because of the high currents and fast rise/fall times encountered. The test for ground clip problems is to touch the probe tip to the alligator clip, with the clip connected to the regulator ground point. Any trace seen on the screen is caused either by circulating currents in a ground loop, or by antenna action of the ground clip.

The fix for ground clip "pickup" is to throw the clip wire away and replace it with a special soldered-in probe terminator which can be obtained from the probe manufacturer. The plastic probe tip cover is pulled off to reveal the naked coaxial metal tube shield which extends to the small needle tip. This tube slips into the terminator to complete the ground connection. This technique will allow you to measure millivolts of output ripple on a switching regulator even in the presence of high magnetic fields.

Rule #3: *Don't make any low level measurements on a switching regulator using a standard ground clip lead.* If an official terminator is not available, solder a solid bare hookup wire to the desired ground point and wrap it around the exposed probe coaxial tube with absolute minimum distance between the ground point and the tube. Position the ground point so that the probe needle tip can touch the desired test point.

Wires are not shorts

A common error in probing switching regulators is to assume that the voltage anywhere on a wire path is the same. A typical example is the ripple voltage measured at the output of a switching regulator. If the regulator delivers square waves of current to the output capacitor, a positive to negative converter for instance, the current rise/fall time will be approximately 10^8A/sec. This dI/dt will generate ≈2V per inch "spikes" in the lead inductance of the output capacitor. The output (load) traces of the regulator should connect directly to the through-hole points where the radial-lead output capacitor leads are soldered in. The oscilloscope probe tip terminator (no ground clips, please) must be tied in directly at the base of the capacitor also.

The 2V/in. number can cause significant measurement errors even at high level points. When the input voltage to a switching regulator is measured across the input bypass capacitor, the spikes seen may be only a few tenths of a volt. If that capacitor is several inches away from the LT1074 though, the spikes "seen" by the regulator may be many volts. This can cause problems, especially at a low input voltage. Probing the "wrong" point on the input wire might mask these spikes.

Rule #4: If you want to know what the voltage is on a high AC current signal path, define exactly which component voltage you are measuring and connect the probe terminator directly across that component. As an example, if your circuit has a snubber to protect against switch overvoltage, connect the probe terminator directly to the IC switch terminals. Inductance in the leads connecting the switch to the snubber may cause the switch voltage to be many volts higher than the snubber voltage.

EMI suppression

Electromagnetic interference (EMI) is a fact of life with switching regulators. Consideration of its effects should occur early in the design so that the electrical, physical, and monetary implications of any required filtering or shielding are understood and accounted for. EMI takes two basic forms; "conducted," which travels down input and output wiring, and "radiated," which takes the form of electric and magnetic fields.

Conducted EMI occurs on input lines because switching regulators draw current from their input supply in pulses, either square wave, or triangular, or a combination of these. This pulsating current can create bothersome ripple voltage on the input supply and it can radiate from input lines to surrounding lines or circuitry.

Conducted EMI on the output of a switching regulator is usually limited to the voltage ripple on the output nodes. Ripple frequencies from buck regulators consist almost entirely of the fundamental switching frequency, whereas boost and inverting regulator outputs contain much higher frequency harmonics if no additional filtering is used.

Electric fields are generated by the fast rise and fall times of the switch node in the regulator. EMI from this source is usually of secondary concern and can be minimized by keeping all connections to this node as short as possible and by keeping this node "internal" to the switching regulator circuitry so that surrounding components act as shields.

The primary source of electric field problems within the regulator itself is coupling between the switching node and the feedback pin. The switching node has a typical slew rate of $0.8 \cdot 10^9$V/sec., and the impedance at the feedback pin is typically 1.2kΩ. Just 1PF coupling between these pins will generate 1V spikes at the feedback pin, creating erratic switching waveforms. Avoid long traces on the feedback pin by locating the feedback resistors immediately adjacent to the pin. When coupling to switching node cannot be avoided, a 1000pF capacitor from the LT1074 ground pin to the feedback pin will prevent most pickup problems.

Magnetic fields are more troublesome because they are generated by a variety of components, including the input and output capacitors, catch diode, snubber networks, the inductor, the LT1074 itself, and many of the wires connecting these components. While these fields do not usually cause regulator problems, they can create problems for surrounding circuitry, especially with low level signals such as disc drives, data acquisition, communication, or video processing. The following guidelines will be helpful in minimizing magnetic field problems.

1. Use inductors or transformers with good EMI characteristics such as toroids or pot cores. The worst offenders from an EMI standpoint are "rod" inductors. Think of them as cannon barrels firing magnetic flux lines in every direction. Their only application in switchers should be in the output filter where ripple current is very low.

2. Route all traces carrying high ripple current over a ground plane to minimize radiated fields. This includes the catch diode leads, input and output capacitor leads, snubber leads, inductor leads, LT1074 input and switch pin leads, and input power leads. Keep these leads short and the components close to the ground plane.
3. Keep sensitive low level circuitry as far away as possible, and use field-cancelling tricks such as twisted-pair differential lines.
4. In critical applications, add a "spike killer" bead on the catch diode to suppress high harmonics. These beads will prevent very high dI/dt signals, but will also make the diode appear to turn on slowly. This can create higher transient switch voltages at switch turn-off, so switch waveforms should be checked carefully.
5. Add an input filter if radiation from input lines could be a problem. Just a few μH in the input line will allow the regulator input capacitor to swallow nearly all the ripple current created at the regulator input.

Troubleshooting hints

Low efficiency

The major contributors here are switch and diode loss. These are readily calculable. If efficiency is abnormally low after factoring in these effects, zero in on the inductor. Core or copper loss may be the problem. Remember that inductor current may be much higher than output current in some topologies. A very handy substitution tool is a 500μH inductor wound with heavy wire on a large molypermalloy core. 100μH and 200μH taps are helpful. This inductor can be substituted for suspect units when inductor losses are suspected. If you read this Application Note, you will know that a large core is used not to reduce core loss, but to allow enough room for large wire that eliminates copper loss.

If inductor losses are not the problem, check all the nickel and dime effects such as quiescent current and capacitor loss to see if the sum is no longer negligible.

Alternating switch timing

Switch on time may alternate from cycle to cycle if excess switching frequency ripple appears on the V_C pin. This can occur naturally because of high ESR in the output capacitor or because of pickup on the FB pin or the V_C pin. A simple check is to put a 3000pF capacitor from V_C pin to the ground pin close to the IC. If the erratic switching improves or is cured, excess V_C pin ripple is the problem. Isolate it by connecting the capacitor from FB to ground pin. If this also makes the problem disappear, V_C pin pickup is eliminated, and FB pickup is the likely culprit. The feedback resistors should be located close to the IC so that connections to the FB pin are short and routed away from switching nodes. A 500pF capacitor from FB to ground pin will usually be sufficient if pickup cannot be eliminated. Occasionally, excess output ripple is the problem. This can be checked by paralleling the output capacitor with a second unit. A 1000pF to 3000pF capacitor on V_C can often be used to stop erratic switching caused by high output ripple, but be sure the ripple current rating of the output capacitor is adequate!

Input supply won't come up

Switching regulators have negative input resistance at DC. Therefore, they draw high current at low V_{IN}. This can latch input supplies low. See the Soft-Start section for details.

Switching frequency is low in current limit

This is normal. See the Frequency Shifting at the Feedback Pin in the Pin Description section.

IC blows up!

Like the LT1070 before it, the only thing that can destroy the LT1074 or LT1076 is excess switch voltage. (I am ignoring obvious stuff like voltage reversal or wiring errors).

Start-up surges can sometimes cause momentary large switch voltages, so check voltages carefully with an oscilloscope. Read the section on oscilloscope techniques.

IC runs hot

A common mistake is to assume that heat sinks are no longer needed with a switching design. This is often true for small load currents, but as load current climbs above 1A, switch loss may increase to the point where a heat sink is needed. A T0-220 package has a thermal resistance of 50°C/W with no heat sink. A 5V, 3A output (15W) with 10% switch loss, will dissipate over 1.5W in the IC. This means a 75°C temperature rise, or 100°C case temperature at room ambient. This is normally referred to as hot! A small heat sink solves the problem. Simply soldering the T0-220 tab to an enlarged copper pad on the PC board will reduce thermal resistance to ≈25°C/W.

High output ripple or noise spikes

First read the Oscilloscope Techniques section to avoid possible embarrassment, then check ESR of the output capacitor. Remember that fast (<100ns) spikes will be greatly attenuated by parasitic supply line inductance and load capacitance even if supply lines are only a few inches long.

Poor load or line regulation

Check in this order:

1. Secondary output filter DC resistance if it is outside the loop.
2. Ground loop error in oscilloscope.
3. Improper connection of output divider resistors to current carrying lines.
4. Excess output ripple. The LT1074 can peak detect ripple voltages on the FB pin if they exceed $50mV_{P-P}$.

See the Reference Shift with Ripple Voltage graph in the Typical Performance Characteristics section.

500kHz-5MHz oscillations, especially at light load

This is discontinuous mode ringing and is quite normal and harmless. See buck converter waveform description for more details.

Section 3

Linear Regulator Design

High efficiency linear regulators (6)

Presents circuit techniques permitting high efficiency to be obtained with linear regulation. Particular attention is given to the problem of maintaining high efficiency with widely varying inputs, outputs and loading. Appendix sections review component characteristics and measurement methods.

High efficiency linear regulators

6

Jim Williams

Introduction

Linear voltage regulators continue to enjoy widespread use despite the increasing popularity of switching approaches. Linear regulators are easily implemented, and have much better noise and drift characteristics than switchers. Additionally, they do not radiate RF, function with standard magnetics, are easily frequency compensated, and have fast response. Their largest disadvantage is inefficiency. Excess energy is dissipated as heat. This elegantly simplistic regulation mechanism pays dearly in terms of lost power. Because of this, linear regulators are associated with excessive dissipation, inefficiency, high operating temperatures and large heat sinks. While linears cannot compete with switchers in these areas they can achieve significantly better results than generally supposed. New components and some design techniques permit retention of linear regulator's advantages while improving efficiency.

One way towards improved efficiency is to minimize the input-to-output voltage across the regulator. The smaller this term is, the lower the power loss. The minimum input/output voltage required to support regulation is referred to as the "dropout voltage." Various design techniques and technologies offer different performance capabilities. Appendix A, "Achieving Low Dropout," compares some approaches. Conventional three terminal linear regulators have a 3V dropout, while newer devices feature 1.5V dropout (see Appendix B, "A Low Dropout Regulator Family") at 7.5A, decreasing to 0.05V at 100μA.

Regulation from stable inputs

Lower dropout voltage results in significant power savings where input voltage is relatively constant. This is normally the case where a linear regulator post-regulates a switching supply output. Figure 6.1 shows such an arrangement. The main output ("A") is stabilized by feedback to the switching regulator. Usually, this output supplies most of the power taken from the circuit. Because of this, the amount of energy in the transformer is relatively unaffected by power demands at the "B" and "C" outputs. This results in relatively constant "B" and "C" regulator input voltages. Judicious design allows the regulators to run at or near their dropout voltage, regardless of loading or switcher input voltage. Low dropout regulators thus save considerable power and dissipation.

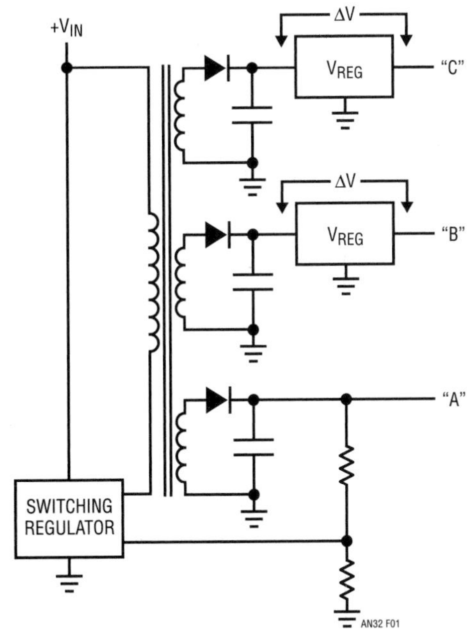

Figure 6.1 • Typical Switching Supply Arrangement Showing Linear Post-Regulators

Analog Circuit and System Design: Immersion in the Black Art of Analog Design. http://dx.doi.org/10.1016/B978-0-12-397888-2.00006-7

Regulation from unstable input—AC line derived case

Unfortunately, not all applications furnish a stable input voltage. One of the most common and important situations is also one of the most difficult. Figure 6.2 diagrams a classic situation where the linear regulator is driven from the AC line via a step-down transformer. A 90VAC (brownout) to 140VAC (high line) line swing causes the regulator to see a proportionate input voltage change. Figure 6.3 details efficiency under these conditions for standard (LM317) and low dropout (LT®1086) type devices. The LT1086's lower dropout improves efficiency. This is particularly evident at 5V output, where dropout is a significant percentage of the output voltage. The 15V output comparison still favors the low dropout regulator, although efficiency benefit is somewhat reduced. Figure 6.4 derives resultant regulator power dissipation from Figure 6.3's data. These plots show that the LT1086 requires less heat sink area to maintain the same die temperature as the LM317.

Both curves show the deleterious effects of poorly controlled input voltages. The low dropout device clearly cuts losses, but input voltage variation degrades obtainable efficiency.

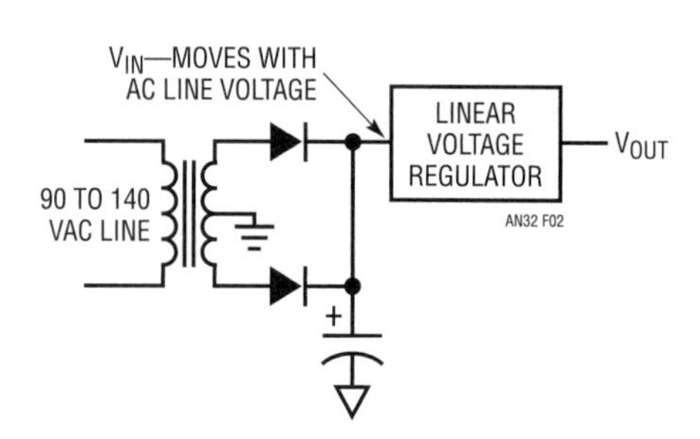

Figure 6.2 • Typical AC Line Driven Linear Regulator

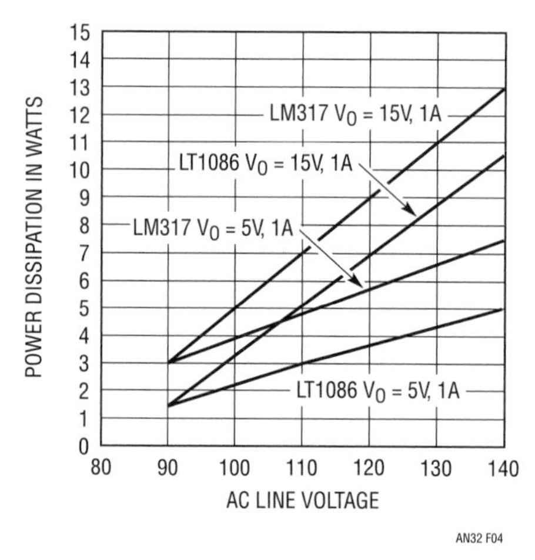

Figure 6.4 • Power Dissipation for Different Regulators vs AC Line Voltage. Rectifier Diode Losses are not Included

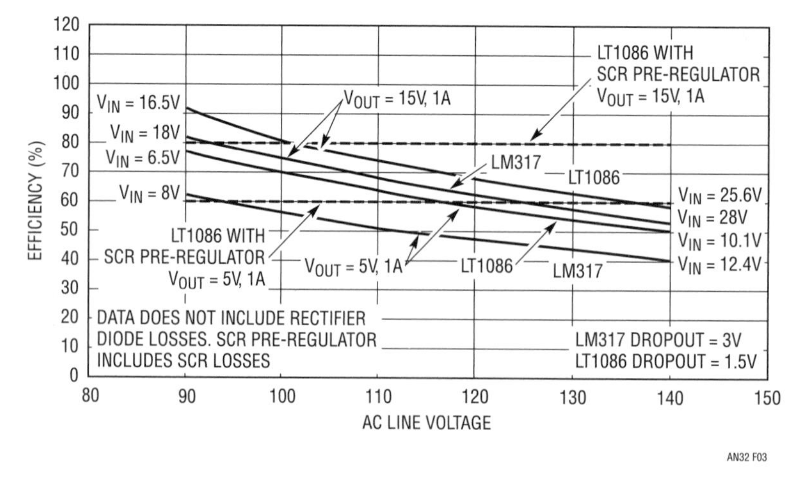

Figure 6.3 • Efficiency vs AC Line Voltage for LT1086 and LM317 Regulators

SCR pre-regulator

Figure 6.5 shows a way to eliminate regulator input variations, even with wide AC line swings. This circuit, combined with a low dropout regulator, provides high efficiency while retaining all the linear regulators desirable characteristics. This design servo controls the firing point of the SCRs to stabilize the LT1086 input voltage. A1 compares a portion of the LT1086's input voltage to the LT1004 reference. The amplified difference voltage is applied to C1B's negative input. C1B compares this to a line synchronous ramp (Trace B, Figure 6.6) derived by C1A from the transformers rectified secondary (Trace A is the "sync" point in the Figure 6.6). C1B's pulse output (Trace C) fires the appropriate SCR and a path from the main transformer to L1 (Trace D) occurs. The resultant current flow (Trace E), limited by L1, charges the 4700μF capacitor. When the transformer output drops low enough the SCR commutates and charging ceases. On the next half-cycle the process repeats, except that the alternate SCR does the work (Traces F and G are the individual SCR currents). The loop phase modulates the SCR's firing point to maintain a constant LT1086 input voltage. A1's 1μF capacitor compensates the loop and its output 10kΩ-diode network ensures start-up. The three terminal regulator's current limit protects the circuit from overloads.

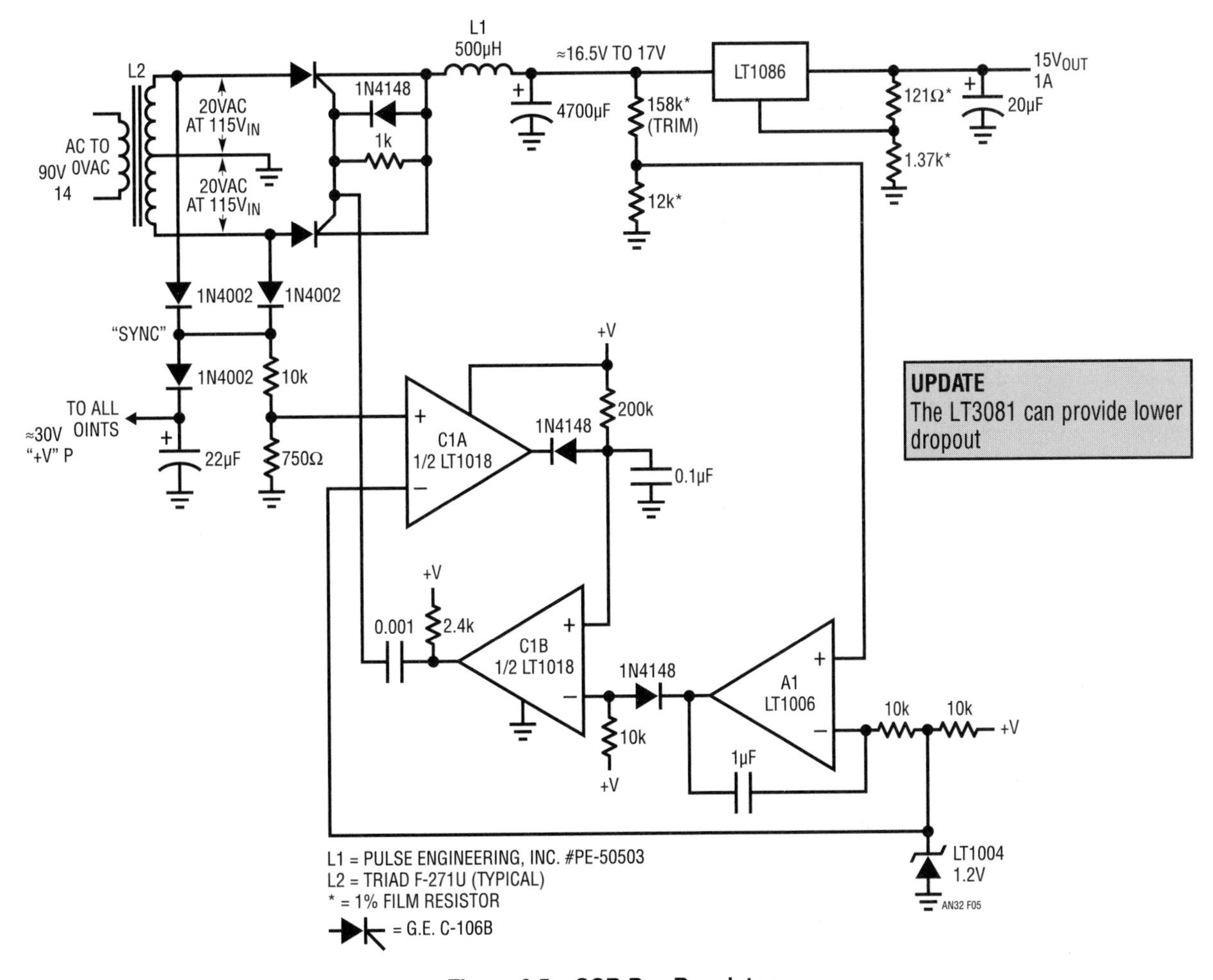

Figure 6.5 • SCR Pre-Regulator

This circuit has a dramatic impact on LT1086 efficiency versus AC line swing[1]. Referring back to Figure 6.3, the data shows good efficiency with no change for 90VAC to 140VAC input variations. This circuit's slow switching preserves the linear regulators low noise. Figure 6.7 shows slight 120Hz residue with no wideband components.

DC input pre-regulator

Figure 6.8a's circuit is useful where the input is DC, such as an unregulated (or regulated) supply or a battery. This circuit is designed for low losses at high currents. The LT1083 functions in conventional fashion,

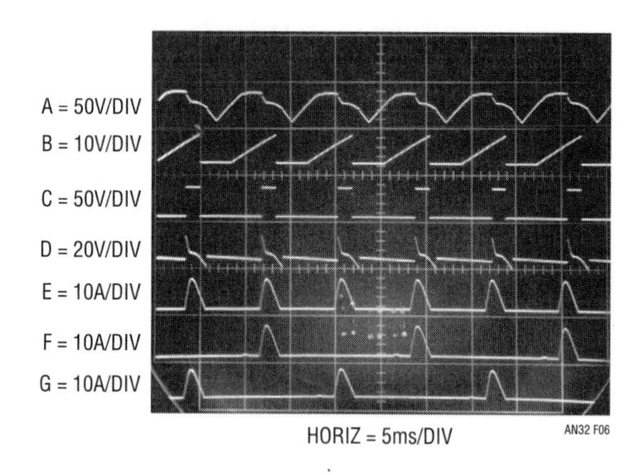

Figure 6.6 • Waveforms for the SCR Pre-Regulator

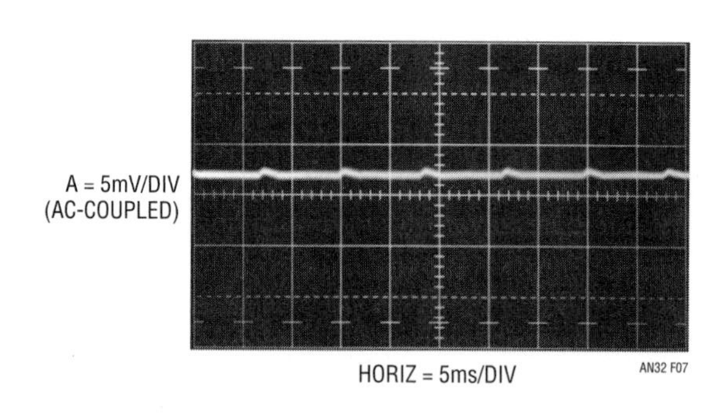

Figure 6.7 • Output Noise for the SCR Pre-Regulated Circuit

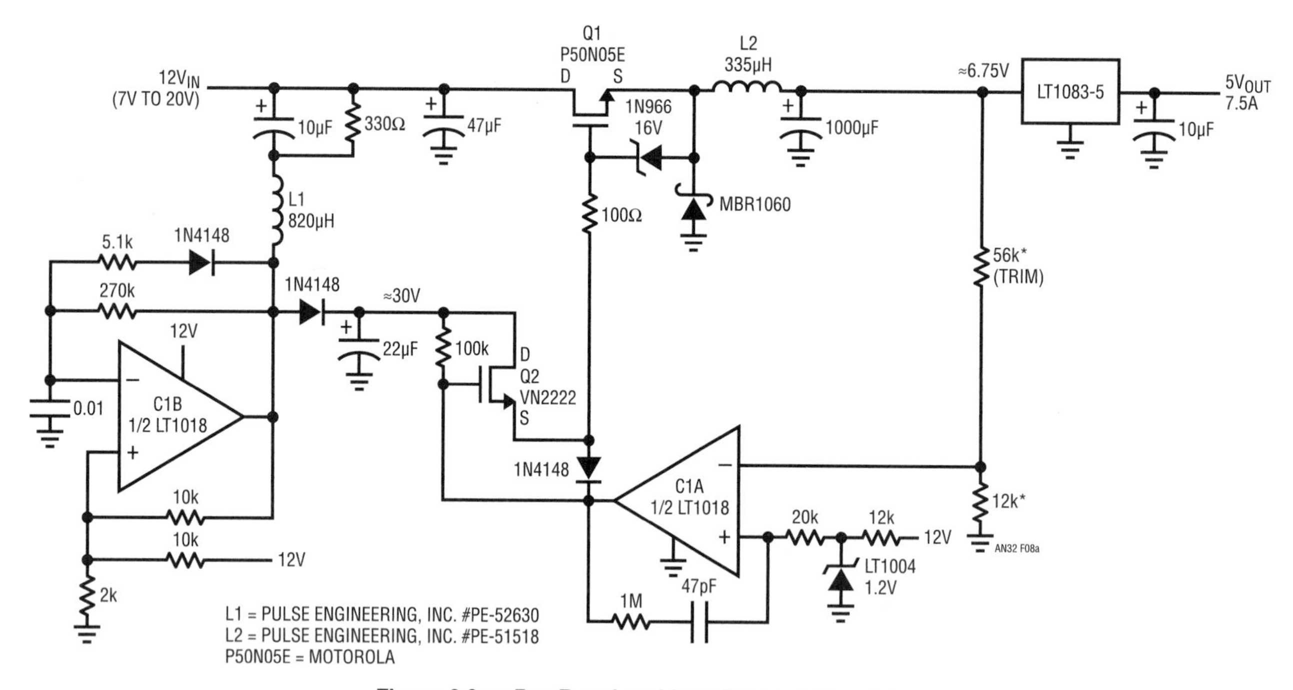

Figure 6.8a • Pre-Regulated Low Dropout Regulator

Note 1: The transformer used in a pre-regulator can significantly influence overall efficiency. One way to evaluate power consumption is to measure the actual power taken from the 115VAC line. See Appendix C, "Measuring Power Consumption."

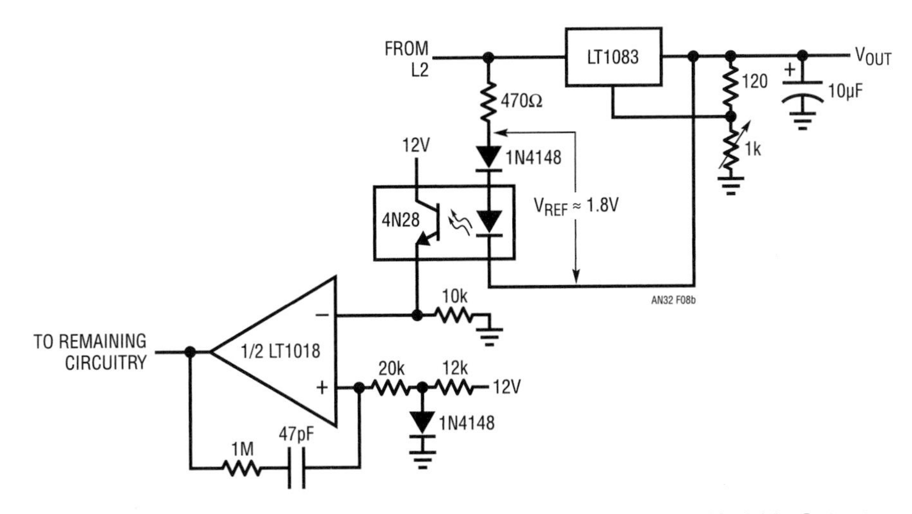

Figure 6.8b • Differential Sensing for the Pre-Regulator Allows Variable Outputs

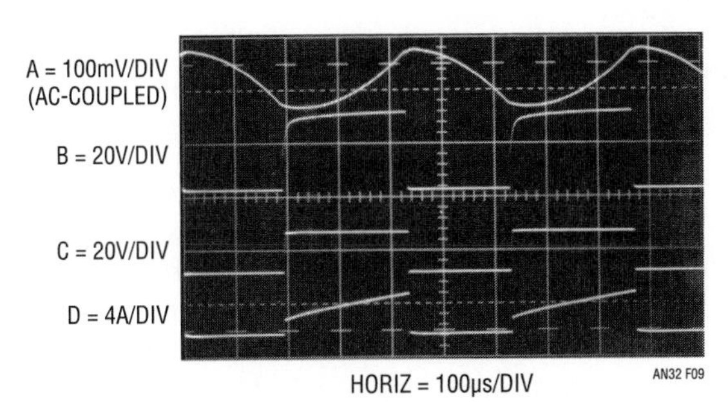

Figure 6.9 • Pre-Regulator Waveforms

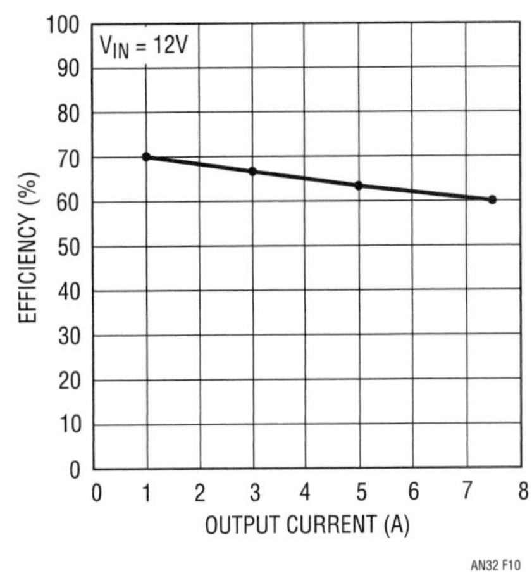

Figure 6.10 • Efficiency vs Output for Figure 6.8a

supplying a regulated output at 7.5A capacity. The remaining components form a switched-mode dissipation regulator. This regulator maintains the LT1083 input just above the dropout voltage under all conditions. When the LT1083 input (Trace A, Figure 6.9) decays far enough, C1A goes high, allowing Q1's gate (Trace B) to rise. This turns on Q1, and its source (Trace C) drives current (Trace D) into L2 and the 1000µF capacitor, raising regulator input voltage. When the regulator input rises far enough C1A returns low, Q1 cuts off and capacitor charging ceases. The MBR1060 damps L2's flyback spike and the 1M-47pF combination sets loop hysteresis at about 100mV.

Q1, an N-channel MOSFET, has only 0.028Ω of saturation loss but requires 10V of gate-source turn-on bias. C1B, set up as a simple flyback voltage booster, provides about 30V DC boost to Q2. Q2, serving as a high voltage pull-up for C1A, provides voltage overdrive to Q1's gate. This ensures Q1 saturation, despite its source follower connection. The Zener diode clamps excessive gate-source overdrives. These measures are required because alternatives are unattractive. Low loss P-channel devices are not currently available, and bipolar approaches require large drive currents or have poor saturation. As before, the linear regulator's current limit protects against overloads. Figure 6.10 plots efficiency for the pre-regulated LT1083 over a range of currents. Results are favorable, and the linear regulator's noise and response advantages are retained.

Figure 6.8b shows an alternate feedback connection which maintains a fixed small voltage across the LT1083 in

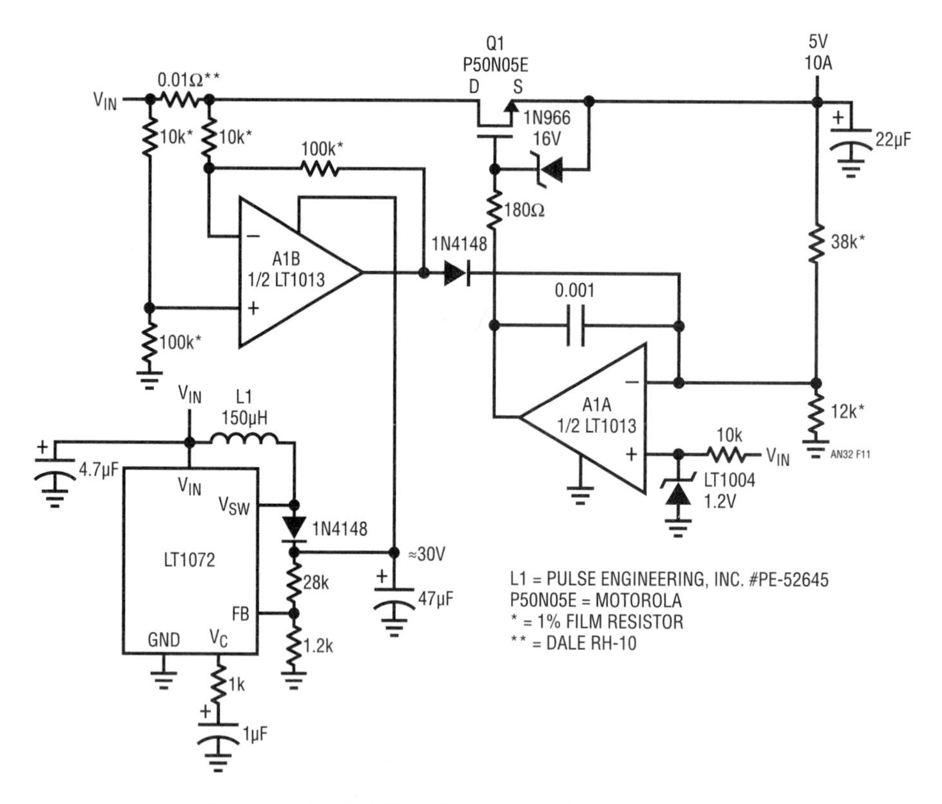

Figure 6.11 • 10A Regulator with 400mV Dropout

applications where variable output is desired. This scheme maintains efficiency as the LT1083's output voltage is varied.

10A regulator with 400mV dropout

In some circumstances an extremely low dropout regulator may be required. Figure 6.11 is substantially more complex than a three terminal regulator, but offers 400mV dropout at 10A output. This design borrows Figure 6.8a's overdriven source follower technique to obtain extremely low saturation resistance. The gate boost voltage is generated by the LT1072 switching regulator, set up as a flyback converter.[2] This configuration's 30V output powers A1, a dual op amp. A1A compares the regulators output to the LT1004 reference and servo controls Q1's gate to close the loop. The gate voltage overdrive allows Q1 to attain its 0.028Ω saturation, permitting the extremely low dropout noted. The Zener diode clamps excessive gate-source voltage and the 0.001µF capacitor stabilizes the loop. A1B, sensing current across the 0.01Ω shunt, provides current limiting by forcing A1A to swing negatively. The low resistance shunt limits loss to only 100mV at 10A output. Figure 6.12 plots current limit performance for the regulator. Roll-off is smooth, with no oscillation or undesirable characteristics.

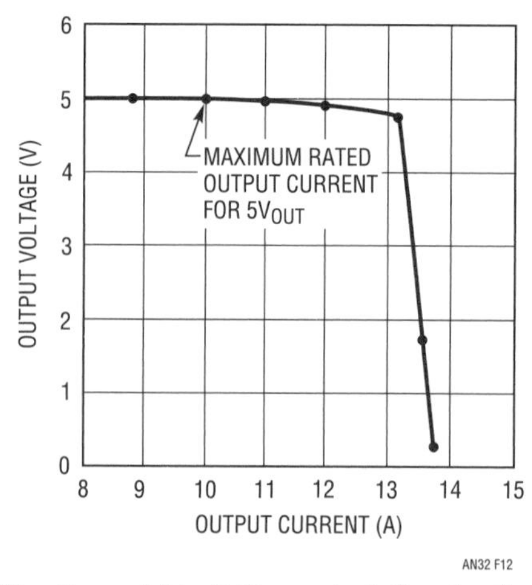

Figure 6.12 • Current Limit Characteristics for the Discrete Regulator

Note 2: If boost voltage is already present in the system, significant circuit simplification is possible. See LTC Design Note 32, "A Simple Ultra-Low Dropout Regulator."

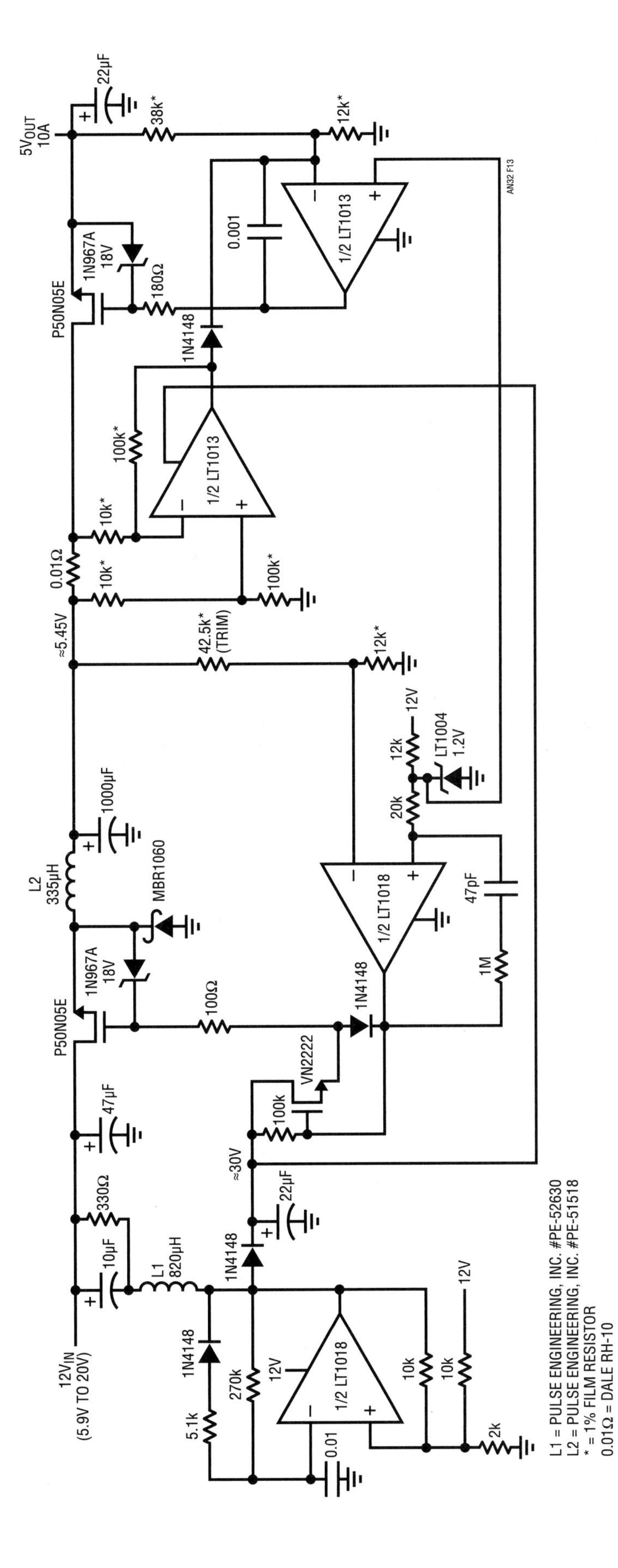

Figure 6.13 • Ultralow Dropout Linear Regulator with Pre-Regulator

Ultrahigh efficiency linear regulator

Figure 6.13 combines the preceding discrete circuits to achieve highly efficient linear regulation at high power. This circuit combines Figure 6.8a's pre-regulator with Figure 6.11's discrete low dropout design. Modifications include deletion of the linear regulators boost supply and slight adjustment of the gate-source Zener diode values. Similarly, a single 1.2V reference serves both pre-regulator and linear output regulator. The upward adjustment in the Zener clamp values ensures adequate boost voltage under low voltage input conditions. The pre-regulator's feedback resistors set the linear regulators input voltage just above its 400mV dropout.

This circuit is complex, but performance is impressive. Figure 6.14 shows efficiency of 86% at 1A output, decreasing to 76% at full load. The losses are approximately evenly distributed between the MOSFETs and the MBR1060 catch diode. Replacing the catch diode with a synchronously switched FET (see Linear Technology AN29, Figure 32) and trimming the linear regulator input to the lowest possible value could improve efficiency by 3% to 5%.

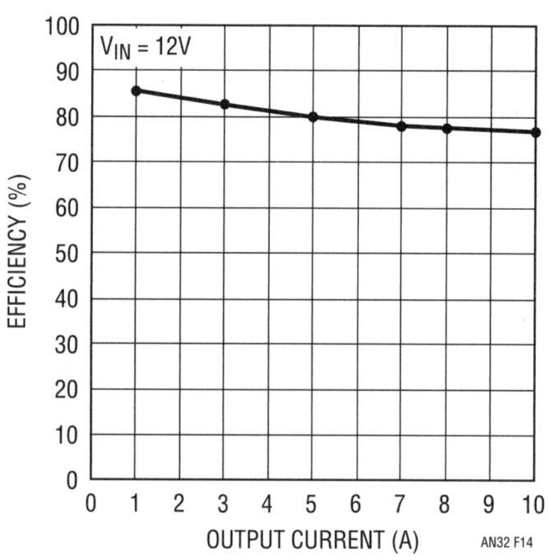

Figure 6.14 • Efficiency vs Output Current for Figure 6.13

Micropower pre-regulated linear regulator

Power linear regulators are not the only types which can benefit from the above techniques. Figure 6.15's pre-regulated micropower linear regulator provides excellent

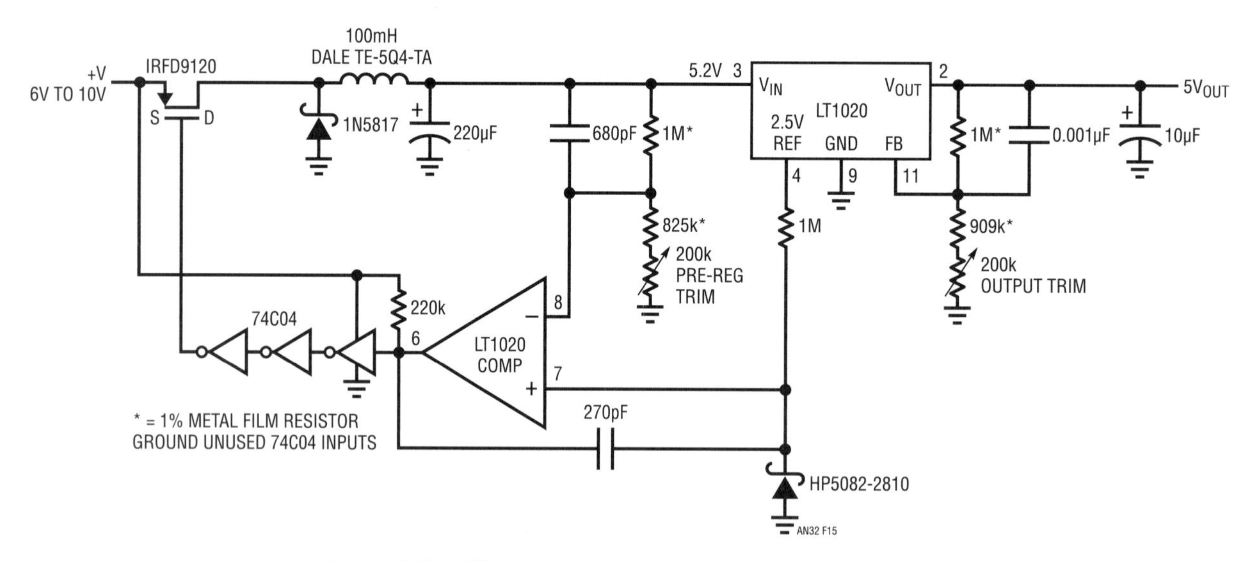

Figure 6.15 • Micropower Pre-Regulated Linear Regulator

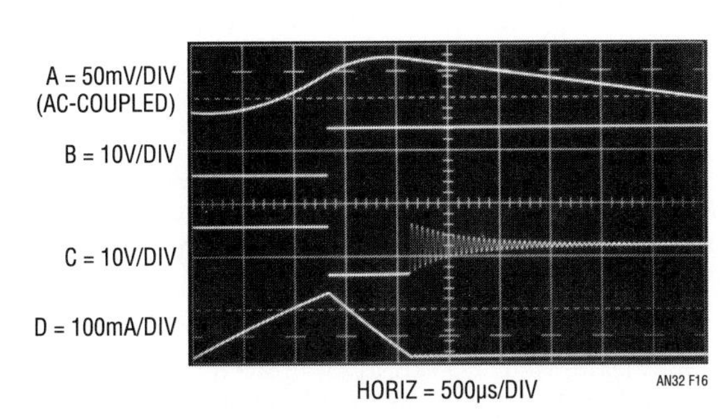

Figure 6.16 • Figure 6.15's Waveforms

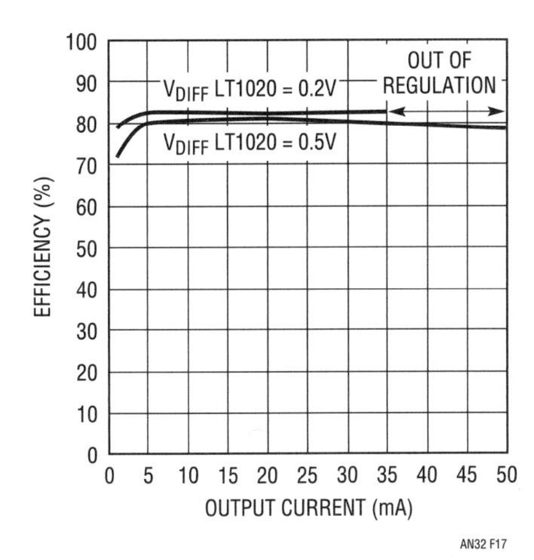

Figure 6.17 • Efficiency vs Output Current for Figure 6.15

efficiency and low noise. The pre-regulator is similar to Figure 6.8a. A drop at the pre-regulator's output (Pin 3 of the LT1020 regulator, Trace A, Figure 6.16) causes the LT1020's comparator to go high. The 74C04 inverter chain switches, biasing the P-channel MOSFET switch's grid (Trace B). The MOSFET comes on (Trace C), delivering current to the inductor (Trace D). When the voltage at the inductor-220μF junction goes high enough (Trace A), the comparator switches high, turning off MOSFET current flow. This loop regulates the LT1020's input pin at a value set by the resistor divider in the comparator's negative input and the LT1020's 2.5V reference. The 680pF capacitor stabilizes the loop and the 1N5817 is the catch diode. The 270pF capacitor aids comparator switching and the 2810 diode prevents negative overdrives.

The low dropout LT1020 linear regulator smooths the switched output. Output voltage is set with the feedback pin associated divider. A potential problem with this circuit is start-up. The pre-regulator supplies the LT1020's input but relies on the LT1020's internal comparator to function. Because of this, the circuit needs a start-up mechanism. The 74C04 inverters serve this function. When power is applied, the LT1020 sees no input, but the inverters do. The 200k path lifts the first inverter high, causing the chain to switch, biasing the MOSFET and starting the circuit. The inverter's rail-to-rail swing also provides good MOSFET grid drive.

The circuit's low 40μA quiescent current is due to the low LT1020 drain and the MOS elements. Figure 6.17 plots efficiency versus output current for two LT1020 input-output differential voltages. Efficiency exceeding 80% is possible, with outputs to 50mA available.

References

1. Lambda Electronics, Model LK-343A-FM Manual
2. Grafham, D.R., "Using Low Current SCRs," General Electric AN200.19. Jan. 1967
3. Williams, J., "Performance Enhancement Techniques for Three-Terminal Regulators," Linear Technology Corporation. AN2
4. Williams, J., "Micropower Circuits for Signal Conditioning," Linear Technology Corporation. AN23
5. Williams, J. and Huffman, B., "Some Thoughts on DC-DC Converters," Linear Technology Corporation. AN29
6. Analog Devices, Inc, "Multiplier Application Guide"

Note: This application note was derived from a manuscript originally prepared for publication in EDN magazine.

Appendix A
Achieving low dropout

Linear regulators almost always use Figure A1a's basic regulating loop. Dropout limitations are set by the pass elements on-impedance limits. The ideal pass element has zero impedance capability between input and output and consumes no drive energy.

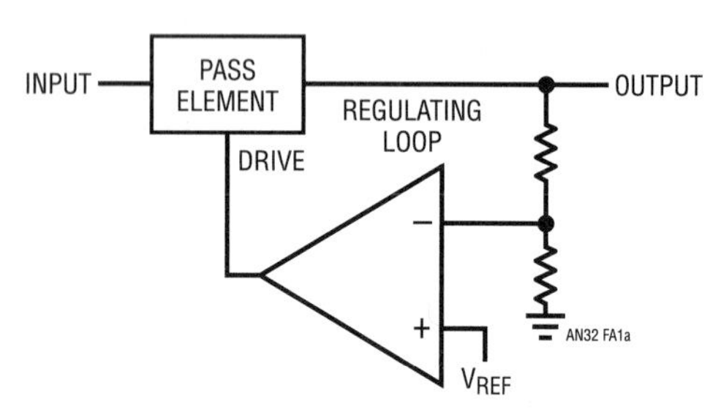

Figure A1a • Basic Regulating Loop

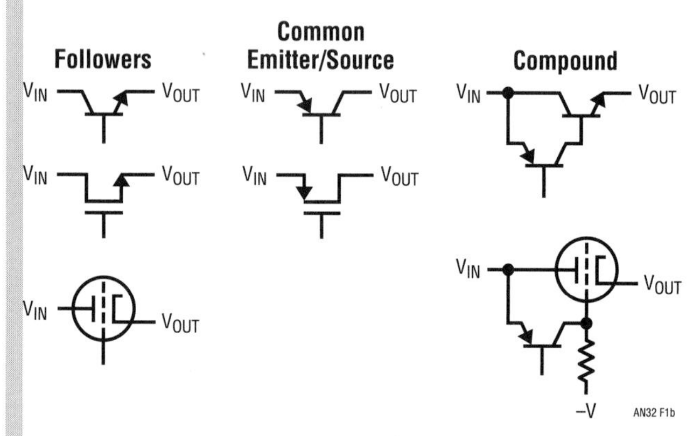

Figure A1b • Linear Regulator with Some Pass Element Candidates

A number of design and technology options offer various trade-offs and advantages. Figure A1b lists some pass element candidates. Followers offer current gain, ease of loop compensation (voltage gain is below unity) and the drive current ends up going to the load. Unfortunately, saturating a follower requires voltage overdriving the input (e.g., base, gate). Since drive is usually derived directly from VIN this is difficult. Practical circuits must either generate the overdrive or obtain it elsewhere. This is not easily done in IC power regulators, but is realizable in discrete circuits (e.g., Figure 6.11). Without voltage overdrive the saturation loss is set by V_{BE} in the bipolar case and channel on-resistance for MOS. MOS channel on-resistance varies considerably under these conditions, although bipolar losses are more predictable. Note that voltage losses in driver stages (Darlington, etc.) add directly to the dropout voltage. The follower output used in conventional three terminal IC regulators combines with drive stage losses to set dropout at 3V.

Common emitter/source is another pass element option. This configuration removes the VBE loss in the bipolar case. The PNP version is easily fully saturated, even in IC form. The trade-off is that the base current never arrives at the load, wasting substantial power. At higher currents, base drive losses can negate a common emitter's saturation advantage. This is particularly the case in IC designs, where high beta, high current PNP transistors are not practical. As in the follower example, Darlington connections exacerbate the problem. At moderate currents PNP common emitters are practical for IC construction. The LT1020/LT1120 uses this approach.

Common source connected P-channel MOSFETs are also candidates. They do not suffer the drive losses of bipolars, but typically require 10V of gate-channel bias to fully saturate. In low voltage applications this usually requires generation of negative potentials. Additionally, P-channel devices have poorer saturation than equivalent size N-channel devices.

The voltage gain of common emitter and source configurations is a loop stability concern, but is manageable.

Compound connections using a PNP driven NPN are a reasonable compromise, particularly for high power (beyond 250mA) IC construction. The trade-off between the PNP VCE saturation term and reduced drive losses over a straight PNP is favorable. Also, the major current flow is through a power NPN, easily realized in monolithic form. This connection has voltage gain, necessitating attention to loop frequency compensation. The LT1083-6 regulators use this pass scheme with an output capacitor providing compensation.

Readers are invited to submit results obtained with our emeritus thermionic friends, shown out of respectful courtesy.

Appendix B
A low dropout regulator family

The LT1083-6 series regulators detailed in Figure B1 feature maximum dropout below 1.5V. Output currents range from 1.5A to 7.5A. The curves show dropout is significantly lower at junction temperatures above 25°C. The NPN pass transistor based devices require only 10mA load current for operation, eliminating the large base drive loss characteristic of PNP approaches (see Appendix A for discussion).

In contrast, the LT1020/LT1120 series is optimized for lower power applications. Dropout voltage is about 0.05V at 100μA, rising to only 400mV at 100mA output. Quiescent current is 40μA.

LT1083 Dropout Voltage vs Output Current

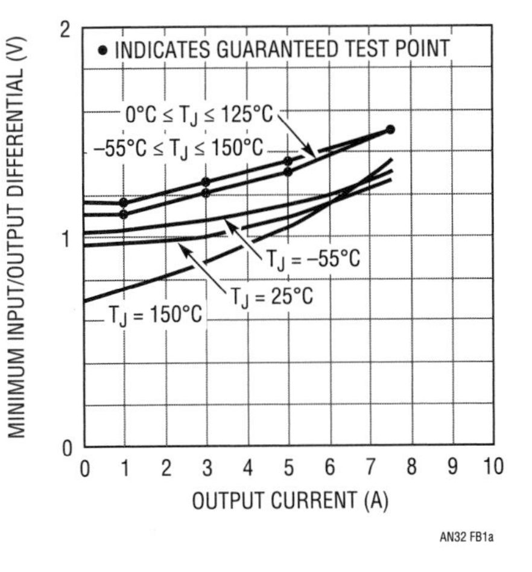

LT1084 Dropout Voltage vs Output Current

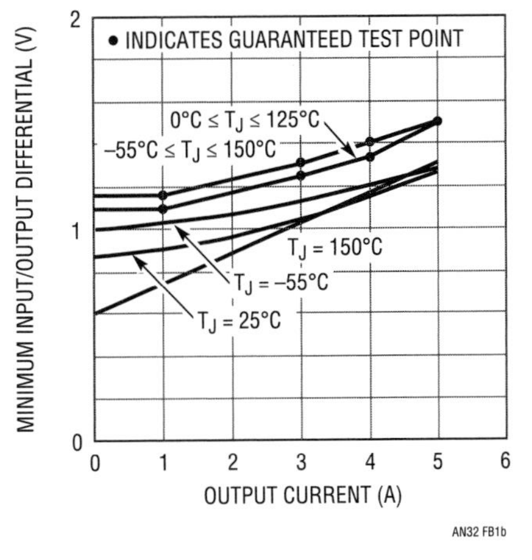

LT1085 Dropout Voltage vs Output Current

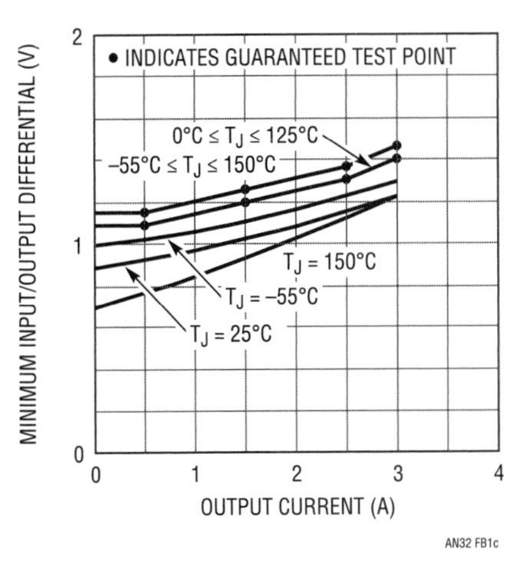

LT1086 Dropout Voltage vs Output Current

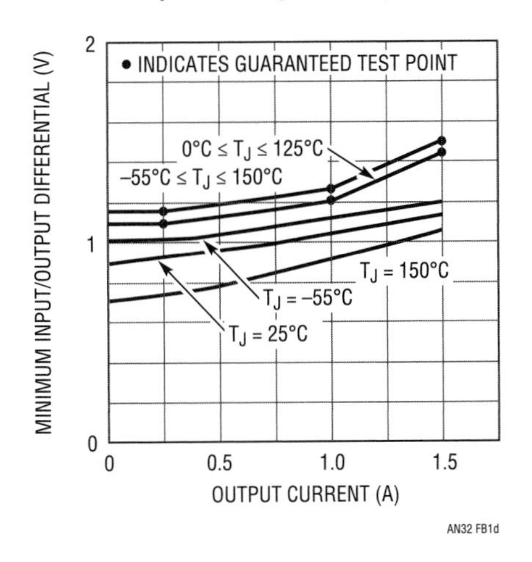

LT1020/LT1120 Dropout Voltage and Supply Current

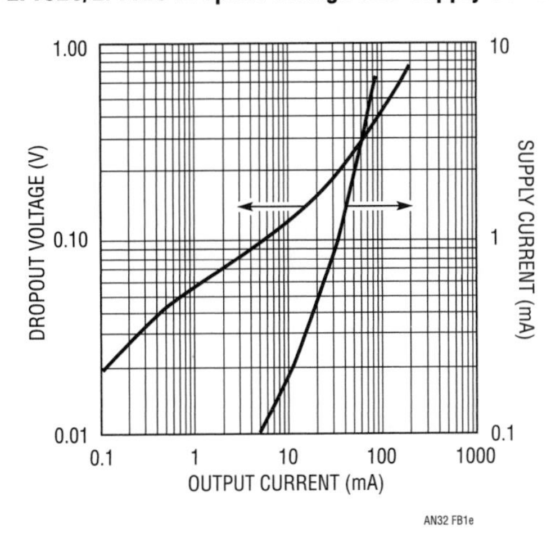

Figure B1 • Characteristics of Low Dropout IC Regulators

Appendix C
Measuring power consumption

Accurately determining power consumption often necessitates measurement. This is particularly so in AC line driven circuits, where transformer uncertainties or lack of manufacturer's data precludes meaningful estimates. One way to measure AC line originated input power (Watts) is a true, real time computation of E-I product. Figure C1's circuit does this and provides a safe, usable output.

> BEFORE PROCEEDING ANY FURTHER, THE READER IS WARNED THAT CAUTION MUST BE USED IN THE CONSTRUCTION, TESTING AND USE OF THIS CIRCUIT. HIGH VOLTAGE, AC LINE-CONNECTED POTENTIALS ARE PRESENT IN THIS CIRCUIT. EXTREME CAUTION MUST BE USED IN WORKING WITH AND MAKING CONNECTIONS TO THIS CIRCUIT. REPEAT: THIS CIRCUIT CONTAINS DANGEROUS, AC LINE-CONNECTED HIGH VOLTAGE POTENTIALS. USE CAUTION.

The AC load to be measured is plugged into the test socket. Current is measured across the 0.01Ω shunt by A1A with additional gain and scaling provided by A1B. The diodes and fuse protect the shunt and amplifier against severe overloads. Load voltage is derived from the 100k-4k divider. The shunts low value minimizes voltage burden error.

The voltage and current signals are multiplied by a 4-quadrant analog multiplier (AD534) to produce the power product. All of this circuitry fl oats at AC line potential, making direct monitoring of the multipliers output potentially lethal. Providing a safe, usable output requires a galvanically isolated way to measure the multiplier output. The 286J isolation amplifier does this, and may be considered as a unity-gain amplifier with inputs fully isolated from its output. The 286J also supplies the floating ±15V power required for A1 and the AD534.

The 286J's output is referred to circuit common (⏚). The 281 oscillator/driver is necessary to operate the 286J (see Analog Devices data sheet for details). The LT1012 and associated components provide a filtered and scaled output. A1B's gain switching provides decade ranging from 20W to 2000W full scale. The signal path's bandwidth permits accurate results, even for nonlinear or discontinuous loads (e.g. SCR choppers). To calibrate this circuit install a known full-scale load, set A1B to the appropriate range, and adjust the trimpot for a correct reading. Typical accuracy is 1%.

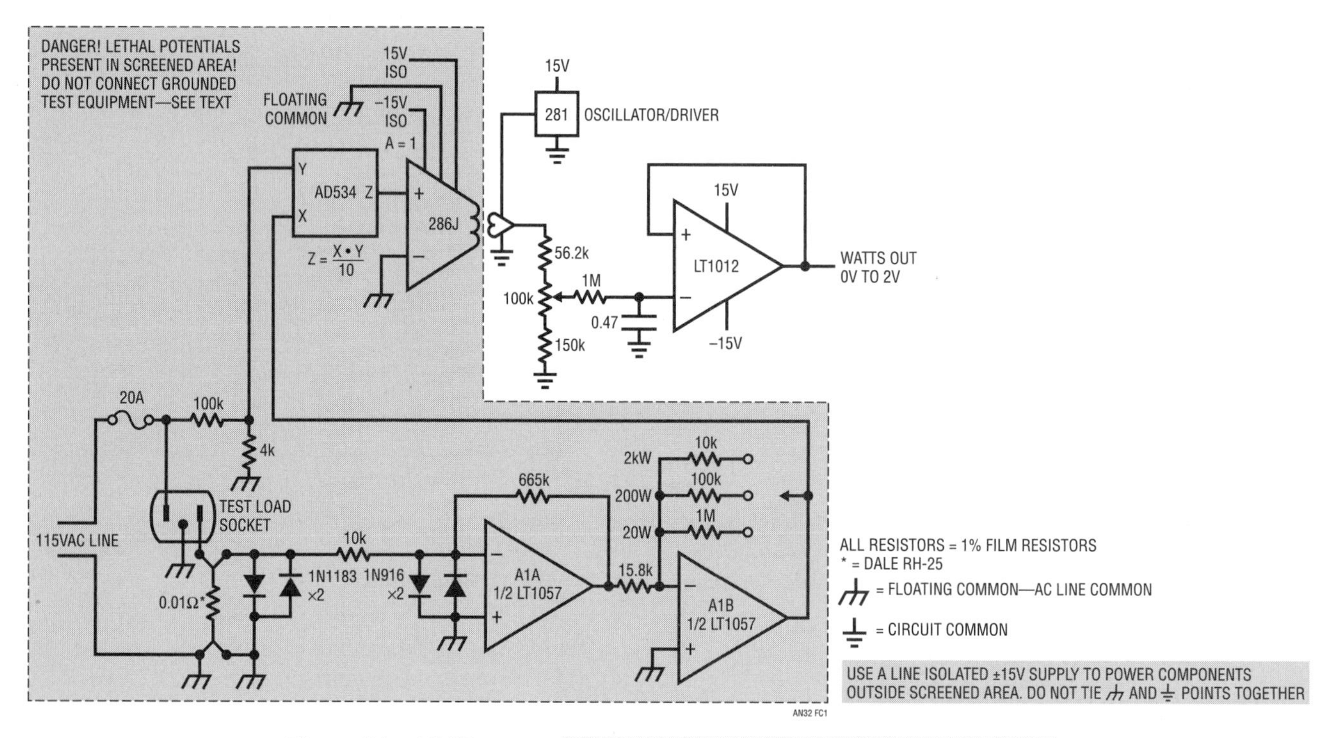

Figure C1 • AC Wattmeter DANGER! Lethal Potentials Present—See Text

Section 4

High Voltage and High Current Applications

High voltage, low noise, DC/DC converters (7)

Photomultipliers (PMT), avalanche photodiodes (APD), ultrasonic transducers, capacitance microphones, radiation detectors and similar devices require high voltage, low current bias. Additionally, the high voltage must be pristinely free of noise; well under a millivolt is a common requirement with a few hundred microvolts sometimes necessary. Normally, switching regulator configurations cannot achieve this performance level without employing special techniques. One aid to achieving low noise is that load currents rarely exceed 5mA. This freedom permits output filtering methods that are usually impractical.

High voltage, low noise, DC/DC converters

A kilovolt with 100 microvolts of noise

7

Jim Williams

Introduction

Photomultipliers (PMT), avalanche photodiodes (APD), ultrasonic transducers, capacitance microphones, radiation detectors and similar devices require high voltage, low current bias. Additionally, the high voltage must be pristinely free of noise; well under a millivolt is a common requirement with a few hundred microvolts sometimes necessary. Normally, switching regulator configurations cannot achieve this performance level without employing special techniques. One aid to achieving low noise is that load currents rarely exceed 5mA. This freedom permits output filtering methods that are usually impractical.

This publication describes a variety of circuits featuring outputs from 200V to 1000V with output noise below 100μV measured in a 100MHz bandwidth. Special techniques enable this performance, most notably power stages optimized to minimize high frequency harmonic content. Although sophisticated, all examples presented utilize standard, commercially available magnetics—no custom components are required. This provision is intended to assist the user in quickly arriving at a produceable design. Circuits and their descriptions are presented beginning with the next ink.

BEFORE PROCEEDING ANY FURTHER, THE READER IS WARNED THAT CAUTION MUST BE USED IN THE CONSTRUCTION, TESTING AND USE OF THE TEXT'S CIRCUITS. HIGH VOLTAGE, LETHAL POTENTIALS ARE PRESENT IN THESE CIRCUITS. EXTREME CAUTION MUST BE USED IN WORKING WITH, AND MAKING CONNECTIONS TO, THESE CIRCUITS. REPEAT THESE CIRCUITS CONTAIN DANGEROUS, HIGH VOLTAGE POTENTIALS. USE CAUTION.

Resonant Royer based converters

The resonant Royer topology is well suited to low noise operation due to its sinosoidal power delivery[1]. Additionally, the resonant Royer is particularly attractive because transformers originally intended for LCD display backlight service are readily available. These transformers are multiply sourced, well proven and competitively priced.

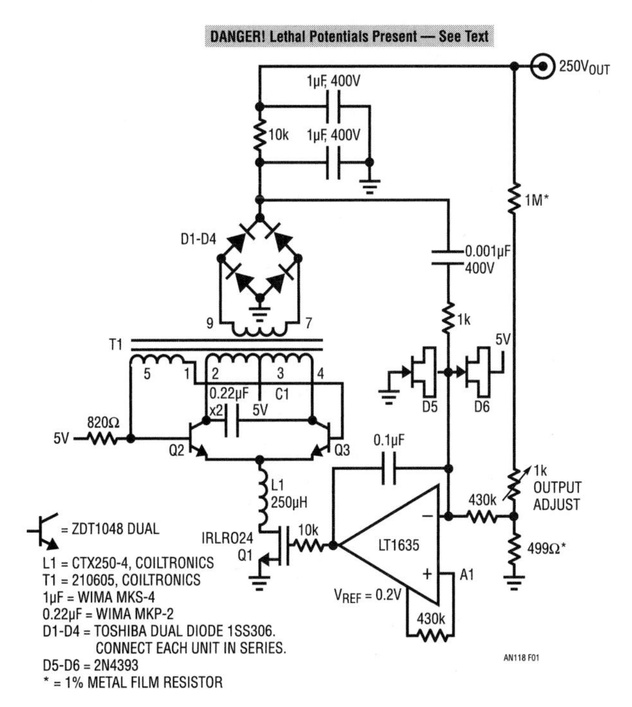

Figure 7.1 • Current Fed Resonant Royer Converter Produces High Voltage Output. A1 Biases Q1 Current Sink, Enforcing Output Voltage Stabilizing Feedback Loop. A1s 0.001μF-1kΩ Network Phase Leads Output Filter, Optimizing Transient Response. D5-D6, Low Leakage Clamps, Protect A1

Note 1: This publication sacrifices academic completeness for focus on the title subject. As such, operating details of the various switching regulator architectures utilized are not covered. Readers desiring background tutorial are directed to the References. Resonant Royer theory appears in Reference 1.

Analog Circuit and System Design: Immersion in the Black Art of Analog Design. http://dx.doi.org/10.1016/B978-0-12-397888-2.00007-9

Figure 7.1's resonant Royer topology achieves 100μVP-P noise at 250V output by minimizing high frequency harmonic in the power drive stage. The self oscillating resonant Royer circuitry is composed of Q2, Q3, C1, T1 and L1. Current flow through L1 causes the T1, Q2, Q3, C1 circuitry to oscillate in resonant fashion, supplying sine wave drive to T1's primary with resultant sine-like high voltage appearing across the secondary.

T1's rectified and filtered output is fed back to amplifier- reference A1 which biases the Q1 current sink, completing a control loop around the Royer converter. L1 ensures that Q1 maintains constant current at high frequency. Milliampere level output current allows the 10k resistor in the output filter. This greatly aids filter performance with minimal power loss.[2] The RC path to A1's negative input combines with the 0.1μF capacitor to compensate A1's loop. D5 and D6, low leakage clamps, protect A1 during start-up and transient events. Although Figure 7.2's collector waveforms are distorted, no high frequency content is present.

The circuit's low harmonic content combined with the RC output filter produces a transcendently clean output. Output noise (Figure 7.3) is just discernible in the monitoring instrumentation's 100μv noise floor[3].

Figure 7.4's variant of Figure 7.1 maintains 100μV output noise while extending input supply range to 32V. Q1 may require heat sinking at high input supply voltage. Converter and loop operation is as before although compensation components are re-established to accommodate the LT1431 control element.

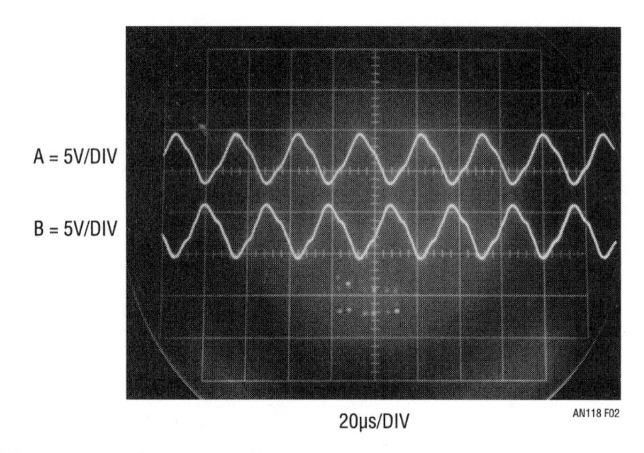

Figure 7.2 • Resonant Royer Collector Waveforms Are Distorted Sinosoids; No High Frequency Content is Present

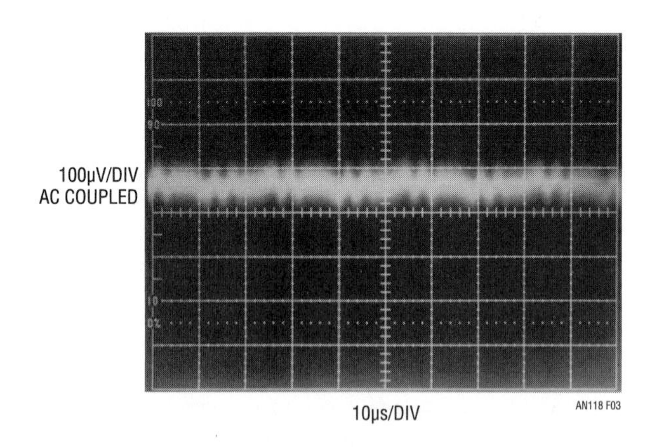

Figure 7.3 • Figure 7.1's is Output Noise is Just Discernable in Monitoring Instrumentation's 100μV Noise Floor

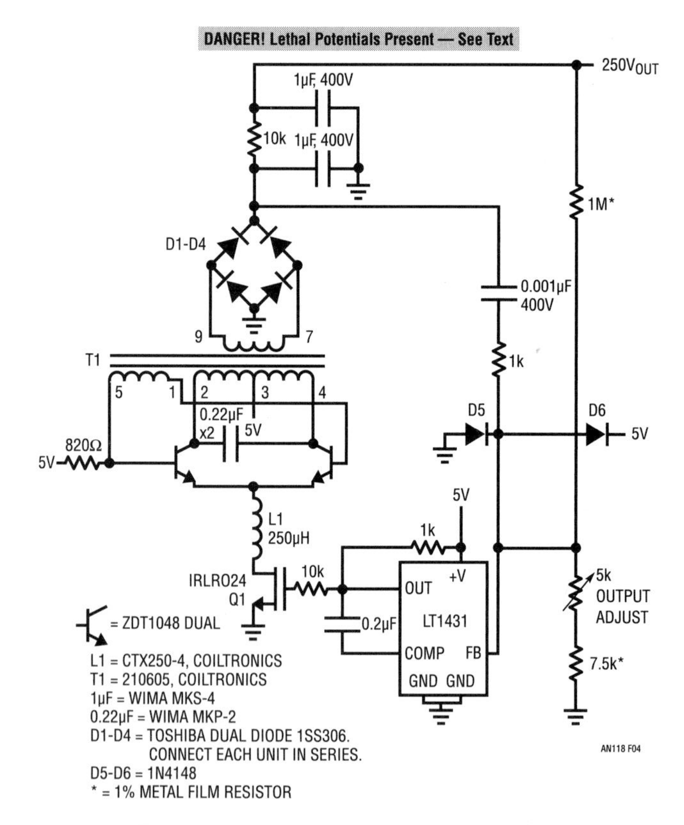

Figure 7.4 • LT1431 Regulator Based Variant of Figure 7.1 Maintains 100μV Output Noise While Extending Input Supply Range to 32V. Q1 May Require Heat Sinking at High Input Supply Voltages

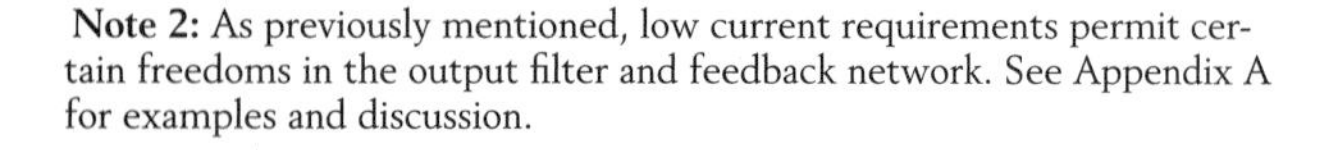
Note 2: As previously mentioned, low current requirements permit certain freedoms in the output filter and feedback network. See Appendix A for examples and discussion.

Note 3: Measurement technique and instrumentation choice for faithful low level noise measurement requires diligence. See Appendices B through E for practical considerations.

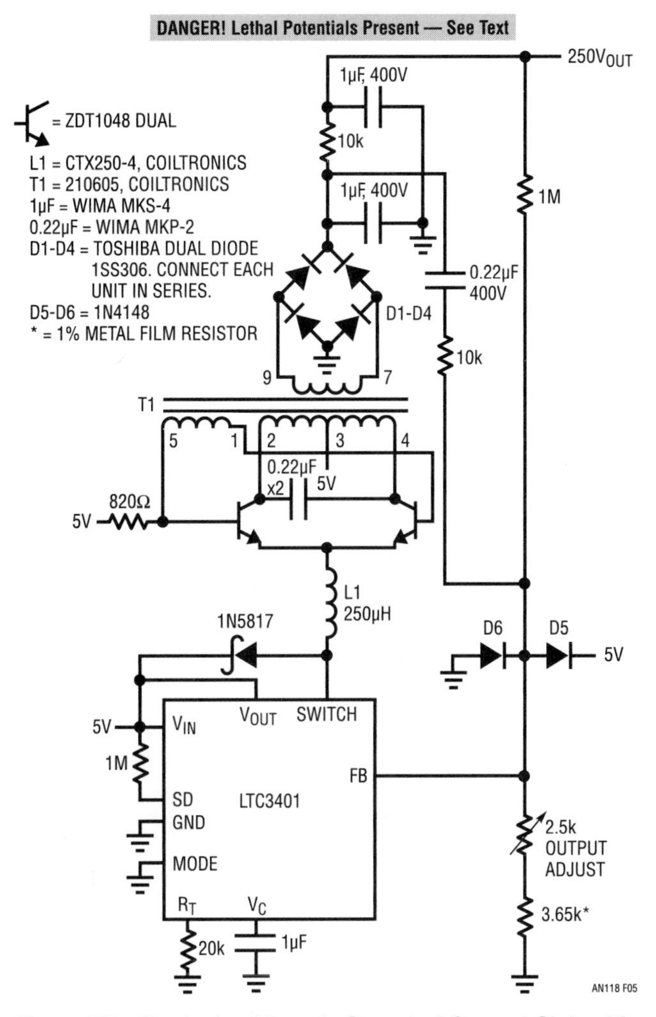

Figure 7.5 • Replacing Linearly Operated Current Sink with Switching Regulator Minimizes Heating Although Output Noise Increases

Switched current source based resonant Royer converters

The previous resonant Royer examples utilize linear control of converter current to furnish harmonic free drive. The trade off is decreased efficiency, particularly as input voltage scales. Improved efficiency is possible by employing switched mode current drive to the Royer converter. Unfortunately, such switched drive usually introduces noise. As will be shown, this undesirable consequence can be countered.

Figure 7.5 replaces the linearly operated current sink with a switching regulator. The Royer converter and its loop are as before; Figure 7.6's transistor collector waveshape (trace A) is similar to the other circuits. The high speed, switch mode current sink drive (trace B) efficiently feeds L1. This switched operation improves efficiency but degrades output noise. Figure 7.7 shows switching regulator harmonic clearly responsible for 3mV peak to peak output noise – about 30 times greater than the linearly operated circuits.

Careful examination of Figure 7.7 reveals almost no Royer based residue. The noise is dominated by switching regulator artifacts. Eliminating this switching regulator originated noise while maintaining efficiency requires special circuitry but is readily achievable.

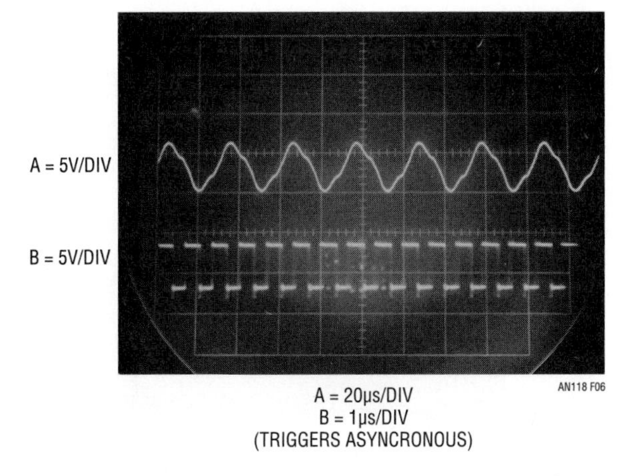

Figure 7.6 • Resonant Royer Collector Waveshape (Trace A) is Similar to Previous Circuits. High Speed, Switched Mode Current Sink Drive (Trace B) Efficiency Feeds L1

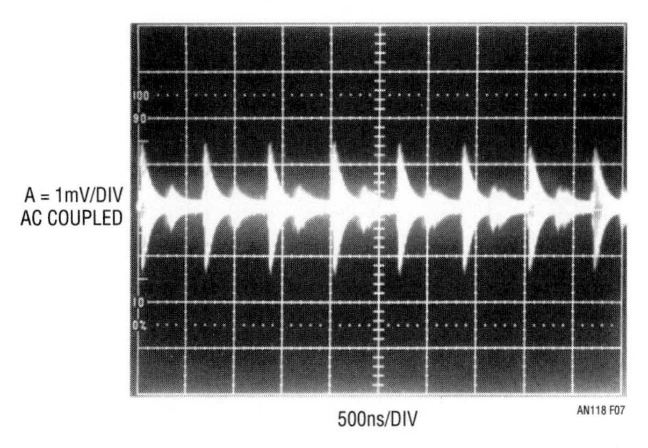

Figure 7.7 • Switching Regulator Harmonic Results in 3mV$_{P-P}$ Output Noise

Low noise switching regulator driven resonant Royer converters

Figure 7.8 examplifies the aforementioned "special circuitry". The resonant Royer converter and its loop are reminiscent of previous circuits. The fundamental difference is the LT1534 switching regulator which utilizes controlled transition times to retard high frequency harmonic while maintaining efficiency. This approach blends switching and linear current sink benefits[4]. Voltage and current transition rate, set by R_V and R_I respectively, is a compromise between efficiency and noise reduction.

Figure 7.9's Royer collector waveshape (trace A) is nearly identical to the one produced by Figure 7.5's circuit. Trace B, depicting LT1534 controlled transition times, markedly departs from its Figure 7.5 counterpart. These controlled transition times dramatically reduce output noise (Figure 7.10) to $150\mu V_{P\text{-}P}$—a 20x improvement vs Figure 7.7's LTC3401 based results.

Note 4: As stated, this forum must suffer brevity to maintain focus. The LT1534's controlled transition time operation mandates further study. See Reference 3.

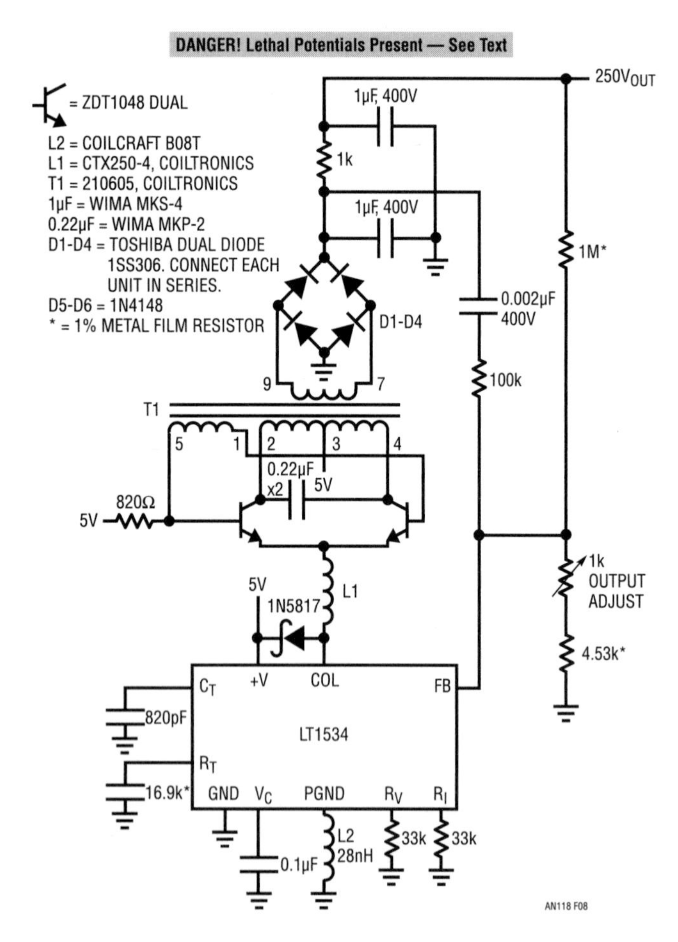

Figure 7.8 • LT1534's Controlled Transition Times Retard High Frequency Harmonic and Maintain Low Heat Dissipation. Approach Blends Switching and Linear Current Sink Benefits

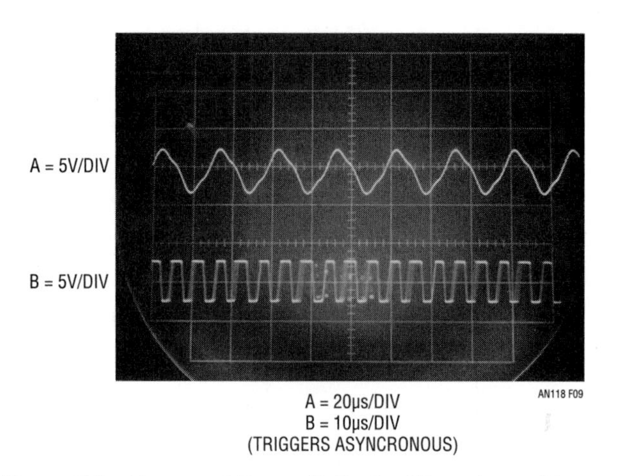

Figure 7.9 • Resonant Royer Collector Waveshape (Trace A) is Identical to Figure 7.5s LT3401 Circuit; LT1534 Current Sink's Controlled Transition Times (Trace B) Attenuate High Frequency Harmonic

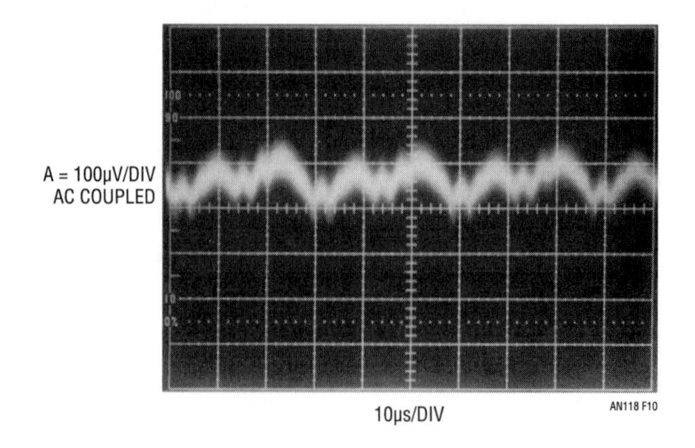

Figure 7.10 • Switched Current Sinks Controlled Transition Times Dramatically Lower Noise to $150pV_{P\text{-}P}$—A 20x Improvement vs Figure 7.7's LTC3401 Results

Figure 7.11 is essentially identical to Figure 7.8 except that it produces a negative 1000V output. A1 provides low impedance, inverting feedback to the LT1534. Figure 7.12a's output noise measures inside 1mV. As before, resonant Royer ripple dominates the noise—no high frequency content is detectable. It is worth noting that this noise figure proportionally improves with increased filter capacitor values. For example, Figure 7.12b indicates only 100μV noise with filter capacitor values increased by 10x, although capacitor physical size is large. The original values selected represent a reasonable compromise between noise performance and physical size.

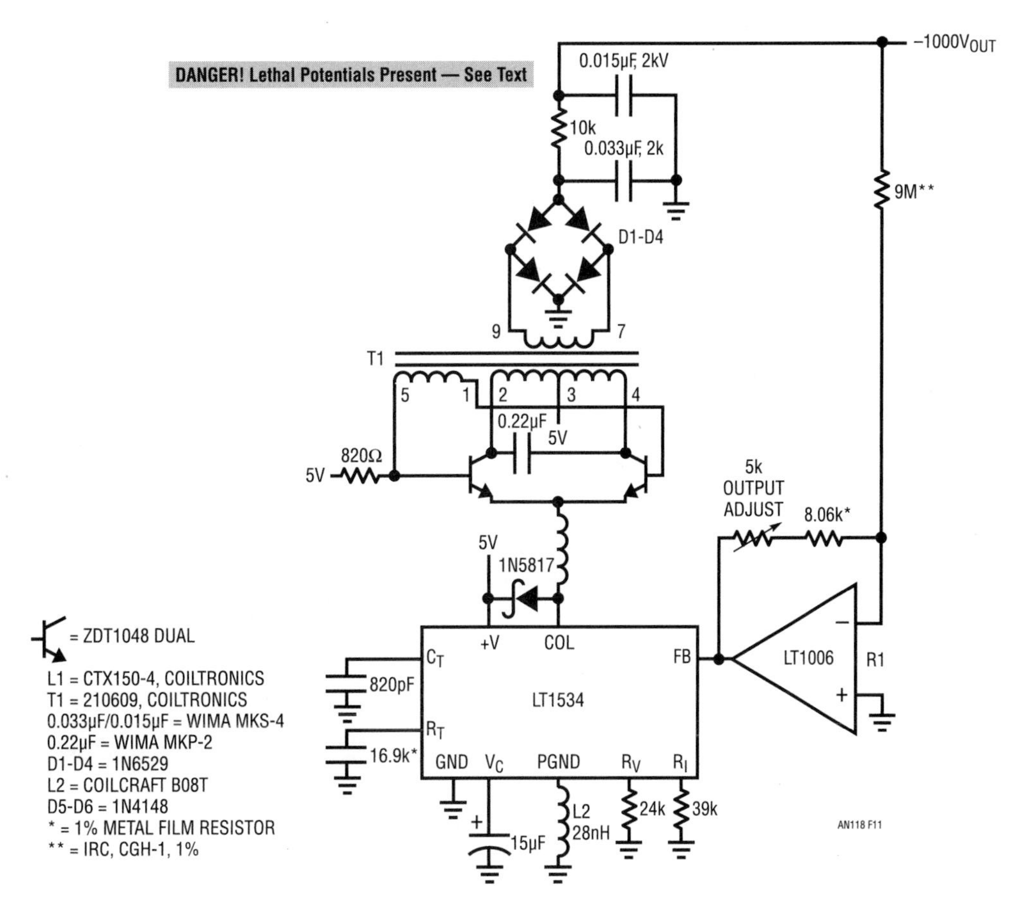

Figure 7.11 • Controlled Transition Time Switching Regulator Applied to a Negative Output, 1000V Converter. A1 Provides Low Impedance, Inverting Feedback to LT1534

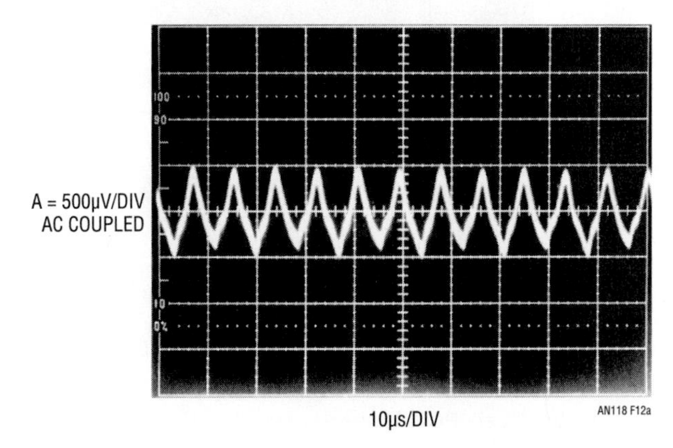

Figure 7.12a • –1000V Converter Output Noise Measures Inside 1mV (1PPM-0.0001%) in 100MHz Bandwidth. Resonant Royer Related Ripple Dominates Residue—No High Frequency Content is Detectable

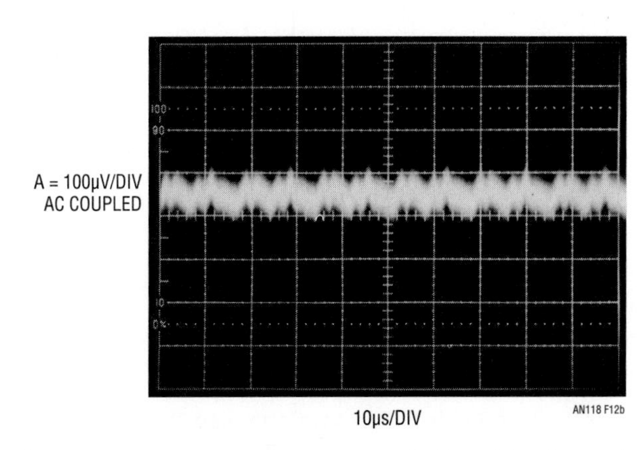

Figure 7.12b • 10x Increase in Figure 7.11's Filter Capacitor Values Reduces Noise to 100μV. Penalty is Capacitor Physical Size

Controlled transition push-pull converters

Controlled transition techniques are also directly applicable to push-pull architectures. Figure 7.13 uses a controlled transition push-pull regulator in a simple loop to control a 300V output converter. Symmetrical transformer drive and controlled switching edge times promote low output noise. The D1-D4 connected damper further minimizes residual aberrations. In this case, inductors are used in the output filter although appropriate resistor values could be employed.

Figure 7.14 displays smooth transitions at the transformer secondary outputs (trace A is T1 Pin 4, trace B, T1 Pin 7). Absence of high frequency harmonic results in extremely low noise. Figure 7.15's fundamental related output residue approaches the 100μV measurement noise floor in a 100MHz bandpass. This is spectacularly low noise performance in any DC/DC converter and certainly in one providing high voltage. Here, at 300V output, noise represents less than 1 part in 3 million.

Figure 7.16 is similar except that output range is variable from 0V to 300V The LT1533 is replaced by an LT3439 which contains no control elements. It simply drives the transformer with 50% duty cycle, controlled switching transitions. Feedback control is enforced by A1-Q1-Q2 driving current into T1's primary center tap. A1 compares a resistively derived portion of the output with a user supplied control voltage. The values shown produce a 0V to 300V output in response to a 0V to 1V control voltage. An RC network from Q2's collector to A1's positive input compensates the loop. Collector waveforms and output noise signature are nearly identical to Figure 7.13. Output noise is 100pV$_{P-P}$ over the entire 0V to 300V output range.

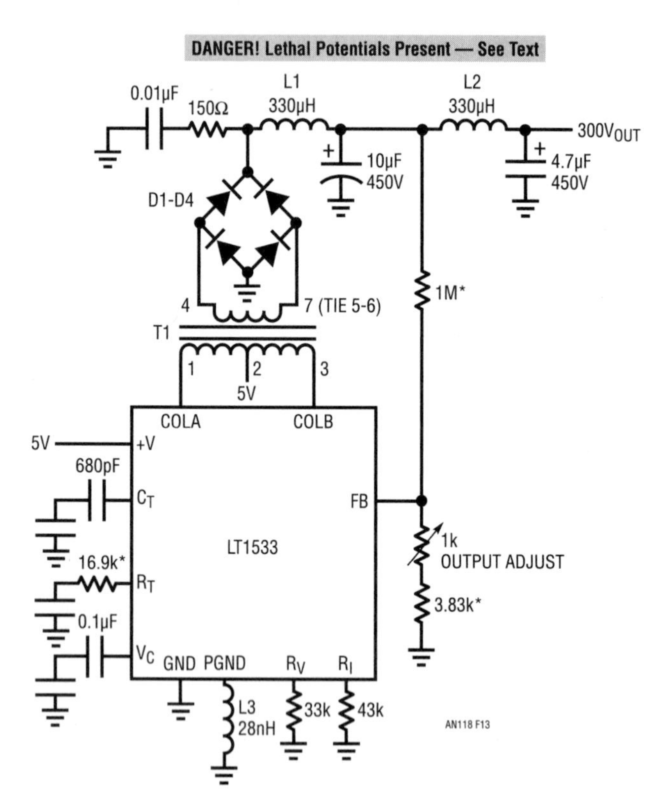

Figure 7.13 • A Push-Pull Drive, Controlled Transition, 300V Output Converter. Symmetrical Transformer Drive and Slow Edges Promote Low Output Noise

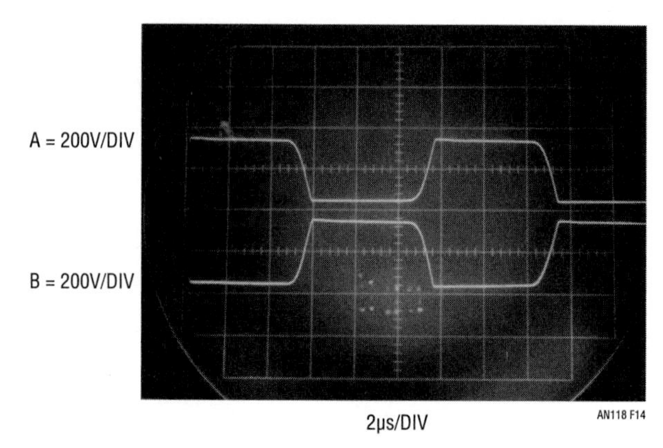

Figure 7.14 • Transformer Secondary Outputs Show No High Frequency Artifacts

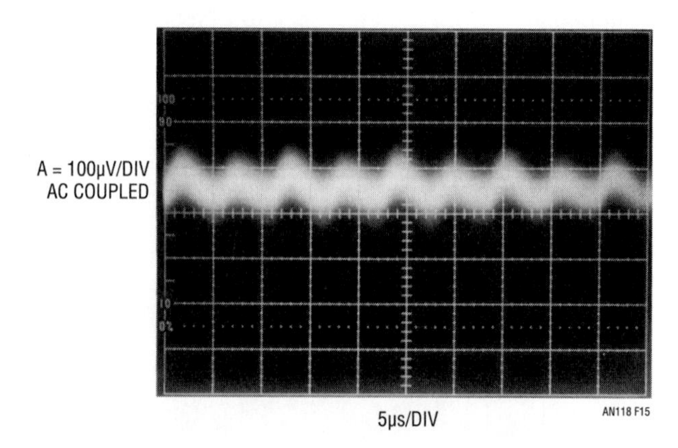

Figure 7.15 • Push-Pull Converter Related Residue Approaches 100μV Measurement Noise Floor. No Wideband Components Appear in 100MHz Measurement Bandpass

Flyback converters

Flyback converters, due to their abrupt, poorly controlled energy delivery, are not usually associated with low noise output. However; careful magnetic selection and layout can provide surprisingly good performance, particularly at low output current.

Figure 7.17's design provides 200V from a 5V input[5]. The scheme is a basic inductor flyback boost regulator with some important deviations. Q1, a high voltage device, has been interposed between the LT1172 switching regulator and the inductor. This permits the regulator to control Q1's high voltage switching without undergoing high voltage stress. Q1, operating as a "cascode" with the LT1172's internal switch, withstands L1's high voltage flyback events[6].

Diodes associated with Q1's source terminal clamp L1 originated spikes arriving via Q1's junction capacitance. The high voltage is rectified and filtered, forming the circuits' output. The ferrite bead, 100Ω and 300Ω resistors aid

Note 5: LTC application note veterans, a weary crew, will recognize material in this section from AN98 and AN113. The original circuits and text have been modified as necessary to suit low noise operation. See References.

Note 6: See References 13-17 for historical perspective and study on cascodes.

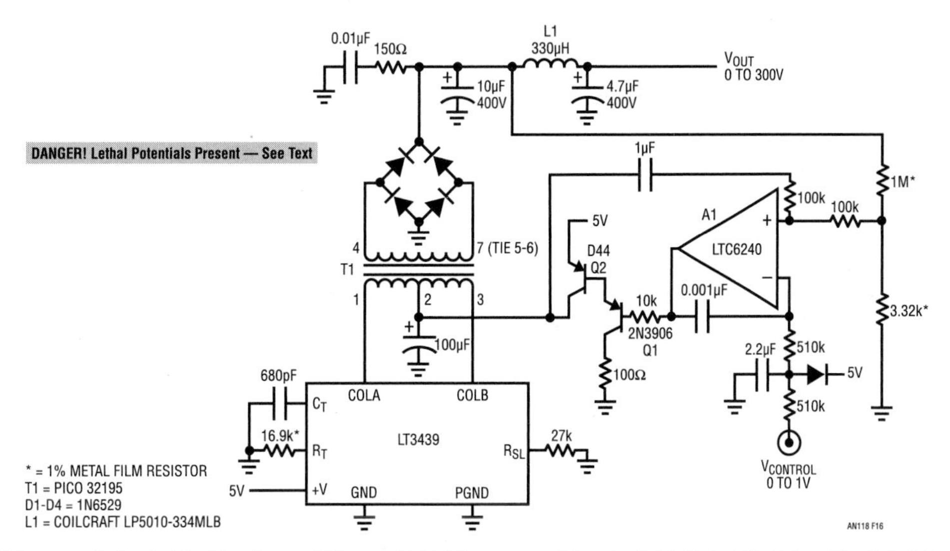

Figure 7.16 • Full Range Adjustable Version of Figure 7.13. $V_{CONTROL}$ Directed A1 Sets T1 Drive Via Q1-Q2. 1M-3.32k Divider Provides Feedback, Stabilized by A1's Input Capacitors. Waveforms Are Similar to Figure 7.13. Output Noise is $100\mu V_{P-P}$

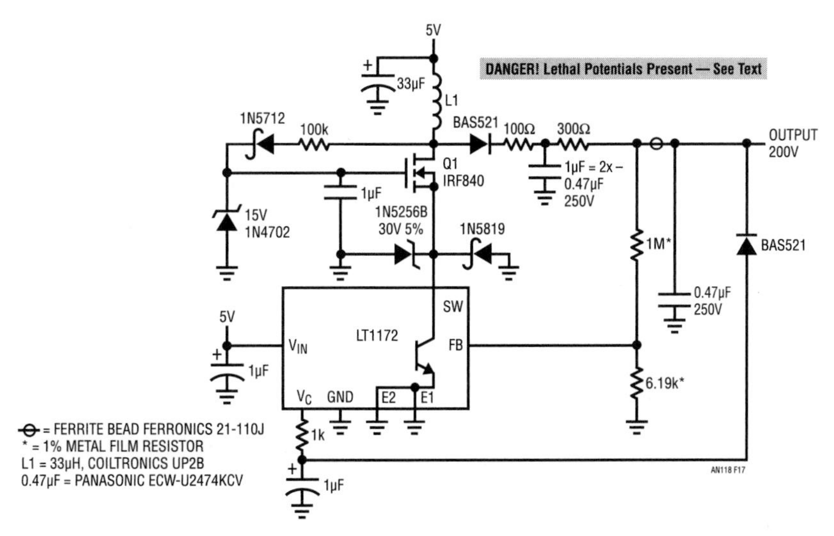

Figure 7.17 • 5V to 200V Output Converter. Cascoded Q1 Switches High Voltage, Allowing Low Voltage Regulator to Control Output. Diode Clamps Protect Regulator from Transients; 100k Path Bootstraps Q1s Gate Drive from L1's Flyback Events. Output Connected 300Ω-Diode Combination Provides Short-Circuit Protection. Ferrite Bead, 100Ω and 300Ω Resistors Minimize High Frequency Output Noise

filter efficiency[7] Feedback to the regulator stabilizes the loop and the V_C pin network provides frequency compensation. A 100k path from L1 bootstraps Q1's gate drive to about 10V ensuring saturation. The output connected diode provides short-circuit protection by shutting down the LT1172 if the output is accidentally grounded.

Figure 7.18's traces A and C are LT1172 switch current and voltage, respectively. Q1's drain is trace B. Current ramp termination results in a high voltage flyback event at Q1's drain. A safely attenuated version of the flyback appears at the LT1172 switch. The sinosoidal signature, due to inductor ring-off between conduction cycles, is harmless.

Figure 7.19, output noise, is composed of low frequency ripple and wideband, flyback related spikes measuring $1mV_{P-P}$ in a 100MHz bandpass.

Figure 7.20, contributed by Albert M. Wu of LTC, is a transformer coupled flyback circuit. The transformer secondary provides voltage step-up referred to the flyback

Note 7: Tutorial on ferrite beads appears in Appendix F

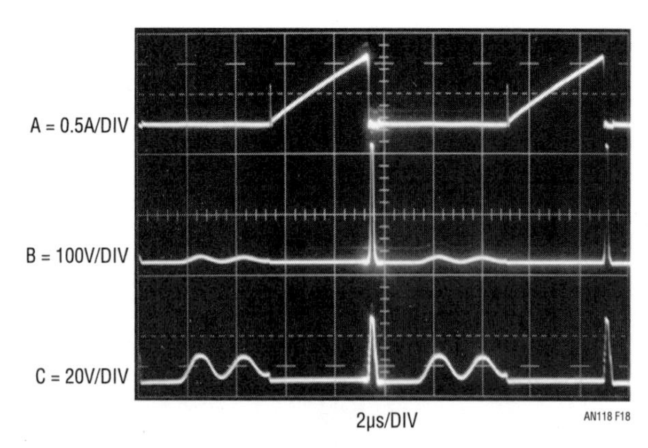

Figure 7.18 • Waveforms for 5V to 200V Converter Include LT1172 Switch Current and Voltage (Traces A and C, Respectively) and Q1's Drain Voltage (Trace B). Current Ramp Termination Results in High Voltage Flyback Event at Q1 Drain. Safely Attenuated Version Appears at LT1172 Switch. Sinosoidal Signature, Due to Inductor Ring-Off Between Current Conduction Cycles, is Harmless. All Traces Intensified Near Center Screen for Photographic Clarity

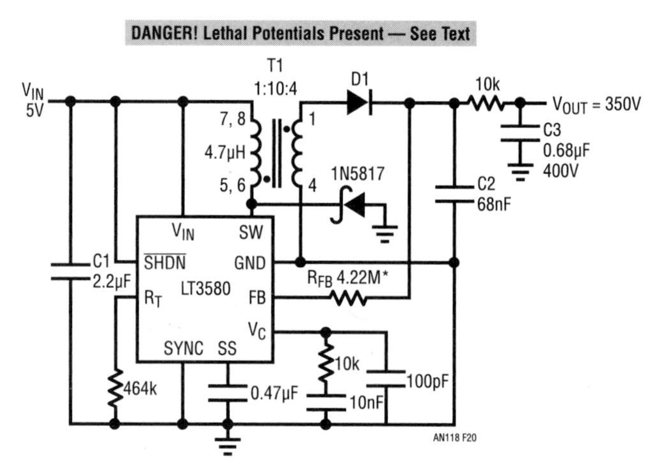

C1: 2.2μF, 25V, X5R, 1206
C2: TDK C3225X7R2J683M
D1: VISHAY GSD2004S DUAL DIODE CONNECTED IN SERIES
T1: TDK LDT565630T-041
C3: WIMA MKS-4
* = IRC-CGH-1, 1%

Figure 7.20 • 5V Powered Transformer Coupled Flyback Converter Produces 350V Output

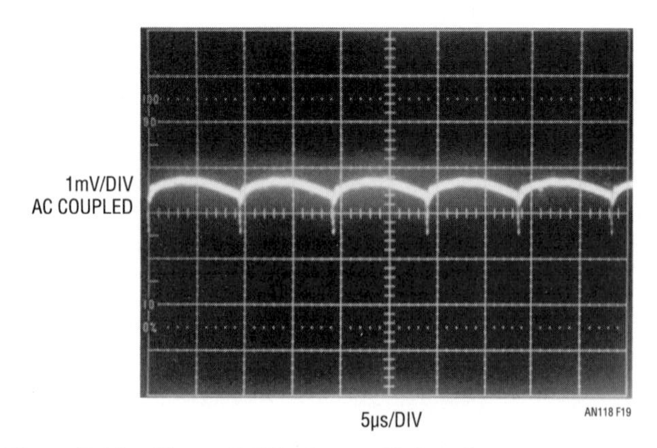

Figure 7.19 • Figure 7.17's Output Noise, Composed of Low Frequency Ripple and Wideband, Flyback Related Spikes, Measures $1mV_{P-P}$ in 100MHz Bandpass

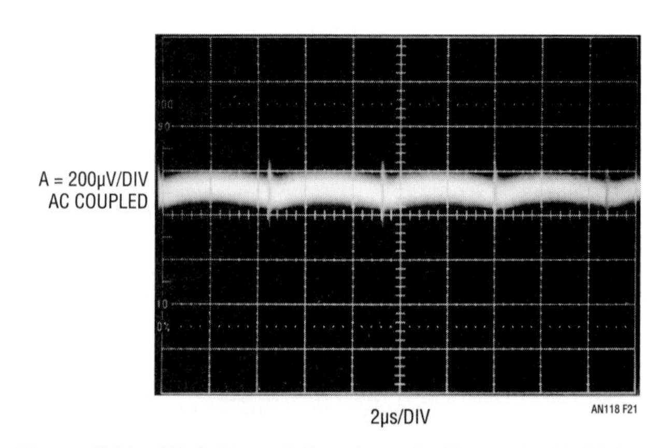

Figure 7.21 • High Speed Transients in Figure 7.20's Noise Signature are Within $300\mu V_{P-P}$

driven primary. The 4.22M resistor supplies feedback to the regulator, closing a control loop. A 10k-0.68μF filter network attenuates high frequency harmonic with minimal voltage drop. Flyback related transients are clearly visible in Figure 7.21's output noise although within 300μV_{P-P}.

Figure 7.22 employs the LT3468 photoflash capacitor charger as a general purpose high voltage DC/DC converter. Normally, the LT3468 regulates its output at 300V by sensing T1's flyback pulse characteristic. This circuit allows the LT3468 to regulate at lower voltages by truncating its charge cycle before the output reaches 300V. A1 compares a divided down portion of the output with the program input voltage. When the program voltage (A1 + input) is exceeded by the output derived potential (A1 − input) A1's output goes low, shutting down the LT3468. The feedback capacitor provides AC hysteresis, sharpening A1's output to prevent chattering at the trip point. The LT3468 remains shut down until the output voltage drops low enough to trip A1's output high, turning it back on. In this way, A1 duty cycle modulates the LT3468, causing the output voltage to stabilize at a point determined by the program input.

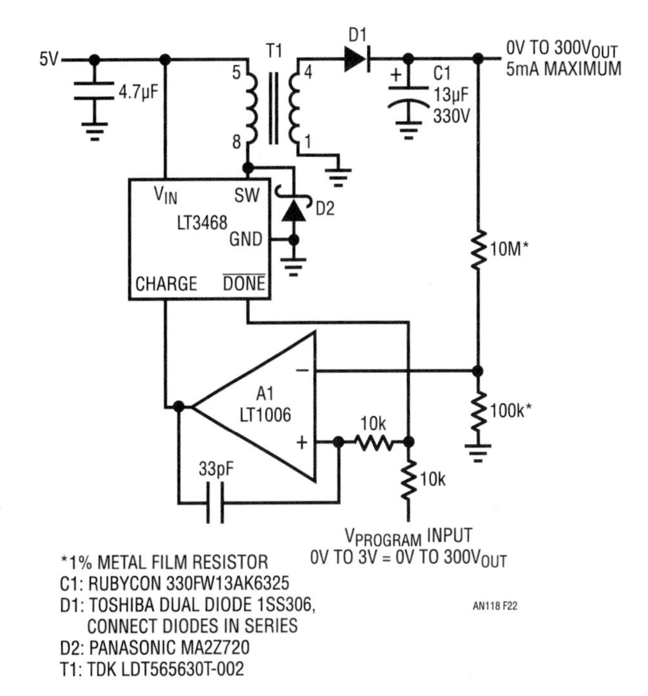

Figure 7.22 • A Voltage Programmable 0V to 300V Output Regulator. A1 Controls Regulator Output by Duty Cycle Modulating LT3468/T1 DC/DC Converter Power Delivery

Figure 7.23's 250V DC output (Trace B) decays down about 2V until A1 (Trace A) goes high, enabling the LT3468 and restoring the loop. This simple circuit works well, regulating over a programmable 0V to 300V range, although its inherent hysteretic operation mandates the (unacceptable) 2V output ripple noted. Loop repetition rate varies with input voltage, output set point and load but the ripple is always present. The following circuit greatly reduces ripple amplitude although complexity increases.

Figure 7.24's post-regulator reduces Figure 7.22's output ripple and noise to only 2mV. A1 and the LT3468 are identical to the previous circuit, except for the 15V zener diode in series with the 10M-100k feedback divider. This component causes C1's voltage, and hence Q1's collector, to regulate 15V above the $V_{PROGRAM}$ input dictated point. The $V_{PROGRAM}$ input is also routed to the A2-Q2-Q1 linear post-regulator. A2's 10M-100k feedback divider does not include a zener so the post-regulator follows the $V_{PROGRAM}$ input with no offset. This arrangement forces 15V across Q1 at all output voltages. This figure is high enough to eliminate undesirable ripple and noise from the output while keeping Q1 dissipation low.

Q3 and Q4 form a current limit, protecting Q1 from overload. Excessive current through the 50Ω shunt turns Q3 on. Q3 drives Q4, shutting down the LT3468. Simultaneously, a portion of Q3's collector current turns Q2 on hard, shutting off Q1. This loop dominates the normal regulation feedback, protecting the circuit until the overload is removed.

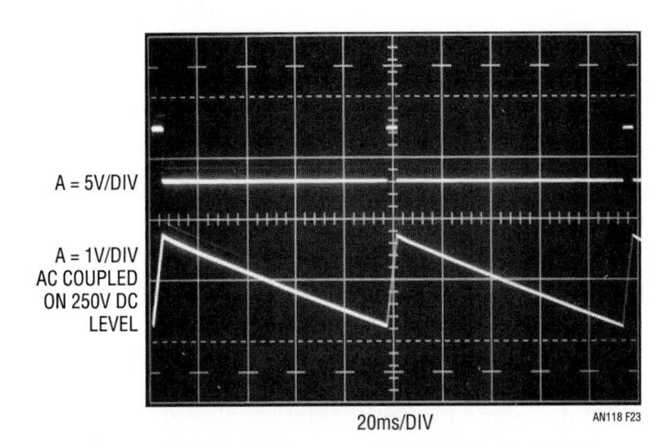

Figure 7.23 • Details of Figure 7.22's Duty Cycle Modulated Operation. High Voltage Output (Trace B) Ramps Down Until A1 (Trace A) Goes High, Enabling LT3468/T1 to Restore Output. Loop Repetition Rate Varies with Input Voltage, Output Set Point and Load

Figure 7.25 shows just how effective the post regulator is. When A1 (trace A) goes high, Q1's collector (trace B) ramps up in response (note LT3468 switching artifacts on ramps upward slope). When the A1-LT3468 loop is

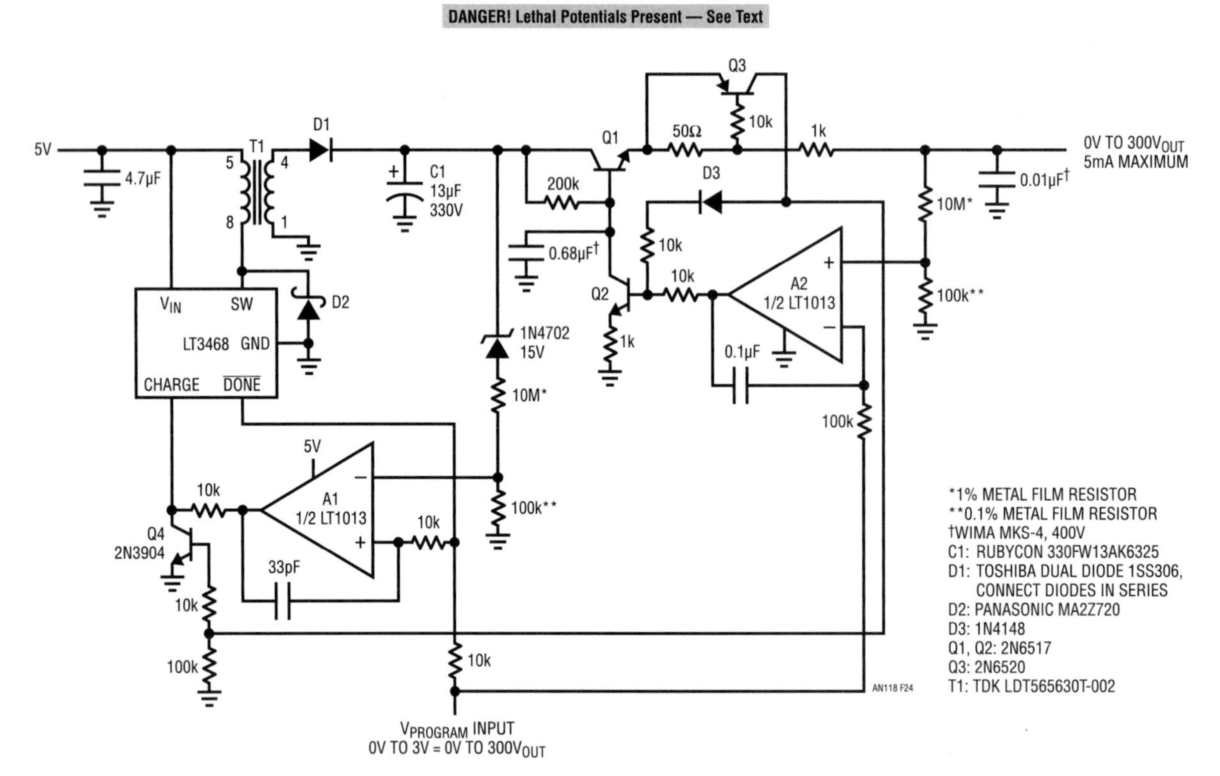

Figure 7.24 • Post-Regulation Reduces Figure 7.22's 2V Output Ripple to 2mV. LT3468-Based DC/DC Converter, Similar to Figure 7.22, Delivers High Voltage to Q1 Collector. A2, Q1, Q2 Form Tracking, High Voltage Linear Regulator. Zener Sets Q1 V_{CE} = 15V, Ensuring Tracking with Minimal Dissipation. Q3-Q4 Limit Short-Circuit Output Current

satisfied, A1 goes low and Q1's collector ramps down. The output post-regulator (trace C), however, rejects the ripple, showing only 2mV of noise. Slight trace blurring derives from A1-LT3468 loop jitter.

Summary of circuit characteristics

Figure 7.26 summarizes the circuits presented with salient characteristics noted. This chart is only a generalized guideline and not an indicator of capabilities or limits. There are too many variables and exceptions to accomodate the categorical statement a chart implies. The interdependence of circuit parameters makes summarizing or rating various approaches a hazardous exercise. There is simply no intellectually responsible way to streamline the selection and design process if optimum results are desired. A meaningful choice *must* be the outcome of laboratory-based experimentation. There are just too many interdependent variables and surprises for a systematic, theoretically based selection. Charts seek authority through glib simplification and simplification is Disaster's deputy. Nonetheless, Figure 7.26, in all its appropriated

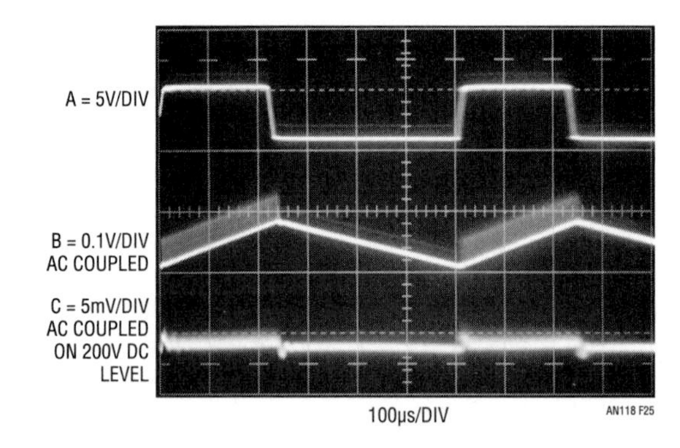

Figure 7.25 • Low Ripple Output (Trace C) is Apparent in Post-Regulator's Operation. Traces A and B are A1 Output and Q1 s Collector, Respectively. Trace Blurring, Right of Photo Center, Derives from Loop Jitter

glory, lists input supply range, output voltage and current along with comments for each circuit[8].

Note 8: Readers detecting author ambivalence at Figure 7.26's inclusion are not hallucinating. Locally based marketeers champion such charts; the writer is less enthusiastic. This application note was derived from a manuscript originally prepared for publication in EDN magazine.

CIRCUIT TYPE	FIGURE NUMBER	SUPPLY RANGE (1mA LOAD)	MAXIMUM OUTPUT CURRENT AT TEST VOLTAGE	COMMENTS
LT1635 - Linear Resonant Royer	1	2.7V to 12V	2mA at 250V	<100μV Wideband Noise. Easily Voltage Controlled. Potential Dissipation Issue at High Supply Voltages.
LT1431 - Linear Resonant Royer	4	2.7V to 32V	2mA at 250V	<100μV Wideband Noise. Wide Supply Range. Potential Dissipation Issue at High Supply Voltages.
LT3401 - Switched Resonant Royer	5	2.7V to 5V	3.5mA at 250V	3mV Wideband Noise. High Output Current, Better Efficiency than Figures 1 and 4.
LT1534 - Switched Resonant Royer	8	2.7V to 15V	2mA at 250V	≈100μV Wideband Noise. Good Trade-Off Between Figures 1, 4 and 5.
LT1534 - Swiched Resonant Royer	11	4.5V to 15V	1.2mA at –1000V	1mV Wideband Noise Reducable to 100μV. Negative 1000V Output Suits Photomultiplier Tubes.
LT1533 Push-Pull	13	2.7V to 15V	2mA at 300V	≈100μV Wideband Noise.
LT3439 Push-Pull	16	4.5V to 6V	2mA at 0V to 300V	Full Range Adjustable Version of Figure 13. ≈100μV Wideband Noise.
LT1172 - Cascode Inductor Flyback	17	3.5V to 30V	2mA at 200V	V_{OUT} Limit ≈200V. ≈1mV Wideband Noise.
LT3580 - XFMR Flyback	20	2.7V to 20V	4mA at 350V	300μV Wideband Noise. Wide Supply Range. High Output Current. Small Transformer.
LT3468 - LT1006 XFMR Flyback	22	3.8V to 12V	5mA at 250V	1.5V Noise. Simple Voltage Control Input $0V_{IN}$ to $3V_{IN}$ = $0V_{OUT}$ – $300V_{OUT}$.
LT3468 - LT1013 XFMR Flyback - Linear	24	3.8V to 12V	5mA at 250V	2mV Wideband Noise. Voltage Control Input $0V_{IN}$ to $3V_{IN}$ = $0V_{OUT}$ to $300V_{OUT}$.

Figure 7.26 • Summarized Characteristics of Techniques Presented. Applicable Circuit Depends on Application Specifics

References

1. Williams, Jim, "A Fourth Generation of LCD Backlight Technology," Linear Technology Corporation, Application Note 65, November 1995, p. 32-34, 119
2. Bright, Pittman and Royer, "Transistors As On-Off Switches in Saturable Core Circuits," Electrical Manufacturing, December 1954. Available from Technomic Publishing, Lancaster, PA
3. Williams, Jim, "A Monolithic Switching Regulator with 100μV Output Noise," Linear Technology Corporation, Application Note 70, October 1997
4. Baxendall, P.J., "Transistor Sine-Wave LC Oscillators," British Journal of IEEE, February 1960, Paper No. 2978E
5. Williams, Jim, "Low Noise Varactor Biasing with Switching Regulators," Linear Technology Corporation, Application Note 85, August 2000, p. 4-6
6. Williams, Jim, "Minimizing Switching Residue in Linear Regulator Outputs". Linear Technology Corporation, Application Note 101, July 2005
7. Morrison, Ralph, "Grounding and Shielding Techniques in Instrumentation," Wiley-Interscience, 1986
8. Fair-Rite Corporation, "Fair-Rite Soft Ferrites," Fair-Rite Corporation, 1998
9. Sheehan, Dan, "Determine Noise of DC/DC Converters," Electronic Design, September 27, 1973
10. Ott, Henry W., "Noise Reduction Techniques in Electronic Systems." Wiley Interscience, 1976
11. Tektronix, Inc. "Type 1A7A Differential Amplifier Instruction Manual," Check Overall Noise Level Tangentially", p. 5-36 and 5-37, 1968
12. Witt, Jeff, "The LT1533 Heralds a New Class of Low Noise Switching Regulators," Linear Technology, Vol. VII, No. 3, August 1997, Linear Technology Corporation
13. Williams, Jim, "Bias Voltage and Current Sense Circuits for Avalanche Photodiodes," Linear Technology Corporation, Application Note 92, November 2002, p. 8
14. Williams, Jim, "Switching Regulators for Poets," Appendix D, Linear Technology Corporation, Application Note 25, September 1987
15. Hickman, R.W. and Hunt, F.V., "On Electronic Voltage Stabilizers," "Cascode," Review of Scientific Instruments, January 1939, p. 6-21, 16
16. Williams, Jim, "Signal Sources, Conditioners and Power Circuitry," Linear Technology Corporation, Application Note 98, (November 2004), p. 20-21
17. Williams, Jim, "Power Conversion, Measurement and Pulse Circuits," Linear Technology Corporation, Application Note 113, August 2007
18. Williams, Jim and Wu, Albert, "Simple Circuitry for Cellular Telephone/Camera Flash Illumination," Linear Technology Corporation, Application Note 95, March 2004
19. LT3580 Data Sheet, Linear Technology Corporation

Appendix A

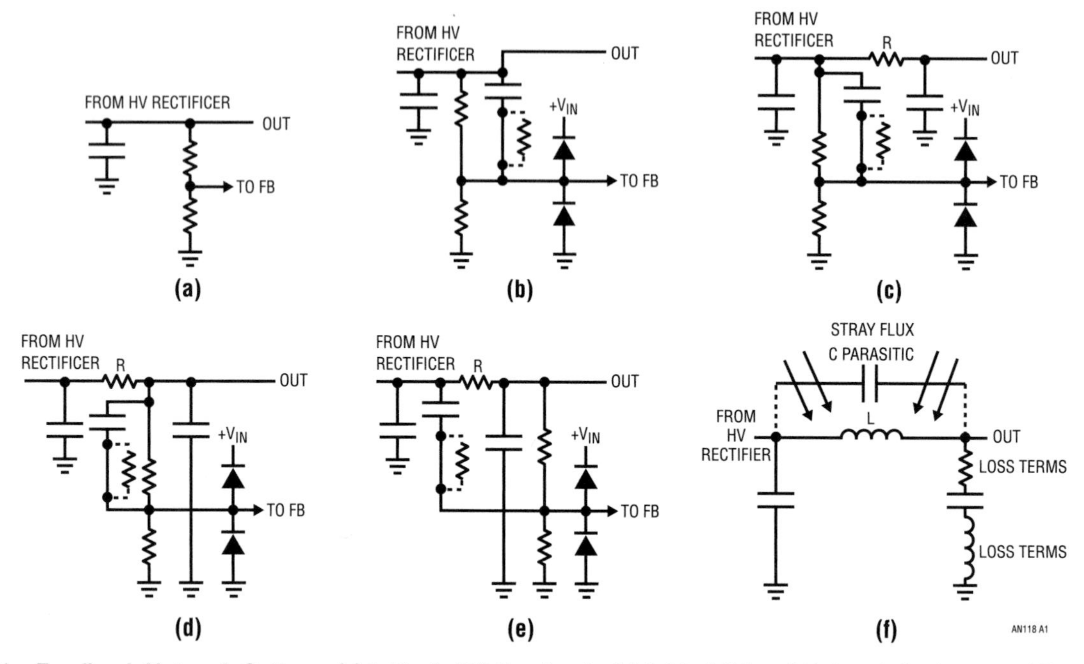

Figure A1 • Feedback Network Options. (a) Is Basic DC Feedback. (b) Adds AC Lead Network for Improved Dynamics. Diode Clamps Protect Feedback Node from Capacitor's Differentiated Response. (c)'s Low Ripple Two Section Filter Slows Loop Transmission but Lead Network Provides Stability. Resistor R Sets DC Output Impedance. (d) Encloses R in DC Loop, Lowering Output Resistance. Feedback Capacitor Supplies Leading Response. (e) Moves Feedback Capacitor to Filter Input, Further Extending (d)s Leading Response. (f), Replacing Filter Resistor (R) with Inductor, Lowers Output Resistance but Introduces Parasitic Shunt Capacitive Path and Stray Flux Sensitivity

Feedback considerations in high voltage DC/DC converters

A high voltage DC/DC converter feedback network is a study in compromise. The appropriate choice is application dependent. Considerations include desired output impedance, loop stability, transient response and high voltage induced overstress protection. Figure A1 lists typical options.

(a) is basic DC feedback and requires no special commentary. (b) adds an AC lead network for improved dynamics. Diode clamps protect the feedback node from the capacitors differentiated response. (c)'s low ripple, two section filter slows transient response but a lead network provides stability. Resistor R, outside the loop, sets DC output impedance. (d) encloses R within the DC loop, lowering output resistance but delaying loop transmission. A feedback capacitor supplies corrective leading response. (e) moves the feedback capacitor to the filter input, further extending (d)'s leading response. (f) replaces filter resistor R with an inductor, lowering output resistance but introducing parasitic shunt capacitance which combines with capacitor loss terms to degrade filtering. The inductor also approximates a transformer secondary, vulnerable to stray flux pick-up with resulting increased output noise.[1]

A common concern in any high voltage feedback network is reliability. Components must be quite carefully chosen. Voltage ratings should be conservative and strictly adhered to. While component ratings are easily ascertained, more subtle effects such as ill-suited board material and board wash contaminants can be reliability hazards. Long term electro-migration effects can have undesirable results. Every potential unintended conductive path should be considered as an error source and layout planned accordingly. Operating temperature, altitude, humidity and condensation effects must be anticipated. In extreme cases, it may be necessary to rout the board under components operating at high voltage. Similarly, it is common practice to use several units in series to minimize voltage across the output connected feedback resistor. Contemporary packaging requirements emphasize tightly packed layout which may conflict with

Note 1: See Appendix G.

high voltage standoff requirements. This tradeoff must be carefully reviewed or reliability will suffer. The potentially deleterious (disastrous) effects of environmental factors, layout and component choice over time cannot be overstated. Clear thinking is needed to avoid unpleasant surprises.

Editor's Note: Appendices B through E are thinly edited and modified versions of tutorials first appearing in AN70. Although originally intended to address controlled transition applications (e.g. LT1533, 4 and LT3439) the material is directly relevant and warrants inclusion here.

Appendix B Specifying and measuring something called noise

Undesired output components in switching regulators are commonly referred to as "noise." The rapid, switched mode power delivery that permits high efficiency conversion also creates wideband harmonic energy. This undesirable energy appears as radiated and conducted components, or "noise." Actually switching regulator output "noise" isn't really noise at all, but coherent, high frequency residue directly related to the regulator's switching. Unfortunately, it is almost universal practice to refer to these parasitics as "noise, " and this publication maintains this common, albeit inaccurate, terminology.[1]

Measuring noise

There are an almost uncountable number of ways to specify noise in a switching regulator's output. It is common industrial practice to specify peak-to-peak noise in a 20MHz bandpass.[2] Realistically, electronic systems are readily upset by spectral energy beyond 20MHz, and this specification restriction benefits no one.[3] Considering all this, it seems appropriate to specify peak-to-peak noise in a verified 100MHz bandwidth. Reliable low level measurements in this bandpass require careful instrumentation choice and connection practices.

Our study begins by selecting test instrumentation and verifying its bandwidth and noise. This necessitates the arrangement shown in Figure B1. Figure B2 diagrams signal flow. The pulse generator supplies a subnanosecond rise time step to the attenuator; which produces a <1mV version of the step. The amplifier takes 40dB of gain (A = 100) and the oscilloscope displays the result. The "front-to-back" cascaded bandwidth of this system should be about 100MHz (t_{RISE} = 3.5ns) and Figure B3 reveals this to be so. Figure B3's trace shows 3.5ns rise time and about 100μV of noise. The noise is limited by the amplifier's 50Ω noise floor.[4]

Figure B4's presentation of output noise shows barely visible switching artifacts (at vertical graticule lines 4, 6 and 8) in the 100MHz bandpass. Fundamental ripple is seen more clearly, although similarly noise floor dominated. Restricting measurement bandwidth to 10MHz (Figure B5) reduces noise floor amplitude, although switching noise and ripple amplitudes are preserved. This indicates that there is no signal power beyond 10MHz. Further measurements as bandwidth is successively reduced can determine the highest frequency content present.

The importance of measurement bandwidth is further illustrated by Figures B6 to B8. Figure B6 measures a commercially available DC/DC converter in a 1MHz bandpass. The unit appears to meet its claimed $5mV_{P-P}$ noise specification. In Figure B7, bandwidth is increased to 10MHz. Spike amplitude enlarges to 6mV $_{P-P}$, about 1mV outside the specification limit. Figure B8's 50MHz viewpoint brings an unpleasant surprise. Spikes measure $30mV_{P-P}$—six times the specified limit![5]

Low frequency noise

Low frequency noise is rarely a concern, because it almost never affects system operation. Low frequency noise is shown in Figure B9. It is possible to reduce low frequency noise by rolling off control loop bandwidth.

Note 1: Less genteelly, "If you can't beat 'em, join 'em."

Note 2: One DC/DC converter manufacturer specifies RMS noise in a 20MHz bandwidth. This is beyond deviousness and unworthy of comment.

Note 3: Except, of course, eager purveyors of power sources who specify them in this manner

Note 4: Observed peak-to-peak noise is somewhat affected by the oscilloscope's "intensity" setting. Reference 11 describes a method for normalizing the measurement.

Note 5: Caveat Emptor.

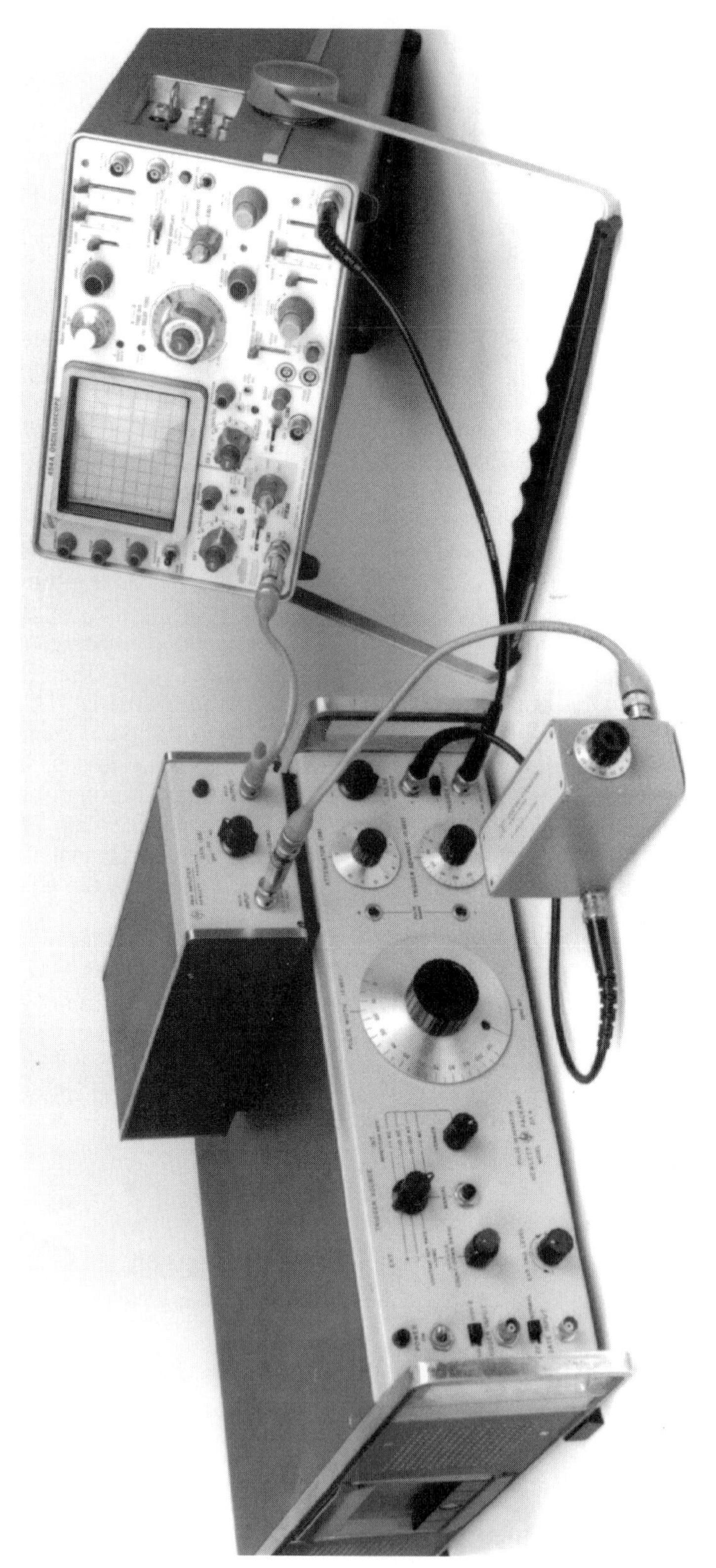

Figure B1 • 100MHz Bandwidth Verification Test Setup. Note Coaxial Connections for Wideband Signal Integrity

Figure B10 shows about a five times improvement when this is done, even with greater measurement bandwidth. A possible disadvantage is loss of loop bandwidth and slower transient response.

Preamplifier and oscilloscope selection

The low level measurements described require some form of preamplification for the oscilloscope. Current generation oscilloscopes rarely have greater than

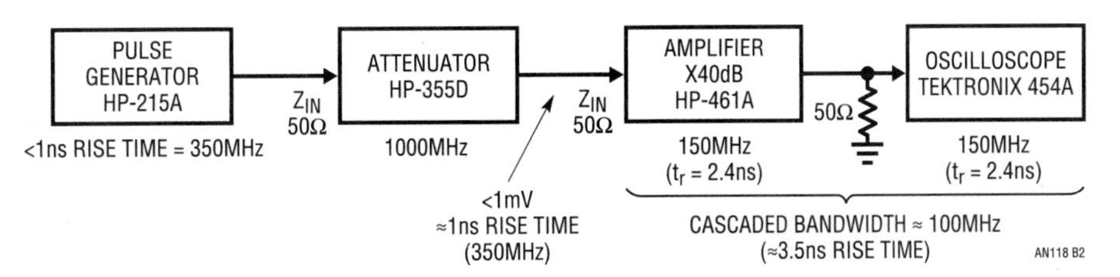

Figure B2 • Subnanosecond Pulse Generator and Wideband Attenuator Provide Fast Step to Verify Test Setup Bandwidth

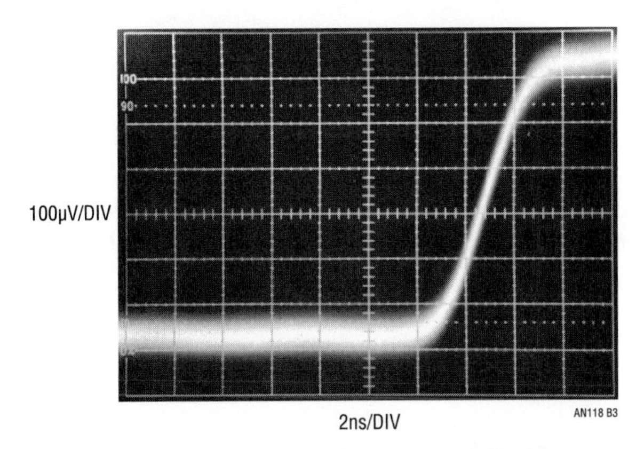

Figure B3 • Oscilloscope Display Verifies Test Setup's 100MHz (3.5ns Rise Time) Bandwidth. Baseline Noise Derives from Amplifier's 50Ω Input Noise Floor

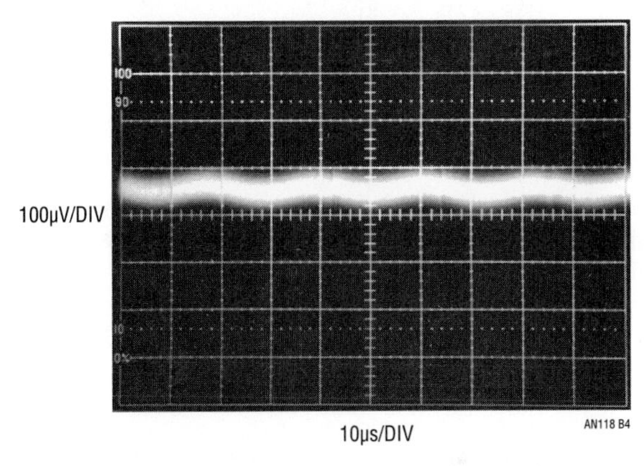

Figure B4 • Output Switching Noise Is Just Discernible in a 100MHz Bandpass

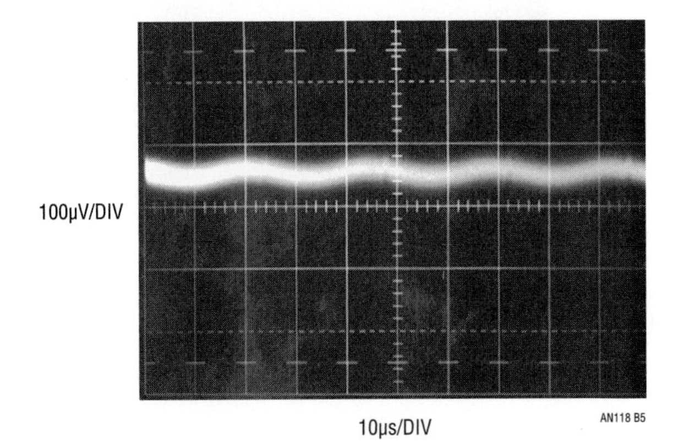

Figure B5 • 10MHz Band Limited Version of Preceding Photo. All Switching Noise Information is Preserved, Indicating Adequate Bandwidth

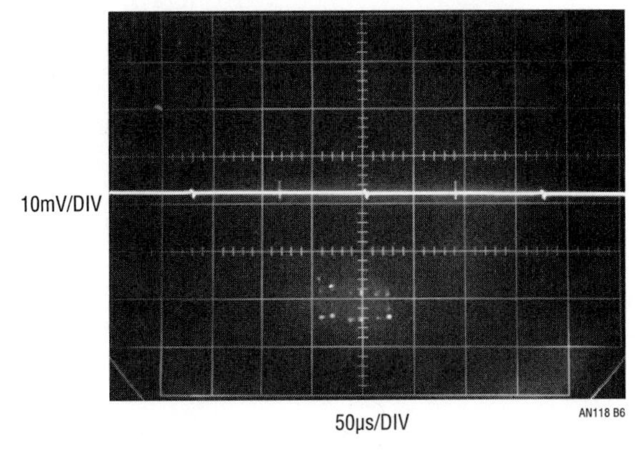

Figure B6 • Commercially Available Switching Regulator's Output Noise in a 1MHz Bandpass. Unit Appears to Meet its $5mV_{P-P}$ Noise Specification

2mV/DIV sensitivity, although older instruments offer more capability. Figure B11 lists representative preamplifiers and oscilloscope plug-ins suitable for noise measurement. These units feature wideband, low noise performance. It is particularly significant that the majority of these instruments are no longer produced. This is in keeping with current instrumentation trends, which emphasize digital signal acquisition as opposed to analog measurement capability.

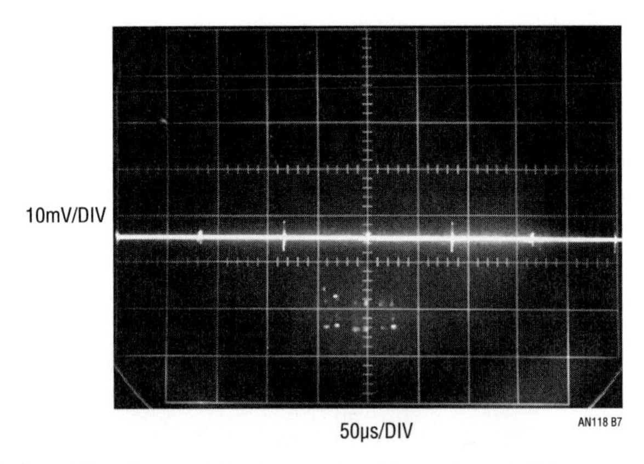

Figure B7 • Figure B6's Regulator Noise in a 10MHz Bandpass. 6mVp-p Noise Exceeds Regulator's Claimed 5mV Specification

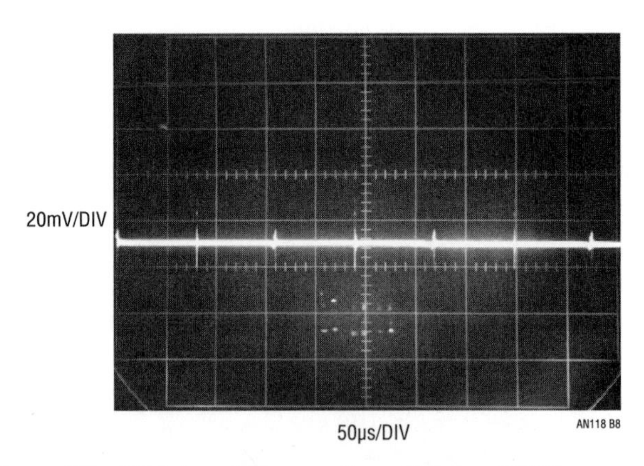

Figure B8 • Wideband Observation of Figure B7 Shows 30mV$_{P-P}$ Noise—Six Times the Regulator's Specification!

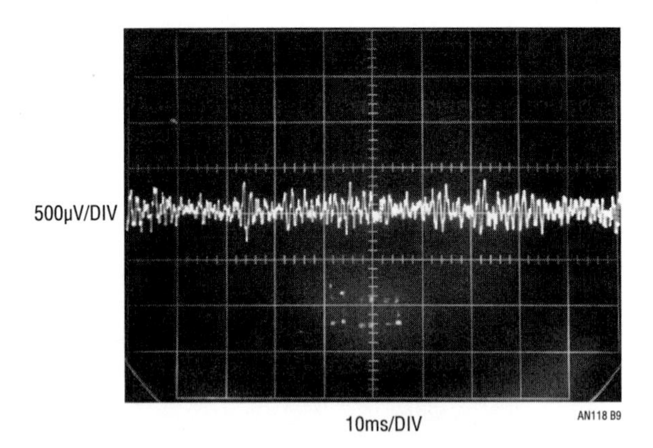

Figure B9 • 1Hz to 3kHz Noise Using Standard Frequency Compensation. Almost All Noise Power is Below 1kHz

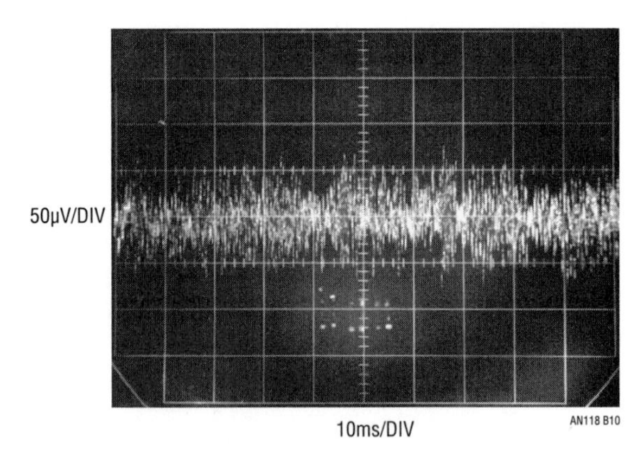

Figure B10 • Feedback Lead Network Decreases Low Frequency Noise, Even as Measurement Bandwidth Expands to 100kHz

The monitoring oscilloscope should have adequate bandwidth and exceptional trace clarity. In the latter regard high quality analog oscilloscopes are unmatched. The exceptionally small spot size of these instruments is well-suited to low level noise measurement.[6] The digitizing uncertainties and raster scan limitations of DSOs impose display resolution penalties. Many DSO displays will not even register the small levels of switching-based noise.

INSTRUMENT TYPE	MANUFACTURER	MODEL NUMBER	−3dB BANDWIDTH	MAXIMUM SENSITIVITY/GAIN	AVAILABILITY	COMMENTS
Amplifier	Hewlett-Packard	461A	175MHz	Gain = 100	Secondary Market	50Ω Input, Stand-Alone. $100\mu V_{P-P}$ (≈20μV RMS) noise in 100MHz bandwidth. Best of this group for noise measurement described in text.
Differential Amplifier	Tektronix	1A5	50MHz	1mV/DIV	Secondary Market	Requires 500 Series Mainframe
Differential Amplifier	Tektronix	7A13	100MHz	1mV/DIV	Secondary Market	Requires 7000 Series Mainframe
Differential Amplifier	Tektronix	11A33	150MHz	1mV/DIV	Secondary Market	Requires 11000 Series Mainframe
Differential Amplifier	Tektronix	P6046	100MHz	1mV/DIV	Secondary Market	Stand-Alone
Differential Amplifier	Preamble	1855	100MHz	Gain = 10	Current Production	Stand-Alone, Settable Bandstops
Differential Amplifier	Tektronix	1A7/1A7A	1MHz	10μV/DIV	Secondary Market	Requires 500 Series Mainframe, Settable Bandstops
Differential Amplifier	Tektronix	7A22	1MHz	10μV/DIV	Secondary Market	Requires 7000 Series Mainframe, Settable Bandstops
Differential Amplifier	Tektronix	5A22	1MHz	10μV/DIV	Secondary Market	Requires 5000 Series Mainframe, Settable Bandstops
Differential Amplifier	Tektronix	ADA-400A	1MHz	10μV/DIV	Current Production	Stand-Alone with Optional Power Supply, Settable Bandstops
Differential Amplifier	Preamble	1822	10MHz	Gain = 100	Current Production	Stand-Alone, Settable Bandstops
Differential Amplifier	Stanford Research Systems	SR-560	1MHz	Gain = 50000	Current Production	Stand-Alone, Settable Bandstops, Battery or Line Operation
Differential Amplifier	Tektronix	AM-502	1MHz	Gain = 100000	Secondary Market	Requires TM-500 Series Power Supply

Figure B11 • Some Applicable High Sensitivity, Low Noise Amplifiers. Trade-Offs Include Bandwidth, Sensitivity and Availability. All Require Protective Input Network to Prevent Catastrophic Failure. See Figure B12 and Associated Text

Note 6: In our work we have found Tektronix types 454, 454A, 547 and 556 excellent choices. Their pristine trace presentation is ideal for discerning small signals of interest against a noise floor limited background.

Auxillary measurement circuits

Figure B12 is the clamp circuit referred to in the preceding figure caption. It must be employed with any of Figure B12's amplifiers to insure protection against catastrophic overloading.[7] The network is simply an AC coupled diode clamp. The coupling capacitor specified withstands the text examples high voltage outputs and the 10M resistors bleed residual capacitor charge. Built into a small BNC equipped enclosure, its output should be *directly* connected to the amplifier. 50Ω inputs may be directly driven; high impedance input amplifiers should be shunted with a coaxial 50Ω terminator.

B13's battery powered, 1MHz, 1mV square wave amplitude calibrator facilitates "end-to-end" amplifier—oscilloscope path gain verification. The 221k resistor associated area is sensitive to variations in stray capacitance and is shielded as per the schematic. A 4.5V reference stabilizes output amplitude against battery voltage change and a peaking trim optimizes front and trailing corner fidelity. Figure B14 shows that the simple peaking network does not quite achieve square corners, but 1mV pulse amplitude is clearly delineated. Trace thickening in the waveform flats indicates amplifier noise floor.

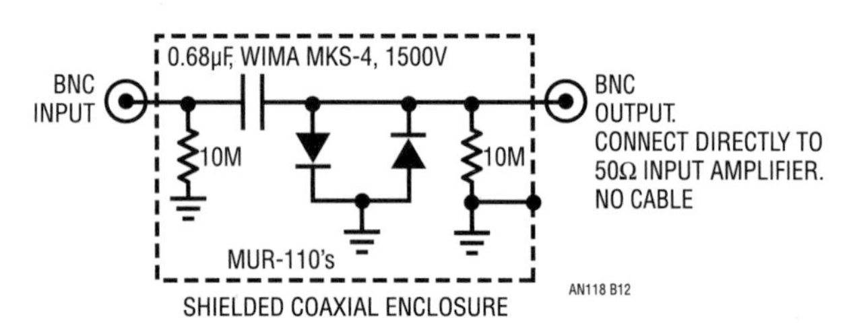

Figure B12 • Coaxially Fixtured Clamp Protects Figure B11's Low Noise Amplifiers From High Voltage Inputs. Resistors Insure Capacitor Discharge

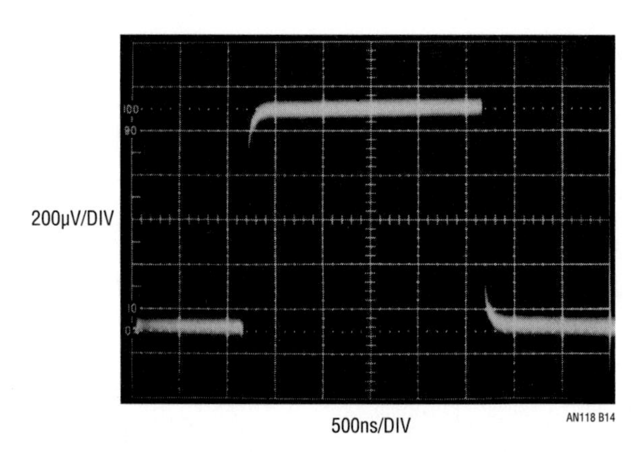

Figure B14 • 1mV Amplitude Calibrator Output Has Minor Corner Rounding but Pulse Flats Indicate Desired Amplitude. Trace Thickening Describes Amplifier Noise Floor

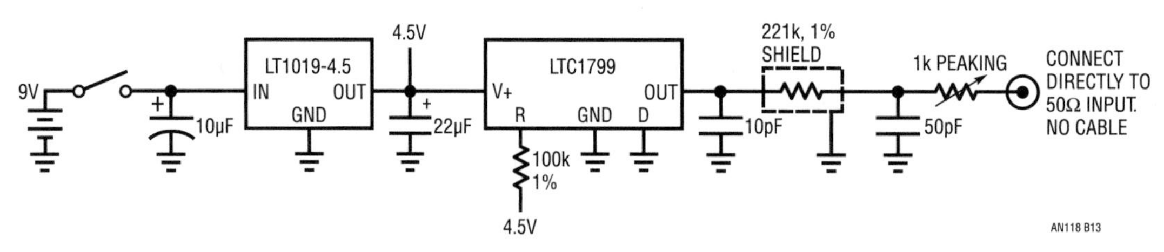

Figure B13 • Battery Powered, 1MHz, 1mV Square Wave Amplitude Calibrator Permits Signal Path Gain Verification. Peaking Trim Optimizes Front and Trailing Corner Fidelity

Note 7: Don't say we didn't warn you.

Appendix C Probing and connection techniques for low level, wideband signal integrity

The most carefully prepared breadboard cannot fulfill its mission if signal connections introduce distortion. Connections to the circuit are crucial for accurate information extraction. The low level, wideband measurements demand care in routing signals to test instrumentation.

Ground loops

Figure C1 shows the effects of a ground loop between pieces of line-powered test equipment. Small current flow between test equipment's nominally grounded chassis creates 60MHz modulation in the measured circuit output.

This problem can be avoided by grounding all line powered test equipment at the same outlet strip or otherwise ensuring that all chassis are at the same ground potential. Similarly, any test arrangement that permits circuit current flow in chassis interconnects must be avoided.

Pickup

Figure C2 also shows 60Hz modulation of the noise measurement. In this case, a 4-inch voltmeter probe at the feedback input is the culprit. *Minimize the number of test connections to the circuit and keep leads short.*

Poor probing technique

Figure C3's photograph shows a short ground strap affixed to a scope probe. The probe connects to a point which provides a trigger signal for the oscilloscope. Circuit output

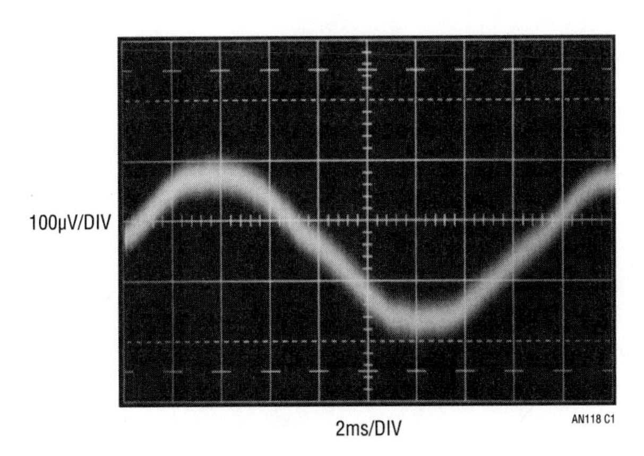

Figure C1 • Ground Loop Between Pieces of Test Equipment Induces 60Hz Display Modulation

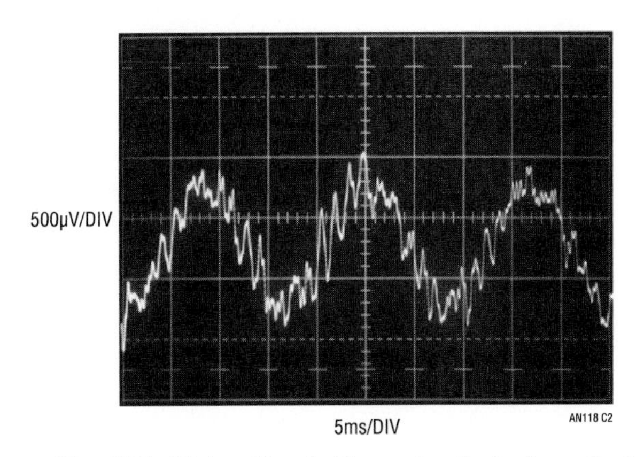

Figure C2 • 60Hz Pickup Due to Excessive Probe Length at Feedback Node

Figure C3 • Poor Probing Technique. Trigger Probe Ground Lead Can Cause Ground Loop-Induced Artifacts to Appear in Display

noise is monitored on the oscilloscope via the coaxial cable shown in the photo.

Figure C4 shows results. A ground loop on the board between the probe ground strap and the ground referred cable shield causes apparent excessive ripple in the display. *Minimize the number of test connections to the circuit and avoid ground loops.*

Violating coaxial signal transmission—felony case

In Figure C5, the coaxial cable used to transmit the circuit output noise to the amplifier-oscilloscope has been replaced with a probe. A short ground strap is employed as the probe's return. The error inducing trigger channel probe in the previous case has been eliminated; the 'scope is triggered by a noninvasive, isolated probe.[1] Figure C6 shows excessive display noise due to breakup of the coaxial signal environment. The probe's ground strap violates coaxial transmission and the signal is corrupted by RF. *Maintain coaxial connections in the noise signal monitoring path.*

Note 1: To be discussed. Read on.

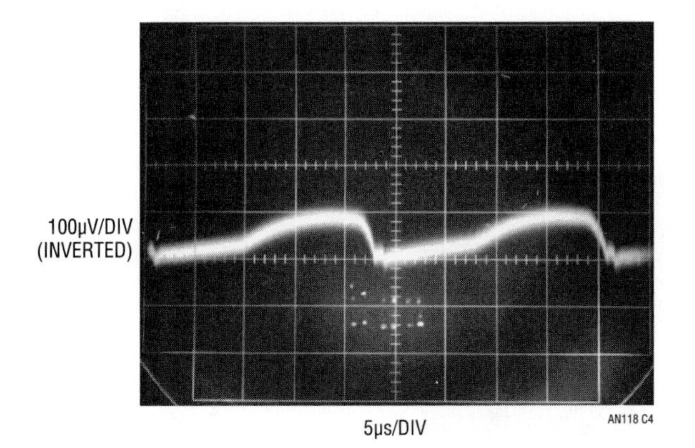

Figure C4 • Apparent Excessive Ripple Results from Figure C3s Probe Misuse. Ground Loop on Board Introduces Serious Measurement Error

Violating coaxial signal transmission—misdemeanor case

Figure C7's probe connection also violates coaxial signal flow, but to a less offensive extent. The probe's ground strap is eliminated, replaced by a tip grounding attachment. Figure C8 shows better results over the preceding case, although signal corruption is still evident. *Maintain coaxial connections in the noise signal monitoring path.*

Proper coaxial connection path

In Figure C9, a coaxial cable transmits the noise signal to the amplifier-oscilloscope combination. In theory, this affords the highest integrity cable signal transmission. Figure C10's trace shows this to be true. The former examples aberrations and excessive noise have disappeared. The switching residuals are now faintly outlined in the amplifier noise floor. *Maintain coaxial connections in the noise signal monitoring path.*

Direct connection path

A good way to verify there are no cable-based errors is to eliminate the cable. Figure C11's approach eliminates all cable between breadboard, amplifier and oscilloscope. Figure C12's presentation is indistinguishable from Figure C10, indicating no cable-introduced infidelity. *When results seem optimal, design an experiment to test them. When results seem poor design an experiment to test them. When results are as expected, design an experiment to test them. When results are unexpected, design an experiment to test them.*

Test lead connections

In theory, attaching a voltmeter lead to the regulator's output should not introduce noise. Figure C13's increased noise reading contradicts the theory. The regulator's output impedance, albeit low, is not zero, especially as frequency scales up. The RF noise injected by the test lead works against the finite output impedance, producing the 200µV of noise indicated in the figure. If a voltmeter lead must be connected to the output during testing, it should be done through a 10kΩ-10µF filter. Such a network eliminates Figure C13's problem while introducing minimal error in the monitoring DVM. *Minimize the number of test lead connections to the circuit while checking noise. Prevent test leads from injecting RF into the test circuit.*

Figure C5 • Floating Trigger Probe Eliminates Ground Loop, but Output Probe Ground Lead (Photo Upper Right) Violates Coaxial Signal Transmission

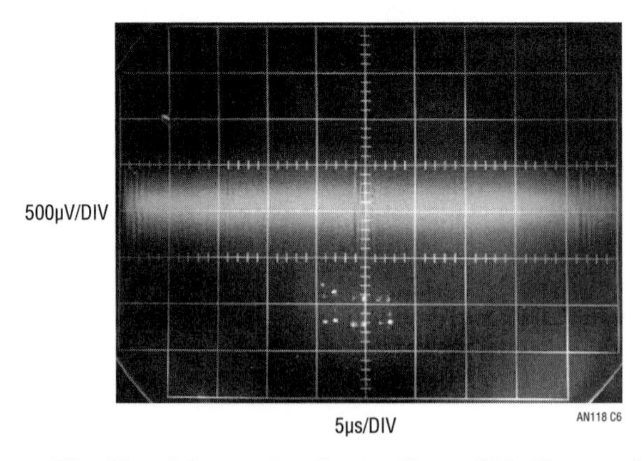

Figure C6 • Signal Corruption Due to Figure C5's Noncoaxial Probe Connection

Figure C7 • Probe with Tip Grounding Attachment Approximates Coaxial Connection

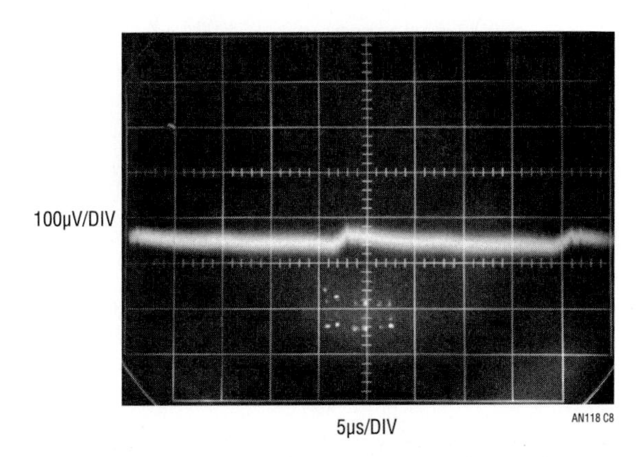

Figure C8 • Probe with Tip Grounding Attachment Improves Results. Some Corruption Is Still Evident

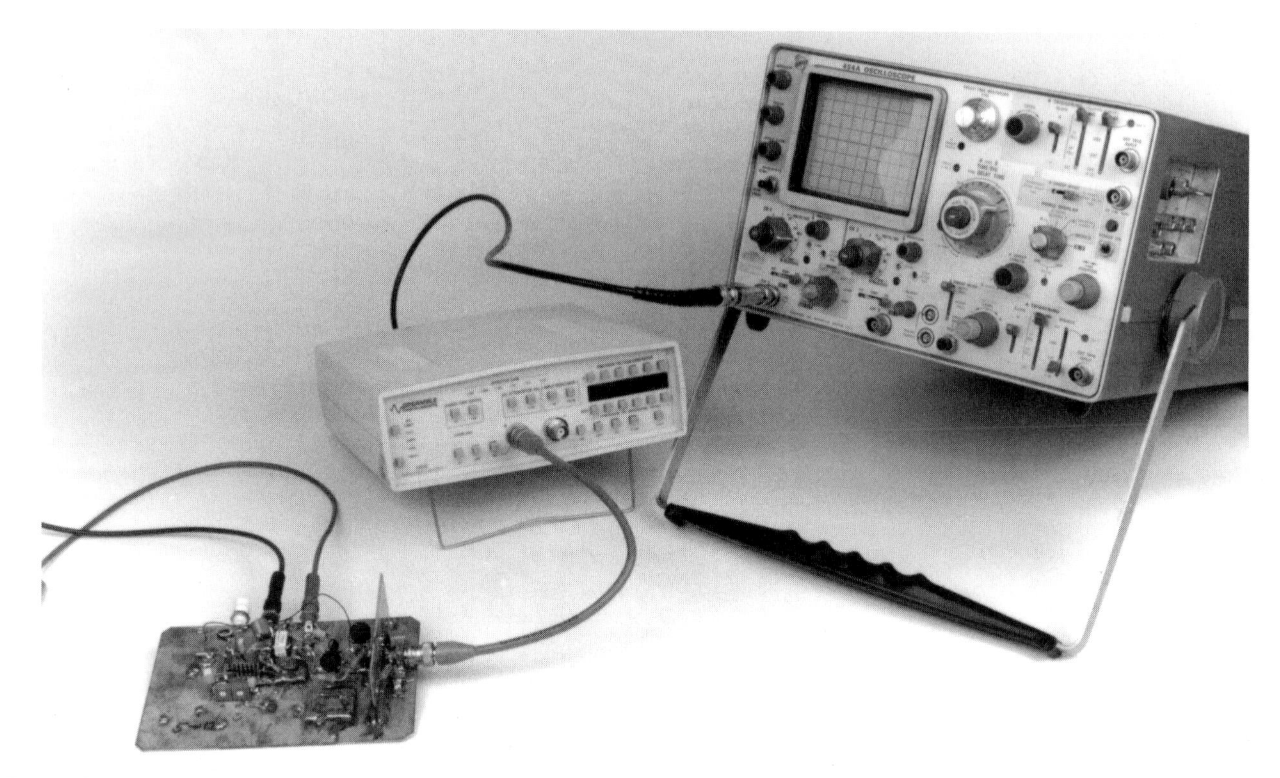

Figure C9 • Coaxial Connection Theoretically Affords Highest Fidelity Signal Transmission

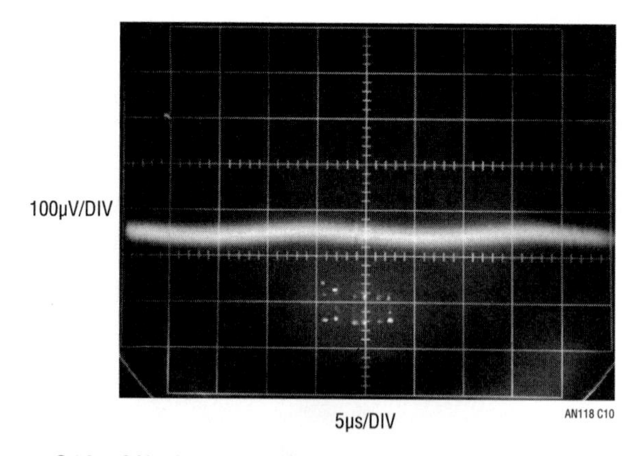

Figure C10 • Life Agrees with Theory. Coaxial Signal Transmission Maintains Signal Integrity. Switching Residuals Are Faintly Outlined in Amplifier Noise

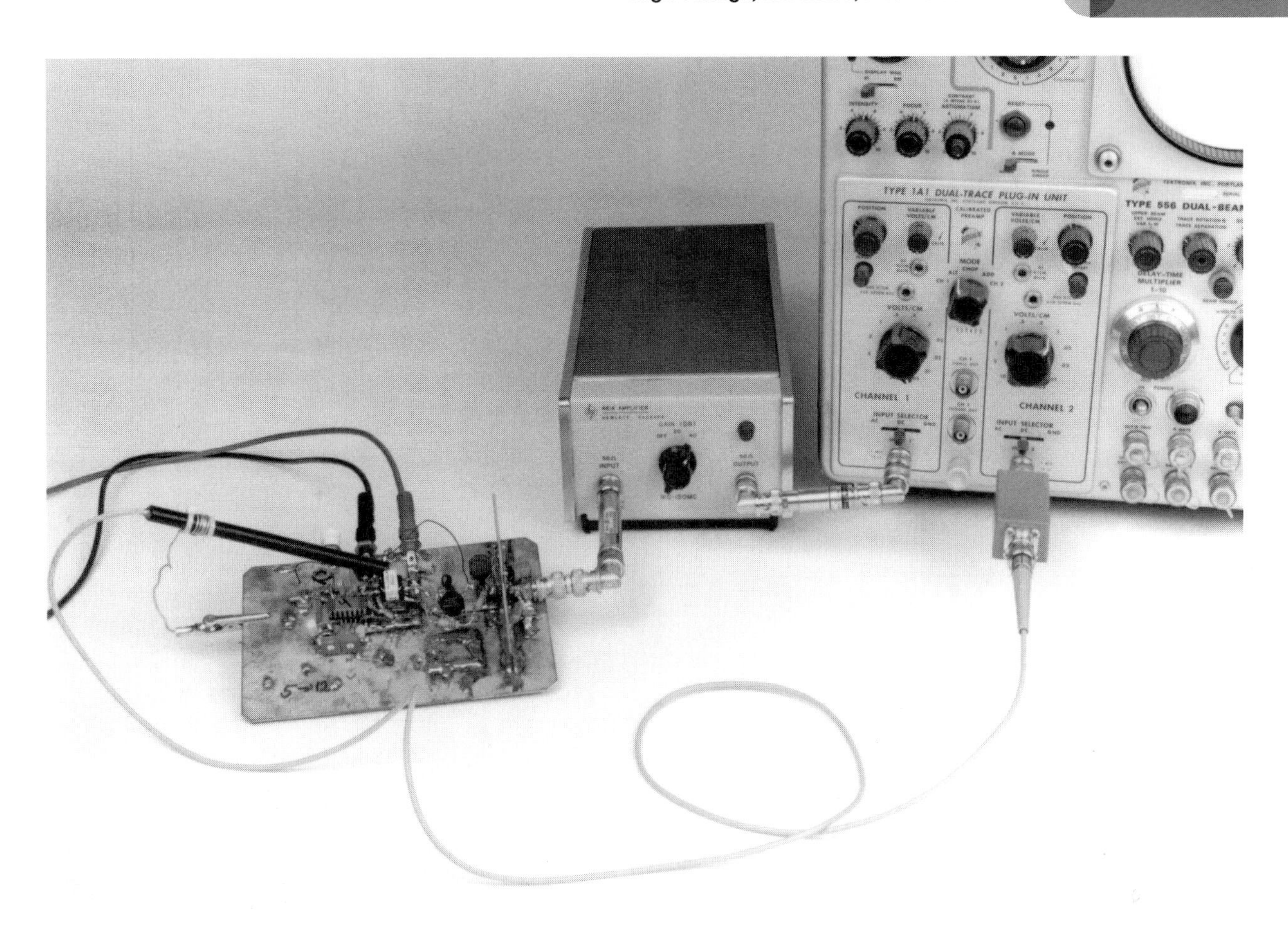

Figure C11 • Direct Connection to Equipment Eliminates Possible Cable-Termination Parasitics, Providing Best Possible Signal Transmission

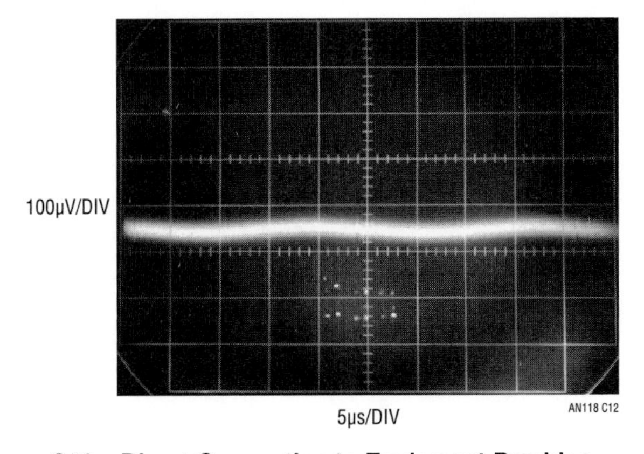

Figure C12 • Direct Connection to Equipment Provides Identical Results to Cable-Termination Approach. Cable and Termination Are Therefore Acceptable

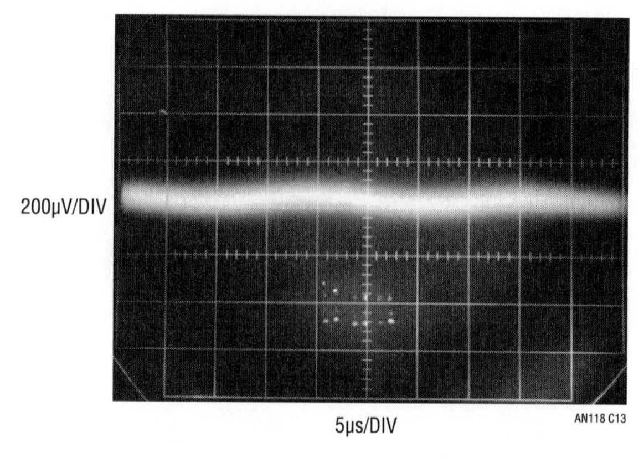

Figure C13 • Voltmeter Lead Attached to Regulator Output Introduces RF Pickup, Multiplying Apparent Noise Floor

Isolated trigger probe

The text associated with Figure C5 somewhat cryptically alluded to an "isolated trigger probe." Figure C14 reveals this to be simply an RF choke terminated against ringing. The choke picks up residual radiated field, generating an isolated trigger signal. This arrangement furnishes a 'scope trigger signal with essentially no measurement corruption. The probe's physical form appears in Figure C15. For good results the termination should be adjusted for minimum ringing while preserving the highest possible amplitude output. Light compensatory damping produces Figure C16's output, which will cause poor 'scope triggering. Proper adjustment results in a more favorable output (Figure C17), characterized by minimal ringing and well-defined edges.

Trigger probe amplifier

The field around the switching magnetics is small and may not be adequate to reliably trigger some oscilloscopes. In such cases, Figure C18's trigger probe amplifier is useful. It uses an adaptive triggering scheme to compensate for variations in probe output amplitude. A stable 5V trigger output is maintained over a 50:1 probe output range. A1, operating at a gain of 100, provides wideband AC gain. The output of this stage biases a 2-way peak detector (Q1 through Q4). The maximum peak is stored in Q2's emitter capacitor, while the minimum excursion is retained in Q4's emitter capacitor. The DC value of the midpoint of A1's output signal appears at the junction of the 500pF capacitor and the 3MΩ units. This point always sits midway between the signal's excursions, regardless of absolute amplitude. This signal-adaptive voltage is buffered by A2 to set the trigger voltage at the LT1394's positive input. The LT1394's negative input is biased directly from A1's output. The LT1394's output, the circuit's trigger output, is unaffected by >50:1 signal amplitude variations. An X100 analog output is available at A1.

Figure C19 shows the circuit's digital output (trace B) responding to the amplified probe signal at A1 (trace A).

Figure C20 is a typical noise testing setup. It includes the breadboard, trigger probe, amplifier, oscilloscope and coaxial components.

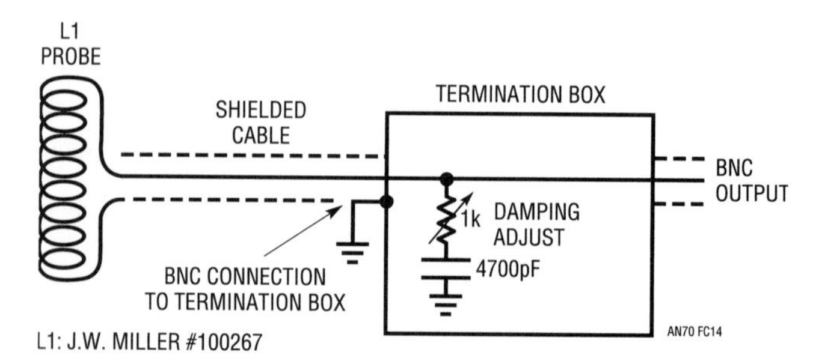

Figure C14 • Simple Trigger Probe Eliminates Board Level Ground Loops. Termination Box Components Damp L1's Ringing Response

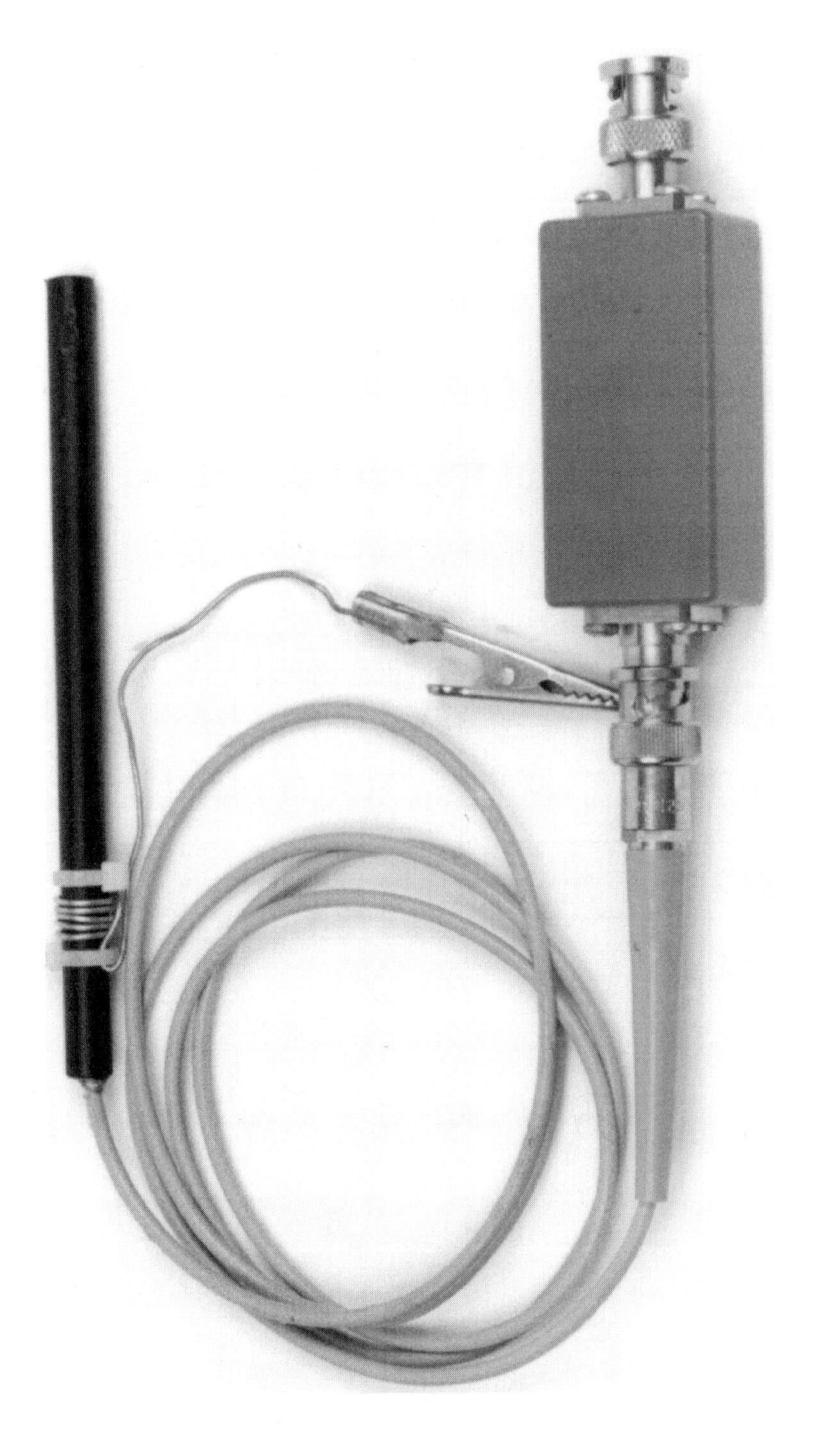

Figure C15 • The Trigger Probe and Termination Box. Clip Lead Facilitates Mounting Probe, Is Electrically Neutral

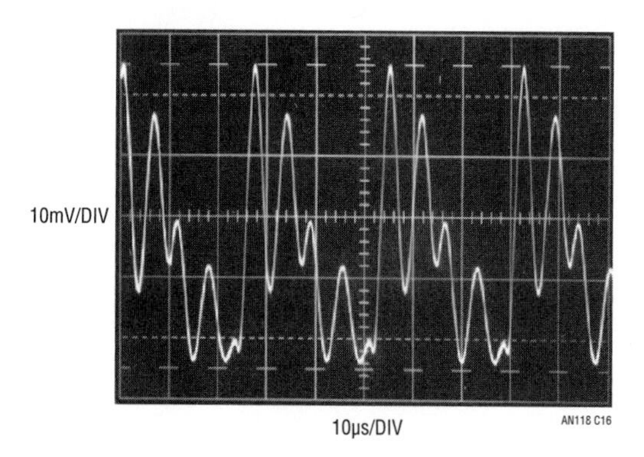

Figure C16 • Misadjusted Termination Causes Inadequate Damping. Unstable Oscilloscope Triggering May Result

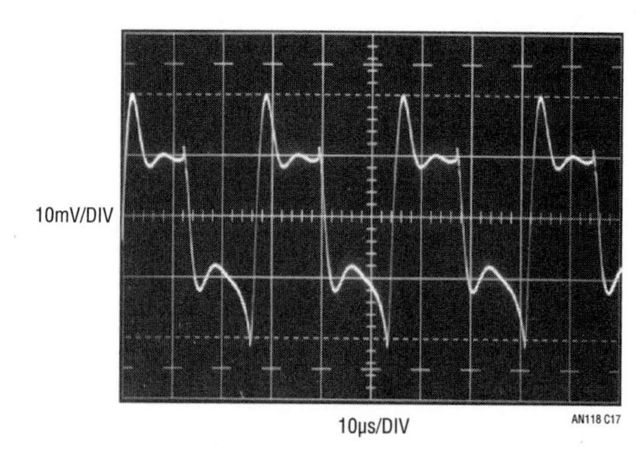

Figure C17 • Properly Adjusted Termination Minimizes Ringing with Small Amplitude Penalty

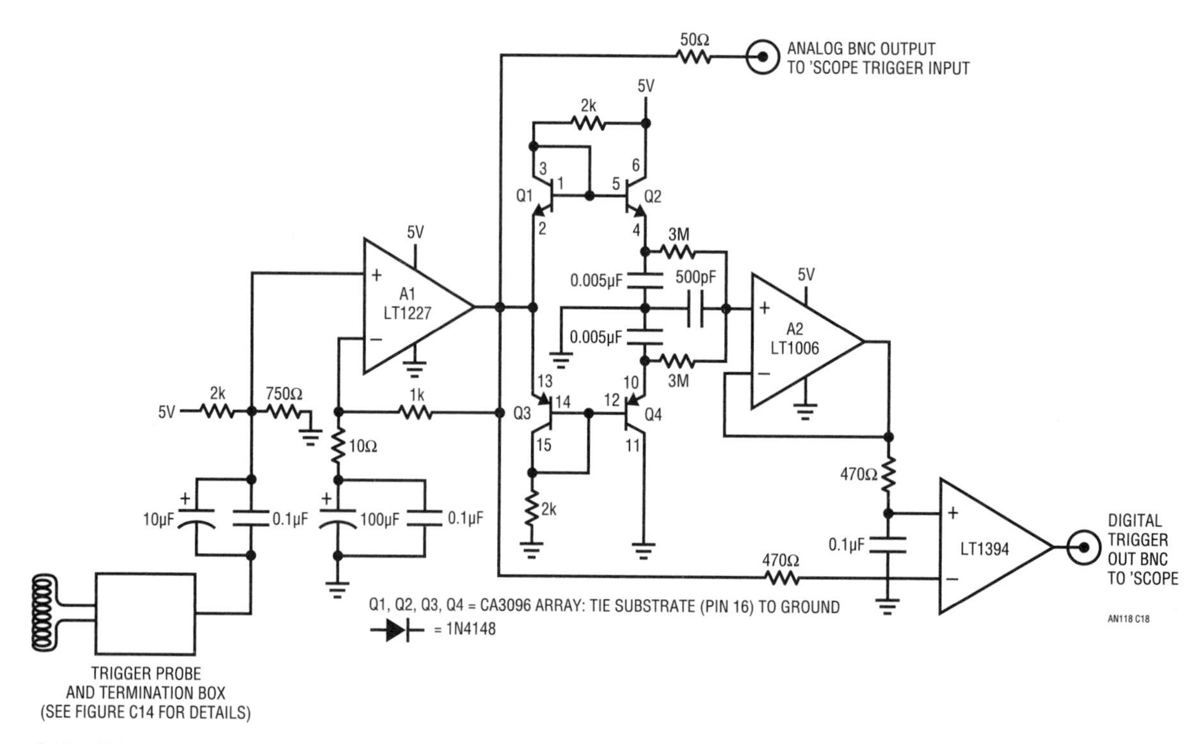

Figure C18 • Trigger Probe Amplifier Has Analog and Digital Outputs. Adaptive Threshold Maintains Digital Output Over 50:1 Probe Signal Variations

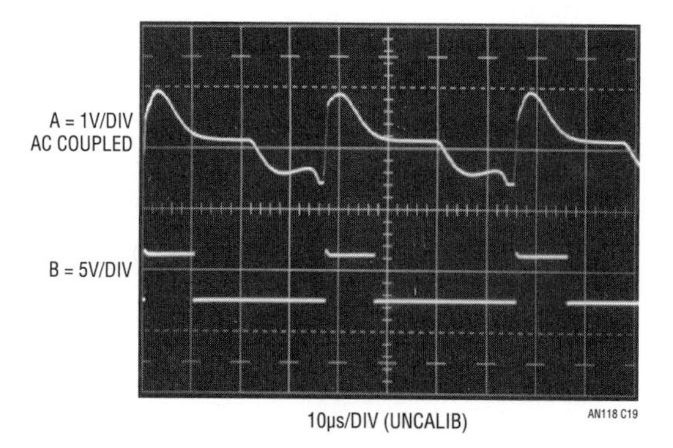

Figure C19 • Trigger Probe Amplifier Analog (Trace A) and Digital (Trace B) Outputs

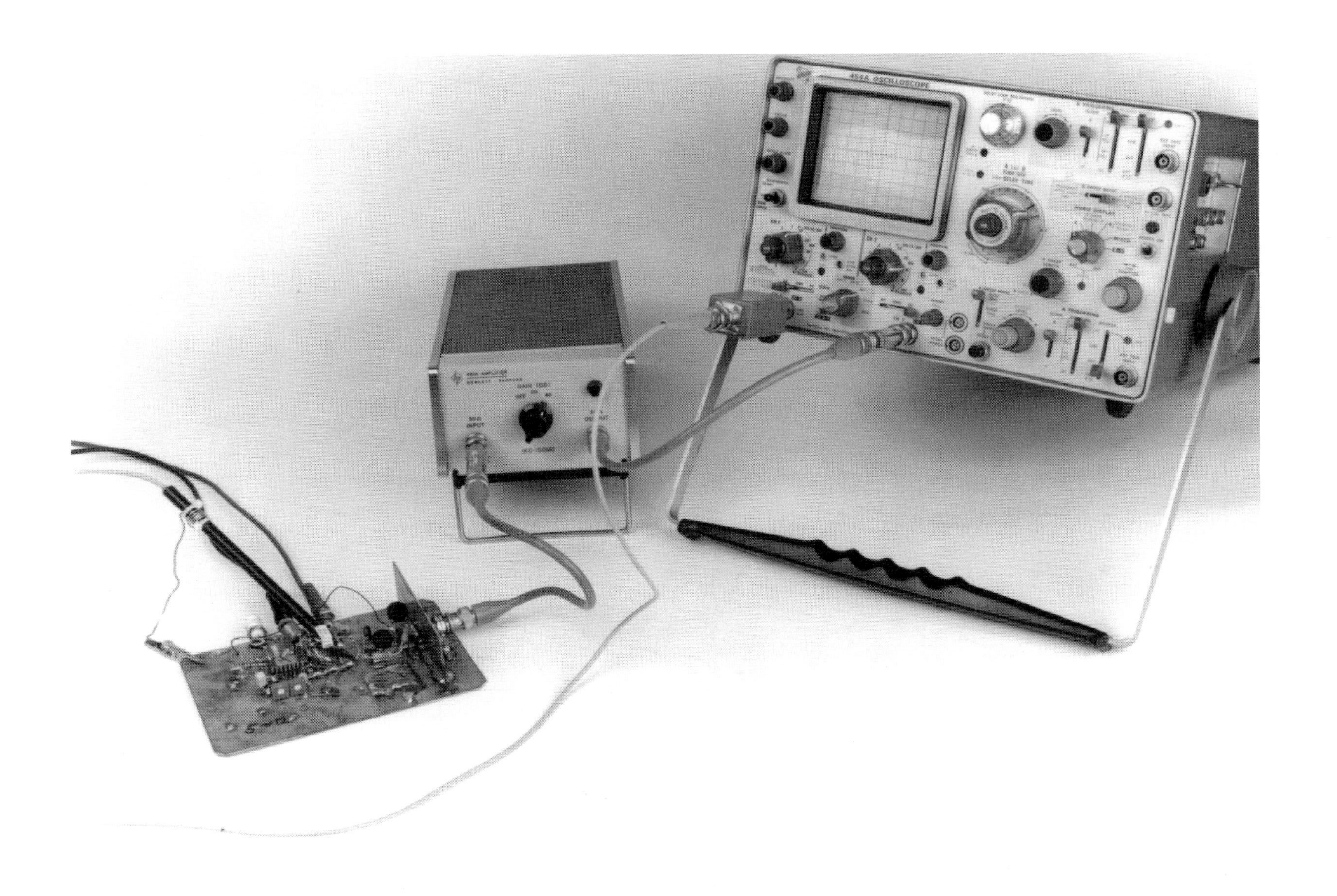

Figure C20 • Typical Noise Test Setup Includes Trigger Probe, Amplifier, Oscilloscope and Coaxial Components

Appendix D Breadboarding, noise minimization and layout considerations

LT1533-based circuit's low harmonic content allows their noise performance to be less layout sensitive than other switching regulators. However, some degree of prudence is in order. As in all things, cavalierness is a direct route to disappointment. Obtaining the absolute lowest noise figure requires care, but performance below 500μV is readily achieved. In general, lowest noise is obtained by preventing mixing of ground currents in the return path. Indiscriminate disposition of ground currents into a bus or ground plane will cause such mixing, raising observed output noise. The LT1533's restricted edge rates mitigate against corrupted ground path-induced problems, but best noise performance occurs in a "single-point" ground scheme. Single-point return schemes may be impractical in production PC boards. In such cases, provide the lowest possible impedance path to the power entry point from the inductor associated with the LT1533's power ground pin. (Pin 16). Locate the output component ground returns as close to the circuit load point as possible. Minimize return current mixing between input and output sections by restricting such mixing to the smallest possible common conductive area.

Noise minimization

The LT1533's controlled switching times allow extraordinarily low noise DC/DC conversion with surprisingly little design effort. Wideband output noise well below 500μV is easily achieved. In most situations this level of performance is entirely adequate. Applications requiring the lowest possible output noise will benefit from special attention to several areas.

Noise tweaking

The slew time versus efficiency trace-off should be weighted towards lowest noise to the extent tolerable. Typically, slew times beyond 1.3μs result in "expensive" noise reduction in terms of lost efficiency, but the benefit is available. The issue is how much power is expendable to obtain incremental decreases in output noise. Similarly, the layout techniques previously discussed should be reviewed. Rigid adherence to these guidelines will result in correspondingly lower noise performance. The text's breadboards were originally constructed to provide the lowest possible noise levels, and then systematically degraded to test layout sensitivity. This approach allows experimentation to determine the best layout without expanding fanatical attention to details that provide essentially no benefit.

The slow edge times greatly minimize radiated EMI, but experimentation with the component's physical orientation can sometimes improve things. Look at the components (yes, literally!) and try and imagine just what their residual radiated field impinges on. In particular, the optional output inductor may pick up field radiated by other magnetics, resulting in increased output noise. Appropriate physical layout will eliminate this effect, and experimentation is useful. The EMI probe described in Appendix E is a useful tool in this pursuit and highly recommended.

Capacitors

The filter capacitors used should have low parasitic impedance. Sanyo OS-CON types are excellent in this regard and contributed to the performance levels quoted in the text. Tantalum types are nearly as good. The input supply bypass capacitor, which should be located directly at the transformer center tap, needs similarly good characteristics. Aluminum electrolytics are not suitable for any service in LT1533 circuits.

Damper network

Some circuits may benefit from a small (e.g., 300Ω-1000pF) damper network across the transformer secondary if the absolutely lowest noise is needed. Extremely small (20μV to 30μV) excursions can briefly appear during the switching interval when no energy is coming through the transformer. These events are so minuscule that they are barely measurable in the noise floor, but the damper will eliminate them.

Measurement technique

Strictly speaking, measurement technique is not a way to obtain lowest noise performance. Realistically, it is essential that measurement technique be trustworthy. Uncountable hours have been lost chasing "circuit problems" that in reality are manifestations of poor measurement technique. Please read Appendices B and C before pursuing solutions to circuit noise that isn't really there.[10]

Appendix E
Application note E101: EMI "sniffer" probe

Bruce Carsten Associates, Inc.
6410 NW Sisters Place, Corvallis, Oregon 97330
541-745-3935

The EMI Sniffer Probe[1E] is used with an oscilloscope to locate and identify magnetic field sources of electromagnetic interference (EMI) in electronic equipment. The probe consists of a miniature 10 turn pickup coil located in the end of a small shielded tube, with a BNC connector provided for connection to a coaxial cable (Figure E1). The Sniffer Probe output voltage is essentially proportional to the rate of change of the ambient magnetic field, and thus to the rate of change of nearby currents.

The principal advantages of the Sniffer Probe over simple pickup loops are:

1. Spatial resolution of about a millimeter.
2. Relatively high sensitivity for a small coil.
3. A 50Ω source termination to minimize cable reflections with unterminated scope inputs.
4. Faraday shielding to minimize sensitivity to electric fields.

The EMI Sniffer Probe was developed to diagnose sources of EMI in switch mode power converters, but it can also be used in high speed logic systems and other electronic equipment.

Sources of EMI

Rapidly changing voltages and currents in electrical and electronic equipment can easily result in radiated and conducted noise. Most EMI in switch mode power converters is thus generated during switching transients when power transistors are turned on or off.

Conventional scope probes can readily be used to see dynamic voltages, which are the principal sources of common mode conducted EMI. (High dV/dt can also feed through poorly designed filters as normal mode voltage spikes and may radiate fields from a circuit without a conductive enclosure.)

Dynamic currents produce rapidly changing magnetic fields which radiate far more easily than electric fields as they are more difficult to shield. These changing magnetic fields can also induce low impedance voltage transients in other circuits, resulting in unexpected normal and common mode conducted EMI.

These high dI/dt currents and resultant fields can not be directly sensed by voltage probes, but are readily detected and located with the Sniffer Probe. While current probes can sense currents in discrete conductors and wires, they are of little use with printed circuit traces or in detecting dynamic magnetic fields.

Probe response characteristics

The Sniffer Probe is sensitive to magnetic fields only along the probe axis. This directionality is useful in locating the paths and sources of high dI/dt currents. The resolution is usually sufficient to locate which trace on a printed circuit board, or which lead on a component package, is conducting the EMI generating current.

For "isolated" single conductors or PC traces, the Probe response is greatest just to either side of the conductor where the magnetic flux is along with probe axis. (Probe response may be a little greater with the axis tilted

Note 1: I do not wax pedantic here. My guilt in this offense runs deep.

Note 1E: The EMI Sniffer Probe is available from Bruce Carsten Associates at the address noted in the title of this appendix.

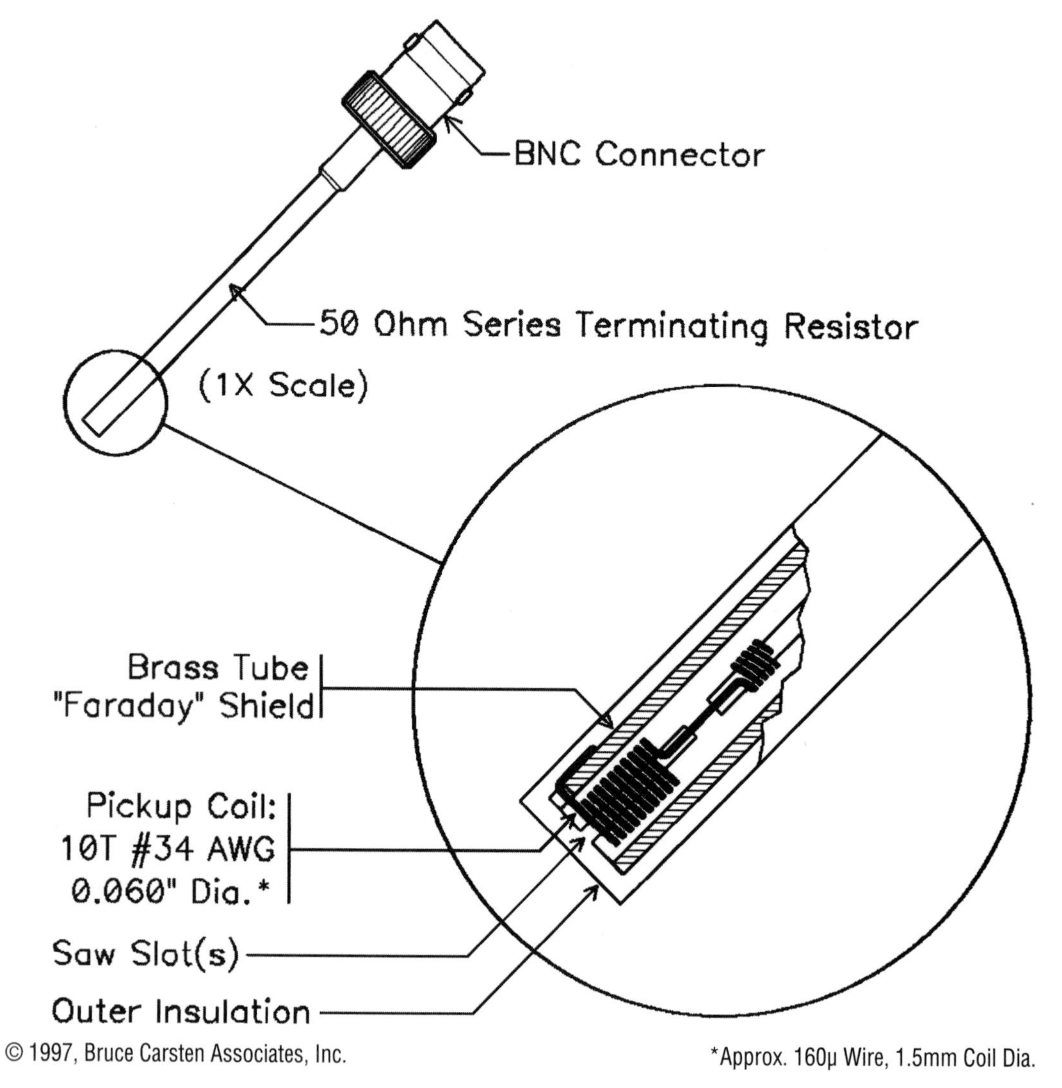

Figure E1 • Construction of the EMI "Sniffer Probe" for Locating and Identifying Magnetic Field Sources of EMI

towards the center of the conductor.) As shown in Figure E2, there is a sharp response null in the middle of the conductor, with a 180° phase shift to either side and a decreasing response with distance. The response will increase on the inside of a bend where the flux lines are crowded together, and is reduced on the outside of a bend where the flux lines spread apart.

When the return current is in an adjacent parallel conductor, the Probe response is greatest between the two conductors as shown in Figure E3. There will be a sharp null and phase shift over each conductor, with a lower peak response outside the conductor pair, again decreasing with distance.

The response to a trace with a return current on the opposite side of the board is similar to that of a single isolated trace, except that the probe response may be greater with the Probe axis tilted away from the trace. A "ground plane" below a trace will have a similar effect, as there will be a counter-flowing "image" current in the ground plane.

The Probe frequency response to a uniform magnetic field is shown in Figure E4. Due to large variations in field

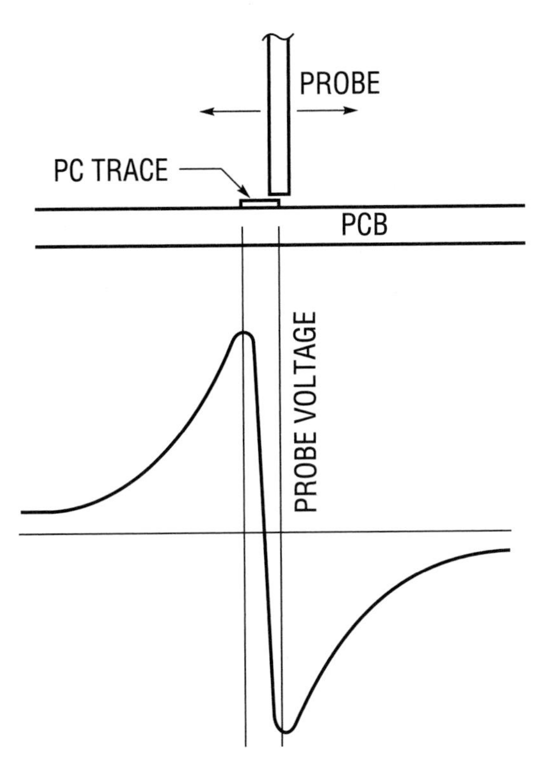

Figure E2 • Sniffer Probe Response to Current in a Physically "Isolated" Conductor

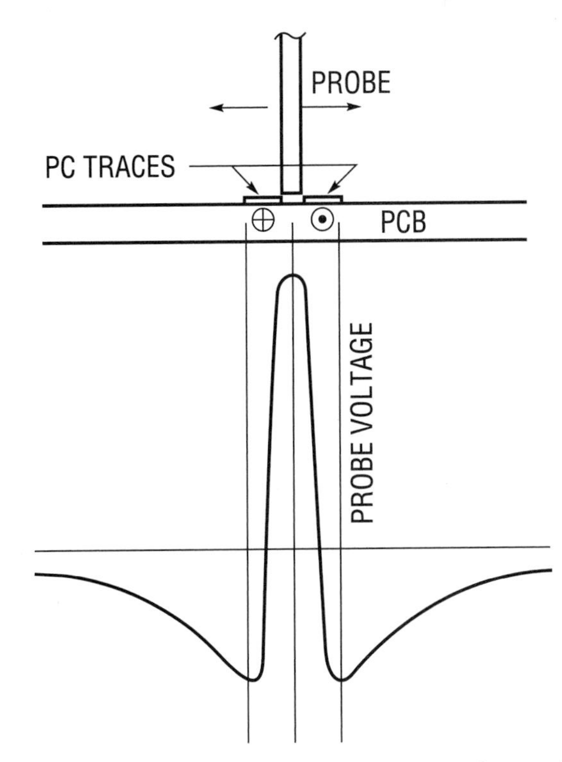

Figure E3 • Sniffer Probe Response with Return Current in a Parallel Conductor

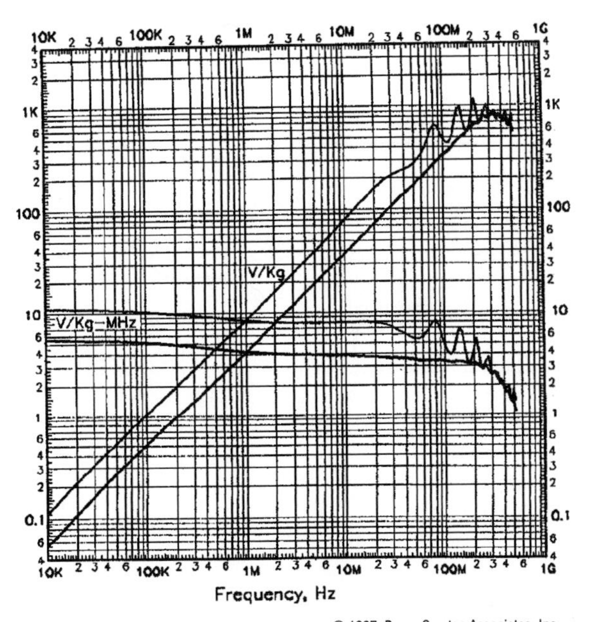

Figure E4 • Typical EMI "Sniffer" Probe Frequency Response Measured with 1.3m (51") of 50Ω Coax to Scope Upper Traces: 1Meg Scope Input Impedance Lower Traces: 50Ω Scope Input Impedance

strength around a conductor, the Probe should be considered as a qualitative indicator only, with no attempt made to "calibrate" it. The response fall-off near 300MHz is due to the pickup coil inductance driving the coax cable impedance, and the mild resonant peaks (with a 1MΩ scope termination) at multiples of 80MHz are due to transmission line reflections.

Principles of probe use

The Sniffer Probe is used with at least a 2-channel scope. One channel is used to view the noise whose source is to be located (which may also provide the scope trigger) and the other channel is used for the Sniffer Probe. The probe response nulls make it inadvisable to use this scope channel for triggering.

A third scope trigger channel can be very useful, particularly if it is difficult to trigger on the noise. Transistor drive waveforms (or their predecessors in the upstream logic) are ideal for triggering; they are usually stable, and allow immediate precursors of the noise to be viewed.

Start with the Probe at some distance from the circuit with the Probe channel at maximum sensitivity. Move the probe around the circuit, looking for "something happening" in the circuit's magnetic fields at the same time as the noise problem. A precise "time domain" correlation between EMI noise transients and internal circuit fields is fundamental to the diagnostic approach.

As a candidate noise source is located, the Probe is moved closer while the scope sensitivity is decreased to keep the probe waveform on-screen. It should be possible to quickly bring the probe down to the PC board trace (or wiring) where the probe signal seems to be a maximum. This may not be near the point of EMI generation, but it should be near a PC trace or other conductor carrying the current from the EMI source. This can be verified by moving the probe back and forth in several directions; when the appropriate PC trace is crossed at roughly right angles, the probe output will go through a sharp null over the trace, with an evident phase reversal in probe voltage on each side of the trace (as noted above).

This EMI "hot" trace can be followed (like a bloodhound on the scent trail) to find all or much of the EMI generating current loop. If the trace is hidden on the back side (or inside) of the board, mark its path with a felt pen and locate the trace on disassembly, on another board

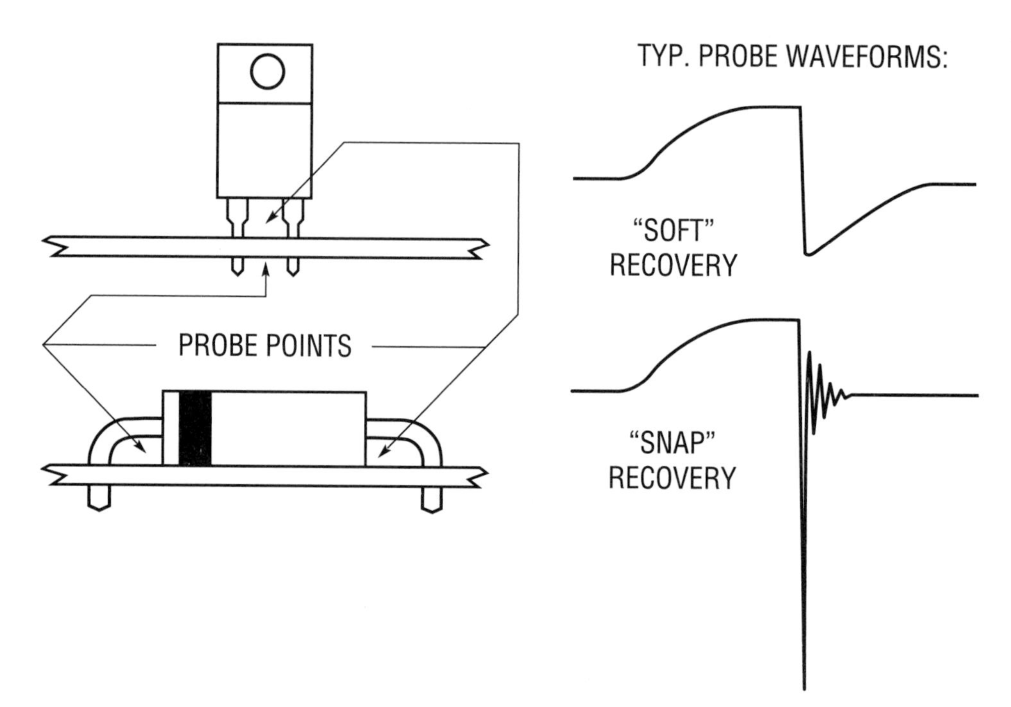

Figure E5 • Rectifier Reverse Recovery Typical Fix: Tightly Coupled R-C Snubber

or on the artwork. From the current path and the timing of the noise transient, the source of the problem usually becomes almost self-evident.

Several not-uncommon problems (all of which have been diagnosed with various versions of the Sniffer Probe) are discussed here with suggested solutions or fixes.

Typical dI/dt EMI problems

Rectifier reverse recovery

Reverse recovery of rectifiers is the most common source of dI/dt-related EMI in power converters; the charge stored in P-N junction diodes during conduction causes a momentary reverse current flow when the voltage reverses. This reverse current may stop very quickly (<1ns) in diodes with a "snap" recovery (more likely in devices with a PIV rating of less than 200V), or the reverse current may decay more gradually with a "soft" recovery. Typical Sniffer Probe waveforms for each type of recovery are shown in Figure E5.

The sudden change in current creates a rapidly changing magnetic field, which will both radiate external fields and induce low impedance voltage spikes in other circuits. This reverse recovery may "shock" parasitic L-C circuits into ringing, which will result in oscillatory waveforms with varying degrees of damping when the diode recovers. A series R-C damper circuit in parallel with the diode is the usual solution.

Output rectifiers generally carry the highest currents and are thus the most prone to this problem, but this is often recognized and they may be well-snubbed. It is not uncommon for unsnubbed catch or clamp diodes to be *more* of an EMI problem. (The fact that a diode in an R-C-D snubber may need its *own* R-C snubber is not always self-evident, for example).

The problem can usually be identified by placing the Sniffer Probe near a rectifier lead. The signal will be strongest on the inside of a lead bend in an axial package, or between the anode and cathode leads in a TO-220, TO-247 or similar type of package, as shown in Figure E5.

Using "softer" recovery diodes is a possible solution and Schottky diodes are ideal in low voltage applications. However, it must be recognized that a P-N diode with soft recovery is also inherently lossy (while a "snap" recovery is not), as the diode simultaneously develops a reverse voltage while still conducting current: The fastest possible diode (lowest recovered charge) with a moderately soft recovery is usually the best choice. Sometimes a faster, slightly "snappy" diode with a tightly coupled R-C snubber works as well or better than a soft but excessively slow recovery diode.

If significant ringing occurs, a "quick-and-dirty" R-C snubber design approach works fairly well: increasingly large damper capacitors are placed across the diode until the ringing frequency is halved. We know that the total ringing capacity is now quadrupled or that the original ringing capacity is 1/3 of the added capacity. The damper resistance required is about equal to the capacitive reactance of the original ringing capacity at the original ringing frequency.

The "frequency halving" capacity is then connected in series with the damping resistance and placed across the diode, as tightly coupled as possible.

Snubber capacitors must have a high pulse current capability and low dielectric loss. Temperature stable (disc or multilayer) ceramic, silvered mica and some plastic film-foil capacitors are suitable. Snubber resistors should be noninductive; metal film, carbon film and carbon composition resistors are good, but wirewound resistors must be avoided. The maximum snubber resistor dissipation can be estimated from the product of the damper capacity, switching frequency and the square of the peak snubber capacitor voltage.

Snubbers on passive switches (diodes) or active switches (transistors) should always be coupled as closely as physically possible, with minimal loop inductance. This minimizes the radiated field from the change in current path from the switch to the snubber. It also minimizes the turn-off voltage overshoot "required" to force the current to change path through the switch-snubber loop inductance.

Ringing in clamp Zeners

A capacitor-to-capacitor ringing problem can occur when a voltage clamping Zener or TransZorb® is placed across the output of a converter for overvoltage protection (OVP). Power Zeners have a large junction capacity, and this can ring in series with the lead ESL and the output capacitors, with some of the ringing voltage showing up on the output. This ringing current can be most easily detected near the Zener leads, particularly on the inside of a bend as shown in Figure E6.

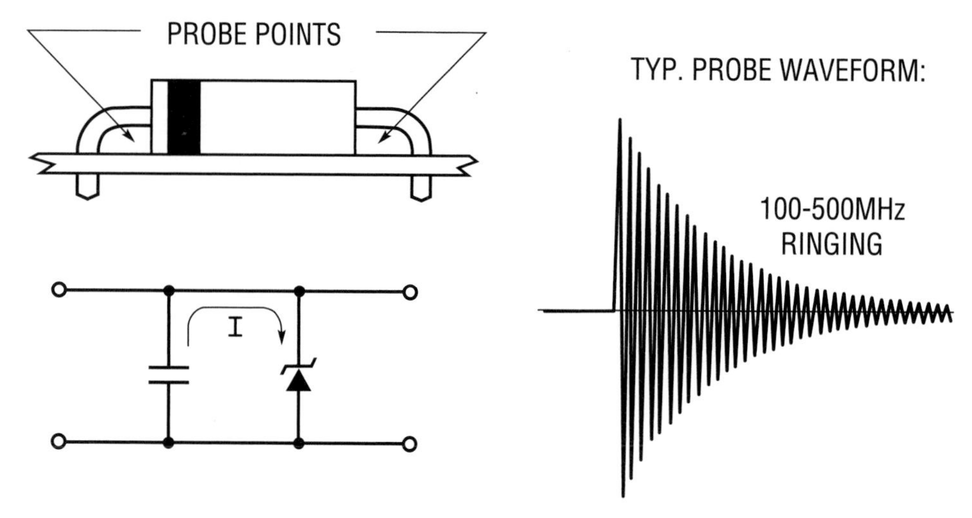

Figure E6 • Ringing Between Clamp Zener and Capacitor Typical Fix: Small Ferrite Bead on Zener Lead(s)

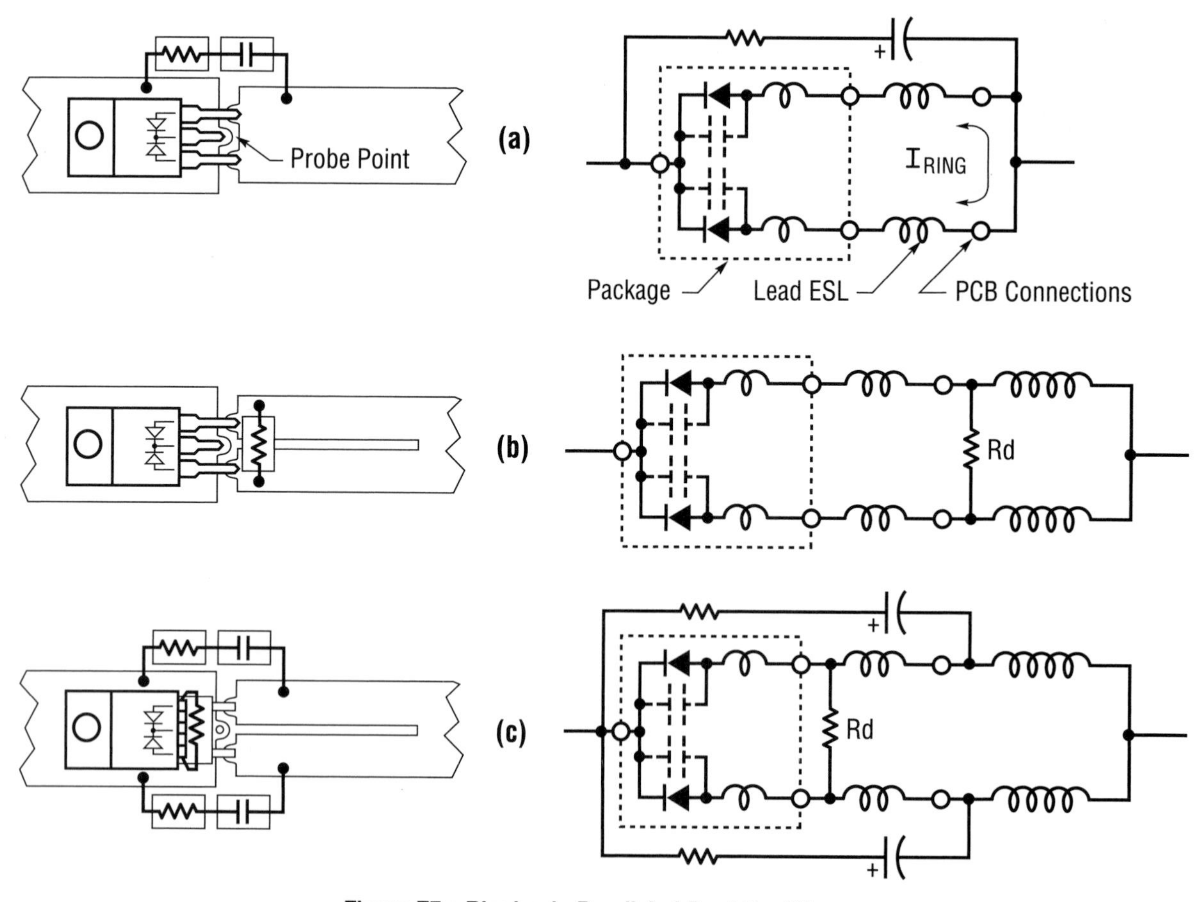

Figure E7 • Ringing in Paralleled Dual Rectifiers

R-C snubbers have not been found to work well in this case as the ringing loop inductance is often as low or lower than the obtainable parasitic inductance in the snubber. Increasing the external loop inductance to allow damping is not advisable as this would limit dynamic clamping capability. In this case, it was found that a small ferrite bead on one or both of the Zener leads dampened the HF oscillations with minimal adverse side effects (a high permeability ferrite bead quickly saturates as soon as the Zener begins to conduct significant current).

Paralleled rectifiers

A less evident problem can occur when dual rectifier diodes in a package are paralleled for increased current capability, even with a tightly coupled R-C snubber. The two diodes seldom recover at exactly the same time, which can cause a very high frequency oscillation (hundreds of MHz) to occur between the capacities of the two diodes in series with the anode lead inductances, as shown in Figure E7.This effect can really only be observed by placing the probe between the two anode leads, as the ringing current exists almost nowhere else (the ringing is nearly "invisible" to a conventional voltage probe, like many other EMI effects that can be easily found with a magnetic field Sniffer Probe).

This "teeter-totter" oscillation has a voltage "null" about where the R-C snubber is connected, so it provides little or no damping (see Figure E7a). It is actually very difficult to insert a suitable damping resistance into this circuit.

The easiest way to dampen the oscillation is to "slit" the anode PC trace for an inch or so and place a damping resistor at the anode leads as shown in Figure E7b. This increases the inductance in series with the diode-diode loop external to the package and leads, while having minimal effect on the effective series inductance. Even better damping is obtained by placing the resistor across the anode leads at the entry point to the case, as shown in Figure E7c, but this violates the mindset of many production engineers.

It is also preferable to split the original R-C damper into two (2R) – (C/2) dampers, one on each side of the dual rectifier (also shown in Figure E7c). In practice, it is always preferable to use dual R-C dampers, one each side of the diode; loop inductance is cut about in half, and the external dI/dt field is reduced even further due to the oppositely "handed" currents in the two snubber networks.

Paralleled snubber or damper caps

A problem similar to that with the paralleled diodes occurs when two or more low loss capacitors are paralleled and driven with a sudden current change. There is a tendency for a current to ring between the two capacitors in series with their lead inductances (or ESL), as shown in Figure E8a. This type of oscillation can usually be detected by placing the Sniffer Probe between the leads of the paralleled capacitors. The ringing frequency is much lower than with the paralleled diodes (due to the larger capacity), and the effect *may be* benign if the capacitors are sufficiently closer together.

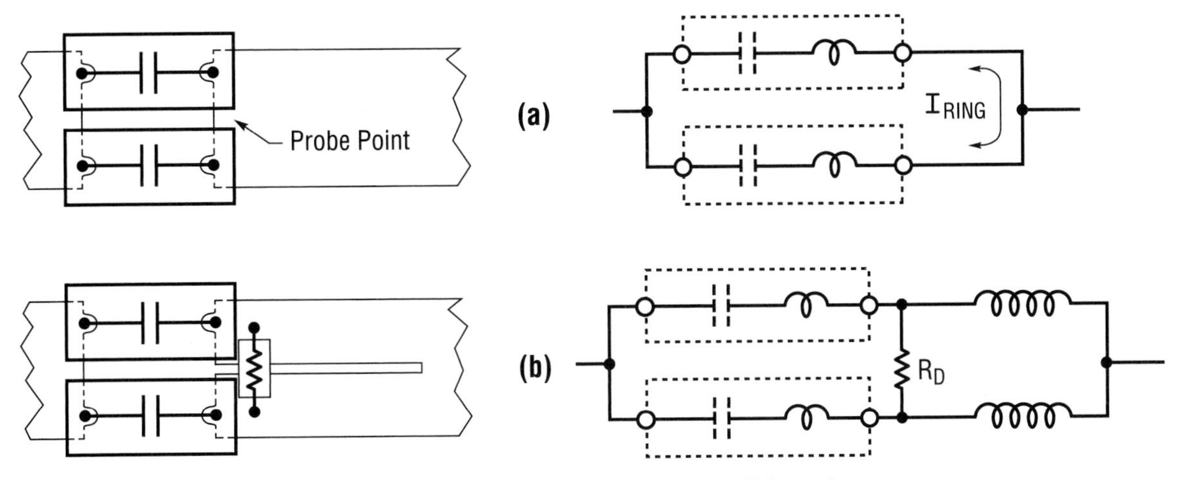

Figure E8 • Ringing in Paralleled "Snubber" Capacitors

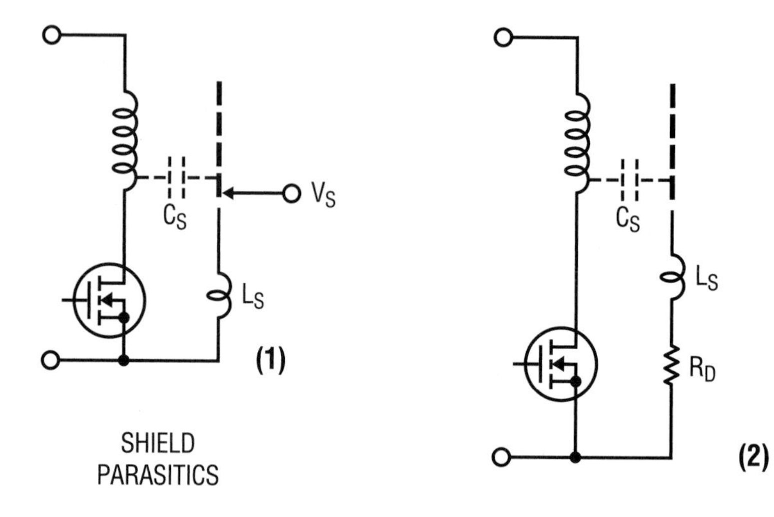

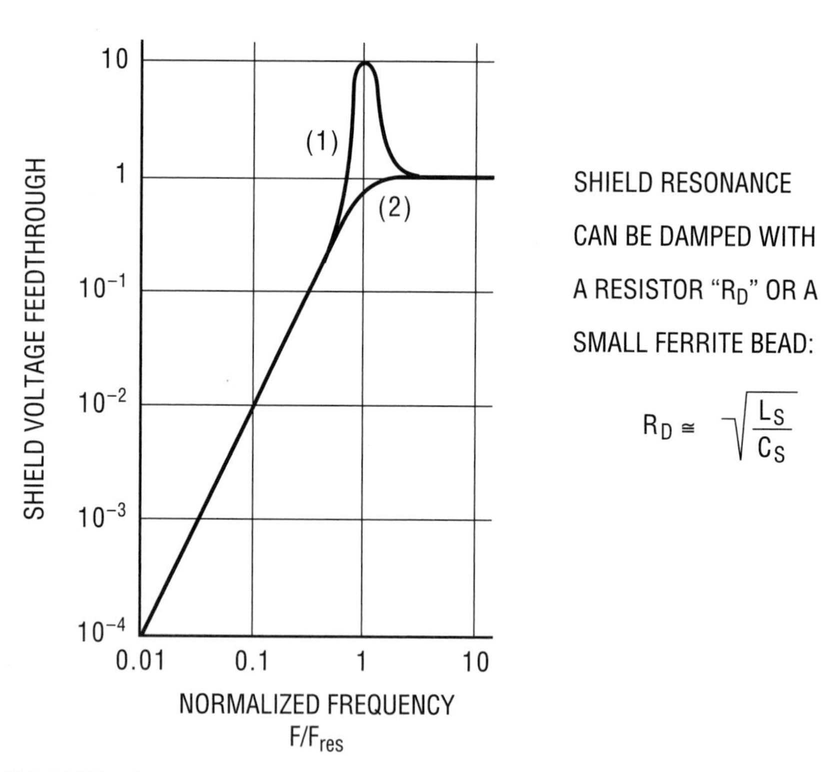

SHIELD RESONANCE CAN BE DAMPED WITH A RESISTOR "R_D" OR A SMALL FERRITE BEAD:

$$R_D \cong \sqrt{\frac{L_S}{C_S}}$$

Figure E9 • Shield Effectiveness at High Frequencies is Limited by Shield Capacity and Lead Inductance

If the resultant ringing *is* picked up externally, it can be damped in a similar way as with the parallel diodes as shown in Figure E8b. In either case, the dissipation in the damping resistor tends to be relatively small.

Ringing in transformer shield leads

The capacity of a transformer shield to other shields or windings (C_S in Figure E9) forms a series resonant circuit with its "drain wire" inductance (L_S) to the bypass point. This resonant circuit is readily excited by typical square wave voltages on windings, and a poorly damped oscillatory current may flow in the drain wire. The shield current may radiate noise into other circuits, and the shield voltage will often show up as common mode conducted noise. The shield voltage is very difficult to detect with a voltage probe in most transformers, but the ringing shield current can be observed by holding the Sniffer Probe near the shield drain wire (Figure E10), or the shield current's return path in the circuit.

This ringing can be dampened by placing a resistor R_D in series with the shield drain wire, whose value is approximately equal to the surge impedance of the resonant circuit, which may be calculated from the formula in Figure E9.

The shield capacitance (C_S) can readily be measured with a bridge (as the capacity from the shield to all facing shields and/or windings), but L_S is usually best calculated from C_S and the ringing frequency (as sensed by the Sniffer Probe). This resistance is typically on the order of tens of ohms.

One or more small ferrite beads can also be placed on the drain wire instead to provide damping. This option may be preferable as a late "fix" when the PC board has already been laid out.

In either case, the damper losses are typically quite small. The damper resistor has a moderately adverse impact on shield effectiveness below the shield and drain wire resonant frequency; damper beads are superior in this respect as their impedance is less at lower frequencies. The drain wire connection should also be as short as possible to the circuit bypass point, both to minimize EMI and to raise the shield's maximum effective (i.e., resonant) frequency.

Leakage inductance fields

Transformer leakage inductance fields emanate from between primary and secondary windings. With a single primary and secondary, a significant dipole field is created, which may be seen by placing the Sniffer Probe near the winding ends as shown in Figure E11a. If this field is generating EMI, there are two principal fixes:

1. Split the Primary *or* Secondary in two, to "sandwich" the other winding, and/or:
2. Place a shorted copper strap "electromagnetic shield" around the *complete-core and winding* assembly as shown in Figure E12. Eddy currents in the shorted strap largely cancel the external magnetic field.

The first approach creates a "quadrupole" instead of a dipole leakage field, which significantly reduces the distant field intensity. It also reduces the eddy current losses in any shorted strap electromagnetic shield used, which may or may not be an important consideration.

External air gap fields

External air gaps in an inductor, such as those in open "bobbin core" inductors or with "E" cores spaced apart (Figure E11b), can be a major source of external magnetic fields when significant ripple or AC currents are present. These fields can also be easily located with the Sniffer Probe; response will be a maximum near an air gap or near the end of an open inductor winding.

"Open" inductor fields are not readily shielded and if they present an EMI problem the inductor must usually be redesigned to reduce external fields. The external field around spaced E cores can be virtually eliminated by placing all of the air gap in the center leg. Fields due to a (possibly intentional) residual or minor outside air gap can be minimized with the shorted strap electromagnetic shield of Figure E12, if eddy current losses prove not to be too high.

A less obvious problem may occur when inductors with "open" cores are used as second stage filter chokes. The minimal ripple current may not *create* a significant field, but such an inductor can "pick up" external magnetic fields and convert them to noise voltages or be an EMI *susceptibility* problem.[2]

Note 2: Ed Note. See Appendix D for additional commentary.

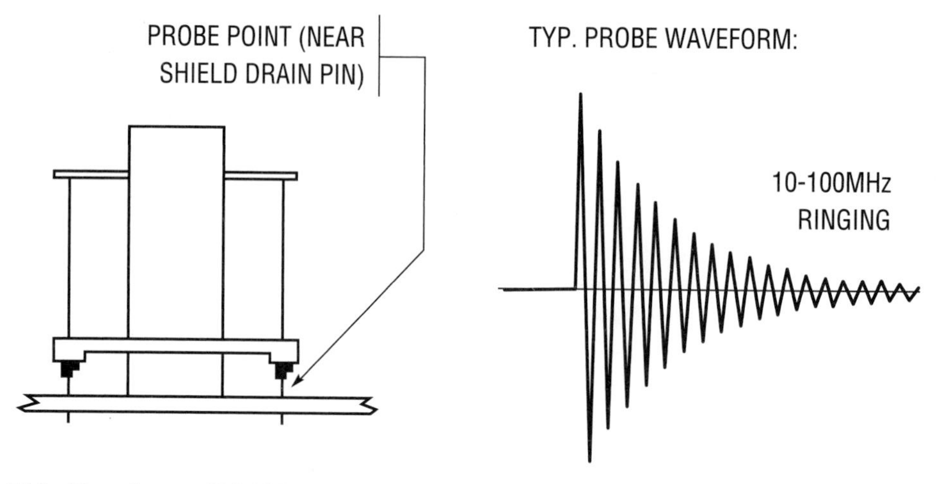

Figure E10 • Transformer Shield Ringing Typical Fix: 10Ω to 100Ω Resistor (or Ferrite Bead in Drain Wire)

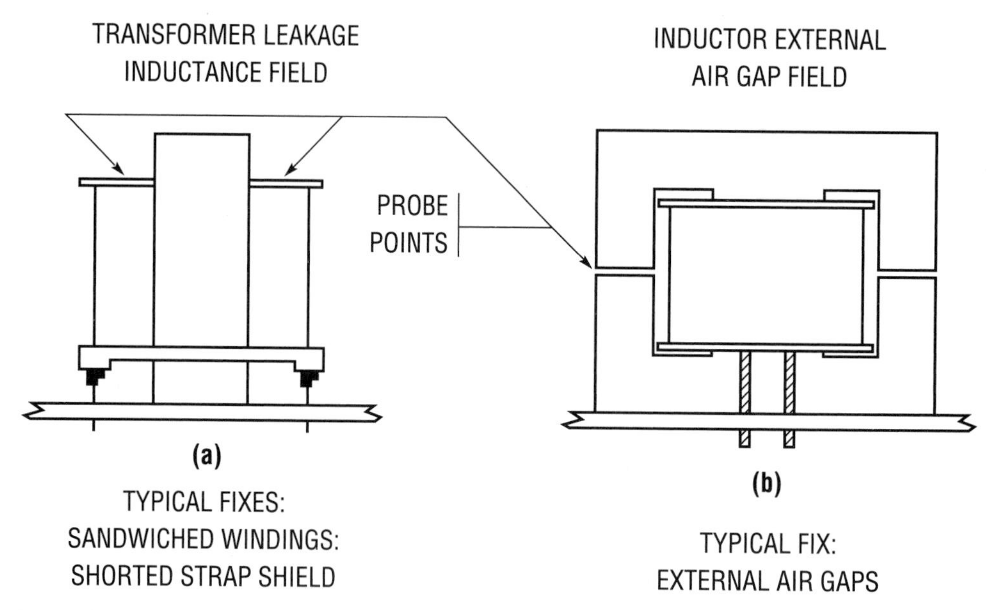

Figure E11 • Probe Voltages Resemble the Transformer and Inductor Winding Waveforms

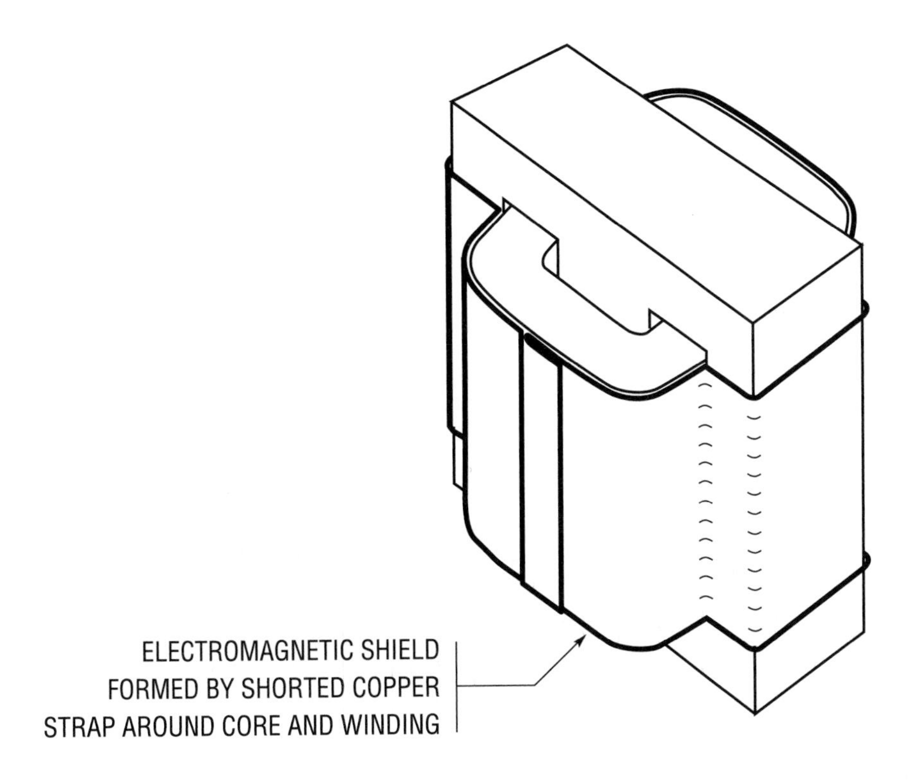

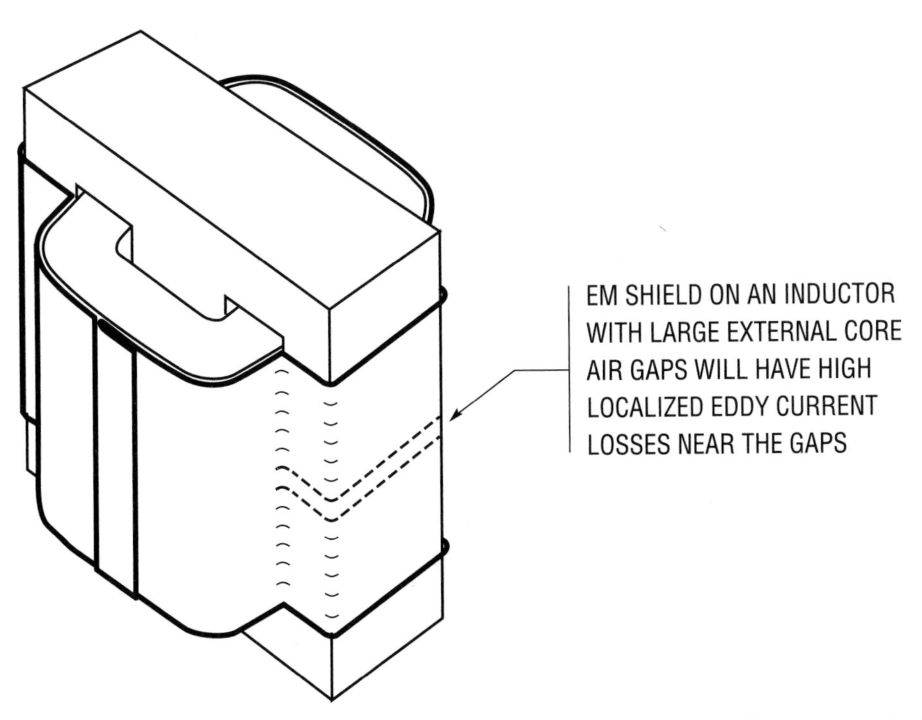

Figure E12 • A "Sandwiched" PRI-SEC Transformer Winding Construction Reduces Electromagnetic Shield Eddy Current Losses

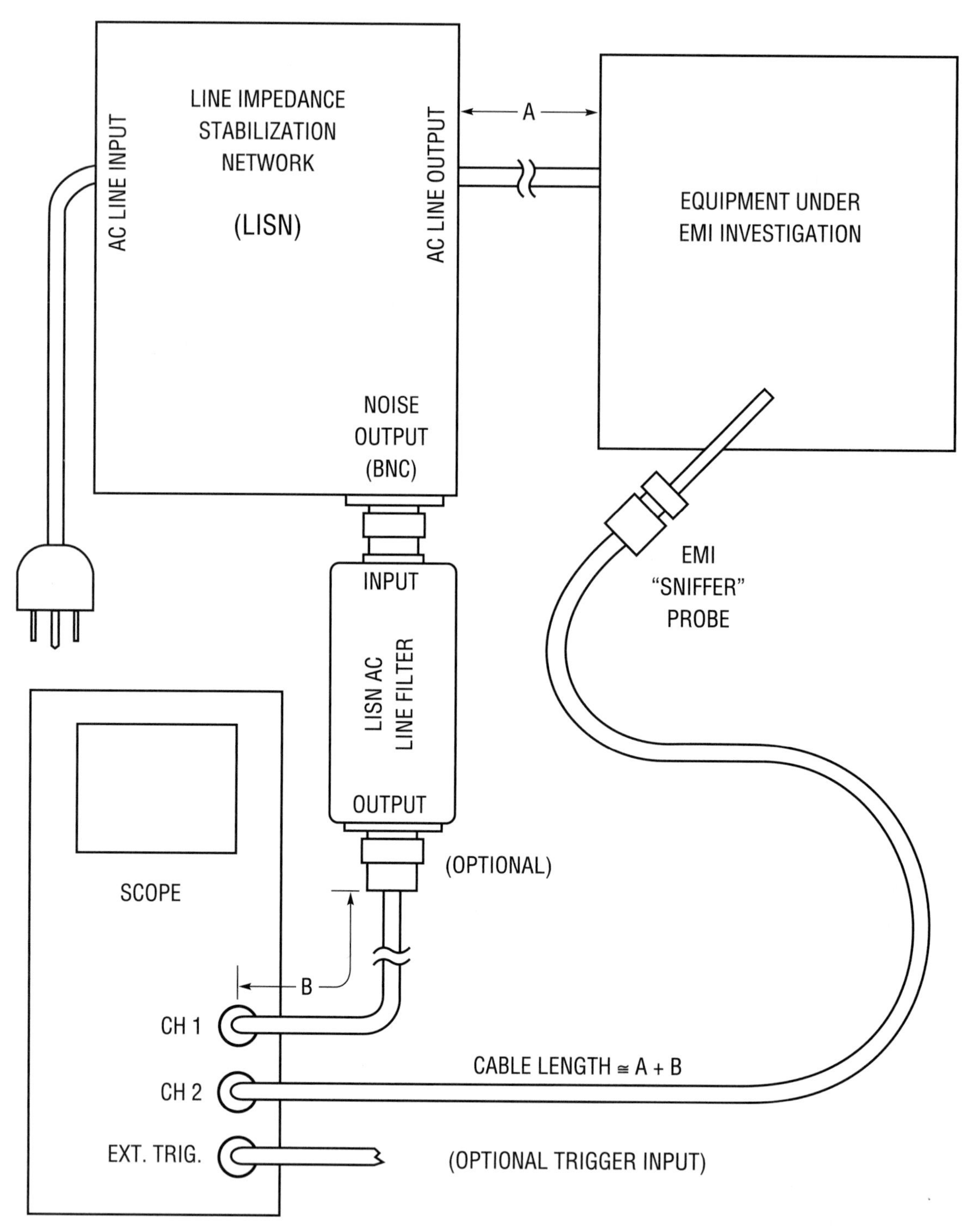

Figure E13 • Using the Probe with a "LISN"

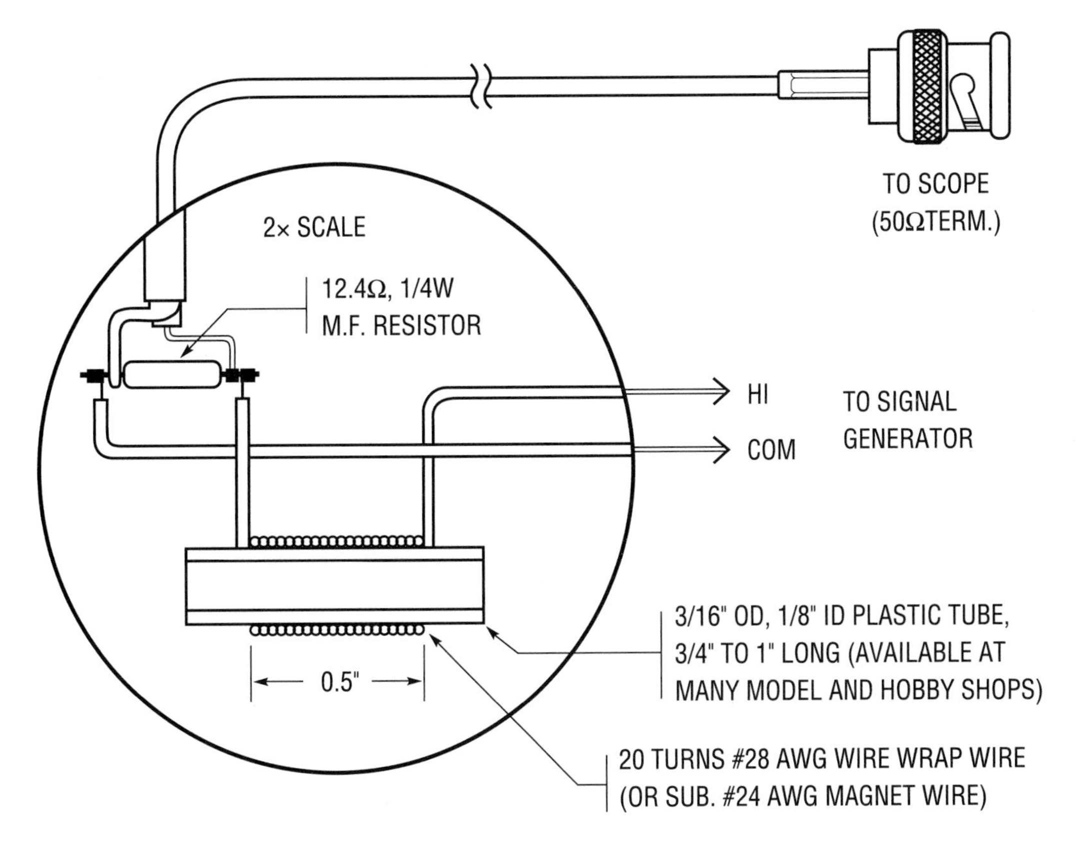

The Sniffer Probe Tip is centered inside the test coil where the Probe voltage is greatest. The approximate flux density in the middle of a coil can be calculated from the formula:

$$B = H = 1.257\ NI/l \quad \text{(CGS Units)}$$

For the 1.27cm long, 20-turn test coil, the flux density is about 20 Gauss per amp. At 1MHz, the Sniffer Probe voltage is $19mV_{P-P}$ (±10%) per $100mA_{P-P}$ for a 1MΩ load impedance, and half that for a 50Ω load.

Figure E14 • EMI "Sniffer" Probe Test Coil

Poorly bypassed high speed logic

Ideally, all high speed logic should have a tightly coupled bypass capacitor for each IC and/or have power and ground distribution planes in a multilayer PCB.

At the other extreme, I have seen *one* bypass capacitor used at the power entrance to a logic board, with power and ground led to the ICs from opposite sides of the board. This created large spikes on the logic supply voltage and produced significant electromagnetic fields around the board.

With a Sniffer Probe, I was able to show which pins of which ICs had the larger current transients in synchronism with the supply voltage transients. (The logic design engineers were accusing the power supply vendor of creating the noise. I found that the supplies were fairly quiet; it was the poorly designed logic power distribution system that was the problem.)

Probe use with a "LISN"

A test setup using the Sniffer Probe with a Line Impedance Stabilization Network (LISN) is shown in Figure E13. The optional "LISN AC LINE FILTER" reduces AC line voltage feedthrough from a few 100mV to microvolt

1) Use a 2-channel scope, preferably one with an external trigger.
2) One scope channel is used for the Sniffer Probe, which is not to be used for triggering.
3) The second channel is used to view the noise transient whose source is to be located, which may also be used for triggering if practical.
4) More stable and reliable triggering is achieved with an "external trigger" (or a 3rd channel) on a transistor drive waveform (or preceding logic transition), allowing immediate precursors to the transient to be viewed. (Nearly all noise transients occur during, or just after, a power transistor turn-on or turn-off.
5) Start with the Probe at some distance from the circuit with maximum sensitivity and "sniff around" for something happening in *precise sync* with the noise transient. The Probe waveform will not be identical to the noise transient, but will usually have a strong resemblance.
6) Move the Probe closer to the suspected source while decreasing sensitivity. The conductor carrying the responsible current is located by the sharp response null on top of the conductor with inverted polarity on each side.
7) Trace out the noise current path as much as possible. Identify the current path on the schematic.
8) The source of the noise transient is usually evident from the current path and the timing information.

Figure E15 • EMI "Sniffer" Probe Procedure Outline

levels, simplifying EMI diagnosis when a suitable DC voltage source is not available or cannot be used.

Testing the sniffer probe

The Sniffer Probe can be functionally tested with a jig similar to that shown in Figure E14, which is used to test probes in production.

Conclusion

The Sniffer Probe is a simple, but very fast and effective means to locate dI/dt sources of EMI. These EMI sources are very difficult to locate with conventional voltage or current probes.

Summary

A summarized procedure for using the EMI "Sniffer" Probe appears in Figure E15.

Sniffer probe amplifier

Figure E16 shows a 40MHz amplifier for the Sniffer Probe. A gain of 200 allows an oscilloscope to display probe output over a wide range of sensed inputs. The

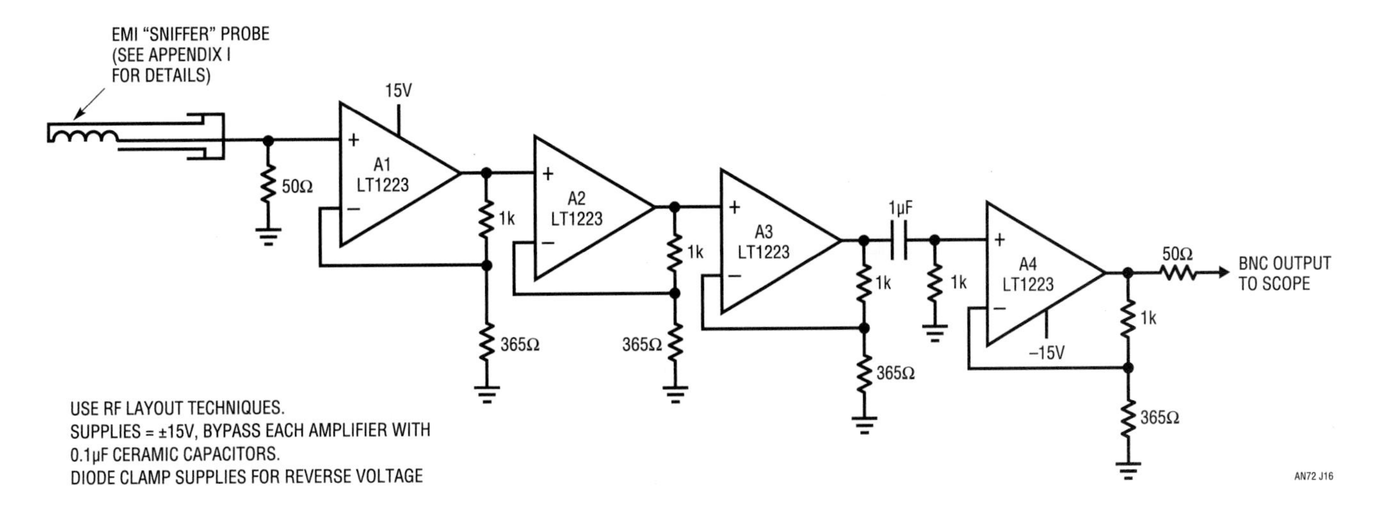

Figure E16 • 40MHz Amplifier for EMI Probe

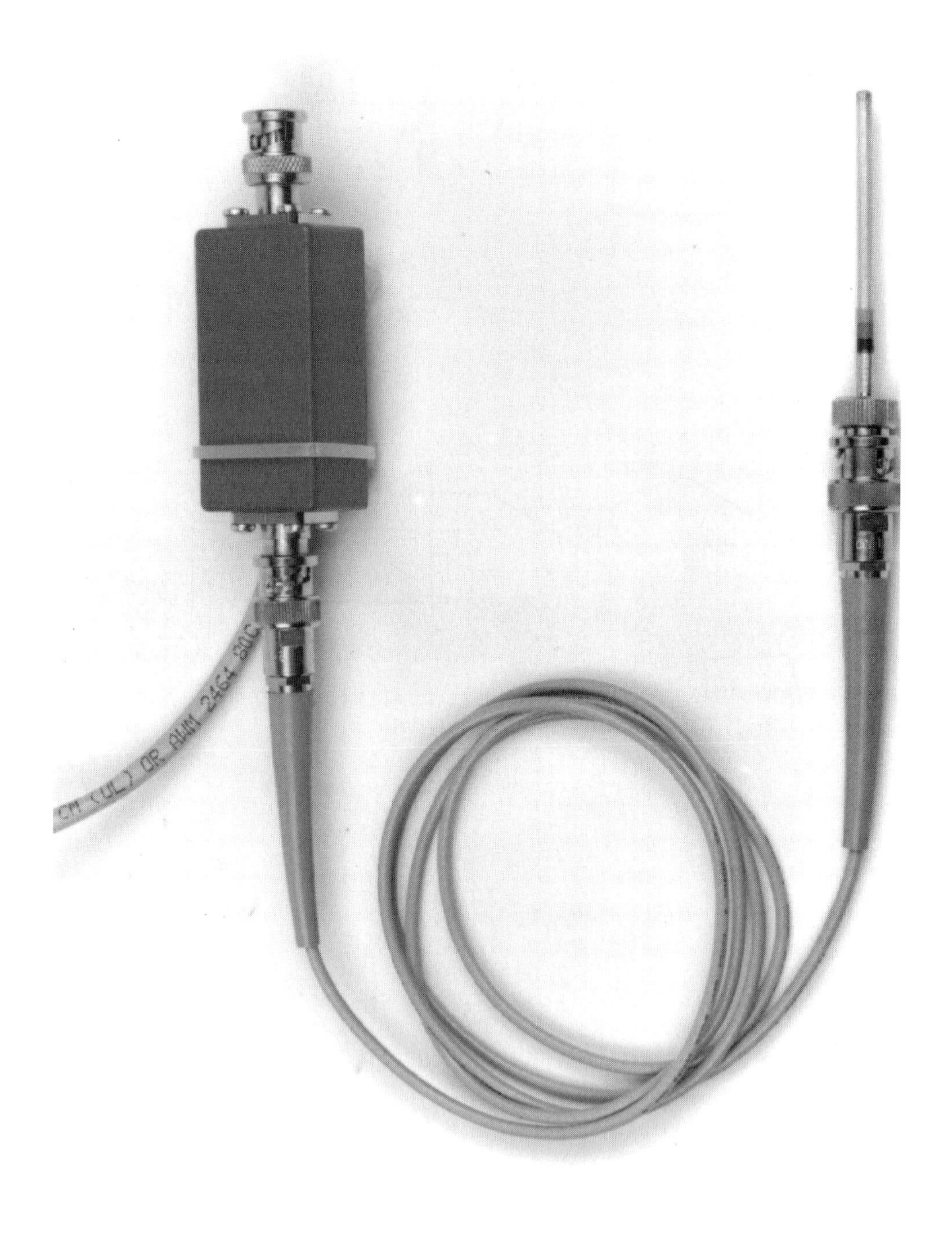

Figure E17 • Sniffer Probe and Amplifier. Note All BNC-Based Signal Transmission. ±15V Power Enters Box via Separate Cable

amplifier is built into a small aluminum box. The probe should connect to the amplifier via BNC cable, although the 50Ω termination does not have to be a high quality coaxial type. The probe's uncalibrated, relative output means high frequency termination aberrations are irrelevant. A simple film resistor, contained in the amplifier box, is adequate. Figure E17 shows the Sniffer Probe and the amplifier.

An alternate approach utilizes Appendix B's (Figure B11) HP-461A 50Ω amplifier.

Appendix F
About ferrite beads

A ferrite bead enclosed conductor provides the highly desirable property of increasing impedance as frequency rises. This effect is ideally suited to high frequency noise filtering of DC and low frequency signal carrying conductors. The bead is essentially lossless within a linear regulator's passband. At higher frequencies the bead's ferrite material interacts with the conductors magnetic field, creating the loss characteristic. Various ferrite materials and geometries result in different loss factors versus frequency and power level. Figure F1's plot shows this. Impedance rises from 0.01Ω at DC to 50Ω at 100MHz. As DC current, and hence constant magnetic field bias, rises, the ferrite becomes less effective in offering loss. Note that beads can be "stacked" in series along a conductor, proportionally increasing their loss contribution. A wide variety of bead materials and physical configurations are available to suit requirements in standard and custom products.

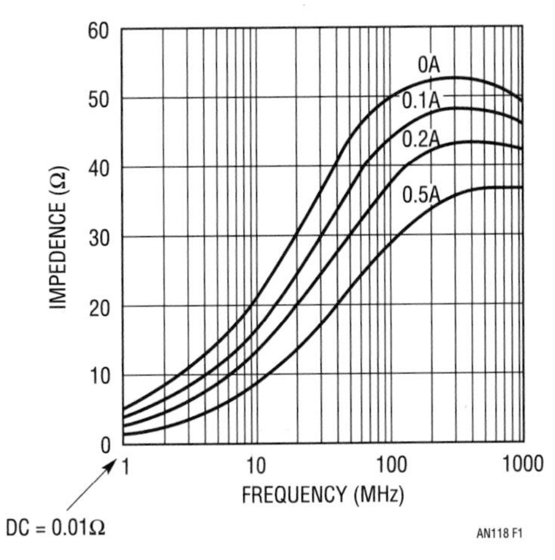

Figure F1 • Impedance vs Frequency at Various DC Bias Currents for a Surface Mounted Ferrite Bead (Fair-Rite 2518065007Y6). Impedance is Essentially Zero at DC and Low Frequency, Rising Above 50Ω Depending on Frequency and DC Current. Source: Fair-Rite 2518065007Y6 Datasheet

Appendix G
Inductor parasitics

Inductors can sometimes be used for high frequency filtering instead of beads but parasitics must be kept in mind. Advantages include wide availability and better effectiveness at lower frequencies, e.g., ≤100kHz. Figure G1 shows disadvantages are parasitic shunt capacitance and potential susceptibility to stray switching regulator radiation. Parasitic shunt capacitance allows unwanted high frequency feedthrough. The inductors circuit board position may allow stray magnetic fields to impinge its winding, effectively turning it into a transformer secondary. The resulting observed spike and ripple related artifacts masquerade as conducted components, degrading performance.

Figure G2 shows a form of inductance based filter constructed from PC board trace. Such extended length traces, formed in spiral or serpentine patterns, look inductive at high frequency. They can be surprisingly effective in some circumstances, although introducing much less loss per unit area than ferrite beads.

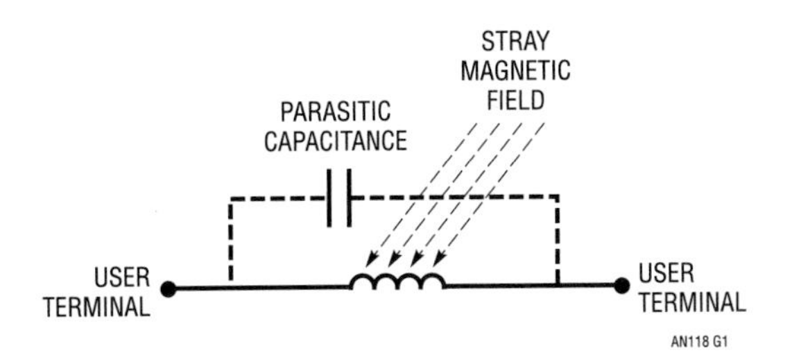

Figure G1 • Some Parasitic Terms of an Inductor. Unwanted Capacitance Permits High Frequency Feedthrough. Stray Magnetic Field Induces Erroneous Inductor Current

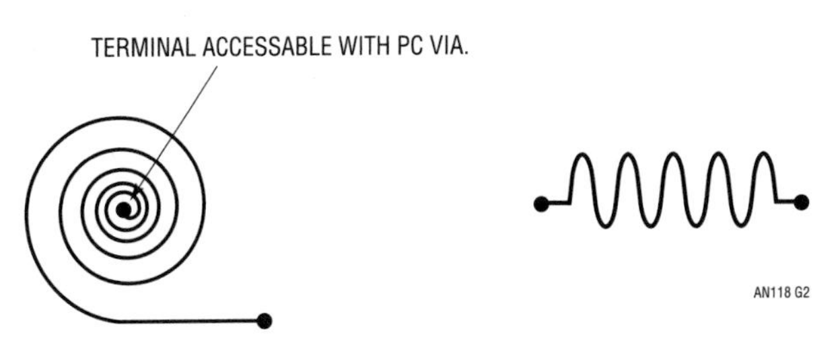

Figure G2 • Spiral and Serpentine PC Patterns are Sometimes Used as High Frequency Filters, Although Less Effective Than Ferrite Beads

Section 5

Powering Illumination Devices

A fourth generation of LCD backlight technology (8)

This publication treats the subject of LCD backlighting comprehensively. The text considers lamps, display and layout induced losses, circuitry, efficiency related issues, optimization and measurement techniques. Twelve appended sections cover lamp types, mechanical design, electrical and photometric measurement, layout, circuitry and related topics.

Simple circuitry for cellular telephone/camera flash illumination (9)

This publication concerns implementation of high quality flash illumination in cellular telephones/cameras. Performance vs LED based illumination is discussed and flashlamp operation reviewed. Considerations for support circuitry are given, and a practical circuit, accompanied by performance data, is described. Layout and RFI issues are treated and a sample layout provided. An appended section details operation of the LT3468 flash capacitor charger used in the text's circuit and lists appropriate magnetic components.

A fourth generation of LCD backlight technology

Component and measurement improvements refine performance

8

Jim Williams

Preface

Current generation portable computers and instruments utilize backlit LCDs (liquid crystal displays). These displays have also appeared in applications ranging from medical equipment to automobiles, gas pumps and retail terminals. Cold cathode fluorescent lamps (CCFLs) provide the highest available efficiency for backlighting the display. These lamps require high voltage AC to operate, mandating an efficient high voltage DC/AC converter. In addition to good efficiency, the converter should deliver the lamp drive in sine wave form. This is desirable to minimize RF emissions. Such emissions can cause interference with other devices, as well as degrading overall operating efficiency. The sine wave excitation also provides optimal current-to-light conversion in the lamp. The circuit should permit lamp control from zero to full brightness with no hysteresis or "pop-on," and must also regulate lamp intensity vs power supply variations.

The small size and battery-powered operation associated with LCD equipped apparatus mandate low component count and high efficiency for these circuits. Size constraints place severe limitations on circuit architecture and long battery life is usually a priority. Laptop and handheld portable computers offer an excellent example. The CCFL and its power supply are responsible for almost 50% of the battery drain. Additionally, these components, including PC board and all hardware, usually must fit within the LCD enclosure with a height restriction of 0.25 inches.

A practical, efficient LCD backlight design is a classic study of compromise in a transduced electronic system. Every aspect of the design is interrelated, and the physical embodiment is an integral part of the electrical circuit. The choice and location of the lamp, wires, display housing and other items have a major effect on electrical characteristics. The greatest care in every detail is required to achieve a practical high efficiency LCD backlight. Getting the lamp to light is just the beginning!

First generation backlights were crude, with poor performance in almost all areas. LTC (Linear Technology Corporation) has introduced feedback stabilization and optimized lamp driving configurations in three successive generations of technology. The effort has culminated in dedicated ICs for backlight driving.

This fourth publication reviews our recent work in components and measurement techniques applicable to LCD backlighting. Theoretical considerations are presented with practical suggestions, remedies and circuits. As always, we welcome reader comments, questions and requests for consultation.

Update

While LED backlights have mostly replaced CCFLs, this Application Note shows proper design and layout for high voltage inverters.

Analog Circuit and System Design: Immersion in the Black Art of Analog Design. http://dx.doi.org/10.1016/B978-0-12-397888-2.00008-0

Introduction

This scribing marks the fourth LTC publication in as many years concerning LCD illumination.[1] The extraordinary user response to previous efforts has resulted in a continuing LCD backlight development effort by our company. This level of interest, along with significant performance advances since the last publication, justifies further discussion of LCD backlighting.

Development of attractive solutions for LCD illumination has necessitated the longest sustained LTC application engineering effort to date. A single circuit in a 1991 publication (Measurement and Control Circuit Collection, LTC Application Note 45, June 1991) has resulted in four years of continuous investigation, summarized in three successive, dedicated publications.

The impetus for all this bustle has been an overwhelming and continuously ascending reader response. Practical, high performance LCD backlighting solutions are needed in a wide range of applications. The optical, transductive and electronic aspects combine (conspire?) to present an extraordinarily challenging problem. The LCD backlight problem's interdisciplinary nature, along with highly interactive effects, provides an exquisitely subtle engineering exercise. Backlights present the most complex set of interdependencies the author has ever encountered. Our academic interest in this challenge is, of course, well patinaed with capitalistic intent. Substantial comfort arrives with the certainty that the audience is similarly acculturated.

This publication includes pertinent information from previous efforts in addition to updated sections and a large body of new material. The partial repetition is a small penalty compared to the benefits of text flow, completeness and time efficient communication. Older material has been altered, abridged or augmented as appropriate, while simultaneously introducing new findings. Previous work has emphasized obtaining and verifying high efficiency. This characteristic is still quite desirable, but other backlight requirements have become evident. These include low voltage operation, improved system interface, minimization of display-induced losses, circuitry compaction and better measurement/optimization techniques. These advances have been enabled by development of new ICs and instrumentation.

Finally, this preamble must appreciate the text's arrangement and review by various LTC personnel and customers. They transmuted a psychotic uproar of a manuscript into this finessed presentation. Hopefully, readers will join the author in applause.

Note 1: Previous publications are annotated in References 1, 18 and 25.

Perspectives on display efficiency

The LCD displays currently available require two power sources, a backlight supply and a contrast supply. The display backlight is the single largest power consumer in a typical portable apparatus, accounting for almost 50% of battery drain with the display at maximum intensity. As such, every effort must be expended to maximize backlight efficiency.

Study of LCD energy management should consider the problem from an interdisciplinary viewpoint. The backlight presents a cascaded energy attenuator to the battery (Figure 8.1). Battery energy is lost in the electrical-to-electrical conversion to high voltage AC to drive the CCFL. This section of the energy attenuator is the most efficient; conversion efficiencies exceeding 90% are possible. The CCFL, although the most efficient electrical-to-light converter available today, has losses exceeding 80%. Additionally, the optical transmission efficiency of present displays is under 10% for monochrome with color types much lower.

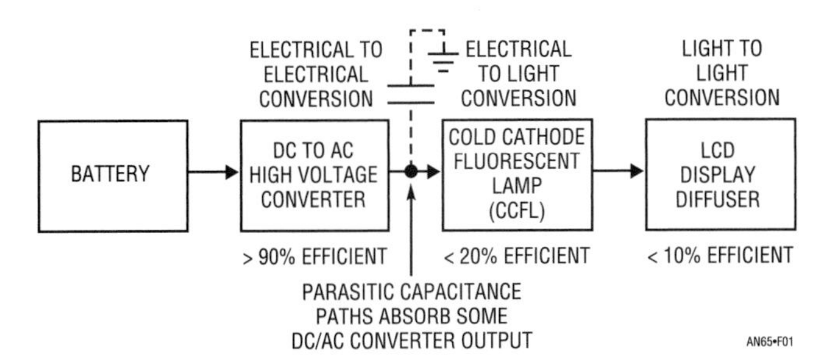

Figure 8.1 • Backlit LCD Display Presents a Cascaded Energy Attenuator to the Battery. DC/AC Conversion is Significantly More Efficient than Energy Conversions in Lamp and Display

The very high DC/AC conversion efficiency highlights some significant issues. Anything that improves energy transfer in the other "attenuator" areas will have greater impact than further electrical efficiency improvements. Additional improvements in electrical efficiency, while certainly desirable, are reaching the point of diminishing returns. Clearly, overall backlight efficiency gains must come from lamp and display improvements.

There is very little electrical workers can do to improve lamp and display efficiency besides call attention to the problems (see the following sections on lamps and displays).[2] Improvements are, however, possible in related areas. In particular, the *form* of drive applied to the lamp is quite critical. The waveshape supplied to the lamp influences its current-to-light conversion efficiency. Thus, dissimilar waveforms containing equivalent power can produce different amounts of lamp light output. This implies that a more *electrically* efficient inverter with a nonoptimal output waveshape could produce less light than a "less efficient" inverter with a more appropriate waveform. Experiment reveals this to be true. As such, distinction between electrical and photometric efficiency is necessary and requires attention.

Another practical area where improvement is possible is transmission of inverter drive to the lamp. The high frequency AC waveform is subject to losses due to parasitic capacitances in the wiring and display. Controlling the parasitic capacitances and the manner in which lamp drive is applied can yield significant efficiency improvement.

Practical methods addressing both aforementioned areas are contained in subsequent sections of this publication.

Cold cathode fluorescent lamps (CCFLs)

Any discussion of CCFL power supplies must consider lamp characteristics. These lamps are complex transducers, with many variables affecting their ability to convert electrical current to light. Factors influencing conversion efficiency include the lamp's current, temperature, drive waveform characteristics, length, width, gas constituents and the proximity to nearby conductors.

These and other factors are interdependent, resulting in a complex overall response. Figures 8.2 through 8.8 show some typical characteristics. A review of these curves hints at the difficulty in predicting lamp behavior as operating conditions vary. The lamp's current, temperature and warm-up time are clearly critical to emission, although electrical efficiency may not necessarily correspond to the best optical efficiency point. Because of this, both electrical and photometric evaluation of a circuit is often required. It is possible, for example, to construct a CCFL circuit with 94% electrical efficiency which produces less light output than an approach with 80% electrical efficiency. (See Appendix L, "A Lot of Cut Off Ears

Note 2: "Call attention to the problems" constitutes a pleasant euphemism for complaining. This publication's section on displays presents such complaints in visual form along with suggested remedies.

and No Van Goghs—Some Not-So-Good Ideas.") Similarly, the performance of a very well matched lamp/circuit combination can be severely degraded by a lossy display enclosure or excessive high voltage wire lengths. Display enclosures with too much conducting material near the lamp have huge losses due to capacitive coupling. A poorly designed display enclosure can easily degrade efficiency by 20%. High voltage wire runs typically cause 1% loss per inch of wire.

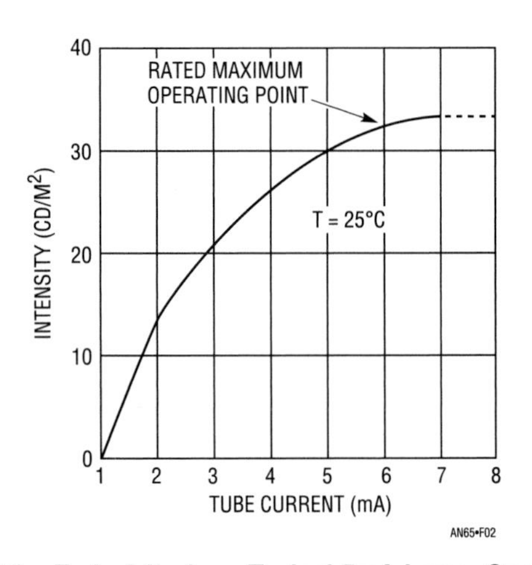

Figure 8.2 • Emissivity for a Typical 5mA Lamp. Curve Flattens Badly Above 6mA

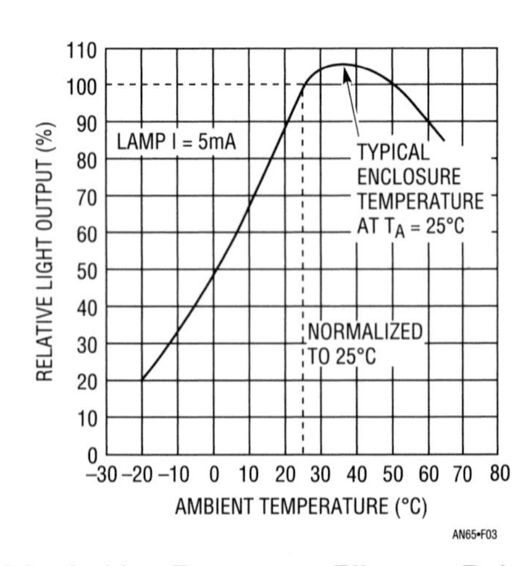

Figure 8.3 • Ambient Temperature Effects on Emissivity of a Typical 5mA Lamp. Lamp and Enclosure Must Come to Thermal Steady State Before Measurements are Made

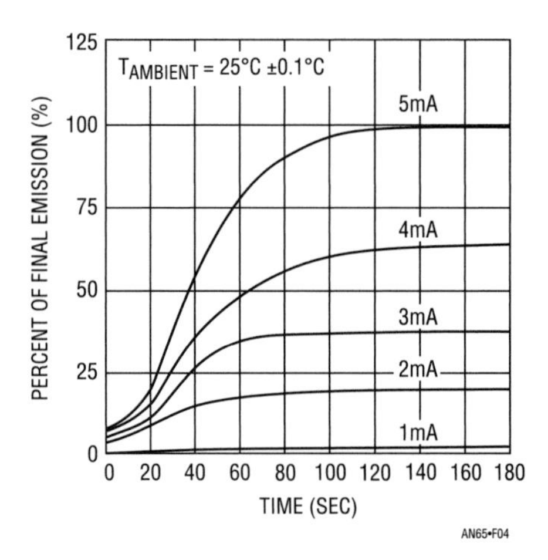

Figure 8.4 • Emissivity vs On-Time for a Typical Lamp in Free Air. Lamp Must Arrive at Temperature Before Emission Stabilizes

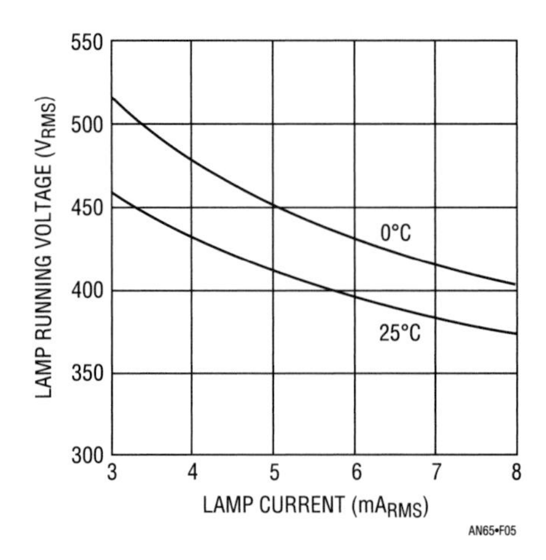

Figure 8.5 • Lamp Current vs Voltage in the Operating Region. Note Large Temperature Coefficient

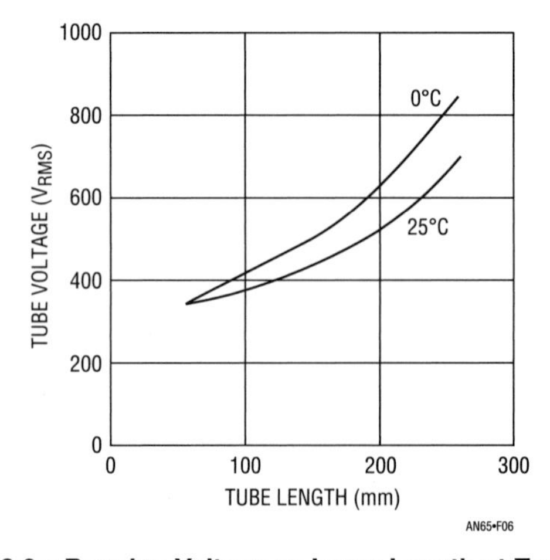

Figure 8.6 • Running Voltage vs Lamp Length at Two Temperatures. Start-Up Voltages are Usually 50% to 200% Higher Over Temperature

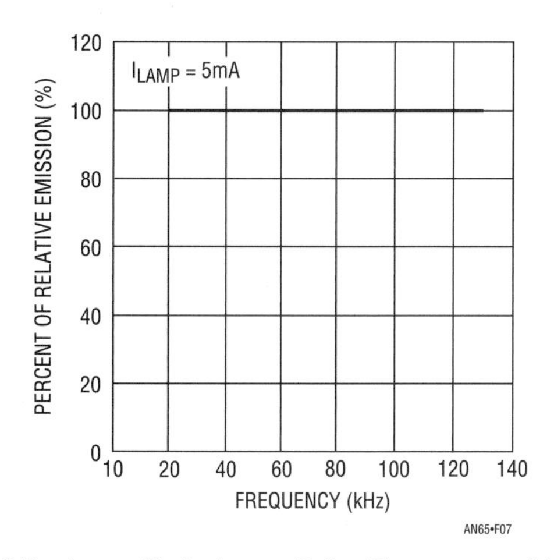

Figure 8.7 • Lamp Emission vs Drive Frequency with Lamp in Free Space. No Change is Measurable from 20kHz to 130kHz, Indicating Lamp Insensitivity to Frequency

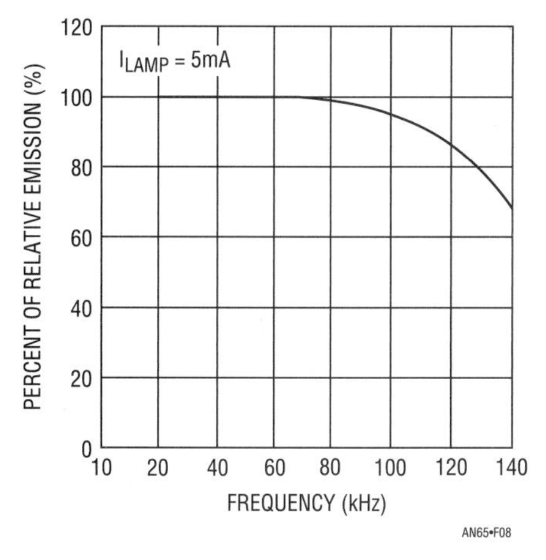

Figure 8.8 • Figure 8.7's Lamp Shows Significant Emission vs Drive Frequency Degradation When Mounted in a Display. Cause is Frequency-Dependent Loss Due to Displays Parasitic Capacitance Paths

The optimum drive frequency is determined by display and wiring losses, not lamp characteristics. Figure 8.7 shows lamp emissivity is essentially flat over a wide frequency range. Figure 8.8 shows results with the same lamp mounted in a typical display.

The apparent emissivity fall-off at high frequencies is caused by reduced lamp current due to parasitic capacitance-induced losses. As frequency increases, the display's parasitic capacitance diverts progressively more energy, lowering lamp current and emission. This effect is sometimes misinterpreted, leading to the mistaken conclusion that lamp emissivity degrades with increasing frequency.

CCFL load characteristics

These lamps are a difficult load to drive, particularly for a switching regulator. They have a "negative resistance" characteristic; the starting voltage is significantly higher than the operating voltage. Typically, the start voltage is about 1000V; although higher and lower voltage lamps are common. Operating voltage is usually 300V to 500V although other lamps may require different potentials. The lamps will operate from DC, but migration effects within the lamp will quickly damage it. As such, the waveform must be AC. No DC content should be present.

Figure 8.9a shows an AC driven lamp's characteristics on a curve tracer. The negative resistance-induced

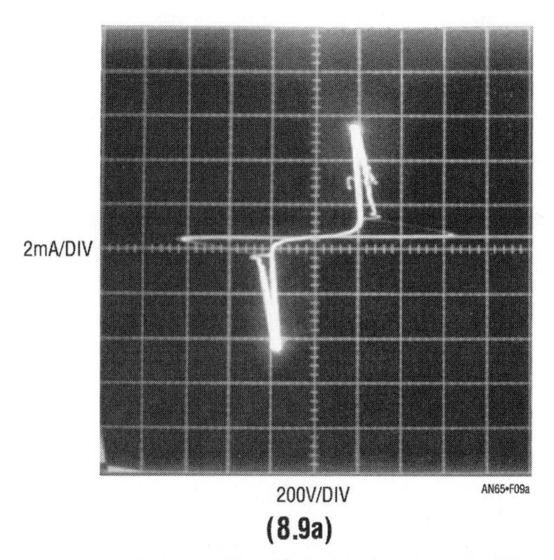

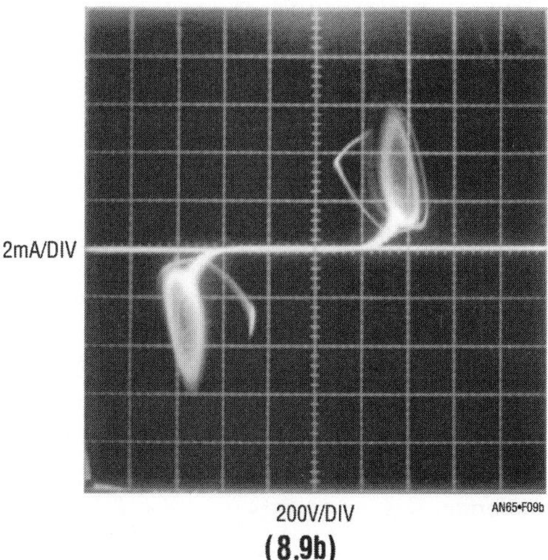

Figure 8.9 • Negative Resistance Characteristic for Two CCFL Lamps. "Snap-Back" is Readily Apparent, Causing Oscillation in 8.9b. These Characteristics Complicate Power Supply Design

"snap-back" is apparent. In Figure 8.9b another lamp, acting against the curve tracer's drive, produces oscillation. These tendencies, combined with the frequency compensation problems associated with switching regulators, can cause severe loop instabilities, particularly on start-up. Once the lamp is in its operating region it assumes a linear load characteristic, easing stability criteria. Lamp operating frequencies are typically 20kHz to 100kHz and a sine-like waveform is preferred. The sine drive's low harmonic content minimizes RF emissions, which could cause interference and efficiency degradation.[3] A further benefit to the continuous sine drive is its low crest factor and controlled rise times, which are easily handled by the CCFL. CCFLs RMS current-to-light output efficiency and lifetime degrades with fast rise, high crest factor drive waveforms.[4]

Display and layout losses

The physical layout of the lamp, its leads, the display housing and other high voltage components are integral parts of the circuit. Placing the lamp into a display introduces pronounced electrical loading effects which must be considered. Poor layout can easily degrade efficiency by 25% and higher layout-induced losses have been observed. Producing an optimal layout requires attention to how losses occur. Figure 8.10 begins our study by examining potential parasitic paths between the transformer's output and the lamp. Parasitic capacitance to AC ground from any point between the power supply output and the lamp creates a path for undesired current flow. Similarly, stray coupling from any point along the lamp's length to AC ground induces parasitic current flow. All parasitic current flow is wasted, causing the circuit to produce more energy to maintain desired current flow in the lamp. The high voltage path from the transformer to the display housing should be as short as possible to minimize losses. A good rule of thumb is to assume 1% efficiency loss per inch of high voltage lead. Any PC board traces, ground or power planes should be relieved by at least 1/4" in the high voltage area. This not only prevents losses but eliminates arcing paths.

Parasitic losses associated with lamp placement within the display housing require attention. High voltage wire length within the housing must be minimized, particularly for displays using metal construction. Ensure that the high voltage is applied to the shortest wire(s) in the display. This may require disassembling the display to verify wire length and layout. Another loss source is the reflective foil commonly used around lamps to direct light into the actual LCD. Some foil materials absorb considerably more field energy than others, creating loss. Finally, displays supplied in metal enclosures tend to be lossy. The metal absorbs significant energy and an AC path to ground is unavoidable. Direct grounding of metal enclosed displays further increases losses. Some display manufacturers have addressed this issue by relieving the metal in the lamp area with other materials. Losses introduced by the display are substantial and vary widely with different displays. These losses not only degrade overall efficiency, but complicate meaningful determination of the lamp current.

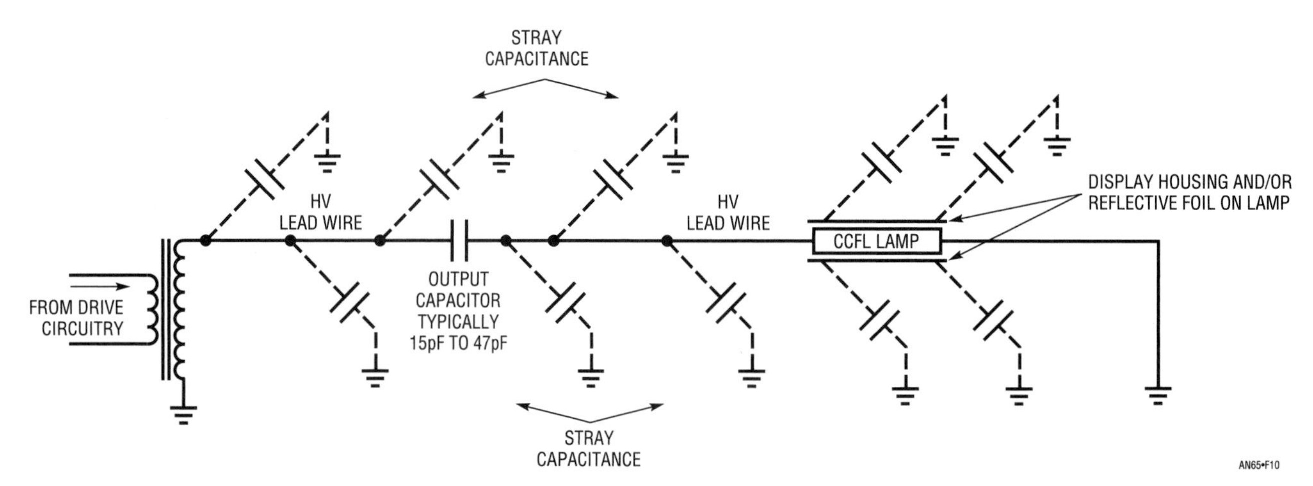

Figure 8.10 • Loss Paths Due to Stray Capacitance in a Practical LCD Installation. Minimizing These Paths is Essential for Good Efficiency

Note 3: Many of the characteristics of CCFLs are shared by so-called "Hot" cathode fluorescent lamps. See Appendix A, "Hot" Cathode Fluorescent Lamps.

Note 4: See Appendix L, "A Lot of Cut Off Ears and No Van Goghs—Some Not-So-Great Ideas."

Figure 8.11 shows effects of distributed parasitic capacitance loss paths on lamp current. The display housing and reflective foil induced loss paths provide a continuous conduit for loss current flow. This results in a continuously varying value of "lamp current" along the lamp's length. In cases where one end of the lamp is at or near ground, the current fall-off is greatest in the lamp's high voltage regions. Although parasitic capacitance is usually uniformly distributed, its effect becomes far greater as voltage scales up.

These effects illustrate why designing around lamp specifications is such a frustrating exercise. Display vendors typically call out lamp operating parameters based on information received from the lamp manufacturer. Lamp vendors often determine operating characteristics in a completely different enclosure, or none at all. This set of uncertainties complicates design effort. The only viable solution is to determine lamp performance with the display of interest. This is the only practical way to maximize performance and ensure against overdriving the lamp, which wastes power and shortens lamp life.

In general, the display introduces parasitics which degrade performance. Latter portions of this text discuss some compensatory techniques, but the deleterious effects of display parasitics dominate practical backlight design.

There are some benefits to lossy displays. One advantage of display parasitics is that they effectively lower lamp breakdown voltage. The parasitic shunt capacitance along the tube's length forms a distributed electrode, effectively shortening the breakdown path, lowering the lamp's turn-on voltage. This accounts for the fact that many display mounted lamps start at lower voltages than the "naked" lamp breakdown voltage specification suggests. This effect aids low temperature start-up (see Figures 8.5 and 8.6).

A second potential advantage of distributed parasitic lamp capacitance is enhancement of low current operation. In some cases extended dimming range is possible because the parasitics provide a more evenly distributed field along the lamp's length. This tends to maintain illumination along the lamp's entire length at low operating currents, allowing low luminosity operation.

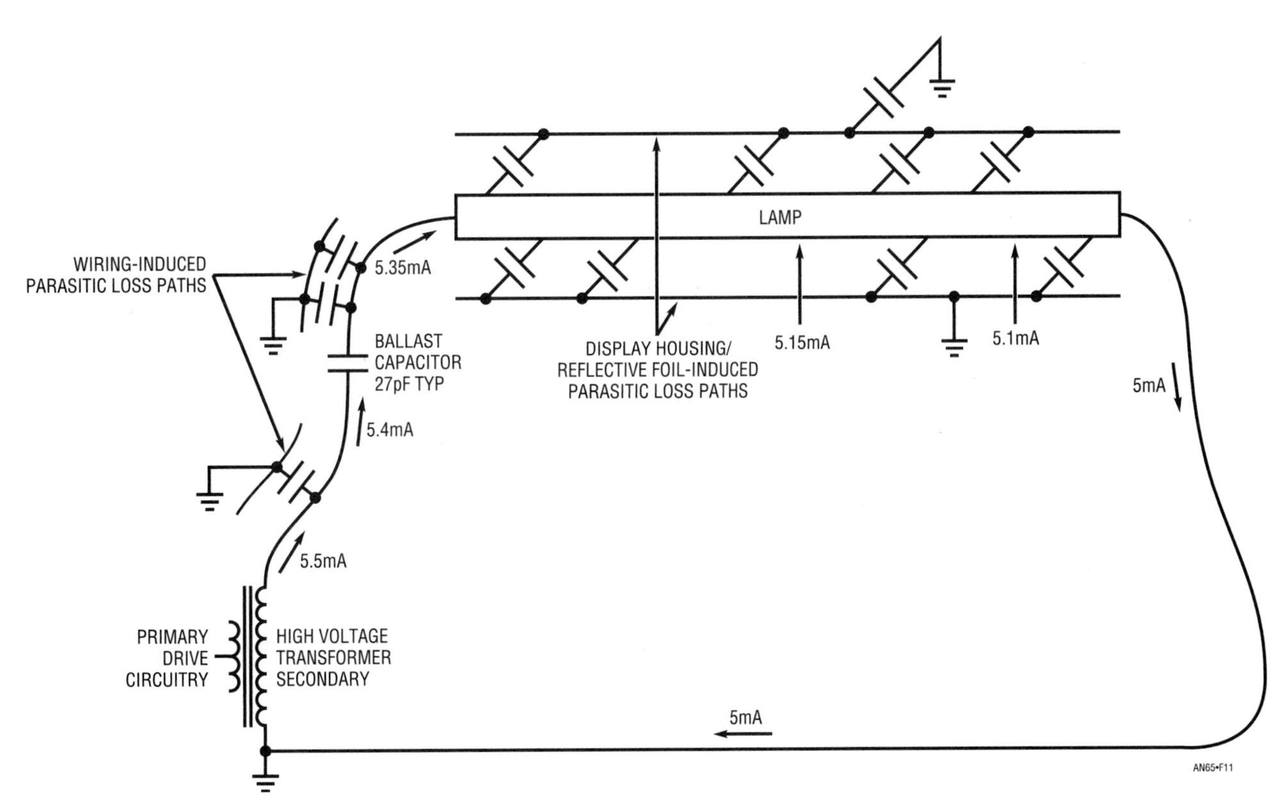

Figure 8.11 • Distributed Parasitic Capacitances in a Practical Situation Cause Continuous Downward Shift in Measured "Lamp Current." In This Case 0.5mA is Lost to Parasitic Paths. Most Loss Occurs in High Voltage Regions

The lessons here are clear. A thorough characterization of lamp/display losses is crucial to understanding trade-offs and obtaining the best possible performance. The highest efficiency "in system" backlights have been produced by careful attention to these issues. In some cases the entire display enclosure was re-engineered for lower losses.

The display loss issue, central to backlight design, merits detailed attention. The following briefly commented photographs (Figures 8.12 through 8.32) illustrate a variety of display situations. Hopefully, this visual tour will alert display users and manufacturers to the problems involved, promoting appropriate action by both.

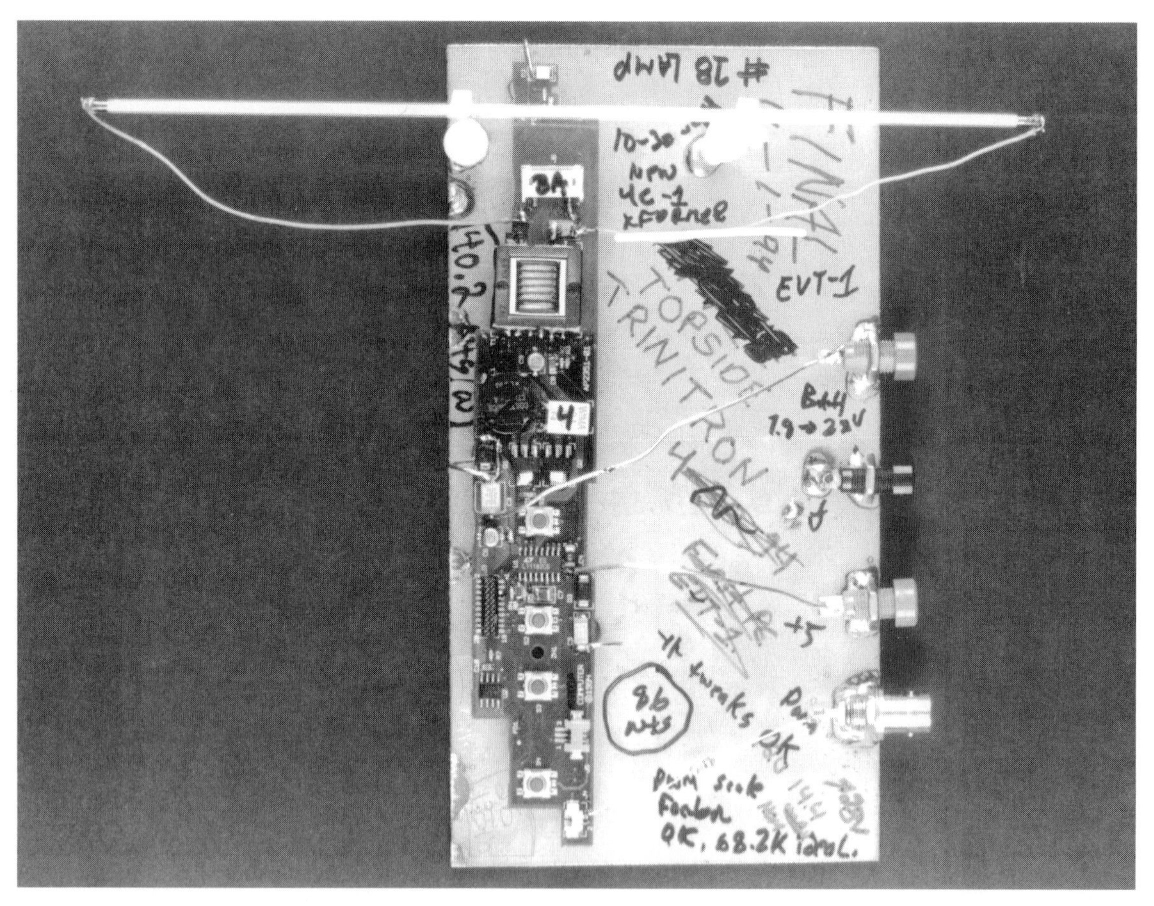

Figure 8.12 • The Ideal Display is No Display. Drive Electronics Connected to a "Naked" Lamp Simulates a Zero Loss Display. Note Nylon Stand-Offs. Results Obtained Have No Relationship to Practical Display Driving

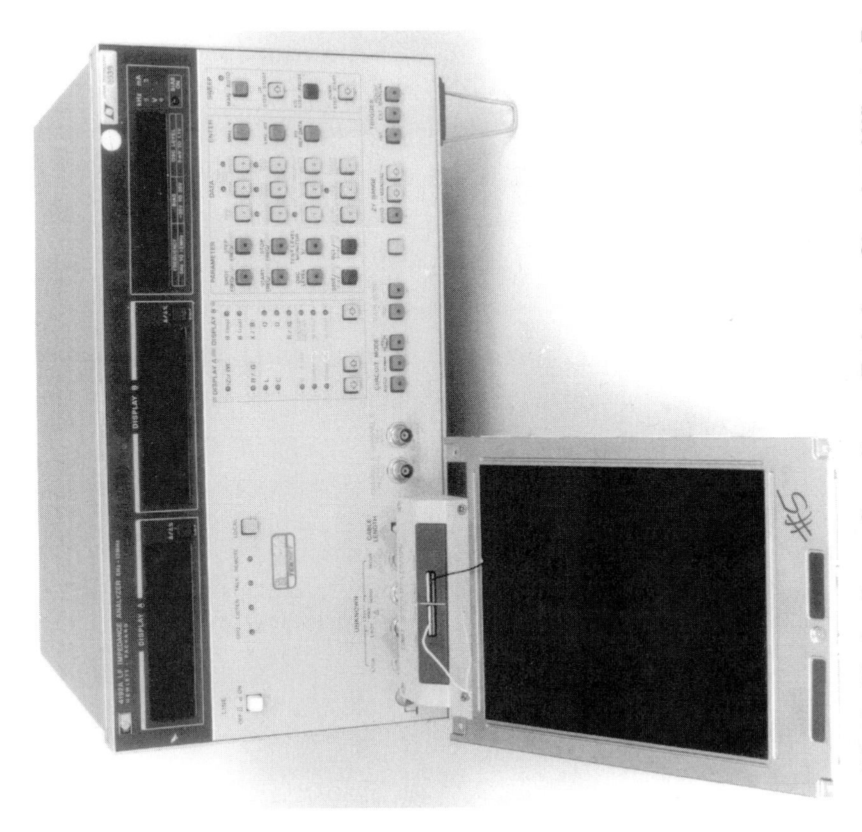

Figure 8.13 • Measuring Lamp Wire to Display Frame Capacitance. Technique Gives Lead Wire-to-Frame Loss Information but Not Lamp-to-Foil-or-Frame Loss Data. Lamp Must Be Energized Before Its Parasitics are Measurable

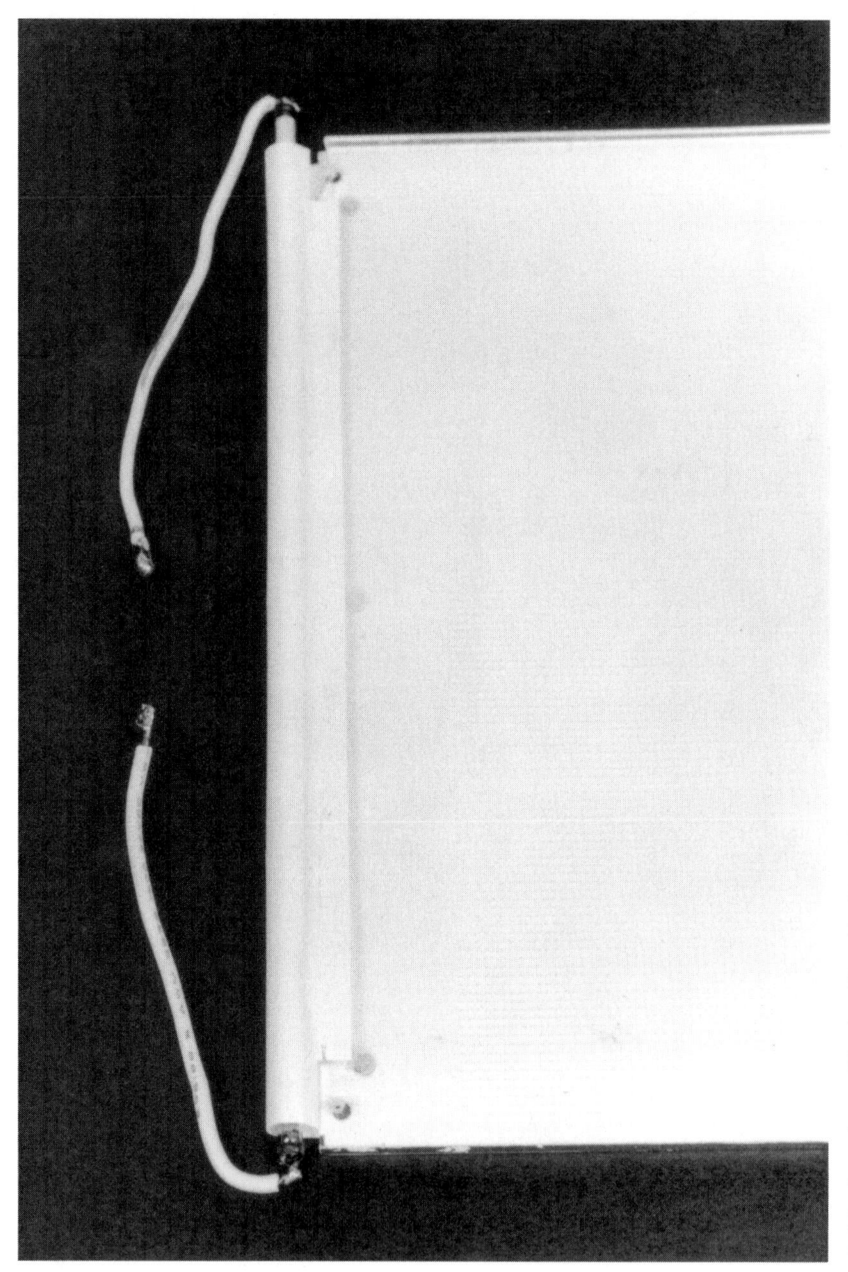

Figure 8.14 • Low Loss Display Has No Metal in Lamp Region. Reflective Foil Floats from Ground and Has Low Absorption. Display Loss About 1.5%

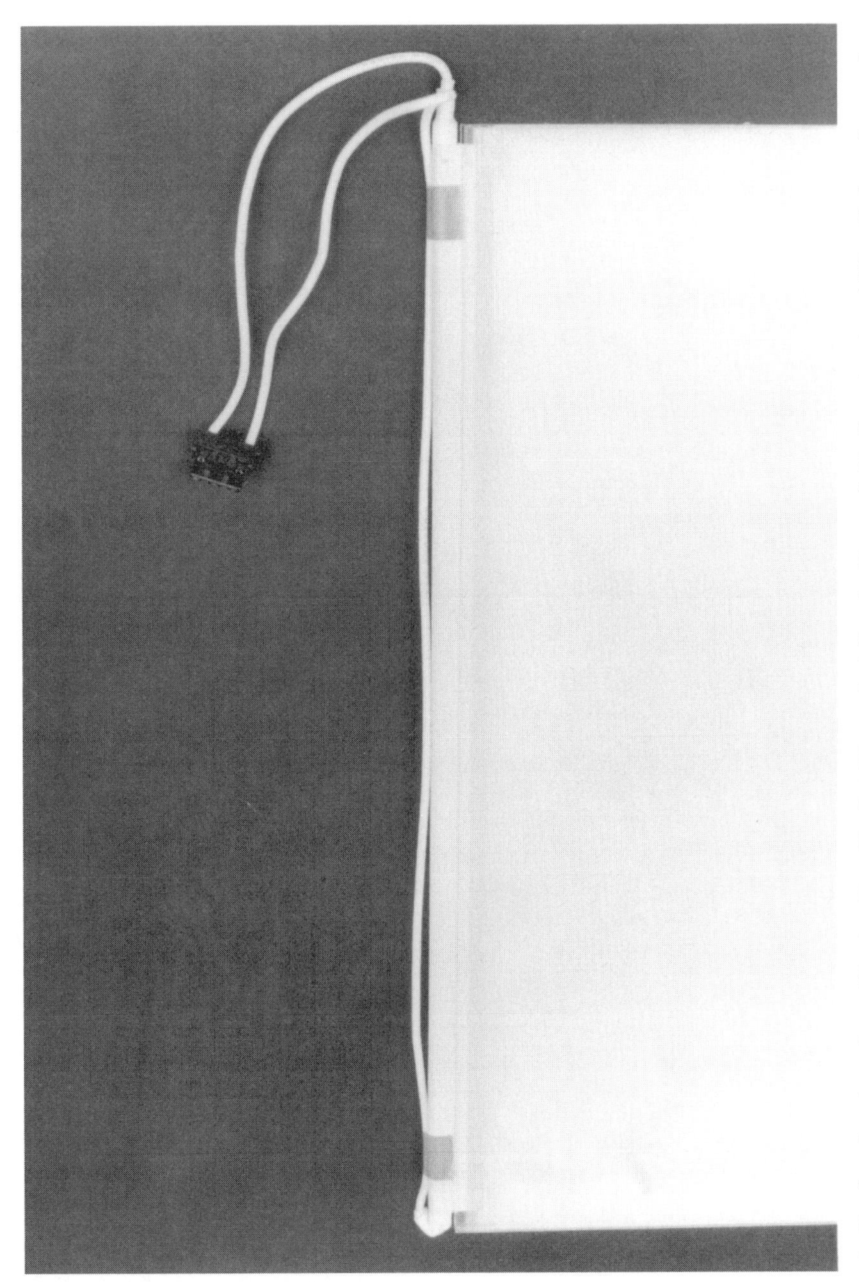

Figure 8.15 • Another Low Loss Display Has Similar Characteristics to Figure 8.14. Running Long Wire Return Across Lamp Length Increases Loss to about 4%. Spacing Wire Away from Lamp Would Cut Loss by Half

Figure 8.16 • A Custom Designed, Extremely Low Loss Display. All Metal Is Eliminated in Lamp Area (Lower Portion of Photo). A Good Compromise Between Mechanical Strength and Loss Control

Figure 8.17 • Figure 8.16's Reverse Side. All Metal is Relieved in Lamp Area, Maintaining Low Losses. An Excellent, Practical Display

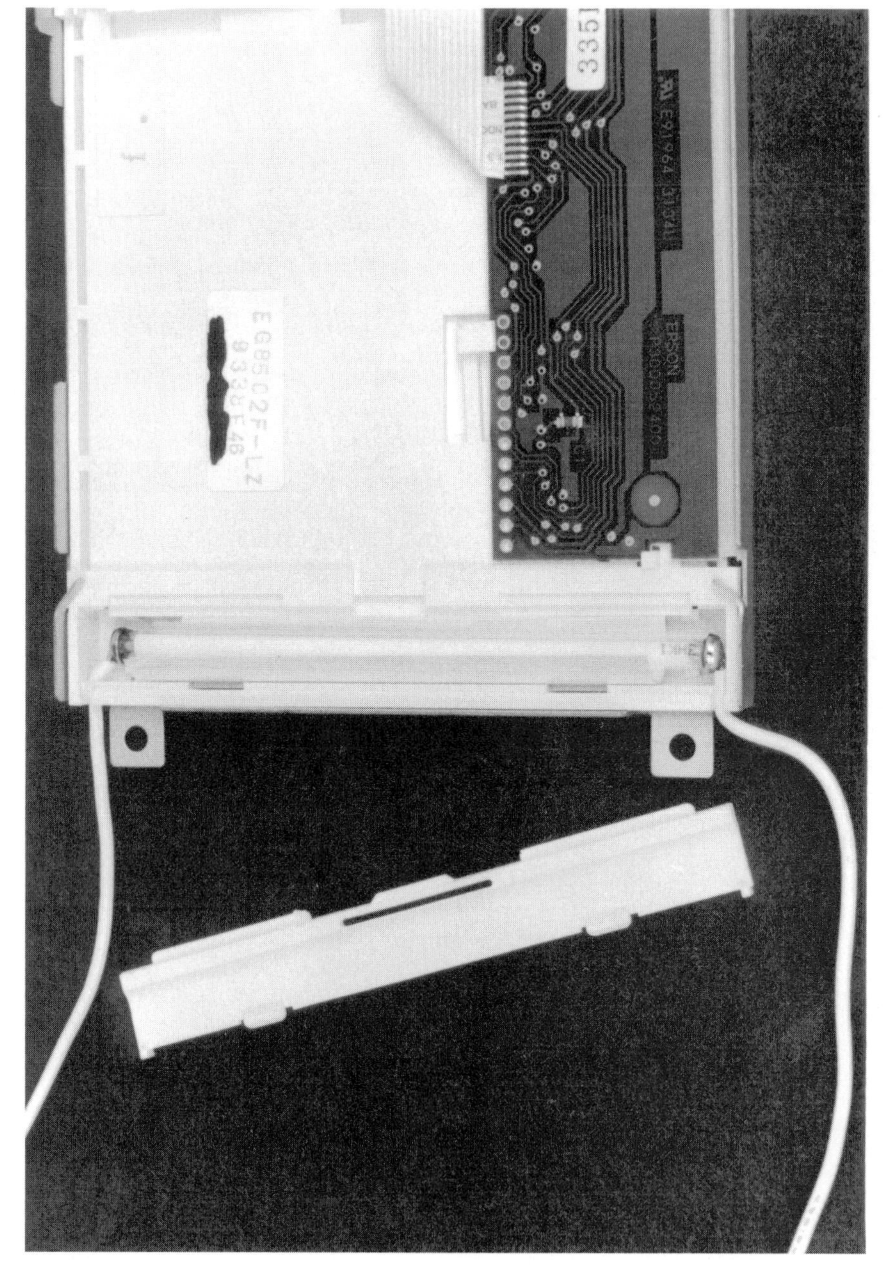

Figure 8.18 • Plastic "Cocoon" Cuts Losses. Metallic Foil is Absorptive but Floats from Grounded Display Frame. A Good Compromise with about 4% Loss

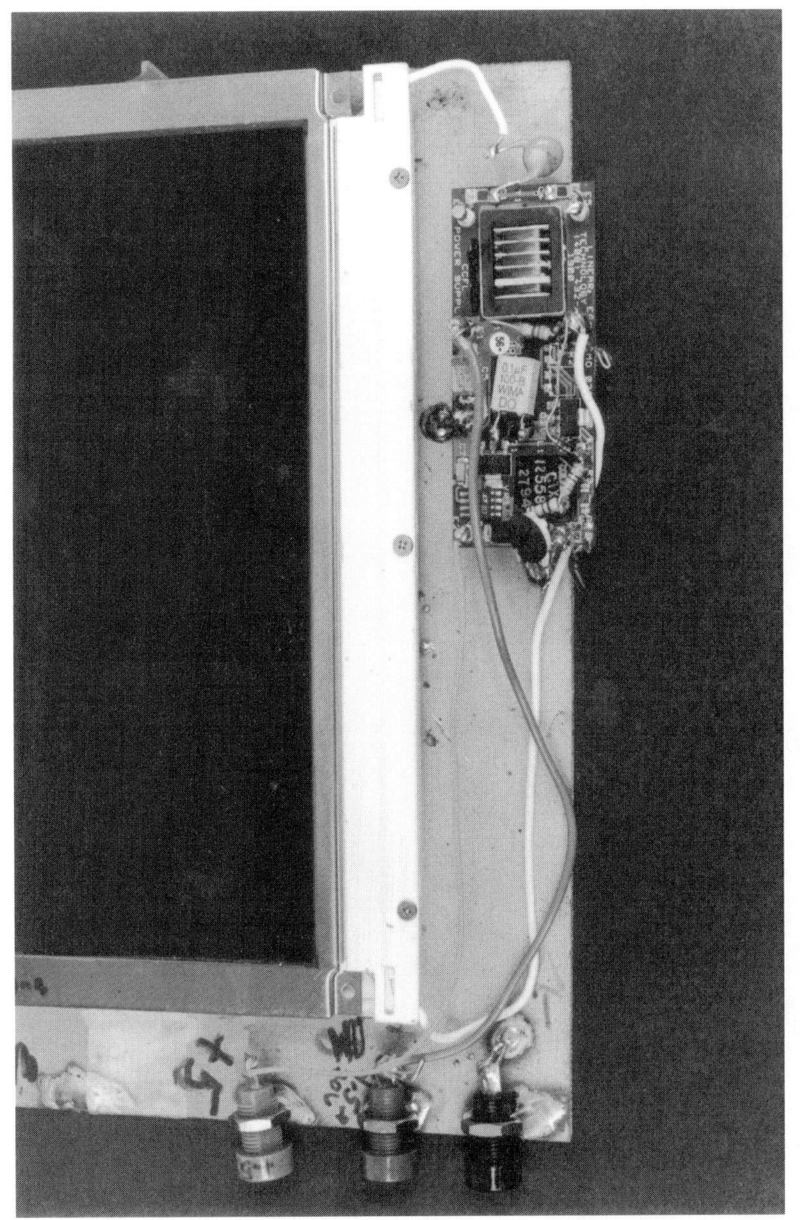

Figure 8.19 • Plastic "Outrigger" Isolates Lamp from Metal Display Frame Loss Path

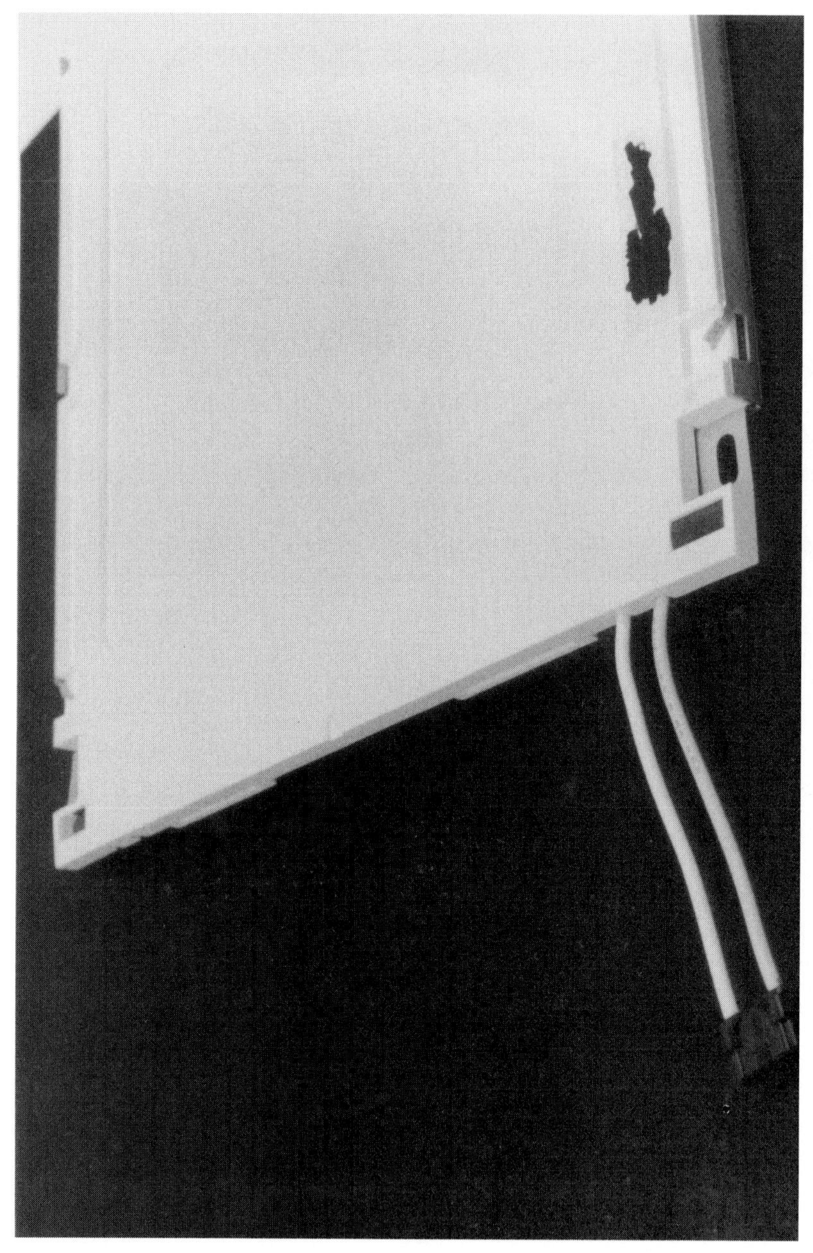

Figure 8.20 • Plastic Isolates Lamp from Metal Frame in This Display's Rear View

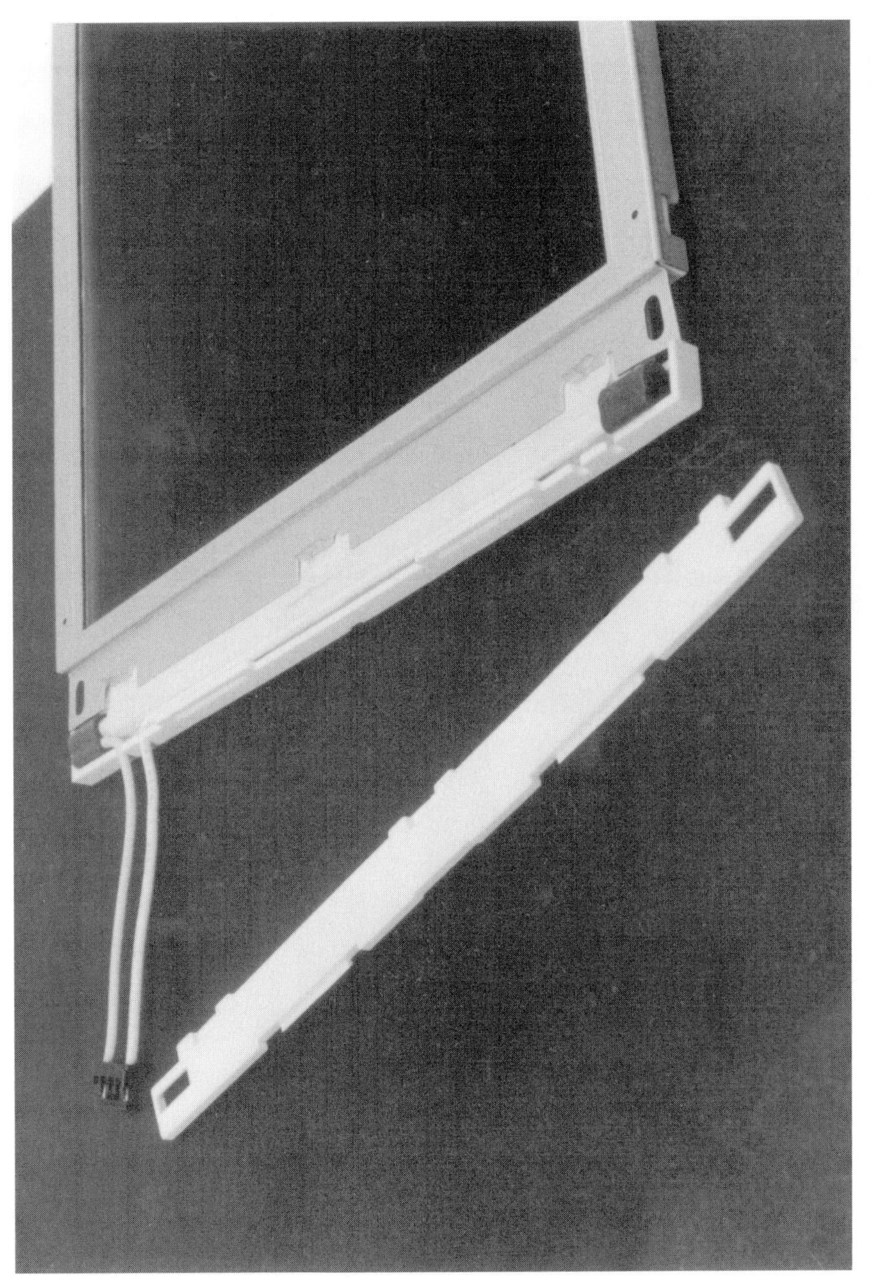

Figure 8.21 • Figure 8.20's Display Front View Continues Plastic Isolation Treatment but Reflective Foil (over Lamp) Contacts Metal Frame. Massive Losses via This Path Cause Overall 12% Loss. Trimming Foil from Metal Cuts Loss to 4%

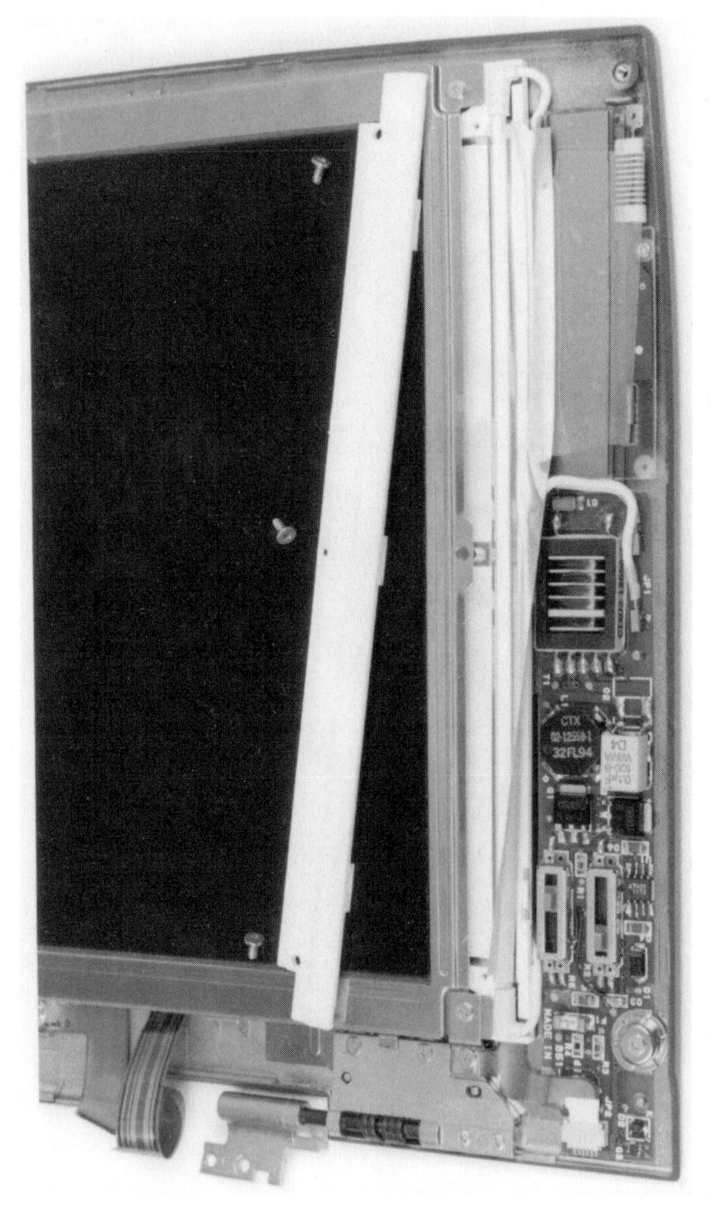

Figure 8.22 • Another "Outrigged" Plastic Enclosure Suffers Foil Contacting Display's Frame Metal. Relieving Foil from Metal Cuts Losses from 13% to 6%. Poor Wire Routing (Lower Right) Causes 3% Loss

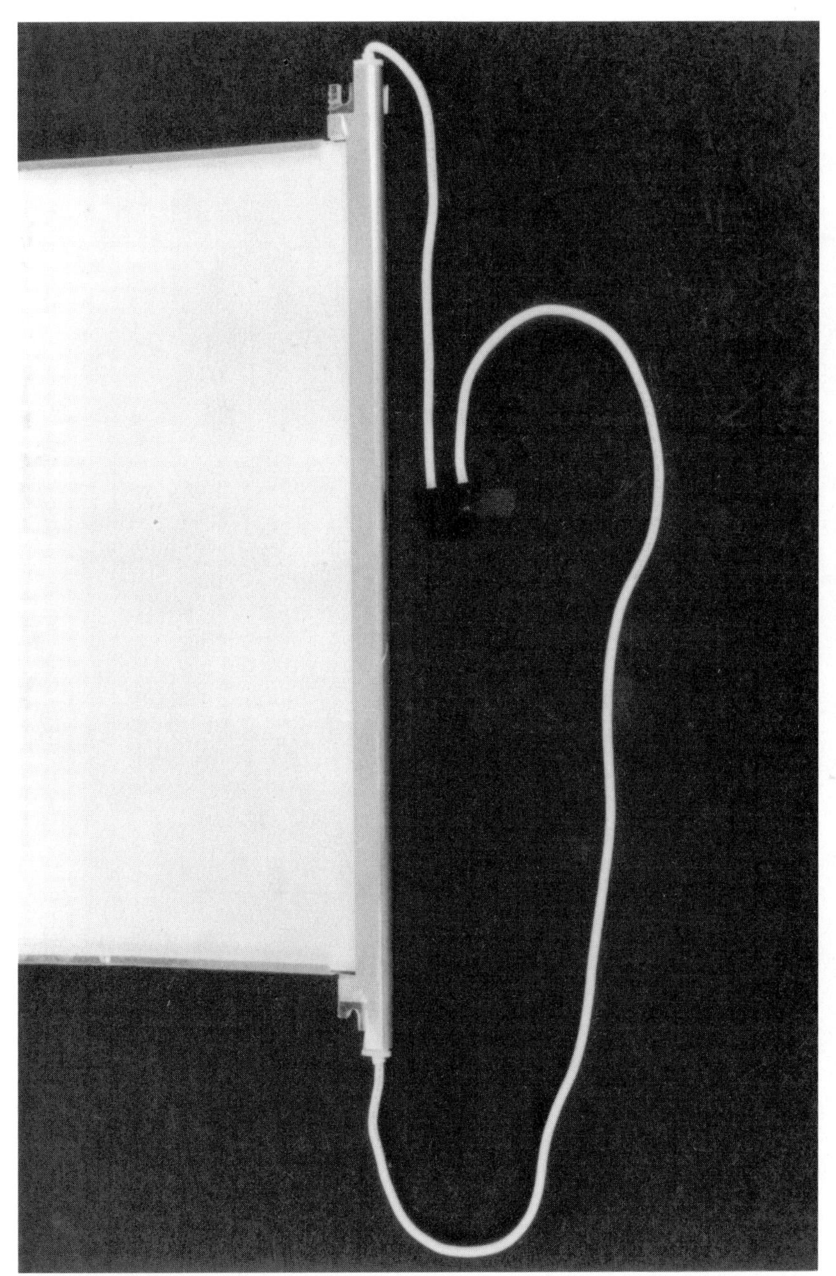

Figure 8.23 • Isolation Slits (Center Right and Left) in Metal Reflector Prevent Losses to Grounded Metal Frame (Upper Right and Left). Overall Losses about 6%

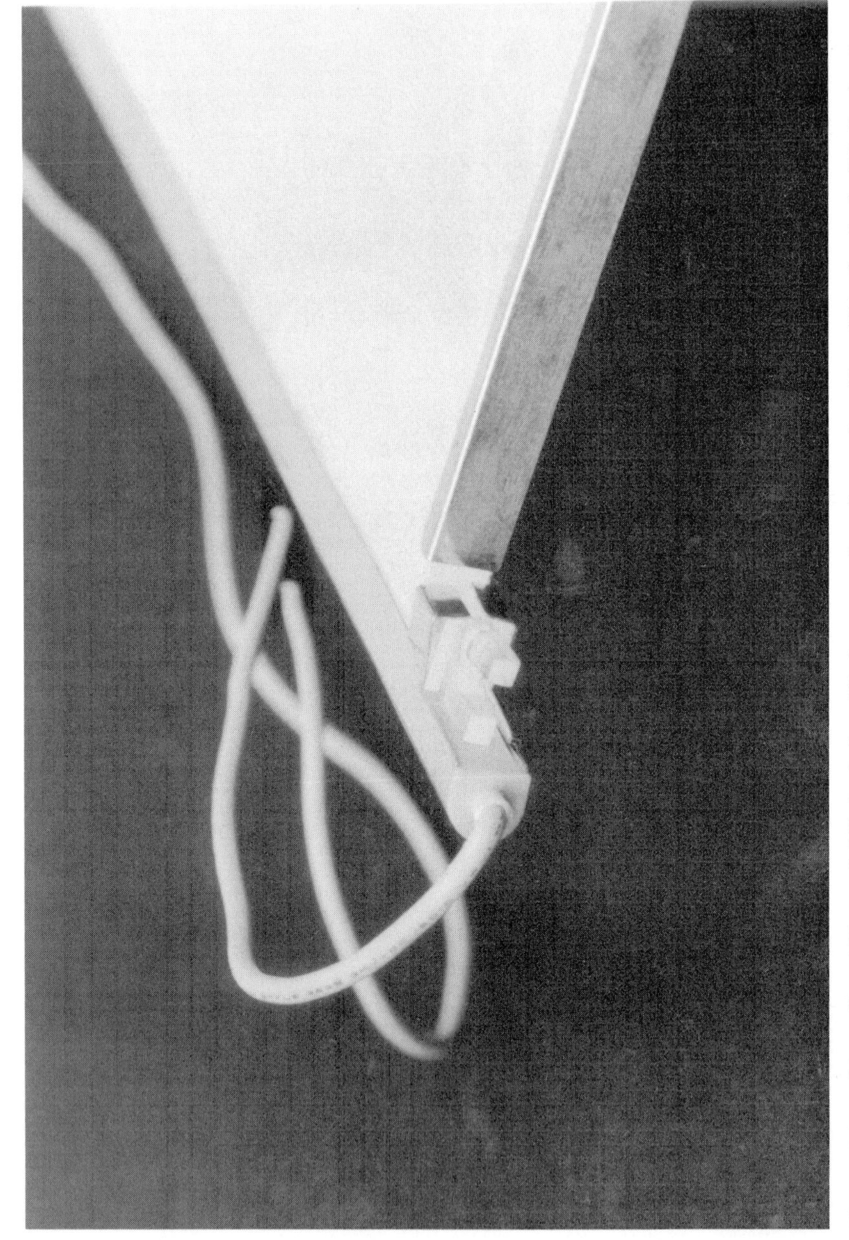

Figure 8.24 • Close-Up of Figure 8.23's Isolation Slit Construction. Secondary Benefit is Control of Reflector-to-Lamp Distance, Minimizing Capacitance

Figure 8.25 • Metal Cover over Lamp Causes 15% Loss. Replacing Cover Securing Screws with Nylon Types Floats Cover from Ground, Dropping Loss to 8%. Replacing Cover with Plastic Improves Loss to Only 3%...a 5X Improvement!

Figure 8.26 • Huge Metal Area over Lamp Causes 14% Loss. Replacing Metal in Lamp Area with Plastic Cuts Loss to 6%

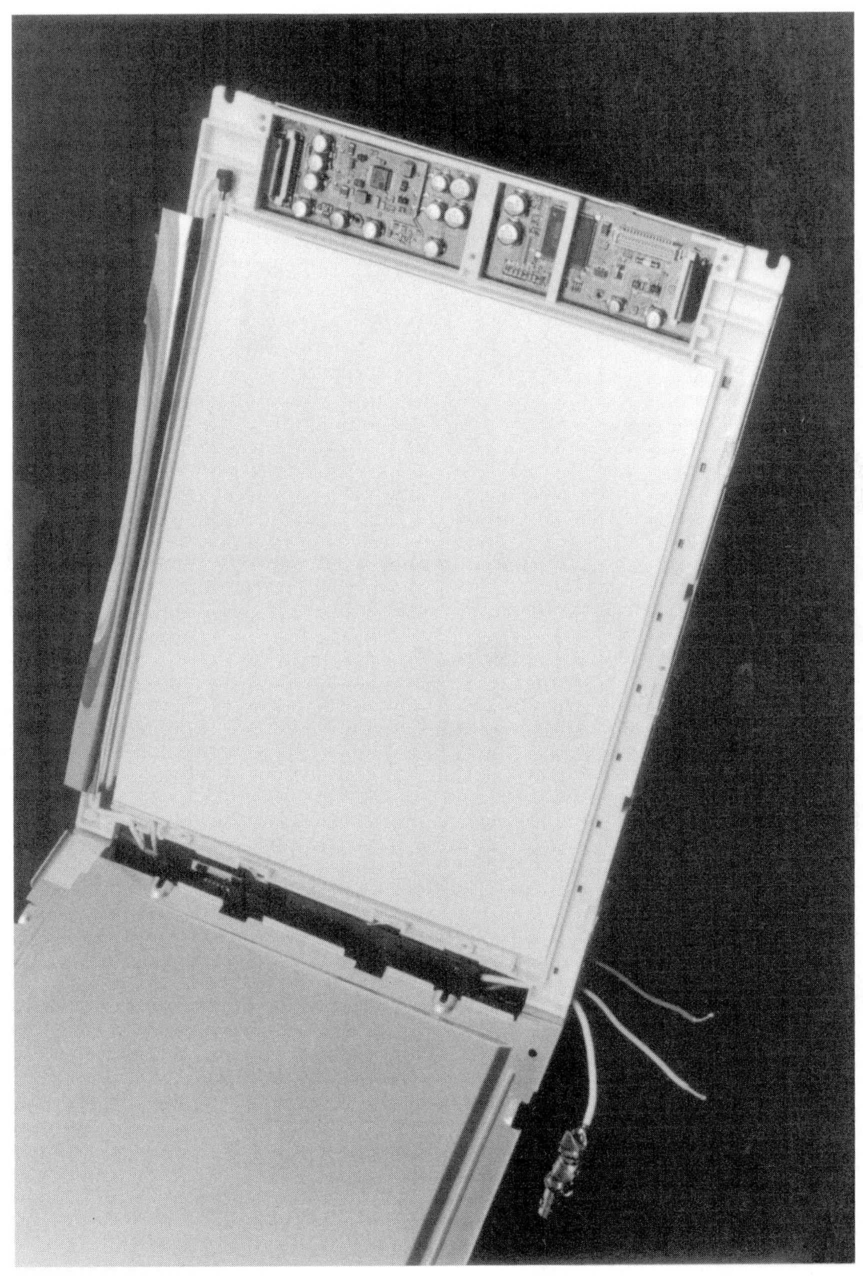

Figure 8.27 • Metallic Foil over Lamp (Upper Center) Dumps Absorbed Energy to Metal Rear Cover. 16% Loss Results

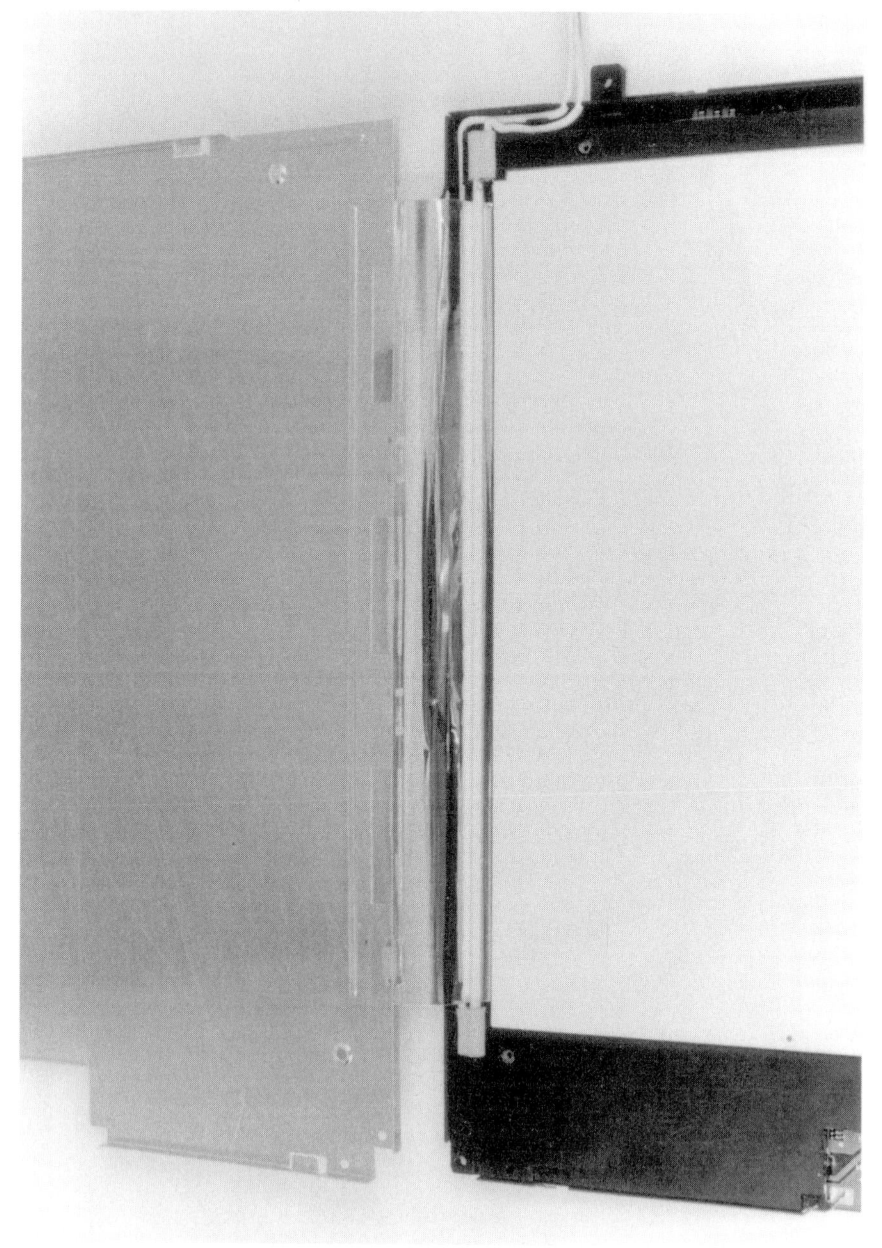

Figure 8.28 • Low Losses of the Display's Nonconductive Frame (Black Plastic) are Thrown Away by Lossy Reflective Foil Contacting Massive Metal Rear Cover. 15% Loss Results

Figure 8.29 • A Similar Situation to Figure 8.28. Large Metal Rear Cover Contacts Lossy Foil (Not Visible), Causing Huge Losses

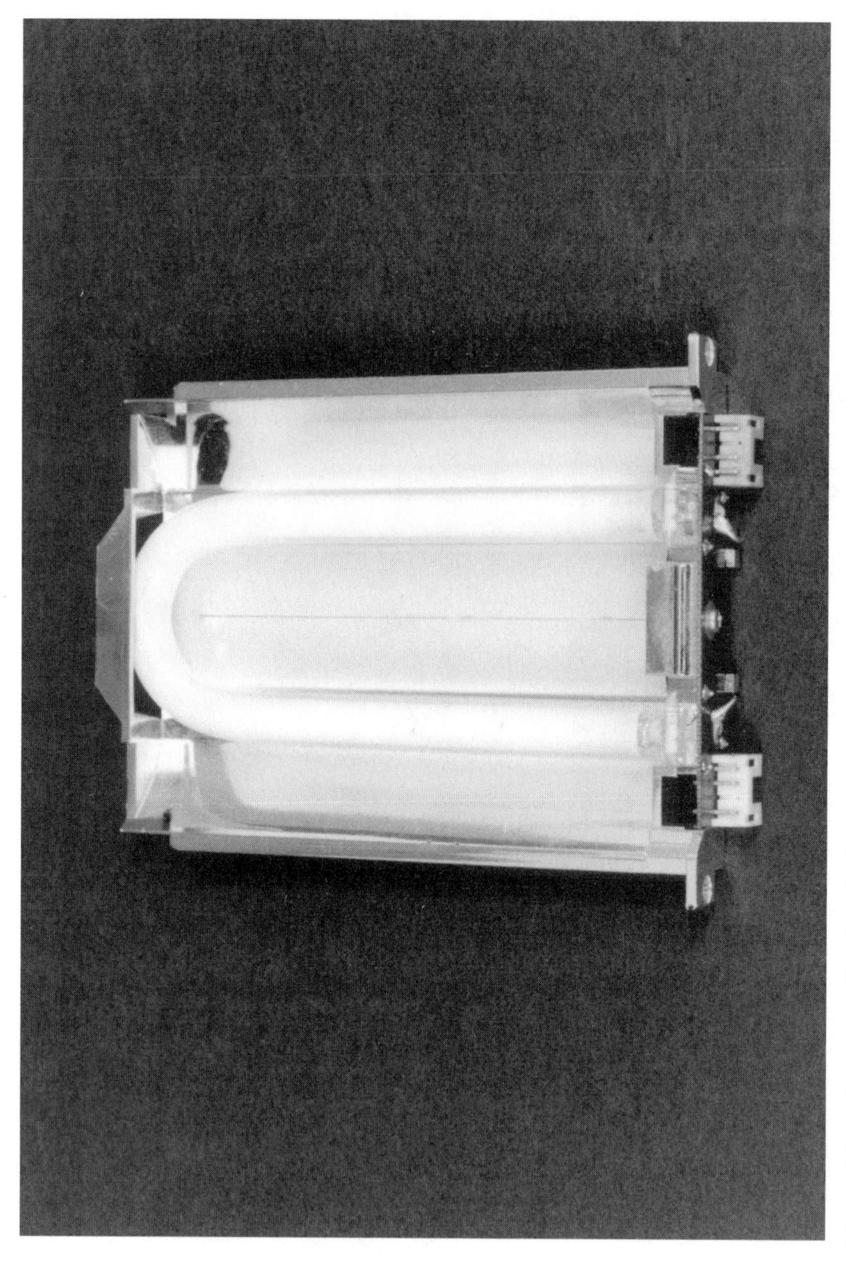

Figure 8.30 • Grounded Metallic Optical Reflector in Automotive Lamp Introduces 18% Loss. Optical Gain Over Nonmetallic Reflector May Justify Large Electrical Loss

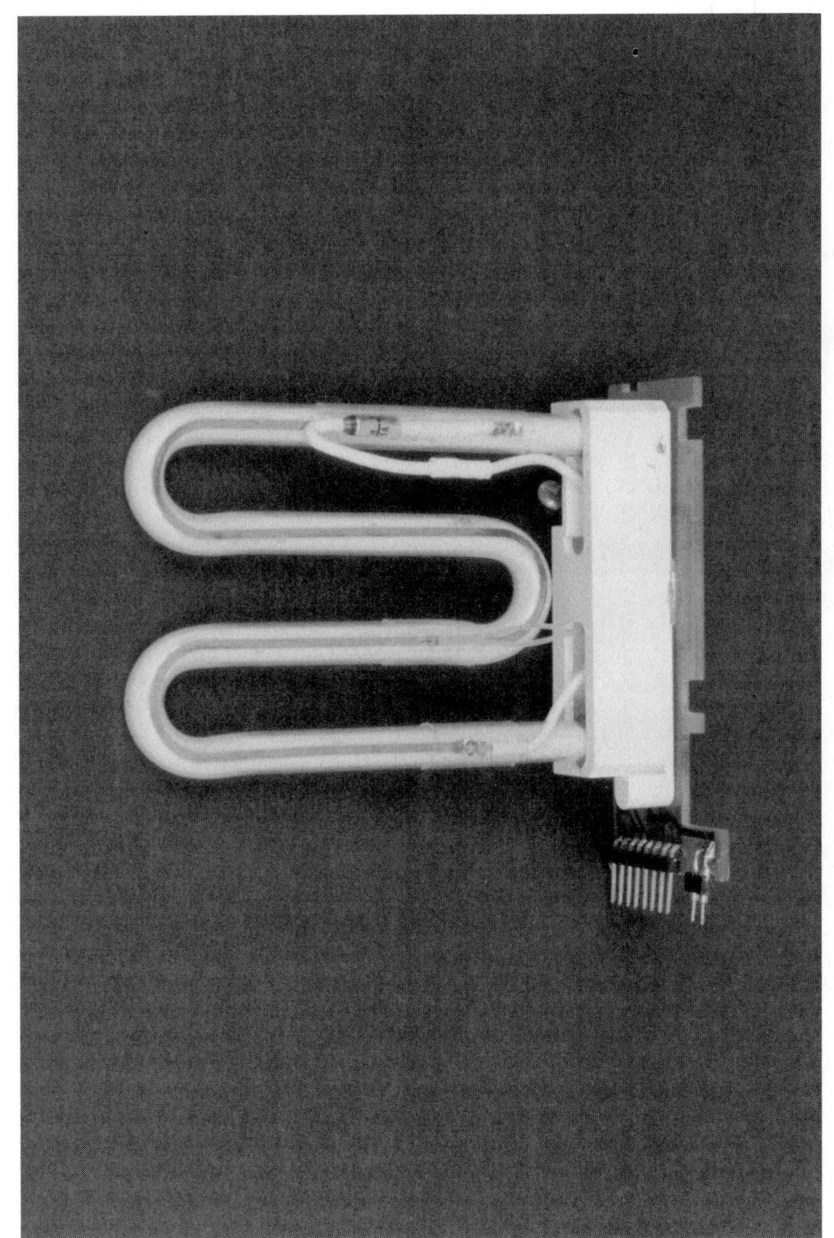

Figure 8.31 • Metallic Heater on Lamp in This Automotive Application Eases Low Temperature Starting but Causes 31% Loss

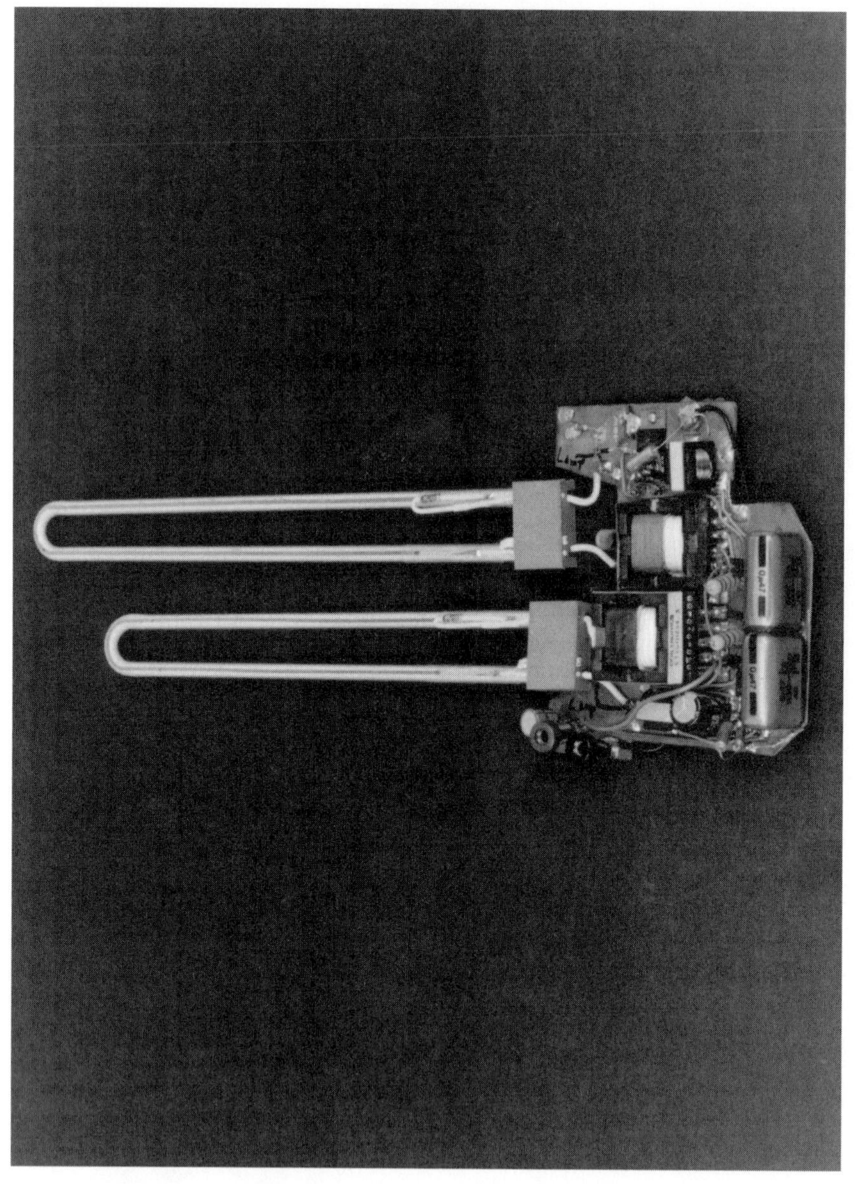

Figure 8.32 • Similar to Figure 8.31. Metallic Cold Start Heaters in Automotive Application Induce 23% Loss

Considerations for multilamp designs

Multiple-lamp designs are not recommended if lamp intensity matching is important. Maintaining emission matching over time, temperature and production variations is quite difficult. In some restricted cases multilamp displays may be a viable option, but a single lamp with good diffuser optics is almost always the better approach. Information on dual-lamp displays is presented here for reference purposes only.[5]

Systems using two lamps have some unique layout problems. Almost all dual-lamp displays are color units. The lower light transmission characteristics of color displays necessitates more light. As such, display manufacturers sometimes use two lamps to produce more light. The wiring layout of these dual-lamp color displays affects efficiency and illumination balance in the lamps. Figure 8.33 shows an "x-ray" view of a typical display. This symmetrical arrangement presents equal parasitic losses. If C1 and C2 and the lamps are well-matched, the circuit's current output splits evenly and equal illumination occurs.

Note 5: The text's tone is intended to convey our distaste for multilamp displays. They are the very soul of heartache.

Figure 8.34's display arrangement is less friendly. The asymmetrical wiring forces unequal losses and the lamps receive imbalanced current. Even with identical lamps, illumination may not be balanced. This condition is partially correctable by skewing values of C1 and C2. C1, because it drives greater parasitic capacitance, should be larger than C2. This tends to equalize the currents, promoting equal lamp drive. It is important to realize that this compensation does nothing to recapture the lost energy—efficiency is still compromised. There is no substitute for minimizing loss paths. Similarly, any change in lamp characteristics (e.g., aging) can cause imbalanced illumination to recur.

In general, imbalanced illumination causes fewer problems than might be supposed at high intensity levels. Unequal illumination is much more noticeable at lower levels. In the worst case the dimmer lamp may only partially illuminate. This phenomenon, sometimes called "Thermometering," is discussed in detail in the text section, "Floating Drive Circuits."

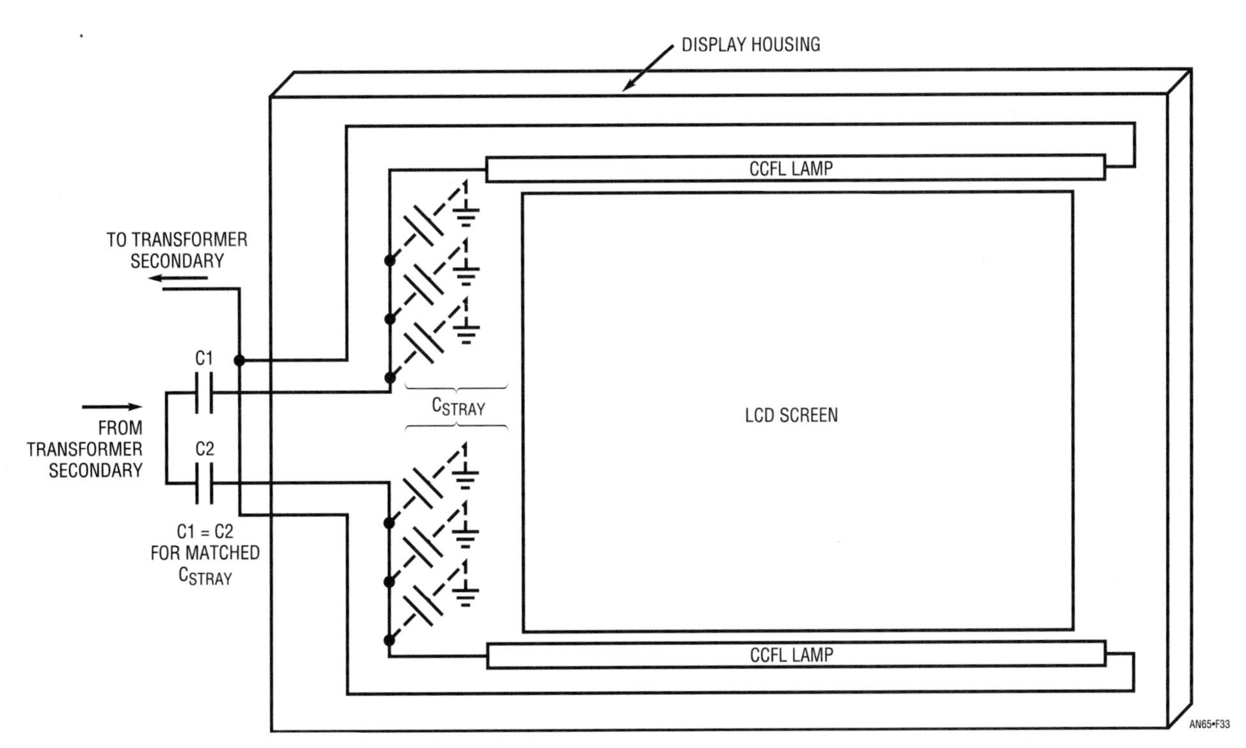

Figure 8.33 • Loss Paths for "Best Case" Dual-Lamp Display. Symmetry Promotes Balanced Illumination, but Lamp Limitations Dominate Achievable Results

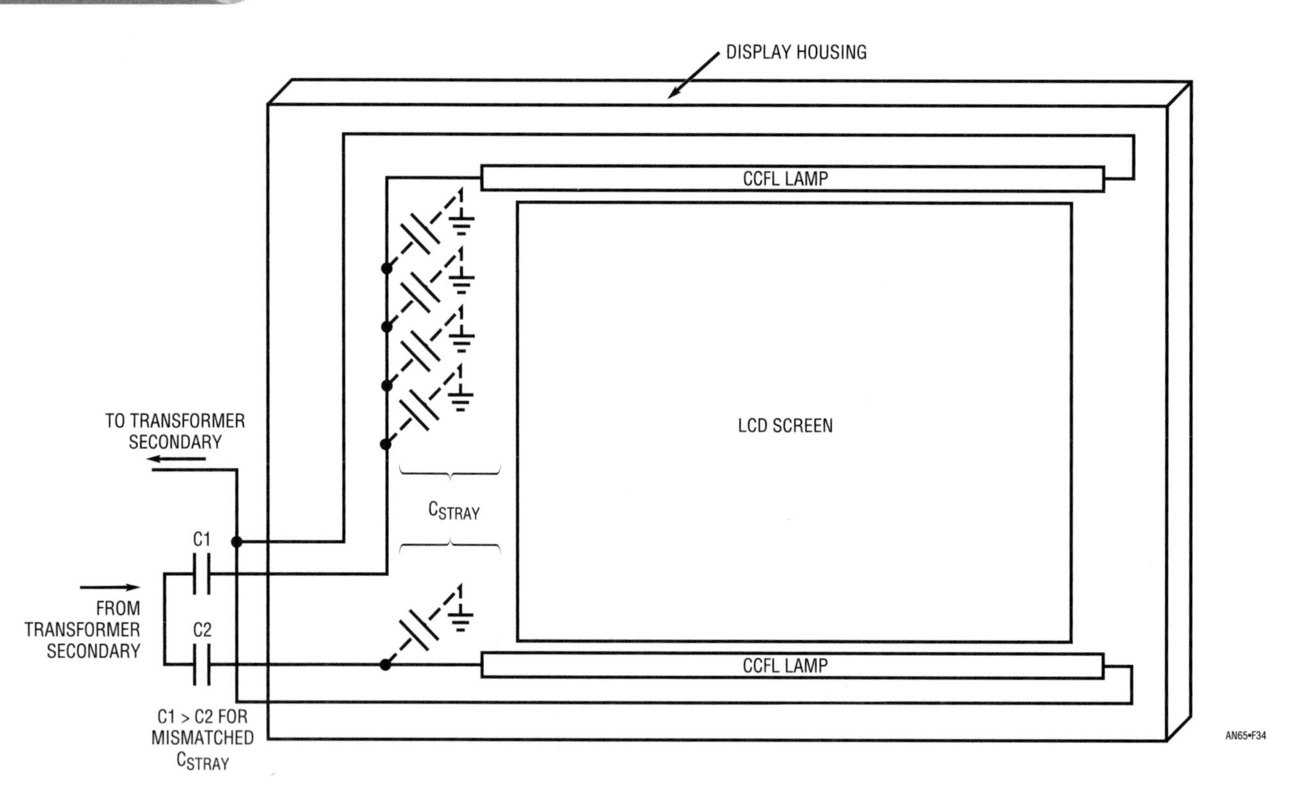

Figure 8.34 • Asymmetric Losses in a Dual-Lamp Display. Skewing C1 and C2 Values Compensates Imbalanced Loss Paths but Not Wasted Energy

CCFL power supply circuits

Choosing an approach for a general purpose CCFL power supply is difficult. A variety of disparate considerations make determining the "best" approach a thoughtful exercise. Above all, the architecture must be extraordinarily flexible. The sheer number and diversity of applications demands this. The considerations take many degrees of freedom. Power supply voltages range from 2V to 30V with output power from minuscule to 50W. The load is highly nonlinear and varies over operating conditions. The backlight is often located some distance from the primary power source, meaning the supply must tolerate substantial supply bus impedances. Similarly, it must not corrupt the supply bus with noise, or introduce appreciable RFI into the system or environment. Component count should be low and the supply must be physically quite small as space is usually extremely limited. Additionally, the circuit must be relatively layout-insensitive because of varying board shape requirements. Interface for shutdown and dimming control should accommodate either digital or analog inputs, including voltage, current, resistive, PWM or serial bit-stream addressing. Finally, lamp current should be predictable and stable with changes in time, temperature and supply voltage.

A current-fed, feedback-controlled resonant Royer converter meets these requirements.[6] This approach, because of its extreme flexibility, is a favorable compromise. It operates over wide supply ranges and scales well over a broad output power range. Current is taken from the supply bus almost continuously, making the circuit tolerate supply bus impedance. This characteristic also means that circuit operation does not corrupt power supply lines. There is no RFI problem and component count is low. It is small, relatively insensitive to layout and easy to interface to. Lastly, lamp current is stable and predictable over operating conditions.

Figure 8.35 is a practical CCFL power supply circuit based on the above discussion. Efficiency is 88% with an input voltage range of 6.5V to 20V. This efficiency figure can be degraded by about 3% if the LT172 V_{IN} pin is powered from the same supply as the main circuit V_{IN}

Note 6: See Appendices K and L for detailed discussion on architecture selection and the Royer configuration.

terminal. Lamp intensity is continuously and smoothly variable from zero to full intensity. When power is applied the LT1172 switching regulator's Feedback pin is below the device's internal 1.2V reference, causing full duty cycle modulation at the V_{SW} pin (Trace A, Figure 8.36). V_{SW} conducts current (Trace B) which flows from L1's center tap, through the transistors, into L2. L2's current is deposited in switched fashion to ground by the regulator's action.

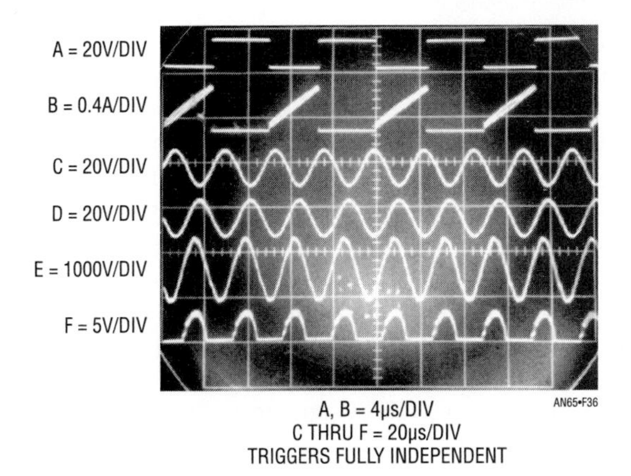

Figure 8.36 • Waveforms for the Cold Cathode Fluorescent Lamp Power Supply. Note Independent Triggering on Traces A and B, and C through F

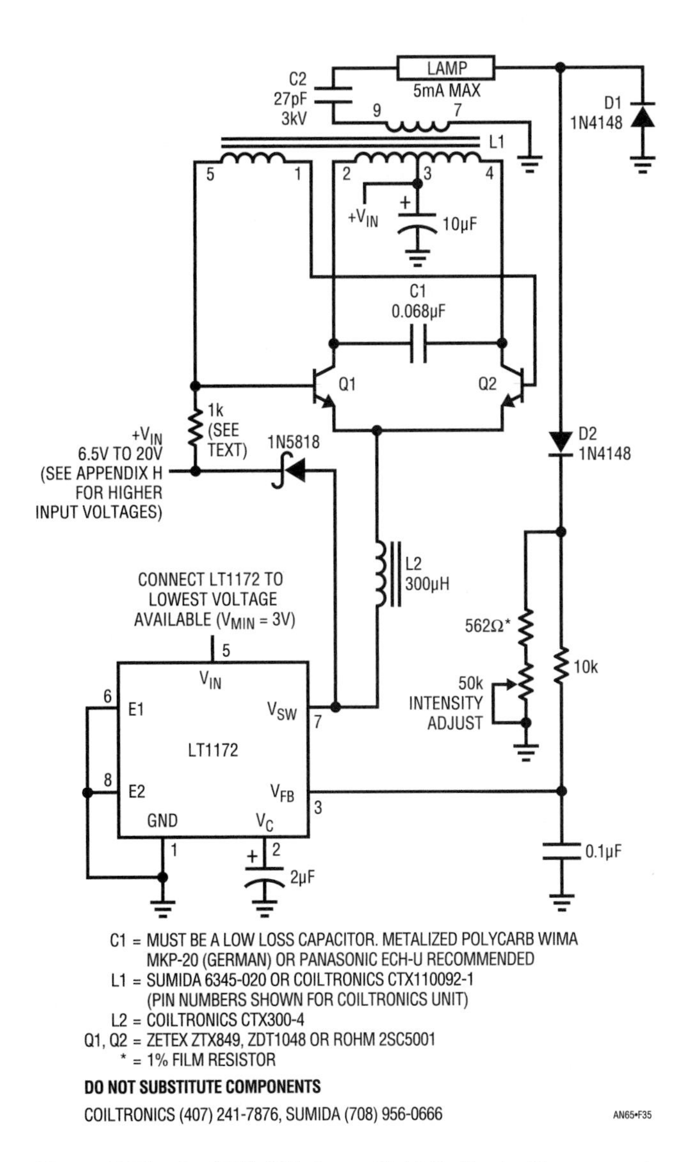

Figure 8.35 • An 88% Efficiency Cold Cathode Fluorescent Lamp Power Supply

L1 and the transistors comprise a current driven Royer class converter[7] which oscillates at a frequency primarily set by L1's characteristics (including its load) and the 0.068µF capacitor. LT1172 driven L2 sets the magnitude of the Q1/Q2 tail current, hence L1's drive level. The 1N5818 diode maintains L2's current flow when the LT1172 is off. The LT1172's 100kHz clock rate is asynchronous with respect to the push/pull converter's (60kHz) rate, accounting for Trace B's waveform thickening.

The 0.068µF capacitor combines with L1's characteristics to produce sine wave voltage drive at the Q1 and Q2 collectors (Traces C and D respectively). L1 furnishes voltage step-up and about $1400V_{P\text{-}P}$ appears at its secondary (Trace E). Current flows through the 27µF capacitor into the lamp. On negative waveform cycles, the lamp's current is steered to ground via D1. Positive waveform cycles are directed via D2 to the ground referred 562Ω/50k potentiometer chain. The positive half-sine appearing across the resistors (Trace F) represents 1/2 the lamp current. This signal is filtered by the 10k/0.1µF pair and presented to the LT1172's Feedback pin. This connection closes a control loop which regulates lamp current. The 2µF capacitor at the LT1172's V_C pin

Note 7: See Appendix K, "Who Was Royer and What Did He Design?" See also Reference 2.

provides stable loop compensation. The loop forces the LT1172 to switch mode modulate L2's average current to whatever value is required to maintain constant current in the lamp. The constant current's value, and hence lamp intensity, may be varied with the potentiometer. The constant current drive allows full 0% to 100% intensity control with no lamp dead zones or "pop-on" at low intensities.[8] Additionally, lamp life is enhanced because current cannot increase as the lamp ages.

The circuit's 0.1% line regulation is notably better than some other approaches. This tight regulation prevents lamp intensity variation when abrupt line changes occur. This typically happens when battery-powered apparatus is connected to an AC-powered charger. The circuit's excellent line regulation derives from the fact that L1's drive waveform never changes shape as input voltage varies. This characteristic permits the simple 10kΩ/0.1µF RC to produce a consistent response. The RC averaging characteristic has serious error compared to a true RMS conversion, but the error is constant and "disappears" in the 562Ω shunt's value.

This circuit is similar to one previously described[9] but its 88% efficiency is 6% higher. The efficiency improvement is primarily due to the transistor's higher gain and lower saturation voltage. The base drive resistor's value (nominally 1k) should be selected to provide full V_{CE} saturation without inducing base overdrive or beta starvation. A procedure for doing this is described in a following section, "General Optimization and Measurement Considerations."

Figure 8.37's circuit is similar, but uses a transformer with lower copper and core losses to increase efficiency to 91%. The trade-off is slightly larger transformer size. Additionally, a higher frequency switching regulator offers slightly lower V_{IN} current, aiding efficiency. L1's smaller value, a result of the higher frequency operation, permits slightly reduced copper loss. The transformer options listed allow efficiency optimization over the supply range of interest. Value shifts in C1, L2 and the base drive resistor reflect different transformer characteristics. This circuit also features shutdown and a DC or pulse width controlled dimming input. Appendix F "Intensity Control and Shutdown Methods, "details operation of these features. Figure 8.38, directly derived from Figure 8.37, produces 10mA output to drive color LCDs at 92% efficiency. The slight efficiency improvement comes from a reduction in regulator "housekeeping" current as a percentage of total current drain. Value changes in components are the result of higher power operation. The most significant change involves driving two lamps. Accommodating two lamps involves separate ballast capacitors but circuit operation is similar. Dual-lamp designs reflect slightly different loading back through the transformer's primary. C2 usually ends up in the 10pF to 47pF range. Note that C2A and B appear with their lamp loads in parallel across the transformer's secondary. As such, C2's value is often

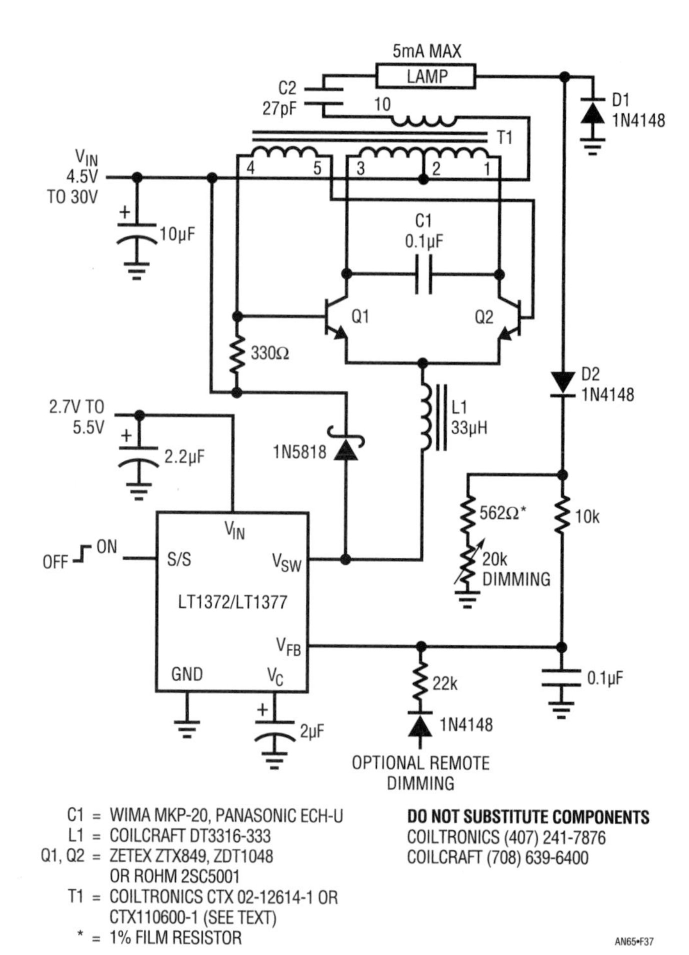

Figure 8.37 • A 91% Efficient CCFL Supply for 5mA Loads Features Shutdown and Dimming Inputs. Higher Frequency Switching Regulator Reduces L1's Size While Requiring Less V_{IN} Current

Note 8: Controlling a nonlinear load's current, instead of its voltage, permits applying this circuit technique to a wide variety of nominally evil loads. See Appendix I, "Additional Circuits."

Note 9: See "Illumination Circuity for Liquid Crystal Displays," Linear Technology Corporation, Application Note 49, August 1992 and "Techniques for 92% Efficient LCD Illumination," Linear Technology Corporation, Application Note 55, August 1993.

smaller than in a single-lamp circuit using the same type lamp. Ideally, the transformer's secondary current splits evenly between the C2-lamp branches, with the total load current being regulated. In practice, differences between C2A and B, and differences in lamps and lamp wiring layout preclude a perfect current split. Practically, these differences are small and the lamps appear to emit an equal amount of light at high intensity. Layout and lamp matching can influence C2's value. Some techniques for dealing with these issues appear in the text section, "Considerations for Multilamp Designs." As previously stated, dual-lamp designs are distinctly not recommended, particularly if balanced illumination over wide dimming ranges is required.

Figure 8.39 uses a dedicated CCFL IC, the LT1183, to enhance circuit performance. The Royer-based high voltage converter portion is recognizable from previous circuits, with the 200kHz LT1183 performing the switching regulator/feedback function. This IC also features open lamp protection circuitry, simplified frequency compensation, a separate regulator providing LCD contrast and other features.[10] The contrast supply is driven by the LT1183 with L3 and associated discrete components completing the function. The CCFL and contrast outputs may be adjusted with DC, PWM or potentiometers.

Note 10: Open lamp protection is often desirable and may be added to the previous circuits at the cost of some discrete components. See Appendix E, "Open Lamp/Overload Protection." Frequency compensation issues are covered in the text section "Feedback Loop Stability Issues." See Appendix J for discussion of LCD contrast supplies.

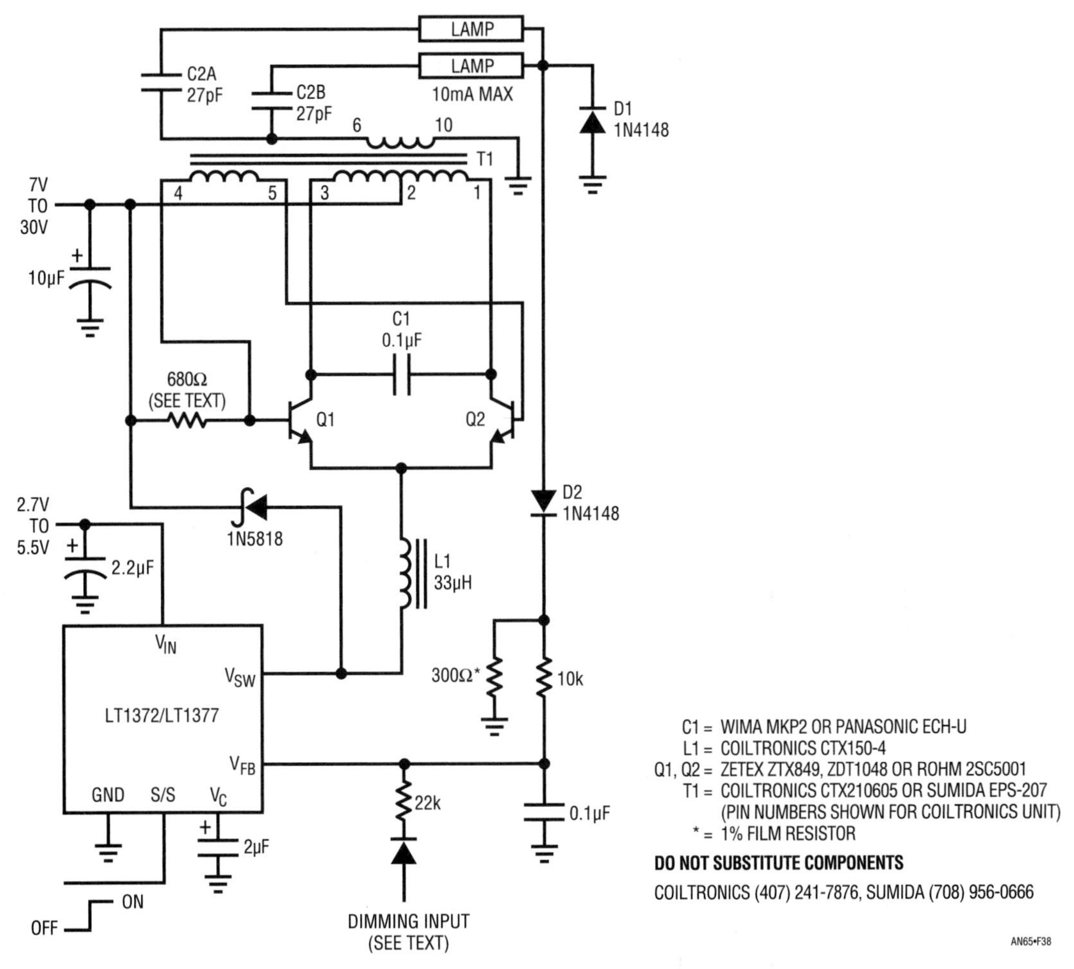

Figure 8.38 • A 92% Efficient CCFL Supply for 10mA Loads Features Shutdown and Dimming Inputs. Dual-Lamp Designs, Typical of Early Color Displays, are Not Recommended

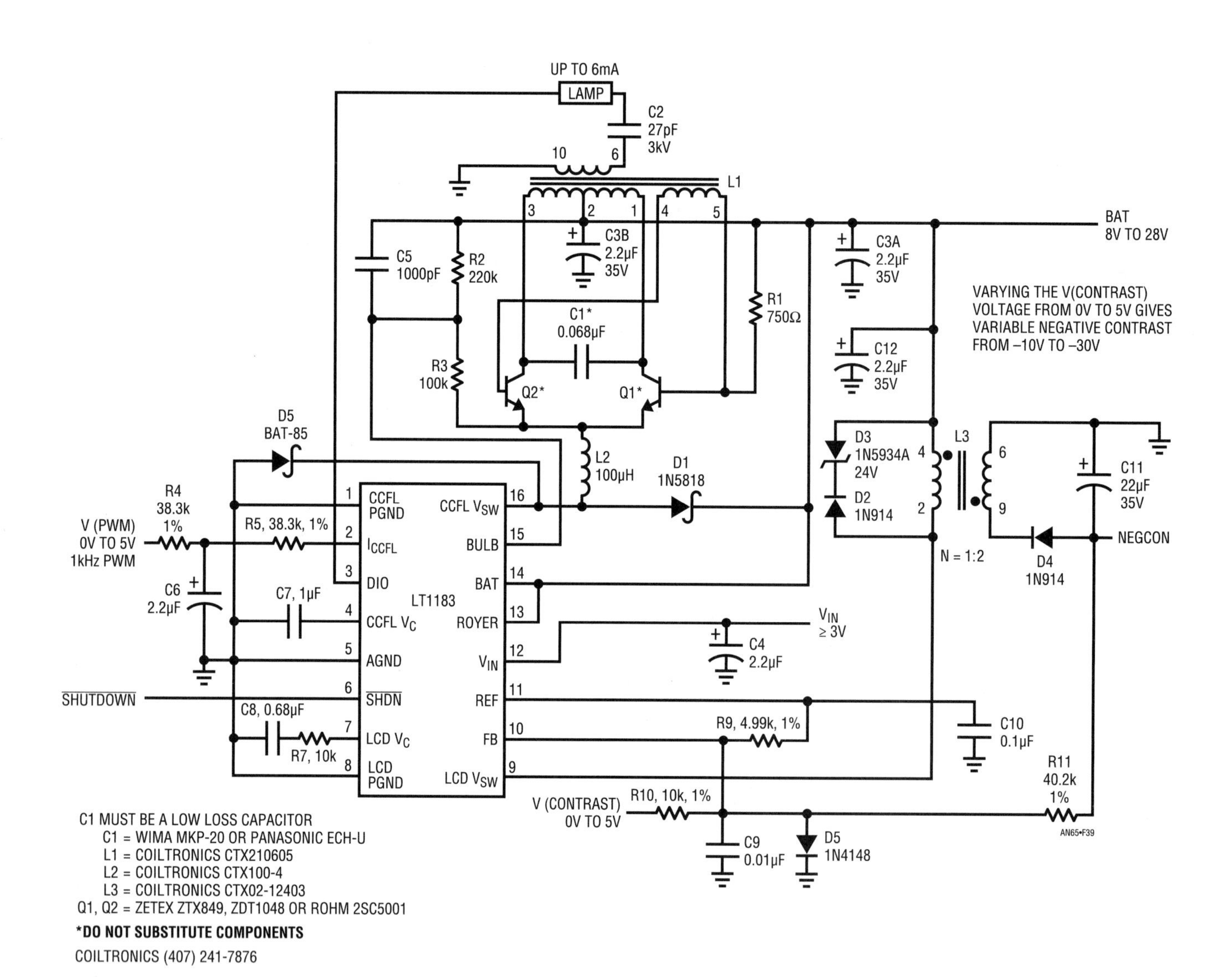

Figure 8.39 • Dedicated Backlight IC Includes Switching Regulator, Open Lamp Protection and LCD Contrast Supply. 200kHz Operation Minimizes L2 Size. Shutdown and Control Inputs Are Simplified

Low power CCFL power supplies

Many applications require relatively low power CCFL backlighting. Figure 8.40's variation, optimized for low voltage inputs, produces 4mA output. Circuit operation is similar to the previous examples. The fundamental difference is L1's higher turns ratio, which accommodates the reduced available drive voltage. The circuit values given are typical, although some variation occurs with various lamps and layouts.

Figure 8.41's design, the so-called "dim backlight," is optimized for very low current lamp operation. The circuit is meant for use at low input voltages, typically 2V to 6V with a 1mA maximum lamp current. This circuit maintains control down to lamp currents of 1μA, a very dim light! It is intended for applications where the longest possible battery life is desired. Primary supply drain ranges from hundreds of microamperes to 100mA with lamp currents of microamps to 1mA. In shutdown the circuit pulls only 100μA. Maintaining high efficiency at low lamp currents requires modifying the basic design.

Achieving high efficiency at low operating current requires lowering quiescent power drain. To do this the previously employed pulse width modulator-based devices are replaced with an LT1173. The LT1173 is a Burst Mode® operation regulator. When this device's Feedback pin is too low it delivers a burst of output current pulses, putting energy into the transformer and restoring the feedback point. The regulator maintains control by appropriately modulating the burst duty cycle. The ground referred diode at the V_{SW} pin prevents substrate turn-on due to excessive L2 ring-off.

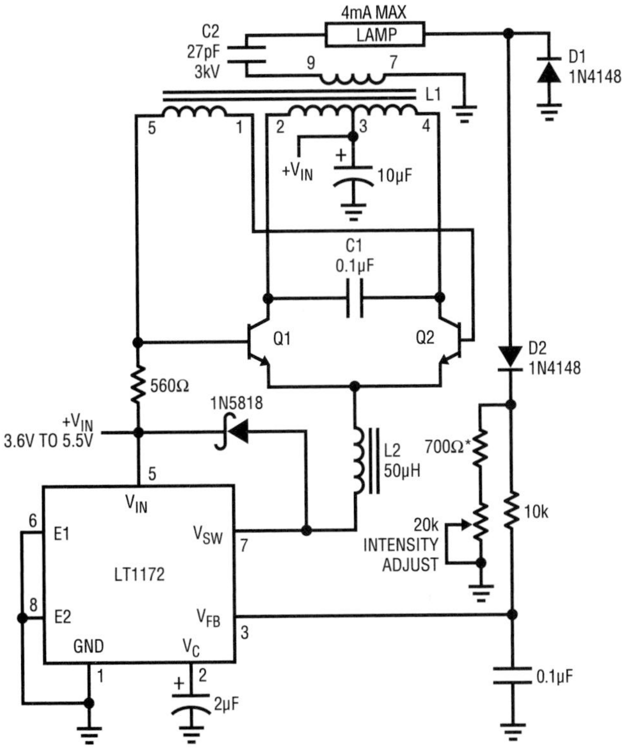

C1 = MUST BE A LOW LOSS CAPACITOR.
METALIZED POLYCARB
WIMA MKP-20 (GERMAN) OR PANASONIC ECH-U RECOMMENDED
L1 = COILTRONICS CTX110654-1
L2 = COILTRONICS CTX50-4
Q1, Q2 = ZETEX ZTX849, ZDT1048 OR ROHM 2SC5001
* = 1% FILM RESISTOR
DO NOT SUBSTITUTE COMPONENTS
COILTRONICS (407) 241-7876, SUMIDA (708) 956-0666

AN65•F40

Figure 8.40 • A 4mA Design Intended for Low Voltage Operation. L1's Modified Turns Ratio Allows Operation Down to 3.6V

C1 = MUST BE A LOW LOSS CAPACITOR.
METALIZED POLYCARB
WIMA FKP2, MKP-20 (GERMAN) OR PANASONIC ECH-U RECOMMENDED
L1 = SUMIDA 6345-020 OR COILTRONICS CTX110092-1
PIN NUMBERS SHOWN FOR COILTRONICS UNIT
L2 = TOKO 262LYF-0091K (408) 432-8251
Q1, Q2 = ZETEX ZTX849, ZDT1048 OR ROHM 2SC5001
DO NOT SUBSTITUTE COMPONENTS

AN65•F41

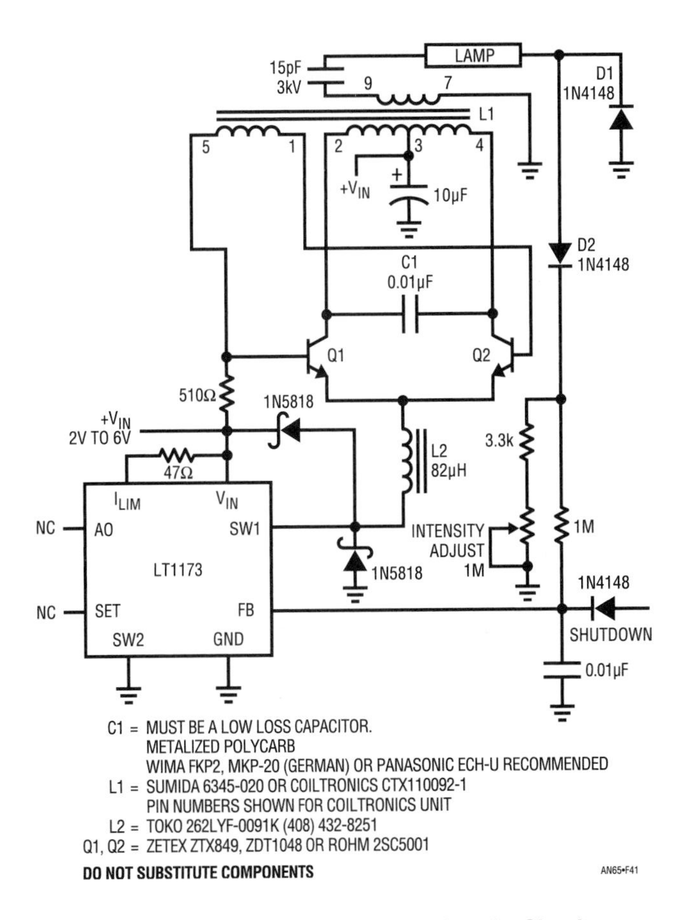

Figure 8.41 • Low Power CCFL Power Supply. Circuit Controls Lamp Current Over a 1μA to 1mA Range

During the off periods the regulator is essentially shut down. This type of operation limits available output power, but cuts quiescent current losses. In contrast, the other circuit's pulse width modulator type regulators maintain "housekeeping" current between cycles. This results in more available output power but higher quiescent currents.

Figure 8.42 shows operating waveforms. When the regulator comes on (Trace A, Figure 8.42) it delivers bursts of output current to the L1/Q1/Q2 high voltage converter. The converter responds with bursts of ringing at its resonant frequency.[11] The circuit's loop operation is similar to the previous designs except that T1's drive waveform varies with supply. Because of this, line regulation suffers and the circuit is not recommended for wide ranging inputs.

Some lamps may display nonuniform light emission at very low excitation currents. See the text section, "Floating Lamp Circuits."

A CCFL power supply that addresses the previous circuit's line regulation problems and operates from 2V to 6V is detailed in Figure 8.43. This circuit, contributed by

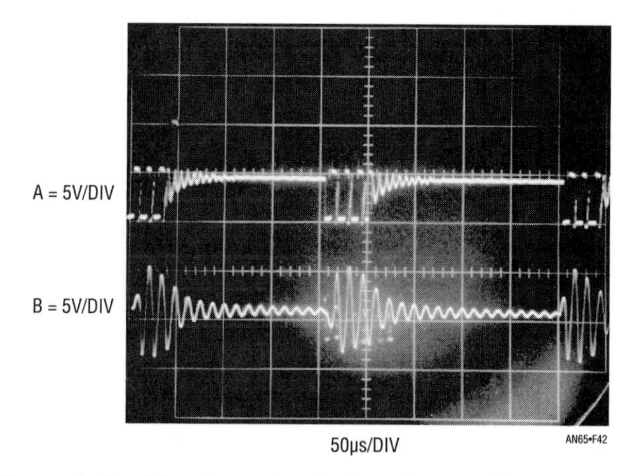

Figure 8.42 • Waveforms for the Low Power CCFL Power Supply. LT1173 Burst Type Regulator (Trace A) Periodically Excites the Resonant High Voltage Converter (Q1 Collector Is Trace B)

Note 11: The discontinous energy delivery to the loop causes substantial jitter in the burst repetition rate, although the high voltage section maintains resonance. Unfortunately, circuit operation is in the "chop" mode region of most oscilloscopes, precluding a detailed display. "Alternate" mode operation causes waveform phasing errors, producing an inaccurate display. As such, waveform observation requires special techniques. Figure 8.42 was taken with a dual-beam instrument (Tektronix 556) with both beams slaved to one time base. Single sweep triggering eliminated jitter artifacts. Most oscilloscopes, whether analog or digital, will have trouble reproducing this display.

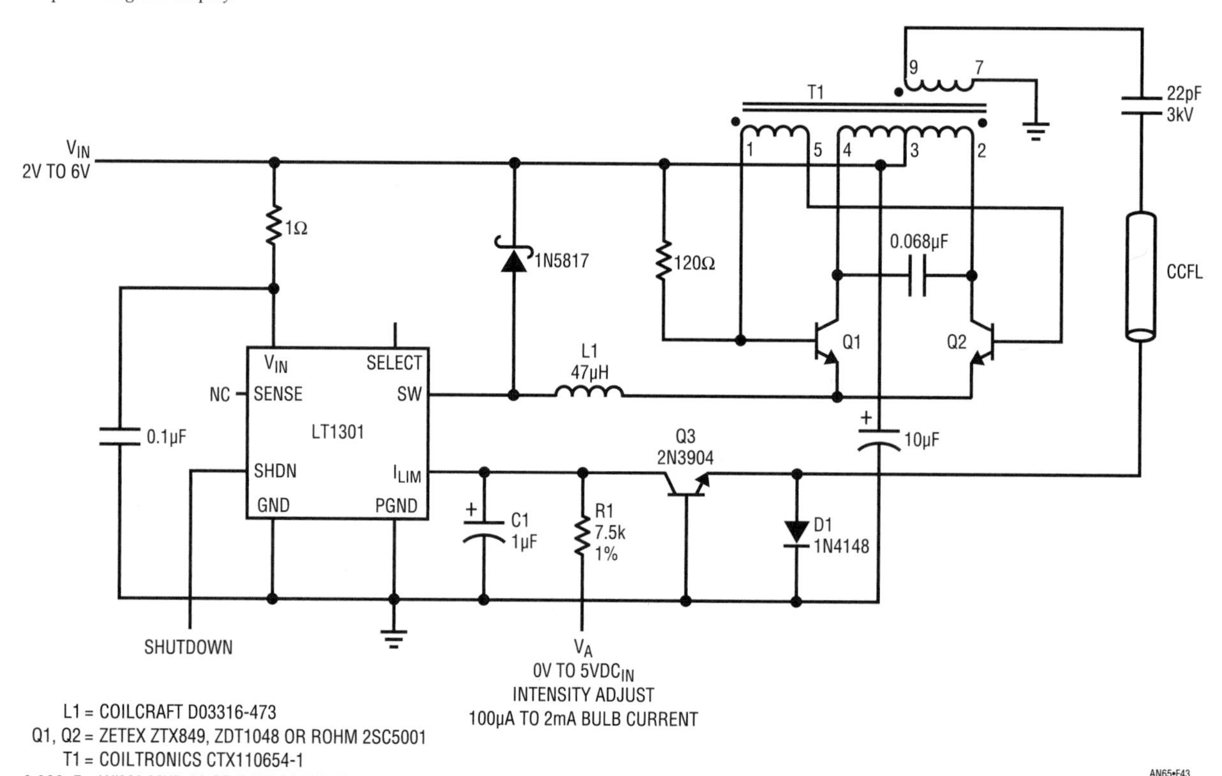

Figure 8.43 • Low Power Cold Cathode Fluorescent Lamp Supply Is Optimized for Low Voltage Inputs and Small Lamps

Steve Pietkiewicz of LTC, can drive a small CCFL over a 100μA to 2mA range.

The circuit uses an LT1301 micropower DC/DC converter IC in conjunction with a current driven Royer class converter comprised of T1, Q1, and Q2. When power and intensity adjust voltage are applied, the LT1301's I_{LIM} pin is driven slightly positive, causing maximum switching current through the IC's internal Switch pin (SW). Current flows from T1's center tap, through the transistors, into L1. L1's current is deposited in switched fashion to ground by the regulator's action.

Circuit efficiency ranges from 80% to 88% at full load, depending on line voltage. Current mode operation combined with the Royer's consistent waveshape vs input results in excellent line rejection. The circuit has none of the line rejection problems attributable to the hysteretic voltage control loops typically found in low voltage micropower DC/DC converters. This is an especially desirable characteristic for CCFL control, where lamp intensity must remain constant with shifts in line voltage.

The Royer converter oscillates at a frequency primarily set by T1's characteristics (including its load) and the 0.068μF capacitor. LT1301 driven L1 sets the magnitude of the Q1/Q2 tail current, hence T1's drive level. The 1N5817 diode maintains L1's current flow when the LT1301's switch is off. The 0.068μF capacitor combines with T1's characteristics to produce sine wave voltage drive at the Q1 and Q2 collectors. T1 furnishes voltage step-up and about 1400$V_{P\text{-}P}$ appears at its secondary. Alternating current flows through the 22pF capacitor into the lamp. On positive half-cycles the lamp's current is steered to ground via D1. On negative half-cycles the lamp's current flows through Q3's collector and is filtered by C1. The LT1301's I_{LIM} pin acts as a 0V summing point with about 25μA bias current flowing out of the pin into C1. The LT1301 regulates L1's current to equalize Q3's average collector current, representing 1/2 the lamp current, and R1's current, represented by $V_A/R1$. C1 smooths all current flow to DC. When V_A is set to zero, the I_{LIM} pin's bias current forces about 100μA bulb current.

High power CCFL power supply

As mentioned, the CCFL circuit approach presented here scales quite nicely over a wide range of output power. Most circuits are in the 0.5W to 3W region due to the application's small size and battery-driven nature. Automotive, aircraft, desktop computer and other displays often require much higher power.

Figure 8.44's arrangement is a scaled-up version of the text's CCFL circuits. This design, similar to ones employed for automotive use, drives a 25W CCFL. There are virtually no configuration changes, although most component power ratings have increased. The transistors can handle the higher currents, but all other power components are higher capacity. Efficiency is about 80%.

Additional high power circuits appear in Appendix I, "Additional Circuits."

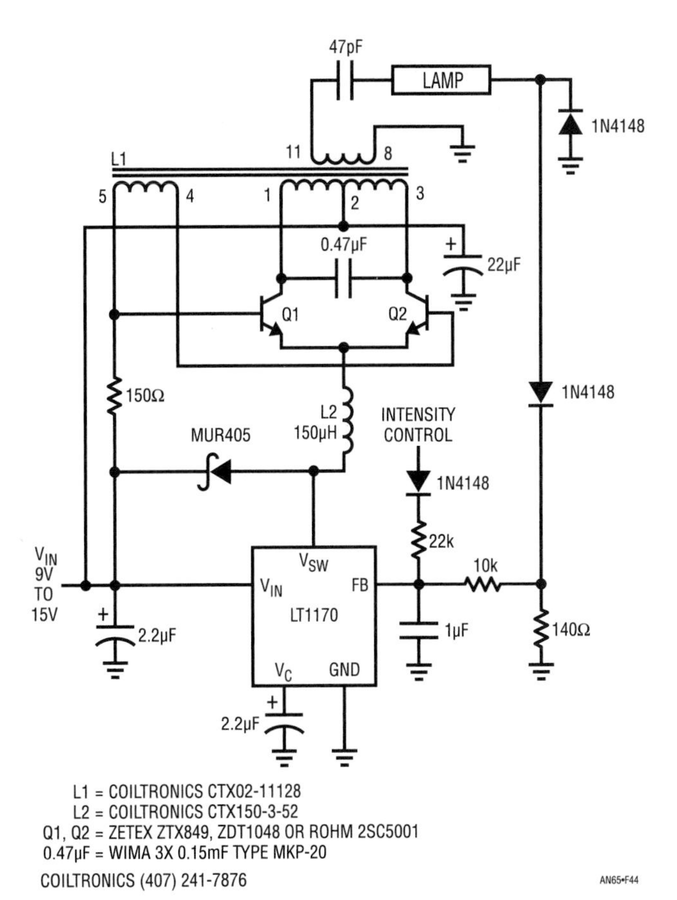

Figure 8.44 • A 25W CCFL Supply Is a Scaled Version of Lower Power Circuits

"Floating" lamp circuits

All circuits presented to this point drive the lamp in single-ended fashion. Similarly, Figure 8.45 shows one lamp electrode receiving drive with the other terminal essentially at ground. This causes significant loss via parasitic paths associated with the lamp's driven end. This is so because of the large voltage swing in this region. The parasitic paths near the lamp's grounded end undergo relatively little swing, contributing small energy loss. Unfortunately, the lost energy is heavily voltage-dependent ($E = 1/2\ CV^2$) and net energy loss is excessive if driven end parasitics are large. Figure 8.46 minimizes the losses by altering the drive scheme. In this case the lamp is driven from both ends instead of grounding one end. This "floating" lamp arrangement requires only half the voltage swing at each lamp end instead of full swing at one end. This introduces more loss in the parasitic paths previously associated with the grounded end. In most cases these increased losses are favorably offset by the reduced swing because of the V^2 loss term associated with voltage amplitude.

The advantage gained varies considerably with display type, although a 10% to 20% reduction in lost energy is common. In some displays loss reduction is not as good, and occasionally improvement is negligible. Heavily asymmetric wiring to or within the display can sometimes make floating drive more lossy than grounded drive. In such cases testing in both modes is necessary to determine which type drive is most efficient.

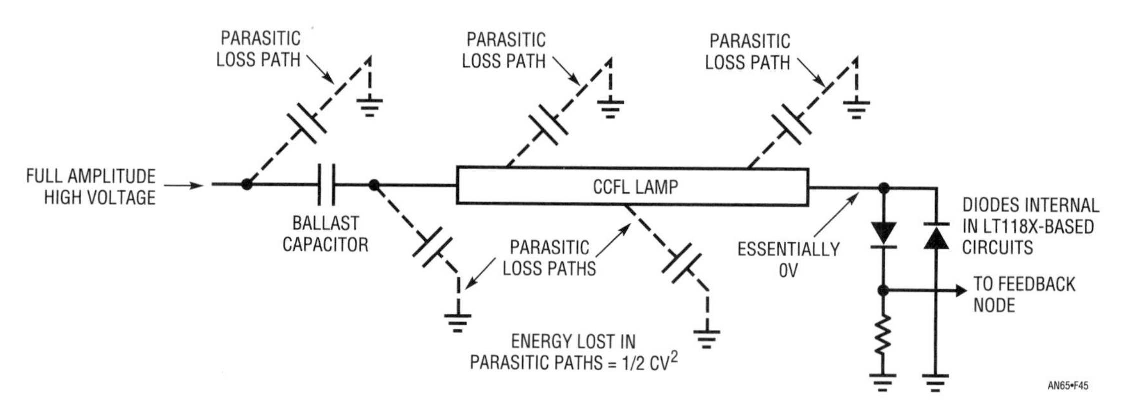

Figure 8.45 • Ground Referred Lamp Drive Has Large Energy Loss in High Voltage Regions Due to Full Amplitude Swing

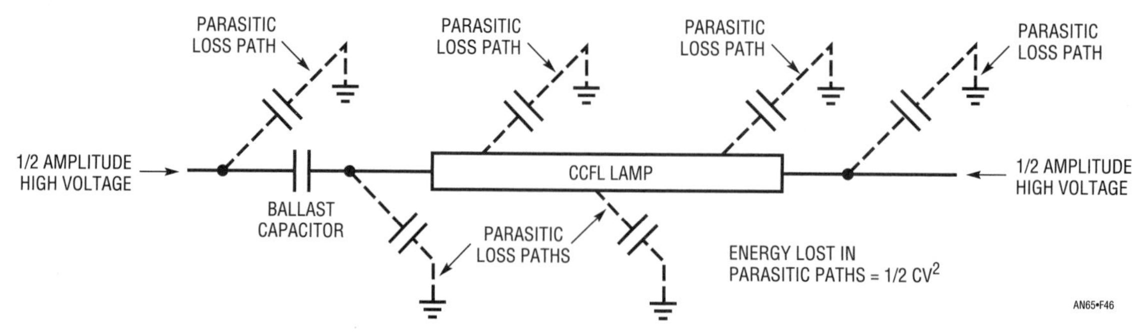

Figure 8.46 • "Floating" Lamp Allows Reduced, Bipolar Drive, Cutting Losses Due to Parasitic Capacitance Paths. Formerly Grounded Lamp End's Paths Absorb More Energy Than Before, but Overall Loss Is Lower Due to Equation's V^2 Term

A second advantage of floating operation is extended illumination range. "Grounded" lamps operating at relatively low currents may display the "thermometer effect," that is, light intensity may be nonuniformly distributed along lamp length.

Figure 8.47 shows that although lamp current density is uniform, the associated field is imbalanced. The field's low intensity, combined with its imbalance, means that there is not enough energy to maintain uniform phosphor glow beyond some point. Lamps displaying the thermometer effect emit most of their light near the driven electrode, with rapid emission fall-off as distance from the electrode increases. Placing a conductor along the lamp's length largely alleviates "thermometering." The trade-off is decreased efficiency due to energy leakage.[12] It is worth noting that various lamp types have different degrees of susceptibility to the thermometer effect.

Some displays require extended illumination range. "Thermometering" usually limits the lowest practical illumination level. One acceptable way to minimize "thermometering" is to eliminate the large field imbalance. The

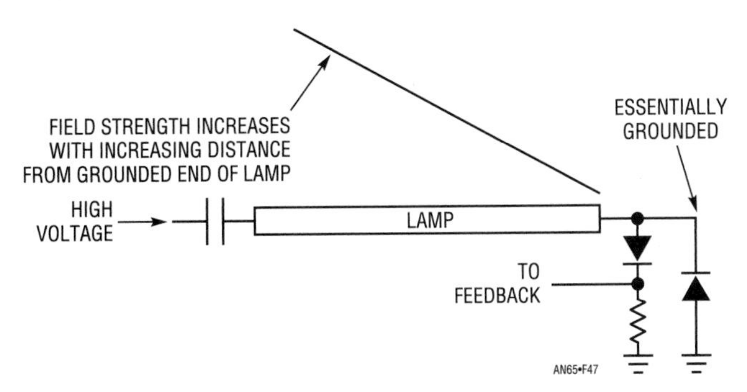

Figure 8.47 • Field Strength vs Distance for a Ground Referred Lamp. Field Imbalance Promotes Uneven Illumination at Low Drive Levels

Note 12: A very simple experiment quite nicely demonstrates the effects of energy leakage. Grasping the lamp at its low voltage end (low field intensity) with thumb and forefinger produces almost no change in circuit input current. Sliding the thumb/forefinger combination towards the high voltage (higher field intensity) lamp end produces progressively greater input currents. Don't touch the high voltage lead or you may receive an electrical shock. Repeat: Do not touch the high voltage lead or you may receive an electrical shock.

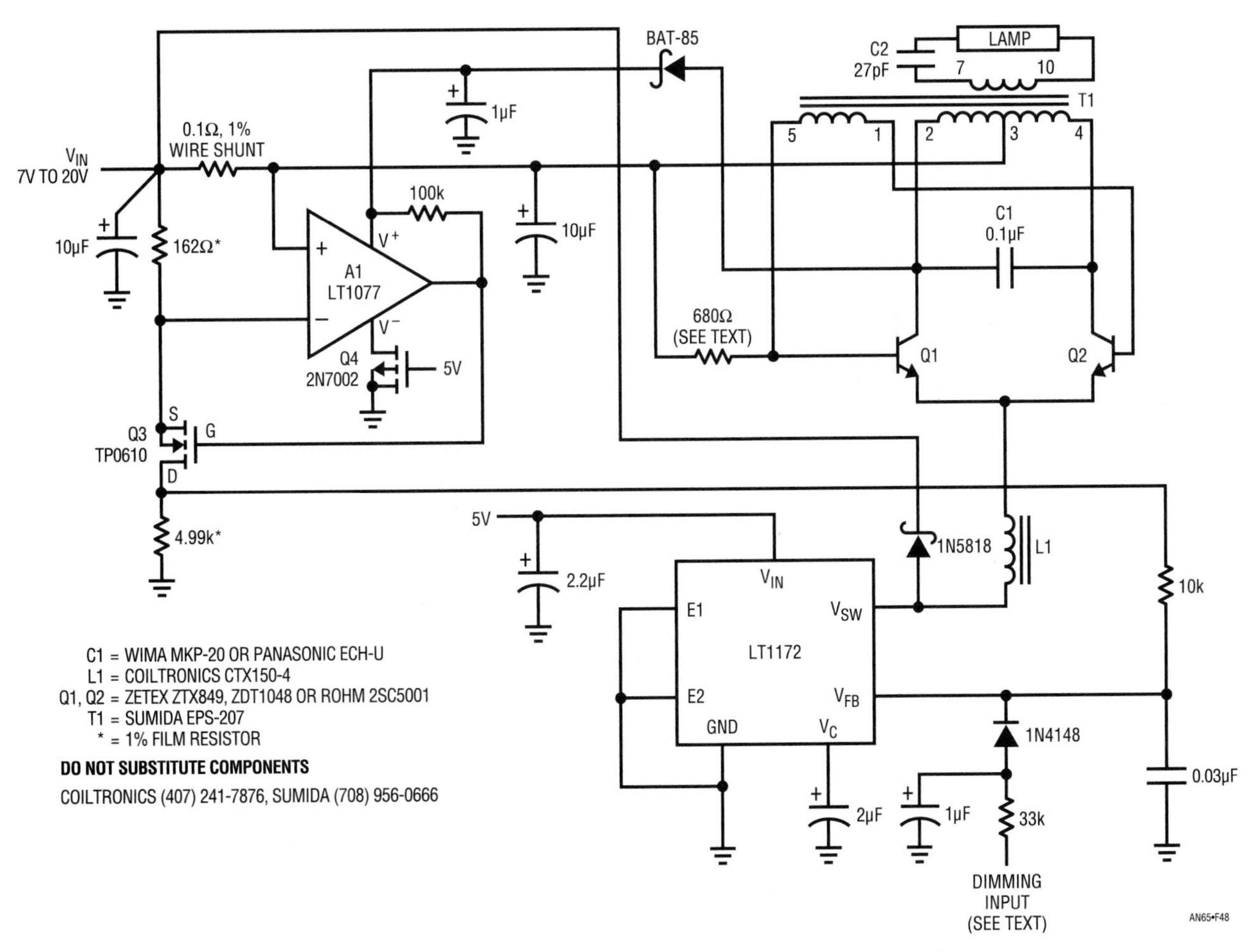

Figure 8.48 • Practical "Floating" Lamp Drive Circuit. A1 Senses Royer Input Current with Q3 Providing Resultant Feedback Information to Switching Regulator. Circuit Reduces Lost Energy Due to Parasitics by 10% to 20%

floating drive used to reduce energy loss also provides a way to minimize "thermometering." Figure 8.48 reviews a circuit originally introduced in a previous publication.[13] The circuit's most significant aspect is that the lamp is fully floating—there is no galvanic connection to ground as in the previous designs. This allows T1 to deliver symmetric, differential drive to the lamp. Such balanced drive eliminates field imbalance, reducing thermometering at low lamp currents. This approach precludes any feedback connection to the now floating output. Maintaining closed-loop control necessitates deriving a feedback signal from some other point. In theory, lamp current proportions to T1's or L1's drive level and some form of sensing this can be used to provide feedback. In practice, parasitics make a practical implementation difficult.[14]

Figure 8.48 derives the feedback signal by measuring Royer converter current and feeding this information back to the LT1172. The Royer's drive requirement closely proportions to lamp current under all conditions. A1 senses this current across the 0.1Ω shunt and biases Q3, closing a local feedback loop. Q3's drain voltage presents an amplified, single-ended version of the shunt voltage to the feedback point, closing the main loop. A1's power supply pin is bootstrapped to T1's boosted swing via the BAT-85 diode, permitting it to sense across the supply-fed shunt resistor. Internal A1 characteristics ensure start-up and substitution of this device is not recommended.[15]

The lamp current is not as tightly controlled as before but 0.5% regulation over wide supply ranges is possible. The dimming in this circuit is controlled by a 1kHz PWM signal. Note the heavy filtering (33k/1μF) outside the feedback loop. This allows a fast time constant, minimizing turn-on overshoot.[16]

In all other respects operation is similar to the previous circuits. This circuit typically permits the lamp to operate with less energy loss and over a 40:1 intensity range without "thermometering." The normal feedback connection is usually limited to a 10:1 range.

Note 13: See Reference 1.

Note 14: See Appendix L, "A Lot of Cut Off Ears and No Van Goghs—Some Not-So-Great Ideas, " for details.

Note 15: See Reference 1, then don't say we didn't warn you.

Note 16: See text section, "Feedback Loop Stability Issues."

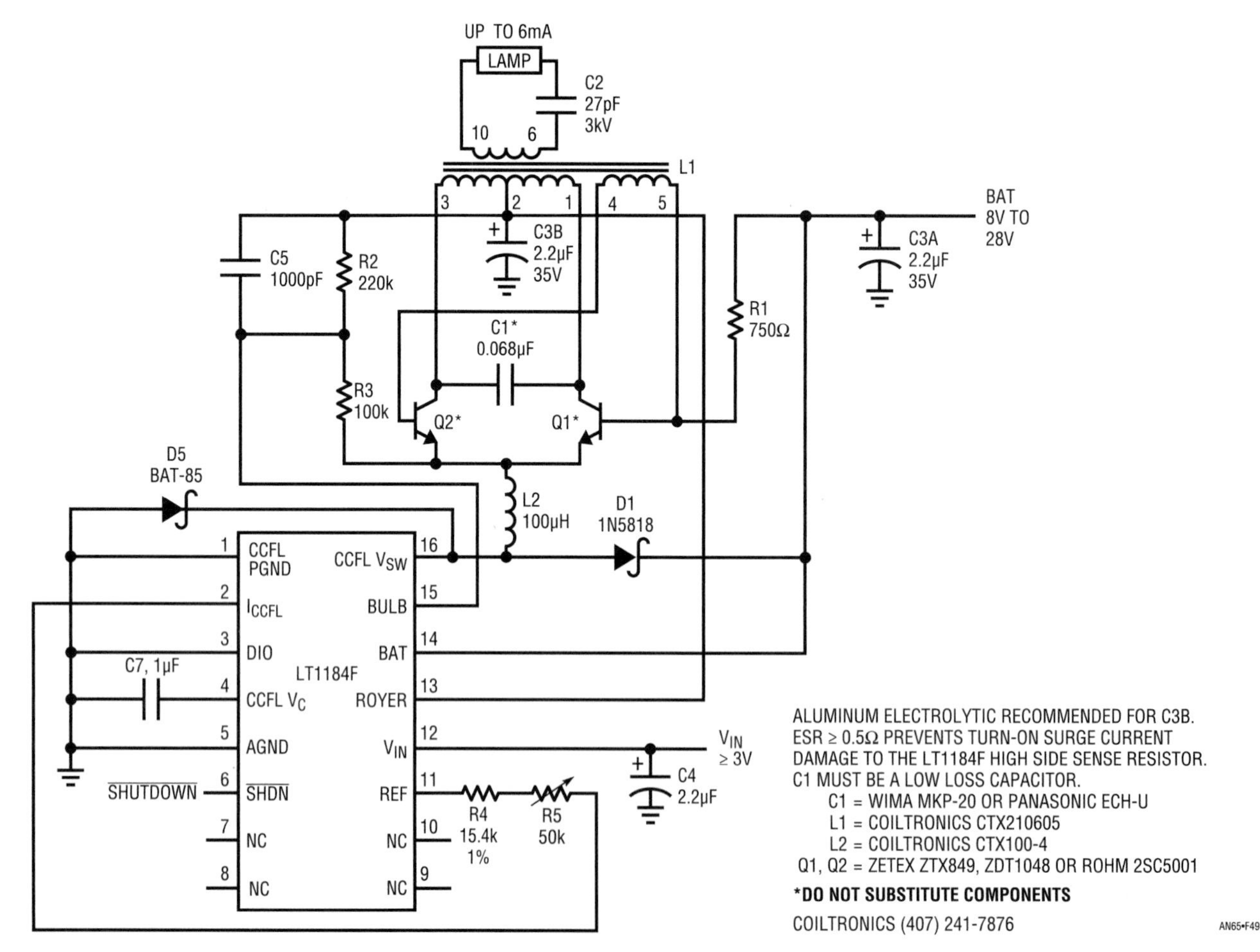

Figure 8.49 • LT1184F IC Version of Figure 8.48's Floating Lamp Circuit Offers Similar Performance with Fewer Components. Open Bulb Protection and Shutdown Are Included

IC-based floating drive circuits

Figure 8.49 compacts Figure 8.48 into a low component count, floating drive circuit. The LT1184F IC contains all functions except the Royer-based high voltage converter. The circuit also has "open lamp" protection and a 1.23V reference for biasing the dimming potentiometer.

Figure 8.50 adds a bipolar LCD contrast supply output to Figure 8.49. The LT1182 allows setting contrast supply polarity by simply grounding the appropriate output terminal. The CCFL portion is similar to the previous circuit, although intensity is controlled with a varying PWM or 0V to 5V input.

Figure 8.51's circuit is similar, although no contrast supply is included. The LT1186 implements a floating lamp drive similar to Figure 8.49. This IC contains an internal D/A converter which may be addressed by accumulating a bit stream or serial protocol. Figure 8.52 shows a typical arrangement using an 80C31 type microcontroller. Figure 8.53 gives the complete software listing which was written by Tommy Wu of LTC.

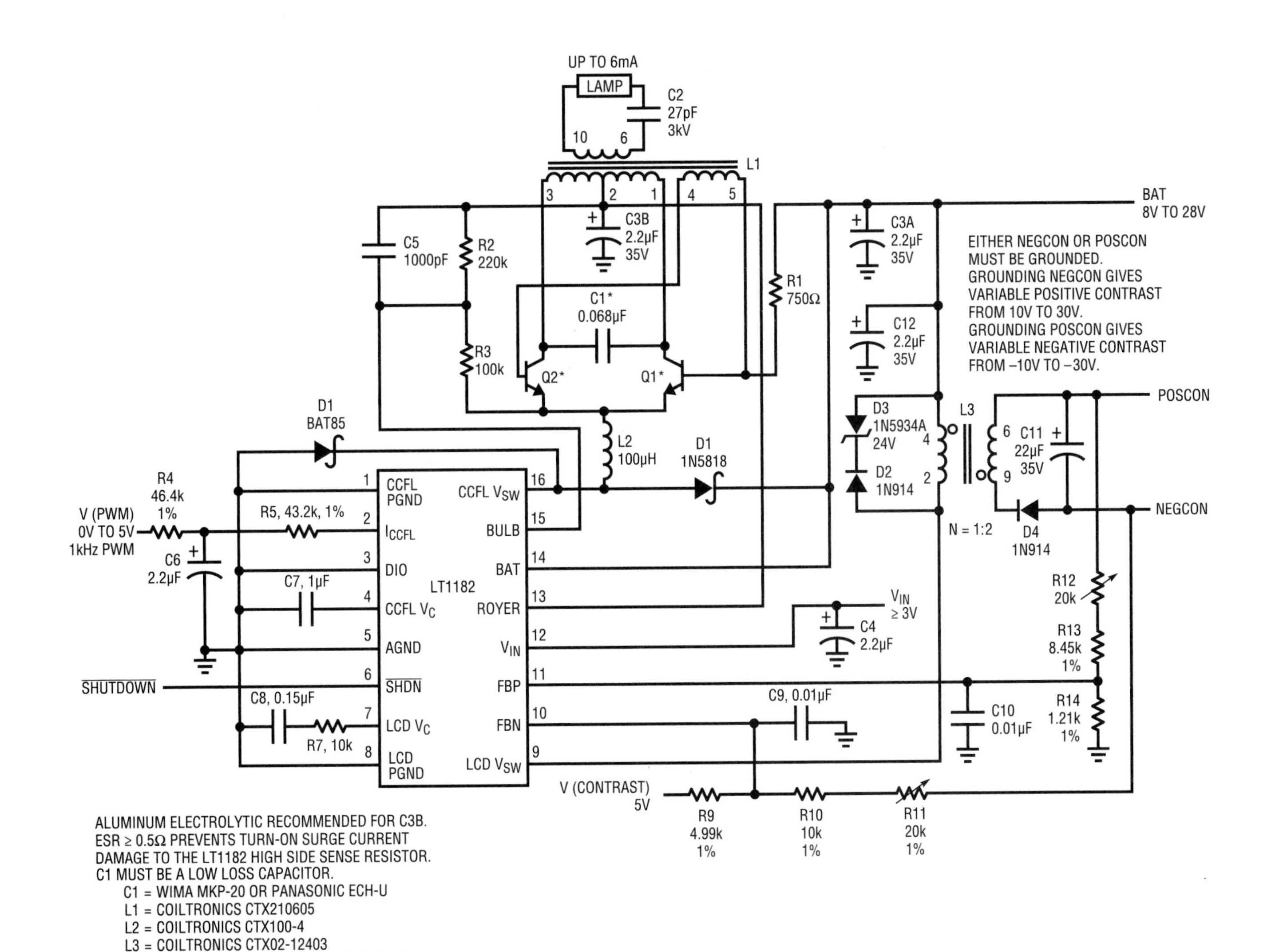

Figure 8.50 • LT1182 Has Bipolar Output Contrast Supply in Addition to Floating Lamp Drive

High power floating lamp circuit

High power floating lamp circuits require more current than the LT118X series can deliver. In such cases the function can be built from discrete components and ICs. Figure 8.54 shows a 30W CCFL circuit used in an automotive application. This 4-lamp circuit uses an LT1269

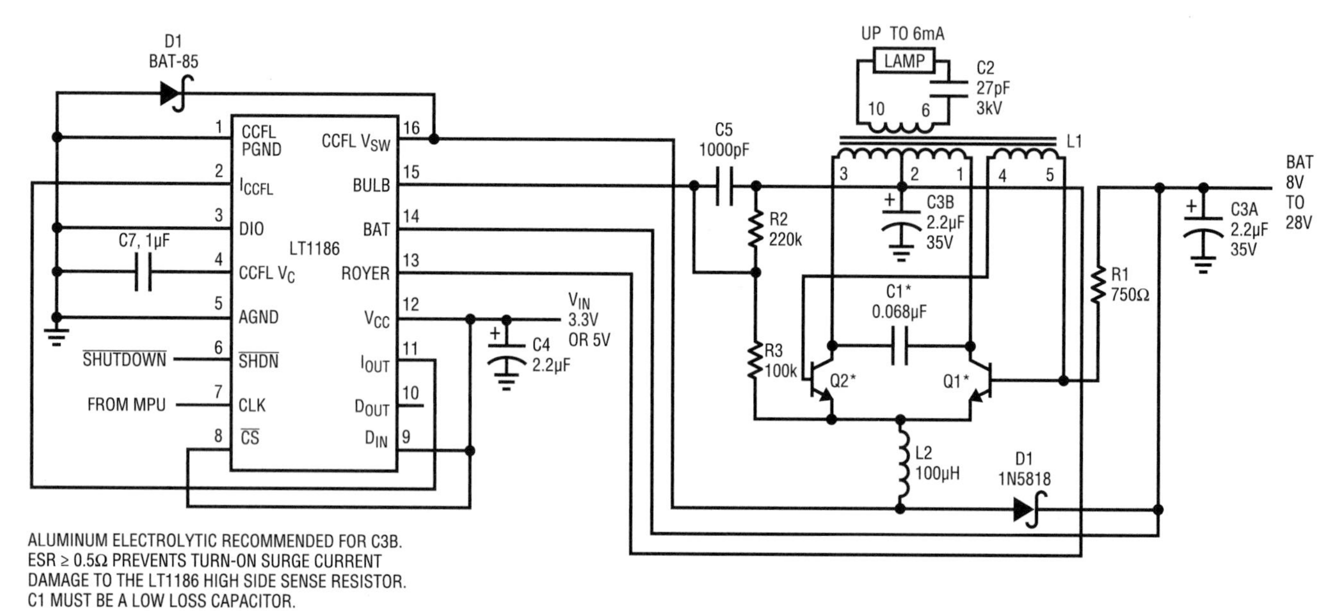

ALUMINUM ELECTROLYTIC RECOMMENDED FOR C3B.
ESR ≥ 0.5Ω PREVENTS TURN-ON SURGE CURRENT
DAMAGE TO THE LT1186 HIGH SIDE SENSE RESISTOR.
C1 MUST BE A LOW LOSS CAPACITOR.

C1 = WIMA MKP-20 OR PANASONIC ECH-U
L1 = COILTRONICS CTX210605
L2 = COILTRONICS CTX100-4
Q1, Q2 = ZETEX ZTX849, ZDT1048 OR ROHM 2SC5001

***DO NOT SUBSTITUTE COMPONENTS**

COILTRONICS (407) 241-7876

Figure 8.51 • LT1186 Permits Serial or Bit Stream Data Addressing to Set Floating Lamp Current

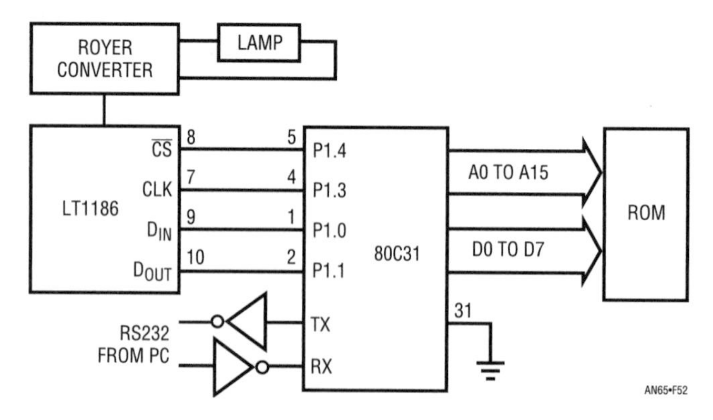

Figure 8.52 • Typical Processor Interface for Figure 8.51

The LT1186 DAC algorithm is written in assembly code in a file named LT1186A.ASM as a function call from the MAIN fuction below.
Note: A user inputs an integer from 0 to 255 on a keyboard and the LT1186 adjusts the IOUT programming current to control the operating lamp current and the brightness of the LCD display.

```
#include <stdio.h>
#include <reg51.h>
#include <absacc.h>

extern char lt1186(char);  /* external assembly function in lt1186a.asm*/
sbit Clock = 0x93;

main()
{
    int number = 0;
    int LstCode;

    Clock = 0;

    TMOD = 0x20;  /* Establish serial communication 1200 baud */
    TH1 = 0xE8;
    SCON = 0x52;
    TCON = 0x69;

    while(1)    /* Endless loop */
    {
        printf("\nEnter any code from 0 - 255:");
        scanf("%d",&number);
        if((0>number)l(number>255))
            {
                number = 0;
                printf("The number exceeds its range. Try again!");
            }
                else
                {
                    LstCode = lt1186(number);
                    printf("Previous # %u",(LstCode&0xFF)); /* AND the previous number with 0xFF to turn off sign
                                                extension */
                }
    number = 0;
    }
}

; The following assembly program named LT1186A.ASM receives the Din word from the main C program,
; lt1186 lt1186(). Assembly to C interface headers, declarations and memory allocations are listed before the
; actual assembly code.
;
; Port p1.4 = CS
; Port p1.3 = CLK
; Port p1.1 = Dout
; Port p1.0 = Din
;
```

Figure 8.53 • Complete Software Listing for Figure 8.52's Processor Interface

```
NAME LT1186_ CCFL
PUBLIC lt1186, ?lt1186?BYTE

?PR?ADC_INTERFACE?LT1186_CCFL SEGMENT CODE
?DT?ADC_INTERFACE?LT1186_CCFL SEGMENT  DATA

            RSEG  ?DT?ADC_INTERFACE?LT1186_CCFL
?lt1186?BYTE:  DS  2

            RSEG  ?PR?ADC_INTERFACE?LT1186_CCFL

CS      EQU  p1.4
CLK     EQU  p1.3
DOUT    EQU  p1.1
DIN     EQU  P1.0

lt1186: setb  CS                ;set CS high to initialize the LT1186
        mov   r7,?lt1186?BYTE   ;move input number(Din) from keyboard to R7
        mov   p1, #01h          ;setup port p1.0 becomes input
        clr   CS                ;CS goes low, enable the DAC
        mov   a, r7             ;move the Din to accumulator
        mov   r4, #08h          ;load counter 8 counts
        clr   c                 ;clear carry before rotating
        rlc   a                 ;rotate left Din bit(MSB) into carry
loop:   mov   DIN, c            ;move carry bit to Din port
        setb  CLK               ;Clk goes high for LT1186 to latch Din bit
        mov   c, DOUT           ;read Dout bit into carry
        rlc   a                 ;rotate left Dout bit into accumulator
        clr   CLK               ;clear clock to shift the next Dout bit
        djnz  r4, loop          ;next data bit loop
        mov   r7, a             ;move previous code to R7 as character return
        setb  CS                ;bring CS high to disable DAC
        ret
        END

Note: When CS goes low, the MSB of the previous code appears at Dout.
```

Figure 8.53 • (Continued) Complete Software Listing for Figure 8.52's Processor Interface

current-fed Royer converter to provide high power Lamp current is sensed in current transformer T2. A1 and associated components form a synchronous rectifier for T2's low level output. A2 provides gain and closes a loop back at the LT1269's feedback terminal. T2's isolated sensing permits the advantages of floating operation with the LT1269 providing high power capability. This circuit has about 83% efficiency at 30W output, a wide dimming range and 0.1% line regulation.

Selection criteria for CCFL circuits

Selecting which CCFL circuit to use for a specific application involves numerous trade-offs. A variety of issues determine which circuit is the "best" approach. At a minimum, the user should consider the following guidelines before committing to any approach. Related discussion to all of the following topics is covered in appropriate text sections.

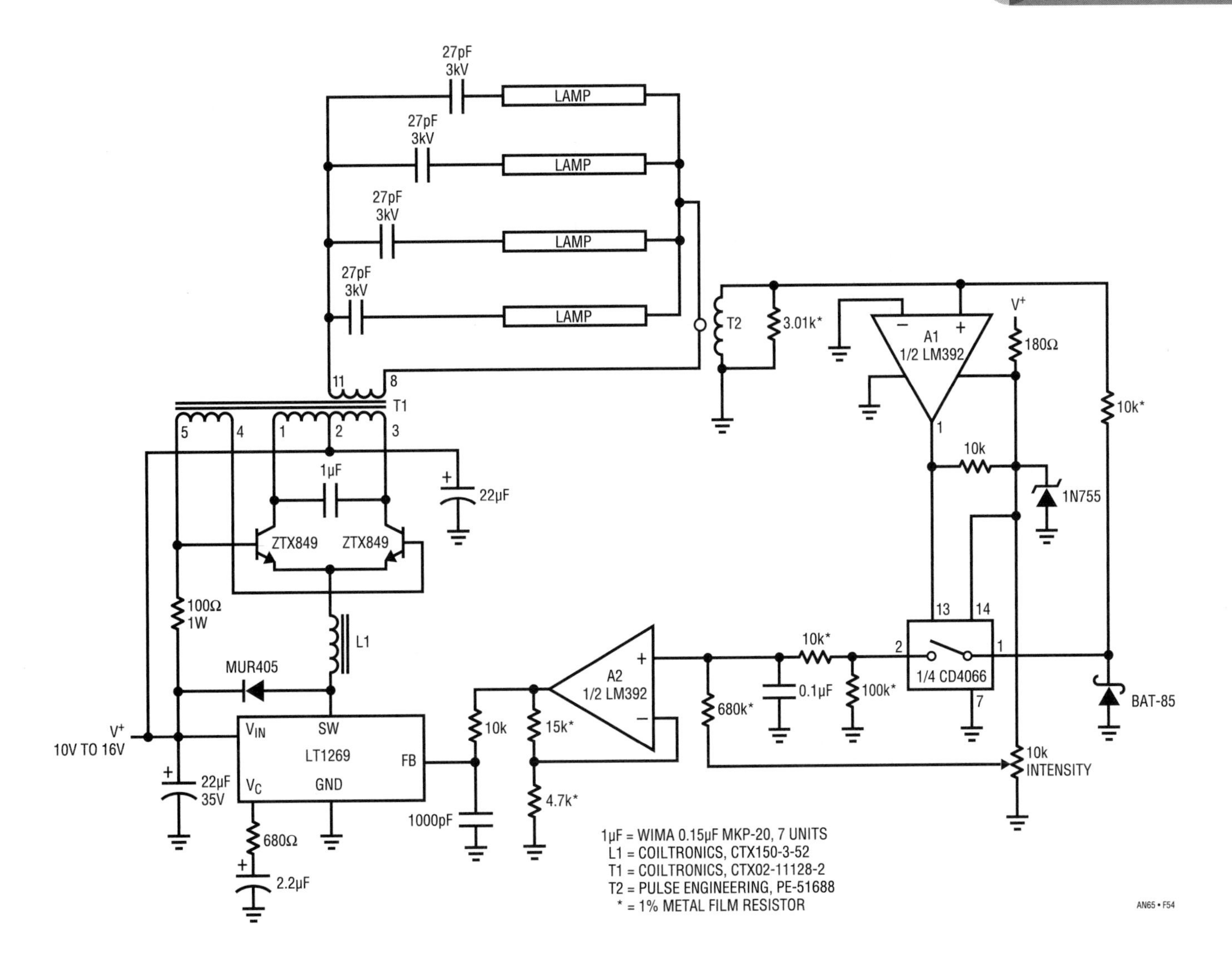

Figure 8.54 • A High Power, Multilamp Display Using the Floating Drive Approach. Power Requirement Necessitates LT1269 Regulator and Discrete Component Approach. Floating Feedback Path Is Via Current Transformer

Display characteristics

The display characteristics (including wiring losses) should be well-understood. Typically, display manufacturers list lamp requirements. These specifications are often obtained from the lamp vendor, who usually tests in free air, with no significant parasitic loss paths. This means that actual required power; start and running voltages may significantly differ from data sheet specifications. The only way to be certain of display characteristics is to measure them. The measured display energy loss can determine if a floating or grounded circuit is applicable. Low loss displays (relatively rare) usually provide better overall efficiency with grounded drive. As losses become worse (unfortunately, relatively common) floating drive becomes a better choice. Efficiency measurements may be required in both modes to determine the best choice. (See "General Optimization and Measurement Considerations.")

Operating voltage range

The operating voltage range includes the minimum to maximum voltages the circuit must operate from. In battery-driven apparatus supply range can easily be 3:1, and sometimes greater. Best backlight performance is usually obtained in the 8V to 28V range. In general, potentials below 7V require some efficiency trade-offs at moderate (1.5W to 3W) power levels. Some systems reduce backlight power when running from the battery, and this can have a pronounced effect on the design. Even seemingly small (e.g., 20%) reductions in power may make painful trade-offs unnecessary. In particular, high turns ratio transformers are required to support low voltage operation at full lamp output. They work well but somewhat less efficiently than lower ratio types due to the higher peak currents characteristic of their operation. Current trends in battery technology encourage system operation at low voltages, necessitating extreme care in transformer selection and Royer circuit design.

Auxiliary operating voltages

Auxiliary, logic supply voltages should be used (if available) to run CCFL "housekeeping" currents, such as IC "V_{IN}" pins. This saves power. Always run switching regulators from the lowest potential available, usually 3.3V or 5V Many systems provide these voltages in switched form, making separate shutdown lines unnecessary. Simply turning off the switching regulator's supply shuts the entire backlight circuit down.

Line regulation

Grounded lamp circuits, by virtue of their true global feedback, provide the best line regulation. For abrupt changes, a user may notice anything beyond 1% regulation. A grounded circuit easily meets this requirement; a floating circuit usually will. Slowly changing line inputs causing excursions outside 1% are not normally a problem because they are not detectable. Rapid line changes, such as plugging in a systems AC line adapter, require good regulation to avoid annoying display flicker.

Power requirements

The CCFL's power requirement, including display and wiring losses, should be well-defined over all conditions, including temperature and lamp specification variations. Usually, IC versions of floating lamp circuits are restricted to 3W to 4W output power while grounded circuit power is easily scaled.

Supply current profile

The backlight is often physically located far "forward" in the system. Impedances in cables, switches, traces and connectors can build up to significant levels. This means that a CCFL circuit should draw operating power continuously, rather than requiring discrete, high current "chunks" from a lossy supply line. Royer-based architectures are nearly ideal in this regard, pulling current smoothly over time and requiring no special bypassing, supply impedance or layout treatment. Similarly, Royer type circuits do not cause significant disturbances to the supply line, preventing noise injection back into the supply.

Lamp current certainty

The ability to predict lamp current at full intensity is important to maintain lamp life. Excessive overcurrent greatly shortens lamp life, while yielding little luminosity benefit (see Figure 8.2). Grounded circuits are excellent in this category with 1% usually achieved. Floating circuits are typically in the 2% to 5% range. Tight current tolerances do not benefit unit/unit display luminosity because lamp emission and display attenuation variations approach ±20% and vary over life.

Efficiency

CCFL backlight efficiency should be considered from two perspectives. The electrical efficiency is the ability of the circuit to convert DC power to high voltage AC and deliver it to the load (lamp and parasitics) with minimum loss. The optical efficiency is perhaps more meaningful to the user. It is simply the ratio of display luminosity to DC power into the CCFL circuit. The electrical and optical losses are lumped together in this measurement to produce a luminosity vs power specification. It is quite significant that the electrical and optical peak efficiency operating points do not necessarily coincide. This is primarily due to the lamp's emissivity dependence on waveshape. The optimum waveshape for emissivity may or may not coincide with the circuit's electrical operating peak. In fact, it is quite possible for "inefficient" circuits to produce more light than "more efficient" versions. The only way to ensure peak efficiency in a given situation is to optimize the circuit to the display.

Shutdown

System shutdown almost always requires turning off the backlight. In many cases the low voltage supply is already available in switched form. If this is so, the CCFL circuits shown go off, absorbing very little power. If switched low voltage power is not available the shutdown inputs may be used, requiring an extra control line.

Transient response

The CCFL circuit should turn on the lamp without attendant overshoot or poor control loop settling characteristics. This can cause objectionable display flicker, and in the worst case result in transformer overstress and failure. Properly prepared floating and grounded CCFL circuits have good transient response, with LT118X-based types inherently easier to optimize.

Dimming control

The method of dimming should be considered early in the design. All of the circuits shown can be controlled by potentiometers, DC voltages and currents, pulse width modulation or serial data protocol. A dimming scheme with high accuracy at maximum current prevents excessive lamp drive and should be employed.

Open lamp protection

The CCFL circuits deliver a current source output. If the lamp is broken or disconnected, compliance voltage is limited by transformer turns ratio and DC input voltage. Excessive voltages can cause arcing and resultant damage. Typically, the transformers withstand this condition but open lamp protection ensures against failures. This feature is built into the LT118X series; it must be added to other circuits.

Size

Backlight circuits usually have severe size and component count limitations. The board must fit within tightly defined dimensions. LT118X series-based circuits offer lowest component count, although board space is usually dominated by the Royer transformer. In extremely tight spaces it may be necessary to physically segment the circuit but this should be considered as a last resort.[17]

Contrast supply capability

Some LT118X parts provide contrast supply outputs. The other circuits do not. The LT118X's onboard contrast supply is usually an advantage but space is sometimes so restricted that it cannot be used. In such cases the contrast supply must be remotely located.

Emissions

Backlight circuits rarely cause emission problems and shielding is usually not required. Higher power versions (e.g., >5W) may require attention to meet emission requirements. The fast rise switching regulator output sometimes causes more RFI than the high voltage AC waveform. If shielding is used, its parasitic effects are part of the inverter load and optimization must be carried out with the shield in place.

Summary of circuits

The interdependence of backlight parameters makes summarizing or rating various approaches a hazardous exercise. There is simply no intellectually responsible way to streamline the selection and design process if optimum results are desired. A meaningful choice must be the outcome of laboratory-based experimentation. There are just too many interdependent variables and surprises for a systematic, theoretically based selection. Pure analytics are pretty; working circuits come from the bench. Some generalizations having limited usefulness are, however, possible. Figures 8.55 and 8.56 attempt to summarize salient characteristics vs part type and may (however cautiously) be considered a beginning point.[18]

Note 17: See Appendix G, "Layout, Component and Emissions Considerations."

Note 18: Readers detecting author ambivalence about inclusion of Figures 8.55 and 8.56 are not hallucinating.

ISSUES	LT118X SERIES	LT117X SERIES	LT137X SERIES
Optical Efficiency	Grounded output versions display dependent. Floating versions usually 5% to 20% better.	Display dependent	Display dependent
Electrical Efficiency	Grounded output versions—75% to 90%, depending on supply voltage and display. Floating output versions slightly lower.	75% to 90%, depending on supply voltage and display	75% to 92%, depending on supply voltage and display
Lamp Current Certainty	1% to 2% for grounded versions, 1% to 4% for floating output types	2% maximum	2% maximum
Line Regulation	0.1% to 0.3% for grounded types, 0.5% to 6% for floating versions	0.1% to 0.3%	0.1% to 3%
Operating Voltage Range	5.3V to 30V, depending on output power, temperature range, display, etc.	4.0V to 30V, depending on output power, temperature range, display etc.	4.0V to 30V, depending on output power, temperature range, display, etc.
Power Range	0.75W to 6W typical	0.75W to 20W typical	0.5W to 6W typical
Supply Current Profile	Continuous—no high current peaks	Continuous—no high current peaks	Continuous—no high current peaks
Shutdown Control	Yes—logic compatible	Requires small FET or bipolar transistor	Yes—logic compatible
Transient Response—Overshoot	Excellent—no optimization required	Excellent—requires optimization in some cases	Excellent—requires optimization in some cases
Dimming Control	Pot., PWM, variable DC voltage or current. LT1186 has serial digital input with data storage.	Pot., PWM, variable DC voltage or current	Pot., PWM, variable DC voltage or current
Emissions	Low	Low	Low, although high power versions may require attention to layout and shielding
Open Lamp Protection	Internal to IC	Requires external small-signal transistor and some discretes at high supply voltages	Requires external small-signal transistor and some discretes at high supply voltages
Size	Low component count, small overall board footprint. 200kHz magnetics.	Small—100kHz magnetics	Small—1MHz magnetics for fastest versions
Contrast Supply Capability	Various contrast supply options available, including bipolar output	No	No

AN65•F55-1

Figure 8.55 • Design Issues vs Typical Part Choice. Chart Makes Simplistic Assumptions and Is Intended As a Guide Only

LT1269/LT1270	LT1301	LT1173
Display dependent	Display dependent	Display dependent
75% to 90%, depending on supply voltage and display	70% to 88%, depending on supply voltage and display	65% to 75%, depending on supply voltage and display
2% maximum	2% typical	5%
0.1% to 0.3%	0.1% to 0.3%	8% to 10%
4.5V to 30V, depending on output power, temperature range, display, etc.	2V to 10V practical	2V to 6V practical
5W to 35W typical	0.02W to 1W practical	Essentially 0W to about 0.6W
Continuous—no high current peaks	Continuous—no high current peaks	Irregular—relatively high current peaking requires attention to supply rail impedance
Requires small FET or bipolar transistor	Yes—logic compatible	Logic compatible shutdown practical
Excellent—requires optimization in some cases	Excellent—no optimization required	Excellent—no optimization required
Pot., PWM, variable DC voltage or current	Pot., PWM, variable DC voltage or current	Pot., PWM, variable DC voltage or current
High power mandates attention to layout and shielding	Real low	Itsy-bitsy
Requires external small-signal transistor and some discretes at high supply voltages	Requires external small-signal tansistor and some discretes, but low supply voltages usually eliminate this consideration	None, but low supply, low power operation usually eliminates this issue
Relatively large due to high power 100kHz magnetics	Very small—low power magnetics cut size	Small—low power magnetics cut size
No	No	No

AN65•F55-2

	LT1182	LT1183	LT1184	LT1184F	LT1186
Floating Lamp Operation	Yes	Yes	No	Yes	Yes
Grounded Lamp Operation	Yes	Yes	Yes	Yes	Yes
Contrast Supply	Bipolar Contrast Outputs	Unipolar Contrast Outputs	No	No	No
Voltage Reference Available	No	Yes	Yes	Yes	No
Internal Control DAC	No	No	No	No	Yes

Figure 8.56 • Features of Various LT118X IC Backlight Controllers

Figure 8.55 summarizes characteristics of all the circuits. Figure 8.56 focuses on the features of the LT118X series parts.

General optimization and measurement considerations

Once a display/lamp combination has been picked, the appropriate circuit can be selected and optimized. "Optimization" implies maximizing performance in those areas most important in a particular application. This may involve trading off characteristics in one area to gain advantage in another. The circuit types described impose mild penalty in this regard because they are quite flexible.

A desirable characteristic is something often loosely referred to as "efficiency." There are really two types of efficiency in a backlight circuit. The optical efficiency measures the circuit/display combination as a transducer. It is the ratio of light output to electrical power input. This ratio lumps the converter's electrical loss with lamp and display losses. A backlight's electrical efficiency measures the converter's electrical input vs output power without regard to optical performance. Obviously, high electrical efficiency is required and a reliable way to measure it is desirable. More subtly, the ability to measure and manipulate purely electrical terms offers a way to influence optical efficiency. This is so because the lamp is sensitive to the drive waveform's shape. Best emissivity and lifetime are usually obtained with low crest factor, sinusoidal waveforms. The Royer circuit's transformer and capacitors can be selected to provide this characteristic for any given display/lamp combination. Doing this optimizes lamp drive but also effects the converter's electrical efficiency. This interaction between the optimum electrical and optical operating points must be accounted for to obtain best optical efficiency. The relationship is quite complex with a number of variables determining just where peak optical efficiency occurs.

Typically, optical output peaking occurs with a fairly clean, low harmonic waveform at the Royer collectors (Figure 8.57). This is usually the result of a relatively large resonating capacitor and a small ballast capacitor. Conversely, the converter's peak electrical efficiency point usually comes just as appreciable second harmonic appears in the Royer collector waveform (Figure 8.58). The peak electrical and optical efficiency points almost never coincide and optical efficiency often occurs 5% or more off the electrical efficiency peak. Happily, this very messy

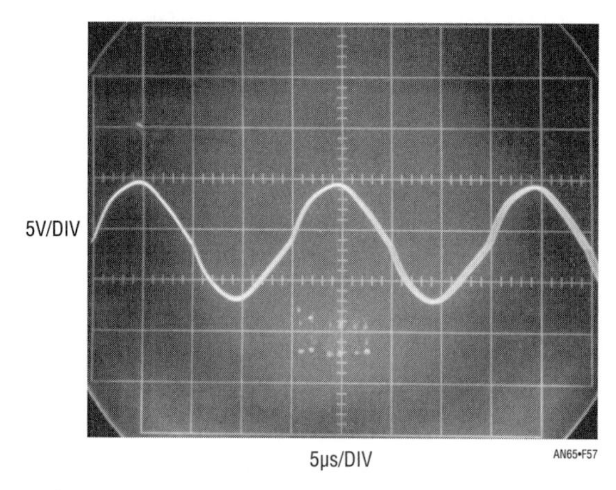

Figure 8.57 • Typical Royer Collector Waveform at the Peak Optical Output Point. Relatively Large Resonating Capacitor May Degrade Electrical Efficiency

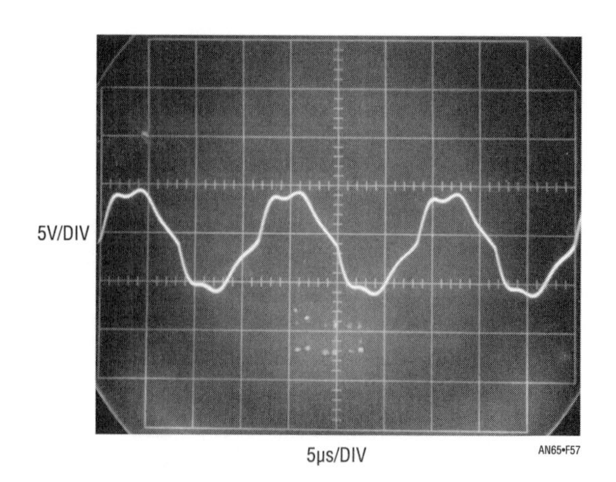

Figure 8.58 • Typical Royer Collector Waveform at the Peak Electrical Efficiency Point. Relatively High Harmonic Content May Degrade Optical Efficiency

situation can be resolved by a relatively simple functional trim. The trimming procedure assumes transformer turns ratio and ballast capacitor values commensurate with the lowest required circuit operating voltage have been chosen. If this factor is not considered, the optical efficiency peak will be realized but the design may not regulate at low supply voltages. Low supply voltage operation mandates high turns ratio and larger ballast capacitor values for a given display loss. If display loss is high, ballast capacitor value generally must rise to offset voltage dividing effects between it and the display's parasitic loss paths. Establish the lowest values of turns ratio and ballast capacitor that maintain regulation at minimum supply voltage before performing the trim.

Achieving peak optical efficiency involves comparing display luminosity to input power for different resonating capacitor values. For a given lamp/transformer/ballast capacitor combination different resonating capacitors produce varying amounts of light. Large values tend to smooth harmonics, peaking optical output but increasing converter circulating losses. Smaller values promote lower circulating currents but less light output. Figure 8.59 shows typical results for five capacitor values at a forced 10V main supply and 5mA lamp current. Large values produce more light but require more supply current. The raw data is expressed as the ratio of light output-per-watt of input power in the right-most column. This Nits-per-Watt ratio peaks at 0.1µF indicating the best optical efficiency.[19]

CAPACITOR (µF)	10V MAIN SUPPLY CURRENT	5V SUPPLY CURRENT	TOTAL SUPPLY WATTS	INTENSITY (NITS)	NITS/WATT
0.15	0.304	0.014	3.11	118	37.9
0.1	0.269	0.013	2.75	112	40.7
0.068	0.259	0.013	2.65	101	38.1
0.047	0.251	0.013	2.57	95	37.3
0.033	0.240	0.013	2.46	88	35.7

Note: Maintain I_{MAIN} Supply = 10.0V and I_{LAMP} = $5mA_{RMS}$ under all conditions.

Figure 8.59 • Typical Data Taken for Optical Efficiency Optimization. Note Emissivity Peak (Nits/Watt) for 0.1µF Resonating Value, Indicating Best Trade-Off Point for Electrical vs Optical Efficiency. Data Should Be Retaken for Several Ballast Capacitor Values to Ensure Maximum Optical Efficiency

This test must be performed in a stable thermal environment because of the lamp's emission sensitivity to temperature (see Figure 8.3). Additionally, some arrangement for rapidly switching the capacitor values is desirable. This avoids power interruptions and the resultant long display warm-up times.

Electrical efficiency optimization and measurement

Several points should be kept in mind when observing operation of these circuits. The high voltage secondary can only be monitored with a wideband, high voltage probe fully specified for this type of measurement. *The vast majority of oscilloscope probes will break down and fail if used for this measurement.*[20] Tektronix probe types P-6007 and P-6009 (acceptable in some cases) or types P6013A and P6015 (preferred) probes must be used to read L1's output.

Another consideration involves observing waveforms. The switching regulator frequency is completely asynchronous from the Royer converter's switching. As such, most oscilloscopes cannot simultaneously trigger and display all the circuit's waveforms. Figure 8.36 was obtained using a dual beam oscilloscope (Tektronix 556). Traces A and B are triggered on one beam while the remaining traces are triggered on the other beam. Single beam instruments with alternate sweep and trigger switching (e.g., Tektronix 547) can also be used but are less versatile and restricted to four traces.

Obtaining and verifying high electrical efficiency[21] requires some amount of diligence. The optimum efficiency values given for C1 and C2 (C1 is the resonating capacitor and C2 is the ballast capacitor) are typical and will vary for specific types of lamps. An important realization is that the term "lamp" includes the total load seen

Note 19: Optical measurement units are beyond arcane; a monument to obscuration. Candela/Meter^2 is a basic unit, and 1 Nit = 1 Candela/Meter^2. "Nit" is a contracted form of the Latin word "Nitere, " meaning "to emit light . . . to sparkle."

Note 20: Don't say we didn't warn you!

Note 21: The term "efficiency" as used here applies to electrical efficiency. In fact, the ultimate concern centers around the efficient conversion of power supply energy into light. Unfortunately, lamp types show considerable deviation in their current-to-light conversion efficiency. Similarly, the emitted light for a given current varies over the life and history of any particular lamp. As such, this text portion treats "efficiency" on an electrical basis; the ratio of power removed from the primary supply to the power delivered to the lamp. When a lamp/display combination has been selected, the ratio of primary supply power to lamp emitted light energy may be measured with the aid of a photometer. This is covered in the immediately preceding text and Appendix D.

by the transformer's secondary. This load, reflected back to the primary, sets transformer input impedance. The transformer's input impedance forms an integral part of the LC tank that produces the high voltage drive. Because of this, circuit efficiency must be optimized with the wiring, efficiency than might otherwise be possible. In practice, a "first cut" efficiency optimization with "best guess" lead lengths and the intended lamp in its display housing usually produces results within 5% of the achievable figure. Final values for C1 and C2 may be established when the physical layout to be used in production has been decided on. C1 sets the circuit's resonance point, which varies to some extent with the lamp's characteristic. C2 ballasts the lamp, effectively buffering its negative resistance characteristic. Small C2 values provide the most load isolation but require relatively large transformer output voltage for loop closure. Large C2 values minimize transformer output voltage but degrade load buffering. C2 values also affect waveform distortion, influencing lamp emissivity and optical efficiency (see previous text discussion). Also, C1's "best" value is somewhat dependent on the lamp type used. Both C1 and C2 must be selected for given lamp types. Some interaction occurs, but generalized guidelines are possible. Typical values for C1 are 0.01µF to 0.15µF C2 usually ends up in the 10pF to 47pF range. C1 must be a low loss capacitor and substitution of the recommended devices is not recommended. A poor quality dielectric for C1 can easily degrade efficiency by 10%. Before capacitor selection the Q1/Q2 base drive resistor should be set to a value which ensures saturation, e.g., 470Ω. Next, C1 and C2 are selected by trying different values for each and iterating towards best efficiency. During this procedure ensure that loop closure is maintained. Several trials usually produce the optimum C1 and C2 values. Note that the highest efficiencies are not necessarily associated with the most esthetically pleasing waveshapes, particularly at Q1, Q2 and output. Finally, the base drive resistor's value should be optimized.

The base drive resistor's value (nominally 1k) should be selected to provide full V_{CE} saturation without inducing base overdrive or beta starvation. This point may be established for any lamp type by determining the peak collector current at full lamp power.

The base resistor should be set at the largest value that ensures saturation for worst-case transistor beta. This condition may be verified by varying the base drive resistor about the ideal value and noting small variations in input supply current. The minimum obtainable current corresponds to the best beta vs saturation trade-off. In practice, supply current rises slightly on either side of this point. This "double value" behavior is due to efficiency degradation caused by either excessive base drive or saturation losses.

Other issues influencing efficiency include lamp wire length and energy leakage from the lamp. The high voltage side(s) of the lamp should have the smallest practical lead length. Excessive length results in radiative losses which can easily reach 3% for a 3-inch wire. Similarly, no metal should contact or be in close proximity to the lamp. This prevents energy leakage which can exceed 10%.[22]

It is worth noting that a custom designed lamp affords the best possible results. A jointly tailored lamp/circuit combination permits precise optimization of circuit operation, yielding highest efficiency.

These considerations should be made with knowledge of other LCD issues. See Appendix B, "Mechanical Design Considerations for Liquid Crystal Displays." This section was guest-written by Charles L. Guthrie of Sharp Electronics Corporation.

Special attention should be given to the circuit board layout since high voltage is generated at the output. The output coupling capacitor must be carefully located to minimize leakage paths on the circuit board. A slot in the board will further minimize leakage. Such leakage can permit current flow outside the feedback loop, wasting power. In the worst case, long term contamination buildup can increase leakage inside the loop, resulting in starved lamp drive or destructive arcing. It is good practice for minimization of leakage to break the silk screen line which outlines the transformer. This prevents leakage from the high voltage secondary to the primary. Another technique for minimizing leakage is to evaluate and specify the silk

Note 22: This footnote annotates similar issues raised in Footnote 12 and associated text. The repetition is based on the necessity for emphasis. A very simple experiment quite nicely demonstrates the effects of energy leakage. Grasping the lamp at its low voltage end (low field intensity) with thumb and forefinger produces almost no change in circuit input current. Sliding the thumb/forefinger combination towards the high voltage (higher field intensity) lamp end produces progressively greater input currents. Don't touch the high voltage lead or your may receive an electrical shock. Repeat: Do not touch the high voltage lead or you may receive an electrical shock.

screen ink for its ability to withstand high voltages. Appendix G, "Layout, Component and Emissions Considerations," details high voltage layout practice.

Electrical efficiency measurement

Once these procedures have been followed efficiency can be measured. Efficiency may be measured by determining lamp current and voltage. Measuring current involves utilization of a wideband, high accuracy clip-on current probe having a true (thermally based) RMS readout. No commercially manufactured current probe will meet the accuracy and bandwidth requirements and the probe must be constructed.[23]

Lamp RMS voltage is measured at the lamp with a wideband, properly compensated high voltage probe.[24] Multiplying these two results gives power in watts, which may be compared to the DC input supply (E)(I) product. In practice, the lamp's current and voltage contain small out-of-phase components but their error contribution is negligible.

Both the current and voltage measurements require a wideband true RMS voltmeter. The meter must employ a thermal type RMS converter—the more common logarithmic computing type-based instruments are inappropriate because their bandwidth is too low.

The previously recommended high voltage probes are designed to see a 1MΩ/10pF-22pF oscilloscope input. The RMS voltmeters have a 10MΩ input. This difference necessitates an impedance matching network between the probe and the voltmeter. Floating lamp circuits require this matching and differential measurement, severely complicating instrumentation design. See Footnote 24.

Feedback loop stability issues

The circuits shown to this point rely on closed-loop feedback to maintain the operating point. All linear closed-loop systems require some form of frequency compensation to achieve dynamic stability. Circuits operating with relatively low power lamps may be frequency compensated by simply overdamping the loop. Text Figures 8.35, 8.37, and 8.38 use this approach. The higher power operation associated with color displays requires more attention to loop response. The transformer produces much higher output voltages, particularly at start-up. Poor loop damping can allow transformer voltage ratings to be exceeded, causing arcing and failure. As such, higher power designs may require optimization of transient response characteristics. LT118X series parts almost never require optimization because their error amplifier's gain/phase characteristics are specially tailored to CCFL load characteristics. The LT1172, LT1372 and other general purpose switching regulators require more attention to ensure proper behavior. The following discussion, applicable to general purpose LTC switching regulators in CCFL applications, uses the LT1172 as an example.

Figure 8.60 shows the significant contributors to loop transmission in these circuits. The resonant Royer converter delivers information at about 50kHz to the lamp. This information is smoothed by the RC averaging time constant and delivered to the LT1172's feedback terminal as DC.

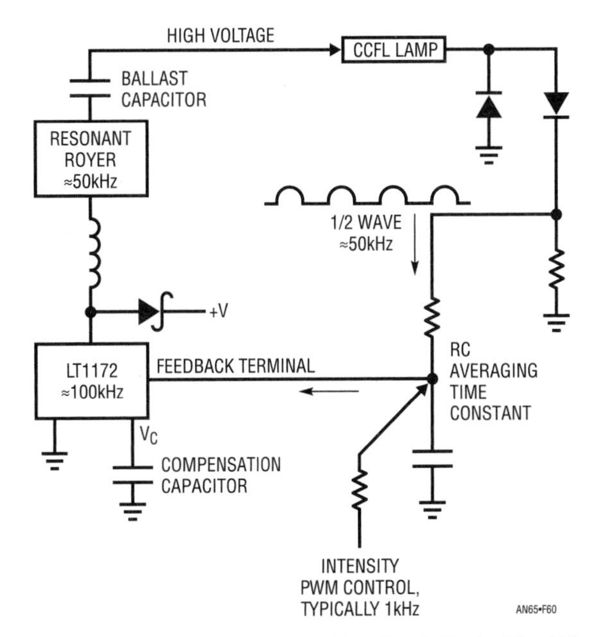

Figure 8.60 • Delay Terms in the Feedback Path. The RC Time Constant Dominates Loop Transmission Delay and Must Be Compensated for Stable Operation

Note 23: Justification for this requirement and construction details appear in Appendix C, "Achieving Meaningful Electrical Measurements."

Note 24: Measuring floating lamp circuit voltages is a particularly demanding exercise requiring a wideband differential high voltage probe. Probe construction details appear in Appendix C.

The LT1172 controls the Royer converter at a 100kHz rate, closing the control loop. The capacitor at the LT1172 rolls off gain, nominally stabilizing the loop. This compensation capacitor must roll off the gain-bandwidth at a low enough value to prevent the various loop delays from causing oscillation.

Which of these delays is the most significant? From a stability viewpoint the LT1172's output repetition rate and the Royer's oscillation frequency are sampled data systems. Their information delivery rate is far above the RC averaging time constants delay and is not significant. The RC time constant is the major contributor to loop delay. This time constant must be large enough to turn the half wave rectified waveform into DC. It also must be large enough to average any intensity control PWM signal to DC. Typically, these PWM intensity control signals come in at a 1kHz rate (see Appendix F "Intensity Control and Shutdown Methods"). The RC's resultant delay dominates loop transmission. It must be compensated by the capacitor at the LT1172. A large enough value for this capacitor rolls off loop gain at low enough frequency to provide stability. The loop simply does not have enough gain to oscillate at a frequency commensurate with the RC delay.[25]

This form of compensation is simple and effective. It ensures stability over a wide range of operating conditions. It does, however, have poorly damped response at system turn-on. At turn-on the RC lag delays feedback, allowing output excursions well above the normal operating point. When the RC acquires the feedback value the loop stabilizes properly. This turn-on overshoot is not a concern if it is well within transformer breakdown ratings. Color displays, running at higher power, usually require large initial voltages. If loop damping is poor, the overshoot may be dangerously high. Figure 8.61 shows such a loop responding to turn-on. In this case the RC values are 10k and 4.7μF with a 2μF compensation capacitor. Turn-on overshoot exceeds 3500V for over 10ms! Ring-off takes over 100ms before settling occurs. Additionally, an inadequate (too small) ballast capacitor and excessively lossy layout force a 2000V output once loop settling occurs. This photo was taken with a transformer rated well below this figure. The resultant arcing caused transformer destruction, resulting in field failures. A typical destroyed transformer appears in Figure 8.62.

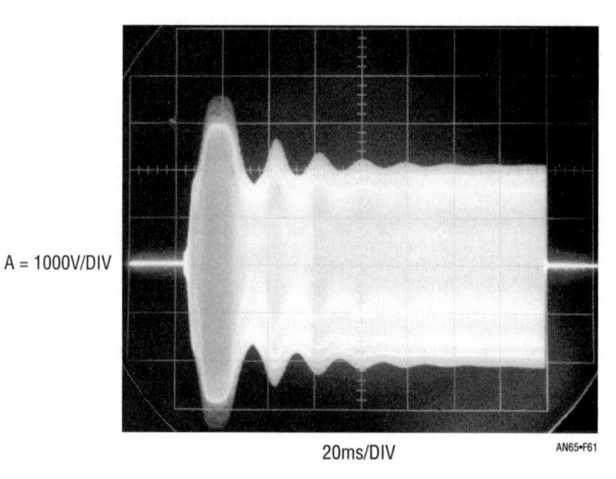

Figure 8.61 • Destructive High Voltage Overshoot and Ring-Off Due to Poor Loop Compensation. Transformer Failure and Field Recall Are Nearly Certain. Job Loss May also Occur

Figure 8.62 • Poor Loop Compensation Caused This Transformer Failure. Arc Occured in High Voltage Secondary (Lower Right). Resultant Shorted Turns Caused Overheating

Figure 8.63 shows the same circuit with the RC values reduced to 10k and 1μF The ballast capacitor and layout have also been optimized. Figure 8.63 shows peak voltage reduced to 2.2kV with duration down to about 2ms (note horizontal scale change). Ring-off is also much quicker

Note 25: The high priests of feedback refer to this as "Dominant Pole Compensation." The rest of us are reduced to more pedestrian descriptives.

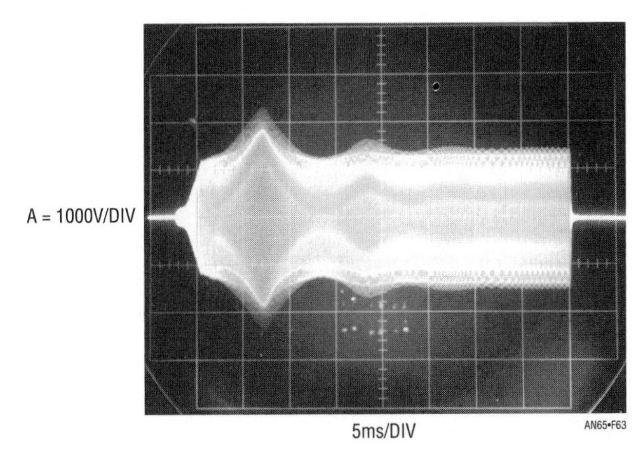

Figure 8.63 • Reducing RC Time Constant Improves Transient Response, Although Peaking, Ring-Off and Run Voltage Are Still Excessive

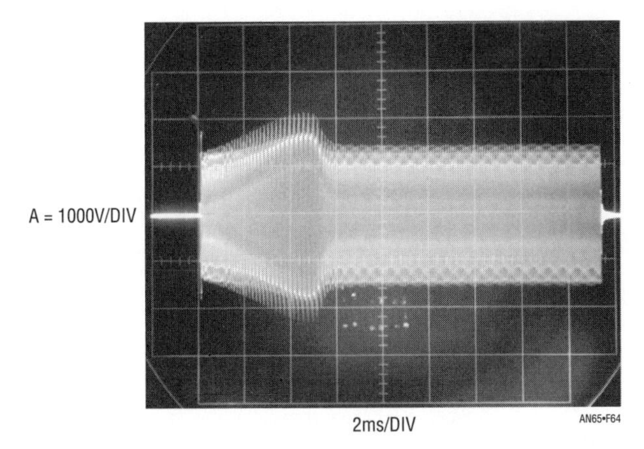

Figure 8.64 • Additional Optimization of RC Time Constant and Compensation Capacitor Reduces Turn-On Transient. Run Voltage Is Large, Indicating Possible Lossy Layout and Display

with lower amplitude excursion. Increased ballast capacitor value and wiring layout optimization reduce running voltage to 1300V Figure 8.64's results are even better. Changing the compensation capacitor to a 3kΩ/2μF network introduces a leading response into the loop, allowing faster acquisition. Now, turn-on excursion is slightly lower, but greatly reduced in duration (again, note horizontal scale change). The running voltage remains the same.

The photos show that changes in compensation, ballast value and layout result in dramatic reductions in overshoot amplitude and duration. Figure 8.62's performance almost guarantees field failures while Figures 8.63 and 8.64 do not overstress the transformer. Even with the improvements, more margin is possible if display losses can be controlled. Figures 8.62, 8.63 and 8.64 were taken with an exceptionally lossy display. The metal enclosure was very close to the metallic foil wrapped lamp, causing large losses with subsequent high turn-on and running voltages. If the display is selected for lower losses, performance can be greatly improved.

Figure 8.65 shows a low loss display responding to turn-on with a 2μF compensation capacitor and 10k/1μF RC values. Trace A is the transformer's output while Traces B and C are the LT1172's $V_{COMPENSATION}$ and Feedback pins, respectively. The output overshoots and rings badly, peaking to about 3000V. This activity is reflected by overshoots at the $V_{COMPENSATION}$ pin (the LT1172's error amplifier output) and the Feedback pin. In Figure 8.66 the RC is reduced to 10kΩ/0.1μF This substantially reduces loop delay. Overshoot goes down to only 800V—a reduction of almost a factor of four. Duration is also much shorter.

The $V_{COMPENSATION}$ and Feedback pins reflect this tighter control. Damping is much better, with slight overshoot induced at turn-on. Further reduction of the RC to

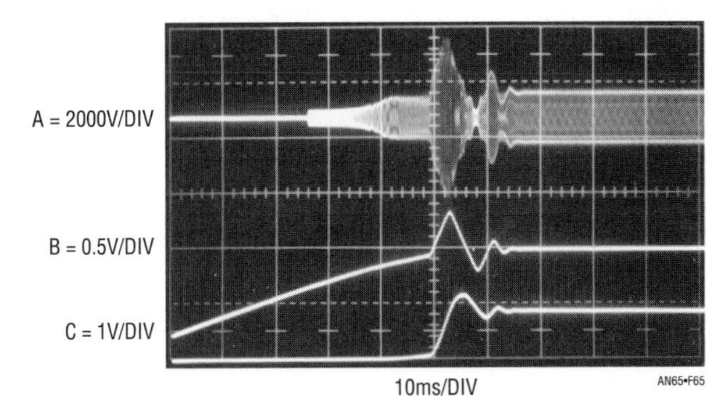

Figure 8.65 • Waveform for a Lower Loss Layout and Display. High Voltage Overshoot (Trace A) Is Reflected at Compensation Node (Trace B) and Feedback Pin (Trace C)

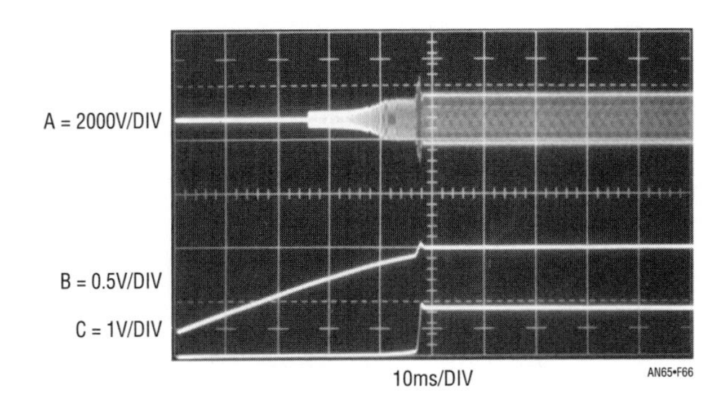

Figure 8.66 • Reducing RC Time Constant Produces Quick, Clean Loop Behavior. Low Loss Layout and Display Result in 650V_{RMS} Running Voltage

10kΩ/0.01µF (Figure 8.67) results in even faster loop capture but a new problem appears. In Trace A lamp turn-on is so fast the overshoot does not register in the photo. The $V_{COMPENSATION}$ (Trace B) and feedback nodes (Trace C) reflect this with exceptionally fast response. Unfortunately, the RC's light filtering causes ripple to appear when the feedback node settles. As such, Figure 8.66's RC values are probably more realistic for this situation.

The lesson from this exercise is clear. The higher voltages involved in color displays mandate attention to transformer outputs. Under running conditions layout and display losses can cause higher loop compliance voltages, degrading efficiency and stressing the transformer. At turn-on improper compensation causes huge overshoots, resulting in possible transformer destruction. Isn't a day of loop and layout optimization worth a field recall?

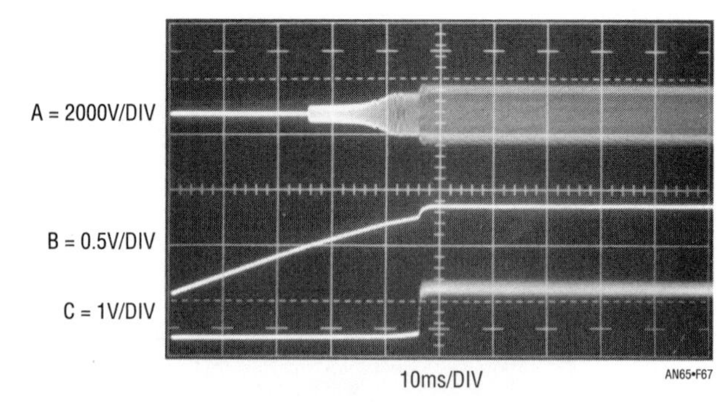

Figure 8.67 • Very Low RC Value Provides Even Faster Response, but Ripple at Feedback Pin (Trace C) Is Too High. Figure 8.66 Is the Best Compromise

References

1. Williams, J., "Techniques for 92% Efficient LCD Illumination," Linear Technology Corporation, Application Note 55, August 1993
2. Bright, Pittman and Royer, "Transistors As On-Off Switches in Saturable Core Circuits," Electrical Manufacturing, December 1954. Available from Technomic Publishing, Lancaster, PA
3. Sharp Corporation, "Flat Panel Displays," 1991
4. C. Kitchen, L. Counts, "RMS-to-DC Conversion Guide," Analog Devices, Inc., 1986
5. Williams, Jim, "A Monolithic IC for 100MHz RMS-DC Conversion," Linear Technology Corporation, Application Note 22, September 1987
6. Hewlett-Packard, "1968 Instrumentation. Electronic-Analytical-Medical," AC Voltage Measurement, p.197-198, 1968
7. Hewlett-Packard, "Model 3400RMS Voltmeter Operating and Service Manual," 1965
8. Hewlett-Packard, "Model 3403C True RMS Voltmeter Operating and Service Manual," 1973
9. Ott, W.E., "A New Technique of Thermal RMS Measurement," IEEE Journal of Solid State Circuits, December 1974
10. Williams, J.M. and Longman, TL., "A 25MHz Thermally Based RMS-DC Converter" 1986 IEEE ISSCC Digest of Technical Papers
11. O'Neill, PM., "A Monolithic Thermal Converter," H.P Journal, May 1980
12. Williams, J., "Thermal Techniques in Measurement and Control Circuitry, 50MHz Thermal RMS-DC Converter," Linear Technology Corporation, Application Note 5, December 1984
13. Williams, J. and Huffman, B., "Some Thoughts on DC-DC Converters," Appendix A, "The +5 to ±15V Converter—A Special Case," Linear Technology Corporation, Application Note 29, October 1988
14. Baxendall, PJ., "Transistor Sine-Wave LC Oscillators," British Journal of IEEE, February 1960, Paper No. 2978E
15. Williams, J., "Temperature Controlling to Microdegrees," Massachusetts Institute of Technology, Education Research Center, 1971 (out of print)
16. Fulton, S.P, "The Thermal Enzyme Probe," Thesis, Massachusetts Institute of Technology, 1975
17. Williams, J., "Designer's Guide to Temperature Measurement," Part II, EDN, May 20, 1977
18. Williams, J., "Illumination Circuitry for Liquid Crystal Displays," Linear Technology Corporation, Application Note 49, August 1992
19. Olsen, J.V, "A High Stability Temperature Controlled Oven," Thesis, Massachusetts Institute of Technology, 1974
20. "The Ultimate Oven," MIT Reports on Research, March 1972
21. McDermott, James, "Test System at MIT Controls Temperature to Microdegrees," Electronic Design, January 6, 1972
22. McAbel, Walter, "Probe Measurements," Tektronix, Inc. Concept Series, 1969
23. Weber, Joe, "Oscilloscope Probe Circuits," Tektronix, Inc. Concept Series, 1969
24. Tektronix, Inc., "P6015 High Voltage Probe Operating Manual"
25. Williams, Jim, "Measurement and Control Circuit Collection," Linear Technology Corporation, Application Note 45, June 1991
26. Williams, J., "High Speed Amplifier Techniques," Linear Technology Corporation, Application Note 47, August 1991
27. Williams, J., "Practical Circuitry for Measurement and Control Problems," Linear Technology Corporation, Application Note 61, August 1994
28. Chadderton, Neil, "Transistor Considerations for LCD Backlighting," Zetex plc. Application Note 14, February 1995

Appendix A
"Hot" cathode fluorescent lamps

Many CCFL characteristics are shared by so-called "hot" cathode fluorescent lamps (HCFLs). The most significant difference is that HCFLs contain filaments at each end of the lamp (see Figure A1). When the filaments are powered they emit electrons, lowering the lamp's ionization potential. This means a significantly lower voltage will start the lamp. Typically, the filaments are turned on, a relatively modest voltage impressed across the lamp and start-up occurs. Once the lamp starts, filament power is removed. Although HCFLs reduce the high voltage requirement they require a filament supply and sequencing circuitry. The CCFL circuits shown in the text will start and run HCFLs without using the filaments. In practice, this involves simply driving the filament connections at the HCFL ends as if they were CCFL electrodes.

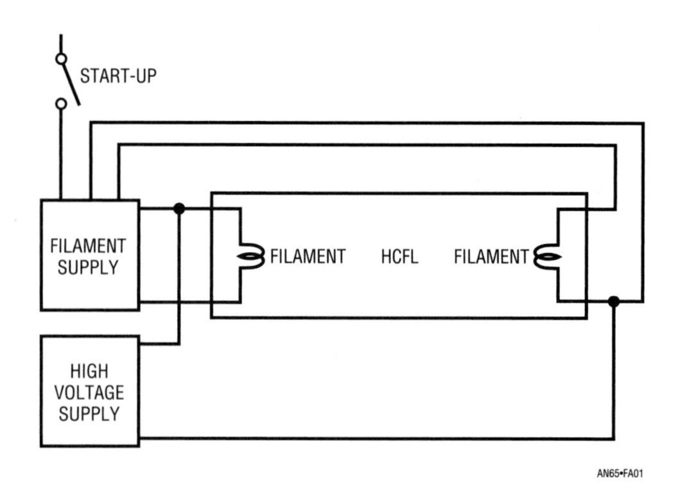

Figure A1 • A Conceptual Hot Cathode Fluorescent Lamp Power Supply. Heated Filaments Liberate Electrons, Lowering the Lamp's Start-Up Voltage Requirement. CCFL Supply Discussed in Text Eliminates Filament Supply

Appendix B
Mechanical design considerations for liquid crystal displays

Charles L. Guthrie, Sharp Electronics Corporation

Introduction

As more companies begin the manufacturing of their next generation of computers, there is a need to reduce the overall size and weight of the units to improve their portability. This has sparked the need for more compact designs where the various components are placed in closer proximity, thus making them more susceptible to interaction from signal noise and heat dissipation. The following is a summary of guidelines for the placement of the display components and suggestions for overcoming difficult design constraints associated with component placement.

In notebook computers the thickness of the display housing is important. The design usually requires the display to be in a pivotal structure so that the display may be folded down over the keyboard for transportation. Also, the outline dimensions must be minimal so that the package will remain as compact as possible. These two constraints drive the display housing design and placement of the display components. This discussion surveys each of the problems facing the designer in detail and offers suggestions for overcoming the difficulties to provide a reliable assembly.

The problems facing the pen-based computer designer are similar to those realized in notebook designs. In addition, however, pen-based designs require protection for the face of the display. In pen-based applications, as the pen is moved across the surface of the display, the pen has the potential for scratching the front polarizer. For this reason the front of the display must be protected. Methods for protecting the display face while minimizing effects on the display image are given.

Additionally, the need to specify the flatness of the bezel is discussed. Suggestions for acceptable construction techniques for sound design are included. Further, display components likely to cause problems due to heat buildup are identified and methods for minimization of the heat's effects are presented.

The ideas expressed here are not the only solutions to the various problems and have not been assessed as to whether they may infringe on any patents issued or applied for.

Flatness and rigidity of the bezel

In the notebook computer the bezel has several distinct functions. It houses the display, the inverter for the back light, and in some instances, the controls for contrast and brightness of the display. The bezel is usually designed to tilt to set the optimum viewing angle for the display.

It is important to understand that the bezel must provide a mechanism to keep the display flat, particularly at the mounting holds. Subtle changes in flatness place uneven stress on the glass which can cause variations in contrast across the display. Slight changes in pressure may cause significant variation in the display contrast. Also, at the extreme, significantly uneven pressures can cause the display glass to fail.

Because the bezel must be functional in maintaining the flatness of the display, consideration must be made for the strength of the bezel. Care must be taken to provide structural members while minimizing the weight of the unit. This may be executed using a parallel grid, normal to the edges of the bezel, or angled about 45° off of the edges of the bezel. The angled structure may be more desirable in that it provides resistance to torquing the unit while lifting the cover with one hand. Again, the display is sensitive to stresses from uneven pressure on the display housing.

Another structure which will provide excellent rigidity, but adds more weight to the computer, is a "honeycomb" structure. This "honeycomb" structure resists torquing from all directions and tends to provide the best protection for the display.

With each of these structures it is easy to provide mounting assemblies for the display. "Blind nuts" can be molded into the housing. The mounting may be done to either the front or rear of the bezel. Attachment to the rear may provide better rigidity for placement of the mounting hardware.

One last caution is worth noting in the development of a bezel. The bezel should be engineered to absorb most of the shock and vibration experienced in a portable computer. Even though the display has been carefully designed, the notebook computer presents extraordinary shock and abuse problems.

Avoiding heat buildup in the display

Several of the display components are sources for heat problems. Thermal management must be taken into account in the design of the display bezel. A heated display may be adversely affected; a loss of contrast uniformity usually results. The cold cathode fluorescent tube (CCFT) itself gives off a small amount of heat relative to the amount of power dissipated in its glow discharge. Likewise, even though the inverters are designed to be extremely efficient, there is some heat generated. The buildup of heat in these components will be aggravated by the typically "tight" designs currently being introduced. There is little ventilation designed into most display bezels. To compound the problem, the plastics used are poor thermal conductors, thus causing the heat to build up which may affect the display.

Some current designs suffer from poor placement of the inverter and/or poor thermal management techniques. These designs can be improved even where redesign of the display housing with improved thermal management is impractical.

One of the most common mistakes in current designs is that there has been no consideration for the buildup of heat from the CCFT. Typically, the displays for notebook applications have only one CCFT to minimize display power requirements. The lamp is usually placed along the right edge of the display. Since the lamp is placed very close to the display glass, it can cause a temperature rise in the liquid crystal. It is important to note that variations in temperature of as little as 5°C can cause an apparent non-uniformity in display contrast. Variations caused by slightly higher temperature variations will cause objectionable variations in the contrast and display appearance.

To further aggravate the situation, some designs have the inverter placed in the bottom of the bezel. This has a tendency to cause the same variations in contrast, particularly when the housing does not have any heat sinking for the inverter. This problem manifests itself as a "blooming" of the display, just above the inverter. This "blooming" looks like a washed out area where, in the worst case, the characters on the display fade completely.

The following section discusses the recommended methods for overcoming these design problems.

Placement of the display components

One of the things that can be done is to design the inverter into the base of the computer with the motherboard. In some applications this is impractical because this requires the high voltage leads to be mounted within the hinges connecting the display bezel to the main body. This causes a problem with strain relief of the high voltage leads and thus with UL certification.

One mistake made most often is placing the inverter at the bottom of the bezel next to the lower edge of the display. It is a fact that heat rises, yet this is one of the most overlooked problems in new notebook designs. Even though the inverters are very efficient, some energy is lost in the inverter in the form of heat. Because of the insulating properties of the plastic materials used in the bezel construction, heat builds up and affects the display contrast.

Designs with the inverter at the bottom can be improved in one of three ways. The inverter can be relocated away from the display, heat sinking materials can be placed between the display and the inverter, or ventilation can be provided to remove the heat.

In mature designs it may be impractical to do what is obvious and move the inverter up to the side of the display towards the top of the housing. In these cases the inverter may be insulated from the display with a "heat dam." One method of accomplishing this would be to use a piece of mica insulator die cut to fit tightly between the inverter and the display. This heat dam would divert the heat around the end of the display bezel to rise harmlessly to the top of the housing. Mica is recommended in this application because of its thermal and electrical insulation properties.

The last suggestion for removing heat is to provide some ventilation to the inverter area. This has to be done very carefully to prevent exposing the high voltage. Ventilation may not be a practical solution because resistance to liquids and dust is compromised.

The best solution for the designer of new hardware is to consider the placement of the inverter to the side of the display and at the top of the bezel. In existing designs of this type the effects of heat from the inverter, even in tight housings, has been minimal or nonexistent.

One problem which is aggravated by the placement of the inverter at the bezel is heat dissipated by the CCFT. In designs where the inverter is placed up and to the side of the display, fading of the display contrast due to CCFT heat is not a problem. However, when the inverter is placed at the bezel bottom, some designs experience a loss of contrast aggravated by the heat from the CCFT and inverter.

In cases where the inverter must be left at the bottom, and the CCFT is causing a loss of contrast, the problem can be minimized by using an aluminum foil heat sink. This does not remove the heat from the display but dissipates it over the entire display area, thus normalizing the display contrast. The aluminum foil is easy to install and in some present designs has successfully improved the display contrast.

Remember that the objection to the contrast variation stems more from nonuniformity than from a total loss of contrast.

Protecting the face of the display

One of the last considerations in the design of notebook and pen-based computers is protection of the display face. The front polarizer is made of a mylar base and thus is susceptible to scratching. The front protection for the display, along with providing scratch protection, may also provide an antiglare surface.

There are several ways that scratch resistance and antiglare surfaces can be incorporated. A glass or plastic cover may be placed over the display, thus providing protection. The material should be placed as close to the display as possible to minimize possible parallax problems due to reflections off of the cover material. With antiglare materials the further the material is from the front of the display the greater the distortion.

In pen applications, the front antiscratch material is best placed in contact with the front glass of the display. The cover glass material normally needs to be slightly thicker to protect the display from distortion when pressure is being exerted on the front.

There are several methods for making the pen input devices. Some use the front surface of the cover glass to provide input data and some use a field effect to a printed wiring board on the back of the display. When the pen input is on the front of the display, the input device is usually on a glass surface.

To limit specular reflection in this application, the front cover glass should be bonded to the display. Care must be taken to ensure that the coefficient of thermal expansion is matched for all of the materials used in the system. Because of the difficulties encountered with the bonding of the cover glass, and the potential to destroy the display through improper workmanship, consulting an expert is strongly recommended.

Appendix C Achieving meaningful electrical measurements

Obtaining reliable efficiency data for the CCFL circuits presents a high order difficulty measurement problem. The accuracy required in the high frequency AC measurements is uncomfortably close to the state-of-the-art. Establishing and maintaining accurate wideband AC measurements is a textbook example of attention to measurement technique. The combination of high frequency, harmonic laden waveforms and high voltage makes meaningful results difficult to obtain. The choice, understanding and use of test instrumentation is crucial. Clear thinking is needed to avoid unpleasant surprises![1]

The lamp's current and voltage waveforms contain energy content over a wide frequency range. Most of this energy is concentrated at the inverter's fundamental frequency and immediate harmonics. However, if 1% measurement uncertainty is desirable, then energy content out to 10MHz must be accurately captured. Figure C1, a spectrum analysis of lamp current, shows significant energy out to 500kHz. Diminished, but still significant,

Note :1 It is worth considering that various constructors of text Figure 8.35 have reported efficiencies ranging from 8% to 115%.

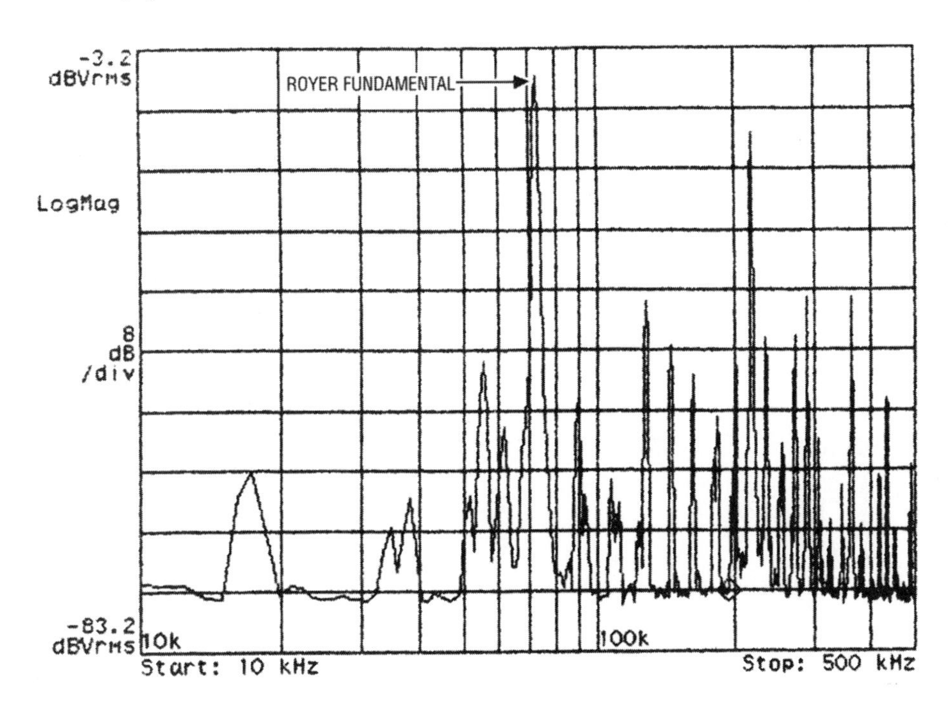

Figure C1 • Hewlett-Packard HP89410A Spectral Plot of Lamp Current Shows Significant Energy Out to 500kHz

content shows up in Figure C2's 6MHz wide plot. This data suggests that monitoring instrumentation must maintain high accuracy over wide bandwidth.

Accurate determination of RMS operating current is important for electrical and emissivity efficiency computations and to ensure long lamp life. Additionally, it is desirable to be able to perform current measurements in the presence of high common mode voltage (>1000V_{RMS}). This capability allows investigation and quantification of display and wiring-induced losses, regardless of their origins in the lamp drive circuitry.

Current probe circuitry

Figure C3's circuitry meets the discussed requirements. It signal-conditions a commercially available "clip-on" current probe with a precision amplifier to provide 1% measurement accuracy to 10MHz. The "clip-on" probe provides convenience, even in the presence of the high common voltages noted. The current probe biases A1, operating at a gain of about 3.75. No impedance matching is required due to the probe's low output impedance termination. Additional amplifiers provide distributed gain, maintaining wide bandwidth with an overall gain of about

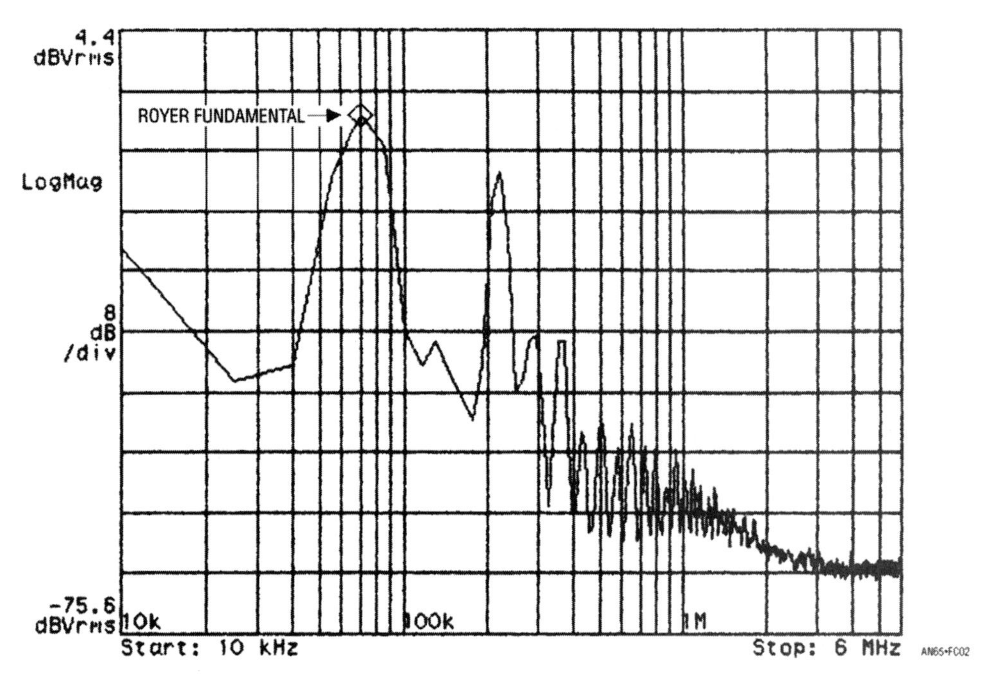

Figure C2 • Extended HP89410A Spectral Plot Shows Lamp Current Has Measurable Energy Well into MHz Range. Data Indicates that Lamp Voltage and Current Instrumentation Must Have Precision, Wideband Response

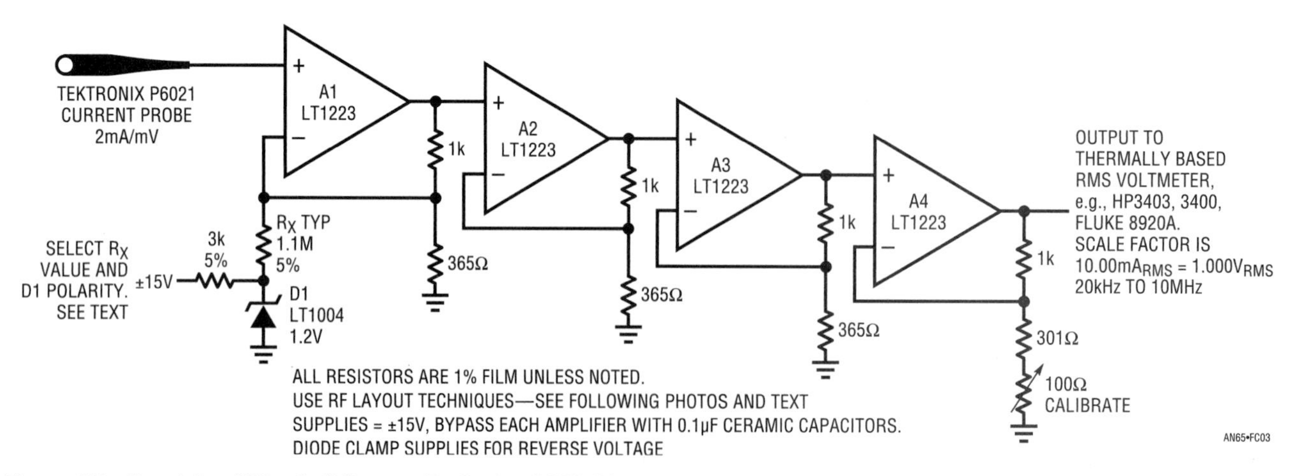

Figure C3 • Precision "Clip-On" Current Probe for CCFL Measurements Maintains 1% Accuracy Over 20kHz to 10MHz Bandwidth

200. The individual amplifiers avoid any possible cross-talk-based error that could be introduced by a monolithic quad amplifier. D1 and R_X are selected for polarity and value to trim overall amplifier offset. The 100Ω trimmer sets gain, fixing the scale factor. The output drives a thermally based, wideband RMS voltmeter. In practice, the circuit is built into a 2.25" × 1" × 1" enclosure which is directly connected via BNC hardware to the voltmeter. No cable is used. Figure C4 shows the probe/amplifier combination. Figure C5 details RF layout techniques used in the amplifier's construction. Figure C6 shows a version of the amplifier, detailing enclosure layout and construction. The result is a "clip-on" current probe with 1% accuracy over a 20kHz to 10MHz bandwidth.

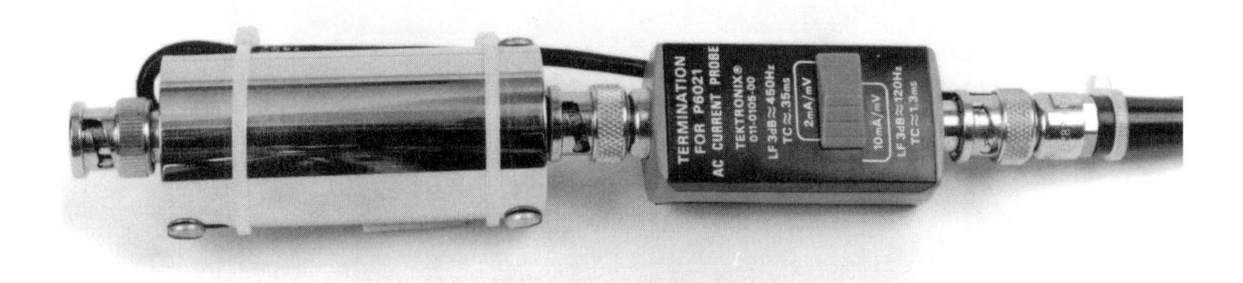

Figure C4 • Current Probe Amplifier Mated to the Current Probe Termination Box

Figure C5 • RF Layout Technique for the Current Probe Amplifier Is Required for Performance Levels Quoted in Text

Figure C6 • A Version of the Current Probe Amplifier in Its Housing. Current Probe Terminator Is at Left

This tool has proven to be indispensable to any rigorously conducted backlight work. Figure C7 shows response for the probe/amplifier as measured on a Hewlett-Packard HP4195A network analyzer.

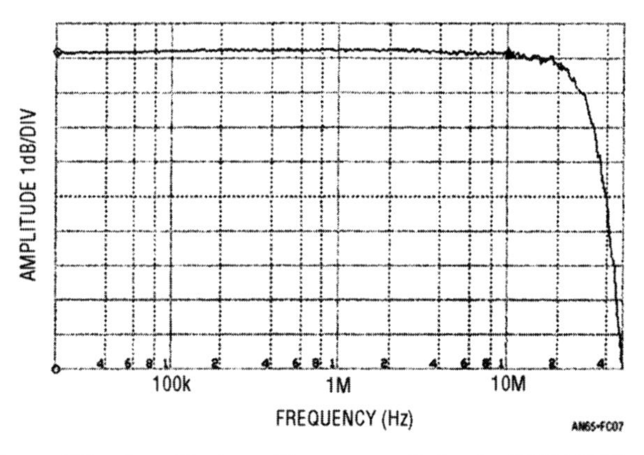

Figure C7 • Amplitude vs Frequency Output of HP4195A Network Analyzer. Current Probe/Amplifier Maintains 1% (0.1dB) Error Bandwidth from 20kHz to 10MHz. Small Aberrations Between 10MHz and 20MHz Are Test Fixture Related

Current calibrator

Figure C8's circuit, a current calibrator, permits calibration of the probe/amplifier and can be used to periodically check probe accuracy. A1 and A2 form a Wein bridge oscillator. Oscillator output is rectified by A4 and A5 and compared to a DC reference at A3. A3's output controls Q1, closing an amplitude stabilization loop. The stabilized amplitude is terminated into a 100Ω, 0.1% resistor to provide a precise 10.00mA, 60kHz current through the series current loop. Trimming is performed by altering the nominal 15k resistor for exactly 1.000V_{RMS} across the 100Ω unit.

In use, this current probe has shown 0.2% baseline stability with 1% absolute accuracy over one year's time. The sole maintenance requirement for preserving accuracy is to keep the current probe jaws clean and avoid rough or abrupt handling of the probe.[2] Figure C9a shows the probe/calibrator used with an RMS voltmeter: Figure C9b shows the current probe in use, in this case determining display frame parasitic loss.

Note 2: Private communication, Tektronix, Inc.

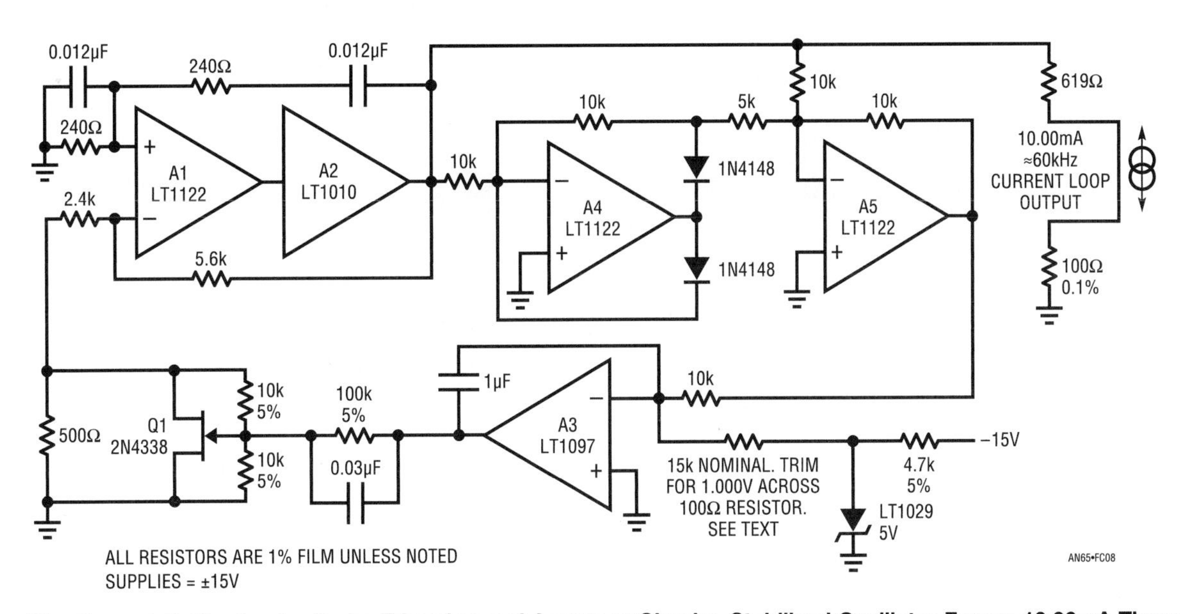

Figure C8 • Current Calibrator for Probe Trimming and Accuracy Checks. Stabilized Oscillator Forces 10.00mA Through Output Current Loop at 60kHz

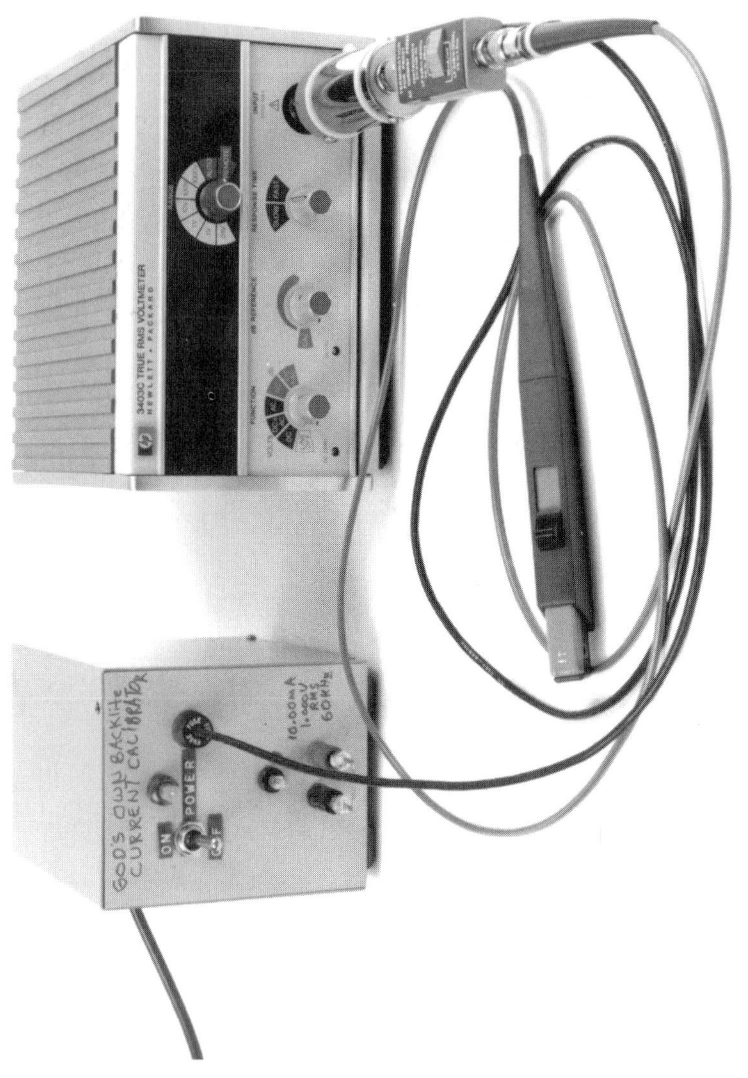

Figure C9a • Complete Current Probe Test Set Includes Probe, Amplifier, Calibrator and Thermally Based RMS Voltmeter. Accuracy Is 1% to 10MHz

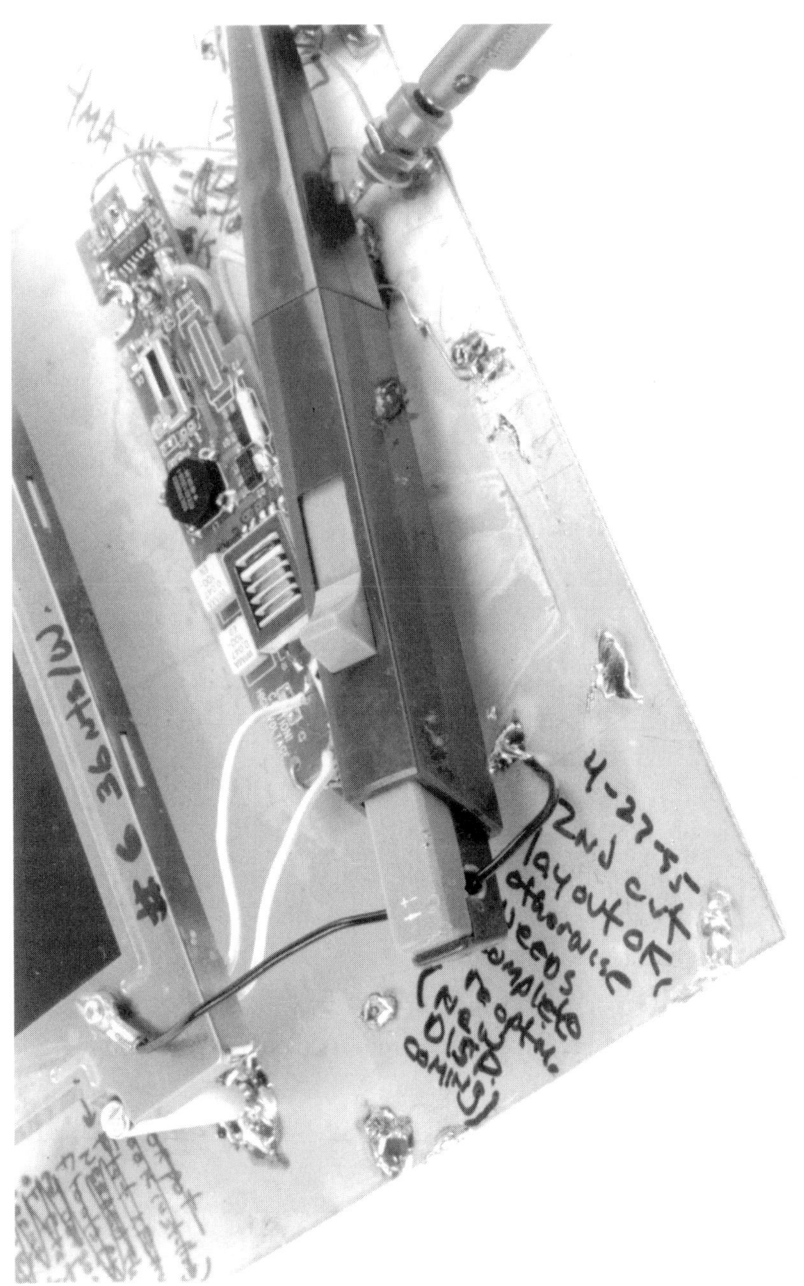

Figure C9b • Current Probe Measuring Display Frame Parasitic Current. "Clip-On" Capability Allows Measurement at Any Point in Lamp Circuit

Voltage probes for grounded lamp circuits

The high voltage measurement across the lamp is quite demanding on the probe. The simplest case is measuring grounded lamp circuits. The waveform fundamental is at 20kHz to 100kHz, with harmonics into the MHz region. This activity occurs at peak voltages in the kilovolt range. The probe must have a high fidelity response under these conditions. Additionally, the probe should have low input capacitance to avoid loading effects which would corrupt the measurement. The design and construction of such a probe requires significant attention. Figure C10 lists some recommended probes along with their characteristics. As stated in the text, almost all standard oscilloscope probes will fail[3] if used for this measurement. Attempting to circumvent the probe requirement by resistively dividing the lamp voltage also creates problems. Large value resistors often have significant voltage coefficients and their shunt capacitance is high and uncertain. As such, simple voltage dividing is not recommended. Similarly, common high voltage probes intended for DC measurement will have large errors because of AC effects. The P6013A and P6015 are the favored probes; their 100MΩ input and small capacitance introduces low loading error.

The penalty for their 1000X attenuation is reduced output, but the recommended voltmeters (discussion to follow) can accommodate this.

All of the recommended probes are designed to work into an oscilloscope input. Such inputs are almost always 1MΩ paralleled by (typically) 10pF to 22pF. The recommended voltmeters, which will be discussed, have significantly different input characteristics. Figure C11's table shows higher input resistances and a range of capacitances. Because of this the probe must be compensated for the voltmeter's input characteristics. Normally, the optimum compensation point is easily determined and adjusted by observing probe output on an oscilloscope. A known amplitude square wave is fed in (usually from the oscilloscope calibrator) and the probe adjusted for correct response. Using the probe with the voltmeter presents an unknown impedance mismatch and raises the problem of determining when compensation is correct.

The impedance mismatch occurs at low and high frequency. The low frequency term is corrected by placing

Note 3: That's twice we've warned you nicely.

TEKTRONIX PROBE TYPE	ATTENUATION FACTOR	ACCURACY	INPUT RESISTANCE	INPUT CAPACITANCE	RISE TIME	BAND-WIDTH	MAXIMUM VOLTAGE	DERATED ABOVE	DERATED TO AT FREQUENCY	COMPENSATION RANGE	ASSUMED TERMINATION RESISTANCE
P6007	100X	3%	10M	2.2pF	14ns	25MHz	1.5kV	200kHz	$700V_{RMS}$ at 10MHz	15pF to 55pF	1M
P6009	100X	3%	10M	2.5pF	2.9ns	120MHz	1.5kV	200kHz	$450V_{RMS}$ at 40MHz	15pF to 47pF	1M
P6013A	1000X	Adjustable	100M	3pF	7ns	50MHz	12kV	100kHz	$800V_{RMS}$ at 20MHz	12pF to 60pF	1M
P6015	1000X	Adjustable	100M	3pF	4.7ns	75MHz	20kV	100kHz	$2000V_{RMS}$ at 20MHz	12pF to 47pF	1M

Figure C10 • Characteristics of Some Wideband High Voltage Probes. Output Impedances Are Designed for Oscilloscope Inputs

MANUFACTURER AND MODEL	FULL-SCALE RANGES	ACCURACY AT 1MHz	ACCURACY AT 100kHz	INPUT RESISTANCE AND CAPACITANCE	MAXIMUM BANDWIDTH	CREST FACTOR
Hewlett-Packard 3400 Meter Display	1mV to 300V, 12 Ranges	1%	1%	0.001V to 0.3V Range = 10M and < 50pF, 1V to 300V Range = 10M and < 20pF	10MHz	10:1 At Full Scale, 100:1 At 0.1 Scale
Hewlett-Packard 3403C Digital Display	10mV to 1000V, 6 Ranges	0.5%	0.2%	10mV and 100mV Range = 20M and 20pF ±10%, 1V to 1000V Range = 10M and 24pF ±10%	100MHz	10:1 At Full Scale, 100:1 At 0.1 Scale
Fluke 8920A Digital Display	2mV to 700V, 7 Ranges	0.7%	0.5%	10M and < 30pF	20MHz	7:1 At Full Scale, 70:1 At 0.1 Scale

Figure C11 • Pertinent Characteristics of Some Thermally Based RMS Voltmeters. Input Impedances Necessitate Matching Network and Compensation for High Voltage Probes

an appropriate value resistor in shunt with the probe's output. For a 10MΩ voltmeter input a 1.1M resistor is suitable. This resistor should be built into the smallest possible BNC equipped enclosure to maintain a coaxial environment. No cable connections should be employed; the enclosure should be placed directly between the probe output and the voltmeter input to minimize stray capacitance. This arrangement compensates the low frequency impedance mismatch. Figure C12 shows the impedance matching box attached to the high voltage probe.

Correcting the high frequency mismatch term is more involved. The wide range of voltmeter input capacitances combined with the added shunt resistor's effects presents problems. How is the experimenter to know where to set the high frequency probe compensation adjustment? One solution is to feed a known value RMS signal to the probe/voltmeter combination and adjust compensation for a proper reading. Figure C13 shows a way to generate a known RMS voltage. This scheme is simply a standard backlight circuit reconfigured for a constant voltage output. The op amp permits low RC loading of the 5.6kΩ feedback termination without introducing bias current error. The 5.6kΩ value may be series or parallel trimmed for a 300V output. Stray parasitic capacitance in the feedback network affects output voltage. Because of this, all feedback associated nodes and components should be rigidly fixed and the entire circuit built into a small metal box. This prevents any significant change in the parasitic terms. The result is a known $300V_{RMS}$ output.

Now, the probe's compensation is adjusted for a 300V voltmeter indication using the shortest possible connection (e.g., BNC-to-probe adapter) to the calibrator box. This procedure, combined with the added resistor completes the probe-to-voltmeter impedance match. If the probe compensation is altered (e.g., for proper response on an oscilloscope) the voltmeter's reading will be erroneous.[4] It is good practice to verify the calibrator box output before and after every set of efficiency measurements. This is done by directly connecting, via BNC adapters, the calibrator box to the RMS voltmeter on the 1000V range.

Voltage probes for floating lamp circuits

Measuring voltage of floating lamp circuits requires a nearly heroic effort. Floating lamp measurement involves all the difficulties of the grounded case but also needs

Note 4: The translation of this statement is to hide the probe when you are not using it. If anyone wants to borrow it, look straight at them, shrug your shoulders and say you don't know where it is. This is decidedly dishonest, but eminently practical. Those finding this morally questionable may wish to re-examine their attitude after producing a day's worth of worthless data with a probe that was unknowingly readjusted.

Figure C12 • The Impedance Matching Box (Extreme Left) Mated to the High Voltage Probe. Note Direct Connection. No Cable Is Used

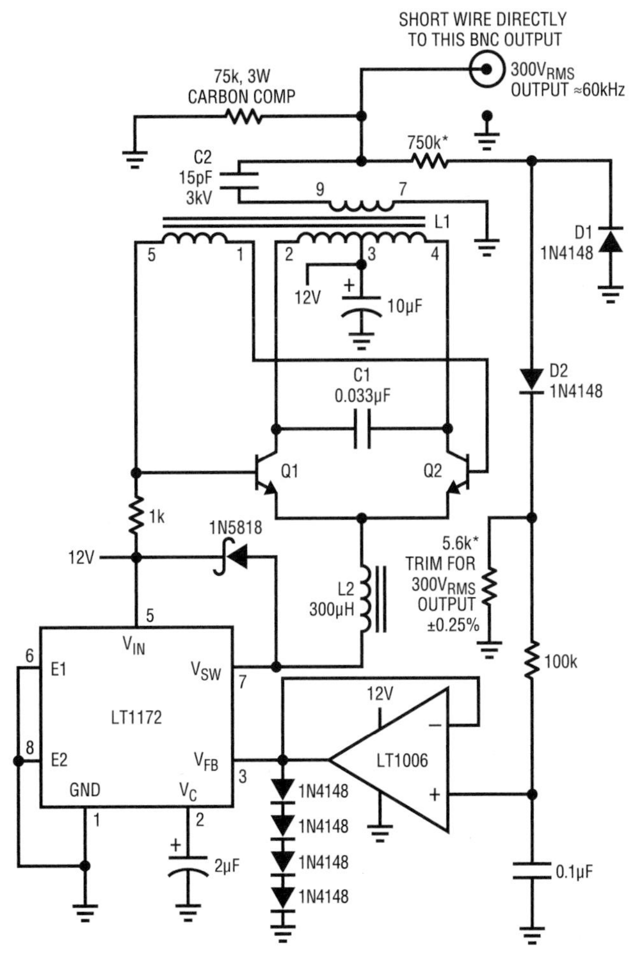

C1 = MUST BE A LOW LOSS CAPACITOR.
METALIZED POLYCARB
WIMA FKP2, MKP-20 (GERMAN) OR PANASONIC ECH-U RECOMMENDED
L1 = SUMIDA 6345-020 OR COILTRONICS CTX110092-1
PIN NUMBERS SHOWN FOR COILTRONICS UNIT
L2 = COILTRONICS CTX300-4
Q1, Q2 = ZETEX ZTX849 OR ZDT1048
* = 1% FILM RESISTOR (TEN 75k RESISTORS IN SERIES)

DO NOT SUBSTITUTE COMPONENTS

COILTRONICS (407) 241-7876, SUMIDA (708) 956-0666

AN65•FC13

Figure C13 • High Voltage RMS Calibrator Is Voltage Output Version of CCFL Circuit

a fully differential input. This is so because the lamp is freely floating from ground. The two probes must not only be properly compensated but matched and calibrated within 1%. Additionally, a fully floating source is required to check calibration instead of Figure C13's simple single ended approach.

Figure C14's differential amplifier converts the differential output of the high voltage probes to a single-ended signal for driving an RMS voltmeter. It introduces less than 1% error in 10MHz bandwidth if probe compensation and calibration are correct (discussion to follow). Both probe inputs feed source followers (Q1-Q4) via RC networks that provide proper probe termination. Q2 and Q4 bias differential amplifier A2, running at a gain of ~2. FET DC and low frequency differential drift is controlled by A1. A1 measures a band limited version of A2's inputs and biases Q4's gate termination resistor. This forces Q4 and Q2 to equal source voltages. This control loop eliminates DC and low frequency error due to FET mismatches.[5] Q1 and Q3 also follow the probe output and feed a small, frequency-dependent, summed signal to A2's auxiliary input. This term is used to correct high frequency common mode rejection limitations of A2's main inputs. A2's output drives the RMS voltmeter via a 20:1 divider. The divider combines with A2's gain-bandwidth characteristics to give 1% accuracy out to 10MHz at the voltmeter input.

To calibrate the amplifier, tie both inputs together and select R_X (shown at Q4) so A1's output is near 0V. It may be necessary to place R_X at Q2 to make this trim. Next, drive the shorted inputs with a 1V 10MHz sine wave. Adjust the "10MHz CMRR trim" for a minimum RMS voltmeter reading, which should be below 1mV Finally,

Note 5: A more obvious and less complex way to control FET mismatch-induced offset would utilize a matched dual monolithic FET Readers are invited to speculate on why this approach has unacceptable high frequency error.

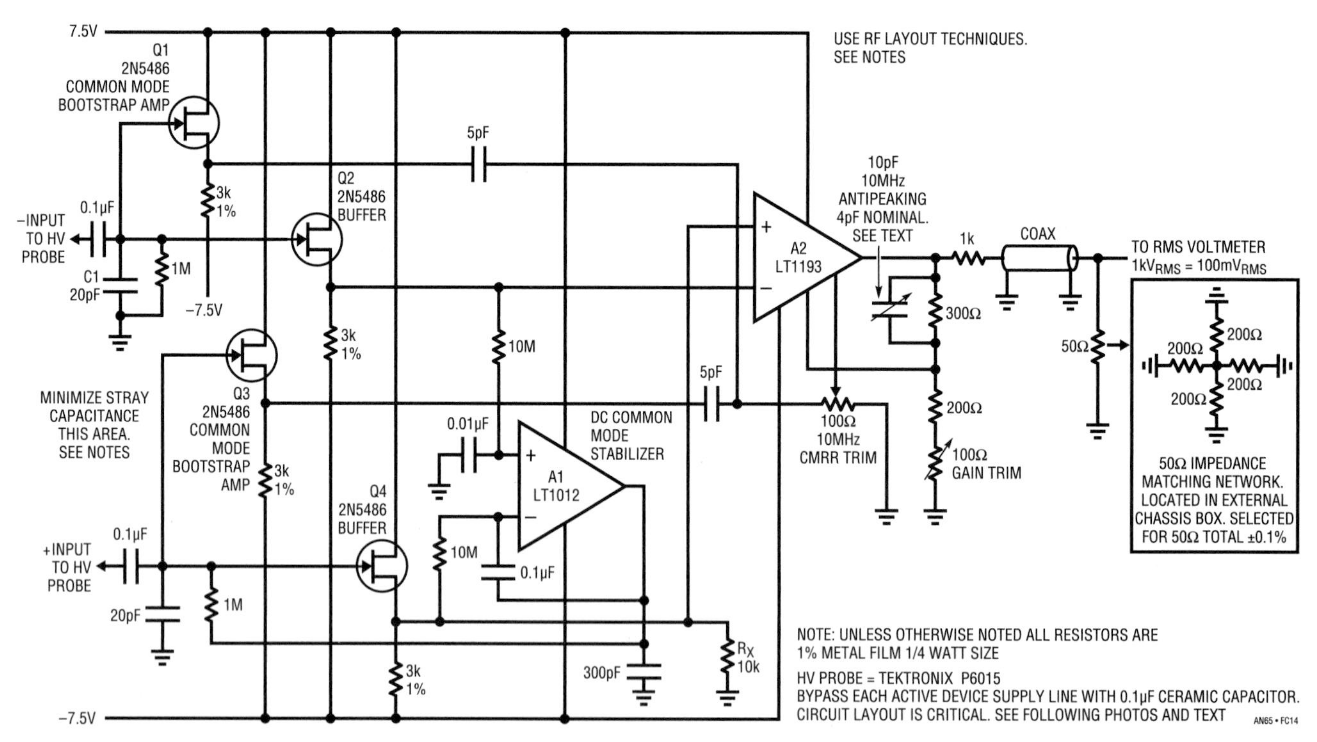

Figure C14 • Precison Wideband Differential Probe Amplifier Permits Floating Lamp Voltage Measurement. Source Followers Combine with Impedance Matching Networks to Unload Probes. A2 Provides Differential-to-Single-Ended Transition

lift the "+" input from ground, apply $1V_{RMS}$ at 60kHz and set A2's gain trim for a 100mV voltmeter reading. As a check, grounding the "+" input and driving the "-" input with the 60kHz signal should produce an identical meter reading. Further, known differential inputs at any frequency from 10kHz to 10MHz should produce corresponding calibrated and stable RMS voltmeter readings within 1%. Errors outside this figure at the highest frequencies are correctable by adjusting the "10MHz antipeaking" trim. This completes amplifier calibration.

The high voltage probes must be properly frequency compensated to give calibrated results with the amplifier. The RC values at the amplifier inputs approximate the termination impedance the probe is designed for Individual probes must, however, be precisely frequency compensated to achieve required accuracy. This is quite a demanding exercise because of probe characteristics.

Figure C15 is an approximate schematic of the Tektronix P6015 high voltage probe. A physically large, 100M resistor occupies the probe head. Although the resistor has repeatable wideband characteristics, it suffers distributed parasitic capacitances. These distributed capacitances combine with similar cable losses, presenting a distorted version of the probed waveform to the terminator box. The terminator box impedance-frequency characteristic, when properly adjusted, corrects the distorted information, presenting the proper waveform at the output. The probe's 1000X attenuation factor, combined with its high impedance, provides a safe, minimally invasive measure of the input waveform.

The large number of parasitic terms associated with the probe head and cable result in a complex, multi-time constant response characteristic. Faithful wideband response requires the terminator box components to separately compensate each of these time constants. As such, no less than seven user adjustments are required to compensate the probe to any individual instrument input. These trims are interactive, requiring a repetitive sequence before the probe is fully compensated. The probe manual describes the trimming sequence, using the intended oscilloscope display as the output. In the present case the ultimate output is an RMS voltmeter via the differential amplifier just described. This complicates determining the probe's proper compensation point but can be accommodated.

To compensate the probes, connect them *directly* to the calibrated differential amplifier (see Figures C16, C17, C18) and ground the probe associated with the "-" input. Drive the "+" input probe with a 100V 100kHz square wave that has a clean 10ns edge with minimal aberrations following the transition.[6] The absolute amplitude of the waveform is unimportant. Monitor this waveform[7] on an oscilloscope. Additionally, monitor A2's output in

Note 6: Suitable instruments include the Hewlett-Packard 214A and the Tektronix type 106 pulse generators.

Note 7: Use a properly compensated probe, please!

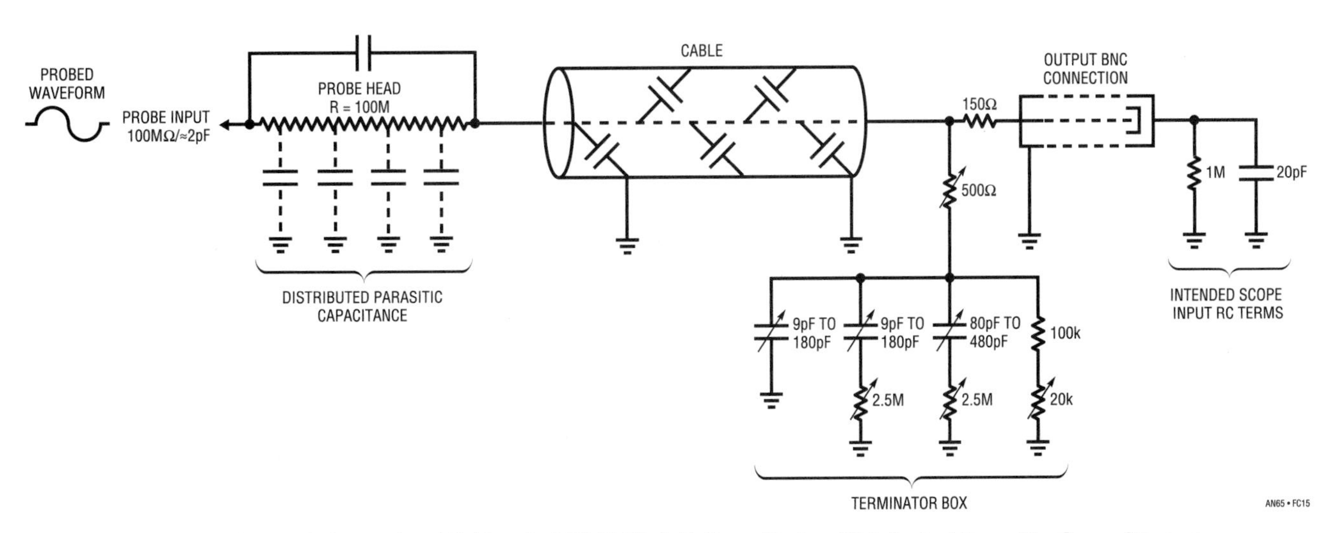

Figure C15 • Approximate Schematic of Tektronix P6015 High Voltage Probe. Distributed Parasitic Capacitances Necessitate Numerous Interactive Trims, Complicating Probe Matching to Voltmeter

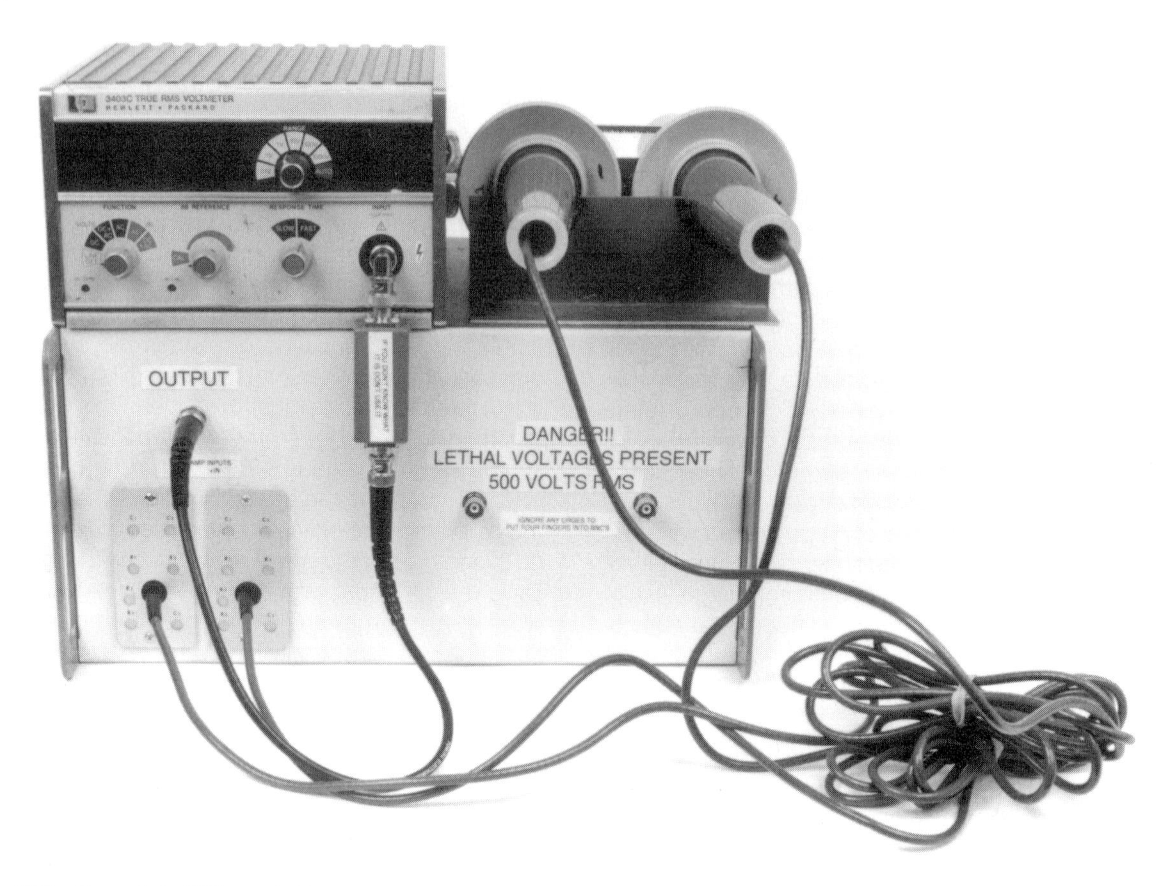

Figure C16 • Complete Differential Probe and Calibrator. BNC Outputs Provide Precision, Floating $500V_{RMS}$ Calibration Source to Check Probe/Amplifier Section

the differential amplifier (see Figure C14) with the oscilloscope.[8] Perform the compensation procedure described in the Tektronix P6015 manual until both waveforms displayed on the oscilloscope have identical shapes. When this state is reached, repeat this procedure with the "-" input probe driven and the "+" input probe grounded. This sequence brings the probe's interactive adjustments reasonably close to the optimum points.

To complete the calibration, connect the 50Ω precision termination (see Figures C14 and C16) and the RMS voltmeter to the differential amplifier's output (see Figure C16). Ground the "-" input probe and drive the "+" probe with a known amplitude high voltage waveform of about 60kHz.[9]

Perform *very slight* readjustments of this probe's compensation trims to get the voltmeter's reading to agree with the calibrated input (account for scaling differences—e.g., ignore the voltmeter's range and decimal point indications). The trim(s) having the greatest influence should be utilized for this adjustment—only a slight adjustment should be required. Upon completing this step repeat the procedure using the 100V 100kHz square wave, verifying input/output waveform edge fidelity. If waveform fidelity has been lost retrim and try again. Several iterations may be necessary until both conditions are met.

Repeat the above procedure for the "-" probe adjustment with the "+" probe grounded.

Note 8: See Footnote 7.

Note 9: Figure C13's calibrator is appropriate.

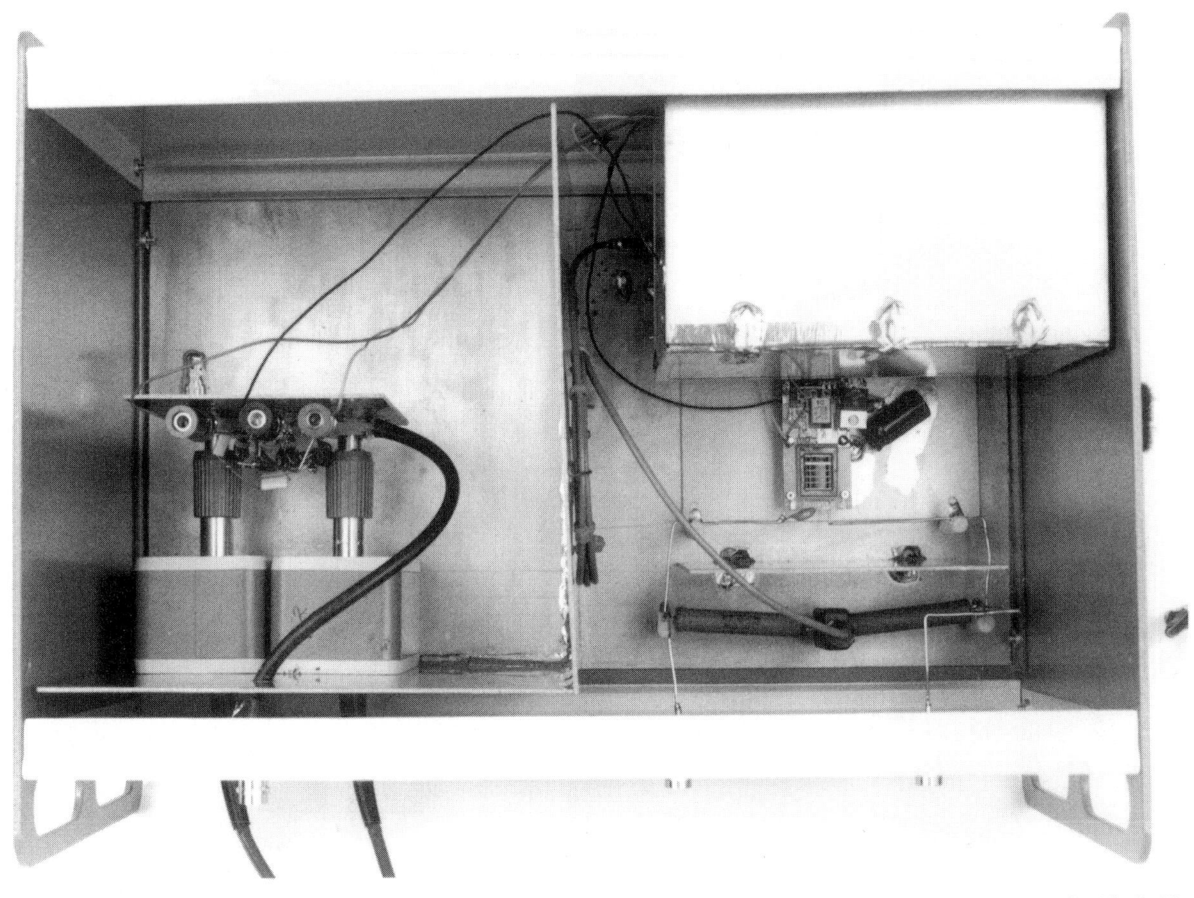

Figure C17 • Top View of Differential Probe/Calibrator. Probes Are Directly Mated to Differential Amplifier (Left). Calibrator Is to Right. Current Transformer Is Located Between Load Resistors

Short both probes together and drive them with the 100V 100kHz square wave. The RMS voltmeter should read (ideally) zero. Typically, it should indicate well below 1% of input. The differential amplifier's "10MHz CMRR trim" (Figure C14) can be adjusted to minimize the voltmeter reading.

Next, with the probes still shorted, apply a swept 20kHz to 10MHz sine wave with the highest amplitude available. Monitor A2's output with an RMS voltmeter, ensuring that it never rises above 1% of input amplitude. Finally,[10] apply the highest available known amplitude, swept 20kHz to 10MHz signal to each probe with the other probe grounded. Verify that the RMS voltmeter indicates correct and flat gain over the entire swept frequency range for each case. If any condition described in this paragraph is not met, the entire calibration sequence must be repeated. This completes the calibration.

Note 10: "Finally" is more than an appropriate descriptive. Achieving a wideband, matched probe response involving 14 interactive adjustments takes time, patience and utter determination. Allow at least six hours for the entire session. You'll need it.

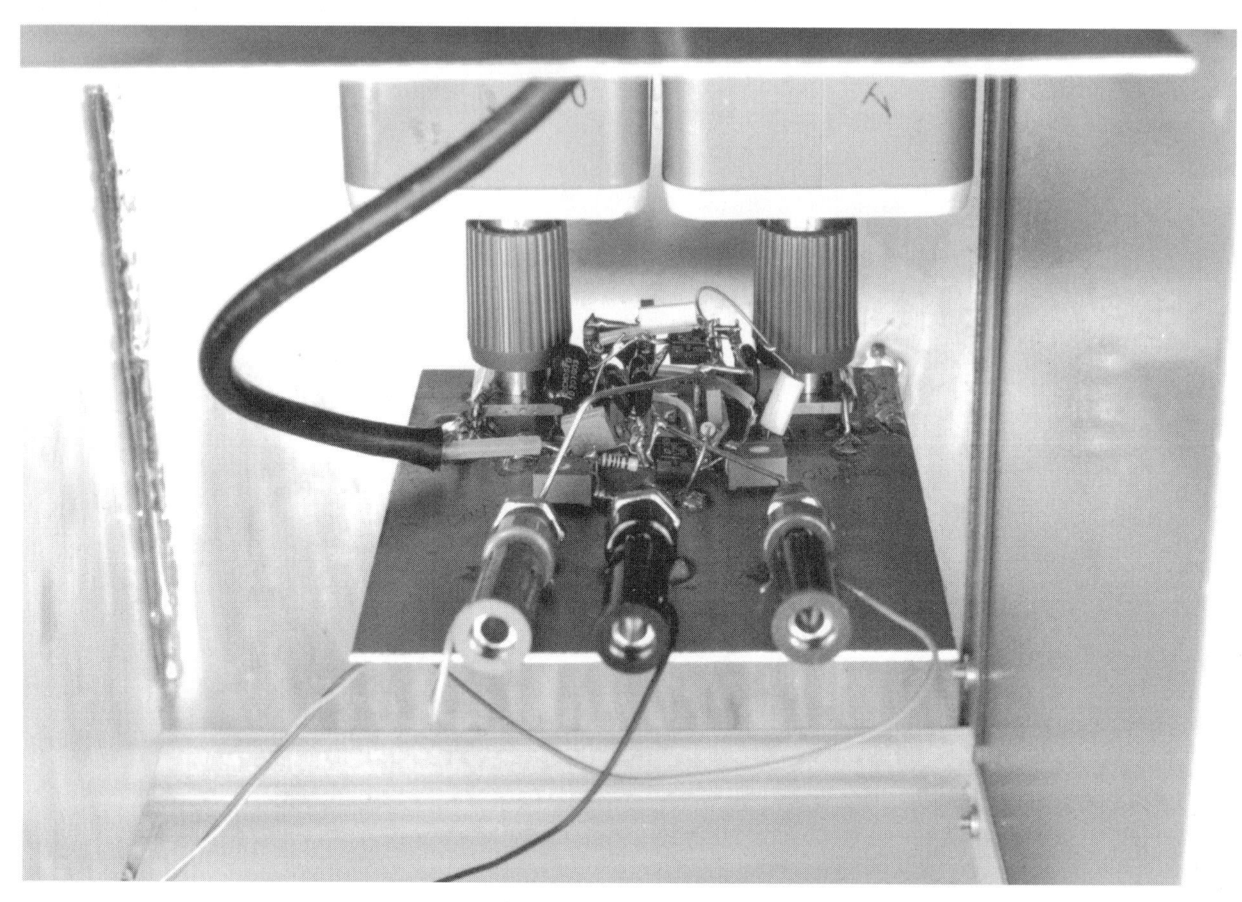

Figure C18 • Detail of Probe/Amplifier Connection Shows Direct, Low Loss BNC Coupling

Differential probe calibrator

A calibrator with a fully floating, differential output allows periodic operational checking of the differential probe's accuracy. This calibrator is built into the same enclosure as the differential probe (Figure C16). Figure C19 is a schematic of the calibrator.

The circuit is a highly modified form of the basic backlight power supply. Here, T1's output drives two precision resistors which are well-specified for high frequency, high voltage operation. The resistor's current is monitored by L2, a wideband current transformer. L2's placement between the resistors combines with T1's floating drive to minimize the effects of L2's parasitic capacitance. Although L2 has parasitic capacitance, it is bootstrapped to essentially 0V negating its effect.

L2's secondary output is amplified by A1 and A2, with A3 and A4 serving as a precision rectifier. A4's output is smoothed by the 10kΩ/0.1µF filter, closing a loop at the LT1172's Feedback pin. Similar to previously described CCFL circuits, the LT1172 controls Royer drive, setting T1's output.

To calibrate this circuit, ground the LT1172's V_C pin, open T1's secondary and select the LT1004's polarity and associated resistor value for 0V at A4's output. Next, put a 5.00mA, 60kHz current through L2.[11] Measure A4's smoothed output (the LT1172's Feedback pin) and adjust the "output trim" for 1.23V Next, reconnect T2's secondary, remove the current calibrator connection and unground the LT1172 V_C pin. The result is $500V_{RMS}$ at

Note 11: Figure C8's output, rescaled for 5.00mA, is a source of calibrated current.

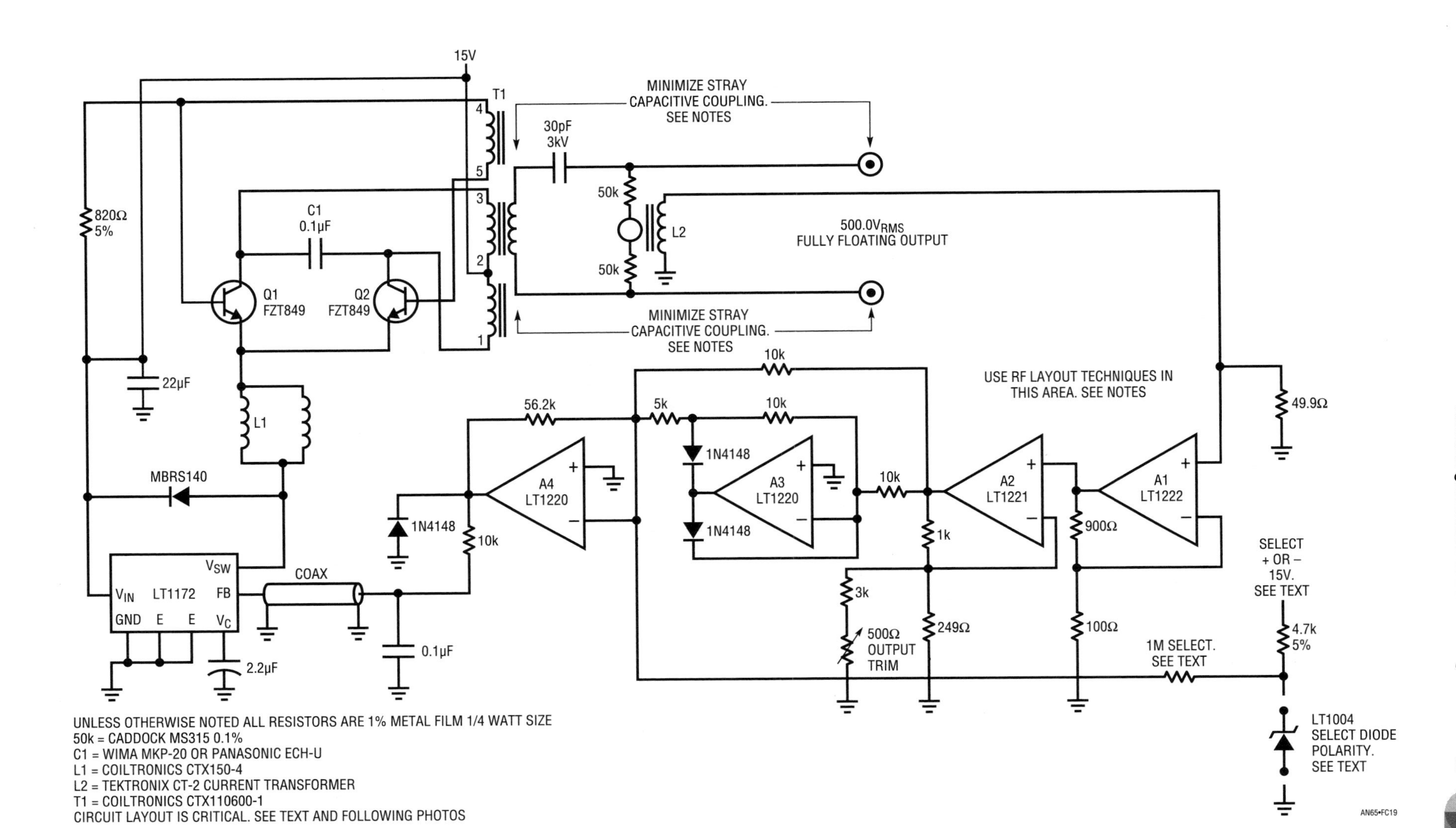

Figure C19 • The Floating Output Calibrator. Current Transformer Permits Floating Output While Maintaining Tight Loop Control. Amplifiers Provide Gain to Inverter Circuit's Feedback Node

the calibrator's differential output. This may be checked with the differential probe. Reversing the probe connections should have no effect, with readings well within 1%.[12]

The differential probe and floating output calibrator require almost fanatical attention to layout to achieve the performance levels noted. The wideband amplifier sections utilize RF layout techniques which are reasonably well-documented.[13] Practical construction considerations for parasitic capacitance related issues are photographically detailed in Figures C17 through C23.

Note 12: Those who construct and trim the differential probe and calibrator will experience the unmitigated joy that breaks loose when they agree within 1%.

Note 13: See Reference 26.

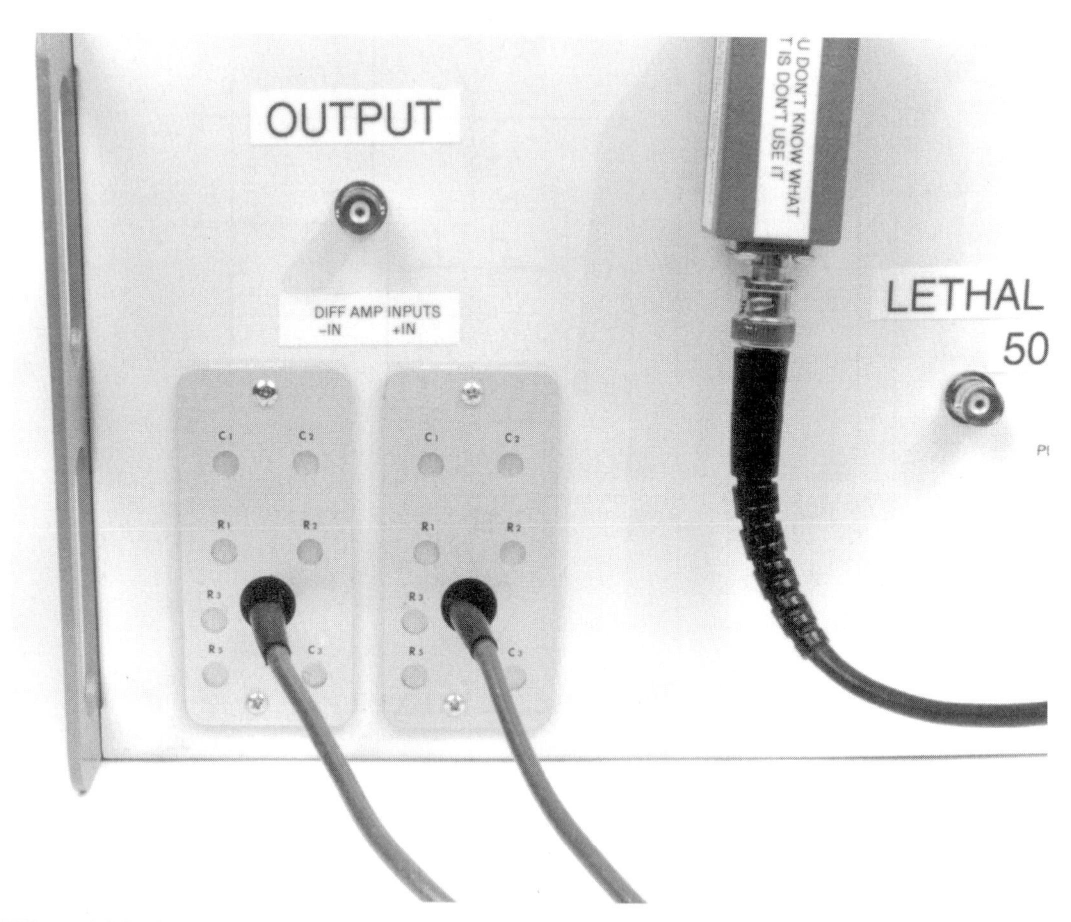

Figure C20 • Differential Probes Are Mechanically Secured to Chassis, Discouraging Unauthorized Removal. All Compensation Access Holes Are Sealed, Preventing Unwanted Adjustment

Figure C21 • Calibrator Section Detail. Inverter in Center with Load Resistors and Current Transformer in Foreground. Note Shield Between Inverter and Load Resistors, and Low Capacitance Layout

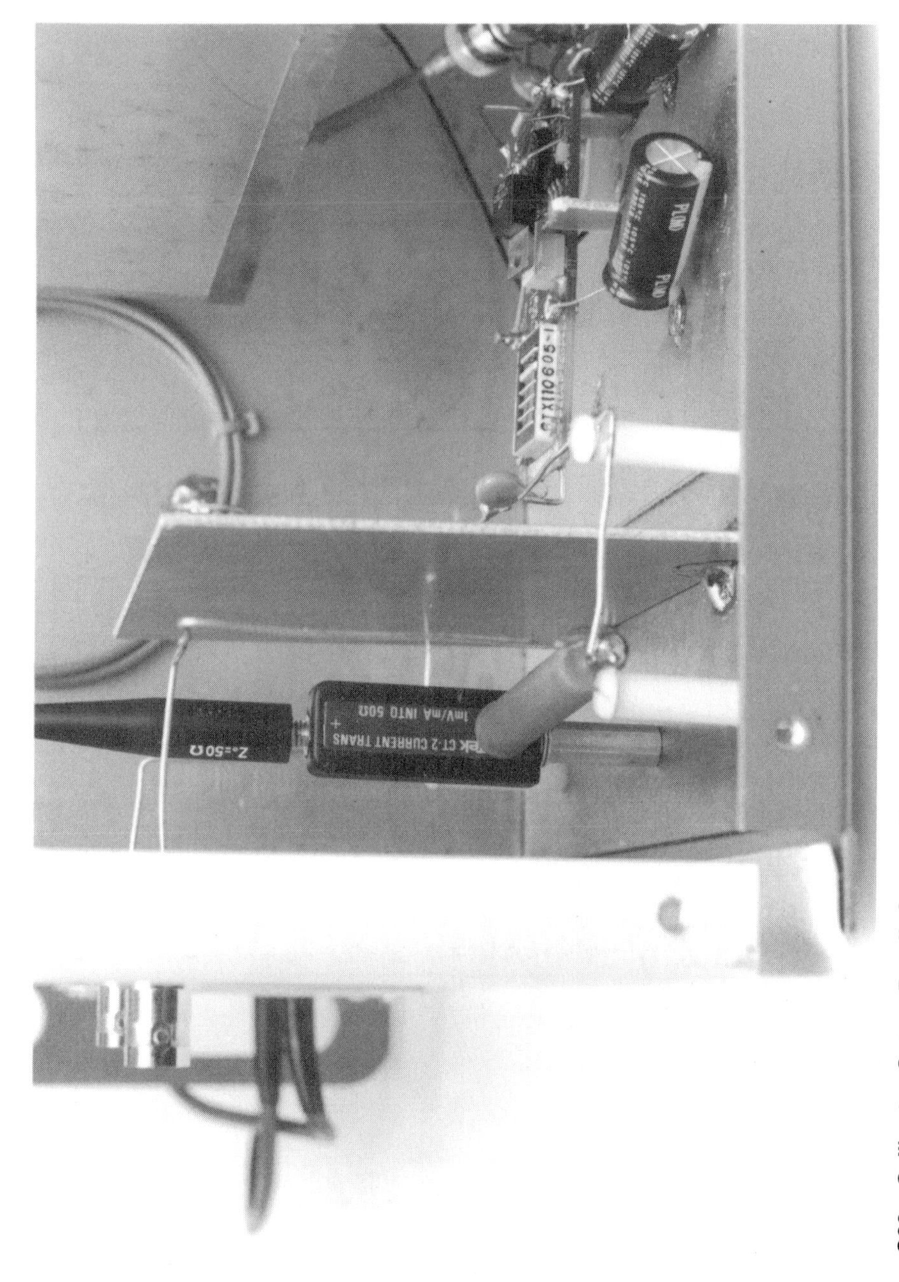

Figure C22 • Calibrator Output Detail. Current Transformer Is Bootstrapped to Load Resistor's OV Midpoint. Shield (Center) Prevents Interaction Between Transformer Field and Load Resistors or Current Transformer. Bus Wire and Nylon Stand-Offs Minimize Stray Capacitance

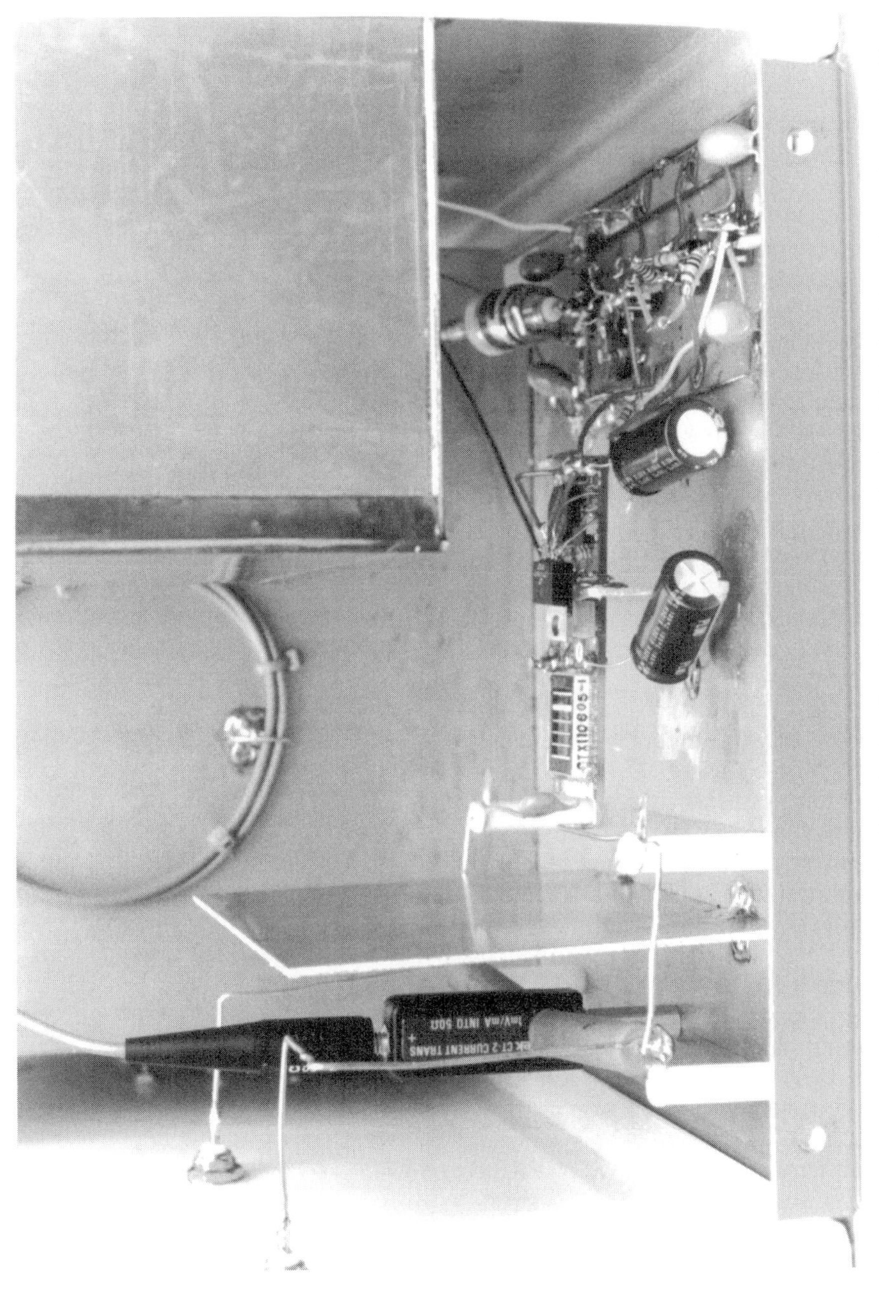

Figure C23 • Calibrator Section Showing Bus Wire/Nylon Post Construction Used to Minimize Stray Capacitance (Left). Inverter in Center; Control Electronics Located at Lower Right. Power Supply Is Enclosed in Shielded Box at Upper Right

RMS voltmeters

The efficiency measurements require an RMS responding voltmeter. This instrument must respond accurately at high frequency to irregular and harmonically loaded waveforms. These considerations eliminate almost all AC voltmeters, including DVMs with AC ranges.

There are a number of ways to measure RMS AC voltage. Three of the most common include *average*, *logarithmic* and *thermally* responding. Averaging instruments are calibrated to respond to the average value of the input waveform, which is almost always assumed to be a sine wave. Deviation from an ideal sine wave input produces errors. Logarithmically based voltmeters attempt to overcome this limitation by continuously computing the input's true RMS value. Although these instruments are "real time" analog computers, their 1% error bandwidth is well below 300kHz and crest factor capability is limited. Almost all general purpose DVMs use such a logarithmically based approach and, as such, are not suitable for CCFL efficiency measurements. Thermally based RMS voltmeters are direct acting thermoelectronic analog computers. They respond to the input's RMS heating value. This technique is explicit, relying on the very definition of RMS (e.g., the heating power of the waveform). By turning the input into heat, thermally based instruments achieve vastly higher bandwidth than other techniques.[14] Additionally, they are insensitive to waveform shape and easily accommodate large crest factors. These characteristics are necessary for the CCFL efficiency measurements.

Figure C24 shows a conceptual thermal RMS/DC converter. The input waveform warms a heater resulting in increased output from its associated temperature sensor. A DC amplifier forces a second, identical, heater/sensor pair to the same thermal conditions as the input driven pair. This differentially sensed, feedback enforced loop makes ambient temperature shifts a common mode term, eliminating their effect. Also, although the voltage and thermal interaction is nonlinear, the input/output RMS voltage relationship is linear with unity gain.

The ability of this arrangement to reject ambient temperature shifts depends on the heater/sensor pairs being isothermal. This is achievable by thermally insulating them with a time constant well below that of ambient shifts. If the time constants to the heater/sensor pairs are matched, ambient temperature terms will affect the pairs equally in phase and amplitude. The DC amplifier rejects this common mode term. Note that although the pairs are isothermal, they are insulated from each other. Any thermal interaction between the pairs reduces the system's thermally based gain terms. This would cause unfavorable signal-to-noise performance, limiting dynamic operating range.

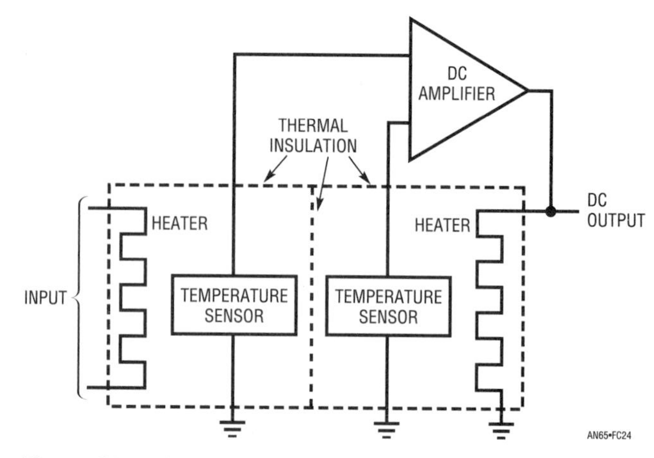

Figure C24 • Conceptual Thermal RMS/DC Converter

Figure C24's output is linear because the matched thermal pair's nonlinear voltage/temperature relationships cancel each other.

The advantages of this approach have made its use popular in thermally based RMS/DC measurements.

The instruments listed in Figure C11, while considerably more expensive than other options, are typical of what is required for meaningful results. The HP3400A and the Fluke 8920A are currently available from their manufacturers. The HP3403C, an exotic and highly desirable instrument, is no longer produced but readily available on the secondary market.

Figure C25 shows an RMS voltmeter which can be constructed instead of purchased.[15] Its small size permits it to be built into bench and production test equipment. As shown, it is designed to be used with Figure C14's differential probe, although the configuration is adaptable to any CCFL-related measurement. It provides a true RMS/DC conversion from DC to 10MHz

Note 14: Those finding these descriptions intolerably brief are commended to References 4–6, 9–12.

Note 15: This circuit derives from Reference 27.

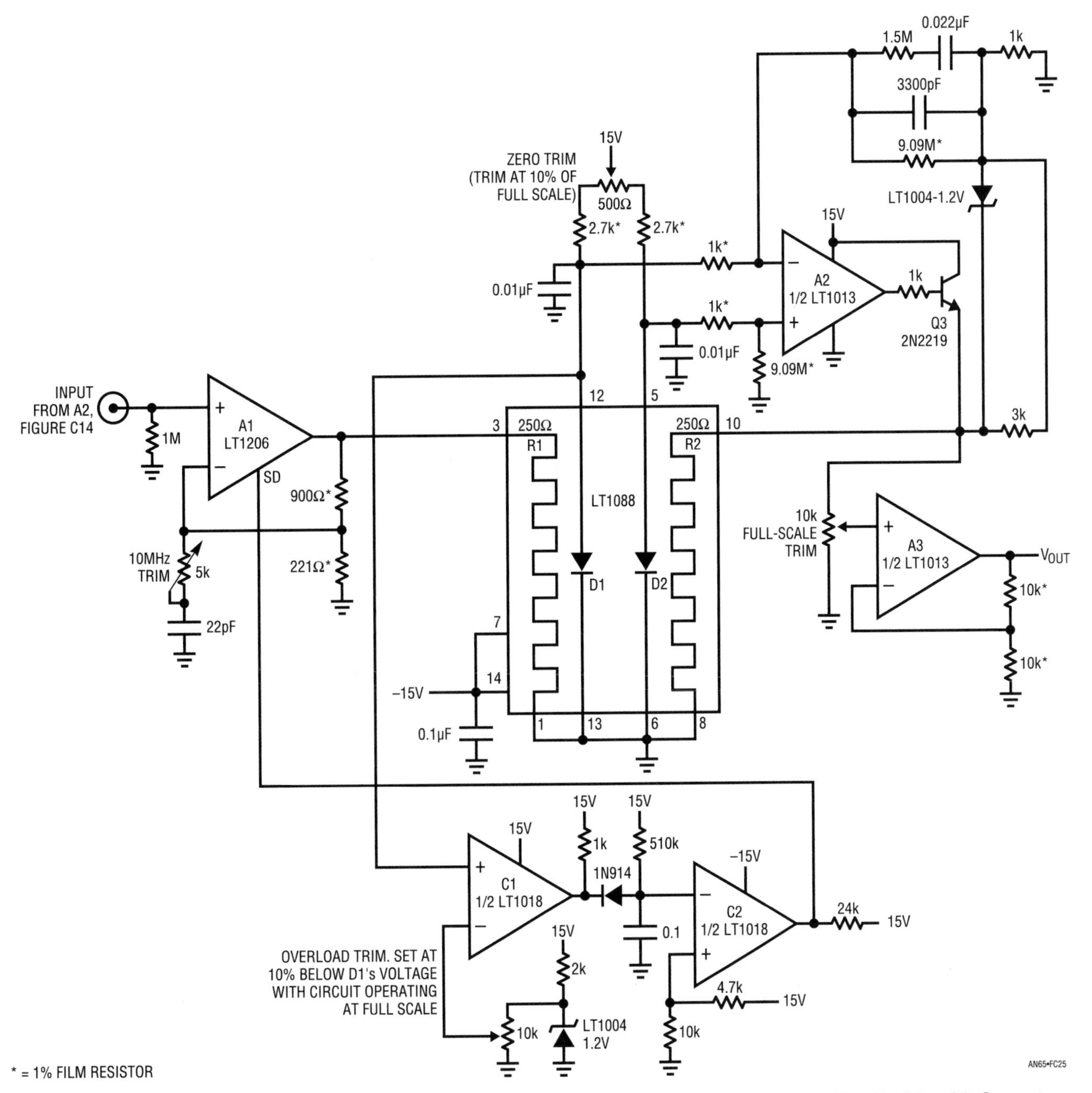

Figure C25 • Wideband RMS/DC Converter for Use with Differential Probe/Amplifier. Circuit Is Also Usable with Current Probe/Amplifier with Appropriate Gain Adjustments

with less than 1% error regardless of input signal waveshape. It also features high input impedance and overload protection.

The circuit consists of three blocks; a wideband amplifier, the RMS/DC converter and overload protection. The amplifier provides high input impedance and gain, and drives the RMS/DC converter's input heater. Input resistance is defined by the 1M resistor with input capacitance about 10pF The LT1206 provides a flat 10MHz bandwidth gain of 5. The 5kΩ/22pF network gives A1 a slight peaking characteristic at the highest frequencies, allowing 1% flatness to 10MHz. A1's output drives the RMS/DC converter.

The LT1088-based RMS/DC converter is made up of matched pairs of heaters and diodes and a control amplifier. The LT1206 drives R1, producing heat which lowers D1's voltage. Differentially connected A2 responds by driving R2 via Q3 to heat D2, closing a loop around the amplifier. Because the diodes and heater resistors are matched, A2's DC output is related to the RMS value of the input, regardless of input frequency or waveshape. In practice, residual LT1088 mismatches necessitate a gain trim, which is implemented at A3. A3's output is the circuit output. The LT1004 and associated components provide loop compensation and good settling time over wide ranges of operating conditions (see Footnote 14).

Start-up or input overdrive can cause A1 to deliver excessive current to the LT1088 with resultant damage. C1 and C2 prevent this. Overdrive forces D1's voltage to an abnormally low potential. C1 triggers low under these conditions, pulling C2's input low. This causes C2's output to go high, putting A1 into shutdown and terminating the overload. After a time determined by the RC at C2's input, A1 will be enabled. If the overload condition still exists the loop will almost immediately shut A1 down again. This oscillatory action will continue, protecting the LT1088 until the overload condition is removed.

Performance for the circuit is quite impressive. Figure C26 plots error from DC to 11MHz. The graph shows 1% error bandwidth of 11MHz. The slight peaking out to 5MHz is due to the gain boost network at A1's negative input. The peaking is minimal compared to the total error envelope, and a small price to pay to get the 1% accuracy to 10MHz.

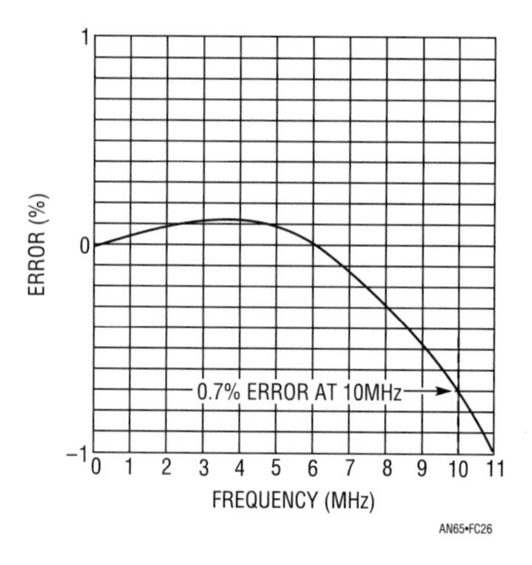

Figure C26 • Error Plot for the RMS/DC Converter. Frequency-Dependent Gain Boost at A1 Preserves 1% Accuracy but Causes Slight Peaking Before Roll-Off

To trim this circuit put the 5kΩ potentiometer at its maximum resistance position and apply a 100mV 5MHz signal. Trim the 500Ω adjustment for exactly $1V_{OUT}$. Next, apply a 5MHz, 1V input and trim the 10kΩ potentiometer for $10.00V_{OUT}$. Finally, put in 1V at 10MHz and adjust the 5kΩ trimmer for $10.00V_{OUT}$. Repeat this sequence until circuit output is within 1% accuracy for DC-10MHz inputs. Two passes should be sufficient.

Calorimetric correlation of electrical efficiency measurements

Careful measurement technique permits a high degree of confidence in the efficiency measurement's accuracy. It is, however, a good idea to check the method's integrity by measuring in a completely different domain. Figure C27 does this by calorimetric techniques. This arrangement, identical to the thermal RMS voltmeter's operation (Figure C24), determines power delivered by the CCFL circuit by measuring its load temperature rise. As in the thermal RMS voltmeter, a differential approach eliminates ambient temperature as an error term. The differential amplifier's output, assuming a high degree of matching in the two thermal enclosures, proportions to load power. The ratio of the enclosure's E • I products yields efficiency information. In a 100% efficient system the amplifier's output energy would equal the power supply's output. Practically it is always less as the CCFL circuit has losses. This term represents the desired efficiency information.

Figure C28 is similar except that the CCFL circuit board is placed within the calorimeter. This arrangement nominally yields the same information, but is a much more demanding measurement because far less heat is generated. The signal-to-noise (heat rise above ambient) ratio is unfavorable, requiring almost fanatical attention to thermal and instrumentation considerations.[16] It is significant that the *total* uncertainty between electrical and both calorimetric efficiency determinations was 3.3%. The two thermal approaches differed by about 2%. Figure C29 shows the calorimeter and its electronic instrumentation. Descriptions of this instrumentation and thermal measurements can be found in the References section following the main text.

Note 16: Calorimetric measurements are not recommended for readers who are short on time or sanity.

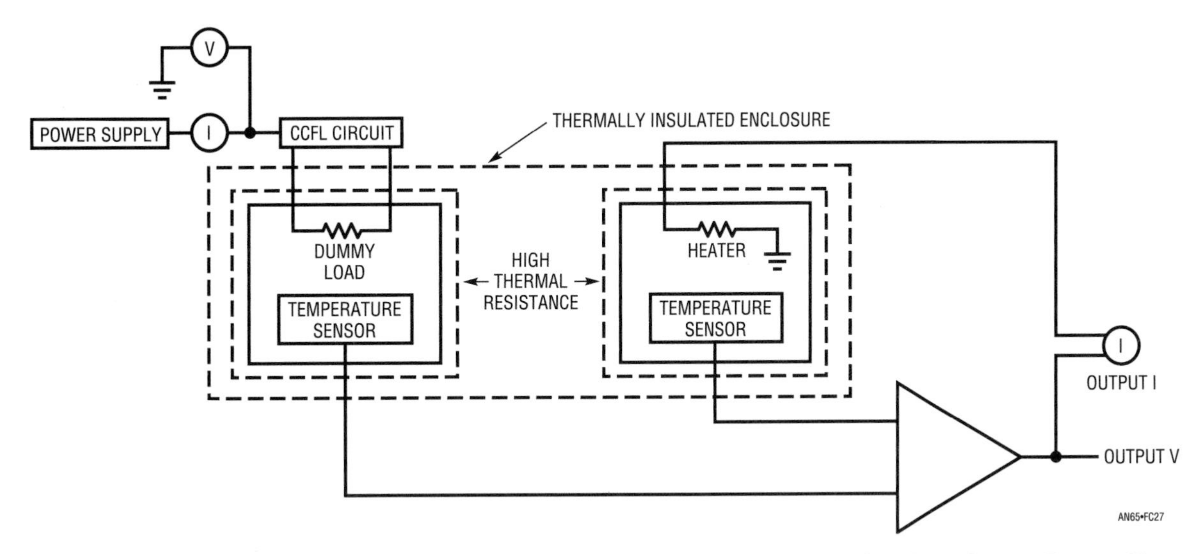

Figure C27 • Efficiency Determination via Calorimetric Measurement. Ratio of Power Supply to Output Energy Gives Efficiency Information

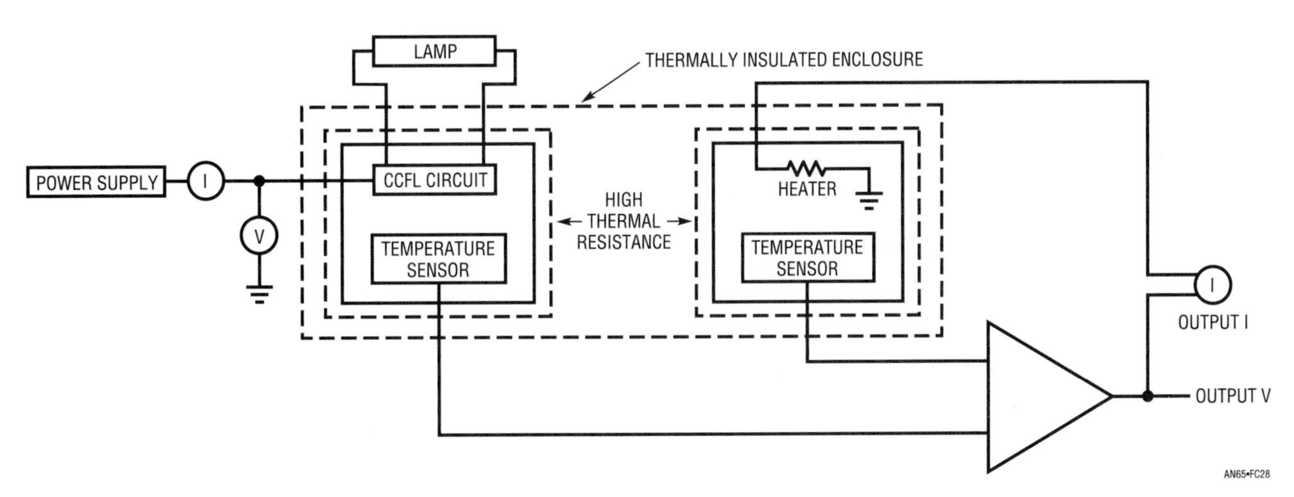

Figure C28 • The Calorimeter Measures Efficiency by Determining Circuit Heating Losses

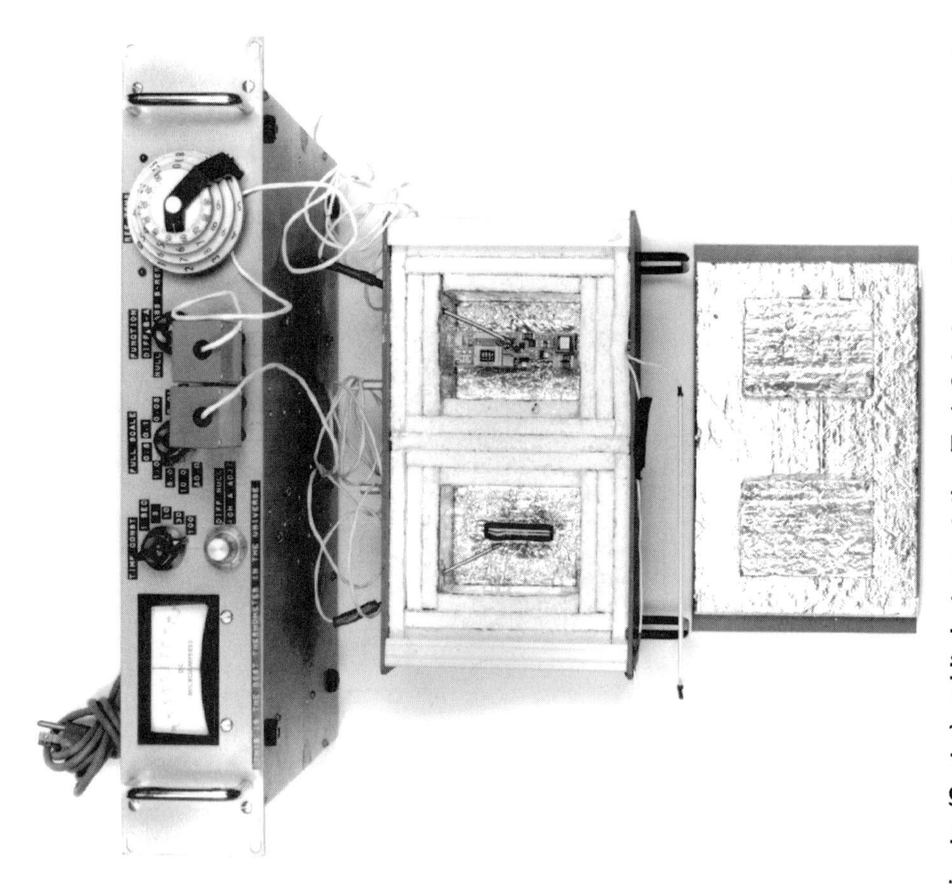

Figure C29 • The Calorimeter (Center) and Its Instrumentation (Top). Calorimeter's High Degree of Thermal Symmetry Combined with Sensitive Servo Instrumentation Produces Accurate Efficiency Measurements. Lower Portion of Photo Is Calorimeter's Top Cover

Appendix D
Photometric measurements

In the final analysis, the ultimate concern centers around the efficient conversion of power supply energy to light. Emitted light varies monotonically with power supply energy,[1] but certainly not linearly. In particular, lamp luminosity may be highly nonlinear, particularly at high power vs drive power. There are complex trade-offs involving the amount of emitted light vs power consumption, drive waveform shape and battery life. Evaluating these trade-offs requires some form of photometer. The relative luminosity of lamps may be evaluated by placing the lamp in a light tight tube and sampling its output with photodiodes. The photodiodes are placed along the lamp's length and their outputs electrically summed. This sampling technique is an uncalibrated measurement, providing relative data only. It is, however, quite useful in determining relative lamp emittance under various drive conditions. Additionally, because the enclosure has essentially no parasitic capacitance, lamp performance may be evaluated under "zero loss" conditions. Figure D1 shows this "glometer," with its uncalibrated output appropriately scaled in "brights." The switches allow various sampling diodes along the lamp's length to be disabled. The photodiode signal-conditioning electronics are mounted behind the switch panel with the drive electronics located to the left.

Figure D2 details the drive electronics. A1 and A2 form a stabilized output. We in bridge sine wave oscillator. A1 is the oscillator and A2 provides gain stabilization in concert with Q1. The stabilizing loop's operating point is derived from the LT1021 voltage reference. A3 and A4 constitute a voltage-controlled amplifier which feeds power stage A5. A5 drives T1, a high ratio step-up transformer. T1's output sources current to the lamp. Lamp current is rectified and its positive portion terminated into the 1k resistor. The voltage appearing across this resistor, indicative of lamp current, biases A6. Band-limited A6 compares the lamp current-derived signal against the LT1021 reference and closes a loop back to A3. This loop's operating point, and hence lamp current, is set by the "current amplitude" adjustment over a 0mA to 6mA range. A1's "Frequency Adjust" control permits a 20kHz to 130kHz frequency operating range. The switch located at A1's output permits external sources of various waveforms and frequencies to drive the amplifier.

Note 1: But not always! It is possible to build highly electrically efficient circuits that emit less light than "less efficient" designs. See previous text and 252.283 ptAppendix L, "A Lot of Cut Off Ears and No Van Goghs—Some Not-So-Great Ideas."

Figure D1 • "Glometer" Measures Relative Lamp Emissivity Under Various Drive Conditions. Test Lamp Is Inside Cylindrical Housing. Photodiodes on Housing Convert Light to Electrical Output (Center) Via Amplifiers (Not Visible in Photo). Electronics (Left) Permit Varying Drive Waveforms and Frequency

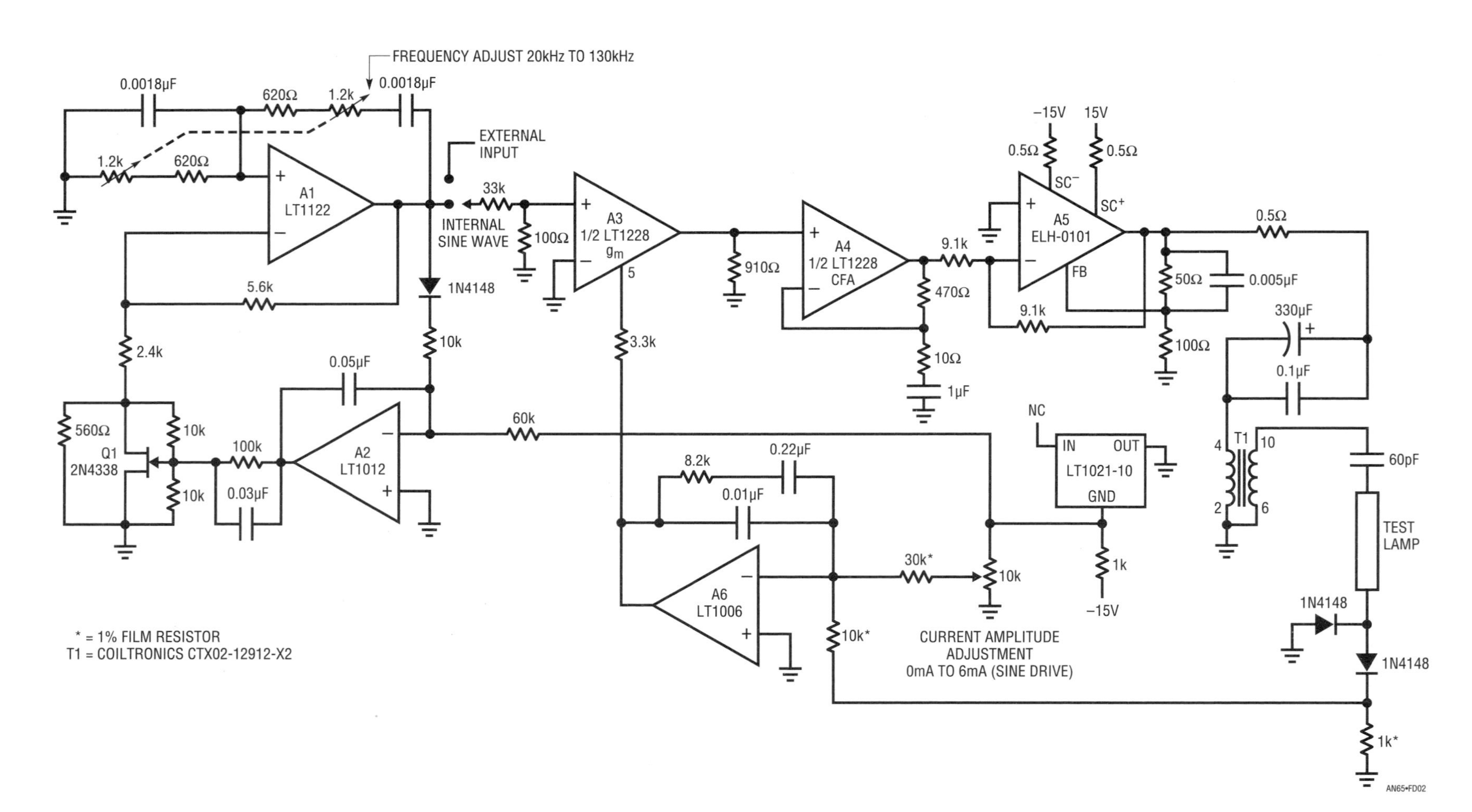

Figure D2 • Glometer Drive Electronics Permits Varying Frequency and Waveshape Applied to Test Lamp. Resultant Data Shows Lamp Sensitivity to These Parameters

The drive scheme and wideband transformer provide extremely faithful response. Figure D3 shows waveform fidelity at 100kHz with a 5mA lamp load. Trace A is T1's primary drive and Trace B is the high voltage output. Figure D4, a horizontal and vertical expansion of D3, indicates well-controlled phase shift. Residual effects cause slight primary impedance variations (note primary drive nonlinearity at the sixth vertical division), although the output remains singularly clean.

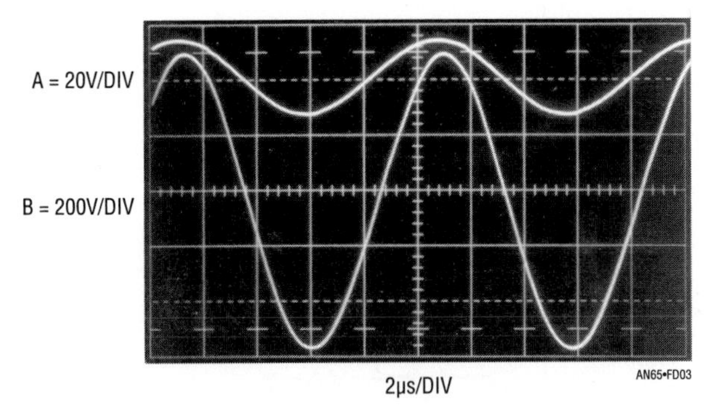

Figure D3 • Wideband Transformers Input (Trace A) and Output (Trace B) Waveforms Indicate Clean Response at 100kHz

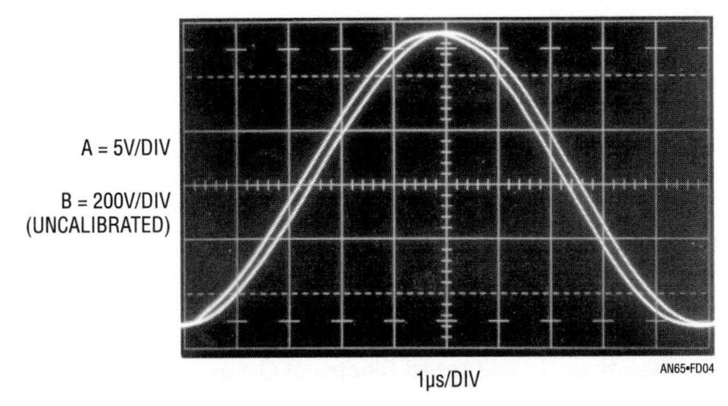

Figure D4 • Magnified Versions of Figure D3s Waveforms. Output (Trace B) Is Undistorted Despite Slight Drive (Trace A) Deformity at Sixth Vertical Division. Transformer Roll-Off Dictates This Desirable Behavior

Figure D5 shows the photodiode signal conditioning. Groups of various photodiodes bias amplifiers A1 through A6. Each amplifier's output is fed via a switch to summing amplifier A7. The switches permit establishment of "dead zones" along the test lamp's length, enhancing ability to study emissivity vs location. A7's output represents the summation of all sensed lamp emission.

The glometer's ability to measure relative lamp emission under controlled settings of frequency, waveshape and drive current in a "lossless" environment is invaluable for evaluating lamp performance. Evaluating display performance and correlating results with customers requires absolute light intensity measurements.

Calibrated light measurements call for a true photometer. The Tektronix J-17/J1803 photometer is such an instrument. It has been found particularly useful in evaluating display (as opposed to simply the lamp) luminosity under various drive conditions. The calibrated output permits reliable correlation with customer results.[2] The light tight measuring head allows evaluation of emittance evenness at various display locations.

Figure D6 shows the photometer in use evaluating a display. Figure D7 is a complete display evaluation setup. It includes lamp and DC input voltage and current instrumentation, the photometer described and a computer (lower right) for calculating optical and electrical efficiency.

Note 2: It is unlikely customers would be enthusiastic about correlating the "brights" units produced by the aforementioned glometer.

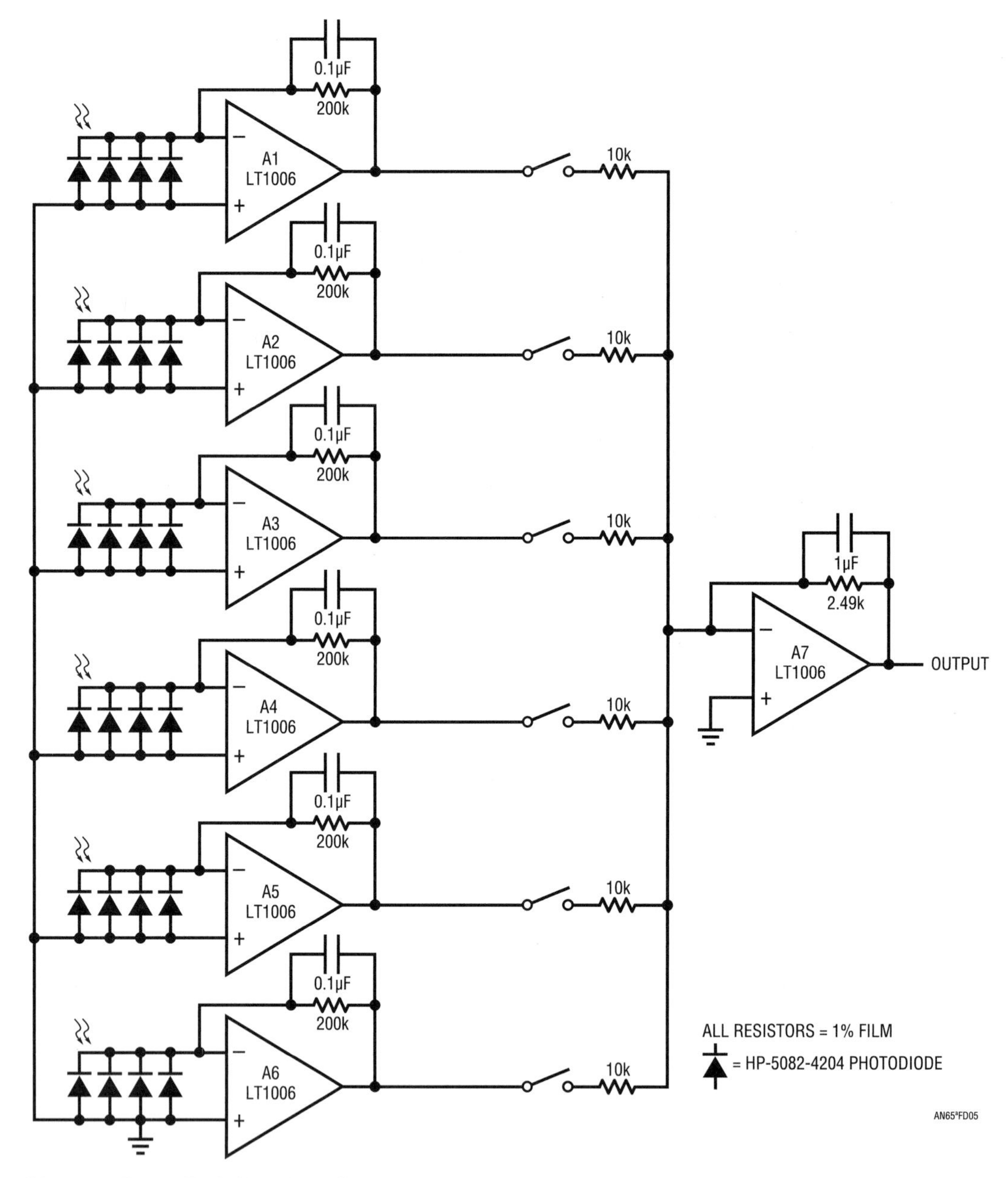

Figure D5 • Glometer Photodiode/Amplifier Converts Lamp Light to Relative, Uncalibrated Electrical Output. Switches Permit Investigation of Individual Portions of Lamp Output

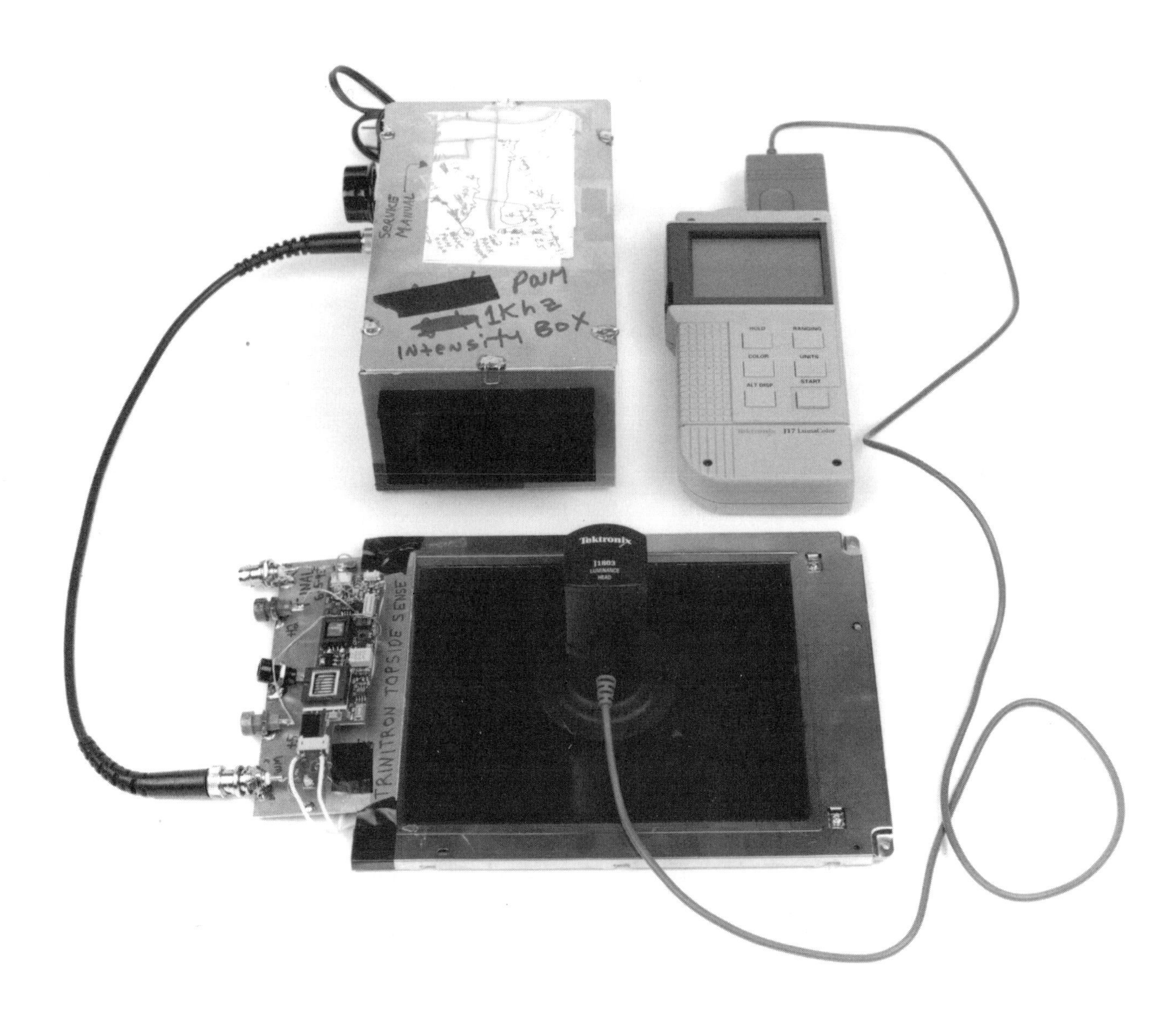

Figure D6 • Apparatus for Calibrated Photometric Display Evaluation. Photometer (Upper Right) Indicates Display Luminosity Via Sensing Head (Center). CCFL Circuit (Left) Intensity Is Controlled by Figure F6's Calibrated Pulse Width Generator (Upper Left)

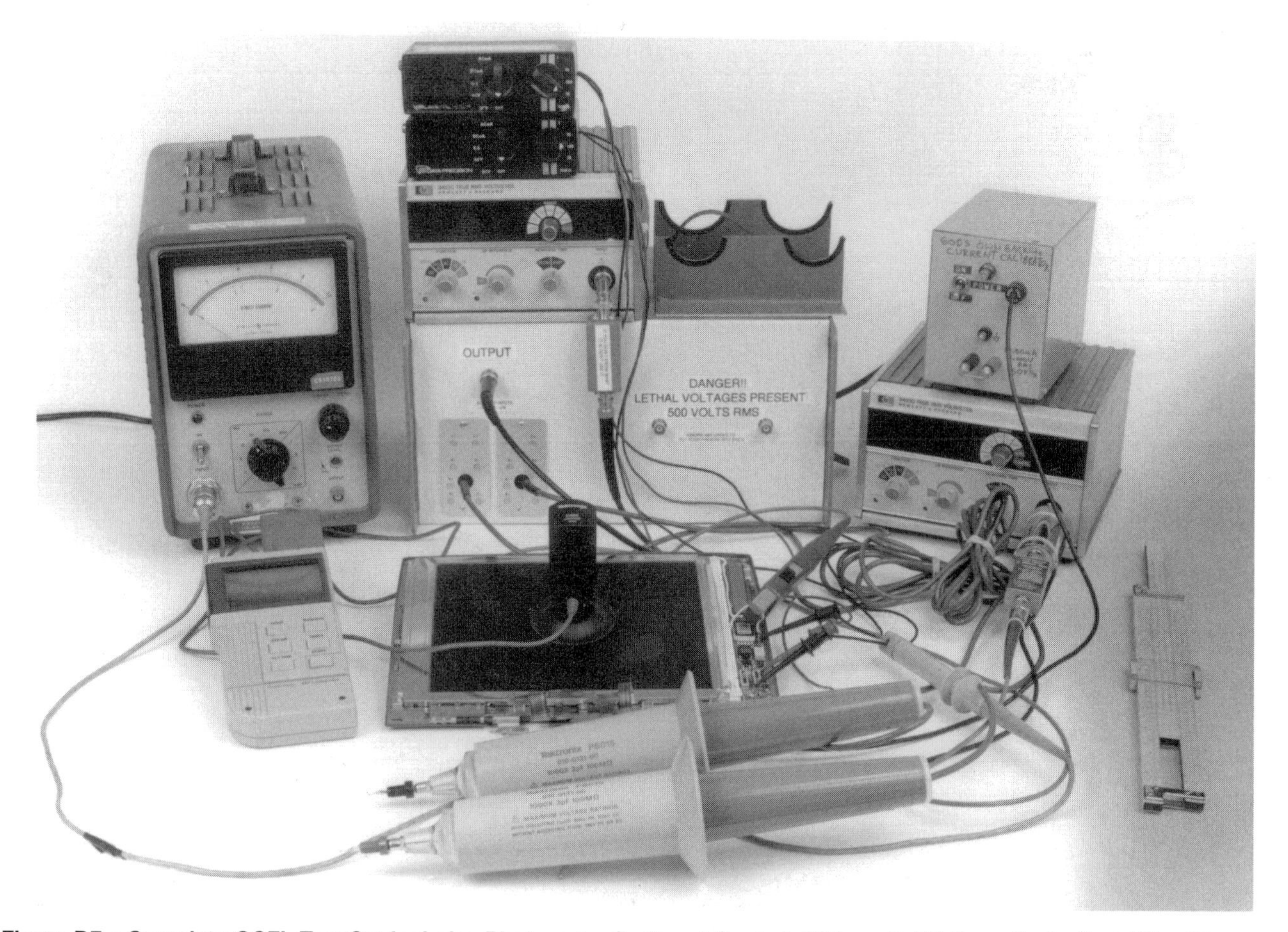

Figure D7 • Complete CCFL Test Set Includes Photometer (Left and Center), Differential Voltage Probe/Amplifier (Upper and Lower Center), Current Probe Electronics (Right) and Input V and I DC Instrumentation (Upper Left). Computer (Lower Right) Permits Calculation of Electrical and Optical Efficiency

Appendix E
Open lamp/overload protection

The CCFL circuit's current source output means that "open" or broken lamps cause full voltage to appear at the transformer output. Safety or reliability considerations sometimes make protecting against this condition desirable. This protection is built into the LT118X series parts. Figure E1 shows a typical circuit. C5, R2 and R3 sense differentially across the Royer converter. Normally, the voltage across the Royer is controlled to relatively small values. An open lamp will cause full duty cycle modulation at the V_{SW} pin, resulting in large current drive through L2. This forces excessive Royer voltage at the C5/R2/R3 network, causing LT1184 shutdown via the "bulb" pin. C5 sets a delay, allowing Royer operation at high drive levels during lamp start-up. This prevents unwarranted shutdown during the lamp's transient high impedance start-up state.

The LT1172 and similar switching regulator parts need additional circuitry for open lamp protection. Figure E2 details the modifications. Q3 and associated components form a simple voltage mode feedback loop that operates if V_Z turns on. If T1 sees no load, there is no feedback and the Q1/Q2 pair receives full drive. Collector voltage rises to abnormal levels, and V_Z biases via Q1's V_{BE} path. Q1's collector current drives the feedback node and the circuit finds a stable operating point. This action controls Royer drive and hence output voltage. Q3's sensing across the Royer provides power supply rejection. V_Z's value should be somewhat above the worst-case Q1/Q2 V_{CE} voltage under running conditions. It is desirable to select V_Z's value so clamping occurs at the lowest output voltage possible while still permitting lamp start-up. This is not as tricky as it sounds because the 10k/1μF RC delays the effects of Q3's turn-on. Usually, selecting V_Z several volts above the worst-case Q1/Q2 V_{CE} will suffice.

Additional protection for all CCFL circuits is possible by fusing the main supply line, typically at a value twice the largest expected DC current. Also, a thermally activated fuse is sometimes mated to Q1 and Q2. Excessive Royer current causes heating in the transistors, activating the fuse.

Overload protection

In certain cases it is desirable to limit output current if either lamp wire shorts to ground. Figure E3 modifies a switching regulator-based circuit to do this. The current sensing network, normally series connected with the lamp, is moved to the transformer. Any overload current must originate from the transformer. Feedback sensing in this path provides the desired protection. This connection measures *total* delivered current, including parasitic terms, instead of lamp return current. Slight line regulation and current accuracy degradation occurs, but not to an objectionable extent.

Floating lamp circuits, because of their isolation, are inherently immune to ground referred shorts. Shorted lamp wires are also tolerated because of the primary side current sensing.

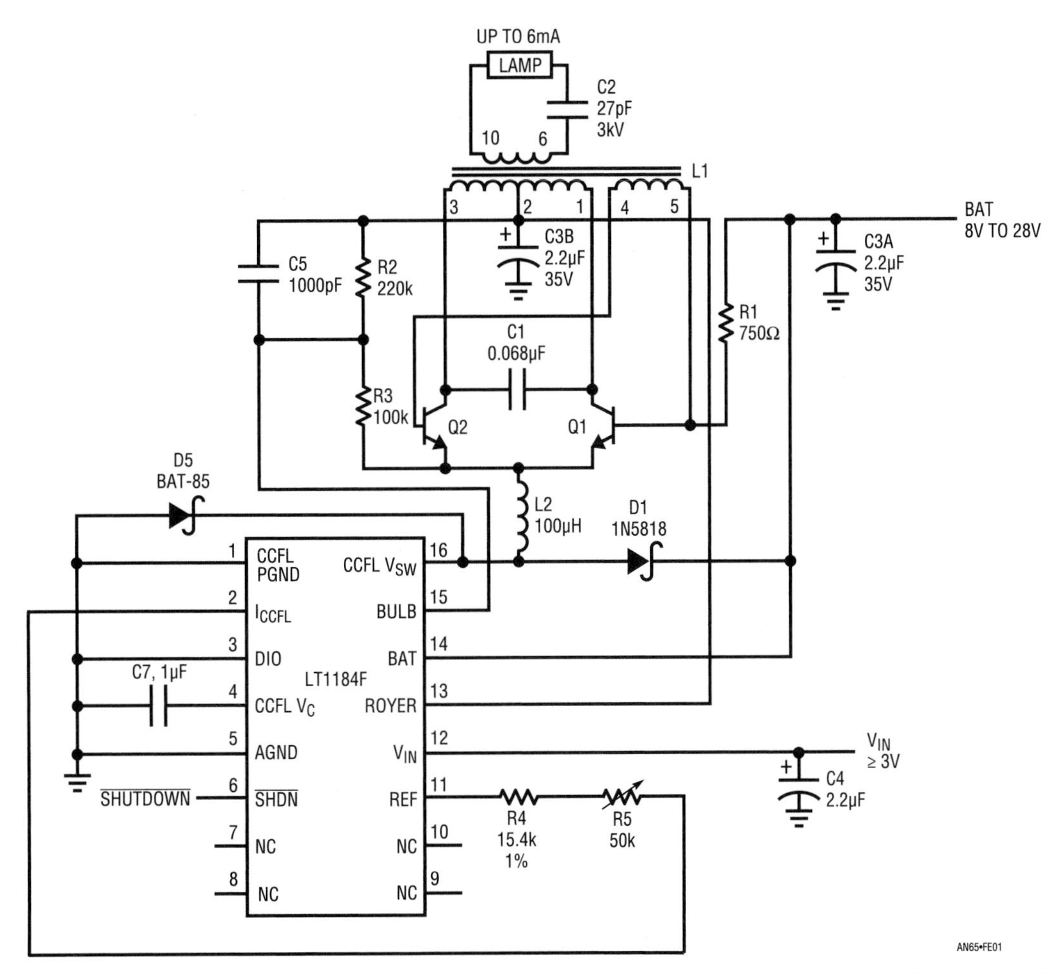

Figure E1 • C5, R2 and R3 Provide Delayed Sensing Across Royer Converter, Protecting Against Open Lamp Conditions in LT118X Series ICs

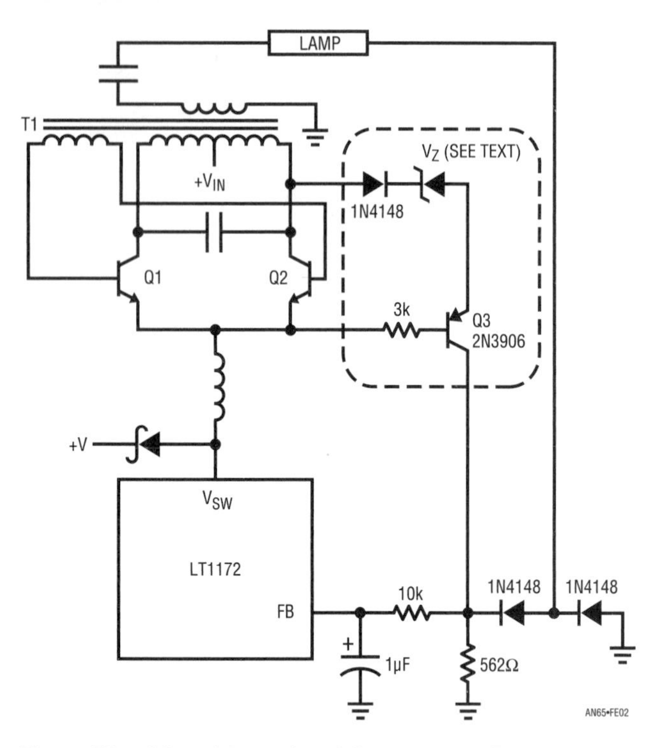

Figure E2 • Q3 and Associated Components Form a Local Regulating Loop to Limit Output Voltage

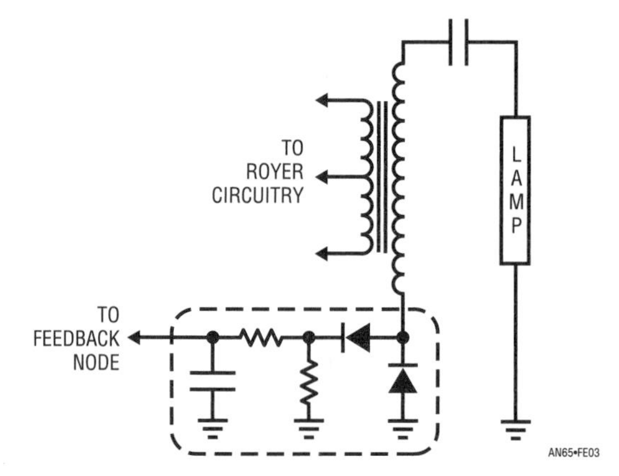

Figure E3 • Relocating Feedback Network (Circled Components) to Transformer Secondary Maintains Current Control When Output Is Shorted. Trade-Off Is Slight Degradation in Line Regulation and Current Accuracy

Appendix F
Intensity control and shutdown methods

The CCFL circuits usually require shutdown capability and some form of intensity (dimming) control. Figure F1 lists various options for the LT118X parts. Control sources include pulse width modulation (PWM), potentiometers and DACs or other voltage sources. The LT1186 (not shown) uses a digital serial-bit stream data input and is discussed in text associated with Figure 51.

In all cases shown the average current into the I_{CCFL} pin sets lamp current. As such, the amplitude and duty cycle must be controlled in cases A and B. The remaining examples use the LT118X's reference to eliminate amplitude uncertainty-induced errors.

Figure F2 shows shutdown options for LT118X parts. The parts have a high impedance Shutdown pin, or power may simply be removed from V_{IN}. Switching V_{IN} power requires a higher current control source but shutdown current is somewhat lower.

Figure F3 shows options for dimming control in LT1172 and similar regulator-based CCFL circuits. Three basic ways to control intensity appear in the figure. The most common intensity control method is to add a potentiometer in series with the feedback termination. When using this method ensure that the minimum value (in this case 562Ω) is a 1% unit. If a wide tolerance resistor is used the lamp current, at maximum intensity setting, will vary appropriately.

Pulse width modulation or variable DC is sometimes used for intensity control. Two interfaces work well. Directly driving the Feedback pin via a diode—22k resistor with DC or PWM produces intensity control. The other method shown is similar, but places the 1µF capacitor

(F1a) LT1182/LT1183 I_{CCFL} PWM Programming

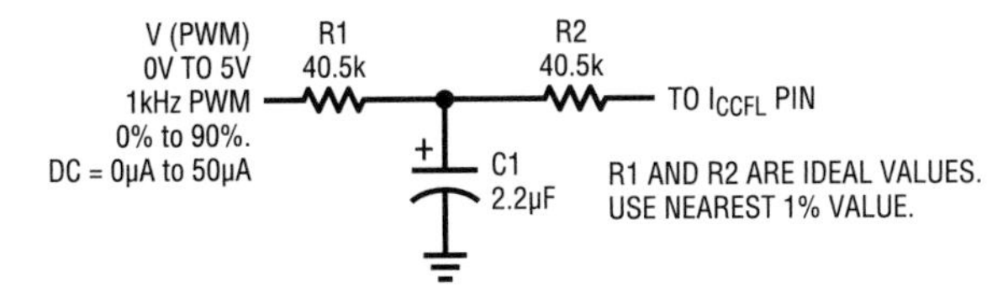

(F1b) LT1184/LT1184F I_{CCFL} PWM Programming

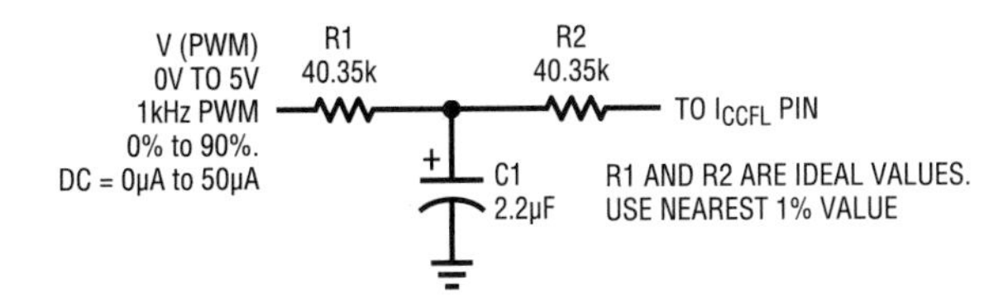

(F1c) LT1183 I_{CCFL} Programming with Potentiometer Control

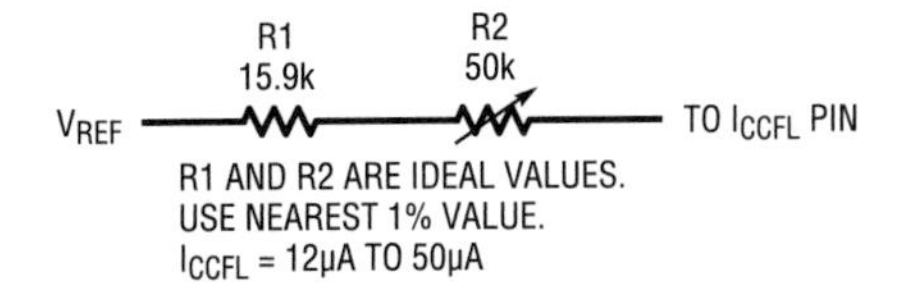

(F1d) LT1184/LT1184F I_{CCFL} Programming with Potentiometer Control

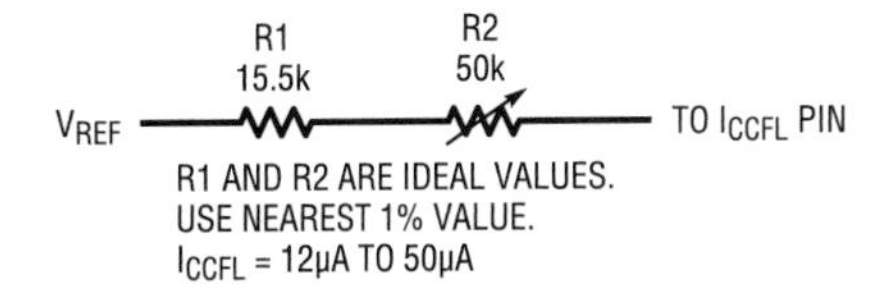

(F1e) LT1182/LT1183/LT1184/LT1184F I_{CCFL} Programming with DAC or Voltage Source Control

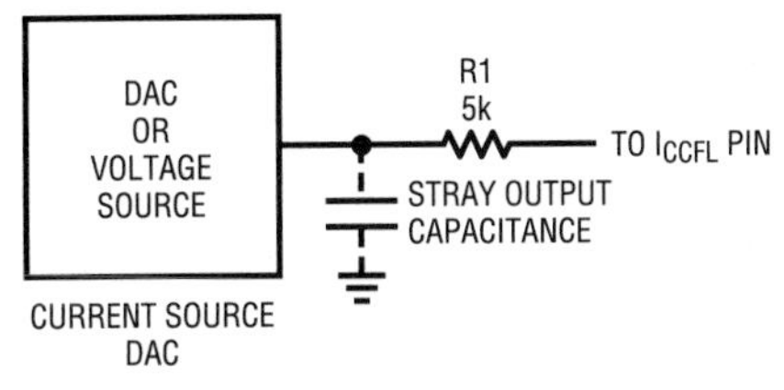

(F1f) LT1183 I_{CCFL} PWM Programming with V_{REF}

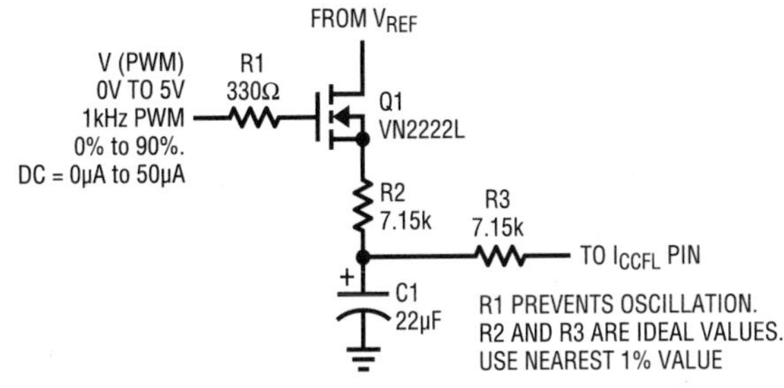

(F1g) LT1184/LT1184F I_{CCFL} PWM Programming with V_{REF}

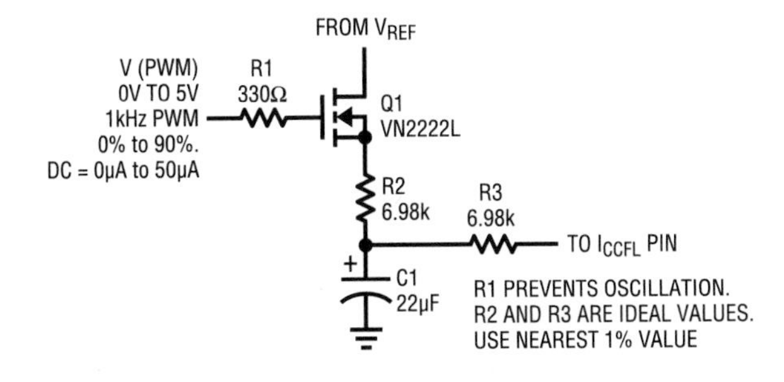

(F1h) LT1183 I_{CCFL} PWM Programming with V_{REF}

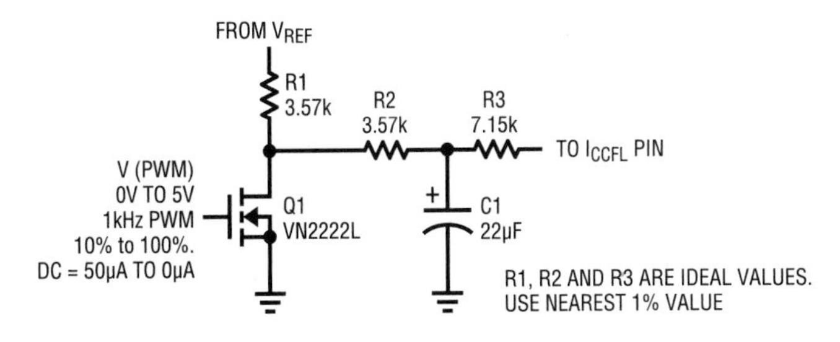

(F1i) LT1184/LT1184F I_{CCFL} PWM Programming with V_{REF}

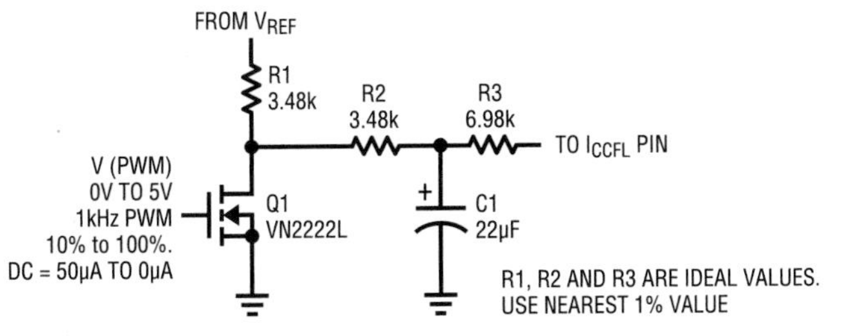

Figure F1 • Various Dimming Options for LT118X Series Parts. LT1186 (Not Shown) Has Serial-Bit Stream Digital Dimming Input

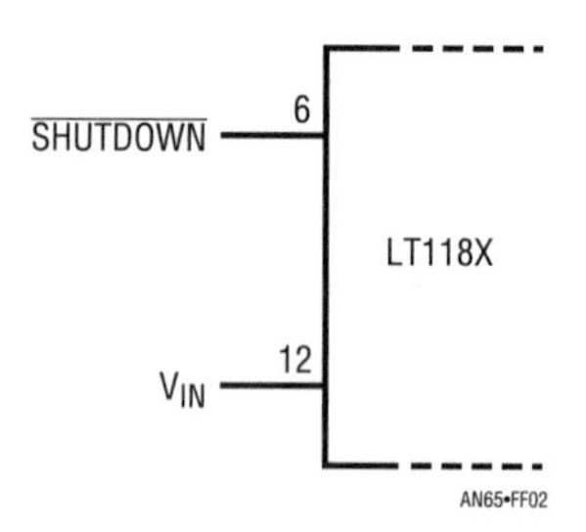

Figure F2 • Shutdown Options for LT118X Series Parts Include Shutdown Pin or Simply Removing V_{IN}

Because of this, it is sometimes advantageous to use the device as a voltage divider instead of a rheostat. The key issue in potentiometer-based dimming is usually ensuring that lamp overdrive cannot occur. This is why arranging dimming schemes for maximum intensity at "shorted" potentiometer positions is preferable. The "end resistance" tolerance, which should be checked, is often less significant and more repeatable than that for maximum resistance or even ratio setting. Other issues include wiper current capability, taper characteristic and circuit sensitivity to "opens." Always review circuit behavior for maximum wiper current demands. CCFL dimming schemes almost never require significant wiper current, but ensure that the particular scheme used doesn't have this problem. The

outside the feedback loop to get best turn-on transient response. This is the best method if output overshoot must be minimized. Note that in all cases the PWM source amplitude at 0% duty cycle does not, by definition, effect full-scale lamp current certainty. See the main text section, "Feedback Loop Stability Issues" for pertinent discussion.

Figure F4 shows methods for shutting down switching regulator-based CCFL circuits. In LT1172 circuits pulling the V_C pin to ground puts the circuit into micropower shutdown. In this mode about 50μA flows into the LT1172 V_{IN} pin with essentially no current drawn from the main (Royer center tap) supply. Turning off V_{IN} power eliminates the LT1172's 50μA drain. Other regulators, such as the LT1372, have a separate Shutdown pin.

About potentiometers

Potentiometers, frequently used in CCFL dimming, require thought to avoid problems. In particular, resistance, ratio tolerances and other issues can upset an ill-prepared design. Keep in mind that ratio tolerances (see Figure F5) are usually better than absolute resistance specifications.

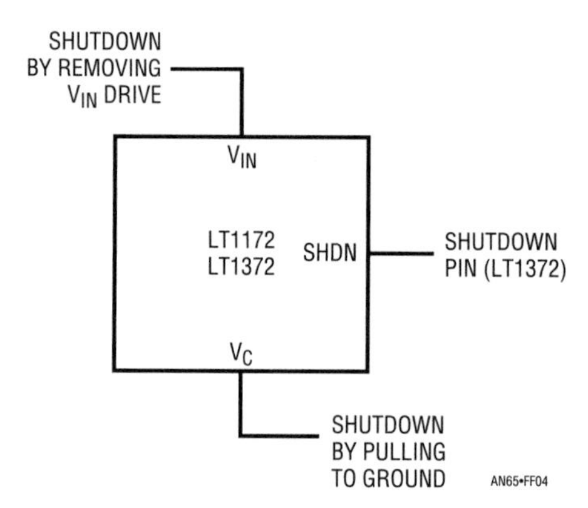

Figure F4 • Various Shutdown Options in LT1172/LT1372 Type CCFL Circuits

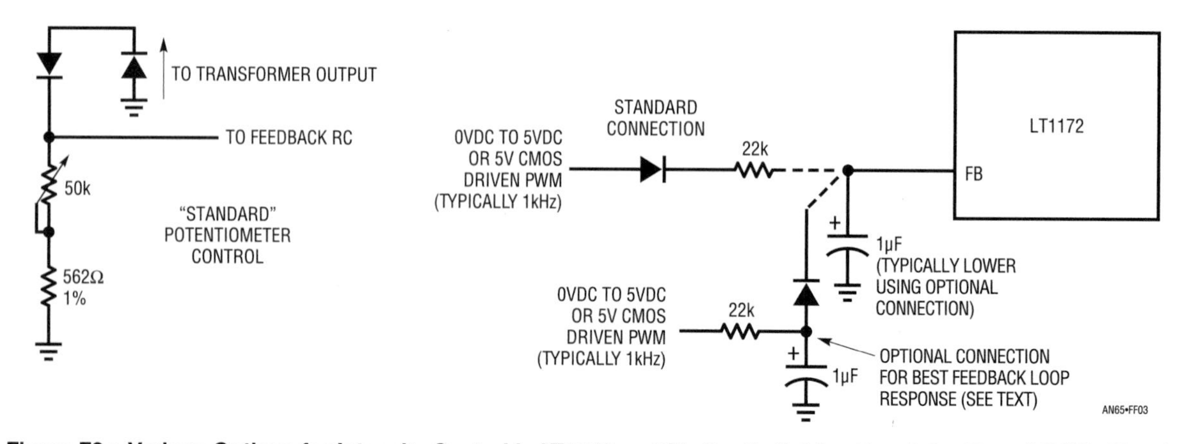

Figure F3 • Various Options for Intensity Control in LT1172 and Similar Switching Regulator-Based CCFL Circuits

potentiometers taper which may be linear or logarithmic, should be matched to the lamp's current-vs-light output characteristic to provide easy user settability. A poorly chosen unit can cause most of the useful dimming range to occur in a small section of potentiometer travel. Finally, always evaluate how the circuit will react if any terminal develops an open condition, which sometimes happens. It is imperative that the circuit have some relatively benign failure mode instead of forcing excessive lamp current or some other regrettable behavior.

Electronic equivalents of potentiometers are monolithic resistor chains tapped by MOS switches. Some devices feature nonvolatile onboard memory. These units have voltage rating restrictions which must be adhered to as with any integrated circuit. Additionally, they have all the limitations discussed in the section on mechanical potentiometers. Their most serious potential difficulty in backlight dimming applications is extremely high end resistance. In the "shorted" position the FET switch's on-resistance is typically 200Ω—much higher than a mechanical unit. Because of this, electronic potentiometers must almost always be set up as 3-terminal voltage dividers. This can usually be accommodated but may eliminate these devices in some applications.

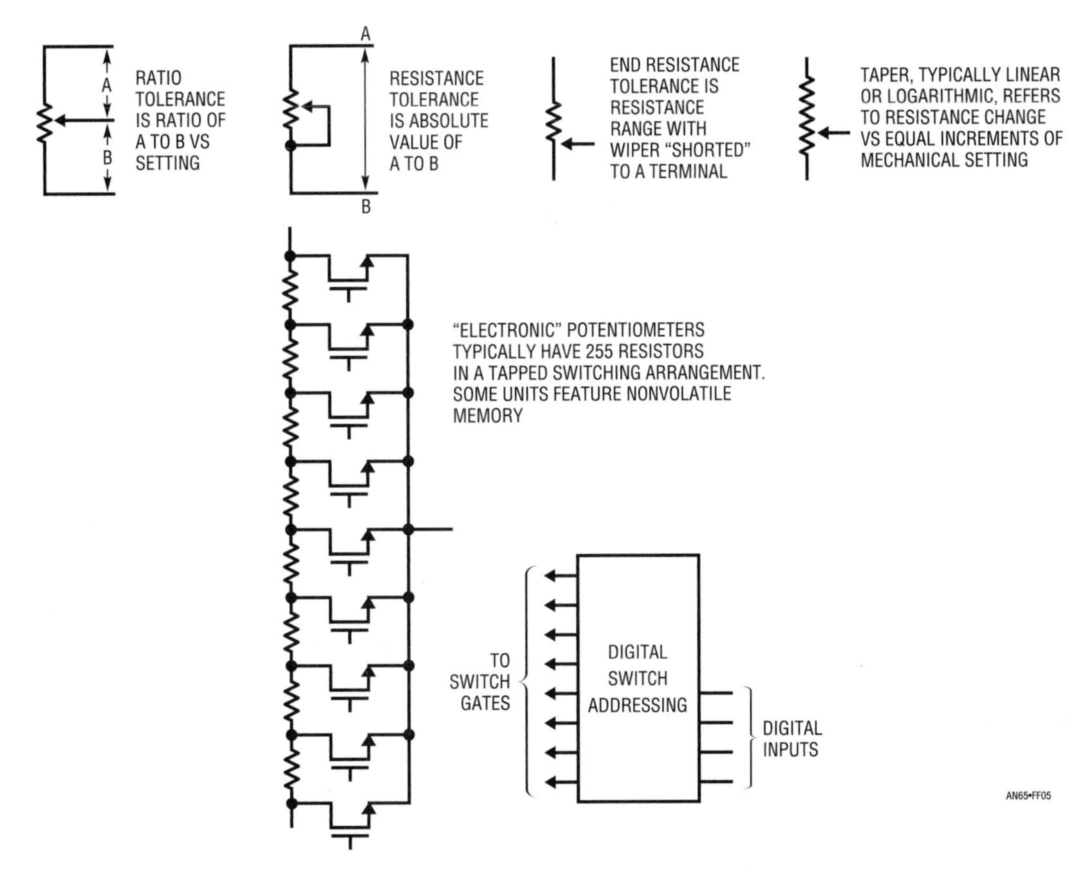

Figure F5 • Relevant Characteristics of Mechanical and Electronic Potentiometers in CCFL Dimming Applications

Precision PWM generator

Figure F6 shows a simple circuit which generates precision variable pulse widths. This capability is useful when testing PWM-based intensity schemes. The circuit is basically a closed-loop pulse width modulator. The crystal controlled 1kHz input clocks the C1/Q1 ramp generator via the differentiator/CMOS inverter network and the LTC201 reset switch. C1's output drives a CMOS inverter, the output of which is resistively sampled, averaged and presented to A1's negative input. A1 compares this signal with a variable voltage from the potentiometer. A1's output biases the pulse width modulator, closing a loop around it. The CMOS inverter's purely ohmic output structure combines with A1's ratiometric operation (e.g., both of A1's input signals derive from the 5V supply) to hold pulse width constant. Variations in time, temperature and supply have essentially no effect. The potentiometer's setting is the sole determinant of output pulse width. Additional inverters provide buffering and furnish the output. The Schottky diodes protect the output from latchup due to cable-induced ESD or accidental events[1] during testing.

The output width is calibrated by monitoring it with a counter while adjusting the 2kΩ trim pot.

As mentioned, the circuit is insensitive to power supply variation. However, the CCFL circuit averages the PWM output. It cannot distinguish between a duty cycle shift and supply variation. As such, the test box's 5V supply should be trimmed ±0.01V. This simulates a "design centered" logic supply under actual operating conditions. Similarly, paralleling additional logic inverters to get lower output impedance should be avoided. In actual use, the CCFL dimming port will be driven from a single CMOS output, and its impedance characteristics must be accurately mimicked.

Note 1: "Accidental events" is a nice way of referring to the stupid things we all do at the bench. Like shorting a CMOS logic output to a -15V supply (then I installed the diodes).

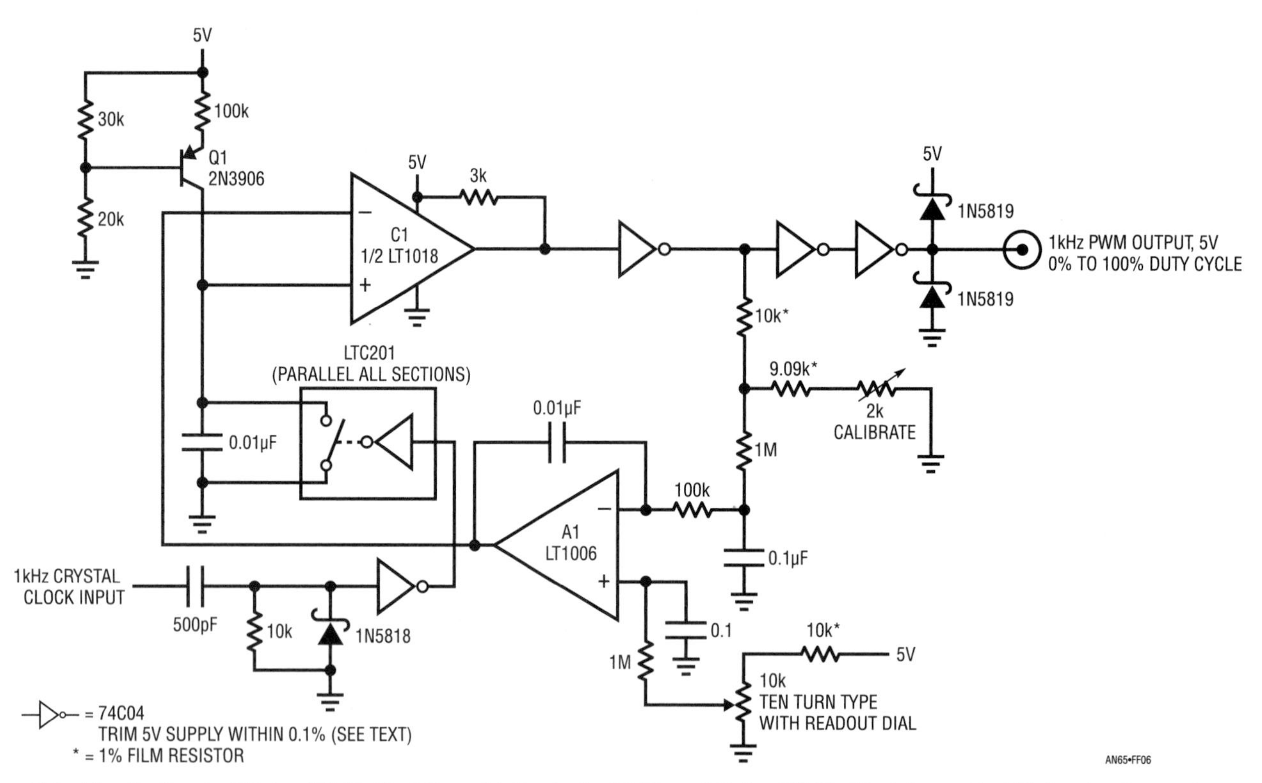

Figure F6 • The Calibrated Pulse Width Test Box. A1 Controls C1-Based Pulse Width Modulator, Stabilizing Its Operating Point

Appendix G
Layout, component and emissions considerations

The CCFL circuits described in the text are remarkably tolerant of layout and impedance in supply lines. This is due to the Royefs relatively continuous current drain over time. Some review of current flow is, however; worthwhile. Figure G1 shows the more critical paths in thick lines for switching regulator-based CCFL circuits. In actual layout, these traces should be reasonably short and thick. The most critical consideration is that C1, T1's center tap and the diode should be connected *directly* together with minimum trace area between them. Similarly, C2 should be near the V_{IN} pin, although this placement is not nearly as critical as C1's.

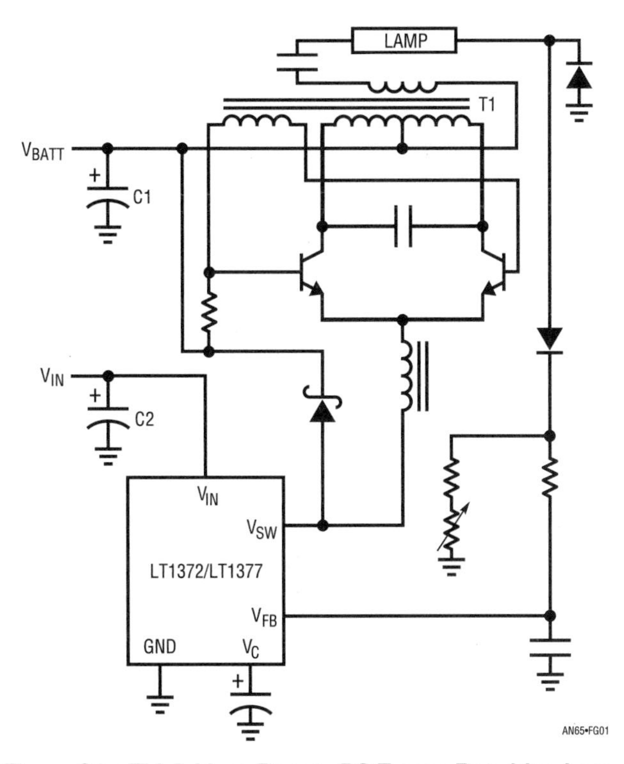

Figure G1 • Thick Lines Denote PC Traces Requiring Low Impedance Layout in LT1172/LT1372 Type CCFL Circuits. Bypass Capacitors Associated with These Paths Should Be Mounted Near Load Point

Figure G2 indicates similar layout treatment for LT118X-based circuits. As before, the Royer and V_{IN} bypass capacitors should be near their respective load points, with the diode in close proximity to the Royer center tap.

Circuit segmenting

In cases where space is extremely limited it may be desirable to physically segment the circuit. Some designs have placed a section of the circuit near the display with another portion remotely located. The best place for segmenting is the junction of the Royer transistor emitters and the inductor (Figure G3). Introducing a long, relatively lossy connection at this point imposes no penalty because signal flow into the inductor closely resembles a constant current source.

There are no wideband components due to the inductor's filtering effect. Figure G4 shows emitter voltage (Trace A) and current (Trace B) waveforms. There is no wideband component or other significant high speed energy movement. The inductor current waveform trace thickening due to Royer and switching regulator frequency mixing is not deleterious.

A very special case of segmentation involves replacing the transformer with two smaller units. Aside from space (particularly height) savings, electrical advantages are also realized. See Appendix I for details.

High voltage layout

Special attention is required for the board's high voltage sections. Board leakage, which can increase dramatically over life due to condensation cycling and particulate matter trapping, must be minimized. If precautions are not taken leakage will cause degraded operation, failures or destructive arcing. The only sure way to eliminate these possibilities is to *completely isolate the high voltage points from the circuit*. Ideally, no high voltage point should be within 0.25" of any conductor: Additionally, moisture trapping due to condensation cycling or improper board washing can be eliminated by routing the area under the transformer. This treatment, standard technique in high voltage layout, is strongly recommended. In general, carefully evaluate all high voltage areas for possible leakage or arcing problems due to layout, board manufacturing or environmental factors. Clear thinking is needed to avoid unpleasant surprises. The following commented photographs, visually summarizing the above discussion, are examples of high voltage layout.

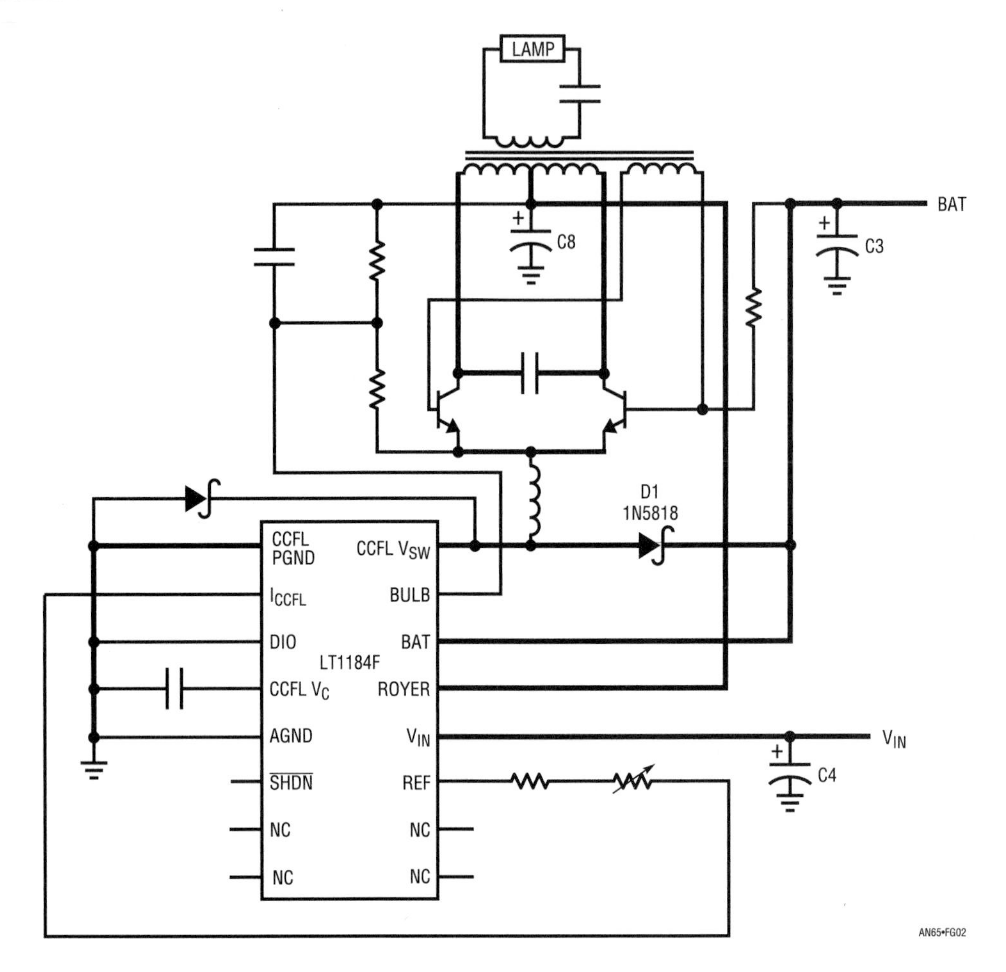

Figure G2 • Critical Current Paths for LT118X Type Circuits. Thick Lines Denote PC Traces Requiring Low Impedance. Bypass Capacitors Associated with These Paths Should Be Near Load Points

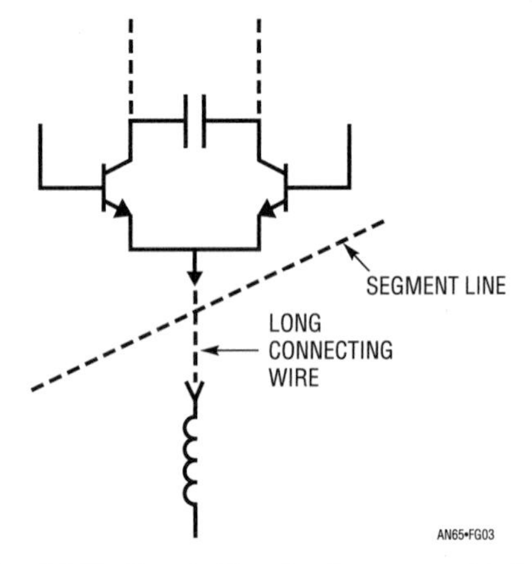

Figure G3 • CCFL Circuit May Be Segmented in Limited Space Applications. Breaking at Emitter/Inductor Junction Imposes No Penalty

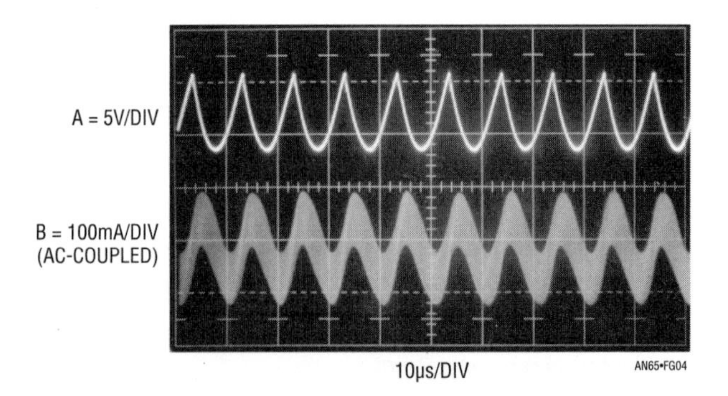

Figure G4 • Royer Emitter/Inductor Junction Is Ideal Point for Segmenting CCFL Circuit. Voltage (Trace A) and Current (Trace B) Waveforms Contain Little High Frequency Content. Trace Thickening of Current Waveform Is Due to Frequency Mixing in Inductor

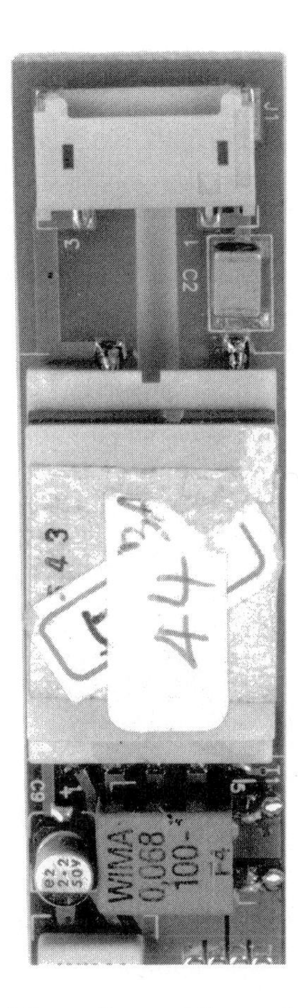

Figure G5 • Transformer Output Terminals, Ballast Capacitor and Connector Are Isolated at End of Board. Slit Prevents Leakage

Figure G6 • Reverse Side of G5. Note Routed Area Under Transformer, Eliminating Possibility of Moisture or Contaminant Trappings

Figure G7 • A Fully Routed Transformer, with High Voltage Capacitor Mounted Well Away from Transformer Ground Terminal (Right). Note Connector "Low Side" Trace Running Directly Away from High Voltage Points

Figure G8 • Routing Detail of Figure G7's Reverse Side. Transformer Header Sits Inside Routed Area, Saving Height Space. Board Markings Are Allowable Because HV Contacts Do Not Plate Through and Board Dielectric Strength Is Known

Figure G9 • Very Thorough Routing Treatment Breaks Up Leakage. Routing Under Ballast Capacitor and Around Connector "Low Side" Pin Allows Tight Layout

Figure G10 • Reverse Side of G9 Shows Offset Transformer Placement Necessitated by Packaging Restrictions. Transformer Header Sits in Routed Area, Minimizing Overall Board Height

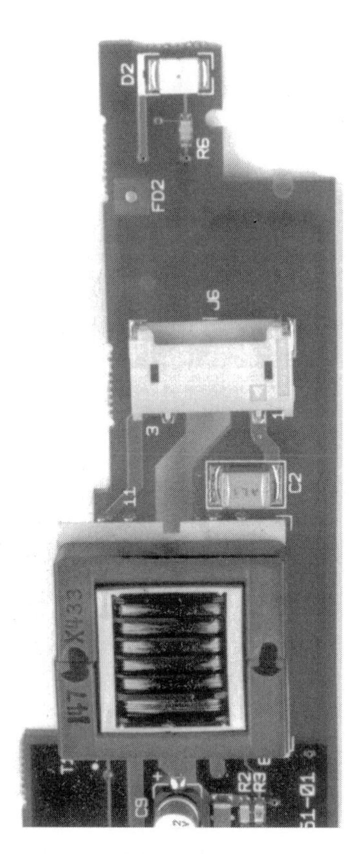

Figure G11 • Topside of Board Shows Isolation Slit Running to the HV Connector

Figure G12 • Bottom of G11 Shows Routed Area. Traces at Board Extreme Left Are Undesirable but Acceptable. Isolation Slit Between Traces and HV Points Would Be Preferable

Figure G13 • A Disaster. Cross-Hatched Ground Plane Surrounds Output Connector and High Voltage Transformer Pins (Upper Center) in This Computer "Aided" Layout. Board Failed Spectacularly at Turn-On

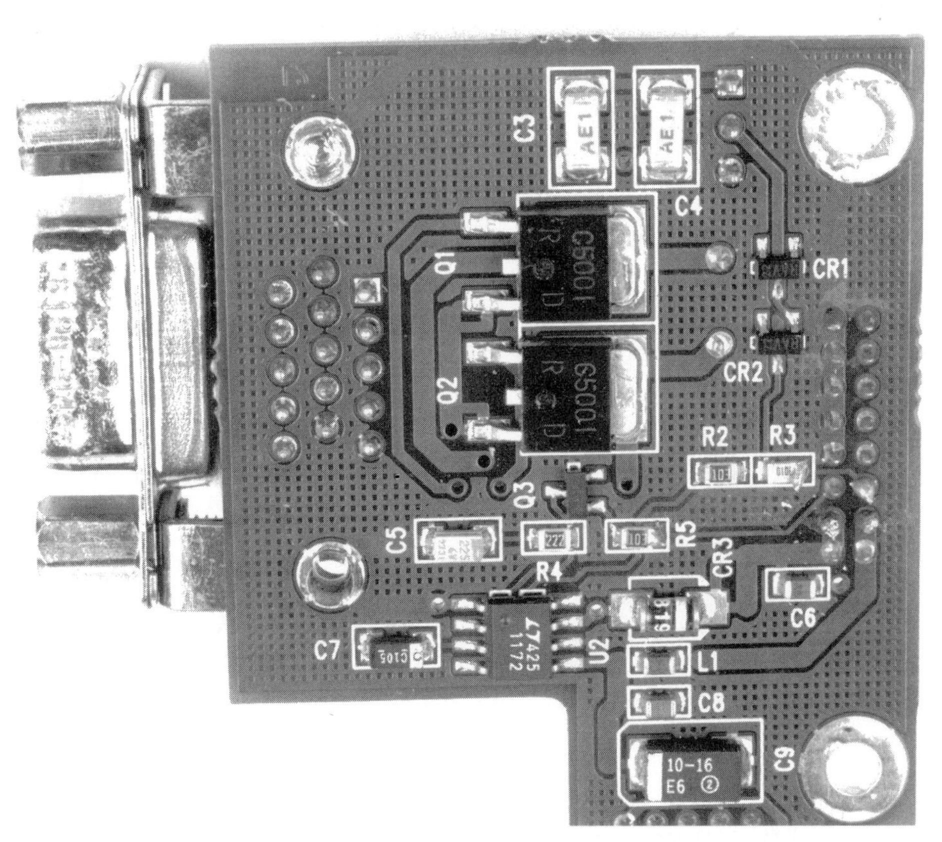

Figure G14 • Bottom Side of G13. Ground Plane in Region of Paralleled Ballast Capacitors (Upper Center) Caused Massive Arcing at Turn-On. Board Needs Complete Re-Layout. Computer Layout Software Package Needs E and M Course

Discrete component selection

Discrete component selection is quite critical to CCFL circuit performance. A poorly chosen dielectric for the collector resonating capacitor can easily degrade efficiency by 5% to 8%. The WIMA and Panasonic types specified are quite good and very few other capacitors perform as well. The Panasonic unit is the only surface mounting type recommended although about 1% more lossy than the "through-hole" WIMA.

The transistors specified are quite special. They feature extraordinary current gain and V_{CE} saturation specifications. The ZDT1048, a dual unit designed specifically for backlight service, saves space and is the preferred device. Figure G15 summarizes relevant characteristics. Substitution of standard devices can degrade efficiency by 10% to 20% and in some cases cause catastrophic failures.[1]

The following section, excerpted with permission from Zetex Application Note 14 (see Reference 28), reviews Royer circuit operation with emphasis on transistor operating conditions and requirements.

Excerpted from "Transistor Considerations for LCD Backlighting," Neil Chadderton, Zetex plc.

Note 1: Don't say we didn't warn you.

ELECTRICAL CHARACTERISTICS (at T_{amb} = 25°C unless otherwise stated).

PARAMETER	SYMBOL	MIN.	TYP.	MAX.	UNIT	CONDITIONS.
Collector-Emitter Breakdown Voltage	V_{CES}	50	85		V	I_C=100μA
Collector-Emitter Breakdown Voltage	V_{CEV}	50	85		V	I_C=100μA, V_{EB}=1V
Collector Cut-Off Current	I_{CBO}		0.3	10	nA	V_{CB}=35V
Collector-Emitter Saturation Voltage	$V_{CE(sat)}$		27 55 120 200 250	45 75 160 240 350	mV mV mV mV mV	I_C=0.5A, I_B=10mA* I_C=1A, I_B=10mA* I_C=2A, I_B=10mA* I_C=5A, I_B=100mA* I_C=5A, I_B=20mA*
Static Forward Current Transfer Ratio	h_{FE}	280 300 300 250 50	440 450 450 300 80	 1200 		I_C=10mA, V_{CE}=2V* I_C=0.5A, V_{CE}=2V* I_C=1A, V_{CE}=2V* I_C=5A, V_{CE}=2V* I_C=20A, V_{CE}=2V*
Transition Frequency	f_T		150		MHz	I_C=50mA, V_{CE}=10V f=50MHz

*Measured under pulsed conditions. Pulse width=300μs. Duty cycle ≤ 2%

ZETEX
U.K. FAX: 0161627-5467
U.S. FAX: 5168647630
HONG KONG FAX: 987 9595

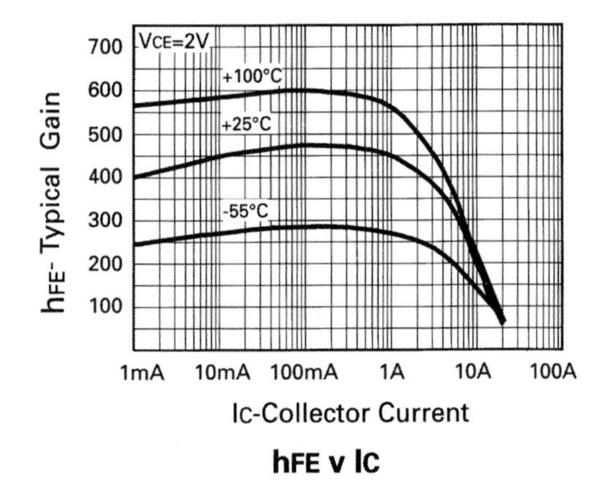

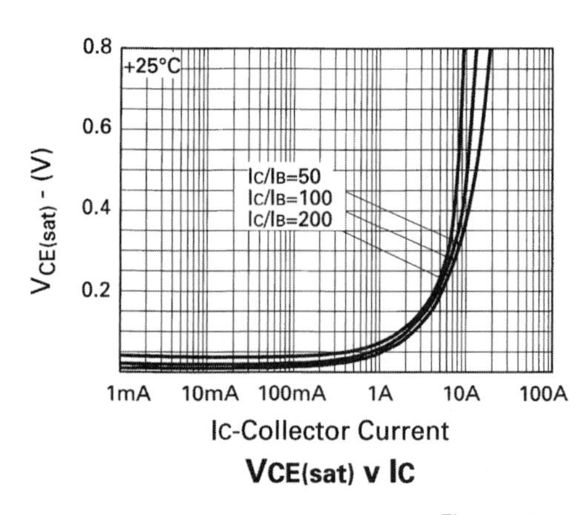

Figures reproduced with permission of Zetex.

Figure G15 • Short Form Specifications for Zetex ZDT1048 Dual Transistor. Extraordinary Beta and Saturation Characteristics Are Ideal for Royer Converter Section of Backlight Circuits

Basic operation of converter

The drive requirements dictated by the CCFL tube's behaviour and preferred operating conditions can be achieved by the resonant push-pull converter shown in FigureG16.This is also referred to as the Royer Converter, after G.H. Royer who proposed the topology in 1954 as a power converter. (Note: Strictly speaking the backlighting converter uses a modified version of the Royer converter-the original used a saturating transformer to fix the operating frequency, and therefore produced a square-wave drive waveform). The circuit looks simple but this is very deceptive: many components interact, and while the circuit is capable of operation with widely varying component values, (useful during development) optimisation is required for each design to achieve the highest possible efficiencies.

Transistors Q1 and Q2 are alternatively saturated by the base drive provided by the feedback winding W4. The base current is defined by resistors R1 and R2. Supply inductor L1 and primary capacitance C1 force the circuit to run sinusoidally thereby minimising harmonic generation and RFI, and providing the preferred drive waveform to the load. Voltage step-up is achieved by the W1:(W2+W3) turns ratio. C2 is the secondary winding ballast capacitor, and effectively sets the tube current.

Prior to the tube striking, or when no tube is connected, the operating frequency is set by the resonant parallel circuit comprising the primary capacitance C1, and the transformer's primary winding W2+W3. Once the tube has struck, the ballast capacitor C2 plus distributed tube and parasitic capacitances are reflected back through the transformer, and the operating frequency is lowered.

The secondary load can become dominant in circuits with a high transformer turns ratio. Eg. those designed to operate from very low DC input voltages.

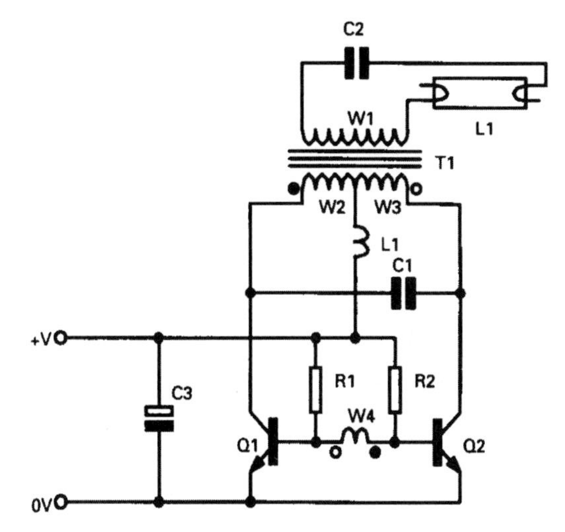

Figure G16 • Generalised Royer Converter

Each transistor's collector is subject to a voltage= 2 x $\pi/2$ x Vs, (or just π x Vs) where Vs is the DC input voltage to the converter. (The $\pi/2$ factor being due to the relationship between average and peak values for a sinewave, and the x2 multiplier being due to the 2:1 autotransformer action of the transformer's centre-tapped primary). This primary voltage is stepped up by the transformer turns ratio Ns:Np, to a high enough level to reliably strike the tube under all conditions:- starting voltage is dependent on display housing, location of ground planes, tube age, and ambient temperature.

The basic converter shown in Figure G16 is a valid and useful circuit that has been utilised for many systems and indeed offered as a sub-system by several manufacturers.

Requisite transistor characteristics

The relatively low operating frequency as required by the backlighting Royer Converter (to minimise HV parasitic capacitance losses), and the ease of transformer drive, makes this circuit particularly suitable for bipolar transistor implementation. This isn't to exclude MOSFET based designs (some IC vendors have specified MOS as this suits their technology) but in terms of equivalent on-resistance and silicon efficiency, the low voltage bipolar device has no equal. For example, the ZETEX ZTX849 E-LINE (TO-92 compatible) transistor exhibits a RcE(sat) of 36MΩ. This can only be matched by a much larger (and expensive) MOSFET die, only available in TO-220, D-PAK, and similar larger packages.

The most important transitor characterstic are voltage rating, VCE(sat), and hFE, and are considered in some detail below.

The voltage rating required deserves some thought with respect to the standard transitor breakdown parameters, as it is possible to over-specify a device on grounds of voltage rating, and thereby incur a reduction in efficency due to unnecessary on-resistance losses. The primary breakdown voltage BVCBO, of a planar bipolar transistor depends on the epitaxial layer-specially it's thickness and resistivity. The breakdown voltage of most interest to the designer is usally that attained across the Collector-Emitter(C-E) terminals. This value can vary between the primary breakdown BVCBO and a much lower voltage dependent on the base terminal bias.

[The breakdown mechanism is caused by the avalanche multiplication effect, whereby free electrons can be imparted with sufficient energy by the reverse bias electric field such that any collisions can lead to ionisation of the lattice atoms. The free electrons thus generated are then accelerated by the field and produce further ionisation. This multiplication of free carriers increases the reverse current dramatically, and so the junction effectively clamps the applied voltage. The base terminal can obviously influence the junction current-thereby modulating the voltage required for a breakdown condition.]

Figure G17 shows how the breakdown characteristic is seen to vary for different circuit conditions. The BV_{CEO} rating (or when the base is open circuit) allows the Collector-Base (C-B) leakage current I_{CBO} to be effectively amplified by the transistor's β thus significantly increasing the leakage component to I_{CEO}. Shorting the base to the Emitter (BV_{CES}) provides a parallel path for the C-B leakage, and so the voltage required for breakdown is higher than the open base condition. BV_{CER} denotes the case between the open and shorted base options:- R indicating an external base-emitter resistance, the value of which is typically 100 to 10kΩ. BV_{CEV} or BV_{CEX} is a special case where the base-emitter is reverse biased; this can provide a better path for the C-B leakage, and so this rating yields a voltage close to, or coincident with the BV_{CBO} value. Figure G18 shows a curve tracer view of the relevant breakdown modes of the ZTX849 transistor, including a curve showing the device in the "on" state. Curves 1 and 2 are virtually coincident and show BV_{CBO} and BV_{CES} respectively. Curve 3 shows the BV_{CEV} case with an applied base bias (V_{EB}) of -1V. Curve 4 shows BV_{CEO} at approximately 36V. Curve 5 is a BV_{CE} curve, showing how the breakdown condition is affected by a positive base bias of 0.5V.

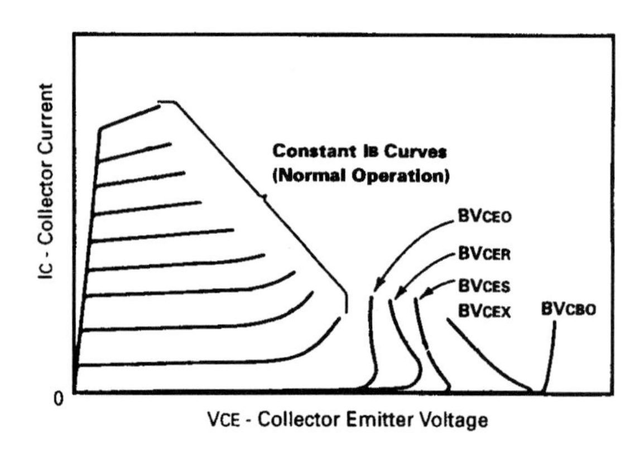

Figure G17 • Voltage Breakdown Modes of Bipolar

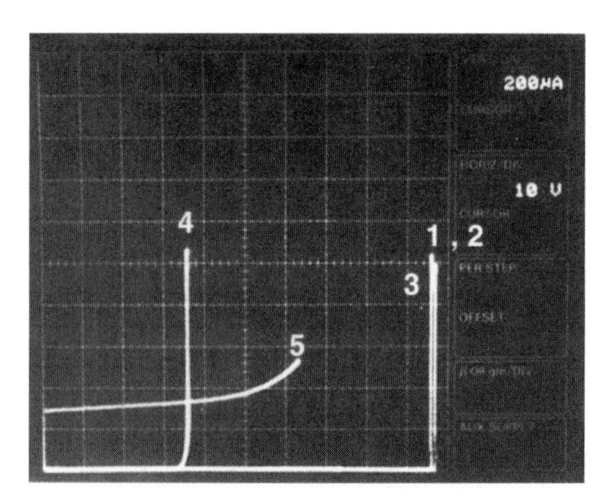

Figure G18 • Breakdown Modes of ZTX849

The BV_{CEV} rating has particular relevance to the Royer Converter, as can be surmised from Figure 19. Examination of this will show that the transistor only experiences the high C-E voltage when the base voltage has been taken negative by the feedback winding, these events of course being in perfect synchronism. An expanded view of the C-E and B-E waveforms is shown in Figure G20.

[Note: The voltage applied by the feedback winding must not exceed the BVebo of the transistor. This is specified at 5V usually, against an actual of 7.5 to 8.5V].

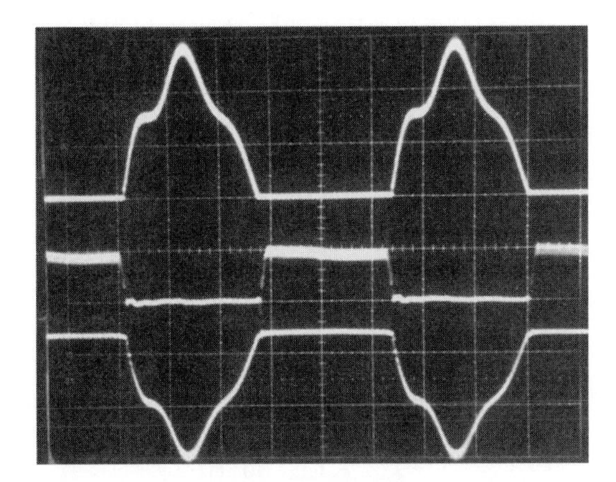

Figure G19 • Royer Converter Operating Waveforms: V_{CE} 10V/div; I_E 0.5A/div; V_{BE} 2V/div Respectively, 2μs/div Horizontal

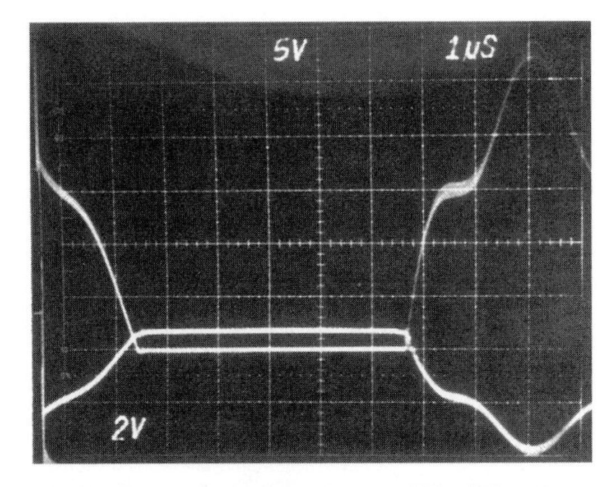

Figure G20 • Royer Converter: V_{CE} and V_{BE} Waveforms: 5V/div and 2V/div Respectively

The $V_{CE(sat)}$ and h_{FE} parameters have a direct bearing on the circuit's electrical conversion efficiency. This is especially true of low voltage battery powered systems, due to the high current levels involved. Selection of standard LF amplifier transistors provides far from ideal results; these parts are for general purpose linear and non-critical switching use only. The high $V_{CE(sat)}$ inherent to these parts, and low current gain could reduce circuit efficiency to less than 50%. For example, the stated $V_{CE(sat)}$ maximum measured at 500mA, for the FZT849 SOT223 transistor, and a LF device sometimes quoted as a suitable Royer Converter transistor are 50mV and 0.5V respectively. Eg.

	VCE(sat)	@Ic	Ib
FZT849	50mV	0.5A	20mA
BCP56	0.5mV	0.5A	50mA

To address the $V_{CE(sat)}$ issue, large power transistors are occasionally specified. Unfortunately their capacitance, and characteristic low base transport factor (a feature of Epitaxial Base devices) can lead to problems with cross-conduction losses due to long storage and switching times. The current gain is also important, as the losses in the base bias can be significant to the overall figure; judicious selection of the bias resistor to ensure a minimum $V_{CE(sat)}$ while preventing base overdrive needs to consider supply variation, maximum lamp current, and transistor h_{FE} minimum value and range.

For the above reasons, transistors designed and optimised for high current switching applications offer the most cost-effective and efficient solutions. Figure G21 shows the $V_{CE(sat)}$ exhibited by the ZTX1048A for a range of forced gain values. This device is one of the ZTX1050 series of transistors that employ a scaled up variant of the highly efficient Matrix geometry, developed for the ZETEX "Super-SOT" series. This enables a $V_{CE(sat)}$ performance similar to the ZTX850 series at the low to moderate currents relevant to this application, though utilising a smaller die, and therefore providing a cost and possibly a space saving advantage.

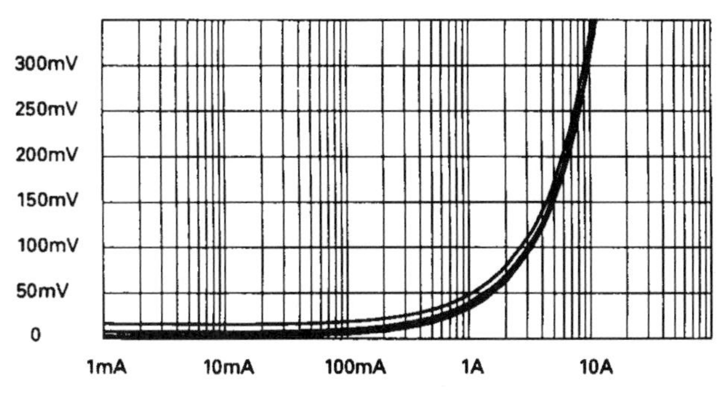

Figure G21 • $V_{CE(sat)}$ vs I_C for the ZTX1048A Forced Gains of 10, 20, 50, 100

Additional discrete component considerations

The magnetics specified have also been carefully selected and substitution can lead to problems ranging from poor efficiency to bad line regulation.

Bypass capacitors can be any type specified for switching regulator service, although tantalum types should be avoided for Royer bypassing if the supply is capable of delivering high current. As of this writing no tantalum supplier can guarantee reliability in the face of high current turn-on. If tantalums must be used, an X2 voltage derating factor must be enforced.[2]

The 2.2μF Royer bypass value used in LT118X-based circuits has been selected to ensure against any possible long-term damage to the IC's internal current shunt. Turn-on current surges can be large and this value limits them to safe excursions.

The high speed catch diode associated with the VSW pin should be capable of handling the fast current spikes encountered. Schottky types offer lower loss than regular high speed units.[3]

Emissions

There are rarely emission problems with the CCFL circuits. The Royer circuit's resonant operation minimizes radiated energy at frequencies of interest. There is often more RF energy associated with the switching regulators V_{SW} node and minimizing exposed trace area eliminates problems. Incidental radiation from magnetics is reasonably low. Some (relatively rare) cases require consideration of magnetics placement to prevent interaction with other circuitry. If shielding is required its effects should be evaluated early in the design. Shielding in the vicinity of the Royer transformer can cause effects ranging from changing the inverter resonance to secondary arcing.

Appendix H
Operation from high voltage inputs

Some applications require higher input voltages. The 20V maximum input specified in the figures is set by the LT1172 going into its isolated flyback mode (see LT1172 data sheet), not breakdown limits. If the LT1172 V_{IN} pin is driven from a low voltage source (e.g., 5V) the 20V limit may be extended by using Figure H1's network. If the LT1172 is driven from the same supply as L1's center tap, the network is unnecessary, although efficiency will suffer. No other switching regulator discussed in the text is subject to this issue. Their operating voltage is set solely by voltage breakdown limits.

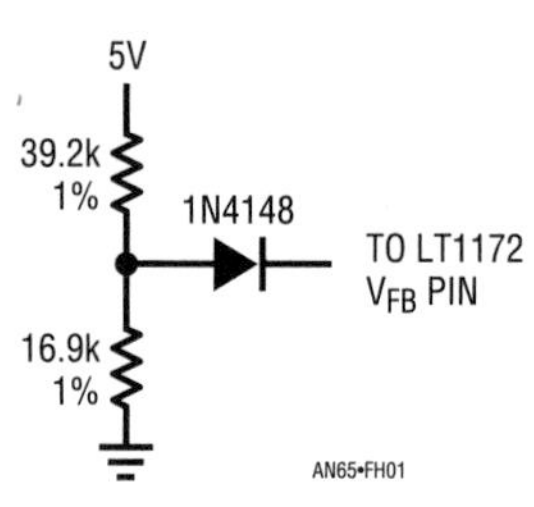

Figure H1 • Network Allows LT1172 Operation Beyond 20V Inputs

Note 2: See Footnote 1. Read it twice.

Note 3: Discussing utilization of 60Hz rectifier diodes (e.g., 1N4002) in this application qualifies as obscene literature. See also Footnote 1.

Appendix I Additional circuits

Desktop computer CCFL power supply

Desktop computers, being line operated, can support higher power displays. High power operation permits high luminosity, enlarged display area or both. Typically, desktop displays absorb 4W to 6W and run from a relatively high voltage, regulated supply. Figure I1 shows such a display.

This "grounded lamp" LT1184-based configuration, similar to previously described versions, requires little comment. The transformer is a high power type, and scaling of the "I_{CCFL}" current programming resistors allows 9mA lamp current. In this case programming is via clamped PWM (see Appendix F), although all other methods described are feasible.

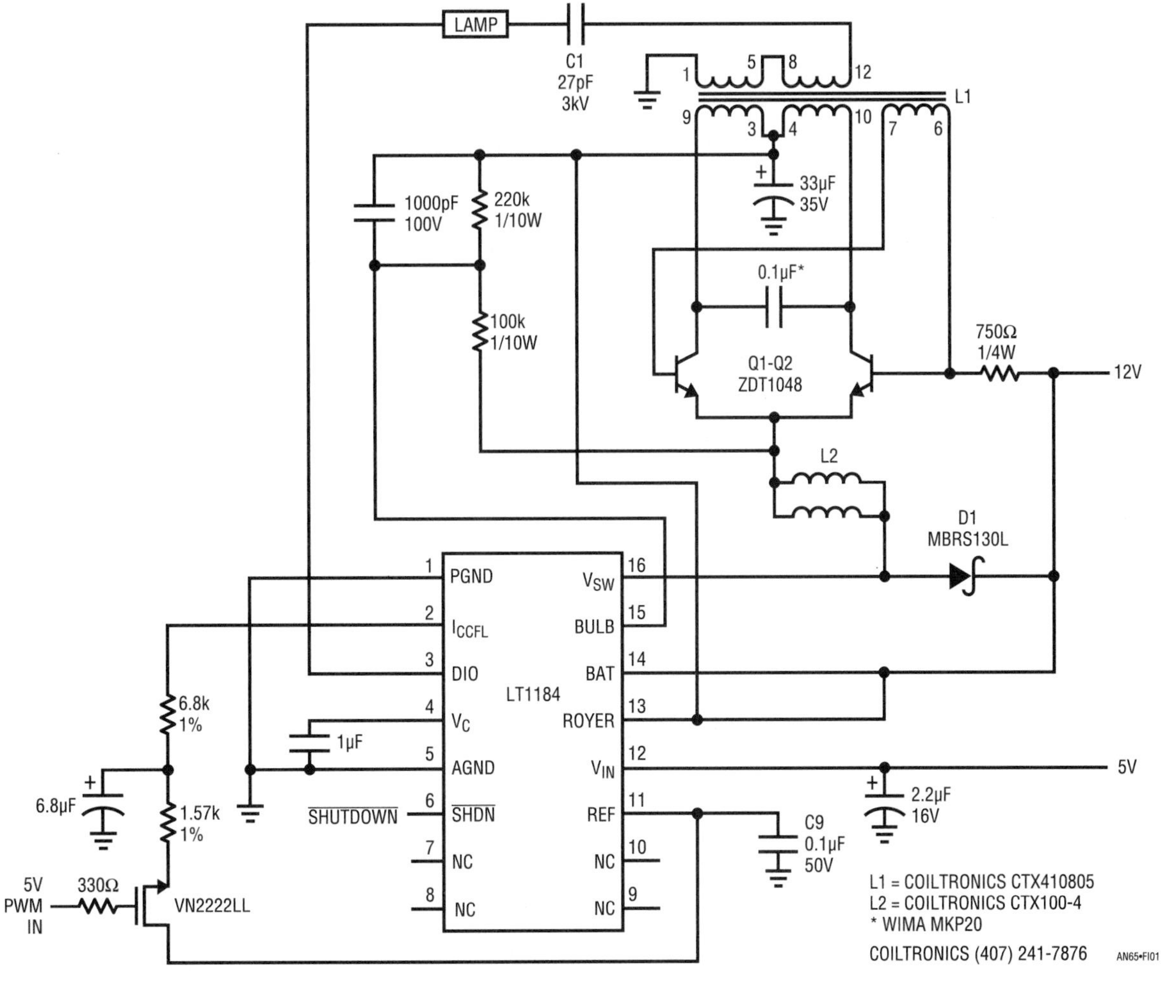

Figure I1 • High Power Transformer and Scaled I_{CCFL} Values Permit Desktop Computer LCD Operation

Dual transformer CCFL power supply

Space constraints may dictate utilization of two small transformers instead of a single, larger unit. Although this approach is somewhat more expensive, it can solve space problems and offers other attractive advantages. Figure I2's approach is essentially a "grounded lamp" LT1184-based circuit. The transistors drive two transformer primaries in parallel. The transformer secondaries, stacked in series, provide the output. The relatively small transformers, each supplying half the load power; may be located directly at the lamp terminals. Aside from the obvious space advantage (particularly height), this arrangement minimizes parasitic wiring losses by eliminating high voltage lead length. Additionally, although the lamp receives differential drive, with its attendant low parasitic losses, the feedback signal is ground referred. Thus, the stacked secondaries afford floating lamp operation efficiency with grounded mode current certainty and line regulation.

L1 is directly driven, with winding 4-5 furnishing feedback in the normal fashion. L3, "slaved" to L1, produces phase-opposed output at its secondary. L1's and L3's interconnects must be laid out for low inductance to maintain waveform purity. The traces should be as wide as possible (e.g., 1/8") and overlaid to cancel inductive effects.

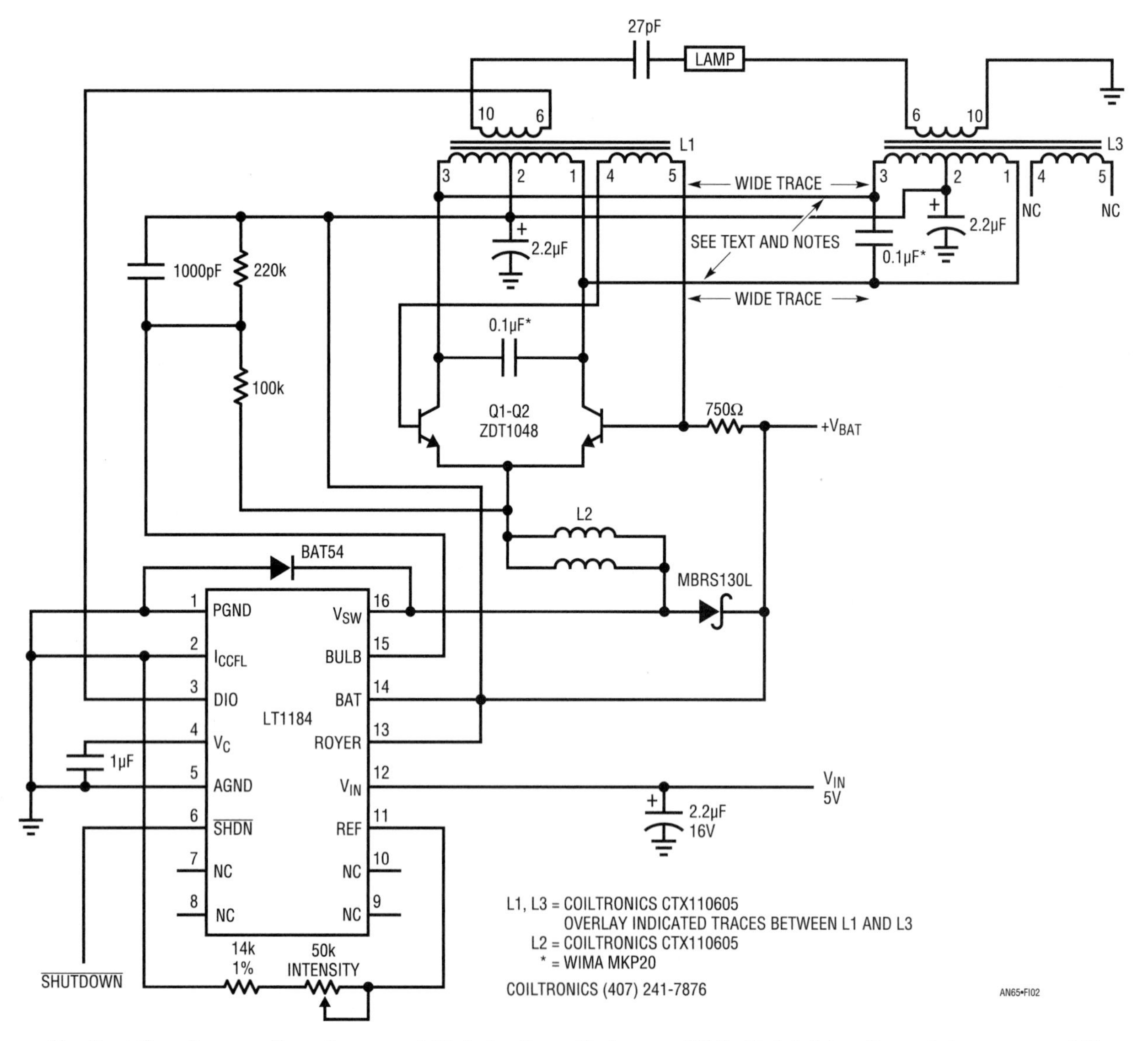

Figure I2 • Dual Transformers Save Space and Minimize Parasitic Losses While Maintaining Current Accuracy and Line Regulation. Trade-Off Is Increased Cost

HeNe laser power supply

Helium-neon lasers, used for a variety of tasks, are difficult loads for a power supply. They typically need almost 10kV to start conduction, although they require only about 1500V to maintain conduction at their specified operating currents. Powering a laser usually involves some form of start-up circuitry to generate the initial breakdown voltage and a separate supply for sustaining conduction. Figure I3's circuit considerably simplifies driving the laser. The start-up and sustaining functions have been combined into a single closed-loop current source with over 10kV of compliance. The circuit is recognizable as a reworked CCFL power supply with a voltage tripled DC output.

When power is applied, the laser does not conduct and the voltage across the 190Ω resistor is zero. The LT1170 switching regulator FB pin sees no feedback voltage, and its Switch pin (V_{SW}) provides full duty cycle pulse width modulation to L2. Current flows from L1's center tap through Q1 and Q2 into L2 and the LT1170. This current flow causes Q1 and Q2 to switch, alternately driving L1. The 0.47μF capacitor resonates with L1, providing boosted sine wave drive. L1 provides substantial step-up, causing about 3500V to appear at its secondary. The capacitors and diodes associated with L1's secondary form a voltage tripler, producing over 10kV across the laser. The laser breaks down and current begins to flow through it. The 47k resistor

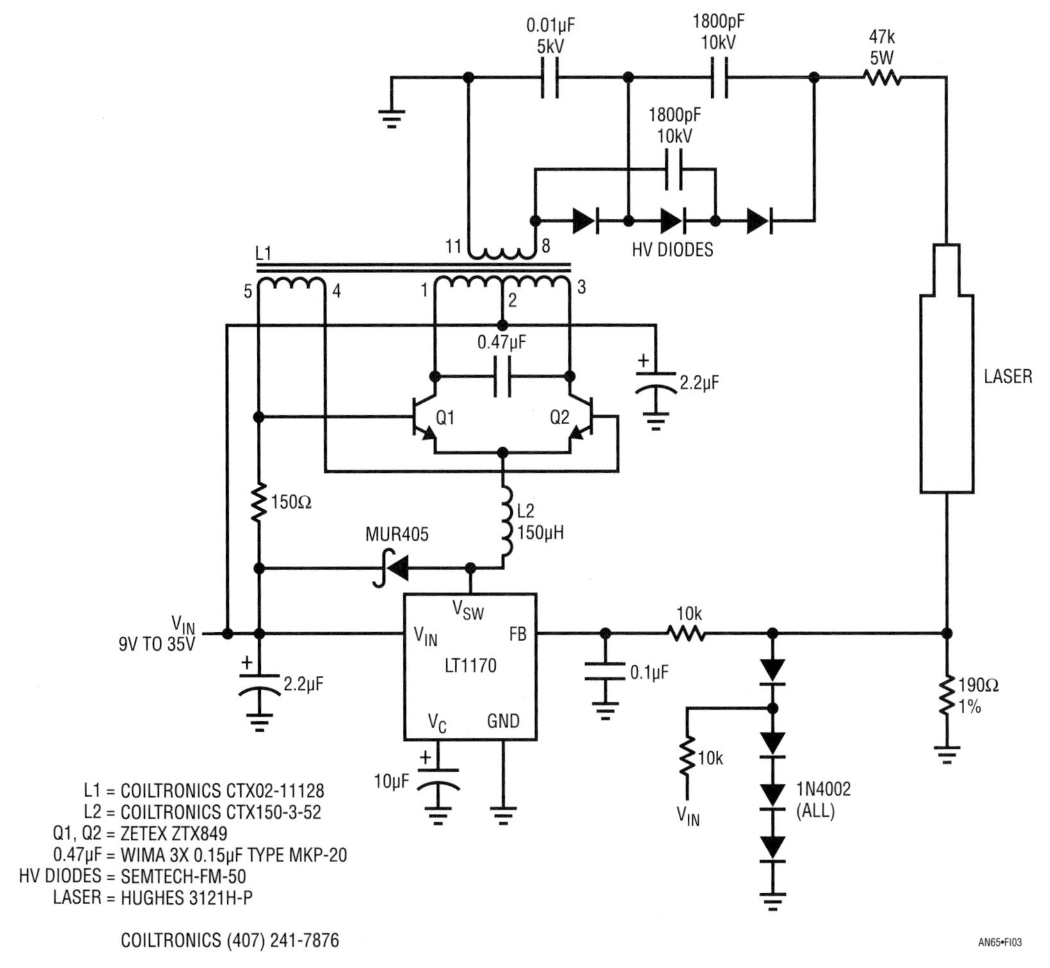

Figure I3 • Laser Power Supply, Based on the CCFL Circuit, Is Essentially a 10, 000V Compliance Current Source

limits current and isolates the laser's load characteristic. The current flow causes a voltage to appear across the 190Ω resistor. A filtered version of this voltage appears at the LT1170 FB pin, closing a control loop. The LT1170 adjusts pulse width drive to L2 to maintain the FB pin at 1.23V regardless of changes in operating conditions. In this fashion, the laser sees constant current drive, in this case 6.5mA. Other currents are obtainable by varying the 190Ω value. The 1N4002 diode string clamps excessive voltages when laser conduction first begins, protecting the LT1170. The 10µF capacitor at the Vq pin frequency compensates the loop and the MUR405 maintains L1's current flow when the LT1170 V_{SW} pin is not conducting. The circuit will start and run the laser over a 9V to 35V input range with an electrical efficiency of about 80%.

Appendix J LCD contrast circuits

LCD panels require variable output contrast control circuits. Contrast power supplies of various capabilities are presented here.

Figure J1 is a contrast supply for LCD panels. It was designed by Steve Pietkiewicz of LTC. The circuit is noteworthy because it operates from a 1.8V to 6V input, significantly lower than most designs. In operation the LT1300/LT1301 switching regulator drives T1 in flyback fashion, causing negative biased step-up at T1's secondary. D1 provides rectification, and C1 smooths the output to DC. The resistively divided output is compared to a command input, which may be DC or PWM, by the IQ's I_{LIM} pin. The IC, forcing the loop to maintain 0V at the I_{LIM} pin, regulates circuit output in proportion to the command input.

Efficiency ranges from 77% to 83% as supply voltage varies from 1.8V to 3V. At the same supply limits, available output current increases from 12mA to 25mA.

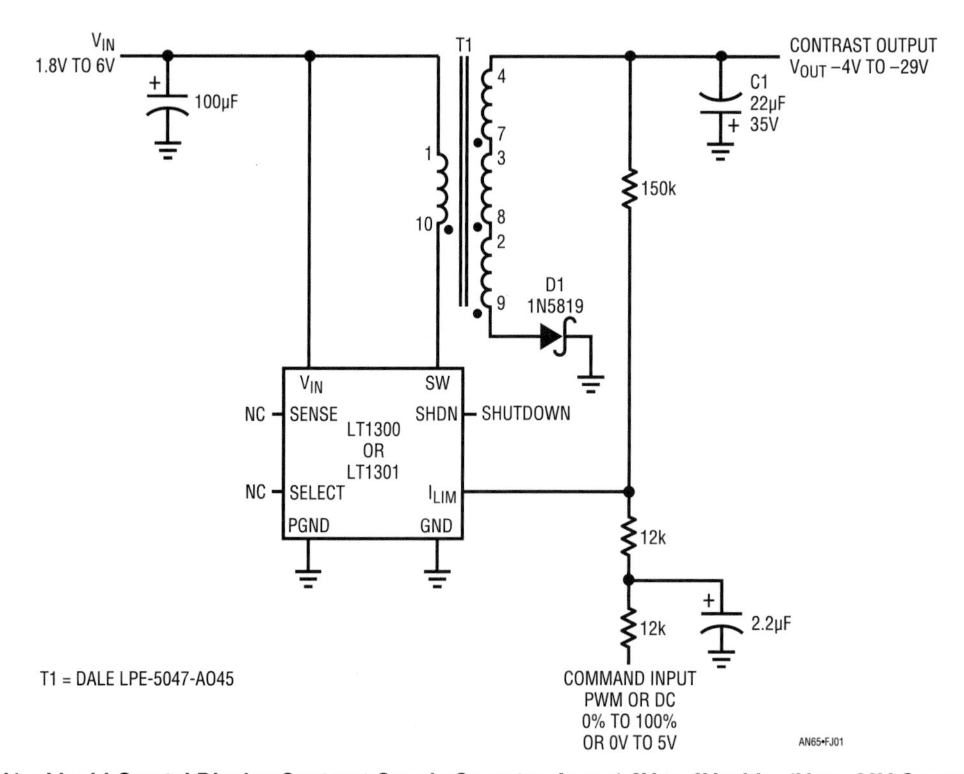

Figure J1 • Liquid Crystal Display Contrast Supply Operates from 1.8V to 6V with -4V to -29V Output Range

Another LCD bias generator, also developed by Steve Pietkiewicz of LTC, is shown in Figure j2. In this circuit U1 is an LT1173 micropower DC/DC converter. The 3V input is converted to 24V by U1's switch, L2, D1 and C1. The Switch pin SW1 also drives a charge pump composed of C2, C3, D2 and D3 to generate -24V Line regulation is less than 0.2% from 3.3V to 2V inputs. Load regulation, although suffering somewhat since the −24V output is not directly regulated, measures 2% from a 1mA to 7mA load. The circuit will deliver 7mA from a 2V input at 75% efficiency.

If greater output power is required, Figure J2's circuit can be driven from a 5V source. R1 should be changed to 47Ω and C3 to 47μF With a 5V input, 40mA is available at 75% efficiency. Shutdown is accomplished by bringing D4's anode to a logic high, forcing the feedback pin of U1 to go above the internal 1.25V reference voltage. Shutdown current is 110μA from the input source and 36μA from the shutdown signal.

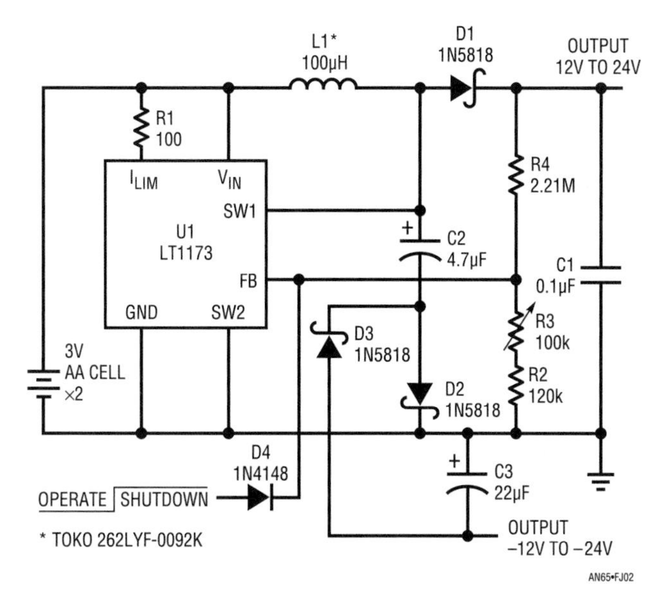

Figure J2 • DC/DC Converter Generates LCD Bias from 3V Supply

Dual output LCD bias voltage generator

The many different kinds of LCDs available make programming LCD bias voltage at the time of manufacture attractive. FigureJ3's Circuit, developed by Jon Dutra of LTC, is an AC-Coupled boost topology. The feedback signal is dervied seperately from the outputs, so loading does not effect loop compensation, althouh, load regulation is comewhat compromised . with 28V out, from 10% to 100% load(4mA to 40mA), the output voltage sags about 0.65V! From 1mA to 40mA load the output voltage drops about 1.4V! This is acceptable for most displays.

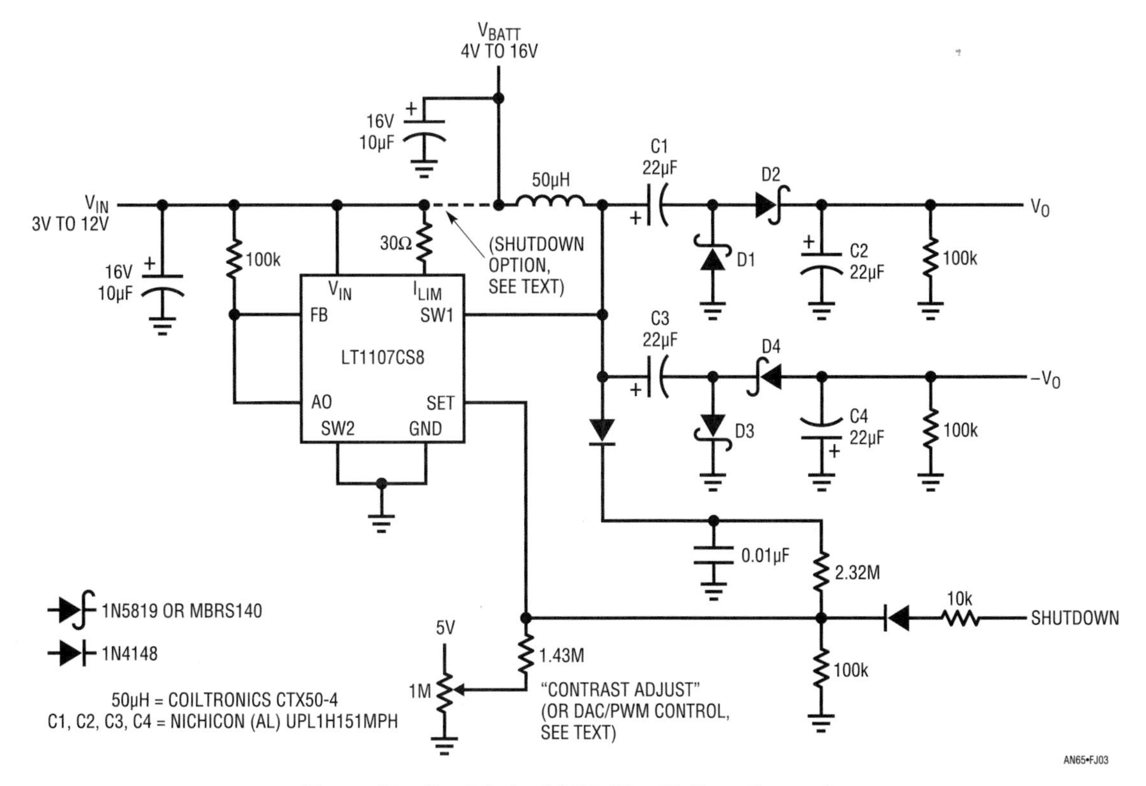

Figure J3 • Dual Output LCD Bias Voltage Generator

Output noise is reduced by using the auxiliary gain block within the LT1107 (see LT1107 data sheet) in the feedback path. This added gain effectively reduces comparator hysteresis and tends to randomize output noise. Output noise is below 30mV over the output load range. Output power increases with V_{BATT}, from about 1.4W with $5V_{IN}$ to about 2W with 8V or more. Efficiency is 80% over a broad output power range. If only a positive or negative output voltage is required, the diodes and capacitors associated with the unused output can be eliminated. The 100k resistor is required on each output to load a parasitic voltage doubler created by D2/D4 shunt capacitance. Without this minimum load, the output voltage can rise to unacceptable levels.

The voltage at the Switch pin SW1 swings from 0V to V_{OUT} plus two diode drops. This voltage is AC-coupled to the positive output through C1 and D1, and to the negative output through C3 and D3. C1 and C3 have the full RMS output current flowing through them. Most tantalum capacitors are not rated for current flow. Use of a rated tantalum or electrolytic is recommended for reliability. At lower output currents monolithic ceramics are also an option.

The circuit may be shut down in several ways. The easiest is to pull the Set pin above 1.25V This approach consumes 200μA in shutdown. A lower power method is to turn off V_{IN} to the LT1107 by a high side switch or simply disable the input supply (see option in schematic). This drops quiescent current from the V_{BATT} input below 10μA. In both cases V_{OUT} drops to 0V In the event $+V_{OUT}$ does not need to drop to zero, C1 and D1 can be eliminated. The output voltage can be adjusted from any voltage above V_{BATT} to 46V Output voltage can be controlled by the user with DAC, PWM or potentiometer control. Summing currents into the feedback node allows downward adjustment of output voltage.

LT118X Series Contrast Supplies

Some LT118X series parts include a contrast supply based on a boost regulator Figure J4 shows a basic positive output circuit. The V_{SW}-driven inductor provides voltage step-up with D5 and C11 rectifying and filtering the output to DC. The R12/R14 divider chain sets feedback ratio and hence output voltage. The connection to the LT1182 Feedback pin closes a control loop with R7 and C8 providing frequency compensation.

Figure J5 is similar, except that it uses charge pump techniques to reduce shutdown current. D4 and C12 are placed in L3's discharge path, AC coupling it to the output. In shutdown, no DC current can flow through L3, reducing battery drain over J4's DC-coupled approach.

Figure J6's transformer-fed output provides negative output voltages with the LT1183's "FBN" pin directly accepting the resultant negatively biased feedback signal. No level shift is required. In this case output voltage is set by a voltage control input, although potentiometer or PWM inputs could be accommodated (see Appendix F).

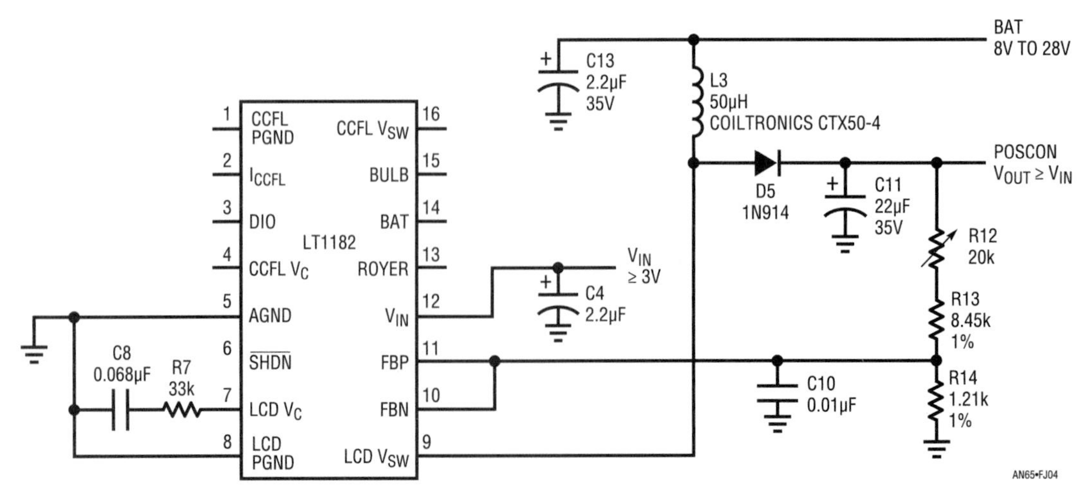

Figure J4 • LT1182 LCD Contrast Positive Boost Converter. CCFL Circuitry Is Omitted for Clarity

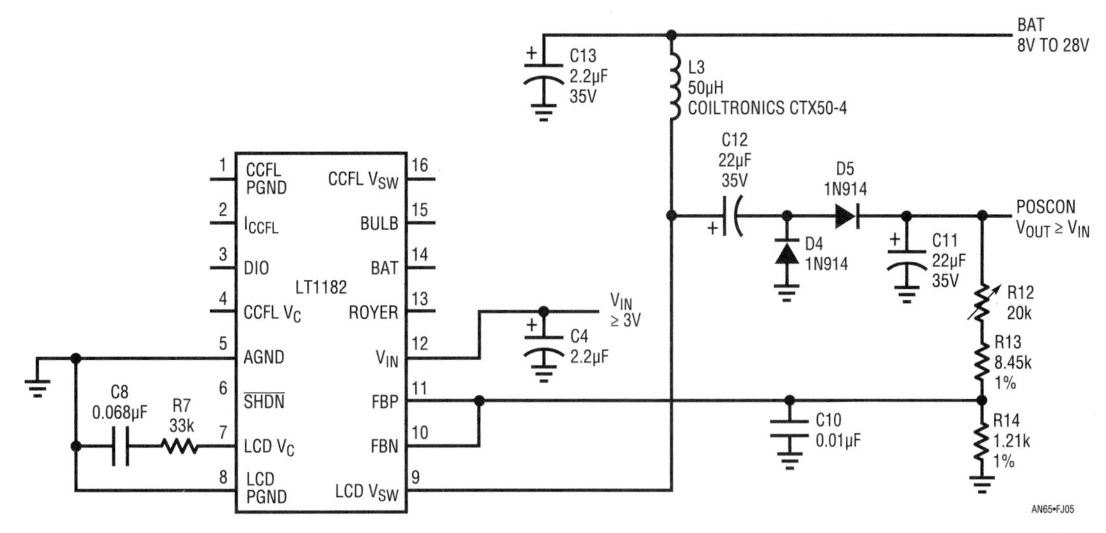

Figure J5 • LT1182 LCD Contrast Positive Boost/Charge Pump Converter Reduces Battery Current in Shutdown

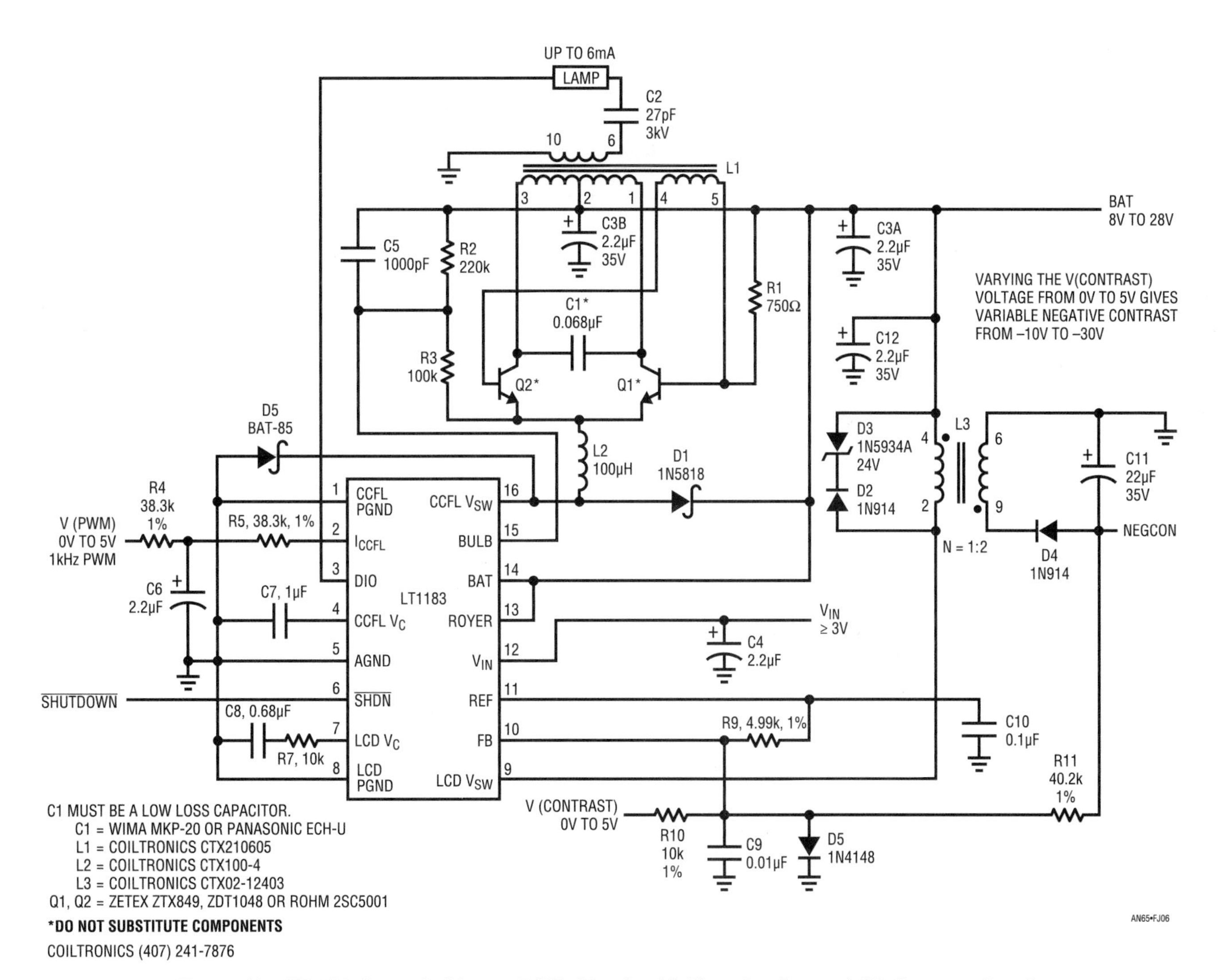

Figure J6 • LT1183 Grounded Lamp CCFL Circuit with Negative Output LCD Contrast Supply

D3 and D2 damp L3 flyback amplitude to safe levels and the isolated secondary permits low shutdown current compared to a simple inductor-based circuit.

Figure J7 takes advantage of the LT1182's bipolar feedback inputs to provide selectable output polarity. This scheme permits the same circuit to be used with LCD's requiring either positive or negative bias. This can be a significant advantage in volume production involving different LCD panels. In operation the circuit is similar to Figure J6, except that L3's secondary winding feeds two separate feedback paths. Output polarity is selected by simply grounding the appropriate L3 secondary terminal.

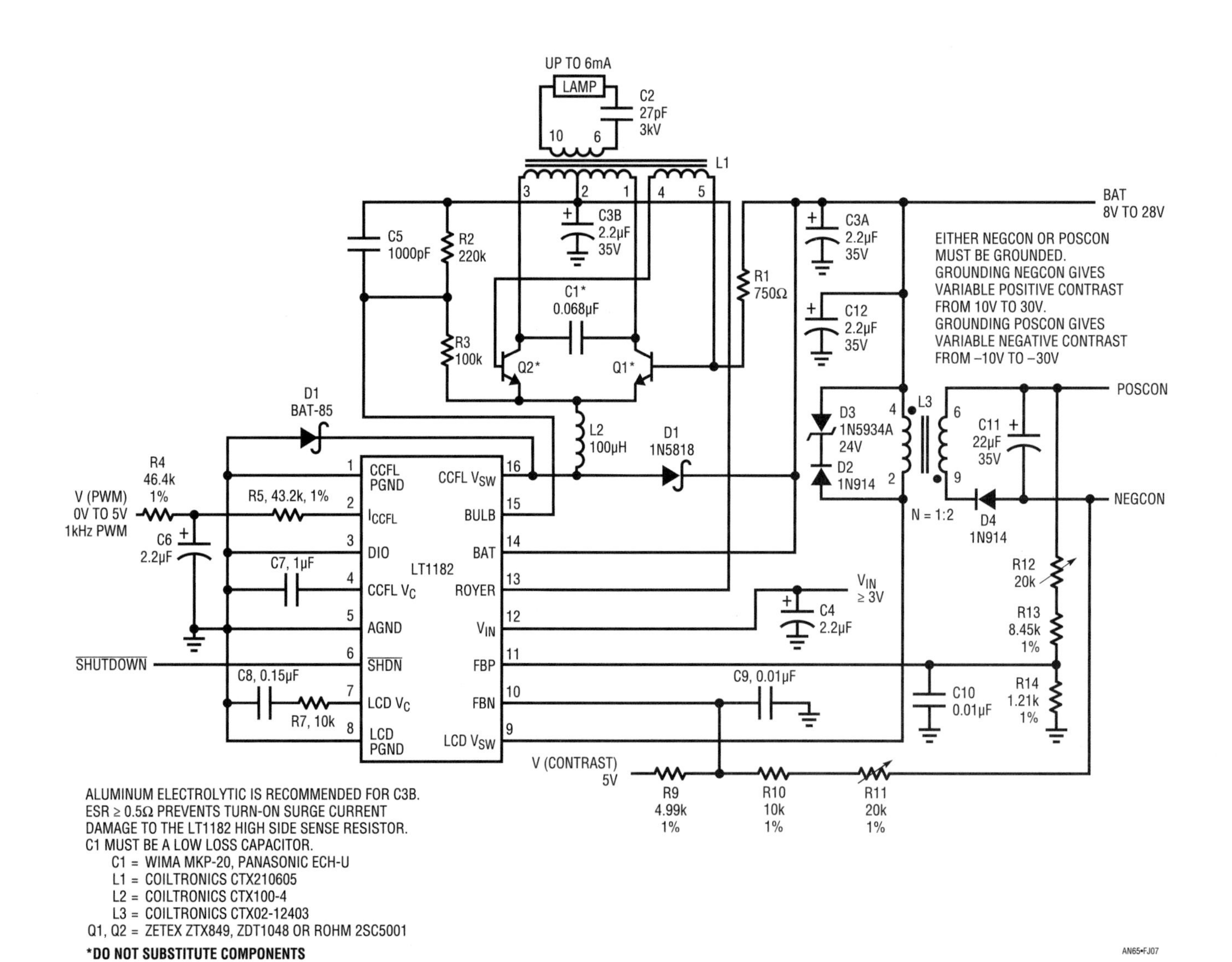

Figure J7 • LT1182 Floating Lamp CCFL Circuit with Positive or Negative LCD Contrast Supply

Appendix K
Who was Royer and what did he design?

In December 1954 the paper "Transistors as On-Off Switches in Saturable-Core Circuits" appeared in Electrical Manufacturing. George H. Royer, one of the authors, described a "d-c to a-c converter" as part of this paper. Using Westinghouse 2N74 transistors, Royer reported 90% efficiency for his circuit. The operation of Royer's circuit is well-described in this paper. The Royer converter was widely adopted and used in designs from watts to kilowatts. It is still the basis for a wide variety of power conversion.

Royer's circuit is not an LC resonant type. The transformer is the sole energy storage element and the output is a square wave. Figure K1 is a conceptual schematic of a typical converter. The input is applied to a self-oscillating configuration composed of transistors, a transformer and a biasing network. The transistors conduct out of phase, switching (Figure K2, Traces A and C are Q1's collector and base, while Traces B and D are Q2's collector and base) each time the transformer saturates. Transformer saturation causes a quickly rising, high current to flow (Trace E).

This current spike, picked up by the base drive winding, switches the transistors. This phase opposed switching causes the transistors to exchange states. Current abruptly drops in the formerly conducting transistor and then slowly rises in the newly conducting transistor until saturation again forces switching. This alternating operation sets transistor duty cycle at 50%.

Figure K3 is a time and amplitude expansion of K2's Traces B and E. It clearly shows the relationship between transformer current (Trace B, Figure K3) and transistor collector voltage (Trace A, Figure K3).[1]

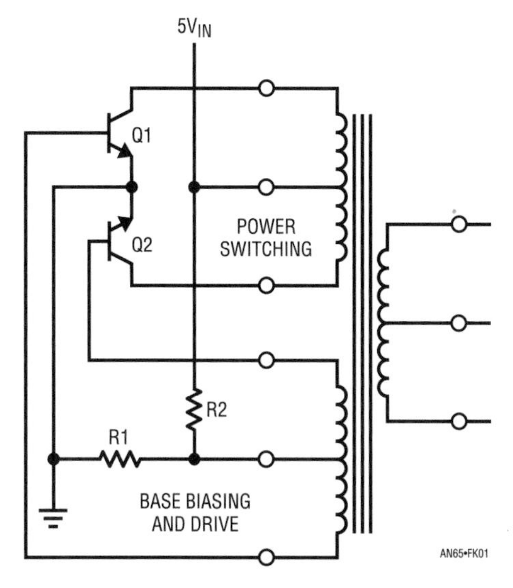

Figure K1 • Conceptual Classic Royer Converter. Transformer Approaching Saturation Causes Switching

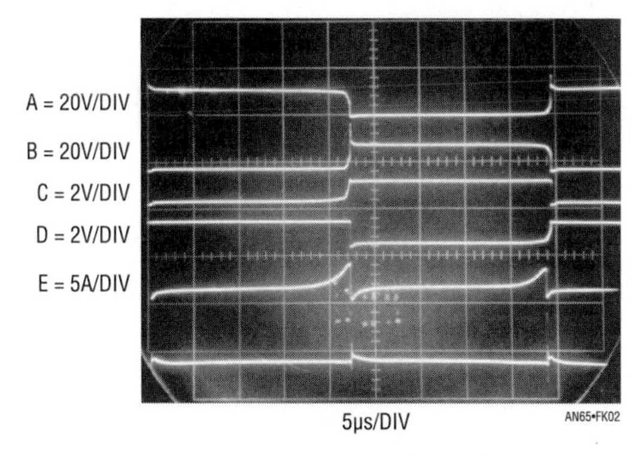

Figure K2 • Waveforms for the Classic Royer Circuit

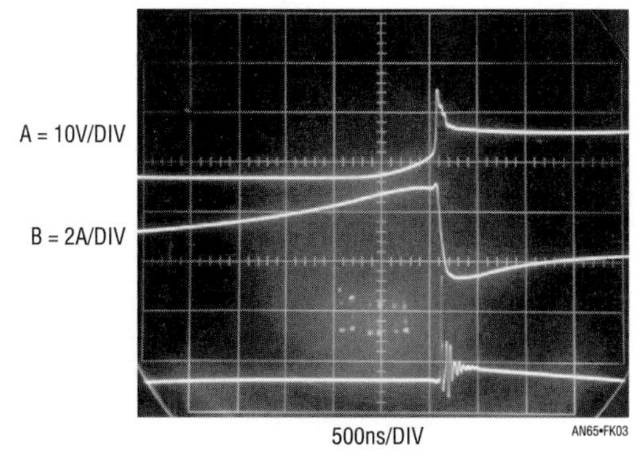

Figure K3 • Detail of Transistor Switching. Turn-Off (Trace A) Occurs Just as Transformer Heads into Saturation (Trace B)

Note 1: The bottom traces in both photographs are not germane and are not referenced in the discussion.

Appendix L
A lot of cut off ears and no Van Goghs

Some not-so-great ideas

The hunt for a practical, broadly applicable and easily utilized CCFL power supply covered (and is still covering) a lot of territory. The wide range of conflicting requirements combined with ill-defined lamp characteristics produces plenty of unpleasant surprises. This section presents a selection of ideas that turned into disappointing breadboards. Backlight circuits are one of the deadliest places the author has ever encountered for theoretically interesting circuits.

Not-so-great backlight circuits

Figure L1 seeks to boost efficiency by eliminating the LT1172's saturation loss. Comparator Cl controls a free running loop around the Royer by on-off modulation of transistor base drive. The circuit delivers bursts of high voltage sine drive to the lamp to maintain the feedback node. The scheme worked, but had poor line rejection due to the varying waveform vs supply seen by the RC averaging pair. Also, the "burst" modulation forces the loop to constantly restart the lamp at the burst rate, wasting energy. Finally, lamp power is delivered by a high crest factor waveform, causing inefficient current-to-light conversion in the lamp and shortening its life.

Figure L2 attempts to deal with some of these issues. It converts the previous circuit to an amplifier-controlled current mode regulator. Also, the Royer base drive is controlled by a clocked, high frequency pulse width modulator. This arrangement provides a more regular waveform to the averaging RC, improving line rejection. Unfortunately, the improvement was not adequate. To avoid annoying flicker, 1% line rejection is required

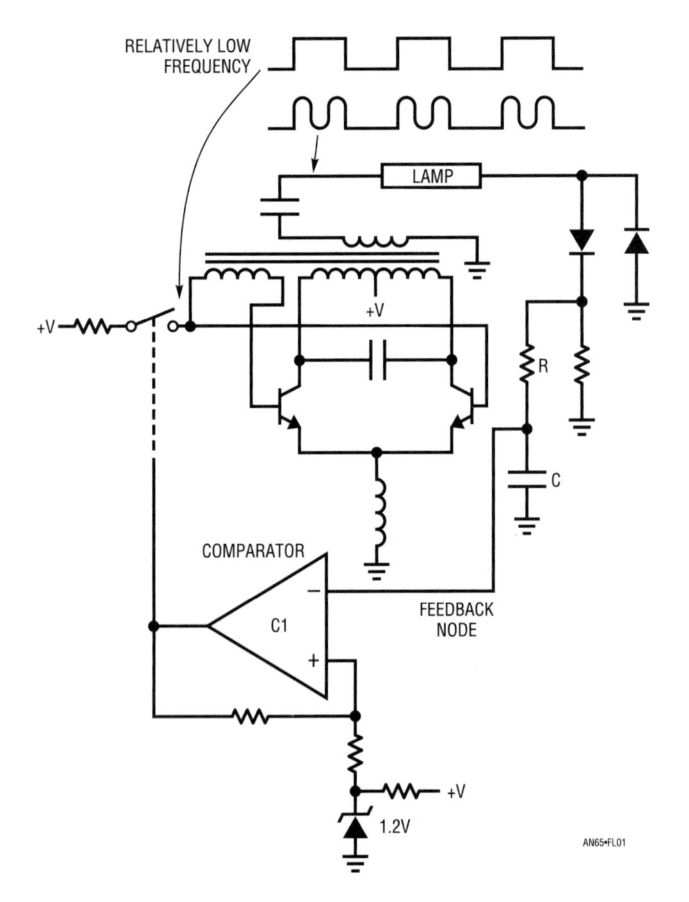

Figure L1 • A First Attempt at Improving the Basic Circuit. Irregular Royer Drive Promotes Losses and Poor Regulation

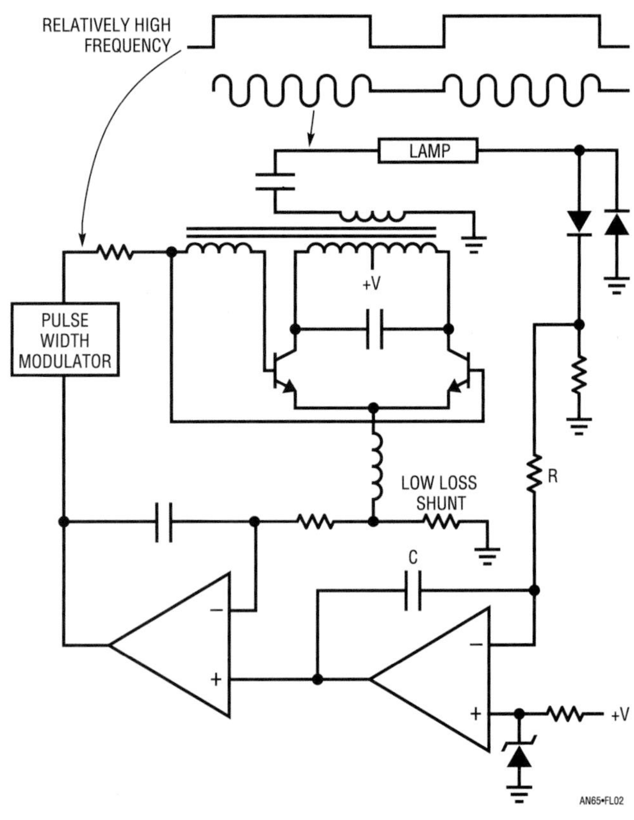

Figure L2 • A More Sophisticated Failure Still Has Losses and Poor Line Regulation

when the line moves abruptly, such as when a charger is activated. Another difficulty is that, although reduced by the higher frequency PWM, crest factor is still nonoptimal with respect to lamp emissivity and life. Finally, the lamp is still forced to restart at each PWM cycle, wasting power.

Figure L3 adds a "keep alive" function to prevent the Royer from turning off. This aspect worked well. When the PWM goes low the Royer is kept running, maintaining low level lamp conduction. This eliminates the continuous lamp restarting, saving power. The "supply correction" block feeds a portion of the supply into the RC averager, improving line rejection to acceptable levels.

This circuit, after considerable fiddling, achieved almost 94% efficiency but produced less output light than a "less efficient" version of text Figure 35! The villain is lamp waveform crest factor. The keep alive circuit helps, but the lamp still cannot handle even moderate crest factors and lamp lifetime is still questionable.

Figure L4 is a very different approach. This circuit is a driven square wave converter: The resonating capacitor is eliminated. The base drive generator shapes the edges, minimizing harmonics for low noise operation. This circuit works well, but relatively low operating frequencies are required to get good efficiency. This is so because the sloped drive must be a small percentage of the fundamental to maintain low losses. This mandates relatively large magnetics—a crucial disadvantage. Also, square waves have a different crest factor and rise time than sines, forcing inefficient lamp transduction.

RMS voltage across the shunt (e.g., the Royer current) is unaffected by this, but the simple RC averager

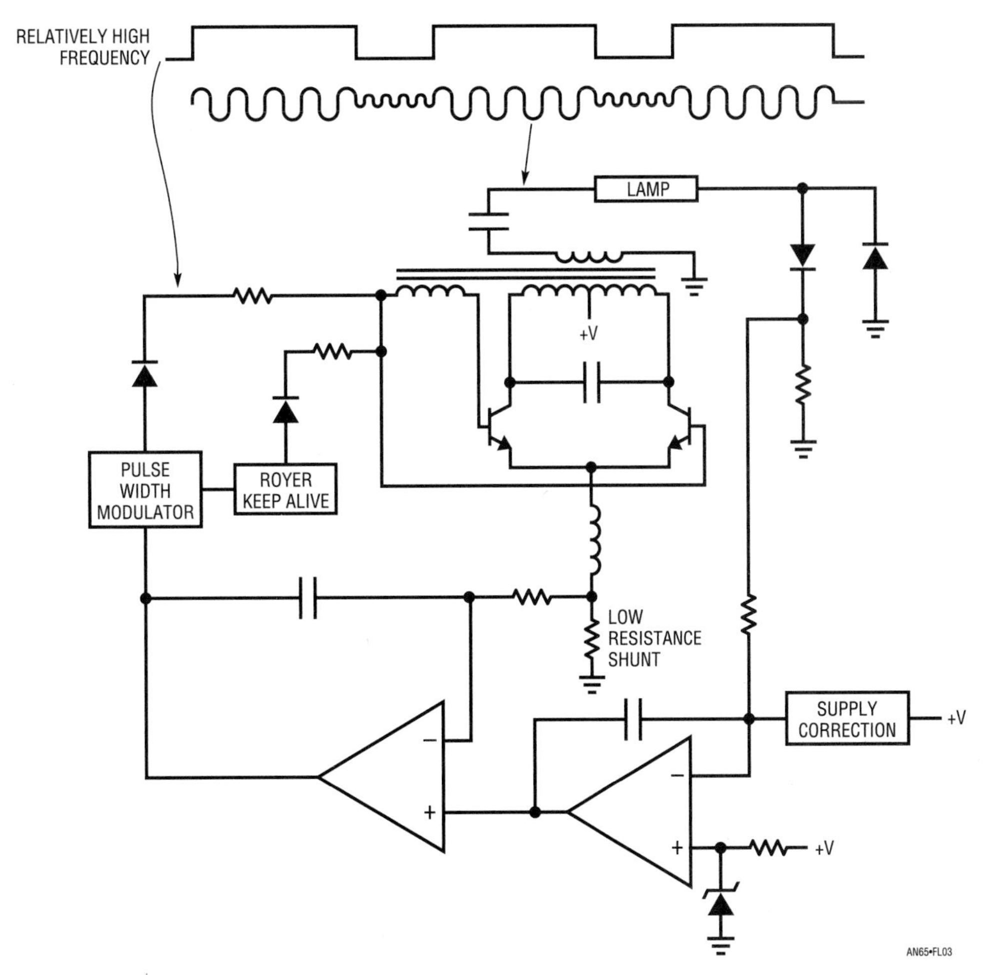

Figure L3 • "Keep Alive" Circuit Eliminates Turn-On Losses and Has 94% Efficiency. Light Emission Is Lower Than "Less Efficient" Circuits

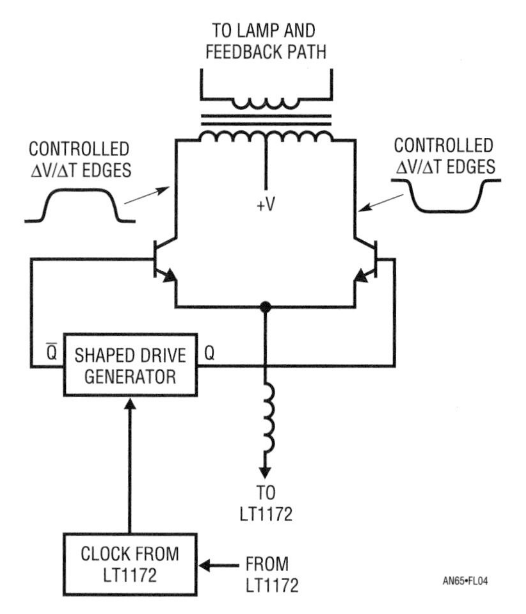

Figure L4 • A Nonresonant Approach. Slew Retarded Edges Minimize Harmonics, but Transformer Size Goes Up. Output Waveform Is Also Nonoptimal, Causing Lamp Losses

produces different outputs for the various waveforms. This causes this approach to have very poor line rejection, rendering it impractical. L6 senses inductor flux, which should correlate with Royer current. This approach promises attractive simplicity. It gives better line regulation but still has some trouble giving reliable feedback as waveshape changes. Also, in keeping with most flux sampling schemes, it regulates poorly under low current conditions.

Not-so-great primary side sensing ideas

Various text figures use primary side current sensing to control lamp intensity. This permits the lamp to fully float, extending its dynamic operating range. A number of primary side sensing approaches were tried before the "top side sense" won the contest.

L5's ground-referred current sensing is the most obvious way to detect Royer current. It offers the advantage of simple signal conditioning—there is no common mode voltage. The assumption that essentially all Royer current derives from the LT1172 emitter pin path is true. Also true, however, is that the waveshape of this path's current varies widely with input voltage and lamp operating current (see Figures L5, L6). The RMS voltage across the shunt (e.g., the Royer current) is unaffected by this, but the simple RC averager produces different outputs for the various waveforms. This causes this approach to have very poor line rejection, rendering it impractical. L6 senses inductor flux, which should correlate with Royer current. This approach promises attractive simplicity. It gives better line regulation but still has some trouble giving reliable feedback as waveshape changes. Also, in keeping with most flux sampling schemes, it regulates poorly under low current conditions.

Figure L7 senses flux in the transformer. This takes advantage of the transformer's more regular waveform. Line regulation is reasonably good because of this, but low current regulation is still poor Figure L8 samples Royer collector voltage capacitively, but the feedback signal does not accurately represent start-up, transient and low current conditions.

Figure L9 is a true, photometrically sensed feedback loop. In theory, it gets around all of the above difficulties by directly sensing lamp emission and feeding back a representative electrical signal. In practice, it introduces severe drawbacks.

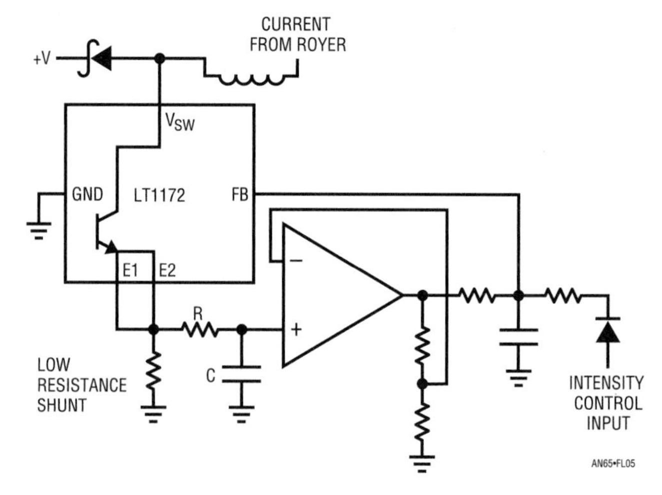

Figure L5 • "Bottom Side" Current Sensing Has Poor Line Regulation Due to RC Averaging Characteristics

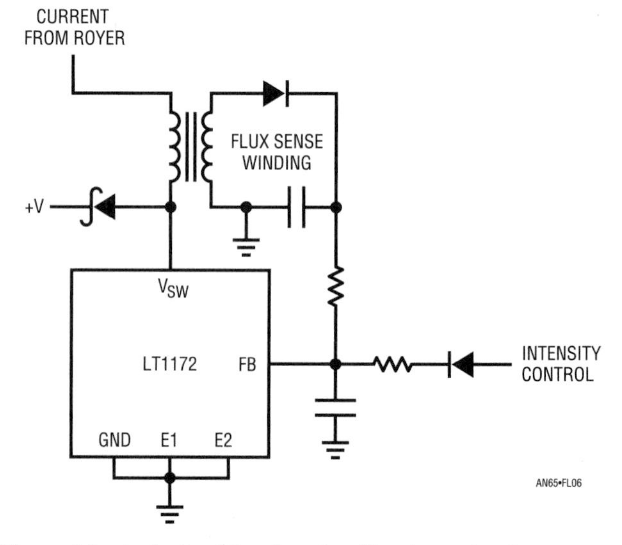

Figure L6 • Inductor Flux Sensing Has Irregular Outputs, Particularly at Low Currents

The loop servo controls current to whatever value is required to force lamp emission to the photodiode determined point. This eliminates the gradually increasing lamp output (see text Figure 8.4) at turn-on. Unfortunately, it also forces huge turn-on currents through the lamp for 10 to 20 seconds, greatly shortening lamp life. Typically, the display immediately settles to the final emission point, but turn-on current peaks at four to six times lamp rating. It is possible to clamp or limit this behavior, but a more insidious problem remains.

As the lamp ages its emissivity drops. Typically, a properly driven lamp will drop to 70% of its original emission level after 10,000 hours. In a photometrically sensed loop, the inverter will continually raise lamp current to counteract decreasing emissivity. Although lamp emission remains constant, life is greatly shortened by the continually increasing overdrive required to maintain output. This positive feedback enforced degenerative spiral assures rapid, systematic lamp destruction. A five to eight times lamp lifetime reduction in this type of loop has been observed. As before, some form of limiting or 2-loop control scheme can mitigate the undesired characteristics, but advantages would be obviated. Finally, an economical photosensor with well-specified response is elusive.

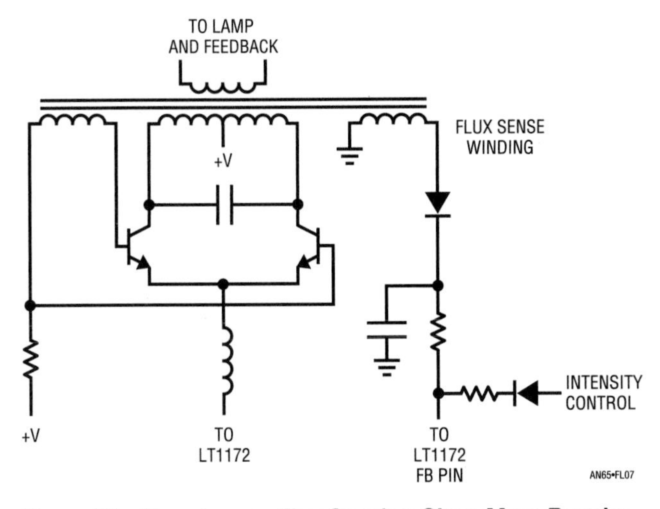

Figure L7 • Transformer Flux Sensing Gives More Regular Feedback, but Not at Low Currents

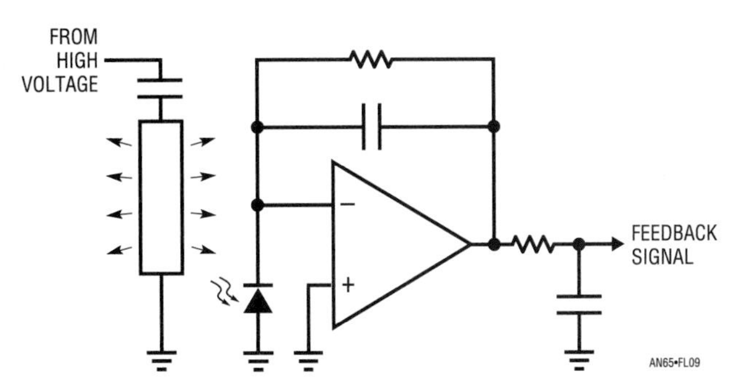

Figure L9 • True Optical Sensing Eliminates Feedback Irregularities, but Introduces Systematic Lamp Degradation

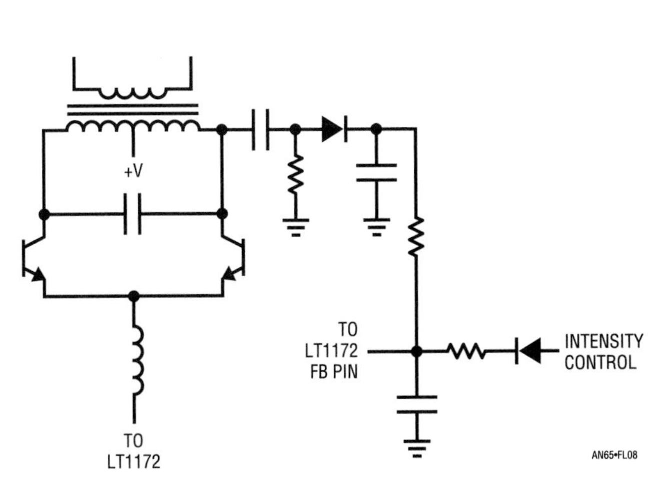

Figure L8 • AC-Coupled Drive Waveform Feedback Is Not Reliable at Low Currents

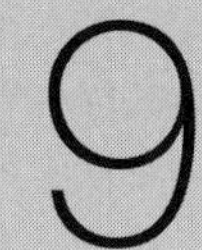

Simple circuitry for cellular telephone/camera flash illumination

A practical guide for successfully implementing flashlamps

Jim Williams Albert Wu

Introduction

Next generation cellular telephones will include high quality photographic capability. Improved image sensors and optics are readily utilized, but high quality "Flash" illumination requires special attention. Flash lighting is crucial for obtaining good photographic performance and must be quite carefully considered.

Flash illumination alternatives

Two practical choices exist for flash illumination—LEDs (Light Emitting Diode) and flashlamps. Figure 9.1 compares various performance categories for LED and flashlamp approaches. LEDs feature continuous operating capability and low density support circuitry among other advantages. Flashlamps, however, have some particularly important characteristics for high quality photography. Their line source light output is hundreds of times greater than point source LEDs, resulting in dense, easily diffused light over a wide area. Additionally, the flashlamp color temperature of 5500°K to 6000°K, quite close to natural light, eliminates the color correction necessitated by a white LED's blue peaked output.

Flashlamp basics

Figure 9.2 shows a conceptual flashlamp. The cylindrical glass envelope is filled with Xenon gas. Anode and cathode electrodes directly contact the gas; the trigger electrode,

PERFORMANCE CATEGORY	FLASHLAMP	LED
Light Output	High—Typically 10 to 400× Higher Than LEDs. Line Source Output Makes Even Light Distribution Relatively Simple	Low. Point Source Output Makes Even Light Distribution Somewhat Difficult
Illumination vs Time	Pulsed—Good for Sharp, Still Picture	Continuous—Good for Video
Color Temperature	5500°K to 6000°K—Very Close to Natural Light. No Color Correction Necessary	8500°K—Blue Light Requires Color Correction
Solution Size	Typically 3.5mm × 8mm × 4mm for Optical Assembly. 27mm × 6mm × 5mm for Circuitry—Dominated by Flash Capacitor (≈6.6mm Diameter; May be Remotely Mounted)	Typically 7mm × 7mm × 2.4mm for Optical Assembly, 7mm × 7mm × 5mm for Circuitry
Support Circuit Complexity	Moderate	Low
Charge Time	1 to 5 Seconds, Dependant Upon Flash Energy	None—Light Always Available
Operating Voltage and Currents	Kilovolts to Trigger, 300V to Flash. I_{SUPPLY} to Charge ≈ 100mA to 300mA, Dependant Upon Flash Energy. Essentially Zero Standby Current	Typically 3.4V to 4.2V at 30mA per LED Continuous—100mA Peak. Essentially Zero Standby Current
Battery Power Consumption	200 to 800 Flashes per Battery Recharge, Dependent Upon Flash Energy	≈120mW per LED (Continuous Light) ≈400mW per LED (Pulsed Light)

Partial Source: Perkin Elmer Optoelectronics

Figure 9.1 • Performance Characteristics for LED and Flashlamp-Based Illumination. LEDs Feature Small Size, No Charge Time and Continuous Operating Capability; Flashlamps are Much Brighter with Better Color Temperature

Analog Circuit and System Design: Immersion in the Black Art of Analog Design. http://dx.doi.org/10.1016/B978-0-12-397888-2.00009-2

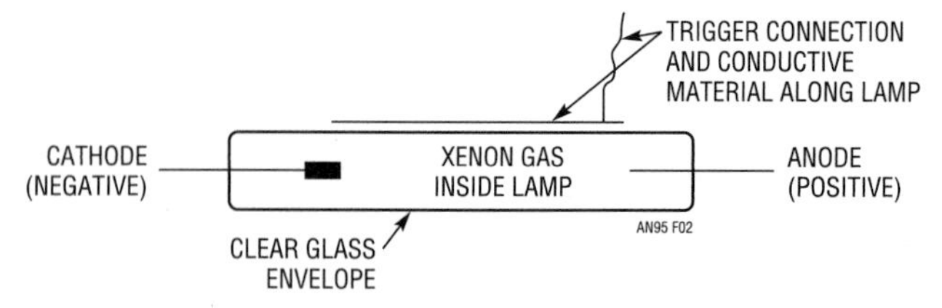

Figure 9.2 • Flashlamp Consists of Xenon Gas-Filled Glass Cylinder with Anode, Cathode and Trigger Electrodes. High Voltage Trigger Ionizes Gas, Lowering Breakdown Potential to Permit Light Producing Current Flow Between Anode and Cathode. Distributed Trigger Connection Along Lamp Length Ensures Complete Lamp Breakdown, Resulting in Optimal Illumination

distributed along the lamp's outer surface, does not. Gas breakdown potential is in the multikilovolt range; once breakdown occurs, lamp impedance drops to $\leq 1\Omega$. High current flow in the broken down gas produces intense visible light. Practically, the large current necessary requires that the lamp be put into its low impedance state before emitting light. The trigger electrode serves this function. It transmits a high voltage pulse through the glass envelope, ionizing the Xenon gas along the lamp length. This ionization breaks down the gas, placing it into a low impedance state. The low impedance permits large current to flow between anode and cathode, producing intense light. The energy involved is so high that current flow and light output are limited to pulsed operation. Continuous operation would quickly produce extreme temperatures, damaging the lamp. When the current pulse decays, lamp voltage drops to a low point and the lamp reverts to its high impedance state, necessitating another trigger event to initiate conduction.

Support circuitry

Figure 9.3 diagrams conceptual support circuitry for flashlamp operation. The flashlamp is serviced by a trigger circuit and a storage capacitor that generates the high transient current. In operation, the flash capacitor is typically charged to 300V. Initially, the capacitor cannot discharge because the lamp is in its high impedance state. A command applied to the trigger circuit results in a multikilovolt trigger pulse at the lamp. The lamp breaks down, allowing the capacitor to discharge[1]. Capacitor, wiring and lamp impedances typically total a few ohms, resulting in transient current flow in the 100A range. This heavy current pulse produces the intense flash of light. The ultimate limitation on flash repetition rate is the lamp's ability to safely dissipate heat. A secondary limitation is the time required for the charging circuit to fully charge the flash capacitor. The large capacitor charging towards a high voltage combines with the charge circuit's finite output impedance, limiting how quickly charging can occur. Charge times of 1 to 5 seconds are realizable, depending upon available input power, capacitor value and charge circuit characteristics.

The scheme shown discharges the capacitor in response to a trigger command. It is sometimes desirable to effect partial discharge, resulting in less intense light flashes. Such operation permits "red-eye" reduction, where the main flash is immediately preceded by one or more reduced intensity flashes[2]. Figure 9.4's modifications provide this operation. A driver and a high current switch have been added to Figure 9.3. These components permit stopping

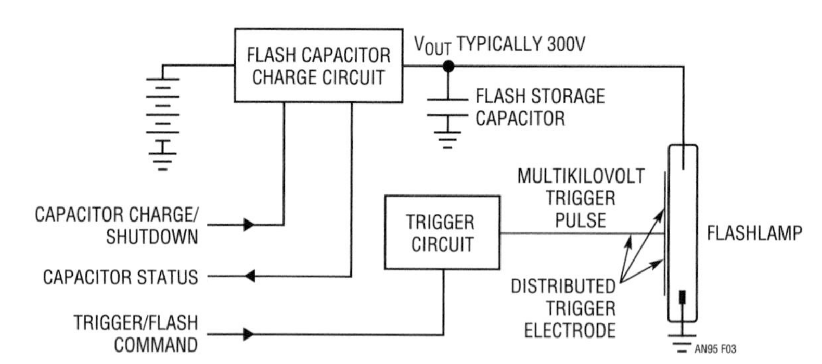

Figure 9.3 • Conceptual Flashlamp Circuitry Includes Charge Circuit, Storage Capacitor, Trigger and Lamp. Trigger Command Ionizes Lamp Gas, Allowing Capacitor Discharge Through Flashlamp. Capacitor Must be Recharged Before Next Trigger Induced Lamp Flash Can Occur

Note 1: Strictly speaking, the capacitor does not fully discharge because the lamp reverts to its high impedance state when the potential across it decays to some low value, typically 50V.

Note 2: "Red-eye" in a photograph is caused by the human retina reflecting the light flash with a distinct red color. It is eliminated by causing the eye's iris to constrict in response to a low intensity flash immediately preceding the main flash.

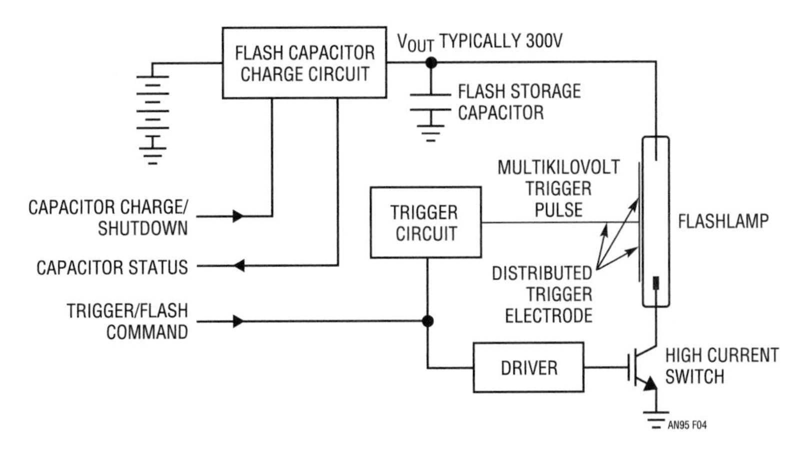

Figure 9.4 • Driver/Power Switch Added to Figure 9.3 Permits Partial Capacitor Discharge, Resulting in Controllable Light Emission. Capability Allows Pulsed Low Level Light Before Main Flash, Minimizing "Red-Eye" Phenomena

flash capacitor discharge by opening the lamp's conductive path. This arrangement allows the "trigger/flash command" control line pulse width to set current flow duration, and hence, flash energy. The low energy, partial capacitor discharge allows rapid recharge, permitting several low intensity flashes in rapid succession immediately preceeding the main flash without lamp damage.

Flash capacitor charger circuit considerations

The flash capacitor charger (Figure 9.5) is basically a transformer coupled step-up converter with some special capabilities[3]. When the "charge" control line goes high, the regulator clocks the power switch, allowing step-up transformer T1 to produce high voltage pulses. These pulses are rectified and filtered, producing the 300V DC output. Conversion efficiency is about 80%. The circuit regulates by stopping drive to the power switch when the desired output is reached. It also pulls the "$\overline{\text{DONE}}$" line low, indicating that the capacitor is fully charged. Any capacitor leakage-induced loss is compensated by intermittent power switch cycling. Normally, feedback would be provided by resistively dividing down the output voltage. This approach is not acceptable because it would require excessive switch cycling to offset the feedback resistor's constant power drain. While this action would maintain regulation, it would also drain excessive power from the primary source, presumably a battery. Regulation is instead obtained by monitoring T1's flyback pulse characteristic, which reflects T1's secondary amplitude. The output voltage is set by T1's turns ratio[4]. This feature permits tight capacitor voltage regulation, necessary to ensure consistent flash intensity without exceeding lamp energy or capacitor voltage ratings. Also, flashlamp energy is conveniently determined by the capacitor value without any other circuit alterations.

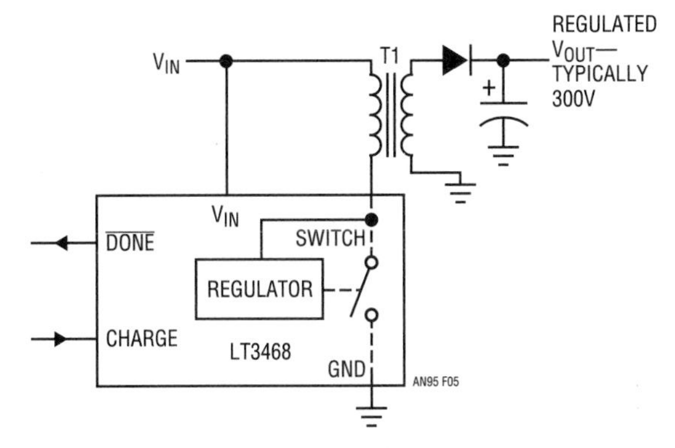

Figure 9.5 • Flash Capacitor Charger Circuit Includes IC Regulator, Step-Up Transformer, Rectifier and Capacitor. Regulator Controls Capacitor Voltage by Monitoring T1's Flyback Pulse, Eliminating Conventional Feedback Resistor Divider's Loss Path. Control Pins Include Charge Command and Charging Complete ("$\overline{\text{DONE}}$") Indication

Note 3: Details on this device's operation appear in Appendix A, "A Monolithic Flash Capacitor Charger."

Note 4: See Appendix A for recommended transformers.

Detailed circuit discussion

BEFORE PROCEEDING ANY FURTHER, THE READER IS WARNED THAT CAUTION MUST BE USED IN THE CONSTRUCTION, TESTING AND USE OF THIS CIRCUIT HIGH VOLTAGE, LETHAL POTENTIALS ARE PRESENT IN THIS CIRCUIT EXTREME CAUTION MUST BE USED IN WORKING WITH, AND MAKING CONNECTIONS TO, THIS CIRCUIT REPEAT: THIS CIRCUIT CONTAINS DANGEROUS, HIGH VOLTAGE POTENTIALS. USE CAUTION.

Figure 9.6 is a complete flashlamp circuit based on the previous text discussion. The capacitor charging circuit, similar to Figure 9.5, appears at the upper left. D2 has been added to safely clamp T1-originated reverse transient voltage events. Q1 and Q2 drive high current switch Q3. The high voltage trigger pulse is formed by step-up transformer T2. Assuming C1 is fully charged, when Ql-Q2 turns Q3 on, C2 deposits current into T2's primary. T2's secondary delivers a high voltage trigger pulse to the lamp, ionizing it to permit conduction. C1 discharges through the lamp, producing light.

Figure 9.7 details the capacitor charging sequence. Trace A, the "charge" input, goes high. This initiates T1 switching, causing C1 to ramp up (trace B). When C1 arrives at the regulation point, switching ceases and the resistively pulled-up "$\overline{\text{DONE}}$" line drops low (trace C), indicating C1's charged state. The "TRIGGER" command (trace D), resulting in C1's discharge via the lamp-Q3 path, may occur any time (in this case ≈600ms) after "$\overline{\text{DONE}}$" goes low. Note that this figure's trigger command is lengthened for photographic clarity; it normally is 500µs to 1000µs in duration for a complete C1 discharge. Low level flash events, such as for "red-eye" reduction, are facilitated by short duration trigger input commands.

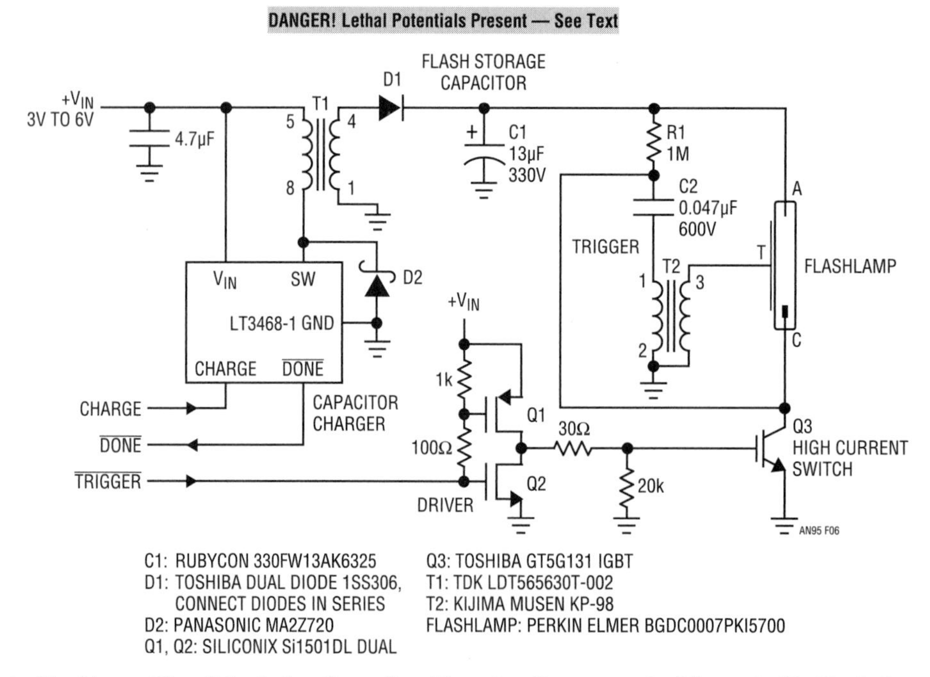

Figure 9.6 • Complete Flashlamp Circuit Includes Capacitor Charging Components (Figure Left), Flash Capacitor C1, Trigger (R1, C2, T2), Q1-Q2 Driver, Q3 Power Switch and Flashlamp. $\overline{\text{TRIGGER}}$ Command Simultaneously Biases Q3 and Ionizes Lamp via T2. Resultant C1 Discharge Through Lamp Produces Light

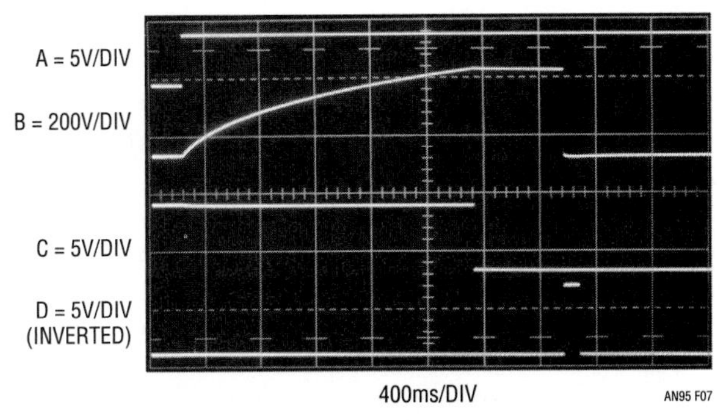

Figure 9.7 • Capacitor Charging Waveforms Include Charge Input (Trace A), C1 (Trace B), $\overline{\text{DONE}}$ Output (Trace C) and $\overline{\text{TRIGGER}}$ Input (Trace D). C1's Charge Time Depends Upon Its Value and Charge Circuit Output Impedance. $\overline{\text{TRIGGER}}$ Input, Widened for Figure Clarity, May Occur any Time After $\overline{\text{DONE}}$ Goes Low

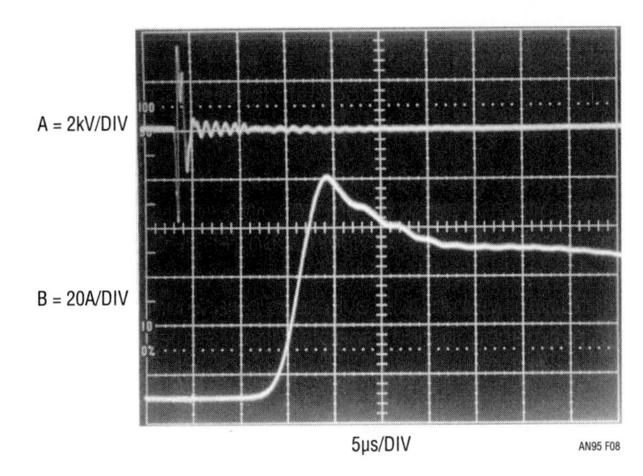

Figure 9.8 • High Speed Detail of Trigger Pulse (Trace A) and Resultant Flashlamp Current (Trace B). Current Approaches 100A After Trigger Pulse Ionizes Lamp

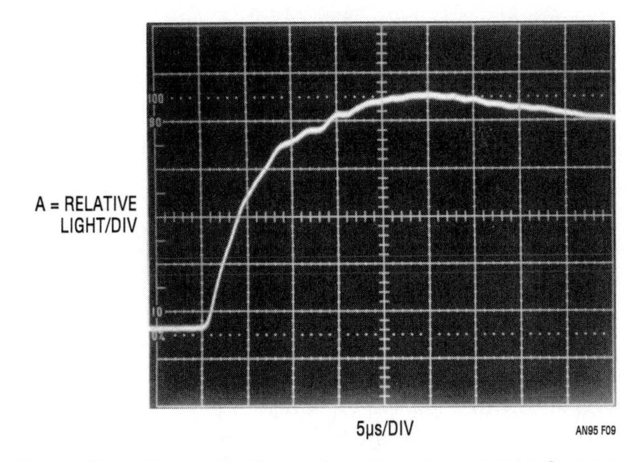

Figure 9.9 • Smoothly Ascending Flashlamp Light Output Peaks in 25μs

Figure 9.8 shows high speed detail of the high voltage trigger pulse (trace A) and resultant flashlamp current (trace B). Some amount of time is required for the lamp to ionize and begin conduction after triggering. Here, 10μs after the $8kV_{P-P}$ trigger pulse, flashlamp current begins its ascent to nearly 100A. The current rises smoothly in 5μs to a well defined peak before beginning its descent. The resultant light produced (Figure 9.9) rises more slowly, peaking in about 25μs before decaying. Slowing the oscilloscope sweep permits capturing the entire current and light events. Figure 9.10 shows that light output (trace B) follows lamp current (trace A) profile, although current peaking is more abrupt. Total event duration is ≈500μs with most energy expended in the first 200μs. The leading edge's discontinuous presentation is due to oscilloscope chopped display mode operation.

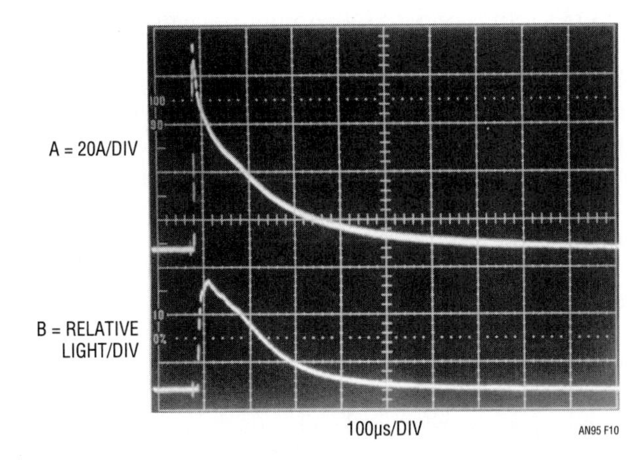

Figure 9.10 • Photograph Captures Entire Current (Trace A) and Light (Trace B) Events. Light Output Follows Current Profile Although Peaking is Less Defined. Leading Edges Dashed Presentation Derives from Oscilloscope's Chopped Display Operation

Lamp layout, RFI and related issues

Lamp considerations

Several lamp related issues require attention. Lamp triggering requirements must be thoroughly understood and adhered to. If this is not done, incomplete or no lamp flash may occur. Most trigger related problems involve trigger transformer selection, drive and physical location with respect to the lamp. Some lamp manufacturers supply the trigger transformer, lamp and light diffuser as a single, integrated assembly[5]. This obviously implies trigger transformer approval by the lamp vendor, assuming it is driven properly. In cases where the lamp is triggered with a user-selected transformer and drive scheme, it is essential to obtain lamp vendor approval before going to production.

The lamp's anode and cathode access the lamp's main discharge path. Electrode polarity must be respected or severe lifetime degradation will occur. Similarly, lamp energy dissipation restrictions must be respected or lifetime will suffer. Severe lamp energy overdrive can result in lamp cracking or disintegration. Energy is easily and reliably controlled by selecting capacitor value and charge voltage and restricting flash repetition rate. As with triggering, lamp flash conditions promoted by the user's circuit require lamp manufacturer approval before production.

Assuming proper triggering and flash energy, lamp lifetimes of ≈5000 flashes may be expected. Lifetime for various lamp types differs from this figure, although all are vendor specified. Lifetime is typically defined as the point where lamp luminosity drops to 80% of its original value.

Layout

The high voltages and currents mandate layout planning. Referring back to Figure 9.6, C1's discharge path is through the lamp, Q3 and back to ground. The ≈100A peak current means this discharge path must be maintained at low impedance. Conduction paths between C1, the lamp and Q3 should be short and well below 1Ω. Additionally, Q3's emitter and C1's negative terminal should be directly connected, the goal being a tight, highly conductive loop between C1's positive terminal, the lamp and Q3's return back to C1. Abrupt trace discontinuities and vias should be avoided as the high current flow can cause conductor erosion in local high resistivity regions. If vias must be employed they should be filled, verified for low resistance or used in multiples. Unavoidable capacitor ESR, lamp and Q3 resistances typically total 1Ω to 2.5Ω, so total trace resistance of 0.5Ω or less is adequate. Similarly, the high current's relatively slow risetime (see Figure 9.8) means trace inductance does not have to be tightly controlled.

C1 is the largest component in the circuit; space considerations may make remotely mounting it desirable. This can be facilitated with long traces or wires so long as interconnect resistance is maintained within the limits stated above.

Capacitor charger IC layout is similar to conventional switching regulator practice. The electrical path formed by the IC's V_{IN} pin, its bypass capacitor, the transformer primary and the switch pin must be short and highly conductive. The IC's ground pin should directly return to a low impedance, planar ground connection. The transformer's 300V output requires larger than minimum spacing for all high voltage nodes to meet circuit board breakdown requirements. Verify board material breakdown specifications and ensure that board washing procedures do not introduce conductive contaminants. T2's multikilovolt trigger winding must connect *directly* to the lamp's trigger electrode, preferably with less than 1/4" of conductor. Adequate high voltage spacing must be employed. In general, what little conductor there is should not contact the circuit board. Excessive T2 output length can cause trigger pulse degradation or radio frequency interference (RFI). Modular flashlamp-trigger transformer assemblies are excellent choices in this regard.

A demonstration layout for Figure 9.6 appears in Figure 9.11. The topside component layer is shown. Power and ground are distributed on internal layers. LT3468 layout is typical of switching regulator practice previously described, although wide trace spacing accomodates T1's 300V output. The ≈100A pulsed current flows in a tight, low resistance loop from C1's postive terminal, through the lamp, into Q3 and back to C1. In this case lamp connections are made with wires, although modular flashlamp-trigger transformers allow trace-based connections[6].

Radio frequency interference

The flash circuit's pulsed high voltages and currents make RFI a concern. The capacitor's high energy discharge is actually far less offensive than might be supposed. Figure 9.12 shows the discharge's 90A current peak confined to a 70kHz bandwidth by its 5μs risetime. This means there is little harmonic energy at radio frequencies, easing this concern. Conversely, Figure 9.13's T2 high voltage output has a 250ns risetime (BW ≈ 1.5MHz), qualifying it as a potential RFI source. Fortunately, the energy involved and the exposed path length (see layout comments) are small, making interference management possible.

The simplest interference management involves placing radiating components away from sensitive circuit nodes or employing shielding. Another option takes advantage of the predictable time when the flash circuit operates. Sensitive circuitry within the telephone can be blanked during flash events, which typically last well under 1ms.

Note 5: See Reference 1.

Note 6: See Reference 1

Note: This application note was derived from a manuscript originally prepared for publication in EDN magazine.

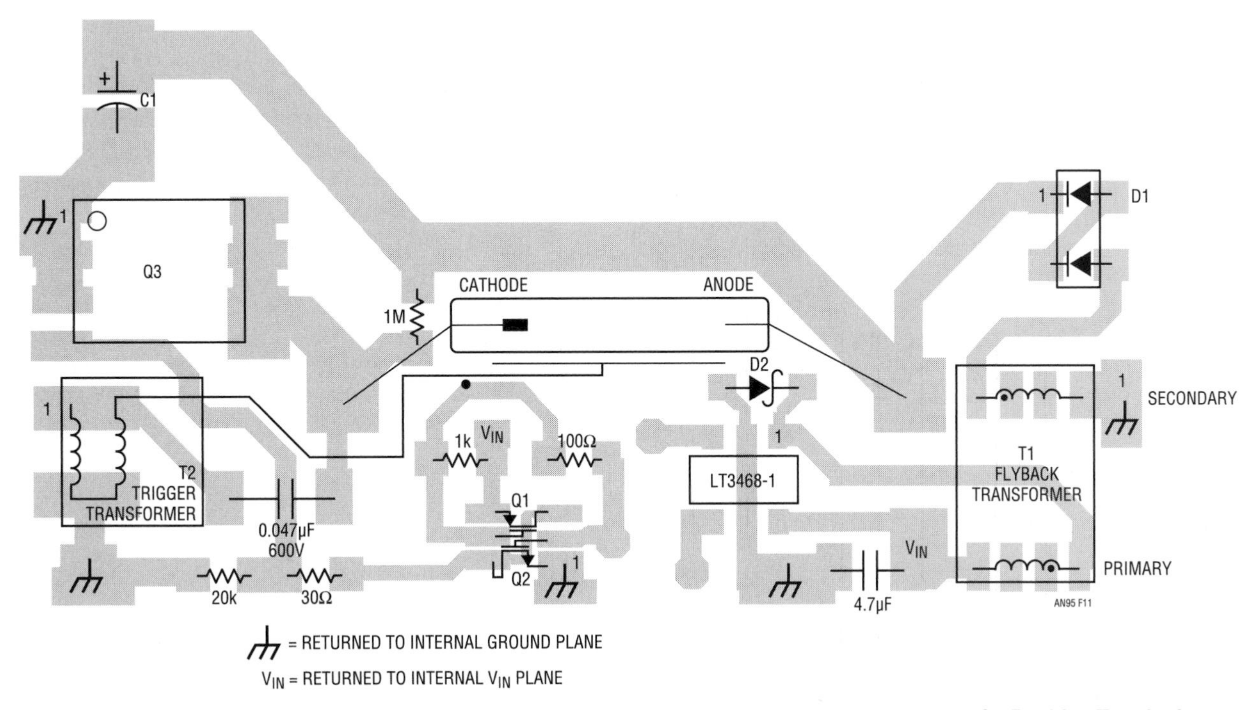

Figure 9.11 • Magnified Demonstration Layout for Figure 9.6. High Current Flows in Tight Loop from C1 Positive Terminal, Through Lamp, Into Q3 and Back to C1. Lamp Connections are Wires, Not Traces. Wide T1 Secondary Spacing Accommodates 300V Output

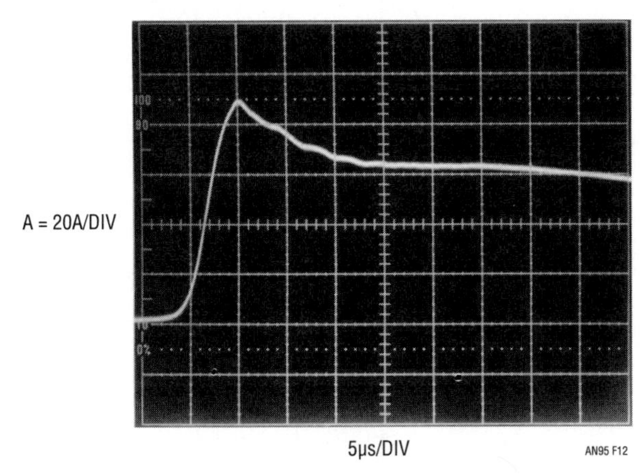

Figure 9.12 • 90A Current Peak is Confined Within 70kHz Bandwidth by 5µs Risetime, Minimizing Noise Concerns

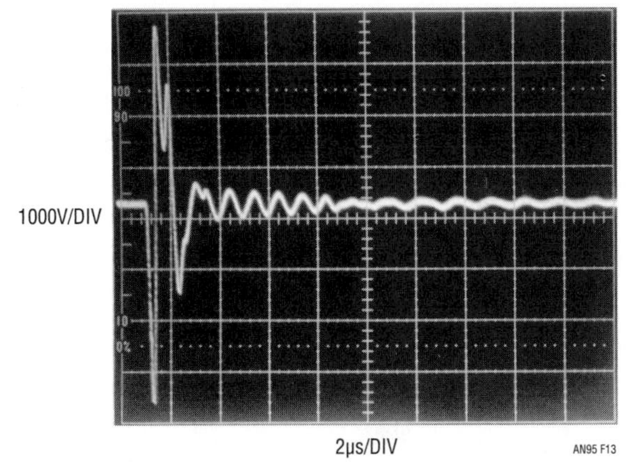

Figure 9.13 • Trigger Pulses High Amplitude and Fast Risetime Promote RFI, but Energy and Path Exposure are Small, Simplifying Radiation Management

References

1. Perkin Elmer, "Flashtubes."
2. Perkin Elmer, "Everything You Always Wanted to Know About Flashtubes"
3. Linear Technology Corporation, "LT®3468/ LT3468-1/ LT3468-2 Data Sheet"
4. Wu, Albert, "Photoflash Capacitor Chargers Fit Into Tight Spots," Linear Technology, Vol. XIII, No. 4, December 2003
5. Rubycon Corporation. Catalog 2004, "Type FW Photoflash Capacitor," Page 187

Appendix A
A monolithic flash capacitor charger

The LT3468/LT3468-1/LT3468-2 charge photoflash capacitors quickly and efficiently. Operation is understood by referring to Figure A1. When the CHARGE pin is driven high, a one shot sets both SR latches in the correct state. Power NPN, Q1, turns on and current begins ramping up in T1's primary. Comparator A1 monitors switch current and when peak current reaches 1.4A (LT3468), 1A(LT3468-2) or 0.7A (LT3468-1), Q1 is turned off. Since T1 is utilized as a flyback transformer, the flyback pulse on the SW pin causes A3's output to be high. SW pin voltage must be at least 36mV above V_{IN} for this to happen.

During this phase, current is delivered to the photoflash capacitor via T1's secondary and D1. As the secondary current decreases to zero, the SW pin voltage begins to collapse. When the SW pin voltage drops to 36mV above V_{IN} or lower, A3's output goes low. This fires a one shot which turns Q1 back on. This cycle continues, delivering power to the output.

Output voltage detection is accomplished via R2, R1, Q2 and comparator A2. Resistors R1 and R2 are sized so that when the SW voltage is 31.5V above V_{IN}, A2's output goes high, resetting the master latch. This disables Q1, halting power delivery. Q3 is turned on, pulling the $\overline{\text{DONE}}$ pin low, indicating the part has finished charging. Power delivery can only be restarted by toggling the CHARGE pin.

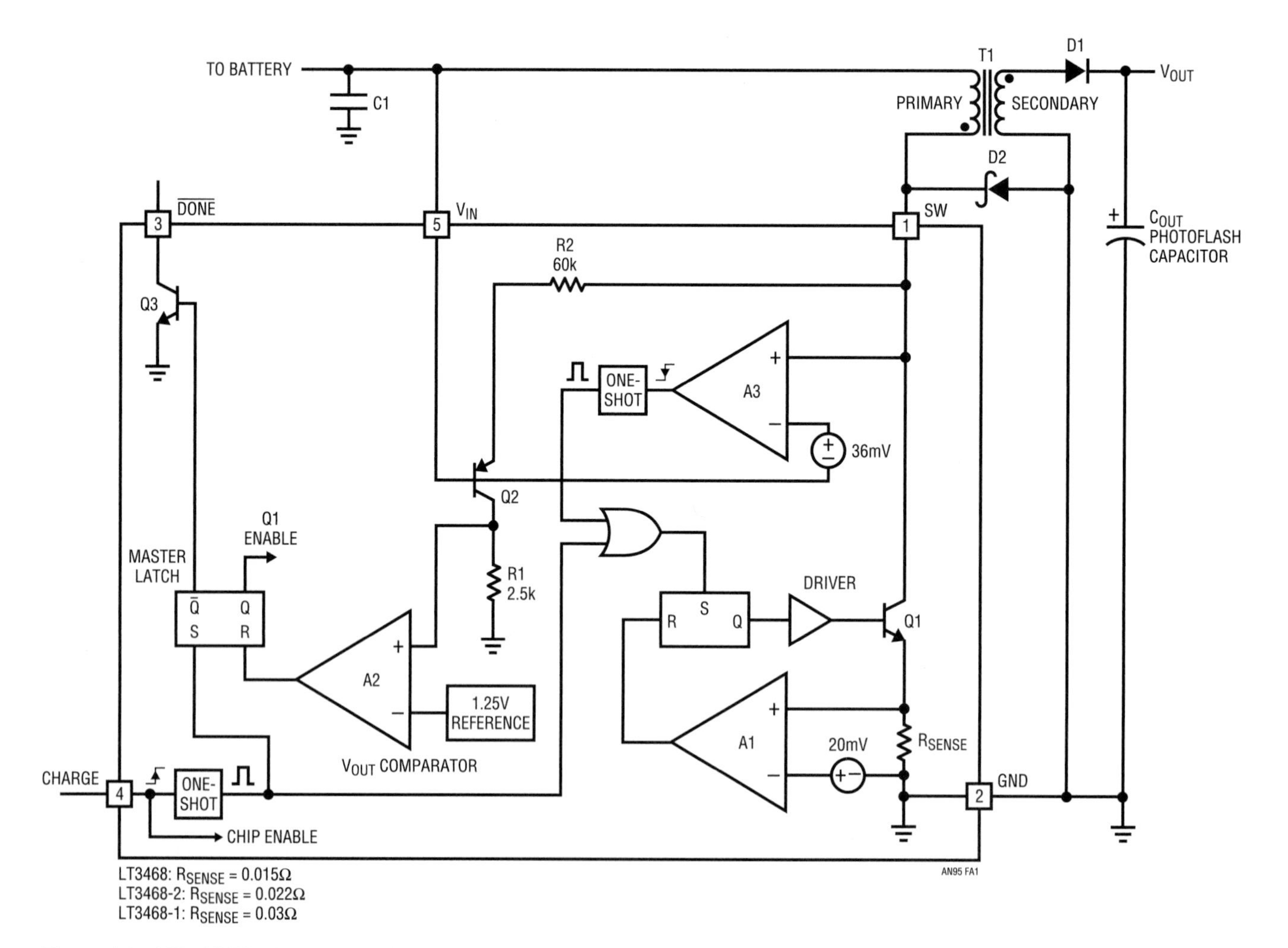

Figure A1 • LT3468 Block Diagram. Charge Pin Controls Power Switching to T1. Photoflash Capacitor Voltage is Regulated by Monitoring T1's Flyback Pulse, Eliminating Conventional Feedback Resistors Loss Path

The CHARGE pin gives the user full control of the part. Charging can be halted at any time by bringing the CHARGE pin low. Only when the final output voltage is reached will the $\overline{DONE}$ pin go low. Figure A2 shows these various modes in action. When CHARGE is first brought high, charging commences. When CHARGE is brought low during charging, the part shuts down and V_{OUT} no longer rises. When CHARGE is brought high again, charging resumes. When the target V_{OUT} voltage is reached, the $\overline{DONE}$ pin goes low and charging stops. Finally, the CHARGE pin is brought low again, the part enters shutdown and the $\overline{DONE}$ pin goes high.

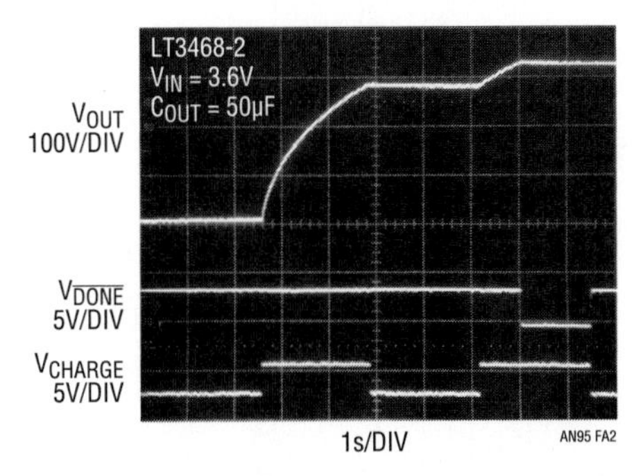

Figure A2 • Halting the Charging Cycle with the CHARGE Pin

The only difference between the three LT3468 versions is the peak current level. The LT3468 offers the fastest charge time. The LT3468-1 has the lowest peak current capability, and is designed for applications requiring limited battery drain. Due to the lower peak current, the LT3468-1 can use a physically smaller transformer. The LT3468-2 has a current limit between the LT3468 and the LT3468-1. Comparative plots of the three versions charge time, efficiency and output voltage tolerance appear in Figures A3, A4 and A5.

Standard off-the-shelf transformers, available for all LT3468 versions, are available and detailed in Figure A6. For transformer design considerations, as well as other supplemental information, see the LT3468 data sheet.

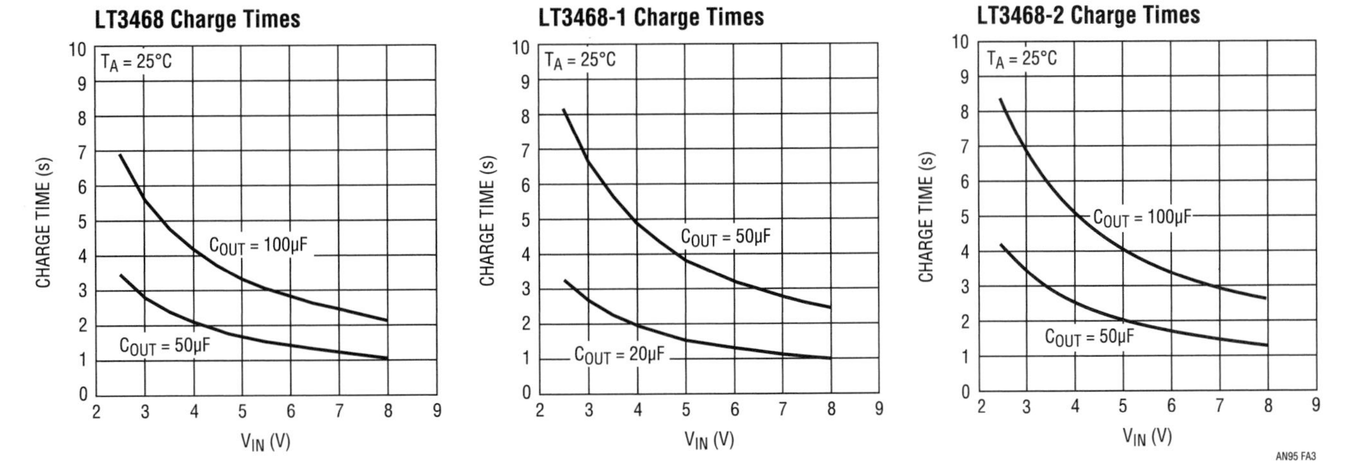

Figure A3 • Typical LT3438 Charge Times. Charge Time Varies with IC Version, Capacitor Size and Input Voltage

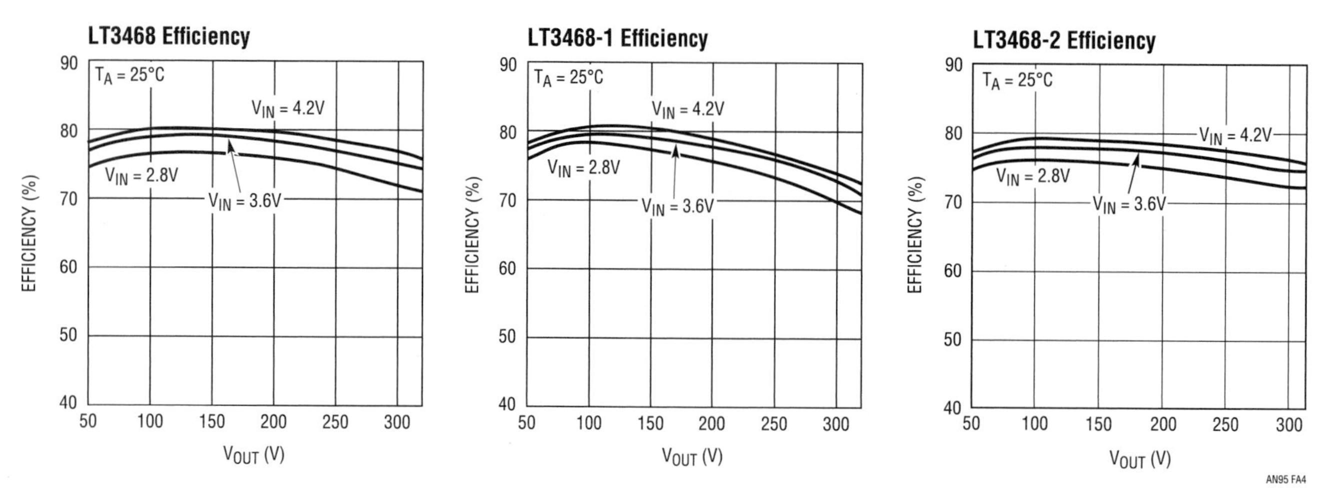

Figure A4 • Efficiency for the Three LT3468 Versions Varies with Input and Output Voltages

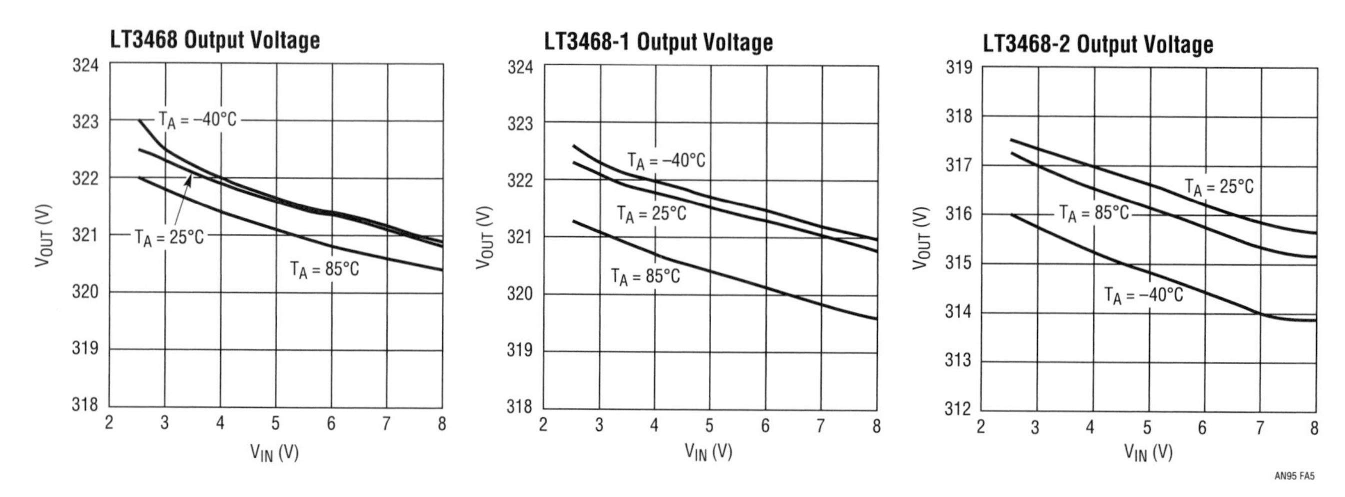

Figure A5 • Typical Output Voltage Tolerance for the Three LT3468 Versions. Tight Voltage Tolerance Prevents Overcharging Capacitor, Controls Flash Energy

FOR USE WITH	TRANSFORMER NAME	SIZE (W × L × H) mm	L_{PRI} (μH)	$L_{PRI-LEAKAGE}$ (nH)	N	R_{PRI} (mΩ)	R_{SEC} (Ω)	VENDOR
LT3468/LT3468-2 LT3468-1	SBL-5.6-1 SBL-5.6S-1	5.6 × 8.5 × 4.0 5.6 × 8.5 × 3.0	10 24	200 Max 400 Max	10.2 10.2	103 305	26 55	Kijima Musen Hong Kong Office 852-2489-8266 (ph) kijimahk@netvigator.com (email)
LT3468 LT3468-1 LT3468-2	LDT565630T-001 LDT565630T-002 LDT565630T-003	5.8 × 5.8 × 3.0 5.8 × 5.8 × 3.0 5.8 × 5.8 × 3.0	6 14.5 10.5	200 Max 500 Max 550 Max	10.4 10.2 10.2	100 Max 240 Max 210 Max	10 Max 16.5 Max 14 Max	TDK Chicago Sales Office (847) 803-6100 (ph) www.components.tdk.com
LT3468/LT3468-1 LT3468-1	T-15-089 T-15-083	6.4 × 7.7 × 4.0 8.0 × 8.9 × 2.0	12 20	400 Max 500 Max	10.2 10.2	211 Max 675 Max	27 Max 35 Max	Tokyo Coil Engineering Japan Office 0426-56-6336 (ph) www.tokyo-coil.co.jp

Figure A6 • Standard Transformers Available for LT3468 Circuits. Note Small Size Despite High Output Voltage

Section 6

Automotive and Industrial Power Design

Extending the input voltage range of PowerPath circuits for automotive and industrial applications (10)

The voltage range of Linear Technology's PowerPath™ circuits can be easily extended with just a few components, thus allowing them to meet the needs of virtually all applications. This application note presents solutions for circuits that must withstand large negative voltages, a reverse adapter input for example, and circuits that must withstand large positive inputs, such as automotive load-dump.

Extending the input voltage range of PowerPath circuits for automotive and industrial applications

10

Greg Manlove

Introduction

The voltage range of Linear Technology's PowerPath® circuits can be easily extended with just a few components, thus allowing them to meet the needs of virtually all applications. This application note presents solutions for circuits that must withstand large negative voltages, a reverse adapter input for example, and circuits that must withstand large positive inputs, such as automotive load-dump.

Extending the voltage range

Any of Linear's PowerPath controller circuits can benefit from an extended voltage range, even those that already have wide operating and absolute maximum voltage ranges. For instance, the LTC4412HV and LTC4414 will each withstand voltages from −14V to 40V, which can be extended further using the techniques described here. Likewise, the LTC4412's range of −14V to 28V can be extended. The voltage ranges of monolithic PowerPath solutions such as the LTC4411, which ranges from −0.3V to 6V, can also be extended, though not as far.

There are two different approaches to extending the voltage range of the PowerPath circuits. The first addresses the negative input voltage requirements with the addition of a Schottky diode. This change assures that the external P-channel pass transistor is held in the off state as the input goes below ground. The second approach allows the ICs to operate both above the specified voltage range and below ground. The external circuit count is still compact, requiring only three additional components.

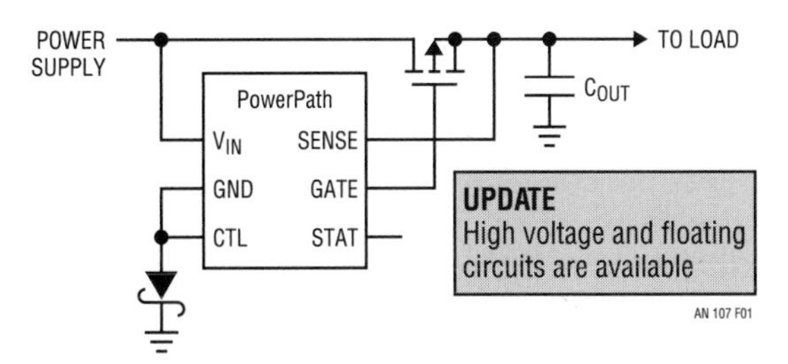

Figure 10.1 • Circuit Capable Of Operation With A Large Negative Input Supply

Circuit for large negative input voltages

Refer to Figure 10.1 for a description of the circuit. The ground and control pins of the PowerPath IC are tied through a Schottky diode to the system ground. When the power supply goes below ground, the diode is reverse biased, blocking the negative supply path to ground. The maximum negative voltage for the circuit is limited by the maximum allowed voltage difference between the SENSE pin and the V_{IN} pin. In the case of both the LTC4412HV and LTC4414, this difference is 40V, so the negative voltage limit is −40V. Likewise, LTC4411 is limited to −6V. These values both assume the SENSE Pin (load side) is 0V. Because the LTC4412HV and LTC4414 are capable of withstanding −14V with no diode present, the reverse breakdown voltage of the Schottky diode must exceed 26V to achieve the −40V capability at the input (40V – 14V = 26V).

During normal operation, when the input supply is positive, the voltage at the ground pin is equal to the forward voltage of the Schottky or approximately 0.2V. In turn, this additional voltage on the ground pin raises the minimum operating supply of the circuit by approximately 0.2V. The control signal input threshold increases by the same amount.

Analog Circuit and System Design: Immersion in the Black Art of Analog Design. http://dx.doi.org/10.1016/B978-0-12-397888-2.00010-9

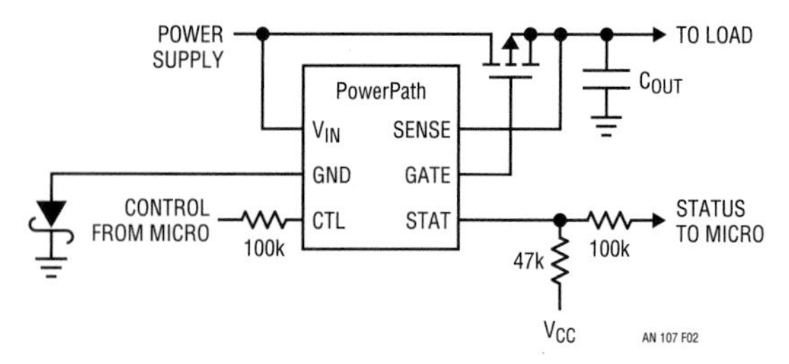

Figure 10.2 • PowerPath Circuit Capable of Operating with a Large Negative Supply with Controls and Status Available

When the input supply goes more negative than the normal operating range of the part (−14V on the LTC4412 and LTC4414), the ground pin begins to go negative. The IC continues to hold the P-channel FET off as the power supply goes further below ground until the maximum V_{SENSE} minus V_{IN} voltage is reached (−40V on the LTC4412 and LTC4414 and −6V on the LTC4411).

The control and status pins also go negative under this large negative supply condition. Refer to Figure 10.2 for a circuit that allows control of the PowerPath IC during normal operation. A 100k series resistor must be added between the micro-processor output and the control input. The series resistor allows the control pin to go below ground without causing excessive current in the microprocessor or other device controlling the part. The status pin also goes below ground under a negative input supply, requiring a 100k resistor in series between the status pin and the microprocessor input. Again, the resistor is added to protect the microprocessor from the negative input signal. Realistically, V_{CC} is not valid if the input power supply is negative, so the part operates for all valid supply conditions. The 100k series resistors have minimal impact on the control threshold or Status output. Both signals have a nominal ground reference at the V_F of the Schottky diode or approximately 0.2V. This is the largest deviation from nominal and should not present a problem in most systems.

Circuit for large positive input voltages

Refer to Figure 10.3 for a description of the circuit. The IC ground and control pins of the PowerPath circuit are wired together and grounded through a resistor. They are also connected through a Zener diode to the input power supply. The breakdown voltage on the Zener must be less than the breakdown voltage of the IC: that is, a 5V Zener for the LTC4411, and a 36V Zener for both the LTC4412HV and the LTC4414.

When a large positive voltage is applied to the system, the Zener diode clamps the voltage between the V_{IN} and ground pins of the IC. The voltage on the resistor connected to system ground rises. The quiescent current of the PowerPath products are typically under 50μA, thus a 2k resistor causes the nominal voltage on the ground line to rise only 0.1V. This increases the minimum operating voltage by the voltage drop across the resistor or approximately 0.1V. The ground resistor must have a high enough power rating (V^2/R) for the circuit. For example, the LTC4412HV with a 36V Zener and an 80V input, produces 44V across the resistor. The resistor power rating is equal to $(44V)^2/2k$ or approximately 1W. If 80V only occurs during a transient, the power rating of the resistor can be reduced.

The ground pin of the PowerPath IC is positive when the input supply exceeds the Zener clamp voltage. This ground signal can be run through a 100k resistor to a microprocessor input to provide a control signal to the system. The voltage on the Overvoltage Status pin can be quite large, injecting too much current into the microprocessor input pin. A Schottky diode can be added between the 100k resistor and system supply to clamp the signal, if required.

When the input supply is a diode voltage below ground, the Zener diode conducts. This pulls the ground resistor terminal to within a diode of the negative power supply. The part sees virtually no external voltage between the ground and input pins. The maximum negative supply is limited by the maximum voltage difference between V_{IN} and the Sense pin. On the LTC4412HV and the LTC4414, the limit is 40V.

The LTC4411 has a negative absolute maximum voltage of −0.3V. The forward voltage of the Zener diode may be too large to assure minimal current in the IC under the negative supply condition. If the current is too large, a Schottky diode can be placed in parallel with the Zener diode. The reverse breakdown of the Schottky must be greater than the Zener breakdown of 5V. The forward voltage of the Schottky is less than 0.3V assuring no excessive current in the IC. Again, the maximum negative voltage allowed is the maximum differential between IN and OUT or 6V.

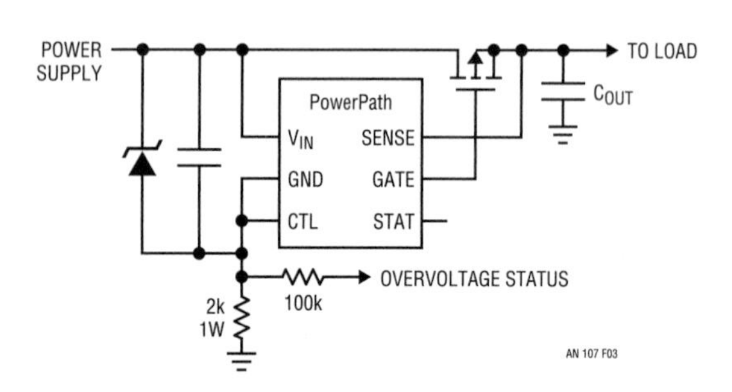

Figure 10.3 • Powerpath Circuit Capable of Operating with a Large Positive Input Supply

Conclusion

The circuit techniques presented here extend the supply voltage ranges of Linear Technology's PowerPath products, thereby extending their applicability beyond their data sheet voltage range.

PART 2

Data Conversion, Signal Conditioning and High Frequency/RF

Section 1

Data Conversion

Circuitry for single cell operation (11)

1.5V powered circuits for complex linear functions are detailed. Designs include a V/F converter, a 10-bit A/D, sample-hold amplifiers, a switching regulator and other circuits. Also included is a section of component considerations for 1.5V powered linear circuits.

Component and measurement advances ensure 16-bit DAC settling time (12)

DAC DC specifications are relatively easy to verify. AC specifications require more sophisticated approaches to produce reliable information. In particular, the settling time of the DAC and its output amplifier is extraordinarily difficult to determine to 16-bit resolution. This application note presents methods for 16-bit DAC settling time measurement and compares results. Appendices discuss oscilloscope overdrive, frequency compensation, circuit and optimization techniques, layout, power stages and an historical perspective on precision DACs.

Fidelity testing for A→D converters (13)

The ability to faithfully digitize a sine wave is a sensitive test of high resolution A→D converter fidelity. This test requires a sine wave generator with residual distortion products approaching one part-per-million. Additionally, a computer-based A→D output monitor is necessary to read and display converter output spectral components. Performing this testing at reasonable cost and complexity requires construction of its elements and performance verification prior to use.

11

Circuitry for single cell operation

Jim Williams

Portable, battery-powered operation of electronic apparatus has become increasingly desirable. Medical, remote data acquisition, power monitoring and other applications are good candidates for battery operation. In some circumstances, due to space, power or reliability considerations, it is preferable to operate the circuitry from a single 1.5V cell. Unfortunately, a 1.5V supply eliminates almost all linear ICs as design candidates. In fact, the LM10 op amp-reference and the LT®1017/LT1018 comparators are the only IC gain blocks fully specified for 1.5V operation. Further complications are presented by the 600mV drop of silicon transistors and diodes. This limitation consumes a substantial portion of available supply

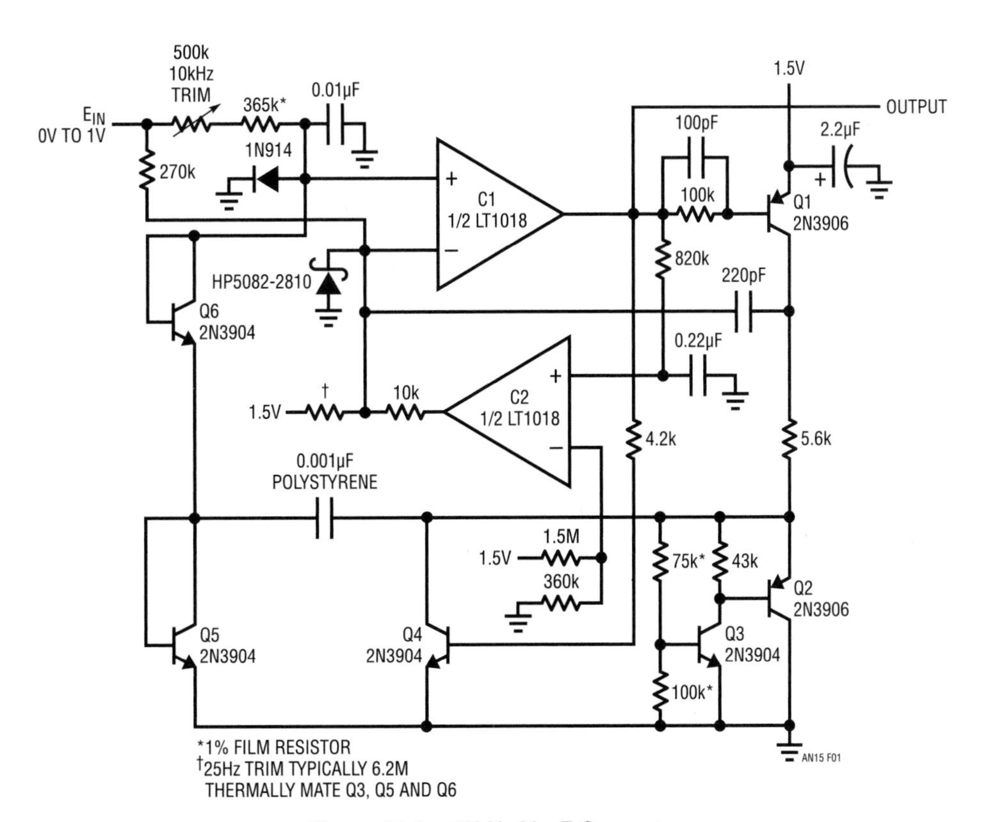

Figure 11.1 • 10kHz V→F Converter

Analog Circuit and System Design: Immersion in the Black Art of Analog Design. http://dx.doi.org/10.1016/B978-0-12-397888-2.00011-0

range, making circuit design difficult. Additionally, any circuit designed for 1.5V operation must function at end-of-life battery voltage, typically 1.3V. (See Box Section, "Components for 1.5V Operation.")

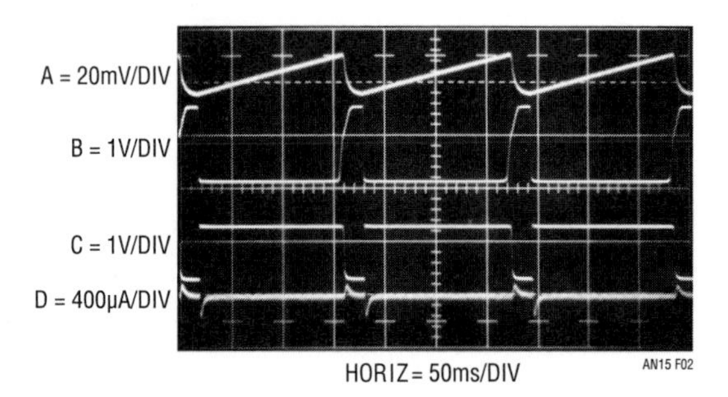

Figure 11.2 • V→F Operating Waveforms

These restrictions are painful, especially if complex linear circuit functions such as data converters and sample-holds are needed. Despite the problems, designing such circuits is possible by combining considerable attention to component characteristics with usual circuit methods.

10kHz V→F converter

Figure 11.1, an example of this approach, is a complete 1.5V powered 10kHz V→F converter. A 0V to 1V input produces a 25Hz to 10kHz output, with a transfer linearity of 0.35%. Gain drift is 250ppm/°C and current consumption about 205μA.

To understand circuit operation, assume C1's positive input is slightly below its negative input (C2's output is low). The input voltage causes a positive going ramp at C1's positive input (Trace A, Figure 11.2). C1's output (Trace B) is low, biasing Q1 on. Q1's collector current drives the Q2-Q3 combination, forcing Q2's emitter (Trace C) to clamp at 1V. The 0.001μF capacitor charges to ground (0.001μF unit's current waveform is Trace D) via Q5. When the ramp at C1's positive input goes high enough, C1's output goes high, cutting off Q1, Q2 and Q3. Q4 conducts, pulling current from C1's positive input capacitor via Q6. This current removal resets C1's positive input ramp to a potential slightly below ground, forcing C1's output low. The 100pF capacitor at Q1's collector furnishes AC positive feedback, ensuring that C1's output remains positive long enough for a complete discharge of the 0.001μF capacitor.

The Schottky diode prevents C1's input from being driven outside its negative common mode limit. This actioncuts off Q4, Q1-Q3 come on and the entire cycle repeats. The oscillation frequency directly depends on

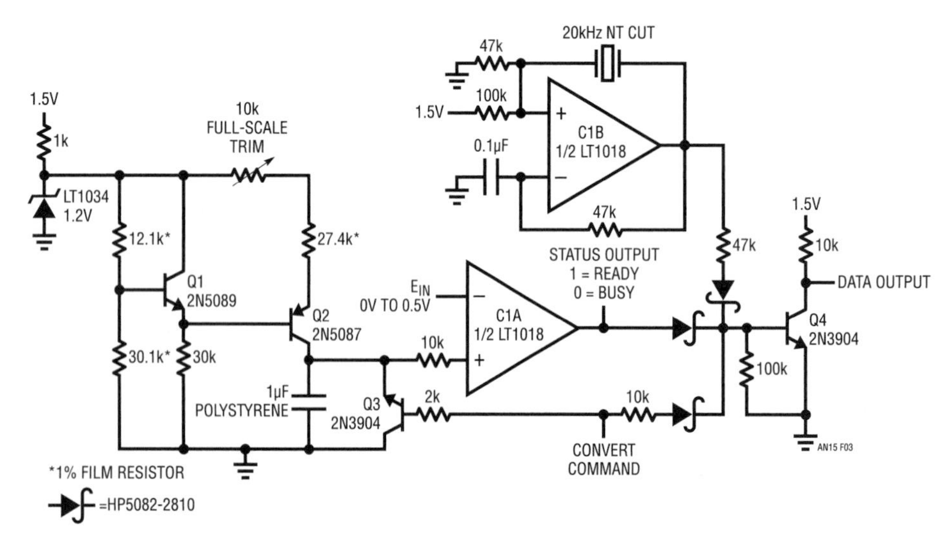

Figure 11.3 • 10-Bit A/D Converter

the input voltage derived current. The temperature coefficient of the Q2-Q3 1V clamp is largely compensated by the junction tempcos of Q5 and Q6, minimizing overall temperature drift. The 270k resistor path provides an input voltage derived trip point for C1, enhancing circuit linearity performance. This resistor should be selected to achieve the quoted linearity.

Circuit start-up or overdrive can cause the circuit's AC-coupled feedback loop to latch. If this occurs, C1's output goes high. C2 detects this, via the 820k-0.22µF lag, and also goes high, lifting C1's negative input towards 1.5V. Because C1's positive input is diode clamped at 600mV, its output switches low, initiating normal circuit behavior.

To calibrate this circuit, select the 100k value for V_{CLAMP} = 1V. Next, apply 2.5mV at the input and select the resistor value indicated at C1's input for a 25Hz output. Then, put in exactly 1V and trim the 500kΩ potentiometer for 10kHz output.

10-bit A/D converter

Figure 11.3 is another data converter circuit. This integrating A/D converter has a 60ms conversion time, consumes 460µA from its 1.5V supply and maintains 10-bit accuracy over a 15°C to 35°C temperature range.

A pulse applied to the convert command line (Trace A, Figure 11.4) causes Q3, operating in inverted mode, to discharge the 1µF capacitor (Trace B). Simultaneously, Q4 is biased through the 10k diode path, forcing its collector (Trace D) low. Q3's inverted mode switching results in a capacitor discharge within 1mV of ground. When the convert command falls low, Q3 goes off, Q4's collector lifts, and the LT1004 stabilized Q1-Q2 current source charges the 1µF unit with a linear ramp. During the time the ramps value is below the input voltage, C1A's output is low (Trace C). This allows pulses from C1B, a quartz stabilized oscillator, to modulate Q4. Output data appears

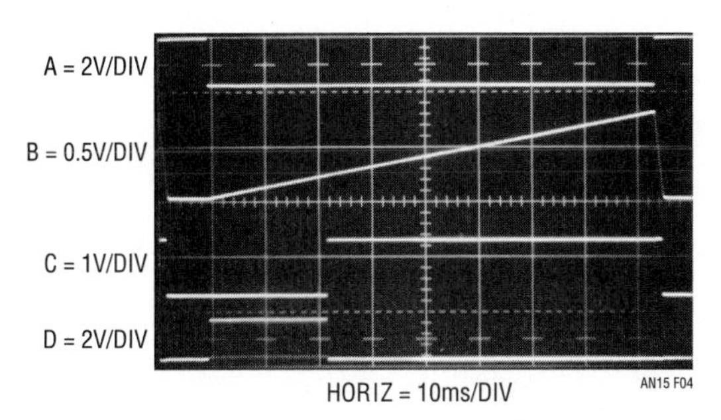

Figure 11.4 • A/D Converter Waveforms

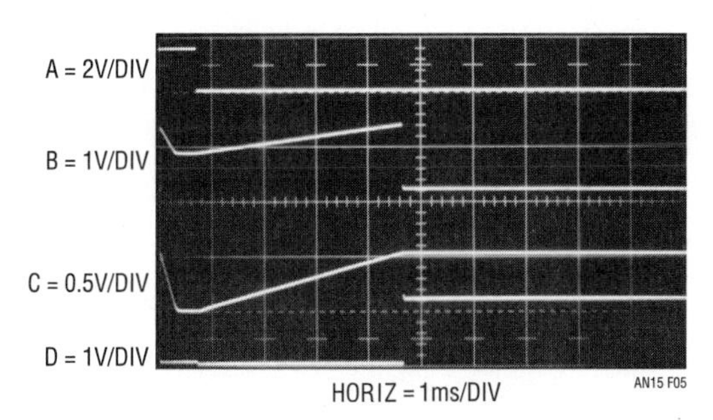

Figure 11.5 • Sample-Hold Waveforms

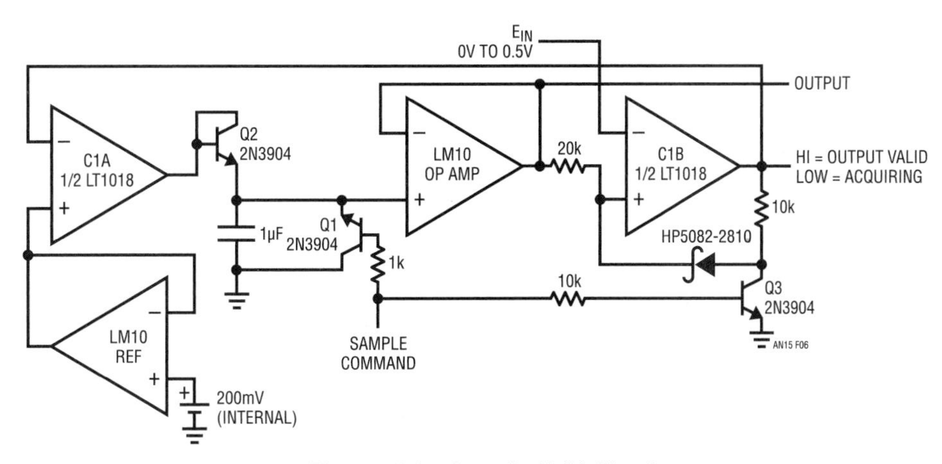

Figure 11.6 • Sample-Hold Circuit

at Q4's collector (Trace D). When the ramp crosses the input voltages value C1A's output goes high, biasing Q4 and output data ceases. The number of pulses appearing at the output is directly proportional to the input voltage. To calibrate this circuit, apply 0.5000V to the input and trim the 10k potentiometer for exactly 1000 pulses out each time the convert command line is pulsed. No zero trim is required, although Q3's inverted 1mV saturation voltage limits zero resolution to 2 LSBs.

Sample-hold amplifier

A logical companion to the A/D converter described is a sample-hold amplifier. A sample-hold is one of the most difficult circuits to design for 1.5V operation, primarily because FET switches with low enough pinch-off voltages are not available. Two methods are presented here. The first circuit gets around the switch problem with an approach that eliminates the switch. Although an unusual way to implement a sample-hold, it requires no special components or trimming, is easy to build and has a 4ms acquisition time to 0.1%. The second circuit, a more conventional design, requires specially selected and matched components and is more complex, but offers 125µs (0.1%) acquisition time—a 30× improvement over the other design.

When a sample command (Figure 11.5, Trace A) is applied to the circuit of Figure 11.6, Q1, operating in inverting mode, discharges the 1µF capacitor (Trace C). When the sample command falls, Q1 goes off and C1A's internal output pull-up current source (Trace B) charges the capacitor via Q2, connected as a low leakage diode. The capacitors charging ramp is followed by the LM10, which biases C1B's positive input. When the ramp potential crosses the circuit's input voltage, applied to C1B's negative input, C1B's output goes high (Trace D).

This forces C1A's output low, and the 1µF capacitor stops charging. Under these conditions, the circuit is in the "hold" mode. The voltage the capacitor sits at is the same as the input voltage, and the circuit output is taken at the LM10. The 10k diode path at C1B provides a latch, preventing input voltage changes or noise from affecting the value stored in the 1µF capacitor. When the next sample command is received, Q3 breaks the latch and circuit action repeats.

Acquisition time is directly proportional to input value, with 4ms required for full-scale (0.5V). Although faster acquisition is possible, the delay in shutting off C1A's output will degrade accuracy. The circuits primary advantages are elimination of the FET switch requirement and relative simplicity. Accuracy is 0.1%, droop rate specs at 10µV/ms and current consumption is 350µA.

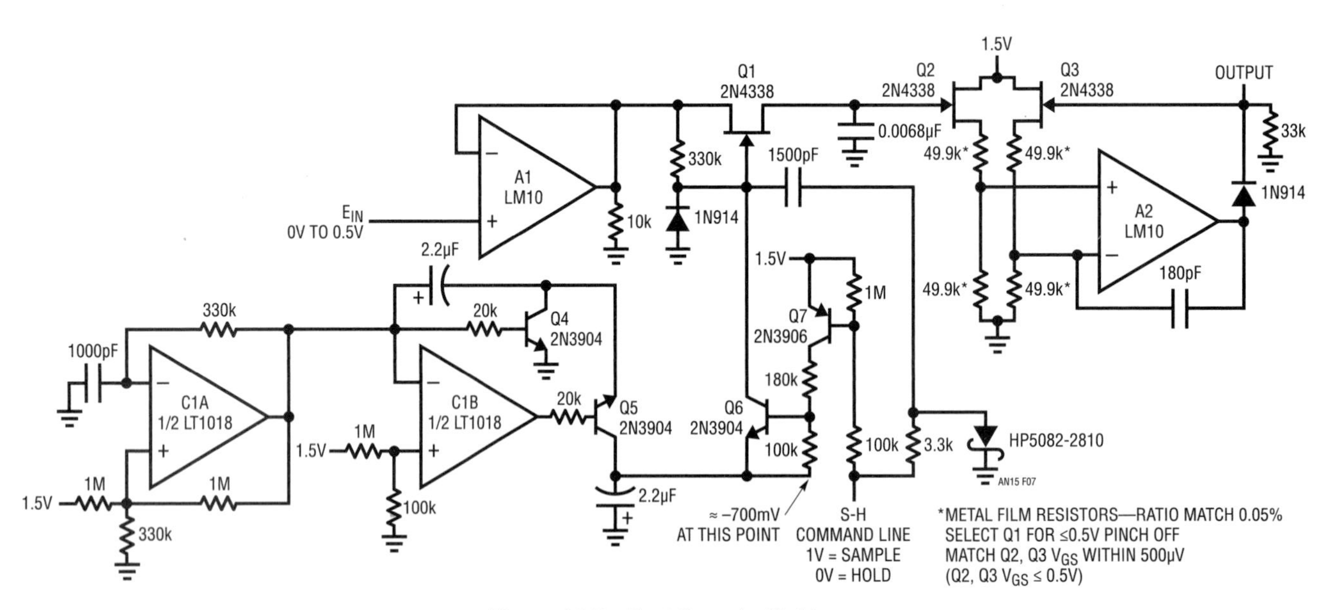

Figure 11.7 • Fast Sample-Hold

Fast sample-hold amplifier

Figure 11.7, a more conventional approach to sample-hold, is significantly faster, but also more complex and has special construction requirements. Q1 serves as the sample-hold switch, with Q6 and Q7 providing a level shift to drive the gate. To minimize power consumption, a 1500pF feedforward path is used for fast gate switching without resorting to high operating currents in Q6 and Q7. C1A, a simple square wave oscillator, drives Q4. C1B inverts C1A's output and biases Q5. The transistors serve as synchronous switches and charge is pumped to the 2.2μF capacitor at Q5's collector, resulting in a negative potential there.

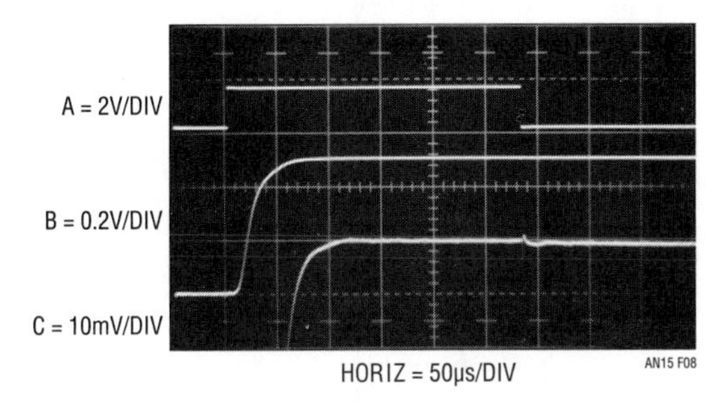

Figure 11.8 • Fast Sample-Hold Waveforms

Q1's low pinch-off voltage is obtained at the expense of on-resistance. The typical R_{ON} of 1.5k to 2k means the circuit's hold capacitor must be small if fast acquisition is desired. This mandates a low bias current output amplifier, or droop rate will suffer. Q2, Q3 and A2 meet this need. Q2 and Q3 are set up as source followers, with the resistors used as level shifters to keep A2's inputs inside the LM10's common mode range. A2's output diode ensures clean dynamic performance for voltages close to zero by setting the LM10's output bias point well above ground. The 180pF capacitor compensates the composite amplifier.

Several special considerations are required to use this circuit. Q1, an extremely low pinch-off device, must be further selected for a pinch-off below 500mV to enable proper turn-off. Also, any V_{GS} mismatch between Q2 and Q3 will contribute offset error, and these devices must be selected for V_{GS} matching within 500μV. Additionally, the Q2-Q3 V_{GS} absolute value must be inside 500mV or A2 may encounter common mode limitations for circuit inputs near full-scale.

Finally, mismatches in the resistor level shift contribute a gain error. To hold 0.1% circuit accuracy, the resistors should be ratio matched with 0.05%.

Once these special provisions have been attended to, the circuit delivers excellent specifications for a 1.5V powered sample-hold. Acquisition time is 125μs to 0.1% with a droop rate of 10μV/ms. Current consumption is inside 700μA.

Figure 11.8 shows the circuit acquiring a full-scale input. Trace A is the sample-hold command, while Trace B is the circuit's output. Trace C, an amplitude expanded version of B, shows acquisition detail. The input is acquired within 125μs and sample-to-hold offset is within a millivolt.

Temperature compensated crystal clock

Many systems require a stable clock source and crystal oscillators which run from 1.5V are relatively easy to construct. However, if good stability over temperature is required, things become more difficult. Ovenizing the crystal is one approach, but power consumption is excessive. An alternate method provides open loop, frequency correcting bias to the oscillator. The bias value is determined by absolute temperature. In this fashion, the oscillator's thermal drift, which is repeatable, is corrected. The simplest way to do this is by slightly varying the crystal's resonance point with a variable shunt or series impedance. Varactor diodes, the capacitance of which varies with reverse voltage, are commonly employed for this purpose. Unfortunately, these diodes require volts of reverse bias to generate significant capacitance shift, making direct 1.5V powered operation impossible.

Figure 11.9's circuit accomplishes the temperature compensation function. The transistor and associated components form a Colpitts class oscillator which runs directly from the 1.5V supply. The varactor diode, in series with the crystal, tunes oscillator frequency as its DC bias varies. An ambient temperature dependent DC bias is generated by the remaining circuitry.

The thermistor network and the LM10 amplifier are arranged to produce a temperature dependent signal which corrects the thermal drift of the crystal type specified. Normally, the 1.5V powered LM10 could not provide the output levels required to bias the varactor. Here, however, a self-exciting switching up-converter (T1 and associated components) is included in the LM10's feedback loop. The LM10 drives the switching converter's input to generate whatever output voltage is required to close the loop. The thermistor-bridge network and amplifier feedback resistor values are scaled to produce appropriate temperature dependent varactor bias. The LM10's reference portion stabilizes the temperature network against 1.5V supply variations. The 100pF positive feedback forces the LM10's output into switched mode operation, conserving power.

Figure 11.10 plots compensated versus uncompensated oscillator drift. The compensation improves drift performance by more than a factor of ten. The residual aberrance

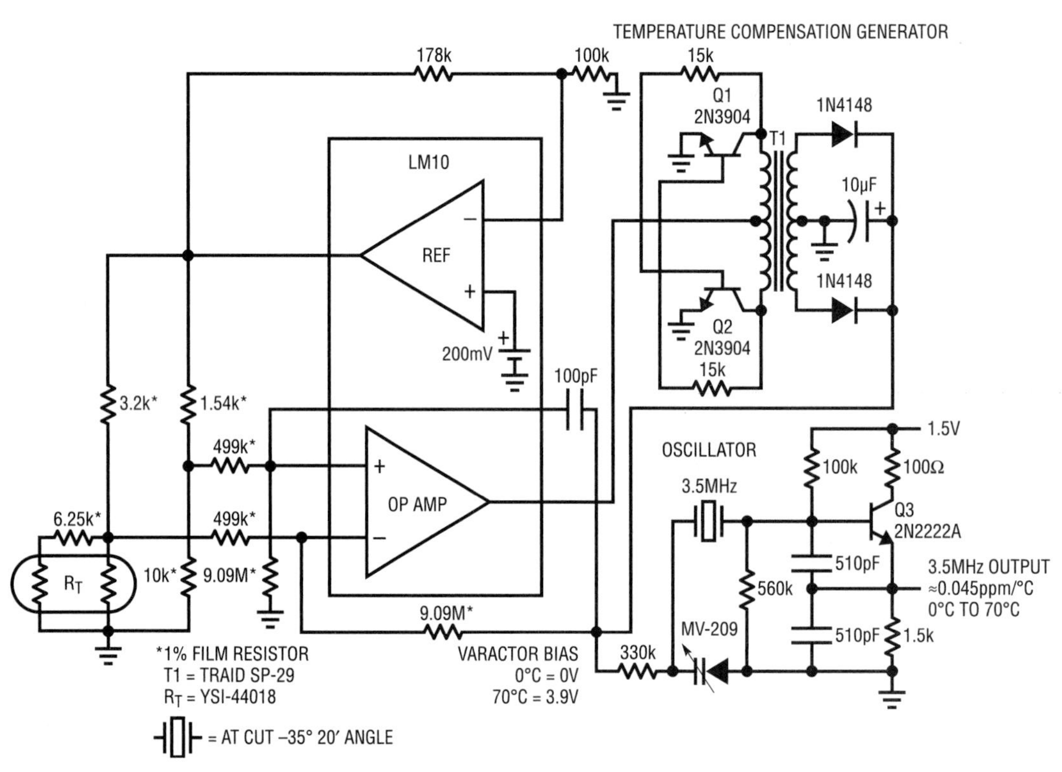

Figure 11.9 • Temperature Compensated Crystal Oscillator

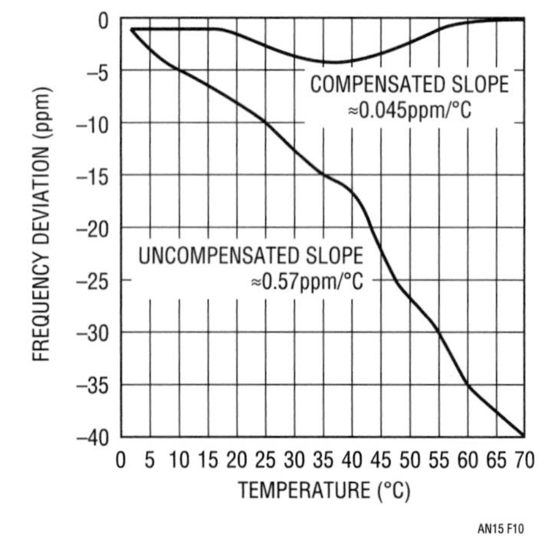

Figure 11.10 • Compensated vs Uncompensated Oscillator Results

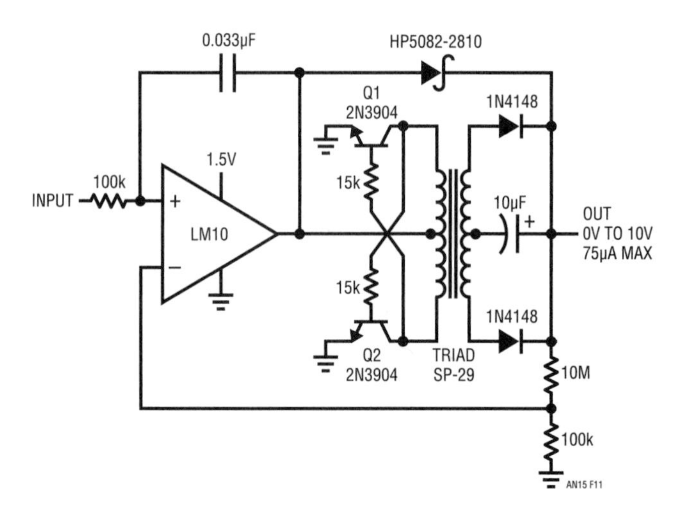

Figure 11.11 • Voltage-Boosted Output Op Amp

in the compensated curve is due to the first-order linear correction used. Current consumption is inside 850μA.

Voltage boosted output amplifier

In many circumstances, it is desirable to have 1.5V powered circuitry interface to higher voltage systems. The most obvious example is 1.5V driven, remote data acquisition apparatus which feeds a line-powered data gathering point. Although the battery-powered portion may locally process signals with 1.5V circuitry, it is useful to address the monitoring high level instrumentation at high voltage. Figure 11.11's design borrows from the method used in Figure 11.9 to generate high voltage outputs. This 1.5V powered amplifier provides 0V to 10V outputs at up to 75μA capacity. The LM10 drives the self-exciting up-converter with whatever energy is required to close the feedback loop. In this case, the amplifier is set up with a gain of 101, although other gains are easily realized. The sole restriction is that the 1.5V powered LM10's common mode input range not be exceeded. The Schottky diode bypasses the up-converter for low voltage outputs, aiding output noise performance. Overlap between the up-converter's turn-on threshold and the diode forward breakdown ensures clean dynamic behavior at the transition point. To increase efficiency the 0.033μF capacitor provides AC positive feedback, forcing the LM10 output to pulse-width modulate the up-converter.

Figure 11.12 details operation. The circuit's output (Trace A) decays until the LM10 switches (Trace B), starting the up converter. The two transistors alternately drive the transformer (transistor collectors are Traces C and D) until the output voltage rises high enough to shut off the LM10 output. This sequence repeats, with repetition rate dependent upon output voltage and loading conditions.

5V output switching regulator

No commercially available logic, processor or memory family will operate from 1.5V. Many of the circuits described previously normally work in logic-driven systems. Because of this, a way to permit use of standard logic functions from a 1.5V battery is necessary. The simplest way to do this is a switching regulator specifically designed for 1.5V input operation.

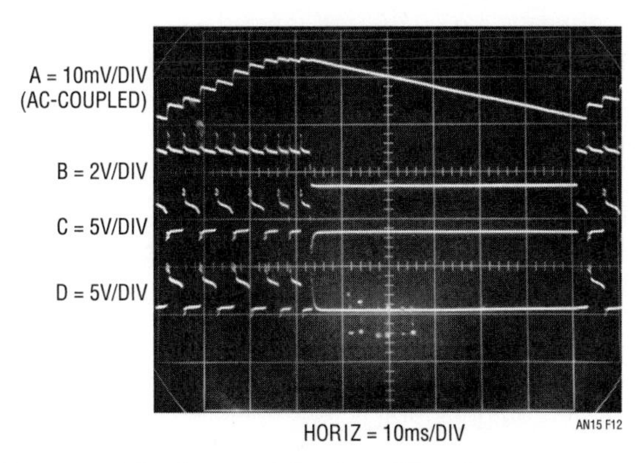

Figure 11.12 • Boosted Op Amp Waveforms

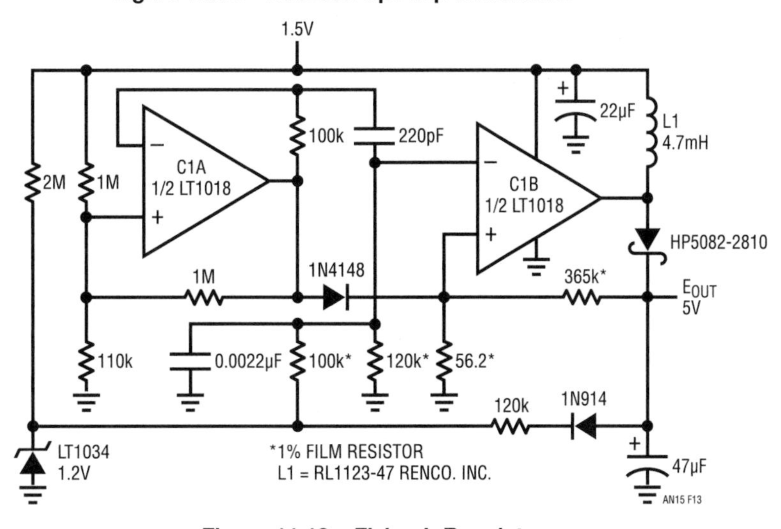

Figure 11.13 • Flyback Regulator

Figure 11.13's flyback configuration, a variant of a design by R. J. Widlar, gives a 5V output. C1A serves as an oscillator, providing a ramp (Trace A, Figure 11.14) at C1B's DC-biased negative input. C1B compares a divided version of the output to a reference point derived from the LT1034. The ramp signal, summed with the reference point, causes C1B's output to width modulate (Trace B). During the time C1B is low, current builds in its output inductor (Trace C). When the ramp at C1B goes low enough, C1B's output goes high, and the inductor discharges into the 47µF capacitor. The diode from C1A's output to C1B's positive input supplies a pulse (Trace D) on each oscillator cycle, ensuring loop start-up. The 120kΩ diode path from the output bootstraps LT1034 bias, aiding overall regulation.

Figure 11.15 plots regulator efficiency. Small loads produce lowest efficiency because of fixed losses in the regulator, although 80% efficiency is achieved above 150µA.

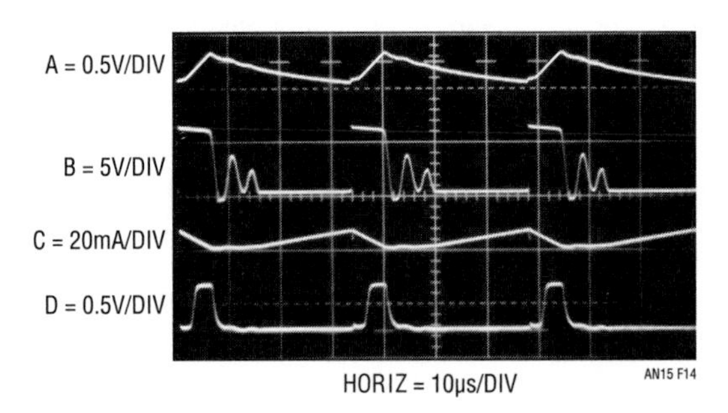

Figure 11.14 • Flyback Regulator Waveforms

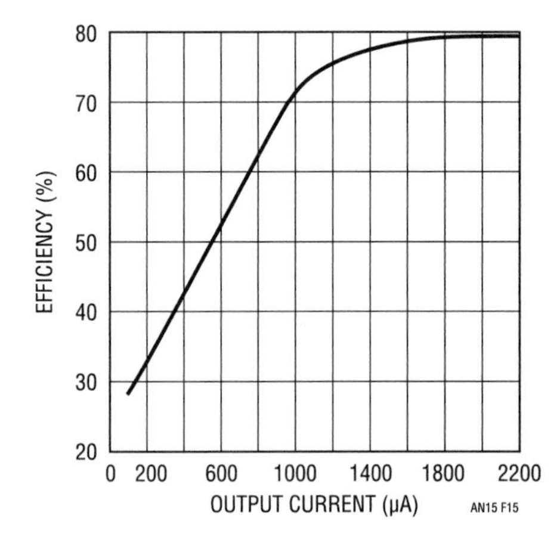

Figure 11.15 • Flyback Regulator Efficiency

Components for 1.5V operation

Almost all commercially available linear ICs are not capable of 1.5V operation. Two that are capable include the LM10 and LT1017/LT1018. The LM10 op amp-reference runs as low as 1.1V; the LT1017/LT1018 comparator goes down to 1.2V. The LM10 provides good DC input characteristics, although speed is limited to 0.1V/µs. The LT1017/LT1018 comparator series features microsecond range response time, high gain and good DC characteristics. Both devices feature low power consumption. The LT1004 and LT1034 voltage references feature 20µA operating currents and 1.2V operation.

Standard PN junction diodes have a 600mV drop, a substantial percentage of available supply range. At currents below 10µA to 20µA, this figure reduces to about 450mV. Schottky diodes typically exhibit only 300mV drop, although reverse leakage is higher than standard diodes. Germanium diodes are lowest, with 150mV to 200mV drop, even at relatively high currents. Often, though, the significant reverse leakage of Germanium precludes its use.

Standard silicon transistors have a 600mV V_{BE}, although this figure comes down somewhat at very low base currents. The V_{CE} saturation of silicon transistors is well below 100mV at reasonable currents, and judicious device selection and use can reduce this figure below 25mV. Inverted mode operation allows V_{CE} saturation losses below 1mV, although beta is often below 0.1, necessitating substantial base drive. Germanium transistors have 2 to 3 times lower V_{BE} and V_{CE} losses, although speed, leakage and beta are generally not as good as silicon types.

Perhaps the most important component is the battery. Many types of cells are available, and the best choice varies with the application. Two common types are Carbon-Zinc and Mercury. Carbon-Zinc offers higher initial voltage, but Mercury units have a much flatter discharge curve (e.g., better supply regulation) if currents are controlled (see Figure 11.16).

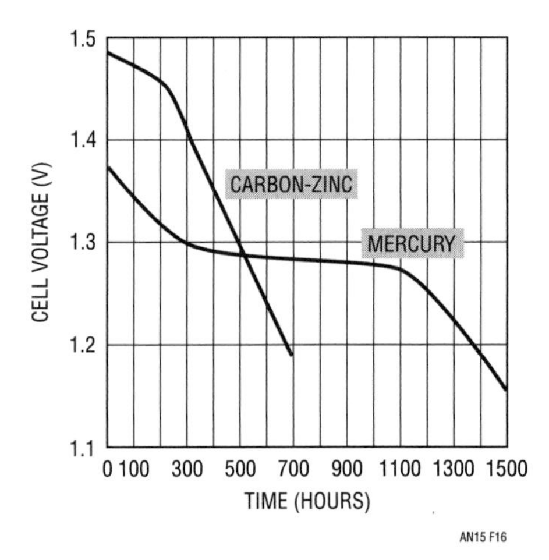

Figure 11.16 • Typical Discharge Curves of Similar Size (AA) Mercury and Carbon Zinc Cells (1mA Load)

Component and measurement advances ensure 16-bit DAC settling time

The art of timely accuracy

12

Jim Williams

Introduction

Instrumentation, waveform generation, data acquisition, feedback control systems and other application areas are beginning to utilize 16-bit data converters. More specifically, 16-bit digital-to-analog converters (DACs) have seen increasing use. New components (see Components for 16-Bit Digital-to-Analog Conversion, page 2) have made 16-bit DACs a practical design alternative.[1] These ICs provide 16-bit performance at reasonable cost compared to previous modular and hybrid technologies. The DC and AC specifications of the monolithic DAC's approach or equal previous converters at significantly lower cost.

DAC settling time

DAC DC specifications are relatively easy to verify. Measurement techniques are well understood, albeit often tedious. AC specifications require more sophisticated approaches to produce reliable information. In particular, the settling time of the DAC and its output amplifier is extraordinarily difficult to determine to 16-bit resolution. DAC settling time is the elapsed time from input code application until the output arrives at and remains within a specified error band around the final value. It is usually specified for a full-scale 10V transition. Figure 12.1 shows that DAC settling time has three distinct components. The *delay time* is very small and is almost entirely due to propagation delay through the DAC and output amplifier: During this interval there is no output movement.

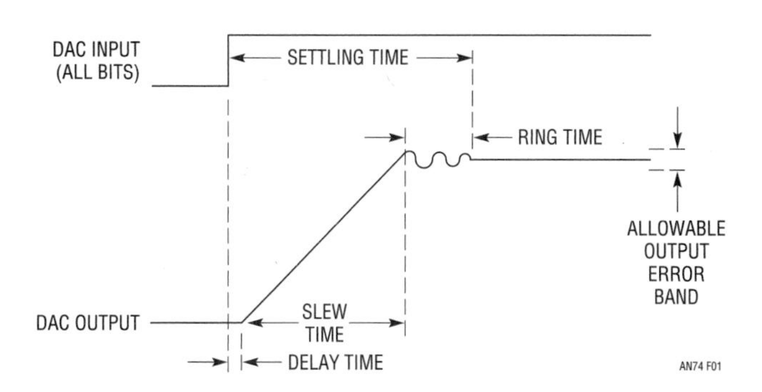

Figure 12.1 • DAC Settling Time Components Include Delay, Slew and Ring Times. Fast Amplifiers Reduce Slew Time, Although Longer Ring Time Usually Results. Delay Time is Normally a Small Term

During *slew time* the output amplifier moves at its highest possible speed towards the final value. *Ring time* defines the region where the amplifier recovers from slewing and ceases movement within some defined error band. There is normally a trade-off between slew and ring time. Fast slewing amplifiers generally have extended ring times, complicating amplifier choice and frequency compensation. Additionally, the architecture of very fast amplifiers usually dictates trade-offs which degrade DC error terms.[2]

Measuring anything at any speed to 16 bits ($\approx$0.0015%) is hard. Dynamic measurement to 16-bit resolution is particularly challenging. Reliable 16-bit settling time measurement constitutes a high order difficulty problem requiring exceptional care in approach and experimental technique.

Note 1: See Appendix A, "A History of High Accuracy Digital-to-Analog Conversion".

Note 2: This issue is treated in detail in latter portions of the text. Also see Appendix D "Practical Considerations for DAC-Amplifier Compensation.

Analog Circuit and System Design: Immersion in the Black Art of Analog Design. http://dx.doi.org/10.1016/B978-0-12-397888-2.00012-2

Components for 16-bit D/A conversion

Components suitable for 16-bit D/A conversion are members of an elite class. 16 binary bits is one part in 65,536—just 0.0015% or 15 parts-per-million. This mandates a vanishingly small error budget and the demands on components are high. The digital-to-analog converters listed in the chart all use Si-Chrome thin-film resistors for high stability and linearity over temperature. Gain drift is typically 1ppm/°C or about 2LSBs over 0°C to 70°C. The amplifiers shown contribute less than 1LSB error over 0°C to 70°C with 16-bit DAC driven settling times of 1.7µs available. The references offer drifts as low as 1LSB over 0°C to 70°C with initial trimmed accuracy to 0.05%

Short Form Descriptions of Components Suitable for 16-Bit Digital-to-Analog Conversion

Component type	Error contribution over 0°C to 70°C	Comments
LTC®1597 DAC	≈2LSB Gain Drift 1LSB Linearity	Full Parallel Inputs Current Output
LTC1595 DAC	≈2LSB Gain Drift 1LSB Linearity	Serial Input 8-Pln Package Current Output
LTC1650 DAC	≈3.5LSB Gain Drift 6LSB Offset 4LSB Linearity	Complete Voltage Output DAC
LT®1001 Amplifier	<1LSB	Good Low Speed Choice 10mA Output Capability
LT1012 Amplifier	<1LSB	Good Low Speed Choice Low Power Consumption
LT1468 Amplifier	<2LSB	1.7µs Settling to 16 Bits Fastest Available
LM199A Reference-6.95V	≈1LSB	Lowest Drift Reference in This Group
LT1021 Reference-10V	≈4LSB	Good General Purpose Choice
LT1027 Reference-5V	≈4LSB	Good General Purpose Choice
LT1236 Reference-10V	≈10LSB	Trimmed to 0.05% Absolute Accuracy
LT1461 Reference-4.096V	≈10LSB	Recommended for LTC1650 DACs (see Above)

Considerations for measuring DAC settling time

Historically, DAC settling time has been measured with circuits similar to that in Figure 12.2. The circuit uses the "false sum node" technique. The resistors and DAC-amplifier form a bridge type network. Assuming ideal resistors, the amplifier output will step to V_{IN} when the DAC inputs move to all ones. During slew, the settle node is bounded by the diodes, limiting voltage excursion. When settling occurs, the oscilloscope probe voltage should be zero. Note that the resistor divider's attenuation means the probe's output will be one-half of the actual settled voltage.

In theory, this circuit allows settling to be observed to small amplitudes. In practice, it cannot be relied upon to produce useful measurements. The oscilloscope connection presents problems. As probe capacitance rises, AC loading of the resistor junction influences observed settling waveforms. A 10pF probe alleviates this problem but its 10× attenuation sacrifices oscilloscope gain. 1× probes are not suitable because of their excessive input capacitance. An active 1× FET probe will work, but another issue remains.

The clamp diodes at the settle node are intended to reduce swing during amplifier slewing, preventing excessive oscilloscope overdrive. Unfortunately, oscilloscope overdrive recovery characteristics vary widely among different types and are not usually specified. The Schottky diodes' 400mV drop

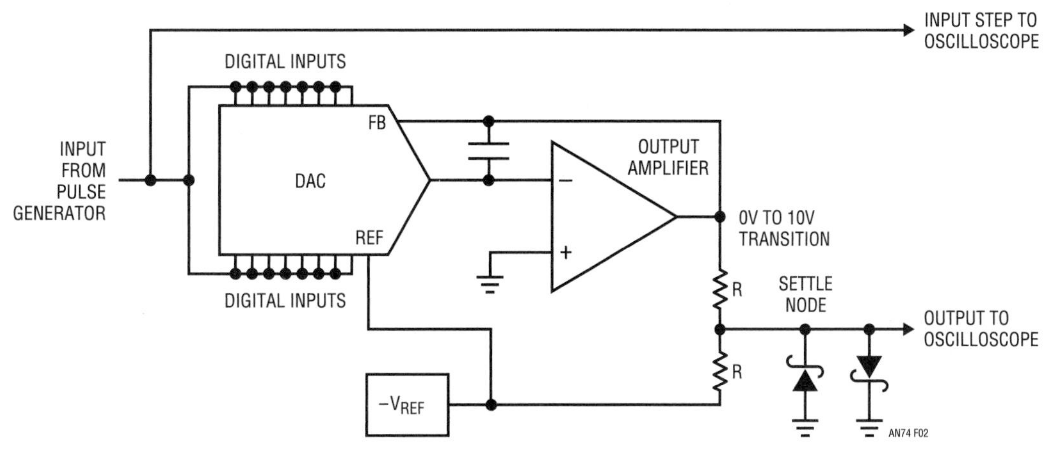

Figure 12.2 • Popular Summing Scheme for DAC Settling Time Measurement Provides Misleading Results. 16-Bit Measurement Causes >200× Oscilloscope Overdrive. Displayed Information is Meaningless

means the oscilloscope may see an unacceptable overload, bringing displayed results into question.[3]

At 10-bit resolution (10mV at the DAC output—5mV at the oscilloscope), the oscilloscope typically undergoes a 2× overdrive at 50mV/DIV and the desired 5mV baseline is just discernible. At 12-bit or higher resolution, the measurement becomes hopeless with this arrangement. Increasing oscilloscope gain brings commensurate increased vulnerability to overdrive induced errors. At 16 bits, there is clearly no chance of measurement integrity.

The preceding discussion indicates that measuring 16-bit settling time requires a high gain oscilloscope that is somehow immune to overdrive. The gain issue is addressable with an external wideband preamplifier that accurately amplifies the diode-clamped settle node. Getting around the overdrive problem is more difficult.

The only oscilloscope technology that offers inherent overdrive immunity is the classical sampling 'scope.[4] Unfortunately, these instruments are no longer manufactured (although still available on the secondary market). It is possible, however, to construct a circuit that borrows the overload advantages of classical sampling 'scope technology. Additionally, the circuit can be endowed with features particularly suited for measuring 16-bit DAC settling time.

Practical DAC settling time measurement

Figure 12.3 is a conceptual diagram of a 16-bit DAC settling time-measurement circuit. This figure shares attributes with Figure 12.2, although some new features appear. In this case, the preamplified oscilloscope is connected to the settle point

Note 3: For a discussion of oscilloscope overdrive considerations, see Appendix B, "Evaluating Oscilloscope Overdrive Performance".

Note 4: Classical sampling oscilloscopes should not be confused with modern era digital sampling 'scopes that have overdrive restrictions. See Appendix B, "Evaluating Oscilloscope Overload Performance" for comparisons of various type 'scopes with respect to overdrive. For detailed discussion of classical sampling 'scope operation see references 14 through 17 and 20 through 22. Reference 15 is noteworthy; it is the most clearly written, concise explanation of classical sampling instruments the author is aware of. A 12 page jewel.

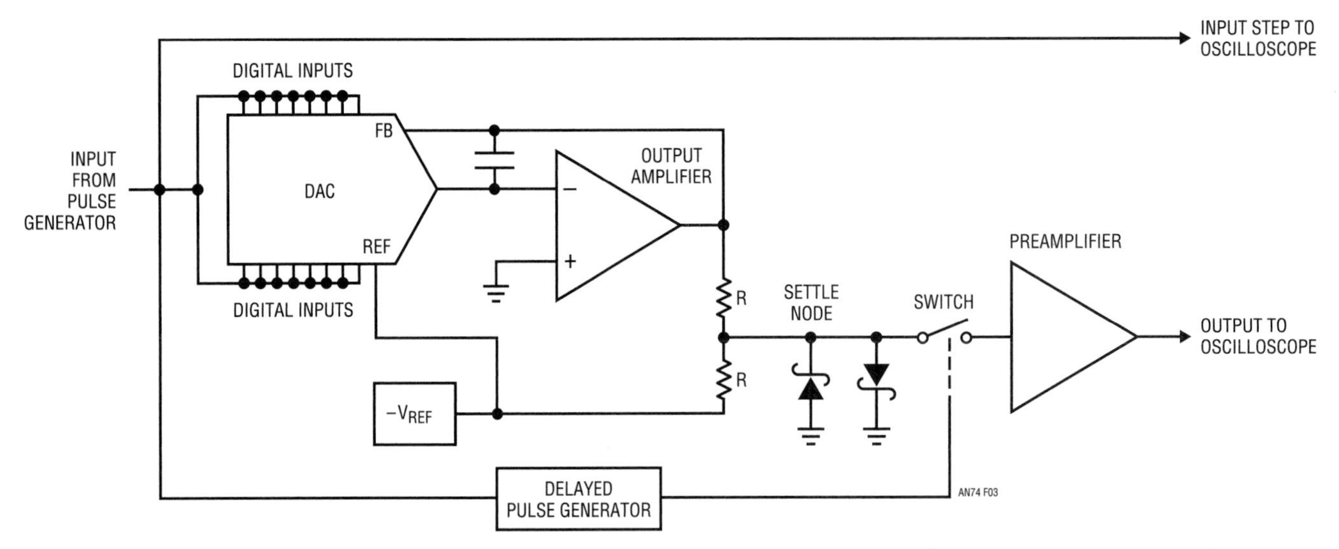

Figure 12.3 • Conceptual Arrangement Eliminates Oscilloscope Overdrive. Delayed Pulse Generator Controls Switch, Preventing Oscilloscope from Monitoring Settle Node Until Settling is Nearly Complete

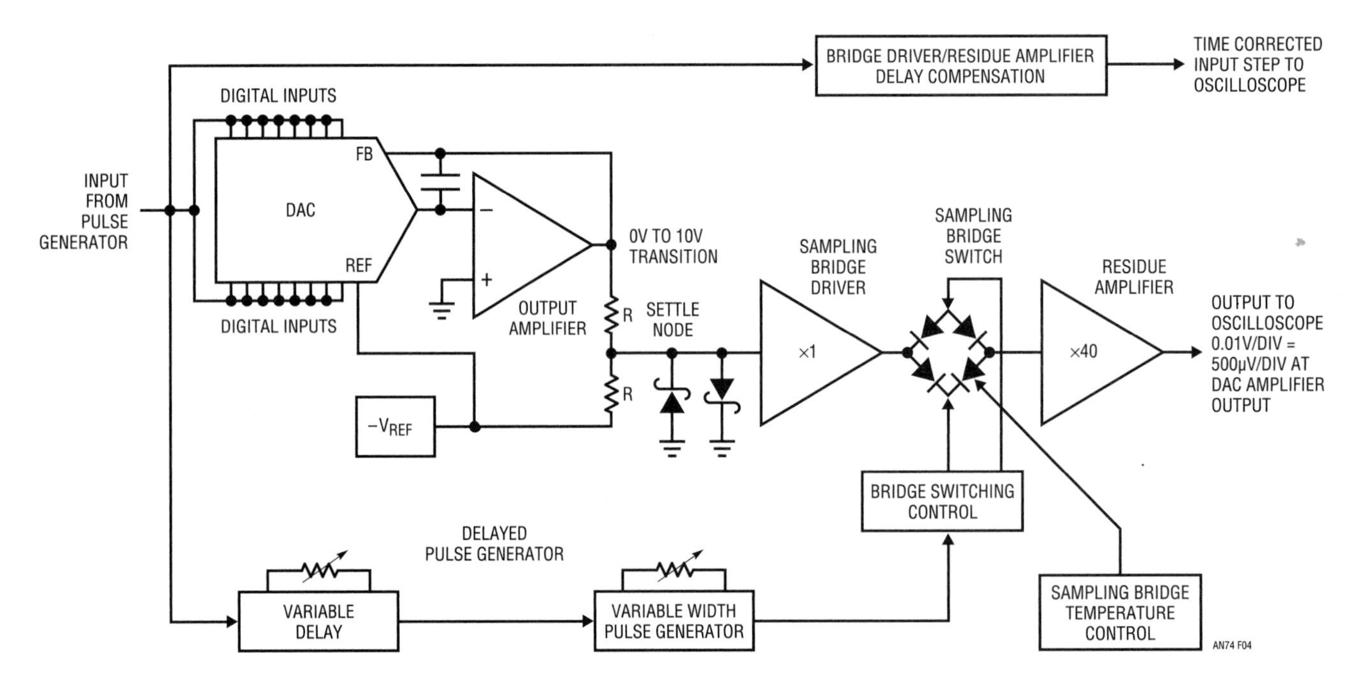

Figure 12.4 • Block Diagram of DAC Settling Time Measurement Scheme. Diode Bridge Switch Minimizes Switching Feedthrough, Preventing Residue Amplifier-Oscilloscope Overdrive. Temperature Control Maintains 10μV Switch Offset Baseline. Input Step Time Reference is Compensated for ×1 and ×40 Amplifier Delays

by a switch. The switch state is determined by a delayed pulse generator, which is triggered from the same pulse that controls the DAC. The delayed pulse generator's timing is arranged so that the switch does not close until settling is very nearly complete. In this way the incoming waveform is sampled in time, as well as amplitude. The oscilloscope is never subjected to overdrive—no off-screen activity ever occurs.

Figure 12.4 is a more complete representation of the DAC settling time scheme. Figure 12.3's blocks appear in greater detail and some new refinements show up. The DAC-amplifier summing area is unchanged. Figure 12.3's delayed pulse generator has been split into two blocks; a delay and a pulse generator, both independently variable. The input step to the oscilloscope runs through a section that compensates for the propagation delay of the settling-time-measurement path. The most striking new aspect of the diagram is the diode bridge switch. Borrowed from classical sampling oscilloscope circuitry, it is the key to the measurement. The diode bridge's inherent balance eliminates charge injection based errors in the output. It is far superior to other electronic switches in this characteristic. Any other high speed switch technology contributes excessive output spikes due to charge-based feedthrough. FET switches are not suitable because their gate-channel capacitance permits such feedthrough. This capacitance allows gate-drive artifacts to corrupt the oscilloscope display, inducing overload and defeating the switches purpose.

The diode bridge's balance, combined with matched, low capacitance monolithic diodes and complementary high speed switching, yields a cleanly switched output. The monolithic diode bridge is also temperature controlled, providing a bridge offset error below 10μV stabilizing the measurement baseline. The temperature control is implemented using uncommitted diodes in the monolithic array as heater and sensor.

Figure 12.5 details considerations for the diode bridge switch. The bridge diodes tend to cancel each other's temperature coefficient—unstabilized bridge drift is about 100μV/°C and the temperature control reduces residual drift to a few microvolts/°C.

Bridge temperature control is achieved by using one diode as a sensor. Another diode, running in reverse breakdown ($V_Z \approx 7V$), serves as the heater. The control amplifier, comparing the sensor diode to a voltage at it's negative terminal, drives the heater diode to temperature stabilize the array.

DC balance is achieved by trimming the bridge on-current for zero input-output offset voltage. Two AC trims are required. The "AC balance" corrects for diode and layout capacitive imbalances and the "skew compensation" corrects for any timing asymmetry in the nominally complementary bridge drive. These AC trims compensate small dynamic imbalances that could result in parasitic bridge outputs.

Detailed settling time circuitry

Figure 12.6 is a detailed schematic of 16-bit DAC settling-time-measurement circuitry. The input pulse switches all DAC bits simultaneously and is also routed to the oscilloscope via a delay-compensation network. The delay network, composed of CMOS inverters and an adjustable RC network, compensates the oscilloscope's input step signal for the 12ns delay through the circuit's measurement path.[5] The DAC amplifier's output is compared against the LT1236-10V reference via the precision 3k summing resistor ratio set. The LT1236 also furnishes the DAC reference, making the measurement ratiometric. The clamped settle node is unloaded by A1, which drives the sampling bridge. Note the additional clamp diodes at A1's output. These diodes prevent any possibility of abnormal A1 outputs (due to lost supply or supply sequencing anomalies) from damaging the diode array.[6] A3 and associated

Note 5: See Appendix C, "Measuring and Compensating Residue Amplifier Delay."

Note 6: This can and did happen. The author was unfit for human companionship upon discovering this mishap. Replacing the sampling bridge was a lengthy and highly emotionally charged task. To see why, refer to Appendix G, "Breadboarding, Layout and Connection Techniques."

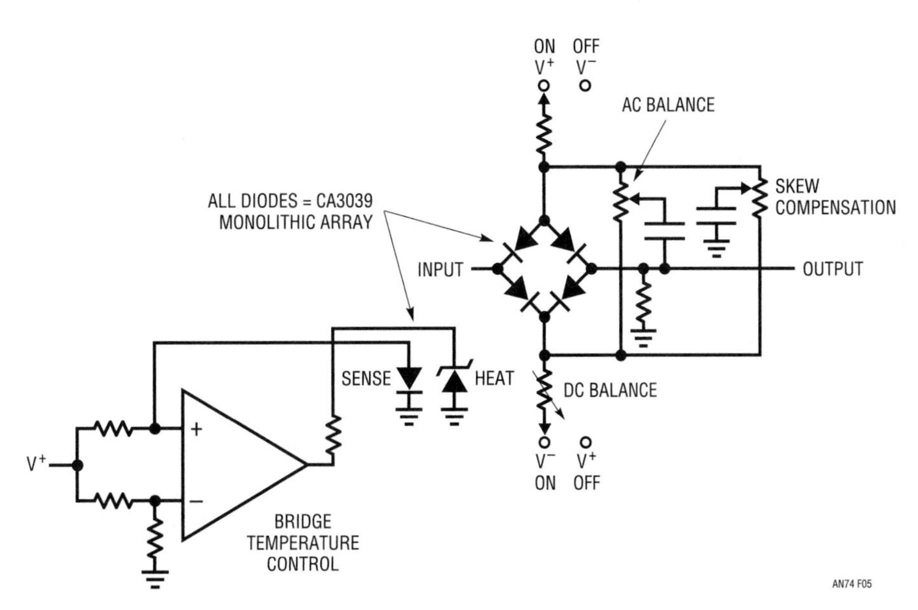

Figure 12.5 • Diode Bridge Switch Trims Include AC and DC Balance and Switch Drive Timing Skew. Remaining Diodes in Monolithic Array are Used for Temperature Control

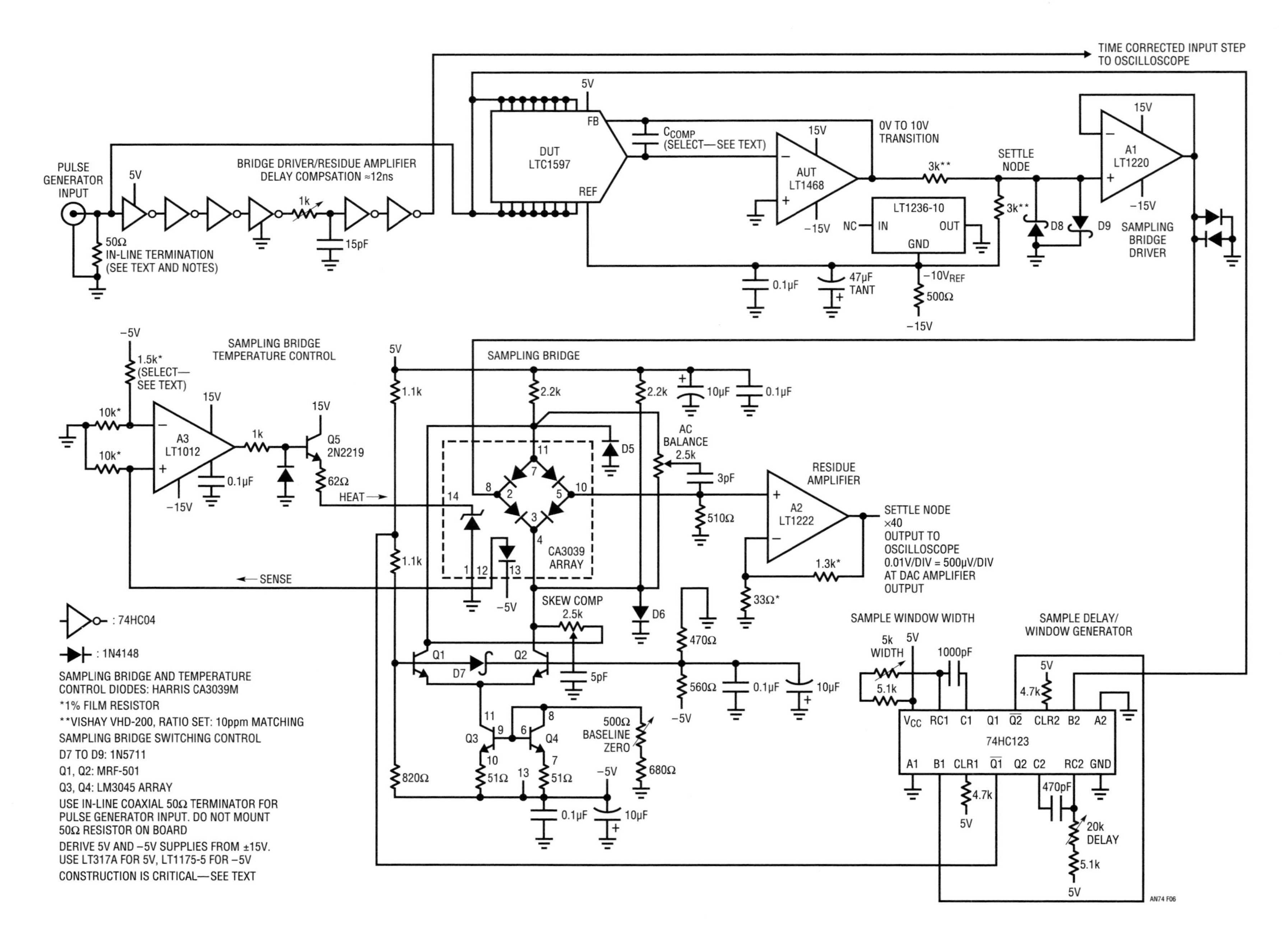

Figure 12.6 • Detailed Schematic of DAC Settling Time Measurement Circuit Closely Follows Block Diagram. Optimum Performance Requires Attention to Layout

components temperature control the sampling diode bridge by comparing a diodes's forward drop to a stable potential derived from the −5V regulator. Another diode, operated in the reverse direction ($V_Z \approx 7V$) serves as a chip heater. The pin connections shown on the schematic have been selected to provide best temperature control performance.

The input pulse triggers the 74HC123 one shot. The one shot is arranged to produce a delayed (controllable by the 20k potentiometer) pulse whose width (controllable by the 5k potentiometer) sets diode bridge on-time. If the delay is set appropriately, the oscilloscope will not see any input until settling is nearly complete, eliminating overdrive. The sample window width is adjusted so that all remaining settling activity is observable. In this way the oscilloscope's output is reliable and meaningful data may be taken. The one shot's output is level shifted by the Q1-Q4 transistors, providing complementary switching drive to the bridge. The actual switching transistors, Q1-Q2, are UHF types, permitting true differential bridge switching with less than 1ns of time skew.[7]

A2 monitors the bridge output, provides gain and drives the oscilloscope. Figure 12.7 shows circuit waveforms. Trace A is the input pulse, trace B the DAC amplifier output, trace C the sample gate and trace D the residue amplifier output. When the sample gate goes low, the bridge switches cleanly, and the last 1.5mV of slew are easily observed. Ring time is also clearly visible, and the amplifier settles nicely to final value. When the sample gate goes high, the bridge switches off, with only 600μV of feedthrough. The 100μV peak before bridge switching (at ≈3.5 vertical divisions) is feedthrough from A1's output, but it is similarly well controlled. Note that there is no off-screen activity at any time—the oscilloscope is never subjected to overdrive.

The circuit requires trimming to achieve this level of performance. The bridge temperature control point is set by grounding Q5's base prior to applying power. Next, apply power and measure A3's positive input with respect to the −5V rail. Select the indicated resistor (1.5k nominal) for a voltage at A3's negative input (again, with respect to −5V) that is 57mV below the positive input's value. Unground Q5's base and the circuit will control the sampling bridge to about 55°C:

$$25°C\ \text{room} + \frac{57\text{mV}}{1.9\text{mV}/°C\ \text{diode drop}} = 30°C\ \text{rise} = 55°C$$

The DC and AC bridge trims are made once the temperature control is functional. Making these adjustments requires disabling the DAC and amplifier (disconnect the input pulse from the DAC and set all DAC inputs low) and shorting the settle node directly to the ground plane. Figure 12.8 shows typical results before trimming. Trace A is the input pulse, trace B the sample gate and trace C the residue amplifier output. With the DAC-amplifier disabled and the settle node grounded, the residue amplifier output should (theoretically) always be zero. The photo shows this is not the case for an untrimmed bridge. AC and DC errors are present. The sample gate's transitions cause large, off-screen residue amplifier swings (note residue amplifier's response to the sample gate's turn-off

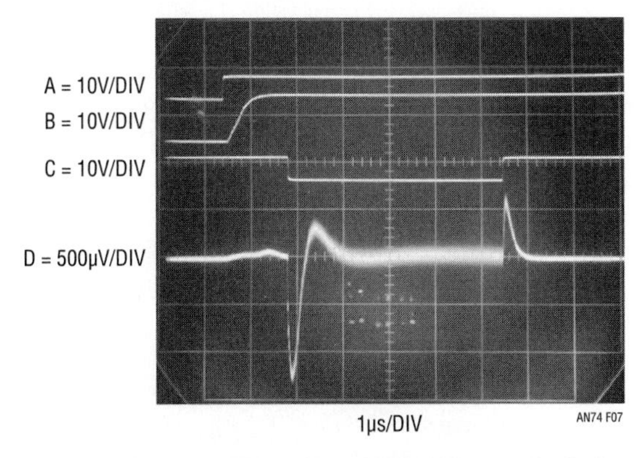

Figure 12.7 • Settling Time Circuit Waveforms Include Time Corrected Input Pulse (Trace A), DAC Amplifier Output (Trace B), Sample Gate (Trace C) and Settling Time Output (Trace D). Sample Gate Window's Delay and Width are Variable

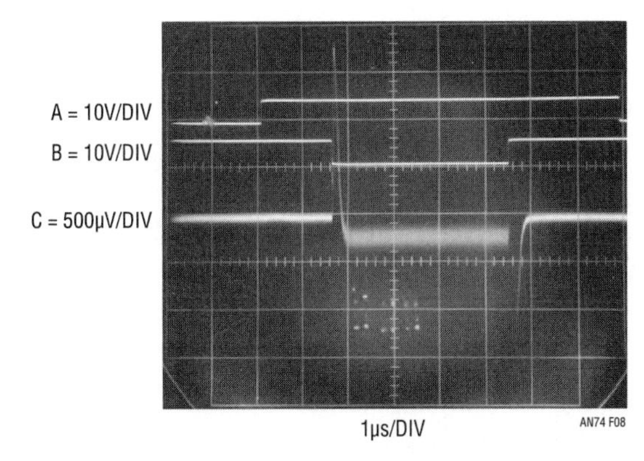

Figure 12.8 • Settling Time Circuit's Output (Trace C) with Unadjusted Sampling Bridge AC and DC Trims. DAC is Disabled and Settle Node Grounded for This Test. Excessive Switch Drive Feedthrough and Baseline Offset are Present. Traces A and B are Input Pulse and Sample Window, Respectively

Note 7: The bridge switching scheme was developed at LTC by George Feliz.

at the ≈8.5 vertical division). Additionally, the residue amplifier output shows significant DC offset error during the sampling interval. Adjusting the AC balance and skew compensation minimizes the switching induced transients. The DC offset is adjusted out with the baseline zero trim. Figure 12.9 shows the results after making these adjustments. All switching related activity is now well on-screen

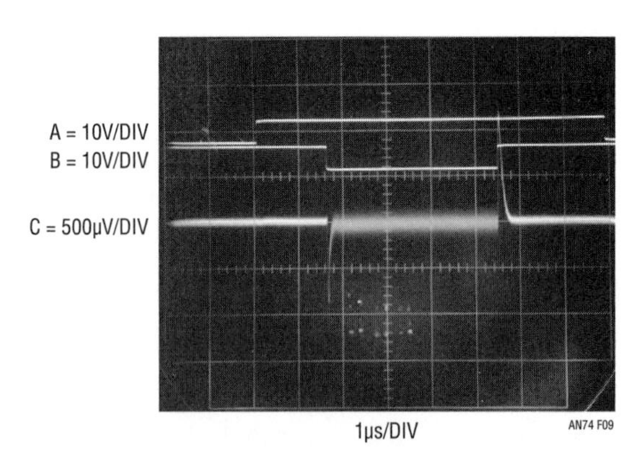

Figure 12.9 • Settling Time Circuit's Output (Trace C) with Sampling Bridge Trimmed. As in Figure 12.8, DAC is Disabled and Settle Node Grounded for This Test. Switch Drive Feedthrough and Baseline Offset are Minimized. Traces A and B are Input Pulse and Sampling Gate, Respectively

and offset error reduced to unreadable levels. Once this level of performance has been achieved, the circuit is ready for use.[8] Unground the settle node and restore the input pulse connection to the DAC.

Using the sampling-based settling time circuit

Figure 12.10 through 12.12 underscore the importance of positioning the sampling window properly in time. In Figure 12.10 the sample gate delay initiates the sample window (trace A) too early and the residue amplifier's output (trace B) overdrives the oscilloscope when sampling commences. Figure 12.11 is better, with only slight off-screen activity. Figure 12.12 is optimal. All amplifier residue is well inside the screen boundaries.

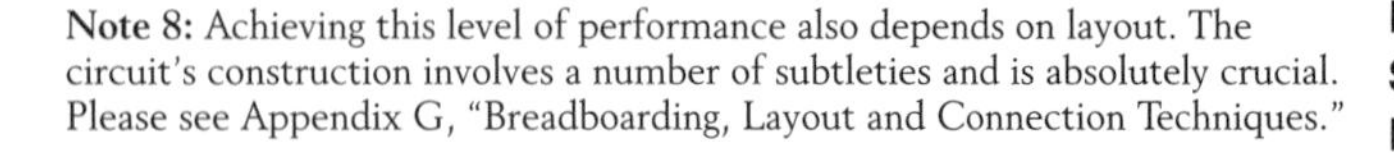

Note 8: Achieving this level of performance also depends on layout. The circuit's construction involves a number of subtleties and is absolutely crucial. Please see Appendix G, "Breadboarding, Layout and Connection Techniques."

In general, it is good practice to "walk" the sampling window up to the last millivolt or so of amplifier slewing so that the onset of ring time is observable. The sampling based approach provides this capability and it is a very powerful measurement tool. Additionally, remember that slower amplifiers may require extended delay and/or

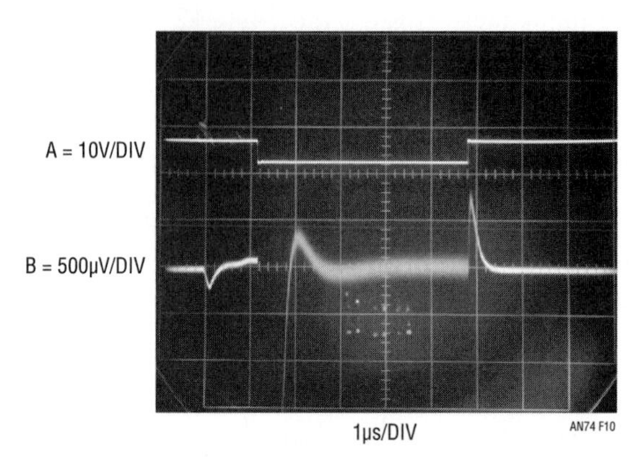

Figure 12.10 • Oscilloscope Display with Inadequate Sample Gate Delay. Sample Window (Trace A) Occurs Too Early, Resulting in Off-Screen Activity in Settle Output (Trace B). Oscilloscope is Overdriven, Making Displayed Information Questionable

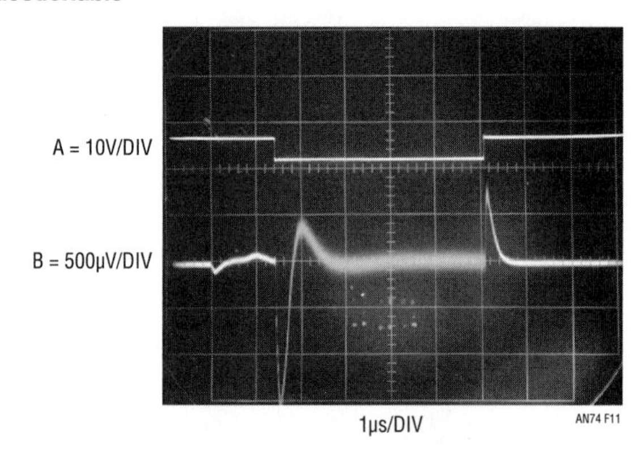

Figure 12.11 • Increasing Sample Gate Delay Positions Sample Window (Trace A) So Settle Output (Trace B) Activity is On-Screen

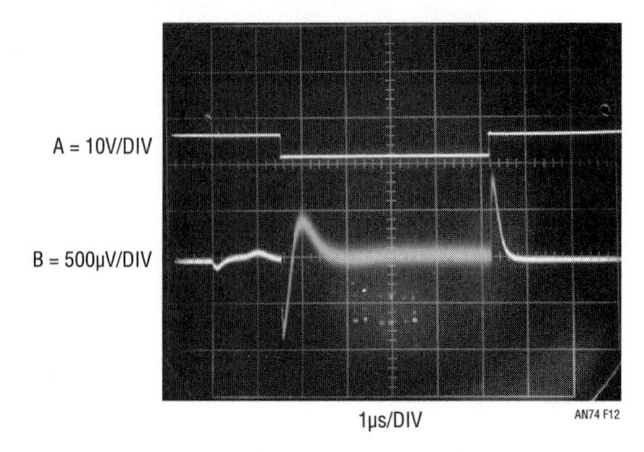

Figure 12.12 • Optimal Sample Gate Delay Positions Sampling Window (Trace A) So All Settle Output (Trace B) Information is Well Inside Screen Boundaries

sampling window times. This may necessitate larger capacitor values in the 74H123 one-shot timing networks.

Compensation capacitor effects

The DAC amplifier requires frequency compensation to get the best possible settling time. The DAC has appreciable output capacitance, complicating amplifier response and making careful compensation capacitor selection even more important.[9] Figure 12.13 shows effects of very light compensation. Trace A is the time corrected input pulse and trace B the residue amplifier output. The light compensation permits very fast slewing but excessive ringing amplitude over a protracted time results. The ringing is so severe that it feeds through during a portion of the sample gate off-period, although no overdrive results. When sampling is initiated (just prior to the sixth vertical division) the ringing is seen to be in its final stages, although still offensive. Total settling time is about 2.8μs. Figure 12.14 presents the opposite extreme. Here a large value compensation capacitor eliminates all ringing but slows down the amplifier so much that settling stretches out to 3.3μs. The best case appears in Figure 12.15. This photo was taken with the compensation capacitor carefully chosen for the best possible settling time. Damping is tightly controlled and settling time goes down to 1.7μs.

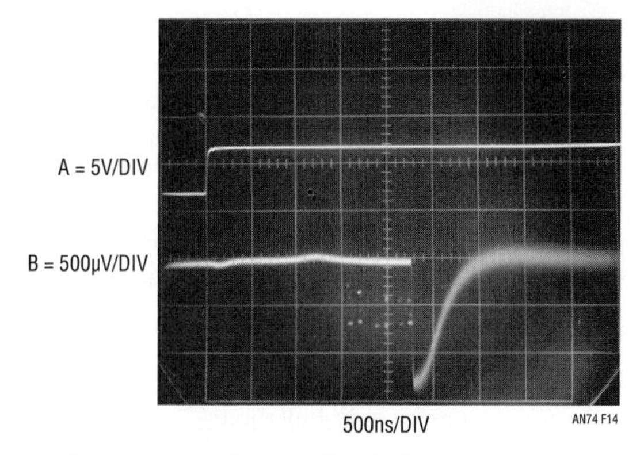

Figure 12.14 • Excessive Feedback Capacitance Overdamps Response. $t_{SETTLE} = 3.3\mu s$

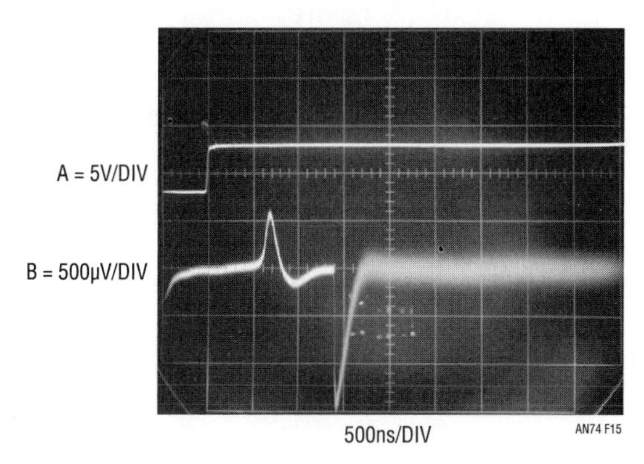

Figure 12.15 • Optimal Feedback Capacitance Yields Tightly Damped Signature and Best Settling Time. $t_{SETTLE} = 1.7\mu s$

Verifying results—alternate methods

The sampling-based settling time circuit appears to be a useful measurement solution. How can its results be tested to ensure confidence? A good way is to make the same measurement with alternate methods and see if results agree. To begin this exercise we return to the basic diode-bounded settle circuit.

Figure 12.16 repeats Figure 12.2's basic settling time measurement, with the same problem. The Schottky-bounded

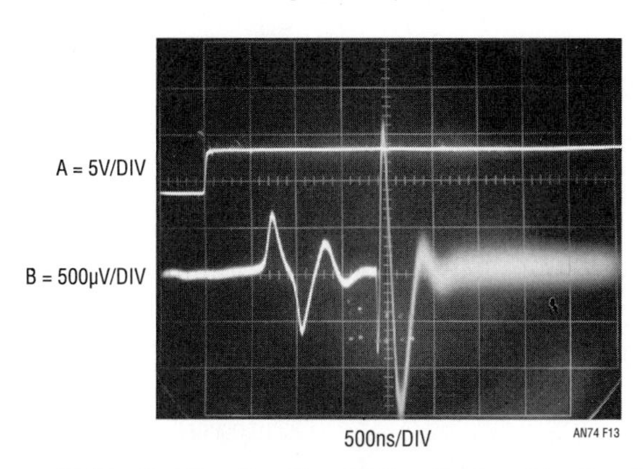

Figure 12.13 • Settling Profile with Inadequate Feedback Capacitance Shows Underdamped Response. Excessive Ringing Feeds Through During Sample Gate Off-Period (Second Through ≈Sixth Vertical Divisions) But is Tolerable. $t_{SETTLE} = 2.8\mu s$

Note 9: This section discusses frequency compensation of the DAC amplifier within the context of sampling-based settling time measurement. As such, it is necessarily brief. Considerably more detail is available later in the text and in Appendix D, "Practical Considerations for DAC-Amplifier Compensation."

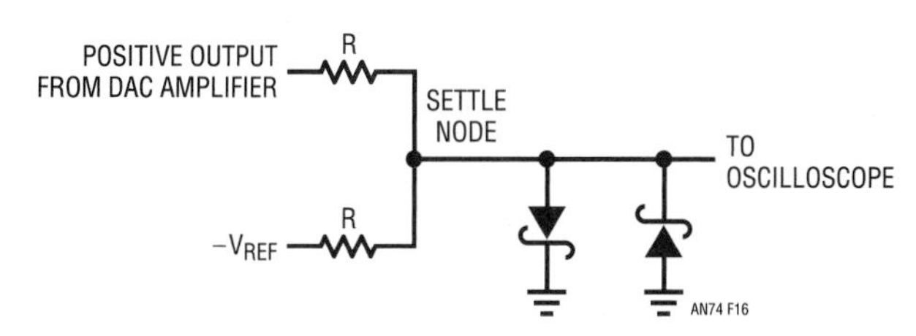

Figure 12.16 • Clamped Settle Node Permits Oscilloscope Overdrive Because Diodes Have 400mV Drop

settle node forces a 400mV overdrive to the oscilloscope, rendering all measurements useless. Now, consider Figure 12.17. This arrangement is similar, but the diodes are returned to bias voltages that are slightly lower than the diode drops. Theoretically, this has the same effect as ground-referred diodes with an inherently lower forward drop, greatly reducing oscilloscope overdrive. In practice, diode V-I characteristics and temperature effects limit achievable performance to uninteresting levels. Clamping reduction is minimal and diode forward leakage when the settle node reaches zero causes signal amplitude errors. Although impractical, this approach does hint at the way to a more useful method.

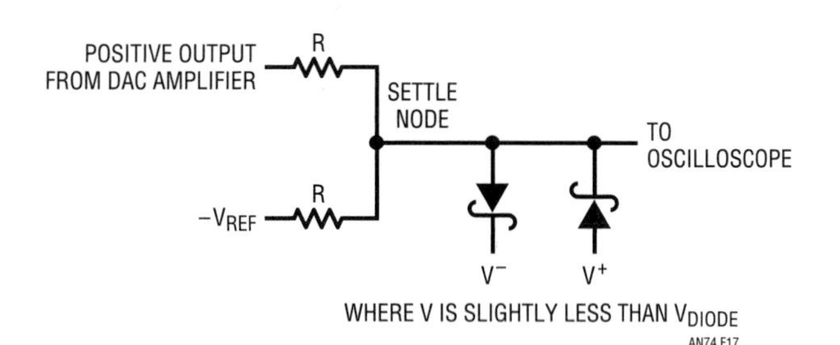

Figure 12.17 • Biasing Diodes Theoretically Lowers Clamp Voltage. In Practice, V-I Characteristics and Temperature Effects Limit Performance

Alternate method I—bootstrapped clamp

Figure 12.18's approach returns the diodes to amplifier- generated voltages bootstrapped from the settle-node input signal. In this way, the diode bias is actively maintained at the optimum point with respect to the signal to be clamped. During DAC amplifier slew, the settle-node signal is large and the amplifiers supply a resultant large bias to the diodes, forcing the desired small clamp voltage. When the DAC amplifier comes out of slew, the settle-node signal very nearly approaches zero, the amplifiers supply almost no diode bias and the oscilloscope monitors the uncorrupted settle node output. Adjustable amplifier gains permit optimal setting of positive and negative bound limits. This scheme offers the possibility of minimizing oscilloscope overdrive while preserving signal-path integrity.

A practical bootstrapped clamp appears in Figure 12.19. The actual clamp circuit, composed of A3 and A4, is nearly identical to the previous figure's theoretical incarnation. A1 and A2 are added, suppling a nonsaturating gain of 80 to the clamp. This permits a 500μV/DIV oscilloscope scale factor with respect to the DAC amplifier output. In Figure 12.20 the amplifier-bound voltages are set equal to the diode drops, and bootstrapping does not occur. The response is essentially identical to that of a simple diode clamp. In Figure 12.21 A4's gain is adjusted, reducing the positive clamp excursion. A3's gain is similarly trimmed in Figure 12.22, producing a corresponding reduction in the negative clamp limit. Note that in both photos, the small amplitude settle signal waveform (beginning about the fifth vertical division) is unaffected. Further refinement of the positive and negative trims produces Figure 12.23. The trims are optimized for minimal peak-to-peak amplitude while maintaining settle-signal waveform fidelity. This permits an oscilloscope running at 20mV/DIV (500μV at the DAC amplifier) to monitor the settle signal with only a 2.5× overdrive. This is not as ideal a situation as

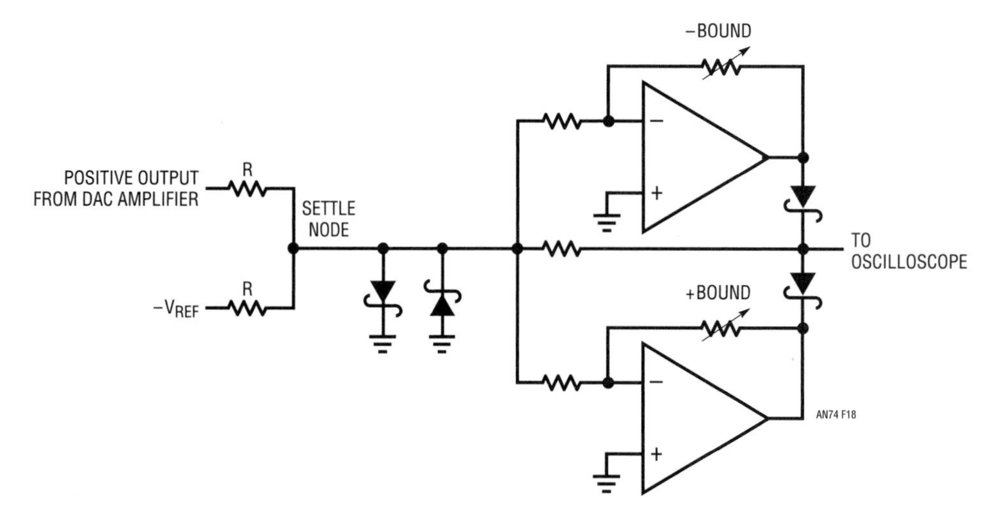

Figure 12.18 • Conceptual Bootstrapped Clamp Biases Diodes From Input Signal, Minimizing Effects of V-I Characteristics and Temperature

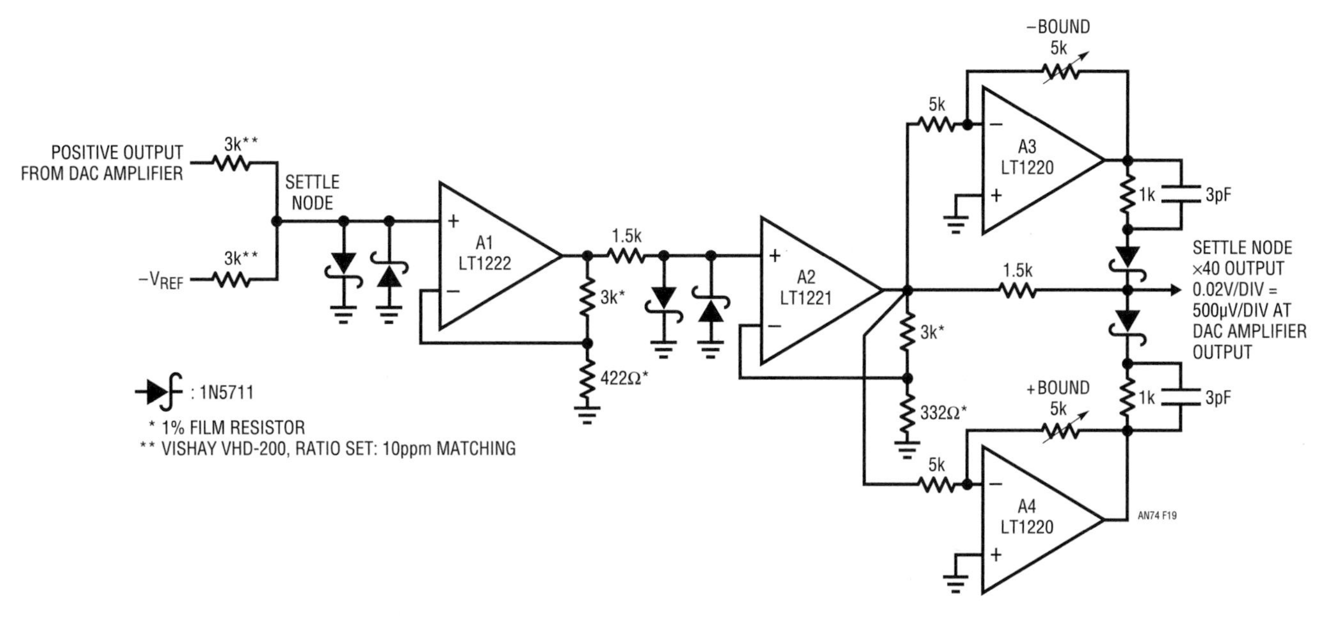

Figure 12.19 • A Practical Bootstrapped Clamp. A1 and A2 Provide Gain to Bootstrapped Section. Positive and Negative Bounds are Adjustable

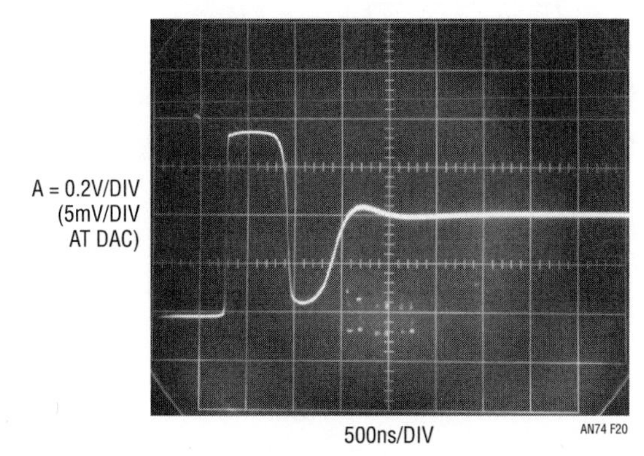

Figure 12.20 • Bootstrapped Clamp Waveform with Bound Limits Equal to Diode Drops. Bootstrap Action Does Not Occur. Response is Identical to Diode Clamp

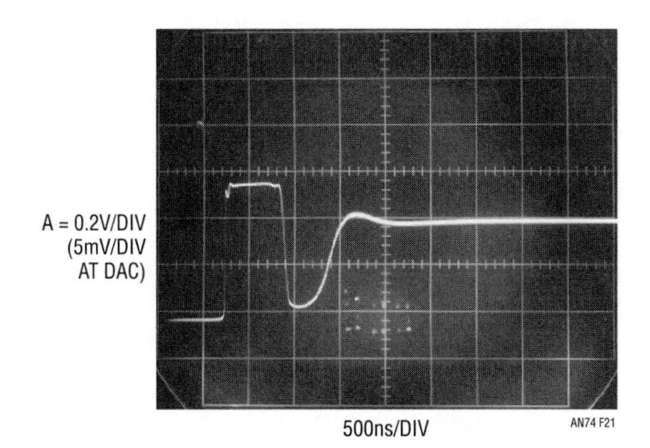

Figure 12.21 • The Positive Bound is Adjusted, Reducing Positive Clamp Excursion

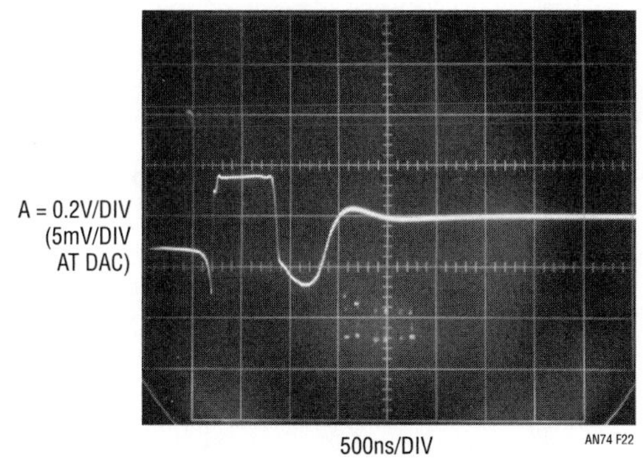

Figure 12.22 • The Negative Bound is Trimmed, Reducing Negative Clamp Limit

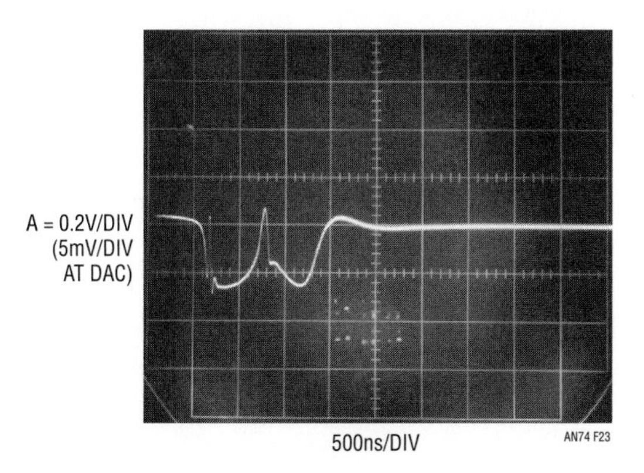

Figure 12.23 • Positive and Negative Bound Adjustments are Optimized for Minimal Peak-to-Peak Amplitude. Waveform Information in Settling Region (Right of Fourth Vertical Division) is Undistorted and Identical to Figure 12.20

the sampling approach, which has no overdrive, but is markedly improved over the simple diode clamp. The monitoring oscilloscope selected must be verified to produce reliable displays while withstanding the 2.5× overdrive.[10]

Figure 12.24 shows the bootstrapped clamp adapted to Figure 12.6's settling-time test circuit. The settle node feeds the residue amplifier, which drives the bootstrapped clamp. As before, the input pulse is time corrected for signal path delays.[11] Additionally, similar type FET probes at the outputs ensure overall delay matching.[12] Figure 12.25 shows the results. Trace A is the time-corrected input step and trace B the settle signal. The oscilloscope undergoes

Note 10: This limitation is surmountable by improving the bootstrapped clamp's dynamic operating range. Future work will be directed towards this end. For the present, the following oscilloscopes have been found to produce faithful results under the 2.5× overdrive conditions noted in the text. The instruments include Tektronix types 547 and 556 (type 1A1 or 1A4 plug-in) and types 453, 454, 453A and 454A. See also Appendix B, "Evaluating Oscilloscope Overload Performance."

Note 11: Characterization of signal path delay is treated in Appendix C, "Measuring and Compensating Residue Amplifier Delay."

Note 12: The bootstrapped clamp's output impedance mandates a FET probe. A second FET probe monitors the input step, but only to maintain channel delay matching.

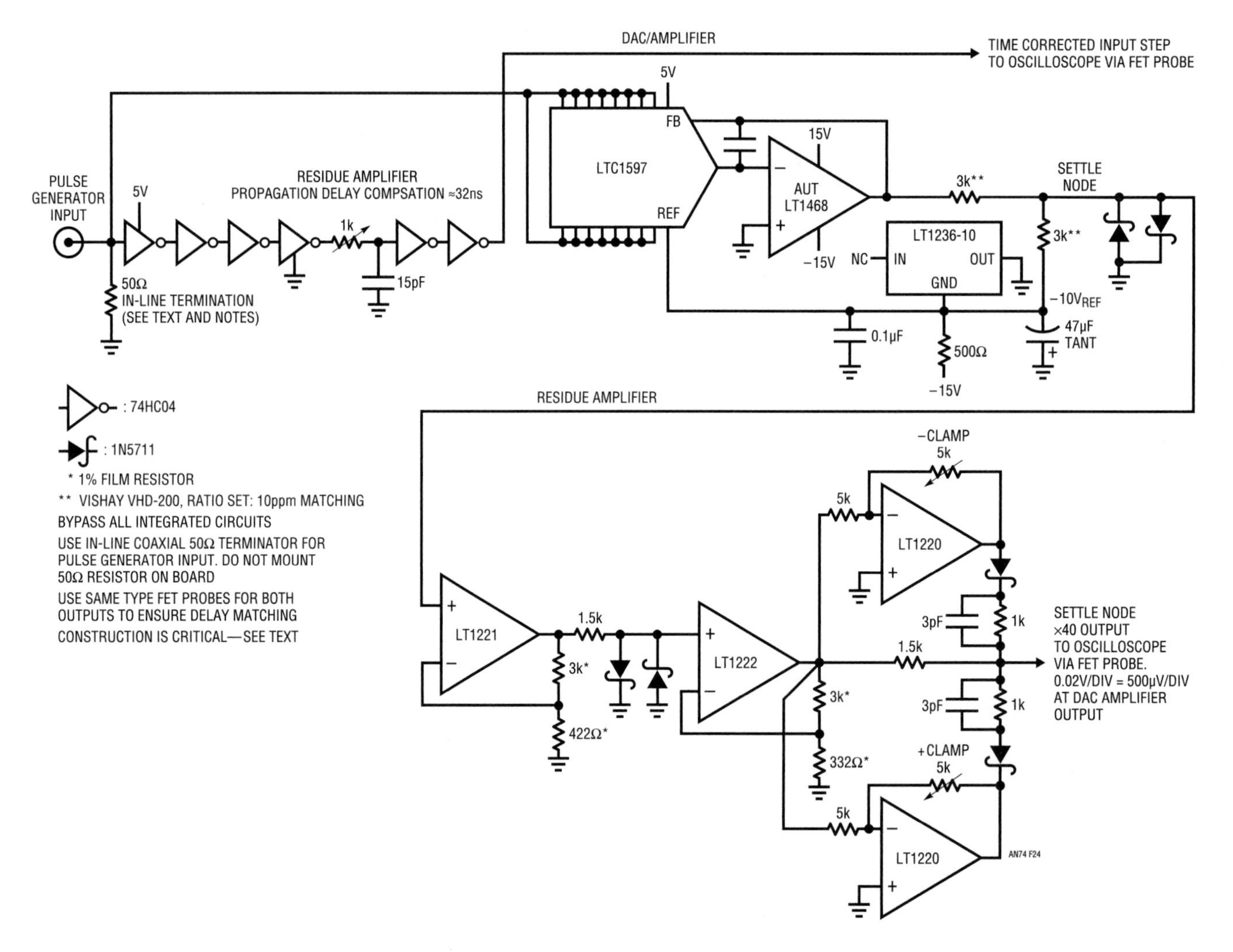

Figure 12.24 • Complete Bootstrapped Clamp-Based DAC Settling Time Measurement Circuit. Overdrive is Substantially Reduced Over Conventional Diode Clamp, But Oscilloscope Must Tolerate ≈2.5× Screen Overdrive

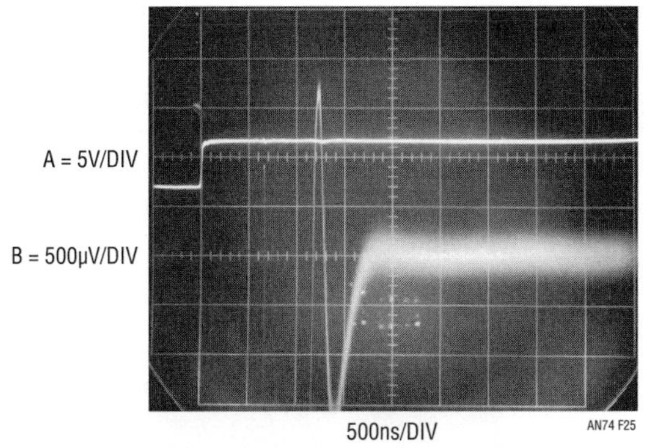

A = 5V/DIV

B = 500μV/DIV

500ns/DIV

AN74 F25

Figure 12.25 • The Bootstrapped Clamp-Amplifier Measuring Settling Time. Oscilloscope Must Tolerate 2.5× Screen Overdrive for Meaningful Results

about a 2.5× overdrive, although the settling signal appears undistorted.

Alternate method II—sampling oscilloscope

It was stated earlier that classical sampling oscilloscopes were inherently immune to overdrive.[13] If this is so, why not utilize this feature and attempt settling time measurement with a simple diode clamp? Figure 12.26 does this. The schematic is identical to Figure 12.24 except that the bootstrapped clamp has been replaced with a simple diode clamp. Under these conditions the sampling 'scope[14] is heavily overdriven, but is ostensibly immune to the insult. Figure 12.27 puts the sampling oscilloscope to the test.

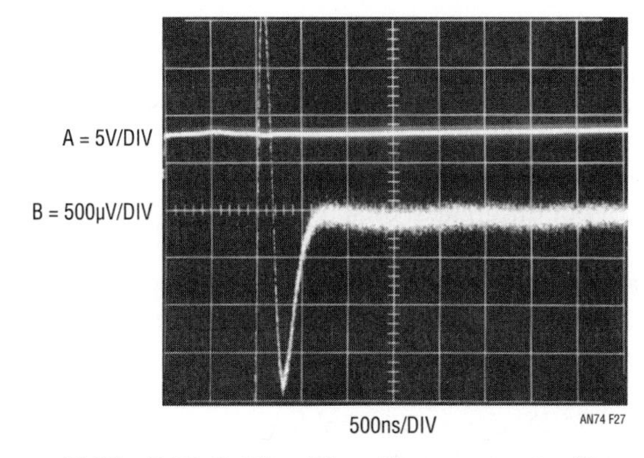

Figure 12.27 • DAC Settling Time Measurement with the Classical Sampling 'Scope. Oscilloscope's Overload Immunity Permits Accurate Measurement Despite Extreme Overdrive

Note 13: See Appendix B, "Evaluating Oscilloscope Overdrive Performance, " for in-depth discussion.

Note 14: Tektronix type 661 with 4S1 vertical and 5T3 timing plug-ins.

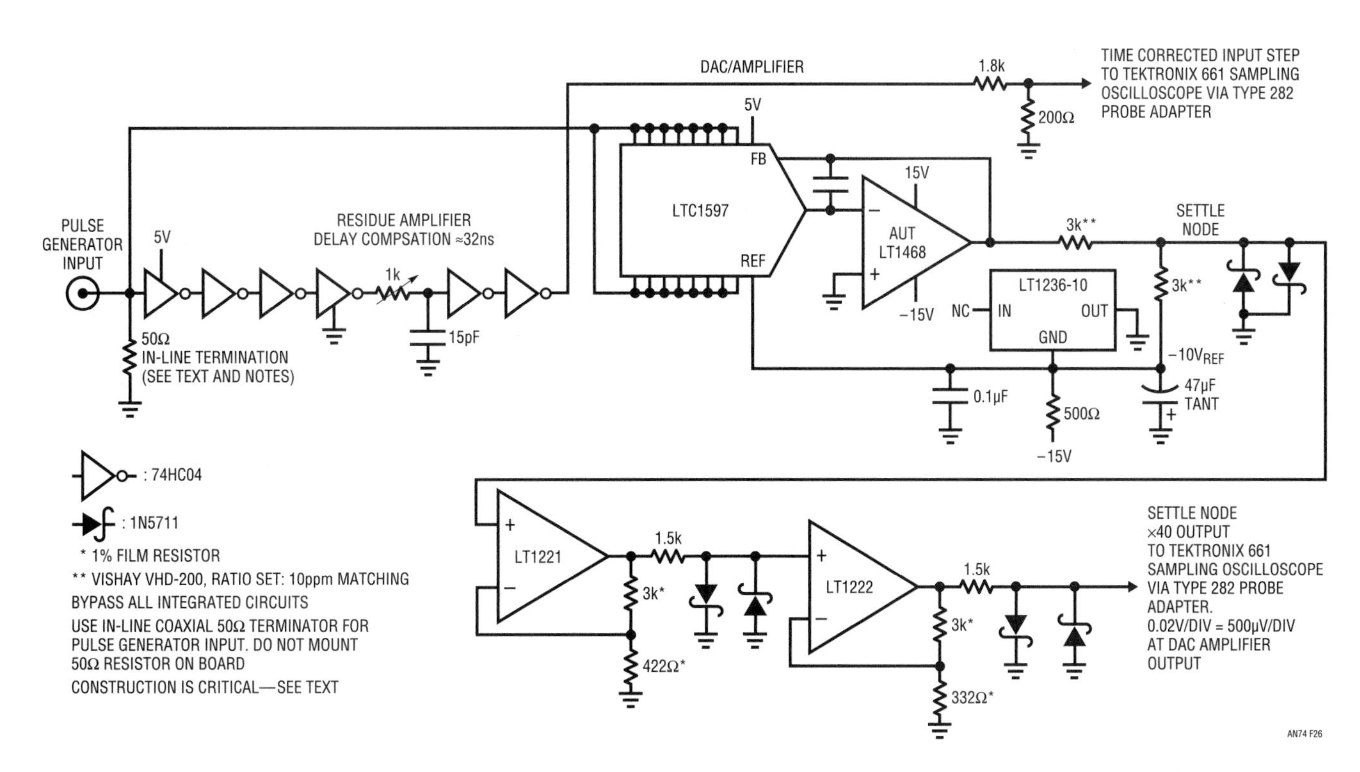

Figure 12.26 • DAC Settling Time Test Circuit Using Classical Sampling Oscilloscope. Circuit is Similar to Figure 12.24. Sampling 'Scope's Inherent Overload Immunity Eliminates Bootstrapped Clamp Requirement

Trace A is the time corrected input pulse and trace B the settle signal. Despite a brutal overdrive, the 'scope appears to respond cleanly, giving a very plausible settle signal presentation.

Alternate method III—differential amplifier

In theory, a differential amplifier with one input biased at the expected settled voltage can measure settling time to 16-bit resolution. In practice, this is an extraordinarily demanding measurement for a differential amplifier. The amplifier's overload recovery characteristics must be pristine. In fact, no commercially produced differential amplifier or differential oscilloscope plug-in has been available that meets this requirement. Recently, an instrument has appeared that, although not fully specified at these levels, appears to have superb overload recovery performance. Figure 12.28 shows the differential amplifier (type and manufacturer appear in the schematic notes) monitoring the DAC output amplifier. The amplifier's negative input is biased from its internal adjustable reference to the expected settled voltage. The diff. amp's clamped output, operating at a gain of 10, feeds A1-A2, a bounded, nonsaturating gain of 40. Note that the monitoring oscilloscope, operating at 0.2V/DIV (500µV/DIV at the DAC amplifier) cannot be overdriven. Figure 12.29 shows the results. Trace A is the time corrected input step and trace B the settle signal. The settle signal is seen to come smoothly out of bound, entering the amplified linear region between the third and fourth vertical divisions. The settling signature appears reasonable and complete settling occurs just beyond the fourth vertical division.

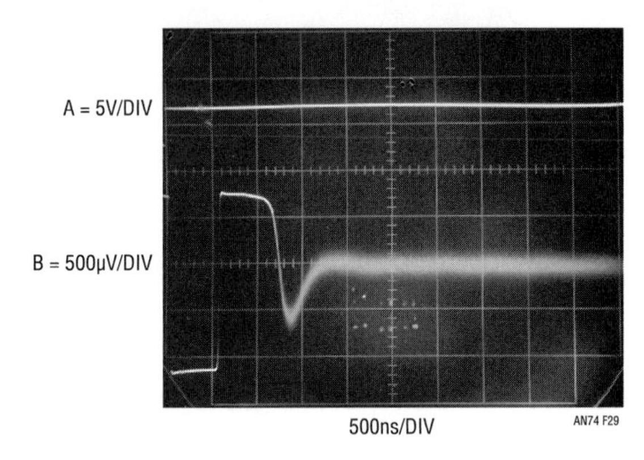

Figure 12.29 • DAC Settling Time Measurement with the Differential/Clamped Amplifiers. All Oscilloscope Input Signal Excursions are On-Screen

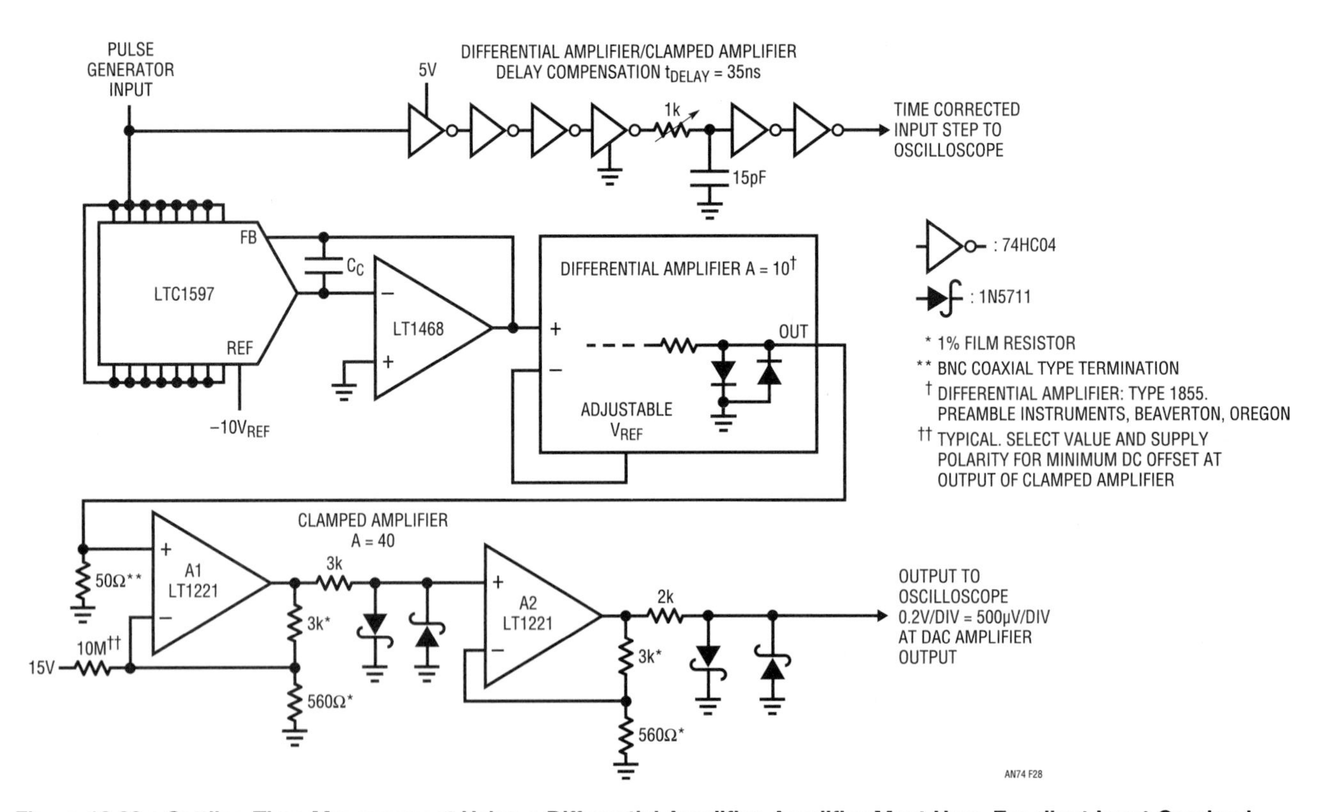

Figure 12.28 • Settling Time Measurement Using a Differential Amplifier. Amplifier Must Have Excellent Input Overload Recovery. Clamped Amplifier's Bounded Gain Stages Limit Amplitude While Maintaining Linear Region Operation. Oscilloscope is Not Overdriven

Summary of results

The simplest way to summarize the four different method's results is by visual comparison. Figures 12.30 through 12.33 repeat previous photos of the four different settling-time methods. If all four approaches represent good measurement technique and are constructed properly, results should be indentical.[15] If this is the case, the identical data produced by the four methods has a high probability of being valid.

Examination of the four photographs shows identical 1.7μs settling times and settling waveform signatures. The shape of the settling waveform, in every detail, is identical in all four photos. This kind of agreement provides a high degree of credibility to the measured results. It also provides the confidence necessary to characterize a wide variety of amplifiers. Figure 12.34 lists various LTC amplifiers and heir measured settling times to 16 bits.

About this chart

The writer despises charts. In their attempt to gain authority they simplify, and glib simplification is the host of mother nature's surprise party. Any topic as complex as DAC-amplifier settling time to 16 bits is a dangerous place for oversimplification. There are simply too many variables and exceptions to accommodate the categorical statement a chart implies. It is with these reservations that Figure 12.34 is presented.[16] The chart

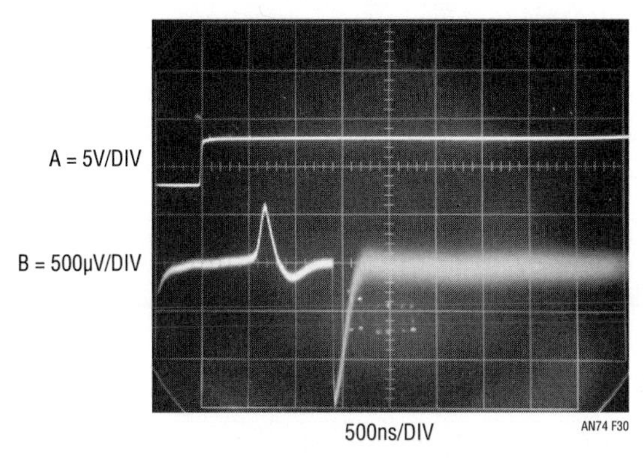

Figure 12.30 • DAC Settling Time Measurement Using the Sampling Bridge Circuit. t_{SETTLE} = 1.7μs

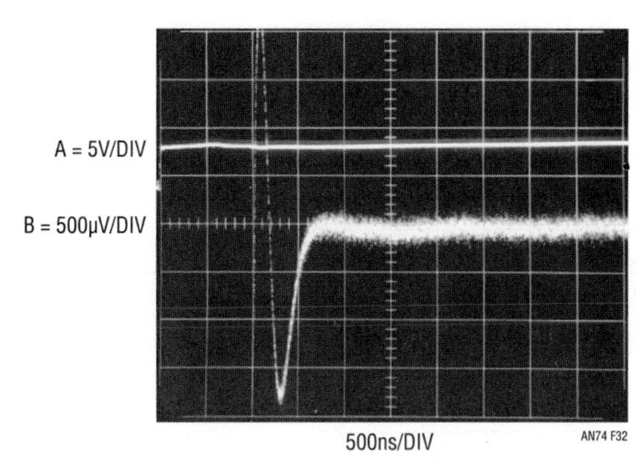

Figure 12.32 • DAC Settling Time Measurement Using the Classical Sampling 'Scope. t_{SETTLE} = 1.7μs

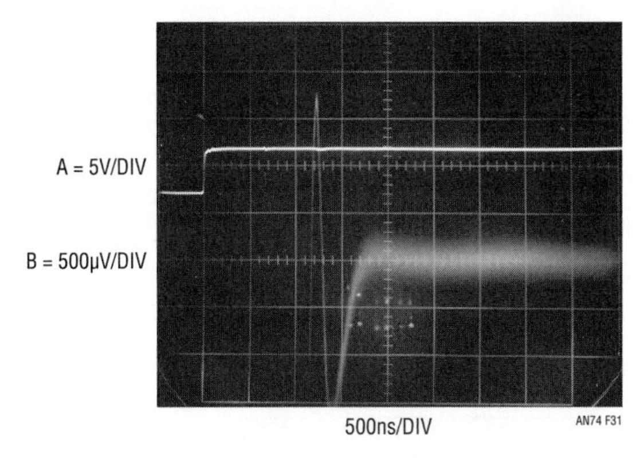

Figure 12.31 • DAC Settling Time Measurement with the Bootstrapped Clamp Method. t_{SETTLE} = 1.7μs

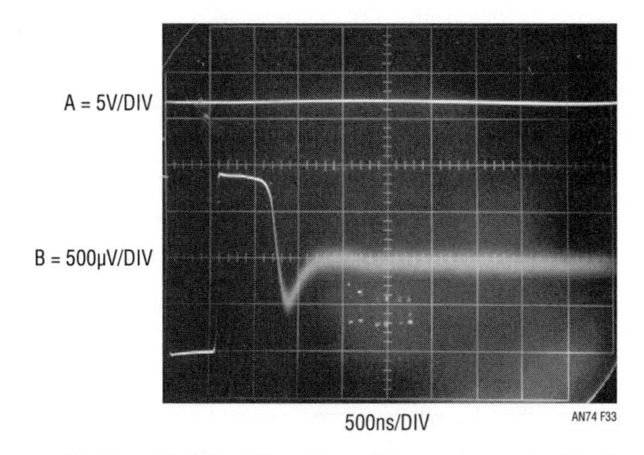

Figure 12.33 • DAC Settling Time Measurement with the Differential Amplifier. t_{SETTLE} = 1.7μs

Note 15: Construction details of the settling time fixtures discussed here appear (literally) in Appendix G, "Breadboarding, Layout and Connection Techniques."

Note 16: Readers detecting author ambivalence about the inclusion of Figure 12.34's chart are not hallucinating.

AMPLIFIER	OPTIMIZED SETTLING TIME AND TYPICAL COMPENSATION VALUE		CONSERVATIVE SETTLING TIME AND COMPENSATION VALUE		COMMENTS
LT1001	65µs	100pF	120µs	100pF	Good Low Speed Choice
LT1006	26µs	66pF	50µs	150pF	
LT1007	17µs	100pF	19µs	100pF	I_B Gives ≈1LSB Error at 25°C
LT1008	64µs	100pF	115µs	100pF	
LT1012	56µs	75pF	116µs	75pF	Good Low Speed Choice
LT1013	50µs	150pF	75µs	150pF	≈1LSB Error Due to V_{OS} over Temperature
LT1055	3.7µs	54pF	5µs	75pF	V_{OS} Gives ≈2 to 3LSB Error over Temperature
LT1077	110µs	100pF	200µs	100pF	
LT1097	60µs	75pF	120µs	75pF	Good Low Speed Choice
LT1122	3µs	51pF	3.5µs	68pF	V_{OS} Induced Errors
LTC1150	7ms	100pF	10ms	100pF	Special Case. See Appendix E. Needs Output Booster, e.g., LT1010
LT1178	330µs	100pF	450µs	100pF	
LT1179	330µs	100pF	450µs	100pF	
LT1211	5.5µs	73pF	6.5µs	82pF	I_B and V_{OS} Based Errors
LT1213	4.6µs	58pF	5.8µs	68pF	I_B and V_{OS} Based Errors
LT1215	3.6µs	53pF	4.7µs	68pF	I_B and V_{OS} Based Errors
LT1218	110µs	100pF	200µs	100pF	≈1.5LSB Error Due to V_{OS}. ≈4 to 5LSB I_B Based Errors
LT1220	2.3µs	41pF	3.1µs	56pF	V_{OS} and I_B Based Errors
LT1366	64µs	100pF	100µs	150pF	V_{OS} and I_B Based Errors
LT1413	45µs	100pF	75µs	120pF	≈2LSB Error Due to V_{OS}
LT1457	7.4µs	100pF	12µs	120pF	5 to 6LSB Error From V_{OS} over Temperature
LT1462	78µs	100pF	130µs	120pF	7 to 8LSB Error Due to V_{OS} over Temperature
LT1464	19µs	90pF	30µs	110pF	See LT1462 Comments Above
LT1468	1.7µs	20pF	2.5µs	30pF	Fastest Settling with 16-Bit Performance
LT1490	175µs	100pF	300µs	100pF	V_{OS} Based Errors
LT1492	7.5µs	80pF	10µs	100pF	V_{OS} and I_B Based Errors
LT1495	10ms	100pF	25ms	100pF	Measured with Hourglass and Differential Voltmeter. Needs Output Booster, e.g., LT1010
LT1498	5µs	60pF	7.3µs	82pF	V_{OS} and I_B Based Errors
LT1630	4.5µs	63pF	6.7µs	82pF	Significant I_B Based Error
LT1632	4µs	55pF	5.2µs	68pF	Significant I_B Based Error
LTC1650	6µs		7.3µs		DAC On-Board. ±4V Step. ≈10LSB V_{OS} Related Error Over Temperature
LT2178	330µs	100pF	450µs	100pF	1 to 2LSB V_{OS} Based Error

Figure 12.34 • 16-Bit Settling Time for Various Amplifiers Driven by the LT1597 DAC. Optimized Settling Times Require Trimming Compensation Capacitor, Conservative Times are Untrimmed. LT1468 (Shaded) Offers Fastest Settling Time While Maintaining Accuracy Over Temperature

lists measured settling times to 16 bits for various LTC amplifiers used with the LTC1595-7 16-bit DACs. A number of conditions and comments apply to interpreting the chart's information. The amplifiers selected are not all accurate to 16 bits over temperature, or (in some cases) even at 25°C. However, many applications, such as AC signal processing, servo loops or waveform generation, are insensitive to DC offset error and, as such, these amplifiers are worthy candidates. Applications requiring DC accuracy to 16 bits (10V full scale) must keep input errors below 15nA and 152µV to maintain performance.

The settling times are quoted for "optimized" and "conservative" cases. The optimized case uses a typical amplifier-DAC combination. This implies "design centered" values for amplifier slew rate and DAC output resistance and capacitance. It also permits trimming the amplifier's feedback capacitor to obtain the best possible settling time. The conservative category assumes worst-case amplifier slew rate, highest DAC output impedances and untrimmed, standard 5% feedback capacitors. This worst-case error summation is perhaps unduly pessimistic; RMS summing may represent a more realistic compromise. However, such a maudlin outlook helps avoid unpleasant surprises in production. Settling times are quoted using ±15V supplies, a −10V DAC reference and a 10V positive output step. The sole exception to this is the LTC1650, a 16-bit DAC with amplifier onboard. This device is powered by ±5V supplies and settling is measured with a 4V reference and a ±4V swing.[17] All feedback capacitances listed were determined with a General Radio model 1422-CL precision variable air capacitor.[18]

In general, the slower amplifiers' extended slew times make their ring times vanishingly small settling-time contributors. This is reflected in identical feedback capacitor values for the optimized and conservative cases. Conversely, faster amplifiers' ring times are significant terms, resulting in different compensation values for the two categories. Additional considerations for compensation are discussed in Appendix D, "Practical Considerations for DAC-Amplifier Compensation."

Thermally induced settling errors

A final category of settling-time error is thermally based. Some poorly designed amplifiers exhibit a substantial "thermal tail" after responding to an input step. This phenomenon, due to die heating, can cause the output to wander outside desired limits long after settling has apparently occurred. After checking settling at high speed it is always a good idea to slow the oscilloscope sweep down and look for thermal tails. Figure 12.35 shows such a tail. The amplifier slowly (note horizontal sweep speed) drifts 200µV after settling has apparently occurred. Often, the thermal tail's effect can be accentuated by loading the amplifier's output. Figure 12.36 doubles the error by increasing amplifier loading.

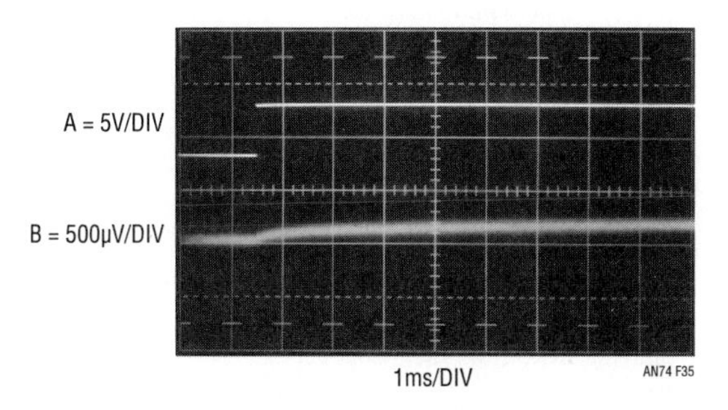

Figure 12.35 • Typical Thermal Tail in a Poorly Designed Amplifier. Device Drifts 200µV (>1LSB) After Settling Apparently Occurs

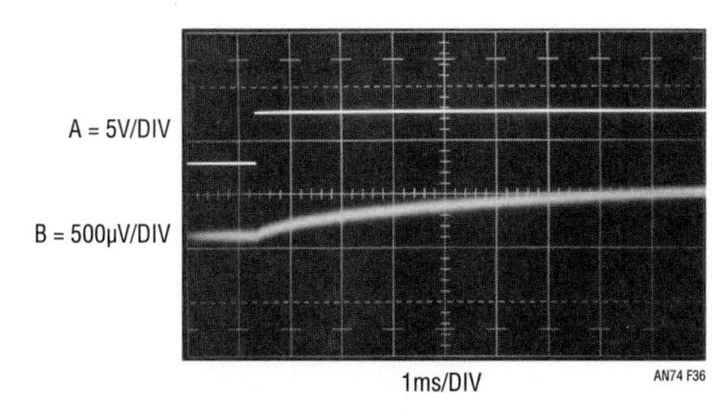

Figure 12.36 • Loading the Amplifier Increases Thermal Tail Error to 400µV (>2.5LSB)

Note 17: See Appendix F "Settling Time Measurement of Serially Loaded DACs."

Note 18: A thing of transcendent beauty. It is worth owning this instrument just to look at it. It is difficult to believe humanity could fashion anything so perfectly gorgeous.

References

1. Williams, Jim, "Methods for Measuring Op Amp Settling Time," Linear Technology Corporation, Application Note 10, July 1985
2. Demerow, R., "Settling Time of Operational Amplifiers, "Analog Dialogue, Volume 4-1, Analog Devices, Inc., 1970
3. Pease, R. A., "The Subtleties of Settling Time," The New Lightning Empiricist, Teledyne Philbrick, June 1971
4. Harvey, Barry, "Take the Guesswork Out of Settling Time Measurements," EDN, September 19, 1985
5. Williams, Jim, "Settling Time Measurement Demands Precise Test Circuitry," EDN, November 15, 1984
6. Schoenwetter H. R., "High-Accuracy Settling Time Measurements," IEEE Transactions on Instrumentation and Measurement, Vol. IM-32. No. 1, March 1983
7. Sheingold, D. H., "DAC Settling Time Measurement," Analog-Digital Conversion Handbook, pg 312–317. Prentice-Hall, 1986
8. Williams, Jim, "Evaluating Oscilloscope Overload Performance," Box Section A, in "Methods for Measuring Op Amp Settling Time," Linear Technology Corporation, Application Note 10, July 1985
9. Orwiler, Bob, "Oscilloscope Vertical Amplifiers," Tektronix, Inc., Concept Series, 1969
10. Addis, John, "Fast Vertical Amplifiers and Good Engineering," Analog Circuit Design; Art, Science and Personalities, Butterworths, 1991
11. W. Travis, "Settling Time Measurement Using Delayed Switch," Private, Communication. 1984
12. Hewlett-Packard, "Schottky Diodes for High-Volume, Low Cost Applications," Application Note 942, Hewlett-Packard Company, 1973
13. Harris Semiconductor, "CA3039 Diode Array Data Sheet, Harris Semiconductor, 1993
14. Carlson, R., "A Versatile New DC-500MHz Oscilloscope with High Sensitivity and Dual Channel Display," Hewlett-Packard Journal, Hewlett-Packard Company, January 1960
15. Tektronix, Inc., "Sampling Notes," Tektronix, Inc., 1964
16. Tektronix, Inc., "Type 1S1 Sampling Plug-In Operating and Service Manual," Tektronix, Inc., 1965
17. Mulvey, J., "Sampling Oscilloscope Circuits," Tektronix, Inc., Concept Series, 1970
18. Addis, John, "Sampling Oscilloscopes," Private Communication, February, 1991
19. Williams, Jim, "Bridge Circuits—Marrying Gain and Balance," Linear Technology Corporation, Application Note 43, June, 1990
20. Tektronix, Inc., "Type 661 Sampling Oscilloscope Operating and Service Manual,' Tektronix, Inc., 1963
21. Tektronix, Inc., "Type 4S1 Sampling Plug-In Operating and Service Manual," Tektronix, Inc., 1963
22. Tektronix, Inc., "Type 5T3 Timing Unit Operating and Service Manual," Tektronix, Inc., 1965
23. Williams, Jim, "Applications Considerations and Circuits for a New Chopper-Stabilized Op Amp," Linear Technology Corporation, Application Note 9, March, 1985
24. Morrison, Ralph, "Grounding and Shielding Techniques in Instrumentation," 2nd Edition, Wiley Interscience, 1977
25. Ott, Henry W., "Noise Reduction Techniques in Electronic Systems," Wiley Interscience, 1976
26. Williams, Jim, "High Speed Amplifier Techniques," Linear Technology Corporation, Application Note 47, 1991
27. Williams, Jim, "Power Gain Stages for Monolithic Amplifiers," Linear Technology Corporation, Application Note 18, March 1986

Appendix A
A history of high accuracy digital-to-analog conversion

People have been converting digital-to-analog quantities for a long time. Probably among the earliest uses was the summing of calibrated weights (Figure A1, left center) in weighing applications. Early electrical digital-to-analog conversion inevitably involved switches and resistors of different values, usually arranged in decades. The application was often the calibrated balancing of a bridge or reading, via null detection, some unknown voltage. The most accurate resistor-based DAC of this type is Lord Kelvin's Kelvin-Varley divider (Figure, large box). Based on switched resistor ratios, it can achieve ratio accuracies of 0.1ppm (23+ bits) and is still widely employed in standards laboratories. High speed digital-to-analog conversion resorts to electronically switching the resistor network. Early electronic DACs were built at the board level using discrete precision resistors and germanium transistors (Figure, center foreground, is a 12-bit DAC from a Minuteman missile D-17B inertial navigation system, circa 1962). The first electronically switched DACs available as standard product were probably those produced by Pastoriza Electronics in the mid 1960s. Other manufacturers followed and discrete- and monolithically-based modular DACs (Figure, right and left) became popular by the 1970s. The units were often potted (Figure, left) for ruggedness, performance or to (hopefully) preserve proprietary knowledge. Hybrid technology produced smaller package size (Figure, left foreground). The development of Si-Chrome resistors permitted precision monolithic DACs such as the LTC1595 (Figure, immediate foreground). In keeping with all things monolithic, the cost-performance trade off of modern high resolution IC DACs is a bargain. Think of it! A 16-bit DAC in an 8-pin IC package. What Lord Kelvin would have given for a credit card and LTC's phone number.

Figure A1 • Historically Significant Digital-to-Analog Converters Include: Weight Set (Center Left), 23+ Bit Kelvin-Varley Divider (Large Box), Hybrid, Board and Modular Types, and the LTC1595 IC (Foreground). Where Will It All End?

Appendix B
Evaluating oscilloscope overdrive performance

Most of the settling-time circuits are heavily oriented towards providing little or no overdrive to the monitoring oscilloscope. This is done to avoid overdriving the oscilloscope. Oscilloscope recovery from overdrive is a grey area and almost never specified. Some of the settling time measurement methods require the oscilloscope to be overdriven. In these cases, the oscilloscope is required to supply an accurate waveform after the display has been driven off screen. How long must one wait after an overdrive before the display can be taken seriously? The answer to this question is quite complex. Factors involved include the degree of overdrive, its duty cycle, its magnitude in time and amplitude and other considerations. Oscilloscope response to overdrive varies widely between types and markedly different behavior can be observed in any individual instrument. For example, the recovery time for a 100× overload at 0.005V/DIV may be very different than at 0.1V/DIV. The recovery characteristic may also vary with waveform shape, DC content and repetition rate. With so many variables, it is clear that measurements involving oscilloscope overdrive must be approached with caution.

Why do most oscilloscopes have so much trouble recovering from overdrive? The answer to this question requires some study of the three basic oscilloscope types' vertical paths. The types include analog (Figure B1A), digital (Figure B1B) and classical sampling (Figure B1C) oscilloscopes. Analog and digital 'scopes are susceptible to overdrive. The classical sampling 'scope is the only architecture that is inherently immune to overdrive.

An analog oscilloscope (Figure B1A) is a real time, continuous linear system.[1] The input is applied to an attenuator, which is unloaded by a wideband buffet: The vertical preamp provides gain, and drives the trigger pick-off, delay line and the vertical output amplifier. The attenuator and delay line are passive elements and require little comment. The buffer, preamp and vertical output amplifier are complex linear gain blocks, each with dynamic operating range restrictions. Additionally, the operating point of each block may be set by inherent circuit balance, low frequency stabilization paths or both. When the input is overdriven, one or more of these stages may saturate, forcing internal nodes and components to abnormal operating points and temperatures. When the overload ceases, full recovery of the electronic and thermal time constants may require surprising lengths of time.[2]

The digital sampling oscilloscope (Figure B1B) eliminates the vertical output amplifier, but has an attenuator buffer and amplifiers ahead of the A/D converter. Because of this, it is similarly susceptible to overdrive recovery problems.

The classical sampling oscilloscope is unique. Its nature of operation makes it inherently immune to overload. Figure B1C shows why. The sampling occurs *before* any gain is taken in the system. Unlike Figure B1B's digitally sampled 'scope, the input is fully passive to the sampling point. Additionally, the output is fed back to the sampling bridge, maintaining its operating point over a very wide range of inputs. The dynamic swing available to maintain the bridge output is large and easily accommodates a wide range of oscilloscope inputs. Because of all this, the amplifiers in this instrument do not see overload, even at 1000× overdrives, and there is no recovery problem. Additional immunity derives from the instrument's relatively slow sample rate—even if the amplifiers were overloaded, they would have plenty of time to recover between samples.[3]

The designers of classical sampling 'scopes capitalized on the overdrive immunity by including variable DC offset generators to bias the feedback loop (see Figure B1C, lower right). This permits the user to offset a large input, so small amplitude activity on top of the signal can be accurately observed. This is ideal for, among other things, settling time measurements. Unfortunately, classical sampling oscilloscopes are no longer manufactured, so if you have one, take care of it!

Note 1: Ergo, the Real Thing. Hopelessly bigoted residents of this locale mourn the passing of the analog 'scope era and frantically hoard every instrument they can find.

Note 2: Some discussion of input overdrive effects in analog oscilloscope circuitry is found in reference 10.

Note 3: Additional information and detailed treatment of classical sampling oscilloscope operation appears in references 14-17 and 20-22.

Although analog and digital oscilloscopes are susceptible to overdrive, many types can tolerate some degree of this abuse. The early portion of this appendix stressed that measurements involving oscilloscope overdrive must be approached with caution. Nevertheless, a simple test can indicate when the oscilloscope is being deleteriously affected by overdrive.

The waveform to be expanded is placed on the screen at a vertical sensitivity that eliminates all off-screen activity. Figure B2 shows the display. The lower right hand portion is to be expanded. Increasing the vertical sensitivity by a factor of two (Figure B3) drives the waveform off-screen, but the remaining display appears reasonable. Amplitude has doubled and waveshape is consistent with the original display. Looking carefully, it is possible to see small amplitude information presented as a dip in the waveform at about the third vertical division. Some small disturbances are also visible. This observed expansion of

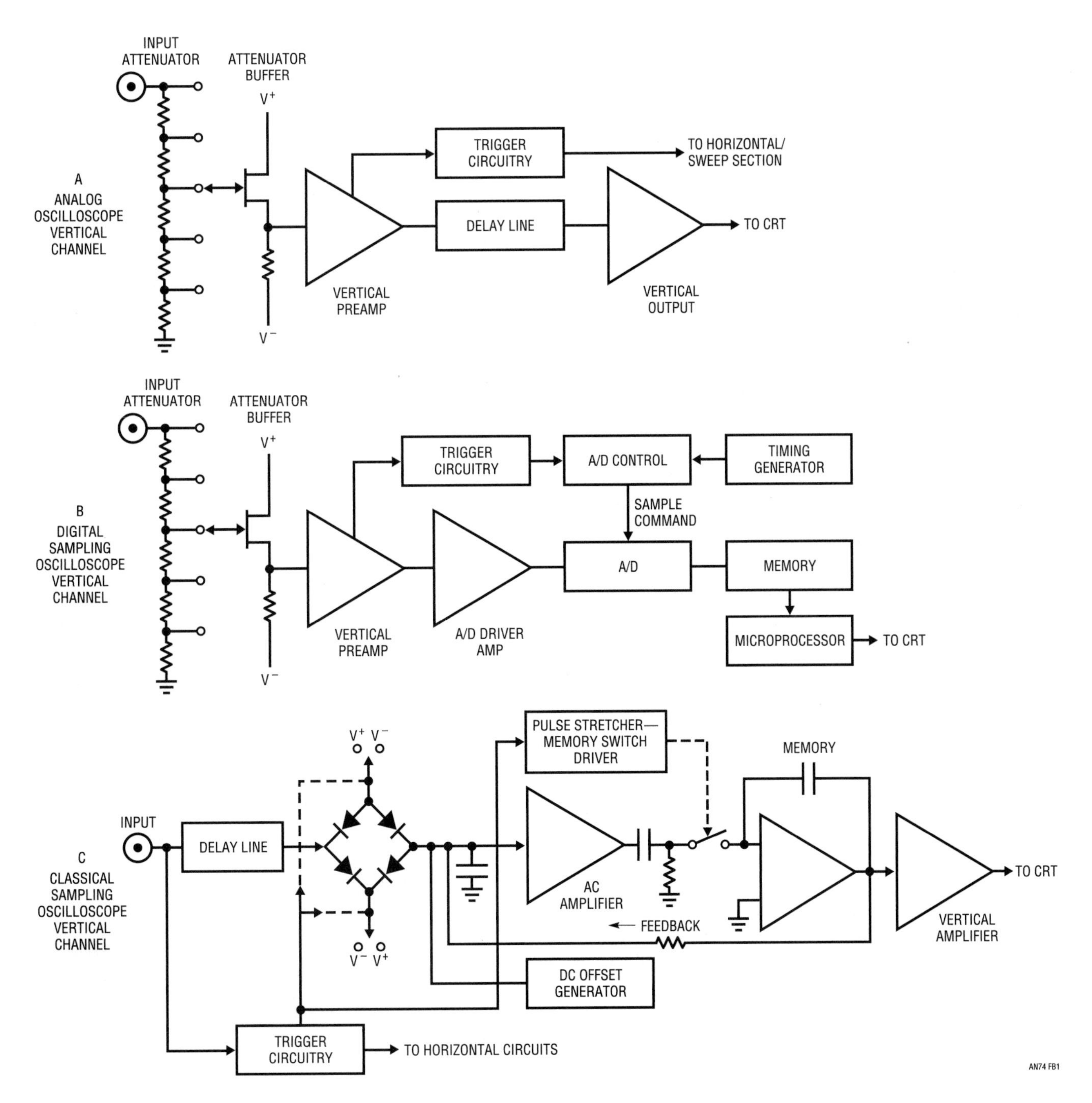

Figure B1 • Simplified Vertical Channel Diagrams for Different Type Oscilloscopes. Only the Classical Sampling 'Scope (C) Has Inherent Overdrive Immunity. Offset Generator Allows Viewing Small Signals Riding On Large Excursions

the original waveform is believable. In Figure B4, gain has been further increased, and all the features of Figure B3 are amplified accordingly. The basic waveshape appears clearer and the dip and small disturbances are also easier to see. No new waveform characteristics are observed. Figure B5 brings some unpleasant surprises. This increase in gain causes definite distortion. The initial negative-going peak, although larger, has a different shape. Its bottom appears less broad than in Figure B4. Additionally, the peak's positive recovery is shaped slightly differently. A new rippling disturbance is visible in the center of the screen. This kind of change indicates that the oscilloscope is having trouble. A further test can confirm that this waveform is being influenced by overloading. In Figure B6 the gain remains the same but the vertical position knob has been used to reposition the display at the screen's bottom. This shifts the oscilloscope's DC operating point which, under normal circumstances, should not affect the displayed waveform. Instead, a marked shift in waveform amplitude and outline occurs. Repositioning the waveform to the screen's top produces a differently distorted waveform (Figure B7). It is obvious that for this particular waveform, accurate results cannot be obtained at this gain.

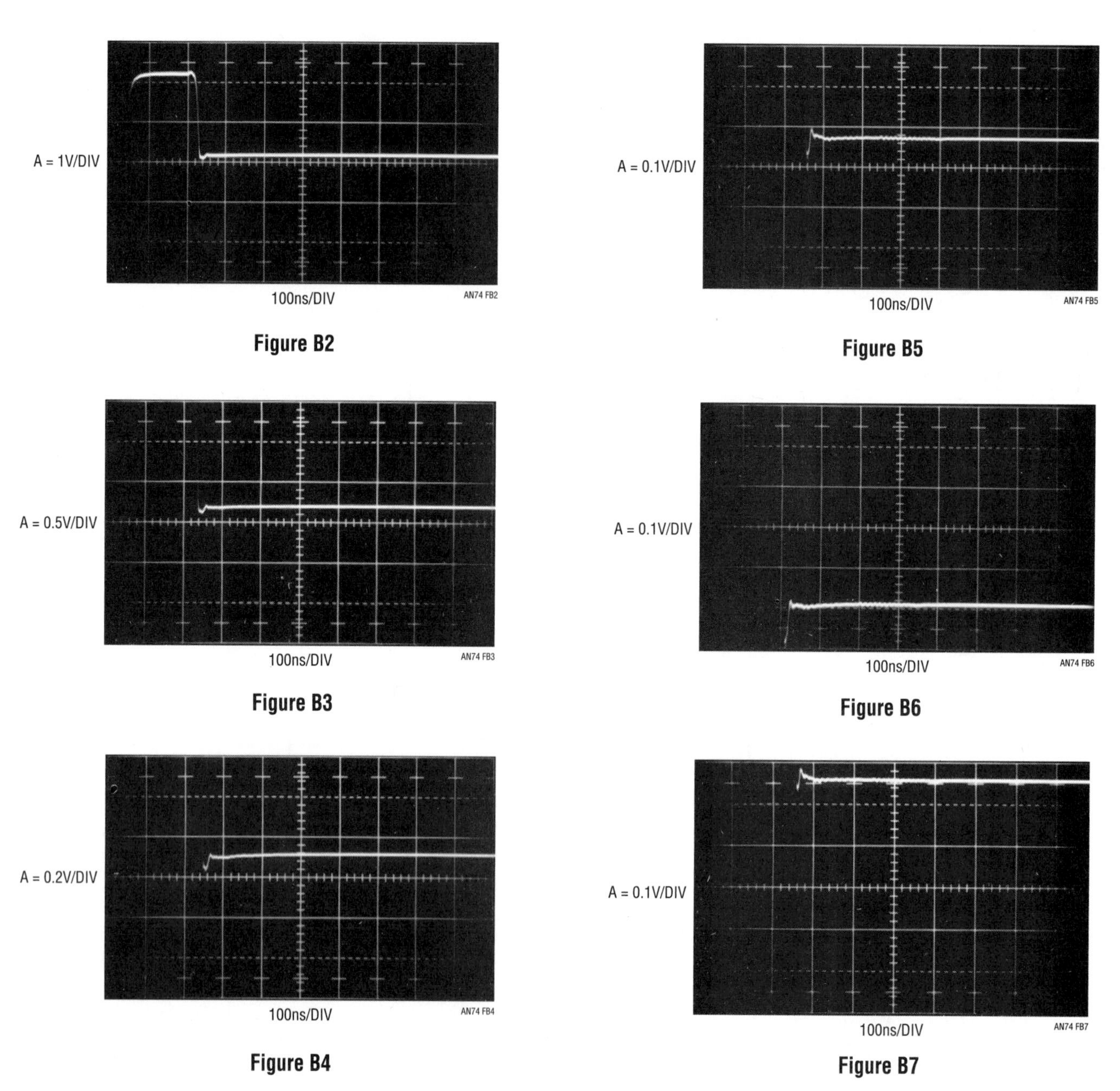

Figure B2–B7 • The Overdrive Limit is Determined by Progressively Increasing Oscilloscope Gain and Watching for Waveform Aberrations

Appendix C
Measuring and compensating residue-amplifier delay

The settling-time circuits utilize an adjustable delay network to time correct the input pulse for delays in the signal processing path. Typically, these delays introduce errors of a few percent, so a first-order correction is adequate. Setting the delay trim involves observing the network's input-output delay and adjusting for the appropriate time interval. Determining the "appropriate" time interval is somewhat more complex. Measuring the sampling bridge- based circuit's signal path delay involves modifications to Figure 12.6, shown in Figure C1. These changes lock the circuit into its "sample" mode, permitting an input-to-output delay measurement under signal-level conditions similar to normal operation. In Figure C2, trace A is the pulse-generator input at 200μV/DIV (note 10k-1Ω divider feeding the settle node). Trace B shows the circuit output at A2, delayed by about 12ns. This delay is a small error, but is readily corrected by adjusting the delay network for the same time lag.

Figure C3 takes a similar approach with Figure 12.26's sampling 'scope-based measurement. Modifications permit a small amplitude pulse to drive the settle node, mimicking normal operation signal level conditions. Circuit output at A2 is monitored for delay with respect to the input pulse. Note that A2's high impedance output requires a FET probe to avoid loading. As such, the input pulse must be routed to the oscilloscope via a similar FET probe to maintain delay matching. Figure C4 shows results. The output (trace B) lags the input by 32ns. This factor is used to calibrate text Figure 12.26's delay network, compensating the circuit's signal path propagation time error. Delay compensation values for the bootstrapped clamp (text Figure 12.24) and differential amplifier (text Figure 12.28) circuits were determined in similar fashion.

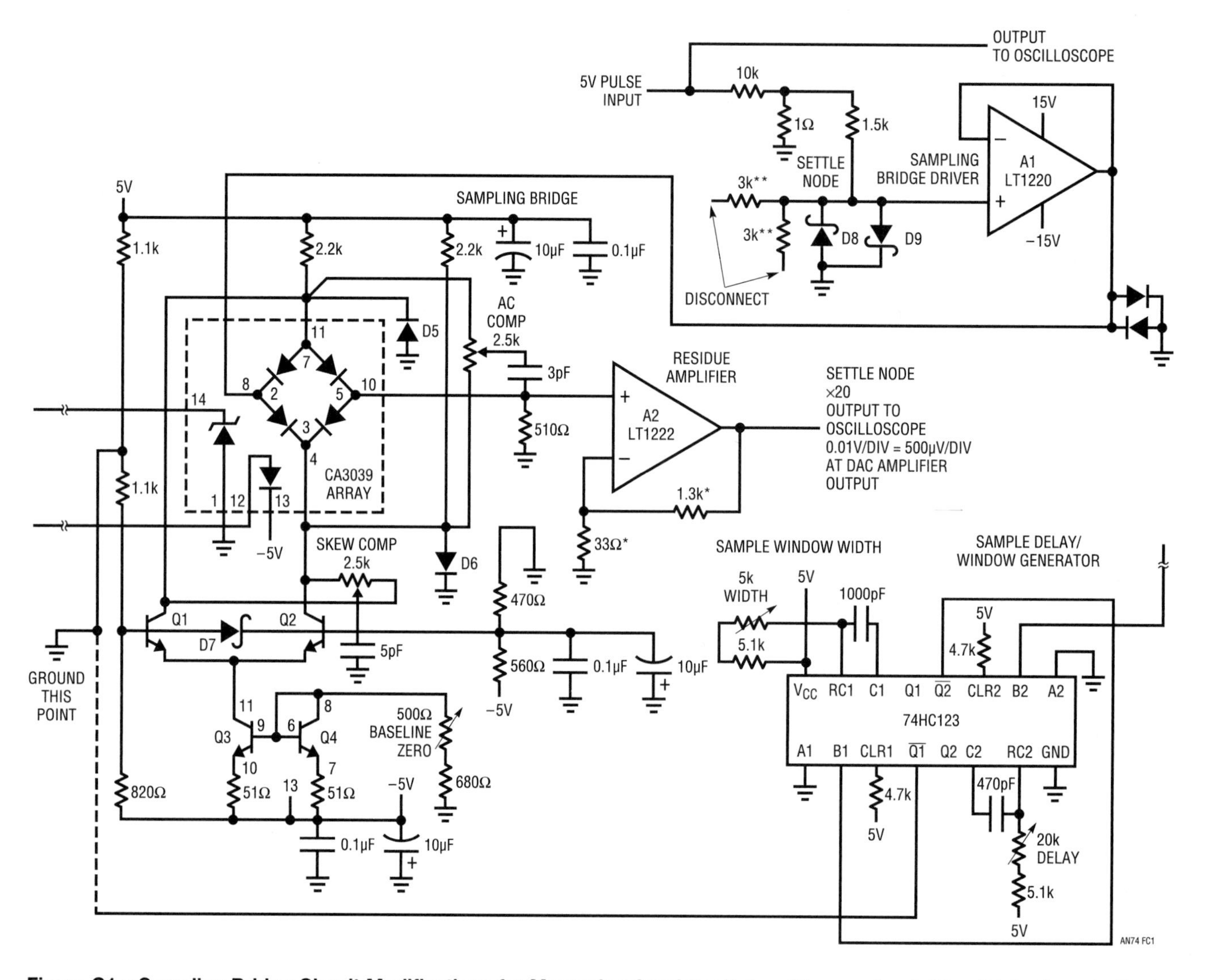

Figure C1 • Sampling Bridge Circuit Modifications for Measuring Amplifier Delay. Changes Lock Circuit into Sample Mode, Permitting Input-to-Output Delay Measurement

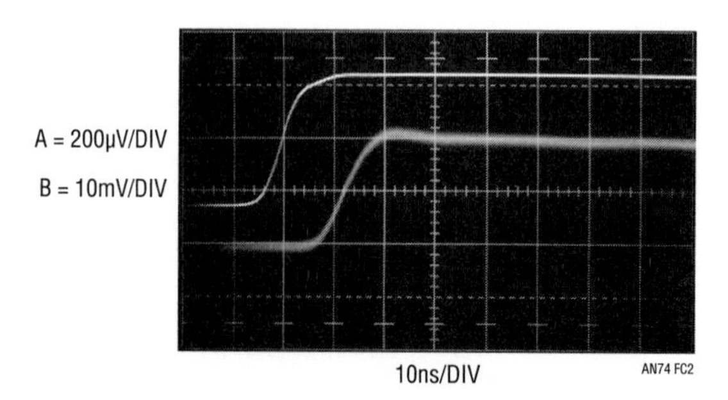

Figure C2 • Input-Output Delay for Sampling Bridge Circuit Measures About 12ns

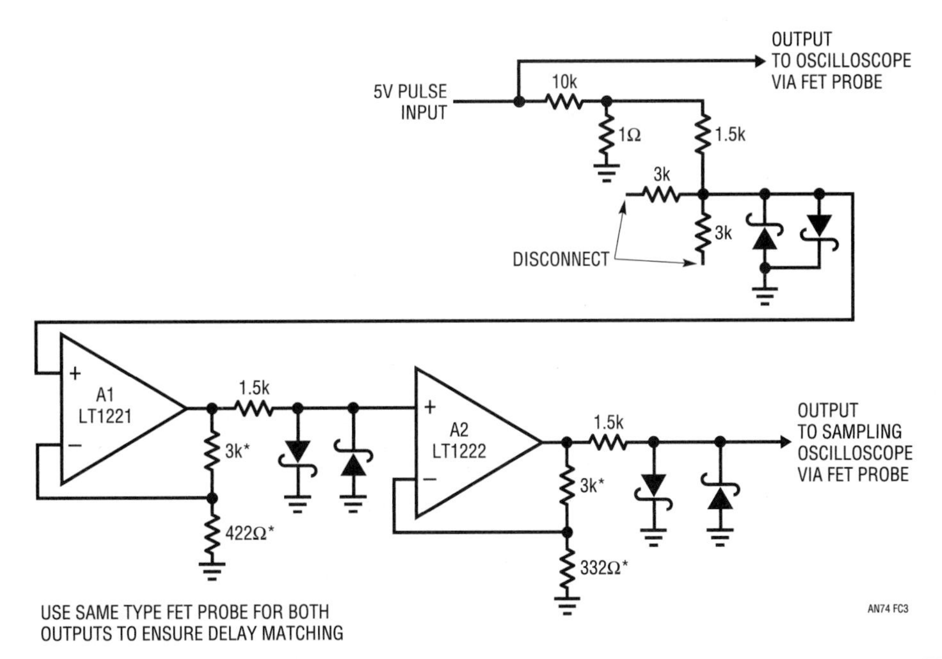

Figure C3 • Partial Version of Figure 12.26 Showing Modifications Permitting Delay Time Measurement. FET Probes are the Same Type, Eliminating Time Skewing Error

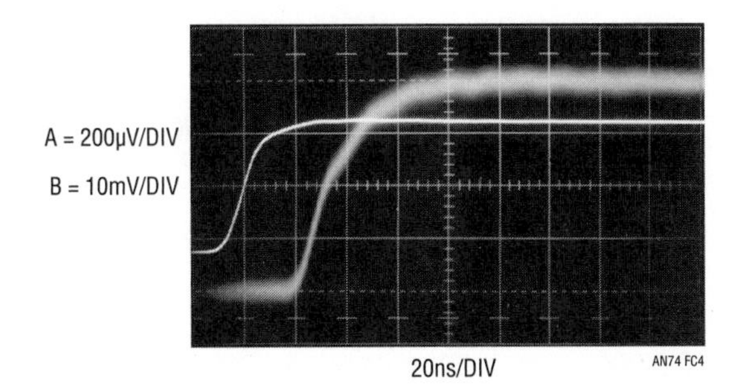

Figure C4 • Delay Measurement Results for Figure C3. Input-Output Time Lag is About 32ns

Appendix D
Practical considerations for DAC-amplifier compensation

There are a number of practical considerations in compensating the DAC-amplifier pair to get fastest settling time. Our study begins by revisiting text Figure 12.1 (repeated here as Figure D1). Settling time components include delay, slew and ring times. Delay is due to propagation time through the DAC-amplifier and is a very small term. Slew time is set by the amplifier's maximum speed. Ring time defines the region where the amplifier recovers from slewing and ceases movement within some defined error band. Once a DAC-amplifier pair have been chosen, only ring time is readily adjustable. Because slew time is usually the dominant lag, it is tempting to select the fastest slewing amplifier available to obtain best settling. Unfortunately, fast slewing amplifiers usually have extended ring times, negating their brute force speed advantage. The penalty for raw speed is, invariably, prolonged ringing, which can only be damped with large compensation capacitors. Such compensation works, but results in protracted settling times. The key to good settling times is to choose an amplifier with the right balance of slew rate and recovery characteristics and compensate it properly. This is harder than it sounds because amplifier settling time cannot be predicted or extrapolated from any combination of data sheet specifications. It must be measured in the intended configuration. In the case of a DAC-amplifier, a number of terms combine to influence settling time. They include amplifier slew rate and AC dynamics, DAC output resistance and capacitance, and the compensation capacitor. These terms interact in a complex manner, making predictions hazardous.[1] If the DAC's parasitics are eliminated and replaced with a pure resistive source, amplifier settling time is still not readily predictable. The DAC's output impedance terms just make a difficult problem more messy. The only real handle available to deal with all this is the feedback compensation capacitor, CF. CF's purpose is to roll off amplifier gain at the frequency that permits best dynamic response. Normally, the DAC's current output is unloaded directly into the amplifier's summing junction, placing the DAC's parasitic capacitance from ground to the amplifier's input. The capacitance introduces feedback phase shift at high frequencies, forcing the amplifier to "hunt" and ring about the final value before settling. Different DACs have different values of output capacitance. CMOS DACs have the highest output capacitance, typically 100pF and it varies with code.

Best settling results when the compensation capacitor is selected to functionally compensate for all the above parasitics. Figure D2 shows results for an optimally

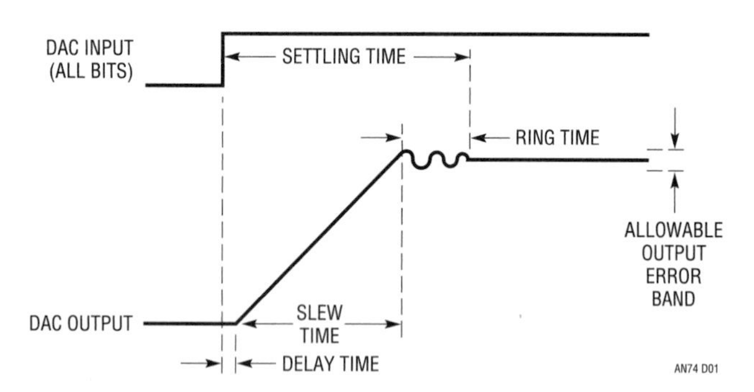

Figure D1 • DAC-Amplifier Settling Time Components Include Delay, Slew and Ring Times. For Given Components, Only Ring Time is Readily Adjustable

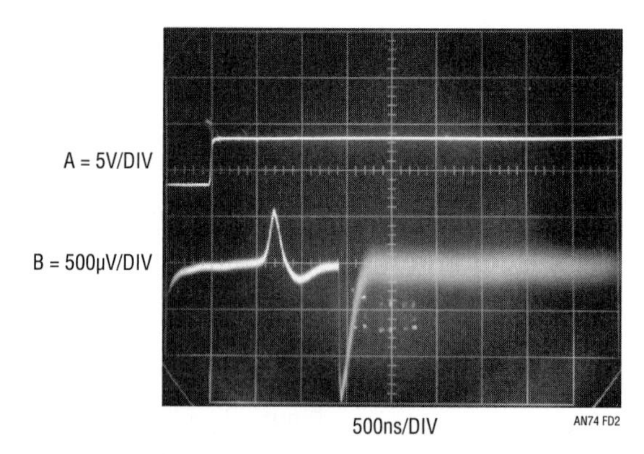

Figure D2 • Optimized Compensation Capacitor Permits Nearly Critically Damped Response, Fastest Settling Time. $t_{SETTLE} = 1.7\mu s$

Note 1: Spice aficionados take notice.

selected feedback capacitor. Trace A is the DAC input pulse and trace B the amplifier's settle signal. The amplifier is seen to come cleanly out of slew (sample gate opens just prior to fifth vertical division) and settle very quickly.

In Figure D3, the feedback capacitor is too large. Settling is smooth, although overdamped, and a 600ns penalty results. Figure D4's feedback capacitor is too small, causing a somewhat underdamped response with resultant excessive ring time excursions. Settling time goes out to 2.3μs.

When feedback capacitors are individually trimmed for optimal response, the DAC, amplifier and compensation capacitor tolerances are irrelevant. If individual trimming is not used, these tolerances must be considered to determine the feedback capacitor's production value. Ring time is affected by DAC capacitance and resistance, as well as the feedback capacitor's value. The relationship is nonlinear, although some guidelines are possible. The DAC impedance terms can vary by ±50% and the feedback capacitor is typically a ±5% component. Additionally, amplifier slew rate has a significant tolerance, which is stated on the data sheet. To obtain a production feedback capacitor value, determine the optimum value by individual trimming *with the production board layout* (board layout parasitic capacitance counts too!). Then, factor in the worst-case percentage values for DAC impedance terms, slew rate and feedback capacitor tolerance. Add this information to the trimmed capacitors measured value to obtain the production value. This budgeting is perhaps unduly pessimistic (RMS error summing may be a defensible compromise), but will keep you out of trouble.[2] Figure 12.34's "conservative" settling time values were arrived at in this manner. Note that the chart's slow slewing amplifiers have the same compensation capacitor for "optimal" and "conservative" cases. This reflects the fact that their ring times are very small compared to their slew intervals.

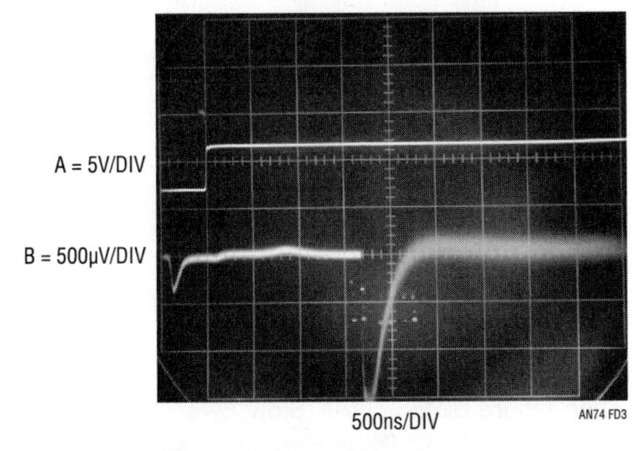

Figure D3 • Overdamped Response Ensures Freedom from Ringing, Even with Component Variations in Production. Penalty is Increased Settling Time. $t_{SETTLE} = 2.3\mu s$

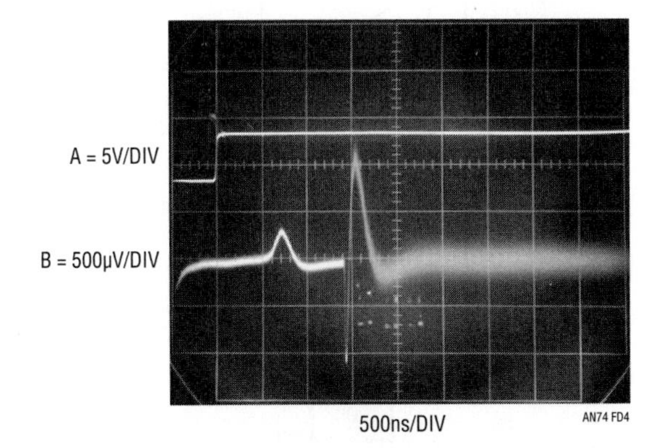

Figure D4 • Underdamped Response Results from Undersized Capacitor. Component Tolerance Budgeting Will Prevent This Behavior. $t_{SETTLE} = 2.3\mu s$

Note 2: The potential problems with RMS error summing become clear when sitting in an airliner that is landing in a snowstorm.

Appendix E
A very special case—measuring settling time of chopper-stabilized amplifiers

Figure 12.34's table lists the LTC1150 chopper-stabilized amplifier. The term "special case" appears in the "comments" column. A special case it is! To see why requires some understanding of how these amplifiers work. Figure E1 is a simplified block diagram of the LTC1150 CMOS chopper-stabilized amplifier. There are actually two amplifiers. The "fast amp" processes input signals directly to the output. This amplifier is relatively quick, but has poor DC offset characteristics. A second, clocked, amplifier is employed to periodically sample the offset of the fast channel and maintain its output "hold" capacitor at whatever value is required to correct the fast amplifier's offset

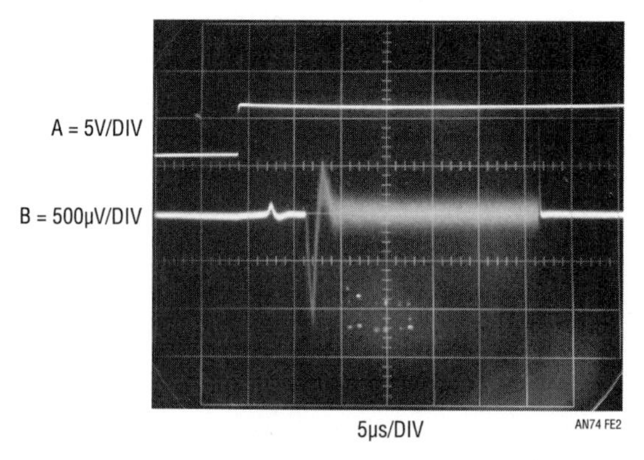

Figure E2 • Short-Term Settling Profile of Chopper-Stabilized Amplifier Seems Typical. Settling Appears to Occur in 10μs

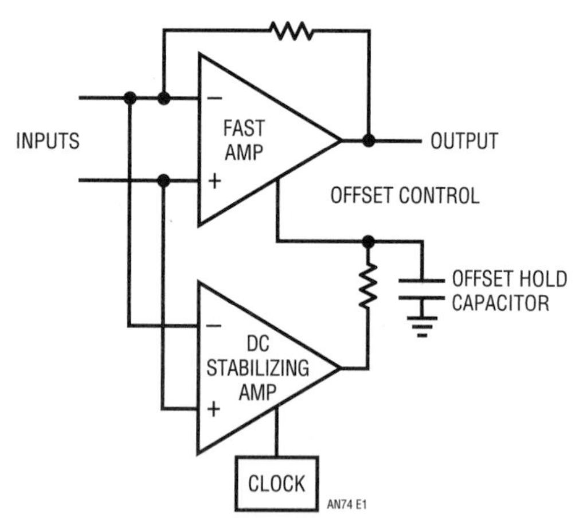

Figure E1 • Highly Simplified Block Diagram of Monolithic Chopper-Stabilized Amplifier. Clocked Stabilizing Amplifier and Hold Capacitor Cause Settling Time Lag

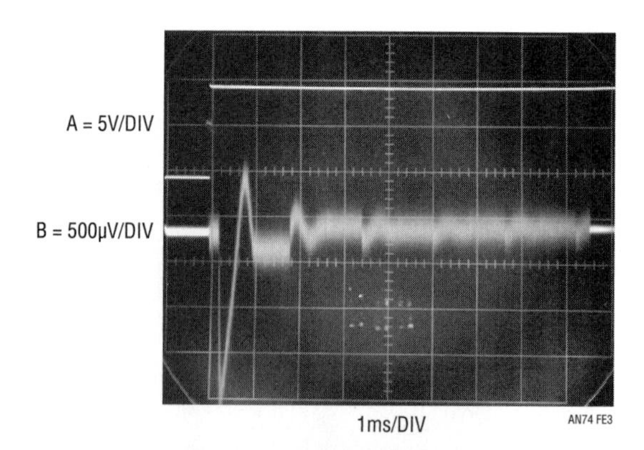

Figure E3 • Surprise! Actual Settling Requires 700× More Time Than Figure E2 Indicates. Slow Sweep Reveals Monstrous Tailing Error (Note Horizontal Scale Change) Due to Amplifier's Clocked Operation. Stabilizing Loop's Iterative Corrections Progressively Reduce Error Before Finally Disappearing Into Noise

errors. The DC stabilizing amplifier is clocked to permit it to operate (internally) as an AC amplifier, eliminating its DC terms as an error source.[1] The clock chops the stabilizing amplifier at about 500Hz, providing updates to the hold capacitor-offset control every 2ms.[2]

The settling time of this composite amplifier is a function of the response of the fast and stabilizing paths. Figure E2 shows short-term settling of the amplifier. Trace A is the DAC input pulse and trace B the settle signal. Damping is reasonable and the 10μs settling time and profile appear typical. Figure E3 brings an unpleasant surprise.

If the DAC slewing interval happens to coincide with the amplifier's sampling cycle, serious error is induced. In Figure E3, trace A is the amplifier output and trace B the settle signal. Note the slow horizontal scale. The amplifier initially settles quickly (settling is visible in the 2nd vertical division region) but generates a huge error 200μs later when its internal clock applies an offset correction. Successive clock cycles progressively chop the error into the noise but 7 *milliseconds* are required for complete recovery. The error derives from the fact that the amplifier sampled its offset when its input was driven well outside its bandpass. This caused the stabilizing amplifier to acquire erroneous offset information. When this "correction" was applied, the result was a huge output error.

This is admittedly a worst case. It can only happen if the DAC slewing interval coincides with the amplifier's internal clock cycle, but it can happen.[3,4]

Appendix F Settling time measurement of serially loaded DACs

LTC models LTC1595 and LTC1650 are serially loaded 16-bit DACs. The LTC1650 includes an onboard output amplifier. Measuring these device's settling time with the methods described in the text requires additional circuitry. The circuitry must provide a "start" pulse to the settling time measurement after serially loading a full-scale step into the DAC. Figure F1 's circuitry, designed and constructed by Jim Brubaker, Kevin Hoskins, Hassan Malik and Tuyet Pham (all of LTC) does this. The "start" pulse is taken from U1B's Q output. The DAC amplifier is monitored in the normal manner. This permits Figure F2 to display settling results in (what should be by now) familiar fashion. Settling time (trace B) is measured from the rising edge of trace A's start pulse. Figure F3 is a similar circuit configuration for the LTC1650 voltage output DAC. Reference voltage and other changes are required to accommodate the DAC's ±4V output swing and different architecture, but overall operation is similar. Figure F4 shows settling results.

Note 1: This AC processing of DC information is the basis of all chopper and chopper-stabilized amplifiers. In this case, if we could build an inherently stable CMOS amplifier for the stabilizing stage, no chopper stabilization would be necessary.

Note 2: Those finding this description intolerably brief are commended to reference 23.

Note 3: Readers are invited to speculate on the instrumentation requirements for obtaining Figure E3's photo.

Note 4: The spirit of Appendix D's footnote 2 is similarly applicable in this instance.

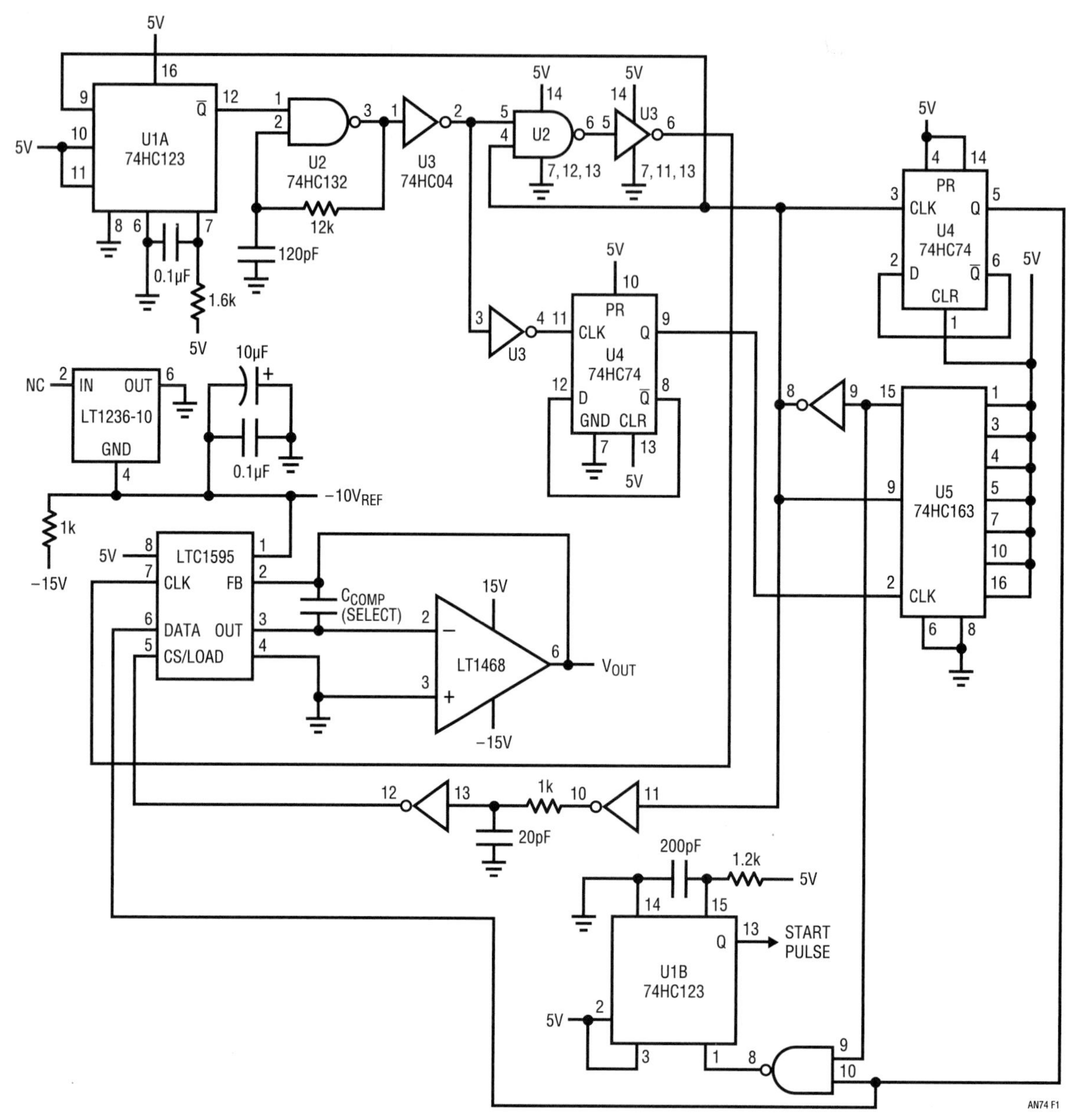

Figure F1 • Logic Circuitry Permits Measuring Settling Time of Serially Loaded LTC1595 DAC

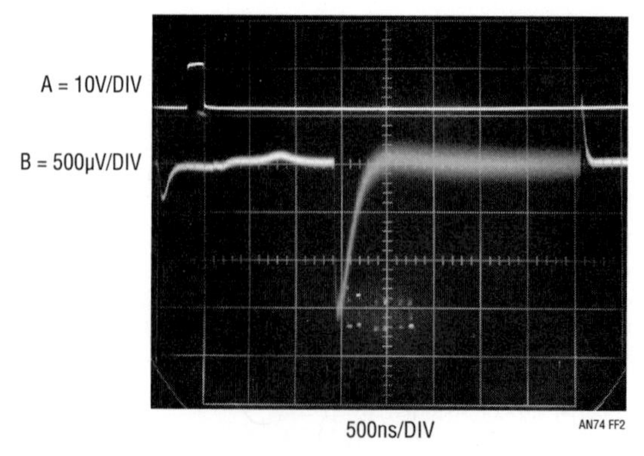

Figure F2 • Oscilloscope Display of Serially Loaded DAC Settling Time Measurement. Settling Time is Measured from "Start Pulse" (See Schematic) Rising Edge (Trace A)

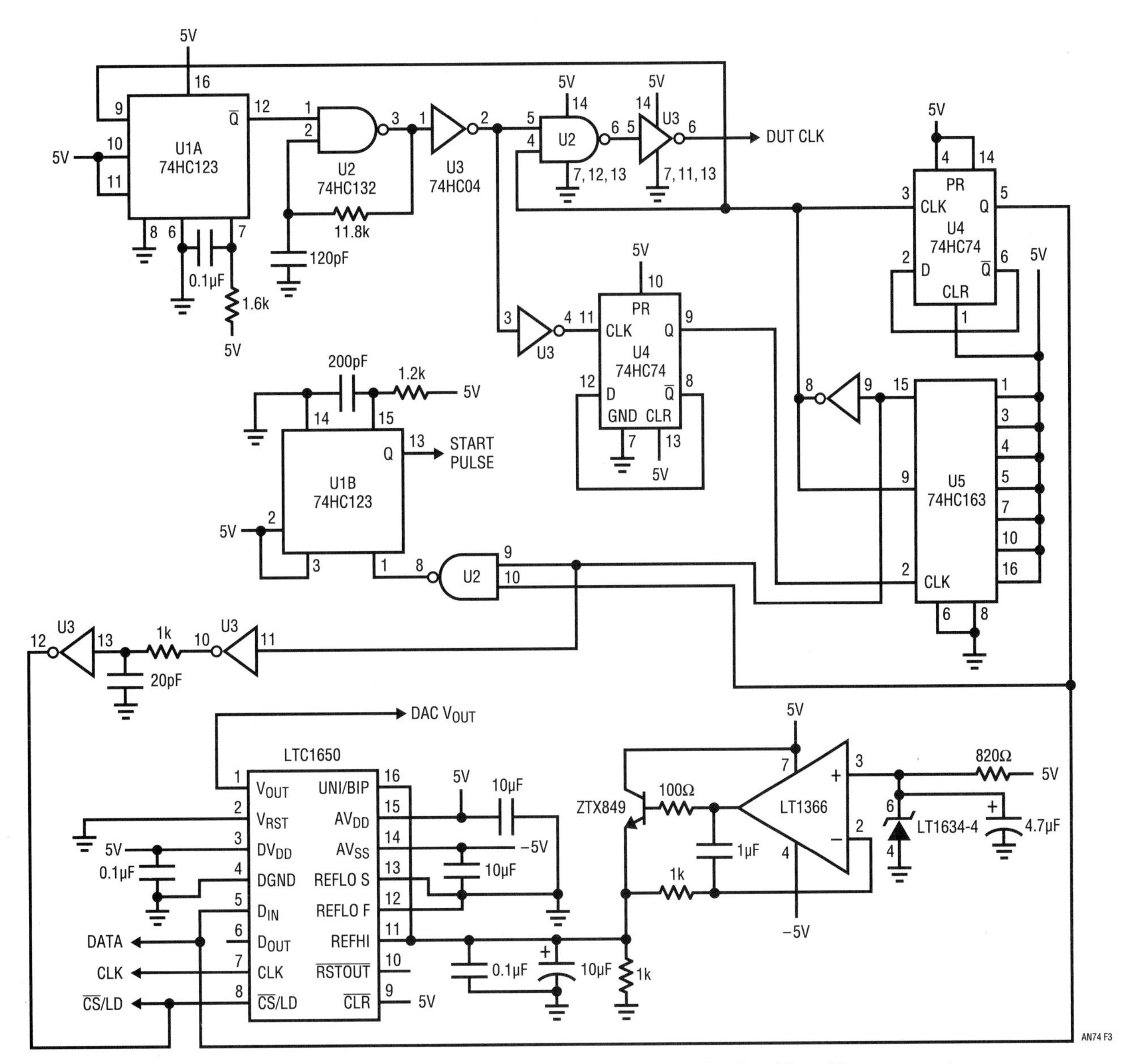

Figure F3 • Reconfiguration Figure F1 Allows LTC1650 Settling Time Measurement

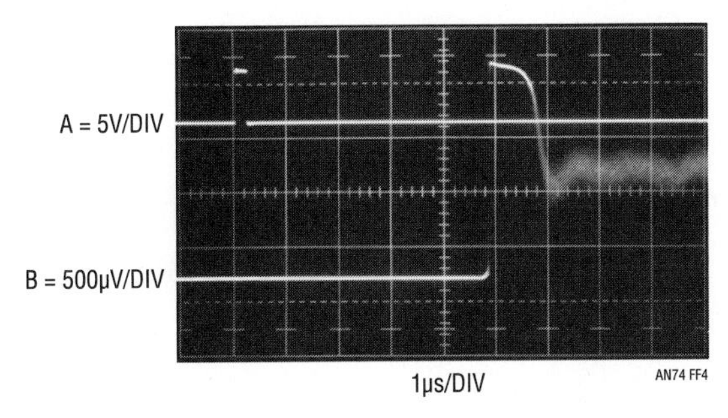

Figure F4 • LTC1650 Voltage Output DAC Shows 6µs Settling Time. A is Start Pulse, B the Settle Signal

Appendix G Breadboarding, layout and connection techniques

The measurement results presented in this publication required painstaking care in breadboarding, layout and connection techniques. Wideband, 100μV resolution measurement does not tolerate cavalier laboratory attitude. The oscilloscope photographs presented, devoid of ringing, hops, spikes and similar aberrations, are the result of an exhaustive (and frustrating) breadboarding exercise.[1] The sampler-based breadboard (Figures G1, G2) was rebuilt six times and required days of layout and shielding experimentation before obtaining a noise/uncertainty floor worthy of 16-bit measurement.

Ohm's law

It is worth considering that Ohm's law is a key to successful layout.[2] Consider that 1mA running through 0.1Ω generates 100μV—almost 1LSB at 16 bits! Now, run that milliampere at 5 to 10 nanosecond rise times (≈75MHz) and the need for layout care becomes clear. A paramount concern is disposal of circuit ground return current and disposition of current in the ground plane. The impedance of the ground plane between any two points is *not* zero, particularly as frequency scales up. This is why the entry point and flow of "dirty" ground returns must be carefully placed within the grounding system. In the sampler-based breadboard, the approach was separate "dirty" and "signal" ground planes (see Figures G1 to G7), tied together at the supply ground origin.

A good example of the importance of grounding management involves delivering the input pulse to the breadboard. The pulse generator's 50Ω termination *must* be an in-line coaxial type, and it cannot be directly tied to the signal ground plane. The high speed, high density (5V pulses through the 50Ω termination generate 100mA current spikes) current flow must return directly to the pulse generator. The coaxial terminator's construction ensures this substantial current does this, instead of being dumped into the signal ground plane (100mA termination current flowing through 1 *milliohm* of ground plane produces ≈1LSB of error!). Figure G3 shows that the BNC shield floats from the signal plane, and is returned to "dirty" ground via RF braid. Additionally, Figure G1 shows the pulse generator's 50Ω termination physically distanced from the breadboard via a coaxial extension tube. This further ensures that pulse generator return current circulates in a tight local loop at the terminator, and does not mix into the signal plane.

It is worth mentioning that every ground return in the entire circuit must be evaluated with these concerns in mind. A paranoiac mindset is quite useful.

Shielding

The most obvious way to handle radiation-induced errors is shielding. Various following figures show shielding. Determining where shields are required should come *after* considering what layout will minimize their necessity. Often, grounding requirements conflict with minimizing radiation effects, precluding maintaining distance[3] between sensitive points. Shielding[4] is usually an effective compromise in such situations.

A similar approach to ground path integrity should be pursued with radiation management. Consider what points are likely to radiate, and try to lay them out at a distance from sensitive nodes. When in doubt about odd effects, experiment with shield placement and note results, iterating towards favorable performance.[5] *Above all, never rely on filtering or measurement bandwidth limiting to "get rid of" undesired signals whose origin is not fully understood*. This is not only intellectually dishonest, but may produce wholly invalid measurement "results," even if they look pretty on the oscilloscope.

Note 1: "War" is perhaps a more accurate descriptive.

Note 2: I do not wax pedantic here. My abuse of this postulate runs deep.

Note 3: Distance is the physicist's approach to reducing radiation induced effects.

Note 4: Shielding is the engineer's approach to reducing radiation induced effects.

Note 5: After it works, you can figure out why.

Connections

All signal connections to the breadboard must be coaxial. Ground wires used with oscilloscope probes are forbidden. A 1" ground lead used with a 'scope probe can easily generate several LSBs of observed "noise"! Use coaxially mounting probe tip adapters![6]

Figures G1 to G10 restate the above sermon in visual form while annotating the text's measurement circuits.

Note 6: See Reference 26 for additional nagging along these lines.

Figure G1 • Overview of Settling Time Breadboard. Pulse Generator Input Enters Top Left—50Ω Coaxial Terminator Mounted On Extension Minimizes Pulse Generator Return Current Mixing Into Signal Ground Plane (Bottom Board Facing Viewer). "Dirty" Ground Paths Return Separately from Signal Ground Plane Via Planed Horizontal Strip (Upper Left of Main Board). DAC-Amplifier and Support Circuitry are at Extreme Left on Vertical Board. Sampler Circuitry Occupies Board Lower Center. Nonsaturating Amplifier-Bootstrapped Clamp is Thin Board, Extreme Right. Note Coaxial Board Signal and Probe Connections

Figure G2 • Settling Time Breadboard Detail. Note Radiation Shielding at Delayed Pulse Generator (Lower Left Center), Sampling Bridge Area (Lower Center Right) and DAC-Amplifier Board (Extreme Left). "Dirty" Grounds Return Via Separate Plane (Horizontal Strip, Photo Center Left). Vertical Shield (Center) Bisects Board, Separating Delayed Pulse Generator's (Lower Left Center) Fast Edges from Sampling Bridge Circuitry. DAC Amplifier Output is Routed Via Thin Copper Strip (Angled, Running Left-Right) to Settle Node (Board Center Right). Shield (Angled, Center) Prevents Radiation Into Bridge Area

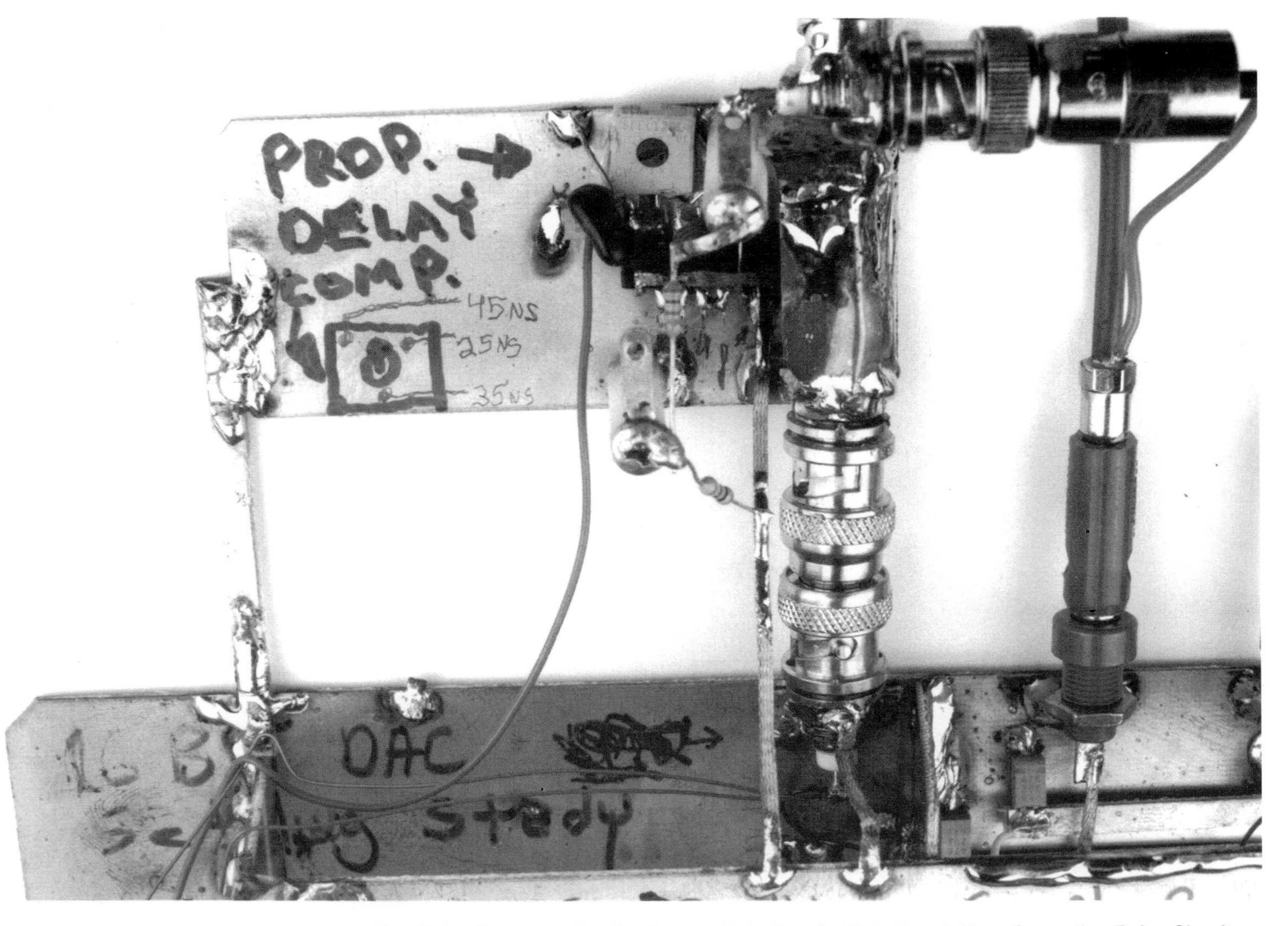

Figure G3 • Detail of Pulse Generator Input—Delay Compensation Section and Interface to Main Board. Time Correction Delay Circuitry is at Photo Upper Center. Coaxial Probe Pick-Off Upper Right. Time Corrected Pulse Enters Main Board at Lower Center Plane (Clad Stand-Off Just Visible at Lower Connector Right). Connector Shell Ground is Tied To "Dirty" Ground Plane Via RF Braid (Lower Right Center), Preventing High Speed Return Currents from Corrupting Main Board Signal Ground Plane. Active Current Returns Flow Through Separate Braid (Long Vertical Run, Photo Center Right) to Main Board "Dirty" Ground Plane

Figure G4 • Delayed Pulse Generator is Fully Shielded (Vertical Shield, Photo Center Right) from Sampling Circuitry (Partial, to Right of Vertical Shield). Delayed Pulse Generators Output Lead (Photo Center) Sneaks Under Main Ground Plane to Minimize Radiation Into Sampling Bridge. Screw Adjustment (Photo Center Left) Sets Delayed Pulse Width, Large Potentiometer (Partial, Photo Upper Left Center) Sets Delay

Figure G5 • Sampling Bridge and Support Circuitry. Delayed Pulse Generator Output Arrives from Under Ground Plane (Photo Center, Just to Right of T0-220 Power Package), Triggers Complementary Level Shifters (Center Left). Sampling Bridge is SOT-16 Package at Photo Center. Skew Compensation and Bridge AC Balance Trimpots are at Photo Center Right. Baseline Zero is Large Knob at Left. Sampling Bridge Temperature Control Circuitry Appears Upper Right Center

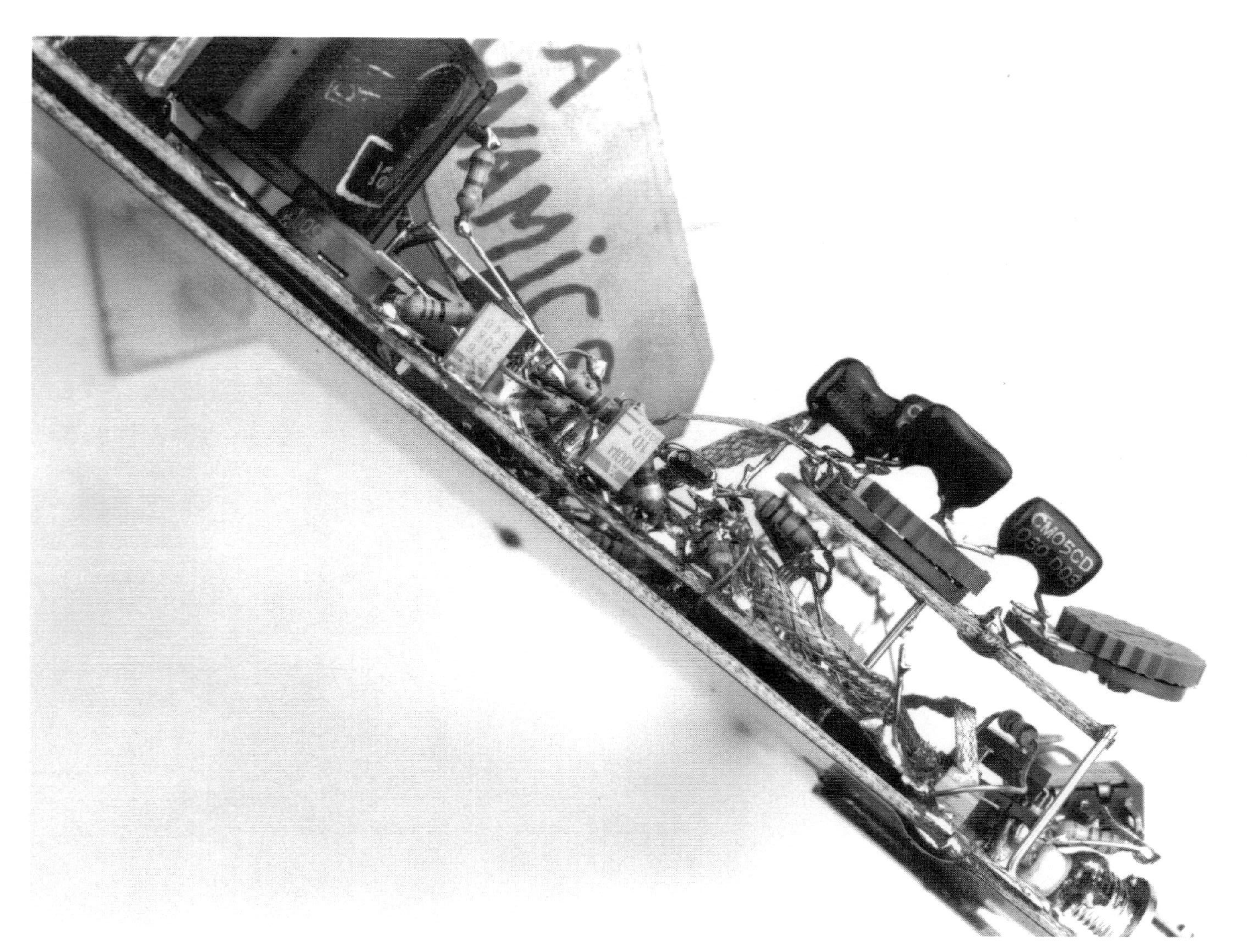

Figure G6 • Side-On View of Sampling Bridge Circuitry. Delayed Pulse Generator Output Line is Just Visible in Shielded Space Under Ground Plane (≈45° Angle, Photo Extreme Left, Running Towards Center). SOT-16 Packaged Sampling Bridge (Exact Photo Center) Floats on Flying Leads to Maximize Thermal Resistance, Aiding Temperature Control. Upper Board (Photo Right) Mechanically Supports AC Bridge Trimpots

Figure G7 • DAC-Ampiifier Board Contains DAC (Right), Amplifier ("W4" at Center Right) and Reference (Lower Right). Digital IC Packages at Left are Serial DAC Interface, Potential Noise Generators. Insulating Strip Across Entire Board Bottom Fully Isolates it from Main Board Signal Ground Plane. Individual Board Returns are Routed Separately to Main Board Power Common

Figure G8 • Gain-of-80 Nonsaturating Amplifier (Right of BNC Adapter) and Bootstrapped Clamp (Left of BNC Adapter). Ground Planed Construction and Minimized Summing Point Capacitances Aid Wideband Response

Figure G9 • Wideband Nonsaturating Gain of 40 for Figure 12.28's Differential Amplifier. Layout Ensures Minimal Feedthrough, Particularly When Amplifier is Outside Gain Region. Note Input Shield (Photo Right)

Figure G10 • Detail of Nonsaturating Gain of 40 Input Shielding (Photo Right). Shield Prevents Input Excursions Outside Amplifier's Gain Region from Feeding Through to Output, Corrupting Data

Appendix H
Power gain stages for heavy loads and line driving

Some applications require driving heavy loads. The load may be static, transient or both. Practical examples of loads include actuators, cables and power voltage/current sources in test equipment. Required load currents may range from tens of milliamperes to amperes, while simultaneously maintaining 16-bit performance. Figure H1 summarizes the system problems involved in applying a power gain stage, sometimes referred to as a booster.

The booster stage's output impedance must be low enough to accurately drive complex loads at all frequencies of interest. "Complex" loads may include interconnecting cable capacitance, pure resistance, capacitive and inductive components. Significant effort should be expended to characterize the load prior to designing the booster. Note that reactive load components will almost certainly add a stability term, complicating wideband loop dynamics. These considerations dictate that the booster's output impedance must be exceptionally low at all frequencies encountered. Also, the booster must be fast to avoid delay-induced stability problems. Any booster included in an amplifier's loop must be transparent to the amplifier to preserve dynamic performance.[1]

Grounding and connection considerations mandate special attention in a high current, 16-bit, DAC driven system. A 1A load current returning through 1mΩ of parasitic resistance causes almost 7LSBs of error. Similar errors occur if feedback sensing is improperly arranged. As such, single-point grounding, in the strictest sense of the word, is mandatory. In particular, the load-return conductor should be thick, short, flat and highly conductive. Feedback sensing should be arranged so that the DAC's R_{FB} terminal is connected *directly* to the load via a low impedance conductor.

Note 1: This discussion must suffer brevity in this forum. For more detail, see References 26 and 27.

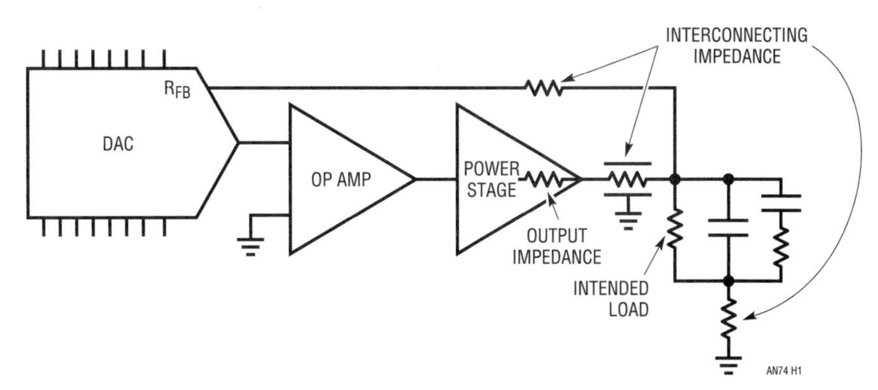

Figure H1 • Conceptual Power Gain Stage for DAC-Amplifier. System Issues Involve Booster Output Impedance, Interconnections, Load Characteristics and Grounding

Booster circuits

Figure H2 is a simple power output stage using the LT1010 IC booster. Output currents to 125mA are practical. Figure H3 is similar, although offering higher speed operation and more output power. The LT1206 op amp, configuration D as a unity-gain follower, permits 250mA outputs; the LT1210 extends current capacity to 1.1A. The indicated optional capacitor improves dynamic performance with capacitive loading (see data sheets). Figure H4 extends current output capacity to 2A by utilizing a wideband discrete state. In the positive signal-path output transistor Q4 is an RF power type, driven by Darlington connected Q3. The diode in Q1's emitter compensates the additional V_{BE} introduced by Q3, preventing crossover distortion.

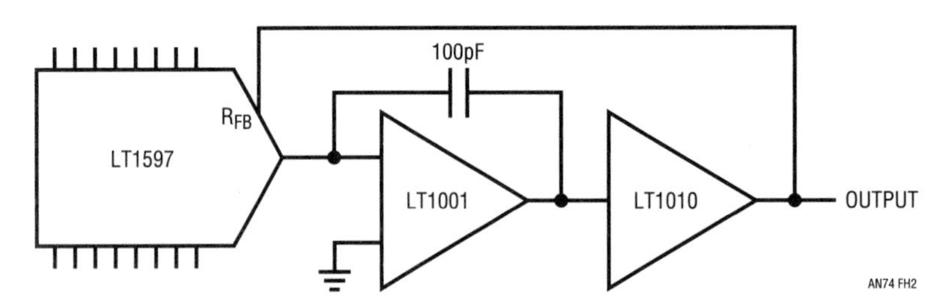

Figure H2 • LT1010 Permits 125mA Output Currents

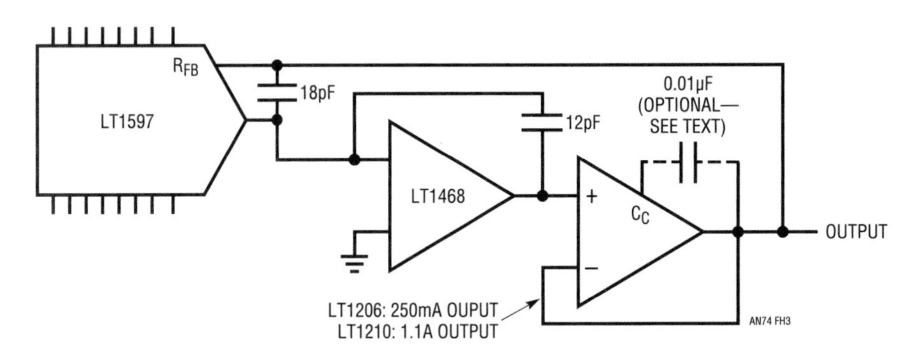

Figure H3 • LT1206/LT1210 Output Stages Supply 250mA and 1.1A Loads, Respectively

The negative signal path substitutes the Q5-Q6 connection to simulate a fast PNP power transistor. This arrangement is necessitated by the lack of availability of wideband PNP power transistors. Although this configuration acts like a fast PNP follower, it has voltage gain and tends to oscillate. The local 2pF feedback capacitor suppresses these parasitic oscillations and the composite transistor is stable.

This circuit also includes a feedback capacitor to optimize AC response. Current limiting is provided by Q7 and Q8, which sense across the 0.2Ω shunts.

Figure H5 is a voltage gain stage. The high voltage stage is driven in closed-loop fashion by A2, instead of being included in the DAC-amplifier (A1) loop. This avoids driving the DAC's monolithic feedback resistor from the

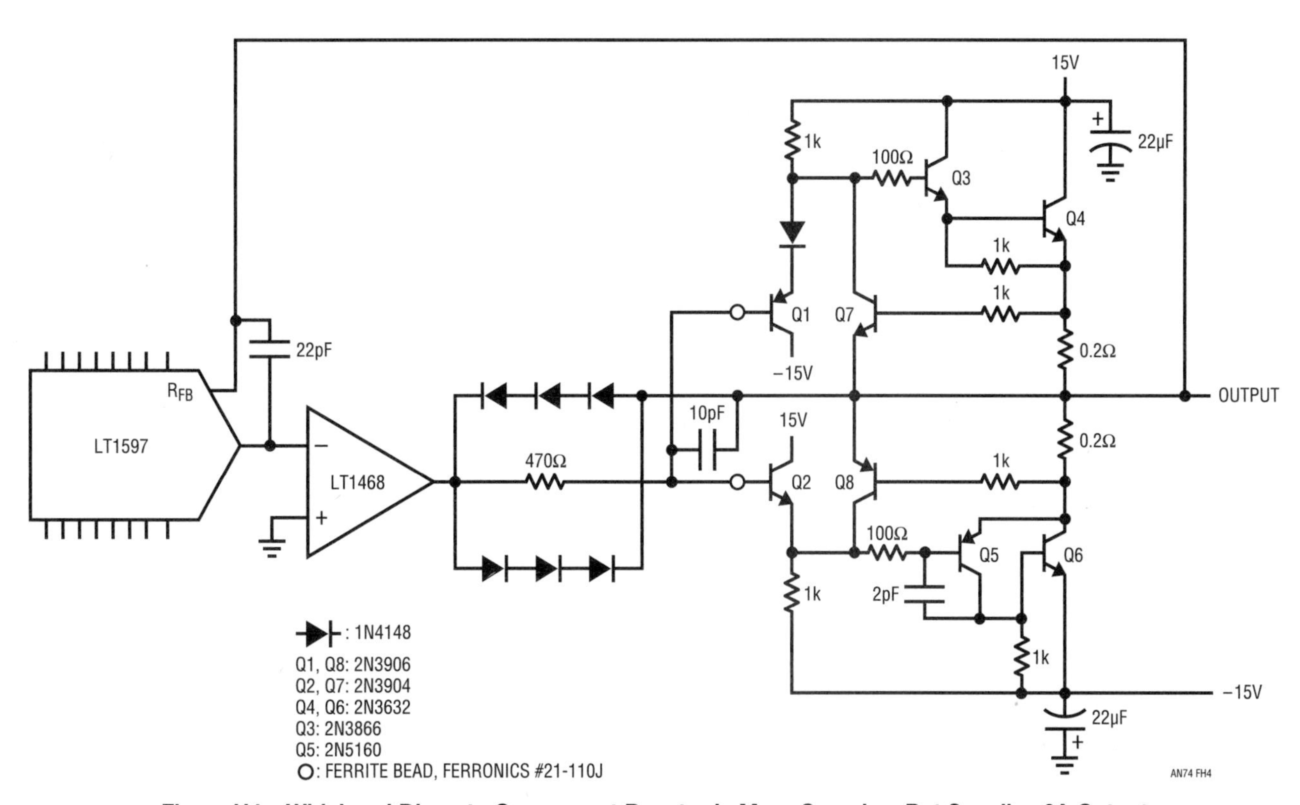

Figure H4 • Wideband Discrete Component Booster is More Complex, But Supplies 2A Output

100V output, preserving DAC temperature coefficient and, not incidentally, the DAC. Q1 and Q2 furnish voltage gain, and feed the Q3-Q4 emitter follower outputs. Q5 and Q6 set current limit at 25mA by diverting output drive when voltages across the 27Ω shunts become too high. The local 1M-50k feedback pairs set stage gain at 20, allowing ±12V A2 drives to cause full ±120V output swing. The local feedback reduces stage gain-bandwidth, making dynamic control easier. This stage is relatively simple to frequency compensate because only Q1 and Q2 contribute voltage gain. Additionally, the high voltage transistors have large junctions, resulting in low f_ts, and no special high frequency roll-off precautions are needed. Because the stage inverts, feedback is returned to A2's positive input. Frequency compensation is achieved by rolling off A2 with the local 330pF-10k pair. The 15pF capacitor in the feedback peaks edge response and is not required for stability. If over compensation is required, it is preferable to increase the 330pF value, instead of increasing the 15pF loop-feedback capacitor.

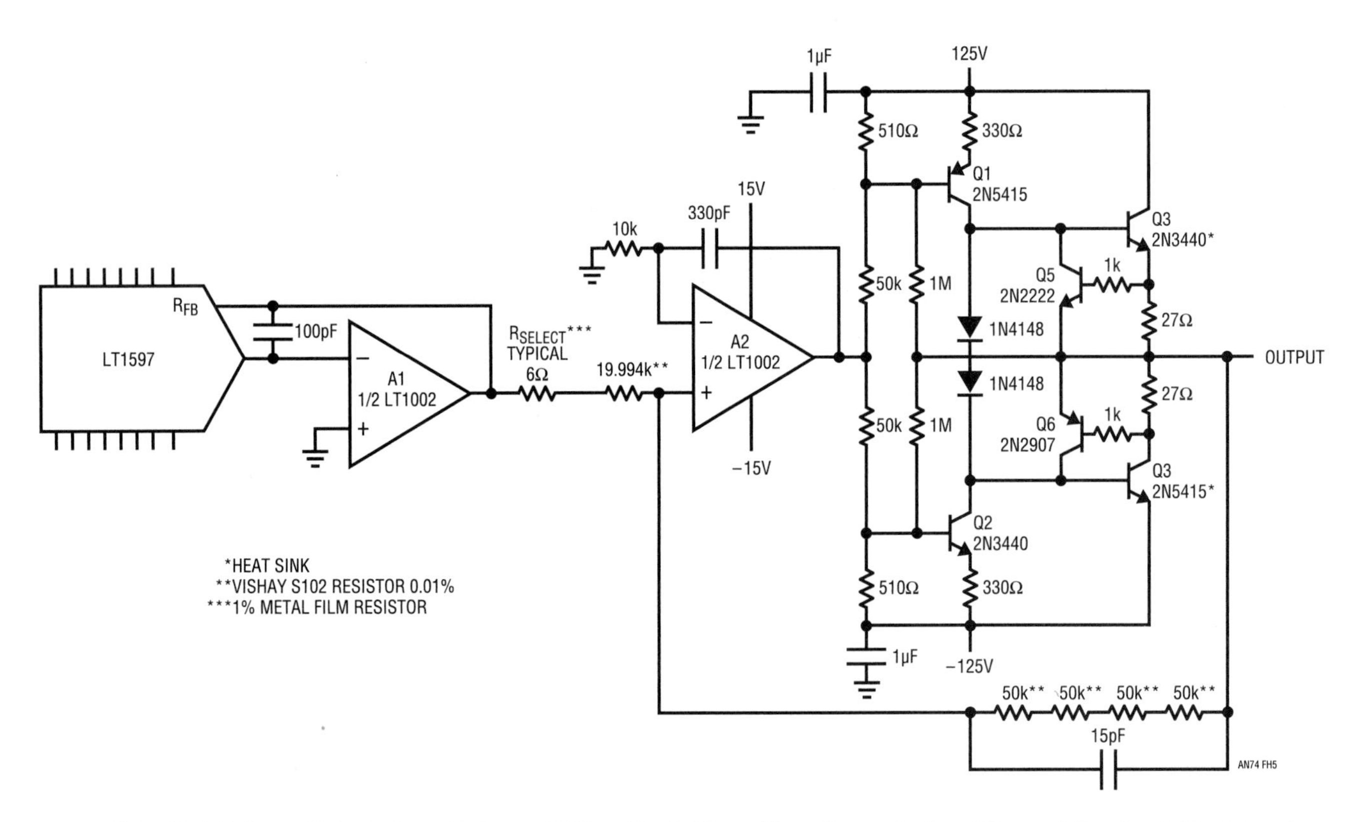

Figure H5 • High Voltage Output Stage Delivers 100V at 25mA. Stage Uses Separate Amplifier and Feedback Resistors to Preserve DAC's Gain Temperature Coefficient

This prevents excessive high voltage energy from coupling to A2's inputs during slew. If it is necessary to increase the feedback capacitor, the summing point should be diode clamped to ground or to the ±15V supply terminals. Trimming involves selecting the indicated resistor for exactly 100.000V output with the DAC at full scale.

The dynamic response issues discussed in the text apply to all the above circuits.

Figure H6 summarizes the booster stage's characteristics. The IC-based stages offer simplicity while the complex discrete designs provide more output power.

FIGURE	VOLTAGE GAIN	CURRENT GAIN	COMMENTS
H2	No	Yes	Simple 125mA Stage
H3	No	Yes	Simple 250mA/1.1A Stage
H4	No	Yes	Complex 2A Output
H5	Yes	Minimal	Complex ±120V Output

Figure H6 • Summary of Booster Stage Characteristics

Fidelity testing for A→D converters

Proving purity

13

Jim Williams Guy Hoover

Introduction

The ability to faithfully digitize a sine wave is a sensitive test of high resolution A→D converter fidelity. This test requires a sine wave generator with residual distortion products approaching one part-per-million. Additionally, a computer-based A→D output monitor is necessary to read and display converter output spectral components. Performing this testing at reasonable cost and complexity requires construction of its elements and performance verification prior to use.

Overview

Figure 13.1 diagrams the system. A low distortion oscillator drives the A→D via an amplifier. The A→D output interface formats converter output and communicates with the computer which executes the spectral analysis software and displays the resulting data.

Oscillator circuitry

The oscillator is the system's most difficult circuit design aspect. To meaningfully test 18-bit A→Ds, the oscillator must have transcendentally low levels of impurity, and these characteristics must be verified by independent means. Figure 13.2 is basically an "all inverting" 2kHz Wien bridge design (A1-A2) adapted from work by Winfield Hill of Harvard University. The original designs J-FET gain control is replaced with a LED driven CdS photocell isolator, eliminating J-FET conductivity modulation introduced errors and the trim required to minimize them. Band limited A3 receives A2 output and DC offset bias, providing output via a 2.6KHz filter which drives the A→D input amplifier. Automatic gain control (AGC) for the A1-A2 oscillator is taken from the circuit output ("AGC sense") by AC-coupled A4 which feeds rectifier A5-A6. A6's DC output represents the AC amplitude of the circuit output sine wave. This value is balanced against the LT®1029 reference by current summing resistors which terminate into AGC amplifier A7. A7,

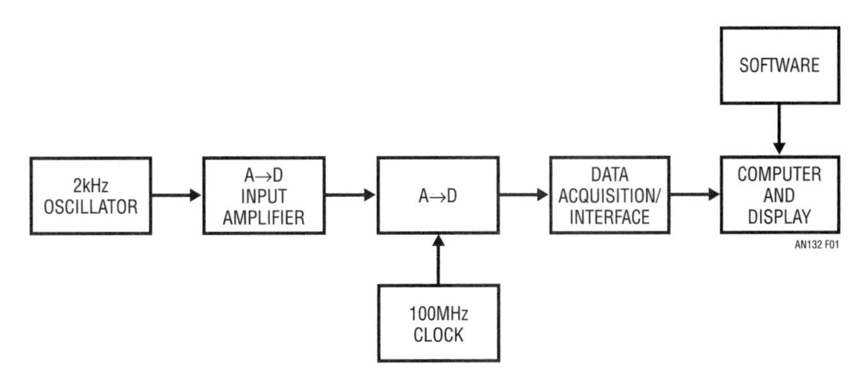

Figure 13.1 • Block Diagram of A→D Spectral Purity Test System. Assuming a Distortionless Oscillator, Computer Displays Fourier Components Due to Amplifier and A→D Infidelities

Analog Circuit and System Design: Immersion in the Black Art of Analog Design. http://dx.doi.org/10.1016/B978-0-12-397888-2.00013-4

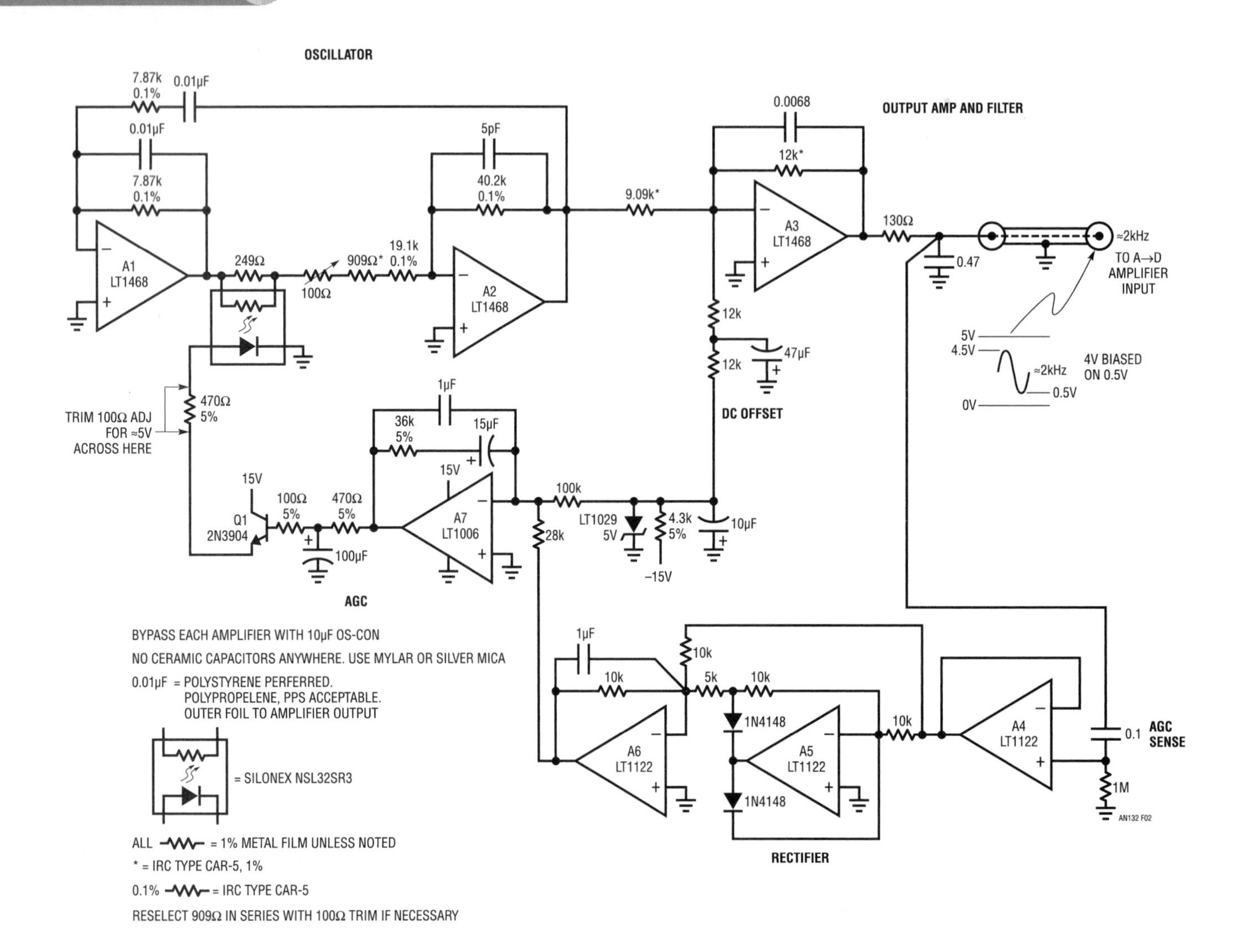

Figure 13.2 • Wien Bridge Oscillator Uses Inverting Amplifiers in Signal Path, Achieves 3ppm Distortion. LED Photocell Replaces Usual J-FET as Gain Control, Eliminating Conductivity Modulation Induced Distortion. A3 Associated Filtering Attenuation is Compensated by Sensing AGC Feedback at Circuit Output. DC Offset Biases Output Into A→D Input Amplifier Range

driving Q1, closes a gain control loop by setting LED current, and hence CdS cell resistance, to stabilize oscillator output amplitude. Deriving gain control feedback from the circuit's output maintains output amplitude despite the attenuating band-limiting response of A3 and the output filter. It also places demands on A7 loop closure dynamics. Specifically, A3's band limiting combines with the output filter, A6's lag and ripple reduction components in Q1's base to generate significant phase delay. A 1µF dominant pole at A7 along with an RC zero accommodates the delay, achieving stable loop compensation. This approach replaces closely tuned high order output filters with simple RC roll-offs, minimizing distortion while maintaining output amplitude.[1]

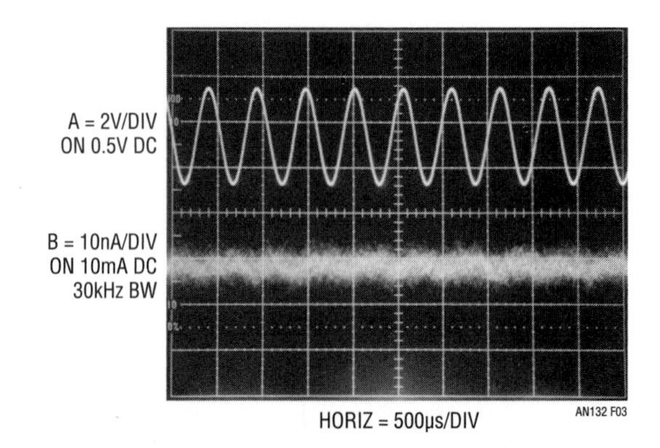

Figure 13.3 • Oscillator (Trace A) Related Residue (Trace B), Just Discernible in Q1 Emitter Noise, ≈1nA, About 0.1ppm of LED Current. Characteristic, Deriving From Heavy AGC Signal Path Filtering, Prevents Modulation Products From Influencing Photocell Response

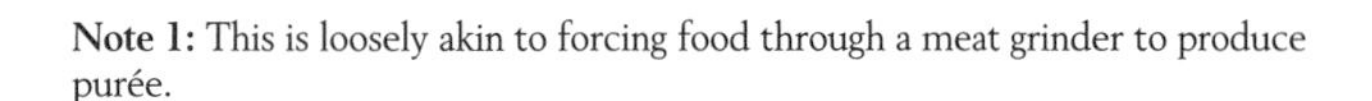

Note 1: This is loosely akin to forcing food through a meat grinder to produce purée.

Eliminating oscillator related components from the LED bias is essential to maintaining low distortion. Any such residue will amplitude modulate the oscillator, introducing impure components. The band-limited AGC signal forward path is well filtered and the heavy RC time constant in Q1's base provides a final, steep roll-off. Figure 13.3, Q1's emitter current, shows about 1nA of oscillator related ripple out of a 10mA total, less than 0.1ppm.

The oscillator achieves its performance using only a single trim. This adjustment, which centers AGC capture range, is set in accordance with the schematic note.

Verifying oscillator distortion

Verifying oscillator distortion necessitates sophisticated measurement techniques. Attempts to measure distortion with a conventional distortion analyzer, even a high grade type, encounter limitations. Figure 13.4 shows oscillator output (Trace A) and its indicated distortion residuals at the analyzer output (Trace B). Oscillator related activity is faintly outlined in the analyzer noise and uncertainty floor. The HP-339A employed specifies a minimum measurable distortion of 18ppm; this photograph was taken with the instrument indicating 9ppm. This is beyond specification, and highly suspect, because of the pronounced uncertainties introduced when measuring distortion at or near equipment limits.[2] Specialized analyzers with exquisitely low uncertainty floors are needed to meaningfully measure oscillator distortion. The Audio Precision 2722, specified at a 2.5ppm Total Harmonic Distortion + Noise (THD + N) limit (1.5ppm typical), supplied Figure 13.5's data. This figure indicates a Total Harmonic Distortion (THD) of -110dB, or about 3ppm. Figure 13.6, taken with the same instrument, shows THD + N of 105dB, or about 5.8ppm. In Figure 13.7's final test, the analyzer determines the oscillator's spectral components with the third harmonic dominating at -112dB, or about 2.4ppm. These measurements provide confidence in applying the oscillator to A→D fidelity characterization.

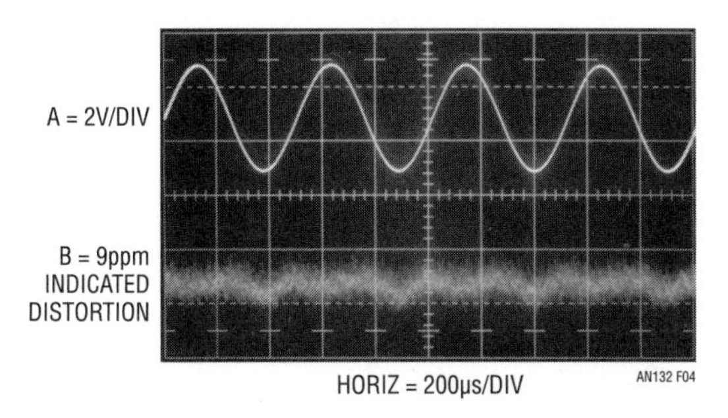

Figure 13.4 • HP-339A Distortion Analyzer Operating Beyond Its Resolution Limit Provides Misleading Distortion Indication (Trace B). Analyzer Output Contains Uncertain Combination of Oscillator and Instrument Signatures and Cannot Be Relied Upon. Trace A Is Oscillator Output

A→D testing

A→D testing routes oscillator output to the A→D via its input amplifier: The test measures distortion products produced by the input amplifier/A→D combination. A→D output is examined by the computer, which quantitatively indicates spectral error components in Figure 13.8's display.[3] The display includes time domain information showing the biased sine wave centered into the converter's operating range, a Fourier transform indicating spectral error components and detailed tabulated readings. The LTC®2379 18-Bit A→D/LT6350 amplifier combination under test produces 2nd harmonic distortion of —111 dB, about 2.8ppm, with higher frequency harmonics well below this level. This indicates the A→D and its input amplifier are operating properly and within specifications. Possible harmonic cancellation between the oscillator and amplifier/ A→D mandates testing several amplifier/A→D samples to enhance measurement confidence.[4]

Note 2: Distortion measurements at or near equipment limits are full of unpleasant surprises. See LTC Application Note 43, "Bridge Circuits", Appendix D, "Understanding Distortion Measurements" by Bruce Hofer of Audio Precision.

Note 3: Input amplifier/A→D converter, computer data acquisition and clock boards, necessary for testing, are available from LTC. Software code may be downloaded at www.linear.com. See Appendix A "Tools for A→D Fidelity Testing" for details.

Note 4: Review text section, "Verifying Oscillator Distortion" and footnote 2 for relevant commentary.

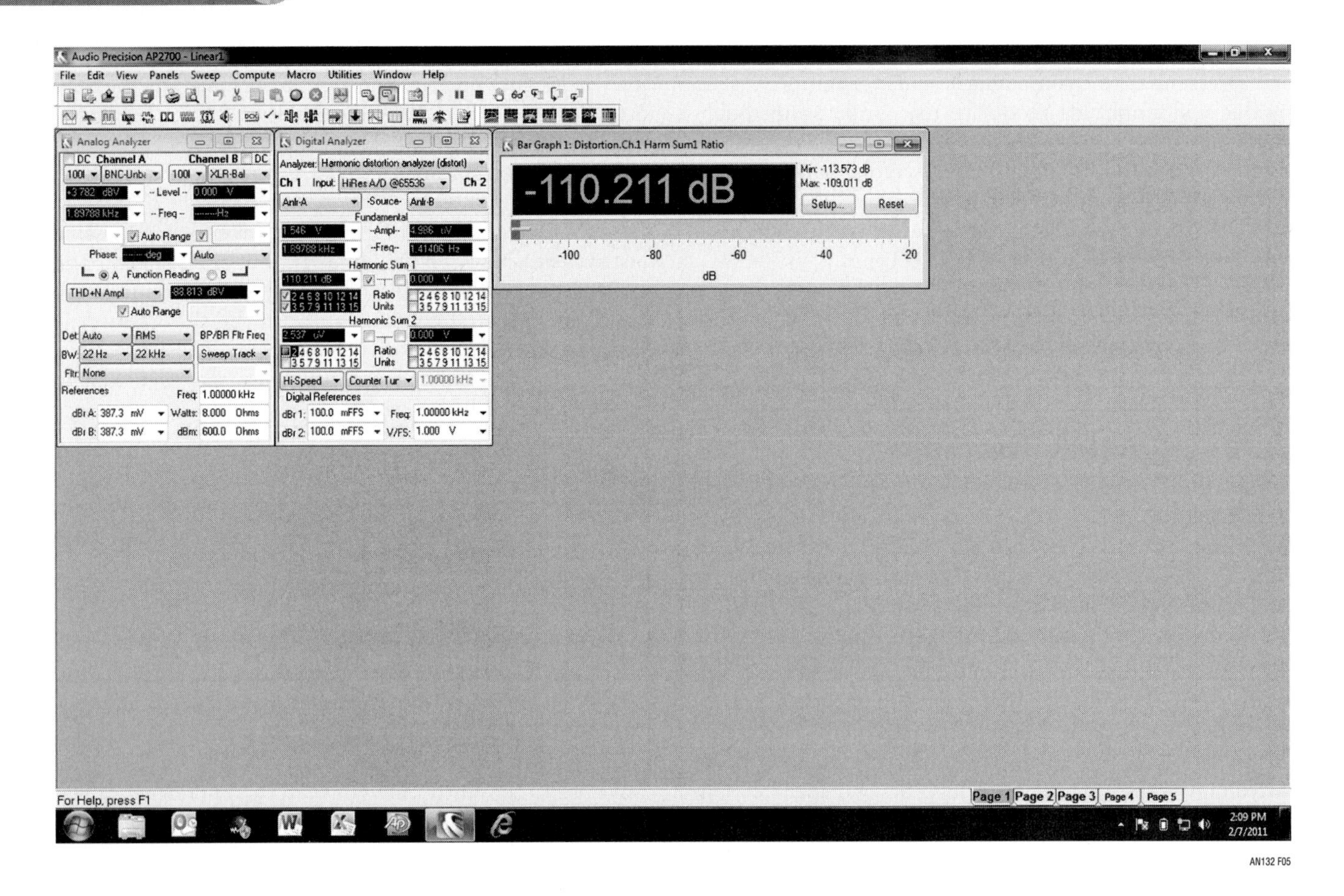

Figure 13.5 • Audio Precision 2722 Analyzer Measures Oscillator THD at –110dB, About 3ppm

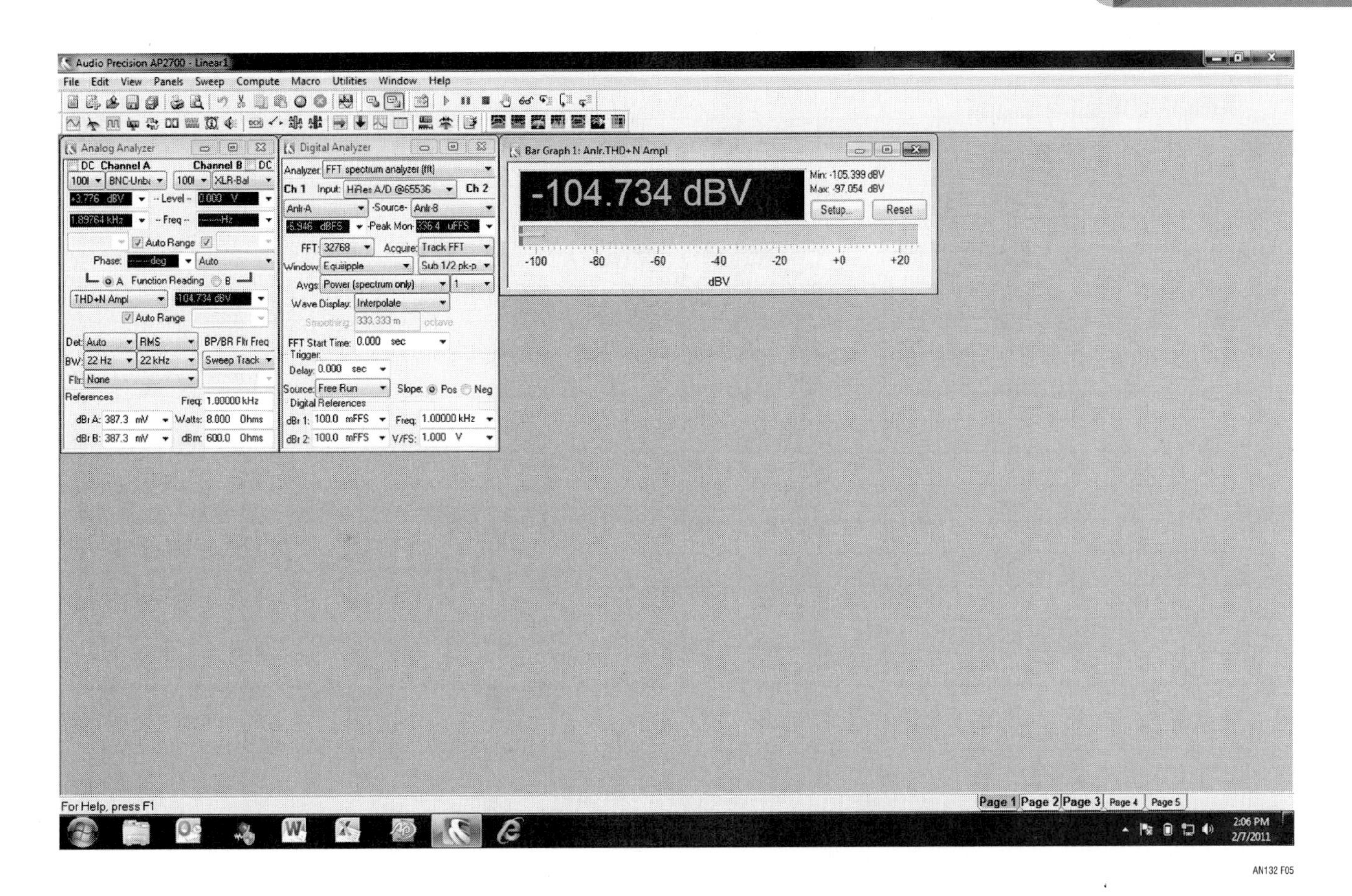

Figure 13.6 • AP-2722 Analyzer Measures Oscillator THD + N at ≈ −105dB, About 5.8ppm

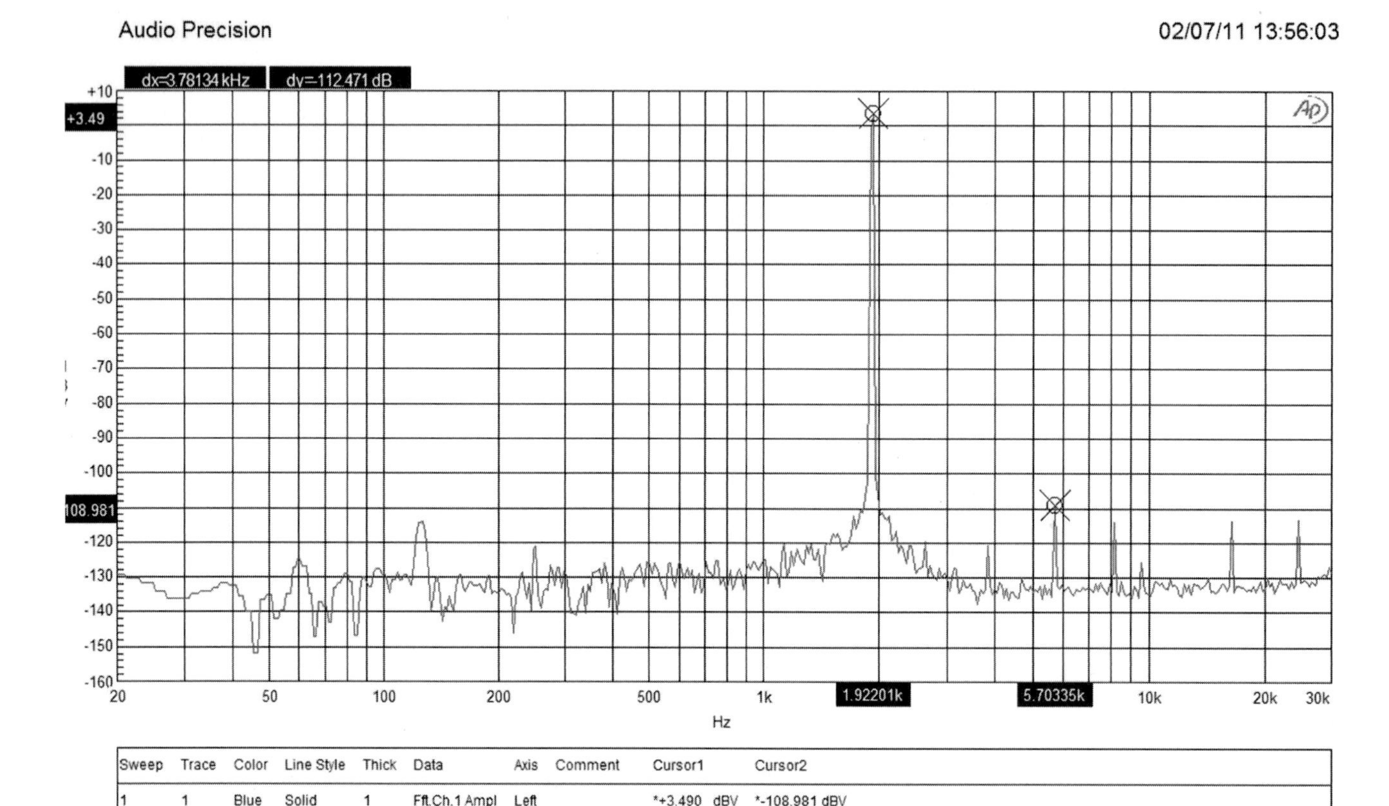

Figure 13.7 • AP-2722 Spectral Output Indicates 3rd Harmonic Peak at –112.5dB, ≈2.4ppm

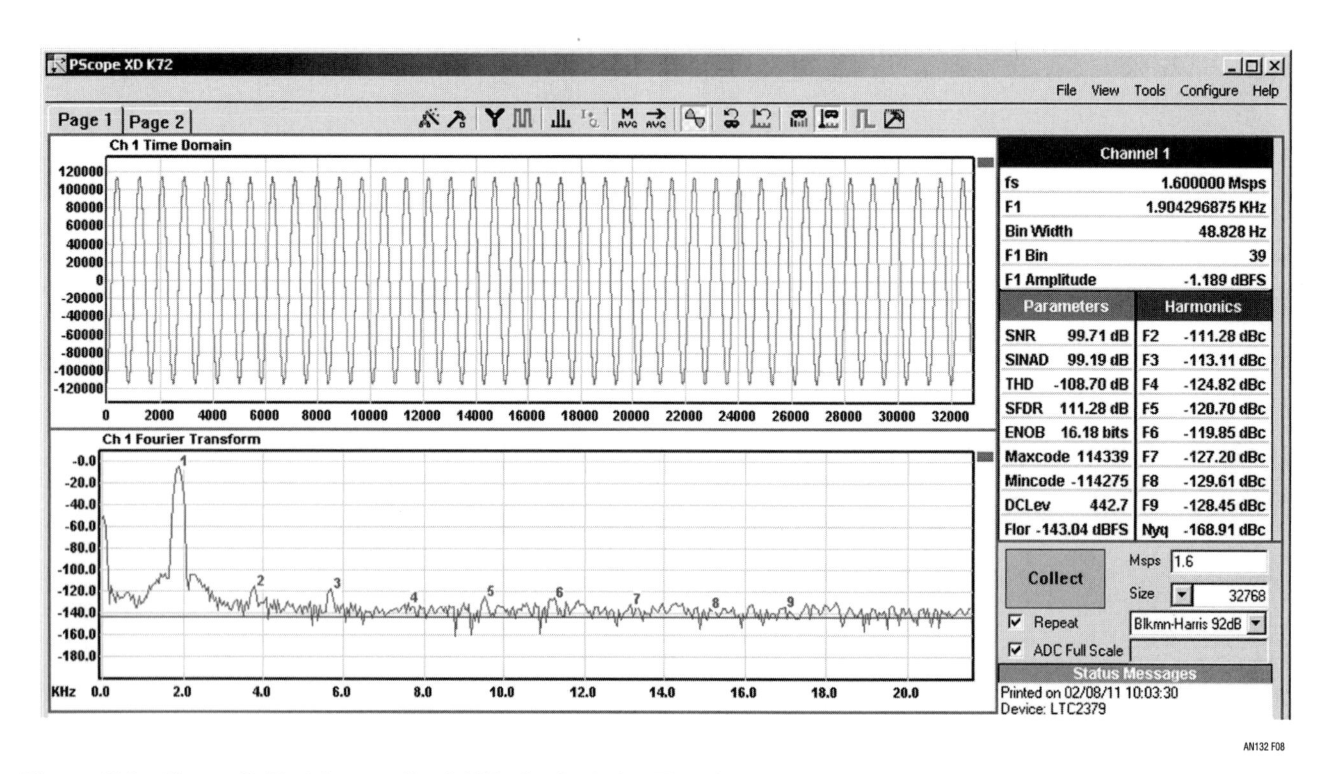

Figure 13.8 • Figure 1's Test System Partial Display Includes Time Domain Information, Fourier Spectral Plot and Detailed Tabular Readings for LTC2379 18-Bit A→D Driven by LT6350 Amplifier

Appendix A
Tools for A→D fidelity testing

Circuit boards for implementing the text's A→D testing are available. Table 13.1 lists the board functions and their part numbers. The computer software, PScope™, is also available from Linear Technology and may be downloaded at www.linear.com.

Table 13.1

BOARD FUNCTION	PART NUMBER
LT6350/LTC1279 Amp/A→D	DC-1783A-E
Interface	DC718
100MHz Clock*	DC1216A-A
Oscillator	To Be Released

* Any stable, low phase noise 3.3V clock capable of driving 50Ω may be used.

Section 2

Signal Conditioning

Applications for a new power buffer (14)

The LT1010 150mA power buffer is described in a number of useful applications such as boosted op amp, a feedforward, wideband DC stabilized buffer, a video line driver amplifier, a fast sample-hold with hold step compensation, an overload protected motor speed controller, and a piezoelectric fan servo.

Thermal techniques in measurement and control circuitry (15)

Six applications utilizing thermally based circuits are detailed. Included are a 50MHz RMS to DC converter, an anemometer, a liquid flowmeter and others. A general discussion of thermodynamic considerations involved in circuitry is also presented.

Methods for measuring op amp settling time (16)

This note begins with a survey of methods for measuring op amp settling time. This commentary develops into circuits for measuring settling time to 0.0005%. Construction details and results are presented. Appended sections cover oscilloscope overload limitations and amplifier frequency compensation.

High speed comparator techniques (17)

This note is an extensive discussion of the causes and cures of problems in very high speed comparator circuits. A separate applications section presents circuits, including a 0.025% accurate 1Hz to 30MHz V/F converter, a 200ns 0.01% sample-hold and a 10MHz fiber-optic receiver. Five appendices covering related topics complete this note.

Designs for high performance voltage-to-frequency converters (18)

A variety of high performance V/F circuits is presented. Included are a 1Hz to 100MHz design, a quartz-stabilized type and a 0.0007% linear unit. Other circuits feature 1.5V operation, sine wave output and nonlinear transfer functions. A separate section examines the trade-offs and advantages of various approaches to V/F conversion.

Unique IC buffer enhances op amp designs, tames fast amplifiers (19)

This note describes some of the unique IC design techniques incorporated into a fast, monolithic power buffer, the LT1010. Also, some application ideas are described such as capacitive load driving, boosting fast op amp output current and power supply circuits.

Power gain stages for monolithic amplifiers (20)

This note presents output state circuits which provide power gain for monolithic amplifiers. The circuits feature voltage gain, current gain, or both. Eleven designs are shown, and performance is summarized. A generalized method for frequency compensation appears in a separate section.

Composite amplifiers (21)

Applications often require an amplifier that has extremely high performance in several areas. For example, high speed and DC precision are often needed. If a single device cannot simultaneously achieve the desired characteristics, a composite amplifier made up of two (or more) devices can be configured to do the job. This note shows examples of composite approaches in designs combining speed, precision, low noise and high power.

A simple method of designing multiple order all pole bandpass filters by cascading 2nd order sections (22)

Presents two methods of designing high quality switched-capacitor bandpass filters. Both methods are intended to vastly simplify the mathematics involved in filter design by using tabular methods. The text assumes no filter design experience but allows high quality filters to be implemented by techniques not presented before in the literature. The designs are implemented by numerous examples using devices from Linear's switched-capacitor filter family: LTC1060, LTC1061, and LTC1064. Butterworth and Chebyshev bandpass filters are discussed.

FilterCAD user's manual, version 1.10 (23)

This note is the manual for FCAD, a computer-aided design program for designing filters with Linear Technology's switched-capacitor filter family. FCAD helps users design good filters with a minimum amount of effort. The experienced filter designer can use the program to achieve better results by providing the ability to play "what if" with the values and configuration of various components.

30 nanosecond settling time measurement for a precision wideband amplifier (24)

This application note permits verification of 30 nanosecond amplifier settling times to 0.1% resolution. The sampling-based technique used is detailed and results presented. Appendices cover oscilloscope overdrive issues, construction of a subnanosecond rise time pulse generator, amplifier compensation, circuit construction and calibration procedures.

Application and optimization of a 2GHz differential amplifier/ADC driver (25)

Modern high speed analog-to-digital converters (ADC), including those with pipeline or successive approximation register (SAR) topologies, have fast switched-capacitor sampling inputs. The switched capacitor inputs result in significant charge injection with every switch movement, which requires a front end that can absorb that charge and settle to the correct voltage by the time the ADC input switches again, in nanoseconds or less. This Application Note discusses the features and boundaries of the LTC6400 family and how to achieve optimal performance from the amplifiers in real applications.

2 nanosecond, 0.1% resolution settling time measurement for wideband amplifiers (26)

Amplifier DC specifications are relatively easy to verify. Measurement techniques are well understood, albeit often tedious. AC specifications require more sophisticated approaches to produce reliable information. In particular, amplifier settling time is extraordinarily difficult to determine. Settling time is the elapsed time from input application until the output arrives at and remains within a specified error band around the final value. Reliable nanosecond region settling time measurement constitutes a high order difficulty problem requiring exceptional care in approach and experimental technique. This publication details a method for making high speed amplifier settling time measurements and evaluates its performance. Eight appendices cover related topics.

An introduction to acoustic thermometry (27)

Acoustic thermometry is an arcane, elegant temperature measurement technique. It utilizes sound's temperature dependent transit time in a medium to measure temperature. The medium may be a solid, liquid or gas. Acoustic thermometers function in environments that conventional sensors cannot tolerate. Examples include extreme temperatures, applications where the sensor would be subjected to destructive physical abuse and nuclear reactors. This application note describes theory, practical acoustic thermometer circuitry and transducer construction details. Two appended sections cover calibration and present a complete software listing.

Applications for a new power buffer

14

Jim Williams

A frequent requirement in systems involves driving analog signals into non-linear or reactive loads. Cables, transformers, actuators, motors and sample-hold circuits are examples where the ability to drive difficult loads is required. Although several power buffer amplifiers are available, none have been optimized for driving difficult loads. The LT®1010 can isolate and drive almost any reactive load. It also offers current limiting and thermal overload protection which protect the device against output fault conditions. The combination of good speed, output protection, and reactive load driving capability (see box section, "The LT1010 at a Glance") make the device useful in a variety of practical situations.

Buffered output line driver

Figure 14.1 shows the LT1010 placed within the feedback loop of an operational amplifier. At lower frequencies, the buffer is within the feedback loop and its offset voltage and gain error are negligible. At higher frequencies, feedback is through C_F so that phase shift from load capacitance acting against the buffer's output resistance does not cause loop instability.

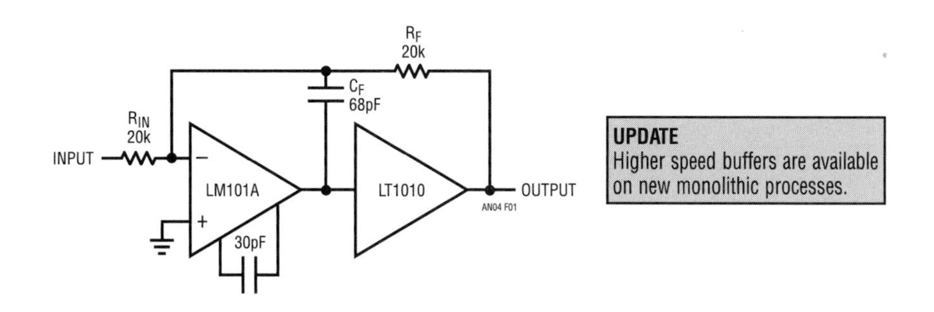

Figure 14.1 • Practical LT1010 Based Boosted Op Amp

Analog Circuit and System Design: Immersion in the Black Art of Analog Design. http://dx.doi.org/10.1016/B978-0-12-397888-2.00014-6

Figure 14.2 shows this configuration driving a 50Ω-0.33µF load. The waveform is clean, with controlled damping. With C load increased to a brutal 2µF, the circuit is still stable (Trace A, Figure 14.3), even though the large capacitance requires substantial current (Trace B) from the LT1010. Adjustment of the R_F-C_F time constant would allow improved damping.

Although this circuit is useful, its speed is limited by the op amp.

Fast, stabilized buffer amplifier

Figure 14.4 shows a way to eliminate this restriction, while maintaining good DC characteristics. Here, the LT1010 is combined with a wideband gain stage, Q1-Q3, to form a fast inverting configuration. The LT1008 op amp DC stabilizes this stage by biasing the Q2-Q3 emitters to force a zero DC potential at the circuit's summing junction. The roll-offs of the fast stage and the op amp are arranged to provide smooth overall circuit response.

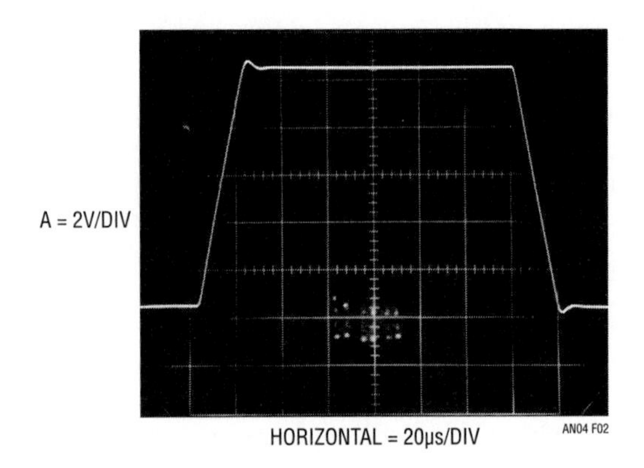

Figure 14.2 •

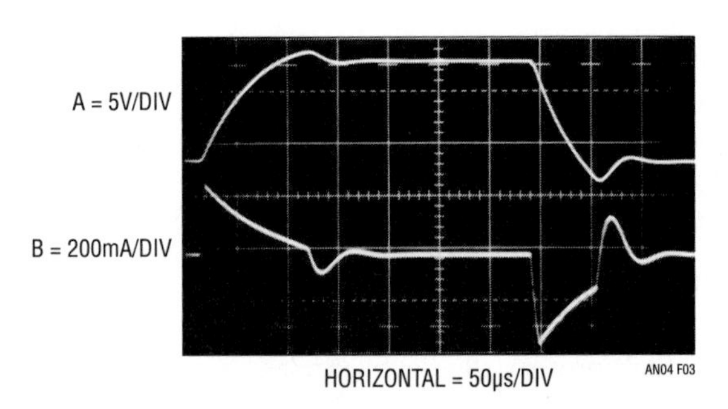

Figure 14.3 •

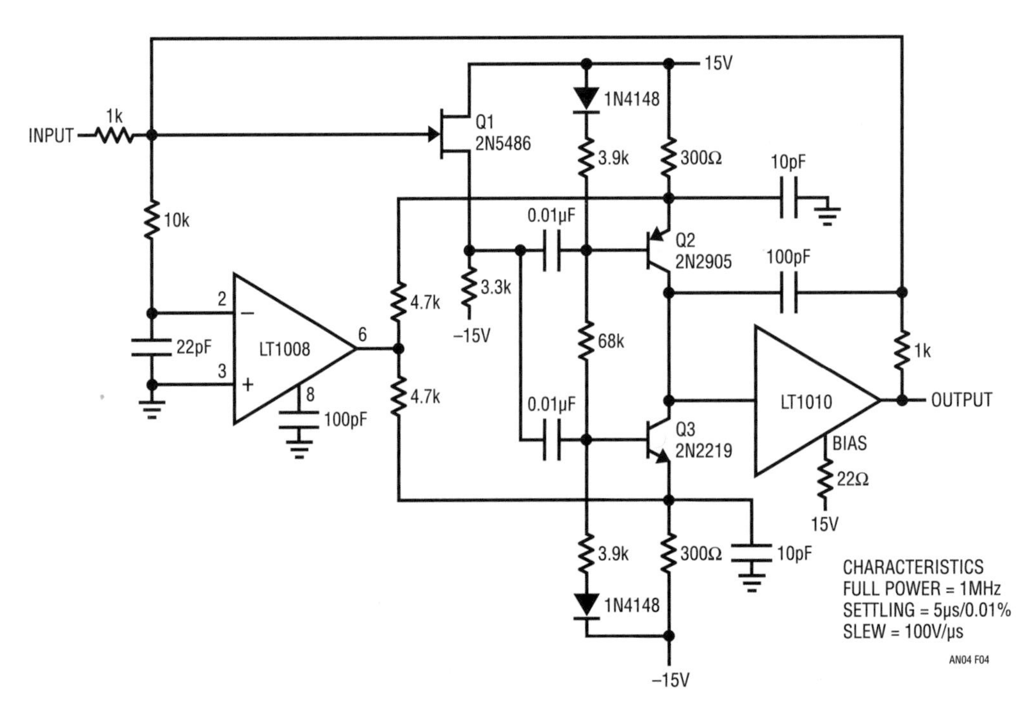

Figure 14.4 • Fed Forward. Wideband DC-Stabilized Buffer

Because the circuit's DC stabilization path occurs in parallel with the buffer, higher speed is obtainable. Figure 14.5 shows the circuit driving a 600Ω-2500pF load. Despite the heavy load, the output (Trace B) does a good job of following the input (Trace A) at a gain of –1.

Video line driving amplifier

In many applications, DC stability is unimportant and AC gain is required. Figure 14.6 shows how to combine the LT1010's load handling capability with a fast, discrete gain stage. Q1 and Q2 form a differential stage which single ends into the LT1010. The capacitively terminated feedback divider gives the circuit a DC gain of 1, while allowing AC gains up to 10. Using a 20Ω bias resistor (see box section), the circuit delivers $1V_{P\text{-}P}$ into a typical 75Ω video load. For applications sensitive to NTSC requirements, dropping the bias resistor value will aid performance.

At A = 2, the gain is within 0.5dB to 10MHz with the –3dB point occurring at 16MHz. At A = 10, the gain is flat (±0.5dB to 4MHz) with a –3dB point at 8MHz. The peaking adjustment should be optimized under loaded output conditions.

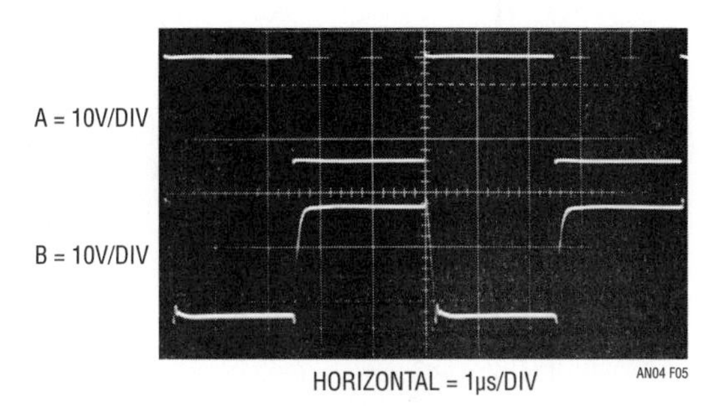

Figure 14.5 •

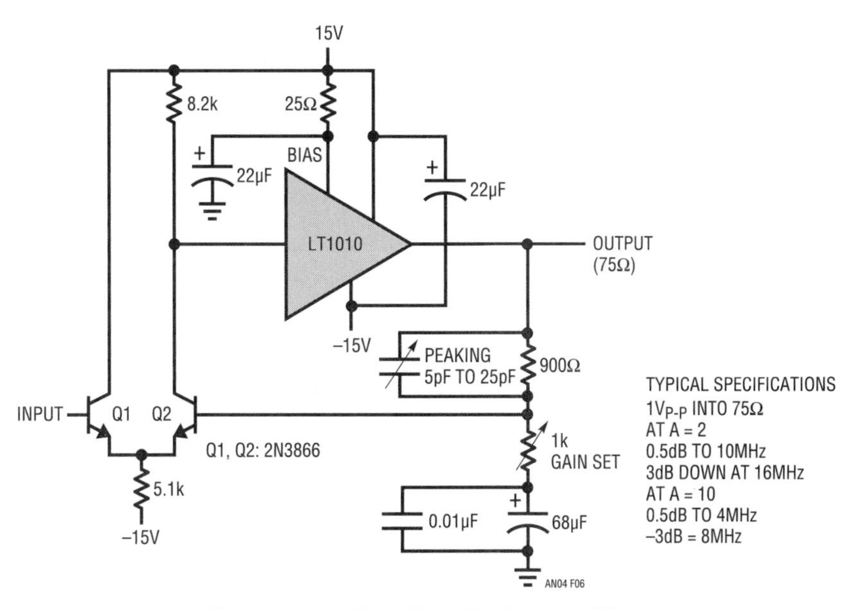

Figure 14.6 • Video Line Driving Amplifier

Figure 14.7 shows a video distribution amplifier. In this example, resistors are included in the output line to isolate reflections from unterminated lines. If the line characteristics are known, the resistors may be deleted. To meet NTSC gain-phase requirements, a small value boost resistor is used. Each $1V_{P-P}$ channel output is essentially flat through 6MHz into a 75Ω load.

Fast, precision sample-hold circuit

Sample-hold circuits require high capacitive load driving capability to achieve fast acquisition times. Additionally, other trade-offs must be considered to achieve a good design. The conceptual circuit of Figure 14.8 illustrates

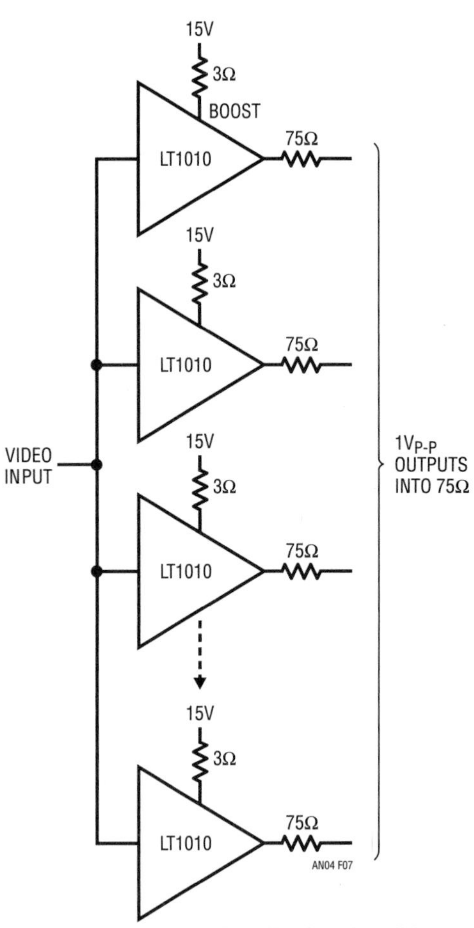

Figure 14.7 • Video Distribution Amplifier

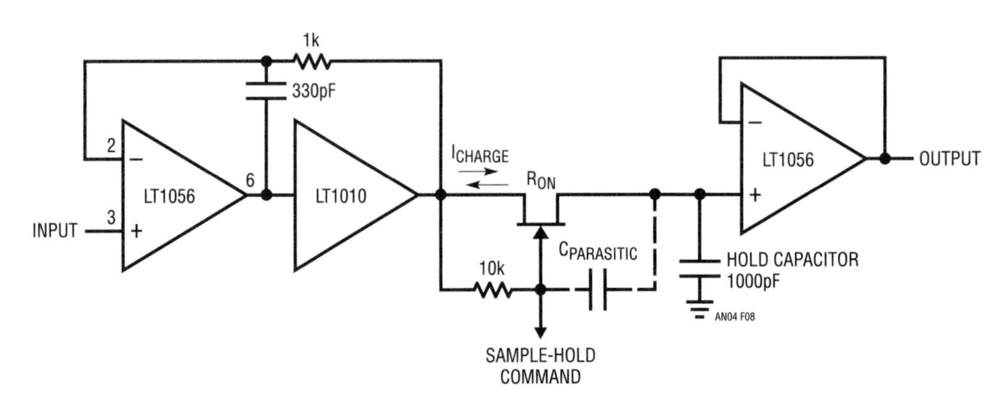

Figure 14.8 • Conceptual Sample-Hold

some of the issues encountered. Fast acquisition requires high charge currents and dynamic stability, which the LT1010 can provide. To get reasonable droop rate, the hold capacitor must be appropriately sized, but too large a value means FET switch on-resistance will effect acquisition time. If very low on-resistance FETs are used, the parasitic gate-source capacitance becomes significant and a substantial amount of charge is removed from the hold capacitor when the gate is switched off. This charge removal causes the stored voltage to abruptly change with the circuit is switched into the hold mode. This phenomenon, called "hold step", limits accuracy. It can be combatted by increasing the hold capacitor's value, but then acquisition time suffers. Finally, since a TTL compatible input is desirable, the FET requires a level shift. This level shift must provide adequate pinch-off voltage over the entire range of circuit inputs and must also be fast. Delays will result in aperture errors, introducing dynamic sampling inaccuracies.

Figure 14.9 shows a circuit which combines the LT1010 with some techniques to produce a fast, precise sample-hold circuit. Q1 through Q4 constitute a very fast TTL compatible level shift. Total delay from the TTL input switching into hold to Q6 turning off is 16ns. Baker clamped Q1 biases Q3's emitter to switch level shifter Q4. Q2 drives a heavy feedforward network, speeding Q4's switching. This stage affords low aperture errors, while providing the necessary level shift for Q6's gate. The hold step error due to Q6's parasitic gate-source capacitance is compensated for by Q5 and the LT318A amplifier (A3).

The amount of charge removed by Q6's parasitic capacitance is signal dependent (Q = CV). To compensate this error, A2 measures the circuit output and biases the Q5 switch. Each time the circuit switches into hold mode, an appropriate amount of charge is delivered through the potentiometer—15pF network in Q5's emitter. The

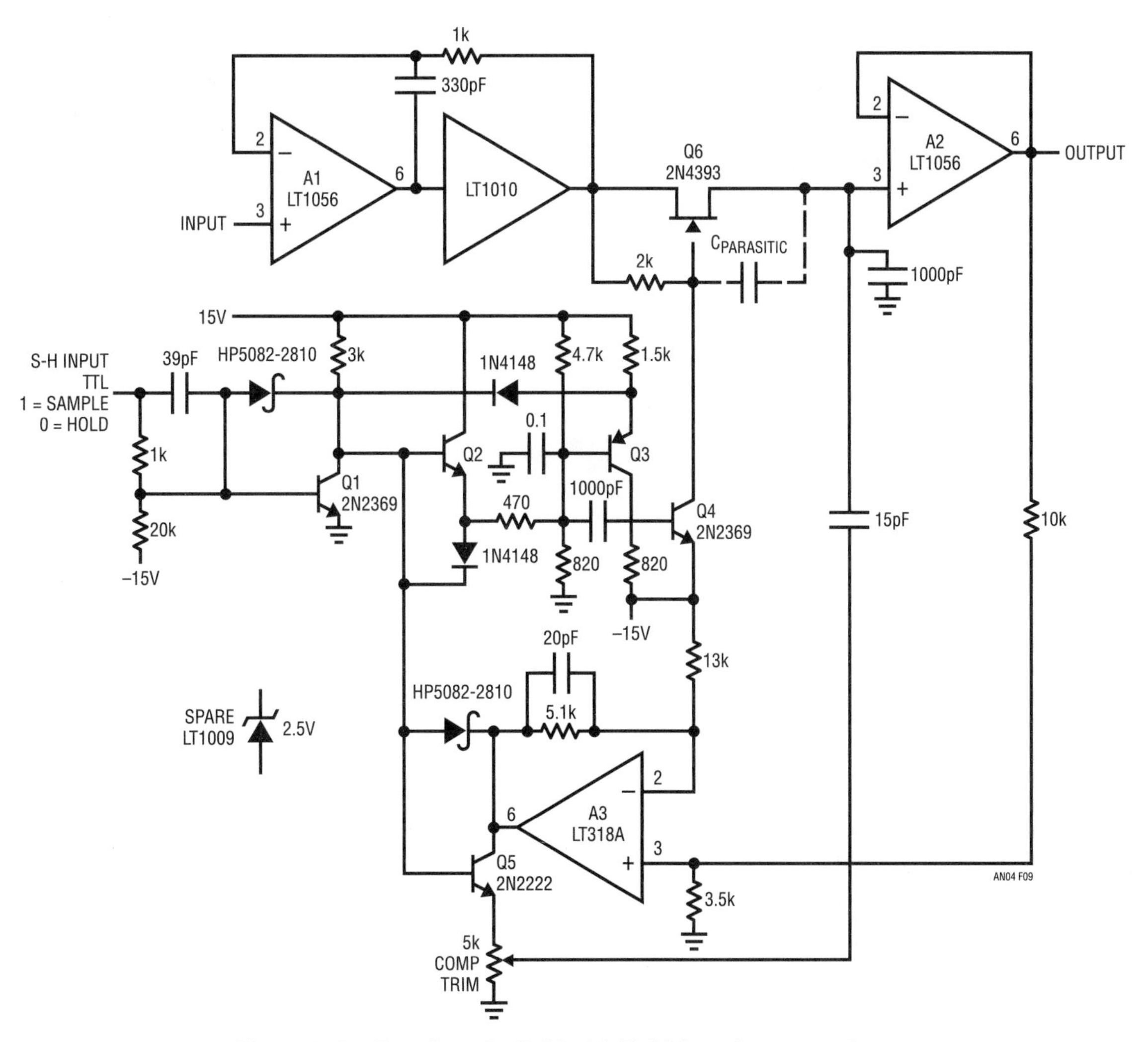

Figure 14.9 • Fast Sample-Hold with Hold Step Compensation

amount of charge is scaled to compensate for charge removal due to Q6's parasitic term. A3's inverting input is biased so that negative supply shifts, which alter the charge removed through C parasitic, are accounted for in the compensating charge. Compensation is set by grounding the signal input, clocking the S-H line and adjusting the potentiometer for minimum disturbance at the circuit's output.

Figure 14.10 shows the circuit at work. When the sample-hold input (Trace A, Figure 14.10) goes into hold, charge cancellation occurs and the output (Trace B) sees less than 250μV of hold step error within 100ns. Without compensation, the error would be 50mV (Trace B, Figure 14.11—Trace A is the sample-hold input).

Figure 14.12 shows the LT1010's contribution to fast acquisition. The circuit acquires a 10V signal in this photograph. Trace A is the sample-hold input. Trace B shows the LT1010 delivering over 100mA to the hold capacitor and Trace C depicts the output value slewing and settling to final value. Note that the acquisition time is limited by amplifier settling time and not capacitor charge time. Pertinent specifications include:

Acquisition time: 2μs to 0.01%
Hold settling time: <100ns to 1mV
Aperture time: 16ns

Motor speed control

The LT1010's ability to drive difficult loads is exploited in Figure 14.13's circuit. Here, the buffer drives a motor-tachometer combination. The tachometer signal is fed back and compared to a reference current and the LM301A amplifier closes a control loop. The 0.47μF capacitor provides stable compensation. Because the tachometer output is bipolar, the speed is controllable in both directions, with clean transitions through zero. The LT1010's thermal protection is particularly useful in this application, preventing device destruction in the event of mechanical overload or malfunction.

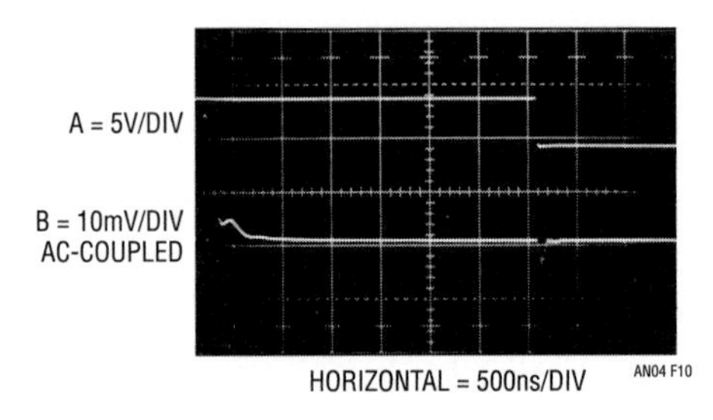

Figure 14.10 •

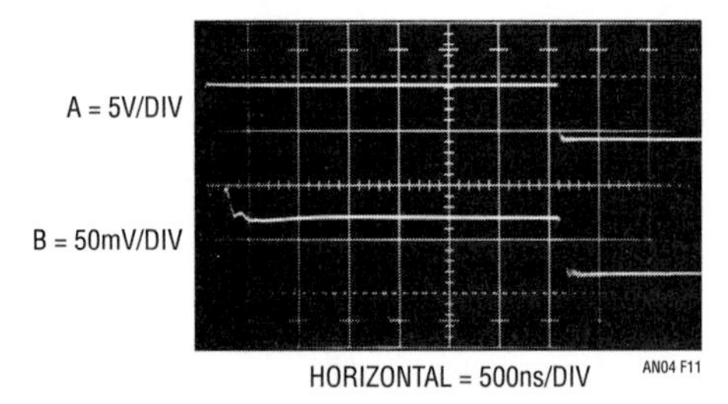

Figure 14.11 •

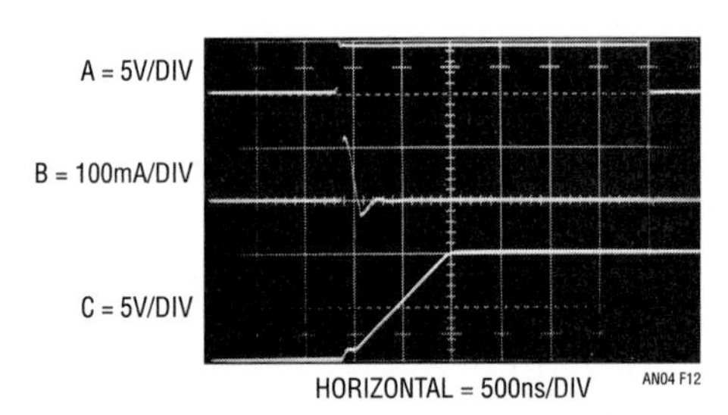

Figure 14.12 •

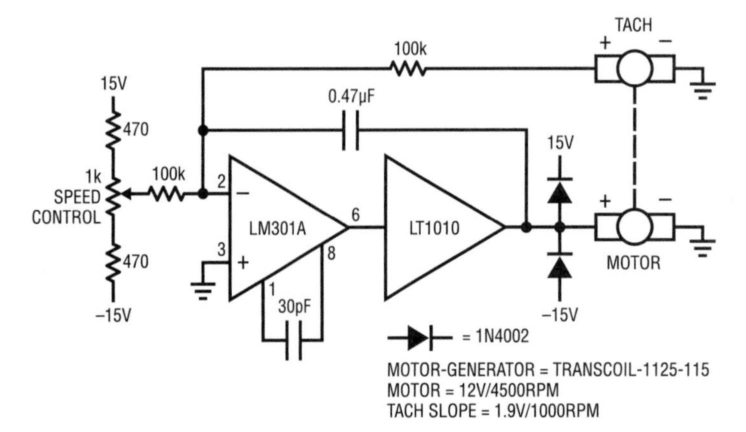

Figure 14.13 • Overload Protected Motor Speed Controller

Fan-based temperature controller

Figure 14.14 shows a way to use the LT1010 to control a fan motor's speed to regulate instrument temperature. The fan employed is one of the new electrostatic types which has very high reliability because it contains no wearing parts. These devices require high voltage drive. When power is applied, the thermistor (located in the fan's exhaust stream) is at a high value. This unbalances the A3 amplifier-driven bridge, A1 receives no power, and the fan does not run. As the instrument enclosure warms, the thermistor value decreases until A3 begins to oscillate. A2 provides isolation and gain and A4 drives the transformer to generate high voltage for the fan. In this fashion, the loop acts to maintain a stable instrument temperature by controlling the fan's exhaust rate. The 100μF time constant across the error amplifier pins is typical of such configurations. Fast time constants will produce audibly annoying "hunting" in the servo. Optimal values for this time constant and gain depend upon the thermal and airflow characteristics of the enclosure being controlled.

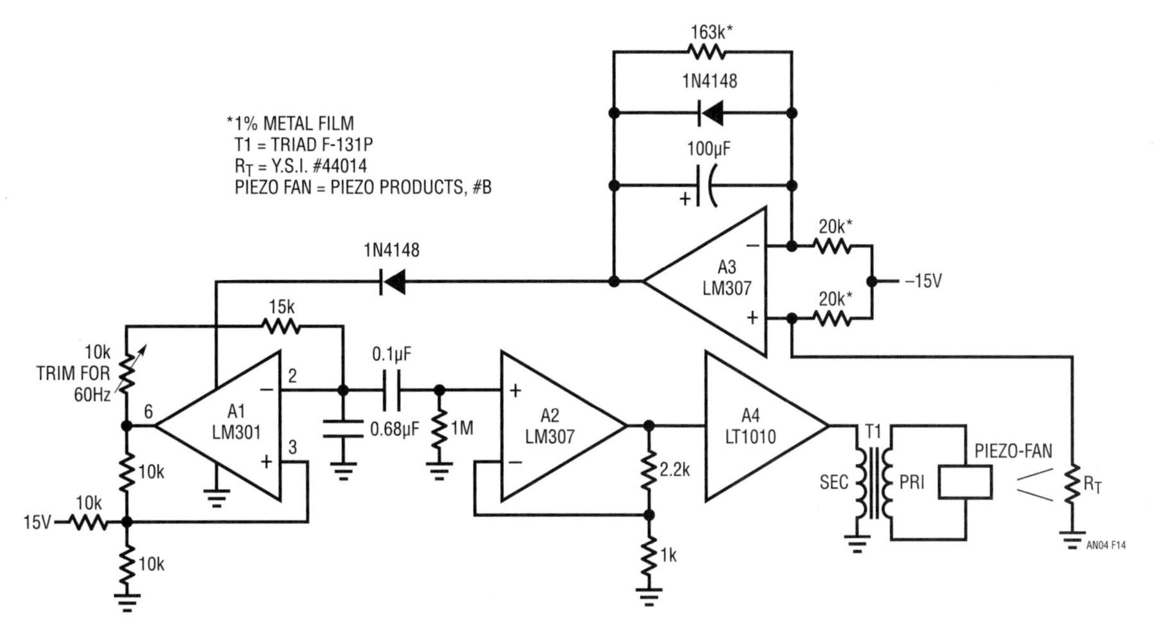

Figure 14.14 • Piezo-Electric Fan Servo

The LT1010 at a glance

R. J. Widlar

The schematic describes the basic elements of the buffer design. The op amp drives the output sink transistor, Q3, such that the collector current of the output follower never drops below the quiescent value (determined by I and the area ratio of D1 and D2). As a result, the high frequency response is essentially that of a simple follower, even when Q3 is supplying the load current. The internal feedback loop is isolated from the effects of capacitive loading in the output lead.

The scheme is not perfect in that the rate of rise of sink current is noticeably less than for source current. This can be mitigated by connecting a resistor between the bias terminal at V^+, raising quiescent current. A feature of the final design is that the output resistance is largely independent of the follower quiescent current or the output load current. The output will also swing to the negative rail, which is particularly useful with single supply operation.

The buffer is no more sensitive to supply bypassing than slower op amps as far as stability is concerned. The 0.1μF disc ceramic capacitors usually recommended for op amps are certainly adequate or low frequency work. As always, keeping the capacitor leads short and using a ground plane are prudent, especially when operating at high frequencies.

The buffer slew rate can be reduced by inadequate supply bypass. With output current changes much above 100mA/μs, using 10μF solid tantalum capacitors on both supplies is good practice, although bypassing from the positive to the negative supply may suffice.

When used in conjunction with an op amp and heavily loaded (resistive or capacitive), the buffer can couple into supply leads common to the op amp, causing stability problems with the overall loop. Adequate bypassing can usually be provided by 10μF solid tantalum capacitors. Alternately, smaller capacitors could be used with decoupling resistors. Sometimes the op amp has much better high frequency rejection on one supply, so bypass requirements are less on this supply.

Power dissipation

In many applications, the LT1010 will require heat sinking. Thermal resistance, junction to still air, is 150°C/W for the TO-39 package and 60°C/W for the TO-3 package. Circulating air, a heat sink or mounting the TO-3 package to a printed circuit board will reduce thermal resistance.

In DC circuits, buffer dissipation is easily computed. In AC circuits, signal wave shape and the nature of the load determine dissipation. Peak dissipation can be several times average with reactive loads. it is particularly important to determine dissipation when driving large load capacitance.

Overload protection

The LT1010 has both instantaneous current limit and thermal overload protection. Foldback current limiting has not been used, enabling the buffer to drive complex loads without limiting. Because of this, it is capable of power dissipation in excess of its continuous ratings.

Normally, thermal overload protection will limit dissipation and prevent damage. However, with more than 30V across the conducting output transistor, thermal limiting is not quick enough to ensure protection in current limit. The thermal protection is effective with 40V across the conducting output transistor as long as the load current is otherwise limited to 150mA.

Drive impedance

When driving capacitive loads, the LT1010 likes to be driven from a low source impedance at high frequencies. Some low power op amps are marginal in this respect. Some care may be required to avoid oscillations, especially at low temperatures.

Bypassing the buffer input with more than 200pF will solve the problem. Raising the operating current also works, but this can be done only on the TO-3 package.

The LT1010 at a Glance

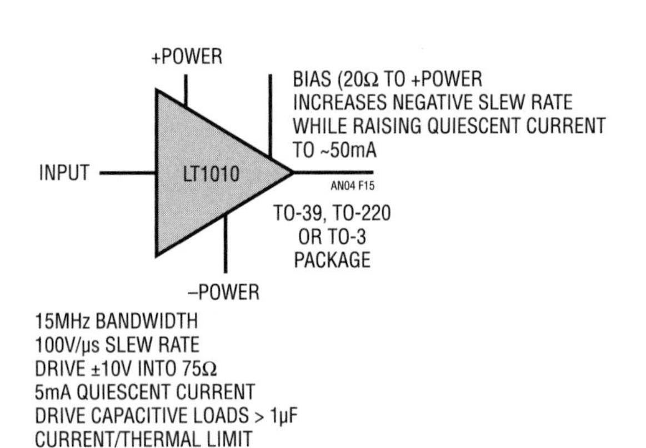

The LT1010 Conceptual Schematic

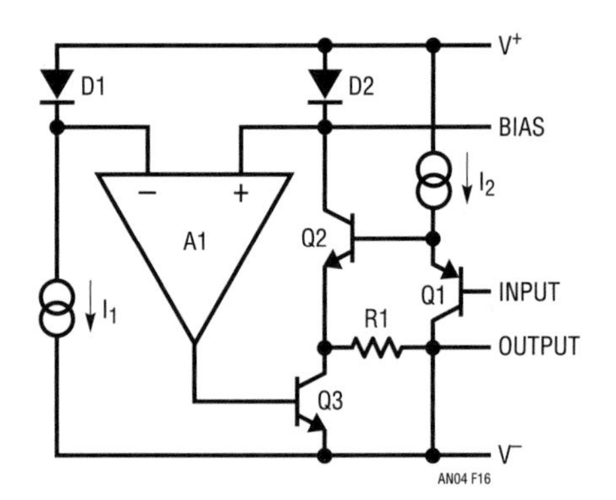

Thermal techniques in measurement and control circuitry

15

Jim Williams

Designers spend much time combating thermal effects in circuitry. The close relationship between temperature and electronic devices is the source of more design headaches than any other consideration.

In fact, instead of eliminating or compensating for thermal parasitics in circuits, it is possible to utilize them. In particular, applying thermal techniques to measurement and control circuits allows novel solutions to difficult problems. The most obvious example is temperature control. Familiarity with thermal considerations in temperature control loops permits less obvious, but very useful, thermally-based circuits to be built.

Temperature controller

Figure 15.1 shows a precision temperature controller for a small components oven. When power is applied, the thermistor, a negative TC device, is at a high value. A1 saturates positive. This forces the LT®3525A switching regulator's output low, biasing Q1. As the heater warms, the thermistor's value decreases. When its inputs finally balance, A1 comes out of saturation and the LT3525A pulse width modulates the heater via Q1, completing a feedback path. A1 provides gain and the LT3523A furnishes high efficiency. The 2kHz pulse width modulated heater power is much faster than the thermal loop's response and the oven sees an even, continuous heat flow.

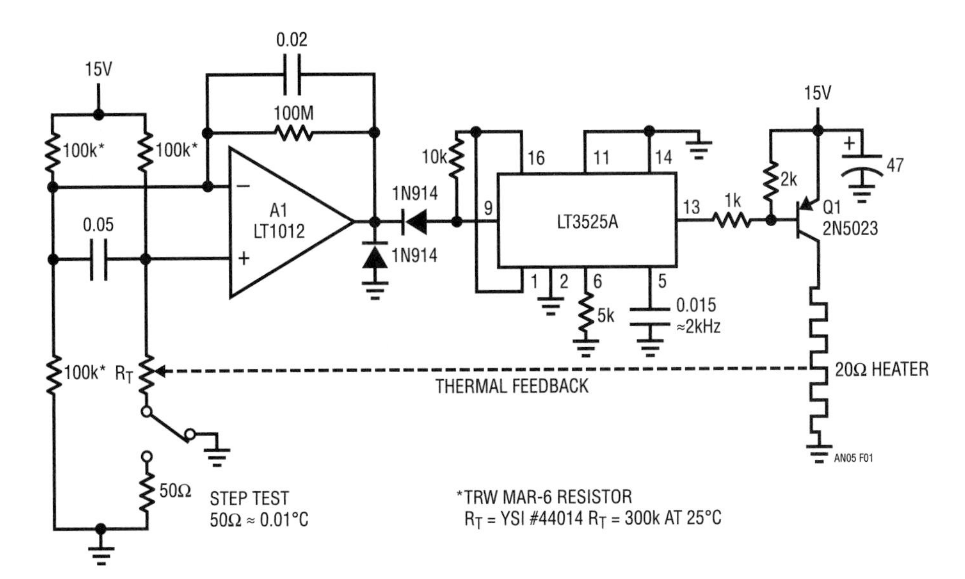

Figure 15.1 • Precision Temperature Controller

Analog Circuit and System Design: Immersion in the Black Art of Analog Design. http://dx.doi.org/10.1016/B978-0-12-397888-2.00015-8

The key to high performance control is matching the gain bandwidth of A1 to the thermal feedback path. Theoretically, it is a simple matter to do this using conventional servo-feedback techniques. Practically, the long time constants and uncertain delays inherent in thermal systems present a challenge. The unfortunate relationship between servo systems and oscillators is very apparent in thermal control systems.

The thermal control loop can be very simply modeled as a network of resistors and capacitors. The resistors are equivalent to the thermal resistance and the capacitors equivalent to thermal capacity. In Figure 15.2 the heater, heater-sensor interface, and sensor all have RC factors that contribute to a lumped delay in the ability of a thermal system to respond. To prevent oscillation, A1's gain bandwidth must be limited to account for this delay. Since high gain bandwidth is desirable for good control, the delays must be minimized. The physical size and electrical resistivity of the heater selected give some element of control over the heater's time constant. The heater-sensor interface time constant can be minimized by placing the sensor in intimate contact with the heater.

The sensor's RC product can be minimized by selecting a sensor of small size relative to the capacity of its thermal environment. Clearly, if the wall of an oven is 6" thick aluminum, the tiniest sensor available is not an absolute necessity. Conversely, if one is controlling the temperature of 1/16" thick glass microscope slide, a very small sensor (i.e., fast) is in order.

After the thermal time constants relating to the heater and sensor have been minimized, some form of insulation for the system must be chosen. The function of insulation is to keep the loss rate down so the temperature control device can keep up with the losses. For any given system, the higher the ratio between the heater-sensor time constants

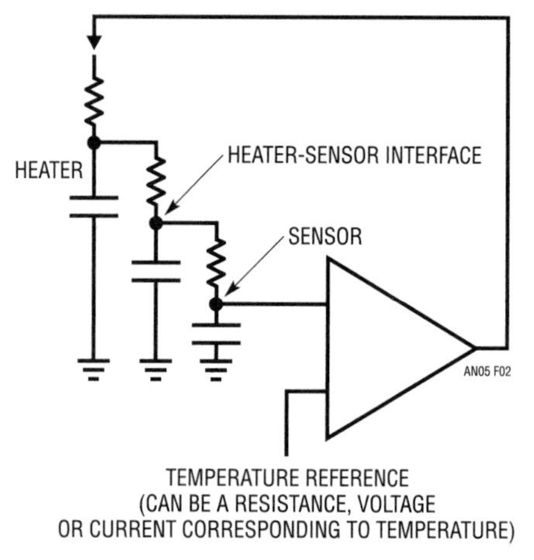

Figure 15.2 • Thermal Control Loop Model

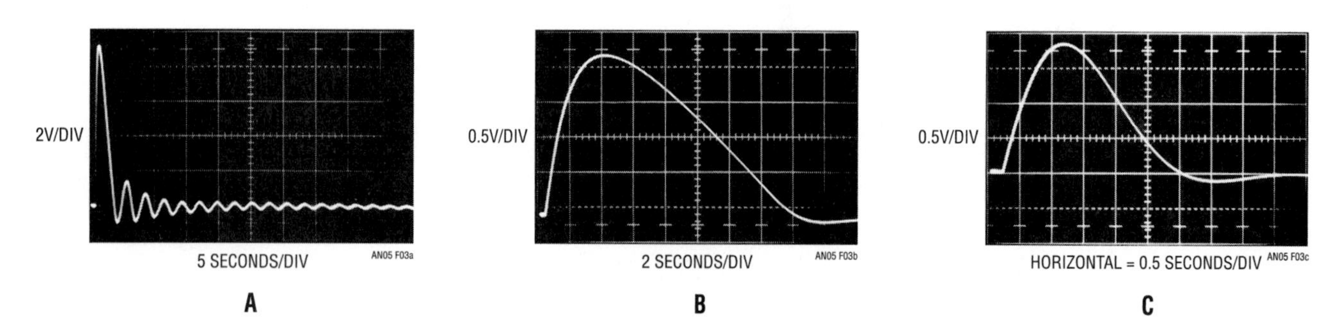

Figure 15.3 • Loop Response for Various Gain Bandwidths

and the insulation time constants, the better the performance of the control loop.

After these thermal considerations have been attended to, the control loop's gain bandwidth can be optimized. Figures 15.3A, 15.3B and 15.3C show the effects of different compensation values at A1. Compensation is trimmed by applying small steps in temperature setpoint and observing the loop response at A1's output. The 50Ω resistor and switch in the thermistor leg of the bridge furnish a 0.01°C step generator. Figure 15.3A shows the effects of too much gain bandwidth. The step change forces a damped, ringing response over 50 seconds in duration! The loop is marginally stable. Increasing A1's gain bandwidth (GBW) will force oscillation. Figure 15.3B shows what happens when GBW is reduced. Settling is much quicker and more controlled. The waveform is overdamped, indicating that higher GBW is achievable without stability compromises. Figure 15.3C shows the response for the compensation values given and is a nearly ideal critically damped recovery. Settling occurs within 4 seconds. An oven optimized in this fashion will easily attenuate external temperature shifts by a factor of thousands without overshoots or excessive lags.

Thermally stabilized pin photodiode signal conditioner

PIN photodiodes are frequently employed in wide range photometric measurements. The photodiode specified in Figure 15.4 responds linearly to light intensity over a 100dB range. Digitizing the diode's linearly amplified output would require an A/D converter with 17 bits of range. This requirement can be eliminated by logarithmically compressing the diode's output in the signal conditioning circuity. Logarithmic amplifiers utilize the logarithmic relationship between V_{BE} and collector current in transistors. This characteristic is very temperature sensitive and requires special components and layout considerations to achieve good results. Figure 15.4's circuit logarithmically signal conditions the photodiode's output with no special components or layout.

A1 and Q4 convert the diode's photocurrent to a voltage output with a logarithmic transfer function. A2 provides offsetting and additional gain. A3 and its associated components form a temperature control loop which maintains Q4 at constant temperature (all transistors in

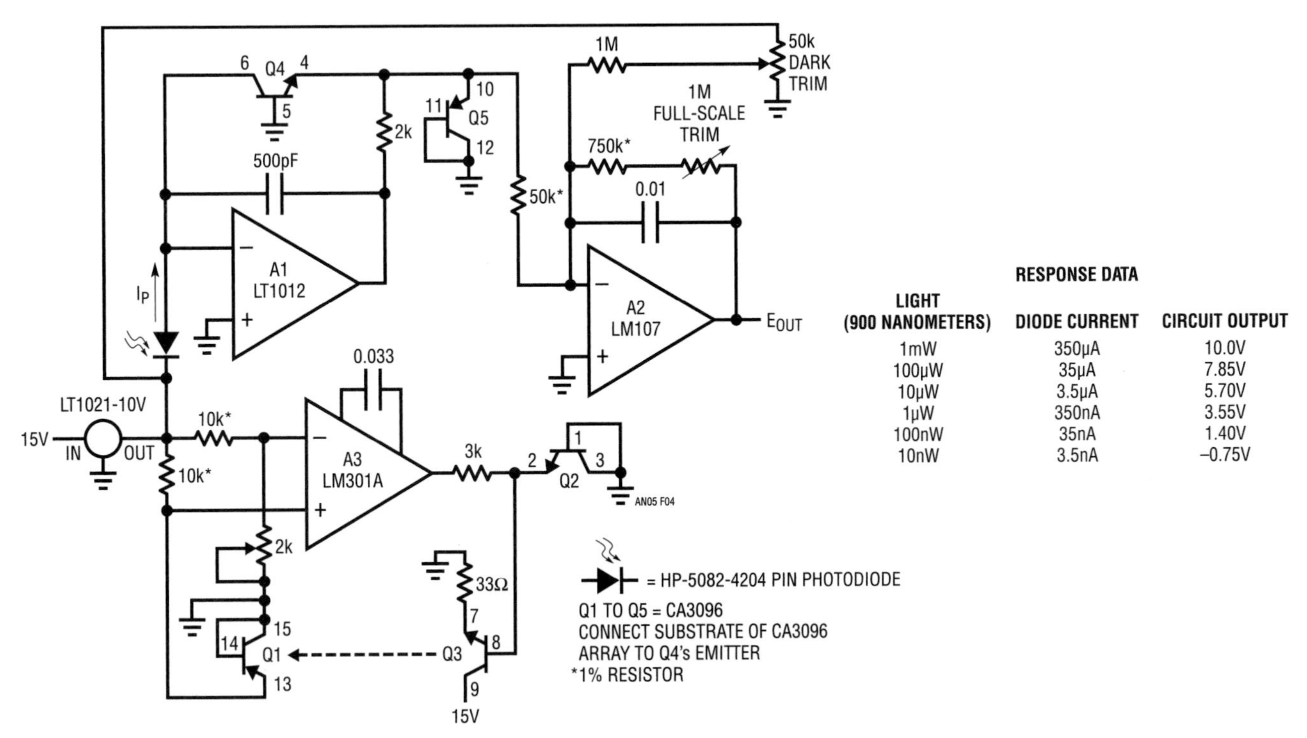

RESPONSE DATA

LIGHT (900 NANOMETERS)	DIODE CURRENT	CIRCUIT OUTPUT
1mW	350μA	10.0V
100μW	35μA	7.85V
10μW	3.5μA	5.70V
1μW	350nA	3.55V
100nW	35nA	1.40V
10nW	3.5nA	–0.75V

Figure 15.4 • 100dB Range Logarithmic Photodiode Amplifier

this circuit are part of a CA3096 monolithic array). The 0.033µF value at A3's compensation pins gives good loop damping if the circuit is built using the array's transistors in the locations shown. These locations have been selected for optimal control at Q4, the logging transistor. Because of the array die's small size, response is quick and clean. A full-scale step requires only 250ms to settle (photo, Figure 15.5) to final value. To use this circuit, first set the thermal control loop. To do this, ground Q3's base and set the 2k pot so A3's negative input voltage is 55mV below its positive input. This places the servo's setpoint at about 50°C (25°C ambient + 2.2mV/°C • 25°C rise = 55mV = 50°C). Unground Q3's base and the array will come to temperature. Next, place the photodiode in a completely dark environment and adjust the "dark trim" so A2's output is 0 V. Finally, apply or electrically simulate (see chart, Figure 15.4) 1mW of light and set the "full-scale" trim to 10V out. Once adjusted, this circuit responds logarithmically to light inputs from 10nW to 1mW with an accuracy limited by the diode's 1% error.

50MHz bandwidth thermal RMS→DC converter

Conversion of AC waveforms to their equivalent DC power value is usually accomplished by either rectifying and averaging or using analog computing methods. Rectification averaging works only for sinusoidal inputs. Analog computing methods are limited to use below 500kHz. Above this frequency, accuracy degrades beyond the point of usefulness in instrumentation

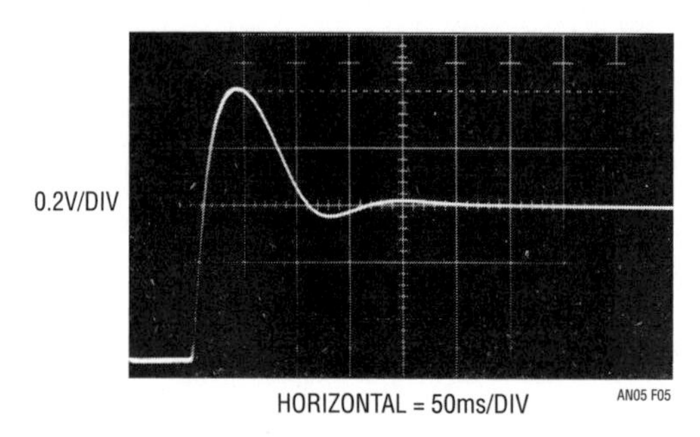

Figure 15.5 • Figure 15.4's Thermal Loop Response

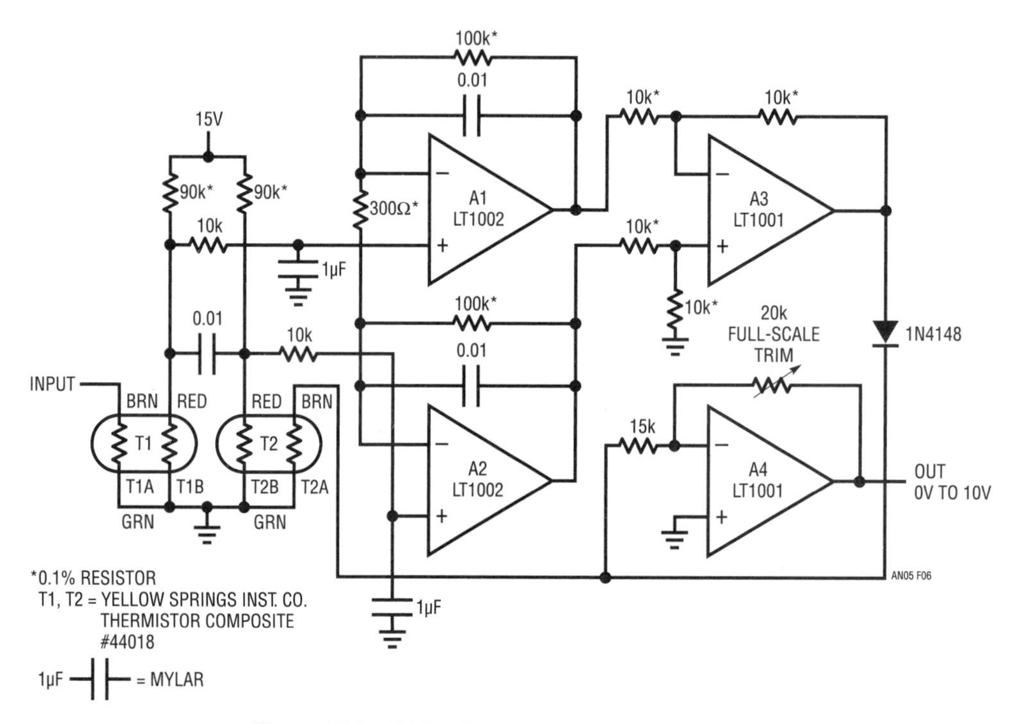

Figure 15.6 • 50MHz Thermal RMS→DC Converter

applications. Additionally, crest factors greater than 10 cause significant reading errors.

A way to achieve wide bandwidth and high crest factor performance is to measure the true power value of the waveform directly. The circuit of Figure 15.6 does this by measuring the DC heating power of the input waveform. By using thermal techniques to integrate the input waveform, 50MHz bandwidth is easily achieved with 2% accuracy. Additionally, because the thermal integrator's output is at low frequency, no wideband circuitry is required. The circuit uses standard components and requires no special trimming techniques. It is based on measuring the amount of power required to maintain two similar but thermally decoupled masses at the same temperature. The input is applied to T1, a dual thermistor bead. The power dissipated in one leg (T1A) of this bead forces the other section (T1B) to shift down in value, unbalancing the bridge formed by the other bead and the 90k resistors. This imbalance is amplified by the A1-A2-A3 combination. A3's output is applied to a second thermistor bead, T2. T2A heats, causing T2B to decay in value. As T2B's resistance drops, the bridge balances. A3's output adjusts drive to T2A until T1B and T2B have equal values. Under these conditions, the voltage at T2A is equal to the RMS value of the circuit's input. In fact, slight mass imbalances between T1 and T2 contribute a gain error, which is corrected at A4. RC filters at A1 and A2 and the 0.01µF capacitor eliminate possible high frequency error due to capacitive coupling between T1A and T1B. The diode in A3's output line prevents circuit latch-up.

Figure 15.7 details the recommended thermal arrangement for the thermistors. The Styrofoam block provides an isothermal environment and coiling the thermistor leads attenuates heat pipe effects to the outside ambient. The 2-inch distance between the devices allows them to see identical thermal conditions without interaction. To calibrate this circuit, apply $10V_{DC}$ to the input and adjust the full-scale trim for 10V out at A4. Accuracy remains within 2% from DC to 50MHz for inputs of 300mV to 10V. Crest factors of 100:1 contribute less than 0.1% additional error and response time to rated accuracy is five seconds.

Low flow rate thermal flowmeter

Measuring low flow rates in fluids presents difficulties. "Paddle wheel" and hinged vane type transducers have low and inaccurate outputs at low flow rates. If small diameter tubing is required, as in medical or biochemical work, such transduction techniques also become mechanically impractical. Figure 15.8 shows a thermally-based flowmeter which features high accuracy at rates as low as 1mL/minute and has a frequency output which is a linear function of flow rate. This design measures the differential temperature between two sensors (Figure 15.9). One sensor, T1, located before the heater resistor, assumes the fluid's temperature before it is heated by the resistor. The second sensor, T2, picks up the temperature rise induced into the fluid by the resistor's heating. The sensor's difference signal appears at A1's output. A2 amplifies this difference with a time constant set by the 10MΩ adjustment. Figure 15.10 shows A2's output versus flow rate. The function has an inverse relationship. A3 and A4 linearize this relationship, while simultaneously providing a frequency output (Figure 15.10). A3 functions as an integrator which is biased from the LT1004 and the 383k input resistor. Its output is compared to A2's output at A4. Large inputs from A2 force the integrator to run for a long time before A4 can go high, turning on Q1 and resetting A3. For small inputs from A2, A3 does not have to integrate very long before resetting action occurs. Thus, the configuration oscillates at a frequency which is inversely proportional to A2's output voltage. Since this voltage is inversely related to flow rate, the oscillation frequency linearly corresponds to flow rate.

Several thermal considerations are important in this circuit. The amount of power dissipated into the stream

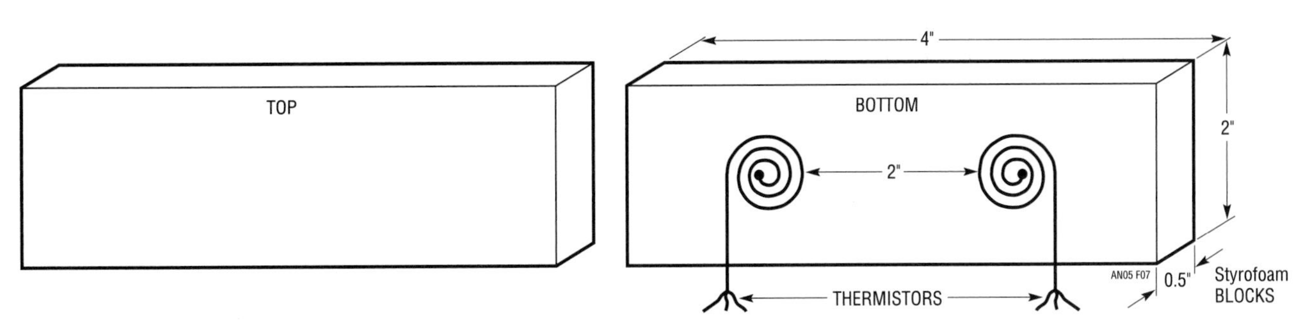

Figure 15.7 • Thermal Arrangement for RMS→DC Converter

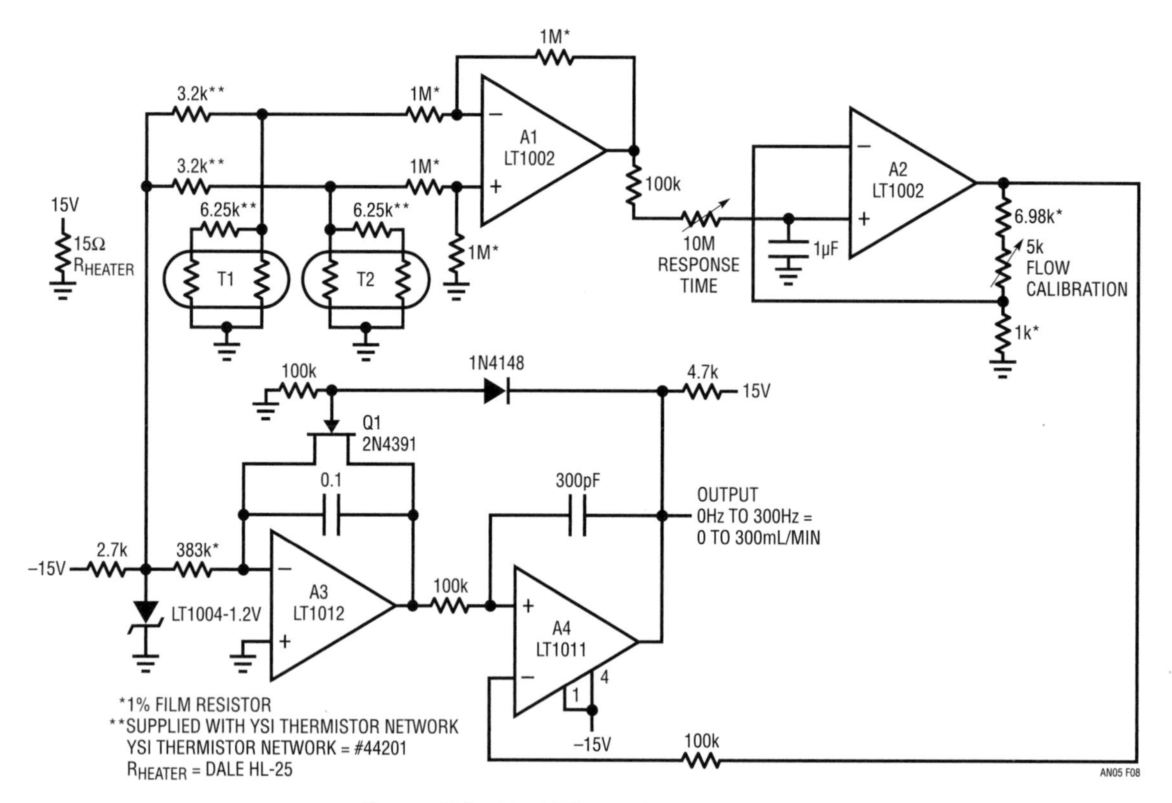

Figure 15.8 • Liquid Flowmeter

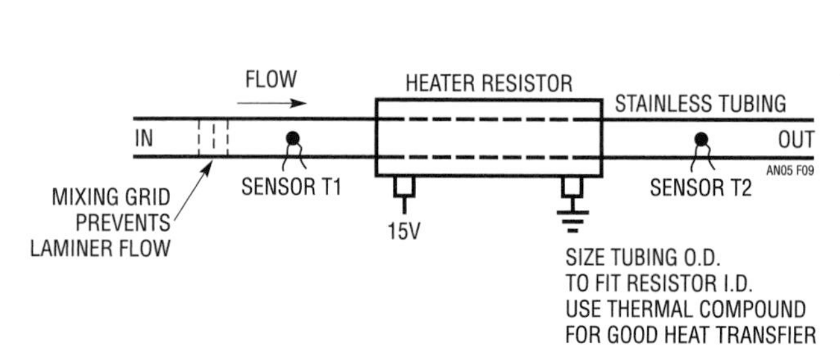

Figure 15.9 • Flowmeter Transducer Details

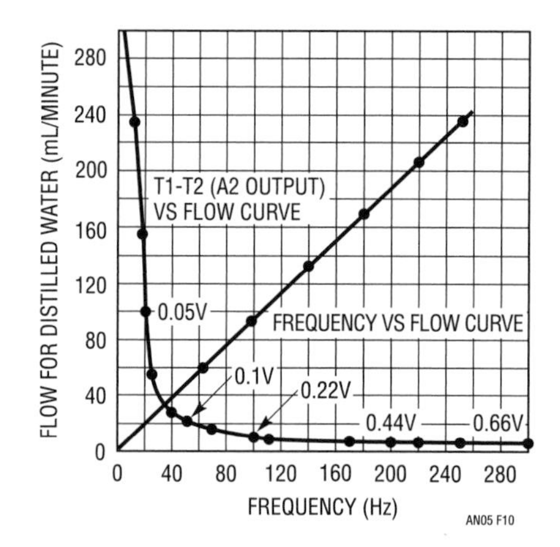

Figure 15.10 • Flowmeter Response Data

should be constant to maintain calibration. Ideally, the best way to do this is to measure the VI product at the heater resistor and construct a control loop to maintain constant wattage dissipation. However, if the resistor specified is used, its drift with temperature is small enough to assume constant dissipation with a fixed voltage drive. Additionally, the fluid's specific heat will affect calibration. The curves shown are for distilled water. To calibrate this circuit, set a flow rate of 10mL/minute and adjust the flow calibration trim for 10Hz output. The response time adjustment is convenient for filtering flow aberrations due to mechanical limitations in the pump driving the system.

Thermally-based anemometer (air flowmeter)

Figure 15.11 shows another thermally-based flowmeter, but this design is used to measure air or gas flow. It works by measuring the energy required to maintain a

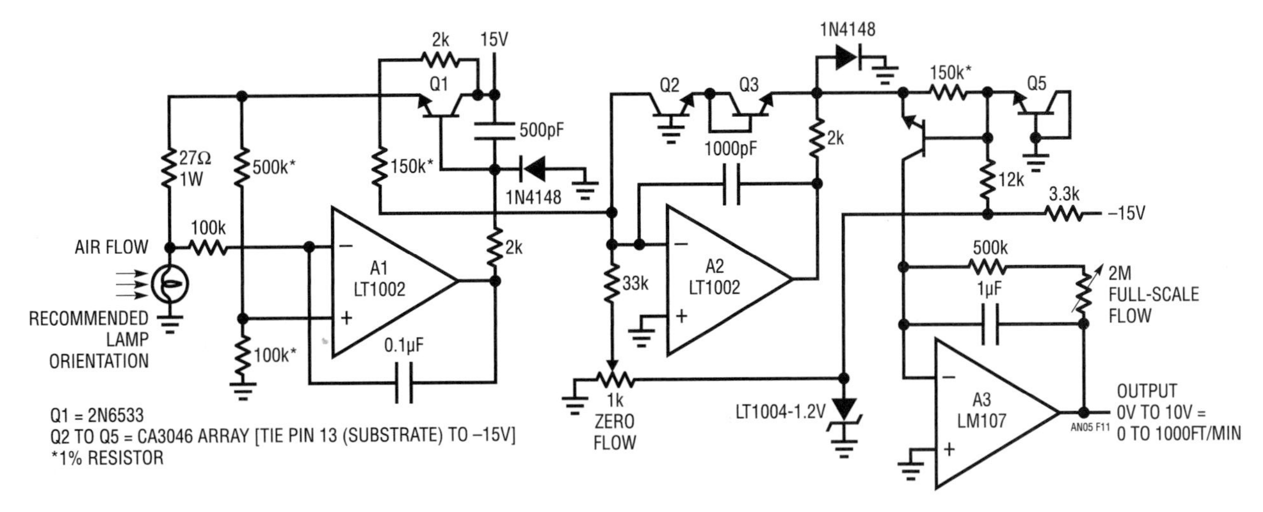

Figure 15.11 • Thermal Anemometer

heated resistance wire at constant temperature. The positive temperature coefficient of a small lamp, in combination with its ready availability, makes it a good sensor. A type 328 lamp is modified for this circuit by removing its glass envelope. The lamp is placed in a bridge which is monitored by A1. A1's output is current amplified by Q1 and fed back to drive the bridge. The capacitors and 220Ω resistor ensure stability. The 2k resistor furnishes start-up. When power is applied, the lamp is at a low resistance and Q1's emitter tries to come full on. As current flows through the lamp, its temperature quickly rises, forcing its resistance to increase. This action increases A1's negative input potential. Q1's emitter voltage decreases and the circuit finds a stable operating point. To keep the bridge balanced, A1 acts to force the lamp's resistance, hence its temperature, constant. The 10k-2k bridge values have been chosen so that the lamp operates just below the incandescence point. This high temperature minimizes the effects of ambient temperature shifts on circuit operation. Under these conditions, the only physical parameter which can significantly influence the lamp's temperature is a change in dissipation characteristic. Air flow moving by the lamp provides this change. Moving air by the lamp tends to cool it and A1 increases Q1's output to maintain the lamp's temperature. The voltage at Q1's emitter is nonlinearly, but predictably, related to air flow by the lamp. A2, A3 and the array transistors form a circuit which squares and amplifies Q1's emitter voltage to give a linear, calibrated output versus air flow rate. To use this circuit, place the lamp in the air flow so that its filament is a 90° angle to the flow. Next, either shut off the air flow or shield the lamp from it and adjust the zero flow potentiometer for a circuit output of 0V. Then, expose the lamp to air flow of 1000 feet/minute and trim the full flow potentiometer for 10V output. Repeat these adjustments until both points are fixed. With this procedure completed, the air flowmeter is accurate within 3% over the entire 0 to 1000 foot/minute range.

Low distortion, thermally stabilized Wien Bridge oscillator

The positive temperature coefficient of lamp filaments is employed in a modern adaptation of a classic circuit in Figure 15.12. In any oscillator it is necessary to control the gain as well as the phase shift at the frequency of interest. If gain is too low, oscillation will not occur. Conversely, too much gain will cause saturation limiting. Figure 15.12 uses a variable Wien Bridge to provide frequency tuning from 20Hz to 20kHz. Gain control comes from the positive temperature coefficient of the lamp. When power is applied, the lamp is at a low resistance value, gain is high and oscillation amplitude builds. As amplitude builds, the lamp current increases, heating occurs and its resistance goes up. This causes a reduction in amplifier gain and the circuit finds a stable operating point. The lamp's gain-regulating behavior is flat within 0.25dB over the 20Hz-20kHz range of the circuit. The smooth, limiting nature of the lamp's operation, in combination with its simplicity, gives good results. Trace A, Figure 15.13 shows circuit output at 10kHz. Harmonic distortion is shown in Trace B and is below 0.003%. The trace shows that most of the distortion is due to second harmonic content and some crossover disturbance is noticeable. The low resistance values in the Wein network and the $3.8\text{nV}\sqrt{\text{Hz}}$ noise specification of the LT1037 eliminate amplifier noise as an error term.

At low frequencies, the thermal time constant of the small normal mode lamp begins to introduce distortion levels above 0.01%. This is due to "hunting" as the oscillator's frequency approaches the lamp thermal time constant. This effect can be eliminated, at the expense of reduced output amplitude and longer amplitude settling time, by

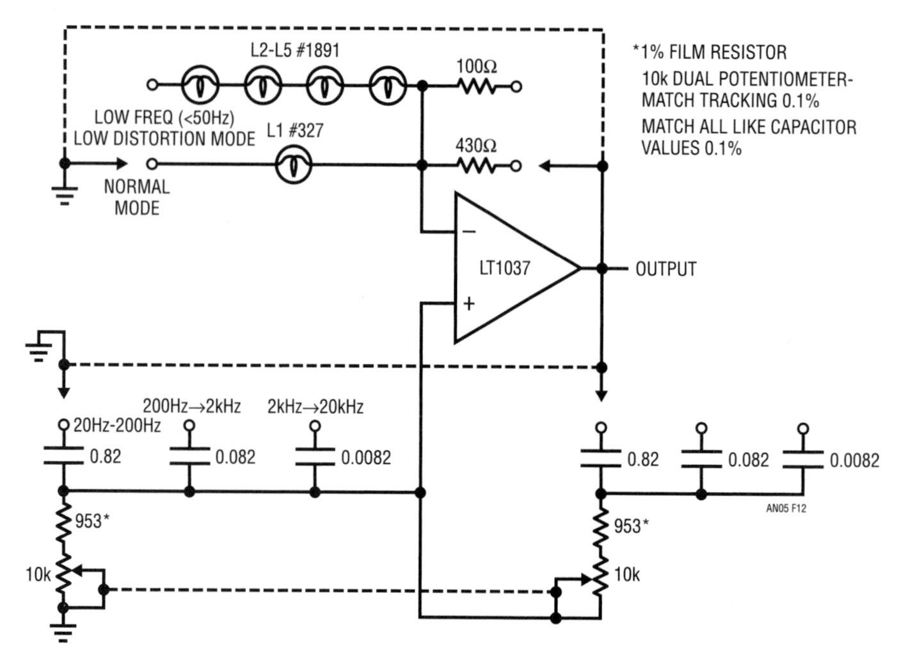

Figure 15.12 • Low Distortion Sinewave Oscillator

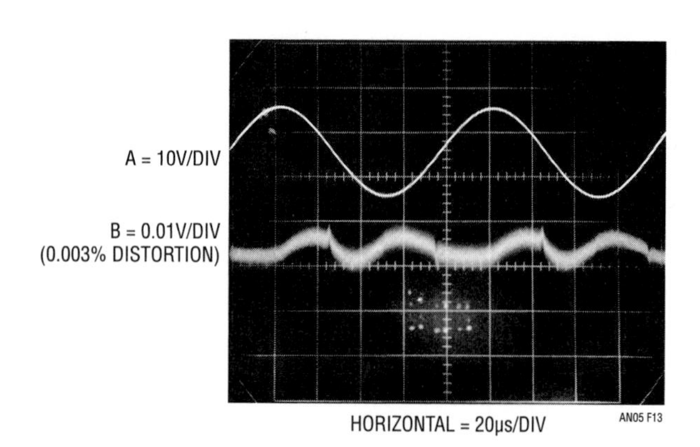

Figure 15.13 • Oscillator Waveforms

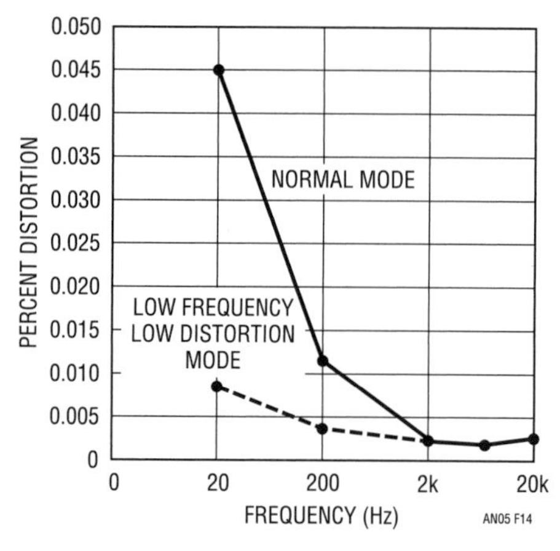

Figure 15.14 • Oscillator Distortion vs Frequency

switching to the low frequency, low distortion mode. The four large lamps give a longer thermal time constant and distortion is reduced. Figure 15.14 plots distortion versus frequency for the circuit

References

1. Multiplier Application Guide, pp. 7–9, "Flowmeter," Analog Devices, Inc., Norwood, Massachusetts
2. Olson, J.V., "A High Stability Temperature Controlled Oven," S.B. Thesis M.I.T., Cambridge, Massachusetts, 1974
3. PIN Photodiodes—5082-4200 Series, pp. 332–335, Optoelectronics Designers' Catalog, 1981, Hewlett Packard Company, Palo Alto, California
4. Y.S.I. Thermilinear Thermistor, #44018 Data Sheet, Yellow Springs Instrument Company, Yellow Springs, Ohio
5. Hewlett, William R., "A New Type Resistance-Capacitor Oscillator," M.S. Thesis, Stanford University, Palo Alto, California, 1939

Methods of measuring op amp settling time

16

Jim Williams

Servo, DAC and data acquisition amplifiers all require good dynamic response. In particular, the time required for an amplifier to settle to final value after an input step is especially important. This specification allows setting a circuit's timing margins with confidence that the data produced is accurate. The settling time is the total length of time from input step application until the amplifier remains within a specified error band around the final value.

Figure 16.1 shows one way to measure amplifier settling time (see References 1, 2, and 3). The circuit uses the "false sum node" technique. The resistors and amplifier form a bridge-type network. Assuming ideal resistors, the amplifier output will step to $-V_{IN}$ when an input pulse is applied. During slew, the oscilloscope probe is bounded by the diodes, limiting voltage excursion. When settling occurs, the oscilloscope probe voltage should be zero. Note that the resistor divider's attenuation means the probe's output will be one-half of the actual settled voltage.

In theory, this circuit allows settling to be observed to small amplitudes. In practice, it cannot be relied upon to produce useful measurements. Several flaws exist. The circuit requires the input pulse to have a flat top within the required measurement limits. Typically, settling within 10mV or less for a 10V step is of interest. No general-purpose pulse generator is meant to hold output amplitude and noise within these limits. Generator output-caused aberration appearing at the oscilloscope probe will be indistinguishable from amplifier output movement, producing unreliable results. The oscilloscope connection presents additional problems. As probe capacitance rises, AC loading of the resistor junction will influence observed settling waveforms. The 20pF probe shown alleviates this problem but its 10X attenuation sacrifices oscilloscope gain. 1X probes are not suitable because of their excessive input capacitance. An active 1X FET probe will work, but another issue remains.

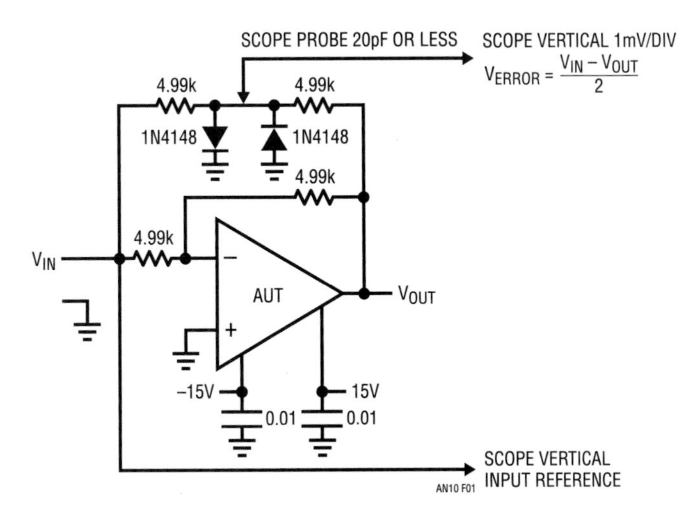

Figure 16.1 • Typical Settling Time Test Circuit

Analog Circuit and System Design: Immersion in the Black Art of Analog Design. http://dx.doi.org/10.1016/B978-0-12-397888-2.00016-x

The clamp diodes at the probe point are intended to reduce swing during amplifier slewing, preventing excessive oscilloscope overdrive. Unfortunately, oscilloscope overdrive recovery characteristics vary widely among different types and are not usually specified. The diodes' 600mV drop means the oscilloscope may see an unacceptable overload, bringing displayed results into question (for a discussion of oscilloscope overdrive considerations, see Box Section A, "Evaluating Oscilloscope Overload Response").

Figure 16.2 shows a practical settling time test circuit that addresses the problems discussed. Combined with a careful evaluation of the test oscilloscope used, it permits reliable settling time measurements in the 0.1% to 0.01% region. The input pulse does not drive the amplifier, but switches a Schottky bridge via a clamp. The bridge is biased from two low noise LT1021-10V references. Depending on input pulse polarity, current flows through the appropriate 10k resistor to bias the amplifier's summing point. The bridge switches cleanly and quickly, producing a flat-topped current pulse into the AUT. The circuit's input pulse characteristics do not influence the measurement. A second clamp-bridge arrangement supplies an opposite polarity signal which is nulled against the amplifier's output at point B. Schottky clamp diodes limit this point's voltage excursion to ±300mV.

The Q1-Q5 configuration forms a low input capacitance, high speed buffer to drive the oscilloscope. Q1A's 1pF to 2pF input capacitance provides very light AC loading, eliminating probe-caused problems. Q1B, running as a current sink, compensates Q1A's V_{GS} drop. Q2-Q5 form a complementary emitter-follower, which can drive substantial cable capacitance without distortion.

The circuit should be built on a ground planed board with particular care taken to ensure low stray capacitance at points A and B. The AUT socket should be selected for short pin lengths. Very high speed amplifiers (t_{SETTLE} < 200ns) should be directly soldered into the circuit.

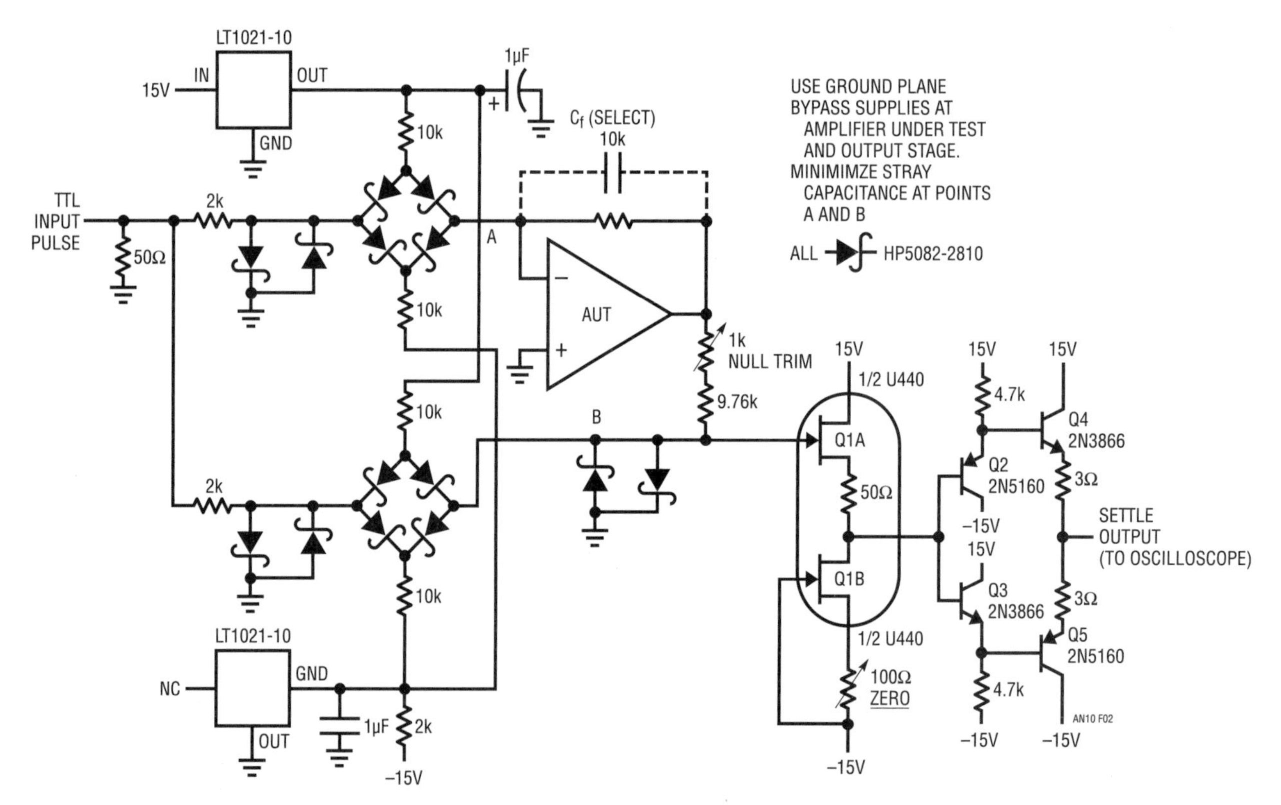

Figure 16.2 • Improved Settling Time Test Circuit

This circuit, combined with a judiciously chosen oscilloscope, allows observation of amplifier settling to a millivolt (0.01%) for a 10V step. A good way to gain confidence in the circuit is to test a very fast UHF amplifier. Figure 16.3 shows response for an amplifier (Teledyne Philbrick 1435) specified to settle in 70ns within a millivolt for a 10V step. Trace A is the input pulse, Trace B is the amplifier output and Trace C is the settle signal. Settling occurs inside 70ns, indicating good agreement between the circuit and the AUT specification. Since most amplifiers are not nearly this fast, it is reasonable to assume that the circuit will always provide reliable results.

Because this circuit works by nulling opposite polarity sources, it seems unable to test followers—but it can. The AUT is battery-powered and completely floated from the circuit's power supply (Figure 16.4). The AUT output is connected to circuit ground and the battery center tap becomes the output. The positive input is driven from the Schottky bridge. The floating power supply lets the follower fool the circuit into thinking it is testing an inverter. The AUT's output appears inverted, but this is not a significant penalty.

To calibrate this circuit, ground point B and adjust the "zero trim" for 0V output. Next, temporarily tie

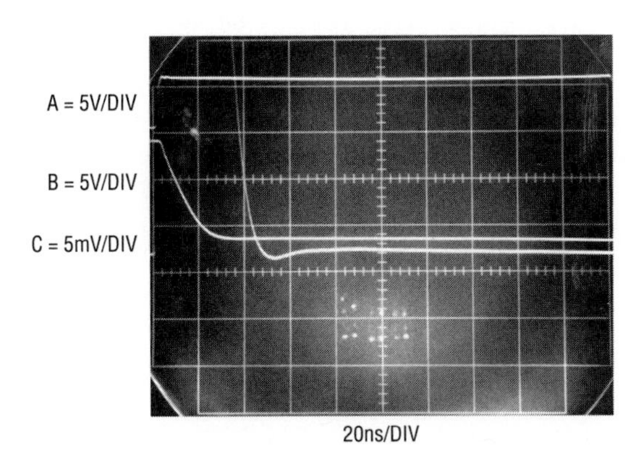

Figure 16.3 • Settling Detail for a Fast Amplifier

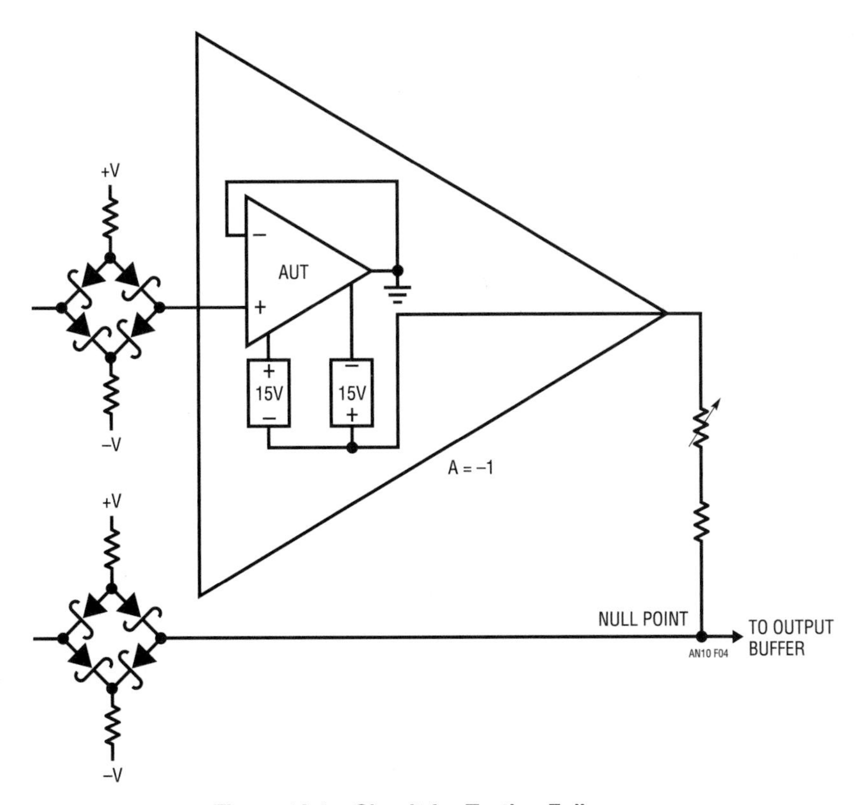

Figure 16.4 • Circuit for Testing Followers

the pulse input to +15V through 680Ω and adjust the "null trim" for 0V output. Remove the 680Ω resistor and the circuit is ready for use. When measuring settling times remember to experiment with the value of C_F to obtain best performance (see Box Section B, "Amplifier Compensation").

In the past, amplifier settling measurements below 1mV were not required. Recently, 16-bit and 18-bit D→A converters have become relatively common, requiring users to consider sub-millivolt settling time performance. Also, the offset specifications of current generation monolithic amplifiers are good enough to make very high precision settling time data worthwhile. Previously, being able to see an amplifier settle within 50μV wasn't interesting because its thermal drift swamped this figure.

The newer amplifier's substantially lower drift means very high precision settling time measurement data is useful. Figure 16.2's circuit is limited to 0.01% (1mV out of 10V) resolution by the 300mV Schottky clamp potential at point B. Simply increasing oscilloscope gain to get higher resolution will not work because of severe overload problems. With the oscilloscope set at 50μV/division, the Schottky bound allows a 6000:1 overdrive. This is much more than any vertical amplifier is designed to accommodate. The oscilloscope's overload recovery will completely dominate the observed waveform and all measurements will be meaningless.

One way to obtain higher precision settling time measurements is to clip the incoming waveform in *time*, as well as amplitude. If the oscilloscope is prevented from seeing the waveform until settling is nearly complete, overload is avoided. Doing this requires placing a switch at the settle circuit's output and controlling it with an input-triggered, variable delay. FET switches are not suitable because of their gate-source capacitance. This capacitance will allow gate drive artifacts to corrupt the oscilloscope display, producing confusing readings. In the worst case, gate drive transients will be large enough to induce overload, defeating the switch's purpose.

Figure 16.5 shows a way to implement the switch, which largely eliminates these problems. This circuit, connected to the basic settle circuit of Figure 16.2, allows settling within 10μV to be observed. The Schottkys sampling bridge is the actual switch. The bridge's inherent balance,

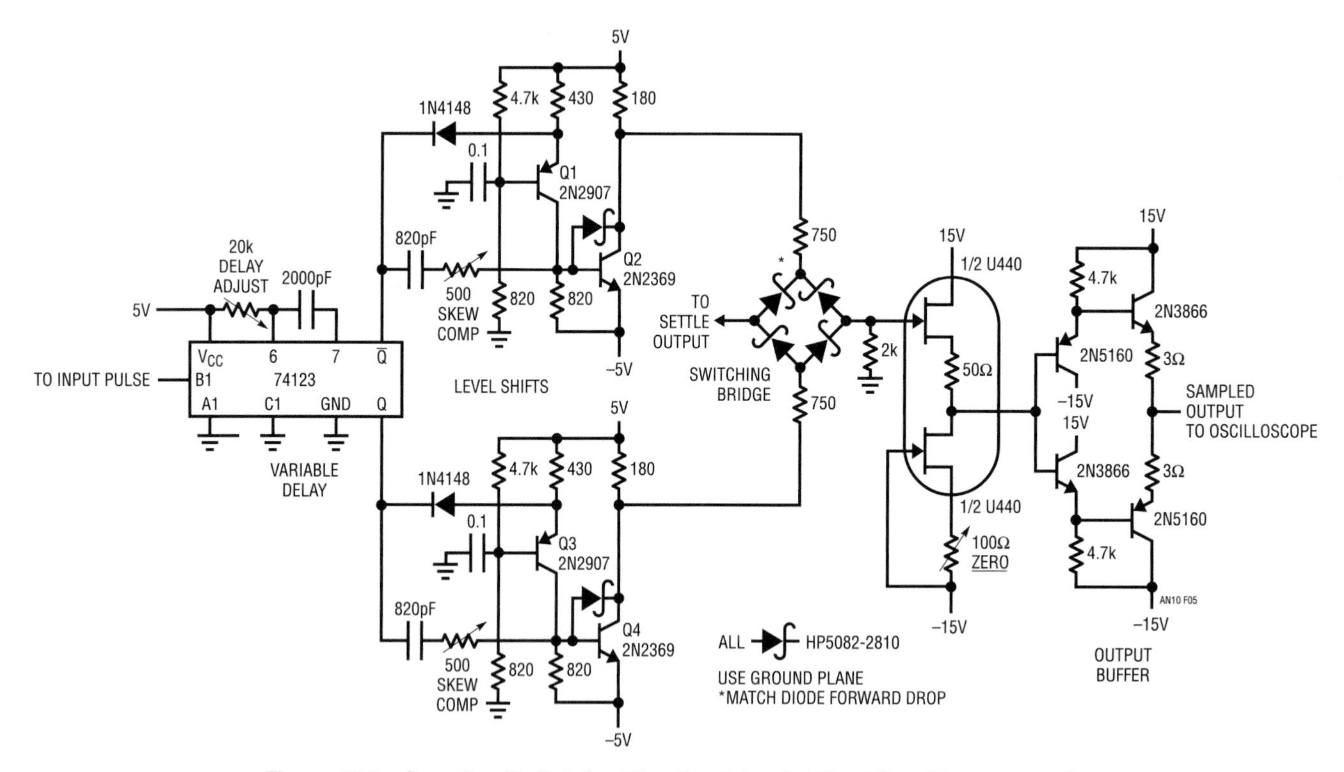

Figure 16.5 • Sampling Switch for Ultra Precision Settling Time Measurement

combined with matched diodes and very high speed complementary bridge switching, yields a clean, switched output. An output buffer stage, identical to the one used in Figure 16.2, unloads the bridge and drives the oscilloscope.

The complementary bridge switching drive is supplied from the Q1-Q2 and Q3-Q4 level shifters. Each circuit converts the delay one-shot's TTL output to ±5V levels. The identical stages are comprised of an emitter-switched current source feeding a Baker-clamped common emitter output. Feedforward capacitance to the output transistor aids speed and overall delays are about 3ns. The level shifters must switch simultaneously to minimize drive-induced disturbances in the bridge's output. The "skew compensation" trims permit very small phasing adjustments in each level shifter, compensating skew in the 74123 one-shot's outputs. To trim this circuit, ground the bridge input and pulse the 74123's C1 input. Next, set the oscilloscope to 100μV/division and adjust the skew trims for minimum indication on the screen. Connect the bridge input back to the settle circuit's output and the circuit is ready for use.

Construction of this circuit requires care. A ground plane is mandatory and all bridge connections should be as short and symmetrical as possible. To maintain low noise, the bridge's output ground return should be routed away from high current returns such as the 74123's ground pin.

This switch circuit, carefully constructed and used with the basic settle circuit, provides good results. Figure 16.6 shows an LT1001 precision op amp as the AUT. Trace A is the input pulse, while Trace B is the AUT output. During the AUT's slewing period the 74123 is fired (Trace C is Q), turning off the bridge. The bridge input appears in Trace D. The 74123 delay is adjusted so the bridge is switched when settling is nearly complete. Trace E is the circuit's final output, showing settling details at 100μV/division. The narrow peaking at the waveform's leading edge is due to switching residue. Figure 16.7 lists measured settling times to 50μV (0.0005% of a 10V step) for a group of precision amplifiers.

Some poorly designed amplifiers exhibit a substantial "thermal tail" after responding to an input step. This phenomenon, due to die heating, can cause the output to wander outside desired limits long after settling has apparently occurred. After checking settling at

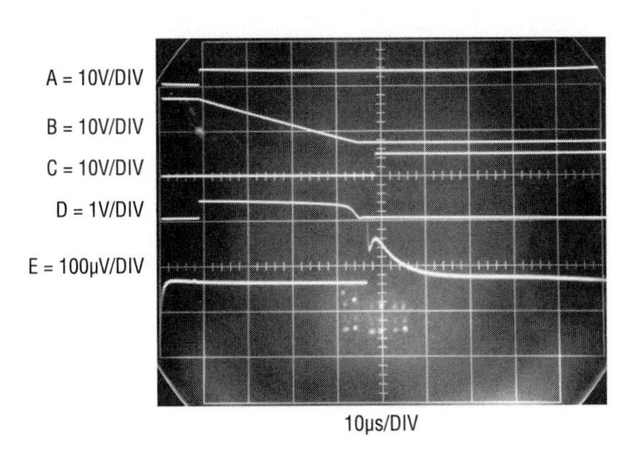

Figure 16.6 • Sampling Switch Waveforms

Amplifier	Settling Time	Remarks
LT1001	65μs	
LT1007	18μs	
LT1008	65μs	Standard Compensation
LT1008	35μs	Feedforward Compensation
LT1012	70μs	
LT1055	6μs	
LT1056	5μs	

Figure 16.7 • Measured Settling Times to 0.0005%

high speed it is always a good idea to slow the oscilloscope sweep down and look for thermal tails. Often the thermal tail's effect can be accentuated by loading the amplifier's output. Figure 16.8 shows the thermal tail of an amplifier, which appears to have settled in a much shorter time than it actually has.

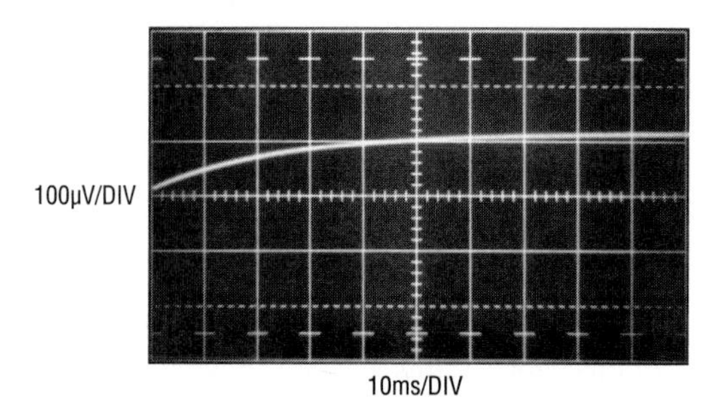

Figure 16.8 • Typical Thermal Tail

References

1. Analog Devices AD544 Data Sheet, "Settling Time Test Circuit"
2. National Semiconductor, LF355/356/357 Data Sheet, "Settling Time Test Circuit"
3. Precision Monolithics, Inc., OP-16 Data Sheet, "Settling Time Test Circuit"
4. R. Demrow, "Settling Time of Operational Amplifiers," *Analog Dialogue*, volume 4–1, 1970 (Analog Devices)
5. R. A. Pease, "The Subtleties of Settling Time," The New *Lightning Empiricist*, June 1971, Teledyne Philbrick
6. W. Travis, "Settling Time Measurement Using Delayed Switch," Private Communication

Box section A
Evaluating oscilloscope overload performance

Settling time measurement relies heavily on the oscilloscope used. In many cases the oscilloscope is required to supply an accurate waveform after the display has been driven off screen. How long must one wait after an overload before the display can be taken seriously? The answer to this question is quite complex. Factors involved include the degree of overload, its duty cycle, its magnitude in time and amplitude and other considerations. Oscilloscope response to overload varies widely between types and markedly different behavior can be observed in any individual instrument. For example, the recovery time for a 100X overload at 0.005V/division may be very different than at 0.1V/division. The recovery characteristic may also vary with waveform shape, DC content and repetition rate. With so many variables, it is clear that measurements involving oscilloscope overload must be approached with caution. Nevertheless, a simple test can indicate when the oscilloscope is being deleteriously affected by overdrive.

The waveform to be expanded is placed on the screen at a vertical sensitivity, which eliminates all off-screen activity. Figure A1 shows the display. The lower right hand portion is to be expanded. Increasing the vertical sensitivity by a factor of two (Figure A2) drives the waveform off-screen, but the remaining display appears reasonable. Amplitude has doubled and waveshape is consistent with the original display. Looking carefully, it is possible to see small amplitude information presented as a dip in the waveform at about the third vertical division. Some small disturbances are also visible. This observed expansion of the original waveform is believable. In Figure A3, gain has been further increased, and all the features of Figure A2 are amplified accordingly. The basic waveshape appears clearer and the dip and small disturbances are also easier to see. No new waveform characteristics are observed. Figure A4 brings some unpleasant surprises. This increase in gain causes definite distortion. The initial negative-going peak, although larger, has a different shape. Its bottom appears less broad than in Figure A3. Additionally, the peak's positive recovery is shaped slightly differently. A new rippling disturbance is visible in the center of the screen. This kind of change indicates that the oscilloscope is having trouble. A further test can confirm that this waveform is being influenced by overloading. In Figure A5 the gain remains the same but the vertical position knob has been used to reposition the display at the screen's bottom. This shifts the oscilloscope's DC operating point, which, under normal circumstances, should not affect the displayed waveform. Instead, a marked shift in waveform amplitude and outline occurs. Repositioning the waveform to the screen's top produces a differently distorted waveform (Figure A6). It is obvious that for this particular waveform, accurate results cannot be obtained at this gain.

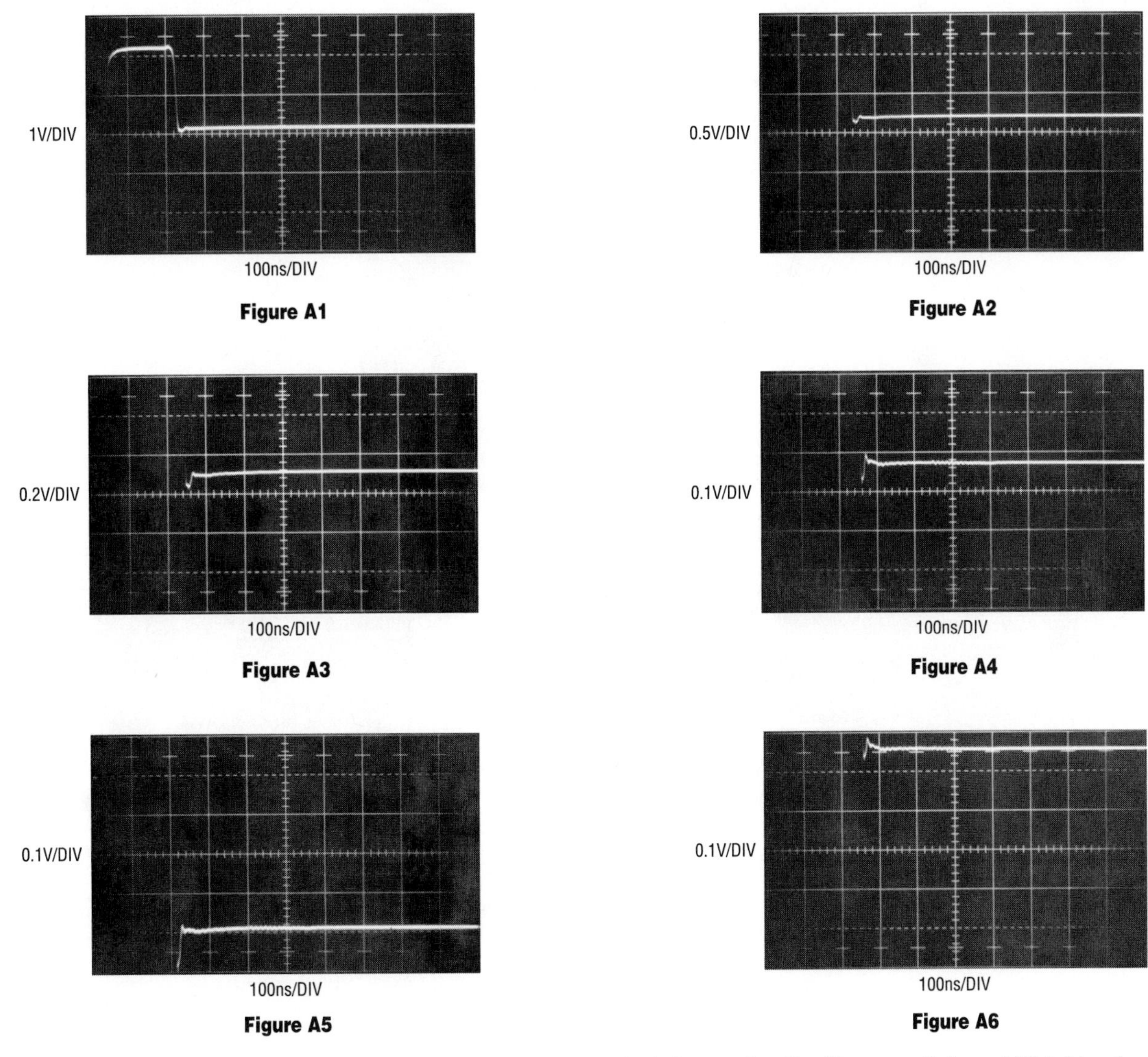

Figures A1-A6 • The Overdrive Limit is Determined by Progressively Increasing Oscilloscope Gain and Watching for Waveform Aberrations

Box section B
Amplifier compensation

To get the best possible settling time from any amplifier, the feedback capacitor, C_F, should be carefully chosen. C_F's purpose is to roll off amplifier gain at the frequency, which permits best dynamic response. The optimum value for C_F will depend on the feedback resistor's value and the characteristics of the source. One of the most common sources is also one of the most difficult. Digital-to-analog converters' current outputs must often be converted to a voltage. Although an op amp can easily do this, care is required to obtain good dynamic performance. A fast DAC can settle to 0.01% in 200ns but its output also includes a parasitic capacitance term, making the amplifier's job more difficult. Normally, the DAC's current output is unloaded directly into the amplifier's summing junction, placing the parasitic capacitance from ground to the amplifier's input. The capacitance introduces feedback phase shift at high frequencies, forcing the amplifier to "hunt" and ring about the final value before settling. Different DACs have different values of output capacitance. CMOS DACs have the highest output capacitance and it varies with code. Bipolar DACs typically have 20pF to 30pF of capacitance, stable over all codes. Because of their output capacitance, DACs furnish an instructive example in amplifier compensation. In practice, the Schottky bridge which feeds the AUT in the settle circuit is replaced with the DAC to be used. Depending on DAC input coding, it may be necessary to use inverters in the DAC input lines to maintain circuit nulling action. Figure B1 shows the response of an industry standard DAC-80 type and an LT1023 op amp, which is optimized for inverting applications. Trace A is the input, while Traces B and C are the amplifier and settle outputs, respectively. In this example no compensation capacitor is used and the amplifier rings badly before settling. In B2, and 82pF unit stops the ringing and settling time goes down to 4μs. The overdamped response means that C_F dominates the capacitance at the AUT's input and stability is assured. If fastest response is desired, C_F must be reduced. B3 shows critically damped behavior obtained with a 22pF unit. The settling time of 2μs is the best obtainable for this DAC-amplifier combination.

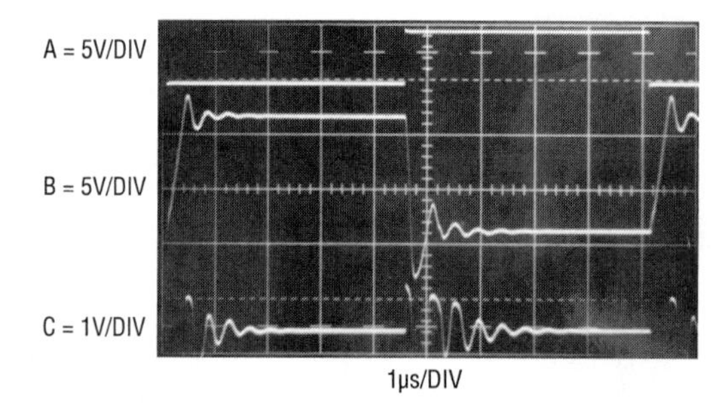

Figure B1

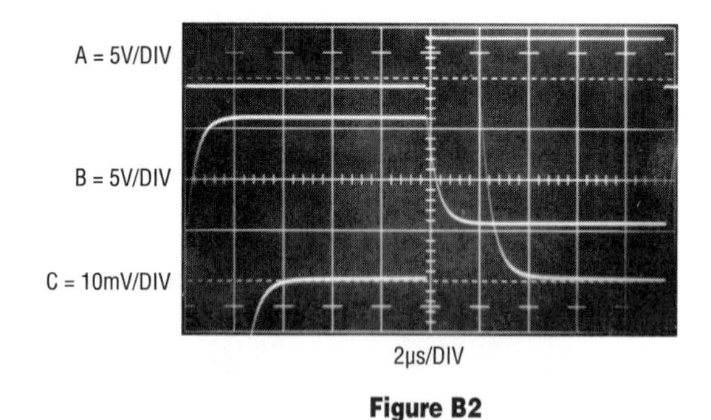

Figure B2

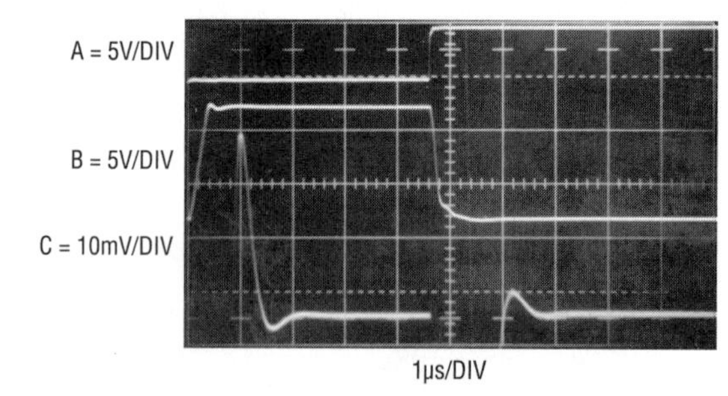

Figure B3

Figures B1-B3 • Effects of Different Feedback Capacitors on DAC-Op Amp Combination

High speed comparator techniques

17

Jim Williams

Introduction

Comparators may be the most underrated and underutilized monolithic linear component. This is unfortunate because comparators are one of the most flexible and universally applicable components available. In large measure the lack of recognition is due to the IC op amp, whose versatility allows it to dominate the analog design world. Comparators are frequently perceived as devices, which crudely express analog signals in digital form—a 1-bit A/D converter. Strictly speaking, this viewpoint is correct. It is also wastefully constrictive in its outlook. Comparators don't "just compare" in the same way that op amps don't "just amplify".

Comparators, in particular high speed comparators, can be used to implement linear circuit functions which are as sophisticated as any op amp-based circuit. Judiciously combining a fast comparator with op amps is a key to achieving high performance results. In general, op amp-based circuits capitalize on their ability to close a feedback loop with precision. Ideally, such loops are maintained continuously over time. Conversely, comparator circuits are often based on speed and have a discontinuous output over time. While each approach has its merits, a fusion of both yields the best circuits.

This effort's initial sections are devoted to familiarizing the reader with the realities and difficulties of high speed comparator circuit work. The mechanics and subtleties of achieving precision circuit operation at DC and low frequency have been well documented. Relatively little has appeared which discusses, in practical terms, how to get fast circuitry to work. In developing such circuits, even the most veteran designers sometimes feel that nature is conspiring against them. In some measure this is true. Like all engineering endeavors, high speed circuits can only work if negotiated compromises with nature are arranged. Ignorance of, or contempt for, physical law is a direct route to frustration. In this regard, much of the text and appendices are directed at developing awareness of and respect for circuit parasitics and fundamental limitations. This approach is maintained in the applications section, where the notion of "negotiated compromises" is expressed in terms of resistor values and compensation techniques. Many of the application circuits use the LT®1016's speed to improve on a standard circuit. Some utilize the speed to implement a traditional function in a non-traditional way, with attendant advantages. A (very) few operate at or near the state-of-the-art for a given circuit type, regardless of approach. Substantial effort has been expended in developing these examples and documenting their operation. The resultant level of detail is justified in the hope that it will be catalytic. The circuits should stimulate new ideas to suit particular needs, while demonstrating the LT1016's capabilities in an instructive manner.

The LT1016—an overview

A new ultra high speed comparator, the LT1016, features TTL-compatible complementary outputs and 10ns response time. Other capabilities include a latch pin and

Analog Circuit and System Design: Immersion in the Black Art of Analog Design. http://dx.doi.org/10.1016/B978-0-12-397888-2.00017-1

good DC input characteristics (see Figure 17.1). The LT1016's outputs directly drive all TTL families, including the new higher speed ASTTL and FAST parts. Additionally, TTL outputs make the device easier to use in linear circuit applications where ECL output levels are often inconvenient.

A substantial amount of design effort has made the LT1016 relatively easy to use. It is much less prone to oscillation and other vagaries than some slower comparators, even with slow input signals. In particular, the LT1016 is stable in its linear region, a feature no other high speed comparator has. Additionally, output stage switching does not appreciably change power supply current, further enhancing stability. These features make the application of the 200GHz gain-bandwidth LT1016 considerably easier than other fast comparators. Unfortunately, laws of physics dictate that the circuit *environment* the LT1016 works in must be properly prepared. The performance limits of high speed circuitry are often determined by parasitics such as stray capacitance, ground impedance, and layout. Some of these considerations are present in digital systems where designers are comfortable describing bit patterns and memory access times in terms of nanoseconds. The LT1016 can be used in such fast digital systems and Figure 17.2 shows just how fast the device is. The simple test circuit allows us to see that the LT1016's (Trace B) response to the pulse generator (Trace A) is faster than a TTL inverter (Trace C)! In fact, the inverter's output never gets to a TTL "0" level. Linear circuits operating with this kind of speed make many engineers justifiably wary. Nanosecond domain linear circuits are widely associated with oscillations, mysterious shifts in circuit characteristics, unintended modes of operation and outright failure to function.

Other common problems include different measurement results using various pieces of test equipment, inability to make measurement connections to the circuit without inducing spurious responses and dissimilar operation

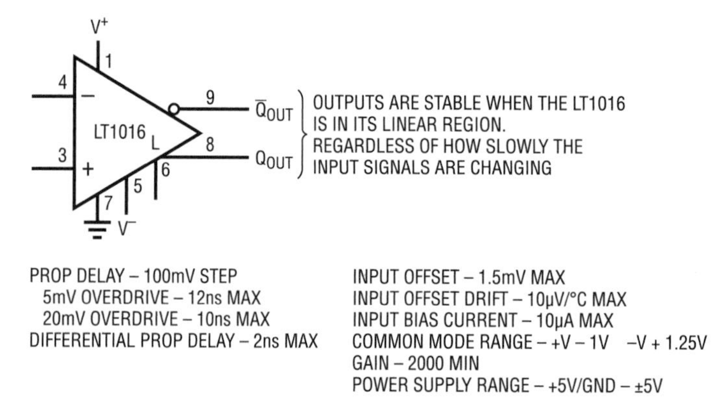

Figure 17.1 • The LT1016 at a Glance

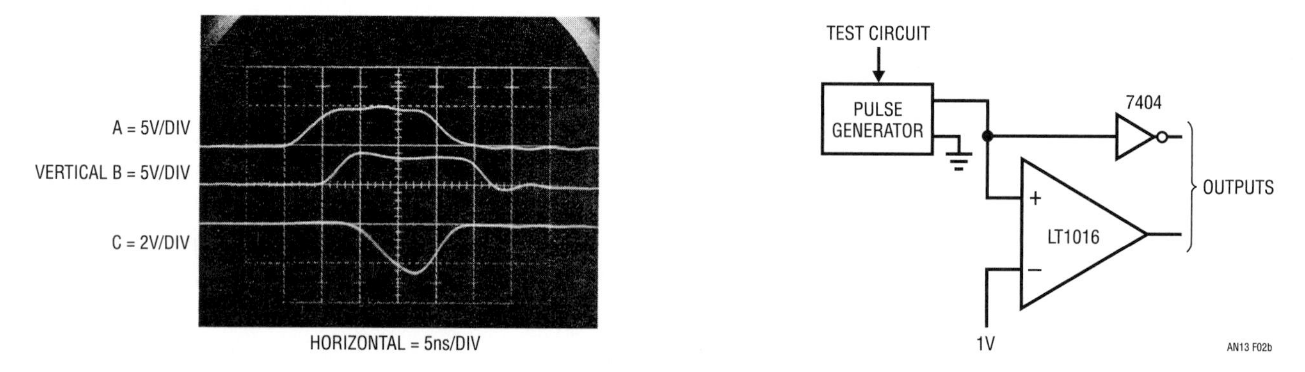

Figure 17.2 • LT1016 vs a TTL Gate

between two "identical" circuits. If the components used in the circuit are good and the design is sound, all of the above problems can usually be traced to failure to provide a proper circuit "environment." To learn how to do this requires studying the causes of the aforementioned difficulties.

The Rogue's gallery of high speed comparator problems

By far the most common error involves power supply bypassing. Bypassing is necessary to maintain low supply impedance. DC resistance and inductance in supply wires and PC traces can quickly build up to unacceptable levels. This allows the supply line to move as internal current levels of the devices connected to it change. This will almost always cause unruly operation. In addition, several devices connected to an unbypassed supply can "communicate" through the finite supply impedances, causing erratic modes. Bypass capacitors furnish a simple way to eliminate this problem by providing a local reservoir of energy at the device. The bypass capacitor acts like an electrical flywheel to keep supply impedance low at high frequencies. The choice of what type of capacitors to use for bypassing is a critical issue and should be approached carefully (see Appendix A, "About Bypass Capacitors"). An unbypassed LT1016 is shown responding to a pulse input in Figure 17.3. The power supply the LT1016 sees at its terminals has high impedance at high frequency. This impedance forms a voltage divider with the LT1016, allowing the supply to move as internal conditions in the comparator change. This causes local feedback and oscillation occurs. Although the LT1016 responds to the input pulse, its output is a blur of 100MHz oscillation. *Always use bypass capacitors.*

In Figure 17.4 the LT1016's supplies are bypassed, but it still oscillates. In this case, the bypass units are either too far from the device or are lossy capacitors. *Use capacitors with good high frequency characteristics and mount them as close as possible to the LT1016. An inch of wire between the capacitor and the LT1016 can cause problems.*

In Figure 17.5 the device is properly bypassed but a new problem pops up. This photo shows both outputs of the comparator. Trace A appears normal, but Trace B shows an excursion of almost 8V—quite a trick for a device running from a +5V supply. This is a commonly reported problem in high speed circuits and can be quite confusing. It is not due to suspension of natural law, but is traceable to a grossly miss-compensated or improperly selected oscilloscope probe. *Use probes which match your oscilloscope's input characteristics and compensate them properly* (for a discussion on probes, see Appendix B, "About Probes and Scopes"). Figure 17.6 shows another probe-induced problem. Here, the amplitude seems correct but the 10ns response time LT1016 appears to have 50ns edges! In this case, the probe used is too heavily compensated or slow for the oscilloscope. Never use 1× or "straight" probes. Their bandwidth is 20MHz or less and capacitive loading is high. *Check probe bandwidth to ensure it is adequate for the measurement. Similarly, use an oscilloscope with adequate bandwidth.*

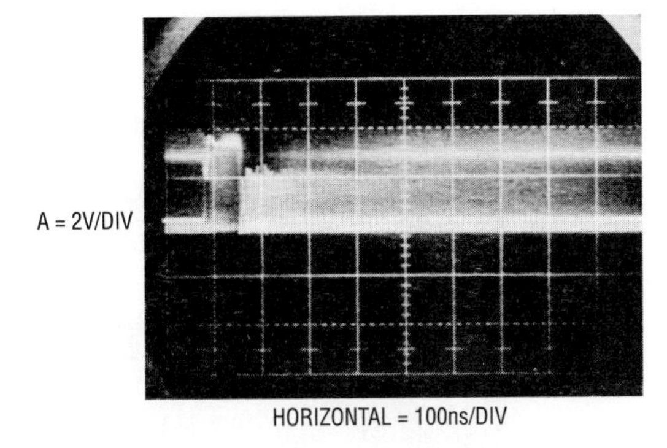

Figure 17.3 • Unbypassed LT1016 Response

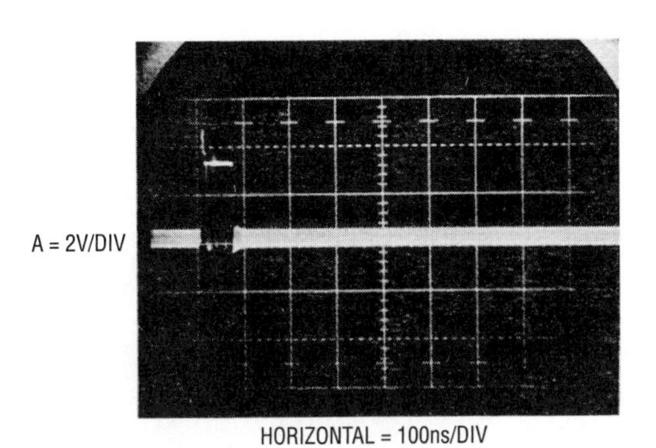

Figure 17.4 • LT1016 Response with Poor Bypassing

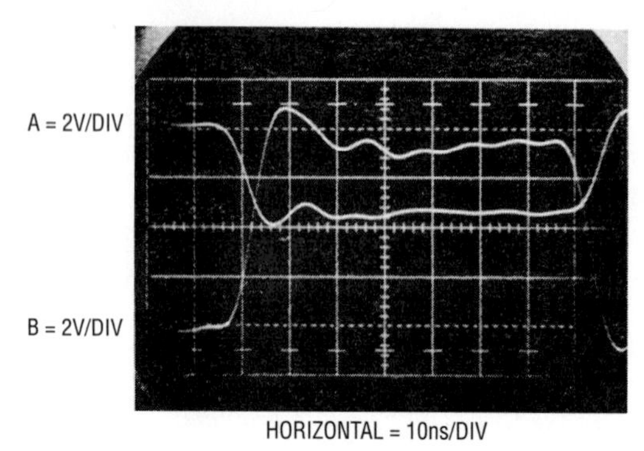

Figure 17.5 • Improper Probe Compensation Causes Seemingly Unexplainable Amplitude Error

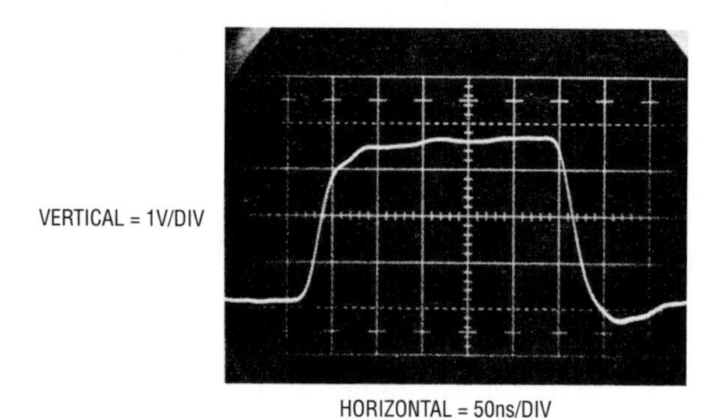

Figure 17.6 • Overcompensated or Slow Probes Make Edges Look Too Slow

In Figure 17.7 the probes are properly selected and applied but the LT1016's output rings and distorts badly. In this case, the probe ground lead is too long. For general purpose work most probes come with ground leads about six inches long. At low frequencies this is fine. At high speed, the long ground lead looks inductive, causing the ringing shown. High quality probes are always supplied with some short ground straps to deal with this problem. Some come with very short spring clips which fix directly to the probe tip to facilitate a low impedance ground connection. For fast work, the ground connection to the probe should not exceed one inch in length. *Keep the probe ground connection as short as possible.*

The difficulty in Figure 17.8 is delay and inadequate amplitude (Trace B). A small delay on the leading edge is followed by a large delay before the falling edge begins. Additionally, a lengthy, tailing response stretches 70ns before finally settling out. The amplitude only rises to 1.5V. A common oversight is responsible for these conditions.

A FET probe monitors the LT1016 output in this example. The probe's common mode input range has been exceeded, causing it to overload and clip the output badly. The small delay on the rising edge is characteristic of active probes and is legitimate. During the time the output is high, the probe is driven deeply into saturation. When the output falls, the probe's overload recovery is lengthy and uneven, causing the delay and tailing.

Know your FET probe. Account for the delay of its active circuitry. Avoid saturation effects due to common mode

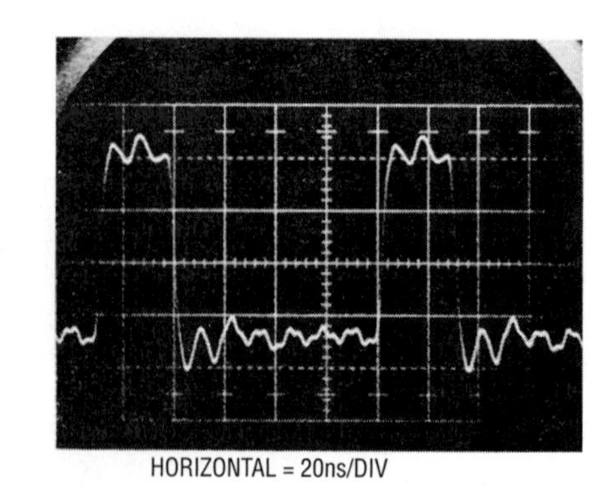

Figure 17.7 • Typical Results Due to Poor Probe Grounding

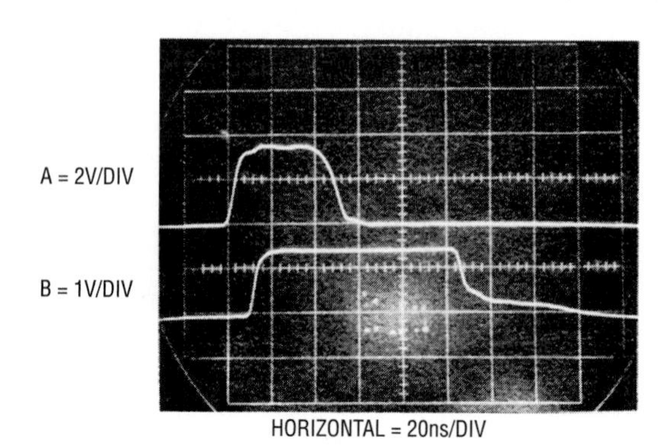

Figure 17.8 • Overdriven FET Probe Causes Delayed Tailing Response

input limitations (typically ±1V). Use 10× and 100× attenuator heads when required.

Figure 17.9 shows the LT1016's output (Trace B) oscillating near 40MHz as it responds to an input (Trace A). Note that the input signal shows artifacts of the oscillation. This example is caused by improper grounding of the comparator. In this case, the LT1016's ground pin connection is one inch long. The ground lead of the LT1016 must be as short as possible and connected directly to a low impedance ground point. Any substantial impedance in the LT1016's ground path will generate effects like this. The reason for this is related to the necessity of bypassing the power supplies. The inductance created by a long device ground lead permits mixing of ground currents, causing undesired effects in the device. The solution here is simple. *Keep the LT1016's ground pin connection as short (typically 1/4 inch) as possible and run it directly to a low impedance ground. Do not use sockets.*

Figure 17.10 addresses the issue of the "low impedance ground", referred to previously. In this example, the output is clean except for chattering around the edges. This photograph was generated by running the LT1016 without a "ground plane". A ground plane is formed by using a continuous conductive plane over the surface of the circuit board (the theory behind ground planes is discussed in Appendix C). The only breaks in this plane are for the circuit's necessary current paths. The ground plane serves two functions. Because it is flat (AC currents travel along the surface of a conductor) and covers the entire area of the board, it provides a way to access a low inductance ground from anywhere on the board. Also, it minimizes the effects of stray capacitance in the circuit by referring them to ground. This breaks up potential unintended and harmful feedback paths. *Always use a ground plane with the LT1016.*

"Fuzz" on the edges is the difficulty in Figure 17.11. This condition appears similar to Figure 17.10, but the oscillation is more stubborn and persists well after the output has gone low. This condition is due to stray capacitive feedback from the outputs to the inputs. A 3kΩ input source impedance and 3pF of stray feedback allowed this

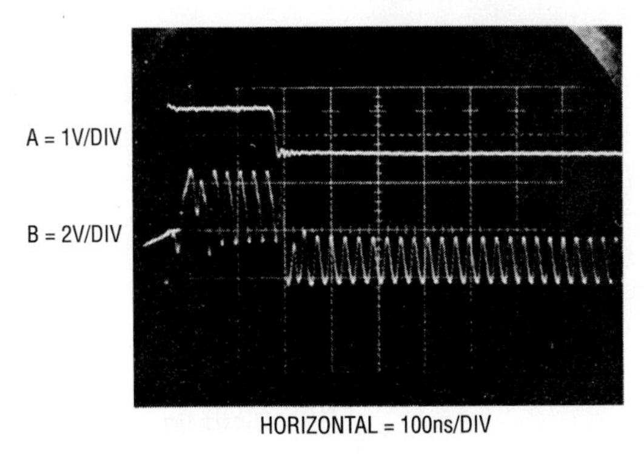

Figure 17.9 • Excessive LT1016 Ground Path Resistance Causes Oscillation

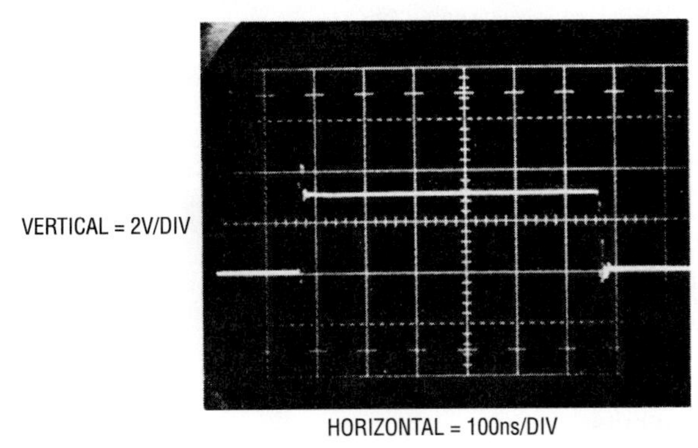

Figure 17.10 • Transition Instabilities Due to No Ground Plane

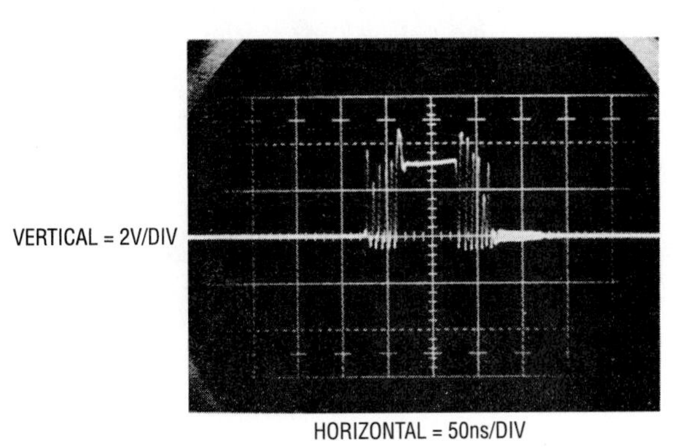

Figure 17.11 • 3pF Stray Capacitive Feedback with 3kΩ Source Can Cause Oscillation

oscillation. The solution for this condition is not too difficult. *Keep source impedances as low as possible, preferably 1kΩ or less. Route output and input pins and components away from each other.*

The opposite of stray-caused oscillations appears in Figure 17.12. Here, the output response (Trace B) badly lags the input (Trace A). This is due to some combination of high source impedance and stray capacitance to ground at the input. The resulting RC forces a lagged response at the input and output delay occurs. An RC combination of 2kΩ source resistance and 10pF to ground gives a 20ns time constant—significantly longer than the LT1016's response time. *Keep source impedances low and minimize stray input capacitance to ground.*

Figure 17.13 shows another capacitance-related problem. Here the output does not oscillate, but the transitions are discontinuous and relatively slow. The villain of this situation is a large output load capacitance. This could be caused by cable driving, excessive output lead length or the input characteristics of the circuit being driven. In most situations this is undesirable and may be eliminated by buffering heavy capacitive loads. In a few circumstances it may not affect overall circuit operation and is tolerable. *Consider the comparator's output load characteristics and their potential effect on the circuit. If necessary, buffer the load.*

Another output-caused fault is shown in Figure 17.14. The output transitions are initially correct but end in a ringing condition. The key to the solution here is the ringing. What is happening is caused by an output lead which is too long. The output lead looks like an unterminated transmission line at high frequencies and reflections occur. This accounts for the abrupt reversal of direction on the leading edge and the ringing. If the comparator is driving TTL this may be acceptable, but other loads may not tolerate it. In this instance, the direction reversal on the leading edge might cause trouble in a fast TTL load. Keep output lead lengths short. *If they get much longer than a few inches, terminate with a resistor (typically 250Ω to 400Ω).*

A final malady is presented in Figure 17.15. These waveforms are reminiscent of the input RC-induced delay of Figure 17.12. The output waveform initially responds to the input's leading edge, but then returns to zero before

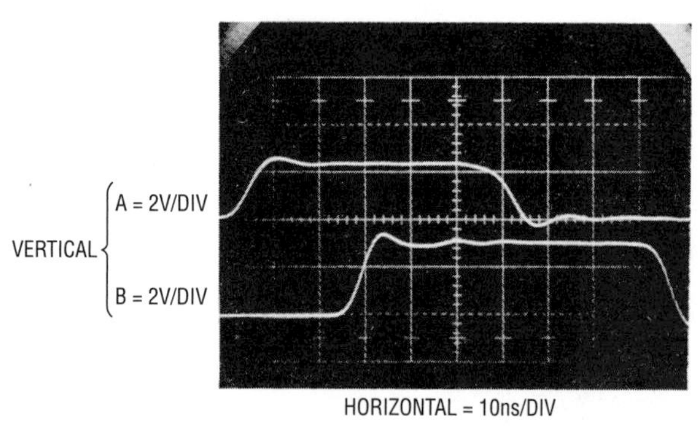

Figure 17.12 • Stray 5pF Capacitance from Input to Ground Causes Delay

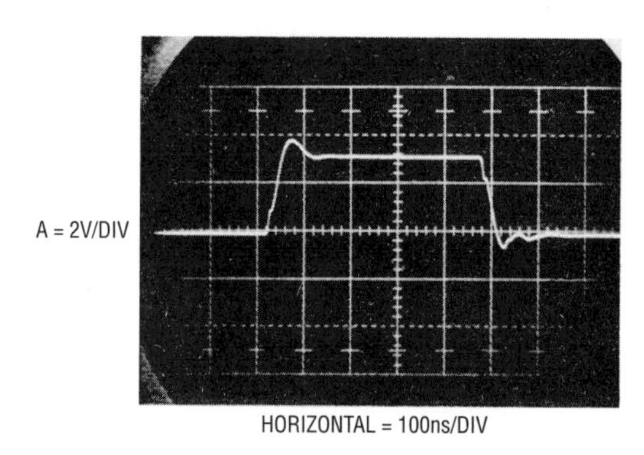

Figure 17.13 • Excessive Load Capacitance Forces Edge Distortion

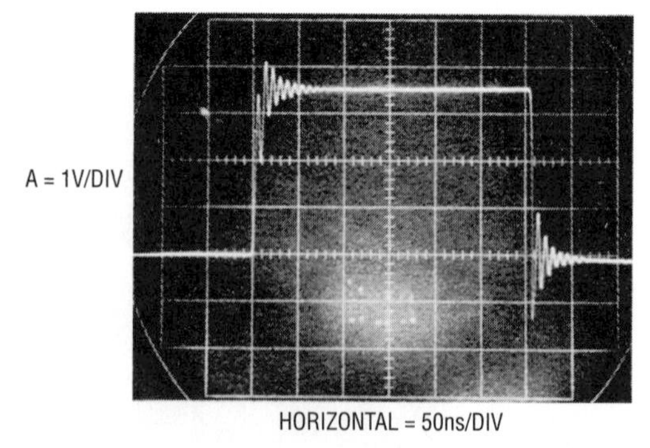

Figure 17.14 • Lengthy, Unterminated Output Lines Ring from Reflections

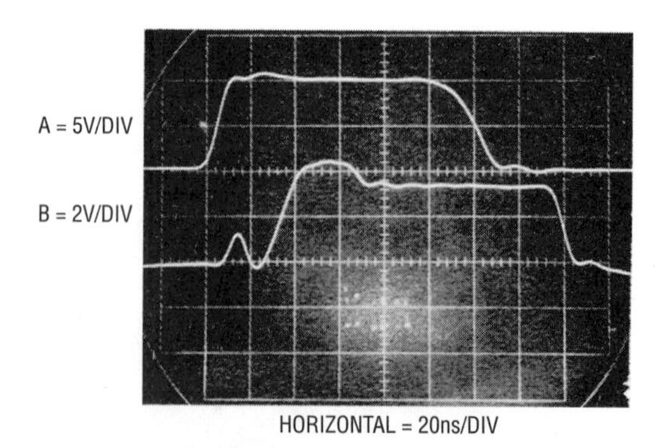

Figure 17.15 • Input Common-Mode Overdrive Generates Odd Outputs

going high again. When it does go high, it slews slowly. Additional odd characteristics include pronounced overshoot and pulse top aberration. The fall time is also slow and well delayed from the input. This is certainly strange behavior from a TTL output. What is going on here? The input pulse is responsible for all these anomalies. Its 10V amplitude is well outside the +5V powered LT1016's common mode input range. Internal input clamps prevent this pulse from damaging the LT1016, but an overdrive of this magnitude results in poor response. *Keep input signals inside the LT1016's common mode range at all times.*

Oscilloscopes

A few of the examples illustrated dealt with probe-caused problems. Although it should be obvious, it is worth mentioning that the choice of oscilloscope employed is crucial. Be certain of the characteristic of the probe-oscilloscope combination you are using. Rise time, bandwidth, resistive and capacitive loading, delay, overdrive recovery and other limitations must be kept in mind. High speed linear circuitry demands a great deal from test equipment and countless hours can be saved if the characteristics of the instruments used are well known (see Appendix C, "Measuring Equipment Response"). In fact, it is possible to use seemingly inadequate equipment to get good results if the equipment's limitations are well known and respected. All of the applications which follow involve rise times and delays well above the 100MHz to 200MHz region, but 90% of the development work was done with a 50MHz oscilloscope. Familiarity with equipment and thoughtful measurement technique permit useful measurements seemingly beyond instrument specifications. A 50MHz oscilloscope cannot track a 5ns rise time pulse, but it can measure a 2ns delay between two such events. Using such techniques, it is often possible to deduce the desired information. There are situations where no amount of cleverness will work and the right equipment, e.g., a faster oscilloscope, must be used.

In general, use equipment you trust and measurement techniques you understand. Keep asking questions and don't be satisfied until everything you see on the oscilloscope is accounted for and makes sense.

The LT1016, combined with the precautionary notes listed above, permits fast linear circuit functions which are difficult or impossible using other approaches. Many of the applications presented represent the state-of-the-art for a particular circuit function. Some show new and improved ways to implement standard functions by utilizing the LT1016's speed. All have been carefully (and painfully) worked out and should serve as good idea sources for potential users of the device.

Applications section

1Hz to 10MHz V→F converter

The LT1016 and the LT1012 low drift amplifier combine to form a high speed V→F converter in Figure 17.16. A variety of circuit techniques is used to achieve a 1Hz to 10MHz output. Overrange to 12MHz (V_{IN} = 12V) is provided. This circuit has a wider dynamic range (140dB, or 7 decades) than any commercially available unit. The 10MHz full-scale frequency is 10 times faster than currently available monolithic V→Fs. The theory of operation is based on the identity Q = CV.

Each time the circuit produces an output pulse, it feeds back a fixed quantity of charge (Q) to a summing node (Σ). The circuit's input furnishes a comparison current at the summing node. The difference signal at the node is integrated in a monitoring amplifier's feedback capacitor.

The amplifier controls the circuit's output pulse generator, completing a feedback loop around the integrating amplifier. To maintain the summing node at zero, the pulse generator runs at a frequency which permits enough charge pumping to offset the input signal. Thus, the output frequency will be linearly related to the input voltage. A1 is the integrating amplifier.

For low bias, high speed operation, a pair of discrete FETs directly drives A1's output stages, replacing A1's monolithic input circuitry. A1's input stage is turned off by connecting the input pins to the negative 15V rail. The FET gates become the "+" and "–" inputs of the amplifier. 0.2μV/°C offset drift performance is obtained by stabilizing the A1-FET combination with A2, a precision op amp. A2 measures the DC value of the negative input, compares it to ground, and forces the positive input to maintain offset balance in the A1-FET combination. Note that A2 is configured as an integrator and cannot see high frequency signals. It functions only at DC and low frequency. The A1-FET combination is arranged as an integrator with a 100pF feedback capacitor. When a positive voltage is applied to the input, A1's output integrates in a negative direction (Trace A, Figure 17.17). During this period, C1's inverting output is low. A very high speed level shifter, Q1-Q2 (see Appendix D, "About Level

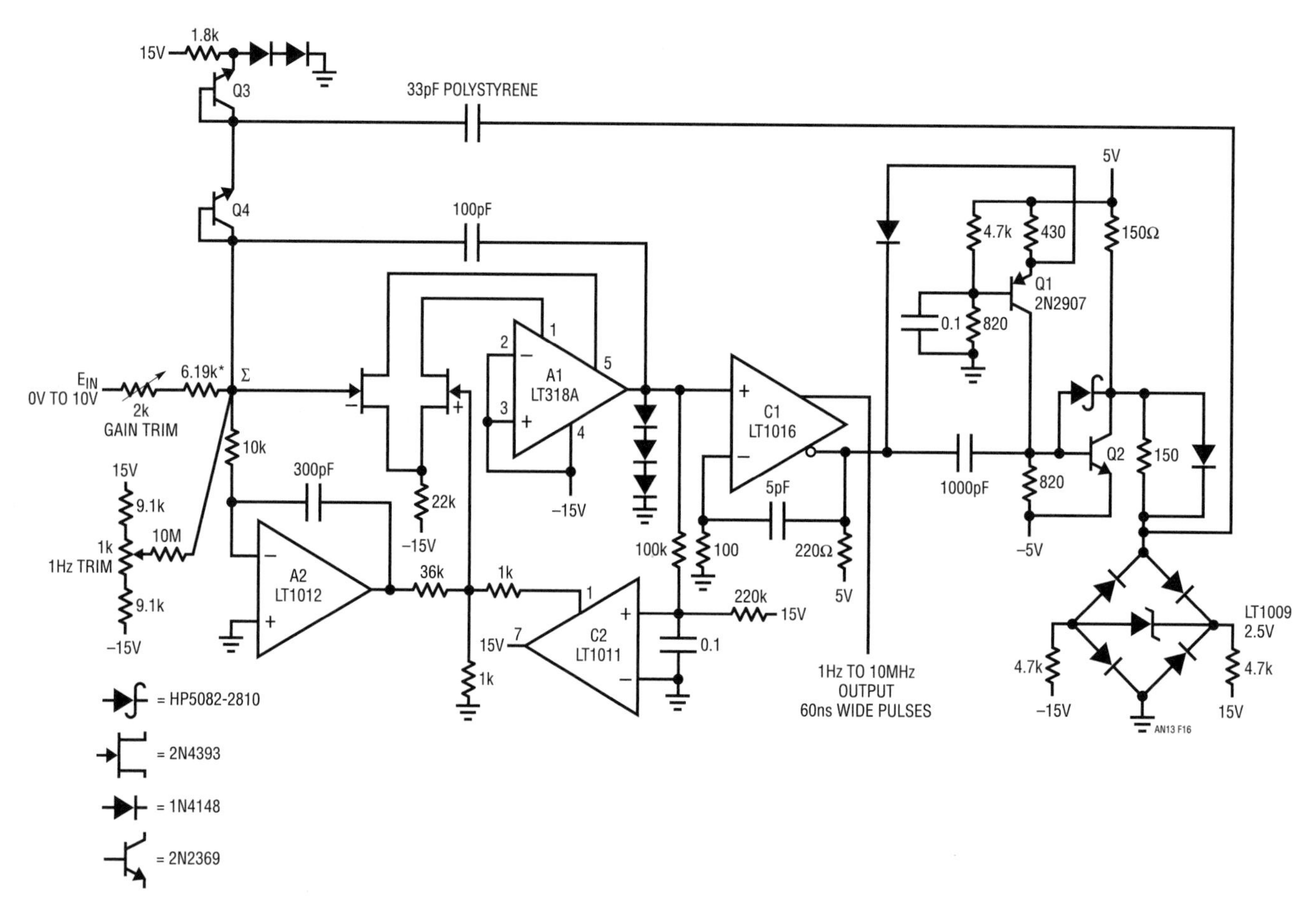

Figure 17.16 • 1Hz to 10MHz V→F Converter

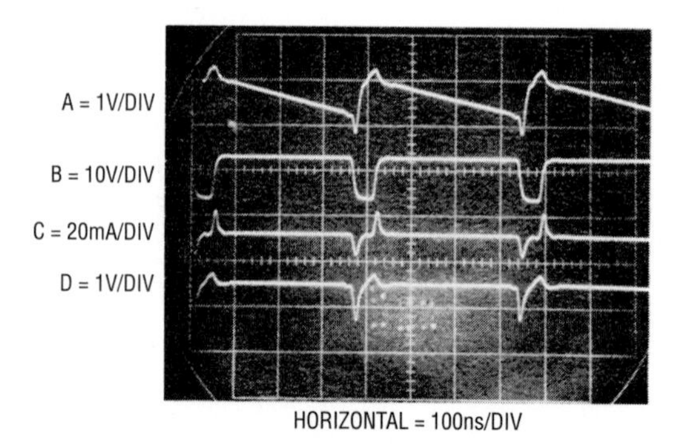

Figure 17.17 • 10MHz V→Fs Operating Waveforms

Shifters"), inverts this output and drives the Zener reference bridge. The bridge's positive output is used to charge the 33pF capacitor. The 1.2V diode string provides cancellation and temperature compensation for the diode drops in the bridge so that the 33pF unit charges to $V_Z + V_{BE}$ Q3.

When A1's output crosses zero, C1's inverting output goes high and Q2's (Trace B) collector goes to −5V. This causes the 33pF unit to dispense charge into the summing node via Q4's V_{BE}. The amount of charge dispensed is a direct function of the voltage that the 33pF unit was charged to ($Q = CV$). Q4's V_{BE} compensates the Q3 V_{BE} term in the capacitor's charge equation. The current, which flows through the 33pF unit (Trace C) reflects this charge pumping action. The removal of current from A1's summing junction (Trace D) causes the junction to be driven very quickly negative. The initial negative-going 20ns transient at A1's output is due to amplifier delay. The input signal feeds directly through the feedback capacitor and appears at the output. When the amplifier finally responds, its output (Trace A) slew limits as it attempts to regain control of the summing node. The amount of time Q2's collector (Trace B) remains at −5V depends on how long it takes A1 to recover and the 5pF-100Ω hysteresis network at C1. This 60ns interval is long enough for the 33pF unit to fully discharge. After this, C1 changes state and Q2's collector swings positive. The capacitor is recharged and the entire cycle repeats. The frequency at which this oscillation occurs is directly related to the voltage-input-derived current into the summing junction. Any input current will require a corresponding oscillation frequency to hold the summing point at an average value of 0V.

Maintaining this relationship at megahertz frequencies places severe restrictions on circuit timing. The key to achieving 10MHz full-scale operating frequency is the ability to transmit information around the loop as quickly as possible. The discharge-reset sequence is particularly critical and is detailed in Figure 17.18. Trace A is the A1 integrator output. Its ramp output crosses 0V at the first left vertical graticule division. A few nanoseconds later, C1's inverting output begins to rise (Trace B), driving the Q1-Q2 level shifter output negative (Trace C). Q2's collector begins to head negative about 12ns after A1's output crosses 0V. 4ns later, the summing point (Trace D) begins to go negative as current is pulled from it through the 33pF capacitor. At 25ns, C1's inverting output is fully up, Q2's collector is at −5V, and the summing point has been pulled to its negative extreme. Now, A1 begins to take control. Its output (Trace A) slews rapidly in the positive direction, restoring the summing point. At 60ns, A1 is in control of the summing node and the integration ramp begins again.

Start-up and overdrive conditions could force A1's output to go to the negative rail and stay there. The AC-coupled nature of the charge dispensing loop can preclude normal operation and the circuit may latch, C2 provides a "watchdog" function for this condition. If A1's output tries to go too far below zero, C2 switches, forcing the "+" input FET gate positive. This causes A1's output to slew positive, initiating normal circuit action. The diode chain at C1's input prevents common mode overdrive at the LT1016. To trim this circuit, ground the input and adjust the 1k pot for 1Hz output. Next, apply 10,000V and set the 2kΩ unit for 10.000MHz output. The transfer linearity of the circuit is 0.06%. Full-scale drift is typically 50ppm/°C and zero point error about 0.2μV/°C (0.2Hz/°C).

Quartz-stabilized 1Hz to 30MHz V→F converter

Figure 17.16's upper limit on operating frequency is imposed by delays in the active elements in the LT1016's feedback path. Higher speed is possible by minimizing these delays. Figure 17.19 shows a way to do this while

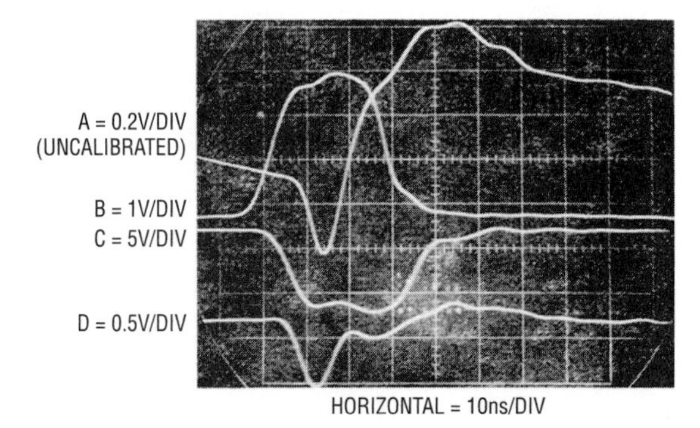

Figure 17.18 • Detail of 60ns Reset Sequence (Whoosh!)

retaining good drift and linearity characteristics. The circuit's untrimmed 150dB dynamic range is 1000 times greater than commercially available V→F converters, whether monolithic, hybrid, or modular.

The technique employed allows the LT1016 to roar along at a 30MHz full-scale output frequency, substantially faster than any commercially available V→F. The actual V→F conversion is performed by the circuit shown inside the dashed lines. This circuit functions similarly to Figure 17.16.

The level shift and Zener bridge are eliminated. Q1 charges the 200pF capacitor, which is unloaded by the Q2-Q3 buffer. When the LT1016's negative input rises above its positive input, its output goes low, pulling charge out of the capacitor via Q4, which serves as a low leakage diode. The 2.7pF capacitor provides positive feedback. If the left end of the 100k input resistor is driven from a voltage source, the LT1016 oscillates over a 1Hz to 30MHz range. Although this simple circuit is fast, its linearity is poor and drift exceeds 5000ppm/°C.

The remaining components in Figure 17.19 form a quartz-locked sampled-data loop to correct these terms without sacrificing speed. The loop works by counting the number of pulses at the LT1016's output during a fixed interval and converting this information to a voltage. The voltage is compared to the circuit's input by an amplifier

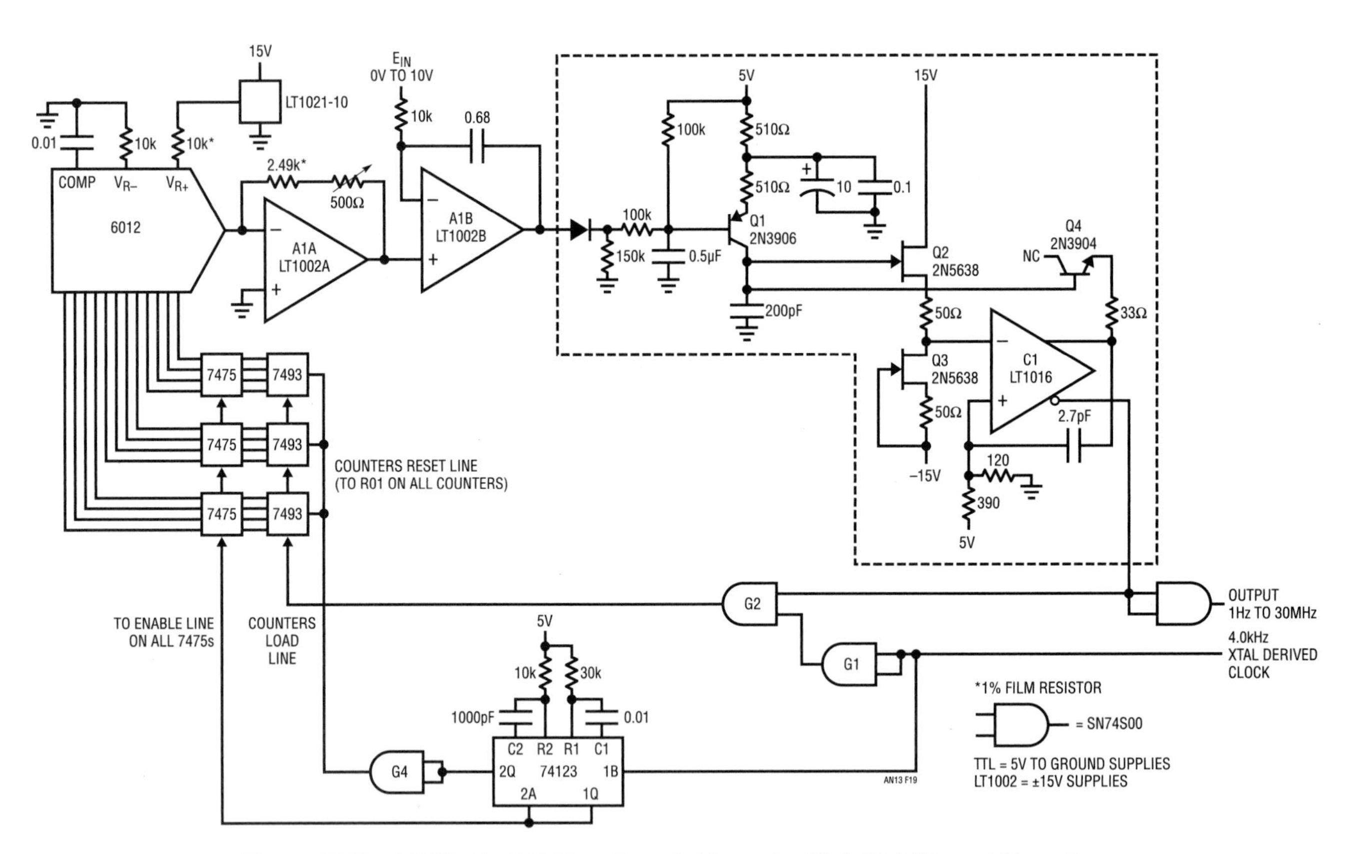

Figure 17.19 • 30MHz V→F Utilizes Sampled Loop for High Stability and Linearity

which drives the LT1016 V→F circuit. This closed loop technique relies on the stability of the time interval and the digital-to-voltage conversion to achieve circuit stability. Frequent updating of the loop ensures long term stability. Figure 17.20 shows how the circuit functions.

Waveforms A, B and C are the LT1016's negative input, output and positive input, respectively. Their similarity to Figure 17.17 (Traces A, B and C) reflects the two circuits' commonality of operation. Trace D shows the quartz-crystal-derived 4kHz clock. During the clock's low portion, the LT1016's gated output appears at G2's output (Trace E). This data is loaded into counters which drive a 12-bit DAC via the 7475 latches. When the clock goes high, one section of the 74123 one-shot generates a pulse (Trace F), allowing the latches to acquire the counter's data. After this pulse goes low, the one-shot's second half pulses (Trace G) the counter's reset line. At the clock's next falling edge the entire cycle repeats. The DAC and its associated output amplifier (A1A) provide a voltage representation of the digital word at the 7475 outputs. This voltage is compared to the circuit's input by A1B, whose output drives the LT1016-based V→F. Any drift or nonlinearity in the V→F will be corrected by the feedback action of this stabilizing loop. The 10k-0.68μF time constant at A1B provides loop compensation.

Although it is not obvious, the frequency setting resolution is much greater than the 12-bit quantization limit of the DAC. This is because the DAC's output dithering around the LSB is integrated to a pure DC level by loop time constants. Once the DAC has settled within an LSB, its output acts like a 4kHz clocked pulse width modulator. The slow loop time constants integrate the width-modulated information to DC, affording smooth, continuous frequency setting capability. The practical limit on resolution is due to the LT1016 oscillator's short term jitter and is about 25ppm of reading.

Although this approach allows higher speeds than Figure 17.16, there are some trade-offs. The loop's sampled nature, combined with its long time constants, limit settling time to about 100ms. Thus, although its output is faster than Figure 17.16's, it cannot track quickly varying inputs. Circuit linearity is DAC limited to 0.025% with full-scale drift of 50ppm/°C. Zero point drift of 1Hz/°C is due to A1B's 0.3μV/°C offset drift.

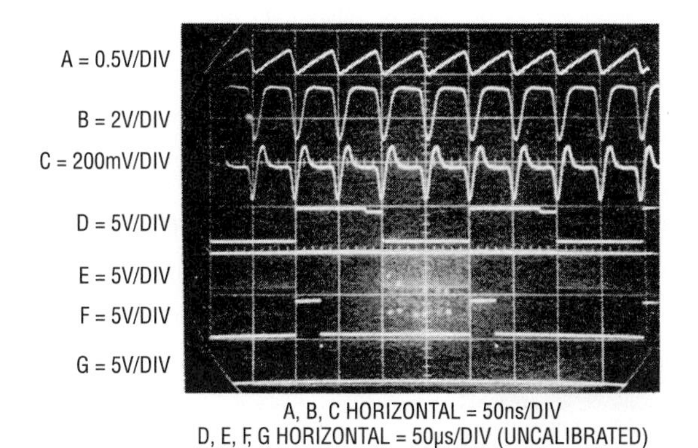

Figure 17.20 • Waveforms for Figure 17.19. Sampled Data Loop (Traces D-G) Stabilizes Basic V→F (Traces A-C)

1Hz to 1MHz voltage-controlled sine wave oscillator

Both V→F converters described have pulse outputs. Many applications such as audio, shaker table driving, and automatic test equipment require voltage-controlled oscillators (VCO) with a sine wave output. The circuit of Figure 17.21 meets this need, spanning a 1Hz to 1MHz range (120dB, or 6 decades) for a 0V to 10V input. It is over 10 times faster than previously published circuits, while maintaining 0.25% frequency linearity and 0.40% distortion specifications.

To understand the circuit, assume Q5 is on and its collector (Trace A, Figure 17.22) is at −15V, cutting off Q1. The positive input voltage is inverted by A3, which biases the summing node of integrator A1 through the 3.6k resistor and the self-biased FETs. A current, −I, is pulled from the summing point. A2, a precision op amp, DC stabilizes A1. A1's output (Trace B, Figure 17.22) integrates positive until C1's input (Trace C) crosses 0V. When this happens, C1's inverting output goes negative, the Q4-Q5 level shifter turns off, and Q5's collector goes to +15V. This allows Q1 to come on. The resistors in Q1's path are scaled to produce a current, +2I, exactly twice the absolute magnitude of the current, −I, being removed from the summing node. As a result, the net current into the junction becomes +I and A1 integrates negatively at the same rate its positive excursion took.

When A1 integrates far enough in the negative direction, C1's "+" input crosses zero and its outputs reverse. This switches the Q4-Q5 level shifter's state, Q1 goes off

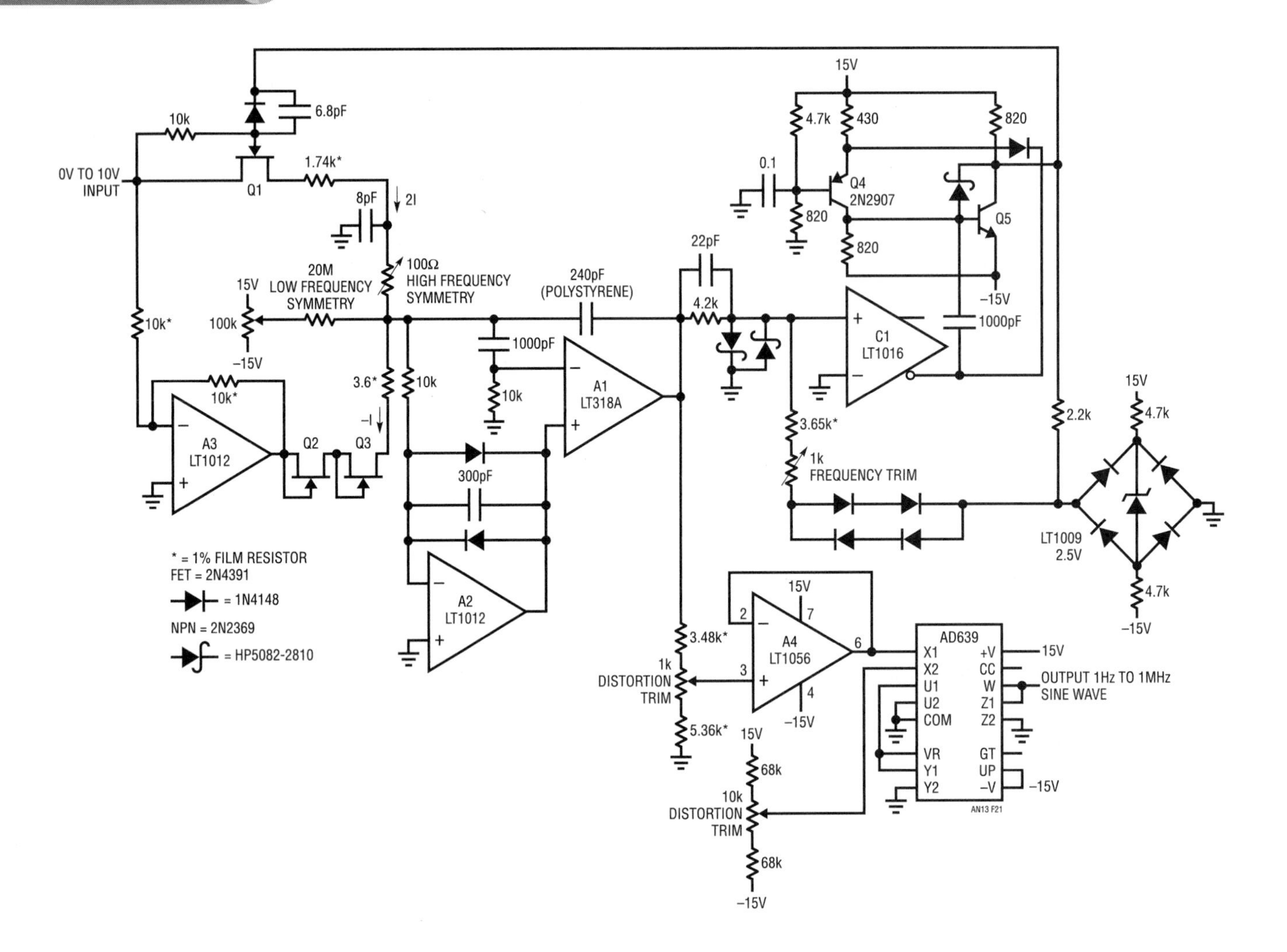

Figure 17.21 • 1Hz to 1MHz Sine Wave Output VCO

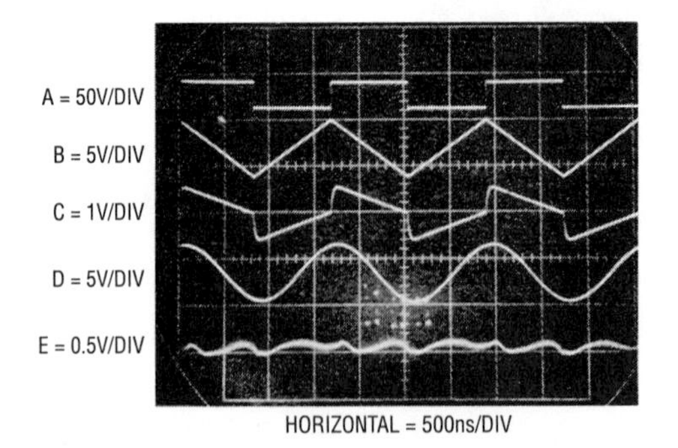

Figure 17.22 • Sine Wave VCO Waveforms

and the entire cycle repeats. The result is a triangle waveform at A1's output. The frequency of this triangle is dependent on the circuit's input voltage and varies from 1Hz to 1MHz with a 0V to 10V input.

The LT1009 diode bridge and the series-parallel diodes provide a stable bipolar reference which always opposes the sign of A1's output ramp. The Schottky diodes bound C1's "+" input, assuring it clean recovery from overdrive.

The AD639 trigonometric function generator, biased via A4, converts A1's triangle output into a sine wave (Trace D).

The AD639 must be supplied with a triangle wave which does not vary in amplitude or output distortion will result. At higher frequencies, delays in the A1 integrator switching loop result in late turn on and turn off of Q1. If these delays are not minimized, triangle amplitude will increase with frequency, causing distortion level to also increase with frequency. The total delay generated by the LT1016, the Q4-Q5 level shifter, and Q1 is 14ns. This small delay, combined with the 22pF feedforward network at the LT1016's input, keeps distortion to just 0.40% over the entire 1MHz range. At 100kHz, distortion is typically inside 0.2%. The effects of gate-source charge transfer, which happens whenever Q1 switches, are minimized by the 8pF unit in Q1's source line. Without this capacitor, a sharp spike would occur at the triangle peaks, increasing distortion. The Q2-Q3 FETs compensate the temperature-dependent on-resistance of Q1, keeping the +2I/−I relationship constant with temperature.

To adjust this circuit, put in 10.00V and trim the 100Ω pot for a symmetrical triangle output at A1. Next, put in 100μV and trim the 100k pot for triangle symmetry. Then, put in 10.00V again and trim the 1k "frequency trim" adjustment for a 1MHz output frequency. Finally, adjust the "distortion trim" potentiometers for minimum distortion as measured on a distortion analyzer (Trace E). Slight readjustment of the other potentiometers may be required to get lowest possible distortion.

200ns-0.01% sample-and-hold circuit

Figure 17.23's circuit uses the LT1016's high speed to improve upon a standard circuit function. The 200ns acquisition time is well beyond monolithic sample-and-hold capabilities and is matched only by hybrid and modular units selling in the $200 range. Other specifications exceed the best commercial unit's performance. This circuit also gets around many of the problems associated with standard sample-and-hold approaches, including FET switch errors and amplifier settling time. To achieve this, the LT1016's high speed is used in a circuit which completely abandons traditional sample-and-hold methods. When the sample-and-hold command line (Trace A, Figure 17.24) is high, Q2 conducts current, biasing Q3 on and forcing the 1000pF capacitor (Trace B) to discharge toward Q4's emitter potential. Q4's emitter, in turn, sits at a potential slightly below Q3's collector voltage. Q5 and the LT1009, biased from the input voltage, drive Q4. Concurrently, the TTL gate at the LT1016 grounds the latch pin (enabling the comparator) and the comparator's inverting output (Trace C) goes high. When the sample-and-hold command line (Trace A) falls, Q2 and Q3 go off and the Q1 current source charges the 1000pF unit (Trace B) with a fast linear ramp. This capacitor is buffered by Q7, a current sink loaded source follower. When Q7's output reaches the circuit's input value, the LT1016's inverting output (Trace C) switches low. The Q1 current source cuts off in about 2ns and capacitor charging ceases. The LT1016's low state also means the NOR gate's output is high, latching the comparator's output. This prevents input line noise or a change in signal from affecting the stored value in the 1000pF hold capacitor.

Ideally, Q7's output now sits at exactly the sampled value of the input voltage. In practice, a slight error exists because of the LT1016's delay and the turn-off time of Q1 (total 12ns). Because of these delays, the capacitor is able to charge to a higher voltage than the input before current stops. This error term is compensated by removing a small quantity of charge from the 1000pF capacitor when the LT1016's inverting output goes low. The charge is removed through the 8pF-1kΩ potentiometer network. Because the charging ramp's slope is fixed, the error term is constant and the compensation works over the circuit's ±3V input common mode range. The lower four traces are expanded to show detail of the compensation and the circuit's critical ramp turn-off sequence. When the LT1016 goes off (Trace D), the ramp is seen to slightly overshoot its final value (Trace E)

The 1kΩ-8pF combination pulls enough charge (Trace F) out of the 1000pF hold capacitor to bring it back to the correct value. Trace G is the $\overline{\text{NOW}}$ line. It falls low two gate delays after the LT1016 inverting output goes low. When this line goes low, the circuit's sampled output has settled from the correction transient and is valid data. The total time from the falling of the sample-and-hold line to the $\overline{\text{NOW}}$ output going low will always be inside 200ns.

The circuit's 200ns acquisition time is due to the high slew rate of the charging ramp and the action of Q4, Q5 and the LT1009. These components form a wideband tracking amplifier whose output is always a fixed amount below the input. Q7's current source load (Q6) ensures that its V_{GS} does not change. Thus, Q3 will always reset the capacitor a small, relatively constant amount below any circuit input. In this way, the ramp does not have to run very long before it crosses the input value, and acquisition time versus input voltage is constant. In Figure 17.25 the circuit is shown sampling a bipolar triangle wave. Trace A is the input and Trace B is the circuit output. Trace C is

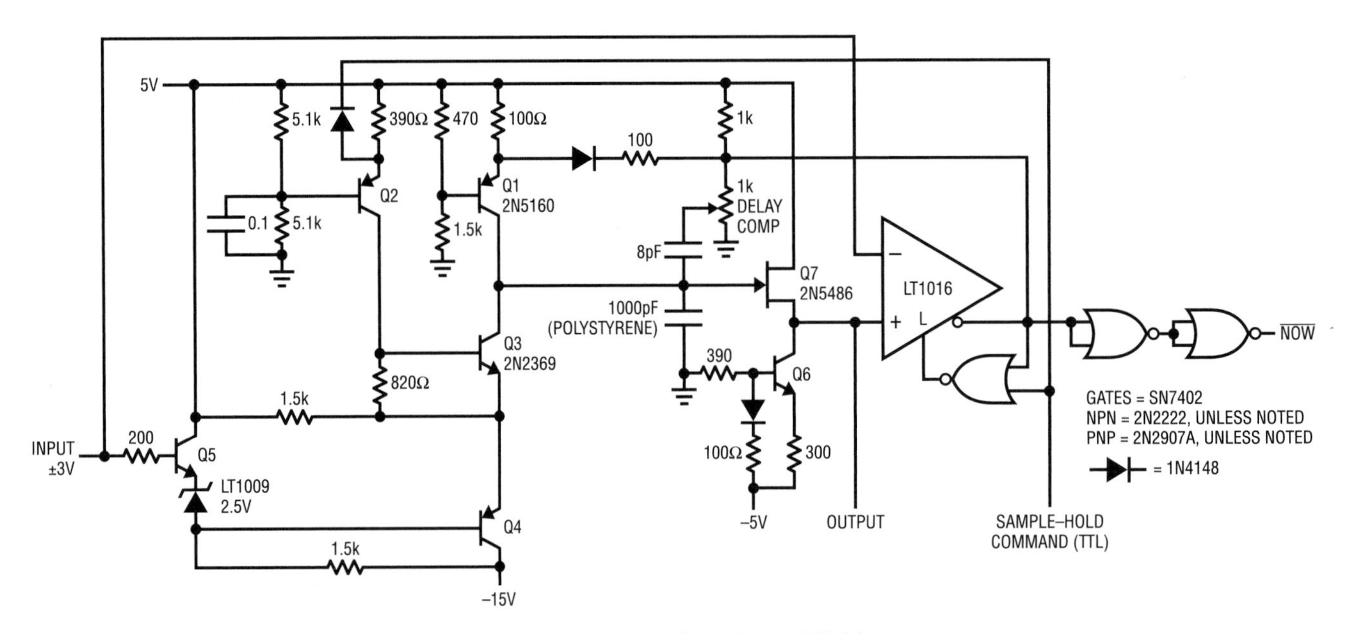

Figure 17.23 • 200ns Sample-and-Hold

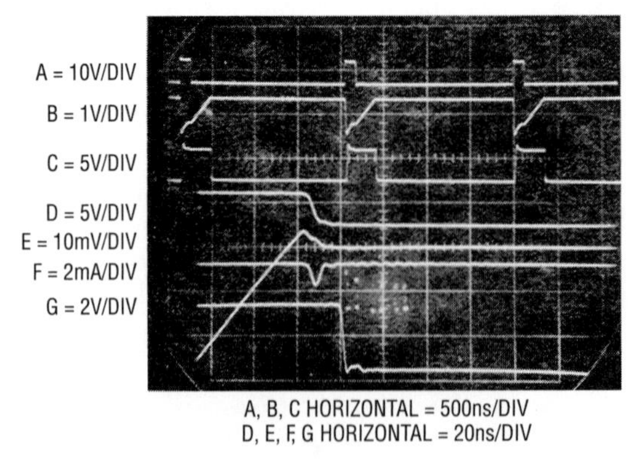

Figure 17.24 • Fast Sample-and-Hold Waveforms. Traces A-C Show Ramp-Compare Action. Traces D-G Detail Delay Compensation

an expansion of Trace B (the "smearing" of the sampled pedestals in Trace C is due to the repetitive asynchronous sampling of the triangle). The action of the tracking amplifier is readily apparent. It always resets the ramp to the same point below the input voltage, regardless of the common mode level. To calibrate the circuit, ground the input, repetitively pulse the sample-and-hold command line, and adjust the 1kΩ pot for 0V output.

Important specifications for this circuit include:

Acquisition Time	<200ns
Common Mode Input Range	±3V
Droop	1μV/μs
Hold Step	2mV
Hold Settling Time	15ns
Feedthrough Rejection	>>100dB

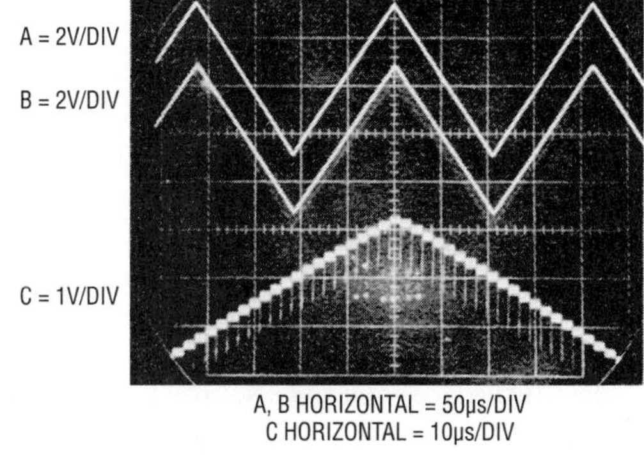

Figure 17.25 • Fast Sample-and-Hold Tracking a Triangle Wave. Trace C is Expanded to Show the Ramping of the Circuit's Output

Fast track-and-hold circuit

The track-and-hold circuit shown in Figure 17.26 is generically related to the sample-and-hold circuit of Figure 17.23. It also forsakes standard techniques in favor of an approach based on the LT1016's speed. This circuit's main

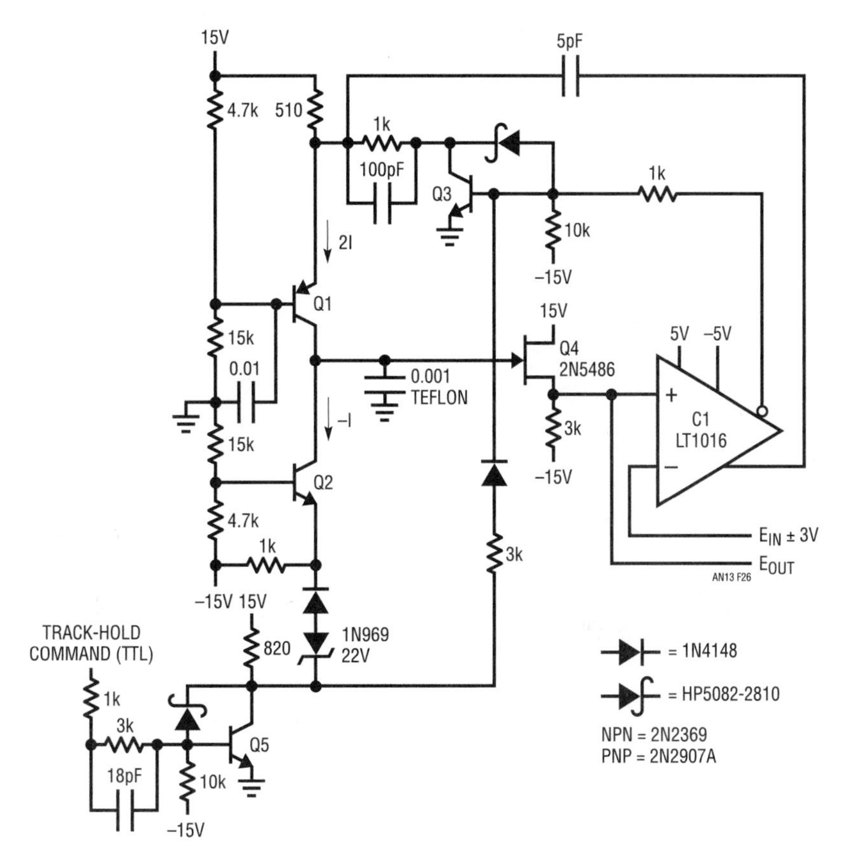

Figure 17.26 • Comparator-Based Track-and-Hold

blocks are a switched current source (Q1-Q3), a current sink (Q2), a FET follower (Q4), and the LT1016. To understand the circuit, assume the voltage stored in the 0.001μF hold capacitor is below the input potential and the track-and-hold command line (Trace A, Figure 17.27) is at a TTL "1" (track mode). Under these conditions Q5 is on and C1's output is positive. C1's inverting output is low and Q3 is off, allowing the Q1 current source to charge the hold capacitor. The Q2 current sink is also operating, but at one-half the current density of Q1. The hold capacitor charges positively. When Q4's source (Trace B, Figure 17.27) ramps to the input voltage's value, C1's outputs reverse state. Q3 comes on, quickly turning off the Q1 current source. The 5pF feedforward capacitor speeds up Q1's turnoff by bypassing Q3. With Q1 off, Q2's sink current discharges the hold capacitor. This causes C1's output to change state and oscillation commences (Trace B, Figure 17.27). This controlled, 10mV-25MHz oscillation centers itself around the input voltage's value. When the track-and-hold line (Trace A) goes low, Q5 ceases conducting, Q1 and Q2 immediately go off, oscillations cease and the circuit's output sits within ±5mV of the input value at the time of turn-off. This 5mV uncertainty, caused by the nature of the circuit's operation, limits accuracy to 8 bits.

Figure 17.28 shows what happens when a square wave is fed into the circuit. Trace A is the input. Trace B is the output. Trace C is the track-and-hold command line and Trace D is the LT1016's output. Note that the controlled oscillation stops cleanly when the track-and-hold line goes low. If the source-sink transistors were run at higher currents, the circuit's output would slew much faster to keep up with the input's transitions. The oscillation's error band would also proportionately enlarge. The 25MHz update rate allows this circuit to track a relatively slow signal very closely with settling time under 10ns when switched into hold.

10ns sample-and-hold

Figure 17.29 shows a 10ns acquisition time sample-and-hold which can be used with repetitive signals only. Here, the LT1016 (C1) drives a differential integrator's (A1) input. Feedback from the integrator back to the LT1016 closes a loop around the circuit. Figure 17.30 shows what happens when a 1MHz sine wave (Trace A, Figure 17.30) is applied to the input. C2 generates a zero crossing signal (Trace B) and one-shot "A" (Trace C) provides an adjustable width. One-shot B's Q output produces a 30ns pulse (Trace D) which is fed into a logic network with the $\overline{Q}$ signal. The two inverter delays in Q's path give its associated gate a shorter duration output (Trace F) than $\overline{Q}$'s gate (Trace E). The last gate subtracts these two signals and generates a 10ns spike. This is inverted (Trace G) and fed to C1's latch pin. Each time the latch is enabled the comparator responds to the condition of the summing junction at its "+" input. If summing error is positive, A1 pulls current. If the error is negative, A1 sources current to the junction. After a number of input cycles, A1's output settles at a DC value which is the same as the level sampled during the time the latch is enabled. The "delay adjust" allows the 10ns sampling "window" to be positioned anywhere on the input sine wave.

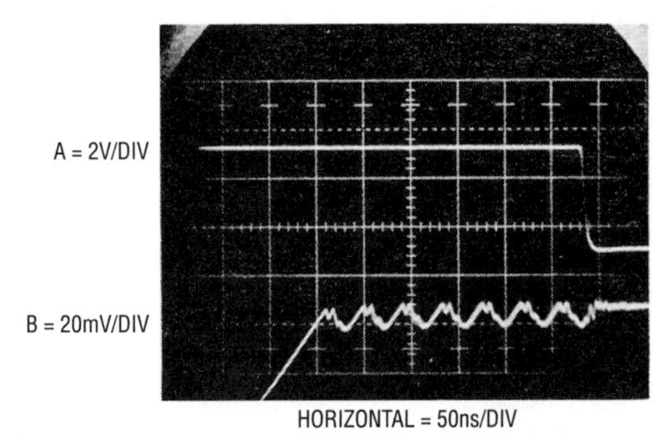

Figure 17.27 • Track-and-Hold Circuit Acquiring an Input

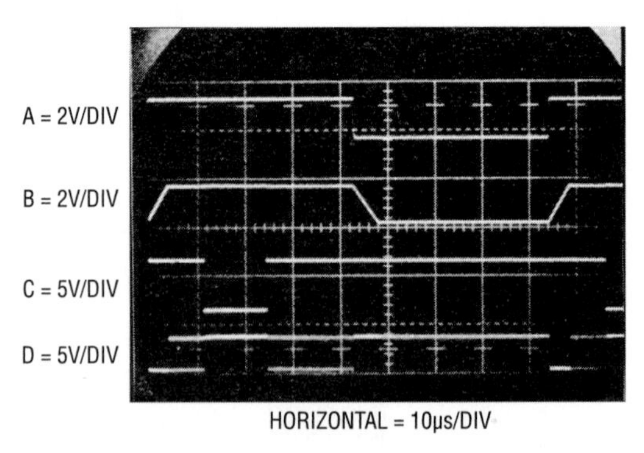

Figure 17.28 • Track-and-Hold Responding to a Square Wave Input

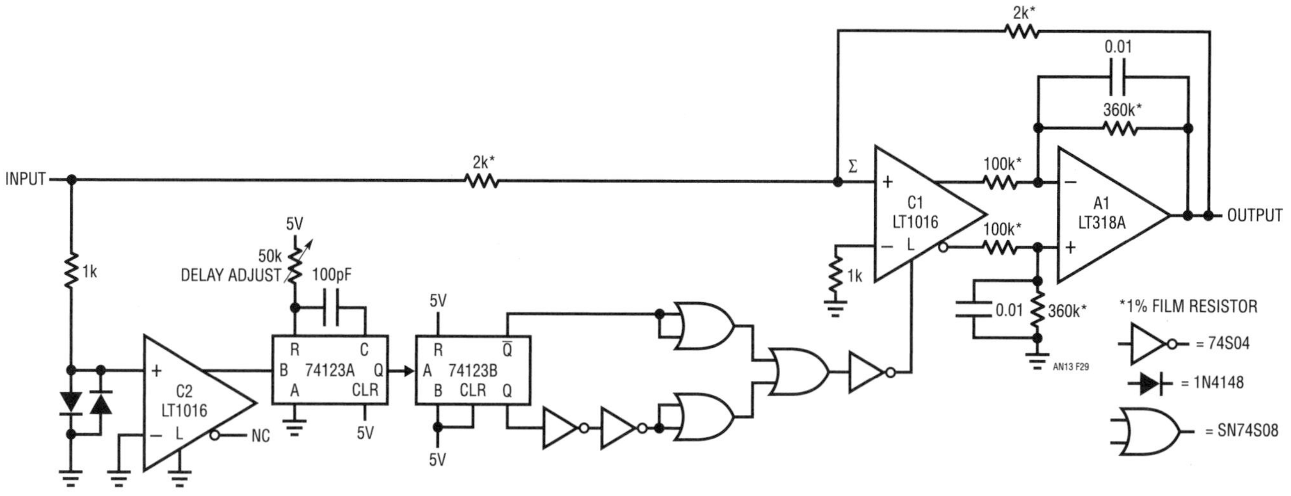

Figure 17.29 • 10ns Sample-and-Hold for Repetitive Signals

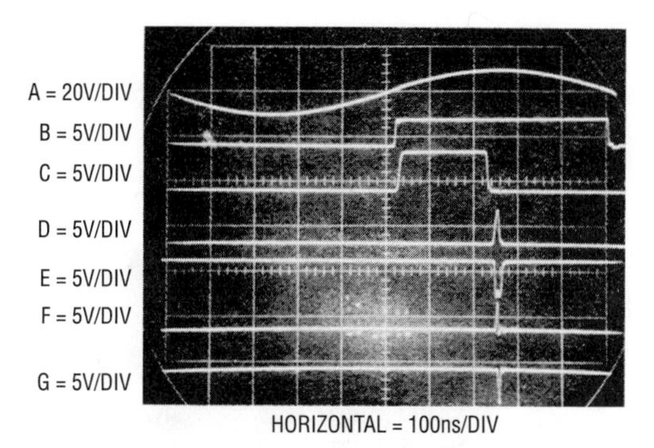

Figure 17.30 • Waveforms for 10ns Sample-and-Hold. 10ns Sampling Window (Trace G) May be Positioned Anywhere on Input (Trace A)

5μs, 12-Bit A/D converter

The LT1016's high speed is used to implement a fast 12-bit A/D converter in Figure 17.31. The circuit is a modified form of the standard successive approximation approach and is faster than most commercial SAR 12-bit units. In this arrangement the 2504 successive approximation register (SAR), A1 and C1 test each bit, beginning with the MSB, and produce a digital word representing V_{IN}'s value. To get faster conversion time, the clock (C2) is sped up after the third MSB is converted. This takes advantage of the segmented DAC used, which has significantly faster settling time for the lower 9 bits.

A1 provides preamplification for C1 while adding only 7ns delay. The preamplification allows clean response to one-half LSB (1.22mV) overdrives at A1's input. Figure 17.32 shows the converter at work. To aid in observing operation, A1 has been eliminated and the DAC-input node drives the LT1016 "+" input directly. A1 should be employed in normal use.

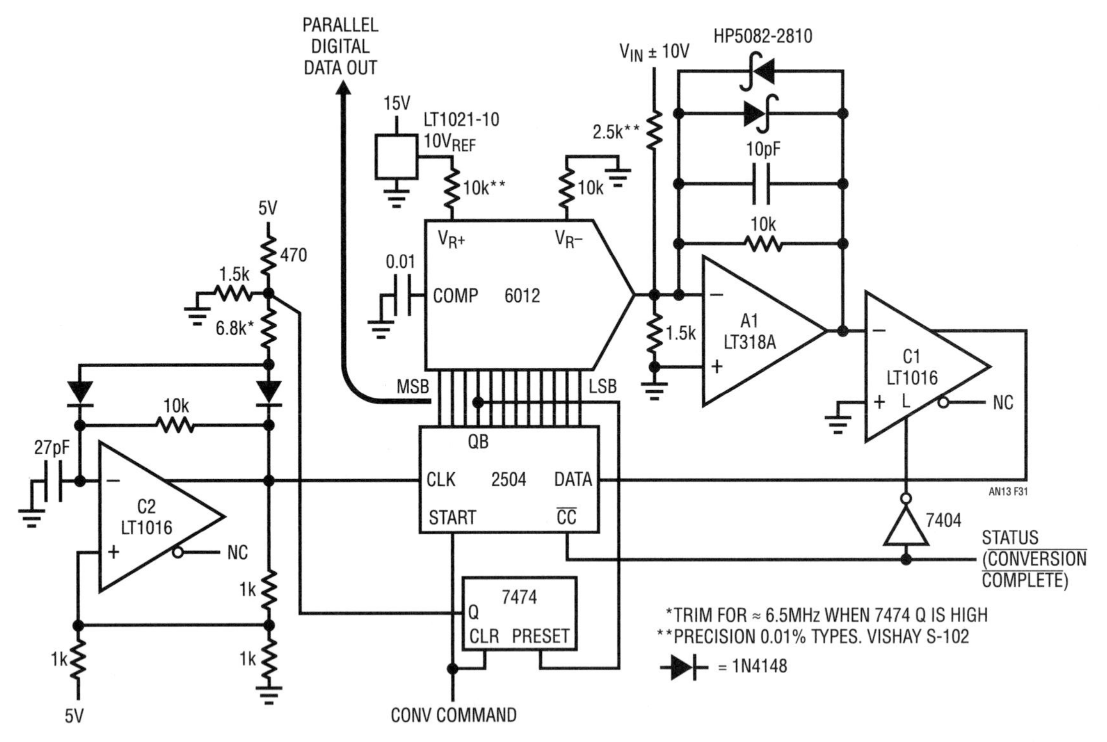

Figure 17.31 • 5µs, 12-Bit SAR Converter. Clock is Sped Up After the Third Bit, Shortening Overall Conversion Time

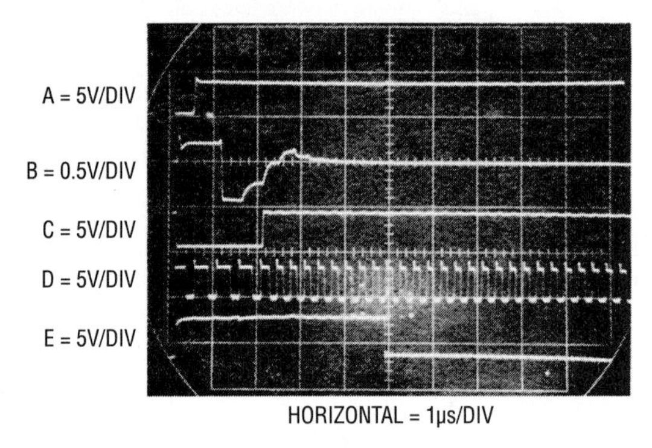

Figure 17.32 • Fast SAR Converter Waveforms. Note Clock (Trace D) Speed-Up After 3rd Bit Conversion

The conversion begins when the "convert command" line (Trace A, Figure 17.32) drops low. When this happens the SAR begins to test each bit. The DAC output (Trace B), fed to the Schottky-clamped C1 input, sequentially converges toward final value. After the third MSB has been established, the 7474 Q line goes high (Trace C), forcing the 2.1MHz clock to shift to 3.2MHz (Trace D). This speeds up conversion of the remaining 9 bits, minimizing overall A/D time. When conversion is complete, the status line (Trace E) drops low and C1's latch is set by the TTL inverter, preventing the comparator from responding to input noise or shifts.

The next convert command reinitiates the entire cycle. Note that on the lowest order bits C1 must accurately respond to small signals without sacrificing speed. The high gain-bandwidth required makes this application one of the most difficult for a comparator. This circuit's 5µs conversion time is fast for a 12-bit A/D. Faster conversion time is possible, although the design becomes more complex. A "stretched" version of this circuit, with 1.8µs conversion time, appears in AN17, "Considerations for Successive Approximation A/D Converters".

Inexpensive, fast 10-bit serial output A/D

Figure 17.33 shows a simple way to build a fast, inexpensive 10-bit A/D converter. This circuit is especially useful where a large number of converters is required and all of them can be serviced by one clock. The design consists of a current source, an integrating capacitor, a comparator and some gates.

Every time a pulse is applied to the convert command input (Trace A, Figure 17.34), Q1 resets the 1000pF capacitor to 0V (Trace B). This resetting action takes 200ns—the minimum acceptable convert command pulse width. On the falling edge of the convert command pulse, the capacitor begins to charge linearly. In precisely 10μs, it charges to 2.5V (over range to 3.0V is provided). Normally, Q1 would not be able to reset the capacitor to zero due to its V_{CE} saturation voltage. This effect is compensated by Q4. This device switches in inverting mode, resulting in a reset within 1mV of ground. Q1 absorbs most of the capacitor's charge and Q4 completes the discharge.

The 10μs ramp is applied to the LT1016's positive input. The LT1016 compares the ramp to Ex, the unknown, at its negative input. For a 0V to 2.5V range, Ex is applied to the 2.5kΩ resistor. For a 0V to 10V range, the 2.5kΩ resistor is grounded and Ex is applied to the 7.5kΩ resistor. The 2.0k resistor at the positive input provides balanced source impedance for C1. The output of the LT1016 is a pulse (Trace C) whose width is directly dependent on the value of Ex. This pulse width is used to gate a 100MHz clock. The 74AS00 gate achieves this function and also gates out the portion of the LT1016 output pulse due to the convert command pulse. Thus, the 100MHz clock pulse bursts that appear at the output

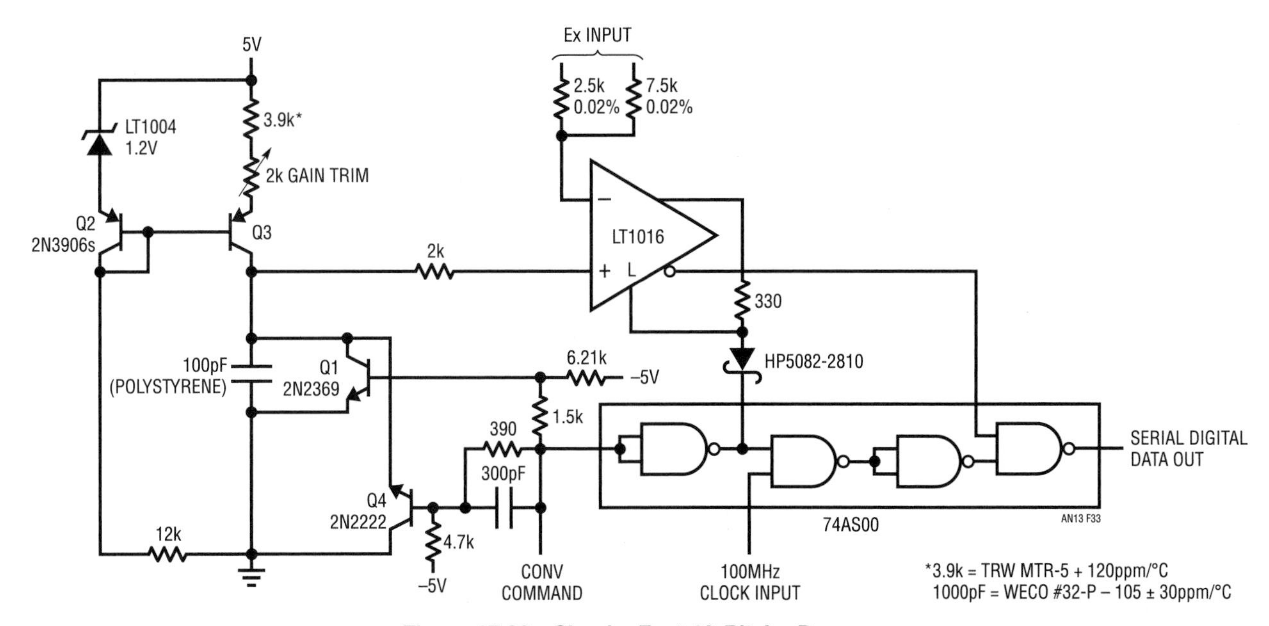

Figure 17.33 • Simple, Fast 10-Bit A→D

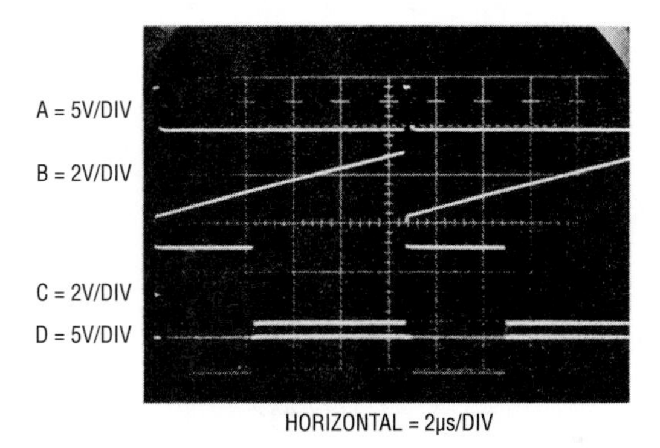

Figure 17.34 • Waveforms for 10-Bit A→D

(Trace D) are proportional to Ex. For a 0V to 10V input, 1024 pulses appear at full-scale, 512 at 5.00V, etc. The resistor-diode network at the LT1016's latch pin ensures clean comparator transitions by locking the LT1016 outputs after the conversion is completed. This latch is broken by the next convert command pulse.

The current source scaling resistor and ramp capacitor specified provide good temperature compensation because of their opposing thermal coefficients. The circuit will typically hold ±1 LSB accuracy over 0°C to 70°C with an additional ±1 LSB uncertainty due to the asynchronous relationship between the clock and the conversion sequence.

Figure 17.35 details the most critical part of the converter's operation, the reset phase. Trace A is the convert command. Trace B is the capacitor (greatly magnified) resetting to zero. The comparator output appears in Trace C and Trace D is the gated serial output. Observe that the output pulses do not appear until the capacitor has started to ramp (just past mid-screen), even though the comparator is high.

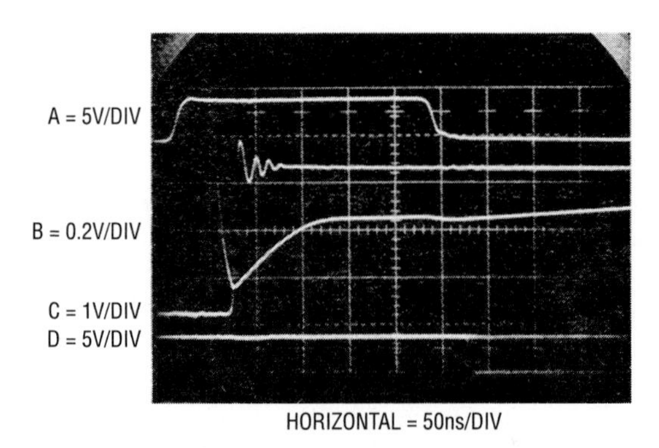

Figure 17.35 • Figure 17.33's Reset Sequence. Q1-Q4 Combination Gives Quick, Low Offset Zero Reset

2.5MHz precision rectifier/AC voltmeter

Most precision rectifier circuits rely on operational amplifiers to correct for diode drops. Although this scheme works well, bandwidth limitations usually restrict these circuits to operation below 100kHz. Figure 17.36 shows the LT1016 in an open-loop, synchronous rectifier configuration which has high accuracy out to 2.5MHz. An input 1MHz sine wave (Trace A, Figure 17.37) is zero cross detected by C1. Both of C1's outputs drive identical level shifters with fast (delay = 2ns to 3ns), ±5V outputs. These outputs bias a Schottky switching bridge (Traces B and C are the switched corners of the bridge). The input signal is fed to the

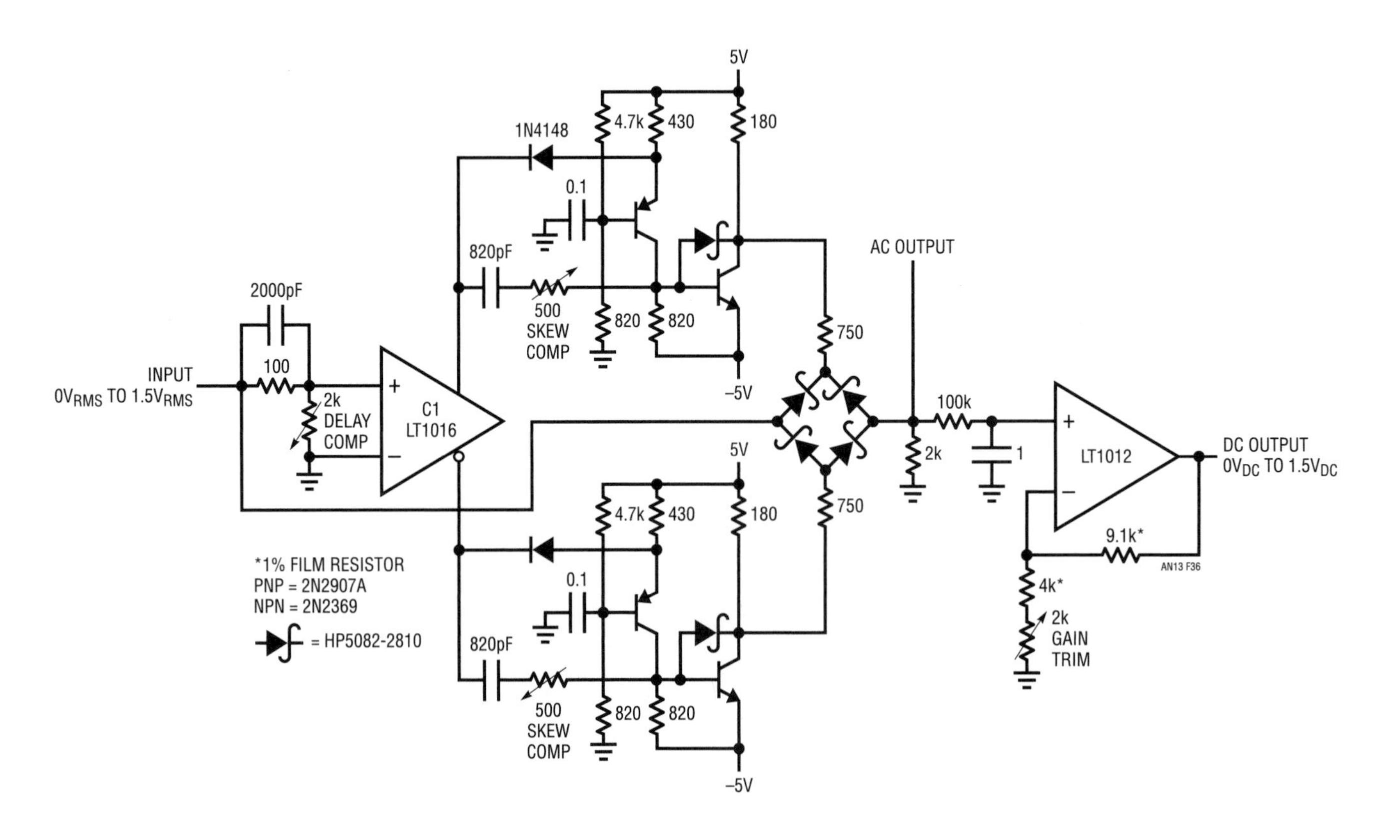

Figure 17.36 • Fast, Synchronous Rectifier-Based AC/DC Converter

left-midsection of the bridge. Because C1 drives the bridge synchronously with the input signal, a half-wave rectified sine appears at the AC output (Trace D). The DC RMS value appears at the DC output. The Schottky bridge gives fast switching and eliminates the charge pump-through that a FET switch would contribute. This is evident in Trace E, which is an expanded version of Trace D. The waveform is clean with the exception of very small disturbances where bridge switching occurs. To calibrate this circuit, apply a 1MHz to 2MHz $1V_{p\text{-}p}$. Sine wave and adjust the delay compensation so bridge switching occurs when the sine crosses zero. This adjustment corrects for the small delays through the LT1016 and the level shifters. Next, adjust the skew compensation potentiometers for minimum aberrations in the AC output signal. These trims slightly shift the phase of the rising output edge of their respective level shifter. This allows skew in the complementary bridge drive signals to be kept within 1ns to 2ns, minimizing output disturbances when switching occurs. A 100mV sine input will produce a clean output with a DC output accuracy of better than 0.25%.

10MHz fiber optic receiver

Reception of high data rate fiber optic data is not easy. The high speed data and uncertain intensity of the light level can cause erroneous results unless the receiver is carefully designed. The fiber optic receiver shown in Figure 17.38 will accurately condition a wide range of light inputs at up to 10MHz data rates. Its digital output features an adaptive threshold trigger which accommodates varying signal intensities due to component aging and other causes. An analog output is also available to monitor the detector output. The optical signal is detected by the PIN photodiode and amplified by a broadband fed-back stage, Q1-Q3. A second, similar, stage gives further amplification. The output of this stage (Q5's collector) biases a 2-way peak detector (Q6-Q7). The maximum peak is stored in Q6's emitter capacitor, while the minimum excursion is retained in Q7's emitter capacitor. The DC value of Q5's output signal's mid-point appears at the junction of the 500pF capacitor and the 22MΩ unit. This point will always sit midway between the signal's excursions, regardless of absolute amplitude. This signal-adaptive voltage is buffered by the low bias LT1012 to set the trigger voltage at the LT1016's positive input. The LT1016's negative input is biased directly from Q5's collector. Figure 17.39 shows the results using the test circuit indicated in Figure 17.38. The pulse generator's output is Trace A, while Q5's collector (analog output monitor) appears in Trace B. The LT1016 output is Trace C. The wideband amplifier responds within 5ns and rises in 25ns. Note that the LT1016's output transitions line up with the mid-point of Trace B, in accordance with the adaptive trigger's operation.

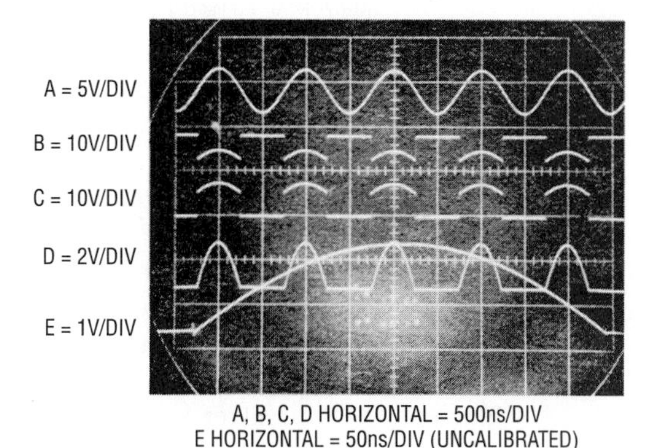

Figure 17.37 • Fast AC/DC Converter Operating at 1MHz. Clean Switching is Due to LT1016's Speed and Compensations for Delay and Switching Skew

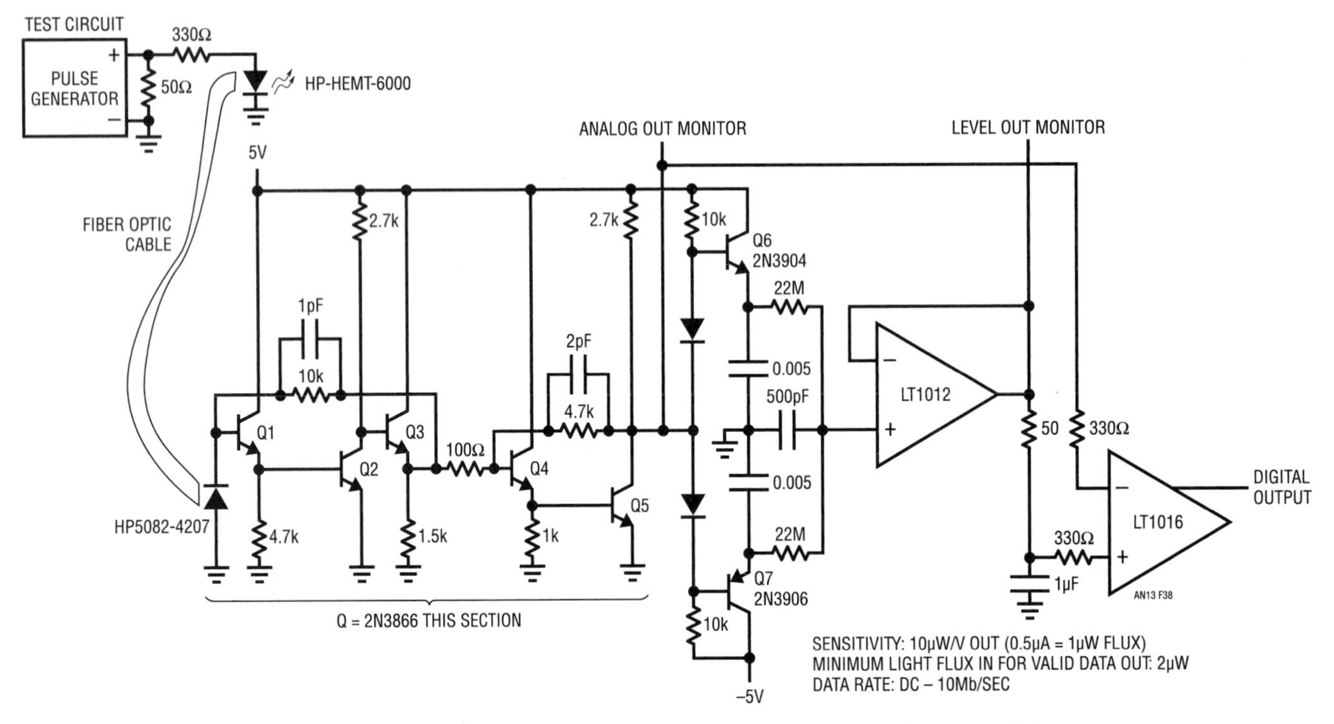

Figure 17.38 • Fast Fiber Optic Receiver is Immune to Shifts in Operating Point

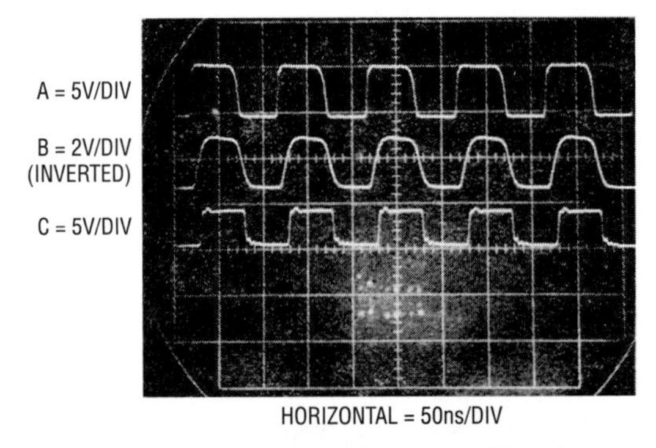

Figure 17.39 • Fiber Optic Receiver Waveforms

12NS circuit breaker

Figure 17.40 shows a simple circuit which will turn off current in a load 12ns after it exceeds a preset value. This circuit has been used to protect integrated circuits during developmental probing and is also useful for protecting expensive loads during trimming and calibration. It is three times faster and less complex than previously published circuits. Under normal conditions the voltage across the 10Ω shunt is smaller than the potential at the LT1016's negative input. This keeps Q1 off and Q2 receives bias, driving the load. When an overload occurs (in this case via a test circuit, chose output is Trace A, Figure 17.41), the current through the 10Ω sense resistor begins to increase (Trace B, Figure 17.41). When this current exceeds the preset value, the LT1016's outputs (non-inverting output shown in Trace C) reverse. This provides ideal turn-on drive for Q1 and it cuts off Q2 (Q2 emitter is Trace D) in 5ns. The delay from the onset of excessive load current to complete shutdown is just 13ns. Once the circuit has triggered, the LT1016 is held in its latched state by feedback from the non-inverting output. When the load fault has been cleared the pushbutton can be used to reset the circuit.

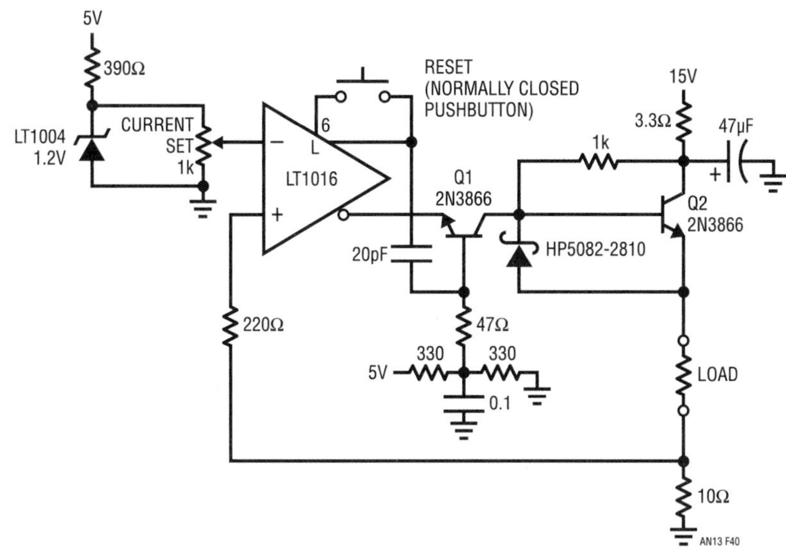

Figure 17.40 • 12ns Circuit Breaker

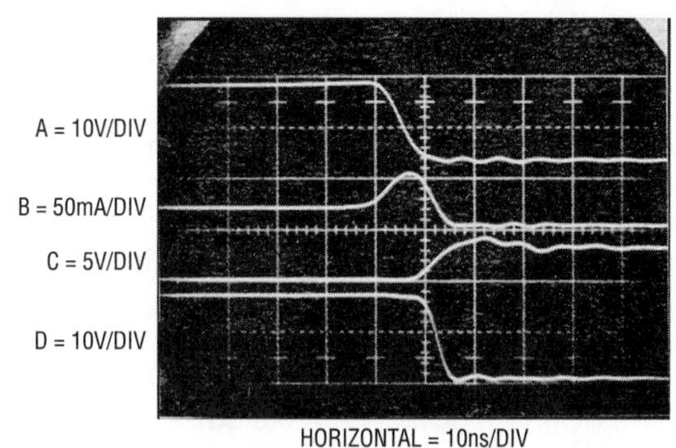

Figure 17.41 • Operating Waveforms for the 12ns Circuit Breaker. Circuit Output (Trace D) Starts to Shut Down 12ns After Output Current (Trace B) Begins to Rise

50MHz trigger

Counters and other instruments require a trigger circuit. Designing a fast, stable trigger is not easy, and often entails a considerable amount of discrete circuitry. Figure 17.42 shows a simple trigger with 100mV sensitivity at 50MHz. The FETs comprise a simple high speed buffer and the LT1016 compares the buffer's output to the potential at the "trigger level" potentiometer, which may be either polarity. The 10k resistor provides hysteresis, eliminating "chattering" caused by noisy input signals. Figure 17.43 shows the trigger's response (Trace B) to a 50MHz sine wave (Trace A). To calibrate this circuit, ground the input and adjust the "input zero" control for 0V at Q2's drain terminal.

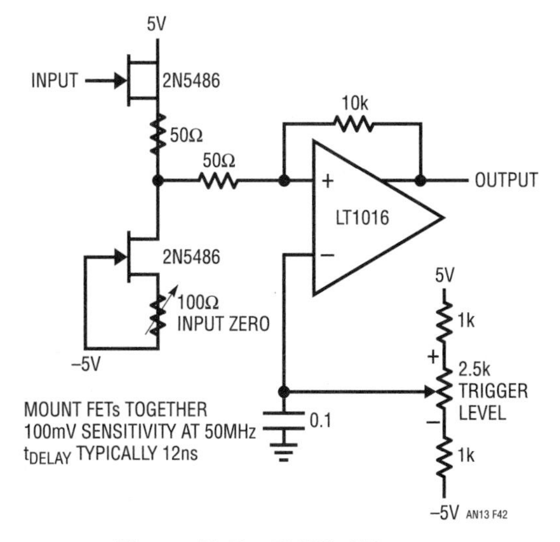

Figure 17.42 • 50MHz Trigger

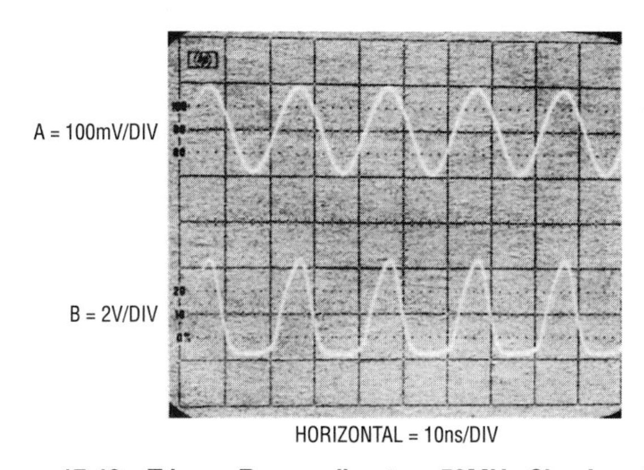

Figure 17.43 • Trigger Responding to a 50MHz Sine Input

References

1. Dendinger, S., "One IC Makes Precision Sample and Hold," EDN, May 20, 1977
2. Pease, R. A., "Amplitude to Frequency Converter," U.S. Patent #3, 746, 968, Filed September, 1972
3. Hewlett-Packard Application Note #915, "Threshold Detection of Visible and Infra-Red Radiation with PIN Photodiodes,"
4. Williams, J., "A Few Proven Techniques Ease Sine-Wave-Generator Design," EDN, November 20, 1980, page 143
5. Williams, J., "Simple Techniques Fine-Tune Sample-Hold Performance," Electronic Design, November 12, 1981, page 235
6. Baker, R. H., "Boosting Transistor Switching Speed," Electronics, Vol. 30, 1957, pages 190 to 193
7. Bunze, V., "Matching Oscilloscope and Probe for Better Measurements,' Electronics, March 1, 1973, pages 88 to 93

Appendix A
About bypass capacitors

Bypass capacitors are used to maintain low power supply impedance at the point of load. Parasitic resistance and inductance in supply lines mean that the power supply impedance can be quite high. As frequency goes up, the inductive parasitic becomes particularly troublesome. Even if these parasitic terms did not exist, or if local regulation is used, bypassing is still necessary because no power supply or regulator has zero output impedance at 100MHz. What type of bypass capacitor to use is determined by the application, frequency domain of the circuit, cost, board space and many other considerations. Some useful generalizations can be made.

All capacitors contain parasitic terms, some of which appear in Figure A1. In bypass applications, leakage and dielectric absorption are second order terms but series R and L are not. These latter terms limit the capacitor's ability to damp transients and maintain low supply impedance. Bypass capacitors must often be large values so they can absorb long transients, necessitating electrolytic types which have large series R and L.

Different types of electrolytics and electrolytic-nonpolar combinations have markedly different characteristics. Which type(s) to use is a matter of passionate debate in some circles and the test circuit (Figure A2) and accompanying photos are useful. The photos show the response of five bypassing methods to the transient generated by the test circuit. Figure A3 shows an unbypassed line which sags and ripples badly Figure A5 at large amplitudes. Figure A4 uses an aluminum 10μF electrolytic to considerably cut the disturbance, but there is still plenty of potential trouble. A tantalum 10μF unit offers cleaner response in A5 and the 10μF aluminum combined with a 0.01μF ceramic type is even better in Figure A6. Combining electrolytics with non-polarized capacitors is a popular way to get good response but beware of picking the wrong duo. The right (wrong) combination of supply line parasitics and paralleled dissimilar capacitors can produce a resonant, ringing response, as in Figure A7. Caveat!

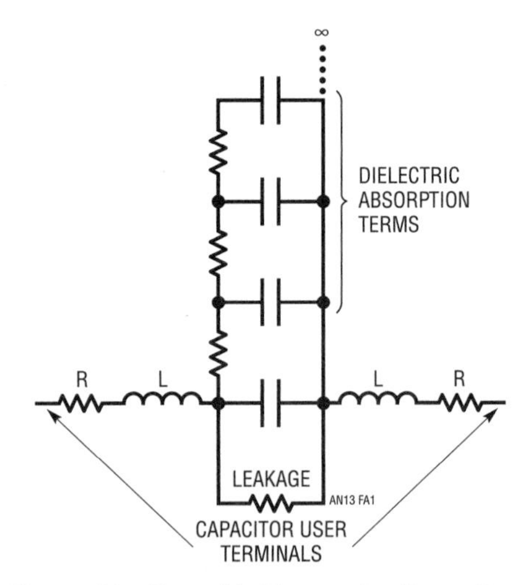

Figure A1 • Parasitic Terms of a Capacitor

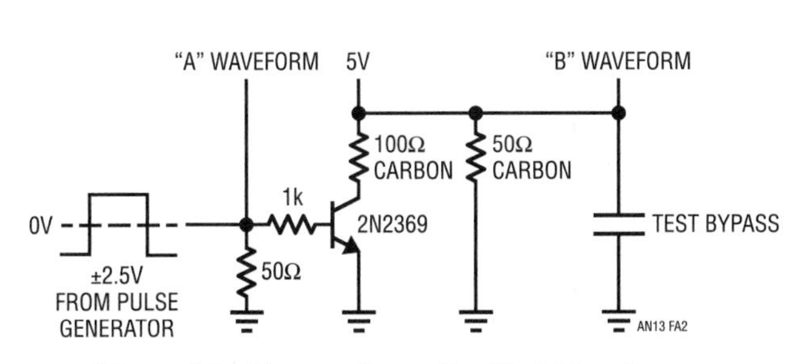

Figure A2 • Bypass Capacitor Test Circuit

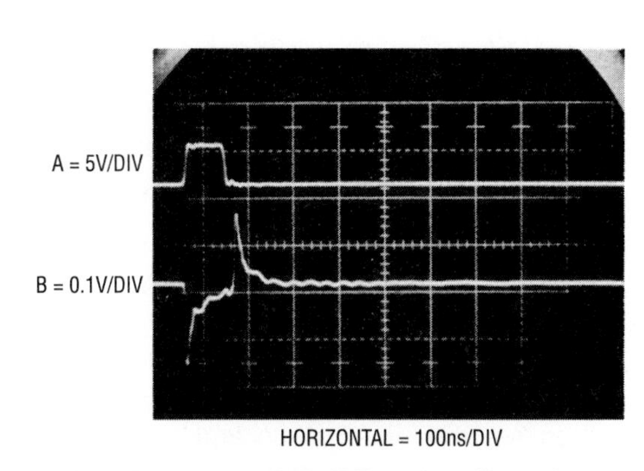

Figure A5 • Response of 10μF Tantalum Capacitor

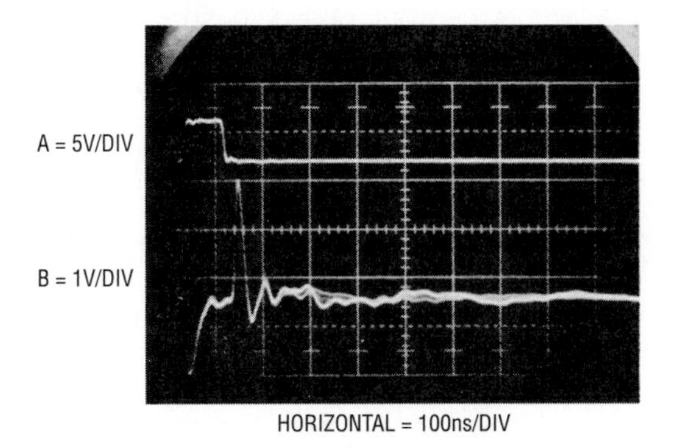

Figure A3 • Response of Unbypassed Line

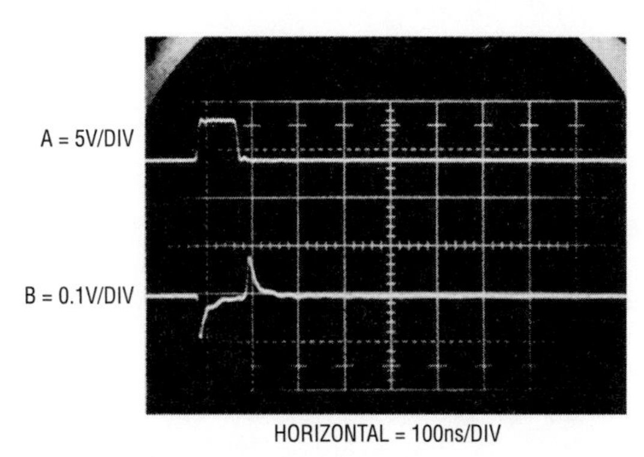

Figure A6 • Response of 10μF Aluminum Paralleled by 0.01μF Ceramic

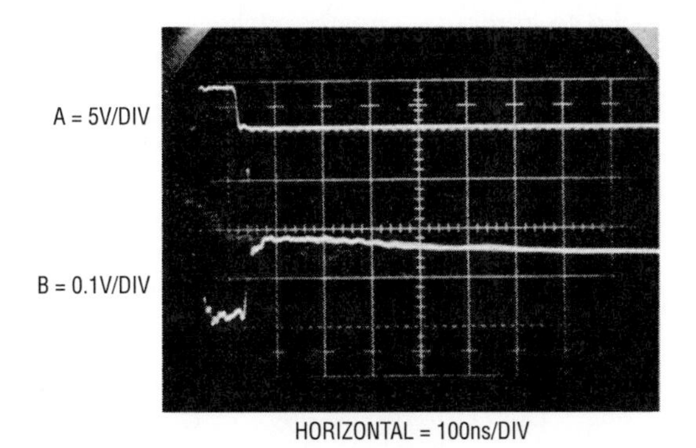

Figure A4 • Response of 10μF Aluminum Capacitor

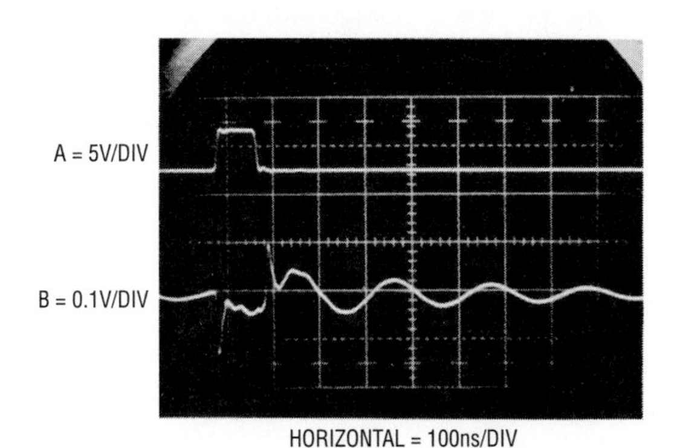

Figure A7 • Some Paralleled Combinations Can Ring. Try Before Specifying!

Appendix B
About probes and oscilloscopes

The oscilloscope-probe combination used in high speed work is the most important equipment decision the designer must make. Ideally, the oscilloscope should have at least 150MHz bandwidth for work with the LT1016, but slower instruments are acceptable if their limitations are well understood. Be aware of your scope's behavior with respect to input impedance, noise, overdrive recovery, sweep nonlinearity, triggering, channel-to-channel feedthrough and other characteristics.

Probes are the most overlooked cause of oscilloscope mismeasurement. All probes have some effect on the point they are measuring. The most obvious is input resistance, but input capacitance usually dominates in a high speed measurement. Much time can be lost chasing circuit events which are actually due to improperly selected or applied probes. An 8pF probe looking at a 1kΩ source impedance will form an 8ns lag—close to the LT1016's response time! Low impedance probes, designed for 50Ω inputs, (with 500Ω to 1kΩ resistance) usually have input capacitance of 1pF or 2pF. They are a very good choice if you can stand the low resistance. FET probes maintain high input resistance and keep capacitance at the 1 pF level but have substantially more delay than passive probes. FET probes also have limitations on input common mode range which must be adhered to or serious measurement errors will result. Contrary to popular belief, FET probes *do not* have extremely high input resistance—some types are as low as 100kΩ.

Current probes are useful and convenient. The passive transformer-based types are fast and have less delay than the Hall effect-based versions. The Hall types, however, respond at DC and low frequency and the transformer types typically roll off around 100Hz to 1kHz. Both types have saturation limitations which, when exceeded, cause odd results on the CRT which will confuse the unwary.

When using different probes remember that they all have different delay times, meaning that apparent timing errors will occur on the CRT. Know what the individual probe delays are and account for them in interpreting the CRT display.

By far the greatest source of error in probe use is grounding. Poor probe grounding can cause ripples and discontinuities in the waveform observed. In some cases the choice and placement of a probe's ground strap will affect waveforms on another channel. In the worst case, connecting the probe's ground wire will virtually disable the circuit being measured. The cause of these problems is due to parasitic inductance in the probe's ground connection. In most oscilloscope measurements this is not a problem, but at nanosecond speeds it becomes critical. Fast probes are always supplied with a variety of spring clips and accessories designed to aid in making the lowest possible inductive connection to ground. Most of these attachments assume a ground plane is in use, which it should be. Always try to make the shortest possible connection to ground—anything longer than one inch may cause trouble.

The simple network of Figure B1 shows just how easy it is for poorly chosen or used probes to cause bad results. A 9pF input capacitance probe with a four inch long ground strap monitors the output (Trace B, Figure B2). Although the input (Trace A) is clean, the output contains ringing. Using the same probe with a one-fourth inch spring tip ground connection accessory seemingly cleans up everything (Figure B3). However, substituting a 1pF FET probe (Figure B4) reveals a 50% output amplitude error in measurement B3! The FET probe's low input capacitance allows a more accurate version of circuit action. The FET probe does, however, contribute its own form of error. Note that the probe's response is tardy by 5ns due to delay in its active circuitry. Hence, separate measurements with each probe are required to determine amplitude and timing parameters of the output.

A final form of probe is the human finger. Probing the circuit with a finger can accentuate desired or undesired effects, giving clues that may be useful. The finger can be used to introduce stray capacitance to a suspected circuit node while observing results on the CRT. Two fingers, lightly moistened, can be used to provide an experimental resistance path. Some high speed engineers are particularly adept at these techniques and can estimate the capacitive and resistive effects created with surprising accuracy.

Examples of the probes discussed, along with different forms of grounding implements, are shown in Figure B5.

Probes A, B, E, and F are standard types equipped with various forms of low impedance grounding attachments. The conventional ground lead used on G is more convenient to work with but will cause ringing and other effects at high frequencies, rendering it useless. H has a very short ground lead. This is better, but can still cause trouble at high speeds. D is a FET probe. The active circuitry in the probe and a very short ground connector ensure low parasitic capacitance and inductance. C is a separated FET probe attenuator head. Such heads allow the probe to be used at higher voltage levels (e.g., ±10V or ±100V). The miniature coaxial connector shown can be mounted on the circuit board and the probe mated with it. This technique provides the lowest possible parasitic inductance in the ground path and is especially recommended.

I is a current probe. A ground connection is not usually required. However, at high speeds the ground connection may result in a cleaner CRT presentation. Because no current flows in the ground lead of these probes, a long strap is usually permissible. J is typical of the finger probes described in the text. Note the ground strap on the third finger.

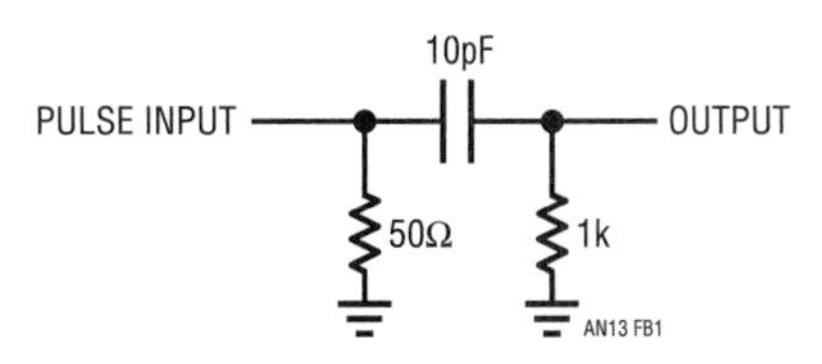

Figure B1 • Probe Test Circuit

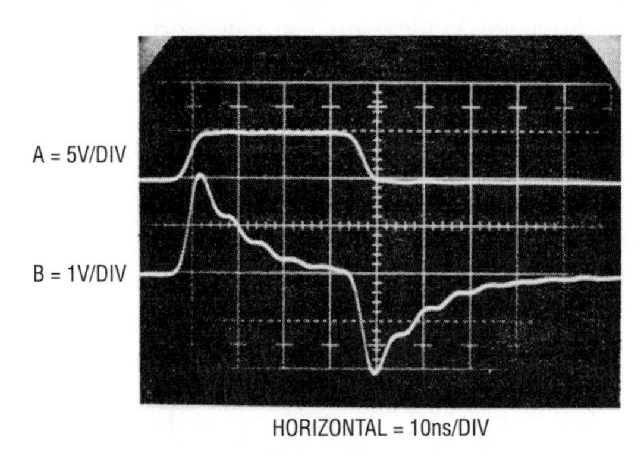

Figure B2 • Test Circuit Output with 9pF Probe and 4 Inch Ground Strap

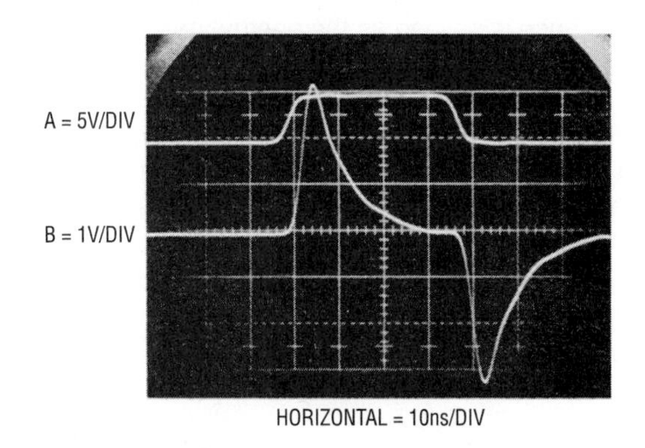

Figure B4 • Test Circuit Output with FET Probe

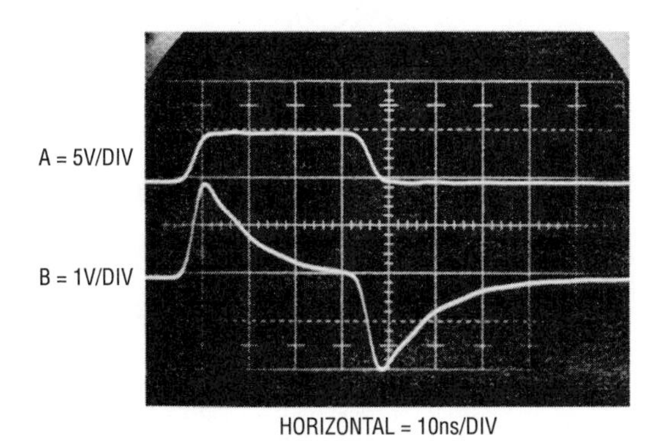

Figure B3 • Test Circuit Output with 9pF Probe and 0.25 Inch Ground Strap

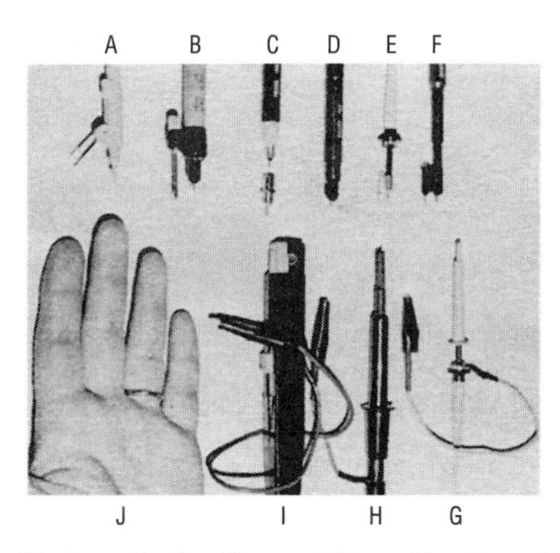

Figure B5 • Various Probe-Ground Strap Configurations

The low inductance ground connectors shown are available from probe manufacturers and are always supplied with good quality, high frequency probes. Because most oscilloscope measurements do not require them, they invariably become lost. There is no substitute for these devices when they are needed, so it is prudent to take care of them. This is especially applicable to the ground strap on the finger probe.

Appendix C
About ground planes

Many times in high frequency circuit layout the term "ground plane" is used, most often as a mystical and ill-defined cure to spurious circuit operation. In fact, there is little mystery to the usefulness and operation of ground planes, and like many phenomena, their fundamental operational principle is surprisingly simple.

Ground planes are primarily useful for minimizing circuit inductance. They do this by utilizing basic magnetic theory. Current flowing in a wire produces an associated magnetic field. The field's strength is proportional to the current and inversely related to the distance from the conductor. Thus, we can visualize a wire carrying current (Figure C1) surrounded by radii of magnetic field. The unbounded field becomes smaller with distance. A wire's inductance is defined as the energy stored in the field set up by the wire's current. To compute the wire's inductance requires integrating the field over the wire's length and the total radial area of the field. This implies integrating on the radius from $R = R_W$ to infinity, a very large number. However, consider the case where we have two wires in space carrying the same current in either direction (Figure C2). The fields produced cancel.

In this case, the inductance is much smaller than in the sample wire case and can be made arbitrarily small by reducing the distance between the two wires. This reduction of inductance between current carrying conductors is the underlying reason for ground planes. In a normal circuit, the path current takes from the signal source, through its conductor and back to ground includes a large loop area. This produces a large inductance for this conductor which can cause ringing due to LRC effects. It is worth noting that 10nH at 100MHz has an impedance of 6Ω. At 10mA a 60mV drop results.

A ground plane provides a return path directly under the signal carrying conductor through which return current can flow. The conductor's small physical separation means the inductance is low. Return current has a direct path to ground, regardless of the number of branches associated with the conductor. Currents will always flow through the return path of lowest impedance. In a properly designed ground plane this path is directly under the signal conductor. In a practical circuit it is desirable to "ground plane" one whole side of the PC card (usually the component side for wave solder considerations) and run the signal conductors on the other side. This will give a low inductance path for all the return currents.

Aside from minimizing parasitic inductance, ground planes have additional benefits. Their flat surface minimizes resistive losses due to AC "skin effect" (AC currents travel along a conductor's surface). Additionally, they aid the circuit's high frequency stability by referring stray capacitances to ground.

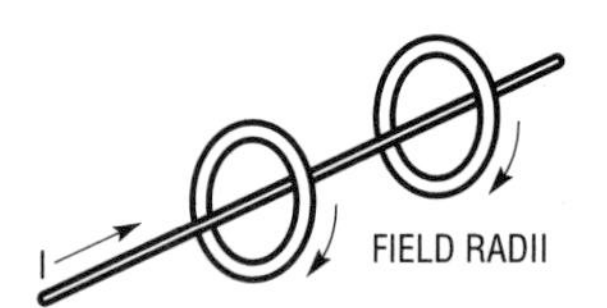

Figure C1 • Single Wire Case

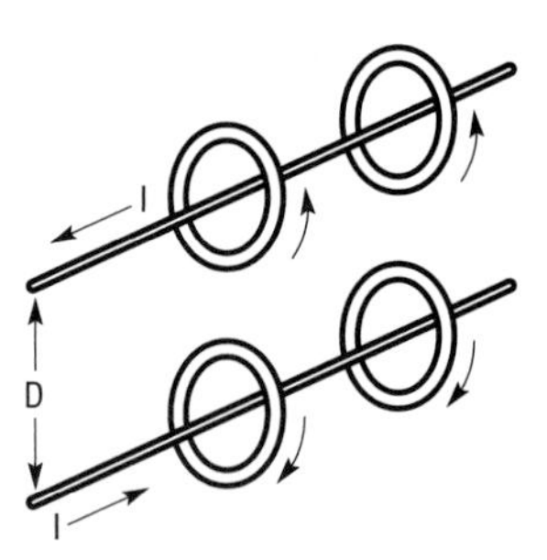

Figure C2 • Two Wire Case

Some practical hints for ground planes are:

1. Ground plane as much area as possible on the component side of the board, especially under traces that operate at high frequency.
2. Mount components that conduct substantial fast rise currents (termination resistors, ICs, transistors, decoupling capacitors) as close to the board as possible.
3. Where common ground potential is important (i.e., at comparator inputs), try to single point the critical components into the ground plane to avoid voltage drops.

   For example, in Figure C3's common A/D circuit, good practice would dictate that grounds 2, 3, 4 and 6 be as close to single point as possible.

   Fast, large currents must flow through R1, R2, D1 and D2 during the DAC settle time. Therefore, D1, D2, R1 and R2 should be mounted close to the ground plane to minimize their inductance. R3 and C1 don't carry any current, so their inductance is less important; they could be vertically inserted to save space and to allow point 4 to be single point common with 2, 3 and 6. In critical circuits the designer must often trade off the beneficial effects of lowered inductance versus the loss of single point ground.
4. Keep trace length short. Inductance varies directly with length and no ground plane will achieve perfect cancellation.

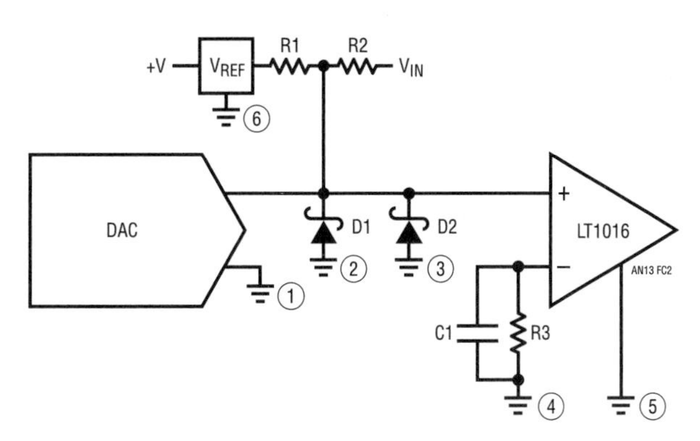

Figure C3 • Typical Grounding Scheme

Appendix D
Measuring equipment response

The 10ns response time of the LT1016 and the circuitry it is used in will challenge the best test equipment. Many of the measurements made utilize equipment near the limit of its capabilities. It is a good idea to verify parameters such as probe and scope rise time and differences in delays between probes and even oscilloscope channels. To do this, a source of very fast, clean pulses is necessary. The circuit shown in Figure D1 uses a tunnel diode to generate a pulse with a rise time well under 1ns.

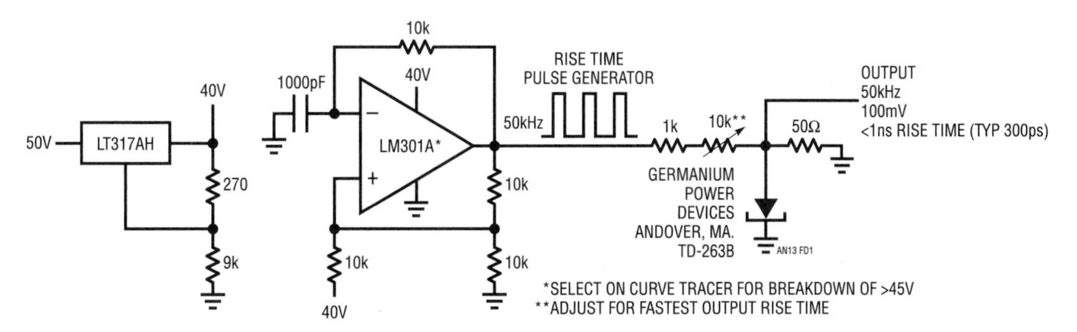

Figure D1 • Tunnel Diode-Based 1ns Rise Time Pulse Generator

Figure C2 shows that the pulse is also very clean, with no attendant ringing or noise. In this photo the pulse is used to check a probe-scope combination with a specified 1.4ns rise time. The display Figure D2 shows that the equipment is being properly used and is in specification. Using the tunnel diode generator to perform tests such as this can save countless hours pursuing "circuit problems," which in reality are caused by misapplied or out of spec equipment.

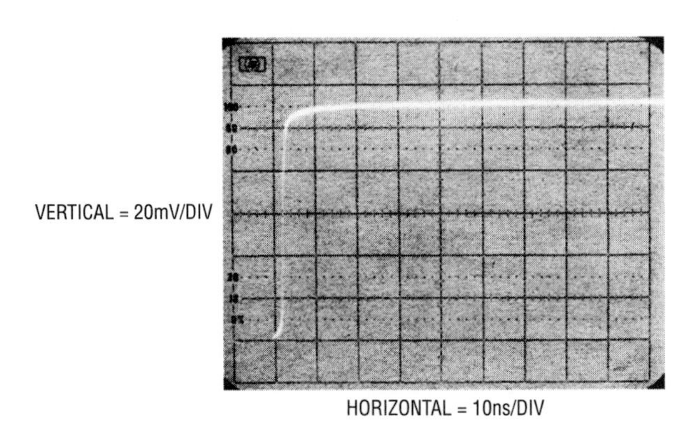

Figure D2 • Figure D1's Output Monitored on a 275MHz Oscilloscope

Appendix E
About level shifts

The TTL output of the LT1016 will interface with many circuits directly. Many applications, however, require some form of level shifting of the output swing. With LT1016-based circuits this is not trivial because it is desirable to maintain very low delay in the level shifting stage. When designing level shifters, keep in mind that the TTL output of the LT1016 is a sink-source pair (Figure E1) with good ability to drive capacitance (such as feedforward capacitors). Figure E2 shows a noninverting voltage gain stage with a 15V output. When the LT1016 switches, the base-emitter voltages at the 2N2369 reverse, causing it to switch very quickly. The 2N3866 emitter-follower gives a low impedance output and the Schottky diode aids current sink capability. Figure E3 is a very versatile stage. It features a bipolar swing

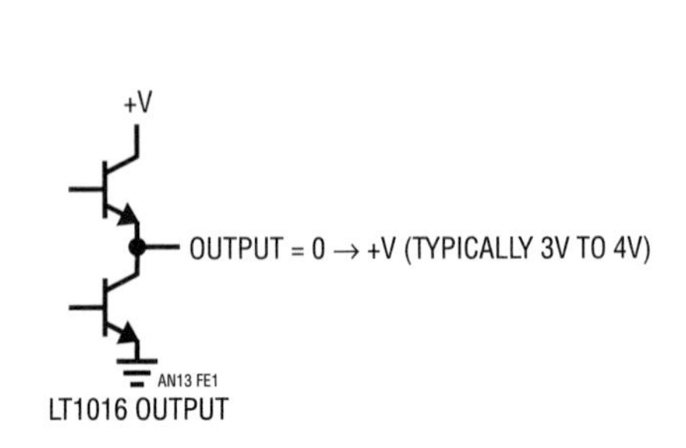

Figure E1 •

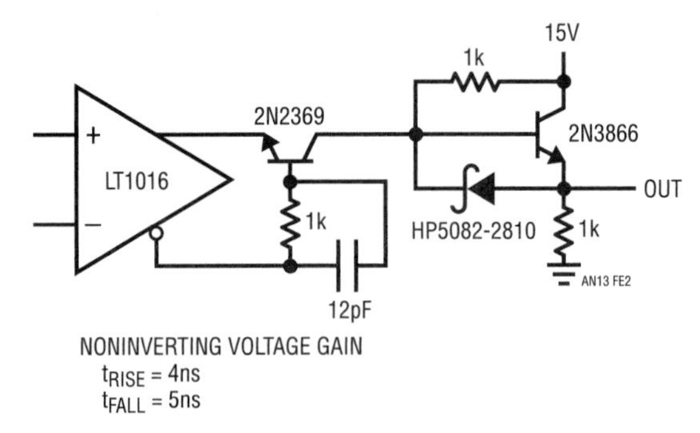

Figure E2 •

which may be programmed by varying the output transistor's supplies. This 3ns delay stage is ideal for driving FET switch gates. Q1, a gated current source, switches the Baker-clamped output transistor, Q2. The heavy feedforward capacitor from the LT1016 is the key to low delay, providing Q2's base with nearly ideal drive. This capacitor loads the LT1016's output transition (Trace A, Figure E5), but Q2's switching is clean (Trace B, Figure E5) with 3ns delay on the rise and fall of the pulse.

Figure E4 is similar to E2 except that a sink transistor has replaced the Schottky diode. The two emitter-followers drive a power MOSFET which switches 1A at 15V. Most of the 7ns to 9ns delay in this stage occurs in the MOSFET and the 2N2369.

When designing level shifters, remember to use transistors with fast switching times and high f_T's. To get the kind of results shown, switching times in the ns range and f_T's approaching 1GHz are required.

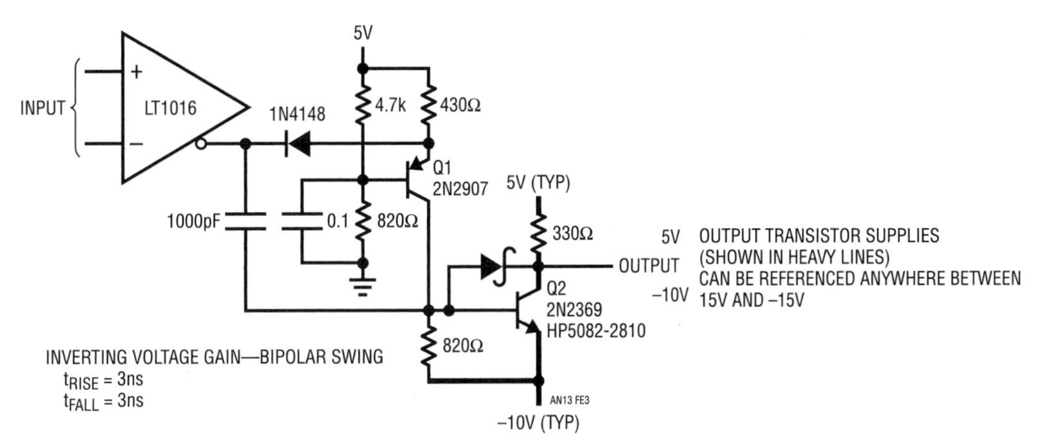

Figure E3 •

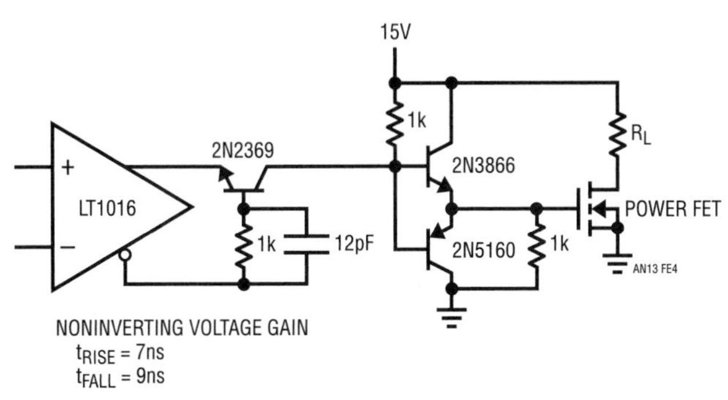

Figure E4 •

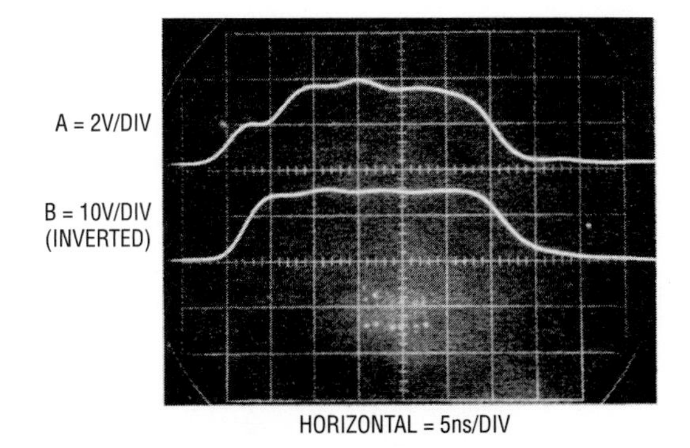

Figure E5 • Figure E3's Waveforms

Designs for high performance voltage-to-frequency converters

18

Jim Williams

Monolithic, modular and hybrid technologies have been used to implement voltage-to-frequency converters. A number of types are commercially available and overall performance is adequate to meet many requirements. In many cases, however, very high performance or special characteristics are required and available units will not work. In these instances V→F circuits specifically optimized for the desired parameters(s) are required. This application note presents examples of circuits which offer substantially improved performance over commercially available V→Fs. Various approaches (see Box Section, "V→F Design Techniques") permit improvements in speed, dynamic range, stability and linearity. Other circuits feature low voltage operation, sine wave output and deliberate nonlinear transfer functions.

Ultra-high speed 1Hz to 100MHz V→F converter

Figure 18.1's circuit uses a variety of circuit methods to achieve wider dynamic range and higher speed than any commercial V→F. Rocketing along at 100MHz full-scale (10% overrange to 110MHz is provided), it leaves all other V→Fs far behind. The circuit's 160dB dynamic range (8 decades) allows continuous operation down to 1Hz. Additional specifications include 0.06% linearity, 25ppm/°C gain temperature coefficient, 50nV/°C (0.5Hz/°C) zero shift and a 0V to 10V input range.

In this circuit an LTC®1052 chopper-stabilized amplifier servo-biases a crude by wide range V→F converter. The V→F output drives a charge pump. The averaged difference between the charge pump's output and the circuit's input biases the servo amplifier, closing a control loop around the wide range V→F. The circuit's wide dynamic range and high speed are derived from the basic V→F's characteristics. The chopper-stabilized amplifier and charge pump stabilize the circuit's operating point, contributing high linearity and low drift. The LTC1052's 50nV/°C offset drift allows the circuit's 100nV/Hz gain slope, permitting operation down to 1Hz.

The positive input voltage causes A1, the servo amplifier, to swing positive. The 2N3904 current sink pulls current (Trace A, Figure 18.2) from the varactor diode, serving as an integrated capacitor. A3 unloads the varactor and biases a trigger made up of the ECL gate and its associated components. This circuit, similar to those employed in oscilloscope triggering applications, features voltage threshold hysteresis and 1ns response time. When A3 ramps to the trigger's lower trip point, its outputs reverse state. The inverting output, operating as an unterminated emitter-follower, deposits a fast positive current spike (Trace B) into the varactor diode integrator. The trigger-gate's complementary output goes low (Trace C), clocking the ECL ÷ 16 counter. This counter's output (Trace D), level shifted by the differential pair of 2N5160s, feeds the 4013 flip-flop. The 4013's square wave drive (Trace E) to the LTC1043 provides charge pump action. The switch-capacitor pairs in the LTC1043 run out of phase and charge is pumped (Trace F) from A1's positive input on each edge of the LTC1043's square wave input. The amount of charge delivered per cycle is primarily dependent on the LT®1009 voltage reference and the 100pF value of the capacitors ($Q = CV$). The slight difference between the charge delivered on the clock's rising and falling edge is due to capacitor tolerances and does not influence circuit operation. The charge pump's overall accuracy is determined by the stability of the LT1009 and the capacitors and the low charge injection of the LTC1043. The ECL counter and the flip-flop divide the trigger's output by 32, setting the LTC1043's maximum switching frequency at about 3MHz (100MHz ÷ 32); within its specified operating range. The 0.22μF capacitor integrates the pumping action to DC. The averaged difference between the positive input-derived current and the charge pump feedback signal is amplified by A1, which servo-controls the circuit's operating point. The compensation capacitor at A1 provides stable loop compensation. Nonlinearity and drift in the basic V→F circuit are compensated by A1's servo action, resulting in the high linearity and low drift previously noted.

Analog Circuit and System Design: Immersion in the Black Art of Analog Design. http://dx.doi.org/10.1016/B978-0-12-397888-2.00018-3

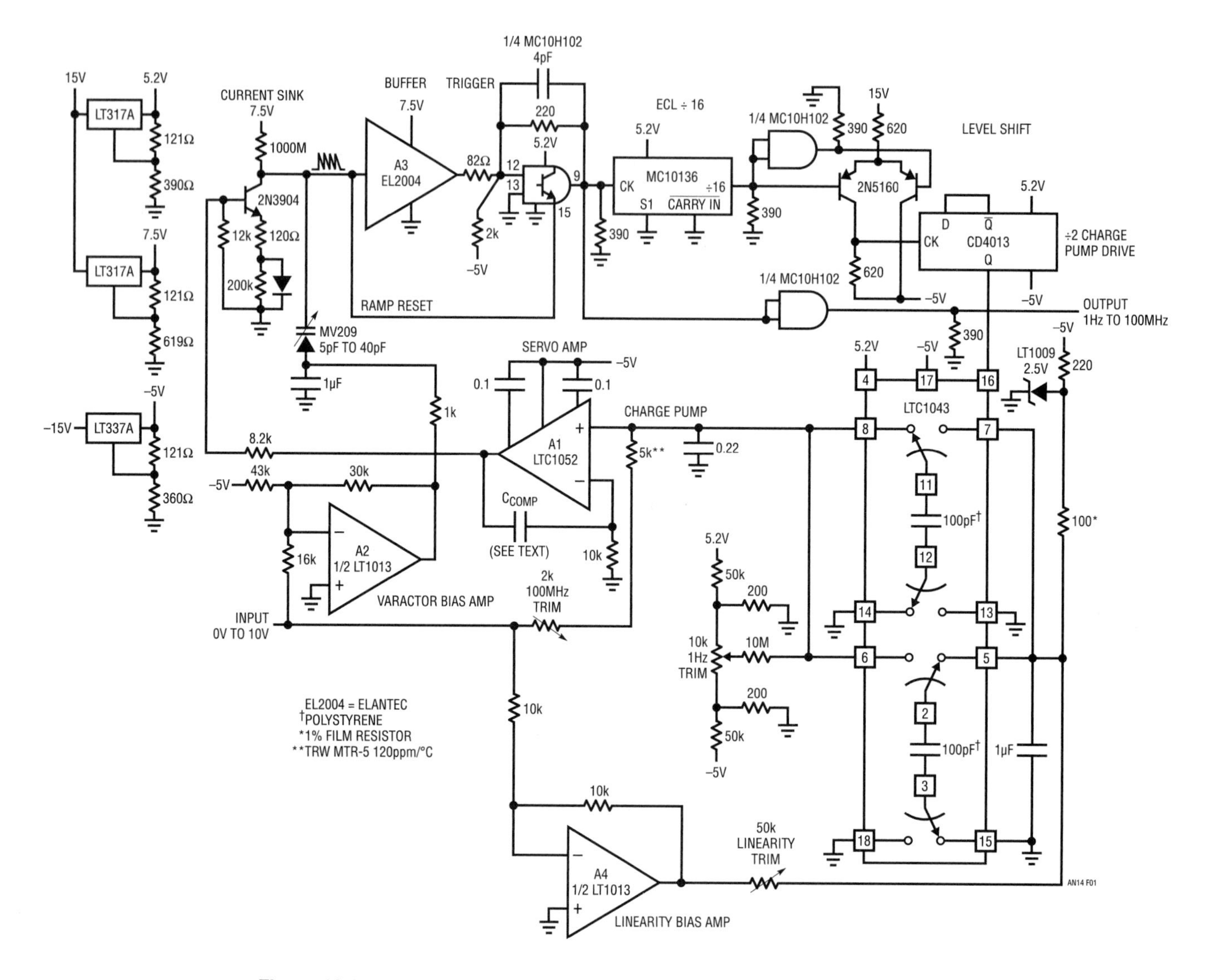

Figure 18.1 • 1Hz to 100MHz Voltage-to-Frequency Converter (King Kong V→F)

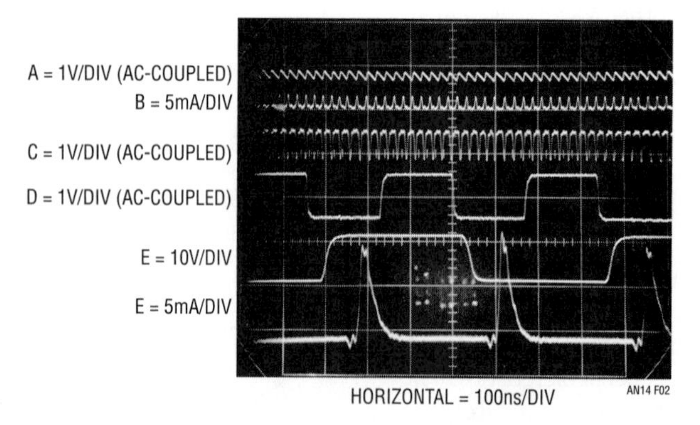

Figure 18.2 • 1Hz to 100MHz V→F Waveforms

Some special techniques are required for this circuit to achieve its specifications. A2, driven from the input voltage, provides DC bias for the varactor diode-integrating capacitor. This DC bias causes the varactor's capacitance to vary inversely with input, helping the circuit achieve its 8-decade dynamic range. The 1μF capacitor, in series with the varactor, gives the relatively large ramp currents a low impedance path to ground. The 1000M resistor in the current sink sources enough current to swamp the effects of all leakages from the 2N3904 collector. This ensures that current must always be sunk from the varactor-integrator to sustain oscillation, even at the very lowest frequencies.

The 200k diode combination in the 2N3904's emitter reduces low frequency jitter. It does this by reducing current sink noise at low frequencies by increasing emitter resistance at low base bias voltages.

The 2k pull-down resistor at the trigger input ensures clean, quick transitions at low ramp slew rates, aiding low frequency jitter performance.

The 5k input resistor specified has a temperature coefficient which opposes that of the polystyrene capacitors in the charge pump. This reduces the effect of their tempco, lowering overall circuit gain drift.

A4 supplies a small, input-related current to the charge pump's voltage reference, correcting nonlinear terms due to residual charge imbalance in the LTC1043. The input-derived correction is effective because the effect of this imbalance varies directly with frequency.

The 100MHz full-scale frequency sets stringent restrictions on oscillator cycle time. At this frequency only 10ns is available for a complete ramp-and-reset sequence. The ultimate limitation on speed in the circuit is the time required to reset the varactor integrator. Figure 18.3 shows high speed details. The combination of a small amplitude ramp and fast ECL switching yields the necessary high speed operation. Trace A is the ramp and Trace B is the reset current from the ECL gate's open emitter. Note that reset occurs in 3.5ns, with little aberration or overshoot.

Figure 18.4 plots output frequency jitter as a function of frequency. At 100MHz, jitter is 0.01%, falling to about 0.002% at 1MHz. In this range the jitter is dominated by noise in the current source and ECL inputs. Below this, jitter slowly rises as operating frequency approaches the servo amplifier's roll off. At 1kHz (10ppm of full scale) jitter is still below 1%, with about 10% jitter at 1Hz (0.01ppm) for $C_{COMP} = 1\mu F$. With $C_{COMP} = 0.1\mu F$, jitter increases below 1kHz and operation below 10Hz is not possible due to loop instability and A1's noise floor. The trade-off is loop settling time. With the larger compensation capacitor the loop settles in 600ms. The 0.1μF value permits 60ms settling.

To calibrate this circuit, apply 10.000V and trim the 100MHz adjustment for 100.00MHz at the output. If a fast enough counter is not available, the ÷32 signal at Pin 16 of the LTC1043 will read 3.1250MHz. Next, ground the input, install $C_{COMP} = 1\mu F$ and adjust the "1Hz trim" until the circuit oscillates at 1Hz. Finally, set the "linearity trim" to 50.00MHz for a 5.000V input. Repeat these adjustments until all three points are fixed.

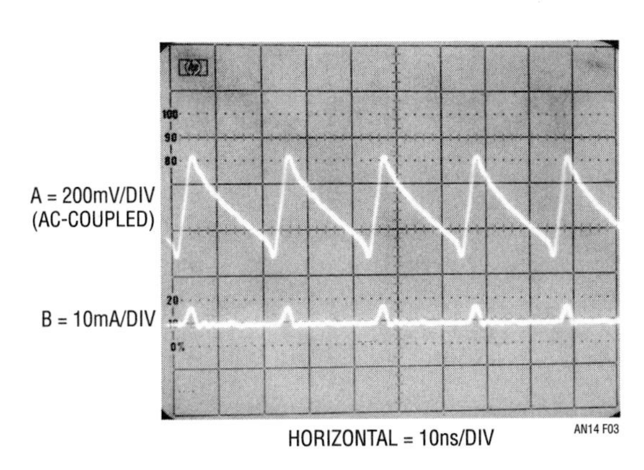

Figure 18.3 • Ramp and Reset Current Detail at 50MHz

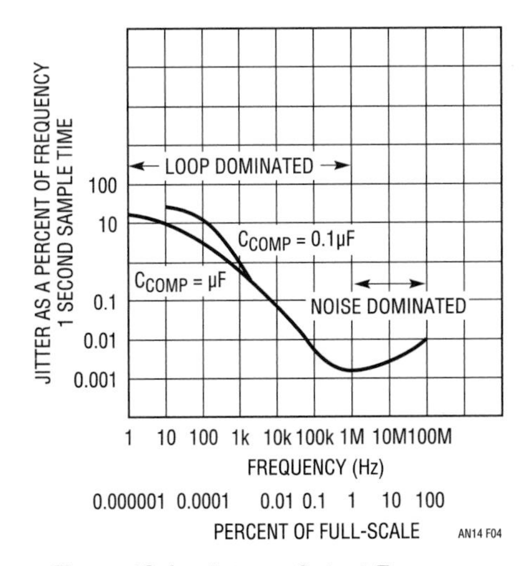

Figure 18.4 • Jitter vs Output Frequency

Fast response 1Hz to 2.5MHz V→F converter

Figure 18.5's circuit is not nearly as fast as Figure 18.1's, but its 2.5MHz output settles from a full-scale input step in only 3μs. This makes the circuit a good candidate for FM applications or any area where fast response to input movement is required. Linearity is 0.05% with a 50ppm/°C gain tempco. A chopper-stabilized correction network holds zero-point error to 0.025Hz/°C. This circuit, a high speed charge-dispensing type (see Box Section) also uses charge feedback. The charge feedback scheme used is a highly modified, high speed variant of the approach originally described by R. A. Pease (see References). A servo amplifier is not used, permitting fast response to input steps. Instead, the charge is fed back directly to the oscillator, which can respond immediately. Although this approach permits fast response, it also requires attention to parasitics to achieve high linearity and low drift.

When an input voltage is applied, A1 integrates in a negative direction (Trace A, Figure 18.6). When its output crosses zero, A2's output switches, causing the paralleled

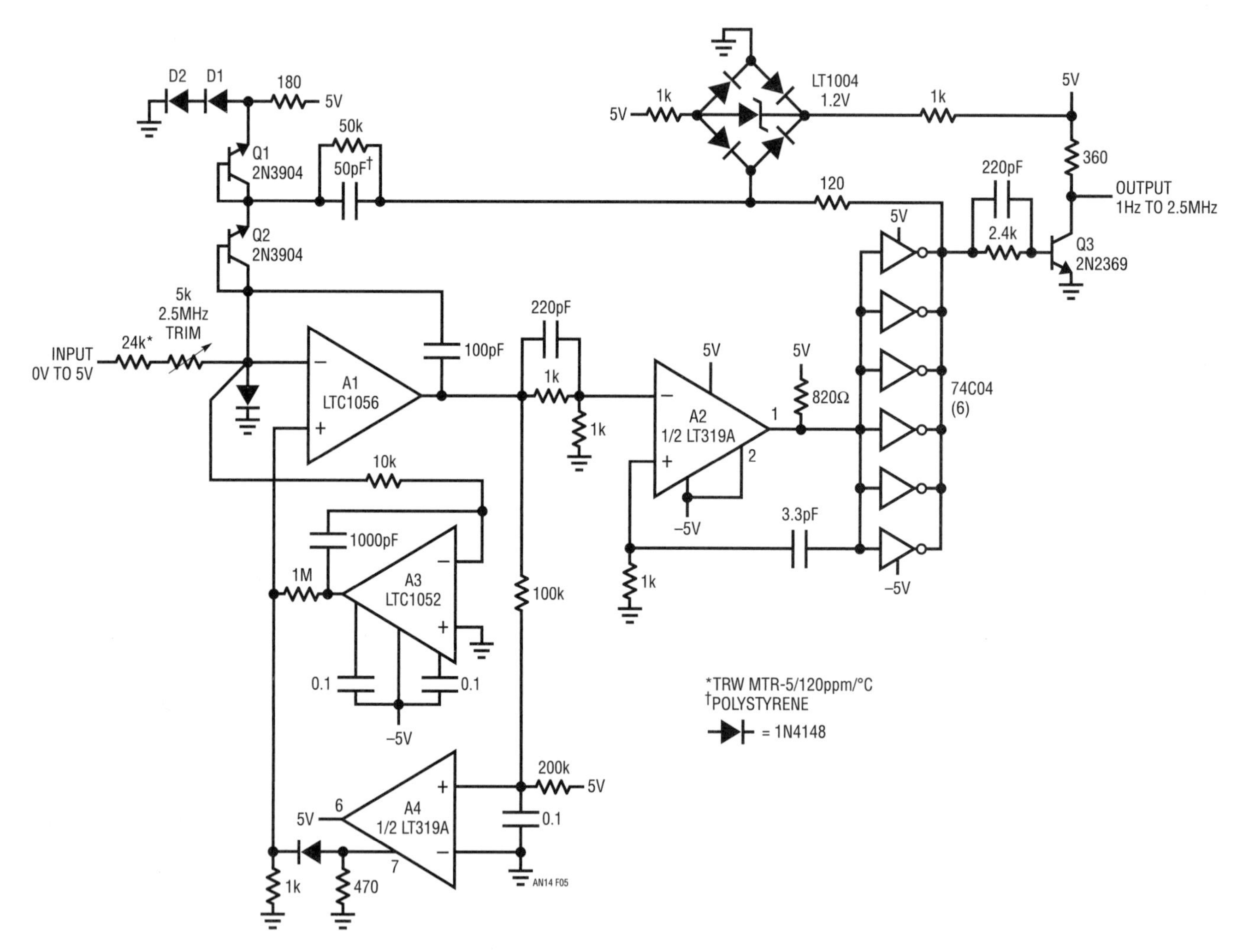

Figure 18.5 • 1Hz to 2.5MHz Fast Response V→F

inverters to go low (Trace B). The feedforward network in A2's negative input aids response. This causes the LT1004 diode bridge to bound at −2.4V ($-V_Z$ LT1004) + ($-2V_{FWD}$). Local positive feedback at A2's positive input (Trace C) reinforces this action. During this interval, charge is pulled (Trace D) from A1's summing junction via the 50pF-50k combination, forcing A1's output to move quickly positive. This causes the A2 inverter combination to switch positive (Trace B), bounding the LT1004 diode bridge at 2.4V. Now the 50pF capacitor receives charge, while A1 again integrates negative, and the entire cycle repeats. The frequency of this action is a linear function of the input voltage.

D1 and D2 compensate the diodes in the bridge. Diode-connected Q1 compensates steering diode Q2. (The diode-connected transistors provide lower leakage from the summing junction than conventional diodes.) A3, a chopper-stabilized op amp, offset stabilizes A1, eliminating the necessity for zero trimming.

A4 guards against circuit latch-up, which can occur due to the AC-coupled feedback loop. If the circuit latches, A1's output goes to the negative rail and stays there. This causes A4's output (A4 is used in emitter-follower output mode) to go high. A1's output now heads positive, initiating normal circuit behavior. The diode at A1's negative input ensures that the start-up loop will dominate over any input condition.

The 50k resistor across the 50pF charge-dispensing capacitor improves linearity by permitting complete discharge on each cycle, despite junction tailing effects in Q2. The input resistor specified has a temperature coefficient opposite that of the capacitor's enhancing circuit gain tempco.

Figure 18.7 shows circuit step response. Trace A is the input, while Trace B is the output. Frequency shift is quick and clean, with no evidence of poor dynamics or time constants.

To trim the circuit, apply 5.000V and adjust the 5k potentiometer for a 2.500MHz output. A3's low offset eliminates the requirement for a zero trim. The circuit maintains 0.05% linearity with 50ppm/°C drift from 1Hz to 2.5MHz. The TTL-compatible output is available at Q3's collector (Trace E). A 10MHz full-scale circuit of this type appears in AN13.

High stability quartz stabilized V→F converter

The gain temperature coefficient of the previous circuits is affected by drift in the charge pumping capacitors. Although compensation schemes were employed in both cases to minimize the effect of this drift, another approach is required to get significantly lower gain drift.

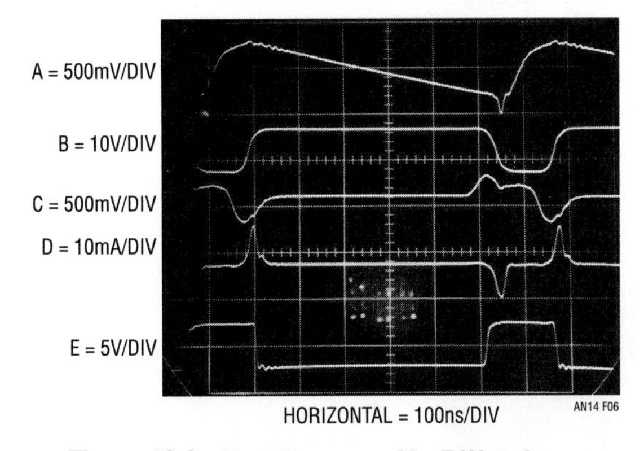

Figure 18.6 • Fast Response V→F Waveforms

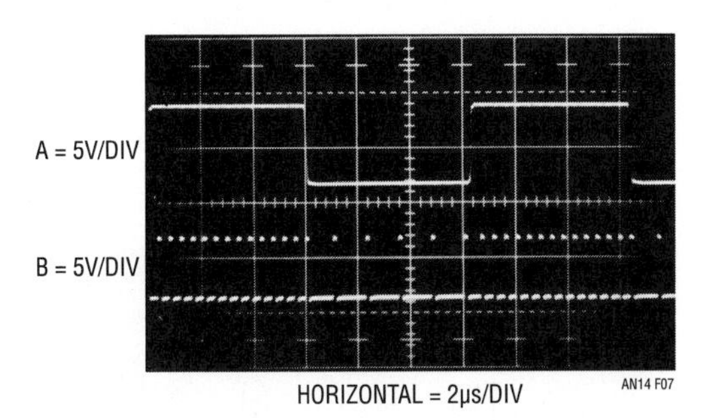

Figure 18.7 • Step Response of 2.5MHz V→F

Figure 18.8's circuit reduces gain TC to 5ppm/°C by replacing the capacitor with a quartz-stabilized clock.

In charge pump-based circuits the feedback is based on Q = CV. In a quartz-stabilized circuit the feedback is based on Q = IT, where I is a stable current source and T is an interval of time derived from the clock.

Figure 18.9. details Figure 18.8's waveforms of operation. A positive input voltage causes A1 to integrate in the negative direction (Trace A, Figure 18.9). The flip-flop's Q1 output (Trace B) changes state at the first positive-going clock edge after A1's output has crossed the D input's switching threshold. The 50kHz clock (Trace C) comes from the flip-flop's other half, which is driven by A2, a quartz-stabilized relaxation oscillator. The flip-flop's Q1 output controls the gating of a precision current sink composed of A3, the LM199 voltage reference, a FET and the LTC1043 switch. When A1 is integrating negative, the Q1 output is high and the LTC1043 directs the current sink's output to ground via Pins 11 and 7. When A1's output crosses the D input's switching threshold, Q1 goes low at the first positive clock edge. LTC1043 Pins 11 and 8 close and a precise, quickly rising current flows out of A1's summing point (Trace D).

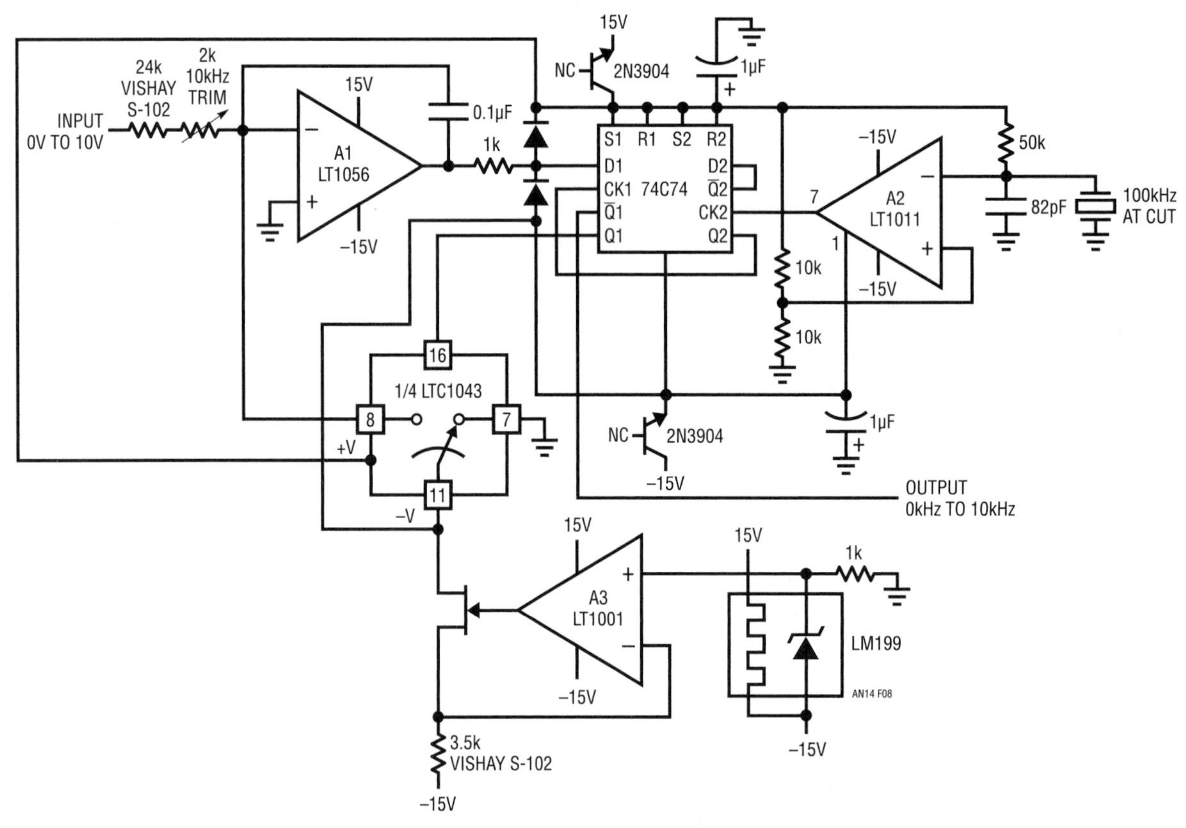

Figure 18.8 • Quartz-Stabilized V→F

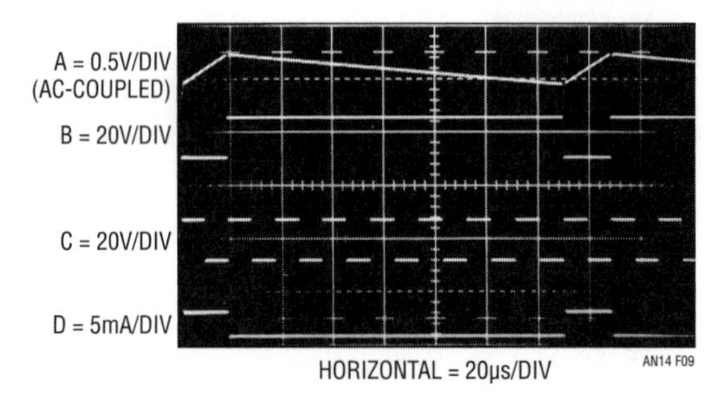

Figure 18.9 • Waveforms for Quartz-Stabilized V→F

This current, scaled to be greater than the maximum signal-derived input current, causes A1's output to reverse direction. At the first positive clock pulse after A1's output crosses the D input's trip point, switching again occurs and the entire process repeats. The repetition frequency depends on the input-derived current, hence the frequency of oscillation is directly related to the input voltage. The circuit's output may be taken from the flip-flop's Q1 or $\overline{Q}1$ outputs. Because this circuit replaces the capacitor with a quartz-locked clock, temperature drift is low, typically 5ppm/°C. The quartz crystal contributes about 0.5ppm/°C, with the remaining drift a function of the current source components, switching time variations and the input resistor.

The reverse-biased 2N3904s serve as Zener diodes, providing about 15V across the CMOS flip-flop. The diodes at the D1 input prevent transient overdrive from A1 during circuit start-up.

A V→F of this type is usually restricted to relatively low full-scale frequencies, e.g., 10kHz to 100kHz, because of speed limitations in accurately switching the current sink.

Additionally, short-term frequency jitter may occur because of the uncertain timing relationship between A1's output switching the flip-flop and the clock phase. This is normally not a problem because the circuit's output is usually read over many cycles, e.g., 0.1 to 1 second.

As shown circuit linearity is 0.005%, gain temperature coefficient is 5ppm/°C and full-scale frequency is 10kHz. The LT1056's low input offset reduces zero point error to 0.005Hz/°C. To trim this circuit, apply exactly 10V in and adjust the 2k potentiometer for 10.000kHz output.

Ultra-linear V→F converter

Figure 18.10 shows a V→F circuit optimized for very high linearity. Although it may be used in a "stand-alone" mode it is specifically intended for processor-driven applications which require 17-bit accuracy, such as weighing scales. This V→F has a resolution of 1ppm, with linearity inside 7ppm (0.0007%). When combined with a processor-driven gain/zero calibration loop it has negligible zero and gain drift. To further ease interface with processor-based systems, the circuit functions from a single 5V power supply.

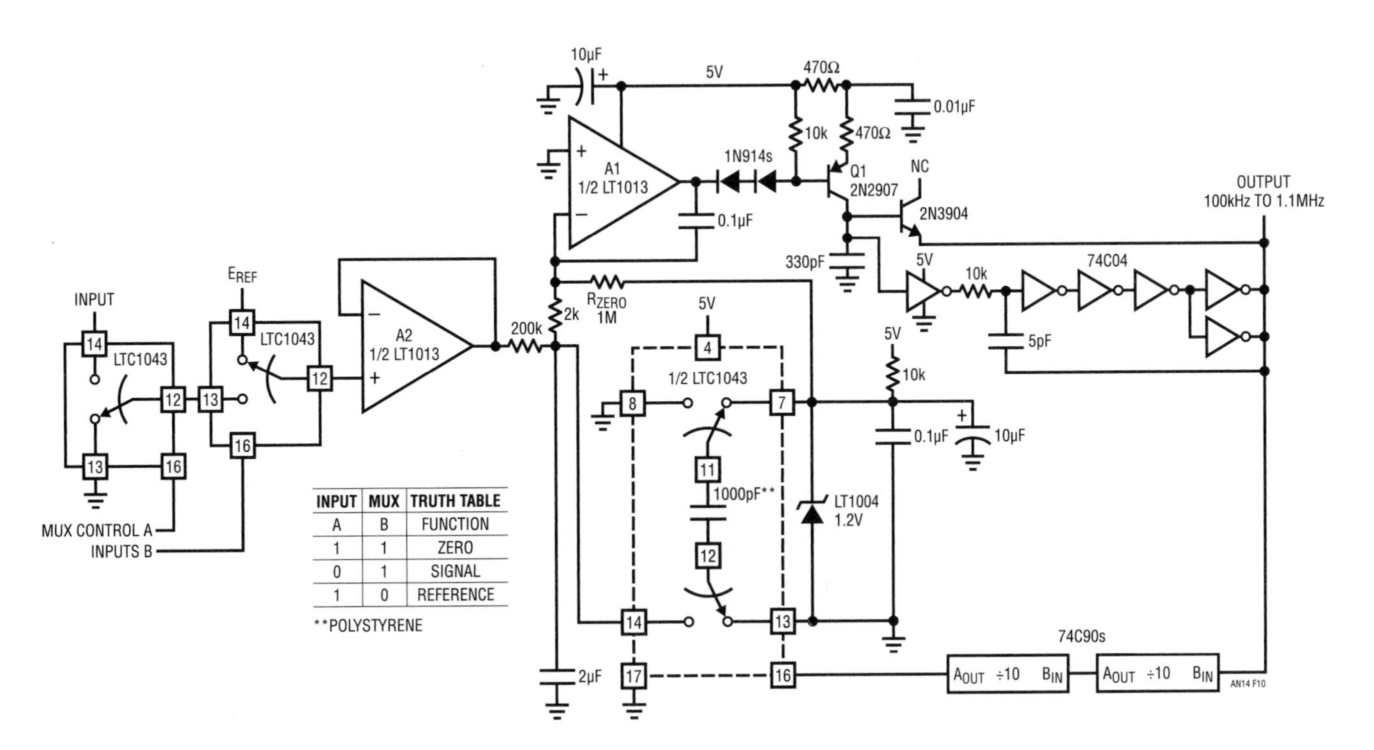

INPUT	MUX	TRUTH TABLE
A	B	FUNCTION
1	1	ZERO
0	1	SIGNAL
1	0	REFERENCE

Figure 18.10 • Ultra-Linear V→F

The circuit is conceptually similar to the 100MHz V→F of Figure 18.1. A1 servo-controls a crude V→F converter composed of Q1, in this case a current source, and the 74C04 gates. The V→F's output is divided digitally and drives a charge pump whose output closes a loop back at A1. In Figure 18.1's case, the crude V→F's output was divided down to permit the LTC1043 to function; it cannot operate at 100MHz toggle rates. Here, the divider's purpose is to lower the toggle frequency, allowing the charge pump to achieve much higher precision than with direct feedback.

Before discussing processor-driven operation it is necessary to understand basic circuit operation. To do this, delete A2 and R_{ZERO}. Assume a positive voltage is applied to the left end of the 200k resistor which was previously connected to A2. This forces A1's output to move negatively, turning on Q1. A1's collector (Trace A, Figure 18.11) ramps the 330pF capacitor positively. When this ramp crosses the 74C04 inverter's threshold, its output moves toward ground, causing the entire chain to switch. AC positive feedback from the paralleled outputs enhances switching. The output inverter's signal (Trace B), the circuit's output, also drives the ÷100 counter chain. The counter's output (Trace C) clocks the LTC1043 which is configured to pump negative charge (Trace D) into the 200k-2k-2μF junction. The 2μF capacitor integrates the discrete charge events to DC, closing a loop around A1. Thus, A1 biases Q1 at whatever point is required to maintain its inputs at balance. This forces the crude V→F's output frequency to be a direct function of the input voltage over a 0MHz to 1MHz output range. The relatively low LTC1043 clock frequency furnished by the dividers permits 0.0007% V→F linearity. For processor-driven auto-zero/gain loop operation, the input multiplexer and R_{ZERO} must be added. With the multiplexer set to the "zero" function (see Truth Table), A2's input is grounded and the 200k resistor receives no drive. A1 receives bias via R_{ZERO}, however, and the circuit oscillates around 100kHz. After the processor has read this frequency it shifts the multiplexer to the "signal" function. Here, A2's output is a buffered version of the signal input. The circuit's output frequency is now determined by this input and the current through R_{ZERO}. Typical outputs will range from 100kHz to 1MHz. After reading this frequency the processor selects the "reference" multiplexer state and determines the frequency produced. The reference voltage must be greater than the largest signal input. It may be either a stable potential or one ratiometrically related to the signal input, as is the case in many transducer-based systems. Typically, it will produce a 1.1MHz output. Once this measurement sequence is completed the processor has enough information to determine the value of the signal input by mathematical manipulation. Additionally, because the multiplexing sequence occurs relatively quickly, drifts in the V→F are cancelled. No precision components are required, although the polystyrene capacitor is needed for high linearity. The circuit's 7ppm linearity and 1ppm resolution will suit almost all applications, although processor techniques could be used to obtain even better linearity.

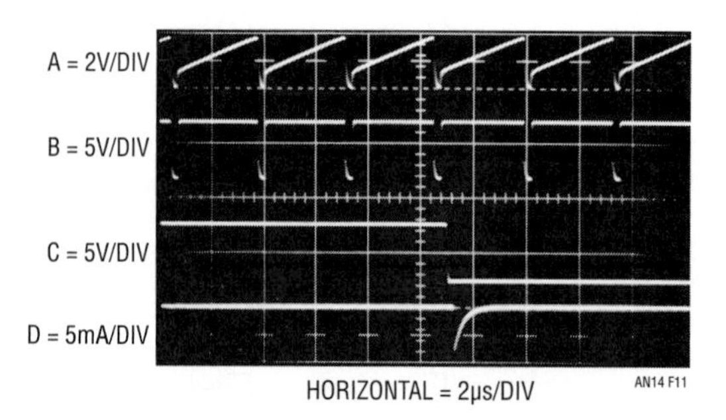

Figure 18.11 • Ultra-Linear V→F Waveforms

Single cell V→F converter

High speed and precision are not the only areas where special V→F circuits are needed. Figure 18.12 shows a circuit which runs from a single 1.5V cell with only 125μA current drain. The circuit uses an LT1017 dual micropower comparator in a servo-controlled charge pump configuration. The input is applied to C1, which is compensated by the 10μF and 1μF capacitors to act as an op amp. C1's output drives the 110k-0.02μF RC, causing the capacitor to ramp (Trace A, Figure 18.13).

During the ramp, C2's output is high, turning off Q1 and biasing Q2 on. The potential across the Q3-Q4 V_{BE} voltage reference (Trace B) is zero. The 0.01μF capacitor receives no charge. When the ramp equals the potential at C2's positive input, switching occurs. C2's output goes low, and the 0.02μF unit discharges. AC positive feedback (Trace C) "hangs up" C2 long enough for a ramp reset of about 80mV. Concurrently, Q1 comes on and Q2 goes off. The Q3-Q4 reference comes on (Trace B) and charges the 0.01μF capacitor via Q6.

When the positive feedback at C2 ceases, its output returns high, cutting off Q1 and biasing Q2. Now, the

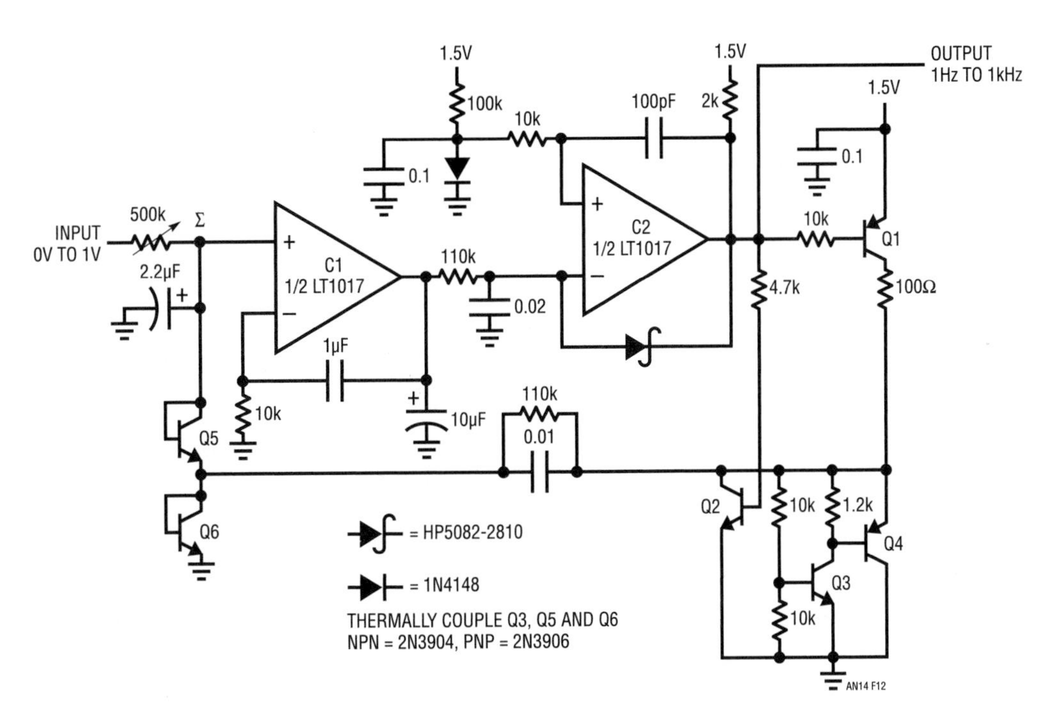

Figure 18.12 • Single Cell V→F

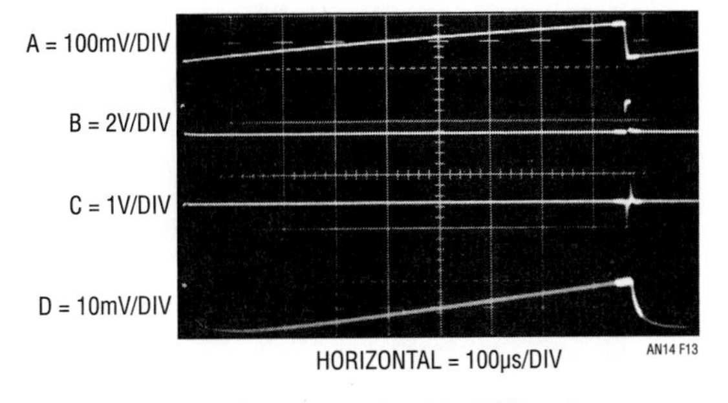

Figure 18.13 • Single Cell V→F Waveforms

0.01μF capacitor discharges, forcing current to flow from C1's 2.2μF summing point capacitor (Trace D) via Q5 and Q2. C1 servo-controls this oscillator to whatever frequency is required to maintain C1's summing point near zero. Since the current into C1's input is a linear function of the input voltage, oscillator frequency is also linear. The 1μF-10k combination at C1 provides loop stability. The 100k resistor across the 0.01μF capacitor influences its discharge characteristic, aiding overall circuit linearity.

The temperature coefficient of the 1.2V Q3-Q4 reference is largely compensated by the junction tempcos of Q5 and Q6, giving the circuit a 250ppm/°C gain drift. Battery discharge introduces less than 1% error over 1000 hours operation.

Sine wave output V→F converter

Almost all V→F converters have a pulse or square wave output. Many applications such as audio, filter testing and automatic test equipment require a sine wave output. The circuit of Figure 18.14 meets this need, spanning a 1Hz to 100kHz range (100dB or 5 decades) for a 0V to 10V input. It is significantly faster than previously published circuits while maintaining 0.1% frequency linearity and 0.2% distortion specifications.

To understand the circuit, assume C1 is low, cutting off Q1. The positive input voltage is inverted by A3, which biases the summing node of integrator A1 through

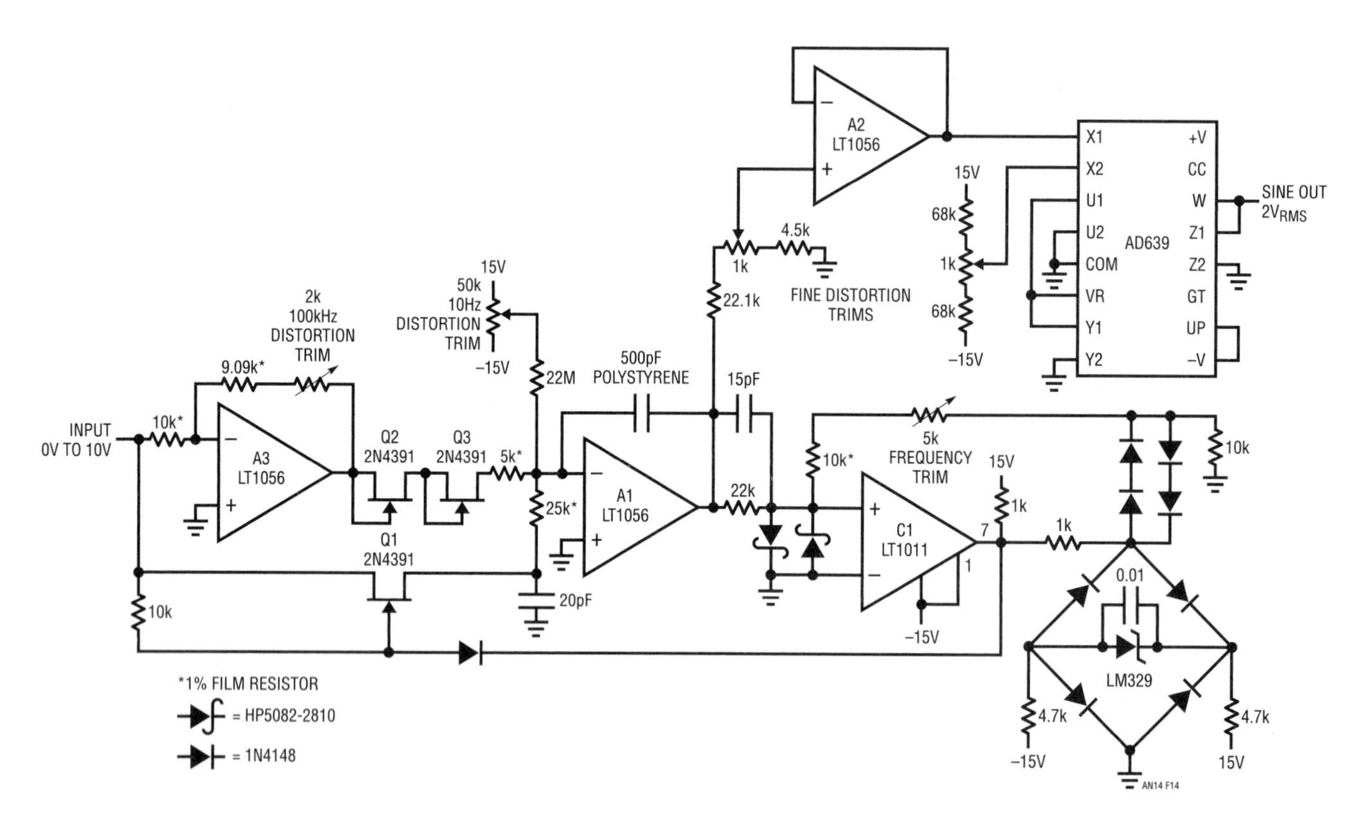

Figure 18.14 • Sine Wave Output 1Hz to 100kHz V→F (VCO)

the 5k resistor and the self-biased FETs. A current, −I, is pulled from the summing point. A1's output (Trace A, Figure 18.15) integrates positive until C1's input crosses 0 V. When this happens, C1's output goes positive (Trace B), allowing Q1 to come on. The resistor in Q1's path is scaled to produce a current, +2I, exactly twice the absolute magnitude of the current, −I, being removed from the summing node. As a result, the net current into the junction becomes +I and A1 integrates negatively at the same rate its positive excursion took. When A1 integrates far enough in the negative direction, C1's positive input crosses zero and it again switches. This turns Q1 off and the entire cycle repeats. The result is a triangle waveform at A1's output. The frequency of this triangle is dependent on the circuit's input voltage and varies from 1Hz to 100kHz with a 0V to 10V input. The LM329 diode bridge and the series-parallel diodes provide a stable bipolar reference which always opposes the sign of A1's output ramp. The Schottky diodes bound C1's positive input, assuring it clean recovery from overdrive. The AD639 trigonometric function generator, biased via A2, converts A1's triangle output into a sine wave (Trace C). The AD639 must be supplied with a triangle wave which does not vary in amplitude or output distortion will result. At high frequencies, delays in the A1 integrator switching loop result in late turn on and turn off of Q1. If the effects of the delays are not minimized, triangle amplitude will increase with frequency, causing distortion level to also increase with frequency. The 15pF feedforward network at C1's input compensates the delay, keeping distortion to just 0.2% over the entire 100kHz range. At 10kHz, distortion is inside 0.07%. The effects of gate-source charge transfer, which happens whenever Q1 switches, are minimized by the 20pF unit in Q1's source line. Without this capacitor, a sharp spike would occur at the triangle peaks, increasing distortion. The Q2-Q3 FETs compensate the temperature dependent on-resistance of Q1, keeping the +2I/−I relationship constant with temperature. Circuit gain TC is 150ppm and zero point drift is 0.1Hz/°C.

This circuit features extremely fast response to input changes, something most sine wave circuits cannot do. Figure 18.16 shows what happens when the input switches between two levels (Trace A). The circuit's output (Trace B) shifts frequency immediately, with no glitching or poor dynamics.

To adjust this circuit, put in 10.00V and trim the 2k pot for a symmetrical triangle output at A1. Next, put in 100μV and trim the 50k pot for triangle symmetry. Then, put in 10.00V again and trim the 5k "frequency trim" adjustment for a 100.0kHz output frequency. Finally, adjust the "distortion trim" potentiometers for minimum distortion as measured on a distortion analyzer (Trace D). Slight readjustment of the other potentiometers may be required to get lowest possible distortion.

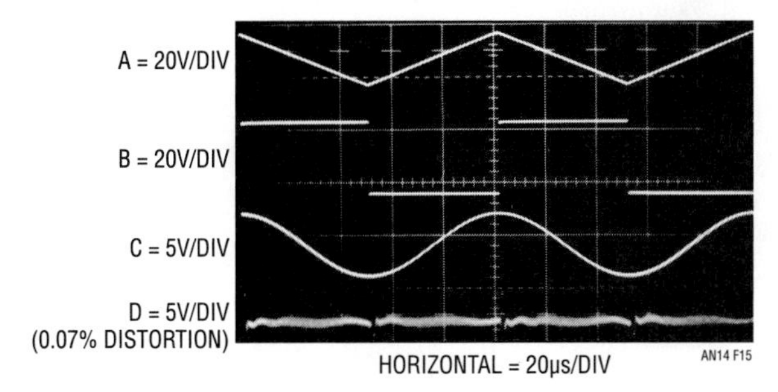

Figure 18.15 • Sine Wave Output V→F Converter Waveforms

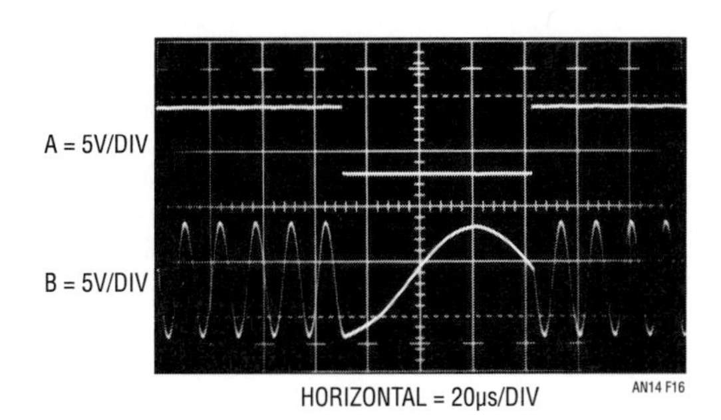

Figure 18.16 • Sine Wave Output V→F Input Step Response

1/X transfer function V→F converters

Another dimension in V→F design is converters which have a deliberate nonlinear transfer function. Such converters are useful in linearizing outputs from transducers such as gas sensors and flow meters. Figure 18.17's circuit converts input voltages of 0V to 10V to an output frequency of 1kHz to 2Hz with a 0.05% accurate 1/X conformity.

A1 integrates current from the LT1009 2.5V reference. A1's negative output ramp (Trace A, Figure 18.18) is compared at C1 to the input voltage via a current summing network. When C1's input goes negative, its output (Trace B) falls, triggering the flip-flop (Trace C) Q output high. This turns on Q1, resetting the ramp. When the ramp reset gets very near ground, C2 triggers low (Trace D), resetting the Q output low. This turns off Q1, allowing the ramp to begin again and the entire cycle repeats. Waveforms E, F, G and H are expanded versions of A through D, respectively, and show detail of the ramp resetting sequence.

In most V→F converters the input signal controls the integrator slope. Here the integrator runs at a fixed slope. The length of time the integrator requires to cross the input voltage is inversely proportional to the input's amplitude and loop oscillation is related by 1/X to the input. The ramp reset time is a first order error term because it is lost in the integration. At low frequencies the ramp reset time is a small term, even though reset takes longer

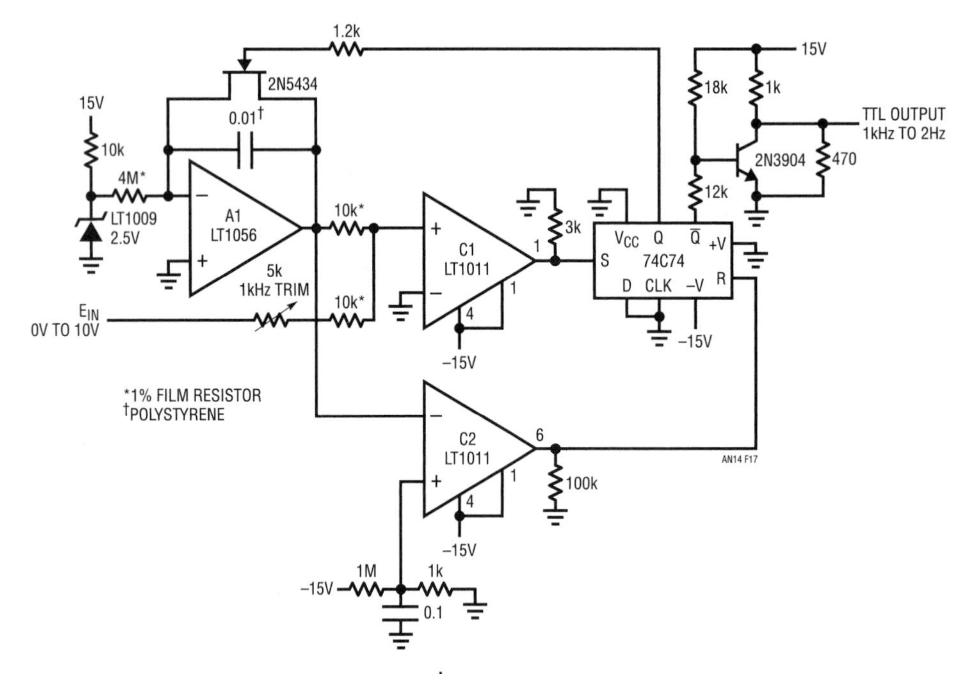

Figure 18.17 • $\frac{1}{E_{IN}}$ → Frequency Converter

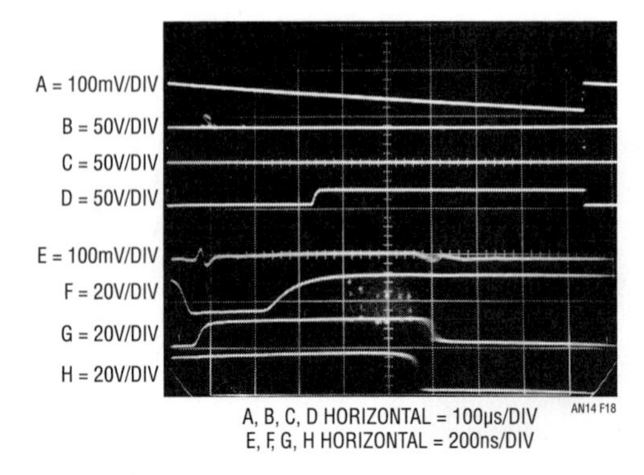

Figure 18.18 • $\frac{1}{E_{IN}}$ → Frequency Converter Waveforms

(because the ramp had to run to a higher amplitude to cross the input). At higher frequencies, even though it is shorter, the reset period becomes significant because its "dead time" is a substantial percentage of the oscillation frequency. The 2-comparator flip-flop reset scheme reduces this error by adaptively controlling and minimizing the ramp reset time, regardless of peak ramp amplitude. A simple fixed AC feedback scheme would not do this because its time constant would have to be long enough to reset the ramp from large peak amplitudes (e.g., at low frequency). Even with this reset arrangement, the circuit's 0.05% 1/X conformity can only be achieved by limiting maximum frequency to about 1kHz. It is worth noting that this circuit has almost ten times the accuracy of analog multipliers and other analog 1/X computing techniques. Circuit drift is about 150ppm/°C. To trim the circuit, put in 50mV and adjust the 5k potentiometer for 1kHz output.

Figure 18.19's 1/X V→F, developed by R. Essaff, provides better performance, although it is somewhat more complex. This charge pump class design gives 0.005% 1/X conformity, 50ppm/°C drift and 10kHz to 50Hz outputs for 0V to 5V in.

A1 and its associated components form an integrator which ramps positive (Trace A, Figure 18.20). When A1's output crosses zero, C1 goes negative (Trace B), triggering

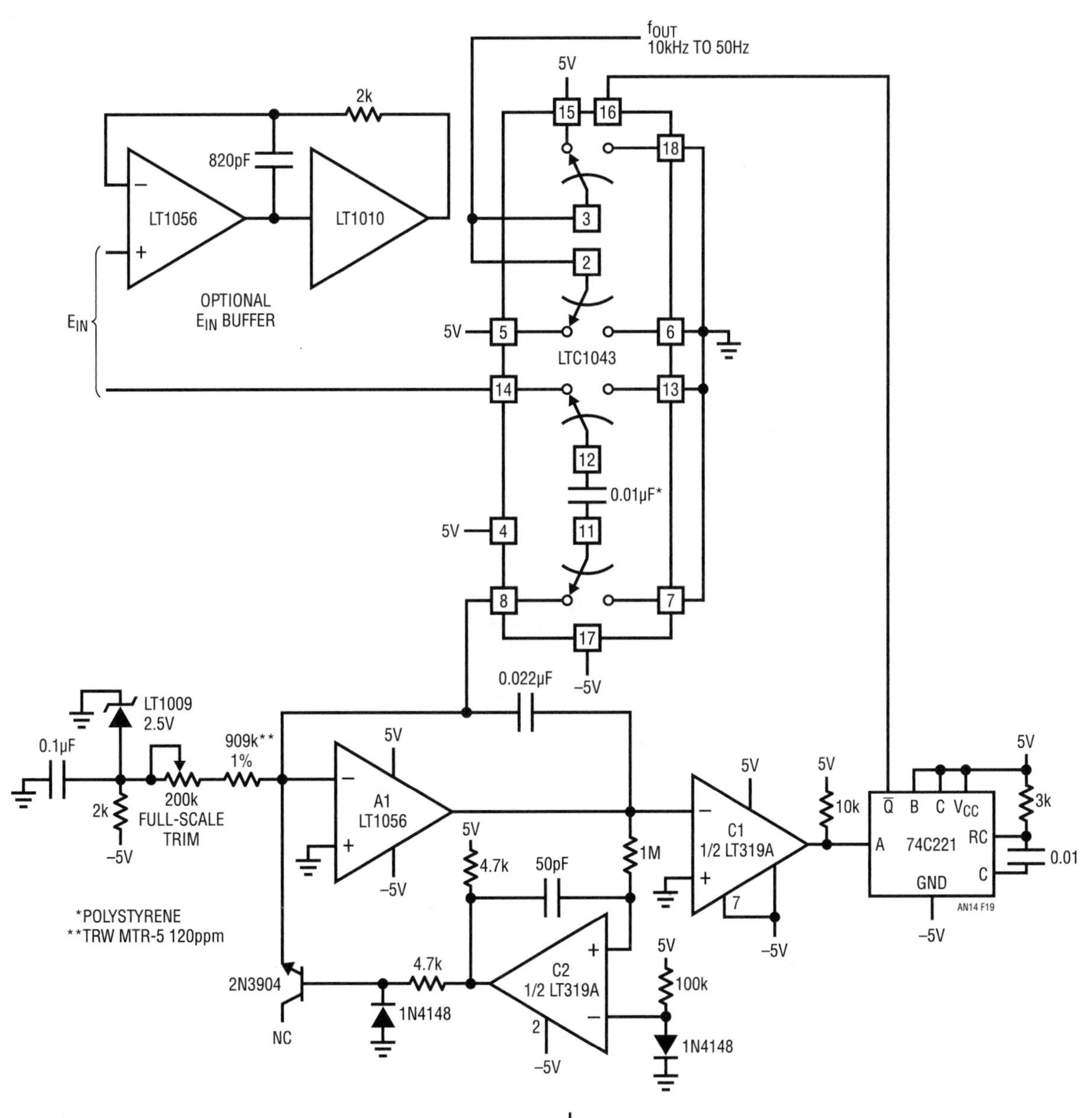

Figure 18.19 • Charge Pump $\frac{1}{E_{IN}}$ → Frequency Converter

the one-shot. The one-shot output (Trace C) toggles the LTC1043 switch, transferring charge from E_{IN} to A1's summing point via the 0.01µF capacitor (Trace D). This forces A1's output negative by an amount related to the charge transferred. When charge transfer ceases, A1 again ramps positively. The depth of A1's negative excursion is directly proportional to E_{IN}, hence loop oscillation frequency is inversely (1/X) related to E_{IN}.

The circuit's output is taken from the paralleled LTC1043 switch sections.

Because this circuit relies on charge feedback, integrator reset time does not influence accuracy. The loop runs at whatever frequency is required to maintain A1's summing point at zero.

If A1's output ever overruns 0V, the oscillator loop will latch. This condition is detected by C2, which goes high, driving current into A1's summing point via the low leakage 2N3904 BE junction. A1's output is forced negative, and normal circuit operation commences.

This circuit's primary disadvantage is that the input signal must be capable of supplying substantial current each time the LTC1043 commutates the 0.01µF capacitor to A1's summing point. The current required varies directly with input voltage, with 25mA drawn at E_{IN} = 5V. The optional input buffer shown will provide the necessary drive, although input voltage range must fall within the buffer's common mode limits.

To calibrate this circuit, apply exactly 5V and trim the 200kΩ potentiometer for 50Hz output.

E^X transfer function V→F converter

Figure 18.21's V→F circuit responds exponentially to its input voltage. It is ideally suited to electronic music synthesizers and, as shown, has a 1V in/octave of frequency out scale factor. Exponential conformity is within 0.13% over a 10Hz to 20kHz range and drift is 150ppm/°C. The circuit has a pulse output and also provides a ramp output for applications which require substantial power at the fundamental frequency.

A1's 1µF input capacitor integrates current from Q4's emitter, forming a ramp at A1's input (Trace A, Figure 18.22). When the ramp crosses zero, A1's output flips

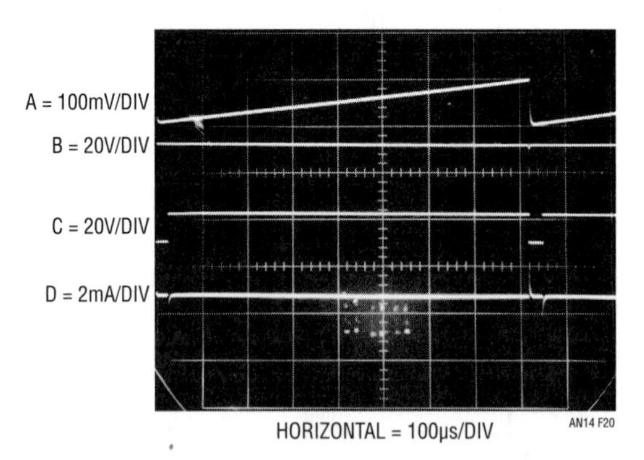

Figure 18.20 • Charge Pump-Based $\frac{1}{E_{IN}}$ → Frequency Converter Waveforms

(Trace B, Figure 18.22), causing the LTC1043 to change states. The 0.0012μF capacitor, charged to the LT1021's 10V potential, is switched to pull current from A1's summing point (Trace C). The 30pF capacitor provides A1's positive input with positive AC feedback (Trace D), insuring enough time for a complete discharge of the 0.0012μF unit. This action forces A1's input ramp to go in a negative direction, resetting it toward zero. When the positive AC feedback around A1 decays, the cycle repeats. Q5 and its associated components form a start-up loop, insuring proper circuit start sequence. Start-up conditions or input overdrive could force A1's output to go to the negative rail and stay there. If this occurs, Q5 comes on, pulling A1's negative input toward −15V and initializing normal circuit operation.

The oscillation frequency of this charge pump class current-to-frequency converter is linearly related to Q4's emitter current. Q4's emitter current, in turn, is exponentially

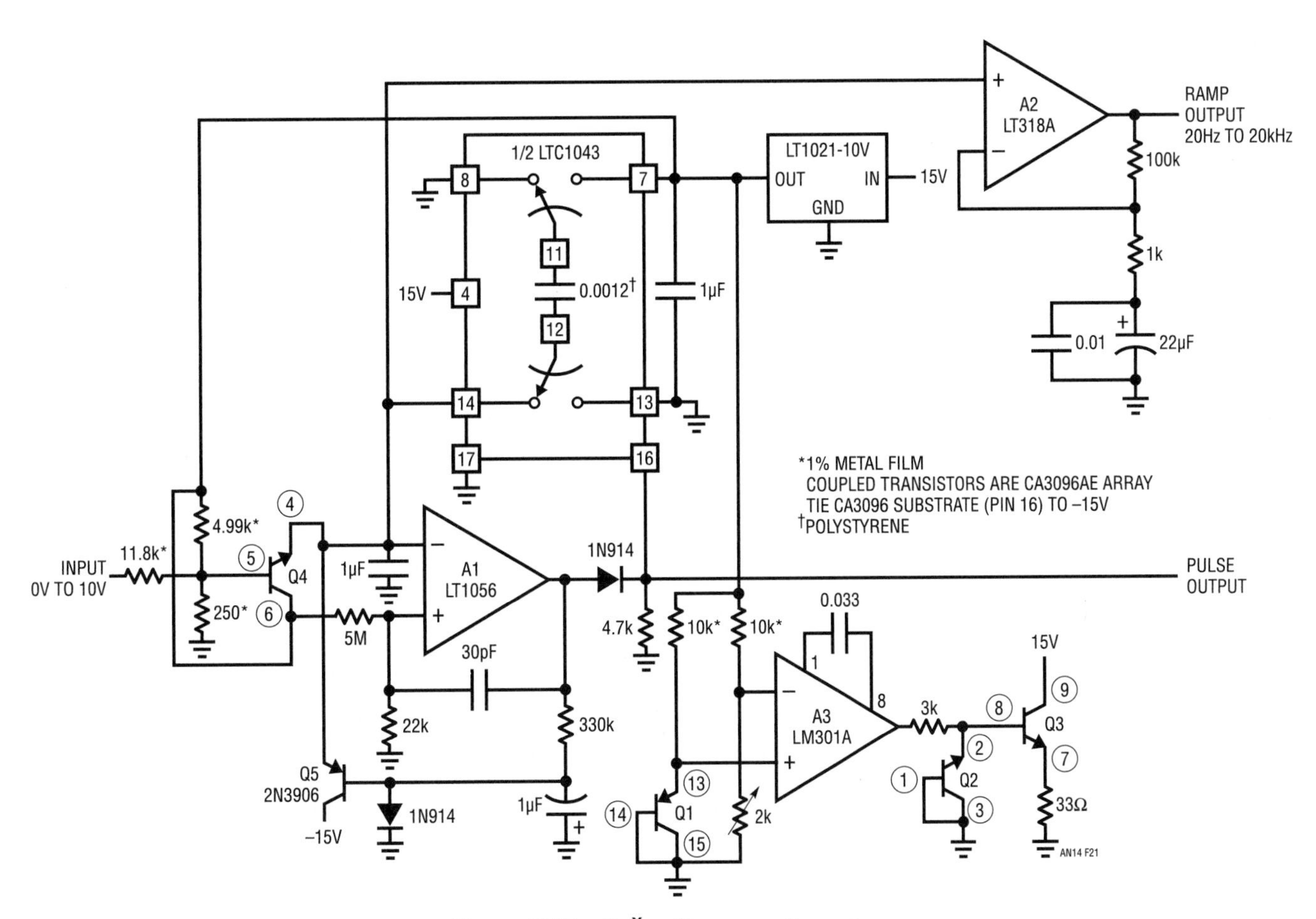

Figure 18.21 • E_{IN}^{X} → Frequency Converter

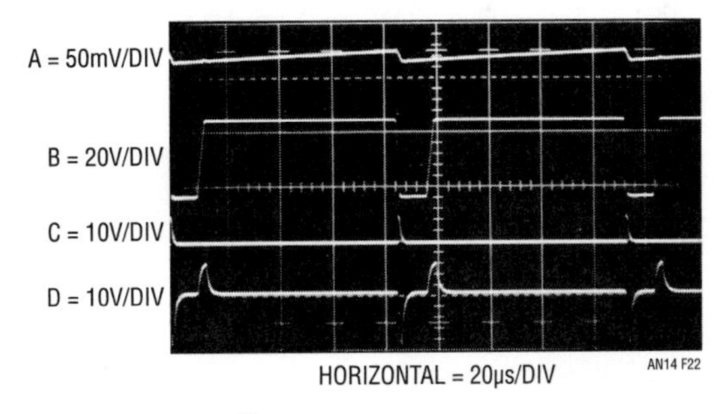

Figure 18.22 • E_{IN}^{X} → Frequency Converter Waveforms

related to its V_{BE}, which is determined by the resistors connected to it and the input voltage. This is in accordance with the well-known relationship between collector current and V_{BE} in transistors. Normally Q4's operating point would be quite sensitive to temperature, but it is part of an array which is temperature-stabilized by the A3 configuration. Q1, also part of the array, senses temperature. A3 compares Q1's V_{BE} with a bridge potential and drives array transistor Q3 to close a thermal control loop. This stabilizes the array, preventing ambient temperature shifts from influencing Q4's operation. Q2, serving as a clamp, ensures against loop lock-up conditions and prevents Q3 from ever becoming reverse biased.

With the thermal loop controlling Q4, the circuit's exponential behavior is stable and repeatable. The 5MΩ value from Q4's collector to A1's positive input introduces a slight shift in A1's operating point at high frequencies (e.g., high Q4 collector currents). This compensates Q4's bulk emitter resistance term, maintaining good exponential performance up to 20kHz. The 4.99k resistor sets the 0V input frequency at about 10Hz, while the 250Ω value establishes circuit k factor, nominally 1V in/octave output, as shown.

To use this circuit, adjust the 2k potentiometer so that A3's negative input is 100mV below its positive input with Q3's base grounded. Next, underground Q3's base and the circuit is ready for use.

$\frac{R1}{R2} = \frac{V1}{V2}$ →frequency converter

Figure 18.23's circuit produces an output frequency proportional to the ratio of the voltages across two externally supplied resistors. This circuit has wide application in transducer signal conditioning. Both R1 and R2 are ground referred, preferable for noise considerations. In this case, R1 is a Platinum resistance sensor, with R2 being set at the sensor's 0°C value. The grounded end of R2 allows fine trimming with decade boxes without excessive noise problems. R1's grounded side allows it to be located at the end of a cable run, with similar noise rejection properties.

The 6012 DAC serves as a simple source of two identical currents. The DAC's MSB is set high and all other bits are low. This sets the DAC's output currents equal. With constant, equal, currents through them, R1 and R2 produce a differential voltage which is sampled by the LTC1043 switch-capacitor configuration. The LTC1043's internal clock continuously switches the 3900pF capacitor across the R1-R2 pair and then dumps the charge into A1's summing point. The quantity of charge delivered per cycle is a direct function of the voltage difference across R1 and R2 ($Q = CV$). A1's output ramps (Trace A, Figure 18.24) negative. The ramp is compared to A2's output at C1. A2's DC output is a function of the 330pF charge pump capacitor at the LTC1043, A2's feedback resistor and the LTC1043 clock frequency. Because A1 and A2 are receiving charge at the same rate, LTC1043 oscillator drift affects each equally and does not contribute error.

When A1's ramp crosses A2's output value, C1 goes high (Trace B), turning on the FET. AC positive feedback to C1's positive input (Trace C) ensures a complete discharge for A1's feedback capacitor. When the feedback ceases, the cycle repeats. The oscillation frequency is a linear function of the R1-R2 ratio.

The two polystyrene capacitors at the LTC1043 provide temperature coefficient cancellation. A2's specified feedback resistor compensates A1's polystyrene feedback capacitor. Overall circuit tempco is about 35ppm/°C. As shown, a 0°C to 100°C excursion at the R1 sensor gives a 0kHz to 1kHz output with an accuracy, limited by the sensor, of 0.35°C. This is well outside the dead time error produced by A1's reset time and the circuit contributes no appreciable measurement error. In practice, slight trimming of R2's value may be required to compensate for individual R1 tolerances at 0°. The 5k potentiometer trims for 1kHz out at a 100°C R1 temperature. This circuit may be used with any resistive based transducer. For negative tempco devices, reverse the positions of R1 and R2.

References

1. "Trigger Circuit" Model 2235 Oscilloscope Service Manual, Tektronix, Inc
2. Pease R. A., "A New Ultra-Linear Voltage-to-Frequency Converter," 1973 NEREM Record, Vol. I, page 167
3. Pease R. A., assignee to Teledyne, "Amplitude to Frequency Converter," U.S. patent 3, 746, 968, filed September 1972
4. Williams, J., "Low Cost A→D Conversion Uses Single-Slope Techniques," EDN, August 5, 1978, pages 101–104
5. Gilbert B., "A Versatile Monolithic Voltage-to-Frequency Converter," IEEE J. Solid State Circuits, Volume SC-11, pages 852–864, December 1976
6. Williams, J., "Applications Considerations and Circuits for a New Chopper Stabilized Op Amp," 1Hz-30MHz V→F, pages 14–15, Linear Technology Corporation, Application Note 9

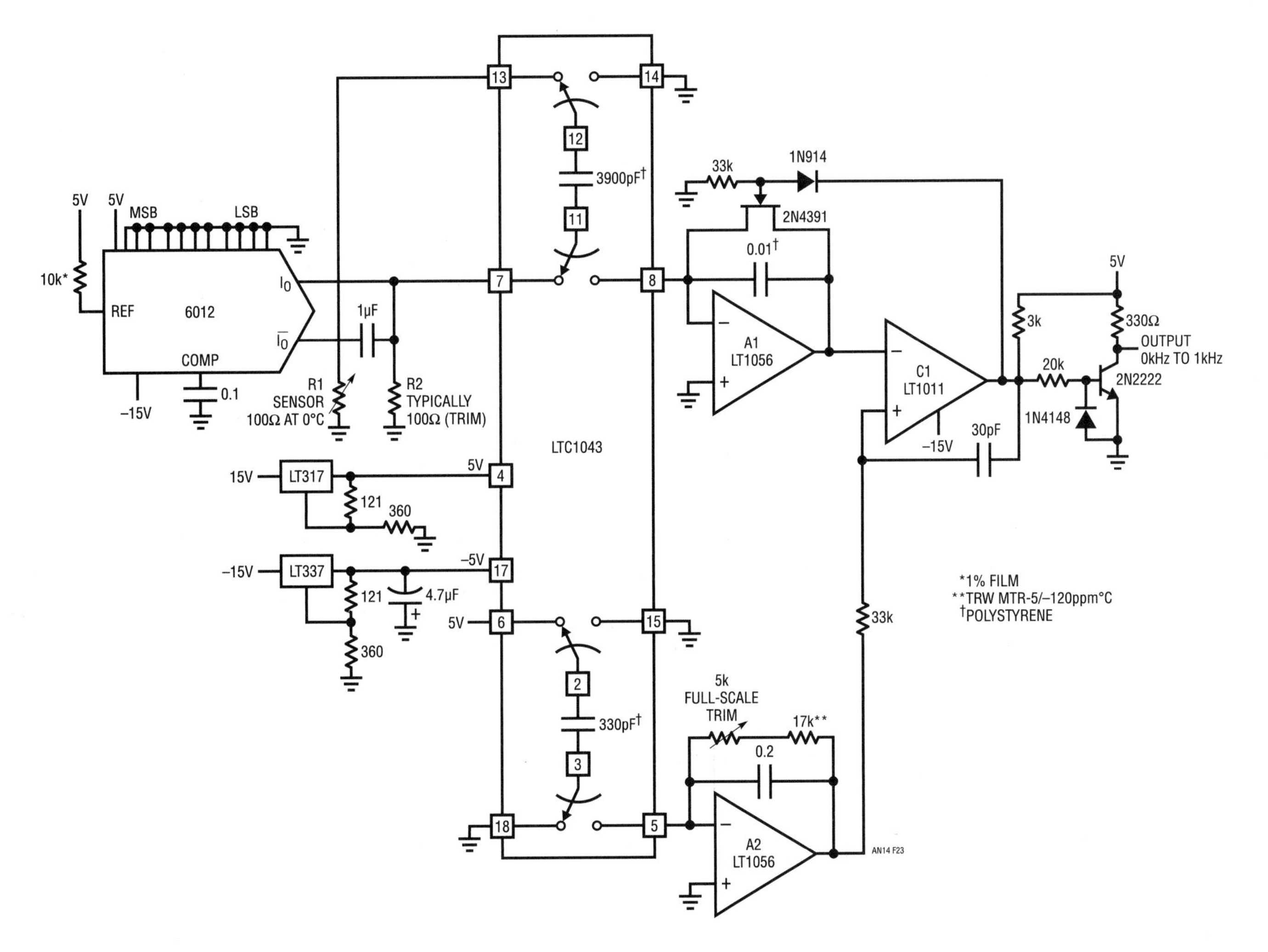

Figure 18.23 • $\frac{R1}{R2} = \frac{V1}{V2}$**→Frequency Converter**

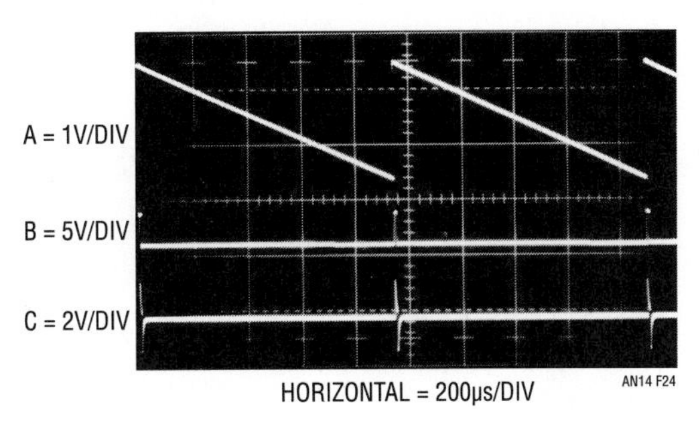

Figure 18.24 • $\frac{R1}{R2} = \frac{V1}{V2}$**→Frequency Converter Waveforms**

Box Section

V→F techniques

There are many ways to convert a voltage to a frequency. The best approach in an application varies with desired precision, speed, response time, dynamic range and other considerations. Figure B1 shows one of the most obvious. The input drives an integrator. The integrator's ramp slope varies with the input-derived current. When the ramp crosses V_{REF}, the comparator turns on the switch, discharging the capacitor and reinitializing the cycle. The frequency of this action directly relates to input voltage. With careful design, one op amp can serve as both integrator and comparator, providing circuit economy.

A serious drawback to this approach is the capacitor's discharge-reset time. This time, "lost" in the integration, 5796 results in significant linearity error as operating frequency approaches it. For example, a 1µs reset interval introduces 0.1% error at 1kHz, rising to 1% at 10kHz. Also, variations in reset time contribute additional errors. Because of this, circuit operation is restricted to relatively low frequencies if good linearity and stability are required. Although various compensation methods can reduce these errors, performance is still limited.

Figure B2 gets around B1's problems by enclosing the integrator in a charge-dispensing loop. In this approach C1 charges to V_{REF} during the integrator's ramping time. When the comparator trips, C1 is discharged into A1's summing point,

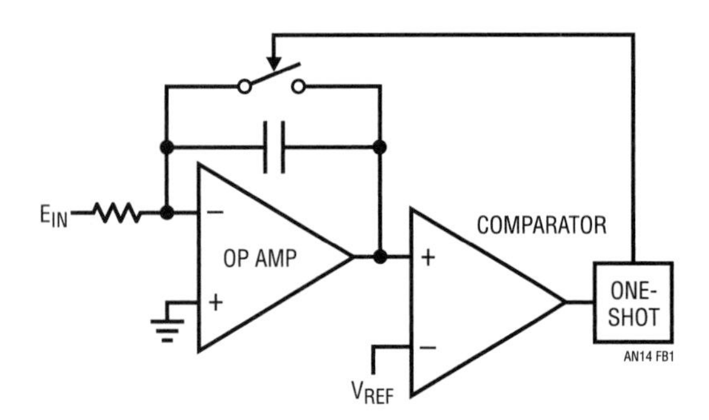

Figure B1 • Ramp-Comparator V→F

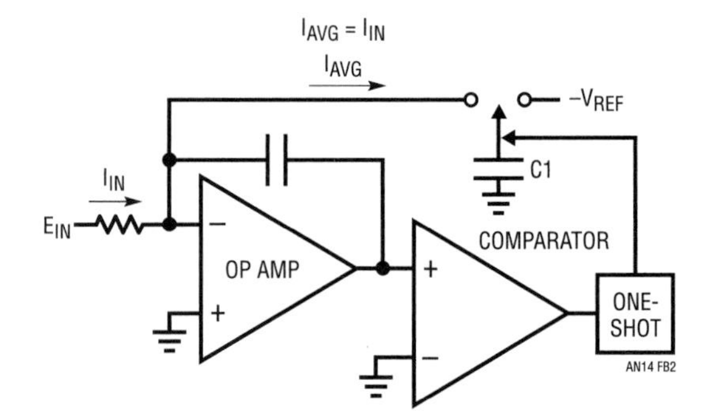

Figure B2 • Charge Pump V→F

forcing its output high. After C1's discharge, A1 begins to ramp and the cycle repeats.

Because the loop acts to force the average summing currents to zero, integrator time constant and reset time do not affect frequency. This approach yields high linearity (typically 0.01%) up to high frequencies. With attention to design, converters of this type can be constructed with a single op amp.

Figure B3 is conceptually similar, except that it uses feedback current instead of charge to maintain the op amp's summing point. Each time the op amp's output trips the comparator, the current sink pulls current from the summing point. Current is pulled from the summing point for the timing reference's duration, forcing the integrator positive. At the end of the current sink's period, the integrators output again heads negative. The frequency of this action is input-related.

Figure B4 uses DC loop correction. This arrangement offers all the advantages of charge and current balancing except that response time is slower. Additionally, it can achieve exceptionally high linearity (0.001%), output speeds exceeding 100MHz and very wide dynamic range (160dB). The DC amplifier controls a relatively crude V→F. This V→F is designed for high speed and wide dynamic range at the expense of linearity and thermal stability. The circuit's output switches a charge pump whose output, integrated to DC, is compared to the input voltage.

The DC amplifier forces V→F operating frequency to be a direct function of input voltage. The DC amplifier's frequency compensation capacitor, required because of loop delays, limits loop response time. Figure B5 is similar, except that the charge pump is replaced by digital counters, a quartz time base and a DAC. Although it is not immediately obvious, this circuit's resolution is not restricted by the DAC's quantizing

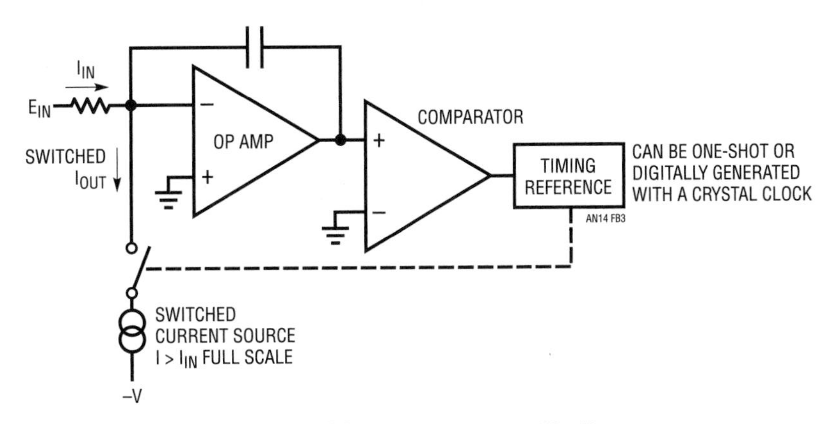

Figure B3 • Current Balance V→F

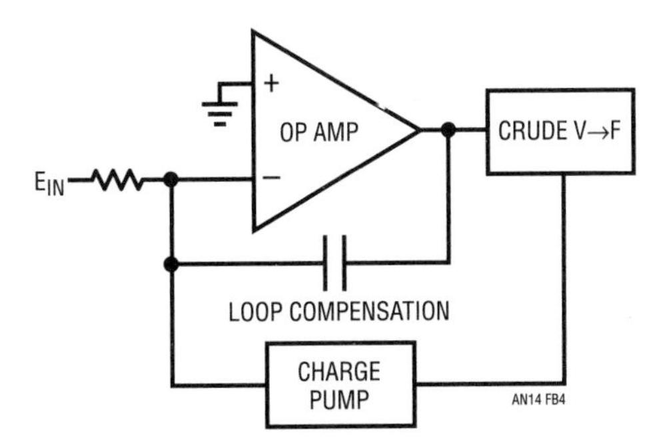

Figure B4 • Loop-Charge Pump V→F

limitations. The loop forces the DAC's LSB to oscillate around the ideal value. These oscillations are integrated to DC in the loop compensation capacitor. Hence, the circuit will track input shifts much smaller than a DAC LSB. Typically, a 12-bit DAC (4096 steps) will yield 1 part in 50,000 resolution. Circuit linearity, however, is set by the DAC's specification. An example of this approach appears in AN-13, "High Speed Comparator Techniques".

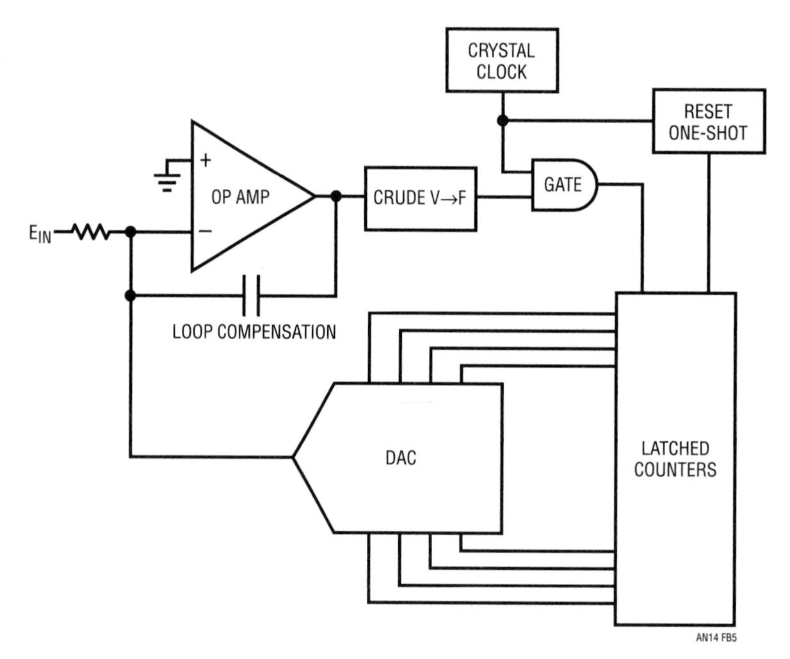

Figure B5 • Loop-DAC V→F

Unique IC buffer enhances op amp designs, tames fast amplifiers

19

Robert J. Widlar

Introduction

An output buffer can do much more than increase the output swing of an op amp. It can also eliminate ringing with large capacitive loads. Fast buffers can improve the performance of high speed followers, integrators and sample/hold circuits, while at the same time making them much easier to work with.

Interest in buffers has been low because a reasonably priced, high performance, general purpose part has not been available. Ideally, a buffer should be fast, have no crossover distortion and drive a lot of current with large output swing. At the same time, the buffer should not eat much power, drive all capacitive loads without stability problems and cost about the same as the op amps it is used with. Naturally, current limiting and thermal overload protection would be nice.

These goals have been a dream for twenty years; but thanks to some new IC design techniques, they have finally been reached. A truly general purpose buffer has been made that is faster than most op amps but not hard to use in slow applications. It is manufactured using standard bipolar processing, and die size is 50 × 82 mils.

The electrical characteristics of the buffer are summarized in Table 19.1. Offset voltage and bias current win no medals; but the buffer will usually be driven from an op amp output and put within the feedback loop, virtually eliminating these terms as errors. Loaded voltage gain is mostly determined by the output resistance. Again, any error is much reduced with the buffer inside a feedback loop.

Unloaded, the output swings within a volt of the positive supply and almost to the negative rail. With ±150mA loading, this saturation voltage increases by 2.2V. Except for output voltage swing, performance is little affected for a total supply voltage between 4V and 40V. This means that it can be powered by a single 5V logic supply or ±20V op amp supplies.

Bandwidth and slew rate decrease somewhat with reduced load resistance. The values given in Table19.1 are for a 100Ω in parallel with 100pF. The speed is quite impressive considering that quiescent current is but 5mA.

Table 19.1 Typical Performance of the Buffer at 25°C. Supply Voltage Range is 4V to 40V

PARAMETER	VALUE
Output Offset Voltage	70mV
Input Bias Current	75μA
Voltage Gain	0.999
Output Resistance	7Ω
Positive Saturation Voltage	0.9V
Negative Saturation Voltage	0.1V
Output Saturation Resistance	15Ω
Peak Output Current	±300mA
Bandwidth	22MHz
Slew Rate	100V/μs
Supply Current	5mA

Analog Circuit and System Design: Immersion in the Black Art of Analog Design. http://dx.doi.org/10.1016/B978-0-12-397888-2.00019-5

Design concept

The functional schematic in Figure 19.1 describes the basic elements of the buffer design. The op amp drives the output sink transistor, Q30, such that the collector current of the output follower, Q29, never drops below the quiescent value (determined by I_1 and the area ratio of Q12 and Q28). As a result, the high frequency response is essentially that of a simple follower even when Q30 is supplying the load current. The internal feedback loop is isolated from the effects of capacitive loading by a small resistor in the output lead.

The scheme is not perfect in that the rate of rise of sink current is noticeably less than for source current. This can be mitigated by connecting a resistor between the bias terminal and V^+, raising quiescent current. A feature of the final design is that the output resistance is largely independent of the follower current, giving low output resistance at low quiescent current. The output will swing to the negative rail, which is particularly useful with single-supply operation.

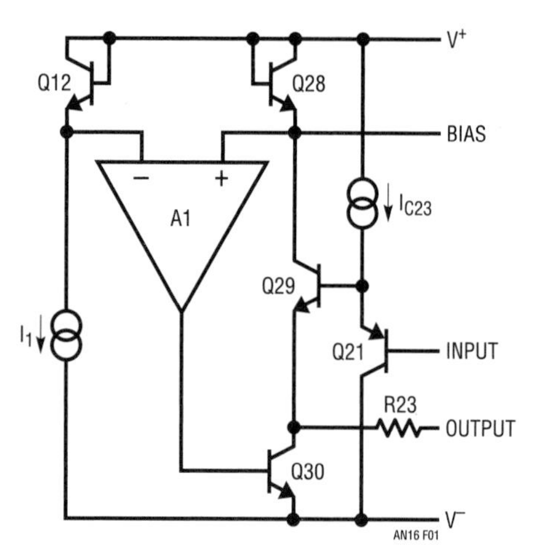

Figure 19.1 • In the Buffer, Main Signal Path Is Through Followers Q21 and Q29. Op Amp Keeps Q29 Turned on Even When Q30 Is Supplying Load Current, So Response Is That of Followers

Basic design

Figure 19.2 shows the essential details of the buffer design using the concept in Figure 19.1 (for clarity, parts common to simplified and developed schematics use the same number). The op amp uses a common base PNP pair, Q10 and Q11, degenerated with R6 and R7 for an input stage. The differential output is converted to single-ended by a current mirror, Q13 and Q14; and this drives the output sink transistor, Q30, through a follower, Q19.

A clamp, Q15, is included to insure that the output sink transistor does not turn off completely. Its biasing circuitry Q6 through Q9, is arranged such that the emitter current of Q15 is about equal to the base current of Q19 with no output load.

The control loop is stabilized with a feedforward capacitor, C1. Above 2MHz, feedback is predominantly through the capacitor. The break frequency is determined by C1 and R7 plus the emitter resistance of Q11. The loop is made stable for capacitive and resonant loading by R23, which limits the phase lag that can be induced at the emitter of Q29.

A resistor, R10, has been added to improve the negative slew response. With a large negative transient, Q29 will cut off. When this happens, R10 pulls stored charge from Q28 and provides enough voltage swing to get Q30 from its clamp level into conduction.

Start-up biasing is done with a collector FET, Q4. Once in operation, the collector current of Q6 is added to the drain current of Q4 to bias Q5. These currents plus the current through Q9 and Q10 flow through Q12 to set the output quiescent current (along with R10).

Follower boost

The boost circuit in Figure 19.3 reduces the buffer standby current by at least a factor of three while improving performance. It does this by increasing the effective current gain of Q29 so that the current source current I_{C23}, can be drastically cut. Secondly, it can give under 0.5Ω follower output resistance at less than 3mA bias, something that normally takes over 40mA. Hard as it may be to believe, the boost does not degrade the high frequency response of the final design.

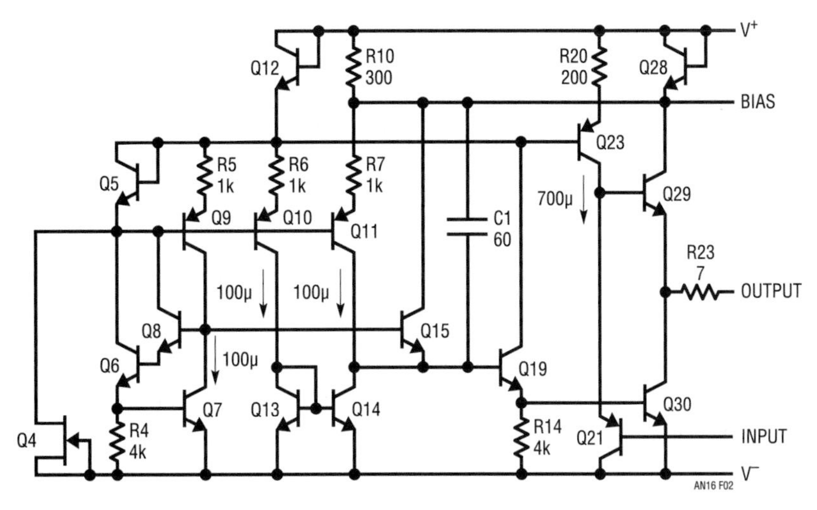

Figure 19.2 • Implementation of the Buffer in Figure 19.1. Simple Op Amp Uses Common Base PNP Input Transistors (Q10 and Q11). Control Loop Is Stabilized with Feedforward Capacitor (C1); and Clamp (Q15) Keeps Q30 from Turning Off Entirely

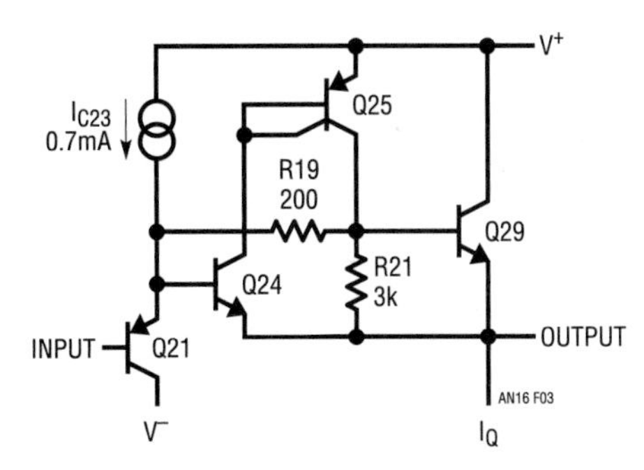

Figure 19.3 • This Boost Circuit Raises Effective Current Gain and Transconductance of the Output Transistor, Giving Low Standby Current Along with Low Output Resistance

If R19 is removed (opened), circuit operation becomes clearer. Output resistance is determined by Q24, with Q25 and Q29 providing current gain. If the current through R21 is larger than the base current of Q29, output resistance is proportionately reduced. Without R21, output resistance depends on Q29 bias, like a simple follower.

The purpose of R19 is to provide a direct AC path at high frequencies and kill unneeded gain in the boost feedback loop. If R21 is properly selected, voltage change across R19 with loading is less than 40mV, so a small value causes no problems (increasing load does cause Q21 bias current to increase). The quiescent drop across R19 is set by sizing Q24, Q25, and Q29 geometries.

Charge storage PNP

At high frequencies, a lateral PNP looks like a low impedance between the base and emitter because charge stored between the emitter and subcollector (the PNP base) has a capacitive effect. The input PNP, Q21, has been designed to have more than 30 times the stored charge of a standard lateral for a given emitter current. This stored charge couples in the input to slew internal stray capacitances and drive the output follower while the boost circuitry is coming into action.

Stored charge can be maximized in a lateral PNP by using large emitter area and wide base spacing. Dimensions of several mils are practical; diffusion lengths are in the order of 6 mils with good processing.

A sketch of a charge storage PNP is shown in Figure 19.4. With the dimensions shown, current gains of 10 can be obtained regularly. A sinker base contact is shown here because a low resistance from the base terminal to the area under the emitter is important.

The charge stored *under* the emitter is most effective in obtaining a fast charge transfer from base to emitter with minimum change of emitter base voltage. Using the notation in Figure 19.4, this charge varies as:

$$Q_E \propto \frac{W_B A_E}{S_E}$$
$$\propto (X_C - X_E)X_E$$

where S_E is the emitter periphery. With X_C fixed, it can be shown that Q_E is maximized for $X_E = 0.5X_C$.

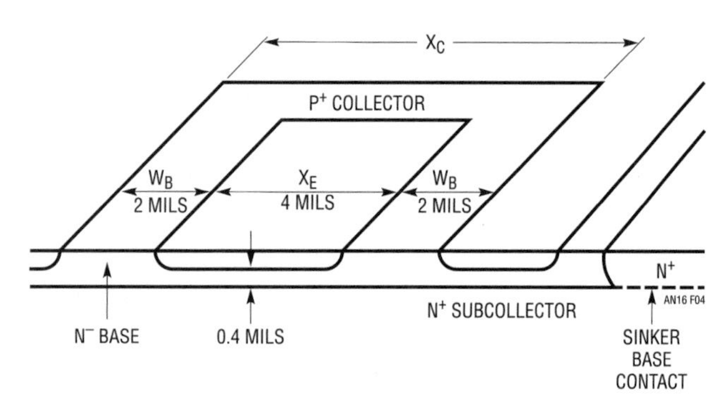

Figure 19.4 • Charge Storage PNP is Lateral Structure with Base and Emitter Dimensions of Several Mils. As Above, Current Gains of 10 are Practical

Isolation-base transistor

Transistors can be made by substituting an isolation diffusion for the normal base diffusion. Figure 19.5 shows the impurity profile of such a transistor. Base doping under the emitter is three orders of magnitude higher than standard transistors, and the base extends all the way to the subcollector. The measured current gains of 0.1 are not lower than might be expected.

The emitter-base voltage of an isolation-base transistor is about 120mV greater than a standard IC transistor when operating at the same emitter current. Production variations in V_{BE} are much less than standard NPNs, probably because net base doping is little affected by anything but the isolation doping.

As will be seen on the complete schematic, the isolation-base transistor is used as a bias diode for current sources because of its high V_{BE}. One (Q28) is also used in the collector of the output follower because the behavior at very high current densities is much better than a standard transistor.

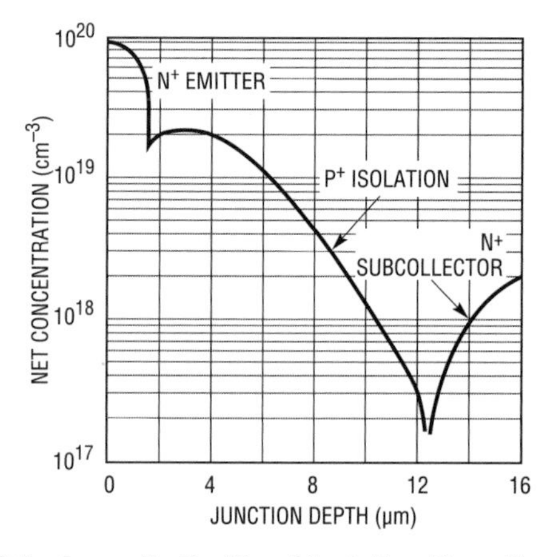

Figure 19.5 • Impurity Profile of Isolation-Base Transistor. In Contrast, Typical Standard NPN Has Peak Base Concentration of 5 x 10^{16} cm^{-3} and Base Width of 1µm

Complete circuit

A complete schematic of the LT1010 buffer is given in Figure 19.6. Component identification corresponds to the simplified schematics. All details discussed thus far have been integrated into the diagram.

Current limiting for the output follower is provided by Q22 and Q31, which serve to clamp the voltage into the follower boost circuitry when the voltage across R22 equals a diode drop.

Negative current limit is less conventional because putting a sense resistor in the emitter of Q30 will seriously degrade negative slew under load. Instead, the sense resistor, R17, is in the collector. When the drop across it turns on Q27, this transistor supplies current directly to the sink current control amplifier, limiting sink current.

Should the output terminal rise above V^+ because of some fault condition, Q27 can saturate, breaking the current limit loop. Should this happen, Q26 (a lateral collector near Q27 base) takes over to control current by removing sink drive through Q16. This reserve current limit oscillates, but in a controlled fashion.

Clamp diodes, from the output to each supply, should be used if the output can be driven beyond the supplies by a high-current source. Unlike most ICs, the LT1010 is designed so that ordinary junction diodes are effective even when the IC is much hotter than the external diodes.

Current limit is backed up by thermal overload protection. The thermal sensor is Q1, with its base biased near 400mV. When Q1 gets hot enough to pull base drive off Q2 (about 160°C), the collector of Q2 will rise, turning on Q16 and Q20. These two transistors then shut down the buffer. Including R2 generates hysteresis to control the frequency of thermal limit oscillation.

Base drive to Q20 is limited by R15, a pinched base resistor. The value of this resistor varies as transistor h_{fe} over temperature and in production, controlling the turn off current near 2mA. An emitter into the isolation wall capacitor, C2, keeps Q20 from turning on with fast signals on its collector.

In current limit or thermal limit, excessive input-output voltage might damage internal circuitry. To avoid this, back-to-back isolation Zeners, Q32 and Q33, clamp the input to the output. They are effective as long as the input current is limited to about 40mA.

Other details include the negative saturation clamp, Q17 and Q18. This clamp allows the output to saturate within 100mV of the negative supply rail without increasing supply current while recovering cleanly from saturation. The base of Q17 is connected internally into Q30 to sense voltage on the internal collector side of the saturation resistance to insure optimum operation at high currents.

When sinking large currents, the base of Q19 loads the control amplifier. This unbalances the control loop and reduces the output follower bias current. To compensate for this, the base current of Q30 is routed to the bias diode, Q12, through Q19. A small resistor, R19, aids compensation. This action raises the bias to Q23 and is responsible for increasing the input PNP bias current with sink current.

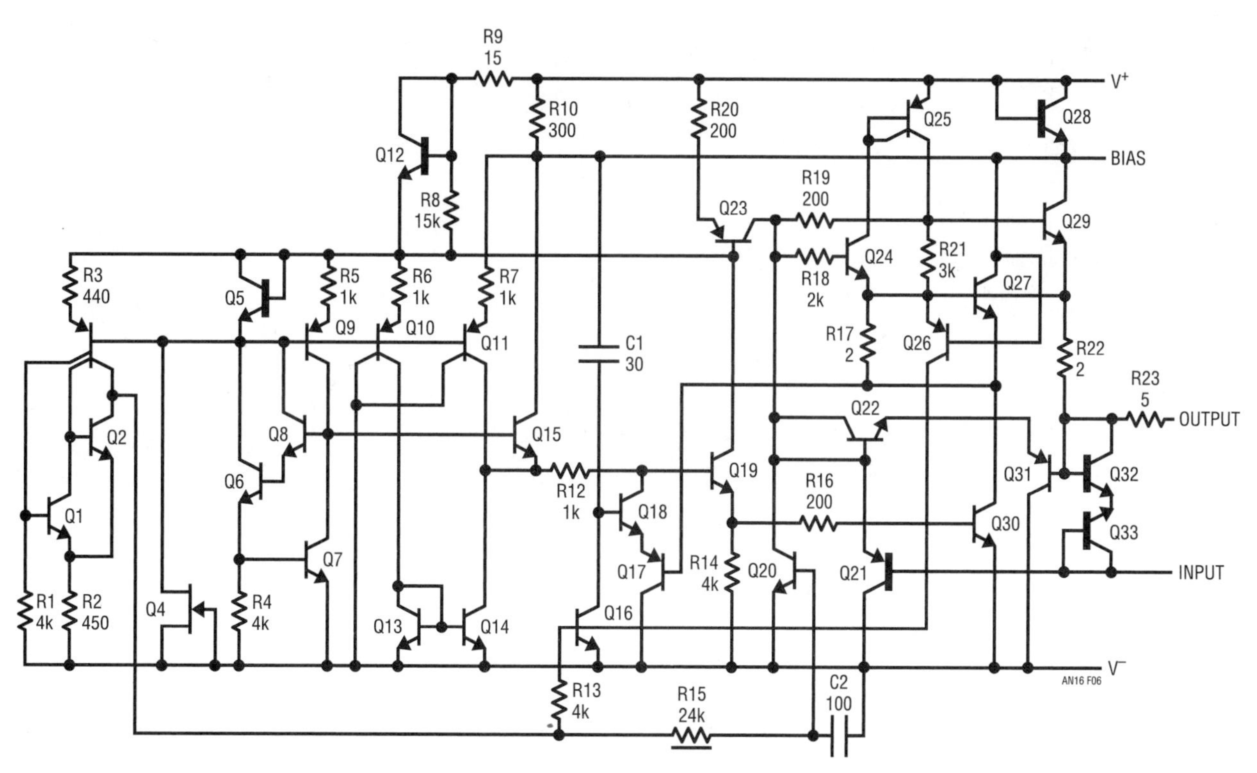

Figure 19.6 • Complete Schematic of the LT1010 Buffer. Component Identification Corresponds to Simplified Schematics. The Isolation-Base Transistors Are Drawn with Heavy Base, as Is the Charge Storage PNP. Follower Drive Boost Has Been Included Along with Negative Saturation Clamp (Q17 and Q18) and Protection Circuitry

Final details of the design are that the collectors of Q10 and Q11 are segmented so that only a fraction of the emitter current is sent to the current mirror, with the rest dumped to V^-. This allows the transistors to be operated at their f_T peak without requiring large C1. Lastly, R8 has been included to shape the temperature characteristics of output stage quiescent current.

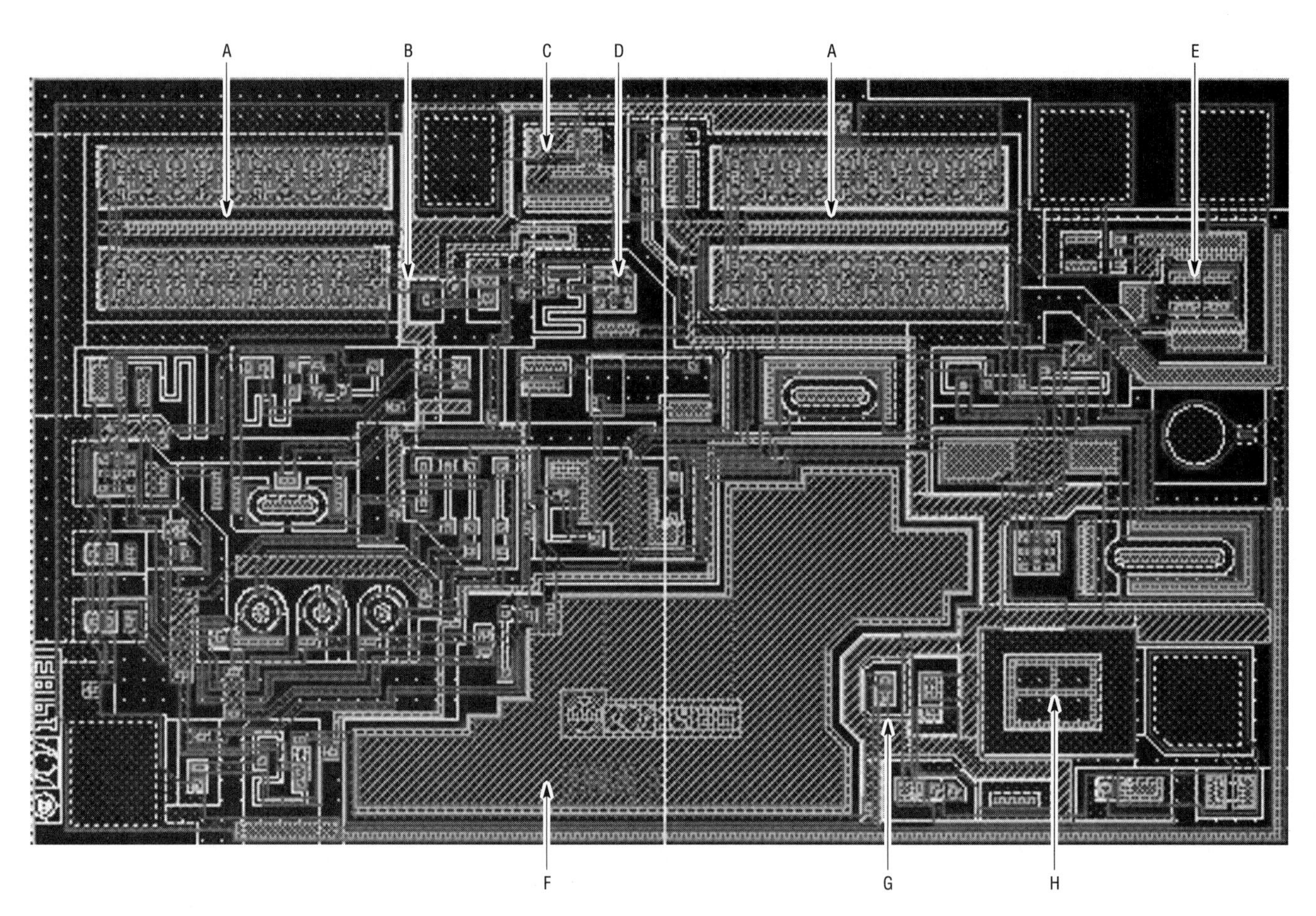

Figure 19.7 • Plot of the LT1010. Die Size Is 50 × 82 Mils

A photomicrograph of the LT1010 die is shown in Figure 19.7. The features pointed out are identified below.

A) Output transistors were designed to maximize high frequency performance, while obtaining some ballasting.
B) Clamp PNP base (Q17) is connected by subcollector stripe to region furthest from Q30 collector contact to isolate saturation resistance.
C) Output resistors are in floating tub so that IC tubs are not forward biased when junction diodes clamp output below V^-.
D) A high f_T, 0.3 mil stripe, cross geometry is used for the sink transistor driver (Q19).
E) Isolation-base transistor (Q28) carries the same 500mA peak current as the output transistor but is much smaller.
F) MOS capacitor (C1) takes up considerable area.
G) Capacitance formed by diffusing emitter into isolation wall takes advantage of unused area.
H) Charge storage PNP.

Buffer performance

Table19.1 in the Introduction summarizes the typical specifications of the LT1010 buffer. The IC is supplied in three standard power packages: the solid kovar base TO-5 (TO-39), the steel TO-3, and the plastic TO-220. The bias terminal is not available in the TO-39 package because it has only four leads, compared to five for the other packages.

The thermal resistance for one output transistor, excluding the package, is 20°C/W because it was kept as small as possible to enhance speed. This explains the junction-to-case thermal resistance of 40°C/W for the TO-39 package and 25°C/W for the TO-3 and TO-220, again for one transistor. With AC loads, both transistors will be conducting; if the frequency is high enough, thermal resistance is reduced by 10°C/W.

The operating case temperature range for the LT1010 is −55°C to 125°C. The maximum junction temperature for the internal power transistors is 150°C. A commercial version, the LT1010C, is also available. It rated for 0°C to 100°C case temperature with a maximum junction temperature of 125°C.

The following curves describe the buffer performance in some detail. The fact that quiescent current boost (5mA – 40mA) is not available on the TO-39 package should be noted.

Bandwidth

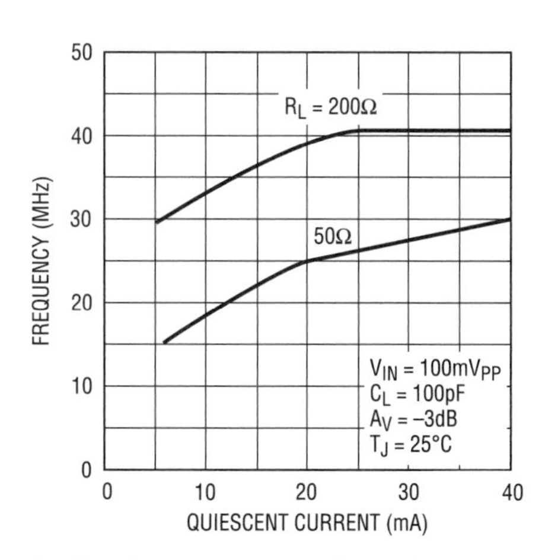

Figure 19.8 • The Dependence of Small Signal Bandwidth on Load Resistance and Quiescent Current Boost Is Shown Here. The 100pF Capacitive Load That Is Specified Limits the Bandwidth That Can Be Obtained with Boost and Light Loads

Phase delay

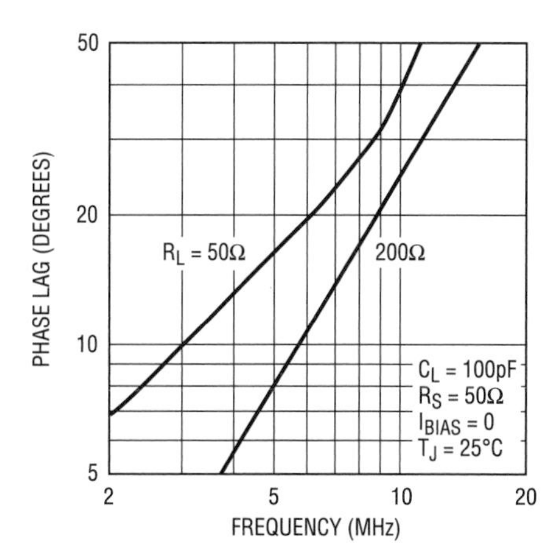

Figure 19.9 • The Phase Delay Gives More Useful Information About High Frequency Performance Than Bandwidth. This Is a Plot of Phase Delay as a Function of Frequency with 50Ω and 100Ω Loads. Capacitive Loading Is 100pF, and Quiescent Current Is Not Boosted

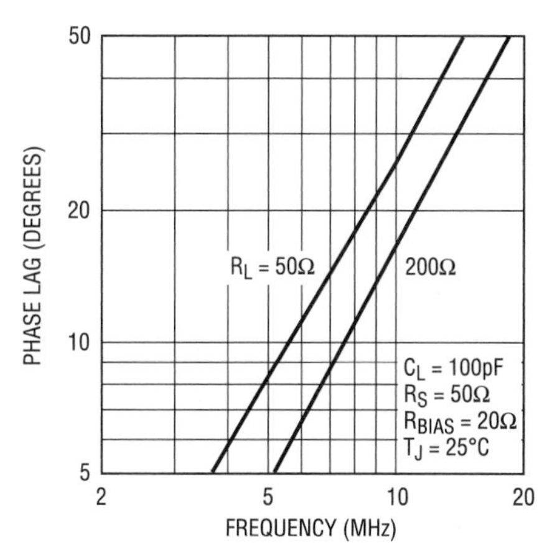

Figure 19.10 • This Shows Reduction in Phase Lag with Quiescent Current Boosted to 40mA (R_{BIAS} = 20Ω)

Step response

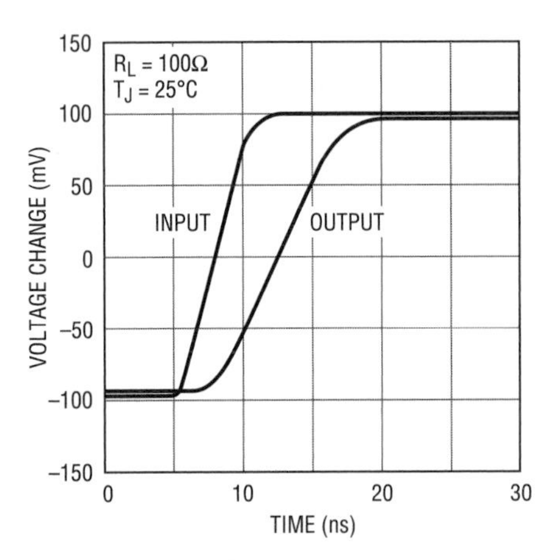

Figure 19.11 • The Small Signal Step Response with 100Ω Load Shows a 2ns Output Delay. This Gives an Excess Phase Delay of 15° at 20MHz, Explaining Why the –3dB Bandwidth Is Greater Than the Frequency for 45° Phase Delay

Capacitive loading

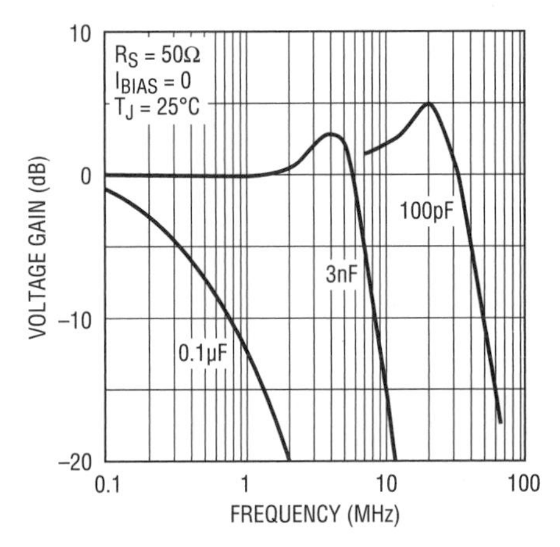

Figure 19.13 • These Frequency Response Plots, with Capacitive Load Only, Show That Nothing Unusual Happens as Load Capacitance Is Varied Over a Wide Range. Minor Peaking Is Reduced with Quiescent Current Boost

Output impedance

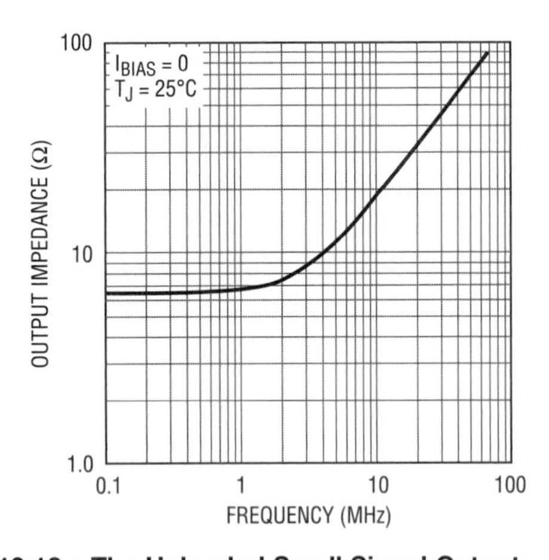

Figure 19.12 • The Unloaded Small Signal Output Impedance Stays Down to 1MHz, Indicating the Frequency Limit of the Follower Boost Circuitry

Slew response

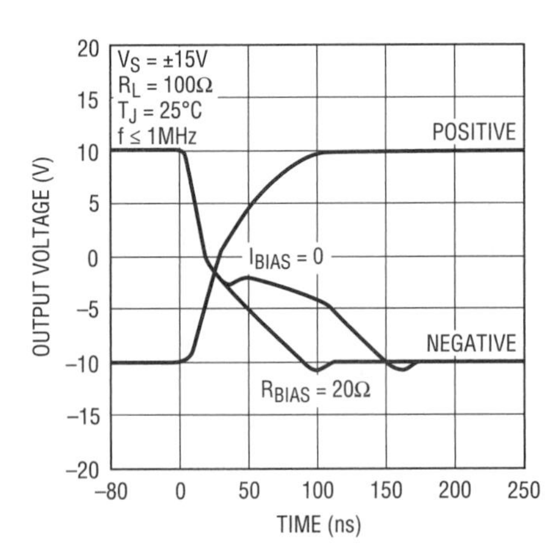

Figure 19.14 • The Negative Slew Delay Is Reduced by Using Quiescent Current Boost (40mA). Positive Slew Is Not Affected by Boost

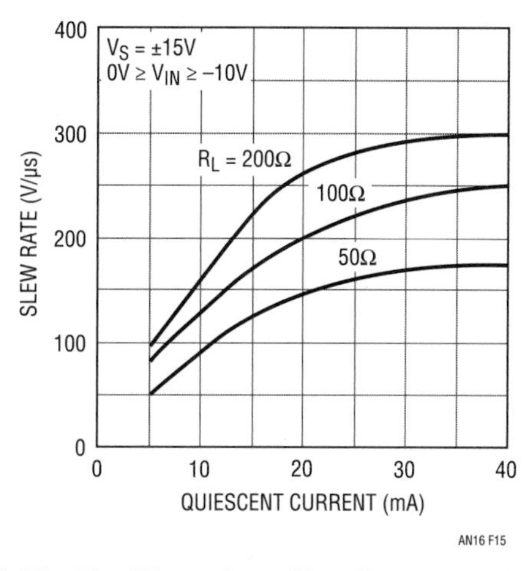

Figure 19.15 • The Worst-Case Slew Response, Going from 0V to −10V, Is Plotted Here. It Is Clear That Substantial Improvement Can be Made with Quiescent Current Boost

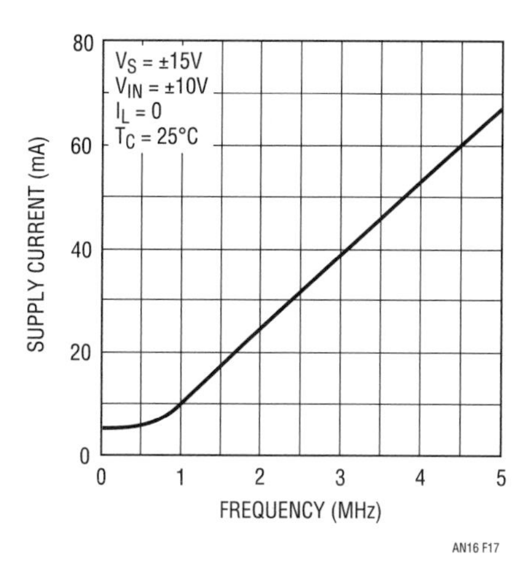

Figure 19.17 • The No Load Supply Current Increases Above 1MHz Under Large Signal Conditions. This Is a Quiescent Current Boost Caused by Charging of Internal Capacitances. It Does Give Very Good Power Bandwidth Even with Load, Although the Excess Dissipation May Cause the IC to Go into Power Limit

Input offset voltage

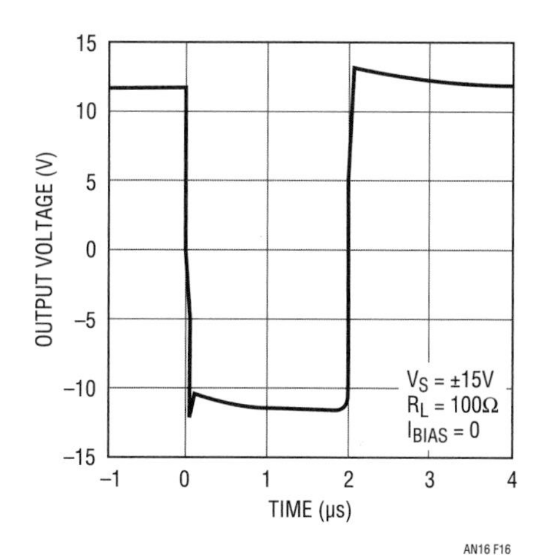

Figure 19.16 • This 500ns Slew Residue Is Caused by Recovery of the Follower Boost Circuitry. For Positive Outputs, the Boost Circuit Is Hit Hard by the Input Through the Charge Storage PNP. For Negative Outputs, it Is Hit by the Leading Edge Overshoot on the Output. Recovery Is from a Positive Boost Overshoot in Both Cases

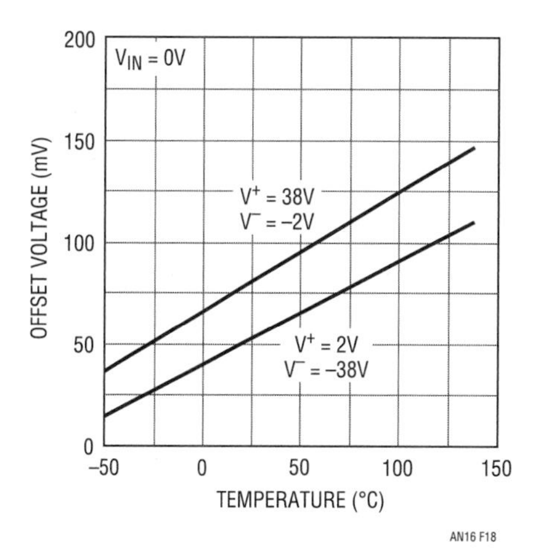

Figure 19.18 • The Offset Voltage is Determined by Matching Between the Output Follower and the Input PNP. The Charge Storage PNP on the Input Is Run at High Injection Levels to Maximize Stored Charge. Therefore, the High Offset Voltage Drift Shown Here Is No Surprise. The Offset Voltage Change with Supply Voltage Shown in the Figure Is Mostly Positive Supply Sensitivity. Changing the Negative Supply by 35V Shifts Offset by 5mV

Input bias current

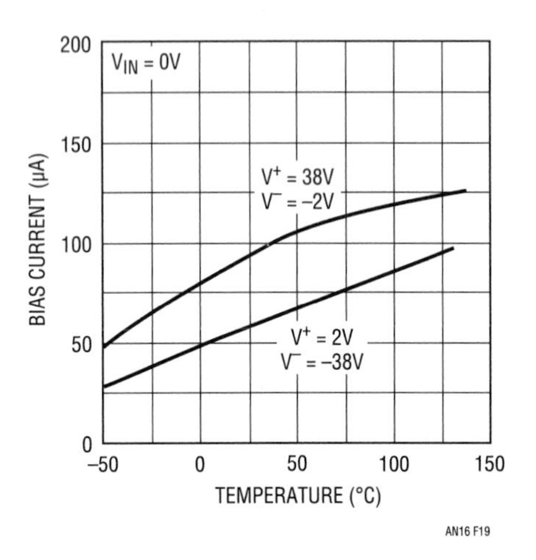

Figure 19.19 • The Increase in Bias Current with Temperature Reflects the Current Gain Characteristics of the Charge Storage PNP. Sensitivity of Bias Current to Supply Voltage Is About Three Times Greater on Positive Supply

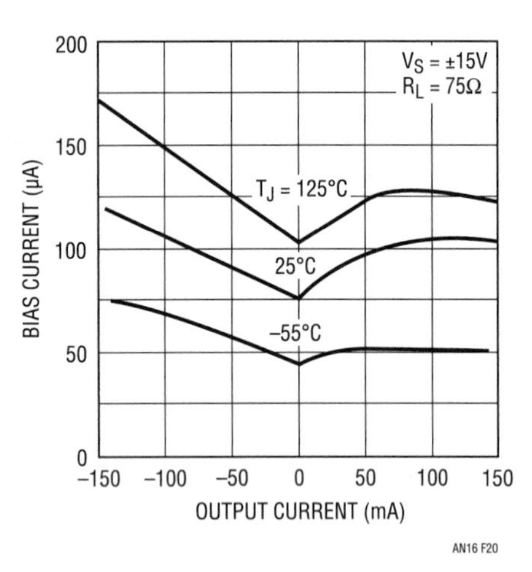

Figure 19.20 • The Change in Input Bias Current with Load Current Is Not Excessive, but it Shows That the Follower Is Not Designed for Working with High Source Resistances. For Positive Output Current, Increase Is Caused by Follower Boost. For Negative Output, It Results from Sink Transistor Base Current Increasing Bias to the Input PNP Current Source

Voltage gain

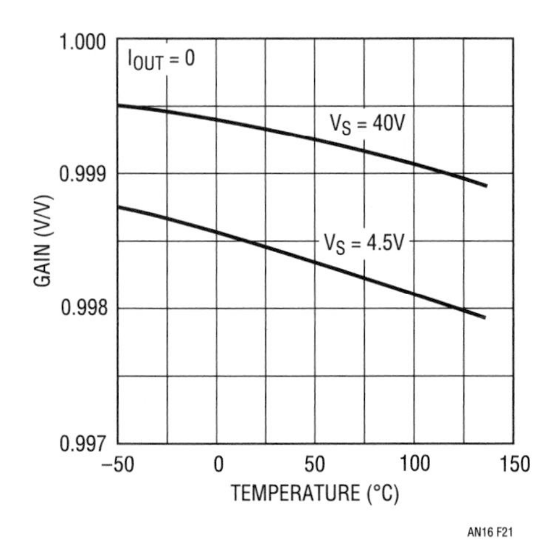

Figure 19.21 • The Unloaded Voltage Gain Is High Enough to Be Ignored in Most Any Application. In Practice, Gain Will Be Determined by the Load Working Against the Output Resistance

Output resistance

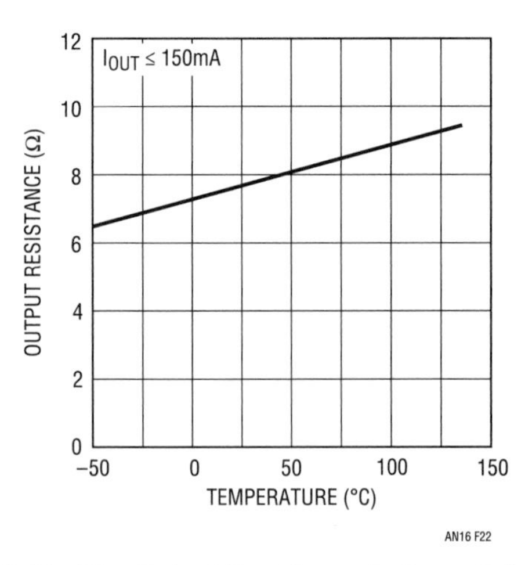

Figure 19.22 • The Output Resistance Is Essentially Independent of DC Output Loading. The Temperature Sensitivity Is Shown Here

Output noise voltage

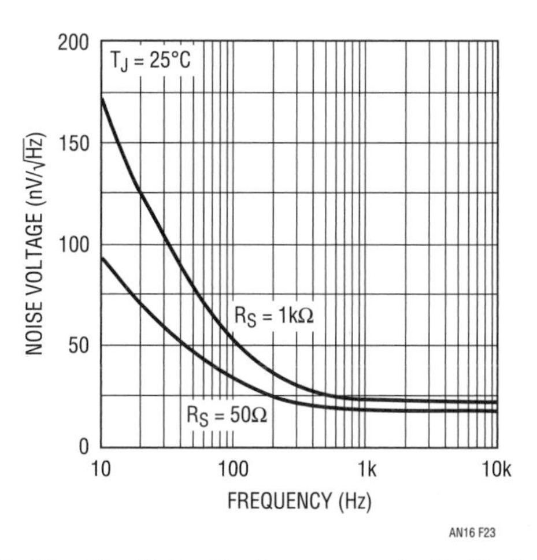

Figure 19.23 • The Noise Performance of a Buffer Is of Small Concern Unless it Is Grossly Bad. This Plot Shows That the Buffer Noise Is Low by Comparison to the Excess Output Noise of Op Amps

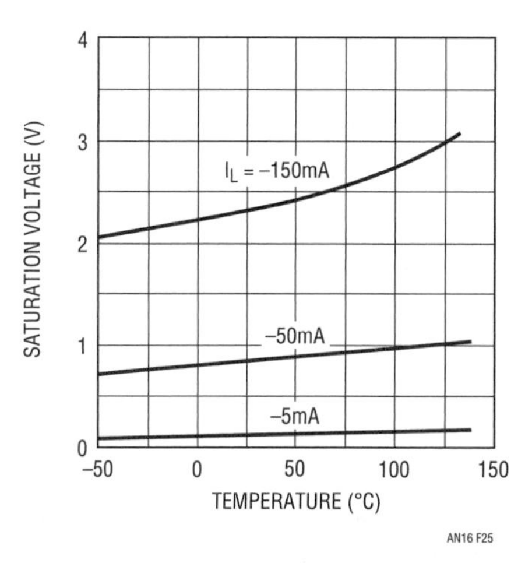

Figure 19.25 • This Curve Gives the Negative Saturation Voltage. Unloaded Saturation Voltage Is <0.1V, Again Increasing Linearly with Current. The Saturation Characteristics Are Negligibly Affected by Supply Voltage and Are Used to Determine Output Swing Under Load

Saturation voltage

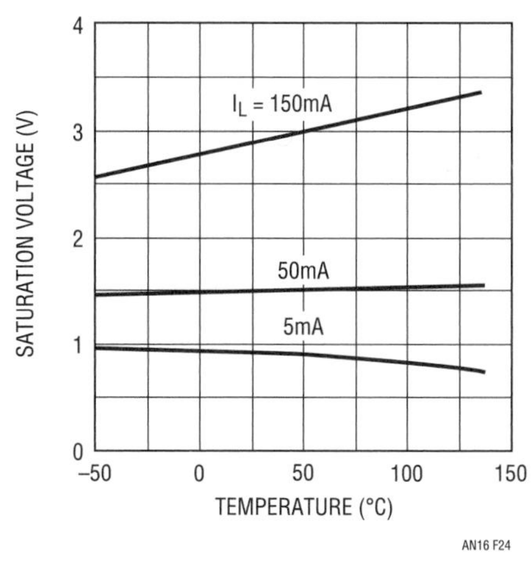

Figure 19.24 • The Positive Saturation Voltage (Referred to the Positive Supply) Is Plotted Here as a Function of Temperature. Unloaded Saturation Voltage Is 0.9V, with the Saturation Voltage Increasing Linearly with Current to 150mA

Supply current

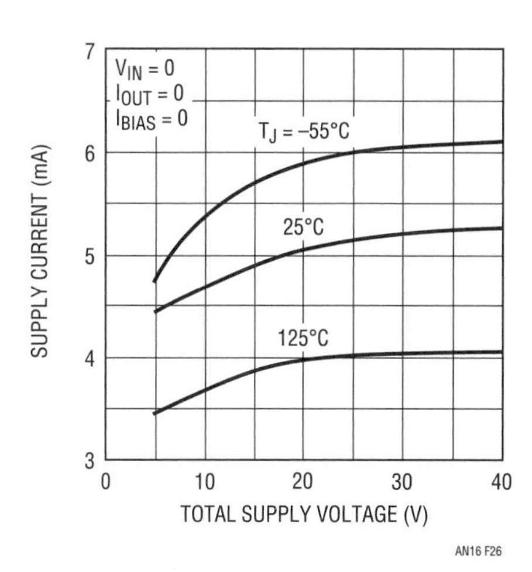

Figure 19.26 • Supply Current Is Not Greatly Affected by Supply Voltage, as Shown in This Expanded-Scale Plot. This Accounts for the 4V to 40V Supply Range with Unchanged Specifications

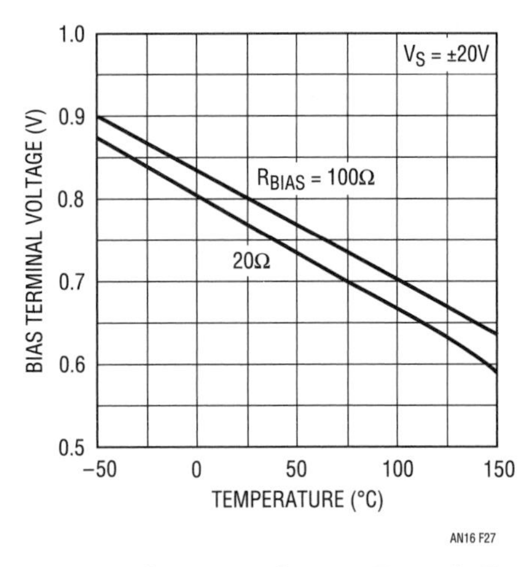

Figure 19.27 • The Quiescent Current Boost Is Determined by the Bias Terminal Voltage Across an External Resistor. This Expanded-Scale Plot Shows the Change in Bias Terminal Voltage with Temperature. The Voltage Increases Less Than 20mV as the Total Supply Voltage Is Raised from 4.5V to 40V

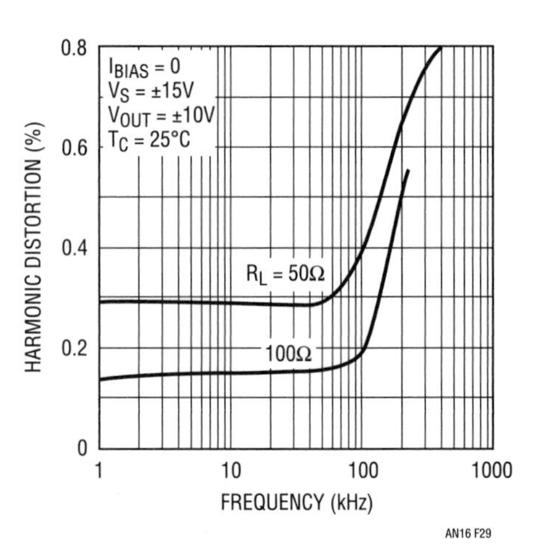

Figure 19.29 • Distortion Is Low to 100kHz, Even without Quiescent Current Boost. The Influence of Load Resistance Is Indicated Here

Total harmonic distortion

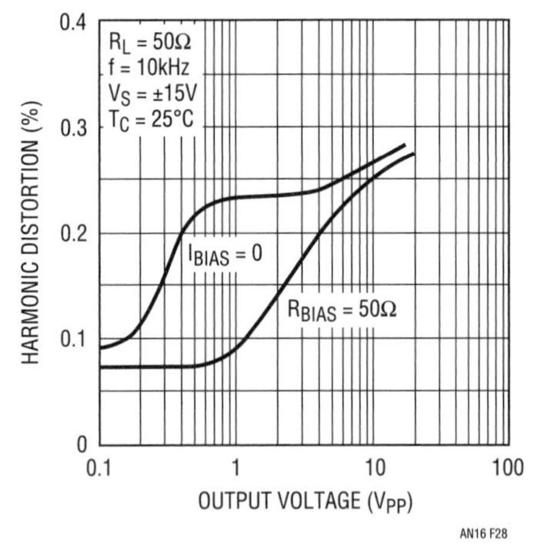

Figure 19.28 • The Buffer Distortion Is Not High, Even When it Is Outside a Feedback Loop, as Shown Here. The Reduced-Distortion Curve Is for 20mA Supply Current

Maximum power

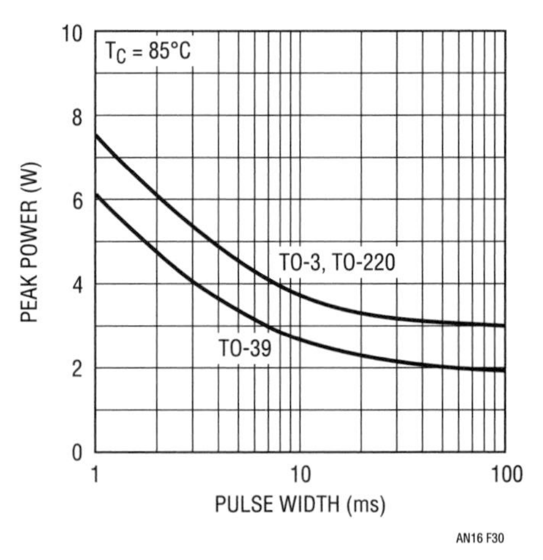

Figure 19.30 • These Curves Indicate the Peak Power Capability of One Output Transistor for T_C = 85°C. With AC Loading, Power Is Divided Between the Two Output Transistors. This Can Reduce Thermal Resistance to 30°C/W for the TO-39 and 15°C/W for the TO-3, as Long as the Frequency Is High Enough That the Peak Rating of Neither Transistor Is Exceeded

Short circuit characteristics

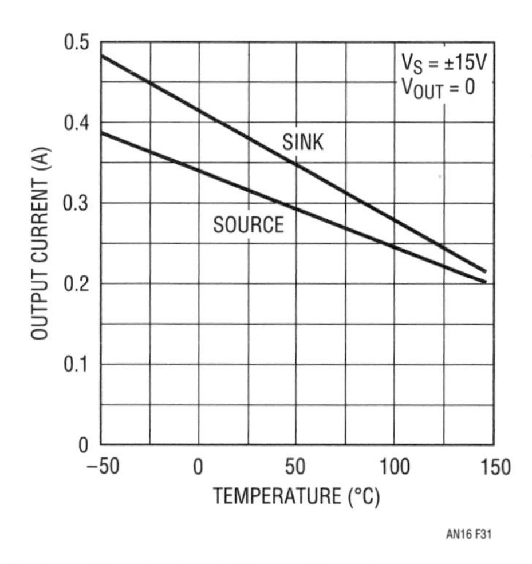

Figure 19.31 • The Output Short Circuit Current Is Plotted Here as a Function of Temperature. Above 160°C it Falls Off Sharply Because of Thermal Limit. The Peak Output Current Is Equal to the Short Circuit Current; with Capacitive Loads Greater Than 1nF, Current Limiting Can Reduce Slew Rate

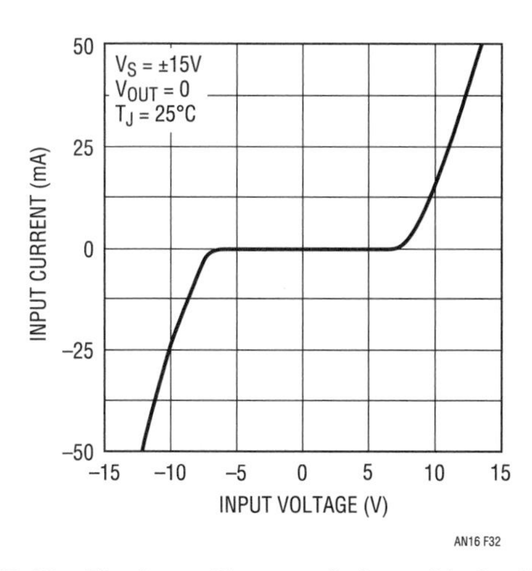

Figure 19.32 • The Input Characteristics, with the Output Shorted, Are Plotted Here. The Input Is Clamped to the Output to Protect Internal Circuitry. Therefore, it Is Necessary to Externally Limit Input Current. The Output-Current Limit of IC Op Amps Is Adequate Protection

Isolating capacitive loads

The buffered follower in Figure 19.33a shows the recommended method of isolating capacitive loads. At lower frequencies, the buffer is within the feedback loop so that offset voltage and gain errors are negligible. At higher frequencies (above 80kHz here) op amp feedback is through C1 so that phase shift from the load capacitance acting against the buffer output impedance does not cause instability.

The initial step response is the same as if the buffer were outside the feedback loop; the gain error of the buffer is then corrected by the op amp with a time constant determined by R1C1. This is shown in Figure 19.33b.

With small load capacitors, the bandwidth is determined by the slower of the two amplifiers. The op amp and the buffer in Figure 19.33 give a bandwidth near 15MHz. This is reduced for capacitive loads greater than 1nF (determined by the output impedance of the buffer).

Feedback-loop stability with large capacitive loads is determined by the ratio of the feedback time constant (R1C1) to that of the buffer output resistance and load capacitance ($R_{OUT}C_L$). A stability factor, m, can be expressed as

$$m = \frac{R1C1}{R_{OUT}C_L}$$

where R_{OUT} is the buffer output resistance.

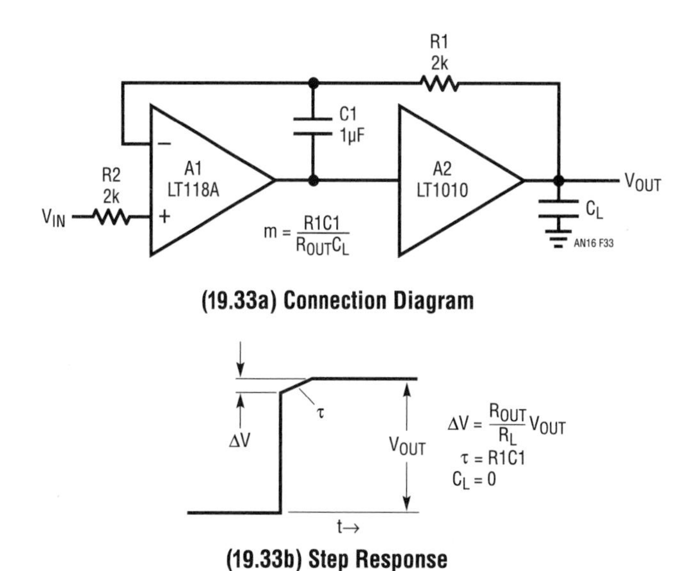

Figure 19.33 • Capacitive Loading on This Buffered Follower Reduces Bandwidth Without Causing Ringing. Step Response with No Capacitive Load Has Residue as Shown Here

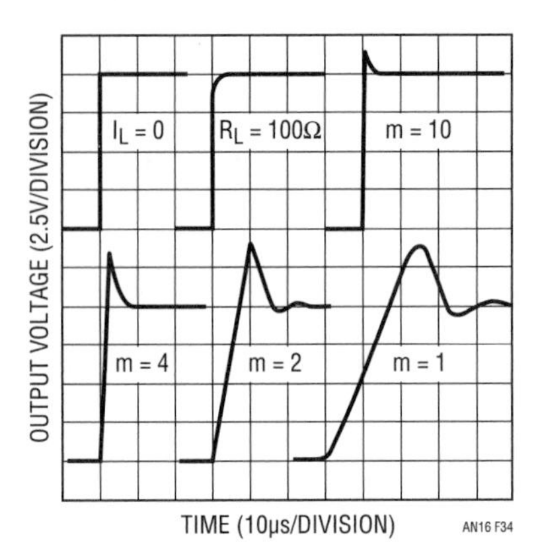

Figure 19.34 • Large Signal Step Response (±5V) of the Buffered Follower in Figure 19.33 for Indicated Loads

The measured large signal step response for the circuit in Figure 19.33a is given in Figure 19.34 for various loads. For $m \geq 4$ ($C_L \leq 0.068\mu F$) there is overshoot but no ringing. For $m < 1$ ($C_L > 0.33\mu F$) ringing becomes pronounced.

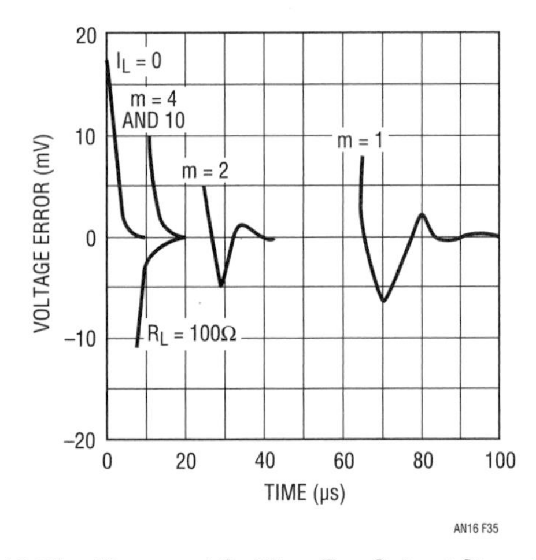

Figure 19.35 • Measured Settling For Output Steps in Figure 19.34. For Capacitive Loads Less Than 0.068µF (m = 4) Settling Is Based on a 2µs Time Constant

The settling time constant is determined by R1C1 for $m \geq 4$. Without capacitive loading, the initial error on the output step is smaller, so time to settle is less. The settling characteristics are shown in Figure 19.35.

With R1C1 as shown in Figure 19.33, any op amp with a bandwidth greater than 200kHz will give the same results on stability. Settling time, however, will be dominated by the slew rate limitations of slow op amps.

Certain op amps, like the LM118, have back-to-back protection diodes across the input terminals. With input rise times in excess of the op amp slew rate, C1 can be charged through these diodes, increasing settling time. Including R2 in series with the input takes care of the problem. Good supply bypass (22µF solid tantalum) should be used because high peak currents are required to drive load capacitors and supply transients can feed into the op amp, increasing settling time.

The same load isolation technique is shown applied to an inverting amplifier in Figure 19.36. The response differs in that the output rise time and bandwidth are limited by R1C1. This does reduce overshoot for $m \geq 4$, as shown in Figure 19.37. For $m < 4$, response approaches that of the follower.

Although the small signal bandwidth is reduced by C1, considerable isolation can be obtained without reducing it

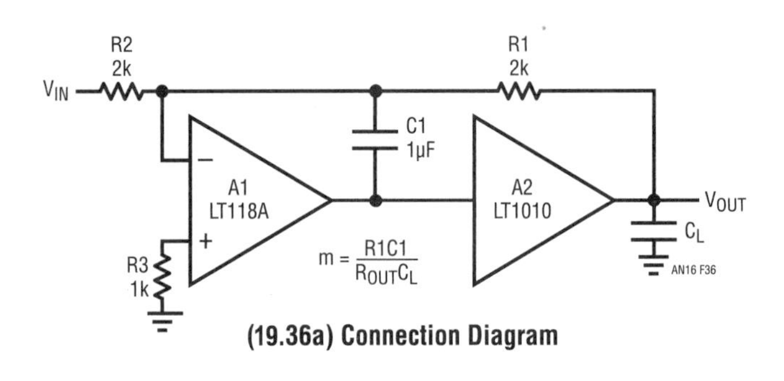

(19.36a) Connection Diagram

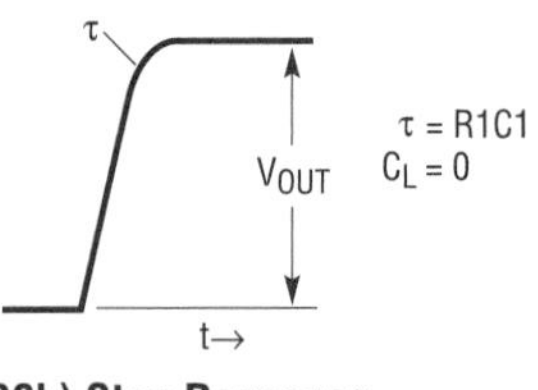

(19.36b) Step Response

Figure 19.36 • With an Inverter, Bandwidth and Rise Time Are Limited by $R1C_L$. For $m \geq 4$, Capacitive Loading Has Little Effect on Bandwidth

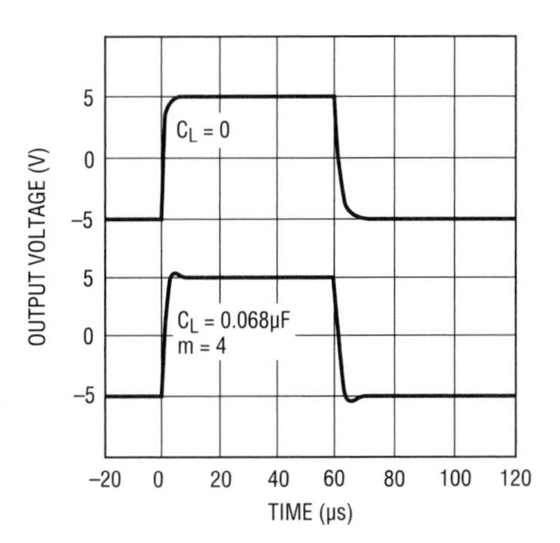

Figure 19.37 • Large Signal Pulse Response of the Inverter in Figure 19.36

below the power bandwidth. Often, bandwidth reduction is desirable to filter high frequency noise or unwanted signals.

An alternate method of isolating capacitive loads is to buffer an inverter output with the follower shown in Figure 19.33.

Capacitive load isolation for non-inverting amplifiers is shown in Figure 19.38, along with the step response for small C_L. Rise time of the initial step is reduced with increasing C_L, and response approaches that of the inverter.

Integrators

A lowpass amplifier can be formed just by using large C1 with the inverter in Figure 19.36, as long as the op amp is capable of supplying the required current to the summing junction and the increase in closed loop output impedance above the cutoff frequency is not a problem (it will never rise above the buffer output impedance).

If the integrating capacitor must be driven from the buffer output, the circuit in Figure 19.39 can be used to provide capacitive load isolation. The method does introduce errors, as is shown in the figure.

The op amp does not respond instantly to an input step, and the input current is supplied by the buffer output. The resulting change in buffer output voltage is seen at the real summing junction and is corrected at an R1C1 time constant. As the output ramps, the voltage change across C1 generates a current through R1, shifting the real summing junction off ground.

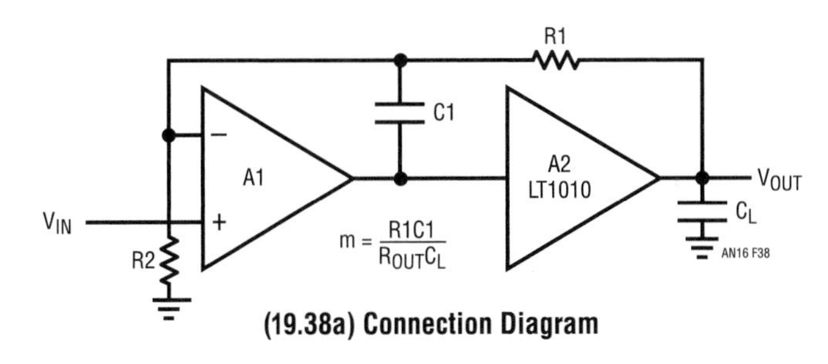

(19.38a) Connection Diagram

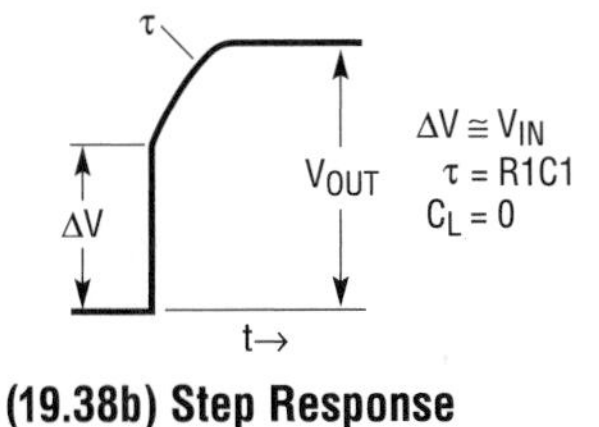

(19.38b) Step Response

Figure 19.38 • With Non-Inverting Amplifier, Rise Time of Initial Step Decreases with Increasing C_L. Stability Requirements Are the Same as for Follower and Inverter

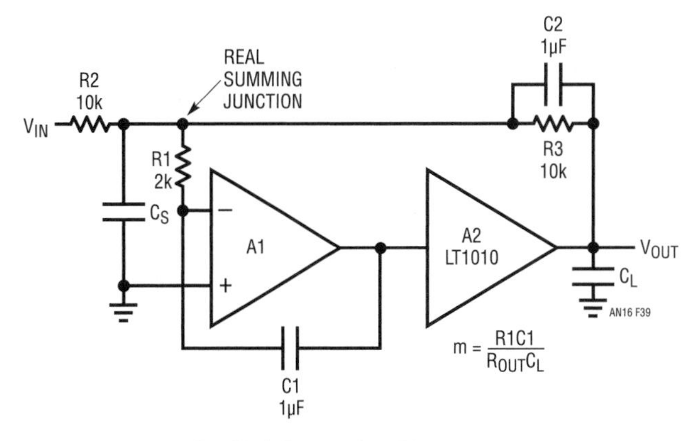

(19.39a) Connection Diagram

$$\Delta V = \left(\frac{R1C1}{R2C2} + \frac{R_{OUT}}{R_{IN}}\right)\Delta V_{IN}$$

$\tau = R1C1$

$C_L = 0$

ΔV τ V_{OUT} t→

(19.39b) Step Response

Figure 19.39 • Capacitive Load Isolation for a Lowpass or Integrating Amplifier When Integrating Capacitor Must Go to Buffer Output. Response Given Is for Negative Input Step

Figure 19.40 shows the voltage on the real summing junction for an input square wave. Both error terms are apparent in the top curve. With $C_L = 0.33\mu F$, response is reasonable. This suggest that m = 1 be used as a stability criterion for this type of circuit if the shift of real summing node voltage with output ramp is a problem. A capacitor can be used on the real summing junction to absorb current transients and reduce spiking, as shown in the lower curve.

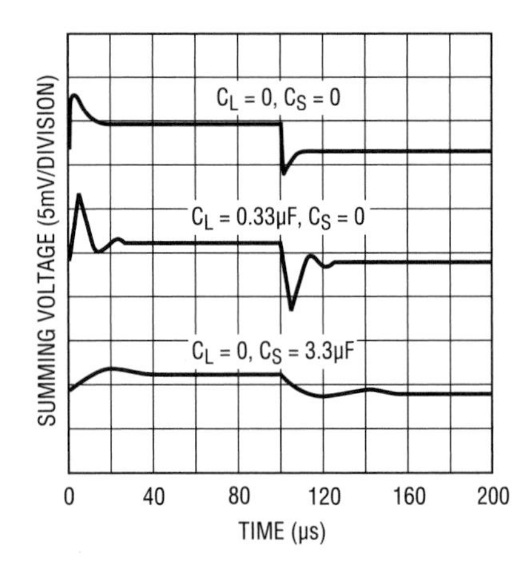

Figure 19.40 • Step Response of the Integrating Amplifier in Figure 19.39. The Real Summing Junction Voltage Is Shown for ±0.5mA Input Change

With large R2 and $C_S = 0$, the output voltage of the integrator will be the response of an ideal integrator plus the voltage of the real summing junction. Large C_S will increase the high frequency loop gain so that this is no longer true.

Impulse integrator

With certain sensors, like radiation detectors, the output is delivered in short, high current bursts. Frequently, it is necessary to integrate these impulses to determine net charge. A complication with some solid-state sensors is that the peak voltage across them must be kept low to avoid error.

The circuit in Figure 19.41 will integrate high current pulses while keeping the summing note under control. Although it increases noise gain, C_S is often required for stability and to absorb the leading edge of fast pulses. The buffer increases the peak current available to the summing node and improves stability by isolating C_f and C_S from the op amp output. Increased output drive capability is a bonus.

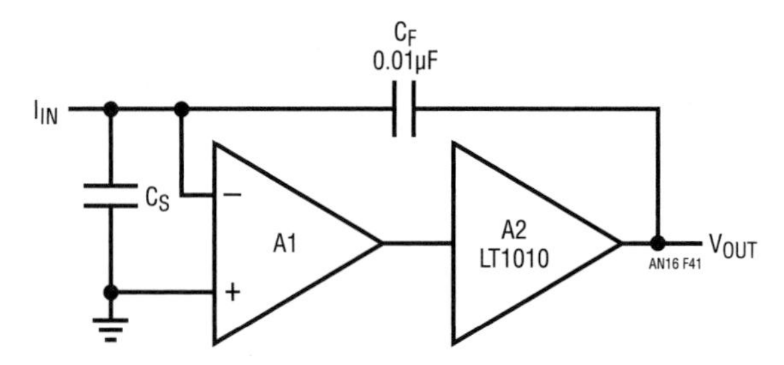

Figure 19.41 • Buffer Increases Current Available to Summing Node. Input Capacitor Absorbs Input Impulses and Raises Loop Gain

The summing node response to a 100mA, 100ns input impulse is shown in Figure 19.42 for three different cases. With $C_S = 0.33\mu F$, the LT118A will settle faster than the LF156 because of its higher gain-bandwidth product; but C_S cannot be made much smaller for $C_f = 0.01\mu F$. The LF156 works with $C_S = 0.02\mu F$ and settles even faster because it goes through unity gain at a frequency where the LT1010 is better able to handle $C_f = 0.01\mu F$ as a load capacitance. However, the smaller C_S does allow the summing node to get further off null during the input impulse.

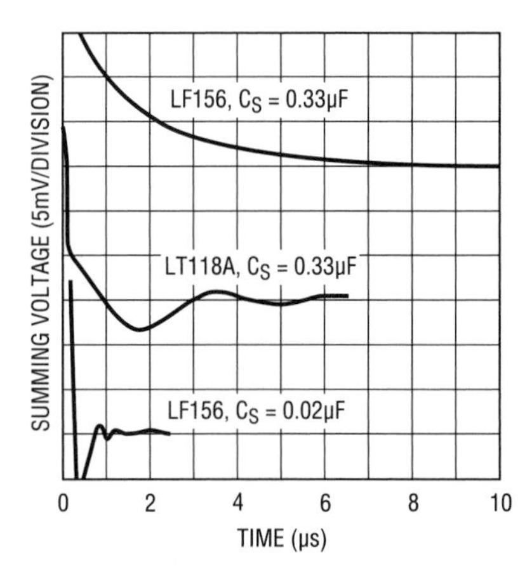

Figure 19.42 • Summing Node Voltage of Impulse Integrator in Figure 19.41 with 100mA, 100ns Input Impulse and −10mA Recovery

Parallel operation

Parallel operation provides reduced output impedance, more drive capability and increased frequency response under load. Any number of buffers can be directly paralleled as long as the increased dissipation in individual units caused by mismatches of output resistance and offset voltage is taken into account.

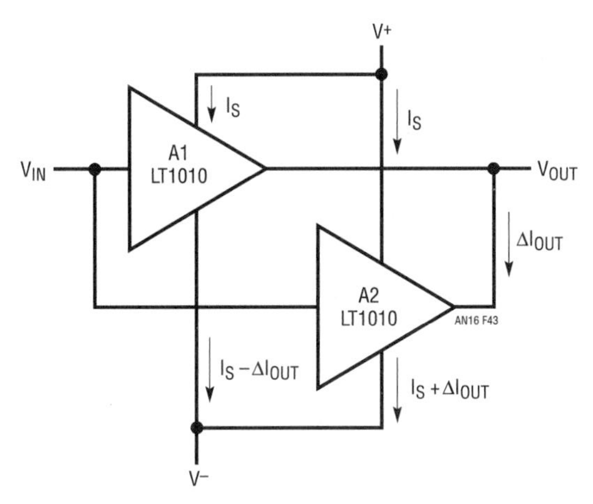

Figure 19.43 • When Two Buffers Are Paralleled, a Current Can Flow Between Outputs, But Total Supply Current Is Not Greatly Affected

When the inputs and outputs of two buffers are connected together as shown in Figure 19.43, a current, ΔI_{OUT}, flows between the output:

$$\Delta I_{OUT} = \frac{V_{OS1} - V_{OS2}}{R_{OUT1} + R_{OUT2}}$$

where V_{OS} and R_{OUT} are the offset voltage and output resistance of the respective buffers.

Normally, the negative supply current of one unit will increase and the other decrease, with the positive supply current staying the same. The worst-case ($V_{IN} \rightarrow V^+$) increase in standby dissipation can be assumed to be $\Delta I_{OUT} V_T$, where V_T is the total supply voltage.

Offset voltage is specified worst-case over a range of supply voltages, input voltage and temperature. It would be unrealistic to use these worst-case numbers above because paralleled units are operating under identical conditions. The offset voltage specified for $V_S = \pm 15V$, $V_{IN} = 0$ and $T_A = 25°C$ will suffice for a worst-case condition.

Output load current will be divided based on the output resistance of the individual buffers. Therefore, the available output current will not quite be doubled unless output resistances are matched. As for offset voltage above, the 25°C limits should be used for worst-case calculations.

Parallel operation is not thermally unstable. Should one unit get hotter than its mates, its share of the output and its standby dissipation will decrease.

As a practical matter, parallel connection needs only some increased attention to heat sinking. In some applications, a few ohms equalization resistance in each output may be wise. Only the most demanding applications should require matching, and then just of output resistance at 25°C.

Wideband amplifiers

Figure 19.44 shows the buffer inside the feedback loop of a wideband amplifier that is not unity gain stable. In this case, C1 is not used to isolate capacitive loads. Instead, it provides an optimum value of phase lead to correct for the buffer phase lag with a limited range of load capacitances.

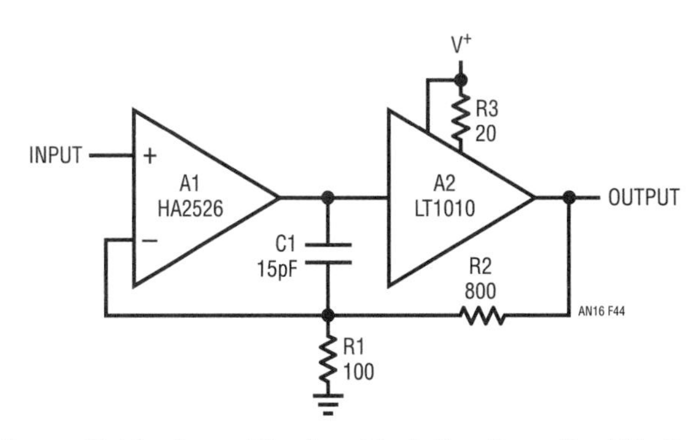

Figure 19.44 • Capacitive Load Isolation Described Earlier Does Not Apply For Amplifiers That Are Not Unity Gain Stable. This 8MHz, $A_V = 9$ Amplifier Handles Only 200pF Load Capacitance

With the TO-3 and TO-220 packages, behavior can be improved by raising the quiescent current with a 20Ω resistor from the bias terminal to V^+. Alternately, devices in the TO-39 package can be operated in parallel.

Putting the buffer outside the feedback loop, as shown in Figure 19.45, will give capacitive load isolation, with large output capacitors only reducing bandwidth. Buffer offset, referred to the op amp input, is divided by the gain. If the load resistance is known, gain error is determined by the output resistance tolerance. Distortion is low.

The 50Ω video line splitter in Figure 19.46 puts feedback on one buffer, with others slaved. Offset and gain accuracy of slaves depends on their matching with master.

When driving long cables, including a resistor in series with the output should be considered. Although it reduces gain, it does isolate the feedback amplifier from the effects of unterminated lines which present a resonant load.

When working with wideband amplifiers, special attention should always be paid to supply bypassing, stray capacitance and keeping leads short. Direct grounding of test probes, rather than the usual ground clip lead, is absolutely necessary for reasonable results.

The LT1010 has slew limitations that are not obvious from standard specifications. Negative slew is subject to glitching, but this can be minimized with quiescent current boost. The appearance is always worse with fast rise signal generators than in practical applications.

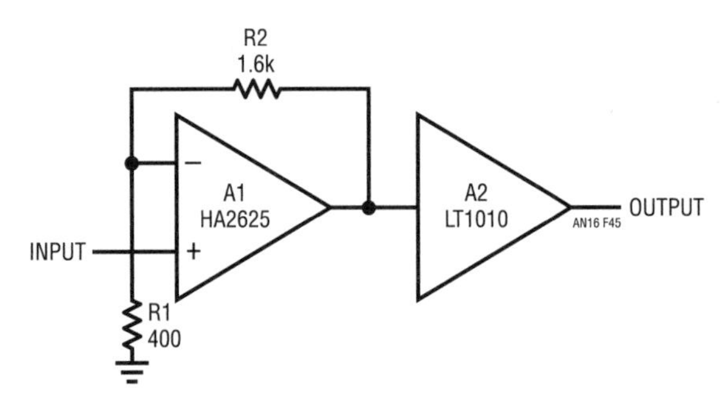

Figure 19.45 • Buffer Outside Feedback Loop Gives Capacitive Load Isolation. Buffer Offset Is Divided by Amplifier Gain, Gain Error Is Determined by Output Resistance Tolerance and Distortion Is Low

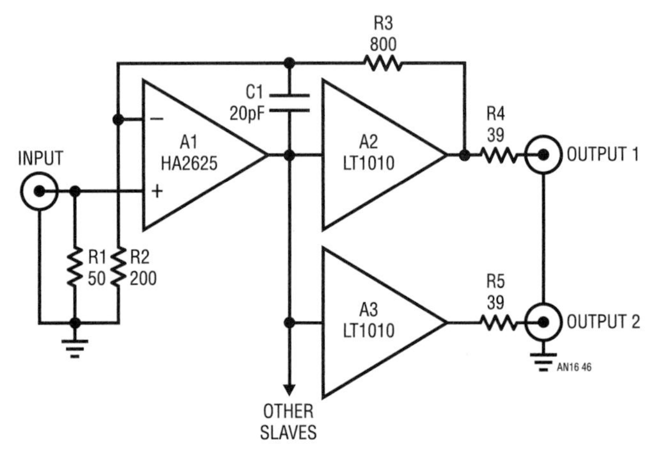

Figure 19.46 • This Video Line Splitter Has Feedback on One Buffer with Others Slaved. Offset and Gain Accuracy of Slaves Depends on Matching with Master

Track and hold

A 5MHz track and hold circuit is shown in Figure 19.47. It has a power bandwidth of 400kHz with a ±10V signal swing.

The buffered input-follower drives the hold capacitor, C4, through Q1, a low resistance (<5Ω) FET switch. The positive hold command is supplied by TTL logic with Q3 level shifting to the switch driver, Q2.

When the FET gate is driven to V^- for hold, it pulls charge that depends upon the input voltage and drain-gate capacitance out of the hold capacitor. A compensating charge is put into the hold capacitor through C3.

Below the FET pinch voltage, the gate capacitance increases sharply. Since the FET will always be pinched off in hold, the turn-off charge from this excess capacitance will be constant over the input voltage range.

Going into hold, the inverting amplifier, A4, makes the positive voltage step into C3 proportional to the negative step on the switch gate, plus a constant to account for the increased capacitance below pinch-off. The step into hold is made independent of the input level with R7 and adjusted to zero with R10 (initially setting up for V_{IN} = ±5V avoids special problems at input voltage extremes). The circuit is brought into adjustment range for a particular design with an appropriate value for C3, although a couple hundred ohms in series with C3 may be advised for larger values to insure the stability of A4.

The positive input voltage range is determined by the common mode range of the op amps. However, if the output of A4 saturates, gate-capacitance compensation will be affected.

The input voltage must be above the negative supply by at least the pinch voltage of the FET to keep it off in hold. In addition, the negative supply must be sufficient to maintain current in D2; or gate-capacitance compensation will suffer. The voltage on the emitter of Q2 can be made more negative than the op amp supplies to extend the operating range.

Since internal dissipation can be quite high when driving fast signals into a capacitive load, using a buffer in a power package is recommended.[1] Raising buffer quiescent current to 40mA with R3 improves frequency response.

Note 1: Overheating of the buffer causes a sharp reduction in slew rate before thermal limit is activated.

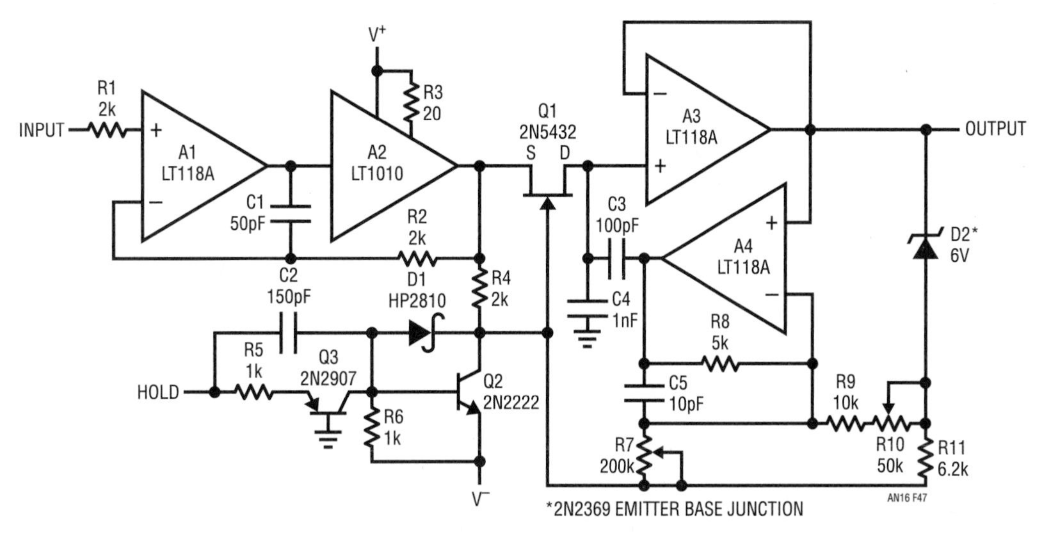

Figure 19.47 • A 5MHz Track and Hold. With Buffer, Bandwidth and Slew Rate Is Little Affected by the Hold Capacitor. Compensation for Gate Capacitance of FET Switch Is Included

This circuit is equally useful as a fast acquisition sample and hold. An LF156 might be used for A3 to reduce drift in hold because its lower slew rate is not usually a problem in this application.

Bidirectional current sources

The voltage-to-current converter in Figure 19.48 uses the standard op amp configuration. It has differential input, so either input can be grounded for the desired output sense. Output is bidirectional.

Maximum output resistance is obtained by trimming the resistors. High frequency output characteristics will depend on the bandwidth and slew rate of the op amp, as well as stray capacitance to the op amp inputs. This ±150mA current source had a measured output resistance of 3MΩ and 48nF equivalent output capacitance.

Using an LT118A and lower feedback resistors would give much lower output capacitance at the expense of output resistance.

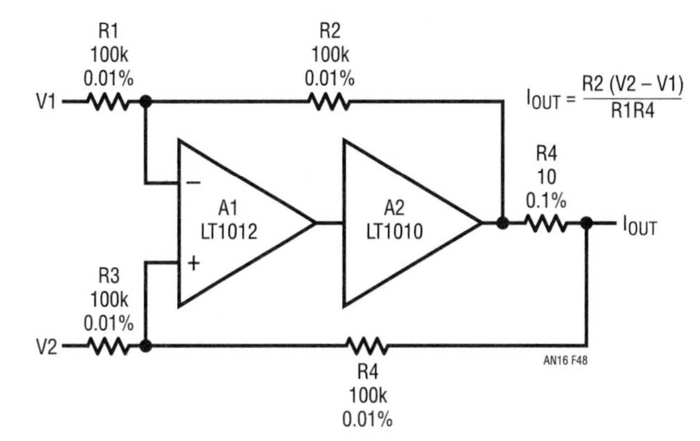

Figure 19.48 • This Voltage/Current Converter Requires Excellent Resistor Matching or Trimming to Get High Output Resistance. Buffer Increases Output Current and Capacitive Load Stability with Small R4

In Figure 19.49, an instrumentation amplifier is used to eliminate the feedback resistors and any sensitivity to stray capacitances. The circuit had a measured output resistance of 6MΩ and an equivalent output capacitance of 19nF. Pins 7 and 8 of the LM163 are differential inputs, but they are loaded internally with 50kΩ to V^-. Either input can be grounded to get the desired output sense. Because of the loading, the input should be driven from a low impedance source like an op amp.

Both circuits are stable for all capacitive loads.

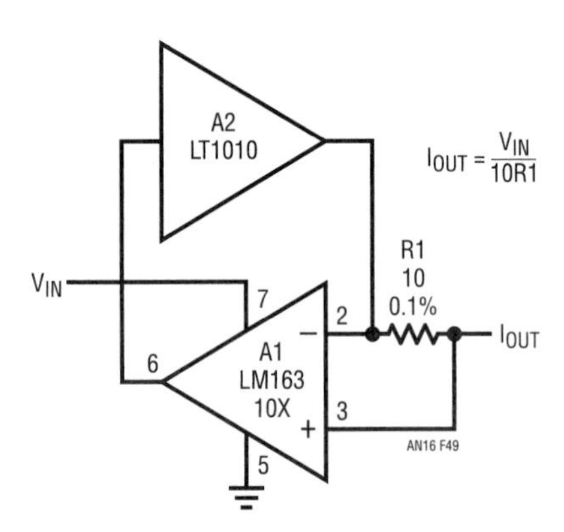

Figure 19.49 • Voltage/Current Converter Using Instrumentation Amplifier Does Not Require Matched Resistors

Voltage regulator

Even though it operates from a single supply, the circuit in Figure 19.50 will regulate voltage down to 200mV. It will also source or sink current.

The circuit's ability to handle capacitive loads is determined by R3 and C1. The values given are optimized for up to 1μF output capacitance, as might be required for an IC test supply.

The purpose of C1 is to lower the drive impedance to the buffer at high frequencies because the high frequency output impedance of the LM10 runs above 1kΩ. Without C1 there could be low level oscillation at certain capacitive loads.

It is important to connect Pin 4 of the LM10 and the bottom of R2 to a common ground point to avoid poor regulation because of ground loop problems.

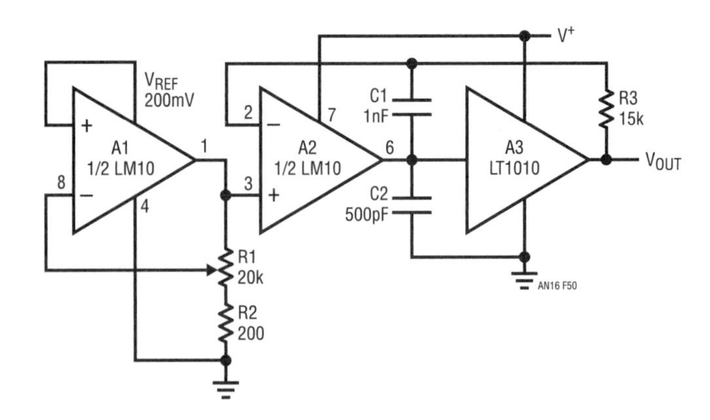

Figure 19.50 • This Voltage Regulator Operates From a Single Supply Yet Is Adjustable Down to 200mV and Can Source or Sink Current

Voltage/current regulator

Figure 19.51 shows a fast power buffer that regulates the output voltage at V_V until the load current reaches a value programmed by V_I. For heavier loads it is a fast, precision current regulator.

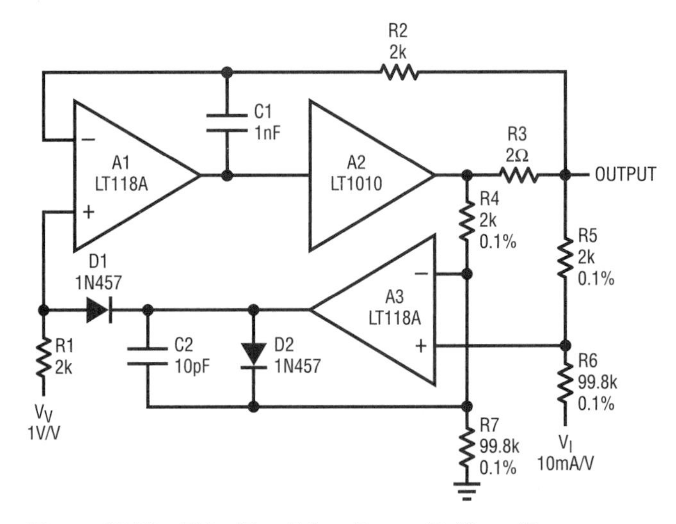

Figure 19.51 • This Circuit Is a Power Buffer with Automatic Transition into Precision, Programmable Current Limit. Fast, Clean Response Into and Out of Current Limit is a Feature of the Design

With output current below the current limit, the current regulator is disconnected from the loop by D1, with D2 keeping its output out of saturation. This output clamp enables the current regulator to get control of the output current from the buffer current limit within a microsecond for an instantaneous short.

In the voltage regulation mode, A1 and A2 act as a fast voltage follower using the capacitive load isolation technique described earlier. Load transient recovery, as well as capacitive load stability, are determined by C1. Recovery from short circuit is clean.

Bidirectional current limit can be provided by adding another op amp connected as a complement to A3. Increased output current and less sensitivity to capacitive loading are obtained by paralleling buffers.

This circuit can be used to make an operational power supply with a bandwidth up to 10MHz that is well suited to IC testing. Output impedance is low without output capacitors and current limit is fast so that it will not damage sensitive circuits. The bandwidth and slew rate are reduced to 2MHz and 15V/μs[2] (without paralleling) by the 0.01μF required for supply bypass on many ICs. Large output capacitors can be accommodated by switching a larger capacitor across C1.

Supply splitter

Dual supply op amps and comparators can be operated from a single supply by creating an artificial ground at half the supply voltage. The supply splitter in Figure 19.52 can source or sink 150mA.

The output capacitor, C2, can be made as large as necessary to absorb current transients. An input capacitor is also used on the buffer to avoid high frequency instability that can be caused by high source impedance.

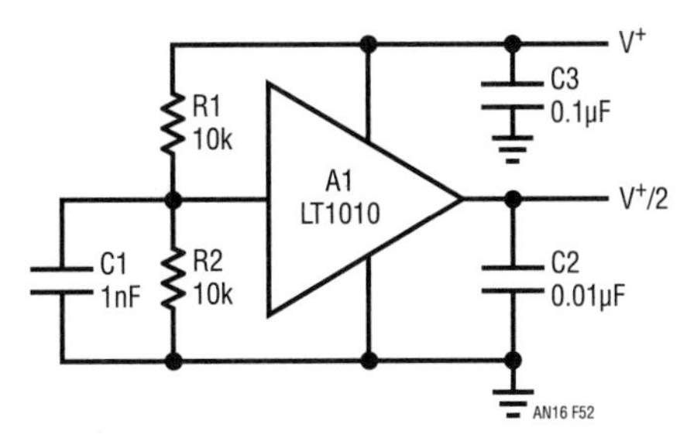

Figure 19.52 • Using the Buffer to Supply an Artificial Ground ($V^+/2$) to Operate Dual Supply Op Amps and Comparators from a Single Supply

Overload clamping

The input of a summing amplifier is at virtual ground as long as it is in the active region. With overloads this is no longer true unless the feedback is kept active.

Figure 19.53 shows a chopper-stabilized current-to-voltage converter. It is capable of 10pA resolution, yet is able to keep the summing node under control with overload currents to ±150mA.

During normal operation, D3 and D4 are not conducting; and R1 absorbs any leakage current from the Zener clamps, D6 and D7. In overload, current is supplied to the summing node through the Zener clamps rather than the scaling resistor, R2. A capacitor on the input absorbs fast transients.

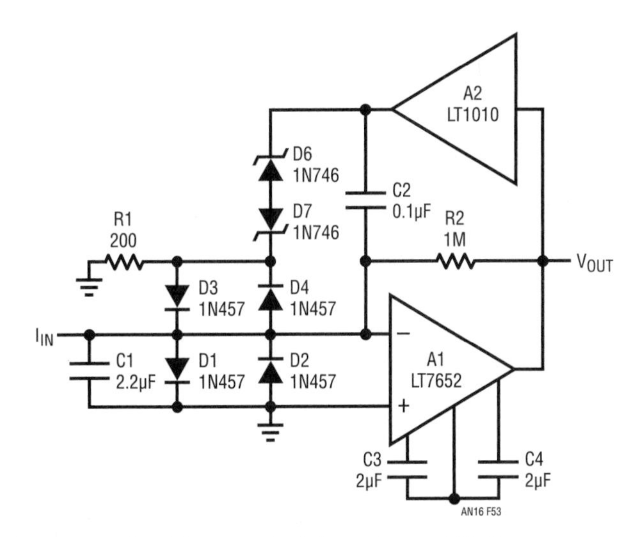

Figure 19.53 • Chopper-Stabilized Current/Voltage Converter Has Picoampere Sensitivity, Yet Is Capable of Keeping Summing Node Under Control with 150mA Input Current

Note 2: Slewing large capacitors causes high buffer dissipation.

Conclusions

A new class-B output stage has been described that is particularly well suited to IC designs. It is fast and avoids the parasitic oscillation problems of the quasi-complementary output. This has been combined with the charge storage transistor, a new diode structure and a novel boost circuit to make a general-purpose buffer that combines speed, large output drive and low standby current. The buffer has been well characterized and shows few disagreeable characteristics.

The applications section has demonstrated that buffers can be quite useful in everyday analog design. They also make touchy wideband amplifiers easy to use. The availability of a low cost, high performance IC buffer should be a stimulus to expanding upon these applications. Buffers no longer need to be considered an exotic component; they will become a standard analog design tool.

Acknowledgement

Thanks are due to Felisa Velasco for special engineering assembly which was key to product development and to Guy Hoover for doing most of the experimental work presented here.

Appendix

The following summarizes some design details that might otherwise be overlooked when first using the buffer. An equivalent circuit is given, and guaranteed electrical characteristics from the data sheet are listed for reference.

Supply bypass

The buffer is no more sensitive to supply bypassing than slower op amps, as far as stability is concerned. The 0.1μF disc ceramic capacitors usually recommended for op amps are certainly adequate for low frequency work. As always, keeping the capacitor leads short and using a ground plane is prudent, especially when operating at high frequencies.

The buffer slew rate can be reduced by inadequate supply bypass. With output current changes much above 100mA/μs, using 10μF solid tantalum capacitors on both supplies is good practice, although bypassing from the positive to the negative supply may suffice.

When used in conjunction with an op amp and heavily loaded (resistive or capacitive), the buffer can couple into supply leads common to the op amp causing stability problems with the overall loop and extended settling time.

Adequate bypassing can usually be provided by 10μF solid tantalum capacitors. Alternately, smaller capacitors could be used with decoupling resistors. Sometimes the op amp has much better high frequency rejection on one supply, so bypass requirements are less on this supply.

Power dissipation

In many applications, the LT1010 will require heat sinking. Thermal resistance, junction to still air is 150°C/W for the TO-39 package, 100°C/W for the TO-220 package and 60°C/W for the TO-3 package. Circulating air, a heat sink or mounting the package to a printed circuit board will reduce thermal resistance.

In DC circuits, buffer dissipation is easily computed. In AC circuits, signal waveshape and the nature of the load determine dissipation. Peak dissipation can be several times average with reactive loads. It is particularly important to determine dissipation when driving large load capacitance.

With AC loading, power is divided between the two output transistors. This reduces the effective thermal resistance, junction to case, to 30°C/W for the TO-39 package and 15°C/W for the TO-3 and TO-220 packages, as long as the peak rating of neither output transistor is exceeded. Figure 19.30 indicates the peak dissipation capabilities of one output transistor.

Overload protection

The LT1010 has both instantaneous current limit and thermal overload protection. Foldback current limiting has not been used, enabling the buffer to drive complex loads without limiting. Because of this, it is capable of power dissipation in excess of its continuous ratings.

Normally, thermal overload protection will limit dissipation and prevent damage. However, with more than 30V across the conducting output transistor, thermal limiting is not quick enough to insure protection in current limit. The thermal protection is effective with 40V across the conducting output transistor as long as the load current is otherwise limited to 150mA.

Drive impedance

When driving capacitive loads, the LT1010 likes to be driven from a low source impedance at high frequencies. Certain low power op amps (e.g., the LM10) are marginal in this respect. Some care may be required to avoid oscillations, especially at low temperatures.

Bypassing the buffer input with more than 200pF will solve the problem. Raising the operating current also works, but this cannot be done with the TO-39 package.

Equivalent circuit

Below 1MHz, the LT1010 is quite accurately represented by the equivalent circuit shown in Figure A for both small and large signal operation. The internal element, A1, is an idealized buffer with the unloaded gain specified for the LT1010. Otherwise, it has zero offset voltage, bias current and output resistance. The output of A1 saturates to its supply terminals.

Loaded voltage gain can be determined from the unloaded gain, A_V, the output resistance, R_{OUT}, and the load resistance, R_L, using

$$A_{VL} = \frac{A_{VL} R_L}{R_{OUT} + R_L}$$

Maximum positive output swing is given by

$$V_{OUT^+} = \frac{(V^+ - V_{SOS^+})R_L}{R_{SAT} + R_L}$$

where V_{SOS} is the unloaded output saturation voltage and R_{SAT} is the output saturation resistance.

The input swing required for this output is

$$V_{IN^+} = V_{OUT^+}\left(1 + \frac{R_{OUT}}{R_L}\right) - V_{OS} + \Delta V_{OS}$$

where ΔV_{OS} is the clipping allowed in making the saturation measurements (100mV).

The negative output swing and input drive requirements are determined similarly. The values given in Figure A are typicals; worst-case numbers are obtained from the data sheet reproduced on the back page.

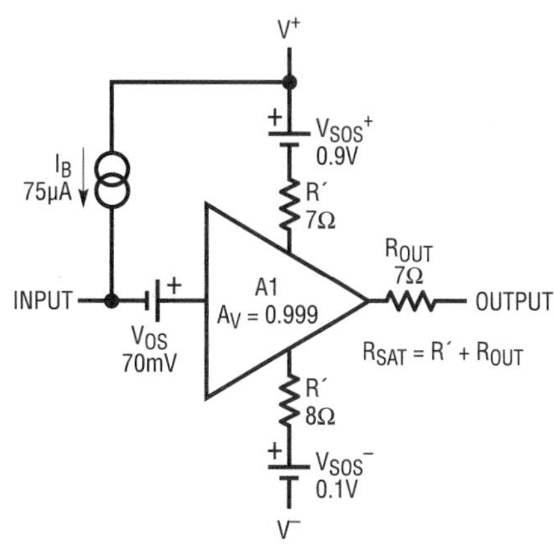

Figure A • An Idealized Buffer, A1, as Modified by This Equivalent Circuit Describes the LT1010 at Low Frequencies

Absolute Maximum Ratings	
Total Supply Voltage	±22V
Continuous Output Current	±150mA
Continuous Power Dissipation (Note 1)	
LT1010MK	5.0W
LT1010CK	4.0W
LT1010CT	4.0W
LT1010MH	3.1W
LT1010CH	2.5W
Input Current (Note 2)	±40mA
Operating Junction Temperature	
LT1010M	−55°C to 150°C
LT1010C	0°C to 125°C
Storage Temperature Range	−65°C to 150°C
Lead Temperature (Soldering, 10 sec)	300°C

Connection diagrams

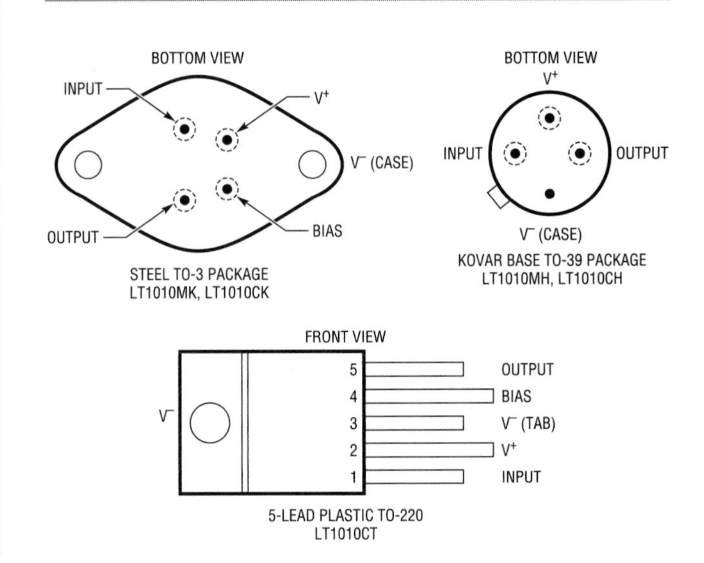

Electrical Characteristics

SYMBOL	PARAMETER	CONDITIONS (Note 4)		LT1010M MIN	LT1010M MAX	LT1010C MIN	LT1010C MAX	Units
V_{OS}	Output Offset Voltage	(Note 3)		20	150	0	150	mV
			•	**−10**	**220**	**−20**	220	mV
		$V_S = \pm 15V$, VIM = 0V		40	90	20	100	mV
I_B	Input Bias Current	$I_{OUT} = 0mA$		0	150	0	250	μA
		$I_{OUT} \leq 150mA$		0	250	0	500	μA
			•	**0**	**300**	**0**	**800**	μA
A_V	Large-Signal Voltage Gain		•	**0.995**	**1.00**	**0.995**	**1.00**	V/V
R_{OUT}	Output Resistance	$I_{OUT} = \pm 1mA$		6	9	5	10	Ω
		$I_{OUT} = \pm 150mA$						
				6	9	5	10	Ω
			•		**12**		**12**	Ω
	Slew Rate	$V_S = \pm 15V$, $V_{IN} = \pm 10V$, $V_{OUT} = \pm 8V$, $R_L = 100\Omega$		75		75		V/μs
V_{SOS}^+	Positive Saturation Offset	Note 4, $I_{OUT} = 0$			1.0		1.0	V
			•		**1.1**		**1.1**	V
V_{SOS}^-	Negative Saturation Offset	Note 4, $I_{OUT} = 0$			0.2		0.2	V
			•		**0.3**		**0.3**	V
R_{SAT}	Saturation Resistance	Note 4, $I_{out} = \pm 150mA$			18		22	Ω
			•		**24**		**28**	Ω
V_{BIAS}	Bias Terminal Voltage	Note 5, $R_{BIAS} = 20\Omega$		750	810	700	840	mV
			•	**560**	**925**	**560**	**880**	mV
I_S	Supply Current	$I_{OUT} = 0$, $I_{BIAS} = 0$			8		9	mV
			•		**9**		**10**	mV

Note 1: For case temperatures above 25°C, dissipation must be derated based on a thermal resistance of 25°C/W with the K and T packages or 40°C/W with the H package. See applications information.
Note 2: In current limit or thermal limit, input current increases sharply with input-output differentials greater than 8V; so input current must be limited. Input current also rises rapidly for input voltages 8V above V^+ or 0.5V below V^-.
Note 3: Specifications apply for $4.5V \leq V_S \leq 40V$, $V^- + 0.5V \leq V_{IN} \leq V^+ - 1.5V$ and $I_{OUT} = 0$, unless otherwise stated. Temperature range is $-55°C \leq T_J \leq 150°C$, $T_C \leq 125°C$, for the LT1010M and $0°C \leq T_J \leq 125°C$, $T_C \leq 100°C$, for the LT1010C. The • and **boldface type** on limits denote the specifications that apply over the full temperature range.
Note 4: The output saturation characteristics are measured with 100mV output clipping. See applications information for determining available output swing and input drive requirements for a given load.
Note 5: With the TO-3 and TO-220 packages, output stage quiescent current can be increased by connecting a resistor between the bias pin and V^+. The increase is equal to the bias terminal voltage divided by this resistance.

Power gain stages for monolithic amplifiers

20

Jim Williams

Most monolithic amplifiers cannot supply more than a few hundred milliwatts of output power. Standard IC processing techniques set device supply levels at 36V, limiting available output swing. Additionally, supplying currents beyond tens of milliamperes requires large output transistors and causes undesirable IC power dissipation.

Many applications, however, require greater output power than most monolithic amplifiers will deliver. When voltage or current gain (or both) is needed, a separate output stage is necessary. The power gain stage, sometimes called a "booster", is usually placed within the monolithic amplifier's feedback loop, preserving the IC's low drift and stable gain characteristics.

Because the output stage resides in the amplifier's feedback path, loop stability is a concern. The output stage's gain and AC characteristics must be considered if good dynamic performance is to be achieved. Overall circuit phase shift, frequency response and dynamic load handling capabilities are issues that cannot be ignored when designing a power gain stage for a monolithic amplifier. The output stage's added gain and phase shift can cause poor AC response or outright oscillation. Judicious application of frequency compensation methods is needed for good results (see box section, "The Oscillation Problem").

The type of circuitry used in an output stage varies with the application, which can be quite diverse. Current and voltage boosting are common requirements, although both are often simultaneously required. Voltage gain stages are usually associated with the need for high voltage power supplies, but output stages which inherently generate such high voltages are an alternative.

A simple, easily used current booster is a good place to begin a study of power gain stages.

150mA output stage

Figure 20.1a shows the LT®1010 monolithic 150mA current booster placed within the feedback loop of a fast FET amplifier. At lower frequencies, the buffer is within the feedback loop so that its offset voltage and gain errors are negligible. At higher frequencies, feedback is through C_f, so that phase shift from the load capacitance acting against the buffer output resistance does not cause loop instability.

Small-signal bandwidth is reduced by C_f, but considerable load isolation can be obtained without reducing it below the power bandwidth. Often a bandwidth reduction is desirable to filter high frequency noise or unwanted signals.

The LT1010 is particularly adept at driving large capacitive loads, such as cables.

The follower configuration (Figure 20.1b) is unique in that capacitive load isolation is obtained without a reduction in small-signal bandwidth, although the output impedance of the buffer has a 10MHz bandwidth without capacitive loading, yet it is stable for all load capacitance to over 0.3μF.

Figure 20.1c shows LT1010s used in a bridge type differential output stage. This permits increased voltage swing across the load, although the load must float.

All of these circuits will deliver 150mA of output current. The LT1010 supplies short-circuit and thermal overload protection. Slew limit is set by the op amp used.

Analog Circuit and System Design: Immersion in the Black Art of Analog Design. http://dx.doi.org/10.1016/B978-0-12-397888-2.00020-1

High current booster

Figure 20.2 uses a discrete stage to get 3A output capacity. The configuration shown provides a clean, quick way to increase LT1010 output power. It is useful for high current loads, such as linear actuator coils in disk drives.

The 33Ω resistors sense the LT1010's supply current, with the grounded 100Ω resistor supplying a load for the LT1010. The voltage drop across the 33Ω resistors biases

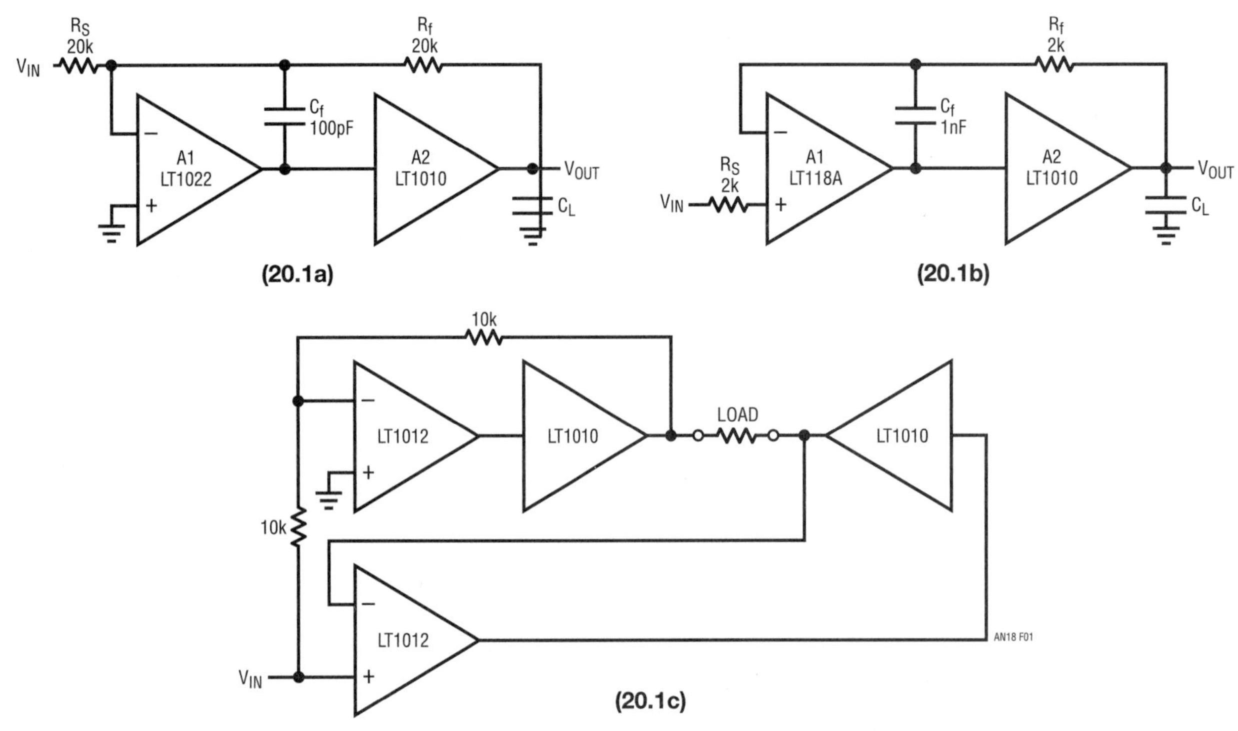

Figure 20.1 • LT1010 Output Stages

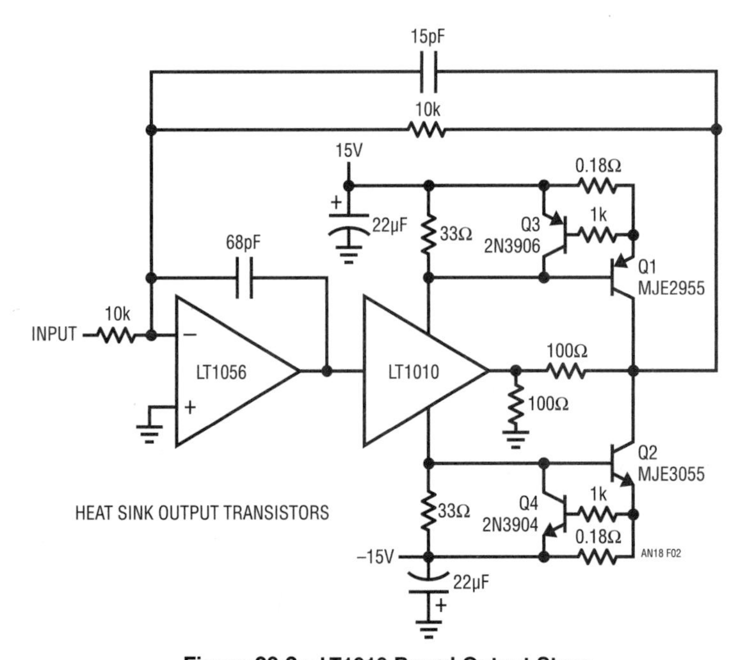

Figure 20.2 • LT1010 Based Output Stage

Q1 and Q2. Another 100Ω value closes a local feedback loop, stabilizing the output stage. Feedback to the LT1056 control amplifier is via the 10k value. Q3 and Q4, sensing across the 0.18Ω units, furnish current limiting at about 3.3A.

The output transistors have low F_t, and no special frequency compensation considerations are required. The LT1056 is rolled off by the 68pF capacitor for dynamic stability, and the 15pF feedback capacitor trims edge response. At full power (±10V, 3A peaks), bandwidth is 100kHz and slew rate about 10V/μs.

UltraFast™ fed–forward current booster

The previous circuits place the output stage booster within the op amp's feedback loop. Although this ensures low drift and gain stability, the op amp's response limits speed. Figure 20.3 shows a very wideband current boost stage. The LT1012 corrects DC errors in the booster stage, and does not see high frequency signals. Fast signals are fed directly to the stage via Q5 and the 0.01μF coupling capacitors. DC and low frequency signals drive the stage via the op amp's output. This parallel path approach allows very broadband performance without sacrificing the DC stability of the op amp. Thus, the LT1012's output is effectively current and speed boosted. The output stage consists of current sources Q1 and Q2 driving the Q3-Q5 and Q4-Q7 complementary emitter followers. The transistors specified have F_t's approaching 1GHz, resulting in a very fast stage. The diode network at the output steers drive away from the transistor bases when output current exceeds 250mA, providing fast short-circuit protection. Net inversion in the stage means the feedback must return to the LT1012's positive input. The circuit's high frequency summing node is the junction of the 1k and 10k resistors at the LT1012. The 10k-39pF pair filters high frequencies, permitting accurate DC summation at the LT1012's positive input. The low frequency roll-off of the fast stage is matched to the high frequency characteristics of the LT1012 section, minimizing aberration in the circuit's AC response. The 8pF feedback capacitor is selected to optimize settling characteristics at the highest speeds.

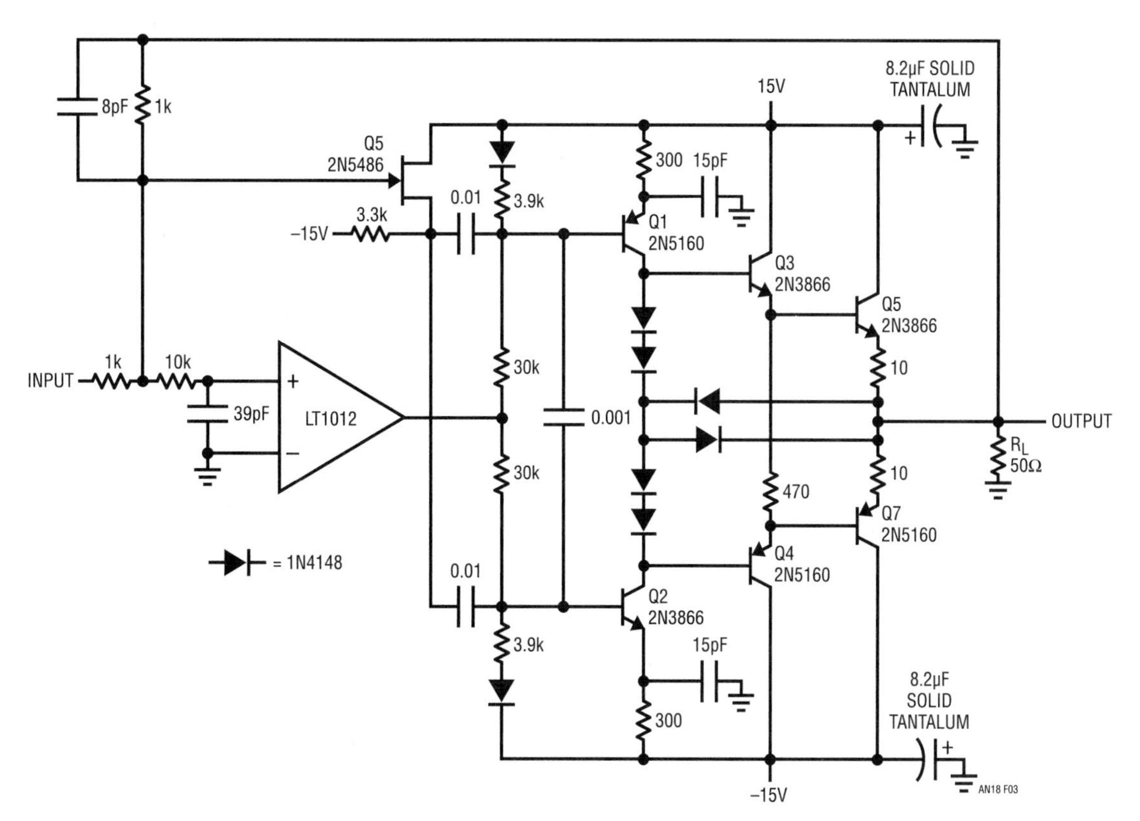

Figure 20.3 • Fed-Forward Wideband Current Booster

This current boosted amplifier features a slew rate in excess of 1000V/µs, a full-power bandwidth of 7.5MHz and a 3dB point of 14MHz. Figure 20.4 shows the circuit driving a 10V pulse into a 50Ω load. Trace A is the input and Trace B is the output.

Slew and settling characteristics are quick and clean, with pulse fidelity approaching the quality of the input pulse generator. Note that this circuit relies on summing action, and cannot be used in the noninverting mode.

Simple voltage gain stages

Voltage gain is another type of output stage. A form of voltage gain stage is one that allows output swing very near the supply rails. Figure 20.5a utilizes the resistive nature of the complementary outputs of a CMOS logic inverter to make such a stage. Although this is an unusual application for a logic inverter, it is a simple, inexpensive way to extend an amplifier's output swing to the supply rails. This circuit is particularly useful in 5V powered analog systems, where improvements in available output swing are desirable to maximize signal processing range.

The paralleled logic inverters are placed within the LT1013's feedback loop. The paralleling drops output resistance, aiding swing capability. The inversion in the loop requires the feedback connection to go to the amplifier's positive input. An RC damper eliminates oscillation in the inverter stage, which has high gain bandwidth when running in its linear region. Local capacitive feedback at the amplifier gives loop compensation. The table provided shows that output swing is quite close to the positive rail, particularly at loads below several milliamperes.

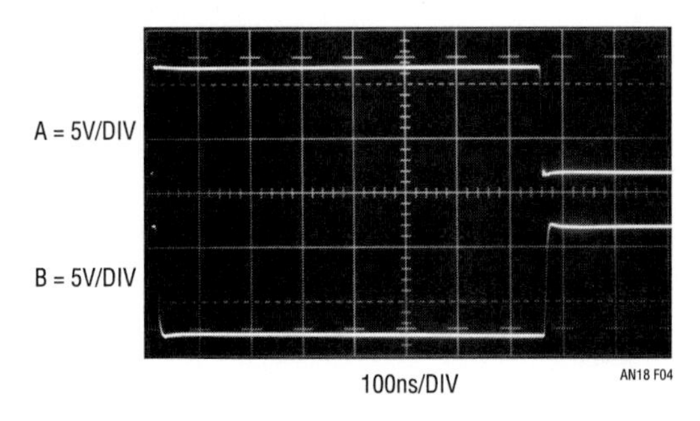

Figure 20.4 • Figure 20.3's Response. Slew Rate Exceeds 1000V/µs at 10V Into 50Ω

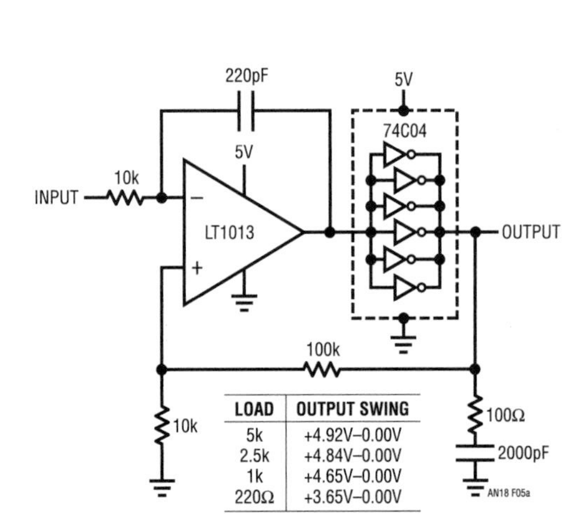

LOAD	OUTPUT SWING
5k	+4.92V–0.00V
2.5k	+4.84V–0.00V
1k	+4.65V–0.00V
220Ω	+3.65V–0.00V

Figure 20.5a • CMOS Inverter-Based Output Stage with Voltage Gain

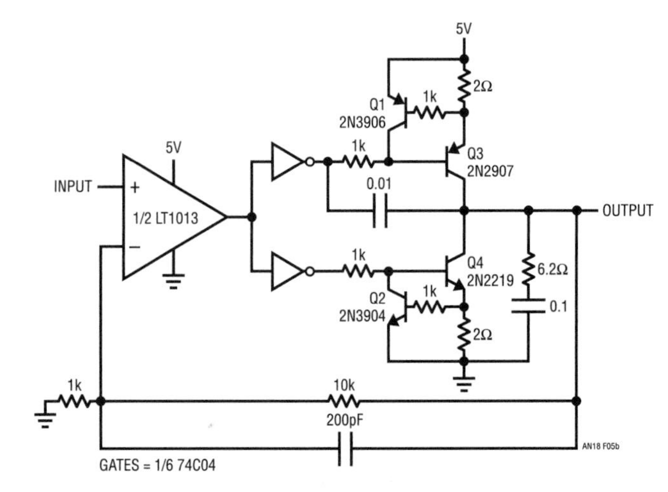

Figure 20.5b • Common Emitter Output Stage with Voltage Gain

Figure 20.5b is similar, except that the CMOS inverters drive bipolar transistors to reduce saturation losses, even at relatively high currents. Figure 20.6a shows Figure 20.5b's output saturation characteristics. Note the extremely low saturation limits below 25mA. Removing the current limit circuitry permits even better performance, particularly at high output currents.

Figure 20.6b shows waveforms of operation for circuit Figure 20.5a. The LT1013's output (Trace B) servos around the 74C04's switching threshold (about 1/2 supply voltage) as it controls the circuit's output (Trace A). This allows the amplifier to operate well within its output swing range while controlling a circuit output with nearly rail-to-rail capability.

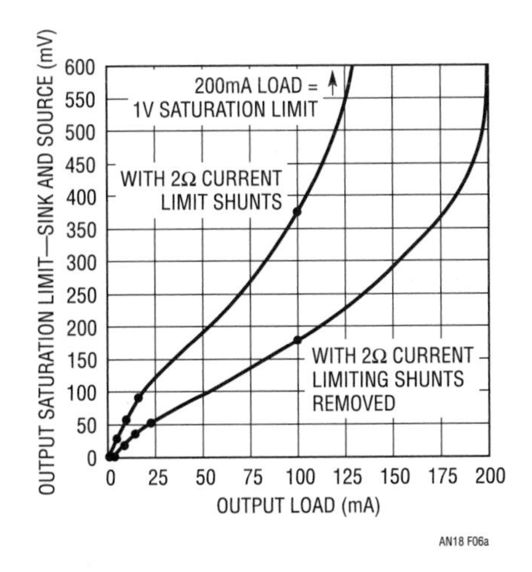

Figure 20.6a • Figure 20.5b's Saturation Characteristics

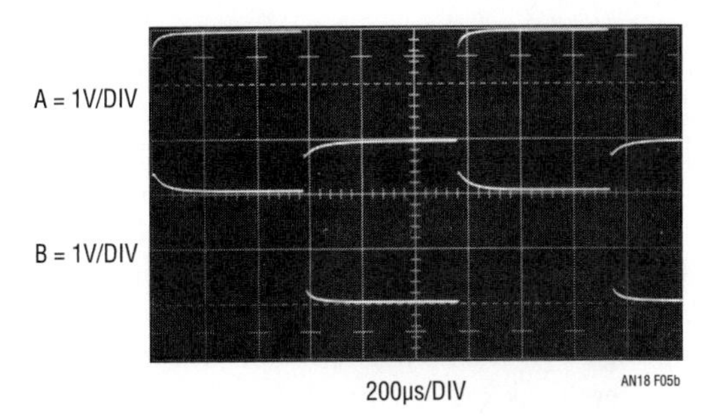

Figure 20.6b • Figure 20.5a's Waveforms

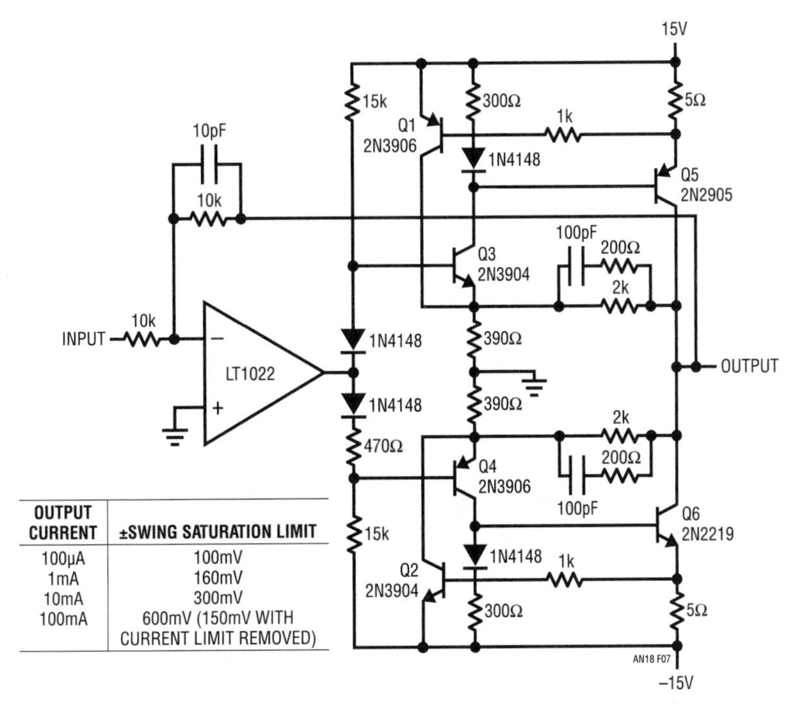

OUTPUT CURRENT	±SWING SATURATION LIMIT
100μA	100mV
1mA	160mV
10mA	300mV
100mA	600mV (150mV WITH CURRENT LIMIT REMOVED)

Figure 20.7 • Complementary Closed-Loop Common Emitter Stages Provide High Current and Good Saturation Perfomance

High current rail-to-rail output stage

Figure 20.7 is another rail-to-rail output stage, but features higher output current and voltage capability. The stage's voltage gain and low saturation losses allow it to swing nearly to the rails while simultaneously supplying current gain.

Q3 and Q4, driven from the op amp, provide complementary voltage gain to output transistors Q5-Q6. In most amplifiers, the output transistors run as emitter followers, furnishing current gain. Their V_{BE} drop, combined with voltage swing limitations of the driving stage, introduces the swing restrictions characteristic of such stages. Here, Q5 and Q6 run common emitter, providing additional voltage gain and eliminating V_{BE} drops as a concern. The voltage inversion of these devices combines with the drive stage inversion to yield overall noninverting operation. Feedback is to the LT1022's negative input. The 2k-390Ω local feedback loop associated with each side of the booster limits stage gain to about 5. This is necessary for stability. The gain bandwidth available through the Q3-Q5 and Q4-Q6 connections is quite high, and not readily controllable. The local feedback reduces the gain bandwidth, prompting stage stability. The 100pF-200Ω damper across each 2k feedback resistor provides heavy gain attenuation at very high frequencies, eliminating parasitic local loop oscillations in the 50MHz to 100MHz range. Q1 and Q2, sensing across the 5Ω shunts, furnish 125mA current limiting. Current flow above 125mA causes the appropriate transistor to come on, shutting off the Q3-Q4 driver stage.

Even with the feedback enforced gain-bandwidth limiting, the stage is quite fast. AC performance is close to the amplifier used to control the stage. Using the LT1022, full-power bandwidth is 600kHz and slew rate exceeds 23V/μs under 100mA output loading. The chart in the figure 20.shows output swing versus loading. Note that, at high current, output swing is primarily limited by the 5Ω current sense resistors, which may be removed. Figure 20.8 shows response to a bipolar input pulse for 25mA loading. The output swings nearly to the rails, with clean dynamics and good speed.

±120V output stage

Figure 20.9 is another voltage gain output stage. Instead of minimizing saturation losses, it provides high voltage outputs from a ±15V powered amplifier. Q1 and Q2 furnish voltage gain, and feed the Q3-Q4 emitter follower outputs. ±15V power for the LT1055 control amplifier is derived from the high voltage supplies via the Zener diodes. Q5 and Q6 set current limit at 25mA by diverting output drive when voltages across the 27Ω shunts become too high. The local 1M-50k feedback pairs set stage gain at 20, allowing ±10V LT1055 drives to cause full ±120V output swing. As in Figure 20.7, the local feedback reduces stage gain bandwidth, making dynamic control easier. This stage is relatively simple to frequency compensate because only Q1 and Q2 contribute voltage gain. Additionally, the high voltage transistors have large junctions, resulting in low F_t's, and no special high frequency roll-off precautions are needed. Because the stage inverts, feedback is returned to the LT1055's positive input. Frequency compensation is achieved by rolling off the LT1055 with the local 100pF-10k pair The 33pF capacitor in the feedback peaks edge response and is not required for stability. Full power bandwidth is 15kHz with a slew limit of about 20V/μs. As shown, the circuit operates in inverting mode, although noninverting operation is possible by exchanging the input and ground assignments at the LT1055's input. Under noninverting conditions, the LT1055's input common mode voltage limits must be observed, setting the minimum noninverting gain at 11. If over compensation is required, it is preferable to increase the 100pF value, instead of increasing the 33pF loop feedback capacitor This prevents excessive

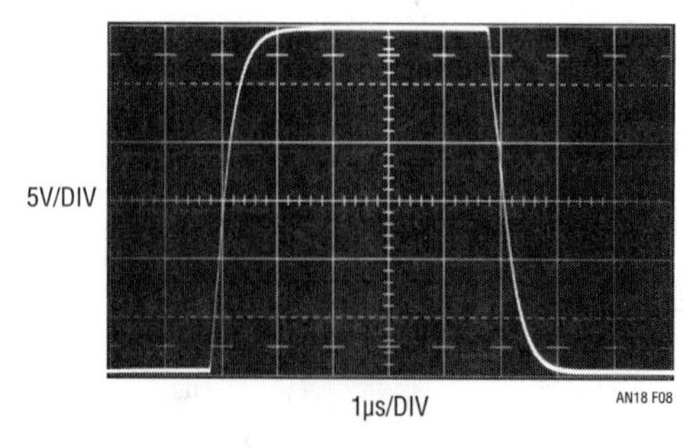

Figure 20.8 • Figure 20.7 Drives ±14.85V Into a 100mA Load

high voltage energy from coupling to the LT1055's inputs during slew. If it is necessary to increase the feedback capacitor, the summing point should be diode clamped to ground or to the LT1055 supply terminals. Figure 20.10 shows results with a ±12V input pulse (Trace A). The output (Trace B) responds with a cleanly damped 240V peak-to-peak pulse.

Figure 20.11 is a similar stage, except that Figure 20.9's output transistors are replaced with vacuum tubes. Most of this stage is conceptually identical to Figure 20.9, but major changes are needed to get the vacuum tube output to swing negatively. Positive swing is readily achieved by simply replacing Figure 20.9's NPN emitter follower with a cathode follower (V1A). Negative outputs require PNP Q3 to drive a Zener biased common cathode configuration. The transistor inverter is necessitated because our thermionic friends have no equivalent to PNP transistors. Zener biasing of V1B's cathode allows Q3's swing to cut off the tube, a depletion mode device.

Without correction, the DC biasing asymmetry caused by the Q3-V1B configuration will force the LT1055 to bias well away from zero. Tolerance stack-up could cause

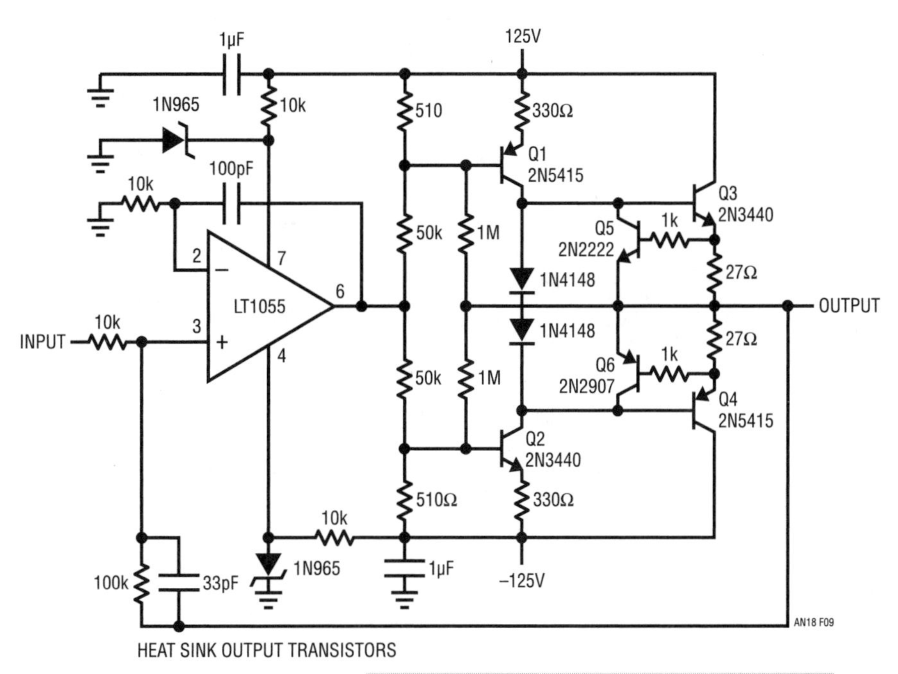

Figure 20.9 • ±120V Output Stage. DANGER! High Voltages Present. Use Caution

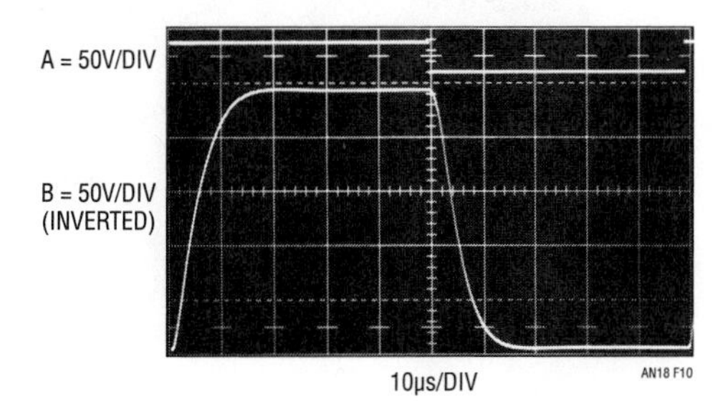

Figure 20.10 • Figure 20.9 Swinging ±120V Into 6kΩ. DANGER! High Voltages Present. Use Caution

saturation limiting in the LT1055's output, reducing overall available swing. This is avoided by skewing the stage's bias string with the potentiometer adjustment. To make this adjustment, ground the input and trim the potentiometer for 0V out at the LT1055.

Figure 20.11's full-power bandwidth is 12kHz, with a slew rate of about 12V/μs. Figure 20.12 shows response to a bipolar input (Trace A). The output responds cleanly, although the slew and settling characteristics reflect the stage's asymmetric gain bandwidth. This stage's output is extremely rugged, due to the inherent forgiving nature of vacuum tubes. No special short-circuit protection is needed, and the output will survive shorts to voltages many times the value of the ±150V supplies.

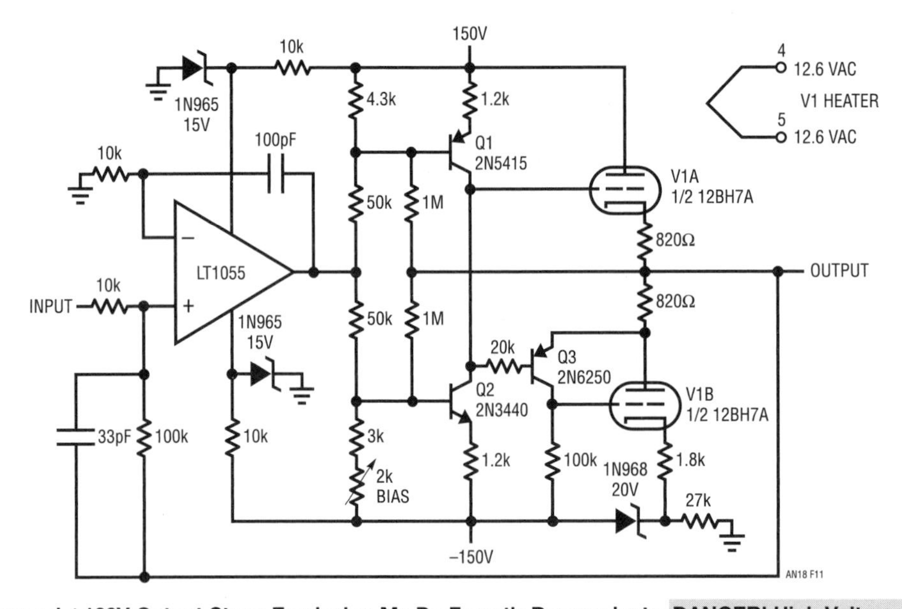

Figure 20.11 • Rugged ±120V Output Stage Employing Mr. De Forest's Descendants. DANGER! High Voltages Present. Use Caution

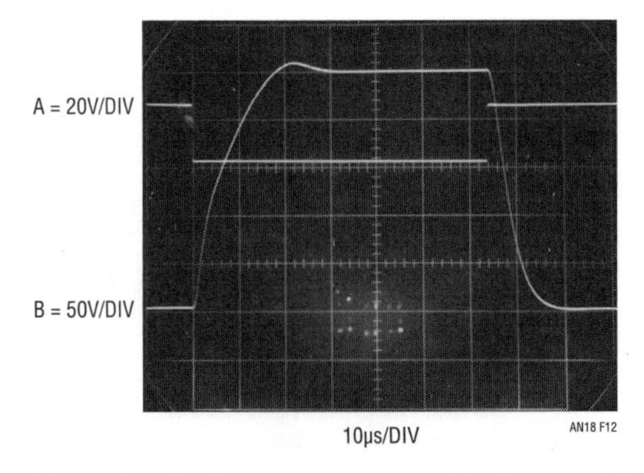

Figure 20.12 • Figure 20.11's Response. Asymmetric Slew and Settling Are Due to Q3-V1B Connection. DANGER! High Voltages Present. Use Caution

Unipolar output, 1000V gain stage

Figure 20.13 shows a unipolar output gain stage which swings 1000V and supplies 15W. This boost stage has the highly desirable property of operating from a single, low voltage supply. It does not require a separate high voltage supply. Instead, the high voltage is directly generated by a switching converter which is an integral part of the gain stage.

A2's output drives Q3, forcing current into T1. T1's primary is chopped by MOSFETs Q1 and Q2, which receive complementary drive from the 74C04 based square wave oscillator. A1 supplies power to the oscillator. T1 provides voltage step-up. Its rectified and filtered output is the boost stage's output. The 1M-10k divider furnishes feedback to A2, closing a loop around it. The 0.01μF capacitor from Q3's emitter to A2's negative input gives loop stability and the 0.002μF unit trims step response damping. C1 is used for short-circuit limiting. Current from Q1 and Q2 passes through the 0.1Ω shunt. Abnormal output currents cause shunt voltage to rise, tripping C1's output low. This simultaneously removes drive from Q3, Q1 and Q2's gates and the oscillator, resulting

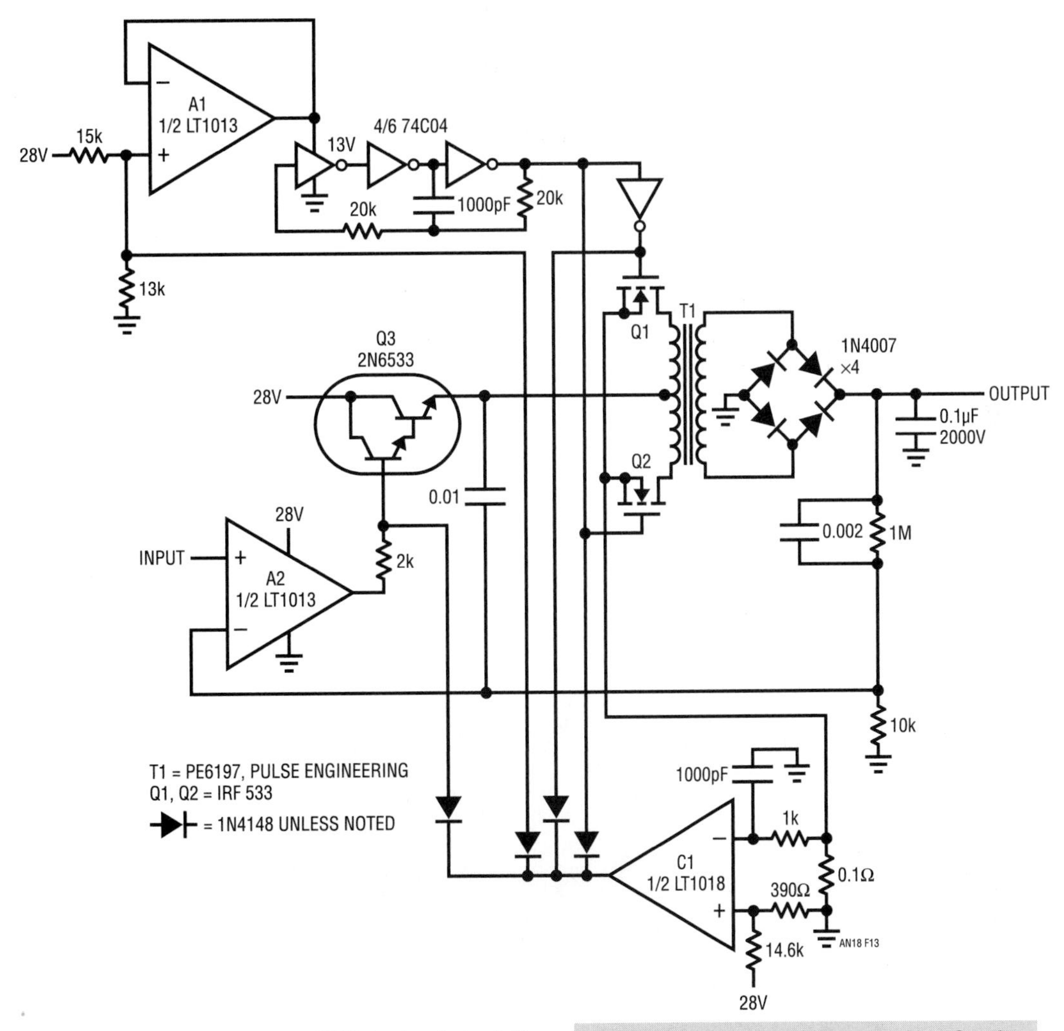

Figure 20.13 • 15 Watt, 1000V Unipolar Output Stage. DANGER! High Voltages Present. Use Caution

in output shutdown. The 1k-1000pF filter ensures that C1 does not trip due to current spikes or noise during normal operation.

A2 supplies whatever drive is required to close the loop, regardless of the output voltage called for. The low, resistive saturation losses of the VMOS FETs combined with A2's servo action allows controlled outputs all the way down to 0 V.

Substituting higher power devices for Q1 and Q2 along with a larger transformer allows more output power, although dissipation in Q3 will become excessive. If higher power is desired, a switched mode stage should be substituted for Q3 to maintain efficiency.

The 0.1µF filter capacitor at the output limits full-power bandwidth to about 60Hz. Figure 20.14 shows dynamic response at full load. Trace A, a 10V input, produces a 1000V output in Trace B. Note that slew is faster on the leading edge, because the stage cannot sink current. The falling edge slew rate is determined by the load resistance.

±15V powered, bipolar output, voltage gain stage

Figure 20.13's output is limited to unipolar operation because the step-up transformer cannot pass DC polarity information.

Obtaining bipolar output from a transformer based voltage booster requires some form of DC polarity restoration at the output. Figure 20.15's ±15V powered circuit does this, using synchronous demodulation to preserve polarity in its ±100V output. This booster features 150mA current output, 150Hz full-power output and a slew rate of 0.1V/µs.

The high voltage output is generated in similar fashion to Figure 20.13's circuit. The 74C04 based oscillator furnishes complementary gate drive to VMOS devices Q1 and Q2, which chop Q3's output into T1, a step-up transformer. In this design, however, a synchronously switched absolute value amplifier is placed between servo amplifier A1 and Q3's drive point. Input signal polarity information, derived from A1's output, causes C1 to switch the LTC1043 section located at A2's positive input. This circuitry is arranged so that A2's output is the positive absolute value of A1's input signal. A second, synchronously switched LTC1043 section gates oscillator pulses to the appropriate SCR trigger transformer at the output. For positive inputs LTC1043 Pins 2 and 6 are connected, as well as Pins 3 and 18. A2, acting as a unity gain follower, passes A1's output directly and drives Q3. Simultaneously, oscillator pulses are conducted through an inverter via LTC1043 Pin 18. The inverter drives trigger transformer T2, turning Q4 on. Q4, biased from the full wave bridge's positive point, supplies positive polarity voltage to the output.

Negative inputs cause the LTC1043 switch positions to reverse. A2, functioning as an inverter, again supplies Q3 with positive voltage drive. The Schottky diode at A2 prevents the LTC1043 from seeing transient negative voltages. Oscillator pulses are directed to SCR Q5 via LTC1043 Pin 15, its associated inverter and T3. This SCR connects the full wave bridge's negative point to the output. Both SCR cathodes are tied together to form the circuit's output. The

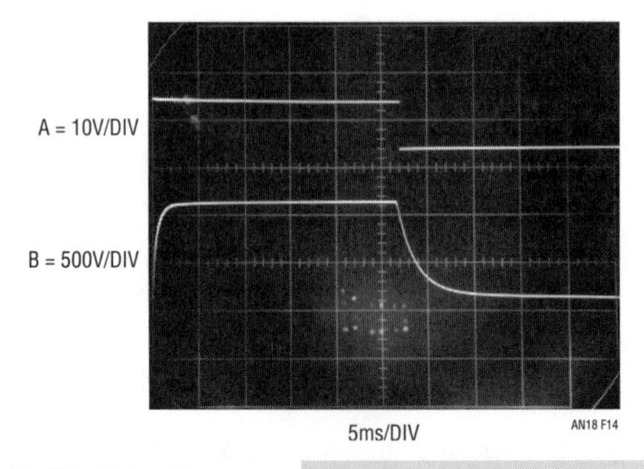

Figure 20.14 • Figure 20.13's Pulse Response. DANGER! High Voltage Present. Use Caution

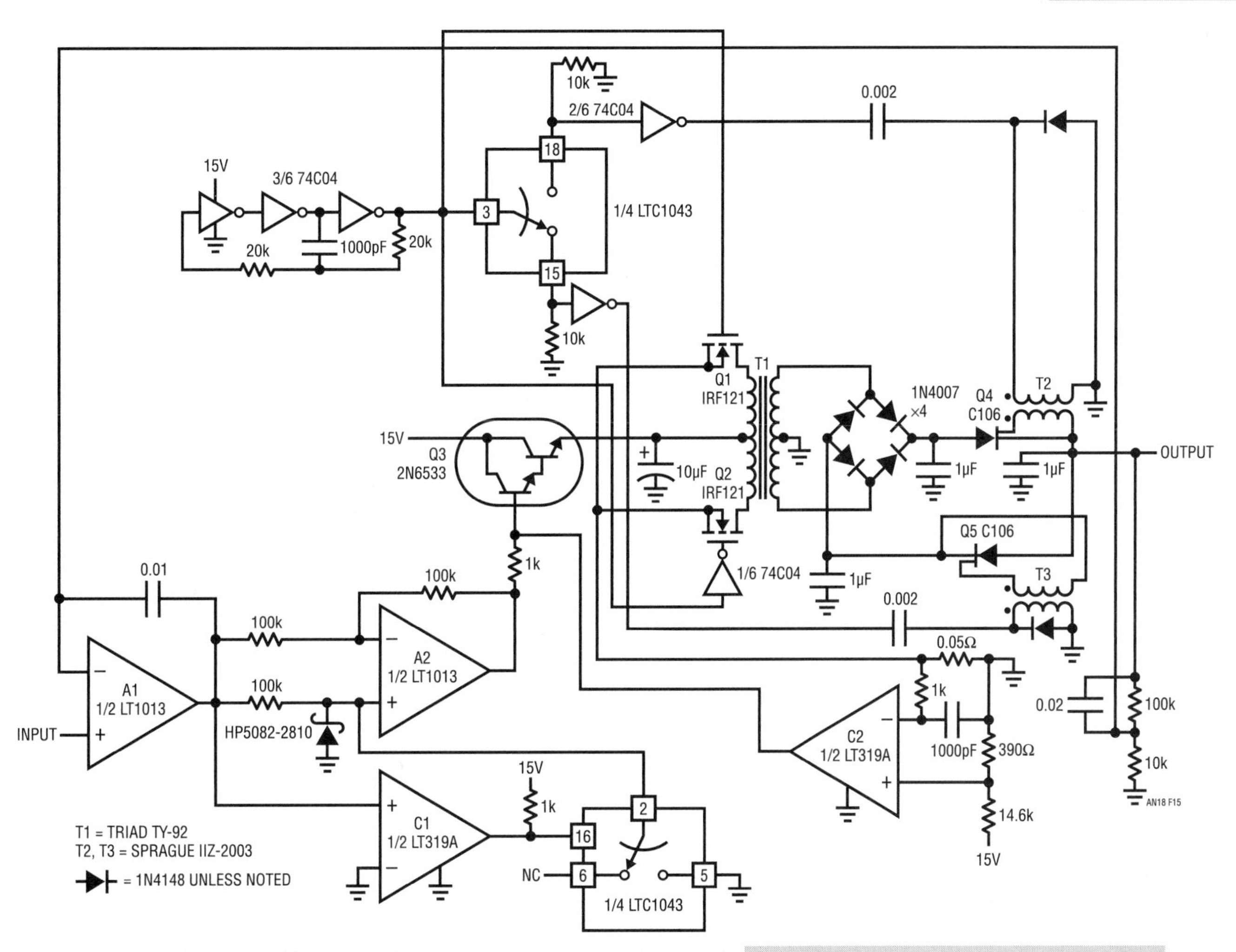

Figure 20.15 • ±100V Output Stage Runs from ±15V Supply. DANGER! High Voltage Present. Use Caution

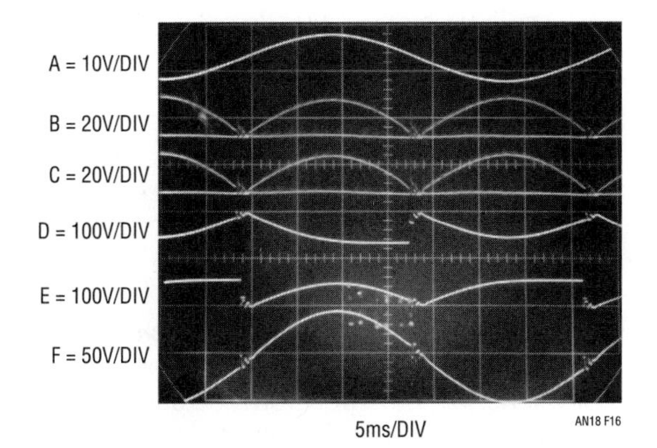

Figure 20.16 • Figure 20.15's Operating Details. DANGER! High Voltages Present. Use Caution

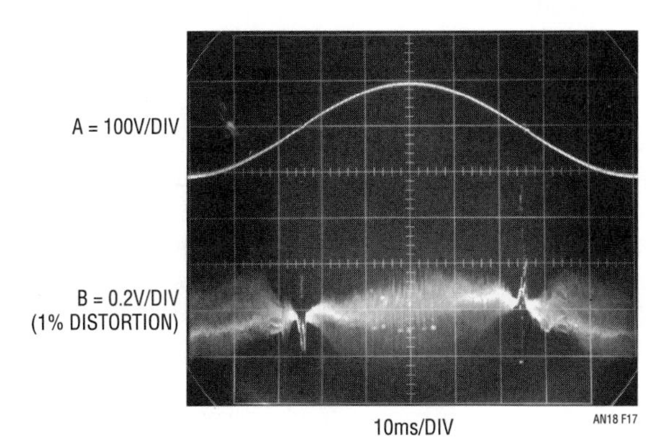

Figure 20.17 • Crossover Residue Reflects Chopping and Zero Cross Switching. DANGER! High Voltages Present. Use Caution

100k-10k divider supplies feedback to A1 in the conventional manner. The synchronous switching allows polarity information to be preserved in the stage's output, permitting full bipolar operation. Figure 20.16 shows waveforms for a sine wave input. Trace A is A1's input. Traces B and C are Q1 and Q2's drain waveforms. Traces D and E are the full wave bridge's negative and positive outputs, respectively. Trace F, the circuit output, is an amplified, reconstructed version of A1's input. Phase skewing between the SCR switching and the carrier borne signal causes some distortion at the zero crossover. The amount of skew is both load and signal frequency dependent, and is not readily compensated. Figure 20.17 shows distortion products (Trace B) at a 10Hz output (Trace A) at full load (±100V at 150mA peak). Residual high frequency carrier components are clearly present, and the zero point SCR switching causes the sharp peaks. RMS distortion measured 1% at 10Hz, rising to 6% at 100Hz.

C2 supplies current limiting in identical fashion to Figure 20.13's scheme. Frequency compensation is also similar. A 0.01µF capacitor at A1 gives loop stability, while the 0.02µF feedback unit sets damping. Figure 20.18 summarizes the capabilities of the power gain stages presented, and should be useful in selecting an approach for a given application.

FIGURE	VOLTAGE GAIN	CURRENT GAIN	FULL-POWER BANDWIDTH	COMMENTS
1a	No	Yes, 150mA Output	600kHz	Simple, Easy
1b	No	Yes, 150mA Output	1.5MHz	Simple, Easy
2	No	Yes, 3A	100kHz	
3	No	Yes, 200mA	7.5MHz	Feedforward Technique Gives High Bandwidth >1000V/µs Slew. Inverting Operation Only
5a, 5b	Yes	No	Depends On Op Amp	Simple Stages Allow Wide Swing, Almost to Rails
7	Yes	Yes, 125mA	600kHz	High Current, Nearly Rail-to-Rail Swing Capability
9	Yes, ±120V	Yes, 25mA	15kHz	Good, General Purpose High Voltage Stage
11	Yes, ±120V	Yes, 25mA	12kHz	Almost Indestructible Output
13	Yes, 1000V	No	60Hz	High Voltage Output with No External High Voltage Supplies Required. Limited Bandwidth with Asymmetrical Slewing. Positive Outputs Only
15	Yes, ±100V	Yes, 150mA	150Hz	High Voltage Outputs with No External High Voltage Supplies. Limited Bandwidth. Full Bipolar Output.

Figure 20.18 • Summary of Circuit Characteristics

The oscillation problem (Frequency compensation without tears)

All feedback systems have the propensity to oscillate. Basic theory tells us that gain and phase shift are required to build an oscillator. Unfortunately, feedback systems, such as operational amplifiers, have gain and phase shift. The close relationship between oscillators and feedback amplifiers requires careful attention when an op amp is designed. In particular, excessive input-to-output phase shift can cause the amplifier to oscillate when feedback is applied. Further, any time delay placed in the amplifier's feedback path introduces additional phase shift, increasing the likelihood of oscillation. This is why feedback loop enclosed power gain stages can cause oscillation.

A large body of complex mathematics is available which describes stability criteria, and can be used to predict stability characteristics of feedback amplifiers. For the most sophisticated applications, this approach is required to achieve optimum performance.

However, little has appeared which discusses, in practical terms, how to understand and address the issues of compensating feedback amplifiers. Specifically, a practical approach to stabilizing amplifier-power gain stage combinations is discussed here, although the considerations can be generalized to other feedback systems.

Oscillation problems in amplifier-power booster stage combinations fall into two broad categories; *local* and *loop* oscillations. *Local* oscillations can occur in the boost stage, but should not appear in the IC op amp, which presumably was debugged prior to sale. These oscillations are due to transistor parasitics, layout and circuit configuration caused instabilities. They are usually relatively high in frequency, typically in the 0.5MHz to 100MHz range. Usually, local booster stage oscillations do not cause loop disruption. The major loop continues to function, but contains artifacts of the local oscillation. Text Figure 20.7 furnishes an instructive example. The Q3-Q5 and Q4-Q6 pairs have high gain bandwidth. The intended resistive feedback loops allow them to oscillate in the 50MHz to 100MHz region without the 100pF-200Ω network shunting the DC feedback. This network rolls off gain bandwidth, preventing oscillation. It is worth noting that a ferrite bead in series with the 2k resistor will give similar results. In this case, the bead raises the inductance of the wire, attenuating high frequencies.

The photo in Figure B1 shows text Figure 20.7 following a bipolar square wave input with the local high frequency RC compensation networks removed. The resultant high frequency oscillation is typical of locally caused disturbances. Note that the major loop is functional, but the local oscillation corrupts the waveform.

Eliminating such local oscillations starts with device selection. Avoid high F_t transistors unless they are needed. When high frequency devices are in use, plan layout carefully. In very stubborn cases, it may be necessary to lightly bypass transistor junctions with small capacitors or RC networks. Circuits which use local feedback can sometimes require careful transistor selection and use. For example, transistors operating in a local loop may require different F_t's to achieve stability. Emitter followers are notorious sources of oscillation, and should never be directly driven from low impedance sources.

Text Figure 20.5 uses an RC damper network from the 74C04 inverters to ground to eliminate local oscillations. In that circuit the 74C04s are forced to run in their linear region. Although their DC gain is low, bandwidth is high. Very small parasitic feedback terms result in high frequency oscillations. The damper network provides a low impedance to ground at high frequency, breaking up the unwanted feedback path.

Loop oscillations are caused when the added gain stage supplies enough delay to force substantial phase shift. This causes the control amplifier to run too far out of phase with the gain stage. The control amplifier's gain combined with the added delay causes oscillation. Loop oscillations are usually relatively low in frequency, typically 10Hz to 1MHz.

A good way to eliminate loop caused oscillations is to limit the gain bandwidth of the control amplifier. If the booster stage has higher gain bandwidth than the control amplifier, its phase delay is easily accommodated in the loop. When control amplifier gain bandwidth dominates, oscillation is assured. Under these conditions, the control amplifier hopelessly tries to servo a feedback signal which consistently arrives "too late." The servo action takes the form of an electronic tail chase, with oscillation centered around the ideal servo point.

Frequency response roll-off of the control amplifier will almost always cure loop oscillations. In many situations it is preferable to "brute force" compensation using large capacitors in the major feedback loop. As a general rule, it is wise to stabilize the loop by rolling off control amplifier gain bandwidth. The feedback capacitor serves to trim step response only and should not be relied on to stop outright oscillation.

Figures B2 and B3 illustrate these issues. The 600kHz gain bandwidth LT1012 amplifier used with the LT1010 current buffer produces the output shown in Figure B2. The LT1010's 20MHz gain bandwidth introduces negligible loop delay, and dynamics are clean. In this case, the LT1012's internal roll-off is well below that of the output stage, and stability is achieved with no external compensation components. Figure B3 uses a 15MHz LT318A as the control amplifier. The associated photo shows the results. Here, the control amplifier's roll-off, close to the output stages, causes problems. The phase shift through the LT1010 is now appreciable and oscillations occur. Stabilizing this circuit requires degenerating the LT318A's gain bandwidth (see text Figure 20.1).

The fact that the slower op amp circuit doesn't oscillate is a key to understanding how to compensate booster loops. With the slow device, compensation is "free". The faster amplifier makes the AC characteristics of the output stage become significant and requires roll-off components for stability.

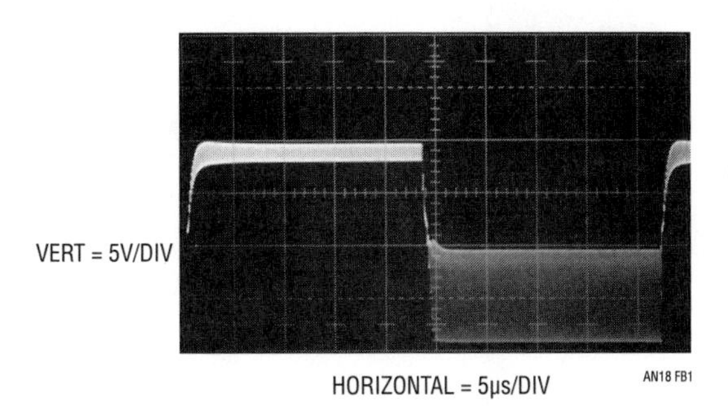

Figure B1 • Typical Local Output Stage Oscillation

Text Figure 20.9's high voltage stage is an interesting case. The high voltage transistors are very slow devices, and the LT1055 amplifier has a much higher gain bandwidth than the output stage. The LT1055 is locally compensated by the 10k-100pF network, giving it an integrator-like response. This compensation, combined with the damping provided by the 33pF feedback capacitor, gives good loop response. The procedure used to compensate this circuit is typical of what is done to stabilize boosted amplifier loops and is worth reviewing.

With no compensation components installed, the circuit is turned on and oscillations are observed (photo, Figure B4). The relatively slow oscillation frequency suggests a loop oscillation problem. The LT1055 gain bandwidth is degenerated with the RC components around the amplifier. The RC time constant is chosen

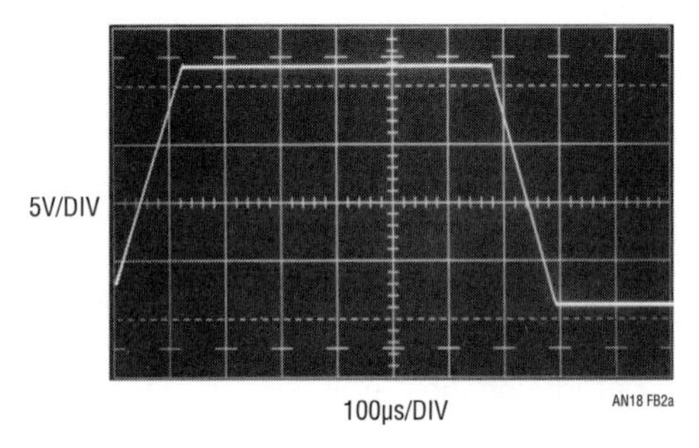

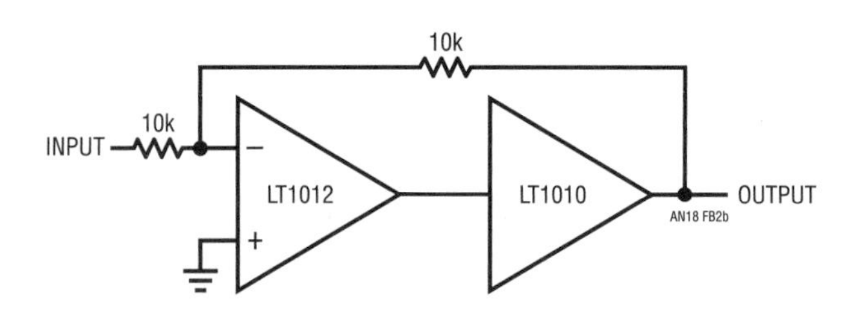

Figure B2 • A Slow Control Amplifier Gives "Free" Loop Compensation

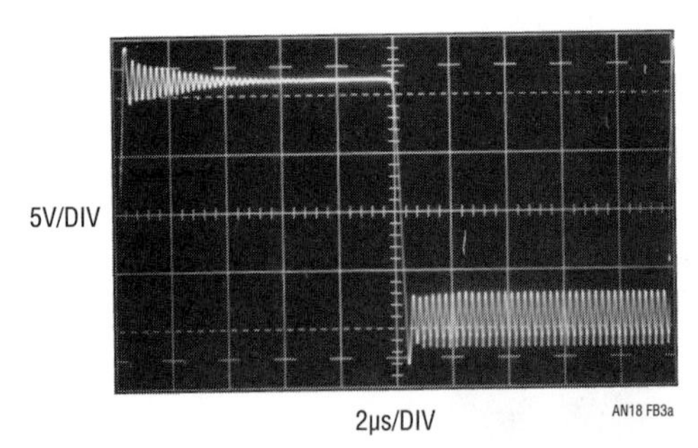

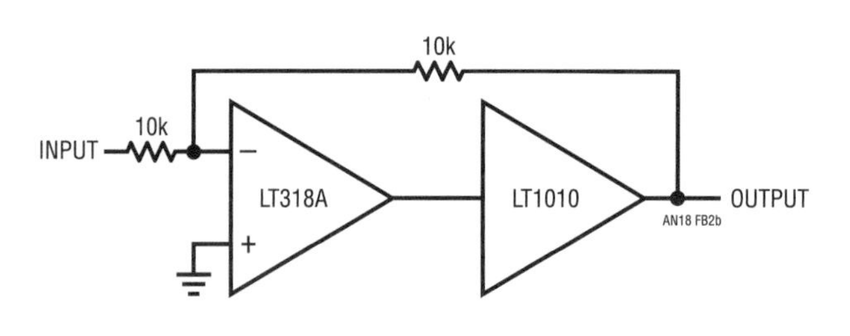

Figure B3 • A Fast Control Amplifier Gives "Free" Loop Oscillation

to eliminate oscillations and give the best possible response (photo, Figure B5) with no loop feedback capacitor in place. Observe that the 1µs time constant selected offers significant attenuation at the oscillation frequency noted in the photo, Figure B4. Finally, the loop feedback capacitor (33pF) is selected to give the optimum damping shown in text Figure 20.10.

When making tests like these, remember to investigate the effects of various loads and output operating voltages. Sometimes a compensation scheme which appears fine gives bad results for some output conditions. For this reason, check the completed circuit over as wide a variety of operating conditions as possible.

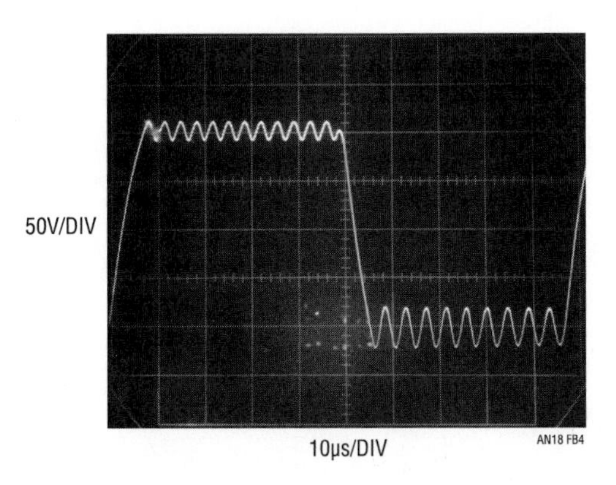

Figure B4 • Oscillations After Slewing Suggest a Loop Problem

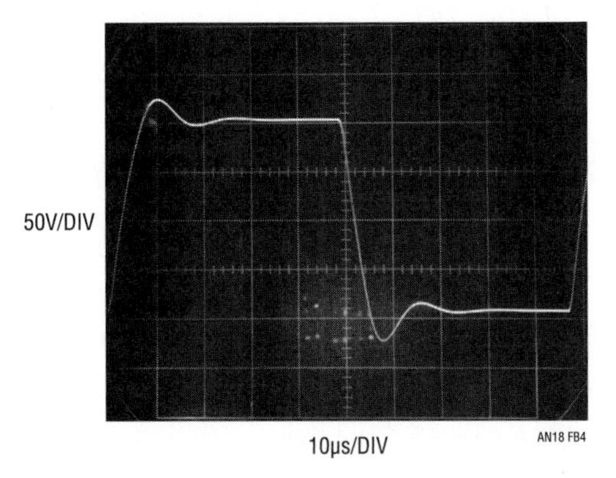

Figure B5 • Control Amplifier Roll-Off Stabilizes B4's Problems

References

1. Roberge, J. K.; Operational Amplifiers: Theory and Practice, Chapters IV and V; Wiley
2. Tobey, Graeme, Huelsman; Operation Amplifiers, Chapter 5; McGraw-Hill
3. Janssen, J. and Ensing, L.; The Electro-Analogue, An Apparatus For Studying Regulating Systems; Philips Technical Review; March 1951
4. Williams, J.; Thermal Techniques in Measurement and Control Circuitry; Application Note 5, pages 1–3; Linear Technology Corporation
5. EICO Corp.; Series-Parallel RC Combination Decade Box; Model 1140A

Composite amplifiers

21

Jim Williams

Amplifier design, regardless of the technology utilized, is a study in compromise. Device limitations make it difficult for a particular amplifier to achieve optimal speed, drift, bias current, noise and power output specifications. As such, various amplifier families emphasizing one or more of these areas have evolved. Some amplifiers are very good attempts at doing everything well, but the best achievable performance figures are limited to dedicated designs.

Practical applications often require an amplifier that has extremely high performance in several areas. For example, high speed and DC precision are often needed. If a single device cannot simultaneously achieve the desired characteristics, a composite amplifier made up of two (or more) devices can be configured to do the job. Composite designs combine the best features of two or more amplifiers to achieve performance unobtainable in a single device. More subtly, composite designs permit circuit approaches which are normally impractical. This is particularly true of high speed stages which may be designed with little attention to DC biasing considerations if a separate stabilizing stage is employed.

Figure 21.1 shows a composite made up of an LT®1012 low drift device and an LT1022 high speed amplifier. The overall circuit is a unity-gain inverter, with the summing node located at the junction of three 10k resistors. The LT1012 monitors this summing node, compares it to ground and drives the LT1022's positive input, completing a DC stabilizing loop around the LT1022. The 10k-300pF time constant at the LT1012 limits its response to low frequency signals. The LT1022 handles high frequency inputs while the LT1012 stabilizes the DC operating point. The 4.7k-220Ω divider at the LT1022 prevents excessive input overdrive during start-up. This circuit combines the LT1012's 35μV offset and 1.5V/°C

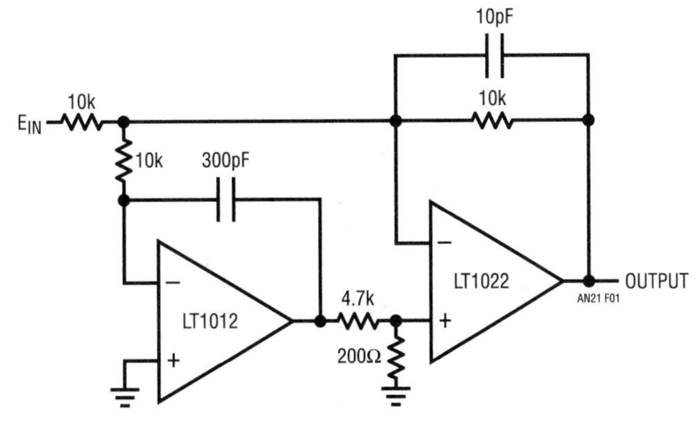

Figure 21.1 • Basic DC-Stabilized Fast Amplifier

Analog Circuit and System Design: Immersion in the Black Art of Analog Design. http://dx.doi.org/10.1016/B978-0-12-397888-2.00021-3

drift with the LT1022's 23V/µs slew rate and 300kHz full-power bandwidth. Bias current, dominated by the LT1012, is about 100pA.

Figure 21.2 is similar, but uses discrete FETs to more than triple the speed. Here A1's input stage is turned off by connecting its inputs to the negative rail. The differentially connected FETs bias the second state via A1's offset pins. This connection replaces A1's input stage, reducing bias current and increasing speed. FET mismatch would normally result in excessive offset and drift. A2 corrects this by monitoring the summing point (the junction of the two 4.7k resistors) and forcing Q2's gate to eliminate overall offset. The 10k-1000pF pair limits A2's response to low frequency, and the 1k divider chain prevents overdrive to Q2 on startup. The 1k-10pF damper at the summing node aids high frequency stability. Figure 21.3 shows pulse response. Trace A is the input and Trace B the output. Slew rate exceeds 100V/µs, with clean damping. Full-power bandwidth is about 1MHz, and the input bias current is in the 100pA range. DC offset and drift are similar to Figure 21.1.

Figure 21.4 shows a highly stable unity-gain buffer with good speed and high input impedance. Q1 and Q2 constitute a simple, high speed FET input buffer. Q1 functions as a source follower, with the Q2 current source load setting the drain-source channel current. The LT1010 buffer provides output drive capability for cables or whatever load is required. Normally, this open-loop configuration would be quite drifty because there is no DC feedback. The LTC ®1052 contributes this function to stabilize the circuit. It does this by comparing the filtered circuit output to a similarly filtered version of the input signal. The amplified difference between these signals is used to set Q2's bias, and hence Q1's channel current. This forces Q1's V_{GS} to whatever voltage is required to match the circuit's input and output potentials. The 2000pF capacitor at A1 provides stable loop compensation. The RC network in A1's output prevents it from seeing high speed edges coupled through Q2's collector-base junction. A2's output is also fed back to the shield around Q1's gate lead, bootstrapping the circuit's effective input capacitance down to less than 1pF.

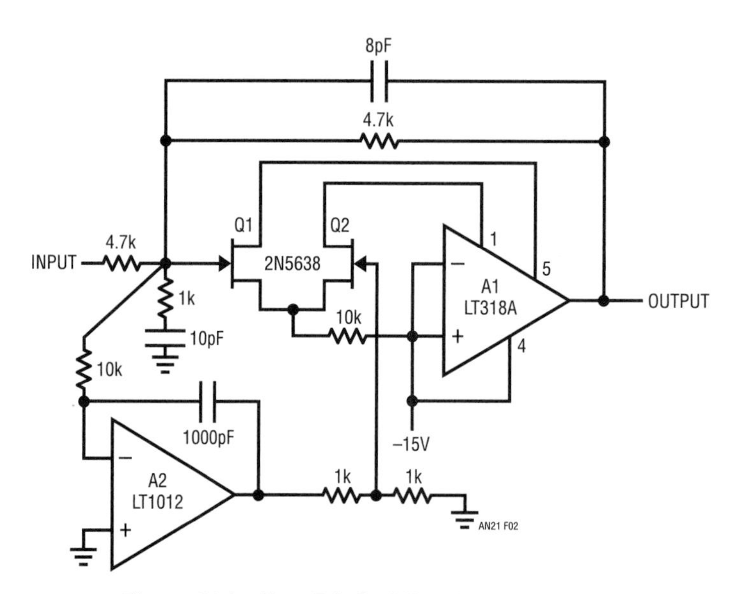

Figure 21.2 • Fast DC-Stabilized FET Amplifier

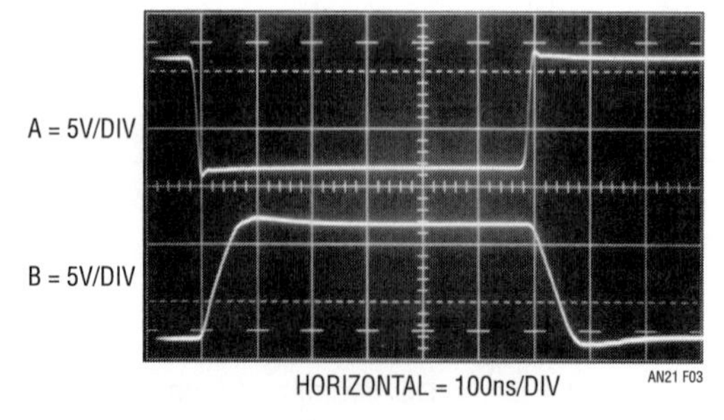

Figure 21.3 • Figure 21.2's Waveforms

The LT1010's 15MHz bandwidth and 100V/μs slew rate, combined with its 150mA output, are fast enough for most circuits. For very fast requirements, the alternate discrete component buffer shown will be useful. Although its output is current limited at 75mA, the GHz range transistors employed provide exceptionally wide bandwidth, fast slewing and very little delay. Figure 21.5 shows the LTC1052 stabilized buffer circuit's response using the discrete stage. Response is clean and quick, with delay inside 4ns. Slew exceeds 2000V/μs with full-power bandwidth approaching 50MHz. Note that rise time is limited by the pulse generator and not the circuit. For either stage, offset is set by the LTC1052 at 5μV, with gain about 0.95.

A potential difficulty with Figure 21.4's circuit is that the gain is not quite unity. Figure 21.6 maintains high speed and low bias while achieving a true unity-gain transfer function.

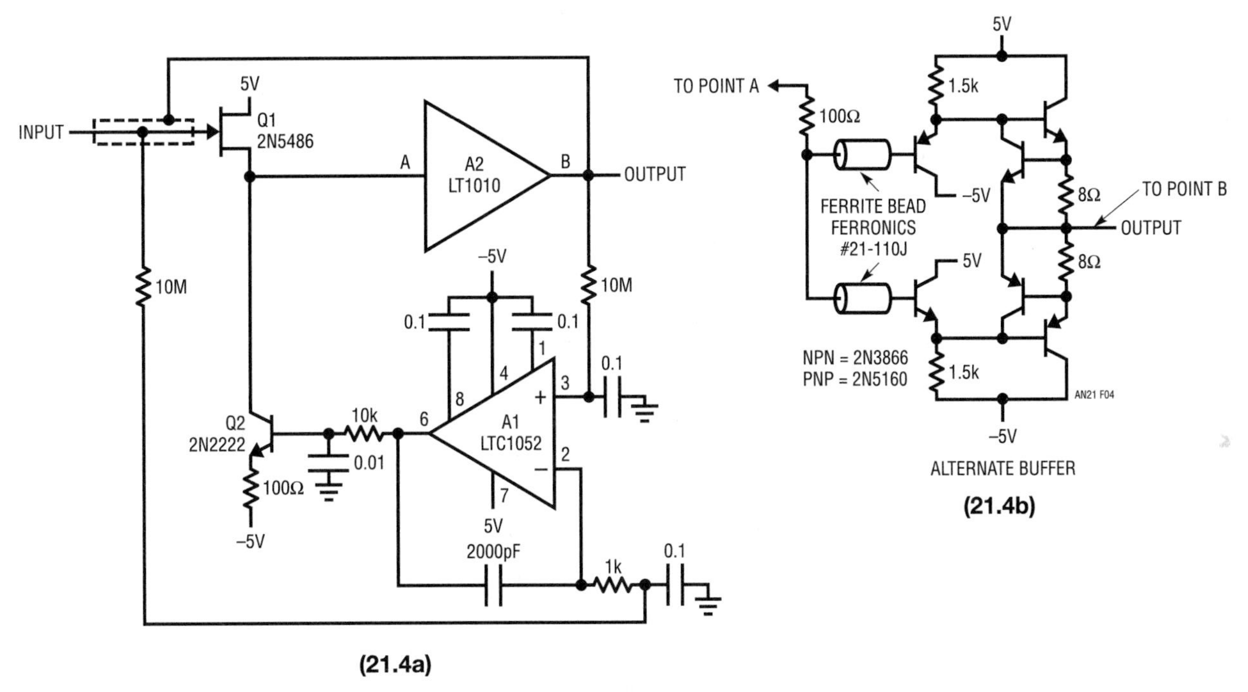

(21.4a)

Figure 21.4 • Wideband FET Input Stabilized Buffer

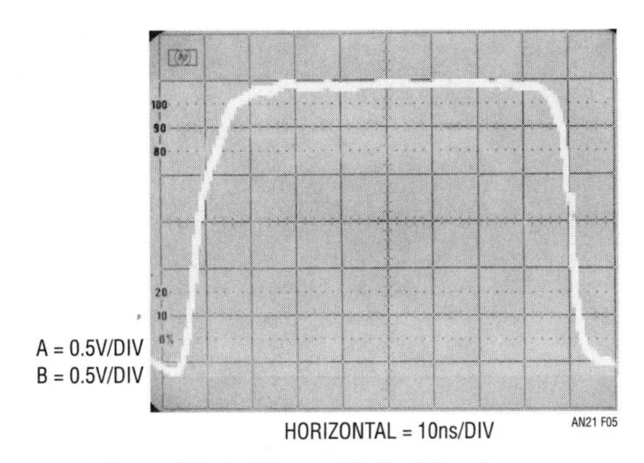

Figure 21.5 • Figure 21.4's Waveforms

This circuit is somewhat similar to Figure 21.4, except that the Q2-Q3 stage takes gain. A2 DC stabilizes the input-output path, and A1 provides drive capability. Feedback is to Q2's emitter from A1's output. The 1k adjustment allows the gain to be precisely set to unity. With the LT1010 output stage slew and full-power bandwidth ($1V_{P-P}$) are 100V/μs and 10MHz, respectively. −3dB bandwidth exceeds 35MHz. At A = 10 (e.g., 1k adjustment set at 50Ω) full-power bandwidth stays at 10MHz while the −3dB point falls to 22MHz.

With the optional discrete stage, slew exceeds 1000V/μs and full-power bandwidth ($1V_{P-P}$) is 18MHz. −3dB bandwidth is 58MHz. At A = 10, full power is available to 10MHz, with the −3dB point at 36MHz.

Figures 21.7a and 12.7b show response with both output stages. The LT1010 is used in Figure 21.7a (Trace A = input, Trace B = output). Figure 21.7b uses the discrete stage and is slightly faster. Either stage provides more than adequate performance for driving video cable or data converters, and the LT1012 maintains DC stability under all conditions.

Figure 21.8 is another DC-stabilized fast amplifier which functions over a wide range of gains (typically 1-10). It combines the LT1010 and a fast discrete

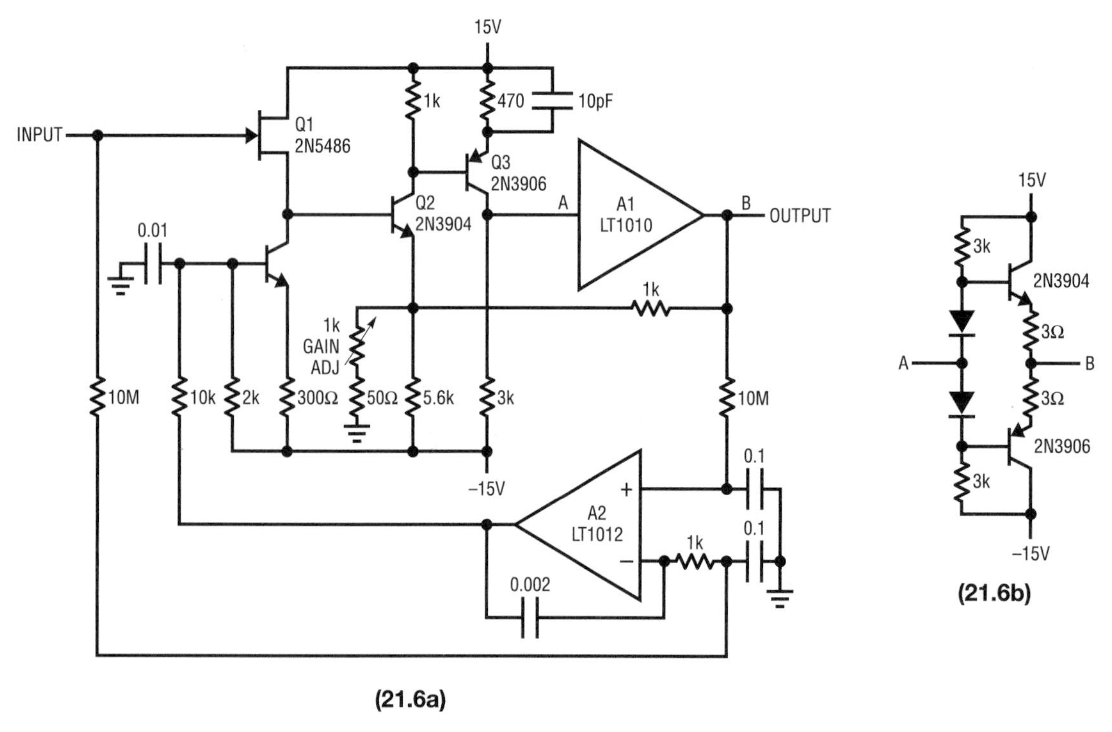

(21.6a)

Figure 21.6 • Gain Trimmable Wideband FET Amplifier

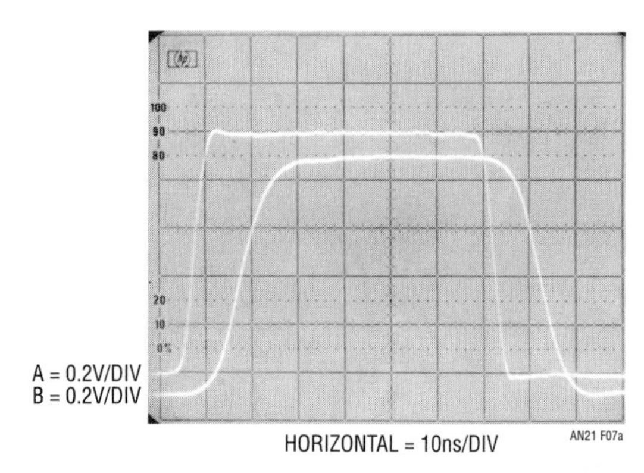

Figure 21.7a • Figure 21.6's Waveforms Using the LT1010

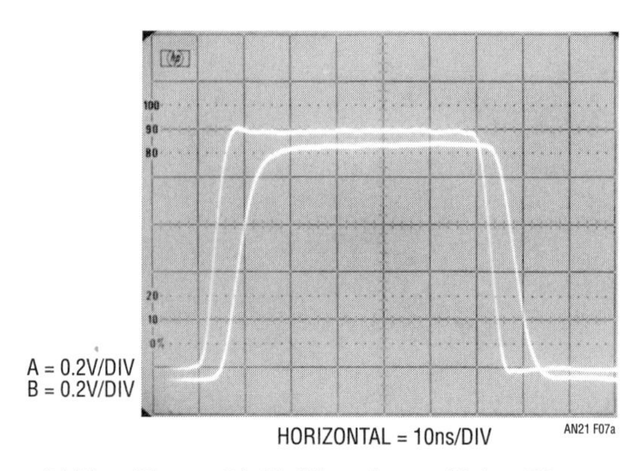

Figure 21.7b • Figure 21.6's Waveforms Using Discrete Stage

stage with an LT1008 based DC stabilizing loop. Q1 and Q2 form a differential stage which single-ends into the LT1010. The circuit delivers $1V_{P-P}$ into a typical 75Ω video load. At A = 2, the gain is within 0.5dB to 10MHz with the −3dB point occurring at 16MHz. At A = 10, the gain is flat (±0.5dB to 4MHz) with a −3dB point at 8MHz. The peaking adjustment should be optimized under loaded output conditions.

Normally, the Q1-Q2 pair would be quite drifty, but the LT1008 corrects for this. This correction stage is similar to the one in Figures 21.4 and 21.6, except that the feedback is taken from a divided down sample of the fast amplifier. The ratio of this divider should be set to the same value as the circuit's closed-loop gain. Frequency roll-off of this stage is set by the 1M-0.022μF filters in the LT1008's input lines. The 0.22μF capacitor at the amplifier eliminates oscillations. The DC loop servo controls drift by biasing the DC operating point of Q2's collector to force zero error between the LT1008's inputs.

This is a simple stage for fast applications where relatively low output swing is required. Its $1V_{P-P}$ output works nicely for video circuits. A possible problem is the relatively high bias current, typically 10μA. Additional swing is possible, but more circuitry is needed.

Figure 21.9's circuit addresses these issues. It trades speed for output swing and reduced bias current. As before, a sepa-

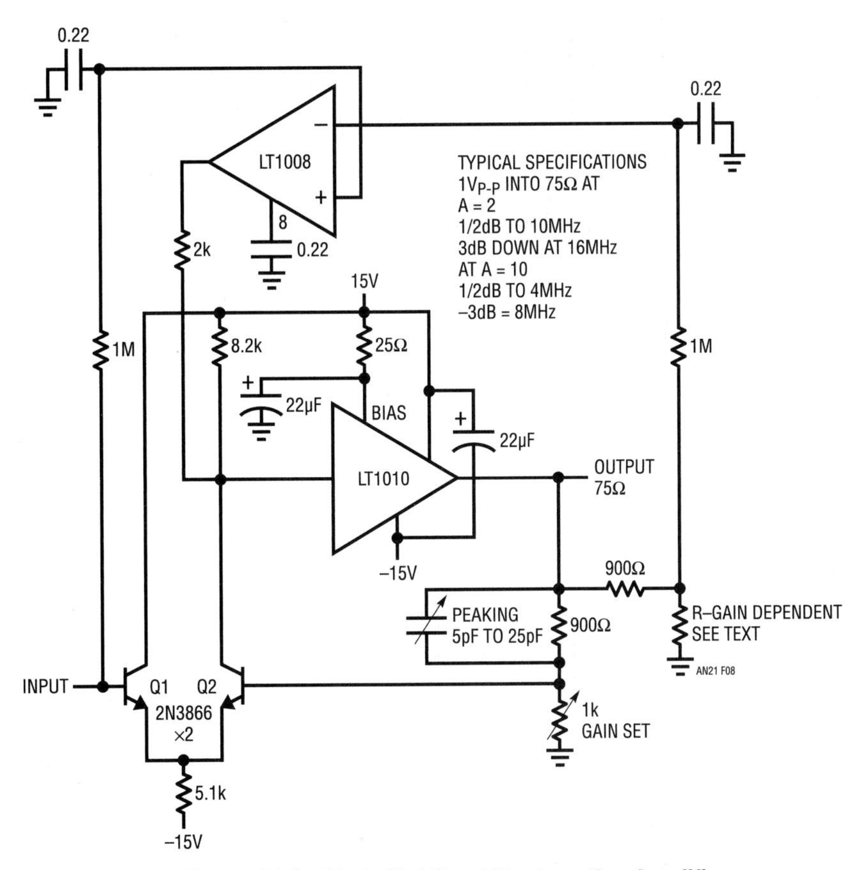

Figure 21.8 • Fast, Stabilized Noninverting Amplifier

rate loop maintains DC stability. This circuit is a good example of an approach made practical by composite techniques. Without the separate stabilizing loop, the DC imbalances in the signal path would preclude any level of operation.

In this arrangement a PNP level-shifting stage (Q4) has been added to Figure 21.8's circuit to increase available swing at the LT1010 output. This is obtained at the expense of available bandwidth and amplifier stability. The 33pF capacitor from Q4's collector to the circuits summing node (Q3's gate) affords stable loop compensation.

Figure 21.8's bias current errors are eliminated by Q3, a FET source follower. This device buffers the summing point from the relatively high bias current required by Q2. Normally, this configuration would cause volts of offset, due to Q3's gate-source voltage. Here, A1 closes a DC restoration loop, forcing Q1's base to whatever point is required to compensate for the offset. Thus, A1's operation not only provides low DC error but permits a simplistic approach to minimizing summing point bias current. Figure 21.10 shows operating waveforms for a 10V output. Trace A is the input, while Trace B is the output.

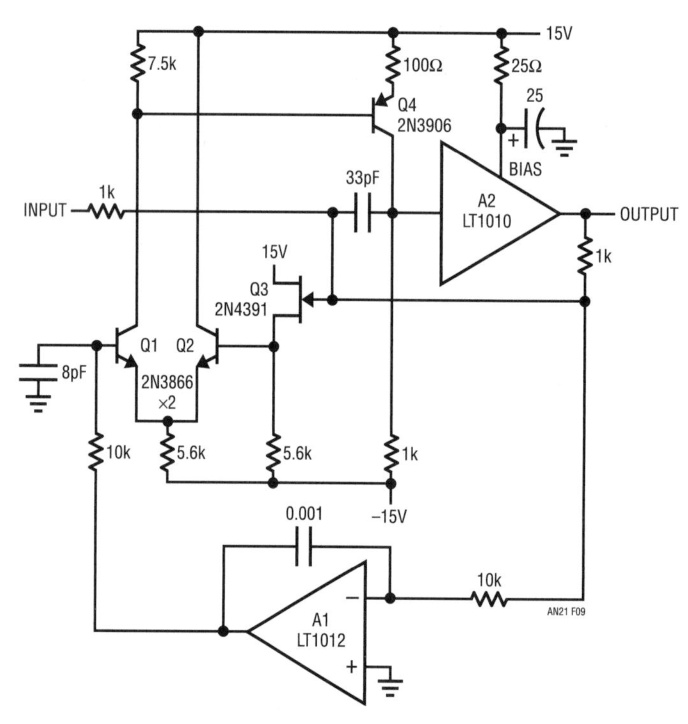

Figure 21.9 • Fast, Stabilized Inverting Amplifier with Low Summing Point Bias Current

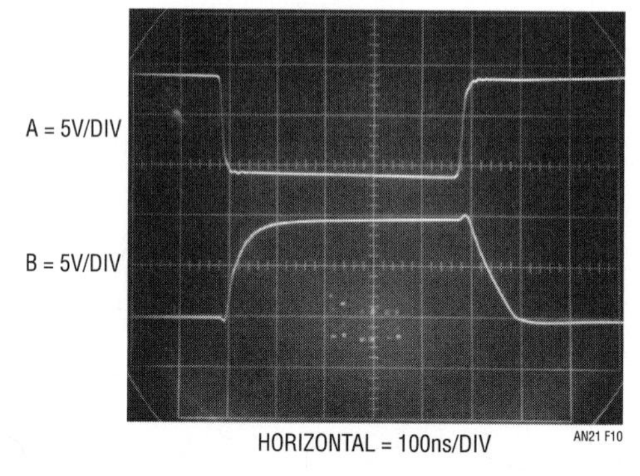

Figure 21.10 • Figure 21.9's Pulse Response

Slew rate is about 100V/μs, with a full-power bandwidth of 1MHz. The LT1010 allows 100mA outputs and makes cable driving practical at these speeds.

Figure 21.11 shows another fast stage with wide output swing. The circuit is noninverting, and has higher input impedance than Figure 21.9. Additionally, it's operation is based on an arrangement commonly referred to as "current mode" feedback. This technique, well established in RF design and also employed in some monolithic instrumentation amplifiers, permits fixed bandwidth over a wide range of closed-loop gains. This contrasts with normal feedback schemes, where bandwidth degrades as closed-loop gain increases.

The overall amplifier is composed of two LT1010 buffers and a gain stage, Q1 and Q2. A3 acts as a DC restoration loop. The 33Ω resistors sense A1's operating current, biasing Q1 and Q2. These devices furnish complementary voltage gain to A2, which provides the circuit's output. Feedback is from A2's output to A1's output, which is a low impedance point.

A3's stabilizing loop compensates large offsets in the signal path, which are dominated by mismatch in Q1 and Q2. Correction is implemented by controlling the current through Q3, which shunts Q2's base bias resistor. Adequate loop capture range is assured by deliberate skewing of Q1's operating point via the 330Ω unit. The 9k-1k feedback divider feeding A3 is selected to equal the gain ratio of the circuit, in this case 10.

The feedback scheme makes A1's output look like the negative input of the amplifier, with closed-loop gain set by the ratio of the 470Ω and 51Ω resistors. The outstanding feature of this connection is that bandwidth becomes relatively independent of closed-loop gain over a reasonable range. For this circuit, full-power bandwidth remains at 1MHz over gains of 1 to about 20. The loop is quite stable, and the 15pF value at A2's input provides good damping over a wide range of gains. The LT1010 buffers limit bandwidth in this circuit. Dramatic speed improvement is possible if they are replaced by discrete stages.

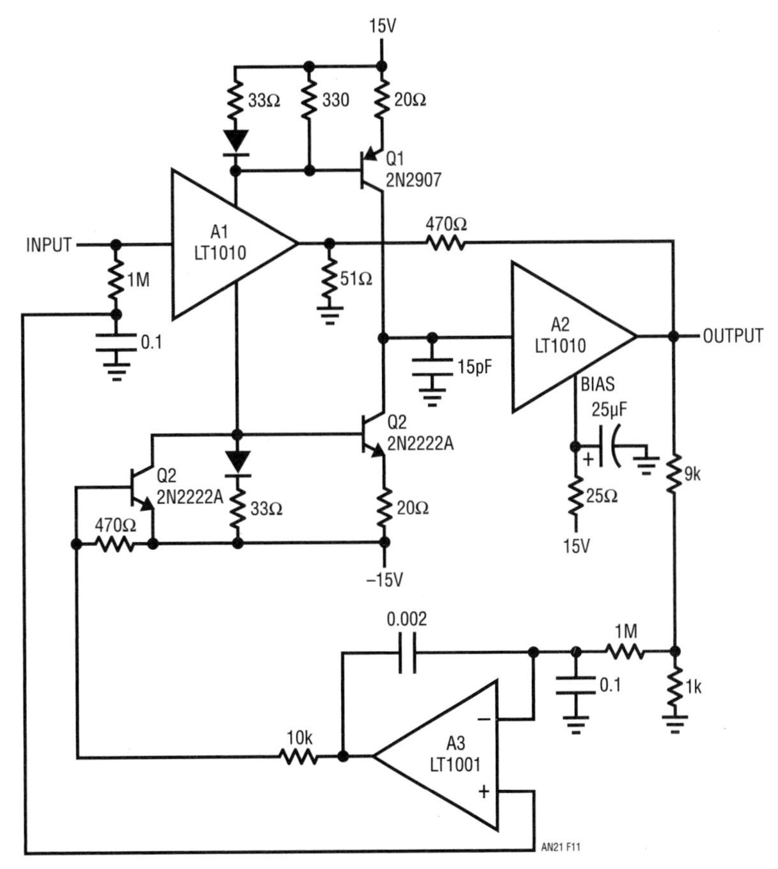

Figure 21.11 • "Current Mode Feedback" Amplifier

Figure 21.12 substitutes discrete elements for Figure 21.11's LT1010s. Although this arrangement is substantially more complex, it provides an extraordinarily wideband amplifier. This composite design is composed of three amplifiers; the discrete wideband stage, a quiescent current control amp and an offset servo. Q1-Q4 replace Figure 21.11's A1, although complementary voltage gain is taken at the collectors Q3 and Q4. Q5 and Q6 provide additional gain, similar to Q1 and Q2 in Figure 21.11. Q7-Q10 form the output buffer stage. The feedback scheme is identical to Figure 21.11's, with summing action at the Q3-Q4 emitter connection. To obtain

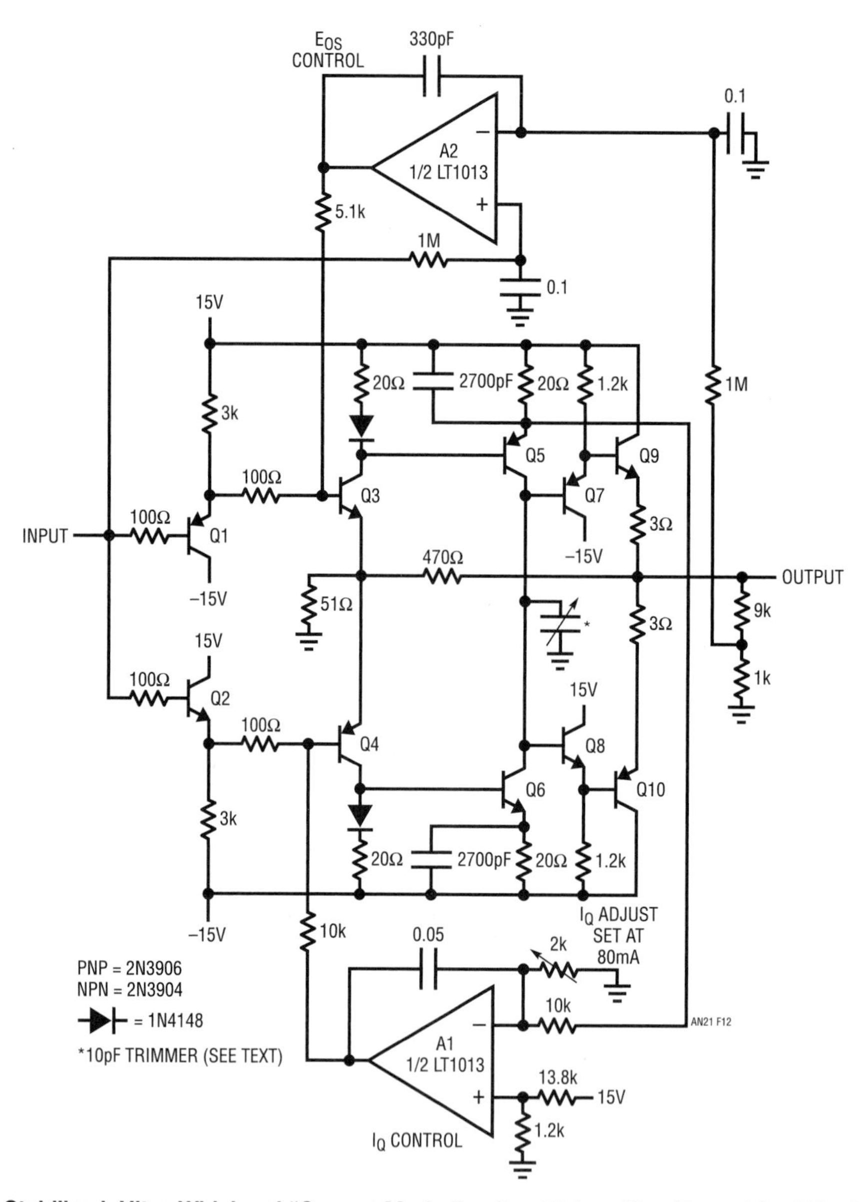

Figure 21.12 • Stabilized, Ultra-Wideband "Current Mode Feedback" Amplifier (Son of Godzilla Amplifier)

maximum bandwidth, quiescent current is quite high. Without closed-loop control, the circuit will quickly go into thermal runaway and destroy itself. A1 provides the required servo control of quiescent current. It downs this by sampling a resistively divided version of the voltage across Q5's emitter resistor and comparing it to a power supply derived reference. A1's output biases Q4, completing a loop which forces fixed current through Q5. This action effectively controls overall quiescent current in the discrete stage. Simultaneously, A2 corrects for offset by forcing Q3's base to equalize the DC input and output values at the discrete stage. Because the closed-loop gain is set at 10 (470Ω and 51Ω ratio), A2 samples the output via the 10:1 divider. Both A1 and A2 have local roll-off, limiting their response to low frequency. Casual consideration of A1 and A2's operation might raise concern about interaction, but detailed analysis shows this is not so. The offset and quiescent current loops do not influence each others operation.

When this circuit is constructed using high frequency layout techniques and a ground plane, performance is quite impressive. For gains of 1 to 20, full-power bandwidth remains at 25MHz, with the −3dB point beyond 110MHz. Slew rate exceeds 3000V/μs. These figures can be improved upon by using RF transistors, although the types shown are inexpensive and readily available. Figure 21.13 shows pulse response for a ±12V output (Trace B) at a gain of 10 (input is Trace A). Delay is about 6ns, with rise time limited by the input pulse generator. Damping is optimized with the 10pF trimmer at the Q5-Q6 collector line. To use this circuit, adjust the I_Q level to 80mA *IMMEDIATELY* after turn on. Next, set A2's input resistor divider to a ratio appropriate to the closed-loop circuit gain. Finally, adjust the 10pF trimmer for best response. Note that, in the interests of speed, this circuit has no output protection.

Although speed and offset combinations are the most common area for composite techniques, other circuits are possible. Figure 21.14 shows a way to combine a low drift chopper-stabilized amplifier with an ultralow noise bipolar amplifier. The LTC1052 measures the DC error at the LT1028's input terminals and

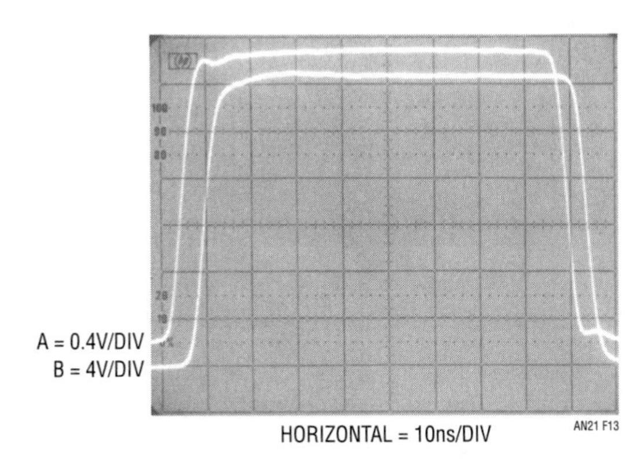

Figure 21.13 • Figure 21.12's Pulse Response (Measurement Limited by Pulse Generator)

biases its offset pins to force offset to a few microvolts. The 1N758 Zeners allow the LTC1052 to function from ±15V rails. The offset pin biasing at the LT1028 is arranged so the LTC1052 will always be able to find the servo point. The 0.01μF capacitor rolls off the LTC1052 at low frequency, and the LT1028 handles high frequency signals. The combined characteristics of these amplifiers yield the following performance:

Offset Voltage	5μV Max
Offset Drift	50nV/°C Max
Noise	$1.1nV\sqrt{Hz}$ Max

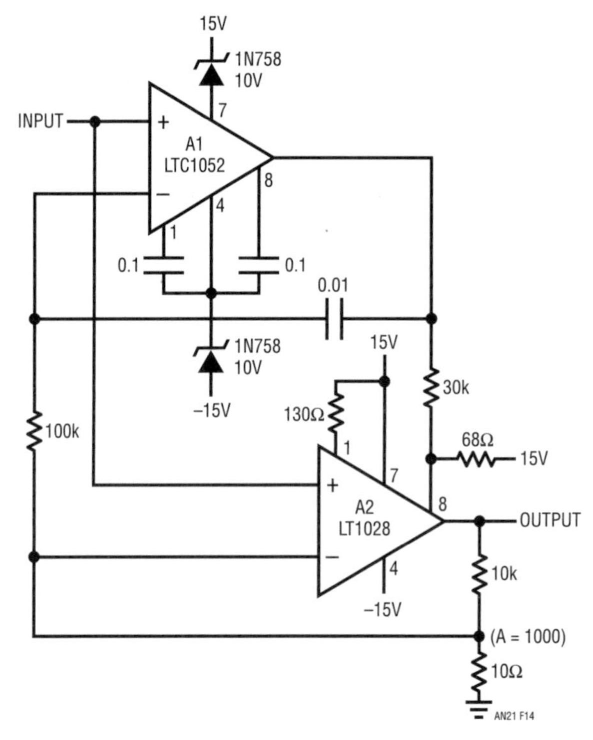

Figure 21.14 • DC-Stabilized, Low Noise Amplifier

Figure 21.15 plots noise amplitude over time in a 0.1Hz to 10Hz bandwidth.

Figure 21.16 uses multiple LT1028 low noise amplifiers in a statistical noise reduction technique. It is based on the fact that noise decreases by the $\sqrt{N}$ of the number of devices in parallel. For example, for nine paralleled amplifiers, noise would decrease by a factor of three, to about $0.33nV\sqrt{Hz}$ at 1kHz. A potential penalty of this connection is that the input current noise increases by $\sqrt{N}$ devices.

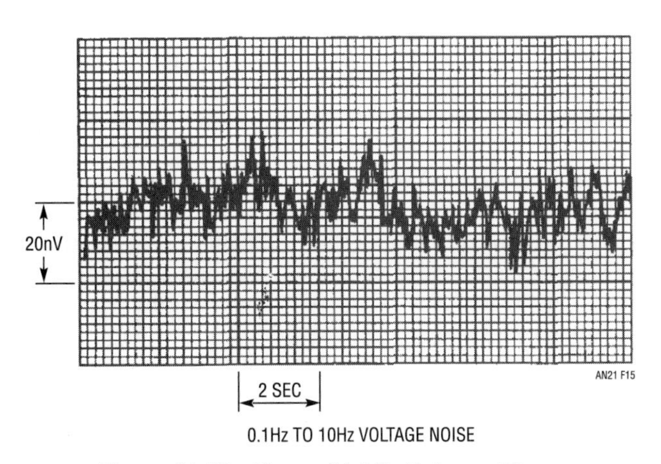

Figure 21.15 • Figure 21.14's Noise vs Time

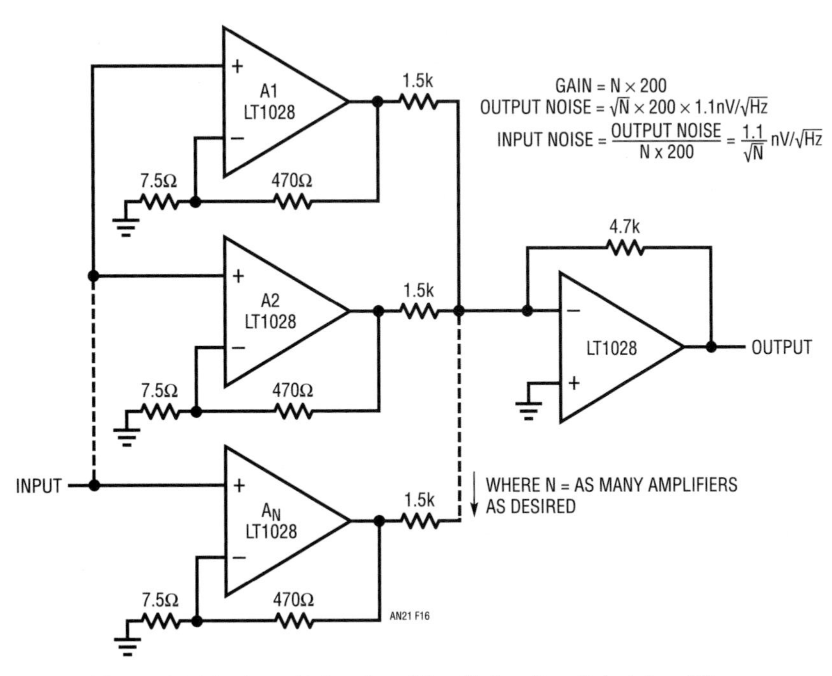

Figure 21.16 • Low Noise Amplifier Using Paralleled Amplifiers

A final circuit, Figure 21.17, uses a composite of paralleled LT1010 buffers to create a simple, high current stage. Parallel operation provides reduced output impedance, more drive capability and increased frequency response under load. Any number of LT1010s can be directly paralleled as long as the increased dissipation in individual units caused by mismatches of output resistance of offset voltage is taken into account.

When the inputs and outputs of two buffers are connected together, a current, ΔI_{OUT}, flows between the outputs:

$$I_{OUT} = \frac{V_{OS1} - V_{OS2}}{R_{OUT1} + R_{OUT2}}$$

where V_{OS} and R_{OUT} are the offset voltage and output resistance of the respective buffers.

Normally, the negative supply current of one unit will increase and the other decrease, with the positive supply current staying the same. The worst-case ($V_{IN} \rightarrow V^+$) increase in standby dissipation can be assumed to be $\Delta I_{OUT}\ V_T$, where V_T is the total supply voltage.

Offset voltage is specified worst-case over a range of supply voltages, input voltage and temperature. It would be unrealistic to use these worst-case numbers above because paralleled units are operating under identical conditions. The offset voltage specified for $V_S = \pm 15V$, $V_{IN} = 0$ and $T_A = 25°C$ will suffice for a worst-case condition.

Output load current will be divided based on the output resistance of the individual buffers. Therefore, the available output current will not quite be doubled unless output resistances are matched. As for offset voltage above, the 25°C limits should be used for worst-case calculations.

Parallel operation is not thermally unstable. Should one unit get hotter than its mates, its share of the output and its standby dissipation will decrease.

As a practical matter, parallel connection needs only some increased attention to heat sinking. In some applications a few ohms equalization resistance in each output may be wise. Only the most demanding applications should require matching, and then just of output resistance at 25°C.

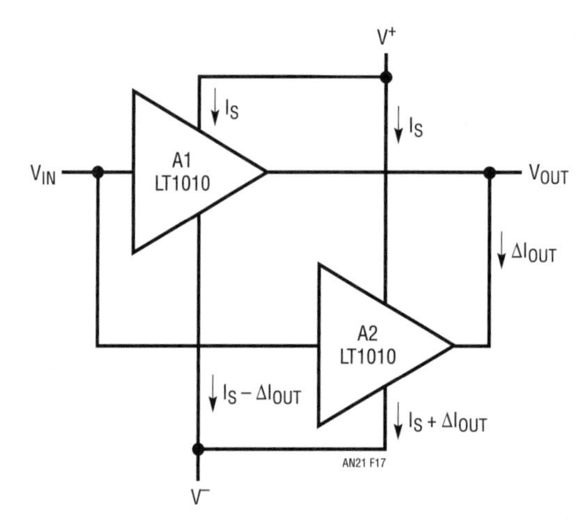

Figure 21.17 • Paralleling Scheme for High Current Output

A simple method of designing multiple order all pole bandpass filters by cascading 2nd order sections

22

Nello Sevastopoulos Richard Markell

Introduction

Filter design, be it active, passive, or switched capacitor, is traditionally a mathematically intensive pursuit. There are many architectures and design methods to choose from. Two methods of high order bandpass filter design are discussed herein. These methods allow the filter designer to simplify the mathematical design process and allow LTC's switched capacitor filters (LTC®1059, LTC1060, LTC1061, LTC1064) to be utilized as high quality bandpass filters.

The first method consists of the traditional cascading of non-identical 2nd order bandpass sections to form the familiar Butterworth and Chebyshev bandpass filters. The second method consists of cascading identical 2nd order bandpass sections. This approach, although "non-textbook," enables the hardware to be simple and the mathematics to be straightforward. Both methods will be described here.

This is the first of a series of application notes from LTC concerning our universal filter family. Additional notes in the series will discuss notch, lowpass and highpass filters implemented with the universal switched capacitor filter. An addition to this note will extend the treatment of bandpass filters to the elliptic or Cauer forms.

This note will first present a finished design example and proceed to present the design methodology, which relies on tabular simplification of traditional filter design techniques.

Designing bandpass filters

Table 22.1 was developed to enable anyone to design Butterworth bandpass filters. We will discuss the tables in more detail later in this paper, but let's first design a filter.

Example 1—design

A 4th order 2kHz Butterworth bandpass filter with a –3dB bandwidth equal to 200Hz is required as shown in Figure 22.1.

Noting that $(f_{0BP}/BW) = 10/1$ we can go directly to Table 22.1 for our normalized center frequencies. From Table 22.1 under 4th order Butterworth bandpass filters, we go to $(f_{0BP}/BW) = 10$.

We find $f_{01} = 0.965$ and $f_{02} = 1.036$ (both normalized to $f_{0BP} = 1$). To find our desired actual center frequencies,

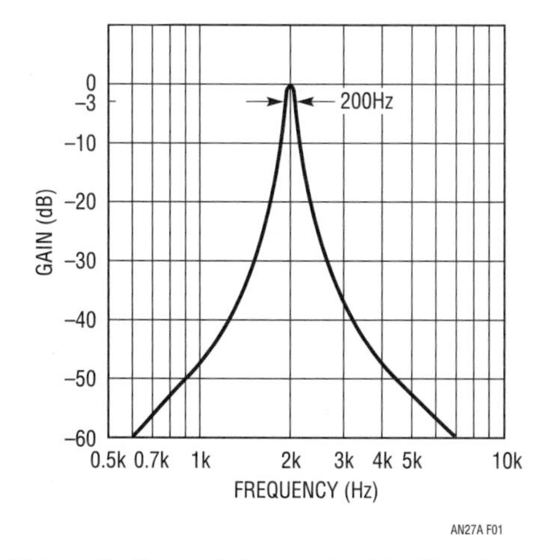

Figure 22.1 • 4th Order Butterworth BP Filter, f_{oPB} = 2kHz

Analog Circuit and System Design: Immersion in the Black Art of Analog Design. http://dx.doi.org/10.1016/B978-0-12-397888-2.00022-5

we must multiply by $f_{OBP} = 2kHz$ to obtain $f_{o1} = 1.930kHz$ and $f_{o2} = 2.072kHz$.

The Qs are $Q_1 = Q_2 = 14.2$ which is read directly from Table 22.1. Also available from the table is K, which is the product of each individual bandpass gain H_{OBP}. To put it another way, the value of K is the gain required to make the gain, H, of the overall filter equal to 1 at f_{OBP}. Our filter parameters are highlighted in the following table:

f_{OBP}	f_{01}	f_{02}	Q_S	K
2kHZ	1.93kHZ	2.072kHZ	Q1=Q2=14.2	2.03

Hardware implementation

Universal switched capacitor filters are simple to implement. A bandpass filter can be built from the traditional state-variable filter topology. Figure 22.2 shows this topology for both switched capacitor and active operational amplifier implementations. Our example requires four resistors for each 2nd order section. So eight resistors are required to build our filter.

We start with two 2nd order sections (1 LTC1060, 2/3 LTC1061 or 1/2 LTC1064), Figure 22.3.

We associate resistors as belonging to 2nd order sections, so R1x belongs to the x section. Thus R12, R22, R33 and R42 all belong to the second of two 2nd order sections in our example.

Our requirements are shown in the following table:

SECTION 1	SECTION 2
$f_{01} = 1.93kHz$	$f_{02} = 2.072kHz$
Q1 = 14.2	Q2 = 14.2
$H_{OBP1} = 1$	$H_{OBP2} = 2.03$

Note that $H_0bp1 \times H_0bp2 = K$ and this is the reason for choosing $H_0bp2 = 2.03$.

For this example we choose the $f_0 = \frac{f_{CLK}}{50}\sqrt{\frac{R2}{R4}}$ mode, so we will tie the 50/100/Hold pin on the SCF chip to V+, generally (5V to 7V). We choose 100kHz as our clock and calculate resistor values. Choosing the nearest 1% resistor values we can implement the filter using Figure 22.3's topology and the resistor values listed below.

R11 = 147k	R12 = 71.5k
R21 = 10k	R22 = 10.7k
R31 = 147k	R32 = 147k
R41 = 10.7k	R42 = 10k

Our design is now complete. We have only to generate a TTL or CMOS compatible clock at 100kHz, which we feed to the clock pin of the switched capacitor filter, and we should be "on the air."

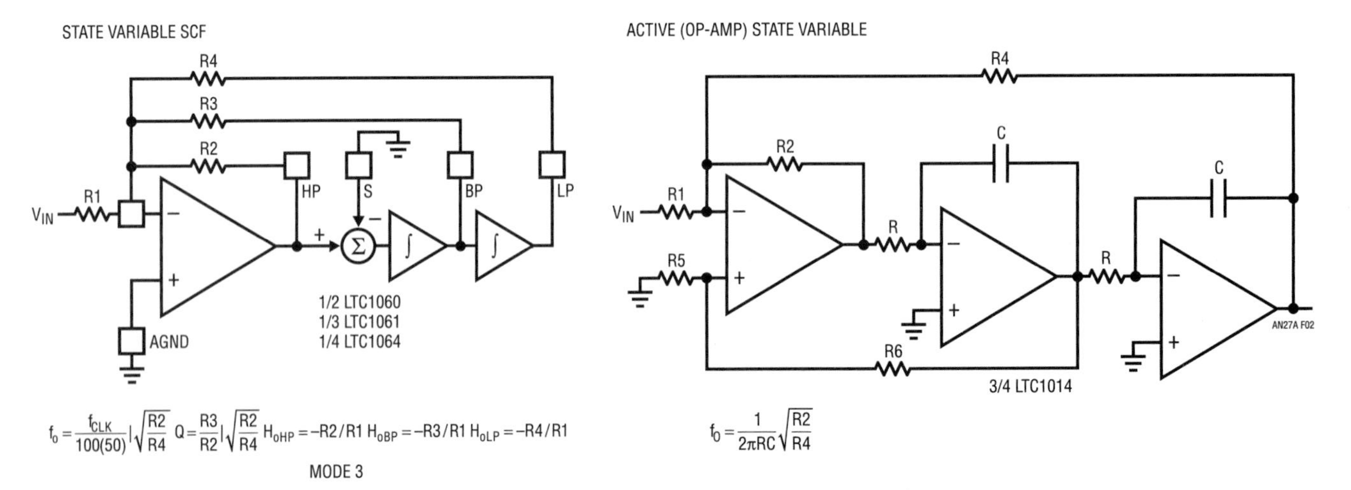

Figure 22.2 • Switched Capacitor vs Active RC State Variable Topology

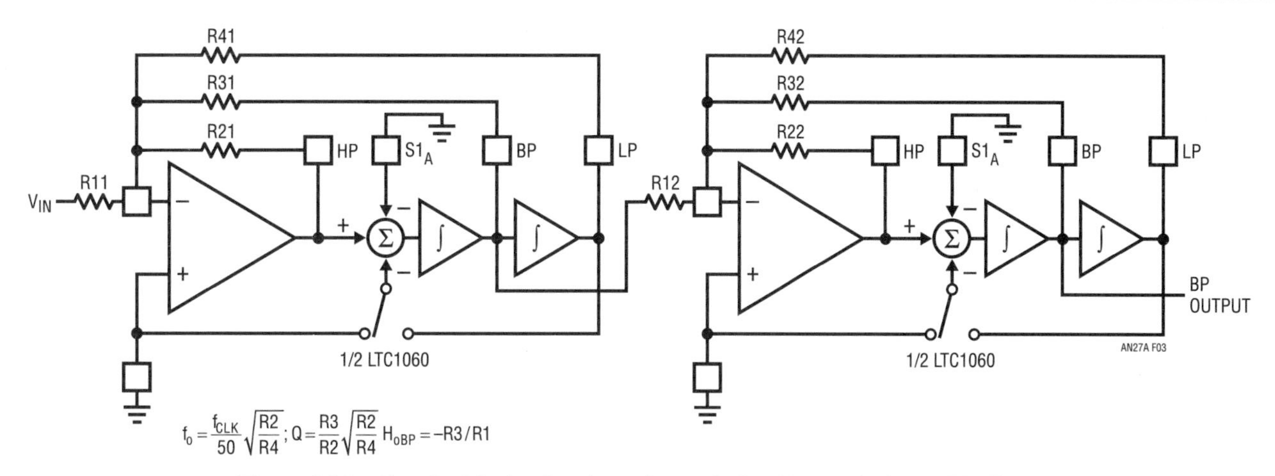

Figure 22.3 • Two 2nd Order Sections Cascaded to Form 4th Order BP Filter

Designing bandpass filters—theory behind the design

Traditionally, bandpass filters have been designed by laborious calculations requiring some time to complete. At the present time programs for various personal or laboratory computers are often used. In either case, no small amount of time and/or money is involved to evaluate, and later test, a filter design.

Many designers have inquired as to the feasibility of cascading 2nd order bandpass sections of relatively low Q to obtain more selective, higher Q, filters. This approach is ideally suited to the LTC family of switched capacitor filters (LTC1059, LTC1060, LTC1061 and LTC1064). The clock to center frequency ratio accuracy of a typical "Mode 1" design with non "A" parts is better than 1% in a design that simply requires three resistors of 1% tolerance or better. Also, no expensive high precision film capacitors are required as in the active op amp state variable design.

We present here an approach for designing bandpass filters using the LTC1059, LTC1060, LTC1061 or the LTC1064 which many designers have "on the air" in days instead of weeks.

Cascading identical 2nd order bandpass sections

When we want to detect single frequency tones and simultaneously reject signals in close proximity, simple 2nd order bandpass filters often do the job. However, there are cases where a 2nd order section cannot be implemented with the required characteristics (generally the Qs are too high). We wish to explore here the use of cascaded identical 2nd order sections for building high Q bandpass filters.

For a 2nd order bandpass filter

$$Q = \frac{\sqrt{1 - G^2}}{G} \times \frac{f/f_0}{|1 - (f/f_0)^2|} \quad (1)$$

Where Q is the required filter quality factor

f is the frequency where the filter should have gain, G, expressed in Volts/V.

f_0 is at the filter center frequency. Unity gain is assumed at f_0.

Example 2—design

We wish to design a 2nd order BP filter to pass 150Hz and to attenuate 60Hz by 50dB. The required Q may be calculated from Equation (1):

$$S_0, Q = \frac{\sqrt{1 - (3.162 \times 10^{-3})^2}}{3.162 \times 10^{-3}} \times \frac{60/150}{|1 - (60/150)^2|} = 150.7$$

This very high Q dictates a –3dB bandwidth of 1Hz.

Although the universal switched capacitor filters can realize such high Qs, their guaranteed center frequency accuracy of ±0.3%, although impressive, is not enough to pass the 150Hz signal without gain error. According to the previous equation, the gain at 150Hz will be 1 ±26%; the rejection, however, at 60Hz will remain at –50dB. The gain inaccuracy can be corrected by tuning resistor R4 when mode 3, Figure 22.2, is used. Also, if only detection of the signal is sought, the gain inaccuracy could be acceptable.

This high Q problem can be solved by cascading two identical 2nd order bandpass sections. To achieve a gain, G, at frequency f the required Q of each 2nd order section is:

$$Q = \frac{\sqrt{1-G}}{\sqrt{G}} \times \frac{f/f_0}{|1-(f/f_0)^2|} \quad (2)$$

The gain at each bandpass section is assumed unity.

In order to obtain 50dB attenuation at 60Hz, and still pass 150Hz, we will use two identical 2nd order sections.

We can calculate the required Q for each of two 2nd order sections from Equation (2):

$$SO, Q = \frac{\sqrt{1-3.162 \times 10^{-3}}}{\sqrt{3.162 \times 10^{-3}}} \times \frac{60/150}{|1-(60/150)^2|} = 8.5!!$$

With two identical 2nd order sections each with a potential error in center frequency, f_0, of ±0.3% the gain error at 150Hz is 1 ±0.26%. If lower cost (non "A" versions of LTC1060 and LTC1064) 2nd order bandpass sections are used with an f_0 tolerance of ±0.8%, the gain error at 150Hz is 1 ±1.8%! The benefits of lower Q sections are therefore obvious.

Hardware implementation

Mode 1 operation of LTC1060, LTC1061, LTC1064

As previously discussed, we associate resistors with each 2nd order section, so R1x belongs to x section. Thus R12, R22 and R23 belong to the second of the two 2nd order sections, Figure 22.4.

Each section has the same requirements as shown:

$$f_{01} = f_{02} = 150\text{Hz}$$
$$Q1 = Q = 8.5$$
$$H_{OBP1} = H_{OBP2} = 1$$

Note that we could get gain out of our BP filter structure by letting the product of the H_{OBP} terms be >1 (within the performance limits of the filter itself).

For our example using the LTC1060 we will use $f_{01} = f_{02} = f_{CLK}/100$. So we input a 15kHz clock and tie the 50/100/ Hold pin to mid-supplies (ground for ±5V supplies).

We can implement this filter using the two sections of an LTC1060 filter operated in mode 1. Mode 1 is the fastest operating mode of the switched capacitor filters. It provides Lowpass, Bandpass and Notch outputs.

Each 2nd order section will perform approximately as shown in Figure 22.5, curve (a).

Implementation in mode 1 is simple as only three resistors are required per section. Since we are cascading *identical* sections, the calculations are also simple.

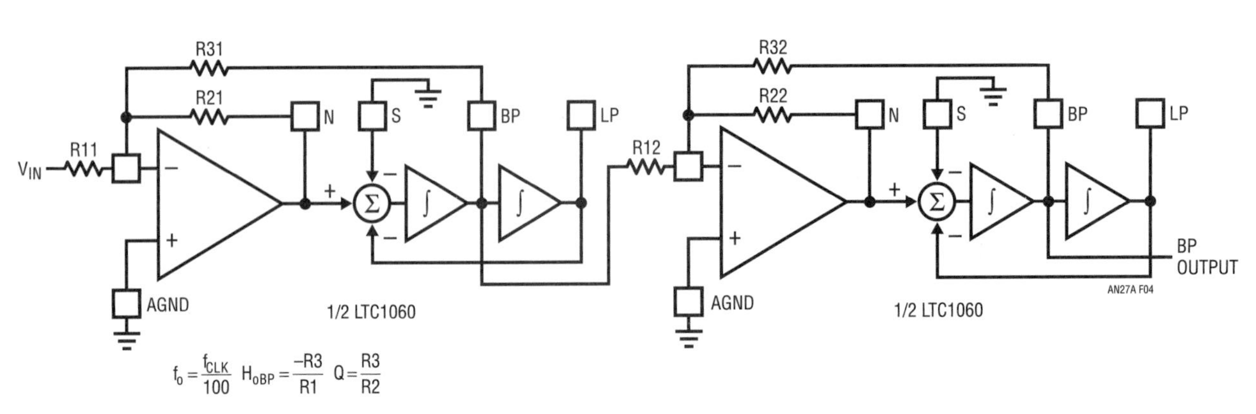

Figure 22.4 • LTC1060 as BP Filter Operating in Mode 1

We can calculate the resistor values from the indicated formulas and then choose 1% values. (Note that we let our minimum value be 20k.) The required values are:

$$R11 = R12 = 169k$$
$$R21 = R22 = 20k$$
$$R31 = R32 = 169k$$

Our design is complete. The performance of two 2nd order sections cascaded versus one 2nd order section is shown in Figure 22.5, curve (b). We must, however, generate a TTL or CMOS clock at 15kHz to operate the filter.

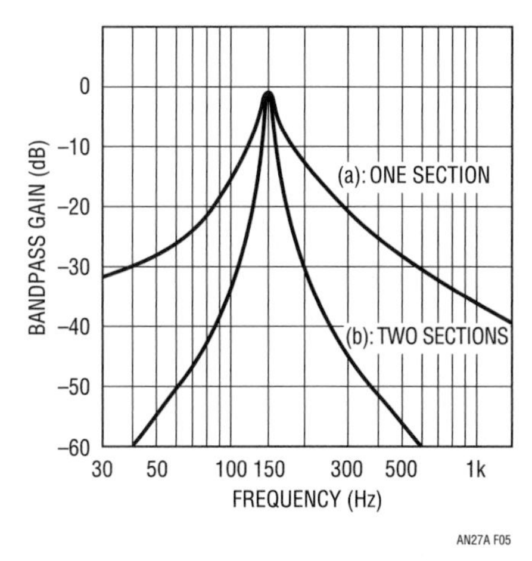

Figure 22.5 • Cascading Two 2nd Order BP Sections for Higher Q Response

Mode 2 operation of LTC1060 family

Suppose that we have no 15kHz clock source readily available. We can use what is referred to as mode 2, which allows the input clock frequency to be less than 50:1 or 100:1 [$f_{CLK}/f_o = 50$ or 100]. This still depends on the connection of the 50/100/Hold pin.

If we wish to operate our previous filter from a television crystal at 14.318MHz we could divide this frequency by 1000 to give us a clock of 14.318kHz. We could then set up our mode 2 filter as shown in Figure 22.6.

We can calculate the resistor values from the formulas shown and then choose 1% values. The required values are:

$$R11, R12 = 162k$$
$$R21, R22 = 20k$$
$$R31, R32 = 162k$$
$$R41, R42 = 205k$$

Cascading more than two identical 2nd order BP sections

If more than two identical bandpass sections (2nd order) are cascaded, the required Q of each section may be shown to be:

$$Q = \frac{\sqrt{1 - G^{2/n}}}{G^{1/n}} \times \frac{(f/f_0)}{|1 - (f/f_0)^2|} \qquad (3)$$

where Q, G, f and f_o are as previously defined and n = the number of cascaded 2nd order sections.

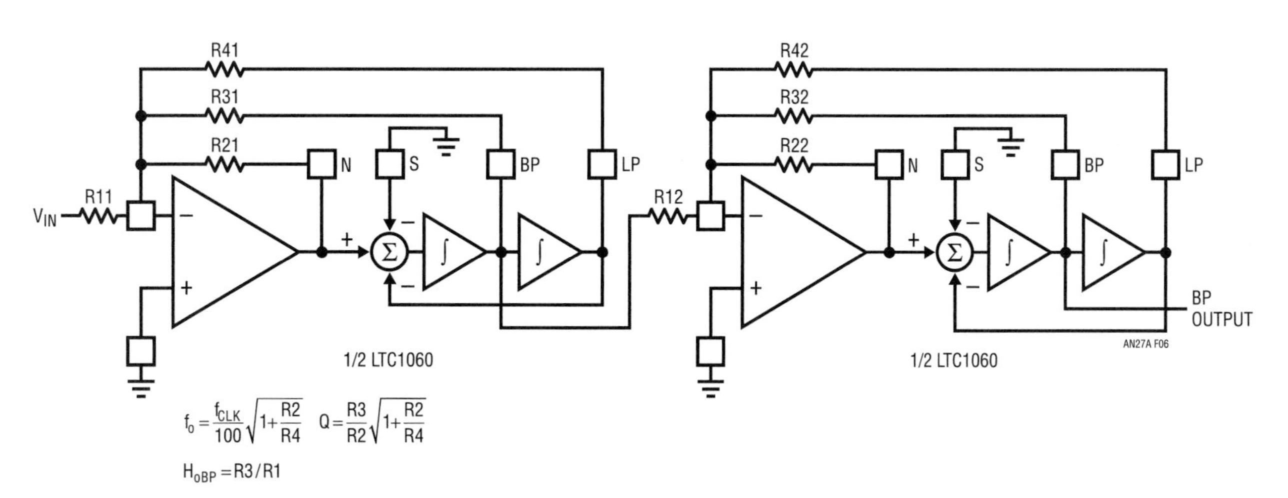

Figure 22.6 • LTC1060 as BP Filter Operating in Mode 2

The equivalent Q of the overall bandpass filter is then:

$$Q_{equiv} = \frac{Q_{(identical\ section)}}{\sqrt{(2^{1/n}) - 1}} \qquad (4)$$

Figure 22.7 shows the passband curves for Q = 2 cascaded bandpass sections where n is the number of 2nd order sections cascaded.

The benefits can be seen for two and three cascaded sections. Cascading four or more sections increases the Q, but not as rapidly. Nevertheless for designers requiring high Q bandpass filters cascading identical sections is a very real option considering the simplicity.

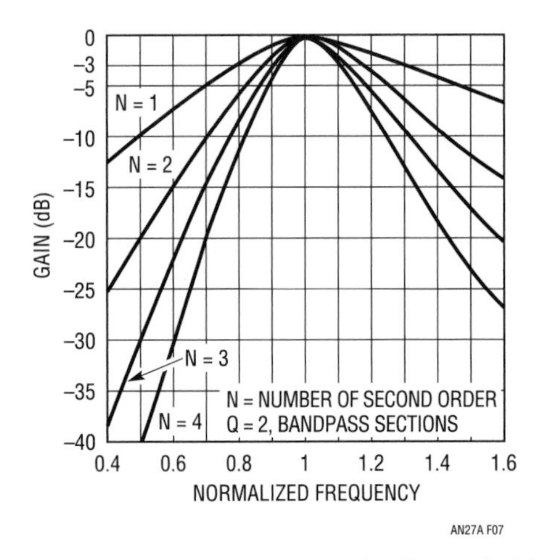

Figure 22.7 • Frequency Response of n Cascaded Identical 2nd Order Bandpass Sections

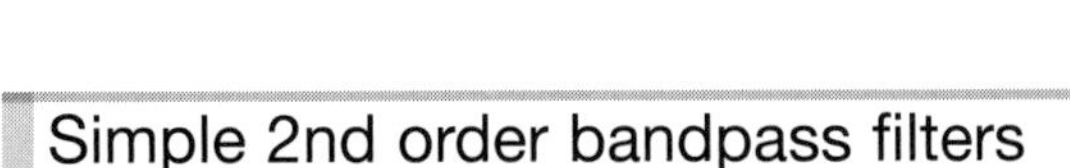

Simple 2nd order bandpass filters

Gain and phase relations

The bandpass output of each 2nd order filter section of the LTC1059, LTC1060, LTC1061 and LTC1064, closely approximates the gain and phase response of an ideal "textbook" filter.

$$G = \frac{(H_{OBP}) \times (ff_O)/Q}{\left[(f_O^2 - f^2)^2 + (ff_O/Q)^2\right]^{1/2}}$$

G = filter gain in Volts/V
f_O = the filter's center frequency
Q = the quality coefficient of the filter
H_OBP = the maximum voltage gain of the filter occuring at f_O
$\frac{f_O}{Q}$ = -3dB bandwidth of the filter

Figure 22.8 illustrates the above definitions. Figure 22.9 illustrates the bandpass gain, G, for various values of Q. This figure is very useful for estimating the filter attenuation when several identical 2nd order bandpass filters are cascaded. High Qs make the filter more selective, and at the same time, more noisy and more difficult to realize. Qs in excess of 100 can be easily realized with the universal switched capacitor filters, LTC1059, LTC1060, LTC1061 and LTC1064, and still maintain low center frequency and Q drift, but for system considerations, this may not be practical.

The phase shift, ϕ, of a 2nd order bandpass filter is:

$$\phi = \arctan\left[\left(\frac{f_O^2 - f^2}{ff_O}\right) \times Q\right]$$

The phase shift at f_O is 0° or, if the filter is inverting, it is −180°.

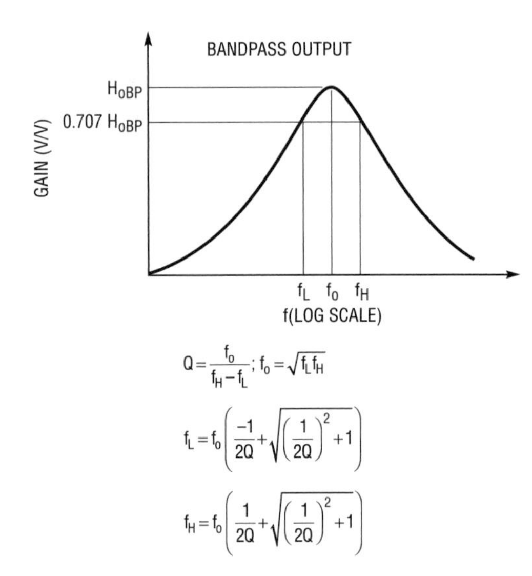

Figure 22.8 • Bandpass Filter Parameters

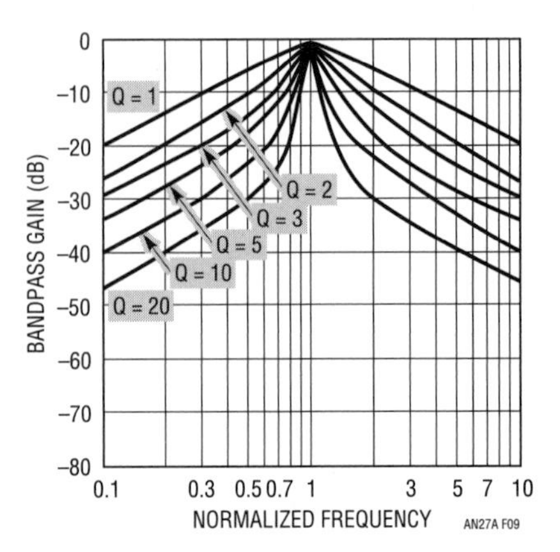

Figure 22.9 • Bandpass Gain as a Function of Q

Simple 2nd order bandpass filters (continued)

All the bandpass outputs of the LTC1059, LTC1060, LTC1061 and LTC1064 universal filters are inverting. The phase shift, especially in the vicinity of f_0, depends on the value of Q, see Figure 22.10. By the same argument, the phase shift at a given frequency varies from device to device due to the fo tolerance. This is true especially for high Qs and in the vicinity of f_0. For instance, an LTC1059A, 2nd order universal filter, has a guaranteed initial center frequency tolerance of ±0.3%. The ideal phase shift at the ideal f_0 should be –180°. With a Q of 20, and without trimming, the worst-case phase shift at the ideal f_0 will be –180° ±6.8°. With a Q of 5 the phase shift tolerance becomes –180° ±1.7°. These are important considerations when bandpass filters are used in multichannel systems where phase matching is required. By way of comparison, a state variable active bandpass filter built with 1% resistors and 1% capacitors may have center frequency variation of ±2% resulting in phase variations of ±2% resulting in phase variations of ±33.8° for Q = 20 and ±11.4° for Q = 5.

Constant Q versus constant BW

The bandpass outputs of the universal filters are "constant Q." For instance, a 2nd order bandpass filter operating in mode 1 with a 100kHz clock (see LTC1060 data sheet) ideally has a 1kHz or 2kHz center frequency, and a –3dB bandwidth equal to (f_0/Q). When the clock frequency varies, the center frequency and bandwidth will vary at the same rate. In a constant bandwidth filter, when the center frequency varies, the Q varies accordingly to maintain a constant (f_0/Q) ratio. A constant bandwidth BP filter could be implemented using 2nd order switched capacitor filters but this is beyond the scope of this paper.

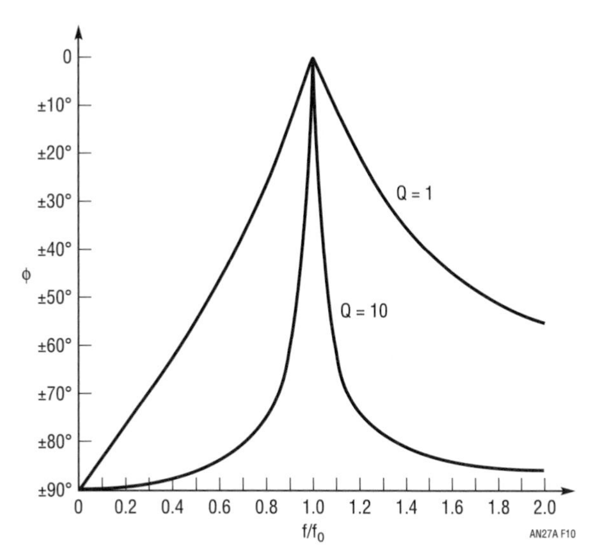

Figure 22.10 • Phase Shift, φ, of a 2nd Order BP Filter Section (LTC1059, 1/2 LTC1060, 1/3 LTC1061)

Using the tables

Tables 22.1 through 22.4 were derived from textbook filter theory. They can be easily applied to the LTC filter family (LTC1059, LTC1060, LTC1061 and LTC1064) if the Qs are kept relatively low (<20) and the tuning resistors are at least 1% tolerance. These lower Q designs provide almost textbook BP filter performance using LTC's switched capacitor filters. For higher Q implementations, tuning should be avoided and the "A" versions of the LTC1059, LTC1060, LTC1061 or LTC1064 should be specified. Also, resistor tolerances of better than 1% are a necessity.

Table 22.1 may be used to find pole positions and Qs for Butterworth bandpass filters. It should be noted that the bandpass filters in these tables are geometrically symmetrical about their center frequencies, f_0BP. Any frequency, f_3, as shown in Figure 22.11 has its geometrical counterpart f_4 such that:

$$f_4 = \frac{f_{0BP^2}}{f_3}$$

Additionally, Table 22.1 illustrates the attenuation at the frequencies f_3, f_5, f_7 and f_9, which correspond to bandwidths 2, 3, 4 and 5 times the passband (see Figure 22.11). These values allow the user to get a good estimate of filter selectivity,

An important approximation can be made for not only the Butterworth filters in Table 22.1, but also for the Chebyshev filter Tables 22.2–22.4. Treating Figure 22.11 (or Figure 22.12) as a generalized bandpass filter, the two corner frequencies f_2 and f_1 can be seen to be nearly arithmetically symmetrical with respect to f_{0BP} provided that:

$$\frac{f_{0PB}}{BW} >> \frac{1}{2}, BW = f_2 - f_1$$

Under this condition, for either Butterworth or Chebyshev bandpass filters:

$$f_{0BP} \cong \frac{f_3 - f_4}{2} + f_3$$

$$f_{0PB} \cong \frac{f_5 - f_6}{2} + f_3$$

•
•
•

This is true for any bandwidth, BW, and any set of frequencies.The tables can now be arithmetically scales as illustrated

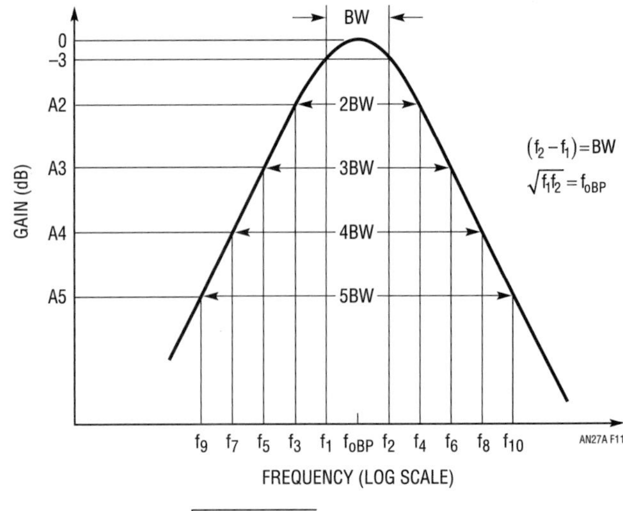

$$(f_1, f_2) = \frac{\pm BW + \sqrt{(BW)^2 + 4(f_{0BP})^2}}{2}$$

MORE GENERALLY $(f_x, f_{x+1}) = \frac{\pm nBW + \sqrt{(nBW)^2 + 4(f_{0BP})^2}}{2}$

(VALID FOR ANY f_x, f_{x+1} PAIR, ANY BW)

Figure 22.11 • Generalized Bandpass Butterworth Response (See Table 22.1)

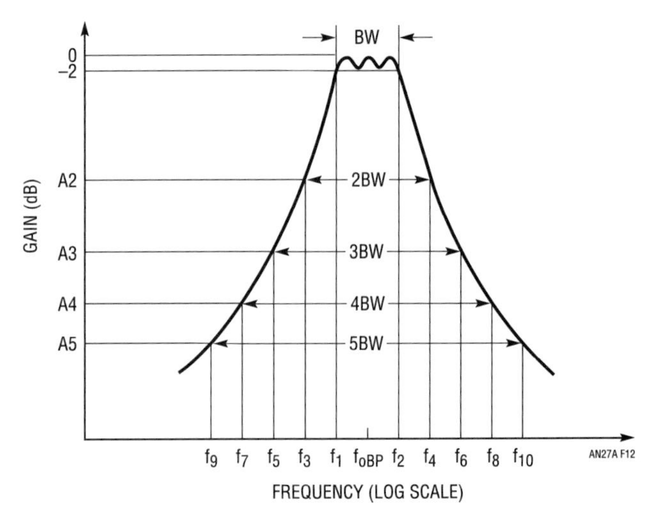

$\sqrt{f_4 f_3} = f_{0PB}$

$$(f_4, f_3) = \frac{\pm 2BW + \sqrt{(2BW)^2 + 4(f_{0BP})^2}}{2}$$

FOR ANY (f_x, f_{x-1}) PAIR AND ANY CORRESPONDING BANDWIDTH (2BW, 3BW, ETC.)

FOR EXAMPLE:

$$(f_6, f_5) = \frac{\pm 3BW + \sqrt{(3BW)^2 + 4(f_{0BP})^2}}{2}$$

Figure 22.12 • Generalized 4th, 6th, and 8th Order Chebyshev Bandpass Filter with 2dB Passband Ripple (A_{MAX})

Table 22.1 Butterworth Bandworth Filters Normalized to $f_{oBP} = 1$

f_{oBP}(Hz)	f_{oBP}/ BW(Hz)	f_{o1} (Hz)	f_{o2} (HZ)	f_{o3} (Hz)	f_{o4} (HZ)	f $_{-3dB}$(Hz)	f_{-3dB} (HZ)	Q1=Q2		K	f_1 (Hz)	f_3 (Hz)	GAIN AT f_3 (dB)-A2	f_5 (Hz)	GAIN AT f_5 (dB)-A3	f_7(dB)	GAIN AT f_7 (dB)-A4	f_9 (Hz)	GAIN AT f_9 (dB)-A5
4th Order Butterworth Bandpass Filter Normalized to its Center Frequency, $f_{OBP} = 1$, and -3dB Bandwidth (BW)																			
1	1	0.693	1.442			0.500	2.000	1.5		2.28	0.500	0.414	-12.3	0.303	-19.1	0.236	-24.0	0.193	-28.0
1	2	0.836	1.195			0.781	1.281	2.9		2.07	0.781	0.618	-12.3	0.500	-19.1	0.414	-24.0	0.351	-28.0
1	3	0.885	1.125			0.847	1.180	4.3		2.07	0.847	0.721	-12.3	0.618	-19.1	0.535	-24.0	0.469	-28.0
1	5	0.932	1.073			0.905	1.105	7.1		2.04	0.905	0.820	-12.3	0.744	-19.1	0.677	-24.0	0.618	-28.0
1	10	0.965	1.036			0.951	1.051	14.2		2.03	0.951	0.905	-12.3	0.861	-19.1	0.820	-24.0	0.781	-28.0
1	20	0.982	1.018			0.975	1.025	28.3		2.03	0.975	0.951	-12.3	0.928	-19.1	0.905	-24.0	0.883	-28.0
6th Order Butterworth Bandpass Filter Normalized to its Center Frequency, $f_{OBP} = 1$, and -3dB Bandwidth (BW)																			
									Q3										
1	1	0.650	1.539	1.000		0.500	2.000	2.2	1.0	4.79	0.500	0.414	-18.2	0.303	-28.6	0.236	-36.1	0.193	-41.9
1	2	0.805	1.242	1.000		0.781	1.281	4.1	2.0	4.18	0.781	0.618	-18.2	0.500	-28.6	0.414	-36.1	0.351	-41.9
1	3	0.866	1.155	1.000		0.847	1.180	6.1	3.0	4.07	0.847	0.721	-18.2	0.618	-28.6	0.535	-36.1	0.469	-41.9
1	5	0.917	1.091	1.000		0.905	1.105	10.0	5.0	4.03	0.905	0.820	-18.2	0.744	-28.6	0.677	-36.1	0.618	-41.9
1	10	0.958	1.044	1.000		0.951	1.051	20.0	10.0	4.01	0.951	0.905	-18.2	0.861	-28.6	0.820	-36.1	0.781	-41.9
1	20	0.979	1.022	1.000		0.975	1.025	40.0	20.0	4.00	0.975	0.951	-18.2	0.928	-28.6	0.905	-36.1	0.883	-41.9
8th Order Butterworth Bandpass Filter Normalized to its Center Frequency, $f_{OBP} = 1$, and -3dB Bandwidth (BW)																			
									Q3 = Q4										
1	1	0.809	1.237	0.636	1.574	0.500	2.000	1.1	2.9	10.14	0.500	0.414	-24.0	0.303	-38.0	0.236	-48.1	0.193	-55.8
1	2	0.907	1.103	0.795	1.259	0.781	1.281	2.2	5.4	8.48	0.781	0.618	-24.0	0.500	-38.0	0.414	-48.1	0.351	-55.8
1	3	0.938	1.066	0.858	1.166	0.847	1.180	3.3	7.9	8.15	0.847	0.721	-24.0	0.618	-38.0	0.535	-48.1	0.469	-55.8
1	5	0.962	1.039	0.912	1.097	0.905	1.105	5.4	13.1	8.05	0.905	0.820	-24.0	0.744	-38.0	0.677	-48.1	0.618	-55.8
1	10	0.981	1.019	0.955	1.047	0.951	1.051	10.8	26.2	8.00	0.951	0.905	-24.0	0.861	-38.0	0.820	-48.1	0.781	-55.8
1	20	0.990	1.010	0.977	1.023	0.975	1.025	21.6	52.3	8.00	0.975	0.951	-24.0	0.928	-38.0	0.905	-48.1	0.883	-55.8

Table 22.2 4th Order Chebyshev Bandpass Filter Normalized to its Center Frequency $f_{OBP} = 1$

f_{oBP}(Hz)	f_{oBp}/BWi* (Hz)	f_{o1} (Hz)	f_{o2} (HZ)	f_{0Bp}/BW_2** (Hz)	f_{-3dB} (Hz)	f_{-3dB} (HZ)	Q1=Q2	K	f_1 (Hz)	f_3 (Hz)	GAIN AT f_3 (dB)-A2	f_5 (Hz)	GAIN AT f_5 (dB)-A3	f_7(Hz)	GAIN AT f_7 (dB)-A4	f_9 (Hz)	GAIN AT f_9 (dB)-A5
Passband Ripple, $A_{MAX} = 0.1$dB																	
1	1	0.488	2.050	0.52	0.423	2.364	1.1	3.81	0.500	0.414	-3.2	0.303	-08.7	0.236	-13.6	0.193	-17.4
1	2	0.703	1.422	1.03	0.626	1.597	1.8	2.66	0.781	0.618	-3.2	0.500	-08.7	0.414	-13.6	0.351	-17.4
1	3	0.793	1.261	1.54	0.727	1.375	2.6	2.48	0.847	0.721	-3.2	0.618	-08.7	0.535	-13.6	0.469	-17.4
1	5	0.871	1.148	2.58	0.825	1.213	4.3	2.38	0.905	0.820	-3.2	0.744	-08.7	0.677	-13.6	0.618	-17.4
1	10	0.933	1.071	5.15	0.908	1.102	8.5	2.38	0.951	0.905	-3.2	0.861	-08.7	0.820	-13.6	0.781	-17.4
1	20	0.966	1.035	10.31	0.953	1.050	16.9	2.37	0.975	0.951	-3.2	0.928	-08.7	0.905	-13.6	0.883	-17.4
Passband Ripple $A_{MAX} = 0.5$dB																	
1	1	0.602	1.660	0.72	0.523	1.912	1.6	3.80	0.500	0.414	-7.9	0.303	-15.0	0.236	-20.2	0.193	-24.1
1	2	0.777	1.287	1.44	0.711	1.406	2.9	3.17	0.781	0.618	-7.9	0.500	-15.0	0.414	-20.2	0.351	-24.1
1	3	0.845	1.182	2.16	0.795	1.258	4.3	3.07	0.847	0.721	-7.9	0.618	-15.0	0.535	-20.2	0.469	-24.1
1	5	0.904	1.106	3.60	0.871	1.149	7.1	3.03	0.905	0.820	-7.9	0.744	-15.0	0.677	-20.2	0.618	-24.1
1	10	0.951	1.051	7.19	0.933	1.072	14.1	2.98	0.951	0.905	-7.9	0.861	-15.0	0.820	-20.2	0.781	-24.1
1	20	0.975	1.025	14.49	0.966	1.035	28.1	2.97	0.975	0.951	-7.9	0.928	-15.0	0.905	-20.2	0.883	-24.1
Passband Ripple $A_{MAX} = 1.0$dB																	
1	1	0.639	1.564	0.82	0.562	1.779	2.0	4.42	0.500	0.414	-10.3	0.303	-17.7	0.236	-23.0	0.193	-27.0
1	2	0.799	1.251	1.64	0.741	1.349	3.7	3.85	0.781	0.618	-10.3	0.500	-17.7	0.414	-23.0	0.351	-27.0
1	3	0.861	1.161	2.47	0.818	1.223	5.5	3.76	0.847	0.721	-10.3	0.618	-17.7	0.535	-23.0	0.469	-27.0
1	5	0.914	1.094	4.12	0.886	1.129	9.2	3.71	0.905	0.820	-10.3	0.744	-17.7	0.677	-23.0	0.618	-27.0
1	10	0.956	1.046	8.20	0.941	1.063	18.2	3.70	0.951	0.905	-10.3	0.861	-17.7	0.820	-23.0	0.781	-27.0
1	20	0.978	1.022	16.39	0.970	1.031	36.5	3.63	0.975	0.951	-10.3	0.928	-17.7	0.905	-23.0	0.883	-27.0
Passband Ripple $A_{max} = 2.0$dB																	
1	1	0.668	1.496	0.93	0.598	1.672	2.7	6.00	0.500	0.414	-12.7	0.303	-20.3	0.236	-25.5	0.193	-29.5
1	2	0.816	1.225	1.86	0.767	1.304	5.1	5.30	0.781	0.618	-12.7	0.500	-20.3	0.414	-25.5	0.351	-29.5
1	3	0.873	1.145	2.79	0.837	1.195	7.5	5.22	0.847	0.721	-12.7	0.618	-20.3	0.535	-25.5	0.469	-29.5
1	5	0.922	1.085	4.65	0.898	1.113	12.5	5.13	0.905	0.820	-12.7	0.744	-20.3	0.677	-25.5	0.618	-29.5
1	10	0.960	1.041	9.35	0.948	1.055	24.9	5.13	0.951	0.905	-12.7	0.861	-20.3	0.820	-25.5	0.781	-29.5
1	20	0.980	1.021	18.87	0.974	1.027	49.8	5.07	0.975	0.951	-12.7	0.928	-20.3	0.905	-25.5	0.883	-29.5

* f_{0Bp}/BW_i -This is the ratio of the bandpass filter center frequency to the **ripple bandwidth** of the filter.

** f_{0BP}/BW_2 -This is the ratio of the bandpass filter center frequency to the **-3dB filter bandwidth.**

Table 22.3 6th Order Chebychev Bandpass Filter Normalized to its Center Frequency $f_{oBp} = 1$

f_{oBp}(Hz)	f_{oBp}/BW_1* (Hz)	f_{01} (Hz)	f_{02} (HZ)	f_{03}(Hz)	f_{oBP}/BW_2** (Hz)	f_{-3dB} (HZ)	f_{-3dB} (HZ)	Q1=Q2	Q = 3	K	f_1 (Hz)	f_3 (Hz)	GAIN AT f_3 (dB)-A2	f_5 (Hz)	GAIN AT f_5 (dB)-A3	f_7 (Hz)	GAIN AT f_7 (dB)-A4	f_9 (Hz)	GAIN AT f_9 (dB)-A5
Passband Ripple, A_{MAX} = 0.1dB																			
1	1	0.558	1.791	1.000	0.72	0.523	1.912	2.4	1.0	9.9	0.500	0.414	-12.2	0.303	-23.6	0.236	-31.4	0.193	-37.3
1	2	0.741	1.349	1.000	1.44	0.711	1.406	4.3	2.1	7.9	0.781	0.618	-12.2	0.500	-23.6	0.414	-31.4	0.351	-37.3
1	3	0.818	1.222	1.000	2.16	0.795	1.258	6.3	3.1	7.5	0.847	0.721	-12.2	0.618	-23.6	0.535	-31.4	0.469	-37.3
1	5	0.886	1.128	1.000	3.60	0.871	1.149	10.4	5.2	7.4	0.905	0.820	-12.2	0.744	-23.6	0.677	-31.4	0.618	-37.3
1	10	0.941	1.062	1.000	7.19	0.933	1.072	20.6	10.3	7.3	0.951	0.905	-12.2	0.861	-23.6	0.820	-31.4	0.781	-37.3
1	20	0.970	1.030	1.000	14.49	0.966	1.035	41.3	20.6	7.3	0.975	0.951	-12.2	0.928	-23.6	0.905	-31.4	0.883	-37.3
Passband Ripple, A_{MAX} = 0.5dB																			
1	1	0.609	1.641	1.000	0.86	0.574	1.741	3.6	1.6	14.8	0.500	0.414	-19.2	0.303	-30.8	0.236	-38.6	0.193	-44.5
1	2	0.776	1.288	1.000	1.72	0.750	1.333	6.6	3.2	12.5	0.781	0.618	-19.2	0.500	-30.8	0.414	-38.6	0.351	-44.5
1	3	0.844	1.185	1.000	2.57	0.824	1.213	9.7	4.8	12.0	0.847	0.721	-19.2	0.618	-30.8	0.535	-38.6	0.469	-44.5
1	5	0.903	1.107	1.000	4.29	0.890	1.123	16.1	8.0	11.8	0.905	0.820	-19.2	0.744	-30.8	0.677	-38.6	0.618	-44.5
1	10	0.950	1.052	1.000	8.55	0.943	1.060	32.0	16.0	11.8	0.951	0.905	-19.2	0.861	-30.8	0.820	-38.6	0.781	-44.5
1	20	0.975	1.026	1.000	16.95	0.971	1.030	63.8	32.0	11.4	0.975	0.951	-19.2	0.928	-30.8	0.905	-38.6	0.883	-44.5
Passband Ripple, A_{MAX} = 1 0dB																			
1	1	0.626	1.598	1.000	0.91	0.593	1.687	4.5	2.0	20.1	0.500	0.414	-22.5	0.303	-34.0	0.236	-41.9	0.193	-47.8
1	2	0.787	1.271	1.000	1.83	0.763	1.310	8.3	4.1	17.1	0.781	0.618	-22.5	0.500	-34.0	0.414	-41.9	0.351	-47.8
1	3	0.852	1.174	1.000	2.74	0.834	1.199	12.3	6.1	16.7	0.847	0.721	-22.5	0.618	-34.0	0.535	-41.9	0.469	-47.8
1	5	0.908	1.101	1.000	4.59	0.897	1.115	20.3	10.1	16.4	0.905	0.820	-22.5	0.744	-34.0	0.677	-41.9	0.618	-47.8
1	10	0.953	1.050	1.000	9.17	0.947	1.056	40.5	20.2	16.4	0.951	0.905	-22.5	0.861	-34.0	0.820	-41.9	0.781	-47.8
1	20	0.976	1.024	1.000	18.18	0.973	1.028	81.0	40.5	16.4	0.975	0.951	-22.5	0.928	-34.0	0.905	-41.9	0.883	-47.8
Passband Ripple, A_{MAX} = 2.0dB																			
1	1	0.639	1.565	1.000	0.97	0.609	1.642	6.0	2.7	31.7	0.500	0.414	-26.0	0.303	-37.5	0.236	-45.4	0.193	-51.3
1	2	0.795	1.257	1.000	1.94	0.775	1.291	11.1	5.4	27.4	0.781	0.618	-26.0	0.500	-37.5	0.414	-45.4	0.351	-51.3
1	3	0.858	1.165	1.000	2.91	0.843	1.187	16.5	8.1	26.7	0.847	0.721	-26.0	0.618	-37.5	0.535	-45.4	0.469	-51.3
1	5	0.912	1.096	1.000	4.83	0.902	1.109	27.2	13.6	26.2	0.905	0.820	-26.0	0.744	-37.5	0.677	-45.4	0.618	-51.3
1	10	0.955	1.047	1.000	9.71	0.950	1.053	54.3	27.1	26.0	0.951	0.905	-26.0	0.861	-37.5	0.820	-45.4	0.781	-51.3
1	20	0.977	1.023	1.000	19.61	0.975	1.026	108.5	54.2	26.0	0.975	0.951	-26.0	0.928	-37.5	0.905	-45.4	0.883	-51.3

* f_{0Bp}/BW_1 -This is the ratio of the bandpass filter center frequency to the **ripple bandwidth** of the filter.

** f_{0Bp}/BW_2 -This is the ratio of the bandpass filter center frequency to the **-3dB filter bandwidth.**

Table 22.4 8th Order Chebychev Bandpass Filter Normalized to its Center Frequency f0BP = 1

f_{oBP}(Hz)	f_{oBp}/BWi* (Hz)	f_{01} (Hz)	f_{02} (HZ)	f_{03} (HZ)	f_{04} (HZ)	f_{oBp}/BW_2** (Hz)	f_{-3dB} (HZ)	f_{-3dB} (HZ)	Q1=Q2	Q3=Q4	K	f_1 (Hz)	f_3 (Hz)	GAIN AT f_3 (dB)-A2	f_5 (Hz)	GAIN AT f_5 (dB)-A3	f_7(Hz)	GAIN AT f_7 (dB)-A4	f_9 (Hz)	GAIN AT f_9 (dB)-A5
Passband Ripple, A_{MAX} = 0.1dB																				
1	1	0.785	1.274	0.584	1.713	0.82	0.563	1.776	1.6	4.4	40.6	0.500	0.414	-23.4	0.303	-38.8	0.236	-49.3	0.193	-57.1
1	2	0.889	1.125	0.757	1.320	1.65	0.742	1.348	3.2	7.9	32.1	0.781	0.618	-23.4	0.500	-38.8	0.414	-49.3	0.351	-57.1
1	3	0.925	1.081	0.830	1.204	2.48	0.818	1.222	4.7	11.6	30.5	0.847	0.721	-23.4	0.618	-38.8	0.535	-49.3	0.469	-57.1
1	5	0.954	1.048	0.894	1.118	4.12	0.886	1.129	7.9	19.1	29.9	0.905	0.820	-23.4	0.744	-38.8	0.677	-49.3	0.618	-57.1
1	10	0.977	1.023	0.945	1.058	8.20	0.941	1.063	15.7	37.9	29.8	0.951	0.905	-23.4	0.861	-38.8	0.820	-49.3	0.781	-57.1
1	20	0.988	1.012	0.972	1.028	16.39	0.970	1.031	31.4	75.7	29.8	0.975	0.951	-23.4	0.928	-38.8	0.905	-49.3	0.883	-57.1
Passband Ripple, A_{MAX} = 0.5dB																				
1	1	0.808	1.238	0.613	1.632	0.91	0.593	1.686	2.4	6.4	90.1	0.500	0.414	-30.2	0.303	-45.5	0.236	-56.0	0.193	-63.9
1	2	0.900	1.111	0.777	1.286	1.83	0.763	1.310	4.8	11.8	74.3	0.781	0.618	-30.2	0.500	-45.5	0.414	-56.0	0.351	-63.9
1	3	0.932	1.073	0.845	1.183	2.74	0.834	1.199	7.1	17.4	71.5	0.847	0.721	-30.2	0.618	-45.5	0.535	-56.0	0.469	-63.9
1	5	0.959	1.043	0.903	1.107	4.59	0.897	1.115	11.8	28.7	70.0	0.905	0.820	-30.2	0.744	-45.5	0.677	-56.0	0.618	-63.9
1	10	0.979	1.021	0.950	1.052	9.17	0.947	1.056	23.6	57.1	70.0	0.951	0.905	-30.2	0.861	-45.5	0.820	-56.0	0.781	-63.9
1	20	0.989	1.010	0.975	1.026	18.18	0.973	1.028	47.2	114.0	70.0	0.975	0.951	-30.2	0.928	-45.5	0.905	-56.0	0.883	-63.9
Passband Ripple, A_{MAX} = 1.0dB																				
1	1	0.814	1.228	0.622	1.607	0.95	0.604	1.656	3.0	8.0	162.8	0.500	0.414	-32.9	0.303	-48.3	0.236	-58.8	0.193	-66.6
1	2	0.903	1.107	0.784	1.275	1.90	0.771	1.297	6.0	14.8	133.2	0.781	0.618	-32.9	0.500	-48.3	0.414	-58.8	0.351	-66.6
1	3	0.934	1.070	0.850	1.177	2.85	0.840	1.191	8.9	21.8	128.1	0.847	0.721	-32.9	0.618	-48.3	0.535	-58.8	0.469	-66.6
1	5	0.960	1.041	0.906	1.103	4.74	0.900	1.111	14.9	36.0	127.7	0.905	0.820	-32.9	0.744	-48.3	0.677	-58.8	0.618	-66.6
1	10	0.980	1.020	0.952	1.050	9.52	0.949	1.054	29.7	71.7	124.0	0.951	0.905	-32.9	0.861	-48.3	0.820	-58.8	0.781	-66.6
1	20	0.990	1.010	0.976	1.025	18.87	0.974	1.027	59.4	143.0	120.0	0.975	0.951	-32.9	0.928	-48.3	0.905	-58.8	0.883	-66.6
Passband Ripple, A_{MAX} = 2.0dB																				
1	1	0.820	1.220	0.629	1.589	0.98	0.613	1.631	4.0	10.6	374.8	0.500	0.414	-35.4	0.303	-50.8	0.236	-61.3	0.193	-69.2
1	2	0.905	1.104	0.789	1.268	1.96	0.777	1.287	7.9	19.6	312.6	0.781	0.618	-35.4	0.500	-50.8	0.414	-61.3	0.351	-69.2
1	3	0.936	1.068	0.853	1.172	2.95	0.845	1.184	11.9	29.0	302.0	0.847	0.721	-35.4	0.618	-50.8	0.535	-61.3	0.469	-69.2
1	5	0.961	1.040	0.909	1.100	4.90	0.903	1.107	19.7	47.9	302.0	0.905	0.820	-35.4	0.744	-50.8	0.677	-61.3	0.618	-69.2
1	10	0.980	1.020	0.953	1.049	9.80	0.950	1.052	39.5	95.4	302.0	0.951	0.905	-35.4	0.861	-50.8	0.820	-61.3	0.781	-69.2
1	20	0.990	1.010	0.976	1.024	19.61	0.975	1.026	79.0	190.0	302.0	0.975	0.951	-35.4	0.928	-50.8	0.905	-61.3	0.883	-69.2

* f_{0Bp}/BW_i -This is the ratio of the bandpass filter center frequency to the **ripple bandwidth** of the filter.

f_{0Bp}/BW_2 -This is the ratio of the bandpass filter center frequency to the **-3dB filter bandwidth.

Chebyshev or Butterworth—a system designer's confusion

The filter designer/mathematician is familiar with terms such as:

$$K_C = \tanh A$$

$$A = \frac{1}{n}\cosh^{-1}\frac{1}{\epsilon}$$

Ripple bandwidth = 1/cosh A and $A_{dB} = 10 \log [1 + \epsilon^2(C_n{}^2(\Omega))]$.

This is all gobbledygook (not to be confused with flooby dust) to the system designer. The system designer is accustomed to –3dB bandwidths and may be tempted to use only Butterworth filters because they have the cherished –3dB bandwidths. But specs are specs and Butterworth bandpass filters are only so good. Chebyshev bandpass filters trade off ripple in the passband for somewhat steeper rolloff to the stopband. More ripple translates to a higher "Q" filter. The pain of the filter designer is sometimes tolerable to the system designer.

Tables 22.1 through 22.4 are unique (we think) in that they present -3dB bandwidths for Chebyshev filters for use by system designers. Nevertheless we would be amiss to Mr. Chebyshev if we did not, at least, explain ripple bandwidth.

Figure 22.13 shows the Chebyshev bandpass filter at frequencies near the passband.

It can be clearly seen that the ripple bandwidth ($f_{1ripple} - f_{2ripple}$) is the band of passband frequencies where the ripple is less than or equal to a specific value (R_{dB}). The –3dB bandwidth is seen to be greater than the ripple bandwidth and that is the

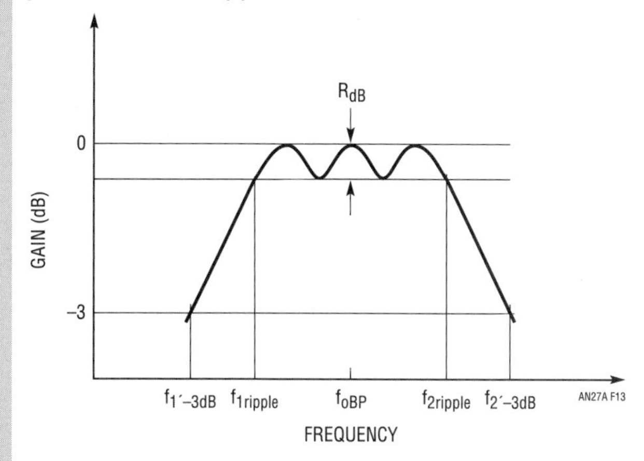

Figure 22.13 • Typical Chebyshev BP Filter—Close-Up of Passband

subject of much confusion on the part of the system designer.

Tables 22.1 through 22.4 allow the system designer to use –3dB bandwidths to specify Chebyshev BP filters. The Chebyshev approximation to the ideal BP filter has many benefits over the Butterworth filter near the cutoff frequency.

YOU CAN DESIGN WITH CHEBYSHEV FILTERS!!!

Example 3—design

Use Table 22.4 to design an 8th order all pole Chebyshev bandpass filter centered at f_0BP = 10.2kHz with a *–3dB bandwidth* equal to 800Hz as shown in Figure 22.14.

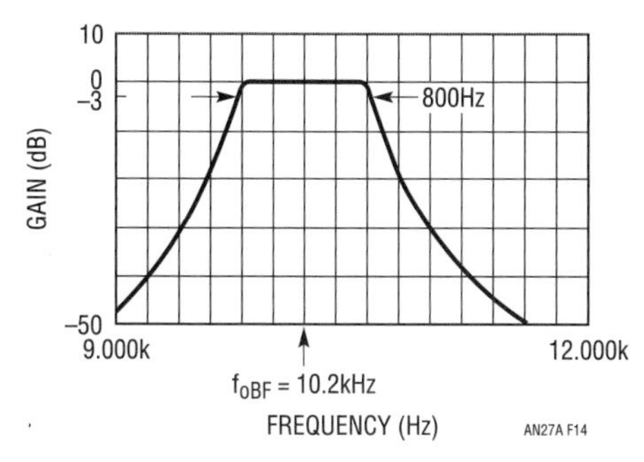

Figure 22.14 • Example 3—8th Order Chebyshev BP Filter f_{OBP} = 10.2kHz, BW = 800Hz

We choose $A_{MAX} = 0.1$dB. Now we calculate:

$$\frac{f_{OBP}}{f_{BW(-3dB)}} = \frac{10.2kHz}{800Hz} = 12.75$$

We can now extract from Table 22.4 the following line:

f_{OBP}	f_{OBP}/ BW_1	f_{o1}(Hz)	f_{o2}(Hz)	f_{o3}(Hz)	f_{o4}(Hz)	f_{OBP}/BW_2	Q1=Q2	Q3=Q4	K
1	10	0.977	1.023	0.945	1.058	8.20	15.7	37.9	29.8

Since our bandwidth ratio f_{OBP}/BW_2 is not exactly on a chart line, but between two lines, we must arithmetically scale to obtain our design parameters. Our f_{OBP}/BW_2 ratio lies between 8.2 and 16.39. (Remember, this is –3dB BW!)

For a symmetrical bandpass filter the poles are symmetrical about f_{0BP}. Then:

$$(f_{02} - f_{01}) = (1.023 - 0.977) \times 10.2\text{kHz} \times \left(\frac{8.2}{12.75}\right) = 302\text{Hz}$$

Note: $\left(\frac{8.2}{12.75}\right) = \frac{f_0BP}{BW}$ Scaling Factor

So our first two poles lie symmetrically about f_o(10.2kHz) and are 302Hz apart:

$f_{o2} = 10200\text{Hz} + 302\text{Hz}/2 = 10351\text{Hz}$

$f_{o1} = 10200\text{Hz} - 302\text{Hz}/2 = 10049\text{Hz}$

The Q of these two poles is equal and is also scaled:

$$Q1 = Q2 = 15.7 \times \frac{12.75}{8.2} = 24.4$$

We calculate the two additional poles:

$$(f_{04} - f_{03}) = (1.058 - 0.945) \times 10.2\text{kHz} \times \frac{8.2}{12.75} = 741\text{Hz}$$

$f_{o3} = 10200\text{Hz} - 741\text{Hz}/2 = 9830\text{Hz}$

$f_{o4} = 10200\text{Hz} + 741\text{Hz}/2 = 10571\text{Hz}$

The Qs are:

$$Q3 = Q4 = 37.9 \times \frac{12.75}{8.2} = 58.9$$

Qs of this magnitude are difficult to realize no matter how the filter is realized. The filter designer should strive for Qs no greater than 20 and perhaps no greater than 10 at frequencies above 20kHz. K, for this example, is not scaled and will be equal to 29.8 from Table 22.4.

Example 3—frequency response estimation

Table 22.4 (and also Tables 22.1–22.3) may be used by the filter designer to obtain a good approximation to the overall shape of the bandpass filter. Referring to Figure 22.12 for Chebyshev filters, we may use the charts to find f_3, f_5, f_7, ... These frequencies define the band edges at 2, 3, 4, ...times the ripple bandwidth of the Chebyshev filter.

Example 3 specified a 10.2kHz bandpass filter with an 800Hz -3dB bandwidth. Our task, if we choose to accept it, is to convert our -3dB bandwidth to the ripple bandwidth of the filter so that we may use the tables.

Recalling that:

$$\frac{f_{0BP}}{BW_2\,(-3\text{dB})} = 12.75 \text{ and that } f_{0BP} = 1,$$

(Because all the tables are normalized), we calculate $BW_{2(-3dB)} = .0784$

Comparing the Table 22.4 values for $A_{MAX} = 0.1\text{dB}$ we note that:

$$\frac{f_{0BP}}{BW_{1(\text{ripple})}} \cong \frac{f_{0BP}}{BW_{2(-3dB)}} \times (\text{Scaling Factor})$$

For Amax = 0.1dB, 8th order Chebyshev, this factor is approximately 0.82. For other order filters and/or different values of A_{MAX} we can examine the corresponding chart values to find our scaling factor.

So our ripple BW is:

$$BW_{2(-3dB)} \times (\text{Scaling Factor}) = BW_{1(\text{ripple})}$$

$$.0784 \times 0.82 = .0643$$

Now we can calculate f_3, f_5, f_7, ...Notice that once we find f_3, f_5, f_7, ...it does not matter where on the table our filter falls. The filter bandwidth determines f_3, f_5, f_7, ...and once we know these frequencies we can directly get our gains at these frequencies.

By formula:

$$(f_x, f_{x+1}) + \frac{\pm nBW + \sqrt{(nBW)^2 + 4(f_{OBP})^2}}{2}$$

for our case $f_0BP = 1$

Calculating:

$$2BW = .1286 \frac{\pm 2BW + \sqrt{(.1286)^2 + 4}}{2} = 1.0664, 0.9378$$

$$3BW = .1929 \frac{\pm 3BW + \sqrt{(1929)^2 + 4}}{2} = 1.1011, 0.9082$$

Then we can denormalize to find points for our Bode plot:

$(f_3, f_4) = 0.9378 \times f_{OBP} = 0.9378 \times 10.2\text{kHz} = 9.566\text{kHz}$
$1.0664 \times f_{OBP} = 1.0664 \times 10.2\text{kHz} = 10.877\text{kHz}$

Gain = –23.4dB both f_3 and f_4
$(f_5, f_6) = 0.9082 \times f_{OBP} = 0.9082 \times 10.2\text{kHz} = 9.264\text{kHz}$
$1.1011 \times f_{OBP} = 1.1011 \times 10.2\text{kHz} = 11.231\text{kHz}$

Gain = –38.8dB both f_5 and f_6

Example 3—implementation

The 10.2kHz (f_{OBP}), 8th order bandpass filter can be implemented with an LTC1064A using three sections in mode 2 and one section in mode 3. The implementation is shown briefly in Figures 22.15 and 22.16. The calculations are not shown here, but are similar to the previous hardware implementations of examples 1 and 2.

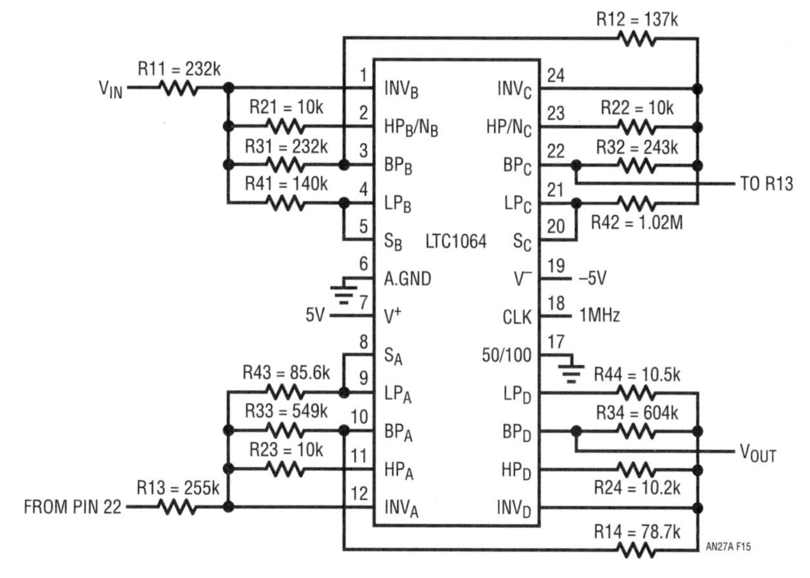

Figure 22.15 • LTC1064 Implementation Pinout—10.2kHz 8th Order BPF

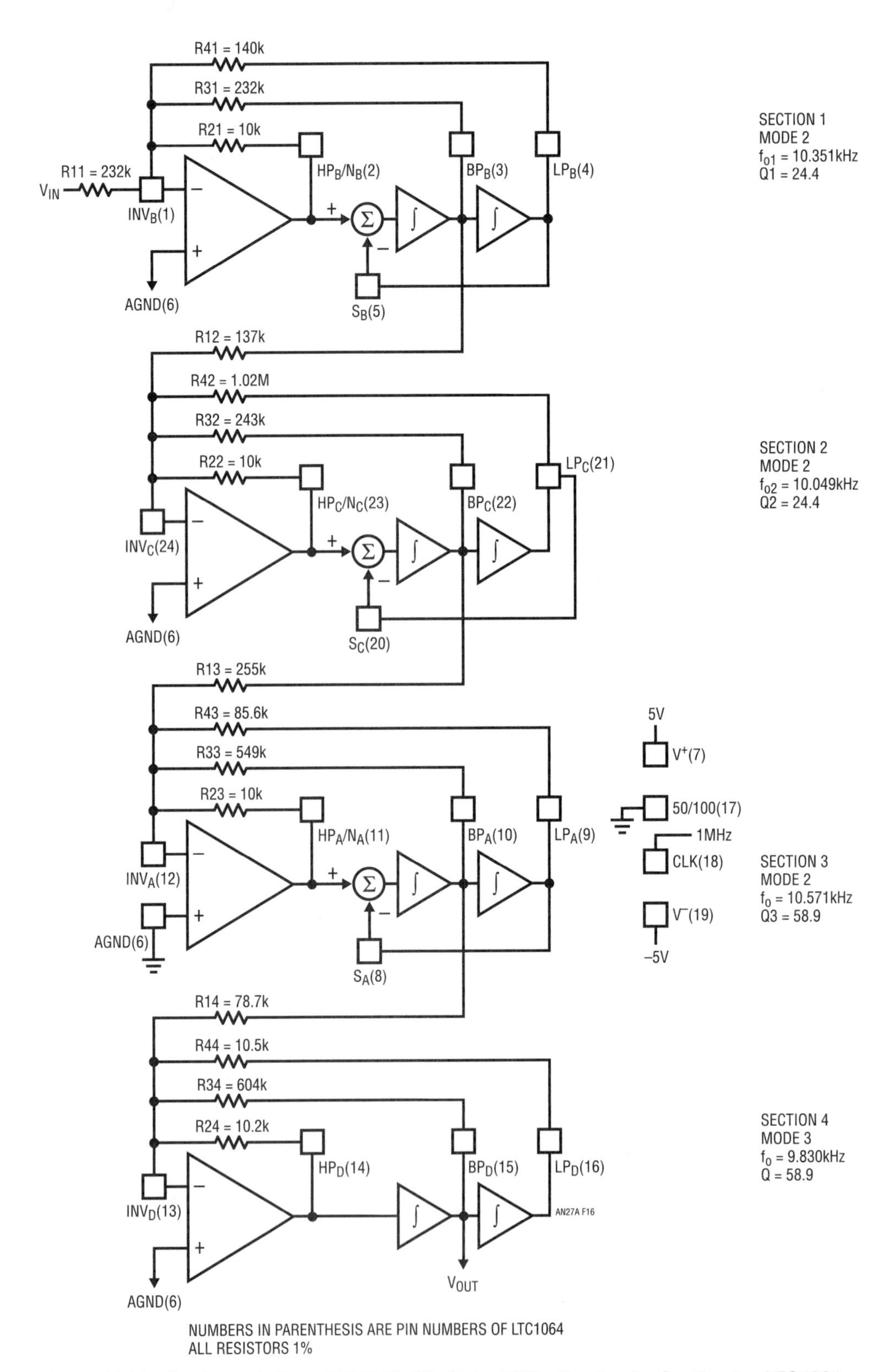

Figure 22.16 • Implementation of 10.2kHz 8th Order BPF—Section by Section for LTC1064

23 FilterCAD user's manual, version 1.10

What is FilterCAD?

FilterCAD is designed to help users without special expertise in filter design to design good filters with a minimum of effort. It can also help experienced filter designers achieve better results by providing the ability to play "what if" with the values and configuration of various components.

With FilterCAD, you can design any of the four major filter types (lowpass, highpass, bandpass and notch), with Butterworth, Chebyshev, Elliptic, or custom-designed response characteristics. (Bessel filters can be realized by manually entering pole and Q values, but FilterCAD cannot synthesize a Bessel response in this version.) FilterCAD is limited to designs which can be achieved by cascading state-variable 2nd order sections. FilterCAD plots amplitude, phase and group-delay graphs, selects appropriate devices and modes, and calculates resistor values. Device selection, cascade order and modes can be edited by the user.

License agreement/disclaimer

This copy of FilterCAD is provided as a courtesy to the customers of Linear Technology Corporation. It is licensed for use in conjunction with Linear Technology Corporation products only. The program is not copy protected and you may make copies of the program as required, provided that you do not modify the program, and that said copies are used only with Linear Technology Corporation products.

While we have made every effort to ensure that FilterCAD operates in the manner described in this manual, we do not guarantee operation to be error free. Upgrades, modifications, or repairs to this program will be strictly at the discretion of Linear Technology Corporation. If you encounter problems in installing or operating FilterCAD, you may obtain technical assistance by calling our applications department at (408) 432-1900, between 8:00 a.m. and 5:00 p.m. Pacific time, Monday through Friday. Because of the great variety of operating-system versions, and peripherals currently in use, we do not guarantee that you will be able to use FilterCAD successfully on all such systems. If you are unable to use FilterCAD, Linear Technology Corporation does guarantee to provide design support for LTC filter products by whatever means necessary.

Linear Technology Corporation makes no warranty, either expressed or implied, with respect to the use of FilterCAD or its documentation. Under no circumstances will Linear be liable for damages, either direct or consequential, arising from the use of this product or from the inability to use this product, even if we have been informed in advance of the possibility of such damages.

FilterCAD download

The FilterCAD tool, although not supported, can be downloaded at www.linear.com. Locate the downloaded file on your computer and manually start installation in that directory.

Your FilterCAD distribution includes the following files. If, after installing the program, you have difficulty in running FilterCAD, check to be sure all of the necessary files are present.

README.DOC	(Optional) if present, includes updated information on FilterCAD not included in this manual
INSTALL.BAT	Automatic installation program—installs FilterCAD on hard drive
FCAD.EXE	Main program file for FilterCAD
FCAD.OVR	Overlay file for FilterCAD—used by FCAD.EXE

Analog Circuit and System Design: Immersion in the Black Art of Analog Design. http://dx.doi.org/10.1016/B978-0-12-397888-2.00023-7

FCAD.ENC	Encrypted copyright protection file—DO NOT TOUCH!
FDPF.EXE	Device-parameter file editor—used to update FCAD.DPF (see Appendix 1)
FCAD.DPF	Device-parameter file—holds data for all device types supported by FilterCAD
ATT.DRV	AT&T graphics adapter driver
CGA.DRV	IBM CGA or compatible graphics driver
EGAVGA.DRV	EGA and VGA graphics drivers
HERC.DRV	Hercules monochrome graphics driver
ID.DRV	Identification file for all driver specifications

Note: *Once you have configured FilterCAD and selected your display type, you can delete unnecessary drivers if you need to conserve disk space. (Be sure not to delete any drivers.)*

Before you begin

Please check the FilterCAD program to see if it contains the README.DOC file. This file, if present, will contain important information about FilterCAD not included in this manual. Please read this file before attempting to install and use FilterCAD. To display the README file on your screen, type:

TYPE README.DOC [Enter]

Press

[Ctrl] S

to pause scrolling. Press any key to resume scrolling. To print a hard copy of the README file on your printer type:

TYPE README.DOC>PRN [Enter]

Procedure for FilterCAD installation in Win7 PC

The FilterCAD installation in Win7 downloads reliably to a target folder.

1. If an LTC program like LTspice® or QuikEval™ has been installed then the following directory folder exists:
   a. C*Program Files**LTC* (in a 32-bit system)
   or
   b. C*Program Files(86)**LTC* (in a 64-bit system).
   If not then create a directory folder as in a. or b.

The FilterCAD download is at: http://www.linear.com/designtools/software/#Filter

2. Start the FilterCAD download and open the FilterCAD.zip to extract the "FilterCADv300.exe."
   "Right Click" on "FilterCAD.exe,"
   and select
   "Run as an administrator"
   then select the following Directory:
   C\Program Files\LTC (in a 32-bit system)
   or
   C\Program Files(86)\LTC (in a 64-bit system).
3. Go to
   C\Program Files\LTC (in a 32-bit system)
   or
   C\Program Files(86)\LTC (in a 64-bit system)
   open
   "OPEN THIS FOLDER TO INSTALL FilterCAD"
   and
   "Run" SETUP.exe
   then FilterCAD is installed in:
   C\Program Files\LTC\FILTERCAD (in a 32-bit system)
   or
   C\Program Files(86)\LTC\FILTERCAD (in a 64-bit system).
   END

Hardware requirements

A list of the graphics adapters and modes supported by FilterCAD will be found in the Configuration section. FilterCAD is a calculation-intensive program, and should therefore, be run on the most powerful system available.

What is a filter?

A filter is a circuit that selectively passes a certain range of the frequencies present at its input to its output, while blocking (attenuating) other frequencies. Filters are normally described in terms of the frequencies that they pass.

Most filters conform to one of four common types. Lowpass filters pass all frequencies below a specified frequency (called the cutoff frequency) and progressively attenuate frequencies above the cutoff frequency. Highpass filters do exactly the opposite; they pass frequencies above the cutoff frequency while progressively attenuating frequencies below the cutoff frequency. Bandpass filters pass a band of frequencies around a specified center frequency,

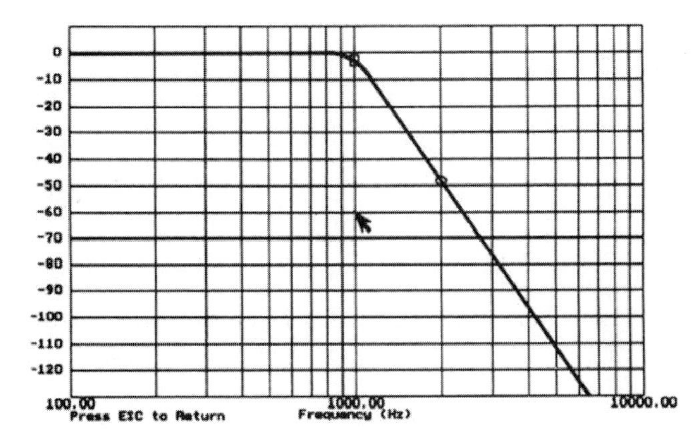

Figure 23.1 • Lowpass Response

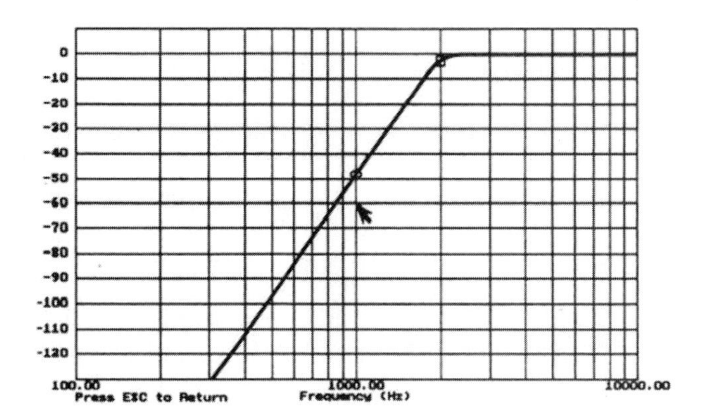

Figure 23.2 • Highpass Response

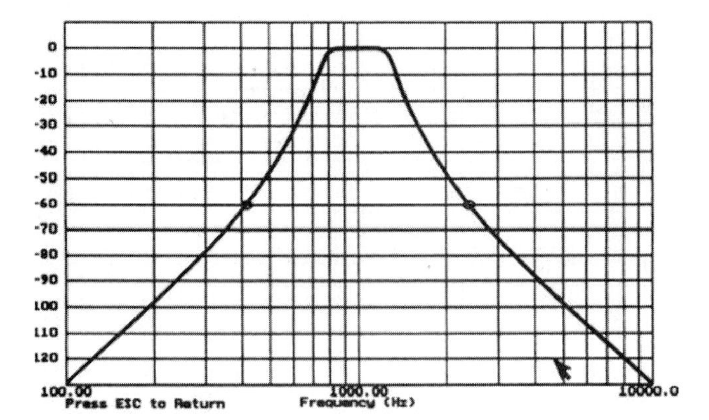

Figure 23.3 • Bandpass Response

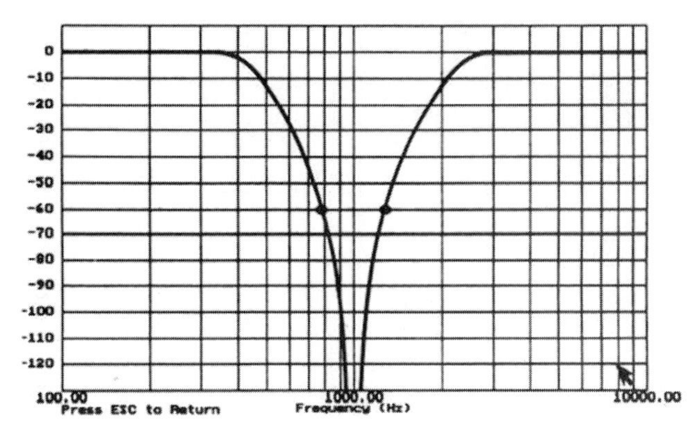

Figure 23.4 • Notch Response

attenuating frequencies above and below. Notch or bandstop filters attenuate the frequencies around the center frequency, passing frequencies above and below. The four basic filter types are illustrated in Figures 23.1 to 23.4. There are also allpass filters, which, not surprisingly, pass all of the frequencies present at their input.[1] In addition, it is possible to create filters with more complex responses which are not easily categorized.

The range of frequencies that a filter passes is known, logically enough as its "passband." The range of frequencies that a filter attenuates is known as its "stopband." Between the passband and stopband is the "transition region." An ideal filter might be expected to pass all of the frequencies in its passband without modification while infinitely attenuating frequencies in its stopband. Such a response is shown in Figure 23.5. Regrettably, real-world filters do not meet these imaginary specifications. Different types of filters have different characteristics, less-than-infinite rates of attenuation versus frequency in the transition region. In other words, the amplitude response of a given filter has a characteristic slope. Frequencies in the passband may also be modified, either in amplitude ("ripple") or in phase.

Note 1: While allpass filters don't affect the relative amplitudes of signals with different frequencies, they do selectively affect the phase of different frequencies. This characteristic can be used to correct for phase shifts introduced by other devices, including other types of filters. FilterCAD cannot synthesize allpass filters.

Real-world filters all represent compromises: steepness of slope, ripple, and phase shift (plus, of course, cost and size).

FilterCAD permits the design of filters with one of three response characteristics (plus custom responses). These three response types, which are known as "Butterworth," "Chebyshev," and "Elliptic," represent three different compromises among the previously described characteristics. Butterworth filters (Figure 23.6) have the optimum flatness in the passband, but have a slope that rolls off more gradually after the cutoff frequency than the other two types. Chebyshev filters (Figure 23.7) can have a steeper initial roll off than Butterworths, but at the expense of more than 0.4dB of ripple in the passband. Elliptic filters (Figure 23.8) have the steepest initial roll off of all. But exhibit ripple in both the passband and the stopband. Elliptic filters have higher Qs, which may (if not carefully implemented) translate to a noisier filter. These high Qs have made elliptic filters difficult to implement with active RC filters because of the increased stability and center frequency accuracy requirements. Elliptic filters can be implemented with SCFs due to their inherently better stabilities and center frequency accuracies when compared to active RC filters. Chebyshev and elliptic designs can achieve greater stopband attenuation for a given number of 2nd order sections than can Butterworths.

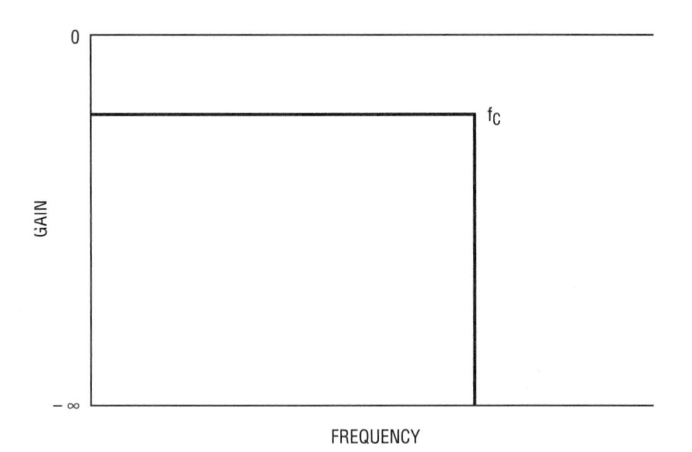

Figure 23.5 • Ideal Lowpass Response

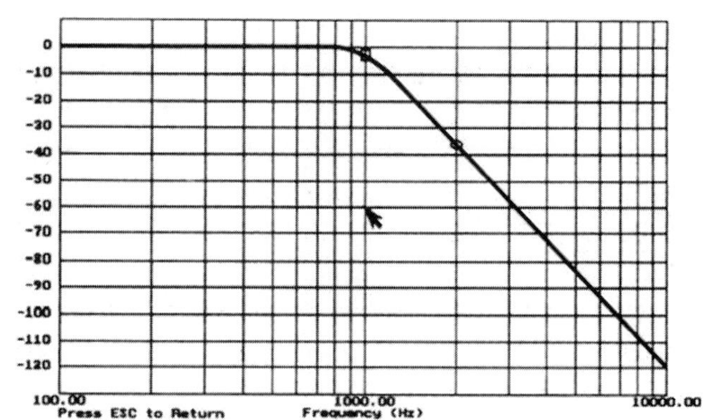

Figure 23.6 • 6th Order Butterworth Lowpass Response

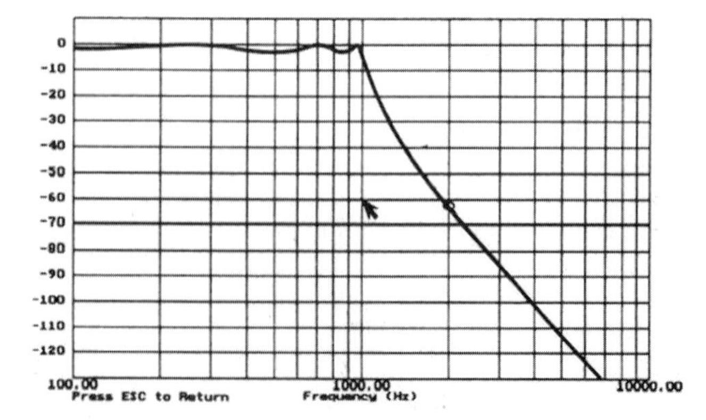

Figure 23.7 • 6th Order Chebyshev Lowpass Response

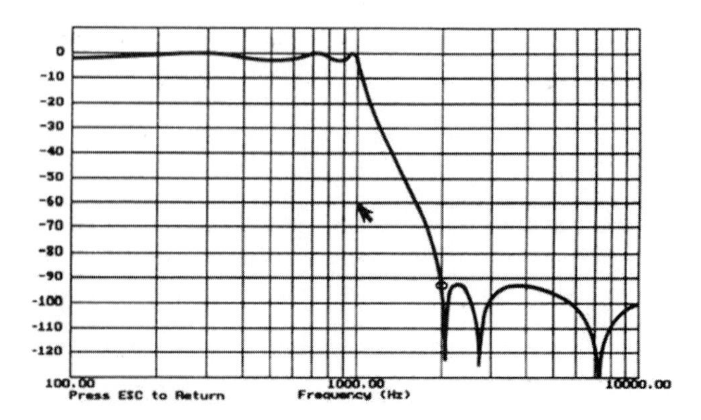

Figure 23.8 • 6th Order Elliptic Lowpass Response

Filters are typically built up from basic building blocks known as 1st order and 2nd order sections. Each LTC filter contains circuitry which, together with an external clock and a few resistors, closely approximates 2nd order filter functions. These are tabulated in the frequency domain.

1. **Bandpass function:** available at the bandpass output pin, refer to Figure 23.9

$$G(S) = H_{OBP}\frac{s\,\omega_0/Q}{s^2 + (s\,\omega_0/Q) + \omega_0^2}$$

H_{OBP} = Gain at $\omega = \omega_0$

$f_0 = \omega_0/2\pi$; f_0 is the center frequency of the complex pole pair. At this frequency, the phase shift between input and output is $-180°$.

Q = Quality factor of the complex pole pair. It is the ration of f_0 to the -3dB bandwidth of the 2nd order bandpass function. The Q is always measured at the filter BP output

2. **Lowpass function:** available at the LP output pin, refer to Figure 23.10

$$G(s) = H_{OLP}\frac{\omega_0^2}{s^2 + s(\omega_0/Q) + \omega_0^2}$$

H_{OLP} = DC gain of the LP output.

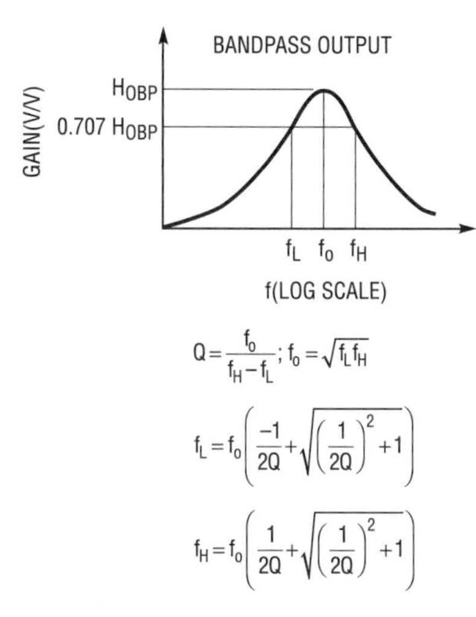

Figure 23.9 • 2nd Order Bandpass Section

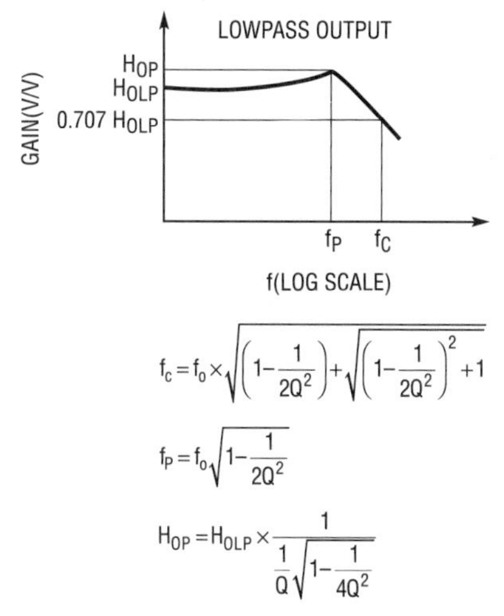

Figure 23.10 • 2nd Order Lowpass Section

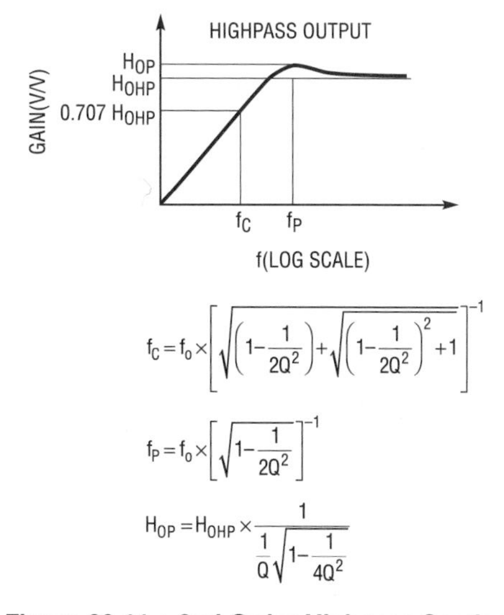

Figure 23.11 • 2nd Order Highpass Section

3. **Highpass function:** available only in mode 3 at the HP output pin, refer to Figure 23.11

$$G(s) = H_{OHP}\frac{s^2}{s^2 + s(\omega_0/Q) + \omega_0^2}$$

H_{OHP} = gain of the HP output for $f \rightarrow \frac{f_{CLK}}{2}$

4. **Notch function:** available at the N output for several modes of operation.

$$G(s) = (H_{ON2})\frac{(s^2 + \omega_n^2)}{s^2 + s(\omega_0/Q) + \omega_0^2}$$

H_{ON2} = gain of the notch output for $f \rightarrow \frac{f_{CLK}}{2}$
H_{ON1} = gain of the notch output for $f \rightarrow 0$
$f_n = \omega_n/2\pi$; f_n is the frequency of the notch occurrence.

These sections are cascaded (the output of one section fed to the input of the next) to produce higher-order filters which have steeper slopes. Filters are described as being of a certain "order," which corresponds to the number and type of cascaded sections they comprise. For example, an 8th order filter would require four cascaded 2nd order sections, whereas a 5th order filter would require two 2nd order sections and one 1st order section. (The order of a filter also corresponds its number of poles, but an explanation of poles is outside the scope of this manual.)

Step one, the basic design

The first item on FilterCAD's MAIN MENU is "DESIGN Filter." To access the DESIGN Filter screen, press

1

On the DESIGN Filter screen, you make several basic decisions about the type of filter you're going to design. First, you must select your basic filter type (lowpass, highpass, bandpass, or notch). Press the

Spacebar

to step through the options. When the filter type that you want is displayed, press

[Enter]

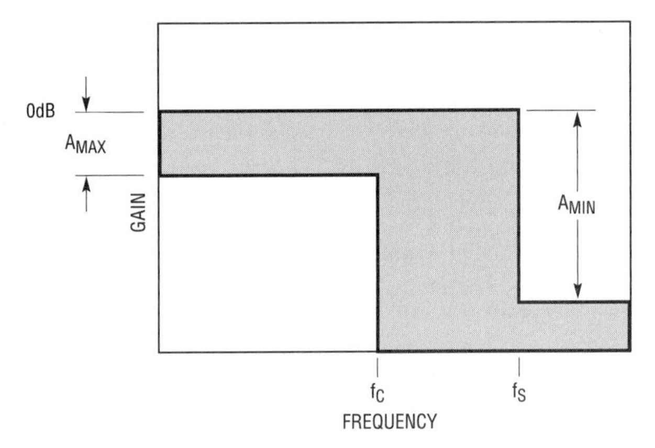

Figure 23.12 • Lowpass Design Parameters: A_{MAX} = Maximum Passband Ripple, f_C = Corner Frequency, f_S = Stopband Frequency, A_{MIN} = Stopband Attenuation

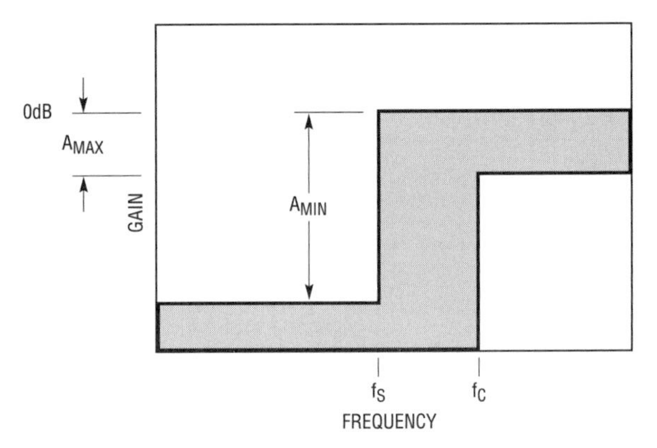

Figure 23.13 • Highpass Design Parameters: A_{MAX} = Maximum Passband Ripple, f_C = Corner Frequency, f_S = Stopband Frequency, A_{MIN} = Stopband Attenuation

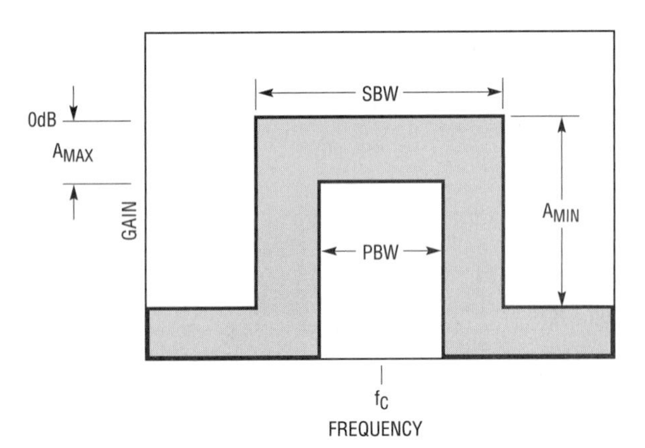

Figure 23.14 • Bandpass Design Parameters: A_{MAX} = Maximum Passband Ripple, f_C = Center Frequency, PBW = Pass Bandwidth, SBW = Stop Bandwidth, A_{MIN} = Stopband Attenuation

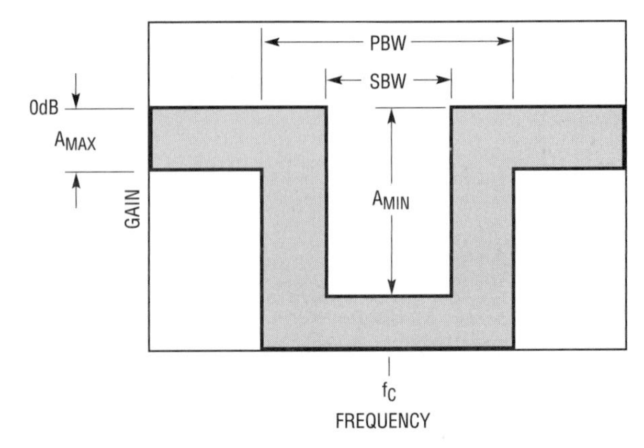

Figure 23.15 • Notch Design Parameters: A_{MAX} = Maximum Passband Ripple, f_C = Center Frequency, PBW = Pass Bandwidth, SBW = Stop Bandwidth, A_{MIN} = Stopband Attenuation

Next, you must select the type of response characteristic you want (Butterworth, Chebyshev, Elliptic, or Custom). Again, use the

Spacebar

to step through the options and press

[Enter]

when the response type you want is displayed.

Next, you will enter the most important parameters for your filter. Exactly what these parameters will be depends on the type of filter you have chosen. If you have selected lowpass or highpass, you must enter the maximum pass-band ripple, in dB (must be greater than zero, or, in the case of Butterworth response, must be 3dB); the stopband attenuation, in dB; the corner frequency (also known as the cutoff frequency), in Hz; and the stopband frequency. If you chose a bandpass or notch filter, you must enter the maximum passband ripple and the stopband attenuation, followed by the center frequency, in Hz; the pass bandwidth, in Hz; and the stop bandwidth, in Hz. (The meanings of these various parameters in the different design contexts are illustrated in Figures 23.12 to 23.15) If you chose a custom response, you're in an entirely different ball game, which will be described later.

Type in the parameters for your chosen filter, pressing

[Enter]

after each parameter. If you want to go back and alter one of the parameters you have previously entered, use the

Up-Arrow

and

Down-Arrow

keys to move to the appropriate location and retype the parameter. When you have entered all of the correct parameters, move the cursor to the last parameter in the list and press

[Enter]

FilterCAD will now calculate and display additional parameters of the filter you have designed, including its order, actual stopband attenuation, and gain, and will display a list of the 2nd order and 1st order sections needed to realize the design, along with their f_0, Q, and f_n values (as appropriate). These numbers will be used later to implement your filter design and calculate resistor values.

FilterCad will, in many cases, prevent you from entering inappropriate values. For instance, in the case of a lowpass filter, the program will not permit you to enter a stopband frequency that is lower than the corner frequency. Similarly, in the case of a highpass filter, the stopband frequency must be lower than the corner frequency. In addition, you cannot enter a maximum passband ripple value that is greater than the stopband attenuation, nor can you enter a set of values that will lead to a filter of an order greater than 28.

Custom filters

The custom-response option on the DESIGN screen can be used in two ways. It can be used to modify filter designs created by the method previously described, or it can be used to create filters with custom responses from scratch, by specifying a normalization value and manually entering the desired f_0, Q, and f_n values for the necessary 2nd order and 1st order sections. To edit the response of a filter that has already been designed, press

ESC

to exit the DESIGN screen, then press

1

to re-enter the DESIGN screen. Move the cursor to "FILTER RESPONSE" and use the

Spacebar

to select "CUSTOM." You can now edit the f_0, Q, and f_n values for the 2nd order and 1st order sections in the window at the bottom of the screen. When you customize an existing design, the normalization frequency is automatically set to the previously-specified corner/center frequency. It can, however, be edited by the user.

To design a custom filter from scratch, simply select "CUSTOM" as your response type upon first entering the DESIGN screen, then type in the appropriate f_0, Q, and f_n values for the desired response. By default, the normalization value for custom filters is 1Hz. Once you have entered your values, you can change the normalization frequency to any desired value and FilterCAD will scale the f_0, and f_n frequencies accordingly. To change the normalization frequency, press

N

type in the new value, and press

[Enter]

By alternately graphing the resulting response and modifying the f_0, Q, and f_n values, almost any kind of response shape can be achieved by successive approximations.

It should be understood that true custom filter design is the province of a small number of experts. If you have a "feel" for the type of pole and Q values that produce a particular response, then FilterCAD will allow you to design by this "seat of the pants" method. If you lack such erudition, however, it is beyond the scope of the program or this manual to supply it. Nevertheless, novice designers can make productive use of FilterCAD's custom-response feature by entering design parameters from published tables. An example of this technique will be found in the following section.

Step two, graphing filter response

After you have designed your filter, the next step is item two on the MAIN MENU, GRAPH Filter Response. You can graph the amplitude, phase, and/or group-delay characteristics of your filter, and you can plot your graph on either a linear or logarithmic scale. The graph also highlights the 3dB-down point(s) (for Butterworth filters only) and the point(s) where the calculated attenuation is achieved. An additional option on the Graph Menu is called "Reduced View." This option displays a reduced view of your graph in a window in the upper right-hand corner of the full-sized graph. This feature is useful in conjunction with the "Zoom" option.

Use the
 Up-Arrow
and
 Down-Arrow
keys to move through the list of graph parameters and press the
 Spacebar
to step through the selections for each parameter that you wish to modify. When all of the graph parameters have been set correctly, press
 [Enter]
to begin plotting the graph.

Plotting to the screen

If you chose the screen as your output device, FilterCAD will immediately begin plotting the graph. Generating the graph is a calculation-intensive process. It is here that the speed and power of your CPU and the presence or absence of a math coprocessor will become evident. The graph will be generated in a matter of seconds. Note, however, that the speed of calculation and plotting can be increased by reducing the number of data points to be plotted. To modify this parameter, use the "Change GRAPH Window" option, found under item 6, "Configure DISPLAY Parameters," on the Configuration Menu. The number of points can range from 50 to 500. Of course, choosing a smaller number of points will result in a courser graph, but this may be an acceptable trade-off for quicker plotting.

The zoom feature

When you display the graph on the screen, you have the additional option of magnifying or "zooming in" on areas of the graph that are of particular interest. (Before using the zoom feature, it is a good idea to enable the "Reduced View" option on the graph menu. When you zoom in, the area of the zoom will be indicated by a box on the reduce view of the full-sized graph.) Note the arrow in the lower right-hand corner of the graph. This arrow can be used to select the region of the graph to zoom in on. It can also be used to pinpoint the frequency and gain values of any given point on the graph. (These values are displayed at the upper right-hand corner of the screen.) The location of the arrow is controlled by the arrow keys (the cursor control keys) on the numeric keypad. The arrow can move in either fine or coarse increments. To select course movement, press the

 +

key. To select fine movement, press the

 –

key. Move the arrow to one corner of the area that you wish to magnify, and press

 [Enter]

Next, move the arrow to the opposite corner of the area of interest. As you move the arrow, you will see a box expand to enclose the area to be magnified. If you want to relocate the box, you must press

 ESC

and restart the selection process. When the box encloses the desired area, press

 [Enter]

and the screen will be redrawn and a new graph will be plotted within the selected range. Note that the program actually calculates new points as appropriate for the higher precision of the magnified section.

It is possible to zoom in repeatedly to magnify progressively smaller areas of the graph. There is a limit to this process, but the available magnification is more than adequate for any practical purpose. If you want to output the magnified graph to the plotter or a disk file, you can do so. Just press

ESC

once to return to the GRAPH MENU screen, change the output device to plotter or disk, press

[Enter]

and then proceed with the plotting process as described later. To zoom out to the previous graph, press

L

(for Large). You can zoom out as many times in succession as you have zoomed in. Note, however, that for each successive zoom out the graph is recalculated and replotted. If you have done several successive zooms and you want to get back to the full-sized graph without the intermediate steps, it will be quicker to press

ESC

twice to return to the MAIN MENU, then re-enter the GRAPH screen.

Printing the Screen—You can dump your screen graph to your printer at any time by pressing

[Alt] P

This same feature can be used from the Device Screen (see the implementation section) to print FilterCAD's mode diagrams. The screen-print routine will check to see whether your printer is connected and turned on, and will warn you if it is not. If your printer is connected and turned on, but off line, FilterCAD will put it on line and begin printing. Once printing begins, however, FilterCAD does no error checking, so turning off your printer or taking it off line to stop printing may cause the program to "hang up."

Plotting to a plotter, HPGL file, or text file

If you choose to send your graph to a plotter or to an HPGL disk file, you will first be shown the PLOTTER STATUS MENU. First, you will be asked "GENERATE CHART (Y/N)." You are not being asked here whether you want to plot a graph, but whether, when you plot to disk or plotter, you want to draw the grid or only plot the data. This may seem like an absurd choice, but it is here for a reason. If you are plotting to a plotter, you could overlay the response graphs of several different filter designs on one sheet of paper for comparison. If you drew the grid every time, you would end up with a mess. This option allows you to draw the grid on the first pass, then plot only the data on successive passes. This same process can be used, albeit with slightly more effort, when plotting to a disk file. The procedure here is to plot two (or more) separate files, the first with the grid turned on, and the remainder with the grid turned off. Then exit FilterCAD and use the DOS COPY command to concatenate the files. For example, if you wanted to concatenate three HPGL files named SOURCE1, SOURCE2, and SOURCE3 into a single file called TARGET, you would use the following syntax:

COPY/B SOURCE1 + SOURCE2 + SOURCE3 TARGET [Enter]

After you have answered the question

GENERATE CHART (Y/N)

the remainder of the PLOTTER STATUS MENU will be displayed. Here, you can select the dimensions of your graph, the pen colors to be used, and whether to print the design parameters below the graph. When you have determined that the plotter options are set correctly, press

P

to begin plotting. If you wish to exit the PLOTTER STATUS MENU without plotting, press

ESC

You can also plot your graph as a set of data points for gain, phase and group delay plotted to disk in ASCII text format. Select

DISK/TEXT

as the output device. Then select which parameters to plot and press

[Enter]

You will then be prompted for a file name into which the data will be placed.

These points may then be imported to a spreadsheet program, for example, for data manipulation.

If your graph consists mostly of straight horizontal lines or slopes, this indicates that the frequency and amplitude ranges of the graph are probably not set appropriately for the particular filter you are graphing (e.g., a graph frequency range of 100Hz to 10, 000Hz for a highpass filter with a corner frequency of 20Hz). To adjust the frequency and gain ranges of your graph, use the "Change GRAPH Window" option, found under item 6, "Configure DISPLAY Parameters, on the Configuration Menu.

Implementing the filter

The third item on the MAIN MENU, "IMPLEMENT Filter," is where we transform the numbers generated in step one into practical circuitry. There are several steps to this process.

Optimization

The first step is to optimize your filter for one of two characteristics.[2] You can optimize for lowest noise or lowest harmonic distortion. When you optimize for noise, the cascaded sections are ordered in such a way as to produce the lowest output noise when the filter input is grounded. In the absence of any other, conflicting design criteria, this is the most obvious characteristic for optimization. When you optimize for harmonic distortion, the sections are cascaded so as to minimize the internal swings of the respective amplifiers, resulting in reduced harmonic distortion. *Note, however, that optimizing for harmonic distortion will result in the worst noise performance.*

The Optimization screen also allows you to select the ratio of the internal clock frequency to f_0 (50:1 or 100:1) and the clock frequency in Hz, and to turn automatic device selection on and off. It should be understood that the clock frequency ratio represents the state of a particular pin on the device, and does not *necessarily* correspond to the actual ratio of the clock frequency to f_0. Thus, if you change the clock frequency to a value other than that automatically selected by FilterCAD, the frequency ratio *will not change accordingly.* Novice filter designers are advised to use the clock frequency selected by FilterCAD unless there is a compelling reason to do otherwise, and to leave automatic device selection "ON." Clock frequencies and frequency ratios are not arbitrary, but have a relationship to the corner or center frequency that depends upon the selected mode. For more information on clock frequencies, frequency ratios, and modes, consult LTC product data sheets.

When you have selected the characteristic for optimization and adjusted the other options to your satisfaction, move the cursor back to "OPTIMIZE FOR" and press

O

FilterCAD will respond by displaying the selected device and mode(s) and its rationale for selecting a particular cascade order. If you want to re-optimize for some other characteristic, press

O

again and repeat the process previously described.

Implementation

When you are satisfied with the optimization, press

I

to proceed with the implementation. This won't do anything obvious except to clear the window where the optimization information is displayed, so press

D

to view the selected device, cascade order, and modes. When the Device screen is first displayed, it will show detailed specifications of the selected part. To see the mode diagrams of the 2nd and 1st order sections that implement the design, press the

Down-Arrow

Note 2: FilterCAD will not optimize custom filter designs, nor will it select the modes for such designs. This is just another indication that custom designs are the province of experienced designers.

key to move the cursor through the cascade-order list on the left-hand side of the screen.

You will have probably observed that, on the implementation menu, this screen is entitled "Edit DEVICE/MODEs," and you can, in fact, manually edit both the device selection and the modes of the 2nd and 1st order sections. This capability, however, is one that the novice user would be best advised to ignore. If you manually edit the device selection or the modes, you are essentially ignoring the expertise that is built into the program in favor of your own judgment. Experienced designers may choose to do this under some circumstances, but if you want to get from a specification to a working design with the least time and effort, accept the device and modes that FilterCAD selects. To complete the implementation process, press

ESC

to exit the device screen, then press

R

to calculate the resistor values. You can calculate absolute values by pressing

A

Or the nearest 1% tolerance values by pressing

P

There is one more option on the implementation menu that we have not yet examined, "Edit CASCADE ORDER." This option allows you to exchange poles and zeros among and edit the cascade order of the 2nd order sections that make up your filter. This represents one of the most arcane and esoteric aspects of filter design—even the experts are sometimes at a loss to explain the benefits to be gained by tweaking these parameters. So, once again, we must recommend that the novice leave this feature alone.[3]

Saving your filter design

Item 4 on the MAIN MENU, "SAVE Current Filter Design," allows you to save your design to disk. Press

4

to save your design. By default, a new file will be saved with the name "NONAME," while a file that was previously saved and loaded will be saved under its previous name. If you want to save the file under a different name, just type it in at the cursor position. Type the file name only, eight characters or fewer; do not type an extension. All files are saved with the extension .FDF (*Filter Design File*).

By default, the file will be stored in the current directory (the directory that was active when FilterCAD was started). This directory will be displayed at the top of the screen. If you want to save the file to a different directory, press

Home

then type in a new path. When the file name and path are correct, press

[Enter]

to save the file. If there is already a file on the disk with the name that you have selected, FilterCAD will ask whether you want to overwrite the file. Press

Y

to overwrite the file or press

N

to save the file under a different name.

Note 3: Of course, there is no harm in the novice experimenting with the advanced features of FilterCAD, provided he or she realizes that the results of such experiments will not necessarily be useful filter designs.

Loading a filter design file

To load a Filter Design File that you have previously saved, select

5

on the main menu. The LOAD FILE MENU screen will display a directory of all of the .FDF files in the current directory. Use the cursor keys to move the pointer to the name of the file you want to load and press

[Enter]

If there are more .FDF files than can be displayed on the screen at one time, press the

PgDn

key to see additional files. If you want to load a file from a different directory, press

P

then type in the new path. You can also enter a mask to restrict the file names which will appear on the screen. This mask can consist of any of the characters that DOS allows for file names, including the DOS wildcards "*" and "?". By default, the mask is "*", allowing all file names to be displayed. To change the mask, press

M

then type in a new mask of up to eight characters. For example, if you named your .FDF files in such a way that the first two letters of the file name represented the filter type (LP for lowpass, HP for highpass, etc.), you could change the mask to LP* to display only the lowpass filter design files.

Note: *If you attempt to load an .FDF file created with an earlier version of FilterCAD, the program will issue a warning and ask you whether you want to abort loading or proceed at your own risk. Differences between .FDF files from different versions of FilterCAD are minor, and you should be able to load and use earlier .FDF files without difficulty.*

Caution: When you load a filter design file, FilterCAD DOES NOT prompt you to save any design currently in memory, so when a new file is loaded, any unsaved work in memory will be lost.

Printing a report

Item 7 on the MAIN MENU, "SYSTEM Status/Reports," displays a varied collection of information on your system, such as the date and time, the presence or absence of a math coprocessor, and the status of your printer and communications ports. It also shows the state of progress of the current design, including the total design time. (Gosh, a 6th order Butterworth lowpass in 00:05:23, only five seconds shy of the record!) The main function of this screen, however, is to print a design report. This report will include all of the information about your filter available on the design, optimization and implementation screens, except for the mode diagrams. You can also print the gain, phase and group delay of your filter design by pressing

P

Note that if you have not completed the design process by implementing the design and calculating resistor values, the "REPORT AVAILABLE" line will say "PARTIAL." A partial report, lacking modes and resistor values, can be printed, but for a complete report you must implement and calculate resistor values.

Quitting FilterCAD

The ninth and final item of FilterCAD's MAIN MENU, "END FilterCAD," is self-explanatory. If you haven't saved your current design before attempting to exit the program, FilterCAD will ask you if you wish to do so. Press

Y

to exit the program or press

N

to remain in FilterCAD.

Now that we have examined the principal features of FilterCAD, let's walk through a few typical filter designs to get a better idea of how the program works.

A Butterworth lowpass example

First, we'll design a Butterworth lowpass filter, one of the most basic filter types. Load FilterCAD, if you haven't already done so, and press

1

to go to the design screen. Select lowpass as the design type and Butterworth as the response type. Now we have four additional parameters to enter. The passband ripple must be specified as 3dB: this places the cutoff frequency –3dB down with respect to the filters DC gain. Should you desire a Butterworth response with other than 3dB passband ripple you can do so by going to the custom menu. Let's select an attenuation of 45dB, a corner frequency of 1000Hz, and a stopband frequency of 2000Hz. Press

[Enter]

after each parameter. When the last parameter is entered, FilterCAD will synthesize the response. We soon see that the result is an 8th order filter with an actual attenuation of 48.1442dB at 2000Hz. It is composed of four 2nd order lowpass sections, all with corner frequencies of 1000Hz and modest Qs. (Having the same corner frequency for all of the cascaded sections is a characteristic of Butterworth filters.) This is a good time to experiment with some of the filter's parameters to see how they affect the resulting design. Try increasing the attenuation or lowering the stopband frequency. You'll discover that any modification that results in a significantly steeper roll-off will increase the order of the filter proportionally. For instance, reducing the stopband frequency to 1500Hz changes results in a filter of order 13! If a very steep roll-off is required and some ripple in the passband is acceptable, a response type other than Butterworth would probably be preferable.

Next, we'll graph the amplitude and phase characteristics of our Butterworth lowpass filter. Press

ESC

to return to the MAIN MENU, then press

2

to go to the graph menu. We're going to output this graph to the screen, so press

[Enter]

to begin graphing immediately. In a few seconds or a few minutes, depending on the type of computer system you're using, you should see a graph very much like the one in Figure 23.16 Amplitude (in dB's) is indicated on the left side of the graph and phase (in degrees) is indicated on the right side. (If you have your graph parameters set differently than FilterCAD's defaults, your graph may show less of the frequency and amplitude range than the figure. If you don't see a graph substantially like the one in Figure 23.16, you may need to adjust the graph's ranges. Exit the graph screen, go to the MAIN MENU, and select 6, CONFIGURE FilterCAD. Next, select item 6, "Configure DISPLAY Parameters," followed by item 2, "Change GRAPH Window.")

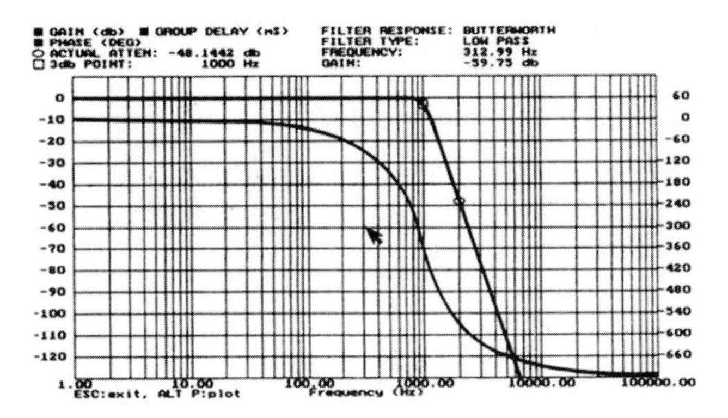

Figure 23.16 • Butterworth Lowpass Filter Response

Observe the characteristic amplitude and phase-response curves of the Butterworth response. (The amplitude curve is the one that begins at 0dB and begins to fall off sharply around 1000Hz—of course, if you have a color display you can make the amplitude and phase curves easily distinguishable by assigning them different colors.) The amplitude in the passband is extremely flat (you could magnify a small segment of the passband many times and still find no observable ripple), and the slope of the roll-off begins just before the corner frequency, reaching the 3dB down point (which, in a Butterworth response is synonymous with the corner frequency) at 1000Hz, and continues to roll-off at the same constant rate to the stopband and beyond. (In theory, the slope will continue to roll-off at this same rate all the way to an infinite attenuation at an infinite frequency.) The phase response begins at 0, slopes exponentially to near −360° as it approaches the corner frequency, then continues down until it asymptotes to −720° in the filter's stopband. Butterworth filters offer the most linear phase response of any type except the Bessel. Figure 23.17 shows the phase response of the Butterworth lowpass filter using a linear phase scale.

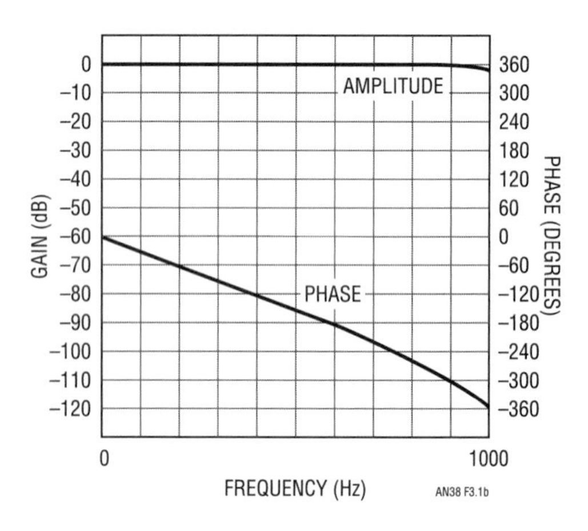

Figure 23.17 • Butterworth Lowpass Phase Response

Having graphed our filter's response, we will next go to the implementation screen to transform it into a practical design. Press

ESC

twice to exit the graph display, then press

3

to go to the implementation screen. The first step is to optimize. Lacking any other pressing need, we'll optimize for noise (the default optimization strategy). We'll use a clock frequency ratio of 50:1 and we'll leave auto device selection ON. Press

O

to execute optimization. FilterCAD selects the LTC1164 and indicates that the Qs have been intermixed for the lowest noise. Next, press

I

for implement and then

D

to display the device screen. This screen shows detailed specs of the LTC1164, and, in the window on the left-hand side of the screen, indicates that **all four of the 2nd order sections in the design will use mode 1.** Press the

Down-Arrow

key and you'll see a diagram of a mode 1 network like the one in Figure 23.18 in place of the LTC1164 specs. Press the

Down-Arrow

three more times, and you'll see three more examples of the same network, differing only in the Q values. This configuration would be fine except for one thing: the LTC1164 has four 2nd order sections, but the fourth lacks an accessible summing node, and therefore cannot be configured in mode 1. You must manually change the mode of the last stage to mode 3. This illustrates the limitations of the present version of FilterCAD.

Each mode 1 section requires three resistors, and the final mode 3 section requires four. To calculate the resistor values, press

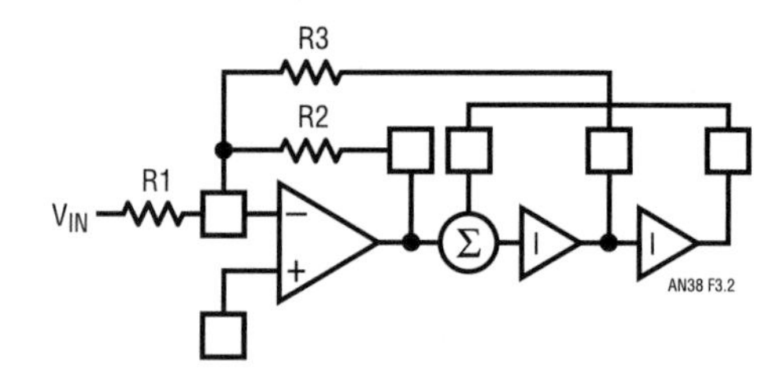

Figure 23.18 • Mode 1 Network

ESC

to exit the Device screen, then press

R

On pressing

P

to select 1% tolerance resistors, FilterCAD displays the values in Table 23.1. This completes the implementation of our Butterworth lowpass example.

Table 23.1 Resistors for Butterworth Lowpass Example

STAGE	R1	R2	R3	R4
1	16.50k	16.50k	10.00k	
2	16.20k	10.00k	25.50k	
3	10.00k	12.10k	10.70k	
4	15.00k	20.50k	10.00k	20.50k

A Chebyshev bandpass example

For our next example, we'll design a bandpass filter with a Chebyshev response. (We'll assume you know your way around the program reasonably well at this point, so we'll dispense with telling you specific keys to press unless we introduce a new feature.) For our Chebyshev design, we'll select a maximum passband ripple of 0.05dB, an attenuation of 50dB, and a center frequency of 5000Hz. We'll specify a pass bandwidth of 600Hz and a stop bandwidth of 3000Hz. This results in another 8th order filter, consisting of four 2nd order bandpass sections, with their corner frequencies staggered around 5000Hz and moderate Qs illustrated in Table 23.2.[4]

The Qs of the sections have been kept within reasonable limits by specifying the very low minimum passband ripple of 0.05dB. That this should be the case may not be obvious until you consider that the passband ripple consists of the product of the resonant peaks of the 2nd order sections. By keeping the passband ripple to a minimum, the Qs of the individual sections are reduced proportionally. You can verify this fact by changing the passband ripple to a higher value and observing the effect on the Qs of the 2nd order sections.

Table 23.2 f_0, Q, and f_n Values for 8th Order Chebyshev Bandpass

STAGE	f_0	Q	fn
1	4657.8615	27.3474	0.0000
2	5367.2699	27.3474	INFINITE
3	4855.1190	11.3041	0.0000
4	5149.2043	11.3041	INFINITE

Next we'll graph the response of our design. The result appears in Figure 23.19 (We have reset the frequency range of the graph to focus on the area of interest.) Observe that the slope of the amplitude response rolls off quite steeply in the transition region and that the slopes become more gradual well into the stopband. In other words, the slope is not constant. This is a characteristic of Chebyshev filters. The characteristic passband ripple is not observable at the current scale, but we can see it by zooming in on the passband.

To zoom, first press

+

to select coarse motion, then use the arrow keys on the cursor keypad to move the arrow to one corner of the rectangle you want to zoom in on (just outside the pass-band), then press

[Enter]

Now move the arrow again and you'll see a box expand to enclose the area to be magnified. When the box encloses the passband, press

[Enter]

again and the new graph will be calculated and plotted. It may require two or three consecutive zooms, but eventually you'll get a close-up of the passband that shows the 0.05dB ripple quite clearly, as in Figure 23.20 (Note that in this figure, the graph style has been reset to "linear.") To return to the full-scale graph, press

L

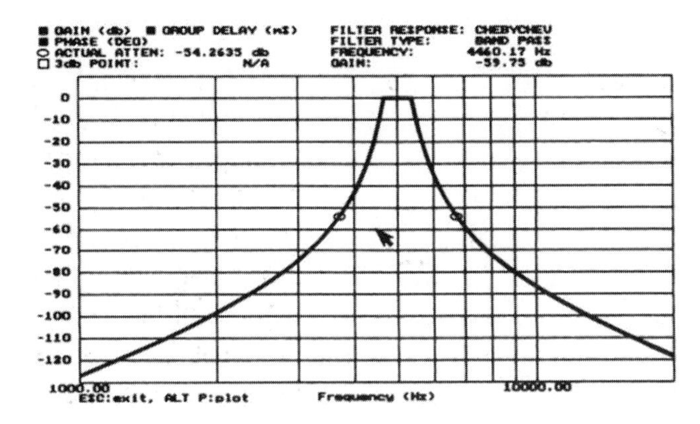

Figure 23.19 • Chebyshev Bandpass Response

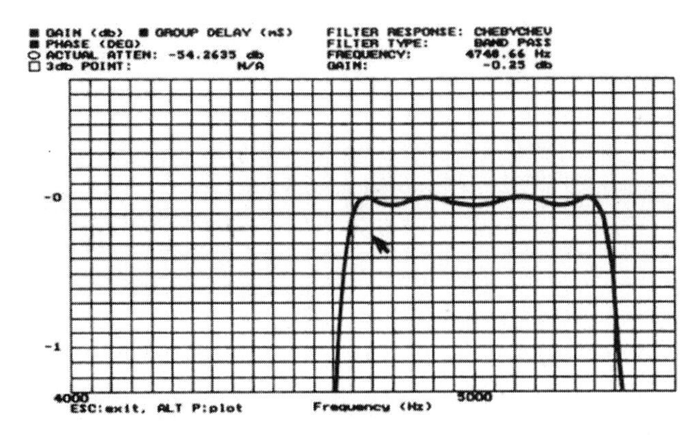

Figure 23.20 • Close-Up of Passband

Note 4: It is possible to design a bandpass filter from a mixture of highpass and lowpass or highpass, lowpass and bandpass sections, but FilterCAD will not do this. When you specify a particular filter type, all of the sections used to realize the design will be of that same type.

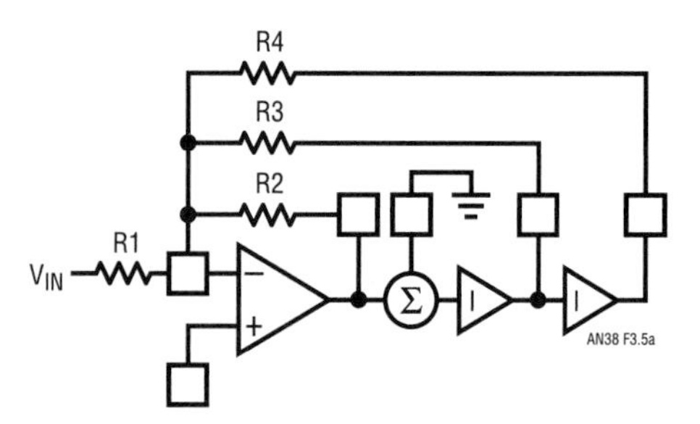

Figure 23.21 • Mode 3 Network

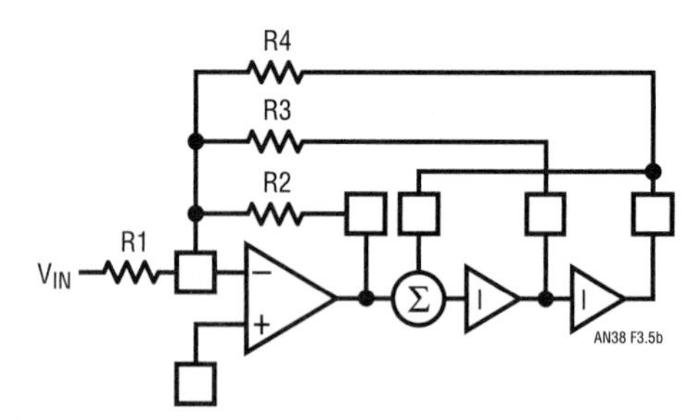

Figure 23.22 • Mode 2 Network

Now, we'll implement our design, optimizing, as before, for lowest noise. We will select clock to f_0 ratio equal to 50:1 and clock frequency equal to 250,000Hz. Once again the LTC1164 is selected and the Qs of the sections are intermixed for the lowest noise. This time mode 3 has been chosen for the first two stages (where $f_0 < f_{CLK}/50$) and mode 2 has been selected for the remaining two stages (where $f_0 > f_{CLK}/50$). The overall gain of 27.29dB has been evenly distributed among stages two through four and the gain of stage one has been set to 1 for improved dynamics. When we view the Device screen, we'll see the specs of the LTC1164 and two diagrams of the mode 3 network followed by two diagrams of the mode 2 network with different f_0, Q, and f_n values (see Figures 23.21 and 23.22). Calculating the 1% resistor values produces the results in Table 23.3.

Table 23.3 Resistors for Chebyshev Bandpass Example

STAGE	R1	R2	R3	R4
1	115.0k	10.00k	68.10k	10.70k
2	215.0k	10.00k	113.0k	11.50k
3	17.80k	10.00k	205.0k	64.90k
4	90.90k	10.00k	105.0k	165.0k

Two elliptic examples

For our next example, we will design a lowpass filter with an elliptic response. We'll specify a maximum passband ripple of 0.1dB, an attenuation of 60dB, a corner frequency of 1000Hz, and a stopband frequency of 1300Hz. In the case of an elliptic response we have one additional question to answer before the response is synthesized. When we have entered the other parameters, FilterCAD asks "Remove highest f_n?" (Y/N). This question requires a bit of explanation. An elliptic filter creates notches by summing the highpass and lowpass outputs of 2nd order stages. To create a notch from the last in a series of cascaded 2nd order stages, an external op amp will be required to sum the highpass and lowpass outputs. Removing the last notch from the series eliminates the need for the external op amp, but does change the response slightly, as we will see.

Note: *The last notch can be removed only from an even-order elliptic filter. If you are synthesizing an elliptic response for the first time and you are uncertain what order of response will result, answer "NO" when asked if you want to remove the last notch. If an even-order response results you can go back and remove the last notch if you wish.*

For comparison, we will synthesize both responses. The f_0, Q, and f_n values for both designs (both are 8th order) are shown in Table 23.4. Observe that the removal of the highest f_n produces slight variations in all of the other values.

Table 23.4 f_0, Q, and f_n Values for Lowpass Elliptic Examples

STAGE	f_0	Q	f_n
Highest f_n Not Removed			
1	478.1819	0.6059	5442.3255
2	747.3747	1.3988	2032.7089
3	939.2728	3.5399	1472.2588
4	1022.0167	13.3902	1315.9606
Highest f_n Removed			
1	466.0818	0.5905	INFINITE
2	723.8783	1.3544	2153.9833
3	933.1712	3.5608	1503.2381
4	1022.0052	13.6310	1333.1141

When we graph our two elliptic examples, (Figures 23.23 and 24) we see that the response of the filter without the highest f_n removed shows four notches in the stopband and a gradual slope after the last notch, whereas the filter with the highest f_n removed exhibits only three notches followed by a steeper slope. Both examples have the steep initial roll-off

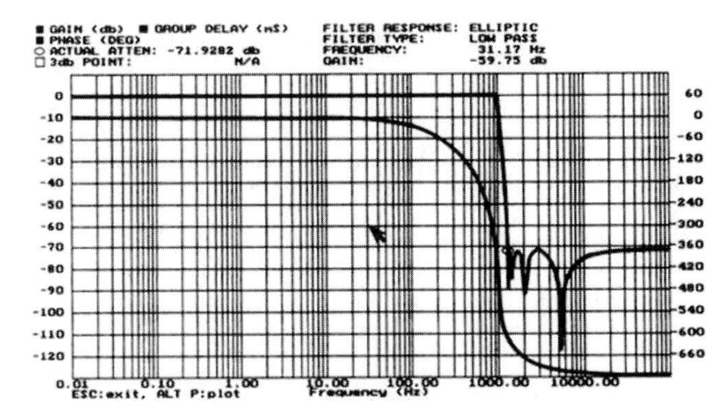

Figure 23.23 • Lowpass Elliptic, Highest f_n Not Removed

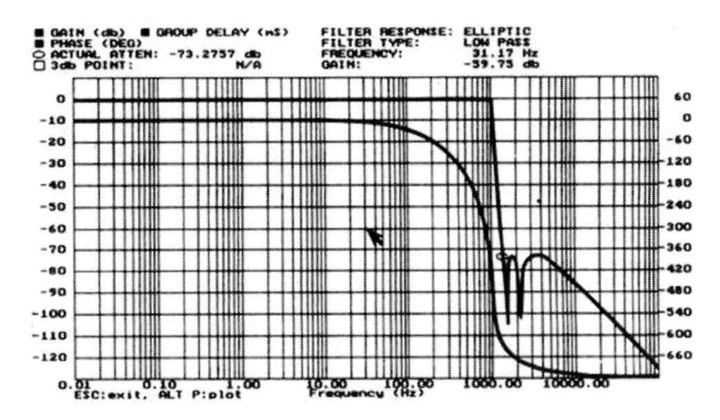

Figure 23.24 • Lowpass Elliptic, Highest f_n Removed

and extremely non-linear phase response in the vicinity of the corner frequency that are essential characteristics of the elliptic response. If your only goal is stopband attenuation greater than 60dB, either implementation would be satisfactory, and the version with the highest f_n removed would probably be selected due to its lower parts count.

When we optimize our two elliptic filters for noise, Filter-CAD selects the LTC1164 and specifies mode 3A for all four stages. Mode 3A is the standard mode for elliptic and notch filters, as it sums the highpass and lowpass outputs of the 2nd order sections as described previously. When we go to the Device screen, we see four mode 3A diagrams, each showing the external op amp, as in Figure 23.25. In practice, this external summing amp is not needed in every case. When cascading sections, the highpass and lowpass outputs of the previous section can be summed into the inverting input of the next section, an external summing amp being required only for the last section. If the highest f_n is removed, external op amps can be dispensed with entirely.

Calculating 1% resistor values (for clock to f_0 ratio 50:1, clock frequency equals 50, 000Hz) for our two elliptic variations yields the results in Table 23.5. R_H and R_L are the resistors, which sum the highpass and lowpass outputs of the successive stages, and R_G is the resistor that sets the gain of the external op amp. Therefore, there is one fewer R_H/R_L pair in the version with the last f_n removed, and R_G is found only in the last stage of the first example. Also, R1, the resistor connected to the inverting input of the input amplifier, is used only for the first stage. **The R_H/R_L pair takes the place of R1 in subsequent stages.**

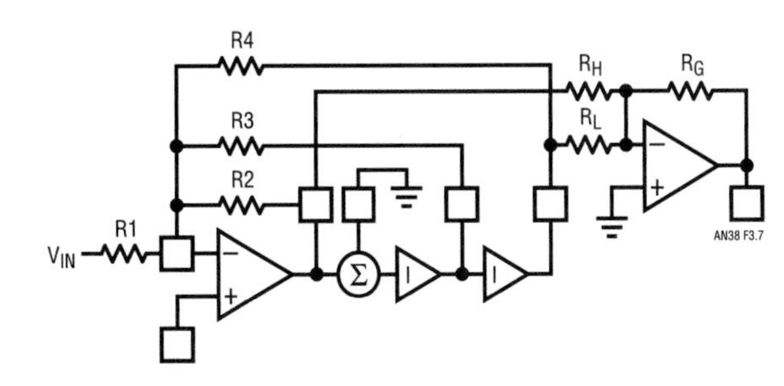

Figure 23.25 • Mode 3A Network

Table 23.5 Resistor Values for Lowpass Elliptic Examples

STAGE	R1	R2	R3	R4	R_G	R_H	R_L
Highest f_n Not Removed							
1	24.90k	10.00k	17.40k	17.80k		237.0k	57.60k
2		10.50k	73.20k	10.00k		21.50k	12.70k
3		10.00k	30.90k	11.30k		21.50k	10.00k
4		19.60k	24.30k	86.60k	10.20k	294.0k	10.00k
Highest f_n Removed							
1	26.10k	10.00k	17.40k	19.10k		261.0k	56.20k
2		10.50k	75.00k	10.00k		23.20k	13.00k
3		10.00k	31.60k	11.50k		22.60k	10.00k
4		18.70k	23.20k	86.60k			

A custom example

For a simple example of how the custom design option works, we'll design a 6th order lowpass Bessel filter by manually entering pole and Q values. When you set the response on the Design screen to "Custom" and press

[Enter]

the usual parameter-entry stage is bypassed and you go directly to the f_0, Q and f_n section, where you can enter any values you want. We'll use values from Table 23.6, for a filter normalized to –3db = 1Hz. The published table from which these values were taken didn't mention f_n values at all, so when the author typed them in initially, he left the f_n values as he found them, as zeros. The result was not the desired lowpass filter, but its highpass mirror image. This shows the kind of trap that awaits the unwary.

Once the values have been entered, they can be re-normalized for any desired corner frequency. Just press

[Enter]

In this case we will re-normalize to 1000Hz, which simply multiplies the f_0 values in the table by 1000.

Looking at the graph of the resulting response (Figure 23.26), we see the characteristic Bessel response, with its droopy passband and very gradual initial roll-off. When we go to the implementation stage, the process is a little different than we are accustomed to. FilterCAD won't optimize a custom design, nor will it specify the mode(s). It will, however, select the device, (the envelope please...) the LTC1164. Now we need to go to the device screen and manually select the mode for each of the three 2nd order sections. We will select mode 3 for all sections, because the three sections each have different corner frequencies, and mode 3 provides for independent tuning of the individual sections by means of the ratio R2/R4. (We've seen what the mode 3 network looks like before, so we won't duplicate it here.) Having selected the mode, we can calculate resistor values. The results are shown in Table 23.7.

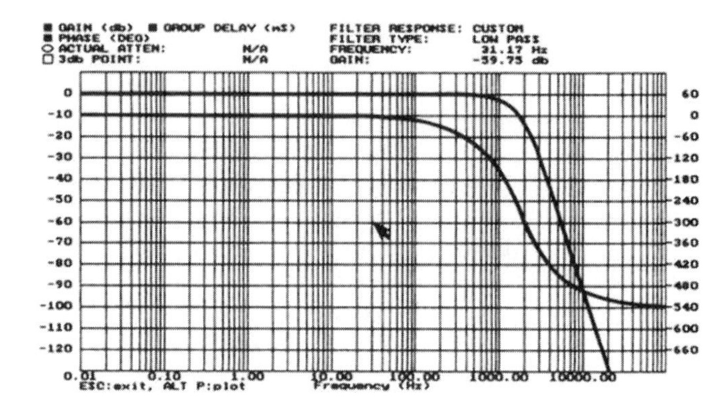

Figure 23.26 • 6th Order Lowpass Bessel Response

Table 23.6 f_0, Q, and f_n Values for 6th Order Lowpass Bessel, Normalized for 1Hz

STAGE	f_0	Q	f_n
1	1.606	0.510	INFINITY
2	1.691	0.611	INFINITY
3	1.907	1.023	INFINITY

Table 23.7 Resistors for 6th Order Bessel Lowpass Example

STAGE	R1	R2	R3	R4
1	13.00k	33.20k	10.00k	13.00k
2	10.50k	29.40k	10.00k	10.50k
3	10.00k	36.50k	17.40k	10.00k

Editing cascade order

As stated earlier in this manual, optimizing performance by editing cascade order and/or swapping pole and Q values is among the most arcane esoteric aspects of active filter design. Although certain aspects of this process are understood by experienced designers, current knowledge is not sufficiently systematic to guarantee the success of algorithmic optimization. Hence the need for manual editing. In the discussion that follows, we will consider briefly the underlying principals of optimization for minimizing noise or harmonic distortion. This will be followed by some concrete examples illustrating the effect of these principles on real-world filter designs. It should be emphasized that the fine-tuning process described here may or may not be necessary for a particular application. If you need assistance in maximizing performance of a filter using LTC parts, do not hesitate to contact our applications department for advice and counsel.

Optimizing for noise

The key to noise optimization is the concept of band limiting. Band limiting of noise is achieved by placing the 2nd order section with the lowest Q and lowest f_0 (in the case of a lowpass filter) *last* in the cascade order. To understand why this works, we must consider the response shapes of 2nd order sections. A 2nd order section with a low Q begins rolling off before f_0 (Figure 23.27). The lower the Q, the farther into the passband the roll-off begins. 2nd order sections with high Qs, on the other hand, have resonance peaks centered at f_0 (Figure 23.28). The higher the Q, the higher the resulting peak. The most noise in cascaded filters is contributed by the stages with the highest Qs, the noise being greatest in the vicinity of the resonance peaks. By placing the stage with the lowest Q and the lowest f_0 last in the cascade order, we place much of the noise contributed by previous stages *outside* the passband of this final stage, resulting in a reduction of the overall noise. Also, because the final stage is the lowest in Q, it contributes relatively little noise of its own. This technique allows the realization of selective elliptic lowpass filters with acceptable noise levels.

Optimizing for harmonic distortion

Distortion in switched-capacitor filters can be caused by three factors. First, distortion can be produced by loading. The CMOS amplifiers that are used in LTC switched-capacitor filter devices are not suited to driving heavy loads. For best results, no node should see an impedance of less than 10kΩ, and you will observe that FilterCAD never calculates a resistor value below this limit. Further, it may be desirable, when trying to obtain optimal distortion performance, to scale up resistor values calculated by FilterCAD by a factor of two or three to minimize loading.

The second factor that affects distortion performance is the clock frequency. Each LTC switched capacitor filter device has an optimum clock frequency range. Using a clock frequency significantly above the optimal range will result in increased distortion. For information on acceptable clock frequency ranges, consult LTC data sheets and application notes. If you do not observe these two design

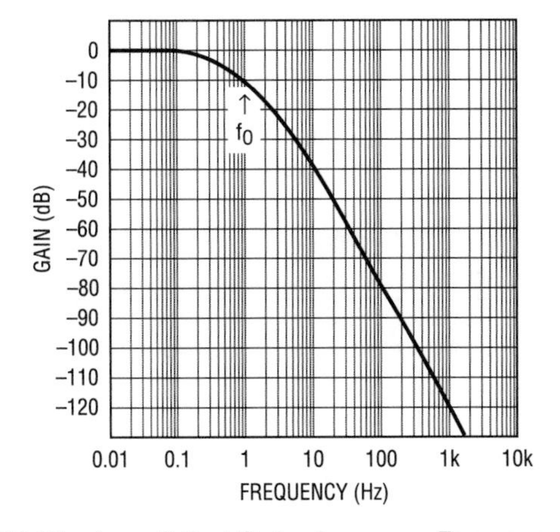

Figure 23.27 • Low Q 2nd Order Lowpass Response

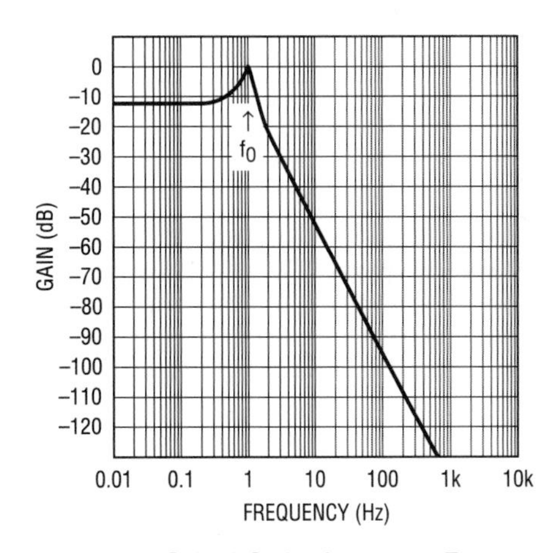

Figure 23.28 • High Q 2nd Order Lowpass Response

factors, any attempt to optimize THD performance by editing cascading order will likely be wasted.

The third factor is distortion introduced by the non-linear effects of the internal op amps when they swing close to their rails.

Both the gain and the position of the highest Q section are significant factors in this process. As previously discussed, high Q 2nd order sections (Q > 0.707) have a resonance peak in the vicinity of f_0. In order to maintain an overall gain of 1 for the circuit, and to minimize distortion, it is necessary to give high Q stages a **DC gain** of less than 1 and proportionally increase the gain of subsequent stages. (Note that FilterCAD automatically performs dynamic optimization for designs based exclusively on mode 3A, independent of the cascading order of the 2nd order sections.) If each stage were given a gain of 1, the overall gain for the circuit would, of course, be 1. However, when a high Q section has a gain of 1 at DC, the frequencies in the vicinity of f_0 will receive an additional boost from the resonance peak, resulting in a gain greater than 1 for those frequencies (see Figure 23.29 and the LTC1060 data sheet). Depending on the strength of the input signal, the output from the high Q stage may saturate the following input stage, driving it into its non-linear region and thereby creating distortion. Setting the gain of the high Q stage so that the peak at f_0 does not exceed 0dB results in a DC gain of less than 1 for the stage. This has the effect of significantly attenuating most of the frequencies in the passband, thereby minimizing the excursions of the input amplifier. Although this strategy reduces harmonic distortion, it can create noise problems, because the noise generated by a 2nd order stage increases with Q. (As a rule of thumb, noise can be regarded as increasing at approximately the square root of Q.) When the output from the high Q stage is amplified by subsequent stages in order to bring the overall passband gain up to 1, its noise component is amplified proportionally (Figure 23.30). Thus, as we stated previously, THD optimization is inimical to noise optimization, so the "best" cascade is a compromise between the two.

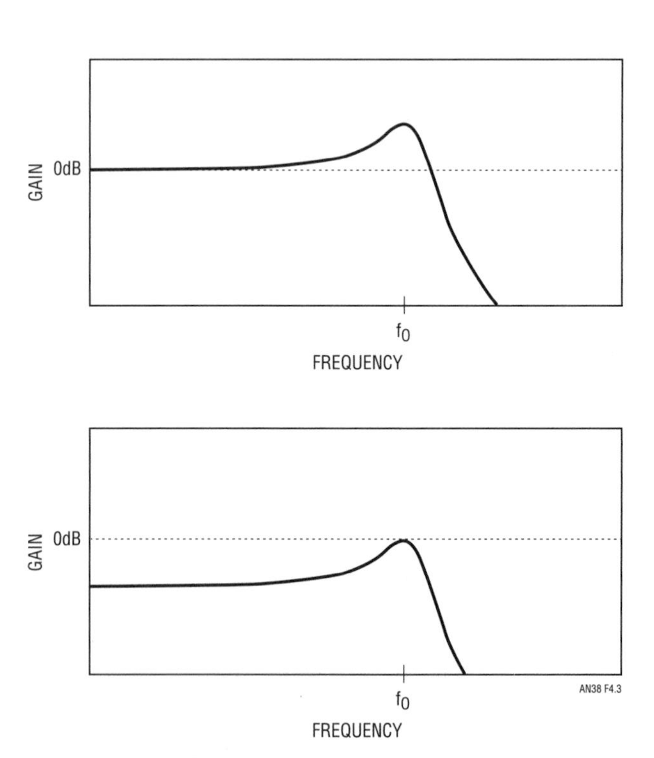

Figure 23.29 • DC Gain of 1 Results in Amplitude Greater Than 0dB at f_0; DC Gain is Reduced, Attenuating Frequencies in the Passband

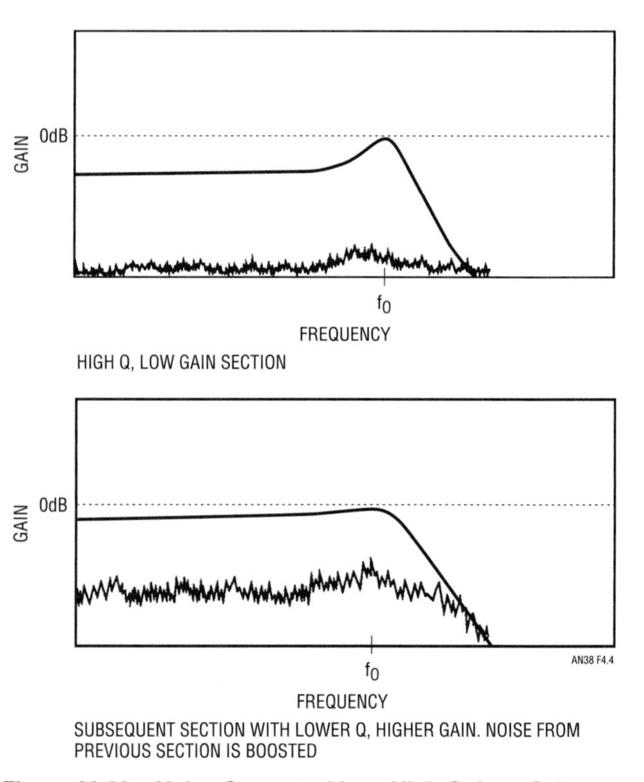

Figure 23.30 • Noise Generated by a High Q, Low Gain Stage Is Amplified by a Subsequent Low Q, High Gain Stage

More practical examples

To illustrate how the sorting of cascade order can affect performance, we will examine two concrete examples. The first is an 8th order Butterworth lowpass filter, normalized to 1Hz. The maximum passband ripple is 3dB, the stopband attenuation is 48dB, the corner frequency is 1Hz, and the stopband frequency is 2Hz. Two different versions of this design were implemented, one with the cascade order sorted for decreased harmonic distortion (THD), and the other sorted for lowest noise. Table 23.8 shows the f_0, Q and f_n values for both versions of our example. Of the four stages, three have Qs of less than 1, and one has a Q greater than 2.5. The two cascade orders differ only in the position of this high Q section. Observe that in the first case, the highest Q stage was placed in the second position, rather than the first, as the previous discussion indicated. This is a compromise to minimize harmonic distortion while maintaining acceptable noise performance. In the second case, the highest Q section is placed in the third position, followed immediately by the section with the lowest Q. Since this is a Butterworth filter, all of the sections have the same f_0. Nevertheless, because the low Q section has a droopy passband (see Figure 23.28) it still has the effect of band limiting the noise from the preceding section.

Mode 3 was selected for all stages because it produces lower harmonic distortion than mode 1. The clock-frequency ratio is 50:1, with an actual clock frequency of 400kHz giving an actual f_0 value of 8kHz. These two designs were breadboarded, using the resistor values given in Table 23.9. All of the resistor values calculated by FilterCAD were multiplied by 3.5 to minimize loading, except for the R1 values, which were selected to set the gains for the various sections so that no node will go above 0dB (lowpass gain for mode 3 = R4/R1).

The harmonic distortion performance was measured, yielding the results in Figures 23.31 and 23.32. The graphs indicate total harmonic distortion as a percentage of the input voltage. Each graph shows THD performance for inputs of 1V and $2.5V_{RMS}$. The difference between the two designs with a $1V_{RMS}$ input is negligible, but with a 2.5V input, a clear improvement in harmonic distortion is visible in Figure 23.31. In both examples, the distortion takes a significant dip as we approach the corner frequency. This is somewhat deceptive, however, since in this region the third and higher harmonics begin to be attenuated by the filter. While giving better harmonic distortion performance, Figure 23.31 has a wideband noise spec of $90\mu V_{RMS}$, whereas Figure 23.32 yielded a wideband noise spec of $80\mu V_{RMS}$.

Our second example is a 6th order elliptic lowpass filter, again normalized to 1Hz and realized in two versions, one optimized for noise, and the other optimized for harmonic distortion. This example has a maximum passband ripple of 1dB, a stopband attenuation of 50dB, a corner frequency of 1Hz, and a stopband frequency of 1.20Hz. The clock to filter cutoff frequency ratio is 100:1. The cascade orders for the two sections are given in Table 23.10. The

Table 23.8 f_0, Q, and f_n Values for 8th Order Butterworth Lowpass

STAGE	f_0	Q	f_n
Sorted for Reduced Harmonic Distortion			
1	1.0000	0.6013	INFINITE
2	1.0000	2.5629	INFINITE
3	1.0000	0.9000	INFINITE
4	1.0000	0.5098	INFINITE
Sorted for Low Noise			
1	1.0000	0.6013	INFINITE
2	1.0000	0.9000	INFINITE
3	1.0000	2.5629	INFINITE
4	1.0000	0.5098	INFINITE

Table 23.9 Resistor Values for 8th Order Butterworth Lowpass

STAGE	R1	R2	R3	R4	DC GAIN
Optimized for Reduced Harmonic Distortion					
1	61.90k	60.90k	35.00k	60.90k	0.98
2	57.60k	35.00k	77.35k	35.00k	0.61
3	43.20k	42.35k	35.00k	42.35k	0.96
4	39.20k	71.75k	35.00k	71.75k	1.83
(Total)					1.05
Optimized for Low Noise					
1	60.20k	60.90k	35.00k	60.90k	1.01
2	41.60k	42.35k	35.00k	42.35k	1.00
3	56.20k	35.00k	77.35k	35.00k	0.60
4	43.20k	71.75k	35.00k	71.75k	1.66
(Total)					1.05

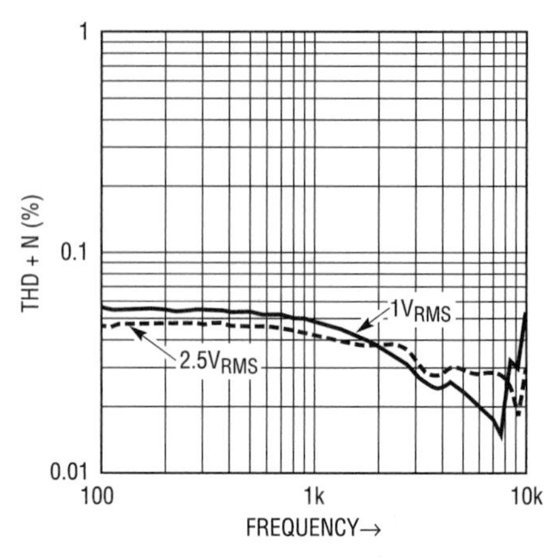

Figure 23.31 • THD Performance, 8th Order Butterworth Sorted for Reduced THD

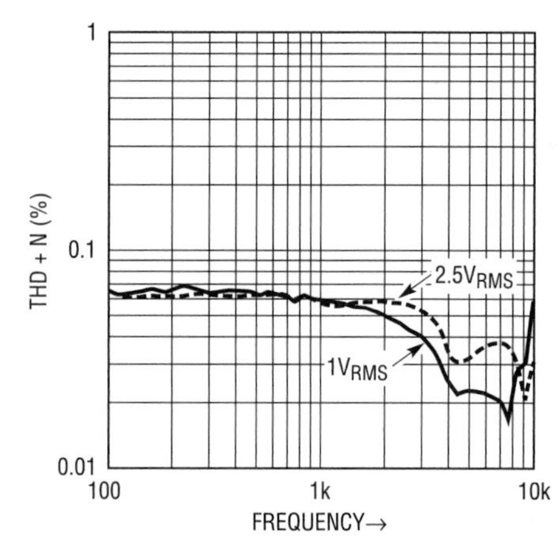

Figure 23.32 • THD Performance, 8th Order Butterworth Sorted for Reduced Noise

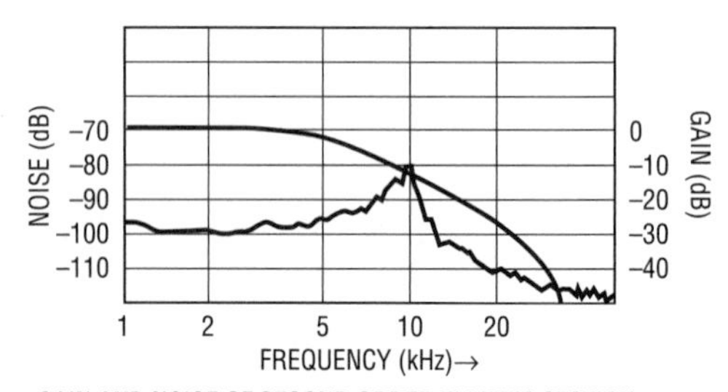

GAIN AND NOISE OF SECOND ORDER ELLIPTIC SECTION BANDLIMITING NOISE FROM PREVIOUS SECTION.
Q = 0.79, f_0 = 4.6kHz, f_n = 36kHz

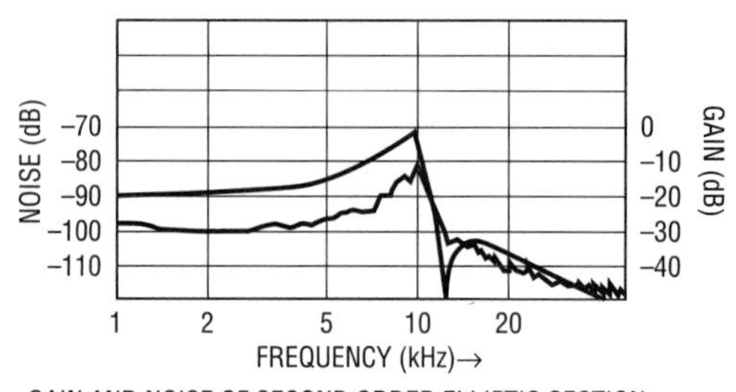

GAIN AND NOISE OF SECOND ORDER ELLIPTIC SECTION
Q = 15, f_0 = 10kHz, f_n = 12.2kHz

Figure 23.33 • Using Cascade-Order to Band-Limit Noise

case of an elliptic filter is more complex than the Butterworth in that the 2nd order responses have f_n values (notches or 0s) as well as f_0s and Qs. The ratio of f_n to f_0 in a particular section affects the height of the resonance peaks resulting from high Qs. The closer f_n is to f_0, the lower the peak.

Table 23.10 f_0, Q, and fn Values for 6th Order Elliptic Lowpass

STAGE	f_0	Q	f_n
Sorted for Reduced Harmonic Distortion			
1	0.9989	15.0154	1.2227
2	0.8454	3.0947	1.4953
3	0.4618	0.7977	3.5990
Sorted for Low Noise			
1	0.8454	3.0947	1.4953
2	0.9989	15.0154	1.2227
3	0.4618	0.7977	3.5990

In the first instance, the section with the highest f_0 and the highest Q is paired with the lowest f_n, and placed first in the cascade order. The second-highest Q is paired with the second-lowest f_n, and so on. This pairing minimizes the difference between the highest peak and the lowest gain in each second order section. Referring to Table 23.11, which gives the resistor values and lowpass gains for the stages, we see that the first stage has a very low gain of 0.067, and that most of the gain is provided by stage three, which has the lowest Q. Thus, the input swings of the individual stages are minimized, input-induced distortion is reduced, and an overall gain of 1 for the circuit is obtained. (In the case of mode 3A sections, lowpass gain for the first section is determined by R4/R1, and lowpass gain for subsequent stages is determined by R4 divided by RL of the previous stage. The final gain stage is provided by the external op amp, and is determined by RG/RL. Highpass gain is not taken into account here.)

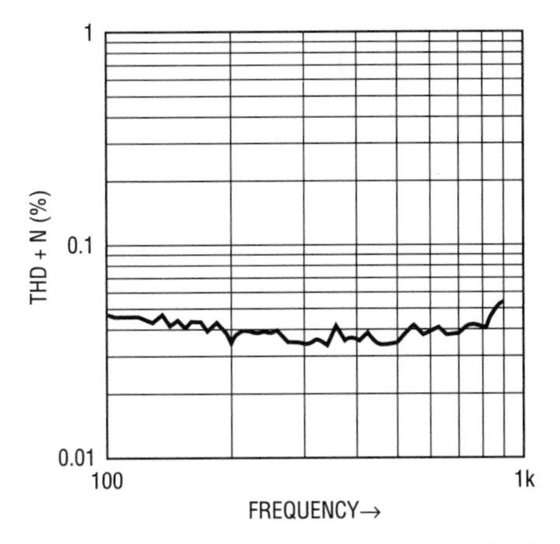

Figure 23.34 • THD Performance, 6th Order Elliptic Sorted for Reduced THD

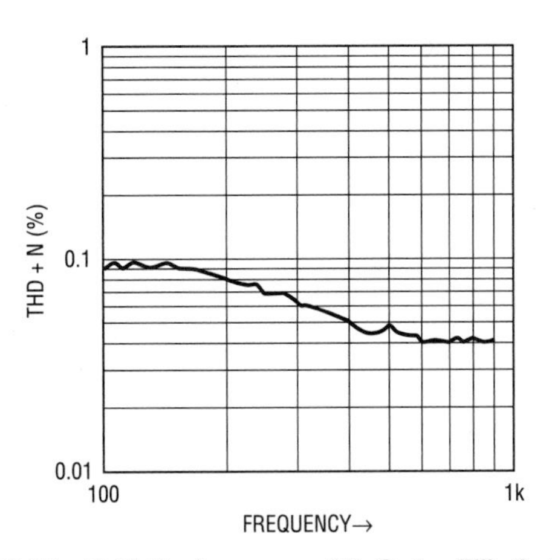

Figure 23.35 • THD Performance, 6th Order Elliptic Sorted for Reduced Noise

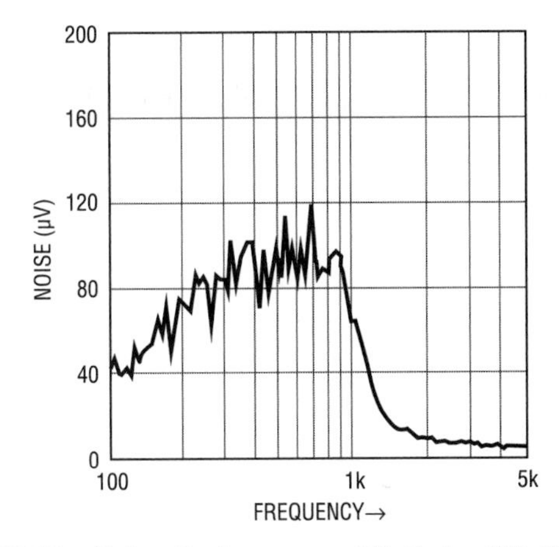

Figure 23.36 • Noise Performance, 6th Order Elliptic Sorted for Reduced THD

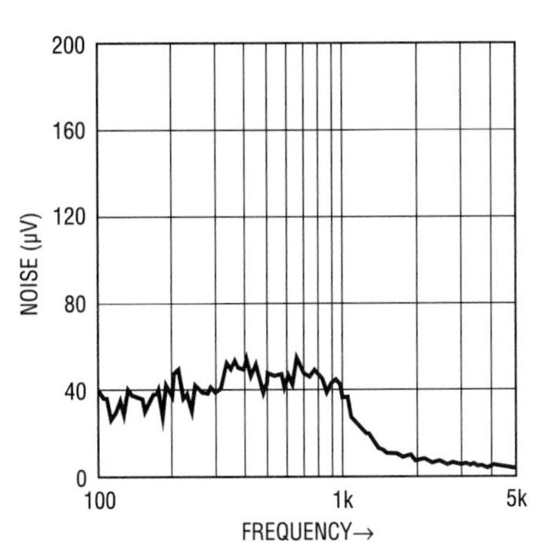

Figure 23.37 • Noise Performance, 6th Order Elliptic Sorted for Reduced Noise

In the second example, 2nd order stages have been sorted for reduced noise. In this case, the stage with the highest Q and f_0 is placed in the middle of the cascade order and is followed immediately by the stage with the lowest Q and f_0. Most of the gain is provided by stage three, which would tend to boost the noise generated by the previous stage, but the greater-than -2:1 ratio between the f_0s of the two sections causes much of the noise generated by stage two to fall outside of stage three's passband (see Figure 23.33). This produces the band-limiting effect described previously, and improves the overall noise performance of the circuit significantly. Figures 23.34 through 23.37 detail noise and THD performance of the two 6th order elliptic examples.

Notches...the final frontier

Notch filters, especially those with high Qs and/or high attenuations, are the most difficult to implement with universal switched-capacitor filter devices. You may design a notch filter with FilterCAD, with specifications that purport to yield a stopband attenuation of greater than 60dB, and find that in practice an attenuation of 40dB or less is the result. This is primarily due to the sampled data nature of the universal filter blocks; signals of equal amplitude and opposite phase do not ideally cancel when summed together as they would do in a purely analog system. Notches of up to 60dB *can* be obtained, but to do so

Table 23.11 Resistor Values for 6th Order Lowpass Elliptic (100:1, f_{CLK} to f_c)

STAGE	R1	R2	R3	R4	R_G	R_H	R_L	LOWPASS GAIN
Sorted for Low THD								
1	150.0k	10.00k	150.0k	10.00k		15.00k	10.00k	0.067
2		11.80k	43.20k	16.50k		22.60k	10.00k	1.650
3		16.90k	28.70k	78.70k	11.50k	130.0k	10.00k	7.870
External Op Amp								1.150
(Total)								1.000
Sorted for Low Noise								
1	43.20k	10.00k	36.50k	14.00k		110.0k	48.70k	0.324
2		10.00k	150.0k	10.00k		15.00k	10.00k	0.205
3		28.00k	48.70k	130.0k	11.50k	130.0k	10.00k	13.00
External Op Amp								1.150
(Total)								0.993

Table 23.12 f_0, Q, and f_n Values for 40kHz, 60dB Notch

STAGE	f_0	Q	f_n
1	35735.6793	3.3144	39616.8585
2	44773.1799	3.3144	40386.8469
3	35242.9616	17.2015	39085.8415
4	45399.1358	17.2105	40935.5393

Table 23.13 f_0, Q, and f_n Values for 40kHz, 60dB Notch

STAGE	f_0 (kHz)	Q	f_n (kHz)	MODE
1	40.000	10.00	40.000	1
2	43.920	11.00	40.000	2
3	40.000	10.00	40.000	1
4	35.920	8.41	40.000	3

requires techniques not covered by this version of FilterCAD. Some of these techniques will be examined here. We will start by using FilterCAD to enter the parameters for an elliptic notch response. We'll specify a maximum passband ripple of 0.1dB, an attenuation of 60dB, a center frequency of 40kHz, a stop bandwidth of 2kHz, and a pass bandwidth of 12kHz. Given these parameters, FilterCAD synthesizes the response shown in Table 23.12. This 8th order filter claims an actual stopband attenuation of greater than 80dB, a level of performance that would be exceedingly difficult to achieve in the real world. A working filter with an attenuation of 60dB can be achieved, but only be deviating significantly from the advice provided by FilterCAD.

Switched-capacitor filter devices give the best performance when certain operating parameters are kept within particular ranges. Those conditions which produce the best results for a particular parameter are called its "figure of merit." For example, in the case of the LTC1064, the best specs for clock to center frequency ratio (f_{CLK}/f_0) accuracy are published for a clock frequency of 1MHz and a Q of 10. As we depart from this "figure of merit" (as we must do to produce the 40kHz notch in our example), performance will gradually deteriorate. One of the problems that we will encounter is "Q-enhancement." That is, the Qs of the stages will appear slightly greater than those set by resistors. (Note that Q-enhancement is mostly a problem in modes 3 and 3A and is *not* limited to notches but occurs in LP, BP and HP filters as well.) This results in peaking above and below the notch. Q-enhancement can be compensated for by placing small capacitors (3pF to 30pF) in parallel with R4 (mode 2 or 3). With this modification, Q-enhancement can be compensated for in notch filters with center frequencies as high as 90kHz. The values suggested here are compromise values for a wide-range clock-tunable notch. If you want to produce a fixed-frequency notch, you can use larger caps at higher frequencies. At least in the case of the LTC1064, Q-enhancement is unlikely to be a problem below 20kHz. Adding capacitors at lower frequencies will have the effect of widening the notch.

As mentioned previously, the other problem in implementing notch filters is inadequate attenuation. For low frequency notches, stopband attenuation may be increased by boosting the clock to notch frequency up to 250:1. Attenuation may also be improved by adding external capacitors, this time in parallel with R2 (modes 1, 2, and 3A). Capacitors of 10pF to 30pF in this position can increase stopband attenuation by 5dB to 10dB. Of course, this capacitor/resistor combination constitutes a passive 1st order lowpass stage with a corner frequency at $1/(2\pi RC)$. In the case of the values indicated above, the corner frequency will appear so far out in the passband that it is unlikely to be significant. However, if the notch is needed at a frequency below 20kHz, the capacitor value will need to be increased and the corner frequency of the 1st order stage will be lowered proportionally. For a capacitor of 100pF and an R2 of 10k, the corner frequency will be 159kHz, a value that is still unlikely to cause problems in most applications. For a capacitor of 500pF (a value that might prove necessary for a deep notch at a low center frequency) and an R2 of 20k, the corner frequency drops down to 15.9kHz. If maximum stopband attenuation is more important than a wide passband, such a solution may prove acceptable. Adding resistors in parallel with R2 produces one additional problem: it increases the Q that we just controlled with the capacitors across R4. Resistor values must be adjusted to bring the Q down again.

Table 23.13 contains the parameters for a *real* notch filter which actually meets our 60dB attenuation spec using the techniques previously outlined. This is essentially the clock-tunable 8th order notch filter described in the LTC1064 data sheet. Note the mixture of modes used. This is a solution that FilterCAD is incapable of proposing.

It should be apparent that the methods for notch filters described here are primarily empirical at this point, and that the account given here is far from comprehensive. We have not even touched on optimizing these filters for noise or distortion, for instance. No simple rules can be given for this process. Such optimization is possible, but must be addressed on a case-by-case basis. If you need to implement a high performance notch filter and the tips above prove inadequate, please call the LTC applications department for additional assistance.

Appendix 1
The FilterCAD device-parameter editor

The FilterCAD Device-Parameter Editor (FDPF.EXE) allows you to modify the FilterCAD Device-Parameter File (FCAD.DPF). This file contains the data about LTC switched-capacitor filter devices which FilterCAD uses in making device selections in its implementation phase. The Device-Parameter Editor is a menu-driven program that has a similar command structure to FilterCAD. Its main menu includes the following entries:

1. ADD New Device
2. DELETE Device
3. EDIT Existing Device
4. SAVE Device Parameter File
5. LOAD Device Parameter File
6. CHANGE Path to Device Parameter File
9. END Device Parameter Editor

The principle reason for editing the Device-Parameter File is to add data for new LTC devices that were released after this revision of FilterCAD. To enter data for a new device, press

1

You will see a blank form with fields for the necessary parameters. Use the arrow keys to move through the fields and type in the appropriate values from the LTC data sheet. Press

[Enter]

to accept the data in each field, then press

ESC

when you have finished entering the data. Don't forget to SAVE the new .DPF file. The program will inform you that the FCAD.DPF already exists and ask you whether you want to overwrite it. Press

Y

to save your file.

Another possible reason for using the Device-Parameter Editor is to delete some devices from the Device-Parameter File so that FilterCAD could only select devices that you have on hand. To delete a device, press

2

When it shows you a device name on the screen, press

Y

to delete the device from the file or press

N

to cycle through the list of devices until it displays the one that you wish to delete.

You can also use the Device-Parameter Editor to edit the data for a device supported by this version of FilterCAD in the event that the specifications for that device are revised. To edit a device already in the Device-Parameter File, press

3

You will see a form, like the one described previously for adding new devices, except the fields will contain data. Use the

PgDn

and

PgUp

keys to page through the devices to find the one you wish to edit, move the cursor to the fields that you want to modify and type in the new data. You must press

[Enter]

to accept the new value in each field. Press

ESC

when you have finished editing.

The remainder of the options on the Device-Parameter File Editor's main menu are self explanatory.

Appendix 2 Bibliography

For more information on the theory of filter design, consult one of the works listed below.

1. Daryanani, Gobind, "Principles of Active Network Synthesis and Design." New York: John Wiley and Sons, 1976.
2. Ghausi, M.S., and K.R. Laker, " Modern Filter Design, Active RC and Switched Capacitor." Englewood Cliffs, New Jersey: Prentice-Hall, Inc., 1981
3. Lancaster, Don, "The Active Filter Cookbook." Indianapolis, Indiana: Howard W. Sams & Co., Inc., 1975.
4. Williams, Arthur B., "Electronic Filter Design Handbook." New York: McGraw-Hill, Inc., 1981.

Note: Applications and algorithms by Nello Sevastopoulos, Philip Karantzalis and Richard Markell.

30 nanosecond settling time measurement for a precision wideband amplifier

Quantifying prompt certainty

24

Jim Williams

Introduction

Instrumentation, waveform generation, data acquisition, feedback control systems and other application areas utilize wideband amplifiers. New components (see page 2 "A Precision Wideband Dual Amplifier with 30ns Settling Time") have introduced precision while maintaining high speed operation. The amplifier's DC and AC specifications approach or equal previous devices at significantly lower cost while saving power.

Settling time defined

Amplifier DC specifications are relatively easy to verify. Measurement techniques are well understood, albeit often tedious. AC specifications require more sophisticated approaches to produce reliable information. In particular, amplifier settling time is extraordinarily difficult to determine. Settling time is the elapsed time from input application until the output arrives at and remains within a specified error band around the final value. It is usually specified for a full-scale transition. Figure 24.1 shows that settling time has three distinct components. The *delay time* is small and is almost entirely due to amplifier propagation delay. During this interval there is no output movement. During *slew time* the amplifier moves at its highest possible speed towards the final value. *Ring time* defines the region where the amplifier recovers from slewing and ceases movement within some defined error band. There is normally a trade-off between slew and ring time. Fast slewing amplifiers generally have extended ring times, complicating amplifier choice and frequency compensation. Additionally, the architecture of very fast amplifiers usually dictates trade-offs which degrade DC error terms.[1]

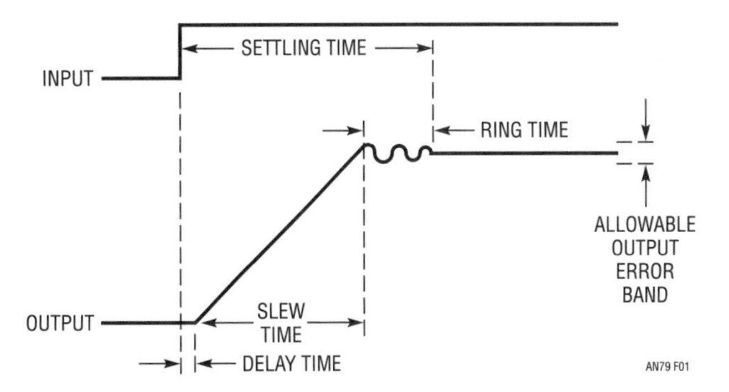

Figure 24.1 • Settling Time Components Include Delay, Slew and Ring Times. Fast Amplifiers Reduce Slew Time, Although Longer Ring Time Usually Results. Delay Time is Normally a Small Term

Measuring anything at any speed requires care. Dynamic measurement is particularly challenging. Reliable nanosecond region settling time measurement constitutes a high order difficulty problem requiring exceptional care in approach and experimental technique.[2]

Note 1: This issue is treated in detail in latter portions of the text. Also see Appendix D "Practical Considerations for Amplifier Compensation".

Note 2: The approach used for settling time measurement and its description borrows heavily from a previous publication. See Reference 1.

Analog Circuit and System Design: Immersion in the Black Art of Analog Design. http://dx.doi.org/10.1016/B978-0-12-397888-2.00024-9

A precision wideband dual amplifier with 30ns settling time

Until recently, wideband amplifiers provided speed, but sacrificed precision, power consumption and, often, settling time. The LT®1813 dual op amp does not require this compromise. It features low offset voltage and bias current and high DC gain while operating at low supply current. Settling time is 30ns to 0.1% for a 5V step. The output will drive a 100Ω load to ±3.5V with ±5V supplies, and up to 100pF capacitive loading is permissible. The table below provides short form specifications.

LT1813 Short Form Specifications

CHARACTERISTIC	SPECIFICATION
Offset Voltage	0.5mV
Offset Voltage vs Temperature	10μV/°C
Bias Current	1.5μA
DC Gain	3000
Noise Voltage	$8nV/\sqrt{Hz}$
Output Current	60mA
Slew Rate	750V/μS
Gain-Bandwidth	100MHz
Delay	2.5ns
Settling Time	30ns/0.1%
Supply Current	3mA per Amplifier

Considerations for measuring nanosecond region settling time

Historically, settling time has been measured with circuits similar to that in Figure 24.2. The circuit uses the "false sum node" technique. The resistors and amplifier form a bridge type network. Assuming ideal resistors, the amplifier output will step to $-V_{IN}$ when the input is driven. During slew, the settle node is bounded by the diodes, limiting voltage excursion. When settling occurs, the oscilloscope probe voltage should be zero. Note that the resistor divider's attenuation means the probe's output will be one-half of the actual settled voltage.

In theory, this circuit allows settling to be observed to small amplitudes. In practice, it cannot be relied upon to produce useful measurements. Several flaws exist. The circuit requires the input pulse to have a flat top within the required measurement limits. Typically, settling within 5mV or less for a 5V step is of interest. No general purpose pulse generator is meant to hold output amplitude and noise within these limits. Generator output-caused aberrations appearing at the oscilloscope probe will be indistinguishable from amplifier output movement, producing unreliable results. The oscilloscope connection also presents problems. As probe capacitance rises, AC loading of the resistor junction influences observed settling waveforms. A 10pF probe alleviates this problem but its 10× attenuation sacrifices oscilloscope gain. 1×probes are not suitable because of their excessive input capacitance. An active 1×FET probe will work, but another issue remains.

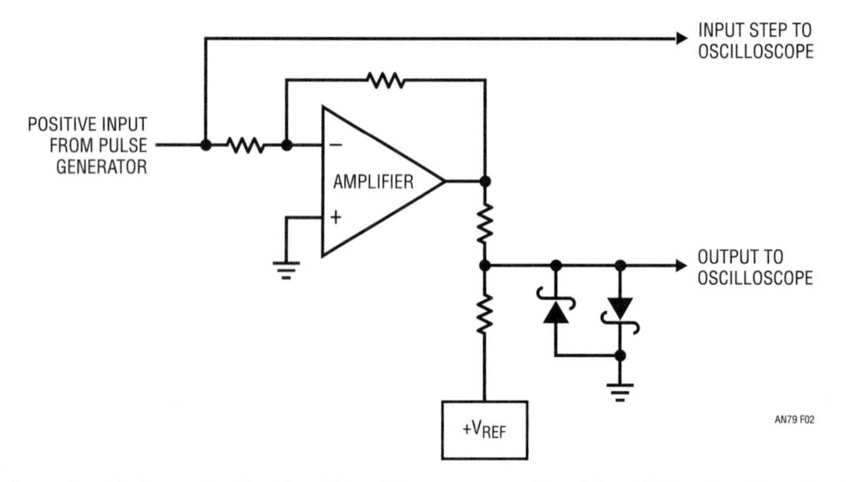

Figure 24.2 • Popular Summing Scheme for Settling Time Measurement Provides Misleading Results. Pulse Generator Posttransition Aberrations Appear at Output. 10× Oscilloscope Overdrive Occurs. Displayed Information Is Meaningless

The clamp diodes at the settle node are intended to reduce swing during amplifier slewing, preventing excessive oscilloscope overdrive. Unfortunately, oscilloscope overdrive recovery characteristics vary widely among different types and are not usually specified. The Schottky diodes' 400mV drop means the oscilloscope will undergo an unacceptable overload, bringing displayed results into question.[3]

At 0.1% resolution (5mV at the output—2.5mV at the oscilloscope), the oscilloscope typically undergoes a 10× overdrive at 10mV/DIV and the desired 2.5mV baseline is unattainable. At nanosecond speeds, the measurement becomes hopeless with this arrangement. There is clearly no chance of measurement integrity.

The preceding discussion indicates that measuring amplifier settling time requires an oscilloscope that is somehow immune to overdrive and a "flat-top" pulse generator. These become the central issues in wideband amplifier settling time measurement.

The only oscilloscope technology that offers inherent overdrive immunity is the classical sampling 'scope.[4] Unfortunately, these instruments are no longer manufactured (although still available on the secondary market). It is possible, however, to construct a circuit that borrows the overload advantages of classical sampling 'scope technology. Additionally, the circuit can be endowed with features particularly suited for measuring nanosecond range settling time.

The "flat-top" pulse generator requirement can be avoided by switching current, rather than voltage. It is much easier to gate a quickly settling current into the amplifier's summing node than to control a voltage. This makes the input pulse generator's job easier, although it still must have a rise time of 1 nanosecond or less to avoid measurement errors.[5]

Practical nanosecond settling time measurement

Figure 24.3 is a conceptual diagram of a settling time measurement circuit. This figure shares attributes with Figure 24.2, although some new features appear. In this case, the oscilloscope is connected to the settle point by a switch. The switch state is determined by a delayed pulse generator, which is triggered from the input pulse. The delayed pulse generator's timing is arranged so that the switch does not close until settling is very nearly

Note 3: For a discussion of oscilloscope overdrive considerations, see Appendix A, "Evaluating Oscilloscope Overdrive Performance."

Note 4: Classical sampling oscilloscopes should not be confused with modern era digital sampling 'scopes that have overdrive restrictions. See Appendix A, "Evaluating Oscilloscope Overload Performance" for comparisons of various type 'scopes with respect to overdrive. For detailed discussion of classical sampling 'scope operation see References 16 through 19 and 22 through 24. Reference 17 is noteworthy; it is the most clearly written, concise explanation of classical sampling instruments the author is aware of—a 12-page jewel.

Note 5: Subnanosecond rise time pulse generators are considered in Appendix B, "Subnanosecond Rise Time Pulse Generators for the Rich and Poor."

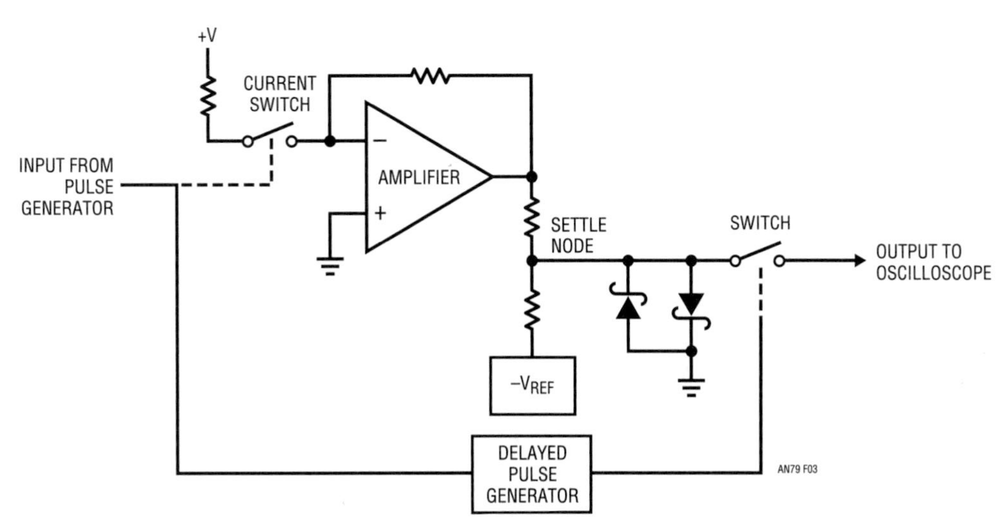

Figure 24.3 • Conceptual Arrangement is Insensitive to Pulse Generator Aberrations and Eliminates Oscilloscope Overdrive. Switch at Input Gates Current Step to Amplifer. Second Switch is Controlled by Delayed Pulse Generator, Preventing Oscilloscope from Monitoring Settle Node Until Settling is Nearly Complete

complete. In this way the incoming waveform is sampled in time, as well as amplitude. The oscilloscope is never subjected to overdrive—no off-screen activity ever occurs.

A switch at the amplifier's summing junction is controlled by the input pulse. This switch gates current to the amplifier via a voltage-driven resistor. This eliminates the "flat-top" pulse generator requirement, although the switch must be fast and devoid of drive artifacts.

Figure 24.4 is a more complete representation of the settling time scheme. Figure 24.3's blocks appear in greater detail and some new refinements show up. The amplifier summing area is unchanged. Figure 24.3's delayed pulse generator has been split into two blocks; a delay and a pulse generator, both independently variable. The input step to the oscilloscope runs through a section that compensates for the propagation delay of the settling time measurement path. The most striking new aspect of the diagram are the diode bridge switches. Borrowed from classical sampling oscilloscope circuitry, they are the key to the measurement. The diode bridge's inherent balance eliminates charge injection based errors. It is far superior to other electronic switches in this characteristic. Any other high speed switch technology contributes excessive output spikes due to charge-based feedthrough. FET switches are not suitable because their gate-channel capacitance permits such feedthrough. This capacitance allows gate-drive artifacts to corrupt switching, defeating the switches purpose.

The diode bridge's balance, combined with matched, low capacitance monolithic diodes and high speed switching, yields clean switching. The input-driven bridge switches current into the amplifier's summing point very quickly, with settling inside a few nanoseconds. The diode clamp to ground prevents excessive bridge drive swings and ensures that input pulse characteristics are irrelevant.

Figure 24.5 details considerations for the output diode bridge switch. This bridge requires considerable attention to achieve desired performance. The monolithic bridge diodes tend to cancel each other's temperature

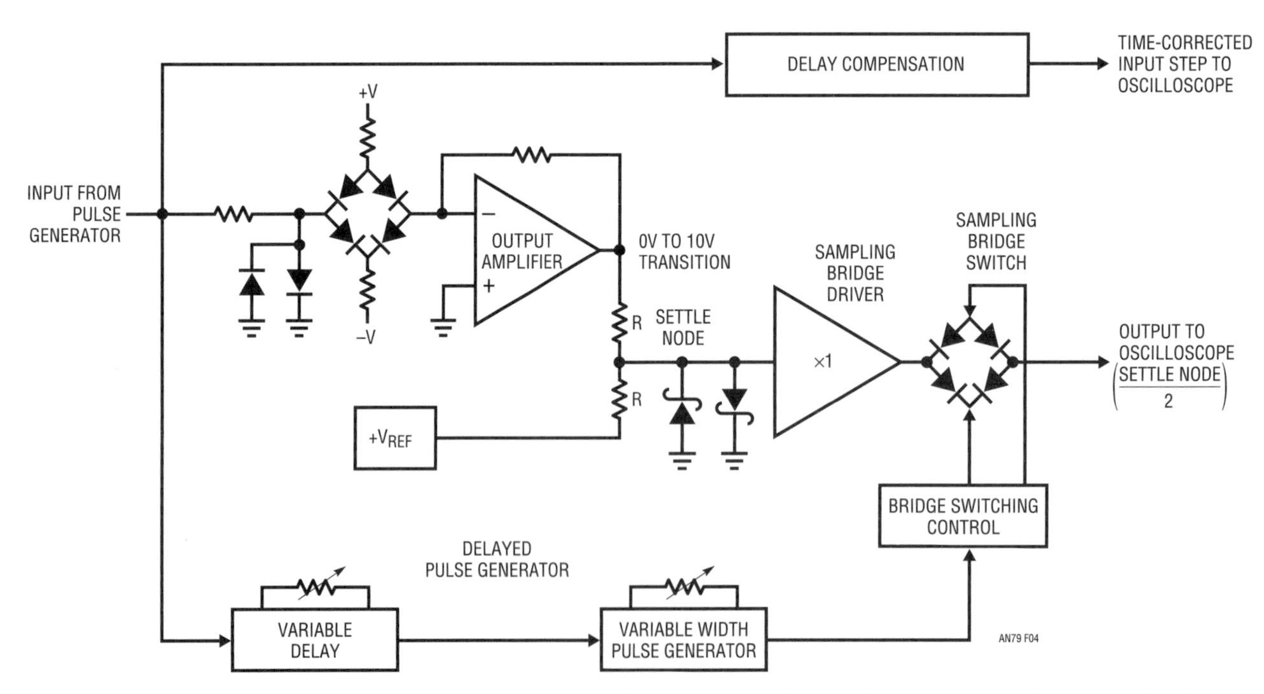

Figure 24.4 • Block Diagram of Settling Time Measurement Scheme. Diode Bridge Switches Input Current to Amplifier. Second Diode Bridge Switch Minimizes Switching Feedthrough, Preventing Oscilloscope Overdrive. Input Step Time Reference is Compensated for Test Circuit Delays

coefficient—drift is only about 100μV/°C—but a DC balance is required to minimize offset.

DC balance is achieved by trimming the bridge on-current for zero input-output offset voltage. Two AC trims are required. The "AC balance" corrects for diode and layout capacitive imbalances and the "skew compensation" corrects for any timing asymmetry in the nominally complementary bridge drive. These AC trims compensate small dynamic imbalances, minimizing parasitic bridge outputs.

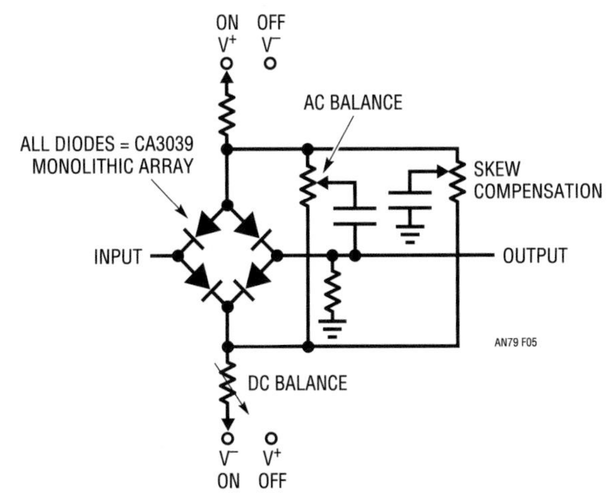

Figure 24.5 • Diode Sampling Bridge Switch Trims Include AC and DC Balance and Switch Drive Timing Skew

Detailed settling time circuitry

Figure 24.6 is a detailed schematic of the settling time measurement circuitry. The input pulse switches the input bridge and is also routed to the oscilloscope via a delay-compensation network. The delay network, composed of a fast comparator and an adjustable RC network, compensates the oscilloscope's input step signal for the 6ns delay through the circuit's measurement path.[6] The amplifier's output is compared against the 5V reference via the summing resistors. The 5V reference also furnishes the bridge input current, making the measurement ratiometric. The −5V reference supply pulls a current from the summing point, allowing the amplifier a 5V step from 2.5V to −2.5V. The clamped settle node is unloaded by A1, which drives the sampling bridge.

The input pulse triggers the C2-C3 based delayed pulse generator. This circuitry is arranged to produce a delayed (controllable by the 10k potentiometer) pulse whose width (controllable by the 2k potentiometer) sets diode bridge on-time. If the delay is set appropriately, the oscilloscope will not see any input until settling is nearly complete, eliminating overdrive. The sample window width is adjusted so that all remaining settling activity is observable. In this way the oscilloscope's output is reliable and meaningful data may be taken. The delayed generator's output is level shifted by the Q1-Q4 transistors, providing complementary switching drive to the bridge. The actual switching transistors, Q1-Q2, are UHF types, permitting true differential bridge switching with less than 1ns of time skew.[7]

Figure 24.7 shows circuit waveforms. Trace A is the time-corrected input pulse, trace B the amplifier output, trace C the sample gate and trace D the settling time output. When the sample gate goes low, the bridge switches cleanly, and the last 10mV of slew are easily observed. Ring time is also clearly visible, and the amplifier settles nicely to final value. When the sample gate goes high, the bridge switches off, with only millivolts of feedthrough. Note that there is no off-screen activity at any time—the oscilloscope is never subjected to overdrive.

Figure 24.8 expands vertical and horizontal scales so that settling detail is more visible.[8] Trace A is the time-corrected input pulse and trace B the settling output. The last 15mV of slew (beginning at the center-screen vertical marker) are easily observed, and the amplifier settles inside 5mV (0.1%) in 30 nanoseconds.

The circuit requires trimming to achieve this level of performance. DC and AC trims are required. Making these adjustments requires disabling the amplifier (disconnect the input current switch and the 1k resistor at the amplifier), and shorting the settle node directly to the ground plane. Figure 24.9 shows typical results before trimming.

Trace A is the input pulse and trace B the settle signal output. With the amplifier disabled and the settle node

Note 6: See Appendix C, "Measuring and Compensating Settling Circuit Delay."

Note 7: The bridge switching scheme was developed at LTC by George Feliz.

Note 8: In this and all following photos, settling time is measured from the onset of the time-corrected input pulse. Additionally, settling signal amplitude is calibrated with respect to the amplifier, not the sampling bridge output. This eliminates ambiguity introduced by the summing resistor's ÷ 2 ratio.

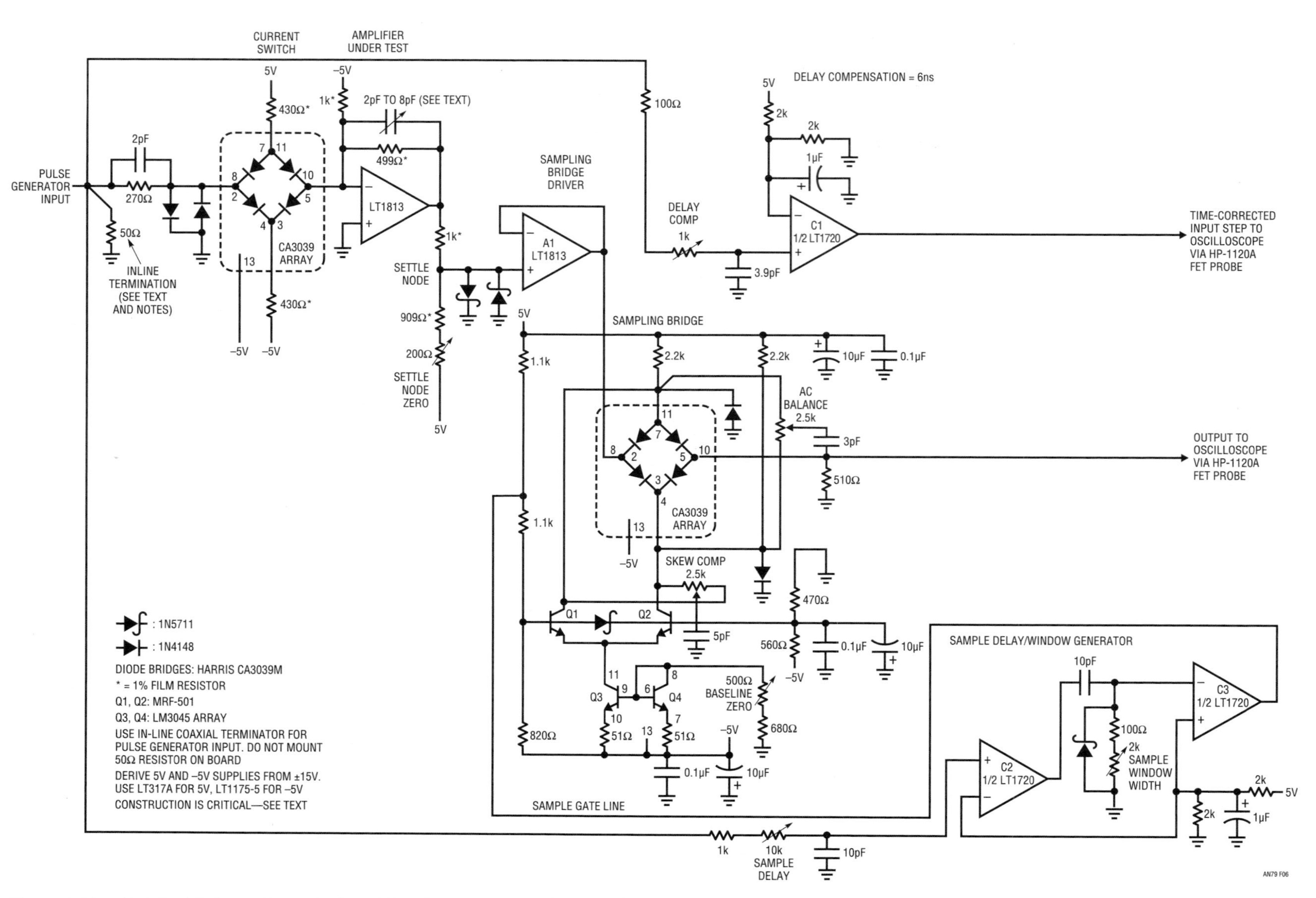

Figure 24.6 • Detailed Schematic of Settling Time Measurement Circuit Closely Follows Block Diagram. Optimum Performance Requires Attention to Layout

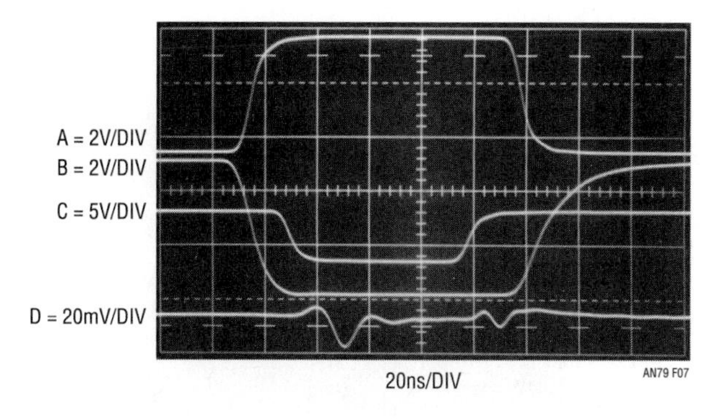

Figure 24.7 • Settling Time Circuit Waveforms Include Time-Corrected Input Pulse (Trace A), Amplifier-Under-Test Output (Trace B), Sample Gate (Trace C) and Settling Time Output (Trace D). Sample Gate Window's Delay and Width are Variable

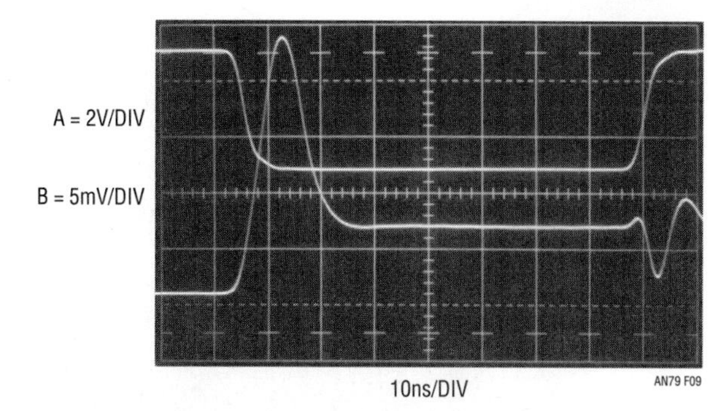

Figure 24.9 • Settling Time Circuit's Output (Trace B) with Unadjusted Sampling Bridge AC and DC Trims. Settle Node is Grounded for This Test. Excessive Switch Drive Feedthrough and Baseline Offset are Present. Trace A is the Sample Gate

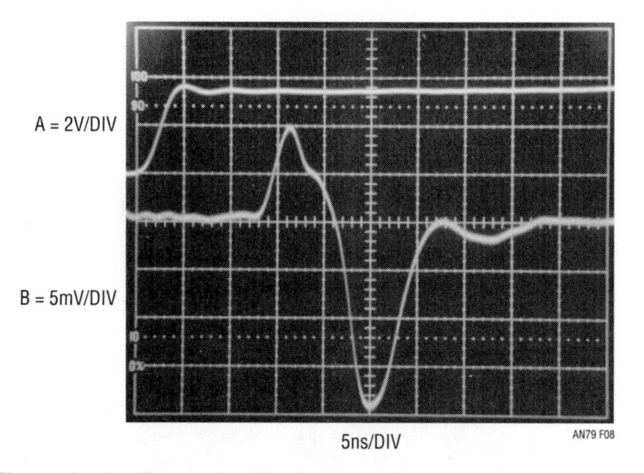

Figure 24.8 • Expanded Vertical and Horizontal Scales Show 30ns Amplifier Settling Within 5mV (Trace B). Trace A is Time-Corrected Input Step

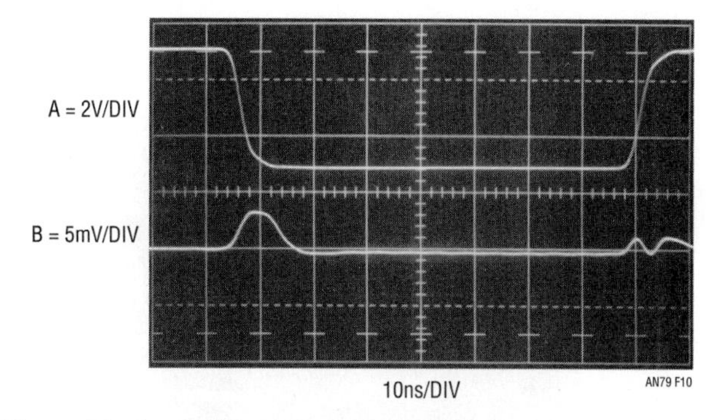

Figure 24.10 • Settling Time Circuit's Output (Trace B) with Sampling Bridge Trimmed. As in Figure 24.9, Settle Node is Grounded for This Test. Switch Drive Feedthrough and Baseline Offset are Minimized. Trace A is the Sample Gate

grounded, the output should (theoretically) always be zero. The photo shows this is not the case for an untrimmed bridge. AC and DC errors are present. The sample gate's transitions cause large swings. Additionally, the output shows significant DC offset error during the sampling interval. Adjusting the AC balance and skew compensation minimizes the switching induced transients. The DC offset is adjusted out with the baseline zero trim. Figure 24.10 shows the results after making these adjustments. All switching related activity is minimized and offset error reduced to unreadable levels. Once this level of performance has been achieved, the circuit is nearly ready for use.[9] Unground the settle node and restore the current switch and resistor connections to the amplifier. Any further differences between pre- and postsettling baseline are corrected with the "settle node zero" trim.

Using the sampling-based settling time circuit

Figures 24.11 and 24.12 underscore the importance of positioning the sampling window properly in time. In Figure 24.10 the sample gate delay initiates the sample window (trace A) too early and the residue amplifier's output (trace B) overdrives the oscilloscope when sampling commences. Figure 24.12 is better, with no off-screen activity.

Note 9: Achieving this level of performance also depends on layout. The circuit's construction involves a number of subtleties and is absolutely crucial. Please see Appendix E, "Breadboarding, Layout and Connection Techniques."

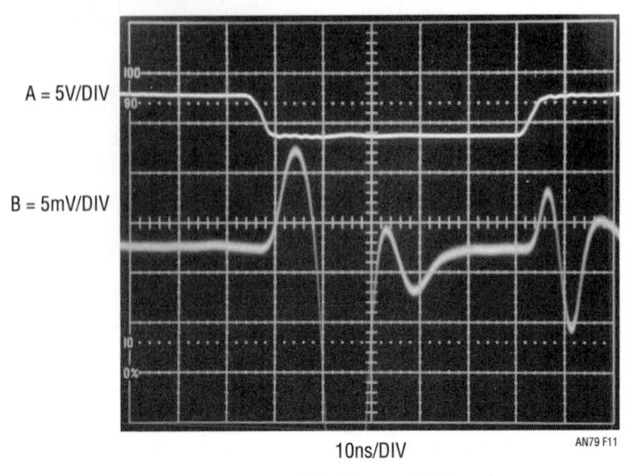

Figure 24.11 • Oscilloscope Display with Inadequate Sample Gate Delay. Sample Window (Trace A) Occurs Too Early, Resulting in Off-Screen Activity in Settle Output (Trace B). Oscilloscope is Overdriven, Making Displayed Information Questionable

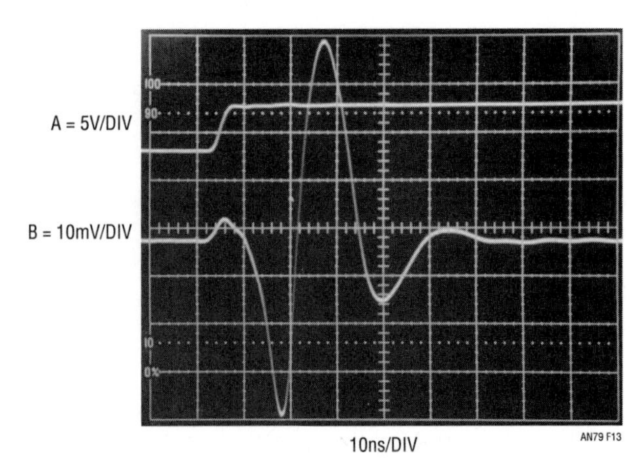

Figure 24.13 • Settling Profile with Inadequate Feedback Capacitance Shows Underdamped Response. Trace A is Time-Corrected Input Pulse. Trace B is Settling Residue Output. $t_{SETTLE} = 43ns$

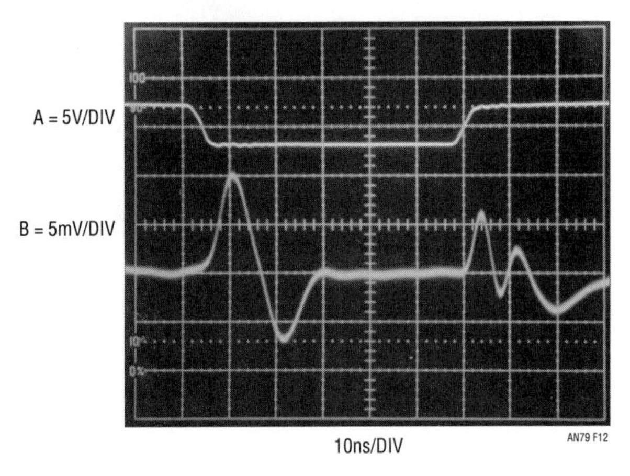

Figure 24.12 • Optimal Sample Gate Delay Positions Sampling Window (Trace A) So All Settle Output (Trace B) Information is Well Inside Screen Boundaries

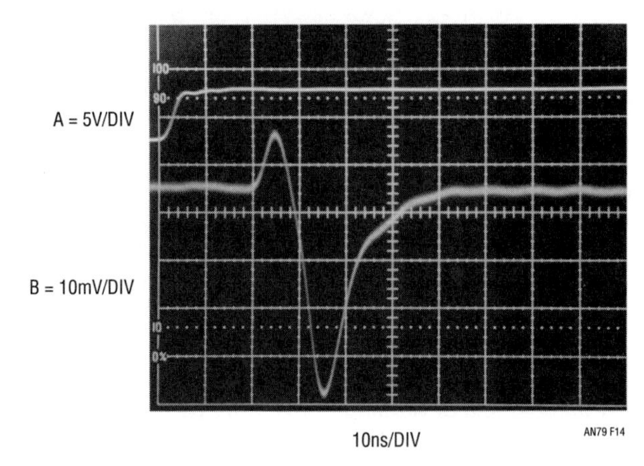

Figure 24.14 • Excessive Feedback Capacitance Overdamps Response. $t_{SETTLE} = 50ns$

All amplifier settling residue is well inside the screen boundaries.

In general, it is good practice to "walk" the sampling window up to the last ten millivolts or so of amplifier slewing so that the onset of ring time is observable. The sampling based approach provides this capability and it is a very powerful measurement tool. Additionally, remember that slower amplifiers may require extended delay and/or sampling window times. This may necessitate larger capacitor values in the delayed pulse generator timing networks.

Compensation capacitor effects

The amplifier requires frequency compensation to get the best possible settling time.[10] Figure 24.13 shows effects of very light compensation. Trace A is the time-corrected input pulse and trace B the settling residue output. The light compensation permits very fast slewing but excessive ringing amplitude over a protracted time results. When sampling is initiated (just prior to the fourth

Note 10: This section discusses frequency compensation of the amplifier within the context of sampling-based settling time measurement. As such, it is necessarily brief. Considerably more detail is available in Appendix D, "Practical Considerations for Amplifier Compensation."

vertical division) the ringing is seen to be in its final stages, although still offensive. Total settling time is about 43ns. Figure 24.14 presents the opposite extreme. Here a large value compensation capacitor eliminates all ringing but slows down the amplifier so much that settling stretches out to 50ns.

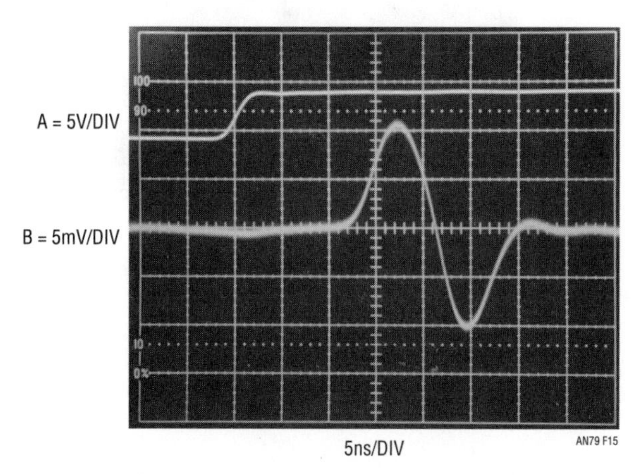

Figure 24.15 • Optimal Feedback Capacitance Yields Tightly Damped Signature and Best Settling Time. Optimum Response Allows Expanded Horizontal and Vertical Scales. $t_{SETTLE} \leq 30ns$

The best case appears in Figure 24.15. This photo was taken with the compensation capacitor carefully chosen for the best possible settling time. Damping is tightly controlled and settling time goes down to 30ns.

Verifying results—alternate method

The sampling-based settling time circuit appears to be a useful measurement solution. How can its results be tested to ensure confidence? A good way is to make the same measurement with an alternate method and see if results agree. It was stated earlier that classical sampling oscilloscopes were inherently immune to overdrive.[11] If this is so, why not utilize this feature and attempt settling time measurement directly at the clamped settle node? Figure 24.16 does this. Under these conditions, the sampling 'scope[12] is heavily overdriven, but is ostensibly immune to the insult. Figure 24.17 puts the sampling oscilloscope to the test. Trace A is the time corrected input

Note 11: See Appendix A, "Evaluating Oscilloscope Overdrive Performance," for in-depth discussion.

Note 12: Tektronix type 661 with 4S1 vertical and 5T3 timing plug-ins.

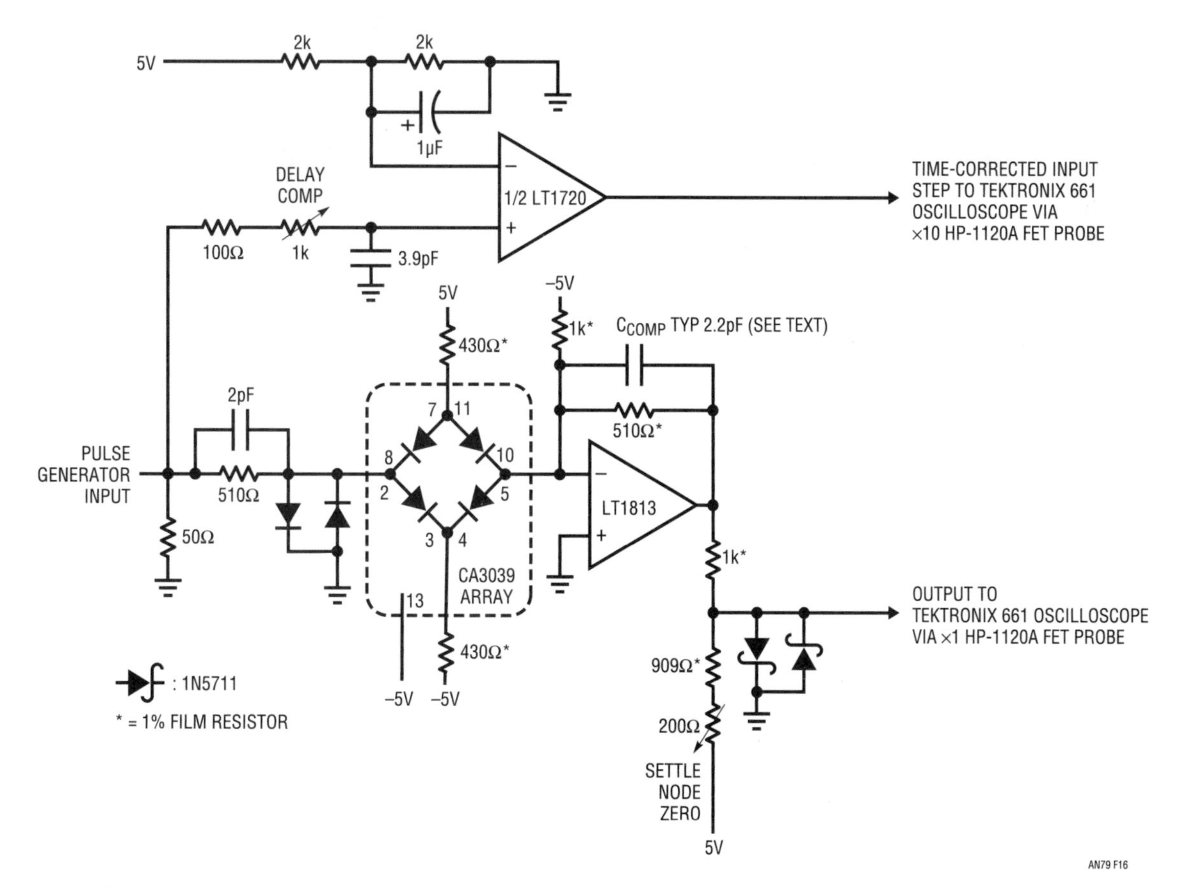

Figure 24.16 • Settling Time Test Circuit Using Classical Sampling Oscilloscope. Sampling 'Scope's Inherent Overload Immunity Permits Large Off-Screen Excursions

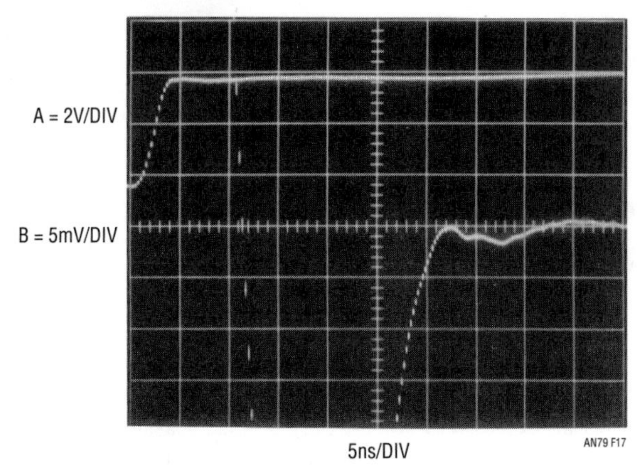

Figure 24.17 • Settling Time Measurement with the Classical Sampling 'Scope. Oscilloscope's Overload Immunity Allows Accurate Measurement Despite Extreme Overdrive

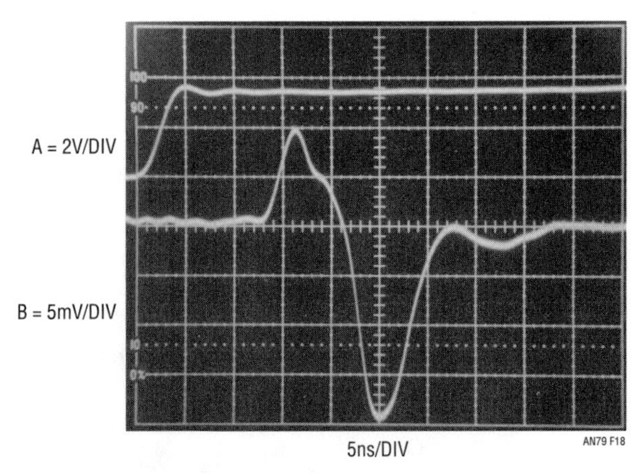

Figure 24.18 • Settling Time Measurement Using the Sampling Bridge Circuit. t_{SETTLE} = 30ns

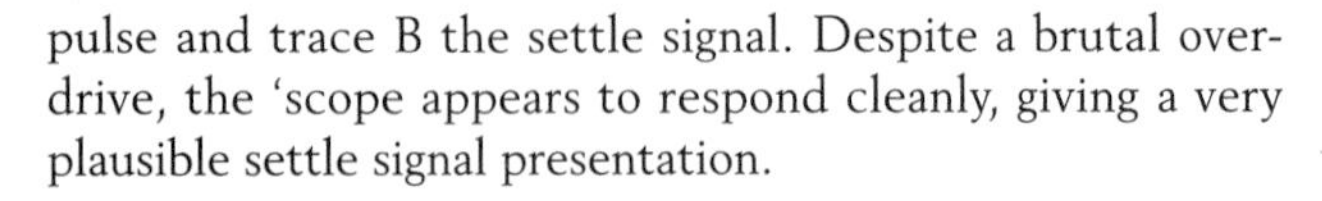

pulse and trace B the settle signal. Despite a brutal overdrive, the 'scope appears to respond cleanly, giving a very plausible settle signal presentation.

Summary and results

The simplest way to summarize the different method's results is by visual comparison. Figures 24.18 and 24.19 repeat previous photos of the two different settling-time methods. If both approaches represent good measurement technique and are properly constructed, results should be indentical.[13] If this is the case, the identical data produced by the two methods has a high probability of being valid.

Examination of the photographs shows nearly identical settling times and settling waveform signatures. The shape of the settling waveform is essentially identical in both photos.[14] This kind of agreement provides a high degree of credibility to the measured results.

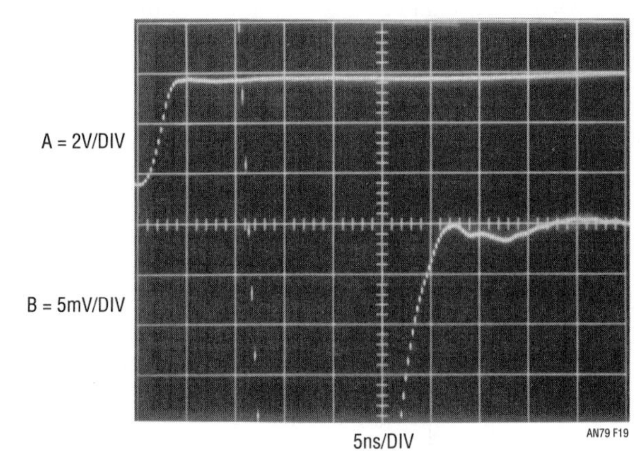

Figure 24.19 • Settling Time Measurement using the Classical Sampling 'Scope. t_{SETTLE} = 30ns

Note 13: Construction details of the settling time fixtures discussed here appear (literally) in Appendix E, "Breadboarding, Layout and Connection Techniques."

Note 14: The slightly rougher appearance of Figure 24.19's final settling movement (7th through 9th vertical divisions) may be due to the sampling 'scope's substantially higher bandwidth. Figure 24.18 was taken with a150MHz instrument; sampling oscilloscope bandwidth is 1GHz.

References

1. Williams, Jim, "Component and Measurement Advances Ensure 16-Bit DAC Settling Time," Linear Technology Corporation, Application Note 74, July 1998
2. Williams, Jim, "Measuring 16-Bit Settling Times: The Art of Timely Accuracy," *EDN*, November 19, 1998
3. Williams, Jim, "Methods for Measuring Op Amp Settling Time," Linear Technology Corporation, Application Note 10, July 1985.
4. Demerow, R., "Settling Time of Operational Amplifiers," Analog Dialogue, Volume 4-1, Analog Devices, Inc., 1970
5. Pease, R. A., "The Subtleties of Settling Time," The New Lightning Empiricist, Teledyne Philbrick, June 1971
6. Harvey, Barry, "Take the Guesswork Out of Settling Time Measurements," *EDN*, September 19, 1985
7. Williams, Jim, "Settling Time Measurement Demands Precise Test Circuitry," *EDN*, November 15, 1984
8. Schoenwetter, H. R., "High-Accuracy Settling Time Measurements," IEEE Transactions on Instrumentation and Measurement, Vol. IM-32. No. 1, March 1983
9. Sheingold, D. H., "DAC Settling Time Measurement," Analog-Digital Conversion Handbook, pg. 312–317. Prentice-Hall, 1986.
10. Orwiler, Bob, "Oscilloscope Vertical Amplifiers," Tektronix, Inc., Concept Series, 1969
11. Addis, John, "Fast Vertical Amplifiers and Good Engineering," Analog Circuit Design; Art, Science and Personalities, Butterworths, 1991
12. W. Travis, "Settling Time Measurement Using Delayed Switch," Private, Communication. 1984
13. Hewlett-Packard, "Schottky Diodes for High- Volume, Low Cost Applications," Application Note 942, Hewlett-Packard Company, 1973
14. Harris Semiconductor, "CA3039 Diode Array Data Sheet," Harris Semiconductor, 1993
15. Korn, G. A. and Korn, T. M., "Electronic Analog and Hybrid Computers," "Diode Switches," pg. 223–226. McGraw-Hill, 1964
16. Carlson, R., "A Versatile New DC-500MHz Oscilloscope with High Sensitivity and Dual Channel Display," Hewlett-Packard Journal, Hewlett-Packard Company, January 1960
17. Tektronix, Inc., "Sampling Notes," Tektronix, Inc., 1964
18. Tektronix, Inc., "Type 1S1 Sampling Plug-In Operating and Service Manual," Tektronix, Inc., 1965
19. Mulvey, J., "Sampling Oscilloscope Circuits," Tektronix, Inc., Concept Series, 1970
20. Addis, John, "Sampling Oscilloscopes," Private Communication, February, 1991
21. Williams, Jim, "Bridge Circuits—Marrying Gain and Balance," Linear Technology Corporation, Application Note 43, June, 1990
22. Tektronix, Inc., "Type 661 Sampling Oscilloscope Operating and Service Manual," Tektronix, Inc., 1963
23. Tektronix, Inc., "Type 4S1 Sampling Plug-In Operating and Service Manual," Tektronix, Inc., 1963
24. Tektronix, Inc., "Type 5T3 Timing Unit Operating and Service Manual," Tektronix, Inc., 1965
25. D. J. Hamilton, F H. Shaver, P G. Griffith, "Avalanche Transistor Circuits for Generating Rectangular Pulses," Electronic Engineering, December 1962
26. R. B. Seeds, "Triggering of Avalanche Transistor Pulse Circuits," Technical Report No. 1653-1, August 5, 1960, Solid-State Electronics Laboratory, Stanford Electronics Laboratories, Stanford University, Stanford, California
27. Haas, Isy, "Millimicrosecond Avalanche Switching Circuit Utilizing Double-Diffused Silicon Transistors," Fairchild Semiconductor, Application Note 8/2 (December 1961)
28. Beeson, R. H. Haas, I., Grinich, V. H., "Thermal Response of Transistors in the Avalanche Mode," Fairchild Semiconductor, Technical Paper 6 (October 1959)
29. Tektronix, Inc., Type 111 Pretrigger Pulse Generator Operating and Service Manual, Tektronix, Inc. (1960)
30. G.B. B. Chaplin, "A Method of Designing Transistor Avalanche Circuits with Applications to a Sensitive Transistor Oscilloscope," paper presented at the 1958 IRE-AIEE Solid State Circuits Conference, Philadelphia, Penn., February 1958
31. Motorola, Inc., "Avalanche Mode Switching," Chapter 9, pp 285–304. Motorola Transistor Handbook, 1963
32. Williams, Jim, "A Seven-Nanosecond Comparator for Single Supply Operation," "Programmable, Sub-Nanosecond Delayed Pulse Generator," pg. 32–34, Linear Technology Corporation, Application Note 72, 1998
33. Morrison, Ralph, "Grounding and Shielding Techniques in Instrumentation," 2nd Edition, Wiley Interscience, 1977
34. Ott, Henry W., "Noise Reduction Techniques in Electronic Systems," Wiley Interscience, 1976
35. Williams, Jim, "High Speed Amplifier Techniques," Linear Technology Corporation, Application Note 47. 1991

Appendix A Evaluating oscilloscope overdrive performance

The sampling bridge-based settling time circuit is heavily oriented towards preventing overdrive to the monitoring oscilloscope. This is done to avoid overdriving the oscilloscope. Oscilloscope recovery from overdrive is a grey area and almost never specified. How long must one wait after an overdrive before the display can be taken seriously? The answer to this question is quite complex. Factors involved include the degree of overdrive, its duty cycle, its magnitude in time and amplitude and other considerations. Oscilloscope response to overdrive varies widely between types and markedly different behavior can be observed in any individual instrument. For example, the recovery time for a 100× overload at 0.005V/DIV may be very different than at 0.1V/DIV. The recovery characteristic may also vary with waveform shape, DC content and repetition rate. With so many variables, it is clear that measurements involving oscilloscope overdrive must be approached with caution.

Why do most oscilloscopes have so much trouble recovering from overdrive? The answer to this question requires some study of the three basic oscilloscope types' vertical paths. The types include analog (Figure A1A), digital (Figure A1B) and classical sampling (Figure A1C) oscilloscopes. Analog and digital 'scopes are susceptible to overdrive. The classical sampling 'scope is the only architecture that is inherently immune to overdrive.

An analog oscilloscope (Figure A1A) is a real time, continuous linear system.[1] The input is applied to an attenuator, which is unloaded by a wideband buffer: The vertical preamp provides gain, and drives the trigger pick-off, delay line and the vertical output amplifier. The attenuator and delay line are passive elements and require little comment. The buffer, preamp and vertical output amplifier are complex linear gain blocks, each with dynamic operating range restrictions. Additionally, the operating point of each block may be set by inherent circuit balance, low frequency stabilization paths or both. When the input is overdriven, one or more of these stages may saturate, forcing internal nodes and components to abnormal operating points and temperatures. When the overload ceases, full recovery of the electronic and thermal time constants may require surprising lengths of time.[2]

The digital sampling oscilloscope (Figure A1B) eliminates the vertical output amplifier, but has an attenuator buffer and amplifiers ahead of the A/D converter. Because of this, it is similarly susceptible to overdrive recovery problems.

The classical sampling oscilloscope is unique. Its nature of operation makes it inherently immune to overload. Figure A1C shows why. The sampling occurs *before* any gain is taken in the system. Unlike Figure A1B's digitally sampled 'scope, the input is fully passive to the sampling point. Additionally, the output is fed back to the sampling bridge, maintaining its operating point over a very wide range of inputs. The dynamic swing available to maintain the bridge output is large and easily accommodates a wide range of oscilloscope inputs. Because of all this, the amplifiers in this instrument do not see overload, even at 1000× overdrives, and there is no recovery problem. Additional immunity derives from the instrument's relatively slow sample rate—even if the amplifiers were overloaded, they would have plenty of time to recover between samples.[3]

The designers of classical sampling 'scopes capitalized on the overdrive immunity by including variable DC offset generators to bias the feedback loop (see Figure A1C, lower right). This permits the user to offset a large input, so small amplitude activity on top of the signal can be accurately observed. This is ideal for, among other things, settling time measurements. Unfortunately, classical sampling oscilloscopes are no longer manufactured, so if you have one, take care of it![4]

Note 1: Ergo, the Real Thing. Hopelessly bigoted residents of this locale mourn the passing of the analog 'scope era and frantically hoard every instrument they can find.

Note 2: Some discussion of input overdrive effects in analog oscilloscope circuitry is found in Reference 11.

Note 3: Additional information and detailed treatment of classical sampling oscilloscope operation appears in References 16–19 and 22–24.

Note 4: Modern variants of the classical architecture (e.g., Tektronix 11801B) may provide similar capability, although we have not tried them.

Although analog and digital oscilloscopes are susceptible to overdrive, many types can tolerate some degree of this abuse. The early portion of this appendix stressed that measurements involving oscilloscope overdrive must be approached with caution. Nevertheless, a simple test can indicate when the oscilloscope is being deleteriously affected by overdrive.

The waveform to be expanded is placed on the screen at a vertical sensitivity that eliminates all off-screen activity. Figure A2 shows the display. The lower right hand portion is to be expanded. Increasing the vertical sensitivity by a factor of two (Figure A3) drives the waveform off-screen, but the remaining display appears reasonable. Amplitude has doubled and waveshape is consistent with the original display. Looking carefully, it is possible to see small amplitude information presented as a dip in the waveform at about the third vertical division. Some small disturbances are also visible. This observed expansion of the original waveform is believable. In Figure A4, gain has been further increased, and all the features of Figure A3 are amplified accordingly. The basic waveshape appears clearer and the dip and small disturbances are also easier to see. No new waveform characteristics are observed. Figure A5 brings some unpleasant surprises. This increase in gain causes definite distortion. The initial negative-going peak, although larger, has a different shape. Its bottom appears less broad than in Figure A4. Additionally, the peak's positive recovery is shaped slightly differently. A new rippling disturbance is visible in the center of the screen. This kind of change indicates that the oscilloscope is having trouble. A further test can confirm that this waveform is being influenced by overloading. In Figure A6 the gain remains the same but the vertical position knob has been used to reposition the display at the screen's bottom. This shifts the oscilloscope's DC operating point which, under normal circumstances, should not affect the displayed waveform. Instead, a marked shift in waveform amplitude and outline occurs. Repositioning the waveform to the screen's top produces a differently distorted waveform (Figure A7). It is obvious that for this particular waveform, accurate results cannot be obtained at this gain.

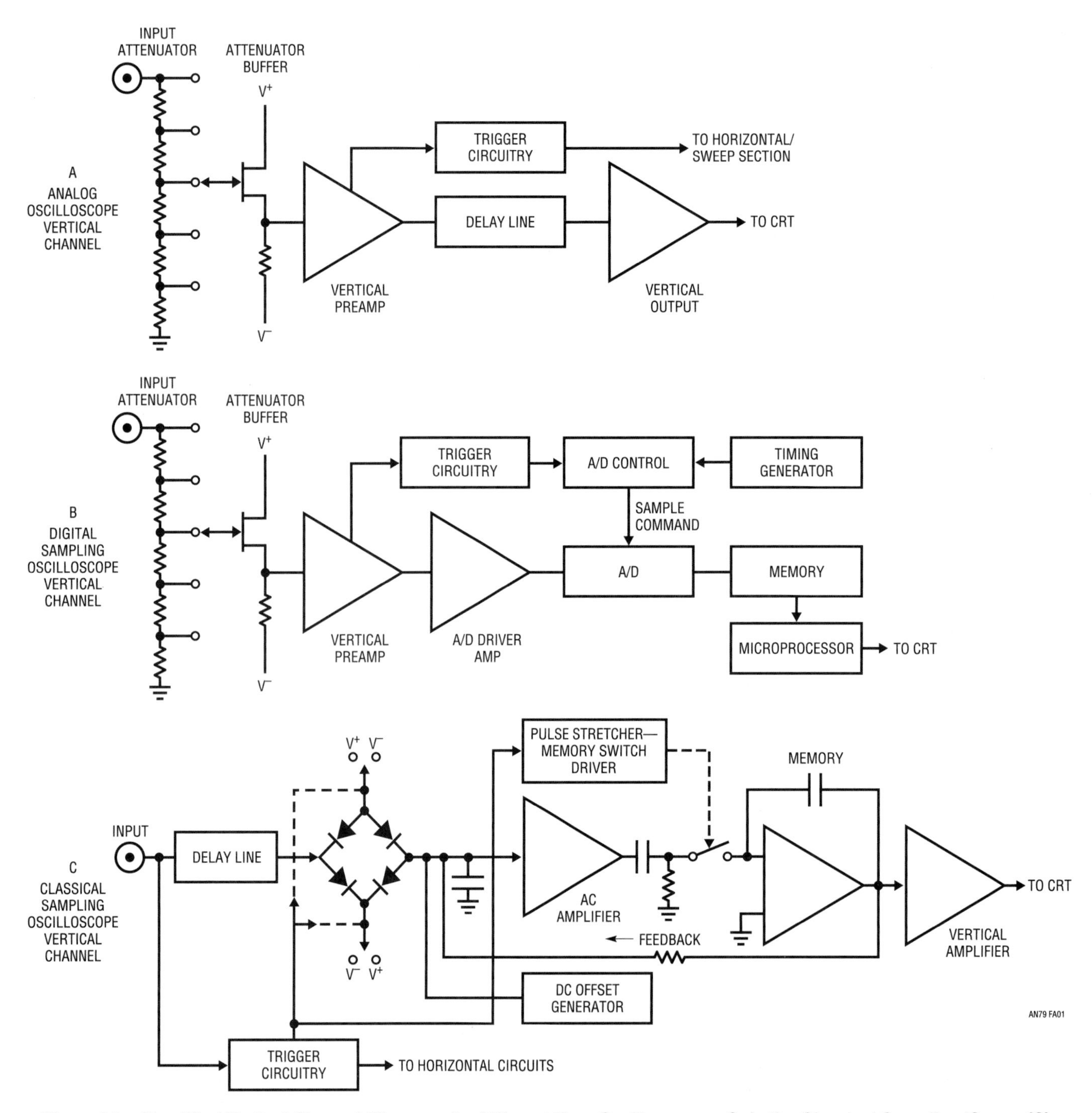

Figure A1 • Simplified Vertical Channel Diagrams for Different Type Oscilloscopes. Only the Classical Sampling 'Scope (C) Has Inherent Overdrive Immunity. Offset Generator Allows Viewing Small Signals Riding On Large Excursions

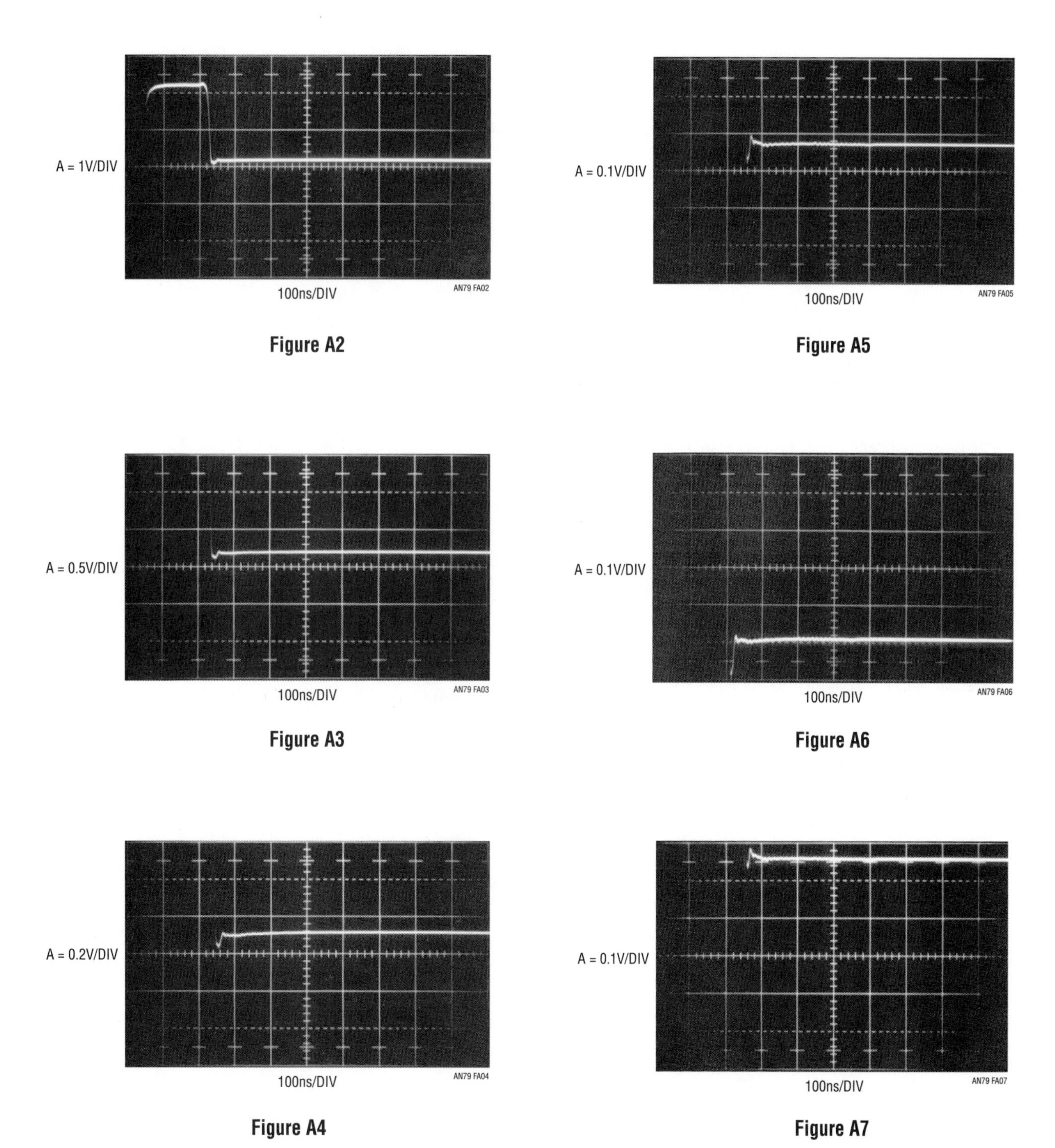

Figures A2–A7 • The Overdrive Limit is Determined by Progressively Increasing Oscilloscope Gain and Watching for Waveform Aberrations

Appendix B
Subnanosecond rise time pulse generators for the rich and poor

The input diode bridge requires a subnanosecond rise time pulse to cleanly switch current to the amplifier under test. The ranks of pulse generators providing this capability are thin. Instruments with rise times of a nanosecond or less are rare, and costs are, in this author's view, excessive. Current production units can easily cost $10,000, with prices rising towards $30,000 depending on features. For bench work, or even production testing, there are substantially less expensive approaches.

The secondary market offers subnanosecond rise time pulse generators at attractive cost. The Hewlett-Packard HP-8082A transitions in under 1ns, has a full complement of controls, and costs about $500. The HP-215A, long out of manufacture, has 800-picosecond edge times and is a clear bargain, with typical price below $50. This instrument also has a very versatile trigger output, which permits continuous time phase adjustment from before to after the main output. External trigger impedance, polarity and sensitivity are also variable. The output, controlled by a stepped attenuator, will put ±10V into 50Ω in 800ps.

The Tektronix type 109 switches in 250 picoseconds. Although amplitude is fully variable, charge lines are required to set pulse width. This reed-relay based instrument has a fixed ≈500Hz repetition rate and no external trigger facility, making it somewhat unwieldy to use. Price is typically $20. The Tektronix type 111 is more practical. Edge times are 500 picoseconds, with fully variable repetition rate and external trigger capabilities. Pulse width is set by charge line length. Price is usually about $25.

A potential problem with older instruments is availability.[1] As such, Figure B1 shows a circuit for producing subnanosecond rise time pulses. Rise time is 500ps, with fully adjustable pulse amplitude. An external input determines repetition rate, and output pulse occurrence is settable from before-to-after a trigger output. This circuit uses an avalanche pulse generator to create extremely fast rise-time pulses.[2]

Q1 and Q2 form a current source that charges the 1000pF capacitor. When the trigger input is high (trace A, Figure B2) both Q3 and Q4 are on. The current source is off and Q2's collector (trace B) is at ground. C1's latch input prevents it from responding and its output remains high. When the trigger input goes low, C1's latch input is disabled and its output drops low. Q4's collector lifts and Q2 comes on, delivering constant current to the 1000pF capacitor (trace B). The resulting linear ramp is applied to C1 and C2's positive inputs. C2, biased from a potential derived from the 5V supply, goes high 30 nanoseconds after the ramp begins, providing the "trigger output" (trace C) via its output network. C1 goes high when the ramp crosses the "delay programming voltage" input, in this case about 250ns. C1 going high triggers the avalanche-based output pulse (trace D), which will be described. This arrangement permits the delay programming voltage to vary output pulse occurrence from 30 nanoseconds before to 300 nanoseconds after the trigger output. Figure B3 shows the output pulse (trace D) occurring 30ns before the trigger output when the delay programming voltage is zero. All other waveforms are identical to Figure B2.

Note 1: Residents of Silicon Valley tend towards inbred techno-provincialism. Citizens of other locales cannot simply go to a flea market, junk store or garage sale and buy a subnanosecond pulse generator.

Note 2: The circuits operation essentially duplicates the aforementioned Tektronix type 111 pulse generator (see Reference 29). Information on avalanche operation appears in References 25–32.

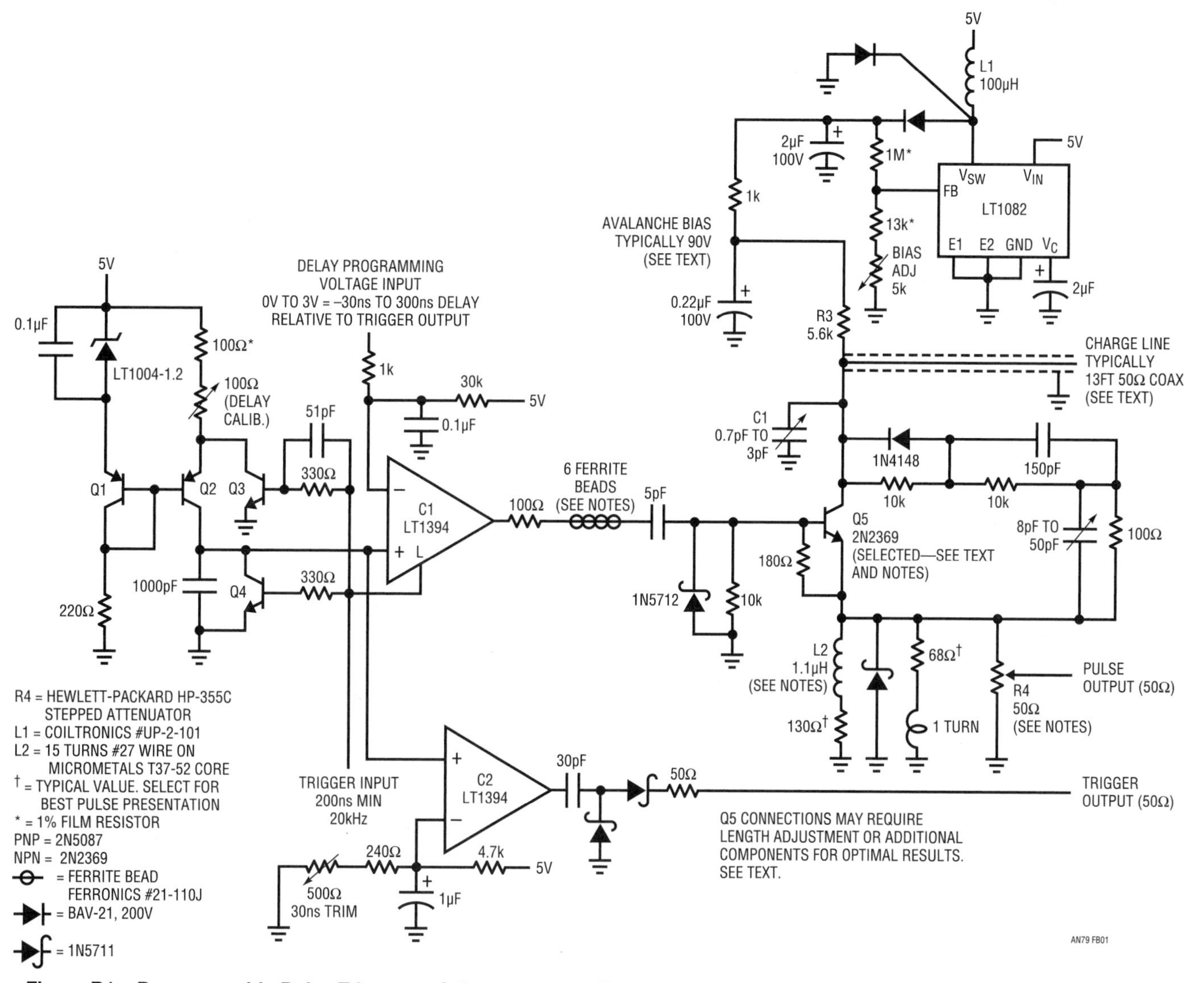

Figure B1 • Programmable Delay Triggers a Subnanosecond Rise Time Pulse Generator. Charge Line at Q5's Collector Determines 40 Nanosecond Output Width. Output Pulse Occurance is Settable from Before-to-After Trigger Output

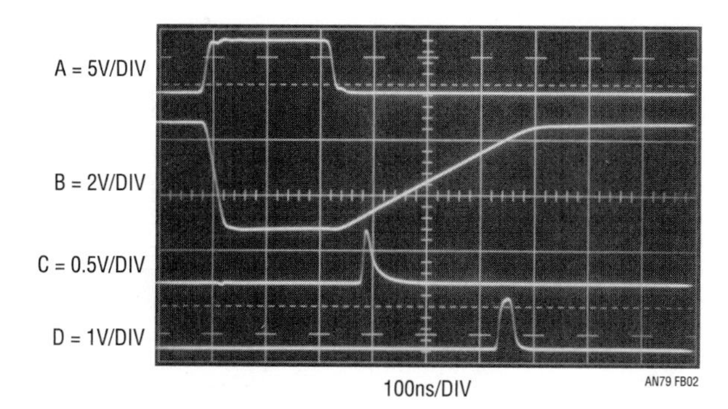

Figure B2 • Pulse Generator's Waveforms Include Trigger Input (Trace A), Q2's Collector Ramp (Trace B), Trigger Output (Trace C) and Pulse Output (Trace D). Delay Sets Output Pulse ≈250ns After Trigger Output

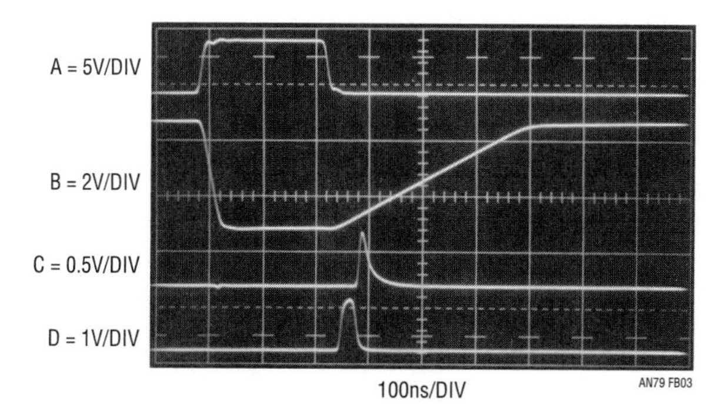

Figure B3 • Pulse Generator's Waveforms with Delay Programmed for Output Pulse Occurence (Trace D) 30ns Before Trigger Output (Trace C). All Other Activity is Identical to Previous Figure

When C1's output pulse is applied to Q5's base, it avalanches. The result is a quickly rising pulse across R4. C1 and the charge line discharge, Q5's collector voltage falls and breakdown ceases. C1 and the charge line then recharge. At C1's next pulse, this action repeats.

Avalanche operation requires high voltage bias. The LT1082 switching regulator forms a high voltage switched mode control loop. The LT1082 pulse width modulates at its 40kHz clock rate. L1's inductive events are rectified and stored in the 2μF output capacitor. The adjustable resistor divider provides feedback to the LT1082. The 1k-0.22μF RC provides noise filtering.

Figure B4, taken with a 3.9GHz bandpass oscilloscope (Tektronix 547 with 1S2 sampling plug-in) shows output pulse purity and rise time. Rise time is 500 picoseconds, with minimal preshoot and pulse top aberrations. This level of cleanliness requires considerable layout experimentation, particularly with Q5's emitter and collector lead lengths and associated components.[3] Additionally, small inductances or RC networks may be required between Q5's emitter and R4 to get best pulse presentation.[4] The charge line sets output pulse width, with 13 feet giving a 40ns wide output.

Q5 may require selection to get avalanche behavior. Such behavior, while characteristic of the device specified, is not guaranteed by the manufacturer. A sample of 50 Motorola 2N2369s, spread over a 12-year date code span, yielded 82%. All "good" devices switched in less than 600ps.

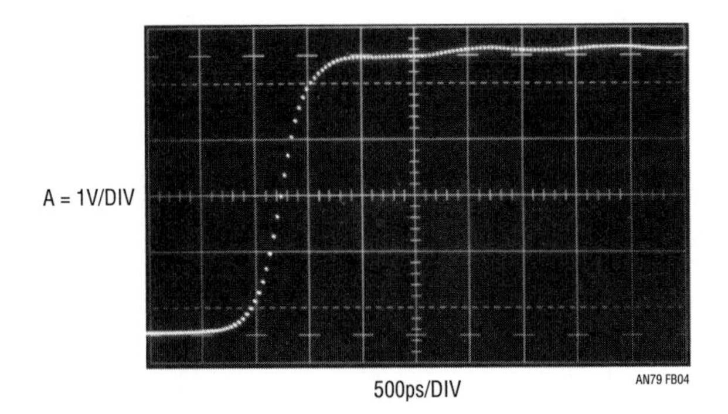

Figure B4 • Pulse Generator Output Shows 500 Picosecond Rise Time with Minimal Pulse-Top Aberrations. Dot Constructed Display is Characteristic of Sampling Oscilloscope Operation

Note 3: See References 29 and 32 for pertinent discussion.

Note 4: Ground plane type construction with high speed layout, connection and termination technique is essential for good results from this circuit. Reference 29 contains extremely useful and detailed procedures for optimizing pulse purity.

Circuit adjustment involves setting the "30ns trim" so C2 goes high 30ns after the trigger input goes low. Next, apply 3V to the delay programming input and set the "delay calibration" so C1 goes high 300ns after the trigger input goes low. Finally, set the high voltage "bias adjust" to the point where free running pulses across R4 *just* disappear with no trigger input applied.

Appendix C Measuring and compensating settling circuit delay

The settling time circuit utilizes an adjustable delay network to time correct the input pulse for delays in the signal-processing path. Typically, these delays introduce errors of 20%, so an accurate correction is required. Setting the delay trim involves observing the network's input-output delay and adjusting for the appropriate time interval. Determining the "appropriate" time interval is somewhat more complex. A wideband oscilloscope with FET probes is required. To ensure accuracy in the following delay measurements probe time skew must be verified. The probes are both connected to a fast rise (<1ns) pulse generator to measure the skew. Figure C1 shows less than 50 picoseconds skewing. This ensures small error for the delay measurements, which will be in the nanosecond range.

Referring to text Figure 24.6, it is apparent that three delay measurements are of interest. The pulse generator to amplifier-under-test, the amplifier-under-test to settle node, and the amplifier-under-test to output. Figure C2 shows 800 picoseconds delay from the pulse generator input to the amplifier-under-test. Figure C3 indicates 2.5 nanoseconds from the amplifier-under-test to the settle node. Figure C4 indicates 5.2 nanoseconds from the amplifier-under-test to the output. In Figure C3's measurement, the probes see severe source impedance mismatch. This is compensated by adding a series 500Ω resistor to the probe monitoring the amplifier-under-test. This provision approximately equalizes probe source impedances, negating the probe's input capacitance (≈1pF) term.

The measurements reveal a circuit input-to-output delay of 6 nanoseconds, and this correction is applied by adjusting the 1k trim at the C1 delay compensation comparator. Similarly, when the sampling 'scope is used, the relevant delays are Figures C2 and C3, a total of 3.3ns. This figure is applied to the delay compensation adjustment when the sampling 'scope-based measurement is taken.

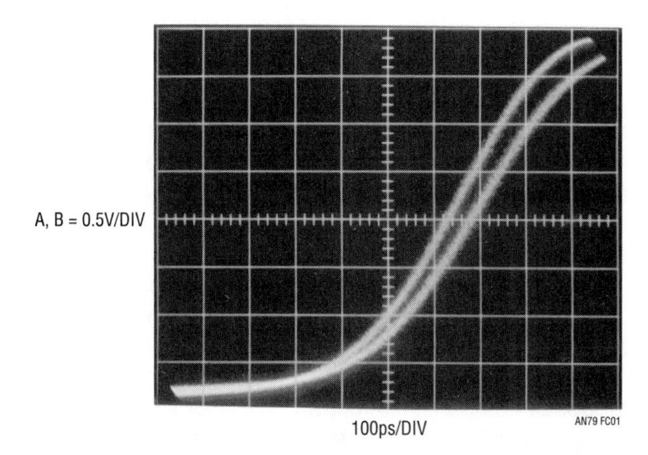

Figure C1 • FET Probe-Oscilloscope Channel-to-Channel Timing Skew Measures 50 Picoseconds

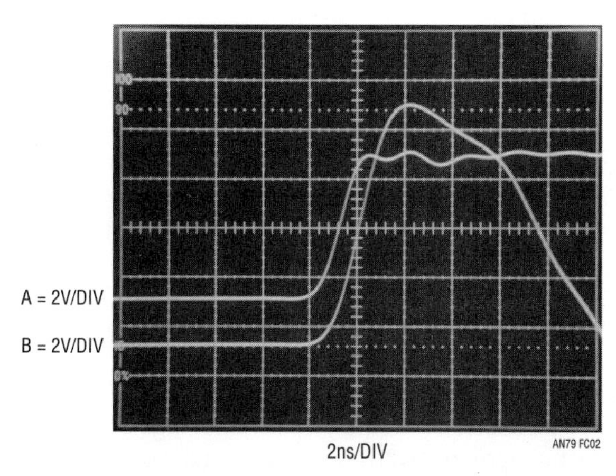

Figure C2 • Pulse Generator (Trace A) to Amplifier-Under-Test Negative Input (Trace B) Delay is 800 Picoseconds

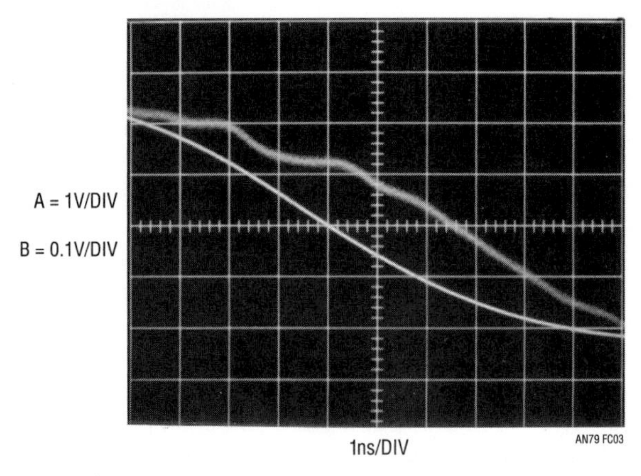

Figure C3 • Amplfier-Under-Test Output (Trace A) to Settle Node (Trace b) Delay is 2.5 Nanoseconds

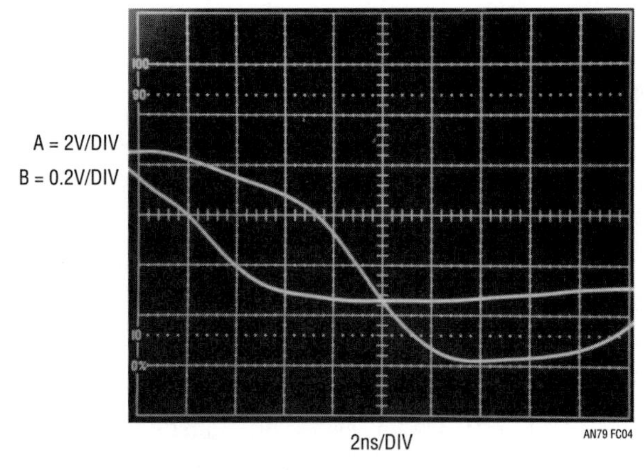

Figure C4 • Amplifier-Under-Test (Trace A) to Output (Trace B) Delay Measures 5.2 Nanoseconds

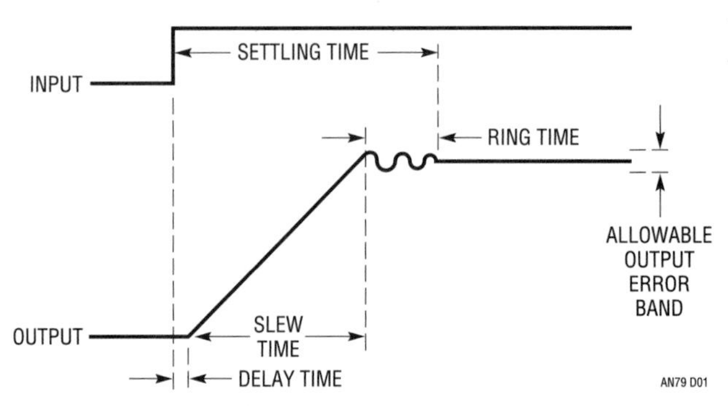

Figure D1 • Amplifier Settling Time Components Include Delay, Slew and Ring Times. For Given Components, Only Ring Time is Readily Adjustable

Appendix D
Practical considerations for amplifier compensation

There are a number of practical considerations in compensating the amplifier to get fastest settling time. Our study begins by revisiting text Figure 24.1 (repeated here as Figure D1). Settling time components include delay, slew and ring times. Delay is due to propagation time through the amplifier and is a relatively small term. Slew time is set by the amplifier's maximum speed. Ring time defines the region where the amplifier recovers from slewing and ceases movement within some defined error band. Once an amplifier has been chosen, only ring time is readily adjustable. Because slew time is usually the dominant lag, it is tempting to select the fastest slewing amplifier available to obtain best settling. Unfortunately, fast slewing amplifiers usually have extended ring times, negating their brute force speed advantage. The penalty for raw speed is, invariably, prolonged ringing, which can only be damped with large compensation capacitors. Such compensation works, but results in protracted settling times. The key to good settling times is to choose an amplifier with the right balance of slew rate and recovery characteristics and compensate it properly. This is harder than it sounds because amplifier settling time cannot be predicted or extrapolated from any combination of data sheet specifications. It must be measured in the intended configuration. A number of terms combine to influence settling time. They include amplifier slew rate and AC dynamics, layout capacitance, source resistance and capacitance, and the compensation capacitor. These terms interact in a complex manner, making predictions hazardous.[1] If the parasitics are eliminated and replaced with a pure resistive source, amplifier settling time is still not readily predictable. The parasitic impedance terms just make a difficult problem more messy. The only real handle available to deal with all this is the feedback compensation capacitor, C_F. C_F's purpose is to roll off amplifier gain at the frequency that permits best dynamic response.

Note 1: Spice aficionados take notice.

Best settling results when the compensation capacitor is selected to functionally compensate for all the above terms. Figure D2 shows results for an optimally selected feedback capacitor. Trace A is the time-corrected input pulse and trace B the amplifier's settle signal. The amplifier is seen to come cleanly out of slew (sample gate opens just prior to sixth vertical division) and settle very quickly.

In Figure D3, the feedback capacitor is too large. Settling is smooth, although overdamped, and a 20ns penalty results. Figure D4's feedback capacitor is too small, causing a somewhat underdamped response with resultant excessive ring time excursions. Settling time goes out to 43ns. Note that Figures D3 and D4 require reduction of vertical and horizontal scales to capture nonoptimal response.

When feedback capacitors are individually trimmed for optimal response, the source, stray, amplifier and compensation capacitor tolerances are irrelevant. If individual trimming is not used, these tolerances must be considered to determine the feedback capacitor's production value. Ring time is affected by stray and source capacitance and output loading, as well as the feedback capacitor's value. The relationship is nonlinear, although some guidelines are possible. The stray and source terms can vary by ±10% and the feedback capacitor is typically a ±5% component.[2] Additionally, amplifier slew rate has a significant tolerance, which is stated on the data sheet. To obtain a production feedback capacitor value, determine the optimum value by individual trimming *with the production board layout* (board layout parasitic capacitance counts too!). Then, factor in the worst-case percentage values for stray and source impedance terms, slew rate and feedback capacitor tolerance. Add this information to the trimmed capacitors measured value to obtain the production value. This budgeting is perhaps unduly pessimistic (RMS error summing may be a defensible compromise), but will keep you out of trouble.[3]

Note 2: This assumes a resistive source. If the source has substantial parasitic capacitance (photodiode, DAC, etc.), this number can easily enlarge to ±50%.

Note 3: The potential problems with RMS error summing become clear when sitting in an airliner that is landing in a snowstorm.

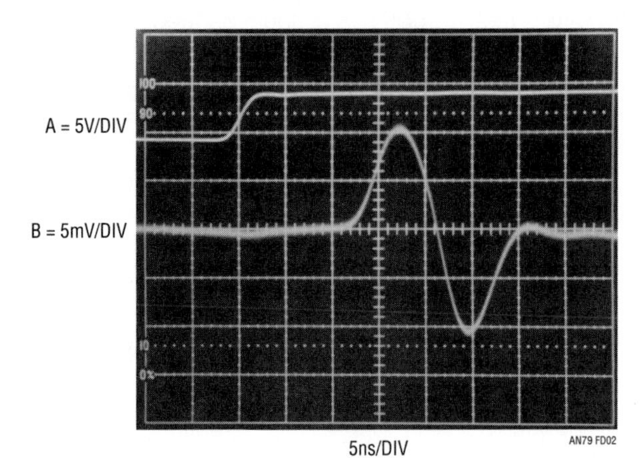

Figure D2 • Optimized Compensation Capacitor Permits Nearly Critically Damped Response, Fastest Settling Time. t_{SETTLE} = 30ns

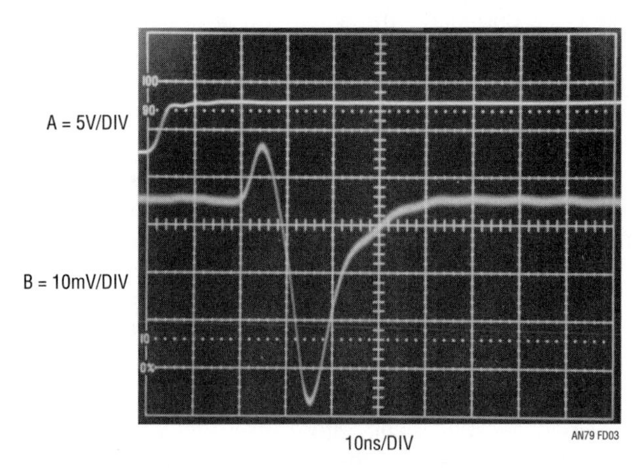

Figure D3 • Overdamped Response Ensures Freedom from Ringing, Even with Component Variations in Production. Penalty is Increased Settling Time. Note Horizontal and Vertical Scale Changes vs Figure D2. t_{SETTLE} = 50ns

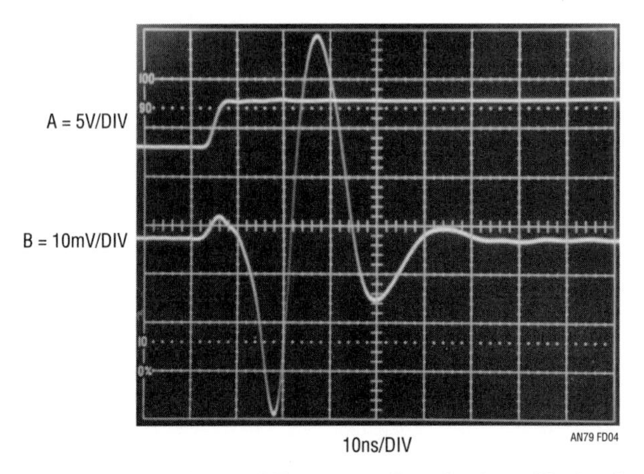

Figure D4 • Underdamped Response Results from Undersized Capacitor. Component Tolerance Budgeting Will Prevent This Behavior. Note Vertical and Horizontal Scale Changes vs Figure D2. t_{SETTLE} = 43ns

Appendix E
Breadboarding, layout and connection techniques

The measurement results presented in this publication required painstaking care in breadboarding, layout and connection techniques. Nanosecond domain, high resolution measurement does not tolerate cavalier laboratory attitude. The oscilloscope photographs presented, devoid of ringing, hops, spikes and similar aberrations, are the result of a careful breadboarding exercise. The sampler-based breadboard required considerable experimentation before obtaining a noise/uncertainty floor worthy of the measurement.

Ohm's law

It is worth considering that Ohm's law is a key to successful layout.[1] Consider that 10mA running through 1Ω generates 10mV—twice the measurement limit! Now, run that current at 1 nanosecond rise times (≈350MHz) and the need for layout care becomes clear. A paramount concern is disposal of circuit ground return current and disposition of current in the ground plane. The impedance of the ground plane between any two points is *not* zero, particularly at nanosecond speeds. This is why the entry point and flow of "dirty" ground returns must be carefully placed within the grounding system. In the sampler-based breadboard, the approach was separate "dirty" and "signal" ground planes tied together at the supply ground origin.

A good example of the importance of grounding management involves delivering the input pulse to the breadboard. The pulse generator's 50Ω termination *must* be an in-line coaxial type, and it cannot be directly tied to the signal ground plane. The high speed, high density (5V pulses through the 50Ω termination generate 100mA current spikes) current flow must return directly to the pulse generator. The coaxial terminator's construction ensures this substantial current does this, instead of being dumped into the signal ground plane (100mA termination current flowing through 50 *milliohms* of ground plane produces ~5mV of error!). Figure E3 shows that the BNC shield floats from the signal plane, and is returned to "dirty" ground via a copper strip. Additionally, Figure E1 shows the pulse generator's 50Ω termination physically distanced from the breadboard via a coaxial extension tube. This further ensures that pulse generator return current circulates in a tight local loop at the terminator, and does not mix into the signal plane.

It is worth mentioning that, because of the nanosecond speeds involved, inductive parasitics may introduce more error than resistive terms. This often necessitates using flat wire braid for connections to minimize parasitic inductive and skin effect-based losses. Every ground return and signal connection in the entire circuit must be evaluated with these concerns in mind. A paranoiac mindset is quite useful.

Note 1: I do not wax pedantic here. My guilt in this matter runs deep.

Shielding

The most obvious way to handle radiation-induced errors is shielding. Various following figures show shielding. Determining where shields are required should come *after* considering what layout will minimize their necessity. Often, grounding requirements conflict with minimizing radiation effects, precluding maintaining distance between sensitive points. Shielding is usually an effective compromise in such situations.

A similar approach to ground path integrity should be pursued with radiation management. Consider what points are likely to radiate, and try to lay them out at a distance from sensitive nodes. When in doubt about odd effects, experiment with shield placement and note results, iterating towards favorable performance.[2] *Above all, never rely on filtering or measurement bandwidth limiting to "get rid of" undesired signals whose origin is not fully understood.* This is not only intellectually dishonest, but may produce wholly invalid measurement "results," even if they look pretty on the oscilloscope.

Connections

All signal connections to the breadboard must be coaxial. Ground wires used with oscilloscope probes are forbidden. A1" ground lead used with a 'scope probe can easily generate large amounts of observed "noise" and seemingly inexplicable waveforms. Use coaxially mounting probe tip adapters![3]

Figures E1 to E6 restate the above sermon in visual form while annotating the text's measurement circuits.

Note 2: After it works, you can figure out why.

Note 3: See Reference 35 for additional nagging along these lines.

Figure E1 • Overview Of Settling Time Breadboard. Pulse Generator Enters Left Side—50Ω Coaxial Terminator Mounted On Extension Tube Minimizes Pulse Generator Return Current Mixing Into Signal Ground Planes (Bottom and Raised Center Boards). Delayed Pulse Generator is Lower Left. Delay Compensation Is Small Board Above Extension Tube (Center Left). Input Bridge-Amplifier-Under-Test Is Between Raised Board (Center) and Delay Pulse Generator (Lower Left). Raised Board Is Sampling Bridge and Drive Circuitry. Note All Coaxial Signal and Probe Connections

Figure E2 • Settling Time Breadboard Detail. Note Radiation Shield (Vertical Board Lower Left) at Delayed Pulse Generator (Lower Left). "Dirty" Ground Return Is Wide Copper Strip Running from Board Lower Center to Banana Jack (Photo Upper Center). Sampling Bridge Circuitry Is Raised Board (Photo Center Right, Foreground). AC Trims (Raised Board Center Right) and DC Adjustment (Raised Board Lower Right) Are Visible

Figure E3 • Detail of Pulse Generator Input and Delay Compensation. Delay Compensation Circuitry Is Small Board Above Pulse Generator Coaxial BNC Fitting (Photo Center Left). Pulse Generator BNC Common Floats from Main Board Via Insulated Vertical Support (Soldered to BNC—Photo Lower Center Left). BNC Is Tied to Ground "Mecca" By Thin Copper Strip (Photo Center Left) Running at Angle to Main Board. Input Bridge and Amplifier-Under-Test Occupy Photo Center Right. "Dirty" Ground Return Bus (Large Rectangular Board) Runs Across Main Board, Ends at Banana Jack

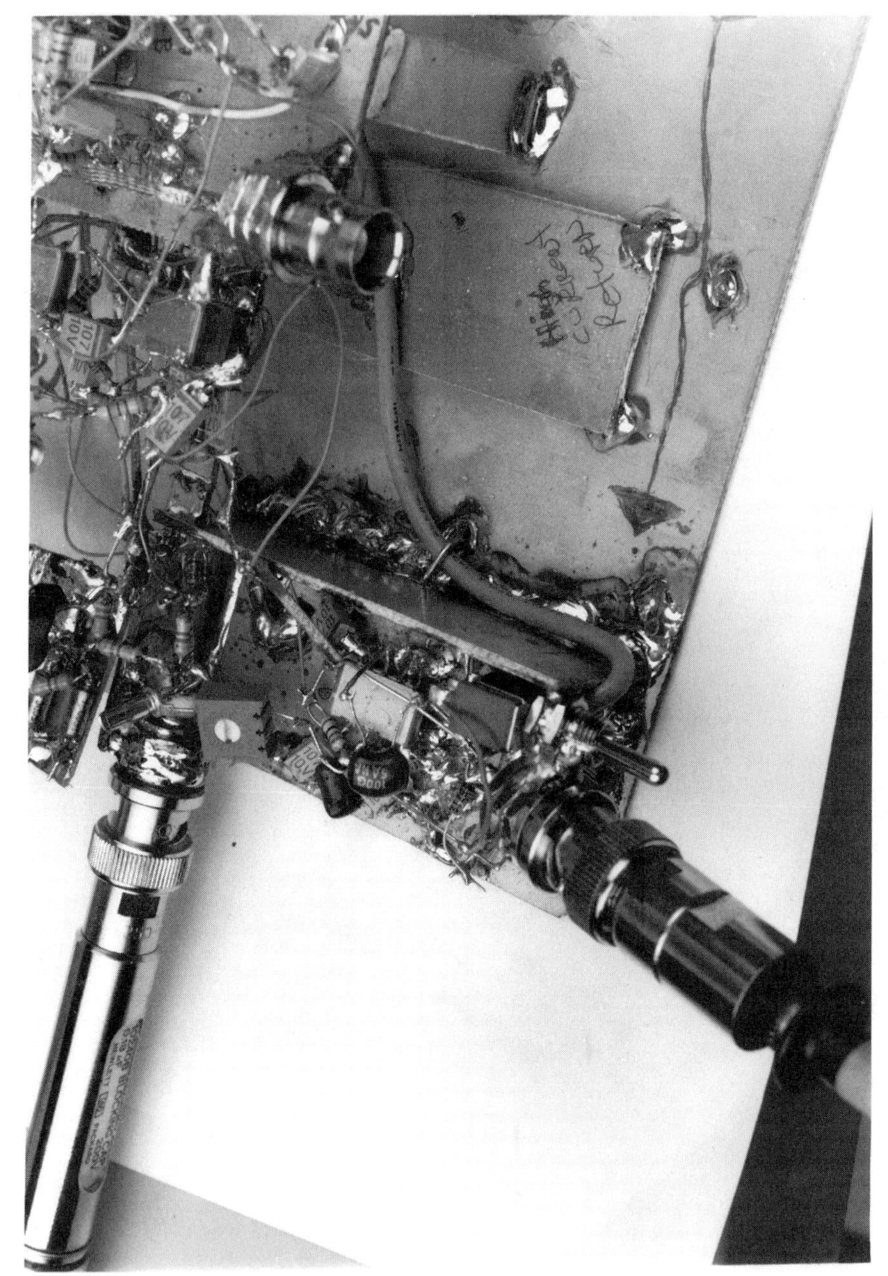

Figure E4 • Delayed Pulse Generator Is Fully Shielded from Input Bridge and Sampler Circuitry (Both Partially Visible, Photo Upper Right). Shield Is Vertical Board (Photo Center). Delayed Pulse Generator Output Routes to Sampling Bridge Via Coaxial Cable (Photo Center Right), Minimizing Radiation

Figure E5 • Input Bridge and Amplifier-Under-Test (AUT) Detail. Pulse Generator Enters Lower Left. Input Bridge Is IC Can (Photo Center); AUT Just Above. AUT Feedback Trim Capacitor Is Upper Center. IC Behind Trim Capacitor Is Bridge Driver Amplifier. Sampling Bridge (Partial) Is Photo Upper. Probe (Photo Extreme Right) Monitors Sampler Input. FET Probe (Photo Extreme Left) Measures Delay Compensated Input Pulse

Figure E6 • Sampling Bridge Viewed from Above. Sample Gate Coaxial Cable Starts at Delayed Pulse Generator (Photo Extreme Upper Left), Goes Under Sampler Board (Photo Center), Reappears at Sampler Board Right Side. Note Vertical Shield Preventing Sample Gate Pulse Radiation from Corrupting Sampler Output. Sampler DC Zero Trim Is Square Potentiometer (Sampler Board Lower Left); Skew and AC Balance Adjustments Are Photo Upper Center. Sampling Bridge Diodes (Not Visible) Are Directly Beneath Shielded Section below Skew and Balance Trims

Application and optimization of a 2GHz differential amplifier/ADC driver

25

Cheng-Wei Pei Adam Shou

Introduction

Modern high speed analog-to-digital converters (ADC), including those with pipeline or successive approximation register (SAR) topologies, have fast switched-capacitor sampling inputs. The highest performance parts maintain low input noise and low signal distortion (high linearity) while consuming less power with each new generation. At the same time, they are sampling at ever increasing rates, which enables wider signal bandwidths and relaxes the requirements for analog antialias filtering. The switched-capacitor inputs result in significant charge injection with every switch movement, which requires a front end that can absorb that charge and settle to the correct voltage by the time the ADC input switches again, in nanoseconds or less.

The task of the ADC driver amplifier is to meet all of these requirements. A good driver must have the ability to output a signal of full-scale amplitude for the ADC with low distortion and low noise to maintain the dynamic range of the ADC. Also, the ADC driver (and any antialias filtering) must be able to withstand the ADC's switch charge injection and recover before the next switch movement to minimize signal degradation. In the ADC driver, this implies good transient response and wide bandwidth relative to the ADC's sampling frequency.

LTC6400 features

Linear Technology's LTC®6400 family of ADC drivers addresses all three issues with a low distortion and low noise ADC driver with reasonable power dissipation. At 70MHz, which is a common IF frequency used in RF/IF signal chains, the LTC6400 family boasts distortion as low as −94dBc (equivalent output IP3 = 51dBm)[1] and input-referred noise density of $1.4\text{nV}/\sqrt{\text{Hz}}$. This allows the LTC6400 family to be used with high performance 14-bit and 16-bit ADCs without compromising their performance. The power draw of the LTC6400 family is as low as 120mW on a single 3V supply.

The LTC6400 comes in two versions, one with the highest performance (LTC6400) and another with half the power draw (LTC6401). To facilitate ease of design and layout, there are four fixed-gain options for each (8dB, 14dB, 20dB and 26dB) for a total of eight parts in the family. The input impedances of the parts vary from 50Ω to 400Ω, which is convenient for impedance matching. The LTC6400 family is unconditionally stable with any input and output termination, and in fact the output does not require any

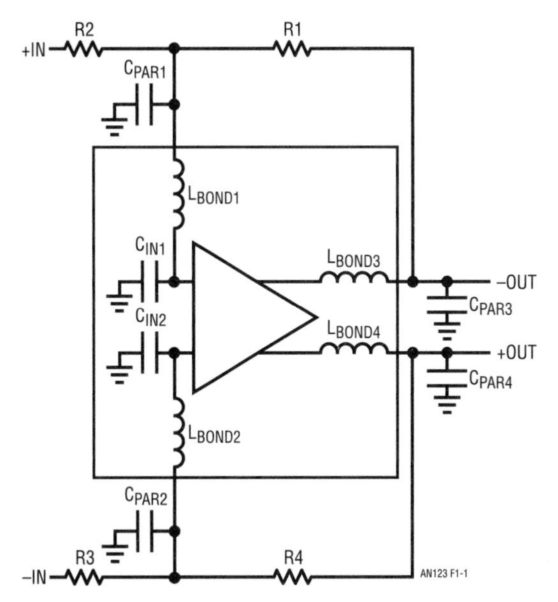

Figure 25.1 • A Typical High Speed Differential Amplifier Circuit with Parasitic Elements that Can Degrade Performance. The Differential Amplifier IC is Shown in the Box. The Bond-Wire Inductance of the Package and the Parasitic Capacitances, Both Internal and External, Will Reduce the Phase Margin of the Amplifier. Notice that the Output Load is in the Feedback Path

Note 1: See "Equivalent Output IP3" section for a definition.

Analog Circuit and System Design: Immersion in the Black Art of Analog Design. http://dx.doi.org/10.1016/B978-0-12-397888-2.00025-0

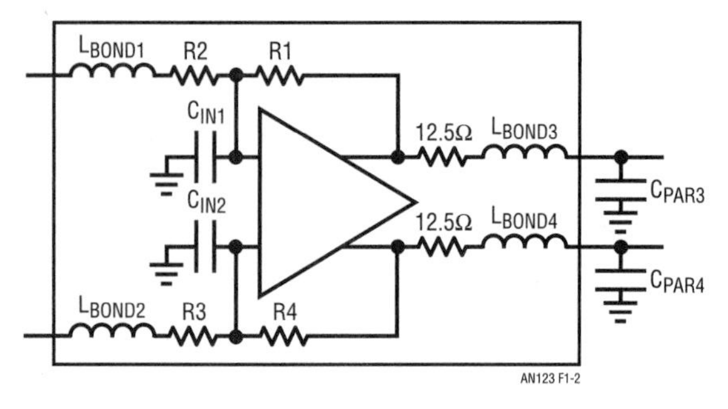

Figure 25.2 • A Differential Amplifier IC with Internal Resistors, Such as the LTC6400. Note that the Bond-Wire Inductances are Not Inside the Feedback Loop, and that the Output Load is Isolated from the Feedback Loop Due to the Resistors and Bond-Wire Inductance. The Input Capacitance of the Differential Amplifier Can Still Affect the Performance, but it is Predictable and Can be Compensated for in the IC Design

impedance-matching components when driving an ADC. The LTC6400 keeps the overall solution size small with its 3mm × 3mm QFN package that requires only a few external power supply bypass capacitors for operation.

This Application Note discusses the features and boundaries of the LTC6400 family and how to achieve optimal performance from the amplifiers in real applications.

Internal gain/feedback resistors

The LTC6400 is manufactured with internal gain and feedback resistors, for a complete differential amplifier solution that is simple to use and less susceptible to parasitic capacitances on the PCB layout than differential amplifiers without internal resistors. This is because the sensitive amplifier feedback loop nodes are contained within the chip.

The benefit of internal resistors can be visualized in Figure 25.1 and Figure 25.2. The first shows a traditional high speed differential amplifier IC, with parasitic inductance and capacitance affecting the stability and frequency response of the amplifier. The second shows the same parasitic elements in an LTC6400 style amplifier with internal resistors. The parasitic elements are outside of the critical feedback loop, and actually help to isolate the output load from the amplifier.

With the LTC6400's internal components, the only required external components are bypass capacitors, which should still be located as close to the amplifier's package as physically possible. For more information and recommendations on PCB layout, see the Layout section of this Application Note.

Low distortion

The LTC6400 family is manufactured on a very high frequency silicon germanium process, and crosses the divide between traditional operational amplifiers (op amps) and RF/IF amplifiers. The LTC6400 maintains very low distortion and low noise with the capability of handling signals

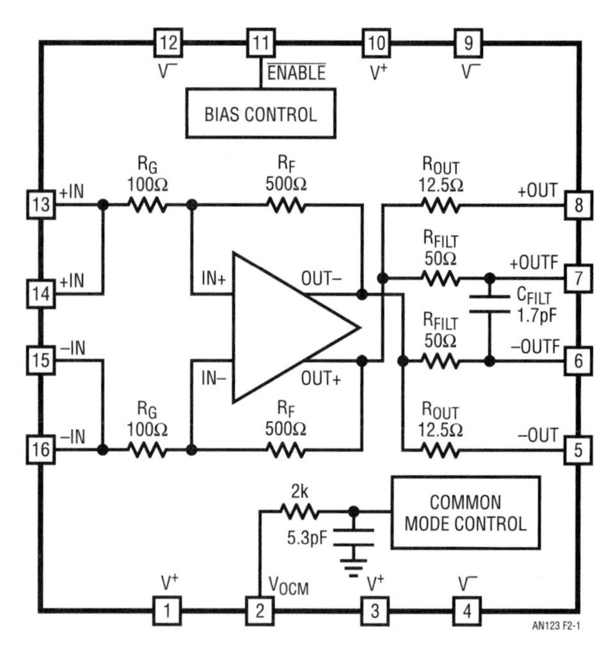

Figure 25.3 • Block Diagram of the LTC6400-14. The LTC6400 Delivers IF Amplifier Performance, but is Topologically Similar to a Differential Op Amp in its use of Feedback

at high intermediate frequencies (IF). This allows the LTC6400 to be used in demanding IF sampling applications of radio receiver signal chains. However, the LTC6400 is still topologically similar to traditional op amps in the use of feedback to achieve its high gain and DC performance.

The block diagram of the LTC6400-14 (14dB fixed gain), shown in Figure 25.3, shows similarities to traditional differential op amps. The main difference lies in the use of internal resistors and other frequency compensation components. By keeping the most sensitive nodes of the amplifier inside the package, the LTC6400 is able to provide almost 2GHz of bandwidth while maintaining good stable performance in real-world layouts. In other words, the user does not need to worry about low product yields due to oscillating high frequency amplifiers. The LTC6400 does still require attention to layout (see the Layout section of this Application Note), but the most difficult part is already done.

Actual bandwidth vs usable bandwidth

The LTC6400 has a very wide bandwidth (approaching 2GHz), but the vast majority of applications will not require frequencies beyond a few hundred Megahertz. The reason lies in the topology. Unlike traditional "open-loop" RF/IF amplifiers, where there is very little or no feedback used in the amplifier circuit, the LTC6400 contains an internal differential op amp with the gain set using a feedback network. The internal amplifier's open-loop gain is much higher than the gain externally, and the amplifier is compensated to push out the overall loop gain roll-off to higher frequencies. The main reason that a "closed-loop" op amp is able to achieve great distortion performance is the combination of feedback and high loop gain, which is able to reduce any distortion created within the amplifier. Once the loop gain begins to roll off at higher frequencies, the distortion performance begins to suffer.

Figure 25.4 shows the 3rd order intermodulation distortion (IMD) products from a 2-tone signal test of the LTC6400-20.[2] At low frequencies, the distortion products approach −100dBc. However, the performance is still good even up to 250MHz-300MHz, which makes the LTC6400 suitable for even mid-to-high IF systems. However, it is important to note that the LTC6400 does not maintain great distortion performance all the way up to its actual −3dB bandwidth.

In other applications, the high bandwidth of the LTC6400 can be a significant advantage. With a slew rate of up to 6700V/μs[3] and a 2V step 1% settling time as fast as 0.8ns, the LTC6400 can be used in high performance video and charge-coupled device (CCD) applications with good results. The wide bandwidth of the LTC6400 results in flat gain to hundreds of Megahertz; Table 25.4 summarizes the gain flatness specifications in the data sheets.

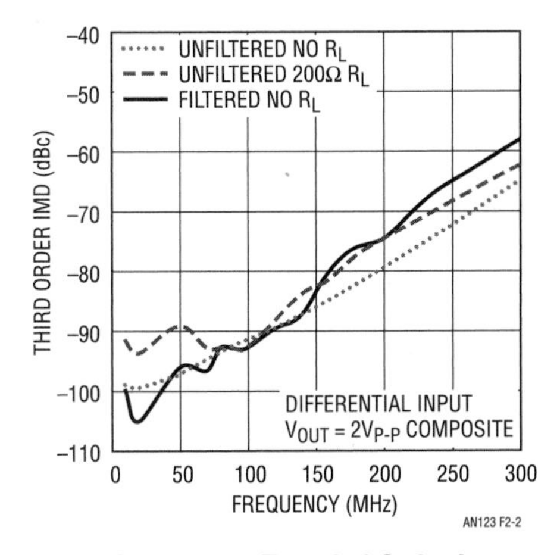

Figure 25.4 • LTC6400-20 2-Tone 3rd Order Intermodulation Distortion. The Feedback Topology of the LTC6400 Means that the Distortion Performance Falls with the Loop-Gain Over Frequency

Low-frequency distortion performance

A standout ability of the LTC6400 among high speed amplifiers is that it can accept inputs down to DC. From Figure 25.4, it can be inferred that the distortion performance approaches or exceeds −100dBc at lower frequencies (10MHz and below). This enables the LTC6400 to be used in very high performance low frequency and baseband systems to provide gain with no measurable degradation in the signal. The low 1/f noise "corner frequency" (on the order of 12kHz) means that the low noise performance of the LTC6400 is maintained well below 1MHz. Although the LTC6400 has low distortion at 100MHz and above, it also has phenomenal distortion performance at 20MHz and below. Figure 25.5 shows measured distortion data with input frequencies as low as 1MHz.

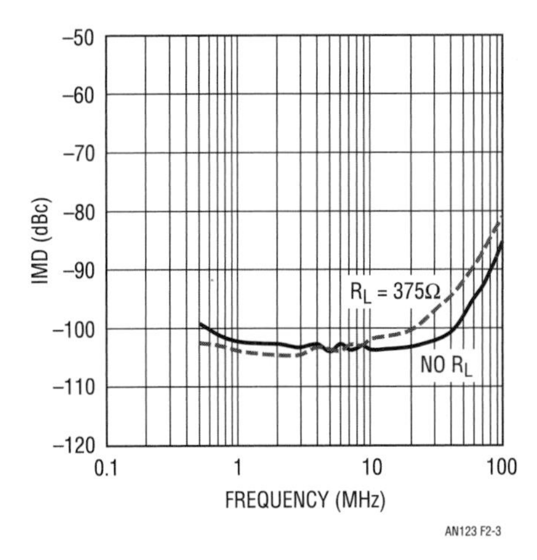

Figure 25.5 • LTC6401-14 Low Frequency Distortion. Results Obtained from Lab Experiments Show that the Distortion of the LTC6400 Family is Better than -100dBc at Frequencies Below 40MHz, Making it an Excellent Choice for Lower Frequency Applications

Distortion performance guaranteed

One of the more unique features of the LTC6400 is the guaranteed distortion specification on the data sheet. Each unit is individually tested in production to meet specifications for functionality, which typically includes gain, offset, supply current, etc. The LTC6400 is unusual among available amplifiers in that each unit is also tested for distortion performance. The production test applies a two-tone input signal and measures the 3rd order intermodulation distortion. Table 25.1 lists the guaranteed distortion specification for each member of the family.

Note 2: For a general discussion on how to measure very low intermodulation distortion products, see (Seremeta 2006).

Note 3: A slew rate of 6700V/μs results in a $2V_{P-P}$ full power bandwidth of 1.066GHz.

Table 25.1 Typical and Guaranteed 2-Tone 3rd Order IMD Specifications for the LTC6400 Family of Products. These Specifications are Measured and Guaranteed at Room Temperature

PART NUMBER	INPUT FREQUENCIES (MHz)	TYPICAL IMD (dBc)	GUARANTEED IMD (dBc)
LTC6400-8	280, 320 (IMD Measured at 360MHz)	-59	-53
LTC6400-14		-63	-57
LTC6400-20		-70	-64
LTC6400-26		-68	-62
LTC6401-8	130, 150 (IMD Measured at 170MHz)	-75	-67
LTC6401-14		-70	-61
LTC6401-20		-69	-61
LTC6401-26		-70	-62

Typically an amplifier's DC specifications are measured, and the AC performance (including distortion) is simply assumed. However, this means that from part to part there can be large variations in actual performance, which can cause the need to design in large performance margins or potentially sacrifice product yield. With a guaranteed specification in the data sheet for distortion, a design can proceed with more confidence knowing exactly what to expect from the amplifier.

Low noise

The challenge of specifying the noise of the LTC6400 family stems from its dual role as a RF/IF signal chain gain block and a traditional voltage-gain differential amplifier. The data sheet specifications of voltage/current noise density ($nV/\sqrt{Hz}$ and $pA/\sqrt{Hz}$) and noise figure (NF) must be correctly interpreted for the user's application. To make things more complicated, the LTC6400 is flexible in source and load impedance, which means that all the noise specifications can change depending on the circuit used. This section expands on the specifications in the data sheet and further explains how to correctly calculate the noise of the LTC6400 family.

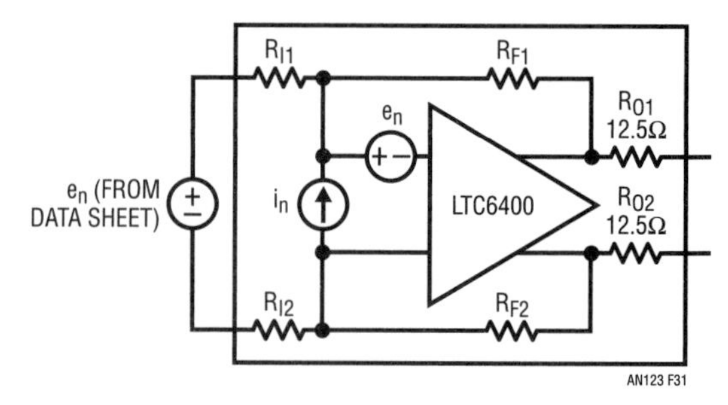

Figure 25.6 • Equivalent Noise Sources of the LTC6400. e_n is the Equivalent Input Voltage Noise Source, and i_n is the Equivalent Differential Input Current Noise Source. In This Example, the Resistors Would Not be Considered Noiseless for Noise Calculation Purposes. The Extra Voltage Source External to the LTC6400 is the e_n Value That is Listed in the Data Sheet, Assuming $R_S = 0\Omega$

Amplifier noise is typically described by an input equivalent noise voltage density, e_n, and noise current density, i_n. Figure 25.6 illustrates the equivalent noise sources. The voltage noise density can be measured by shorting the amplifier inputs, measuring the output voltage noise density, subtracting the effects of the resistors, and dividing by the amplifier's 'noise gain' for $Z_S = 0$. Note that in the case of a feedback amplifier, the noise gain may not equal the input/output signal gain[4]. The current noise density can be determined by connecting the amplifier inputs together with a resistor or capacitor (Z_S), measuring the output voltage noise density, subtracting the noise contribution due to e_n and Z_S, and dividing by the noise gain. For simplicity, the subtraction of the e_n and Z_S effects is done in an RMS fashion (square root of the difference of squares), with the assumption that e_n and i_n are uncorrelated noise sources. Through this procedure, input and output equivalent noise are obtained as shown in Table 25.2.

Notice that the e_n values in Table 25.2 differ from those in the LTC6400 data sheet. This is because the data sheet specifies a voltage noise density referred to the resistor inputs with the assumption that the source impedance is zero. In other words, the output voltage noise density from the data sheet is simply divided by the signal gain of the amplifier. This approach makes part-to-part comparison easier, but it does not lend itself well to general noise calculations with differing source and load impedances, etc. Establishing the noise sources in Figure 25.6 supplements the data sheet numbers and allows more general noise calculations to be made.

Note 4: See (Rich 1988) and (Brisebois 2005) for more background on amplifier noise analysis.

Table 25.2 Equivalent Input and Output Noise at 100MHz Based on the Internal Noise Sources in Figure 25.5. The First Two Rows, e_n and i_n, are Calculated Input-Referred Voltage and Current Noise Components of the Amplifier Alone, Excluding the Noise of the Internal Resistors

PART NUMBER	LTC6400-8	LTC6400-14	LTC6400-20	LTC6400-26
e_n (nV/√Hz)	1.12	1.15	1.03	1.01
i_n (pA/√Hz)	4.00	4.02	2.34	2.57
$e_{n(OUT)}$ (nV/√Hz) (R_S = 0, No R_L)	9.4	12.7	22.7	28.2

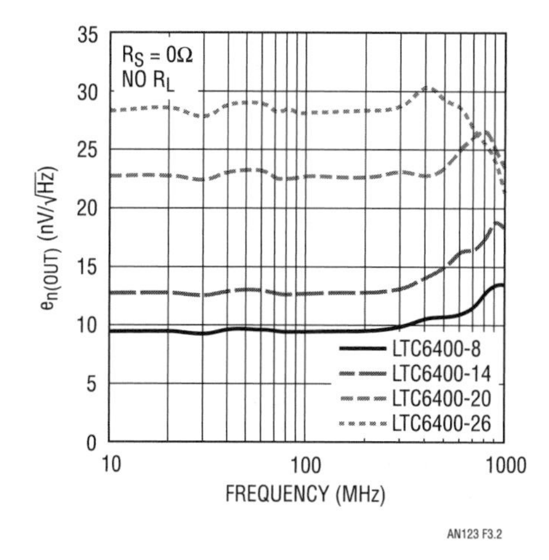

Figure 25.7 • Total Output Noise Voltage Density vs Frequency. The Output Noise is Measured with the Differential Inputs Shorted Together and No Resistive Load on the Amplifiers. The Noise Does Not Scale Linearly with the Gain Values Because the Internal Resistor Values are not the Same

For source impedance, Z_S, the total output noise can be estimated by power superposition of contributions from resistive part of Z_S, e_n and i_n. However this result could be misleading when Z_S is considerably high and/or frequency of interest is high.

There are some limitations to the data in Table 25.2 and its usefulness. Most importantly, e_n and i_n will have significant correlation, since they originate from the same physical noise sources inside the circuit. The RMS subtraction used here neglects this fact. The correlation of the voltage and current noise sources will have an impact on the accuracy of the calculations when $R_S > 100\Omega$ and i_n contributes a higher portion of the total noise at the output. This effect is examined in more detail in Appendix C.

Another limitation is that the e_n and i_n values obtained are only valid at 100MHz or below (down to the flicker noise corner frequency, approximately 12kHz), which is more than an order of magnitude lower than the −3dB bandwidth of the LTC6400. In a typical feedback amplifier, the output voltage noise density at higher frequencies can deviate significantly from the low frequency values (not considering 1/f noise). Figure 25.7 shows the output noise voltage of the LTC6400 family, which increase as the frequency approaches the −3dB bandwidth of the part. This increase in noise density is caused by the reduction in amplifier loop gain with frequency, which is affected by the internal amplifier gain, the compensation network, and the source and load impedances seen by the amplifier. At frequencies above the −3dB bandwidth, the output voltage noise density will fall with the gain of the amplifier.

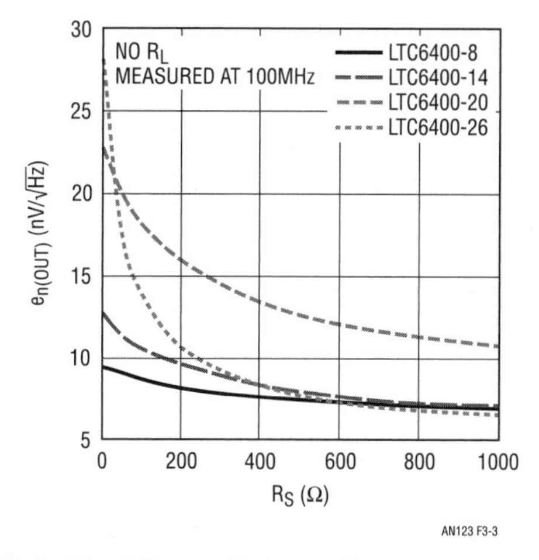

Figure 25.8 • Total Output Noise vs Source Resistance at 100MHz. The Source Resistance is Installed Differentially Across the Two Inputs, and the Output Voltage Noise Density is Measured Without Any Load Resistance on the Amplifier. Note the Trend of Decreasing Output Noise with Increasing Source Resistance

Noise and nf vs source resistance

Figure 25.8 illustrates the measured total output noise voltage for various source resistors at 100MHz. Notice that the LTC6400-8/LTC6400-14/LTC6400-26 output noise curves converge as R_S approaches 1k. This is because all three versions have 500Ω feedback resistors and very similar internal amplifiers. On the other hand, the LTC6400-20 presents a much higher output noise as R_S approaches 1k. This is because the LTC6400-20 has a 1k feedback resistor, which means the effective noise gain is larger than that of the other parts. In addition, the larger feedback resistor contributes more noise direct to the output.

When the output noise curves in Figure 25.8 are translated to noise figure, producing Figure 25.9 (Equation (A2), a slightly different story emerges. The amplifiers with a lower total output noise density (i.e., lower gains) do not necessarily have a lower noise figure. This is because NF is a measure of signal-to-noise ratio degradation, not absolute noise (see Appendix A). Also, the noise Figure curves shown in Figure 25.9 are not monotonic as R_S increases, but instead have a local minimum. There are two effects that are contributing to this result. First, the output voltage noise density of the amplifier levels off as R_S increases, but the source resistor noise continues to increase as R_S does. Second, Figure 25.10 and Figure 25.11 illustrate that the noise gain of the amplifier also decreases as the R_S value increases. So there are two terms increasing the noise figure and one term decreasing it, and the combination of these effects causes the NF to have a local minimum.

Another conclusion from Figure 25.11 is that low NF can be achieved with a certain range of R_S, but at the cost of overall voltage gain. The addition of R_{S1} and R_{S2} in Figure 25.10 contributes to additional input resistance, which lowers the gain. The conclusion is that it is not always desirable to use a source termination that yields the absolute lowest noise figure.

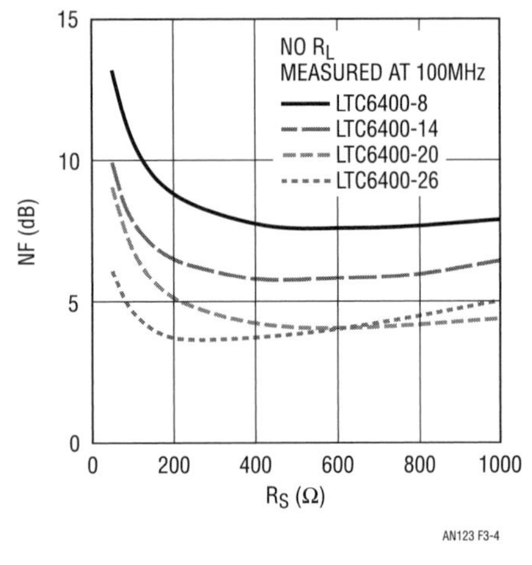

Figure 25.9 • Noise Figure vs Source Resistance of the LTC6400. For Each Member of the Family, There is a Range of Source Resistances That Provide the Lowest Noise Figure Values

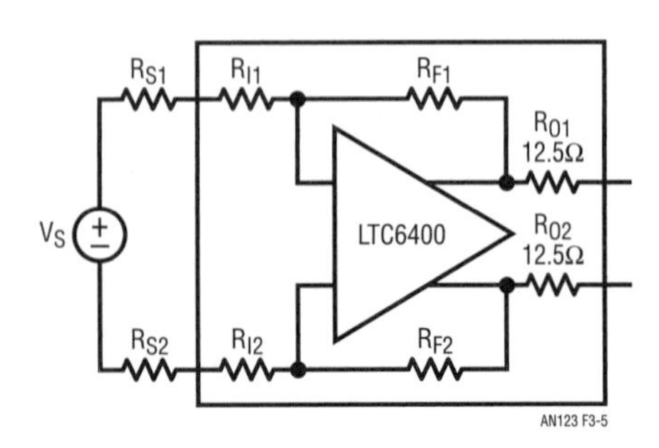

Figure 25.10 • Diagram Showing the Effect of Increased Source Resistance on Overall Voltage Gain. The Source Resistance Adds to the Input Resistance, Which Decreases the Gain of the Amplifier

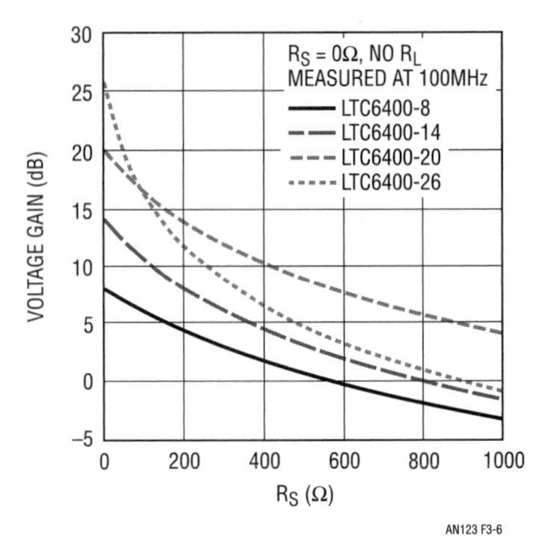

Figure 25.11 • Voltage Gain vs Source Resistance for the LTC6400 Family. The Voltage Gain Defined Here is Calculated from the Source Voltage, V_S, in Figure 25.10 and Assumes No Resistive Load at the Output of the Amplifier

Noise and gain circles

The curves in Figure 25.9 show the effect of varying a real source resistance on noise and noise figure, but what happens if a complex source impedance is used? Noise circles are related to the concept of "minimum noise figure," where there is a certain complex input impedance that will yield the minimum noise figure for a given device at a given frequency, temperature and bias. It is also possible to plot circles of constant noise figure on the same Smith chart, so that any complex source impedance on each circle will yield the same NF For a given source impedance Z_S, the noise factor of a 2-port system can be expressed relative to the minimum noise factor (Fukui 1981):

$$F = F_{MIN} + \frac{G_N}{R_S} \mid Z_S - Z_{OPT} \mid^2 \quad (1)$$

where F_{MIN} is minimum achievable noise factor, G_N is the device's equivalent noise conductance, R_S is the resistive part of Z_S, and Z_{OPT} is the source impedance required to achieve the minimum noise factor. Figure 25.12 shows the noise circles for the LTC6400-8, and Table 25.3 lists the noise parameters for the entire LTC6400 family. The values indicate that inductive source impedances (with positive reactance) yield the optimal NFs for the LTC6400 family. However, when applying noise matching, it is important to remember that other performance factors such as gain, bandwidth and impedance mismatch are also affected by Z_S.

Table 25.3 LTC6400 Complex Noise Parameters Measured at 100MHz. These Parameters are the Basis for the Noise Figure Circles in Figure 25.12, and Allow the Precise Calculation of NF with Any Complex Source Impedance

	LTC6400-8	LTC6400-14	LTC6400-20	LTC6400-26
NF_{MIN} (dB)	7.12	5.6	4.01	3.55
G_N (mS)	1.13	2.45	2.86	7.23
Z_{OPT} (Ω)	531 + j306	440 + j131	516 + j263	272 + j86

One of the other key factors affected by source impedance is overall gain, and Figure 25.13 shows a set of gain circles for the LTC6400-8 plotted on the Smith chart. Transducer gain (G_T) is simply the ratio between power delivered to the load and power available from the source.

Figures 25.12 and 25.13 illustrate the trade-off between noise figure and transducer gain. Looking at the two plots, the minimum NF and maximum G_T cannot be achieved simultaneously, since they are not located in the same spot on the Smith chart. One common strategy is to connect the two optimal points ($\Gamma_{S,OPT}$ and $\Gamma_{S,GT(MAX)}$) with a straight line and pick a source termination on that line, which comes close to optimizing both NF and G_T. However, in many cases the absolute minimum NF or maximum G_T is not truly necessary. Looking at Figures 25.12 and 25.13, there is a large area of the input reflection Smith chart that will achieve within 1dB of both optimal points. There is even a significant portion of the real axis available (including 400Ω), where reactive elements are not necessary, that would provide good performance as well as a wideband impedance match.

Signal-to-noise ratio vs bandwidth

Amplifiers such as the LTC6400 family are typically specified in noise voltage density, with units of nanovolts per root Hertz. When used in practical applications, the question often arises about how to interpret this noise metric to figure out exactly how much performance can be achieved in a system design. Depending on the system, there are many ways to characterize the effect of noise on a signal, but one of the most widely used is signal-to-noise ratio (SNR). SNR is simply a ratio of the maximum possible signal to the total amount of noise present, which determines the dynamic range of the system. The implication is that a

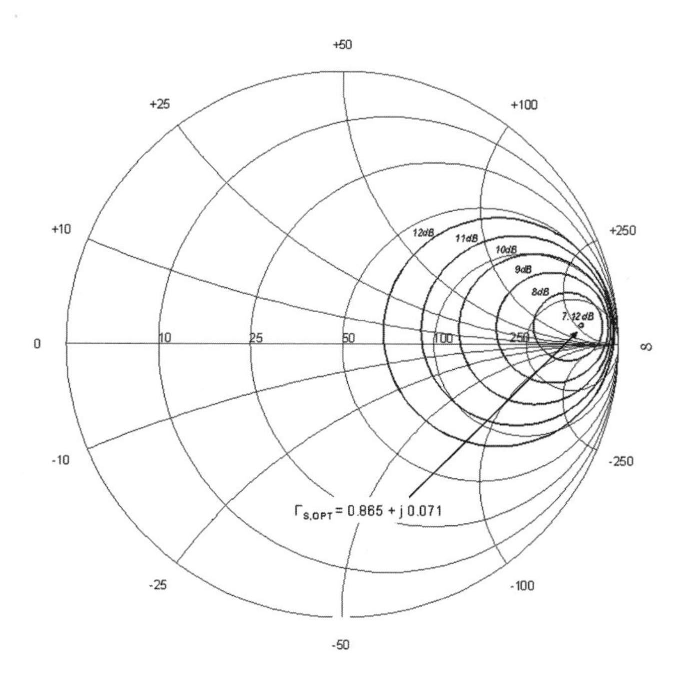

Figure 25.12 • LTC6400-8 Noise Circles on the S11 Reflection Plane at 100MHz

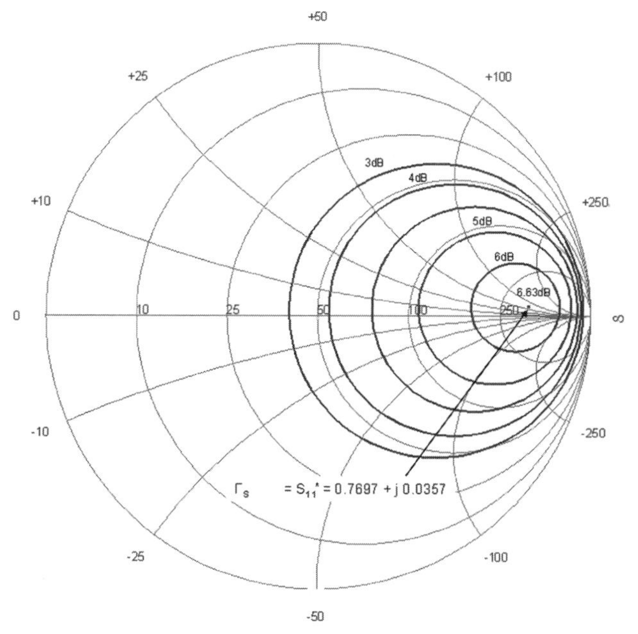

Figure 25.13 • LTC6400-8 Gain Circles on the S11 Reflection Plane at 100MHz with R_L = 375Ω

signal small enough to be "in the noise floor" is not detectable, which is not always true, but SNR allows different designs to be compared side by side.

The LTC6400 family consists of very broadband amplifiers, with −3dB bandwidths over 1GHz. Making the assumption that all the noise is white noise, meaning the noise power density is constant with frequency, the total amount of integrated noise can be calculated by multiplying the noise voltage density by the square root of the total bandwidth.

$$E_{n,TOTAL} = e_n(nV/\sqrt{Hz}) \bullet \sqrt{\alpha \bullet BW}\ nV_{RMS} \quad (2)$$

where α *= Scale factor to include the noise outside the BW, approximately 1.57 for a 1st order roll off and 1.11 for a 2nd order roll off.*

Wide bandwidth amplifiers like the LTC6400 can have a significant impact on SNR. With a −3dB bandwidth of 1GHz, e_n is multiplied by 40,000 in Equation (2), so a $10nV/\sqrt{Hz}$ amplifier output would contribute $400\mu V_{RMS}$ of integrated noise. If the maximum output signal is $2V_{P-P}$ ($0.71V_{RMS}$), as is the case for a typical high speed ADC input, then the maximum theoretical SNR is 1768, or 65dB. This is the SNR of an ADC with 10.5 bits of resolution.

As shown in Table 25.2, the higher gain versions of the LTC6400 have output noise voltage densities up to $28nV/\sqrt{Hz}$. Although the high gain implies that the input-referred noise is very low, the output noise can be significant. The trend in signal conditioning is toward lower supply voltages, and so the maximum signal level (and thus the achievable SNR) is shrinking. For this reason, when driving a 16-bit (or even a 14-bit) ADC, it is almost always desirable to add some filtering at the output of the LTC6400. The design of this external filter depends on the desired input signal bandwidth and the target SNR that is required.

$$SNR_{TARGET} = 20 \bullet \log\left(\frac{V_{SIGNAL}}{E_{n,TOTAL}}\right) dB \quad (3)$$

where V_{SIGNAL} *= maximum input signal in* V_{RMS}*, which can be up to 0.88V for a high performance 3V ADC (2.5Vp-p).*

This SNR calculation ignores the contribution of noise from the ADC. The SNR from the amplifier can be one of the dominant factors in determining the effective ADC resolution. Due to the LTC6400's excellent distortion performance, it is often paired with high performance 14-bit and 16-bit ADCs, and so the bandwidth should be limited in order to achieve the best overall performance.

An important phenomenon that occurs in ADCs is aliasing, which effectively "folds" frequency bands on top of each other, making them indistinguishable in the digital domain. Aliasing does not change the SNR calculation above, but it does affect the ability of any additional digital filtering to improve the SNR. If all of the noise is contained in one Nyquist bandwidth (the width of which is $f_{SAMPLE}/2$), then additional digital filtering can remove noise around the desired signal. However, if many Nyquist bandwidths of noise are aliased together with the desired signal band, then the beneficial effect of digital filtering would be diminished.

A sample calculation along with some further explanation of the noise aliasing phenomenon appears in the SNR Calculation and Aliasing Example section.

Gain and power options

For application flexibility, the LTC6400 and LTC6401 product families come in four different gain versions: 8dB (2.5V/V), 14dB (5V/V), 20dB (10V/V) and 26dB (20V/V). The voltage gain is specified in decibels to maintain consistency with other RF/HF components in the signal chain. Figure 25.14 shows the block diagram of the LTC6400/LTC6400-1.

The gains specified are voltage gains from the input pins to the output pins. Unlike some amplifiers that specify power gain with a known load impedance value, the LTC6400 specifies a voltage gain that does not imply any specific input or output termination.

In addition to the choice of voltage gain, the user also has the choice between the LTC6400 and LTC6401 parts. This choice is mainly a speed/power trade-off; the LTC6400 is the fastest, lowest distortion amplifier and the LTC6401 is a lower power version optimized for lower input frequencies. The LTC6400 maintains good distortion performance out to 300MHz, and the LTC6401 does so out to 140MHz.

Gain, phase and group delay

Providing the gain and feedback resistors internally also enables optimization in the compensation networks within each amplifier in the product family. Therefore, each gain variation of the LTC6400 has similar bandwidth (>1GHz) and minimal peaking in the gain response. Although the usable low distortion bandwidth of the products are much lower in frequency than the −3dB bandwidths, in certain situations gain flatness and phase linearity can be important. Figure 25.15 shows the gain of the LTC6400-20 to be flat within 0.1dB out to approximately 300MHz, and the phase to be linear out to beyond 1GHz. Table 25.4 summarizes the bandwidths of each LTC6400 and LTC6401 version.

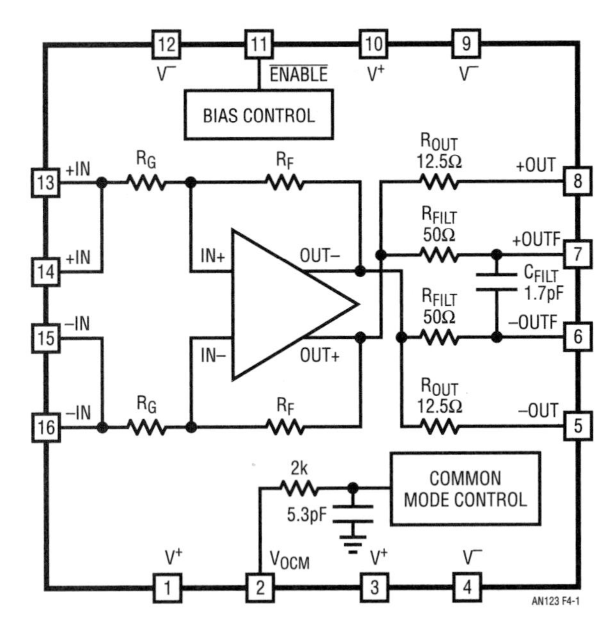

VOLTAGE GAIN	R_G	R_F
8dB	200Ω	500Ω
14dB	100Ω	500Ω
20dB	100Ω	1kΩ
26dB	25Ω	500Ω

FAMILY	SUPPLY CURRENT	INPUT FREQUENCY FOR LOW DISTORTION
LTC6400	85mA TO 90mA AT 3V	DC – 300MHz
LTC6401	45mA TO 50mA AT 3V	DC – 140MHz

Figure 25.14 • LTC6400/LTC6401 Block Diagram and Product Information Tables for the Various Gain and Speed/Power Options Available

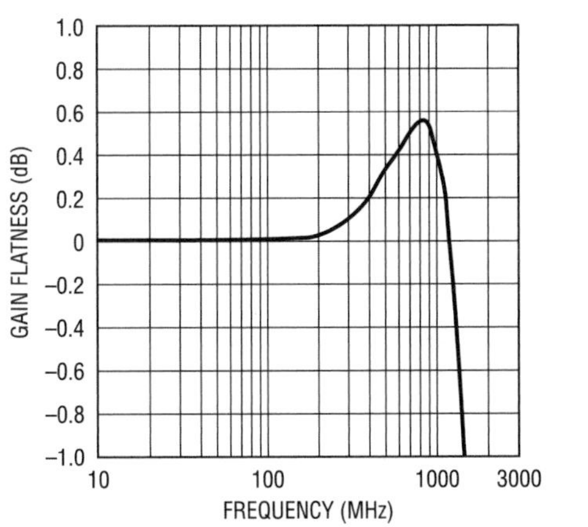

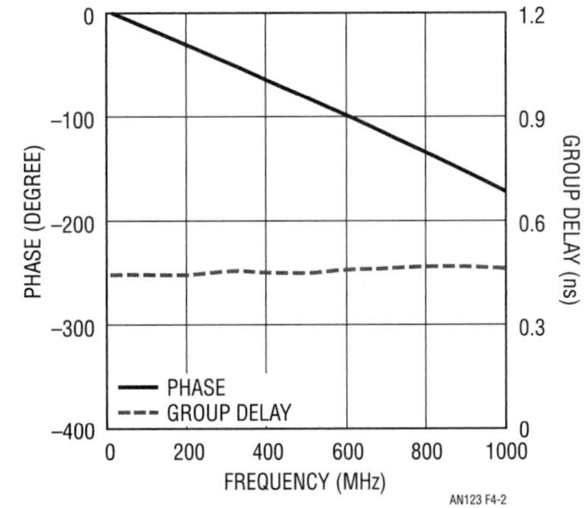

Figure 25.15 • Gain Flatness and Phase/Group Delay Plots of the LTC6400-20

Gain of 1 configuration

In some situations, a simple buffer may be required to drive an ADC with no gain or even attenuation. In this special case, series resistors can be used in front of the LTC6400-8 or LTC6401-8 to lower the gain to 0dB (1V/V). Due to the unconditional stability of the LTC6400, these resistors at the input will not cause instability or oscillation.

Figure 25.16 demonstrates the configuration. When combined with the internal gain resistors of the LTC6400-8, the amplifier becomes a unity-gain buffer with 1kΩ differential input impedance.

When combining external series resistors to lower the gain, there will be some temperature and initial accuracy limitations due to the characteristics of the LTC6400 family's internal resistors. While the internal resistors (200Ω

Table 25.4 Typical Bandwidth and 0.1dB/0.5dB Gain Flatness Frequencies for the LTC6400/LTC6401 Product Families

Product	-3dB Bandwidth	0.1dB Bandwidth	0.5dB Bandwidth
LTC6400-8	2.2GHz	200MHz	430MHz
LTC6400-14	2.37GHz	200MHz	377MHz
LTC6400-20	1.84GHz	300MHz	700MHz
LTC6400-26	1.9GHz	280MHz	530MHz
LTC6401-8	2.22GHz	220MHz	430MHz
LTC6401-14	1.95GHz	230MHz	470MHz
LTC6401-20	1.25GHz	130MHz	250MHz
LTC6401-26	1.6GHz	220MHz	500MHz

and 500Ω in Figure 25.16) are designed to match well over process variations and temperature, their absolute values can vary as much as ±15% (as guaranteed by the LTC6400-8 data sheet). A ±15% variation in the internal resistors, when combined with an ideal 301Ω resistor in our example, will result in less than ±10% variation in the desired gain of the amplifier over temperature and process variations. This does not take into account temperature effects in the external resistor, which should not be expected to track with the internal resistors.

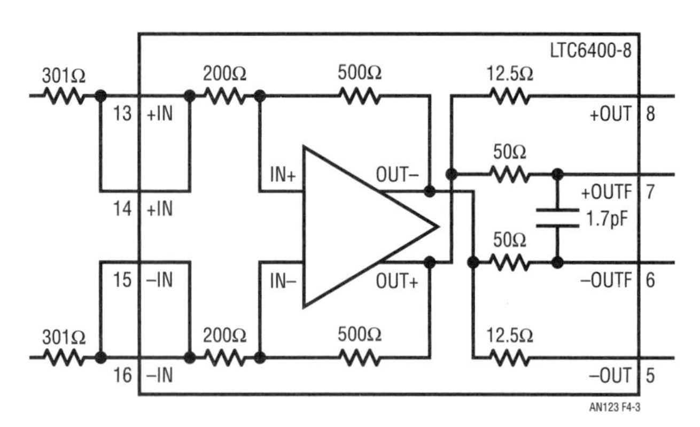

Figure 25.16 • LTC6400-8 with Series Resistors to Lower the Gain to 1V/V (0dB)

Input considerations

Input impedance

The input impedance of a differential amplifier circuit depends on whether it is driven single-ended or differentially. This is significant when performing an impedance match to the amplifier for optimal power transfer or to maximize the voltage signal at the input.

Figure 25.17 shows the LTC6400 with a differential input. To calculate the input impedance, we need to calculate the input current I_{IN} flowing through the input voltage source, V_{IN}. With a fully-differential input and a high gain amplifier, there is very little voltage developed across the internal nodes (INT^+, INT^-). Therefore these nodes act as a differential "virtual short", in a similar fashion to the inverting node of a traditional op amp. Thus, the input impedance is simply the sum of R_{I1} and R_{I2}.

Figure 25.18 shows the LTC6400 with a single-ended input. Assuming that the input signals are high enough in frequency that the DC blocking capacitors C1 and C2 look effectively like short circuits (these are necessary because of the input common mode voltage requirements of the LTC6400), the assumption of no voltage at the INT^+ and INT^- nodes is no longer valid. The *differential* voltage across (INT^+, INT^-) will still be small, but there will be a *common mode* voltage at the two nodes that is proportional to the input voltage.

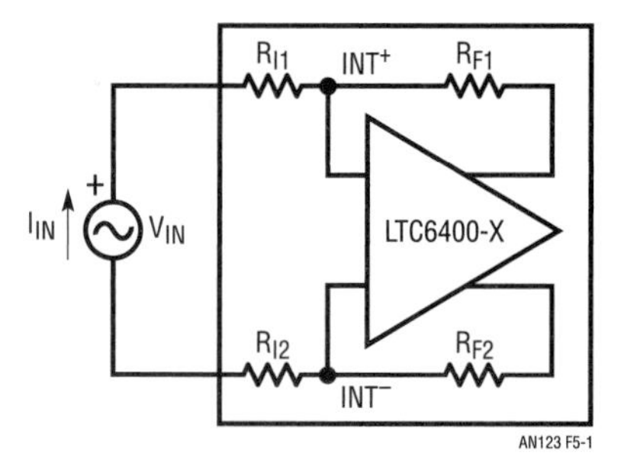

Figure 25.17 • LTC6400 with a Fully Differential Input

$$V_{INT^+} \approx V_{INT^-} \approx V_{OUT} + \bullet \frac{R_{I2}}{R_{F2} + R_{I2}} \tag{4}$$

$$= V_{IN} \bullet \frac{R_{F2}}{2R_{I2}} \bullet \frac{R_{I2}}{R_{F2} + R_{I2}} = V_{IN} \bullet \frac{R_{F2}}{R_{F2} + R_{I2}}$$

$$I_{IN} = \frac{V_{IN} - V_{INT+}}{R_{I1}} = \frac{V_{IN}}{R_{I1}} \bullet \frac{R_{F2} + 2R_{I2}}{2\left(R_{F2} + R_{I2}\right)} \tag{5}$$

$$\text{Input Impedance} = V_{IN}/I_{IN} = \frac{2R_{I1}\left(R_{F2} + R_{I2}\right)}{R_{F2} + 2R_{I2}} \tag{6}$$

As an example, with the LTC6400-20, which has 100Ω input resistors and 1k feedback resistors, the single-ended input impedance is 183Ω. This would be the input impedance seen by V_{IN} in the example. Note that in the case of a V_{IN} with non-zero source impedance, such as a 50Ω signal source, it is beneficial to terminate the unused input with the same effective impedance to maintain balance. This will change the input impedance, and an updated formula can be found in the Resistor Termination section.

Ac coupling vs DC coupling

The LTC6400 is tolerant of AC coupling or DC coupling at both the input and the output, but it does have a range of inputs and output voltages that must be adhered to for best performance. At the input, the common mode input voltage range is defined in the data sheet to be approximately 1V to 1.6V (with $V^+ = 3V$). The common mode input voltage is defined as the *average* voltage of the two inputs, and is separate from the *difference* voltage between the inputs. Input voltages that are centered at ground or at

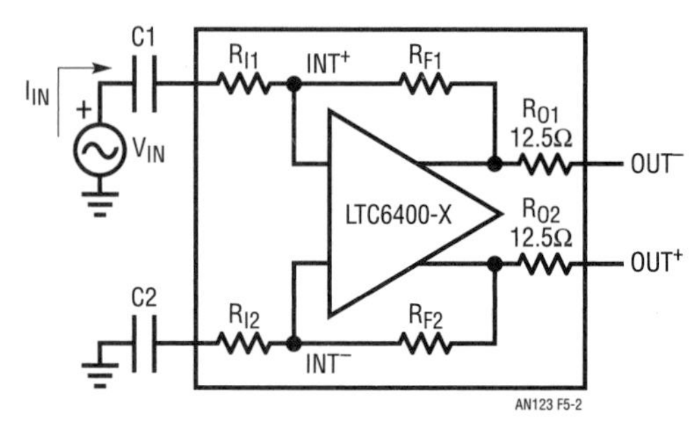

Figure 25.18 • LTC6400 with Single-Ended Input

V_{CC} must therefore be level-shifted before being applied to the LTC6400. As a point of clarification, the common mode input voltage refers to the voltage at input pins of the IC (Pins 13 to 16), not the *internal* input pins of the op amp which are not accessible. Table 25.5 shows the differences between the data sheet common mode limits and the resultant internal node voltages (see Figure 25.17) that result. The data sheet limits are published with a set V_{OCM} of 1.25V, and assuming no additional source resistance. Changing either would shift the DC voltage at the internal nodes, which are the real determinant of the input common mode limits. However, following the common mode limits established in the data sheet only ensure that the input stage operates in the linear region; it does not imply the same performance will be achieved across the entire voltage range.

The V_{OCM} Requirements section addresses the change in distortion for different common mode bias voltages, both at the input and at the output.

The limits in Table 25.5 are based on saturation limits of the input stage, and can be interpreted for the general case of an arbitrary V^+ and V_{OCM}. Changing the V^+ voltage increases the input stage headroom, and changing V_{OCM} changes the bias at the internal nodes, so both will affect the input common mode voltage limits. Referring to the values in Tables 25.5 and 25.6, the input common mode voltage limits can be calculated as:

$$V_{CM,IN(MIN)} = \beta(V_N + V^- - \alpha V_{OCM} - \delta V^+) \quad (7)$$

$$V_{CM,IN(MAX)} = \beta(V_P + (1 - \delta)V^+ - 3.0 - \alpha V_{OCM}) \quad (8)$$

Table 25.5 Limits of Input Common Mode Voltage as Published in the Data Sheet, and the Translated Common Mode Voltage Limits at the Amplifier's Internal Nodes. Knowing the Amplifier's Internal Node Voltages Allows a Design to Stay Within Data Sheet Limits Even with Alternative Configurations. The Internal Node Voltages V_N and V_P Include the Effect of a Small Pull-Up Current in the Amplifier.

		LTC6400-8	LTC6400-14	LTC6400-20	LTC6400-26
Specification from Data Sheet	Input Common Mode Voltage Minimum (I_{VRMIN})	V– + 1.0	V– + 1.0	V– + 1.0	V– + 1.0
	Input Common Mode Voltage Minimum (I_{VRMAX})	1.8 (V+ – 1.2V)	1.8 (V+ – 1.2V)	1.6 (V+ – 1.4V)	1.6 (V+ – 1.4V)
Internal Node Voltage Limits (INT+, INT–)	Common Mode Voltage Minimum (V_N)	1.24	1.14	1.05	1.02
	Common Mode Voltage Maximum (V_P)	1.76	1.78	1.59	1.59
		LTC6401-8	**LTC6401-14**	**LTC6401-20**	**LTC6401-26**
Specification from Data Sheet	Input Common Mode Voltage Minimum (I_{VRMIN})	V– + 1.0	V– + 1.0	V– + 1.0	V– + 1.0
	Input Common Mode Voltage Minimum (I_{VRMAX})	1.6 (V+ – 1.4)	1.6 (V+ – 1.4)	1.6 (V+ – 1.4)	1.6 (V+ – 1.4)
Internal Node Voltage Limits (INT+, INT–)	Common Mode Voltage Minimum (V_N)	1.16	1.09	1.05	1.03
	Common Mode Voltage Maximum (V_P)	1.57	1.58	1.59	1.59

Table 25.6 Constants Used to Calculate Input Common Mode Voltage Limits, Based on the Information Provided in the Data Sheet. Changing the Supply Voltage V^+ and the V_{OCM} Bias Voltage will Affect the Limits Published in the Data Sheet Electrical Tables. These Constants Include the Effect of a Small Pull-Up Current at the Internal Nodes of the Amplifier.

	LTC6400-8	LTC6400-14	LTC6400-20	LTC6400-26	LTC6401-8	LTC6401-14	LTC6401-20	LTC6401-26
Alpha (α)	0.261	0.158	0.090	0.047	0.273	0.162	0.090	0.047
Beta (β)	1.533	1.267	1.117	1.054	1.467	1.233	1.117	1.058
Delta (δ)	0.087	0.053	0.015	0.004	0.045	0.027	0.015	0.008

When calculating the input minimum and maximum voltages with the provided equations, it is important to keep the graphs in Figure 25.30 handy, as a cross-reference. The distortion performance of the LTC6400 family is much less sensitive to the input common mode voltage than the output common mode voltage, but the distortion performance will not be constant over the entire range.

If the inputs are AC-coupled with series capacitors, the inputs of the LTC6400 will self-bias to approximately the same voltage as V_{OCM},[5] and there is no need to apply an external bias voltage to the part. It is only when DC coupling the input that the design needs to address the biasing of the inputs. For more information about performance optimization and the input common mode voltage, see the V_{OCM} Requirements section.

When changing the input and output common mode voltages, it is important to be aware of changing input bias currents. Referring to the block diagram in Figure 25.3, DC bias currents will flow from the outputs back to the inputs through the gain and feedback resistors. The source must be able to source or sink this additional current, which can exceed a milliamp, when setting the DC bias of the LTC6400's input.

The outputs of the LTC6400 will automatically bias to the voltage at the V_{OCM} pin due to an internal common mode loop. This simplifies the task of mating the LTC6400 to Linear Technology's high performance 14-bit and 16-bit ADCs, because the common mode voltage requirements of the ADCs are typically very stringent, and the common mode rejection of the ADC inputs is often not very good.

Ground-referenced inputs

A common case of the DC-coupled application includes having a single-ended ground-referenced input for the ADC driver, where the DC voltage level of the input is at 0V The LTC6400 has an input common mode range that does not include ground, so level-shifting of some sort is necessary to bring the voltages within the limits shown in Table 25.5 and Table 25.6.

If the input source is 50Ω terminated, one possible solution is shown in Figure 25.19. The inputs are pulled up with 75Ω resistors to the supply. This circuit takes advantage of the source's 50Ω termination to create a voltage divider and effectively level shift the input to within the optimal range for the LTC6400. The 75Ω resistors also act as impedance match resistors, transforming the single-ended input impedance of the LTC6400 to a value close to 50Ω.[6] Therefore, the extra pull-up resistors do not attenuate the input signal any more than necessary. Due to the LTC6400-26's lower intrinsic input impedance, the pull-up resistors should be changed to 100Ω, which will yield an input impedance of 42Ω.

Another method for level shifting employs two LTC6400's in series: one to level shift the signal, and another to add gain. Figure 25.20 shows an LTC6400-8 with 1.1k series input resistors to level shift and attenuate the signal. The internal DC bias level is within the LTC6400's range. At the output, an LTC6400-20 amplifies the signal for the ADC. The 200Ω input impedance is not a heavy load for the LTC6400-8. The total gain from the two amplifiers together is 10.7dB.

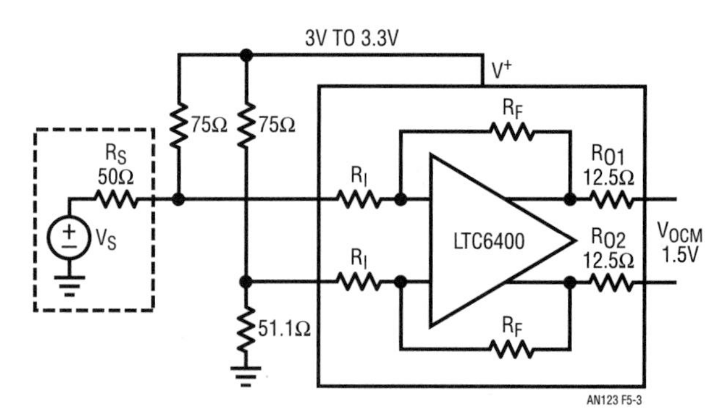

Figure 25.19 • Using Pull-Up Resistors at the Input to Level-Shift a Ground-Referenced Signal and Provide Input Impedance Matching at the Same Time. The Unused Input is Similarly Terminated. The Drawback with this Approach is that a DC Bias Current Will Flow Through the Pull-Up Resistors, Dissipating Extra Power and Requiring the Input Source to Sink the Current

Note 5: In reality, the input common mode voltage is increased slightly by internal pull-up resistors to match the optimal values of input and output bias.

Note 6: The input impedance will be 61Ω with the LTC6400-8, 53Ω with the LTC6400-14 and 55Ω with the LTC6400-20. The impedance can be lowered by lowering the 75Ω resistor values at the expense of power dissipation.

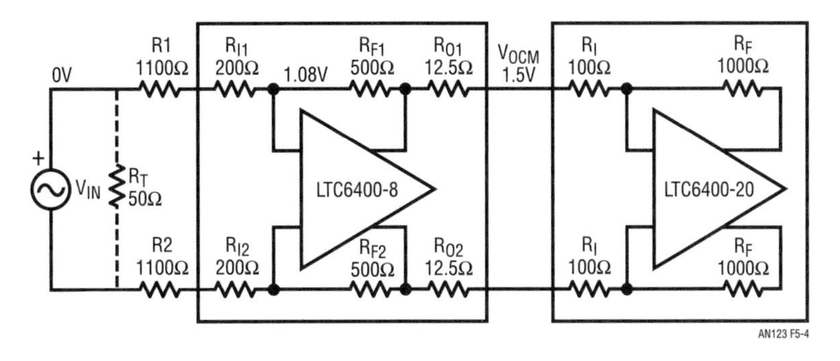

Figure 25.20 • Use Two LTC6400 Amplifiers to Level Shift and Amplify an Input Signal. The LTC6400-8 has Series Input Resistors to Attenuate and Level Shift the Input, and the LTC6400-20 Adds Additional Gain at the Output. The Total Gain of this Circuit is 10.7dB, Including the Attenuation from the LTC6400-8 Input Resistors and the Effect of the 12.5Ω Output Resistors

Impedance matching

Impedance matching is the art of achieving maximum power transfer from a source to a load. Achieving a good impedance match is beneficial for many reasons: maximum signal reception, no reflections to cause signal distortion, and good predictable behavior from the system. In this section, the LTC6400 family's input and output characteristics are discussed to provide the information necessary for impedance matching. A primer on impedance matching is outside of the scope of this note, but there are myriad texts and resources available that discuss it at length.[7]

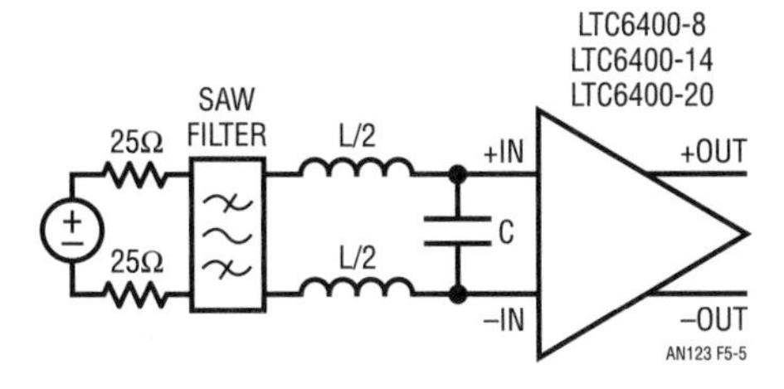

Figure 25.21 • Example of Series-Shunt Input Impedance Matching to a SAW Filter. The LTC6400-26 has an Input Impedance of 50Ω, and Typically does not Require any External Elements for a 50Ω Impedance Match. Using an LC Network Provides an Impedance Match at a specific Frequency; for a Wideband Impedance Match, other Methods Such as Transformers and Resistors are Necessary

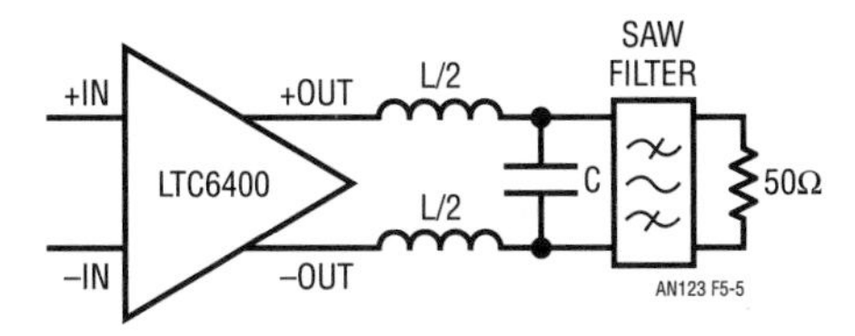

Figure 25.22 • Example of Series-Shunt Output Impedance Matching to a SAW Filter. All Members of the LTC6400 Family have a 25Ω Resistive Output Impedance at Lower Frequencies

Note 7: For impedance matching and general RF design, see (Bowick 1982).

The LTC6400 family provides differential input impedances ranging from 50Ω to 400Ω, and a differential output impedance of 25Ω. In applications where impedance matching is desired, such as when receiving or driving a SAW filter, a simple series L and shunt C network will often suffice. Figure 25.21 presents an example circuit to interface to a SAW filter. At operating frequencies below 100MHz, the LTC6400's input and output impedances are almost purely resistive. At frequencies above that, the reactance must be taken into account. Table 25.7 lists the relevant parameters for impedance matching, assuming the circuits in Figure 25.21 and Figure 25.22. To calculate the C and L values, divide the ωC and ωL values by ω (or 2πf), where ω is the frequency in radians/sec and f is the frequency in Hertz. Figure 25.23 shows the impedance match on the Smith chart.

Figure 25.24 and Figure 25.25, which show the input and output reflection coefficients from 10MHz to 1GHz, can provide the right impedance matching circuit to obtain a better impedance match at high frequencies. Note that the input impedance (of the LTC6400-8/LTC6400-14/LTC6400-20) becomes capacitive and the output impedance becomes inductive as the frequency increases above 100MHz.

Table 25.7 Matching Circuit Parameters for the Circuits in Figure 25.21 and Figure 25.22. To Obtain the Capacitor and Inductor Values from this Table, Which are the Frequency-Independent ωC/ωL Values, Simply Divide the Number by ω (or 2πf). Since the Series Inductors are Differential, Each Inductor Will be Half of the Resultant Value. The LTC6400-26 Does Not Require Matching Elements, as its Input Impedance is 50Ω

	INPUT		OUTPUT
	LTC6400-8	LTC6400-14/20	LTC6400 (all) Unfiltered Output
ωC	0.00661	0.00866	0.02
ωL	132	86.6	25

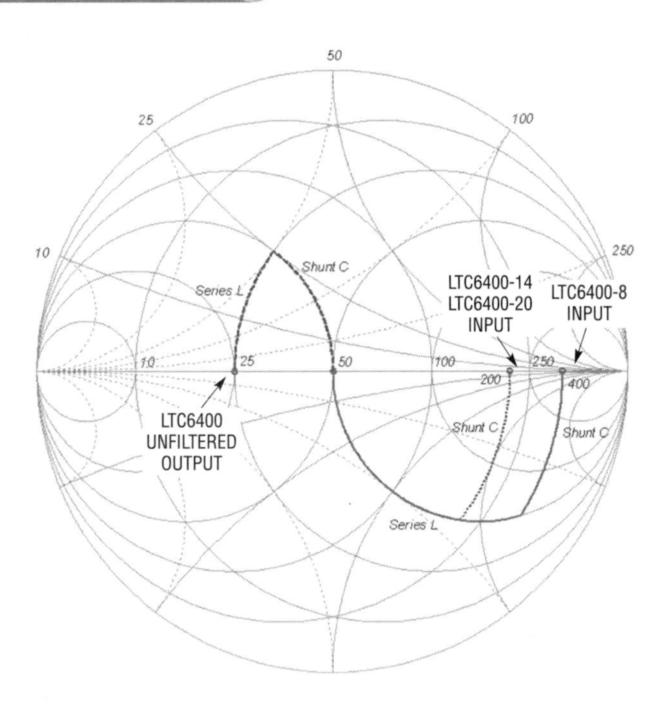

Figure 25.23 • Differential Input and Output Impedance Matching to 50Ω. In Most Matches, a Series Inductor/Shunt Capacitor Network will Yield a Satisfactory Impedance Match

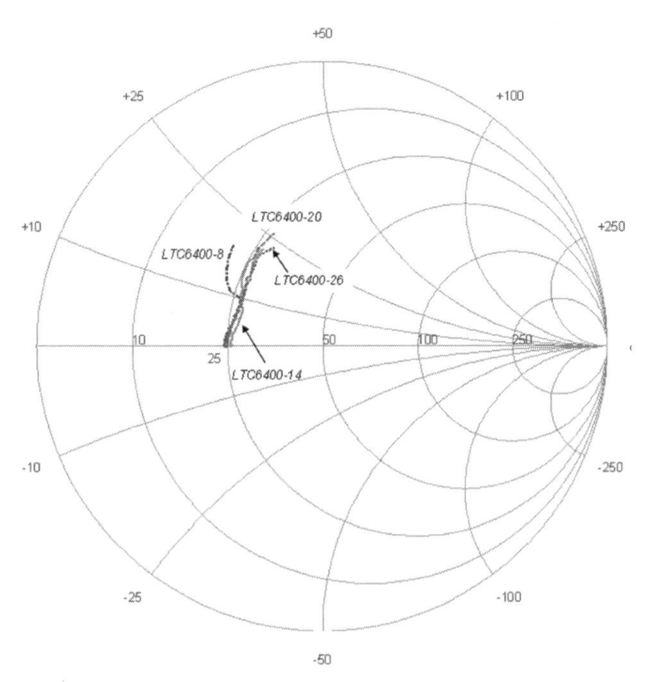

Figure 25.25 • Output Reflection Coefficients (S22) of the LTC6400 Family from 10MHz to 1GHz. Below 100MHz or so, the Impedances are Almost Purely Resistive. Above that, the Reactances Must be Considered for Impedance Matching

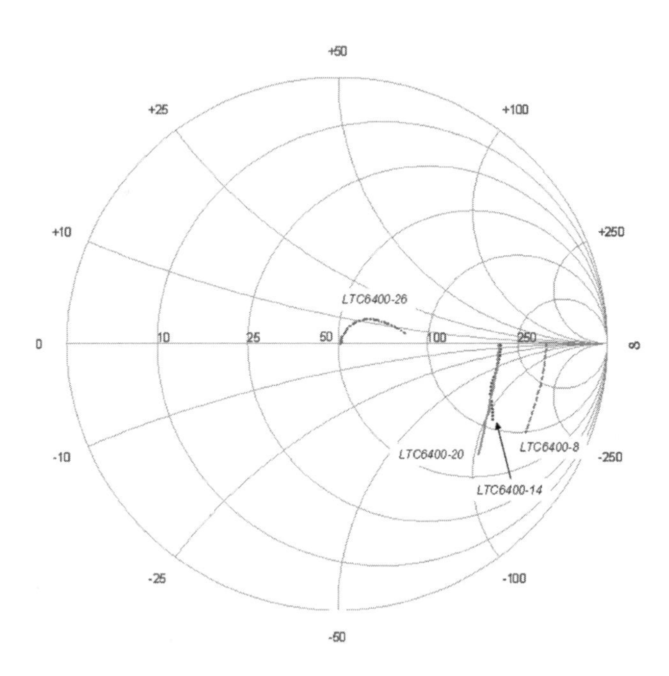

Figure 25.24 • Input Reflection Coefficients (S11) of the LTC6400 Family from 10MHz to 1GHz. Below 100MHz or so, the Impedances are Almost Purely Resistive. Above that, the Reactances Must be Considered for Impedance Matching

Input transformers

The LTC6400 works well with a variety of RF transformers at the input. Transformers are commonly used on LTC6400 demo boards as baluns and impedance-match elements for ease of evaluation. Figure 25.26 shows the schematic of the DC987B standard demo board of the LTC6400. At the input, a 4:1 wideband transmission-line transformer (TCM4-19+) is used for two purposes: to match a 50Ω signal source to the 200Ω input impedance of the IC, and also to convert the single-ended input signal into a differential signal for evaluation. The performance of the LTC6400 as a single-ended-to-differential converter is very good, but for even better performance it should be used with differential inputs.

At the output, another TCM4-19 transformer is used for converting the differential output to a single-ended output and impedance matching a 50Ω load (such as a network or spectrum analyzer). The load will see a 50Ω source impedance, and the amplifier sees a benign 400Ω load impedance. The LTC6400 is designed to drive higher impedance loads, and will not exhibit the same low distortion performance when driving heavy loads (e.g., 50Ω).

For ADC-driving applications, the LTC6400's differential output will connect to the ADC input either directly or through a discrete filter, but typically will not require an output transformer. At the input of the LTC6400, it will still be desirable in many applications to use input

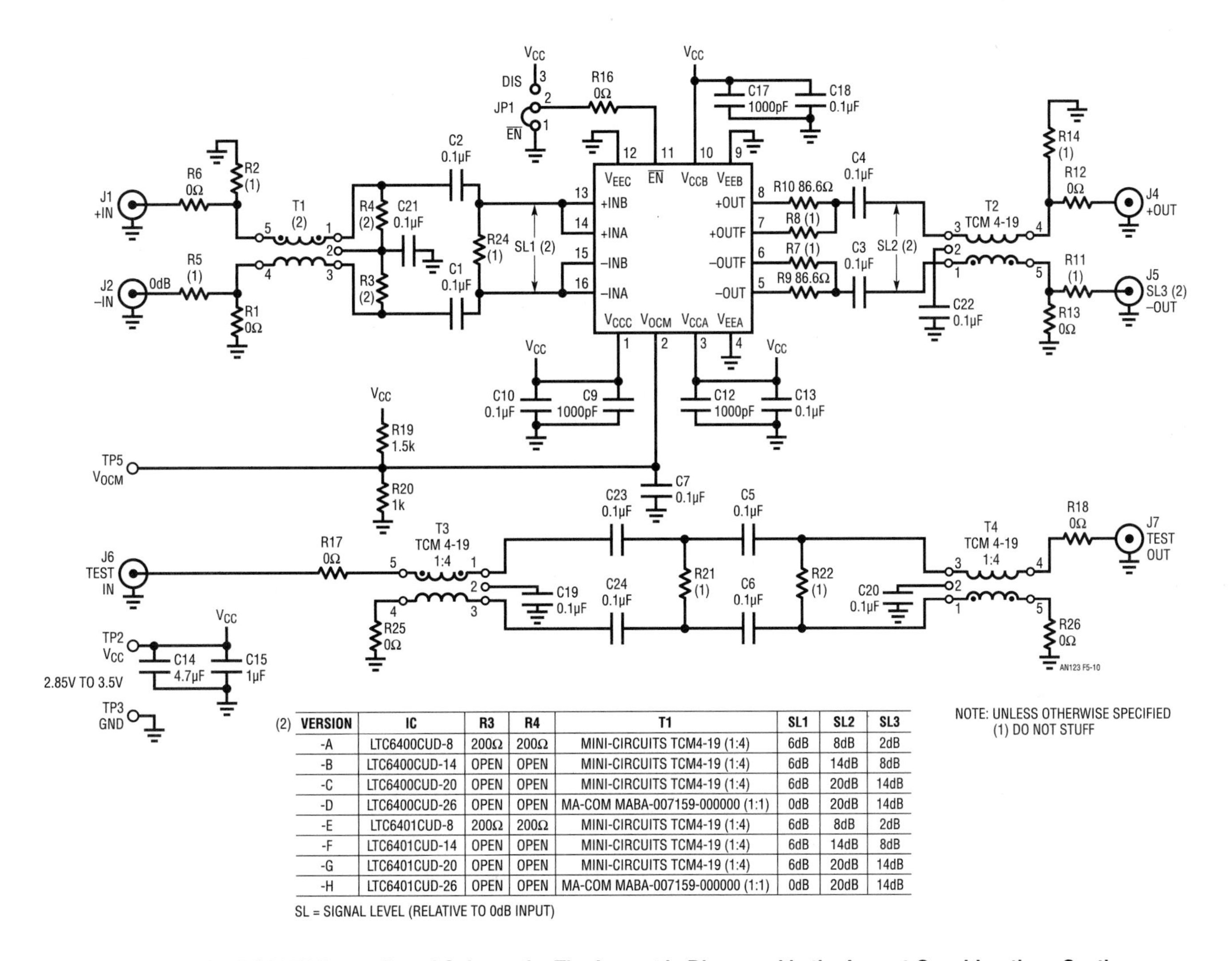

(2) VERSION	IC	R3	R4	T1	SL1	SL2	SL3
-A	LTC6400CUD-8	200Ω	200Ω	MINI-CIRCUITS TCM4-19 (1:4)	6dB	8dB	2dB
-B	LTC6400CUD-14	OPEN	OPEN	MINI-CIRCUITS TCM4-19 (1:4)	6dB	14dB	8dB
-C	LTC6400CUD-20	OPEN	OPEN	MINI-CIRCUITS TCM4-19 (1:4)	6dB	20dB	14dB
-D	LTC6400CUD-26	OPEN	OPEN	MA-COM MABA-007159-000000 (1:1)	0dB	20dB	14dB
-E	LTC6401CUD-8	200Ω	200Ω	MINI-CIRCUITS TCM4-19 (1:4)	6dB	8dB	2dB
-F	LTC6401CUD-14	OPEN	OPEN	MINI-CIRCUITS TCM4-19 (1:4)	6dB	14dB	8dB
-G	LTC6401CUD-20	OPEN	OPEN	MINI-CIRCUITS TCM4-19 (1:4)	6dB	20dB	14dB
-H	LTC6401CUD-26	OPEN	OPEN	MA-COM MABA-007159-000000 (1:1)	0dB	20dB	14dB

SL = SIGNAL LEVEL (RELATIVE TO 0dB INPUT)

Figure 25.26 • DC987B Demo Board Schematic. The Layout is Discussed in the Layout Considerations Section

transformers for impedance conversion and single-ended-to-differential conversion. In the case of the LTC6400-8, LTC6400-14 or LTC6400-20, the amplifiers have 200Ω or 400Ω input impedances. If the signal source has a 50Ω characteristic impedance, then the use of a 4:1 or 8:1 transformer for impedance matching has an important benefit. The transformers will provide some "free" voltage gain compared to other methods of impedance matching, such as a shunt resistor. Since a 4:1 impedance-ratio transformer has a 2:1 voltage gain, the input voltage of the LTC6400 is twice as large. This gain comes without any significant noise or power penalty, and in fact, the gain can actually improve the effective noise figure of the LTC6400 (as compared to using a shunt resistor for impedance matching). Although the LTC6400's voltage noise density does not change, the extra gain means that the input-referred voltage noise is reduced by the extra gain factor.

Resistor termination

RF transformers are excellent in many impedance-matching situations, and they make convenient baluns for using the LTC6400 differentially. However, their frequency response lower limit is largely determined by the size of the transformer, and there is no hope of achieving consistent frequency response all the way down to DC. Resistors used for impedance matching do not have this limitation. Resistors can yield a much more broadband impedance match than even the best RF transformers, and the frequency response of resistors extends down to DC. There is a penalty to pay in noise figure as compared to the transformer method, but using resistors also saves significant cost over transformers. Resistors can also be used to impedance match a single-ended input.

The circuit shown in Figure 25.27 uses a resistor to terminate a differential 50Ω input source. Since the system is fully differential, the input impedance of the amplifier by itself is calculated by adding the two input resistances together (in this case 400Ω). The shunt input resistor matches this impedance to 50Ω, from DC to high frequency. Note that the 400Ω low frequency input impedance is only true as long as the amplifier maintains its internal "virtual ground" node, and as the amplifier's loop gain decreases with frequency, the input impedance will also change. The LTC6400 data sheet contains graphs of input impedance versus frequency.

The two downsides of using a resistor for termination are power/signal attenuation and the resultant increase in noise figure (i.e., degradation in noise performance). For the same input power level, a 50Ω input impedance will result in less voltage swing than a 400Ω input impedance, thereby introducing an effective voltage attenuation. By using a transformer, the impedances are matched losslessly, and no attenuation occurs. Since the input noise power density of the LTC6400 remains the same and the input signal is smaller, the noise figure increases proportionally.

Figure 25.28 shows resistor termination used with a single-ended input source. Notice the extra resistor R_{T2}, which balances the source impedances seen by the two inputs. Balancing the input impedances is desirable because the distortion performance of the LTC6400 can be affected by imbalance in the source impedances. Also, the inequality of the feedback factors will also cause a portion of the LTC6400's common mode noise to become differential mode noise. The best practice is to always balance the source impedances when working with single-ended inputs.

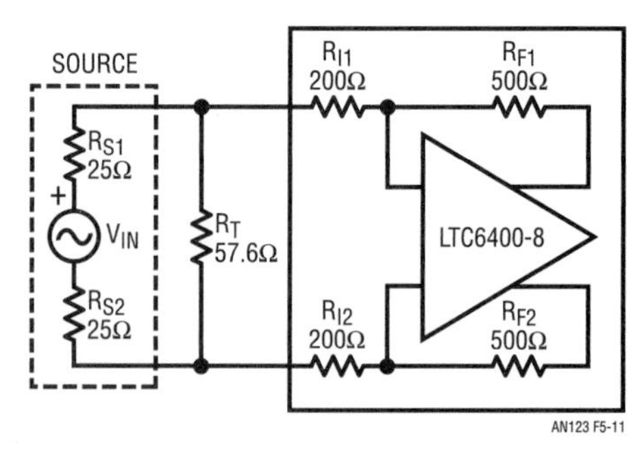

Figure 25.27 • Resistor Termination with a Differential Input Source. A Single Shunt Resistor Transforms the 400Ω Input Impedance into a 50Ω Input Impedance. The Benefit of Resistor Termination is Wideband Performance, from DC to the Maximum Bandwidth of the Amplifier. The Drawback is Power Attenuation and the Resultant Increase in Noise Figure

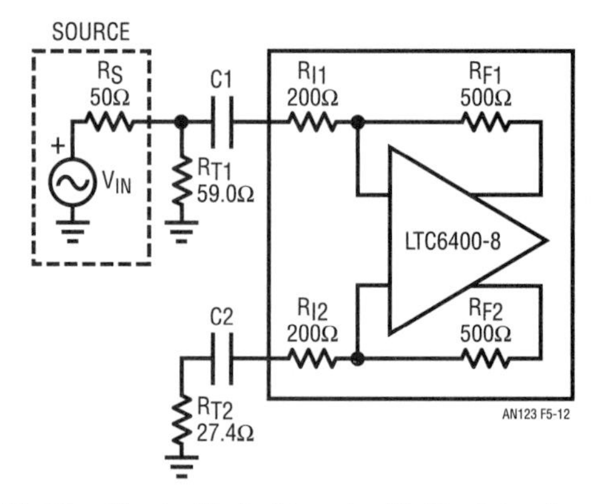

Figure 25.28 • Single-Ended Input with Resistor Impedance Match and Balanced Input Impedances

The addition of R_{T1} and R_{T2} creates additional terms that are not covered in Equation (6). The value of R_{T2} is simply the parallel impedance value of R_{T1} together with Rs (the source impedance). The new values of termination resistors can be calculated as shown:

$$R_{T1} = \frac{1}{2}R_S \bullet \frac{R_SR_F + 2R_SR_I + \sqrt{R_S^2R_F^2 + 4R_F^2R_I^2 + 8R_FR_I^3 + 4R_I^4}}{R_I^2 + R_FR_I - R_S^2} \tag{9}$$

$$R_{T2} = \frac{R_{T1}R_S}{R_{T1} + R_S} \tag{10}$$

In Equations (9) and (10), R_I and R_F are the values of the internal gain and feedback resistors, 200Ω and 500Ω respectively in Figure 25.28. Table 25. 8 lists the values for R_{T1} and R_{T2} that apply in the case of a single-ended 50Ω input source (R_S = 50Ω), calculated using Equation (9) and Equation (10).

The extra source impedance added by R_S and R_{T1}/R_{T2} changes the feedback factor of the differential amplifier, and therefore alters the gain. The overall voltage gain from the ungrounded side of R_{T1} to the differential output can be calculated as follows:

$$\text{Gain} = \frac{2R_F\left(R_{T2}+R_F+R_I\right)}{2R_I^2+2R_I\left(R_F+R_{T2}\right)+R_F R_{T2}} \quad (11)$$

Table 25.8 Termination and Balancing Resistors Used to Match the LTC6400 Family to a 50Ω Single-Ended Input Source. The Resistance Values are Rounded to the Nearest 1% Standard Value

	TERMINATION RESISTOR (R_{T1})	BALANCING RESISTOR (R_{T2})
LTC640x-8	59.0Ω	27.4Ω
LTC640x-14	68.1Ω	28.7Ω
LTC640x-20	66.5Ω	28.7Ω
LTC640x-26	150Ω	37.4Ω

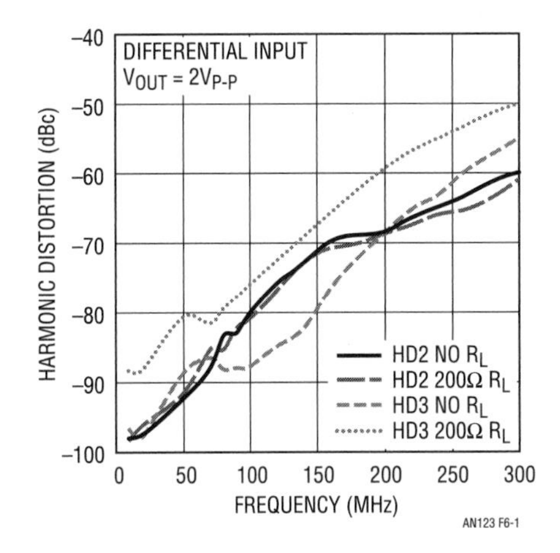

Figure 25.29 • Harmonic Distortion Performance Versus Frequency of the LTC6400-20 with Two Different Resistive Loads. Loads of 200Ω or Greater are Recommended for Optimal Performance

Dynamic range and output networks

The LTC6400 family is fundamentally a very high speed version of the traditional feedback differential amplifier. Therefore, the distortion performance can change dramatically with the choice of output load. The LTC6400 was optimized to drive a high impedance ADC load such as the switched-capacitor inputs of Linear Technology's high performance pipelined ADC families. It was not designed to directly drive a 50Ω load, a characteristic that distinguishes the LTC6400 from other high frequency amplifiers.

Resistive loads

Figure 25.29 shows the distortion versus frequency of the LTC6400-20 with two resistive loads: open (no load) and 200Ω. The distortion with a 200Ω load is measurably higher (worse) than with no resistive load, although the performance is still reasonable. However, it is not recommended to use a 50Ω or 100Ω load with the LTC6400, because the distortion performance would be significantly degraded. Note that the HD2 changes very little with R_L changes; this is because the symmetry of the differential outputs tends to cancel even-order harmonics.

Another important consideration is the output current of the LTC6400. Resistive loads from the outputs directly to ground or to the V^+ supply would cause the LTC6400 outputs to sink or source DC bias currents, which is unnecessary for operation and may impair performance.

V_{OCM} requirements

The LTC6400 is designed for direct DC-coupled driving of Linear Technology's high speed ADCs. The existing families of 3V and 3.3V ADCs have an optimal input common mode DC bias of 1.25V to 1.5V. Therefore, the LTC6400 is optimized for this range of V_{OCM} voltages. Figure 25.30 shows the distortion performance while varying V_{OCM}. If an ADC requires a common mode bias outside of the optimal range for the LTC6400, the output of the LTC6400 can be AC-coupled with capacitors. The downside of this approach is the limited low frequency response.

The distortion performance of the LTC6400 family, like most amplifiers, is dependent on the input and output common mode voltage bias. Figure 25.30 shows the 2-tone 3rd order intermodulation distortion (IMD3) of the LTC6400 at 100MHz (with 1MHz tone spacing). The graphs demonstrate that there is an optimal range for both input and output common mode bias that provides the best overall distortion performance, and that the distortion performance is more dependent on the output common mode voltage (V_{OCM}) than on the input common mode voltage (V_{ICM}).

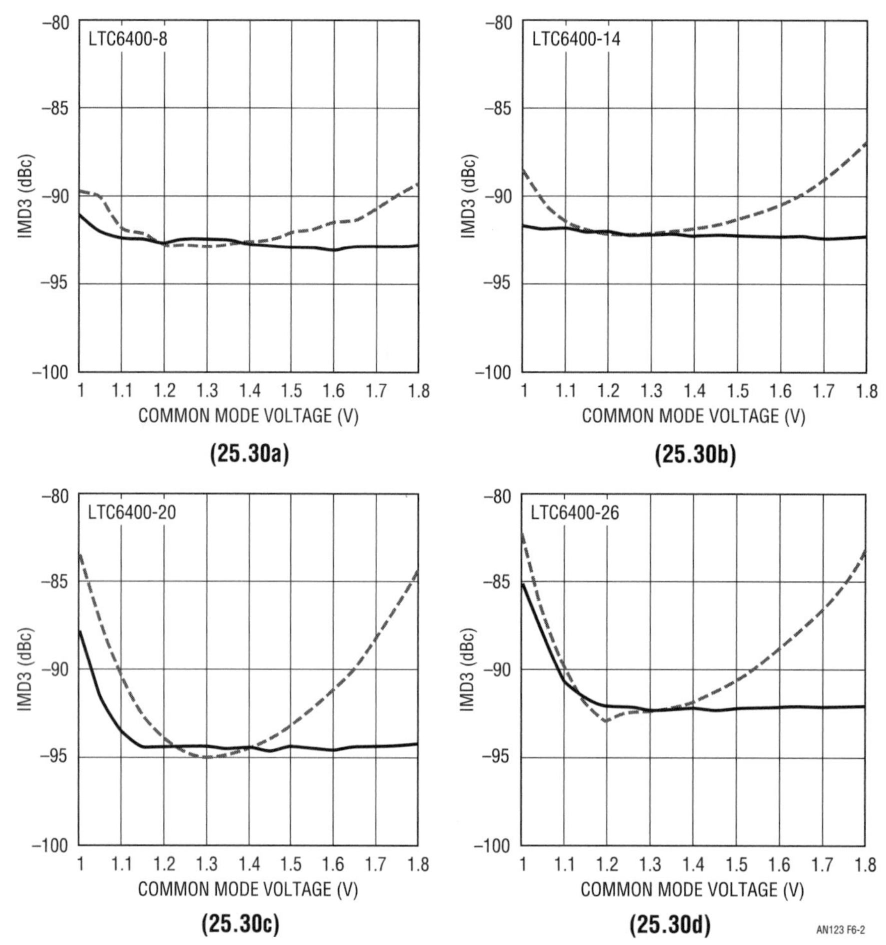

Figure 25.30 • 2-Tone 3rd Order Intermodulation Distortion (IMD3) of the LTC6400 Family Versus Input and Output Common Mode Bias Voltages at 100MHz. Conditions: Differential Input, No R_{LOAD}, V_{OUT} = $2V_{P-P}$ Composite. The Solid Line Shows IMD3 Versus V_{ICM} with V_{OCM} = 1.25V. The Dashed Line Shows IMD3 Versus V_{OCM} with an AC-Coupled (Self-Biased) Input. The Distortion Performance Exhibits More Dependence on V_{OCM} than on V_{ICM}

Unfiltered and filtered outputs

The LTC6400 family includes two sets of parallel outputs for added flexibility in applications. The outputs are not independently buffered, and should not be considered as multiple outputs. In order to help ensure unconditional stability, the LTC6400's unfiltered (normal) outputs include 12.5Ω series resistors on the chip. These resistors must be accounted for in the calculation of voltage drop when driving a resistive load and when designing an antialias filter at the LTC6400 output. Additionally, there is approximately 1nH in bond-wire inductance leading from the series resistors to the package pin, which should not have a significant impact on LTC6400 circuits, but may be important for high order LC filter design.

The filtered outputs of the LTC6400 are designed to potentially save some space in the design of antialias filters. While the unfiltered outputs have 12.5Ω series resistors, vvthe filtered outputs have 50Ω series resistors and 2.7pF of shunt capacitance (including package parasitic capacitance) to form a lowpass filter at the output of the LTC6400. This structure can be used as is or as part of an external antialias filter. By itself, the filter would limit the effective noise bandwidth to under 500MHz. This will prevent the LTC6400's full 1.8GHz noise bandwidth from aliasing and reducing the SNR of the system. When using high performance 14-bit or 16-bit ADCs, it may be necessary to limit the noise bandwidth further to prevent degradation of the ADC's SNR performance.

Output filters and ADC driving networks

It is often desirable to design an output filter for the LTC6400, for antialias or selectivity purposes (or both). Lowpass and bandpass filters are both practical, and the desired bandwidth will usually be determined by the input signal and/or the Nyquist bandwidth of the ADC (half the sample rate). Since the LTC6400 is unconditionally stable with any output load, it is possible to design both RC and LC circuits for this purpose.

Driving a high speed 14-bit or 16-bit ADC that samples at 100 Megasamples per second (and above) is a challenging task. As discussed in the Signal-to-Noise Ratio vs Bandwidth section, the wide bandwidth of the LTC6400-20 means that the noise of the LTC6400-20 can dominate that of a low noise 14-bit or 16-bit ADC. This implies that the configuration shown in Figure 25.31, while highlighting the ease of use of the LTC6400-20, is not sufficient in SNR-critical applications. In order to take advantage of both the low distortion and the low noise of the 16-bit ADC, the noise bandwidth must be limited by a lowpass or bandpass filter.

An ideal drive network for driving a high performance, fast sampling ADC would have low noise, low distortion and low output impedance over frequency. This would allow the network to absorb the charge injection from the ADC's sampling switches and settle in time for the next sample. The frequency content of the charge injection extends out to beyond a GHz, due to the speed with which the sampling switches transition.

The LTC6400 by itself does have low output impedance and the ability to settle relatively quickly after a charge injection event. However, at sampling speeds well above 100Msps, the reduced time for settling may result in incomplete settling from charge injection impulses, which may lead to in increased distortion and noise sampled by the ADC. Fortunately, drive networks can be designed that help to absorb the charge and reduce the impact on the LTC6400. The most basic circuits to do this would be a 1-pole RC lowpass filter or a 2-pole RLC bandpass filter.

The lowpass filter shown in Figure 25.32 is the basic configuration. R_{O1} and R_{O2} represent the frequency dependent output impedance of the LTC6400. R3 and R4 help to absorb the sampling glitches from the ADC input, and are usually sized from 5Ω to 15Ω. The ADC charge injection is typically a common mode event, so C2 and C3 are the dominant charge "reservoirs" that help to absorb the sampling glitches, much like bypass capacitors on a power supply. C1 is a purely differential capacitor, and does not have much effect on the common mode charge injection.

The size of R1 and R2 is not largely constrained, except for in the case of lower frequency cutoff filters. If the total

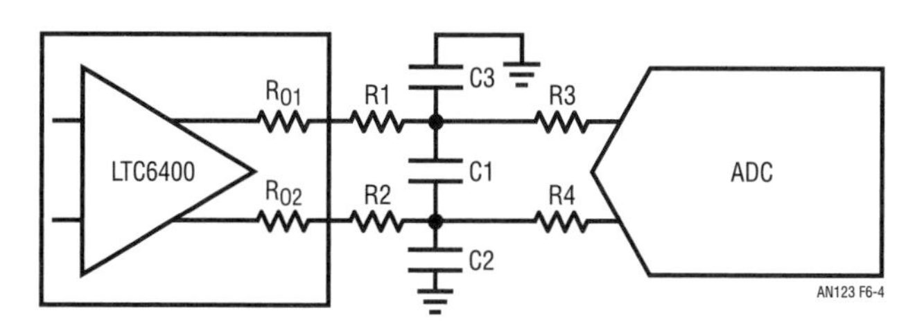

Figure 25.32 • Simple RC Lowpass Filter ADC Drive Network

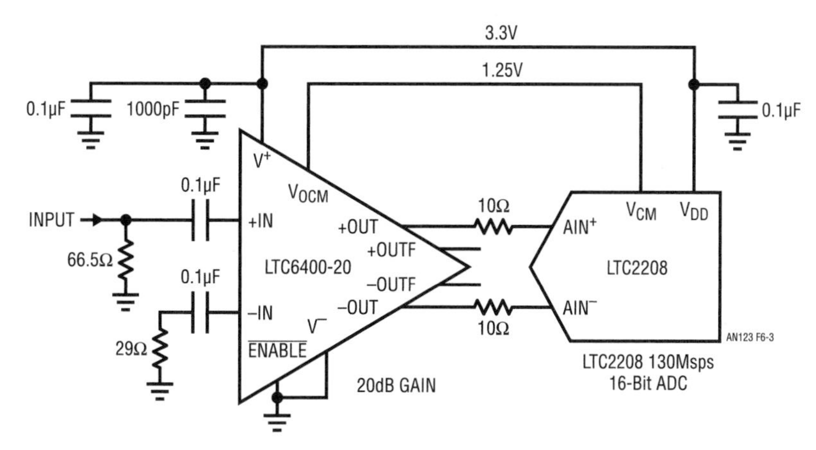

Figure 25.31 • Example of Direct ADC Connection from the LTC6400-20 Data Sheet. This Figure Shows the Flexibility and Simplicity Possible in Interfacing the LTC6400 with a High Speed ADC, but Does Not Address the Issue of SNR and Band Limiting the Amplifier's Output. In Most Applications Where SNR is Critical, the Bandwidth of the LTC6400-20 Output Should be Limited by a Lowpass or Bandpass Filter

resistance seen by the ADC is too high, the linearity of the ADC will typically suffer. This characteristic is different with different ADC families, and is difficult to generalize. At the other extreme, if the resistance R1 and R2 are too small then C1-C3 may be large, and the amplifier loses loop gain when presented with too large of a capacitive load. Overall an R1/R2 value between 10Ω and 100Ω will usually be a good starting point for the filter design.

A simple bandpass filter is shown in Figure 25.33 with the addition of L1. Many of the same considerations apply as for the RC lowpass network. The bandwidth of the bandpass filter is determined by the ratio of the inductance to the total parallel capacitance (C1-C3), as well as the values of R1 and R2. Increasing R1/R2 will make the bandwidth narrower and increase the insertion loss of the filter, so a smaller R1/R2 is desirable, as low as 0Ω. The only time when R1 and R2 may be desired is for improved distortion performance for input signal frequencies close to the passband edges of the filter. The impedance of an LC bandpass filter is at its maximum at the center frequency, but drops quickly in the transition to the stopband. If the bandwidth of the input signal extends out to the edges of the passband, the LTC6400 may be driving a low effective impedance, and the intermodulation distortion may be unacceptably high. In this case, it would be better to increase the bandwidth of the filter (by changing the L/C ratio) and also increase the values of R1/R2. This trades off some insertion loss in the filter for more consistent distortion performance across the passband.

The relationship of the RLC passband frequencies to the sample rate of the ADC is also important. If the ADC's sampling frequency or its harmonic multiples are within the passband of the RLC filter, the network will not attenuate the charge injection of the sampling inputs effectively. If the network resonates at the sampling charge injection frequencies, it will not settle completely between samples at higher sample rates.

An important design consideration is that feedback amplifiers such as op amps typically exhibit poor stability with large capacitive loads, and the LTC6400 is no exception. The phase shift and low impedance presented by a capacitive load, though partially isolated from the main feedback loop,[8] will result in greater gain peaking above 1GHz. Although the LTC6400 is designed to be unconditionally stable, the penalty for a capacitive load will show up as inferior distortion performance. When designing higher order LC filters as antialias filters after the LTC6400, they should ideally be designed to have a series inductor in the first section, or if a shunt capacitive section must be first, the capacitor should have as small of a value as possible. Figure 25.34 shows the difference in distortion performance when a differential capacitive load is presented to the LTC6400-20, and also the mitigating effect of some small series resistors prior to the load.

Note 8: See the Internal Gain/Feedback Resistors section for further explanation of the LTC6400's internal feedback loop.

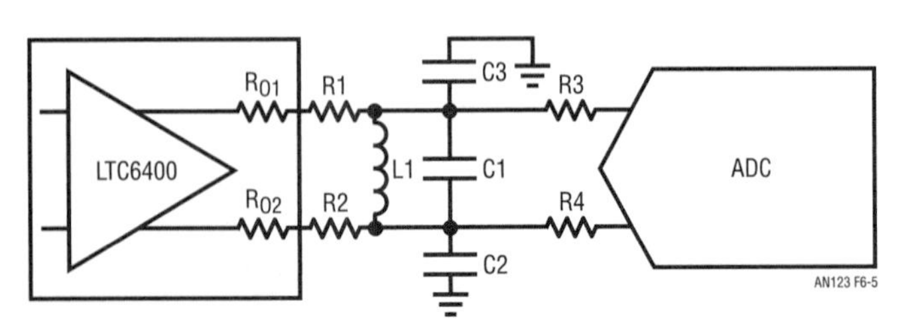

Figure 25.33 • Simple RLC Bandpass Filter ADC Drive Network

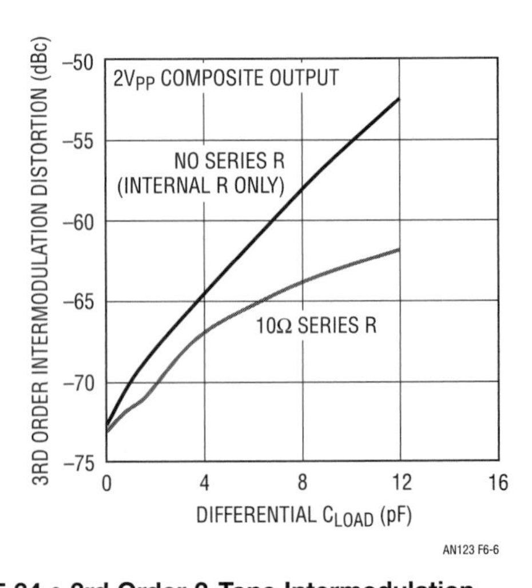

Figure 25.34 • 3rd Order 2-Tone Intermodulation Distortion at 240MHz (1MHz Tone Separation) with a Differential Capacitive Load at the Unfiltered Outputs. The Degradation in Performance is Mitigated by an Additional Series Resistor at Each Output Prior to the Load

Output recovery and line driving

Many feedback amplifiers have difficulty recovering from large input and/or output excursions, which limit their use in high crest factor systems where the output may occasionally be driven to saturation or clipping. In digital drivers and receivers, the output will almost always be at one or the other extreme, which makes the recovery time even more crucial. The LTC6400 family exhibits rapid recovery from an input or output overdrive condition, which enables it to be used in unconventional ways. Figure 25.35 shows a measurement of the LTC6400's propagation delay when used as a digital driver with the inputs and outputs overdriven. The LTC6400 is driving a total 200Ω load, which simulates a terminated 100Ω differential transmission line. The graph demonstrates that even with the inputs overdriven by 1V and the outputs fully clipped, the propagation delay remains less than 3ns. Figure 25.36 is a schematic of the circuit used to measure the data in Figure 25.35.

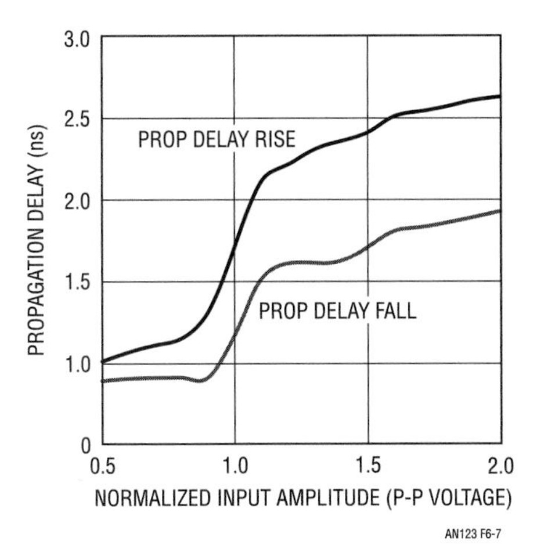

Figure 25.35 • Propagation Delay for the LTC6400-20 with Varying Amounts of Input Overdrive. The X-Axis is the Input Pulse Amplitude (Peak-to-Peak), Normalized to the Pulse Amplitude that Causes Output Clipping. The Y-Axis is Propagation Delay in Nanoseconds, Measured from the 50% Transition of the Input to the 50% Transition of the Outputs

Stability

Stability poses a significant challenge in applications where wideband amplifiers operate at multi-GHz frequencies. Rigorous layout rules, specific feedback components and carefully chosen source/load terminations are often required to guarantee circuit stability. The LTC6400 achieves extraordinary robustness by optimizing the internal compensation networks and isolating sensitive nodes from external parasitic elements. It dramatically reduces the restrictions on user system design and board layout.

If we treat the LTC6400 as a 2-port network, Rollett's stability factor (K factor) can be used as a measurement of overall stability. A circuit is unconditionally stable if both the K-factor is greater than 1 and $|\Delta| < 1$, where (for a 2-port system):

$$K = \frac{1 - |S11|^2 - |S22|^2 + |\Delta|^2}{2|S12||S21|} \tag{12}$$

$$\Delta = S11 \bullet S22 - S12 \bullet S21 \tag{13}$$

Figure 25.37 shows a measurement of the K factor over frequency for LTC6400 (all four gain options), which is based on measured 4-port S parameters. Note that the LTC6400 includes two feedback loops: one for normal differential signals and one to stabilize the common mode bias and reject common mode input signals. Both the differential mode and common mode loops need to be stable to avoid oscillations, and therefore both are presented in Figure 25.37 and Figure 25.38. The requirements of $K > 1$ and $|\Delta| < 1$ are both met by each member of the LTC6400 family.

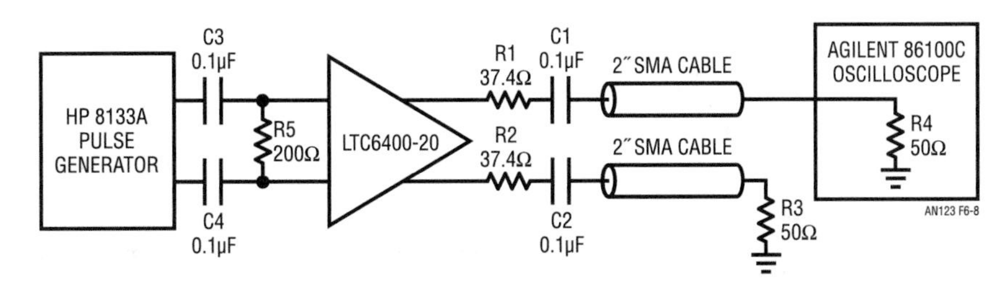

Figure 25.36 • Circuit Used to Measure the Propagation Delay of the LTC6400-20. The Amplifier Receives a Differential Input from the HP 8133A Pulse Generator, and its Output is Measured by an Agilent 86100C Sampling Oscilloscope. The Inputs and Outputs are Impedance Matched with Series or Shunt Resistors for Minimum Line Reflections. The Circuit was Built on the LTC6400-20 Demo Board, DC987B-C

Limitations of stability analysis

The measurements for Figure 25.37 and Figure 25.38 were made at room temperature on a calibrated network analyzer using a PCB with good high frequency layout. As Equations (12) and (13) show, the stability factors K and Δ are calculated from the 2-port S-parameters of the LTC6400, which can be affected by various factors like temperature, bias, and layout. Similar experiments have shown that the LTC6400 family remains stable over temperature and bias conditions, but care must still be taken in layout to ensure good performance. The sensitivity of LTC6400 stability to various layout issues is difficult to quantify, but various best practices have been identified through experience. See the Layout section for more information and recommendations.

Another limitation of this stability calculation method is that the S-parameters are made with symmetrical inputs and outputs. Due to the fact that the LTC6400 is a fully differential amplifier, there are actually four ports to analyze; for the sake of our analysis, we simplify the model down to two independent 2-port networks. In extreme cases of input and output asymmetry, for example, due to layout and/or differences in termination, the differential mode and common mode loops of the amplifier will interact, creating a "mixed mode" situation. If a situation arises where the LTC6400 will be used in this manner, this stability analysis may not apply. In order to ensure unconditional stability in this new mode, the K and Δ analysis should be repeated under the new conditions.

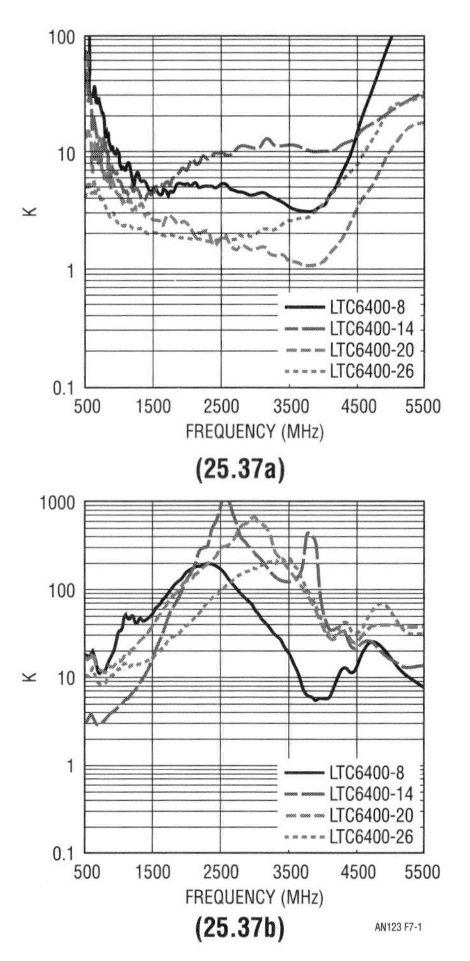

Figure 25.37 • Rollett's Stability Factor (K factor) for the LTC6400 Family of Amplifiers. The K Factor is a Measure of Overall Stability of an Amplifier. (25-37a) is the Differential K factor, and (25.37b) is the K Factor for the Amplifier's Common Mode Loop. A K Factor Value of Greater than 1 at All Relevant Frequencies Implies that the Amplifier is Unconditionally Stable with Any Input or Output Termination

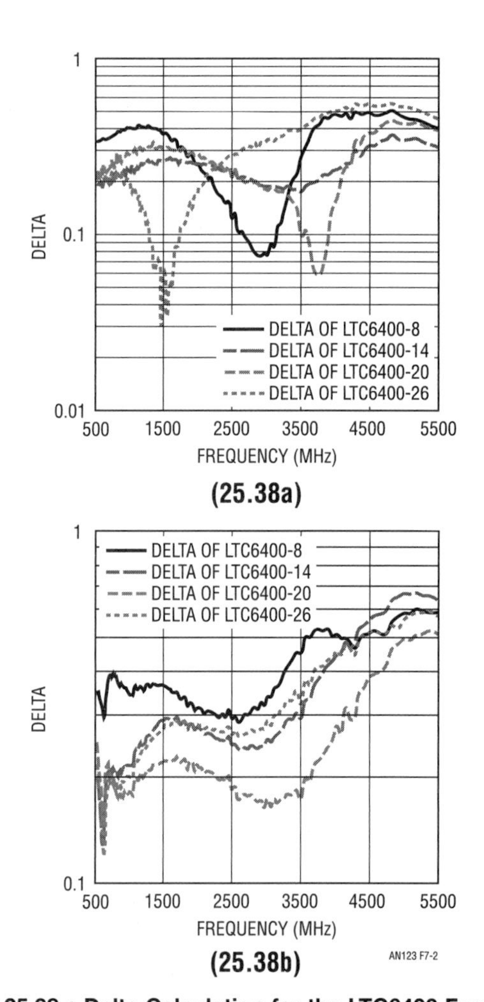

Figure 25.38 • Delta Calculation for the LTC6400 Family's (25.38a) Differential Mode Loop and (25.38b) Common Mode Loop. Delta is a Part of the Rollett's Stability Factor Calculation. A Delta of Less than 1 (Absolute Value) is Part of the Requirement for an Unconditionally Stable System

Layout considerations

The LTC6400 is a high speed fully differential amplifier with almost 2GHz of small-signal bandwidth. This means that as far as the printed circuit board (PCB) is concerned, the LTC6400 family requires the same care in layout design as do sensitive radio frequency (RF) circuits. Poor layout can and will lead to increased distortion, gain and noise peaking, unpredictable signal integrity issues, and in the extreme case, oscillation. Fortunately, the LTC6400 has some features that help to simplify the design of the PCB layout.

First, the resistors for setting the voltage gain are inside the IC, which makes the LTC6400 a "fixed gain" amplifier. Looking at the block diagram, reproduced in Figure 25.39, the critical feedback loop of the amplifier is contained within the chip and is not made available to the user: Traditionally, this is one of the areas where the layout is most critical, and mistakes are most common. Even a very small amount of capacitance at the feedback nodes would significantly affect the frequency response and stability of the amplifier. By keeping the resistors internal, bond-wire inductance and parasitic reactance on the board will not impair the frequency response of the IC as much as it would if the resistors were external.

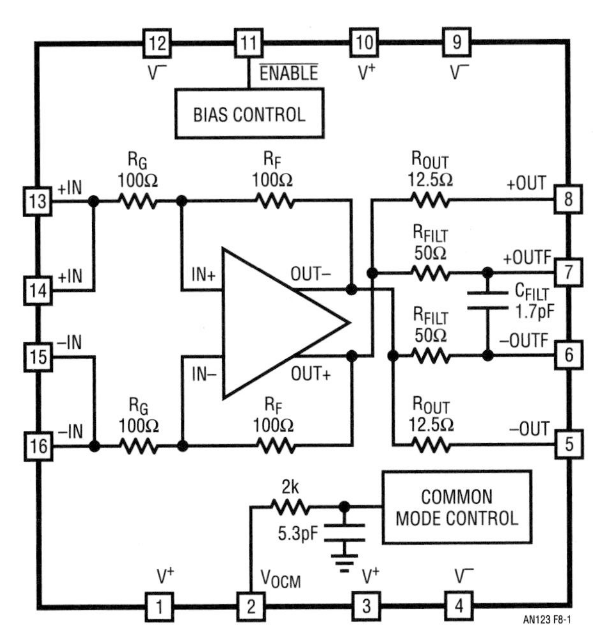

Figure 25.39 • Block Diagram of the LTC6400-20

Second, the LTC6400 has a "flow-through" pinout that is designed for ease of layout. The inputs and outputs are located on opposite sides of the chip, and the power and control pins are located on the remaining two sides. This makes the inputs and outputs easy to route on the board, without the need to route anything around the chip or on internal layers.

One critical aspect of every high speed amplifier layout is the location of the bypass capacitors. The current path from the amplifier's supply pins through the capacitors to the board return and back to the amplifier is critical, because excess inductance or resistance in this path will cause voltage bounce. On the LTC6400, the V^+ and V^- pins are strategically located to allow the customer to place the bypass capacitance in close proximity to the chip, because these components will not hinder the layout of the signal path. Additionally, the LTC6400 family includes approximately 150pF of bypass capacitance on-chip, which makes the layout of the bypass capacitance slightly easier. The external bypass capacitors simply act as a charge reservoir for the on-chip bypass capacitance. Bypass capacitors of 0.1µF are the recommended value to do the job, and are currently available in the 0402 size or smaller.

Note the RC lowpass filter at the input of the V_{OCM} pin, Pin 2 in Figure 25.39. This internal filter makes the layout of the recommended V_{OCM} bypass capacitor less critical. The internal common mode control loop has a bandwidth of over 300MHz, but the V_{OCM} pin itself is much less sensitive to high frequency interference due to this 15MHz lowpass filter. The V_{OCM} bypass capacitor can be located further away from the chip to make room for the more critical V^+ bypass capacitors.

Figure 25.40 shows the layout of the LTC6400 demo board, DC987B. The differential inputs are on the left, and the differential outputs on the right. Users can select either the filtered or unfiltered outputs by changing the resistors on the board. Notice the bypass capacitor groups on the top and bottom sides of the LTC6400—they are located as close as possible with ground vias near the capacitors so that the overall current loop through the capacitors is minimized. Also, the Exposed Pad of the LTC6400 is connected directly to a solid ground with four ground vias, which provides a good thermal and electrical path to board ground.

There are three V^+ (positive voltage supply) pins on the LTC6400, and each of them has its own set of bypass capacitors, as recommended in the product data sheet.

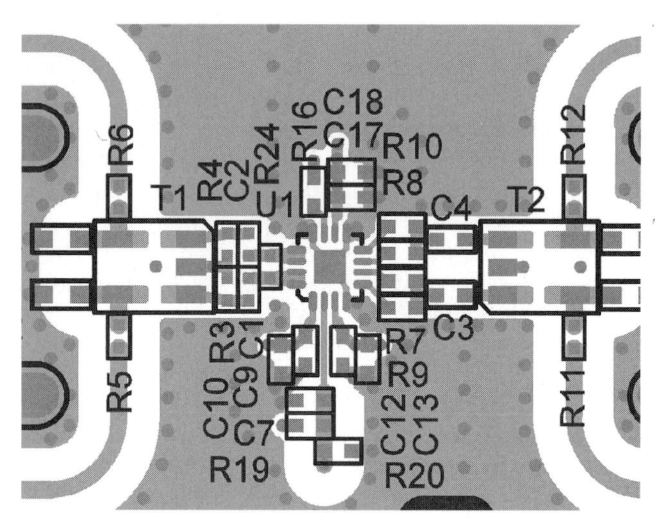

Figure 25.40 • DC987B Demo Board Layout, Layer 1. The Schematic is Shown in Figure 25.26

There is a small capacitor, typically around 1000pF located close to the part. Located further away is a larger (0.1μF to 1.0μF) bypass capacitor. There are also three V^- (negative power supply) pins, which are normally connected to board ground and to the Exposed Pad. It's important that the V^+ pins all be connected to the exact same supply, and that the V^- pins and Exposed Pad are all connected together and tied to board ground.

Thermal layout considerations

The LTC6400 dissipates 250mW and the LTC6401 dissipates 130mW, so thermal issues are not as prevalent as with higher power devices; however, if the part is being operated in a high temperature environment, good thermal layout can protect the silicon from overheating. Good thermal layout consists of providing a square of copper on the board to mount the Exposed Pad, and using four vias underneath the Exposed Pad to help conduct heat to the PCB. The inner ground layers of the LTC6400 (layer 2 should be ground, for best high frequency performance) should have as much unbroken ground plane as possible in the vicinity of the IC, because ground plane copper conducts heat much more efficiently than standard PCB dielectric materials.

Operating with a negative voltage supply

In certain situations, it may be desirable to operate the LTC6400 with a V^- supply that is not the board ground. For example, if operating the LTC6400 with dual ±1.5V supplies, the inputs and outputs can operate at board ground without AC coupling. As long as the V^- pins and Exposed Pad are tied together to the same potential and the absolute voltage from V^+ to V^- does not exceed the data sheet maximum, there is no fundamental problem with this configuration. From a layout point of view, however, it presents a greater challenge to achieve optimal performance.

Besides bypassing the V^+ pins to board ground, it is also necessary to bypass the V^- pins to board ground (or bypass V^+ to V^-). The first step is to squeeze bypass capacitors for V^- onto the top layer in Figure 25.40. There isn't a lot of room for additional capacitors on the topside; however, the close proximity of the V^+ and V^- pins allows for the placement of bypass capacitors directly from V^+ to V^-. This makes for the shortest possible current path, and the use of small-valued 0201 or 0402-sized capacitors (on the order of 1000pF) can serve this function effectively in the layout. In fact, some of the V^+/GND bypass capacitors can be changed to V^+/V^- bypass capacitors without sacrificing performance. Larger-valued bypass capacitors for V^- can be located on the back side of the PCB. Since the LTC6400 layout does not normally require many components on the back of the PCB, there may be ample space to place four or more bypass capacitors around the perimeter of the LTC6400's Exposed Pad area. Figure 25.41 shows a sample top and bottom layout with dual voltage supplies.

Also of concern is thermal layout when the exposed pad is not tied to ground. Ideally, there would be significant V^- copper on one of the internal layers to help spread the heat generated by the part. With the V^- disconnected from board ground, the thermal resistance presented by the PCB will likely be much larger, and even the 250mW dissipated by the LTC6400 may start to present thermal issues at high temperature operation.

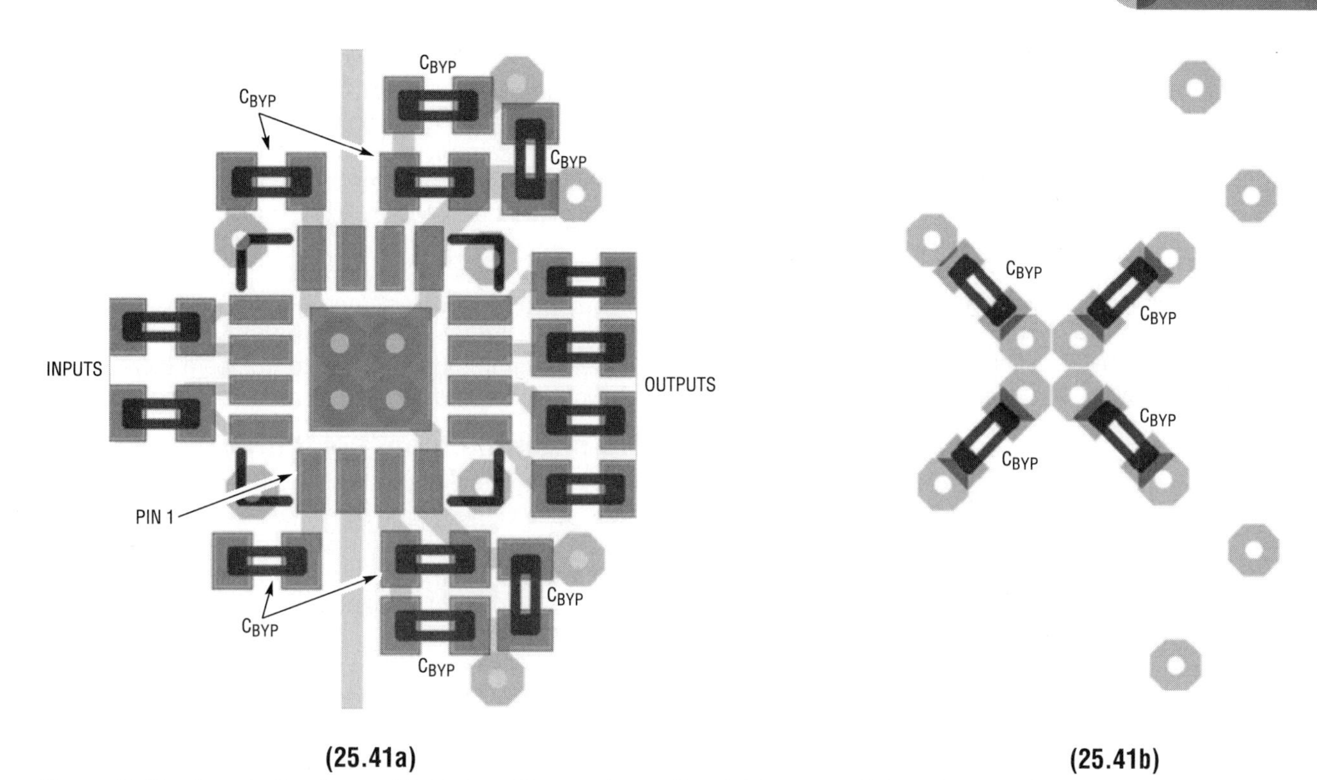

Figure 25.41 • Sample Top Layer (25.41a) and Bottom Layer (25.41b) Layout with Dual Voltage Supplies. All Power Supply Bypass Capacitors are Labeled as C_{BYP}

Conclusion

The LTC6400 takes advantage of a very high speed semiconductor manufacturing process to achieve all of the characteristics necessary for high speed ADC driving, including low distortion, low noise and several gain options up to 26dB. In addition, the LTC6400 includes various ease-of-use features designed to assist in layout and manufacturing: unconditional stability with any input and output terminations, a flow-through layout, and internal resistors to reduce the sensitivity of the layout to parasitic elements.

References

Bowick, Chris. RF Circuit Design. Burlington, MA: Elsevier, 1982.

Brisebois, Glen. Op Amp Selection Guide for Optimum Noise Performance. Design Note 355, Milpitas, CA: Linear Technology, 2005.

Friis, H.T. "Noise Figures in Radio Receivers." Proc. of IRE, July 1944.

Fukui, H. Low-Noise Microwave Transistors and Amplifiers. New York: IEEE Press, 1981.

Gilmore, Rowan, and Les Besser. Practical RF Circuit Design for Modern Wireless Systems Volume II: Active Circuits and Systems. Boston, MA: Artech House, 2003.

Linear Technology. LTC6400-20: 1.8GHz Low Noise, Low Distortion Differential ADC Driver for 300MHz IF. Datasheet, Milpitas, CA: Linear Technology, 2007.

Ludwig, Reinhold, Pavel Bretchko, and Gene Bogdanov. RF Circuit Design: Theory and Applications. Pearson Prentice Hall, 2008.

National Semiconductor. Noise Specs Confusing? Application Note 104, Santa Clara, CA: National Semiconductor, 1974.

Pei, Cheng-Wei. Signal Chain Noise Analysis for RF-to-Digital Receivers. Design Note 439, Milpitas, CA: Linear Technology, 2008.

Rich, Alan. Noise Calculations in Op Amp Circuits. Design Note 15, Milpitas, CA: Linear Technology, 1988.

Sayre, Cotter W. Complete Wireless Design. New York: McGraw-Hill, 2001.

Seremeta, Dorin. Accurate Measurement of LT5514 Third Order Intermodulation Products. Application Note 97, Milpitas, CA: Linear Technology, 2006.

Appendix A

Terms and definitions

This section details some of the more commonly used (and/or misunderstood) terms in the specification and application of the LTC6400 family.

Noise figure (NF)

Noise figure (NF) and noise factor (F) are ratiometric calculations that are useful in RF system design. The fundamental idea is that in an electronic system at a given temperature, there is a certain amount of noise due to random thermal motion. This noise is constant for a given system impedance and comes out to −174 dBm/Hz at room temperature. Whenever an amplifier or any other active element is placed into a system, there is an additional noise power added above and beyond the thermal noise floor. Noise factor is defined as the ratio between the input signal noise ratio (SNR) and output signal noise ratio,[1] or in other words the SNR with the additional amplifier and the SNR without it. Noise figure is simply noise factor in decibels:

$$NF = 10\log\frac{SNR_{IN}}{SNR_{OUT}} = 10\log F \qquad (A1)$$

The concept behind NF is to quantify the amount of additive noise introduced when adding a device to the system. Note that the input SNR only counts the noise contribution from the resistive part of the source impedance Z_S, as ideal capacitors and inductors are noiseless. So substituting R_S for the more general Z_S, we define noise figure as:[2]

$$NF = 10\log\frac{e_{n(OUT)}^2}{e_{n(Z_S)}^2 \bullet G^2} \qquad (A2)$$

where $e_{n(Z_S)}$ is the thermal noise of source resistance and G is the voltage gain of the amplifier for given Z_S.

$$e_{n(Z_s)}^2 = 4kTR_s \qquad (A3)$$

where k is the Boltzmann constant ($1.3806503 \bullet 10^{-23}$ J/K)

For a fixed-impedance system, such as RF systems that operate in 50Ω, using NF is a simple way to compare the noise performance of devices and calculate how the noise of a new device will affect the system-level noise performance. For voltage measurement systems like an ADC, NF is not always as convenient because the denominator of the ratio changes with impedance. This means that the relatively constant voltage noise density of the ADC will result in a different NF for different source impedances. The amplifier inputs of the LTC6400, as another example, have varying impedances from 50Ω to 400Ω, and the outputs may be driving the ADC's high impedance input. So to make system-level noise calculations may require some conversion of noise terms, from the voltage noise density ($V/\sqrt{Hz}$) specified in the data sheets to an 'equivalent' noise figure that is consistent throughout the system.[3] When comparing two similar ADC drivers, it is often easier to compare them in their "native" noise specification, which is voltage noise density.

3rd order intercept point (IP3)

Signal distortion is present in all amplifiers to some extent, and in many amplifier data sheets it will be specified as a harmonic distortion. If a single sine wave tone is generated at the amplifier's output, there will also be a certain amount of unwanted distortion at all multiples of the sine wave frequency. The dominant tones will be the second and 3rd order harmonics. There is also another method of characterizing the 3rd order distortion, which is more useful for narrow-band systems: if two sine waves are generated at the output, spaced closely together in frequency, there will be additional spurious tones generated at the same spacing distance on either side of the two tones. The two closest tones are the 3rd order intermodulation distortion (IMD), and they are due to the mixing that occurs from the amplifier's nonlinearity.

The 3rd order intercept point (IP3) is a useful metric for measuring IMD performance in an amplifier. At a given frequency, set of bias conditions and temperature, the IP3 approximately defines the distortion performance

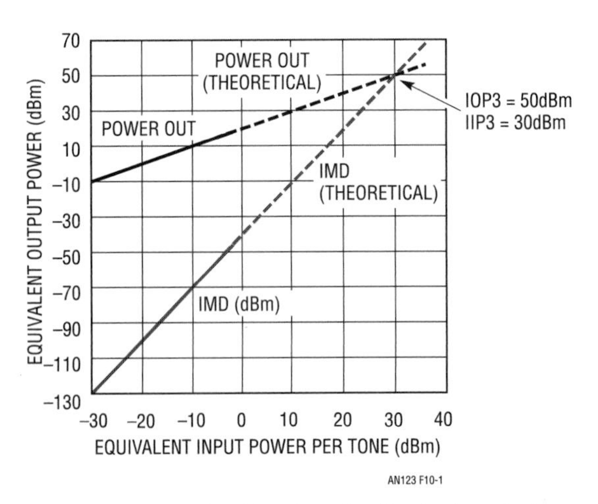

Figure A1 • Graphically Showing the Equivalent 3rd Order Intercept Point (IP3) for a 20dB Gain Amplifier. The Plotted Curves of Output Power (dBm) and 3rd Order Intermodulation Distortion Power (dBm) Theoretically Meet at a Certain Output Power Which is Termed the IP3. This Power Level is Unattainable in Practice Because the Output of the Amplifier Saturates at Lower Power Levels

Note 1: A discussion of noise figure analysis can be found in (Gilmore and Besser 2003).
Note 2: An example of calculating of noise figure for amplifiers can be found in Appendix B or (National Semiconductor 1974).
Note 3: See (Pei 2008) for an example calculation of equivalent NF.

at any power level where the amplifier's output is not saturated (Sayre 2001). IP3 can be referred either to the output (OIP3) or the input (IIP3, output IP3 minus the conversion gain). If you plot the output power versus the input power of an amplifier and overlay the 3rd order intermodulation distortion (IMD) curve versus input power on the same graph, as in Figure A1, the extrapolated curves would meet at the IP3 point. IP3 is typically specified in dBm, which is a logarithmic decibel unit referred to 1mW of power.

In order to apply the output IP3 metric to a voltage feedback amplifier like the LTC6400, it is necessary to define some specialized terminology. Because the LTC6400 is not designed to drive a resistive load of 50Ω, it is not appropriate to define input and output power in the standard way. The LTC6400 amplifies voltage, not power; the current gain through the device is typically very small. When driving a high speed ADC, the LTC6400 may not have a resistive load at all. This leads us to define an ***equivalent output powe***r level, which uses the voltage-swing equivalence into a non-existent 50Ω load. For example, a $1V_{P-P}$ voltage swing into 50Ω would equal 4dBm of RMS output power,[4] so we define $1V_{P-P}$ as 4dBm for the purposes of calculating the OIP3. This allows the LTC6400 to be used in the same signal-chain calculations as other devices, since OIP3 (like NF) can be used for cascaded system-level distortion analysis.

Using this equivalent power terminology, refer back to Figure A1 which shows the theoretical IMD and output power levels for a 20dB amplifier with a 50dBm OIP3 point, matching the typical specification of the LTC6400-20 at 100MHz. Extrapolating the IMD curve back to a 4dBm equivalent power level, which is defined as $1V_{P-P}$ ($2V_{P-P}$ with both tones together), that the IMD level is shown to be −88dBm, or −92dBc.[5]

1dB compression point (P1dB)

Strictly speaking, the "1dB compression point" of an amplifier is the input power level that causes the gain to deviate by 1dB from the ideal linear gain (Sayre 2001). This textbook definition does not include any restrictions on what type of amplifier or the root cause of the gain non-ideality. For an RF gain block, mixer or other "typical RF device" with a 50Ω impedance, the dominant source of this gain compression at high output power levels is the saturation of the output transistor. This "soft saturation" characteristic causes the gain to roll off as the signal swing of the output limits its ability to produce an ever-increasing output power. Consequently, the distortion of the output signal is also increasing in a predictable way as the input power increases. So in some cases, the P1dB of an RF device has a direct relationship with the 3rd order intercept (IP3) of that device, in which case the P1dB can be used to compare one device with another.

This situation does not apply to the LTC6400 family of amplifiers. Due to the large amount of feedback employed in the amplifier topology, the linearity of the LTC6400 (i.e., the amount of distortion) does not follow exactly the same trend as for an RF device. The feedback and loop gain of the LTC6400 circuit linearizes the amplifier's output, and the gain does not significantly deviate from the ideal value until the amplifier's output is actually saturated. So the signal out versus signal in of the LTC6400 exhibits much higher linearity until the output is up against the limit, at which point the distortion increases dramatically. Knowing the P1dB point of the LTC6400 does not give any inferred knowledge of the distortion performance or IP3.

Appendix B

Sample noise calculations

Calculating noise and noise figure for the LTC6400 family is challenging, and there is no one data sheet specification that can completely characterize the noise performance of this family. Due to the gain and feedback resistors being included inside the package, the noise of the amplifier in one configuration (e.g., with the inputs shorted together) will yield different results than with the inputs terminated resistively. This section walks through

Note 4: In 2-tone tests, if each tone has $1V_{P-P}$ amplitude, then the combination of the two will produce a $2V_{P-P}$ composite amplitude, which is the standard amplitude for evaluating the LTC6400.

Note 5: dBc means "decibels referred to the carrier," and uses the carrier's power level as the 0dB reference. In this case, the distortion levels are measured relative to the power level of a single tone in the 2-tone test: −92dBc = −88dBm −4dBm.

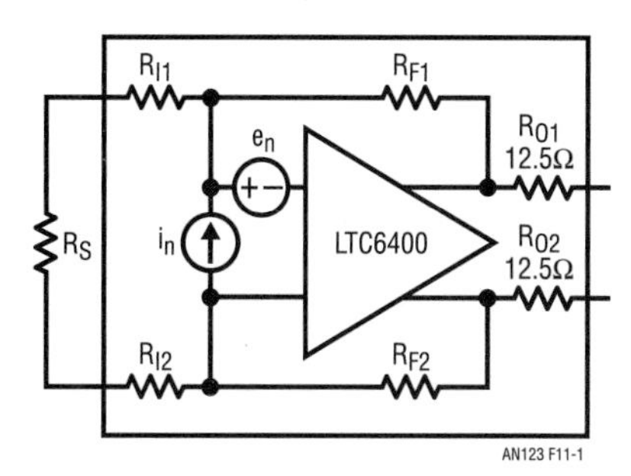

Figure B1 • Schematic Drawing for Noise Analysis with an Arbitrary Source Resistance R_S. This Analysis Ignores the Noise of the Output 12.5Ω Resistors

a few noise calculations in a further attempt to shed light on this topic.

Noise analysis for arbitrary source resistance

Using the noise parameters found in Table 25.2, the differential output noise of the LTC6400 can be calculated for the most generic case, when the gain option and the source resistance are unknown. It is assumed that the source impedance is purely resistive for this exercise.

$$V_{N(OUT)} = \sqrt{\left[e_n \bullet \left(\frac{1+2R_F}{2R_I+R_s}\right)\right]^2 + (2R_F \bullet i_n)^2 + \beta \bullet (2R_I+R_s) \bullet \left(\frac{2R_F}{2R_I+R_S}\right)^2 + \beta \bullet 2R_F}\ (\text{nV}/\sqrt{\text{Hz}}) \quad \text{(B1)}$$

where:

$$\beta = 4kT = 1.6008 \bullet 10^{-20}(\text{J}), R_F = R_{F1} = R_{F2}(\Omega),$$

$$R_I = R_{I1} = R_{I2}(\Omega)$$

The first term in the equation multiplies the input-referred voltage noise density of the LTC6400's internal amplifier with the noise gain. The second term is the noise gain of the input-referred current noise, which is equal to the feedback resistance. The third term deals with the noise of the input and source resistance, which is multiplied by the signal gain of the amplifier. The fourth term is the feedback resistance noise, which has a unity-gain factor.

To use a numeric example, examine the configuration of Figure 25.16, which shows the LTC6400-8 with extra source resistance to lower the effective voltage gain. The terms in the noise equation are: $R_F = 500\Omega$, $R_I = 280\Omega$, $Rs = 602\Omega$. Equation (B1) and the values in Table 25.2 are used to calculate the estimated output noise at 100MHz (Equation (B2).

The result in Equation (B2) matches well with the curve in Figure 25.8. To convert this result to an equivalent noise figure, we refer back to Equation (A2), with the assumption that the 602Ω resistors are the source resistance, with a voltage noise of $\sqrt{X_N} \bullet 602\ \text{nV}/\sqrt{\text{Hz}}$:

$$NF = 10\log\frac{(7.3 \bullet 10^{-9})^2}{(\beta \bullet 602) \bullet 1^2} = 7.4\text{dB} \quad \text{(B3)}$$

The resultant NF matches well with the curve in Figure 25.9. The gain used in the above equation is the signal gain of the LTC6400-8 circuit, which was reduced to 1V/V by the series resistors.

DC987B demo board noise analysis

This section extends the noise calculations to the LTC6400 demo board, DC987B. A good example is the LTC6400-20, which has 200Ω differential input impedance and 1k feedback resistors. Figure B2 contains a noise representation of the board. The transmission line transformers (used mainly for impedance matching) are modeled here as ideal 1:4 impedance transformers together with a −1dB block. This allows the separation of the insertion loss of the transformer from its ideal behavior.

To calculate the noise figure of this system, calculate the noise in the chain for two situations: a noiseless LTC6400-20, which simply amounts to a noiseless gain block, and the real LTC6400-20. That way, the signal gain throughout the chain remains the same, and the difference in the noise will reveal the ratio of the signal-to-noise ratios in these two situations, which is by definition the noise figure. Table 25.9 shows the step-by-step noise calculations that match the locations shown in Figure B2.

Use Equation (A2) to calculate the overall NF from these two cases:

$$NF = 10\log\left(\frac{3.61^2}{1.78^2}\right) = 6.14\text{dB} \quad \text{(B4)}$$

This is the overall noise figure for the demo board DC987B.[1] However, there is still the −1dB loss from the transformer at the input of the LTC6400. For this, use the Friis formula (Friis 1944):

$$F_{TOT} = F1 + \frac{F2-1}{G1} \quad \text{(B5)}$$

To apply the formula, translate the decibel NFs back to the linear noise factor (F). In our case, 6.14dB becomes 4.113, −1dB loss becomes 0.7943, and 1dB noise

$$V_{N(OUT)} = \sqrt{(1.12 \bullet 2)^2 + \left[1000 \bullet (4.00 \bullet 10^{-3})\right]^2 + \beta \bullet 1000 \bullet 1^2 + \beta \bullet 1000} = 7.3\text{nV}/\sqrt{\text{Hz}} \quad \text{(B2)}$$

Note 1: This value approximates the value of NF published in the LTC6400-20 data sheet. To avoid confusion, the measured demo board NF values were used instead of the true amplifier NF value, as calculated in Equation B9.

figure becomes 1.259 (the NF of an attenuator is just the inverse of the attenuation).

$$F_{TOTAL} = F_{LOSS} + \frac{F_{AMP} - 1}{G_{LOSS}} \quad (B6)$$

$$4.113 = 1.259 + \frac{F_{AMP} - 1}{0.7943} \quad (B7)$$

$$F_{AMP} = 3.267 \quad (B8)$$

$$NF_{AMP} = 10\log(F_{AMP}) = 5.14dB \quad (B9)$$

This new calculation subtracts out the effect of the attenuation from the input of the LTC6400, which will give us the true amplifier noise figure. Note that this result is consistent with the data in Figure 25.9.

SNR calculation and aliasing example

This example attempts to shed light on the tradeoffs between amplifier bandwidth and SNR, including the effects of ADC aliasing. Table 25.10 lists the important specifications for this example, for both the amplifier and the ADC.

The first step is to calculate the noise floor of the ADC. Since the ADC samples at 100Msps, its noise can be represented as a constant noise floor extending from DC-50MHz, which is the Nyquist bandwidth. The industry standard for ADC specification is to list SNR with a full-scale sine wave tone. Using the maximum input sine

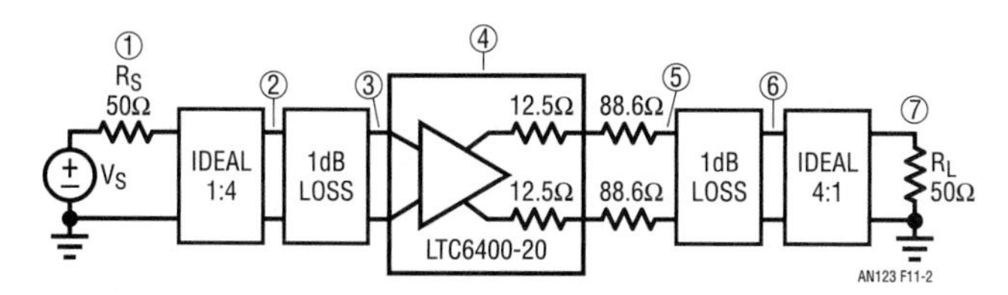

Figure B2 • Equivalent Demo Board Schematic for Noise Analysis. This View Ignores the DC-Blocking Capacitors and Bypass Capacitors of the LTC6400, Which are Inconsequential to the Noise Analysis. The 88.6Ω Resistor at the LTC6400 Output Creates an Approximate 100Ω Source Resistance (or a Differential 200Ω), Which is an Impedance Match for the Reflected R_L

Table 25.10 Specifications for SNR Calculation Example

Amplifier Specifications	Output Noise Density	$8nV/\sqrt{Hz}$
	Effective Noise BW	100MHz
ADC Specifications	Signal-to-Noise Ratio	80dB
	Effective Resolution	(80dB – 1.76)/6.02 = 13 bits
	Sample Rate	100Msps
	Input Span Voltage	2VP-P (0.707V_{RMS} Sine Wave)

Table 25.9 Noise Calculations for the DC987B Demo Board, According to Figure B2. The First Column Corresponds to the Numbered Locations in the Figure. The Second Column Assumes a Noiseless LTC6400-20, and the Third Column Includes the Noise of the LTC6400-20. The End Values of the Two Columns are Compared to Calculate the Noise Figure of the LTC6400-20

LOCATION (Fig B2)	NOISE WITH NOISELESS LTC6400-20 ($nV/\sqrt{Hz}$)	NOISE WITH ACTUAL LTC6400-20 ($nV/\sqrt{Hz}$)	COMMENT/DESCRIPTION
1	0.894	N/A	Noise of Source Resistance at 290K (17°C)
2	0.894	N/A	The 1:4 Transformer Results in a Voltage Doubling, but the Resistive Divider Formed by 200Ω Input Impedance Results in the Voltage Being Halved Again
3	0.80	N/A	After Subtracting the 1dB Loss
4	8.0	16.2	After Gain of 20dB, or 10V/V For the Noisy Case, Look to Figure 25.8 for the Total Output Noise of the LTC6400-20 with a 200Ω Source Resistance, Which Matches the Situation in Figure B2. The Noise Output in the Figure Includes the Source Resistance Noise, Which is Reduced by the 1dB Loss in the Transformer
5	4.0	8.1	Reflected RL Across Transformer is 200Ω, Which Creates a 1:1 Voltage Divider with the Resistors at the LTC6400 Output. For the Sake of Calculations, Treat This Resistance as 200Ω Instead of 101.1Ω
6	3.57	7.22	Subtracting Another 1dB Loss
7	1.78	3.61	Total Noise at Load (Ignoring RL Noise). Transformer Reflects Voltage as 2:1, so the Voltage is Halved

wave amplitude of $0.707V_{RMS}$, and the 80dB SNR, calculate the ADC's intrinsic noise floor:

$$SNR_{LIN} = 10^{80/20} = 10,000 \quad \text{(B10)}$$

$$NOISE_{ADC} = \frac{0.707V}{10,000} = 70.7\mu V_{RMS} \quad \text{(B11)}$$

$$e_{n(ADC)} = \frac{70.7\mu V}{\sqrt{50MHz}} = \frac{10nV}{\sqrt{Hz}} \quad \text{(B12)}$$

The next step is to look at the amplifier's contribution to the overall noise. Since the amplifier's total noise bandwidth is wider than the Nyquist bandwidth of the ADC, then aliasing will occur. See Figure B3 for a visualization of what happens when the ADC samples the amplifier's noise. The ADC has its own noise floor (flat from DC-Nyquist), and the amplifier has flat wideband noise from DC-f_{SAMPLE}, which is two full Nyquist bandwidths. Since the amplifier noise is uncorrelated with the ADC noise, both bands of amplifier noise can be added in an RMS fashion to the ADC's noise floor.

Adding the ADC's $10nV/\sqrt{Hz}$ to the amplifier's $8nV/\sqrt{Hz}$ (twice due to the aliasing), we arrive at the final noise floor of the combined circuit:

$$e_{n(TOTAL)} = \sqrt{10^2 + 8^2 + 8^2} = 15.1nV/\sqrt{Hz} \quad \text{(B13)}$$

Working backward through the previous equations, we arrive back at the new SNR for the system:

$$SNR_{NEW} = 20\log\frac{0.707}{15.1 \bullet 10^{-9} \bullet \sqrt{50MHz}} = 76.4dE(11-14) \quad \text{(B14)}$$

Adding an amplifier with less noise than the ADC but more bandwidth resulted in an overall SNR 3.6dB less than the capability of the ADC. In order for the amplifier to be "transparent," meaning the system SNR is approximately the same as the ADC's SNR, the amplifier needs to have lower output noise density and/or lower bandwidth.

The preceding analysis and Figure B3 assume the alias bands are the same width as the Nyquist bandwidth. If this is not the case, then the resulting noise floor will have a multi-tiered shape, with a rise in the noise floor where the extra aliased noise occurs in frequency. This would also be the case if the amplifier's bandwidth or the anti-alias filter bandwidth was less than one Nyquist band, where the wideband noise rolls off before the Nyquist frequency of the ADC. However, an anti-alias filter would not affect the above calculations, except to change the amplifier's effective noise.

The visual analysis above indicates that once noise is "aliased" into the original Nyquist bandwidth, it is indistinguishable from lower-frequency noise. The two solutions presented so far focus on reducing the noise presented to the ADC input. If digital filtering is possible, there is a third solution, which would be to increase the ADC's sample rate. If the sample rate is increased so that the Nyquist bandwidth exceeds the input bandwidth, there is extra bandwidth that can be removed through digital filtering. Also, for a given analog bandwidth, there would be fewer bands of noise that are aliased, so the resulting noise floor would be lower as well.

Appendix C

Optimizing noise performance by calculation of voltage and current noise correlation

In order to understand the true interaction between current noise and voltage noise at the input of the LTC6400, it is necessary to consider the correlation between the two types of noise. Since the current and voltage noise source elements in the LTC6400 are largely the same, there should be some level of correlation between the two. If there is significant correlation, then it should be possible to find source impedances that would cause the current noise to partially cancel out the voltage noise (or add to it). The current noise source of the LTC6400 can be split into two separate noise sources: i_{nu} for the uncorrelated current noise, and i_{nc} for the correlated current noise. The uncorrelated current noise can be defined by an equivalent noise conductance, and is independent of the voltage noise. The correlated current noise has a complex (vector) relationship with the voltage noise.

$$i_{nu} = \sqrt{4kTG_U} \quad \text{(C1)}$$

$$i_{nc} = Y_C \bullet e_n \quad \text{(C2)}$$

G_U is a conductance that converts voltage to current, k is the Boltzmann constant, Y_c is the complex admittance defined as $Y_C = G_C + jB_C$, and e_n is the total input voltage noise density from the data sheet. The complex admittance (Y) is the inverse of the complex impedance $Z = R + jX$. From the Smith Chart and the values in Table 25.3, there is a relationship between the optimal noise factor and the desired values:[1]

$$F_{MIN} = 1 + 2R_N(G_{OPT} + G_C) \quad \text{(C3)}$$

$$B_C = -B_{OPT} \quad \text{(C4)}$$

$$G_U = R_N \bullet G_{OPT}{}^2 - R_N \bullet G_C^2 \quad \text{(C5)}$$

Note 1: The mathematical derivation for F_{MIN} can be found in (Ludwig, Bretchko and Bogdanov 2008). BC is also derived from the same formulas, based on a minimization of the noise factor equation.

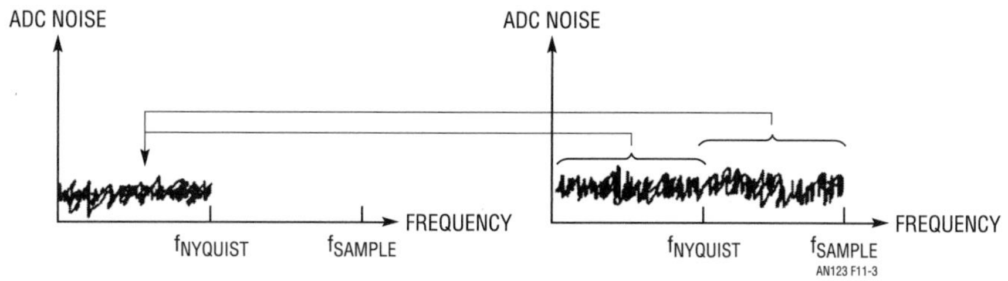

Figure B3 • Representative ADC and Amplifier Noise Floors. Any Amplifier Noise that Exceeds the ADC's Nyquist Bandwidth Will be Aliased Back into the Nyquist Band and Combined with That Noise. In This Example, There are Two Full Nyquist Bands of Amplifier Noise That Will be Summed with the ADC's Noise. Since the Noise is Uncorrelated, the RMS Method (Square Root of Sum of Squares) Can Be Used to Add Them Together

Since F_{MIN} is the linear form of NF_{MIN} and R_N is the inverse of G_N, G_C can be found with the values in Table 25.3. As an example, the LTC6400-8 values will be used to solve for the current noise components.

$$Y_{OPT} = 1/Z_{OPT} = 0.001414 - j0.0008147 \quad (C6)$$

$$F_{MIN} = 5.152 = 1 + 2 \bullet 885 \bullet (0.001414 + G_C) \quad (C7)$$

$$G_C = 0.0009318 \quad (C8)$$

$$Y_C = G_C + B_C = 0.0009318 + j0.0008147 \quad (C9)$$

$$GU = 885 \bullet 0.0014142 - 885 \bullet 0.00093182 = 0.001 \quad (C10)$$

$$i_{nc} = Y_C \bullet e_n = Y_C \bullet 3.7nV/\sqrt{HZ} = 3.45 + j3.01pA/\sqrt{Hz} \quad (C11)$$

$$i_{nu} = \sqrt{(4 \bullet K \bullet 290 \bullet 0.001)} = 4.00\ pA/\sqrt{Hz} \quad (C12)$$

Equation (C11) shows that the correlated current noise has a significant reactive component. If voltage noise were ignored, the total current noise is simply the RMS sum of inu and the magnitude of inc (they are not correlated to each other). However, a complex source impedance Z_S could result in the current noise adding to or partially cancelling the voltage noise (depending on the phase of $i_{nc} \bullet Z_S$), so the total noise varies with source impedance. This happens even if Z_S has no reactive component. The source impedance for optimal noise figure is $\Gamma_{S,OPT}$ in Figure 25.12, where the imaginary part of i_{nc} is cancelled to create a noise minimum.

2 nanosecond, 0.1% resolution settling time measurement for wideband amplifiers

Quantifying quick quiescence

26

Jim Williams

Introduction

Instrumentation, waveform synthesis, data acquisition, feedback control systems and other application areas utilize wideband amplifiers. Current generation components (see box section, next page, "A Precision Wideband Amplifier with 9ns Settling Time") feature good DC precision while maintaining high speed operation. Verifying precision operation at high speed is essential, and presents a high order measurement challenge.

Settling time defined

Amplifier DC specifications are relatively easy to verify. Measurement techniques are well understood, albeit often tedious. AC specifications require more sophisticated approaches to produce reliable information. In particular, amplifier settling time is extraordinarily difficult to determine. Settling time is the elapsed time from input application until the output arrives at and remains within a specified error band around the final value. It is usually specified for a full-scale transition. Figure 26.1 shows that settling time has three distinct components. The *delay time* is small and almost entirely due to amplifier propagation delay. During this interval there is no output movement.

During *slew time* the amplifier moves at its highest possible speed towards the final value. *Ring time* defines the region where the amplifier recovers from slewing and ceases movement within some defined error band. There is normally a trade-off between slew and ring time. Fast slewing amplifiers generally have extended ring times, complicating amplifier choice and frequency compensation. Additionally, the architecture of very fast amplifiers usually dictates trade-offs which degrade DC error terms[1].

Measuring anything at any speed requires care. Dynamic measurement is particularly challenging. Reliable nanosecond region settling time measurement constitutes a high order difficulty problem requiring exceptional care in approach and experimental technique[2].

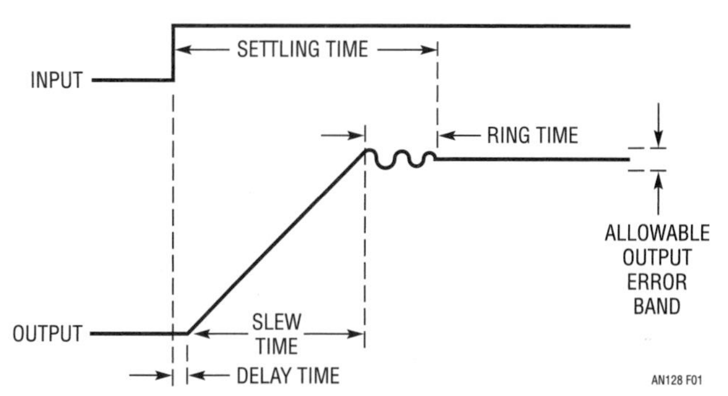

Figure 26.1 • Settling Time Components Include Delay, Slew and Ring Times. Fast Amplifiers Reduce Slew Time, Although Longer Ring Time Usually Results. Delay Time Is Normally a Small Term

Considerations for measuring nanosecond region settling time

Historically, settling time has been measured with circuits similar to that in Figure 26.2. The circuit uses the "false sum node" technique. The resistors and amplifier form a bridge type network. Assuming ideal resistors,

Note 1: This issue is treated in detail in latter portions of the text. Also see Appendix B, "Practical Considerations for Amplifier Compensation."
Note 2: The approach used for settling time measurement and its description, while new, borrows from previous publications. See References 1–5, and Reference 9.

Analog Circuit and System Design: Immersion in the Black Art of Analog Design. http://dx.doi.org/10.1016/B978-0-12-397888-2.00026-2

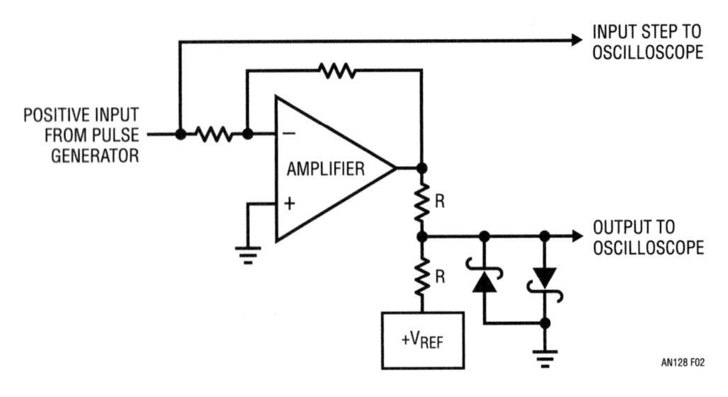

Figure 26.2 • Popular Summing Scheme for Settling Time Measurement Provides Misleading Results. Pulse Generator Post-Transition Aberrations Appear at Output. Large Oscilloscope Overdrive Occurs. Displayed Information Is Meaningless

the amplifier output will step to $-V_{IN}$ when the input is driven. During slew, the settle node is bounded by the diodes, limiting voltage excursion. When settling occurs, the oscilloscope probe voltage should be zero. Note that the resistor divider's attenuation means the probe's output will be one-half of the actual settled voltage.

A precision wideband amplifier with 9ns settling time

Historically, wideband amplifiers provided speed, but sacrificed precision and, often, settling time. The LT1818 op amp does not require this compromise. It features low offset voltage and bias current with adequate gain for 0.1% accuracy. Settling time is 9ns to 0.1% for a 5V step. The output will drive a 100Ω load to ±3.75V with ±5V supplies, and up to 20pF capacitive loading is permissible at unity gain. The table below provides short form specifications.

LT1818 Short Form Specifications

CHARACTERISTIC	SPECIFICATION
Offset Voltage	0.2mV
Offset Voltage vs Temperature	10μV/°C
Bias Current	2μA
DC Gain	2500
Noise Voltage	$6nV/\sqrt{Hz}$
Output Current	70mA
Slew Rate	2500V/μs
Gain-Bandwidth	400MHz
Delay	1ns
Settling Time	9ns/0.1%
Supply Current	9mA

In theory, this circuit allows settling to be observed to small amplitudes. In practice, it cannot be relied upon to produce useful measurements. Several flaws exist. The circuit requires the input pulse to have a flat top within the required measurement limits. Typically, settling within 5mV or less for a 5V step is of interest. No general purpose pulse generator is meant to hold output amplitude and noise within these limits. Generator output-caused aberrations appearing at the oscilloscope probe will be indistinguishable from amplifier output movement, producing unreliable results. The oscilloscope connection also presents problems. As probe capacitance rises, AC loading of the resistor junction influences observed settling waveforms. 1× probes are not suitable because of their excessive input capacitance. A 10× probe's attenuation sacrifices oscilloscope gain and its 10pF capacitance still introduces significant lag at nanosecond speeds. An active 1×, 1pF FET probe largely alleviates the problem but a more serious issue remains.

The clamp diodes at the settle node are intended to reduce swing during amplifier slewing, preventing excessive oscilloscope overdrive. Unfortunately, oscilloscope overdrive recovery characteristics vary widely among different types and are not usually specified. The Schottky diodes' 400mV drop means the oscilloscope will undergo an unacceptable overload, bringing displayed results into question[3].

At 0.1% resolution (5mV at the amplifier output −2.5mV at the oscilloscope), the oscilloscope typically undergoes a 10× overdrive at 10mV/DIV and the desired 2.5mV baseline is unattainable. At nanosecond speeds, the measurement becomes hopeless with this arrangement. There is clearly no chance of measurement integrity.

The preceding discussion indicates that measuring amplifier settling time requires an oscilloscope that is somehow immune to overdrive and a "flat top" pulse generator. These become the central issues in wideband amplifier settling time measurement.

The only oscilloscope technology that offers inherent overdrive immunity is the classical sampling 'scope[4]. Unfortunately, these instruments are no longer manufactured (although still available on the secondary market). It is possible, however; to construct a circuit that borrows the overload advantages of classical sampling 'scope technology. Additionally, the circuit can be endowed with features particularly suited for measuring nanosecond range settling time.

Note 3: For a discussion of oscilloscope overdrive considerations, see Appendix C, "Evaluating Oscilloscope Overdrive Performance".

Note 4: Classical sampling oscilloscopes should not be confused with modern era digital sampling 'scopes that have overdrive restrictions. See Appendix C, "Evaluating Oscilloscope Overload Performance" for comparisons of various type 'scopes with respect to overdrive. For detailed discussion of classical sampling 'scope operation, see References 23–26 and 29–31. Reference 24 is noteworthy; it is the most clearly written, concise explanation of classical sampling instruments the author is aware of—a 12-page jewel.

The flat-top pulse generator requirement can be avoided by switching current, rather than voltage. It is much easier to gate a quickly settling current into the amplifier's summing node than to control a voltage. This makes the input pulse generator's job easier, although it still must have a rise time of about 1 nanosecond to avoid measurement errors.

Practical nanosecond settling time measurement

Figure 26.3 is a conceptual diagram of a settling time measurement circuit. This figure shares attributes with Figure 26.2, although some new features appear. In this case, the oscilloscope is connected to the settle point by a switch. The switch state is determined by a delayed pulse generator, which is triggered from the input pulse. The delayed pulse generator's timing is arranged so that the switch does not close until settling is very nearly complete. In this way, the incoming waveform is sampled in time, as well as amplitude. The oscilloscope is never subjected to overdrive—no off-screen activity ever occurs.

A switch at the amplifier's summing junction is controlled by the input pulse. This switch gates current to the amplifier via a voltage-driven resistor. This eliminates the "flat-top" pulse generator requirement, although the switch must be fast and devoid of drive artifacts.

Figure 26.4 is a more complete representation of the settling time scheme. Figure 26.3's blocks appear in greater detail and some new refinements show up. The amplifier summing area is unchanged. Figure 26.3's delayed pulse generator has been split into two blocks; a delay and a pulse generator, both independently variable. The input step to the oscilloscope runs through a section that compensates for the propagation delay of the settling time measurement path. Similarly, another delay compensates sample gate pulse generator propagation delay. This delay causes the sample gate pulse generator to be driven with a phase-advanced version of the pulse which triggers the amplifier under test. This considerably improves minimum measurable settling time by making sample gate pulse generator propagation delay irrelevant.

The most striking new aspects of the diagram are the diode bridge switch and the multiplier. The diode bridge's balance, combined with matched, low capacitance Schottky diodes and high speed drive, yields clean switching. The bridge switches current into the amplifier's summing point very quickly, with settling inside a nanosecond. The diode clamp to ground prevents excessive bridge drive swings and ensures that non-ideal input pulse characteristics are nearly irrelevant.

Requirements for Figure 26.4's sample gate are stringent. It must faithfully pass wideband signal path information without introducing alien components, particularly those deriving from the switch command channel ("sample gate pulse")[5].

The sample gate multiplier functions as a wideband, high resolution, extremely low feedthrough switch. The great advantage of this approach is that the switch control channel can be maintained in-band; that is, its transition rate is held within the multipliers 250MHz bandpass. The multipliers wide bandwidth means the switch command transition is under control at all times. There are no out-of-band responses, greatly reducing feedthrough and parasitic artifacts.

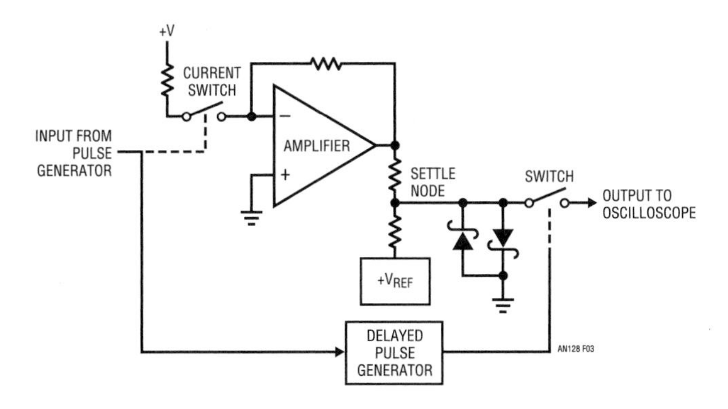

Figure 26.3 • Conceptual Arrangement Is Insensitive to Pulse Generator Aberrations and Eliminates Oscilloscope Overdrive. Input Switch Gates Current Step to Amplifier. Second Switch, Controlled by Delayed Pulse Generator, Prevents Oscilloscope from Monitoring Settle Node Until Settling Is Nearly Complete

Note 5: Conventional choices for the sample gate switch include FET's and the sampling diode bridge. FET parasitic gate to channel capacitances result in large gate drive originated feedthrough into the signal path. For almost all FETs, this feedthrough is many times larger than the signal to be observed, inducing overload and obviating the switches' purpose. The diode bridge is better; its small parasitic capacitances tend to cancel and the symmetrical, differential structure results in very low feedthrough. Practically, the bridge requires DC and AC trims and complex drive and support circuitry. LTC Application Note 74, "Component and Measurement Advances Ensure 16-bit DAC Settling Time" utilized such a sampling bridge and it is detailed in that text. See Reference 3. References 2,9,11 describe a similar sampling bridge based approach.

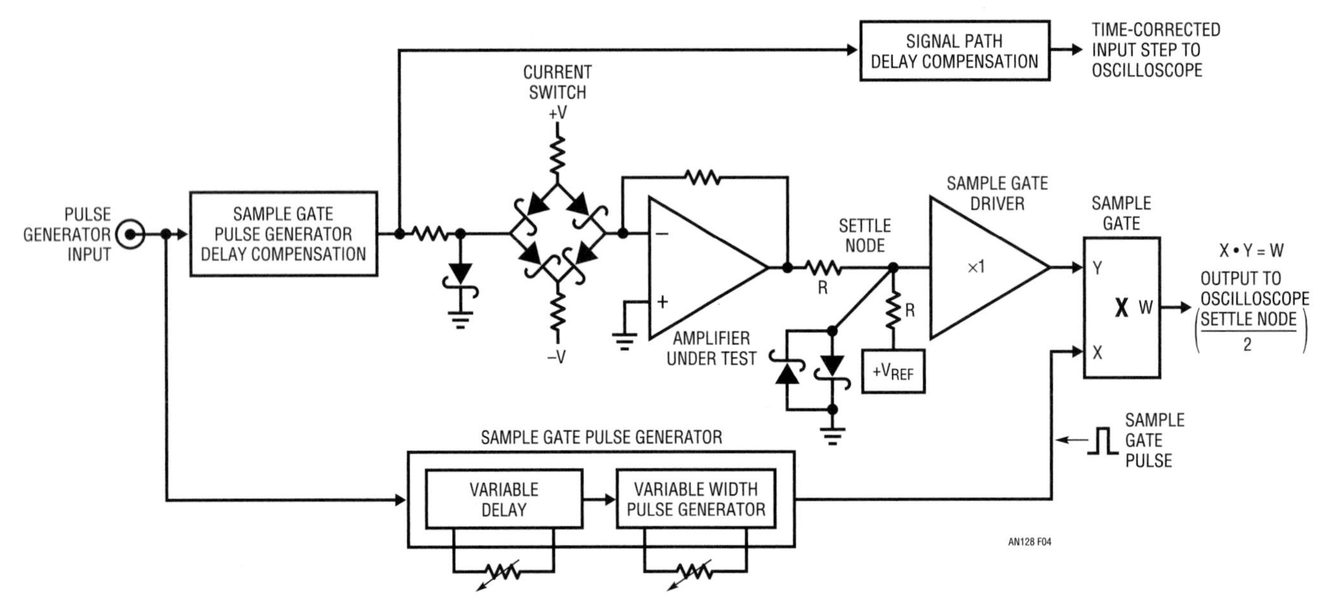

Figure 26.4 • Block Diagram of Settling Time Measurement Scheme. Diode Bridge Cleanly Switches Input Current to Amplifier. Multiplier Based Sampling "Switch" Eliminates Signal Paths Pre-Settling Excursion, Preventing Oscilloscope Overdrive. Input Step Time Reference and Sample Gate Pulse Generator Are Compensated for Test Circuit Delays

Detailed settling time circuitry

Figure 26.5 is a detailed schematic of the settling time measurement circuitry. The input pulse switches the input bridge via a delay network ("A" inverters) and a driver stage ("C" inverters). The delay compensates the sample gate pulse generator's delayed response, ensuring that the sample gate pulse can occur immediately after the amplifier-under-tests' slew time ends. The delay range is chosen so that the sample gate pulse can be adjusted to occur *before* the amplifier slews. This capability is obviously unused in operation although it guarantees that the settling interval will always be capturable.

The "C" inverters form a non-inverting driver stage to switch the diode bridge. Various trims optimize driver output pulse shape, providing a clean, fast impulse to the diode bridge[6]. The high fidelity pulse, devoid of undamped components, prevents radiation and disruptive ground currents from degrading the measurement noise floor. The driver also activates the "B" inverters, which supply a time corrected input step to the oscilloscope.

The driver output pulse transitions through the 1N5712 diode clamp potential in under a nanosecond, causing essentially instantaneous diode bridge switching. The resultant cleanly settling current into the amplifier under tests' summing point causes proportionate amplifier output movement. The negative bias current at the amplifiers summing point combined with the current step produces a +2.5V to −2.5V amplifier output transition. The amplifier's output is compared against a 5V supply derived reference via the summing resistors. The clamped "settle node" is unloaded by A1, which feeds the sample gate signal path information.

The comparator based sample gate pulse generator produces a delayed (controllable by the 20k potentiometer) pulse whose width (controllable by the 2k potentiometer) sets sample gate on-time. The Q1 stage forms the sample gate pulse into a fast rise, exceptionally clean event, furnishing high purity, calibrated amplitude, "on-off" switching instruction to the sample gate multiplier. If the sample gate pulse delay is set appropriately, the oscilloscope will not see any input until settling is nearly complete, eliminating overdrive. The sample window width is adjusted so that all remaining settling activity is observable. In this way, the oscilloscope's output is reliable and meaningful data may be taken.

Figure 26.6 shows circuit waveforms. Trace A is the time-corrected input pulse, Trace B the amplifier output,

Note 6: To maintain text flow and focus, trimming procedures are not presented here. Detailed trimming information appears in Appendix A, "Measuring and Compensating Settling Circuit Delay and Trimming Procedures."

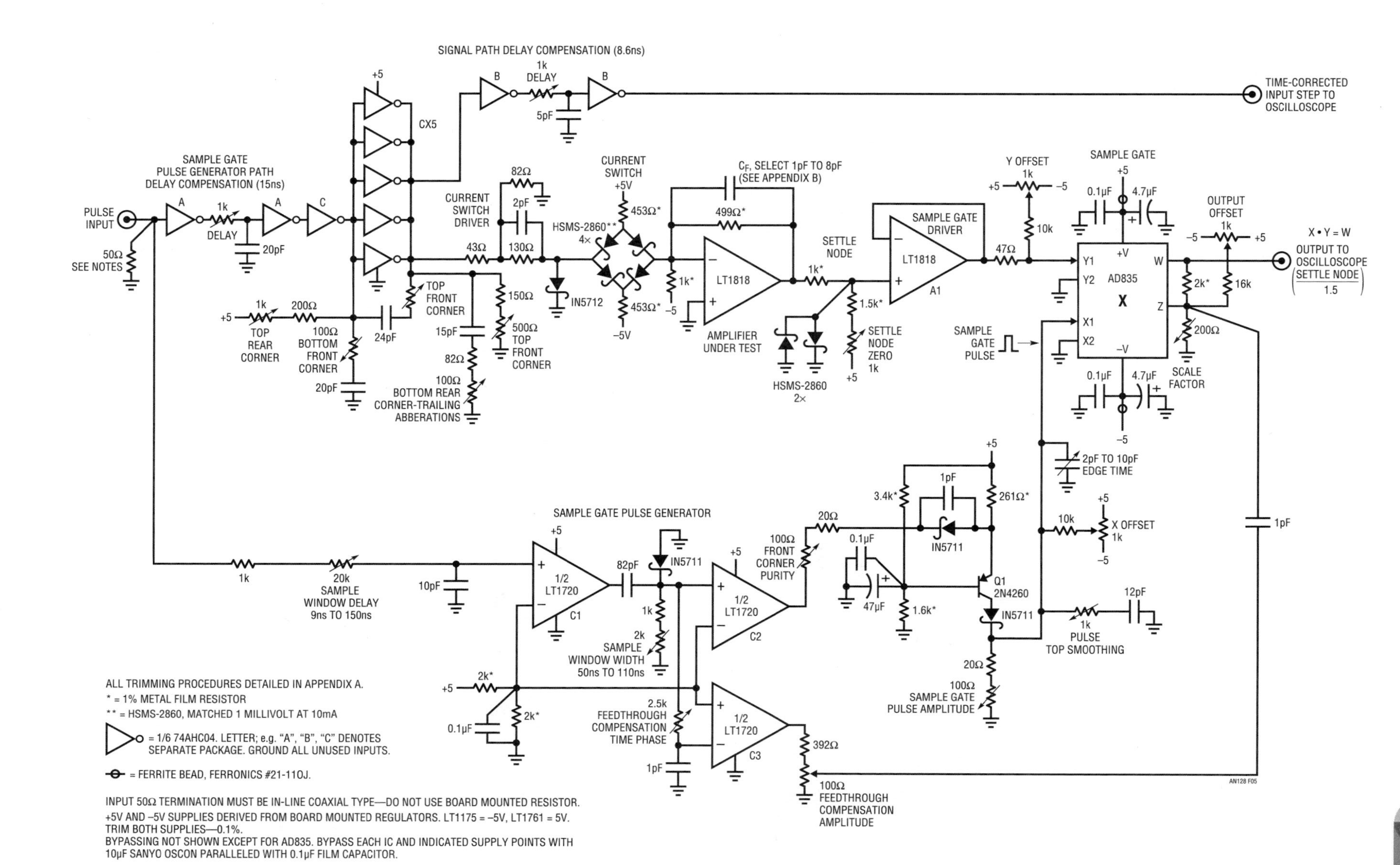

Figure 26.5 • Detailed Schematic of Settling Time Measurement Circuitry Follows Block Diagram. Trimmed, Paralleled Logic Inverters Provide High Speed Drive to Current Switch Bridge. Additional Inverters Form Delay Compensation Networks for Signal Path and Sample Gate Pulse Generator. Transistor Stage Shapes Edges and Amplitude of Sample Gate Pulse Supplied to Multiplier. Multiplier, Functioning as Sample Gate, Passes Settling Time Signal when Sample Gate Pulse Is High

Trace C the sample gate pulse and Trace D the settling time output[7]. When the sample gate pulse goes high, the sample gate switches cleanly, and the last 20mV of slew are easily observed. Ring time is also clearly visible, and the amplifier settles nicely to final value. When the sample gate pulse goes low, the sample gate switches off with only 2mV of feedthrough. Note that there is no off-screen activity at any time—the oscilloscope is never subjected to overdrive.

Figure 26.7 expands vertical and horizontal scales so that settling detail is more visible[8]. Trace A is the time-corrected input pulse and Trace B the settling output. The last 50mV of slew are easily observed, and the amplifier settles inside 5mV (0.1%) in 9 nanoseconds when CF (see Figure 26.5) is optimized[9].

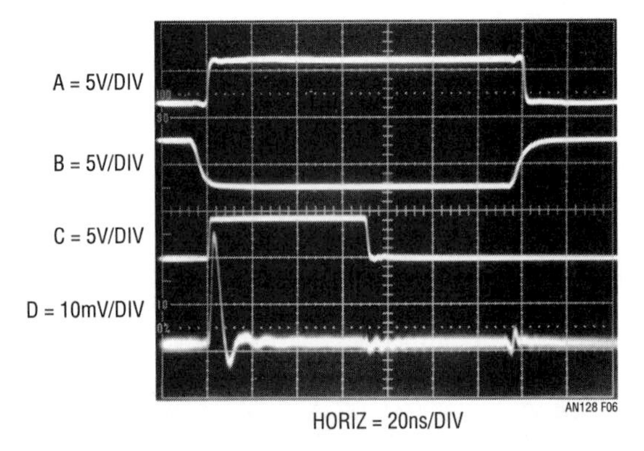

Figure 26.6 • Settling Time Circuit Waveforms Include Time-Corrected Input Pulse (Trace A), Amplifier-Under-Test Output (Trace B), Sample Gate (Trace C) and Settling Time Output (Trace D). Sample Gate Window's Delay and Width Are Variable. Trace B Appears Time Skewed Relative to Time Corrected Trace A

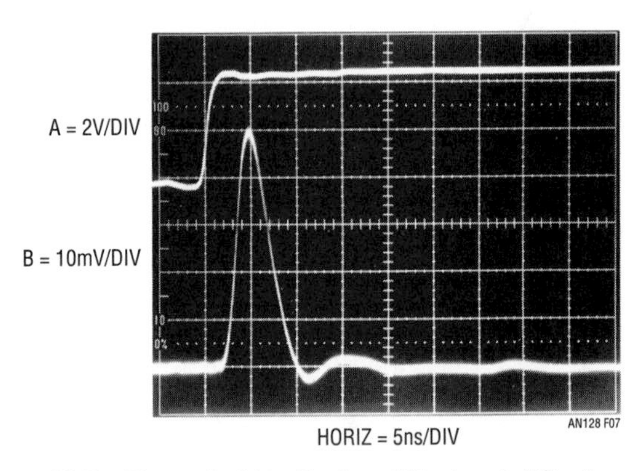

Figure 26.7 • Expanded Vertical and Horizontal Scales Show 9ns Amplifier Settling within 5mV (Trace B). Trace A Is Time Corrected Input Step

Using the sampling-based settling time circuit

In general, it is good practice to walk the sampling window "backwards" in time up to the last 50mV or so of amplifier slewing so that the onset of ring time is observable without encountering oscilloscope overdrive. The sampling based approach provides this capability and it is a very powerful measurement tool. Slower amplifiers may require extended delay and/or sampling window times, necessitating larger capacitor values in the delayed pulse generator timing networks.

Verifying results—alternate method

The sampling-based settling time circuit appears to be a useful measurement solution. How can its results be tested to ensure confidence? A good way is to make the same measurement with an alternate method and see if results agree. It was stated earlier that classical sampling oscilloscopes were inherently immune to overdrive[10]. If this is so, why not utilize this feature and attempt settling time measurement directly at the clamped settle node? Figure 26.8 does this. Under these conditions, the sampling scope[11] is heavily overdriven, but is ostensibly immune to the insult. Figure 26.9 puts the sampling oscilloscope to the test. Trace A is the time corrected input pulse and Trace B the settle signal. Despite a brutal overdrive, the 'scope appears to respond cleanly, giving a very plausible settle signal presentation.

Summary of results and measurement limits

The simplest way to summarize the different method's results is by visual comparison. Ideally, if both approaches represent good measurement technique and are properly constructed, results should be identical. If this is the case, the data produced by the two methods has a high probability of being valid. Examination of Figures 26.9 and 26.10

Note 7: When interpreting waveform placement note that trace B appears time skewed relative to time corrected trace A. This accounts for trace B's falsely apparent movement before trace A's ascent.

Note 8: In this and all following photos, settling time is measured from the onset of the time-corrected input pulse. Additionally, settling signal amplitude is calibrated with respect to the amplifier not the settle node. This eliminates ambiguity due to the settle node's resistance ratio.

Note 9: This section mentions amplifier frequency compensation within the context of sampling-based settling time measurement. As such, it is necessarily brief. Considerably more detail is available in Appendix B, "Practical Considerations for Amplifier Compensation."

Note 10: See Appendix C, "Evaluating Oscilloscope Overdrive Performance" for in-depth discussion.

Note 11: Tektronix type 661 with 4S1 vertical and 5T3 timing plug-ins.

shows nearly identical settling times and highly similar settling waveform signatures. This kind of agreement provides a high degree of credibility to the measured results.

Close observation of settling time circuit operation indicates a noise floor/feedthrough imposed amplitude resolution limit of 2mV. The time resolution limit is about 2 nanoseconds to 5mV settling. For details, see the section "Measurement Limits and Uncertainties", in Appendix A, "Measuring and Compensating Settling Circuit Delay and Trimming Procedures."

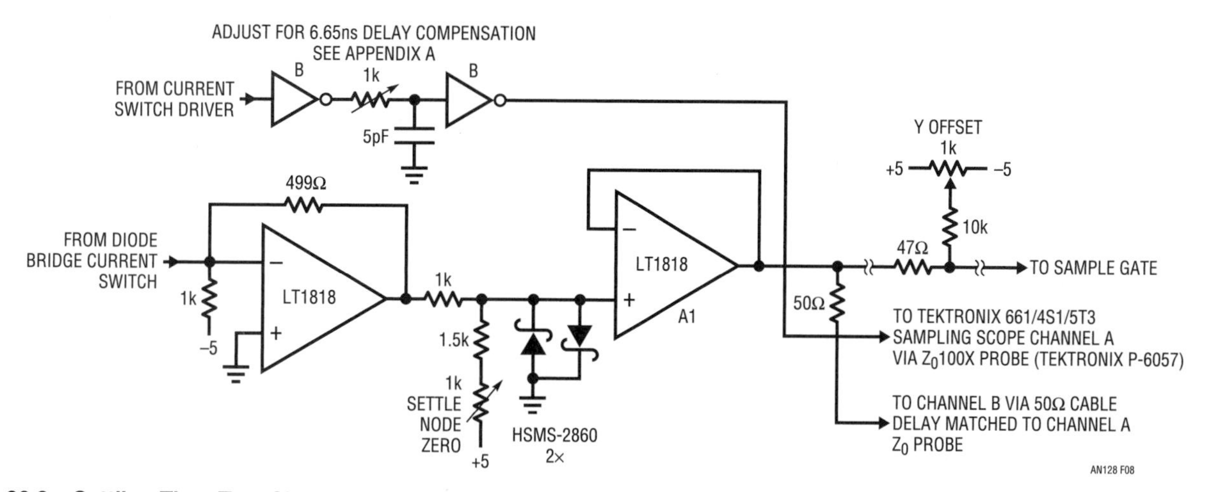

Figure 26.8 • Settling Time Test Circuit Modifications Using Classical Tektronix 661/4S1/5T3 1GHz Sampling Oscilloscope. Sampling 'Scope's Inherent Overload Immunity Permits Large Off-Screen Excursions without Degrading Measurement Fidelity

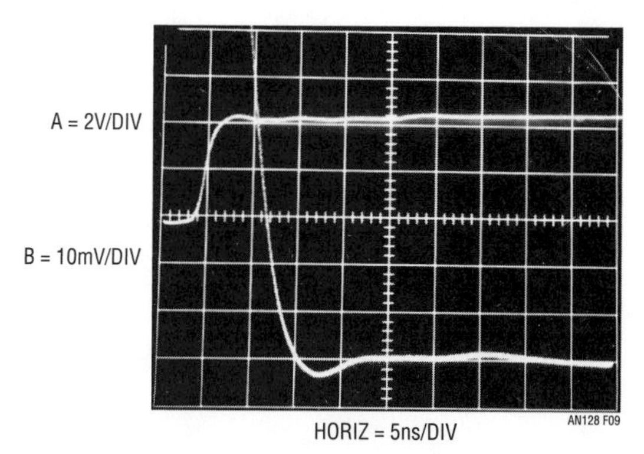

Figure 26.9 • Settling Time Measurement with Classical Sampling 'Scope. Oscilloscope's Overload Immunity Allows Accurate Measurement Despite Extreme Overdrive. 9ns Settling Time and Waveform Profile Are Consistent with Figure 26.7

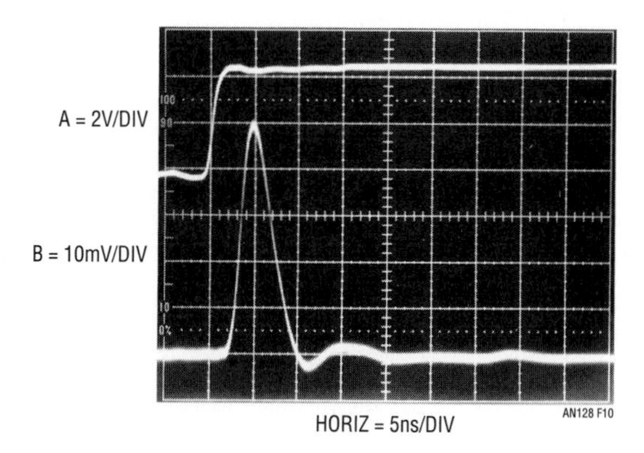

Figure 26.10 • Settling Time Measurement Using Figure 26.5's Circuit. T_{SETTLE} = 9ns. Results Correlate with Figure 26.9

References

1. Williams, Jim, "1ppm Settling Time Measurement for a Monolithic 18-Bit DAC," Linear Technology Corporation, Application Note 120, February 2010
2. Williams, Jim, "30 Nanosecond Settling Time Measurement for a Precision Wideband Amplifier," Linear Technology Corporation, Application Note 79, September 1999
3. Williams, Jim, "Component and Measurement Advances Ensure 16-Bit DAC Settling Time," Linear Technology Corporation, Application Note 74, July 1998
4. Williams, Jim, "Measuring 16-Bit Settling Times: The Art of Timely Accuracy," EDN, November 19, 1998
5. Williams, Jim, "Methods for Measuring Op Amp Settling Time," Linear Technology Corporation, Application Note 10, July 1985
6. LT1818 Data Sheet, Linear Technology Corporation
7. AD835 Data Sheet, Analog Devices, Inc
8. Elbert, Mark, and Gilbert, Barrie, "Using the AD834 in DC to 500MHz Applications: RMS-to-DC Conversion, Voltage-Controlled Amplifiers, and Video Switches", p. 6–47. "The AD834 as a Video Switch", "Applications Reference Manual", Analog Devices, Inc., 1993
9. Kayabasi, Cezmi, "Settling Time Measurement Techniques Achieving High Precision at High Speeds," MS Thesis, Worcester Polytechnic Institute, 2005
10. Demerow, R., "Settling Time of Operational Amplifiers," Analog Dialogue, Volume 4-1, Analog Devices, Inc., 1970
11. Pease, R.A., "The Subtleties of Settling Time," The New Lightning Empiricist, Teledyne Philbrick, June 1971
12. Harvey, Barry, "Take the Guesswork Out of Settling Time Measurements," EDN, September 19, 1985
13. Williams, Jim, "Settling Time Measurement Demands Precise Test Circuitry," EDN, November 15, 1984
14. Schoenwetter, H.R., "High Accuracy Settling Time Measurements," IEEE Transactions on Instrumentation and Measurement, Vol. IM-32. No.1, March 1983
15. Sheingold, D.H., "DAC Settling Time Measurement," Analog-Digital Conversion Handbook, pg. 312–317. Prentice Hall, 1986.
16. Orwiler, Bob, "Oscilloscope Vertical Amplifiers," Tektronix, Inc., Concept Series, 1969
17. Addis, John, "Fast Vertical Amplifiers and Good Engineering," Analog Circuit Design; Art, Science and Personalities, Butterworths, 1991
18. Travis, W., "Settling Time Measurement Using Delayed Switch," Private Communication, 1984
19. Hewlett-Packard, "Schottky Diodes for High Volume, Low Cost Applications," Application Note 942, Hewlett-Packard Company, 1973

20. Jim Williams, "Signal Sources, Conditioners and Power Circuitry," Linear Technology Corporation, Application Note 98 (November 2004) 26–27
21. Williams, Jim and Beebe, David, "Diode Turn-On Induced Failures in Switching Regulators", Linear Technology Corporation, Application Note 122, January 2009, p. 14–19
22. Korn, G.A. and Korn, TM., "Electronic Analog and Hybrid Computers," "Diode Switches," p. 223–226. McGraw-Hill, 1964.
23. Carlson, R., "A Versatile New DC-500MHz Oscilloscope with High Sensitivity and Dual Channel Display," Hewlett-Packard Journal, Hewlett-Packard Company, January 1960
24. Tektronix, Inc. "Sampling Notes," Tektronix, Inc., 1964
25. Tektronix, Inc. "Type 1S1 Sampling Plug-In Operating and Service Manual," Tektronix, Inc. 1965
26. Mulvey, J. "Sampling Oscilloscope Circuits," Tektronix, Inc., Concept Series, 1970
27. Addis, John, "Sampling Oscilloscopes," Private Communication, February 1991
28. Williams, Jim, "Bridge Circuits-Marrying Gain and Balance," Linear Technology Corporation, Application Note 43, June 1990
29. Tektronix, Inc., "Type 661 Sampling Oscilloscope Operating and Service Manual," Tektronix, Inc., 1963
30. Tektronix, Inc., "Type 4S1 Sampling Plug-In Operating and Service Manual," Tektronix, Inc., 1963
31. Tektronix, Inc., "Type 5T3 Timing Unit Operating and Service Manual," Tektronix, Inc., 1965
32. Morrison, Ralph, "Grounding and Shielding Techniques in Instrumentation," 2nd Edition, Wiley Interscience, 1977
33. Ott, Henry W., "Noise Reduction Techniques in Electronic Systems," Wiley Interscience, 1976
34. Williams, Jim, "High Speed Amplifier Techniques," Linear Technology Corporation, Application Note 47, 1991
35. Weber, Joe, "Oscilloscope Probe Circuits," Tektronix, Inc., Concept Series, 1969
36. Ott, Henry, "Electromagnetic Compatibility Engineering," Wiley and Sons, 2009
37. Bogatin, Eric, "Signal and Power Integrity-Simplified," 2nd Edition, Prentice Hall, 2009

Appendix A

Measuring and compensating settling circuit delay and trimming procedures

The settling time circuit requires trimming to achieve quoted performance. The trims fall into four loosely defined categories including current switch bridge drive pulse shaping, circuit delays, sample gate pulse purity and sample gate feedthrough/DC adjustments[1].

Bridge drive trims

The current switch bridge drive is trimmed first. Disconnect all 5 bridge drive related trims and apply a 5V, 1MHz, 10 to 15 nanosecond wide pulse at the circuit input. The paralleled "C" inverter output viewed at the 43Ω back termination's undriven end should resemble Figure A1. Waveform edge times are fast but poorly controlled parasitic excursions risk corrupting the measurement noise floor and must be eliminated. Reconnect all 5 trims and adjust them according to their titles for Figure A2's much improved presentation. There is some interaction between the adjustments but it is limited and favorable results are easily attained. Figure A2's edge times are slightly slower than Figure A1's, but still pass through the 1N5712 clamp level in <1 nanosecond.

Delay determination and compensation

Circuit delay related trims come next. Before making these measurements and adjustments, probe/oscilloscope channel-to-channel time skewing must be corrected. Figure A3 shows 40 picosecond time skew error with both channel probes connected to a 100 picosecond rise time pulse source[2]. The error is corrected in Figure A4 by utilizing the oscilloscopes vertical amplifier variable delay feature (Tektronix 7A29, option 04, installed in a Tektronix 7104 mainframe). This correction permits high accuracy delay measurements to be made[3].

Note 1: The trims require considerable care in instrumentation selection as well as thoughtful wideband probing and oscilloscope measurement technique. See Appendixes D through H for tutorial guidance before proceeding.

Note 2: See Appendix H, "Verifying Rise Time and Delay Measurement Integrity" for fast pulse source recommendations.

Note 3: This assumes the oscilloscope time base has been verified for accuracy. For recommendations, see Appendix H, "Verifying Rise Time and Delay Measurement Integrity".

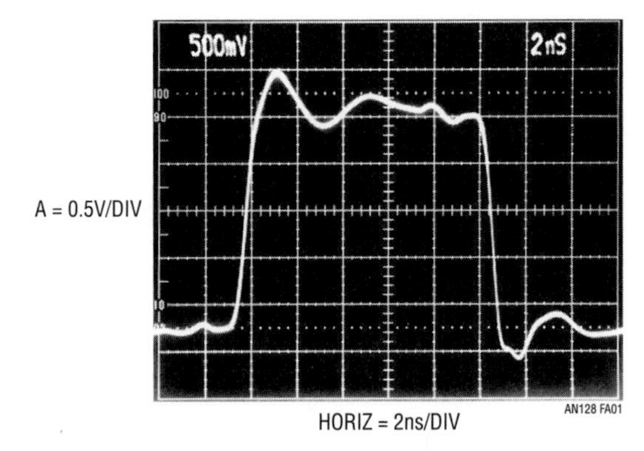

Figure A1 • Untrimmed Current Switch Driver Response at 43Ω Back Termination Viewed in 1GHz, Real Time Bandwidth. Edge Times Are Fast, But Poorly Controlled. Undamped Waveform Artifacts Risk Corrupting Signal Path Noise Floor Via Radiation and Ground Current Disruption Induced Errors

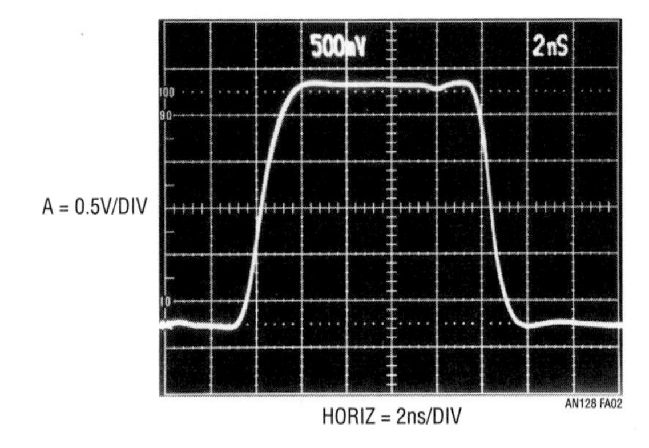

Figure A2 • Trimmed Current Switch Driver Output at 43Ω Back Termination Passes Through 0.6V Diode Clamp Potential in <1 Nanosecond. AC Trims Promote Clean, Well Controlled Waveform

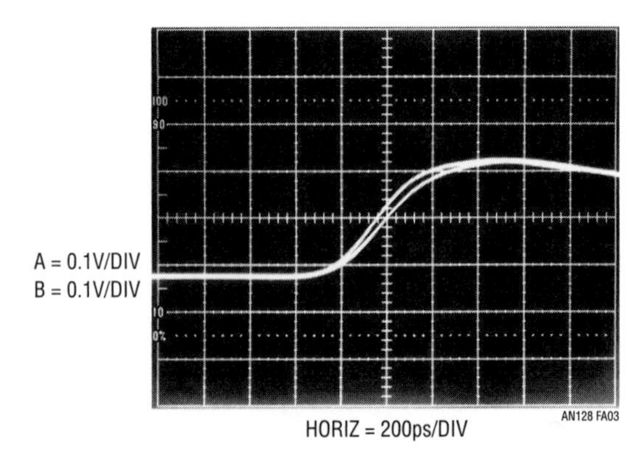

Figure A3 • Probe-Oscilloscope Channel-to-Channel Timing Skew Measures 40 Picoseconds

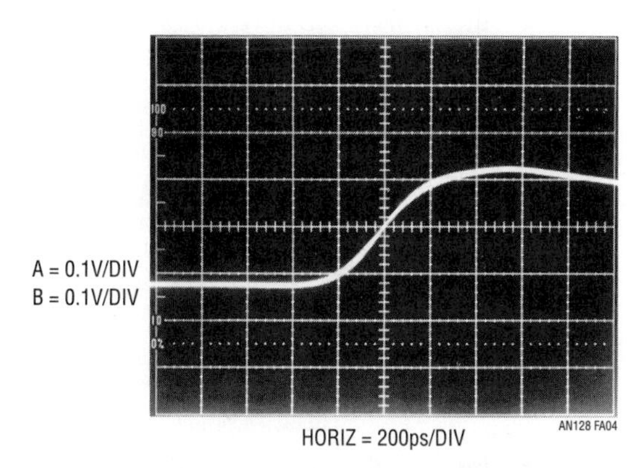

Figure A4 • Corrected Probe/Channel/Skew Shows Nearly Identical Time and amplitude Response

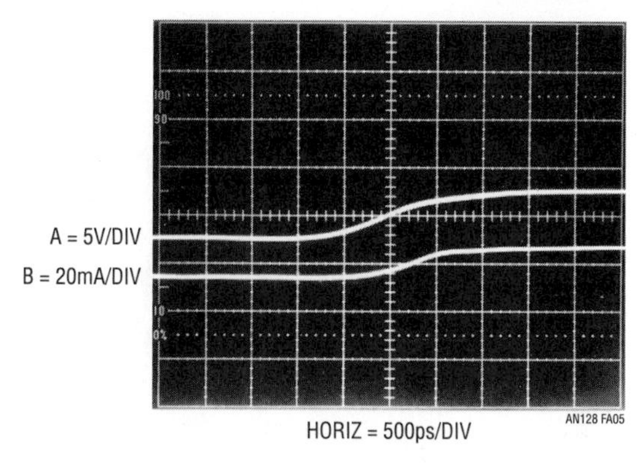

Figure A5 • Current Switch Driver (Trace A) to Amplifier-Under-Test Negative Input (Trace B) Delay Is 250 Picoseconds

The settling time circuit utilizes an adjustable delay network to time correct the input pulse for delays in the signal processing path. Typically, these delays introduce errors approaching 10 nanoseconds, so an accurate correction is required. Setting the delay trim involves observing the network's input-output delay and adjusting for the appropriate time interval. Determining the "appropriate" time interval is somewhat more complex.

Referring to Figure 26.5, it is apparent that three delay measurements are of interest. The current switch driver to amplifier-under-test negative input, the amplifier-under-test output to circuit output and the sample gate multiplier delay. Figure A5 indicates 250 picoseconds delay from the current switch driver to the amplifier-under-test input. Figure A6 reveals 8.4 nanoseconds delay from the amplifier-under-test output to circuit output and Figure A7 shows sample gate multiplier delay of 2 nanoseconds. The measurements indicate a current switch driver-to-circuit output delay of 8.65 nanoseconds; the correction is implemented by adjusting the 1k trim in the "Signal Path Delay Compensation" network for that amount. Similarly, when the sampling 'scope is used, the relevant delays are Figure A5 plus A6 minus Figure A7, a total of 6.65 nanoseconds. This factor is adjusted into the signal path delay compensation network when the sampling 'scope-based measurement is taken.

The "Sample Gate Pulse Generator Path Delay Compensation" trim is less critical. The sole requirement is that it overlap the sample gate pulse generator's delay. Setting the 1k potentiometer in the "A" inverter chain to 15 nanoseconds satisfies this criteria, completing the delay related trims.

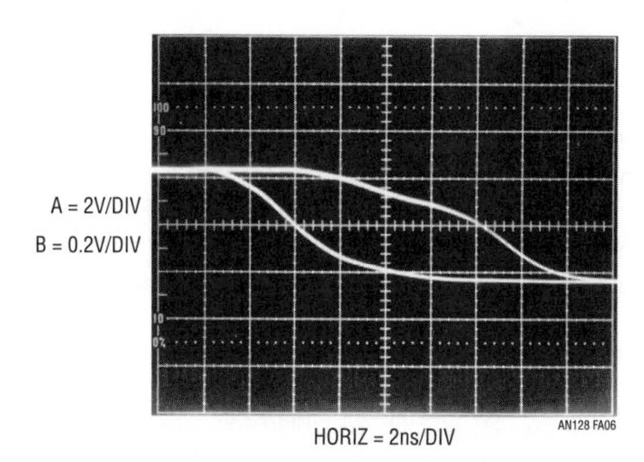

Figure A6 • Amplifier-Under-Test (Trace A) to Circuit Output (Trace B) Delay Measures 8.4 Nanoseconds. Multiplier X Input Held at 1V DC for This Test

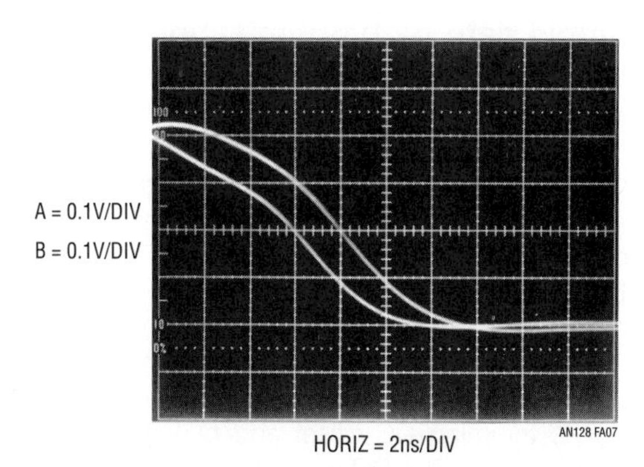

Figure A7 • Multiplier Delay with X Input Held at 1V DC Measures 2ns

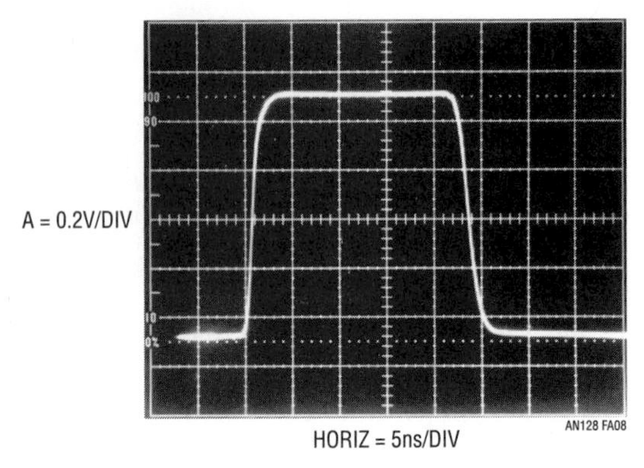

Figure A8 • Sample Gate Pulse Characteristics, Controlled by Edge Shaping, Circuit Configuration and Transistor Choice, Are Kept Within Multiplier's 250MHz (T_{RISE} = 1.4ns) Bandwidth. Accurate, Low Feedthrough, Y Input Signal Path Switching Results

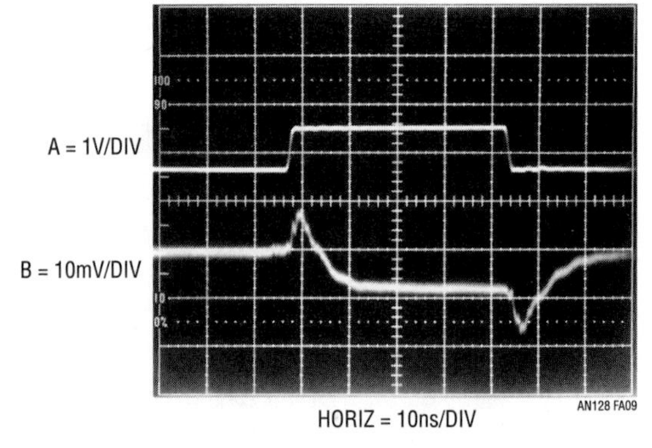

Figure A9 • Settling Time Circuit's Output (Trace B) with Unadjusted Sample Gate Feedthrough and DC Offset. A1's Input Grounded for This Test. Excessive Switch Drive Feedthrough and Baseline Offset Are Present. Trace A Is Sample Gate Pulse

Sample gate pulse purity adjustment

The Q1 sample gate pulse edge shaping stage is adjusted for an optimized front corner, minimum rising edge time, pulse top smoothing and 1V amplitude with the indicated trims. The mildly interactive adjustments converge to Figure A8's display, taken at the sample gate multiplier's X input. The pulse's 2 nanosecond rise time promotes rapid sample gate acquisition but remains within the multipliers 250MHz (t_{RISE} = 1.4ns) bandwidth, assuring freedom from out-of-band parasitic responses. The clean, 1V amplitude pulse top provides calibrated, consistent multiplier output devoid of aberrations which would masquerade as settling signal artifacts. Pulse fall time is irrelevant; it is not germaine to the measurement and its clean falling transition assures controlled multiplier turn-off, precluding off-screen excursions.

Sample gate path optimization

The sample gate path adjustments are the final trims. First, put in 5V DC to the pulse generator input to lock the amplifier-under-test into its −2.5V output state. Adjust the "settle node zero" trim for zero volts within 1mV at A1's output. Next, restore the pulsed circuit input, disconnect the settle node from A1 and ground A1's input with a 750Ω resistor. Figure A9 is typical of the resultant untrimmed response. Ideally, the circuit output (trace B) should be static during sample gate (trace A) switching. The photo reveals errors; correction requires trimming DC offset and dynamic feedthrough related residue. The DC errors are eliminated by adjusting the "X" and "Y" offset trims for a continuous trace B baseline regardless of trace A's sample gate pulse state. Additionally, set the output offset adjustment for minimum multiplier baseline offset voltage. Sample gate gain is set to unity by shutting off the input pulse generator, applying 5V DC to C2's "+" input and forcing 1.00V DC at the previously inserted 750Ω resistor. Under these conditions, adjust "scale factor" for 1.00V DC output. After completing this step, remove the DC bias voltages and the 750Ω resistor; reconnect the settle node and restore the pulsed input.

Feedthrough compensation is accomplished via feedthrough "time phase" and "amplitude" trims. These adjustments set timing and amplitude of the feedthrough correction applied at the multiplier "Z" input. Optimal adjustment results in Figure A10's presentation. This photograph shows the DC and feedthrough trims dramatic effect on Figure A9's pre-trim errors[4].

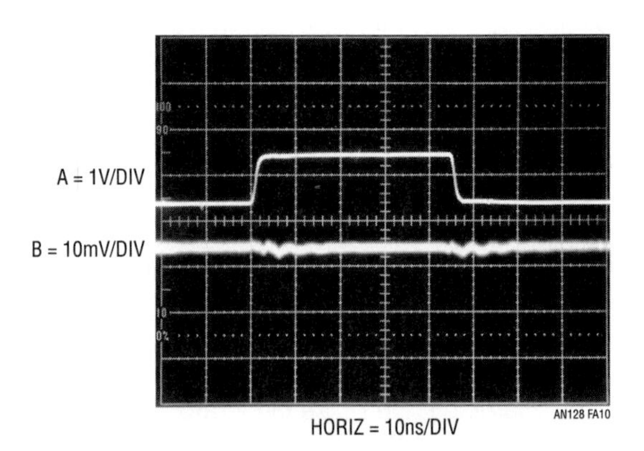

Figure A10 • Settling Time Circuit's Output (Trace B) with Sample Gate Trimmed. As in Figure A9, Al's Input Is Grounded for This Test. Switch Drive Feedthrough and Baseline Offset Are Minimized. Trace A Is Sample Gate Pulse. Measurement Defines Circuit's 2mV Minimum Amplitude Resolution Limit

Note 4: The writer is not much for Hollywood's offerings, but does find drama in feedthrough trims.

Measurement Limits and Uncertainties

Figure A10's post trim response includes a flat baseline and greatly attenuated feedthrough. The measurement defines the circuit's minimum amplitude resolution at 2mV. In another test, A1's input is disconnected from the settle node and biased at 20mV DC via a 750Ω resistor to simulate an infinitely fast settling amplifier. Figure A11 shows circuit output (trace B) settling within 5mV in 2 nanoseconds, arriving inside the 2mV baseline noise limit in 3.6 nanoseconds. This data, taken with sample gate conduction beginning immediately after the time corrected input (trace A) rises, defines the circuit's minimum time resolution limit. Uncertainties in the quoted time and amplitude resolution limits are primarily due to delay compensation limitations, noise and residual feedthrough. Considering likely delay and measurement errors, a time uncertainty of ±500 picoseconds and a 2mV resolution limit is probably realistic. Noise averaging would not improve the amplitude resolution limit because it is imposed by feedthrough residue, a coherent term.

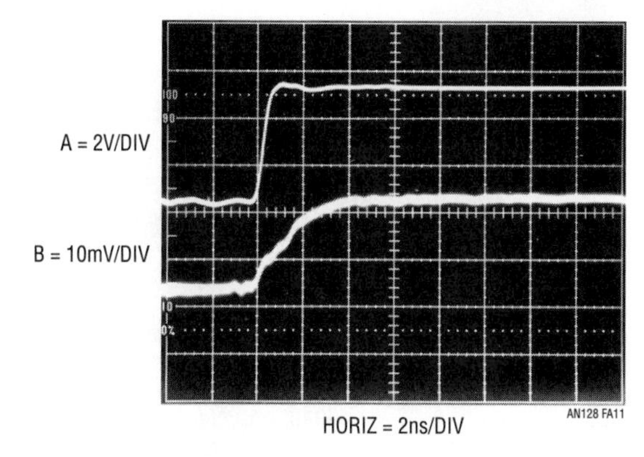

Figure A11 • Circuit Response with 20mV DC Forced at A1's Input. Output (Trace B) Is within 5mV in 2ns, Arriving Inside 2mV Baseline Noise in 3.6ns. Measurement Defines Circuits Minimum Time Resolution Limit. Trace A Is Time Corrected Input Pulse

Appendix B
Practical considerations for amplifier compensation

There are a number of practical considerations in compensating the amplifier to get fastest settling time. Our study begins by revisiting text Figure 26.1 (repeated here as Figure B1). Settling time components include delay, slew and ring times. Delay is due to propagation time through the amplifier and is a relatively small term. Slew time is set by the amplifier's maximum speed. Ring time defines the region where the amplifier recovers from slewing and ceases movement within some defined error band. Once an amplifier has been chosen, only ring time is readily adjustable. Because slew time is usually the dominant lag, it is tempting to select the fastest slewing amplifier available to obtain best settling. Unfortunately, fast slewing amplifiers usually have extended ring times, negating their brute force speed advantage. The penalty for raw speed is, invariably, prolonged ringing, which can only be damped with large compensation capacitors. Such compensation works, but results in protracted settling times. The key to good settling times is to choose an amplifier with the right balance of slew rate and recovery characteristics and compensate it properly. This is harder than it sounds because amplifier settling time cannot be predicted or extrapolated from any combination of data sheet specifications. It must be measured in the intended configuration. A number of terms combine to influence settling time. They include amplifier slew rate and AC dynamics, layout capacitance, source resistance and capacitance, and the compensation capacitor. These terms interact in a complex manner, making predictions hazardous[1]. If the parasitics are eliminated and replaced with a pure resistive source, amplifier settling time is still not readily predictable. The parasitic impedance terms just make a difficult problem more messy. The only real handle available to deal with all this is the feedback compensation capacitor, C_F. C_F's purpose is to roll off amplifier gain at the frequency that permits best dynamic response.

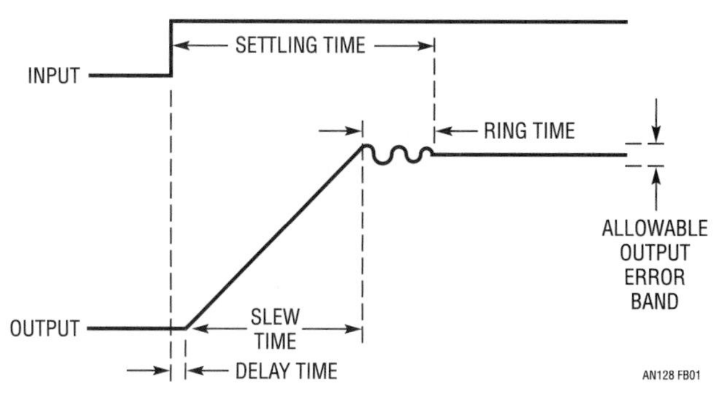

Figure B1 • Settling Time Components Include Delay, Slew and Ring Times. For Given Components, Only Ring Time Is Readily Adjustable

Note 1: Spice aficionados take notice.

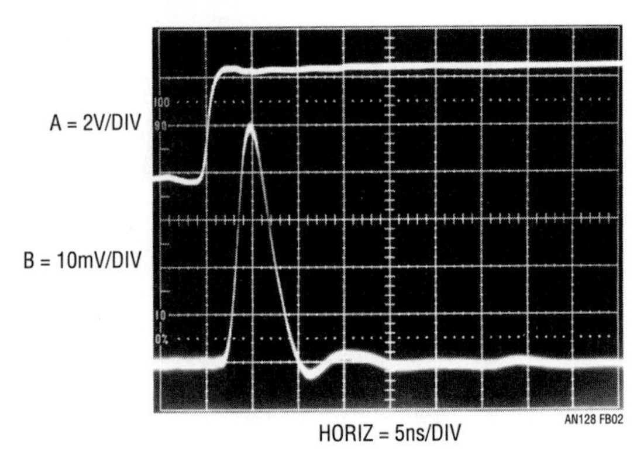

Figure B2 • Optimized Compensation Capacitor Permits Tight Waveform Signature, Nearly Critically Damped Response and Fastest Settling Time. T_{SETTLE} = 9ns. Trace A Is Time Corrected Input Step, Trace B, the Settle Signal

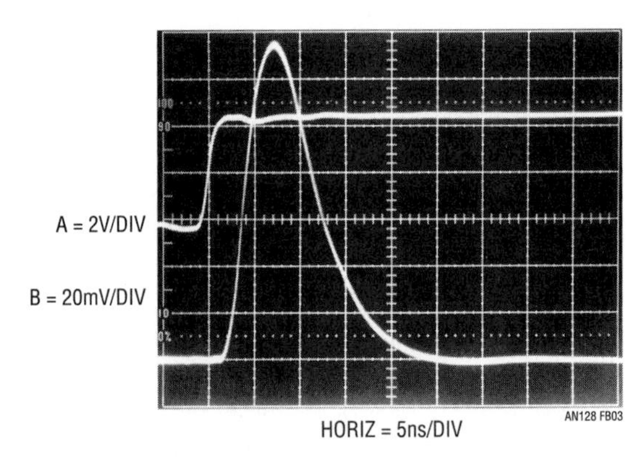

Figure B3 • Overdamped Response Ensures Freedom from Ringing, Even with Component Variations in Production. Penalty Is Increased Settling Time. Note 2X Vertical Scale Change vs. Figure B2. T_{SETTLE} = 22ns. Trace Assignments Same as Previous Figure

Best settling results when the compensation capacitor is selected to functionally compensate for all the above terms. Figure B2 shows results for an optimally selected feedback capacitor Trace A is the time-corrected input pulse and trace B the amplifier's settle signal. The amplifier comes cleanly out of slew (sample gate opens just after the second vertical division) and settles to 5mV in 9 nanoseconds. Waveform signature is tight and nearly critically damped.

In Figure B3, the feedback capacitor is too large. Settling is smooth, although overdamped; a 13 nanosecond penalty results in 22 nanosecond settling. Figure B4 has no feedback capacitor, causing severely underdamped response with resultant excessive ring time excursions. Settling time goes out to 33 nanoseconds. B5 improves on B4 by restoring the feedback capacitor, but the value is too small, resulting in an underdamped response requiring 27 nanoseconds to settle. Note that Figures B3 to B5 require vertical scale reduction to capture non-optimal response.

When feedback capacitors are individually trimmed for optimal response, the source, stray, amplifier and compensation capacitor tolerances are irrelevant. If individual trimming is not used, these tolerances must be considered to determine the feedback capacitor's production value. Ring time is affected by stray and source capacitance and output loading, as well as the feedback capacitor's value. The relationship is nonlinear, although some guidelines are possible. The stray and source

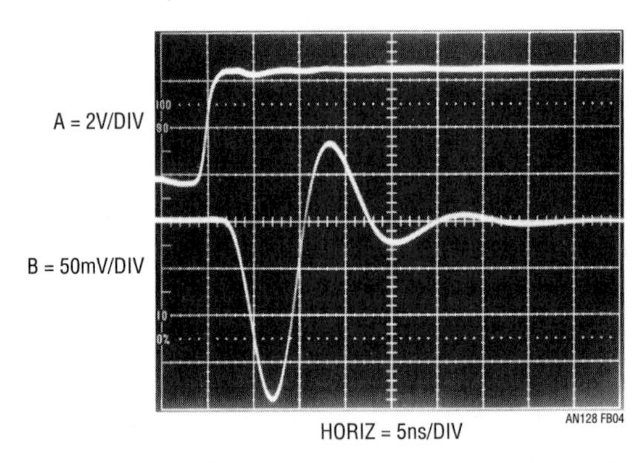

Figure B4 • Severely Underdamped Response Due to No Feedback Capacitor. Note 5X Vertical Scale Change vs Figure B2. T_{SETTLE} = 33ns. Trace Assignments as in Figure B2

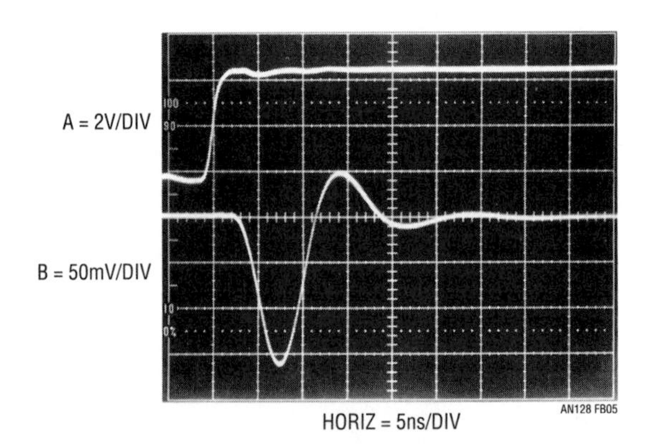

Figure B5 • Underdamped Response Results from Undersized Capacitor. Component Tolerance Budgeting Will Prevent This Behavior. Note 5x Vertical Scale Change vs. Figure B2. T_{SETTLE} = 27ns. Trace Assignments as in Figure B2

terms can vary by ±10% and the feedback capacitor is typically a ±5% component[2]. Additionally, amplifier slew rate has a significant tolerance, which is stated on the data sheet. To obtain a production feedback capacitor value, determine the optimum value by individual trimming with the production board layout (board layout parasitic capacitance counts too!). Then, factor in the worst-case percentage values for stray and source impedance terms, slew rate and feedback capacitor tolerance. Add this information to the trimmed capacitors measured value to obtain the production value. This budgeting is perhaps unduly pessimistic (RMS error summing may be a defensible compromise), but will keep you out of trouble[3].

Note 2: This assumes a resistive source. If the source has substantial parasitic capacitance (photodiode, DAC, etc.), this number can easily enlarge to ±50%

Note 3: The potential problems with RMS error summing become clear when sitting in an airliner that is landing in a snowstorm.

Appendix C

Evaluating oscilloscope overdrive performance

The sampling based settling time circuit is heavily oriented towards preventing overdrive to the monitoring oscilloscope. This is done to avoid overdriving the oscilloscope. Oscilloscope recovery from overdrive is a grey area and almost never specified. How long must one wait after an overdrive before the display can be taken seriously? The answer to this question is quite complex. Factors involved include the degree of overdrive, its duty cycle, its magnitude in time and amplitude and other considerations. Oscilloscope response to overdrive varies widely between types and markedly different behavior can be observed in any individual instrument. For example, the recovery time for a 100× overload at 0.005V/DIV may be very different than at 0.1V/DIV The recovery characteristic may also vary with waveform shape, DC content and repetition rate. With so many variables, it is clear that measurements involving oscilloscope overdrive must be approached with caution.

Why do most oscilloscopes have so much trouble recovering from overdrive? The answer to this question requires some study of the three basic oscilloscope types' vertical path. The types include analog (Figure C1A), digital (Figure C1B) and classical sampling (Figure C1C) oscilloscopes. Analog and digital 'scopes are susceptible to overdrive. The classical sampling 'scope is the only architecture that is inherently immune to overdrive.

An analog oscilloscope (Figure C1A) is a real time, continuous linear system[1]. The input is applied to an attenuator; which is unloaded by a wideband buffer. The vertical preamp provides gain, and drives the trigger pick-off, delay line and the vertical output amplifier. The attenuator and delay line are passive elements and require little comment although they can display reactive behavior at speed and resolution extremes. The buffer, preamp and vertical output amplifier are complex linear gain blocks, each with dynamic operating range restrictions. Additionally, the operating point of each block may be set by inherent circuit balance, low frequency stabilization paths or both. When the input is overdriven, one or more of these stages may saturate, forcing internal nodes and components to abnormal operating points and temperatures. When the overload ceases, full recovery of the electronic and thermal time constants may require surprising lengths of time[2].

The digital sampling oscilloscope (Figure C1B) eliminates the vertical output amplifier, but has an attenuator buffer and amplifiers ahead of the A/D converter. Because of this, it is similarly susceptible to overdrive recovery problems.

The classical sampling oscilloscope is unique. Its nature of operation makes it inherently immune to overload. Figure C1C shows why. The sampling occurs before any gain is taken in the system. Unlike Figure C1B's digitally sampled 'scope, the input is fully passive to the sampling point. Additionally, the output is fed back to the sampling bridge, maintaining its operating point over a very wide range of inputs. The dynamic swing available to maintain the bridge output is large and easily accommodates a wide range of oscilloscope inputs. Because of all this, the amplifiers in this instrument do not see overload, even at 1000× overdrives, and there is no recovery problem. Additional immunity derives from

Note 1: Ergo, the Real Thing. Hopelessly bigoted residents of this locale mourn the passing of the analog 'scope era and frantically hoard every instrument they can find.

Note 2: Some discussion of input overdrive effects in analog oscilloscope circuitry is found in Reference 17.

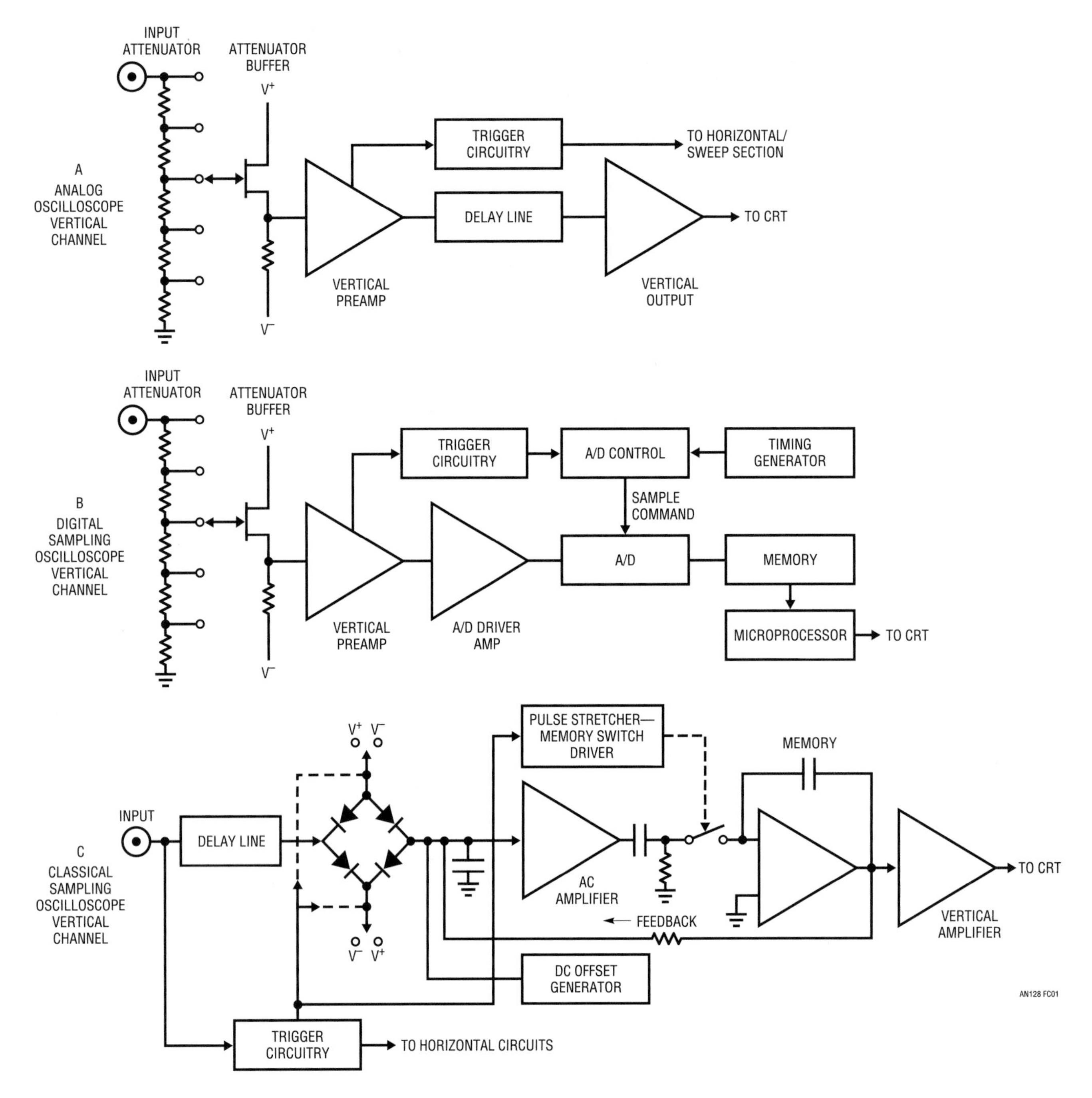

Figure C1 • Simplified Vertical Channel Diagrams for Different Type Oscilloscopes. Only the Classical Sampling 'Scope (C) Has Inherent Overdrive Immunity. Offset Generator Allows Viewing Small Signals Riding On Large Excursions

the instrument's relatively slow sample rate—even if the amplifiers were overloaded, they would have plenty of time to recover between samples.[3]

The designers of classical sampling 'scopes capitalized on the overdrive immunity by including variable DC offset generators to bias the feedback loop (see Figure C1C, lower right). This permits the user to offset a large input, so small amplitude activity on top of the signal can be accurately observed. This is ideal for, among other things, settling time measurements. Unfortunately, classical sampling oscilloscopes are no longer manufactured, so if you have one, take care of it![4]

Note 3: Additional information and detailed treatment of classical sampling oscilloscope operation appears in References 23–26 and 29–31.

Note 4: Modern variants of the classical architecture (e.g., Tektronix 11801B) may provide similar capability, although we have not tried them.

Although analog and digital oscilloscopes are susceptible to overdrive, many types can tolerate some degree of this abuse. The early portion of this appendix stressed that measurements involving oscilloscope overdrive must be approached with caution. Nevertheless, a simple test can indicate when the oscilloscope is being deleteriously affected by overdrive.

The waveform to be expanded is placed on the screen at a vertical sensitivity that eliminates all off-screen activity. Figure C2 shows the display. The lower right hand portion is to be expanded. Increasing the vertical sensitivity by a factor of two (Figure C3) drives the waveform off-screen, but the remaining display appears reasonable. Amplitude has doubled and waveshape is consistent with the original display. Looking carefully, it is possible to see small amplitude information presented as a dip in the waveform at about the third vertical division. Some small disturbances are also visible. This observed expansion of the original waveform is believable. In Figure C4, gain has been further increased, and all the features of Figure C3 are amplified accordingly. The basic waveshape appears clearer and the dip and small disturbances are also easier to see. No new waveform characteristics are observed. Figure C5 brings some unpleasant surprises.

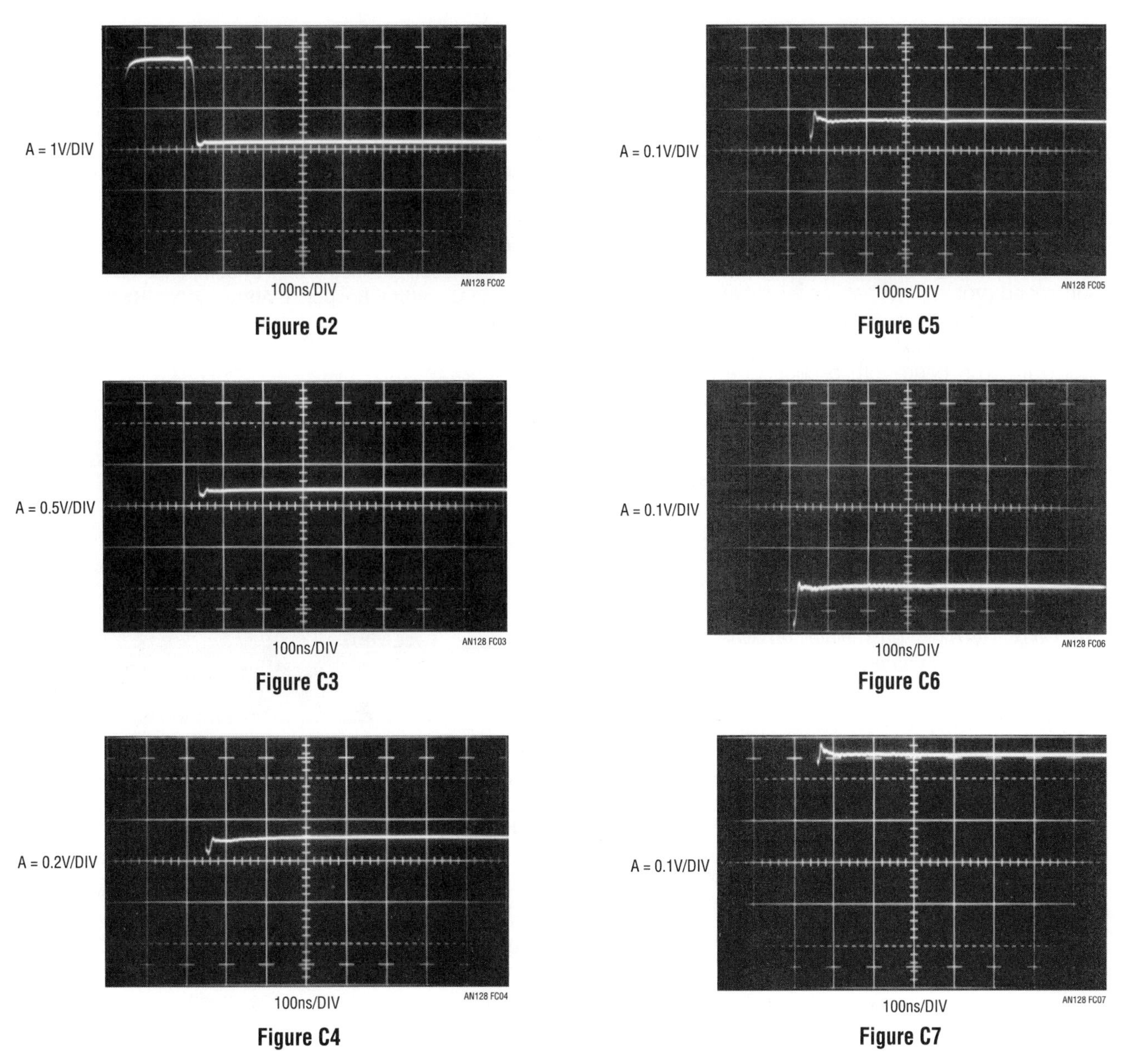

Figure C2 to C7 • The Overdrive Limit Is Determined by Progressively Increasing Oscilloscope Gain and Watching for Waveform Aberrations

This increase in gain causes definite distortion. The initial negative-going peak, although larger, has a different shape. Its bottom appears less broad than in Figure C4. Additionally, the peak's positive recovery is shaped slightly differently. A new rippling disturbance is visible in the center of the screen. This kind of change indicates that the oscilloscope is having trouble. A further test can confirm that this waveform is being influenced by overloading. In Figure C6 the gain remains the same but the vertical position knob has been used to reposition the display at the screen's bottom.[5] This shifts the oscilloscope's DC operating point which, under normal circumstances, should not affect the displayed waveform. Instead, a marked shift in waveform amplitude and outline occurs. Repositioning the waveform to the screen's top produces a differently distorted waveform (Figure C7). It is obvious that for this particular waveform, accurate results cannot be obtained at this gain.

Note 5: Knobs (derived from Middle English, "knobbe", akin to Middle Low German, "knubbe"), cylindrically shaped, finger rotatable panel controls for controlling instrument functions, were utilized by the ancients.

Appendix D
About Z_O probes

When to roll your own and when to pay the money

Z_O (e.g. low impedance) probes provide the most faithful high speed probing mechanism available for low source impedances. Their sub-picofarad input capacitance and near ideal transmission characteristic make them the first choice for high bandwidth oscilloscope measurement. Their deceptively simple operation invites "do-it-yourself" construction but numerous subtleties mandate difficulty for prospective constructors. Arcane parasitic effects introduce errors as speed increases beyond about 100MHz (t_{RISE} = 3.5ns). The selection and integration of probe materials and the probes physical incarnation require extreme care to obtain high fidelity at high speed. Additionally, the probe must include some form of adjustment to compensate small, residual parasitics. Finally, true coaxiality must be maintained when fixturing the probe at the measurement point, implying a high grade, readily disconnectable, coaxial connection capability.

Figure D1 shows that a Z_O probe is basically a voltage divided input 50Ω transmission line. If R1 equals 450Ω, 10× attenuation and 500Ω input resistance result. R1 of 4950Ω causes a 100× attenuation with 5k input resistance. The 50Ω line theoretically constitutes a distortionless transmission environment. The apparent simplicity seemingly permits "do-it-yourself" construction but this section's remaining figures demonstrate a need for caution.

Figure D2 establishes a fidelity reference by measuring a clean 700ps rise time pulse using a 50Ω line terminated via a 10× coaxial attenuator—no probe is employed.

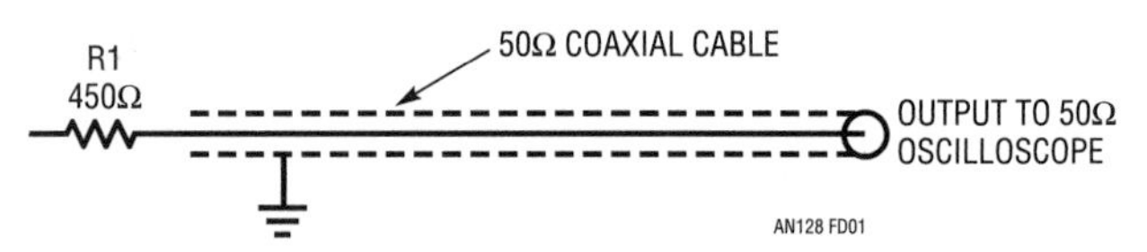

Figure D1 • Conceptual 500Ω, "Z_O", 10x Oscilloscope Probe. If R1 = 4950Ω, 5k Input Resistance with 100x Signal Attenuation Results. Terminated Into 50Ω, Probe Theoretically Constitutes Distortionless Transmission Line. "Do-It-Yourself" Probes Suffer Uncompensated Parasitics, Causing Unfaithful Response Above ≈100MHz (t_{RISE} = 3.5ns)

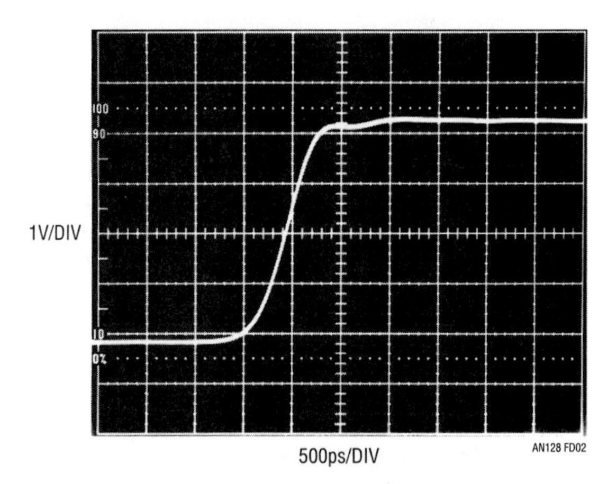

Figure D2 • 700ps Rise Time Pulse Observed via 50Ω Line and 10x Coaxial Attenuator Has Good Pulse Edge Fidelity with Controlled Post-Transition Events

The waveform is singularly clean and crisp with minimal edge and post-transition aberrations. Figure D3 depicts the same pulse with a commercially produced 10× Z_O probe in use. The probe is faithful and there is barely discernible error in the presentation. Photos D4

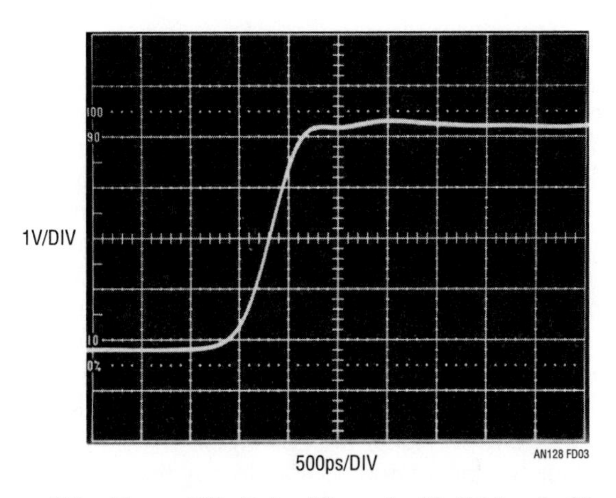

Figure D3 • Figure D2's Pulse Viewed with Tektronix 10x, Z_0 500Ω Probe (P-6056) Introduces Barely Discernible Error

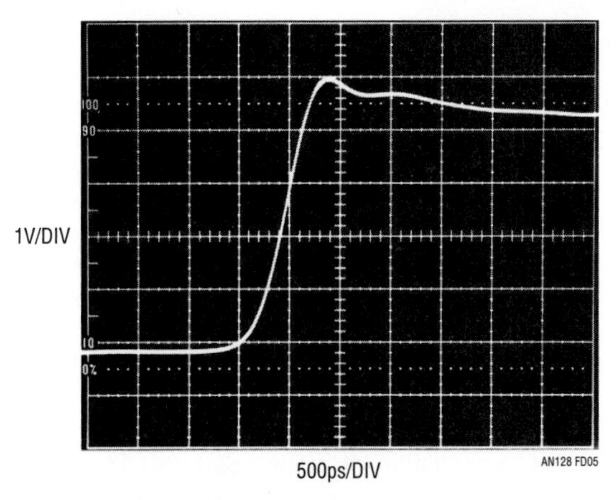

Figure D5 • "Do-It-Yourself" Z_0 Probe #2 Has Overshoot, Again Likely Due to Resistor/Cable Parasitic Terms or Incomplete Coaxiality. Lesson: At These Speeds, Don't "Do-It-Yourself"

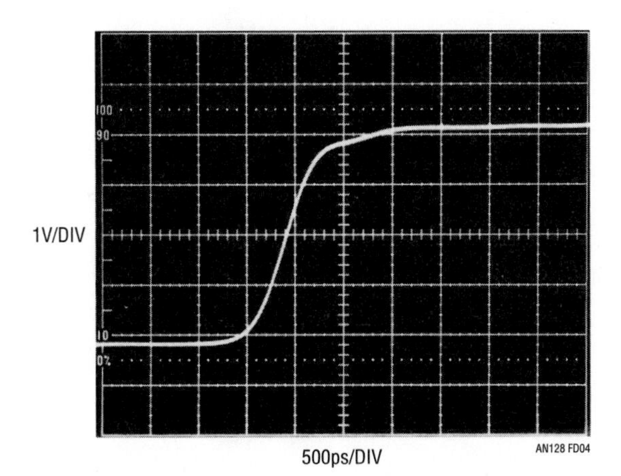

Figure D4 • "Do-It-Yourself" Z_0 Probe #1 Introduces Pulse Corner Rounding, Likely Due to Resistor/Cable Parasitic Terms or Incomplete Coaxiality. "Do-It-Yourself" Z_0 Probes Typically Manifest This Type of Error at Rise Times $\leq$2ns

and D5, taken with two separately constructed "do-it-yourself" Z_0 probes, show errors. In D4, Probe #1 introduces pulse front corner rounding; Probe #2 in D5 causes pronounced corner peaking. In each case, some combination of resistor/cable parasitics and incomplete coaxiality are likely responsible for the errors. In general, "do-it-yourself" Z_0 probes cause these types of errors beyond about 100MHz (t_{RISE} 3.5ns). At higher speeds, if waveform fidelity is critical, it's best to pay the money. For additional terror and wisdom along these lines, see pg. 2-4 → 2-8 of Reference 30 and Reference 35. Both are excellent, informative and, hopefully, sobering to still undaunted do-it-yourselfers.

Appendix E

Connections, cables, adapters, attenuators, probes and picoseconds

Subnanosecond rise time signal paths must be considered as a transmission line. Connections, cables, adapters, attenuators and probes represent discontinuities in this transmission line, deleteriously affecting its ability to faithfully transmit desired signal. The degree of signal corruption contributed by a given element varies with its deviation from the transmission lines nominal impedance. The practical result of such introduced aberrations is degradation of pulse rise time, fidelity, or both. Accordingly, introduction of elements or connections to the signal path should be minimized and necessary connections and elements must be high grade components. Any form of connector, cable, attenuator or probe must be fully specified for high frequency use. Familiar BNC hardware becomes lossy at rise times much faster than 350ps. SMA components are preferred for the rise times described in the text. Additionally, cable should be 50Ω "hard line" or, at least, Teflon-based coaxial cable fully specified for high frequency operation. Optimal connection practice eliminates any cable by coupling the signal output *directly* to the measurement input.

Mixing signal path hardware types via adapters (e.g. BNC/SMA) should be avoided. Adapters introduce significant parasitics, resulting in reflections, rise time degradation, resonances and other degrading behavior. Similarly, oscilloscope connections should be made directly to the instrument's 50Ω inputs, avoiding probes. If probes must be used, their introduction to the signal path mandates attention to their connection mechanism and high frequency compensation. Passive Z_0 types, commercially available in 500Ω (10×) and 5kΩ (100×) impedances, have input capacitance below 1pf.[1] Any such probe must be carefully frequency compensated before use or misrepresented measurement will result. Inserting the probe into the signal path necessitates some form of signal pick-off which nominally does not influence signal transmission. In practice, some amount of disturbance must be tolerated and its effect on measurement results evaluated. High quality signal pick-offs always specify insertion loss, corruption factors and probe output scale factor.

The preceding emphasizes vigilance in designing and maintaining a signal path. Skepticism, tempered by enlightenment, is a useful tool when constructing a signal path and no amount of hope is as effective as preparation and directed experimentation.

Note 1: See Appendix D, "About Z_0 Probes".

Appendix F

Breadboarding, layout and connection techniques

The measurement results presented in this publication required painstaking care in breadboarding, layout and connection techniques. Nanosecond domain, high resolution measurement does not tolerate cavalier laboratory attitude. The oscilloscope photographs presented, devoid of ringing, hops, spikes and similar aberrations, are the result of a careful breadboarding exercise. The breadboard required considerable experimentation before obtaining a noise/uncertainty floor worthy of the measurement.

Ohm's Law

It is worth considering that Ohm's law is a key to successful layout.[1] Consider that 10mA running through 1Ω generates 10mV—twice the measurement limit! Now, run that current at 1 nanosecond rise times (≈350MHz) and the need for layout care becomes clear. A paramount concern is disposal of circuit ground return current and disposition of current in the ground plane. The impedance of the ground plane between any two points is *not* zero, particularly at nanosecond speeds. This is why the entry point and flow of "dirty" ground returns must be carefully placed within the grounding system.

A good example of the importance of grounding management involves delivering the input pulse to the breadboard. The pulse generator's 50Ω termination *must* be an in-line coaxial type. This ensures that pulse generator return current circulates in a tight local loop at the terminator, and does not mix into the signal plane. It is worth mentioning that, because of the nanosecond speeds involved, inductive parasitics may introduce more error than resistive terms. This often necessitates using flat wire braid for connections to minimize parasitic inductive and skin effect-based losses. Every ground return and signal connection in the entire circuit must be evaluated with these concerns in mind. A paranoiac mindset is quite useful.

Shielding

The most obvious way to handle radiation-induced errors is shielding. Determining where shields are required should come *after* considering what layout will minimize their necessity. Often, grounding requirements conflict with minimizing radiation effects, precluding maintaining distance between sensitive points. Shielding is usually an effective compromise in such situations.

A similar approach to ground path integrity should be pursued with radiation management. Consider what points are likely to radiate, and try to lay them out at a distance from sensitive nodes. When in doubt about odd effects, experiment with shield placement and note results, iterating towards favorable performance.[2] *Above all, never rely on filtering or measurement bandwidth limiting to "get rid of" undesired signals whose origin is not fully understood.* This is not only intellectually dishonest, but may produce wholly invalid measurement "results," even if they look pretty on the oscilloscope.

Connections

All signal connections to the breadboard must be coaxial. Ground wires used with oscilloscope probes are forbidden. A 1" ground lead used with a 'scope probe can easily generate large amounts of observed "noise" and seemingly inexplicable waveforms. Use coaxially mounting probe tip adapters![3]

Note 1: I do not wax pedantic here. My guilt in this matter runs deep.

Note 2: After it works, you can figure out why.

Note 3: See Reference 34 for additional nagging along these lines.

Appendix G
How much bandwidth is enough?

Accurate wideband oscilloscope measurements require bandwidth. A good question is just how much is needed. A classic guideline is that "end-to-end" measurement system rise time is equal to the root-sum-square of the system's individual component's rise times. The simplest case is two components; a signal source and an oscilloscope.

Figure G1's plot of $\sqrt{\text{signal}^2 + \text{oscilloscope}^2}$ rise time versus error is illuminating. The figure plots signal-to-oscilloscope rise time ratio versus observed rise time (rise time is bandwidth restated in the time domain, where:

$$\text{Rise Time (ns)} = \frac{350}{\text{Bandwidth (MHz)}})$$

The curve shows that an oscilloscope 3 to 4 times faster than the input signal rise time is required for measurement accuracy inside about 5%. This is why trying to measure a 1ns rise time pulse with a 350MHz oscilloscope (t_{RISE} = 1ns) leads to erroneous conclusions. The curve indicates a monstrous 41% error. Note that this curve does not include the effects of passive probes or cables connecting the signal to the oscilloscope. Probes do not necessarily follow root-sum-square law and must be carefully chosen and applied for a given measurement. For details, see Appendix D. Figure G2, included for reference, gives 10 cardinal points of rise time/bandwidth equivalency between 1MHz and 5GHz.

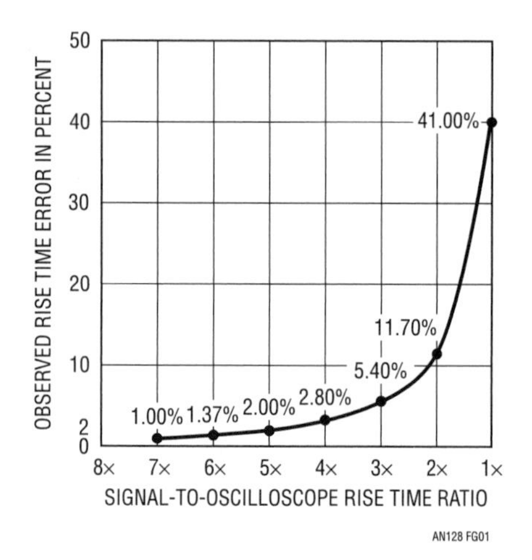

Figure G1 • Oscilloscope Rise Time Effect on Rise Time Measurement Accuracy. Measurement Error Rises Rapidly as Signal-to-Oscilloscope Rise Time Ratio Approaches Unity. Data, Based on Root-Sum-Square Relationship, Does Not Include Passive Probe, Which Does Not Follow Root-Sum-Square Law

RISE TIME	BANDWIDTH
70ps	5GHz
350ps	1GHz
700ps	500MHz
1ns	350MHz
2.33ns	150MHz
3.5ns	100MHz
7ns	50MHz
35ns	10MHz
70ns	5MHz
350ns	1MHz

Figure G2 • Some Cardinal Points of Rise Time/Bandwidth Equivalency. Data Is Based on Text's Rise Time/Bandwidth Formula

Appendix H

Verifying rise time and delay measurement integrity

Any measurement requires the experimenter to insure measurement confidence. Some form of calibration check is always in order. High speed time domain measurement is particularly prone to error and various techniques can promote measurement integrity.

Figure H1's battery-powered 200MHz crystal oscillator produces 5ns markers, useful for verifying oscilloscope time base accuracy. A single 1.5V AA cell supplies the LTC3400 boost regulator, which produces 5V to run the oscillator. Oscillator output is delivered to the 50Ω load via a peaked attenuation network. This provides well defined 5ns markers (Figure H2) and prevents overdriving low level sampling oscilloscope inputs.

Once time base accuracy is confirmed it is necessary to check rise time. The lumped signal path rise time, including attenuators, connections, cables, probes, oscilloscope and anything else, should be included in this measurement. Such "end-to-end" rise time checking is an effective way to promote meaningful results. A guideline for insuring accuracy is to have 4× faster measurement path rise time than the rise time of interest. Thus, verifying the sample gate multipliers 250MHz (1.4ns risetime) bandwidth requires 1GHz (t_{RISE} = 350ps) oscilloscope bandwidth. Verifying the oscilloscope's 350 picosecond rise time, in turn, necessitates a 90 picosecond rise time step to ensure the 'scope is driven to its rise time limit. Figure H3 lists some very fast edge generators for rise time checking.[1] The Tektronix 284, specified at 70ps rise time, was used to check 'scope rise time. Figure H4 indicates 350ps rise time, promoting measurement confidence.

Note 1: This is a fairly exotic group, but equipment of this caliber really is necessary for rise time verification.

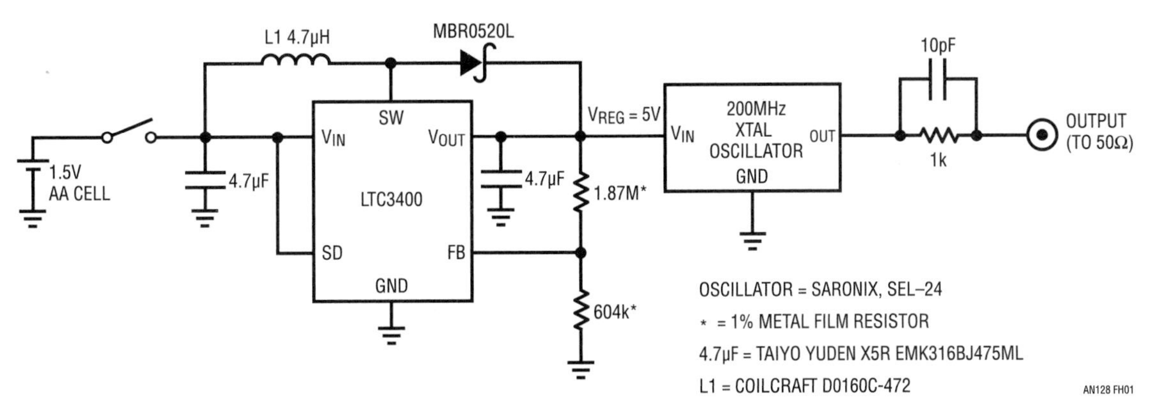

Figure H1 • 1.5V Powered, 200MHz Crystal Oscillator Provides 5ns Time Markers. 1.5V to 5V Switching Regulator Powers Oscillator

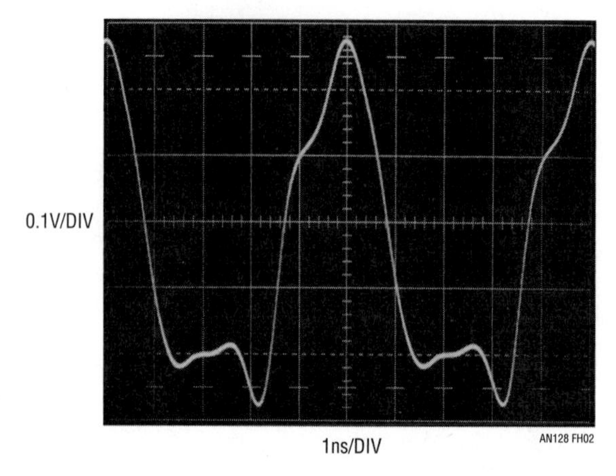

Figure H2 • Time Mark Generator Output Terminated into 50Ω. Peaked Waveform Is Optimal for Verifying Time Base Calibration

MANUFACTURER	MODEL NUMBER	RISE TIME	AMPLITUDE	AVAILABILITY	COMMENTS
Avtech	AVP2S	40ps	0V to 2V	Current Production	Free Running or Triggered Operation, 0MHz to 1MHz
Hewlett-Packard	213B	100ps	≈175mV	Secondary Market	Free Running or Triggered Operation to 100kHz
Hewlett-Packard	1105A/1108A	60ps	≈200mV	Secondary Market	Free Running or Triggered Operation to 100kHz
Hewlett-Packard	1105A/1106A	20ps	≈200mV	Secondary Market	Free Running or Triggered Operation to 100kHz
Picosecond Pulse Labs	TD1110C/TD1107C	20ps	≈230mV	Current Production	Similar to Discontinued HP1105/1106/8A. See above.
Stanford Research Systems	DG535 OPT 04A	100ps	0.5V to 2V	Current Production	Must be Driven with Stand-alone Pulse Generator
Tektronix	284	70ps	≈200mV	Secondary Market	50kHz Repetition Rate. Pre-trigger 5ns, 75ns or 150ns Before Main Output. Calibrated 100MHz and 1GHz Sine Wave Auxiliary Outputs.
Tektronix	111	500ps	≈±10V	Secondary Market	10kHz to 100kHz Repetition Rate. Positive or Negative Outputs. 30ns to 250ns Pre-trigger Output. External Trigger Input. Pulse Width Set with Charge Lines
Tektronix	067-0513-00	30ps	≈400mV	Secondary Market	60ns Pre-trigger Output. 100kHz Repetition Rate
Tektronix	109	250ps	0V to ±55V	Secondary Market	≈600Hz Repetition Rate (High Pressure Hg Reed Relay Based). Positive or Negative Outputs. Pulse Width Set by Charge Lines

Figure H3 • Picosecond Edge Generators Suitable for Rise Time Verification. Considerations Include Speeds, Features and Availability

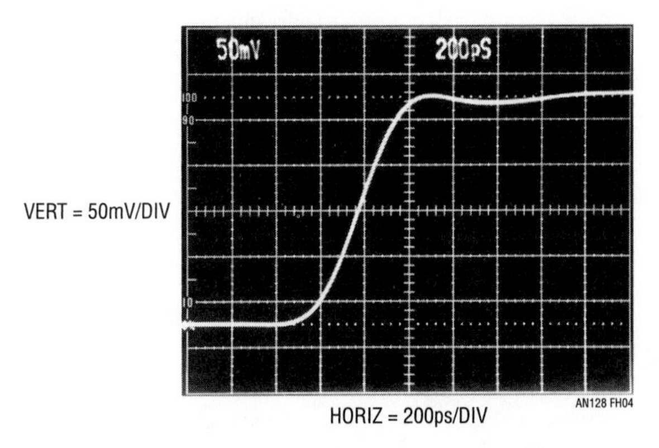

Figure H4 • 70 Picosecond Edge Drives Oscilloscope to its 350 Picosecond Rise Time Limit, Verifing 1GHz Bandwidth

An introduction to acoustic thermometry

An air filled olive jar teaches signal conditioning

27

Jim Williams Omar Sanchez-Felipe

Introduction

We occasionally lecture to university engineering students. A goal of these lectures is to present technology in a novel, even charming, way. This hopefully entices the student towards the topic; an aroused curiosity is fertile ground for education. One such lecture investigates acoustic thermometry as an example of signal conditioning techniques. This subject has drawn enough interest that it is presented here for wider dissemination and as supplementary material for future acoustic thermometry lectures.

Acoustic thermometry

Acoustic thermometry is an arcane, elegant temperature measurement technique. It utilizes sound's temperature dependent transit time in a medium to measure temperature. The medium may be a solid, liquid or gas. Acoustic thermometers function in environments that conventional sensors cannot tolerate. Examples include extreme temperatures, applications where the sensor would be subjected to destructive physical abuse and nuclear reactors. Gas path acoustic thermometers respond very quickly to temperature changes because they have essentially no thermal mass or lag. An acoustic thermometer's "body" ***is*** the measurand. Additionally, an acoustic thermometer's reported "temperature" represents the total measurement path transit time as opposed to a conventional sensor's single point determination. As such, an acoustic thermometer is blind to temperature variations within the measurement path. It reports the measurement path's delay as its "temperature," whether or not the path is iso-thermal.

A pleasant surprise is that the sonic transit time in a gas path thermometer is almost entirely insensitive to pressure and humidity, leaving temperature as the sole determinant. Additionally, sonic speed in air varies predictably as the square root of temperature.

Practical considerations

A practical acoustic thermometer demonstration begins with selecting a sonic transducer and a dimensionally stable measurement path. A wideband ultrasonic transducer

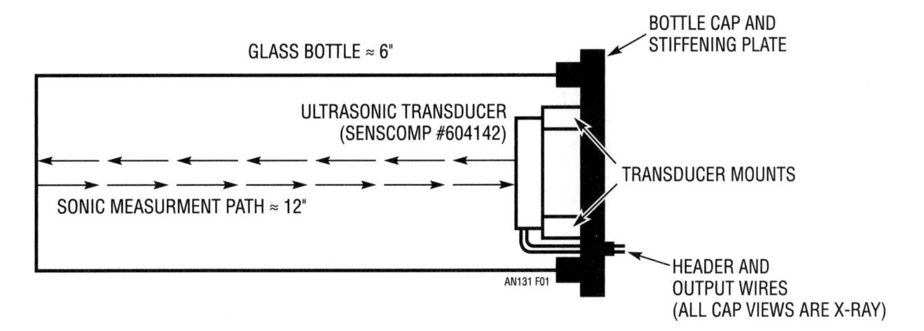

Figure 27.1 • Ultrasonic Transducer Rigidly Mounts Within Stiffened Cap Affixed to Bottle. Structure Defines Fixed Length Measurement Path Essentially Independent of Physical Variables. Sonic Transit Time at 75°F ~ 900μs with ≈1μs/°F Variation

Analog Circuit and System Design: Immersion in the Black Art of Analog Design. http://dx.doi.org/10.1016/B978-0-12-397888-2.00027-4

is desirable to promote fast, low jitter, hi-fidelity response free of resonances and other parasitics. The electrostatic type specified in Figure 27.1 meets these requirements. A single transducer serves as both transmitter and receiver.

The device is rigidly mounted within the stiffened metal cap of a glass enclosure, promoting measurement path dimensional stability. The enclosure and its cap are conveniently furnished by a bottle of "Reese" brand "Cannonball" olives (Figure 27.2). After removing the olives and their residue, the bottle and cap are baked out at 100°C prior to joining. The transducer leads pass through the cap via a coaxial header. This arrangement, along with the glass enclosure's relatively small thermal expansion coefficient, yields a path distance stable against temperature, pressure and mechanically induced changes. Path length, including enclosure bottom bounce and return to transducer, is about 12". This sets a two-way trip time of around 900μs (speed of sound in air ≈ 1.1 ft/ms). At 75°F this path's temperature dependent variation is approximately 1μs/°F. A desired 0.1°F resolution mandates mechanical and electronic induced path length uncertainty inside 100ns; approximately 0.001" dimensional stability referred to the 12" path length. Considering likely error sources, this is a realistic goal.

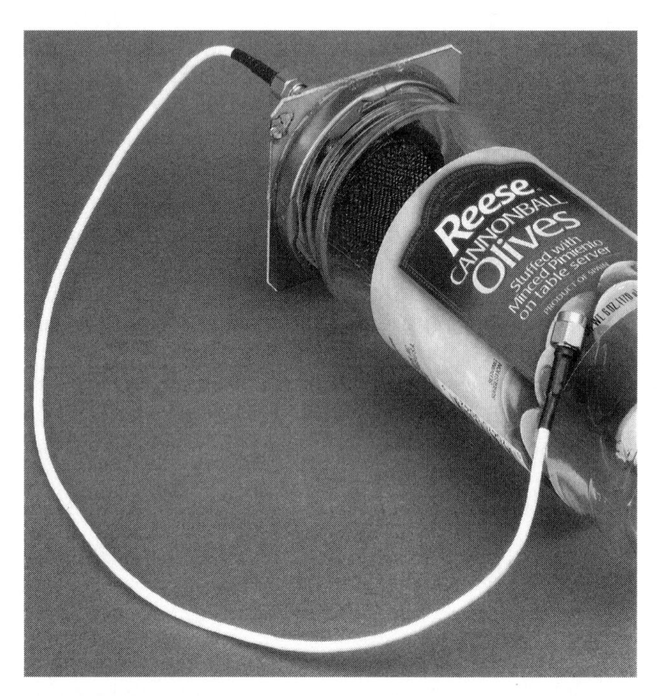

Figure 27.2 • Photograph Details Cap Assembly, Partially Views Glass Enclosed Measurement Path. Ultrasonic Transducer Visible Within Cap. Stiffening Plate, Bonded to Cap Top, Prevents Ambient Pressure or Temperature Changes from Deforming Thin Metal Cap, Promoting Measurement Path Length Stability. Coaxial Header Provides Transducer Connections

Overview

Figure 27.3 is a simplified overview of the acoustic thermometer. The transducer, which can be considered a capacitor, is biased at $150V_{DC}$. The start pulse clock drives it with a short impulse, launching an ultrasonic event into the measurement path. Simultaneously, the width decoding flip-flop is set high. The sonic impulse bounces off the enclosure bottom, returns to the transducer and impinges on it. The resulting minuscule mechanical displacement causes the transducer to give up charge ($Q = \Delta C \bullet V$), which appears as a voltage at the receiver amplifier input. The trigger converts the amplifier's output excursion into a logic compatible level which resets the flip-flop. The flip-flop output width represents the measurement path's temperature dependent sonic transit time. The microprocessor, equipped with the measurement path's temperature/delay calibration constants, calculates the temperature and supplies this information to the display.[1] The start pulse generator has a second output which gates the trigger output off during nearly the entire measurement cycle. The trigger output only passes during the immediate time vicinity when a return pulse is expected. This discriminates against unwanted sonic events originating outside the measurement path, eliminating false triggers. A second gating, sourced from the width decoding flip-flop, shuts down the 150V bias supply switching regulator during the measuring interval. Return pulse amplitude is under 2mV at the transducer and the high gain, wideband receiver amplifier is vulnerable to parasitic inputs. Shutting down the 150V bias supply during the measurement prevents its switching harmonics from corrupting the amplifier. Figure 27.4 describes the system's event sequence. A measurement cycle begins with a start pulse (A) driving the transducer and setting the flip-flop (B) high. After the sonic impulse's transit time, the amplifier responds (C, diagram right), tripping the trigger (D, diagram right) which resets the flip-flop. Gate signals (E and F) protect the trigger from unwanted sonic events and start pulse artifacts and shut off the high voltage regulator during measurement.

Note 1: See Appendix A, "Measurement Path Calibration" for details on determining calibration constants.

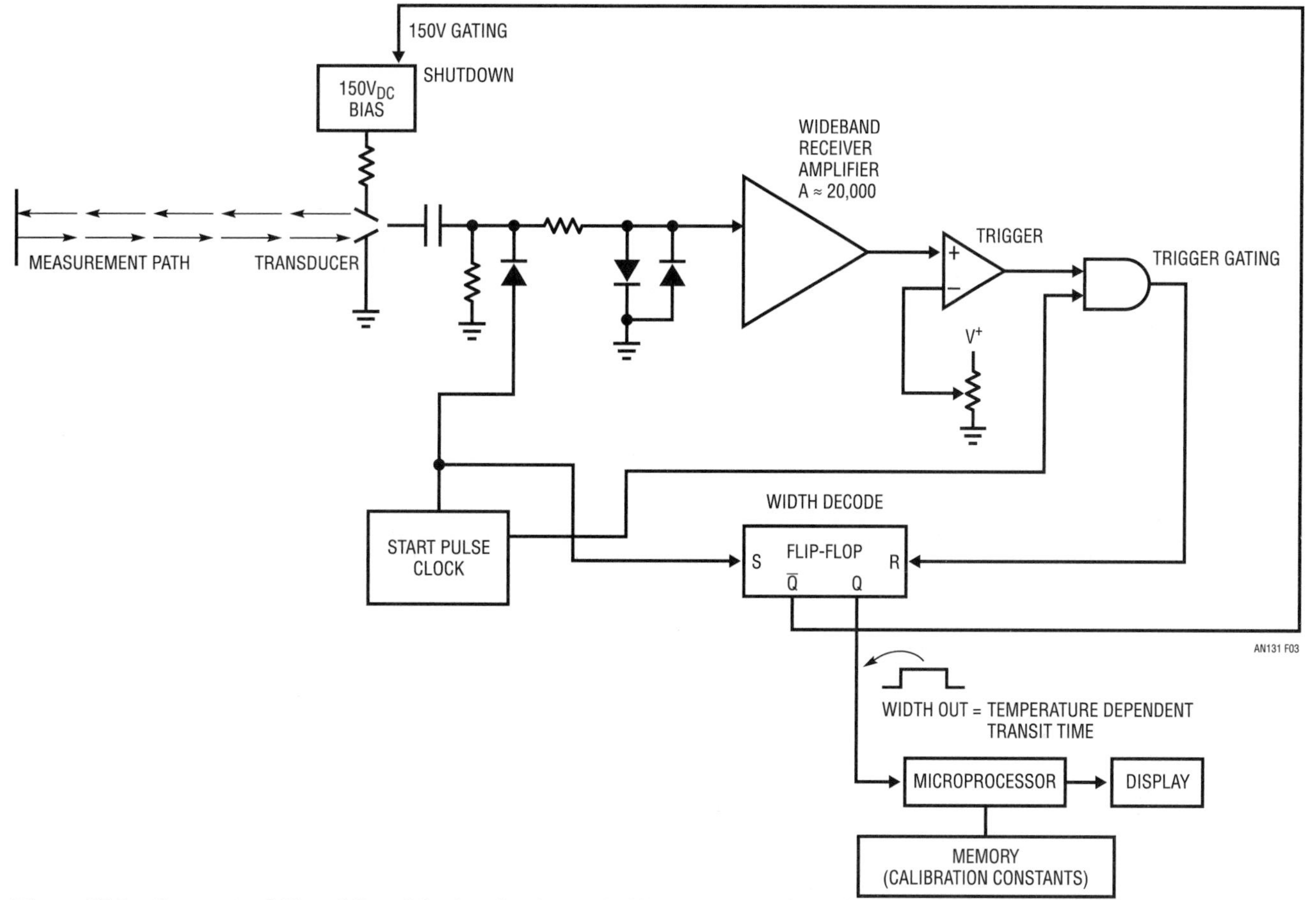

Figure 27.3 • Conceptual Signal Conditioning for Acoustic Thermometer. Start Clock Launches Acoustic Pulse into Measurement Path, Sets Width Decode Flip-Flop High. Acoustic Paths Return Pulse, Amplified by Receiver, Trips Trigger, Resetting Flip-Flop. Resultant "Q" Width Output, Representing Path's Temperature Dependent Transit Time, is Converted to Temperature Reading by Microprocessor. Gating High Voltage Supply and Trigger Prevents Spurious Outputs

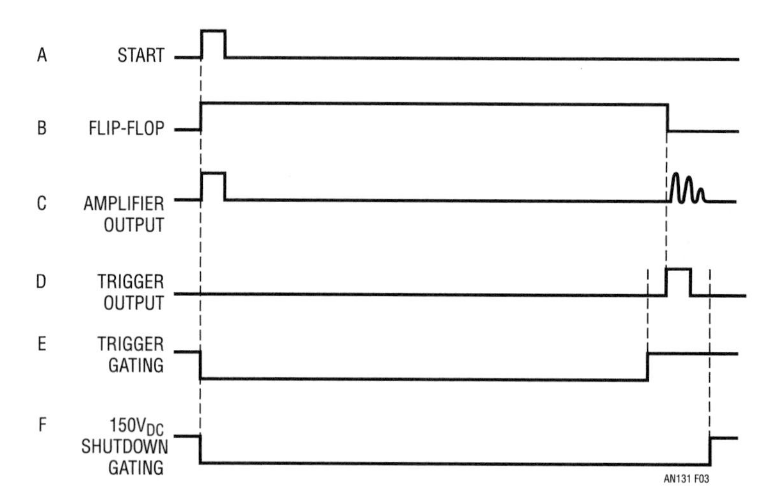

Figure 27.4 • Figure 27.3's Event Sequence. Start Pulse (A) Drives Transducer, Sets Flip-Flop (B) High. Sonic Pulse Return Activates Amplifier (C, Extreme Right), Causing Trigger Output (D) to Reset Flip-Flop (B). Trigger Gating (E) Prevents Erroneous Trigger Response to Start Pulse Caused Amplifier Output (C, Extreme Left) and External Sonic Events. Gating (F) Turns Off 150V Switching Converter During Measurement, Precluding Amplifier Output Corruption

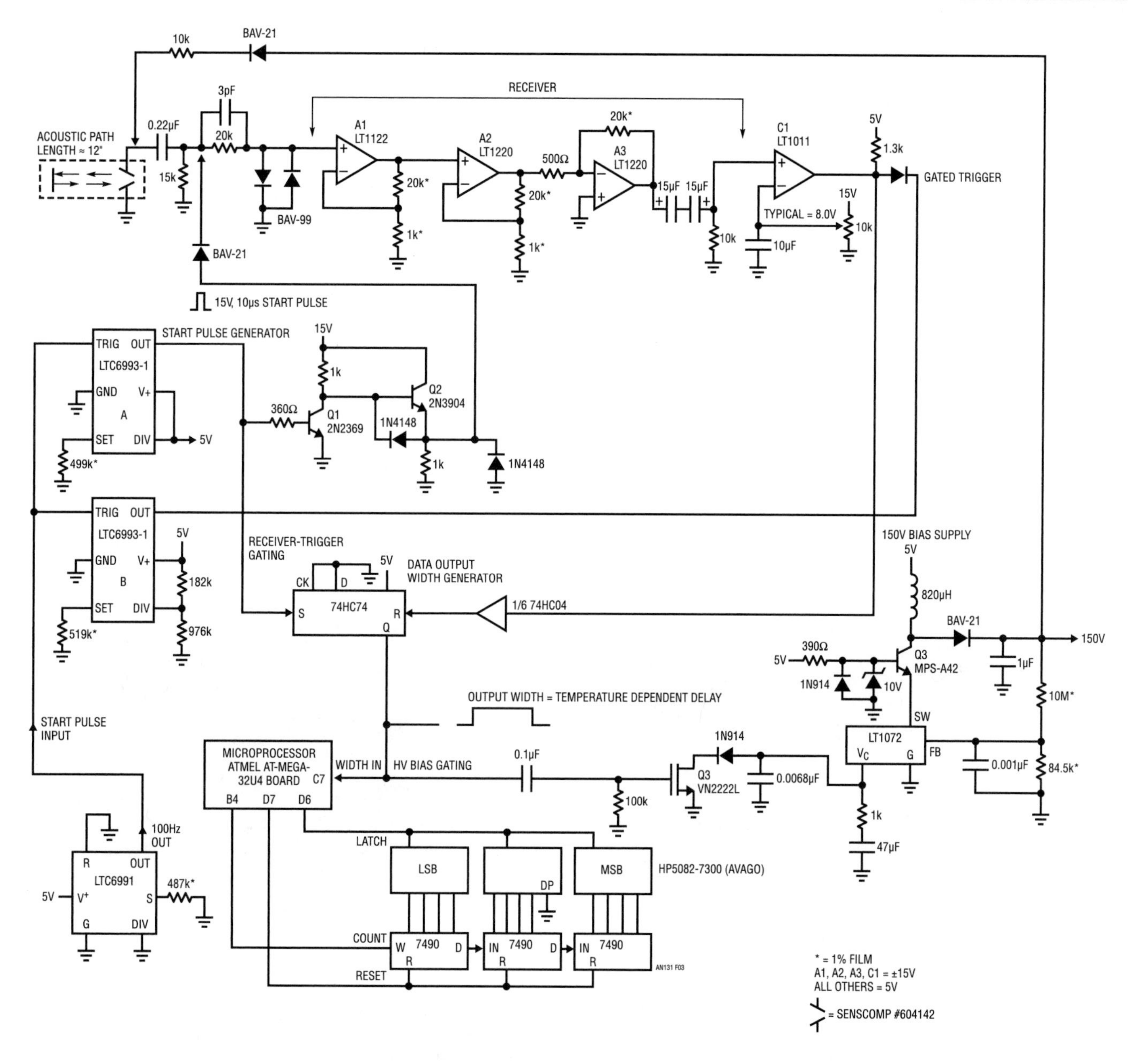

Figure 27.5 • Detailed Circuitry Closely Follows Figure 27.3's Concept. Start Pulse Generator is Comprised of 100Hz Clock, One-Shot Multivibrators and Q1-Q2 Driver Stage. A ≈20,000 Receiver Amplifier Splits Gain Among Three Stages, Biases Trigger Comparator. Flip-Flop Output Width Feeds Microprocessor Which Calculates and Displays Temperature. Capacitive Coupling Isolates High Voltage DC Transducer Bias, Diode Clamping Prevents Destructive Overloads. Switching Regulator Controls High Voltage Via Cascode. Gating Obviates Switching Regulator Noise Originated Interference, Minimizes External Sonic Corruption

Detailed circuitry

Figure 27.5's detailed schematic closely follows Figure 27.3's concepts. An LTC®6991 oscillator furnishes the 100Hz clock. LTC®6993-1 monostable "A" provides a 10μs width to the Q1-Q2 driver, which capacitively couples the start pulse (Trace A, Figure 27.6) to the transducer. Simultaneously, the monostable sets the flip-flop (E) high. The flip-flop high output shuts down the LT®1072 based high voltage converter during the measurement. Monostable "B" produces a pulse (B) which gates off C1's trigger output for a time just shorter than the fastest expected sonic return.

The launched sonic pulse travels down the measurement path, bounces, returns, and impinges on the transducer. The transducer, biased at $150V_{DC}$, releases charge ($Q = \Delta C \bullet V$) which appears as a voltage at the receiver amplifier.

The cascaded amplifier, with an overall gain of ≈17,600, produces A2's output (C) and a further amplified version at A3. C1 triggers (D) at the first event that exceeds its negative input threshold, resetting the flip-flop. The resultant flip-flop width, representing the temperature dependent transit time, is read by the microprocessor which determines the temperature and displays it.[2]

Figure 27.7 studies receiver amplifier operation at the critical return impulse trip point. The returning sonic pulse is viewed at A2 (Trace A). A3 (B) adds gain, softly saturating the signals leading response. C1's trigger output (C) responds to multiple triggers but the flip-flop output (D) remains high after the initial trigger, securing transit time data.

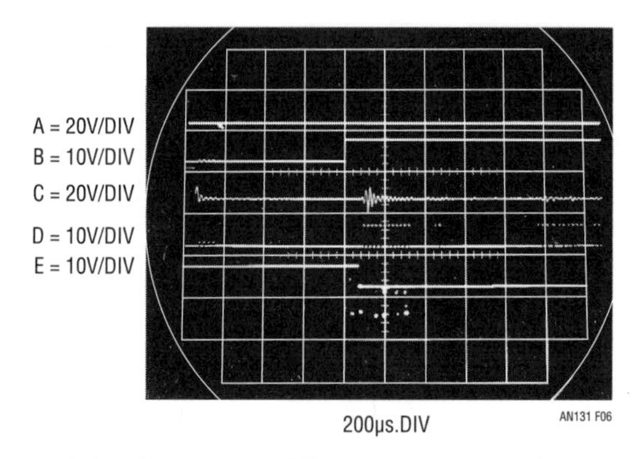

Figure 27.6 • Figure 27.5's Waveforms Include Start Pulse (A), Trigger Gate (B), A2 Output (C), Trigger (D) and Flip-Flop Q Output Width (E). Amplifier Output Causes Multiple Trigger Transitions, But Flip-Flop Maintains Pulse Width Integrity. A2-Trigger Outputs at Photo Right, Due to Acoustic Pulse Second Bounce, Are Inconsequential

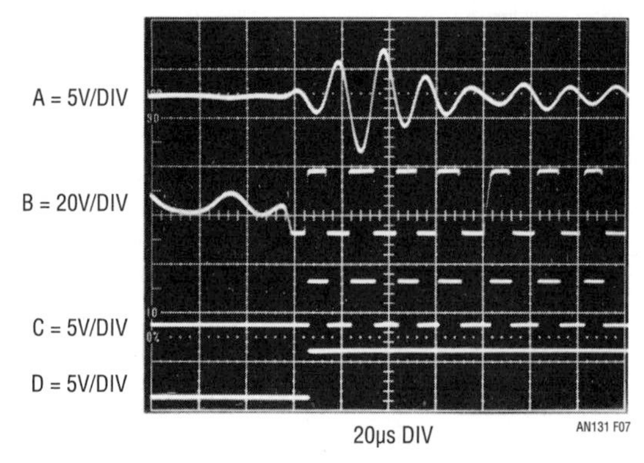

Figure 27.7 • Receiver Amplifier—Trigger Operating Detail at Trip Point. Returning Sonic Pulse Viewed at A2 Output (Trace A) After X440 Gain. A3's Output (B) Adds X40, Softly Saturating Signals Leading Response. C1's Output (C) Responds to Multiple Triggers But Flip-Flop Output (D) Remains High After Initial Trigger

Note 2: Complete processor software code appears in Appendix B, "Software Code."

Gating prevents high voltage supply switching harmonics from producing spurious amplifier-trigger outputs. Figure 27.8 shows gate-off detail. The flip-flop output (Trace A) going high shuts down high voltage switching (B) at the measurement onset. This state persists during the entire transit time, preventing erroneous amplifier-trigger outputs. Figure 27.9's flip-flop fall (Trace A) combines with LT1072 V_C pin associated components, producing delayed high voltage turn-on (B) after the vulnerable, small amplitude return pulse trip point. This assures a clean, noise-free trigger.

Several circuit attributes aid performance. As mentioned, gating the trigger output prevents sonic interference from outside sources. Similarly, gating the 150V converter off prevents its harmonics from corrupting the high gain, wideband receiver amplifier. Additionally, the 150V supply value is a gain term, making its regulation loss during the measurement a potential concern. Practically, the 1μF output capacitor decays only 30mV in this time, or about 0.02%. This small variation is constant, insignificant and may be ignored. Deriving the trigger trip point and the start pulse from the same supply allows trigger voltage to vary ratiometrically with received signal amplitude, enhancing stability. Finally, the transducer used is wideband, highly sensitive and free from resonances, promoting repeatable, jitter-free operation.[3] All of the above directly contribute to the circuit's <100ns (0.1°F) resolution of the ≈ 1ms path length–less than 100ppm uncertainty. Absolute accuracy from 60°F to 90°F is within 1°F referenced to Appendix A's calibration.

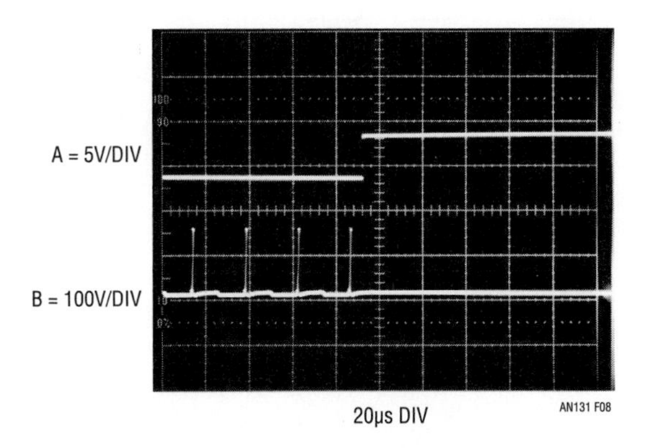

Figure 27.8 • High Voltage Bias Supply Gate-Off Detail. Flip-Flop Output (Trace A) Going High Shuts Down LT1072 Switching Regulator, Turning Off 150V Flyback Events (B). Off-Time Extends During Measurement Interval, Precluding Receiver Amplifier Corruption

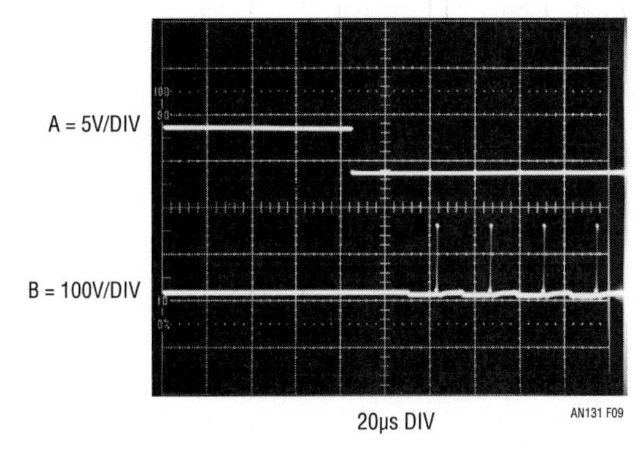

Figure 27.9 • High Voltage Bias Supply Gate-On Detail. Flip-Flop Output (Trace A) Drops Low at Sonic Pulse Return. Components at LT1072 Switching Regulator V_C Pin Delay Bias Supply Turn-On (B). Sequencing Ensures Bias Supply is Off During Amplifier-Trigger's Vulnerable Response to Sonic Pulse Arrival

Note 3: Readers rich in years will recognize the specified transducer descends from 1970's era Polaroid SX-70 automatic focus cameras (every baby boomer had to have one).

Triggering on later bounces offers the potential benefit of easing timing tolerances and merits consideration. Figure 27.10 shows multiple sonic bounces, discernible in A2's output, decaying into acoustic dispersion induced noise contained within the glass enclosure. Triggering on a later bounce would relax timing margins, but incurs unfavorable signal-to-noise characteristics. Signal processing techniques could overcome this, but the increased resolutions utility would have to justify the effort.

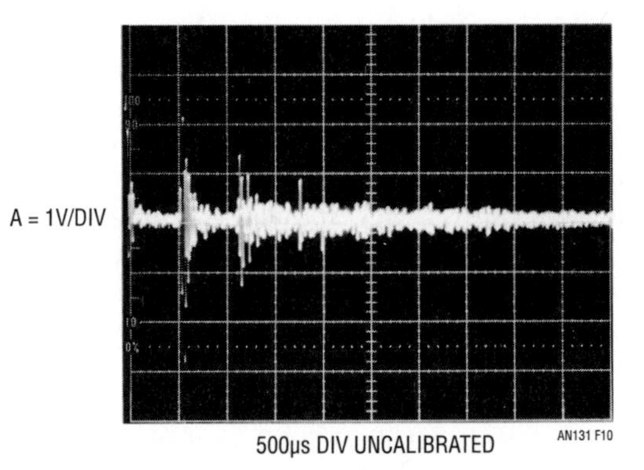

Figure 27.10 • Multiple Sonic Bounces, Discernable in A2's Output, Decay into Acoustic Dispersion Induced Noise Contained Within Glass Enclosure. Triggering on Later Bounce Would Ease Timing Tolerances, But Incur Signal-to-Noise Degradation

References

1. Lynnworth, L.C. and Carnevale, E.H., "Ultrasonic Thermometry Using Pulse Techniques." Temperature: Its Measurement and Control in Science and Industry, Volume 4, p. 715-732, Instrument Society of America (1972)
2. Mi, X.B. Zhang, S.Y. Zhang, J.J. and Yang, Y.T., "Automatic Ultrasonic Thermometry." Presented at Fifteenth Symposium on Thermophysical Properties, June 22-27, 2003, Boulder, Colorado, U.S.A
3. Williams, Jim, "Some Techniques For Direct Digitization of Transducer Outputs," Linear Technology Corporation, Application Note 7, p.4-6, Feb. 1985
4. Analog Devices Inc., "Multiplier Applications Guide," "Acoustic Thermometer," Analog Devices Inc., p. 11-13, 1978

Appendix A
Measurement path calibration

Theoretically, temperature calibration constants can be calculated from measurement path length. In practice, it is difficult to determine path length to the required accuracy. Enclosure, transducer and mounting dimensional uncertainties necessitate calibration vs known temperatures. Figure A1 shows the calibration arrangement. The enclosure is placed in a controllable thermal chamber equipped with an accurate thermometer iso-thermal with the enclosure. Ten evenly spaced temperature points from 60°F to 90°F are generated by stepping chamber set-point. The enclosure has a 30 minute time constant to 0.25°F settling, so adequate time must be allowed for each step to stabilize before readings are taken. Readings consist of noting the counter indicated pulse width at each temperature and recording the data. This information is then loaded into the microprocessor memory. See Appendix B, "Software Code."

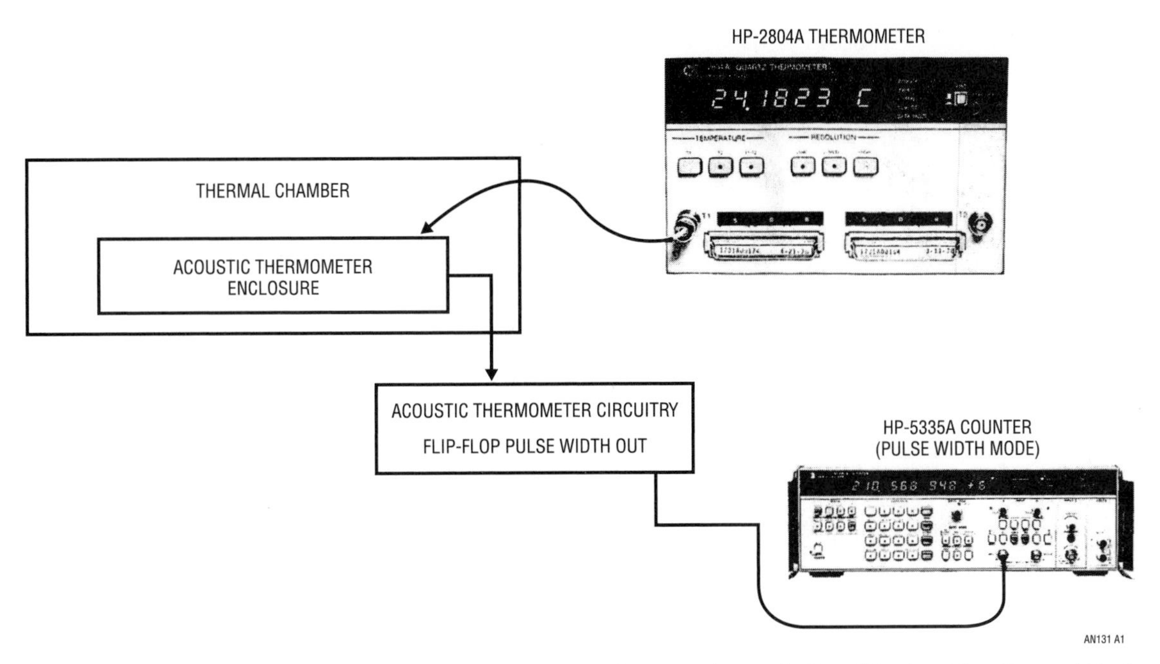

Figure A1 • Calibration Arrangement Consists of Thermometer, Counter, Acoustic Conditioning Circuitry and Enclosure in Thermal Chamber. Temperature is Stepped Every 3°F Between 60°F and 90°F Allowing 30 Minute Stabilization Time Per Point

Appendix B

The software code for the Atmel AT-Mega 32U4 microprocessor, combined with the calibration constants stored in its memory (see Appendix A), enables the processor to calculate and display the sensed temperature. This code, written by Omar Sanchez-Felipe of LTC, appears below.

```
//      OLIVER.C:
//
//      Hot temps from a jar of olives.
//
//      The microprocessor is the Atmel ATmega32U4. See 'atmel.com' for its
//      datasheet. The code is compiled with the current version of the
//      'avr-gcc' compiler obtainable at 'winavr.sourceforge.net'.
//
//      Calibration data:
//         Pulse len (usecs)    Temp (deg)
//           877.40               84.2
//           884.34               76.0
//           887.40               72.4
//           892.56               66.4
//           897.60               60.7
//
#include <avr/io.h>
#include <avr/wdt.h>
//      Some useful defs
#define BYTE            unsigned char
#define WORD            unsigned short
#define DWORD               unsigned long
#define BOOL            unsigned char
#define NULL            (void *)0
#define TRUE            1
#define FALSE           0
#define BSET(r, n)      r |= (1<<n)         // set/clr/tst nth bit of reg
#define BCLR(r, n)      r &= ~(1<<n)
#define BTST(r, n)      (r & (1<<n))
#define DDOUT(r, v)     (r |= (v))          // config (set) bits for OUT
#define DDIN(r, v)      (r &= ~(v))         // config (clear) bits for IN
#define SETBITS(r, v)   (r |= (v))          // set multiple bits on register
#define CLRBITS(r, v)   (r &= ~(v))         // clear ""
#define SYS_CLK         16000000L           // clock (Hz)
#define CLKPERIOD 625
#define POSEDGE         1                   // edge definitions for timerwait
#define NEGEDGE         0
#define TMRGATE         PORTC7              // I/O pins
#define BCDRSET         PORTD7
#define BCDPULS         PORTB4
#define DPYLATCH        PORTD6
```

```
//      Calibration table and entry.
//      Using small units of time and temp allows all calcs
//      to be done in fixed point.
struct calpoint
{
        DWORD   pulse;          // pulse duration, in tenths of nsecs
        WORD    temp;           // temperature, in tenths of degrees
        WORD    slope;          // slope (in tenths of nsecs/tenths of degree)
};
struct  calpoint caltab[] =     // the calibration table
{       {8774000L,      842, 0},
        {8843400L,      760, 0},
        {8874000L,      724, 0},
        {8925600L,      664, 0},
        {8976000L,      607, 0}
};
#define NCALS           (sizeof(caltab)/sizeof(struct calpoint))
#define TEMPERR         999
void    spin(WORD);
BOOL    timerwait(BYTE v, long tmo, WORD *p);
void    setdpy(WORD);
void    dobackgnd(void);
WORD dotemp();

int     main()
{
        WORD temp, i;

        // set master clock divisor
        CLKPR = (1<<CLKPCE); CLKPR = 0;

        // clear WDT
        MCUSR = 0;
        WDTCSR |= (1<<WDCE) | (1<<WDE);
        WDTCSR = 0x00;

        // init the display I/O pins
        BCLR(PORTD,     BCDRSET);
        BCLR(PORTD,     DPYLATCH);
        DDOUT(DDRD,     ((1<<BCDRSET) | (1<<DPYLATCH)));
        BCLR(PORTB,     BCDPULS);
        DDOUT(DDRB,     (1<<BCDPULS));

        // init pulse-width counter (timer #3)
        TCCR3A = 0x00;
        TCCR3B = 0x01;          // no clk prescaling
        DDIN(DDRC, (1<<DDC7));          // PC7 is gate
        BSET(PORTC, PORTC7); // enable pullup
```

```
		// compute the slope for each entry in the cal table
		// the first slope is never used, so we leave it at 0
		// note we're inverting the sign of the slope
		for (i = 1; i < NCALS; i++)
			caltab[i].slope =	(caltab[i].pulse - caltab[i-1].pulse) /
						(caltab[i-1].temp - caltab[i].temp);
		setdpy(0);
for (; ;)
		{
		temp = dotemp();		// compute temperature
		setdpy(temp);			// set the display
		spin(1000);			// spin a second and repeat
		}
} // main
//	Set the display LED's to specified count by brute-force
//	incrementing the BCD counter that feeds it.
//
void	setdpy(WORD cnt)
{
		// reset BCD counter
		BSET(PORTD, BCDRSET);
		asm("nop"); asm("nop");
		BCLR(PORTD, BCDRSET);
		while (cnt--)
			{
			BSET(PORTB, BCDPULS);
			asm("nop"); asm("nop");
			BCLR(PORTB, BCDPULS);
			asm("nop"); asm("nop");
			}
		// latch current val (to avoid flicker)
		BCLR(PORTD, DPYLATCH);
		asm("nop"); asm("nop");
		BSET(PORTD, DPYLATCH);
}
//	Set up the edge detector to trigger on either a positive or
//	negative edge, depending on the 'edge' flag, and then wait for
//	it to happen. See defines for POSEDGE/NEGEDGE above.
//	if param 'tmo' is TRUE, a timeout is set so that we dont wait
//	forever if no pulse materializes. Returns the counter value
//	at which the edge occurs via pointer 'p'.
//
```

```
#definetcnTMO (SYS_CLK/20000L) // approx 1 msecs
BOOL  timerwait(BYTE edge, long tmo, WORD *p)
{
        if (edge == NEGEDGE)
              BCLR(TCCR3B, ICES3); // falling edge
        else BSET(TCCR3B, ICES3); // rising edge
        BSET(TIFR3, ICF3);
        tmo *= tcnTMO;
        while (!BTST(TIFR3, ICF3))
              {if (tmo-- < 0)
        return FALSE;
              }
        *p = ICR3; // return current counter
        return TRUE;
}
//      Measure the next POSITIVE pulse and map into temperature.
//
WORD dotemp()
{
        WORD strt, end, i;
        DWORD dur, temp;
        strt = end = dur = 0;
        // wait for any ongoing pulse to complete
        if (!timerwait(NEGEDGE, 5000, &strt))
           return 0;
        // now catch the first positive pulse
        if (!timerwait(POSEDGE, 5000, &strt))
           return 0;
        // and wait for it to complete
        if (!timerwait(NEGEDGE, 5000, &end))
           return 0;
        dur = (end - strt);
        dur *= CLKPERIOD; // duration now in tenths of nsecs
        // compute temp. If warmer than highest calibrated temp
        // or cooler than lowest calibrated temp, return TEMPERR
        //
```

```
        if (dur < caltab[0].pulse || dur > caltab[NCALS-1].pulse)
            temp = TEMPERR;
        else { // an entry will always be found, but we (re)init
             // temp to avoid complaints from the compiler
            temp = TEMPERR;
            for (i = 1; i < NCALS; i++)
                {if (dur <= caltab[i].pulse)
                    {temp = caltab[i].temp +
                            (caltab[i].pulse - dur) /
                                caltab[i].slope;
                    break;
                    }
                }
            }
        return temp;
}
//      - - - - - - - - - - - - - - - - - - - - - - - - - - - - - - - - - - -
        // Spin for 'ms' millisecs. Spin constant is empirically determined.
        //
#defineSPINC (SYS_CLK / 21600L)
volatile WORD spinx;
void    spin(WORD ms)
{
        WORD i;
        while (ms--)
                for (i = 0; i < SPINC; i++) spinx = i*i;
}
//              END
#               GCC MAKEFILE:
#
#               GMAKE file for the "oliver" code running on the ATmega32U4.
#
CC              = avr-gcc.exe
MCU             = atmega32u4
#               These are common to compile, link and assembly rules
#               The 'no-builtin' opt keeps gcc from assuming defs for putchar() and
#               others
#
COMMON                  = -mmcu=$(MCU) -fno-builtin
CF              = $(COMMON)
CF +            = -Wall -gdwarf-2
##CF +                  = -Wall -gdwarf-2 -O0
AF              = $(COMMON)
AF              += $(CF)
AF              += -x assembler-with-cpp -Wa,-gdwarf2
LF              = $(COMMON)
LF              += -Wl,-Map=$(TMP)$(APP).map
#               weird intel flags
```

```
#
HEX_FLASH_FLAGS = -R .eeprom
HEX_EEPROM_FLAGS = -j .eeprom
HEX_EEPROM_FLAGS += --set-section-flags=.eeprom="alloc,load"
HEX_EEPROM_FLAGS += --change-section-lma .eeprom=0 --no-change-warnings
#               - - - - - - - - - - - - - - - - - - - - - - - - - - - - - - - - - - - -
#               MAKE DIRECTIVES
#
#               The "target directory" TMP is standard for the intermediate files
#               (.obj) and the final product.
#               - - - - - - - - - - - - - - - - - - - - - - - - - - - - - - - - - - - -
TMP             = ./tmp/
APP             = oliver
ELF             = $(TMP)$(APP).elf
OBJS            = $(TMP)oliver.o
all:            $(TMP) $(ELF) $(TMP)$(APP).hex $(TMP)$(APP).eep size
$(TMP):
                rm -rf .\tmp
                mkdir .\tmp
$(TMP)oliver.o: oliver.c
                $(CC) $(INCLUDES) $(CF) -O1 -c $< -o $*.o
oliver.asm: oliver.c
                $(CC) $(INCLUDES) $(CF) -O1 -S $< -o oliver.asm
#               Linker -------------------------------------------------------
$(ELF): $(OBJS)
                $(CC) $(LF) $(OBJS) $(LINKONLYOBJS) $(LIBDIRS) $(LIBS) -o $(ELF)
%.hex: $(ELF)
                avr-objcopy -O ihex $(HEX_FLASH_FLAGS) $< $@
%.eep: $(ELF)
                -avr-objcopy $(HEX_EEPROM_FLAGS) -O ihex $< $@ || exit 0
%.lss: $(ELF)
                avr-objdump -h -S $< > $@
size: $(ELF)
                @echo
                @avr-size -C --mcu=${MCU} $(ELF)
# Misc - - - - - - - - - - - - - - - - - - - - - - - - - - - - - - - - - - - - - -
clean:
                -rm -rf $(OBJS) $(TMP)$(APP).elf ./dep/* $(TMP)$(APP).hex $(TMP)$(APP).eep
                $(TMP)$(APP).map $(TMP)$(APP).d
#               end
```

Section 3

High Frequency/RF Design

Low noise varactor biasing with switching regulators (28)

Telecommunication, satellite links and set-top boxes all require tuning a high frequency oscillation. The actual tuning element, a varactor diode, requires high voltage bias for operation. The high voltage bias must be free of noise to prevent unwanted oscillator outputs. This publication details a method for generating noise-free high voltage from low voltage inputs using switching regulators. Spurious oscillator outputs are below –90dBc. Suggested circuit and layout information is included. Appendices cover varactor diode theory and performance verification techniques.

Low cost coupling methods for RF power detectors replace directional couplers (29)

This note describes an RF feedback coupling method which eliminates the directional coupler. Instead, a 0.4pF ±0.05pF capacitor and 50Ω resistor are used to feed RF signals back to the Linear Technology power controller. This method reduces coupling loss variations, cost and lead time.

Improving the output accuracy over temperature for RMS power detectors (30)

Stable temperature performance is extremely important in base station designs because the ambient temperature can vary widely depending on the surroundings and the location. Using high accuracy over temperature RMS detectors can improve the power efficiency of the base station designs. The LTC5582 and the dual-channel LTC5583 are a family of RMS detectors that offer excellent stable temperature performance (from –40°C to 85°C) at any frequency up to 10GHz for LTC5582, and 6GHz for LTC5583. This application note introduces techniques to improve the temperature stability of these parts.

Low noise varactor biasing with switching regulators

Vanquishing villainous vitiators vis-à-vis vital varactors

28

Jim Williams David Beebe

Introduction

Telecommunication, satellite links and set-top boxes all require tuning a high frequency oscillator. The actual tuning element is a varactor diode, a 2-terminal device that changes capacitance as a function of reverse bias voltage.[1] The oscillator is part of a frequency synthesizing loop, as detailed in Figure 28.1. A phase locked loop (PLL) compares a divided down representation of the oscillator with a frequency reference. The PLL's output is level shifted to provide the high voltage necessary to bias the varactor, which closes a feedback loop by voltage tuning the oscillator. This loop forces the voltage controlled oscillator (VCO) to operate at a frequency determined by the frequency reference and the divider's division ratio.

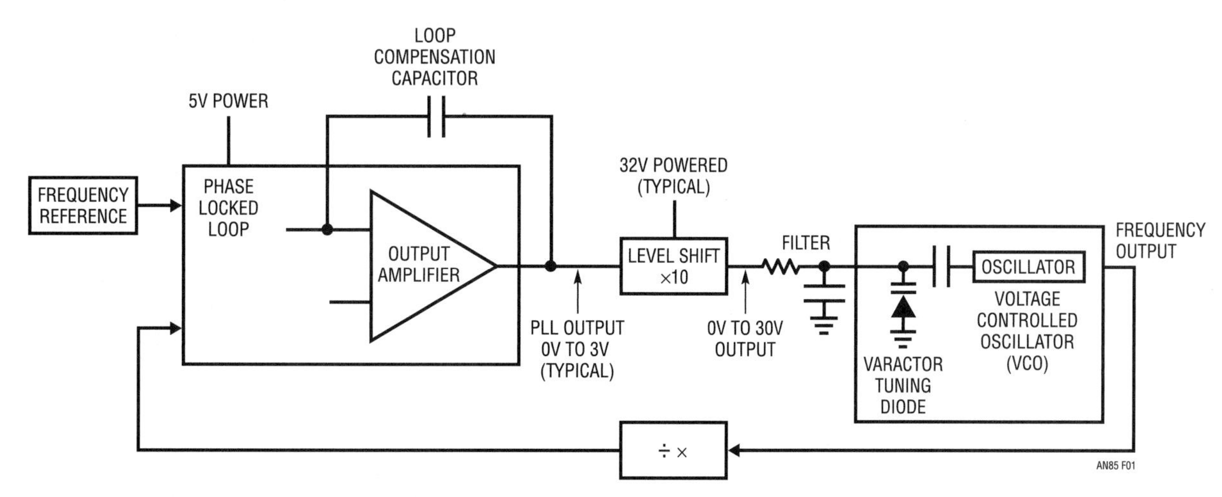

Figure 28.1 • Typical Phase Lock Loop-Based Frequency Synthesizer. Level Shift Furnishes 0V to 30V Bias to VCO Varactor Diode, Although a 32V Supply is Required

Note 1: Theoretical considerations of varactor diodes are treated in Appendix A, "Zetex Variable Capacitance Diodes," guest written by Neil Chadderton of Zetex.

Analog Circuit and System Design: Immersion in the Black Art of Analog Design. http://dx.doi.org/10.1016/B978-0-12-397888-2.00028-6

Varactor biasing considerations

The high voltage bias is required to achieve wide-range varactor operation. Figure 28.2 shows varactor capacitance vs reverse voltage curves for a family of devices. A 10:1 capacitance shift is available, although a 0.1V to 30V swing is required. The curves shown are characteristic of typical "hyperabrupt" devices. Response modification is possible, with compromises in performance, particularly with regard to linearity and sensitivity.[2]

The bias voltage requirement has traditionally been met by utilizing existing high voltage rails. The current trend towards low voltage powered systems means the high voltage bias must be locally generated. This implies some form of voltage step-up switching regulator. This is certainly possible, but varactor noise sensitivity complicates design. In particular, the varactor responds to any form of amplitude variation of its bias, resulting in an undesired capacitance shift. Such a shift causes VCO frequency movement, resulting in spurious oscillator outputs. DC and low frequency shifts are removed by PLL loop action, but activity outside the loop's passband causes undesired outputs. Most applications require spurious oscillator output content to be 80dB or more below the nominal output frequency.[3] This implies a low noise, high voltage supply, mandating caution in the switching regulator design. Switching regulators are often associated with noisy operation, making a varactor bias application seem hazardous. Careful preparation can eliminate this concern, allowing a practical switching regulator-based approach to varactor biasing.

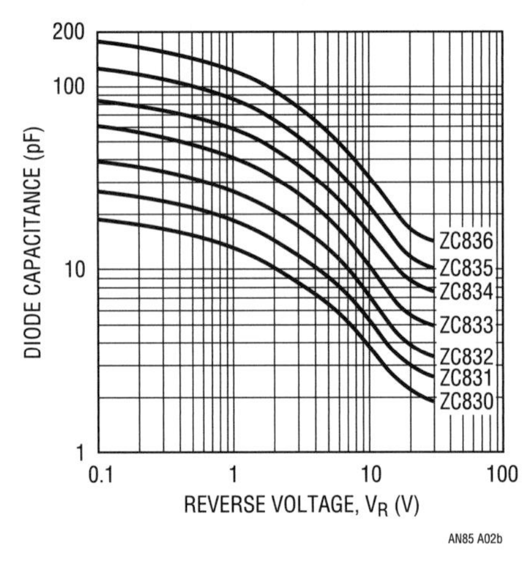

Figure 28.2 • Typical Capacitance Voltage Characteristics for the Zetex ZC830-6 Range. 0.1V to 30V Swing Results in ≈10× Capacitance Shift

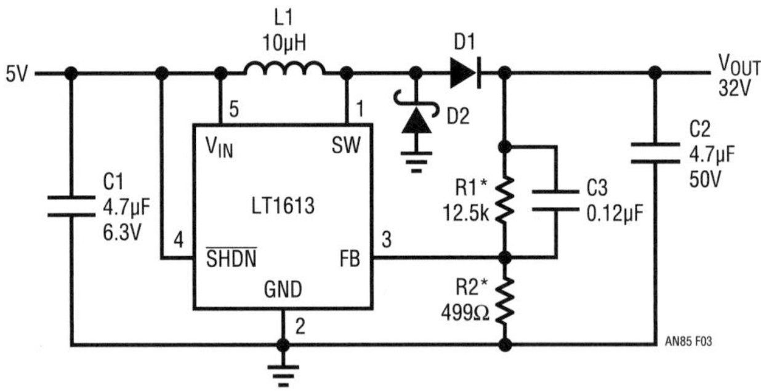

*1% METAL FILM RESISTORS
C1: TAIYO YUDEN JMK212BJ475MG
C2: MURATA GRM235Y5V475Z50
D1: 1N4148
D2: ON SEMICONDUCTOR MBR0540 OR LITE ON/DIODES INC. B0540W
L1: MURATA LQH3C100

Figure 28.3 • LT1613-Based Boost Regulator with Appropriate Component Selection and Layout Has Low Noise Characteristics Needed for Varactor Biasing

Low noise switching regulator design

In theory, a simple flyback regulator will work, but component choice and attention to layout are critical to achieving low noise. Additionally, component count, size and cost are usually considerations in varactor bias applications. Figure 28.3 shows a step-up switching regulator that, properly incarnated, permits low noise varactor biasing. The circuit

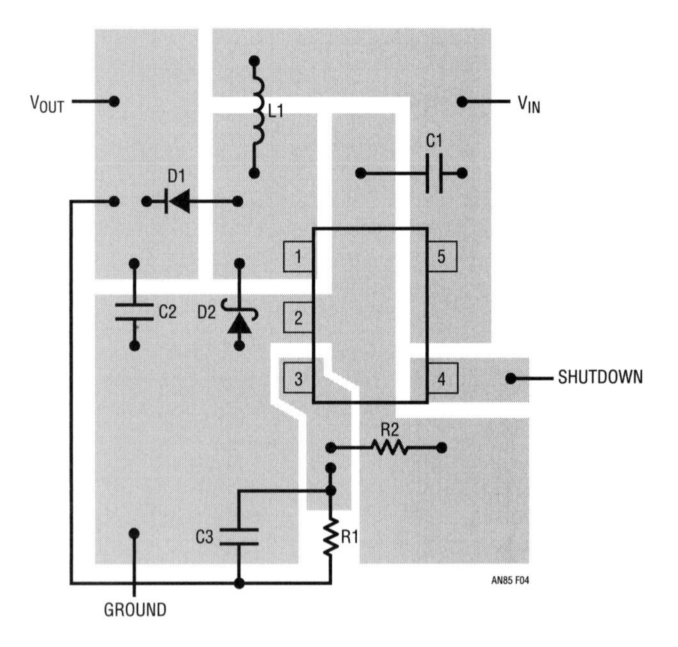

Figure 28.4 • Layout Requires Attention to Component Placement and Ground Current Flow Management. Compact Layout Reduces Parasitic Inductance, Radiation and Crosstalk. Grounding Scheme Minimizes Return Current Mixing

Note 2: The reader is again referred to Appendix A for in-depth discussion of varactor diodes.

Note 3: Spurious oscillator outputs are referred to as "spurs" in RF parlance.

is a simple boost regulator. L1, in conjunction with the SW pin's ground-referred switching, provides voltage step-up. D1 and C2 filter the output to DC, D2 clips possible L1 negative excursions and the feedback resistor ratio sets the loop servo point, and hence, the output voltage. C3 tailors loop frequency response, minimizing switching-frequency ripple components at the output. C1 and C2 are specified for low loss dynamic characteristics and the LT®1613's 1.7MHz switching frequency allows miniature, small value components. This relatively high switching frequency also means that ancillary "downstream" filtering is possible with similarly miniature, small value components.

Layout issues

Layout is the most crucial design aspect for obtaining low noise. Figure 28.4 shows a suggested layout. Ground, V_{IN} and V_{OUT} are distributed in planes, minimizing impedance. The LT1613 GND pin (Pin 2) carries high speed, switched current; its path to the circuit's power exit should be direct and highly conductive at all frequencies. R2's return current, to the extent possible, should not mix with Pin 2's large dynamic currents. C1 and C2 should be located close to Pin 5 and D1 respectively. Their grounded ends should tie directly to the ground plane. L1 has a low impedance path to V_{IN}; its driven end returns directly to LT1613 Pin 1. D1 and D2 should have short, low inductance runs to C2 and Pin 2, respectively; their common connection mating tightly with Pin 1 and L1. Pin 1 has a small area, minimizing radiation. Note that this point is enclosed by planes operating at AC ground, forming a shield. The feedback node (Pin 3) is further shielded from switching radiation, preventing unwanted interaction. Finally, L1 should be oriented so its radiation causes minimal circuit disruption.

Level shifts

The low voltage PLL output (see Figure 28.1) requires an analog level shift to bias the varactor. Figure 28.5 shows some alternatives. Figure 28.5a is an amplifier powered from the LT1613's 32V output. The feedback ratio sets a gain of 10, resulting in a 0V to 30V output for a 0V to 3V input. Figure 28.5b is a noninverting common base stage. Gain is less well controlled than in Figure 28.5a, but overall frequency synthesizer loop action obviates this concern. Figure 28.5c's common emitter circuit is similar except that it inverts.

Test circuit

Figure 28.6 combines the above considerations into a realistic test circuit. The 5V powered design is composed of the LT1613 regulator, an amplifier-based level shift and a GHz range VCO. The amplifier is biased by a filtered LT1004 reference to a 12V output, simulating a typical varactor bias point. The LT1613 configuration's low noise output receives additional filtering via the 100Ω-0.1μF network at the amplifier power pin and by the amplifier's power supply rejection ratio (PSRR). The RC combination provides a theoretical (unloaded) break below 20kHz; the amplifier's PSRR benefit is derived from Figure 28.7. This graph shows PSRR vs frequency for a typical amplifier. There is a steep roll-off beyond 100Hz, although almost 20dB attenuation is available in the MHz region. This implies that the amplifier provides some beneficial filtering of the LT1613's residual 1.7MHz switching components.

A final RC filter section is placed directly at the VCO varactor bias input. Ideally this filter's break frequency is far removed from the 1.7MHz switching rate for maximum ripple attenuation. In practice, the filter is within the PLL loop, placing restrictions on how much delay it can introduce. A PLL loop bandwidth of 5kHz is usually desirable, dictating a filter point of about 50kHz to ensure closed-loop stability. As such, the final RC filter (1.6k-0.002μF) is set at this frequency. It is worth noting that the varactor's input resistance is quite high—essentially that of a reverse biased diode—and no filter buffering is required to drive it.

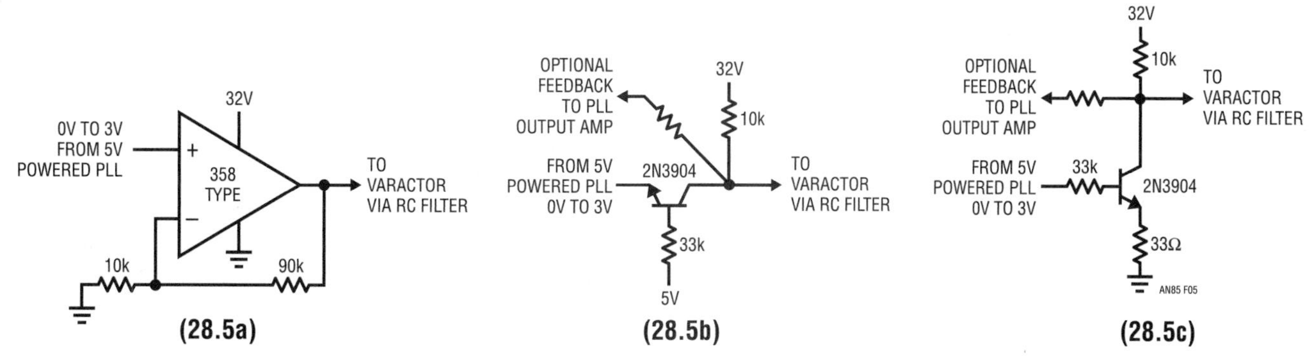

Figure 28.5 • Level Shift Options Include Op Amp (28.5a), Noninverting Common Base (28.5b) and Inverting Common Emitter (28.5c). Op Amp's Operating Point is Inherently Stable; 28.5b and 28.5c Rely on PLL Closed-Loop Action Unless Optional Feedback is Used

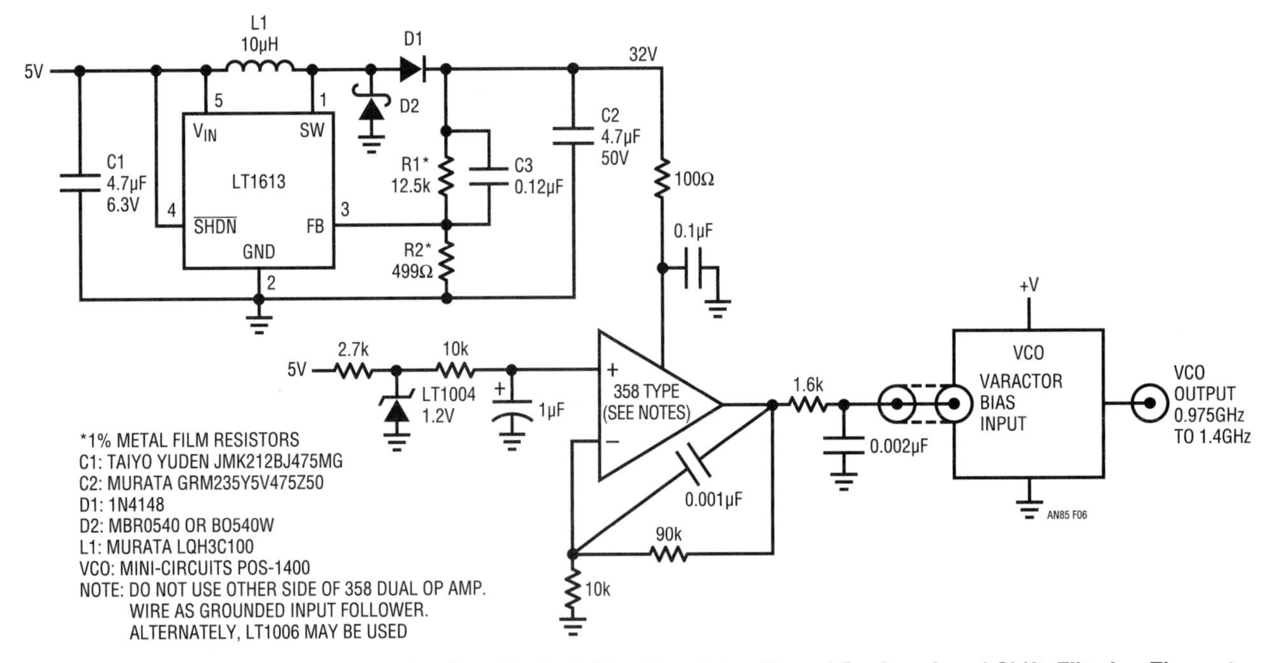

Figure 28.6 • Noise Test Circuit Includes Step-Up Switching Regulator, Biased Op Amp Level Shift, Filtering Elements and GHz Range VCO. Switching Regulator-Associated L1 is the Only Inductor Required

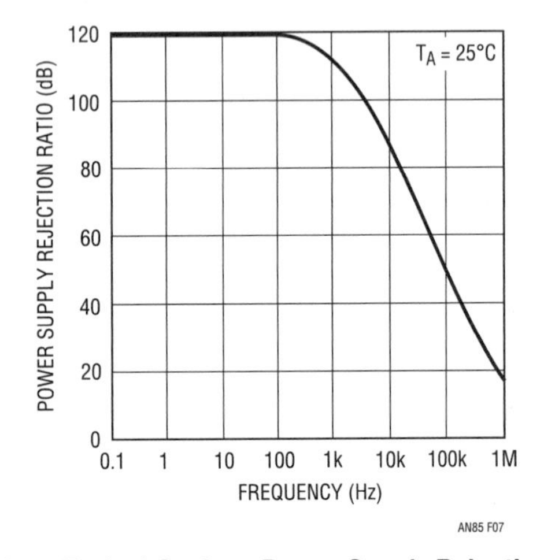

Figure 28.7 • Typical Op Amp Power Supply Rejection Ratio Degrades with Frequency, Although Nearly 20dB is Available in LT1613s MHz Switching Range

Noise performance

Careful measurements permit verification of circuit noise performance.[4] Figure 28.8 shows about 2mV ripple at the LT1613's 32V output. Figure 28.9, taken at the amplifier power pin, shows the effect of the 100Ω-0.1pF filter. Ripple and noise are reduced to about 500µV. Figure 28.10, recorded at the amplifier output, shows the influence of amplifier PSRR. Ripple and noise are further reduced to about 300µV. The actual ripple component is about 100µV. The final RC filter, located directly at the VCO varactor input, gives about 20dB further attenuation. Figure 28.11 shows ripple and noise inside 20µV with a ripple component of about 10µV.

Effects of poor measurement technique

The above results require good measurement technique. The measurements were taken utilizing a purely coaxial probing environment. Deviations from this regime will produce misleading and unduly pessimistic indications.[5] For example, Figure 28.12 shows a 50% amplitude error over Figure 28.8, even though it nominally monitors the same point. The difference is that Figure 28.12 utilizes a 3" probe ground lead instead of Figure 28.8's coaxial ground tip adapter. Similarly, Figure 28.9's amplifier power pin 500µV measurement degrades to Figure 28.13's indicated

Note 4: See Appendix B, "Preamplifier and Oscilloscope Selection," for equipment recommendations to make the high sensitivity oscilloscope measurements described in this section. See also Appendix C, "Probing and Connection Techniques for Low Level, Wideband Signal Integrity."

Note 5: Additional discourse along these lines is presented in Appendix C, "Probing and Connection Techniques for Low Level, Wideband Signal Integrity." See also Reference 2–5.

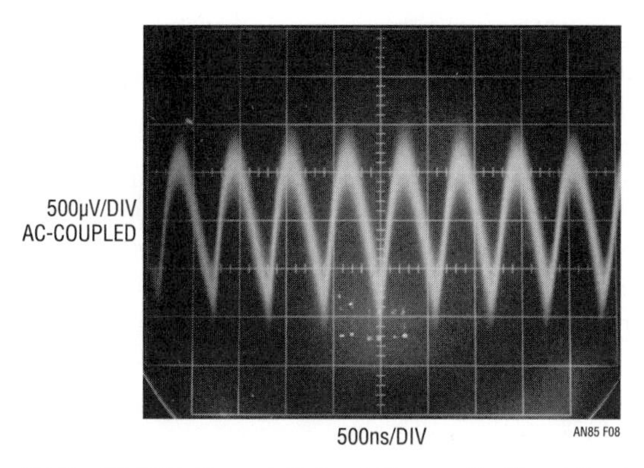

Figure 28.8 • LT1613-Based Output Shows 2mV$_{P-P}$ Ripple and Noise

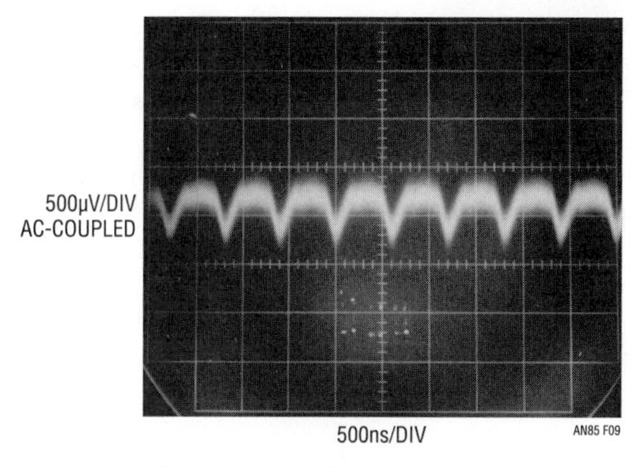

Figure 28.9 • RC Filter at Amplifier's Power Input Pin Reduces Ripple and Noise to 500pV$_{P-P}$

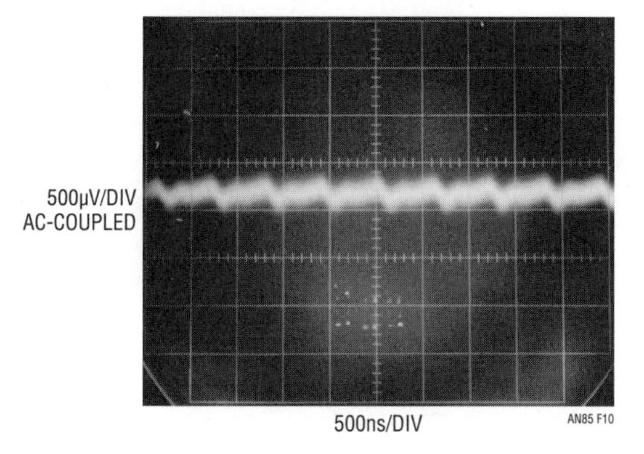

Figure 28.10 • Amplifier Output Shows Additional Filtering Due to Amplifier PSRR. Aberrations Are Inside 300µV

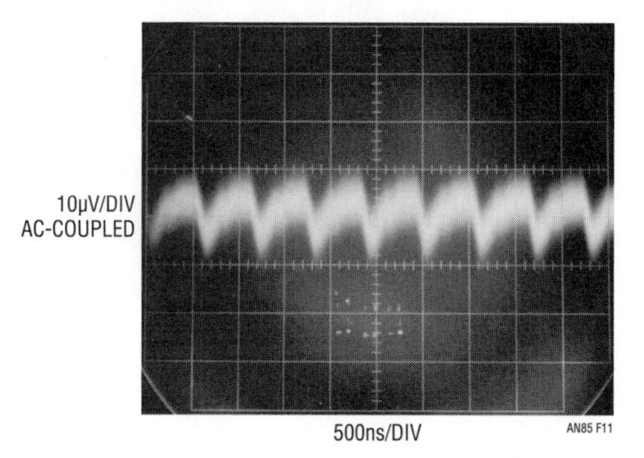

Figure 28.11 • VCO Varactor Bias Input, After 50kHz RC Filter, Displays Less Than 20µV Ripple and Noise. Content Coherent with LT1613's 1.7MHz Switching is Inside 10µV

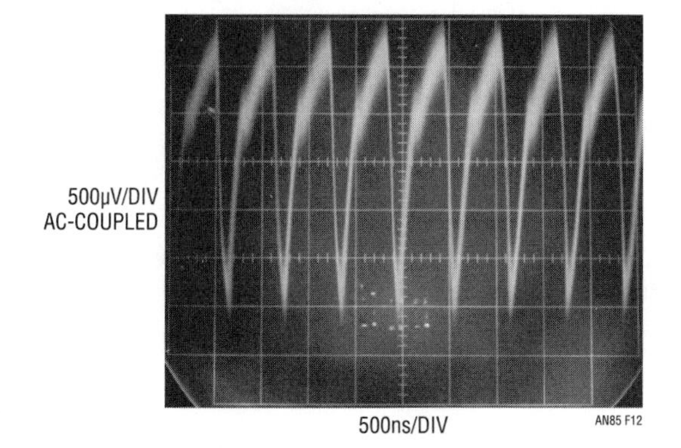

Figure 28.12 • Improper Probing Technique. 3" Ground Lead Causes 50% Display Error vs Figure 28.8's Purely Coaxial Measurement

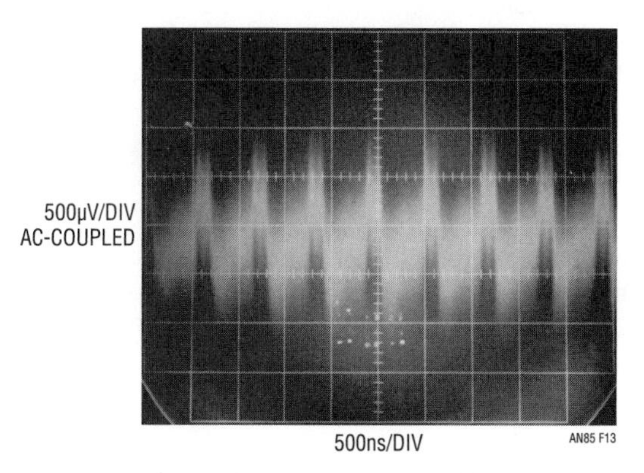

Figure 28.13 • 3" Ground Lead Degrades Figure 28.9s' 500µV Reading to 2mV

2mV representation using the 3" probe ground strap. The same ground strap causes pronounced error in Figure 28.14's apparent 2mV amplifier output vs Figure 28.10's correct 300μV excursion. Figure 28.15 shows a 70μV indication at the VCO varactor input using the 3" ground strap. That's a long way from Figure 28.11's 20μV data taken with the coaxial ground tip adapter![6]

In Figure 28.16 the coaxial ground tip adapter is used but the VCO varactor input shows a blizzard of noise compared to Figure 28.11's orderly trace. The reason is that a 12" voltmeter lead was connected to the point. Pickup and stray RF act against the node's finite output impedance, corrupting the measurement. Figure 28.17, also taken at the VCO input, is better but still shows greater than 50% error. The culprit here is a second probe, located at the LT1613 V_{SW} pin, used to trigger the oscilloscope. Even with coaxial techniques in use at both probe points, the trigger probe dumps transient currents into the ground plane. This introduces small common mode voltages, resulting in the apparent noise increase. The cure is to trigger the oscilloscope with a noninvasive probe.[7]

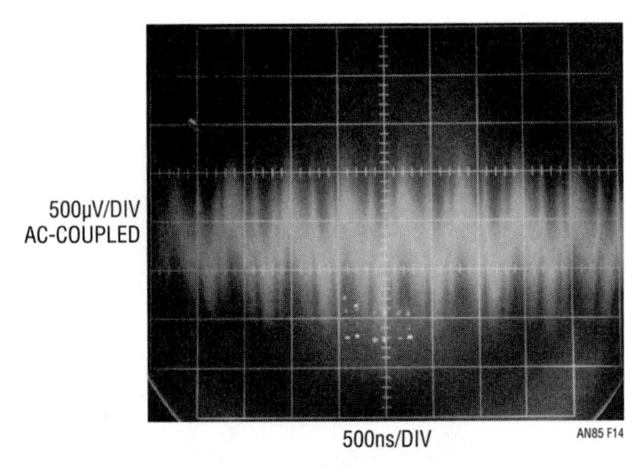

Figure 28.14 • Probe Ground Strap Causes Erroneous 2mV Indication. Actual Value is Figure 28.10's 300μV Reading.

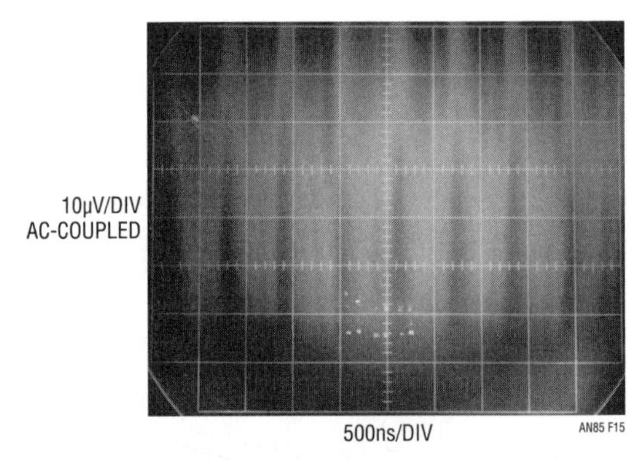

Figure 28.15 • Probe Ground Strap Causes 3.5× Readout Error vs Figure 28.11's Correctly Measured 20μV

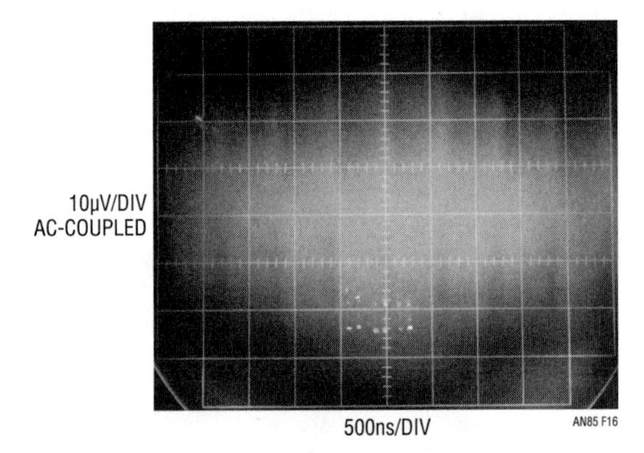

Figure 28.16 • Effect of 12" Voltmeter Probe on VCO Varactor Input. Coaxially Connected 'Scope Probe is in Use. 2.5× Measurement Error Referred to Figure 28.11 Results

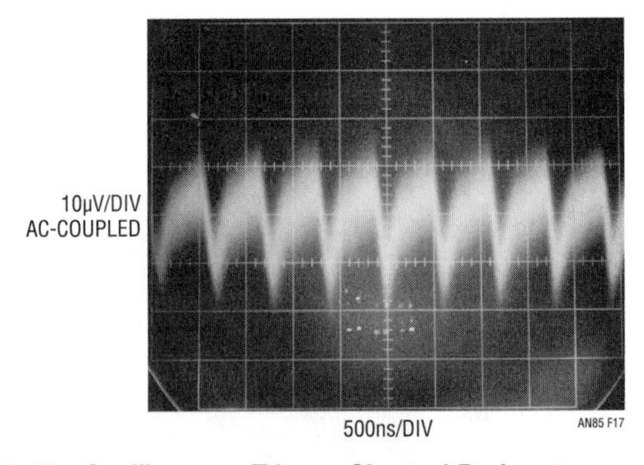

Figure 28.17 • Oscilloscope Trigger Channel Probe at LT1613 SW Pin Causes 50% Measurement Error vs Figure 28.11

Note 6: If you don't think 70μV is a "long way" from 20μV, consider your reaction to a 3.5× income tax reduction.

Note 7: The reader is not being requested to indulge wishful thinking. Such a probe is more easily realized than might be supposed. See Appendix C, "Probing and Connection Techniques for Low Level, Wideband Signal Integrity."

Frequency-domain performance

Although the varactor bias noise amplitude measurements are critical, it is difficult to correlate them with frequency-domain performance. Varactor bias noise amplitude translates into spurious VCO outputs and that is the measurement of ultimate concern. Although it is possible to view the GHz range VCO on an oscilloscope (Figure 28.18), this time domain measurement lacks adequate sensitivity to detect spurious activity. A spectrum analyzer is required. Figure 28.19, a spectral plot of VCO output, shows a center frequency of 1.14GHz, with no apparent spurious activity within the ≈90dB measurement noise floor. A marker has been placed at 1.7MHz (3.5 divisions from center), corresponding to the LT1613's switching frequency. No readily distinguishable activity is apparent at about −90dBc. Succeeding figures "sanity check" this performance by systematically degrading the circuit and noting results. In Figure 28.20, the VCO varactor input's RC filter has been replaced with a direct connection. Now the 1.7MHz spurious outputs are easily seen, about −62dBc.

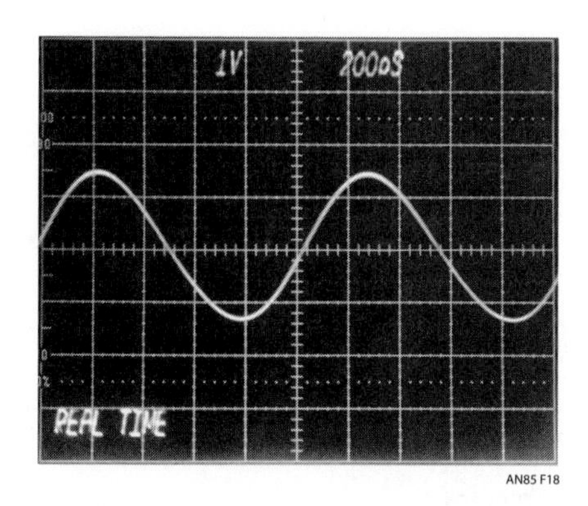

Figure 28.18 • GHz Range VCO Output is Viewable on Oscilloscope, But Spurious Activity is Undetectable. Spectral Measurements Are Required

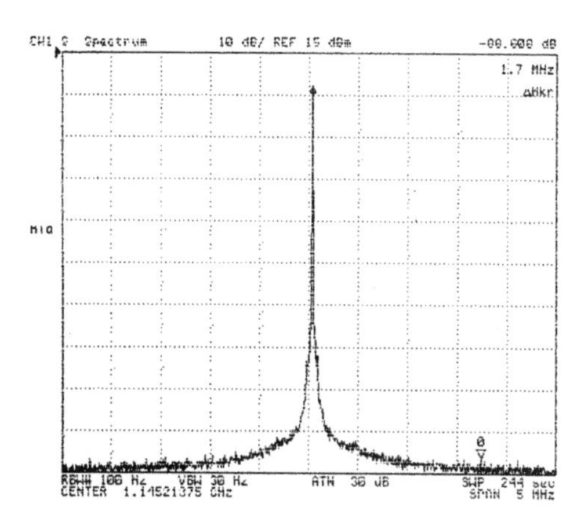

Figure 28.19 • HP-4396B Spectrum Analyzer Indicates Spurious Outputs at Least -90dBc Referred to 1.14GHz VCO Center Frequency

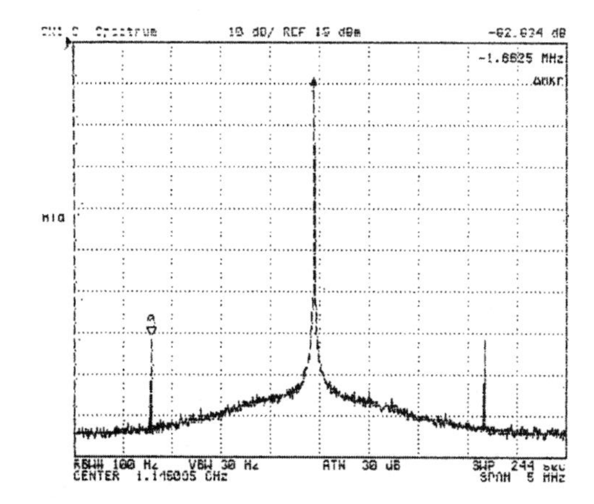

Figure 28.20 • "Sanity Checking" Figure 28.19's Results by Replacing RC Filter at VCO Varactor Input with Direct Connection. LT1613 1.7MHz Switching Frequency Related Activity Appears at -62dBc

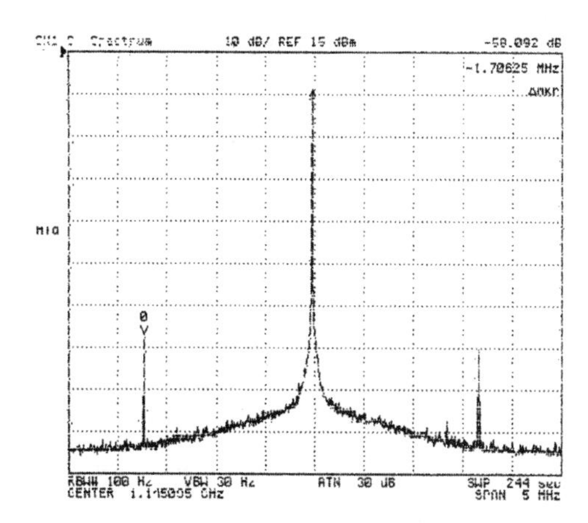

Figure 28.21 • Similar Measurement Conditions to Previous Figure with 12" Voltmeter Probe Added. "Spurs" Increase by 4dB to -58dBc

In Figure 28.21, a 12" voltmeter lead as been connected to the measurement point, resulting in a 4dB degradation, to about −58dBc. Figure 28.22 shows pronounced effects due to poor LT1613 layout (power ground pin routed circuitously, rather than directly, back to input common) and component choice (lossy capacitor substituted for C2). Spurious activity jumps to −48dBc. In Figure 28.23 proper layout and components are used, but the varactor bias line has been placed in close proximity to switching inductor L1. Additionally, the bias line and RC filter components have been distanced from the ground plane. The resultant electromagnetic pickup and increase in bias line effective inductance cause 1.7MHz "spurs" at −54dBc. Additional harmonically related activity, although less severe, is also apparent. Figure 28.24 indicates favorable results when the bias line and RC filter are restored to their proper orientation. The plot is essentially identical to Figure 28.19. The lesson here is clear. Layout and measurement practice are at least as important as circuit design. As always, the "hidden schematic" dominates performance.[8]

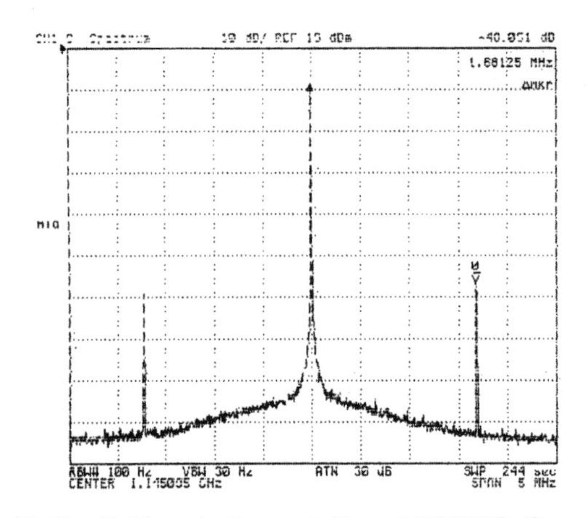

Figure 28.22 • Deliberate Degradation of LT1613's Grounding Scheme and Output Capacitor Raise Spurious Outputs to -48dBc

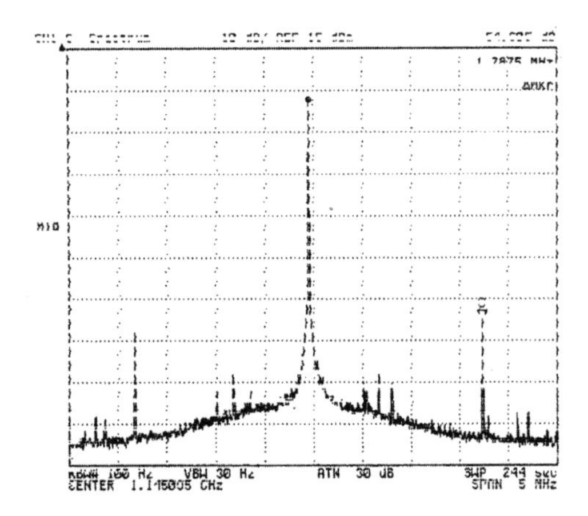

Figure 28.23 • Results with Varactor Bias Line Deliberately Routed Near LT1613's Switching Inductor, and RC Filter Components Lifted from Ground Plane. 1.7MHz "Spurs" at -54dBc; Other Harmonically Related Components Also Appear

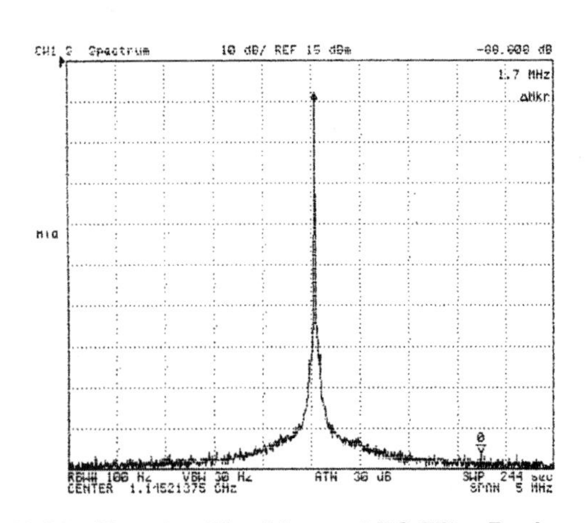

Figure 28.24 • Varactor Bias Line and RC Filter Replaced in Proper Orientation. Figure 28.19's Silence is Restored

Note 8: Charly Gullett of Intel Corporation originated the quoted descriptive, an author favorite.

References

1. Chadderton, Neil, "Zetex Variable Capacitance Diodes," Application Note 9, Issue 2, January 1996. Zetex Applications Handbook, 1998. Zetex plc. UK
2. Williams, Jim, "A Monolithic Switching Regulator with 100pV Output Noise," Linear Technology Corporation, Application Note 70, October 1997
3. Williams, Jim, "High Speed Amplifier Techniques," Linear Technology Corporation, Application Note 47, August 1991
4. Hurlock, Les, "ABCs of Probes," Tektronix, Inc., 1990
5. McAbel, W. E., "Probe Measurements," Tektronix, Inc., 1971

Appendix A

The following section, excerpted with permission from Zetex Application Note 9 (see Reference 1), reviews theoretical considerations of varactor diodes.

Zetex variable capacitance diodes

Neil Chadderton, Zetex plc

Background

The varactor diode is a device that is processed so to capitalise on the properties of the depletion layer of a P-N diode. Under reverse bias, the carriers in each region (holes in the P type and electrons in the N type) move away from the junction, leaving an area that is depleted of carriers. Thus a region that is essentially an insulator has been created, and can be compared to the classic parallel plate capacitor model. The effective width of this depletion region increases with reverse bias, and so the capacitance decreases. Thus the depletion layer effectively creates a voltage dependent junction capacitance, that can be varied between the forward conduction region and the reverse breakdown voltage (typically +0.7V to −35V respectively for the ZC830 and ZC740 series diodes).

Different junction profiles can be produced that exhibit different Capacitance-Voltage (C-V) characteristics. The Abrupt junction type of example, shows a small range of capacitance due to its diffusion profile, and as a consequence of this is capable of high Q and low distortion, while the Hyperabrupt variety allows a larger change in capacitance for the same range of reverse bias. So called Hyper-hyperabrupt, or octave tuning variable capacitance diodes show a large change in capacitance for a relatively small change in bias voltage. This is particularly useful in battery powered systems where the available bias voltage is limited.

The varactor can be modelled as a variable capacitance (Cjv), in series with a resistance (Rs). (Please refer to Figure A1.)

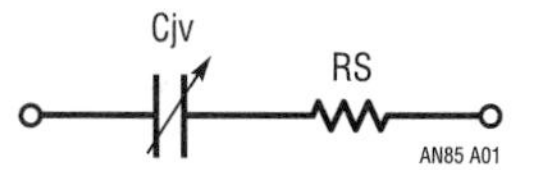

Figure A1 • Common Model for the Varactor Diode

The capacitance, Cjv, is dependent upon the reverse bias voltage, the junction area, and the doping densities of the semiconductor material, and can be described by:

$$\mathrm{Cjv} = \frac{\mathrm{Cj\,0}}{(1 + \mathrm{Vr}/\varphi)^{\mathrm{N}}}$$

where:

Cj0 = Junction capacitance at 0V

Cjv = Junction capacitance at applied bias voltage Vr

Vr = Applied bias voltage

ϕ = Contact potential

N = Power law of the junction or slope factor

The series resistance exists as a consequence of the remaining undepleted semiconductor resistance, a contribution due to the die substrate, and a small lead and package component, and is foremost in determining the performance of the device under RF conditions.

This follows, as the quality factor, Q, is given by:

$$Q = \frac{1}{2\pi fCjvR_S}$$

Where:

Cjv = Junction capacitance at applied bias voltage Vr

R_S = Series Resistance

f = Frequency

So, to maximise Q, R_S must be minimised. This is achieved by the use of an epitaxial structure so minimising the amount of high resistivity material in series with the junction.

Note: Zetex has produced a set of SPICE models to enable designers to simulate their circuits in SPICE, PSPICE and similar simulation packages. The models use a version of the above capacitance equation and so the model parameters may also be of interest for other software packages. Information is also provided to allow inclusion of parasitic elements to the model. These models are available on request, from any Zetex sales office.

Important parameters

This section reviews the important characteristics of varactor diodes with particular reference to the Zetex range of variable capacitance diodes.

The characteristic of prime concern to the designer is the Capacitance-Voltage relationship, illustrated by a C-V curve, and expressed at a particular voltage by Cx, where x is the bias voltage. The C-V curve summarises the range of useful capacitance, and also shows the shape of the relationship, which may be relevant when a specific response is required. Figures A2a, A2b and A2c show families of C-V curves for the ZC740-54 (Abrupt), ZC830-6 (Hyperabrupt), and ZC930 (Hyper-hyperabrupt) ranges respectively. Obviously, the choice of device type depends upon the application, but aspects to consider include: the range of frequencies the circuit must operate with, and hence an appropriate capacitance range; the available bias voltage; and the required response.

The capacitance ratio, commonly expressed as Cx/Cy (where x and y are bias voltages), is a useful parameter that shows how quickly the capacitance changes with applied bias voltage. So, for an Abrupt junction device, a

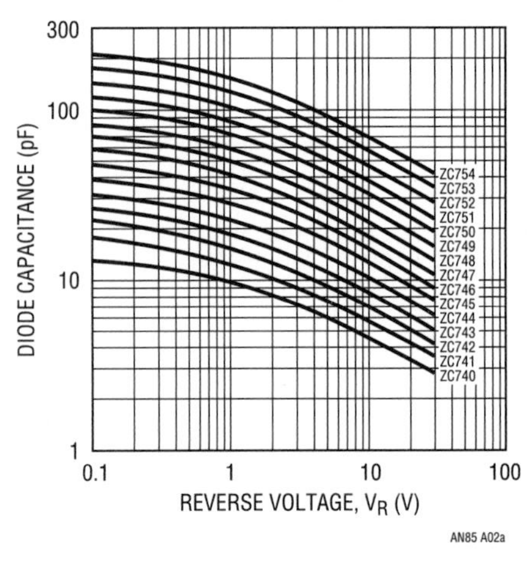

Figure A2a • Typical Capacitance-Voltage Characteristics for the ZC740-54 Range

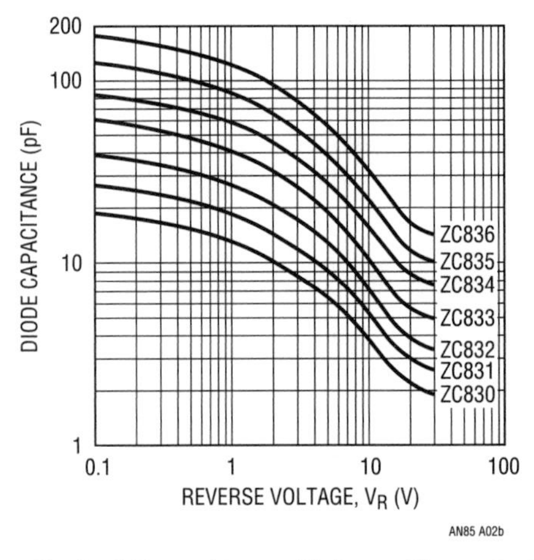

Figure A2b • Typical Capacitance-Voltage Characteristics for the ZC830-6 Range

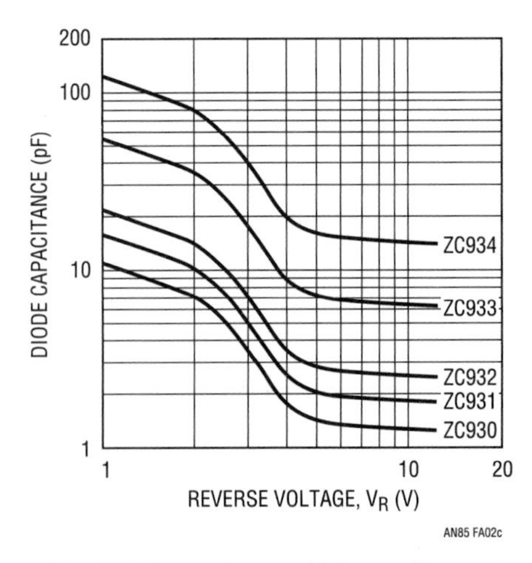

Figure A2c • Typical Capacitance-Voltage Characteristics for the ZC930-4 Range

C2/C20 figure of 2.8 may be typical, whereas a C2/C20 ration of 6 may be expected for a Hyperabrupt device. This feature of the Hyperabrupt variety can be particularly important when assessing devices for battery-powered applications, where the bias voltage range may be limited. In this instance, the ZC930 series that feature a better than 2:1 tuning range for a 0 to 6V bias may be of particular interest.

The quality factor Q, at a particular condition is a useful parameter in assessing the performance of a device with respect to tuned circuits, and the resulting loaded Q.

Zetex guarantees a minimum Q at test conditions of 50MHz, and a relatively low V_R of 3 or 4V and ranges 100 to 450 depending on device type.

The specified V_R is very important in assessing this parameter, because as well as the C-V dependence as detailed previously, a significant part of the series resistance (R_S), is due to the remaining undepleted epitaxial layer, which is also dependant upon V_R. This R_S-V_R relationship is shown in Figure A3 for the ZC830, ZC833 and ZC836 Hyperabrupt devices, measured at frequencies of 470MHz, 300MHz and 150MHz respectively, and also serves to illustrate the excellent performance of Zetex Variable Capacitance Diodes at VHF and UHF.

Also of interest, with respect to stability, is the temperature coefficient of capacitance, as capacitance changes with V_R, and is shown for the three ranges in Figures A4a, A4b and A4c respectively.

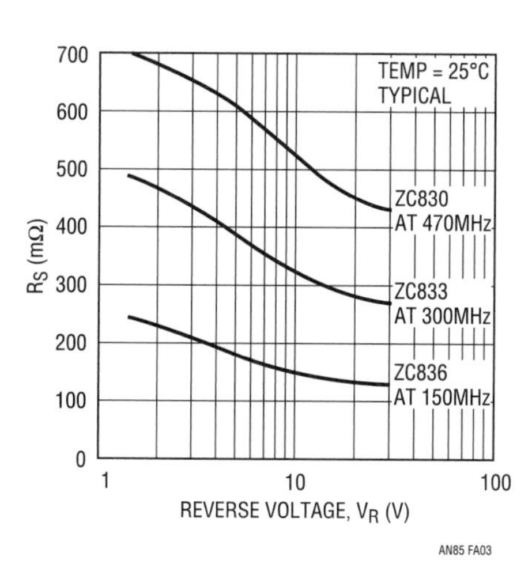

Figure A3 • Typical R_S vs V_R Relationship for ZC830 Series Diodes

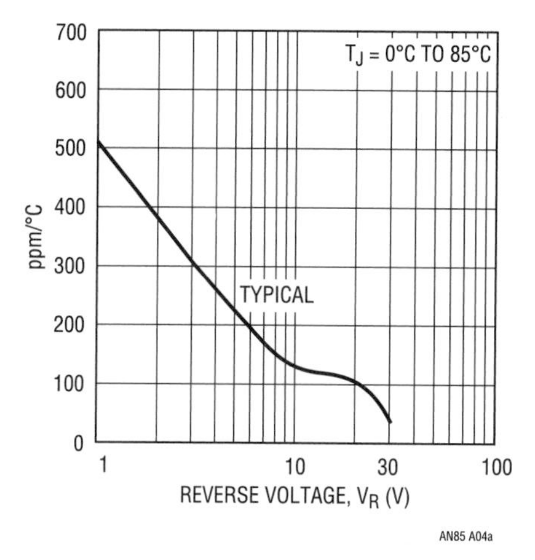

Figure A4a • Temperature Coefficient of Capacitance vs V_R for the ZC740 Series

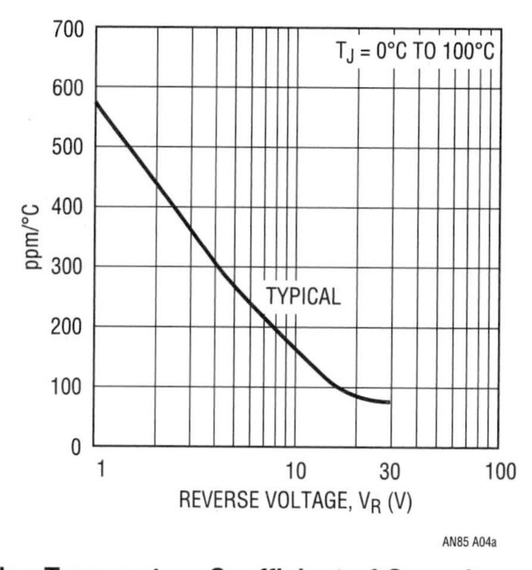

Figure A4b • Temperature Coefficient of Capacitance vs V_R for the ZC830 Series

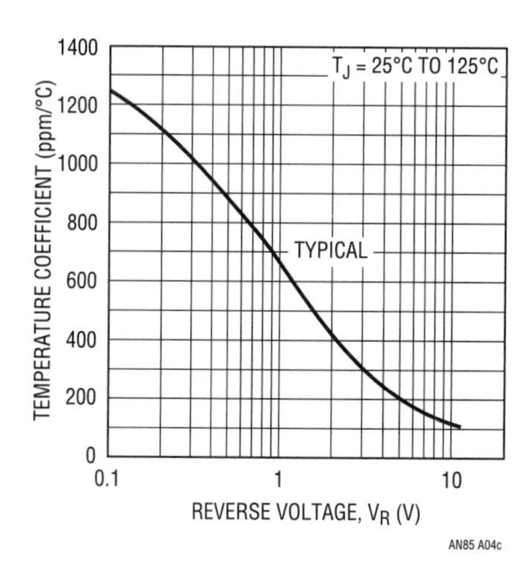

Figure A4c • Temperature Coefficient of Capacitance vs V_R for the ZC930 Series

The reverse breakdown voltage, V(BR) also has a bearing on device selection, as this parameter limits the maximum V_R that may be used when biasing for minimum capacitance. Zetex variable capacitance diodes typically possess a V(BR) of 35V.

The maximum frequency of operation will depend on the required capacitance and the series resistance (and hence useful Q), that is possessed by a particular device type, but also of consequence are the parasitic components exhibited by the device package. These depend on the size, material and construction of the package. For example, the Zetex SOT-23 package has a typical stray capacitance of 0.08pF and a total lead inductance of 2.8nH, while the E-line package shows less than 0.2pF and 5nH respectively. These low values allow a wide frequency application, for example, the ZC830 and ZC930 series, configured as series pairs are ideal for low cost microwave designs extending to 2.5GHz and above.

Appendix B
Preamplifier and oscilloscope selection

The low level measurements described require some form of preamplification for the oscilloscope. Current generation oscilloscopes rarely have greater than 2mV/DIV sensitivity, although older instruments offer more capability. Figure B1 lists representative preamplifiers and oscilloscope plug-ins suitable for noise measurement. These units feature wideband, low noise performance. It is particularly significant that many of these instruments are no longer produced. This is in keeping with current instrumentation trends, which emphasize digital signal acquisition as opposed to analog measurement capability.

The monitoring oscilloscope should have adequate bandwidth and exceptional trace clarity. In the latter regard high quality analog oscilloscopes are unmatched. The exceptionally small spot size of these instruments is well-suited to low level noise measurement.[1] The digitizing uncertainties and raster scan limitations of DSOs impose display resolution penalties. Many DSO displays will not even register the small levels of switching-based noise.

INSTRUMENT TYPE	MANUFACTURER	MODEL NUMBER	BANDWIDTH	MAXIMUM SENSITIVITY/GAIN	AVAILABILITY	COMMENTS
Amplifier	Hewlett-Packard	461A	150MHz	Gain = 100	Secondary Market	50Ω Input, Stand-Alone
Differential Amplifier	Preamble	1855	100MHz	Gain = 10	Current Production	Stand-Alone, Settable Ban(
Differential Amplifier	Tektronix	1A7/1A7A	1MHz	10μV/DIV	Secondary Market	Requires 500 Series Mainf Settable Bandstops
Differential Amplifier	Tektronix	7A22	1MHz	10μV/DIV	Secondary Market	Requires 7000 Series Mair Settable Bandstops
Differential Amplifier	Tektronix	5A22	1MHz	10μV/DIV	Secondary Market	Requires 5000 Series Mair Settable Bandstops
Differential Amplifier	Tektronix	ADA-400A	1MHz	10μV/DIV	Current Production	Stand-Alone with Optional Supply, Settable Bandstop:
Differential Amplifier	Preamble	1822	10MHz	Gain = 1000	Current Production	Stand-Alone, Settable Ban(
Differential Amplifier	Stanford Research Systems	SR-560	1MHz	Gain = 50000	Current Production	Stand-Alone, Settable Ban(Battery or Line Operation

Figure B1 • Some Applicable High Sensitivity, Low Noise Amplifiers. Trade-Offs Include Bandwidth, Sensitivity and Availability

Note 1: In our work we have found Tektronix types 454, 454A, 547 and 556 excellent choices. Their pristine trace presentation is ideal for discerning small signals of interest against a noise floor limited background.

Appendix C
Probing and connection techniques for low level, wideband signal integrity[1]

The most carefully prepared breadboard cannot fulfill its mission if signal connections introduce distortion. Connections to the circuit are crucial for accurate information extraction. The low level, wideband measurements demand care in routing signals to test instrumentation.

Ground loops

Figure C1 shows the effects of a ground loop between pieces of line-powered test equipment. Small current flow between test equipment's nominally grounded chassis creates 60Hz modulation in the measured circuit output. *This problem can be avoided by grounding all line powered test equipment at the same outlet strip or otherwise ensuring that all chassis are at the same ground potential. Similarly, any test arrangement that permits circuit current flow in chassis interconnects must be avoided.*

Pickup

Figure C2 also shows 60Hz modulation of the noise measurement. In this case, a 4-inch voltmeter probe at the feedback input is the culprit. *Minimize the number of test connections to the circuit and keep leads short.*

Poor probing technique

Figure C3's photograph shows a short ground strap affixed to a scope probe. The probe connects to a point which provides a trigger signal for the oscilloscope. Circuit output noise is monitored on the oscilloscope via the coaxial cable shown in the photo.

Figure C4 shows results. A ground loop on the board between the probe ground strap and the ground referred cable shield causes apparent excessive ripple in the display. *Minimize the number of test connections to the circuit and avoid ground loops.*

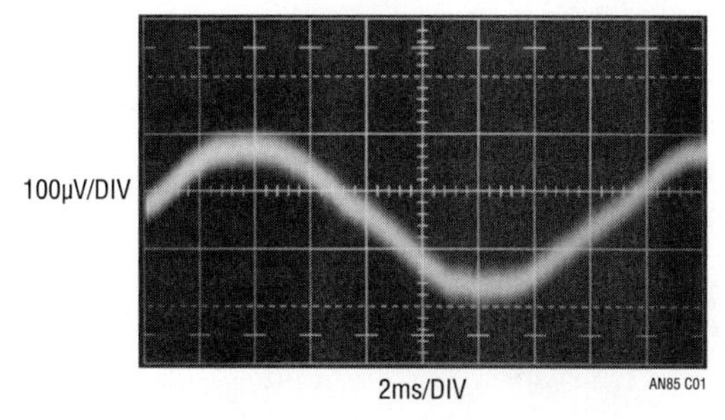

Figure C1 • Ground Loop Between Pieces of Test Equipment Induces 60Hz Display Modulation

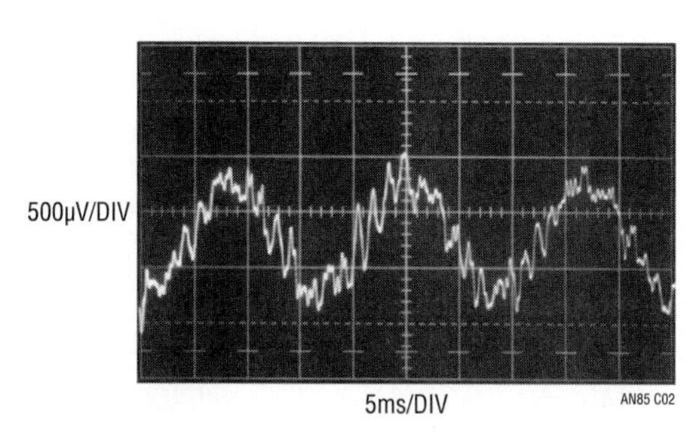

Figure C2 • 60Hz Pickup Due to Excessive Probe Length at Feedback Node

Note 1: Veterans of LTC Application Notes, a hardened crew, will recognize this Appendix from AN70 (see Reference 2). Although that publication concerned considerably more wideband noise measurement, the material is directly applicable to this effort. As such, it is reproduced here for reader convenience.

Figure C3 • Poor Probing Technique Trigger Probe Ground Lead Can Cause Ground Loop-Induced Artifacts to Appear in Display

Violating coaxial signal transmission—felony case

In Figure C5, the coaxial cable used to transmit the circuit output noise to the amplifier-oscilloscope has been replaced with a probe. A short ground strap is employed as the probe's return. The error inducing trigger channel probe in the previous case has been eliminated; the 'scope is triggered by a noninvasive, isolated probe.[2] Figure C6 shows excessive display noise due to breakup of the coaxial signal environment. The probe's ground strap violates coaxial transmission and the signal is corrupted by RF. *Maintain coaxial connections in the noise signal monitoring path.*

Violating coaxial signal transmission—misdemeanor case

Figure C7's probe connection also violates coaxial signal flow, but to a less offensive extent. The probe's ground strap is eliminated, replaced by a tip grounding attachment. Figure C8 shows better results over the preceding case, although signal corruption is still evident. *Maintain coaxial connections in the noise signal monitoring path.*

Proper coaxial connection path

In Figure C9, a coaxial cable transmits the noise signal to the amplifier-oscilloscope combination. In theory, this affords the highest integrity cable signal transmission.

Figure C10's trace shows this to be true. The former example's aberrations and excessive noise have disappeared. The switching residuals are now faintly outlined in the amplifier noise floor. *Maintain coaxial connections in the noise signal monitoring path.*

Direct connection path

A good way to verify there are no cable-based errors is to eliminate the cable. Figure C11's approach eliminates all cable between breadboard, amplifier and oscilloscope. Figure C12's presentation is indistinguishable from Figure C10, indicating no cable-introduced infidelity. When results seem optimal, design an experiment to test them. *When results seem poor, design an experiment to test them. When results are as expected, design an experiment to test them. When results are unexpected, design an experiment to test them.*

Test lead connections

In theory, attaching a voltmeter lead to the regulator's output should not introduce noise. Figure C13's increased noise reading contradicts the theory. The regulator's output impedance, albeit low, is not zero, especially as frequency scales up. The RF noise injected by the test lead works against the finite output impedance, producing the 200μV of noise indicated in the figure. If a voltmeter lead must be connected to the output during testing, it should be done through a 10kΩ-10μF filter: Such a network eliminates Figure C13's problem while introducing minimal error in the monitoring DVM. *Minimize the number of test lead connections to the circuit while checking noise. Prevent test leads from injecting RF into the test circuit.*

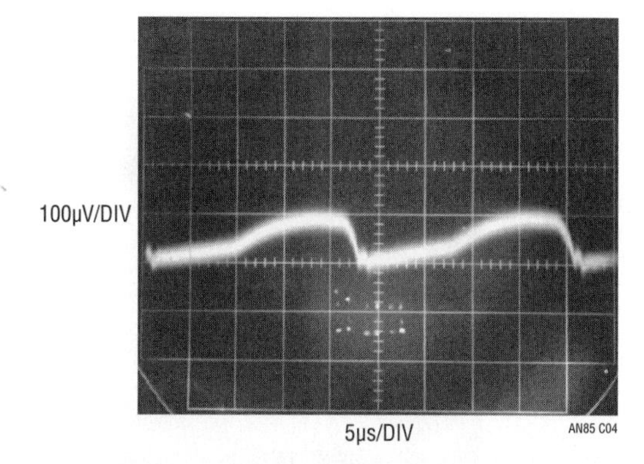

Figure C4 • Apparent Excessive Ripple Results from Figure C3's Probe Misuse. Ground Loop on Board Introduces Serious Measurement Error

Note 2: To be discussed. Read on.

Figure C5 • Floating Trigger Probe Eliminates Ground Loop, But Output Probe Ground Lead (Photo Upper Right) Violates Coaxial Signal Transmission

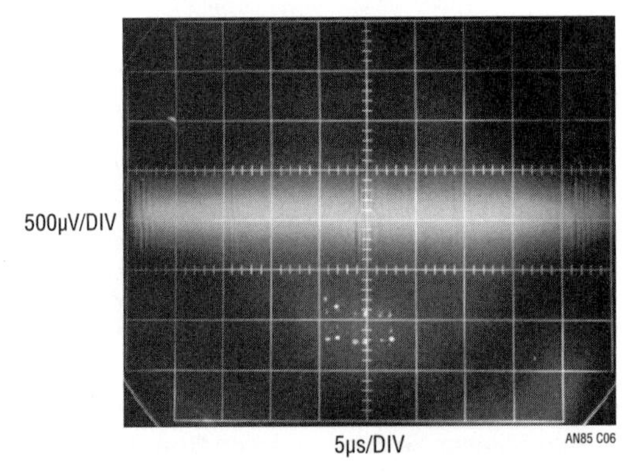

Figure C6 • Signal Corruption Due to Figure C5's Noncoaxial Probe Connection

Figure C7 • Probe with Tip Grounding Attachment Approximates Coaxial Connection

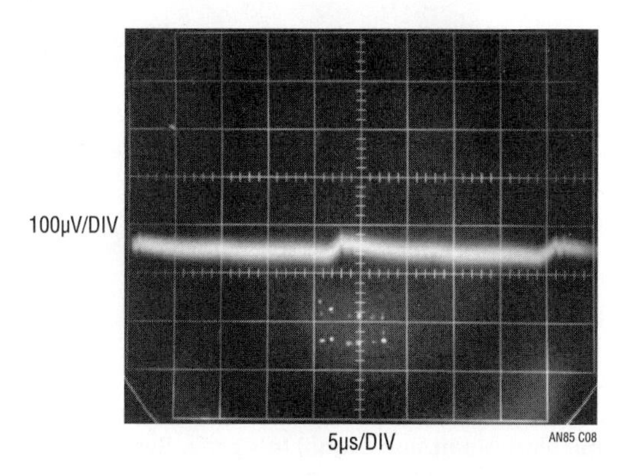

Figure C8 • Probe with Tip Grounding Attachment Improves Results. Some Corruption is Still Evident

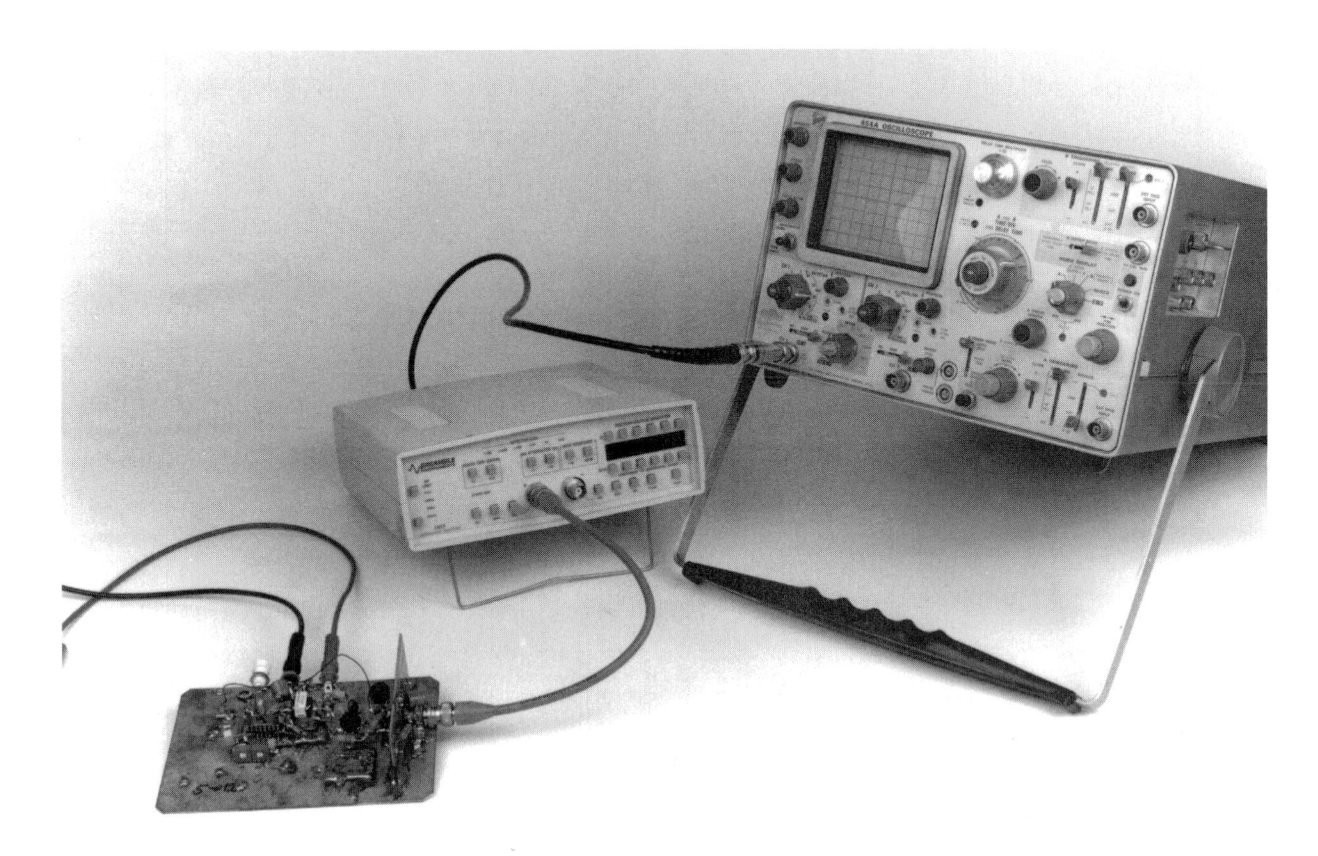

Figure C9 • Coaxial Connection Theoretically Affords Highest Fidelity Signal Transmission

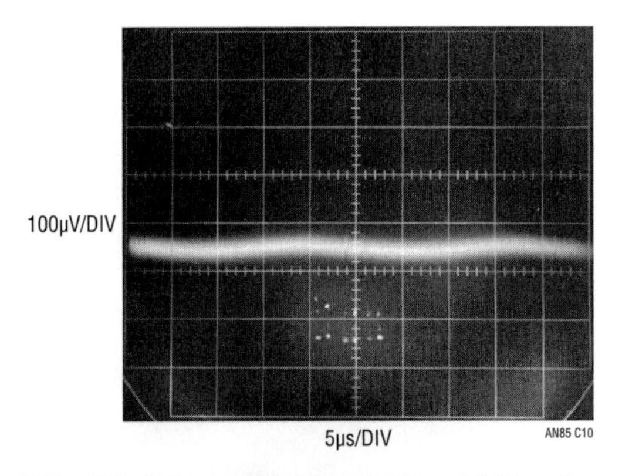

Figure C10 • Life Agrees with Theory. Coaxial Signal Transmission Maintains Signal Integrity. Switching Residuals Are Faintly Outlined in Amplifier Noise

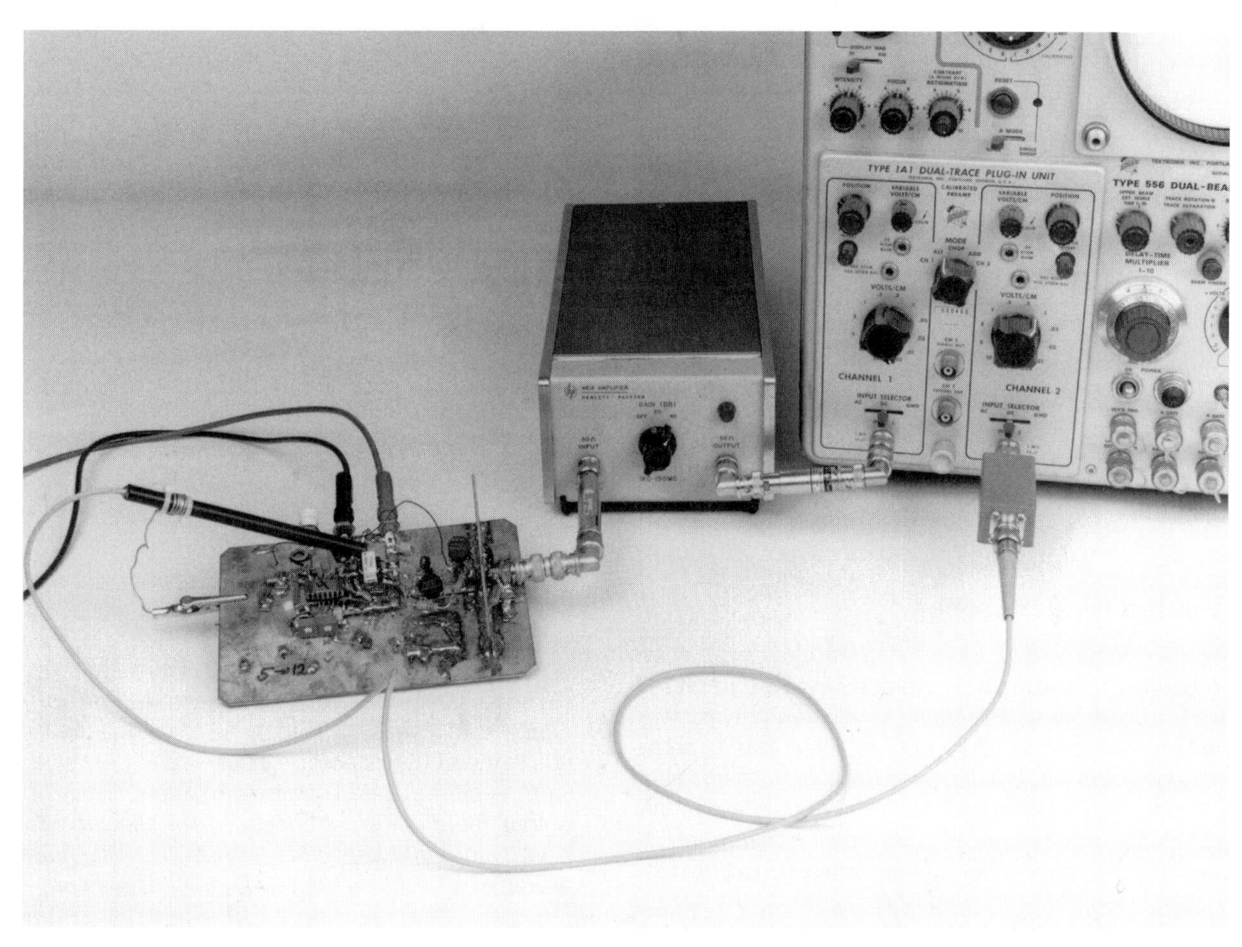

Figure C11 • Direct Connection to Equipment Eliminates Possible Cable-Termination Parasitics, Providing Best Possible Signal Transmission

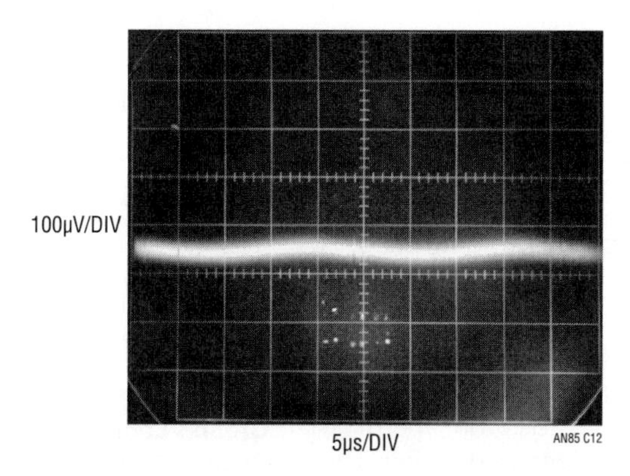

Figure C12 • Direct Connection to Equipment Provides Identical Results to Cable-Termination Approach. Cable and Termination Are Therefore Acceptable

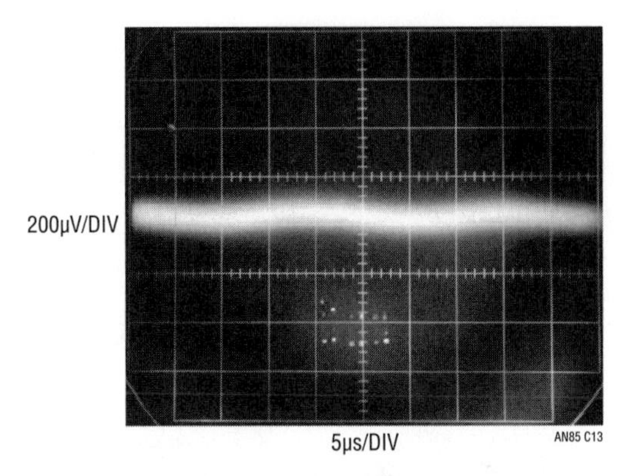

Figure C13 • Voltmeter Lead Attached to Regulator Output Introduces RF Pickup, Multiplying Apparent Noise Floor

Isolated trigger probe

The text associated with Figure C5 somewhat cryptically alluded to an "isolated trigger probe." Figure C14 reveals this to be simply an RF choke terminated against ringing. The choke picks up residual radiated field, generating an isolated trigger signal. This arrangement furnishes a 'scope trigger signal with essentially no measurement corruption. The probe's physical form appears in Figure C15. For good results, the termination should be adjusted for minimum ringing while preserving the highest possible amplitude output. Light compensatory damping produces Figure C16's output, which will cause poor 'scope triggering. Proper adjustment results in a more favorable output (Figure C17), characterized by minimal ringing and well-defined edges.

Trigger probe amplifier

The field around the switching magnetics is small and may not be adequate to reliably trigger some oscilloscopes. In such cases, Figure C18's trigger probe amplifier is useful. It uses an adaptive triggering scheme to compensate for variations in probe output amplitude. A stable 5V trigger output is maintained over a 50:1 probe output range. A1, operating at a gain of 100, provides wideband AC gain. The output of this stage biases a 2-way peak detector (Q1 through Q4). The maximum peak is stored in Q2's emitter capacitor, while the minimum excursion is retained in Q4's emitter capacitor. The DC value of the midpoint of A1's output signal appears at the junction of the 500pF capacitor and the 3MΩ units. This point always sits midway between the signal's excursions, regardless of absolute amplitude. This signal-adaptive voltage is buffered by A2 to set the trigger voltage at the LT1394's positive input. The LT1394's negative input is biased directly from A1's output. The LT1394's output, the circuit's trigger output, is unaffected by >50:1 signal amplitude variations. An X100 analog output is available at A1.

Figure C19 shows the circuit's digital output (Trace B) responding to the amplified probe signal at A1 (Trace A).

Figure C20 is a typical noise testing setup. It includes the breadboard, trigger probe, amplifier, oscilloscope and coaxial components.

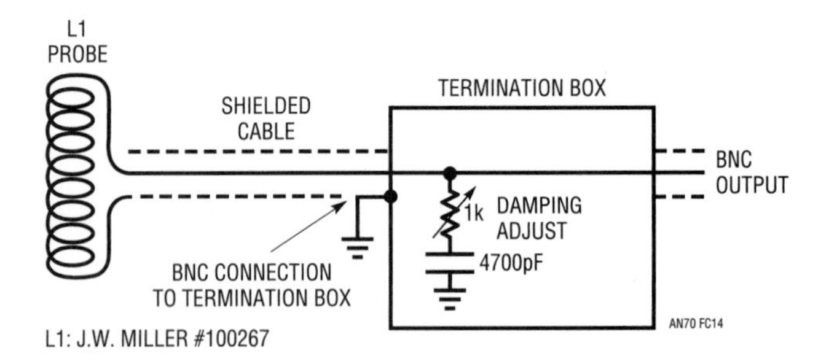

Figure C14 • Simple Trigger Probe Eliminates Board Level Ground Loops. Termination Box Components Damp L1's Ringing Response

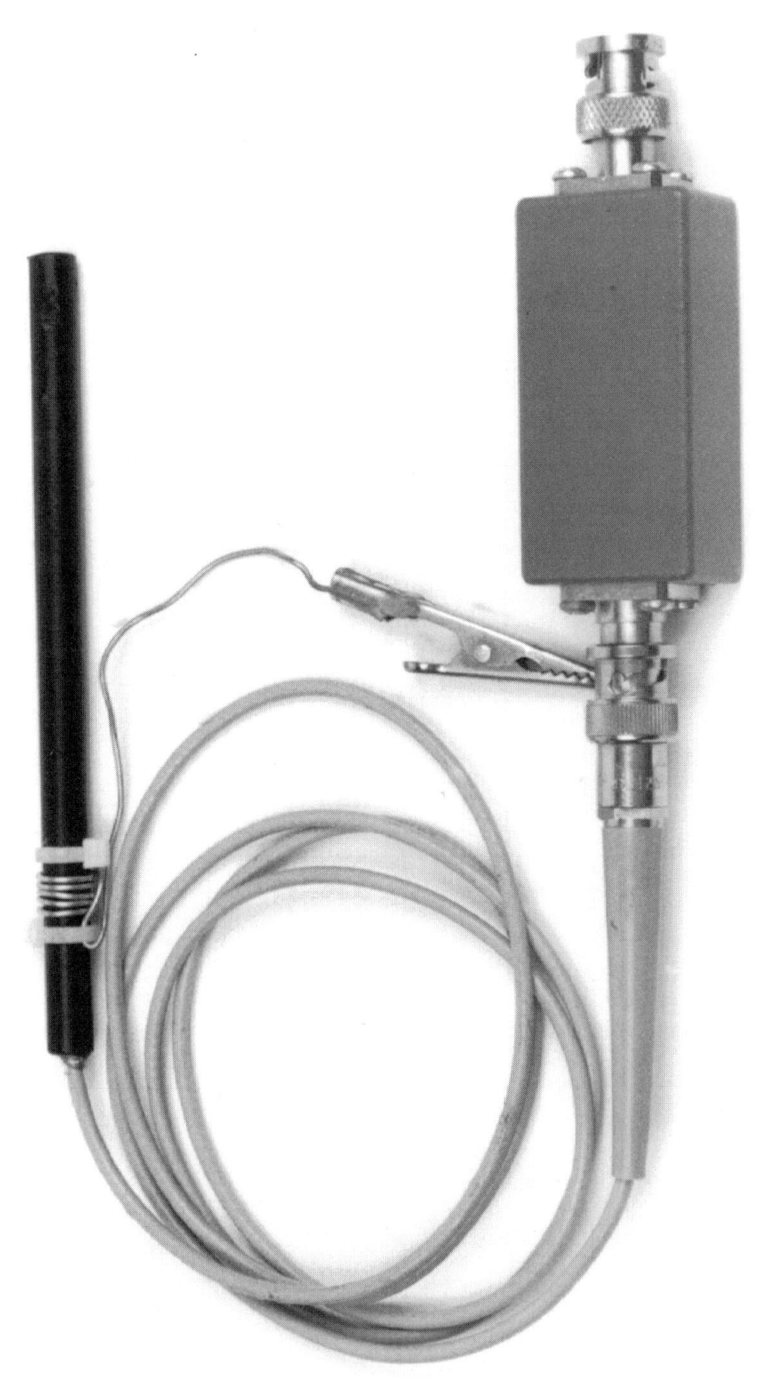

Figure C15 • The Trigger Probe And Termination Box. Clip Lead Facilitates Mounting Probe, Is Electrically Netutral

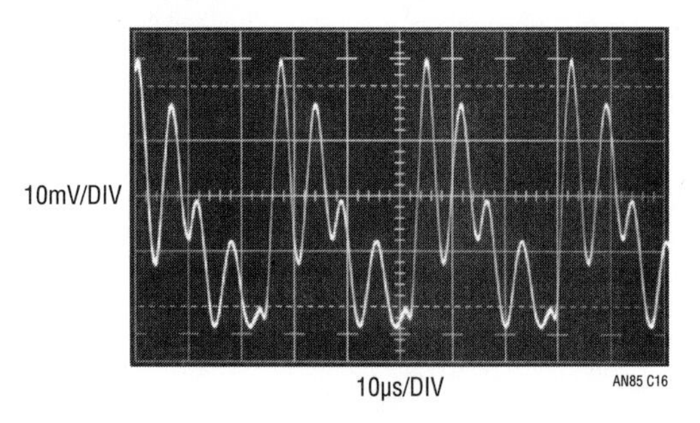

Figure C16 • Misadjusted Termination Causes Inadequate Damping. Unstable Oscilloscope Triggering May Result

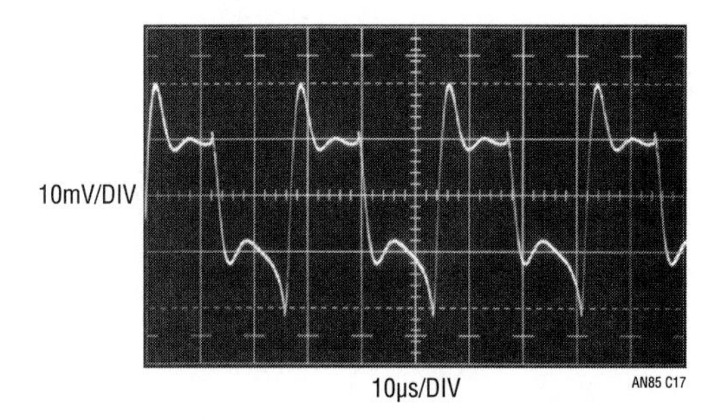

Figure C17 • Properly Adjusted Termination Minimizes Ringing with Small Amplitude Penalty

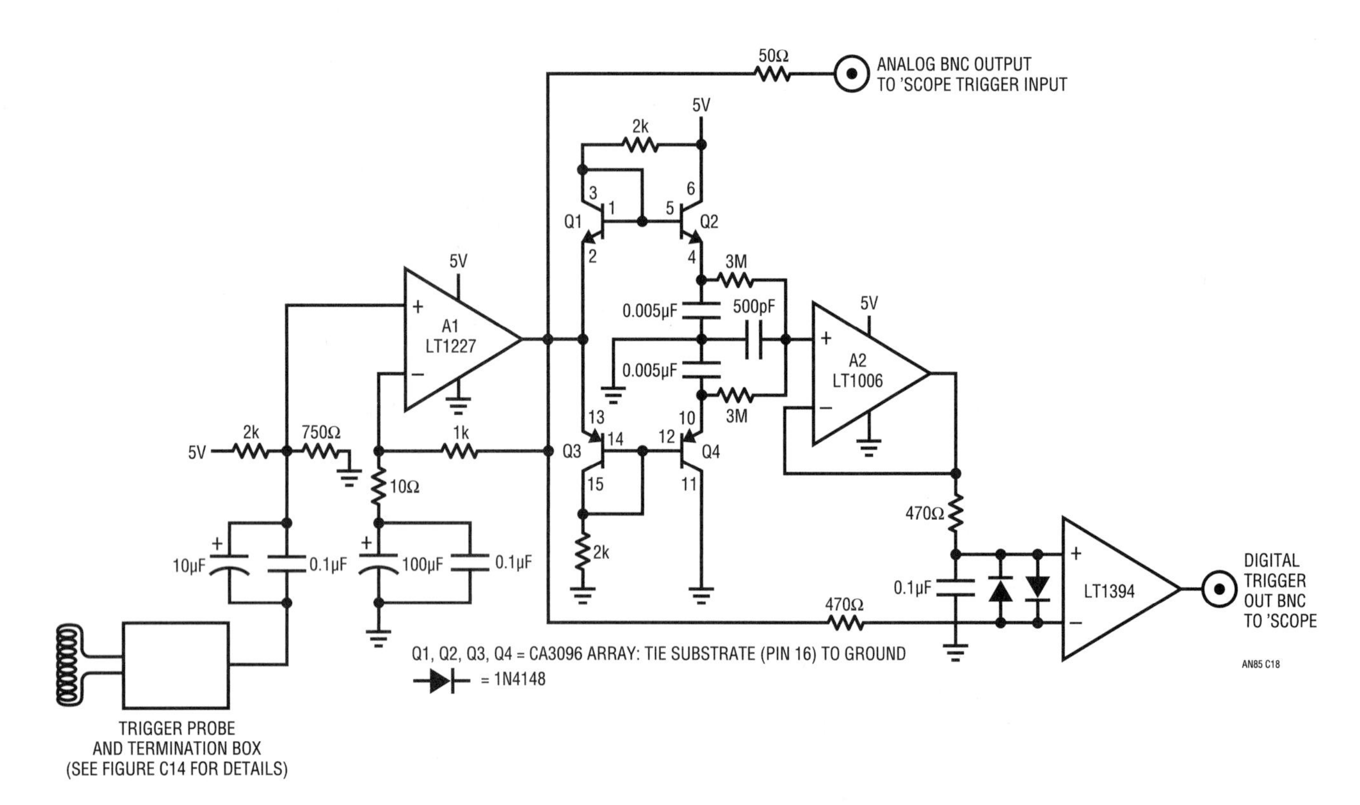

Figure C18 • Trigger Probe Amplifier Has Analog and Digital Outputs. Adaptive Threshold Maintains Digital Output Over 50:1 Probe Signal Variations

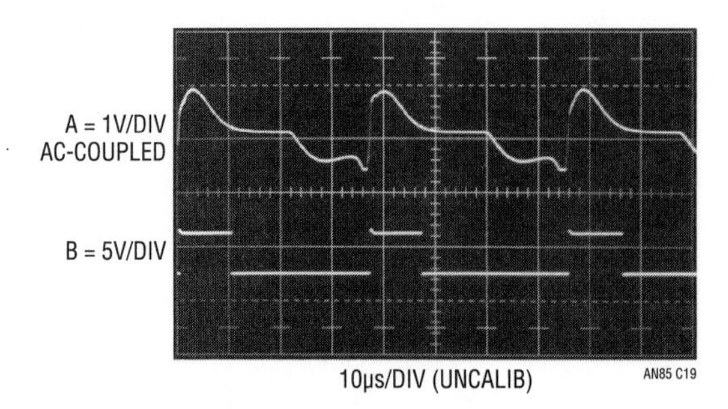

Figure C19 • Trigger Probe Amplifier Analog (Trace A) and Digital (Trace B) Outputs

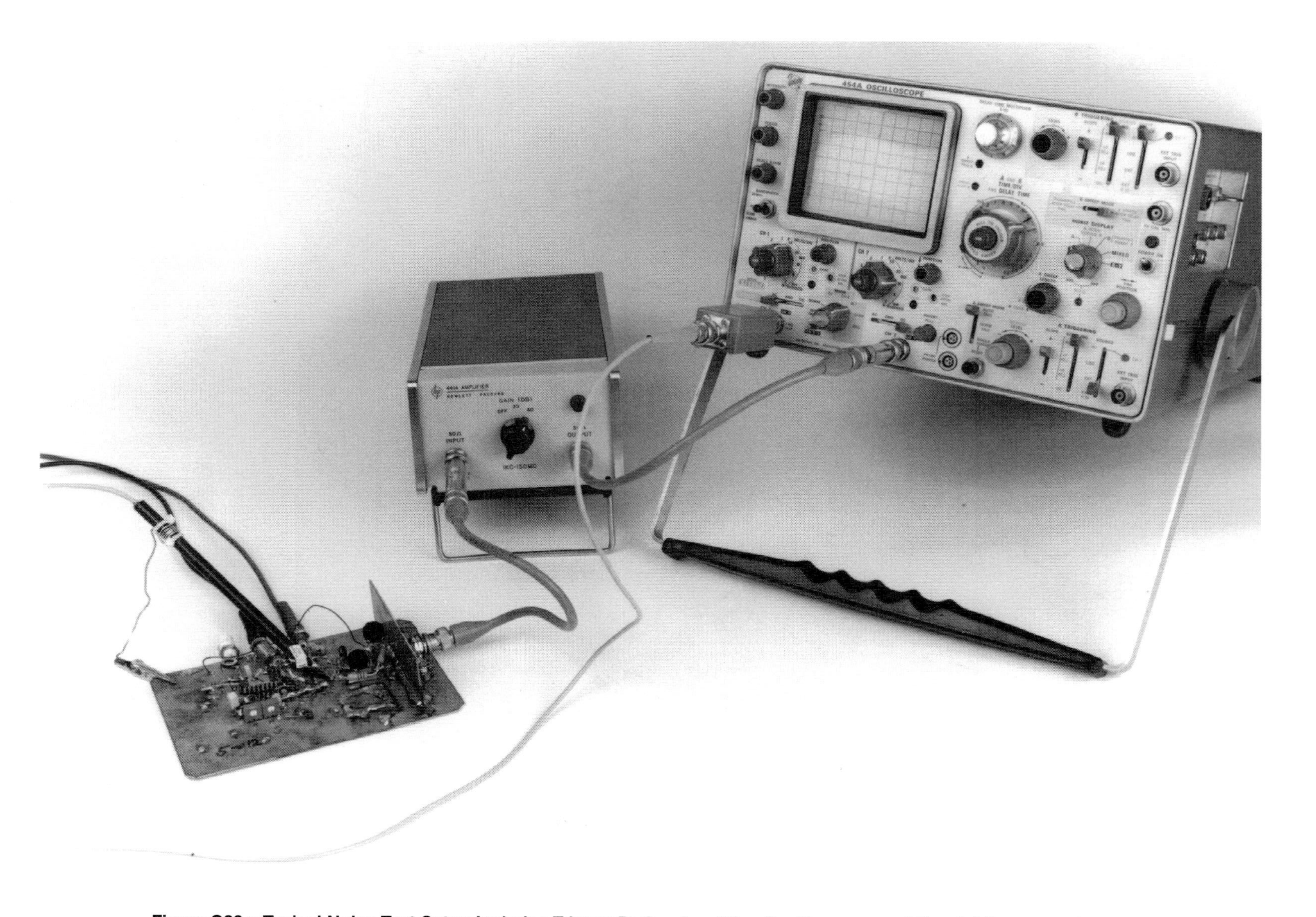

Figure C20 • Typical Noise Test Setup Includes Trigger Probe, Amplifier, Oscilloscope and Coaxial Components

Low cost coupling methods for RF power detectors replace directional couplers

29

Shuley Nakamura   Vladimir Dvorkin

Introduction

Minimizing size and cost is crucial in wireless applications such as cellular telephones. The key components in a typical GSM cellular telephone RF transmit channel consist of an RF power amplifier, power controller, directional coupler and diplexer. Some of the more recent RF power amplifiers incorporate a directional coupler in their module, reducing component count and board area. Most power amplifiers, however, require an external directional coupler. Unfortunately, directional couplers come at a price and sometimes a performance loss. While cost is an issue, long lead-time and wide variations in coupling loss are other concerns facing cell phone designers.

The directional coupler commonly used (Murata LDC21897M190-078) is unidirectional (forward) and dual band. One input is for low frequency signals (897.6MHz ±17.5MHz) and has a coupling factor of 19dB ±1dB. The second input is for higher frequency signals (1747.5MHz ±37.5MHz) and has a coupling factor of 14dB ±1.5dB. The Murata LDC21897M190-078 directional coupler is housed in a 0805 package and requires an external 50Ω termination resistor.

When a signal is passed through one of the inputs, a small portion of RF signal, equal to the difference between P_{OUT} and the coupling factor appears at the coupling output. The remainder of the signal goes to the corresponding signal output. In typical RF feedback configurations, the coupled RF output is passed through a 33pF coupling capacitor and 68Ω shunt resistor (Figure 29.1a).

Linear Technology has developed a coupling scheme for LTC RF power controllers and RF power detectors which is lower cost, more readily available and features tighter tolerance. This coupling method eliminates the 50Ω termination resistor, 68Ω shunt resistor and 33pF coupling capacitor used in traditional coupling schemes. Instead, a 0.4pF capacitor and 50Ω series resistor replace the directional coupler and its external components (Figure 29.1b).[1]

Alternate coupling solutions for use with an LTC power controller

Method 1

The DC401B demo board was designed to demonstrate the performance of the tapped capacitor coupling method (Figure 29.2). RF signal is coupled back to the LTC4401-1 RF input through a 0.4pF capacitor and 50Ω series resistor as shown in Figure 29.1b. The RF signal is fed directly to the diplexer from the power amplifier. The component count is reduced by two.

The 0.4pF series capacitor must have a tolerance of ±0.05pF or less. The tolerance directly affects how much RF signal is coupled back to the power controller RF input. ATC has ultralow ESR, high Q microwave capacitors with the tight tolerances desired. The ATC 600S0R4AW250XT is a 0.4pF capacitor with ±0.05pF tolerance. This capacitor comes in a small 0603 package. The series resistor is 49.9Ω (AAC CR16-49R9FM) with 1% tolerance.

Method 2

The second solution implements a 4.7nH shunt inductor. The inductor compensates for the parasitic shunt capacitance associated with the RF input on the power controller. Consequently, it improves the power control voltage range and sensitivity. In dual-band applications, the inductor value is chosen to increase the sensitivity of one frequency band over the other. Using an inductor requires that a capacitor be placed between the RF input pin and

Note 1: This method has been tested with the LTC4401-1 and the following Hitachi power amplifiers: PF08107B, PF08122B, PF08123B.

Analog Circuit and System Design: Immersion in the Black Art of Analog Design. http://dx.doi.org/10.1016/B978-0-12-397888-2.00029-8

the inductor. This capacitor provides a low impedance path for the RF signal. A 33pF capacitor is used as shown in Figure 29.1c. At each of the frequencies tested, the reactance of the 33pF capacitor is lower than the inductor's.

This method uses the same 0.4pF capacitor and 50Ω resistor implemented in Method 1. The Murata film type inductor, LQP15MN4N7B00D, comes in a 0402 package and has ±1nH tolerance. The 33pF capacitor is an AVX 06035A330JAT1A, comes in a 0603 package and has 5% tolerance. Tight tolerance for the shunt inductor and 33pF capacitor is not critical.

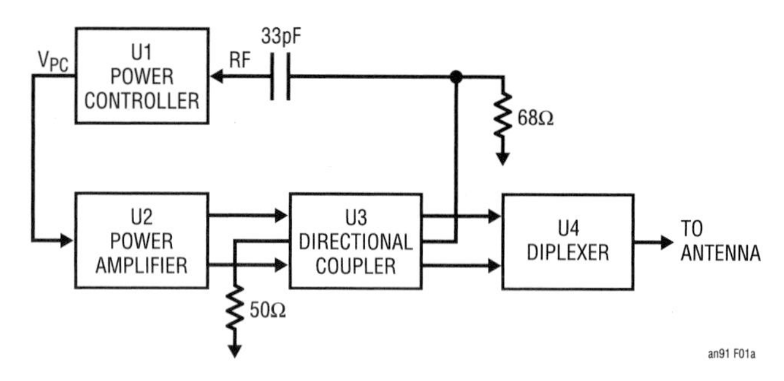

Figure 29.1a • Typical Cellular Phone Coupling Solution

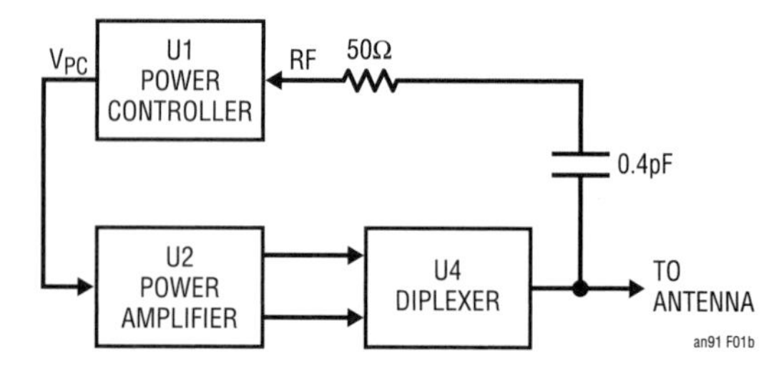

Figure 29.1b • Capacitive Coupling Method 1

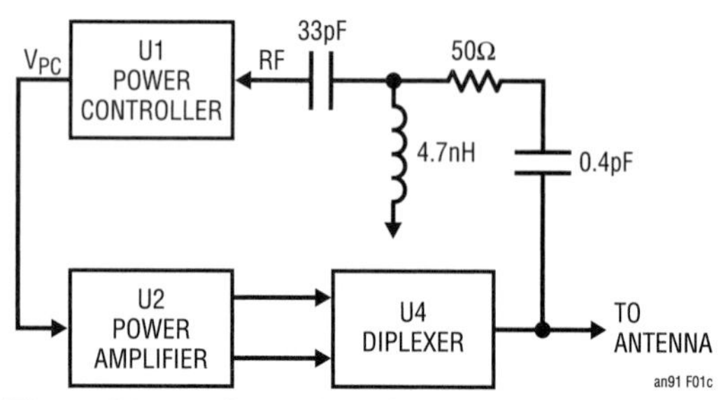

Figure 29.1c • Capacitive Coupling Method 2

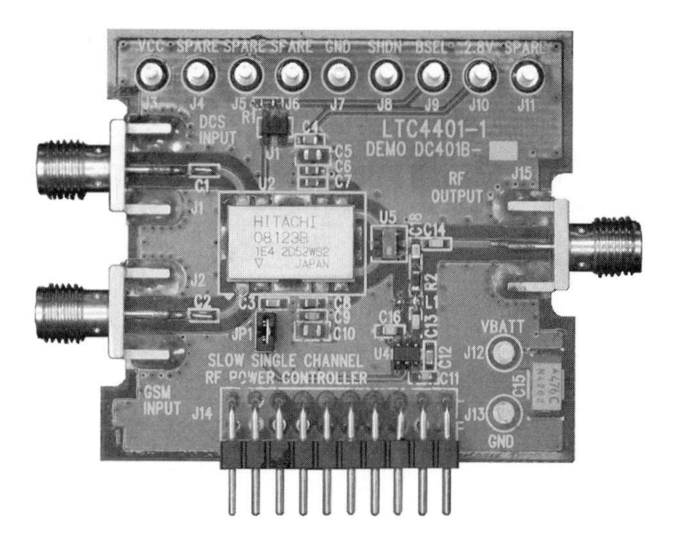

Figure 29.2 • DC401B Demo Board

Theory of operation

The 0.4pF capacitor and 50Ω resistor form a voltage divider with the input impedance of the LTC power controller. The voltage divider ratio varies over frequency. Reactance for capacitors is inversely proportional to frequency. Thus, as frequency increases, the reactance decreases for a fixed capacitance. Similarly, reactance increases as capacitance decreases. A tenth of a picofarad greatly impacts the reactance because the value of the coupling capacitor is so small. This is why tight tolerance is absolutely crucial. Small changes in capacitance will change the reactance and consequently, the voltage divider ratio. Table 29.1 shows the reactance of various components at 900MHz, 1800MHz, and 1900MHz.

Table 29.1 Reactance Variations over Frequency

Frequency (MHz)		900	1800	1900
Component Value	0.3pF	590Ω	295Ω	279Ω
	0.4pF	442Ω	221Ω	210Ω
	0.5pF	354Ω	177Ω	167Ω
	33pF	5.4Ω	2.7Ω	2.5Ω
	4.7nH	27Ω	53Ω	55Ω

The resistor value is determined by the series capacitor value and additional shunt and placement parasitics. When a shunt inductor is utilized, a smaller capacitor can be used, yielding less loss in the main line. The shunt inductor method is tuned to a particular frequency band at the expense of other frequency bands. The second coupling

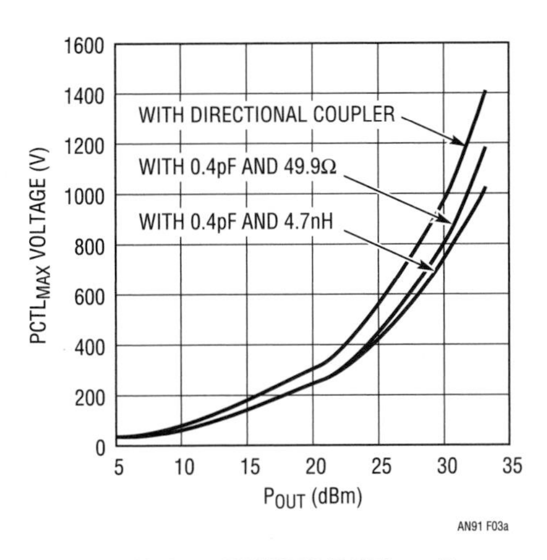

Figure 29.3a • GSM900 PCTL vs P_{OUT}

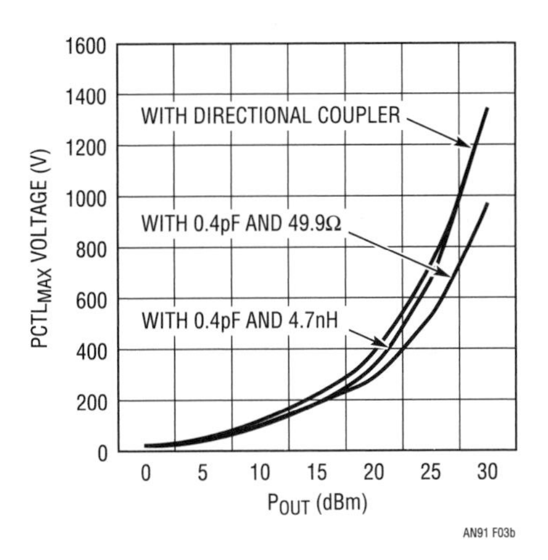

Figure 29.3b • DCS1800 PCTL vs P_{OUT}

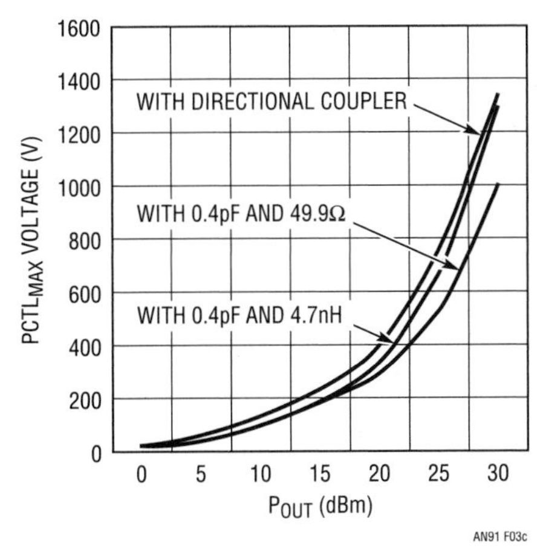

Figure 29.3c • PCS1900 PCTL vs P_{OUT}

method, for example, is tuned to DCS band frequencies. The coupling loss for this method closely resembles the coupling loss of the directional coupler (Figure 29.3b).

Considerations

There are several factors to consider when using either coupling method, such as board layout and loading in the main line. Conservative parts placement is necessary in order to minimize the distance between the TX output 50Ω line and the RF input pin on the power controller. Parasitic effects can also greatly alter the feedback network characteristics. With good layout techniques and use of tight tolerance components, this directional coupler substitute can be used over GSM, DCS and PCS band frequencies.

Test setup and measurement

Three different coupling methods were tested using the DC401A and the DC401B demo boards. The DC401A RF demo board has a triple-band directional coupler and served as the control board. The coupling factor is 19dB at 900MHz and 14dB at 1800MHz and 1900MHz. The DC401B was used to test the two capacitive coupling methods described earlier (Figure 29.8).

Each of these demo boards contains an LTC4401-1 power controller and a Hitachi PF08123B triple-band power amplifier. The component layout of the two boards is identical, except for the components that make up the coupling scheme.

A key measurement of interest is coupling loss. One method of measuring coupled RF signal is to select an RF output power level and compare the PCTL voltages applied in each of the three coupling methods. Figure 29.4 shows what a typical PCTL waveform looks like. Only the maximum level amplitude (maximum PCTL voltage) is adjusted for each measurement. The PCTL waveform is generated by Linear Technology's ramp shaping program, LTRSv2.vxe and is programmed onto the DC314A demo board. The DC314A digital demo board provides regulated power supplies, control logic and a 10-bit DAC to generate the $\overline{SHDN}$ signal and the power control PCTL signal. Input power applied to each power amplifier channel is 0dBm. A nominal battery voltage of 3.6V is used. Figure 29.7 illustrates the test setup.

A higher PCTL voltage indicates less coupling loss (i.e., more RF signal is being coupled back). Having too little coupling loss can be a problem at higher power levels because the PCTL value may exceed the maximum voltage that the DAC can output. Having too much coupling loss can make achieving lower output power levels difficult. Using a PCTL voltage less than 18mV is not recommended, since the RF output will be unstable. Thus, the minimum output power, P_{out}, is limited by PCTL = 18mV.

At 900MHz (GSM900), PCTL voltage measurements were taken at the following output power levels: 5dBm, 10dBm, 13dBm, 20dBm, 23dBm, 30dBm and 33dBm. At 1800MHz (DCS1800) and 1900MHz (PCS1900), PCTL measurements were recorded for the following output powers: 0dBm, 5dBm, 10dBm, 15dBm, 20dBm, 25dBm and 30dBm. Figures 29.3a, 29.3b, and 29.3c relate the output power to the applied PCTL voltage for each coupling method. In general, the capacitive coupling solutions have more coupling loss than the directional coupler. The full output range was achieved using both coupling methods.

Coupling solution for LTC5505 power detector[2]

The tapped capacitor method can also be utilized in systems using the LTC5505 power detector. For example, in the circuit in Figure 29.5, a shunt inductor is implemented at the RF input pin to tune out the parasitic shunt capacitance of the power detector package (5-pin ThinSOT™) and the PCB at the actual operating frequency. Using a shunt inductor improves the sensitivity of the LTC5505-2 by a factor of 2dB to 4dB. If operating between 3GHz to 3.5GHz, the shunt inductor is not recommended because the bond wire inductance compensates for the input parasitic capacitance. A DC blocking capacitor (C4) is needed, because Pin 1 of the LTC5505-2 is internally DC biased.

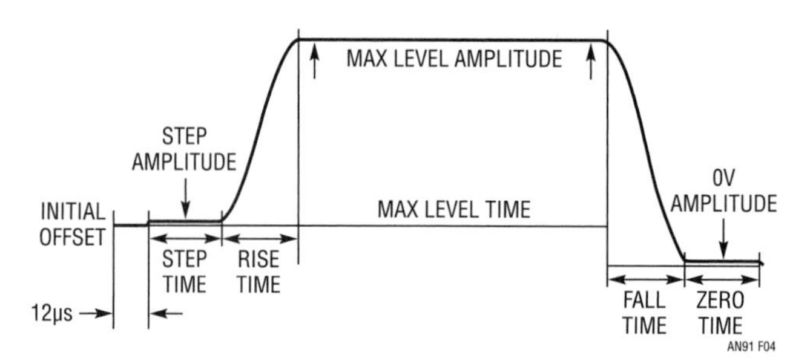

Figure 29.4 • Typical PCTL Ramp Waveform

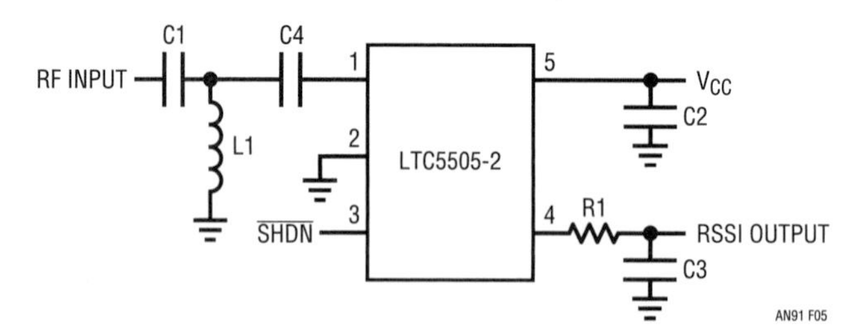

Figure 29.5 • LTC5505-2 Application Diagram with a Shunt Inductor

Note 2: Consult factory for more applications information on LTC power detectors.

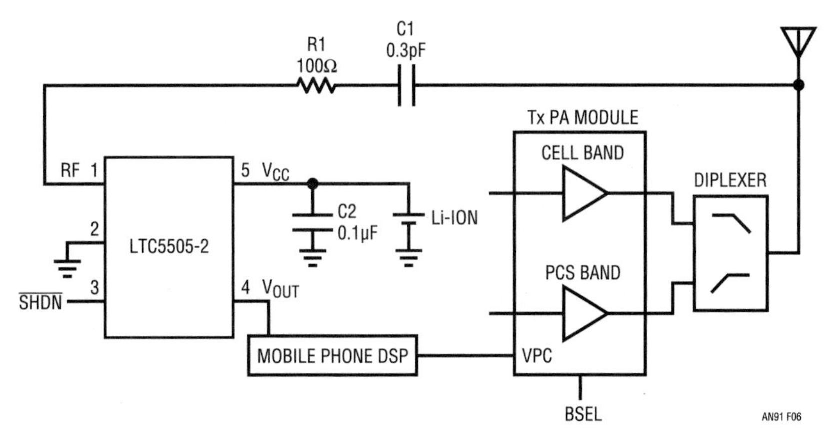

Figure 29.6 • LTC5505-2 Tx Power Control Application Diagram with a Capacitive Tap

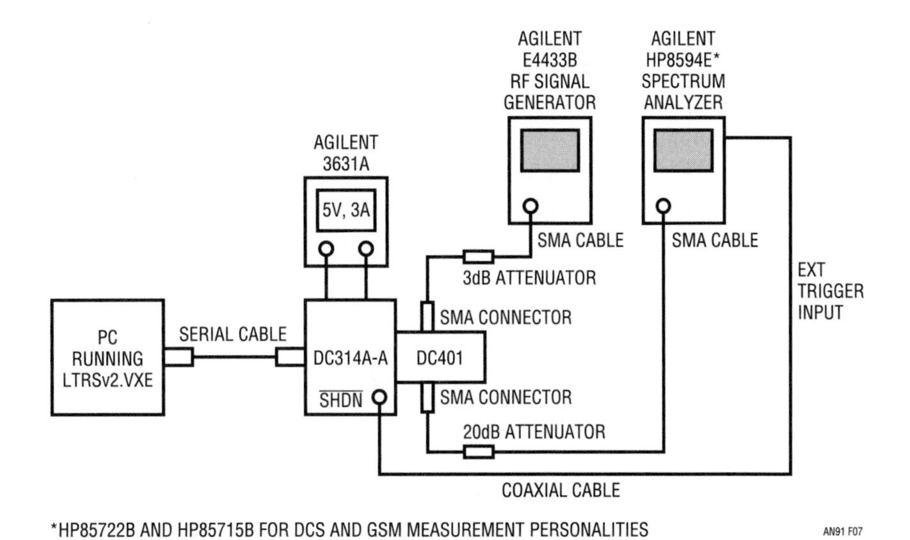

Figure 29.7 • PCTL Measurement Test Setup

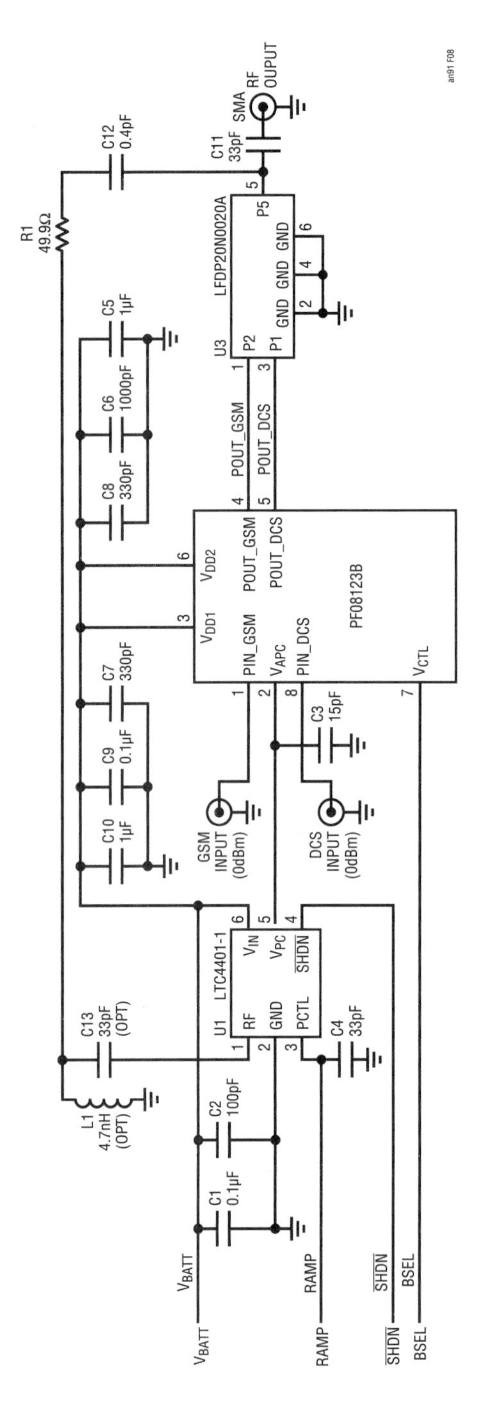

Figure 29.8 • DC401B Schematic

Figure 29.6 illustrates an example of dual band mobile phone transmitter power control with an LTC5505-2 and a capacitive tap instead of a directional coupler. A 0.3pF capacitor (C1) followed by a 100Ω resistor (R1) forms a tapping circuit with about 20dB loss at cellular band (900MHz) and 18dB loss at PCS (1900MHz) band referenced to the LTC5505-2 RF input pin. For best coupling accuracy, C1 should have tight tolerance (±0.05 pF).

Conclusion

Laboratory measurements have shown that the capacitive coupling method is an effective means of coupling the RF output signal. If coupling capacitors with tight tolerances are used, the coupling factor will be consistent. On the other hand, a directional coupler's coupling factor can vary up to 1.5dB. The total number of components decreases if the series resistor and capacitor are used. Cost will also be reduced.

The capacitive coupling scheme has been shown to work with the LTC4401-1 power controller and Hitachi PF08123B power amplifier. This scheme can be applied to all LTC power controllers (LTC1757A, LTC1758, LTC1957, LTC4400, LTC4401, LTC4402, and LTC4403) and supported power amplifiers, as well as LTC power detectors. When used with different power controller and power amplifier combinations, the capacitor and resistor values may need to be adjusted. Decreasing the coupling capacitor or increasing the series resistor will increase the coupling loss. Linear Technology currently supports Anadigics, Conexant, Hitachi, Philips, and RFMD power amplifiers. DC401B demo boards are available upon request.

PARTS LIST (Demo Board DC401B)				
REFERENCE	**QUANTITY**	**PART NUMBER**	**DESCRIPTION**	**VENDOR**
C1, C9	2	0603YC104MAT1A	0.1 mF 16V 20% X7R Capacitor	AVX
C2	1	06035A101JAT1A	100pF 50V 5% NPO Capacitor	AVX
C3	1	06035A150JAT1A	15pF 50V 5% NPO Capacitor	AVX
C4, C11, C13 (OPT)	2	06035A330JAT1A	33pF 50V 5% NPO Capacitor	AVX
C5, C10	2	EMK212BJ105MG-T	1mF 16V 20% X5R Capacitor	Taiyo Yuden
C6	1	06033C102KAT1A	1000pF 25V 10% X7R Capacitor	AVX
C7, C8	2	06035A331JAT1A	330pF 50V 5% NPO Capacitor	AVX
C12	1	600S0R4AW 250 XT	0.4pF ±0.5pF NPO Capacitor	ATC
L1 (OPT)	1	LQP15MN4N7B00	4.7nH 0402 ±0.1nH Inductor	Murata
R1	1	CR16-49R9FM	49.9Ω 1/16W 1% Chip Resistor	AAC
U1	1	LTC4401-1	SOT-23-6 RF Power Control IC	LTC
U2	1	PF08123B	Power Amplifier SMT IC	Hitachi
U3	1	LFDP21920MDP1A048	Dual Wideband Diplexer SMT IC	Murata

Improving the output accuracy over temperature for RMS power detectors

30

Andy Mo

Introduction

Stable temperature performance is extremely important in base-station designs because the ambient temperature can vary widely depending on the surroundings and the location. Using high accuracy over temperature RMS detectors can improve the power efficiency of the base-station designs. The LTC5582 and the dual-channel LTC5583 are a family of RMS detectors that offer excellent stable temperature performance (from −40°C to 85°C) at any frequency up to 10GHz for LTC5582, and 6GHz for LTC5583. However, their temperature coefficients vary with frequency, and without temperature compensation, the error over temperature can be greater than 0.5dB. As a result, sometimes it is necessary to optimize the temperature compensation at different frequencies to improve the accuracy to <0.5dB of error In addition, the temperature compensation can be implemented using only two off chip resistors, with no external circuitry required.

The change in output voltage is governed by the following equation:

$$\Delta V_{OUT} = TC1 \bullet (T_A - t_{NOM}) + TC2 \bullet (T_A - t_{NOM})^2 + detV1 + detV2 \quad (1)$$

Where TC1 and TC2 are the 1st and 2nd order temperature coefficients respectively. TA is the actual ambient temperature, and t_{NOM} is the reference room temperature, 25°C. detV1 and detV2 are output voltage variation when R_{T1} and R_{T2} are not set to zero.

The method to calculate the resistor values for temperature compensation is the same for both LTC5582, and LTC5583. The two control pins are RT1, which sets TC1 (the 1st order temperature compensation coefficient), and RT2 which sets TC2 (the 2nd order temperature compensation coefficient). Shorting RT1 and RT2 to ground conveniently turns off the temperature compensation feature if not needed.

LTC5583 temperature compensation design

LTC5583 includes two additional pins, RP1 that controls the polarity of TC1, and RP2 which controls the polarity of TC2. However; the magnitude of the temperature coefficients are the same with a fixed R_{T1}, or R_{T2} value, only polarity is flipped. Both channel A and channel B share the compensation circuitry, therefore both channels are controlled together.

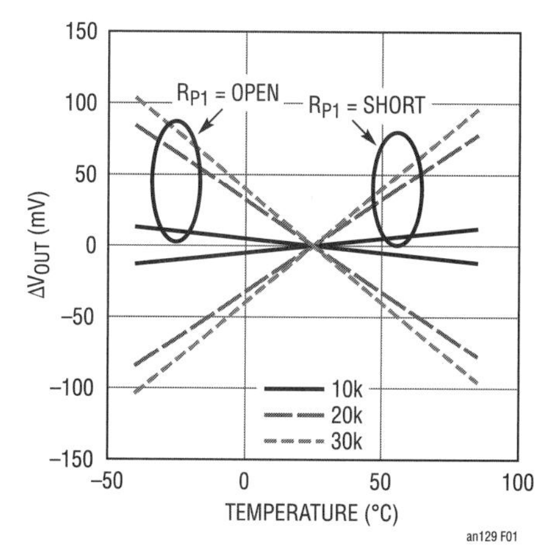

Figure 30.1 • 1st Order ΔV_{OUT} vs Temperature

Figure 30.1 illustrates the change in V_{OUT} as a function of temperature from the 1st order temperature compensation. Only three resistor values are shown to illustrate that increasing resistor values causes an increase in the magnitude of the slope. The polarity of the slope is controlled by the RP1 pin.

Analog Circuit and System Design: Immersion in the Black Art of Analog Design. http://dx.doi.org/10.1016/B978-0-12-397888-2.00030-4

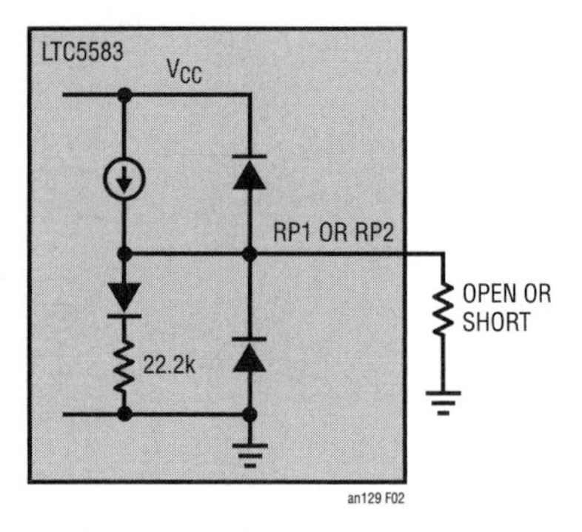

Figure 30.2 • Simplified Schematic of Pins RP1 and RP2

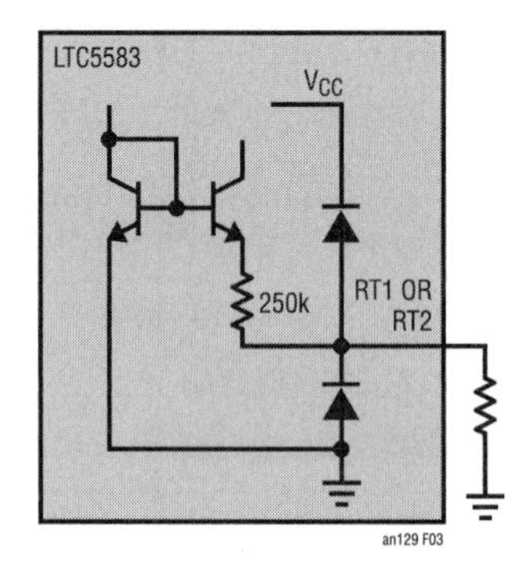

Figure 30.3 • Simplified Schematic of Pins RT1 and RT2

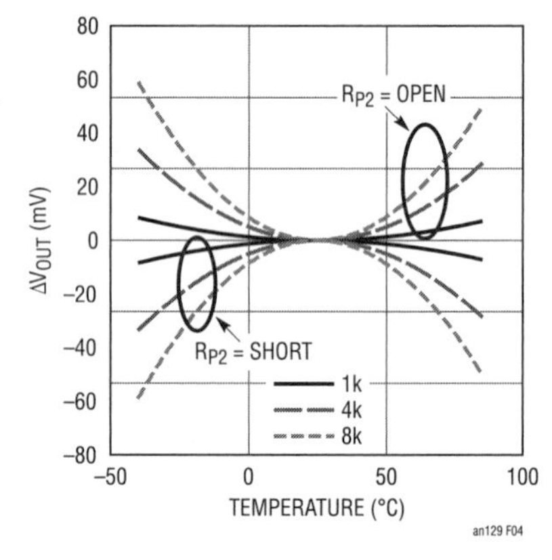

Figure 30.4 • 2nd Order V_{OUT} vs Temperature

Figure 30.4 illustrates the effect of 2nd order temperature compensation on V_{OUT}. The polarity of the curves is controlled by RP2. The curvature depends on the resistor values. The overall effect is the summation of the 1st order and 2nd order temperature compensation given by equation 1.

Take an LTC5583 as an example at 900MHz input. The first step is to measure the V_{OUT} over temperature without compensation. Figure 30.5 shows the uncompensated V_{OUT}. The Linearity error over temperature is referenced to the slope and intercept point at 25°C. To minimize the output voltage change with temperature, the linearity curve in red (85°C) needs to be shifted down, and the linearity curve in blue (−40°C) needs to be shifted up to align with the room temperature in green, and overlap as much as possible. What follows is a step by step design procedure.

Step 1. Estimate the temperature compensation needed in dB, from Figure 30.5. For example, read the plot at an input power of −25dBm, which is the middle of the dynamic range. Multiply the linearity error in dB by 30mV/dB(typical V_{OUT} slope) to convert to mV.

Cold (−40°C) = 13mV or 0.43dB

Hot (85°C) = −20mV or −0.6dB

This is the amount of output voltage adjustment required over temperature.

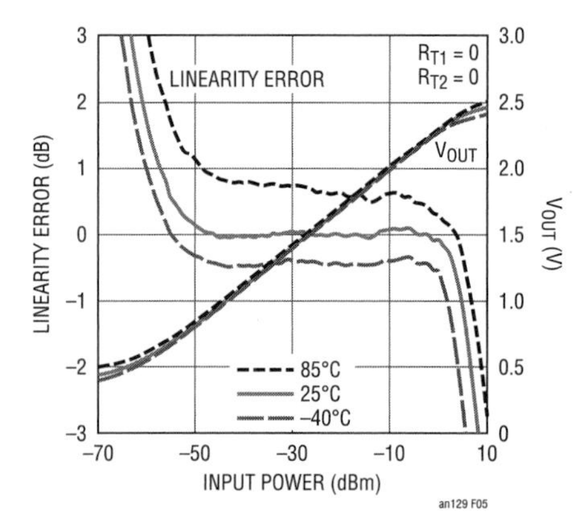

Figure 30.5 • Uncompensated LTC5583 at 900MHz

Step 2. Determine RP1 and RP2, and the solutions for 1st and 2nd order compensation. To find the solutions, let a = 1st order term, and b = 2nd order term. Set them up so they satisfy the temperature compensation at −40°C and 85°C.

$$a - b = 13\text{mV} \quad (2)$$

$$-a - b = -20\text{mV} \quad (3)$$

$$a = 16.5 \quad (1\text{st})$$

$$b = 3.5 \quad (2\text{st})$$

The polarity of a and b in equation 2 and equation 3 are determined by the polarity of the 1st order term and the 2nd order term, such that their summation satisfy the 13mV at cold (−40°C), and −20mV at hot (85°C) adjustment. Refer to Figure 30.6. 1st order term and 2nd order term can be either positive or negative. So there are total of 4 combinations possible. In this case, only when both terms are negative will their sum satisfy the required compensation.

Figure 30.7 shows the 1st and 2nd order compensation required at −40°C and 85°C. Notice the polarity of the 1st order and 2nd order compensation are negative such that when both curves are added, their sum produces

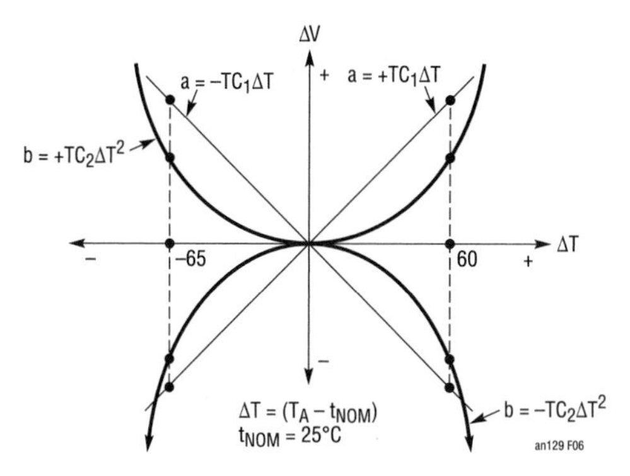

Figure 30.6 • Polarity of 1st and 2nd Order Solutions

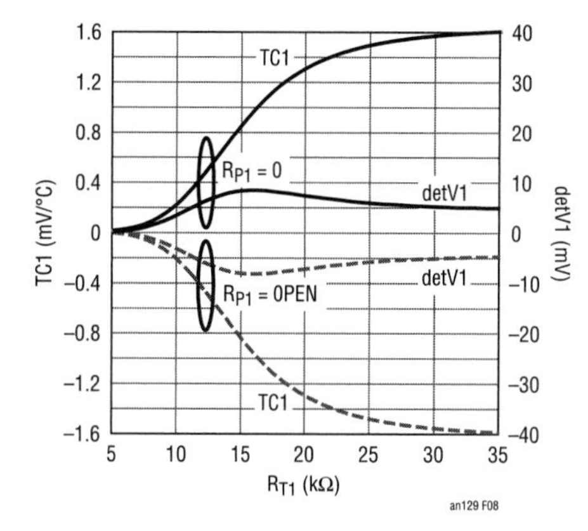

Figure 30.8 • 1st Order Temperature Compensation Coefficient TC1 vs External RT1 Values

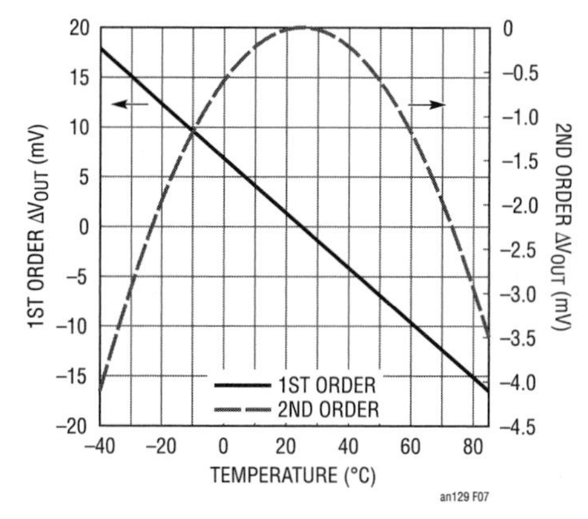

Figure 30.7 • Solutions for the Temperature Compensation

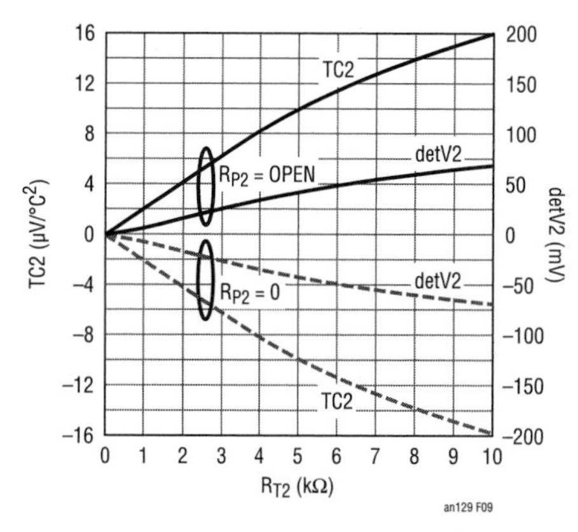

Figure 30.9 • 2nd Order Temperature Compensation Coefficient TC2 vs External RT2 Values

the required adjustment to V_{OUT}. Consequently, TC1 and TC2 are negative, and RP1 and RP2 are determined from Figure 30.8 and Figure 30.9. Notice the values of the two solutions add up to approximately 13mV at −40°C, and −20mV at 85°C.

R_{P1} = open

R_{P2} = short

Step 3. Calculate the temperature coefficients at one of the temperature extremes and determine resistor values R_{T1} and R_{T2}, using Figures 30.8 and 30.9.

$a = 16.5 = TC1 \bullet (85 - 25)$; $TC1 = 0.275mV/°C$

R_{T1} = 11k (from Figure 30.8)

$b = 3.5 = TC2 \bullet (85 - 25)^2$; $TC2 = 0.972\mu V/°C^2$

R_{T2} = 499Ω (from Figure 30.9)

Figure 30.10 shows the LTC5583 performance over temperature for one of the two output channels. Notice an improvement to the temperature performance from uncompensated V_{OUT}, from Figure 30.5. This may be satisfactory for most applications. However, for some applications where even better accuracy is needed, a 2nd iteration can be performed to further improve the temperature performance. To simplify the calculation, detV1 and detV2 terms are ignored because they are not dependent on temperature. As a result, the solutions are not precise. However, it is very helpful in improving the accuracy over temperature, as shown here.

2nd Iteration calculation

Step 1. Find the compensation needed from Figure 30.10, using the same method in first iteration.

Cold (−40°C) = −3mV or −0.1dB

Hot (85°C) = −3mV or −0.1dB

Add the new values to the 1st iteration

Cold (−40°C) = −3mV + 13mV = 10mV

Hot (85°C) = −3mV − 20mV = −23mV

Repeat steps 2 and 3 to calculate the R_{T1} and R_{T2} values.

R_{T1} = 11k

R_{T2} = 953Ω

R_{P1} = open

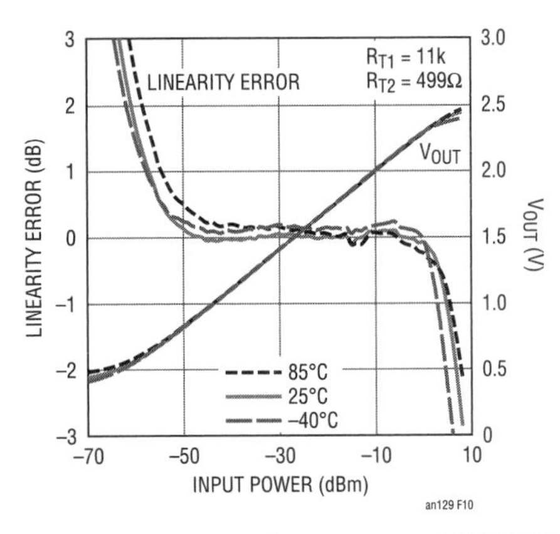

Figure 30.10 • Temperature Compensated LTC5583 Output After 1st Iteration

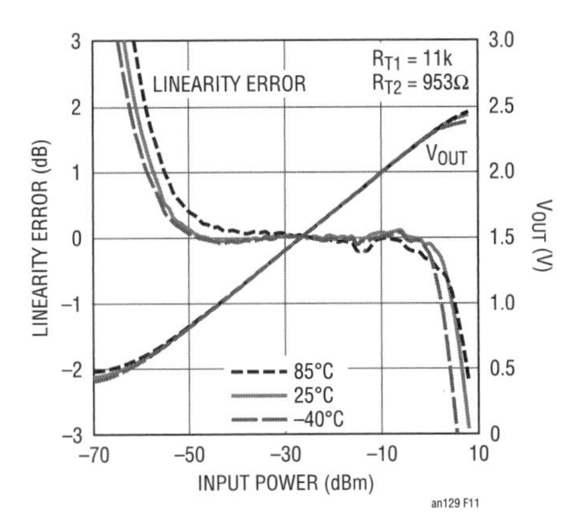

Figure 30.11 • Temperature Compensated LTC5583 Output After 2 Iterations.

R_{P2} = short

The performance results are shown in Figure 30.11 after two iterations. Over temperature, the dynamic range is 50dB with 0.2dB of linearity error, and 56dB of dynamic range with 1.0dB of linearity error. Refer to Table 30.1 for temperature compensation values at other frequencies.

Table 30.1 Recommended Settings and Resistor Values for LTC5583 for Optimal Temperature Performance at Various Frequencies

FREQUENCY (MHz)	R_{P1}	R_{P2}	R_{T1} (kΩ)	R_{T2} (kΩ)
450	Open	0	11.5	1.13
880	Open	0	11.5	1.13
900	Open	0	11	0.953
1800	Open	0	12.1	1.5
2140	Open	0	9.76	1.1
2300	Open	0	10.5	1.43
2500	Open	0	10.5	1.43
2700	Open	0	8.87	1.21

This iteration process can be repeated over and over again to further increase the accuracy. This will allow the designer to dial in the compensation as accurately as needed for most applications.

LTC5582 single detector

The method to calculate the LTC5582 compensation values for R_{T1} and R_{T2} is the same, only easier because the polarity has been predetermined. Both TC1 and TC2 are negative. Refer to Table 30.2 for R_{T1} and R_{T2} values at other frequencies. The compensation coefficients shown in Figure 30.8 and Figure 30.9 are different for LTC5582. Refer to data sheet for additional information.

Table 30.2 Recommended R_{T1} and R_{T2} Values of LTC5582 for Optimal Temperature Performance at Various Frequencies

FREQUENCY (MHz)	R_{T1} (kΩ)	R_{T2} (kΩ)
450	12	2
800	12.4	1.4
880	12	2
2000	0	2
2140	0	2
2600	0	1.6
2700	0	1.6
3000	0	1.6
3600	0	1.6
5800	0	3
7000	10	1.43
8000	10	1.43
10000	10	3

Conclusion

LTC5582 and LTC5583 offer excellent temperature performance with only two external compensation resistors. The procedure to calculate the compensation resistors is simple, and can be reiterated for even better performance. The example shown here is for LTC5583 at 900MHz RF input, but the method can be applied to LTC5582 and LTC5583 at any frequency within the limits of the IC. The performance over temperature is fairly consistent from part to part. The resulting performance provides accuracy over temperature with less than 1% of output voltage.

PART 3

Circuit Collections

Circuit techniques for clock sources (31)

Circuits for clock sources are presented. Special attention is given to crystal-based designs including TXCOs and VXCOs.

Measurement and control circuit collection (32)

A variety of measurement and control circuits is included in this application note. Eighteen circuits, including ultralow noise amplifiers, current sources, transducer signal conditioners, oscillators, data converters and power supplies are presented. The circuits emphasize precision specifications with relatively simple configurations.

Circuit collection, volume I (33)

Presented in the note are a variety of circuits ranging from a 50W high efficiency (>90%) switching regulator to steep roll-off filter circuits with low distortion to 12-bit differential temperature-measurement systems.

Video circuit collection (34)

The Video Circuit Collection features a variety of Linear's video circuits. The LT1204 70MHz multiplexer is featured in a number of circuits which require excellent video isolation from channel to channel. High speed voltage and current feedback amplifiers are highlighted throughout the section on video processing circuits. There is a section on applying current feedback amplifiers (CFAs).

Practical circuitry for measurement and control problems (35)

Most of these circuits were designed at customer request or are derivatives of such efforts. Types of circuits include power converters, transducer signal conditioners, amplifiers and signal generators. Specific circuits include low noise amplifiers, high power single cell DC/DC converters, portable high accuracy barometers, a 10mHz 1% accuracy RMS/DC converter, and random noise generators. Appended sections cover noise theory and present an historical perspective on wideband amplifiers.

Circuit collection, volume III: data conversion, interface and signal processing (36)

This note is a collection of circuits for data conversion, interface and signal processing. The note includes circuits such as fast video multiplexers for high speed video, an ultraselective bandpass filter circuit with adjustable gain, and a fully differential, 8-channel, 12-bit A/D system. The categories included in this application note are data conversion, interface, filters, instrumentation, video/op amps and miscellaneous circuits.

Circuitry for signal conditioning and power conversion (37)

This publication includes designs for data converters and signal conditioners, transducer circuits, crystal oscillators and power converters. Wideband and micropower circuitry receive special attention. Tutorials on micropower design techniques and parasitic effects of test equipment are included.

Circuit collection, volume V: data conversion, interface and signal conditioning products (38)

This note collects useful circuits, covering data conversion, interface and signal conditioning circuits. It includes circuits for high speed video, interface and Hot Swap™ circuits, active RC and switched capacitor filter circuitry and a variety of data conversion and instrumentation circuits. All circuits are conveniently indexed by type.

Signal sources, conditioners and power circuitry (39)

Eighteen circuits are presented in this compilation. Signal sources include a voltage controlled current source, an amplitude/frequency stabilized sine wave oscillator, a versatile, 0V to 50V wideband level shift and four sub-nanosecond pulse generators with rise times as low as 20ps. Five signal conditioners appear: a unique, single positive rail powered amplifier with output to (and below) zero volts, a milliohmmeter, a 0.02% accurate instrumentation amplifier with 120dB CMRR at $125V_{CM}$, a 100MHz switch with 5mV control channel feedthrough and a 5V powered, 15ppm linearity quartz stabilized V→F converter. The power circuits section features a Xenon flashlamp supply, two 5V powered, 0V to 300V DC/DC converters, a fixed 200V output circuit for APD bias, a 100W 0V to 500V, 28V powered converter and a high current paralleling scheme for linear regulators. Two appended sections consider measurement technique and connection practice in sub-nanosecond circuits.

Current sense circuit collection (40)

Sensing and/or controlling current flow is a fundamental requirement in many electronics systems, and the techniques to do so are as diverse as the applications themselves. This note compiles solutions to current sensing problems and organizes the solutions by general application type.

Power conversion, measurement and pulse circuits (41)

This ink marks Linear's eighth circuit collection publication. We are continually surprised, to the point of near mystification, by these circuit amalgams seemingly limitless appeal. Reader requests ascend rapidly upon publication, remaining high for years, even decades. All Linear circuit collections, despite diverse content, share this popularity, although just why remains an open question. Why is it? Perhaps the form; compact, complete, succinct and insular. Perhaps the freedom of selection without commitment, akin to window shopping. Or, perhaps, simply the pleasure of new recruits for the circuit aficionados intellectual palate. Locally based electrosociologists, spinning elegantly contrived theories, offer explanation, but no convincing evidence is at hand. What is certain is that readers are attracted to these compendiums and that calls us to attention. As such, in accordance with our mission to serve customer preferences, this latest collection is presented. Enjoy.

Circuit techniques for clock sources

31

Jim Williams

Almost all digital or communication systems require some form of clock source. Generating accurate and stable clock signals is often a difficult design problem.

Quartz crystals are the basis for most clock sources. The combination of high Q, stability vs time and temperature, and wide available frequency range make crystals a price-performance bargain. Unfortunately, relatively little information has appeared on circuitry for crystals and engineers often view crystal circuitry as a black art, best left to a few skilled practitioners (see box, "About Quartz Crystals").

In fact, the highest performance crystal clock circuitry does demand a variety of complex considerations and subtle implementation techniques. Most applications, however, don't require this level of attention and are relatively easy to serve. Figure 31.1 shows five (5) forms of simple crystal clocks. Types 1a through 1d are commonly referred to as gate oscillators. Although these types are popular, they are often associated with temperamental operation, spurious modes or outright failure to oscillate. The primary reason for this is the inability to reliably identify the analog characteristics of the gates used as gain elements. It is not uncommon in circuits of this type for gates from different manufacturers to produce markedly different circuit operation. In other cases, the circuit works, but is influenced by the status of other gates in the same package. Other circuits seem to prefer certain gate locations within the package. In consideration of these difficulties, gate oscillators are generally not the best possible choice in a production design; nevertheless, they offer low discrete component count, are used in a variety of situations, and bear mention. Figure 31.1a shows a CMOS Schmitt trigger biased into its linear region. The capacitor adds phase shift and the circuit oscillates at the crystal resonant frequency. Figure 31.1b shows a similar version for higher frequencies. The gate gives inverting gain, with the capacitors providing additional phase shift to produce oscillation. In Figure 31.1c, a TTL gate is used to allow the 10MHz operating frequency. The low input resistance of TTL elements does not allow the high value, single resistor biasing method. The R-C-R network shown is a replacement for this function. Figure 31.1d is a version using two gates. Such circuits are particularly vulnerable to spurious operation but are attractive from a component count standpoint. The two linearly biased gates provide 360 degrees of phase shift with the feedback path coming through the crystal. The capacitor simply blocks DC in the gain path. Figure 31.1e shows a circuit based on discrete components. Contrasted against the other circuits, it provides a good example of the design flexibility and certainty available with components specified in the linear domain. This circuit will oscillate over a wide range of crystal frequencies, typically 2MHz to 20MHz.

The 2.2k and 33k resistors and the diodes compose a pseudo current source which supplies base drive.

At 25°C the base current is:

$$\frac{1.2V - 1V_{BE}}{33K} = 18\mu A$$

To saturate the transistor, which would stop the oscillator, requires V_{CE} to go to near zero. The collector current necessary to do this is:

$$IC(sat) = \frac{5V}{1k}(\text{delete } V_{CE} \text{ sat}) = 5mA$$

with 18μA of base drive a beta of:

$$\frac{5mA}{18\mu A} = 278 \text{ is required}$$

At 1mA the DC beta spread of 2N3904's is 70 to ≅210.

The transistor should not saturate…even at supply voltages below 3V.

In similar fashion, the effects of temperature may also be determined.

V_{BE} vs temperature over 25°C – 70°C is:

$$-2.2mV/°C \bullet 45° = -99mV.$$

Analog Circuit and System Design: Immersion in the Black Art of Analog Design. http://dx.doi.org/10.1016/B978-0-12-397888-2.00031-6

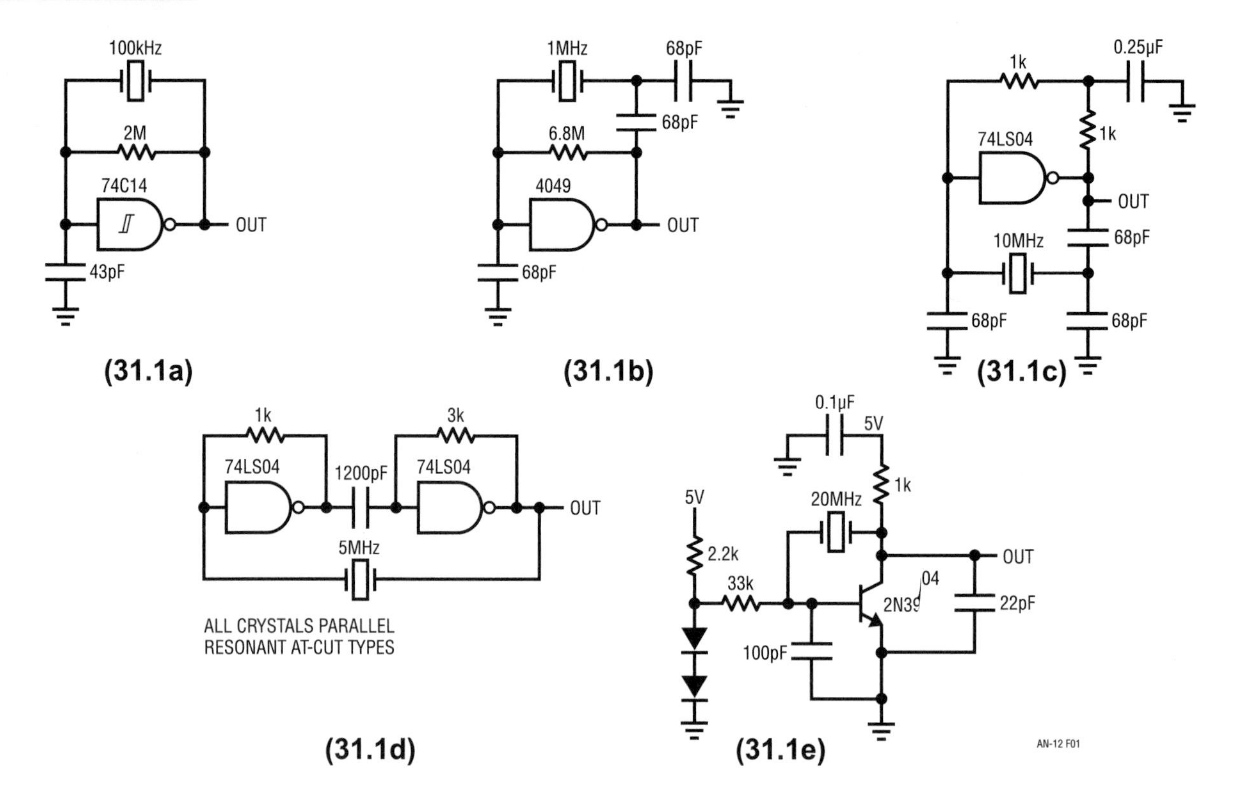

Figure 31.1 • Typical Gate Oscillators and the Preferred Discrete Unit

The compliance voltage of the current source will move:

$$2 \bullet -2.2\text{mV/°C} \bullet 45\text{°C} = -198\text{mV}$$

Hence, a first order compensation occurs:

$$-198\text{mV} - 99\text{mV} = -99\text{mV total shift.}$$

This remaining −99mV over temperature causes a shift in base current:

$$25\text{°C current} = \frac{0.56\text{V}}{33\text{k}} = 18\mu\text{A}$$
$$70\text{°C current} = \frac{0.5\text{V}}{33\text{k}} = 15\mu\text{A}$$
$$18\mu\text{A} - 15\mu\text{A} = 3\mu\text{A}$$

This 3μA shift (about 16%) provides a compensation for transistor h_{FE} shift with temperature, which moves about 20% from 25°C to 70°C. Thus the circuit's behavior over temperature is quite predictable. The resistor, diode and V_{BE} tolerances mean that only first order compensations for V_{BE} and h_{FE} over temperature are appropriate.

Figure 31.2 shows another approach. This circuit uses a standard RC-comparator multivibrator circuit with the crystal connected directly across the timing capacitor. Because the free running frequency of the circuit is close to the crystal's resonance, the crystal "steals" energy from the RC, forcing it to run at the crystal's frequency. The crystal activity is readily apparent in Trace A of Figure 31.3, which is the LT®1011's "–" input. Trace B is the LT1011's output. In circuits of this type, it is important to ensure that enough current is available to quickly start the crystal resonating while simultaneously maintaining an RC time constant of appropriate frequency. Typically, the free running frequency should be set 5% to 10% above crystal resonance with a resistor feedback value calculated to allow about 100μA into the capacitor-crystal network. This type of circuit is not recommended for use above a few hundred kHz because of comparator delays.

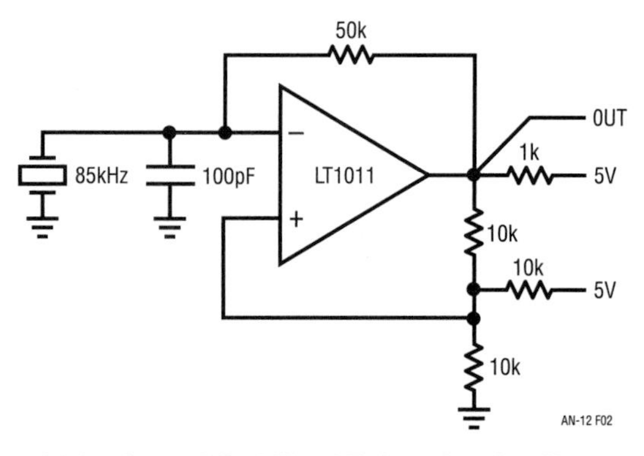

Figure 31.2 • Crystal Stabilized Relaxation Oscillator

Figures 31.4a and 31.4b use another comparator based approach. In Figure 31.4a, the LT1016 comparator is set up with DC negative feedback. The 2k resistors set the common mode level at the device's positive input. Without the crystal, the circuit may be considered as a very wideband (50GHz GBW) unity gain follower biased at 2.5V. With the crystal inserted, positive feedback occurs and oscillation commences. Figure 31.4a is useful with AT-cut fundamental mode crystals up to 10MHz. Figure 31.4b is similar, but supports oscillation frequencies to 25MHz. Above 10MHz, AT-cut crystals operate in overtone mode. Because of this, oscillation can occur at multiples of the desired frequency. The damper network rolls off gain at high frequency, insuring proper operation.

All of the preceding circuits will typically provide temperature coefficients of 1ppm/°C with long term (1 year) stability of 5ppm to 10ppm. Higher stability is achievable with more attention to circuit design and control of

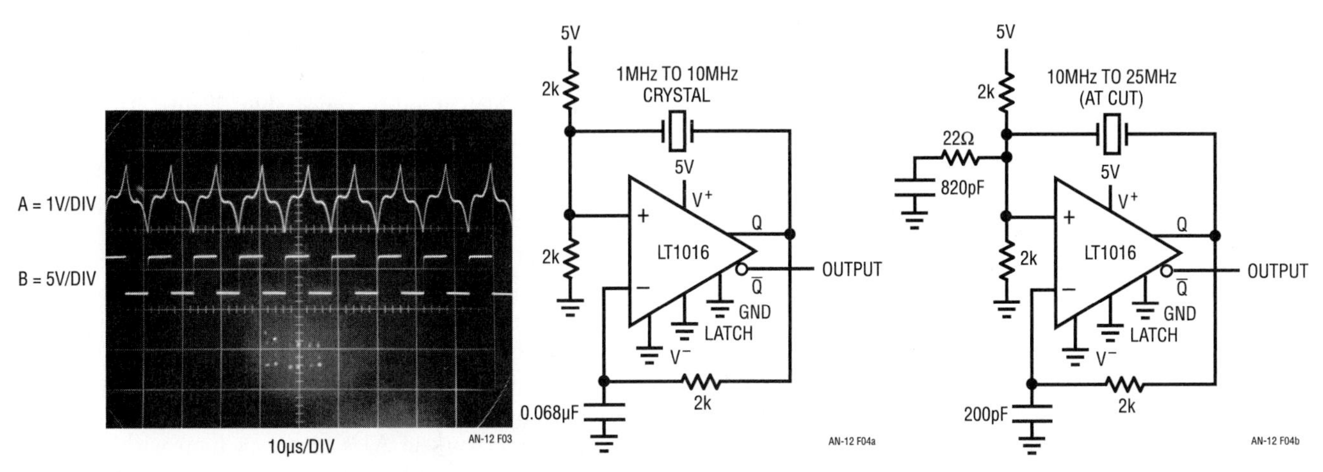

Figure 31.3 • Figure 31.2's Waveforms

Figure 31.4a • 1MHz to 10MHz Crystal Oscillator

Figure 31.4b • 10MHz to 25MHz Crystal Oscillator

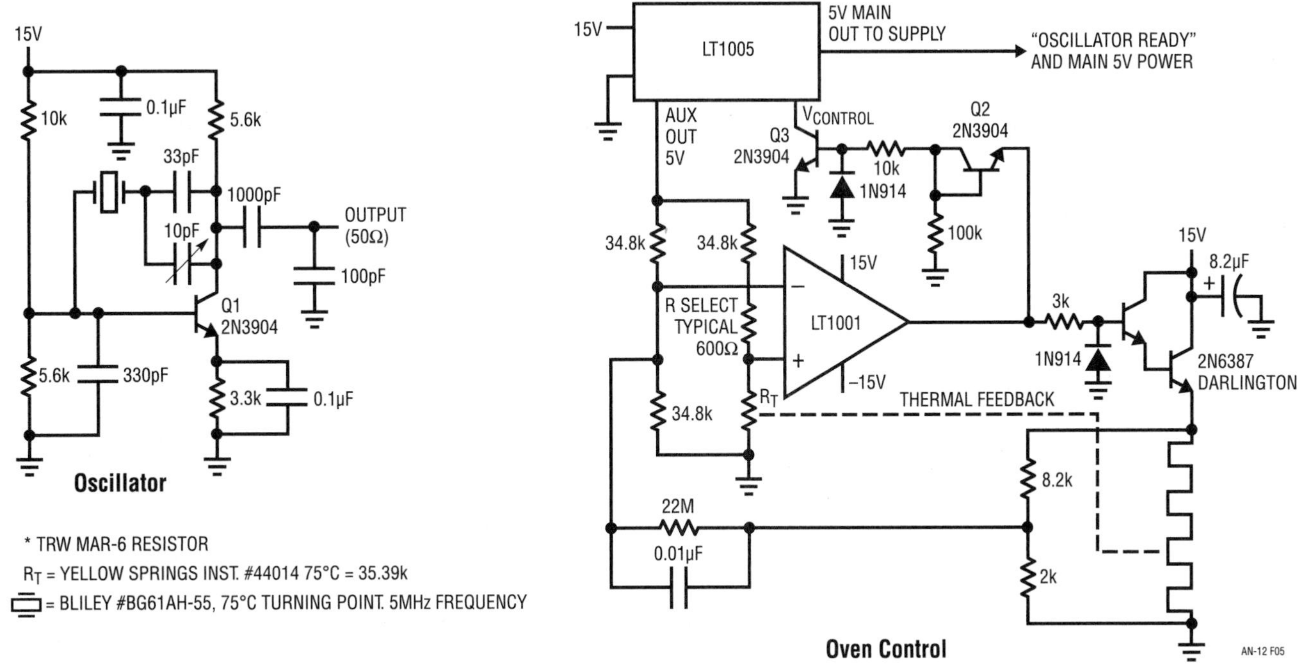

Figure 31.5 • Ovenized Oscillator

temperature. Figure 31.5 shows a Pierce class circuit with fine frequency trimming provided by the paralleled fixed and variable capacitors. The transistor provides 180° of phase shift with the loop components adding another 180°, resulting in oscillation. The LT1005 voltage regulator and the LT1001 op amp are used in a precision temperature servo to control crystal temperature. The LT1001 extracts the differential bridge signal and drives the Darlington stage to power the heater, which is monitored by the thermistor. In practice, the sensor is tightly coupled to the heater. The RC feedback values should be optimized for the thermal characteristics of the oven. In this case, the oven was constructed of aluminum tube stock 3" long × 1" wide × 1/8" thick. The heater windings were distributed around the cylinder and the assembly placed within a small insulating Dewar flask. This allows 75°C setpoint (the zero TC or "turnover" temperature of the crystal specified) control of 0.05°C over 0°C to 70°C. The LT1005 regulator sources bridge drive from its auxiliary output and also keeps system power off until the crystal's temperature (hence, its frequency) is stabilized. When power is applied the negative TC thermistor is high in value, causing the LT1001 to saturate positive. This turns on zener-connected Q2, biasing Q3. Q3's collector current pulls the regulator's control pin low, disabling its output. When the oven arrives at its control point, the LT1001's output comes out of saturation and servo controls the oven at a point well below Q2's zener value. This turns off Q3, enabling the regulator to source power to whatever system the clock is associated with. For the crystal and circuit values specified, this clock will drift less than 1×10^{-9} over 0°C to 70°C with a time drift of 1 part 10^{-9} week.

The oven approach to removing temperature effects of crystal clock frequency is the most effective and in wide use. Ovens do, however, require substantial power and warm-up time. In some situations, this is unacceptable. Another approach to offsetting temperature effects is to measure ambient temperature and insert a scaled compensation factor into the crystal clock's frequency trimming network. This open loop correction technique relies on matching the clock frequency vs temperature characteristic, which is quite repeatable. Figure 31.6 shows a temperature compensated crystal oscillator (TXCO) which uses a first order linear fit to correct for temperature. The oscillator is a Colpitts type, with a capacitive tapped tank network. The LT319A picks off the output and the RC network at the LT319's "–" input provides a signal adaptive trip threshold. The LT1005 regulator's auxiliary output buffers supply variations and the main regulator output control pin allows the system to be shut down without removing power from the oscillator, aiding overall stability. The ambient temperature is sensed by the linear thermistor network in A1's feedback loop with A2 used for scaling and offsetting. A2's voltage output expresses

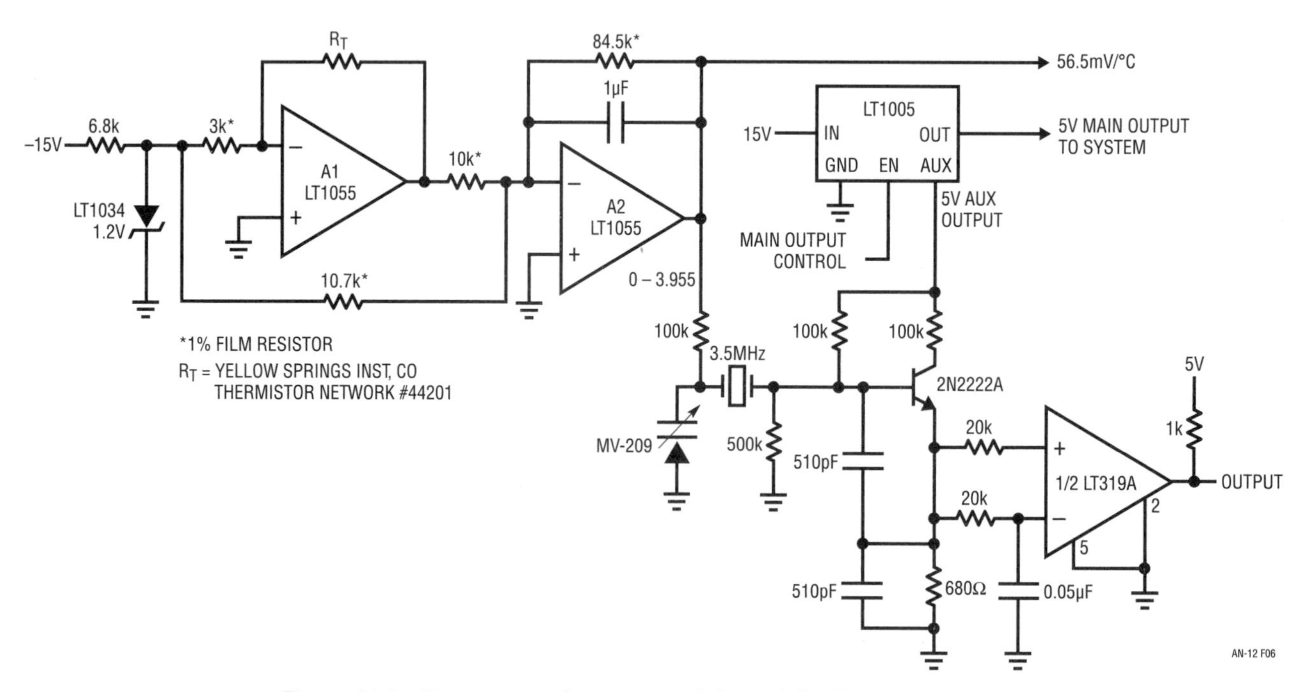

Figure 31.6 • Temperature Compensated Crystal Oscillator (TXCO)

the ambient temperature information required to compensate the clock. The correction is implemented by biasing the varactor diode (a varactor diode's capacitance varies with reverse bias) which is in series with the crystal. The varactor's shift in capacitance is used to pull the crystal's frequency in a complementary fashion to the circuit's temperature error. If the thermistor is maintained isothermally with the circuit, compensation is very effective. Figure 31.7 shows the results. The −40ppm frequency shift over 0°C to 70°C is corrected to within 2ppm. Better compensation is achievable by including 2nd and 3rd order terms in the temperature to voltage conversion to more accurately complement the nonlinear frequency drift characteristic.

Figure 31.8 is another voltage-varactor tuned circuit but is configured to allow frequency shift instead of opposing it. This voltage controlled crystal oscillator (VXCO) has a clean 20MHz sine wave output (Figure 31.9) suitable for communications applications. The curve of Figure 31.10 shows a 7kHz shift from 20MHz over the 10V tuning range. The 25pF trimmer sets the 20MHz zero bias frequency. In many applications, such as phase-locking and narrow bandwidth FM secure communications, the nonlinear response is irrelevant. Improved linearity will require conditioning the tuning voltage or the varactor network's response. In circuits of this type it is important to remember that the limit on pulling frequency is set by the crystals Q, which is high. Achieving wide dynamic "pull" range without stopping the oscillator or forcing it into abnormal modes is difficult. Typical circuits, such as this one, offer pull ranges of several hundred ppm. Larger shifts (e.g., 2000ppm to 3000ppm) are possible without losing crystal lock, although clock output frequency stability suffers somewhat.

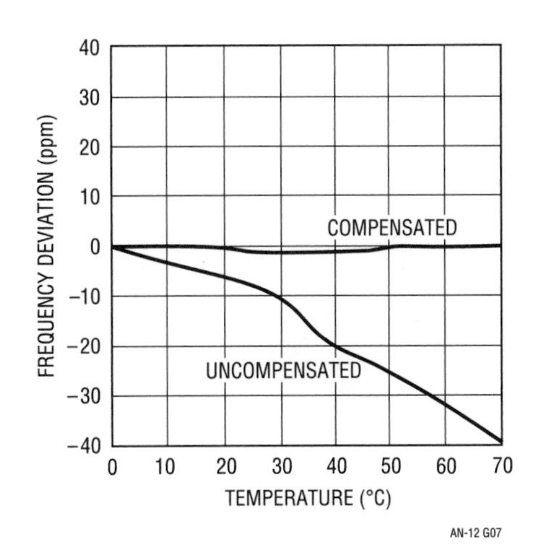

Figure 31.7 • TXCO Drift performance

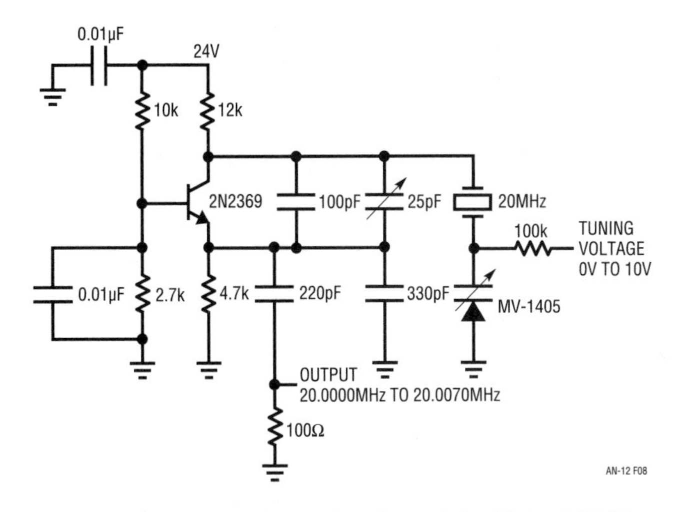

Figure 31.8 • Voltage Controlled Crystal Oscillator (VCXO)

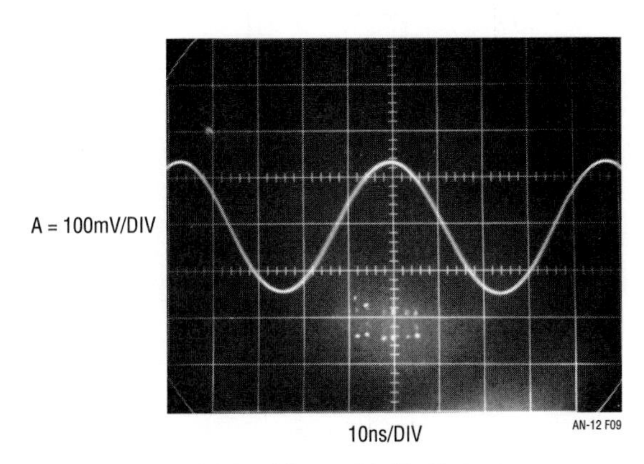

Figure 31.9 • Figure 31.8's Output

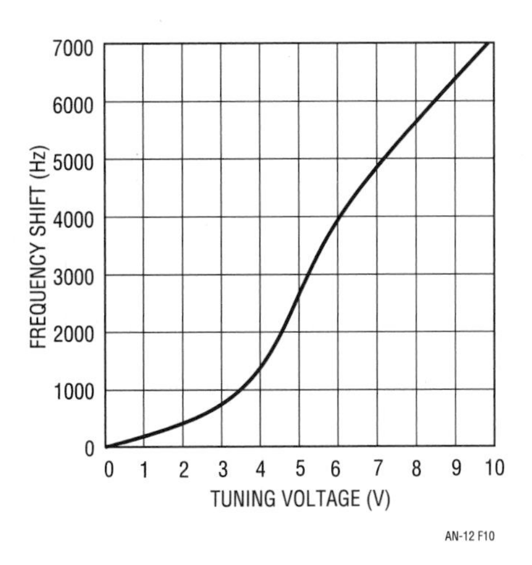

Figure 31.10 • Figure 31.8's Tuning Characteristics

Noncrystal clock circuits

Although crystal based circuits are universally applied, they cannot serve all clock requirements. As an example, many systems require a reliable 60Hz line synchronous clock. Zero crossing detectors or simple voltage level detectors are often employed, but have poor noise rejection characteristics. The key to achieving a good line clock under adverse conditions is to design a circuit which takes advantage of the narrow bandwidth of the 60Hz fundamental. Approaches utilizing wide gain bandwidth, even if hysteresis is applied, invite trouble with noise. Figure 31.11 shows a line synchronous clock which will not lose lock under noisy line conditions. The basic RC multivibrator is tuned to free run near 60Hz, but the AC-line derived synchronizing input forces the oscillator to lock to the line. The circuit derives its noise rejection from the integrator characteristics of the RC network. As Figure 31.12 shows, noise and fast spiking on the 60Hz input (Trace A, Figure 31.12) has little effect on the capacitor's charging characteristics (Trace B, Figure 31.12) and the circuit's output (Trace C, Figure 31.12) is stable.

Figure 31.13 is another synchronous clock circuit. In this instance, the circuit output locks at a higher frequency than the synchronizing input. Circuit operation is the time domain equivalent of a reset stabilized DC amplifier. The LT1055 and its associated components form a stable oscillator. The LM329 diode bridge and compensating diodes provide a stable bipolar charging source for the RC located at the amplifier's negative input. The synchronizing pulse (Trace A, Figure 31.14) is level shifted by the LT1011 comparator to drive the FET. When the synchronizing pulse appears, the FET turns on, grounding the capacitor (Trace B, Figure 31.14). This interrupts normal oscillator action, but only for a small fraction of a cycle. When the sync pulse falls, the capacitor's charge cycle, which has been reset to 0V, starts again. This resetting action forces the frequency of the RC charging to be synchronous and stabilized by the sync pulse. The only evidence of this operation at the output is an occasional, slightly enlarged pulse width (Trace C, Figure 31.14), which is caused by

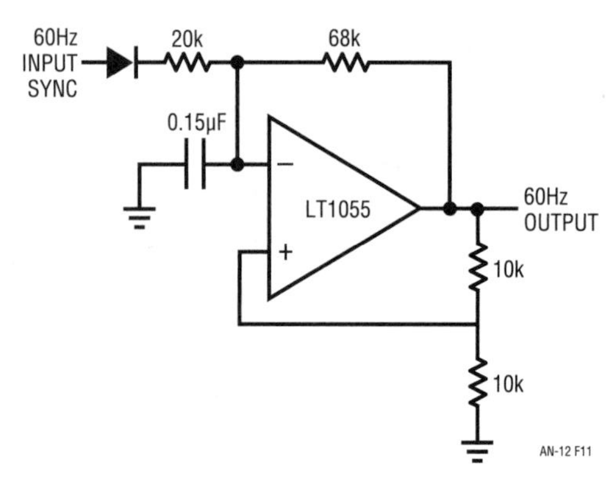

Figure 31.11 • Synchronized Oscillator

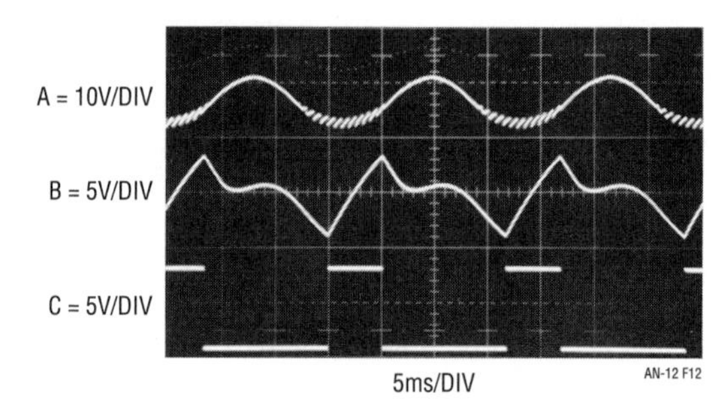

Figure 31.12 • Figure 31.11's Waveforms

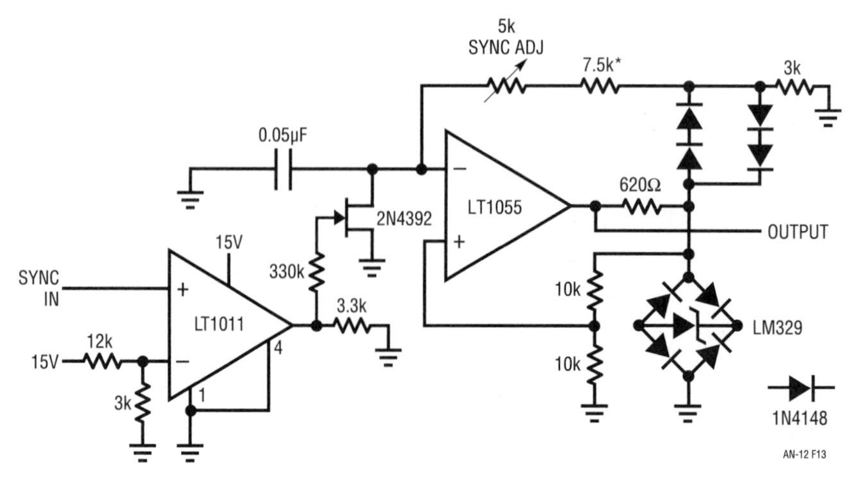

Figure 31.13 • Reset Stabilized Oscillator

the synchronizing interval. The sync adjust potentiometer should be trimmed so the sync pulse appears when the capacitor is near 0V. This minimizes output waveform width deviation and allows maximum protection against losing lock due to RC drift over time and temperature. The maximum practical output frequency to sync frequency ratio is about 50×.

Pure RC oscillators are a final form of clock circuit. Although this class of circuit cannot achieve the stability of a synchronized or crystal based approach, it offers simplicity, economy and direct low frequency output. As such they are used in baud rate generators and other low frequency applications. The key to designing a stable RC oscillator is to make output frequency insensitive to drift in as many circuit elements as possible. Figure 31.15 shows an RC clock circuit which depends primarily on the RC elements for stability. All other components contribute very low order error terms, even for substantial shifts. In addition, the RC components have been chosen for opposing temperature coefficients, further aiding stability. The circuit is a standard comparator-multivibrator with parallel CMOS inverters interposed between the comparator output and the feedback resistors. This replaces the relatively large and unstable bipolar V_{CE} saturation losses of the LT1011 output with the superior ON characteristics of MOS. Not only are the MOS switching losses to the rails low and resistive, but they tend to cancel. The paralleling of inverters further reduces errors to insignificant levels. With this arrangement, the charge and discharge time constant of the capacitor is almost totally immune from supply and temperature shifts. The 10k units need not be precision types, because shifts in them will cancel. In addition, the effect of the comparator's DC input errors is also negated because of the symmetrical nature of the oscillator. This leaves only the RC network as a significant error term. The nominal—120ppm/°C temperature coefficient of the polystyrene capacitor is partially offset by the opposing positive temperature coefficient designed into the specified resistor. In practice, only a first order compensation is achievable because of the uncertainty of the capacitor's exact TC. For the test circuit, 0°C to 70°C temperature excursion showed a 15ppm/°C TC with a power supply rejection factor of less than 20ppm/V. In contrast, a clock constructed from the popular 555 timer, using the compensating RC network, showed 95ppm/°C and 1050ppm/V of supply shift. Because of comparator propagation delays, circuits of this type are less stable above a 5kHz to 10kHz operating frequency.

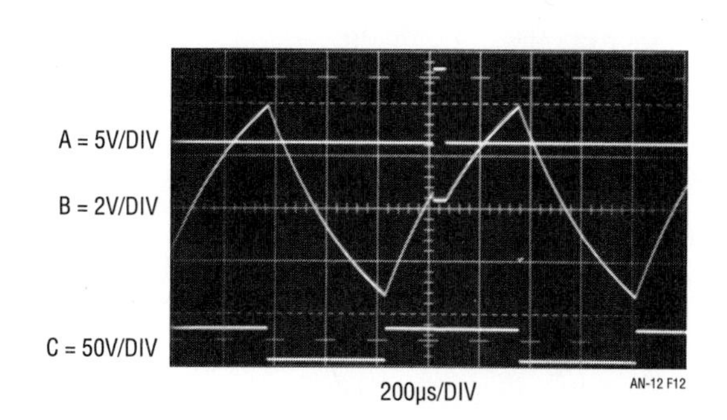

Figure 31.14 • Figure 31.13's Waveforms

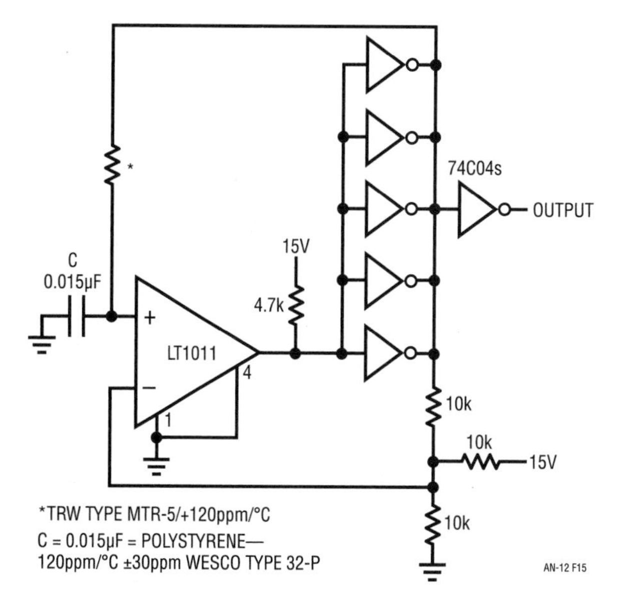

Figure 31.15 • Stable RC Oscillator

About quartz crystals

The frequency stability and repeatability of quartz crystals represent one of nature's best bargains for the circuit

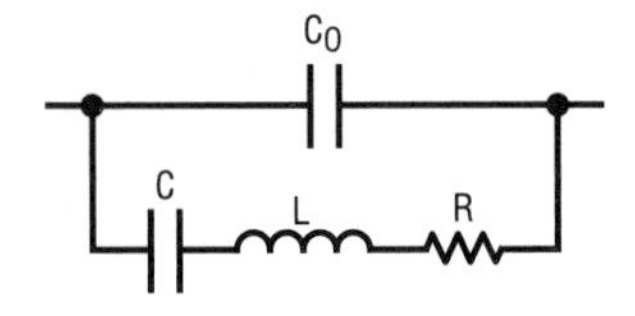

designer. The equivalent circuit of a crystal looks like a series-parallel combination of elements.

Typical Values:

R = 100Ω
L = 500μH
C = 0.01pF
C_O = 5pF
Q = 50,000

C_O is the static capacitance produced by the contact wires, crystal electrodes and the crystal holder. The RLC term is called the motional arm. C is the mechanical mass. R includes all electrical losses in the crystal and L is the reactive component of the quartz. Different angles of cut from the mother crystal produce different electrical characteristics in individual crystals. Cuts can be optimized for temperature coefficient, frequency range and other parameters. The basic "AT" cut used in most crystals in the 1MHz to 150MHz range is a good compromise between temperature coefficient, frequency range, ease of manufacture and other considerations. Other factors affecting resonator performance include the method of lead attachment, package sealing method and internal environment (e.g., vacuum, partial pressure, etc.). Some circuit considerations when using crystals include:

Load capacitance—The reactance the crystal must present to the circuit. Some circuits use the crystal in the parallel resonant mode (e.g., the crystal looks inductive). Other circuits are specified as series resonant and the crystal appears resistive. In this mode, the circuit's load capacitance, including all parasitics, must be specified. A typical number is around 30pF.

Resistance—The impedance the crystal presents when it is resonating.

Drive level—How much power may be dissipated in the crystal and still maintain all specifications. 10mW is typical. Excessive levels can fracture the crystal.

Temperature coefficient/turning point—The tempco of the crystal is usually specified near the "turning point." This is the temperature at which the crystal tempco is zero. Typically the tempco will be below 1ppm/°C over the operating range and the turning point around 75°C, although different cuts can considerably alter these numbers.

Frequency tolerance—The deviation from ideal frequency when used under specified circuit conditions at a defined temperature. Tolerances vary from 50ppm to less than 1ppm.

Measurement and control circuit collection

Diapers and designs on the night shift

32

Jim Williams

Introduction

During my wife's pregnancy I wondered what it would *really* be like when the baby was finally born. Before that time, there just wasn't much mothering and fathering to do. As a consolation, we busied ourselves watching the baby's heartbeat (Figure 32.1) on a thrown-together fetal heart monitor (see References).

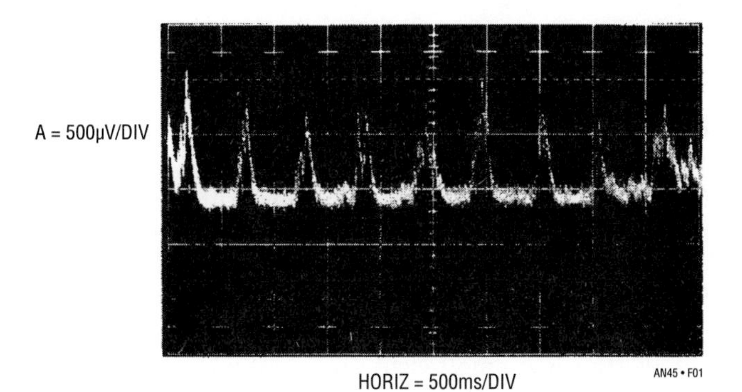

Figure 32.1 • Michael's Fetal Heartbeat 4 1/2 Months into Pregnancy

When Michael was born things got noticeably busier in a hurry. My wife and I split up the evening duties. I got the night shift, 2 am to 7 am. After a few weeks, Michael and I got the hang of it and things began to go (relatively) smoothly. The two of us had mastered feedings, naps, crying jags, bottles, diapers and such and we began looking around for something to do. I decided to introduce Michael to the glories of late night circuit hacking. I first learned about wee hours circuit design at MIT in the 1970s. There was a subculture there that loaded up on pizza, soft drinks, and junk food, took it all into the lab, and closed the door until long after daylight. I was an enthusiastic convert.

Michael and I changed the rules just a bit. We loaded up on formula, diapers, and bottles and went into the lab.

The circuits in this collection represent our efforts, which stopped when he (more or less) began sleeping through the night. Most of the breadboarding occurred between feedings, with design reviews and discussions during feedings. As such, the circuits are annotated with the number of feedings required for their completion; e.g., a "3-bottle circuit" took three feedings. The circuit's degree of difficulty, and Michael's degree of cooperation, combined to determine the bottle rating, which is duly recorded in each figure.

Low noise and drift chopped bipolar amplifier

Figure 32.2's circuit combines the low noise of an LT®1028 with a chopper based carrier modulation scheme to achieve an extraordinarily low noise, low drift DC amplifier. DC drift and noise performance exceed any currently available monolithic amplifier. Offset is inside 1μv with drift less than 0.05μV/°C. Noise in a 10Hz bandwidth is less than 40nV far below monolithic chopper-stabilized amplifiers.

Bias current, set by the bipolar LT1028 input, is about 25nA. These specifications suit demanding transducer signal conditioning situations such as high resolution scales and magnetic search coils.

The 74C04 inverters form a simple 2-phase square wave clock running at about 350Hz. The oscillator provides complementary drive to S1 and S2, causing A1 to see a chopped version of the input voltage. A1 amplifies this AC signal. A1's square wave output is synchronously demodulated by S3 and S4. Because these switches are synchronously driven with the input chopper; proper amplitude and polarity

Analog Circuit and System Design: Immersion in the Black Art of Analog Design. http://dx.doi.org/10.1016/B978-0-12-397888-2.00032-8

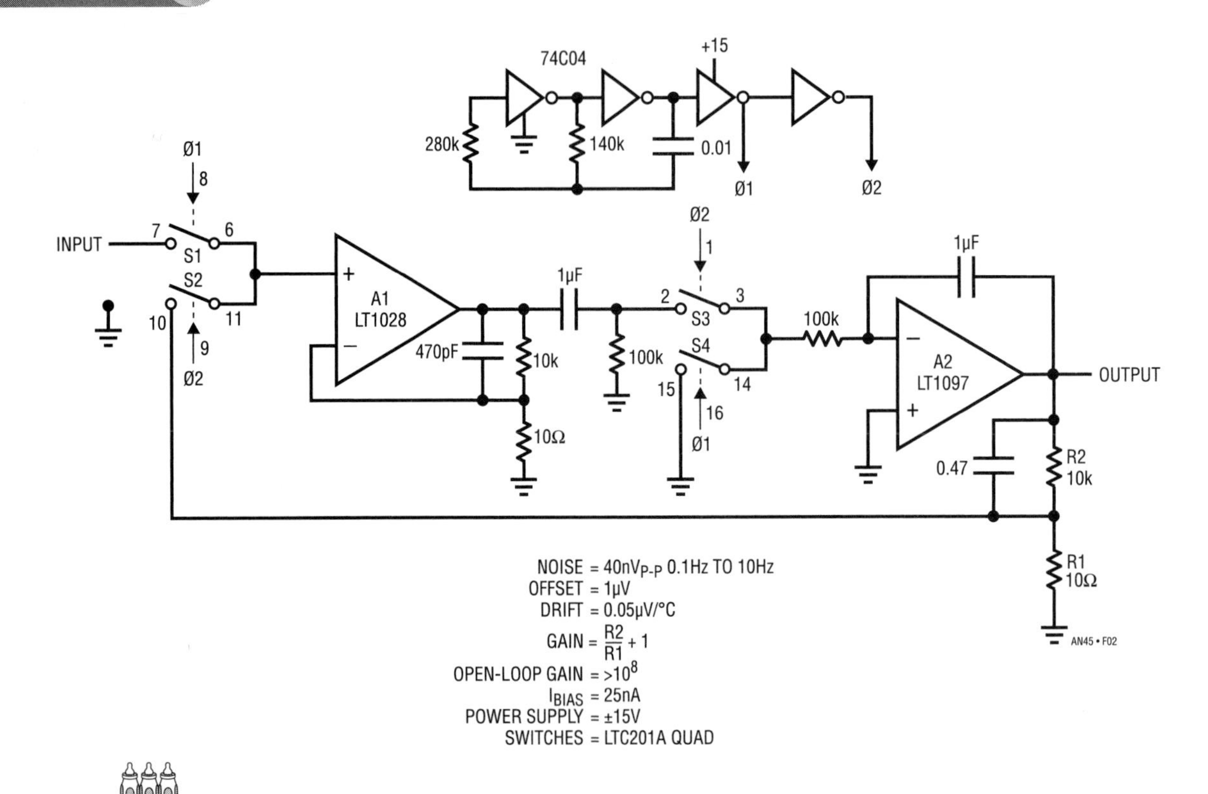

Figure 32.2 • The Chopped Bipolar Amplifier. Noise Is Inside 40nV with 0.05μpV/°C Drift

information is presented to A2, the DC output amplifier. This stage integrates the square wave into a DC voltage, providing the output. The output is divided down (R2 and R1) and fed back to the input chopper where it serves as a zero signal reference. Gain, in this case 1000, is set by the R1-R2 ratio. Because A1 is AC-coupled, its DC offset and drift do not affect overall circuit offset, resulting in the extremely low offset and drift noted.

Figure 32.3, a noise plot of the amplifier in a 0.1Hz to 10Hz bandwidth, shows less than 40nV of peak-to-peak noise. A1 and the 60Ω resistance of S1-S2 contribute about equally to form this noise. When using this amplifier it is important to realize that A1's bias current flowing through the input source impedance causes additional noise. In general, to maintain low noise performance, source resistance should be kept below 500Ω. Fortunately, transducers such as strain gauge bridges, RTDs, and magnetic detectors are well below this figure.

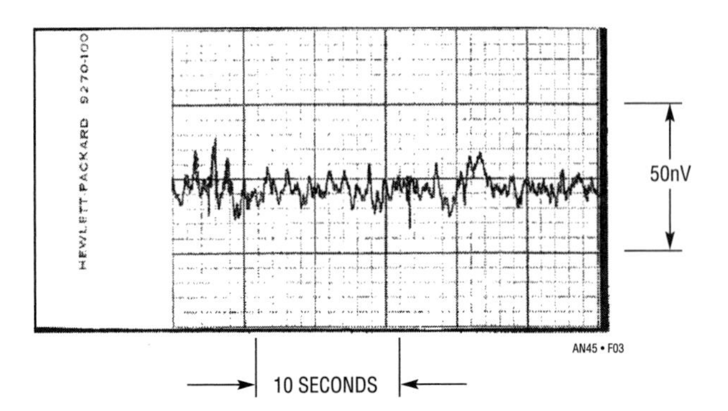

Figure 32.3 • Noise in a 0.1Hz to 10Hz Bandwidth in Less Than 40nV with 0.05μV/°C Drift

Low noise and drift-chopped FET amplifier

Figure 32.4's circuit combines the low drift of a chopper-stabilized amplifier with a pair of low noise FETs. The result is an amplifier with 0.05μV/°C drift, offset within 5μV, 50pA bias current, and 200nV noise in a 0.1Hz to 10Hz bandwidth. The noise performance is especially noteworthy; it is almost eight times better than monolithic chopper-stabilized amplifiers.

FET pair Q1 differentially feeds A2 to form a simple low noise op amp. Feedback, provided by R1 and R2, sets closed-loop gain (in this case 1000) in the usual fashion. Although Q1 has extraordinarily low noise characteristics, its 15mV offset and 25μV/°C drift are poor. A1, a chopper- stabilized amplifier, corrects these deficiencies. It does this by measuring the difference between the amplifier's inputs and adjusting Q1A's channel current to minimize the difference. Q1's skewed drain values ensure that A1 will be able to capture the offset. A1 supplies whatever current is required into Q1A's channel to force offset within 5μV. Additionally, A1's low bias current does not appreciably add to the overall 50pA amplifier bias current.

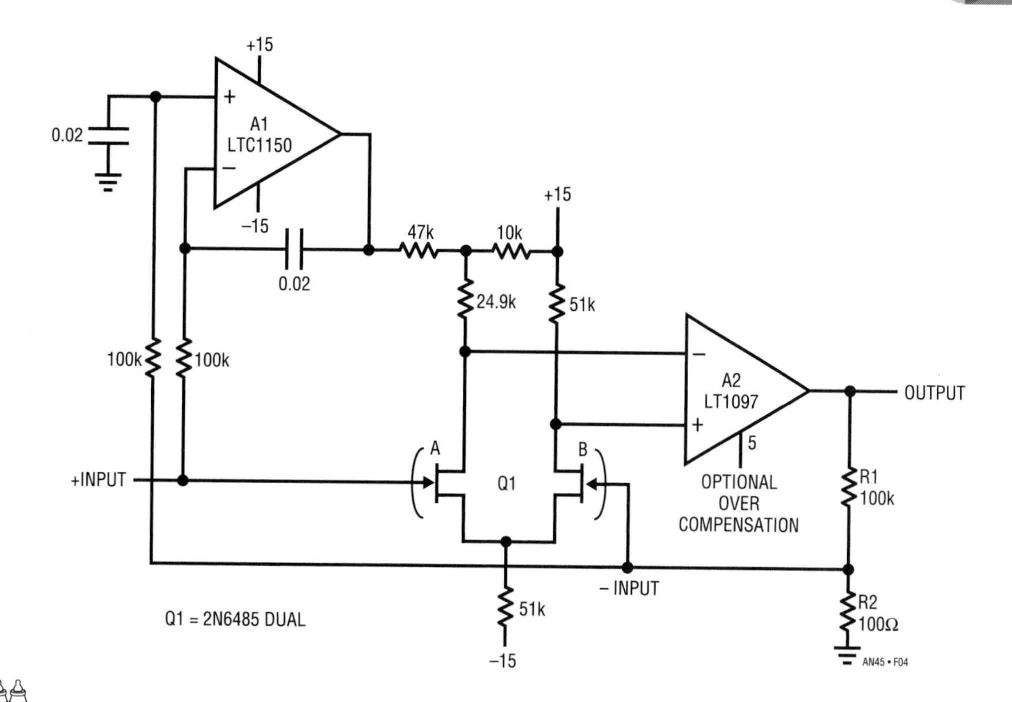

Figure 32.4 • Chopper-Stabilized FET Pair Combines Low Bias, Offset and Drift with 200nV Noise

As shown, the amplifier is set up for a noninverting gain of 1000, although other gains and inverting operation are possible. Figure 32.5 is a plot of noise measured in a 0.1Hz to 10Hz bandwidth. The performance obtained is almost an order of magnitude better than any monolithic chopper-stabilized amplifier, while retaining low offset and drift.

A2's optional overcompensation can be used (capacitor to ground) to optimize damping for low closed-loop gains.

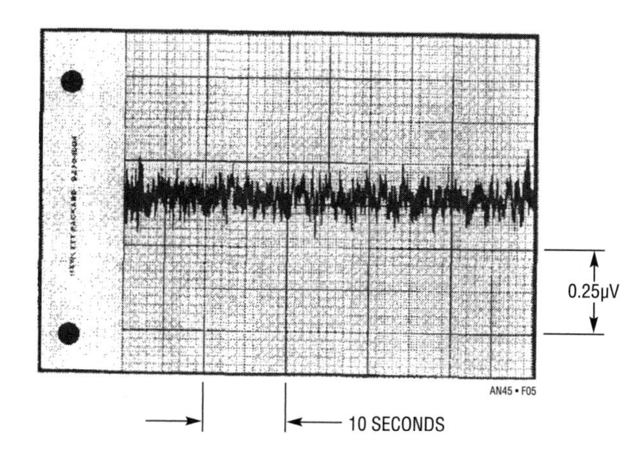

Figure 32.5 • Noise Performance for Figure 32.4. A1's Low Offset and Drift are Retained, but Noise Is Almost Ten Times Better

Stabilized, wideband cable driving amplifier with low input capacitance

Figure 32.6's amplifier has over 20MHz of small-signal bandwidth driving 100mA loads, capacitance or cable. Input capacitance is below 1.5pF and bias current about 100pA. The output is fully protected. These features make this amplifier ideal as an ATE pin amplifier, video A/D input buffer, or cable driver: The amplifier also permits wideband probing when oscilloscope probe loading is not tolerable. The overall amplifier is composed of a low input capacitance FET two LT1010 buffers, and a discrete gain stage. A3 acts as a DC restoration loop. The 33Ω resistors sense A1's operating current, biasing Q3 and Q4. These devices furnish complementary voltage gain to A2, which provides the circuit's output. Feedback is from A2's output to A1's output, which is a low impedance point. This "current mode" feedback permits fixed bandwidth over a wide range of closed-loop gains. This contrasts with normal feedback schemes where bandwidth degrades as closed-loop gain increases.

A3's stabilizing loop compensates large offsets in the signal path, which are dominated by mismatch in Q3 and Q4. A3 measures the DC difference between the amplifier's input and output and biases the signal path to correct for offset. Correction is implemented by controlling Q1's channel current via Q2. The channel current sets Q1's V_{GS}, allowing A3 to control overall circuit offset. The 9k to 1k feedback divider feeding A3 is selected to equal the gain ratio of the circuit, in this case 10.

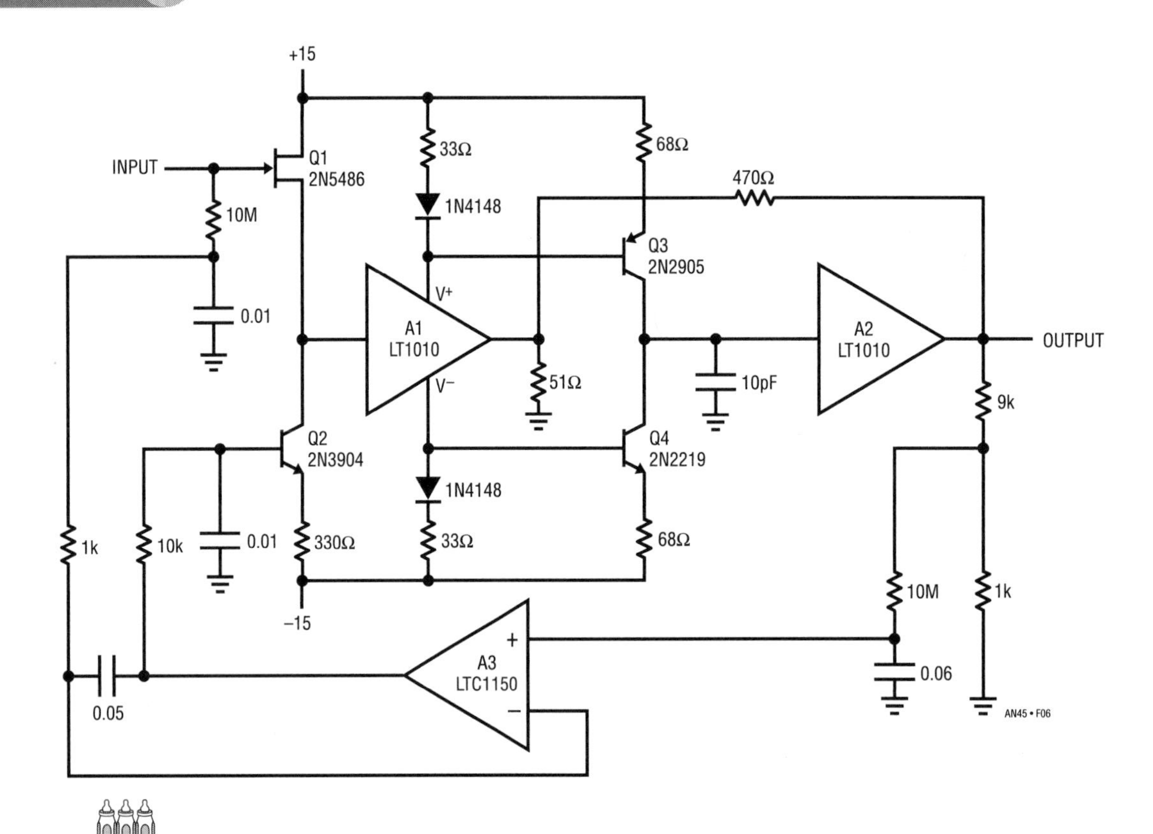

Figure 32.6 • Stabilized, Wideband Cable Driving Amplifier with Low Input Capacitance

The feedback scheme makes A1's output look like the negative input of the amplifier, with closed-loop gain set by the ratio of the 470Ω and 51Ω resistors. The outstanding feature of this connection is that the bandwidth becomes relatively independent of closed-loop gain over a reasonable range. For this circuit, small-signal bandwidth exceeds 20MHz over gains of 1 to 20. The loop is quite stable, and the 10pF value at A2's input provides good damping over a wide range of gains.

Figure 32.7 shows large-signal performance at a gain of 10 driving 10 feet of cable. A fast input pulse (trace A) produces the output shown (trace B). Response is quick and clean with no slew residue or poor dynamics.

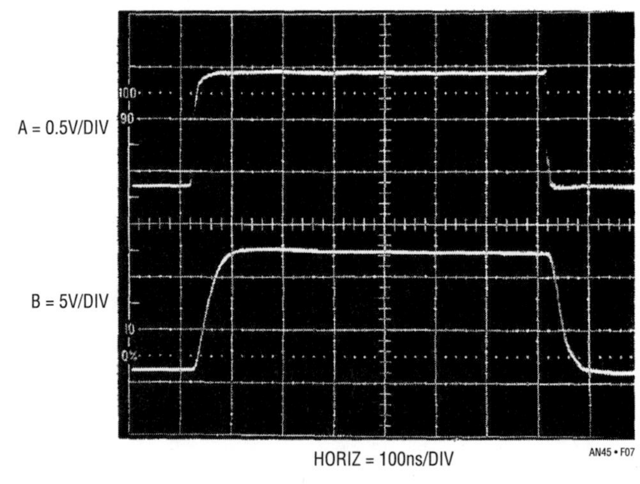

Figure 32.7 • Wideband Amplifier's Response Driving A 10 Foot Cable

Voltage programmable, ground referred current source

Precise, voltage programmable, ground referred current sources are usually complex and require trimming. Figure 32.8's simple, powerful configuration produces output current in strict accordance with the sign and magnitude of the control voltage. Dynamic response is well controlled, and no trimming is required. The circuit's accuracy and stability are almost entirely dependent upon resistor R.

A1, biased by V_{IN}, drives current through R (in this case 10Ω) and the load. Instrumentation amplifier A2, operating at a gain of 100, senses across R. A2's output closes a

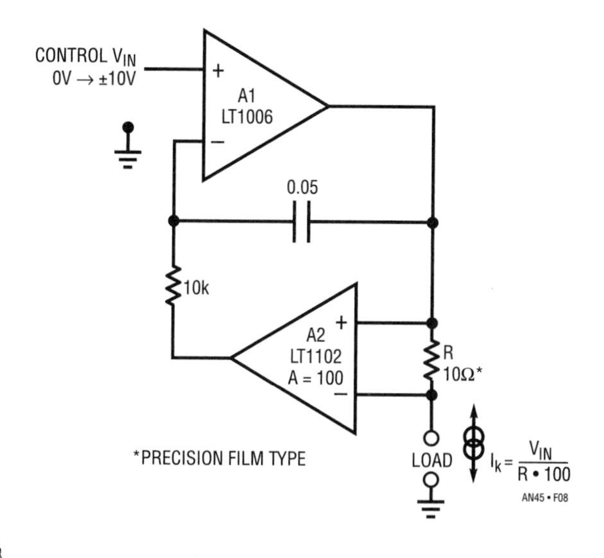

Figure 32.8 • Voltage Programmable Current Source Is Simple and Precise

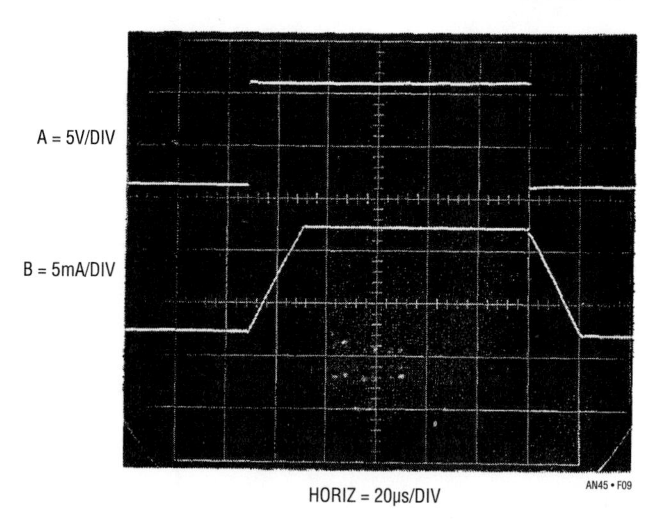

Figure 32.9 • Current Source Dynamics Are Clean, with No Slew Residue or Aberrations

loop back to A1. Because A1's loop forces a fixed voltage across R, the current through the load is constant. The 10k-0.05μF combination sets A1's roll-off, and the circuit is stable.

Assuming an errorless component for R, the circuit's initial error is dominated by A2's 0.05% gain specification and its 5ppm/°C temperature coefficient. High grade film or wirewound resistors will maintain this level of performance.

Figure 32.9 shows dynamic response for a full-scale input step. Trace A is the voltage control input while trace B shows the output current. Response is clean, with no slew residue or aberrations.

5V Powered, fully floating 4mA to 20mA current loop transmitter

4mA to 20mA current-loop transmitters are frequently required in industrial process control. Often, because of uncertain or dangerous common mode voltages, it is desirable that the generated 4mA to 20mA current be completely galvanically isolated from the transmitter's input. Figure 32.10's circuit does this while operating from a single 5V supply.

A2's positive input assumes a bias dependent upon the input and the 4mA trim setting. Under these conditions A2's output heads positive, turning on Q1 and Q2. Q2's collector drives T1's primary, which is chopped by Q3 and Q4. Complementary chopper drive comes from the 74C74 flip-flop outputs, with oscillator I_1 setting a 25kHz clock rate. T1's output, producing voltage step-up, is rectified, filtered, and applied to the load. A3 senses load current across the 16Ω shunt and drives T2's center tap. Q9 and Q10, receiving complementary drive picked off from T1's secondary, modulate T2's DC center-tap voltage. T2's secondary receives this information, with flip-flop driven Q6-Q7 demodulating it back to DC at T2's center tap. T2's center tap voltage is fed A2, completing an isolated control loop. Changes in the circuit's input voltage cause this loop to adjust the load current accordingly. Conversely, load resistance changes have no effect, because the loop forces whatever voltage is necessary to maintain a constant 16Ω shunt voltage. Because T1 can supply up to 50V, load current remains fixed over load resistance swings from 0Ω to 2500Ω. Power supply shifts are similarly rejected by the loop, and the transformer modulation-demodulation scheme permits 0.05% accuracy and stability over temperature and a 250V common mode range. Greater common mode voltages are possible with increased transformer breakdown ratings.

Several subtleties aid circuit performance. I_2-I_3 and I_4-I_5 provide drive delays to Q6 and Q7. These delays approximate the delay through T1 to modulator pair Q9/Q10. This helps the four transistors switch simultaneously, aiding modulator-demodulator accuracy. Zener connected Q5 ensures that T1 produces enough voltage to power A3 and Q9/Q10, even when the load is 0Ω. Q8, similarly Zener connected, clamps gate drive to Q9 and Q10, improving modulator linearity by preventing excessive gate drive variations over operating conditions. The diodes in A3's output ensure proper loop start-up. They prevent T2's center tap from receiving any bias until A3

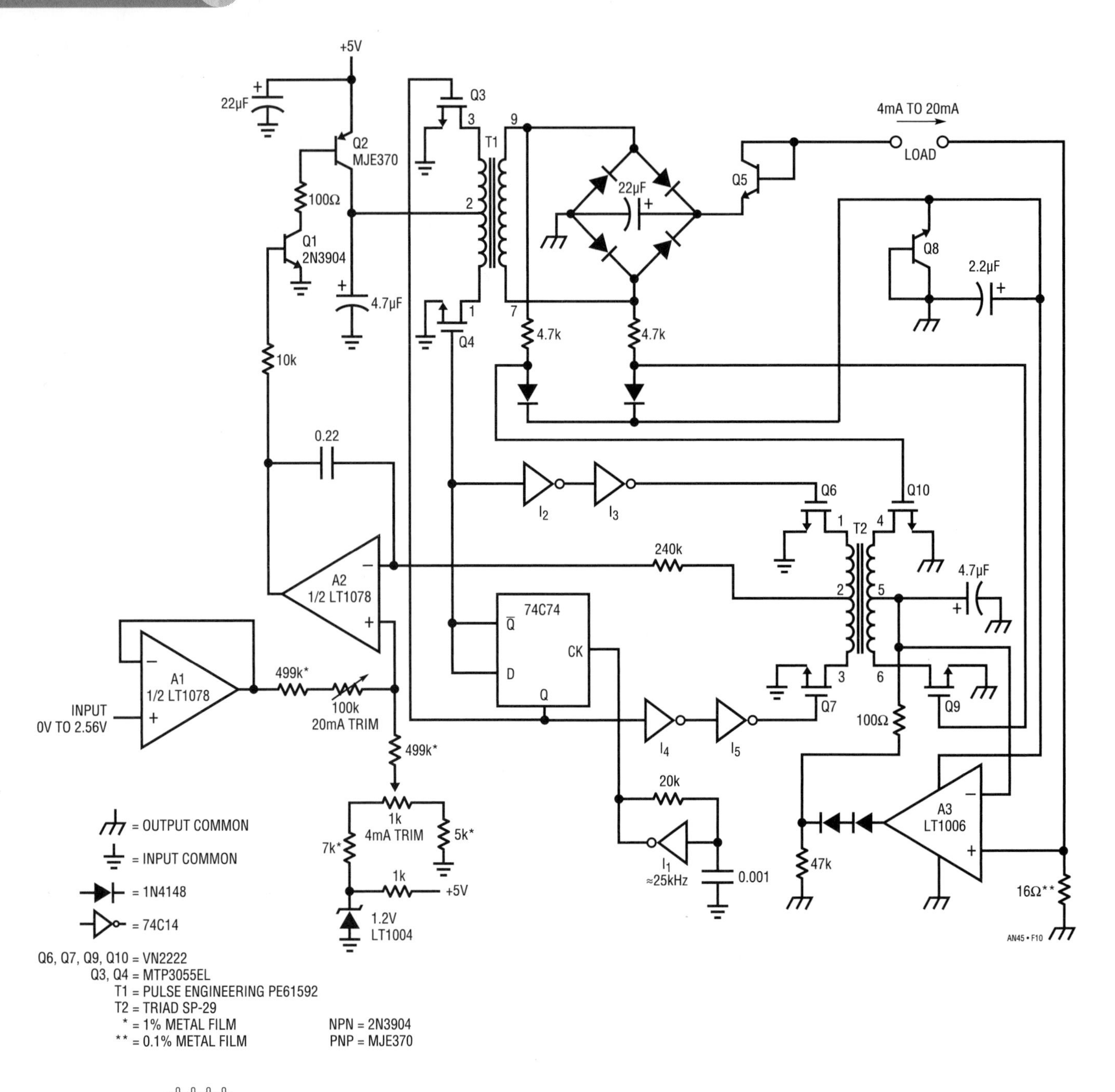

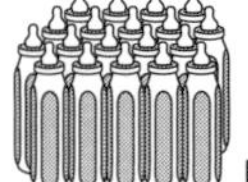

Figure 32.10 • 5V Powered, Fully Floating 4mA to 20mA Current Loop Transmitter

has enough power supply voltage to function normally. To calibrate this circuit apply 0V input and adjust the 4mA trim for 4.00mA output (0.064V across the 16Ω shunt). Next, apply 2.56V input and set the 20mA trimmer for 20.00mA output (0.3200V across the 16Ω shunt). Repeat this procedure until both points are fixed. Note that the 2.56V input range is directly compatible with D/A converter outputs, permitting digital control.

Transistor ΔV_{BE} based thermometer

Low cost makes transistors potentially attractive as temperature sensors. Almost all transistor-sensed thermometer circuits utilize the base-emitter diode voltage shift with temperature as the sensing mechanism. Unfortunately, the absolute diode voltage is unpredictable, necessitating circuit calibration. Additionally, if the transistor sensor ever requires replacement, the calibration must be repeated. This constraint often negates the transistor sensor's cost and convenience advantages.

Figure 32.11's transistor sensor thermometer overcomes this difficulty. The circuit provides a 0V to 10V output corresponding to a 0°C to 100°C temperature excursion at the sensor transistor. Accuracy is ±1°C. No calibration is required, and any common small-signal NPN transistor can serve as the sensor. The circuit is based on the predictable relationship between current and voltage in a transistor V_{BE} junction.[1] At room temperature, the V_{BE} junction diode shifts 59.16mV per decade of current. The temperature dependence of this constant is 0.33%/°C, or 198µV/°C. This ΔV_{BE} versus current relationship holds, regardless of the V_{BE} diode's absolute value.

Note 1: See References 1 through 4.

The LTC®1043 contains switches whose state is controlled by an on-chip oscillator: The 0.01 µF capacitor at Pin 16 sets oscillator frequency at about 500Hz. Q1 operates as a switched-value current source, alternating between about 10µA and 100µA (trace A, Figure 32.12) as the LTC1043 commutates switch Pin 12 and Pin 14. The two currents' exact value is unimportant, so long as their ratio remains constant. Because of this, Q1 requires no reference, although its emitter resistor's ratio is precise. The alternating 10µA to 100µA stepped current to the sensor transistor (Q2) causes the

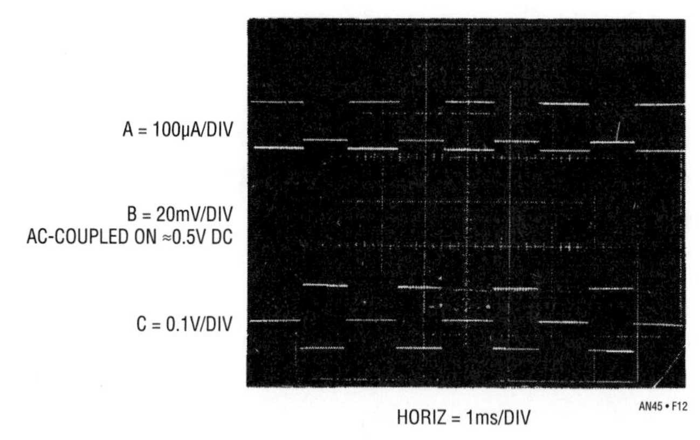

Figure 32.12 • Waveforms for the ΔV_{BE} Based Thermometer

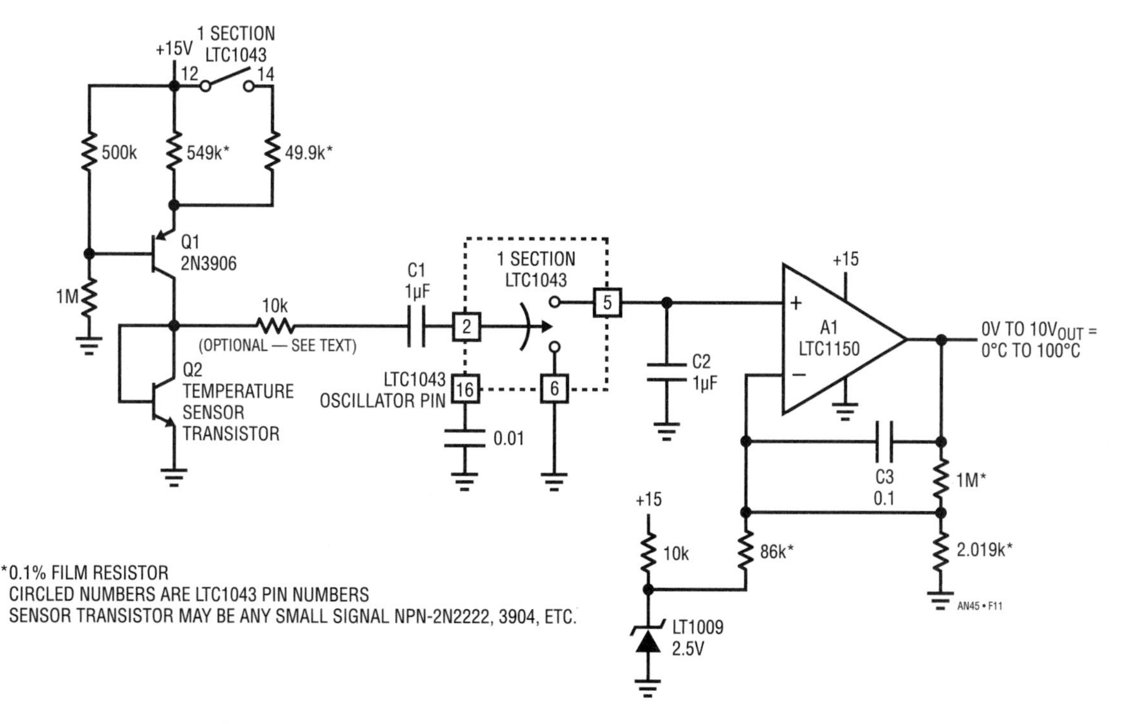

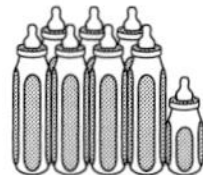

Figure 32.11 • ΔV_{BE} Based Thermometer Does Not Require Calibration

theoretical 59.16mV (25°C) excursion (trace B) to appear across the V_{BE} junction. This signal is coupled to a switched demodulator via C1, which strips off Q2's DC bias. LTC1003 switch Pin 2 (trace C) sees only the 59mV waveform, which is referenced to ground via demodulator action at Pin 5 and Pin 6. Pin 5, connected to capacitor C2, sits at Pin 2's DC peak value. A1 amplifies this DC signal, with the LT1004 providing offset so 0°C equals 0V. The optional 10k resistor protects against ESD events, which may occur if Q2 is located at the end of a cable.

Using the components shown, the circuit achieves ±1°C accuracy over a sensed 0°C to 100°C range. Substituting randomly selected 2N3904s and 2N2222s for Q2 showed less than 0.4°C spread over 25 devices from various manufacturers.

Micropower, cold junction compensated thermocouple-to-frequency converter

Figure 32.13 is a complete, digital output, thermocouple signal conditioner. The circuit produces a 0kHz to 1kHz output in response to a sensed 0°C to 100°C temperature excursion. Cold junction compensation is included, and accuracy is within 1°C with stable 0.1°C resolution. Additionally, the circuit functions from a single supply, which may range from 4.75V to 10V. Maximum current consumption is 360μA.

The LT1025 provides an appropriately scaled cold junction compensation voltage to the type K thermocouple. As a result, the voltage at schematic point "A" varies from 0mV to 4.06mV over a sensed 0°C to 100°C range (type K slope is 40.6μV/°C). The remaining components form a voltage- to-frequency converter that directly converts this millivolt level signal without the usual DC gain stage. A1's negative input is biased by the thermocouple. A1's output drives a crude V-F converter, comprised of Q2, the 74C14 inverters, and associated components. Each V-F output pulse causes a fixed quantity of charge to be dispensed into C3 from C2 via the LTC201 based charge pump. C3 integrates the charge packets, producing a voltage at A1's positive input. A1's output forces the V-F converter to run at whatever frequency is required to balance the amplifier's inputs. This feedback action eliminates drift and nonlinearities in the V-F converter as an error term and the output frequency is solely a function of the DC conditions at A1's inputs. The 0.02μF capacitor forms

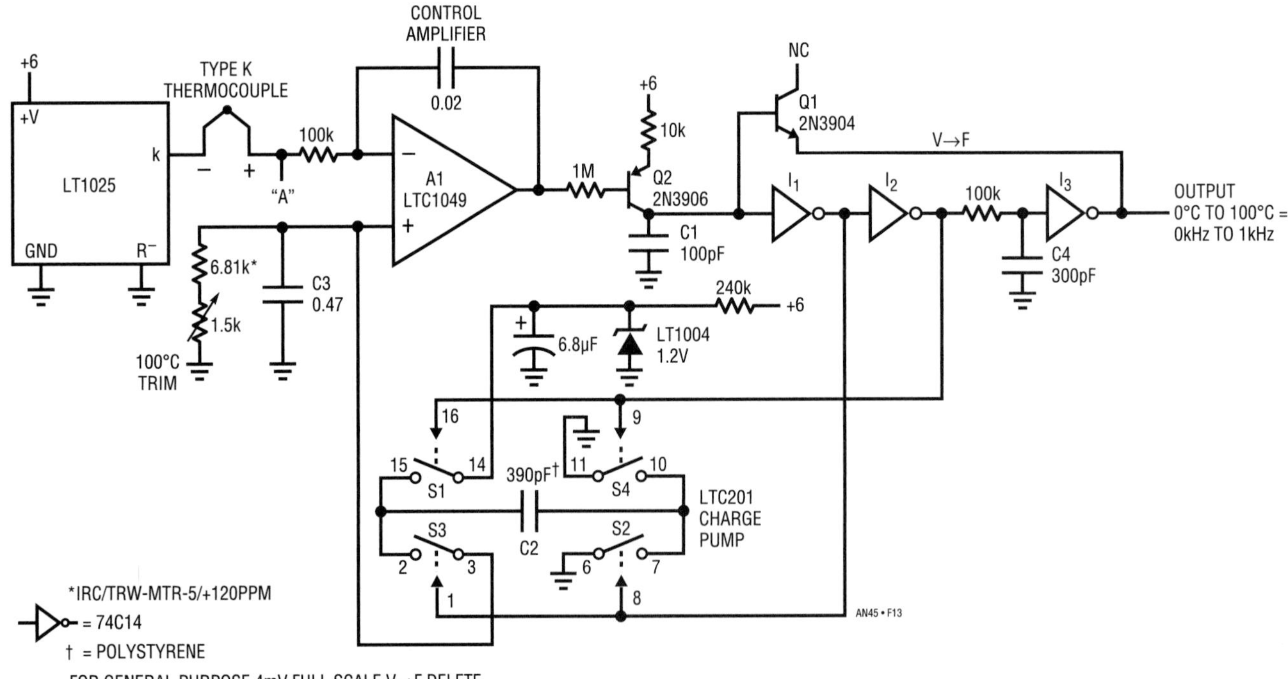

Figure 32.13 • Thermocouple Sensed Temperature-to-Frequency Converter

a dominant response pole at A1, stabilizing the loop. Chopper stabilized A1's low V_{OS} offset and drift eliminate offset error in the circuit, despite an output LSB value of only 4.06µV (0.1°C).

Figure 32.14 details circuit operation. A1's output biases current source Q2, producing a ramp (trace A, Figure 32.14) across C1. When the ramp crosses I_1's threshold, the cascaded inverter chain switches, producing complementary outputs at I_1 (trace B) and I_2 (trace C). I_3's RC delayed response (trace D) turns on diode connected Q1, discharging C1 and resetting the ramp. The ramp aberrations before the reset are due to transient I_1 input currents during switching (near top of ramp). Q1's V_{BE} diode rounding and reverse charge transfer (bottom of ramp) account for the discontinuities during the ramp's low point.

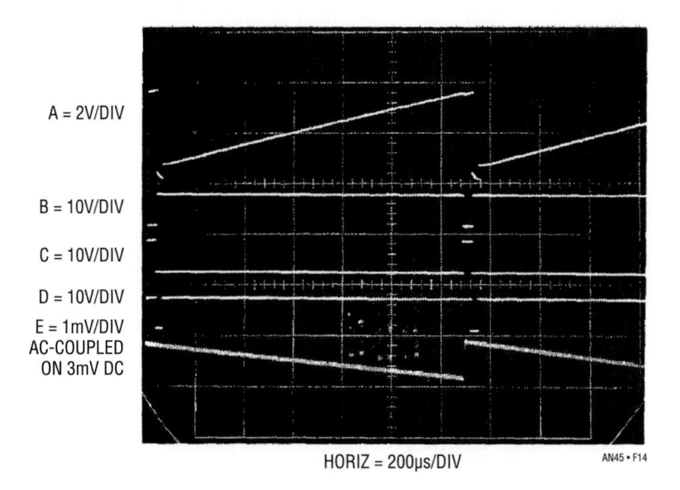

Figure 32.14 • Waveforms for the Thermocouple-to-Frequency Converter

The complementary I_1-I_2 outputs clock the LTC201 switch based charge pump. C2 is alternately charged to the LT1004's reference voltage via S1 and S4 and discharged into C3 through S2 and S3. Each time this cycle occurs, C3's voltage is forced up (trace E). C3's average voltage is set by the 6.81k to 1.5k trimmer resistance across it. A1 servo controls the repetition rate of the V-F to bring its inputs to the same value, closing a control loop. The 0.02µF capacitor smooths A1's response to DC.

To calibrate this circuit, disconnect the thermocouple and drive point "A" with 4.06mV. Next, set the 1.5k trimmer for exactly 1000Hz output. Connect the thermocouple and the circuit is ready for use. Recalibration is not required if the thermocouple is replaced.

It is worth noting that this circuit can directly digitize any millivolt level signal by deleting the LT1025 thermocouple pair and directly driving point "A."

Relative humidity signal conditioner

Relative humidity is a difficult physical parameter to transduce, and most transducers require fairly complex signal conditioning circuitry. Figure 32.15 combines simple circuitry with a capacitively based transducer to achieve good results. This circuit, which runs from a 9V battery, is accurate within 2% in the 5% to 90% relative humidity range.

The sensor specified has a nominal 500pF capacitance at RH=76%, with a slope of 1.7pF/% RH. The average voltage across the device must be zero. This prevents deleterious electrochemical migration in the sensor. LTC1043 section "A," driven by an internal oscillator, alternately charges the sensor from a resistively scaled portion of the LT1004 reference and discharges it into A1's summing point. Note that the switching is arranged so that sensor related current flows out of A1's summing point. The 0.1 µF series capacitor ensures the sensor sees the required zero average voltage, with the 22MΩ resistor preventing charge accumulation, which would stop current flow. The average current out of A1's summing point is balanced by packets of charge delivered by the LTC1043 switched capacitor section "C" in A1's feedback loop. The 0.1µF feedback capacitor gives A1 an integrator-like response, and its output is DC. As such, changes in sensor capacitance are seen as DC shifts in A1's output. A1 responds by raising its output positive to whatever DC potential is required to maintain its summing point at zero.

To allow 0% RH to equal 0V offsetting is required. The signal and feedback terms biasing A1's summing point are expressed in charge form. Because of this, the offset must also be delivered to the summing point as charge, instead of a simple DC current. If this is not done, the circuit will be affected by drift in the LTC1043's internal oscillator. LTC1043 section "B" serves this function, delivering LT1004 referenced offsetting charge to A1.

Drift terms in this circuit include the LT1004 and the ratio stability of the sensor and the polystyrene capacitors. These terms are well within the sensor's 2% accuracy specification, and temperature compensation is not required.

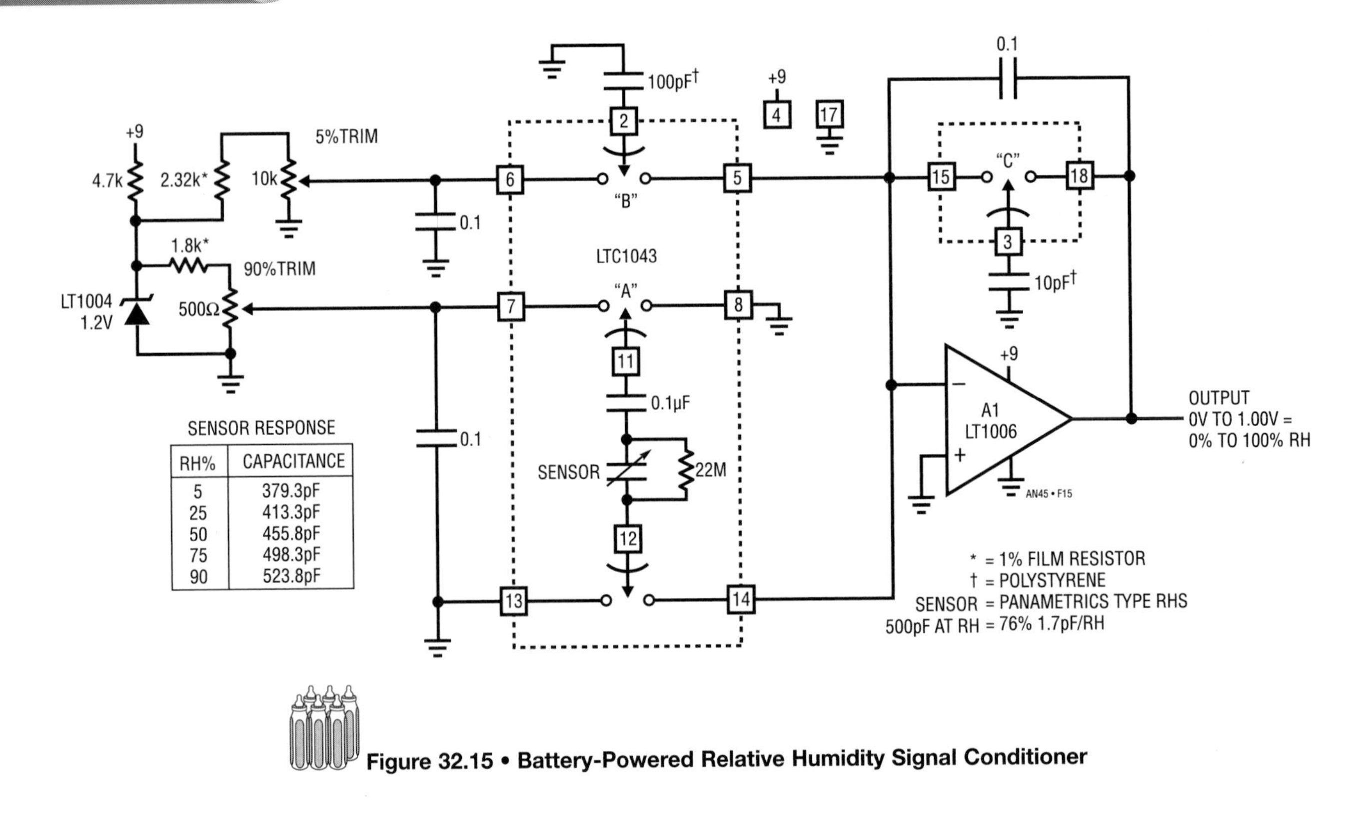

Figure 32.15 • Battery-Powered Relative Humidity Signal Conditioner

To calibrate this circuit, place the sensor in a 5% RH environment and set the "5% RH trim" for 50mV output. Next, place the sensor in a 90% RH environment and set the "90% trim" for 900mV output. Repeat this procedure until both points are fixed. If known RH environments are unavailable, the capacitance versus RH table in Figure 32.15 may be utilized, although it applies for an ideal sensor. The capacitor values may be built-up or directly dialed out on a precision variable air capacitor (General Radio #722D).

Inexpensive precision electronic barometer

Until recently, precision electronically based pressure measurements required expensive transducers. Capacitive and bonded strain gauge based approaches provide unmatched results, but costs are often prohibitive. Additionally, if low power operation is desired, signal conditioning for these devices can become complex.

Semiconductor based pressure transducers becoming available offer significant improvement over earlier devices. Figure 32.16's circuit utilizes such a device to form a low cost barometer. The LT1027 reference and A1 form a current source to put precisely 1.5mA through transducer T1, in accordance with the manufacturer's specifications. Instrumentation amplifier A3 takes a differential gain of 10 from T1's bridge output. A2 provides additional gain to yield a calibrated output directly in inches of mercury.

T1's manufacturer specifies a nominal 115mV at full scale, although each device is supplied with precise calibration data. This information considerably simplifies calibration. To calibrate the circuit, simply adjust the potentiometer at A1 until the output corresponds to the scale factor supplied with the unit.

This circuit, compared to a long column mercury barometer, tracked ambient pressure variations from 29.75" to 30.32" over three months with only two counts of uncertainty. Additionally, over 50 turn-on/turn-off cycles had no measurable effect. Changes in pressure, particularly rapid ones, correlated quite nicely to changing weather conditions.

1.5V Powered radiation detector

Figure 32.17's circuit provides an audible "tick" signal each time radiation or a cosmic ray passes through the detector. The LT1073 switching regulator pulses T1. T1 takes

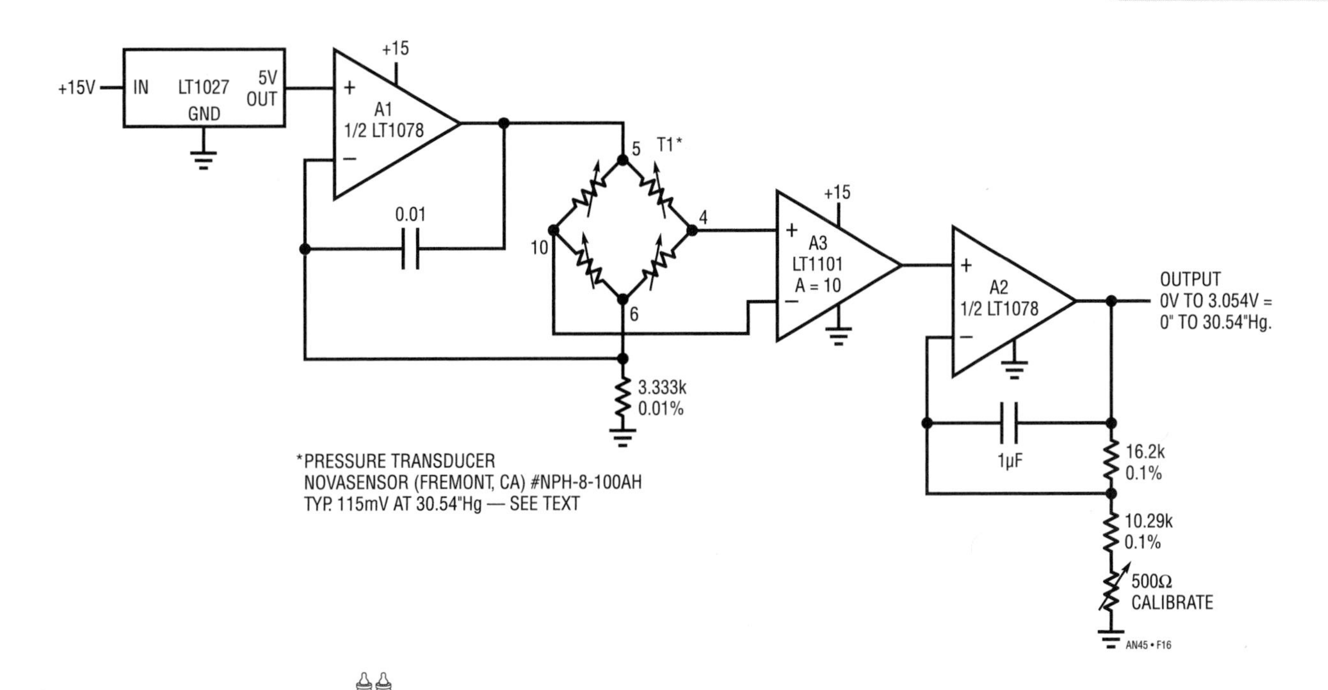

Figure 32.16 • A Simple, Inexpensive Precision Barometer

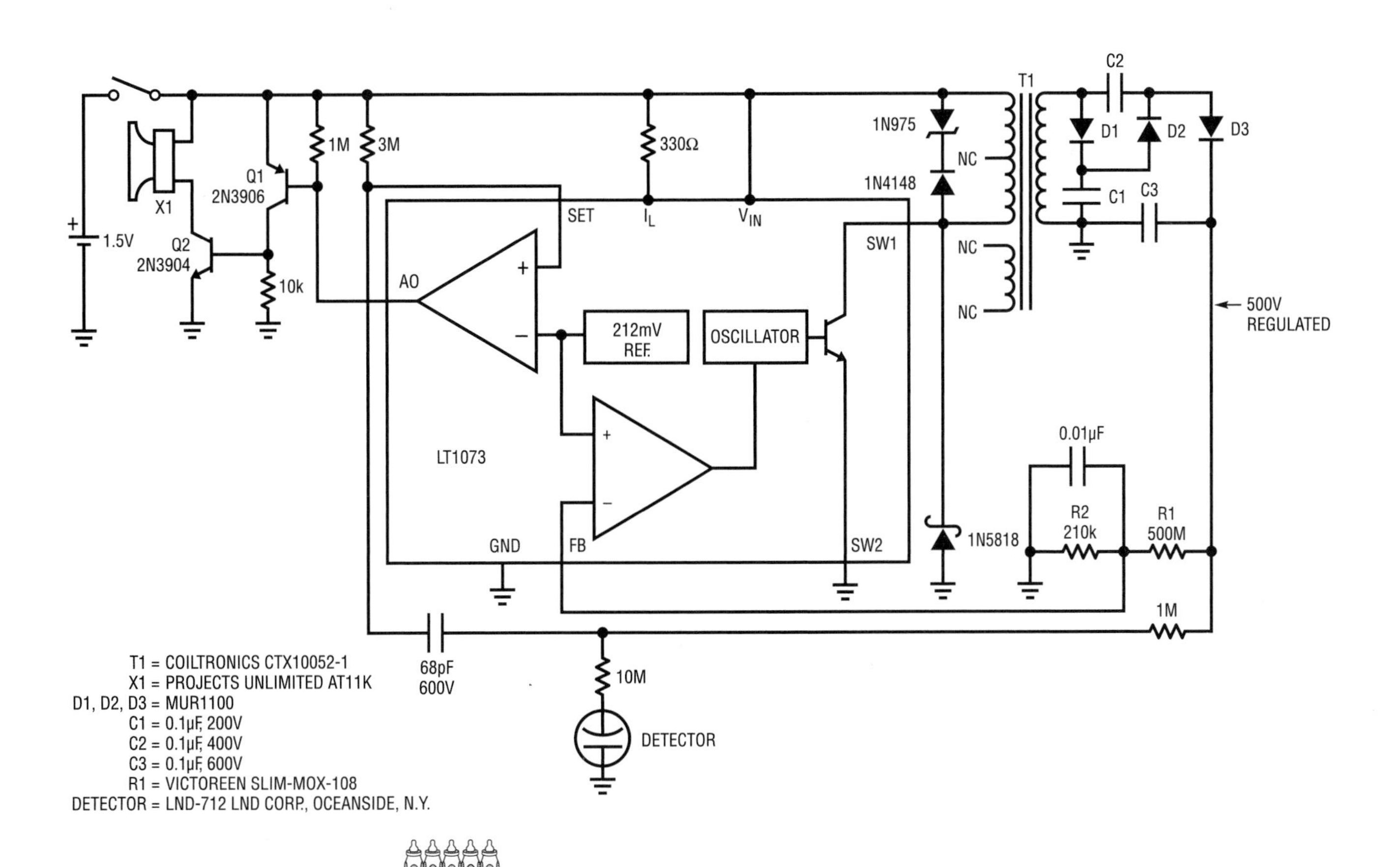

Figure 32.17 • 1.5V Powered Radiation Detector

gain via its turns ratio and drives a voltage tripler, providing 500V bias to the detector: R1 and R2 provide scaled feedback to the LT1073, closing a control loop. The 0.01 μF lag adds AC hysteresis and the Schottky diode clamps negative going T1 excursions. When radiation or a cosmic ray strikes the detector, impedance drops briefly, transferring a quick negative going spike through the 68pF capacitor This spike triggers the LT1073's auxiliary gain block, configured here as a comparator. Q1 and Q2 provide additional gain to drive the audible beeper. About 10 to 15 cosmic rays per minute are recorded in a normal environment.

9ppm Distortion, quartz stabilized oscillator

A spectrally pure sine wave oscillator is required for data converter, filter and audio testing. Figure 32.18 provides a stable frequency output with extremely low distortion. This quartz stabilized 4kHz oscillator has less than 9ppm (0.0009%) distortion in its $10V_{P-P}$ output.

To understand circuit operation, temporarily assume A2's output is grounded. With the crystal removed, A1 and the A3 power buffer form a noninverting amplifier with a grounded input. The gain is set by the ratio of the 47k resistor to the 50k potentiometer—opto-isolator pair. Inserting the crystal closes a positive feedback path at the crystal's resonant frequency, and oscillations occur. A4 compares A3's positive peaks with the LT1004 2.5V negative reference. The diode in series with the LT1004 provides temperature compensation for A3's rectifier diode. A4 biases the LED portion of the opto-isolator, controlling the photoresistor's resistance. This sets loop gain to a

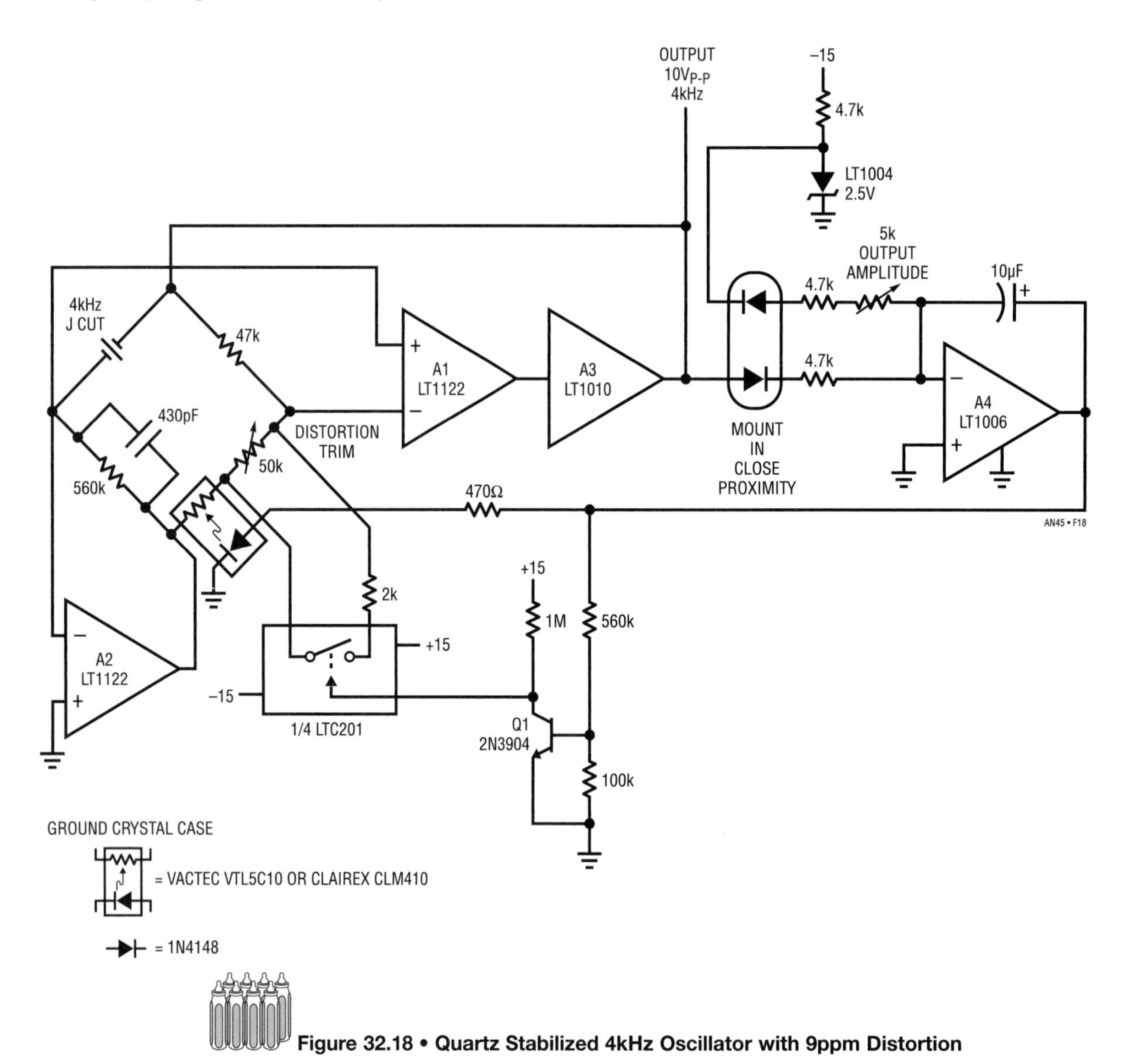

Figure 32.18 • Quartz Stabilized 4kHz Oscillator with 9ppm Distortion

value permitting stable amplitude oscillations. The 10μF capacitor stabilizes this amplitude control loop.

A2's function is to eliminate the common mode swing seen by A1. This dramatically reduces distortion due to A1's common mode rejection limitations. A2 does this by servo controlling the 560kΩ-photocell junction to maintain its negative input at 0V. This action eliminates common mode swing at A1, leaving only the desired differential signal.

Q1 and the LTC201 switch form a start-up loop. When power is first applied oscillations may build very slowly. Under these conditions A4's output saturates positive, turning on Q1. The LTC201 switch turns on, shunting the 2kΩ resistor across the 50KΩ potentiometer. This raises A1's loop gain, forcing a rapid build-up of oscillations. When oscillations rise high enough A4 comes out of saturation, Q1 and the switch go off and the loop functions normally.

The circuit is adjusted for minimum distortion by adjusting the 50kΩ potentiometer while monitoring A3's output with a distortion analyzer. This trim sets the voltage across the photocell to the optimum value for lowest distortion. The circuit's power supply should be well regulated and bypassed to ensure the distortion figures quoted.

After trimming, A3's output (trace A, Figure 32.19) contains less than 9ppm (0.0009%) distortion. Residual distortion components (trace B) include noise and second harmonic residue. Oscillation frequency, set by crystal tolerance, is typically within 50ppm with less than 2.5ppm/°C drift.

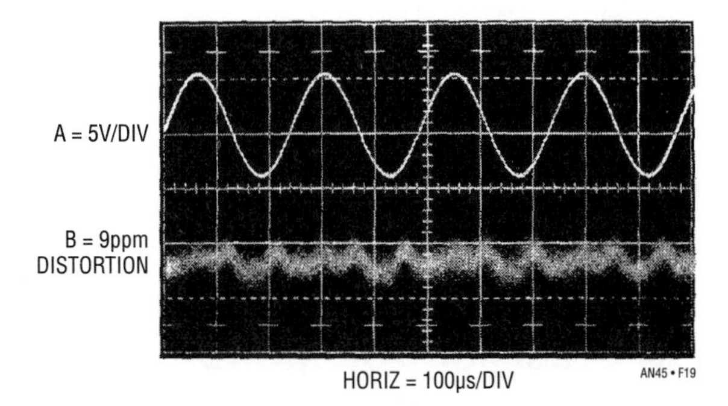

Figure 32.19 • Oscillator Output and its 9ppm Distortion Residue

1.5V Powered temperature compensated crystal oscillator

Many single cell systems require a stable clock source. Crystal oscillators which run from 1.5V are relatively easy to construct. However; if good stability over temperature is required, things become more difficult. Ovenizing the crystal is one approach, but power consumption is excessive. An alternate method provides open loop, frequency correcting bias to the oscillator. The bias value is determined by absolute temperature. In this fashion, the oscillator's thermal drift, which is repeatable, is corrected. The simplest way to do this is by slightly varying the crystal's resonance point with a variable shunt or series impedance. Varactor diodes, the capacitance of which varies with reverse voltage, are commonly employed for this purpose. Unfortunately, these diodes require volts of reverse bias to generate significant capacitance shift, making direct 1.5V powered operation impossible.

Figure 32.20 improves the temperature stability of a 1.5V powered crystal oscillator by a factor of 20. It does this by slightly tuning the crystal's resonance as ambient temperature varies. Q1 and associated components form a 1MHz Colpitts oscillator which normally has a temperature coefficient of about 1ppm/°C. The remainder of the circuit implements the temperature correction. The LM134 senses ambient temperature, converting it to a current which flows through the 30.1k resistor. This resistor's voltage is subtracted from a reference potential by A1. The stable subtraction voltage is derived from the LT1073's 212mV reference via Q2 and the 73.2k to 27.4k resistors. Feedback from Q2's collector to the LT1073's auxiliary amplifier closes the reference loop, which also powers the Colpitts oscillator. The 47μF capacitor frequency compensates the loop.

A1's output controls the remaining portion of the LT1073, which is configured as a voltage step-up switching regulator. L1's high voltage inductive events are rectified and stored in the 47μF output capacitor, resulting in a stepped-up DC potential. This potential is fed back to A1, closing a control loop. Because A1 is biased by the temperature sensitive LM134, the loop's output varies with ambient temperature in a controlled manner. Q3's drop forces the step-up converter to always run, regardless of the loop's required output voltage. This permits smooth and continuous varactor bias from 0V to 3.9V over a 0°C to 70°C ambient operating environment. This output is applied to the varactor diode in the oscillator circuit. The varactor's capacitance, a function of its DC bias, thus varies with ambient temperature. This change in capacitance shifts the crystal's resonant frequency, opposing

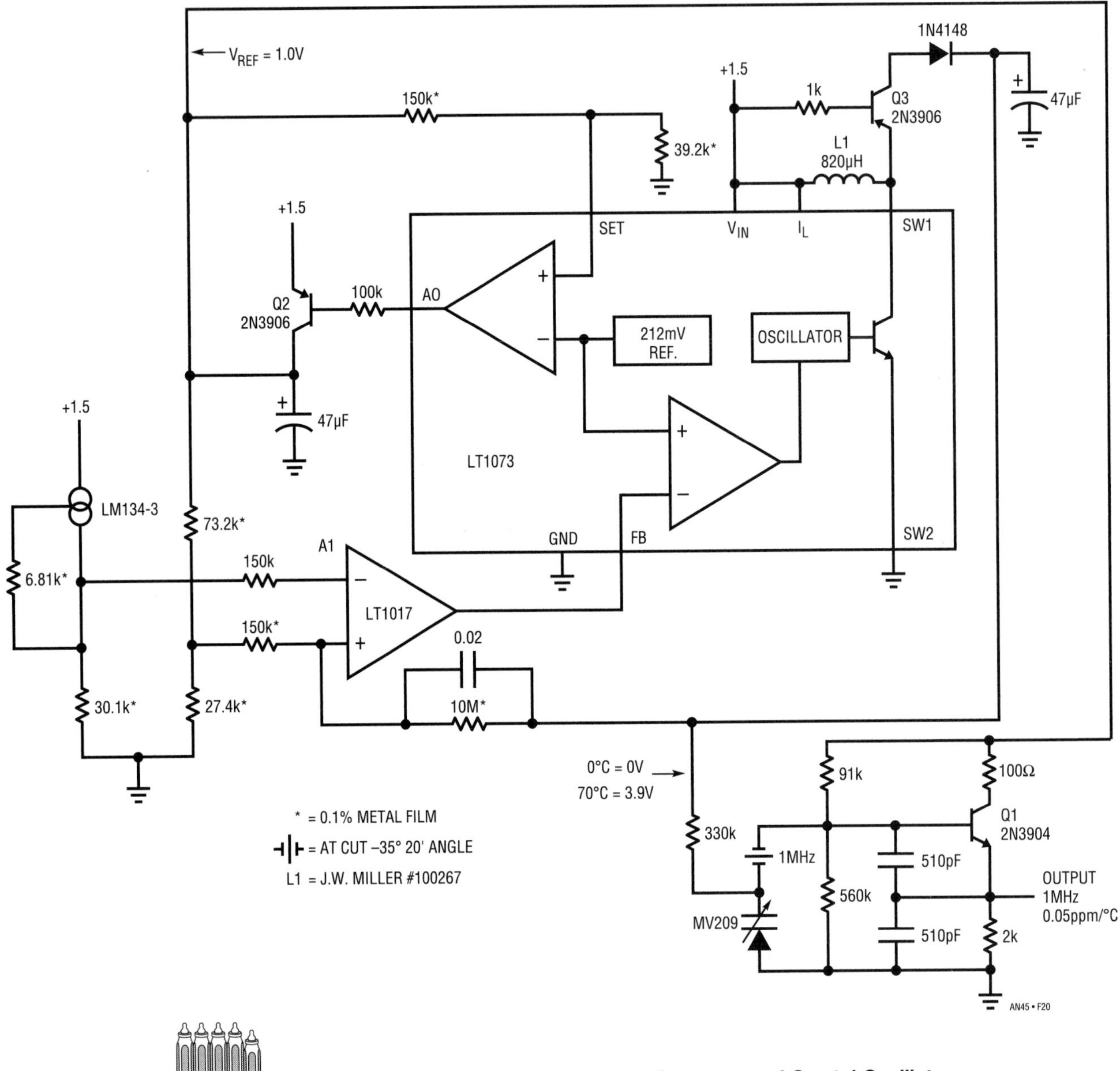

Figure 32.20 • 1.5V Powered Temperature Compensated Crystal Oscillator

temperature induced crystal drift. For the values given in the circuit and the crystal cut specified, residual oscillator drift is only 0.05ppm/°C. This compares favorably with 1ppm/°C drift with no compensation used. The circuit functions from 1.7V down to 1.1V with no specification degradation. Current drain is only 230μA. Applications include portable high accuracy clocks, survival radios, and secure communications.

90μA Precision voltage-to-frequency converter

Figure 32.21 is a micropower voltage-to-frequency converter. A 0V to 5V input produces a 0kHz to 10kHz output with a linearity of 0.05%. Gain drift is 80ppm/°C. Maximum current consumption is only 90μA, almost 30 times lower than currently available V-F converters. To understand circuit operation, assume C1's positive input is slightly below its negative input (C2's output is low). The input voltage causes a positive going ramp at C1's positive input (trace A, Figure 32.22). C1's output is low, biasing

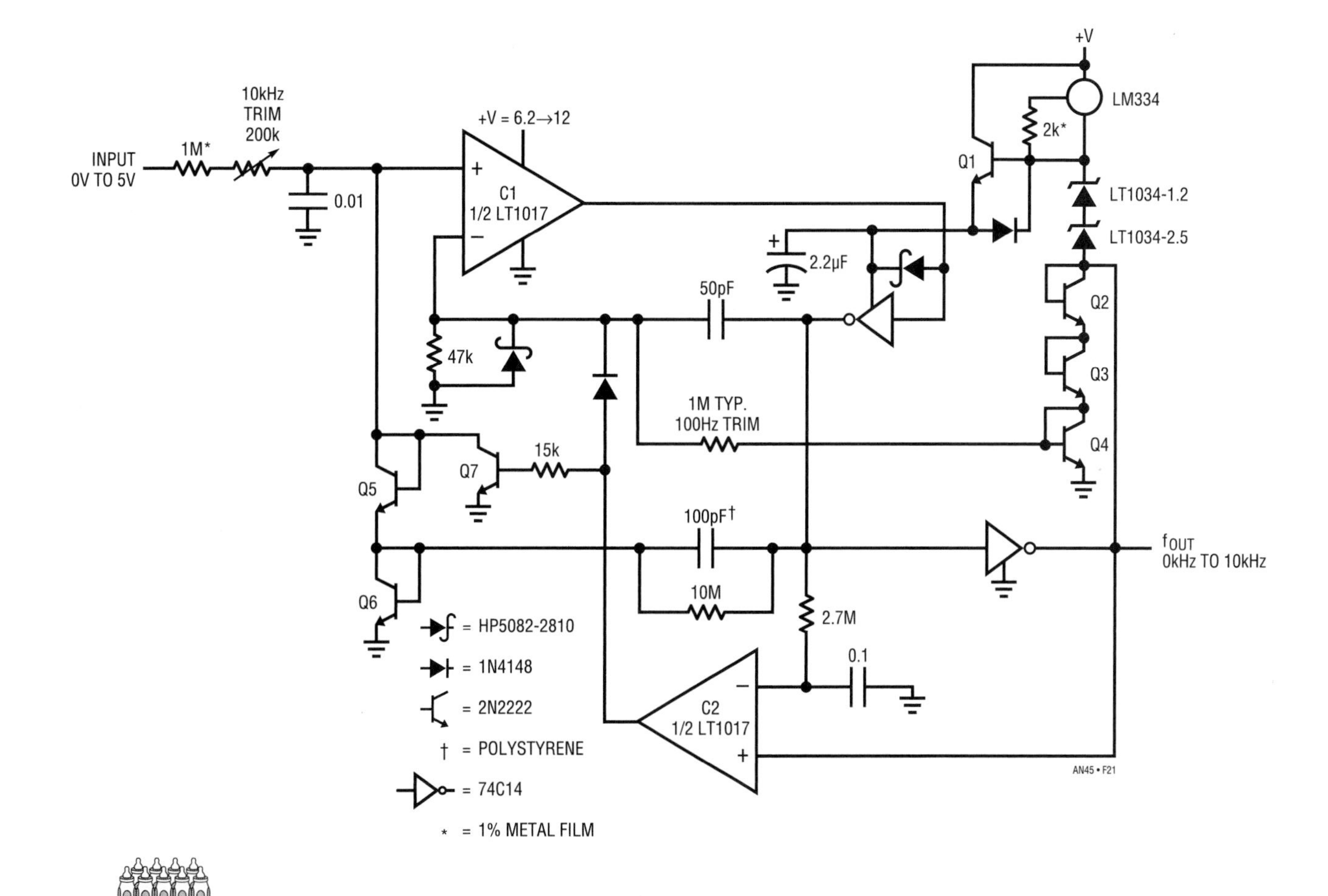

Figure 32.21 • V-to-F Converter Achieves 0.05% Linearity While Requiring Only 90µA Supply Current

the CMOS inverter output high. This allows current to flow from Q1's emitter, through the inverter supply pin to the 100pF capacitor. The 2.2µF capacitor provides high frequency bypass, maintaining low impedance at Q1's emitter. Diode connected Q6 provides a path to ground. The 100pF unit charges to a voltage that is a function of Q1's emitter potential and Q6's drop. When the ramp at C1's positive input goes high enough, C1's output goes high (trace B) and the inverter switches low (trace C). The Schottky clamp prevents CMOS inverter input overdrive. This action pulls current from C1's positive input capacitor via the Q5-100pF route (trace D). This current removal resets C1's positive input ramp to a potential slightly below ground, forcing C1's output to go low. The 50pF capacitor furnishes AC positive feedback, ensuring that C1's output remains positive long enough for a complete discharge of the 100pF capacitor. The Schottky diode prevents C1's input from being driven outside its negative common mode limit. When the 50pF unit's feedback decays, C1 again switches low and the entire cycle repeats. The oscillation frequency depends directly on the input voltage derived current.

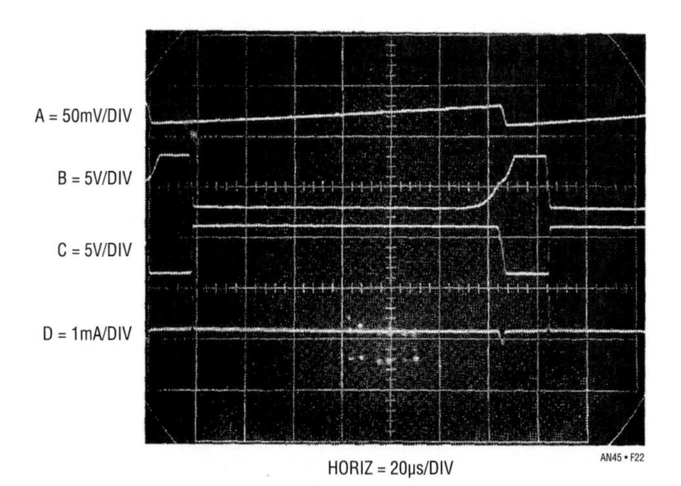

Figure 32.22 • Micropower V-to-F Converter's Waveforms

Q1's emitter voltage must be carefully controlled to get low drive. Q3 and Q4 temperature compensate Q5 and Q6 while Q2 compensates Q1's V_{BE}. The two LT1034s are the actual voltage reference and the LM334 current source provides 35μA bias to the stack. The current drive provides excellent supply immunity (better than 40ppm/V) and also aids circuit temperature coefficient. It does this by utilizing the LM334's 0.3%/°C temperature coefficient to slightly temperature modulate the voltage drop in the Q2-Q4 trio. This correction's sign and magnitude directly oppose that of the −120ppm/°C, 100pF polystyrene capacitor aiding overall circuit stability.

The Q1 emitter-follower efficiently delivers charge to the 100pF capacitor. Both base and collector current end up in the capacitor. The CMOS inverter provides low loss SPDT reference switching without significant drive losses. The 100pF capacitor draws only small transient currents during its charge and discharge cycles. The 50pF-47k positive feedback combination draws insignificantly small switching currents. Figure 32.23, a plot of supply current versus operating frequency, reflects the low power design. At zero frequency, the LT1017's quiescent current and the 35μA reference stack bias account for all current drain. There are no other paths for loss. As frequency scales up, the charge/discharge cycle of the 100pF capacitor introduces the 1.5μA/kHz increase shown.

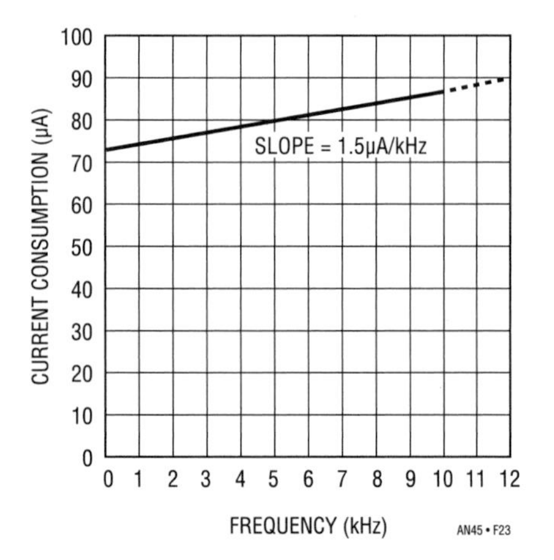

Figure 32.23 • Current Consumption vs Frequency for the V-to-F Converter

Circuit start-up or overdrive can cause the circuit's AC-coupled feedback to latch. If this occurs, C1's output goes high. C2, detecting this via the inverter and the 2.7M-0.1μF lag, also goes high. This lifts C1's negative input and grounds the positive input with Q7, initiating normal circuit action.

Because the charge pump is directly coupled to C1's output, response is fast. The output settles within one cycle for a fast input step. To calibrate this circuit, apply 50mV and select the value at C1's input for a 100Hz output. Then, apply 5V and trim the input potentiometer for a 10kHz output.

Bipolar (AC) input V-F converter

No currently available V-F converter will accept bipolar (AC) inputs. This feature is desirable in power line monitoring and other applications. Figure 32.24's V-F converter accepts ±10V inputs, producing a 0kHz to 10kHz output. Linearity is 0.04%, and temperature coefficient measures about 50ppm/°C. To understand circuit operation, assume a bipolar square wave (trace A, Figure 32.25) is applied to the input. During the input's positive phase, A1's output (trace B) swings negative, driving current through C1 via the full wave diode bridge. A1's current causes C1 to ramp linearly. Instrumentation amplifier A2, operating at a gain of 10, looks differentially across C1. A2's output (trace C) biases comparator A3's negative input. When A2's output crosses zero, A3 fires (trace D). AC positive feedback to A3's positive input (trace E) "hangs up" A3's output for about 20μs. The Q1 level shifter drives ground referred inverters I_1 and I_2 to deliver biphase drive (traces G and H) to the LTC201 switch. The LTC201, set up as a charge pump, places C2 across C1 each time the inverters switch, resetting C1 to a lower voltage. The LT1004 reference, along with C2's value, determines how much charge is removed from C1 each time the charge pump cycles. Thus, each time A2's output tries to cross zero, C2 is switched across C1, resetting it to a small negative voltage and forcing A1 to begin recharging it. The frequency of this oscillatory behavior is directly proportional to the input derived current into A1. During the time C1 is ramping toward zero the LTC201 switches C2 across the LT1004, preparing it for the next discharge cycle. The action is the same for negative input excursions (see Figure 32.25), except that A1's output phasing is reversed. A2, looking differentially across A1's diode bridge, sees the same signal as for positive inputs and circuit action is identical. A4, detecting A1's output polarity, provides a sign bit output (trace F).

Figure 32.26, an amplitude expanded version of A1 and A2's outputs, shows detail. Trace A is the input, while trace B and trace C are A1 and A2's outputs, respectively. Complementary bias points and ramping action are clearly visible in A1's output, while A2 responds identically for

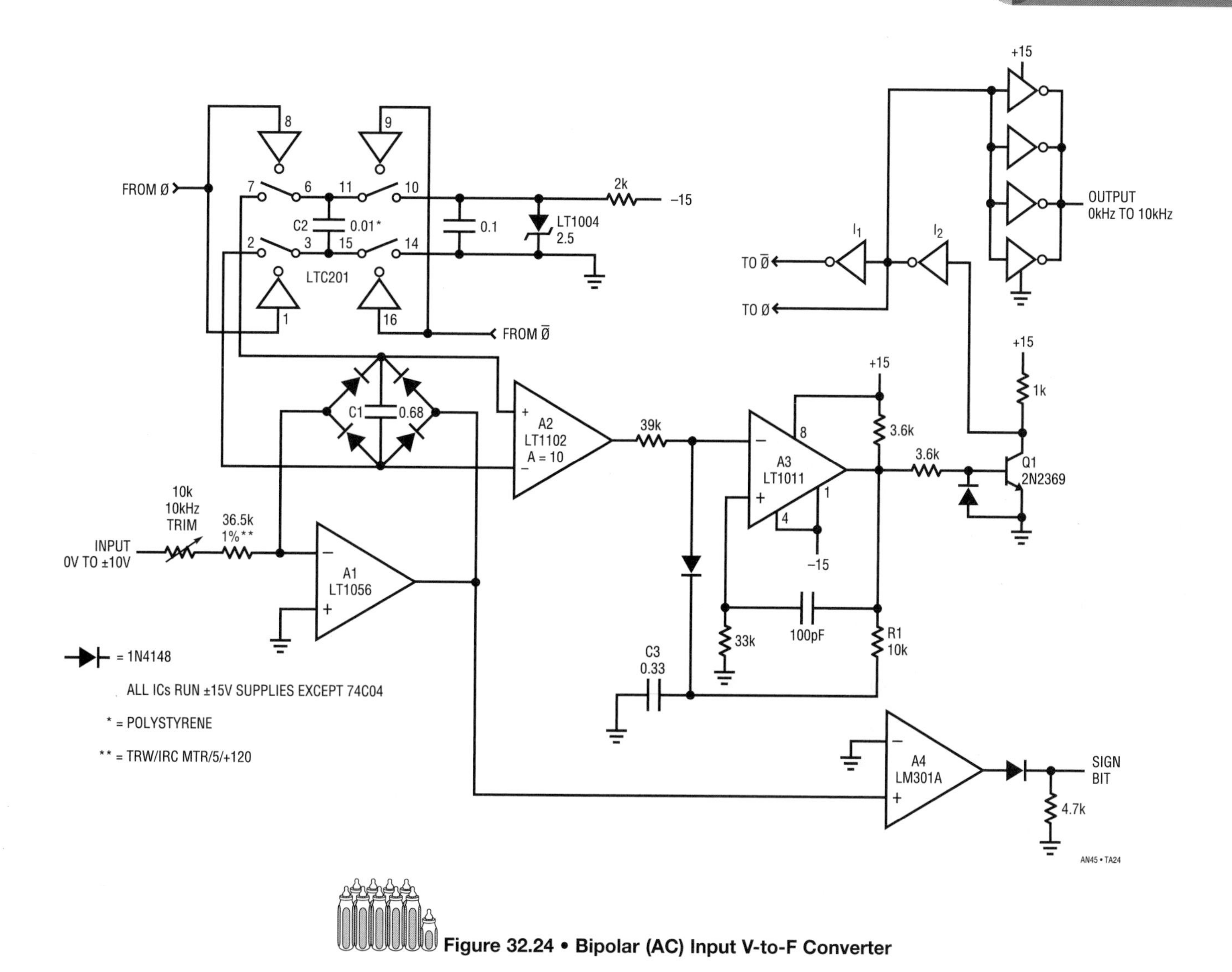

Figure 32.24 • Bipolar (AC) Input V-to-F Converter

A = 20V/DIV
B = 1V/DIV
C = 0.5V/DIV
D = 100V/DIV
E = 50V/DIV
F = 50V/DIV
G = 50V/DIV
H = 50V/DIV
HORIZ = 500µs/DIV
AN45 • F25

Figure 32.25 • Waveforms for the Bipolar Input V-to-F Converter

both input phases. A1's output bias points are established by the two conducting bridge diodes. When the input switches polarity, A1 responds immediately and oscillation frequency settles within 1 to 2 cycles of final value.

Start-up or overdrive conditions could cause this loop to latch. A start-up mechanism, adapted from oscilloscope trigger circuitry, precludes latch-up.[2] If C1 charges past the point where C2 can reset it, loop closure ceases. A2's output saturates positive, causing A3 to go negative. A3's prolonged negative state, detected by the R1-C3 filter pulls its negative input toward −15V. When A3's negative input crosses zero, its output changes state and charges R1-C3 positively. A3's input rises above zero, causing output reversal and free-running oscillation commences.

Note 2: See References 5 and 6.

As in normal mode, the 100pF-33k RC aids transitions. A3's oscillations are transmitted to the LTC201 based charge pump via A1 and the inverters. C2 pumps charge out of C1, driving the voltage across it toward zero. A2 comes out of positive saturation and heads negative, eliminating positive bias at A3's input. A3's free-running oscillation stops, and normal loop action begins.

To calibrate this circuit apply either a -10V or a +10V input and set the 10kΩ trimmer for exactly 10kHz output.

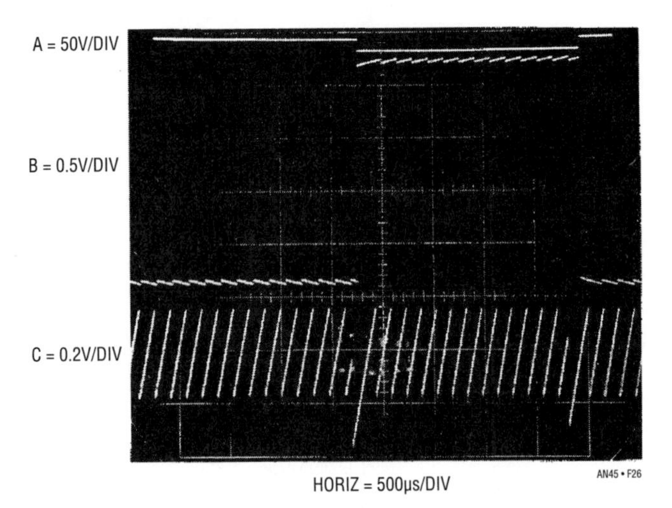

Figure 32.26 • Detail of Integrator and Differential Amplifier Outputs

The low offsets of A1 and A2 permit operation down to a few hertz with no zero trim required.

1.5V Powered, 350ps rise time pulse generator

Verifying the rise time limit of wideband test equipment setups is a difficult task. In particular; the "end-to-end" rise time of oscilloscope-probe combinations is often required to assure measurement integrity. Conceptually, a pulse generator with rise times substantially faster than the oscilloscope-probe combination can provide this information. Figure 32.27's circuit does this, providing a 1ns pulse with rise and fall times inside 350ps. Pulse amplitude is 10V with a 50Ω source impedance. This circuit, built into a small box and powered by a 1.5V battery, provides a simple, convenient way to verify the rise time capability of almost any oscilloscope-probe combination.

The LT1073 switching regulator and associated components supply the necessary high voltage. The LT1073 forms a flyback voltage boost regulator. Further voltage step-up is obtained from a diode-capacitor voltage doubler network. L1 periodically receives charge, and its flyback discharge delivers high voltage events to the doubler network. A portion of the doubler network's DC output is fed back to the LT1073 via the R1, R2 divider, closing a control loop.

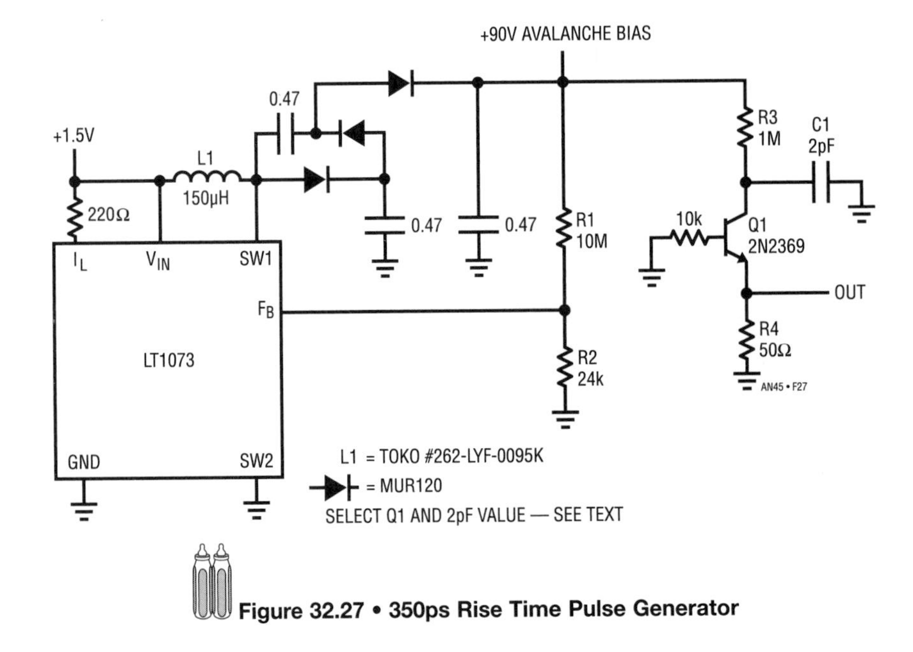

Figure 32.27 • 350ps Rise Time Pulse Generator

The regulator's 90V output is applied to Q1 via the R3-C1 combination. Q1, a 40V breakdown device, non-destructively avalanches when C1 charges high enough.[3] The result is a quickly rising, very fast pulse across R4. C1 discharges, Q1's collector voltage falls and breakdown ceases. C1 then recharges until breakdown again occurs. This action causes free-running oscillation at about 200kHz. Figure 32.28 shows the output pulse. A 1GHz sampling oscilloscope (Tektronix 556 with 1S1 sampling plug-in) measures the pulse at 10V high with about a 1ns base. Rise time is 350ps, with fall time also indicating 350ps. The figures may actually be faster, as the 1S1 is specified with a 350ps rise time limit.[4]

Q1 may require selection to get avalanche behavior. Such behavior; while characteristic of the device specified, is not guaranteed by the manufacturer. A sample of 50 Motorola 2N2369s, spread over a 12 year date code span, yielded 82%. All "good" devices switched in less than 600ps. C1 is selected for a 10V amplitude output. Value spread is typically 2pF to 4pF Ground plane type construction with high speed layout techniques are essential for good results from this circuit. Current drain from the 1.5V battery version is about 5mA.

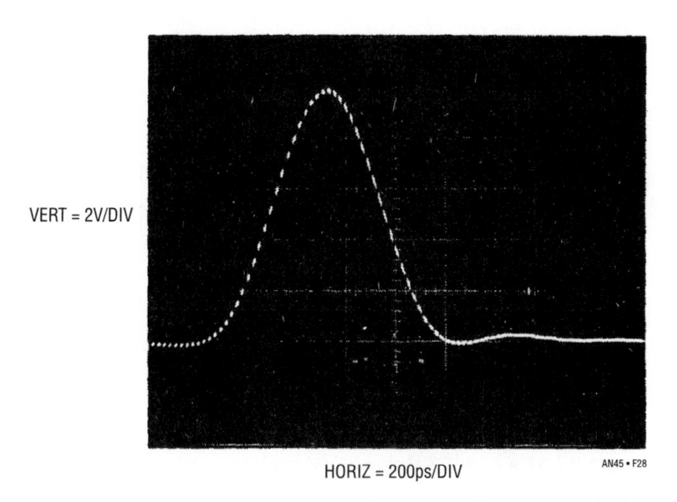

Figure 32.28 • Avalanche Pulse Generator Output Pulse. Waveform Has 350ps Rise and Fall Times. Slightly Under Damped Turn-Off Is Probably Due to Test Fixture Limitations

Figure 32.29 • Alternate 90V DC-DC Converter

For those applications which must run from higher voltage inputs, Figure 32.29 is included. This circuit, which operates from inputs of 4V to 20V will also power the avalanche stage. Cascoded high voltage transistor Q1 combines with the LT1072 switching regulator to form a high voltage switched mode control loop. The LT1072 pulse width modulates Q1 at its 40kHz clock rate. L1's inductive events are rectified and stored in the 2μF output capacitor. The 1MΩ to 12kΩ divider provides feedback to the LT1072. The diode and RC at Q1's base damp inductor related parasitic behavior. The circuit's output drives the avalanche stage in similar fashion to the LT1073 based circuit.

A simple ultralow dropout regulator

Switching regulator post regulators, battery-powered apparatus, and other applications frequently require low dropout linear regulators. Often, battery life is significantly affected by the regulator's dropout performance. Figure 32.30 simple circuit offers lower dropout voltage than any monolithic regulator. Dropout is below 50mV at 1A, increasing to only 450mA at 5A. Line and load regulation

Note 3: See References 7.
Note 4: I'm sorry, but 1GHz is the fastest scope in my house.

are within 5mV and initial output accuracy is inside 1%. Additionally, the regulator is fully short-circuit protected, and has a no load quiescent current of 600μA.

Circuit operation is straightforward. The 3-pin LT1123 regulator (TO-92 package) servo controls Q1's base to maintain its feedback pin (FB) at 5V The 10μF output capacitor provides frequency compensation. If the circuit is located more than six inches from the input source, the optional 10μF capacitor should bypass the input. The optional 20Ω resistor limits LT1123 power dissipation and is selected based upon the maximum expected input voltage (see Figure 32.31).

Normally, configurations of this type offer unpredictable short-circuit protection. Here, the MJE1123 transistor shown has been specially designed for use with the LT1123. Because of this, beta based current limiting is practical. Excessive output current causes the LT1123 to pull down harder on Q1 until beta limiting occurs. Under these conditions the controlled pull-down current combines with Q1's beta and safe operating area characteristics to provide reliable short-circuit limiting. Figure 32.32 details current limit characteristics for 30 randomly selected transistors.

Figure 32.33 shows dropout characteristics. Even at 5A, dropout is about 450mV decreasing to only 50mV at 1A. Monolithic regulators cannot approach these figures, primarily because monolithic power transistors do not offer Q1's combination of high beta and excellent saturation.

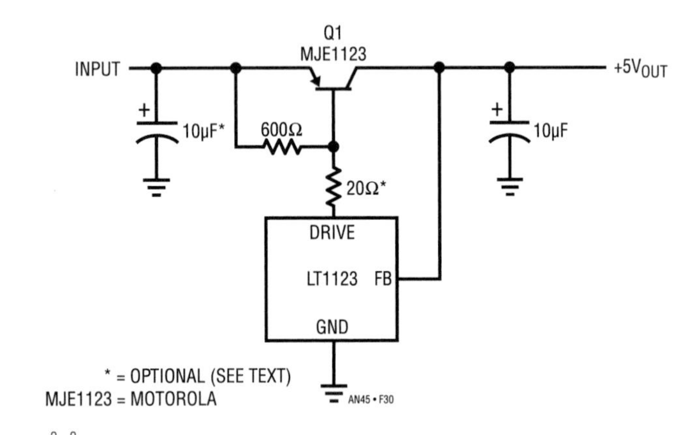

Figure 32.30 • The Ultralow Dropout Regulator. LT1123 Combines with Specially Designed Transistor for Low Dropout and Short-Circuit Protection

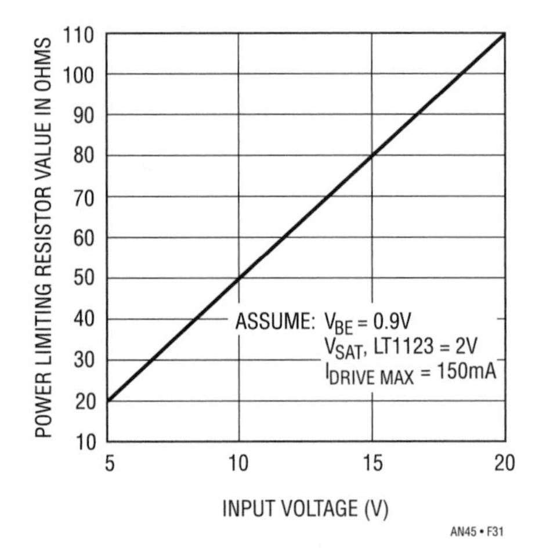

Figure 32.31 • LT1123 Power Dissipation Limiting Resistor Value vs Input Voltage

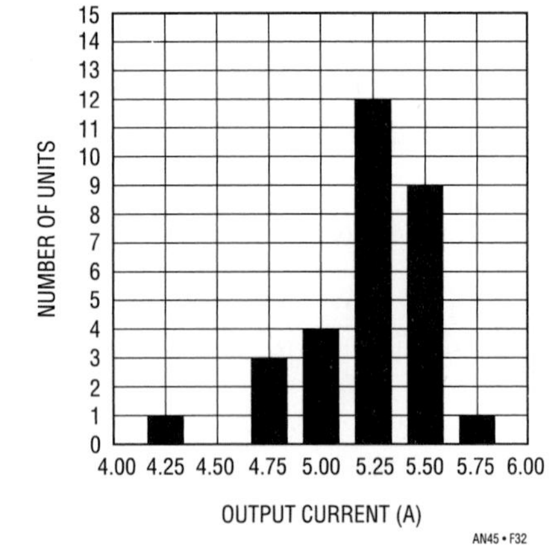

Figure 32.32 • Short Circuit Current for 30 Randomly Selected MJE1123 Transistors at $V_{IN} = 7V$

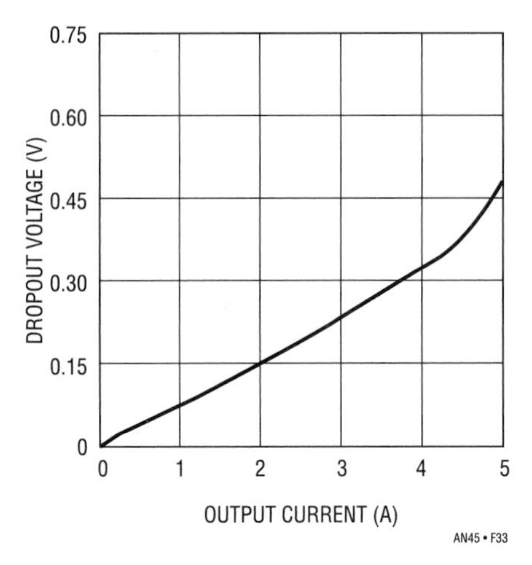

Figure 32.33 • Dropout Voltage vs Output Current

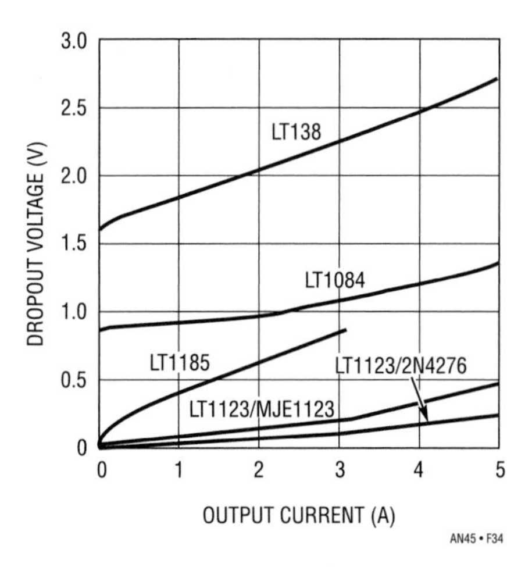

Figure 32.34 • Dropout Voltage vs Output Current for Various Regulators

For comparison, Figure 32.34 compares the circuit's performance against some popular monolithic regulators. Dropout is 10 times better than 138 types, and significantly better than the other types shown. Because of Q1's high beta, base drive loss is only 1% to 2% of output current, even at full 5A output. This maintains high efficiency under the low $V_{IN} - V_{OUT}$ conditions the circuit will typically operate at. As an exercise, the MJE1123 was replaced with a 2N4276, a Germanium device. This combination provided even lower dropout performance, although current limit characteristics cannot be guaranteed.

Figure 32.35 shows a simple way to add shutdown to the regulator. A CMOS inverter or gate biases Q2 to control LT1123 bias. When Q2's base is driven, the loop functions normally. With Q2 unbiased, the circuit goes into shutdown and pulls no current.

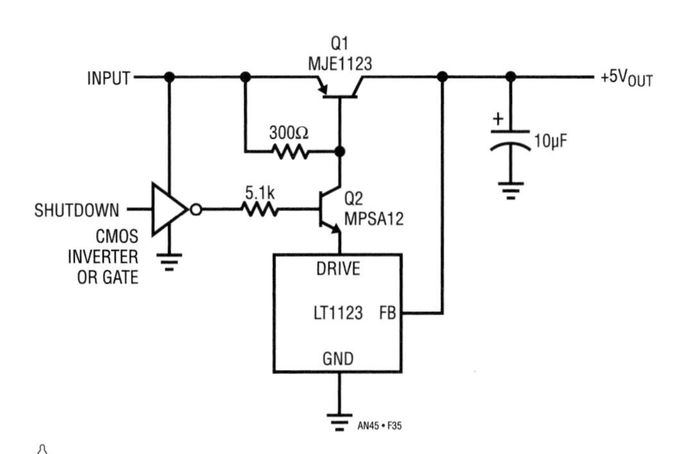

Figure 32.35 • Shutdown for the Low Dropout Regulator

Cold cathode fluorescent lamp power supply

Current generation portable computers utilize back-lit LCD displays. Cold Cathode Fluorescent Lamps (CCFL) provide the highest available efficiency for back lighting the display. These lamps require high voltage AC to operate, mandating an efficient, high voltage DC/AC converter. In addition to good efficiency, the converter should deliver the lamp drive in the form of a sine wave. This is desirable to minimize RF emissions. Such emissions can cause interference with other devices, as well as degrading overall operating efficiency.

Figure 32.36 meets these requirements. Efficiency is 78%, with an input voltage range of 4.5V to 20V. 82% efficiency is possible if the LT1072 is driven from a low

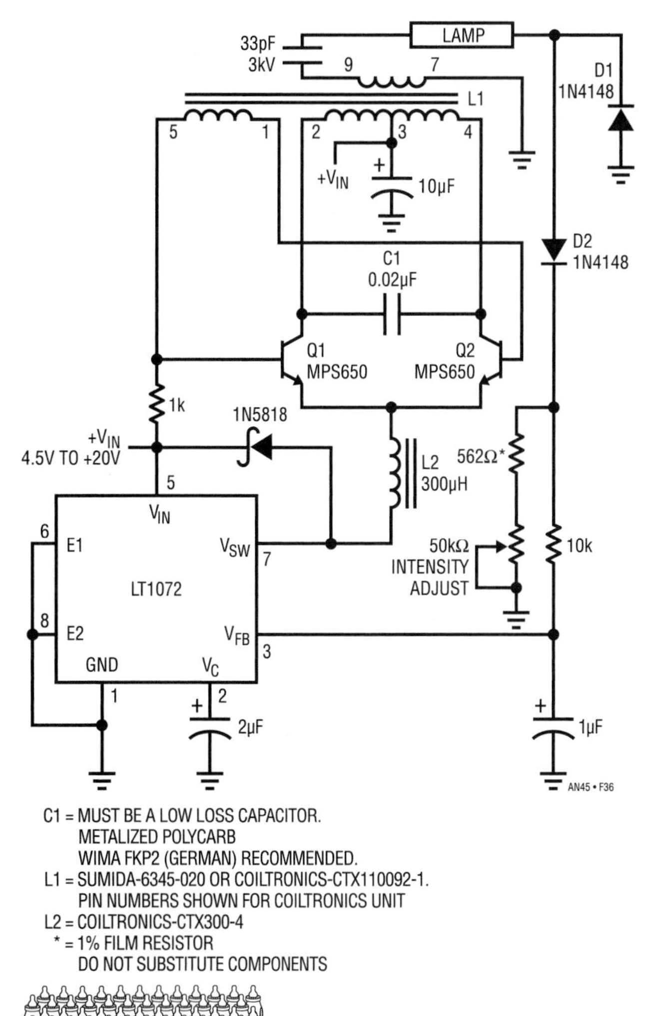

Figure 32.36 • Cold Cathode Fluorescent Lamp Power Supply

voltage (e.g., 3V to 5V) source. Additionally, lamp intensity is continuously and smoothly variable from zero to full intensity.

When power is applied the LT1072 switching regulator's feedback pin is below the devices internal 1.23V reference, causing full duty cycle modulation at the V_{SW} pin (trace A, Figure 32.37). L2 conducts current (trace B), which flows from L1's center tap, through the transistors, into L2. L2's current is deposited in switched fashion to ground by the regulator's action.

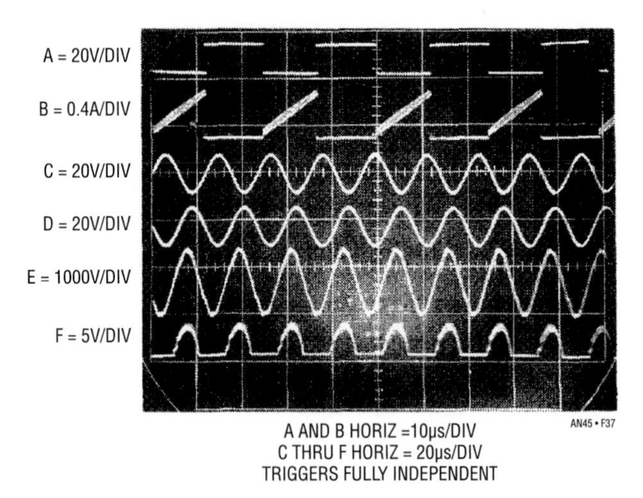

Figure 32.37 • Waveforms for the Cold Cathode Fluorescent Lamp Power Supply. Note Independent Triggering on Traces A and B and C Through F.

L1 and the transistors comprise a current driven Royer class converter[5] which oscillates at a frequency primarily set by L1's characteristics and the 0.02µF capacitor. LT1072 driven L2 sets the magnitude of the Q1-Q2 tail current, and hence L1's drive level. The 1N5818 diode maintains L2's current flow when the LT1072 is off. The LT1072's 40kHz clock rate is asynchronous from the Royer converters (≈60kHz) rate, accounting for trace B's waveform thickening.

The 0.02µF capacitor combines with L1's characteristics to produce sine wave voltage drive at the Q1 and Q2 collectors (traces C and D, respectively). L1 furnishes voltage step-up, and about $1400V_{P\text{-}P}$ appears at its secondary (trace E). Current flows through the 33pF capacitor into the lamp. On negative waveform cycles the lamp's current is steered to ground via D1. Positive waveform cycles are directed, via D2, to the ground referred 562Ω-50k potentiometer chain. The positive half-sine appearing across these resistors (trace F) represents 1/2 the lamp current. This signal is filtered by the 10k-1µF pair and presented to the LT1072's feedback pin. This connection closes a control loop which regulates lamp current. The 2µF capacitor at the LT1072's V_C pin provides stable loop compensation. The loop forces the LT1072 to switch-mode modulate L2's average current to whatever value is required to maintain a constant current in the lamp. The constant current's value, and hence lamp intensity, may be varied with the potentiometer. The constant current drive allows full 0% to 100% intensity control with no lamp dead zones or "pop-on" at low intensities. Additionally, lamp life is enhanced because current cannot increase as the lamp ages.

Several points should be kept in mind when observing this circuit's operation. L1's high voltage secondary can only be monitored with a wideband, high voltage probe fully specified for this type of measurement. *The vast majority of oscilloscope probes will break down and fail if used for this measurement.*[6] Tektronix probe type P-6009 (acceptable) or types P6013A and P6015 (preferred) probes must be used to read L1's output.

Another consideration involves observing waveforms. The LT1072's switching frequency is completely asynchronous from the Q1-Q2 Royer converter's switching. As such, most oscilloscopes cannot simultaneously trigger and display all the circuit's waveforms. Figure 32.37 was obtained using a dual beam oscilloscope (Tektronix 556). LT1072 related traces A and B are triggered on one beam, while the remaining traces are triggered on the other beam. Single beam instruments with alternate sweep and trigger switching (e.g., Tektronix 547) can also be used, but are less versatile and restricted to four traces.

Note 5: See References 8.
Note 6: Don't say we didn't warn you!

References

1. Verster, TC., "P-N Junction as an Ultralinear Calculable Thermometer," Electronic Letters, Vol. 4, pg. 175, May, 1968
2. Verster, TC., "The Silicon Transistor as a Temperature Sensor," International Symposium on Temperature, 1971, Washington, D.C
3. Type 7D13 Plug-In Operating and Service Manual, Tektronix, Inc., 1971
4. Sheingold, D.H., "Nonlinear Circuits Handbook," Chapter 3–1, "Basic Considerations," pgs. 165–166, Analog Devices, Inc., 1974
5. Oscilloscope Trigger Circuits, "Automatic Trigger," pgs. 39–49, Tektronix Concept Series, 1969
6. Type 547 Oscilloscope Operating and Service Manual, "Automatic Stability Circuit," pgs. 3–8, Tektronix, Inc., 1964
7. Type 111 Pretrigger Pulse Generator Operating and Service Manual, Tektronix, Inc., 1960
8. Bright, Pittman and Royer, "Transistors As On-Off Switches in Saturable Core Circuits," Electrical Manufacturing, December, 1954
9. Morrison, John C., MD, Editor. "Antepartal Fetal Surveillance," Obstetrics and Gynecology Clinics of North America, Volume 17:1, March, 1990, W.B. Saunders Co
10. Atkinson, P. Woodcock, J.P., "Doppler Ultrasound," London, Academic Press, 1982
11. Doppler, J.C., "Uber das farbigte Licht der Dopplersterne und einigr anderer Gestirne des Himmels," Abhandl d Konigl Bomischen Gesellschaft der Wissenschaften 2:466, 1843
12. FitzGerald, D.E., Drumm, J.E., "Noninvasive measurement of fetal circulation using ultrasound: A new method," Br Med J 2:1450, 1977
13. Hata, T., Aoki, S., Hata, K., et al, "Intracardiac blood flow velocity waveforms in normal fetuses in utero," Am J Cardiol 58:464, 1987
14. Pourcelot, L., "Applications clinique de l'examen Doppler transcutane," *In* Pourcelot, L. (ed), Velocimetric Ultrasonore Doppler, INSERM 34:213, 1974
15. Shung, K.K., Sigelman, R.A., Reid, J.M., "Scattering of ultrasound by blood," IEEE Trans Biomed Eng BME-23:460, 1976
16. Stuart, B., Drumm, J., FitzGerald, D.E., et al, "Fetal blood velocity waveforms in normal pregnancy," Br J Obstet Gynaecol 87:780, 1980
17. Stabile, I., Bilardo, C., Panella, M., et al, "Doppler measurement of uterine blood flow in the first trimester of normal and complicated pregnancies," Trophoblast Res 3:301, 1988

Circuit collection, volume I

33

Richard Markell Editor

Introduction

Over the past several years *Linear Technology*, the magazine, has come of age. From nothing, the publication has come into its own, as has its subscriber list. Many innovative circuits have seen the light of day in the pages of our now hallowed publication.

This Application Note is meant to consolidate the circuits from the first few years of the magazine in one place. Circuits herein range from laser diode driver circuits to data acquisition systems to a 50W high efficiency switcher circuit. Enough said. I'll stand aside and let the authors explain their circuits.

Analog Circuit and System Design: Immersion in the Black Art of Analog Design. http://dx.doi.org/10.1016/B978-0-12-397888-2.00033-X

A-to-D converters

LTC1292: 12-bit data acquisition circuits

by Sammy Lum

Temperature-measurement system

The circuit in Figure 33.1 shows how a transducer output, such as a platinum RTD bridge, can be digitized with one op amp. This circuit is a modification of that found in Application Note 43.[1] The differential input of the LTC1292 removes the common mode voltage. The LT1006 is used for amplification. The resistor tied between the + input of the LT1006 and the +IN input of the LTC1292 is to compensate for the loading of the bridge by resistor R_S. Full scale can be adjusted by the 500kΩ trim pot and offset can be adjusted by the 100Ω trim pot in series with Rs. A lower R_{PLAT} value than that in AN43 is used here to improve dynamic range. The signal voltage on the +IN pin must not exceed V_{REF}. The differential voltage range is V_{REF} minus approximately 100mV. This is enough range to measure 0°C to 400°C with 0.1°C resolution.

Floating, 12-bit data acquisition system

The circuit in Figure 33.2 demonstrates how to float the LTC1292 to make a differential measurement. This circuit will digitize a 5V range from 10V to 15V with 12 bits of resolution. The digital I/O has been level translated. The LT1019-5 is used in shunt mode to create the floating analog ground for the LTC1292. The digital I/O lines make use of 4.3V Zeners to clamp the single-transistor inverters. Opto-isolators can also be used. The floating analog ground should be laid out as a ground plane for the LTC1292. The 47µF bypass capacitor should be tied from the V_{CC} pin to the floating ground plane with minimum lead length and placed as close to the device as possible. Likewise, keep the lead length from the GND pin to the floating ground plane at a minimum (a low-profile socket is acceptable).

Differential temperature measurement system

The circuit in Figure 33.3 digitizes the difference in temperature between two locations. The two LM134s are used as temperature sensors. These are ideally suited for remote applications because they are current output devices. This allows long wires to run from the sensor back to the LTC1292 without any degradation to the signal from the sensor. Resistor R_{SET} sets the current to 1µA/°K. The current is converted to a voltage by the resistor R1 connected from V^- to ground. The reference voltage and resistor were selected to give a change of 0.05°C/LSB. The resolution is given by °C/LSB = V_{REF} / ((4096)(1mA) (R1)). The maximum temperature at each input is 125°C. Note that if the temperature on the +IN pin is

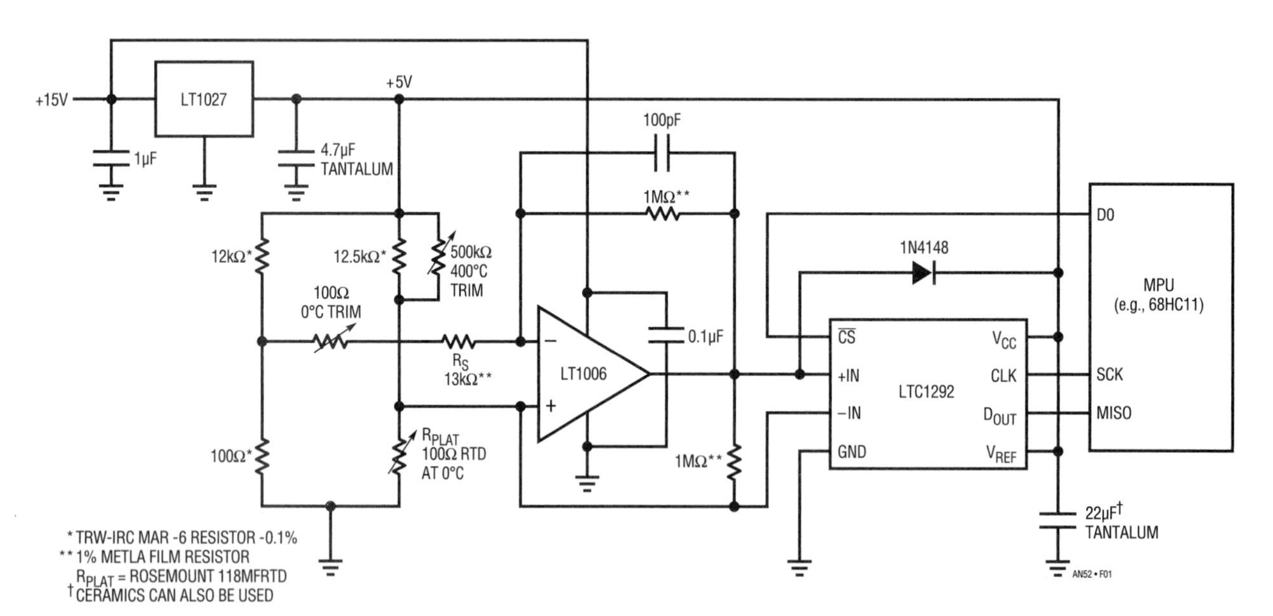

Figure 33.1 • 0° to 400°C Temperature-Measurement System

Note1: Williams, Jim, "Bridge Circuits, Marrying Gain and Balance," Application Note 43, Linear Technology Corp.

less than the temperature on the −IN pin, the output will be zero. Because the LTC1292 is being driven from a high source impedance, you should limit the CLK frequency to 100kHz or less.

The software code for interfacing the LTC1292 to the Motorola MC68HC11 or the Intel 8051 is found in the LTC1292 data sheet. The code needs to be modified for the circuit in Figure 33.2 to account for the inversion introduced by the digital level translators.

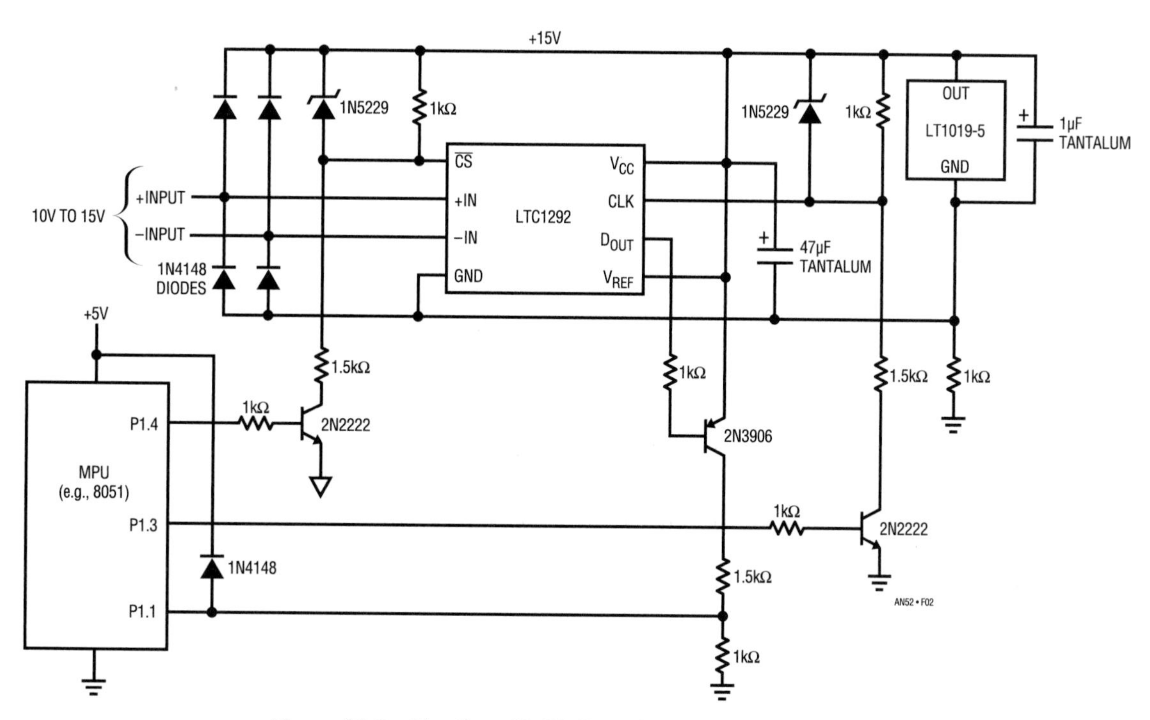

Figure 33.2 • Floating, 12-Bit Data Acquisition System

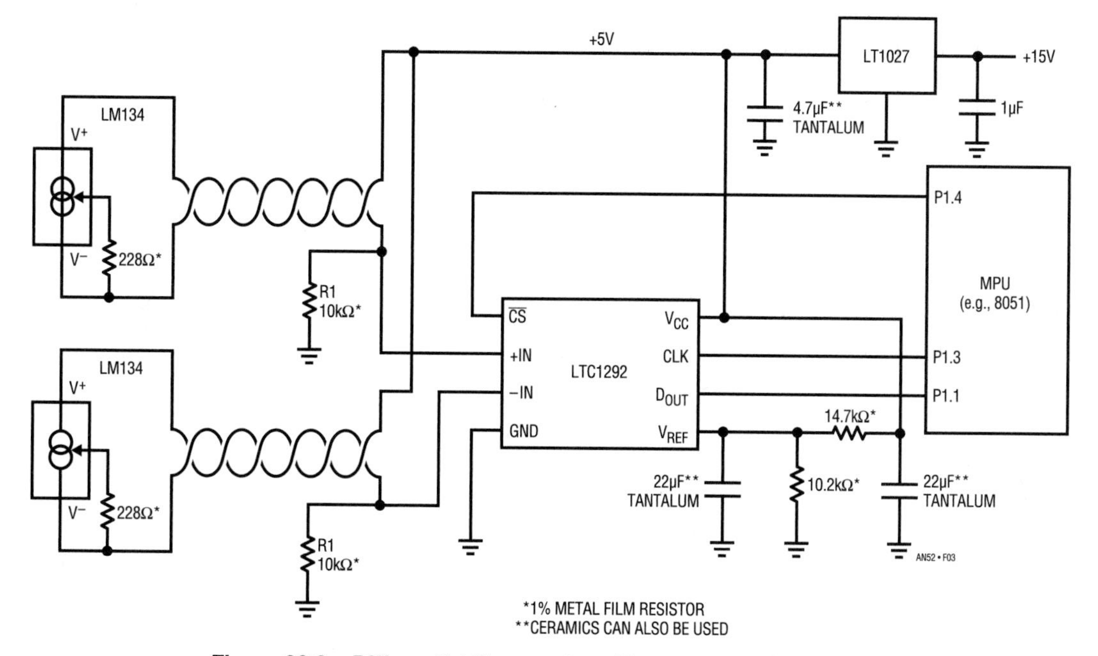

Figure 33.3 • Differential Temperature-Measurement System

Micropower SO8 packaged ADC circuits

by William Rempfer

Floating 8-bit data acquisition system

Figure 33.4 shows a floating system that sends data to a grounded host system. The floating circuitry is isolated by two opto-isolators and powered by a simple capacitor-diode charge pump. The system has very low power requirements because the LTC1096 shuts down between conversions and the opto-isolators draw power only when data is being transferred. The system consumes only 50μA at a sample rate of 10Hz (1ms on-time and 99ms off-time). This is easily within the current supplied by the charge pump running at 5MHz. If a truly isolated system is required, the system's low power simplifies generating an isolated supply or powering the system from a battery.

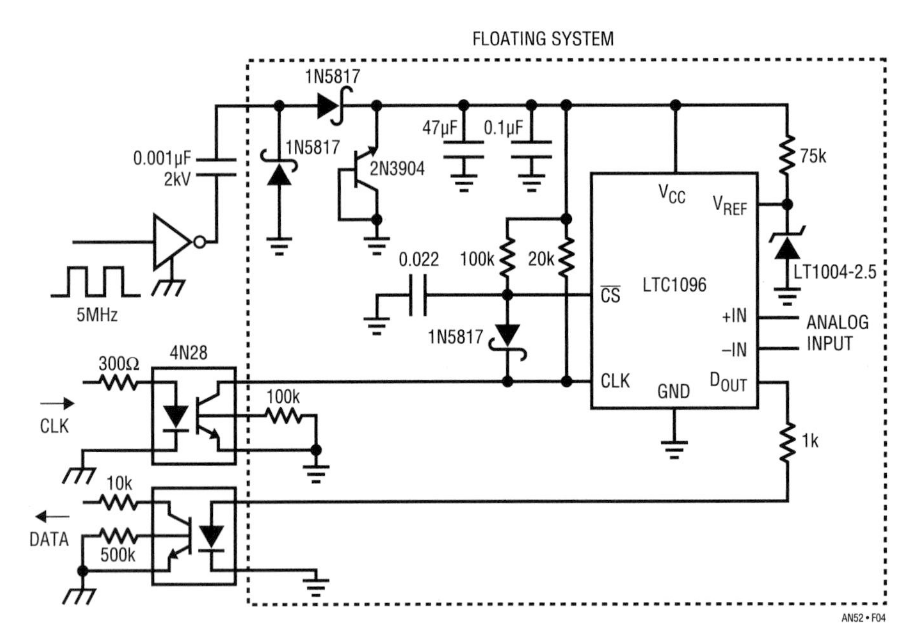

Figure 33.4 • Power for this Floating ADC System Is Provided by a Simple Capacitor-Diode Charge Pump. The Two Opto-Isolators Draw No Current Between Samples, Turning on Only to Send the Clock and Receive Data

0°C–70°C thermometer

Figure 33.5 shows a temperature-measurement system. The LTC1096 is connected directly to the low cost silicon temperature sensor. The voltage applied to the V_{REF} pin adjusts the full scale of the ADC to the output range of the sensor. The zero point of the converter is matched to the zero output voltage of the sensor by the voltage on the LTC1096's negative input.

Operating the ADC directly off batteries can eliminate the space taken by a voltage regulator. Connecting the ADC directly to sensors can eliminate op amps and gain stages. The LTC1096/LTC1098 can operate with small, 0.1μF or 0.01μF chip bypass capacitors.

Figure 33.6 shows the operating sequence of the LTC1096. The converter draws power when the $\overline{CS}$ pin is low and shuts itself down when that pin is high. In systems that convert continuously, the LTC1096/LTC1098 will draw its normal operating power continuously. A 10μs wake up time must be provided to the LTC1096 after each falling $\overline{CS}$.

In systems that have significant time between conversions, lowest power drain will occur with the minimum $\overline{CS}$ low time. Bringing $\overline{CS}$ low, waiting 10μs for the wake up time, transferring data as quickly as possible, and then bringing it back high will result in the lowest current drain.

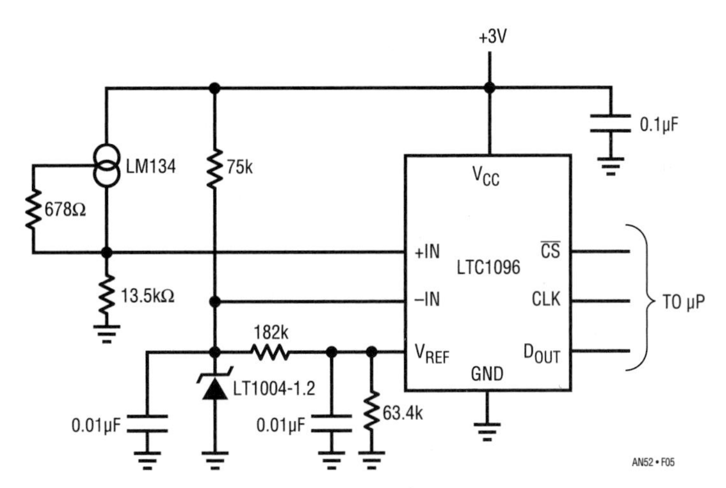

Figure 33.5 • The LTC1096's High-Impedance Input Connects Directly to This Temperature Sensor, Eliminating Signal Conditioning Circuitry in This 0°C–70°C Thermometer

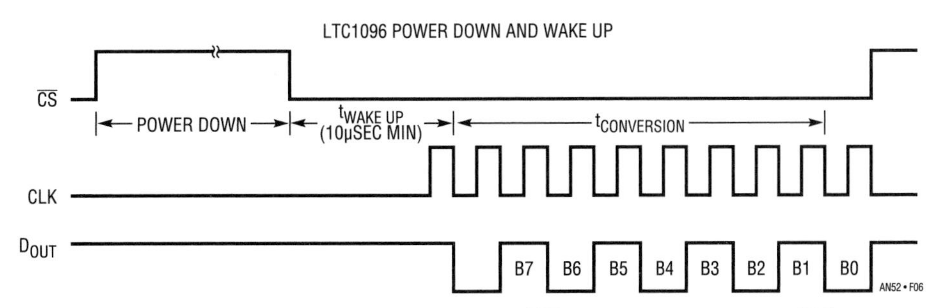

Figure 33.6 • The ADC's Power Consumption Drops to Zero When $\overline{CS}$ Goes High. 10μs After $\overline{CS}$ Goes Low, the ADC Is Ready to Convert. For Minimum Power Consumption Keep $\overline{CS}$ High for as Much Time as Possible Between Conversions

Interface

Low dropout regulator simplifies active SCSI terminators

by Sean Gold

The circuit shown in Figure 33.7 uses an LT1117 low dropout three terminal regulator to control the terminator's local logic supply. The LT1117's line regulation makes the output immune to variations in TERMPWR. After accounting for resistor tolerances and variations in the LT1117's reference voltage, the absolute variation in the 2.85V output is only 4% over temperature. When the regulator drops out at TERMPWR-2.85, or 1.25V the output linearly tracks the input with a 1V/V slope. The regulator provides effective signal termination because the 110Ω series resistor closely matches the transmission line's characteristic impedance, and the regulator provides a good AC ground.

In contrast to a passive terminator, two LT1117s require half as many termination resistors, and operate at 1/15 the quiescent current or 20mA. At these power levels, PC traces provide adequate heat sinking for the LT1117's SOT-223 package. Beyond solving basic signal conditioning problems, this LT1117 terminator handles fault conditions with short circuit current limiting, thermal shutdown, and on-chip ESD protection.

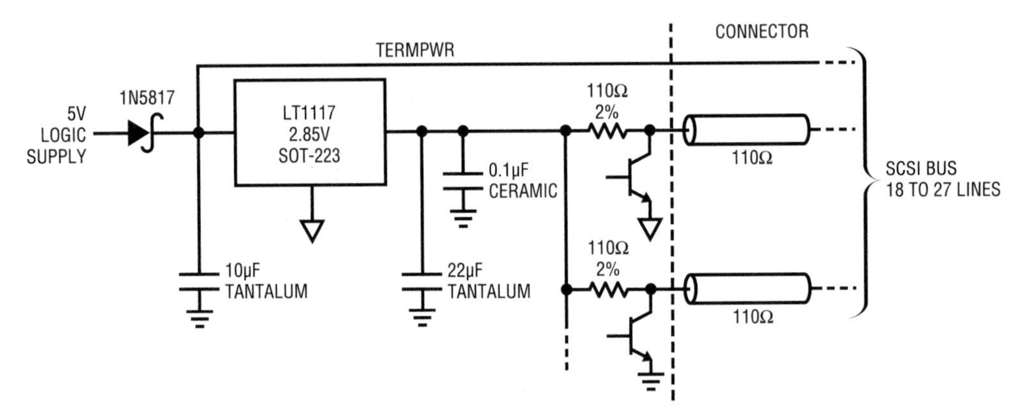

Figure 33.7 • SCSI Active Termination

Power

LT1110 supplies 6 volts at 550mA from 2 AA NiCad cells

by Steve Pietkiewicz

The LT1110 micropower DC-DC converter can provide 5V at 150mA when operating from two AA alkaline cells. The internal switch $V_{CE(SAT)}$ sets this power limit. Even with an external low drop switch, more power is not realistically possible. The internal impedance (typically 200mΩ fresh and 500mΩ at end-of-life) of alkaline AA cells limits peak obtainable battery power: Conversely, nickel-cadmium cells have a constant internal impedance (35mΩ to 50mΩ) for AA size) that increases only when the cell is completely discharged. This allows power to be drawn from the cell at a far greater rate. The circuit in Figure 33.8 uses two AA NiCad cells to supply 6 volts at 550mA. The circuit, developed for pagers with transmit capability, runs at full output current for 15 minutes with two Gates Millennium AA NiCad cells. With a 250mA load, the circuit runs for 36 minutes (see Figure 33.9). Less heat is generated with a reduced load, resulting in the watt-hour difference observable above.

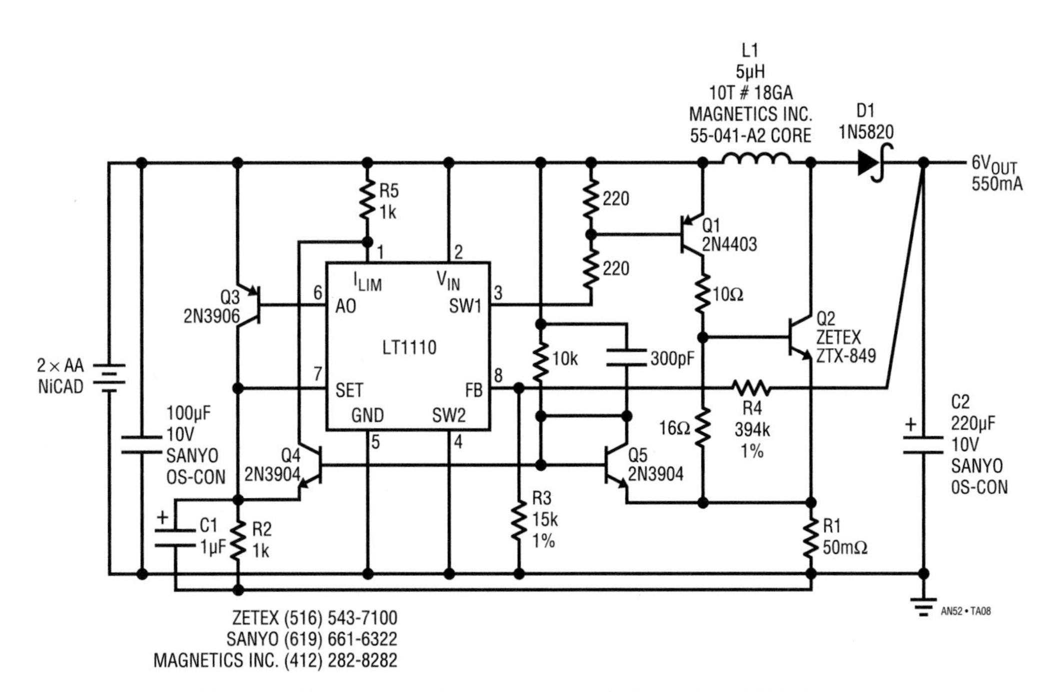

Figure 33.8 • Schematic Diagram, 2 AA NiCad to +6 Volt Converter

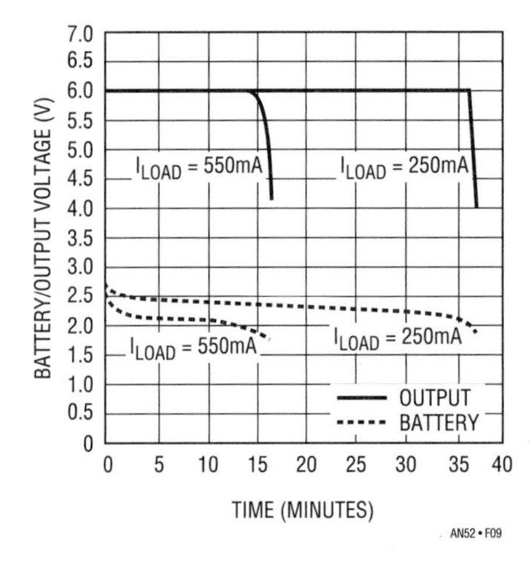

Figure 33.9 • Operating Time at I_{LOAD} = 550mA and 250mA

The circuit uses a micropower LT1110 switching-regulator IC as a controller. The internal switch of the LT1110 furnishes base drive to Q1 through the 220Ω resistors. Q1, in turn, supplies base drive to the power switch Q2. The Zetex ZTX849 NPN device is rated at 5A current and comes in a TO-92 package. For surface-mount fans, the FZT-849, also from Zetex, provides the same performance in an SOT-223 package. The 16Ω resistor provides a turn off path for Q2's stored charge. When Q2 is on, current builds in L1. As Q2 turns off, its collector flies positive until D1 turns on. L1's built-up current discharges through D1 into C2 and the load. The voltage at V_{OUT} is divided by R4 and R3 and fed back into the FB pin of the LT1110, which controls Q2's cycling action. Switch current limit, which is necessary to ensure saturation over supply variations, is implemented by Q3–Q5. Q3, C1, R2, and the auxiliary gain block inside the LT1110, form a 220mV reference point at the LT1110's SET pin. Transistors Q4 and Q5 form a common-base differential amplifier. Q5's emitter monitors the voltage across 50mΩ resistor R1. When the voltage across R1 exceeds 220mV, Q4 turns on hard, pulling current through R5. When the voltage at the I_{LIM} pin of the LT1110 reaches a diode drop below the V_{IN} pin, the internal switch turns off. Thus, maximum switch current is maintained at 220mV/50mΩ, or 4.4A, over input variations and manufacturing spread in the LT1110's on time and frequency.

The circuit's output ripple measures 200mV_{P-P}, and efficiency is 78% at full load with a 2.4V input. Output power can be scaled down for less demanding requirements. To reduce peak current, increase the value of R1. A 100mΩ resistor will limit current to 2.2A. L1 should be increased in value linearly as current is reduced. For a current limit of 2.2A, L1 should be 10μH. Base drive for Q2 can also be reduced by increasing the value of the 10Ω resistor. These lower peak currents are much easier on alkaline cells and will dramatically increase alkaline battery life.

50 watt high efficiency switcher

by Milton Wilcox

The high efficiency 10A step-down (buck) switching regulator shown in Figure 33.10 illustrates how different sized MOSFETs can be driven by the LT1158 without having to worry about shoot-through currents. Since 24V is being dropped down to 5V the duty cycle for the switch (top MOSFET) is only 5/24 or 21%. This means that the bottom MOSFET will dominate the $R_{DS(ON)}$ efficiency losses, because it is turned on nearly four times as long as the top. Therefore a smaller MOSFET is used on the top, and the bottom MOSFET is doubled up, all without having to worry about dead time.

The LT1158 uses an adaptive system that maintains dead time independent of the type, the size, and even the number of MOSFETs being driven. It does this by monitoring the gate turn-off to see that it has fully discharged before allowing the opposite MOSFET to turn on. During turn-on, the hold-off capability of the opposing driver is boosted to prevent transient shoot-through. In this way, cross conduction is completely eliminated as a design constraint.

The non-critical Schottky diode across the bottom MOSFETs reduces reverse-recovery losses. Figure 33.11 shows the operating efficiency for the Figure 33.10 circuit.

Switching regulator applications can take advantage of an important protection feature of the LT1158: remote fault sensing. By sensing the current on the output side of the inductor and returning the LT1158 $\overline{\text{FAULT}}$ pin to the PWM soft-start pin, a true current-mode loop is formed. The Figure 33.10 circuit regulates maximum current in the inductor to 15A with no output voltage overshoot upon recovery from a short circuit.

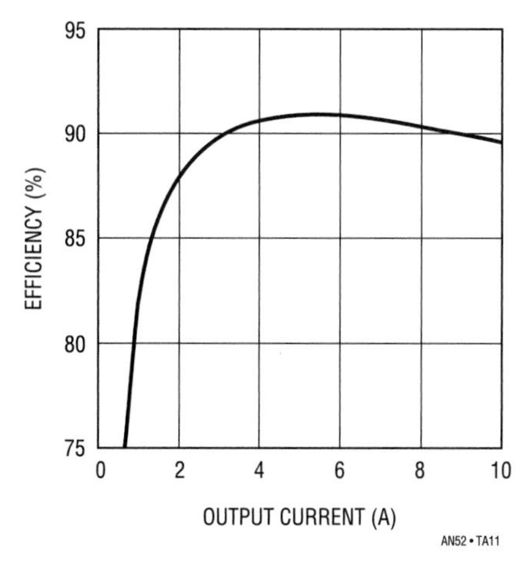

Figure 33.11 • Operating Efficiency for Figure 33.10 Circuit. Current Limit is Set at 15A

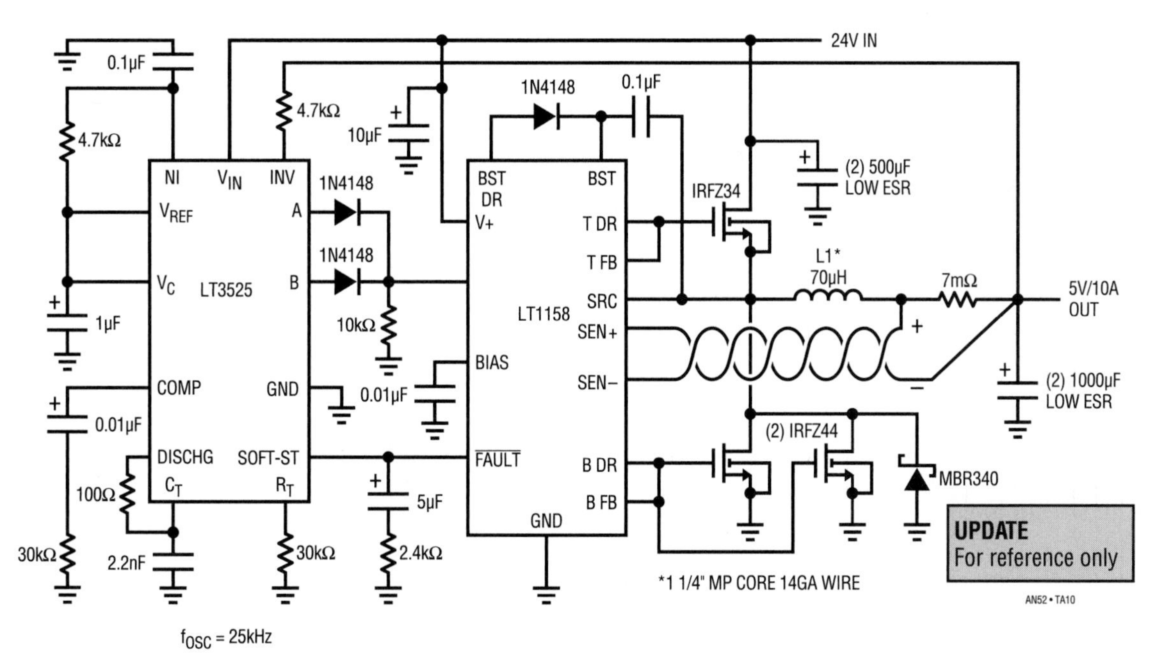

Figure 33.10 • 50W High Efficiency Switching Regulator Illustrates the Design Ease Afforded by Adaptive Dead Time Generation

Filters

Cascaded 8th-order Butterworth filters provide steep roll-off lowpass filter

by Philip Karantzalis and Richard Markell

Sometimes a design requires a filter that exceeds the specifications of the standard "dash-number" filter. In this case, the requirement was a low-distortion (−70dB) filter with roll-off faster than that of an 8th-order Butterworth. An elliptic filter was ruled out because its distortion specifications are too high. Two low power LTC1164-5s were wired in cascade to investigate the specifications that could be achieved with this architecture. The LTC1164-5 is a low power (4 milliamperes with ±5 volt supplies), clock-tunable, 8th-order filter, which can be configured for a Butterworth or Bessel response by strapping a pin. Figure 33.12 shows the schematic diagram of the two-filter system. The frequency response is shown in Figure 33.13, where it can be seen that the filter's attenuation is 80dB at 2.3 times the cutoff frequency. The distortion, as shown in Figure 33.14, is nothing less than spectacular. From 100Hz to 1kHz, the two filters have less than −74dB distortion specifications. At the standard measurement frequency of 1kHz, the specification is −78dB.

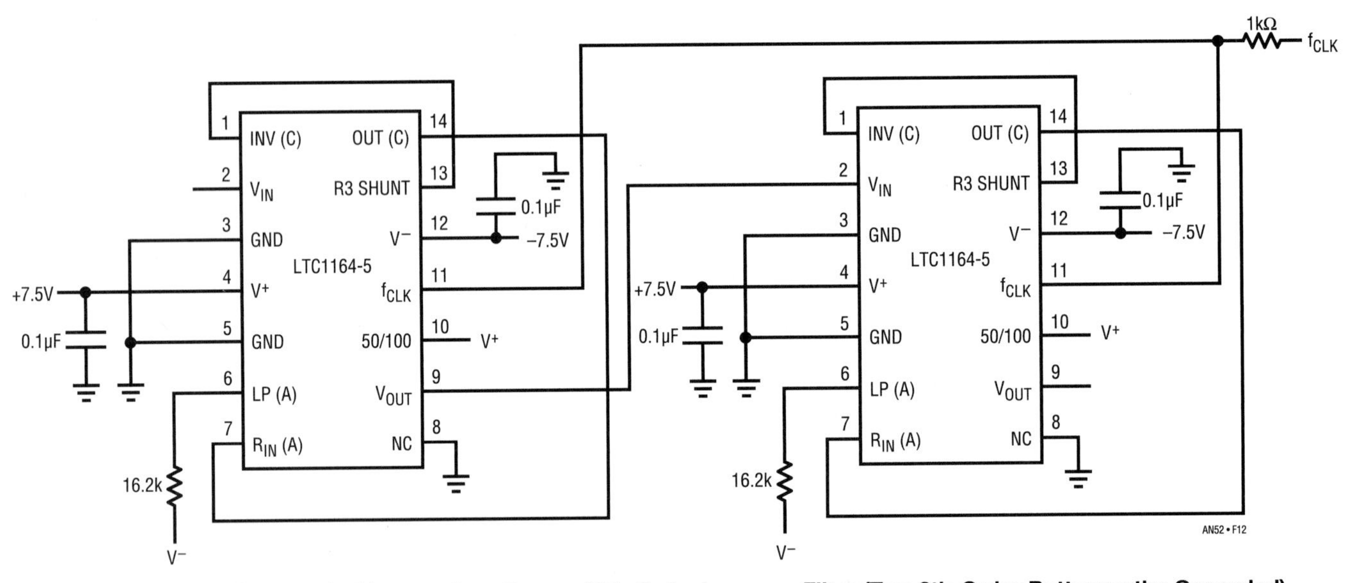

Figure 33.12 • Schematic Diagram: Low Power, 16th-Order Lowpass Filter (Two 8th-Order Butterworths Cascaded)

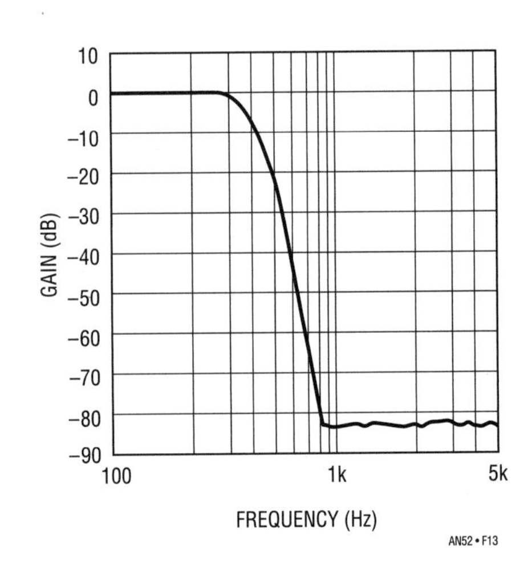

Figure 33.13 • Frequency Response for f_{CLK} = 20kHz

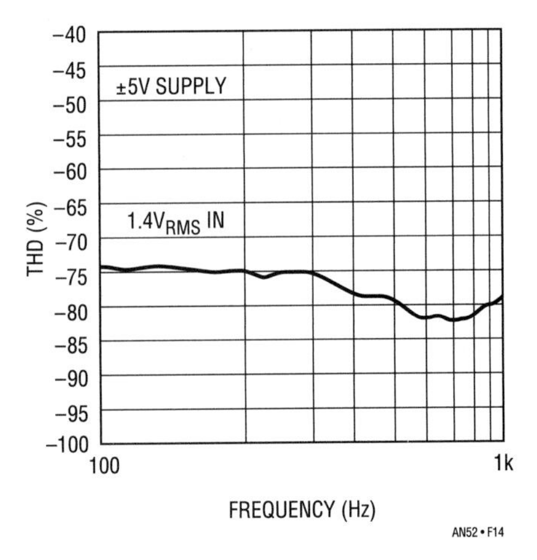

Figure 33.14 • Distortion Performance: Two LTC1164-5s, f_{CLK} = 60kHz (57:1) Pin 10 Connected to V^+

DC-Accurate, programmable-cutoff, fifth-order Butterworth lowpass filter requires no on-board clock

by Richard Markell

The new LTC1063 is a clock-tunable, monolithic filter with low-DC output offset (1mV typical with ±5V supplies). The frequency response of the filter closely approximates a fifth-order Butterworth polynomial.

Most users choose to tune the filter with an on-board microprocessor and/or timer. This is quite convenient if these components are available. If a clock is not available, the LTC1063 can be tuned with an external resistor and capacitor. The scheme shown here allows the filter's cutoff frequency to be programmed using an external microprocessor or the parallel port of a personal computer. This allows the cutoff frequency of the filter to be set before the product is shipped.

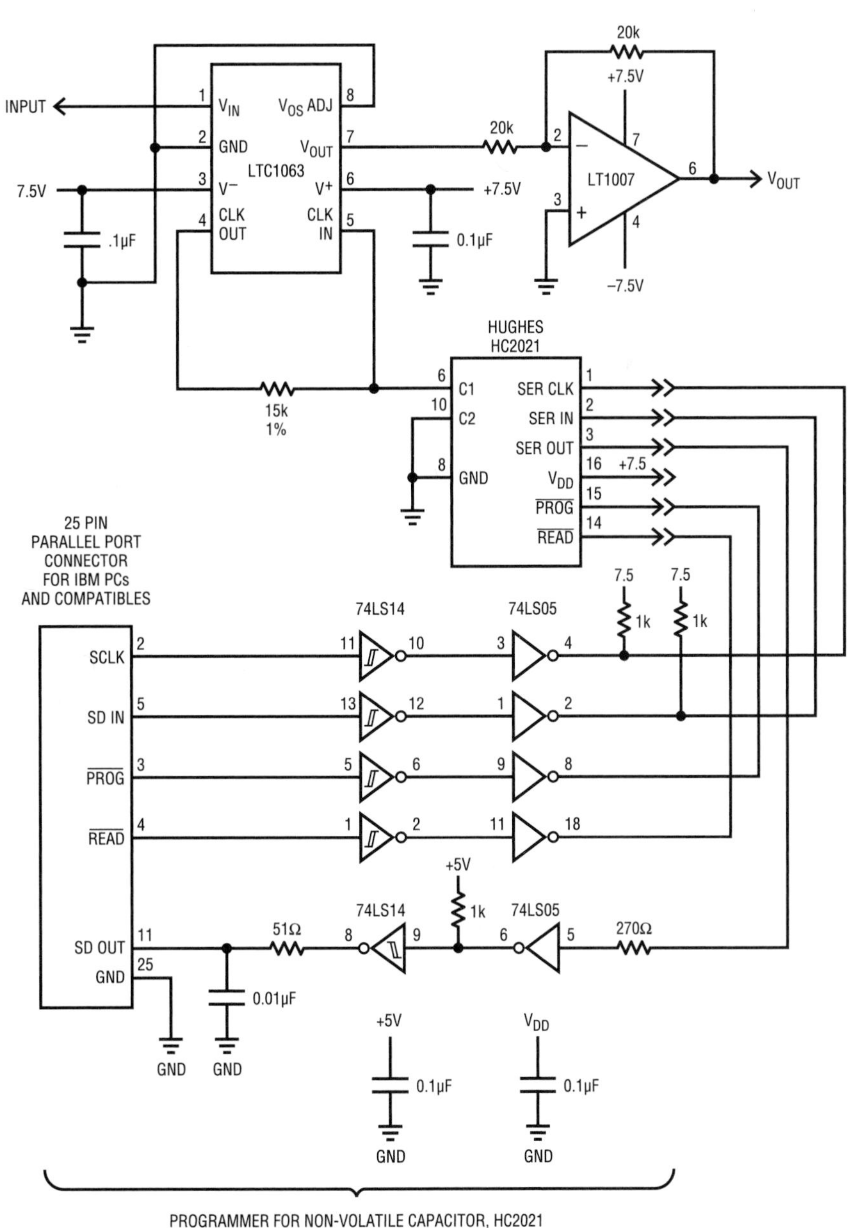

NOTES:
1. THE HC2021 SHOULD BE LOCATED CLOSE TO THE LTC1063 FOR BEST RESULTS.
2. $+3.5 \leq V_{DD} \leq +18.8$.
3. POSITIVE POWER SUPPLY FOR DEVICES 74LS05 AND 74LS14 IS +5V.
4. HUGHES TELEPHONE NUMBER (714) 759-2665.

AN52 • TA15

Figure 33.15 • Schematic Diagram of LTC1063 with Programmable Cutoff Frequency

The tuning scheme makes use of non-volatile, tunable capacitors available from Hughes Semiconductor. These capacitors allow approximately a decade of tuning range. More range could be obtained by using dual devices. Figure 33.15 shows the schematic diagram of the application. Be sure to place the variable capacitor as close as possible to the LTC1063 to minimize parasitic elements. Figure 33.16 shows the frequency response of the filter when the capacitor is varied from minimum to half-value, and then to maximum capacitance. The programming part of the circuit may be disconnected once the variable capacitor is set. The capacitor will remember its value until it is reprogrammed.

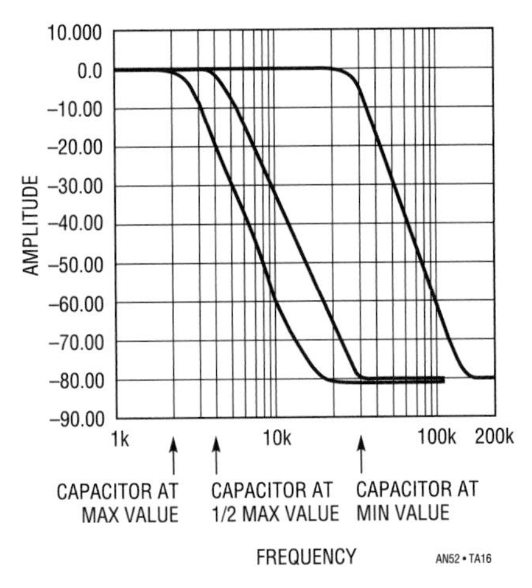

Figure 33.16 • LTC1063 Frequency Response

Miscellaneous circuits

A single cell laser diode driver using the LT1110

by Steve Pietkiewicz

Recently available visible lasers can be operated from 1.5V supplies, given appropriate drive circuits. Because these lasers are exceptionally sensitive to overdrive, power to the laser must be carefully controlled lest it be damaged. Overcurrents as brief as 2 microseconds can cause damage.

In the circuit of Figure 33.17, an LT1110 switching regulator serves as the controller within the single cell powered laser diode driver. The LT1110 regulator is a high speed LT1073.

The LT1110 is used here as an FM controller, driving a PNP power switch Q2, with a typical "ON" time of 1.5 microseconds. Current in L1 reaches a peak value of about 1.0A. The output capacitor C2 has been specified for low ESR, and should not be substituted (damage to the laser diode may result).

The Gain Block output of the LT1110 functions with Q1 as an error amplifier. The differential inputs compare the photodiode current developed as a voltage across R2 to the 212mV reference. The amplifier drives Q1, which modulates current into the I_{LIM} pin. This varies oscillator frequency to control average current.

Overall frequency compensation is provided by R1 and C1, values carefully chosen to eliminate power-up overshoot. The value of current sense resistor R2 determines the laser diode power, as shown the 1000 ohms results in about a 0.8 milliwatt output.

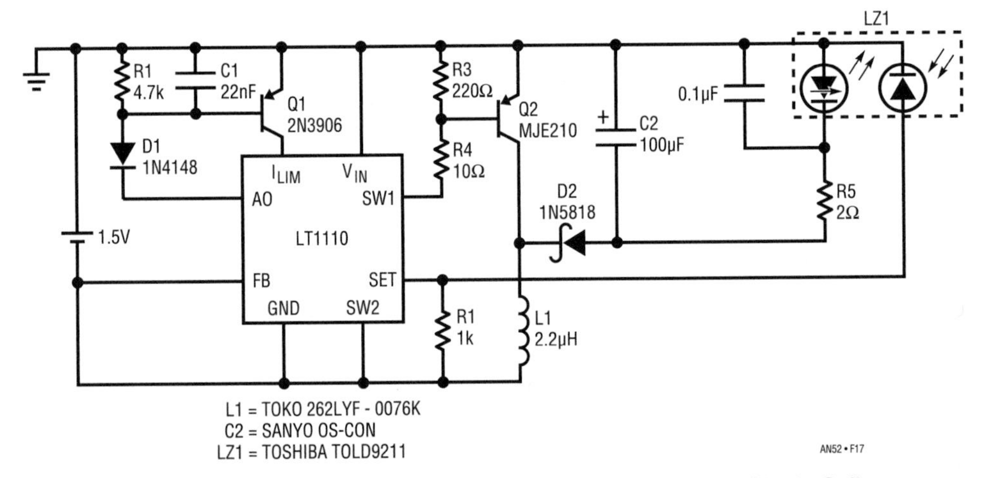

Figure 33.17 • LT1110 Laser Diode Driver Operating from a Single Cell

LT1109 generates VPP for flash memory

by Steve Pietkiewicz

Flash memory chips such as the Intel 28F020 2Megabit device require a VPP program supply of 12 volts at 30mA. A DC-DC converter may be used to generate 12 volts from the 5 volt logic supply. The converter must be physically small, available in surface-mount packaging, and have logic-controlled shutdown. Additionally, the converter must have carefully controlled rise time and zero overshoot. VPP excursions beyond 14 volts for 20ns or longer will destroy the ETOX-process based device.

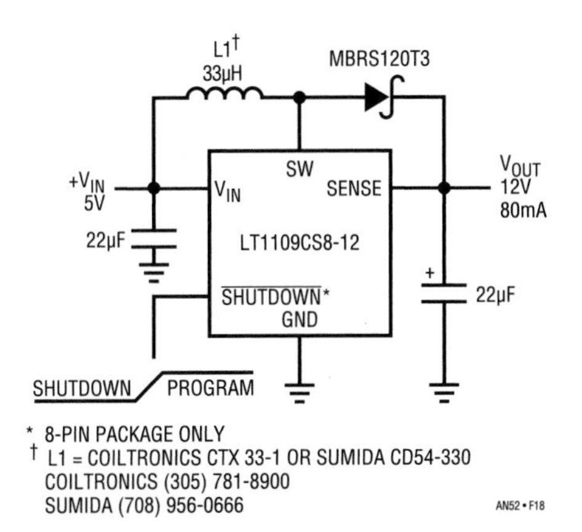

Figure 33.18 • All Surface Mount Flash Memory VPP Generator

Figure 33.18's circuit is well suited for providing VPP power for single or multiple flash memory chips. All associated components, including the inductor, are surface mount devices. The $\overline{\text{SHUTDOWN}}$ input turns off the converter, reducing quiescent current to 300μA when pulled to a logic 0. VPP rises in a controlled fashion, reaching 12 volts ±5% in under 4ms. Output voltage goes to V_{CC} minus a diode drop when the converter is in shutdown mode. This is an acceptable condition for Intel flash memories and does not harm the memory.

RF leveling loop

by Jim Williams

Leveling loops are often a requirement for RF transmission systems. More often than not, low cost is more important than absolute accuracy. Figure 33.19 shows such a circuit.

The RF input is applied to A1, an LT1228 operational transconductance amplifier. A1's output feeds A2, the LT1228's current feedback amplifier. A2's output, the output of the circuit, is sampled by the A3-based gain control configuration. This arrangement closes a gain control loop back at A1. The 4pF capacitor compensates rectifier diode capacitance, enhancing output flatness vs frequency. A1's I_{SET} input current controls its gain, allowing overall output level control. This approach to RF leveling is simple and inexpensive, and provides low output drift and distortion.

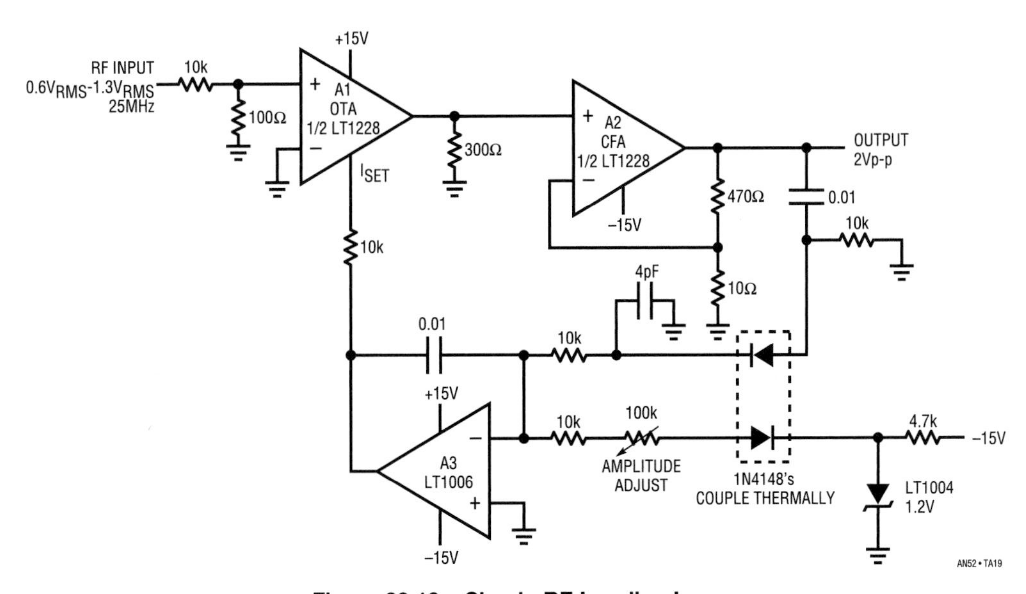

Figure 33.19 • Simple RF Leveling Loop

High accuracy instrumentation amplifier

by Dave Dwelley

The LTC1043 and the LTC1047 combine to make a high performance low frequency instrumentation amplifier as shown in Figure 33.20. The LTC1043 switched capacitor block is configured as a sampling front end, providing exceptional CMRR and rail-to-rail input operation. It works by attaching a 1 μF capacitor across the two inputs, letting it charge to the input voltage. Once charged, the capacitor is disconnected from the input terminals and reconnected to the output terminals, where it transfers its charge to the 1μF capacitor at the LTC1047's input. Any common mode voltage present at the inputs is subjected to a capacitive divider between the 1μF flying cap and the IC's parasitic capacitance. With the LTC1043's parasitics typically below 1μF, this gives AC CMRR above 120dB. The analog switches in the LTC1043 are purely resistive, so they add no DC offset to the signal.

The output signal (with the common mode stripped off) is then amplified by the LTC1047, a precision, micropower zero-drift op amp. The LTC1047 amplifies the signal by the desired amount, adding less than 10μV offset and 0.05μV/°C drift. The sampling frequency of the LTC1043 with single 5V supply is about 400Hz, allowing differential signals below 200Hz to be amplified with no aliasing. Note that common mode signals are not sampled; thus they will not alias regardless of frequency until the common mode/differential mode signal ratio approaches 120dB! The entire system draws 60μA with a single 5V supply and provides two independent channels.

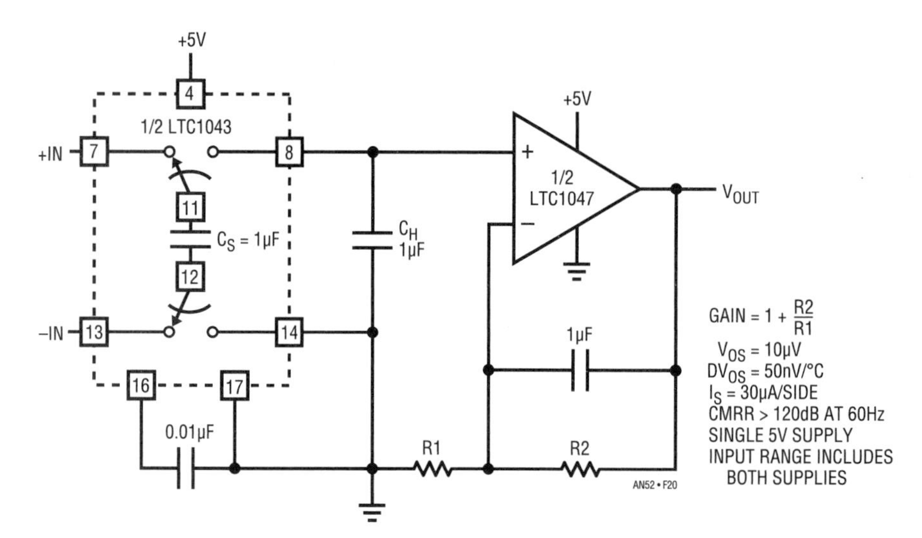

Figure 33.20 • High Accuracy Instrumentation Amplifier

A fast, linear, high current line driver

by Walt Jung and Rich Markell

Among linear applications not usually seen are those which require high speed combined with either very low DC error, or high load current. Such applications can be solved by combining the best attributes of two ICs, either one of which may not be capable by itself of the entire requirement.

A case in point is the line driver of Figure 33.21, which uses an LT1122 JFET input op amp as the gain element combined with an LT1010 buffet: This provides the output current of the LT1010 (typically 150mA) but with the basic DC and low level AC characteristics of the LT1122. The circuit is capable of driving loads as low as 100Ω with very low distortion. The input referred DC error is the low DC offset of the LT1122, typically 0.5mV or less. Large signal characteristics are also very good, due to the 80V/μs symmetrical SR of the LT1122.

The circuit as shown is configured as a precise gain of 5 non-inverting amplifier by gain set resistors R2 and R1, with the LT1010 unity gain voltage follower inside the overall feedback loop. This provides current buffering to the op amp, allowing it to operate most linearly. Small signal bandwidth is set by the time constant of R2 and C1, and is 1MHz as shown, with a corresponding risetime of about 400ns.

Performance with ±18V supplies is shown in Figures 33.22a and 33.22b, with output generally $5V_{RMS}$ or equivalent, driving 100Ω directly. THD is shown in Figure 33.22a, with input level swept up to output clipping level, at a fixed 10kHz frequency. The distortion is generally well below 0.01%, and improves substantially for lower frequencies.

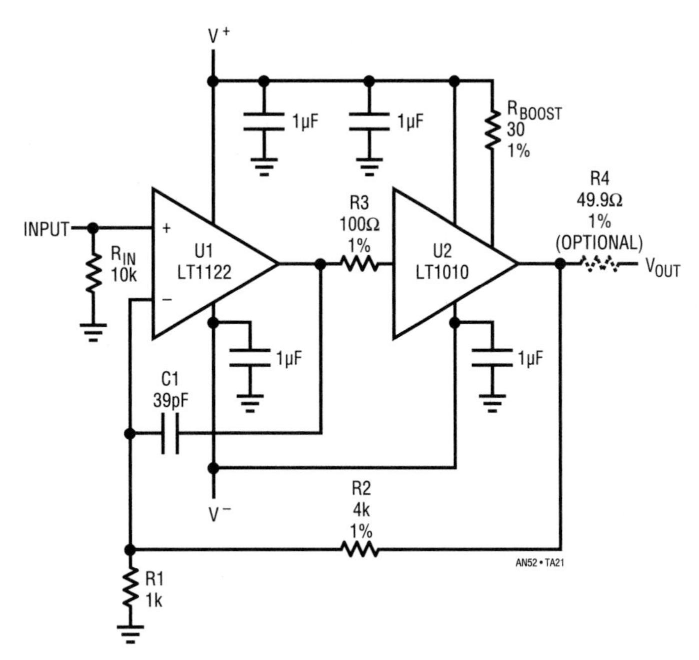

Figure 33.21 • Line Driver

CCIF IM distortion performance of the circuit for similar loading is shown in Figure 33.22b, driving a load of 100Ω at a swept level, again up to output clipping. The LT1122 amplifier is represented by the lower of the two curves, with distortion around the 0.0001% level. Also shown for comparison in this plot is the distortion of a type 156 JFET op amp (also driving the LT1010 buffer with other conditions the same). The 156 op amp uses a design topology with an intrinsically *asymmetric* SR. This manifests itself as rising even order distortion for methods such as this CCIF test. For this example, the distortion is more than an order of magnitude higher than that of the faster, symmetric slewing LT1122 for the same conditions.

Applications of this circuit include low offset linear buffers such as for A/D inputs, line drivers for instrumentation use, and audio frequency range buffers such as very high quality headphone use.

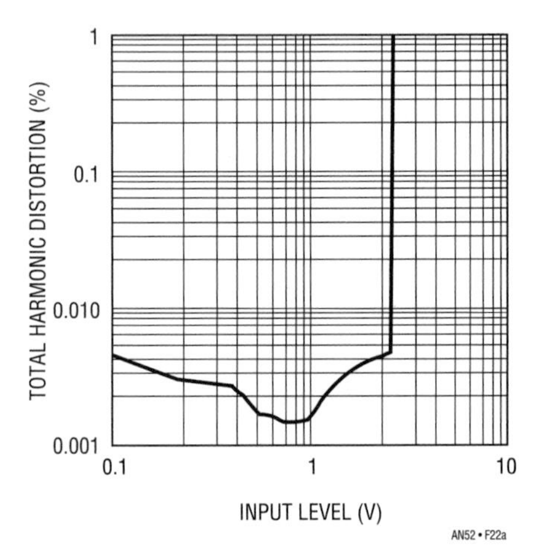

Figure 33.22a • THD vs Input Level

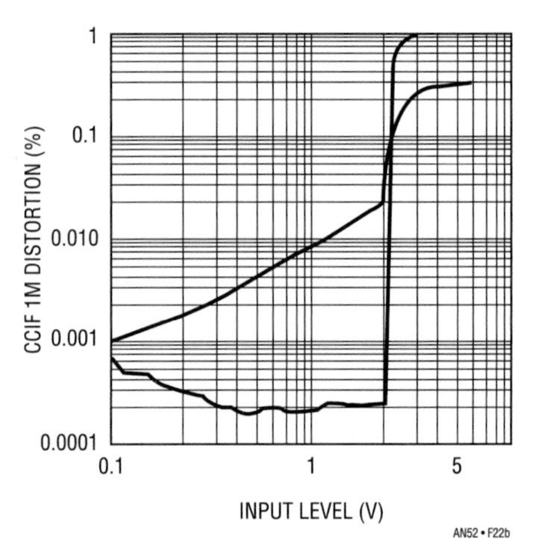

Figure 33.22b • CCIF IM Distortion vs Input Level

Video circuit collection

34

Jon Munson Frank Cox

Introduction

Even in a time of rapidly advancing digital image processing, analog video signal processing still remains eminently viable. The video A/D converters need a supply of properly amplified, limited, DC restored, clamped, clipped, contoured, multiplexed, faded and filtered analog video before they can accomplish anything. After the digital magic is performed, there is usually more amplifying and filtering to do as an adjunct to the D/A conversion process, not to mention all those pesky cables to drive. The analog way is often the most expedient and efficient, and you don't have to write all that code.

The foregoing is only partly in jest. The experienced engineer will use whatever method will properly get the job done; analog, digital or magic (more realistically, a combination of all three). Presented here is a collection of analog video circuits that have proven themselves useful.

Analog Circuit and System Design: Immersion in the Black Art of Analog Design. http://dx.doi.org/10.1016/B978-0-12-397888-2.00034-1

Video Amplifier Selection Guide			
PART	**GBW (MHz)**	**CONFIGURATION**	**COMMENTS**
LT6553	1200 (A = 2)	T	A = 2 (Fixed), 6ns Settling Time
LT6555	1200 (A = 2)	T	2:1 MUX, A = 2 (Fixed)
LT1226	1000 ($A_V \geq 25$)	S	400V/µs SR, Good DC Specs
LT6557	1000 (A = 2)	T	A = 2 (Fixed), Automatic Bias for Single Supply
LT6554	650 (A = 1)	T	A = 1 (Fixed), 6ns Settling Time
LT6556	650 (A = 1)	T	2:1 MUX, A = 1 (Fixed)
LT1222	500 ($A_V \geq 10$)	S	12-Bit Accurate
LT1395/LT1396/LT1397	400	S, D, Q	CFA, DG = 0.02%, DP = 0.04%, 0.1dB Flat to 100MHz
LT1818/LT1819	400	S, D	900V/µs SR, DG = 0.07%, DP = 0.02%
LT1192	350 ($A_V \geq 5$)	S	Low Voltage, ±50mA Output
LT1194	350 ($A_V = 10$)	S	Differential Input, Low Voltage, Fixed Gain of 10
LT6559	300	T	CFA, Independant Enable Controls, Low Cost
LT1398/LT1399	300	D, T	CFA, Independant Enable Controls
LT1675-1/LT1675	250 (A = 2)	S, T	2:1 MUX, A = 2 (Fixed)
LT1815/LT1816/LT1817	220	S, D, Q	750V/µs SR, DG = 0.08%, DP = 0.04%
LT6210/LT6211	200	S, D	CFA, Adjustable Speed and Power
LT1809/LT1810	180	S, D	Low Voltage, Rail-to-Rail Input and Output
LT1203/LT1205	170	D, Q	MUX, 25ns Switching, DG = 0.02%, DP = 0.04°
LT1193	160 ($A_V \geq 2$)	S	Low Voltage, Differential Input, Adjustable Gain, ±50mA Output
LT1221	150 ($A_V \geq 4$)	S	250V/µs SR, 12-Bit Accurate
LT1227	140	S	CFA, 1100V/µs is SR, DG = 0.01%, DP = 0.01°, Shutdown
LT1259/LT1260	130	D, T	RGB CFA, 0.1dB Flat to 30MHz, DG = 0.016%, DP = 0.075°, Shutdown
LT6550/LT6551	110 (A = 2)	T, Q	Low Voltage, Single Supply, A = 2 (Fixed)
LT1223	100	S	CFA, 12-Bit Accurate, Shutdown, 1300V/µs SR, Good DC Specs, DG = 0.02%, DP = 0.12°
LT1229/LT1230	100	D, Q	CFA, 1000V/µs SR, DG = 0.04%, DP = 0.1°
LT1252	100	S	CFA, DG = 0.01%, DP = 0.09°, Low Cost
LT1812	100	S	Low Power, 200V/µs SR
LT6205/LT6206/LT6207	100	S, D, Q	3V Single Supply
LT1191	90	S	Low Voltage, ±50mA Output
LT1253/LT1254	90	D, Q	CFA, DG = 0.03%, DP = 0.28°, Flat to 30MHz, 0.1dB
LT1813/LT1814	85	D, Q	Low Power, 200V/µs SR
LT1228	80 (gm = 0.25)	S	Transconductance Amp + CFA, Extremely Versatile
LT6552	75 ($A_V \geq 2$)	S	Differential Input, Low Power, Low Voltage
LT1204	70	S	CFA, 4-Input Video MUX Amp, 1000V/µs SR, Superior Isolation

(Continued)

Video Amplifier Selection Guide

PART	GBW (MHz)	CONFIGURATION	COMMENTS
LT1363/LT1364/LT1365	70	S, D, Q	1000V/µs SR, $I_S = 7.5mA$ per Amp, Good DC Specs
LT1206	60	S	250mA Output Current CFA, 600V/µs SR, Shutdown
LT1187	50 ($A_V \geq 2$)	S	Differential Input, Low Power
LT1190	50	S	Low Voltage
LT1360/LT1361/LT1362	50	S, D, Q	600V/µs SR, $I_S = 5mA$ per Amp, Good DC Specs
LT1208/LT1209	45	D, Q	400V/µs SR
LT1220	45	S	250V/µs, Good DC Specs, 12-Bit Accurate
LT1224	45	S	400V/µs SR
LT1189	35 (A_V $A_V \geq 10$)	S	Differential Input, Low Power, Decompensated
LT1995	32 (A = 1)	S	Internal Resistor Array
LT1213/LT1214	28	D, Q	Single Supply, Excellent DC Specs
LT1358/LT1359	25	D, Q	600V/µs SR, $I_S = 2.5mA$ per Amp, Good DC Specs
LT1215/LT1216	23	D, Q	Single Supply, Excellent DC Specs
LT1211/LT1212	14	D, Q	Single Supply, Excellent DC Specs
LT1355/LT1356	12	D, Q	400V/µs SR, $I_S = 1.25mA$ per Amp, Good DC Specs
LT1200/LT1201/LT1202	11	S, D, Q	$I_S = 1mA$ per Amp, Good DC Specs
LT1217	10	S	CFA, $I_S = 1mA$, Shutdown

Key to Abbreviations:

CFA = Current Feedback Amplifier	S = Single
DG = Differential Gain	D = Dual
DP = Differential Phase	Q = Quad
MUX = Multiplexer	T = Triple
SR = Slew Rate	

Note: Differential gain and phase is measured with a 150Ω load, except for the LT1203/LT1205 in which case the load is 1000Ω.

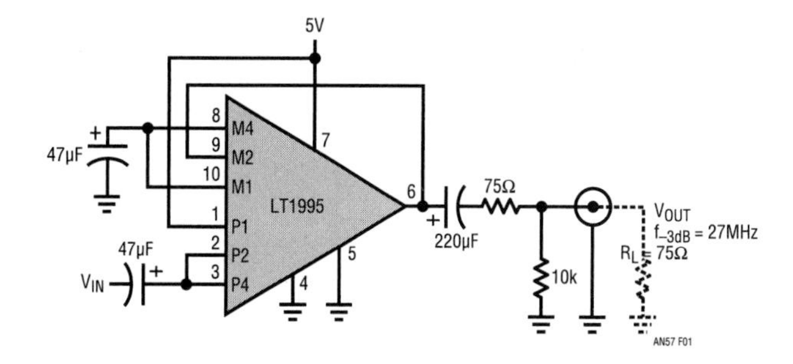

Figure 34.1 • Single Supply Video Line Driver

Video cable drivers

AC-coupled video drivers

When AC-coupling video, the waveform dynamics change with respect to the bias point of the amplifier according to the scene brightness of the video stream. In the worst case, $1V_{P-P}$ video (composite or Luminance + Sync in Y/C or YP_BP_R format) can exhibit a varying DC content of 0.56V, with the dynamic requirement being +0.735V/−0.825V about the nominal bias level. When this range is amplified by two to properly drive a back-terminated cable, the amplifier output must be able to swing $3.12V_{P-P}$, thus a 5V supply is generally required in such circuits, provided the amplifier output saturation voltages are sufficiently small. The following circuits show various realizations of AC-coupled video cable drivers.

Figure 34.1 shows the LT1995 as a single-channel driver. All the gain-setting resistors are provided on-chip to minimize part count.

Figure 34.2 shows an LT6551 quad amplifier driving two sets of "S-video" (Y/C format) output cables from a single Y/C source. Internal gain-setting resistors within the LT6551 reduce part-count.

Figure 34.3 shows the LT6553 ultra-high-speed triple video driver configured for single-supply AC-coupled operation. This part is ideal for HD or high-resolution workstation applications that demand high bandwidth and

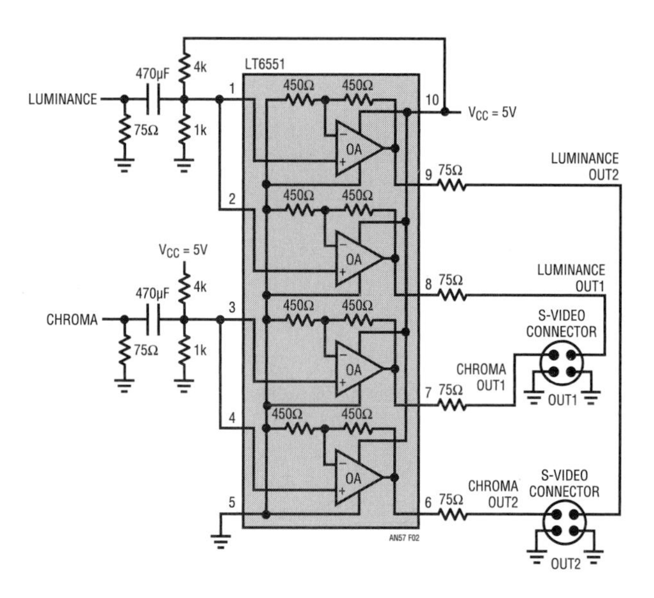

Figure 34.2 • S-Video Splitter

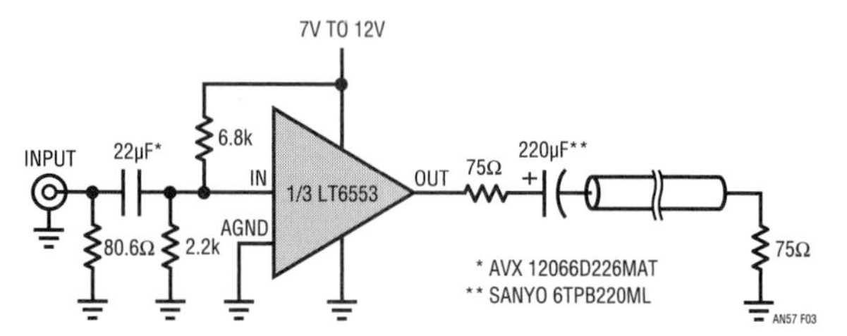

Figure 34.3 • Single Supply Configuration, One Channel Shown

fast settling. The amplifier gains are factory-set to two by internal resistors.

The LT6557 400MHz triple video driver is specifically designed to operate in 5V single supply AC-coupled applications as shown in Figure 34.4. The input biasing circuitry is contained on-chip for minimal external component count. A single resistor programs the biasing level of all three channels.

DC-coupled video drivers

The following circuits show various DC-coupled video drivers. In DC-coupled systems, the video swings are fixed in relation to the supplies used, so back-terminated cable-drivers need only provide 2V of output range when optimally biased. In most cases, this permits operation on lower power supply potential(s) than with AC-coupling (unclamped mode). Generally DC-coupled circuits use split supply potentials since the waveforms often include or pass through zero volts. For single supply operation, the inputs need to have an appropriate offset applied to preserve linear amplifier operation over the intended signal swing.

For systems that lack an available negative supply, the LT1983-3 circuit shown in Figure 34.5 can be used to easily produce a local-use −3V that can simplify an overall cable-driving solution, eliminating large output electrolytics, for example.

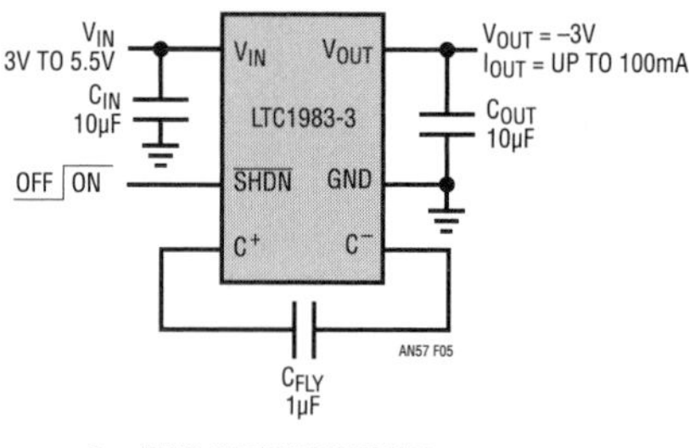

Figure 34.5 • −3V at 100mA DC/DC Converter

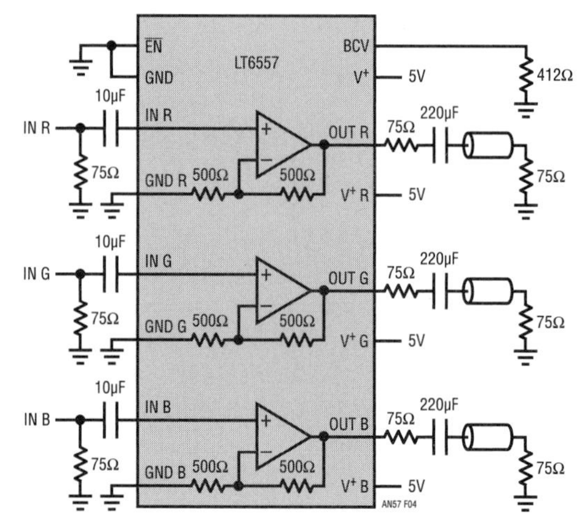

Figure 34.4 • 400MHz, AC-Coupled, 5V Single Supply Video Driver

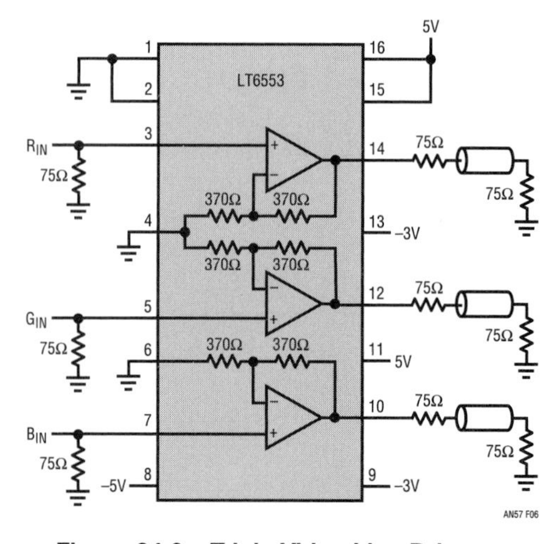

Figure 34.6 • Triple Video Line Driver

Figure 34.6 shows a typical 3-channel video cable driver using an LT6553. This part includes on-chip gain-setting resistors and flow-through layout that is optimal for HD and RGB wideband video applications. This circuit is a good candidate for the LT1983-3 power solution in systems that have only 5V available.

Figure 34.7 shows the LT6551 driving four cables and operating from just 3.3V. The inputs need to have signals centered at 0.83V for best linearity. This application would be typical of standard-definition studio-environment signal distribution equipment (RGBS format).

Figure 34.8 shows a simple video splitter application using an LT6206. Both amplifiers are driven by the input signal and each is configured for a gain of two, one for driving each output cable. Here again careful input biasing is required (or a negative supply as suggested previously).

Figure 34.9 shows a means of providing a multidrop tap amplifier using the differential input LT6552. This circuit taps the cable (loop-through configuration) at a high impedance and then amplifies the signal for transmission to a standard 75Ω video load (a display monitor for example). The looped-through signal would continue on to other locations before being terminated. The exceptional common mode rejection of the LT6552 removes any stray noise pickup on the distribution cable from corrupting the locally displayed video. This method is also useful for decoupling of ground-loop noise between equipment, such as in automotive entertainment equipment. To operate on a single supply, the input signals shown (shield and center of coax feed) should be non-negative, otherwise a small negative supply will be needed, such as the local −3V described earlier.

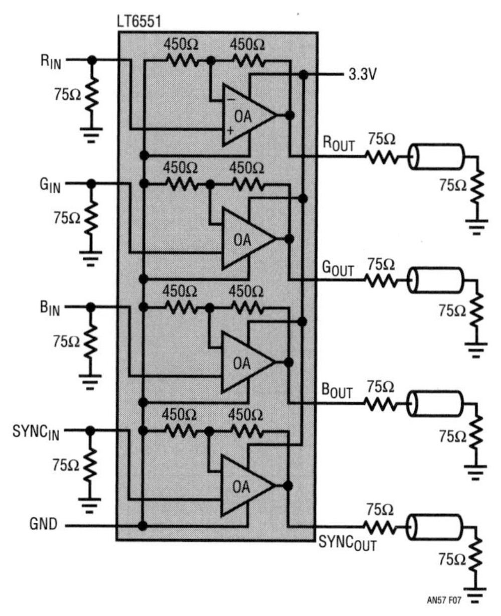

Figure 34.7 • 3.3V Single Supply LT6551 RGB Plus SYNC Cable Driver

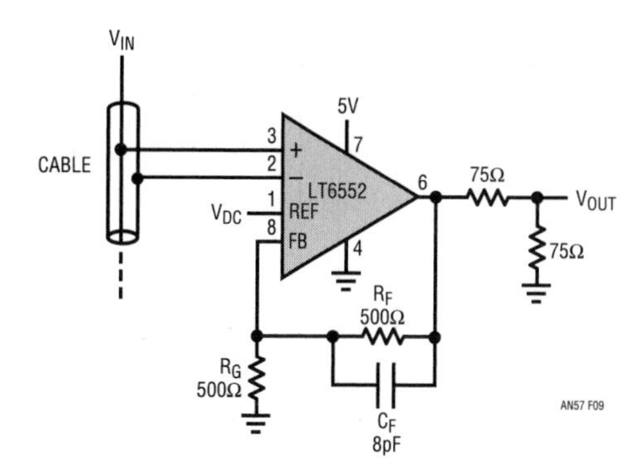

Figure 34.9 • Cable Sense Amplifier for Loop Through Connections with DC Adjust

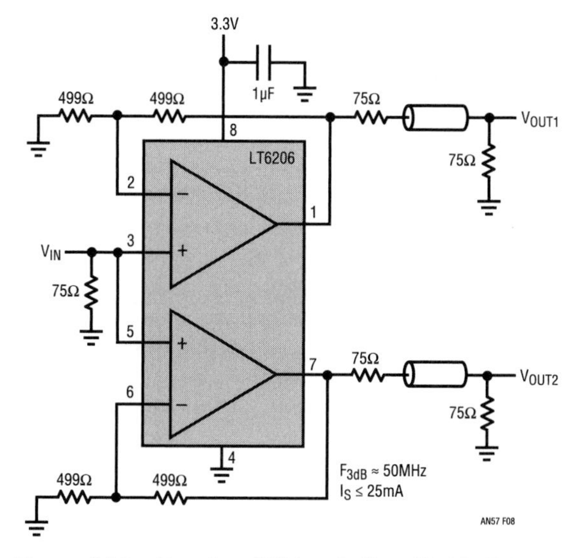

Figure 34.8 • Baseband Video Splitter/Cable Driver

Clamped AC-input video cable driver

The circuit in Figure 34.10 shows a means of driving composite video on standard 75Ω cable with just a single 3.3V power supply. This is possible due to the low output saturation levels of the LT6205 and the use of input clamping to optimize the bias point of the amplifier for standard $1V_{P\text{-}P}$ source video. The circuit provides an active gain of two and 75Ω series termination, thus yielding a net gain of one as seen by the destination load (e.g. display device). Additional detail on this circuit and other low-voltage considerations can be found in Design Note 327.

Twisted-pair video cable driver and receiver

With the proliferation of twisted-pair wiring practices for in-building data communication, video transmission on the same medium offers substantial cost savings compared to

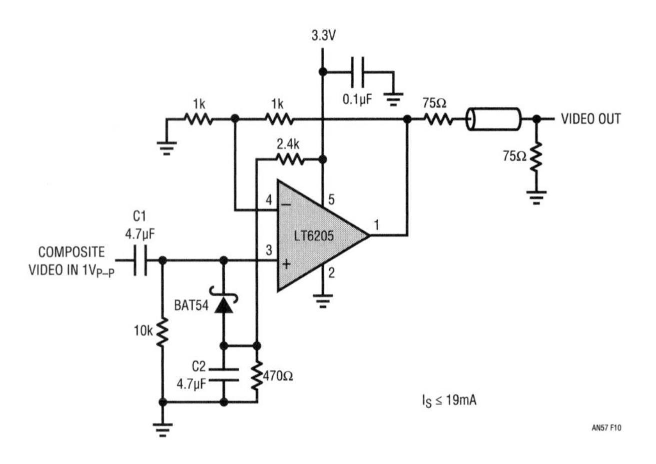

Figure 34.10 • Clamped AC-Input Video Cable Driver

conventional coaxial-cable. Launching a baseband camera signal into twisted pair is a relatively simple matter of building a differential driver such as shown in Figure 34.11. In this realization one LT6652 is used to create a gain of +1 and another is used to make a gain of −1. Each output is series terminated in half the line impedance to provide a balanced drive condition. An additional virtue of using the LT6552 in this application is that the incoming unbalanced signal (from a camera for example) is sensed differentially, thereby rejecting any ground noise and preventing ground loops via the coax shield.

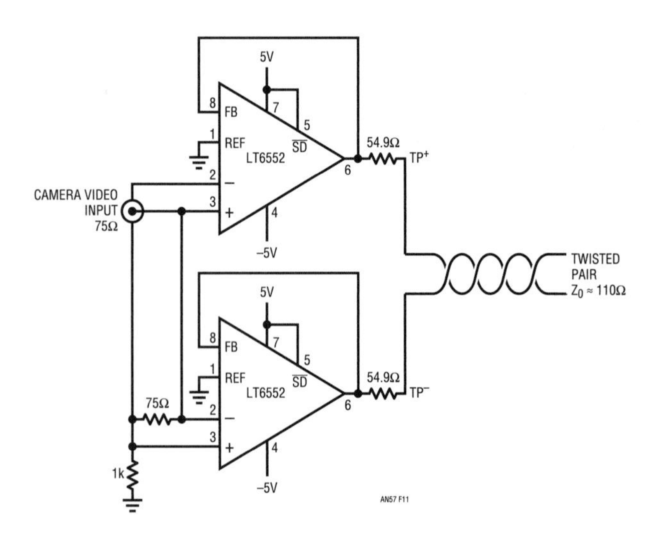

Figure 34.11 • Super-Simple Coax to Twisted-Pair Adapter

At the receiving end of the cable, the signal is terminated and re-amplified to re-create an unbalanced output for connection to display monitors, recorders, etc. The amplifier not only has to provide the 2x gain required for the output drive, but must also make up for the losses in the cable run. Twisted pair exhibits a rolloff characteristic that requires equalization to correct for, so the circuit in Figure 34.12 shows a suitable feedback network that accomplishes this. Here again the outstanding common mode rejection of the LT6552 is harnessed to eliminate stray pickup that occurs in long cable runs.

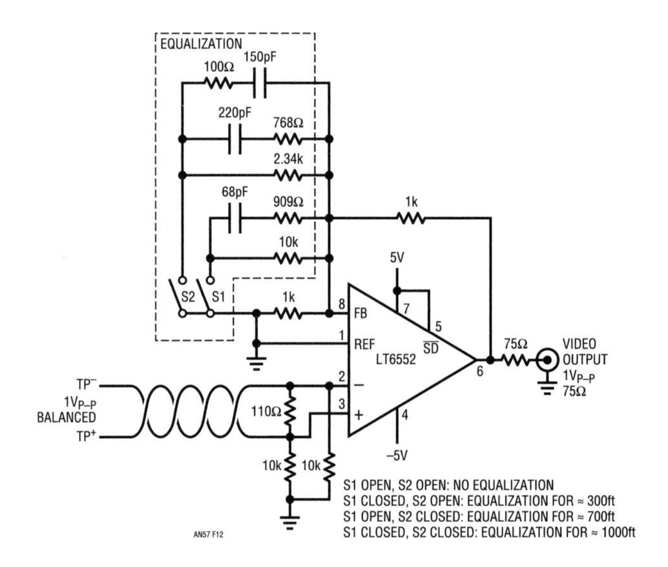

Figure 34.12 • All-In-One Twisted-Pair Video Line Receiver, Cable Equalizer, and Display Driver

Video processing circuits

ADC driver

Figure 34.13 shows the LT6554 triple video buffer. This is a typical circuit used in the digitization of video within high resolution display units. The input signals (terminations not shown) are buffered to present low source impedance and fast settling behavior to ADC inputs that is generally required to preserve conversion linearity to 10 bits or better. With high resolution ADCs, it is typical that the settling-time requirement (if not distortion performance) will call for buffer bandwidth that far outstrips the baseband signals themselves in order to preserve the effective number of [conversion] bits (ENOBs). The 1kΩ loads shown are simply to represent the ADC input for characterization purposes, they are not needed in the actual use of the part.

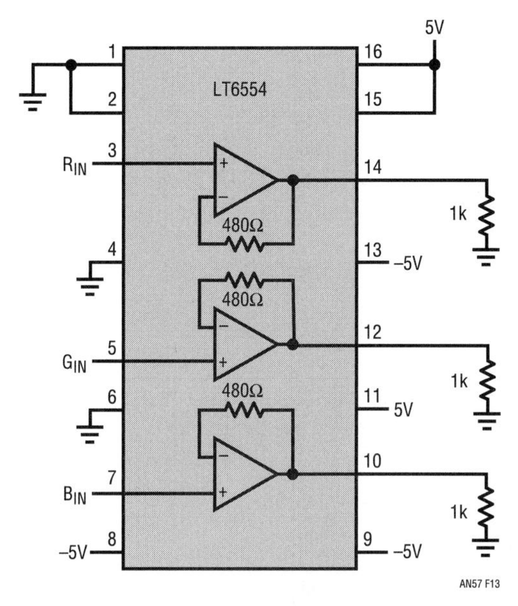

Figure 34.13 • Triple Video Buffer and A/D Driver

Video fader

In some cases it is desirable to adjust amplitude of a video waveform, or cross-fade between two different video sources. The circuit in Figure 34.14 provides a simple means of accomplishing this. The 0V to 2.5V control voltage provides a steering command to a pair of amplifier input sections; at each extreme, one section or the other takes complete control of the output. For intermediate control voltages, the inputs each contribute to the output with a weighting that follows a linear function of control voltage (e.g. at $V_{CONTROL}$ = 1.25V, both inputs contribute at 50%). The feedback network to each input sets the maximum gain in the control range (unity gain is depicted in the example), but depending on the application, other gains or even equalization functions can be voltage controlled (see datasheet and Application Note 67 for additional examples). In the fader example below, it should be noted that both input streams must be gen-locked for proper operation, including a black signal (with sync) if fading to black is intended.

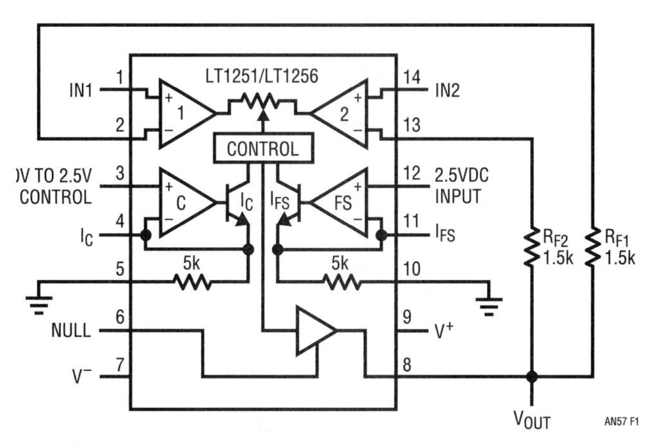

Figure 34.14 • Two-Input Video Fader

Color matrix conversion

Depending on the conventions used by video suppliers in products targeting specific markets, various standards for color signaling have evolved. Television studios have long used RGB cameras and monitor equipment to maximize signal fidelity through the equipment chain. With computer displays requiring maximum performance to provide clear text and graphics, the VESA standards also specify an RGB format, but with separate H and V syncs sent as logic signals. Video storage and transmission systems, on the other hand, seek to minimize information content to the extent that perceptual characteristics of the eye limit any apparent degradation. This has led to utilizing color-differencing approaches that allowed reducing bandwidth on the color information channels without noticeable loss in image sharpness. The consumer 3-channel "component" video connection (YP_BP_R) has a luma + sync (Y) plus blue and red axis color-space signals (P_B and P_R, respectively) that are defined as a matrix multiplication applied to RGB raw video. The color difference signals are typically half the spatial resolution of the luma according to the compression standards defined for DVD playback and digitally broadcast source material, thus lowering "bandwidth" requirements by some 50%. The following circuits show methods of performing color-space mappings at the physical layer (analog domain).

Figure 34.15 shows a method of generating the standard-definition YP_BP_R signals from an RGB source using a pair of LT6550 triple amplifiers. It should be noted that to ensure Y includes a correct sync, correct syncs should be present at all three inputs or else added directly at the Y output (gated 8.5mA current sink or 350Ω switched to -3.3V). This circuit does not deliberately reduce bandwidth on the color component outputs, but most display devices will nonetheless apply a Nyquist filter at the digitizer section of the "optical engine" in the display unit. The circuit is shown as DC-coupled, so ideally black level is near ground for best operation with the low-voltage supplies shown. Adding input coupling capacitors will allow processing source video that has substantial offset.

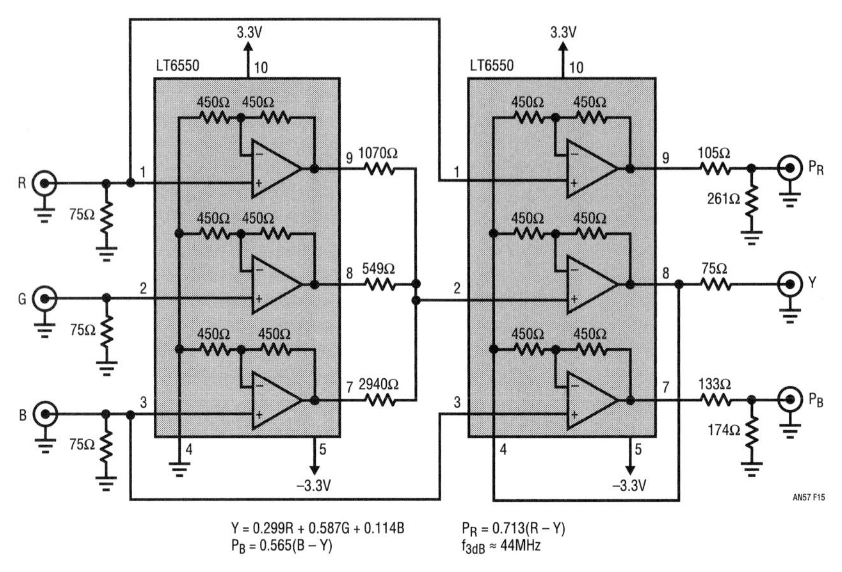

Figure 34.15 • RGB to YP_BP_R Component-Video Conversion

An LT6559 and an LT1395 can also be used to map RGB signals into YP_BP_R "component" video as shown in Figure 34.16. The LT1395 performs a weighted inverting addition of all three inputs. The LT1395 output includes an amplification of the R input by −324/1.07k = −0.3. The amplification of the G input is by −324/549 = −0.59. Finally, the B input is amplified by −324/2.94k = −0.11. Therefore the LT1395 output is −0.3R, −0.59G, −0.11B = −Y. This output is further scaled and inverted by −301/150 = −2 by LT6559 section A2, thus producing 2Y. With the division by two that occurs due to the termination resistors, the desired Y signal is generated at the load. The LT6559 section A1 provides a gain of 2 for the R signal, and performs a subtraction of 2Y from the section A2 output. The output resistor divider provides a scaling factor of 0.71 and forms the 75Ω back-termination resistance. Thus the signal seen at the terminated load is the desired 0.71(R–Y) = P_R. The LT6559 section A3 provides a gain of 2 for the B signal, and also performs a subtraction of 2Y from the section A2 output. The output resistor divider provides a scaling factor of 0.57 and forms the 75Ω back-termination resistance. Thus the signal seen at the terminated load is the desired 0.57(B–Y) = P_B. As with the previous circuit, to develop a normal sync on the Y signal, a normal sync must be inserted on each of the R, G, and B inputs or injected directly at the Y output with controlled current pulses.

Figure 34.17 shows LT6552 amplifiers connected to convert component video (YP_BP_R) to RGB. This circuit maps the sync on Y to all three outputs, so if a separate sync connection is needed by the destination device (e.g. studio monitor), any of the R, G, or B channels may be simply looped-through the sync input (i.e. set Z_{IN} for sync input to unterminated). This particular configuration takes advantage of the unique dual-differential inputs of the LT6552 to accomplish multiple arithmetic functions in each stage, thereby minimizing the amplifier count. This configuration also processes the wider-bandwidth Y signal through just a single amplification level, maximizing the available performance. Here again, operation on low supply voltages is predicated on the absence of substantial input offset, and input coupling capacitors may be used if needed (220µF/6V types for example, polarized according to the input offset condition).

Another realization of a component video (YP_BP_R) to RGB adapter is shown in Figure 34.18 using an LT6207. Amplifier count is minimized by performing passive arithmetic at the outputs, but this requires higher gains, thus a higher supply potential is needed for this (for at least the positive rail). One small drawback to this otherwise compact solution is that the Y channel amplifier must single-handedly drive all three outputs to produce white, so the helper current source shown is needed to increase available drive current. As with the previous circuit, the sync on Y is mapped to all outputs and input coupling-capacitors can be added if the input source has significant offset.

Two LT6559s can also be used to map YP_BP_R "component" video into RGB color space as shown in Figure 34.19. The Y input is properly terminated with 75Ω and buffered with a gain of 2 by amplifier A2. The P_R input is terminated and buffered with a gain of 2.8 by amplifier A1. The P_B input is terminated and buffered with a gain of 3.6 by amplifier A3. Amplifier B1 performs an equally weighted addition of amplifiers A1 and A2

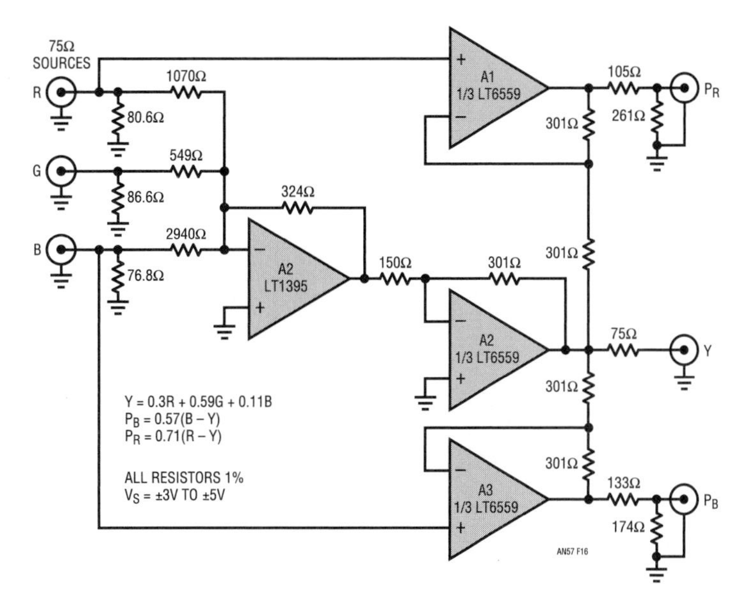

Figure 34.16 • High Speed RGB to YP_BP_R Converter

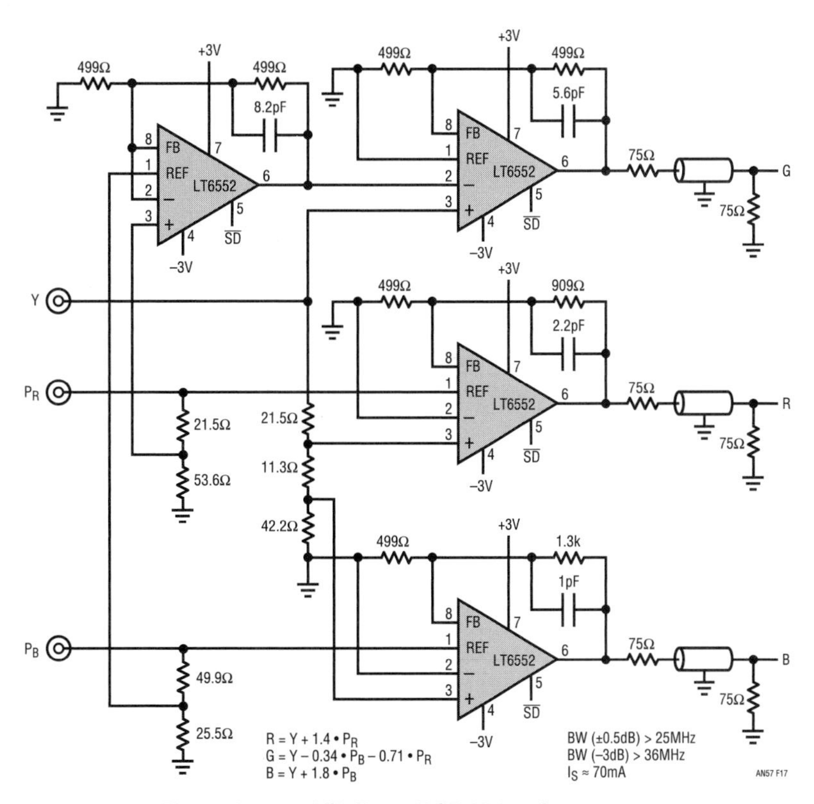

Figure 34.17 • YP_BP_R to RGB Video Converter

outputs, thereby producing $2(Y + 1.4P_R)$, which generates the desired R signal at the terminated load due to the voltage division by 2 caused by the termination resistors. Amplifier B3 forms the equally weighted addition of amplifiers A1 and A3 outputs, thereby producing $2(Y + 1.8P_B)$, which generates the desired B signal at the terminated load. Amplifier B2 performs a weighted summation of all three inputs. The P_B signal is amplified overall by $-301/1.54k \bullet 3.6 = 2(-0.34)$. The P_R signal is amplified overall by $-301/590 \bullet 2.8 = 2(-0.71)$. The Y signal is amplified overall by $1k/(1k + 698) \bullet (1 + [301/(590||1.54k)]) \bullet 2 = 2(1)$. Therefore the amplifier B2

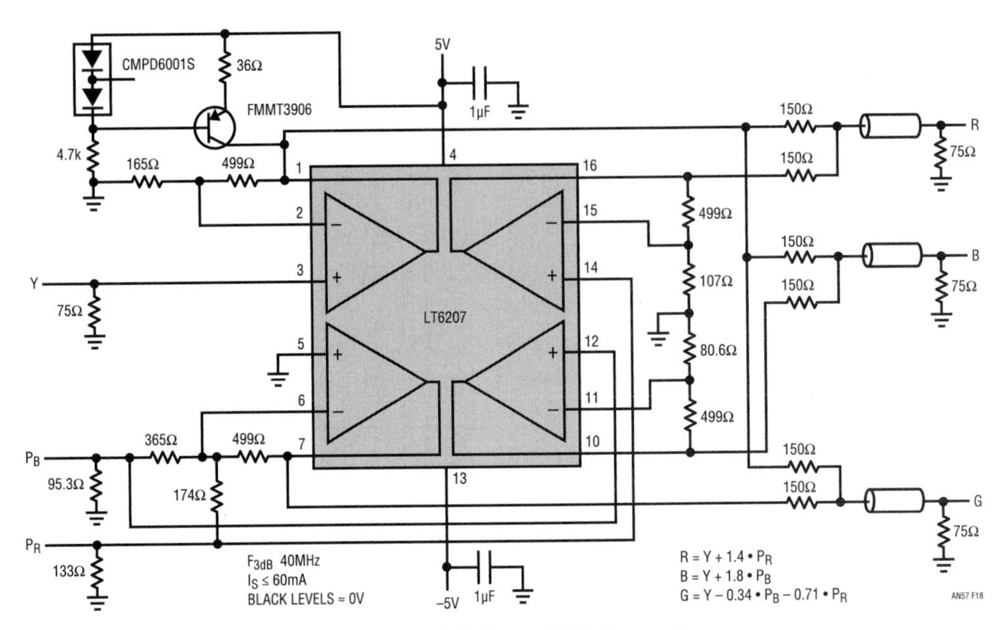

Figure 34.18 • YP_BP_R to RGB Converter

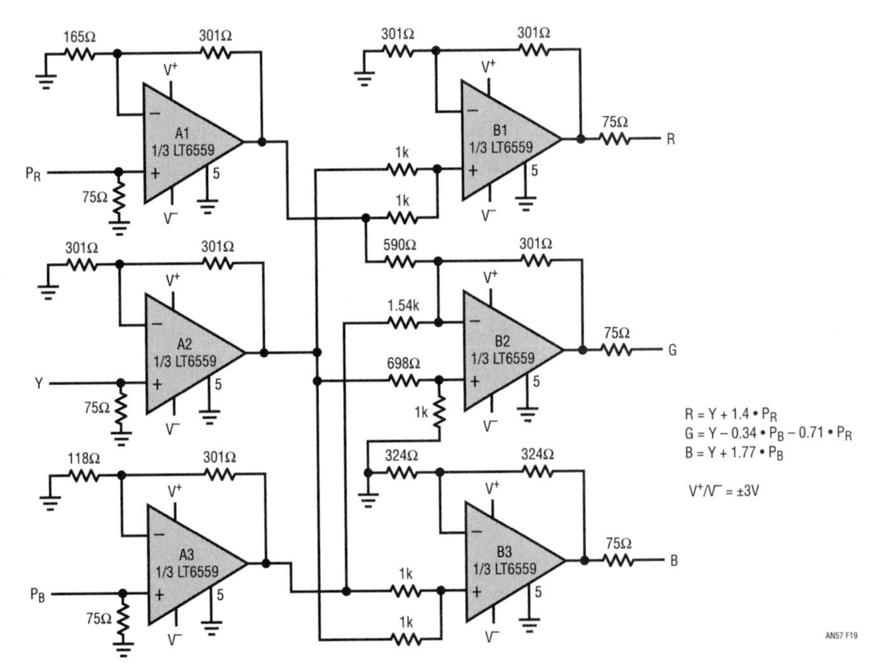

Figure 34.19 • High Speed YP_BP_R to RGB Converter

output is $2(Y–0.34P_B–0.71P_R)$, which generates the desired G signal at the terminated load. Like the previous circuits shown, sync present on the Y input is reconstructed on all three R, G, and B outputs.

Video inversion

The circuit in Figure 34.20 is useful for viewing photographic negatives on video. A single channel can be used for composite or monochrome video. The inverting amplifier stages are only switched in during active video so the blanking, sync and color burst (if present) are not disturbed. To prevent video from swinging negative, a voltage offset equal to the peak video signal is added to the inverted signal.

Graphics overlay adder

Multiplexers that provide pixel-speed switching are also useful in providing simple graphics overlay, such as timestamps or logo "bugs". Figure 34.21 shows an LT1675 pair

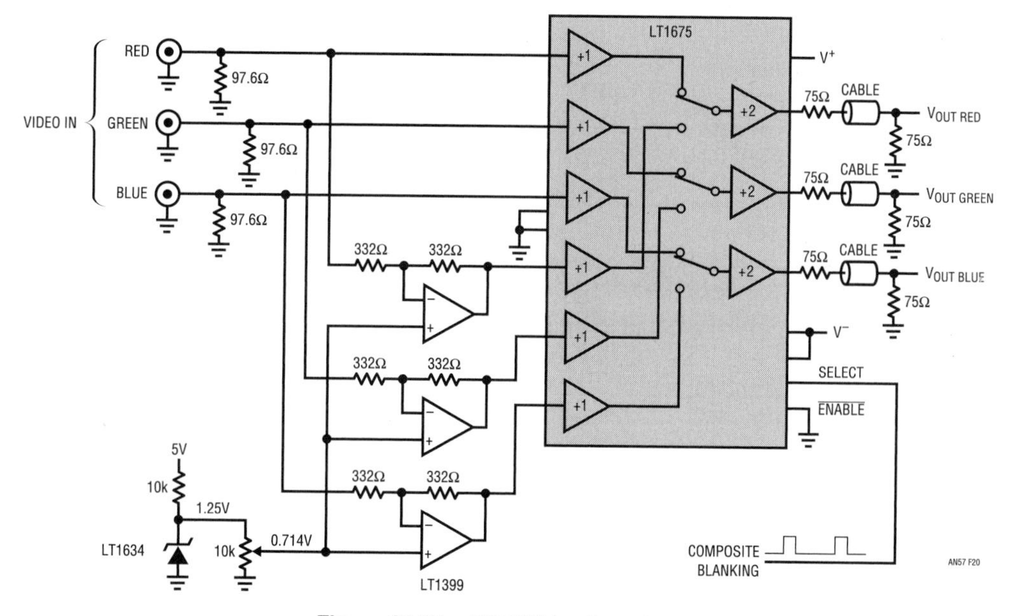

Figure 34.20 • RGB Video Inverter

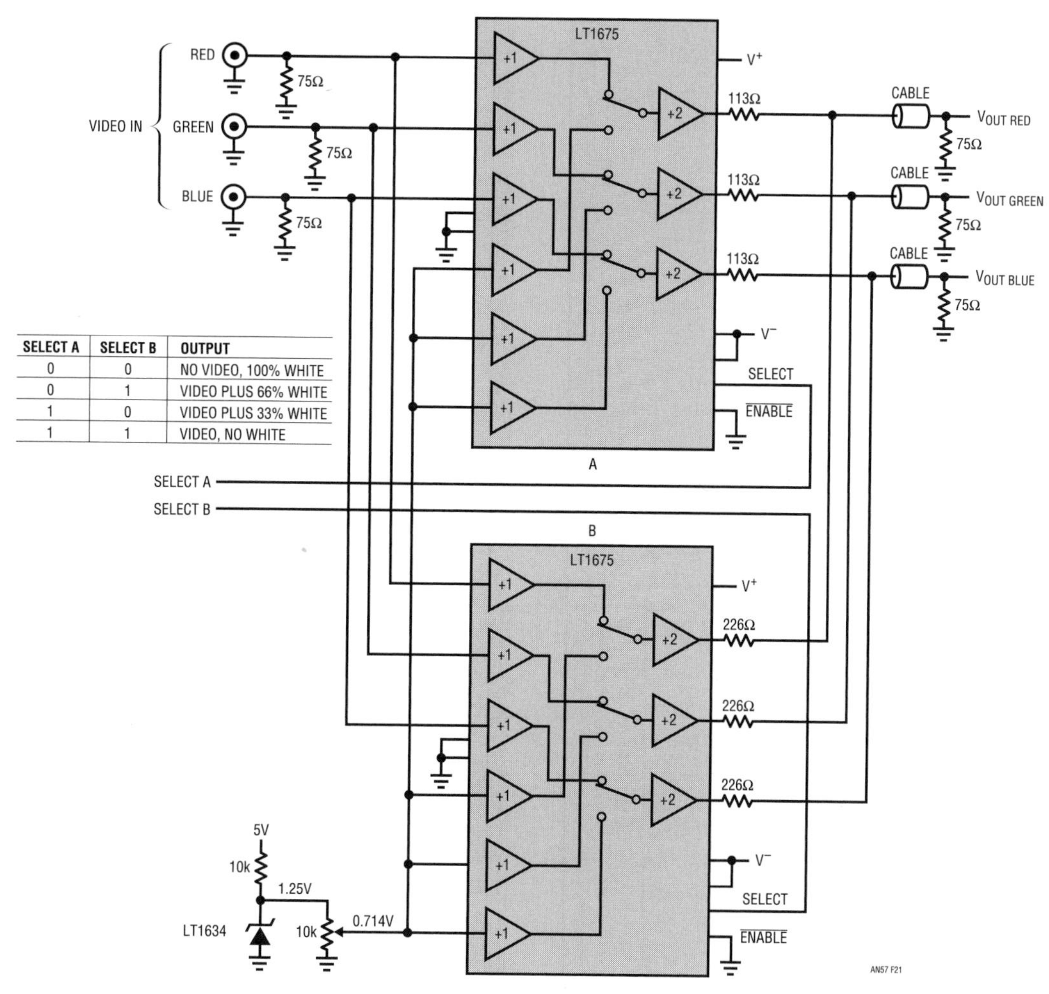

SELECT A	SELECT B	OUTPUT
0	0	NO VIDEO, 100% WHITE
0	1	VIDEO PLUS 66% WHITE
1	0	VIDEO PLUS 33% WHITE
1	1	VIDEO, NO WHITE

Figure 34.21 • Logo or "Bug" Inserter

used to insert multilevel overlay content from a digital generator. The instantaneous state of the two input control lines selects video or white in each device and combines their outputs with the resistor-weighted summing networks at the output. With the four combinations of control line states, video, white, and two differing brightening levels are available.

Variable gain amplifier has ±3dB range while maintaining good differential gain and phase

The circuit in Figure 34.22 is a variable gain amp suitable for composite video use. Feedback around the transcon-duct-ance amp (LT1228) acts to reduce the differential input voltage at the amplifier's input, and this reduces the differential gain and phase errors. Table 34.1 shows the differential phase and gain for three gains. Signal-to-noise ratio is better than 60dB for all gains.

Table 34.1

INPUT (V)	I_{SET} (mA)	DIFFERENTIAL GAIN	DIFFERENTIAL PHASE
0.707	4.05	0.4%	0.15°
1.0	1.51	0.4%	0.1°
1.414	0.81	0.7%	0.5°

Black clamp

Here is a circuit that removes the sync component of the video signal with minimal disturbance to the luminance (picture information) component. It is based on the classic op amp half-wave-rectifier with the addition of a few refinements.

The classic "diode-in-the-feedback-loop" half-wave-rectifier circuit generally does not work well with video frequency signals. As the input signal swings through zero volts, one of the diodes turns on while the other is turned off, hence the op amp must slew through two diode drops. During this time the amplifier is in slew limit and the output signal is distorted. It is not possible to entirely prevent this source of error because there will always be some time when the amp will be open-loop (slewing) as the diodes are switched, but the circuit shown here in Figure 34.23 minimizes the error by careful design.

The following techniques are critical in the design shown in Figure 34.23:

1. The use of diodes with a low forward voltage drop reduces the voltage that the amp must slew.
2. Diodes with a low junction capacitance reduce the capacitive load on the op amp. Schottky diodes are a good choice here as they have both low forward voltage and low junction capacitance.
3. A fast slewing op amp with good output drive is essential. An excellent CFA like the LT1227 is mandatory for good results.
4. Take some gain. The error contribution of the diode switch tends to be constant, so a larger signal means a smaller percentage error.

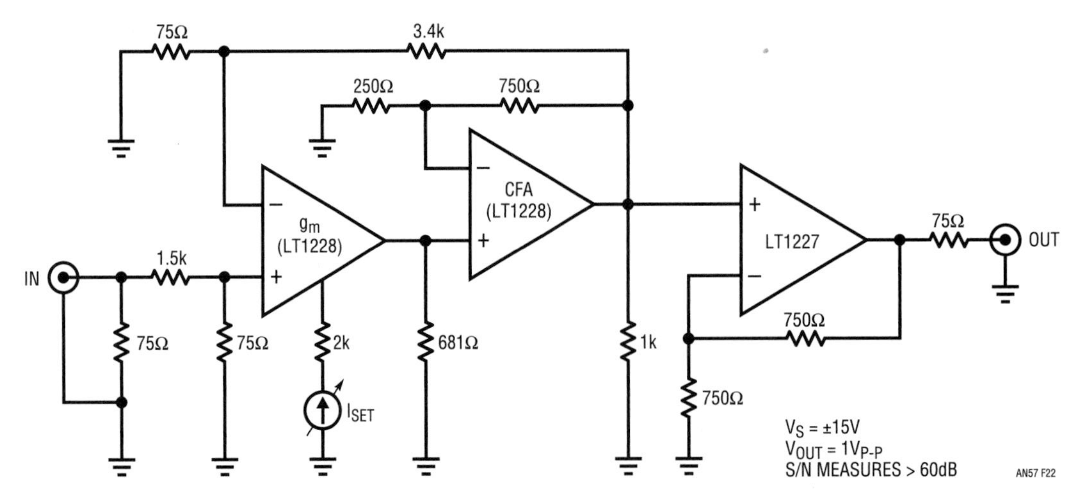

Figure 34.22 • ±3dB Variable Gain Video Amp Optimized for Differential Gain and Phase

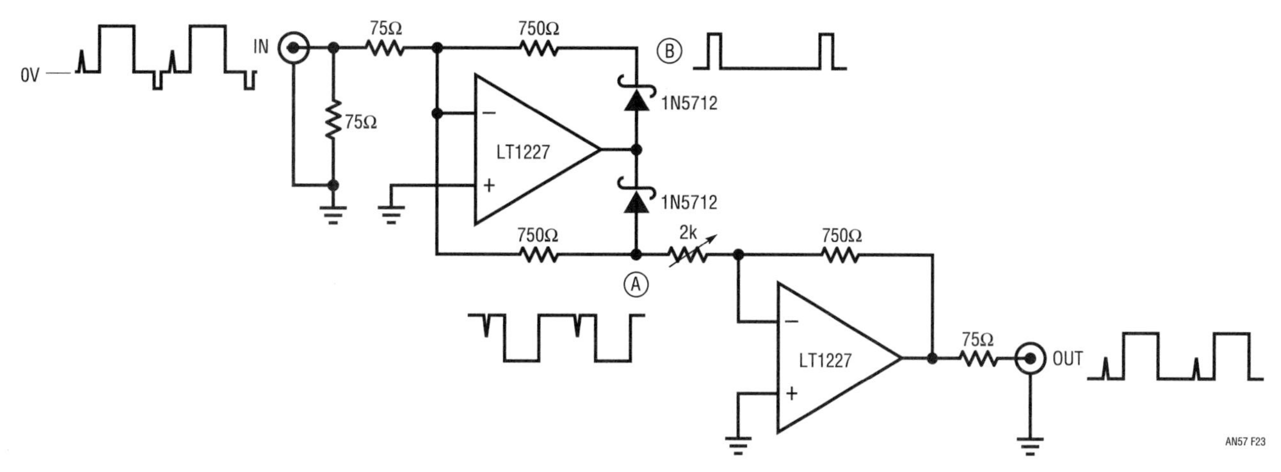

Figure 34.23 • Black Clamp Circuit

Since this circuit discriminates between the sync and video on the basis of polarity, it is necessary to have an input video signal that has been DC restored (the average DC level is automatically adjusted to bring the blanking level to zero volts). Notice that not only is the positive polarity information (luminance: point A in the schematic) available, but that the negative polarity information (sync: point B in the schematic) is also. Circuits that perform this function are called "black clamps." The photograph (Figure 34.24) shows the circuit's clean response to a 1T[1] pulse (some extra delay is added between the input and output for clarity).

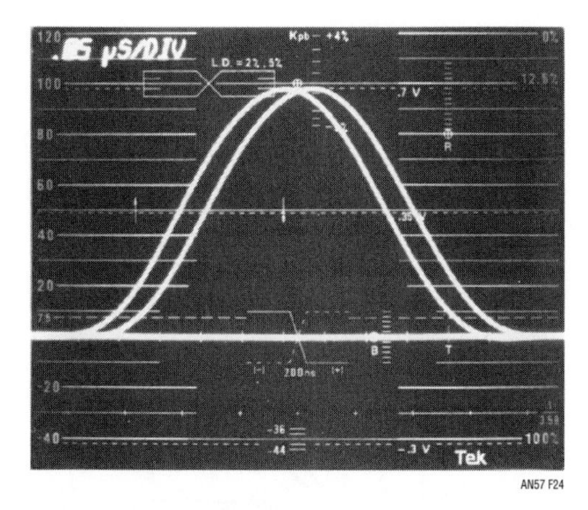

Figure 34.24 • Black Clamp Circuit Response to a "1T" Pulse (±15V Power Supplies)

Note 1: A 1T pulse is a specialized video waveform whose salient characteristic is a carefully controlled bandwidth which is used to quickly quantify gain and phase flatness in video systems. Phase shift and/or gain variations in the video system's passband result in transient distortions which are very noticeable on this waveform (not to mention the picture). [For you video experts out there, the K factor was 0.4% (the TEK TSG120 video signal generator has a K factor of 0.3%)].

Video limiter

Often there is a need to limit the amplitude excursions of the video signal. This is done to avoid exceeding luminance reference levels of the video standard being used, or to avoid exceeding the input range of another processing stage such as an A/D converter. The signal can be hard limited in the positive direction, a process called "white peak clipping," but this destroys any amplitude information and hence any scene detail in this region. A more gradual limiting ("soft limiter") or compression of the peak white excursion is performed by elements called "knee" circuits, after the shape of the amplifier transfer curve.

A soft limiter circuit is shown in Figure 34.25 which uses the LT1228 transconductance amp. The level at which the limiting action begins is adjusted by varying the set current into pin 5 of the transconductance amplifier. The LT1228 is used here in a slightly unusual, closed-loop configuration. The closed-loop gain is set by the feedback and gain resistors (R_F and R_G) and the open-loop gain by the transconductance of the first stage times the gain of the CFA.

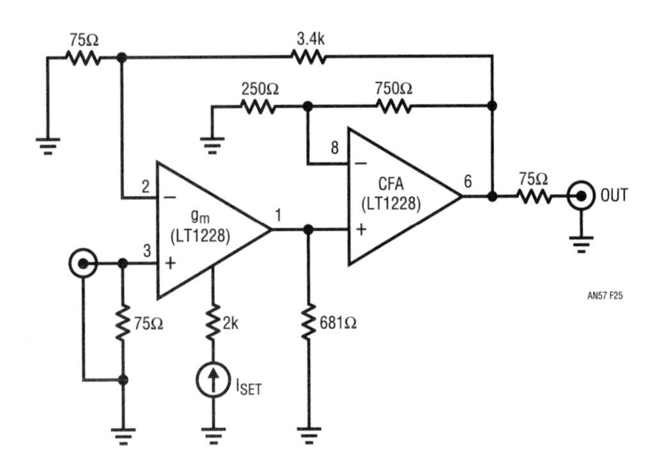

Figure 34.25 • LT1228 Soft Limiter

As the transconductance is reduced (by reducing the set current), the open-loop gain is reduced below that which can support the closed-loop gain and the amp limits. A family of curves which show the response of the limiting amplifier subject to different values of set current with a ramp input is shown in Figure 34.26. Figure 34.27 shows the change in limiting level as I_{SET} is varied.

Circuit for gamma correction

Video systems use transducers to convert light to an electric signal. This conversion occurs, for example, when a camera scans an image. Video systems also use transducers to convert the video signal back to light when the signal is sent to a display, a CRT monitor for example. Transducers often have a transfer function (the ratio of *signal in to light out)* that is unacceptably nonlinear.

The newer generation of camera transducers (CCDs and the improved versions of vidicon-like tubes) are adequately linear, however, picture monitor CRTs are not. The transfer functions of most CRTs follow a power law. The following equation shows this relation:

$$\text{Light Out} = \text{k} \bullet V_{\text{SIG}}^{\gamma}$$

where k is a constant of proportionality and gamma (γ) is the exponent of the power law (gamma ranges from 2.0 to 2.4).

This deviation from nonlinearity is usually called just gamma and is reported as the exponent of the power law. For instance, "the gamma of this vidicon is 0.43." The correction of this effect is *gamma correction.*

In the equation above, notice that a gamma value of 1 results in a linear transfer function. The typical CRT will have a transfer function with a gamma from about 2.0 to 2.4. Such values of gamma give a nonlinear response which compresses the blacks and stretches the whites. Cameras usually contain a circuit to correct this nonlinearity. Such a circuit is a *gamma corrector or simply a gamma circuit.*

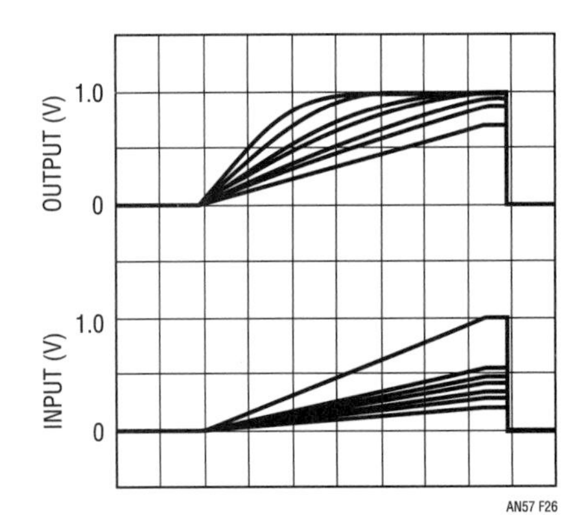

Figure 34.26 • Output of the Limiting Amp (I_{SET} = 0.68mA), with a Ramp Input. As the Input Amplitude Increases from 0.25V to 1V, the Output is Limited to 1V

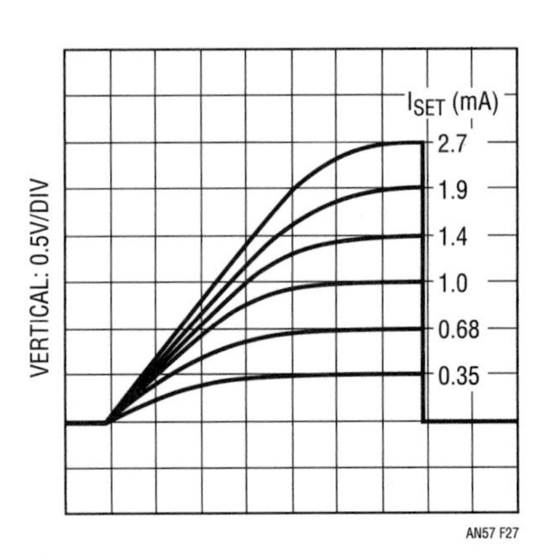

Figure 34.27 • The Output of the Limiting Amp with Various Limiting Levels (I_{SET}). The Input is a Ramp with a Maximum Amplitude of 0.75V

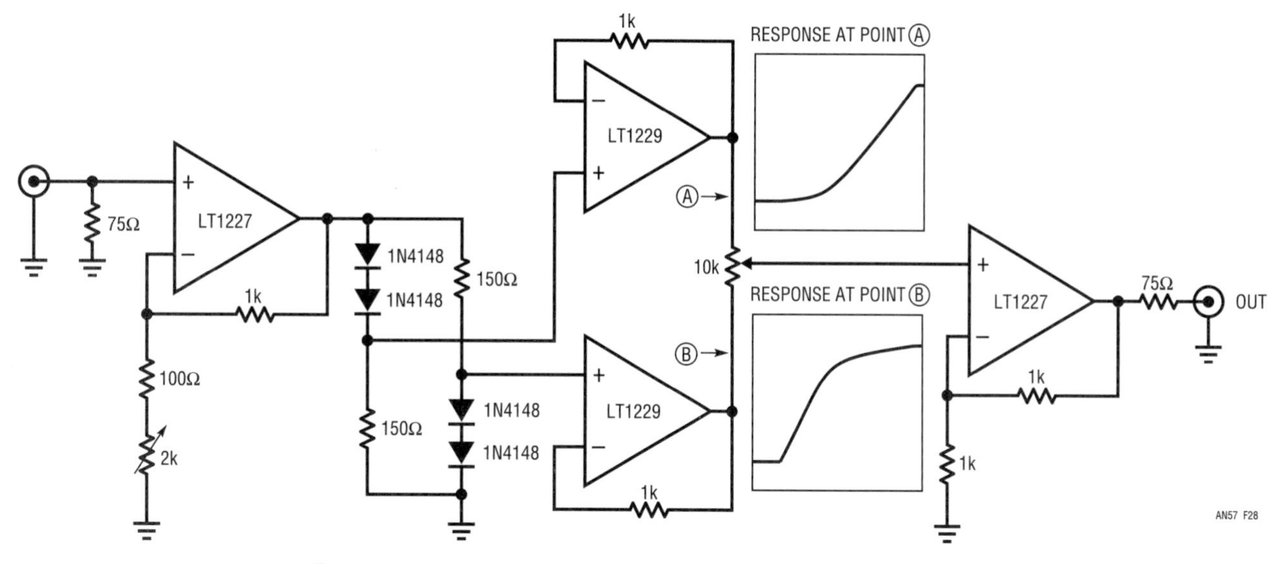

Figure 34.28 • Gamma Amp (Input Video Should Be Clamped)

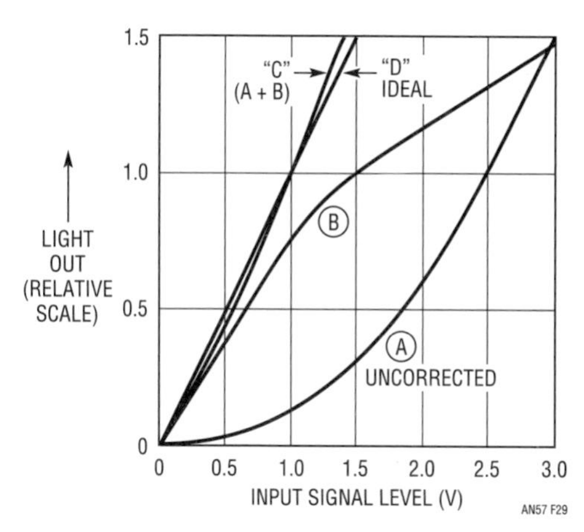

Figure 34.29 • Uncorrected CRT Transfer Function

Figure 34.28 shows a schematic of a typical circuit which can correct for positive or negative gamma. This is an upgrade of a classic circuit which uses diodes as the nonlinear elements. The temperature variation of the diode junction voltages is compensated to the first order by the balanced arrangement. LT1227s and LT1229s were used in the prototype, but a quad (LT1230) could save some space and work as well.

Figure 34.29, curve A, shows a response curve (transfer function) for an uncorrected CRT. To make such a response linear, the gamma corrector must have a gamma that is the reciprocal of the gamma of the device being linearized. The response of a two diode gamma circuit like that in Figure 34.28 is shown in Figure 34.29, curve B. Summing these two curves together, as in Figure 34.29, curve C, demonstrates the action of the gamma corrector. A straight line of appropriate slope, which would be an ideal response, is shown for comparison in Figure 34.29, curve D. Figure 34.30 is a triple exposure photograph of the gamma corrector circuit adjusted for gammas of −3, 1 and +3 (approximately). The input is a linear ramp of duration 52μs which is the period of an active horizontal line in NTSC video.

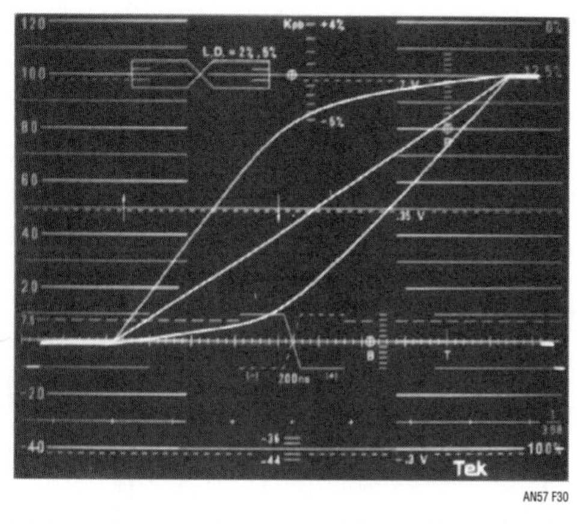

Figure 34.30 • Gamma Corrector Circuit Adjusted for Three Gammas: −3, 1, +3 (Approximately). The Input is a Linear Ramp

LT1228 sync summer

The circuit shown in Figure 34.31a restores the DC level and adds sync to a video waveform. For this example the video source is a high speed DAC with an output which is referenced to −1.2V. The LT1228 circuit (see the LT1228 data sheet for more details) forms a DC restore[2] that maintains a zero volt DC reference for the video. Figure 34.31b shows the waveform from the DAC, the DC restore pulse, and composite sync. The LT1363 circuit sums the video and composite sync signals. The 74AC04 CMOS inverters are used to buffer the TTL composite sync signal. In addition they drive the shaping network and, as they are mounted on the same ground plane as the analog circuitry, they isolate the ground noise from the digital system used to generate the video timing signals. Since the sync is directly summed to the video, any ground bounce or noise gets added in too. The shaping network is simply a third order Bessel lowpass filter with a bandwidth of 5MHz and an impedance of 300Ω. This circuitry slows the edge rate of the digital composite sync signal and also attenuates the noise. The same network, rescaled to an impedance of 75Ω, is used on the output of the summing amp to attenuate the switching noise from the DAC and to remove some of the high frequency components of the waveform. A more selective filter is not used here as the DAC has low glitch energy to start with and the signal does not have to meet stringent bandwidth requirements. The LT1363 used for the summing amp has excellent transient characteristics with no overshoot or ringing. Figure 34.31c shows two horizontal lines of the output waveform

Note 2: This is also referred to as "DC clamp" (or just clamp) but, there is a distinction. Both clamps and DC restore circuits act to maintain the proper DC level in a video signal by forcing the blanking level to be either zero volts or some other appropriate value. This is necessary because the video signal is often AC coupled as in a tape recorder or a transmitter. The DC level of an AC coupled video signal will vary with scene content and therefore the black referenced level must be "restored" in order for the picture to look right. A clamp is differentiated from a DC restore by its speed of response. A clamp is faster, generally correcting the DC error in one horizontal line (63.5μs for NTSC). A DC restore responds slower, more on the order of the frame time (16.7ms for NTSC). If there is any noise on the video signal the DC restore is the preferred method. A clamp can respond to noise pulses that occur during the blanking period and as a result give an erroneous black level for the line. Enough noise causes the picture to have an objectionable distortion called "piano keying." The black reference level and hence the luminance level change from line to line.

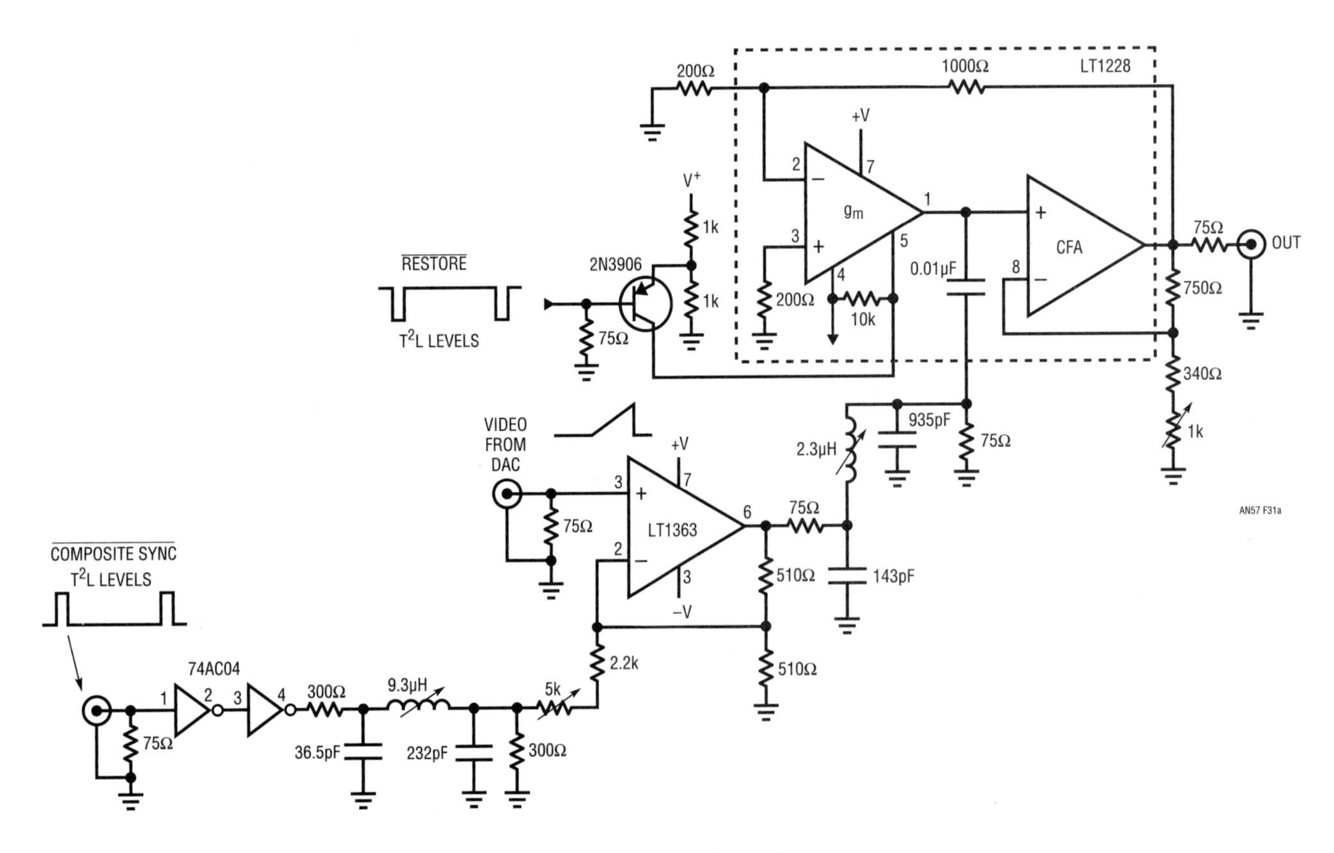

Figure 34.31a • Simple Sync Summer

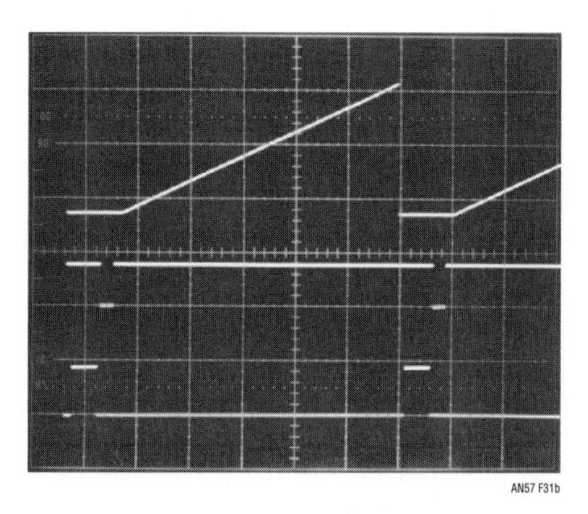

Figure 34.31b • Video Waveform from DAC; Clamp Pulse and Sync Pulse Used as Inputs to Sync Summer

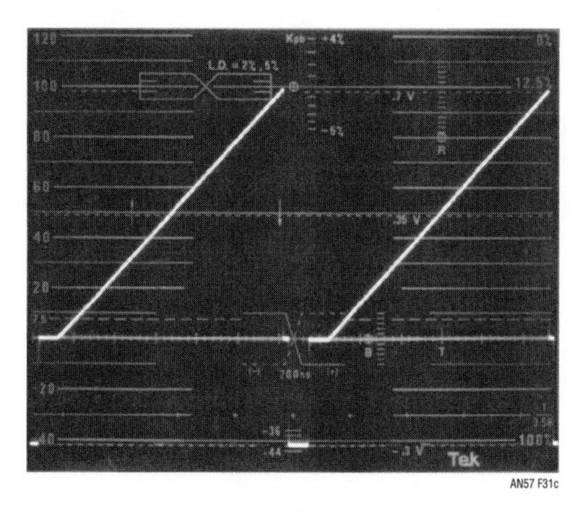

Figure 34.31c • Reconstructed Video Out of Sync Summer

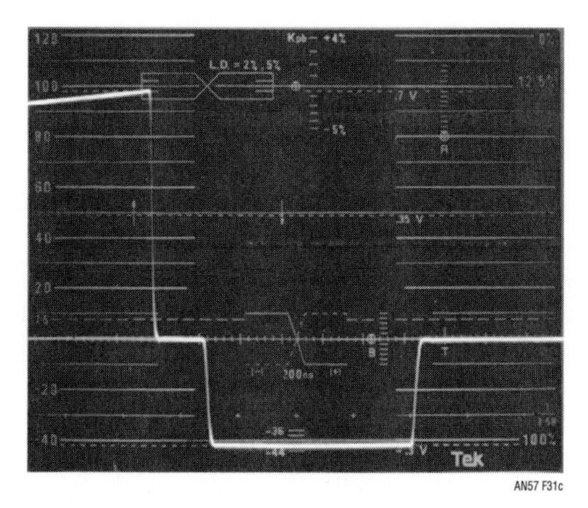

Figure 34.31d • Close-Up of Figure 34.31c, Showing Sync Pulse

with the DC restored and the sync added. Figure 34.31d is an expanded view of the banking interval showing a clean, well formed sync pulse.

Multiplexer circuits

Integrated three-channel output multiplexer

The LT6555 is a complete 3-channel wideband video 2:1 multiplexer with internally set gain of two. This part is ideal as an output port driver for HD component or high-resolution RGB video products. The basic application circuit is shown in Figure 34.32 with terminations shown on all ports, though in many applications the input loading may not be required. One thing this diagram does not reflect is the convenient flow-through pin assignments of the part, in which no video traces need cross in the printed-circuit layout. This maximizes isolation between channels and sources for best picture quality.

Since the LT6555 includes an enable control line, it is possible to extend the selection range of the multiplexer. Figure 34.33 shows two LT6555 devices in a configuration that provides 4:1 selection of RGB sources to an RGB output port (these could also be YP_BP_R signals as well, depending

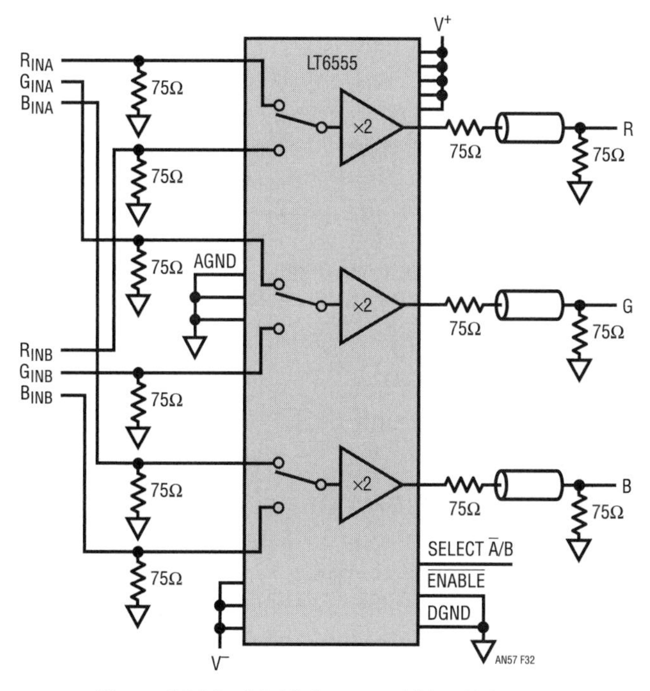

Figure 34.32 • Multiplexer and Line Driver

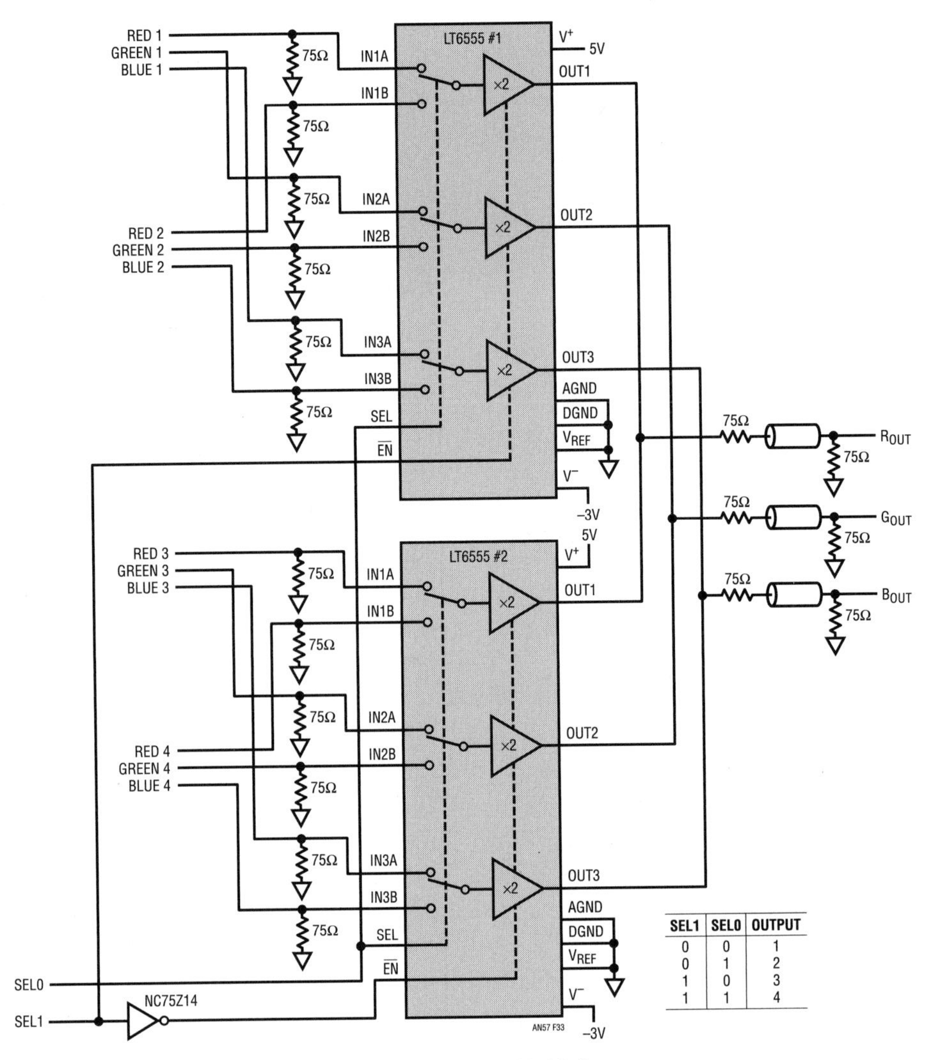

SEL1	SEL0	OUTPUT
0	0	1
0	1	2
1	0	3
1	1	4

Figure 34.33 • 4:1 RGB Multiplexer

on the source). To avoid frequency response anomalies, the two devices should be closely located so that the output lines between parts are as short as possible.

The LT1675 is also an integrated 3-channel 2:1 multiplexer that includes gain of two for cable-driving applications. The basic configuration is shown in Figure 34.34. A single channel version for composite video applications is available as an LT1675-1.

Integrated three-channel input multiplexer

The LT6556 is a complete 3-channel wideband video 2:1 multiplexer with internally set gain of one. This part is ideal as an input port receiver for HD component or high-resolution RGB video products. The basic application circuit is shown in Figure 34.35, with 1kΩ output loads to represent subsequent processing circuitry (the 1kΩ resistors aren't needed, but part characterization was performed with that loading). One thing this diagram does not reflect is the convenient flow-through pin assignments of the part, in which no video traces need cross in the printed-circuit layout. This maximizes isolation between channels and sources for best picture quality.

As with the LT6555, the LT6556 includes an enable control line, so it is possible to extend the selection range of this multiplexer as well. Figure 34.36 shows two LT6556 devices in a configuration that provides 4:1 selection of RGB sources to an RGB signal processing function, such as a digitizer in a projection system (these could be YP_BP_R signals just as well). To avoid frequency response anomalies, the two devices should be closely located so that the output lines between parts are as short as possible.

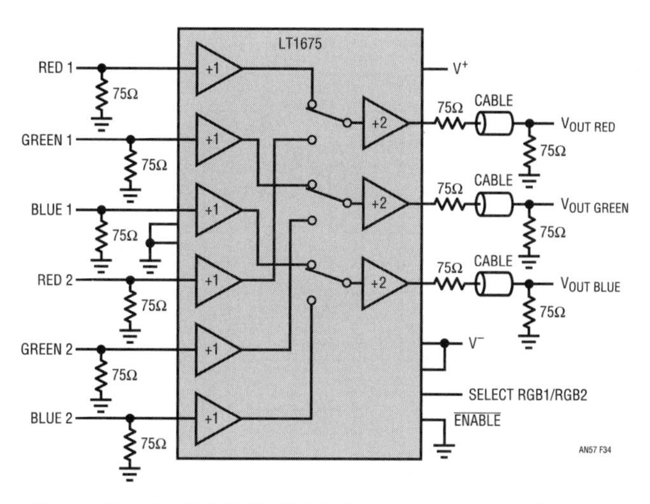

Figure 34.34 • 2:1 RGB Multiplexer and Cable Driver

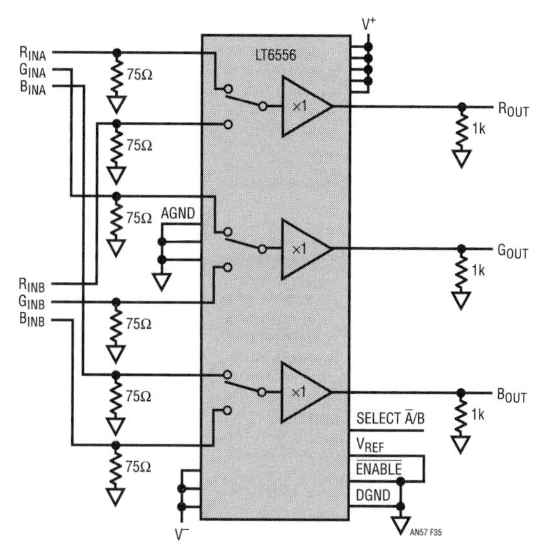

Figure 34.35 • Buffered Input Multiplexer/ADC Driver

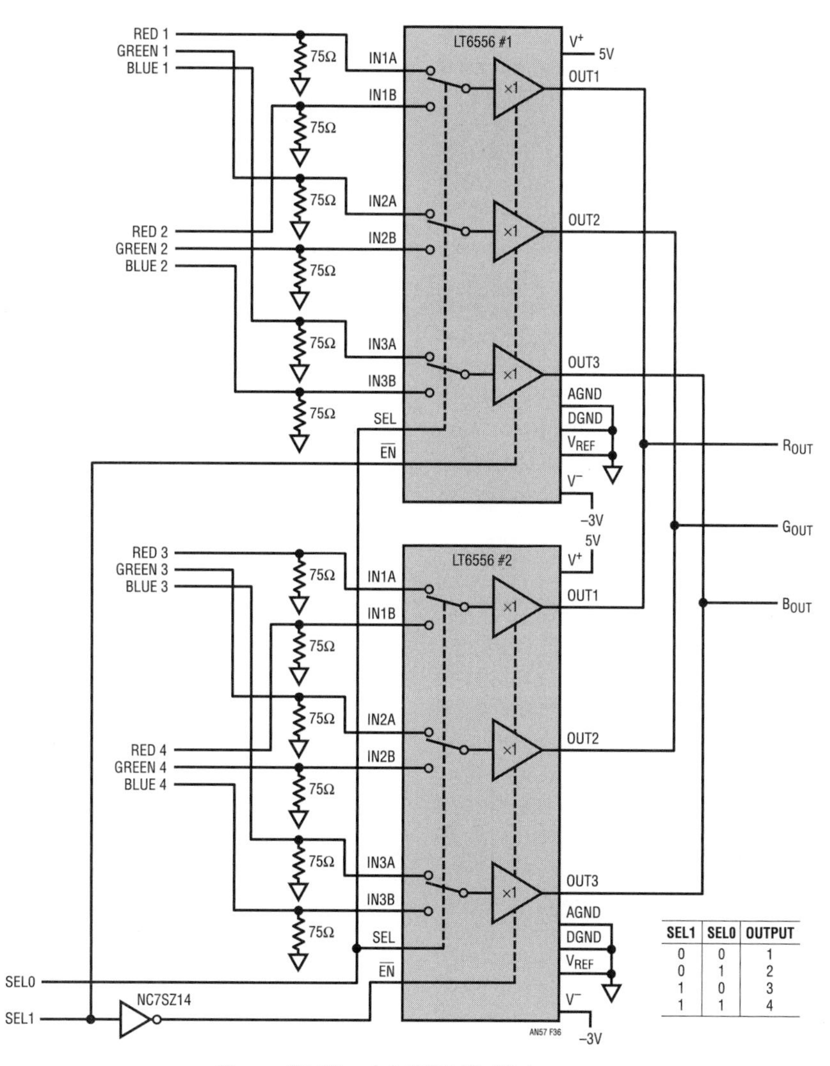

SEL1	SEL0	OUTPUT
0	0	1
0	1	2
1	0	3
1	1	4

Figure 34.36 • 4:1 RGB Multiplexer

A 3:1 cable-driving multiplexer for composite video can be formed from a single LT1399 as shown in Figure 34.37. The LT1399 has the unusual feature of having independent enable controls for each of the three sections. The gain of the amplifiers is set to compensate for passive loss in the loading associated with the off-section feedback networks.

Forming RGB multiplexers from triple amplifiers

The LT6553 triple cable driver and LT6554 triple buffer amp each provide an enable pin, so these parts can be used to implement video multiplexers. Figure 34.38 shows a pair of LT6553 devices configured as a 2:1 output multiplexer and cable driver. Similarly, Figure 34.39 shows a pair of LT6554 devices forming a 2:1 input mux, suitable as an ADC driver. These circuits are functionally similar to the LT6555 and LT6556 integrated multiplexers, but offer the flexibility of providing the mux feature as a simple stuffing option to a single printed circuit design, possibly reducing production costs when multiple product grades are being concurrently manufactured. For best results the two devices should be closely located and use minimal trace lengths between them for the shared output signals.

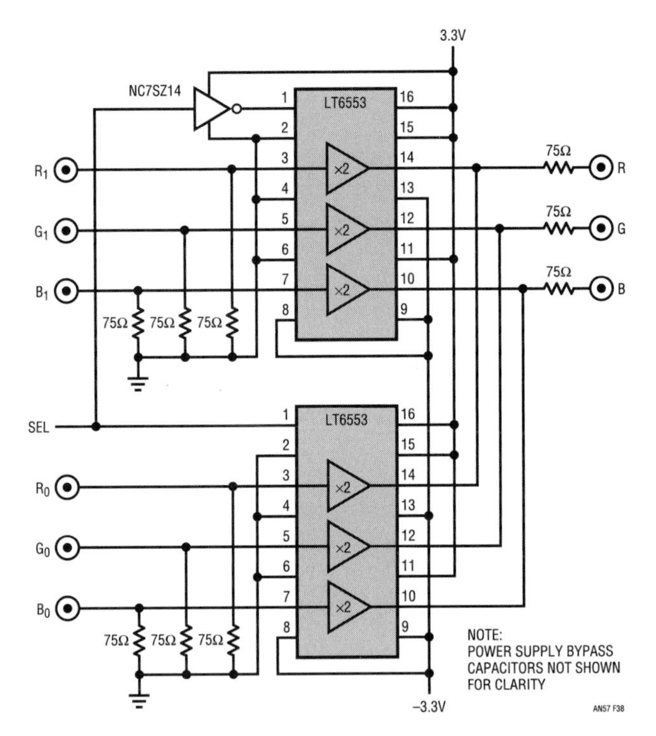

Figure 34.38 • RGB Video Selector/Cable Driver

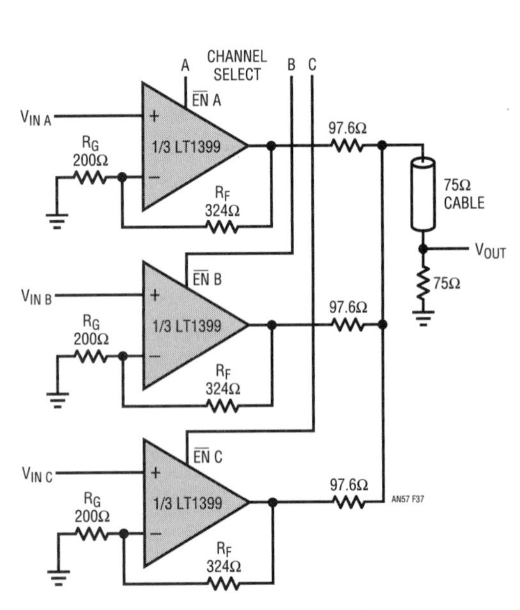

Figure 34.37 • 3-Input Video MUX Cable Driver

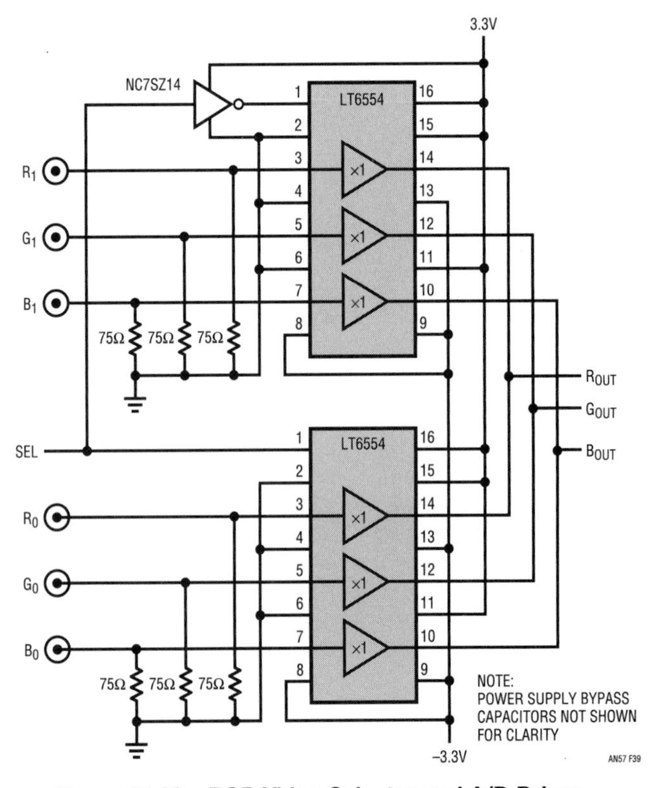

Figure 34.39 • RGB Video Selector and A/D Driver

Stepped gain amp using the LT1204

This is a straightforward approach to a switched-gain amp that features versatility. Figures 34.40 and 34.41 show circuits which implement a switched-gain amplifier; Figure 34.40 features an input Z of 1000Ω, while Figure 34.41's input Z is 75Ω. In either circuit, when LT1204 amp/MUX #2 is selected the signal is gained by one, or is attenuated by the resistor divider string depending on the input selected. When LT1204 amp/MUX #1 is selected there is an additional gain of sixteen. Consult the table in Figure 34.40. The gain steps can be either larger or smaller than shown here.

The input impedance (the sum of the divider resistors) is also arbitrary. Exercise caution in taking large gains however, because the bandwidth will change as the output is switched from one amp to another. Taking more gain in the amp/MUX #1 will lower its bandwidth even though it is a current feedback amplifier (CFA). This is less true for a CFA than for a voltage feedback amp.

LT1204 amplifier/multiplexer sends video over long twisted pair

Figure 34.42 is a circuit which can transmit baseband video over more than 1000 feet of very inexpensive twisted pair

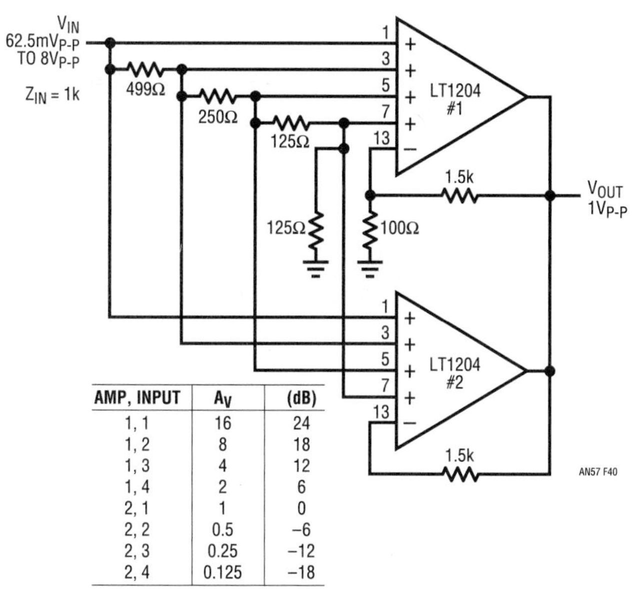

AMP, INPUT	A_V	(dB)
1, 1	16	24
1, 2	8	18
1, 3	4	12
1, 4	2	6
2, 1	1	0
2, 2	0.5	−6
2, 3	0.25	−12
2, 4	0.125	−18

Figure 34.40 • Switchable Gain Amplifier Accepts Inputs from 62.5mV$_{P-P}$ to 8V$_{P-P}$

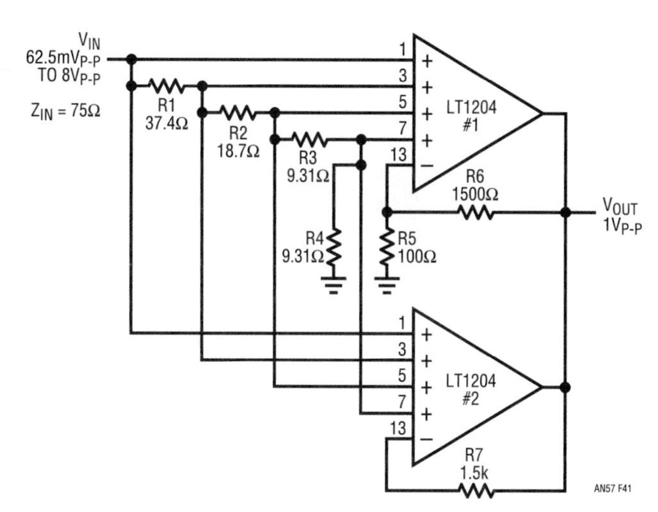

Figure 34.41 • Switchable Gain Amplifier, Z$_{IN}$ = 75Ω Same Gains as Figure 34.37

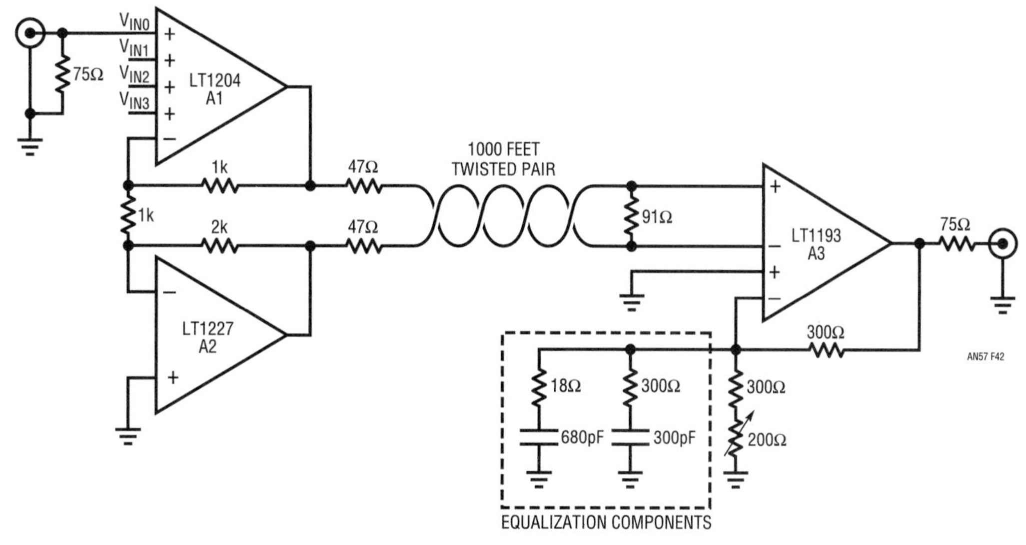

Figure 34.42 • Twisted Pair Driver/Receiver

wire and allow the selection of one-of-four inputs. Amp/MUX A1 (LT1204) and A2 (LT1227) form a single differential driver. A3 is a variable gain differential receiver built using the LT1193. The rather elaborate equalization (highlighted on the schematic) is necessary here as the twisted pair goes self-resonant at about 3.8MHz.

Figure 34.43 shows the video test signal before and after transmission but without equalization. Figure 34.44 shows before and after with the equalization connected. Differential gain and phase are about 1% and 1°, respectively.

Fast differential multiplexer

This circuit (Figure 34.45) takes advantage of the gain node on the LT1204 to make a high speed differential MUX for receiving analog signals over twisted pair. Common-mode noise on loop-through connections is reduced because of the unique differential input. Figure 34.45's circuit also makes a robust differential to single-ended amp/MUX for high speed data acquisition.

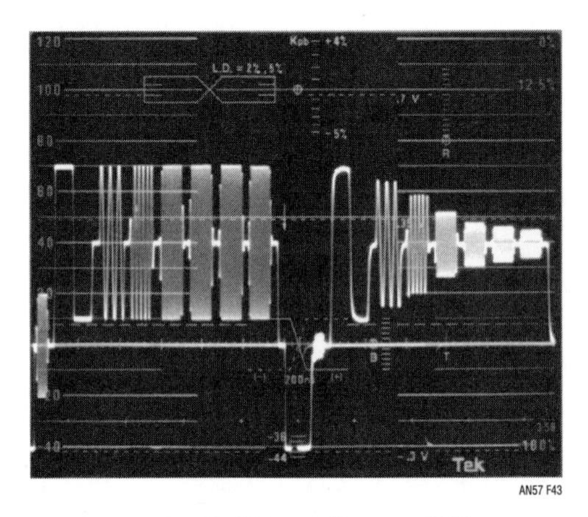

Figure 34.43 • Multiburst Pattern Without Cable Compensation

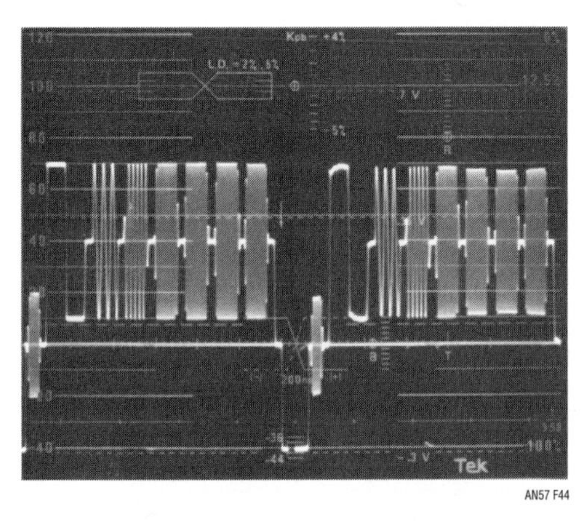

Figure 34.44 • Multiburst Pattern with Cable Compensation

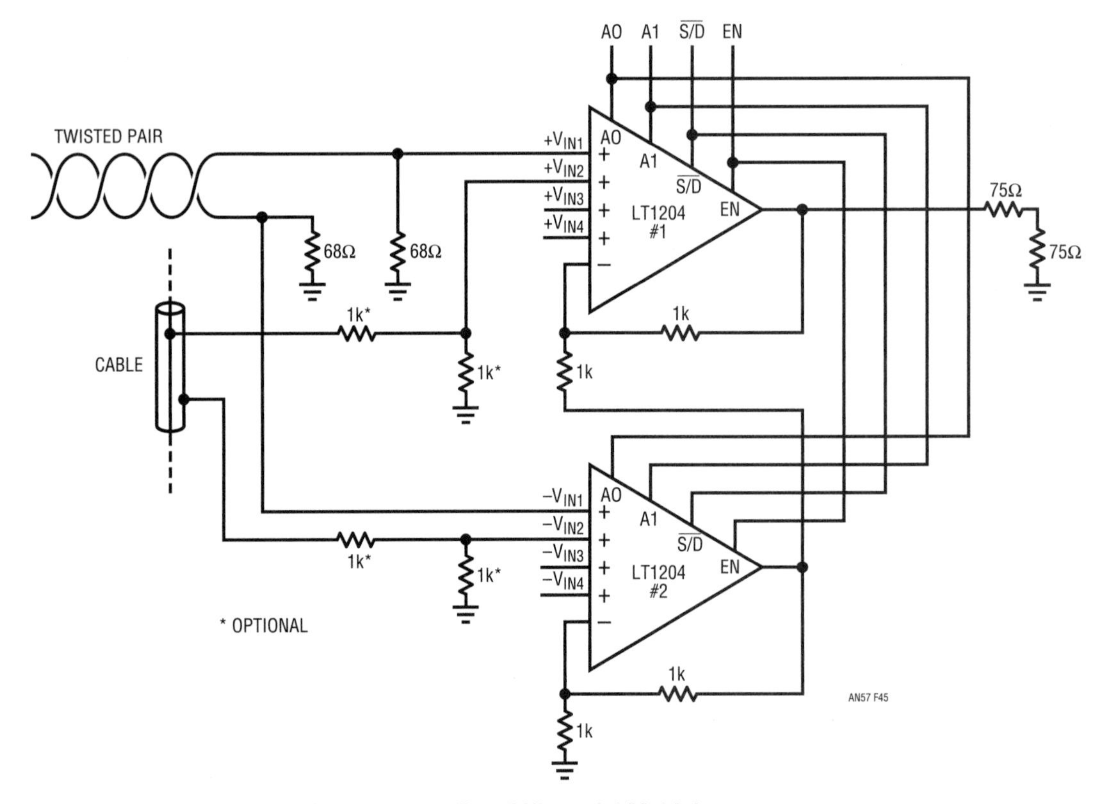

Figure 34.45 • Fast Differential Multiplexer

Signals passing through LT1204 #1 see a noninverting gain of two. Signals passing through LT1204 #2 also see a noninverting gain of two and then an inverting gain of one (for a resultant gain of minus two) because this amp drives the gain resistor on amp #1. The result is differential amplification of the input signal.

The optional resistors on the second input are for input protection. Figure 34.46 shows the differential mode response versus frequency. The limit to the response (at low frequency) is the matching of the gain resistors. One percent resistors will match to about 0.1% (60dB) if they are from the same batch.

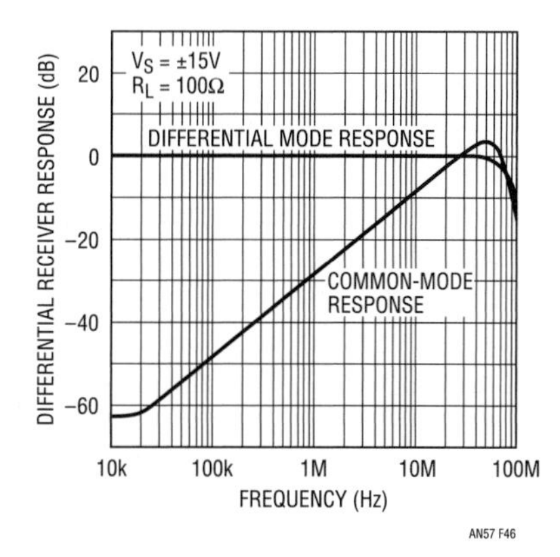

Figure 34.46 • Differential Receiver Response vs Frequency

Misapplications of CFAs

In general the current feedback amplifier (CFA) is remarkably docile and easy to use. These amplifiers feature "real, "usable gain to 100MHz and beyond, low power consumption and an amazingly low price. However, CFAs are still new enough so that there is room for breadboard adventure. Consult the diagrams and the following list for some of the pitfalls that have come to my attention[3].

Note 3: All the usual rules for any high speed circuit still apply, of course. A partial list:
a. Use a ground plane.
b. Use good RF bypass techniques. Capacitors used should have short leads, high self-resonant frequency, and be placed close to the pin.
c. Keep values of resistors low to minimize the effects of parasitics. Make sure the amplifier can drive the chosen low impedance.
d. Use transmission lines (coax, twisted pair) to run signals more than a few inches.
e. Terminate the transmission lines (back terminate the lines if you can).
f. Use resistors that are still resistors at 100MHz.
Refer to AN47 for a discussion of these topics.

1. Be sure there is a DC path to ground on the noninverting input pin. There is a transistor in the input that needs some bias current.
2. Don't use pure reactances for a feedback element. This is one sure way to get the CFA to oscillate. Consult the amplifier data sheet for guidance on feedback resistor values. Remember that these values have a direct effect on the bandwidth. If you wish to tailor frequency response with reactive networks, put them in place of R_G, the gain setting resistor.
3. Need a noninverting buffer? Use a feedback resistor!
4. Any resistance between the inverting terminal and the feedback node causes loss of bandwidth.
5. For good dynamic response, avoid parasitic capacitance on the inverting input.
6. Don't use a high Q inductor for power supply decoupling (or even a middling Q inductor for that matter). The inductor and the bypass capacitors form a tank circuit, which can be excited by the AC power supply currents, causing just the opposite of the desired effect. A lossy ferrite choke can be a very effective way to decouple power supply leads without the voltage drop of a series resistor. For more information on ferrites call Fair Rite Products Corp. (914) 895-2055.

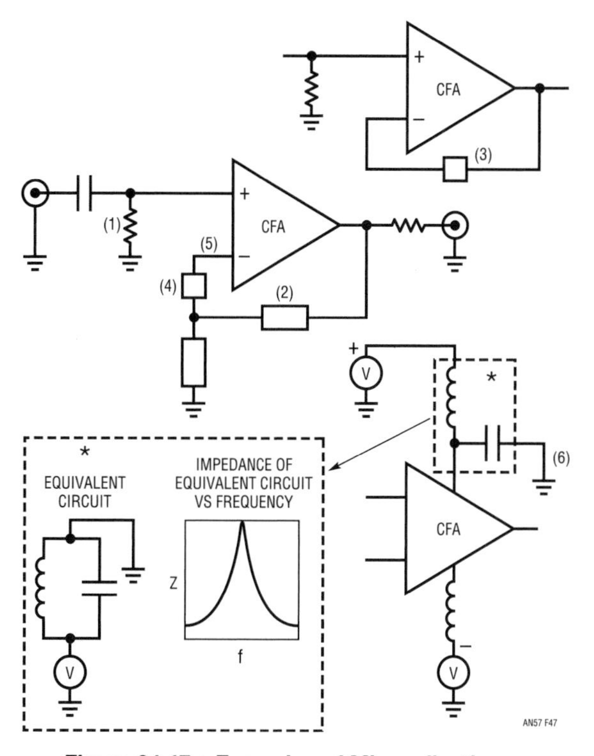

Figure 34.47 • Examples of Misapplications

Appendix A
A temperature-compensated, voltage-controlled gain amplifier using the LT1228

It is often convenient to control the gain of a video or intermediate frequency (IF) circuit with a voltage. The LT1228, along with a suitable voltage-to-current converter circuit, forms a versatile gain-control building block ideal for many of these applications.

In addition to gain control over video bandwidths, this circuit can add a differential input and has sufficient output drive for 50Ω systems.

The transconductance of the LT1228 is inversely proportional to absolute temperature at a rate of −0.33%/°C. For circuits using closed-loop gain control (i.e., IF or video automatic gain control) this temperature coefficient does not present a problem. However, open-loop gain-control circuits that require accurate gains may require some compensation. The circuit described here uses a simple thermistor network in the voltage-to-current converter to achieve this compensation. Table A1 summarizes the circuit's performance.

Figure A1 shows the complete schematic of the gain-control amplifier. Please note that these component choices are not the only ones that will work nor are they necessarily the best. This circuit is intended to demonstrate one approach out of many for this very versatile part and, as always, the designer's engineering judgment must be fully engaged. Selection of the values for the input attenuator, gain-set resistor, and current feedback amplifier resistors is relatively straightforward, although some iteration is usually necessary. For the best bandwidth, remember to keep the gain-set resistor R1 as small as possible and the set current as large as possible with due regard for gain compression. See the "Voltage-Controlled Current Source" (I_{SET}) box for details.

Several of these circuits have been built and tested using various gain options and different thermistor values. Test results for one of these circuits are shown in Figure A2. The gain error versus temperature for this circuit is

Table A1 Characteristics of Exam

Input Signal Range	0.5V to 3.0V_{PK}
Desired Output Voltage	1.0V_{PK}
Frequency Range	0Hz to 5MHz
Operating Temperature Range	0°C to 50°C
Supply Voltages	±15V
Output Load	150Ω (75Ω + 75Ω)
Control Voltage vs Gain Relationship	0V to 5V Min to Max Gain
Gain Variation Over Temperature	±3% from Gain at 25°C

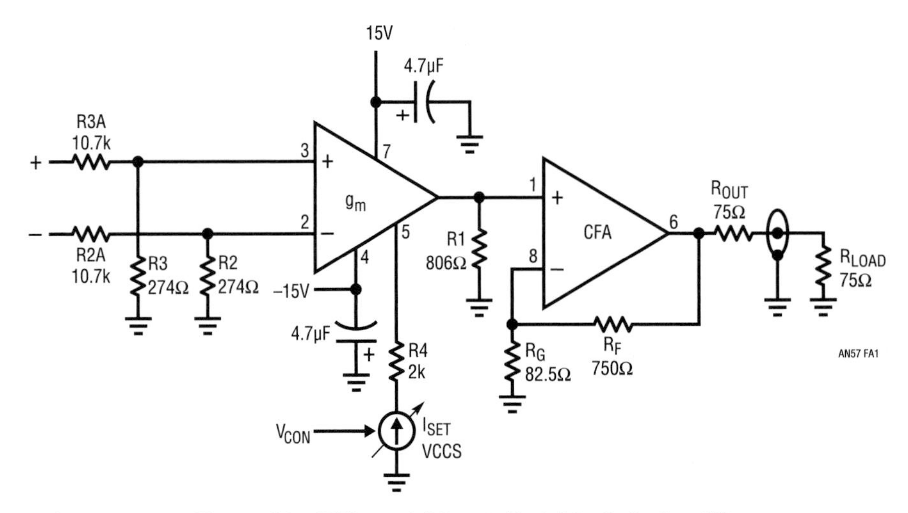

Figure A1 • Differential-Input, Variable-Gain Amplifier

well within the limit of ±3%. Compensation over a much wider range of temperatures or to tighter tolerances is possible, but would generally require more sophisticated methods, such as multiple thermistor networks.

The VCCS is a standard circuit with the exception of the current-set resistor R5, which is made to have a temperature coefficient of −0.33%/°C. R6 sets the overall gain and is made adjustable to trim out the initial tolerance in the LT1228 gain characteristic. A resistor (R_P) in parallel with the thermistor will tend, over a relatively small range, to linearize the change in resistance of the combination with temperature. R_S trims the temperature coefficient of the network to the desired value.

This procedure was performed using a variety of thermistors. BetaTHERM Corporation is one possible source, phone 508-842-0516. Figure A3 shows typical results reported as errors normalized to a resistance with a −0.33%/°C temperature coefficient. As a practical matter, the thermistor need only have about a 10% tolerance for this gain accuracy. The sensitivity of the gain accuracy to the thermistor tolerance is decreased by the linearization network in the same ratio as is the temperature coefficient. The room temperature gain may be trimmed with R6. Of course, particular applications require analysis of aging stability, interchangeability, package style, cost, and the contributions of the tolerances of the other components in the circuit.

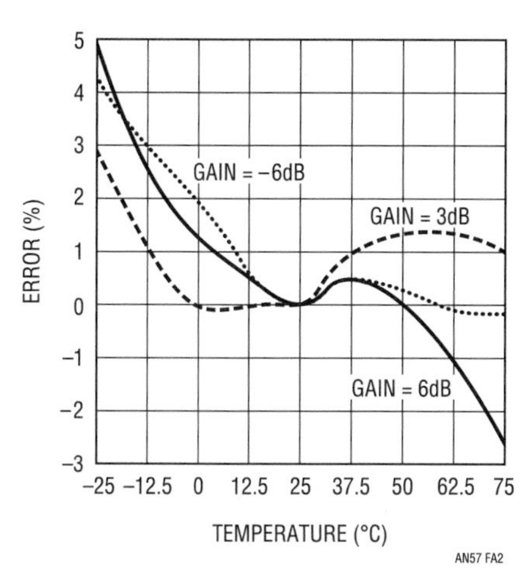

Figure A2 • Gain Error for Circuit in Figure A2 Plus Temperature Compensation Circuit Shown in Figure A4 (Normalized to Gain at 25°C)

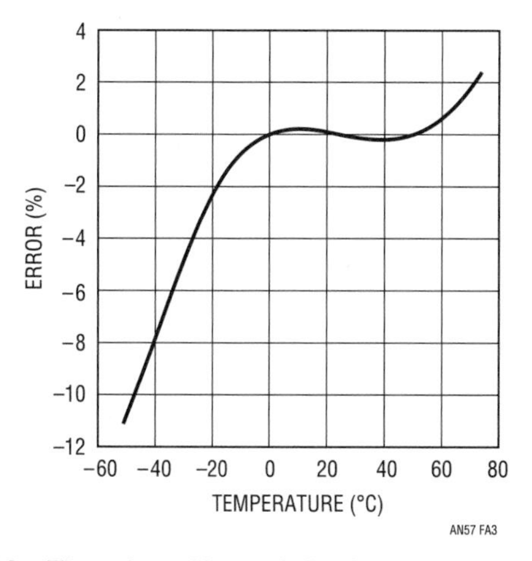

Figure A3 • Thermistor Network Resistance Normalized to a Resistor with Exact −0.33%/°C Temperature Coefficient

Voltage-controlled current source (VCCS) with a compensating temperature coefficient

VCCS design steps

1. Measure, or obtain from the data sheet, the thermistor resistance at three equally spaced temperatures (in this case 0°C, 25°C, and 50°C). Find R_P from:

$$R_p = \frac{(R0 \times R25 + R25 \times R50 - 2 \times R0 \times R50)}{(R0 + R50 - 2 \times R25)}$$

where R0 = thermistor resistance at 0°C
R25 = thermistor resistance at 25°C
R50 = thermistor resistance at 50°C

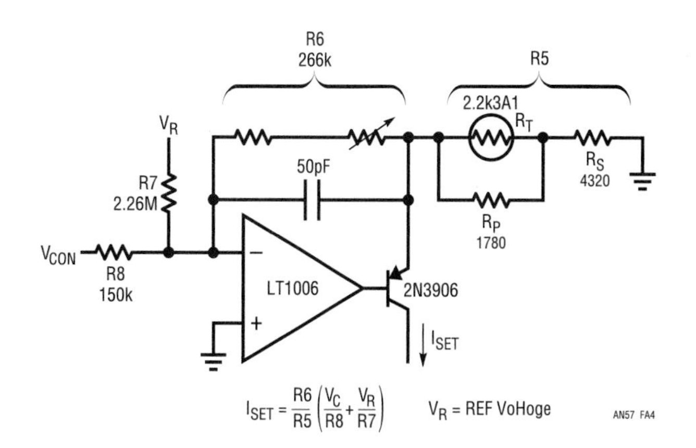

Figure A4 • Voltage-Controlled Current Source (VCCS) with a Compensating Temperature Coefficient

2. Resistor R_P is placed in parallel with the thermistor. This network has a temperature dependence that is approximately linear over the range given (0°C to 50°C).
3. The parallel combination of the thermistor and R_P ($R_P||R_T$) has a temperature coefficient (TC) of resistance given by:

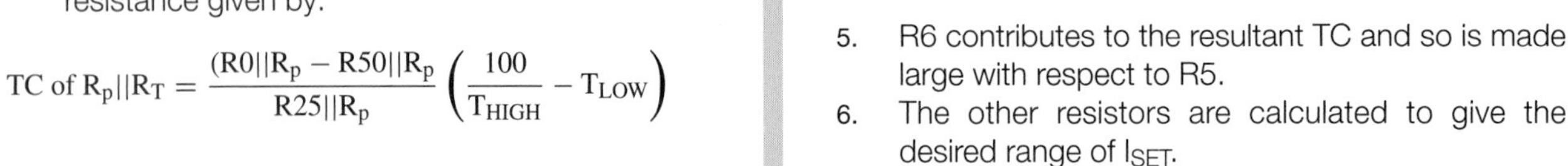

$$\text{TC of } R_p||R_T = \frac{(R0||R_p - R50||R_p}{R25||R_p}\left(\frac{100}{T_{HIGH}} - T_{LOW}\right)$$

4. The desired tempco to compensate the LT1228 gain temperature dependence is −0.33%/°C. A series resistance (R_S) is added to the parallel network to trim its tempco to the proper value. R_S is given by:

$$\frac{\text{TC of } R_p||R_T}{-0.33} \times (R_p||R_{25}) - (R_p||R_{25})$$

5. R6 contributes to the resultant TC and so is made large with respect to R5.
6. The other resistors are calculated to give the desired range of I_{SET}.

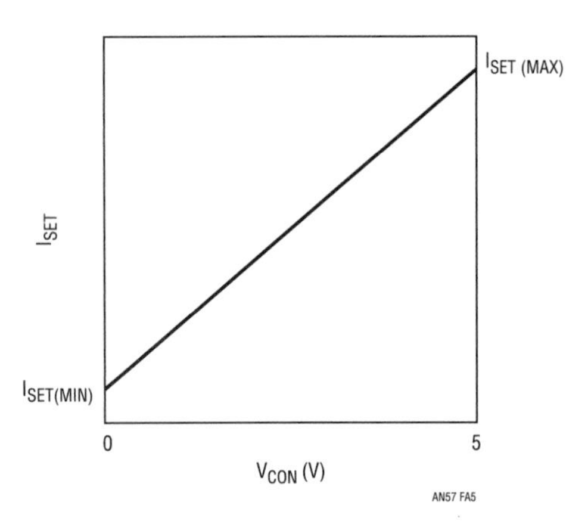

Figure A5 • Voltage Control of I_{SET} with Temperature Compensation

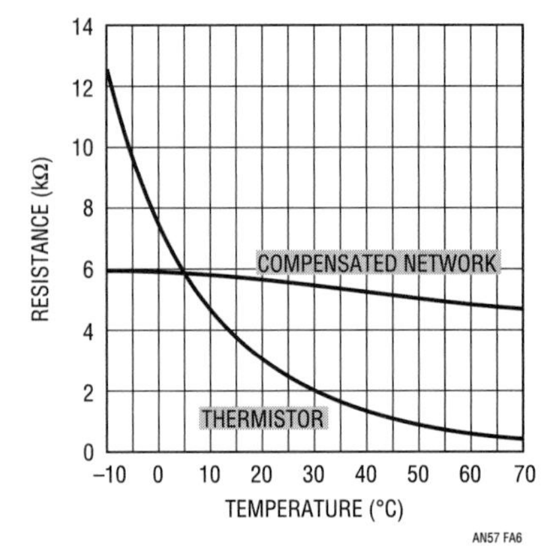

Figure A6 • Thermistor and Thermistor Network Resistance vs Temperature

Appendix B
Optimizing a video gain-control stage using the LT1228

Video automatic-gain-control (AGC) systems require a voltage- or current-controlled gain element. The performance of this gain-control element is often a limiting factor in the overall performance of the AGC loop. The gain element is subject to several, often conflicting restraints. This is especially true of AGC for composite color video systems, such as NTSC, which have exacting phase- and gain-distortion requirements. To preserve the best possible signal-to-noise ratio (S/N),[1] it is desirable for the input signal level to be as large as practical. Obviously, the larger the input signal the less the S/N will be degraded by the noise contribution of the gain-control stage. On the other hand, the gain-control element is subject to dynamic range constraints; exceeding these will result in rising levels of distortion.

Note 1: Signal-to-noise ratio, S/N = 20 × log(RMS signal/RMS noise).

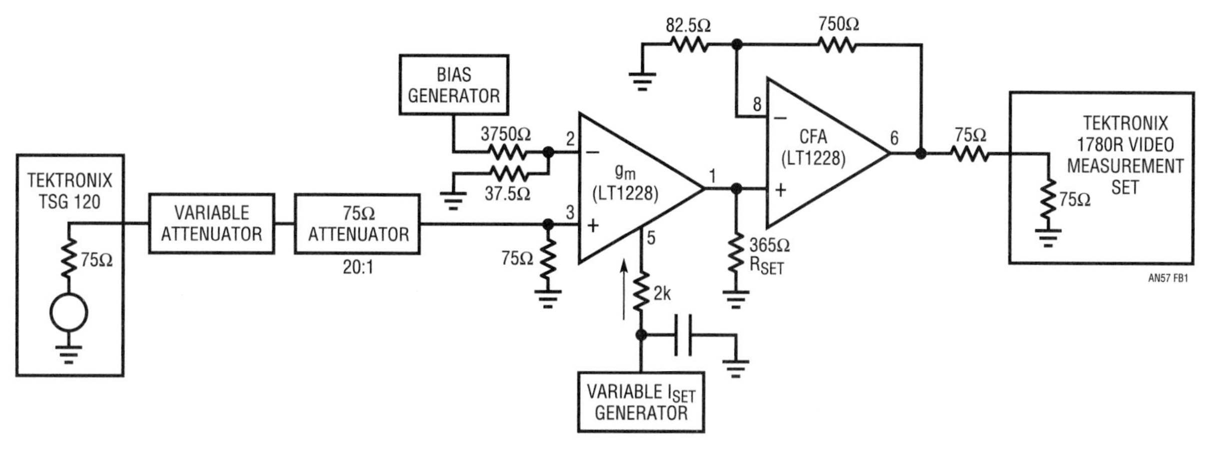

Figure B1 • Schematic Diagram

Table B1 Measured Performance Data (Uncorrected)

INPUT (V)	I_{SET} (mA)	DIFFERENTIAL GAIN	DIFFERENTIAL PHASE	S/N
0.03	1.93	0.5%	2.7°	55dB
0.06	0.90	1.2%	1.2°	56dB
0.09	0.584	10.8%	3.0°	57dB

Table B2 Measured Performance Data (Corrected)

INPUT (V)	BIAS VOLTAGE	I_{SET} (mA)	DIFFERENTIAL GAIN	DIFFERENTIAL PHASE	S/N
0.03	0.03	1.935	0.9%	1.45°	55dB
0.06	0.03	0.889	1.0%	2.25°	56dB
0.09	0.03	0.584	1.4%	2.85°	57dB

Linear Technology makes a high speed transconductance (g_m) amplifier, the LT1228, which can be used as a quality, inexpensive gain-control element in color video and some lower frequency R_F applications. Extracting the optimum performance from video AGC systems takes careful attention to circuit details.

As an example of this optimization, consider the typical gain-control circuit using the LT1228 shown in Figure B1. The input is NTSC composite video, which can cover a 10dB range from 0.56V to 1.8V. The output is to be $1V_{P-P}$ into 75Ω. Amplitudes were measured from peak negative chroma to peak positive chroma on an NTSC modulated ramp test signal. See "Differential Gain and Phase" box.

Notice that the signal is attenuated 20:1 by the 75Ω attenuator at the input of the LT1228, so the voltage on the input (pin 3) ranges from 0.028V to 0.090V. This is done to limit distortion in the transconductance stage. The gain of this circuit is controlled by the current into the I_{SET} terminal, pin 5 of the IC. In a closed-loop AGC system, the loop-control circuitry generates this current by comparing the output of a detector[2] to a reference voltage, integrating the difference and then converting to a suitable current. The measured performance for this circuit is presented in tables B1 and B2. Table B1 has the uncorrected data and Table B2 shows the results of the correction.

All video measurements were taken with a Tektronix 1780R video-measurement set, using test signals generated by a Tektronix TSG 120. The standard criteria for characterizing NTSC video color distortion are the differential gain and the differential phase. For a brief explanation of these tests see the box "Differential Gain and Phase."

For this design exercise the distortion limits were set at a somewhat arbitrary 3% for differential gain and 3° for differential phase. Depending on conditions, this should be barely visible on a video monitor.

Figures B2 and B3 plot the measured differential gain and phase, respectively, against the input signal level (the curves labeled "A" show the uncorrected data from Table B1). The plots show that increasing the input signal

Note 2: One way to do this is to sample the colorburst amplitude with a sample- and-hold and peak detector. The nominal peak-to-peak amplitude of the colorburst for NTSC is 40% of the peak luminance.

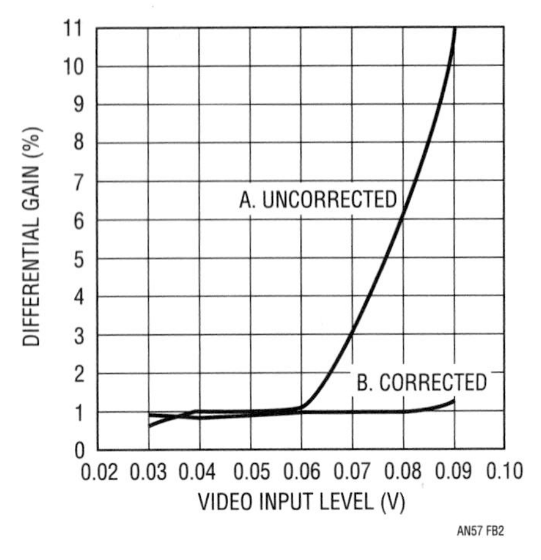

Figure B2 • Differential Gain vs Input Level

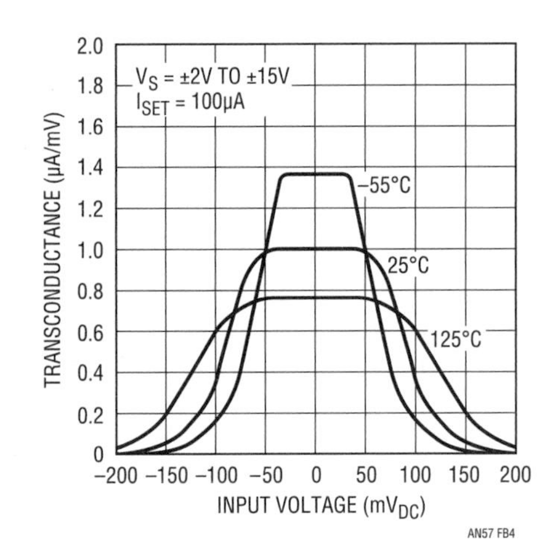

Figure B4 • Small-Signal Transconductance vs DC Input Voltage

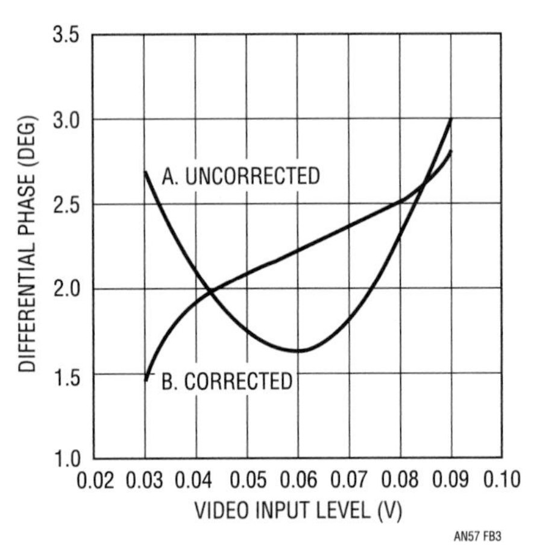

Figure B3 • Differential Phase vs Input Level

level beyond 0.06V results in a rapid increase in the gain distortion, but comparatively little change in the phase distortion. Further attenuating the input signal (and consequently increasing the set current) would improve the differential gain performance but degrade the S/N. What this circuit needs is a good tweak!

Optimizing for differential gain

Referring to the small signal transconductance versus DC input voltage graph (Figure B4), observe that the transconductance of the amplifier is linear over a region centered around zero volts.[3] The 25°C g_m curve starts to become quite nonlinear above 0.050V. This explains why the differential gain (see Figure B2, curve A) degrades so quickly with signals above this level. Most RF signals do not have DC bias levels, but the composite video signal is mostly unipolar.

Video is usually clamped at some DC level to allow easy processing of sync information. The sync tip, the chroma reference burst, and some chroma signal information swing negative, but 80% of the signal that carries the critical color information (chroma) swings positive. Efficient use of the dynamic range of the LT1228 requires that the input signal have little or no offset. Offsetting the video signal so that the critical part of the chroma waveform is centered in the linear region of the transconductance amplifier allows a larger signal to be input before the onset of severe distortion. A simple way to do this is to bias the unused input (in this circuit the inverting input, Pin 2) with a DC level.

In a video system it might be convenient to clamp the sync tip at a more negative voltage than usual. Clamping the signal prior to the gain-control stage is good practice because a stable DC reference level must be maintained.

Note 3: Notice also that the linear region expands with higher temperature. Heating the chip has been suggested.

The optimum value of the bias level on Pin 2 used for this evaluation was determined experimentally to be about 0.03V. The distortion tests were repeated with this bias voltage added. The results are reported in Table B2 and Figures B2 and B3. The improvement to the differential phase is inconclusive, but the improvement in the differential gain is substantial.

Differential gain and phase

Differential gain and phase are sensitive indications of chroma signal distortion. The NTSC system encodes color information on a separate subcarrier at 3.579545MHz. The color subcarrier is directly summed to the black and white video signal. The black and white information is a voltage proportional to image intensity and is called luminance or luma. Each line of video has a burst of 9 to 11 cycles of the subcarrier (so timed that it is not visible) that is used as a phase reference for demodulation of the color information of that line. The color signal is relatively immune to distortions, except for those that cause a phase shift or an amplitude error to the subcarrier during the period of the video line.

Differential gain is a measure of the gain error of a linear amplifier at the frequency of the color subcarrier. This distortion is measured with a test signal called a modulated ramp (shown in Figure B5). The modulated ramp consists of the color subcarrier frequency superimposed on a linear ramp or sometimes on a stair step. The ramp has the duration of the active portion of a horizontal line of video. The amplitude of the ramp varies from zero to the maximum level of the luminance, which, in this case, is 0.714V. The gain error corresponds to compression or expansion by the amplifier (sometimes called "incremental gain") and is expressed as a percentage of the full amplitude range. An appreciable amount of differential gain will cause the luminance to modulate the chroma causing visual chroma distortion. The effect of differential gain errors is to change the saturation of the color being displayed. Saturation is the relative degree of dilution of a pure color with white. A 100% saturated color has 0% white, a 75% saturated color has 25% white, and so on. Pure red is 100% saturated whereas, pink is red with some percentage of white and is therefore less than 100% saturated.

Differential phase is a measure of the phase shift in a linear amplifier at the color subcarrier frequency when the modulated ramp signal is used as an input.

The phase shift is measured relative to the colorburst on the test waveform and is expressed in degrees. The visual effect of the distortion is a change in hue. Hue is the quality of perception which differentiates the

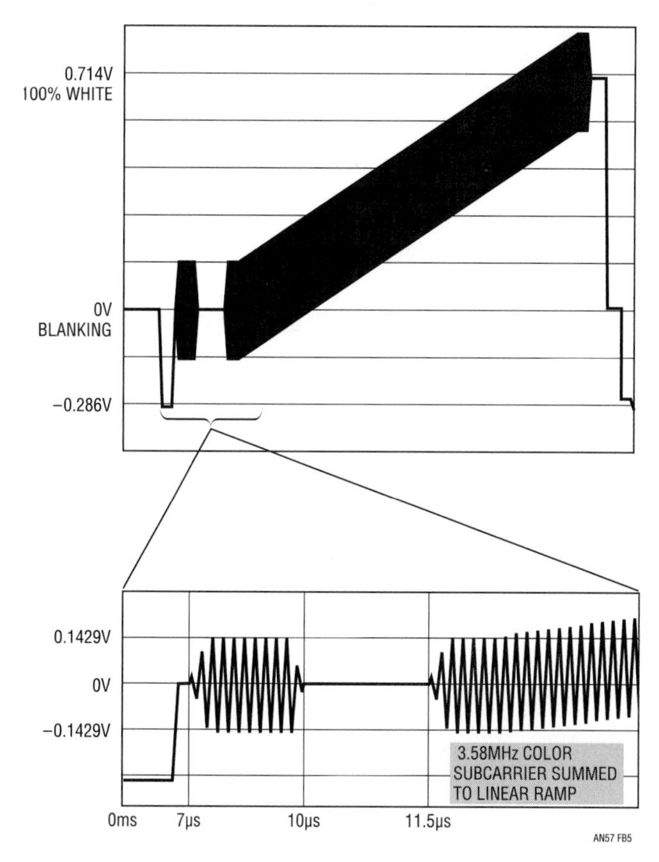

Figure B5 • NTSC Test Signal

frequency of the color, red from green, yellow-green from yellow, and so forth.

Three degrees of differential phase is about the lower limit that can unambiguously be detected by observers. This level of differential phase is just detectable on a video monitor as a shift in hue, mostly in the yellow-green region. Saturation errors are somewhat harder to see at these levels of distortion—3% of differential gain is very difficult to detect on a monitor. The test is performed by switching between a reference signal, SMPTE (Society of Motion Picture and Television Engineers) 75% color bars, and a distorted version of the same signal with matched signal levels. An observer is then asked to note any difference.

In professional video systems (studios, for instance) cascades of processing and gain blocks can reach hundreds of units. In order to maintain a quality video signal, the distortion contribution of each processing block must be a small fraction of the total allowed distortion budget[4]

Note 4: From the preceding discussion, the limits on visibility are about 3° differential phase, 3% differential gain. Please note that these are not hard and fast limits. Tests of perception can be very subjective.

because the errors are cumulative. For this reason, high-quality video amplifiers will have distortion specifications as low as a few thousandths of a degree for differential phase and a few thousandths of a percent for differential gain.

Appendix C
Using a fast analog multiplexer to switch video signals for NTSC "picture-in-picture" displays

The majority of production[1] video switching consists of selecting one video source out of many for signal routing or scene editing. For these purposes the video signal is switched during the vertical interval in order to reduce visual switching transients. The image is blanked during this time, so if the horizontal and vertical synchronization and subcarrier lock are maintained, there will be no visible artifacts. Although vertical-interval switching is adequate for most routing functions, there are times when it is desirable to switch two synchronous video signals during the active (visible) portion of the line to obtain picture-in-picture, key, or overlay effects. Picture-in-picture or active video switching requires signal-to-signal transitions that are both clean and fast. A clean transition should have a minimum of pre-shoot, overshoot, ringing, or other aberrations commonly lumped under the term "glitching."

Using the LT1204

A quality, high speed multiplexer amplifier can be used with good results for active video switching. The important specifications for this application are a small, controlled switching glitch, good switching speed, low distortion, good dynamic range, wide bandwidth, low path loss, low channel-to-channel crosstalk, and good channel-to-channel offset matching. The LT1204 specifications match these requirements quite well, especially in the areas of bandwidth, distortion, and channel-to-channel crosstalk which is an outstanding −90dB at 10MHz. The LT1204 was evaluated for use in active video switching with the test setup shown in Figure C1. Figure C2 shows the video waveform of a switch between a 50% white level and a 0% white level about 30% into the active interval and back again at about 60% of the active interval. The switch artifact is brief and well controlled. Figure C3 is an expanded view of the same waveform.

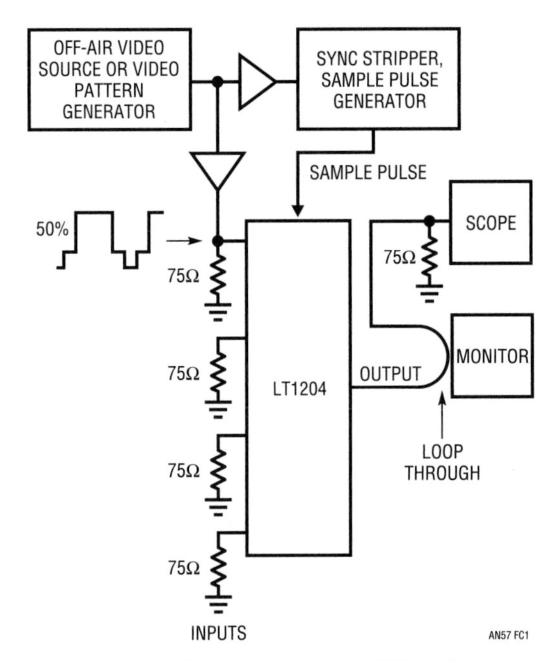

Figure C1 • "Picture-in-Picture" Test Setup

Note 1: Video production, in the most general sense, means any purposeful manipulation of the video signal, whether in a television studio or on a desktop PC.

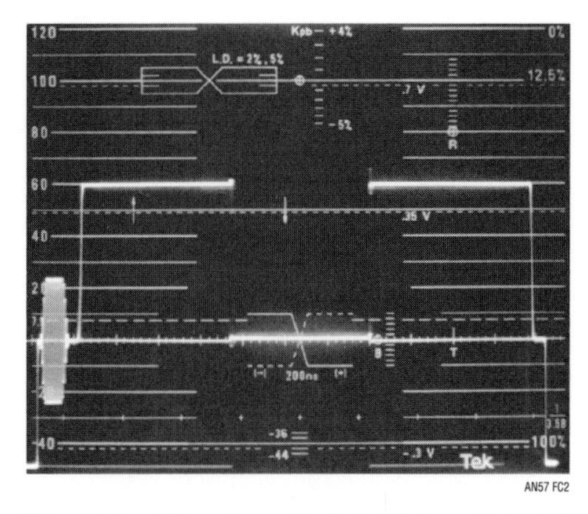

Figure C2 • Video Waveform Switched from 50% White Level to 0% White Level and Back

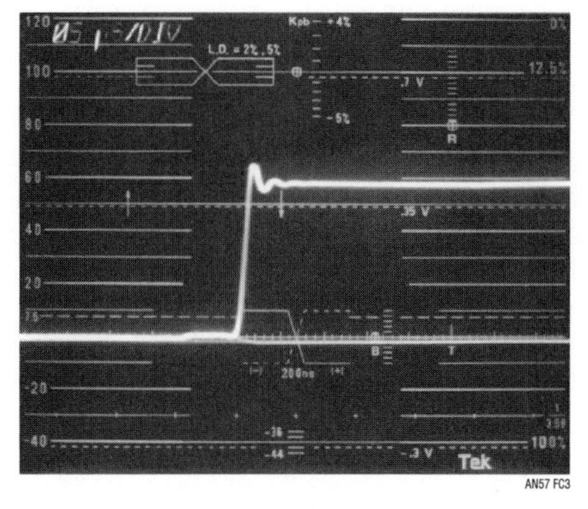

Figure C3 • Expanded View of Rising Edge of LT1204 Switching from 0% to 50% (50ns Horizontal Division)

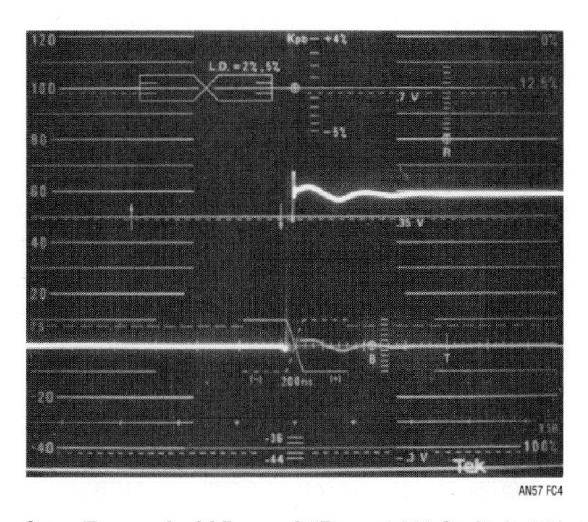

Figure C4 • Expanded View of "Brand-X" Switch 0% to 50% Transition

When viewed on a monitor, the switch artifact is just visible as a very fine line. The lower trace is a switch between two black level (0V) video signals showing a very slight channel-to- channel offset, which is not visible on the monitor. Switching between two DC levels is a worst-case test as almost any active video will have enough variation to totally obscure this small switch artifact.

Video-switching caveats

In a video processing system that has a large bandwidth compared to the bandwidth of the video signal, a fast transition from one video level to another with a low-amplitude glitch will cause minimal visual disturbance. This situation is analogous to the proper use of an analog oscilloscope. In order to make accurate measurement of pulse waveforms, the instrument must have much more bandwidth than the signal in question (usually five times the highest frequency). Not only should the glitch be small, it should be otherwise well controlled. A switching glitch that has a long settling "tail" can be more troublesome (that

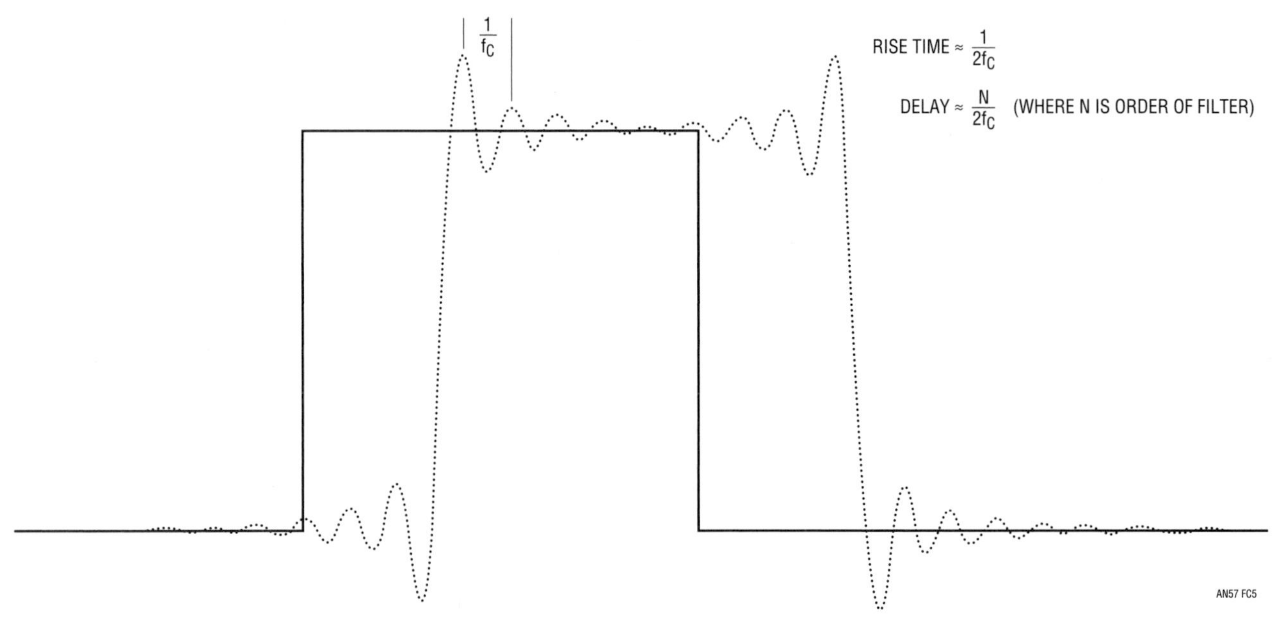

Figure C5 • Pulse Response of an Ideal Sharp-Cutoff Filter at Frequency f_C

is, more visible) than one that has more amplitude but decays quickly. The LT1204 has a switching glitch that is not only low in amplitude but well controlled and quickly damped. Refer to Figure C4 which shows a video multiplexer that has a long, slow-settling tail. This sort of distortion is highly visible on a video monitor.

Composite video systems, such as NTSC, are inherently band-limited and thus edge-rate limited. In a sharply band-limited system, the introduction of signals that contain significant energy higher in frequency than the filter cutoff will cause distortion of transient waveforms (see Figure C5). Filters used to control the bandwidth of these video systems should be group-delay equalized to minimize this pulse distortion. Additionally, in a band-limited system, the edge rates of switching glitches or level-to-level transitions should be controlled to prevent ringing and other pulse aberrations that could be visible. In practice, this is usually accomplished with pulse-shaping networks. Bessel filters are one example. Pulse-shaping networks and delay-equalized filters add cost and complexity to video systems and are usually found only on expensive equipment. Where cost is a determining factor in system design, the exceptionally low amplitude and brief duration of the LT1204's switching artifact make it an excellent choice for active video switching.

Conclusion

Active video switching can be accomplished inexpensively and with excellent results when care is taken with both the selection and application of the high speed multiplexer. Both fast switching and small, well-controlled switching glitches are important. When the LT1204 is used for active video switching between two flat-field video signals (a very critical test) the switching artifacts are nearly invisible. When the LT1204 is used to switch between two live video signals, the switching artifacts are invisible.

Some definitions—

"Picture-in-picture" refers to the production effect in which one video image is inserted within the boundaries of another. The process may be as simple as splitting the screen down the middle or it may involve switching the two images along a complicated geometric boundary. In order to make the composite picture stable and viewable, both video signals must be in horizontal and vertical sync. For composite color signals, the signals must also be in subcarrier lock.

"Keying" is the process of switching among two or more video signals triggering on some characteristic of one of the signals. For instance, a chroma keyer will switch on the presence of a particular color. Chroma keyers are used to insert a portion of one scene into another. In a commonly used effect, the TV weather person (the "talent") appears to be standing in front of a computer generated weather map. Actually, the talent is standing in front of a specially colored background; the weather map is a separate video signal, which has been carefully prepared to contain none of that particular color. When the chroma keyer senses the keying color, it switches to the weather map background. Where there is no keying color, the keyer switches to the talent's image.

Practical circuitry for measurement and control problems

Circuits designed for a cruel and unyielding world

35

Jim Williams

Introduction

This collection of circuits was worked out between June 1991 and July of 1994. Most were designed at customer request or are derivatives of such efforts. All represent substantial effort and, as such, are disseminated here for wider study and (hopefully) use.[1] The examples are roughly arranged in categories including power conversion, transducer signal conditioning, amplifiers and signal generators. As always, reader comment and questions concerning variants of the circuits shown may be addressed directly to the author.

Note 1: "Study" is certainly a noble pursuit but we never fail to emphasize use.

Clock synchronized switching regulator

Gated oscillator type switching regulators permit high efficiency over extended ranges of output current. These regulators achieve this desirable characteristic by using a gated oscillator architecture instead of a clocked pulse width modulator. This eliminates the "housekeeping" currents associated with the continuous operation of fixed frequency designs. Gated oscillator regulators simply self-clock at whatever frequency is required to maintain the output voltage. Typically, loop oscillation frequency ranges from a few hertz into the kilohertz region, depending upon the load.

In most cases this asynchronous, variable frequency operation does not create problems. Some systems, however; are sensitive to this characteristic. Figure 35.1 slightly modifies a gated oscillator type switching regulator

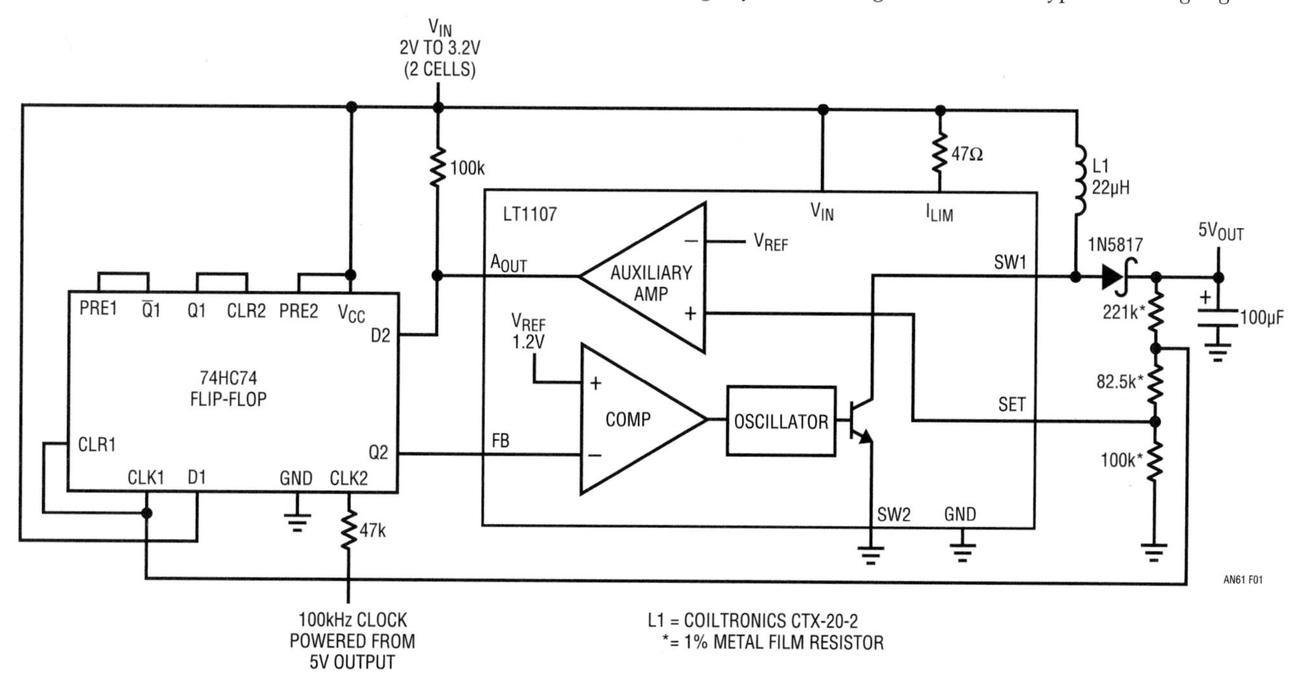

Figure 35.1 • A Synchronizing Flip-Flop Forces Switching Regulator Noise to Be Coherent with the Clock

Analog Circuit and System Design: Immersion in the Black Art of Analog Design. http://dx.doi.org/10.1016/B978-0-12-397888-2.00035-3

by synchronizing its loop oscillation frequency to the systems clock. In this fashion the oscillation frequency and its attendant switching noise, albeit variable, become coherent with system operation.

Circuit operation is best understood by temporarily ignoring the flip-flop and assuming the LT®1107 regulator's A_{OUT} and FB pins are connected. When the output voltage decays the set pin drops below V_{REF}, causing A_{OUT} to fall. This causes the internal comparator to switch high, biasing the oscillator and output transistor into conduction. L1 receives pulsed drive, and its flyback events are deposited into the 100μF capacitor via the diode, restoring output voltage. This overdrives the set pin, causing the IC to switch off until another cycle is required. The frequency of this oscillatory cycle is load dependent and variable. If, as shown, a flip-flop is interposed in the A_{OUT}-FB pin path, synchronization to a system clock results. When the output decays far enough (trace A, Figure 35.2) the A_{OUT} pin (trace B) goes low. At the next clock pulse (trace C) the flip-flop Q2 output (trace D) sets low, biasing the comparator-oscillator. This turns on the power switch (V_{SW} pin is trace E), which pulses L1. L1 responds in flyback fashion, depositing its energy into the output capacitor to maintain output voltage. This operation is similar to the previously described case, except that the sequence is forced to synchronize with the system clock by the flip-flops action. Although the resulting loops oscillation frequency is variable it, and all attendant switching noise, is synchronous and coherent with the system clock.

A start-up sequence is required because this circuit's clock is powered from its output. The start-up circuitry was developed by Sean Gold and Steve Pietkiewicz of LTC. The flip-flop's remaining section is connected as a buffer. The CLR1-CLK1 line monitors output voltage via the resistor string. When power is applied Q1 sets CLR2 low. This permits the LT1107 to switch, raising output voltage. When the output goes high enough Q1 sets CLR2 high and normal loop operation commences.

The circuit shown is a step-up type, although any switching regulator configuration can utilize this synchronous technique.

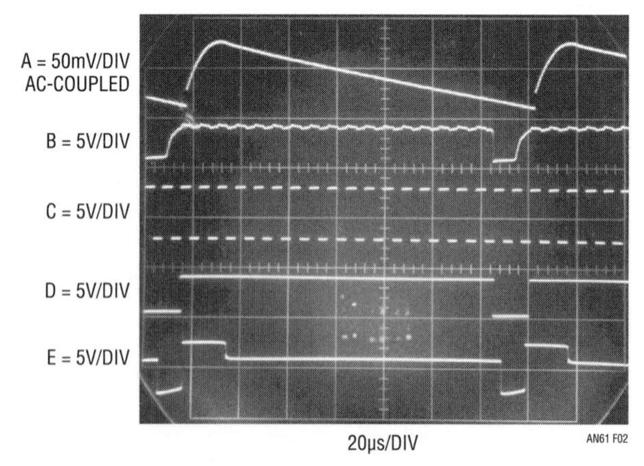

Figure 35.2 • Waveforms for the Clock Synchronized Switching Regulator. Regulator Only Switches (Trace E) on Clock Transitions (Trace C), Resulting in Clock Coherent Output Noise (Trace A)

High power 1.5V to 5V converter

Some 1.5V powered systems (survival 2-way radios, remote, transducer-fed data acquisition systems, etc.) require much more power than stand-alone IC regulators can provide. Figure 35.3's design supplies a 5V output with 200mA capacity.

The circuit is essentially a flyback regulator. The LT1170 switching regulator's low saturation losses and ease of use permit high power operation and design simplicity. Unfortunately this device has a 3V minimum supply requirement. Bootstrapping its supply pin from the 5V output is possible, but requires some form of start-up mechanism.

The 1.5V powered LT1073 switching regulator forms a start-up loop. When power is applied the LT1073 runs,

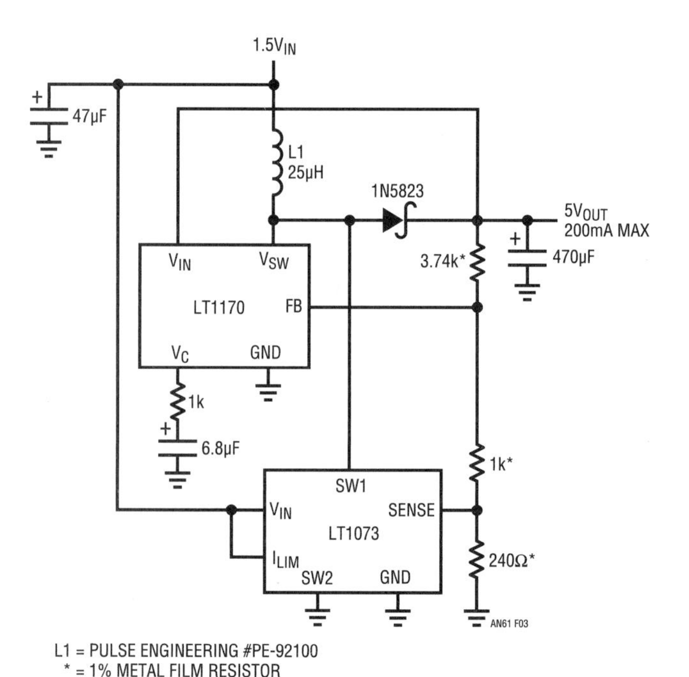

Figure 35.3 • 200mA Output 1.5V to 5V Converter. Lower Voltage LT1073 Provides Bootstrap Start-Up for LT1170 High Power Switching Regulator

causing its V_{SW} pin to periodically pull current through L1. L1 responds with high voltage flyback events. These events are rectified and stored in the 470µF capacitor, producing the circuits DC output. The output divider string is set up so the LT1073 turns off when circuit output crosses about 4.5V. Under these conditions the LT1073 obviously can no longer drive L1, but the LT1170 can. When the start-up circuit goes off, the LT1170 V_{IN} pin has adequate supply voltage and can operate. There is some overlap between start-up loop turn-off and LT1170 turn-on, but it has no detrimental effect.

The start-up loop must function over a wide range of loads and battery voltages. Start-up currents approach 1A, necessitating attention to the LT1073's saturation and drive characteristics. The worst case is a nearly depleted battery and heavy output loading.

Figure 35.4 plots input-output characteristics for the circuit. Note that the circuit will start into all loads with V_{BAT} = 1.2V Start-up is possible down to 1.0V at reduced loads. Once the circuit has started, the plot shows it will drive full 200mA loads down to V_{BAT} = 1.0V. Reduced drive is possible down to V_{BAT} = 0.6V (a very dead battery)! Figure 35.5 graphs efficiency at two supply voltages over a range of output currents. Performance is attractive, although at lower currents circuit quiescent power degrades efficiency. Fixed junction saturation losses are responsible for lower overall efficiency at the lower supply voltage.

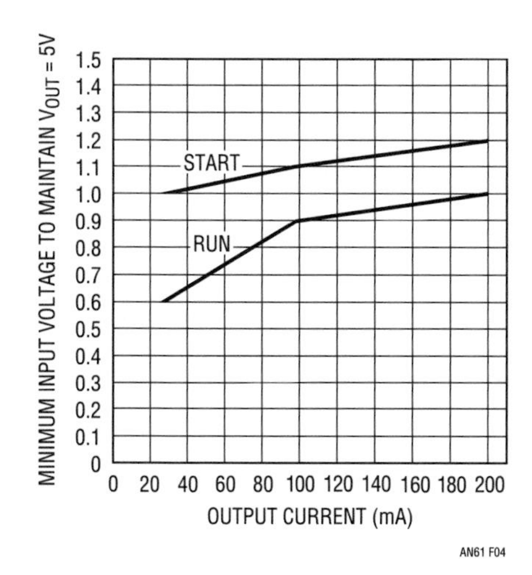

Figure 35.4 • Input-Output Data for the 1.5V to 5V Converter Shows Extremely Wide Start-Up and Running Range into Full Load

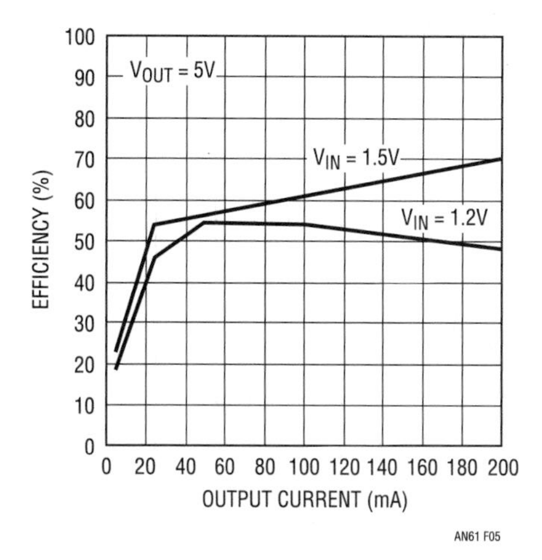

Figure 35.5 • Efficiency vs Operating Point for the 1.5V to 5V Converter. Efficiency Suffers at Low Power Because of Relatively High Quiescent Currents

Low power 1.5V to 5V converter

Figure 35.6, essentially the same approach as the preceding circuit, was developed by Steve Pietkiewicz of LTC. It is limited to about 150mA output with commensurate restrictions on start-up current. It's advantage, good efficiency at relatively low output currents, derives from its low quiescent power consumption.

The LT1073 provides circuit start-up. When output voltage, sensed by the LT1073's "set" input via the resistor divider, rises high enough Q1 turns on, enabling the LT1302. This device sees adequate operating voltage and responds by driving the output to 5V satisfying its feedback node. The 5V output also causes enough overdrive at the LT1073 feedback pin to shut the device down.

Figure 35.7 shows maximum permissible load currents for start-up and running conditions. Performance is quite good, although the circuit clearly cannot compete with the previous design. The fundamental difference between the two circuits is the LT1170's (Figure 35.3) much larger power switch, which is responsible for the higher available power. Figure 35.8, however reveals another difference. The curves show that Figure 35.6 is significantly more efficient than the LT1170 based approach at output currents below 100mA. This highly desirable characteristic is due to the LT1302's much lower quiescent operating currents.

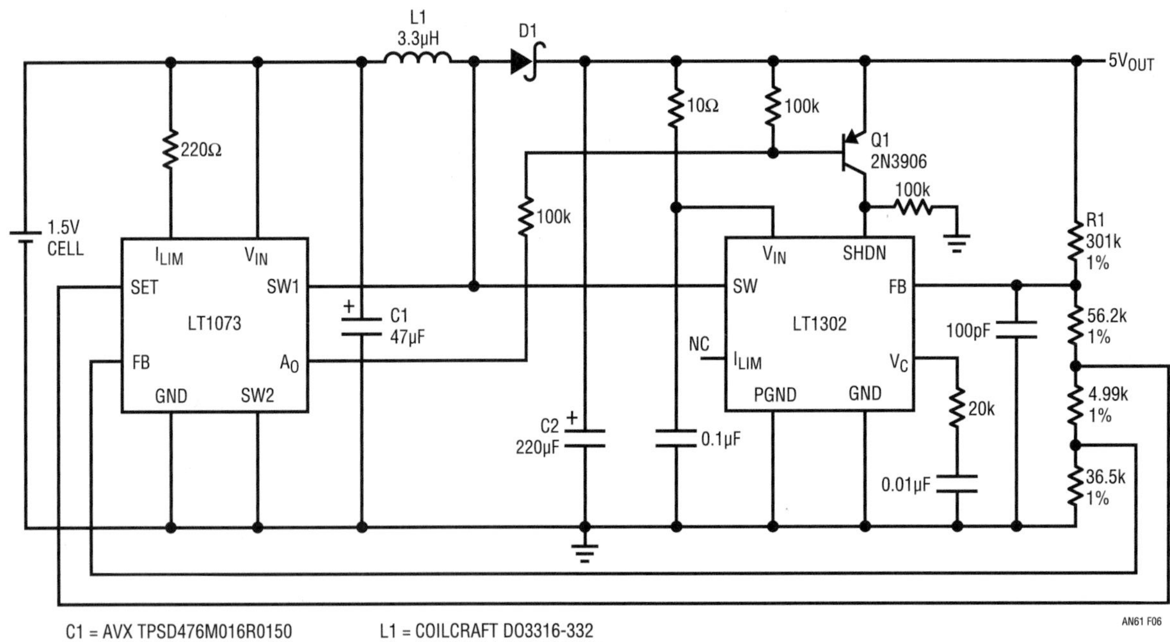

Figure 35.6 • Single-Cell to 5V Converter Delivers 150mA with Good Efficiency at Lower Currents

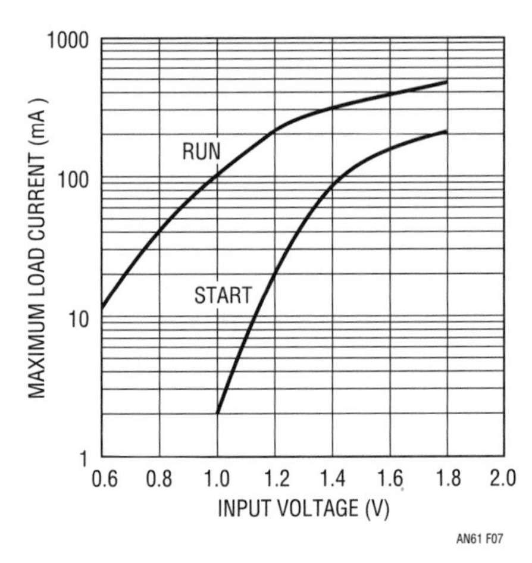

Figure 35.7 • Maximum Permissible Loads for Start-Up and Running Conditions. Allowable Load Current During Start-Up Is Substantially Less Than Maximum Running Current.

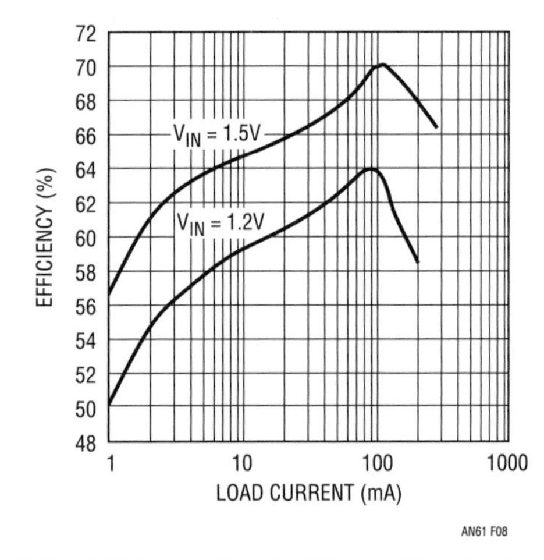

Figure 35.8 • Efficiency Plot for Figure 35.6. Performance Is Better Than the Previous Circuit at Lower Currents, Although Poorer at High Power

Low power, low voltage cold cathode fluorescent lamp power supply

Most Cold Cathode Fluorescent Lamp (CCFL) circuits require an input supply of 5V to 30V and are optimized for bulb currents of 5mA or more. This precludes lower power operation from 2- or 3-cell batteries often used in palmtop computers and portable apparatus. A CCFL power supply that operates from 2V to 6V is detailed in Figure 35.9. This circuit, contributed by Steve Pietkiewicz of LTC, can drive a small CCFL over a 100μA to 2mA range.

The circuit uses an LT1301 micropower DC/DC converter IC in conjunction with a current driven Royer class converter comprised of T1, Q1 and Q2. When power and intensity adjust voltage are applied the LT1301's I_{LIM} pin is driven slightly positive, causing maximum switching current through the IC's internal switch pin (SW). Current flows from T1's center tap, through the transistors, into L1. L1's current is deposited in switched fashion to ground by the regulator's action.

The Royer converter oscillates at a frequency primarily set by T1's characteristics (including its load) and the 0.068μF capacitor. LT1301 driven L1 sets the magnitude of the Q1-Q2 tail current, hence T1's drive level. The 1N5817 diode maintains L1's current flow when the LT1301's switch is off. The 0.068μF capacitor combines with T1's characteristics to produce sine wave voltage drive at the Q1 and Q2 collectors. T1 furnishes voltage step-up and about 1400Vp-p appears at its secondary. Alternating current flows through the 22pF capacitor into the lamp. On positive half-cycles the lamp's current is steered to ground via D1. On negative half-cycles the lamp's current flows through Q3's collector and is filtered by C1. The LT1301's I_{LIM} pin acts as a 0V summing point with about 25μA bias current flowing out of the pin into C1. The LT1301 regulates L1's current to equalize Q3's average collector current, representing 1/2 the lamp current, and R1's current, represented by V_A/R1. C1 smooths all current flow to DC. When V_A is set to zero, the I_{LIM} pin's bias current forces about 100μA bulb current.

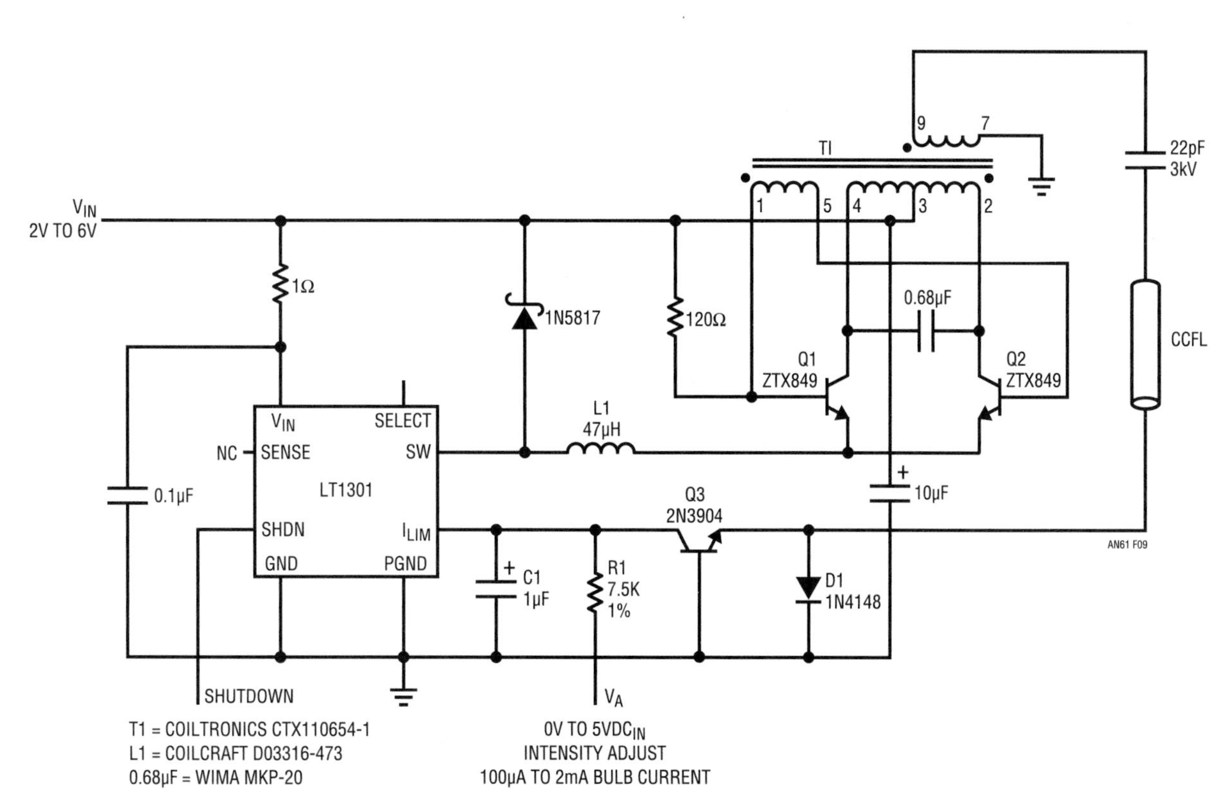

Figure 35.9 • Low Power Cold Cathode Fluorescent Lamp Supply Is Optimized for Low Voltage Inputs and Small Lamps

Circuit efficiency ranges from 80% to 88% at full load, depending on line voltage. Current mode operation combined with the Royer's consistent waveshape vs input results in excellent line rejection. The circuit has none of the line rejection problems attributable to the hysteretic voltage control loops typically found in low voltage micropower DC/DC converters. This is an especially desirable characteristic for CCFL control, where lamp intensity must remain constant with shifts in line voltage. Interaction between the Royer converter, the lamp and the regulation loop is far more complex than might be supposed, and subject to a variety of considerations. For detailed discussion see Reference 3.

Low voltage powered LCD contrast supply

Figure 35.10, a companion to the CCFL power supply previously described, is a contrast supply for LCD panels. It was designed by Steve Pietkiewicz of LTC. The circuit is noteworthy because it operates from a 1.8V to 6V input, significantly lower than most designs. In operation the LT1300/LT1301 switching regulator drives T1 in flyback fashion, causing negative biased step-up at T1's secondary. D1 provides rectification, and C1 smooths the output to DC. The resistively divided output is compared to a command input, which may be DC or PWM, by the IC's "I_{LIM}" pin. The IC, forcing the loop to maintain 0V at the I_{LIM} pin, regulates circuit output in proportion to the command input.

Efficiency ranges from 77% to 83% as supply voltage varies from 1.8V to 3V. At the same supply limits, available output current increases from 12mA to 25mA.

HeNe laser power supply

Helium-Neon lasers, used for a variety of tasks, are difficult loads for a power supply. They typically need almost 10kV to start conduction, although they require only about 1500V to maintain conduction at their specified operating currents. Powering a laser usually involves some form of start-up circuitry to generate the initial breakdown voltage and a separate supply for sustaining conduction. Figure 35.11's circuit considerably simplifies driving the laser.

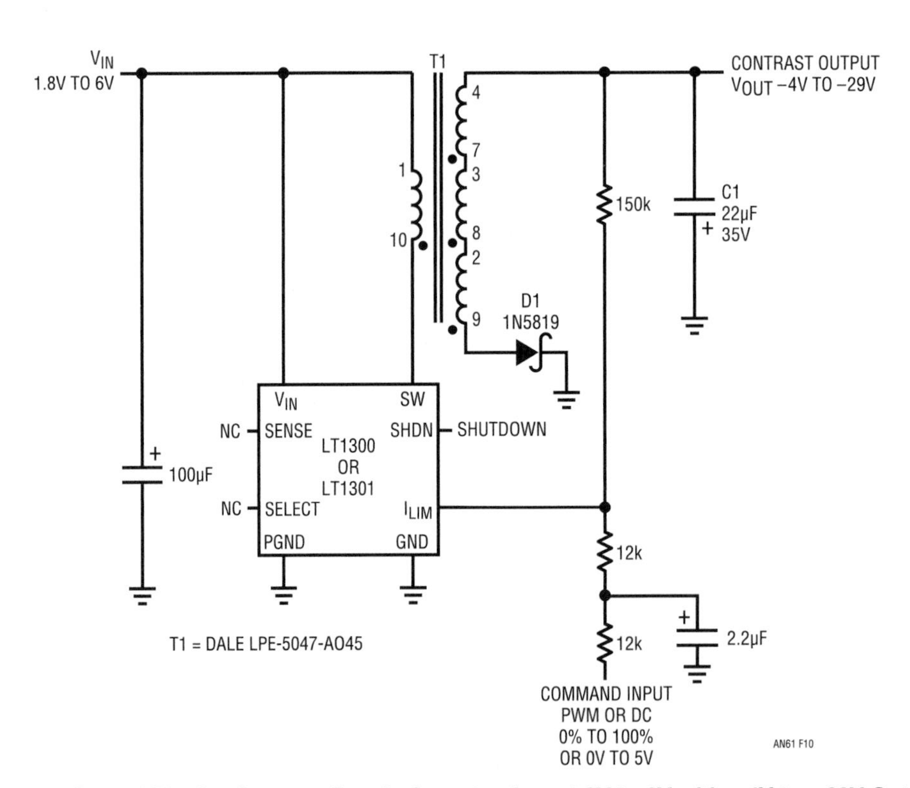

Figure 35.10 • Liquid Crystal Display Contrast Supply Operates from 1.8V to 6V with −4V to −29V Output Range

The start-up and sustaining functions have been combined into a single, closed-loop current source with over 10kV of compliance. The circuit is recognizable as a reworked CCFL power supply with a voltage tripled DC output.[2]

When power is applied, the laser does not conduct and the voltage across the 190Ω resistor is zero. The LT1170 switching regulator FB pin sees no feedback voltage, and its switch pin (V_{SW}) provides full duty cycle pulse width modulation to L2. Current flows from L1's center tap through Q1 and Q2 into L2 and the LT1170. This current flow causes Q1 and Q2 to switch, alternately driving L1. The 0.47µF capacitor resonates with L1, providing boosted sine wave drive. L1 provides substantial step-up, causing about 3500V to appear at its secondary. The capacitors and diodes associated with L1's secondary form a voltage tripler, producing over 10kV across the laser. The laser breaks down and current begins to flow through it. The 47k resistor limits current and isolates the laser's load characteristic. The current flow causes a voltage to appear across the 190Ω resistor. A filtered version of this voltage appears at the LT1170 FB pin, closing a control loop. The LT1170 adjusts pulse width drive to L2 to maintain the FB pin at 1.23V regardless of changes in operating conditions.

Note 2: See References 2 and 3 and this text's Figure 35.9.

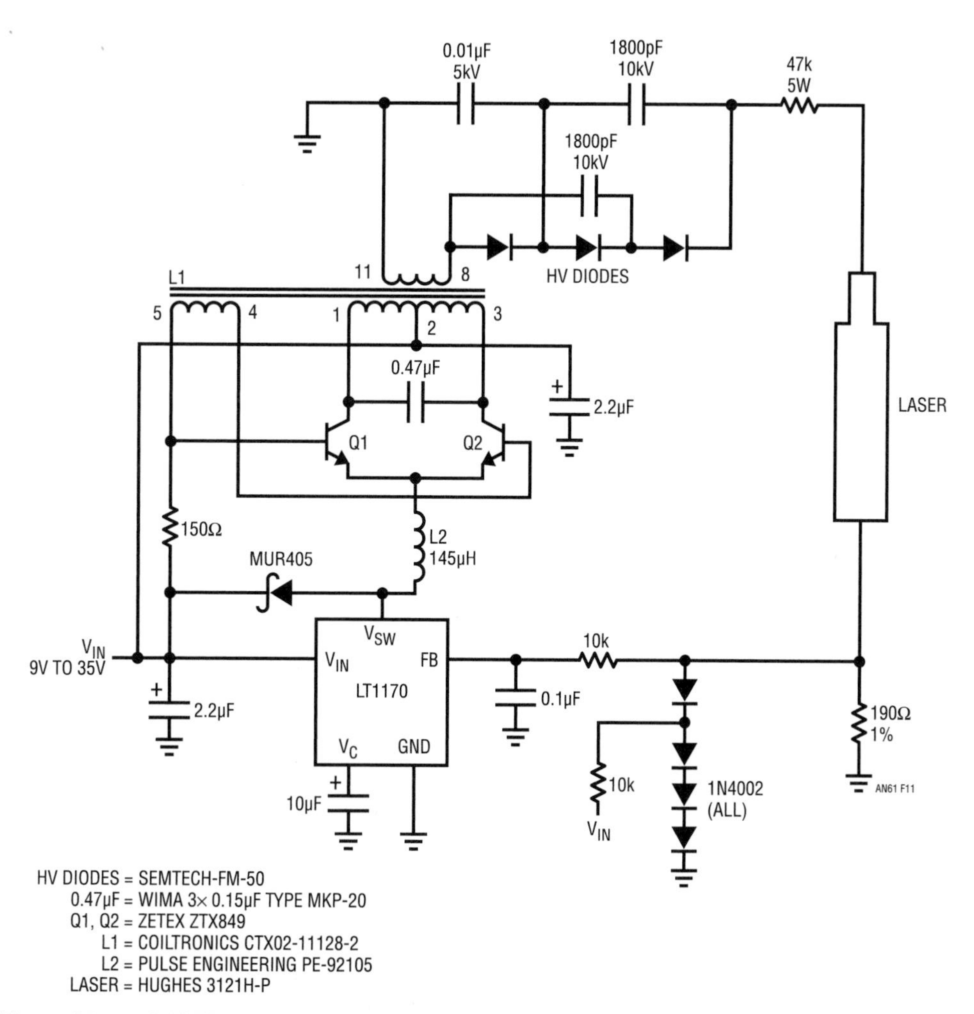

Figure 35.11 • LASER Power Supply Is Essentially A 10,000V Compliance Current Source

In this fashion, the laser sees constant current drive, in this case 6.5mA. Other currents are obtainable by varying the 190Ω value. The 1N4002 diode string clamps excessive voltages when laser conduction first begins, protecting the LT1170. The 10μF capacitor at the V_C pin frequency compensates the loop and the MUR405 maintains L1's current flow when the LT1170 V_{SW} pin is not conducting. The circuit will start and run the laser over a 9V to 35V input range with an electrical efficiency of about 80%.

Compact electroluminescent panel power supply

Electroluminescent (EL) panel LCD backlighting presents an attractive alternative to fluorescent tube (CCFL) backlighting in some portable systems. EL panels are thin, lightweight, lower power; require no diffuser and work at lower voltage than CCFLs. Unfortunately, most EL DC/AC inverters use a large transformer to generate the 400Hz 95V square wave required to drive the panel. Figure 35.12's circuit, developed by Steve Pietkiewicz of LTC, eliminates the transformer by employing an LT1108 micropower DC/DC converter IC. The device generates a 95VDC potential via L1 and the diode-capacitor doubler network. The transistors switch the EL panel between 95V and ground. C1 blocks DC and R1 allows intensity adjustment. The 400Hz square wave drive signal can be supplied by the microprocessor or a simple multivibrator. When compared to conventional EL panel supplies, this circuit is noteworthy because it can be built in a square inch with a 0.5 inch height restriction. Additionally, all components are surface mount types, and the usual large and heavy 400Hz transformer is eliminated.

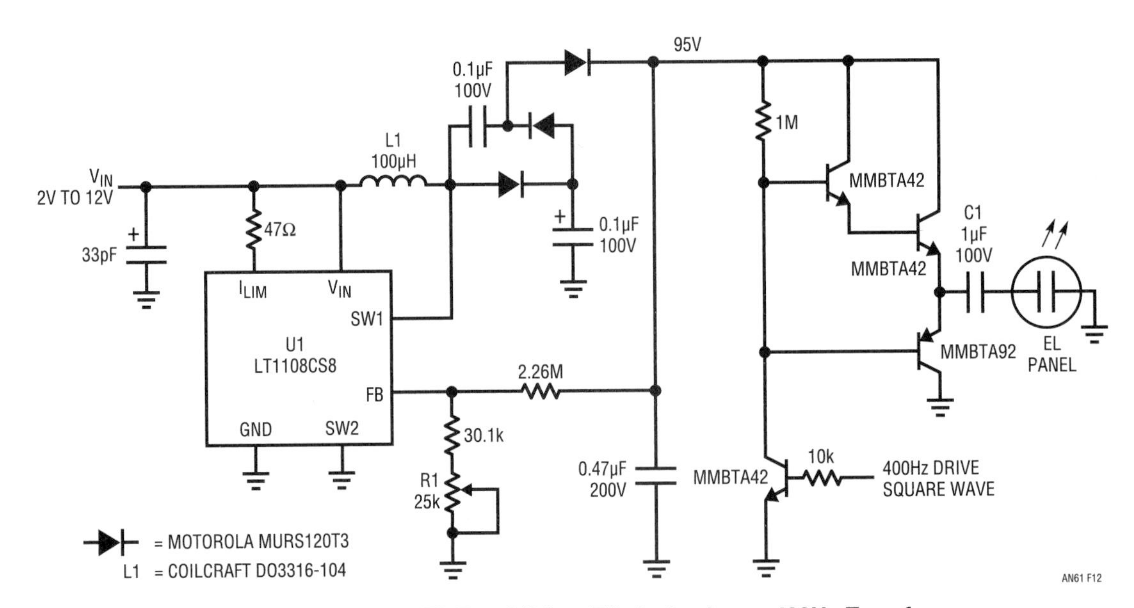

Figure 35.12 • Switch Mode EL Panel Driver Eliminates Large 400Hz Transformer

3.3V powered barometric pressure signal conditioner

The move to 3.3V digital supply voltage creates problems for analog signal conditioning. In particular, transducer based circuits often require higher voltage for proper transducer excitation. DC/DC converters in standard configurations can address this issue but increase power consumption. Figure 35.13's circuit shows a way to provide proper transducer excitation for a barometric pressure sensor while minimizing power requirements.

The 6kΩ transducer T1 requires precisely 1.5mA of excitation, necessitating a relatively high voltage drive. A1 senses T1's current by monitoring the voltage drop across the resistor string in T1's return path.

A1's output biases the LT1172 switching regulator's operating point, producing a stepped up DC voltage which appears as T1's drive and A2's supply voltage. T1's return current out of pin 6 closes a loop back at A1 which is slaved to the 1.2V reference. This arrangement provides the required high voltage drive (≈10V) while minimizing power consumption. This is so because the switching regulator produces only enough voltage to satisfy T1's current requirements. Instrumentation amplifier A2 and A3 provide gain and LTC®1287 A/D converter gives a 12-bit digital output. A2 is bootstrapped off the transducer supply, enabling it to accept T1's common-mode voltage. Circuit current consumption is about 14mA. If the shutdown pin

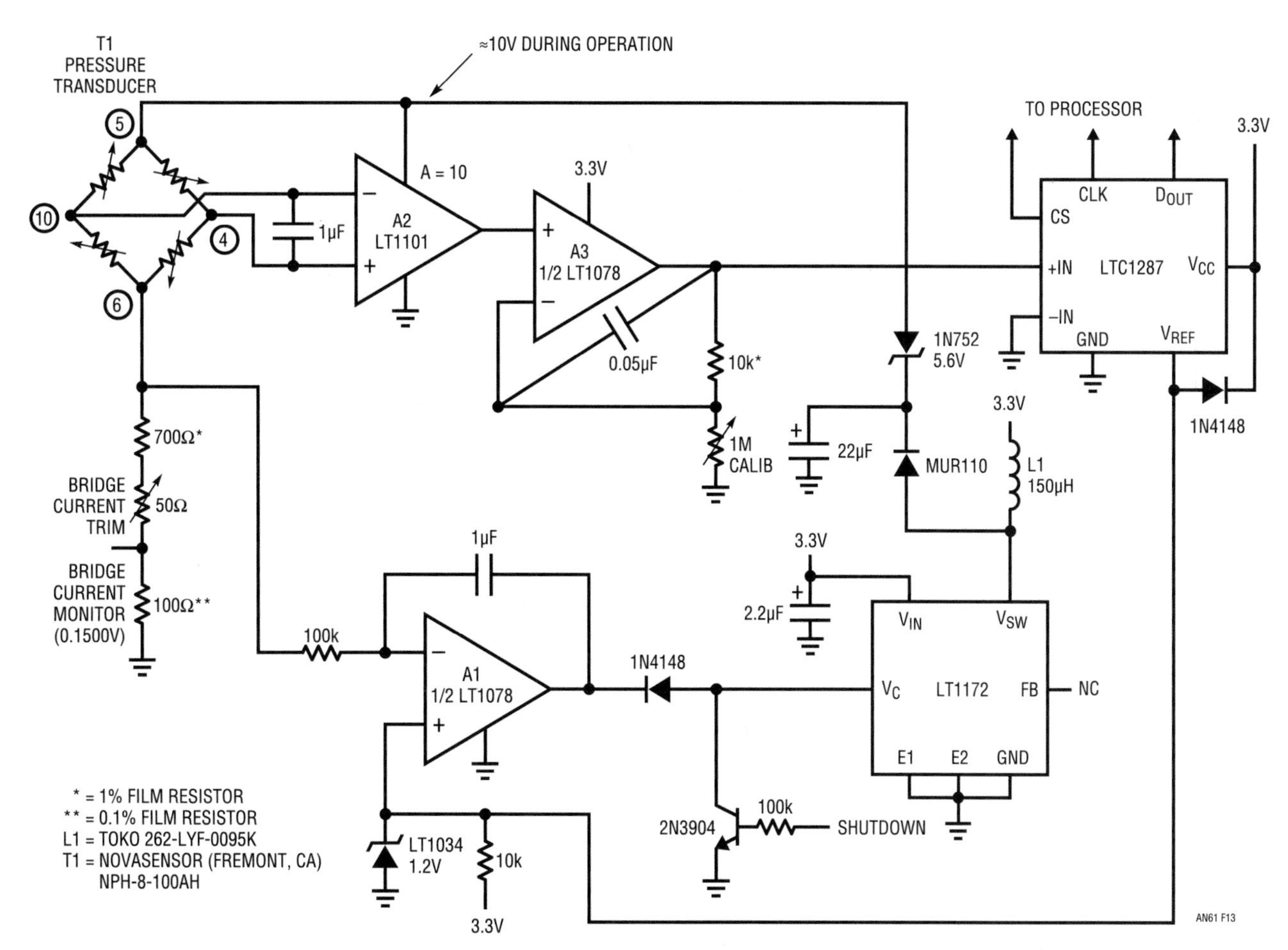

Figure 35.13 • 3.3V Powered, Digital Output, Barometric Pressure Signal Conditioner

is driven high the switching regulator turns off, reducing total power consumption to about 1mA. In shutdown the 3.3V powered A/D's output data remains valid. In practice, the circuit provides a 12-bit representation of ambient barometric pressure after calibration. To calibrate, adjust the "bridge current trim" for exactly 0.1500V at the indicated point. This sets T1's current to the manufacturers specified point. Next, adjust A3's trim so that the digital output corresponds to the known ambient barometric pressure. If a pressure standard is not available the transducer is supplied with individual calibration data, permitting circuit calibration.

Some applications may require operation over a wider supply range and/or a calibrated analog output. Figure 35.14's circuit is quite similar, except that the A/D converter is eliminated and a 2.7V to 7V supply is acceptable. The calibration procedure is identical, except that A3's analog output is monitored.

Single cell barometers

It is possible to power these circuits from a single cell without sacrificing performance. Figure 35.15, a direct extension of the above approaches, simply substitutes a switching regulator that will run from a single 1.5V battery. In other respects loop action is nearly identical.

Figure 35.16, also a 1.5V powered design, is related but eliminates the instrumentation amplifier: As before, the 6kΩ transducer T1 requires precisely 1.5mA of excitation, necessitating a relatively high voltage drive. A1's positive input senses T1's current by monitoring the voltage drop across the resistor string in T1's return path. A1's negative input is fixed by the 1.2V LT1004 reference. A1's output biases the 1.5V powered LT1110 switching regulator. The LT1110's switching produces two outputs from L1. Pin 4's rectified and filtered output powers A1 and T1.

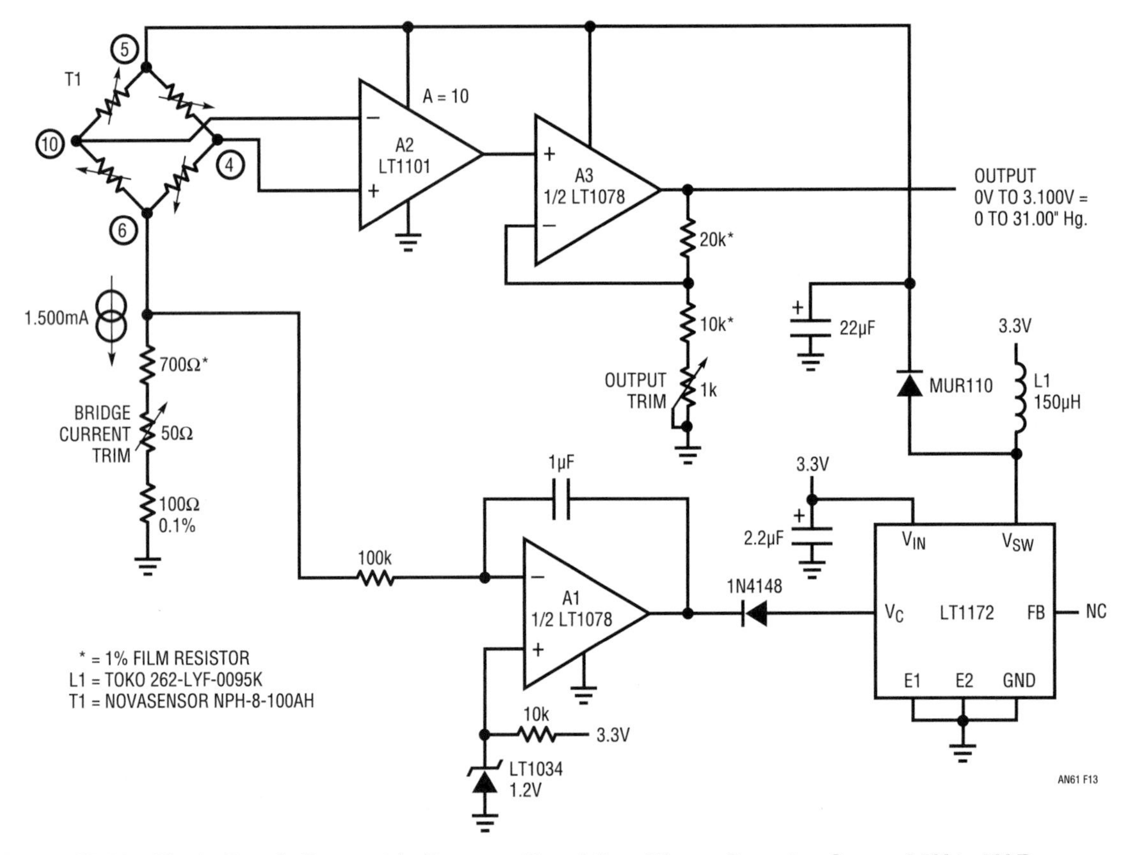

Figure 35.14 • Single Supply Barometric Pressure Signal Conditioner Operates Over a 2.7V to 7V Range

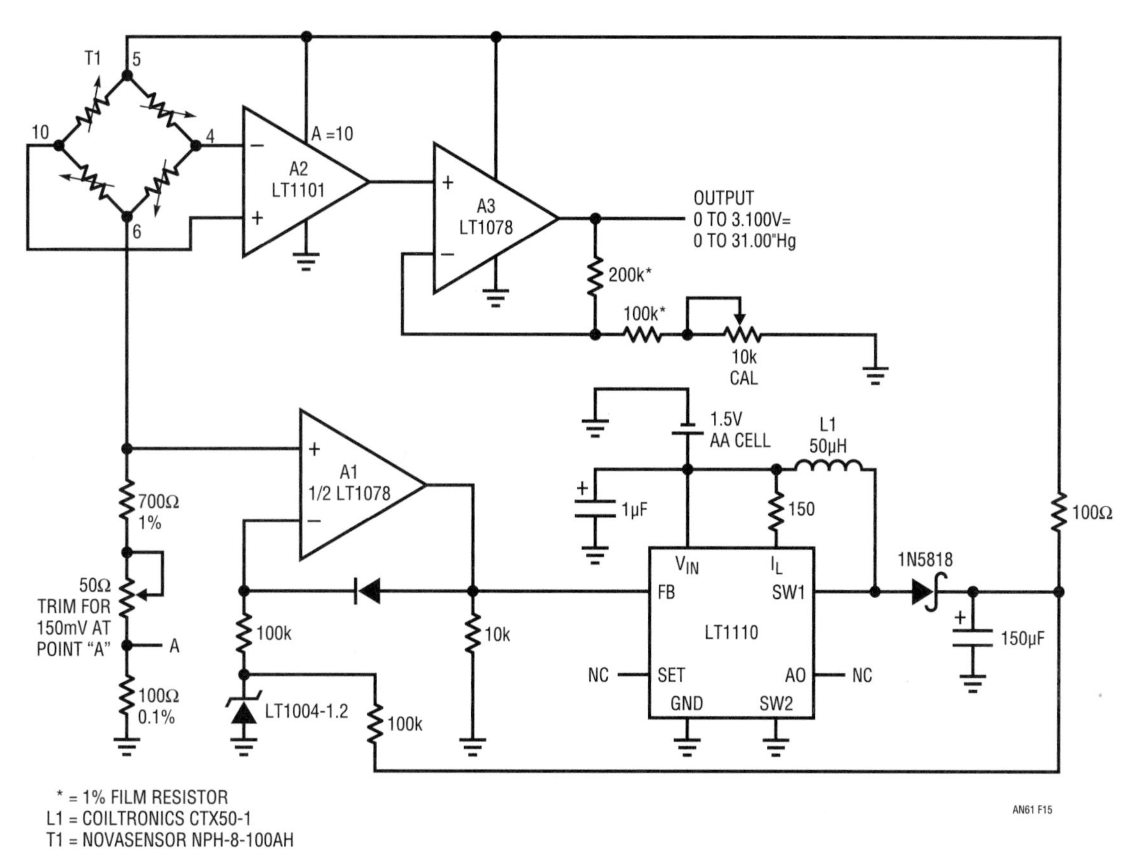

Figure 35.15 • 1.5V Powered Barometric Pressure Signal Conditioner Uses Instrumentation Amplifier and Voltage Boosted Current Loop

A1's output, in turn, closes a feedback loop at the regulator. This loop generates whatever voltage step-up is required to force precisely 1.5mA through T1. This arrangement provides the required high voltage drive while minimizing power consumption. This occurs because the switching regulator produces only enough voltage to satisfy T1's current requirements.

L1 pins 1 and 2 source a boosted, fully floating voltage, which is rectified and filtered. This potential powers A2. Because A2 floats with respect to T1, it can look differentially across T1's outputs, pins 10 and 4. In practice, pin 10 becomes "ground" and A2 measures pin 4's output with respect to this point. A2's gain-scaled output is the circuit's output, conveniently scaled at 3.000V=30.00"Hg. A2's floating drive eliminates the requirement for an instrumentation amplifier, saving cost, power, space and error contribution.

To calibrate the circuit, adjust R1 for 150mV across the 100Ω resistor in T1's return path. This sets T1's current to the manufacturer's specified calibration point. Next, adjust R2 at a scale factor of 3.000V = 30.00"Hg. If R2 cannot capture the calibration, reselect the 200k resistor in series with it. If a pressure standard is not available, the transducer is supplied with individual calibration data, permitting circuit calibration.

This circuit, compared to a high-order pressure standard, maintained 0.01"Hg accuracy over months with widely varying ambient pressure shifts. Changes in pressure, particularly rapid ones, correlated quite nicely to changing weather conditions. Additionally, because 0.01"Hg corresponds to about 10 feet of altitude at sea level, driving over hills and freeway overpasses becomes quite interesting.

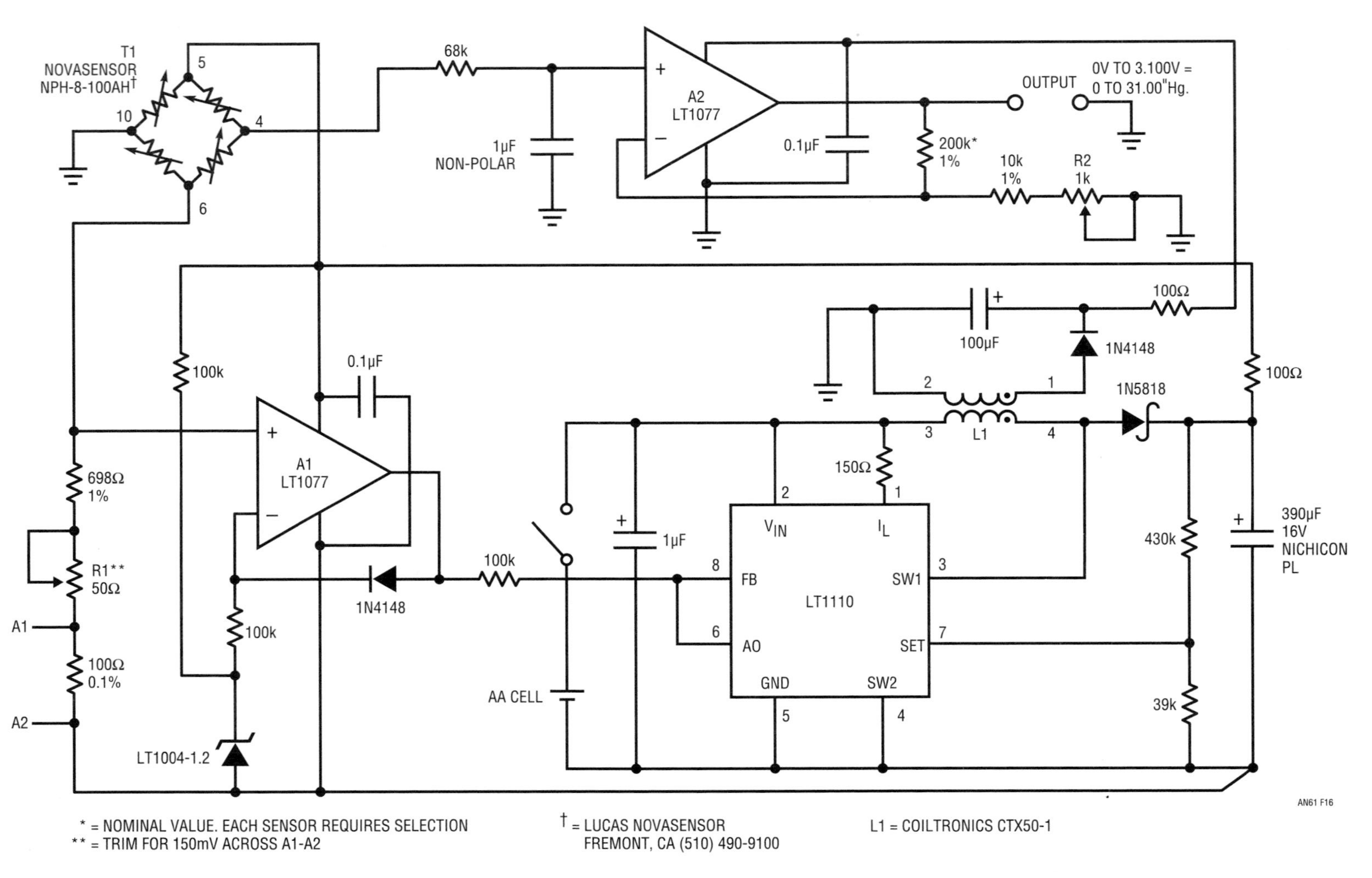

Figure 35.16. • 1.5V Powered Barometric Pressure Signal Conditioner Floats Bridge Drive to Eliminate Instrumentation Amplifier. Voltage Boosted Current Loop Drives Transducer

Until recently, this type of accuracy and stability has only been attainable with bonded strain gauge and capacitively-based transducers, which are quite expensive. As such, semiconductor pressure transducer manufacturers whose products perform at the levels reported are to be applauded. Although high quality semiconductor transducers are still not comparable to more mature technologies, their cost is low and they are vastly improved over earlier devices.

The circuit pulls 14mA from the battery, allowing about 250 hours operation from one D cell.

Quartz crystal-based thermometer

Although quartz crystals have been utilized as temperature sensors (see Reference 5), there has been almost no widespread adaptation of this technology. This is primarily due to the lack of standard product quartz-based temperature sensors. The advantages of quartz-based sensors include simple signal conditioning, good stability and a direct, noise immune digital output almost ideally suited to remote sensing.

Figure 35.17 utilizes an economical, commercially available (see Reference 6) quartz-based temperature sensor in a thermometer scheme suited to remote data collection. The LTQ485 RS485 transceiver is set up in the transmit mode. The crystal and discrete components combine with the IC's inverting gain to form a Pierce type oscillator. The LTC485's differential line driving outputs provide frequency coded temperature data to a 1000-foot cable run. A second RS485 transceiver differentially receives the data and presents a single-ended output. Accuracy depends on the grade of quartz sensor specified, with 1°C over 0°C to 100°C achievable.

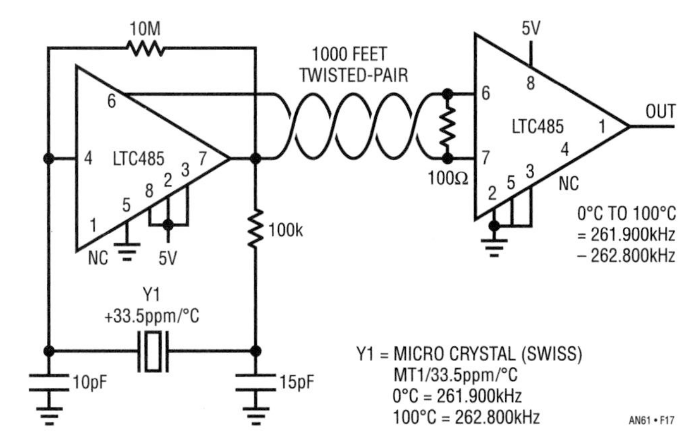

Figure 35.17 • Quartz Crystal Based Circuit Provides Temperature-to-Frequency Conversion. RS485 Transceivers Allow Remote Sensing

Ultra-low noise and low drift chopped-FET amplifier

Figure 35.18's circuit combines the extremely low drift of a chopper-stabilized amplifier with a pair of low noise FETs. The result is an amplifier with 0.05μV/°C drift, offset within 5μV 100pA bias current and 50nV noise in a 0.1Hz to 10Hz bandwidth. The noise performance is especially noteworthy; it is almost 35 times better than monolithic chopper-stabilized amplifiers and equals the best bipolar types.

FETs Q1 and Q2 differentially feed A2 to form a simple low noise op amp. Feedback, provided by R1 and R2, sets closed-loop gain (in this case 10,000) in the usual fashion. Although Q1 and Q2 have extraordinarily low noise characteristics, their offset and drift are uncontrolled. A1, a chopper-stabilized amplifier, corrects these deficiencies. It does this by measuring the difference between the amplifier's inputs and adjusting Q1's channel current via Q3 to minimize the difference. Q1's skewed drain values ensure that A1 will be able to capture the offset. A1 and Q3 supply whatever current is required into Q1's channel to force offset within 5μV Additionally, A1's low bias current does not appreciably add to the overall 100pA amplifier bias current. As shown, the amplifier is set up for a noninverting gain of 10,000 although other gains and inverting operation are possible. Figure 35.19 is a plot of the measured noise performance.

The FETs' V_{GS} can vary over a 4:1 range. Because of this, they must be selected for 10% V_{GS} matching. This matching allows A1 to capture the offset without introducing any significant noise.

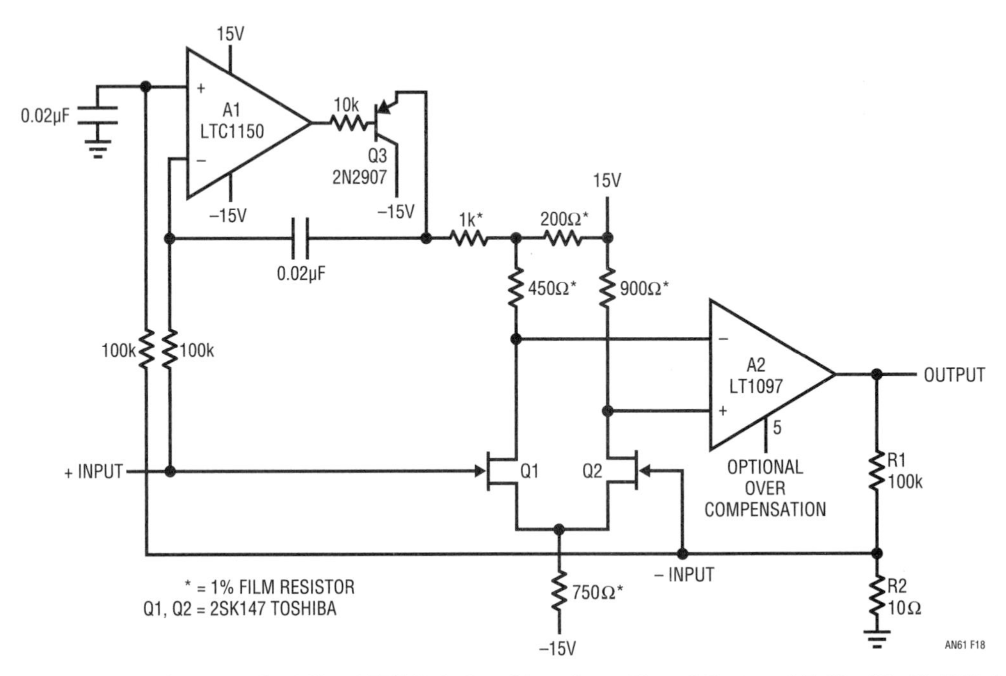

Figure 35.18 • Chopper-Stabilized FET Pair Combines Low Bias, Offset and Drift with 45nV Noise

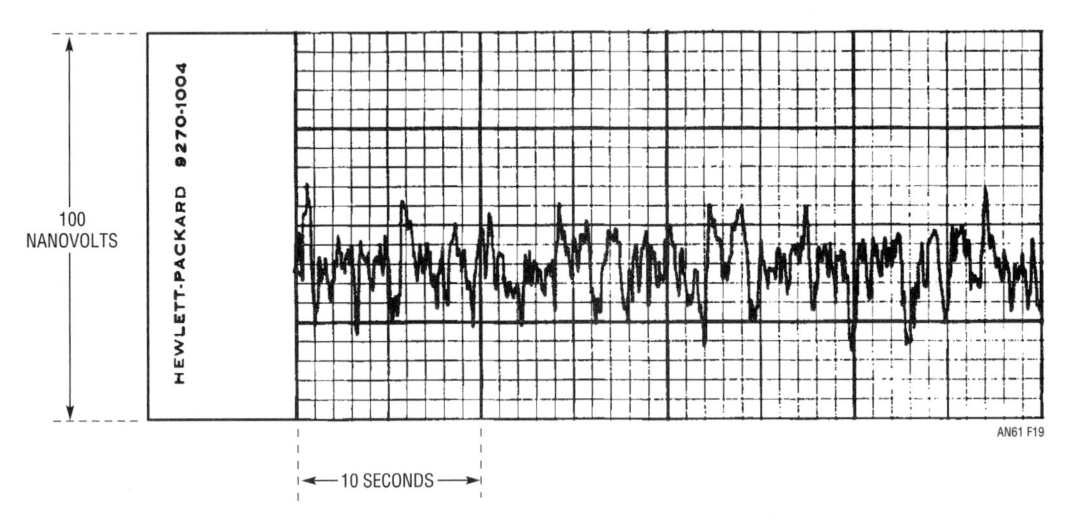

Figure 35.19 • Figure 35.18's 45nV Noise Performance in a 0.1Hz to 10Hz Bandwidth. A1's Low Offset and Drift Are Retained, But Noise Is Almost 35 Times Better

Figure 35.20 shows the response (trace B) to a 1mV input step (trace A). The output is clean, with no overshoots or uncontrolled components. If A2 is replaced with a faster device (e.g., LT1055) speed increases by an order of magnitude with similar damping. A2's optional overcompensation can be used (capacitor to ground) to optimize response for low closed-loop gains.

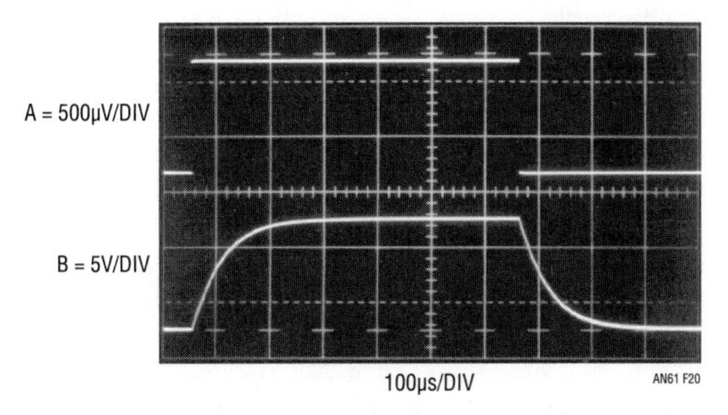

Figure 35.20 • Step Response for the Low Noise x 10,000 Amplifier. A 10x Speed Increase Is Obtainable by Replacing A2 with a Faster Device

High speed adaptive trigger circuit

Line receivers often require an adaptive trigger to compensate for variations in signal amplitude and DC offsets. The circuit in Figure 35.21 triggers on 2mV to 100mV signals from 100Hz to 10MHz while operating from a single 5V rail. A1, operating at a gain of 20, provides wideband AC gain. The output of this stage biases a 2-way peak detector (Q1-Q4). The maximum peak is stored in Q2's emitter capacitor, while the minimum excursion is retained in Q4's emitter capacitor. The DC value of A1's output signal's midpoint appears at the junction of the 500pF capacitor and the 10MΩ units. This point always sits midway between the signal's excursions, regardless of absolute amplitude. This signal-adaptive voltage is buffered by A2 to set the trigger voltage at the LT1116's positive input. The LT1116's negative input is biased directly from A1's output. The LT1116's output, the circuit's output, is unaffected by 50:1 signal amplitude variations. Bandwidth limiting in A1 does not affect triggering because the adaptive trigger threshold varies ratiometrically to maintain circuit output.

Split supply versions of this circuit can achieve bandwidths to 50MHz with wider input operating range (See Reference 7).

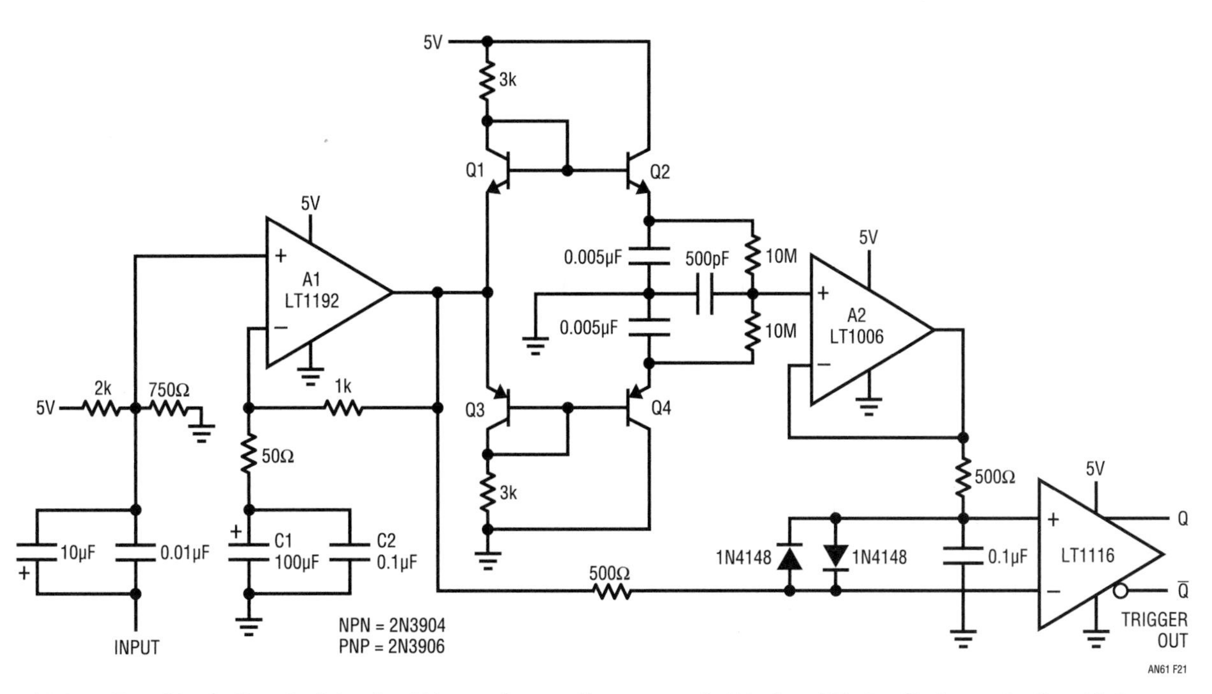

Figure 35.21 • Fast Single Supply Adaptive Trigger. Output Comparator's Trip Level Varies Ratiometrically with Input Amplitude, Maintaining Data Integrity Over 50:1 Input Amplitude Range

Wideband, thermally-based RMS/DC converter

Applications such as wideband RMS voltmeters, RF leveling loops, wideband AGC, high crest factor measurements, SCR power monitoring and high frequency noise measurements require wideband, true RMS/DC conversion. The thermal conversion method achieves vastly higher bandwidth than any other approach. Thermal RMS/DC converters are direct acting, thermoelectronic analog computers. The thermal technique is explicit, relying on "first principles, "e.g., a waveforms RMS value is defined as its heating value in a load.

Figure 35.22 is a wideband, thermally-based RMS/DC converter.[3] It provides a true RMS/DC conversion from DC to 10MHz with less than 1% error, regardless of input signal waveshape. It also features high input impedance and overload protection.

The circuit consists of three blocks; a wideband FET input amplifier, the RMS/DC converter and overload protection. The amplifier provides high input impedance, gain and drives the RMS/DC converters input heater. Input resistance is defined by the 1M resistor with input capacitance about 3pF. Q1 and Q2 constitute a simple, high speed FET input buffer. Q1 functions as a source follower, with the Q2 current source load setting the drain-source channel current. The LT1206 provides a flat 10MHz bandwidth gain of ten. Normally, this open-loop configuration would be quite drifty because there is no DC feedback. The LT1097 contributes this function to stabilize the circuit. It does this by comparing the filtered circuit output to a similarly filtered version of the input signal. The amplified difference between these signals is used to set Q2's bias, and hence Q1's channel current. This forces Q1's V_{GS} to whatever voltage is required to match the circuit's input and output potentials. The capacitor at A1 provides stable loop compensation.

The RC network in A1's output prevents it from seeing high speed edges coupled through Q2's collector-base junction. Q4, Q5 and Q6 form a low leakage clamp which precludes A1 loop latch-up during start-up or overdrive conditions. This can occur if Q1 ever forward biases. The 5k-50pF network gives A2 a slight peaking characteristic at the highest frequencies, allowing 1% flatness to 10MHz. A2's output drives the RMS/DC converter.

The LT1088 based RMS/DC converter is made up of matched pairs of heaters and diodes and a control amplifier. The LT1206 drives R1, producing heat which lowers D1's voltage. Differentially connected A3 responds by driving R2, via Q3, to heat D2, closing a loop around the amplifier. Because the diodes and heater resistors are matched, A3's DC output is related to the RMS value of the input, regardless of input frequency or waveshape. In practice, residual LT1088 mismatches necessitate a gain trim, which is implemented at A4. A4's output is the circuit output. The LT1004 and associated components frequency compensate the loop and provide good settling time over wide ranges of operating conditions (see Footnote 3).

Start-up or input overdrive can cause A2 to deliver excessive current to the LT1088 with resultant damage. C1 and C2 prevent this. Overdrive forces D1's voltage to an abnormally low potential. C1 triggers low under these conditions, pulling C2's input low. This causes C2's output to go high, putting A2 into shutdown and terminating the overload. After a time determined by the RC at C2's input, A2 will be enabled. If the overload condition still exists the loop will almost immediately shut A2 down again. This oscillatory action will continue, protecting the LT1088 until the overload condition is removed.

Note 3: Thermally based RMS/DC conversion is detailed in Reference 9.

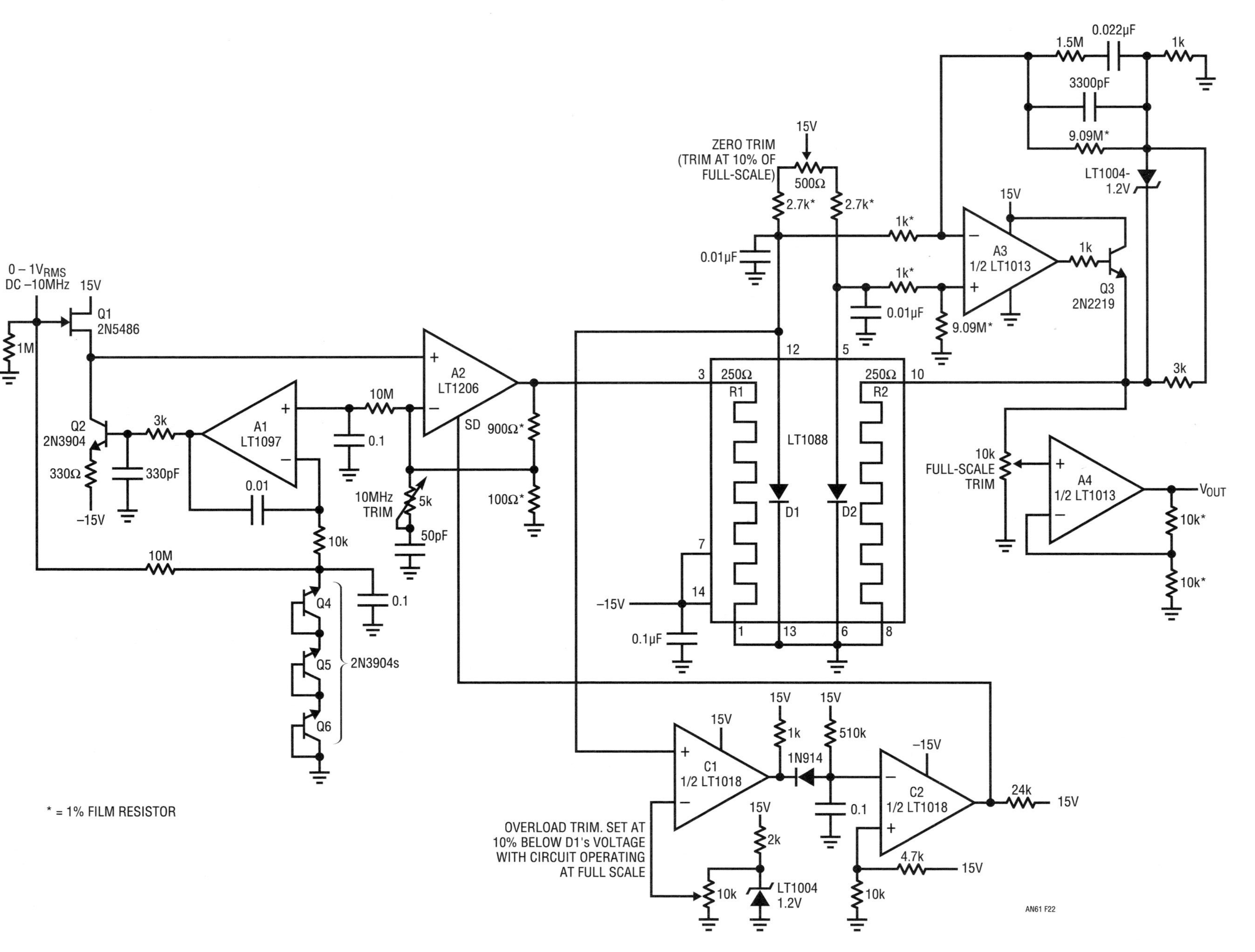

Figure 35.22 • Complete 10MHz Thermally-Based RMS/DC Converter Has 1% Accuracy, High Input Impedance and Overload Protection

Performance for the circuit is quite impressive. Figure 35.23 plots error from DC to 11MHz. The graph shows 1% error bandwidth of 11MHz. The slight peaking out to 5MHz is due to the gain boost network at A2's negative input. The peaking is minimal compared to the total error envelope, and a small price to pay to get the 1% accuracy to 10MHz.

To trim this circuit put the 5kΩ potentiometer at its maximum resistance position and apply a 100mV 5MHz signal. Trim the 500Ω adjustment for exactly $1V_{OUT}$. Next, apply a 5MHz 1V input and trim the 10k potentiometer for $10.00V_{OUT}$. Finally, put in 1V at 10MHz and adjust the 5kΩ trimmer for $10.00V_{OUT}$. Repeat this sequence until circuit output is within 1% accuracy for DC-10MHz inputs. Two passes should be sufficient.

It is worth considering that this circuit performs the same function as instruments costing thousands of dollars.[4]

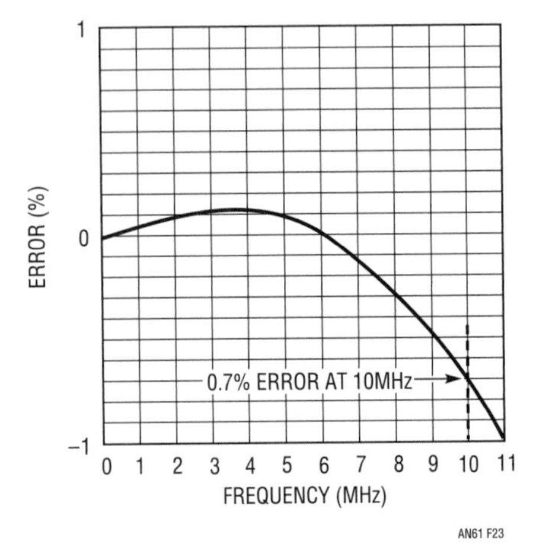

Figure 35.23 • Error Plot for the RMS/DC Converter. Frequency Dependent Gain Boost at A2 Preserves 1% Accuracy, But Causes Slight Peaking Before Roll-Off

Hall effect stabilized current transformer

Current transformers are common and convenient. They permit wideband current measurement independent of common-mode voltage considerations. The most convenient current transformers are the "clip-on" type, commercially sold as "current probes." A problem with all simple current transformers is that they cannot sense DC and low frequency information. This problem was addressed in the mid-1960's with the advent of the Hall effect stabilized current probe. This approach uses a Hall effect device within the transformer core to sense DC and low frequency signals. This information is combined with the current transformers output to form a composite DC-to-high frequency output. Careful roll-off and gain matching of the two channels preserves amplitude accuracy at all frequencies.[5] Additionally, the low frequency channel is operated as a "force-balance," meaning that the low frequency amplifier's output is fed back to magnetically bias the transformer flux to zero. Thus, the Hall effect device does not have to respond linearly over wide ranges of current and the transformer core never sees DC bias, both advantageous conditions. The amount of DC and low frequency information is obtained at the amplifier's output, which corresponds to the bias needed to offset the measured current.

Figure 35.24 shows a practical circuit. The Hall effect transducer lies within the core of the clip-on current transformer specified. A very simplistic way to model the Hall generator is as a bridge, excited by the two 619Ω resistors. The Hall generator's outputs (the midpoints of the "bridge") feed differential input transconductance amplifier A1, which takes gain, with roll-off set by the 50Ω, 0.02μF RC at its output. Further gain is provided by A2, in the same package as A1. A current buffer provides power gain to drive the current transformers secondary. This connection closes a flux nulling loop in the transducer core. The offset adjustments should be set for 0V output with no current flowing in the clip-on transducer. Similarly, the loop gain and bandwidth trims should be set so that the composite output (the combined high and low frequency output across the grounded 50Ω resistor) has clean step response and correct amplitude from DC to high frequency.

Note 4: Viewed from a historical perspective it is remarkable that so much precision wideband performance is available from such a relatively simple configuration. For perspective, see Appendix A, "Precision Wideband Circuitry…Then and Now."

Note 5: Details of this scheme are nicely presented in Reference 15. Additional relevant commentary on parallel path schemes appears in Reference 7.

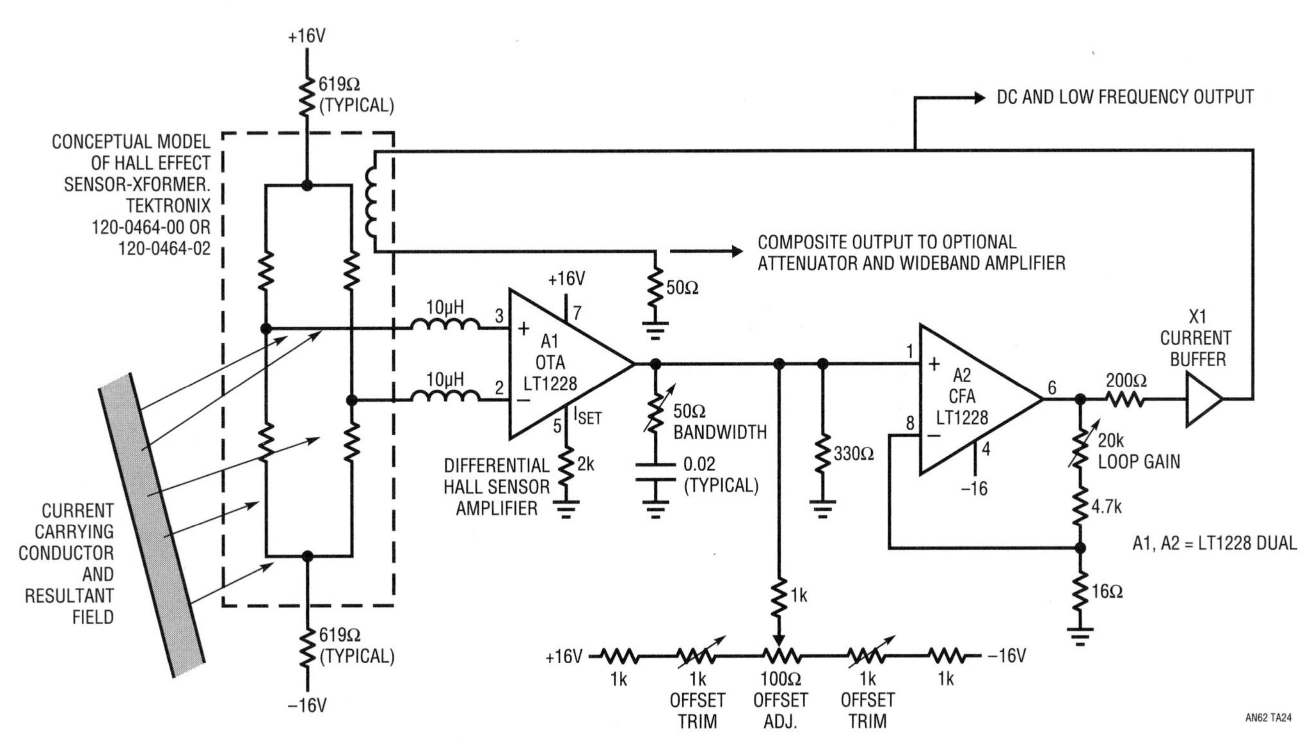

Figure 35.24 • Hall Effect Stabilized Current Transformer (DC → High Frequency Current Probe)

Figure 35.25 shows a practical way to conveniently evaluate this circuits performance. This partial schematic of the Tektronix P-6042 current probe shows a similar signal conditioning scheme for the transducer specified in Figure 35.24. In this case Q22, Q24 and Q29 combine with differential stage M-18 to form the Hall amplifier. To evaluate Figure 35.24's circuit remove M-18, Q22, Q24 and Q29.

Next, connect LT1228 pins 3 and 2 to the former M-18 pins 2 and 10 points, respectively. The ±16V supplies are available from the P-6042's power bus. Also, connect the right end of Figure 35.24's 200Ω resistor to what was Q29's collector node. Finally, perform the offset, loop gain and bandwidth trims as previously described.

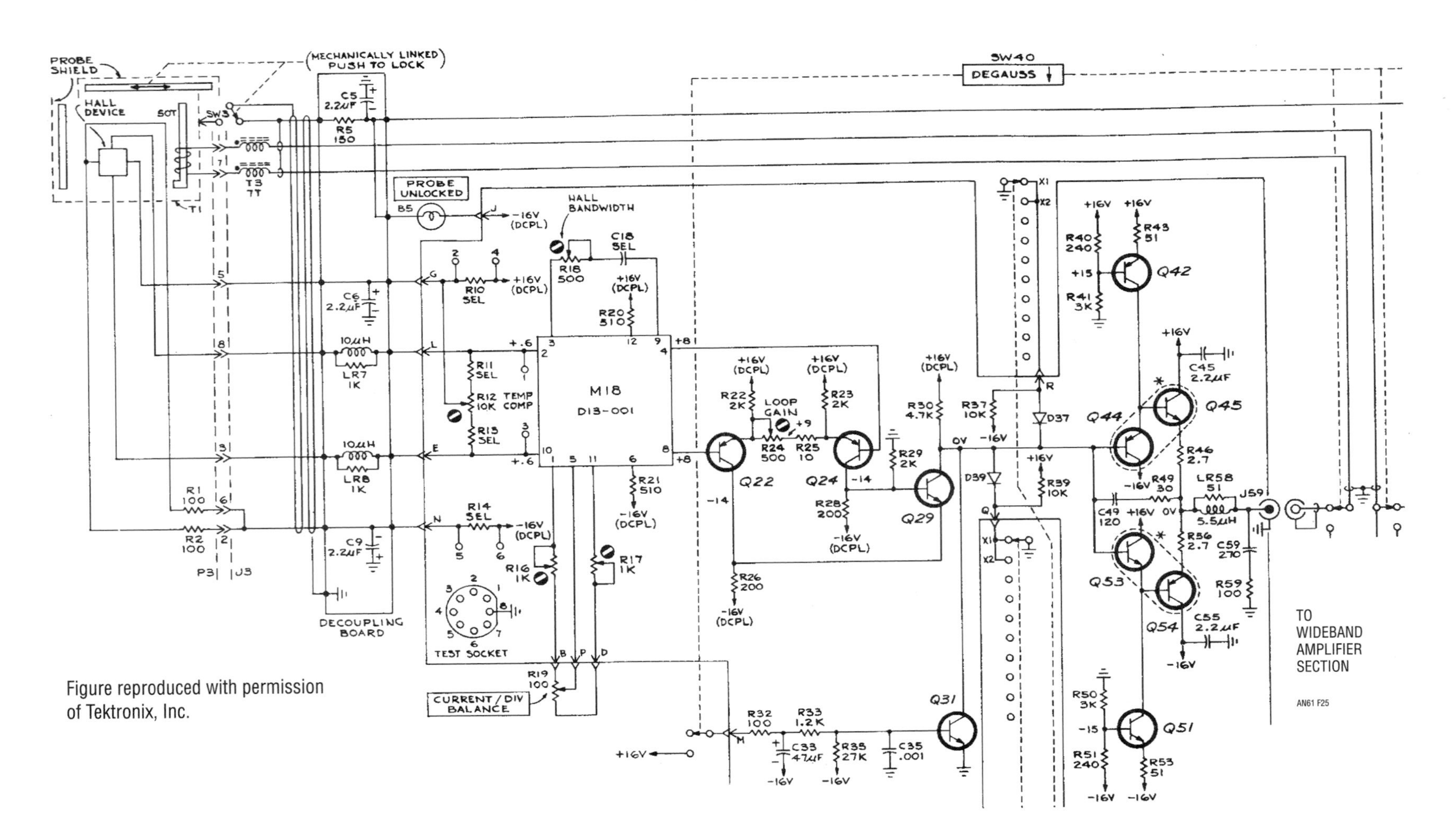

Figure reproduced with permission of Tektronix, Inc.

Figure 35.25 • Tektronix P-6042 Hall Effect Based Current Probe Servo Loop. Figure 35.24 Replaces M18 Amplifier and Q22, Q24 and Q29

Triggered 250 picosecond rise time pulse generator

Verifying the rise time limit of wideband test equipment setups is a difficult task. In particular; the "end-to-end" rise time of oscilloscope-probe combinations is often required to assure measurement integrity. Conceptually, a pulse generator with rise times substantially faster than the oscilloscope-probe combination can provide this information. Figure 35.26's circuit does this, providing an 800ps pulse with rise and fall times inside 250ps. Pulse amplitude is 10V with a 50Ω source impedance. This circuit has similarities to a previously published design (see Reference 7) except that it is triggered instead of free running. This feature permits synchronization to a clock or other event. The output phase with respect to the trigger is variable from 200ps to 5ns.

The pulse generator requires high voltage bias for operation. The LT1082 switching regulator to forms a high voltage switched mode control loop. The LT1082 pulse width modulates at its 40kHz clock rate. L1's inductive events are rectified and stored in the 2μF output capacitor. The adjustable resistor divider provides feedback to the LT1082. The 10k-1μF RC provides noise filtering.

The high voltage is applied to Q1, a 40V breakdown device, via the R3-C1 combination. The high voltage "bias adjust" control should be set at the point where free running pulses across R4 *just* disappear. This puts Q1 slightly below its avalanche point. When an input trigger pulse is applied Q1 avalanches. The result is a quickly rising, very fast pulse across R4. C1 discharges, Q1's collector voltage falls and breakdown ceases. C1 then recharges to just below the avalanche point. At the next trigger pulse this action repeats.[6]

Figure 35.27 shows waveforms. A 3.9GHz sampling oscilloscope (Tektronix 661 with 4S2 sampling pug-in) measures the pulse (trace B) at 10V high with an 800ps base. Rise time is 250ps, with fall time indicating 200ps. The times are probably slightly faster, as the oscilloscope's 90ps rise time influences the measurement.[7] The input trigger pulse is trace A. Its amplitude provides a convenient way to vary the delay time between the trigger and output pulses. A 1V to 5V amplitude setting produces a continuous 5ns to 200ps delay range.

Note 6: This circuit is based on the operation of the Tektronix Type 111 Pulse Generator. See Reference 16.

Note 7: I'm sorry, but 3.9GHz is the fastest 'scope in my house (as of September, 1993).

Figure 35.26 • Triggered 250ps Rise Time Pulse Generator. Trigger Pulse Amplitude Controls Output Phase

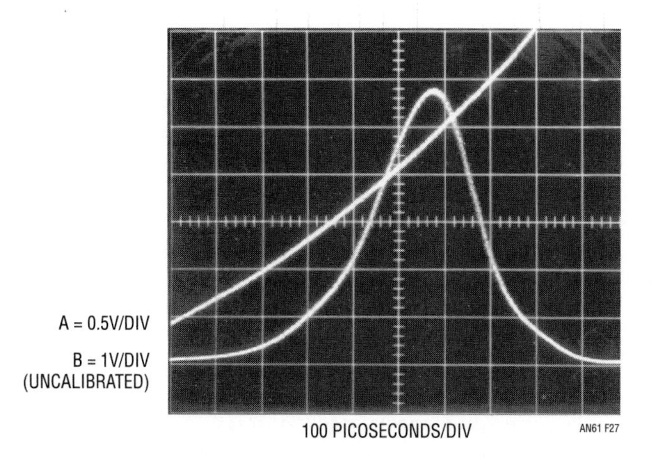

Figure 35.27 • Input Pulse Edge (Trace A) Triggers the Avalanche Pulse Output (Trace B). Display Granularity Is Characteristic of Sampling Oscilloscope Operation

Some special considerations are required to optimize circuit performance. L2's very small inductance combines with C2 to slightly retard the trigger pulse's rise time. This prevents significant trigger pulse artifacts from appearing at the circuit's output. C2 should be adjusted for the best compromise between output pulse rise time and purity. Figure 35.28 shows partial pulse rise with C2 properly adjusted. There are no discernible discontinuities related to the trigger event.

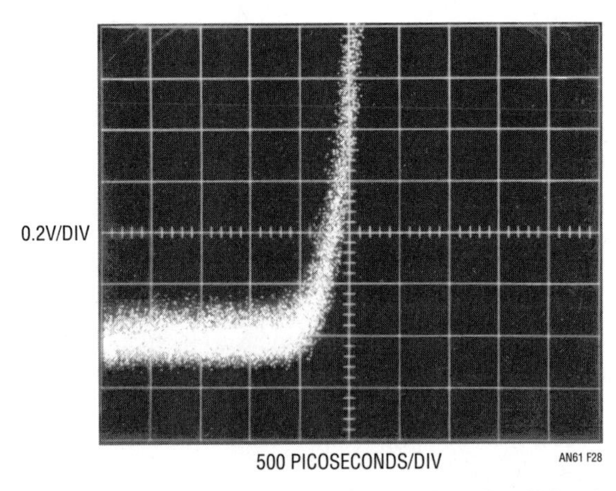

Figure 35.28 • Expanded Scale View of Leading Edge Is Clean with No Trigger Pulse Artifacts. Display Granularity Derives from Sampling Oscilloscope Operation

Q1 may require selection to get avalanche behavior. Such behavior; while characteristic of the device specified, is not guaranteed by the manufacturer. A sample of 50 Motorola 2N2369s, spread over a 12 year date code span, yielded 82%. All "good" devices switched in less than 600ps. C1 is selected for a 10V amplitude output. Value spread is typically 2pF to 4pF Ground plane type construction with high speed layout, connection and termination techniques are essential for a good results from this circuit.

Flash memory programmer

Although "Flash" type memory is increasingly popular, it does require some special programming features. The 5V powered memories need a carefully controlled 12V "VPP" programming pulse. The pulse's amplitude must be within 5% to assure proper operation. Additionally, the pulse must not overshoot, as memory destruction may occur for VPP outputs above 14V.[8] These requirements usually mandate a separate 12V supply and pulse forming circuitry. Figure 35.29's circuit provides the complete flash memory programming function with a single IC and some discrete components. All components are surface mount types, so little board space is required. The entire function runs off a single 5V supply.

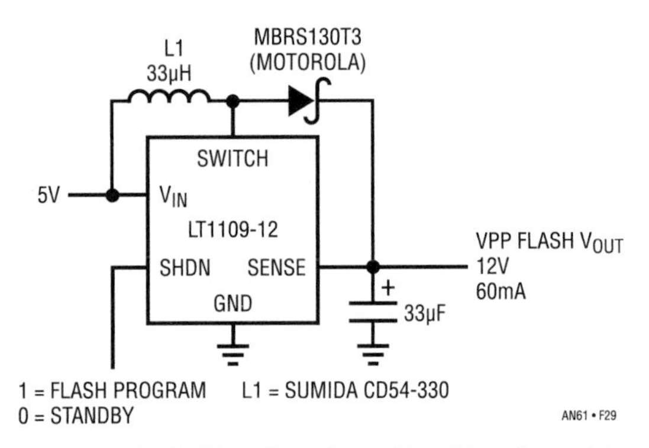

Figure 35.29 • Switching Regulator Provides Complete Flash Memory Programmer

The LT1109-12 switching regulator functions by repetitively pulsing L1. L1 responds with high voltage flyback events, which are rectified by the diode and stored in the 10μF capacitor. The "sense" pin provides feedback, and the output voltage stabilizes at 12V within a few percent. The regulator's "shutdown" pin provides a way to control the VPP programming voltage output. With a logical zero applied to the pin the regulator shuts down, and no VPP programming voltage appears at the output. When the pin goes high (trace A, Figure 35.30) the regulator is activated, producing a cleanly rising, controlled pulse at the output (trace B). When the pin is returned to logical zero, the output smoothly decays off. The switched mode delivery of power combined with the output capacitor's filtering prevents overshoot while providing the required pulse amplitude accuracy. Trace C, a time and amplitude expanded version of trace B, shows this. The output steps up in amplitude each time L1 dumps energy into the output capacitor. When the regulation point is reached the amplitude cleanly flattens out, with only about 75mV of regulator ripple.

Note 8: See Reference 17 for detailed discussion.

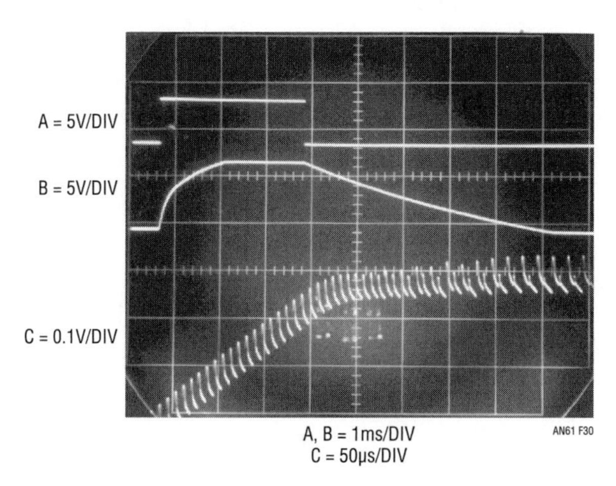

Figure 35.30 • Flash Memory Programmer Waveforms Show Controlled Edges. Trace C Details Rise Time Settling

3.3V powered V/F converter

Figure 35.31 is a "charge pump" type V/F converter specifically designed to run from a 3.3V rail.[9] A 0V to 2V input produces a corresponding 0kHz to 3kHz output with linearity inside 0.05%. To understand how the circuit works assume that A1's negative input is just below 0V. The amplifier output is positive. Under these conditions, LTC1043's pins 12 and 13 are shorted as are pins 11 and 7, allowing the 0.01µF capacitor (C1) to charge to the 1.2V LT1034 reference. When the input-voltage-derived current ramps A1's summing point (negative input-trace A, Figure 35.32) positive, its output (trace B) goes low. This reverses the LTC1043's switch states, connecting pins 12 and 14, and 11 and 8. This effectively connects C1's positively charged end to ground on pin 8, forcing current to flow from A1's summing junction into C1 via LTC1043 pin 14 (pin 14's current is trace C). This action resets A1's summing point to a small negative potential (again, trace A). The 120pF-50k-10k time constant at A1's positive input ensures A1 remains low long enough for C1 to completely discharge (A1's positive input is trace D). The Schottky diode prevents excessive negative excursions due to the 120pF capacitors differentiated response.

When the 120pF positive feedback path decays, A1's output returns positive and the entire cycle repeats. The oscillation frequency of this action is directly related to the input voltage.

This is an AC coupled feedback loop. Because of this, start-up or overdrive conditions could force A1 to go low

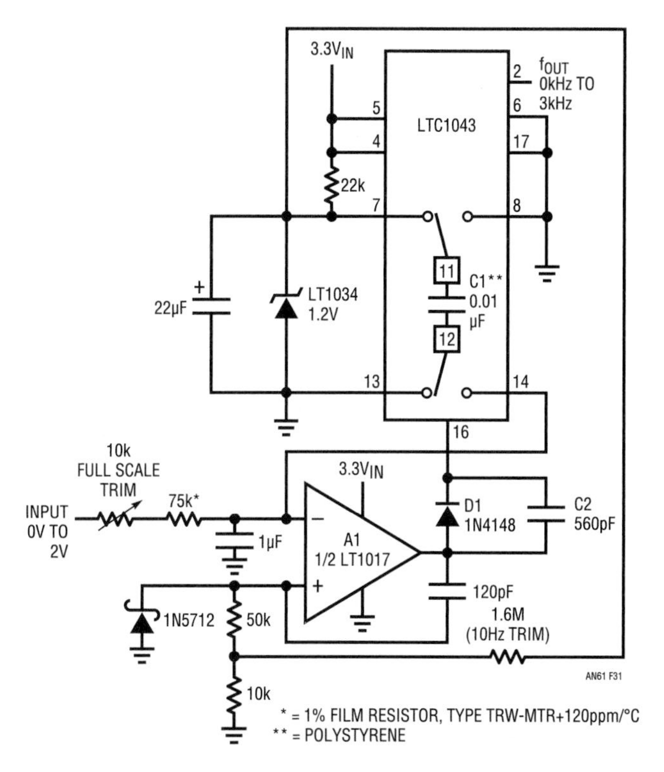

Figure 35.31 • 3.3V Powered Voltage-to-Frequency Converter. Charge Pump Based Feedback Maintains High Linearity and Stability

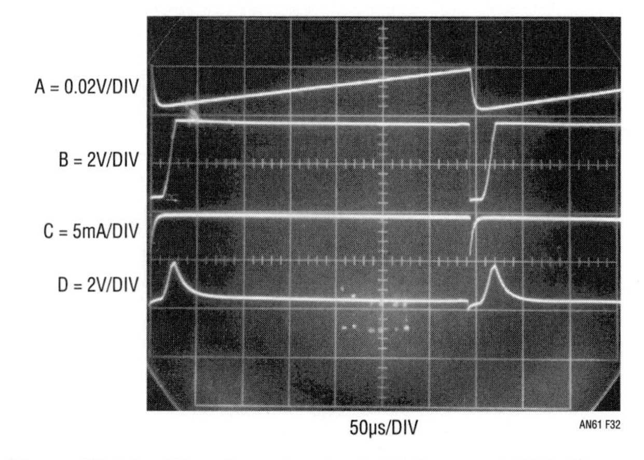

Figure 35.32 • Waveform for the 3.3V Powered V/F. Charge Pump Action (Trace C) Maintains Summing Point (Trace A), Enforcing High Linearity and Accuracy

Note 9: See Reference 20 for a survey of V/F techniques. The circuit shown here is derived from Figure 8 in LTC Application Note 50, "Interfacing to Microprocessor Based 5V Systems" by Thomas Mosteller.

and stay there. When A1's output is low the LTC1043's internal oscillator sees C2 and will begin oscillation if A1 remains low long enough. This oscillation causes charge pumping action via the LTC1043-C1-A1 summing junction path until normal operation commences. During normal operation A1 is never low long enough for oscillation to occur, and controls the LTC1043 switch states via D1.

To calibrate this circuit apply 7mV and select the 1.6M (nominal) value for 10Hz out. Then apply 2.000V and set the 10k trim for exactly 3kHz output. Pertinent specifications include linearity of 0.05%, power supply rejection of 0.04%/V temperature coefficient of 75ppm/°C of scale and supply current of about 200μA. The power supply may vary from 2.6V to 4.0V with no degradation of these specifications. If degraded temperature coefficients are acceptable, the film resistor specified may be replaced by a standard 1% film resistor. The type called out has a temperature characteristic that opposes C1's –120ppm/°C drift, resulting in the low overall circuit drift noted.

Broadband random noise generator

Filter, audio, and RF-communications testing often require a random noise source.[10] Figure 35.33's circuit provides an RMS-amplitude regulated noise source with selectable bandwidth. RMS output is 300mV with a 1kHz to 5MHz bandwidth, selectable in decade ranges.

Noise source D1 is AC coupled to A2, which provides a broadband gain of 100. A2's output feeds a gain control stage via a simple, selectable lowpass filter. The filter's output is applied to A3, an LT1228 operational

Note 10: See Appendix B, "Symmetrical White Gaussian Noise," guest written by Ben Hessen-Schmidt of Noise Com, Inc. for tutorial on noise.

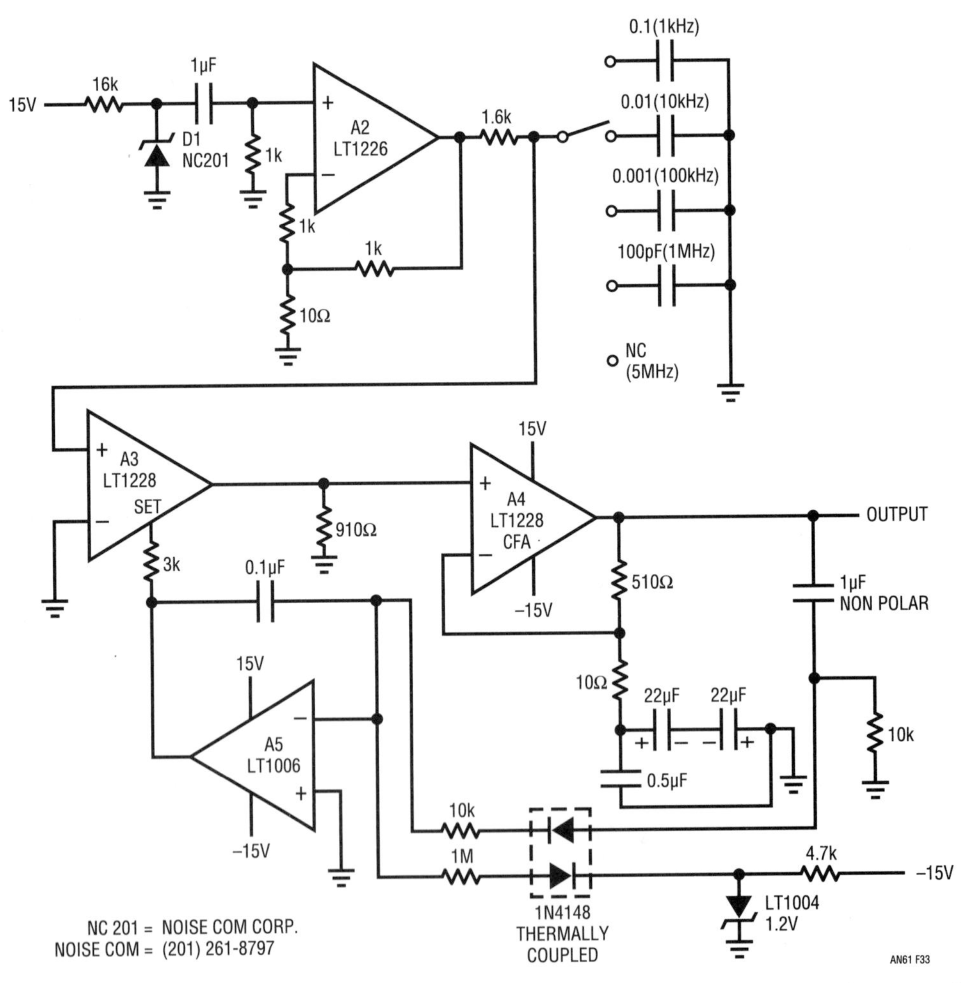

Figure 35.33 • Broadband Random Noise Generator Uses Gain Control Loop to Enhance Noise Spectrum Amplitude Uniformity

transconductance amplifier. A3's output feeds LT1228 A4, a current feedback amplifier. A4's output, also the circuit's output, is sampled by the A5-based gain control configuration. This closes a gain control loop to A3. A3's set current controls gain, allowing overall output level control.

Figure 35.34 shows noise at 1MHz bandpass, with Figure 35.35 showing RMS noise versus frequency in the same bandpass. Figure 35.36 plots similar information at full bandwidth (5MHz). RMS output is essentially flat to 1.5MHz with about ±2dB control to 5MHz before sagging badly.

Figure 35.37's similar circuit substitutes a standard zener for the noise source but is more complex and requires a trim. A1, biased from the LT1004 reference, provides

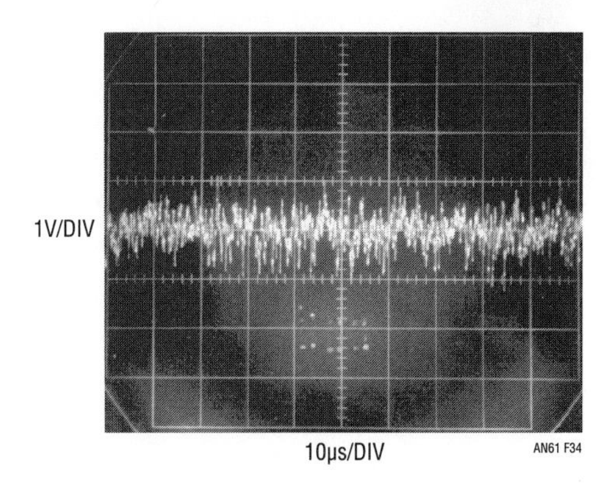

Figure 35.34 • Figure 35.33's Output in the 1MHz Filter Position

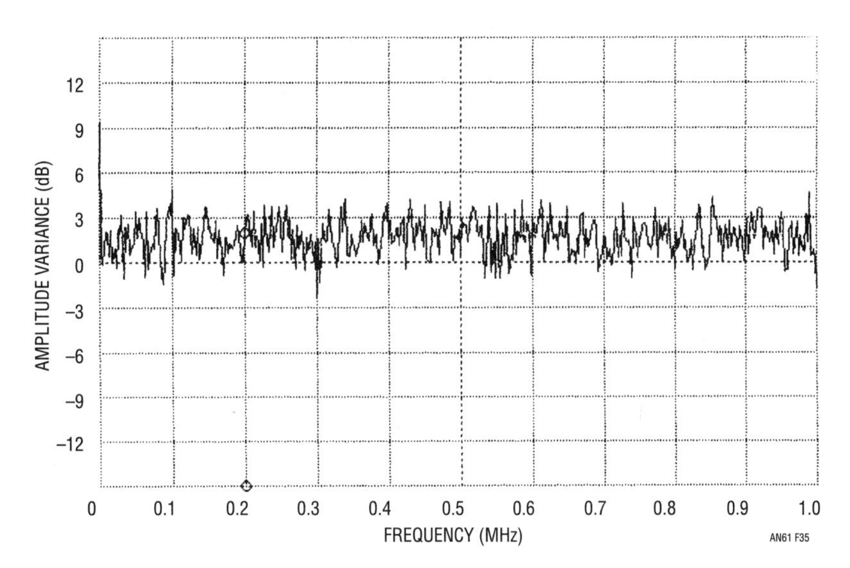

Figure 35.35 • Amplitude vs Frequency for the Random Noise Generator Is Essentially Flat to 1MHz

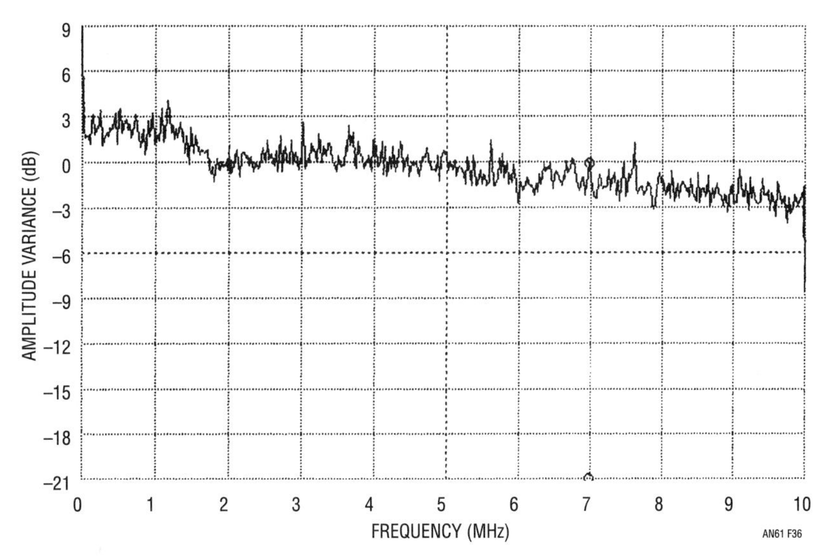

Figure 35.36 • RMS Noise vs Frequency at 5MHz Bandpass Shows Slight Fall-Off Beyond 1MHz

optimum drive for D1, the noise source. AC coupled A2 takes a broadband gain of 100. A2's output feeds a gain-control stage via a simple selectable lowpass filter. The filter's output is applied to LT1228 A3, an operational transconductance amplifier. A3's output feeds LT1228 A4, a current feedbacks amplifier. A4's output, the circuit's output, is sampled by the A5-based gain control configuration. This closes a gain control loop back at A3. A3's set input current controls its gain, allowing overall output level control.

To adjust this circuit, place the filter in the 1kHz position and trim the 5k potentiometer for maximum negative bias at A3, pin 5.

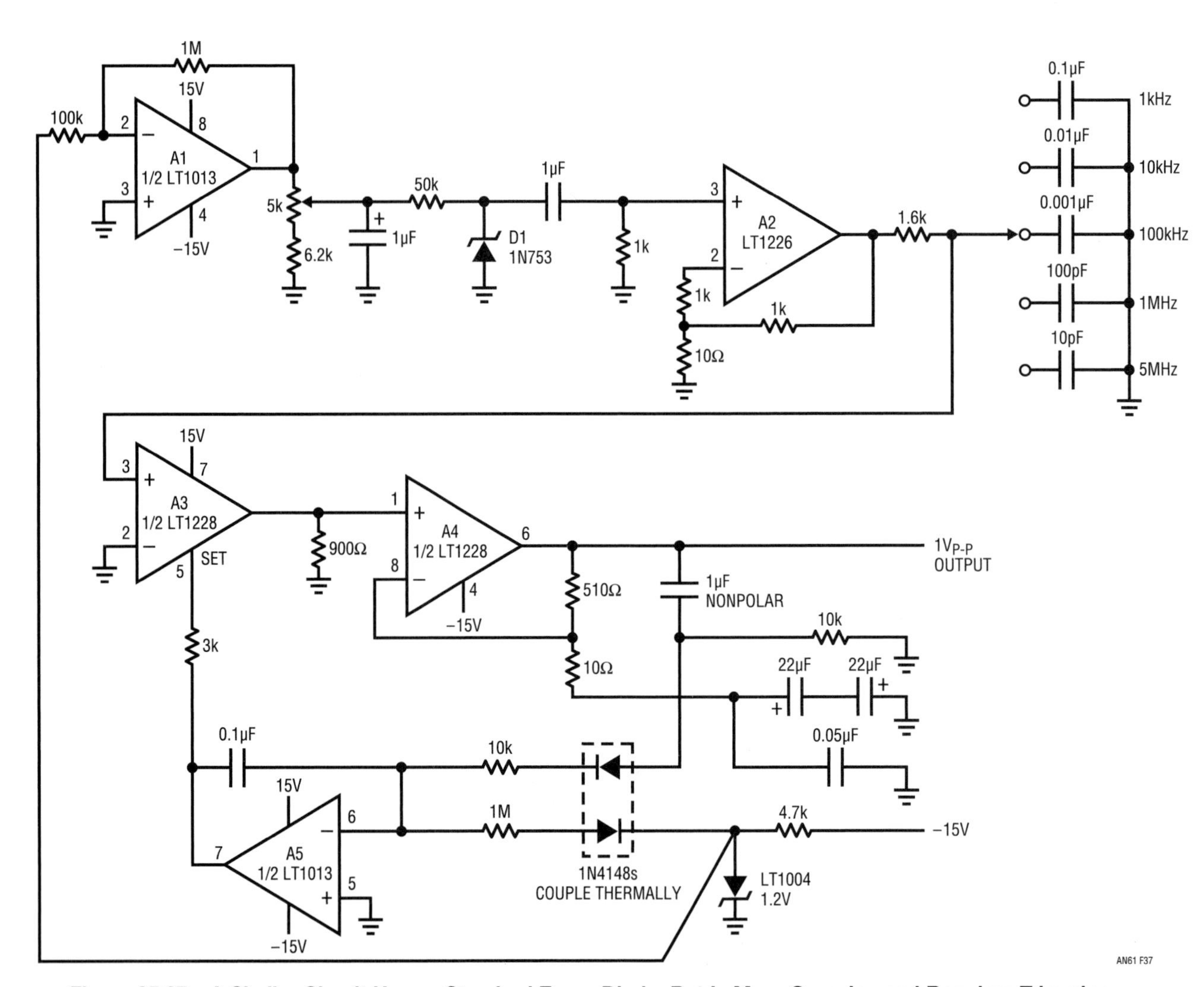

Figure 35.37 • A Similar Circuit Uses a Standard Zener Diode, But Is More Complex and Requires Trimming

Switchable output crystal oscillator

Figure 35.38's simple crystal oscillator circuit permits crystals to be electronically switched by logic commands. The circuit is best understood by initially ignoring all crystals. Further, assume all diodes are shorts and their associated 1k resistors open. The resistors at the LT1116's positive input set a DC bias point. The 2k-25pF path sets up phase shifted feedback and the circuit looks like a wideband unity gain follower at DC. When "Xtal A" is inserted (remember, D1 is temporarily shorted) positive feedback occurs and oscillation commences at the crystals resonant frequency. If D1 and its associated 1k value are realized, oscillation can only continue if logic input A is biased high. Similarly, additional crystal-diode-1k branches permit logic selection of crystal frequency.

For AT cut crystals about a millisecond is required for the circuit output to stabilize due to the high Q factors involved. Crystal frequencies can be as high as 16MHz before comparator delays preclude reliable operation.

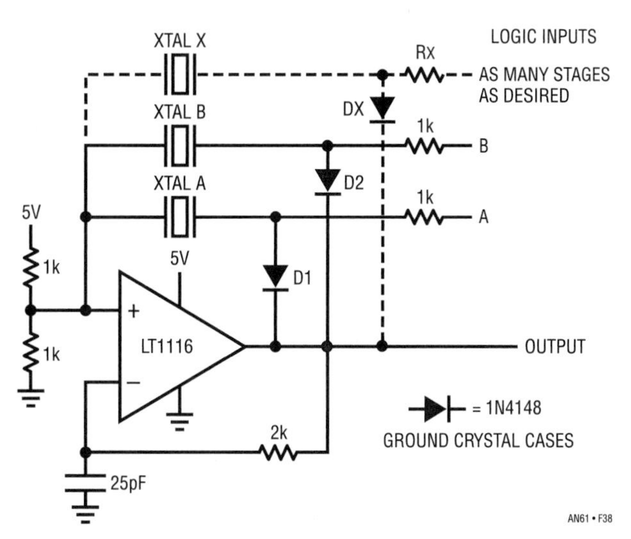

Figure 35.38 • Switchable Output Crystal Oscillator. Biasing A or B High Places the Associated Crystal in the Feedback Path. Additional Crystal Branches Are Permissible

References

1. Williams, Jim and Huffman, Brian. "Some Thoughts on DC-DC Converters," pages 13-17, "1.5V to 5V Converters." Linear Technology Corporation, Application Note 29, October 1988
2. Williams, J., "Illumination Circuitry for Liquid Crystal Displays," Linear Technology Corporation, Application Note 49, August 1992
3. Williams, J., "Techniques for 92% Efficient LCD Illumination," Linear Technology Corporation, Application Note 55, August 1993
4. Williams, J., "Measurement and Control Circuit Collection," Linear Technology Corporation, Application Note 45, June 1991
5. Benjaminson, Albert, "The Linear Quartz Thermometer—a New Tool for Measuring Absolute and Difference Temperatures," Hewlett-Packard Journal, March 1965
6. Micro Crystal-ETA Fabriques d'Ebauches., "Miniature Quartz Resonators - MT Series" Data Sheet. 2540 Grenchen, Switzerland
7. Williams, J., "High Speed Amplifier Techniques," Linear Technology Corporation, Application Note 47, August 1991
8. Williams, Jim, "High Speed Comparator Techniques," Linear Technology Corporation, Application Note 13, April 1985
9. Williams, Jim, "A Monolithic IC for 100MHz RMS-DC Conversion," Linear Technology Corporation, Application Note 22, September 1987

10. Ott, W.E., "A New Technique of Thermal RMS Measurement," IEEEJournal of Solid State Circuits, December 1974
11. Williams, J.M. and Longman, TL., "A 25MHz Thermally Based RMS-DC Converter," 1986 IEEE ISSCC Digest of Technical Papers
12. O'Neill, PM., "A Monolithic Thermal Converter," H.P. Journal, May 1980
13. C. Kitchen, L. Counts, "RMS-to-DC Conversion Guide," Analog Devices, Inc. 1986
14. Tektronix, Inc. "P6042 Current Probe Operating and Service Manual," 1967
15. Weber Joe, "Oscilloscope Probe Circuits," Tektronix, Inc., Concept Series. 1969
16. Tektronix, Inc., Type 111 Pretrigger Pulse Generator Operating and Service Manual, Tektronix, Inc. 1960
17. Williams, J., "Linear Circuits for Digital Systems," Linear Technology Corporation, Application Note 31, February 1989
18. Williams, J., "Applications for a Switched-Capacitor Instrumentation Building Block," Linear Technology Corporation, Application Note 3, July 1985
19. Williams, J., "Circuit Techniques for Clock Sources," Linear Technology Corporation, Application Note 12, October 1985
20. Williams, J. "Designs for High Performance Voltage- to-Frequency Converters," Linear Technology Corporation, Application Note 14, March 1986

Appendix A
Precision wideband circuitry...then and now

Text Figure 35.22's relatively straightforward design provides a sensitive, thermally-based RMS/DC conversion to 10MHzwith less than 1% error. Viewed from a historical perspective it is remarkable that so much precision wideband performance is so easily achieved.

Thirty years ago these specifications presented an extremely difficult engineering challenge, requiring deep-seated knowledge of fundamentals, extraordinary levels of finesse and an interdisciplinary outlook to achieve success.

The Hewlett-Packard model HP3400A (1965 price $525...about 1/3 the yearly tuition at M.I.T.) thermally based RMS voltmeter included all of Figure 35.22's elements, but considerably more effort was required in its execution.[1] Our comparative study begins by considering H-P's version of Figure 35.22's FET buffer and precision wideband amplifier. The text is taken directly from the *HP3400A Operating and Service Manual*.[2]

If that's not enough to make you propose marriage to modern high speed monolithic amplifiers, consider the design heroics spent on the thermal converter.

Note 1: We are all constantly harangued about the advances made in computers since the days of the IBM360. This section gives analog aficionados a stage for their own bragging rights. Of course, an HP3400A was much more interesting than anIBM360 in 1965. Similarly, Figure 35.22's capabilities are more impressive than any contemporary computer I'm aware of.

Note 2: All Hewlett-Packard text and figures used here are copyright 1965 Hewlett-Packard Company. Reproduced with permission.

4-15. IMPEDANCE CONVERTER ASSEMBLY A2.

4-17. The ac signal input to the impedance converter is RC coupled to the grid of cathode follower V201[3] through C201 and R203. The output signal is developed by Q201 which acts as a variable resistance in the cathode circuit of V201. The bootstrap feedback from the cathode of V201 to R203 increases the effective resistance of R203 to the input signal. This prevents R203 from loading the input signal and preserves the high input impedance of the Model 3400A. The gain compensating feedback from the plate of V201 to the base of Q201 compensates for any varying gain in V201 due to age or replacement.

4-18. Breakdown diode CR201 controls the grid bias voltage on V201 thereby establishing the operating point of this stage. CR202 and R211 across the base-emitter junction of Q201 protects Q201 in the event of a failure in the +75 volt power supply. Regulated dc is supplied to V201 filaments to avoid inducing ac hum in the signal path. This also prevents the gain of V201 changing with line voltage variations.

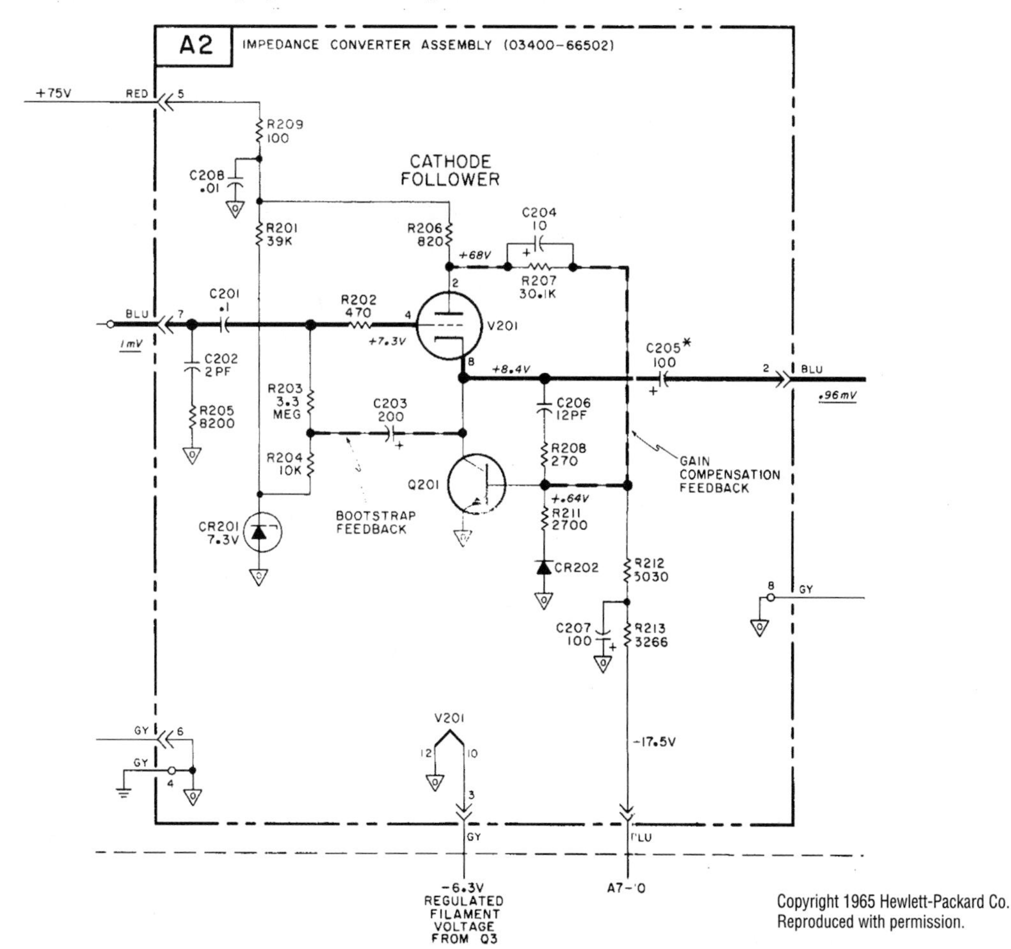

Figure A1 • The "Impedance Converter Assembly," H-Ps Equivalent of Figure 35.22's Wideband FET Buffer

Note 3: Although JFETs were available in 1965 their performance was inadequate for this design's requirements. The only available option was the Nuvistor triode described.

Figure A2 • The Hewlett-Packard 3400A's Wideband Input Buffer. Nuvistor Triode (Upper Center) Provided Speed, Low Noise, and High Impedance. Circuit Required 75V, –17.5V and –6.3V Supplies. Regulated Filament Supply Stabilized Follower Gain While Minimizing Noise

4-22. VIDEO AMPLIFIER ASSEMBLY A4.

4-23. The video amplifier functions to provide constant gain to the ac signal being measured over the entire frequency range of Model 3400A. See video amplifier assembly schematic diagram illustrated on Figure 6-2.

4-24. The ac input signal from the second attenuator is coupled through C402 to the base of input amplifier Q401. Q401, a class A amplifier, amplifies and inverts the signal which is then direct coupled to the base of bootstrap amplifier Q402. The output, taken from Q402 emitter is applied to the base of Q403 and fed back to the top of R406 as a bootstrap feedback. This positive ac feedback increases the effective ac resistance of R406 allowing a greater portion of the signal to be felt at the base of Q402. In this manner, the effective ac gain of Q401 is increased for the midband frequencies without disturbing the static operating voltages of Q401.

4-25. Driver amplifier Q403 further amplifies the ac signal and the output at Q403 collector is fed to the base circuit emitter follower Q404. The feedback path from the collector of Q403 to the base of Q402 through C405 (10 MHz ADJ) prevents spurious oscillations at high input frequencies. A dc feedback loop exists from the emitter circuit of Q403, to the base of Q401 through R425. This feedback stabilizes the Q401 bias voltage. Emitter follower Q404 acts as a driver for the output amplifier consisting of Q405 and Q406; a complimentary pair operating as a push-pull amplifier. The video amplifier output is taken from the collectors of the output amplifiers and applied to thermocouples TC401. A gain stabilizing feedback is developed in the emitter circuits of the output amplifiers. This negative feedback is applied to the emitter of input amplifier Q401 and establishes the overall gain of the video amplifier.

4-26. Trimmer capacitor C405 is adjusted at 10 MHz for frequency response of the video amplifier. Diodes CR402 and CR406 are protection diodes which prevent voltage surges from damaging transistors in the video amplifier. CR401, CR407, and CR408 are temperature compensating diodes to maintain the zero signal balance condition in the output amplifier over the operating temperature range. CR403, a breakdown diode, establishes the operating potentials for the output amplifier.

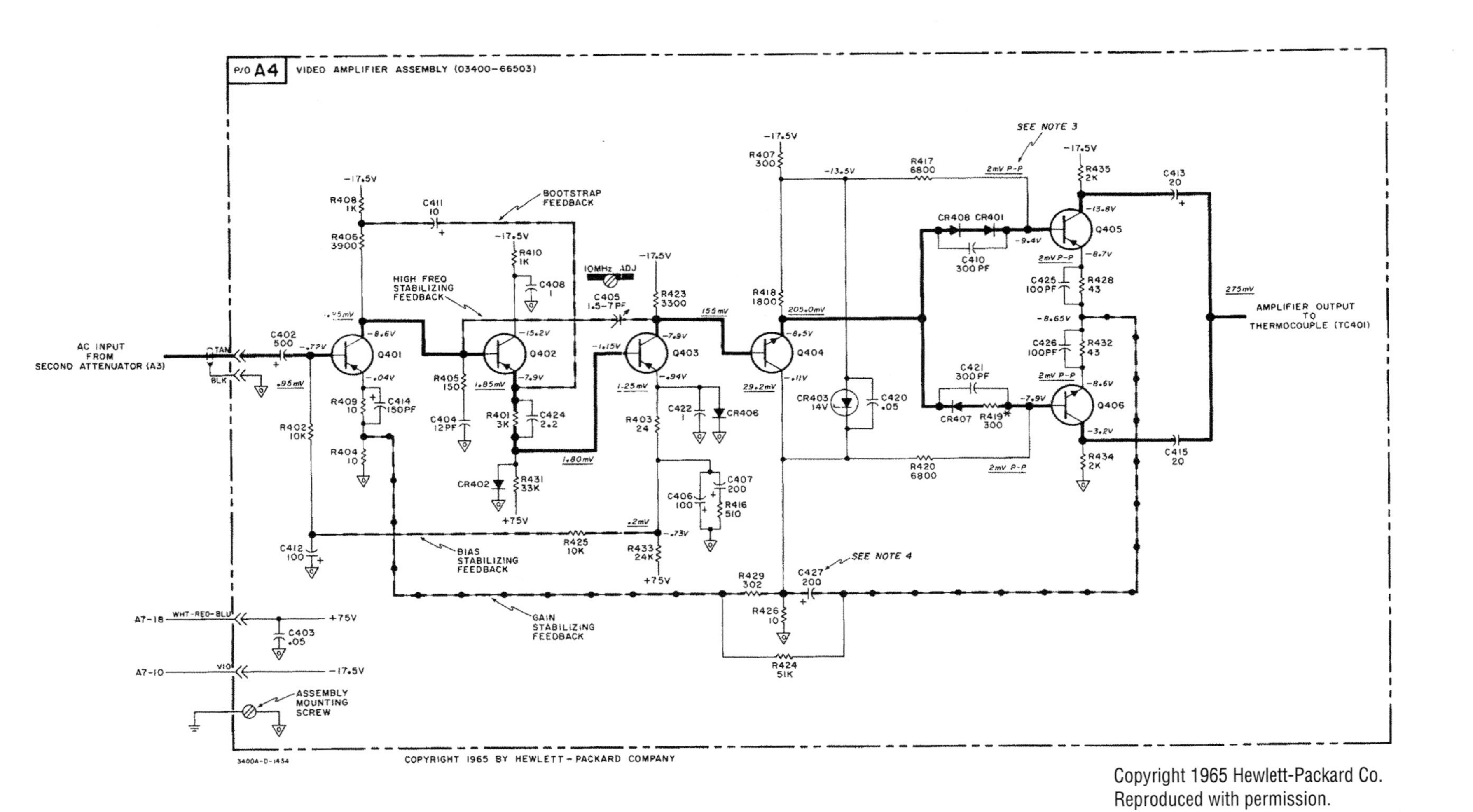

Figure A3 • H-P's Wideband Amplifier, the "Video Amplifier Assembly" Contained DC and AC Feedback Loops, Peaking Networks, Bootstrap Feedback and Other Subtleties to Equal Figure 35.22's Performance

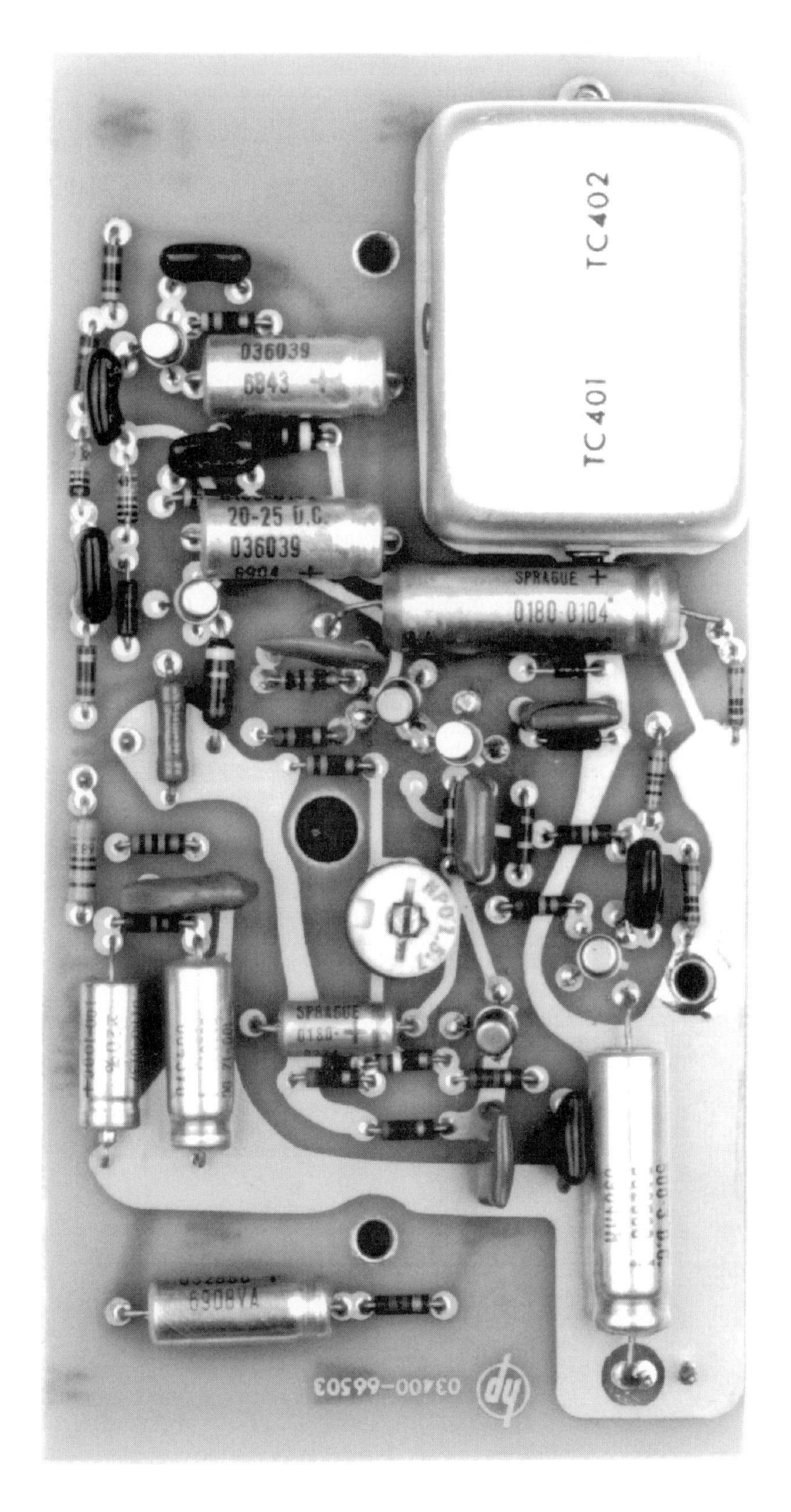

Figure A4 • The Voltmeters "Video Amplifier" Received Input at Board's Left Side. Amplifier Output Drove Shrouded Thermal Converter at Lower Right. Note High Frequency Response Trimmer Capacitor at Left Center

4-27. PHOTOCHOPPER ASSEMBLY A5, CHOPPER AMPLIFIER ASSEMBLY A6, AND THERMOCOUPLE PAIR (PART OF A4).

4-28. The modulator/demodulator, chopper amplifier, and thermocouple pair form a servo loop which functions to position the direct reading meter M1 to the rms value of the ac input signal.[4] See modulator/demodulator, chopper amplifier, and thermocouple pair schematic diagram illustrated in Figure 6-3.

4-29. The video amplifier output signal is applied to the heater of thermocouple TC401. This ac voltage causes a dc voltage to be generated in the resistive portion of TC401 which is proportional to the heating effect (rms value) of the ac input. The dc voltage is applied to photocell V501.

4-30. Photocells V501 and V502 in conjunction with neon lamps DS501 and DS502 form a modulator circuit.[5] The neon lamps are lighted alternately between 90 and 100 Hz. Each lamp illuminates one of the photocells. DS501 illuminates V501;DS502 illuminates V502. When a photocell is illuminated it has a low resistance compared to its resistance when dark. Therefore, when V501 is illuminated, the output of thermocouple TC401 is applied to the input of the chopper amplifier through V501. When V502 is illuminated, a ground signal is applied to the chopper amplifier. The alternate illumination of V501 and V502 modulates the dc input at a frequency between 90 and 100 Hz. The modulator output is a square wave whose amplitude is proportional to the dc input level.

4-31. The chopper amplifier, consisting of Q601 through Q603, is a high gain amplifier which amplifies the square wave developed by the modulator. Power supply voltage variations are reduced by diodes CR601 thru CR603. The amplified output is taken from the collector of Q603 and applied to the demodulator through emitter follower Q604.

4-32. The demodulator comprises two photocells, V503 and V504, which operate in conjunction with DS501 and DS502; the same neon lamps used to illuminate the photocells in the modulator. Photocells V503 and V504 are illuminated by DS501 and DS502, respectively.

4-33. The demodulation process is the reverse of the modulation process discussed in Paragraph 4-30. The output of the demodulator is a dc level which is proportional to the demodulator input. The magnitude and phase of the input square wave determines the magnitude and polarity of the dc output level. This dc output level is applied to two emitter follower output stages.

4-34. The emitter follower is needed to match the high output impedance of the demodulator to the low input impedance of the meter and thermocouple circuits. The voltage drop across CR604 in the collector circuit of Q605 is the operating bias for Q604. This fixed bias prevents Q605 failure when the base voltage is zero with respect to ground.

Note 4: In 1965 almost all thermal converters utilized matched pairs of discrete heater resistors and thermocouples. The thermocouples' low level output necessitated chopper amplifier signal conditioning, the only technology then available which could provide the necessary DC stability.

Note 5: The low level chopping technology of the day was mechanical choppers, a form of relay. H-P's use of neon lamps and photocells as microvolt choppers was more reliable and an innovation. Hewlett-Packard has a long and successful history of using lamps for unintended purposes.

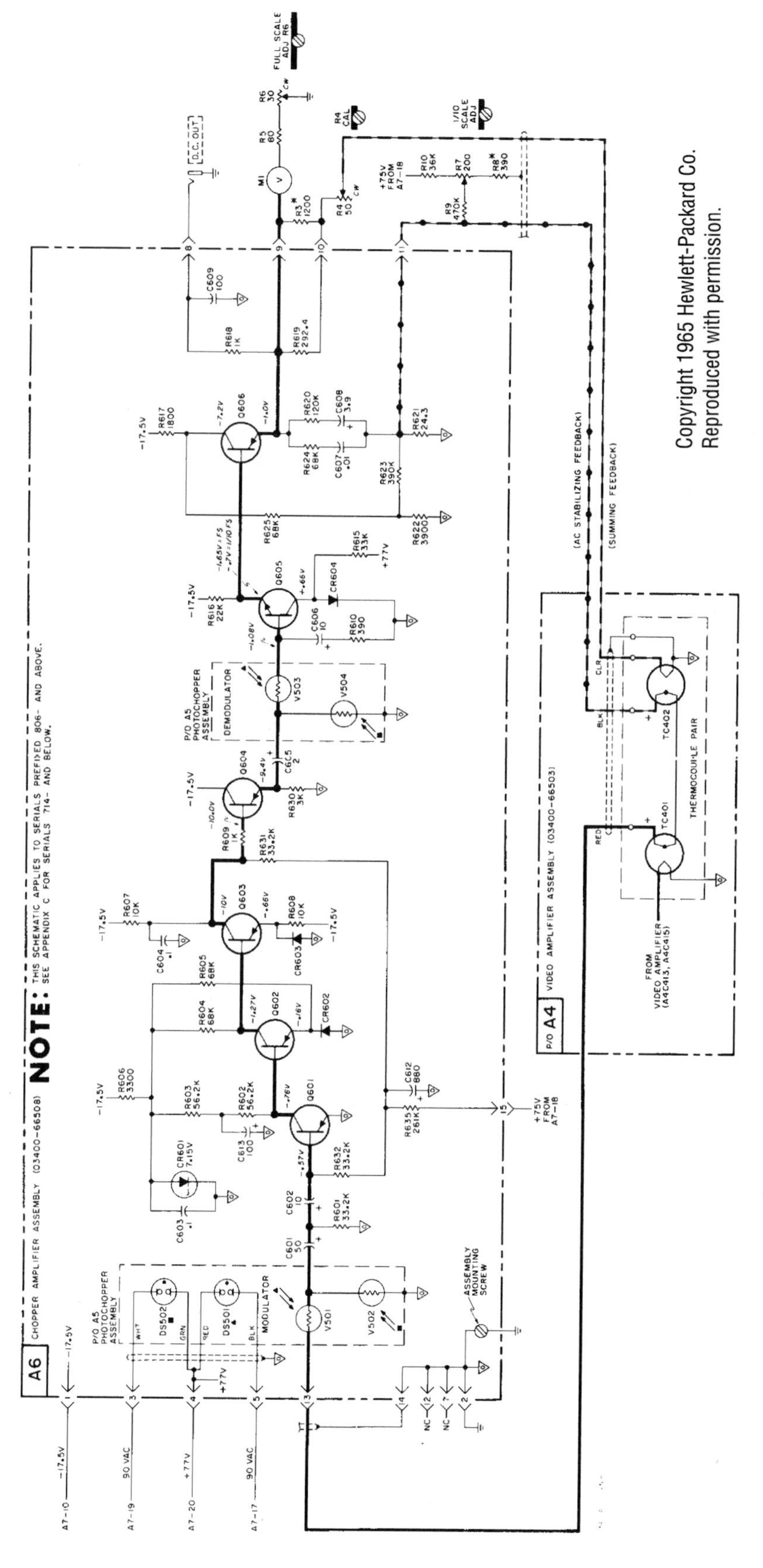

Figure A5 • H-P's Thermal Converter ("A4") and Control Amplifier ("A6") Perform Similarly to Text Figure 35.22's Dual Op Amp and LT1088. Circuit Realization Required Far More Attention to Details

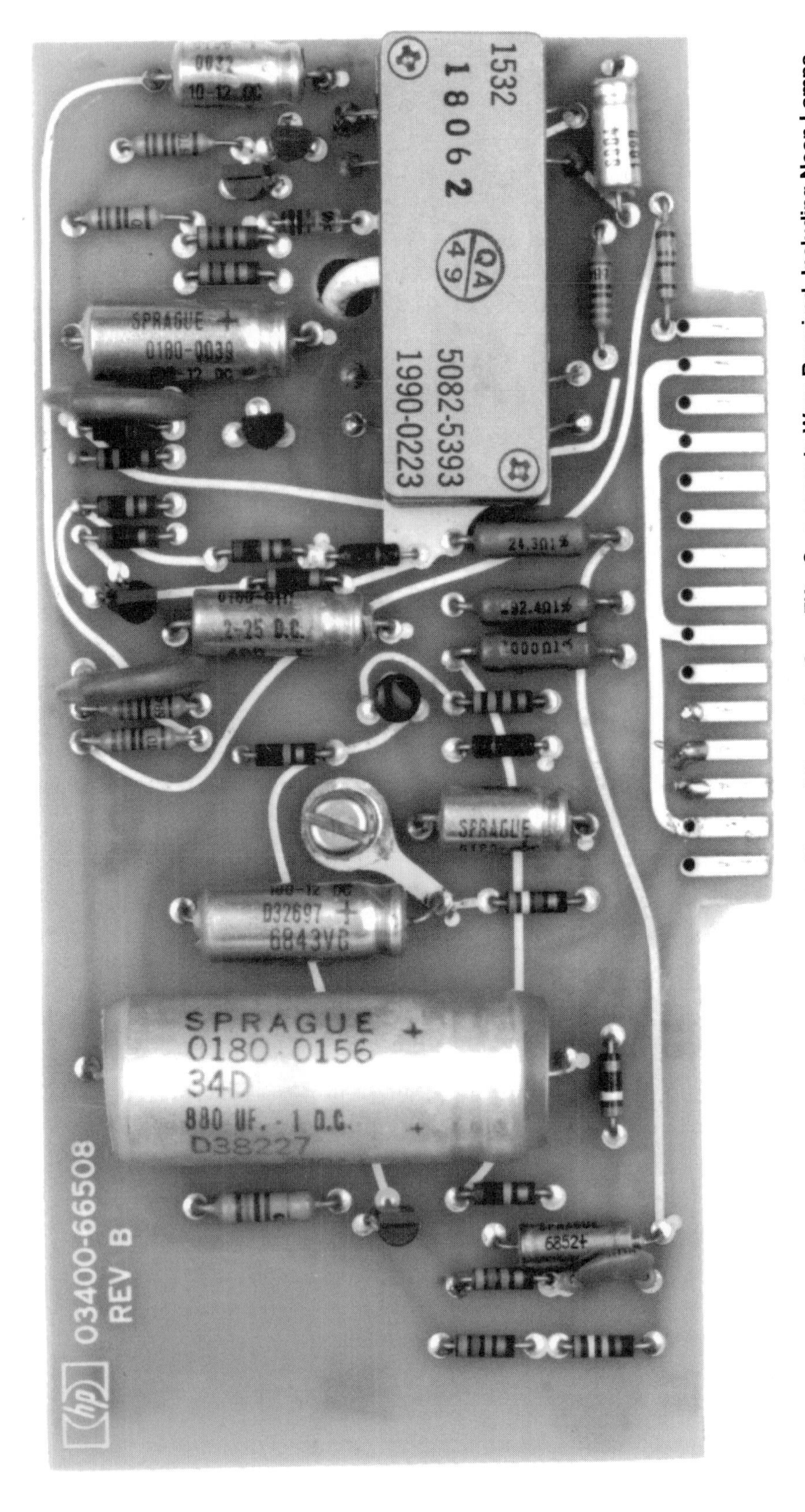

Figure A6 • Chopper Amplifier Board Feedback Controlled the Thermal Converter. Over Fifty Components Were Required, Including Neon Lamps, Photocells and Six Transistors. Photo-Chopper Assembly Is at Board's Lower Right

Figure A7 • Figure 35.22's Circuit Puts Entire HP3400 Electronics on One Small Board. FET Buffer-LT1206 Amplifier Appear Left Center Behind BNC Shield. LT1088 IC (Upper Center) Replaces Thermal Converter. LT1013 (Upper Right) Based Circuitry Replaces Photo-Chopper Board. LT1018 and Components (Lower Right) Provide Overload Protection. Ain't Modern ICs Wonderful?

4-35. The dc level output, taken from the emitter of Q606, is applied to meter M1 and to the heating element of thermocouple TC402. The dc voltage developed in the resistive portion of TC402 is effectively subtracted from the voltage developed by TC401. The input signal to the modulator then becomes the difference in the dc outputs of the two thermocouples. When the difference between the two thermocouples becomes zero the dc from the emitter followers (driving the meter) will be equal to the ac from the video amplifier.

4-36. Noise on the modulated square wave is suppressed by feedback from emitter of Q606 through C607 and C608 to the resistive element of TC402.

When casually constructing a wideband amplifier with a few mini-DIPs, the reader will do well to recall the pain and skill expended by the HP3400A's designers some 30 years ago.

Incidentally, what were *you* doing in 1965?

Appendix B
Symmetrical white Gaussian noise

by Ben Hessen-Schmidt,
NOISE COM, INC.

White noise provides instantaneous coverage of all frequencies within a band of interest with a very flat output spectrum. This makes it useful both as a broadband stimulus and as a power-level reference.

Symmetrical white Gaussian noise is naturally generated in resistors. The noise in resistors is due to vibrations of the conducting electrons and holes, as described by Johnson and Nyquist.[1] The distribution of the noise voltage is symmetrically Gaussian, and the average noise voltage is:

$$\overline{V}_n = 2\sqrt{KT\int R(f)p(f)df} \qquad (1)$$

Where:
k = 1.38E–23 J/K (Boltzmann's constant)
T = temperature of the resistor in Kelvin
f = frequency in Hz
h = 6.62E–34 Js (Planck's constant)
R(f) = resistance in ohms as a function of frequency

$$P(f) = \frac{hf}{kT[\exp(hf/kT) - 1]} \qquad (2)$$

p(f) is close to unity for frequencies below 40GHz when T is equal to 290°K. The resistance is often assumed to be independent of frequency, and ∫df is equal to the noise bandwidth (B). The available noise power is obtained when the load is a conjugate match to the resistor, and it is:

$$N = \frac{\overline{V}_n^2}{4R} = kTB \qquad (3)$$

where the "4" results from the fact that only half of the noise voltage and hence only 1/4 of the noise power is delivered to a matched load.

Equation 3 shows that the available noise power is proportional to the temperature of the resistor; thus it is often called thermal noise power, Equation 3 also shows that white noise power is proportional to the bandwidth.

An important source of symmetrical white Gaussian noise is the noise diode. A good noise diode generates a high level of symmetrical white Gaussian noise. The level is often specified in terms of excess noise ratio (ENR).

$$ENR(\text{in dB}) = 10\text{Log}\frac{(Te - 290)}{290} \qquad (4)$$

Note 1: See Additional Reading at the end of this section.

Te is the physical temperature that a load (with the same impedance as the noise diode) must be at to generate the same amount of noise.

The ENR expresses how many times the effective noise power delivered to a non-emitting, nonreflecting load exceeds the noise power available from a load held at the reference temperature of 290°K (16.8°C or 62.3°F).

The importance of high ENR becomes obvious when the noise is amplified, because the noise contributions of the amplifier may be disregarded when the ENR is 17dB larger than the noise figure of the amplifier (the difference in total noise power is then less than 0.1dB). The ENR can easily be converted to noise spectral density in dBm/Hz or $\mu V/\sqrt{Hz}$ by use of the white noise conversion formulas in Table 35.1.

Table 35.1 Useful White Noise Conversion

dBm = dBm/Hz + 10log (BW)
dBm = 20log ($\bar{V}$n) – 10log(R) + 30dB
dBm = 20log($\bar{V}$n) +13dB for R = 50Ω
dBm/Hz = 20log(μ$\bar{V}$n$\sqrt{Hz}$) – 10log(R) – 90dB
dBm/Hz = –174dBm/Hz + ENR for ENR > 17dB

When amplifying noise it is important to remember that the noise voltage has a Gaussian distribution. The peak voltages of noise are therefore much larger than the average or RMS voltage. The ratio of peak voltage to RMS voltage is called crest factor, and a good crest factor for Gaussian noise is between 5:1 and 10:1 (14 to 20dB). An amplifier's 1dB gain-compression point should therefore be typically 20dB larger than the desired average noise-output power to avoid clipping of the noise.

For more information about noise diodes, please contact NOISE COM, INC. at (201) 261-8797.

Additional Reading

1. Johnson, J.B, "Thermal Agitation of Electricity in Conductors," Physical Review, July 1928, pp. 97-109.
2. Nyquist, H. "Thermal Agitation of Electric Charge in Conductors," Physical Review, July 1928, pp. 110-113.

Circuit collection, volume III

Data conversion, interface and signal processing

36

Richard Markell Editor

Introduction

Application Note 67 is a collection of circuits from the first five years of *Linear Technology*, targeting data conversion, interface and signal processing applications. This Application Note includes circuits such as fast video multiplexers for high speed video, an ultraselective bandpass filter circuit with adjustable gain and a fully differential, 8-channel, 12-bit A/D system. The categories included herein are data conversion, interface, filters, instrumentation, video/op amps and miscellaneous circuits. Application Note 66, which covers power products and circuits from *Linear Technology*'s first five years, is also available from LTC.

Analog Circuit and System Design: Immersion in the Black Art of Analog Design. http://dx.doi.org/10.1016/B978-0-12-397888-2.00036-5

Data conversion

Fully differential, 8-channel, 12-bit A/D system using the LTC1390 and LTC1410

by Kevin R. Hoskins

The LTC1410's fast 1.25Msps conversion rate and differential ±2.5V input range make it ideal for applications that require multichannel acquisition of fast, wide bandwidth signals. These applications include multitransducer vibration analysis, race vehicle telemetry data acquisition and multichannel telecommunications. The LTC1410 can be combined with the LTC1390 8-channel serial interfaced analog multiplexer to create a differential ADC system with conversion throughput rates up to 625ksps. This rate applies to situations where the selected channel changes with each conversion. The conversion rate increases to 1.25Msps if the same channel is used for consecutive conversions.

Figure 36.1 shows the complete differential, 8-channel A/D circuit. Two LTC1390s, U1 and U2, are used as noninverting and inverting input multiplexers. The outputs of the noninverting and inverting multiplexers are applied to

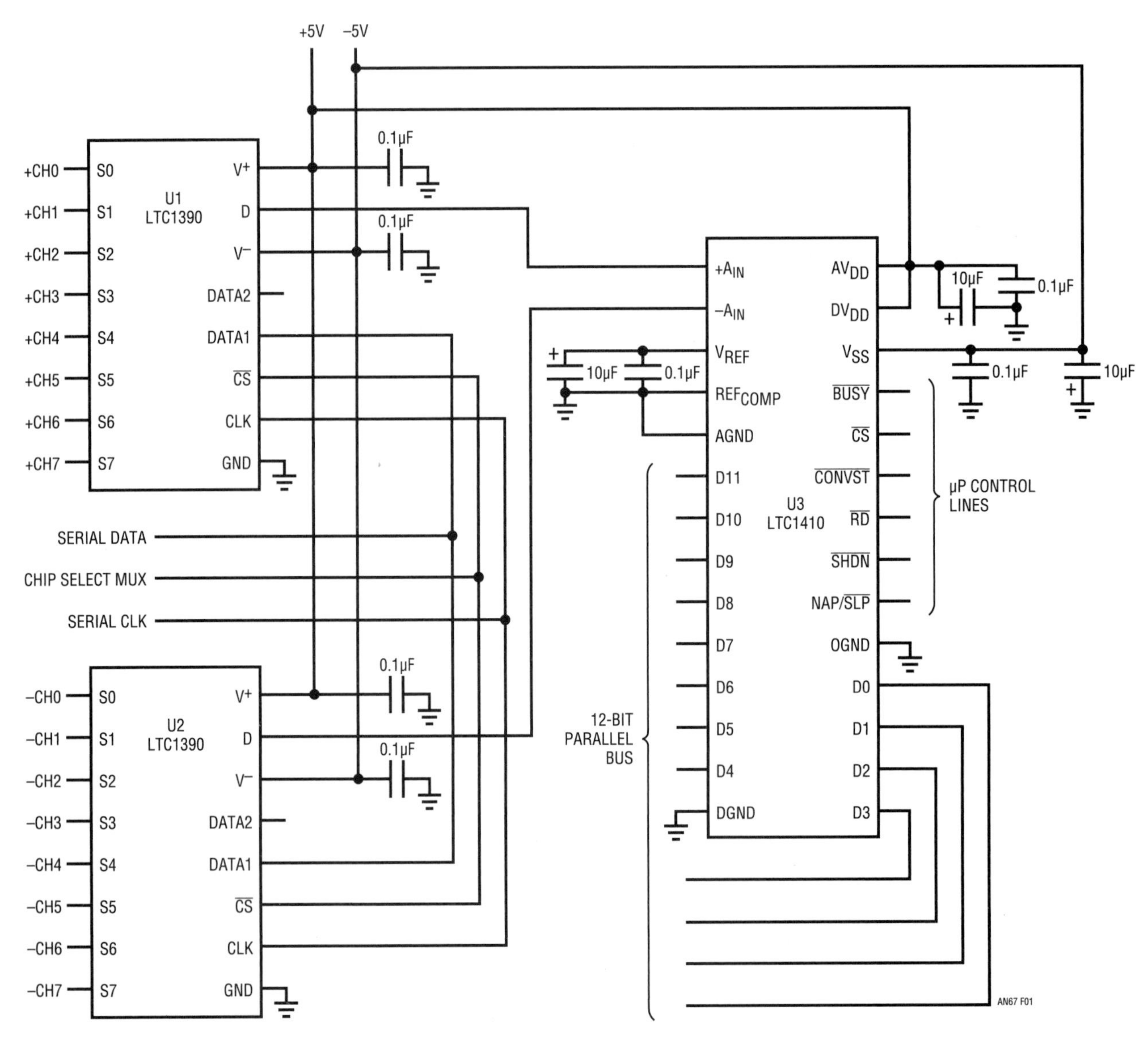

Figure 36.1 • Fully Differential 8-Channel Data Acquisition System Achieves 625ksps Throughput

the LTC1410's $+A_{IN}$ and $-A_{IN}$ inputs, respectively. The LTC1390 share the Chip Select MUX, Serial Data and Serial Clock control signals. This arrangement simultaneously selects the same channel on each multiplexer: S0 for both +CH0 and −CH0, S1 for both +CH1 and −CH1, and so on.

As shown in the timing diagram (Figure 36.2), MUX channel selection and A/D conversion are pipelined to maximize the converter's throughput. The conversion process begins with selecting the desired multiplexerchannel pair. With a logic high applied to the LTC1390's $\overline{CS}$ input, the channel pair data is clocked into each Data 1 input on the rising edge of the 5MHz clock signal. Chip Select MUX is then pulled low, latching the channel pair selection data. The signals on the selected MUX inputs are then applied to the LTC1410's differential inputs. Chip Select MUX is pulled low 700ns before the LTC1410's conversion start input, $\overline{CONVST}$, is pulled low. This corresponds to the maximum time needed by the LTC1390's MUX switches to fully turn on. This ensures that the input signals are fully settled before the LTC1410's S/H captures its sample.

The LTC1410's S/H acquires the input signal and begins conversion on $\overline{CONVST}$'s falling edge. During the conversion, the LTC1390's $\overline{CS}$ input is pulled high and the data for the next channel pair is clocked into Data 1. This pipelined operation continues until a conversion sequence is completed. When a new channel pair is selected for each conversion, the sampling rate of each channel is 78ksps, allowing an input signal bandwidth of 39kHz for each channel of the LTC1390/LTC1410 system.

To maximize the throughput rate, the LTC1410's $\overline{CS}$ input is pulled low at the beginning of a series of conversions. The LTC1410's data output drivers are controlled by the signal applied to $\overline{RD}$. The conversion's results are available 20ns before the rising edge of Busy. The rising edge of the Busy output signal can be used to notify a processor that a conversion has ended and data is ready to read.

This circuit takes advantage of the LTC1410's very high 1.25Msps conversion rate and differential inputs and the LTC1390's ease of programming to create an A/D system that maintains wide input signal bandwidth while sampling multiple input signals.

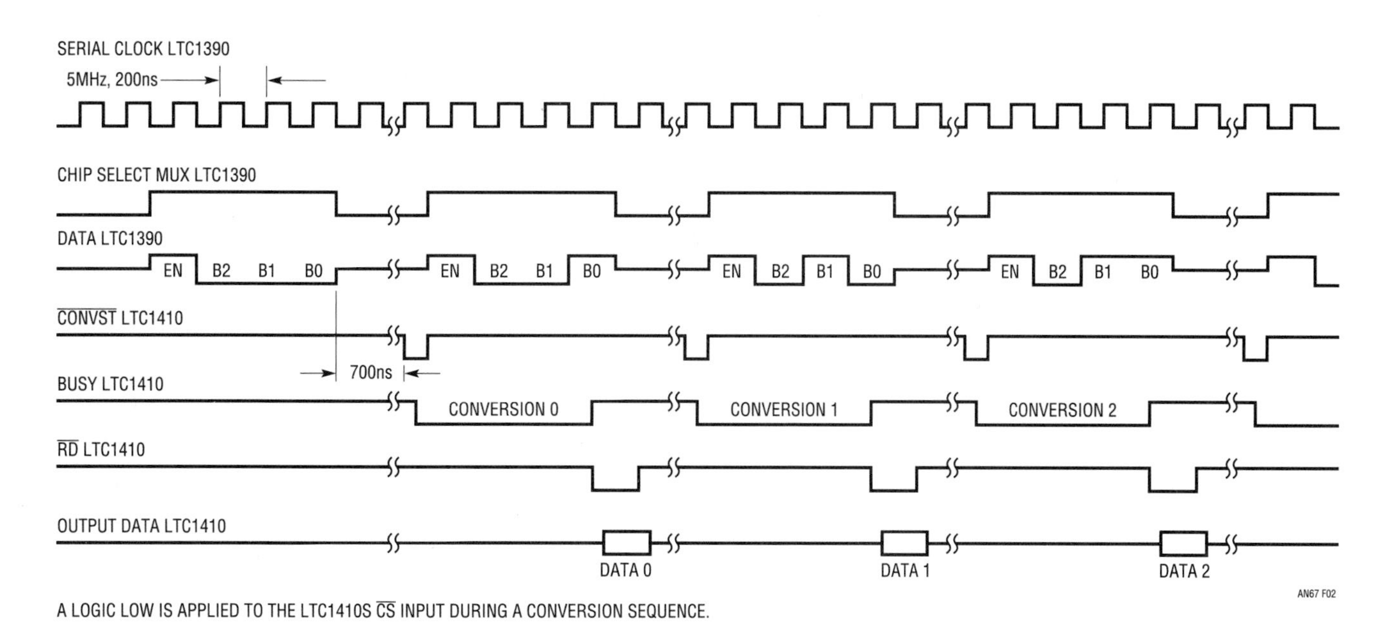

Figure 36.2 • The Figure 36.1 Circuit Timing Diagram

12-bit DAC applications

by Kevin R. Hoskins

System autoranging

System autoranging, adjusting an ADC's full-scale range, is an application area for which the LTC1257 is appropriate. Autoranging is especially useful when using an ADC with multiplexed inputs. Without autoranging only two reference values are used: one to set the full-scale magnitude and another to set the zero scale magnitude. Since it is common to have input signals with different zero scale and full-scale magnitude requirements, fixed reference voltages present a problem. Although the ranges selected for some of the inputs may take advantage of the full range of ADC output codes, inputs that do not span the same range will not generate all codes, reducing the ADC's effective resolution. One possible solution is to match the reference voltage span to each multiplexer input.

The circuit shown in Figure 36.3 uses two LTC1257s to set the full-scale and zero operating points of the LTC1296 12-bit, 8-channel ADC. The ADC shares its serial interface with the DACs. To further simplify bus connections, the DACs' data is daisy-chained. Two chip selects are used, one to select the LTC1296 when programming its multiplexer and the other to select the DACs when setting their output voltages.

During the conversion process, U2 and U3 receive the full and zero scale codes, respectively, that correspond to a selected multiplexer channel. For example, let channel 2's span begin at 2V and end at 4.5V. When a host processor wants a conversion of channel 2's input signal, it first sends code that sets the output of U2 to 2V and U3 to 4.5V fixing the span to 2.5V. The processor then sends data to the LTC1296 selecting channel 2. The processor next clocks the LTC1296 and reads the data generated during the conversion of the $3.5V_{P\text{-}P}$ signal applied to channel 2. As other multiplexer channels are selected the DAC outputs are changed to match their spans.

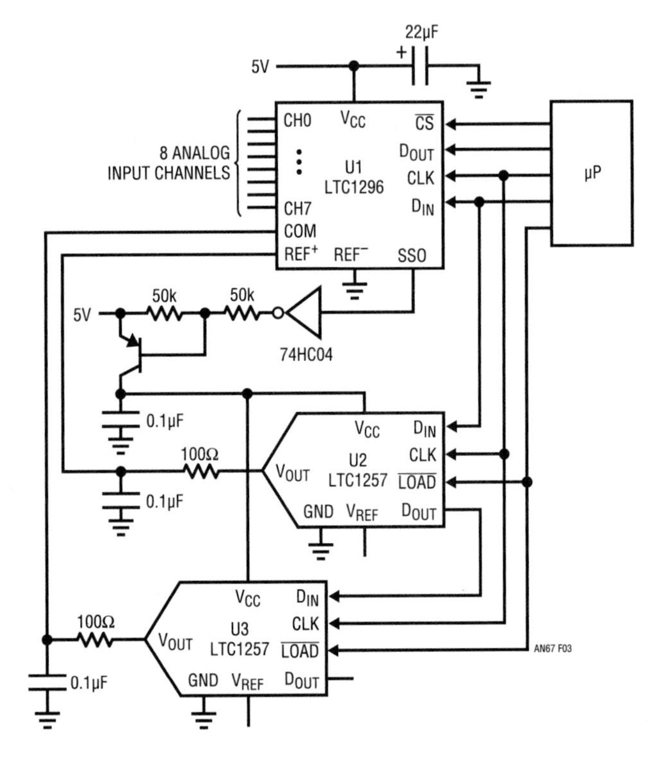

Figure 36.3 • Using Two LTC1257 12-Bit Voltage Output DACs to Set the Input Span of the 12-Bit 8-Channel LTC1296

Computer-controlled 4 – 20mA current loop

A common and useful circuit is the 4 – 20mA current loop. It is used to transmit information over long distances using varying current levels. The advantage of using current over voltage is the absence of IR losses and the transmission errors and signal losses they can create.

The circuit in Figure 36.4 is a computer-controlled 4 – 20mA current loop. It is designed to operate on a single supply over a range of 3.3V to 30V. The circuit's zero output reference signal, 4mA, is set by R1 and calibrated using R2, and its full-scale output current is set by R3 and calibrated using R4. The zero and full-scale output currents are set as follows: with a zero input code applied to the LTC1453, the output current, I_{OUT}, is set to 4mA by adjusting R2; next, with a full-scale code applied to the DAC the full-scale output current is set to 20mA by adjusting R4.

The circuit is self-regulating, forcing the output current to remain stable for a fixed DAC output voltage. This self-regulation works as follows: starting at t=0, the LTC1453's fixed output (in this example, 2.5V) is applied to the left side of R3; instantaneously, the voltage applied to the LT1077's input is 1.25V; this turns on Q1 and the voltage across R_S starts increasing beginning from 1.25V; as the

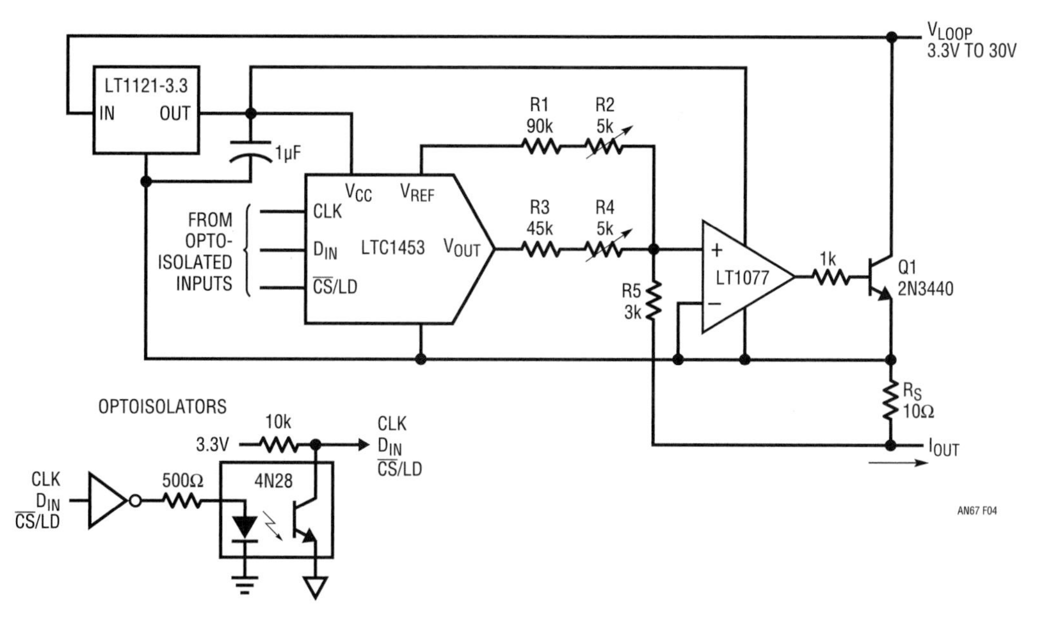

Figure 36.4 • The LTC1453 Forms the Heart of This Isolated 4 – 20mA Current Loop

voltage across R_S increases it lifts the LTC1453's GND pin above 0V; the voltage across R_S continues to increase until it equals the DAC's output voltage.

Once the circuit reaches this stable condition, the constant DAC output voltage sets a constant current through R3 + R4 and R5. This constant current fixes a constant voltage across R5 that is also applied to the LT1077's noninverting input. Feedback from the top of R_S is applied to the inverting input. As the op amp forces its inputs to the same voltage, it will fix the voltage at the top of R_S. This in turn fixes the output current to a constant value.

Optoisolated serial interface

The serial interface of the LTC1451 family and the LTC1257 make optoisolated interfaces very easy and cost effective. Only three optoisolators are needed for serial data communications. Since the inputs of the LTC1451, LTC1452 and LTC1453 have generous hysteresis, the switching speed of the optoisolators is not critical. Further, because each of these DACs can be daisy-chained to others, only three optoisolators are required.

LTC1329 micropower, 8-bit, current output DAC used for power supply adjustment, trimmer pot replacement

by K.S. Yap

Power supply voltage adjustment

Figure 36.5 is a schematic of a digitally controlled power supply voltage adjustment circuit using a 2-wire interface. The LT1107 is configured as a step-up DC/DC converter, with the output voltage (V_{OUT}) determined by the values of the feedback resistors. The LTC1329's DAC current output is connected to the feedback node of these resistors, and an 8051 microprocessor is used to interface to the LTC1329.

By simply clocking the LTC1329, the DAC current output is decreased or increased (decreased if $D_{IN} = 0$, increased if $D_{IN} = 1$), causing V_{OUT} to change accordingly.

Trimmer pot replacement

Figure 36.6 is a schematic of a digitally controlled offset voltage adjustment circuit using a 1-wire interface. By clocking the LTC1329, the DAC current output is increased, causing V_{R2} to increase accordingly. When the DAC current output reaches full scale it will roll over to zero, causing V_{R2} to change from the maximum offset trim voltage to the minimum offset trim voltage.

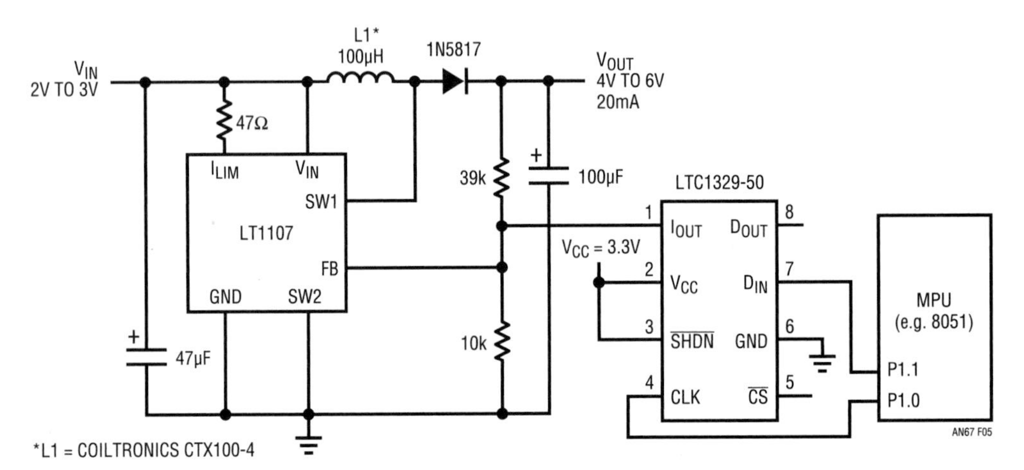

Figure 36.5 • LTC1329 Digitally Controls the Output Voltage of a Power Supply

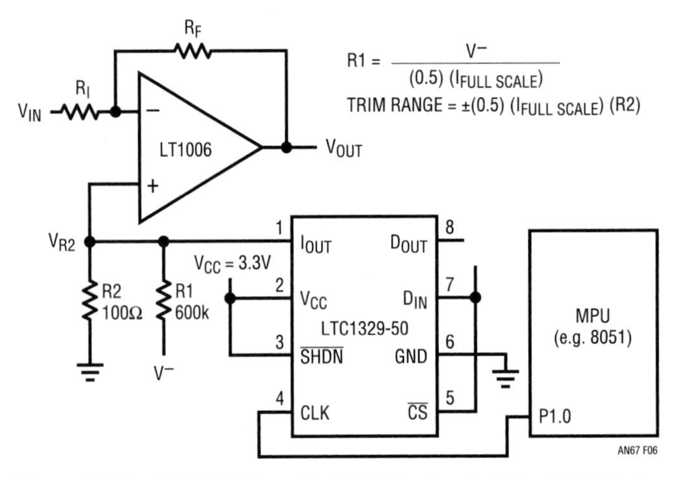

Figure 36.6 • LTC1329 Used to Null Op Amp's Offset Voltage

12-bit cold junction compensated, temperature control system with shutdown

by Robert Reay

The circuit in Figure 36.7 is a 12-bit, single 5V supply temperature control system with shutdown. An external temperature is monitored by a J-type thermocouple. The LT1025A provides the cold junction compensation for the thermocouple and the LTC1050 chopper op amp provides signal gain. The 47kΩ, 1µF RC network filters the chopping noise before the signal is sent to the A/D converter. The LTC1297 A/D converter uses the reference of the LTC1257 after it has been filtered to set full scale. After the A/D measurement is taken $\overline{CS}$ is pulled high and everything except the LTC1257 is powered down, reducing the system supply current to about 350µA. A word can then be written to the LTC1257 and its output can be used as a temperature control signal for the system being monitored.

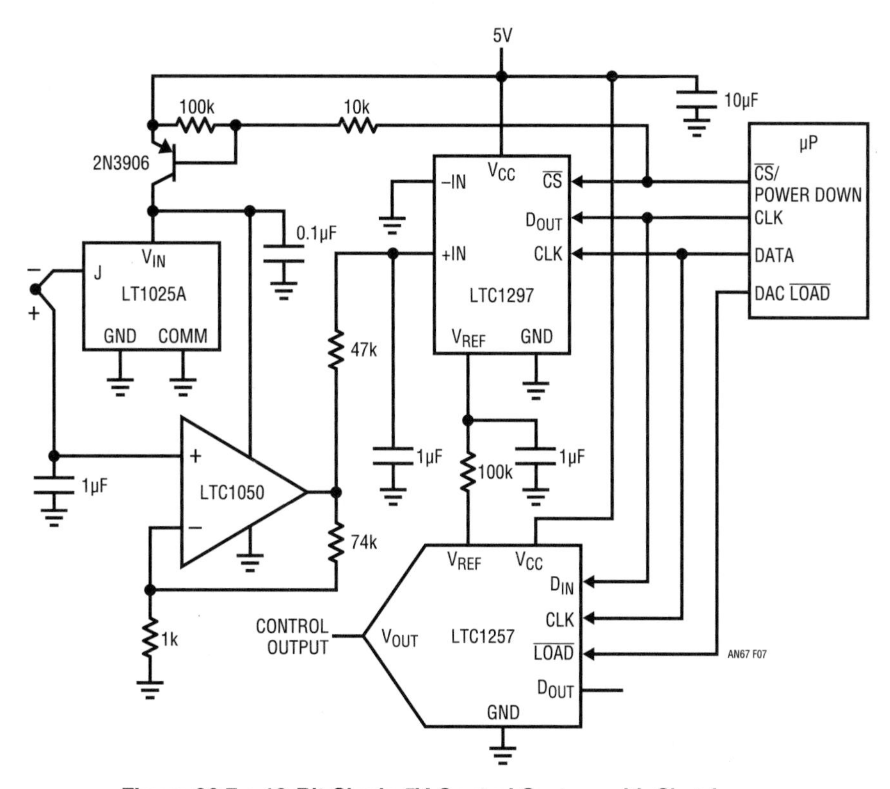

Figure 36.7 • 12-Bit Single 5V Control System with Shutdown

A 12-bit micropower battery current monitor

by Sammy Lum

Introduction

The LTC1297 forms the core of the micropower battery current monitor shown in Figure 36.8. This 12-bit data acquisition system features an automatic power shutdown that is activated after each conversion. In shutdown the supply current is reduced to 6μA, typically. As shown in Figure 36.9, the average power supply current of the LTC1297 varies from milliamperes to a few microamperes as the sampling frequency is reduced. This circuit draws only 190μA from a 6V to 12V battery when the sampling frequency is less than 10 samples per second. Wake-up time is limited by that required by the LTC1297 (5.5μs). For long periods of inactivity, the circuit's supply current can be further reduced to 20μA by using the shutdown feature on the LT1121. More wake-up time is required when using this mode of shutdown. It is usually determined by the amount of capacitance in the circuit and the available charging current from the regulator.

The battery current monitor

The battery voltage of 6V to 12V is regulated down to 5V by the LT1121 micropower regulator. A sense resistor of 0.05Ω is placed in series with the battery to convert the battery current to a voltage. Full scale is designed for 2A, giving a resolution of 0.5mA with the 12-bit ADC. The LTC1047 amplifies the voltage across the sense

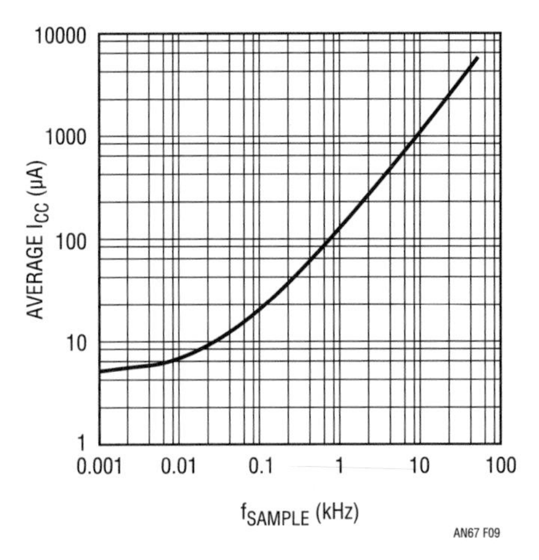

Figure 36.9 • Power Supply Current vs Sampling Frequency for the LTC1297

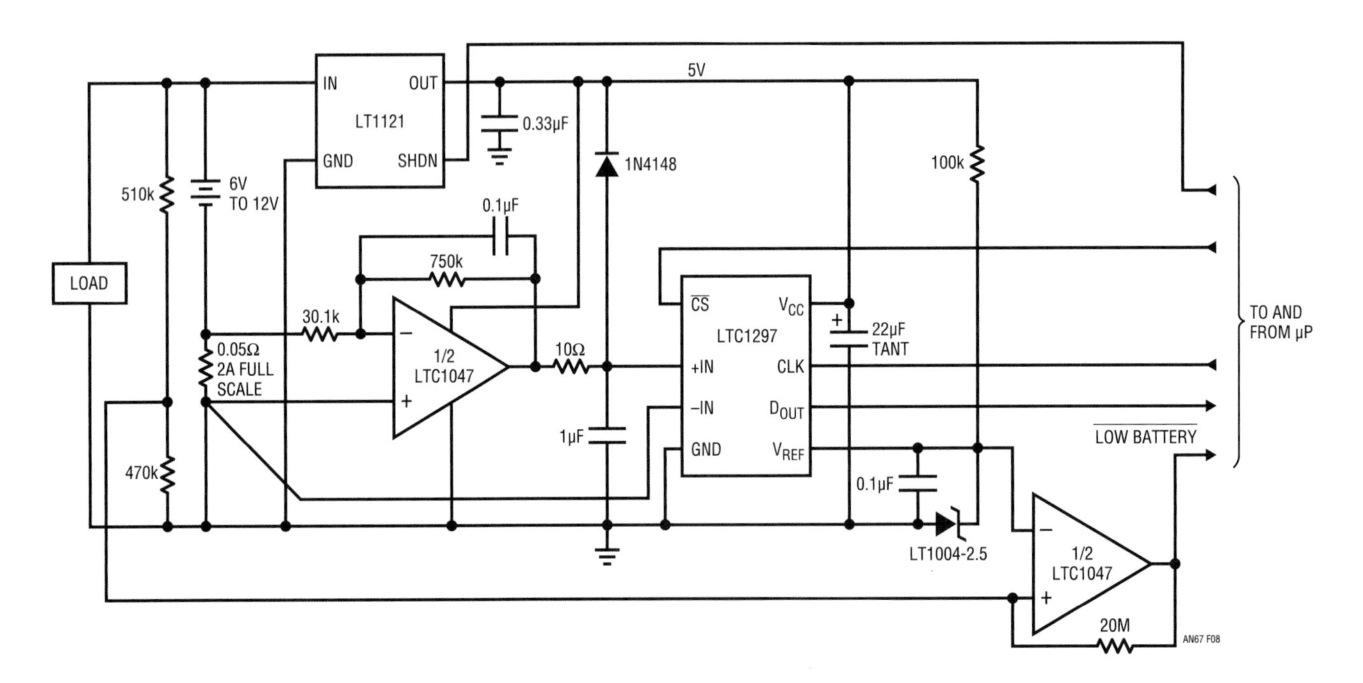

Figure 36.8 • A Micropower Battery Current Monitor Using the LTC1297 12-Bit Data Acquisition System

resistor by 25 V/V. This goes through an RC lowpass filter before being fed into the input of the LTC1297. The RC filter serves two functions. First it helps band limit the input noise to the ADC. Second the capacitor helps the LTC1047 recover from transients due to the switching input capacitor of the LTC1297. The LT1004 provides the full-scale reference for the ADC. A low-battery detection circuit has been created by using the other half of the LTC1047 as a comparator. Its trip point has been set to 5V plus the dropout voltage of the LT1121. Because data is transmitted serially to and from the microprocessor or microcontroller, this current monitor circuit can be located close to the battery.

Interface

V.35 transceivers allow 3-chip V.35 port solution

by Y.K. Sim

Two new LTC interface devices, the LTC1345 and the LTC1346, provide the differential drivers and receivers needed to implement a V.35 interface. When used in conjunction with an RS232 transceiver like the LT1134A, they allow a complete V.35 interface to be implemented with just two transceiver chips and one resistor termination chip. The LTC1345 and LTC1346 provide the three differential drivers and receivers necessary to implement the high speed path and the LT1134A provides the four RS232 drivers and receivers required for the handshaking interface. Both the LTC1345 and the LT1134A provide onboard charge pump power supplies allowing a complete V.35 interface to be powered from a single 5V supply. For systems where ±5V supplies are present the LTC1346 is offered without charge pumps, representing a 30% power savings.

The differential transceivers are capable of operating at data rates above 10MBd in nonreturn-to-zero (NRZ) format.

The RS232 handshaking lines can be implemented with standard RS232 transceivers. The LT1134A provides four RS232 drivers and four receivers, enough to implement the extended 8-line handshaking protocol specified in V.35. The LT1134A also includes an onboard charge pump to generate the higher voltages required by RS232 from a single 5V supply, making it an ideal companion to the LTC1345. These two chips, together with the BI Technologies termination resistor network chip provide a complete surface mountable 5V-only V.35 data port. Systems that have multiple power supply voltages available and use only the simpler 5-signal V.35 handshaking protocol can use the LTC1346 with the LT1135A or LT1039 RS232 transceivers; this combination provides a complete port while saving board space and complexity. Figure 36.10 shows a typical LTC1345/LT1134A V.35 implementation with five differential signals and five basic handshaking signals with an option for three additional handshaking signals.

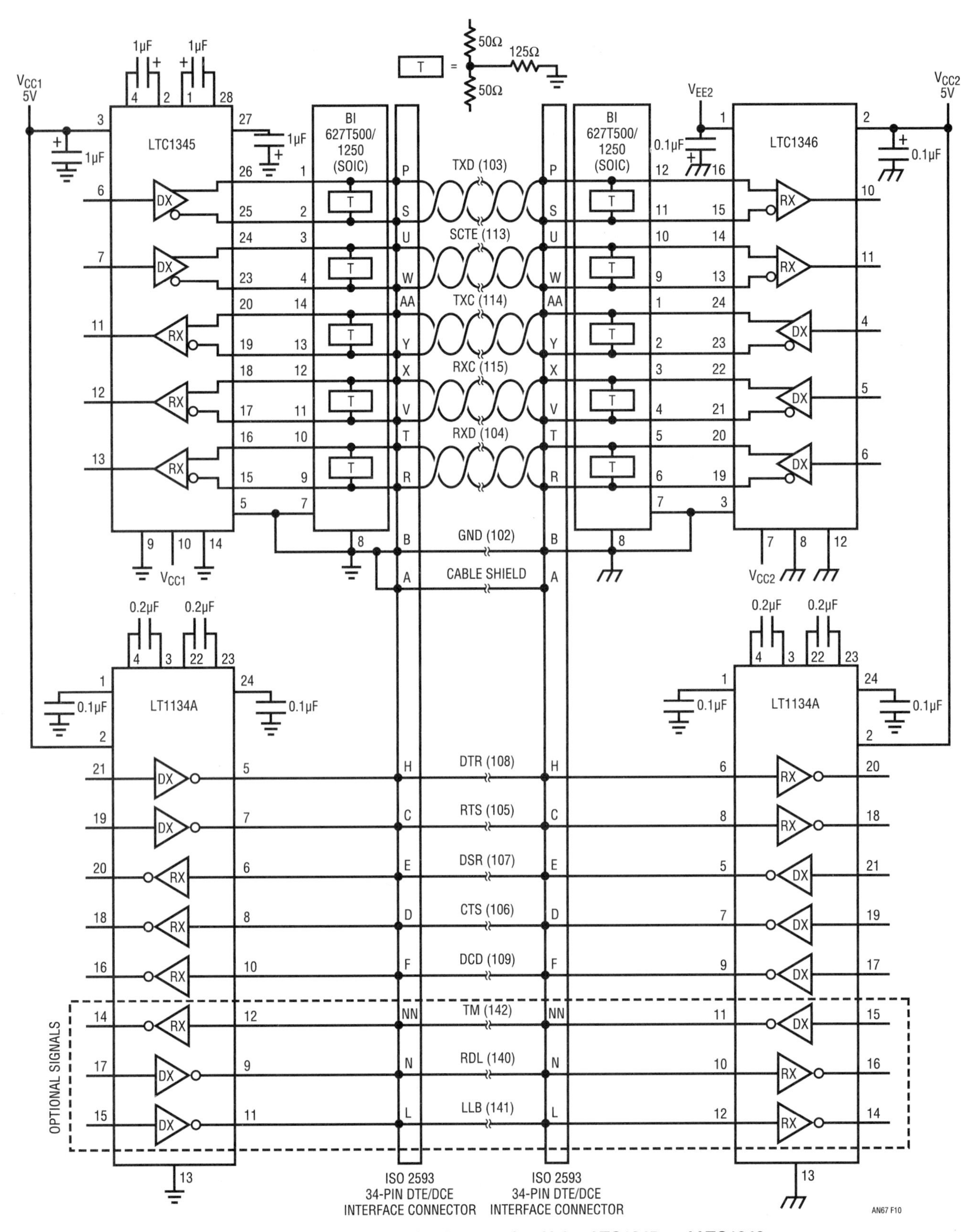

Figure 36.10 • Typical V.35 Implementation Using LTC1345 and LTC1346

Switching, active GTL terminator

by Dale Eagar

Introduction

New high speed microprocessors, especially those used in multiprocessor workstations and video graphics terminals, require high speed backplanes that support peak data rates of up to 1Gbps. The backplane is a passive component, whereas all drivers and receivers are implemented in low voltage swing CMOS (also referred to as GTL logic). These applications require bidirectional terminators, terminators that will either source or sink current (in this case, at 1.55V). The current requirements of the terminator depended on the number of terminations on the backplane. Present applications may require up to 10A. This specification may, of course, be reduced if required.

Circuit operation

The complete schematic of the terminator is shown in Figure 36.11. The circuit is based on the LT1158 half-bridge, N-channel, power MOSFET driver. The LT1158 is configured to provide bidirectional synchronous switching to MOSFETs Q1 through Q6. VR1, an LT1004-1.2, R1 and C1 generate a 1.25V reference voltage that programs the terminator's output voltage. U1A, an LT1215, is a moderate speed (23MHz GBW) precision operational amplifier that subtracts the error voltage at its inverting input from the 1.25V reference. U1A is also used to amplify this error signal. Components R3 and C2 tailor the phase and gain of this section and are selected when evaluating the system's load step response.

U1B and part of U2 provide the gain and the phase inversion necessary to form an oscillator. C3 and C4 provide positive feedback at high frequencies, which is necessary for the system to oscillate in a controlled manner while keeping the voltage excursions within the common mode range of U1B. R8, U2 and C6 provide phase inversion and negative feedback at the middle frequencies, causing U1B to oscillate at a frequency much higher than the feedback loop's response. The DC path for the oscillator is closed through the power MOSFETs Q1 to Q6, the output choke L1, the output capacitor C11 and through the feedback path with the error amplifier. R4 and R7 set the center of the common mode voltage of U1B and are selected to limit the maximum duty factor the oscillator can achieve.

R9, R10, R12 and C9 provide output current sense to U2, allowing it to shut down the oscillator via the Fault pin (Pin 5) to prevent catastrophic or even cataclysmic events from occurring. D2, C8 and the circuitry behind the Boost pin (Pin 16) of U2 work together to provide more than sufficient gate drive for the N-channel FETs Q1-3. D3, R11 and C7 allow the oscillator to start up regardless of the state of the oscillator on powerup.

Performance

The circuit provides excellent transient response, efficiencies in the source mode of better than 80% and efficiencies in the sink mode of better than 90%. Figure 36.12 shows the step response of the terminator.

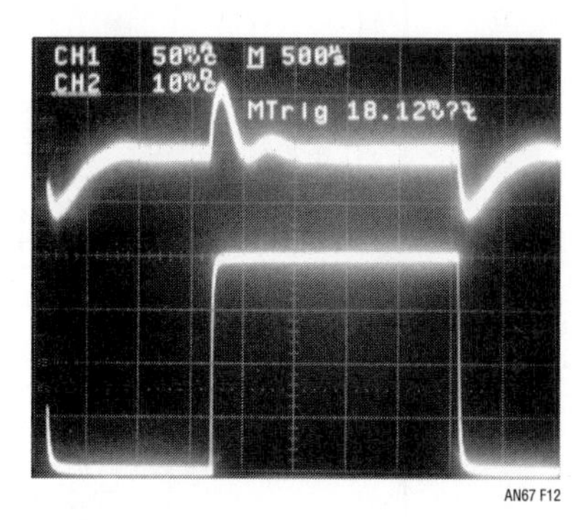

Figure 36.12 • Step Response of LT1158-Based Terminator

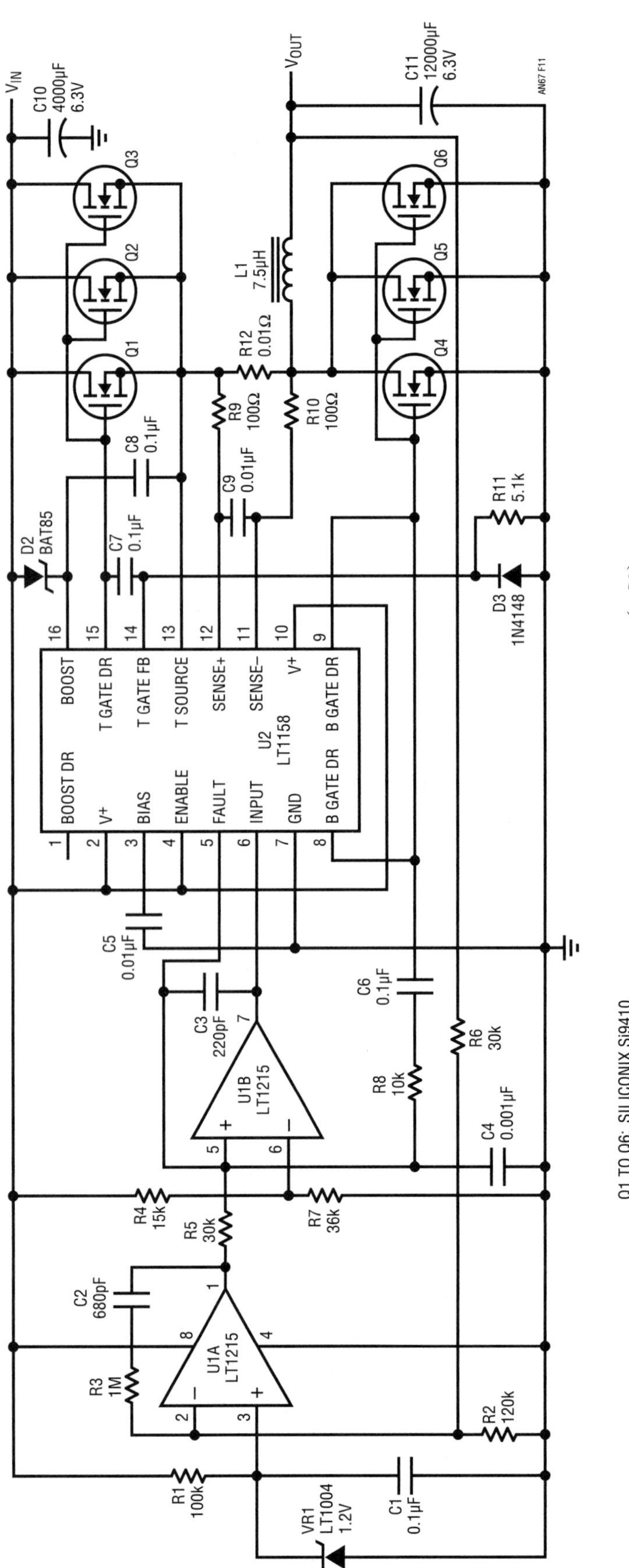

Figure 36.11 • GTL 1.55V Terminator Provides 10A Max Current

RS232 transceivers for DTE/DCE switching

by Gary Maulding

Switched DTE/DCE port

There are situations where a data port is required to act alternately as either a DTE or a DCE. Examples include test equipment and data multiplexers. Figure 36.13 shows a circuit that can switch from a 9-pin DTE to a 9-pin DCE configuration while maintaining full compliance with the RS232 standards.

The circuit uses an LT1137A DTE transceiver and an LT1138A DCE transceiver. A DTE/DCE select logic signal alternately activates or shuts down one of the two transceivers. In addition to drawing no power, the OFF transceiver's drivers achieve a high impedance state, removing themselves from the data line. The receiver inputs will continue to load the line, but this presents no operational problem and does not violate the RS232 standard. The drivers on the activated transceiver can easily drive the extra load of the companion transceiver's inputs along with the termination at the opposite end of the cable. The scope photograph (Figure 36.14) shows the signal outputs of the DTE/DCE switched circuit driving 3k || 1000pF at 120kBd.

To the transceiver at the opposite end of the data line the data port always appears to be a normal fixed port. All signals into the port are properly terminated in 5k.

The schematic in Figure 36.13 shows the essential features needed to implement DTE/DCE switching but other features can be easily included. Shutdown of both transceivers could be implemented by adding an additional logic control signal. Multiplexing of the logic level signals is also possible since receiver outputs remain in a high impedance state when the transceivers are shut down. Two capacitors can be saved by sharing the V^+ and V^- filter capacitors between the two transceivers, but the charge pump capacitors must not be shared.

The circuits used in the demonstration circuit are bipolar, but Linear Technology's CMOS transceivers, such as the LTC1327 and 1328 could be substituted where the absolute minimum power dissipation is required.

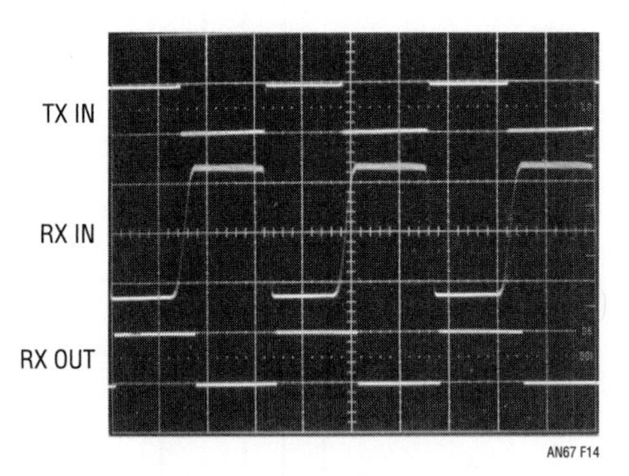

Figure 36.14 • Oscillograph Showing Signal Outputs of the DTE/DCE Circuit of Figure 36.13 Driving 3k || 1000pF at 120kBd

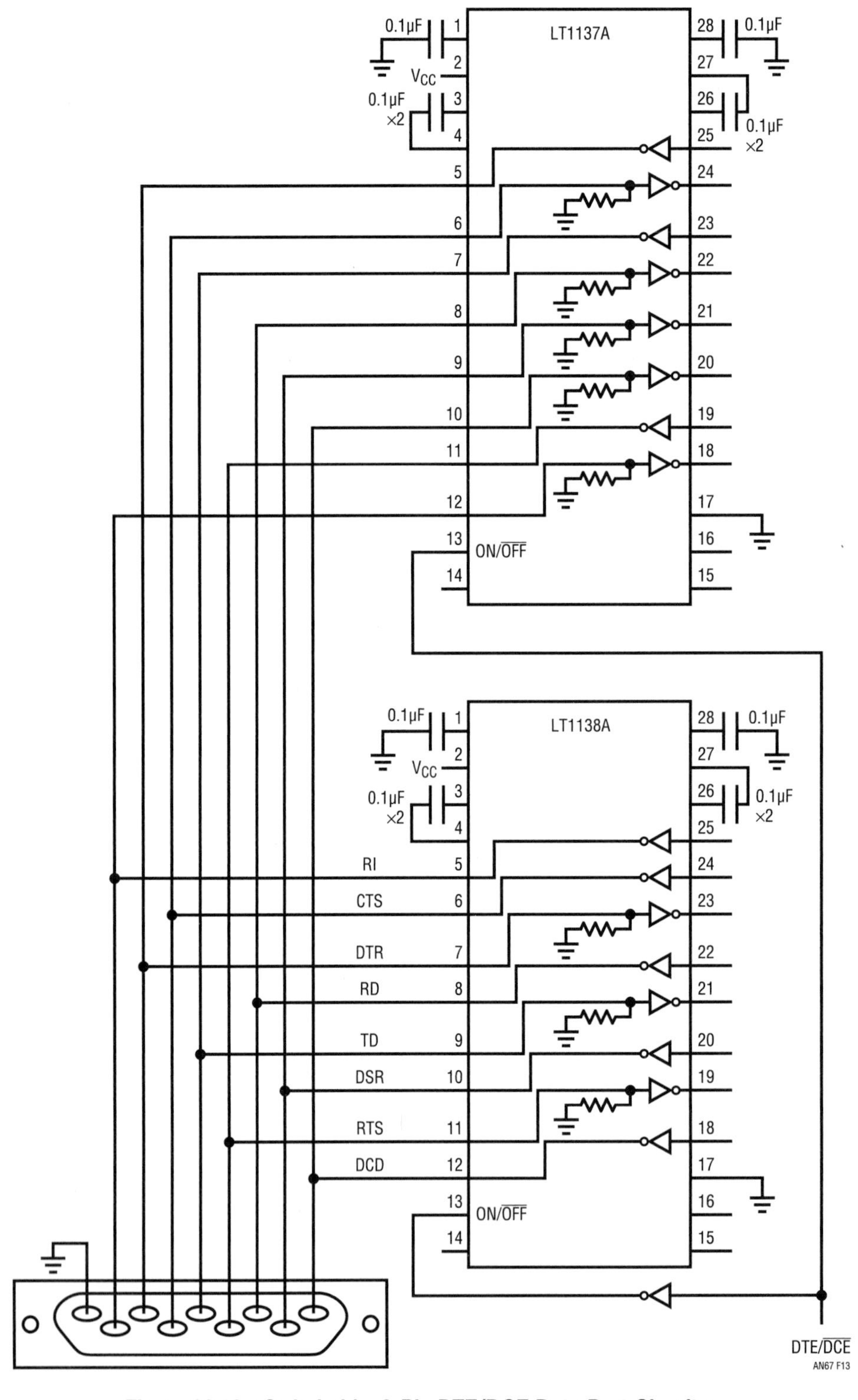

Figure 36.13 • Switchable, 9-Pin DTE/DCE Data Port Circuitry

Active negation bus terminators

by Dale Eagar

High speed data buses require transmission line techniques, including termination, to preserve signal integrity. Lost data on a bus can be attributed to reflections of the signals from the discontinuities of the bus. The solution to this problem is proper termination of the bus.

Early designs of bus terminators were passive (see Figure 36.15). Passive termination works great but wastes lots of precious power; especially when the bus is not being used.

The ideal solution is a voltage source capable of both sourcing and sinking current. Such a voltage source, with termination resistors, is shown in Figure 36.16. This is called active negation. Active negation uses minimal quiescent current, essentially providing only the power needed to properly terminate the bus.

Active negation bus terminator using linear voltage regulation

The active negation circuit shown in Figure 36.17 provides the power to the output at an efficiency of about 50%; the rest of the power is dissipated in either Q1 or U1 depending on the polarity of the output current.

The circuit will source or sink current. Current is sourced from the 5V supply through Q1, an NPN Darlington, to the output. The sink current flows through CR1 into the collector (Pin 1) of the LT1431, and to ground. The LT1431 regulates a scaled version of the output voltage against the

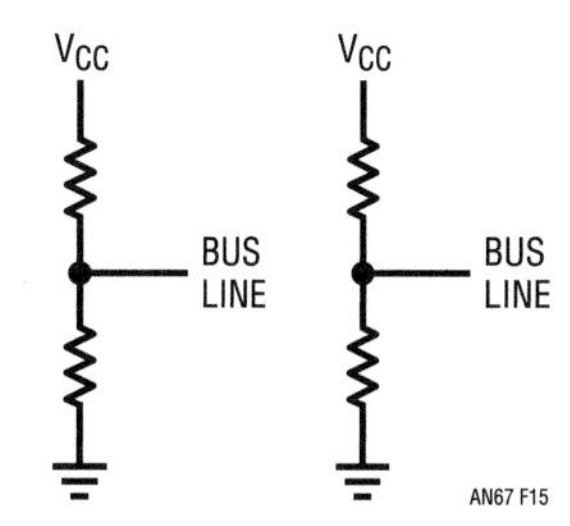

Figure 36.15 • Passive Termination Technique

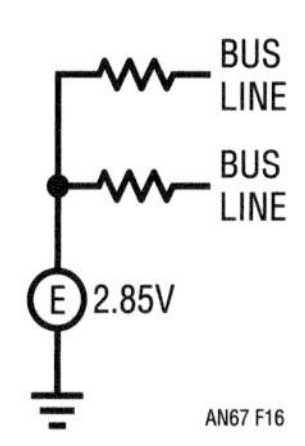

Figure 36.16 • Active Negation Termination Technique

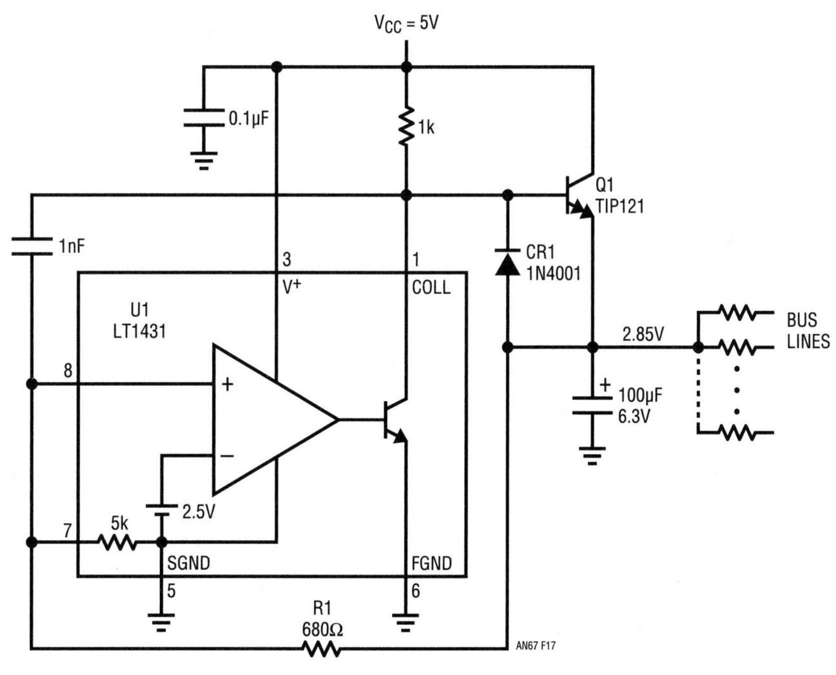

Figure 36.17 • Linear Active Negation Voltage Source

internal 2.5V bandgap reference, driving the base of Q1 or drawing current through CR1 to regulate the output voltage. R1 and the internal 5k resistor of the LT1431 scale the output voltage.

Switching power supply, active negation network

The switching active negation terminator shown in Figure 36.18 is a synchronous switcher. This solution further reduces dissipation and therefore achieves higher efficiency. This type of switcher can both source and sink current.

The switching power supply operates as follows. The 74AC04 hex inverters (U1 and U2) form a 1MHz variable duty factor oscillator. The duty factor is controlled by the output of the regulator, U3, and is maintained at the ratio of $2.85V/V_{IN}$. V_{IN} is the 5V supply that powers U1, U2 and U3. The output voltage is the average voltage of the square wave (V_{IN})(duty factor) from the outputs of U1B–U1F and U2A–U2F. L1 and C2 filter the AC component of the 0V to 5V signal yielding a DC output voltage of 2.85V.

CR1 is added to prevent latchup of U1 and U2 during adverse conditions.

A logic gate could easily be added to the oscillator to add a disable function to this terminator, further lowering the quiescent power when termination is not needed.

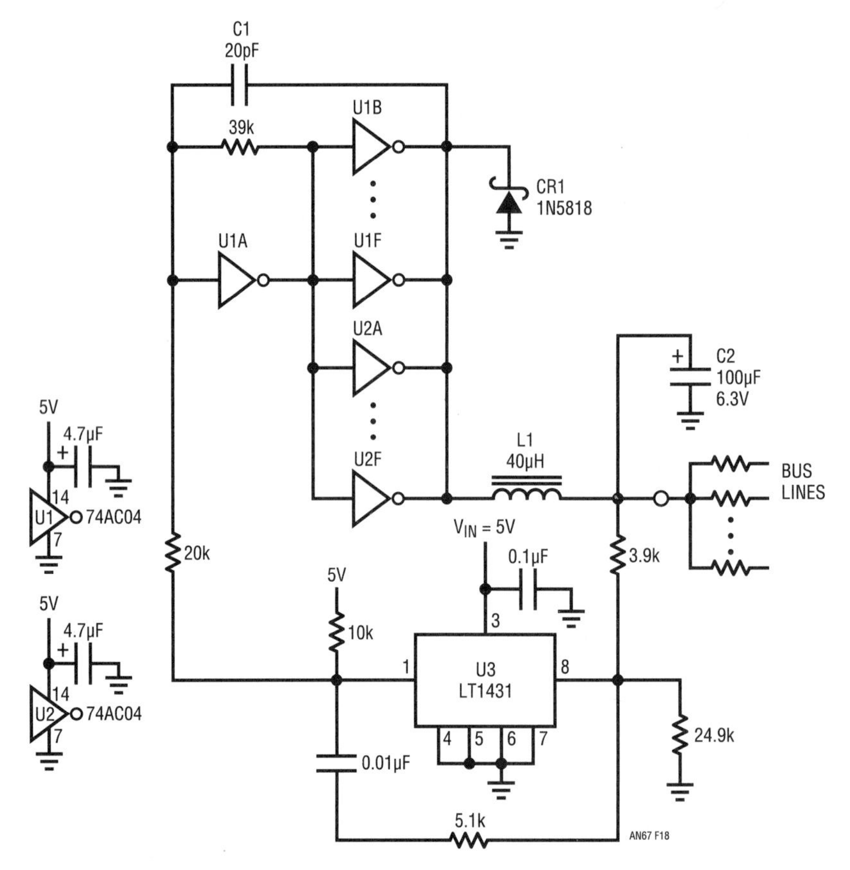

Figure 36.18 • Switching Active Negation Termination

RS485 repeater extends system capability

by Mitchell Lee

RS485 data communications are specified for distances of up to 4000 feet. This limit is the consequence of losses in the twisted pair used to carry the data signals. Beyond 4000 feet, skin effect and dielectric losses take their toll, attenuating the signal beyond use.

If greater distances must be covered some means of repeating the data is necessary. One method is to terminate a long run of cable with a microprocessor-based node capable of relaying data to yet another length of cable.

A more simple solution[1] is shown in Figure 36.19. Two RS485 transceivers are connected back-to-back so as to relay incoming data from either side to the other. A pair of cross coupled one-shots furnish a means of "flow control" so that one and only one transmitter is turned on at any given time. Incoming data is sensed by detecting a 1-0 transition at the output of either idling receiver. The first receiver to spot such a transition triggers its associated one-shot, which, in turn, activates the opposite transmitter and ensures smooth data flow from one side to the other. At the same time the one-shot locks out the other receiver/transmitter/one-shot combination so that only one data path is open.

The one-shot is retriggered by successive 1-0 transitions and start bits, holding the data path in this configuration. The one-shot time constant is set slightly greater than the interval between any two start bits. When the received data stops, the line idles high, producing a 1 at the receiver's output. The one-shot resets, returning the opposite transceiver to the receive mode—ready for any subsequent data flow.

In order to allow adequate time for the one-shot to reset, the software protocol must wait one word length after the end of any data transmission before responding to a call or initiating a new conversation. As shown, the repeater is set up for 100kBd data rates and an 8-bit word length (plus start and stop bits).

Note 1: Honeywell Inc. patent 4,670,886 may apply.

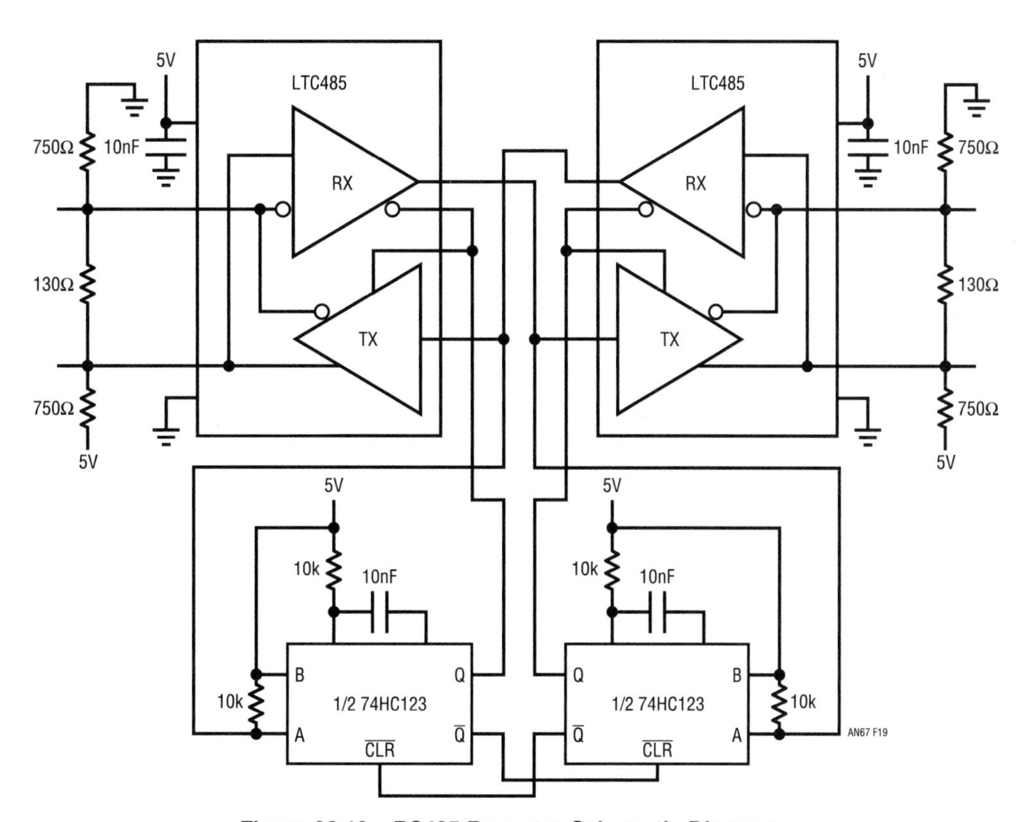

Figure 36.19 • RS485 Repeater Schematic Diagram

An LT1087-based 1.2V GTL terminator

by Mitchell Lee

A recent development in high speed digital design has resulted in a new family of logic chips called Gunning Transition Logic (GTL). Because of the speeds involved, careful attention must be paid to the transmission line characteristics of the interconnections between these chips; active termination is required.

The termination voltage is 1.20V and currents of several amperes are common in a complete system. One method of generating 1.2V is to use a linear regulator operating from 3.3V or 5V. Unfortunately, this method suffers from two major drawbacks. First, the minimum adjust voltage, without the aid of a negative supply, is 1.25V for most adjustable linear regulators. Second, most low voltage linear regulators do not feature low dropout characteristics, rendering them unusable on a 3.3V input. The LT1087 solves both of these problems with an output that can be adjusted to less than the reference voltage and a low dropout architecture.

Figure 36.20 shows the complete circuit. The LT1087 features feedback sense, which, in its original application, was used for remote Kelvin sensing. In the GTL terminator circuit the Sense pins are used to adjust the internal 1.25V reference downward. The result is a 1.20V, 5A regulator with 2% output tolerance over all conditions of line, load and temperature. To minimize power dissipation a 3.3V input source is recommended.

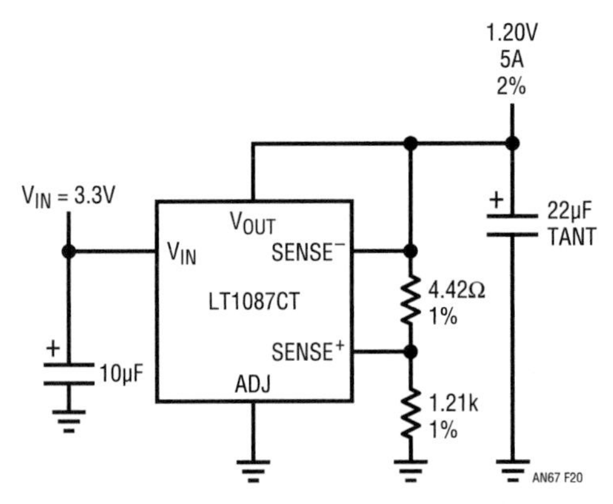

Figure 36.20 • 1.2V GTL Termination Voltage Schematic Diagram

LTC1145/LTC1146 achieve low profile isolation with capacitive lead frame

by James Herr

The LTC1145 and LTC1146 are a new generation of signal isolators. Previously, signal isolation was accomplished by means of optoisolators. Light from an LED was detected across a physical isolation barrier by either a photo diode or transistor and converted to an electrical signal. Isolation levels up to thousands of volts were easily achieved. Attempts have been made to provide signal isolation on a single silicon die. Problems arose due to reliability constraints of damage from ESD or overvoltage. A new technique, using a capacitive lead frame, overcomes the problems associated with single package signal isolation. Further, this technique is suitable for use in thin surface mount packages—a solution not available with optoisolators. The data rates are 200kbps for the LTC1145 and 20kbps for the LTC1146. Both parts can sustain over 1000V across their isolation barriers.

Applications

The LTC1145/LTC1146 can be used in a wide range of applications where voltage transients, differential ground potentials or high noise may be encountered, such as isolated serial data interfaces, isolated analog-to-digital converters for process control, isolated FET drivers and low power optoisolator replacement. One possible application is an isolated RS232 receiver. The D_{IN} pin of the LTC1145 is driven by an RS232 signal through a 5.1k resistor (Figure 36.21). The D_{OUT} pin of the LTC1145 presents isolated

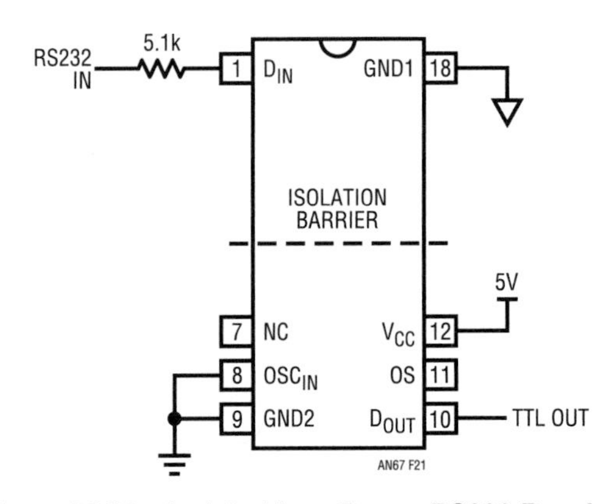

Figure 36.21 • Isolated Low Power RS232 Receiver

TTL-compatible output signals. The GND2 pin of the LTC1145 is connected to the same ground potential as the receiving end of the link. The isolator can accommodate differences of up to 1kV between GND1 and GND2.

Another application is an isolated, thermocouple-sensed temperature-to-frequency converter (see Figure 36.22). The output of I3 produces a 0kHz to1kHz pulse train in response to a 0°C to 100°C temperature excursion (see LTC Application Note 45 for the details). The pulses from I3 drive the D_{IN} pin of LTC1146. The GND1 pin is connected to the same ground potential as I3. The D_{OUT} pin of LTC1146 presents isolated, TTL-compatible output signals. The circuit consumes only 460μA maximum, allowing it to operate from a 9V battery.

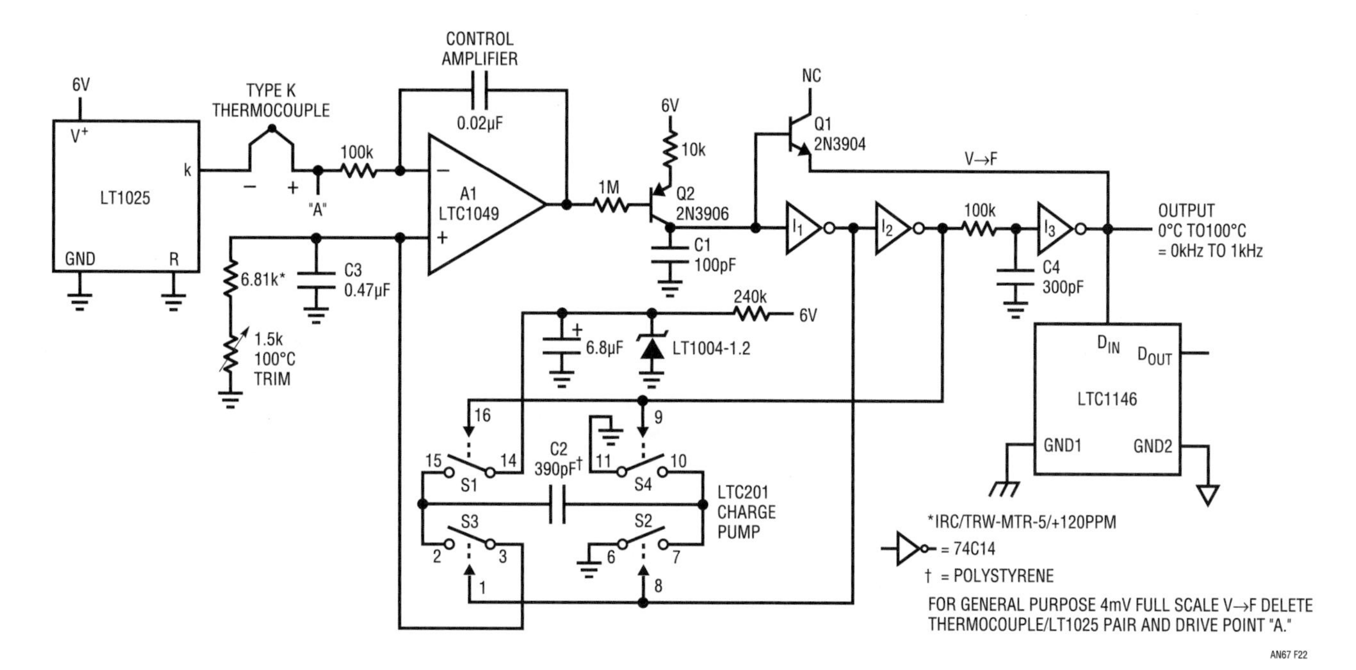

Figure 36.22 • Isolated Temperature-to-Frequency Converter

LTC485 line termination

by Bob Reay

The termination of the data line connecting LTC485 transceivers is very important because an improperly terminated line can cause data errors. The data line is usually a 120Ω shielded twisted pair of wires that is terminated at each end with a 120Ω resistor (Figure 36.23). For some applications a problem occurs when the output of the drivers is forced into a high impedance state because the termination resistors short the inputs to the receivers. Since the receivers are differential comparators with built-in hysteresis, their output will remain in the last logic state.

For the applications that must force the outputs of the receivers to a known state, but still maintain low power consumption, the cable can be terminated as in Figure 36.24. A capacitor (typically 0.1µF) has been connected in series with the 120Ω termination resistor R2 and two bias resistors (R1 and R3) have been added. When data is being transmitted the capacitor looks basically like a short circuit and a differential signal is developed across the termination resistor. When the drivers are forced into a high impedance state, the bias resistors force the receiver into a logic 1 state. The receiver inputs can be reversed when the output must be a logic 0.

Because the capacitor is in series with the bias string, no DC current flows when data is not being transmitted. Care must be taken to transmit data at a high enough rate to prevent the bias resistors from charging the capacitor to the wrong state before the next data bit arrives. Also note that differences in the V^+ supplies or grounds will cause DC current to flow in the cable, but this can be kept to a minimum by using high value bias resistors.

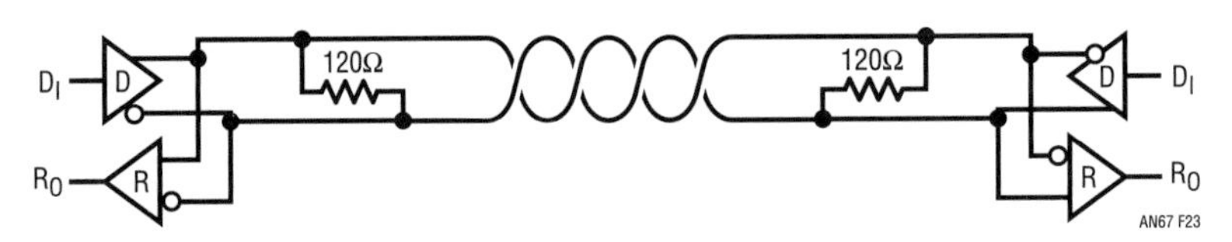

Figure 36.23 • DC Coupled Termination

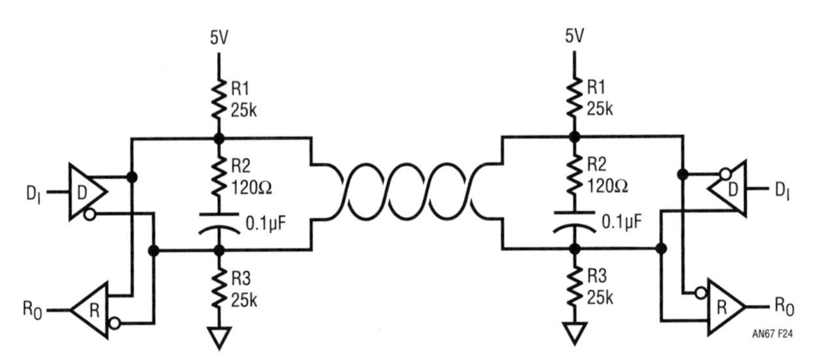

Figure 36.24 • AC Coupled Termination

Filters

Sallen and key filters use 5% values

by Dale Eagar

Lowpass filters designed after Sallen and Key usually take the form shown in Figure 36.25. In the classic Sallen and Key circuit, resistors R1, R2 and R3 are set to the same value to simplify the design equations.

When the three resistors are the same value, the pole placement, and thus the filter characteristics, are set by the capacitor values (C1, C2 and C3). This procedure, although great for the mathematician, can lead to problems. The problem is that, in the real world, the resistors, not the capacitors, are available in a large selection of values.

Taking advantage of the wider range of resistor values is not altogether trivial; the mathematics can be quite cumbersome and time consuming.

This Design Idea includes tables of resistor and capacitor values for third-order Sallen and Key lowpass filters. The resistor values are selected from the standard 5% value pool, and the capacitor values are selected from the standard 10% value pool. Frequencies are selected from the standard 5% value pool used for resistors. Frequencies are in Hertz, capacitance in Farads and resistance in Ohms.

Figure 36.26 details the PSpice simulation of a 1.6kHz Butterworth filter designed from these tables.

How to design a filter from the tables

Pick a cutoff frequency in Hertz as if it were a standard 5% resistor value in Ohms. (that is, if you want a cutoff frequency of 1.7kHz you must choose between 1.6k and 1.8k)

Select the component values from Table 36.1 or Table 36.2 as listed for the frequency (think of the first two color bands on a resistor).

Select a scale factor for the resistors and capacitors from Table 36.3 by the following method:

1. Select a diagonal that represents the frequency multiplier (think of the third color band on a 5% resistor).
2. Choose a particular diagonal box by either choosing a capacitor multiplier from the rows of the table that give you a desired capacitor value or by choosing a resistor multiplier from the columns of the table that gives you a desired resistance value.

Multiply the resistors and capacitors by the scale factors for the rows and columns that intersect at the chosen frequency multiplier box. (for example, 0.68 • 1µF=0.68µF 0.47 • 1kΩ = 470Ω).

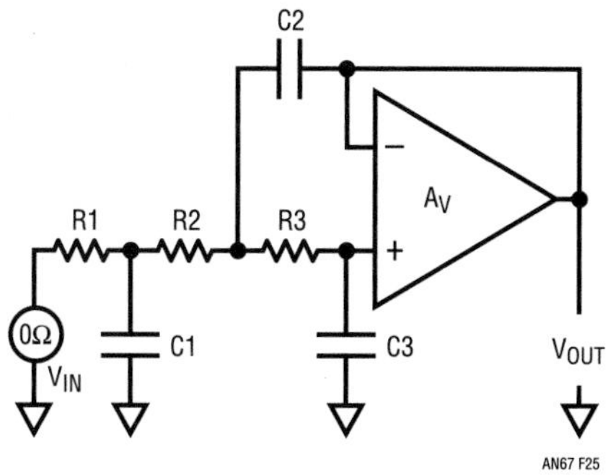

Figure 36.25 • Sallen and Key Lowpass Filter

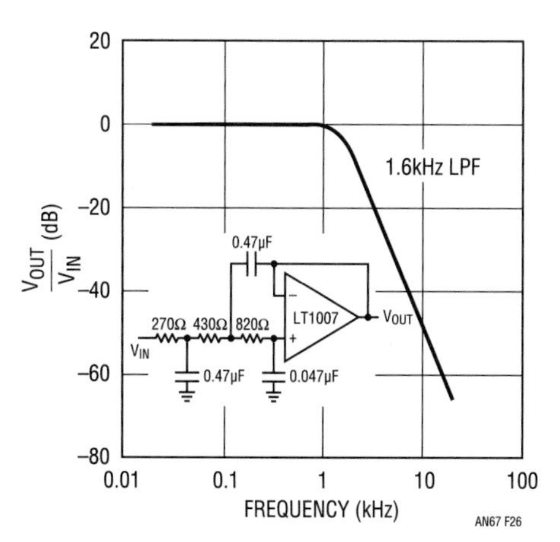

Figure 36.26 • PSpice Simulation of 1.6kHz Butterworth Filter

Table 36.1 Bessel Lowpass Filter

FREQ	R1	R2	R3	C1	C2	C3
1.0	0.39	0.43	8.20	0.47	0.22	0.01
1.1	0.36	0.39	7.50	0.47	0.22	0.01
1.2	0.33	0.36	6.80	0.47	0.22	0.01
1.3	0.36	2.40	0.033	0.22	2.20	0.047
1.5	0.33	4.70	0.012	0.22	4.70	0.022
1.6	0.30	0.10	0.240	0.47	2.20	0.047
1.8	0.30	3.30	5.10	0.22	0.022	0.010
2.0	0.27	0.51	0.027	0.22	2.20	0.100
2.2	0.24	2.70	0.43	0.22	0.10	0.022
2.4	0.22	2.70	3.60	0.22	0.022	0.010
2.7	0.27	0.43	1.30	0.22	0.10	0.022
3.0	0.18	0.82	0.16	0.22	0.22	0.047
3.3	0.15	0.056	1.00	0.47	1.00	0.010
3.6	0.18	0.16	0.022	0.22	2.20	0.100
3.9	0.15	1.50	2.20	0.22	0.022	0.010
4.3	0.13	0.22	0.013	0.22	2.20	0.100
4.7	0.20	0.12	1.20	0.22	0.22	0.010
5.1	0.18	0.068	0.039	0.22	2.20	0.047
5.6	0.20	1.10	0.036	0.10	0.47	0.022
6.2	0.15	0.091	0.91	0.22	0.22	0.010
6.8	0.16	0.91	0.03	0.10	0.47	0.022
7.5	0.15	1.80	0.27	0.10	0.047	0.010
8.2	0.10	0.12	1.00	0.22	0.10	0.010
9.1	0.13	0.56	0.12	0.10	0.10	0.022

Table 36.2 Butterworth Lowpass Filter

FREQ	R1	R2	R3	C1	C2	C3
1.0	0.36	3.3	3.3	0.47	0.10	0.022
1.1	0.47	0.47	6.2	0.47	0.47	0.010
1.2	0.36	0.62	1.0	0.47	0.47	0.047
1.3	0.27	2.00	0.33	0.47	0.47	0.047
1.5	0.24	1.60	0.3	0.47	0.47	0.047
1.6	0.27	0.43	0.82	0.47	0.47	0.047
1.8	0.43	1.20	0.13	0.22	1.00	0.047
2.0	0.36	7.50	0.18	0.22	0.47	0.010
2.2	0.24	0.24	3.00	0.47	0.47	0.010
2.4	0.33	0.91	0.043	0.22	2.20	0.047
2.7	0.27	5.60	0.062	0.22	1.00	0.010
3.0	0.24	5.10	0.056	0.22	1.00	0.010
3.3	0.22	1.60	0.30	0.22	0.22	0.022
3.6	0.22	0.56	0.068	0.22	1.00	0.047
3.9	0.24	0.39	0.68	0.22	0.22	0.022
4.3	0.18	0.51	0.024	0.22	2.20	0.047
4.7	0.16	1.30	0.039	0.22	1.00	0.022
5.1	0.16	0.36	0.051	0.22	1.00	0.047
5.6	0.13	1.10	0.033	0.22	1.00	0.022
6.2	0.13	0.36	0.016	0.22	2.20	0.047
6.8	0.24	1.60	0.33	0.10	0.10	0.010
7.5	0.12	0.30	1.20	0.22	0.10	0.010
8.2	0.12	0.11	0.024	0.22	2.20	0.047
9.1	0.18	1.50	0.091	0.10	0.22	0.010

Table 36.3 Frequency Multipliers

	0.1Ω	1Ω	10Ω	100Ω	1k	10k	100k	1M	10M	100M
1F	10	1	0.1	0.001	—	—	—	—	—	—
0.1F	100	10	1	0.1	0.01	0.001	—	—	—	—
10,000µF	1k	100	10	1	0.1	0.01	0.001	—	—	—
1,000µF	10k	1k	100	10	1	0.1	0.01	0.001	—	—
100µF	100k	10k	1k	100	10	1	0.1	0.01	0.001	—
10µF	1M	100k	10k	1k	100	10	1	0.1	0.01	0.001
1µF	10M	1M	100k	10k	1k	100	10	1	0.1	0.01
0.1µF	100M	10M	1M	100k	10k	1k	100	10	1	0.1
0.01µF	1G	100M	10M	1M	100k	10k	1k	100	10	1
1,000pF	—	1G	100M	10M	1M	100k	10k	1k	100	10
100pF	—	—	1G	100M	10M	1M	100k	10k	1k	100

Low power signal detection in a noisy environment

by Philip Karantzalis and Jimmylee Lawson

Introduction

In signal detection applications where a small narrowband signal is to be detected in the presence of wideband noise, one can design an asynchronous (nonphase sensitive) tone detector using an ultraselective bandpass filter, such as the LTC1164-8. The ultranarrow passband of the LTC1164-8 filter band limits any random noise and increases the detector's signal sensitivity.

The LTC1164-8 is an eighth-order, elliptic bandpass filter with the following features: the filter's f_{CENTER} (the center frequency of the filter's passband) is clock tunable and is equal to the clock frequency divided by 100; the filter's passband is from $0.995f_{CENTER}$ to $1.005f_{CENTER}$ ($\pm0.5\%$ from f_{CENTER}). Figure 36.27 shows a typical LTC1164-8 passband response and the area of passband gain variation.

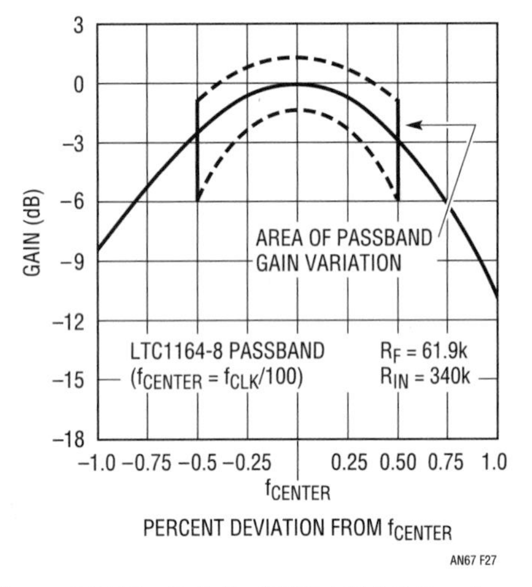

Figure 36.27 • Detail of LTC1164-8 Passband

Outside the filter's passband, signal attenuation increases to more than 50dB for frequencies between $0.96f_{CENTER}$ and $1.04f_{CENTER}$. Quiescent current is typically 2.3mA with a single 5V power supply.

An ultraselective bandpass filter and a dual comparator build a high performance tone detector

The LTC1164-8 has excellent selectivity, which limits the noise that passes from the input to the output of the filter. As a result, one can build a tone detector that can extract small signals from the "mud." Figure 36.28 shows the block diagram of such a tone detector. The detector's input is an LTC1164-8 bandpass filter whose output is AC coupled to a dual comparator circuit. The first comparator converts the filter's output to a variable pulsewidth signal. The pulsewidth varies depending on the signal amplitude. The average DC value of the pulse signal is extracted by a lowpass RC filter and applied to the second comparator. The identification of a tone is indicated by a logic high at the output of the second comparator.

One of the key benefits of using a high selectivity bandpass filter for tone detection is that when wideband noise (white noise) appears at the input of the filter, only a small amount of input noise will reach the filter's output. This results in a dramatically improved signal-to-noise ratio at the output of the filter compared to the signal-to-noise at the input of the filter. If the output noise of the LTC1164-8 is neglected, the signal-to-noise ratio at the output of the filter divided by the signal-to-noise ratio at the input of the filter is:

$$\frac{(S/N)_{OUT}}{S/N)_{IN}} = 20\ \mathrm{Log}\sqrt{\frac{(BW)_{IN}}{(BW)_{f}}}$$

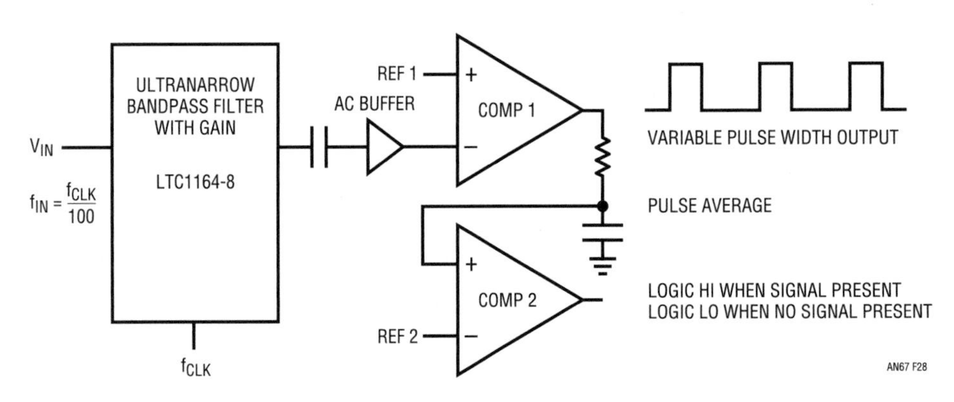

Figure 36.28 • Tone Detector Block Diagram

where: (BW)IN = the noise bandwidth at the input of the filter and (BW)f = (0.01)(fCENTER) is the filter's noise equivalent bandwidth.

For example, a small 1kHz signal is sent through a cable that is also conducting random noise with a 3.4kHz bandwidth. An LTC1164-8 is used to detect the 1kHz signal. The signal-to-noise ratio at the output of the filter is 25.3db larger than the signal-to-noise ratio at the input of the filter:

$$\sqrt{\frac{(BW)_{IN}}{(BW)_f}} = 20\ \text{Log}\sqrt{\frac{3.4\text{kHz}}{(0.01)(1\text{kHz})}} = 25.3\text{dB}$$

Figure 36.29 shows the complete circuit for a 1kHz tone detector operating with a single 5V supply. An LTC1164-8 with a clock input set at 100kHz sets the tone detector's frequency at 1kHz ($f_{CENTER}=f_{CLK}/100$). A low frequency op amp (LT1013) and resistors R_{IN} and R_F set the filter's gain. In order to minimize the filter's output noise and maintain optimum dynamic range, the output feedback resistor R_F should be 61.9k. Capacitor C_F across resistor R_F is added to reduce the clock feedthrough at the filter's output.

To set the gain for the LTC1164-8, R_{IN} should be calculated by the equation:

$$R_{IN} = 340k/\text{Gain}$$

In Figure 36.29, the filter's gain is 10 (R_{IN} = 34k). Capacitor C1 and a unity-gain op amp (LT1013) AC couple the signal at the filter's output to an LTC1040 dual low power comparator. AC coupling is required to eliminate any DC offset caused by the LTC1164-8.

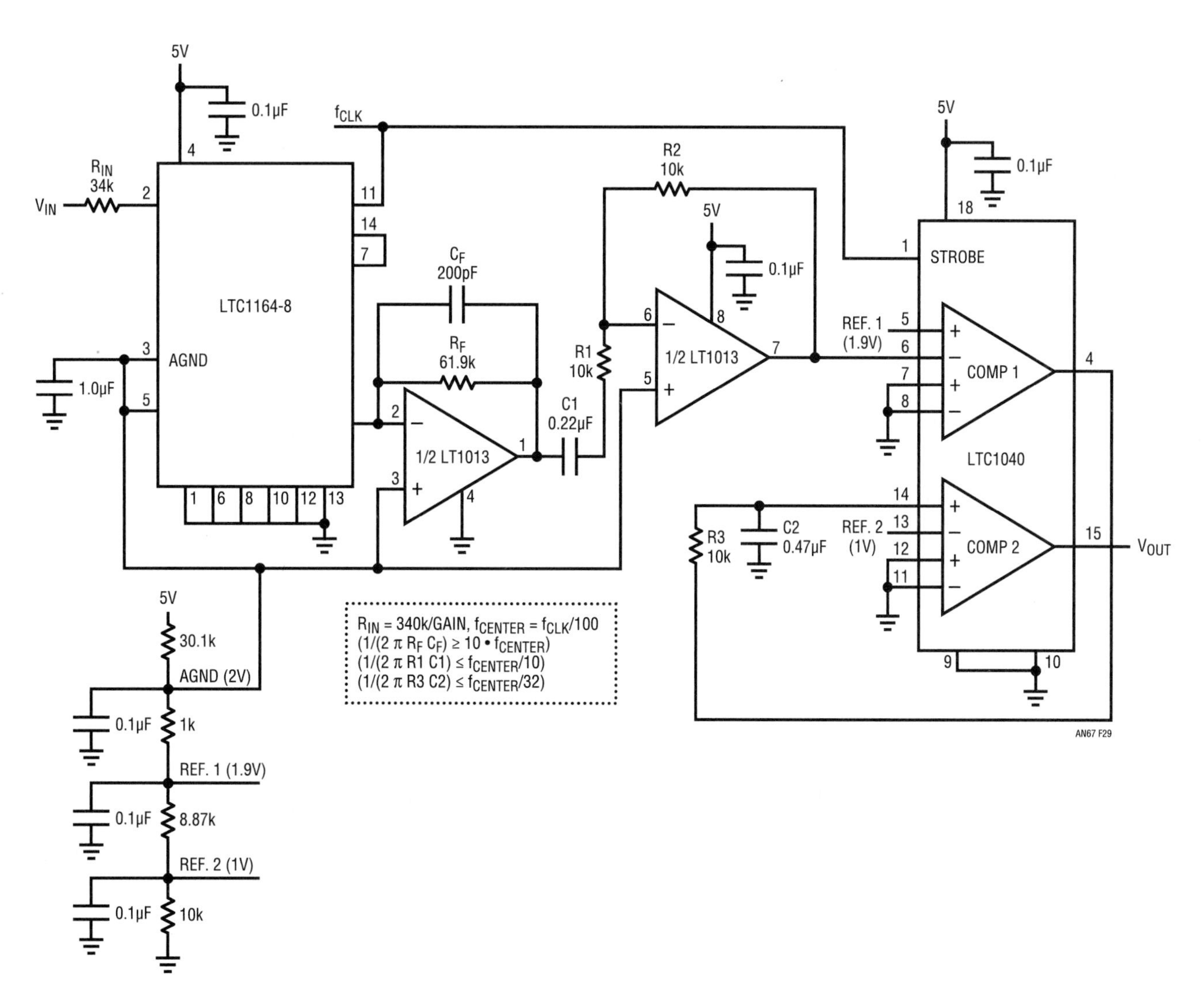

Figure 36.29 • 1kHz Tone Detector with Gain of 10

A resistive divider generates a 2V bias for the LTC1164-8 "ground" (Pins 3 and 5) and the positive input of the LT1013 dual op amps. For single 5V operation the output swing of the LTC1164-8 is from 0.5V to 3.5V centered at 2V. The divider also provides the reference voltages for the LTC1040 dual comparators (Ref. 1 = 1.9V and Ref. 2 = 1V). Power supply variations do not affect the performance of this circuit because all DC reference voltages are derived from the same resistor divider and will track any changes in the 5V power supply.

Theory of operation

The tone detector works by looking at the negative peaks at the output of the filter. Signals below 1.9V at the output of the filter trip the first comparator. The second comparator has a 1V reference and detects the average value of the output of the first comparator. The R3/C2 time constant is set to allow detection only if the duty cycle of the first comparator's output exceeds 25%. Waveforms with duty cycles below 25% are arbitrarily assumed to carry false information.

The circuitry is designed so that two or more negative signal peaks of 160mV at the filter's output produce a 25% duty cycle pulse waveform at the output of the first detector (the 1.9V and 1V references for comparators 1 and 2 respectively, set the $160mV_{PEAK}$ and the 25% duty cycle). The 25% duty cycle requirement establishes an operating point or "minimum detectable signal" for the detector circuit. Thus, the circuitry outputs a "tone present" condition only when the duty cycle is greater than or equal to 25%. The 25% duty cycle requirement sets two conditions for optimum tone detection at the detector's input.

The first input condition is the maximum input noise spectral density that will not trigger the detector's output to indicate the presence of a tone. When only noise is present at the filter's input, the maximum input noise spectral density is conservatively defined as the amount required to produce noise peaks at the filter's output of 160mV or lower amplitude. The 160mV maximum noise peak specification at the filter's output can be converted to output noise in mV_{RMS} by using a crest factor of 5 (the crest factor of a signal is the ratio of its peak value to its RMS value—a theoretical crest factor of 5 predicts 99.3% of the maximum peaks of wideband noise with uniform spectral density). Therefore, the maximum allowable noise at the filter's output is $32mV_{RMS}$ ($160mV_{PEAK}/5$). The noise at the filter's output depends on the filter's gain and noise equivalent bandwidth and the spectral density of the noise at the filter's input. Therefore, the maximum input noise spectral density for Figure 36.3's circuit is:

$$e_{IN} \leq 32mV_{RMS}/(\text{Gain} \bullet \sqrt{(BW)_f})\frac{V_{RMS}}{\sqrt{Hz}}$$

where: Gain is the filter's gain at its center frequency and $(BW)_f$ is the filter's noise equivalent bandwidth.

Note: Compared to $32mV_{RMS}$ the $270mV_{RMS}$ output noise of the LTC1164-8 is negligible. The output noise of the LTC1164-8 is independent of the chosen filter signal gain.

The second input condition is the minimum input signal required so that a tone can be detected when it is buried by the maximum noise, as defined by the first input condition. When a tone plus noise is present at the filter's input, the output of the filter will be a tone whose amplitude is modulated by the bandlimited noise at the filter's output. If a maximum noise peak of 160mV modulates the tone's amplitude, a 320mV tone peak at the filter's output can be detected because the product of the noise and the tone crosses the (negative) $160mV_{PEAK}$ detection threshold and the 25% duty cycle requirement is exceeded. Therefore, a conservative value for the minimum signal at the filter's output can be set to $320mV_{PEAK}$ or $226mV_{RMS}$, but a value of $200mV_{RMS}$ was established experimentally. Therefore, the minimum input signal for reliable tone detection in the presence of the maximum input noise spectral density is:

$$V_{IN(MIN)} = 200mV_{RMS}/\text{Gain}$$

For optimum tone detection, the signal's frequency should be in the filter's passband, within ±0.1% of f_{CENTER}.

Conclusion

A very selective bandpass filter, the LTC1164-8, can be configured as a nonphase-sensitive tone detector. This allows signals to be detected in the presence of comparatively large amounts of noise or signal-to-noise ratios that are less than unity.

Bandpass filter has adjustable Q

by Frank Cox

The bandpass filter circuit shown in Figure 36.30 features an electronically controlled Q. Q for a bandpass filter is defined as the ratio of the 3dB pass bandwidth to the stop bandwidth at some specified attenuation. The center frequency of the bandpass filter in this example is 3MHz, but this can be adjusted with appropriate LC tank components. The upper limit of the usable frequency range is about 10MHz. The width of the passband is adjusted by the current into Pin 5 (set current or I_{SET}) of the transconductance amplifier segment of IC1, an LT1228. Figure 36.31 is a network analyzer plot of frequency response versus set current. This plot shows the variation in Q while the center frequency and the passband gain remain relatively constant.

The circuit's operation is best understood by analyzing the closed-loop transfer function. This can be written in the form of the classic negative feedback equation:

$$H(s) = \frac{A(s)}{1 + A(s)B(s)}$$

where A(s) is the forward gain and B(s) is the reverse gain. The forward gain is the product of the transconductance stage gain (g_m) and the gain of the CFA (A_{CFA}). For this circuit, g_m is ten times the product of I_{SET} and the impedance of the tank circuit as a function of frequency.

This gives the complete expression for the forward gain as a function of frequency:

$$A(s) = 10 I_{SET} A_{CFA} \left(\frac{sL}{1 + s^2 LC} \right)$$

The reverse gain is simply:

$$B(s) = \frac{R7}{R6 + R7}$$

$$\textit{and } A_{CFA} = \frac{R4 + R5}{R4}$$

$$\textit{Setting } B(s) = \frac{1}{A_{CFA}} R_{RATIO}$$

and substituting these expressions into the first equation gives:

$$H(s) = \frac{1}{R_{RATIO}} \frac{10\ I_{SET}\left(\frac{sL}{1+s^2LC}\right)}{1 + 10\ I_{SET}\left(\frac{sL}{1+s^2LC}\right)}$$

The last equation can be rewritten as:

$$H(s) = \frac{1}{R_{RATIO}} \frac{s\left[\frac{1}{\sqrt{LC}}\left(\frac{10\ I_{SET}\sqrt{LC}}{C}\right)\right]}{s^2 + s\left[\frac{1}{\sqrt{LC}}\left(\frac{10\ I_{SET}\sqrt{LC}}{C}\right)\right] + \frac{1}{LC}}$$

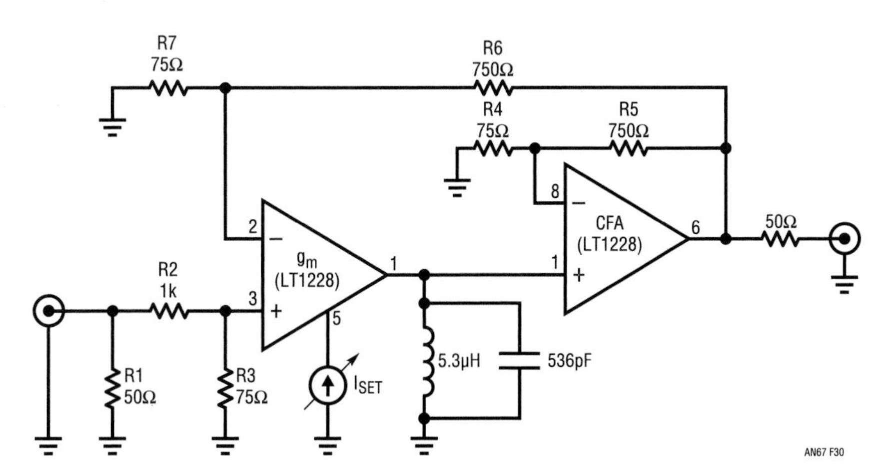

Figure 36.30 • LT1228 Bandpass Filter Circuit Diagram

The transfer function of a second order bandpass filter can be expressed in the form[2]:

$$H(s) = H_{BP}\frac{S(\omega_0/Q)}{S^2 + S(\omega_0/Q) + \omega_0^2}$$

Comparing the last two equations note that

$$\omega_0 = \frac{1}{\sqrt{LC}} \text{and} \frac{1}{Q} = \frac{10\, I_{SET}\sqrt{LC}}{C}$$

$$\text{and therefore } Q = \frac{C}{10\, I_{SET}\sqrt{LC}}$$

Note 2: Thanks to Doug La Porte for this equation hack.

It can be seen from the last equation that the Q is inversely proportional to the set current.

Many variations of the circuit are possible. The center frequency of the filter can be tuned over a small range by the addition of a varactor diode. To increase the maximum realizable Q, add a series LC network tuned to the same frequency as the LC tank on Pin 1 of IC1. To lower the minimum obtainable Q, add a resistor in parallel with the tank circuit. To create a variable Q notch filter, connect the inductor and capacitor at Pin 1 in series rather than in parallel.

A variable Q bandpass filter can be used to make a variable bandwidth IF or RF stage. Another application for this circuit is as a variable-loop filter in a phase locked loop phase demodulator. The variable Q bandpass filter is set for a wide bandwidth while the loop acquires the signal and is then adjusted to a narrow bandwidth for best noise performance after lock is achieved.

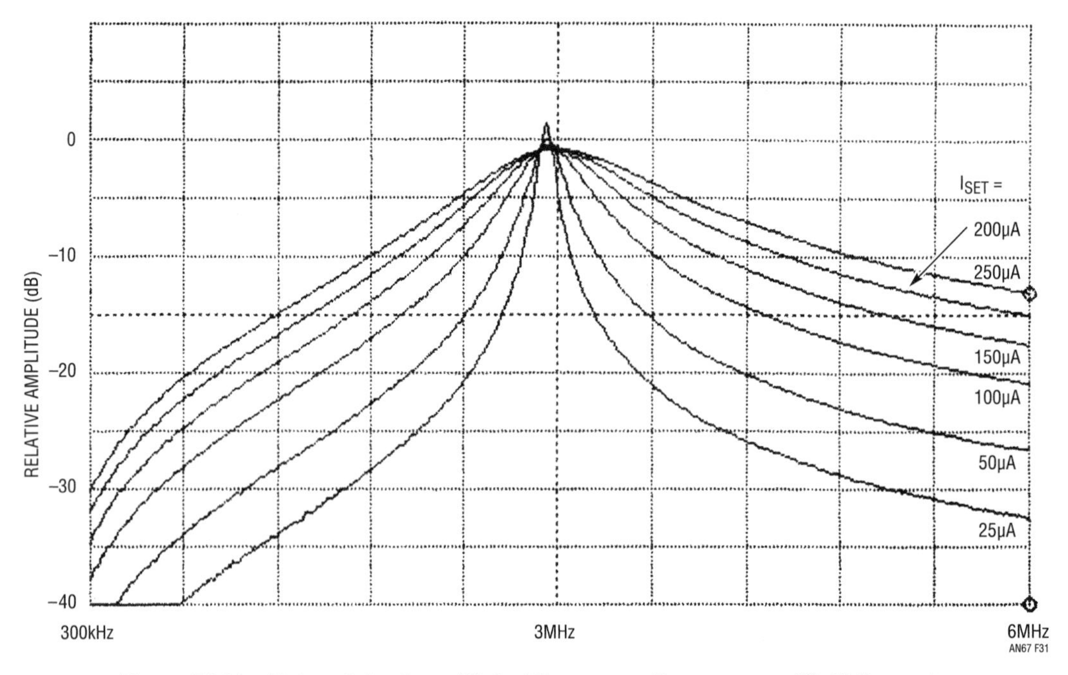

Figure 36.31 • Network Analyzer Plot of Frequency Response vs "Set" Current

An ultraselective bandpass filter with adjustable gain

by Philip Karantzalis

Introduction

The LTC1164-8 is a monolithic, ultraselective, eighth order elliptic bandpass filter: The passband of the LTC1164-8 is tuned with an external clock; the clock-to-center-frequency ratio is 100:1. The stopband attenuation of the LTC1164-8 is greater than 50dB for input frequencies outside a narrow band defined as ±4% of the center frequency of the filter (see Figure 36.32).

One op amp and two resistors build an ultraselective filter

The LTC1164-8 requires an external op amp and two external resistors. The filter's gain at its center frequency is equal to $3.4R_F/R_{IN}$. For optimum dynamic range with a gain equal to one, the external resistor R_F should be 90.9k and the external resistor R_{IN} should be 340k. For gains other than 1, $R_{IN} = 340k/gain$. Gains of up to 1000 are possible. The complete configuration is shown in Figure 36.33. Note that programming the filter's gain with input resistor R_{IN} is equivalent to providing the LTC1164-8 with noiseless preamplification, since the filter's internal noise is not amplified. The wideband noise of the LTC1164-8 measures $400\mu V_{RMS}$ at ±5V and is independent of the filter's gain and center frequency. A capacitor, C_F, across resistor R_F reduces clock feedthrough and provides a smooth sine wave output.

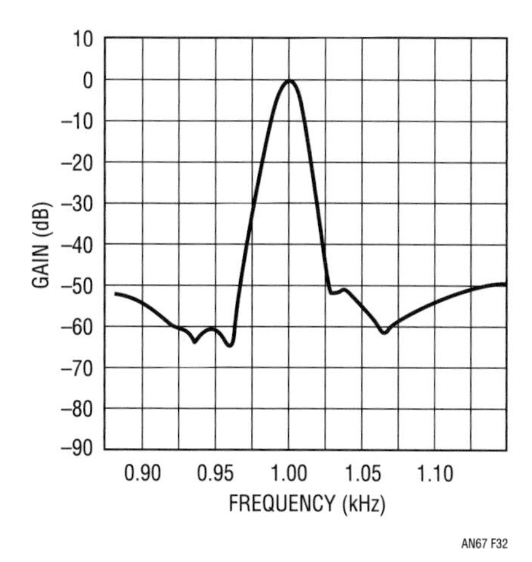

Figure 36.32 • LTC1164-8 Gain vs Frequency Response

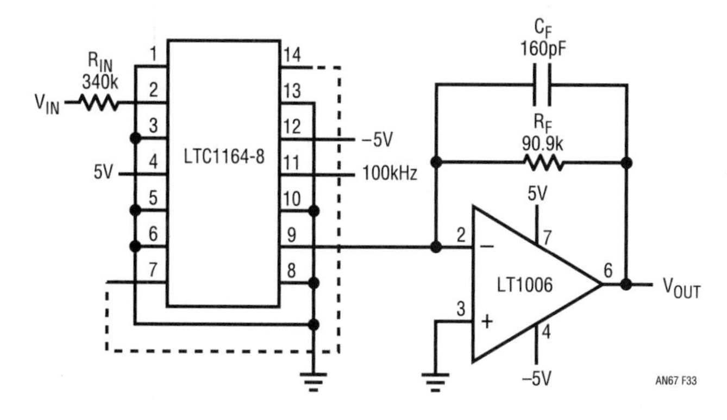

Figure 36.33 • LTC1164-8 Ultranarrow, 1kHz Bandpass Filter with Gain (Gain = $340k/R_{IN}$, $1/2\pi R_F C_F = 10\ f_{CENTER}$)

Signal detection in a hostile environment

An outstanding feature of the LTC1164-8 is its ultraselectivity. A bandpass filter with ultraselectivity is ideal for signal detection applications. One signal detection application occurs when two signals are very closely spaced in the frequency spectrum and only one of the signals has useful information. The LTC1164-8 can extract the signal of interest and suppress its unwanted neighbor. For example, a small 1kHz, $10mV_{RMS}$ signal is combined with an unwanted 950Hz, $40mV_{RMS}$ signal. The two signals differ in frequency by only 5% and the 950Hz signal is four times larger than the 1kHz signal. To detect the 1kHz signal, the LTC1164-8 is set to a gain of 100 and the clock frequency is set to 100kHz. At the filtered output of the LTC1164-8 the following signals will be present: an extracted 1kHz, $1V_{RMS}$ signal and a rejected 950Hz, $2.7mV_{RMS}$ signal, as shown in Figure 36.34. In a narrowband signal separation and extraction application, as described previously, the LTC1164-8 provides a simple and reliable detection circuit solution.

A second signal detection application occurs when a small signal is to be detected in the presence of noise. For example, a 1kHz, $10mV_{RMS}$ signal is mixed with a wideband noise signal that measures $5mV_{RMS}$ in a 400Hz frequency band. The signal-to-noise ratio is just 6dB. With the LTC1164-8 set for a center frequency of 1kHz (f_{CLK} is equal to 100kHz) and a gain of 100, the 1kHz, $10mV_{RMS}$ signal will be detected and amplified. The wideband noise will be band limited by the very narrow band gain response of the LTC1164-8. At the output of a LTC1164-8, the 1kHz signal will be $1V_{RMS}$ as shown in Figure 36.35.

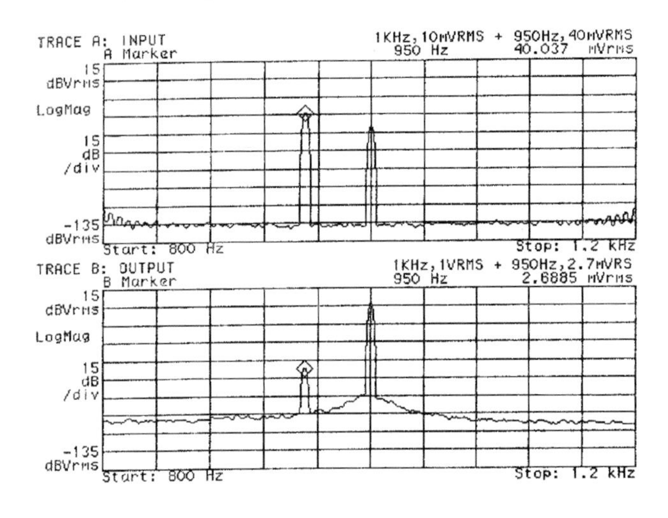

Figure 36.34 • Narrow Band Signal Extraction Showing Input to and Output from the LTC1164-8 Filter. Filter f_{CENTER} set to 1kHz with Gain = 100

The total band limited noise will be $70mV_{RMS}$ with a signal-to-noise ratio of more than 20dB, as shown in Figure 36.36. In applications of signal detection in the presence of noise, the LTC1164-8 provides asynchronous detection. Signal detection circuits such as synchronous demodulators and lock-in amplifiers require the presence of a reference or carrier signal to provide phase and frequency information of the signal to be detected. With an LTC1164-8, signal detection is accomplished by selecting a very narrow signal detection band around the frequency of the desired signal, which is defined as f_{CLK} divided by 100 (f_{CLK} is the clock frequency of the LTC1164-8), and by selecting the filter gain by choosing the value of a resistor.

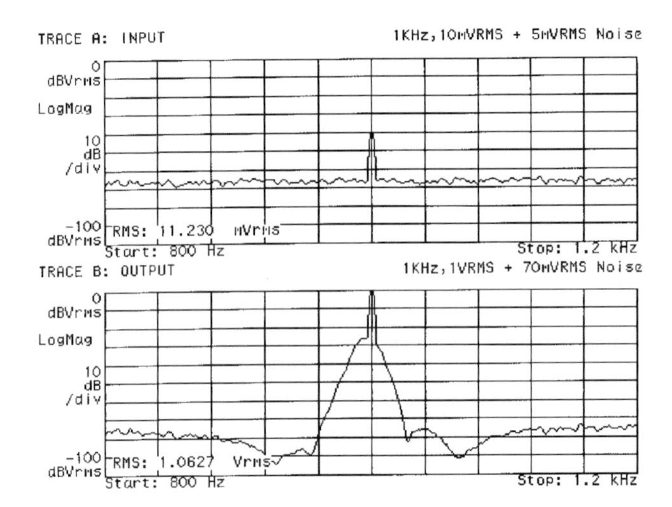

Figure 36.35 • Signal Detection in the Presence of Noise Example Showing Input to and Output from the LTC1164-8 Filter. Filter f_{CENTER} set to 1kHz with Gain = 100

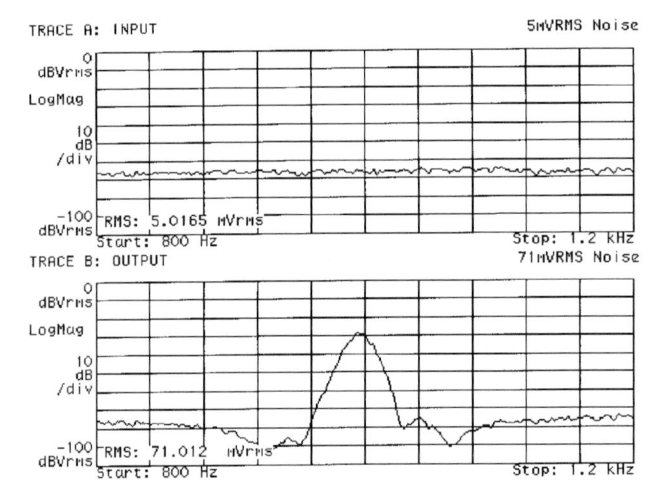

Figure 36.36 • Wideband Noise Input to LTC1164-8 Filter. Plots Show Input to and Output from the Filter. Filter f_{CENTER} set to 1kHz with Gain = 100

LT1367 builds rail-to-rail Butterworth filter

by William Jett and Sean Gold

Single supply 1kHz, 4th order Butterworth filter

The circuit shown in Figure 36.37 takes advantage of all four op amps in the LT1367 to form a 4th order Butterworth filter. The filter is a simplified state-variable architecture consisting of two cascaded second order sections. Each section uses the 360 degree phase shift around the two op amp loop to create a negative summing junction at A1's positive input.[3] The circuit has two-thirds the power dissipation and component count as the classic three op amp biquad,[4] yet it has the same low component sensitivities for center frequency, ω_0 and Q. For cutoff frequencies other than the 1kHz example shown, use the following formula for each section:

$$\omega_0{}^2 = 1/(R1\ C1\ R2\ C2),$$
$$\text{where } R1 = 1/(\omega_0\ QC1) \text{ and } R2 = Q/(\omega_0 C2)$$

The DC bias applied to A2 and A4 for single supply operation is not needed when split supplies are available. The circuit's output can swing rail-to-rail and displays the maximally flat amplitude response with a 1kHz cutoff frequency with 80dB/decade rolloff (Figure 36.38).

Note 3: Hahn, James. 1982. State Variable Filter Trims Predecessor's Component Count. Electronics, April 21, 1982.

Note 4: Thomas, L.C. 1971. The Biquad: Part I—Some Practical Design Considerations. EEE Transactions on Circuit Theory, 3:350-357, May 1971.

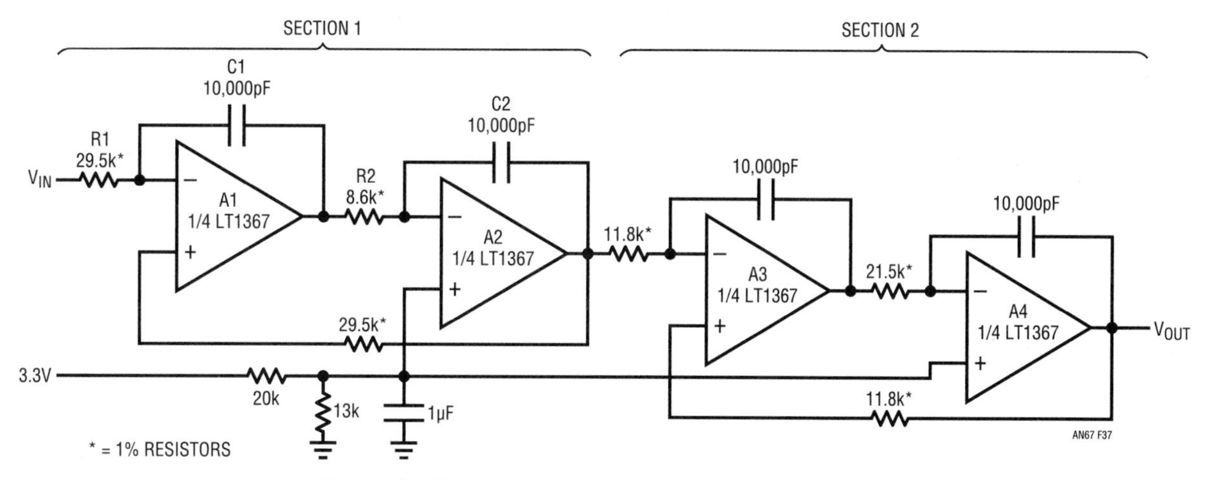

Figure 36.37 • 1kHz 4th Order Butterworth Filter

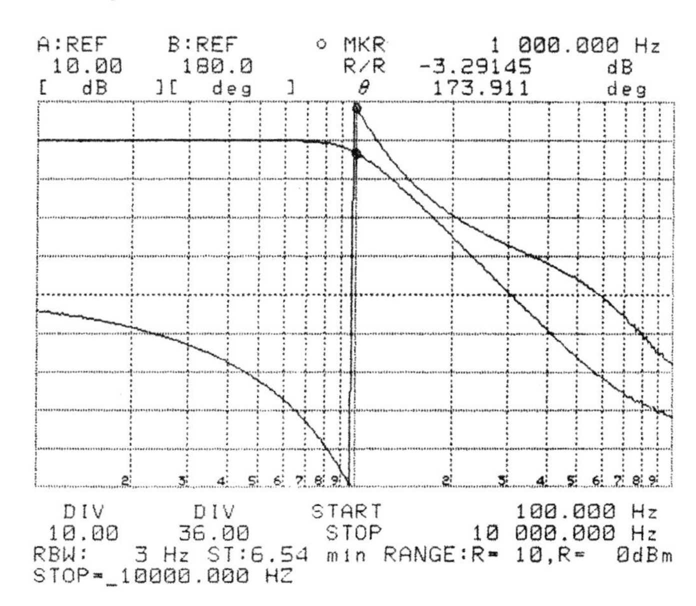

Figure 36.38 • Frequency Response of 4th Order Butterworth Filter

DC accurate, clock tunable lowpass filter with input antialiasing filter

by Philip Karantzalis

In a sampled data system, the sampling theorem says that if an input signal has any frequency components greater than one half the sampling frequency, aliasing errors will appear at the output. In practice aliasing is not always a serious problem. High order switched capacitor lowpass filters are band limited and significant aliasing occurs only for input signals centered around the clock frequency and its multiples.

Figure 36.39 shows the LTC1066-1 aliasing response when operated with a clock-to-f_C ratio of 50:1. With a 50:1 ratio, the LTC1066-1 samples its input twice during one clock period and the effective sampling frequency is twice the clock frequency. Figure 36.39 shows that the maximum aliased output is generated for inputs in the range of $2(f_{CLK} \pm f_C)$. (f_C is the cutoff frequency of the LTC1066-1.) For instance, if the LTC1066-1 is programmed to produce a cutoff frequency of 20kHz with a 1MHz clock, maximum aliasing will occur only for input signals in the narrow range of 2MHz ±20kHz and its multiples.

The simplest antialiasing filter is a passive 1st order lowpass RC filter. The −3dB frequency of the RC filter should be chosen so that the passband of the RC filter does not influence the passband of the LTC1066-1. When the LTC1066-1 clock frequency is 500kHz, an RC filter with the −3dB frequency set at 50kHz attenuates by 26dB any possible aliasing inputs in the range 1MHz ±10kHz. The passband shape of the 50kHz RC filter does not degrade the flat passband of the LTC1066-1 at 10kHz (the passband attenuation of the 50kHz RC filter for frequencies less than 10kHz is less than 0.2dB). If the LTC1066-1 is clock tuned to a cutoff frequency of 5kHz (with a clock frequency of 250kHz), the 50kHz RC filter will provide 20dB attenuation for aliasing inputs in the range of 500kHz ±5kHz. Therefore, a 1st order lowpass RC filter will attenuate all aliasing signals to the LTC1066-1 by a minimum of 20dB for a clock tunable range of one octave.

For added antialiasing bandwidth, a 1st order lowpass RC filter can be tuned by the clock signal of LTC1066-1 to follow the cutoff frequency of the higher order filter. The circuit is shown in Figure 36.40. The circuit operation is as follows. The six comparators inside the LTC1045 detect the clock frequency. The clock signal of the LTC1066-1 is converted to a pulse output whose duty cycle changes with clock frequency. The average voltage of the pulse signal is delivered to a 4-window comparator whose outputs drive the four analog switches of the LTC202. When the LTC1066-1 clock frequency increases or decreases by more than one octave (2x or x/2), a capacitor is switched in or out of the 1st order lowpass filter formed by resistor R1 (1k) and capacitor C1. The −3dB frequency of the lowpass RC filter is therefore doubled or halved if the cutoff frequency of the LTC1066-1 is doubled or halved. Resistor R1 and capacitors C1 through C5 allow the lowpass RC filter to be tuned over a range of five octaves, providing at least 20dB attenuation to any LTC1066-1 input signals in the range $2(f_{CLK} \pm f_C)$ (the RC filter also attenuates all aliasing signals near any multiples of the clock frequency).

The circuit in Figure 36.40 can be used for any clock tunable, 5-octave range for cutoff frequencies from 10Hz to 80kHz (with ±5V supplies for LTC1066-1) or for cutoff frequencies as high as 100kHz (with ±8V supplies for the LTC1066-1). For cutoff frequencies greater than 50kHz, a 15pF capacitor in series with a 30k resistor should be connected between Pins 11 and 13 of the LTC1066-1 to minimize passband gain peaking.

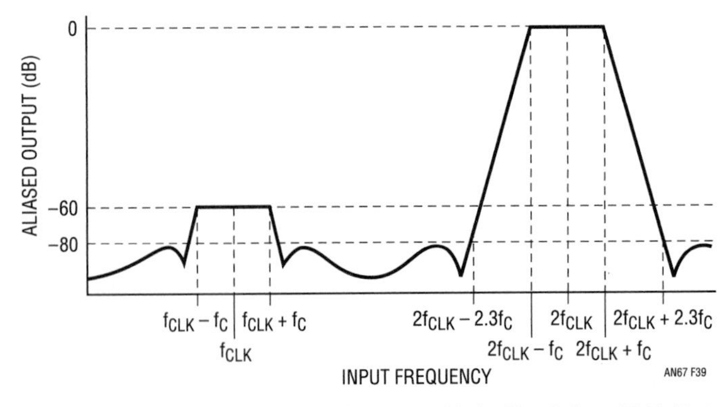

Figure 36.39 • Aliasing vs Frequency f_{CLK}/f_C = 50:1 (Pin 8 to V+); Clock is a 50% Duty Cycle Square Wave

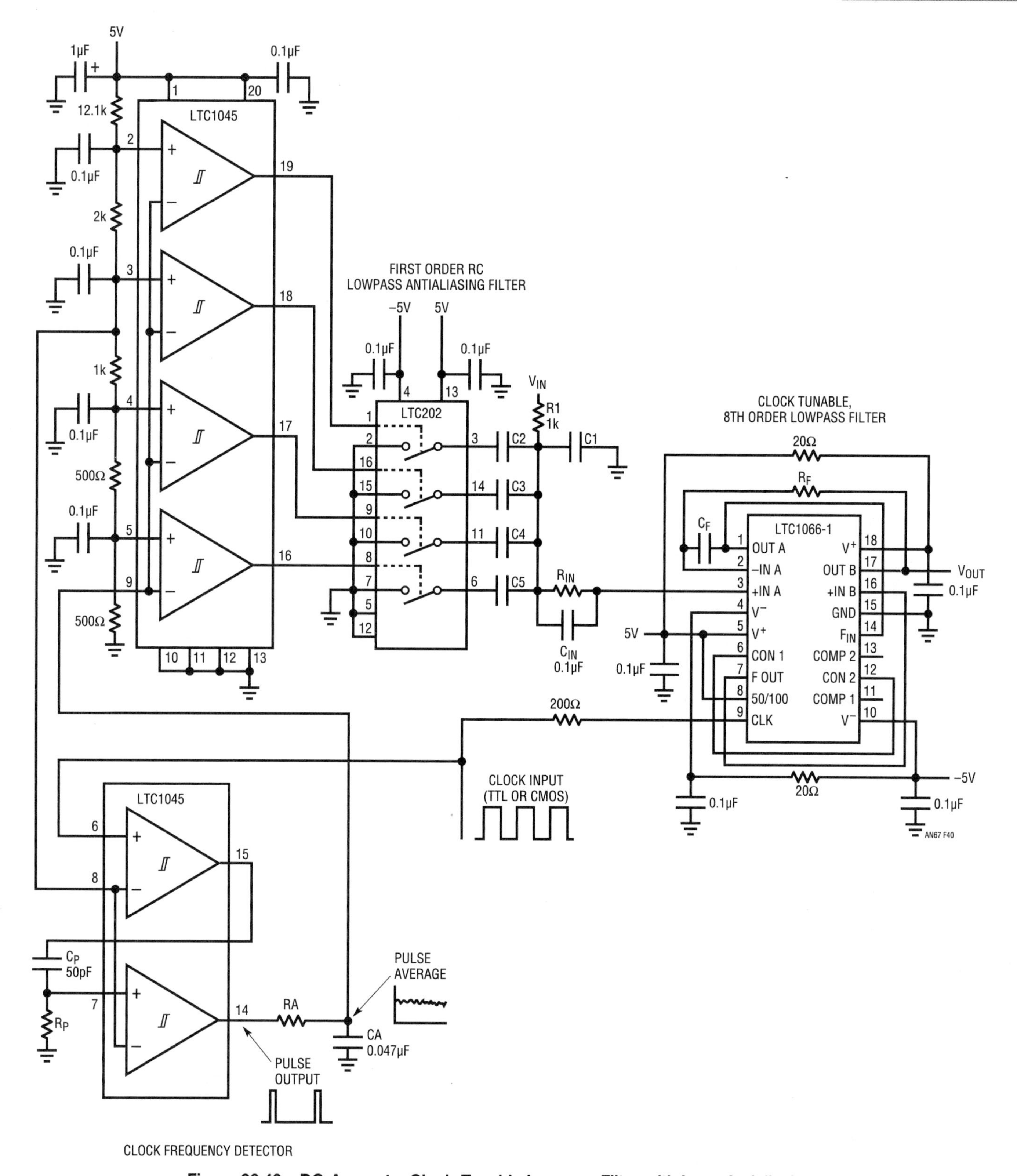

Figure 36.40 • DC-Accurate, Clock-Tunable Lowpass Filter with Input Antialiasing

Use the following design guide for choosing the component values of R_A, R_P, R_F, R_{IN}, C_F, C1 through C5, C_P and C_A.

Definitions

1. The cutoff frequency of the LTC1066-1 is abbreviated as f_C.
2. $f_{C(LOW)}$ is the lowest cutoff frequency of interest
3. A range of five octaves is from $f_{C(LOW)}$ to $32 \bullet f_{C(LOW)}$

Component Calculations

$\frac{1}{2\pi R_F C_F} \frac{f_{C(LOW)}}{250}$ $R_{IN} = R_F$ can be chosen as 20k, R_{IN} and C_{IN} are not needed)

$$C1 = \frac{1}{f_{c(LOW)}} \quad (f_{c(LOW)} \text{ inHz}); R1 = 1K$$

$$C2 = C1 \pm 5\%, C3 = 2(C1) \pm 5\%,$$
$$C4 = 4(C1) \pm 5\%, C5 = 8(C1) \pm 5\%$$

$$C_P = 50pF, R_P = \frac{10^5}{50f_{c(LOW)}}k$$

$$C_A = 0.047\mu F, R_A = \frac{5(10^5)}{50f_{c(LOW)}}k$$

Example

For a five octave range from 1kHz to 32kHz:

$$f_{c(LOW)} = 1KHz$$

$$\text{Let } C_F = 1\mu F \pm 20\%, \text{then } R_F = 40.2k \pm 1\%.$$
$$R_{IN} = R_F = 40.2k \pm 1\%, C_{IN} = 0.1\mu F$$

$$C1 = 0.001\mu F \pm 5\%, C2 = 0.001\mu F \pm 5\%,$$
$$C3 = 0.0022\mu F \pm 5\%$$

$$C4 = 0.0039\mu F \pm 55, C5 = 0.0082\mu F \pm 5\%$$

$$C_P = 50pF, R_p = 2k, C_A = 0.047\mu F, R_A = 10k$$

The LTC1066-1 DC accurate elliptic lowpass filter

by Nello Sevastopoulos

Figure 36.41 shows an application allowing clock tunability from 10Hz to 100kHz. The R_CC_C frequency compensating components (needed only for cutoff frequencies above 60kHz) maintain a flat passband for cutoff frequencies between 50kHz and 100kHz. The input resistor, R_I, reduces the output DC offset caused by the op amp bias current through the 100k feedback resistor, R_F. The measured DC offset and the gain nonlinearity are 4mV and ±0.0063% (84dB), respectively. The 0.1µF bypass capacitor, C_B, helps keep the total harmonic distortion of the filter from being degraded by the 100k input resistor.

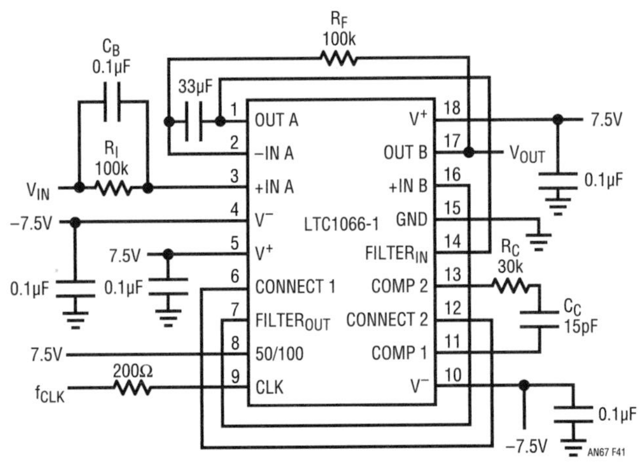

MAXIMUM OUTPUT VOLTAGE OFFSET = 4mV, DC LINEARITY = ±0.0063%, T_A = 25°C.
THE PIN 6 TO 12 CONNECTION SHOULD BE UNDER THE IC AND SHIELDED BY AN ANALOG SYSTEM GROUND PLANE.
RC COMPENSATION BETWEEN PINS 11 AND 13 REQUIRED ONLY FOR f_{CUTOFF} > 50kHz.
THE 33µF CAPACITOR IS A NONPOLARIZED, ALUMINUM ELECTROLYTIC, ±20%, 16V (NICHICON UUPIC 330MCRIGS OR NIC NACEN 33M16V 6.3 × 5.5 OR EQUIVALENT).

Figure 36.41 • DC Accurate, 10MHz to 100kHz 8th Order Elliptic Lowpass Filter, $f_{CLK}/f_C = 50:1$

Clock tunability

An external clock tunes the cutoff frequency of the internal switched capacitor network. The device has been optimized for a clock-to-cutoff-frequency ratio of 50:1. The internal double sampling greatly reduces the risk of aliasing.

The maximum obtainable cutoff frequency, $f_{CUTOFF(MAX)}$, depends on power supply, clock duty cycle and temperature; $f_{CUTOFF(MAX)}$ does not depend on the value of the external resistor/capacitor combination R_FC_F. The R_CC_C compensation is shown in Figure 36.41. The data detailed in Figure 36.42 reveals the important fact that for a cutoff frequency of 100kHz, the stopband attenuation still remains greater than 70dB for input frequencies up to 1MHz.The minimum obtainable cutoff frequency depends on the R_FC_F time constant of the servo loop. For a given R_FC_F time constant, the minimum obtainable cutoff frequency of the LTC1066-1 is:

$$f_{CUTOFF(MIN)} = 250(1/2\pi R_FC_F).$$
$$f_{CUTOFF(MAX)} = 100kHz$$

For instance, if $R_f = 20k, C_f = 1\mu F, f_{CUTOFF(MIN)} = 2kHz$, and $f_{CLOCK(MIN)} = 100kHz$.

Under these conditions, a clock frequency below 100kHz will "warp" the passband gain by more than 0.1dB. Please see the LTC1066-1 data sheet for more details.

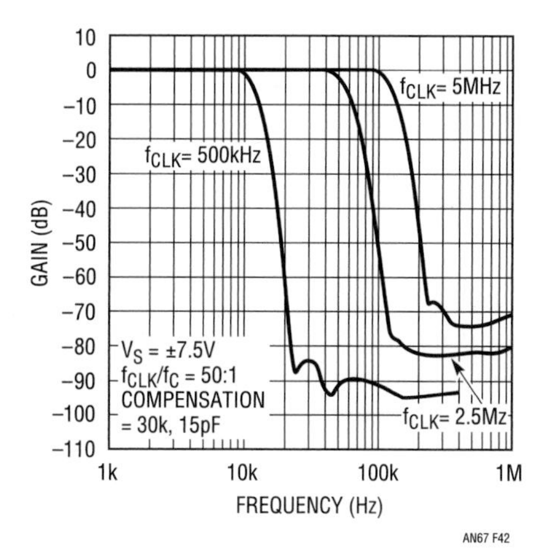

Figure 36.42 • LTC1066-1 Amplitude vs Frequency

Dynamic range

The LTC1066-1 wideband noise is $100\mu V_{RMS}$. Figure 36.43 shows the noise plus distortion versus RMS input voltage at 1kHz. With a ±5V supply, the filter can swing ±2.5V (5V full scale) with better than 0.01% distortion plus noise. The maximum signal-to-noise ratio, in excess

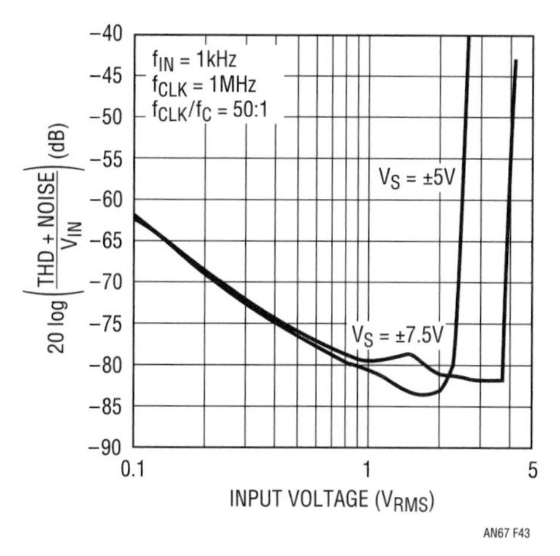

Figure 36.43 • LTC1066-1 Dynamic Range

of 90dB, is achieved with ±7.5V supplies. Unlike previous monolithic filters the data shown in Figure 36.43 is taken without using any input or output op amp buffers. The output buffer of the LTC1066-1 can drive a 200Ω load without dynamic range degradation.

Aliasing and antialiasing

All sampled data systems will alias if their input signals exceed half the sampling rate, but aliasing for high order, band limited, switched capacitor filters need not be a serious problem. The LTC1066-1, when operating with a 50:1 clock-to-cutoff-frequency ratio, will have significant aliasing only for input signals centered around twice the clock frequency and its even multiples. Figure 36.44 shows the input frequencies that will generate aliasing at the filter output. For instance, if the filter is tuned to a 50kHz cutoff frequency using a 2.5MHz clock, significant aliasing will occur only for input frequencies of 5MHz ±50kHz. The filter user should be aware of the spectrum at the input to the filter. Next, an assessment should be made as to whether a simple, continuous-time antialiasing filter in front of the LTC1066-1 is required. The antialiasing filter should do precisely what it is meant to do, that is, provide

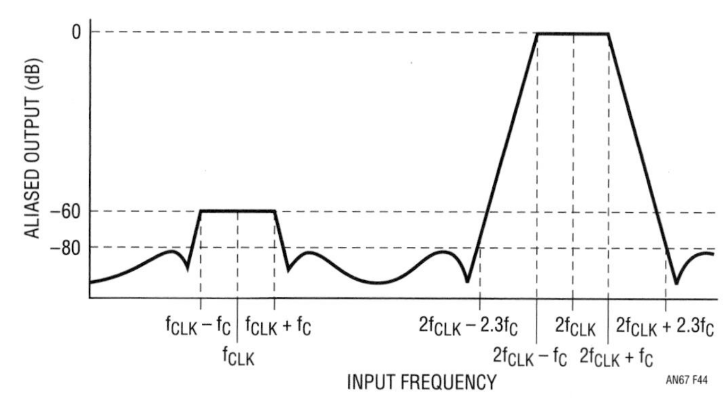

Figure 36.44 • Aliasing vs Frequency f_{CLK}/f_C = 50:1 (Pin 8 to V+). Clock is a 50% Duty Cycle Square Wave

band limiting. The antialiasing filter should not degrade the DC or AC performance of the LTC1066-1.

For fixed cutoff frequency applications, the antialiasing function is quite trivial. Figure 36.45 shows the internal precision input op amp configuration used to perform both the DC accurate function of the LTC1066-1 and the input antialiasing configuration. The cutoff frequency of the RC antialiasing filter is set three times higher than the cutoff frequency of the LTC1066-1. For the example shown in Figure 36.45 the input antialiasing filter provides a 62dB attenuation at twice the clock frequency of the switched capacitor filter.

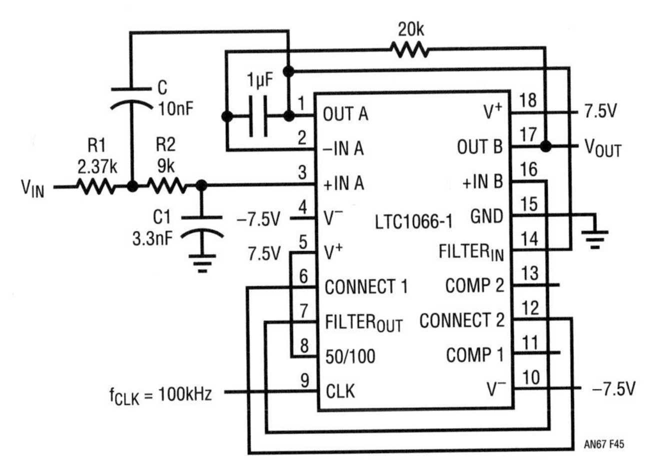

Figure 36.45 • Adding a 2-Pole Butterworth Input Antialiasing Filter. Set C1 = 0.33C, R2 = 3.8(R1); f–3dB (Input Antialiasing) = 0.8993/(2πR1C)

Clock tunable bandpass filter operates to 160kHz in single supply systems

by Philip Karantzalis

When the only available power supply in a system is 5V or 12V and a precision bandpass filter is needed at cutoff frequencies greater than 20kHz, the LTC1264 switched capacitor active filter building block can be configured to realize an 8th order bandpass filter accurate to ±1% or better over temperature (–40°C to 85°C). Figure 36.46 is a schematic diagram of an 8th order bandpass filter tunable with a TTL clock signal to any center frequency up to 70kHz with a 5V supply or to 100kHz with a 12V supply. The clock frequency-to-center frequency ratio is 20:1. The gain response for a 50kHz bandpass filter is shown in Figure 36.47 and the input dynamic range with a 5V supply is shown in Figure 36.48.

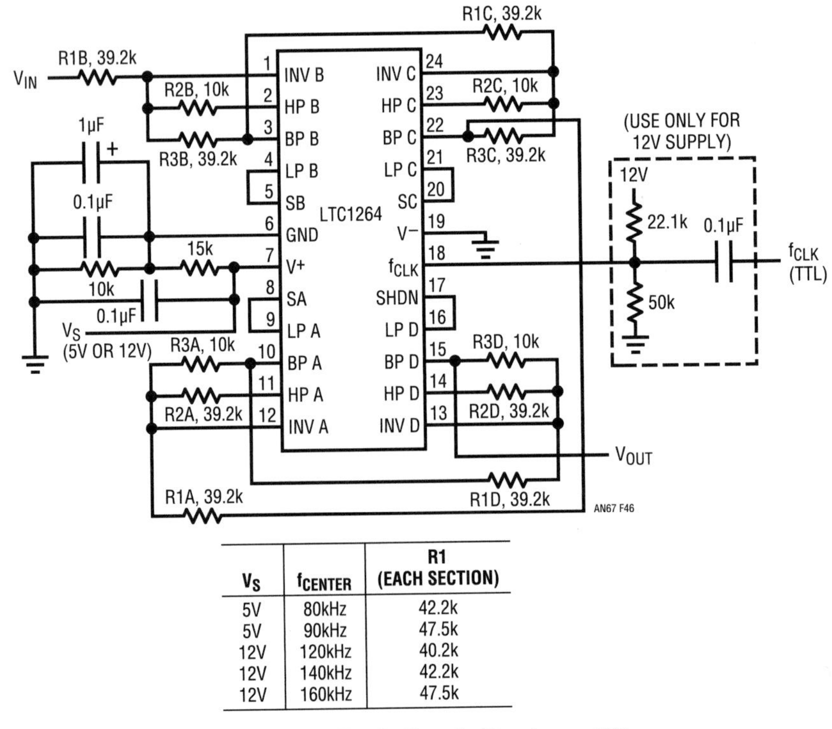

V_S	f_{CENTER}	R1 (EACH SECTION)
5V	80kHz	42.2k
5V	90kHz	47.5k
12V	120kHz	40.2k
12V	140kHz	42.2k
12V	160kHz	47.5k

Figure 36.46 • Single Supply Bandpass Filter

The passband frequency range (the frequency range where the filter's attenuation is 3dB or less) is equal to the center frequency divided by ten. The stopband attenuation reaches 60dB at twice the center frequency and at one-half the center frequency. The typical gain variation at the center frequency is ±0.5dB at 25°C and ±1.5dB over temperature. (Note that an additional ±0.4dB should be added to account for the gain variation due to the 1% resistors). If the operating temperature range is 25°C (±20°C) and the power supply voltage can be controlled to ±2%, the center frequency can be extended to 90kHz for a 5V supply or 160kHz for a 12V supply. Note that the gain error for center frequencies greater than 70kHz with a 5V supply and greater than 100kHz with a 12V supply increases from 1dB to 7dB. Therefore, the value of resistor R1 for each LTC1264 section should be increased to reduce the error to ±1dB (see the table in Figure 36.46).

If the power supply for this filter is a switching regulator, the regulator's output noise can appear at the filter's output if the center frequency of the filter is tuned to the noise frequency of the regulator. This is due to the filter's low power supply rejection near its center frequency. The LTC1264 is not a low power device. The typical quiescent current is 11mA with a 5V supply or 18mA with a 12V supply.

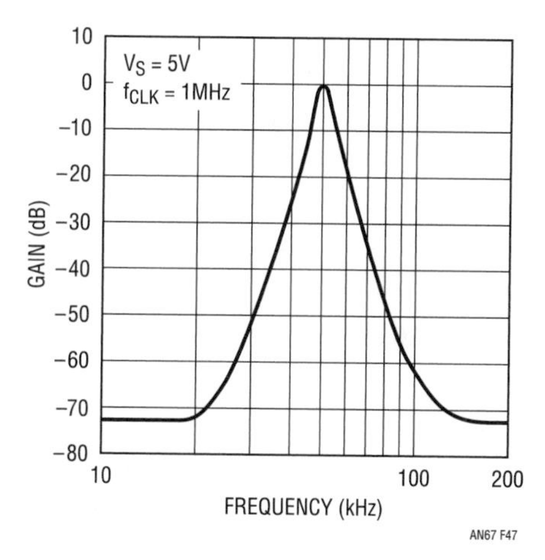

Figure 36.47 • LTC1264 Single 5V Supply, 50kHz Bandpass Response

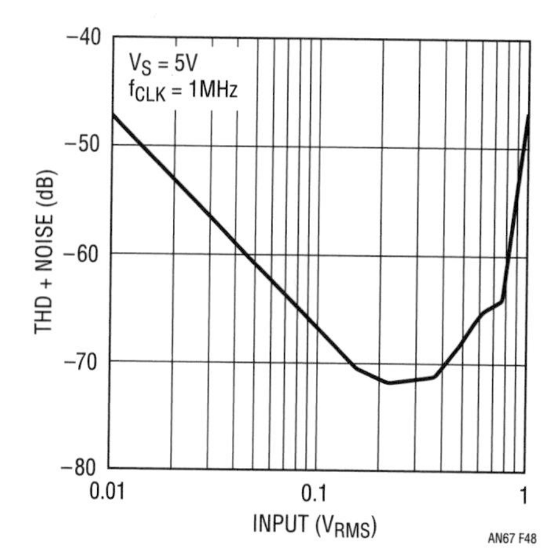

Figure 36.48 • Dynamic Range vs Input Signal. LTC1264 Single 5V Supply, 50kHz Bandpass Filter

A linear-phase bandpass filter for digital communications

by Philip Karantzalis

Bandpass filters with linear passband phase are useful for a variety of data-communications tasks, the most noteworthy of which may be in modulation-demodulation (modem) circuitry. Modems generate signals that must be processed without phase distortion to allow error free transmission and reception of information (or the closest approach to that ideal we can achieve).

Figure 36.49 shows a linear-phase bandpass filter using the LTC1264 high frequency, universal switched capacitor filter building block. This filter is an 8th order narrow bandpass filter, centered at 50kHz for a 1MHz clock input, with flat group delay in its passband. The f_{CLK}-to-f_{CENTER} frequency ratio is 20:1. Figure 36.50 shows the filter's narrowband gain response and Figure 36.51 shows the passband group delay. An interesting feature of linear-phase bandpass filters is that their response to a step input produces a short transient sine wave burst with a symmetrical envelope. Figure 36.52 shows a comparison of the transient responses to a step input for the linear-phase bandpass filter of Figure 36.49 and a bandpass filter with a similar

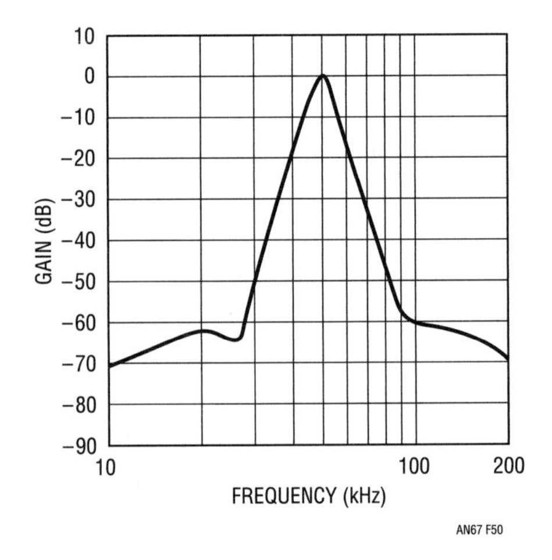

Figure 36.50 • Filter Gain vs Frequency

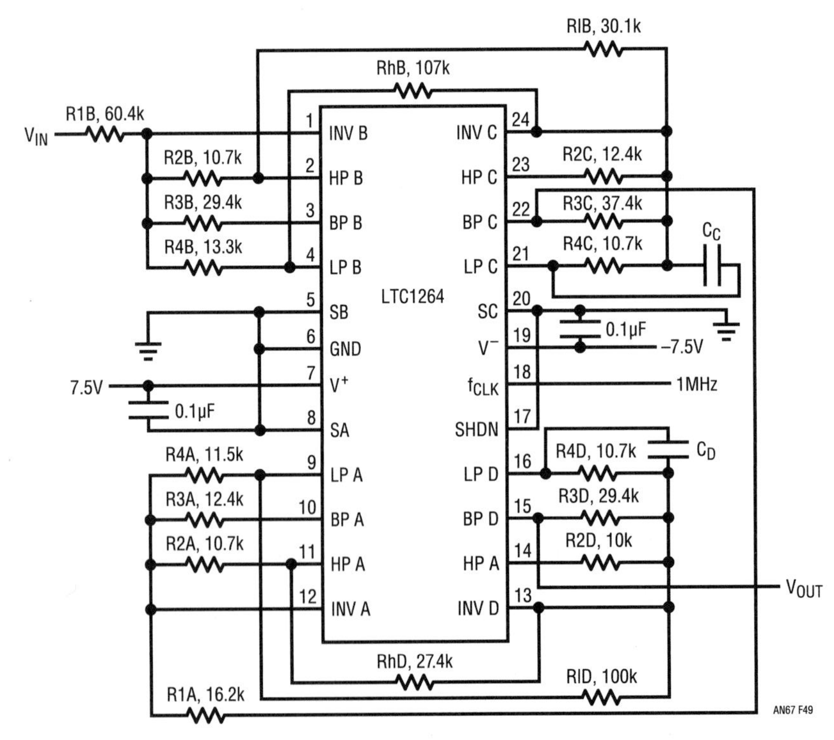

Figure 36.49 • LTC1264 Linear Phase, 8th Order Bandpass Filter

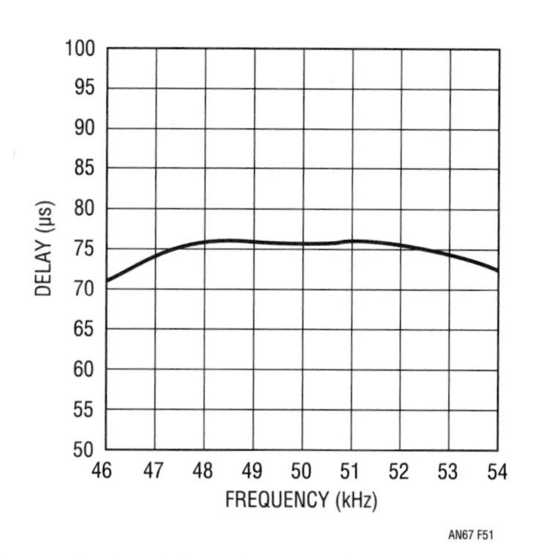

Figure 36.51 • Filter Group Delay vs Frequency

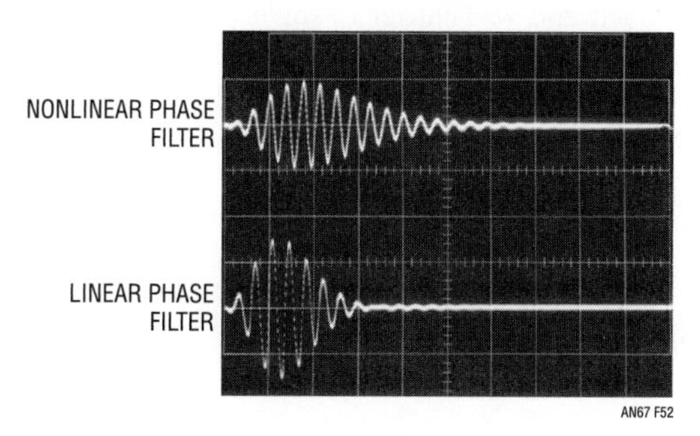

Figure 36.52 • Step Response

passband and nonlinear phase response. The response of a bandpass filter to a step input is a simple qualitative test for determining the linearity of its phase response, although in data transmission systems the measurements are usually made with eye diagrams and constellation displays.

The maximum clock frequency for the filter is 2MHz with ±7.5V supplies. This allows bandpass filters with center frequencies up to 100kHz to be realized without significant phase distortion in the passband.

Capacitor C, across R4 in sections C and D, minimizes gain and phase variations when the filter is used with clock frequencies greater than 1.4MHz. For ±5V supplies the maximum clock frequency is 1.6MHz. Use the Table 36.4 as a guide for the selection of capacitor C.

Table 36.4. Capacitor Selection Guide

Vs	f_{CLK}	$C_C = C_D$
±7.5V	1.8MHz	3pF
	2.0MHz	5pF
±5V	1.6MHz	5pF
	1.4MHz	3pF

Instrumentation

Wideband RMS noise meter

by Mitchell Lee

Recently, I needed to measure and optimize the wideband RMS noise of a power supply over about a 40MHz bandwidth. A quick calculation showed that the 12nV to $15nV/\sqrt{Hz}$ noise floor of my spectrum analyzer would come up short—my circuit was predicted to exhibit a spot noise of perhaps 8nV to $10nV/\sqrt{Hz}$. In fact, I didn't have a single instrument in my lab that would measure $50\mu V_{RMS}$ to $60\mu V_{RMS}$.

For the 40MHz bandwidth, the HP3403C RMS voltmeter is a good choice but its most sensitive range is 100mV about 66dB shy of my requirement. This obsolete instrument today carries a hefty price on the used market. The fact that here in the Silicon Valley HP3403Cs are a common sight at flea markets is of little consolation to most customers wishing to reproduce my measurements. We have several of these meters in the LTC design lab but they are in constant use and closely guarded by "The Keepers of the Secret RMS Knowledge." I resolved to build my own meter using an LT1088 thermal RMS converter.

Full scale on the LT1088 is $4.25V_{RMS}$. To measure $50\mu V$ full scale, I'd need an amplifier with a gain of 100,000. At 40MHz bandwidth, this didn't sound like it would have a good chance of working first time—built by hand and without benefit of a custom casting.

Rather than build a circuit with 40MHz bandwidth and a gain of 100dB, I decided to use just enough gain to put my desired noise performance around twice minimum scale. Aside from gain, this amplifier would also need less than $5nV/\sqrt{Hz}$ input noise, and the output stage would have to drive the 50Ω load presented by the LT1088.

It wasn't hard to find an appropriate output stage. The LT1206 (see Figure 36.53) can easily drive the required 120mA peak current into the LT1088 converter and there's plenty left over for handling noise spikes. To preserve 40MHz bandwidth, the LT1206 was set to run at a gain of 2.

The front end was harder to solve. I needed a low noise, high speed amplifier that could give me plenty of gain. Here I selected the LT1226. This is a 1GHz GBW op amp with only $2.6nV/\sqrt{Hz}$ input noise. It has a minimum stable gain of 25 but in this circuit high gain is an advantage.

Cascading two LT1226s on the front end gives a gain of 625, a little shy of the 5,000 to 10,000 required. Another gain of 5, plus the gain of 2 in the LT1206 adds up to a gain of 6,250—just about right.

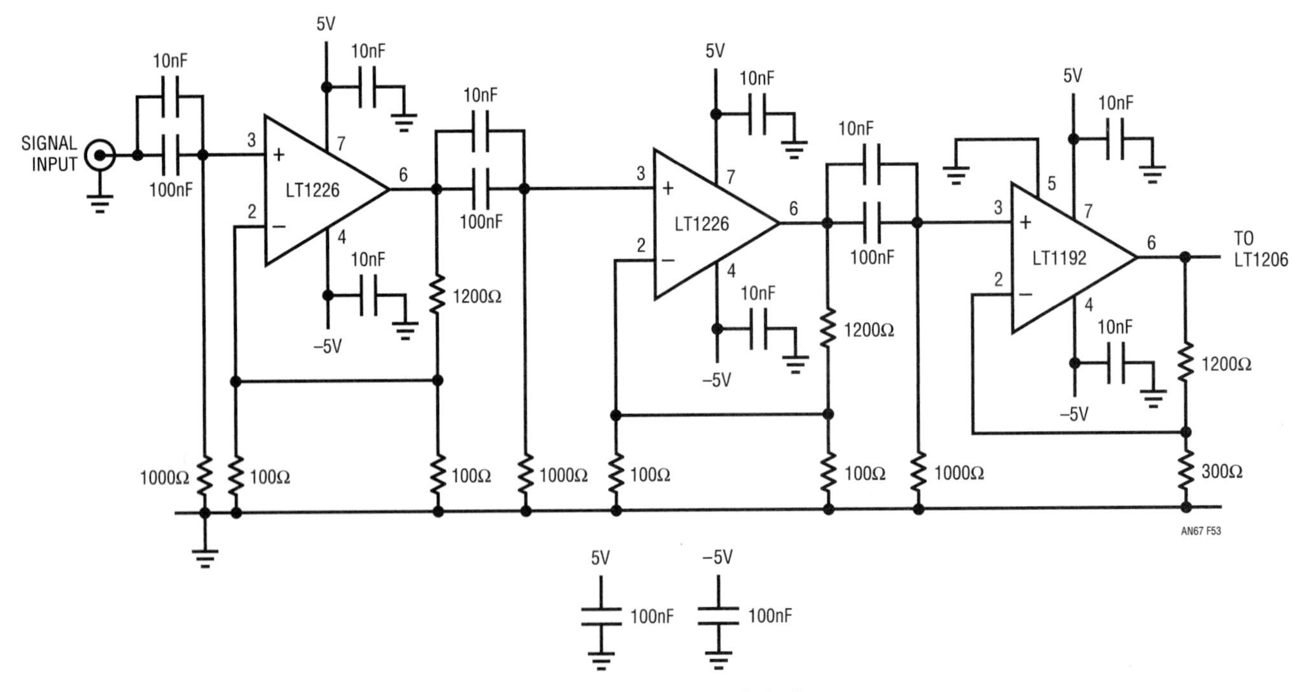

Figure 36.53 • Noise Meter Gain Stage

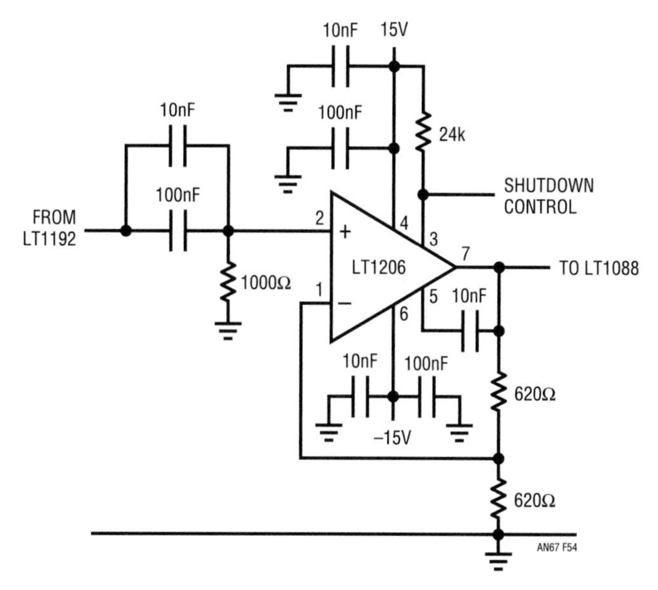

Figure 36.54 • LT1206 Buffer/Driver Section

There are several ways to get 40MHz bandwidth at a gain of 5, including the LT1223 and LT1227 current feedback amplifiers, but I settled on the LT1192 voltage amplifier because it is the lowest cost solution. This brings the gain up to 6250, for a minimum scale sensitivity of $34\mu V_{RMS}$ and a full-scale sensitivity of $680\mu V_{RMS}$.

My advice-filled coworkers assured me that there was no way I could build a wideband amplifier with a gain of 6250 and make it stable. Nevertheless, I built my amplifier on a 1.5" × 6" copperclad board, taking care to maintain a linear layout. The finished circuit was stable provided that a coaxial connection was made to the input. The amplifier was flat with 3dB points at 4kHz and 43MHz and some peaking at high frequencies.

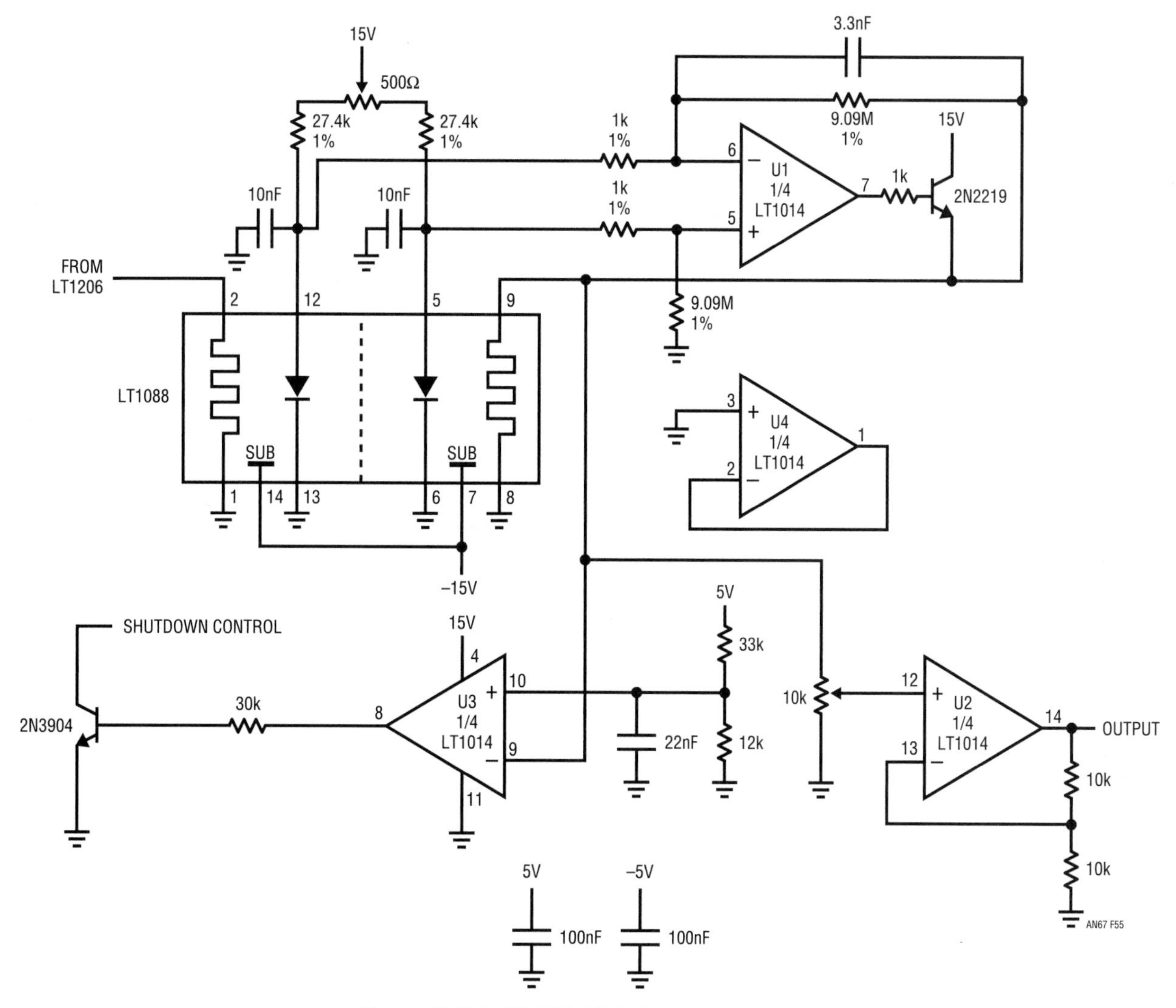

Figure 36.55 • LT1088 RMS Detector Section

Coaxial measurements

When measuring low level signals it is difficult to get a clean, accurate result. Scope probes have two problems. 10× probes attenuate the already small signal, and both 1× and 10× probes suffer from circuitous grounds. Coaxial adapters are a partial solution but these are expensive. They make for a lot of wear and tear on the probes and, without a little forethought, they can be a bear to attach to the circuit under test. My favorite way to get clean measurements of small signals is to directly attach a short length of coaxial cable as shown in Figure 36.56.

I use the good part of a damaged BNC cable, cutting away the shorter portion to leave at least 18" of RG-58/U and one good connector. At the cut, or as I call it, "real world" end of the cable, I unbraid, twist and tin a very small amount of outer conductor to form a stub 1/4" to 3/8" long. Next, I cut away the dielectric, exposing a similar length of center conductor, which I also tin. Now the probe is ready for use. It can be soldered directly to a circuit or breadboard, eliminating any lead length that might otherwise pick up stray noise, or worse, act as an antenna in a sensitive high gain circuit.

Small signals aren't the only beneficiaries of this technique. This works great for looking at ripple on the outputs of switching supplies. Ripple measurements are simplified because the large voltage swings associated with the switch node are completely isolated and no loop is formed where di/dt could inject magnetically coupled noise.

In some instances, I've found it important to retain the 50Ω termination impedance on the cable, but it is rarely possible to place a terminator at the "BNC" end of the cable, since this creates a DC path directly across the circuit under test. There is, however, another way as shown in Figure 36.57. Here, a technique known as back termination is used. No termination is used at the far end of the cable, but a 51Ω resistor is connected in series with the measurement end. Signals sent down the cable reach the BNC connector without attenuation, and fast edges that bounce off the unterminated end are absorbed back into the 51Ω source resistor. I've found this especially useful for measuring fast switch signals, or when measuring the RMS value of small signals, for ensuring that the amplifier input sees a properly terminated source. The resistor trick does not work if the node under test is high impedance; a FET probe is a better choice for high impedance measurements.

While I'm at it, I might as well give away my only other secret. We've all encountered ground loop problems, giving rise to 60Hz (50Hz for my friends overseas) injection into sensitive circuits. Every lab is replete with isolation transformers and "controlled substance" line cords with missing ground prongs for battling ground loops. There are similar problems at high frequencies, but the victim is wave fidelity, not AC pick up. To determine whether or not high frequency grounding, ground loops or common mode rejection is a problem for your oscilloscope, simply clamp a small ferrite E core around the probe lead while observing any effects on the waveform (see Figure 36.58). Sometimes the news is bad; the waveform really is messed up and there is some work to be done on the circuit. But occasionally the circuit is exonerated, the unexplainable aberrations disappear; proving that high frequency gremlins are at work. If necessary, several passes of the probe cable can be made through the E-Core and it can be taped together for as long as needed.

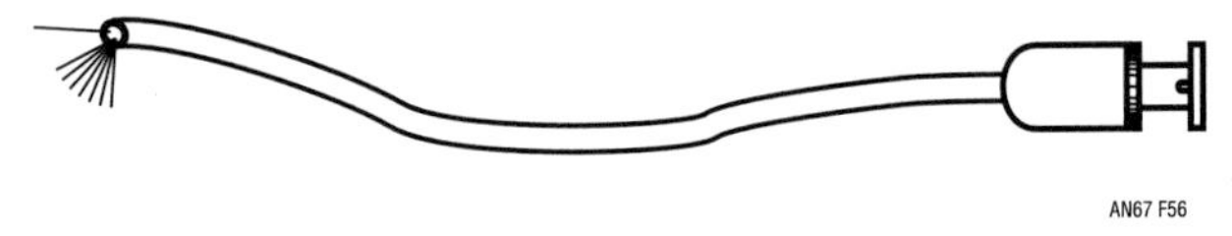

Figure 36.56 • BNC Cable Used as a "Probe"

Figure 36.57 • BNC Cable Probe Back Terminated

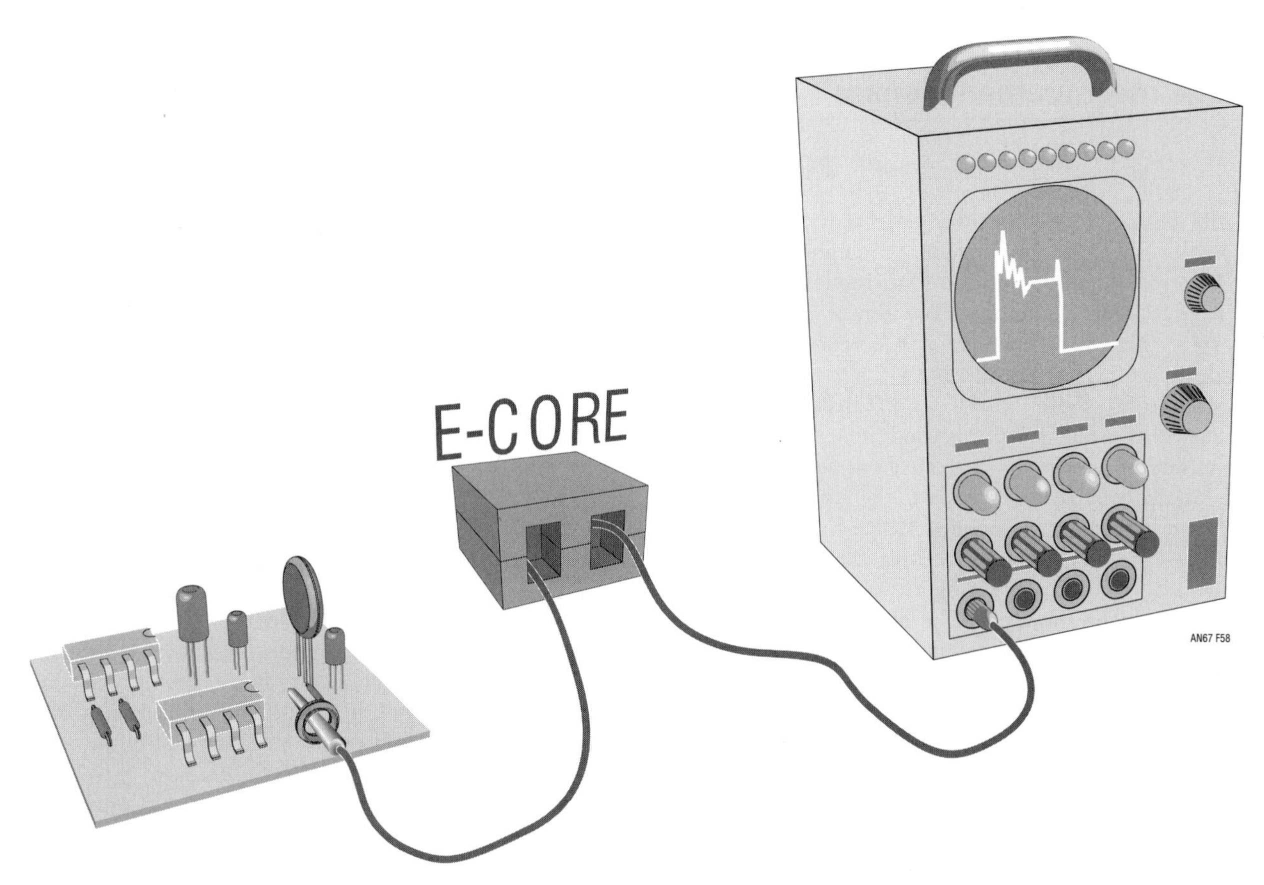

Figure 36.58 • An E-Core Serves to Attenuate High Frequency Common Mode Currents from a Scope Probe

The performance of the amplifier and thermal converter can be optimized by adjusting the value of the feedback and gain setting resistors around the LT1206. Slightly more bandwidth can be achieved at the expense of higher peaking by reducing the resistor values 10%. Reducing the resistor values will decrease peaking effects at the expense of bandwidth. A good compromise value is 680Ω.

I've shown the LT1226 amplifiers operating from ±5 supplies, which puts their bandwidth on the edge at 40MHz. Their bandwidth can be improved by operating at ±15V. Because the LT1206 operates on 15V rails, it is possible to overdrive the LT1088 and possibly cause permanent damage. One section of the LT1014 (U3) is used to sense an overdrive condition on the LT1088 and shut down the LT1206. Sensing the feedback heater instead of the input heater allows the LT1088 to accommodate high crest factor waveforms, shutting down only when the average input exceeds maximum ratings.

By the way, my power supply noise measured 200μV; filtering brought it down to less than 60μV.

LTC1392 micropower temperature and voltage measurement sensor

by Ricky Chow and Dave Dwelley

The LTC1392 is a micropower data acquisition system designed to measure temperature, on-chip supply voltage and differential rail-to-rail common mode voltage. The device incorporates a temperature sensor, a 10-bit A/D converter, a high accuracy bandgap reference and a 3-wire half-duplex serial interface.

Figure 36.59 shows a typical LTC1392 application. A single point "star" ground is used along with a ground plane to minimize errors in the voltage measurements. The power supply is bypassed directly to the ground plane with a 1μF tantalum capacitor in parallel with an 0.1μF ceramic capacitor.

The conversion time is set by the frequency of the signal applied to the CLK pin. The conversion starts when the $\overline{CS}$ pin goes low. The falling edge of $\overline{CS}$ signals the LTC1392 to wake up from micropower shutdown mode. After the LTC1392 recognizes the wake-up signal, it requires an additional 80μs delay for a temperature measurement, or a 10μs delay for a voltage measurement, followed by a 4-bit configuration word shifted into D_{IN} pin. This word configures the LTC1392 for the selected measurement and initiates the A/D conversion cycle. The D_{IN} pin is then disabled and the D_{OUT} pin switches from three-state mode to an active output. A null bit is then shifted out of the D_{OUT} pin on the falling edge of the CLK, followed by the result of the selected conversion. The output data can be formatted as an MSB-first sequence or as an MSB-first followed by an LSB-first sequence, providing easy interface to either LSB-first or MSB-first serial ports. The minimum conversion time for the LTC1392 is 142μs in temperature mode or 72μs in the voltage conversion modes, both at the maximum clock frequency of 250kHz.

Conclusion

The LTC1392 provides a versatile data acquisition and environmental monitoring system with an easy-to-use interface. Its low supply current, coupled with space saving SO-8 or PDIP packaging, makes the LTC1392 ideal for systems that require temperature, voltage and current measurement while minimizing space, power consumption and external component count. The combination of temperature and voltage measurement capability on one chip makes the LTC1392 unique in the market, providing the smallest, lowest power multifunction data acquisition system available.

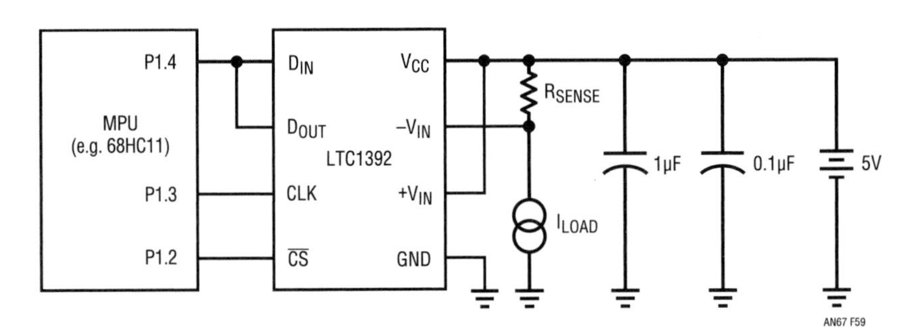

Figure 36.59 • Typical LTC1392 Application

Humidity sensor to data acquisition system interface

by Richard Markell

Introduction

It can be difficult to interface humidity sensors to data acquisition systems because of the sensors' drive requirements and their wide dynamic range. By carefully selecting the devices that comprise the analog front end, users can customize the circuit to meet their humidity sensing requirements and achieve reasonable accuracy throughout the chosen range. This article details the analog front end interface between a Phys-Chem Scientific Corp.[5] model EMD-2000 humidity sensor and a user selected (probably microprocessor-based) data acquisition system.

Design considerations

The Phys-Chem humidity sensor is a small, low cost, accurate resistance-type relative humidity (RH) sensor. This sensor has a well-defined stable response curve and can be replaced in circuit without system recalibration.

The design criteria call for a low cost, high precision analog front end that requires few calibration "tweaks" and operates on a single 5V supply. The sensor requires a square wave or sine wave excitation with no DC component. The sensor reactance varies over an extremely wide range (approximately 700Ω to 20MΩ). The wide dynamic range (approximately 90dB) required to obtain the full RH range of the sensor results in some challenges for the designer.

The circuit shown in the schematic features zero drift operational amplifiers (LTC1250 and LTC1050) and a precision instrumentation switched capacitor block (LTC1043). This design will maintain excellent DC accuracy down to microvolt levels. This method was chosen over the use of a true RMS-to-DC or log converter because of the expense and temperature sensitivity of these parts.

Circuit description

Figure 36.60 is a schematic diagram of the circuit. Only a single 5V power supply is required. Integrated circuit U1, an LTC1046, converts the 5V supply to −5V to supply power to U2, U3 and U4. U2A, part of an LTC1043 switched capacitor building block, provides the excitation for the sensor, switching between 5V and −5V at a rate of approximately 2.2kHz. This rate can be varied, but we recommended that it be kept below approximately 2.4kHz, which is one-half the auto zero rate of U3. We believe the

Note 5: Phys-Chem Scientific Corporation, 26 West 20th Street, New York, NY 10011. (212) 924-2070 Phone, (212) 243-7352 FAX

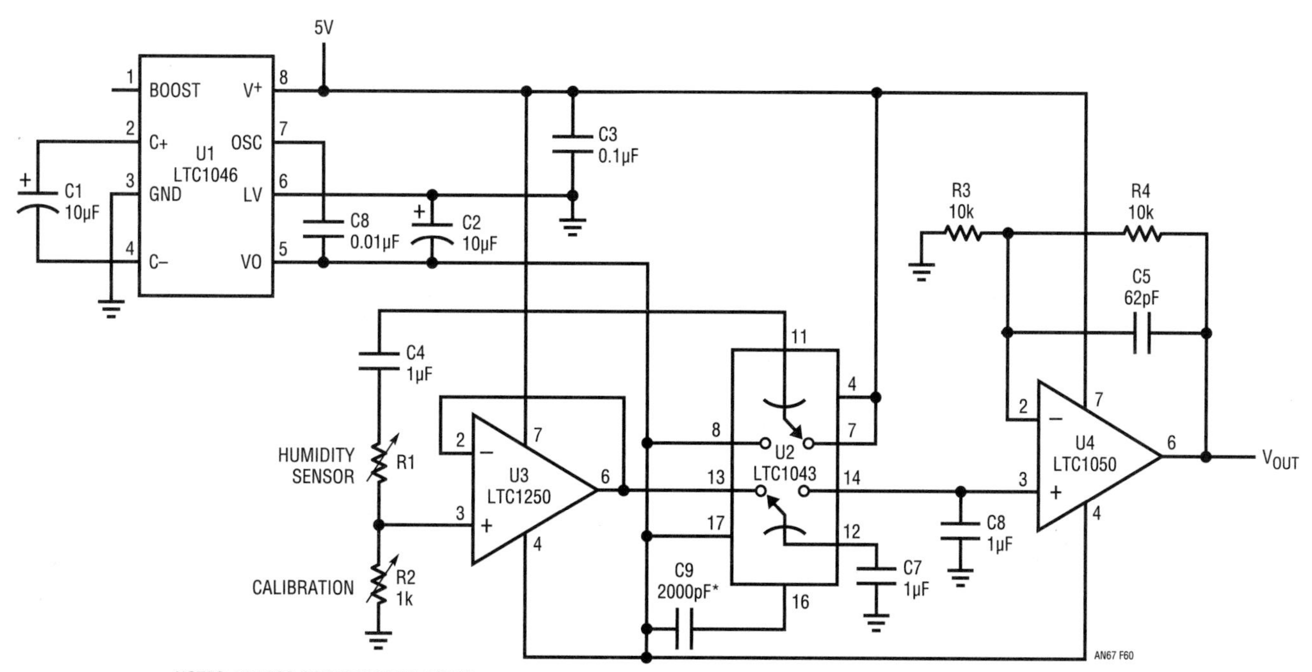

Figure 36.60 • Schematic Diagram of Humidity Sensor Circuit

deviation from the Phys-Chem response curves taken at 5kHz is insignificant.

Variable resistor R2 sets the full-scale output. Since the sensor resistance is 700Ω at approximately 90% humidity, setting R2 at 700Ω will provide a 2:1 voltage divider that, when combined with the gain of U4 (×2), results in an overall gain of one. U3 must be included in order for the circuit to function properly; otherwise C4 and C7 form a voltage divider that is dependent on the resistance of the RH sensor. U3 is a precision auto zero operational amplifier with an auto zero frequency of approximately 4.75kHz. U2B (the "lower" switch) samples the output of U3 and provides this sample to the input of U4. U4 is set to provide a gain of 2.

It is easy to digitize the output of U4. Figure 36.61 is the schematic of a 12-bit converter that can be used for this purpose. The range of humidity that can be sensed depends on the resolution of the converter. The full-scale output (which is equivalent to approximately 90% humidity) is essentially independent of the number of bits in the A/D converter, but the dry (low RH) end of the scale is dependent on the A/D resolution. As an example, the above referenced 12-bit converter will process humidity signals that translate to approximately 20% RH, since the voltage output at this humidity is approximately 2.3mV while 0.5LSB is 1.2mV Digitization down to 10% RH requires the conversion of 350μV signals or a 16-bit converter. From a cost standpoint this seems unwieldy. It is much more economical to use a 2-channel 12-bit converter that changes ranges somewhere in the humidity range.

All of the above solutions measure output voltage from a voltage divider consisting of the RH sensor and a fixed "calibration" resistor. The resistance of the sensor at a fixed output voltage can be calculated from the formula

$$R(\Omega) = \frac{R2V_{FULLSCALE}}{V_{OUT}/2} - R2$$

In this case, if R2 is set to 700Ω and $V_{FULL\ SCALE} =$ 5.00V, then

$$R(\Omega) = \frac{3500}{V_{OUT/2}} - 700$$

Once R is calculated (probably by the microprocessor), the humidity can be calculated from the quadratic approximation in the Phys-Chem literature:

$$RH = \frac{LnR - 13.95 - \sqrt{(13.95 - LnR)^2 + 24.288}}{-0.184}$$

If a suitable humidity chamber is not available, the sensor can be removed and fixed resistors substituted. The circuit should then be calibrated from the EMD-2000 "typical response curve." This should provide approximately 2% accuracy.

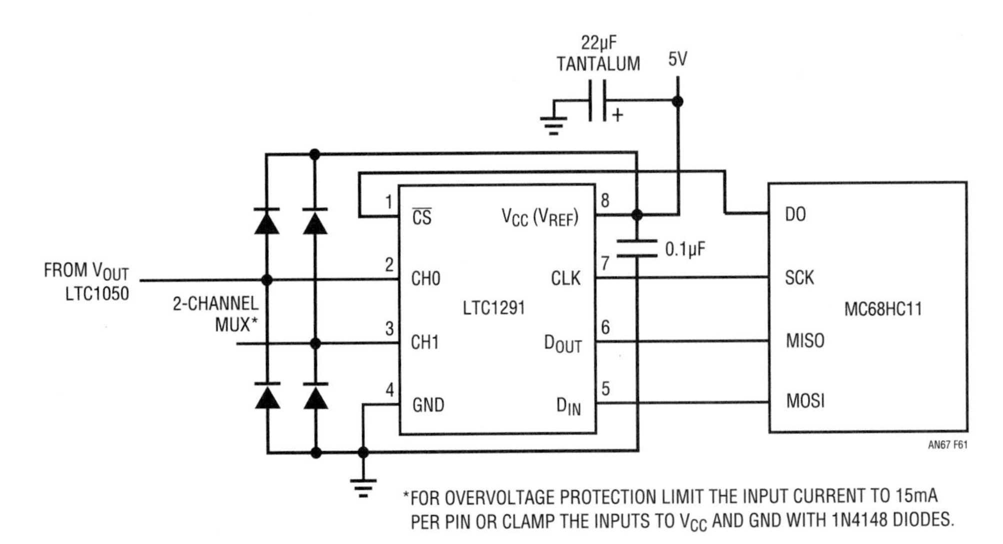

Figure 36.61 • LTC1291 12-Bit A/D Converter Interfaced to MC68HC11

A single cell barometer

by Jim Williams and Steve Pietkiewicz

Figure 36.62, a complete barometric pressure signal conditioner, operates from a single 1.5V battery. Until recently, high accuracy and stability have been obtainable only with bonded strain gauge and capacitively-based transducers, which are quite expensive. This design, using a recently introduced semiconductor transducer, achieves 0.01"Hg (inches of mercury) uncertainty over time and temperature. The 1.5V powered operation permits portable application.

The 6kΩ transducer (T1) requires precisely 1.5mA of excitation, necessitating a relatively high voltage drive. A1's positive input senses T1's current by monitoring the voltage drop across the resistor string in T1's return path. A1's negative input is fixed by the 1.2V LT1004 reference. A1's output biases the 1.5V powered LT1110 switching regulator. The LT1110's switching produces two outputs from L1. Pin 4's rectified and filtered output powers A1 and T1. A1's output, in turn, closes a feedback loop at the regulator. This loop generates whatever voltage step-up is required to force precisely 1.5mA through T1. This arrangement provides the required high voltage drive while minimizing power consumption. This occurs because the switching regulator produces only enough voltage to satisfy T1's current requirements.

L1 Pins 1 and 2 source a boosted, fully floating voltage, which is rectified and filtered. This potential powers A2. Because A2 floats with respect to T1, it can look differentially across T1's outputs, Pins 10 and 4. In practice, Pin 10 becomes "ground" and A2 measures Pin 4's output with respect to this point. A2's gain scaled output is the circuit's output, conveniently scaled at 3.000V=30.00"Hg.

To calibrate the circuit, adjust R1 for 150mV across the 100Ω resistor in T1's return path. This sets T1's current to the manufacturer's specified calibration point. Next, adjust R2 at a scale factor of 3.000V = 30.00"Hg. If R2 cannot capture the calibration, reselect the 200k resistor in series with it. If a pressure standard is not available, the transducer is supplied with individual calibration data permitting circuit calibration.

This circuit, compared to a high order pressure standard, maintained 0.01"Hg accuracy over months with widely varying ambient pressure shifts. Changes in pressure, particularly rapid ones, correlated quite nicely to changing weather conditions. Additionally, because 0.01"Hg corresponds to about 10 feet of altitude at sea level, driving over hills and freeway overpasses becomes quite interesting. The circuit pulls 14mA from the battery, allowing about 250 hours operation from one D cell.

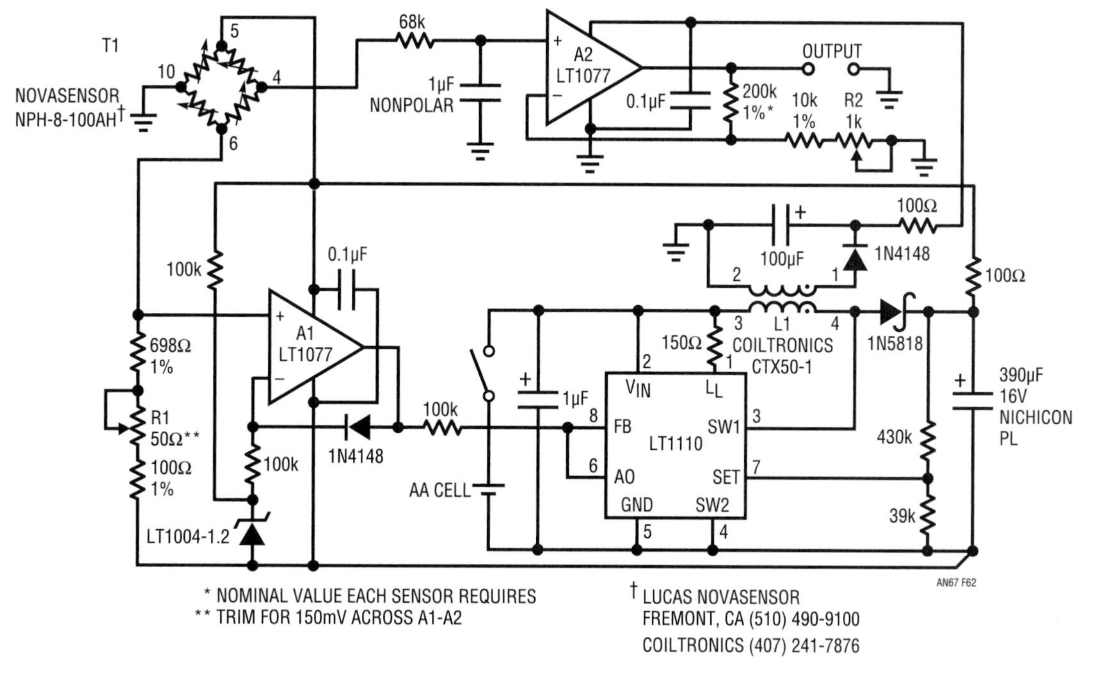

Figure 36.62 • Single Cell Barometer Schematic Diagram

Noise generators for multiple uses

A broadband random noise generator

by Jim Williams

Filter, audio and RF communications testing often require a random noise source. Figure 36.63's circuit provides an RMS amplitude regulated noise source with selectable bandwidth. RMS output is 300mV with a 1kHz to 5MHz bandwidth, selectable in decade ranges.

Noise source D1 is AC coupled to A2, which provides a broadband gain of 100. A2's output feeds a gain control stage via a simple, selectable lowpass filter. The filter's output is applied to A3, an LT1228 operational transconductance amplifier. A1's output feeds LT1228 A4, a current feedback amplifier. A4's output, which is also the circuit's output, is sampled by the A5-based gain control configuration. This closes a gain control loop to A3. A3's I_{SET} current controls gain, allowing overall output level control.

Figure 36.64 plots noise at a 1MHz bandpass, whereas Figure 36.65 shows RMS noise versus frequency in the same bandpass. Figure 36.66 plots similar information at full bandwidth (5MHz). RMS output is essentially flat to 1.5MHz, with about ±2dB control to 5MHz before sagging badly.

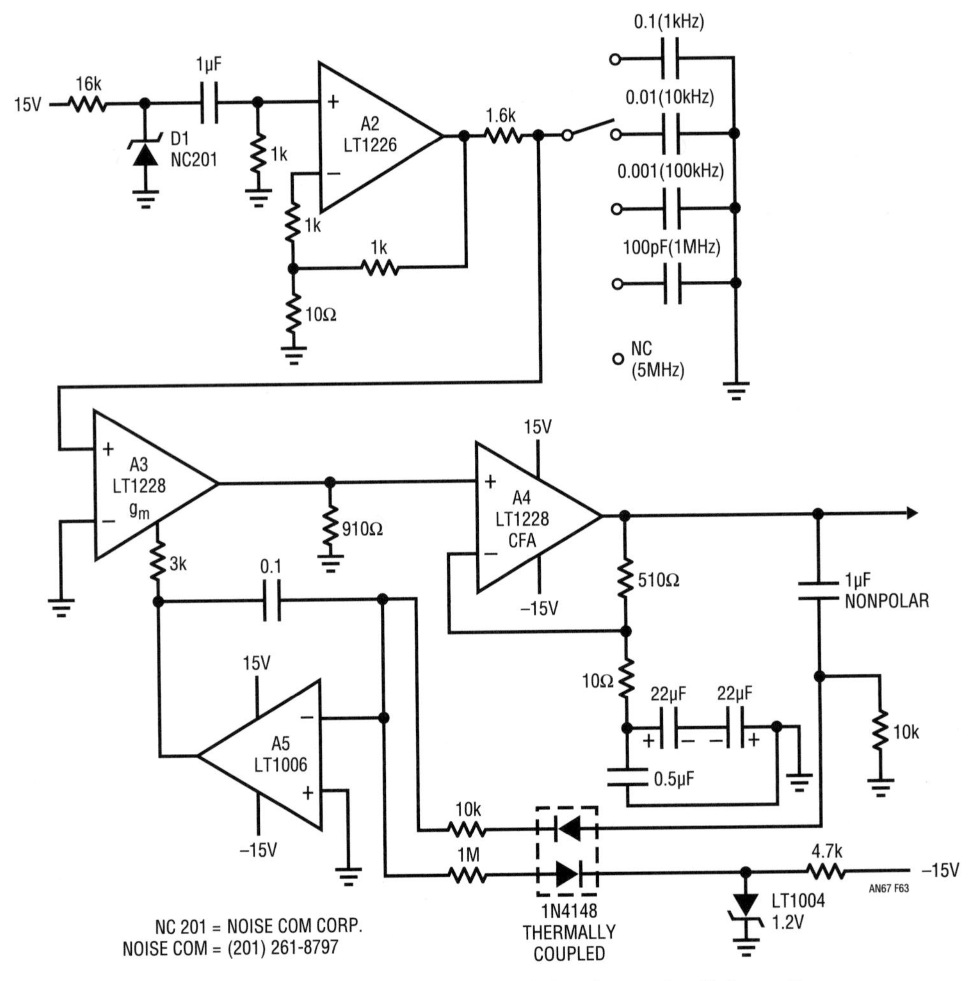

Figure 36.63 • Broadband Random Noise Generator Schematic

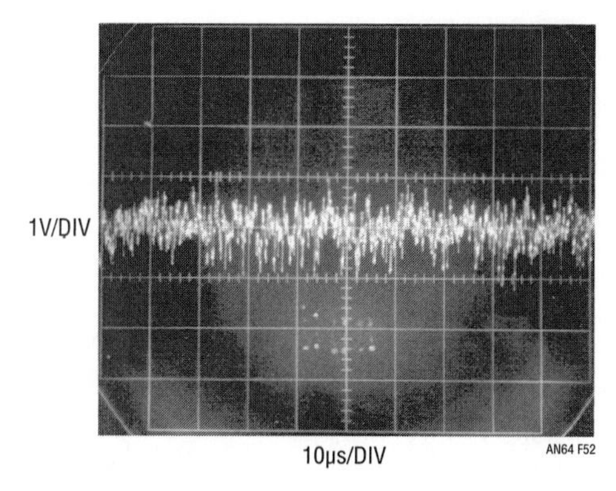

Figure 36.64 • Figure 36.63's Output In the 1MHz Filter Position

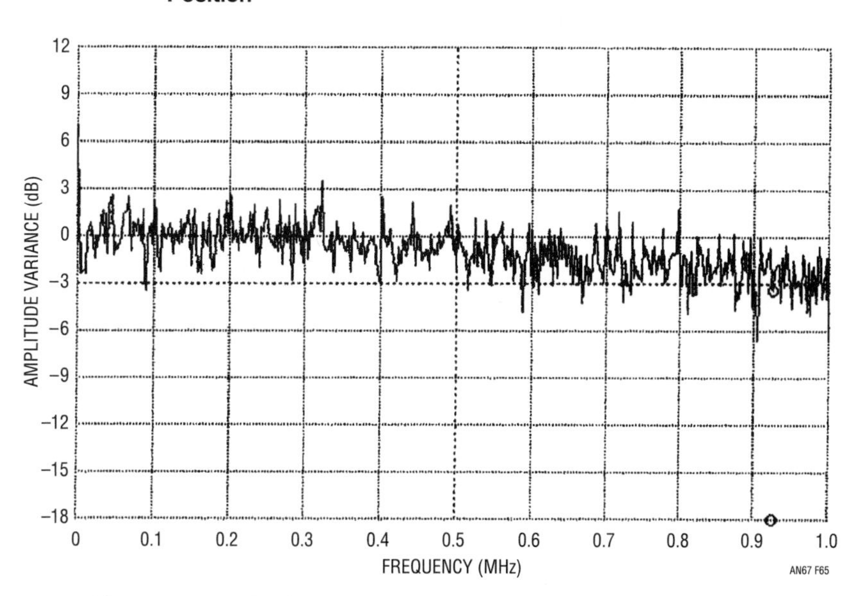

Figure 36.65 • RMS Noise vs Frequency at 1MHz Bandpass

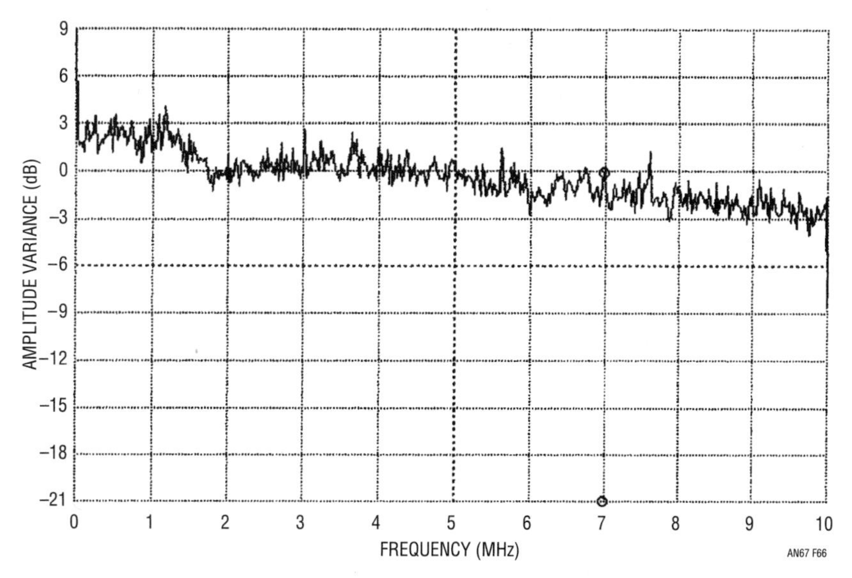

Figure 36.66 • RMS Noise vs Frequency at 5MHz Bandpass

Symmetrical white Gaussian noise

by Bent Hessen-Schmidt, NOISE COM, INC.

White noise provides instantaneous coverage of all frequencies within a band of interest with a very flat output spectrum. This makes it useful both as a broadband stimulus and as a power level reference.

Symmetrical white Gaussian noise is naturally generated in resistors. The noise in resistors is due to vibrations of the conducting electrons and holes as described by Johnson and Nyquist.[6,7] The distribution of the noise voltage is symmetrically Gaussian, and the average noise voltage is:

$$\bar{V}_n = 2\sqrt{kTfR(f)p(f)df} \qquad (1)$$

where:
K = 1.38E-23 J/K (Boltzmann's constant)
T = tempetrature of the resistor in Kelvin
f = frequency in Hz
h = 6.62E-34 Js (Planck's constant)
R(f) = resistance in ohms as a function of frequency

$$P(f) = \frac{hf}{kT[\exp(hf/kT) - 1]} \qquad (2)$$

p(f) is close to unity for frequencies below 40GHz when T is equal to 290°K. The resistance is often assumed to be independent of frequency, and *f*df is equal to the noise bandwidth (B). The available noise power is obtained when the load is a conjugate match to the resistor, and it is:

$$N = \frac{\bar{V}_{n^2}}{4R} = kTB \qquad (3)$$

where the "4" results from the fact that only half of the noise voltage and hence only 1/4 of the noise power is delivered to a matched load.

Equation 3 shows that the available noise power is proportional to the temperature of the resistor; thus it is often called thermal noise power. Equation 3 also shows that white noise power is proportional to the bandwidth. An important source of symmetrical white Gaussian noise is the noise diode. A good noise diode generates a high level of symmetrical white Gaussian noise. The level is often specified in terms of excess noise ratio (ENR).

$$\text{ENR(in dB)} = 10\text{Log}\frac{(Te - 290)}{290} \qquad (4)$$

Te is the physical temperature that a load (with the same impedance as the noise diode) must be at to generate the same amount of noise.

The ENR expresses how many times the effective noise power delivered to a nonemitting, nonreflecting load exceeds the noise power available from a load held at the reference temperature of 290°K (16.8°C or 62.3°F).

The importance of a high ENR becomes obvious when the noise is amplified, because the noise contributions of the amplifier may be disregarded when the ENR is 17dB larger than the noise figure of the amplifier (the difference in total noise power is then less than 0.1dB). The ENR can easily be converted to noise spectral density in dBm/Hz or $\mu V/\sqrt{Hz}$ by use of the white noise conversion formulas in Table 36.5.

Table 36.5. Useful White Noise Conversion
dBm = dBm/Hz + 10log(BW)
dBm = 20log($\bar{V}$n) − 10log(R) + 30dB
dBm = 20log($\bar{V}$n) + 13dB, for R = 50Ω
dBm/Hz = 20log(μVn$\sqrt{Hz}$ − 10log(R) − 90dB
dBm/Hz = −174dBm/Hz + ENR, for ENR > 17dB

When amplifying noise it is important to remember that the noise voltage has a Gaussian distribution. The peak voltages of noise are therefore much larger than the average or RMS voltage. The ratio of peak voltage to RMS voltage is called crest factor, and a good crest factor for Gaussian noise is between 5:1 and 10:1 (14dB to 20dB). An amplifier's 1dB gain compression point should therefore be typically 20dB larger than the desired average noise output power to avoid clipping of the noise.

For more information about noise diodes, please contact NOISE COM, INC. at (201) 261-8797.

Note 6: Johnson, J.B. "Thermal Agitation of Electricity in Conductors," Physical Review, July 1928, pp. 97-109.

Note 7: Nyquist, H. "Thermal Agitation of Electric Charge in Conductors," Physical Review, July 1928, pp. 110-113.

Noise generators for multiple uses

A diode noise generator for "eye diagram" testing

by Richard Markell

The circuit that Jim Williams describes evolved from my desire to build a circuit for testing communications channels by means of "eye diagrams." (See Linear Technology, Volume I, Number 2 for a short explanation of the eye diagram.) I wanted to replace my pseudorandom code generator circuit, which used a PROM, with a more "analog" design—one that more people could build without specialized components. What evolved was a noise source sampled by a very fast comparator (see Figure 36.67). The comparator outputs a random pattern of 1's and 0's.

The noise diode (an NC201) is filtered and amplified by the LT1190 high speed operational amplifier (U1). The output feeds the LT1116 (U2), a 12ns single supply, ground-sensing comparator. The 2kΩ pot at the inverting input of the LT1116 sets the threshold to the comparator so that a quasiequal number of 1's and 0's are output. U3 latches the output from U2 so that the output from the comparator remains latched throughout one clock period. The two-level output is taken from U3's Q0 output.

The additional circuitry shown in the schematic diagram allows the circuit to output four-level data for PAM (pulse amplitude modulation) testing. The random data from the two-level output is input to a shift register, which is reset on every fourth clock pulse. The output from the shift register is weighted by the three 5k resistors and summed into the LT1220 operational amplifier from which the output is taken. The filter network between the 74HC74 output and the 74HC4094 strobe input is necessary to ensure that the output data is correct.

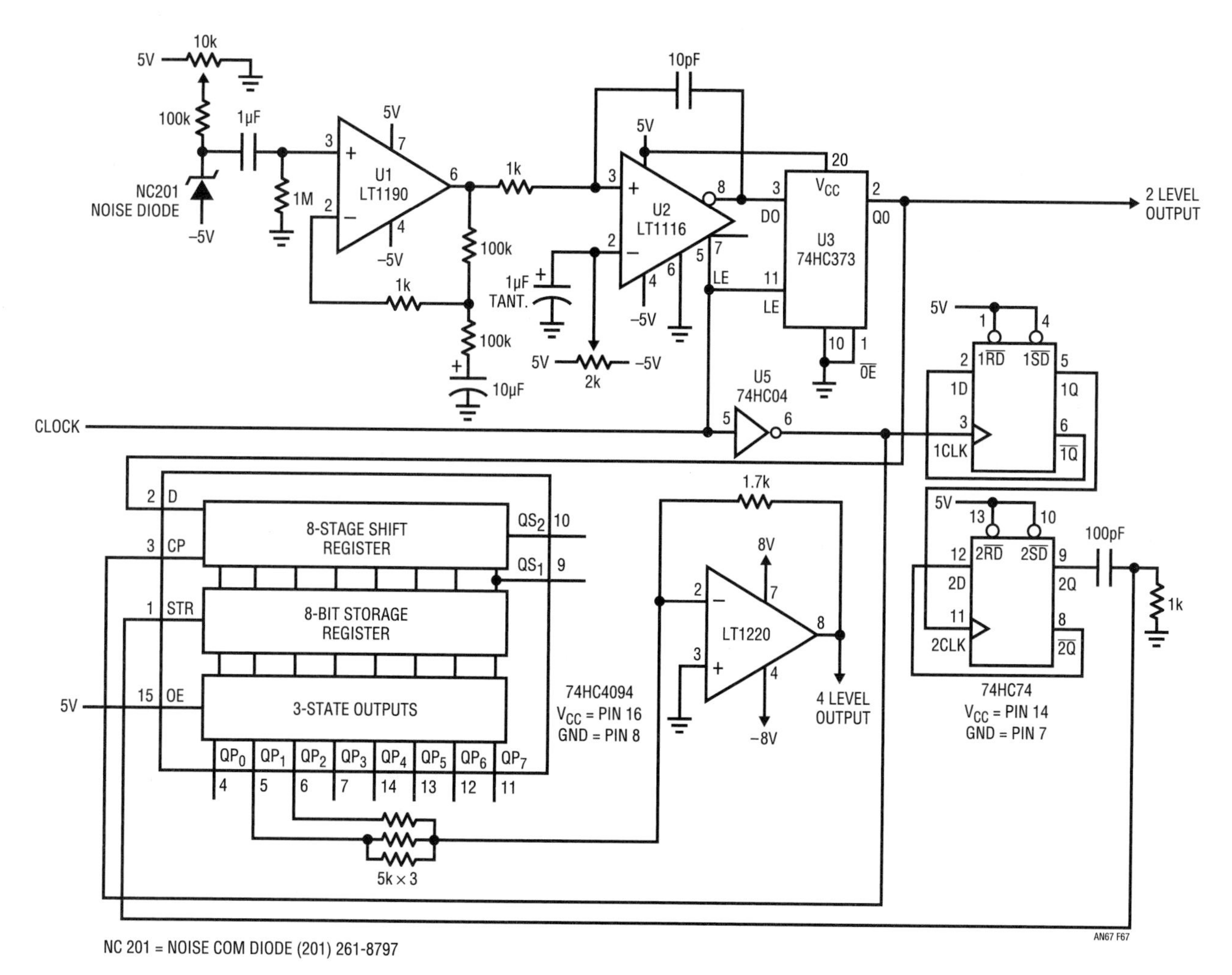

Figure 36.67 • Pseudorandom Code Generator Schematic Diagram

Video/op amps

LT1251 circuit smoothly fades video to black

by Frank Cox

When a video signal is attenuated, there is a point were the sync amplitude is too small for a monitor to process properly. Instead of making a smooth transition to black, the picture rolls and tears. One solution to this problem is to run a separate sync signal into the monitor. This may not be a viable solution in a system where cost and complexity are the prime concerns. What is needed is a simple video "volume control."

The circuit in Figure 36.68 can perform a smooth fade to black, while maintaining good video fidelity. U1, an LT1360 op amp, and its associated components form an elementary sync separator. C1, R1 and D1 clamp the composite video. D2 biases the input of U1 to compensate for the drop across D1. When D1 conducts, the most negative portion of the waveform containing the sync information is amplified by U1. The clamp circuit in the feedback network of U1 (D4 to D8) prevents the amplifier from saturating. D3 and the CMOS inverter U4 complete the shaping of the sync waveform. This sync separator works with most video signals but, because of its simplicity, will not work with very noisy or distorted video. The remainder of the circuit is an LT1251 video fader (U2) configured to fade between the original video and the sync stripped from that video. Thus, the video fades to black.

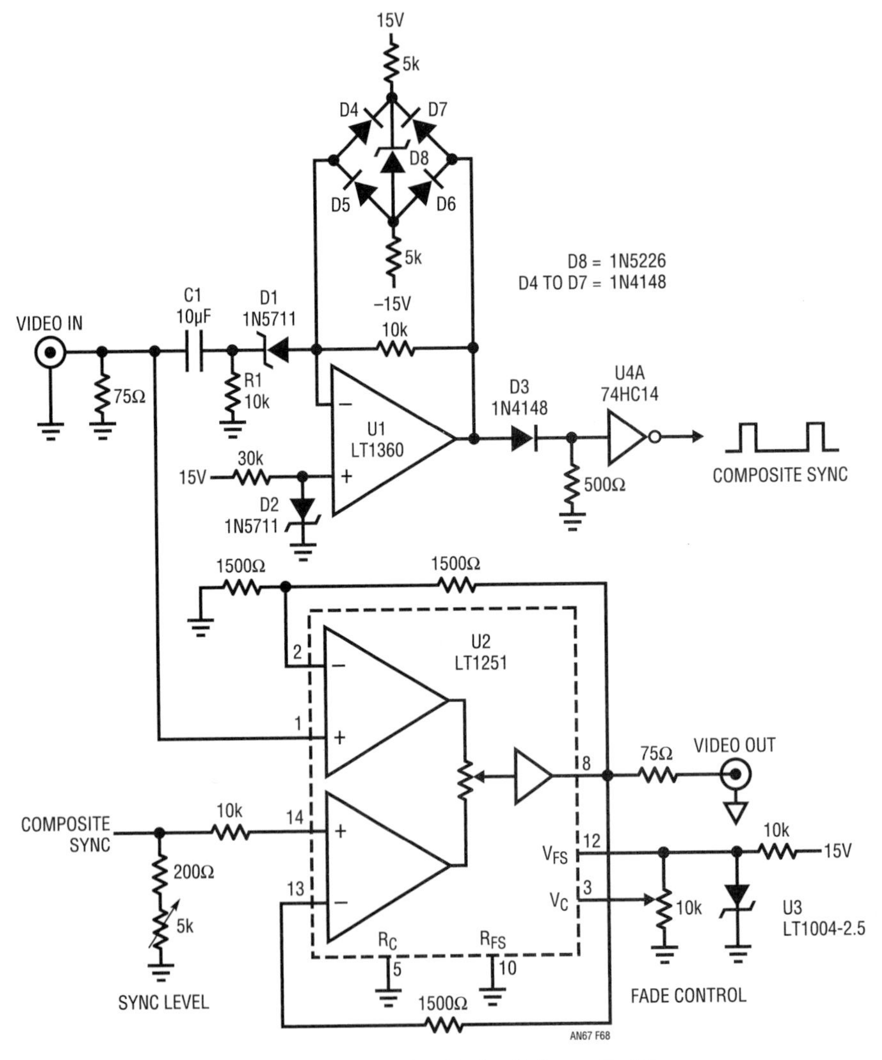

Figure 36.68 • LT1251 Video Fader Schematic Diagram

The control voltage for the fader is generated by a voltage reference and a 10k variable resistor. If this control potentiometer is mounted an appreciable distance from the circuit or if the control generates any noise when adjusted, this node should be bypassed.

Figure 36.69 is a multiple exposure waveform photograph that shows the action of this circuit. Two linear ramp video test signals are shown in this photograph. The video is faded from full amplitude to zero amplitude in six steps. The sync waveform (lower center) remains unchanged. In Figure 36.70, a single video line modulated with color subcarrier is faded from full video amplitude to zero video amplitude. The monitor will eventually lose color lock and shut the color off as the amplitude of the color subcarrier is reduced. This is not a problem in this application because the color decoding circuits in the monitor are designed to work with a variety of signals from tape or broadcast, and so have a large dynamic range. Color portions of the picture will remain after the luminance portion is completely black.

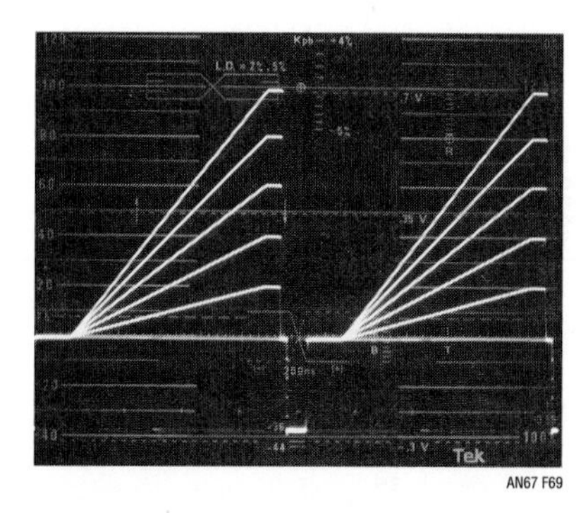

Figure 36.69 • Multiple Exposure Photo Showing Circuit Operation

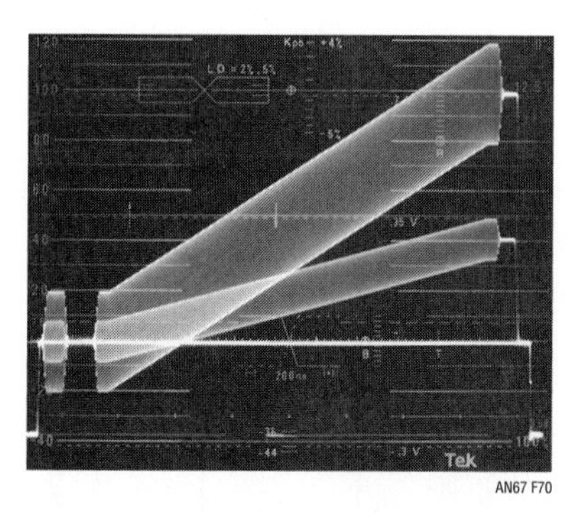

Figure 36.70 • Photo Detailing a Single Video Line with Color Subcarrier Faded to Zero Amplitude

Luma keying with the LT1203 video multiplexer

by Frank Cox

In video systems, the action of switching between two or more active video sources is referred to as a "wipe" or a "key." When the decision to switch video sources is based on an attribute of the active video itself, the action is called keying. A wipe is controlled by a nonvideo signal such as a ramp. The circuit presented in Figure 36.71 is referred to as a "luma key" because it switches between two sources when the luminance ("luma") of a monochrome key signal reaches a set level. It is also possible to key on the color of the video source and this, not surprisingly, is called "chroma keying."

Figure 36.71's operation is very straightforward. A monochrome video source is used to generate the key signal. The LT1363 is used as a buffer and may not be needed in all applications. If the key signal is to be used as one of the switched signals, it is convenient to "loop through" the input of this buffer. The LT1016 comparator switches when the video level exceeds the DC reference on its inverting input, which is controlled by the "key sensitivity" control. The TTL key signal controls an LT1203 video multiplexer. Any two video sources may be connected to the inputs of the LT1203, as long as they are gen locked and within the common mode range (on ±5V supplies this is ±3V over 0°C to 70°C) of the multiplexer The LT1203's fast switching speed, low offset and clean switching make it a natural for an active video switching application like this one. Composite color signals can be used, but the best results will be obtained if the key signal's horizontal sync is phase coherent with the color reference of the sources. The key source video should be monochrome to prevent the key comparator from switching on the color subcarrier.

Nonstandard video signals can be used for the inputs to the LT1203. For instance, it is possible to select between two DC input levels to construct a two-level image. Figure 36.72 is an example of an image constructed this way. A monochrome video signal is sliced and used to key between black (0V) and gray (approximately 0.5V) to generate this image of a famous linear IC designer. An image formed in this way is not a standard video output until the blanking and sync intervals are reconstructed. The second LT1203 blanks the video and an LT1363 circuit sums composite sync to the video and drives a cable. For more information on this part of the circuit, see AN57, page 7. A clamp is not used since the DC levels are arbitrarily set by the inputs, but one could be used, as in the figure on page 7 of AN57, if the sources were video. As another option, Figure 36.73 shows the same key signal used as one of the inputs to the multiplexer.

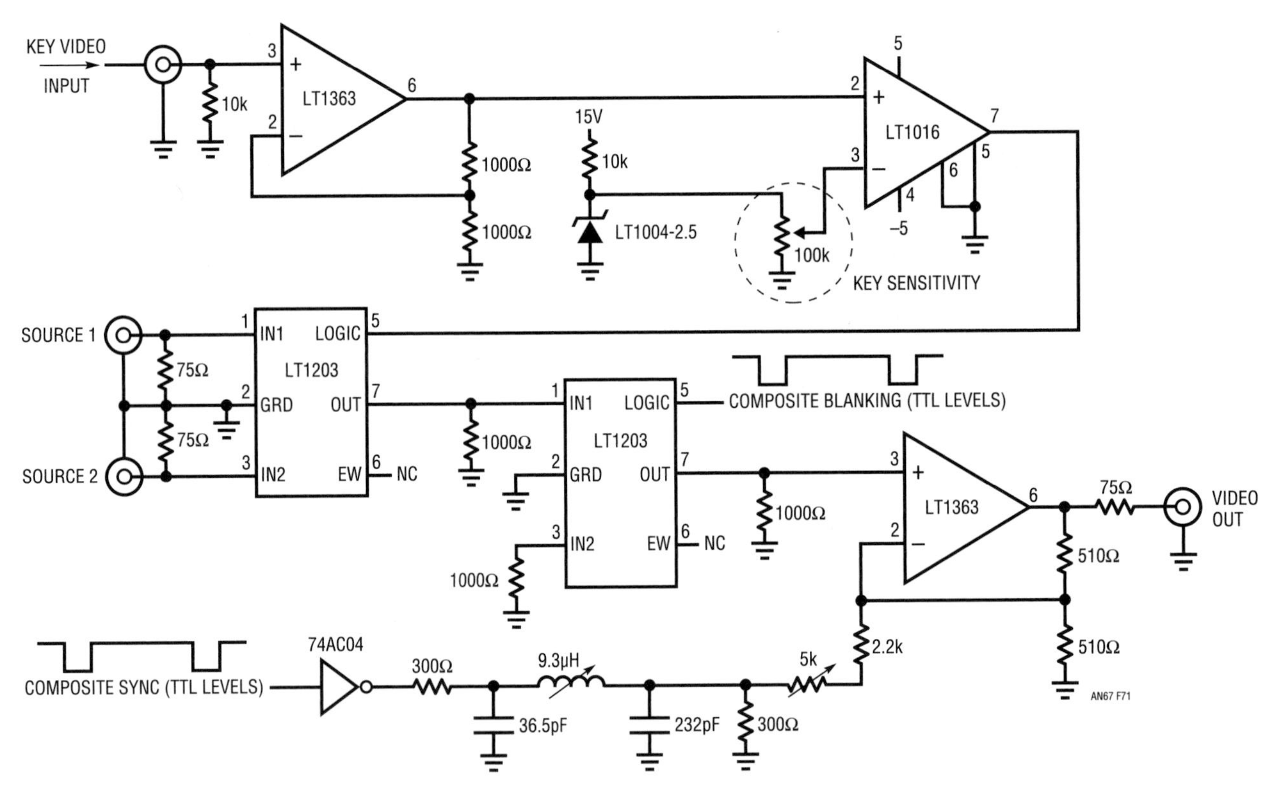

Figure 36.71 • Luma Keyer Schematic Diagram

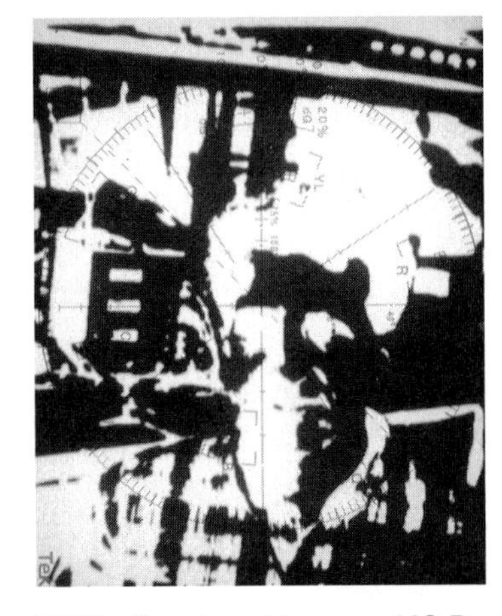

Figure 36.72 • Two Level Image of IC Designer

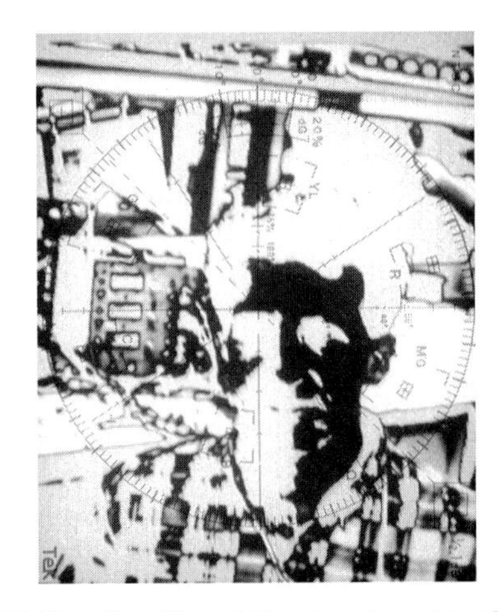

Figure 36.73 • Key Signal Used as Input to the MUX

LT1251/LT1256 video fader and DC gain controlled amplifier

by William H. Gross

The video fader

Figure 36.74 shows the LT1251/LT1256 configured as a fader with unity gain. A full-scale voltage of 2.5V is applied to Pin 12 and the control input drives Pin 3.

Figure 36.75 shows the true response of the control path. The control path is fast enough for quick switching between signals, as when keying on a color or luminance level. The control path introduces only a small (50mV), short (50ns) glitch when switched quickly.

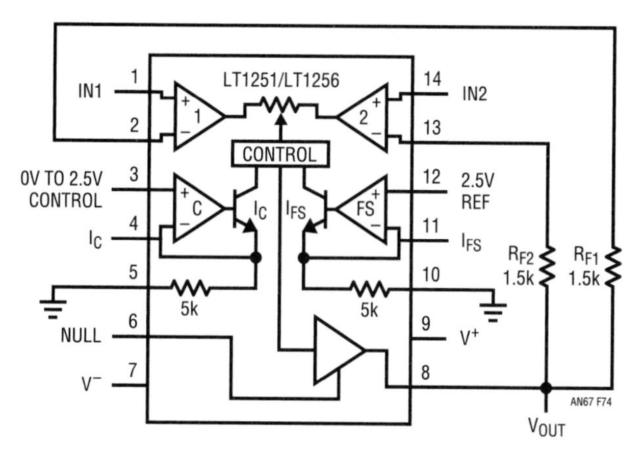

Figure 36.74 • Two Input Video Fader

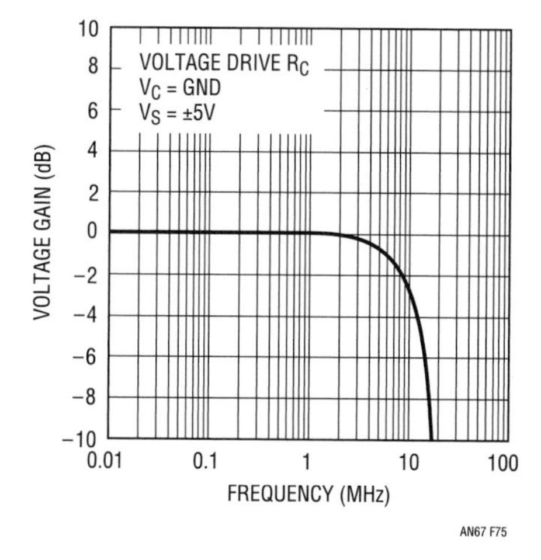

Figure 36.75 • LT1251/56 Control Path Bandwidth

A summary of the LT1251/LT1256 performance operating on ±5V supplies in the configuration shown in Figure 36.74 is given in Table 36.6.

Table 36.6. LT1251/LT1256 Performance Summary

Parameter	Value
Slew Rate (at ±2V R_L = 150Ω)	300V/μs
Full-Power Bandwidth (1V_{RMS})	30MHz
Small-Signal Bandwidth	30MHz
Differential Gain (NTSC, R_L = 150Ω)	0.1%
Differential Phase (NTSC, R_L = 150Ω)	0.1°
Total Harmonic Distortion (1kHz, K = 1)	0.001%
(1kHz, K = 0.5)	0.01%
(1kHz, K = 0.1)	0.4%
Rise Time, Fall Time	11ns
Overshoot	3%
Propagation Delay	10ns
Settling Time (0.1%, Vo = 2V)	65ns
Quiescent Supply Current	13.5mA

Applications

Grounding IN2 of the LT1256 in Figure 36.74 results in a 2-quadrant multiplier. Figure 36.76 shows the 2-quadrant multiplier being used as an AM modulator. The output will deliver 10dBm into 50Ω. The LT1077 op amp senses the LT1256 output DC and drives the Null pin, eliminating any DC at the output. The Null pin voltage is nominally 100mV above the negative supply and therefore the op amp output must be able to swing within a few millivolts of the negative supply. Without the LT1077, the worst-case DC output voltage is 50mV.

By operating one input stage in an inverting configuration and the other in a noninverting configuration and driving both inputs, the LT1256 becomes a 4-quadrant multiplier. Figure 36.77 shows the 4-quadrant multiplier being used as a double-sideband, suppressed-carrier modulator. The LT1077 DC output nulling circuit could be added if necessary.

The LT1251/LT1256 can be used to implement numerous other functions, including voltage controlled filters, phase shifters and oscillators. Squaring and limiting circuits can be designed by feeding the output or input into the Control pins. Gamma correction and other compression circuits are created in a similar manner. The applications are limited only by the designer's imagination.

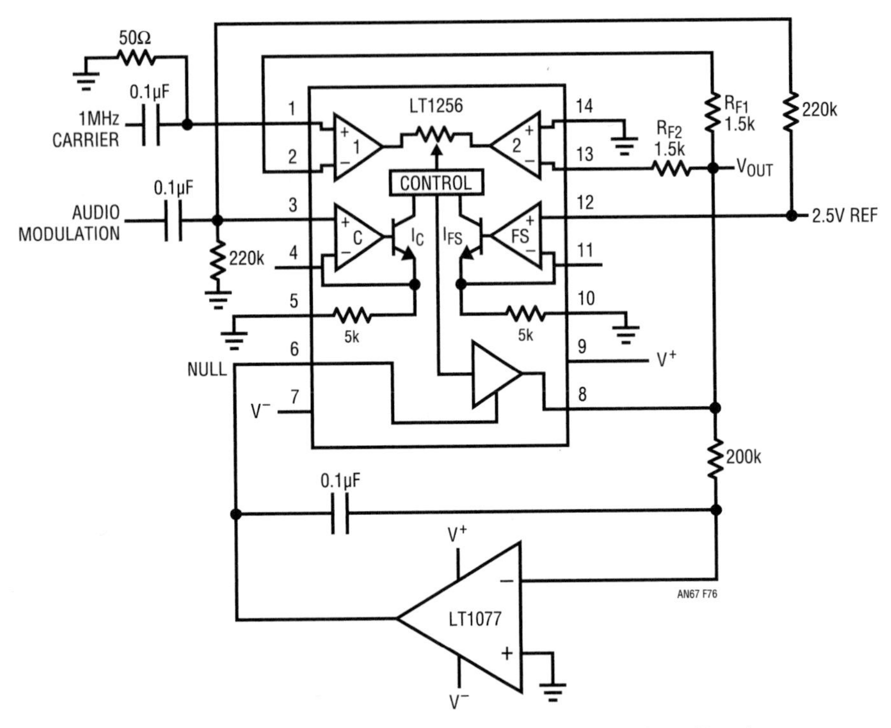

Figure 36.76 • AM Modulator with DC Output Nulling Circuit

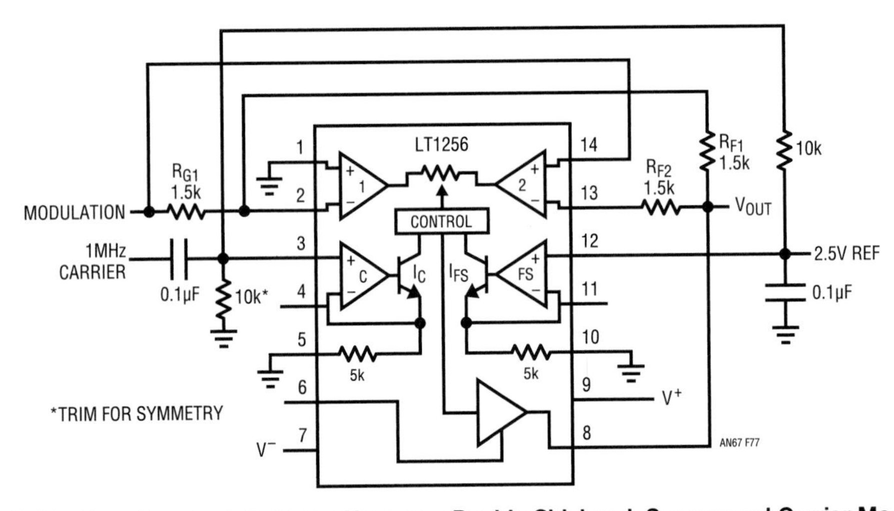

Figure 36.77 • Four Quadrant Multiplier Uses as a Double Sideband, Suppressed Carrier Modulator

Extending op amp supplies to get more output voltage

by Dale Eagar

We often hear of applications that require high output voltage, low output impedance amplifiers. Here is a topology that allows you to extend an op amp's output voltage swing while still maintaining its short-circuit protection. The trick is to suspend the op amp between two MOSFET source followers so that the supply voltages track the op amp's output voltage (see Figure 36.78). The circuit shown in Figure 36.78 will perform very nicely with any run-of-the-mill ideal op amp. The problem is in the lead times of ideal op amps—they just keep getting pushed out to later dates.

Nonideal op amps have realistic lead times and can be made to work in the extended supply mode. They have bandwidth limitations in both CMRR and PSRR. The circuit shown in Figure 36.79 implements the extended supply as shown in Figure 36.78 and has several additional components: C1 is added to decouple the supply, improving high frequency PSRR; R3 and R5 decouple the gates of Q1 and Q2 from AC ground, preventing Q1 and Q2 from running off together to redirect local air traffic; R1, R2 and C4 form a snubber to de-Q the 2-pole system formed by the Miller capacitance of Q1 and Q2 and the high frequency CMRR of IC1; additionally, R4, R6, C2, C3, Z1 and Z2 form the two 15V voltage sources (E1 and E2 in Figure 36.78); CR1 and CR2 are protection diodes that allow the output to be instantaneously shorted to ground when the output is at any output voltage.

The values of R1, R2 and C4 vary with the MOSFETs' Miller capacitance and with the high frequency CMRR of the op amp used. They are selected to minimize the overshoot in the step response of the amplifier.

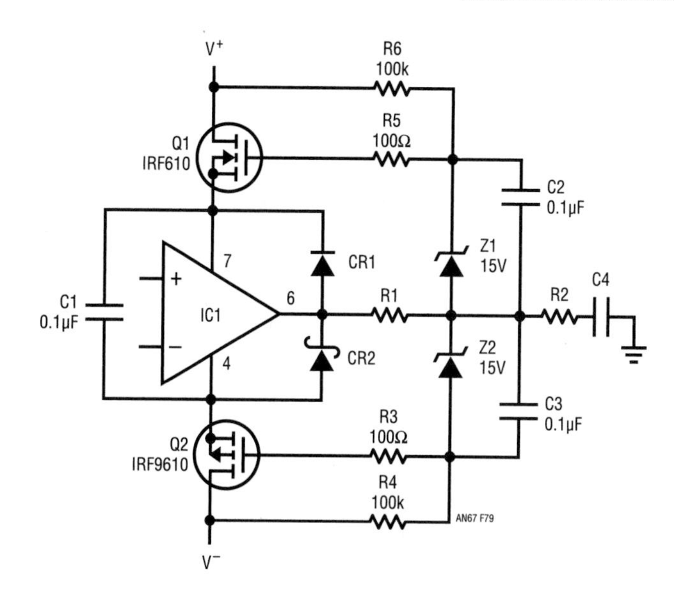

Figure 36.79 • Detailed High Voltage Op Amp

High voltage, high frequency amplifier

Using the LT1227 current feedback amplifier (CFA) in the extended supply mode as shown in Figure 36.79, it is relatively easy to get a 1MHz power bandwidth at $100V_{P-P}$ (see Figure 36.80 for component values). This circuit has short-circuit protection and is stable into all capacitive loads.

If one is good, are two better?

Dual and quad op amps can also be configured with extended supplies, although the design gets just a wee bit tricky. When extending supplies of multiple stages and/or complete circuits, some design rules need to change. Op amp circuits generally require a ground against which to reference all signals. The problem encountered when using extended supply mode is that "ground" is swinging through the common mode range of the op amp and beyond.

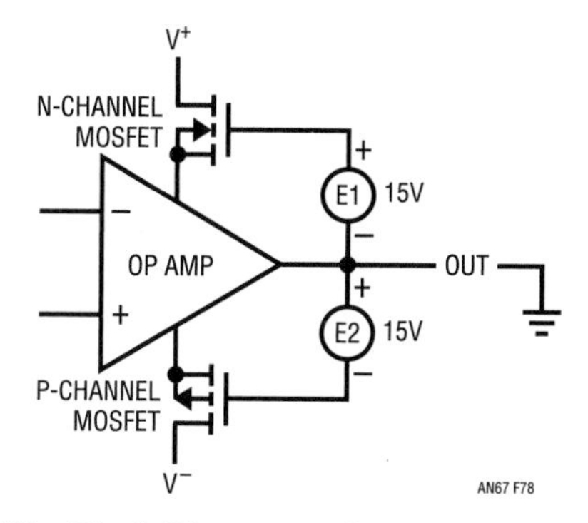

Figure 36.78 • Block Diagram of Suspended Supply Op Amp

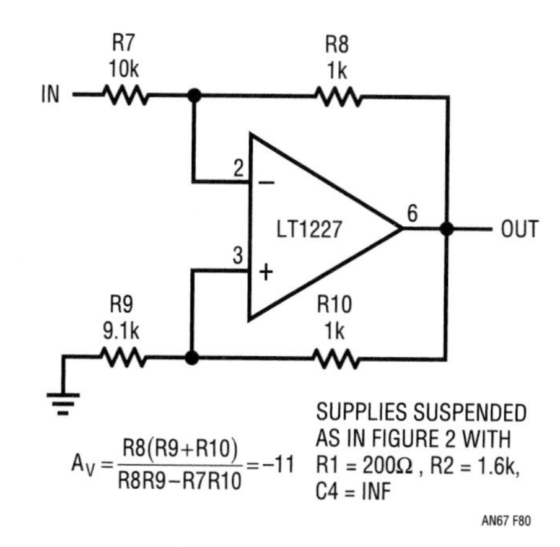

Figure 36.80 • High Speed Suspended Op Amp

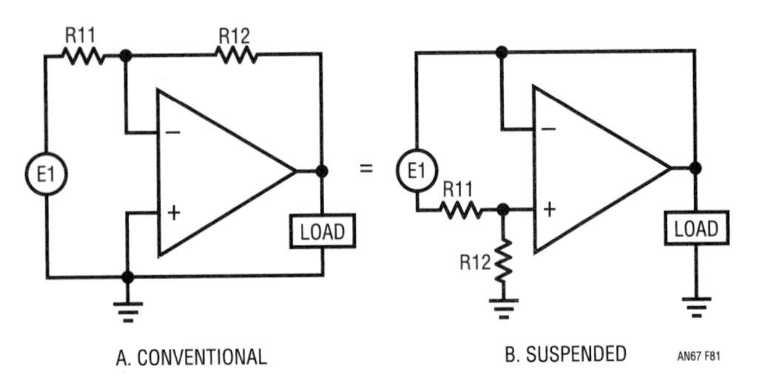

Figure 36.81 • Inverting Amplifiers (A. Conventional, B. Suspended)

This raises the following question: "If I cannot reference the signals to ground, to what can I reference them?" The answer? "Use the output as the signal reference." This works for all stages except the last stage, where using the output as the reference would simply discard the signal. In the last stage, ground is effectively the output and the feedback resistor is R12. This is shown in Figures 36.81a and 36.81b. Figure 36.81a shows a conventional inverting amplifier where the input and output signals are referred to ground. Figure 36.81b shows the equivalent circuit implemented in the extended supply mode.

Here are two rules for design in the extended supply mode, which will be demonstrated in the next application:

Rule 1: When designing multiple stages in the extended supply mode, reference the signals of all stages except the last to the output of the last stage.

Rule 2: Invert the signal using the circuit in Figure 36.81b at the last stage.

Ring-tone generator

Ring-tone generators are sine wave output, high voltage inverters for the specific purpose of ringing telephone bells. In decades past, the phone company generated their ring tones with motor generator sets with the capacity to ring numerous phones simultaneously. Often, ring tones are 20Hz at 90V with less than 10mA per bell output current capability. Since the power supplied is low one would think that the task is minimal. This is not always so. "It's simple—no problem, " is often heard in response to queries about ring tone generators. "Just hook a couple of logic level FETs to two spare output bits of the microprocessor and hook their drains to the primary side of a transformer, with the center tap hooked to 5V or 12V or whatever." At this point everyone is happy until the transformer comes in. After a few phone calls to make sure that the transformer maker shipped the right one, the engineer (face covered with egg) asks if anyone needs a rather large paperweight. The engineer (still wiping egg from his face) then decides to use switching power supply technology to solve this "simple" problem.

Here is a simple ring-tone generator that can be turned on and off with a logic signal. It has a fully isolated output, is short-circuit protected and can be powered by any input voltage from 3V to 24V.

How it works

Suspended along with the dual op amp in Figure 36.82 are two voltage references and an oscillator. Keep in mind when referring to Figure 36.82 that the node labeled "A" is the output; this is the reference common for the references, the oscillator and the first lowpass filter (U1a). The two references, VR1 and VR2, produce ±2.5V. The oscillator U2, running on the ±2.5V references, produces a 20Hz square wave rail-to-rail. U1a is a 2nd order Sallen and Key lowpass filter that knocks off the sharp edges, presenting the somewhat smoothed signal at point "B."

Next comes the tricky stuff. U1b is a 2nd order, multiple- feedback (MFB) lowpass filter/amplifier that performs four functions: first, it subtracts the voltage at point "A" (its own output voltage) from the voltage at point "B" (the incoming signal), forming a difference that is the signal; second, it filters the difference signal with a 2-pole lowpass filter, smoothing out the last wrinkles in the signal; third, it amplifies the filtered difference signal with a gain of 34; and fourth, it references the amplified signal to ground, forming the output.

Note that R99 shown in Figure 36.82 is there to protect the input of U1b in the event that the output is shorted when the output voltage is very high. This measure is necessary because the bottom end of C99 is connected to ground, and C99 could have up to 100V across it. When the output is shorted to ground from a high voltage, R99 limits the current into the input of U1b to an acceptable level.

This circuit, when coupled with the switching power supply shown in Figure 36.83, implements a fully isolated sine wave ring tone generator.

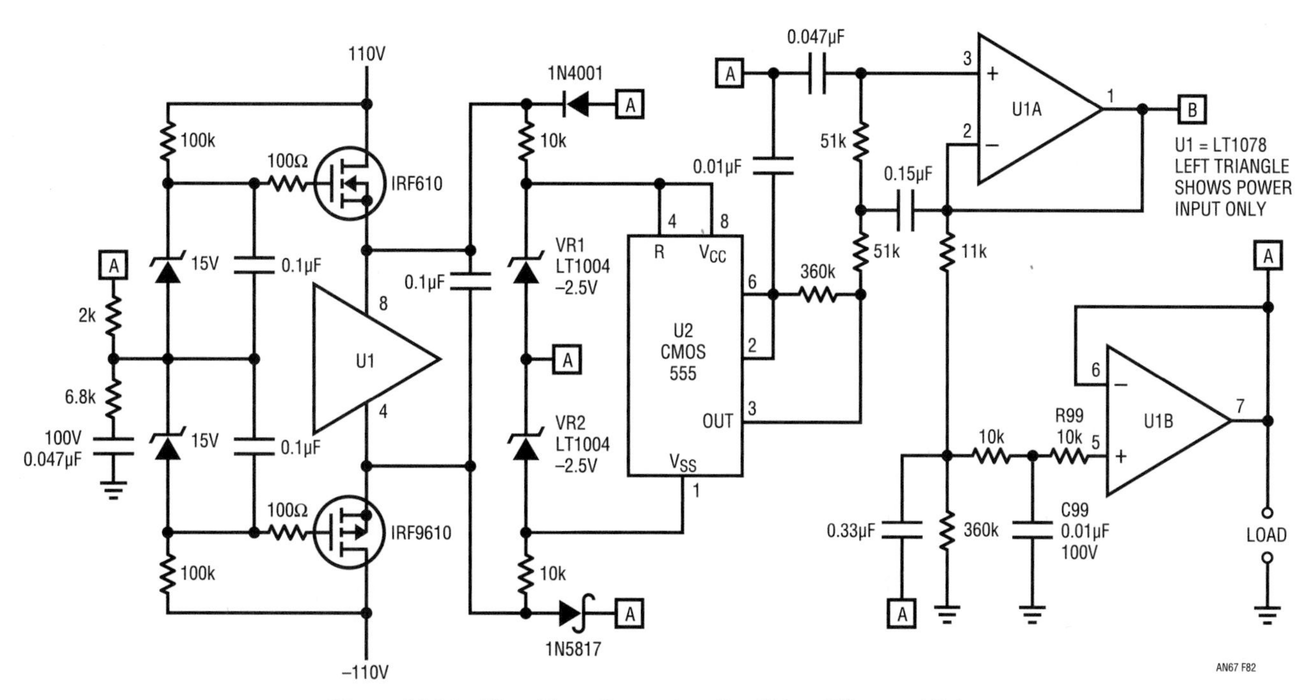

Figure 36.82 • Ring Tone Generator: Oscillator, Filter and Driver

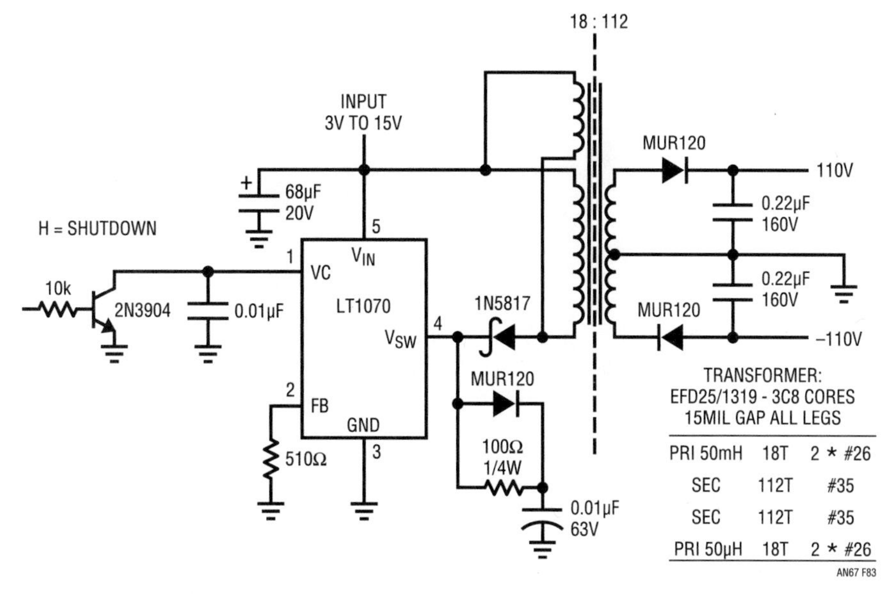

Figure 36.83 • High Voltage Power Supply for Ring Tone Generator

The input current and power versus input voltage for the combination ring-tone generator (Figures 36.82 and 36.83) are shown in Figure 36.84. The output waveform (loaded with one bell) is shown in Figure 36.85 and the harmonic distortion is shown in Figure 36.86.

Although somewhat tricky at first, extended supply mode is a valuable tool to get out of many tight places. There is also a great deal of satisfaction to be gathered when making it work, for those of you who love a technical challenge.

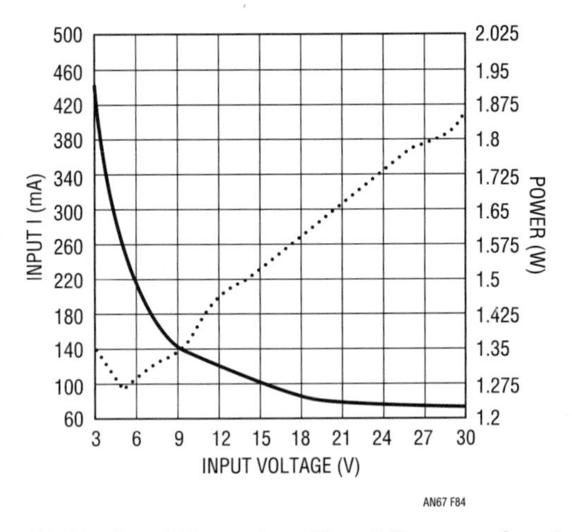

Figure 36.84 • Input Current and Input Power vs Input Voltage While Ringing One Bell for Figure 36.82 Circuit

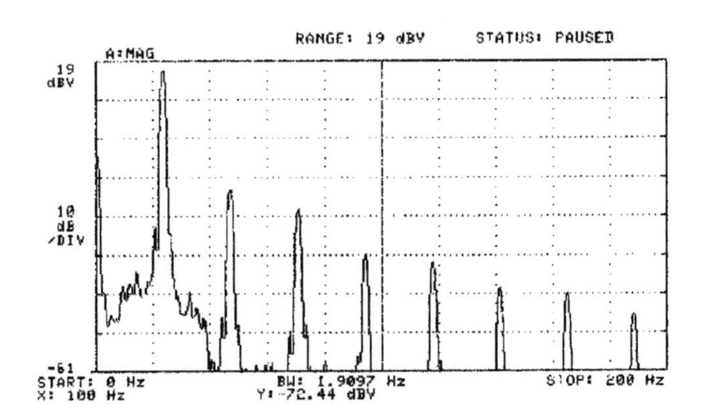

Figure 36.85 • Ring-Tone Generator Frequency Spectrum Plot

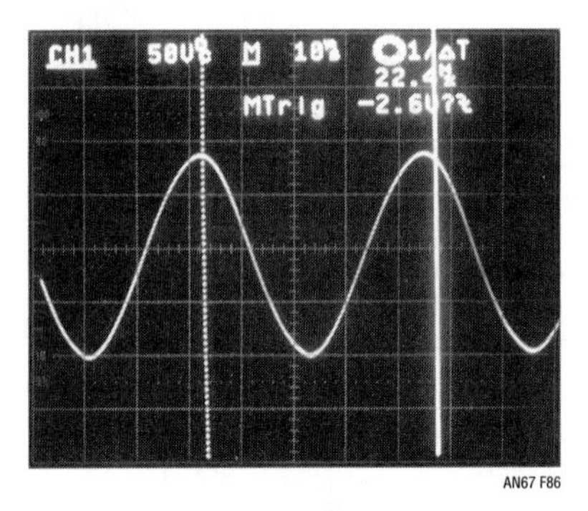

Figure 36.86 • Sine Wave Output from Ring-Tone Generator

Using super op amps to push technological frontiers: an ultrapure oscillator

by Dale Eagar

The advent of high speed op amps allows the implementation of circuits that were impossible just a few years ago. This article describes a new topology that makes use of these new high speed circuits and makes astounding improvements in its performance. An oscillator using such op amps has distortion limits beyond our ability to measure.

An ultralow distortion, 10kHz sine wave source for calibration of 16-bit or higher A/D converters

The path to low distortion in an amplifier or an oscillator begins with amplifiers with the lowest possible open-loop distortion and lots of excess open-loop gain in the frequency band of interest. The next step is closing the loop, thereby reducing open-loop distortion by an amount approximately equal to the loop gain. This is not easy, as certain stability criteria must be met by an amplifier that isn't an oscillator or by an oscillator that oscillates at a specified frequency.

The trick used in this circuit is to build an amplifier that has excessive gain where it is needed but no excess gain or phase shift where it isn't. In many applications the band from DC to 100kHz requires the above mentioned high gain; the gain should fall off when the open-loop gain falls through unity (around 5MHz). How this is done in the flesh (silicon) is shown here.

Circuit operation and circuit evolution

A standard inverting amplifier topology, as shown in Figure 36.87, has a finite open-loop gain in the frequency band of interest (see Figure 36.88), with some open-loop harmonic distortion (about −60dB) and an open-loop output impedance of about 70Ω.

The amplifier shown in Figure 36.87 can achieve low distortion, but since the circuit has a limited loop gain, the curative effects of feedback can only be taken so far The designer must also be careful to ensure that R_L is many times higher than the open-loop output impedance of U1.

Figure 36.89's circuit makes several improvements over the circuit of Figure 36.87. First, the open-loop gain of U1 is multiplied by $A_V(f)$, the gain of the composite amplifier stage A1. Second, the input impedance of A1 can be made very high, further improving both open-loop gain of U1 and the open-loop harmonic distortion of U1. Third, the output voltage swing of U1 is decreased, keeping its output circuitry in its lowest distortion area.

The composite circuit, A1, consists of three sections. The first section, as seen in Figure 36.90, has the gain/phase plot shown in Figure 36.92. Note the high gain at 10kHz (60dB) and the gain of 6dB at 5MHz, with only 17 degrees of phase contribution. In fact, this looks so nice that you might ask, "why not use two?" and thus reduce your distortion by an additional 60dB?

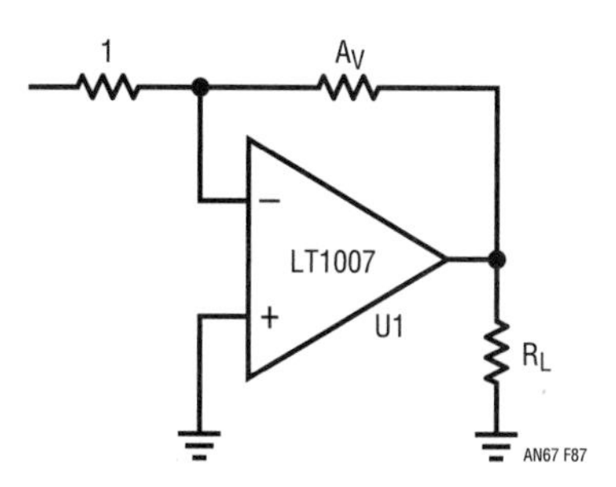

Figure 36.87 • Conventional Inverting Op Amp Topology

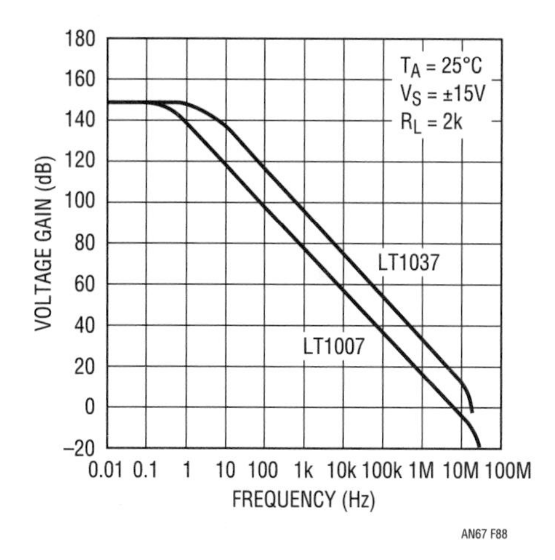

Figure 36.88 • Voltage Gain vs Frequency

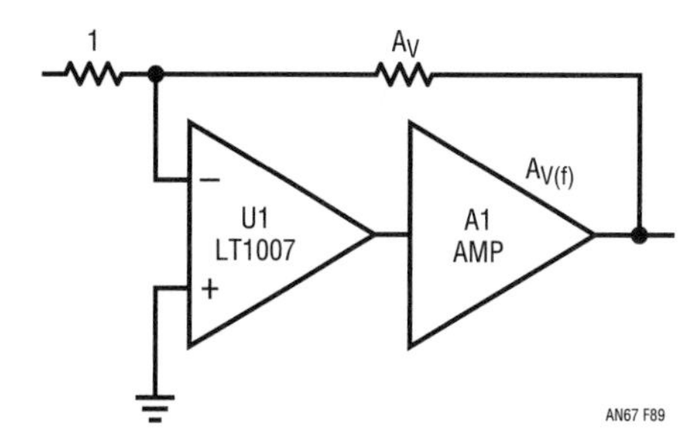

Figure 36.89 • LT1007 Followed by Composite Amplifier A1

The second section, shown in Figure 36.91, has the gain/phase plot shown in Figure 36.93. Note that here the gain doesn't change significantly, but the phase is positive just where we want it (1MHz to 5MHz) to allow a very stable system to be built.

The third section, as you might guess, is the same as the first. In sum, the gain/phase plot of the composite amplifier A1 is shown in Figure 36.94. Note the gain, which is in

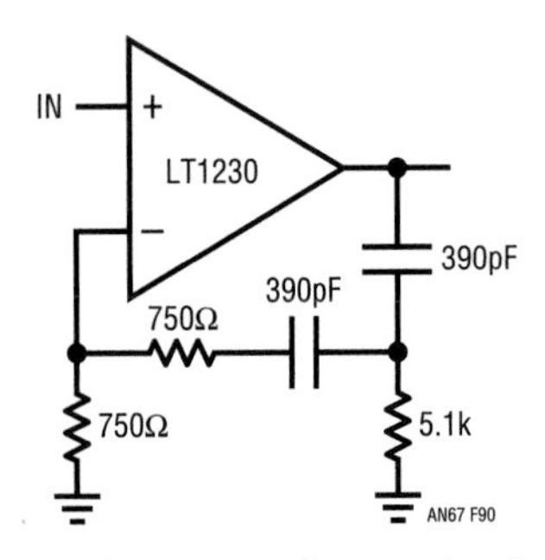

Figure 36.90 • First Section of Composite Amplifier A1

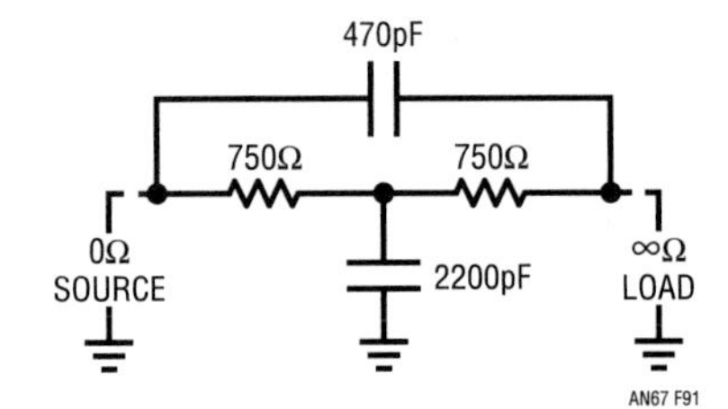

Figure 36.91 • Second Section of Composite Amplifier A1

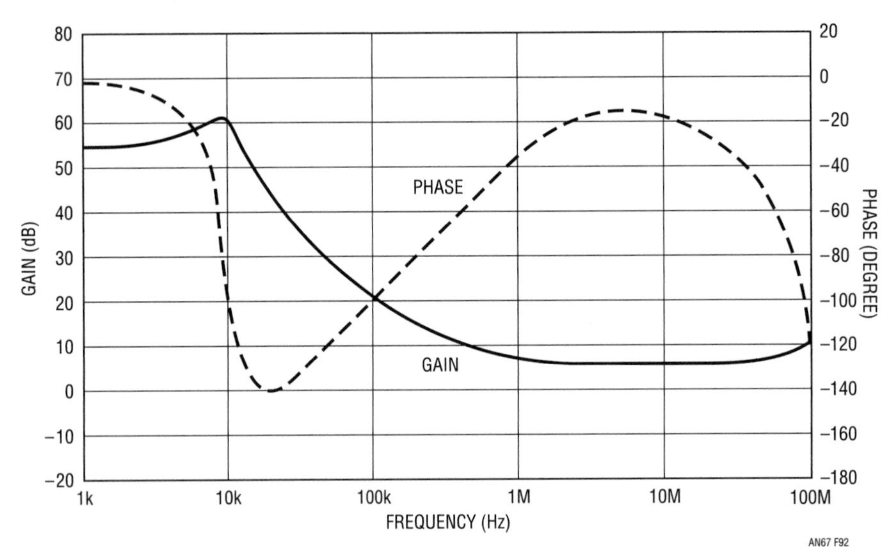

Figure 36.92 • Gain/Phase Response of Circuit Shown in Figure 36.90

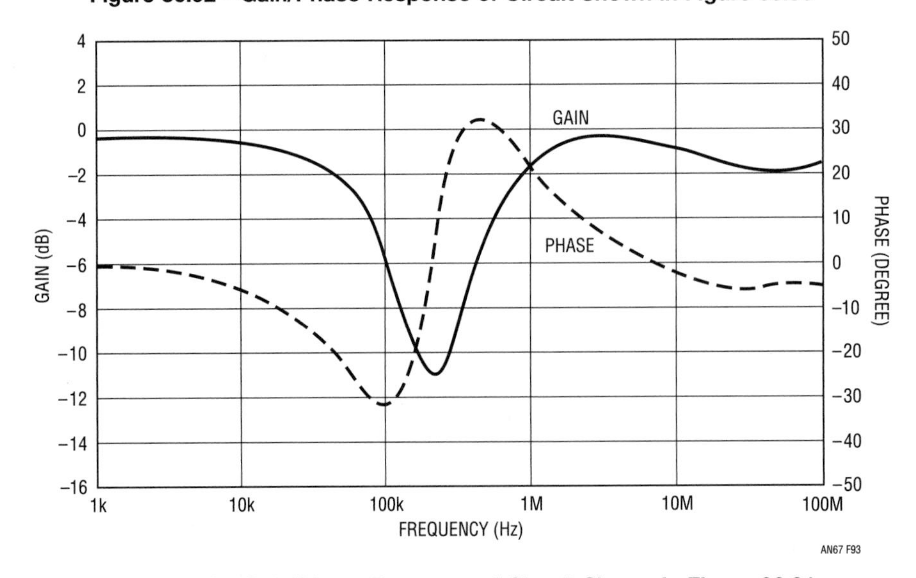

Figure 36.93 • Gain/Phase Response of Circuit Shown in Figure 36.91

excess of 120db at 10kHz and the total phase contribution of about −20 degrees at 5MHz. The complete gain block is shown in Figure 36.95.

Super gain block oscillator circuitry

When A1, as described above is connected with U1, as shown in Figure 36.89, the resulting circuit is not only unity-gain stable but has open-loop gain of 180dB at 10kHz (yes, 1 billion). This means that the closed-loop harmonic distortion can easily be kept in the region of "parts per billion."

A Wien bridge oscillator with harmonic distortion in the parts per billion is shown in Figure 36.96. The super op amps S1 and S2 are the previously described composite amplifiers as shown in Figure 36.95. Note that the output is taken between the two outputs of S1 and S2. This topology gives the best signal-to-noise ratio, in addition to

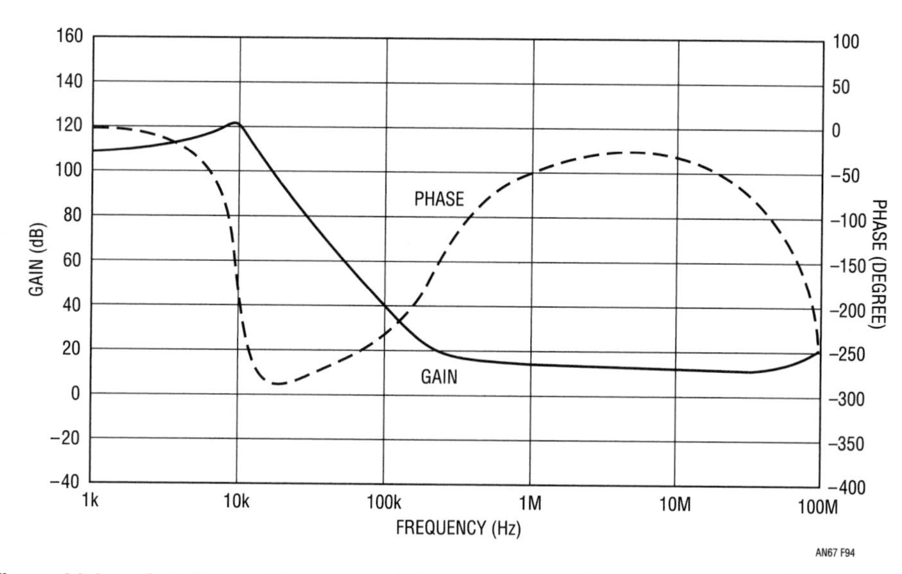

Figure 36.94 • Gain/Phase Response of Composite Amplifier A1 (Shown in Figure 36.89)

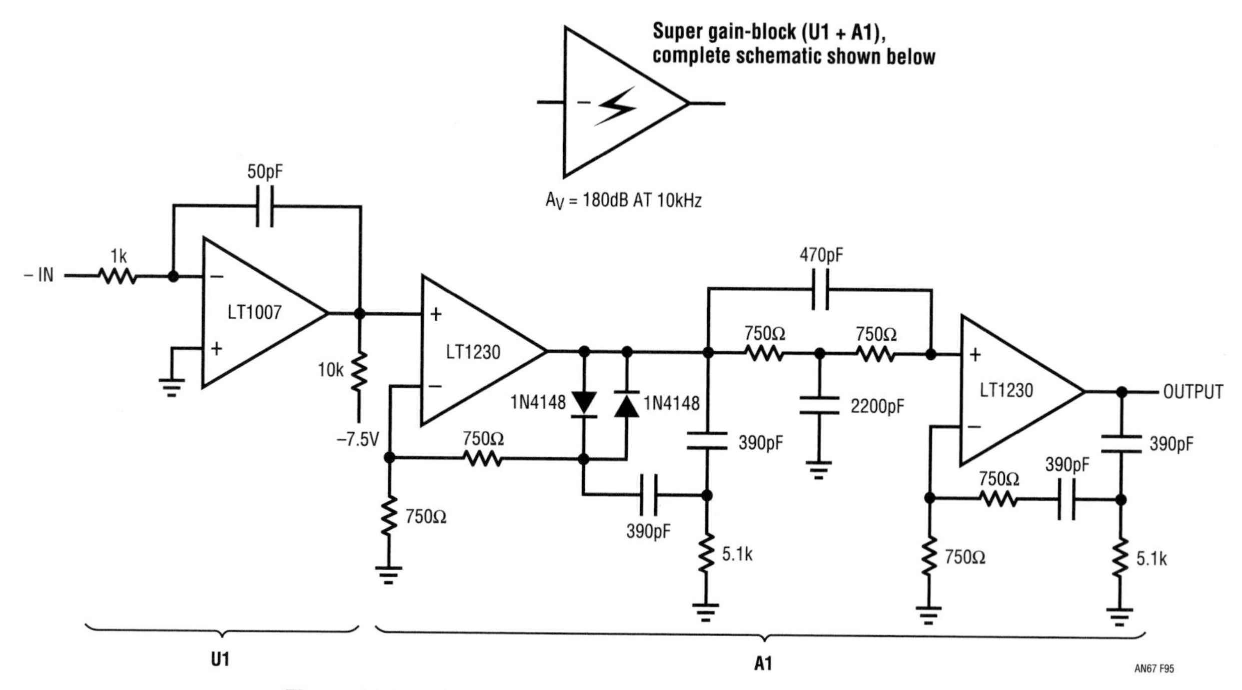

Figure 36.95 • Super Gain Block S1 and S2 Schematic Diagram

balancing the power supply currents and their harmonics. Taking the output from one amplifier's output to ground is also valid.

To align the circuit, first center the output amplitude adjustment potentiometer. Next, adjust the gain trim for oscillation while also adjusting the output amplitude for $5V_{P-P}$ output (single ended). Next, adjust the gain trim to $1V_{P-P}$ at the output of the LT1228. Finally, connect a spectrum analyzer to the output of the LT1228 and adjust the second harmonic trim potentiometer for a null in the second harmonic of the oscillator frequency. The measurement of the harmonic distortion of this oscillator defies all of our resources, but appears to be well into the parts-per-billion range.

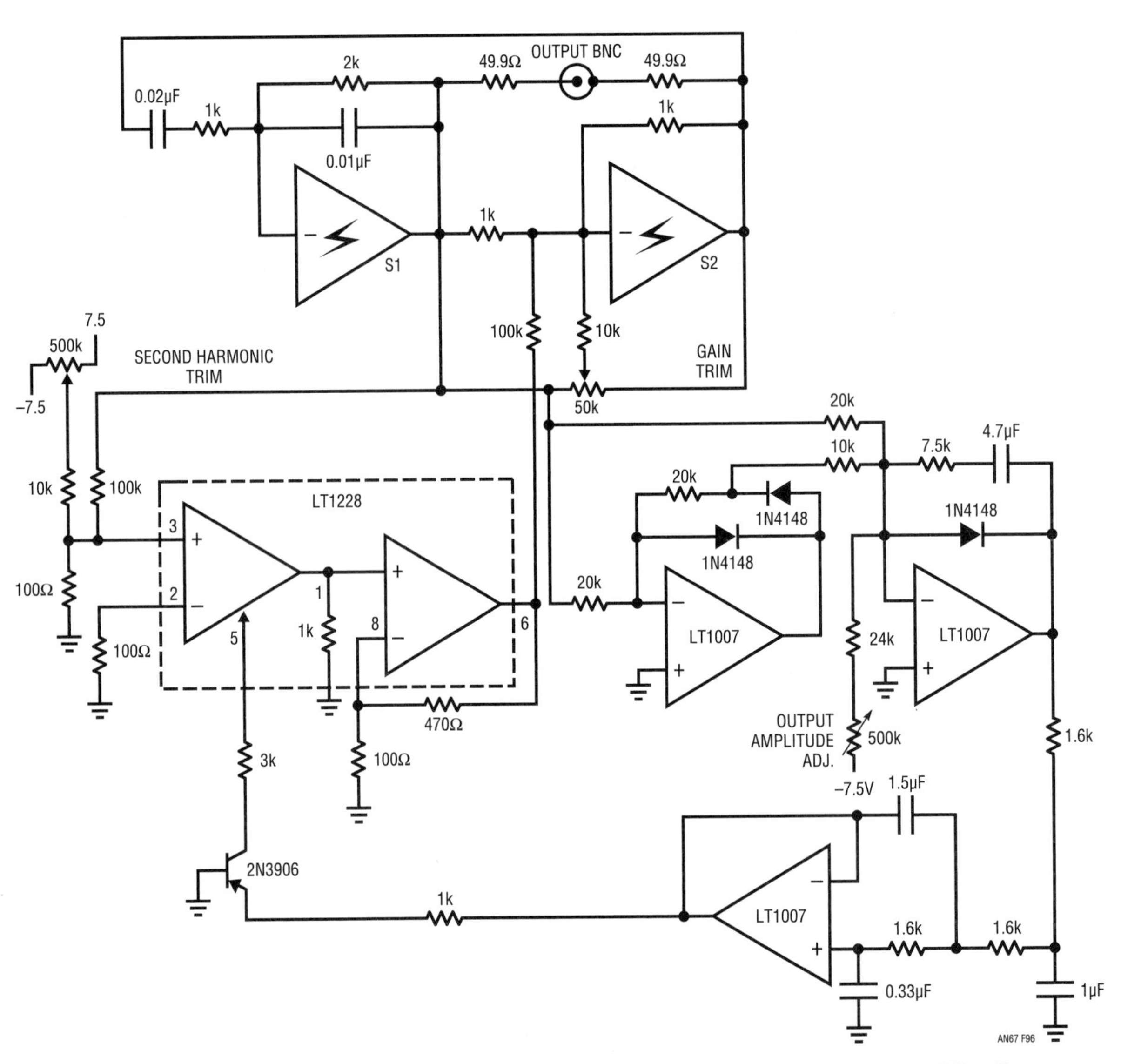

Figure 36.96 • Schematic Diagram: Wien Bridge Oscillator with Distortion in the Parts-per-Billion Range

Fast video MUX uses LT1203/LT1205

by Frank Cox

To demonstrate the switching speed of the LT1203/LT1205, the RGB MUX of Figure 36.97 is used to switch the inputs of an RGB workstation with a 22ns pixel width. Figure 36.98a is a photo showing the workstation output and RGB MUX output. The slight rise time degradation at the RGB MUX output is due to the bandwidth of the LT1260 current feedback amplifier used to drive the 75Ω cable. In Figure 36.98b the LT1203 switches at the end of the first pixel to an input at zero and removes the following pixels.

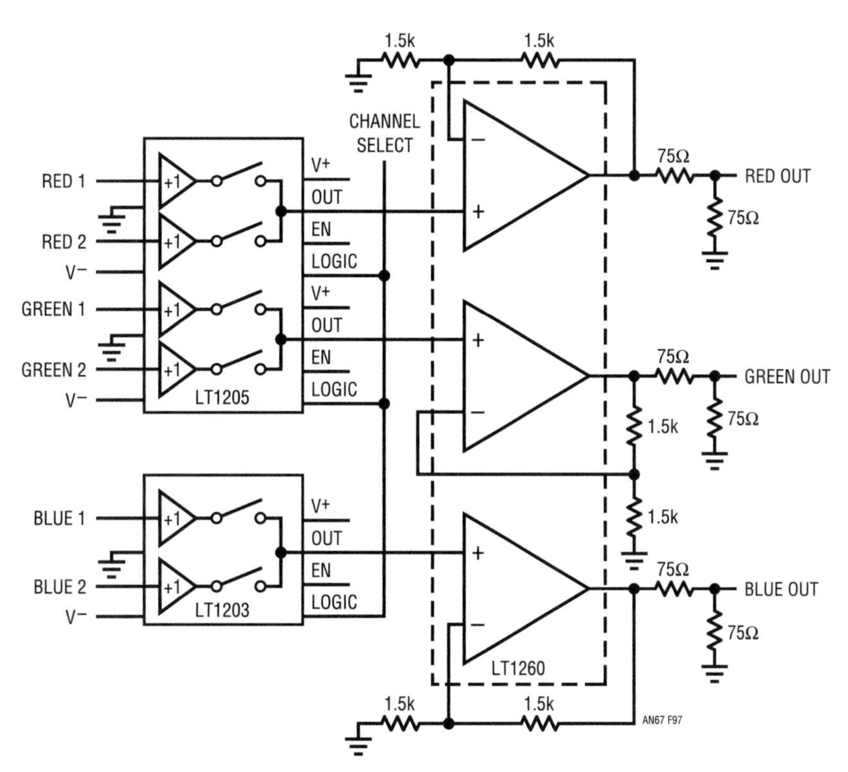

Figure 36.97 • Fast RGB MUX

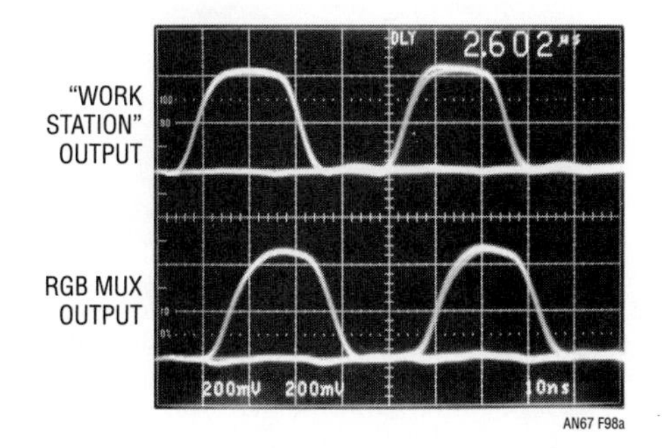

Figure 36.98a • Workstation and RGB MUX Output

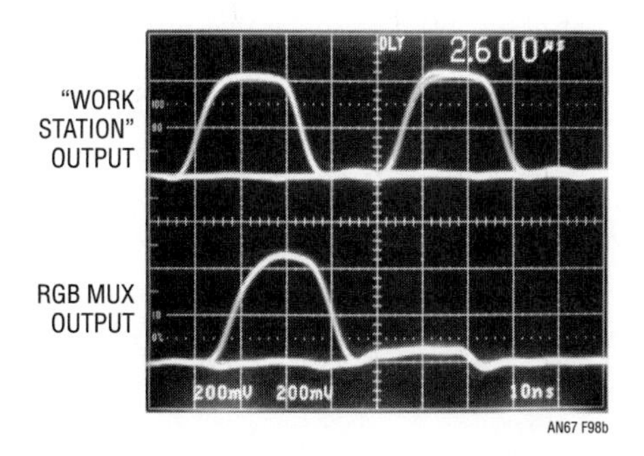

Figure 36.98b • RGB MUX Output Switched to Ground After One Pixel

Using a fast analog multiplexer to switch video signals for NTSC "picture-in-picture" displays

by Frank Cox

Introduction

The majority of production[8] video switching consists of selecting one video source out of many for signal routing or scene editing. For these purposes, the video signal is switched during the vertical interval in order to reduce visual switching transients. The image is blanked during this time, so if the horizontal and vertical synchronization and subcarrier lock are maintained, there will be no visible artifacts. Although vertical interval switching is adequate for most routing functions, there are times when it is desirable to switch two synchronous video signals during the active (visible) portion of the line to obtain picture-in-picture, key or overlay effects. Picture-in-picture or active video switching requires signal-to-signal transitions that are both clean and fast. A clean transition should have a minimum of preshoot, overshoot, ringing or other aberrations commonly lumped under the term "glitching."

Using the LT1204

A quality high speed multiplexer amplifier can be used with good results for active video switching. The important specifications for this application are small, controlled switching glitch, good switching speed, low distortion, good dynamic range, wide bandwidth, low path loss, low channel-to-channel crosstalk and good channel-to-channel offset matching. The LT1204 specifications match these requirements quite well, especially in the areas of bandwidth, distortion and channel-to-channel crosstalk (which is an outstanding 90dB at 10MHz). The LT1204 was evaluated for use in active video switching with the test setup shown in Figure 36.99. Figure 36.100 shows the video waveform of a switch between a 50% white level and a 0% white level about 30% into the active interval and back again at about 60% of the active interval. The switch artifact is brief and well controlled. Figure 36.101 is an expanded view of the same waveform. When viewed on a monitor, the switch artifact is just visible as a very fine line. The lower trace is a switch between two black level

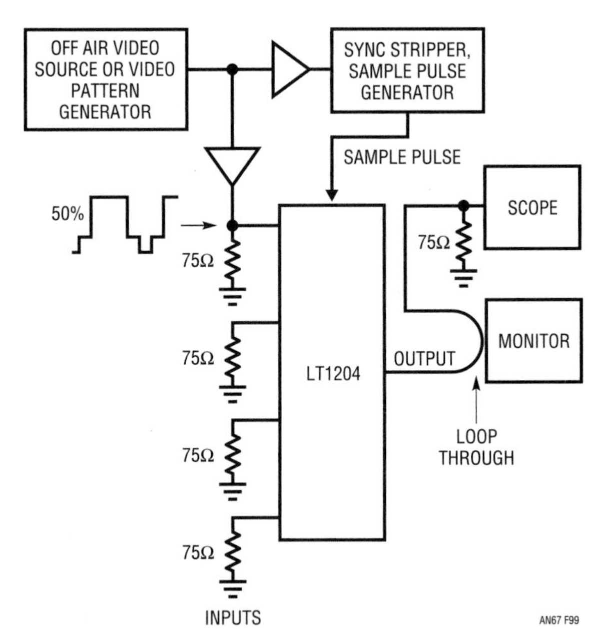

Figure 36.99 • "Picture-in-Picture" Test Setup

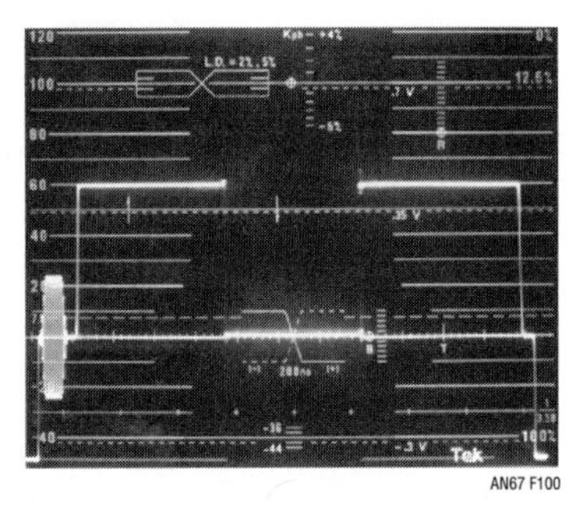

Figure 36.100 • Video Waveform Switched from 50% White Level to 0% White Level and Back

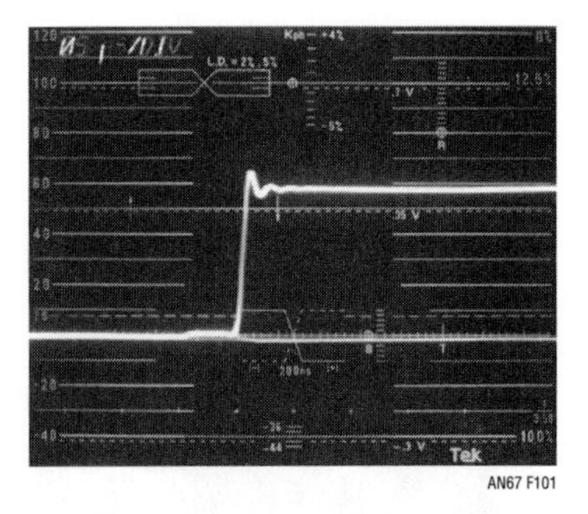

Figure 36.101 • Expanded View of Rising Edge of LT1204 Switching from 0% to 50% (50ns Horizontal Division)

Note 8: Video production, in the most general sense, means any purposeful manipulation of the video signal, whether in a television studio or on a desktop PC.

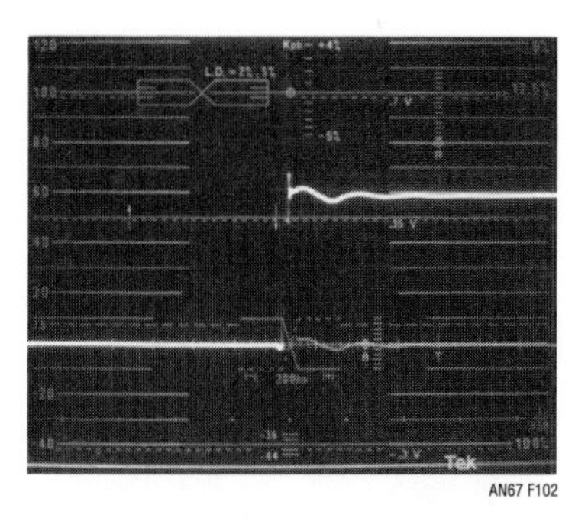

Figure 36.102 • Expanded View of "Brand-X" Switch 0% to 50% Transition

(0V) video signals showing a very slight channel-to-channel offset, which is not visible on the monitor. Switching between two DC levels is a worst-case test, as almost any active video will have enough variation to totally obscure this small switch artifact.

Video switching caveats

In a video processing system that has a large bandwidth compared to the bandwidth of the video signal, a fast transition from one video level to another (with a low amplitude glitch) will cause minimal visual disturbance. This situation is analogous to the proper use of an analog oscilloscope. In order to make accurate measurement of pulse waveforms, the instrument must have much more bandwidth than the signal in question (usually five times the highest frequency). Not only should the glitch be small, it should be otherwise well controlled. A switching glitch that has a long settling "tail" can be more troublesome (that is, more visible) than one that has more amplitude but decays quickly. The LT1204 has a switching glitch that is not only low in amplitude but well controlled and quickly damped. Refer to Figure 36.102, which shows a video multiplexer that has a long, slow-settling tail. This sort of distortion is highly visible on a video monitor.

Composite video systems, such as NTSC, are inherently band limited and thus edge-rate limited. In a sharply band limited system, the introduction of signals that contain significant energy higher in frequency than the filter cutoff will cause distortion of transient waveforms (see Figure 36.103). Filters used to control the bandwidth of these video systems should be group-delay equalized to minimize this pulse distortion. Additionally, in a band limited system, the edge rates of switching glitches or level-to-level transitions should be controlled to prevent ringing and other pulse aberrations that could be visible. In practice, this is usually accomplished with pulse-shaping networks (Bessel filters are one example). Pulse-shaping networks and delay equalized filters add cost and complexity to video systems and are usually found only on expensive equipment. Where cost is a determining factor in system design, the exceptionally low amplitude and brief duration of the LT1204's switching artifact make it an excellent choice for active video switching.

Some definitions—

"Picture in picture" refers to the production effect in which one video image is inserted within the boundaries of another. The process may be as simple as splitting the screen down the middle or it may involve switching the two images along a complicated geometric boundary. In order to make the composite picture stable and viewable, both video signals must be in horizontal and vertical sync. For composite color signals the signals must also be in subcarrier lock.

"Keying" is the process of switching among two or more video signals, triggering on some characteristic of one of the signals. For instance, a chroma keyer will switch on the presence of a particular color. Chroma keyers are used to insert a portion of one scene into another. In a commonly used effect, the TV weather person (the "talent") appears to be standing in front of a computer generated weather map. Actually, the talent is standing in front of a specially colored background; the weather map is a separate video signal which has been carefully prepared to contain none of that particular color When the chroma keyer senses the keying color, it switches to the weather map background. Where there is no keying color the keyer switches to the talent's image.

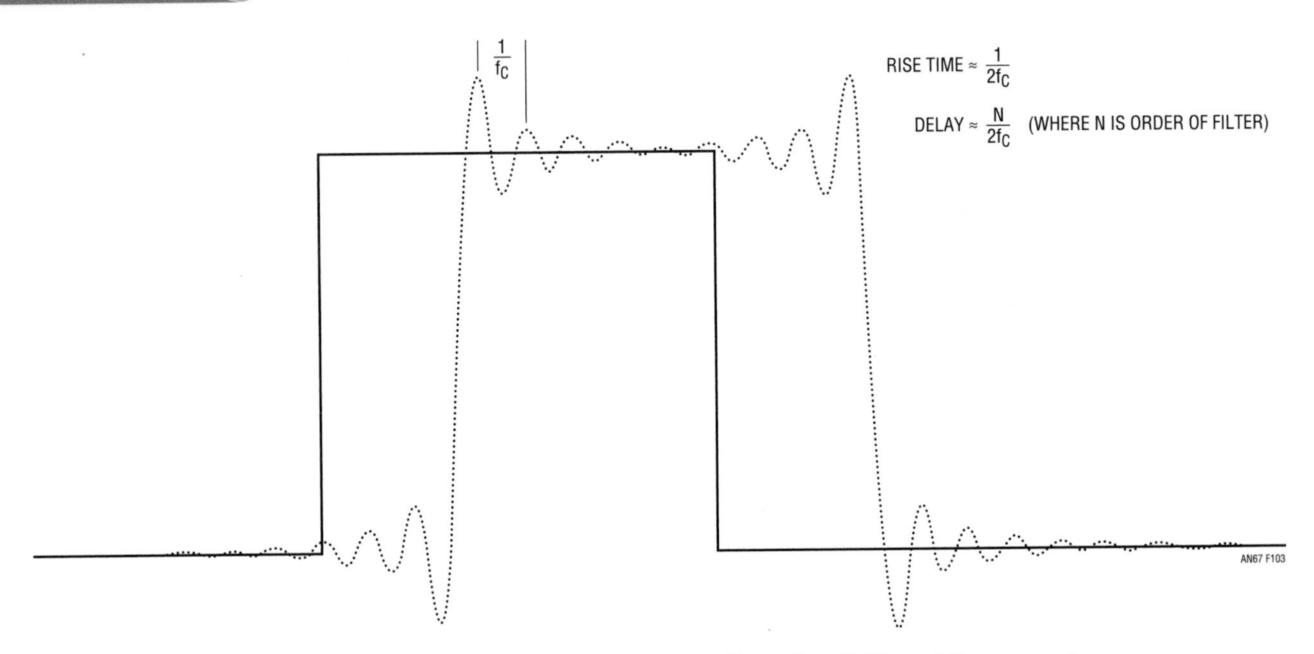

Figure 36.103 • Pulse Response of an Ideal Sharp Cutoff Filter at Frequency fc

Applications for the LT1113 dual JFET op amp

by Alexander Strong

Figure 36.104 shows a low noise hydrophone amplifier with a DC servo. Here one half (A) of the LT1113 is configured in the noninverting mode to amplify a voltage signal from the hydrophone, and the other half (B) of the LT1113 nulls errors due to voltage and current offsets of amplifier A and to null out DC errors of the hydrophone. The value of C1 depends on the capacitance of the hydrophone, which can range from 200pF to 8000pF The time constant of the servo should be larger than the time constant of the hydrophone capacitance and the 100M source resistance. This will prevent the servo from canceling the low frequency signals from the hydrophone.

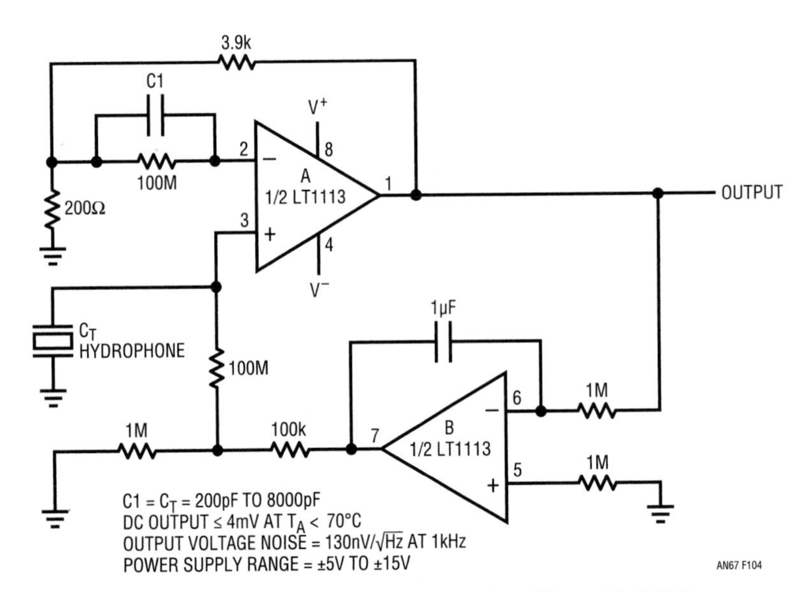

Figure 36.104 • Low Noise Hydrophone Amplifier with DC Servo

Another popular charge-output transducer is the accelerometer Since precision accelerometers are charge-output devices, the inverting mode is used to convert the transducer charge to an output voltage. Figure 36.105 is an example of an accelerometer with a DC servo. The charge from the transducer is converted to a voltage by C1, which should equal the transducer capacitance plus the input capacitance of the op amp. The noise gain will be $1 + C1/C_T$. The low frequency bandwidth of the amplifier will depend on the value of R1 • C1 (or R1 (1 + R2/R3) for a Tee network). As with the hydrophone example, the time constant of the servo (1/R5C5) should be larger than the time constant of the amplifier (1/R1C1).

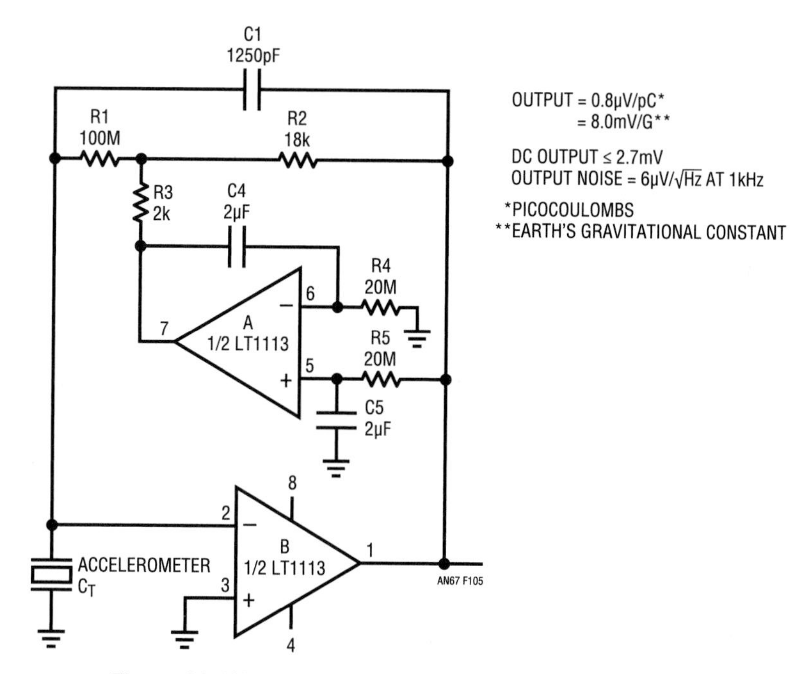

Figure 36.105 • Accelerometer Amplifier with DC Servo

LT1206 and LT1115 make low noise audio line driver

by William Jett

Although the wide bandwidth and high output drive capabilities of the LT1206 make it a natural for video circuits, these characteristics are also useful for audio applications. Figure 36.106 shows the LT1206 combined with the LT1115 low noise amplifier to form a very low noise, low distortion audio buffer with a gain of 10. With a 32Ω load and a $5V_{RMS}$ output level (780mW), the THD + noise for the circuit is 0.0009% at 1kHz, rising to 0.004% at 20kHz. The frequency response is flat to 0.1dB from DC to 600kHz, with a −3dB bandwidth of 4MHz. The circuit is stable with capacitive loads of 250pF or less.

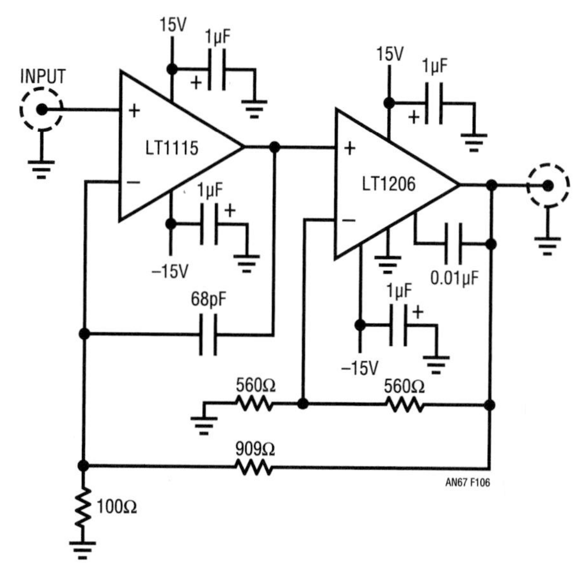

Figure 36.106 • Low Noise x 10 Buffered Line Driver

Driving mulitple video cables with the LT1206

by William Jett

The combination of a 60MHz bandwidth, 250mA output current capability and low output impedance makes the LT1206 ideal for driving multiple video cables. One concern when driving multiple transmission lines is the effect of an unterminated (open) line on the other outputs. Since the unterminated line creates a reflected wave that is incident on the output of the driver, a nonzero amplifier output impedance will result in crosstalk to the other lines. Figure 36.107 shows the LT1206 connected as a distribution amplifier. Each line is separately terminated to minimize the effect of reflections. For systems using composite video, the differential gain and phase performance are also important and have been considered in the internal design of the device. The differential phase and differential gain performance versus supply is shown in Figures 36.108 and 36.109 for 1, 3, 5 and 10 cables. Figure 36.110 shows the output impedance versus frequency. Note that at 5MHz the output impedance is only 0.6Ω.

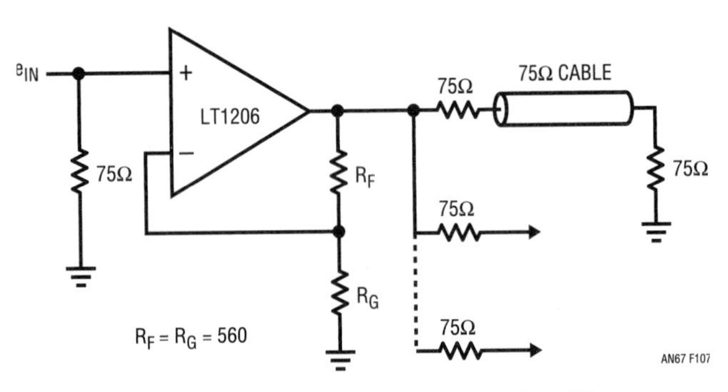

Figure 36.107 • LT1206 Distribution Amplifier

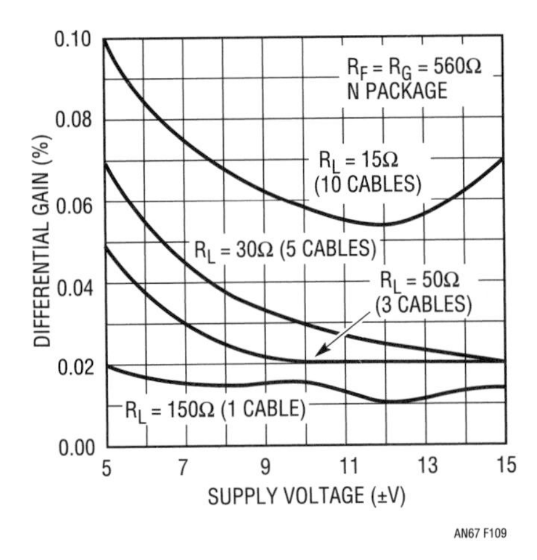

Figure 36.109 • Differential Gain vs Supply Voltage

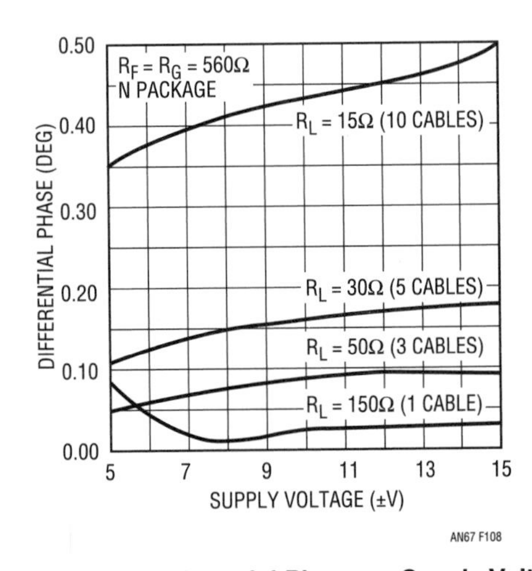

Figure 36.108 • Differential Phase vs Supply Voltage

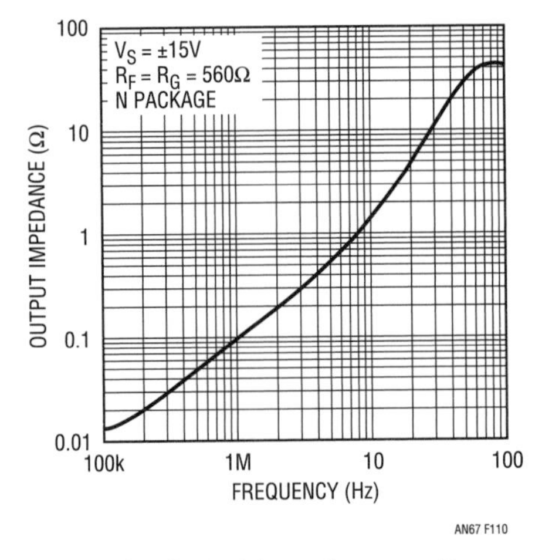

Figure 36.110 • Output Impedance vs Frequency

Optimizing a video gain control stage using the LT1228

by Frank Cox

Video automatic gain control (AGC) systems require a voltage- or current-controlled gain element. The performance of this gain-control element is often a limiting factor in the overall performance of the AGC loop. The gain element is subject to several, often conflicting, restraints. This is especially true of AGC for composite color video systems such as NTSC, which have exacting phase and gain distortion requirements. To preserve the best possible signal-to-noise ratio (S/N),[9] it is desirable for the input signal level to be as large as practical. Obviously, the larger the input signal the less the S/N will be degraded by the noise contribution of the gain control stage. On the other hand, the gain control element is subject to dynamic range constraints and exceeding these will result in rising levels of distortion.

Linear Technology makes a high speed transconductance (g_m) amplifier, the LT1228, which can be used as a quality, inexpensive gain control element in color video and some lower frequency RF applications. Extracting the optimum performance from video AGC systems takes careful attention to circuit details.

As an example of this optimization, consider the typical gain control circuit using the LT1228 shown in Figure 36.111. The input is NTSC composite video, which can cover a 10dB range, from 0.56V to 1.8V. The output is to be 1V peak-to-peak into 75Ω. Amplitudes were measured from peak negative chroma to peak positive chroma on an NTSC modulated ramp test signal (see Differential Gain and Phase Section).

Note 9: Signal to noise ratio, S/N = 20 × log(RMS signal/RMS noise).

Notice that the signal is attenuated 20:1 by the 75Ω attenuator at the input of the LT1228, so the voltage on the input (Pin 3) ranges from 0.028V to 0.090V. This is done to limit distortion in the transconductance stage. The gain of this circuit is controlled by the current into the I_{SET} terminal, Pin 5 of the IC. In a closed-loop AGC system the loop control circuitry generates this current by comparing the output of a detector[10] to a reference voltage, integrating the difference and then converting to a suitable current. The measured performance for this circuit is presented in Table 36.7.

Table 36.7. Measured Performance Data (Uncorrected)

INPUT (V)	I_{SET} (mA)	DIFFERENTIAL GAIN	DIFFERENTIAL PHASE	S/N
0.03	1.93	0.5%	2.7°	55dB
0.06	0.90	1.2%	1.2°	56dB
0.09	0.584	10.8%	3.0°	57dB

All video measurements were taken with a Tektronix 1780R video measurement set, using test signals generated by a Tektronix TSG 120. The standard criteria for characterizing NTSC video color distortion are the differential gain and the differential phase. For a brief explanation of these tests see the "Differential Gain and Phase" section. For this design exercise the distortion limits were set at a somewhat arbitrary 3% for differential gain and 3° for differential phase. Depending on conditions, this should be barely visible on a video monitor.

Note 10: One way to do this is to sample the colorburst amplitude (the nominal peak-to-peak amplitude of the colorburst for NTSC is 40% of the peak luminance) with a sample-and-hold and peak detector.

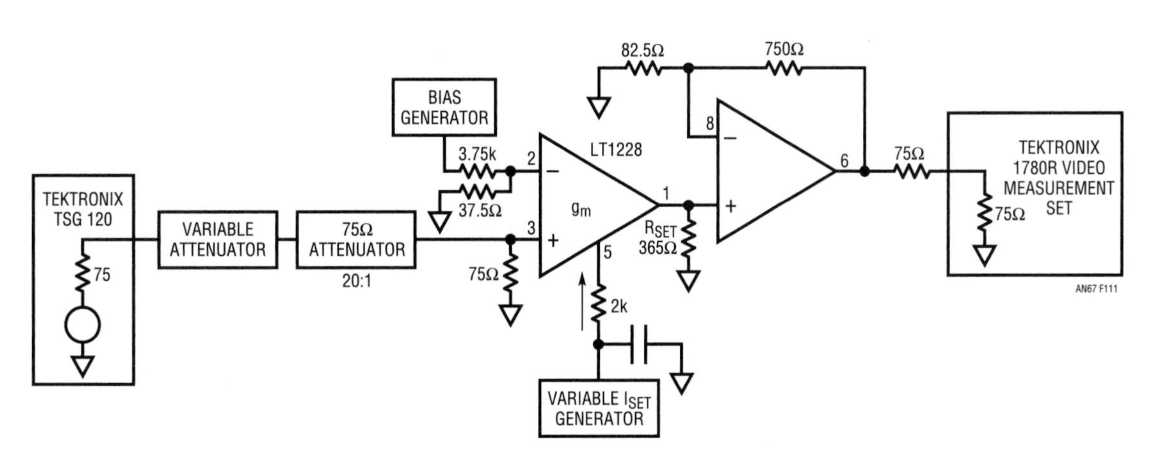

Figure 36.111 • Schematic Diagram

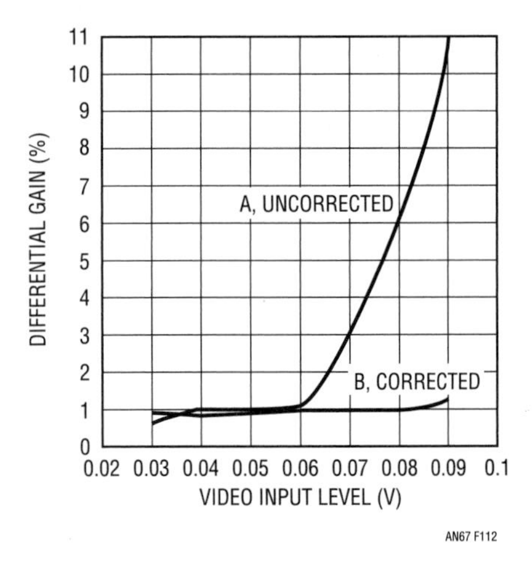

Figure 36.112 • Differential Gain vs Input Level

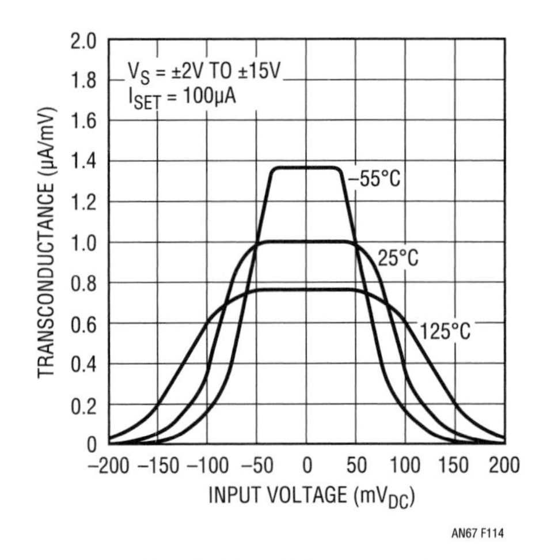

Figure 36.114 • Small-Signal Transconductance vs DC Input Voltage

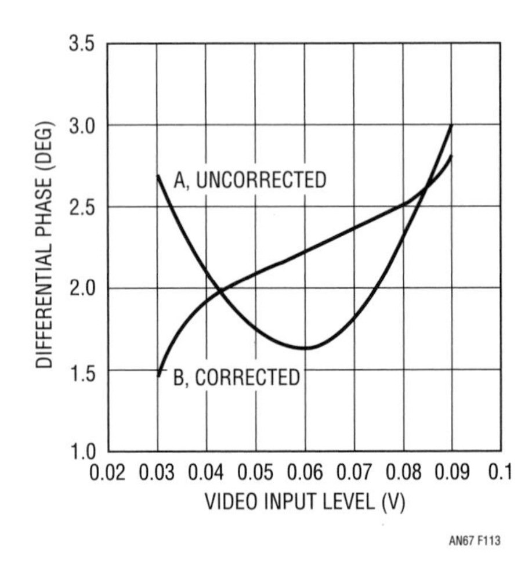

Figure 36.113 • Differential Phase vs Input Level

Figures 36.112 and 36.113 plot the measured differential gain and phase, respectively, against the input signal level (the curves labeled "A" show the uncorrected data from Table 36.7). The plots show that increasing the input signal level beyond 0.06V results in a rapid increase in the gain distortion, but comparatively little change in the phase distortion. Further attenuating the input signal (and consequently increasing the set current) would improve the differential gain performance but degrade the S/N. What this circuit needs is a good tweak.

Optimizing for differential gain

Referring to the small-signal transconductance versus DC input voltage graph (Figure 36.114), observe that the transconductance of the amplifier is linear over a region centered around 0V.[11] The 25°C g_m curve starts to become quite nonlinear above 0.050V. This explains why the differential gain (see Figure 36.112, curve A) degrades so quickly with signals above this level. Most RF signals do not have DC bias levels but the composite video signal is mostly unipolar.

Video is usually clamped at some DC level to allow easy processing of sync information. The sync tip, the chroma reference burst and some chroma signal information swing negative, but 80% of the signal that carries the critical color information (chroma) swings positive. Efficient use of the dynamic range of the LT1228 requires that the input signal have little or no offset. Offsetting the video signal so that the critical part of the chroma waveform is centered in the linear region of the transconductance amplifier allows a larger signal to be input before the onset of severe distortion. A simple way to do this is to bias the unused input (in this circuit the inverting input, Pin 2) with a DC level.

In a video system, it might be convenient to clamp the sync tip at a more negative voltage than usual. Clamping the signal prior to the gain-control stage is good practice because a stable DC reference level must be maintained.

The optimum value of the bias level on Pin 2 used for this evaluation was determined experimentally to be about 0.03V. The distortion tests were repeated with this bias voltage added. The results are reported in Table 36.8 and Figures 36.112 and 36.113 (curves B). The improvement to the differential phase is inconclusive, but the improvement in the differential gain is substantial.

Table 36.8. Measured Performance Data (Corrected)

INPUT (V)	BIAS VOLTAGE	I_{SET} (mA)	DIFFERENTIAL GAIN	DIFFERENTIAL PHASE	S/N
0.03	0.03	1.935	0.9%	1.45°	55dB
0.06	0.03	0.889	1.0%	2.25°	56dB
0.09	0.03	0.584	1.4%	2.85°	57dB

Note 11: Notice also that the linear region expands with higher temperature. Heating the chip has been suggested.

Differential gain and phase

Differential gain and phase are sensitive indications of chroma signal distortion. The NTSC system encodes color information on a separate subcarrier at 3.579545MHz. The color subcarrier is directly summed to the black and white video signal. (The black and white information is a voltage proportional to image intensity and is called luminance or luma.) Each line of video has a burst of 9 to 11 cycles of the subcarrier (so timed that it is not visible) that is used as a phase reference for demodulation of the color information of that line. The color signal is relatively immune to distortions, except for those that cause a phase shift or an amplitude error to the subcarrier during the period of the video line.

Differential gain is a measure of the gain error of a linear amplifier at the frequency of the color subcarrier. This distortion is measured with a test signal called a modulated ramp (shown in Figure 36.115). The modulated ramp consists of the color subcarrier frequency superimposed on a linear ramp (or sometimes on a stair step). The ramp has the duration of the active portion of a horizontal line of video. The amplitude of the ramp varies from zero to the maximum level of the luminance, which in this case, is 0.714V. The gain error corresponds to compression or expansion by the amplifier (sometimes called "incremental gain") and is expressed as a percentage of the full amplitude range. An appreciable amount of differential gain will cause the luminance to modulate the chroma, producing visual chroma distortion. The effect of differential gain errors is to change the saturation of the color being displayed. Saturation is the relative degree of dilution of a pure color with white. A 100% saturated color has 0% white, a 75% saturated color has 25% white, and so on. Pure red is 100% saturated, whereas pink is red with some percentage of white and is therefore less than 100% saturated.

Differential phase is a measure of the phase shift in a linear amplifier at the color subcarrier frequency when the modulated ramp signal is used as an input.

The phase shift is measured relative to the colorburst on the test waveform and is expressed in degrees. The visual effect of the distortion is a change in hue. Hue is that quality of perception which differentiates the frequency of the color, red from green, yellow-green from yellow, and so forth.

Three degrees of differential phase is about the lower limit that can unambiguously be detected by observers. This level of differential phase is just detectable on a video monitor as a shift in hue, mostly in the yellow-green region. Saturation errors are somewhat harder to see at these levels of distortion—3% of differential gain is very difficult to detect on a monitor. The test is performed by switching between a reference signal, SMPTE (Society of Motion Picture and Television Engineers) 75% color bars and a distorted version of the same signal, with matched signal levels. An observer is then asked to note any difference.

In professional video systems (studios, for instance) cascades of processing and gain blocks can reach hundreds of units. In order to maintain a quality video signal, the distortion contribution of each processing block must be a small fraction of the total allowed distortion budget[12] (the errors are cumulative). For this reason, high quality video amplifiers will have distortion specifications as low as a few thousandths of a degree for differential phase and a few thousandths of a percent for differential gain.

Note 12: From the preceding discussions, the limits on visibility are about 3° differential phase, 3% differential gain. Please note that these are not hard and fast limits. Tests of perception can be very subjective.

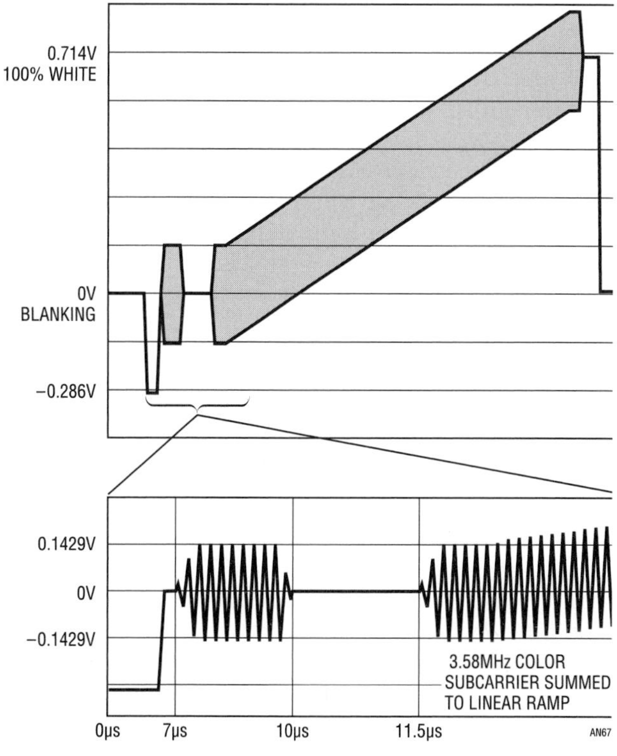

Figure 36.115 • NTSC Test Signal

LT1190 family ultrahigh speed op amp circuits

by John Wright and Mitchell Lee

Introduction

The LT1190 series op amps combine bandwidth, slew rate and output drive capability to satisfy the demands of many high speed applications. This family offers up to 350MHz gain bandwidth product and slew rates of 450V/μs while driving 150Ω (75Ω, double terminated) loads. In 50Ω systems, the LT1190 family can deliver 13.5dBm to a double terminated load. These parts are based on the familiar, easy-to-use, voltage mode feedback topology.

Small-signal performance

Figures 36.116 and 36.117 show the small-signal performance of the LT1190 and LT1191 when configured for gains of+1 and −1. The noninverting plots show peaking at 130MHz, which is characteristic of the socketed test fixture and supply bypass components. A tight PC board layout would reduce the LT1190 peaking to 2dB. The small-signal performance of an LM118 is shown for comparison.

Fast peak detectors

Fast peak detectors place unusual demands on amplifiers. The output stage must have a high slew rate in order to keep up with the intermediate stages of the amplifier. This condition causes either a long overload or DC accuracy errors. To maintain a high slew rate at the output, the amplifier must deliver large currents into the capacitive load of the detector. Other problems include amplifier instability with large capacitive loads and preservation of output voltage accuracy.

The LT1190 is the ideal candidate for this application, with a 450V/μs slew rate, 50mA output current and 70° phase margin. The closed-loop peak detector circuit of Figure 36.118 uses a Schottky diode inside the feedback loop to obtain good accuracy. A 20Ω resistor (Ro) isolates the 10nF load and prevents oscillation.

DC error with a sine wave input is plotted in Figure 36.119 for various input amplitudes. The DC value is read with a DVM. At low frequencies, the error is small and is dominated by the decay of the detector capacitor between cycles. As frequency rises, the error increases because capacitor charging time decreases. During this time the overdrive becomes a very small portion of a sine wave cycle. Finally, at approximately 4MHz, the error rises rapidly owing to the slew-rate limitation of the op amp. For comparison purposes, the error of an LM118 is also plotted for $V_{IN} = 2V_{P-P}$.

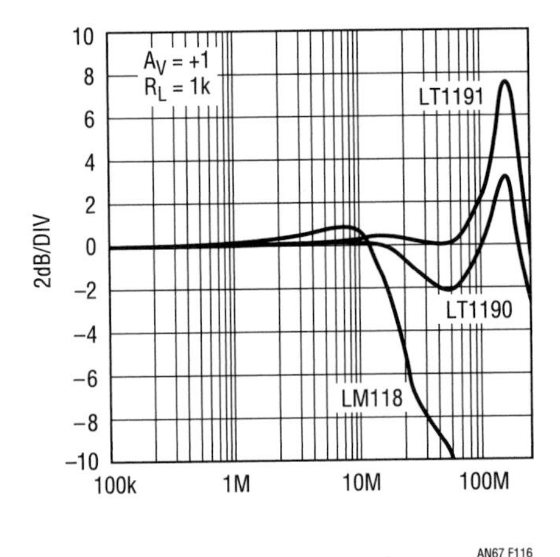

Figure 36.116 • Small-Signal Response $A_V = +1$. 130MHz Peaking Due to Socket and Bypass Components

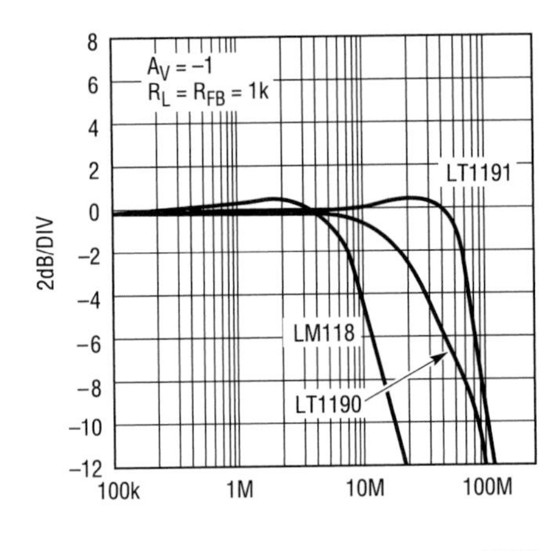

Figure 36.117 • Small-Signal Response $A_V = -1$

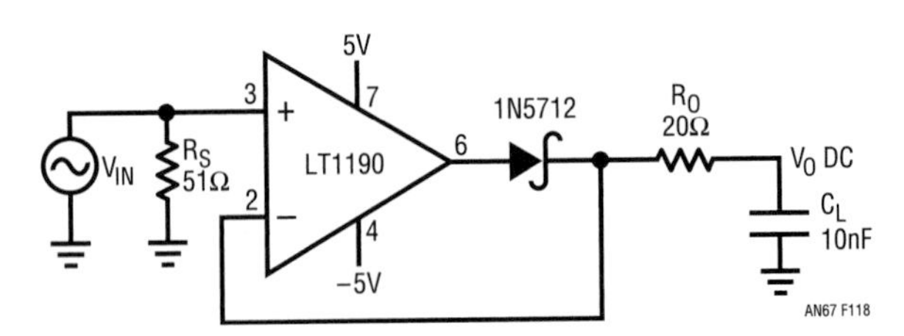

Figure 36.118 • Closed-Loop Peak Detector

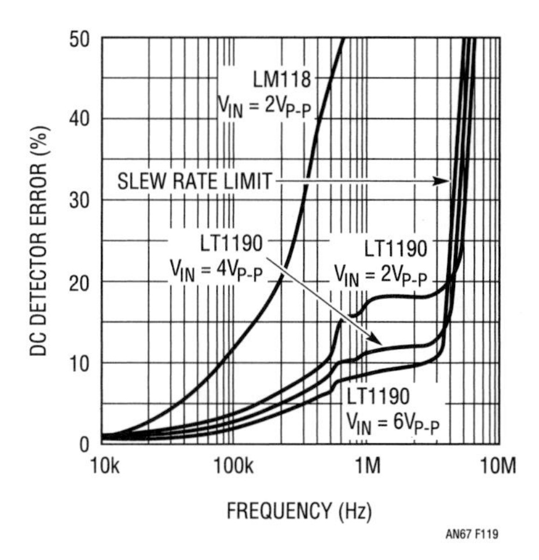

Figure 36.119 • Closed-Loop Peak Detector Error vs Frequency

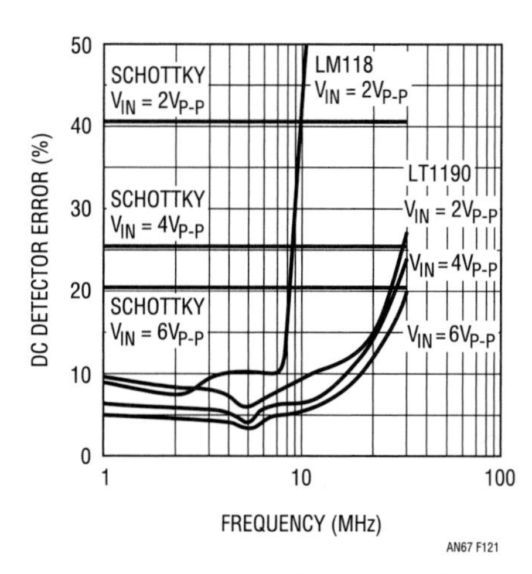

Figure 36.121 • Open-Loop Peak Detector Error vs Frequency

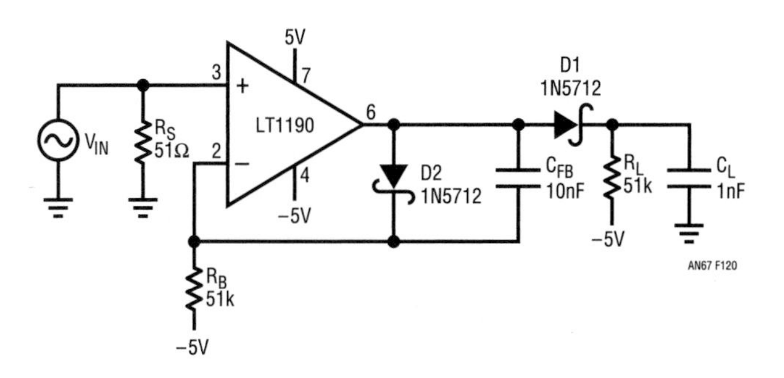

Figure 36.120 • Open-Loop, High Speed Peak Detector

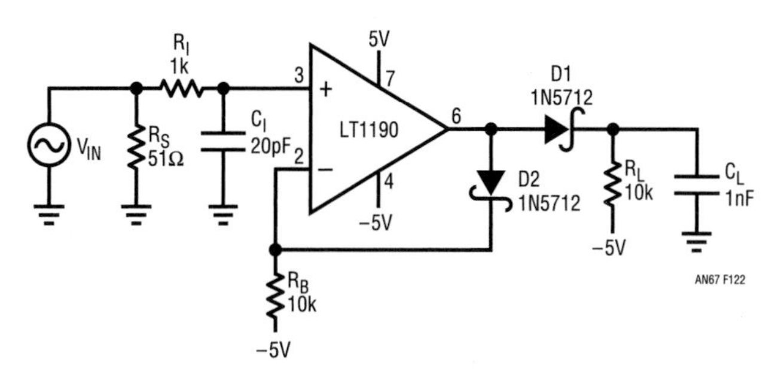

Figure 36.122 • Fast Pulse Detector

A Schottky diode peak detector can be built with a 1nF capacitor and a 10kΩ pulldown. Although this simple circuit is very fast, it has limited usefulness because of the error of the diode threshold and its low input impedance. The accuracy of this simple detector can be improved with the LT1190 circuit of Figure 36.120.

In this open loop design, D1 is the detector diode and D2 is a level shifting or compensating diode. A load resistor, R_L, is connected to −5V and an identical bias resistor, R_B, is used to bias the compensating diode. Equal value resistors ensure that the diode drops are equal. Low values of R_L and R_B (1k to 10k) provide fast response, but at the expense of poor low frequency accuracy. High values of R_L and R_B provide good low frequency accuracy but cause the amplifier to slew rate limit, resulting in poor high frequency accuracy. A good compromise can be made by adding a feedback capacitor, C_{FB}, which enhances the negative slew rate on the (−) input.

The DC error with a sine wave input, as read with a DVM, is plotted in Figure 36.121. For comparison purposes the LM118 error is plotted as well as the error of the simple Schottky detector.

Pulse detector

A fast pulse detector can be made with the circuit of Figure 36.122. A very fast input pulse will exceed the amplifier's slew rate and cause a long overload recovery time. Some amount of dV/dt limiting on the input can help this overload condition; however, it will delay the response.

Figure 36.123 shows the detector error versus pulse width. Figure 36.124 is the response to a $4V_{P\text{-}P}$ input pulse that is 80ns wide. The maximum output slew rate in the photo is 70V/µs. This rate is set by the 70mA current limit driving 1nF As a performance benchmark, the LM118 takes 1.2µs to peak detect and settle, given the

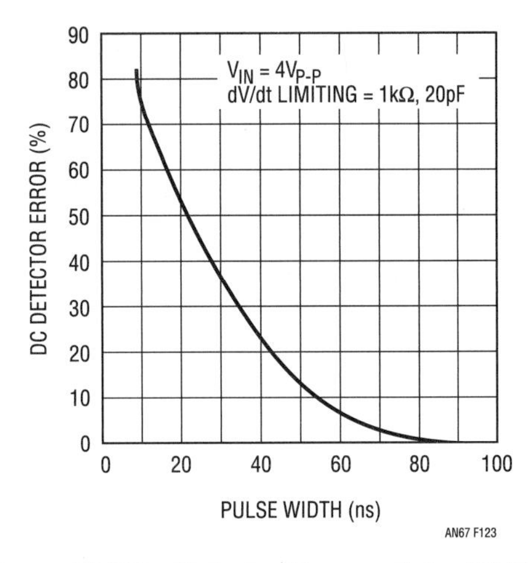

Figure 36.123 • Detector Error vs Pulse Width

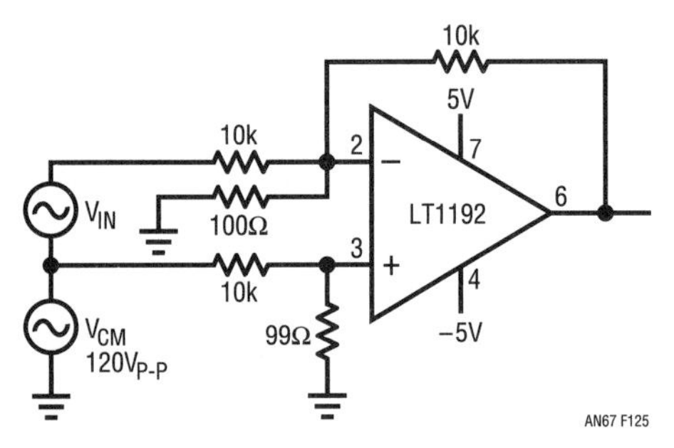

Figure 36.125 • 3.5MHz Instrumentation Amplifier Rejects $120V_{P-P}$

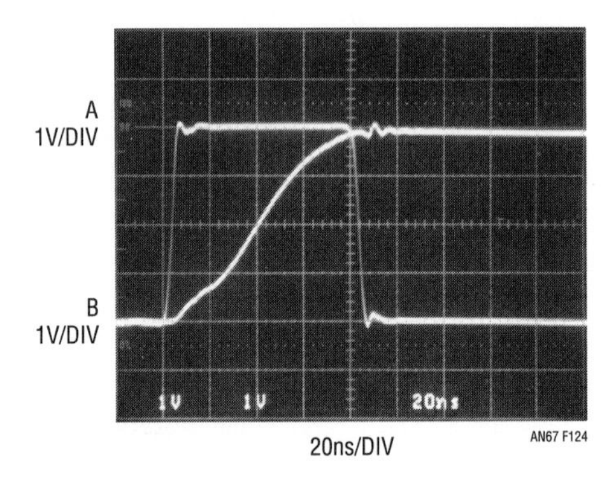

Figure 36.124 • Open-Loop Peak Detector Response

same amplitude input. This slower response is due in part to the much lower slew rate and lower phase margin of the LM118.

Instrumentation amplifier rejects high voltage

Instrumentation amplifiers are normally used to process slowly varying outputs from transducers, rather than fast signals. However, it is possible to make an instrumentation amplifier that responds very quickly, with good common mode rejection. For the circuit of Figure 36.125, an LT1192 is used to obtain 50dB of CMRR from a $120V_{P-P}$ signal. In this application, the CMRR is limited by the matching of the resistors, which should match to better than 0.01%.

An LT1192 is used in this application because the circuit has a noise gain of 100 and because the higher gain bandwidth of the LT1192 allows a −3dB bandwidth of 3.5MHz. Note also that the 100:1 attenuation of the common mode signal presents a common mode voltage to the amplifier of only $1.2V_{P-P}$. Figure 36.126 shows the amplifier output for a 1MHz square wave riding on a $120V_{P-P}$, 60Hz signal. The circuit exhibits 50dB rejection of the common mode signal.

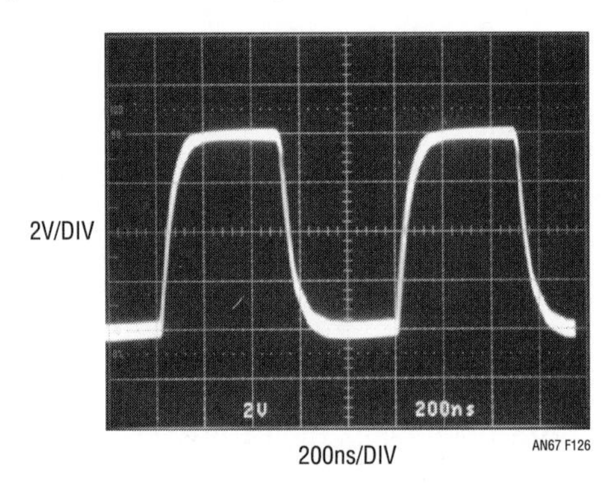

Figure 36.126 • Open-Loop Peak Detector Response

Crystal oscillator

Op amps have found wide use in low frequency (≤100kHz) crystal oscillator circuits, but just haven't had the bandwidth to operate successfully at higher frequencies. The LT1190 and LT1191 make excellent gain stages for high-frequency Colpitts oscillators. A practical implementation is shown in Figure 36.127.

Gain limiting is provided by two Schottky diodes, which maintain the output at approximately+11dBm—sufficient to directly drive +7 or+10dBm diode-ring mixers. Output-stage clipping is not recommended as a means of gain limiting, as this increases distortion and allows internal nodes to be overdriven. The recovery time would add excessive

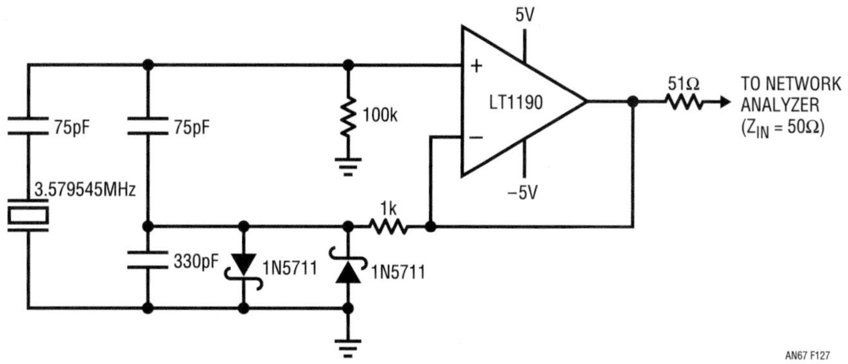

Figure 36.127 • High Frequency Colpitts Oscillator

phase shift in the oscillator loop, degrading frequency stability.

Distortion performance is good, considering that the oscillator consists of one stage and can deliver useful output power. Figure 36.128 shows a spectral plot of the oscillator's output. The second harmonic is approximately 37dB down, limited primarily by the clipping action of the Schottky diodes. Power supply rejection is excellent, showing a frequency sensitivity of approximately 0.1 ppm/V The LT1190 gives acceptable performance to 10MHz, while the LT1191 extends the circuit's operating range to 20MHz.

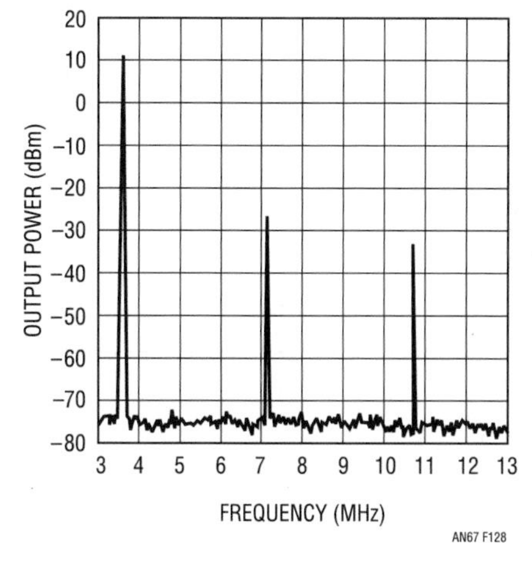

Figure 36.128 • Oscillator Output Spectrum

An LT1112 dual output buffered reference

by George Erdi

A dual output buffered reference application is shown in Figure 36.129.

Figure 36.129 works on two AA batteries, which can be discharged to ±1.3V With two equal 20k resistors, two equal but opposite sign reference voltages are available. Changing the ratio of the two 0.1% resistors allows for other values: one positive and one negative.

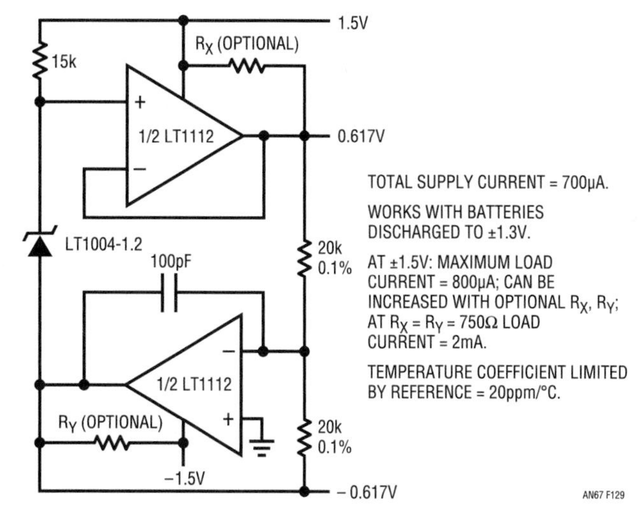

Figure 36.129 • Dual Output Reference Operates on Two AA Cells

Three op amp instrumentation amp using the LT1112/LT1114

by George Erdi

The LT1112/LT1114 are dual and quad universal precision op amps. All important precision specifications have been maintained:

1. Microvolt offset voltage; the low cost grades (including the small outline, 8-pin surface mount package) are guaranteed to 75μV
2. Drift guaranteed to 0.5μV/°C (0.75μV/°C low cost grades)
3. Bias and offset currents are in the picoampere range, even at 125°C
4. Low noise: 0.32μV peak-to-peak, 0.1Hz to 10Hz
5. Supply current is 400μA max per amplifier
6. Voltage gain is in excess of one million

The LT1112/LT1114 also provide a full set of matching specifications, facilitating their use in such matching dependent applications as the three op amp instrumentation amplifier shown in Figure 36.130. The performance of this instrumentation amplifier depends only on the matching parameters not the specifications of the individual amplifiers.

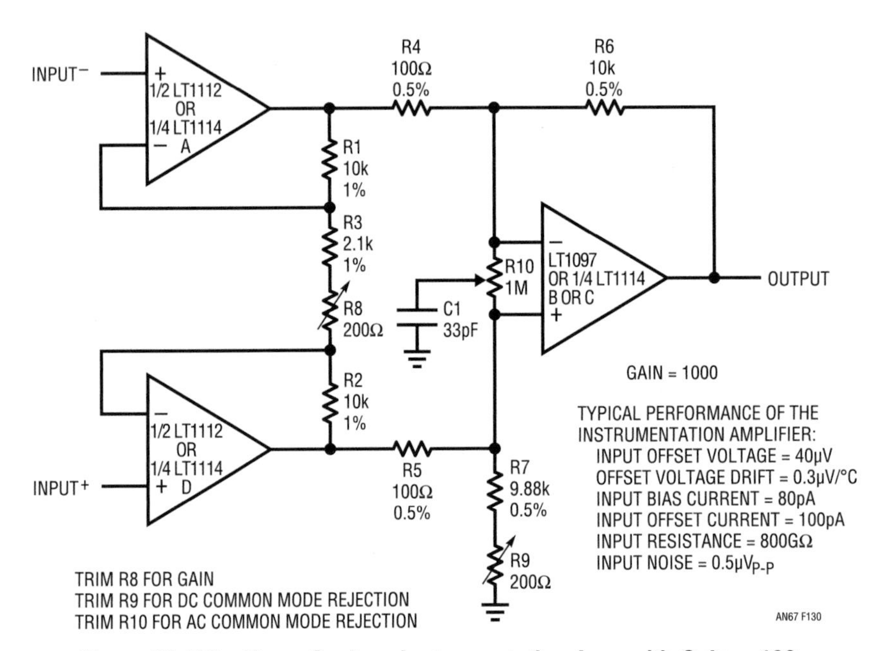

Figure 36.130 • Three Op Amp Instrumentation Amp with Gain = 100

Ultralow noise, three op amp instrumentation amplifier

by George Erdi and Alexander Strong

Op amp instrumentation amplifiers usually have op amps with a fixed gain greater than one at the input stage (Figure 36.131). At low frequencies, decompensated op amps work well, but at high frequencies and with one input grounded, the virtual ground begins to lose its integrity. As the frequency of the input signal increases, the amplitude at the virtual ground increases, making the virtual ground look inductive, eventually requiring a unity-gain stable amplifier. The LT1028 can be made stable under these conditions with bypass capacitors and a little experimenting, but the LT1128 is unconditionally stable.

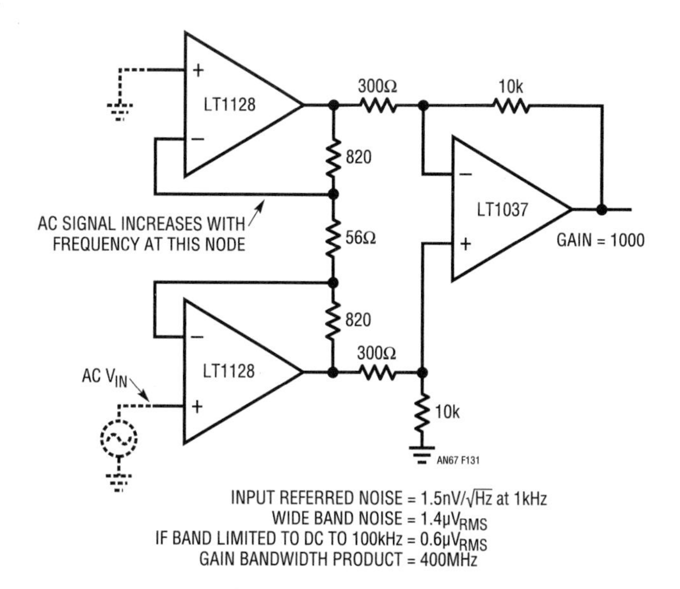

Figure 36.131 • Three Op Amp, Ultralow Noise Instrumentation Amplifier

A temperature compensated, voltage-controlled gain amplifier using the LT1228

by Frank Cox

It is often convenient to control the gain of a video or intermediate frequency (IF) circuit with a voltage. The LT1228, along with a suitable voltage-to-current converter circuit, forms a versatile gain control building block ideal for many of these applications. In addition to gain control over video bandwidths this circuit can add a differential input and has sufficient output drive for 50Ω systems.

The transconductance of the LT1228 is inversely proportional to absolute temperature at a rate of −0.33%/°C. For circuits using closed-loop gain control (i.e., IF or video automatic gain control) this temperature coefficient does not present a problem. However, open-loop gain control circuits that require accurate gains may require some compensation. The circuit described here uses a simple thermistor network in the voltage-to-current converter to achieve this compensation. Table 36.9 summarizes the circuit's performance.

Table 36.9. Characteristics of Example

Input Signal Range	0.5V to 3.0V pk
Desired Output Voltage	1.0V pk
Frequency Range	0Hz to 5MHz
Operating Temperature Range	0°C to 50°C
Supply Voltages	±15V
Output Load	150Ω (75Ω + 75Ω)
Control Voltage vs Gain Relationship	0V to 5V Min to Max Gain
Gain Variation Over Temperature	±3% from Gain at 25°C

Figure 36.132 shows the complete schematic of the gain control amplifier. Please note that these component choices are not the only ones that will work nor are they necessarily the best. This circuit is intended to demonstrate one approach out of many for this very versatile part and, as always, the designer's engineering judgment must be fully engaged. Selection of the values for the input attenuator, gain-set resistor and current feedback amplifier

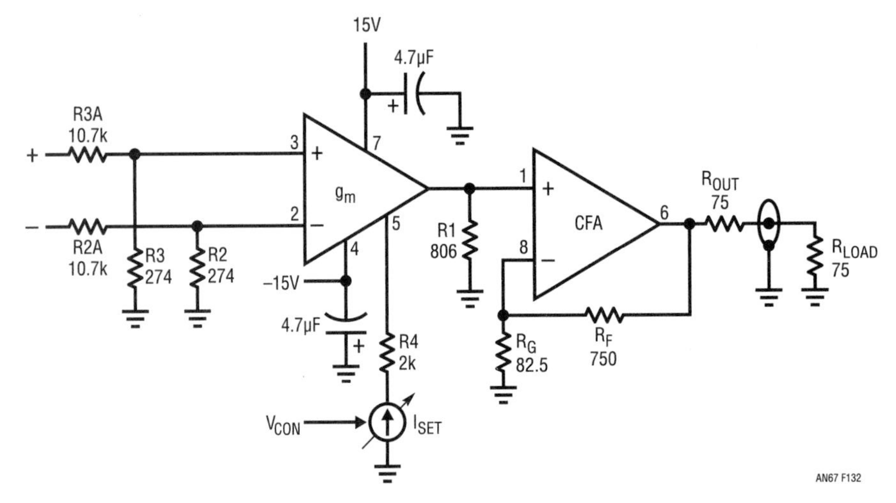

Figure 36.132 • Differential Input, Variable Gain Amplifier

resistors is relatively straightforward, although some iteration is usually necessary. For the best bandwidth, remember to keep the gain-set resistor, R1, as small as possible and the set current as large as possible (with due regard for gain compression). The voltage-controlled current source (I_{SET}) is detailed in the boxed section.

Several of these circuits have been built and tested using various gain options and different thermistor values. Test results for one of these circuits are shown in Figure 36.133. The gain error versus temperature for this circuit is well within the limit of ±3%. Compensation over a much wider range of temperatures or to tighter tolerances is possible, but would generally require more sophisticated methods, such as multiple thermistor networks.

The VCCS is a standard circuit with the exception of the current set resistor R5, which is made to have a temperature coefficient of −0.33%/°C. R6 sets the overall gain and is made adjustable to trim out the initial tolerance in the LT1228 gain characteristic. A resistor (R_P) in parallel with the thermistor will tend, over a relatively small range, to linearize the change in resistance of the combination with temperature. R_S trims the temperature coefficient of the network to the desired value.

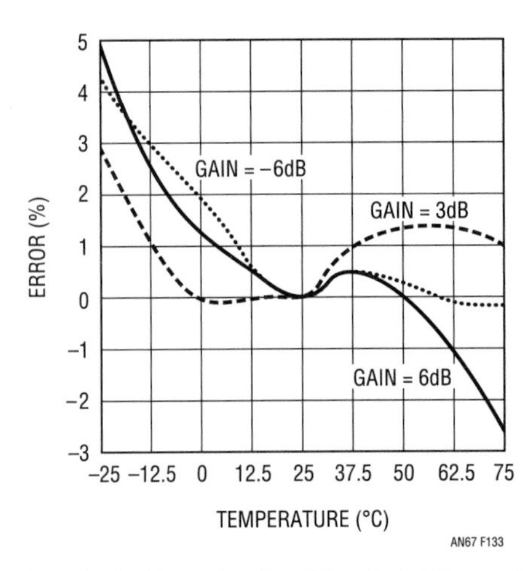

Figure 36.133 • Gain Error for the Circuit in Figure 36.132 Plus the Temperature Compensation Circuit Shown in Figure 36.134 (Normalized to Gain at 25°C)

Voltage controlled current source (VCCS) with a compensating temperature coefficient

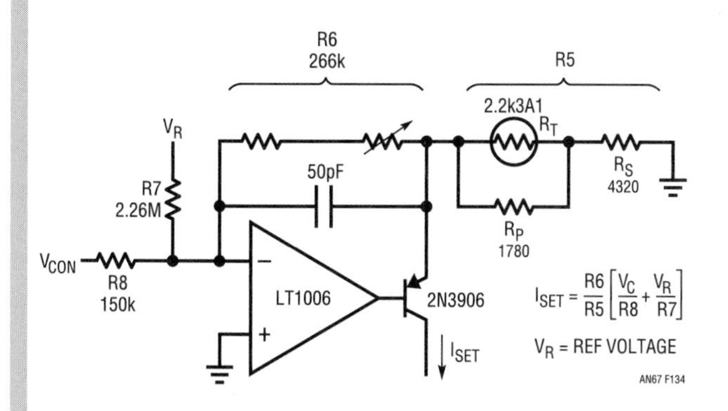

Figure 36.134 • Voltage Controlled Current Source (VCCS) with a Compensating Temperature Coefficient

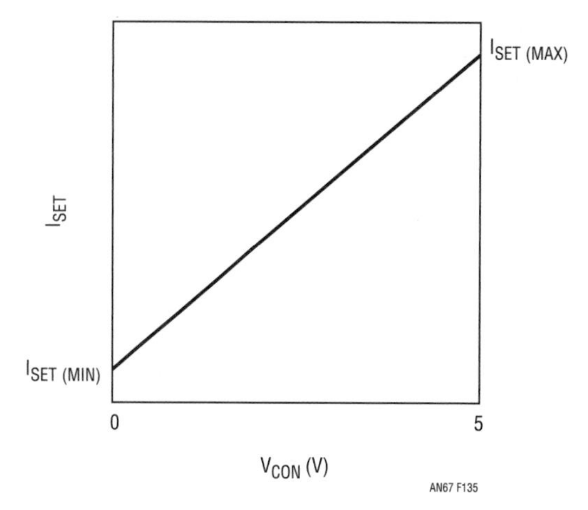

Figure 36.135 • Voltage Control of I_{SET} with Temperature Compensation

VCCS design steps

1. Measure or obtain from the data sheet the thermistor resistance at three equally spaced temperatures (in this case 0°C, 25°C and 50°C). Find R_P from:

$$RP = \frac{(R0 \bullet R25 + R25 \bullet R50 - 2 \bullet R0 \bullet R50)}{R0 + R50 - 2 \bullet R25}$$

where: R0 = thermistor resistance at 0°C
R25 = thermistor resistance at 25°C
R50 = thermistor resistance at 50°C

2. Resistor R_P is placed in parallel with the thermistor. This network has a temperature dependence that is approximately linear over the range given (0°C to 50°C).
3. The parallel combination of the thermistor and R_P ($R_P||R_T$) has a temperature coefficient of resistance (TC) given by:

$$TC\ R_P \,||\, R_T = \left(\frac{R0 \,||\, R_P - R50 \,||\, R_P}{R25 \,||\, R_P}\right)\left(\frac{100}{T_{HIGH} - T_{LOW}}\right)$$

where: T_{HIGH} = the high temperature
T_{LOW} = the low temperature
R_T = the thermistor

4. The desired temp. co. to compensate the LT1228 gain temperature dependence is -0.33%/°C. A series resistance (R_S) is added to the parallel network to trim its TC to the proper value. R_S is given by:

$$\left(\frac{TCR_P \,||\, R_T}{-0.33}\right)(R_P \,||\, R25) - (R_P \,||\, R_T)$$

5. R6 contributes to the resultant temperature and so is made large with respect to R5.
6. The other resistors are calculated to give the desired range of I_{SET}.

This procedure was performed using a variety of thermistors (one possible source is BetaTHERM Corporation—phone 508-842-0516). Figure 36.5 shows typical results reported as errors normalized to a resistance with a −0.33%/°C temperature coefficient. As a practical matter, the thermistor need only have about a 10% tolerance for this gain accuracy. The sensitivity of the gain accuracy to the thermistor tolerance is decreased by the linearization network, in the same ratio as is the temperature coefficient; the room temperature gain may be trimmed with R6. Of course, particular applications require analysis of aging stability, interchangeability, package style, cost, and the contributions of the tolerances of the other components in the circuit.

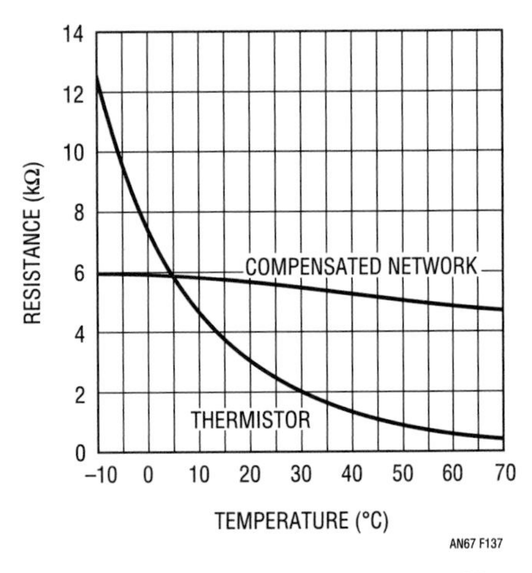

Figure 36.136 • Thermistor and Thermistor Network Resistance vs Temperature

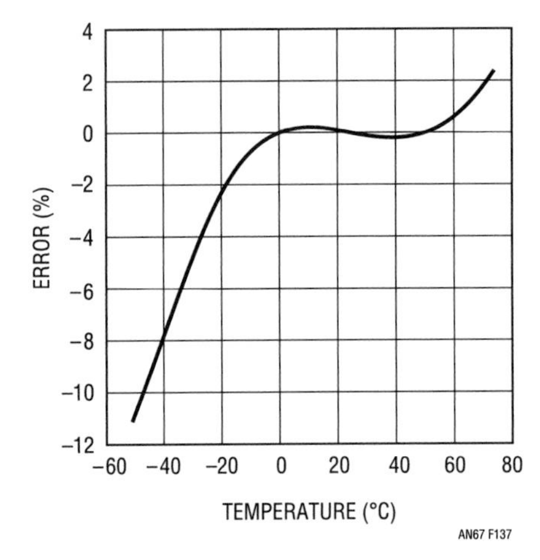

Figure 36.137 • Thermistor Network Resistance Normalized to a Resistor with Exact –0.33%/°C Temperature Coefficient

The LTC1100, LT1101 and LT1102: a trio of effective instrumentation amplifiers

by George Erdi

Next to the universally used op amp, perhaps the most useful linear IC building block is the instrumentation amp, or "IA." Using IAs effectively can in some ways be more challenging than selecting op amps, because IAs have different specs and can also use different topologies. However, the basic task is a fixed gain, differential input, single-ended output amplifier, the definition of an IA. The differential signal typically rides on top of a common mode signal; the differential input is amplified and the common mode voltage is rejected by the IA.

The instrumentation amplifier can be implemented with dedicated IA designs, or with one to three op amps to realize the gain function, and a minimum of four ratio-matched precision resistors configured as two like ratio pairs.

The most familiar IA type is the single op amp variety, usually called a difference amplifier and shown in Figure 36.138. Using just two parts (one op amp and one resistor network), this IA is the height of simplicity and utility. For modest requirements it is built with just a general purpose op amp and four precision resistors. A drawback to this type of IA is that the resistor bridge loads the source. The three op amp configuration uses seven resistors and has high input impedance. It is obviously more difficult to implement than the single op amp version. A nice compromise between these two approaches is illustrated in Figure 36.139. This IA design uses two op amps to buffer the signal inputs and requires only four resistors. The use of two op amps with modern dual devices causes no penalty, and in fact this arrangement has real virtues over the more basic setup of Figure 36.138.

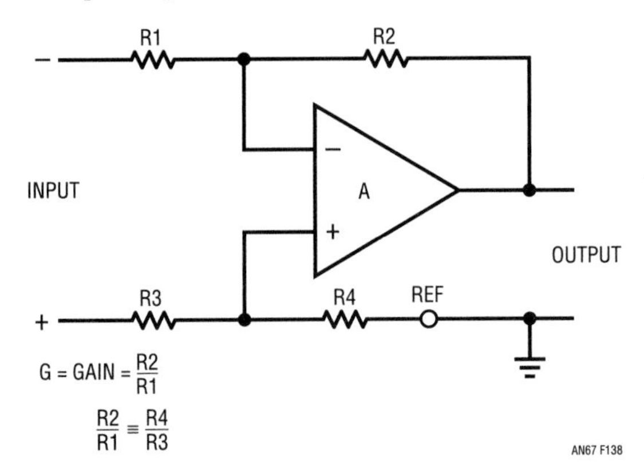

Figure 36.138 • Basic Single Op Amp Instrumentation Amplifier

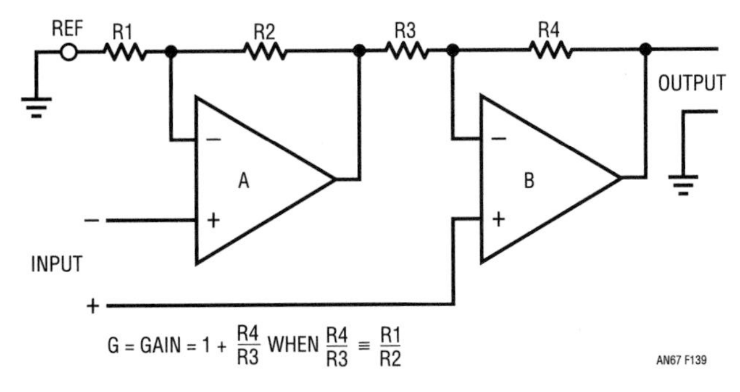

Figure 36.139 • Buffered Dual Op Amp Instrumentation Amplifier

This IA architecture presents minimum loading to the differential source, namely the bias current of the op amp used, which is balanced between the two inputs. The resistor network needs very precise trimming for high common mode rejection (CMRR) and gain accuracy. The trimming is noninteractive; first the R4/R3 ratio is trimmed for gain accuracy then the R1/R2 ratio is trimmed for high CMRR. Trimming compensates not only for resistor inaccuracies, but also for the finite gain and CMRR of the op amps. The amplified difference appears between the output terminal and the voltage applied to the REF terminal (normally grounded).

As a basic building block, this IA can be performance optimized for various applications by a choice of op amps. LTC has taken this step with the LTC1100, LT1101 and LT1102, an instrumentation amplifier series offered in an 8-pin footprint with connections as shown in Figure 36.140. As illustrated, the gain of these IAs is user programmed by taps on the resistor array, for pre-trimmed precision gains of either 10 or 100 for the LT1101 and LT1102. The 8-pin LTC1100 has a fixed gain of 100, but makes the summing points available for user connections. The key specifications of these three devices are summarized in Table 36.10.

Table 36.10. LTC Instrumentation Amplifier Specifications[1]

	LTC1100C	LT1101C/I/M	LT1102C/I/M
Available Gains	100[2]	10/100	10/100
Gain Error (%)	0.01	0.01	0.01
Gain Nonlinearity (ppm)	3	3	7
Gain Drift (ppm/°C)	2	2	10
Vos (μV)	1	60	200
V_{OS} Drift (μV/°C)	0.005	0.5	3
I_B (PA)	2.5	6000	4
I_{OS} (pA)	10	150	4
e_n	1.9μV_{P-P} (DC to 10Hz)	0.9μV_{P-P} (0.1Hz to 10Hz)	20nV/(Hz)1/2 (at 1kHz)
CMRR (dB)	110	112	98
PSRR (dB)	130	114	102
V_S (Total, Mode)	4V to 18V (Single/Dual)	1.8V to 44V (Single/Dual)	10V to 44V (Dual)
I_S (mA)	2.4	0.09	3.4
Gain Bandwidth (MHz)	2	0.37	35
SR (V/μs)	4	0.1	30

[1]Unless otherwise stated all specifications are typical at $T_A = 25°C$. $V_S = \pm 15V$ for LT1101/LT1102 and ±5V for LTC1100.[2]A gain option of 10/100 is available in LTC1100CS (16-Lead SW)

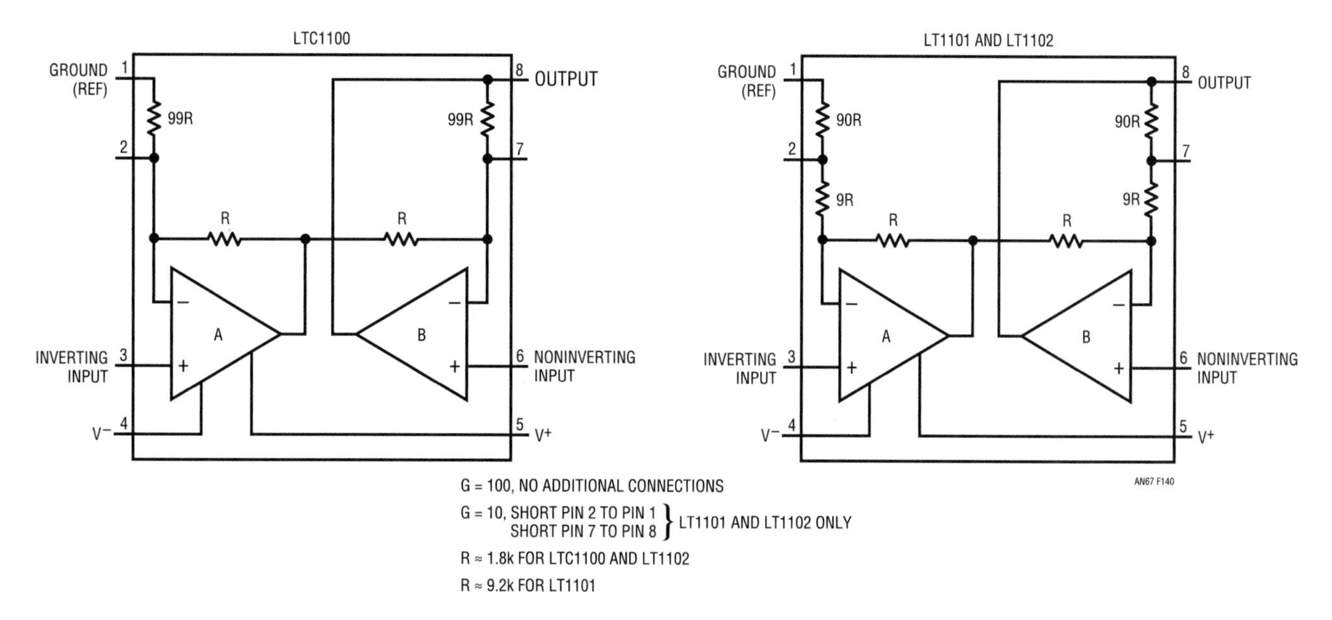

Figure 36.140 • Instrumentation Amplifiers in 8-Pin Packages

It is apparent from Table 36.10 that for these three IAs, there are no output contributions to input errors. With dedicated IA's or with the three op amp configuration there are separate specifications for input and output offset voltage, input and output drift and noise, and input and output power supply rejection ratio. To calculate system errors these input and output terms must be combined. With the LTC1100/01/02 these error calculations are simple.

With these three IA choices, the user can optimize performance for a variety of factors. The LTC1100 operates with dual or single supplies ranging from 4V to 18V whereas the LTC1101 accepts a supply range of from 1.8V to 40V. In addition, the LT1101 consumes only 100μA standby current. For applications that require very low offset voltage and drift, the LTC1100 excels with 1μV of offset and 5nV/°C drift. Where both high speed and low bias current are important, the LT1102 is the IA of choice, albeit at a cost of slightly higher power consumption and dual supplies. As can be seen from the table, all of these devices are outstanding with regard to gain accuracy, linearity and stability. The LTC1100, which is based on a dual chopper amplifier prototype (the LTC1051), is by far the best in terms of offset and drift. Either the LTC1100 or the LT1102 could be the unit of choice in terms of lowest bias current, with the LT1102 gaining an edge at higher temperatures.

Applications considerations

While this IA type is generally outstanding in terms of performance and simplicity, independent of the op amps, some caveats apply to using it most effectively. One concern is AC CMRR. As noted in Figure 36.140, the first op amp (A) is configured for unity gain, and the second op amp (B) provides all of the voltage gain. This has the effect of making the respective CMRR's frequency mismatched, since the CMRR of the higher gain, "B" side, corners at a much lower frequency. The resulting differential CMRR will therefore degrade more quickly with frequency than that of a topology with better AC balance. On the LT1102 this problem is resolved by decompensating amplifier B to gain-of-ten stability. This increases slew rate and bandwidth and also matches the CMRR rolloff with the frequencies of the two op amps when G = 10. At a gain of 100, this rolloff match no longer holds. However, connecting an 18pF capacitor between Pins 1 and 2 matches the CMRRs of the two sides and improves CMRR by an order of magnitude in the 300Hz to 30kHz range (Figure 36.141). As shown on the LTC1100 and LTC1101 data sheets, similar improvements can be obtained from those devices by connecting external capacitors.

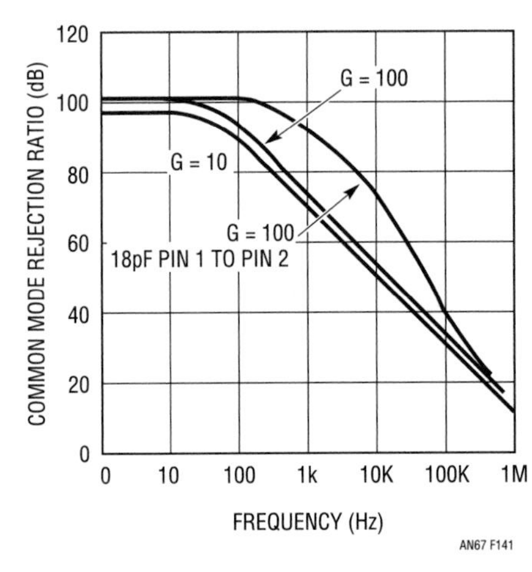

Figure 36.141 • LT1102 Common Mode Rejection Ratio vs Frequency

The LTC1100 and LT1101 also present some important usage considerations because of their single supply abilities, i.e., when operating with the V⁻terminal tied to ground. In this configuration, these devices handle CM inputs near ground and voltage swings to ground and their reference terminals can be tied to ground. One of the most common uses of these two IAs is as bridge amplifiers in conjunction with single supply powered DC strain gauges. As such, these IAs have a unique ability to deliver high gain with precision, while operating with a 1/2 supply voltage CM input. At first glance, it appears that a dual supply IA could operate, for example, on a 9V battery supply with 4.5V common mode input, but its output will not swing to ground and its reference terminal cannot be tied to ground.

For SPICE simulation purposes, a model for the LT1101 is included in the LTC macromodel library. The model is configured as the resistor network shown for the LT1101, combined with a model for the LT1078. A similar model for the LTC1100 can be made by scaling the four resistors appropriately, and using an LTC1051 model from the same library. A close model approximation for the LT1102 can be made with the LT1102 resistor values, combined with an LT1057 model for the "A" side, and a LT1022 model for the "B" side (both also in the library).

Miscellaneous circuits

Driving a high level diode ring mixer with an operational amplifier

by Mitchell Lee

One of the most popular RF building blocks is the diode ring mixer. Consisting of a diode ring and two coupling transformers, this simple device is a favorite with RF designers anywhere a quick multiplication is required, as in frequency conversion, frequency synthesis or phase detection. In many applications these mixers are driven from an oscillator. Rarely does anyone try building an oscillator capable of delivering 7dBm for a "minimum geometry" mixer, let alone one of higher level. One or more stages of amplification are added to achieve the drive level required by the mixer. The new LT1206 high speed amplifier makes it possible to amplify an oscillator to 27dBm in one stage.

Figure 36.143 shows the complete circuit diagram for a crystal oscillator, LT1206 op amp/buffer and diode-ring mixer. Most of the components are used in the oscillator itself, which is of the Colpitts class. Borrowing from a technique used in Hewlett Packard's Unit Oscillator, the current of the crystal is amplified rather than the voltage. There are several advantages to this method, the most important of which is low distortion. Although the voltages present in this circuit have poor wave shape and are sensitive to loading, the crystal current represents essentially a filtered version of the voltage waveform and is relatively tolerant of loading effects.

The impedance, and therefore the voltage at the bottom of the crystal, is kept low by injecting the current into the summing node of an LT1206 current feedback amplifier. Loop gain reduces the input impedance to well under 1Ω. Oscillator bias is adjustable, allowing control of the mixer drive. This also provides a convenient point for closing an output power servo loop.

Operating from ±15V supplies, the LT1206 can deliver 32dBm to a 50Ω load, and with a little extra headroom (the absolute maximum supply voltage is ±18V), it can reach 2W output power into 50Ω. Peak guaranteed output current is 250mA.

Shown in Figures 36.144 to 36.148 are spectral plots for various combinations of single and double termination at power levels ranging from +17dBm to +27dBm—not bad for an inductorless circuit. Double termination may be used to present a 50Ω source impedance to the mixer, or to isolate two or more mixers driven simultaneously from one LT1206 amplifier.

Although a 10MHz example has been presented here, the LT1206's 65MHz bandwidth makes it useful in circuits up to 30MHz. In addition, the shutdown feature can be used to interrupt drive to the mixer. When the LT1206 is shut down, the oscillator will likely stop, since the crystal

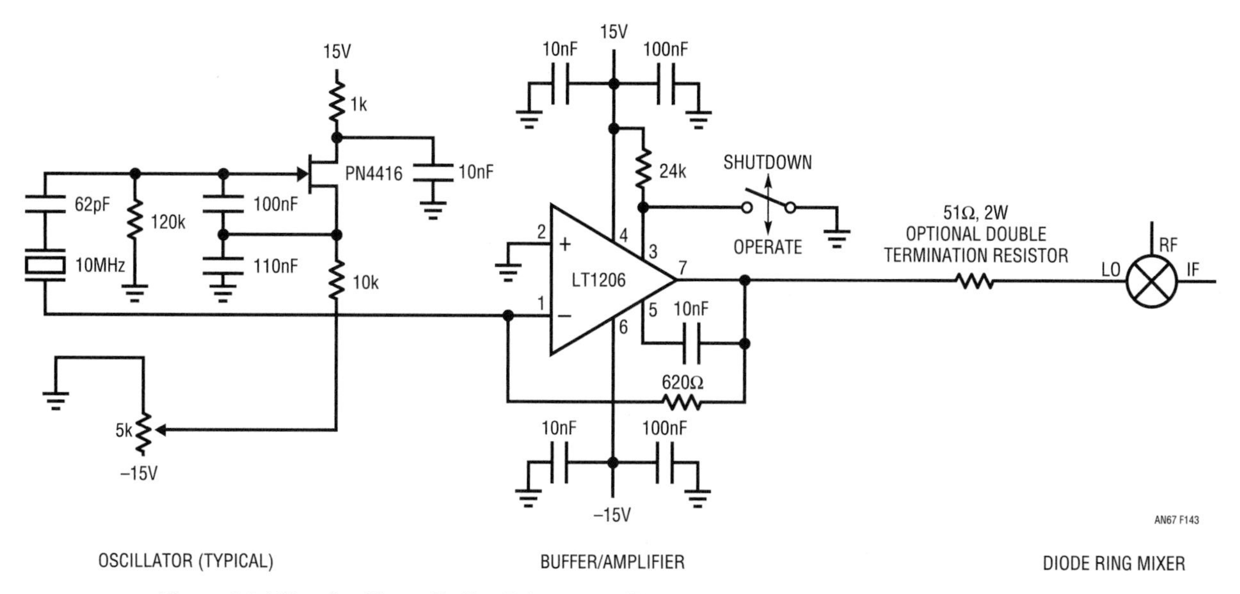

Figure 36.143 • Oscillator Buffer Drives +17dBm to +27dBm Double Balanced Mixers

then sees a series impedance of 620Ω and the mixer itself. Upon re-enabling the LT1206 there will be some time delay before the oscillator returns to full power The circuit works equally well with an LC version of the oscillator.

Note that the current feedback topology is inherently tolerant of stray capacitive effects at the summing node, making it ideal for this application. Another nice feature is the LT1206's ability to drive heavy capacitive loads while remaining stable and free of spurious oscillations.

For mixers below +17dBm, the LT1227 is a lower cost alternative, featuring 140MHz bandwidth in combination with the shutdown feature of the LT1206.

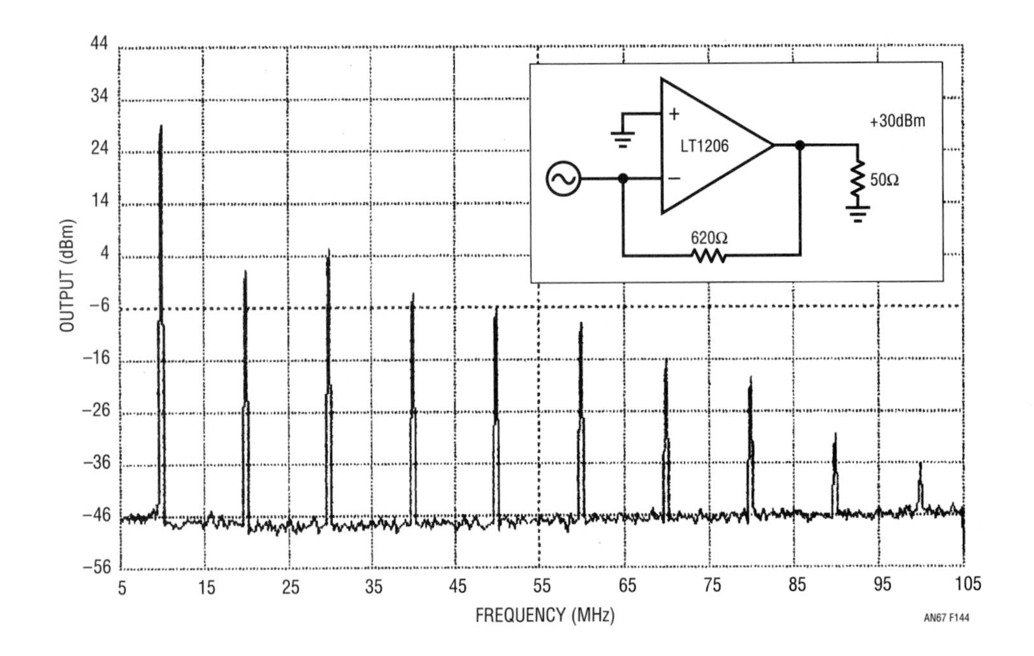

Figure 36.144 • Spectrum Plot of Figure 36.143's Circuit Driving +30dBm into a 50Ω Load (Single Termination)

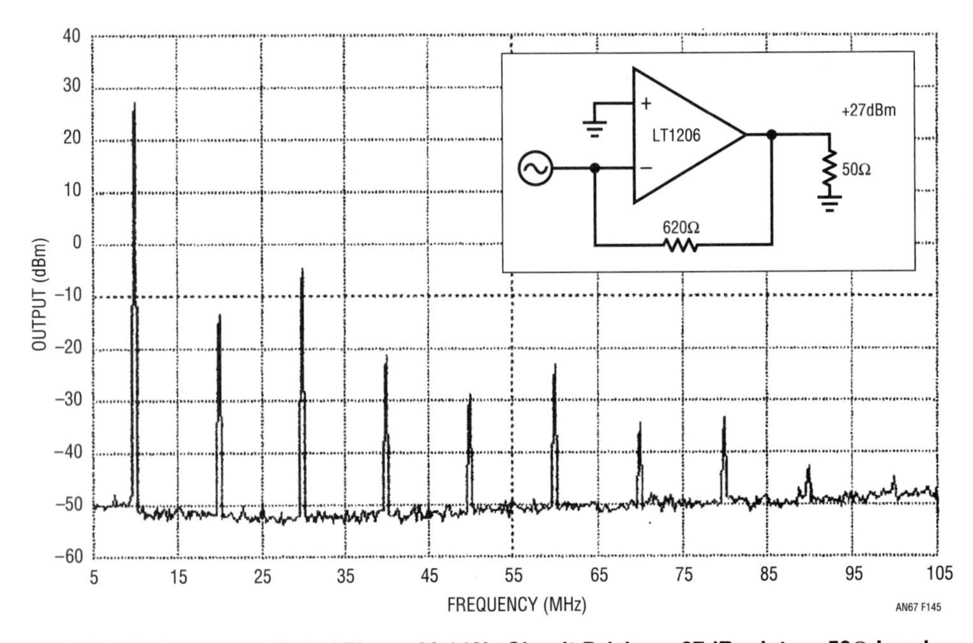

Figure 36.145 • Spectrum Plot of Figure 36.143's Circuit Driving +27dBm into a 50Ω Load

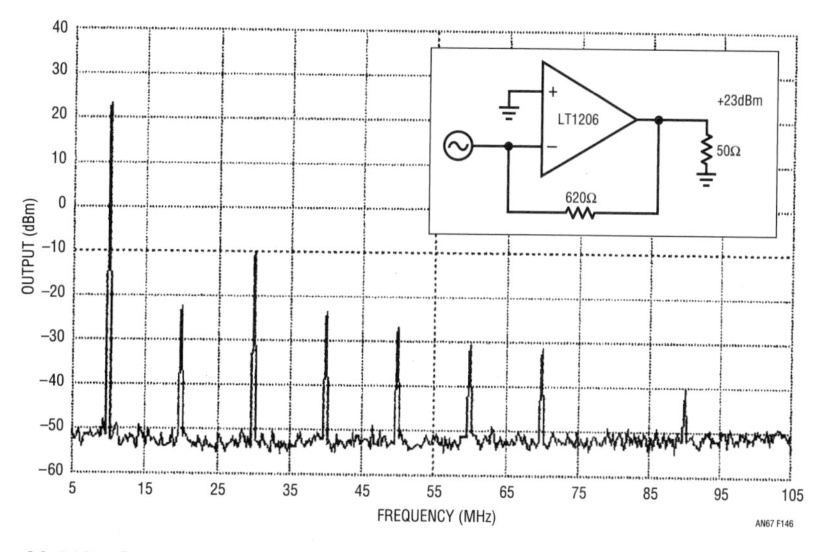

Figure 36.146 • Spectrum Plot of Figure 36.143's Circuit Driving +23dBm into a 50Ω Load

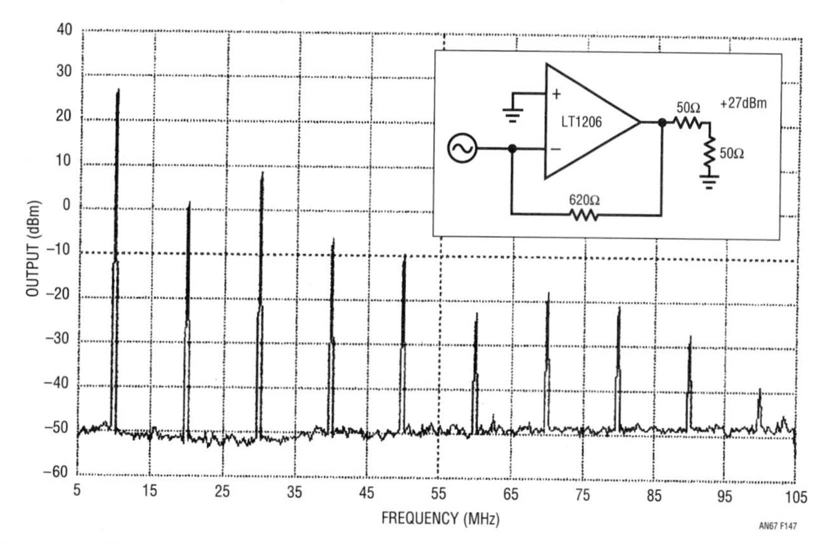

Figure 36.147 • Spectrum Plot of Figure 36.143's Circuit Driving +27dBm into a 50Ω, Double Terminated Load

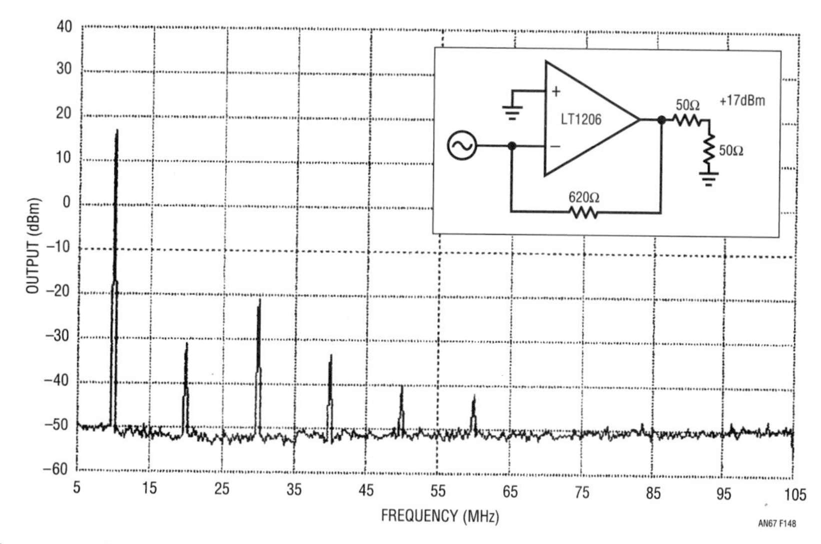

Figure 36.148 • Spectrum Plot of Figure 36.143's Circuit Driving +17dBm into a 50Ω, Double Terminated Load

Circuitry for signal conditioning and power conversion

Designs from a once lazy sabbatical

37

Jim Williams

Introduction

Linear Technology has a sabbatical program. Every five years employees are granted sabbatical leave, which may last up to six weeks. You have 18 months from each five year employment anniversary to take the leave. Sabbatical is fully company paid and has no restrictions. The time is yours to do with as you please.

People exercise all degrees of freedom with their sabbaticals. They go sailing, they go to South Sea islands, they ski some mountain nobody ever heard of, they trek in Nepal. Houses get fixed, cars restored and children played with.

For my third sabbatical I resolved to do absolutely nothing. For the first time in my life I was really tired, and I knew it. A six week rest sounded just fine. I'd walk the dog and spend time with my wife and son. That's it. No transistors, no resistors, no op amps and, above all, no writing. I was so written out the thought of picking up a pencil produced an instant headache.

The first week I really did do nothing but sleep, walk the dog, read and hang around with my wife and kid. Later, on the weekend, I went for a long, cold (top down) ride in the countryside, which, via some convoluted route, ended up at an electronic junk store. There I found a wonderfully pristine, albeit nonfunctional, Hewlett-Packard 215A pulse generator. This instrument, utilizing an exotic, step recovery diode based output stage, has clean, sub-nanosecond transitions. After the requisite economic arm wrestling at the counter, I bought the thing for twenty-five bucks.

I took it home, repaired it, and used it to characterize a fast coincidence detector (Figs 14-18 and associated text) I had previously abandoned. This exercise proved fatally catalytic. Things rapidly proceeded in a predictable direction. The result was a three week binge in the middle of my formerly restful sabbatical. Many of the circuits presented here are refinements or adaptations of previous efforts, although some are new. Also included, and annotated as such, are other authors' works that seemed appropriate.

This publication's title is cursorily descriptive of its contents. A more studied accounting includes categories of data converters and signal conditioners, transducer circuits, oscillators and power converters. They begin immediately.

Micropower voltage-to-frequency converters

Figure 37.1 is a voltage-to-frequency converter. A 0V to 5V input produces a 0Hz to 10kHz output, with a linearity of 0.02%. Gain drift is 60ppm/°C. Maximum current consumption is only 21μA, over 100 times lower than currently available monolithic ICs.

To understand circuit operation, assume that C1's negative input is slightly below its positive input (C2's output is low). The input voltage causes a positive-going ramp at C1's input (trace A, Figure 37.2). C1's output is high, allowing current flow from Q1's emitter, through C1's output stage to the 100pF capacitor: The 2.2μF capacitor provides high frequency bypass, maintaining low impedance at Q1's emitter. Diode connected Q6 provides a path to ground. The voltage to which the 100pF unit charges is a function of Q1's emitter potential and Q6's drop. C1's CMOS output, purely ohmic, contributes no voltage error. When the ramp at C1's negative input goes high enough, C1's output (trace B) goes low and the inverter switches high (trace C). This action pulls current from C1's negative input capacitor via the Q5 route (trace D). This current removal resets C1's negative input ramp to a potential slightly below ground. The 50pF capacitor furnishes AC positive feedback (C1's positive input is trace E) ensuring that C1's output remains negative long enough for a complete discharge of the 100pF capacitor. The Schottky diode prevents C1's input from being driven outside its

Analog Circuit and System Design: Immersion in the Black Art of Analog Design. http://dx.doi.org/10.1016/B978-0-12-397888-2.00037-7

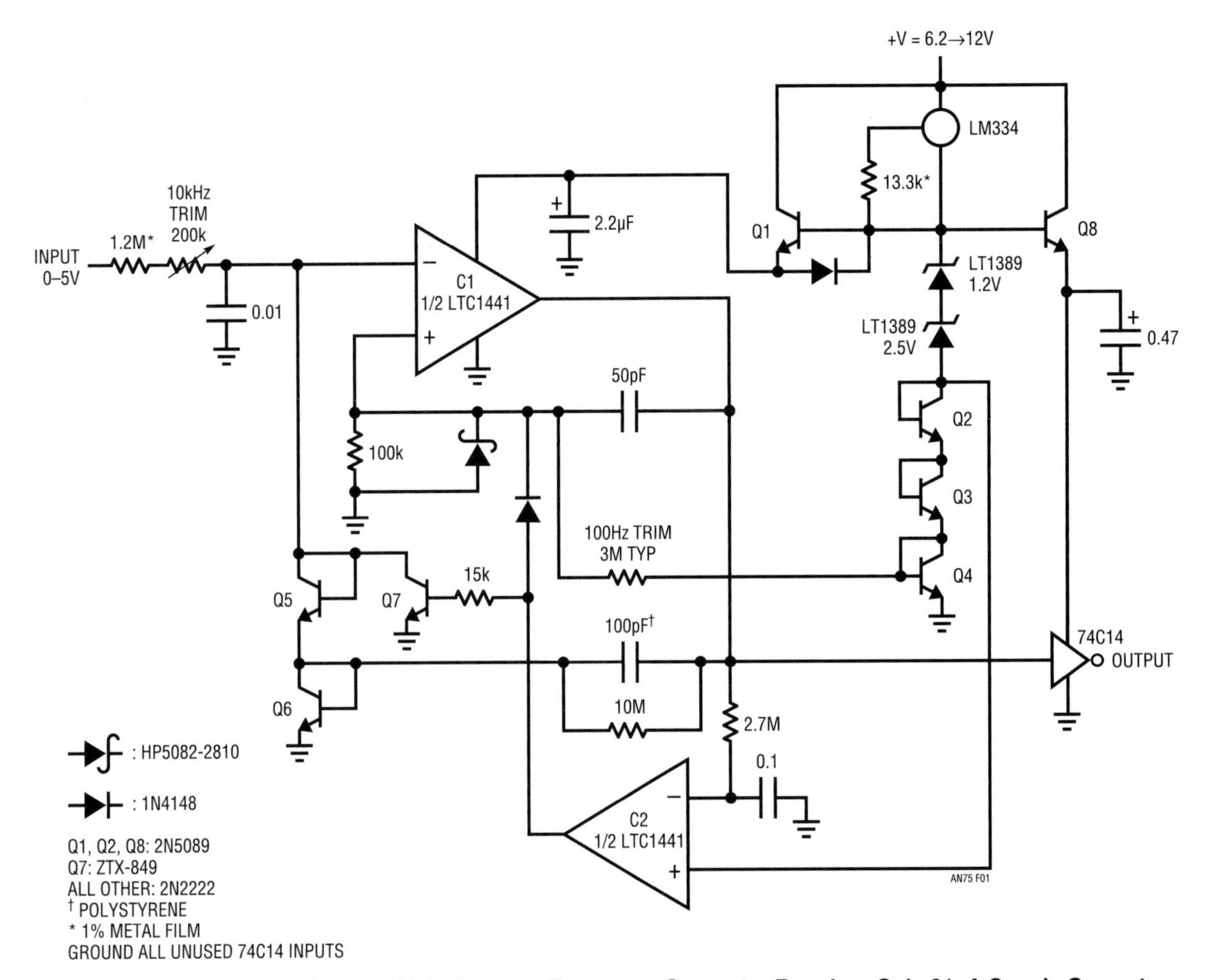

Figure 37.1 • 0.02%, 0Hz to 10kHz Voltage-to-Frequency Converter Requires Only 21µA Supply Current

negative common mode limit. When the 50pF units' feedback decays, C1 again switches high and the entire cycle repeats. The oscillation frequency depends directly on the input-voltage-derived current.

Q1's emitter voltage must be carefully controlled to get low drift. Q3 and Q4 temperature compensate Q5 and Q6 while Q2 compensates Q1's V_{BE}. The two LT®1389s are the actual voltage reference and the LM334 current source provides 5µA bias to the stack. The current drive provides excellent supply immunity (better than 40ppm/V) and also aids circuit temperature coefficient. It does this by using the LM334's 0.3%/°C tempco to slightly temperature modulate the voltage drop in the Q2-Q4 trio. This correction's sign and magnitude directly oppose the -120ppm/°C 100pF polystyrene capacitor's drift, aiding overall circuit stability. Q8's isolated drive to the CMOS inverter prevents output loading from influencing Q1's operating point. This makes circuit accuracy independent of loading.

The Q1 emitter-follower delivers charge to the 100pF capacitor efficiently. Both base and collector current end up in the capacitor. The 100pF capacitor, as small as desired performance permits, draws only small transient currents during its charge and discharge cycles. The 50pF-100k positive feedback combination draws insignificantly small switching currents. Figure 37.3, a plot of supply current vs operating frequency, reflects the low power design. At zero frequency, comparator quiescent current and the 5µA reference stack bias account for all current drain. There are no other paths for loss. As frequency scales up, the 100pF capacitor's charge-discharge cycle introduces the 1.1µA/kHz increase shown. A smaller value capacitor would cut power, but effects of stray capacitance and charge imbalance would introduce linearity and drift errors. Similarly, reduced reference stack drive would save current at the expense of drift.

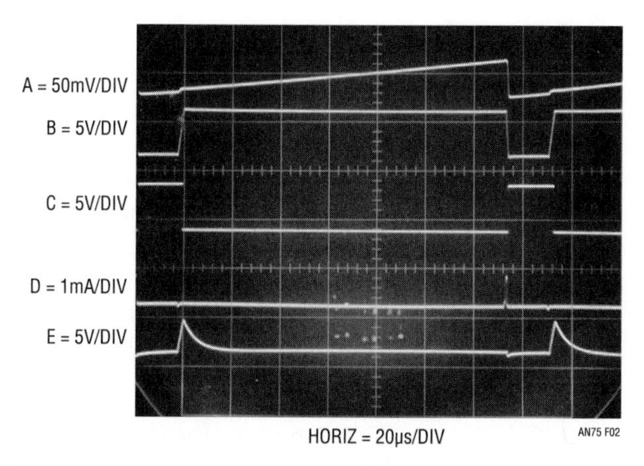

Figure 37.2 • Waveforms for the Micropower Voltage-to-Frequency Converter. Charge-Based Feedback Provides Precision Operation with Extremely Low Power Consumption

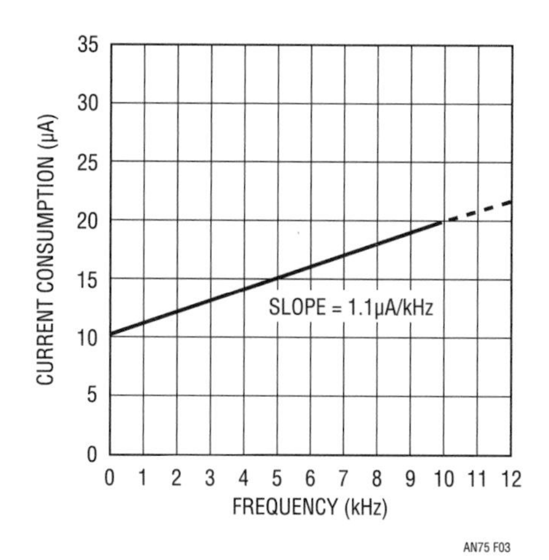

Figure 37.3 • Current Consumption vs Frequency for the Voltage-to-Frequency Converter. Charge Dispensing Cycles Dominate 1.1μA/kHz Current Drain Increase

Circuit start-up or overdrive can cause the circuit's AC-coupled feedback to latch. If this occurs, C1's output goes low; C2, detecting this via the 2.7M-0.1μF lag, goes high. This lifts C1's positive input and grounds the negative input with Q7, initiating normal circuit action.

To calibrate this circuit, apply 50mV and select the indicated resistor at C1's positive input for a 100Hz output. Complete the calibration by applying 5V and trimming the input potentiometer for a 10kHz output.

Figure 37.4's circuit is quite similar, although a reworked reference cuts current drain to just 8.8μA and permits operation from a 5V supply (V_{IN} 3.4V to 36V). The penalty is degraded linearity and drift performance. A 0V to 2.5V input produces a 0Hz to 10kHz output, with 0.03% linearity, 250ppm/°C drift and 10ppm/V supply rejection. Maximum current consumption is only 8.8μA, 300 times lower than currently available ICs. Circuit AC operation is nearly identical to Figure 37.1, although a brief description follows.

Comparator C1 switches a charge pump comprising D1, D2 and the 33pF capacitor to maintain its negative input at 0V A1 and associated components form a temperature compensating reference supply for the C1 based charge pump[1].

Note 1: Okay all you SPICE types out there, start your computers and model the charge pump drift and the reference compensation mechanism.

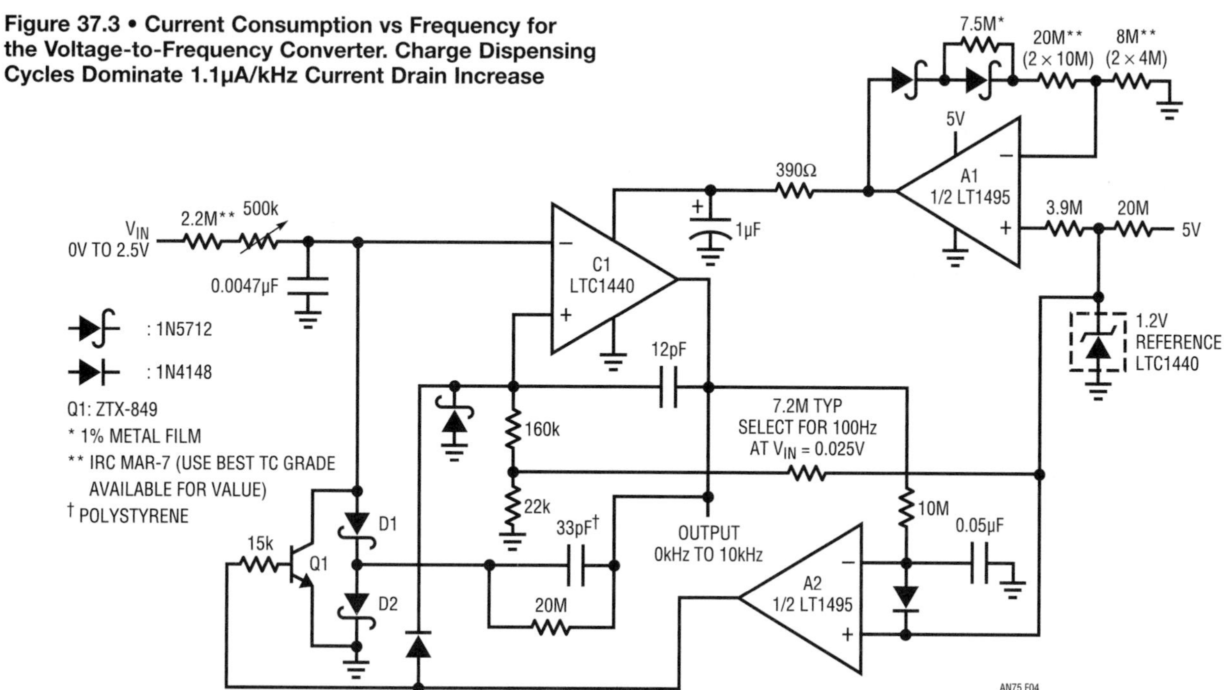

Figure 37.4 • 0Hz to 10kHz Voltage-to-Frequency Converter Consumes Only 8.8μA. Circuit Is Similar to Figure 37.1. Reworked Reference Cuts Power, Although Linearity and Drift Degrade

The 1.2V reference biasing A1 is contained within C1's package. As such, a bootstrapped start-up is required. The 20M resistor provides this, while wasting less than 200nA.

The 33pF capacitor charges to a fixed voltage; hence, the switching repetition rate is the circuit's only degree of freedom to maintain feedback. Comparator C1 pumps uniform packets of charge to its negative input at a repetition rate precisely proportional to the input voltage derived current. This action ensures that circuit output frequency is strictly and solely determined by the input voltage.

Current consumption is extraordinarily low. Figure 37.4 shows only 5.4μA quiescent current, rising to 8.8μA at 10kHz. The 340nA/kHz slope directly relates to the charge dispensing losses.

Start-up or input overdrive can cause the circuit's AC-coupled feedback to latch. If this occurs, C1's output goes low; A2, detecting this via the 10M/0.05μF lag, goes high. This lifts C1's positive input and grounds the negative input with Q1, initiating normal circuit action.

It is worth noting that these voltage-to-frequency circuits are the beneficiaries of considerable attention over a protracted period of time. The evolution of these designs is detailed in Appendix A, "Some Guidelines for Micropower Design and an Example." (see Figure 37.5)

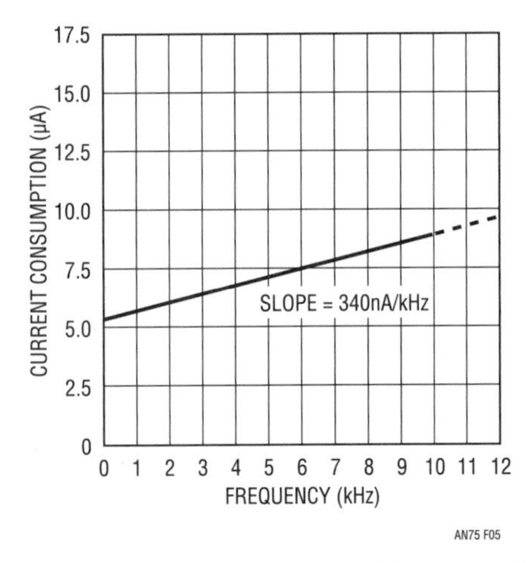

Figure 37.5 • Current Consumption vs Frequency for Figure 37.4. Charge Dispensing Cycles Dictate 340nA/kHz Current Drain Increase

Micropower A/D converters

In general, monolithic A/D converters have replaced discrete types. Occasionally, specific desirable circuit characteristics dictate a discrete design. Examples of such special cases include the need for a passive analog input, output data format, control protocol or economic constraints. Figure 37.6's 8-bit design consumes 12μA maximum, has 70ppm/°C drift (<1LSB 0°C to 70°C) and converts in 90ms. The circuit consists of a switched current source, an integrating capacitor, a comparator and a synchronized clock. When a pulse is applied to the convert command input (trace A, Figure 37.7) Q6 resets the 0.22μF capacitor to zero (trace B). Simultaneously, C1A goes low and Q5 conducts, biasing the LM334 based current source on. Additionally, Q4 conducts, causing the C1B based clock (trace D) to stop oscillating. During this interval the current source stabilizes, delivering its output to ground via Q6. When the convert command pulse falls the 0.22μF capacitor begins to ramp linearly. Concurrently, Q4 goes off, allowing the C1B clock to produce data output pulses (trace D). When the ramp voltage equals the E_X input, C1A switches high (trace C), biasing Q3 to stop the C1A clock. C1A's high state also cuts off Q5, shutting down the current source. Q5's gate going high bleeds a sub-microampere current through the 20M-Q1 path, maintaining ramp charging, but at a greatly reduced rate (this action is not readily discernible in Figure 37.7, but will be detailed). This insures overdrive for C1A while minimizing current source on-time, saving power. C1A's output pulse width (again, trace C) varies linearly with E_X's value. The Q3-Q4 gating of C1B prevents the convert command induced portion of C1A's output from allowing clock pulses. Thus, the clock bursts appearing at the data output (trace D) are directly and solely proportional to E_X. For the arrangement shown, 256 pulses appear for a 2.5V full-scale input.

Some subtleties are involved to achieve stated circuit performance. Q2 and associated biasing values combine with the LM334's inherent 3300ppm/°C temperature coefficient to keep current source drift inside 60ppm/°C. Q2, lacking gold doping, temperature tracks the LM334 more closely than a small signal diode would. The 0.22μF integrating and 220pF clock capacitors, both polystyrene, ratiometrically cancel their temperature coefficients to within 5ppm/°C. The specified resistors in the current source and clock have very low drift. The biasing at C1B's

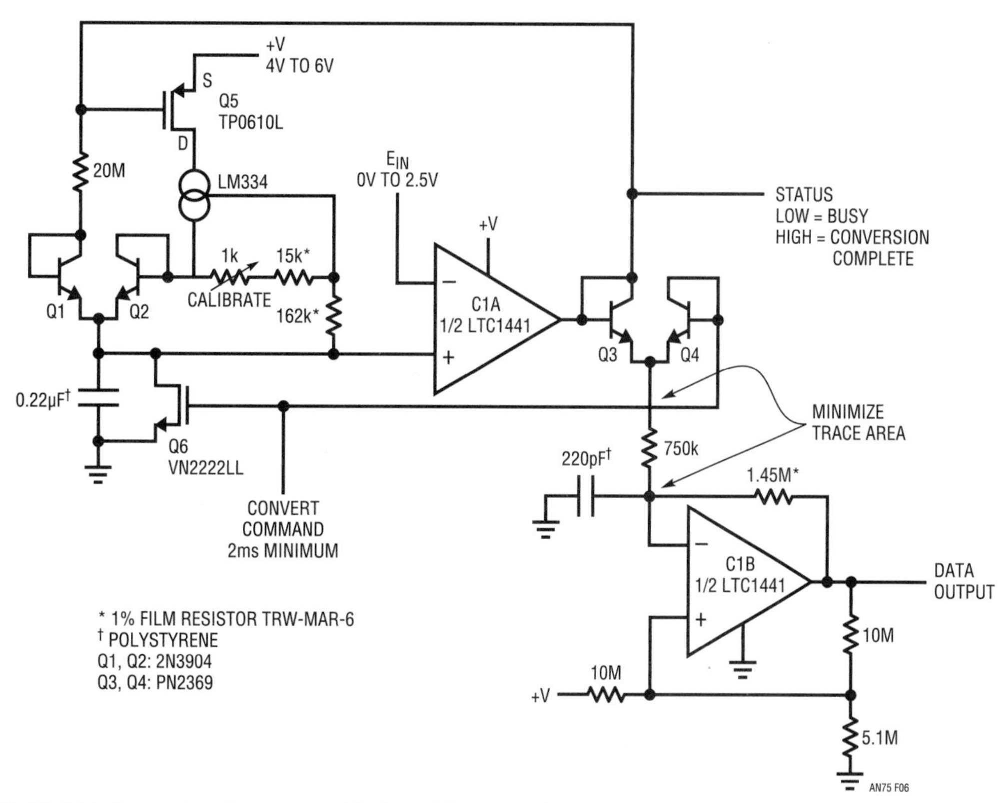

Figure 37.6 • 8-Bit A/D Converter Consumes 12μA and Has Passive Input. Additional Features Include 7μA Quiescent Current, 70ppm/°C Drift and 90ms Conversion Time

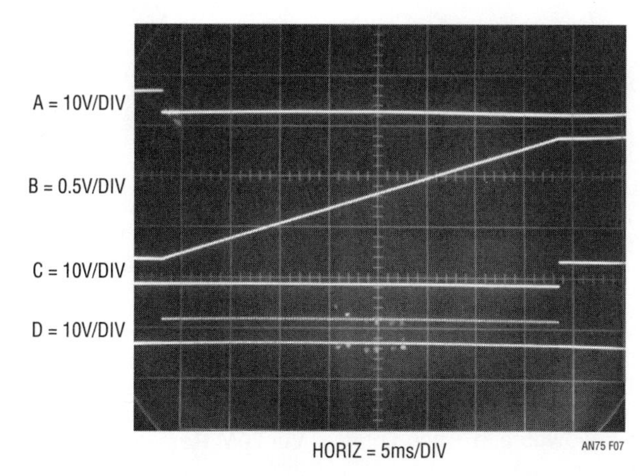

Figure 37.7 • Waveforms for the 12μA A/D Converter (E_{IN} = 1.25V) Include Convert Command (A), Reference Ramp (B), Status Output (C) and Data Output (D). Segmented Ramp Slope Characteristic Is Not Discernable in Trace B

negative input synchronizes the clock oscillator to the conversion sequence, eliminating a ±1 count error source. It also enforces predictable, optimum oscillator start-up, minimizing data jitter. Q3 and Q4 provide lower AC parasitics than diodes, enhancing clean oscillator gating. The converter typically holds 1LSB accuracy over 0°C to 70°C. Achievable conversion time varies with input. At tenth scale 16ms is possible, decreasing to 90ms at full scale.

Figure 37.8 details operation at E_X = 80mV The segmented slope operation due to current source switching is easily seen under these conditions. Trace A is convert command, trace B Q5's drain, trace C the ramp, trace D C1A's output ("status" line), trace E C1B's negative input[2] and trace F the data output. Trace E shows the benefit of the aforementioned optimized biasing at C1B's negative input. Clock oscillations start immediately, with no untoward dynamics.

Figure 37.9 is a study of the segmented slope operation. The photograph, taken at a 180mV input, shows the ramp zero reset and clean switching. When Q5 is on, its drain (trace A, Figure 37.9) is high, turning on the current source. The current source linearly ramps the 0.22μF capacitor (trace B) until C1A switches Q5 off. The current source then goes off, leaving the 20M-Q1 path to continue the charging at a sub-microampere rate. This continued charging ensures that C1A is overdriven, preventing spurious outputs.

The current source operates at almost 5μA. Turning it off after C1A switches saves considerable power particularly at modest E_X values at high conversion rates. When Q5 switches the current source off, charging continues via the 20M-Q1 path, but far less than a microampere is lost.

A/D power consumption is extremely low, due to the low power components and circuit configuration. Current consumption is 12μA for E_X = 2.5V at a 10Hz conversion rate. Intermediate values of E_X and conversion rate result in less current drain, down to a minimum of 7μA at quiescence. Additional power savings are theoretically possible by running a lower current source value, but dynamics and temperature coefficient suffer. Further power economy is possible by shutting off the current source during capacitor reset, but accuracy degrades due to current source settling time requirements.

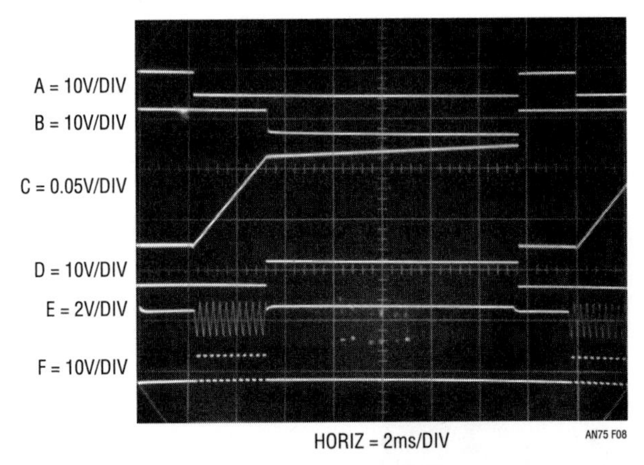

Figure 37.8 • Detailed Operating Waveforms at E_{IN} = 80mV. Trace A Is Convert Command, B, Q5 Drain, C, Ramp, D, Status, E, Clock Capacitor and F, Data Out. Optimized Capacitor Biasing Ensures Immediate, Predictable Clock Start-Up. Segmented Ramp Slope is Viewable in Trace C

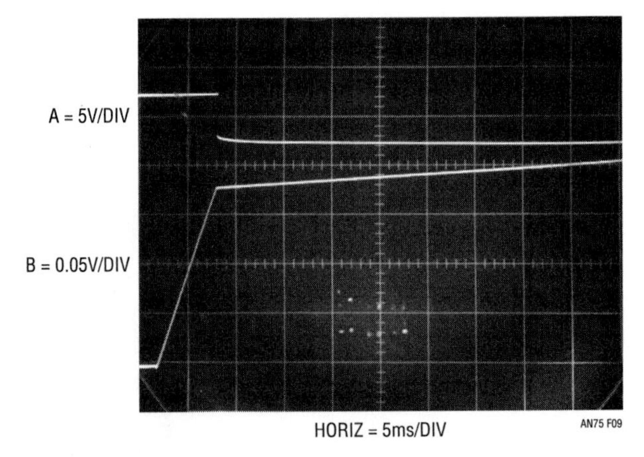

Figure 37.9 • Expanded Detail of Segmented Slope Ramp (B) and Q5 Drain (A) at E_{IN} = 180mV. When Q5 Goes Off, Ramp Current Source Ceases, Saving Power. Ramp Capacitor Charging Continues at Greatly Reduced Rate Via 20M Resistor, Insuring Comparator Overdrive

Note 2: Monitoring this high impedance AC node without incurring probe induced error involves special considerations. See Appendix B, "Parasitic Effects of Test Equipment in Micropower Circuits."

10-bit, micropower A/D converter

Figure 37.10 extends accuracy to 10μ bits, while increasing conversion speed to 35ms. The trade off is current consumption, which increases to 29μA. The circuit's operation is nearly identical to the 8-bit version, although the current source and clock are redesigned for higher accuracy. The LT1389-2N3809 combination is the current source, with the 301k resistor specified to oppose the integrating capacitor's −120ppm/°C temperature coefficient. The clock employs a 32.768kHz "watch" crystal for stability[3]. The quartz crystal's high Q, resonant characteristic precludes direct oscillator gating as was done in the previous circuit. Instead, the clock is synchronized to the conversion sequence with a flip-flop, which in turn transmits the convert command to the converter.

These stability improvements allow 10-bit resolution with 1LSB of drift over 0°C to 70°C. At a 10Hz conversion rate with $E_{IN} = 3V$ current drain is 29μA, decreasing to 21μA at quiescence. As in the previous circuit, different values of E_{IN} and conversion rate result in intermediate amounts of current consumption.

Note 3: A detailed description of this clock circuit appears in the text associated with Figure 37.27.

Differential input, 10MHz RMS/DC converter

Wideband, thermally based RMS/DC conversion has previously been described, utilizing single-ended inputs[4]. Figure 37.11's 10MHz RMS/DC converter has differential inputs while maintaining 1% accuracy beyond 10MHz. A $1V_{RMS}$ differential input produces 10V DC at the output.

The wideband LT1207 dual power op amp receives the differential inputs. The amplifiers, connected for a differential gain of 10, feed the LT1088 RMS/DC converter. The 24pF-5k trim provides a high frequency gain boost, preserving accuracy at the highest frequencies.

The LT1088 based RMS/DC converter is made up of matched pairs of heaters and diodes and a control amplifier. The A1-A2 amplifiers drive R1, producing heat which lowers D1's voltage. Differentially connected A3 responds by driving R2, via Q3, to heat D2, closing a loop around the amplifier. Because the diodes and heater resistors are matched, A3's DC output is related to the RMS value of the input, regardless of input frequency or waveshape. In practice, residual LT1088 mismatches necessitate a gain trim, which is implemented at A4.

Note 4: For examples, see references 10 through 13.

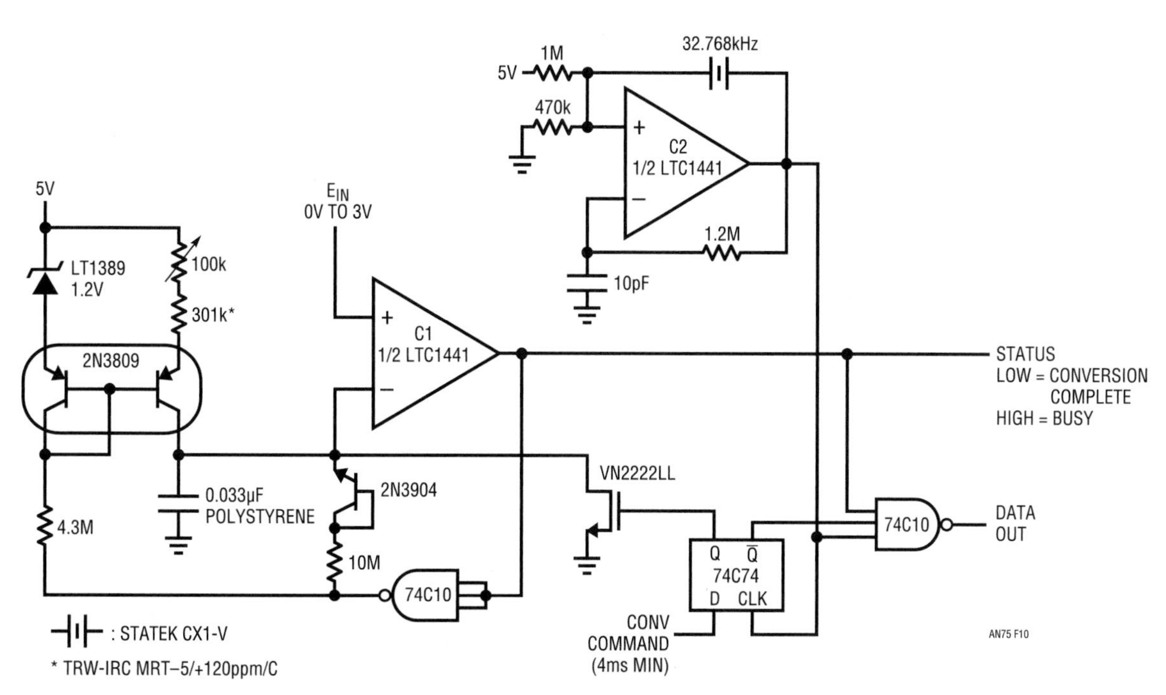

Figure 37.10 • A 10-Bit Version of Figure 37.6. Improvements Include Higher Stability Clock and Current Source. Modifications Permit 1LSB Drift (0°C to 70°C) and 35ms Conversion Speed, Although Current Drain Increases to 29μA

A4's output is the circuit output. The LT1004 and associated components frequency compensate the loop and provide good settling time over wide ranges of operating conditions.

Start-up or input overdrive can cause A2 to deliver excessive current to the LT1088 with resultant damage. C1 and C2 prevent this. Overdrive forces D1's voltage to an abnormally low potential. C1 triggers low under these conditions, pulling C2's input low. This causes C2's output to go high, putting A1 and A2 into shutdown and terminating the overload. After a time determined by the RC at C2's input, A1 and A2 will be enabled. If the overload condition still exists the loop will almost immediately shut A1 and A2 down again. This oscillatory action will continue, protecting the LT1088 until the overload condition is removed.

Performance for the circuit is quite impressive. Figure 37.12 plots error with one input driven at two different gain boost network trims. The graph (B) shows 1% error bandwidth to 11MHz. The slight peaking out to 5MHz is due to the A1-A2 gain boost network. This peaking is minimal compared to the total error envelope, and a small price to pay to get 1% accuracy to 11MHz. One percent accuracy to 14MHz is available if the gain trim and boost network are set to accentuate peaking at the expense of flatness (A).

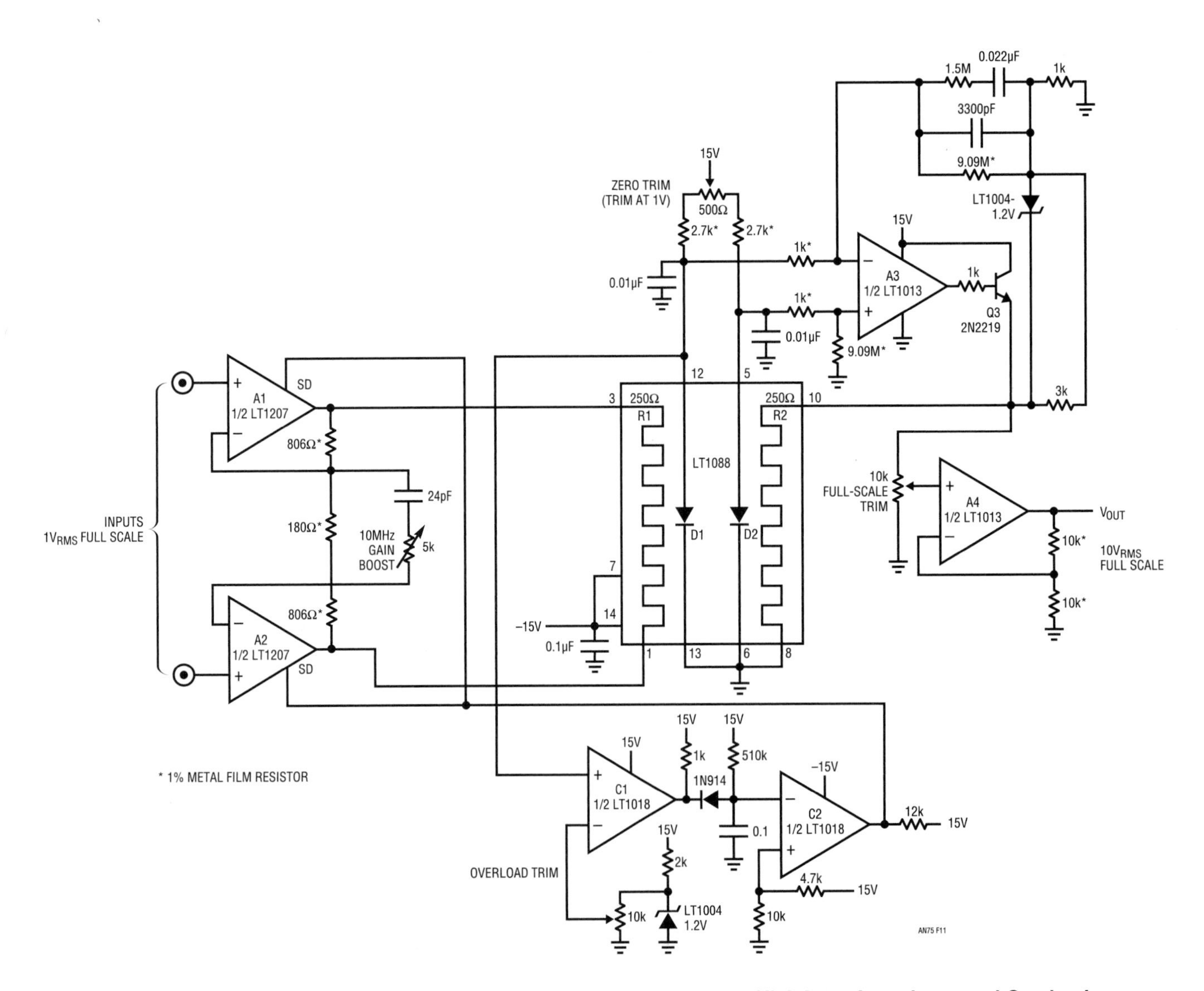

Figure 37.11 • Differential Input 10MHz RMS/DC Converter Has 1% Accuracy, High Input Impedance and Overload Protection. Single-Ended Operation Extends 1% Error Bandwidth to 14MHz

Figure 37.13 shows effects of common mode signals on accuracy. This data was taken with a well shielded, carefully layed out breadboard. Common mode rejection ratio remains high as frequency scales, contributing negligible error until 10.2MHz. The indicated $5V_{RMS}$ common mode drive is a demanding test, with smaller values permitting better performance.

To trim this circuit put the 5kΩ potentiometer at its maximum resistance position and apply a 100mV 5MHz signal. Trim the 500Ω adjustment for exactly $1V_{OUT}$. Next, apply a 5MHz 1V input and trim the 10k potentiometer for $10.00V_{OUT}$. Finally, put in 1V at 10MHz and adjust the 5kΩ trimmer for $10.00V_{OUT}$. Repeat this sequence until circuit output is within 1% accuracy for DC-10MHz inputs. Two passes should be sufficient. The overload trim is set at 10% below D1's voltage with the circuit operating at full scale.

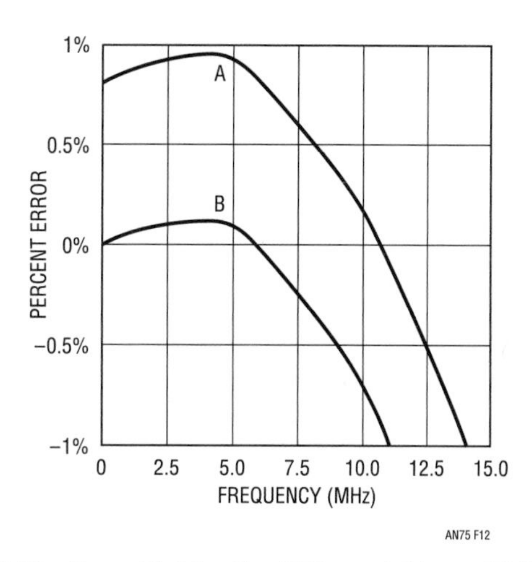

Figure 37.12 • Error Plot for the Differential Input RMS/DC Converter with One Input Driven. Frequency Dependent Gain Boost Preserves 1% Accuracy, But Causes Slight Peaking Before Roll-Off. Boost Is Settable for Maximum Bandwidth (A) or Minimum Error (B)

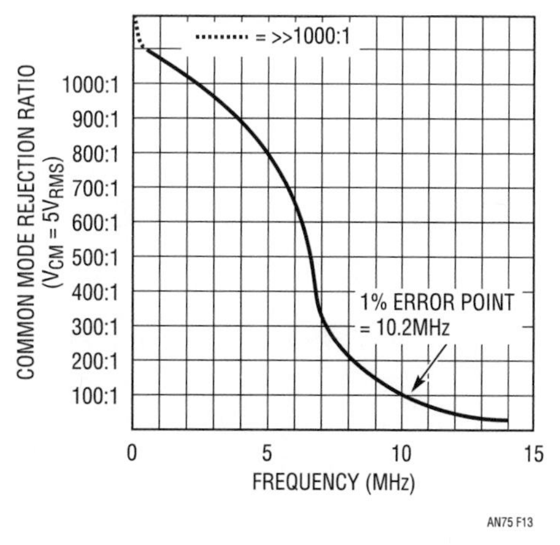

Figure 37.13 • Common Mode Rejection Ratio vs Frequency for the Differential Input RMS/DC Converter. Layout, Amplifier Bandwidth and AC Matching Characteristics Determine Curve

Nanosecond coincidence detector

Figure 37.14's circuit, detecting coincident voltage levels at its inputs, responds with a logical high at its output. The detection trigger level is settable between zero and 4.0V. The circuit will resolve coincidence down to 3ns and has a decision delay time of 4.5ns. The circuit is composed of a pair of fast level discrimination comparators and a sub-nanosecond AND gate. The comparators balance each input against a level threshold, in this case about 1V. The comparator outputs feed Q1 and associated components, which form a 300ps AND gate. Figure 37.15's waveforms show circuit operation. Trace A is one input, while trace B is the remaining input. Trace C is the circuit output. When trace B crosses the 1V recognition threshold the output goes high, remaining high until either input (in this case trace B) drops below 1V The key to this circuit's speed is the fast comparators and the discrete AND gates extremely low delay.

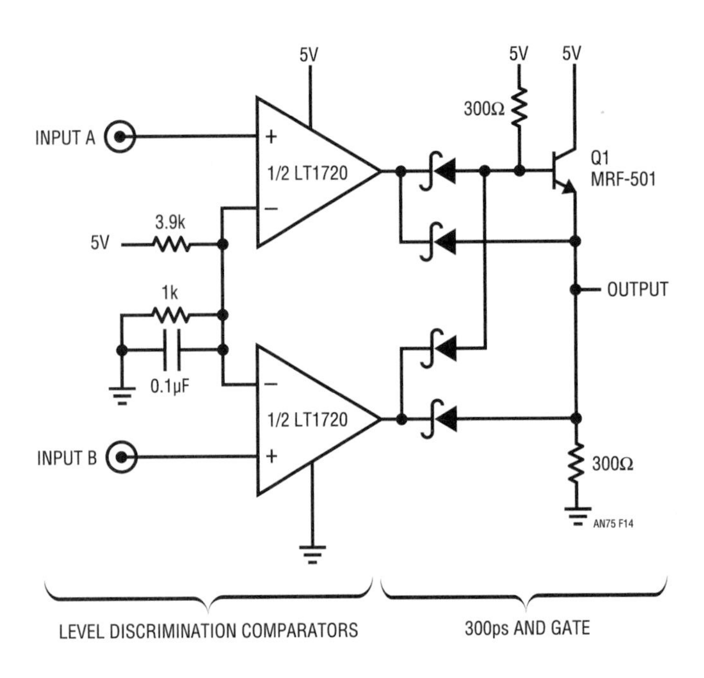

Figure 37.14 • Coincidence Detector Has 3ns Recognition Threshold. Discrete Components Form 300ps AND Gate, Maintaining High Speed Signal Path

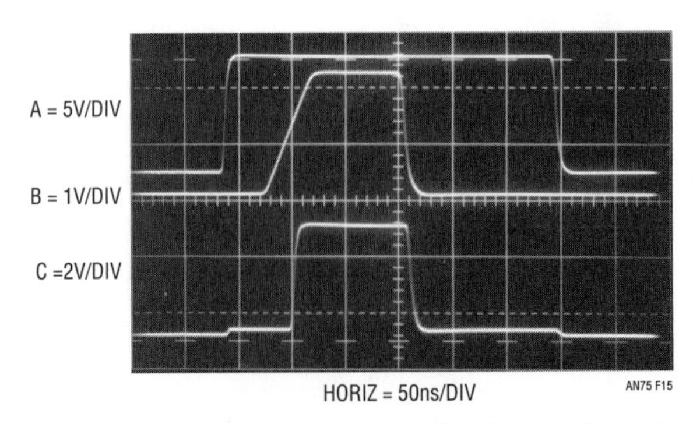

Figure 37.15 • Coincidence Detector Waveforms. Trace A Is Input A, B Is Input B. Trace C Indicates Coincidence When A and B Are Both >1V

Evaluating circuit performance requires a sub-nanosecond rise-time pulse generator and a very fast oscilloscope[5]. Figure 37.16, taken in a 3.9GHz sampled bandpass, shows a comparator output (trace A) and the resultant circuit output (trace B). The Schottky diodes and gigahertz range transistor provide very fast response, and delay is inside 300ps.

Figure 37.17 shows circuit response in a 3.9GHz sampled bandpass with the inputs simultaneously driven by a 3ns, 2V pulse (trace A). This pulse width is just inside the recognition limit, and the output (trace B) responds cleanly. The 4.5ns decision delay characteristic is also readily apparent. Further input pulse width reduction has dramatic results[6]. In Figure 37.18 input width (trace A) is shortened by 600ps. T output (trace B) is caught not quite fully responding. It rises about 2V before falling back in a noisy but controlled decay. The rise slope, degraded from Figure 37.17's, is additional evidence of circuit gain- bandwidth limitations.

Note 5: Refer to this publication's introduction.

Note 6: I offer no apology for the choice of verbiage. Nerds like me find drama in these things.

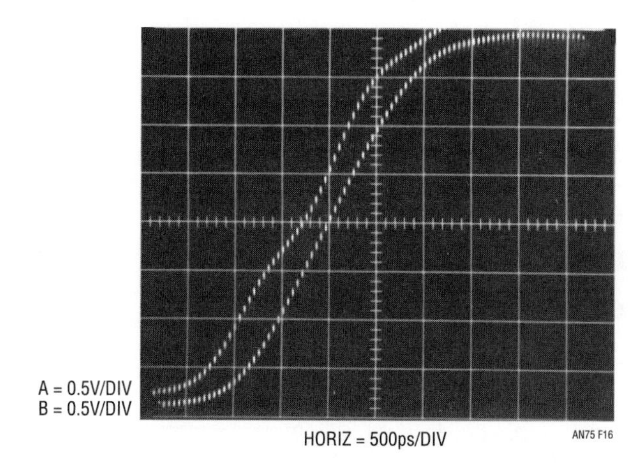

Figure 37.16 • 300ps AND Gate Delay Measured from Comparator Output (A) to Q1 Emitter (B). Fall Time Delay Is Similar. Sampling Oscilloscope Display Presents as a Series of Dots

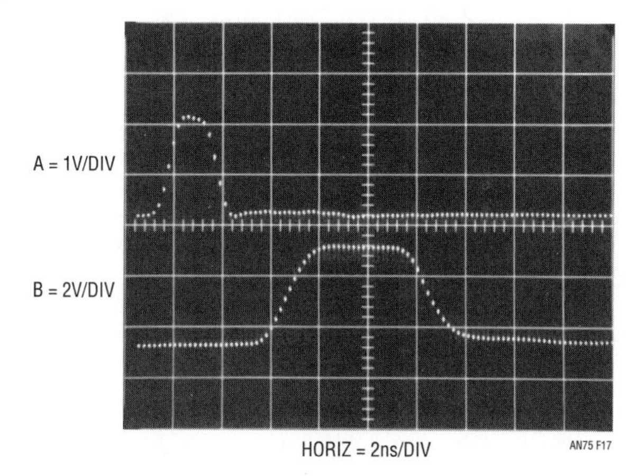

Figure 37.17 • Output (B) Recognizing a 3ns Coincident Pulse (A) at Inputs. Response Is Clean, with Decision Delay of 4.5ns. Segmented Display Is Characteristic of Sampling 'Scope Operation

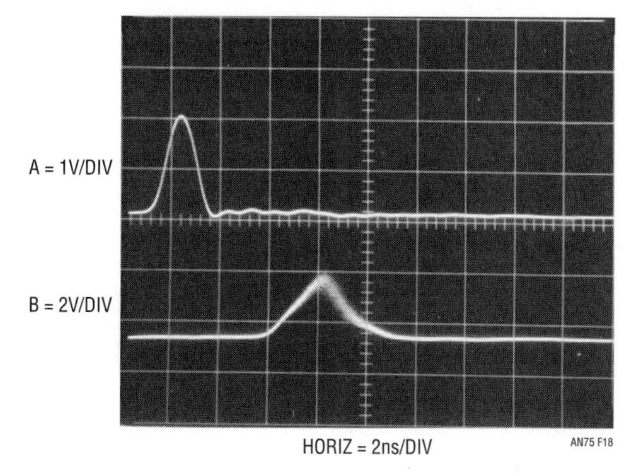

Figure 37.18 • An Unrecognized Coincidence. Output (B) Cannot Fully Respond to ≈2.5ns Coincident Pulse (A). Additional 500ps of Coincidence Would Permit Valid Recognition (See Previous Figure)

15 nanosecond waveform sampler

Figure 37.19 is another high speed circuit. This waveform sampler has 15ns response and a gain of 10. The circuit is made up of a fast, low parasitic switch, its drive components and an output amplifier The switch is formed by the diode bridge. Borrowed from classical sampling oscilloscope circuitry, it is the key to circuit performance[7]. The diode bridge's inherent balance eliminates charge injection based errors in the output. It is far superior to other electronic switches in this characteristic. Any other high speed switch technology contributes excessive output spikes due to charge-based feedthrough. FET switches are not suitable because their gate-channel capacitance permits such feedthrough. This capacitance allows gate-drive artifacts to corrupt switch output. The diode bridge's balance, combined with matched, low capacitance monolithic diodes and complementary high speed switching, yields a cleanly switched output. Trims optimize switch performance. DC balance is achieved by trimming the bridge on-current for zero input-output offset voltage. Two AC trims are required. The "AC balance" corrects for diode and layout capacitive imbalances and the "skew compensation" corrects for any timing asymmetry in the nominally complementary bridge drive. These AC trims compensate small dynamic imbalances that could result in parasitic switch outputs.

Note 7: See references 14 and 15 for design details of diode bridge switches.

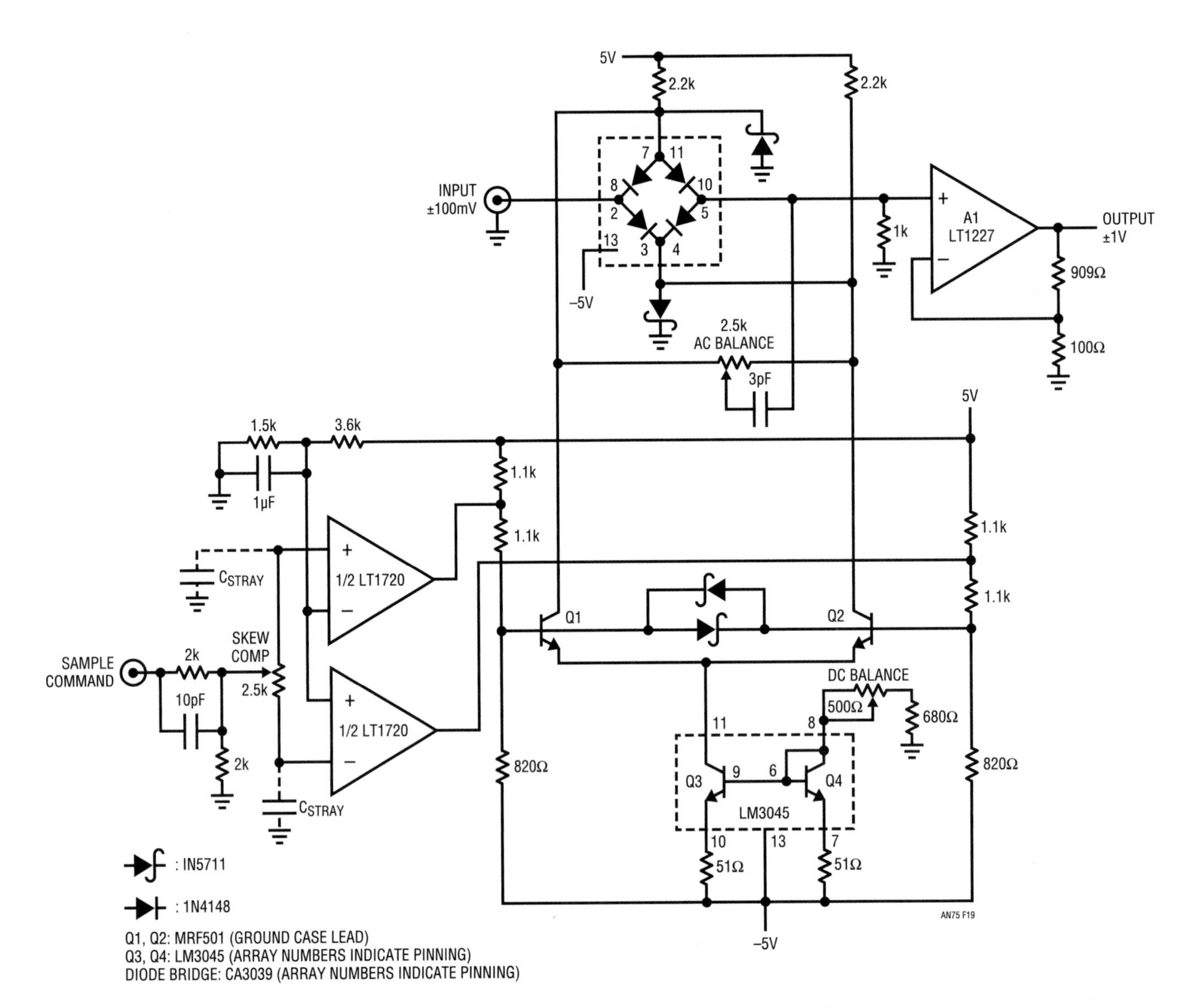

Figure 37.19 • 15ns Waveform Sampler Utilizes Diode Bridge Switch and Wideband x10 Amplifier. Comparators and Associated Components Provide Optimized Diode Bridge Switching

The sample command biases the LT1720 comparators, which furnishes complementary levels to the Q1-Q2 switch drivers. The "skew compensation" trim, working differentially against stray and device capacitance, provides a way to slightly time skew the comparators response. The comparator outputs bias current sink loaded Q1-Q2[8]. These devices provide level shifted drive to the bridge. Bridge output feeds A1, a wideband amplifier; operating at a gain of 10. Figure 37.20 presents waveforms. Trace A is the sample command, trace B and C complementary bridge drives at the Q1-Q2 collectors and trace D the output.

Figure 37.21, an amplitude and time expanded view, shows more detail. Trace assignments are identical, although scale factors are changed. A small delay occurs between the sample command (trace A) and the complementary bridge drives (traces B and C), although no drive time skewing is evident. Trace D, the output, responds cleanly, with some switch induced pre-shoot before falling.

Trimming is required to optimize sampler performance. DC balance is adjusted first. Ground the input and connect the sample command to the 5V supply. Monitor the output and adjust the "DC balance" for 0V The AC trims are made dynamically. Connect the input to a *well bypassed* 50mV DC source and drive the sample command with a 1MHz square wave. A typical pre-trim sampler output appears in Figure 37.22. The pre-shoot (waveform bottom) is due to poor AC balance. The mid-transition discontinuity is characteristic of untrimmed skew compensation. In general, poor AC balance shows up as pronounced pre or post transition events, while unadjusted skew compensation causes distortion during the transition. When properly trimmed, circuit output should be devoid of all such behavior. Figure 37.23 shows this; only very slight disturbances (probably due to residual AC imbalance) are visible.

Pertinent performance specifications include 100μV/°C drift, 15ns delay time, 10MHz full-power bandwidth and a minimum sample window for full-power response of 30ns.

Note 8: The bridge drive scheme presented here is variant of a circuit developed by George Feliz (LTC). See LTC Application Note 74, "Component and Measurement Advances Ensure 16-Bit DAC Settling Time."

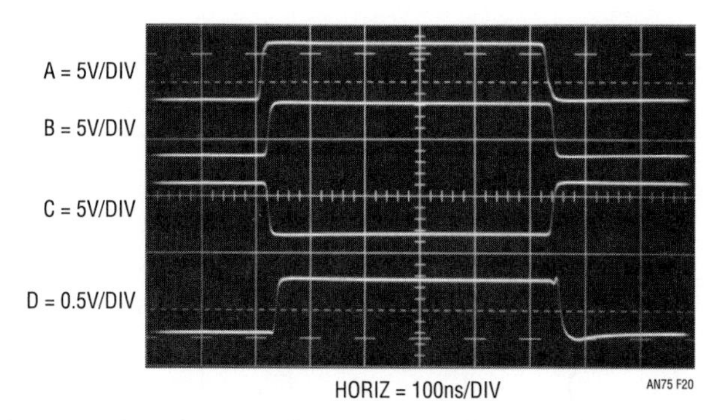

Figure 37.20 • Sampler Operation at 50mV Input. Trace A Is Sample Command, B and C Complementary Bridge Drives. Trace D Is Output

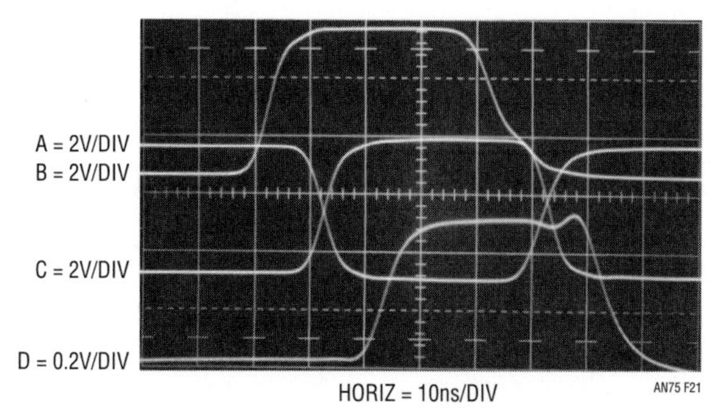

Figure 37.21 • Highly Expanded View of Figure 37.20 Has Same Trace Assignments. Bridge Switching Appears Unskewed and Output Responds Cleanly

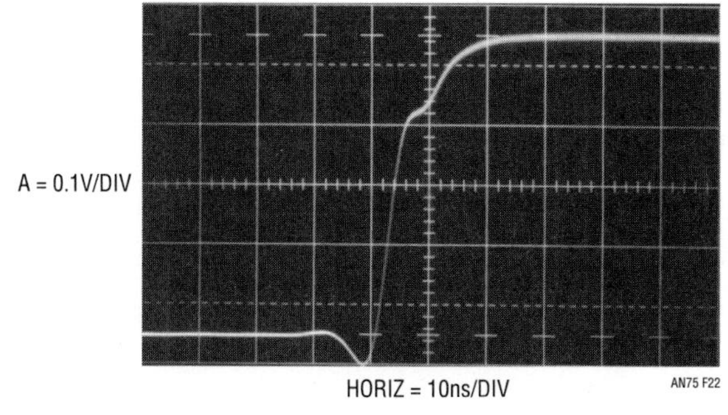

Figure 37.22 • Sampler Output Before Trimming. Aberration at Bottom Is Due to Misadjusted AC Balance. Mid-Transition Discontinuity Derives from Untrimmed Skew Compensation

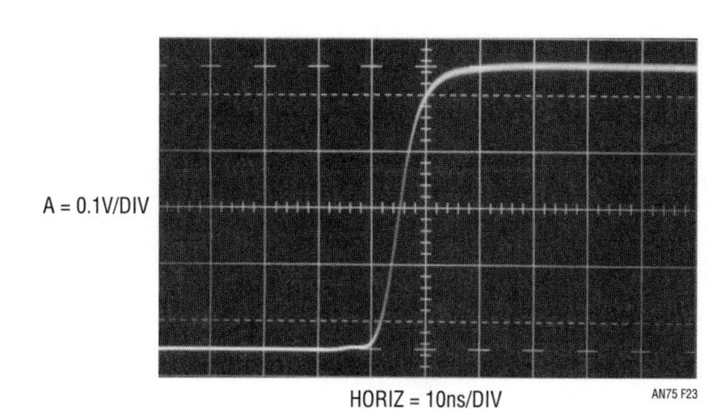

Figure 37.23 • Sampler Output After Optimizing AC Balance and Skew Compensation

5.5μA powered, 0.05μV/°C chopped amplifier

Figure 37.24 shows a chopped amplifier that requires only 5.5μA supply current. Offset voltage is 5μV with 0.05μV/°C drift. A gain of 10^8 affords high accuracy, even at large closed-loop gains.

The micropower comparators form a biphase 5Hz clock. The clock drives the input related switches, causing an amplitude-modulated version of the DC input to appear at A1A's input. AC-coupled A1A takes a gain of 1000, presenting its output to a switched demodulator similar to the aforementioned modulator.

The demodulator output, a reconstructed, DC-amplified version of the circuit's input, is fed to A1B, a DC gain stage. A1B's output is fed back, via gain setting resistors, to the input modulator, closing a feedback loop around the entire amplifier The configuration's DC gain is set by the feedback resistor's ratio, in this case 1000.

The circuit's internal AC coupling prevents A1's DC characteristics from influencing overall DC performance, accounting for the extremely low offset uncertainty noted. The high open-loop gain permits 10ppm gain accuracy at a closed-loop gain of 1000.

The desired micropower operation and A1's bandwidth dictate the 5Hz clock rate. As such, resultant overall bandwidth is *low*. Full-power bandwidth is 0.05Hz with a slew rate of about 1V/s. Clock-related noise, about 5μV can be reduced by increasing C_{COMP}, with commensurate bandwidth reduction.

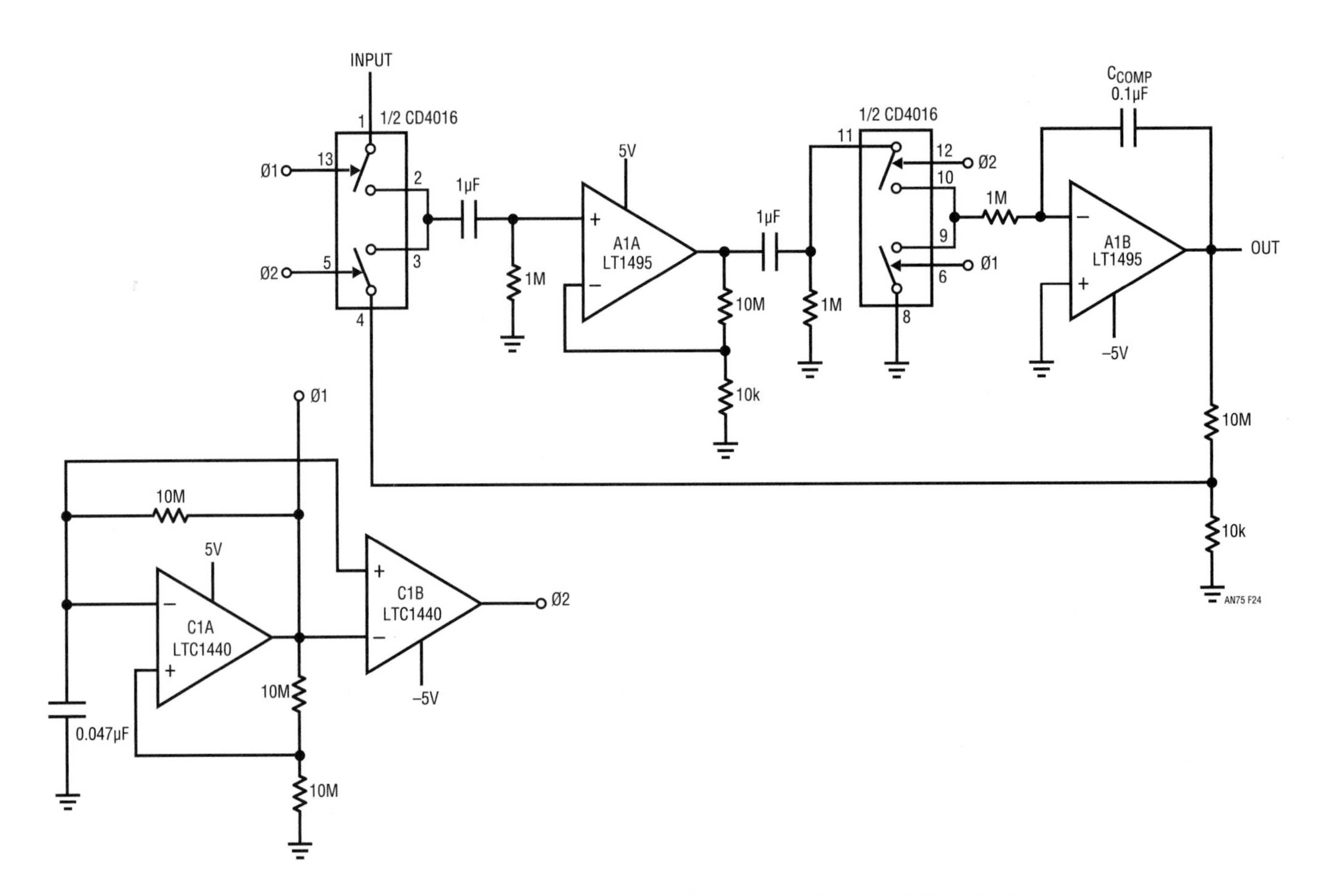

Figure 37.24 • 0.05μV/°C Chopped Amplifier Consumes Only 5.5μA Supply Current

Pilot light flame detector with low-battery lockout

Figure 37.25 shows a pilot light flame detector with low-battery lockout. The amplifier ("A"), running open loop, compares a small portion of the reference with the thermocouple- generated voltage. When the thermocouple is hot, the amplifier's output swings high, biasing Q1 on. Hysteresis, provided by the 10M resistor, ensures clean transitions, while the diodes clamp static generated voltages to the rails. The 100k-2.2μF RC filters the signal to the amplifier.

The comparator ("C") monitors the battery voltage via the 2M-1M divider and compares it to the 1.2V reference. A battery voltage above 3.6V holds C's output high, biasing Q2 on and maintaining the small potential at A's negative input. When the battery voltage drops too low, C goes low, signaling a low-battery condition. Simultaneously, Q2 goes off, causing A's negative input to move to 1.2V. This biases A low, shutting off Q1. The low outputs alert downstream circuitry to shut down gas flow.

Tip-acceleration detector for shipping containers

Figure 37.26's circuit is a tip-acceleration detector for shipping containers. It detects if a shipping container has been subjected to excessive tipping or acceleration and retains the detected output. The sensitivity and frequency response are adjustable. A potentiometer with a small pendulous mass biases the amplifier ("A"), operating at a gain of 12. Normally, A's output is below C's trip point and circuit output is low. Any tip-acceleration event that causes A's output to swing beyond 1.2V will trip C high. Positive feedback around C will latch it in this high state, alerting the receiving party that the shipped goods have been mishandled. Sensitivity is variable with potentiometer mechanical or electrical biasing or A's gain. Bandwidth is settable by selection of the capacitor at A's input. The circuit is prepared for use by applying power and pushing the button in C's output.

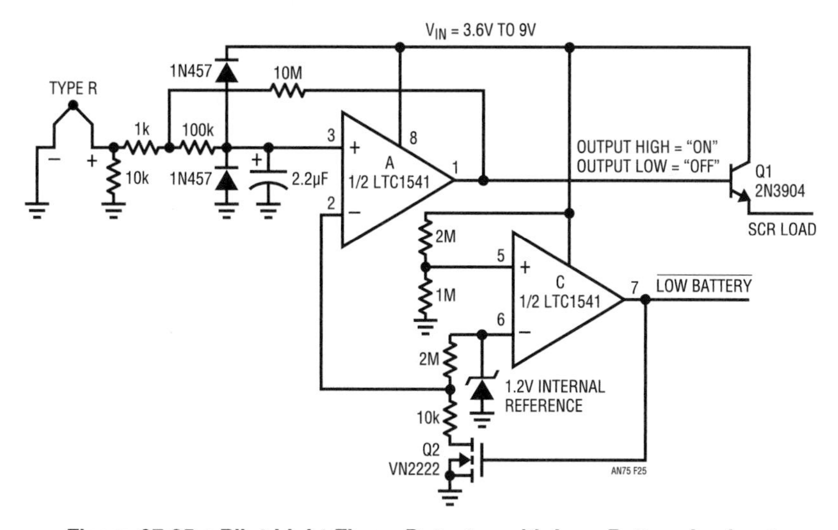

Figure 37.25 • Pilot Light Flame Detector with Low-Battery Lockout

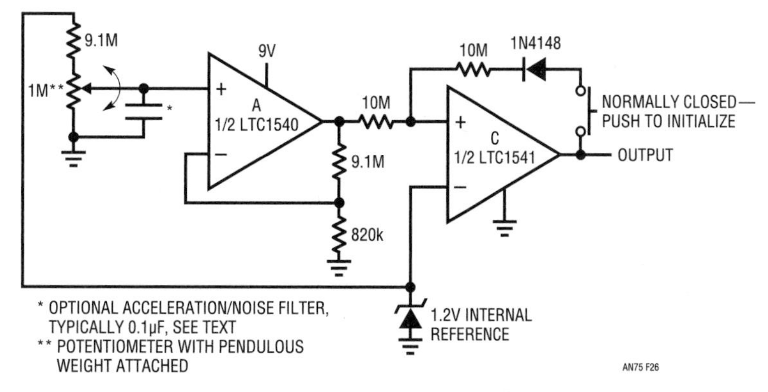

Figure 37.26 • Tip-Acceleration Detector for Shipping Containers Retains Output If Triggered. Sensitivity Is Adjustable Via Amplifier Feedback Values. Capacitor Sets Acceleration Response Bandwidth

32.768kHz "watch crystal" oscillator

Figure 37.27's quartz oscillator, using a standard 32.768kHz "watch crystal, " starts under all conditions with no spurious modes. Current draw is only 9µA at a 2V supply.

The circuit is best understood by initially ignoring the crystal. Resistors at the positive input establish a DC bias point. The 1.2M–10pF path sets up phase shifted negative feedback and the circuit looks like a marginally stable unity gain follower at DC. When the crystal is realized, positive feedback occurs and oscillation commences at the crystal's resonant frequency.

Power consumption is low. The LTC1441's output stage design eliminates "totem" currents, maintaining low drain even as supply increases. Figure 37.28's plot shows 9µA drain at 2V supply, increasing linearly to 18µA at 5V supply. Current drain is reducible by altering component values, but erratic crystal start-up or parasitic modes may result. This is particularly the case if various brands of crystal are employed. The values given represent a compromise between minimized current drain and assured operation.

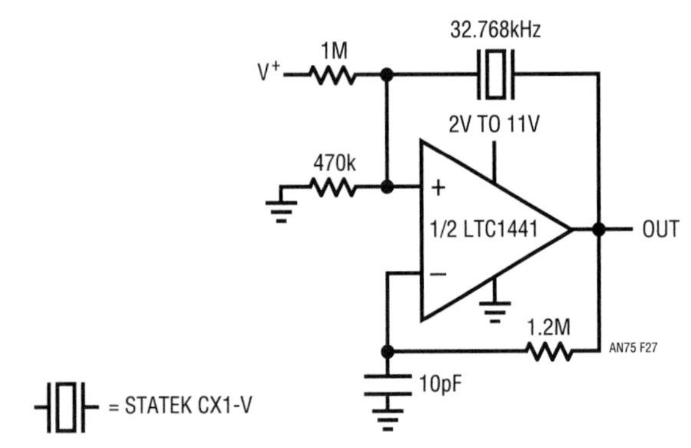

Figure 37.27 • 32.768kHz "Watch Crystal" Oscillator Has No Spurious Modes. Circuit Pulls 9µA at V_S = 2V

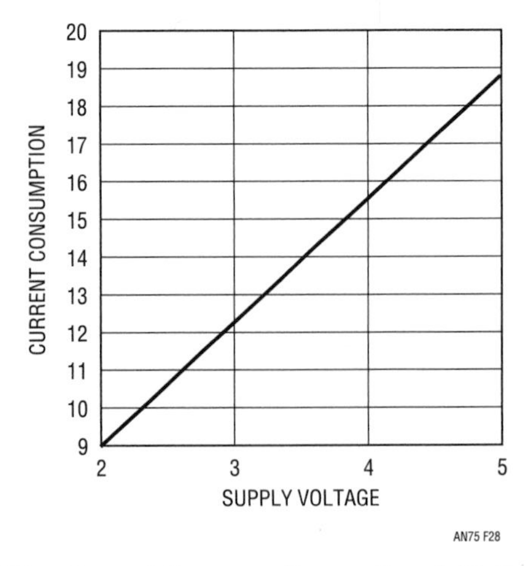

Figure 37.28 • Current Consumption vs Supply Voltage for the 32.768kHz Crystal Oscillator. Characteristic Is Essentially Linear

Complementary output, 50% duty cycle crystal oscillator

Figure 37.29's circuit, developed jointly by Joe Petrofsky (LTC) and the author, uses the LT1720 dual comparator in a 50% duty cycle crystal oscillator. Output frequencies to 10MHz are practical.

Resistors at C1's positive input set a DC bias point. The 2k-0.068μF path furnishes phase-shifted feedback and C1 acts like a wideband, unity-gain follower at DC. The crystal's path provides resonant positive feedback and stable oscillation occurs. C2, sensing C1's input, provides a delay matched, low skew, complementary output. A1 compares band limited versions of the outputs and biases C1's negative input.

Because frequency is fixed, C1's only degree of freedom to respond is variation of pulse width; hence, the outputs are forced to 50% duty cycle.

The circuit operates with AT-cut fundamental crystals from 1MHz to 10MHz, over a 2.7V to 6V power supply range. All biasing is supply derived, and hence ratiometric. As such, 50% duty cycle is maintained at all supply voltages, with output skew below 800ps. Figure 37.30 plots skew, which is seen to vary by about 800ps over a 2.7V to 6V supply excursion.

It is noteworthy that any desired duty cycle may be obtained by summing current into either of A1's inputs. If this is done, the current should derive directly from the supply or supply rejection will be compromised.

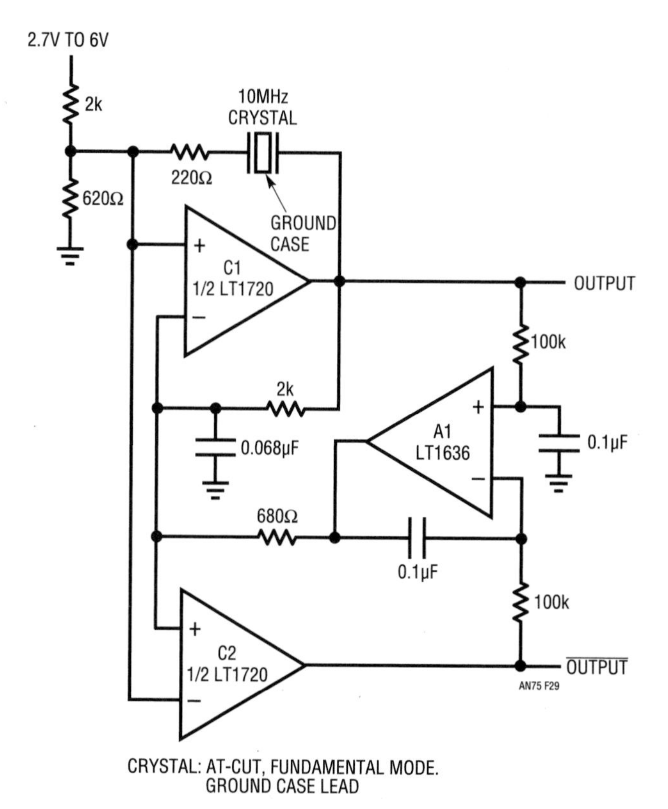

Figure 37.29 • Crystal Oscillator Has Complementary Outputs and 50% Duty Cycle. A1's Feedback Maintains Output Duty Cycle Despite Supply Variations

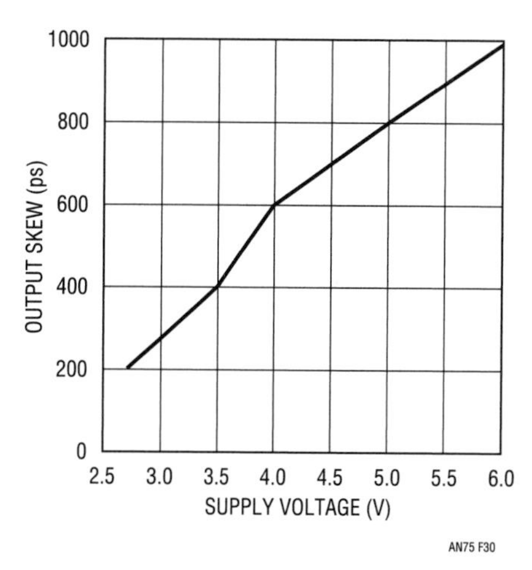

Figure 37.30 • Output Skew vs Supply for 10MHz Clock. Skew Varies Only 800ps Over 2.7V to 6V Supply Excursion

Nonoverlapping, complementary output crystal oscillator

Figure 37.31, an extension of the previous design, generates a nonoverlapping, complementary output crystal clock. The circuit is essentially identical to Figure 37.29, with the exception that C2 receives attenuated bias. This causes the outputs to have a nonoverlapping characteristic.

Under these conditions, the only way A1 can balance its inputsis if the circuit outputs have identical output duty. The nonoverlapping operation is verified in Figure 37.32, which shows the circuit's output. The outputs transition crisply, with no detectable overlap. This circuit shares the previous version's supply immunity due to ratiometric biasing. If the A1 network is deleted output duty will be unequal, but nonoverlapping operation retained.

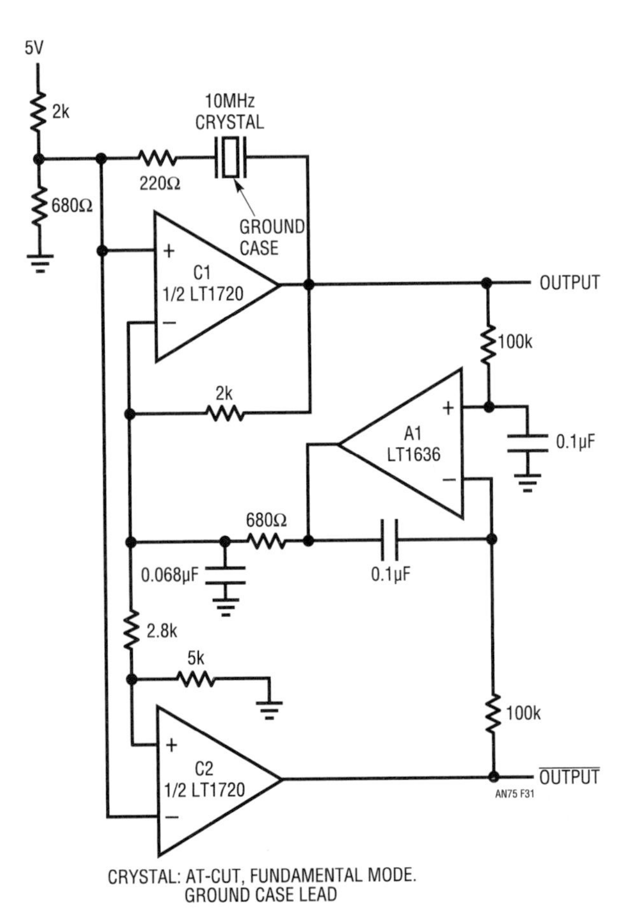

Figure 37.31 • Rearranging Figure 37.29's Comparator Biasing Provides Nonoverlapping Complementary Outputs

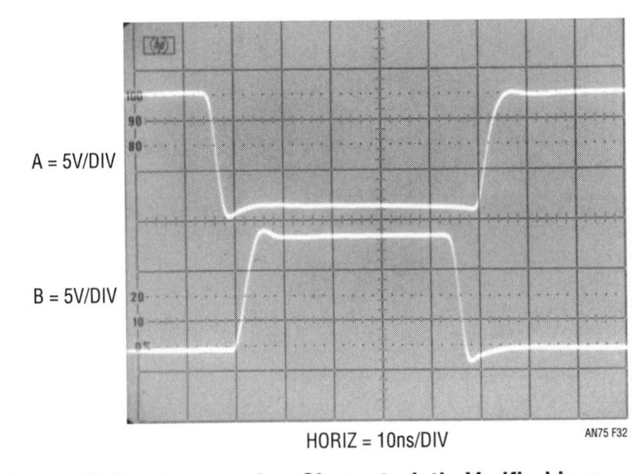

Figure 37.32 • Nonoverlap Characteristic Verified in a 275MHz Bandpass

High power CCFL backlight inverter for desktop displays

Large LCD (liquid crystal display) displays designed to replace CRTs (cathode ray tubes) in desktop computer applications are becoming available. The LCD's reduced size and power requirements allow much smaller product size, a highly desirable feature.

CRT replacement requires a 10W to 20W inverter to drive the CCFL (cold cathode fluorescent lamp) that illuminates the LCD. Additionally, the inverter must provide the wide dimming range associated with CRTs, and it must have safety features to prevent catastrophic failures.

Figure 37.33's circuit meets these requirements. It is a modified, high power variant of an approach employed in laptop computer displays[9]. T1, Q1, Q2 and associated components form a current fed, resonant Royer converter that produces high voltage at T1's secondary. Current flows through the CCFL tubes and is summed, rectified and filtered, providing a feedback signal to the LT1371 switching regulator The LT1371 delivers switched mode power to the L1-D1 node, closing a control loop around the Royer converter The 182Ω resistor provides current-to-voltage conversion, setting the lamp current operating point. The loop stabilizes lamp current against variations in time, supply, temperature and lamp characteristics. The LT1371's frequency compensation is set by C1 and C2.

Note 9: See reference 21.

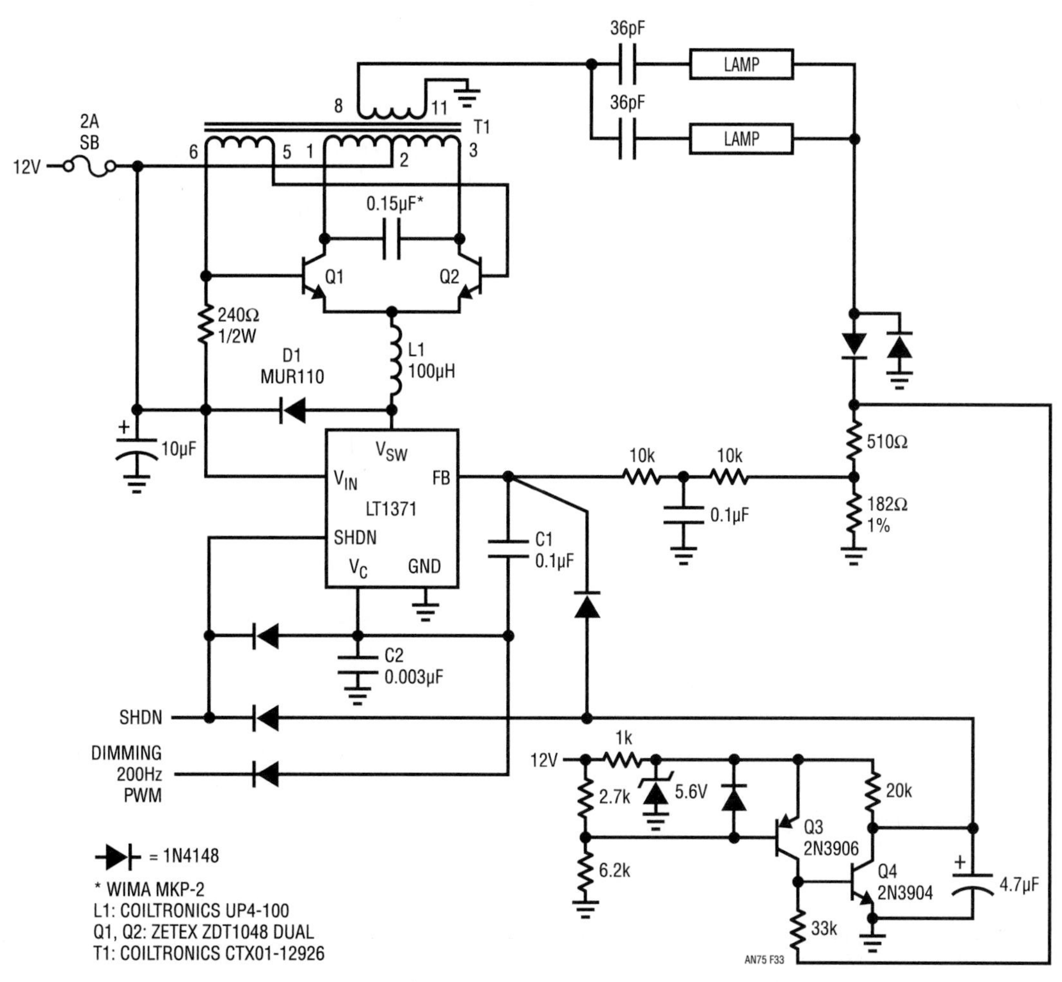

Figure 37.33 • 12W CCFL Backlight Inverter for Desktop Displays Provides Wide Range Dimming and Safety Features

The compensation responds quickly enough to permit the 200Hz PWM input to control dimming over a 30:1 range with no degradation in loop regulation. Applicable waveforms appear in Figure 37.34.

Q3 and Q4 shut down the circuit if lamp current ceases (open or shorted lamps or leads, T1 failure or similar malfunction). Normally, Q4's collector is held near ground by the lamp-current-derived base biasing. If lamp current ceases, Q4's collector voltage increases, overdriving the feedback node and shutting down the circuit. Q3 prevents unwanted shutdown during power supply turn-on by driving Q4's base until supply voltage is above about 7V.

Figure 37.35 shows the shutdown circuit reacting to the loss of lamp feedback. When lamp feedback ceases, the voltage across the 182Ω current sense resistor drops to zero (visible between Figure 37.35's third and fourth vertical graticule lines, trace A). The LT1371 responds to this open- loop condition by driving the Royer converter to full power (Q1's collector is trace B). Simultaneously, Q4's collector (trace C) ramps up, overdriving the LT1371's feedback node in about 50ms. The LT1371 stops switching, shutting off the Royer converter drive. The circuit remains in this state until the failure has been rectified.

This circuit's combination of features provides a safe, simple and reliable high power CCFL lamp drive. Efficiency is in the 85% to 90% range. The closed-loop operation ensures maximum lamp life while permitting extended dimming range. The safety feature prevents excessive heating in the event of malfunction and the use of off-the-shelf components allows ease of implementation.

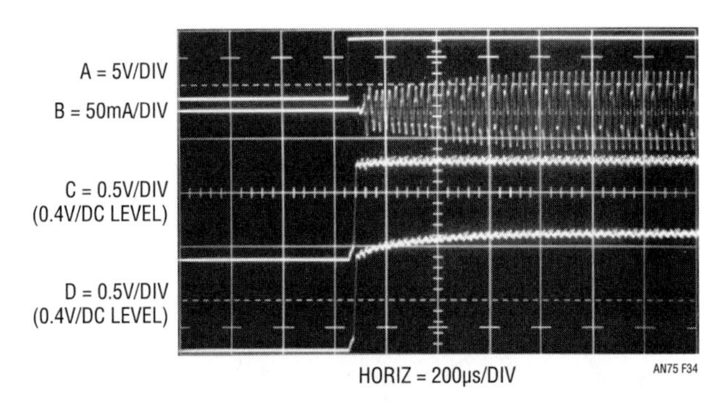

Figure 37.34 • Fast Loop Response Maintains Regulation at 200Hz PWM Rate. Waveforms Include PWM Command (A), Lamp Current (B), LT1371 Feedback (C) and Error Amplifier V_C (D) Pins. Loop Settling Occurs in 500μs

Ultralow noise power converters[10]

Today's circuit designer is often challenged to assemble a high performance system by combining sensitive analog electronics with potentially noisy power converters. Requirements for a small, efficient, cost effective solution are in conflict with acceptable noise performance—noisy switching regulators call for filtering, shielding and layout revisions that add bulk and expense. Most electromagnetic interference (EMI) problems associated with DC/DC converters are due to high speed switching of large currents and voltages. To maintain high efficiency, these switch transitions are designed to occur as quickly as possible. The result is input and output ripple that contains very high harmonics of the switching frequency. These fast edges also couple through stray magnetic and electric fields to nearby signal lines, making efforts to filter the supply lines ineffective.

The LT1534 ultralow noise switching regulator provides an effective and flexible solution to this problem. Using two external resistors, the user can program the slew rates of the current through the internal 2A power switch and the voltage on it. Noise performance can be evaluated and improved with the circuit operating in the final system. The system designer need sacrifice only as much efficiency as is necessary to meet the required noise performance. With the controlled slew rates, system performance is less sensitive to layout, and shielding requirements can be greatly reduced; expensive layout and mechanical revisions can be avoided.

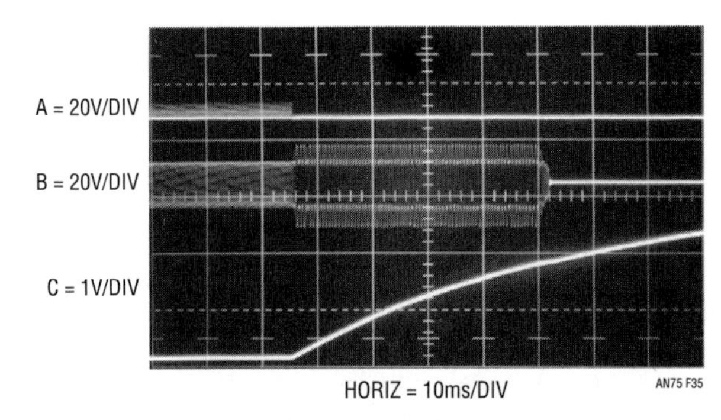

Figure 37.35 • Safety Feature Reacts to Lamp Feedback Loss by Shutting Down Power. Lamp Current Dropout (A) Allows Monitoring Circuit to Ramp Up (C), Shutting Off Drive (B)

Note 10: Figures 37.36 to 37.39 and all associated text are authored by Jeff Witt of LTC. Their original presentation is annotated in reference 22.

The LT1534's internal oscillator can be programmed over a broad frequency range (20kHz to 250kHz) with good initial accuracy. It can also be synchronized to an external signal placing the switching frequency and its harmonics away from sensitive system frequencies.

Low noise boost regulator

In Figure 37.36, the LT1534 boosts 3.3V to supply 650mA at 5V with its oscillator synchronized to an external 50kHz clock. The circuit relies on the low ESR of capacitor C2 to keep the output ripple low at the fundamental frequency; slew rate control reduces the high frequency ripple. Figure 37.37 shows waveforms of the circuit as it delivers 500mA. The top trace shows the voltage on the collector of the internal bipolar power switch (the COL pins), and the middle trace shows the switch current. The lowest trace is the output ripple. The slew rates are programmed to their fastest here, resulting in good efficiency (83%), but also generating excessive high frequency ripple. Figure 37.38 shows the same waveforms with the slew rates reduced. The large high frequency transients have been eliminated.

Low noise bipolar supply

Many high performance analog systems require quiet bipolar supplies. This circuit (Figure 37.39) will generate ±5V from a wide input range of 3V to 12V with a total output power of 1.5W. By using a 1:1:1 transformer, the primary and secondary windings can be coupled using capacitors C2 and C3, allowing the LT1534 to control the switch transitions at the output rectifiers as well as at the switch collector. Secondary damping networks are not required.

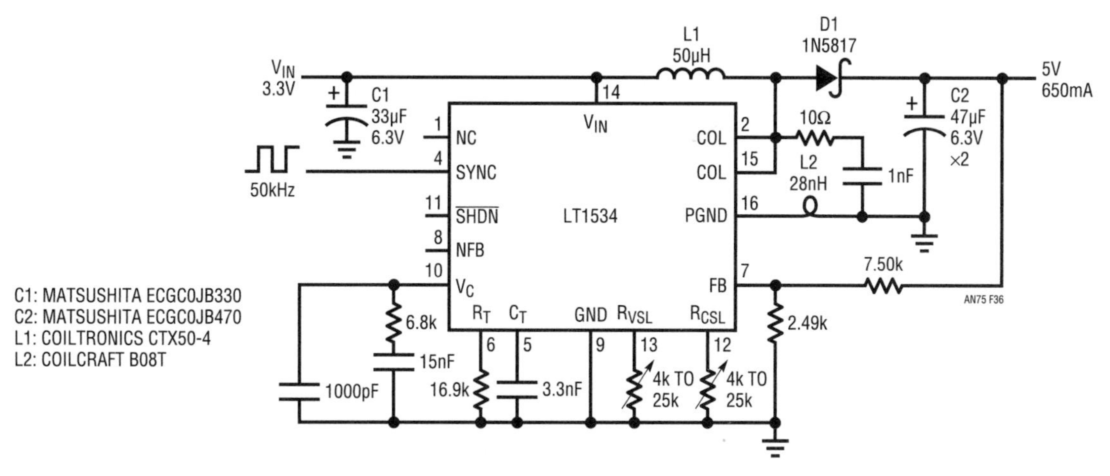

Figure 37.36 • The LT1534 Boosts 3.3V to 5V. The Resistors On the R_{VSL} and R_{CSL} Pins Program the Slew Rates of the Voltage On the Power Switch (COL Pins) and the Current Through It

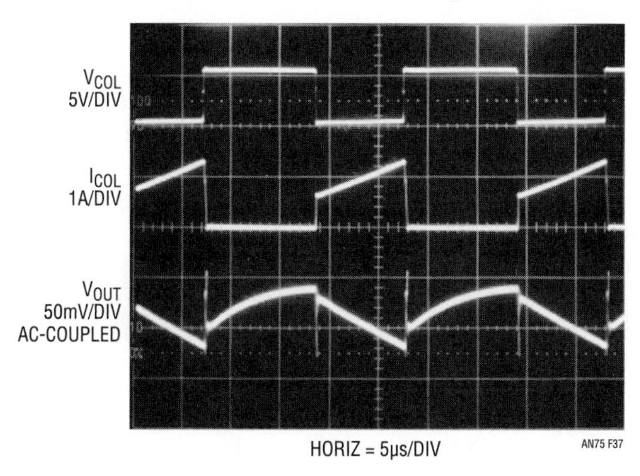

Figure 37.37 • High Slew Rates (R_{CSL} = R_{VSL} = 4k) Result in Good Efficiency But Excess High Frequency Ripple

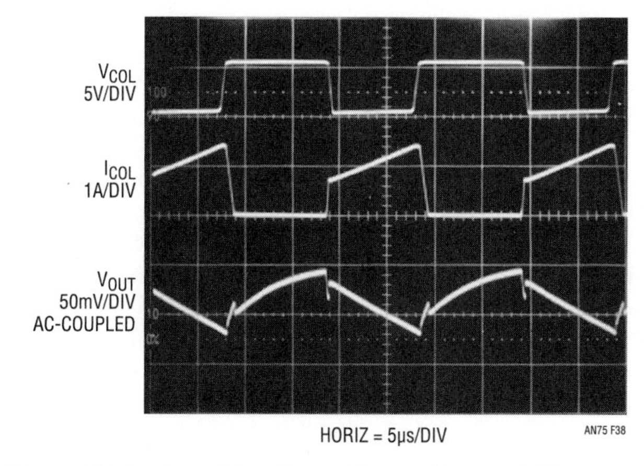

Figure 37.38 • Low Slew Rates (R_{CSL} = R_{VSL} = 24k) Result in an Output Without Troublesome High Frequency Transients

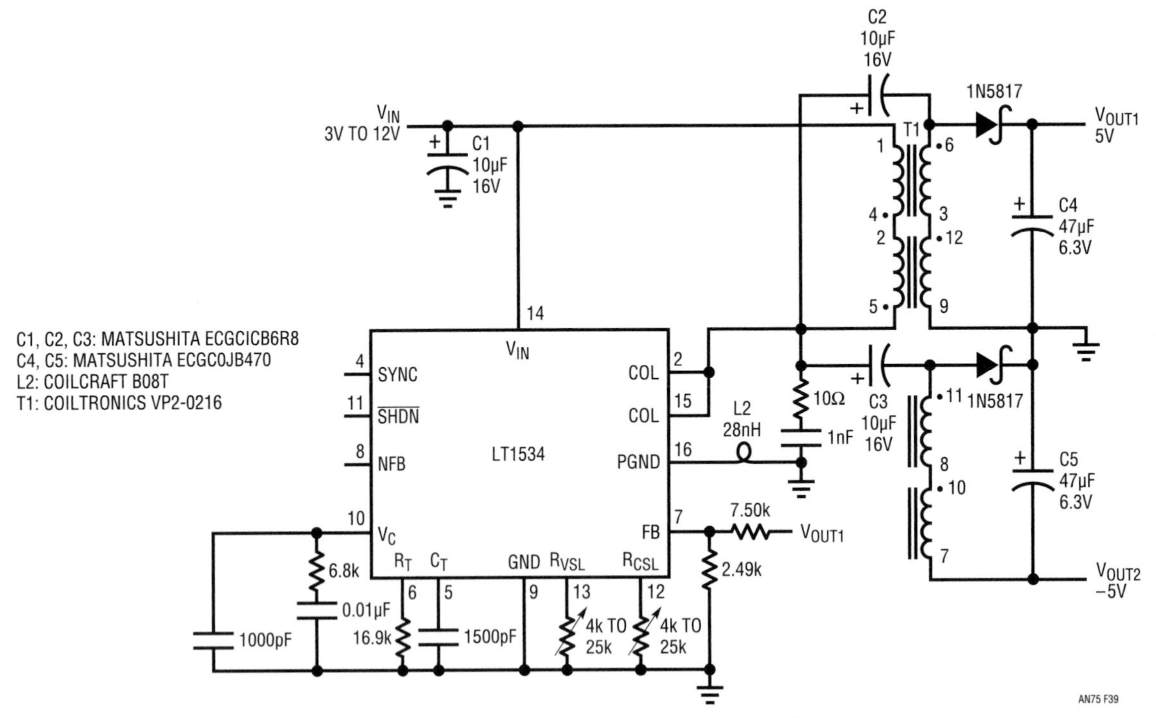

Figure 37.39 • A Low Noise, Wide Input Range ±5V Supply

Ultralow noise off-line power supply

Off-line power supplies require input filtering components to meet FCC emission requirements. Additionally, board layout is usually quite critical, requiring considerable experimentation even for experienced off-line supply designers. These considerations derive from the wideband harmonic energy generated by the fast switching of traditional off-line supplies. A new device, the LT1533 low noise switching regulator, eliminates these issues by continuous, closed-loop control of voltage and current switching times.[11] Additionally, the device's push-pull output drive eliminates the flyback interval of conventional approaches. This further reduces harmonics and smoothes input current drain characteristics. Although intended for DC/DC conversion, the LT1533 adapts nicely to off-line service, while eliminating emission, filtering, layout and noise concerns.

Figure 37.40 shows the supply. Q5 and Q6 drive T1, with a rectifier-filter, the LT1431 and the optocoupler closing an isolated loop back to the LT1533. The LT1533 drives Q5 and Q6 in cascode fashion to achieve high voltage switching capability. It also continuously controls their current and voltage switching times, using the resistors at the I_{SLEW} and V_{SLEW} pins to set transition rates. FET current information is directly available, although FET voltage status is derived via the 360k-10k dividers and routed to the gates via the NPN-PNP followers. The source wave shapes, and hence the voltage slewing information at the LT1533 collector terminals, are nearly identical in shape to the drain waveforms.

Q1, Q2 and associated components provide a bootstrapped bias supply, with start-up transistor Q1 turning off once T1 begins supplying power to Q2. The resistor string at Q2's emitter furnishes various "housekeeping" bias potentials. The LT1533's internal 1A current limit is too high for effective overcurrent protection. Instead, current is sensed via the 0.8Ω shunt at the LT1533's emitter pin (E). C1, monitoring this point, goes low when current limit is exceeded. This pulls the V_C pin low and also accelerates voltage slew rate, resulting in fast limiting while minimizing instantaneous FET stress. Prolonged short-circuit conditions result in C2 going low, putting the circuit into shutdown. Once this occurs, the C1-C2 loop oscillates in a controlled manner, sampling current for about a millisecond every second or so. This action forms a power limit, preventing FET heating and eliminating heat sink requirements.

Note 11: In depth coverage of this device, its use and performance verification appears in reference 23.

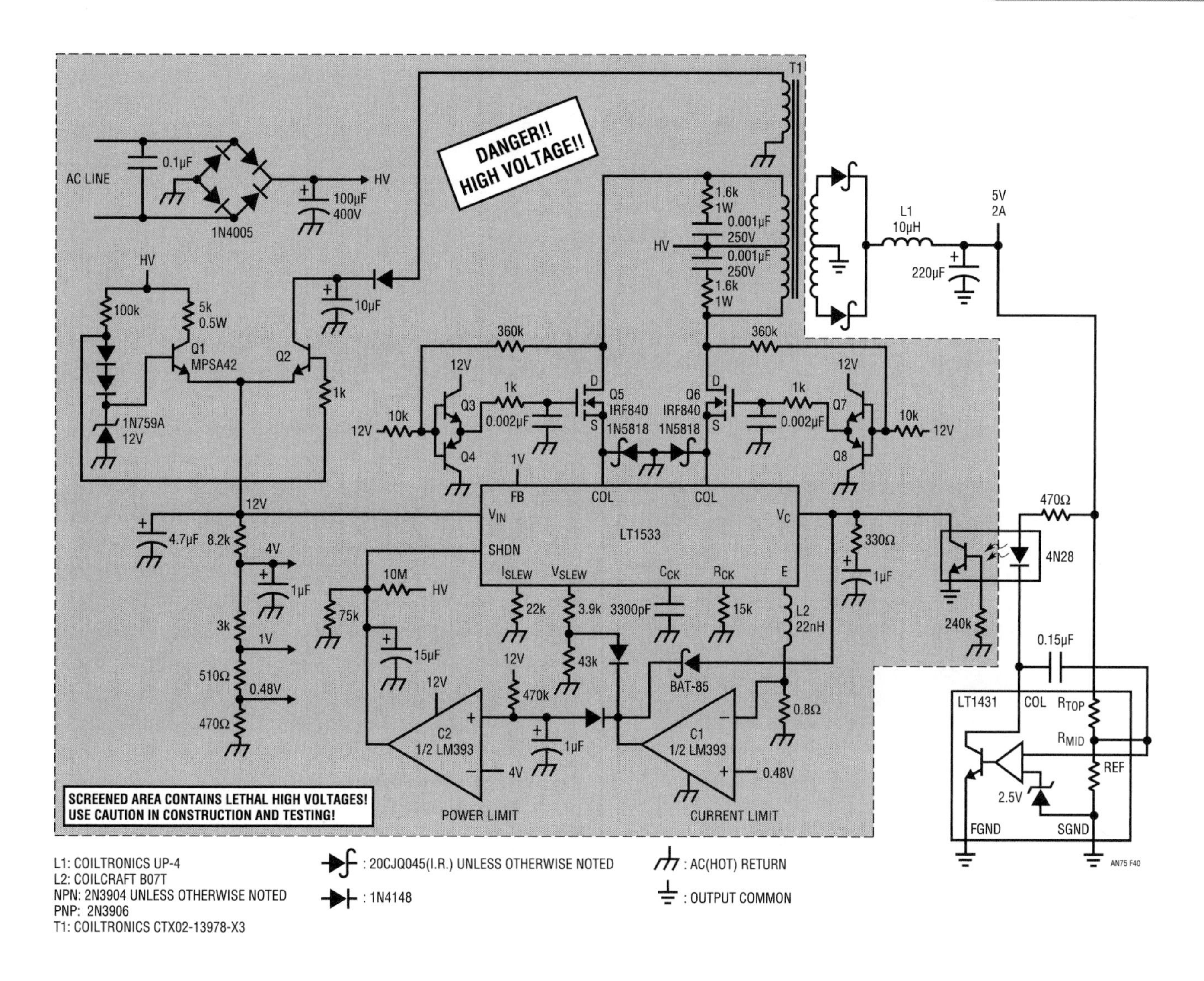

Figure 37.40 • 10W Off-Line Power Supply Passes FCC Emission Requirements Without Filter Components

Figure 37.41 shows waveforms for the power supply. Trace A is one FET source; traces B and C are its gate and drain waveforms, respectively. FET current is trace D. The cascoded drive maintains waveshape fidelity, even as the LT1533 tightly regulates voltage and current transition rates. The wideband harmonic activity typical of off-line supply waveforms is entirely absent. Power delivery to T1 (center screen, trace C) is particularly noteworthy. The waveshapes are smoothly controlled, and no high frequency content is observable. Figure 37.42 increases sweep speed by a factor of 5, but high frequency components are still undetectable. Figure 37.43 shows supply input monitored with a wideband current probe at the "HV" node. The current drain profile is smooth, with complete absence of high frequency content.

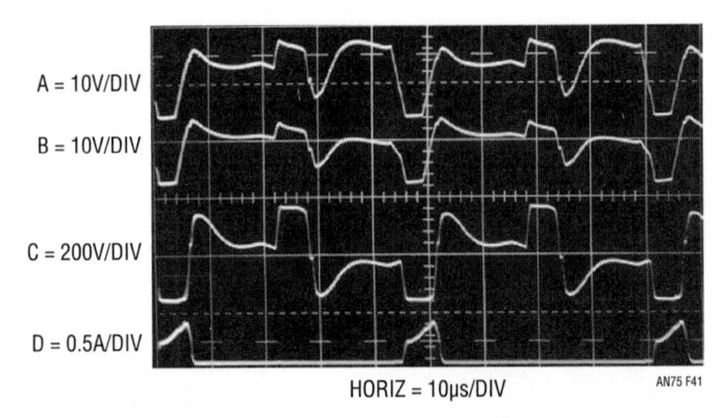

Figure 37.41 • Waveforms for One of the Power Supplies' FETs Show No Wideband Harmonic Activity. LT1533 Provides Continuous Control of Voltage and Current Slewing. Result Is Smoothly Controlled Waveshapes for FET Source (A), Gate (B) and Drain (C). FET Current is Trace D

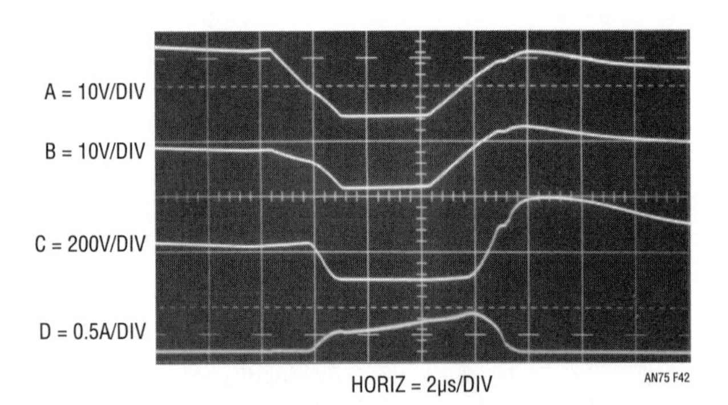

Figure 37.42 • Time Expanded Version of Figure 37.41, with Same Trace Assignments. No Wideband Components Are Detectable

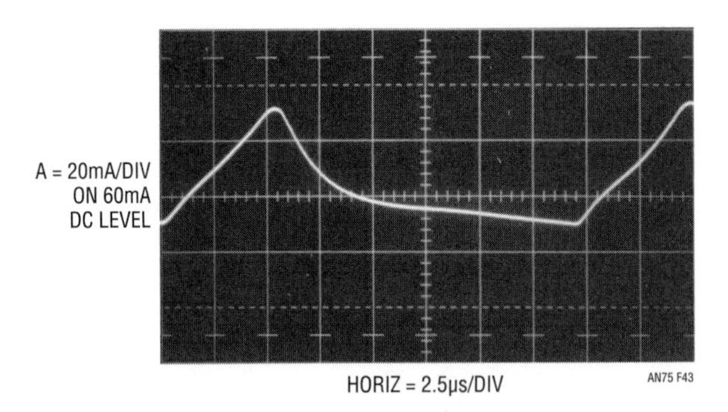

Figure 37.43 • Circuit's Input Current Drain Profile Is Smooth, with No High Frequency Content

Figure 37.44, a 30MHz wide spectral plot, shows circuit emissions well below FCC requirements. This data was taken with no input filtering LC components and a nominally nonoptimal layout.

Output noise is composed of fundamental ripple residue, with essentially no wideband components. Typically, the low frequency ripple is below 50mV If additional ripple attenuation is desired a 100μH-100μF LC section permits <100μV output noise. Figure 37.45 shows this in a 100MHz bandpass. Ripple and noise are so low that the oscilloscope requires a 40dB low noise preamplifier to even register a display (see Note 11).

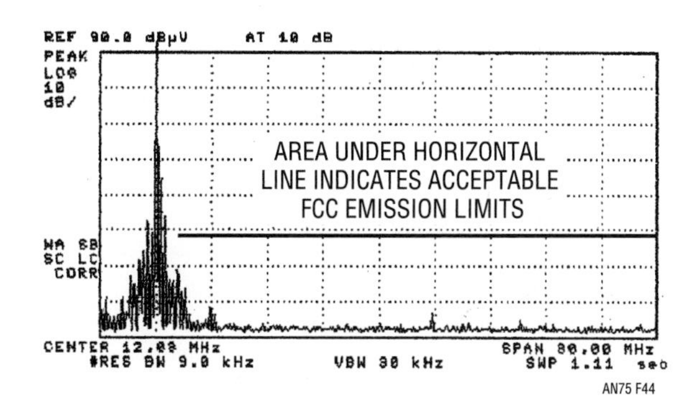

Figure 37.44 • 30MHz Wide Spectral Plot Shows Circuit Emissions Well Below FCC Requirements Despite Lack of Conventional Filter Components

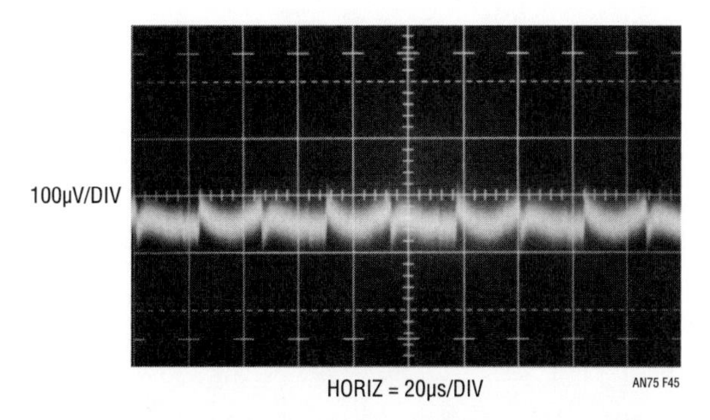

Figure 37.45 • Power Supply Output Noise Below 100μV (100MHz Measurement Bandwidth) Is Obtainable Using Additional Output LC Section. Without LC Section Wideband Harmonic Is Still Absent, Although Fundamental Ripple Is 50mV

References

1. Sylvan, T. P, "Voltage-to-Frequency Converter," Transistor Manual, General Electric Co., 1964, Figure 13.63, p. 346.
2. Pease, R. A., "A new Ultralinear Voltage-to-Frequency Converter," 1973 NEREM Record, Vol. 1, p. 167
3. Pease, R. A., assignee to Teledyne, "Amplitude to Frequency Converter," U.S. patent 3, 746, 968, filed September 1972
4. Williams, J., "Low Cost A/D Conversion Uses Single- Slope Techniques," EDN, August 5, 1978, pp. 101–104
5. Williams, J., "Designs for High Performance Voltage- to-Frequency Converters," Linear Technology Corporation, Application Note 14, March 1986
6. Wilkinson, D. H., "A Stable Ninety-Nine Channel Pulse Amplitude Analyzer for Slow Counting," Proceedings of the Cambridge Philosophical Society, Cambridge, England 46, 508. (1950)
7. Hewlett-Packard Company, "Electronic Test Instruments," Catalog No. 25, Digital, Differential Voltmeters, Ramp (Voltage-to-Time) DVM, Hewlett-Packard Company, 1965, pp. 142–143
8. Hewlett-Packard Company, "Operating and Service Manual—HP3440 DVM," Hewlett-Packard Company, 1961
9. Williams, J., "Micropower Circuits for Signal Conditioning," Linear Technology Corporation, Application Note 23, April 1987
10. Hewlett-Packard Company, "1968 Instrumentation. Electronic—Analytical—Medical," AC Voltage Measurement, Hewlett-Packard Company, 1968, pp. 197–198
11. Gregory Justice, "An RMS-Responding Voltmeter with High Crest Factor Rating," Hewlett-Packard Journal, Hewlett-Packard Company, January, 1964
12. Hewlett-Packard Company, "Model HP3400A RMS Voltmeter Operating and Service Manual," Hewlett-Packard Company, 1965
13. Williams, J., "A Monolithic IC for 100MHz RMS/DC Conversion," Linear Technology Corporation, Application Note 22, September 1987
14. Hewlett-Packard Company, "Schottky Diodes for High Volume, Low Cost Applications," Application Note 942, Hewlett-Packard Company, 1973
15. Tektronix, Inc., "Sampling Notes," Tektronix, Inc., 1964
16. Goldberg, E. A., "Stabilization of Wideband Amplifiers for Zero and Gain," RCA, Review, June 1950, p. 298
17. Williams, J., "Applications Considerations and Circuits for a New Chopper-Stabilized Op Amp," Linear Technology Corporation, Application Note 9, March 1985
18. Mattheys, R. L., "Crystal Oscillator Circuits," Wiley, New York, 1983
19. Frerking, M. E., "Crystal Oscillator Design and Temperature Compensation," Van Nostrand Reinhold, New York, 1978
20. Williams, J., "Circuit Techniques for Clock Sources," Linear Technology Corporation, Application Note 12, October 1985
21. Williams, J., "A Fourth Generation of LCD Backlight Technology," Linear Technology Corporation, Application Note 65, November 1995
22. Witt, J., "LT1534 Ultralow Noise Switching Regulator Controls EMI," Linear Technology Corporation, Design Note 178, April 1998
23. Williams, J., "A Monolithic Switching Regulator with 100pV Output Noise," Linear Technology Corporation, Application Note 70, October 1997
24. Hunt, F V, and Hickman, R. W., "On Electronic Voltage Stabilizers," "Cascode," Review of Scientific Instruments, January 1939, pp. 6–21, p. 16.[12]
25. J. Williams, "High Speed Amplifier Techniques," Linear Technology Corporation, Application Note 47 (1991) 96–97

Note 12: Veterans of LTC Application Notes, a weary brigade, may recognize this reference as the object of Application Note 70's (Footnote 14) champagne prize offer. The mystery solved, the messenger was compensated as specified (Veuve Clicquot Ponsardin).

Appendix A
Some guidelines for micropower design and an example

As with all engineering, micropower circuitry requires attention to detail, awareness of trade-offs and an opportunistic bent towards achieving the design goal.

The most obvious way to save power is to choose components which require little energy. Additional savings require more effort.

Circuits should be examined in terms of current flow. Consider such flow in all DC and AC paths. For example, do DC base currents go where they can do some useful work, or are they thrown away? Try to keep AC signal swings down, particularly if capacitors (parasitic or intended) must be continually charged and discharged. Examine the circuit for areas where power strobing may be allowable.

Consider quiescent vs dynamic power requirements of components to avoid unpleasant surprises. Data sheets usually specify quiescent power because the manufacturer doesn't know what the user's circuit conditions are. For example, everyone "knows" that "MOS devices draw no current." Unfortunately, Mother Nature dictates that as frequency and signal swings go up, the capacitances associated with MOS devices begin to require more power. It is often a mistake to automatically associate low power operation with a process technology. While it's likely that CMOS will provide lower power operation for a given function than 12AX7s, a bipolar approach may be even better. Consider individual situations on the basis of their specific requirements before committing to a technology. Very often, circuits require several technologies (i.e., CMOS, bipolar and discrete) for best results.

Usually, achieving low power operation requires performance trade-offs. Minimizing signal swings and current saves power, but moves circuit operation closer to the noise floor. Offsets, drift, bias currents and noise become increasingly significant error factors as signal amplitudes are constricted to save power. This is a fundamental tradeoff and must be carefully considered. Circuits employing power strobing can sometimes get around this problem by utilizing low duty cycles.

Text Figures 37.1 and 37.4, voltage-to-frequency converters, furnish an example of the evolution of a low power design. Design goals included a 10kHz maximum output, low drift, fast step response, linearity inside 0.05% and minimum supply current. Other specifications appear in the text.

Figure A1 shows an early (1986) version of this circuit. Operation is similar to the text described for Figure 37.1, but a brief description follows. When the input current-derived ramp at C1's negative input crosses zero, C1's output drops low, pulling charge through C1. This forces the negative input below zero. C2 provides positive feedback, allowing a complete discharge for C1. When C2 decays, C1A's output goes high, clamping at the level set by D1, D2 and V_{REF}. C1 receives charge and recycling occurs when C1A's negative input again arrives at zero. The frequency of this action is related to the input voltage. Diodes D3 and D4 provide steering, and are temperature compensated by D1 and D2. C1A's sink saturation voltage is uncompensated, but small. C1B is a start-up loop.

Although the LT1017 and LT1034 have low operating currents, this circuit pulls almost 400μA. The AC current paths include C1's charge-discharge cycle, and C2's branch. The DC path through D2 and V_{REF} is particularly costly. C1's charging must occur quickly enough for 10kHz operation, meaning the clamp seen by C1A's output must have low impedance at this frequency. C3 helps, but significant current still must come from somewhere to keep impedance low. C1A's current limited output cannot do the job unaided, and the resistor from the supply is required. Even if C1A could supply the necessary current, V_{REF}'s settling time would be an issue. Dropping C1's value will reduce impedance requirements proportionally, and would seem to solve the problem. Unfortunately, such reduction magnifies the effects of stray capacitance at the D3-D4 junction. It also mandates increasing R_{IN}'s value to keep scale factor constant. This lowers operating currents at C1A's negative input, making bias current and offset more significant error sources.

Figure A2 shows an initial attempt at dealing with these issues. This scheme is similar to Figure A1, except that Q1 and Q2 appear. V_{REF} receives switched bias via Q1, instead of being on all the time. Q2 provides the sink path for C1. These transistors invert C1A's output, so its input pin assignments are exchanged. R1 provides a light current from the supply, improving reference settling time. This arrangement decreases supply current to about 300μA, a significant improvement. Several problems do exist, however. Q1's switched operation is really effective only at higher frequencies. In the lower ranges, C1A's output is low most of the time, biasing Q1 on and wasting power. Additionally, when C1A's output switches, Q1 and Q2 simultaneously conduct during the transition, effectively shunting R2 across the supply. Finally, the base currents of both transistors flow to ground and are lost. The basic temperature compensation is as before, except that Q2's saturation term replaces the comparator's.

Figure A3 is better. Q1 is gone, Q2 remains but Q3, Q4 and Q5 have been added. V_{REF} and its associated diodes are biased from R1. Q3, an emitter-follower, is used to source current to C1. Q4 temperature compensates Q3's V_{BE}, and Q5 switches Q3.

This method has some distinct advantages. The V_{REF} string can operate at greatly reduced current because of Q3's current gain. Also, Figure A2's simultaneous conduction problem is largely alleviated because Q5 and Q2 are switched at the same voltage threshold out of C1A. Q3's base and emitter currents are delivered to C1. Q5's currents are wasted, although they are much smaller than Q3's. Q2's small base current is also lost. The values for C2 and R3 have been changed. The time constant is the same, but some current reduction occurs due to R3's increase.

If C1 cannot be reduced, then its AC currents cannot be avoided. This leaves only the aforementioned Q5 and Q2 currents as significant wasted terms, along with R3's now smaller loss. Current drain for this circuit is about 200μA maximum.

Figure A4 (1987) is very similar, but eliminates Q5 and Q2's losses to get maximum operating current below 150μA with quiescent current under 80μA. The basic improvement is the use of CMOS inverters for reference switching—the inverters supply pin is driven by the reference buffer NPN and their paralleled outputs switch between V_{REF} and ground. Other enhancements provide better temperature compensation and improved power supply rejection. The modified LM334 driven reference stack begins to look very similar to Figure 37.1's arrangement. This circuit provided excellent precision—0.02% linearity, 40ppm/°C drift and 40ppm/V PSRR.

A variant (1991) of this circuit, Figure A5, reduced supply current to only 90μA maximum, by minimizing the number of CMOS inverters, eliminating their AC input currents. The charge dispensing capacitor was also reduced to 100pF necessitating a larger input resistance value. The price for the current saving was degradation of drift and linearity by factors of 2 and 3, respectively.

Text Figures 37.1 and 37.4 (1997 and 1999, respectively) are direct extensions of the last two circuits. Their markedly decreased operating currents are obtained with minimal performance compromises by utilizing contemporary components. The LTC1440/LTC1441 comparators and the LT1389 reference are the heroes. Some other refinements are involved, but the text's voltage-to-frequency circuits are the final (for now) iteration of the five versions shown here.

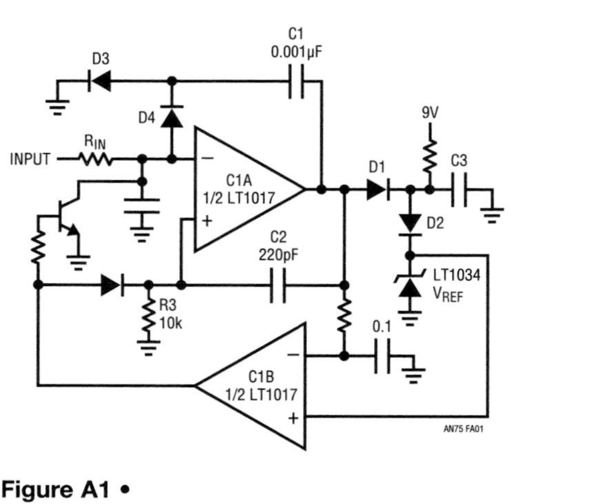

Figure A1 •

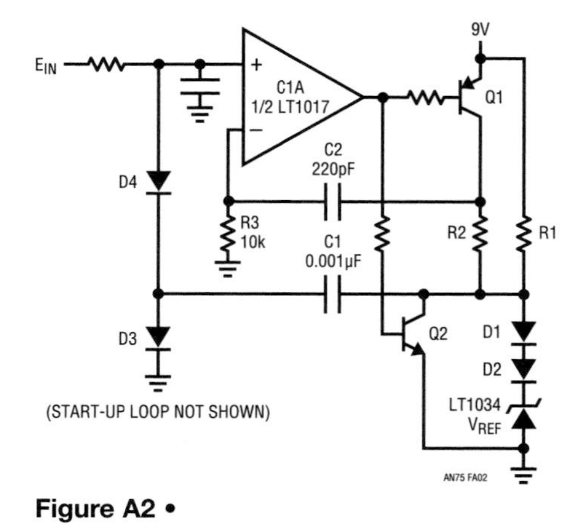

Figure A2 •

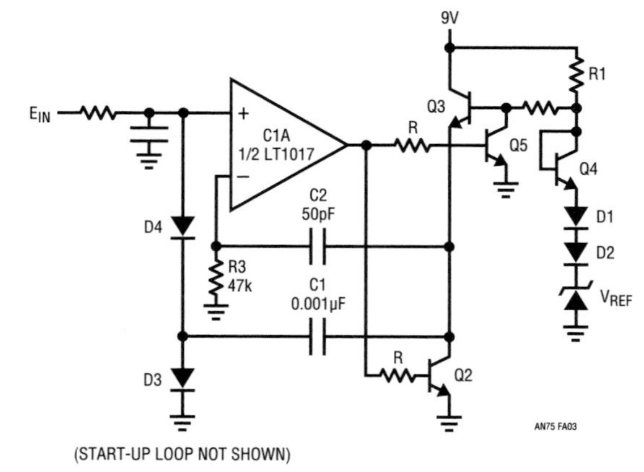

Figure A3 •

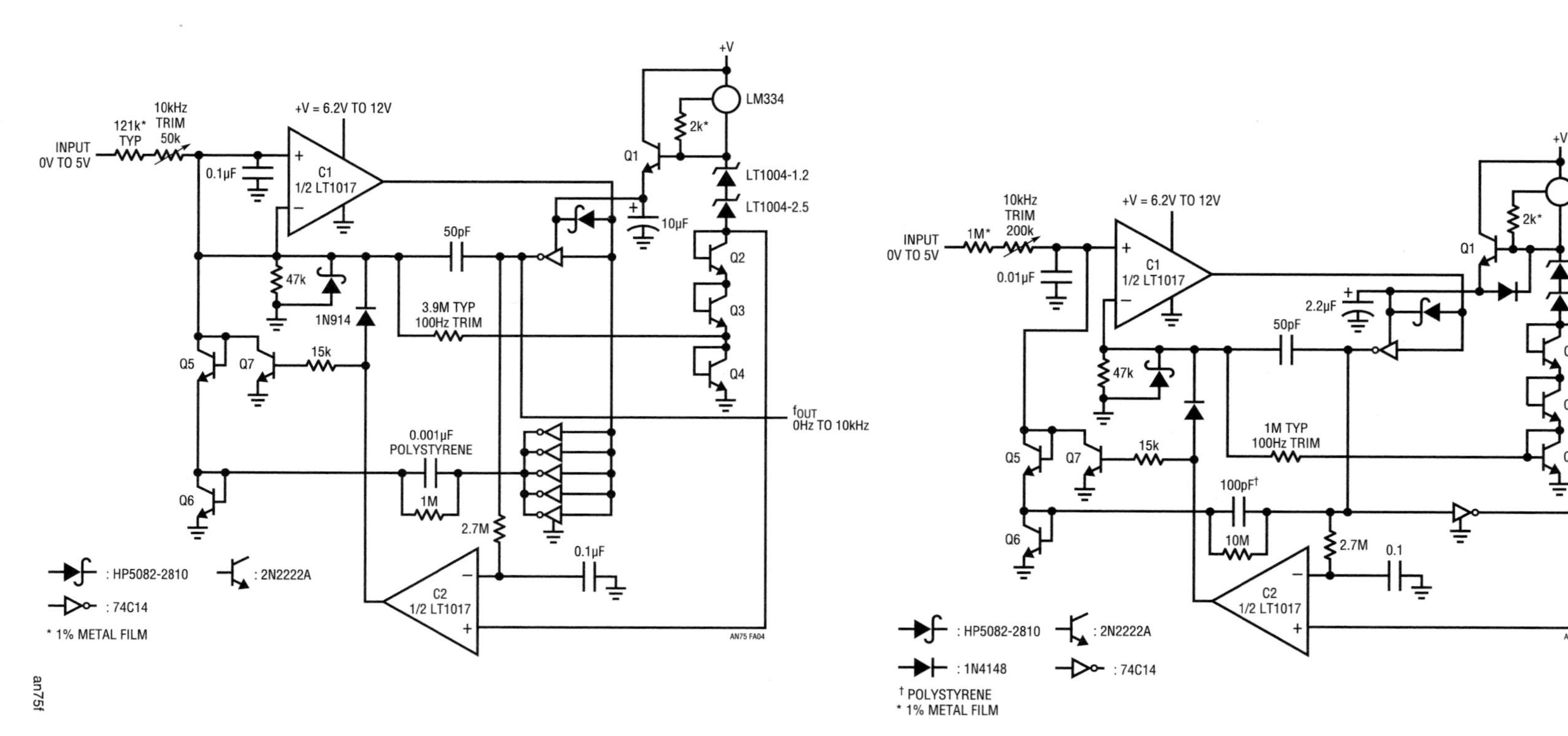

Figure A4 •

Figure A5 •

Appendix B
Parasitic effects of test equipment on micropower circuits

The energy absorbed by test equipment connections to micropower circuits can be significant. Under normal circumstances test equipment and probes have negligible power drain, but microampere level operating currents mandate care. Test instrumentation should be regarded as an integral part of the circuit. DC and AC loading and parasitic effects must be kept in mind to avoid unpleasant surprises. Such instrument connection errors can make the circuit under test look unfairly bad or good.

The DC resistance of oscilloscope probes varies from hundreds of ohms (1X types) to 10MΩ (10X), with some 10X types as low as 1MΩ. Contrary to some expectations, FET probes do not have high input resistance—some types are as low as 100kΩ, although most are about 10MΩ. The DC loading of a 10X 1M probe could introduce as much as 5μA of loss, almost 60% of Figure 37.4's total! The AC loading of a 10pF probe looking at Figure 37.27's 30kHz clock will cause apparent circuit consumption of 1μA, a significant loss in a low power circuit. 1X type probes present about 50pF of loading, with 1MΩ DC resistance when connected to the 'scope. This kind of probe loading can cause large errors in micropower circuits, while virtually disabling some. Such a probe, introduced at C1B's negative input in text Figure 37.6, would stop the circuit's oscillator. If placed across the supply of Figure 37.6 it would consume almost as much current as the circuit.

Probe AC and DC loading are not the only effects. Some DVMs produce "charge spitting" at their inputs. Such parasitic charge, introduced into high impedance nodes, can cause substantial errors. It's also worth remembering that DVM DC loading may change with range. Lower ranges may have very high input impedance, but higher ranges are typically 10MΩ. A 10MΩ DVM reading Figure 37.6's supply introduces almost 10% supply current error.

Figure B1 shows a way test equipment can make the circuit look too good, instead of too bad. If the pulse generator is adjusted more than a diode drop above the regulator's output, the bypass capacitor peak detects the charge delivered through the IC's internal diode. The regulator can't sink current, and with its output forced high it won't source anything. Under these conditions the circuit functions while the current meter reads zero.. .a very low power circuit indeed[1]!

Figure B2 shows a very simple, but useful circuit which greatly aids probe loading problems in micropower circuits. The LT1022 high speed FET op amp drives an LT1010 buffer. The LT1010's output allows DVM cable and probe driving and also biases the circuit's input shield. This bootstraps the input capacitance, reducing its effect. DC and AC errors of this circuit are low enough for almost all work, with enough bandwidth for just about any low power circuit. Built into a small enclosure with its own power supply, it can be used ahead of a 'scope or DVM with good results. Pertinent specifications appear in the diagram.

Figure B3 is a very fast high impedance probe for those occasions which require it. A1, a hybrid FET buffer, forms the electrical core of the probe. This device is a low input capacitance, wideband FET source follower driving a fast bipolar output stage. The input of the probe goes to this device via a 51Ω resistor, reducing the possibility of oscillations in the follower input stage when the probe sees low AC impedance. A1's output drives a guard shield around the probe's input line, reducing effective input capacitance to about 4pF A ground referred shield encircles the guard shield, reducing pickup and making high quality ground connections to the circuit under test easy. Back-terminated A1 drives the output BNC cable, feeding a 50Ω termination at the oscilloscope. Specifications are noted in the figure. Note that the back termination mandates an attenuation of 2, while the buffer's open-loop architecture introduces a small gain error. The probe's physical construction is critical to achieving stated performance. See reference [25] for details.

Note 1: Practically speaking, most regulators and power supplies can sink small amounts of current. Because of this, the current meter may actually read negative.

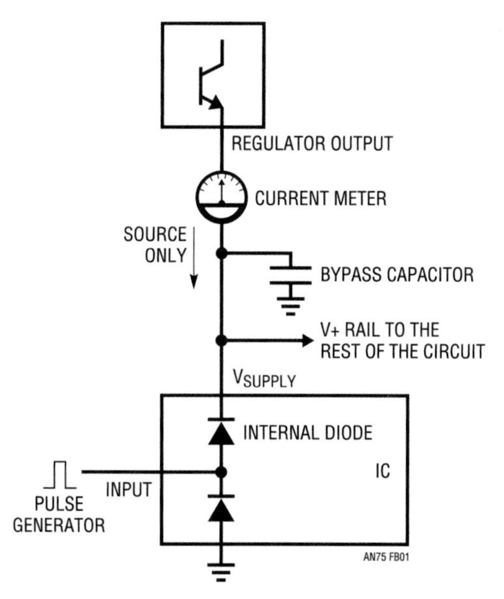

Figure B1 • Parasitic Currents Flowing Into Circuit From Pulse Generator Produce Misleading Current Meter Indications

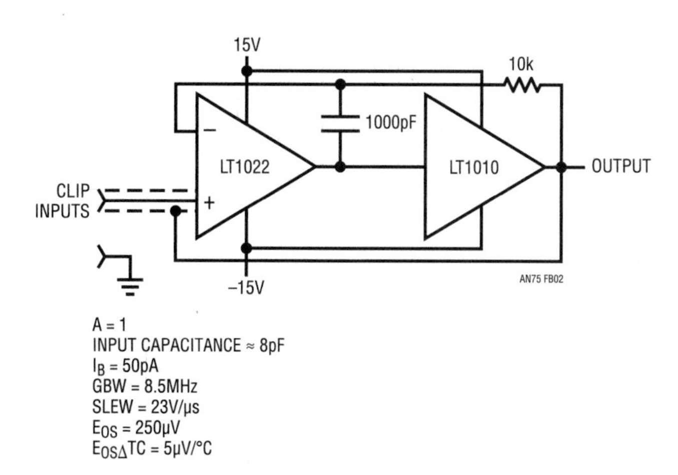

Figure B2 • High Impedance Probe Introduces Minimal Loading. Speed Is Adequate for Most Micropower Circuits

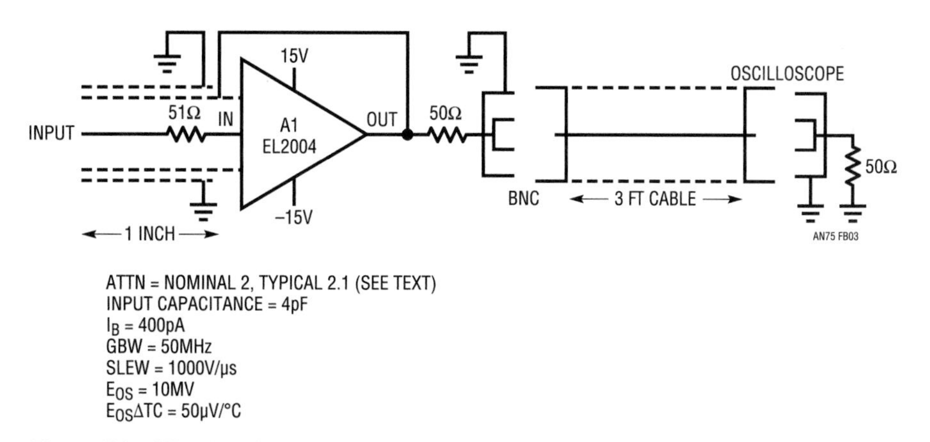

Figure B3 • Ultrafast Buffer Probe Maintains Minimal Loading with 50MHz Bandwidth

Circuit collection, volume V

Data conversion, interface and signal conditioning products

38

Richard Markell, Editor

Introduction

This is the fifth in a series that excerpts useful circuits from *Linear Technology* magazine to preserve them for posterity. This application note highlights data conversion, interface and signal conditioning circuits from issue VI:1 through issue VIII:4. Like its predecessor, AN67, this Application Note includes circuits for high speed video, interface and hot swap circuits, active RC and switched capacitor filter circuitry and a variety of data conversion and instrumentation circuits. There are also several circuits that cannot be so neatly categorized. So, without further ado, I'll let the authors describe their circuits.

Note: Article Titles appear in this application note exactly as they originally appeared in *Linear Technology* magazine. This may result in some inconsistency in the usage of terminology.

Analog Circuit and System Design: Immersion in the Black Art of Analog Design. http://dx.doi.org/10.1016/B978-0-12-397888-2.00038-9

Data converters

The LTC1446 and LTC1446L: world's first dual 12-bit DACs in SO-8 packages

by Hassan Malik and Jim Brubaker

Dual 12-bit rail-to-rail performance in a tiny SO-8

The LTC1446 and LTC1446L are dual 12-bit, single-supply, rail-to-rail voltage output digital-to-analog converters. Both of these parts include an internal reference and two DACs with rail-to-rail output buffer amplifiers, packed in a small, space-saving 8-pin SO or PDIP package. A power-on reset initializes the outputs to zero-scale at power-up.

The LTC1446 has an output swing of 0V to 4.095V making each LSB equal to 1mV It operates from a single 4.5V to 5.5V supply, dissipating 3.5mW (I_{CC} typical = 700μA). The LTC1446L has an output swing of 0V to 2.5V It can operate on a single supply with a wide range of 2.7V to 5.5V It dissipates 1.35mW (ICC typical = 450μA) at a 3V supply.

An autoranging 8-channel ADC with shutdown

Figure 38.1 shows how to use an LTC1446 to make an autoranging ADC. The microprocessor sets the reference span and the common pin for the analog input by loading the appropriate digital code into the LTC1446. V_{OUTA} controls the common pin for the analog inputs to the LTC1296 and V_{OUTB} controls the reference span by setting the REF_+ pin on the LTC1296. The LTC1296 has a shutdown pin that goes low in shutdown mode. This will turn off the PNP transistor supplying power to the LTC1446. The resistor and capacitor on the LTC1446 outputs act as a lowpass filter for noise.

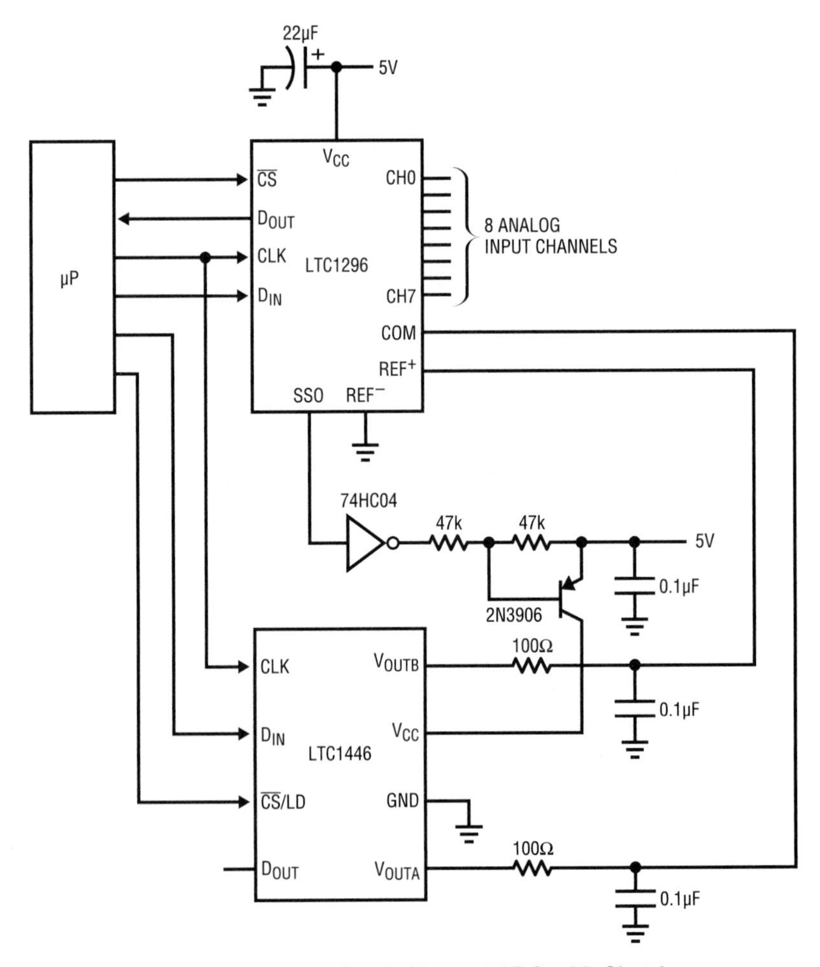

Figure 38.1 • An Autoranging 8-Channel ADC with Shutdown

A wide-swing, bipolar-output DAC with digitally controlled offset

Figure 38.2 shows how to use an LTC1446 and an LT1077 to make a wide bipolar-output-swing 12-bit DAC with an offset that can be digitally programmed. V_{OUTA}, which can be set by loading the appropriate digital code for DAC A, sets the offset. As this value changes, the transfer curve for the output moves up and down, as shown in the figure.

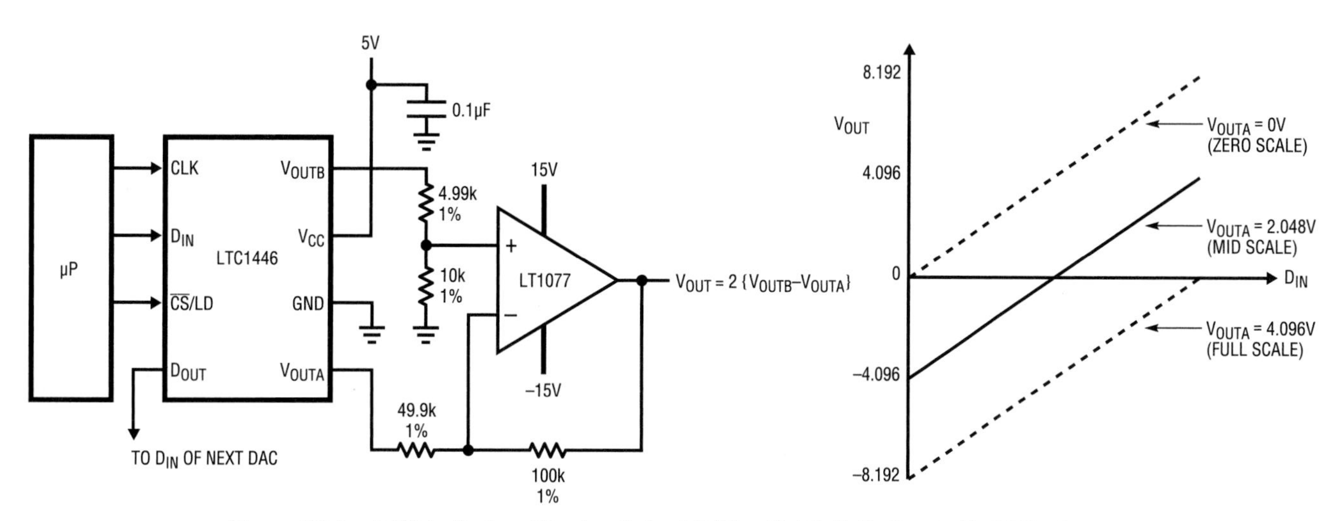

Figure 38.2 • A Wide-Swing, Bipolar Output DAC with Digitally Controlled Offset

Multichannel A/D uses a single antialiasing filter

by LTC Applications Staff

The circuit in Figure 38.3 demonstrates how the LTC1594's independent analog multiplexer can simplify the design of a 12-bit data acquisition system. All four channels are MUXed into a single 1kHz, fourth-order Sallen-Key antialiasing filter, which is designed for single-supply operation. Since the LTC1594's data converter accepts inputs from ground to the positive supply, rail-to-rail op amps were chosen for the filter to maximize dynamic range. The LT1368 dual rail-to-rail op amp is compensated for the 0.1μF load capacitors (C1 and C2) that help reduce the amplifier's output impedance and improve supply rejection at high frequencies. The filter contributes less than 1LSB of error due to offsets and bias currents. The filter's noise and distortion are less than −72dB for a 100Hz, $2V_{P-P}$ offset sine input.

The combined MUX and A/D errors result in an integral nonlinearity error of ±3LSB (maximum) and a differential nonlinearity error of ±0.75LSB (maximum). The typical signal-to-noise plus distortion ratio is 68dB, with approximately −78dB of total harmonic distortion. The LTC1594 is programmed through a 4-wire serial interface that allows efficient data transfer to a wide variety of microprocessors and microcontrollers. Maximum serial clock speed is 200kHz, which corresponds to a 10.5kHz sampling rate.

The complete circuit consumes approximately 800μA from a single 5V supply. For ratiometric measurements, the A/D's reference can also be taken from the 5V supply. Otherwise, an external reference should be used.

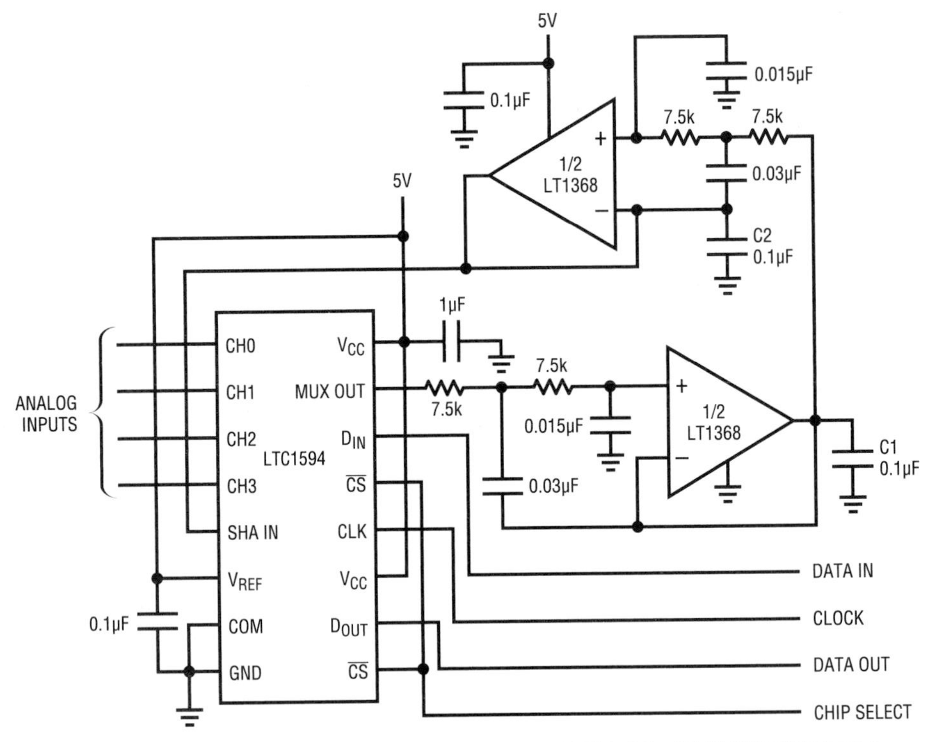

Figure 38.3 • Simple Data Acquisition System Takes Advantage of the LTC1594's MUX OUT/SHA IN Loop to Filter Analog Signals Prior to A/D Conversion

LTC1454/54L and LTC1458/58L: dual and quad 12-bit, rail-to-rail, micropower DACs

by Hassan Malik and Jim Brubaker

Dual and quad rail-to-rail DACs offer flexibility and performance

The LTC1454 and LTC1454L are dual 12-bit, single supply, rail-to-rail voltage-output digital-to-analog converters. The LTC1458 and LTC1458L are quad versions of this family. These DACs have an easy-to-use, SPI-compatible interface. A $\overline{CLR}$ pin and power-on-reset both reset the DAC outputs to zero scale. DNL is guaranteed to be less than 0.5LSB. Each DAC has its own rail-to-rail voltage output buffer amplifier. The onboard reference is brought out to a separate pin and can be connected to the REFHI pins of the DACs. There is also a REFLO pin that can be used to offset the DAC range. For further flexibility the $\times 1/\times 2$ pin for each DAC allows the user to select a gain of either 1 or 2. The LTC1454/54L are available in 16-pin PDIP and SO packages, and the LTC1458/58L are available in 28-pin SO or SSOP packages.

5V and 3V single supply and micropower

The LTC1454 and LTC1458 operate from a single 4.5V to 5.5V supply. The LTC1454 dissipates 3.5mW (I_{CC} typical = 700μA), whereas the LTC1458 dissipates 6.5mW (I_{CC} typical = 1.3mA). There is an onboard reference of 2.048V and a nominal full scale of 4.095V when using the onboard reference and a gain-of-2 configuration.

The LTC1454L and LTC1458L operate on a single supply with a wide range of 2.7V to 5.5V The LTC1454L dissipates 1.35mW (I_{CC} typical = 450μA), whereas the LTC1458L dissipates 2.4mW (I_{CC} typical = 800μA) from a 3V supply. There is a 1.22V onboard reference and a convenient full scale of 2.5V when using the onboard reference and a gain-of-2 configuration.

Flexibility allows a host of applications

These products can be used in a wide range of applications, including digital calibration, industrial process control, automatic test equipment, cellular telephones and portable, battery-powered systems.

A 12-bit DAC with digitally programmable full scale and offset

Figure 38.4 shows how to use one LTC1458 to make a 12-bit DAC with a digitally programmable full scale and offset. DAC A and DAC B are used to control the offset and full scale of DAC C. DAC A is connected in a ×1 configuration and controls the offset of DAC C by moving $REFLO_C$ above ground. The minimum value to which this offset can be programmed is 10mV DAC B is connected in a×2 configuration and controls the full scale of DAC-C by driving $REFHI_C$. Note that the voltage at $REFHI_C$ must be less than or equal to $V_{CC}/2$, corresponding to DAC B's code <2, 500 for $V_{CC} = 5V$ since DAC-C is being operated in ×2 mode for full rail-to-rail output swing.

The transfer characteristic is:

$$V_{OUTC} = 2 \times [D_c \times (2 \times D_B - D_A) + D_A] \times REFOUT$$

Where REFOUT = The reference output
D_A = (DAC A digital code)/4096 this sets the offset
D_B = (DAC B digital code)/4096 this sets the full scale
D_C = (DAC C digital code)/4096

A single-supply, 4-quadrant multiplying DAC

The LTC1454L can also be used for four-quadrant multiplying with an offset signal ground of 1.22V This application is shown in Figure 38.5. The inputs are connected to $REFHI_B$ or $REFHI_A$ and have a 1.22V amplitude around a signal ground of 1.22V The outputs will swing from 0V to 2.44V as shown by the equation with the figure.

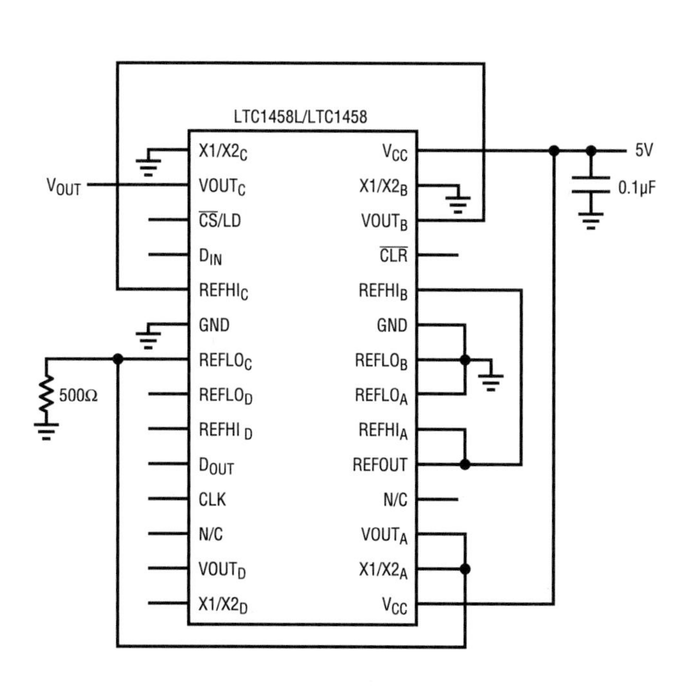

Figure 38.4 • A 12-Bit DAC with Digitally Controlled Zero Scale and Full Scale

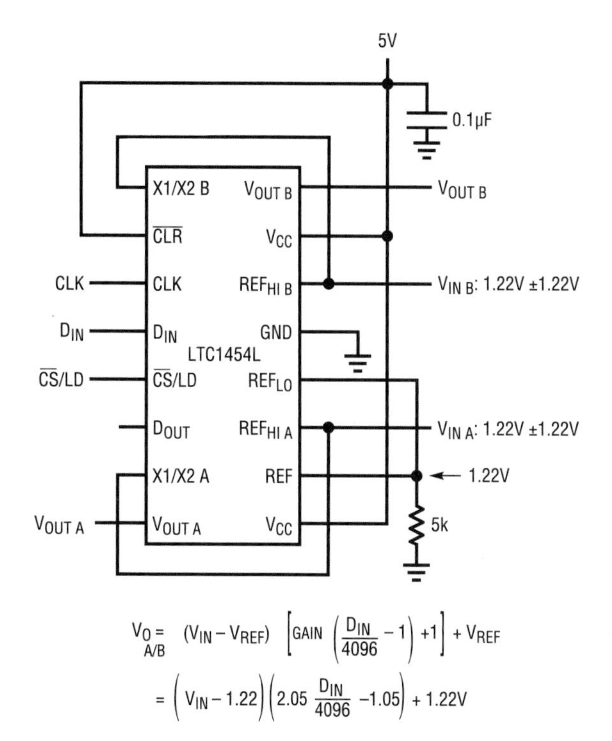

Figure 38.5 • Single-Supply, 4-Quadrant Multiplying DAC

Micropower ADC and DAC in SO-8 give PC 12-bit analog interface

by LTC Applications Staff

Needing to add two channels of simple, inexpensive, low powered, compact analog input/output to a PC computer, The LTC1298 ADC and LTC1446 DAC were chosen. The LTC1298 and the LTC1446 are the first SO-8 packaged 2-channel devices of their kind. The LTC1298 draws just 340μA. A built-in auto shutdown feature further reduces power dissipation at reduced sampling rates (to 30μA at 1ksps). Operating on a 5V supply, the LTC1446 draws just 1mA (typ). Although the application shown is for PC data acquisition, these two converters provide the smallest, lowest power solutions for many other analog I/O applications.

The circuit shown in Figure 38.6 connects to a PC's serial interface using four interface lines: DTR, RTS, CTS and TX. DTR is used to transmit the serial clock signal, RTS is used to transfer data to the DAC and ADC, CTS is used to receive conversion results from the LTC1298 and the signal on TX selects either the LTC1446 or the LTC1298 to receive input data. The LTC1298's and LTC1446's low power dissipation allows the circuit to be powered from the serial port. The TX and RTS lines charge capacitor C4 through diodes D3 and D4. An LT1021-5 regulates the voltage to 5V Returning the TX and RTS lines to a logic high after sending data to the DAC or completion of an ADC conversion provides constant power to the LT1021-5.

Using a 486-33 PC, the throughput was 3.3ksps for the LTC1298 and 2.2ksps for the LTC1446. Your mileage may vary.

Listing 1 is C code that prompts the user to either read a conversion result from the ADC's CH0 or write a data word to both DAC channels.

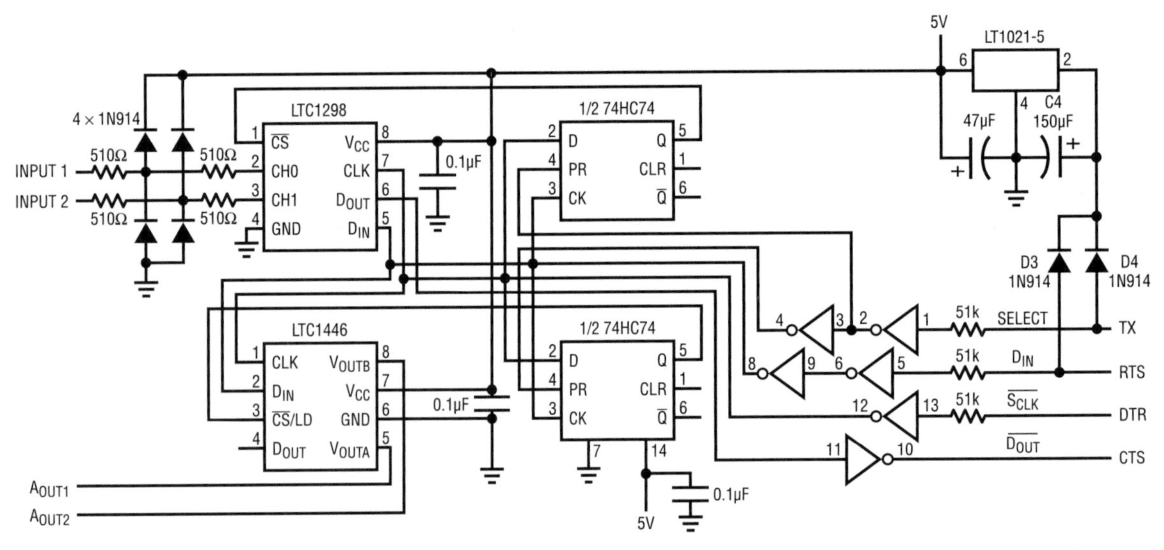

Figure 38.6 • Communicating Over the Serial Port, the LTC1298 and LTC1446 in SO-8 Create a Simple, Low Power, 2-Channel Analog Interface for PCs

Listing 1. C Code to Configure the Analog Interface

```
#define port 0x3FC            /* Control register, RS232 */
#define inprt 0x3FE           /* Status reg. RS232 */
#define LCR 0x3FB         /* Line Control Register */
#define high 1
#define low 0
#define Clock 0x01        /* pin 4, DTR */
#define Din 0x02          /* pin 7, RTS */
#define Dout 0x10         /* pin 8, CTS input */
#include<stdio.h>
#include<dos.h>
#include<conio.h>
```

```
/* Function module sets bit to high or low */
void set_control(int Port,char bitnum,int flag)
{
   char temp;
   temp = inportb(Port);
   if (flag==high)
     temp |= bitnum;     /* set output bit to high */
     else
     temp &= ~bitnum;    /* set output bit to low */
   outportb(Port,temp);
}
            /* This function brings CS high or low (consult the schematic) */
void CS_Control(direction)
{
if (direction)
     {
     set_control(port,Clock,low);          /* set clock high for Din to be read */
     set_control(port,Din,low);            /* set Din low */
set_control(port,Din,low);                 /* set Din high to make CS goes high */
     }
     else {
          outportb(port, 0x01);            /* set Din & clock low */
          Delay(10);
          outportb(port, 0x03);            /* Din goes high to make CS go low */
          }
}
         /* This function outputs a 24-bit (2x12) digital code to LTC1446L */
void Din_(long code,int clock)
{
     int x;
     for(x = 0; x<clock; ++x)
     {
     code <<= 1;                           /* align the Din bit */
     if (code & 0x1000000)
          {
          set_control(port,Clock,high);    /* set Clock low */
          set_control(port,Din,high);      /* set Din bit high */
          }
     else {
          set_control(port,Clock,high);    /* set Clock low */
          set_control(port,Din,low);       /* set Din low */
          }
     set_control(port,Clock,low);          /* set Clock high for DAC to latch */
     }
}
```

```
                                        /* Read bit from ADC to PC */
Dout_()
{
int temp, x, volt =0;
for(x = 0; x<13; ++x)
     {
     set_control(port,Clock,high);
     set_control(port,Clock,low);
     temp = inportb(inprt);          /* read status reg. */
     volt <<= 1;                     /* shift left one bit for serial transmission
*/
     if(temp & Dout)
     volt += 1;                              /* add 1 if input bit is high */
     }
return(volt & 0xfff);
}
/* menu for the mode selection */
char menu()
{
printf("Please select one of the following:\na: ADC\nd: DAC\nq: quit\n\n");
return (getchar());
}
void main()
{
long code;
char mode_select;
int temp,volt=0;
/* Chip select for DAC & ADC is controlled by RS232 pin 3 TX line. When LCR's bit 6 is
       set the DAC is selected and the reverse is true for the ADC. */
outportb(LCR,0x0);                /* initialize DAC */
outportb(LCR,0x64);               /* initialize ADC */
while((mode_select = menu()) != 'q')
     {
     switch(mode_select)
        {
        case 'a':
              {
              outportb(LCR,0x0);                   /* selecting ADC */
              CS_Control(low);                /* enabling the ADC CS */
              Din_(0x680000, 0x5);                          /* channel selection */
              volt = Dout_();
              outportb(LCR,0x64);                  /* bring CS high */
              set_control(port,Din,high);    /* bring Din signal high */
              printf("\ncode: %d\n",volt);
              }
```

```
            break;
        case 'd':
          {
          printf("Enter DAC input code (0 - 4095):\n");
          scanf("%d", &temp);
          code = temp;
          code += (long)temp << 12;        /* converting 12-bit to 24-bit word */
          outportb(LCR,0x64)               /* selecting DAC */
          CS_Control(low);                 /* CS enable */
          Din_(code,24);                   /* loading digital data to DAC */
          outportb(LCR,0x0);               /* bring CS high */
          outportb(LCR,0x64);              /* disabling ADC */
          set_control(port,Din,high);      /* bring Din signal high */
          }
        break;
        }
    }
}
```

The LTC1594 and LTC1598: micropower 4- and 8-channel 12-bit ADCs

by Marco Pan

Micropower ADCs in small packages

The LTC1594 and LTC1598 are micropower 12-bit ADCs that feature a 4- and 8-channel multiplexer, respectively. The LTC1594 is available in a 16-pin SO package and the LTC1598 is available in a 24-pin SSOP package. Each ADC includes a simple, efficient serial interface that reduces interconnects and, thereby, possible sources of corrupting digital noise. Reduced interconnections also reduce board size and allow the use of processors having fewer I/O pins, both of which help reduce system costs.

The LTC1594 and LTC1598 include an auto shutdown feature that reduces power dissipation when the converter is inactive (whenever the CS signal is a logic high).

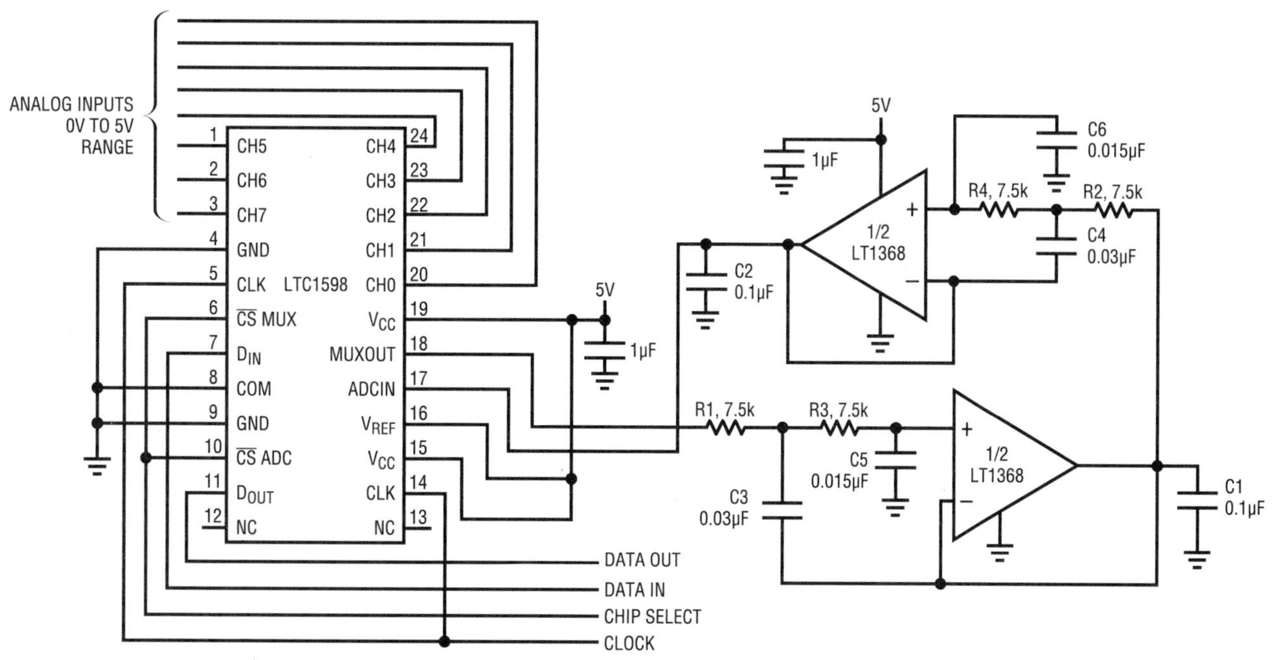

Figure 38.7 • Simple Data Acquisition System Takes Advantage of the LTC1598's MUX OUT/ADCIN Pins to Filter Analog Signals Prior to A/D Conversion

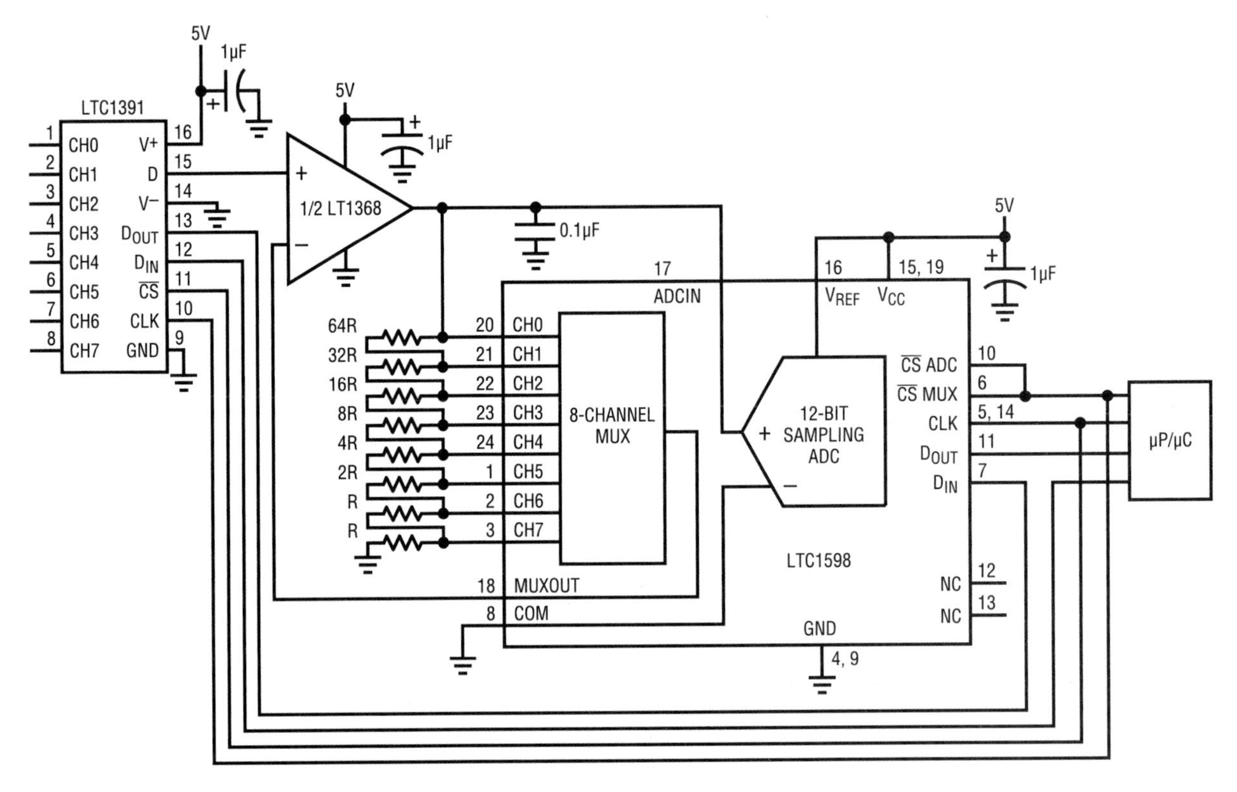

Figure 38.8 • Using the MUXOUT/ADCIN Loop of the LTC1598 to Form a PGA with Eight Gains in a Noninverting Configuration

MUXOUT/ADCIN loop economizes signal conditioning

The MUXOUT and ADCIN pins form a very flexible external loop that allows PGA and/or processing analog input signals prior to conversion. This loop is also a cost effective way to perform the conditioning, because only one circuit is needed instead of one for each channel. Figure 38.7 shows the loop being used to antialias filter several analog inputs. The output signal of the selected MUX channel, present on the MUXOUT pin, is applied to R1 of the Sallen-Key filter. The filter bandlimits the analog signal and its output is applied to ADCIN. The LT1368 rail-to-rail op amps used in the filter will, when lightly loaded as in this application, swing to within 8mV of the positive supply voltage. Since only one circuit is used for all channels, each channel sees the same filter characteristics.

Table 38.1 PGA Gain for Each MUX Channel of Figures 38.8 and 38.9

MUX Channel	Noninverting Gain	Inverting Gain
0	1	−1
1	2	−2
2	4	−4
3	8	−8
4	16	−16
5	32	−32
6	64	−64
7	128	−128

Using MUXOUT/ADCIN loop as PGA

Combined with the LTC1391 (as shown in Figure 38.8) the LTC1598's MUXOUT/ADCIN loop and an LT1368 can be used to create an 8-channel PGA with eight noninverting gains for each channel. The output of the LT1368 drives the ADCIN and the resistor ladder. The resistors above the selected MUX channel form the feedback for the LT1368. The loop gain for this amplifier is $(R_{S1}/R_{S2}) + 1$. R_{S1} is the summation of the resistors above the selected MUX channel and R_{S2} is the summation of the resistors below the selected MUX channel. If CH0 is selected, the loop gain is 1 since R_{S1} is 0. Table 38.1 shows the gain for each MUX channel. The LT1368 dual rail-to-rail op amp is designed to operate with 0.1μF load capacitors. These capacitors provide frequency compensation for the amplifiers, help reduce the amplifiers' output impedance and improve supply rejection at high frequencies. Because the LT1368's I_B is low, the R_{ON} of the selected channel will not affect the loop gain given by the formula above. In the case of the inverting configuration of Figure 38.9, the selected channel's R_{ON} will be added to the resistor that sets the loop gain.

8-Channel, differential, 12-bit A/D system using the LTC1391 and LTC1598

The LTC1598 can be combined with the LTC1391 8-channel, serial-interface analog multiplexer to create a differential A/D system. Figure 38.10 shows the complete 8-channel, differential A/D circuit. The system uses the LTC1598's MUX as the noninverting input multiplexer and the LTC1391 as inverting input multiplexer. The LTC1598's MUXOUT drives the ADCIN directly. The inverting multiplexer's output is applied to the LTC1598's COM input. The LTC1598 and LTC1391 share the CS, D_{IN}, and CLK control signals. This arrangement simultaneously selects the same channel on each multiplexer and maximizes the system's throughput. The dotted-line connection daisy-chains the MUXes of the LTC1391 and LTC1598 together. This configuration provides the flexibility to select any channel in the noninverting input MUX with respect to any channel in the inverting input MUX. This allows any combination of signals applied to the inverting and noninverting MUX inputs to be routed to the ADC for conversion.

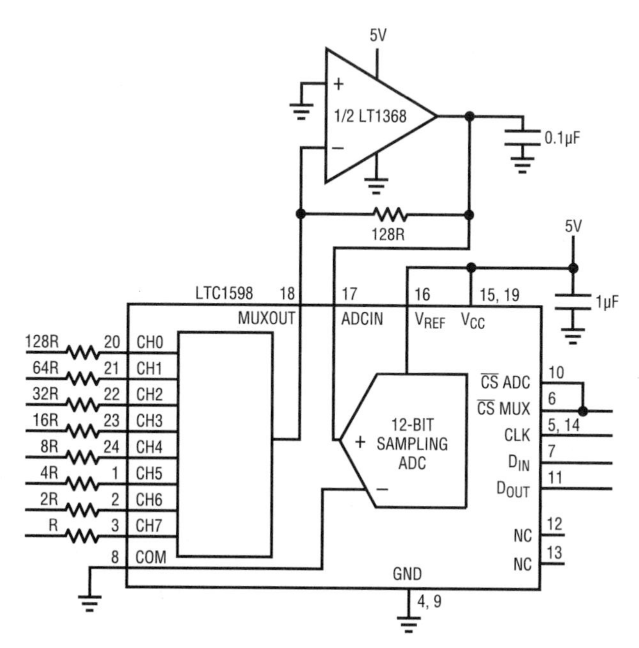

Figure 38.9 • Using the MUXOUT/ADCIN Loop of the LTC1598 to Form a PGA with Eight Inverting Gains

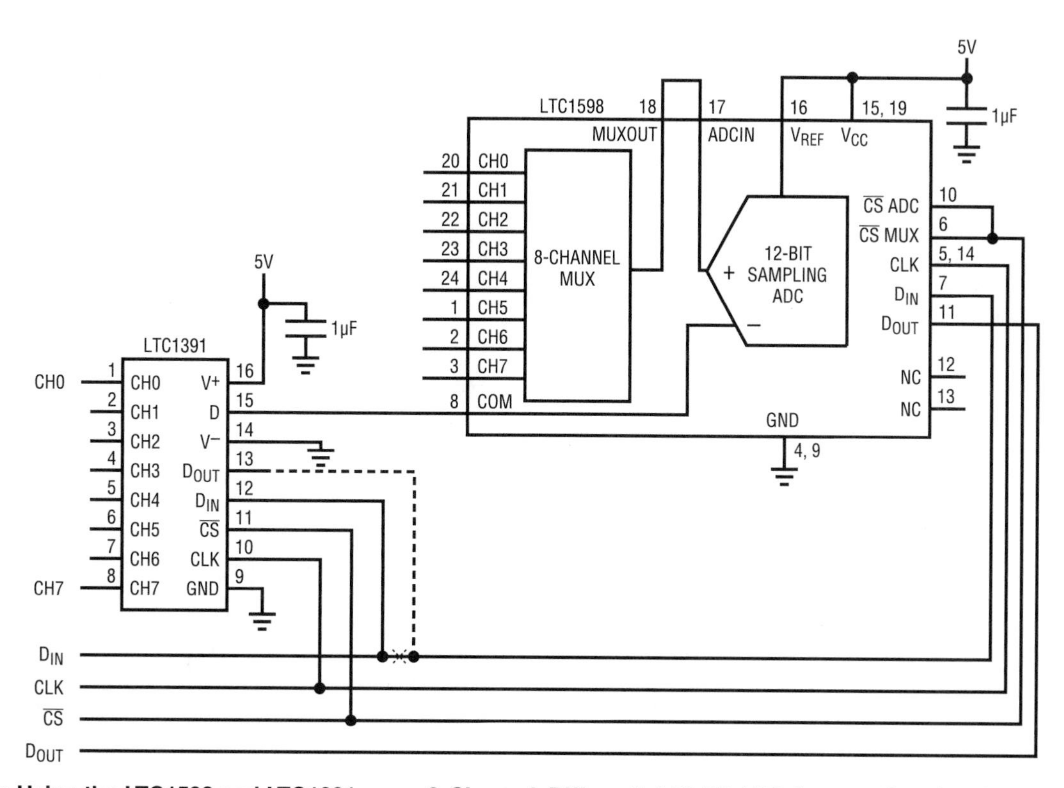

Figure 38.10 • Using the LTC1598 and LTC1391 as an 8-Channel, Differential 12-Bit ADC System: Opening the Indicated Connection and Shorting the Dashed Connection Daisy-Chains the External and Internal MUXes, Increasing Channel-Selection Flexibility

MUX the LTC1419 without software

by LTC Applications Staff

The circuit shown in Figure 38.11 uses hardware instead of software routines to select multiplexer channels in a data acquisition system. The circuit features the LTC1419 800ksps 14-bit ADC. It receives and converts signals from a 74HC4051 8-channel multiplexer. Three of the four output bits from an additional circuit, a 74HC4520 dual 4-bit binary counter, are used to select a multiplexer channel. A logic high power-on or processor-generated reset is applied to the counter's pin 7.

After the counter is cleared, the multiplexer's channel selection input is 000 and the input to channel 0 is applied to the LTC1419's S/H input. The channel-selection counter is clocked by the rising edge of the convert start (CONVST) signal that initiates a conversion. As each CONVST pulse increments the counter from 000 to 111, each multiplexer channel is individually selected and its input signal is applied to the LTC1419. After each of the eight channels has been selected, the counter rolls over to zero and the process repeats. At any time, the input multiplexer channel can be reset to 0 by applying a logic- high pulse to pin 7 of the counter.

This data acquisition circuit has a throughput of 800ksps or 100ksps/channel. As shown in Figure 38.12, the SINAD is 76.6dB for a full-scale ±2.5V 1.19kHz sine wave input signal.

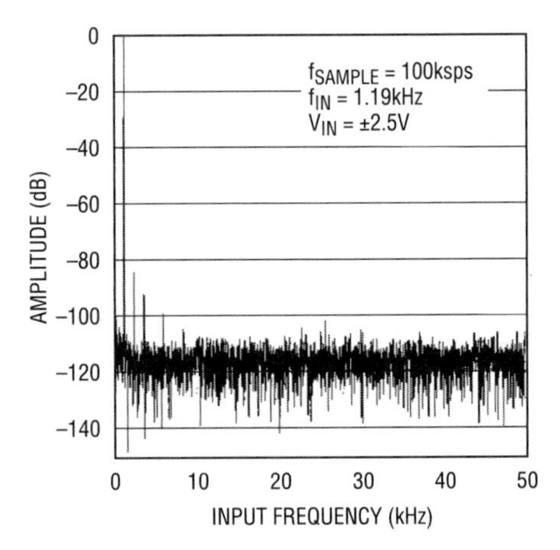

Figure 38.12 • FFT of the MUXed LTC1419's Conversion of a Full-Scale 1.19kHz Sine Wave

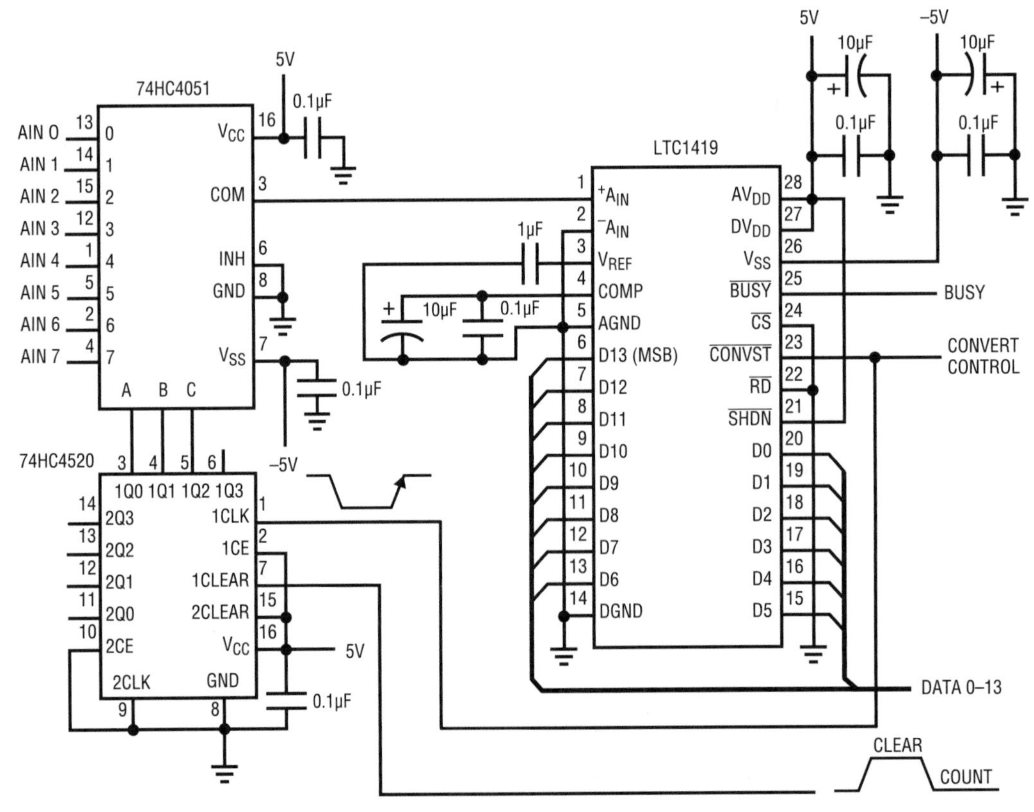

Figure 38.11 • This Simple Stand-Alone Circuit Requires no Software to Sequentially Sample and Convert Eight Analog Signal Channels at 14-bit Resolution and 100ksps/Channel

The LTC1590 dual 12-bit DAC is extremely versatile

by LTC Applications Staff

CMOS multiplying DACs make versatile building blocks that go beyond their basic function of converting digital data into analog signals. This article details some of the other circuits that are possible when using the LTC1590 dual, serially interfaced 12-bit DAC.

The circuit shown in Figure 38.13 uses the LTC1590 to create a digitally controlled attenuator using DAC_A and a programmable gain amplifier (PGA) using DAC_B. The attenuator's gain is set using the following equation:

$$V_{OUT} = -V_{IN}\frac{D}{2^n}$$

Where V_{OUT} =output voltage

V_{IN} = output voltage

n = DAC resolution in bits

D = value of code applied to DAC (min code = 000H)

The attenuator's gain varies from 4095/4096 to 1/4096. A code of 0 can be used to completely attenuate the input signal.

The PGA's gain is set using the following equation:

$$V_{OUT} = -V_{IN}\frac{2^n}{D}$$

Where V_{OUT} = output voltage

V_{IN} = output voltage

n = DAC resolution in bits

D = value of code applied to DAC (min code = 000H)

The gain is adjustable from 4096/4095 to 4096/1. A code of 0 is meaningless, since this results in infinite gain and the amplifier operates open loop. With either configuration, the attenuator's and PGA's gain are set with 12 bits accuracy.

A further modification to the basic attenuator and PGA is shown in Figure 38.14. In this circuit, DAC_A's attenuator circuit is modified to give the output amplifier a gain set by the ratio of resistors R3 and R4. The equation for this attenuator with output gain is

$$V_{OUT} = -V_{IN}\frac{16D}{2^n}$$

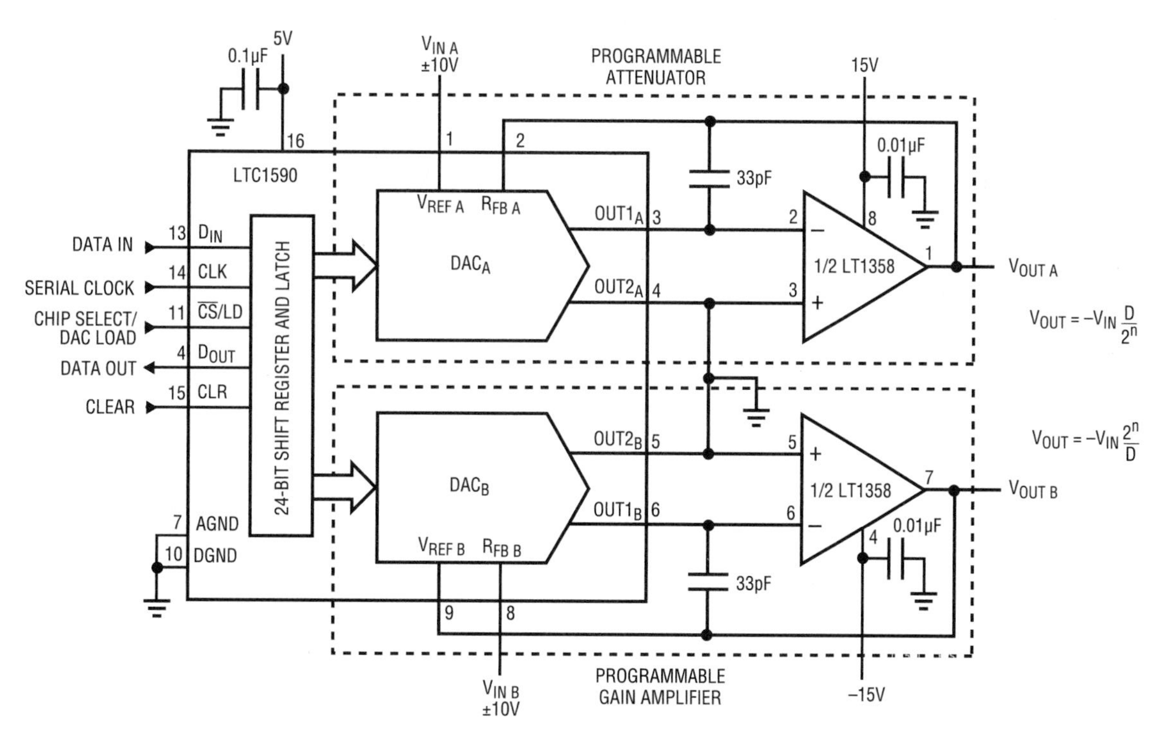

Figure 38.13 • Driving DAC_A's Reference Input (V_{REF}) and Tying the Feedback Resistor (R_{FB}) to the Op Amps Output Creates a 12-Bit- Accurate Attenuator. Reversing the V_{REF} and R_{FB} Connections Configures DAC_B as a Programmable-Gain Amplifier

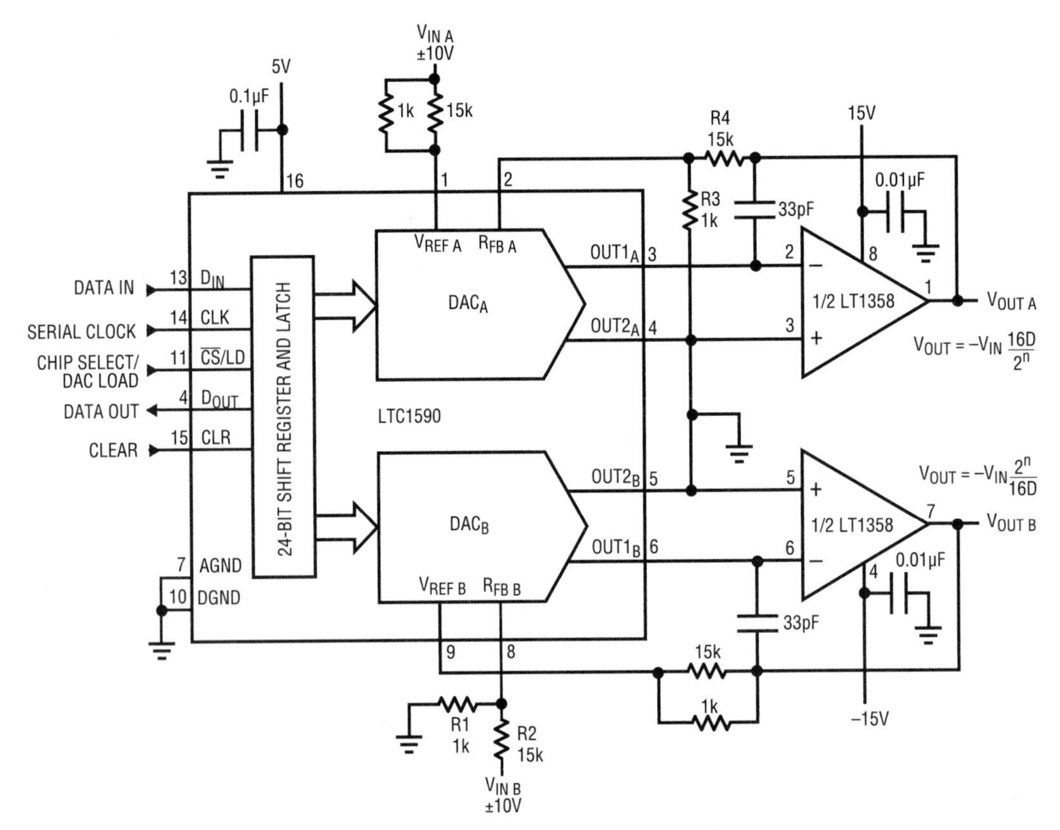

Figure 38.14 • Modifying the Basic Attenuator and PGA Creates Gain for the Attenuator (R3 and R4) and Attenuation at the PGA's Input (R1 and R2)

With the values shown, the attenuator's gain has a range of $-1/256$ to -16. This range is easily modified by changing the ratio of R3 and R4. In the other half of the circuit, an attenuator has been added to the input of DAC_B, configured as a PGA. The equation for this PGA with input attenuation is

$$V_{OUT} = -V_{IN}\frac{2^n}{16D}$$

This sets the gain range from effectively $-1/16$ to -256. Again, this range can be modified by changing the ratio of R1 and R2.

The LTC1590 can also be used as the control element that sets a lowpass filter's cutoff frequency. This is shown in Figure 38.15. The DAC becomes an adjustable resistor that sets the time constant of the integrator formed by U4 and C_I. With the integrator enclosed within a feedback loop, a lowpass filter is created.

The cutoff frequency range is a function of the DAC's resolution and the digital data that sets the effective resistance. The effective resistance is

$$R_{REF} = R_I\frac{2^n}{D}$$

Using this effective resistance, the cutoff frequency is

$$f_c = \frac{D}{2^{n+1} \bullet \pi \bullet R_I \bullet C_I}$$

The cutoff frequency range varies from 0.0000389/RC to 0.159/RC. As an example, to set the minimum cutoff frequency to 10Hz, make $R_I = 8.25k$ and $C_I = 470pF$ At an input code of 1, the cutoff frequency is 10Hz. The cutoff frequency increases linearly with increasing code, becoming 40.95kHz at a code of 4095. Generally, as the code changes by ±1 bit, the cutoff frequency changes by an amount equal to the frequency at $D = 1$. In this example, the cutoff frequency changes in 10Hz steps.

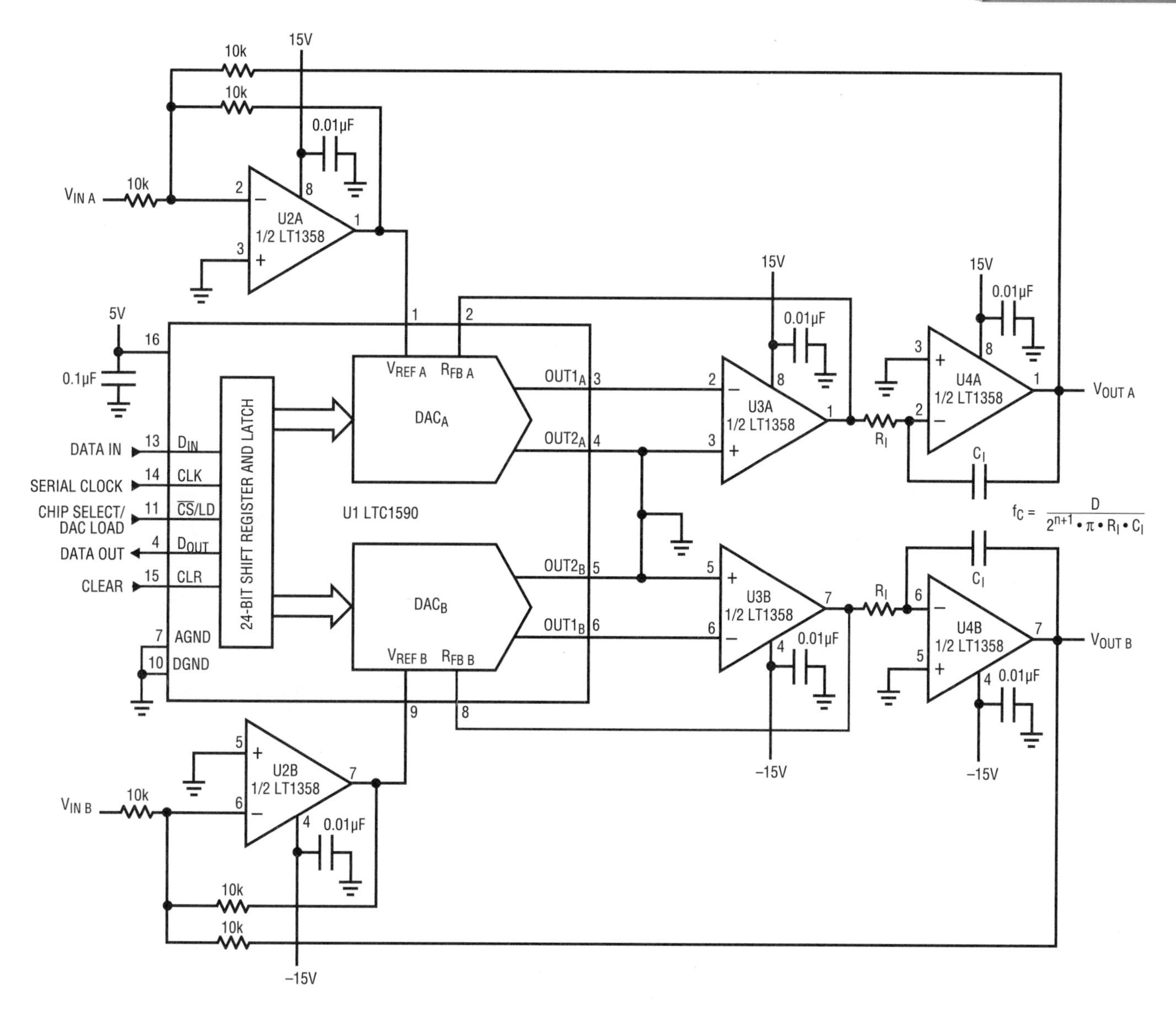

Figure 38.15 • This LTC1590-Controlled Dual Single-Pole Lowpass Filter Uses R_I and the DAC's Input Code to Create an Effective Resistance that Sets the Integrator's Time Constant and, Therefore, the Circuit's Cutoff Frequency

New 16-bit SO-8 DAC has 1LSB max INL and DNL over industrial temperature

by Jim Brubaker and William C. Rempfer

New generations of industrial systems are moving to 16 bits and hence require high performance 16-bit data converters. The new LTC1595/LTC1596 16-bit DACs provide the easiest to use, most cost effective, highest performance solution for industrial and instrumentation applications. The LTC1595/LTC1596 are serial input, 16-bit, multiplying current output DACs. Features of the new DACs include:

- ±1LSB maximum INL and DNL over the industrial temperature range
- Ultralow, 1nV-s glitch impulse
- ±10V output capability
- Small SO-8 package (LTC1595)
- Pin-compatible upgrade for industry-standard 12-bit DACs (DAC8043/8143 and AD7543)

0V-10V and ±10V output capability

Precision 0V-10V outputs with one op amp

Figure 38.16 shows the circuit for a 0V-10V output range. The DAC uses an external reference and a single op amp in this configuration. This circuit can also perform 2-quadrant multiplication where the reference input is driven by a ±10V input signal and V_{OUT} swings from 0V to $-V_{REF}$. The full-scale accuracy of the circuit is very precise because it is determined by precision-trimmed internal resistors. The power dissipation of the circuit is set by the op amp dissipation and the current drawn from the DAC reference input (7k nominal). The supply current of the DAC itself is less than 10μA.

An advantage of the LTC1595/LTC1596 is the ability to choose the output op amp to optimize the accuracy, speed, power and cost of the application. Using an LT1001 provides excellent DC precision, low noise and low power dissipation (90mW total for Figure 38.16's circuit). For higher speed, an LT1007, LT1468 or LT1122 can be used. The LT1122 will provide settling to 1LSB in 3μs for a full-scale transition. Figure 38.17 shows the 3μs settling performance obtained with the LT1122. The feedback capacitor in Figure 38.16 ensures stability. In higher speed applications, it can be used to optimize transient response. In slower applications, the capacitor can be increased to reduce glitch energy and provide filtering.

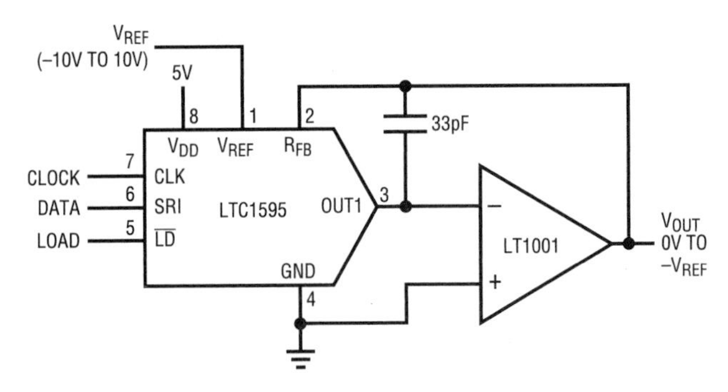

Figure 38.16 • With a Single External Op Amp, the DAC Performs 2-Quadrant Multiplication with ±10V Input and 0V to $-V_{REF}$ Output. With a Fixed −10V Reference, it Provides a Precision 0V–10V Unipolar Output

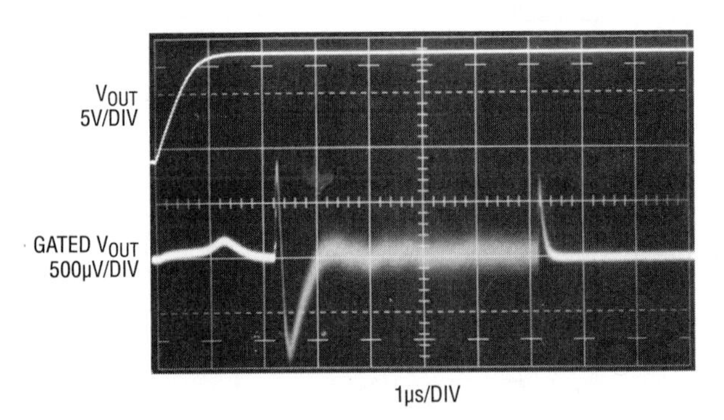

Figure 38.17 • When Used with an LT1122 (in the Circuit of Figure 38.16), the LTC1595/LTC1596 Can Settle in 3μs to a Full-Scale Step. The Top Trace Shows the Output Swinging from 0V to 10V. The Bottom Trace Shows the Gated Settling Waveform Settling to 1LSB (1/3 of a Division) in 3μs

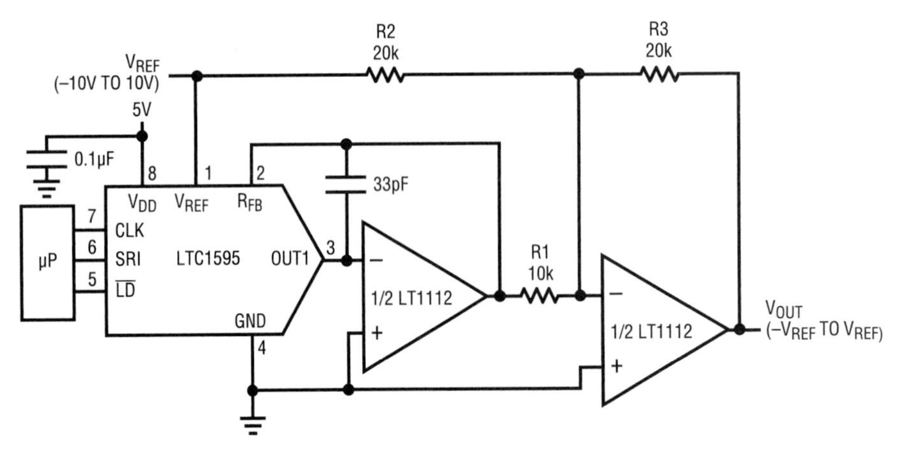

Figure 38.18 • With a Dual Op Amp, the DAC Performs 4-Quadrant Multiplication. With a Fixed 10V Reference, it Provides a ±10V Bipolar Output

Precision ±10V outputs with a dual op amp

Figure 38.18 shows a bipolar, 4-quadrant multiplying application. The reference input can vary from −10V to 10V and V_{OUT} swings from -V_{REF} to +V_{REF}. If a fixed 10V reference is used, a precision ±10V bipolar output will result.

Unlike the unipolar circuit of Figure 38.16, the bipolar gain and offset will depend on the matching of the external resistors. A good way to provide good matching and save board space is to use a pack of matched 20k resistors (the 10k unit is formed by placing two 20k resistors in parallel).

The LT1112 dual op amp is an excellent choice for high precision, low power applications that do not require high speed. The LT1469 or LT1124 will provide faster settling. Again, with op amp selection the user can optimize the speed, power, accuracy, and cost of the application.

LTC1659, LTC1448: smallest rail-to-rail 12-bit DACs have lowest power

by Hassan Malik

In this age of portable electronics, power and size are the primary concerns of most designers. The LTC1659 and the LTC1448 are rail-to-rail, 12-bit, voltage output DACs that address both of these concerns. The LTC1659 is a single DAC in an MSOP-8 package that draws only 250μA from a 3V or 5V supply, whereas the LTC1448 is a dual DAC in an SO-8 package that draws 450μA from a 3V or 5V supply.

Figure 38.19 shows a convenient way to use the LTC1659 in a digital control loop where 12-bit resolution is required. The output of the LTC1659 will swing from 0V to V_{REF}, because there is a gain of one from the REF pin to V_{OUT} at full-scale. Because the output can only swing up to V_{CC}, V_{REF} should be less than or equal to V_{CC} to prevent the loss of codes and degradation of PSRR near full-scale.

To obtain full dynamic range, the REF pin can be connected to the supply pin, which can be driven from a reference to guarantee absolute accuracy (see Figure 38.20). The LT1236 is a precision 5V reference with an input range of 7.2V to 40V In this configuration, the LTC1659 has a wide output swing of 0V to 5V The LTC1448 can be used in a similar configuration where dual DACs are needed.

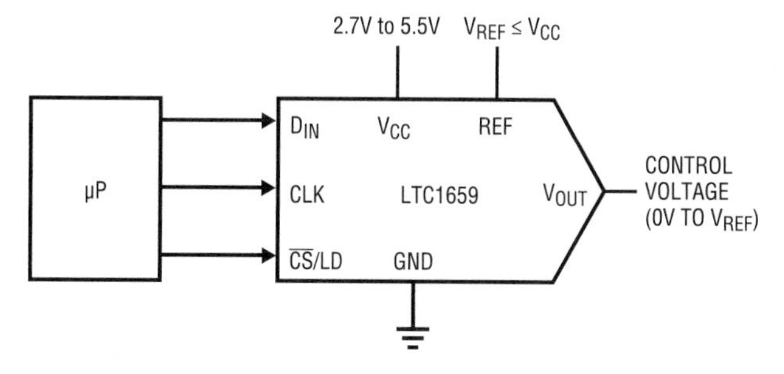

Figure 38.19 • 12-Bit DAC for Digital Control Loop

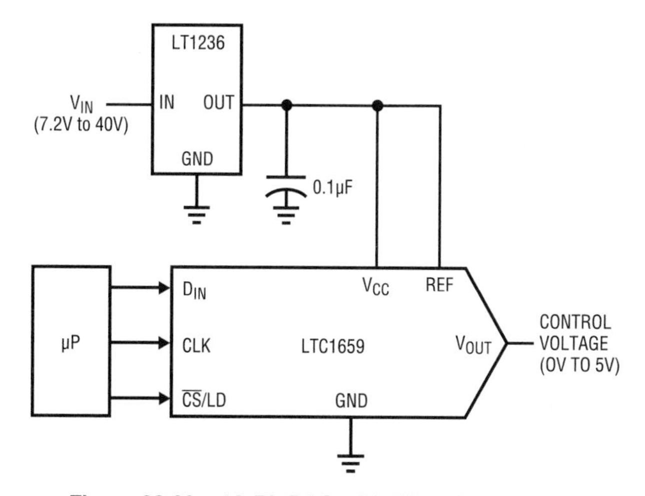

Figure 38.20 • 12-Bit DAC with Wide Output Swing

An SMBus-controlled 10-bit, current output, 50μA full-scale DAC

by Ricky Chow

The LTC1427-50 is a 10-bit, current-output DAC with an SMBus interface. This device provides precision, full-scale current of 50μA $\pm$1.5% at room temperature ($\pm$2.5% over temperature), wide output voltage DC compliance (from −15V to (Vcc − 1.3V)) and guaranteed monotonicity over a wide supply-voltage range. It is an ideal part for applications in contrast/brightness control or voltage adjustment in feedback loops.

Digitally controlled LCD bias generator

Figure 38.21 is a schematic of a digitally controlled LCD bias generator using a standard SMBus 2-wire interface. The LT1317 is configured as a boost converter, with the output voltage (V_{OUT}) determined by the values of the feedback resistors, R1 and R2. The LTC1427-50's DAC current output is connected to the feedback node of the LT1317. The LTC1427-50's DAC current output increases or decreases according to the data sent via the SMBus. As the DAC output current varies from 0μA to 50μA, the output voltage is controlled over the range of 12.7V to 24V. A 1LSB change in the DAC output current corresponds to an 11mV change in the output voltage.

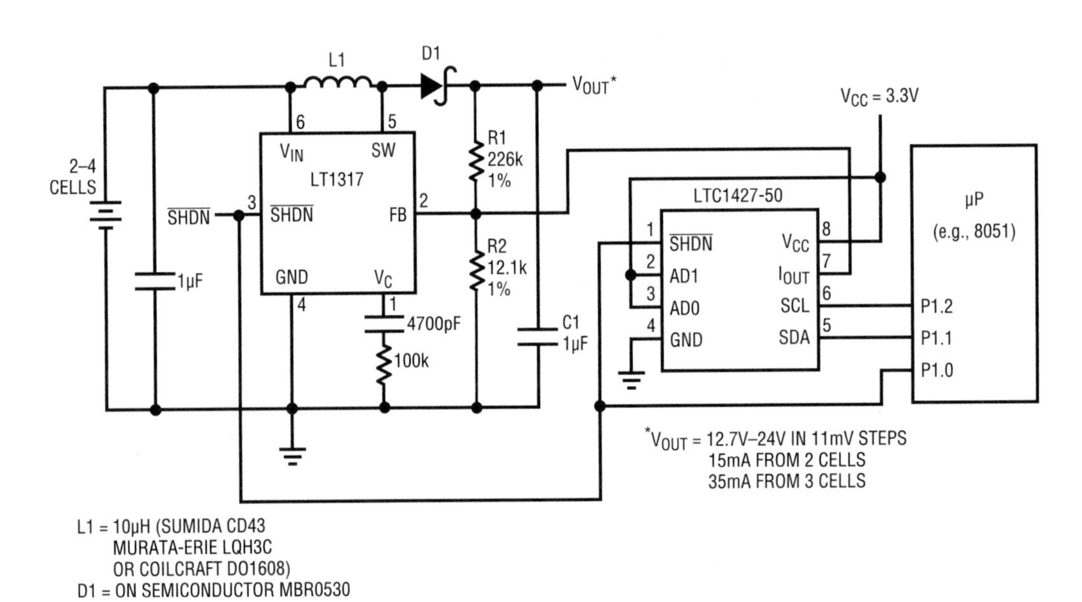

Figure 38.21 • Digitally Controlled LCD Bias Generator

Interface circuits

Simple resistive surge protection for interface circuits

by LTC Applications Staff

Surges and circuits

Many interface circuits must survive surge voltages such as those created by lightning strikes. These high voltages cause the devices within the IC to break down and conduct large currents, causing irreversible damage to the IC. Engineers must design circuits that tolerate the surges expected in their environments. They can quantify the surge tolerance of circuitry by using a surge standard. Standards differ mainly in their voltage levels and wave forms. At LTC, we test surge resistance using the circuit of Figure 38.22. We describe the voltage wave form (Figure 38.23) by its peak value V_P, the "front time" T_F (roughly, the rise time), and the "time to half-value" $T_{1/2}$ (roughly, the time from the beginning of the pulse to when the pulse decays to half of V_P). Surges are similar to ESD, but challenge circuits in a different way. A surge may rise to 1kV in 10ms, whereas an ESD pulse might rise to 15kV in only a few ns. However, the surge lasts for more than 100ms, whereas the ESD pulse decays in about 50ns. Thus, the surge challenges the power dissipation ability of the protection circuitry, whereas the ESD challenges the turn-on time and peak current handling. The Linear Technology LT1137A has on-chip circuitry to withstand ESD pulses up to 15kV (IEC 801-2). This circuitry also increases the surge tolerance of the LT1137A relative to a standard 1488/1489.

Designing for surge tolerance

Many designers enhance the surge tolerance of a circuit by placing a transient voltage suppressor (TVS) in parallel with the vulnerable IC pins, as shown in Figure 38.24. The TVS contains Zener diodes, which break down at a certain voltage and shunt the surge current to ground. Thus, the TVS clamps the voltage at a level safe for the IC. The TVS, like any protection circuitry, increases the manufacturing cost and complexity of the circuit. Alternately, designers can use a series resistor to protect the vulnerable pins, as shown in Figure 38.25. The resistor

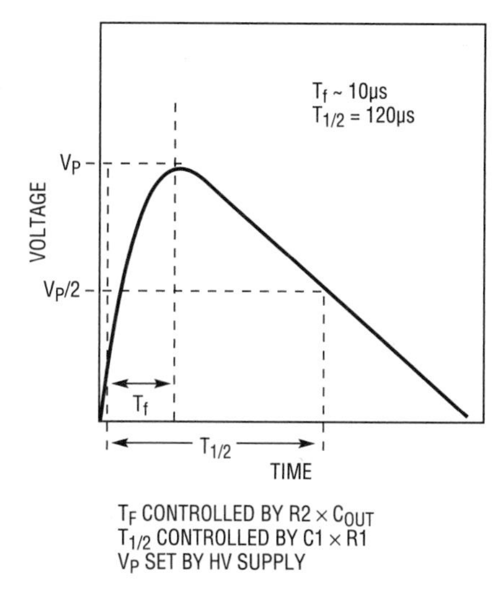

Figure 38.23 • LTC Surge-Test Waveform

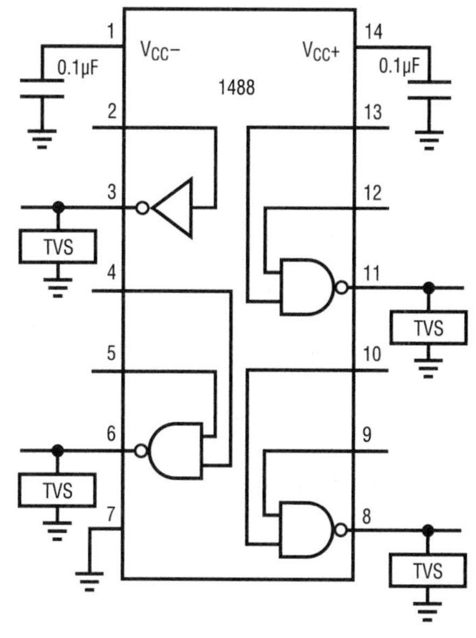

Figure 38.24 • 1488 Line Driver with TVS Surge Protection

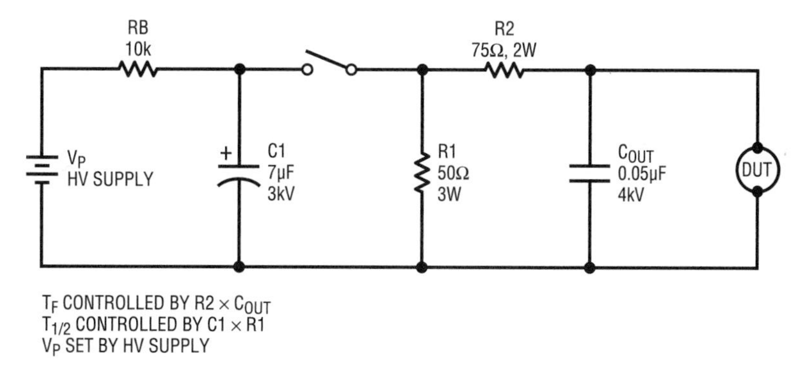

Figure 38.22 • LTC Surge-Test Circuit: TF Controlled by R2 • C_{OUT}; $T_{1/2}$ Controlled by C1 • R1; V_P Set by HV Supply

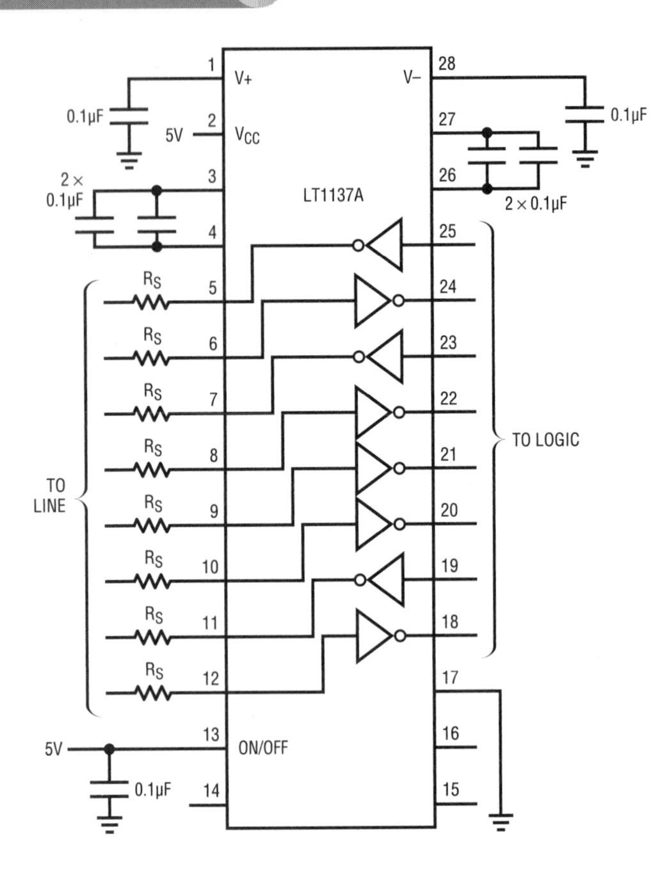

Figure 38.25 • LT1137A with Resistive Surge Protection

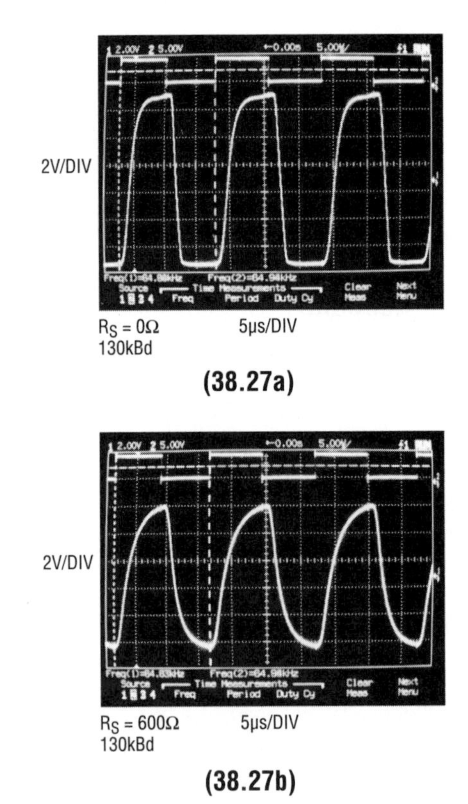

Figure 38.27 • Output Waveforms with Series Resistor

reduces the current flowing into the IC to a safe level. Resistive protection simplifies design and inventory and may offer lower cost. The resistance must be large enough to protect the IC, but not so large that it degrades the frequency performance of the circuit. Larger surge amplitudes require increased resistance to protect the IC. More robust ICs need less resistance for protection against a given surge amplitude. Linear's LT1137A is protected by a much smaller resistor than a 1488, as shown in Figure 38.26.

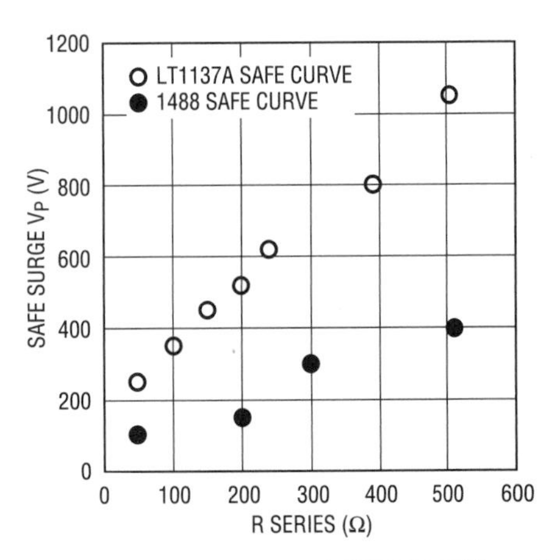

Figure 38.26 • Safe Curves for 1488 (SN75188N) and LT1137A. Safe Curves Represent the Highest V_P for Which No IC Damage Occurred After 10 Surges

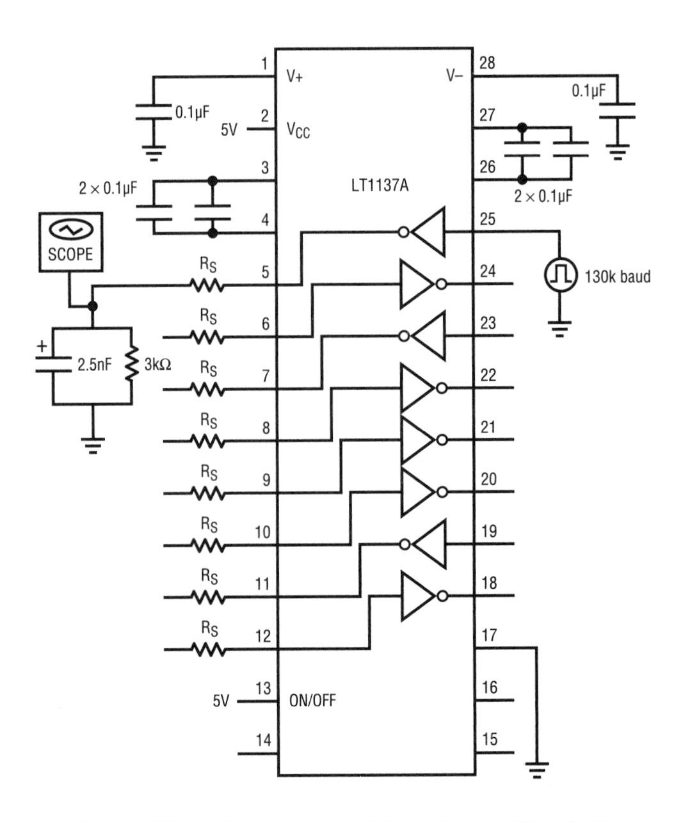

Figure 38.28 • Testing Line Driver Output Waveform

These curves are empirical "rules of thumb." You should test actual circuits.

The series resistor may have an adverse effect on the frequency performance of the circuit. When protecting a receiver; the resistor has little effect. Figures 38.27a and 38.27b show the effect of a 600Ω resistor on the driver-output wave form. These waveforms were obtained with the test circuit of Figure 38.28. A 600Ω resistor is adequate for 1kV surges, but has minimal effect on the driver wave form up to 130kbaud, even with a worst-case load of 3kΩ||2.5nF.

You must choose the series resistor carefully to withstand the surge. Unfortunately, neither voltage ratings nor power ratings provide an adequate basis for choosing surge-tolerant resistors. Usually, through-hole resistors will withstand much larger surges than surface mount resistors of the same value and power rating. Typical 1/8 Watt surface mount resistors are not suitable for protecting the LT1137A. If you use surface mount components, you may need ratings of 1W or more. With the LT1137A, you can use carbon film 1/4W through-hole resistors against surges up to about 900V and 1/2W carbon film resistors against surges up to about 1200V Unfortunately, using series or parallel combinations of resistors *does not* increase the surge handling as one would expect.

Resistive surge protection

The LT1137A has proprietary circuitry that makes it more robust against ESD and surges than the standard 1488/1489. The greater surge tolerance of the LT1137A makes it practical to use resistive surge protection, reducing inventory and component cost relative to TVS surge protection. The major considerations are the surge tolerance required, the resulting resistor value needed, resistor robustness and frequency performance.

The LTC1343 and LTC1344 form a software-selectable multiple-protocol interface port using a DB-25 connector

by Robert Reay

Introduction

With the explosive growth in data networking equipment has come the need to support many different serial protocols using only one connector. The problem facing interface designers is to make the circuitry for each serial protocol share the same connector pins without introducing conflicts. The main source of frustration is that each serial protocol requires a different line termination that is not easily or cheaply switched.

With the introduction of the LTC1343 and LTC1344, a complete software-selectable serial interface port using an inexpensive DB-25 connector becomes possible. The chips form a serial interface port that supports the V.28 (RS232), V.35, V.36, RS449, EIA-530, EIA-530A or X.21 protocols in either DTE or DCE mode and is both NET1 and NET2 compliant. The port runs from a single 5V supply and supports an echoed clock and loop-back configuration that helps eliminate glue logic between the serial controller and the line transceivers.

A typical application is shown in Figure 38.29. Two LTC1343s and one LTC1344 form the interface port using a DB-25 connector, shown here in DTE mode.

Each LTC1343 contains four drivers and four receivers and the LTC1344 contains six switchable resistive terminators. The first LTC1343 is connected to the clock and data signal lines along with the diagnostic LL (local loop- back) and TM (test mode) signals. The second LTC1343 is connected to the control-signal lines along with the diagnostic RL (remote loop-back) signal. The single-ended driver and receiver could be separated to support the RI (ring-indicate) signal. The switchable line terminators in the LTC1344 are connected only to the high speed clock and data signals. When the interface protocol is changed via the digital mode selection pins (not shown), the drivers and receivers are automatically reconfigured and the appropriate line terminators are connected.

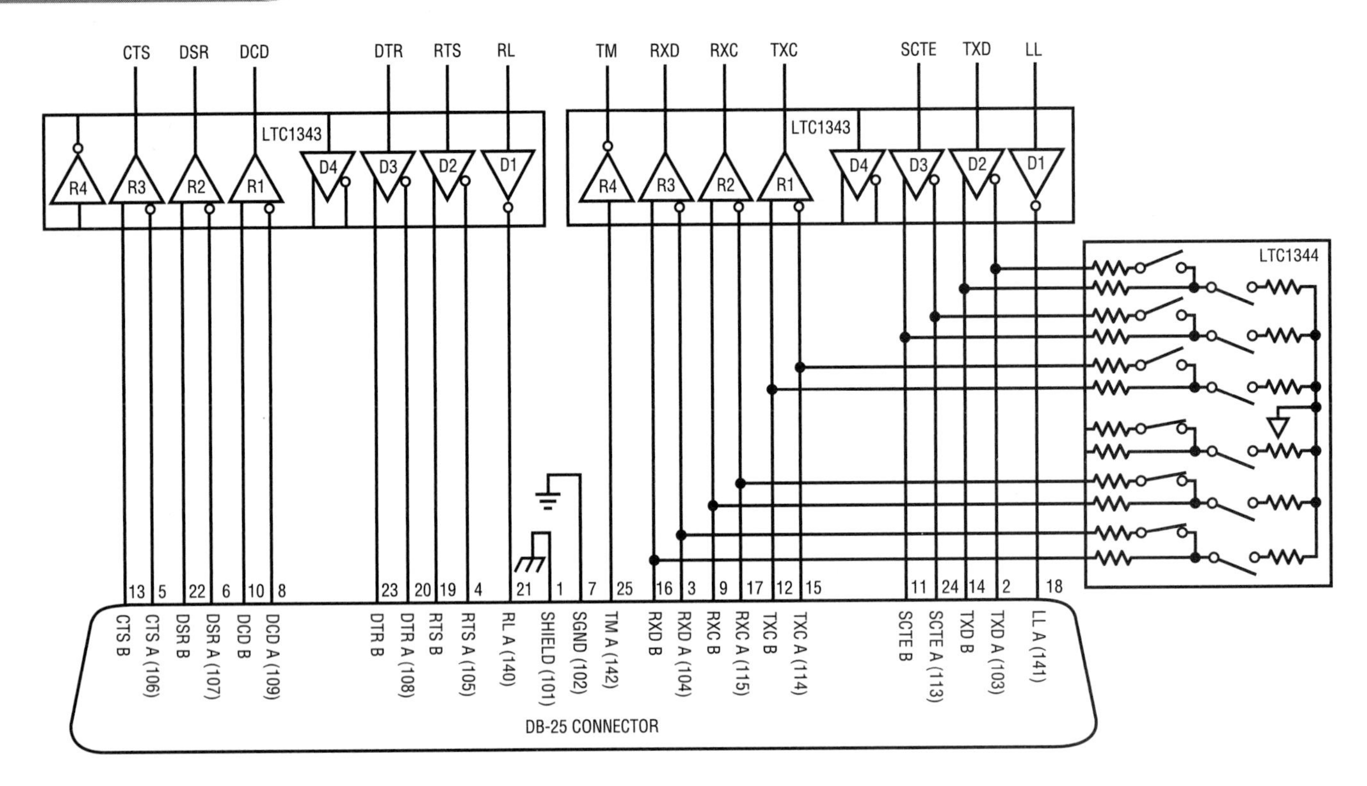

Figure 38.29 • LTC1343/LTC1344 Typical Application

Review of interface standards

The serial interface standards RS232, EIA-530, EIA-530A, RS449, V.35, V.36 and X.21 specify the function of each signal line, the electrical characteristics of each signal, the connector type, the transmission rate and the data exchange protocols. The RS422 (V.11) and RS423 (V.10) standards merely define electrical characteristics. The RS232 (V.28) and V.35 standards also specify their own electrical characteristics. In general, the US standards start with RS or EIA, and the equivalent European standards start with V or X. The characteristics of each interface are summarized in Table 38.2.

Table 38.2 shows only the most commonly used signal lines. Note that each signal line must conform to only one of four electrical standards, V.10, V.11, V.28 or V.35.

V.10 (RS423) interface

A typical V.10 unbalanced interface is shown in Figure 38.30. A V.10 single-ended generator (output A with ground C) is connected to a differential receiver with input A' connected to A and input B' connected to the signal-return ground C.

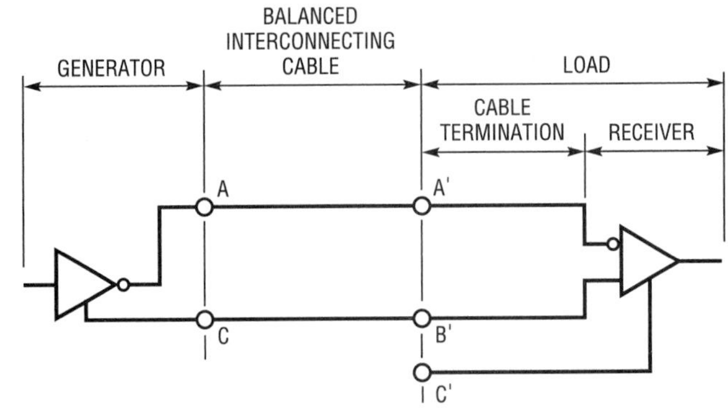

Figure 38.30 • Typical V.10 Interface

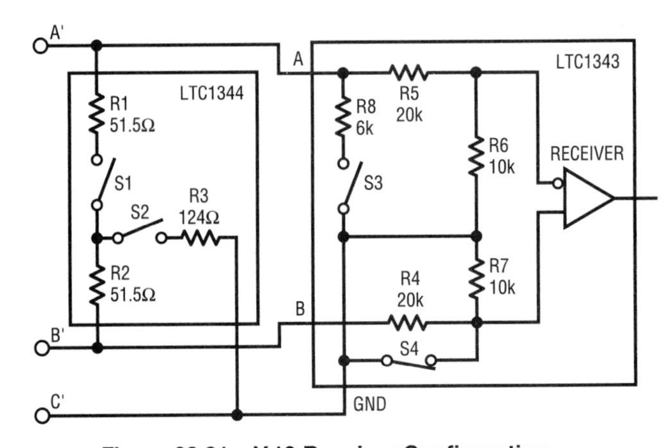

Figure 38.31 • V.10 Receiver Configuration

Table 38.2 Interface Summary

	Clock and Data Signals					Control Signals					Test Signals			
	TXD	SCTE	TXC	RXC	RXD	RTS	DTR	DSR	DCD	CTS	RI	LL	RL	TM
CCITT#	(103)	(113)	(114)	(115)	(104)	(105)	(108)	(107)	(109)	(106)	(125)	(141)	(140)	(142)
RS232	V.28	V.28	V.28	V.28	V.28	V.28	V.28	V.28	V.28	V.28	V.28	V.28	V.28	V.28
EIA-530	V.11	V.11	V.11	V.11	V.11	V.11	V.11	V.11	V.11	V.11	—	V.10	V.10	V.10
EIA-530A	V.11	V.11	V.11	V.11	V.11	V.11	V.10	V.10	V.11	V.11	V.10	V.10	V.10	V.10
RS449	V.11	V.11	V.11	V.11	V.11	V.11	V.11	V.11	V.11	V.11	V.10	V.10	V.10	V.10
V.35	V.35	V.35	V.35	V.35	V.35	V.28	—	V.28	V.28	V.28	—	—	—	—
V.36	V.11	V.11	V.11	V.11	V.11	V.11	—	V.11	V.11	V.11	—	V.10	V.10	V.10
X.21	V.11	V.11	V.11	V.11	V.11	V.11	—	—	V.11	—	—	—	—	—

The receiver's ground C' is separate from the signal return. Usually, no cable termination between A' and B' is required for V.10 interfaces. The V.10 receiver configuration for the LTC1343 and LTC1344 is shown in Figure 38.31.

In V.10 mode, switches S1 and S2 inside the LTC1344 and S3 inside the LTC1343 are turned off. Switch S4 inside the LTC1343 shorts the noninverting receiver input to ground so the B input at the connector can be left floating. The cable termination is then the 30k input impedance to the ground of the LTC1343 V.10 receiver.

V.11 (RS422) interface

A typical V.11 balanced interface is shown in Figure 38.32. A V.11 differential generator with outputs A and B and ground C is connected to a differential receiver with ground C', input A' connected to A and input B' connected to B. The V.11 interface has a differential termination at the receiver end with a minimum value of 100Ω. The termination resistor is optional in the V.11 specification, but for the high speed clock and data lines, the termination is required to prevent reflections from corrupting the data. In V.11 mode, all switches are off except S1 inside the LTC1344, which connects a 103Ω differential termination impedance to the cable, as shown in Figure 38.33.

V.28 (RS232) interface

A typical V.28 unbalanced interface is shown in Figure 38.34. A V.28 single-ended generator (output A with ground C) is connected to a single-ended receiver with input A' connected to A and ground C' connected via the signal return ground to C. In V.28 mode, all switches are off except S3 inside the LTC1343, which connects a 6k impedance (R8) to ground in parallel with 20k (R5) plus 10k (R6), for an combined impedance of 5k, as shown in Figure 38.35. The noninverting input is disconnected inside the LTC1343 receiver and connected to a TTL level reference voltage for a 1.4V receiver trip point.

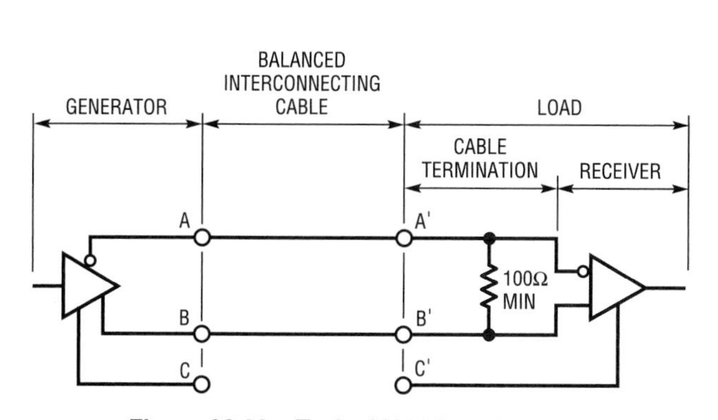

Figure 38.32 • Typical V.11 Interface

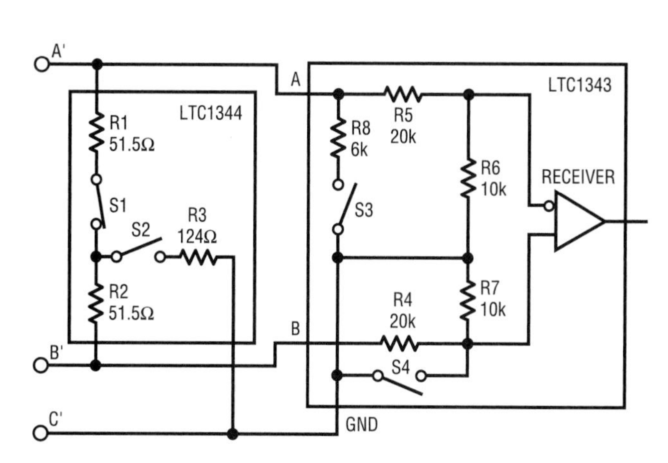

Figure 38.33 • V.11 Receiver Configuration

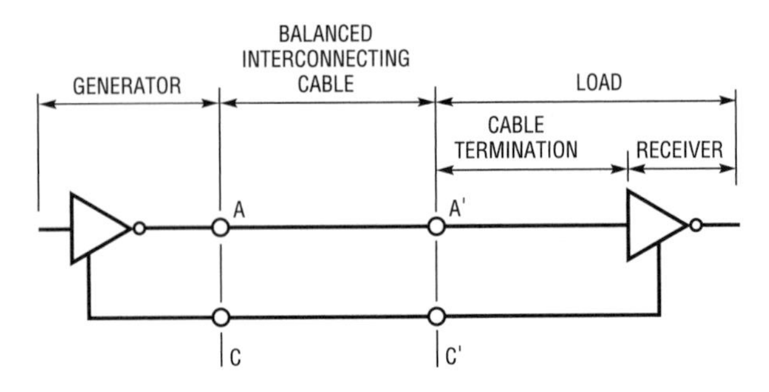

Figure 38.34 • Typical V.28 Interface

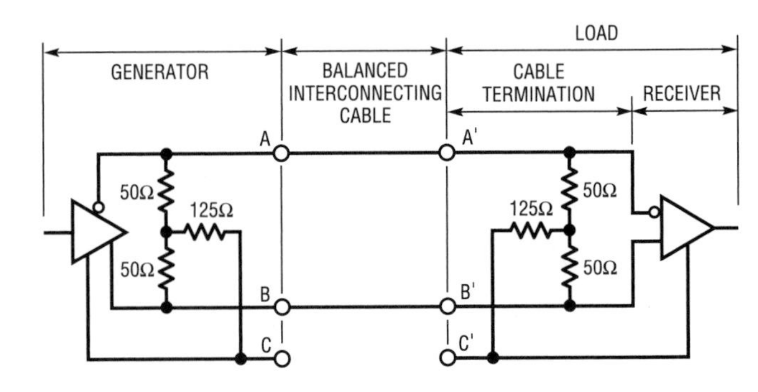

Figure 38.36 • Typical V.35 Interface

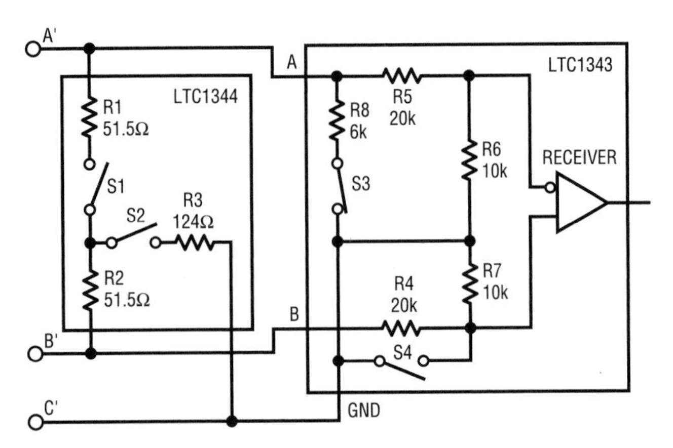

Figure 38.35 • V.28 Receiver Configuration

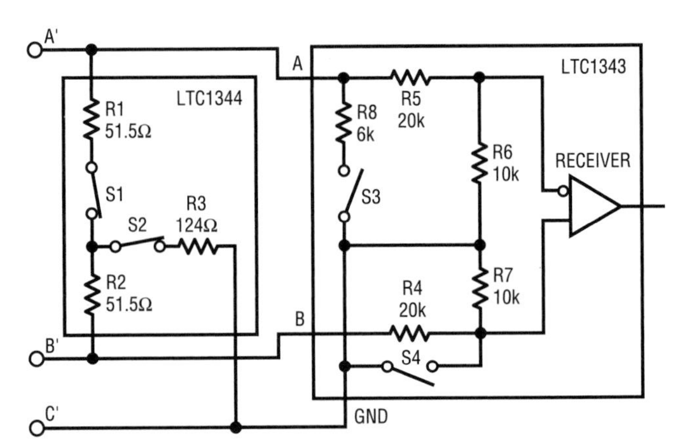

Figure 38.37 • V.35 Receiver Configuration

V.35 interface

A typical V.35 balanced interface is shown in Figure 38.36. A V.35 differential generator with outputs A and B and ground C is connected to a differential receiver with ground C', input A' connected to A and input B' connected to B. The V.35 interface requires T or delta network termination at the receiver end and the generator end. The receiver differential impedance measured at the connector must be 100 ±10Ω, and the impedance between shorted terminals (A' and B') and ground (C') is 150 ±15Ω.

In V.35 mode, both switches S1 and S2 inside the LTC1344 are on, connecting the T-network impedance, as shown in Figure 38.37. Both switches in the LTC1343 are off. The 30k input impedance of the receiver is placed in parallel with the T-network termination, but does not affect the overall input impedance significantly.

The generator differential impedance must be 50Ω to 150Ω, and the impedance between shorted terminals (A and B) and ground (C) is 150Ω ±15Ω. For the generator termination, switches S1 and S2 are both on and the top side of the center resistor is brought out to a pin so it can be bypassed with an external capacitor to reduce common mode noise, as shown in Figure 38.38.

Any mismatch in the driver rise and fall times or skew in driver propagation delays will force current through the center termination resistor to ground, causing a high frequency common mode spike on the A and B terminals. This spike can cause EMI problems that are reduced by capacitor C1, which shunts much of the common mode energy to ground rather than down the cable.

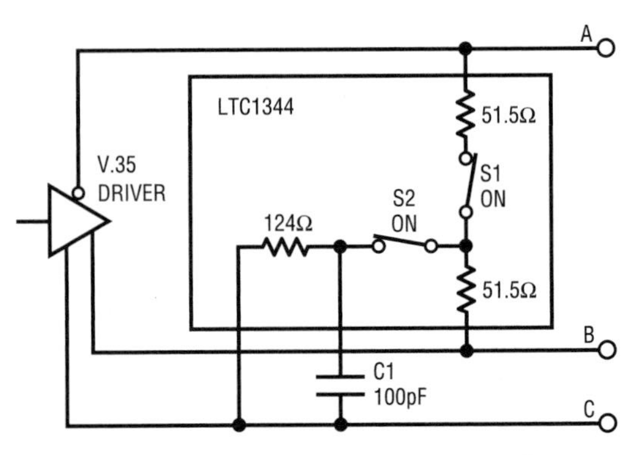

Figure 38.38 • V.35 Driver Using the LTC1344

Table 38.3 LTC1343/LTC1344 Mode Selection

LTC1343 Mode Name	M2	M1	M0	CTRL/ $\overline{CLK}$	D1	D 2	D 3	D 4	R1	R 2	R 3	R4
V.10/RS423	0	0	0	X	V.10	V.10	V.10	V.10	V.10	V.10	V.10	V.10
RS530A clock & data	0	0	1	0	V.10	V.11	V.11	V.11	V.11	V.11	V.11	V.10
RS530A control	0	0	1	1	V.10	V.11	V.10	V.11	V.11	V.10	V.11	V.10
Reserved	0	1	0	X	V.10	V.11	V.11	V.11	V.11	V.11	V.11	V.10
X.21	0	1	1	X	V.10	V.11	V.11	V.11	V.11	V.11	V.11	V.10
V.35 clock & data	1	0	0	0	V.28	V.35	V.35	V.35	V.35	V.35	V.35	V.28
V.35 control	1	0	0	1	V.28	V.28	V.28	V.28	V.28	V.28	V.28	V.28
RS530/RS449/V.36	1	0	1	X	V.10	V.11	V.11	V.11	V.11	V.11	V.11	V.10
V.28/RS232	1	1	0	X	V.28	V.28	V.28	V.28	V.28	V.28	V.28	V.28
No Cable	1	1	1	X	Z	Z	Z	Z	Z	Z	Z	Z

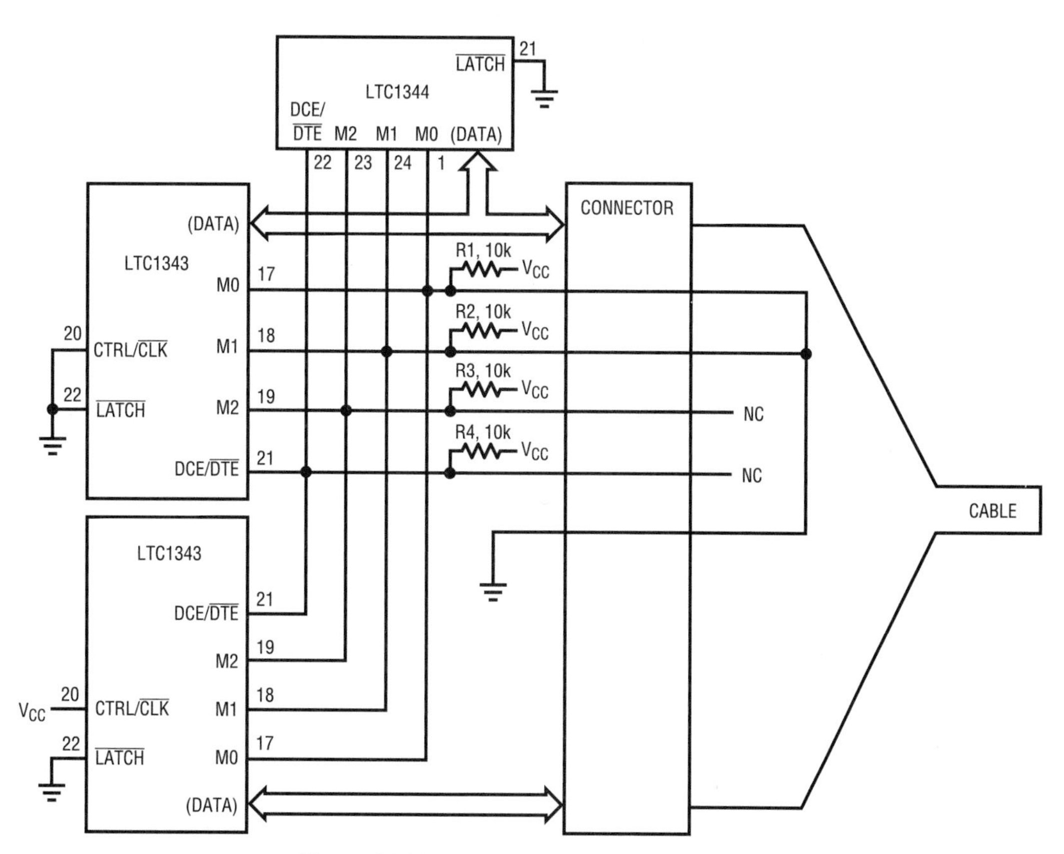

Figure 38.39 • Mode Selection by Cable

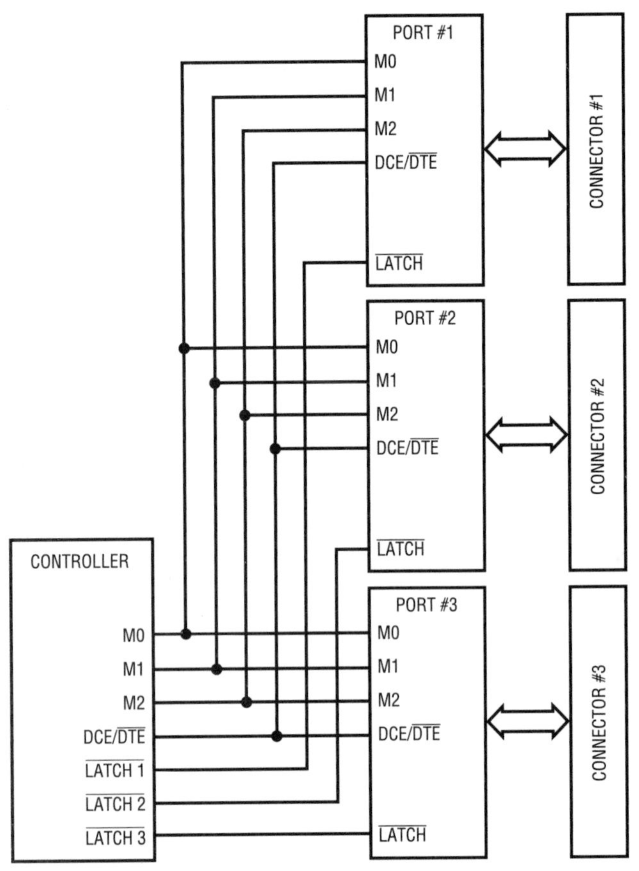

Figure 38.40 • Mode Selection by Controller

LTC1343/LTC1344 mode selection

The interface protocol is selected using the mode select pins M0, M1, M2 and CTRL/CLK, as summarized in Table 38.3. The CTRL/CLK pin should be pulled high if the LTC1343 is being used to generate control signals and pulled low if used to generate clock and data signals.

For example, if the port is configured as a V35 interface, the mode selection pins shouldbe M2 = 1, M1 = 0, M0 = 0. For the control signals, CTRL/CLK = 1 and the drivers and receivers will operate in RS232 (V.28) electrical mode. For the clock and data signals, CTRL/CLK = 0 and the drivers and receivers will operate in V.35 electrical mode, except for the single-ended driver and receiver, which will operate in the RS232 (V.28) electrical mode.

The DCE/CLK pin will configure the port for DCE mode when high, and DTE when low.

The interface protocol may be selected by simply plugging the appropriate interface cable into the connector. The mode pins are routed to the connector and are left unconnected (1) or wired to ground (0) in the cable, as shown in Figure 38.39.

The pull-up resistors R1–R4 ensure a binary 1 when a pin is left unconnected and also ensure that the two LTC1343s and the LTC1344 enter the no-cable mode when the cable is removed. In the no-cable mode, the LTC1343 power supply current drops to less than 200μA and all LTC1343 driver outputs and LTC1344 resistive terminators are forced into a high impedance state. Note that the data latch pin, LATCH, is shorted to ground for all chips.

The interface protocol may also be selected by the serial controller or host microprocessor, as shown in Figure 38.40.

The mode selection pins M0, M1, M2 and DCE/DTE can be shared among multiple interface ports, while each port has a unique data-latch signal that acts as a write enable. When the LATCH pin is low, the buffers on the MO, M1, M2, CTRL/CLK, DCE/DTE, LB and EC pins are transparent. When the LATCH pin is pulled high, the buffers latch the data, and changes on the input pins will no longer affect the chip.

The mode selection may also be accomplished by using jumpers to connect the mode pins to ground or V_{CC}.

Loop-back

The LTC1343 contains logic for placing the interface into a loop-back configuration for testing. Both DTE and DCE loop-back configurations are supported. Figure 38.41 shows a complete DTE interface in the loop-back configuration and Figure 38.42 the DCE loop-back configuration. The loop- back configuration is selected by pulling the LB pin low.

Enabling the single-ended driver and receiver

When the LTC1343 is being used to generate the control signals (CTRL/CLK = high) and the EC pin is pulled low, the DCE/DTE pin becomes an enable for driver 1 and receiver 4 so their inputs and outputs can be tied together, as shown in Figure 38.43.

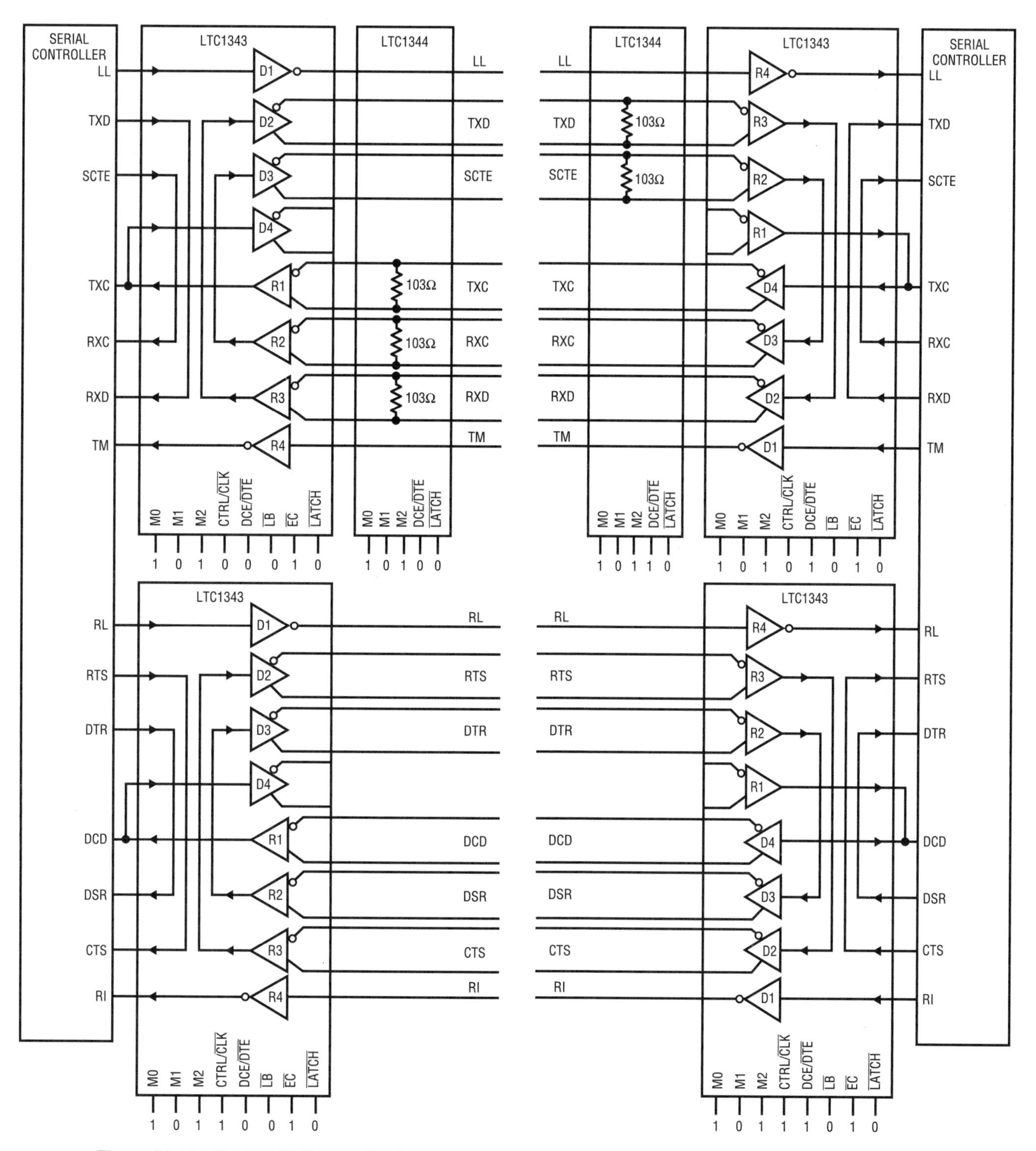

Figure 38.41 • Normal DTE Loop-Back

Figure 38.42 • Normal DCE Loop-Back

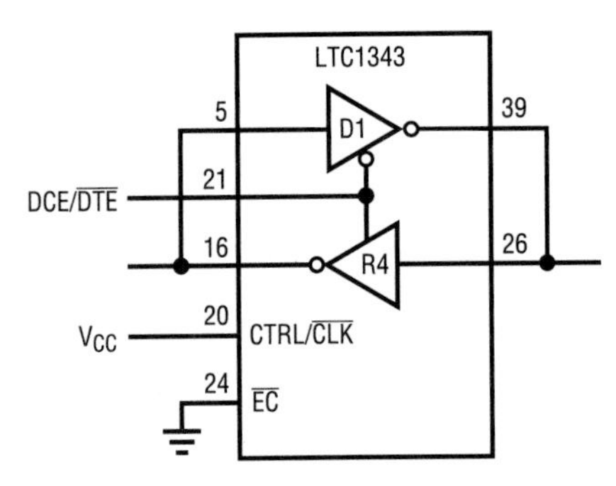

Figure 38.43 • Single-Ended Driver and Receiver Enable

The $\overline{EC}$ pin has no affect on the configuration when CTRL/ $\overline{CLK}$ is high except to allow the DCE/$\overline{DTE}$ pin to become an enable. When DCE/$\overline{DTE}$ is low, the driver 1 output is enabled. The receiver 4 output goes into three-state, and the input presents a 30k load to ground.

When DCE/$\overline{DTE}$ is high, the driver 1 output goes into three- state, and the receiver 4 output is enabled. The receiver 4 input presents a 30k load to ground in all modes except when configured for RS232 operation, when the input impedance is 5k to ground.

Multiprotocol interface with DB-25 or μDB-26 connectors

A multiprotocol serial interface with a standard DB-25 connector EIA-530 pin configuration is shown in Figure 38.44. (Figures 38.44–38.47 follow). The signal lines must be reversed in the cable when switching between DTE and DCE using the same connector. For example, in DTE mode, the RXD signal is routed to receiver 3, but in DCE mode, the TXD signal is routed to receiver 3. The interface mode is selected by logic outputs from the controller or from jumpers to either V_{CC} or GND on the mode-select pins.

The single-ended driver 1 and receiver 4 of the control chip share the RL signal on connector pin 21. With EC low and CTRL/$\overline{CLK}$ high, the DCE/$\overline{DTE}$ pin becomes an enable signal.

Single-ended receiver 4 can be connected to pin 22 to implement the RI (ring indicate) signal in RS232 mode (see Figure 38.45). In all other modes, pin 22 carries the DSR(B) signal.

A cable selectable multiprotocol interface is shown in Figure 38.46. Control signals LL, RL and TM are not implemented. The Vcc supply and select lines M0 and M1 are brought out to the connector. The mode is selected in the cable by wiring M0 (connector pin 18) and M1 (connector pin 21) and DCE/$\overline{DTE}$ (connector pin 25) to ground (connector pin 7) or letting them float. If M0, M1 or DCE/$\overline{DTE}$ are floating, pull-up resistors R3, R4 and R5 will pull the signals to Vcc. The select bit M1 is hard wired to Vcc. When the cable is pulled out, the interface goes into the no-cable mode.

A cable-selectable multiprotocol interface found in many popular data routers is shown in Figure 38.47. The entire interface, including the LL signal, can be implemented using the tiny μDB-26 connector.

Conclusion

The LTC1343 and LTC1344 allow the designer of a multiprotocol serial interface to spend all of his time on the software rather than the hardware. Simply drop the chips down on the board, hook them up to the connector and a serial controller, apply the 5V supply voltage and you're off and running. In addition, the chip set's small size and unique termination topology allow many ports to be placed on a board using inexpensive connectors and cables.

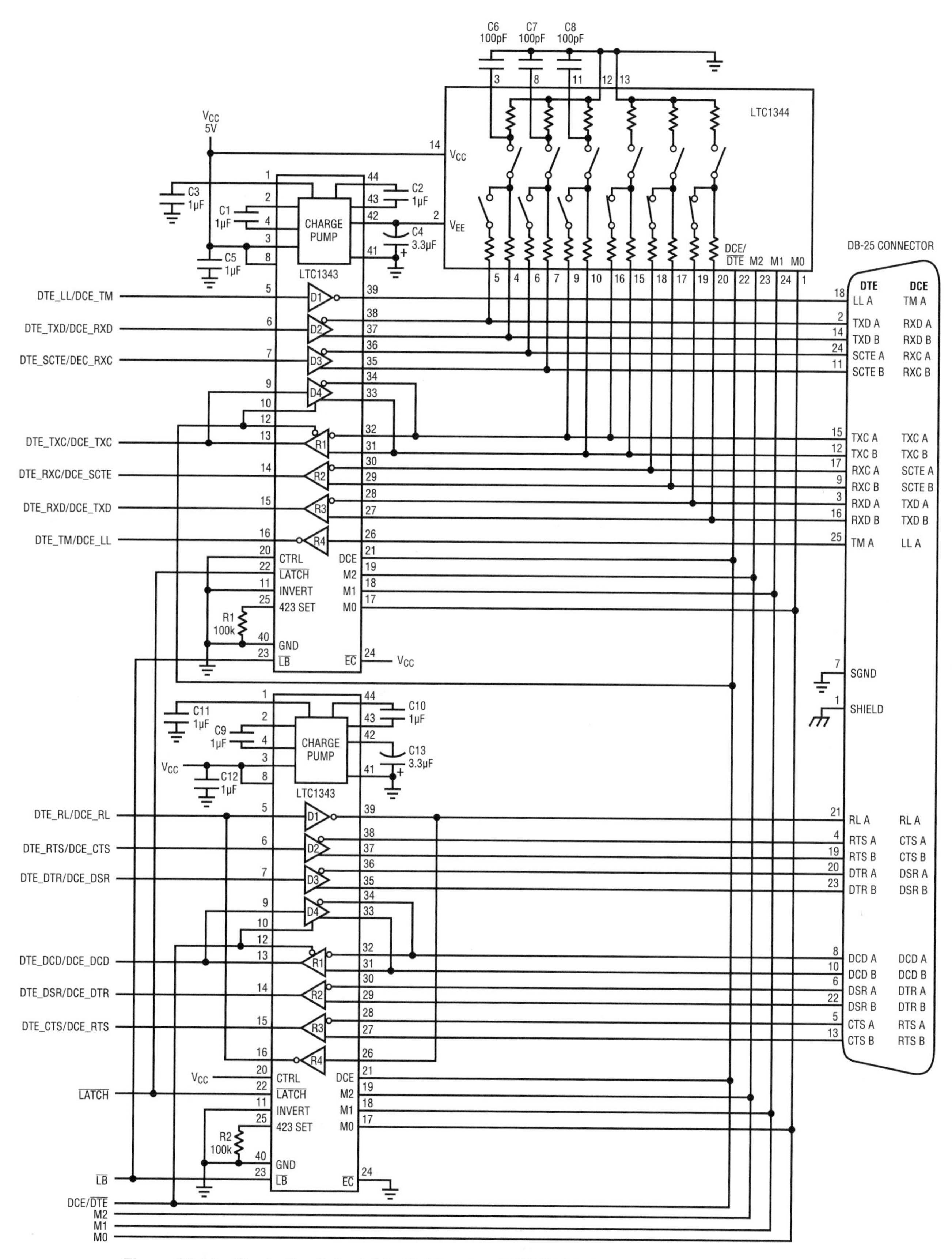

Figure 38.44 • Controller-Selectable Multiprotocol DTE/DCE Port with DB-25 Connector

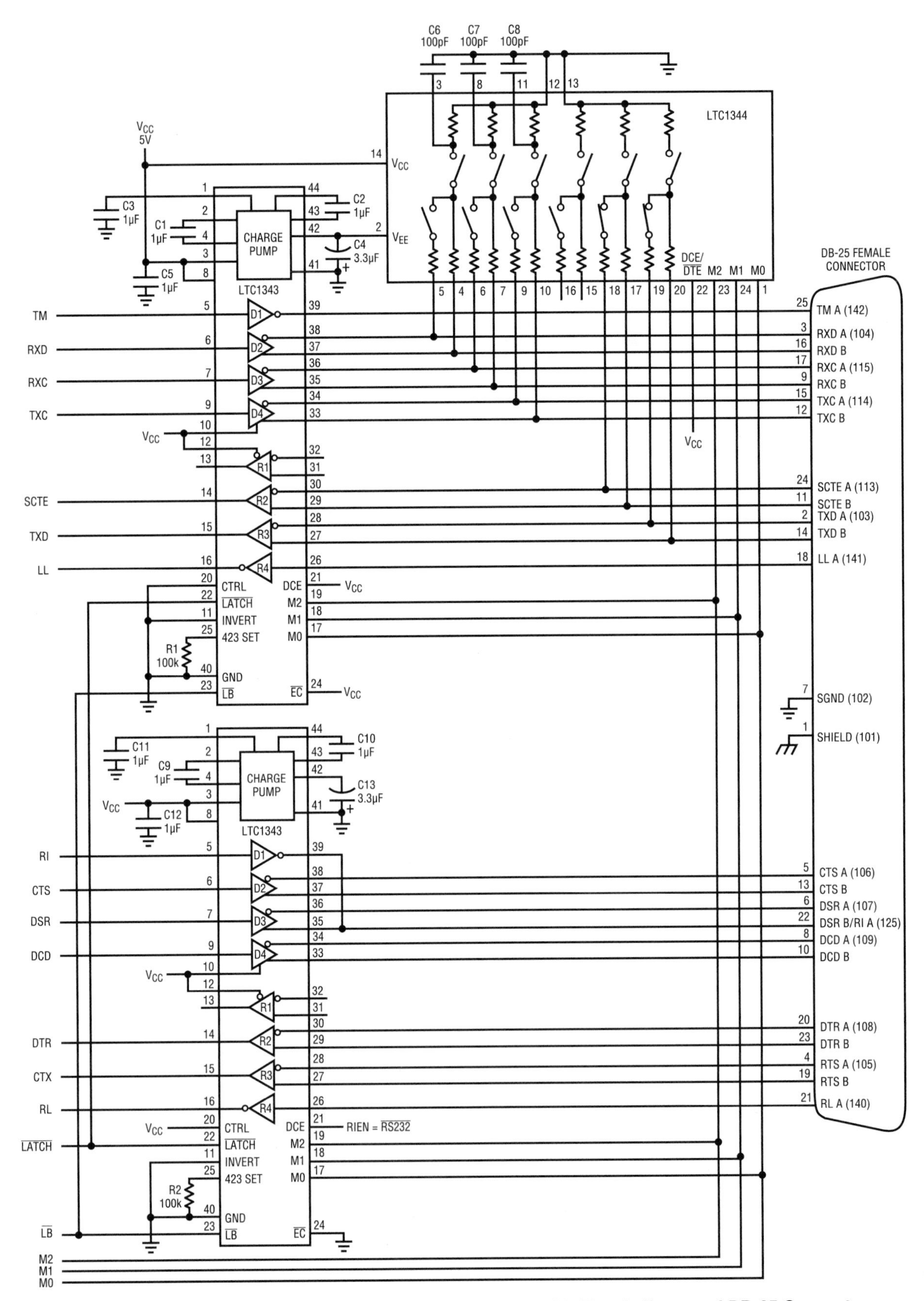

Figure 38.45 • Controller-Selectable Multiprotocol DCE Port with Ring-Indicate and DB-25 Connector

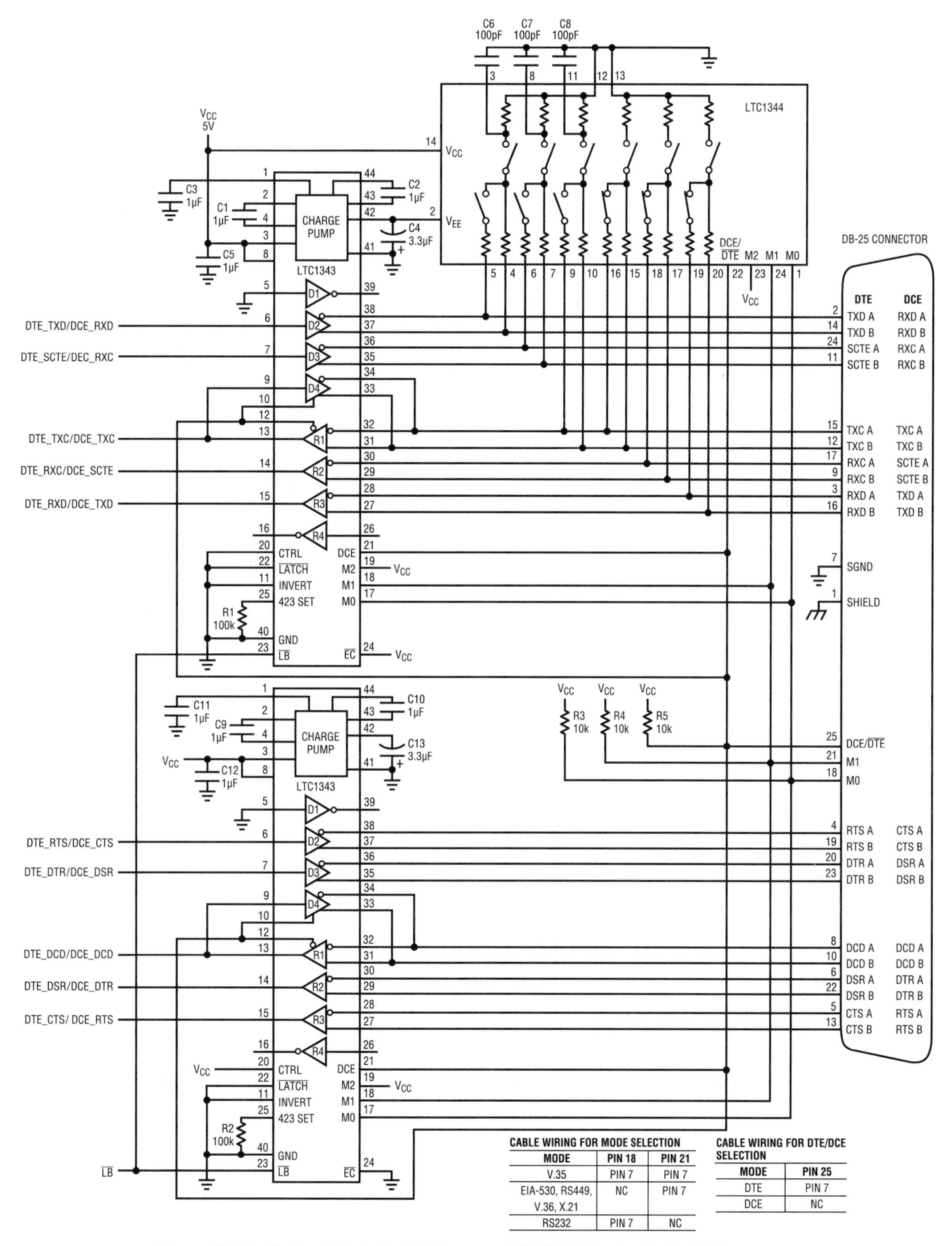

CABLE WIRING FOR MODE SELECTION

MODE	PIN 18	PIN 21
V.35	PIN 7	PIN 7
EIA-530, RS449, V.36, X.21	NC	PIN 7
RS232	PIN 7	NC

CABLE WIRING FOR DTE/DCE SELECTION

MODE	PIN 25
DTE	PIN 7
DCE	NC

Figure 38.46 • Cable-Selectable Multiprotocol DTE/DCE Port with DB-25 Connector

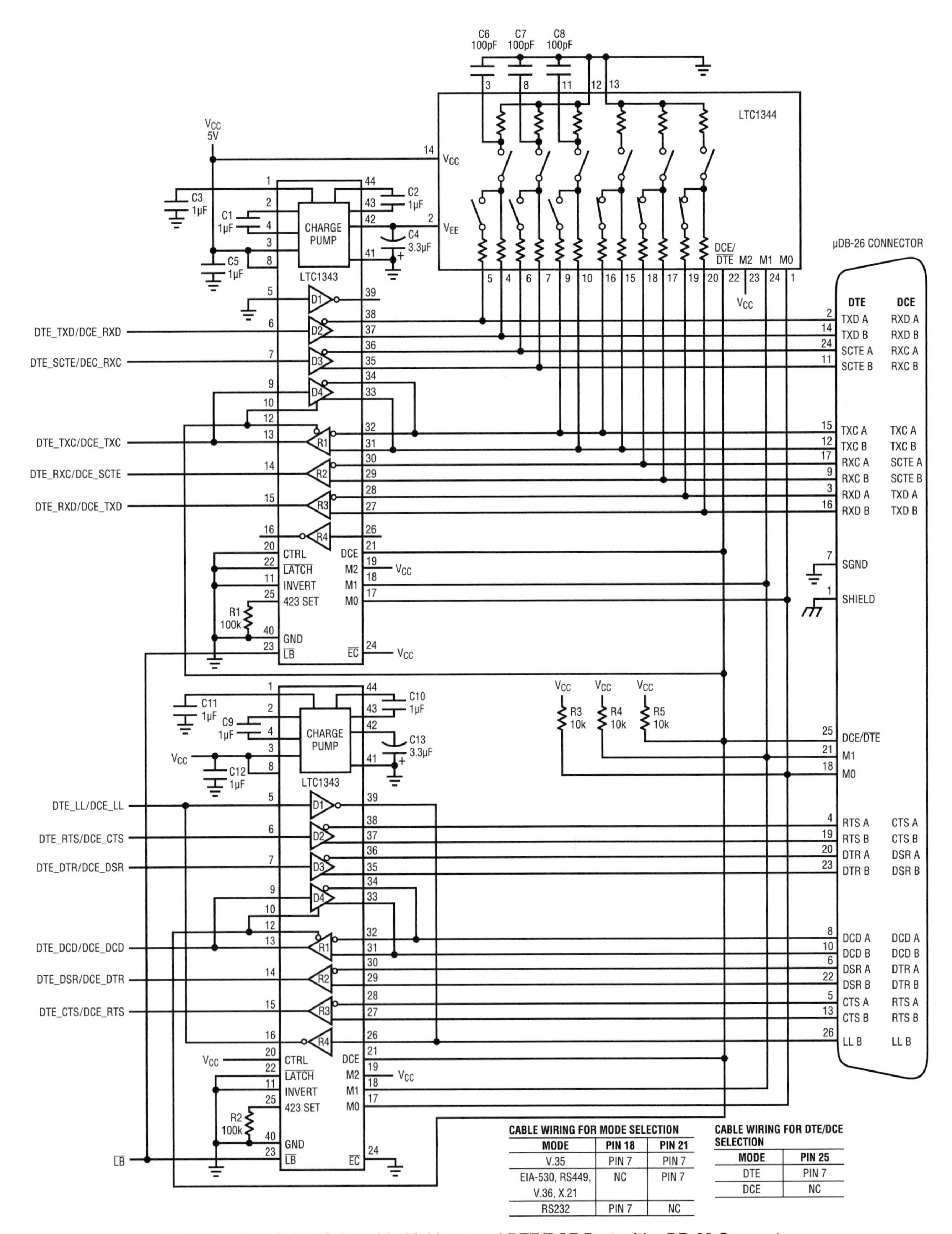

CABLE WIRING FOR MODE SELECTION

MODE	PIN 18	PIN 21
V.35	PIN 7	PIN 7
EIA-530, RS449, V.36, X.21	NC	PIN 7
RS232	PIN 7	NC

CABLE WIRING FOR DTE/DCE SELECTION

MODE	PIN 25
DTE	PIN 7
DCE	NC

Figure 38.47 • Cable-Selectable Multiprotocol DTE/DCE Port with µDB-26 Connector

The LT1328: A low cost IrDA receiver solution for data rates up to 4Mbps

by Alexander Strong

IrDA SIR

The LT1328 circuit in Figure 38.48 operates over the full 1cm to 1 meter range of the IrDA standard at the stipulated light levels. For IrDA data rates of 115kbps and below, a 1.6μs pulse width is used for a zero and no pulse for a one. Light levels are 40mW/sr (milliwatts per steradian) to 500mW/sr. Figure 38.49 shows a scope photo for a transmitter input (top trace) and the LT1328 output (bottom trace). Note that the input to the transmitter is inverted; that is, transmitted light produces a high at the input, which results in a zero at the output of the transmitter. The Mode pin (pin 7) should be high for these dataa rates.

An IrDA-compatible transmitter can also be implemented with only six components, as shown in Figure 38.50. Power requirements for the LT1328 are minimal: a single 5V supply and 2mA of quiescent current.

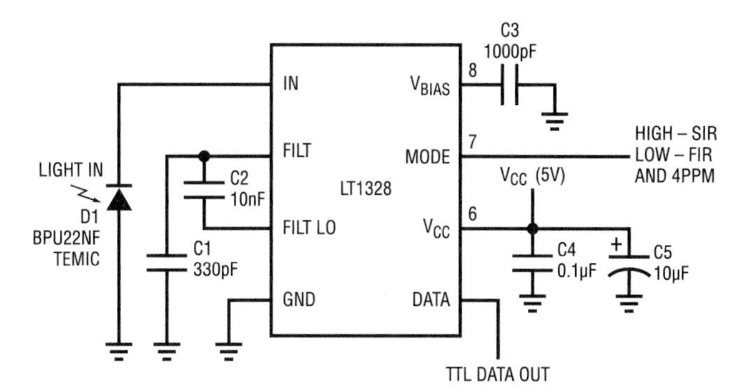

Figure 38.48 • LT1328 IrDA Receiver—Typical Application

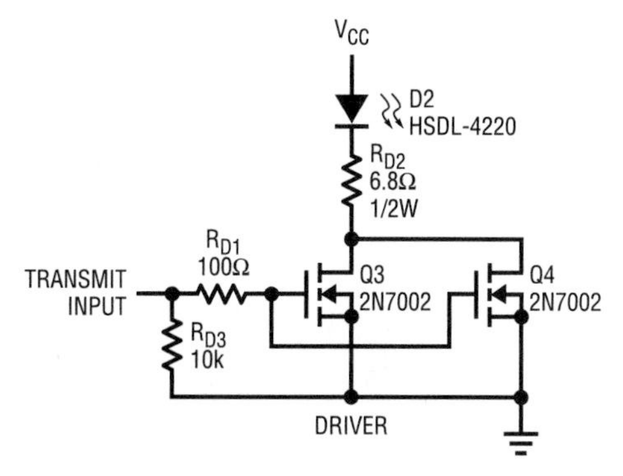

Figure 38.50 • IrDA Transmitter

IrDA FIR

The second fastest tier of the IrDA standard addresses 576kbps and 1.152Mbps data rates, with pulse widths of 1/4 of the bit interval for zero and no pulse for one. The 1.152Mbps rate, for example, uses a pulse width of 217ns; the total bit time is 870ns. Light levels are 100mW/sr to 500mW/sr over the 1cm to 1 meter range. A photo of a transmitted input and LT1328 output is shown in Figure 38.51. The LT1328 output pulse width will be less than 800ns wide over all of the above conditions at 1.152Mbps. Pin 7 should be held low for these data rates and above.

4ppm

The last IrDA encoding method is for 4Mbps and uses pulse position modulation, thus its name: 4ppm. Two bits are encoded by the location of a 125ns wide pulse at one of the four positions within a 500ns interval (2 bits • 1/500ns=4Mbps). Range and input levels are the same as for 1.152Mbps. Figure 38.52 shows the LT1328 reproduction of this modulation.

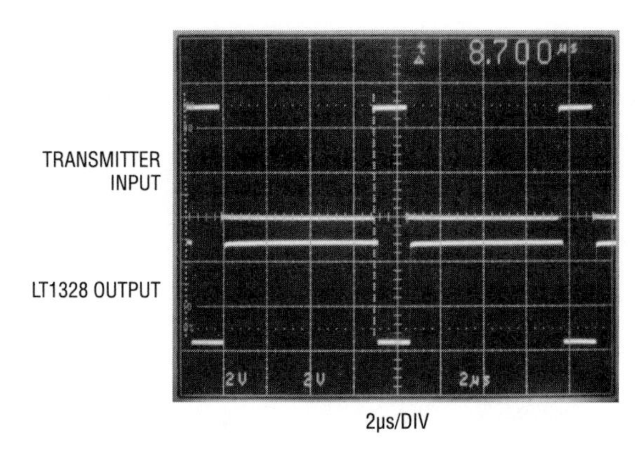

Figure 38.49 • IrDA 115kbps Modulation

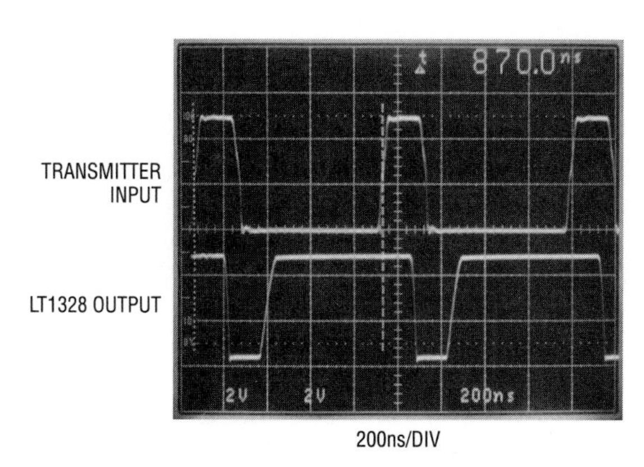

Figure 38.51 • IrDA 1.152Mbps Modulation

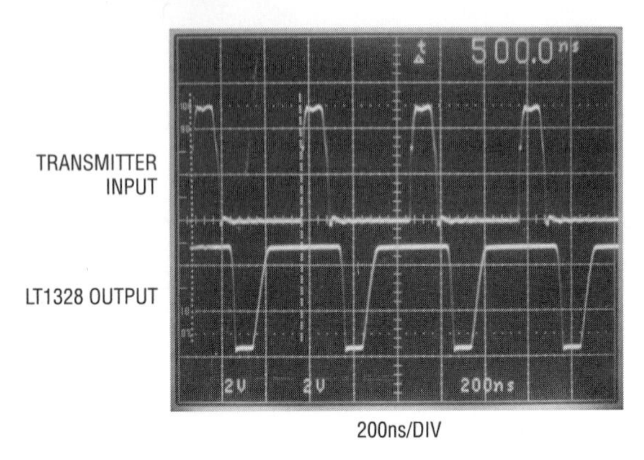

Figure 38.52 • IrDA 4ppm Modulation

LT1328 functional description

Figure 38.53 is a block diagram of the LT1328. Photodiode current from D1 is transformed into a voltage by feedback resistor R_{FB}. The DC level of the preamp is held at V_{BIAS} by the servo action of the transconductance amplifier's gm. The servo action only suppresses frequencies below the R_{gm}/C_{FILT} pole. This highpass filtering attenuates interfering signals, such as sunlight or incandescent or fluorescent lamps, and is selectable at pin 7 for low or high data rates. For high data rates, pin 7 should be held low. The highpass filter breakpoint is set by the capacitor C1 at $f = 25/(2\pi \bullet R_{gm} \bullet C)$, where $R_{gm} = 60k$. The 330pF capacitor (C1) sets a 200kHz corner frequency and is used for data rates above 115kbps. For low data rates (115kbps and below), the capacitance at pin 2 is increased by taking pin 7 to a TTL high. This switches C2 in parallel with C1, lowering the highpass filter breakpoint. A 10nF cap (C2) produces a 6.6kHz corner. Signals processed by the preamp/g_m amplifier combination cause the comparator output to swing low.

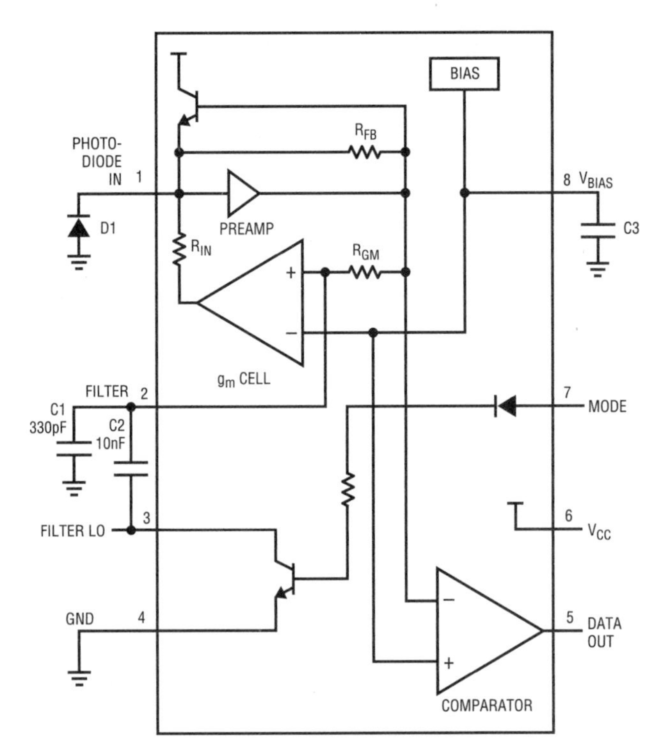

Figure 38.53 • LT1328 Block Diagram

Conclusion

In summary, the LT1328 can be used to build a low cost receiver compatible with IrDA standards. Its ease of use and flexibility also allow it to provide solutions to numerous other photodiode receiver applications. The tiny MSOP package saves on PC board area.

LTC1387 single 5V RS232/RS485 multiprotocol transceiver

by Y.K. Sim

Introduction

The LTC1387 is a single 5V supply, logic-configurable, single-port RS232 or RS485 transceiver. The LTC1387 offers a flexible combination of two RS232 drivers, two RS232 receivers, an RS485 driver, an RS485 receiver and an onboard charge pump to generate boosted voltages for true RS232 levels from a single 5V supply. The RS232 transceivers and RS485 transceiver are designed to share the same port I/O pins for both single-ended and differential signal communication modes. The RS232 transceiver supports both RS232 and EIA562 standards, whereas the RS485 transceiver supports both RS485 and RS422 standards. Both half-duplex and full-duplex communication are supported.

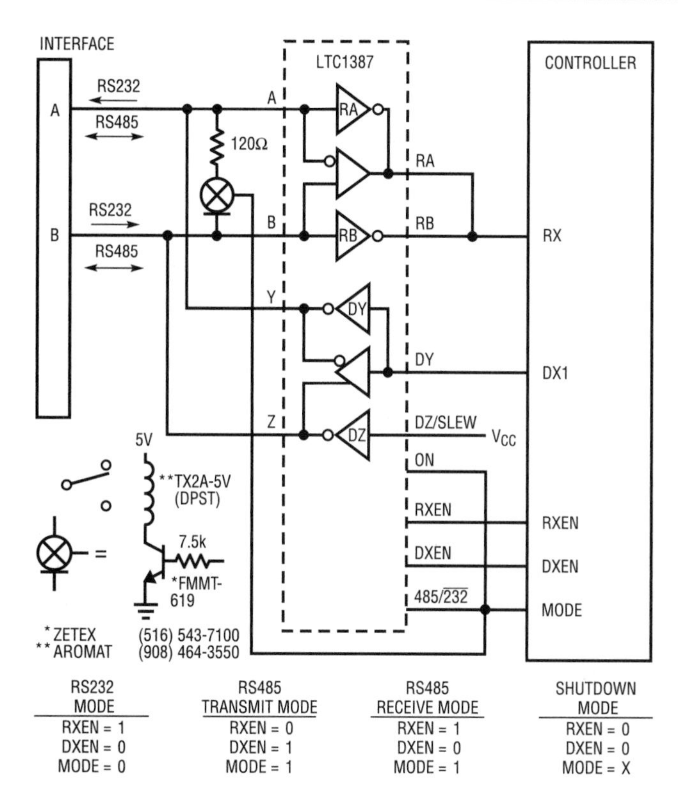

Figure 38.55 • Full-Duplex RS232, Half-Duplex RS485

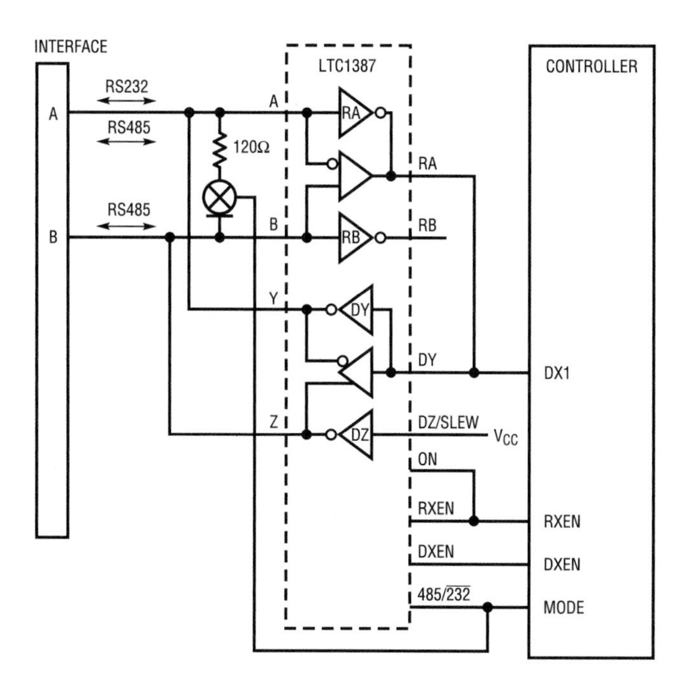

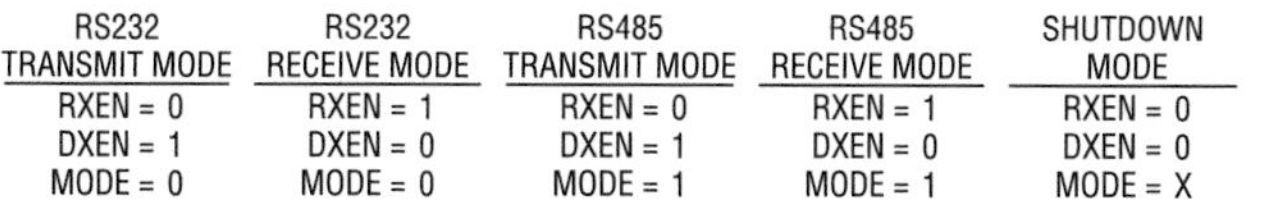

RS232 TRANSMIT MODE	RS232 RECEIVE MODE	RS485 TRANSMIT MODE	RS485 RECEIVE MODE	SHUTDOWN MODE
RXEN = 0	RXEN = 1	RXEN = 0	RXEN = 1	RXEN = 0
DXEN = 1	DXEN = 0	DXEN = 1	DXEN = 0	DXEN = 0
MODE = 0	MODE = 0	MODE = 1	MODE = 1	MODE = X

Figure 38.54 • Half-Duplex RS232, Half-Duplex RS485

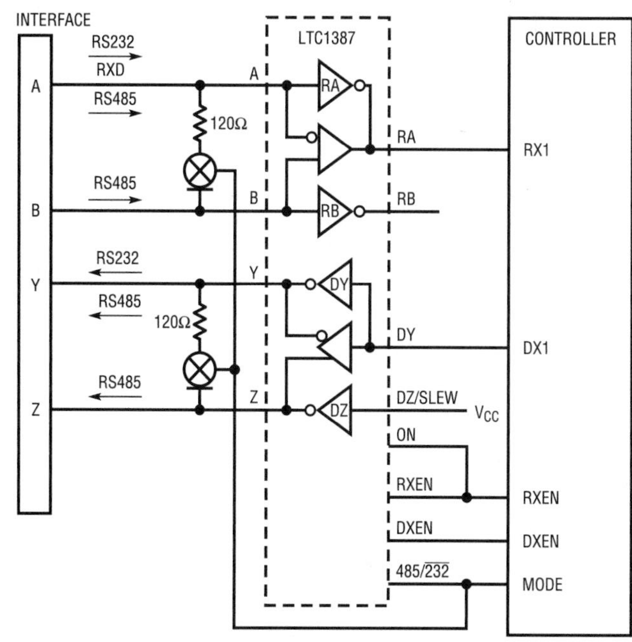

RS232 MODE	RS485 MODE	SHUTDOWN MODE
RXEN = 1	RXEN = 1	RXEN = 0
DXEN = 1	DXEN = 1	DXEN = 0
MODE = 0	MODE = 1	MODE = X

Figure 38.56 • Full-Duplex RS232 (1-Channel), Full-Duplex RS422

A logic input selects between RS485 and RS232 modes. Three additional control inputs allow the LTC1387 to be reconfigured easily via software to adapt to various communication needs, including a one-signal line RS232 I/O mode (see function tables in figures). Four examples of interface port connections are shown in Figs 38.54–38.57.

A SLEW input pin, active in RS485 mode, changes the driver transition between normal and slow slew-rate modes. In normal RS485 slew mode, the twisted pair cable must be terminated at both ends to minimized signal reflection. In slow-slew mode, the maximum signal bandwidth is reduced, minimizing EMI and signal reflection problems. Slow-slew-rate systems can often use improperly terminated or even unterminated cables with acceptable results. If cable termination is required, external termination resistors can be connected through switches or relays.

The LTC1387 features micropower shutdown mode, loop-back mode for self-test, high data rates (120kbaud for RS232 and 5Mbaud for RS485) and 7kV ESD protection at the driver outputs and receiver inputs.

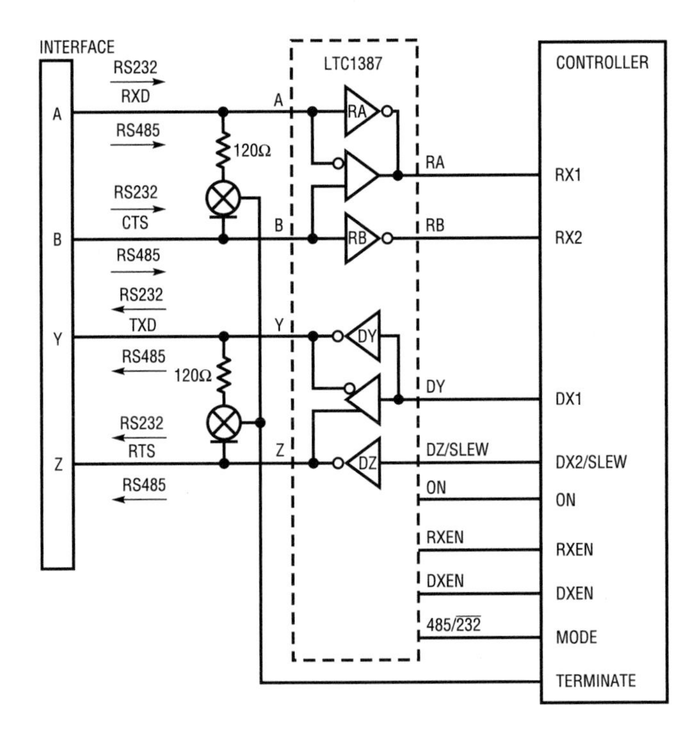

RS232 MODE	RS485 MODE	SHUTDOWN MODE
ON = 1	ON= 1	ON = 0
RXEN = 1	RXEN = 1	RXEN = 0
DXEN = 1	DXEN = 1	DXEN = 0
MODE = 0	MODE = 1	MODE = X

Figure 38.57 • Full-Duplex RS232 (2-Channel), Full-Duplex RS485 with Slew and Termination Control

A 10MB/s multiple-protocol chip set supports Net1 and Net2 standards

by David Soo

Introduction

Typical application

Like the LTC1343 software-selectable multiprotocol transceiver, introduced in the August, 1996 issue of *Linear Technology*, the LTC1543/LTC1544/LTC1344A chip set creates a complete software-selectable serial interface using an inexpensive DB-25 connector. The main difference between these parts is the division of functions: the LTC1343 can be configured as a data/clock chip or as a control-signal chip using the CTRL/$\overline{CLK}$ pin, whereas the LTC1543 is a dedicated data/clock chip and the LTC1544 is a control-signal chip. The chip set supports the V.28 (RS232), V.35, V.36, RS449, EIA-530, EIA-530A and X.21 protocols in either DTE or DCE mode.

Figure 38.58 shows a typical application using the LTC1543, LTC1544 and LTC1344A. By just mapping the chip pins to the connector, the design of the interface port is complete. The figure shows a DCE mode connection to a DB-25 connector.

The LTC1543 contains three drivers and three receivers, whereas the LTC1544 contains four drivers and four receivers. The LTC1344A contains six switchable resistive terminators that are connected only to the high

Table 38.4 Mode-Pin Functions

LTC1543/LTC1544 Mode Name	M2	M1	M0
Not Used	0	0	0
EIA-530A	0	0	1
EIA-530	0	1	0
X.21	0	1	1
V.35	1	0	0
RS449/V.36	1	0	1
RS232/V.28	1	1	0
No Cable	1	1	1

speed clock and data signals. When the interface protocol is changed via the mode selection pins, M2, M1 and M0, the drivers, receivers and line terminators are placed in their proper configuration. The mode pin functions are summarized in Table 38.4. There are internal 50μA pull-up current sources on the mode select pins, DCE/$\overline{\text{DTE}}$ and the INVERT pins.

DTE vs DCE operation

The LTC1543/LTC1544/LTC1344A chip set can be configured for either DTE or DCE operation in one of two ways. The first way is when the chip set is a dedicated DTE or DCE port with a connector of appropriate gender. The second way is when the port has one connector that can be configured for DTE or DCE operation by rerouting the signals to the chip set using a dedicated DTE or DCE cable.

Figure 38.58 is an example of a dedicated DCE port using a female DB-25 connector: The complement to this port is the DTE-only port using a male DB-25 connector, as shown in Figure 38.59.

If the port must accommodate both DTE and DCE modes, the mapping of the drivers and receivers to connector pins must change accordingly. For example, in Figure 38.58, driver 1 in the LTC1543 is connected to pin 3 and pin 16 of the DB-25 connector. In DTE mode, as shown in Figure 38.59, driver 1 is mapped to pins 2 and 14 of the DB-25 connector. A port that can be configured for either DTE or DCE operation is shown in Figure 38.60. This configuration requires separate cables for proper signal routing.

Cable-selectable multiprotocol interface

The interface protocol may be selected by simply plugging the appropriate interface cable into the connector. A cable-selectable multiprotocol DTE/DCE interface is shown in Figure 38.61. The mode pins are routed to the connector and are left unconnected (1) or wired to ground (0) in the cable. The internal pull-up current sources ensure a binary 1 when a pin is left unconnected and also ensure that the LTC1543/LTC1544/LTC1344A enter the no-cable mode when the cable is removed. In the no-cable mode, the LTC1543/LTC1544 power supply current drops to less than 200μA and all of the LTC1543/LTC1544 driver outputs will be forced into the high impedance state.

Adding optional test signal

In some cases, the optional test signals local loopback (LL), remote loopback (RL) and test mode (TM) are required but there are not enough drivers and receivers available in the LTC1543/LTC1544 to handle these extra signals. The solution is to combine the LTC1544 with the LTC1343. By using the LTC1343 to handle the clock and data signals, the chip set gains one extra single-ended driver/receiver pair. This configuration is shown in Figure 38.62.

Compliance testing

A European standard EN 45001 test report is available for the LTC1543/LTC1544/LTC1344A chip set. The report provides documentation on the compliance of the chip set to Layer 1 of the NET1 and NET2 standard. A copy of this test report is available from LTC or from Detecon, Inc. at 1175 Old Highway 8, St. Paul, MN 55112.

Conclusion

In the world of network equipment, the product differentiation is mostly in the software and not in the serial interface. The LTC1543, LTC1544 and LTC1344A provide a simple yet comprehensive solution to standards compliance for multiple-protocol serial interface.

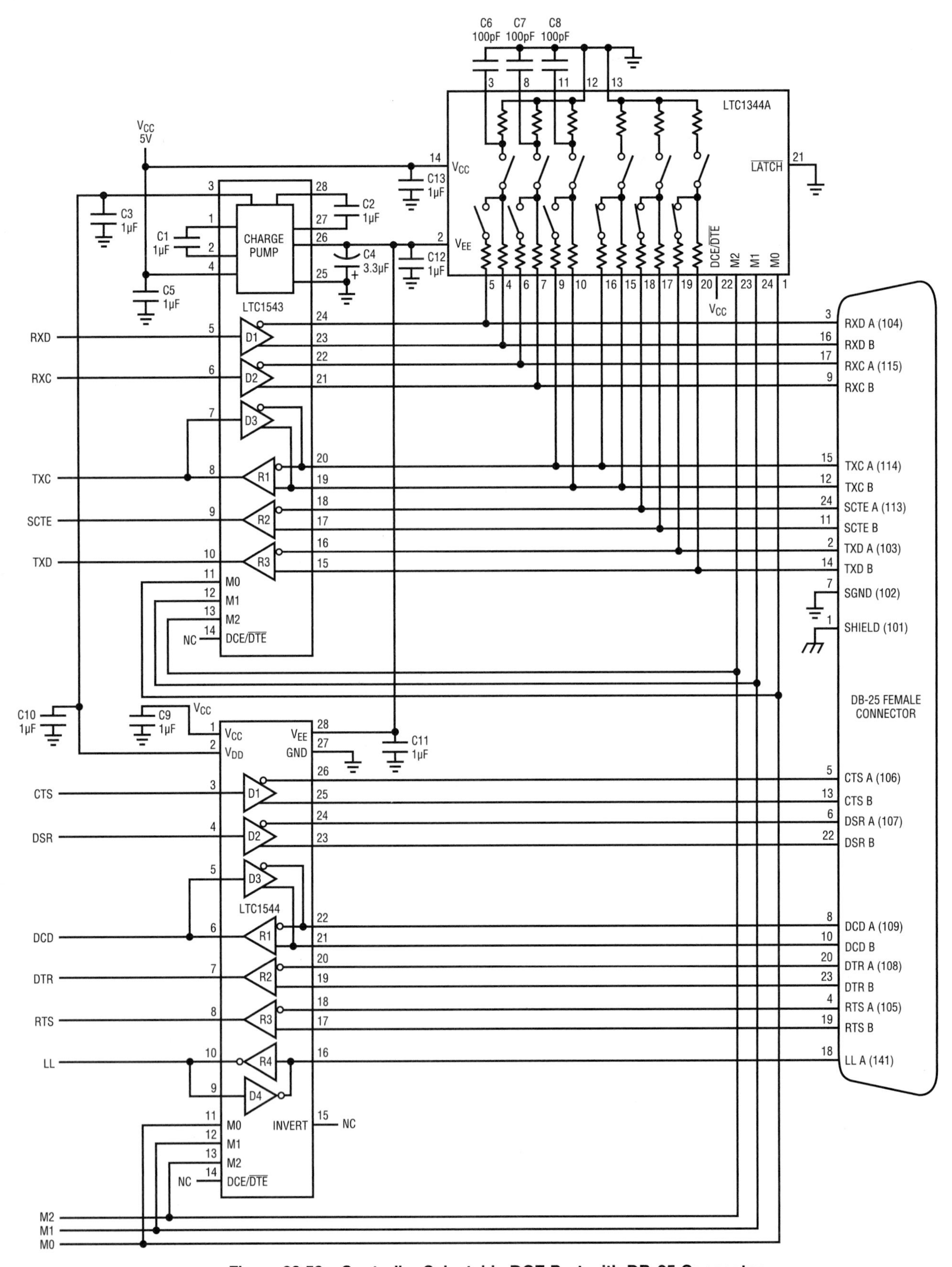

Figure 38.58 • Controller-Selectable DCE Port with DB-25 Connector

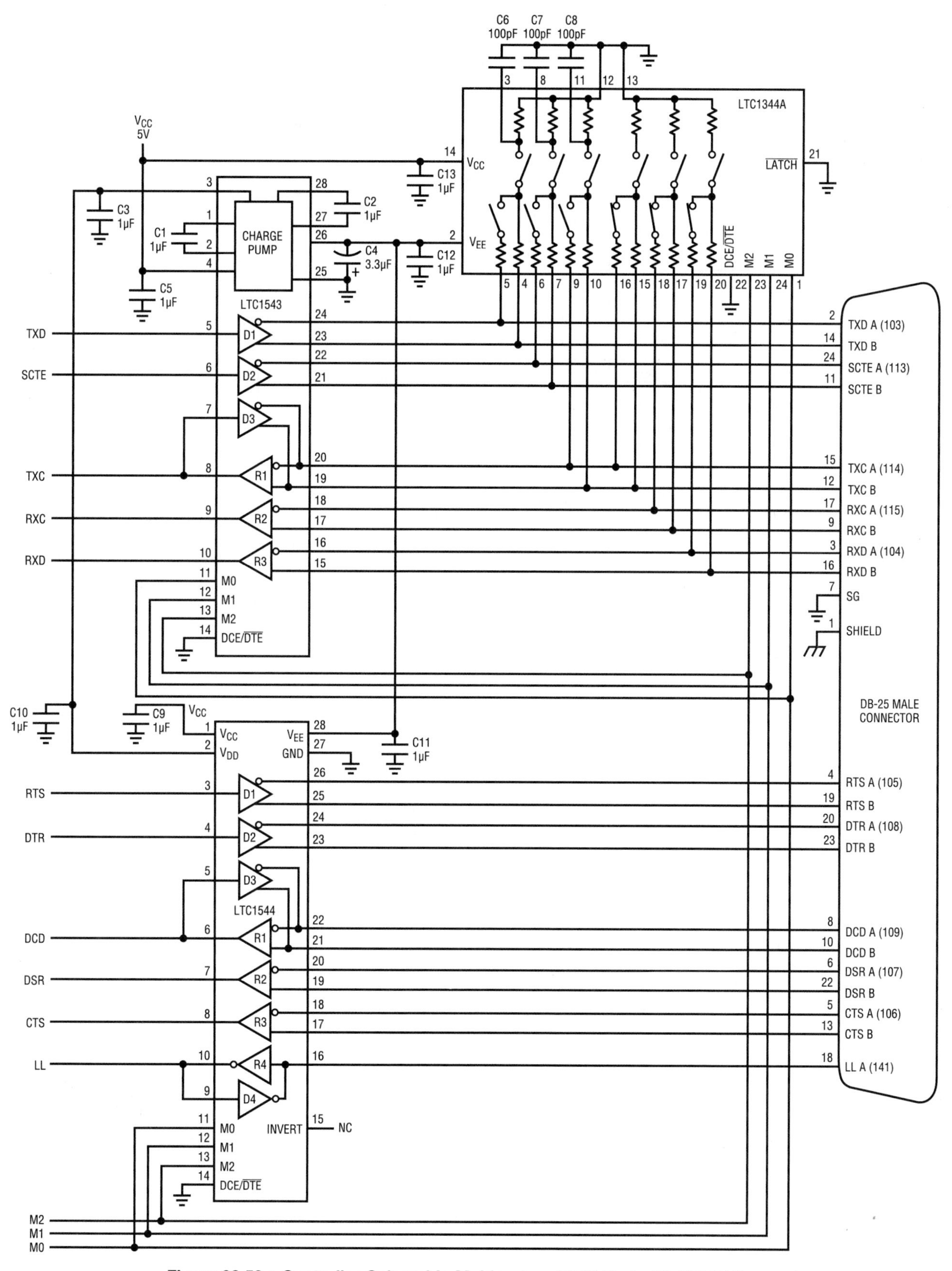

Figure 38.59 • Controller-Selectable Multiprotocol DTE Port with DB-25 Connector

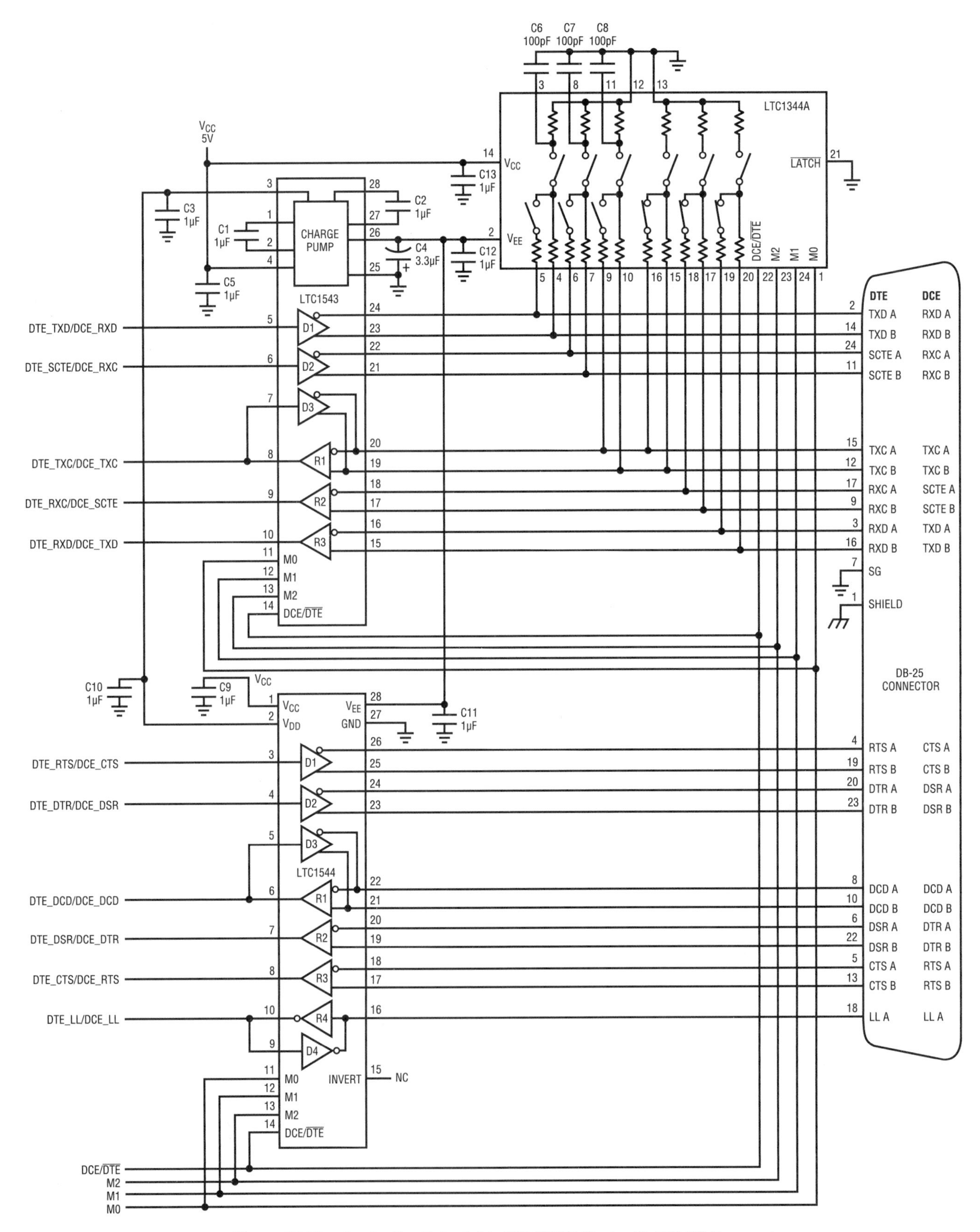

Figure 38.60 • Controller-Selectable DTE/DCE Port with DB-25 Connector

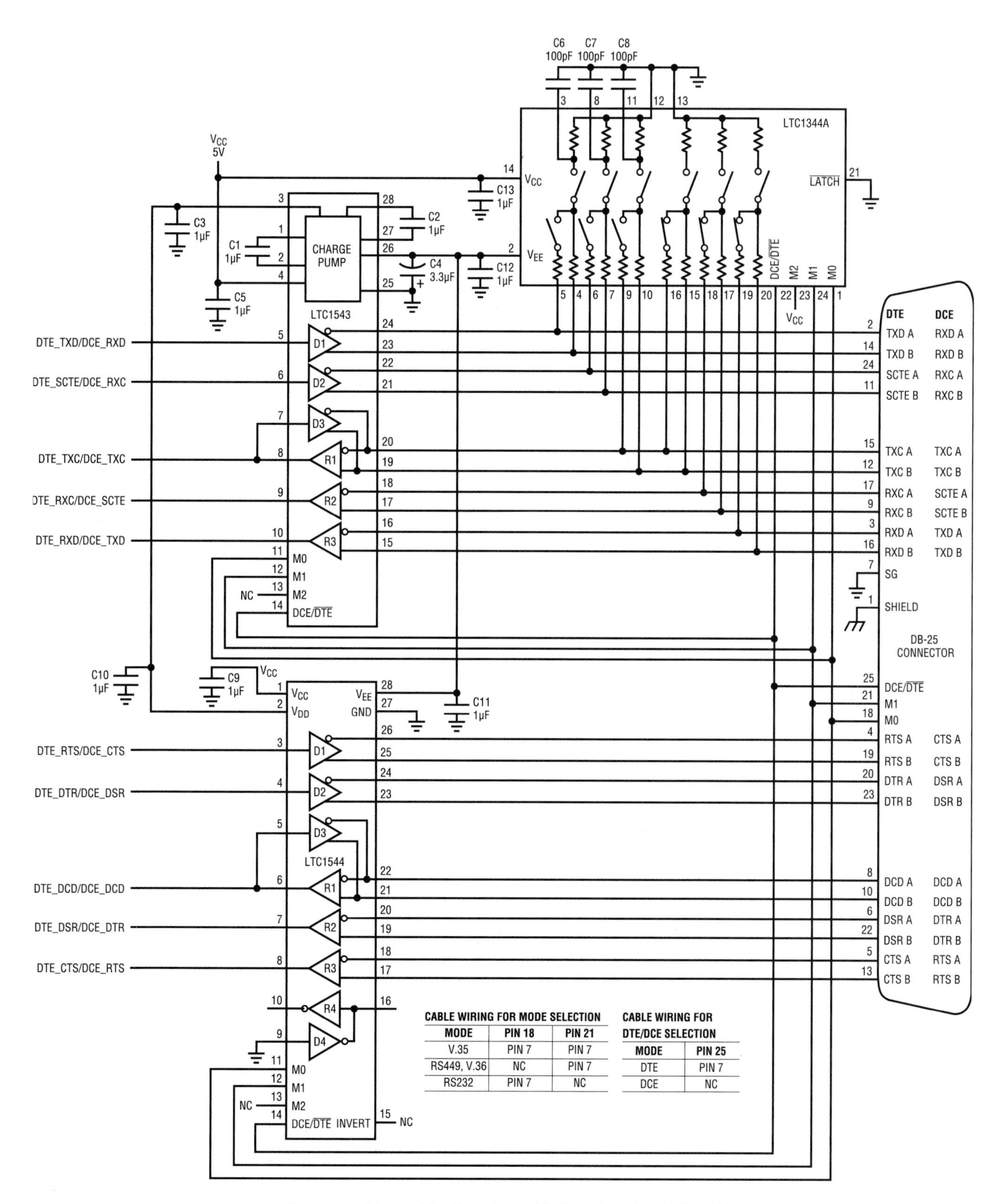

CABLE WIRING FOR MODE SELECTION

MODE	PIN 18	PIN 21
V.35	PIN 7	PIN 7
RS449, V.36	NC	PIN 7
RS232	PIN 7	NC

CABLE WIRING FOR DTE/DCE SELECTION

MODE	PIN 25
DTE	PIN 7
DCE	NC

Figure 38.61 • Cable-Selectable Multiprotocol DTE/DCE Port

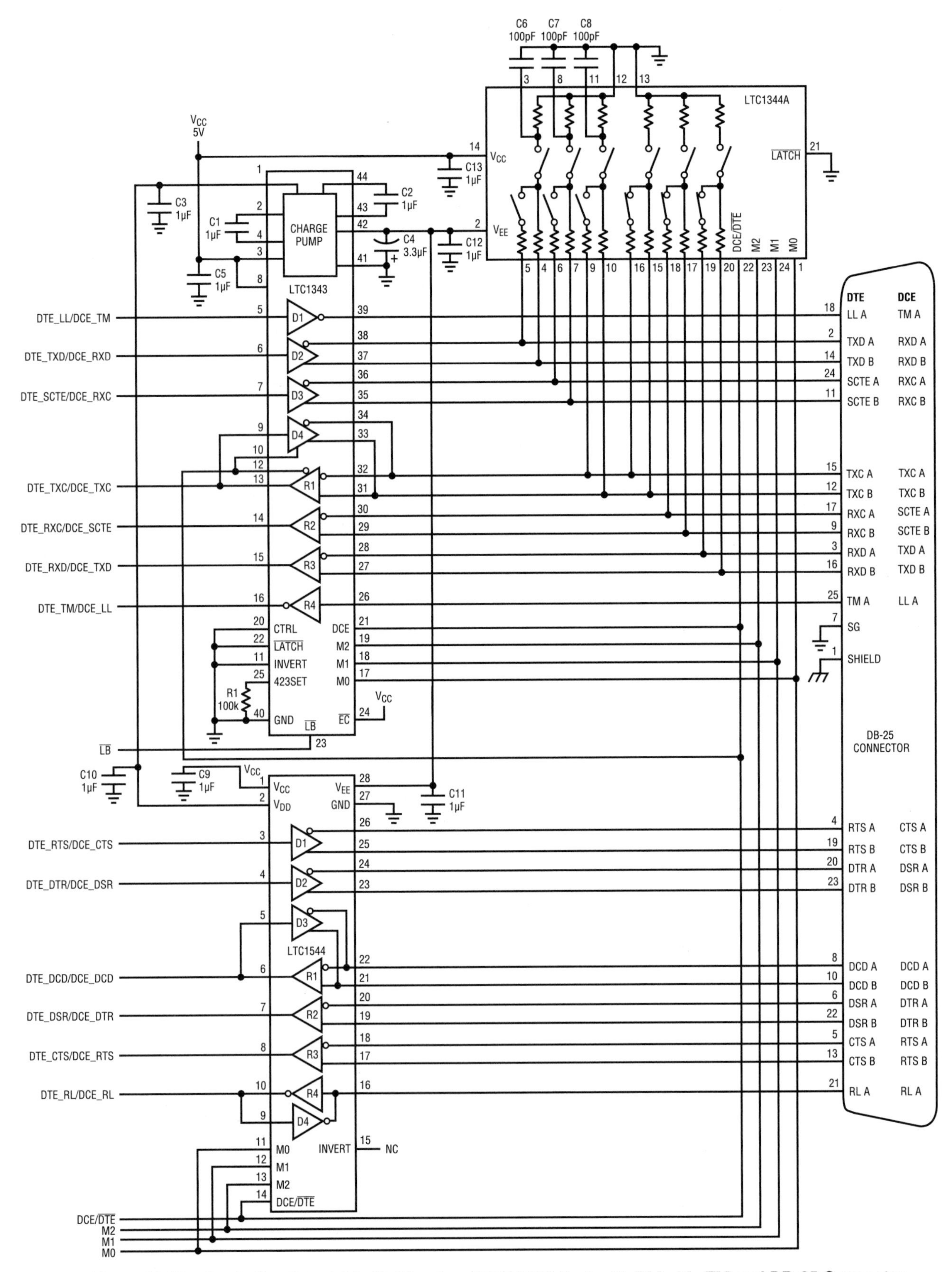

Figure 38.62 • Controller-Selectable Multiprotocol DTE/DCE Port with RLL, LL, TM and DB-25 Connector

Net1 and Net2 serial interface chip set supports test mode

by David Soo

Some serial networks use a test mode to exercise all of the circuits in the interface. The network is divided into local and remote data terminal equipment (DTE) and data-circuit-terminating equipment (DCE), as shown in Figure 38.63. Once the network is placed in a test mode, the local DTE will transmit on the driver circuits and expect to receive the same signals back from either a local or remote DCE. These tests are called local or remote loopback.

The LTC1543/LTC1544/LTC1344A chip set has taken the integrated approach to multiple protocol. By using this chip set, the Net1 and Net2 design work is done. The LTC1545 extends the family by offering test mode capability. By replacing the 6-circuit LTC1544 with the 9-circuit LTC1545, the optional circuits TM (Test Mode), RL (Remote Loop- back) and LL (Local Loopback) can now be implemented.

Figure 38.64 shows a typical application using the LTC1543, LTC1545 and LTC1344A. By just mapping the

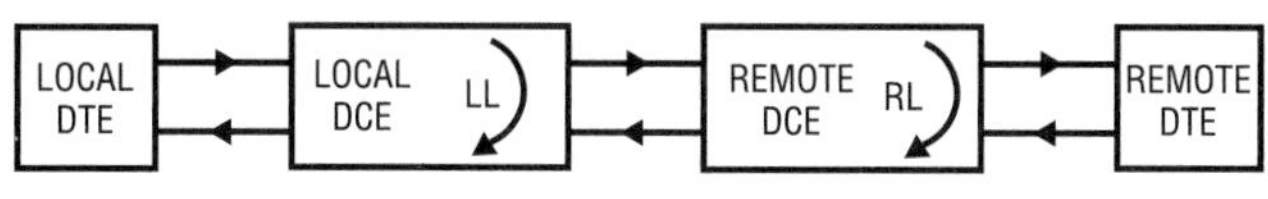

Figure 38.63 • Serial Network

chip pins to the connector, the design of the interface port is complete. The chip set supports the V.28, V.35, V.36, RS449, EIA-530, EIA-530A or X.21 protocols in either DTE or DCE mode. Shown here is a DCE mode connection to a DB-25 connector. The mode-select pins, M0, M1 and M2, are used to select the interface protocol, as summarized in Table 38.5.

Table 38.5 Mode-Pin Functions

LTC1543/ LTC1544 Mode Name	M2	M1	M0
Not Used	0	0	0
EIA-530A	0	0	1
EIA-530	0	1	0
X.21	0	1	1
V.35	1	0	0
RS449/V.36	1	0	1
RS232/V.28	1	1	0
No Cable	1	1	1

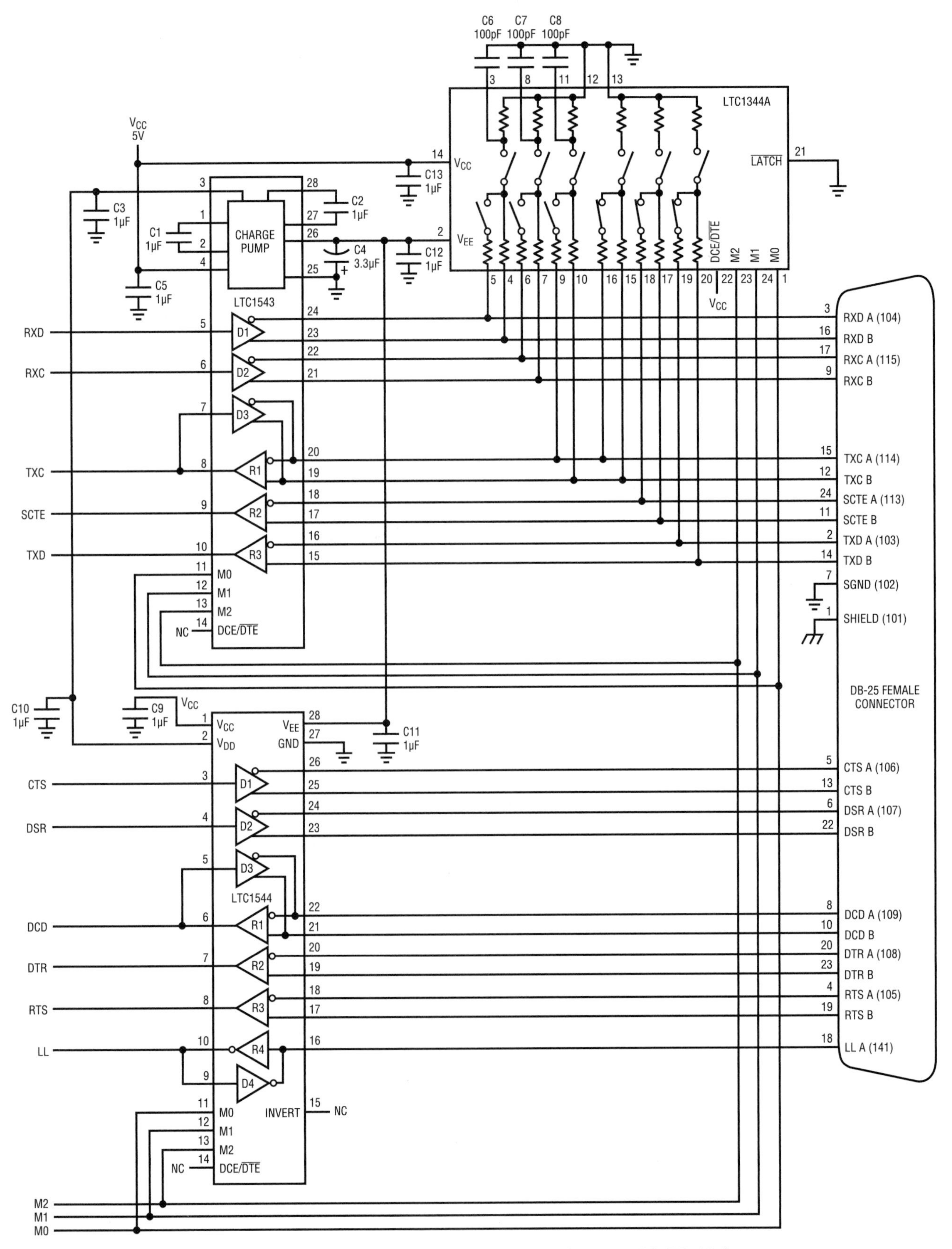

Figure 38.64 • Typical Application: Controller-Selectable DCE Port with DB-25 Connector

Operational amplifiers/video amplifiers

LT1490/LT1491 over-the-top dual and quad micropower rail-to-rail op amps

by Jim Coelho-Sousae

Introduction

The LT1490 is Linear Technology's lowest power, lowest cost and smallest dual rail-to-rail input and output operational amplifier: The ability to operate with its inputs above V_{CC}, its high performance-to-price ratio and its availability in the MSOP package, sets the LT1490 apart from other amplifiers.

An Over-The-Top® application

The battery current monitor circuit shown in Figure 38.65 demonstrates the LT1491's ability to operate with its inputs above the positive supply rail. In this application, a conventional amplifier would be limited to a battery voltage between 5V and ground, but the LT1491 can handle battery voltages as high as 44V The LT1491 can be shut down by removing V_{cc}. With V_{cc} removed the input leakage is less than 0.1nA. No damage to the LT1491 will result from inserting the 12V battery backward.

When the battery is charging, Amp B senses the voltage drop across RS. The output of Amp B causes QB to drain sufficient current through RB to balance the inputs of Amp B. Likewise, Amp A and QA form a closed loop when the battery is discharging. The current through QA or QB is proportional to the current in RS; this current flows into RG, which converts it back to a voltage. Amp D buffers and amplifies the voltage across RG. Amp C compares the output of Amp A and Amp B to determine the polarity of the current through RS. The scale factor for V_{OUT} with S1 open is 1V/A. With S1 closed the scale factor is 1V/100mA, and current as low as 5mA can be measured.

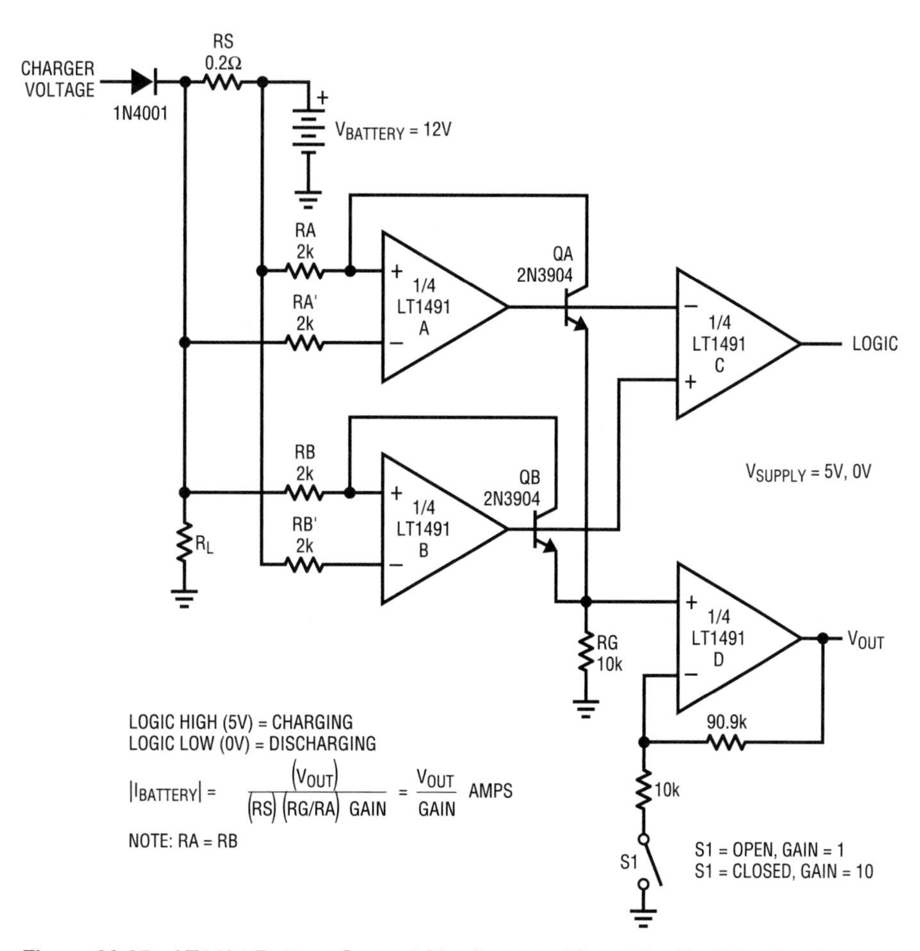

Figure 38.65 • LT1491 Battery Current Monitor—an "Over-The-Top" Application

The LT1210: a 1-ampere, 35MHz current feedback amplifier

by William Jett and Mitchell Lee

Introduction

The LT1210 current feedback amplifier extends Linear Technology's high speed driver solutions to the 1 ampere level. The device combines a 35MHz bandwidth with a guaranteed 1A output current, operation with ±5V to ±15V supplies and optional compensation for capacitive loads, making it well suited for driving low impedance loads. Short circuit protection and thermal shutdown ensure the device's ruggedness. A shutdown feature allows the device to be switched into a high impedance, low current mode, reducing dissipation when the device is not in use. The LT1210 is available in the 7-pin TO-220 package, the 7-pin DD surface mount package and the 16-pin SO-16 surface mount package.

Twisted pair driver

Figure 38.66 shows a transformer-coupled application of the LT1210 driving a 100Ω twisted pair This surge impedance is typical of PVC-insulated, 24 gauge, telephone-grade twisted pair wiring. The 1:3 transformer ratio allows just over 1W to reach the twisted pair at full output. Resistor R_T acts as a primary side back-termination. The overall

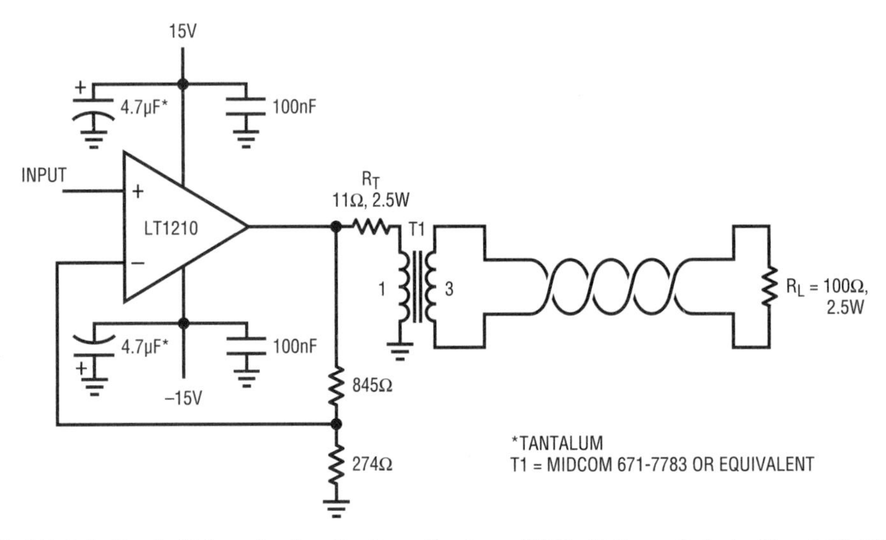

Figure 38.66 • Twisted Pair Is Easily Driven for Applications Such as ADSL. Voltage Gain is About 12. 5V_{P-P} Input Corresponds to Full Output

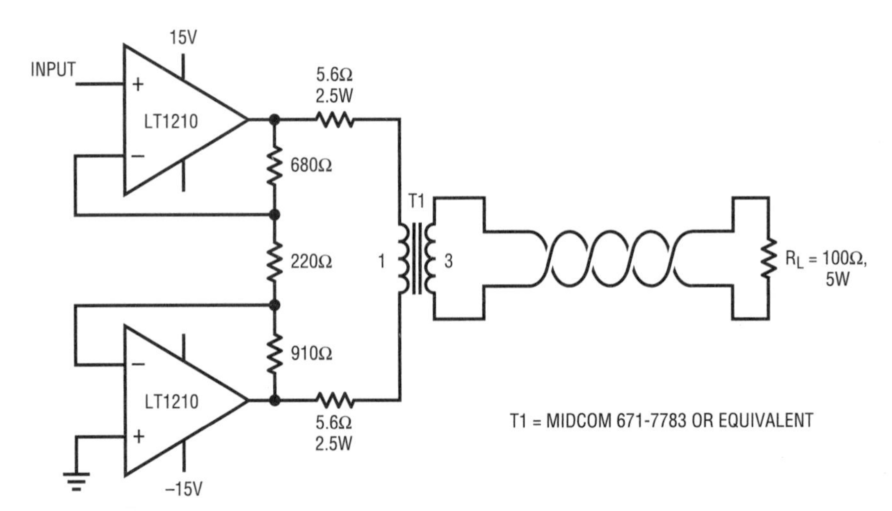

Figure 38.67 • In a Bridge Configuration, the LT1210 Can Deliver Almost 5W to a Twisted Pair (and Another 5W to the Back Termination)

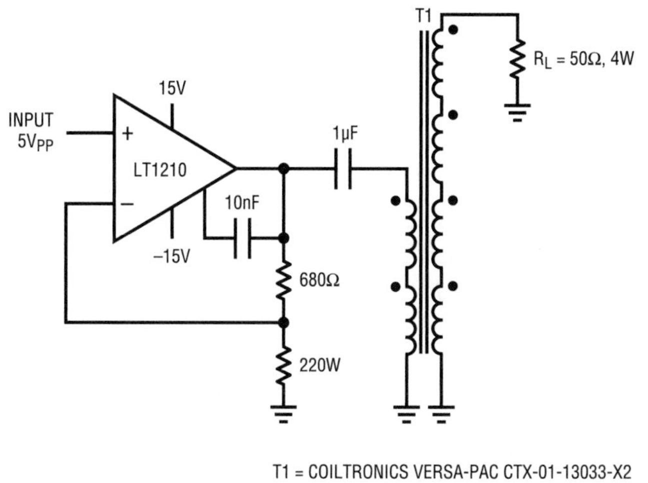

T1 = COILTRONICS VERSA-PAC CTX-01-13033-X2

Figure 38.68 • Matched to a 50Ω Load with a Balun-Mode Transformer, this Circuit Delivers a Measured 35.6dBm (almost 4W). Full-Power Band Limits are 15kHz to Slightly Over 10MHz

frequency response is flat to within 1dB from 500Hz to 2MHz. Distortion products at 1MHz are below −70dBc at a total output power of 560mW (load plus termination), rising to −56dBc at 2.25W.

On a ±15V supply, a maximum output power of 5W is available when a 10Ω load is presented to the LT1210. With the transformer shown in Figure 38.66, a total load impedance of approximately 22Ω limits the output to 2.25W. Bridging allows nearly maximum output power to be delivered into standard 1:3 data communications transformers. Figure 38.67 shows a bridged application with two LT1210s, delivering approximately 9W maximum into the load and termination.

At first glance the resistor values would suggest a gain imbalance between the inverting and noninverting sides of the bridge. On close inspection, however, it is apparent that both sides operate at a closed loop gain of 4 relative to the input signal. This ensures symmetric swing and maximum undistorted output.

Matching 50Ω systems

Few practical systems exhibit a 10Ω impedance, so a matching transformer is necessary for applications driving other loads, such as 50Ω. Multifilar winding techniques exhibit the best high frequency characteristics. Suitable

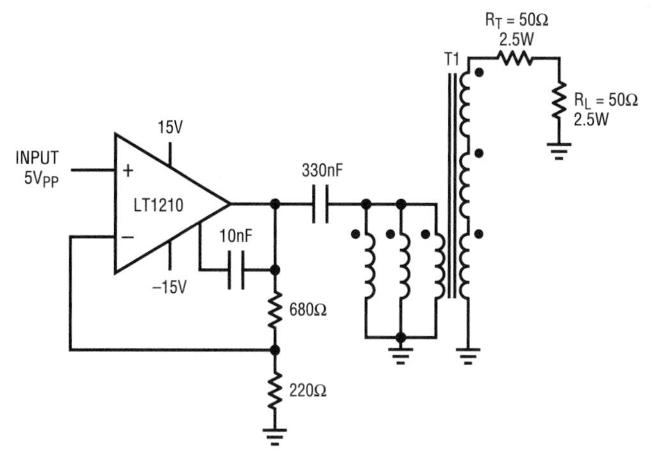

T1 = COILTRONICS VERSA-PAC CTX-01-13033-X2

Figure 38.69 • Wide Bandwidth can be Obtained with Even Higher Impedance Transformations. Here, a 1:3 Step-Up Matches 100Ω and Develops Nearly 4.5W. A Measured +33dBm Reaches the 50Ω Load. Full-Power Band Limits Are 80kHz to 18MHz

off-the- shelf components are available, such as the Coiltronics Versa-Pac™ series. These are hexafilar wound and give power bandwidths in excess of 10MHz. One disadvantage is that using a limited number of 1:1 windings makes it impossible to exactly transform 50Ω to the optimum 10Ω load. Nevertheless, there are several useful connections.

In Figure 38.68 the windings are configured for a 2:4 step-up, reflecting 12.5Ω into the LT1210. The circuit exhibits 18dB gain and drives 50Ω to nearly +36dBm. The large-signal, low frequency response is limited by the magnetizing inductance of the transformer to about 15kHz. The high frequency response is limited to 10MHz by the stack of four secondary windings.

Reconfiguring the transformer windings allows double termination at full power (Figure 38.69). Here the transformer reflects 11.1Ω and the amplifier delivers over +33dBm to the load. Paralleled input windings limit the low frequency response to 80kHz, but fewer series secondary windings extend the high frequency corner to 18MHz.

The coupling capacitor shown in these examples is added to block current flow through the transformer primary, arising from amplifier offsets. The capacitor value is based on setting X_C equal to the reflected load impedance at the frequency where X_L of the primary is also equal to

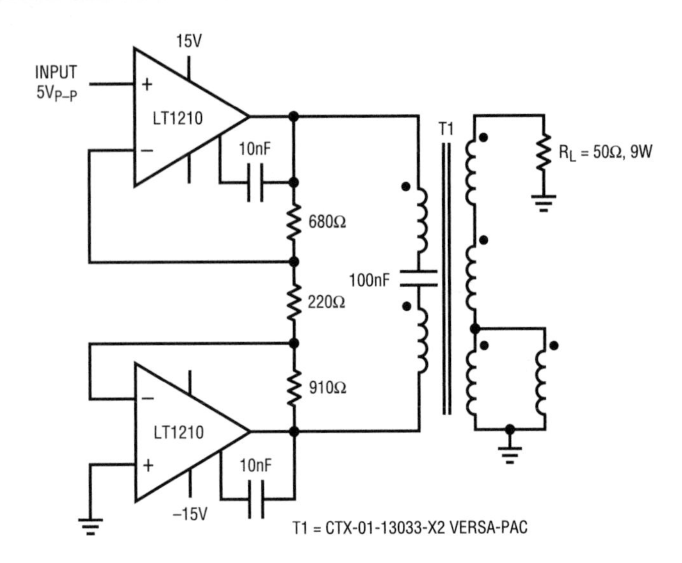

Figure 38.70 • In this Bridge Amplifier, the LT1210 Delivers +39.5dBm (9W) to a 50Ω Load. Power Band Limits Range from 40kHz to 14.5MHz. The Sixth, Otherwise-Unused Winding is Connected in Parallel with One Secondary Winding to Avoid Parasitic Effects Arising from a Floating Winding

the reflected load. This isolates the amplifier from a low impedance short at frequencies below transformer cutoff. In applications where a termination resistor is positioned between the LT1210 amplifier and the transformer, no coupling capacitor is necessary. Note that a low frequency signal, well below the transformer's cutoff frequency, could result in high dissipation in the termination resistor.

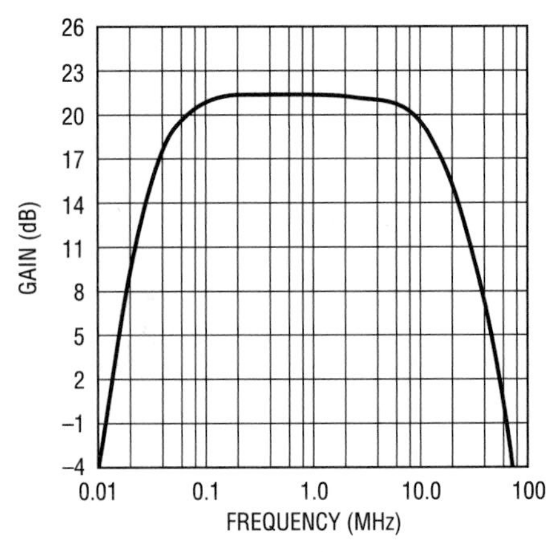

Figure 38.71 • Frequency Response of Figure 38.70's Circuit

Another useful connection for the Versa-Pac transformer is shown in Figure 38.70. A 2:3 transformation presents 11.1Ω to each LT1210 in a bridge, delivering a whopping 9W into 50Ω. In this circuit the lower frequency cutoff was limited by the choice of coupling capacitor to approximately 40kHz (the transformer is capable of 15kHz). The frequency response is shown in Figure 38.71.

Conclusion

The LT1210 combines high output current with a high slew rate to form an effective solution for driving low impedance loads. Power levels of up to 5W can be supplied to a load at frequencies ranging from DC to beyond 10MHz.

The LT1207: an elegant dual 60MHz, 250mA current feedback amplifier

by LTC Applications Staff

Introduction

The LT1207 is a dual version of Linear Technology's LT1206 current feedback amplifier. Each amplifier has 60MHz bandwidth, guaranteed 250mA output current, operates on ±5V to ±15V supply voltages and offers optional external compensation for driving capacitive loads. These features and capabilities combine to make it well suited for such difficult applications as driving cable loads, wide-bandwidth video and high speed digital communication.

LT1088 differential front end

Using thermal conversion, the LT1088 wideband RMS/DC converter is an effective solution for applications such as RMS voltmeters, wideband AGC, RF leveling loops and high frequency noise measurements. Its thermal conversion method achieves vastly wider bandwidth than any other approach. It can handle input signals that have a 300MHz bandwidth and a crest factor of at least 40:1. The thermal technique employed relies on first principles: a wave form's RMS value is defined as its heating value in a load. Another characteristic of the LT1088 is its low impedance inputs (50Ω and 250Ω), common to thermal converters. Though this low impedance represents a difficult load to most drive circuits, the LT1207 can handle it with ease.

Featuring high input impedance and overload protection, the differential input, wideband thermal RMS/DC converter in Figure 38.72 performs true RMS/DC conversion over a 0Hz to 10MHz bandwidth with less than 1% error, independent of input-signal wave shape. The circuit consists of a wideband input amplifier, RMS/DC converter and overload protection.[1] The LT1207 provides high input impedance, gain and output current capability necessary to drive the LT1088's input heater. The 5k/24pF network across the LT1207's 180Ω gain-set resistor is used to adjust a slight peaking characteristic at high frequencies, ensuring 1% flatness at 10MHz. The converter uses matched pairs of heaters and diodes and a control amplifier. R1 produces heat when the LT1207 drives it differentially. This heat lowers D1's voltage. Differentially connected A3 responds by driving R2, heating D2 and closing the loop. A3's DC output directly relates to the input signal's RMS value, regardless of input frequency or wave shape. A4's gain trim compensates residual LT1088 mismatches. The RC network around A3 frequency compensates the loop, ensuring good settling time.

The LT1088 can suffer damage if the 250Ω input is driven beyond $9V_{RMS}$ at 100% duty cycle. An easy remedy to this possibility is to reduce the driver supply voltage. This, however, sacrifices crest factor. Instead, a means of overload protection is included. The LT1018 monitors D1's anode voltage. Should this voltage become abnormally low, A5's output goes low and pulls A6's input low. This causes A6's output to go high, shutting down the LT1207 and eliminating the overload condition. The RC network on A6's input delays the LT1207's reactivation. If the overload condition remains, shutdown is reinstated. This oscillatory action continues, protecting the LT1088 until the overload is corrected. The RMS/DC circuit's 1% error bandwidth and CMRR performance are shown in Figures 38.73 and 38.74, respectively.

CCD clock driver

Charge-coupled-devices (CCDs) are used in many imaging applications, such as surveillance, hand-held and desktop computer video cameras, and document scanners. Using a "bucket-brigade," CCDs require a precise multiphase clock signal to initiate the transfer of light-generated pixel charge from one charge reservoir to the next. Noise, ringing or overshoot on the clock signal must be avoided, since they introduce errors into the CCD output signal. These errors cause aberrations and perturbations in a displayed or printed image.

Two challenges surface in the effort to avoid these error sources when driving a CCD's input. First, CCDs have an input capacitance that varies over a range of 100pF to 2000pF and varies directly with the number of sensing elements (pixels). This presents a high capacitive load to the clock-drive circuitry. Second, CCDs typically require a clock signal whose magnitude is greater than the output capabilities of 5V interfaces and control circuitry. An amplifying filter built around the LT1207 will meet both challenges.

Note 1: Thanks to Jim Williams for this Circuit.

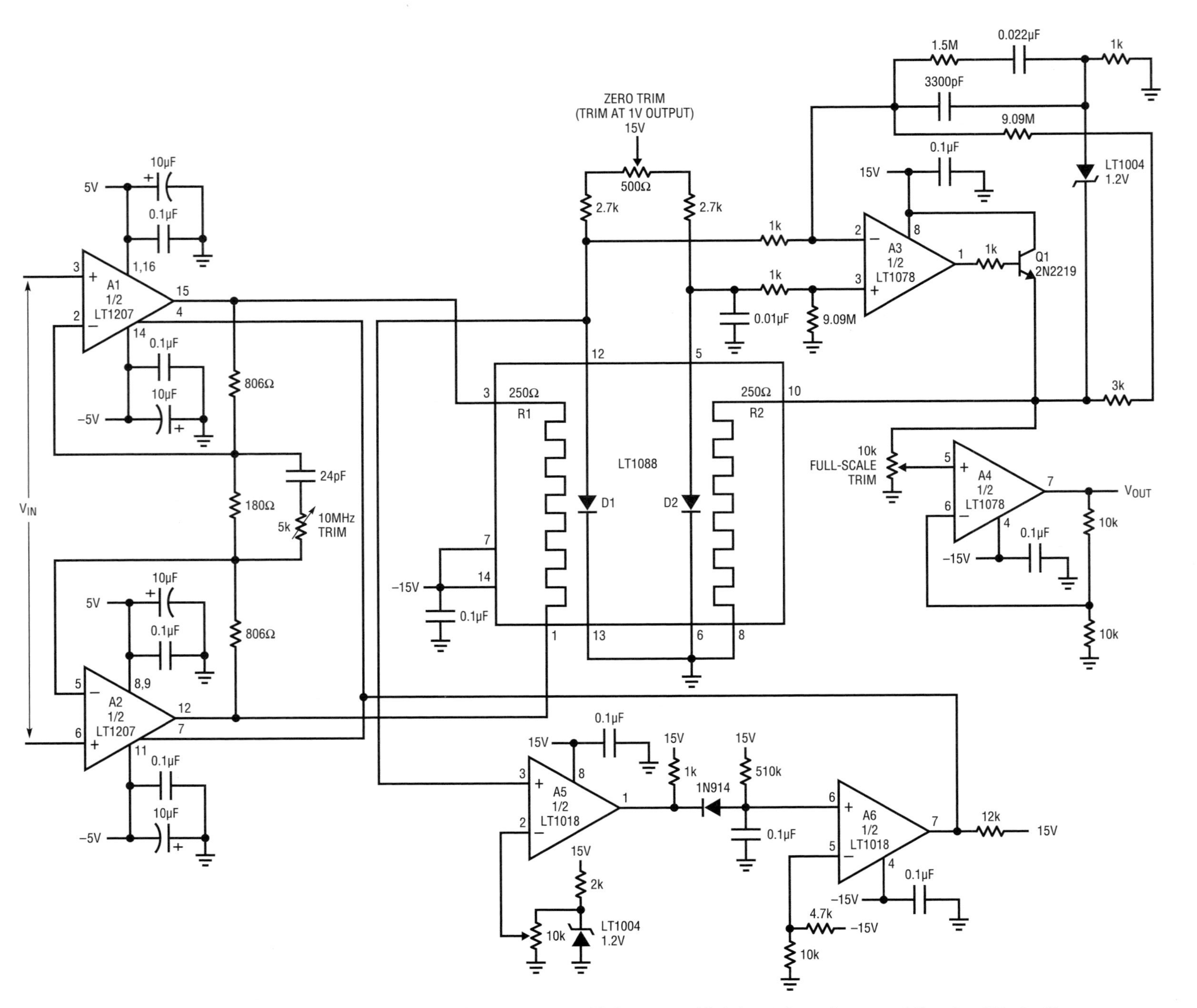

Figure 38.72 • Differential Input 10MHz RMS/DC Converter has 1% Accuracy, High Input Impedance and Overload Protection

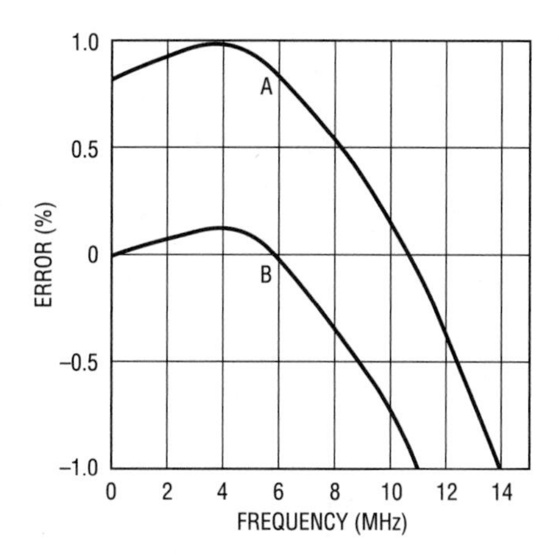

Figure 38.73 • Error Plot for the Differential-Input RMS/DC Converter. Gain Boost at A2 Preserves 1% Accuracy but Causes Slight Peaking before Roll-Off. Boost Can be Set for Maximum Bandwidth (A) or Minimum Error (B)

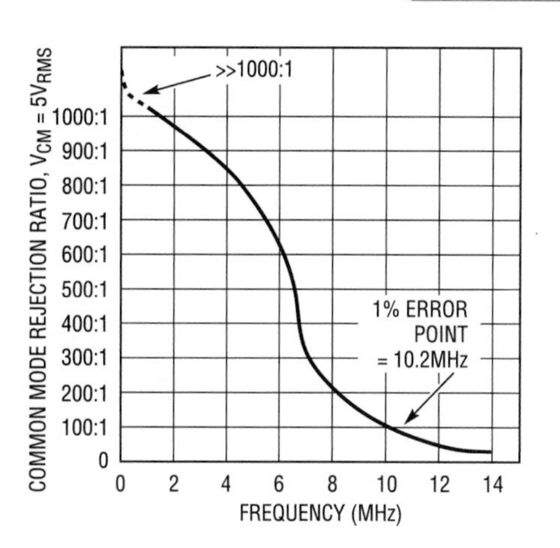

Figure 38.74 • Common Mode Rejection Ratio vs Frequency for the Differential-Input RMS/DC Converter. Layout, Amplifier Bandwidth and AC Matching Characteristics Determine the Curve

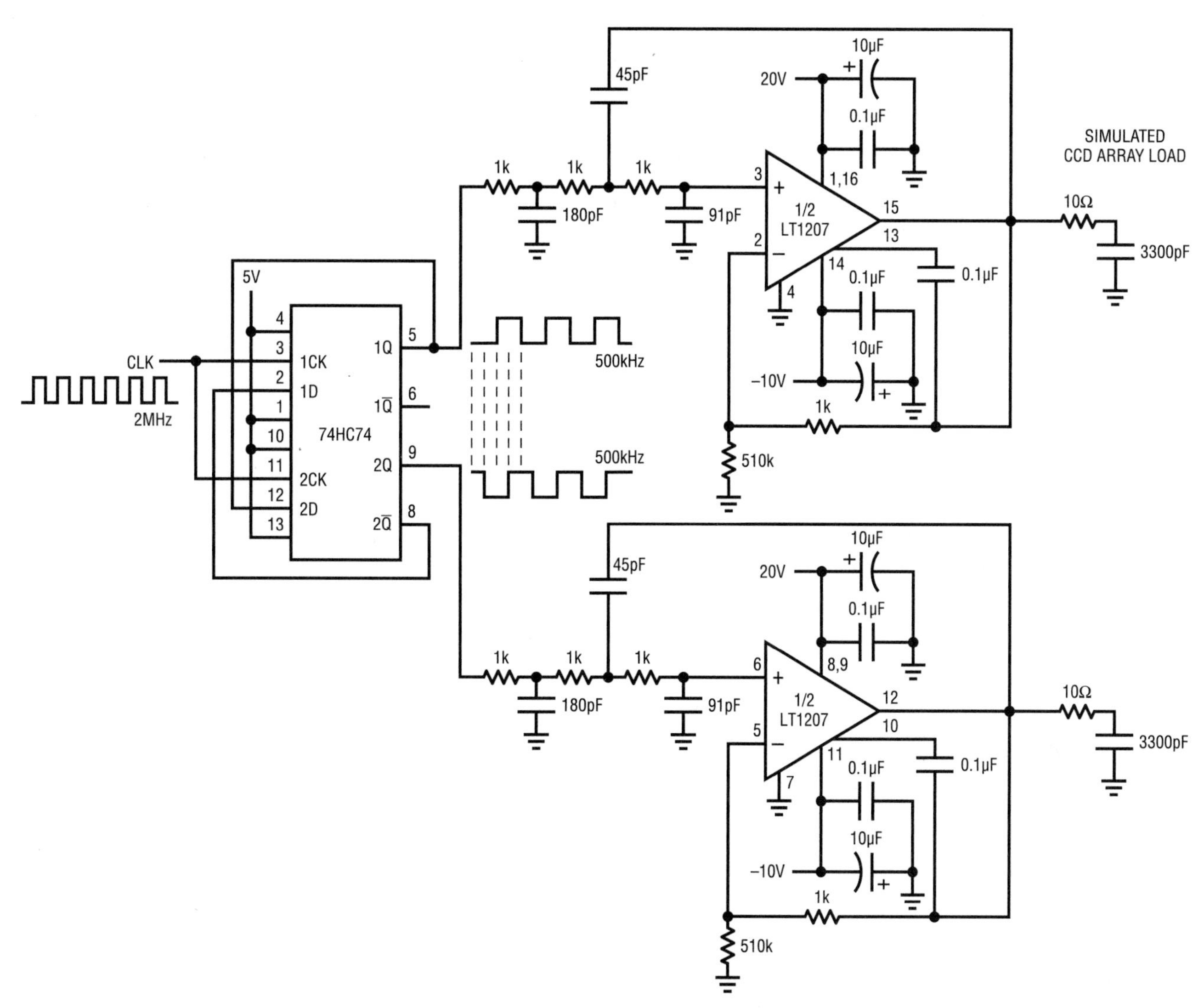

Figure 38.75 • The LT1207 Easily Tames the High Capacitance Loads of CCD Clock Inputs without Ringing or Overshoot

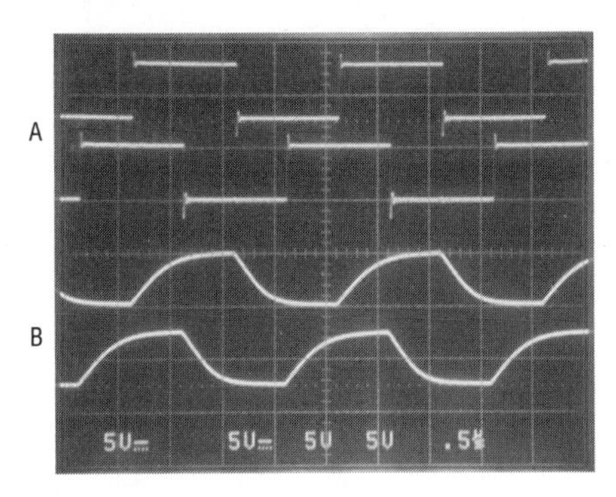

Figure 38.76a • Trace A is the Quadrature Drive Signals. Trace B. is the Voltage at the Input of the Simulated CCD of Figure 38.75, Driven by HC Logic

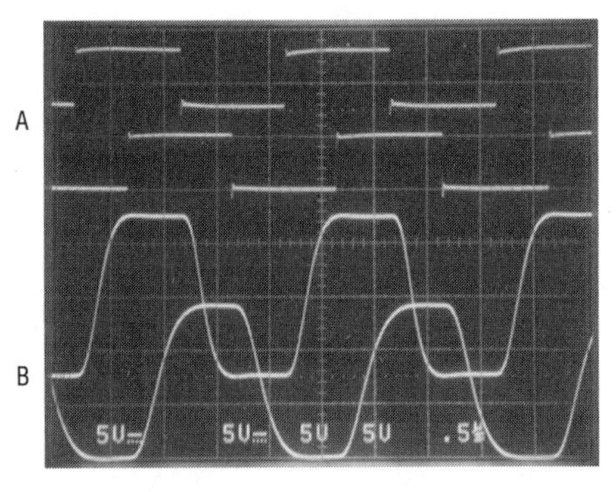

Figure 38.76b • Trace A is the Quadrature Signals. Trace B Shows the Voltage at the Input of the Simulated CCD of Figure 38.75, Driven by the LT1207

Controlling clock signal rise and fall times is one way to avoid ringing or overshoot. This is done by conditioning the clock signal with a nonringing Gaussian filter: The circuit shown in Figure 38.75 uses the LT1207 to filter and amplify control circuitry clock output signals. To reduce ringing and overshoot, each amplifier is configured as a third-order Gaussian lowpass filter with a 1.6MHz cutoff frequency.

Figures 38.76a and 38.76b compare the response of a digital 5V clock-drive signal and the output of the LT1207, each driving a 3300pF load. The digital clock circuit has two major weaknesses that lead to jitter and image distortion.

The CCD's output is changing during charge transfer, producing glitches that decay exponentially. Conversely, the LT1207 circuit's output has a flat top and controlled rise and fall. If an ADC is used to sample a CCD output, the conversion will be much more accurate when the LT1207 circuit is used to clock the pixel changes. With the LT1207's filter configuration, the output has a controlled rise and fall time of approximately 300ns. Ringing and overshoot are absent from the LT1207's output. Wide bandwidth, high output current capability and external compensation allow the LT1207 to easily drive the difficult load of a CCD's clock input.

Micropower, dual and quad JFET op amps feature C-load™ capability and picoampere input bias currents

by Alexander Strong

Introduction

The LT1462/LT1464 duals and the LT1463/LT1465 quads are the first micropower op amps (30μA typical, 40μA maximum per amp for the LT1462; 140μA typical, 200μA maximum per amp for the LT1464) to offer both pico ampere input bias currents (500fA typical) and unity-gain stability for capacitive loads up to 10nF. The outputs can swing a 10k load to within 1.5 volts of either supply. Just like op amps that require an order of magnitude more supply current, the LT1462/LT1463 and the LT1464/LT1465 have open loop gains of 600,000 and 1,000,000, respectively. These unique features, along with a 0.8mV offset, have not been incorporated into a single monolithic amplifier before.

Applications

Figure 38.77 is a track-and-hold circuit that uses a low cost optocoupler as a switch. Leakages for these parts are usually in the nano amp region with 1 to 5 volts across the output. Since there is less than 2mV across the junctions, less than 0.5pA leakage can be achieved for both optocouplers. The input signal is buffered by one op amp while the other buffers the stored voltage; this results in a droop of 50μV/s with a 10nF cap.

Figure 38.78 is a logging photodiode sensor using two LT1462 duals or an LT1463 quad. The low input bias current of the LT1462/LT1463 makes it a natural for amplifying low level signals from high impedance transducers. The 500fA of input bias current contributes only $0.4fA/\sqrt{Hz}$ of current noise. For example, a 1M input impedance converts the noise current to a noise voltage of only $0.4nV/\sqrt{Hz}$. Here, a photodiode converts light to a current, which is converted to a voltage by the first op amp. The first, second and third gain stages are logarithmic amplifiers that perform a logarithmic compression. A DC

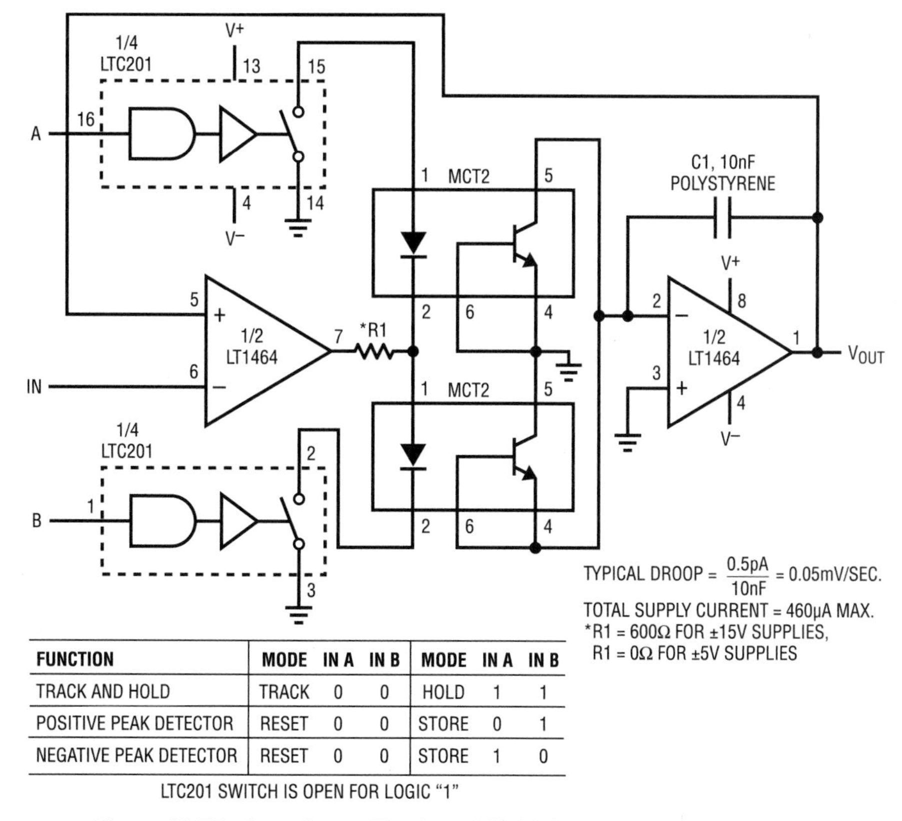

FUNCTION	MODE	IN A	IN B	MODE	IN A	IN B
TRACK AND HOLD	TRACK	0	0	HOLD	1	1
POSITIVE PEAK DETECTOR	RESET	0	0	STORE	0	1
NEGATIVE PEAK DETECTOR	RESET	0	0	STORE	1	0

LTC201 SWITCH IS OPEN FOR LOGIC "1"

Figure 38.77 • Low-Droop Track-and-Hold Circuit/Peak Detector

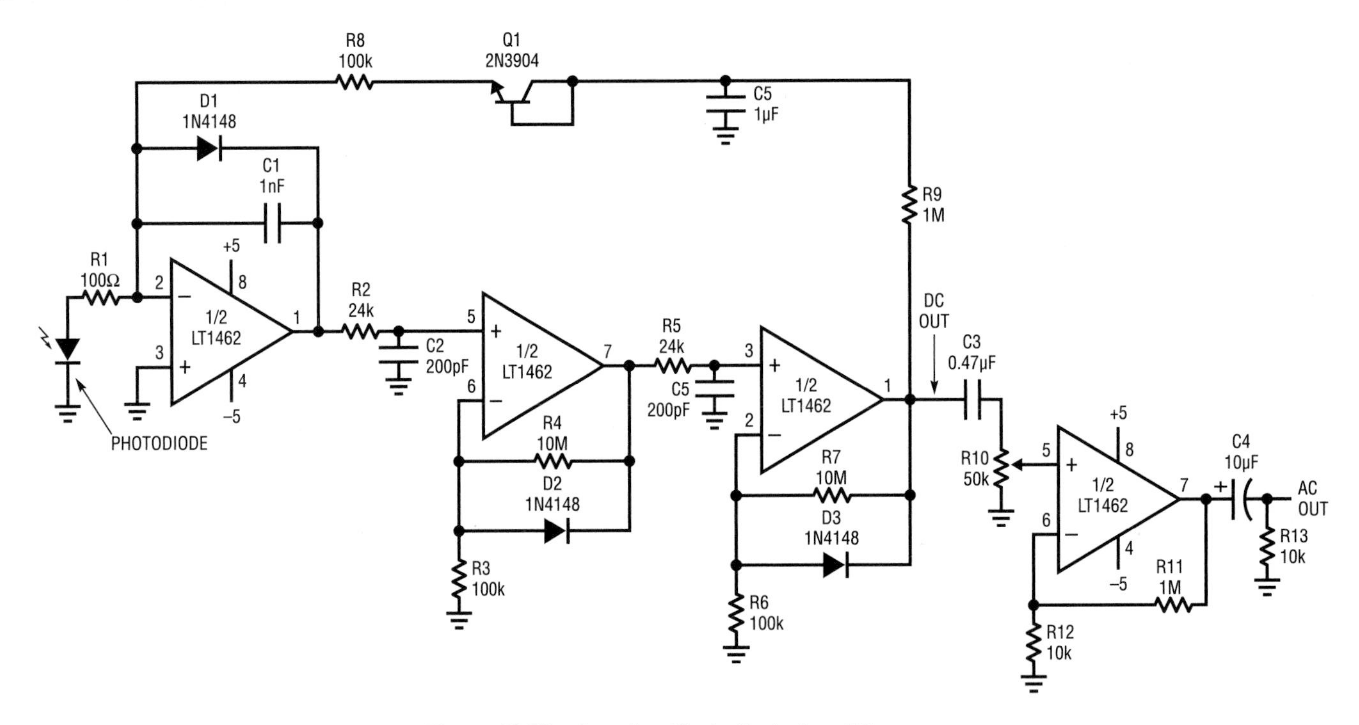

Figure 38.78 • Logging Photodiode Amplifier

feedback path comprising R8, R9, C5 and Q1 is active only for no-light conditions, which are very rare, due to the picoampere sensitivity of the input. Q1 is off when light is present, isolating the photodiode from C5. When the feedback path is needed, a small filtered current through R8 keeps the output of the third op amp within an acceptable range. The third op amp's output voltage, which is proportional to the photodiode current, can serve as a logarithmic DC light meter. Figure 38.79 shows the relationship between DC output voltage and photodiode current. The AC component of the output of third op amp is compressed logarithmically and passed through capacitor C3 and pot R10 for amplitude control. The fourth op amp amplifies this AC signal which is generated across R13. The logarithmic compression of the AC photodiode current allows the user to examine the AC signals for a wide range of input currents.

Conclusion

The LT1462/LT1464 duals and the LT1463/LT1465 quads combine many advantages found in many different op amps, such as low power, (LT1464/LT1465 are 140μA, LT1462/LT1463 are 30μA typical per amplifier), wide input common mode range that includes the positive rail and pico ampere input bias currents. Not only is the output swing specified with 2k and 10k loads, gain is also specified for the same load conditions, which is unheard-of

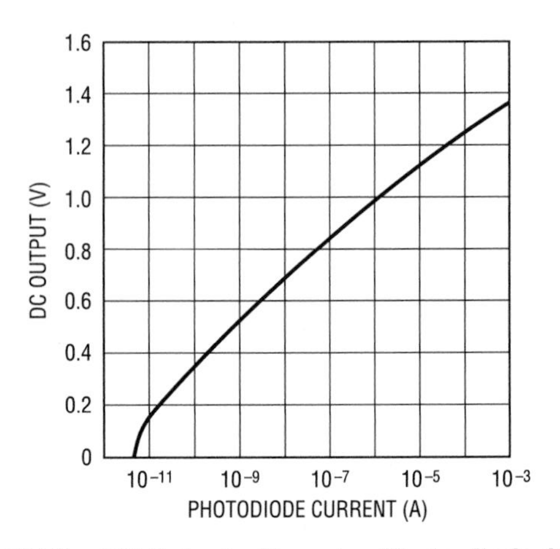

Figure 38.79 • DC Output of Logging Photo diode Amplifier

for micropower op amps. The 1MHz (LT1464/LT1465) or 250kHz (LT1462/LT1463) bandwidth self adjusts to maintain stability for capacitive loads up to 10nF. And don't forget the low 0.8mV offset voltage and DC gains of 1 million (LT1464/LT1465) or 600,000 (LT1462/LT1463) even with 10k loads.

The LT1210: high power op amp yields higher voltage and current

by Dale Eagar

Introduction

The LT1210, a 1 amp current feedback operational amplifier, opens up new frontiers. With 30MHz bandwidth, operation on ±15V supplies, thermal shutdown and 1 amp of output current, this amplifier single-handedly tackles many tough applications. But can it handle output voltages higher than ±15V or currents greater than 1 ampere? This Design Idea features a collection of circuits that open the door to high voltage and high current for the LT1210.

Fast and sassy—telescoping amplifiers

Need ±30V? Cascading LT1210's will get you there. This circuit (Figure 38.80) will provide the ±30V at ±1A and has 13MHz of full-power bandwidth (see Figure 38.81). How does it work? The first LT1210 drives the "ground" of the second LT1210 subcircuit, effectively raising and lowering it while the second LT1210 further amplifies the input signal. This telescoping arrangement can be cascaded with additional stages to get more than ±30V This amplifier is stable into capacitive loads, is short-circuit protected and thermally shuts down when overheated.

Extending power supply voltages

Another method of getting high voltage from an amplifier is the extended-supply mode (see "Extending Op Amp Supplies to Get More Voltage"; *Linear Technology* Volume IV Number 2 (June 1994), pp. 20-22). This involves steering two external regulators with the power supply pins of an op amp to get a high voltage amplifier.

Figure 38.82 shows the LT1210 connected in the extended- supply mode. Placing an amplifier in the extended-supply mode requires changing the return of the compensation node from the power supply pins to system ground. R9 and C5 are selected for clean step response. The process of relocating the return of the compensation node slows the amplifier down to approximately 1MHz (see Figure 38.83).

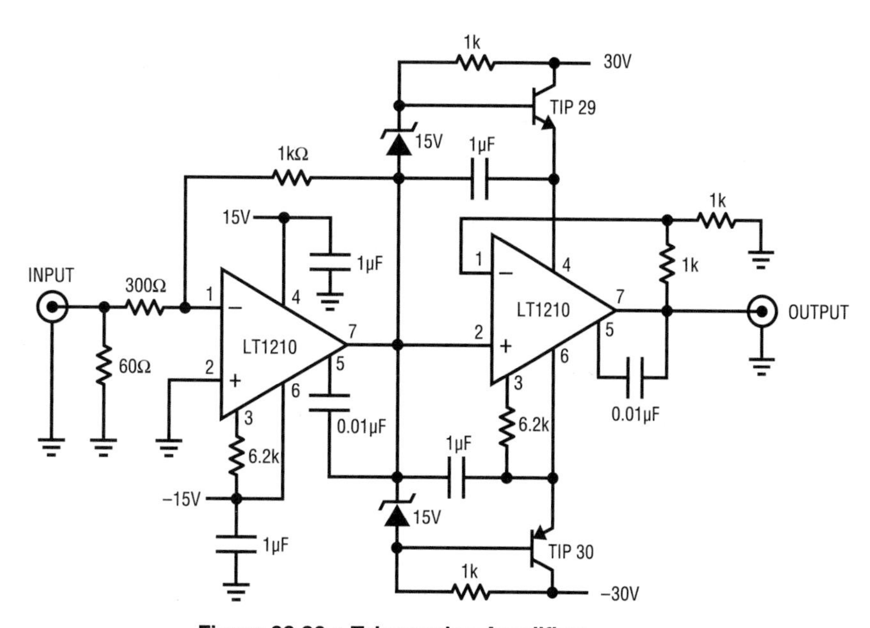

Figure 38.80 • Telescoping Amplifiers

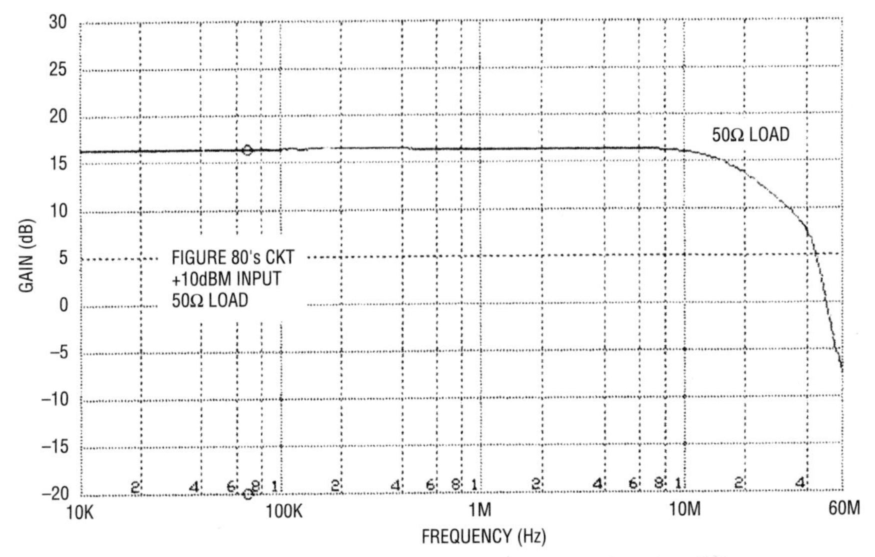

Figure 38.81 • Gain vs Frequency Plot of Telescoping Amplifier

Figure 38.82's circuit will provide ±1A at ±100V is stable into capacitive loads and is short-circuit protected. The two external MOSFETs need heat sinking.

Gateway to the stars

The circuit of Figure 38.82 can be expanded to yield much higher voltages; the first and most obvious way is to use higher voltage MOSFETs. This causes two problems: first, high voltage P-channel MOSFETs are hard to get; second, and more importantly, at ±1A the power dissipated by the MOSFETs is too high for single packages. The solution is to build telescoping regulators, as shown in Figure 38.84. This circuit can provide ±1A of current at ±200V and has the additional power-dissipation ability of four MOSFETs.

Boosting output current

The current booster detailed in Figure 38.85 illustrates a technique for amplifying the output current capability of an op amp while maintaining speed. Among the many niceties of this topology is the fact that both Q1 and Q2 are normally off and thus consume no quiescent current. Once the load current reaches approximately 100mA, Q1 or Q2 turns on, providing additional drive to the output. This transition is seamless to the outside world and takes advantage of the full speed of Q1 and Q2. This circuit's small-signal bandwidth and full-power bandwidth are shown in Figure 38.86.

Boosting both current and voltage

The current-boosted amplifier shown in Figure 38.85 can be used to replace the amplifiers in Figure 38.80, yielding ±10A at ±30V Placing the boosted amplifier in the circuits shown in Figures 38.82 or 38.84 will yield peak powers into the kilowatts.

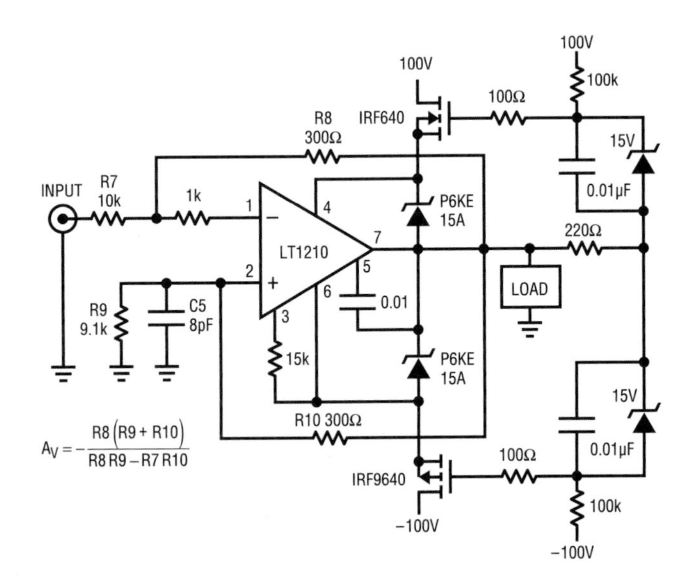

Figure 38.82 • ±100V, ±1A Power Driver

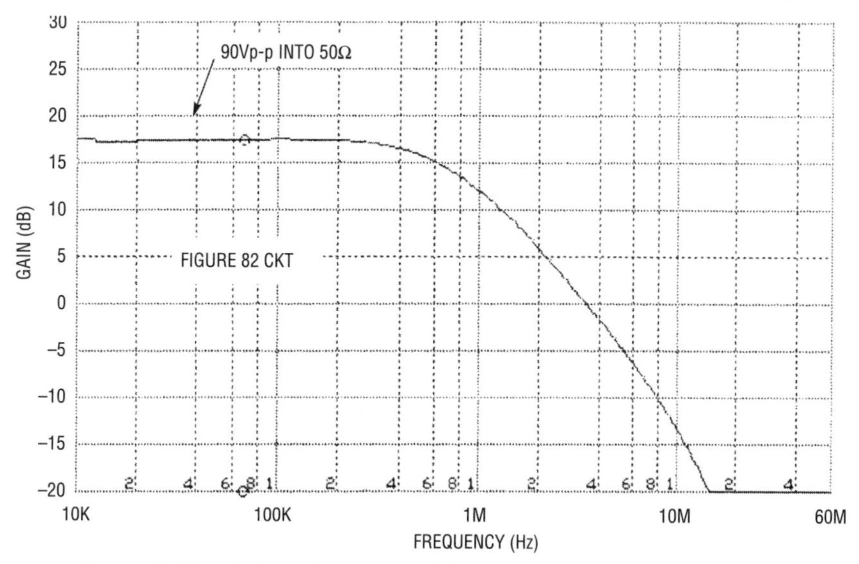

Figure 38.83 • Gain vs Frequency Plot of Extended-Supply Amplifier

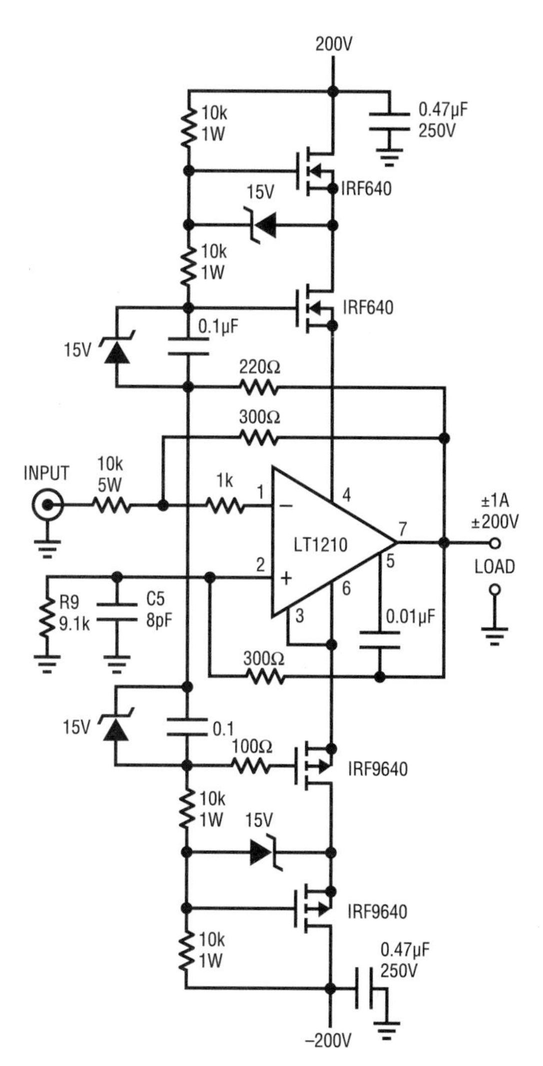

Figure 38.84 • Cascode Power Amplifier

Thermal management

When the LT1210 is used with external transistors to increase its output voltage and/or current range an additional benefit can often be realized: system thermal shutdown. Careful analysis of the thermal design of the system can coordinate the overtemperature shutdown of the LT1210 with the junction temperatures of the external transistors. This essentially extends the umbrella of protection of the LT1210's thermal shutdown to cover the external transistors. The thermal shutdown of the LT1210 activates when the junction temperature reaches 150°C and has about 10°C hysteresis. The thermal resistance $R_{\theta JC}$ of the TO-220 package (LT1210CY) is 5°C/Watt).

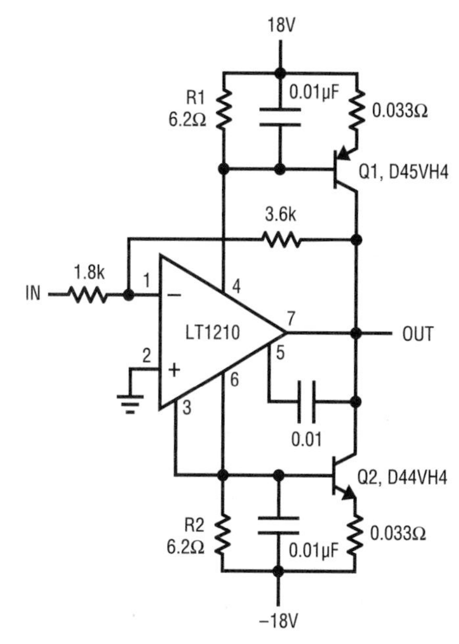

Figure 38.85 • ±10A/1MHz Current-Boosted Power Op Amp

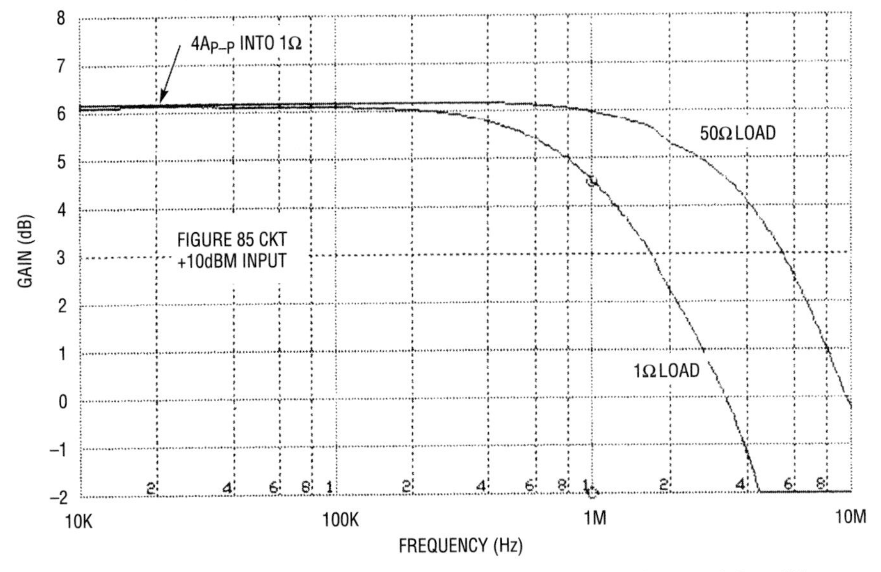

Figure 38.86 • Gain vs Frequency Response of Current-Boosted Amplifier

Summary

The LT1210 is a great part; its performance in terms of speed, output current and output voltage is unsurpassed. Its C-Load™ output drive and thermal shutdown allow it to take its place in the real world—no kid gloves are required here. If the generous output specification of the LT1210 isn't big enough for your needs, just add a couple of transistors to dissipate the additional power and you are on your way. Only the worldwide supply of transistors limits the amount of power you could command with one of these parts.

New rail-to-rail amplifiers: precision performance from micropower to high speed

by William Jett and Danh Tran

Introduction

Linear Technology's latest offerings expand the range of rail-to-rail amplifiers with precision specifications. Rail-to-rail amplifiers present an attractive solution for signal conditioning in many applications. For battery-powered or other low voltage circuitry, the entire supply voltage can be used by both input and output signals, maximizing the system's dynamic range. Circuits that require signal sensing near the positive supply are straightforward using a rail-to-rail amplifier.

Applications

The ability to accommodate any input or output signal that falls within the amplifier supply range makes these amplifiers very easy to use. The following applications demonstrate the versatility of the family of amplifiers.

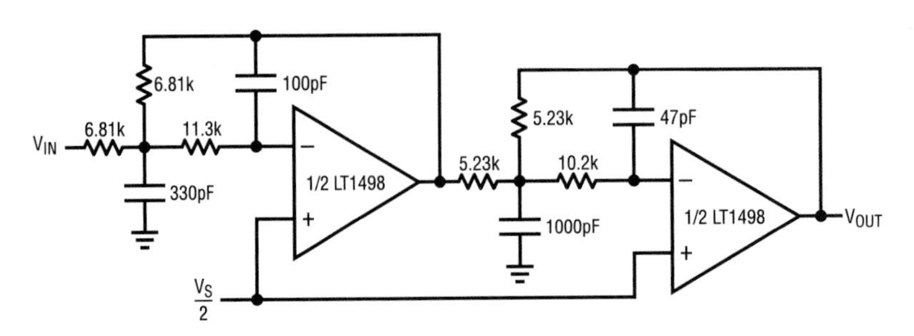

Figure 38.87 • 100kHz 4th Order Butterworth Filter

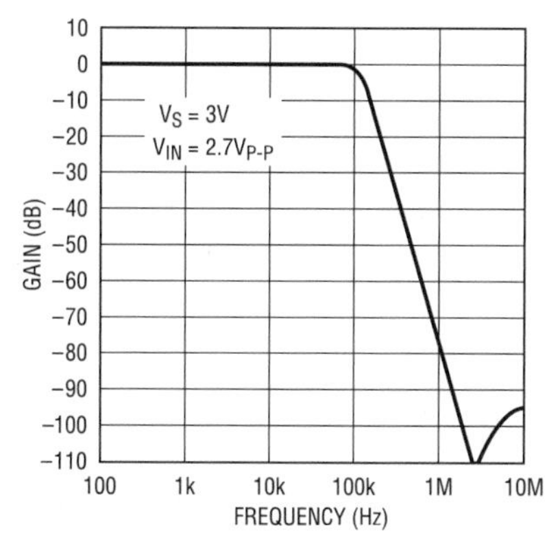

Figure 38.88 • Filter Frequency Response

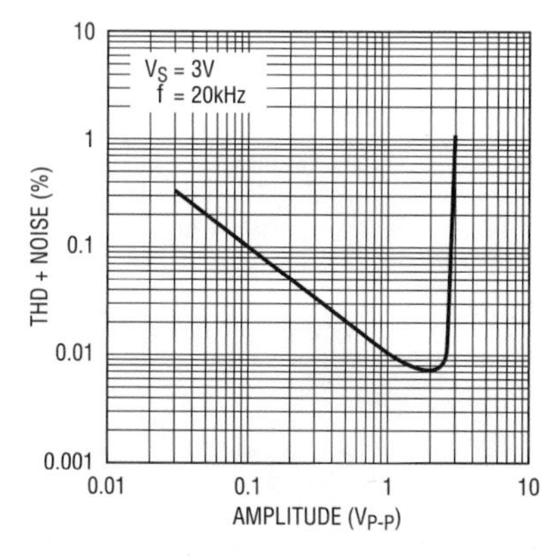

Figure 38.89 • Filter Distortion vs Amplitude

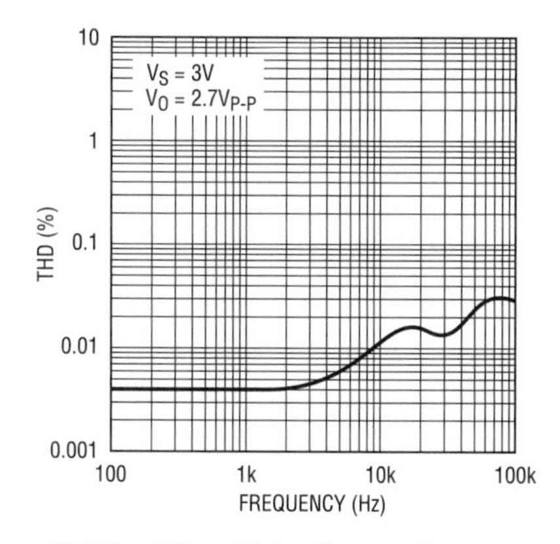

Figure 38.90 • Filter Distortion vs Frequency

100kHz 4th order Butterworth filter for 3V operation

The filter shown in Figure 38.87 uses the low voltage operation and wide bandwidth of the LT1498. Operating in the inverting mode for lowest distortion, the output swings rail-to-rail. The graphs in Figures 38.88– 38.90 display the measured lowpass and distortion characteristics with a 3V power supply. As seen from the graphs, the distortion with a $2.7V_{P\text{-}P}$ output is under 0.03% for frequencies up to the cutoff frequency of 100kHz. The stop band attenuation of the filter is greater than 90dB at 10MHz.

Multiplexer

A buffered MUX with good offset characteristics can be constructed using the shutdown feature of the LT1218. In shutdown, the output of the LT1218 assumes a high impedance, so the outputs of two devices can be tied together (wired OR, as they say in the digital world). As shown in Figure 38.91, the shutdown pins of each LT1218 are driven by a 74HC04 buffet: The LT1218 is active with the shutdown pin high. The photo in Figure 38.92 shows the switching characteristics with a 1kHz sine wave applied to one input and the other input tied to ground. As shown, each amplifier is connected for unity gain, but either amplifier or both could be configured for gain.

Conclusion

The latest members of LTC's family of rail-to-rail amplifiers expand the versatility of rail-to-rail operation to micropower and high speed applications. The devices maintain precision

Vos specifications over the entire rail-to-rail input range and have open loop gains of one million or more. These characteristics, combined with low voltage operation, makes for truly versatile amplifiers.

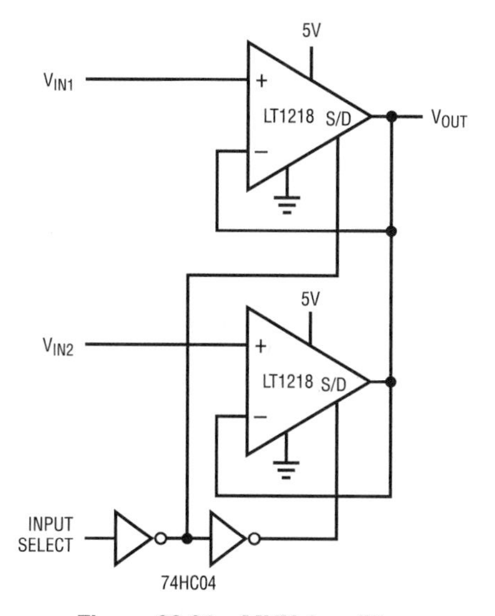

Figure 38.91 • MUX Amplifier

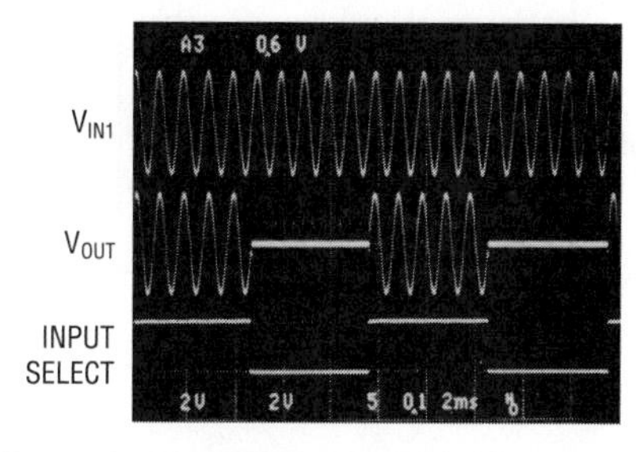

Figure 38.92 • MUX Amplifier Waveforms

LT1256 voltage-controlled amplitude limiter

by Frank Cox

Amplitude-limiting circuits are useful where a signal should not exceed a predetermined maximum amplitude, such as when feeding an A/D or a modulator. A clipper, which completely removes the signal above a certain level, is useful for many applications, but there are times when it is not desirable to lose information. For instance, when video signals have amplitude peaks that exceed the dynamic range of following processing stages, simply clipping the peaks at the maximum level will result in the loss of all detail in the areas where clipping takes place. Often these well illuminated areas are the primary subject of the scene. Because these peaks usually correspond to the highest level of luminosity, they are referred to as "highlights." One way to preserve some of the detail in the highlights is to automatically reduce the gain (compress) at high signal levels.

The circuit in Figure 38.93 is a voltage-controlled breakpoint amplifier that can be used for highlight compression. When the input signal reaches a predetermined level (the breakpoint), the amplifier gain is reduced. As both the breakpoint and the gain for signals greater than the breakpoint are voltage programmable, this circuit is useful for systems that adapt to changing signal levels. Adaptive highlight compression finds use in CCD video cameras, which have a very large dynamic range. Although this circuit was developed for video signals, it can be used to adaptively compress any signal within the 40MHz bandwidth of the LT1256.

The LT1256 video fader is connected to mix proportional amounts of input signal and clipped signal to provide a voltage-controlled variable gain. The clipped signal is provided by a discrete circuit consisting of three transistors. Q1 acts as an emitter follower until the input voltage exceeds the voltage on the base of Q2 (the breakpoint voltage or V_{BP}). When the input voltage is greater than V_{BP}, Q1 is off and Q2 clamps the emitters of the two transistors to V_{BP} plus a V_{BE}. Q3, an NPN emitter follower, buffers the output and drops the voltage a V_{BE} and thus the DC level of the input signal is preserved to the extent allowed by the V_{BE} matching and temperature tracking of

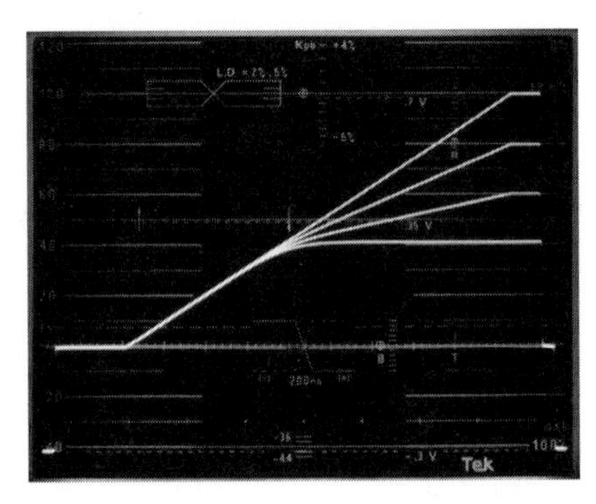

Figure 38.94 • Multiple-Exposure Photograph of a Single Line of Monochrome Video, Showing Four Different Levels of Compression

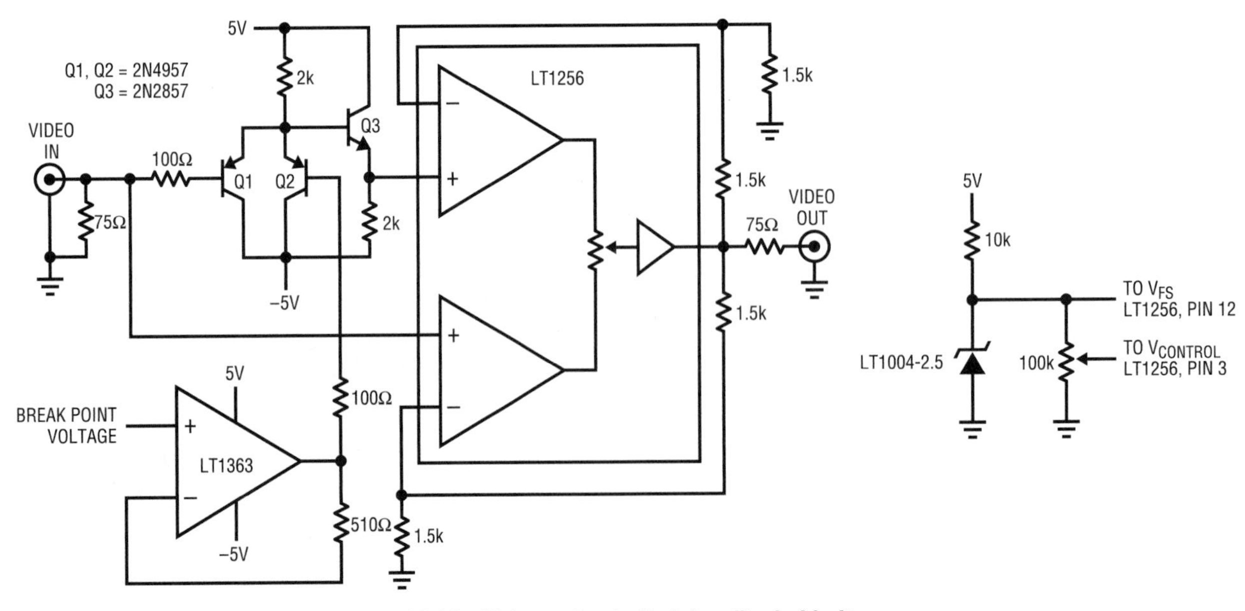

Figure 38.93 • Voltage-Controlled Amplitude Limiter

the transistors used. The breakpoint voltage at the base of Q2 must remain constant when this transistor is turning on or the signal will be distorted. The LT1363 maintains a low output impedance well beyond video frequencies and makes an excellent buffer.

Figure 38.94 is a multiple-exposure photograph of a single line of monochrome video, showing four different levels of compression ranging from fully limited signal to unprocessed input signal. The breakpoint is set to 40% of the peak amplitude to clearly show the effect of the circuit; normally only the top 10% of video would be compressed.

The LT1495/LT1496: 1.5μA rail-to-rail op amps

by William Jett

Introduction

Micropower rail-to-rail amplifiers present an attractive solution for battery-powered and other low voltage circuitry. Low current is always desirable in battery-powered applications, and a rail-to-rail amplifier allows the entire supply range to be used by both the inputs and the output, maximizing the system's dynamic range. Circuits that require signal sensing near either supply rail are easier to implement using rail-to-rail amplifiers. However; until now, no amplifier combined precision offset and drift specifications with a maximum quiescent current of 1.5μA.

Operating on a minuscule 1.5μA per amplifier, the LT1495 dual and LT1496 quad rail-to-rail amplifiers consume almost no power while delivering precision performance associated with much higher current amplifiers.

The LT1495/LT1496 feature "Over-The-Top" operation: the ability to operate normally with the inputs above the positive supply. The devices also feature reverse-battery protection.

Applications

The ability to accommodate any input or output signal that falls within the amplifier supply range makes the LT1495/LT1496 very easy to use. The following applications highlight signal processing at low currents.

Nanoampere meter

A simple 0nA-200nA meter operating from two flashlight cells or one lithium battery is shown in Figure 38.95. The readout is taken from a 0μA-200μA, 500Ω analog meter; the LT1495 supplies a current gain of 1000 in this application. The op amp is configured as a floating I-to-I converter: It consumes only 3μA when not in use, so there is no need for an on/off switch. Resistors R1, R2 and R3 set the current gain. R3 provides a ±10% full-scale adjust for the meter movement. With a 3V supply, maximum current in the meter is limited by R2 + R3 to less than 300μA, protecting the movement. Diodes D1 and D2 and resistor R4 protect the inputs from faults up to 200V Diode currents are below 1nA in normal operation, since the maximum voltage across the diodes is 375μV the V_{OS} of the LT1495. C1 acts to stabilize the amplifier, compensating for capacitance between the inverting input and ground. The unused amplifier should be connected as shown for minimum supply current. Error terms from the amplifier (base currents, offset voltage) sum to less than 0.5% over the operating range, so the accuracy is limited by the analog meter movement.

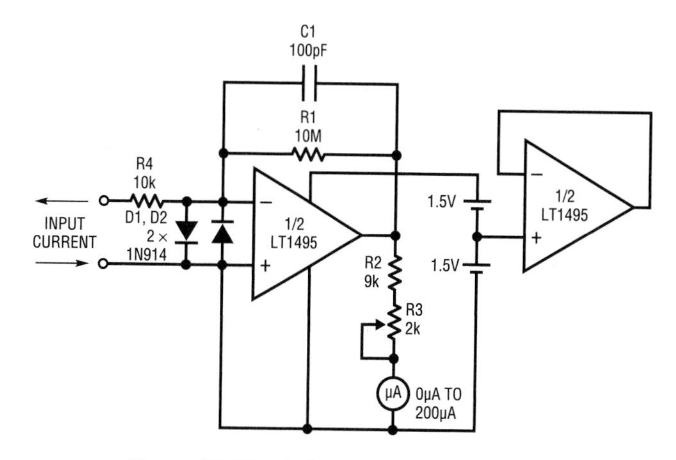

Figure 38.95 • 0nA–200nA Current Meter

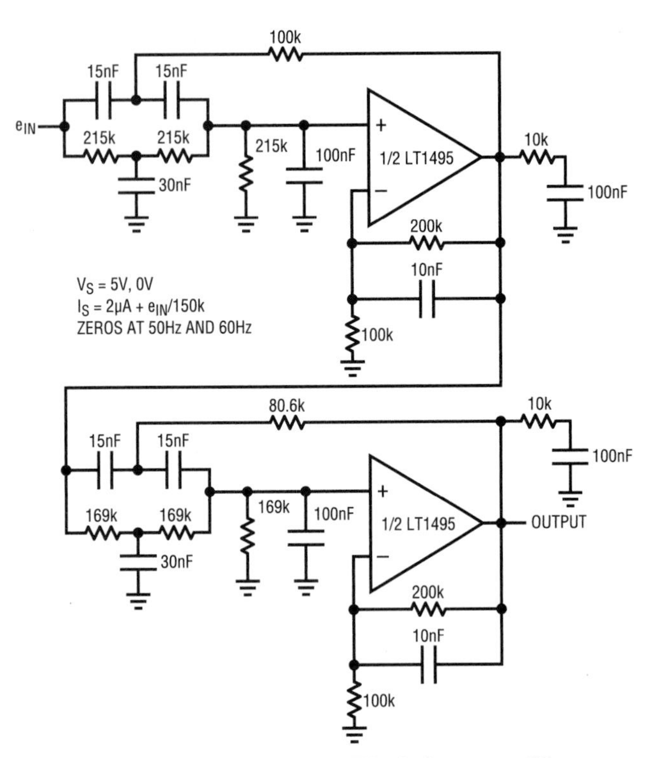

Figure 38.96 • 6th Order 10Hz Elliptic Lowpass Filter

6th order, 10Hz elliptic lowpass filter

Figure 38.96 shows a 6th order, 10Hz elliptic lowpass filter with zeros at 50Hz and 60Hz. Supply current is primarily determined by the DC load on the amplifiers and is approximately $2\mu A + V_O/150k$ ($9\mu A$ for $V_O = 1V$). The overall frequency response is shown in Figure 38.97. The notch depth of the zeros at 50Hz and 60Hz is nearly 60dB and the stopband attenuation is greater than 40dB out to 1kHz. As with all RC filters, the filter characteristics are determined by the absolute values of the resistors and

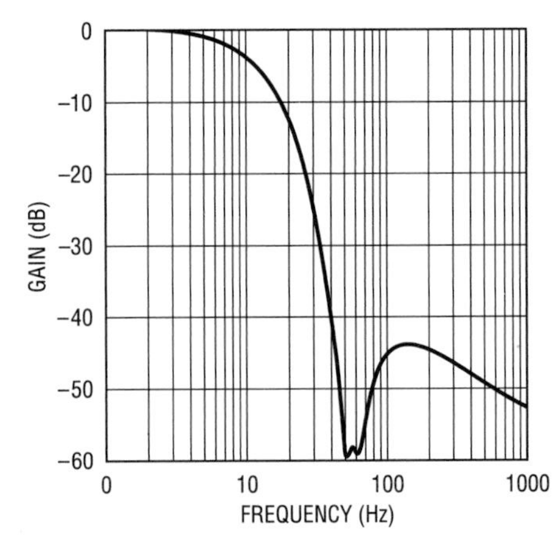

Figure 38.97 • Frequency Response of Figure 38.96's 6th Order Elliptic Lowpass Filter

capacitors, so resistors should have a 1% tolerance or better and capacitors a 5% tolerance or better.

Battery-current monitor with Over-The-Top operation

The bidirectional current sensor shown in Figure 38.98 takes advantage of the extended common mode range of the LT1495 to sense currents into and out of a 12V battery while operating from a 5V supply. During the charge cycle, op amp A1 controls the current in Q1 so that the voltage drop across R_A is equal to $I_L \cdot R_{SENSE}$. This voltage is then amplified at the charge output by the ratio of R_A to R_b. During this cycle, amplifier A2 sees a negative offset, which keeps Q2 off and the discharge output low. During the discharge cycle, A2 and Q2 are active and operation is similar to that during the charge cycle.

Conclusion

The LT1495/LT1496 extends Linear Technology's range of rail-to-rail amplifier solutions to a truly micropower level.

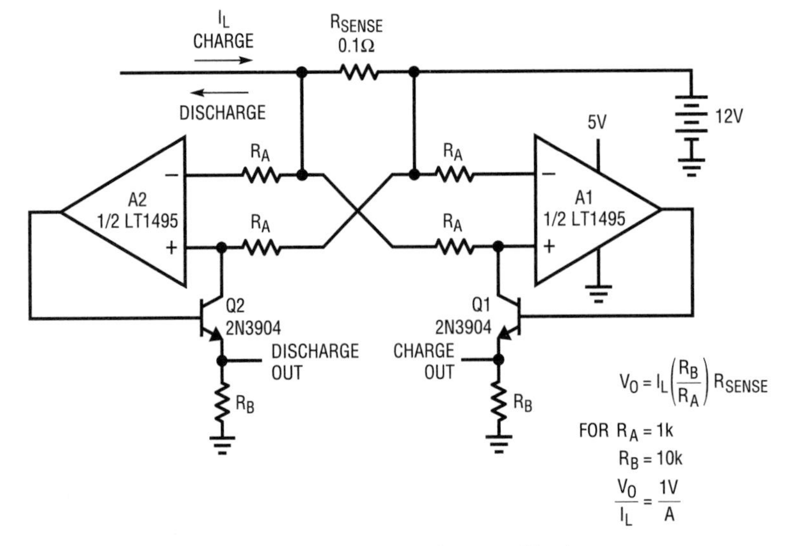

Figure 38.98 • Battery-Current Monitor

The combination of extremely low current and precision specifications provides designers with a versatile solution for battery-operated devices and other low power systems.

Send camera power and video on the same coax cable

by Frank Cox

Because remotely located video surveillance cameras do not always have a ready source of power, it is convenient to run both the power and the video signal through a single coax cable. One way to do this is to use an inductor to present a high impedance to the video and a low impedance to the DC. The difficulty with this method is that the frequency spectrum of a monochrome video signal extends down to at least 30Hz. The composite color video spectrum goes even lower, with components at 15Hz. This implies a rather large inductor. For example, a 0.4H inductor has an impedance of only 75Ω at 30Hz, which is about the minimum necessary. Large inductors have a large series resistance that wastes power. More importantly, large inductors can have a significant amount of parasitic capacitance and stand a good chance of going into self resonance below the 4MHz video bandwidth and thus corrupting the signal. The circuit shown in Figure 38.99 takes a different approach to the problem by using all active components.

The circuitry at the monitor end of the coax cable supplies all the power to the system. U1, an LT1206 current feedback amplifier, forms a gyrator or synthetic inductor. The gyrator isolates the low impedance power supply from the cable by maintaining a reasonably high impedance over the video bandwidth while, at the same time, contributing only 0.1Ω of series resistance. This op amp needs to have enough bandwidth for video and sufficient output drive to supply 120mA to the camera. The selected part has a guaranteed output current of 250mA and a 3dB bandwidth of 60MHz, making it a good fit. Because the video needs to be capacitively coupled, there is no need for split supplies; hence a single 24V supply is used. The 24V supply also gives some headroom for the voltage drop in long cable runs.

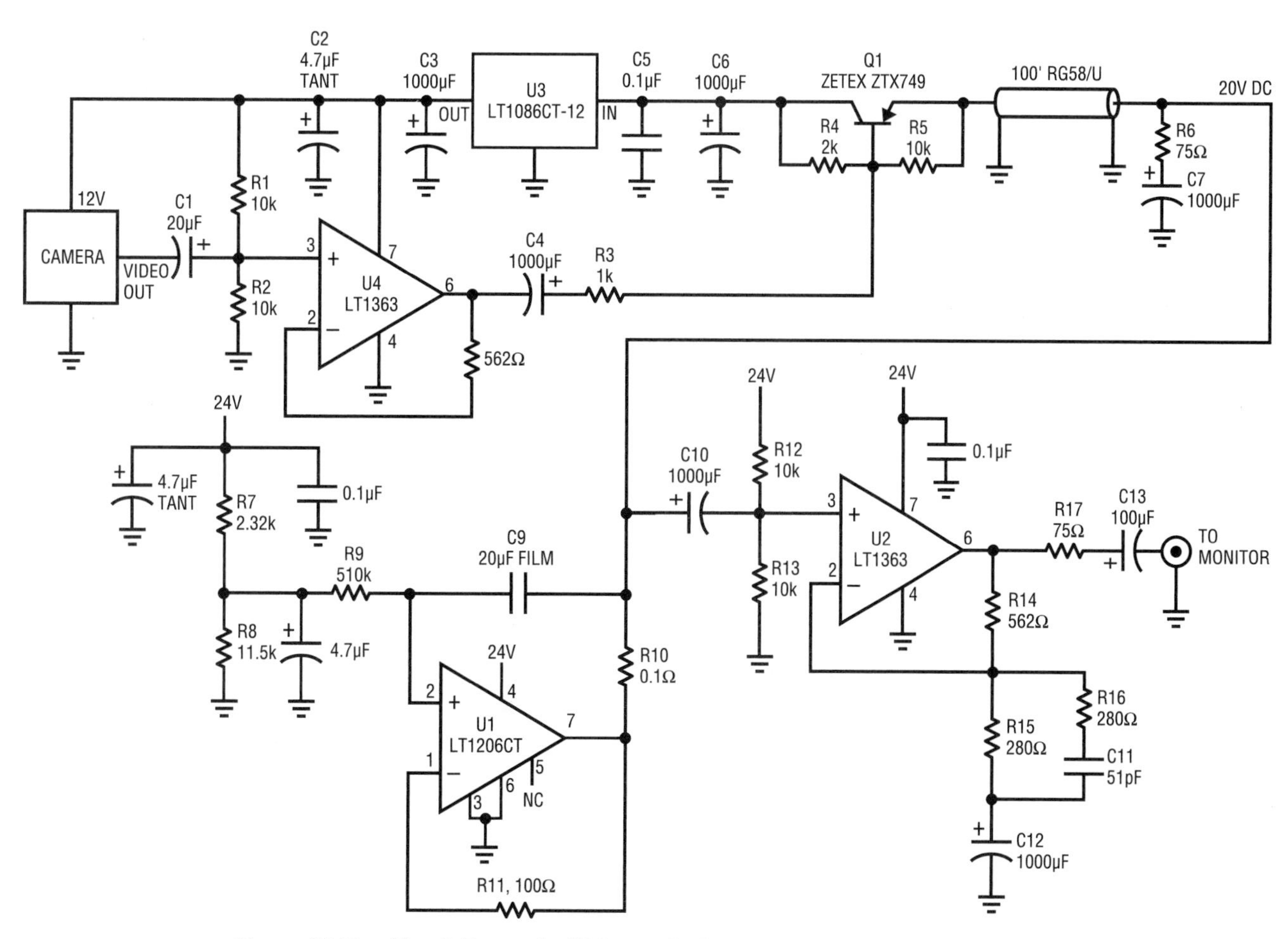

Figure 38.99 • Circuit Transmits Video and 12V Power on the Same Coax Cable

The camera end has an LT1086 fixed 12V regulator (U3) to supply 12V to a black and white CCD video camera. U4, an LT1363 op amp, supplies the drive for Q1, a fast, high current transistor. Q1, in turn, modulates the video on the 20V DC. The collector of Q1 is the input to the 12V regulator. This point is AC ground because it is well bypassed as required by U3. U1 is set up to deliver 20V to the cable. Because the 12V regulator in the camera end needs 1.5V of dropout voltage, the balance of 6.5V can be dropped in the series resistance of the cable. The output of the LT1206 is set to 20V to give headroom between the supply and the video.

U2, another LT1363 video-speed op amp, receives video from the cable, supplies some frequency equalization and drives the cable to the monitor. Equalization is used to compensate for high frequency roll off in the camera cable. The components shown (R16, C11) gave acceptable monochrome video with 100 feet of RG58/U cable.

200μA, 1.2MHz rail-to-rail op amps have Over-The-Top inputs

by Raj Ramchandani

Introduction

The LT1638 is Linear Technology's latest general-purpose, low power dual rail-to-rail operational amplifier; the LT1639 is a quad version. The circuit topology of the LT1638 is based on Linear Technology's popular LT1490/LT1491 op amps, with substantial improvements in speed. The LT1638 is five times faster than the LT1490.

Battery current monitor

The battery-current monitor shown in Figure 38.100 demonstrates the LT1639's ability to operate with its inputs above the positive rail. In this application, a conventional amplifier would be limited to a battery voltage between 5V and ground, but the LT1639 can handle battery voltages as high as 44V The LT1639 can be shut down by removing V_{CC}. With V_{CC} removed, the input leakage is less then

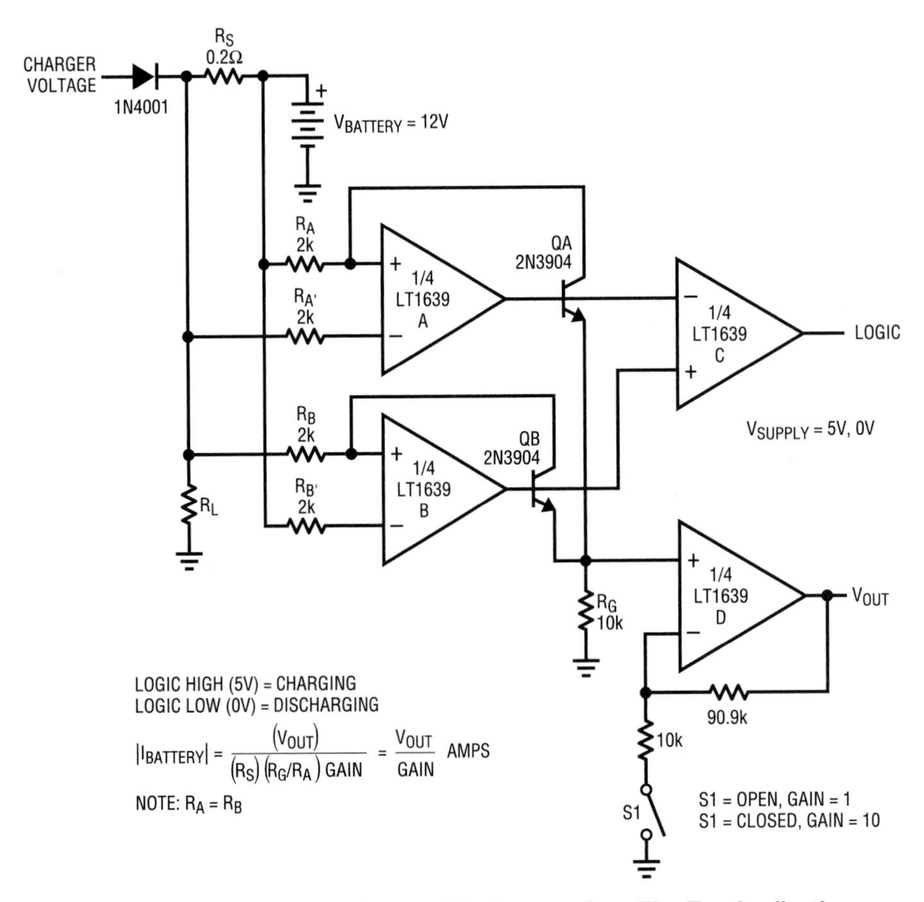

Figure 38.100 • LT1639 Battery Current Monitor—an Over-The-Top Application

0.1nA. No damage to the LT1639 will result from inserting the 12V battery backward.

When the battery is charging, amplifier B senses the voltage drop across R_S. The output of amplifier B causes Q_B to drain sufficient current through R_B to balance the inputs of amplifier B. Likewise, amplifier A and Q_A form a closed loop when the battery is discharging. The current through Q_A or Q_B is proportional to the current in R_S. This current flows into RG and is converted into a voltage. Amplifier D buffers and amplifies the voltage across R_G. Amplifier C compares the outputs of amplifier A and amplifier B to determine the polarity of current through R_S. The scale factor for V_{OUT} with S1 open is 1V/A. With S1 closed the scale factor is 1V/100mA and currents as low as 5mA can be measured.

Low distortion rail-to-rail op amps have 0.003% THD with 100kHz signal

by Danh Tran

Introduction

The LT1630/LT1632 duals and LT1631/LT1633 quads are the newest members of Linear Technology's family of rail-to-rail op amps, which provide the best combination of Ac performance and Dc precision over the widest range of supply voltages. The LT1630/LT1631 deliver a 30MHz gain-bandwidth product, a 10V/μs slew rate and $6nV/\sqrt{Hz}$ input-voltage noise. Optimized for higher speed applications, the LT1632/LT1633 have a 45MHz gain-bandwidth product, a 45V/μs slew rate and $12nV/\sqrt{Hz}$ input voltage noise.

Applications

The ability to accommodate any input and output signals that fall within the device' s supplies makes these amplifiers very easy to use. They exhibit a very good transient response and can drive low impedance loads, which makes them suitable for high performance applications. The following applications demonstrate the versatility of these amplifiers.

400kHz 4th order Butterworth filter for 3V operation

The circuit shown in Figure 38.101 makes use of the low voltage operation and the wide bandwidth of the LT1630 to create a 400kHz 4th order lowpass filter with a 3V supply. The amplifiers are configured in the inverting mode for the lowest distortion and the output can swing rail-to-rail for the maximum dynamic range. Figure 38.102 displays the frequency response of the filter Stopband attenuation is greater than 85dB at 10MHz. With a $2.25V_{P-P}$, 100kHz input signal, the filter has harmonic distortion products of less than −87dBc.

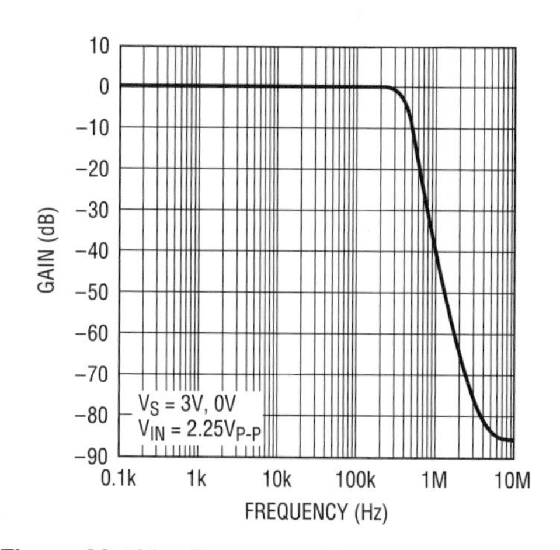

Figure 38.102 • Frequency Response of Filter in Figure 38.101

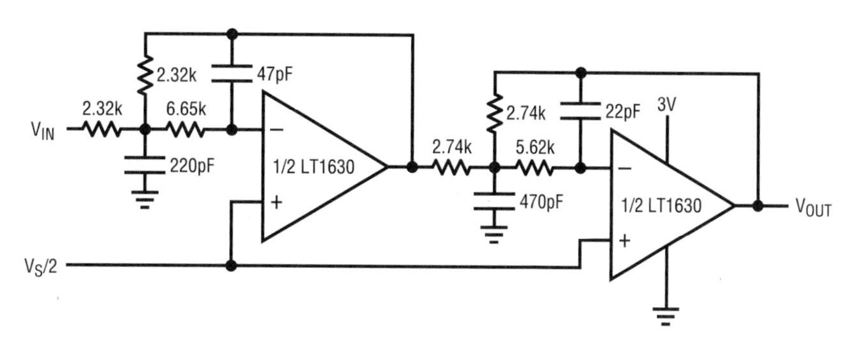

Figure 38.101 • Single-Supply, 400kHz, 4th Order Butterworth Filter

40dB gain, 550kHz instrumentation amplifier

An instrumentation amplifier with a rail-to-rail output swing, operating from a 3V supply, can be constructed with the LT1632, as shown in Figure 38.103. The amplifier has a nominal gain of 100, which can be adjusted with resistor R5. The DC output level is equal to the input voltage (V_{IN}) between the two inputs multiplied by the gain of 100. Common mode range can be calculated by the equations shown with Figure 38.103. For example, the common mode range is from 0.15V to 2.65V if the output voltage is at one-half of the 3V supply. The common mode rejection is greater than 110dB at 100Hz when trimmed with resistor R1. Figure 38.103 shows the amplifier's cutoff frequency of 550kHz.

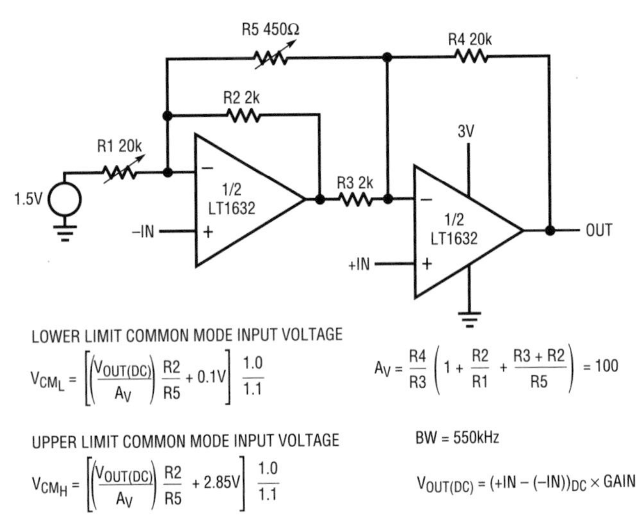

$$V_{CML} = \left[\left(\frac{V_{OUT(DC)}}{A_V}\right)\frac{R2}{R5} + 0.1V\right]\frac{1.0}{1.1}$$

$$V_{CMH} = \left[\left(\frac{V_{OUT(DC)}}{A_V}\right)\frac{R2}{R5} + 2.85V\right]\frac{1.0}{1.1}$$

$$A_V = \frac{R4}{R3}\left(1 + \frac{R2}{R1} + \frac{R3 + R2}{R5}\right) = 100$$

$$V_{OUT(DC)} = (+IN - (-IN))_{DC} \times GAIN$$

Figure 38.103 • Single-Supply Instrumentation Amplifier

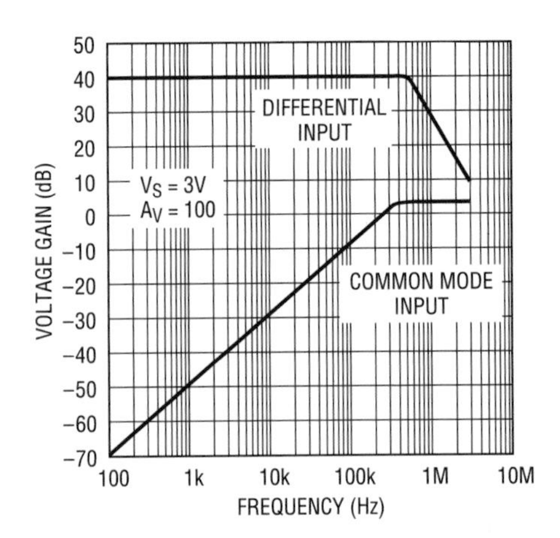

Figure 38.104 • Frequency Response of Figure 38.103's Instrumentation Amplifier

The LT1167: precision, low cost, low power instrumentation amplifier requires a single gain-set resistor

by Alexander Strong

Introduction

The LT1167 is the next-generation instrumentation amplifier designed to replace the previous generation of monolithic instrumentation amps, as well as discrete, multiple op amp solutions. Instrumentation amplifiers differ from operational amplifiers in that they can amplify input signals that are not ground referenced. The output of an instrumentation amplifier is referenced to an external voltage that is independent of the input. Conversely, the output voltage of an op amp, due to the nature of its feedback, is referenced to the differential and common mode input voltage.

Applications

Single-supply pressure monitor

The LT1167's low supply current, low supply voltage operation and low input bias current (350pA max) allow it to fit nicely into battery powered applications. Low overall power dissipation necessitates using higher impedance bridges. Figure 38.105 shows the LT1167 connected to a 3kΩ bridge's differential output. The picoampere input bias currents will still keep the error caused by offset current to a negligible level. The LT1112 level shifts the LT1167's reference pin and the ADC's analog ground pins above ground. This is necessary in single-supply applications because the output cannot swing to ground. The LT1167's and LT1112's combined power dissipation is still less than the bridge's. This circuit's total supply current is just 3mA.

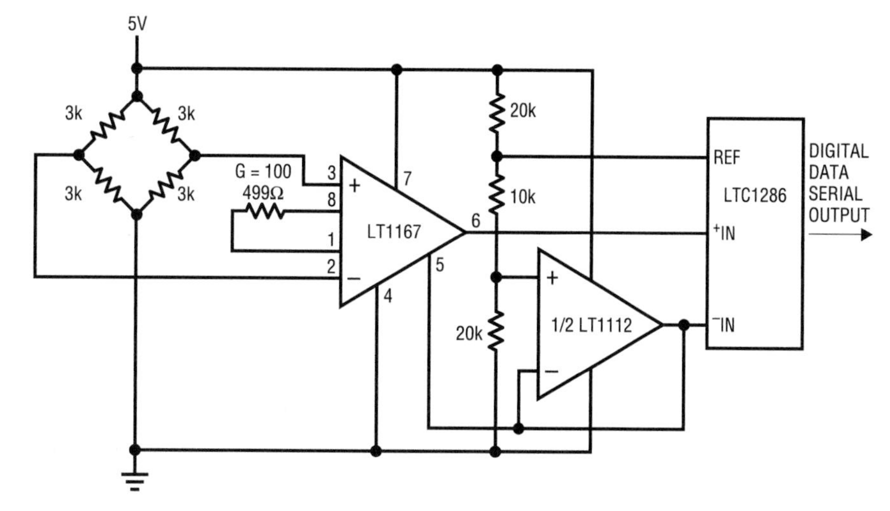

Figure 38.105 • Single-Supply Pressure Monitor

ADC signal conditioning

The LT1167 is shown in Figure 38.106 changing a differential signal into a single-ended signal. The single-ended signal is then filtered with a passive 1st order RC lowpass filter and applied to the LTC1400 12-bit analog-to-digital converter (ADC). The LT1167's output stage can easily drive the ADC's small nominal input capacitance, preserving signal integrity. Figure 38.107 shows two FFTs of the amplifier/ADC's output. Figures 38.107a and 38.107b show the results of operating the LT1167 at unity gain and a gain of ten, respectively. This results in a typical SINAD of 70.6dB.

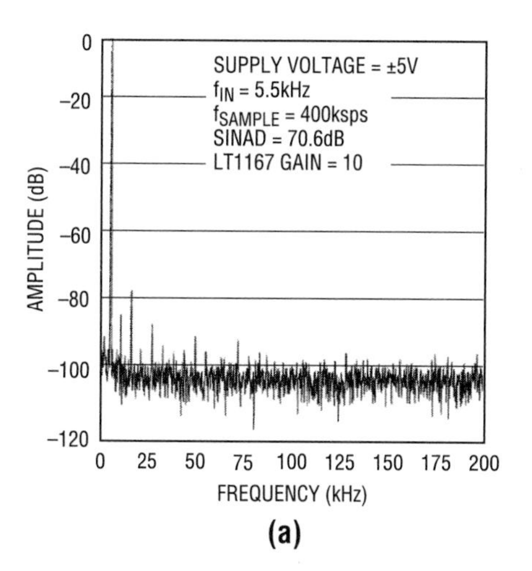

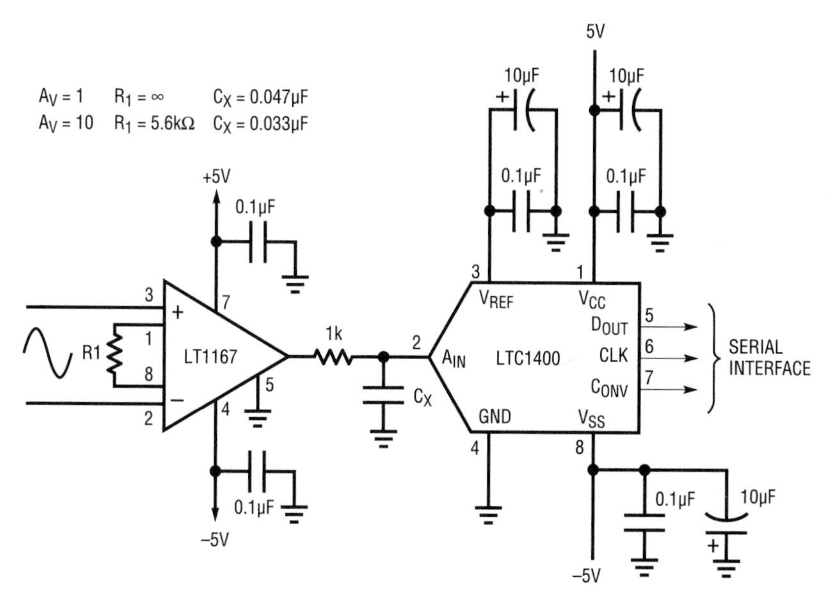

Figure 38.106 • The LT1167 Converting Differential Signals to Single- Ended Signals; the LT1167 is Ideal for Driving the LTC1400

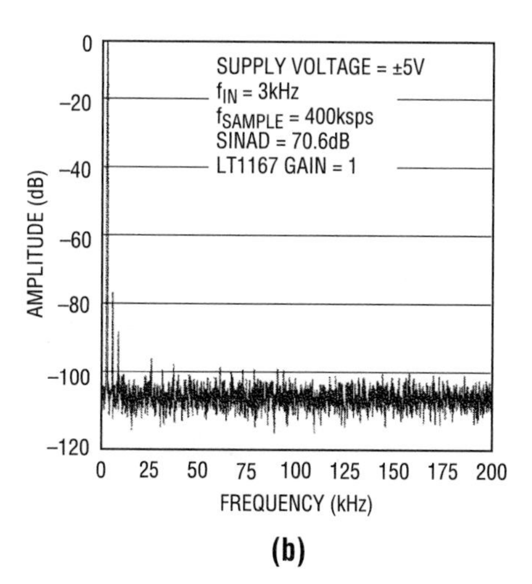

Figure 38.107 • Operating at a Gain of One (a) or Ten (b), Figure 38.106's Circuit Achieves 12-Bit Operation with a SINAD of 70.6dB

Current source

Figure 38.108 shows a simple, accurate, low power programmable current source. The differential voltage across pins 2 and 3 is mirrored across R_G. The voltage across R_G is amplified and applied across R1, defining the output current. The 50μA bias current flowing from pin 5 is buffered by the LT1464 JFET operational amplifier, which increases the resolution of the current source to 3pA.

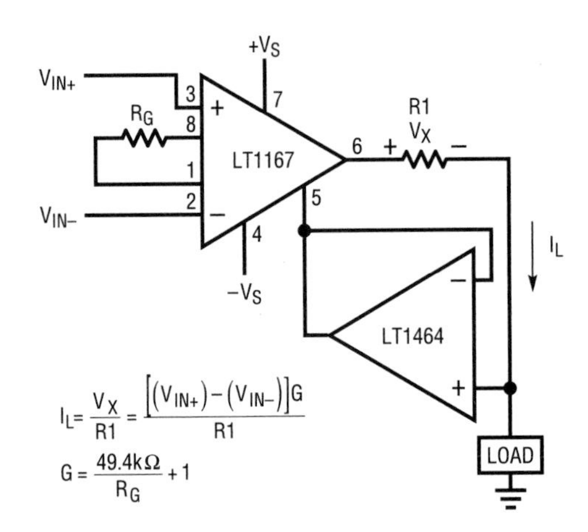

Figure 38.108 • Precision Current Source

Nerve-impulse amplifier

The LT1167's low current noise makes it ideal for ECG monitors that have MΩ source impedances. Demonstrating the LT1167's ability to amplify low level signals, the circuit in Figure 38.109 takes advantage of the amplifier's high gain and low noise operation. This circuit amplifies the low level nerve impulse signals received from a patient at pins 2 and 3 of the LT1167. R_G and the parallel combination of R3 and R4 set a gain of ten. The potential on LT1112's pin 1 creates a ground for the common mode signal. The LT1167's high CMRR of 110db ensures that the desired differential signal is amplified and unwanted common mode signals are attenuated. Since the DC portion of the signal is not important, R6 and C2 make up a 0.3Hz highpass filter. The AC signal at LT1112's pin 5 is amplified by a gain of 101 set by R7/R8 + 1. The parallel combination of C3 and R7 forms a lowpass filter that decreases this gain at frequencies above 1kHz.

The ability to operate at ±3V on 0.9mA of supply current makes the LT1167 ideal for battery-powered applications. Total supply current for this application is 1.7mA. Proper safeguards, such as isolation, must be added to this circuit to protect the patient from possible harm.

Conclusion

The LT1167 instrumentation amplifier delivers the best precision, lowest noise, highest fault tolerance, plus the ease of use provided by single-resistor gain setting. The LT1167 is offered in 8-pin PDIP and SO packages. The SO uses significantly less board space than discrete designs.

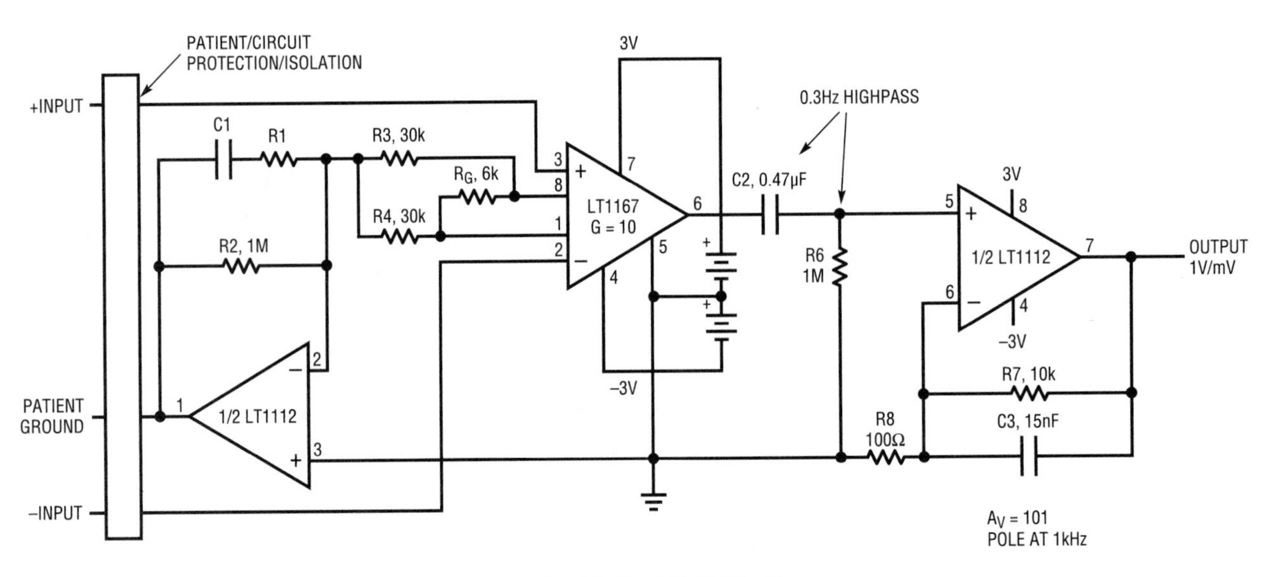

Figure 38.109 • Medical ECG Monitor

Level shift allows CFA video amplifier to swing to ground on a single supply

by Frank Cox

A current feedback (CFA) video amplifier can be made to run off a single supply and still amplify ground-referenced video with the addition of a simple and inexpensive level shifter. The circuit in Figure 38.110 is an amplifier and cable driver for a current output video DAC. The video can be composite or component but it must have sync. The single positive supply is 12V but could be as low as 6V for the LT1227.

The output of the LT1227 CFA used here can swing to within 2.5V of the negative supply with a 150Ω load over the commercial temperature range of 0°C to 70°C. Five diodes in the feedback loop are used, in conjunction with C5, to level shift the output to ground. The video from the output of the LT1227 charges C5 and the voltage across it allows the output to swing to ground or even slightly negative. However, the level of this negative swing will depend on the video signal and so will be unpredictable. When the scene is black, there must be sync on the video for C5 to remain charged. A zero-level component video signal with no sync will not work with this circuit. The CFA output will try to go to zero, or as low as it can, and the diodes will turn off. The load will be disconnected from the CFA output and connected through the feedback resistor to the network of R6 and R7. This causes about 150mVDC to appear at the output, instead of the 0V that should be there.

The ground-referenced video signal at the input needs to be level shifted into the input common mode range of the LT1227 (3V above negative supply). R4 and R5 shift the input signal to 3V In the process, the input video is attenuated by a factor of 2.5. For correct gain, no offset and with a zero source impedance, R4 would be 1.5k. To compensate for the presence of R3, R4 is made 1.5k minus R3, or 1.46k. The trade off is a gain error of about 1.5%. If R4 is left 1.5k, the gain is correct, but there is an offset error of 75mV. R6, R7 and R8 set the gain and the output offset of the amplifier. A noninverting gain of five is taken to compensate for the attenuation in the input level shifter and the cable termination.

The voltage offset on the output of this circuit is a rather sensitive function of the value of the input resistors. For instance, an error of 1% in the value of R6 will cause an offset of 30mV (1% of 3V) on the output. This is in addition to the offset error introduced by the op amp. Precision resistor networks are available (BI Technologies, 714-447-2345) with matching specifications of 0.1% or

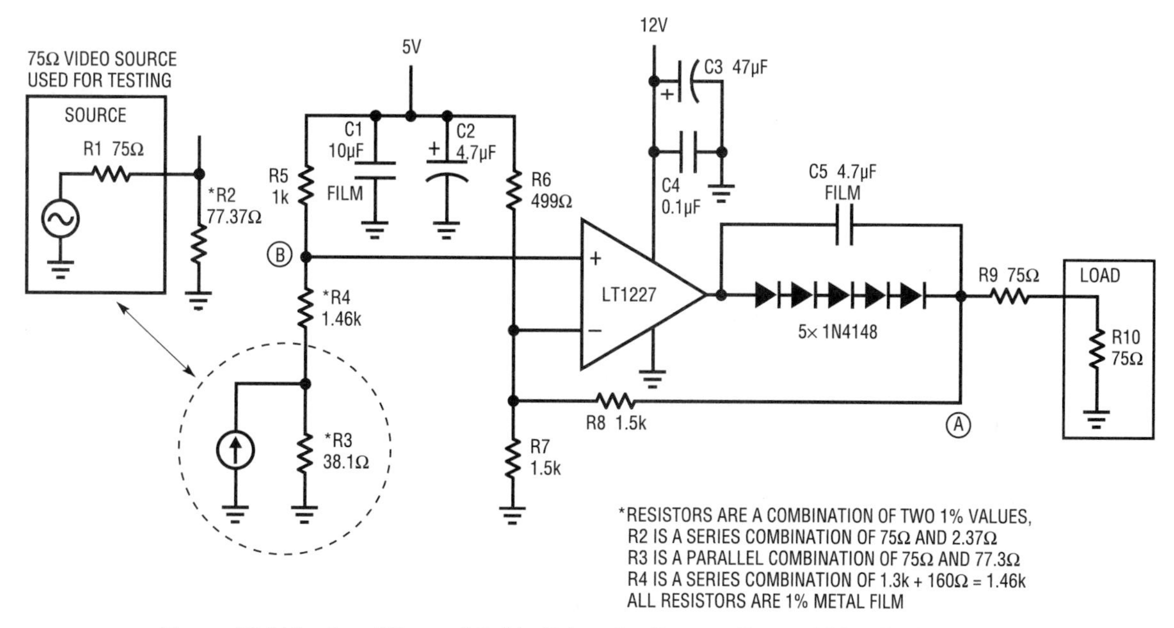

Figure 38.110 • Amplifier and Cable Driver for Current-Output Video DAC

better. These could be used for the level shifting resistors, although this would make adjustments like the one made to R4 difficult.

Fortunately, there is always synchronization information associated with video. A simple circuit can be used to DC restore voltage offsets produced by resistor mismatch, op amp offset or DC errors in the input video. Figure 38.111 shows the additional circuitry needed to perform this function. The LTC201A analog switch and C1 store the offset error during blanking. The clamp pulse should be 3μs or wider and should occur during blanking. It can conveniently be made by delaying the sync pulse with one shots. If the sync tip is clamped, the clamp pulse must start after and end before the sync pulse or offset errors will be introduced. The integrator made with the LT1632 adjusts the voltage at point B (see Figure 38.110) to correct the offset.

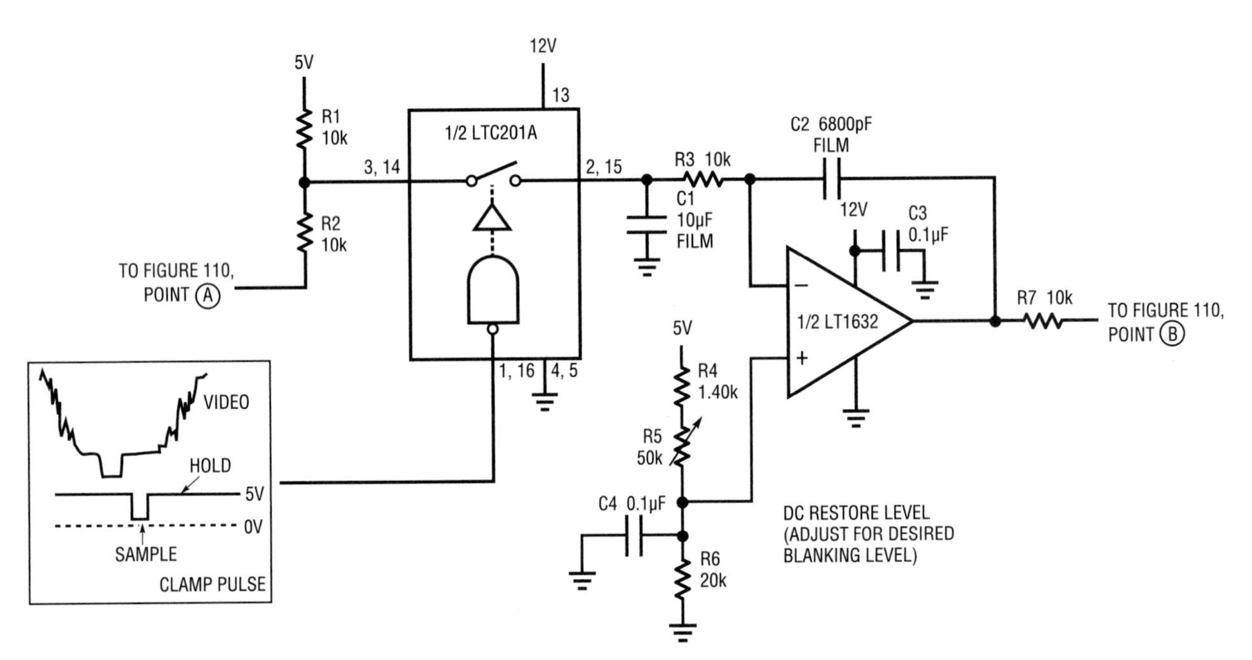

Figure 38.111 • DC Restore Subcircuit

LT1468: an operational amplifier for fast, 16-bit systems

by George Feliz

Introduction

The LT1468 is a single operational amplifier that has been optimized for accuracy and speed in 16-bit systems. Operating from ±15V supplies, the LT1468 in a gain of −1 configuration will settle in 900ns to 150μV for a 10V step. The LT1468 also features the excellent DC specifications required for 16-bit designs. Input offset voltage is 75μV max, input bias current is 10nA maximum for the inverting input and 40nA maximum for the noninverting input and DC gain is 1V/μV minimum.

16-bit DAC current-to-voltage converter with 1.7µs settling time

The key AC specification of the circuit of Figure 38.112 is settling time as it limits the DAC update rate. The settling time measurement is an exceptionally difficult problem that has been ably addressed by Jim Williams, in Linear Technology Application Note 74. Minimizing settling time is limited by the need to null the DAC output capacitance, which varies from 70pF to 115pF depending on code. This capacitance at the amplifier input combines with the feedback resistor to form a zero in the closed-loop frequency response in the vicinity of 200kHz–400kHz. Without a feedback capacitor, the circuit will oscillate. The choice of 20pF stabilizes the circuit by adding a pole at 1.3MHz to limit the frequency peaking and is chosen to optimize settling time. The settling time to 16-bit accuracy is theoretically bounded by 11.1 time constants set by the 6kΩ and

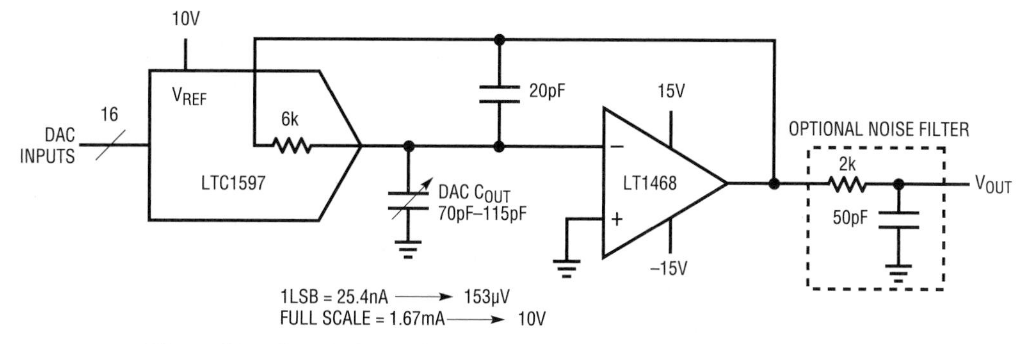

Figure 38.112 • 16-Bit DAC I/V Converter with 1.7µs Settling Time

20pF Figure 38.112's circuit settles in 1.7μs to 150μV for a 10V step. This compares favorably with the 1.33μs theoretical limit and is the best result obtainable with a wide variety of LTC and competitive amplifiers. This excellent settling requires the amplifier to be free of thermal tails in its settling behavior.

The LTC1597 current output DAC is specified with a 10V reference input. The LSB is 25.4nA, which becomes 153μV after conversion by the LT1468, and the full-scale output is 1.67mA, which corresponds to 10V at the amplifier output. The zero-scale offset contribution of the LT1468 is the input offset voltage and the inverting input current flowing through the 6k feedback resistor. This worst-case total of 135μV is less than one LSB. At full-scale there is an insignificant additional 10μV of error due to the 1V/μV minimum gain of the amplifier. The low input offset of the amplifier ensures negligible degradation of the DAC's outstanding linearity specifications.

With its low $5nV/\sqrt{Hz}$ input voltage noise and $0.6pA/\sqrt{Hz}$ input current noise, the LT1468 contributes only an additional 23% to the DAC output noise voltage. As with any precision application, and particularly with wide bandwidth amplifiers, the noise bandwidth should be minimized with an external filter to maximize resolution.

ADC buffer

The important amplifier specifications for an analog-to-digital converter buffer application (Figure 38.113) are low noise and low distortion. The LTC1604 16-bit ADC signal-to-noise ratio (SNR) of 90dB implies $56\mu V_{RMS}$ noise at the input. The noise for the amplifier, 100Ω/3000pF filter and a high value 10kΩ source is $15\mu V_{RMS}$, which degrades the SNR by only 0.3dB. The LTC1604 total harmonic distortion (THD) is a low −94dB at 100kHz. The buffer/filter combination alone has 2nd and 3rd harmonic distortion better than −100dB for a $5V_{P-P}$, 100kHz input, so it does not degrade the AC performance of the ADC.

The buffer also drives the ADC from a low source impedance. Without a buffer, the LTC1604 acquisition time increases with increasing source resistance above 1k and therefore the maximum sampling rate must be reduced. With the low noise, low distortion LT1468 buffer, the ADC can be driven at maximum speed from higher source resistances without sacrificing AC performance.

The DC requirements for the ADC buffer are relatively modest. The input offset voltage, CMRR (96dB minimum) and noninverting input bias current through the source resistance, R_S, affect the DC accuracy, but these errors are an insignificant fraction of the ADC offset and full- scale errors.

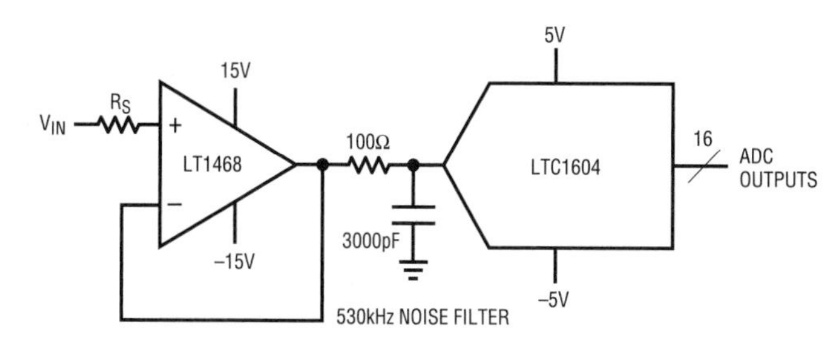

Figure 38.113 • ADC Buffer

Telecommunications circuits

How to ring a phone with a quad op amp

by Dale Eagar

Requirements

When your telephone rings, exactly what is the phone company doing? This question comes up frequently, as it seems everyone is becoming a telephone company. Deregulation opens many new opportunities, but if you want to be the phone company you must ring bells. The voltage requirement for ringing a telephone bell is a $87V_{RMS}$ 20Hz sine wave superimposed on −48VDC.

An open-architecture ring-tone generator

What the module makers offer is a solution to a problem that, by its nature, calls for unusual design techniques. What we offer here is a design that you can own, tailor to your specific needs, lay out on your circuit board and put on your bill of materials. Finally, you will be in control of the black magic (and high voltages) of ring-tone generation.

Not your standard bench supply

Ring-tone generation requires two high voltages, 60VDC and −180VDC. Figure 38.114 details the switching power supply that delivers the volts needed to run the ring-tone circuit. This switcher can be powered from any voltage from 5V to 30V and is shut down when not in use, conserving power. The transformer and optocoupling yield a fully floating output. Faraday shields in the transformer eliminate most switcher noise, preventing mystery system noise problems later. Table 38.6 is the build diagram of the transformer used in the switching power supply.

Quad op amp rings phones

When a phone rings, it rings with a cadence, a sequence of rings and pauses. The standard cadence is one second ringing followed by two seconds of silence. We use the first 1/4 of the LT1491 as a cadence oscillator (developed in Figures 38.115 and 38.116) whose output is at Vcc for one second and then at V_{EE} for two seconds (see Figure 38.120).

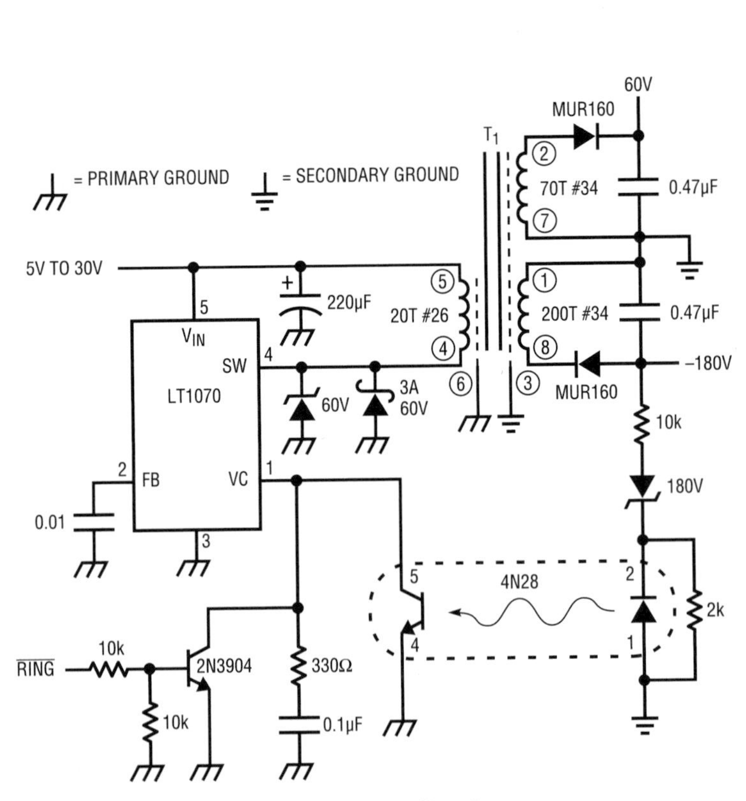

Figure 38.114 • The Switching Power Supply

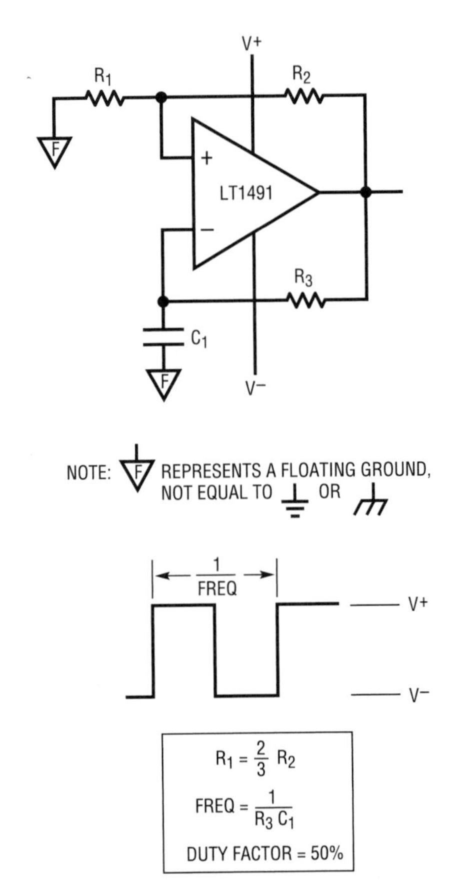

Figure 38.115 • Op Amp Intentionally Oscillates

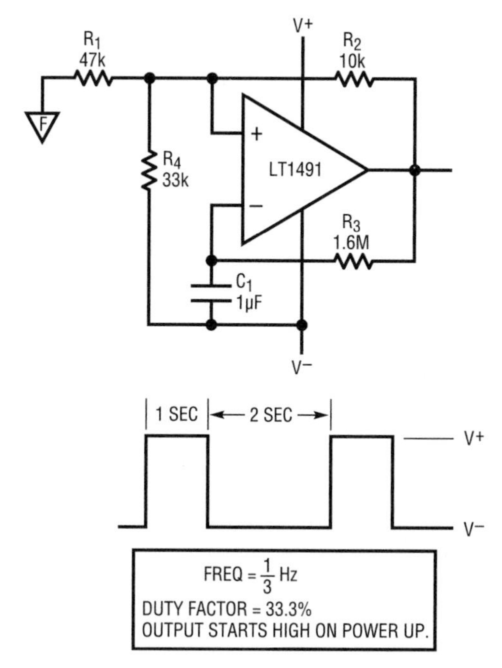

Figure 38.116 • Duty Factor is Skewed

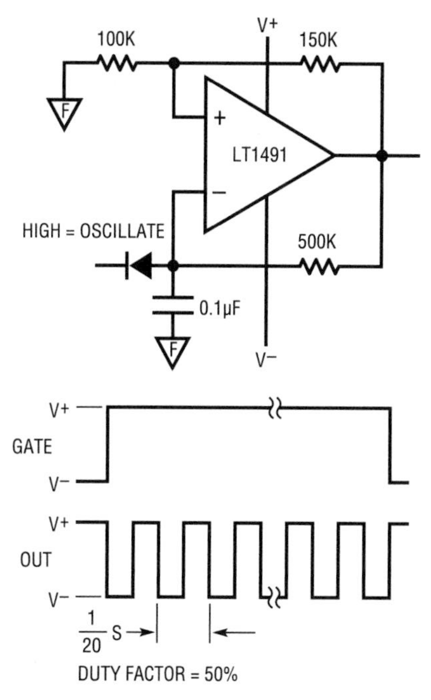

Figure 38.117 • Gated 20Hz Oscillator

This sequence repeats every three seconds, producing the all-too-familiar pattern.

The actual ringing of the bell is performed by a 20Hz AC sine wave signal at a level of $87V_{RMS}$, superimposed on −48VDC. The 20Hz signal is implemented with the second amplifier in the LT1491 (Figure 38.117) which acts as a gated 20Hz oscillator. Connecting the circuit shown in Figure 38.116 to the circuit shown in Figure 38.117 and adding three resistors yields the sequencer as shown in Figure 38.118. The waveform, labled "Square Out," is the fourth trace in Figure 38.120. This waveform is the output of Figure 38.121.

Square wave plus filter equals sine wave

Thevenin will tell you that the output impedance of the sequencer shown in Figure 38.118 is 120kΩ. This impedance can be recycled and used as the input resistance of the filter that follows. The filter detailed in Figure 38.119 uses the Thevenin resistor on its input, yielding a slick, compact design while distorting the nice waveform on the node labeled "square out" to a half sine wave, half square wave.

Appending the filter to the waveform sequencer creates the waveform engine detailed in Figure 38.119. The output of this waveform engine is shown in the bottom

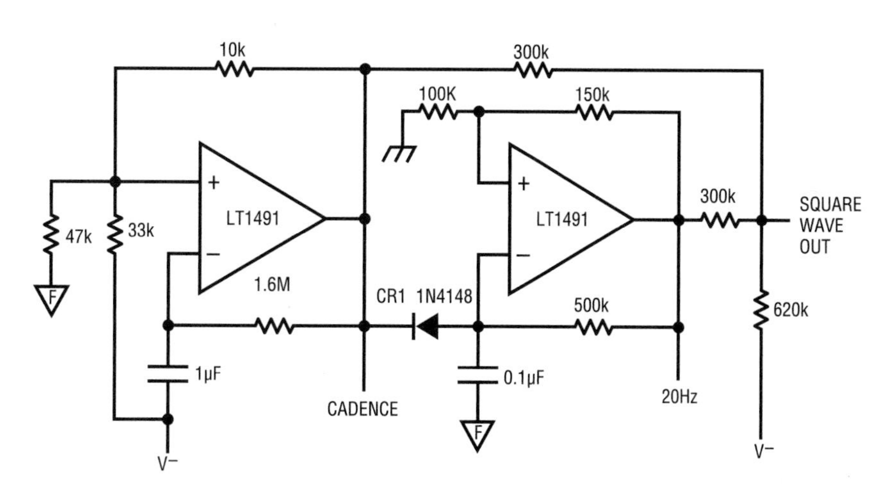

Figure 38.118 • Sequencer: Cadenced 20Hz Oscillator

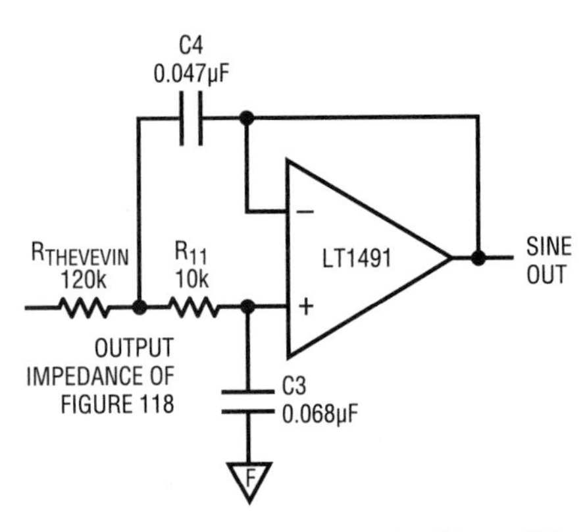

Figure 38.119 • Filter to Remove the Sharp Edges

trace in Figure 38.120. This waveform engine is shown in block form in Figure 38.122.

Mapping out the ring-tone generator in block form

We now build a system-level block diagram of our ring tone generator. We start with the waveform engine of Figure 38.122, add a couple of 15V regulators and a DC offset (47k resistor), then apply some voltage gain with a high voltage amplifier to ring the bell. This hypothetical system-level block diagram is detailed in Figure 38.123. Figure 38.124 shows the output waveform of the ring tone generator; the sequenced ringing starts when the high voltage supply (Figure 38.114) is turned on, and continues as long as the power supply is enabled.

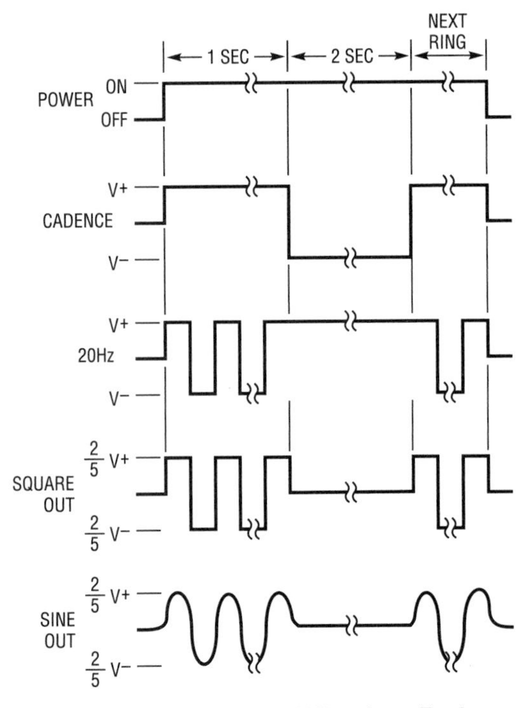

Figure 38.120 • Timing of Waveform Engine

Table 38.6 Ring-Tone High Voltage Transformer Build Diagram

Materials	
2	EFD 20-15-3F8 Cores
1	EFD 20-15-8P Bobbin
2	EFD 20- Clip
2	0.007" Nomex Tape for Gap
Winding 1	Start Pin 1 200T #34
	Term Pin 8
	1 Wrap 0.002" Mylar Tape
Winding 2	Start Pin 2 70T #34
	Term Pin 7
	1 Wrap 0.002" Mylar Tape
Shields	Connect Pin 3 1T Foil Tape Faraday Shield
	1 Wrap 0.002" Mylar Tape
	Connect Pin 6 1T Foil Tape Faraday Shield
	1 Wrap 0.002" Mylar Tape
Winding 3	Start Pin 4 20T #26
	Term Pin 5
	Finish with Mylar Tape

What's wrong with this picture (Figure 38.123)

Careful scrutiny of Figure 38.123 reveals an inconsistency: even though the three fourths of the LT1491 in the waveform engine block are powered by ±15V the final amplifier is shown as powered from 60V and −180V; this poses two problems: first the LT1491 is a quad op amp and all four sections have to share the same supply pins, and second, the LT1491 will not meet specification when powered from 60V and −180V This is because 240V is greater than the **absolute maximum rating** of 44V (V+ to V−). Linear Technology products are noted for their robustness and conservative "specmanship," but this is going too far. It is time to apply some tricks of the trade.

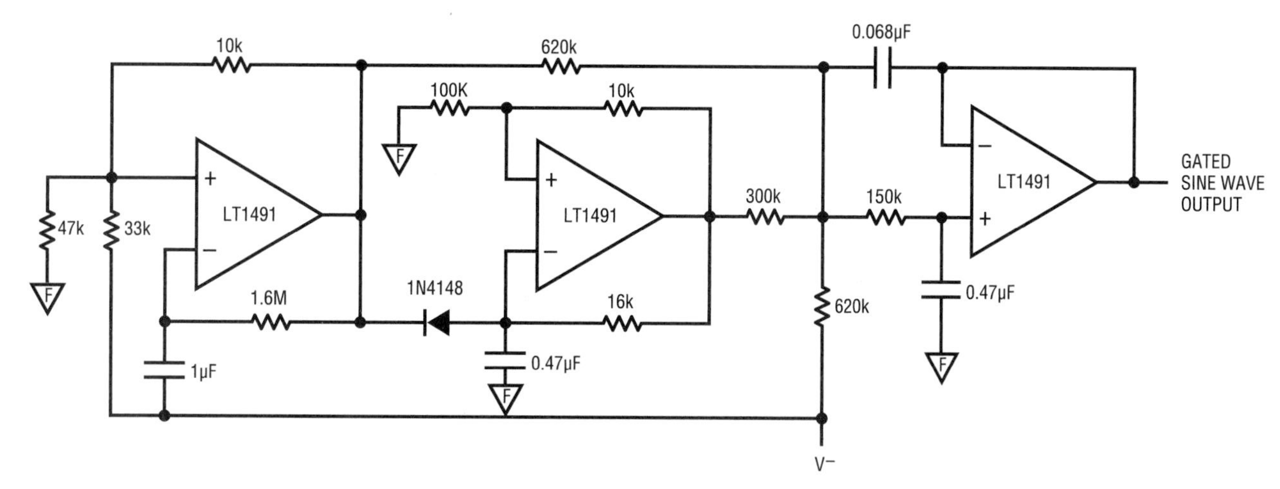

Figure 38.121 • Waveform synthesizer

Building high voltage amplifiers

Setting aside the waveform engine for a moment, we will develop a high voltage amplifier. We start with the ±15V regulators shown in Figure 38.123; these are not your run-of-the-mill regulators, these are high differential voltage regulators, constructed as shown in Figure 38.125. Using these regulators and the final section of the LT1491 quad op amp, we can build a high voltage amplifier. We will use the ±15V regulators as the "output transistors" of our amplifier, because they can both take the voltage and dissipate the power required to provide the ring voltage and current. By connecting the op amp to the regulators, one gets a free cascode high voltage amplifier. This is because the supply current for the op amp is also the regulator current. The trouble one encounters when so doing is that the input common mode range of the op amp is not wide enough to accommodate the full output voltage range of the composite amplifier. This would not be a problem if the amplifier were used as a unity-gain noninverting amplifier, but in this system we need gain to get from our $12V_{P-P}$ to $87V_{RMS}$.

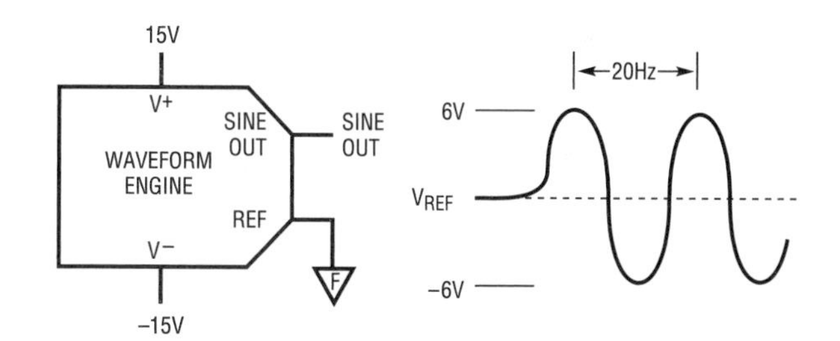

Figure 38.122 • Waveform Engine

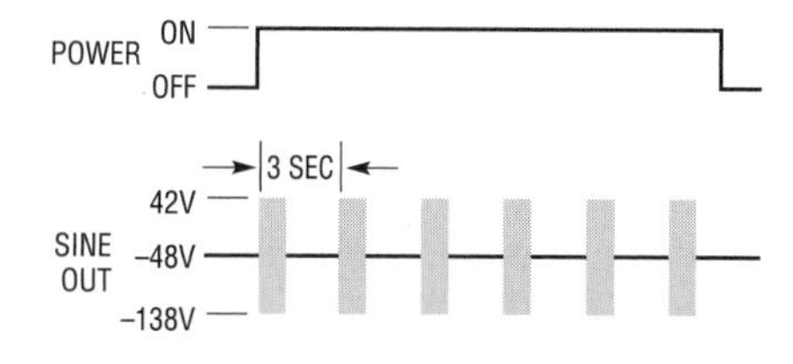

Figure 124 • System Output

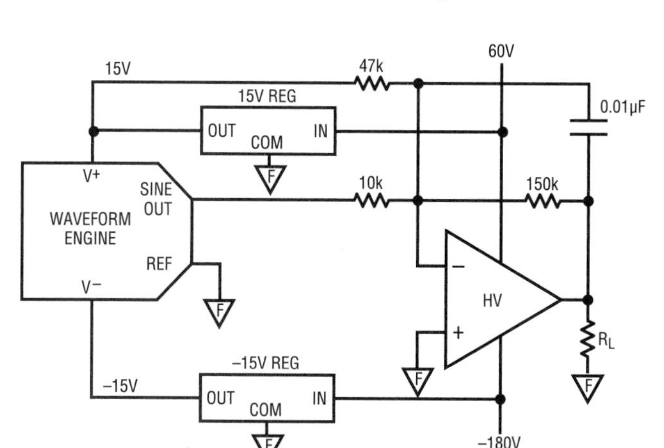

Figure 38.123 • High Voltage Amplifier

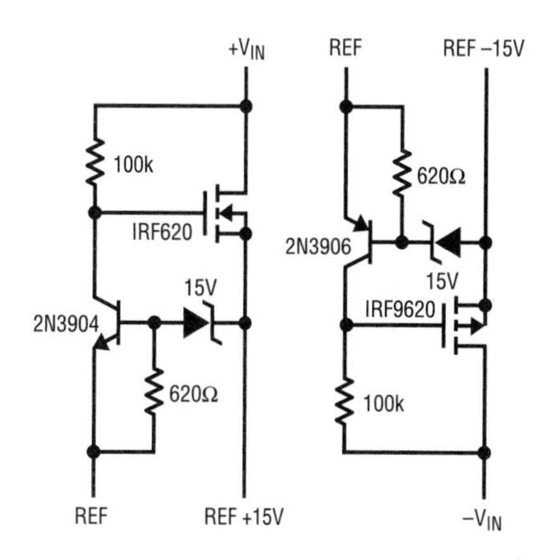

Figure 38.125 • High Differential Voltage Regulators

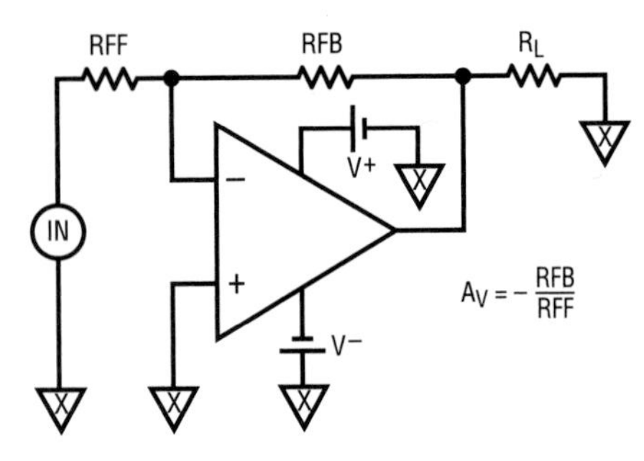

Figure 38.126 • Standard Op Amp Form

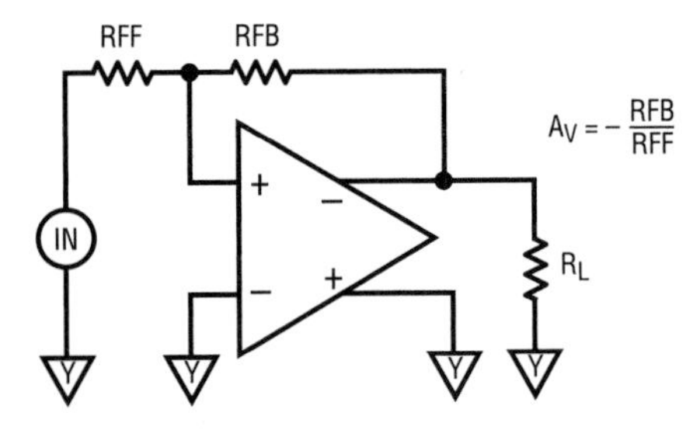

Figure 38.128 • Trade Inputs and Outputs

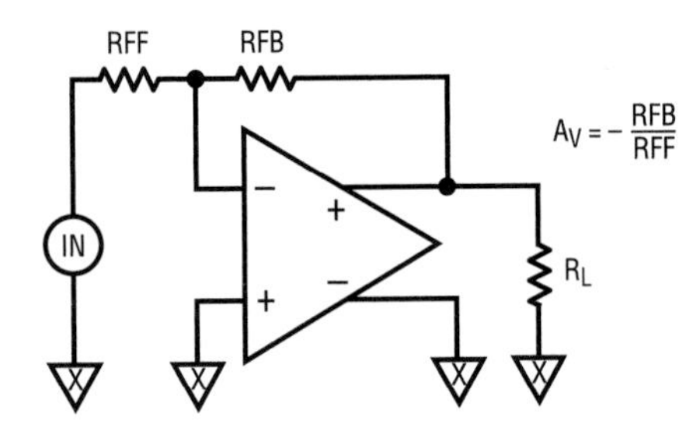

Figure 38.127 • Hide the Batteries Inside the Op Amp

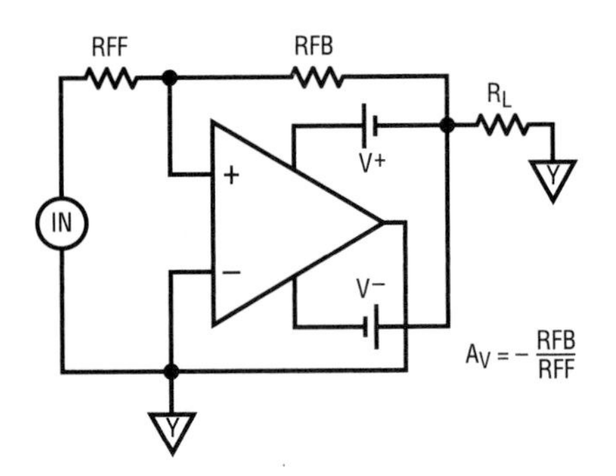

Figure 38.129 • Pull the Batteries Back out of the Amplifier

Moving the amplifier's output transistor function out of the op amp and into the ±15V regulators moves the effective amplifier output from the op amp output to the center of the two supplies sourcing the ±15V regulators. This is a transformative step in the evolution of amplifiers from low voltage op amps to high voltage, extended supply amplifiers.

Inverting op amp circuit gets morphed

Let's focus on this transformative step as it relates to the simple inverting amplifier shown in Figure 38.126.* Were we to look at the amplifier in Figure 38.126 in some strange Darwinistic mood, we might see that the power supplies (batteries) are in fact an integral part of our amplifier. Such an observation would lead us to redraw the circuit to look like Figure 38.127 where the center of the two batteries are brought out of the amplifier as the negative terminal of the output.

Once that is done, one is free to swap the polarities of the inputs and outputs, yielding the circuit shown in Figure 38.128. Finally we pull the two batteries back out of the amplifier to get our morphed inverting amplifier (Figure 38.129). Isn't assisting evolution fun?†

Editor's Notes:

* The grounds X and Y, shown in Figures 38.126– 38.129, are for illustrating the effects of "evolution." Ground X may be regarded as "arbitrary exemplary ground, " and ground Y as "postmetamorphic exemplary ground." Ground X and ground Y are not the same.

† Evolutionary theory invloves pure, random chance. What you have done here requires purposeful thought and design.

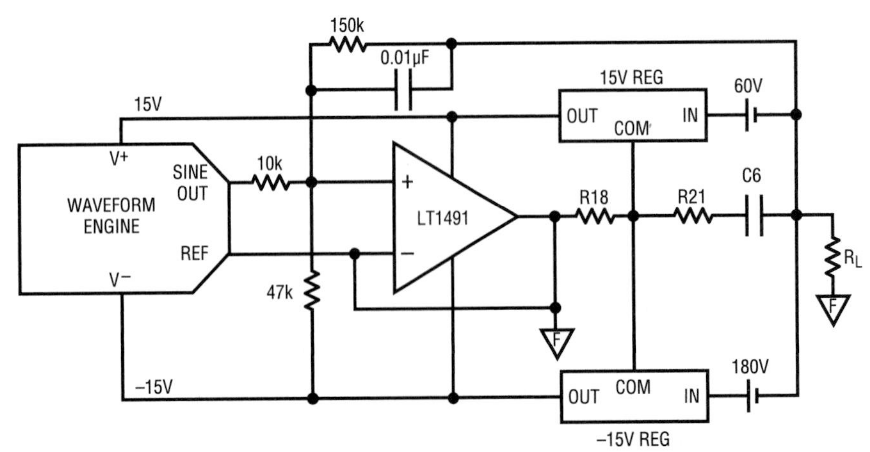

Figure 38.130 • Post-Evolution Block Diagram

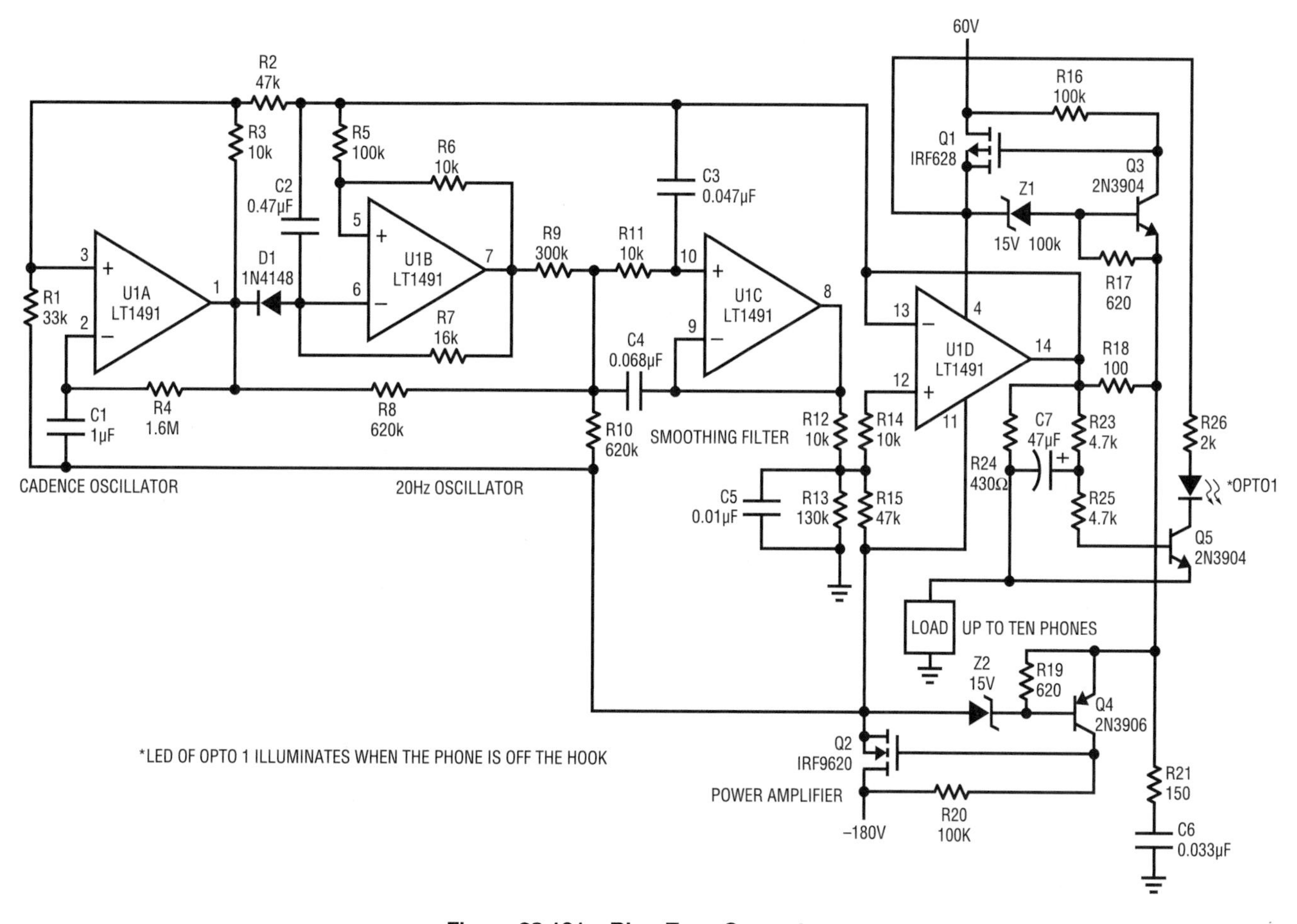

Figure 38.131 • Ring-Tone Generator

Applying the evolutionary forces just described to the block diagram in Figure 38.123, we get the block diagram in Figure 38.130. Actually Figure 38.130 contains three strangers, R18, R21 and C6, parts not predicted by our evolutionary path (unless R18 = 0Ω and R21 is open) These parts are needed because, in our metamorphosis going from Figure 38.127 to Figure 38.128, the amplifier's internal compensation node was moved from ground to the amplifier's output. These parts correct the compensation for the new configuration.

Ring-trip sense

Now that we can ring the telephone, we must sense when the phone is picked up. This is done by sensing the DC current flowing to the phone while it is ringing, using the ring-trip sense circuit comprising R23-R26, C7, Q5 and Opto1 of Figure 38.131, the complete ring-tone generator. This circuit will ring more than ten phones at once, and is protected on its output from shorts to ground or to either the +60V or the -180V supply.

Conclusion

Here is a ring tone generator you can own, a robust circuit that is stable into any load. If your system design requires a circuit with different specifications, you can easily tailor this circuit to meet your needs. Don't hesitate to call us if we can help you with your design.

A low distortion, low power, single-pair HDSL driver using the LT1497

by George Feliz and Adolfo Garcia

Introduction

High speed digital subscriber line (HDSL) interfaces support full-duplex data rates up to 1.544Mbps over 12,000 feet using two standard 135Ω twisted-pair telephone wires. The high data rate is achieved with a combination of encoding 2 bits per symbol using two-binary, one-quaternary (2B1Q) modulation, and sophisticated digital signal processing to extract the received signal. This performance is possible only with low distortion line drivers and receivers. In addition, the power dissipation of the transceiver circuitry is critical because it may be loop-powered from the central office over the twisted pair: Lower power dissipation also increases the number of transceivers that can placed in a single, non-forced-air enclosure. Single-pair HDSL requires the same performance as two-pair HDSL over a single twisted pair and operates at twice the fundamental 2B1Q symbol rate. In HDSL systems that use 2B1Q line coding, the signal passband necessary to carry a data rate of 1.544Mbps is 392kHz. This signal rate will be used to quantify the performance of the LT1497 in this article.

Low distortion line driver

The circuit of Figure 38.132 transmits signals over a 135Ω twisted pair through a 1:1 transformer. The LT1497 dual 125mA, 50MHz current feedback amplifier was chosen for its ability to cleanly drive heavy loads, while consuming a modest 7mA maximum supply current per amplifier in a thermally enhanced SO-8 package. The driver amplifiers are configured in gains of two (A1) and minus one (A2) to compensate for the attenuation inherent in the back-termination of the line and to provide differential drive to the transformer. The transmit power requirement for HDSL is 13.5dBm (22.4mW) into 135Ω, corresponding to a 1.74V_{RMS} signal. Since 2B1Q modulation is a 4-level pulse amplitude modulated signal, the crest factor (peak to RMS) of this signal is 1.61. Thus, a 13.5dBm, 2B1Q modulated signal yields 5.6V_{P-P} across the 135Ω load. The corresponding output signal current is ±20.7mA peak. This modest drive level increases for varying line conditions and is tested with a standardized collection of test loops that can have line impedances as low as 25Ω. The LT1497's high output current and voltage swing drive the 135Ω line at the required distortion level of −72dBc. For a data rate of 1.544Mbps and 2-bit-per-symbol encoding, the fundamental frequency of operation is 392kHz.

The LT1497 provides such low distortion because it operates at only a fraction of its output current capability and is well within its voltage swing limitations. There are other LTC amplifiers that can achieve this performance, but at the expense of higher power dissipation or a larger package.

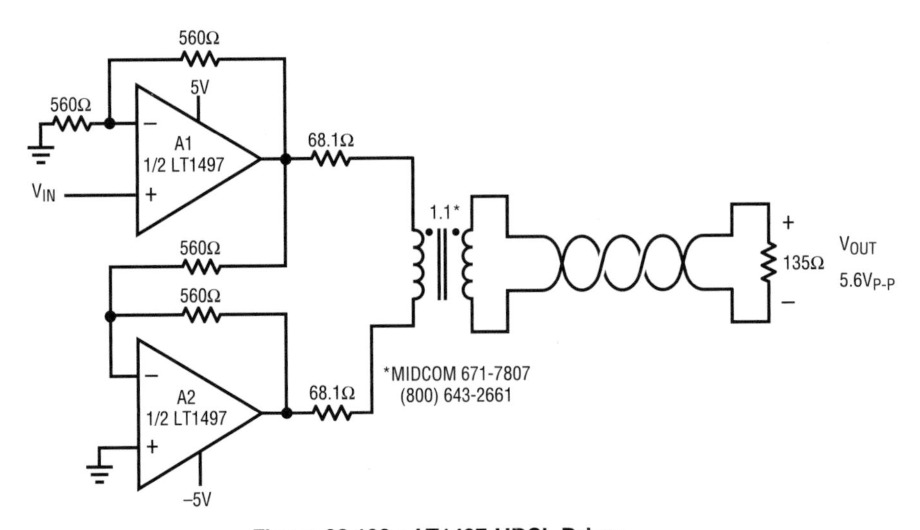

Figure 38.132 • LT1497 HDSL Driver

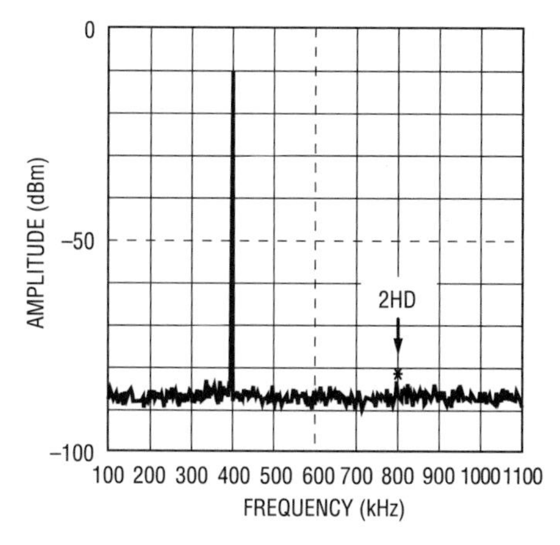

Figure 38.133 • Harmonic Distortion of Figure 38.132's Circuit with a 400kHz Sine Wave and an Output Level of 5.6V_{P-P} into 135Ω

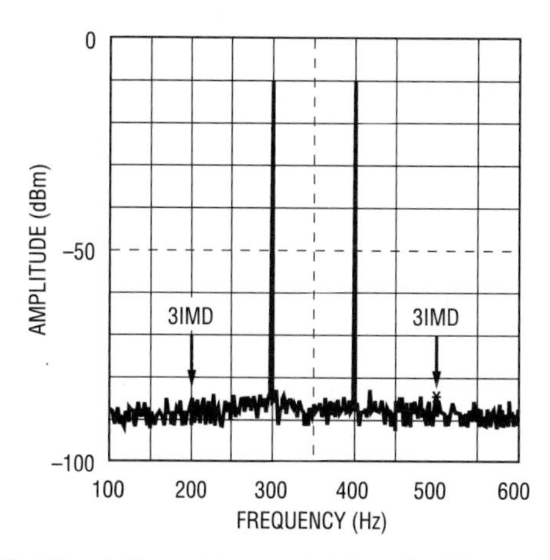

Figure 38.134 • 2-Tone Intermodulation for Figure 38.132's Circuit

Performance

The circuit of Figure 38.132 was evaluated for harmonic distortion with a 400kHz sine wave and an output level of 5.6V_{P-P} into 135Ω. Figure 38.133 shows that the second harmonic is −72.3dB relative to the fundamental for the 135Ω load. Third harmonic distortion is not critical, because received signals are heavily filtered before being digitized by an A/D converter. Performance with a 50Ω load (to simulate more challenging test loops) is slightly better at −75dB. The output signal was attenuated to obtain maximum sensitivity of the HP4195A network analyzer used for the measurements.

With multicarrier applications such as discrete multitone modulation (DMT) becoming as prevalent as single-carrier applications, another important measure of amplifier dynamic performance is 2-tone intermodulation. This evaluation is a valuable tool to gain insight to amplifier linearity when processing more than one tone at a time.

For this test, two sine waves at 300kHz and 400kHz were used with levels set to obtain 5.6V_{P-P} across the 135Ω load. Figure 38.134 shows that the third-order intermodulation products are well below −72dB. With a 50Ω load, performance is within 1dB-2dB of that with the 135Ω load.

Conclusion

The circuit presented provides outstanding distortion performance in an SO-8 package with remarkably low power dissipation. It is ideally suited for single pair digital subscriber line applications, especially for remote terminals.

Comparators

Ultralow power comparators include reference

by James Herr

The LTC1440-LTC1445 family features 1μA comparators with adjustable hysteresis and TTL/CMOS outputs that sink and source current and a 1μA reference that can drive a bypass capacitor of up to 0.01μF without oscillation. The parts operate from a 2V to 11V single supply or a ±1V to ±5V dual supply.

Undervoltage/overvoltage detector

The LTC1442 can be easily configured as a window detector, as shown in Figure 38.135. R1, R2 and R3 form a resistive divider from V_{CC} so that comparator A goes low when V_{CC} drops below 4.5V and comparator B goes low when V_{CC} rises above 5.5V. A 10mV hysteresis band is set by R4 and R5 to prevent oscillations near the trip points.

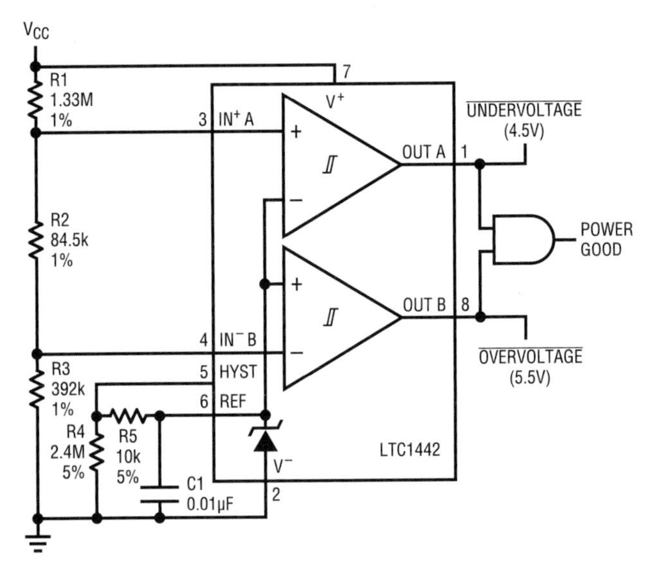

Figure 38.135 • Window Detector

Single-cell lithium-ion battery supply

Figure 38.136 shows a single cell lithium-ion battery to 5V supply with the low-battery warning, low-battery shutdown and reset functions provided by the LTC1444. The LT1300 micropower step-up DC/DC converter boosts the battery voltage to 5V using L1 and D1. Capacitors C2 and C3 provide input and output filtering.

The voltage-monitoring circuitry takes advantage of the LTC1444's open-drain outputs and low supply voltage operation. Comparators A and B, along with R1, R2 and R3, monitor the battery voltage. When the battery voltage drops below 2.65V comparator A's output pulls low to generate a nonmaskable interrupt to the microprocessor to warn of a low-battery condition. To protect the battery from over discharge, the output of comparator B is pulled high by R7 when the battery voltage falls below 2.45V P-channel MOSFET Q1 and the LT1300 are turned off, dropping the quiescent current to 20μA. Q1 is needed to prevent the load circuitry from discharging the battery through L1 and D1.

Comparators C and D provide the reset input to the microprocessor As soon as the boost converter output rises above the 4.65V threshold set by R8 and R9, comparator C turns off and R10 starts to charge C4. After 200ms, comparator D turns off and the Reset pin is pulled high by R12.

Conclusion

With their built-in references, low supply current requirements and variety of configurations, Linear Technology's LTC1440-45 family of micropower comparators is ideal for system monitoring in battery-powered devices such as PDAs, laptop and palmtop computers and hand-held instruments.

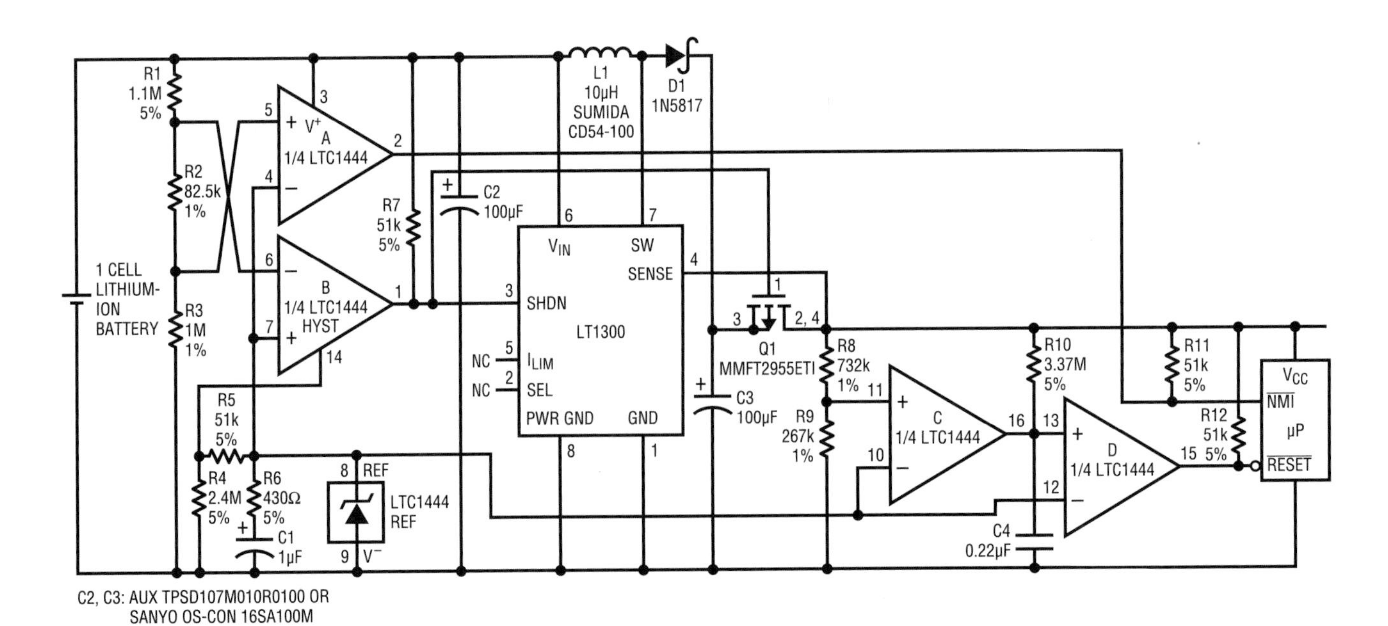

Figure 38.136 • Single-Cell to 5V Supply

A 4.5ns, 4mA, single-supply, dual comparator optimized for 3V/5V operation

by Joseph G. Petrofsky

Introduction

The LT1720 is an UltraFast™ (4.5ns), low power (4mA/comparator), single-supply, dual comparator designed to operate on a single 3V or 5V supply. These comparators feature internal hysteresis, making them easy to use, even with slowly moving input signals. The LT1720 is fabricated in Linear Technology's 6GHz complementary bipolar process, resulting in unprecedented speed for its low power consumption.

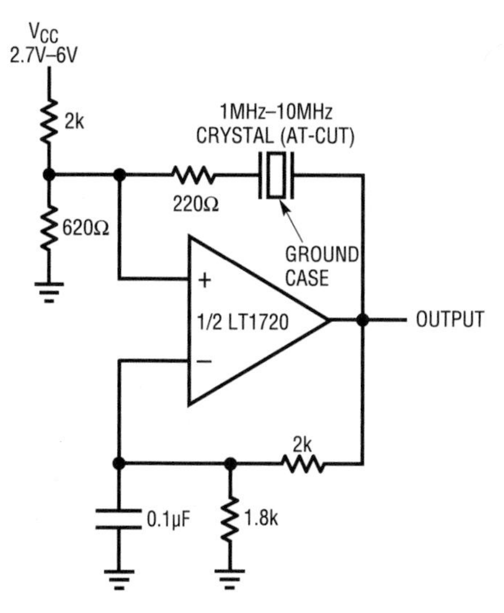

Figure 38.137 • Simple 1MHz to 10MHz Crystal Oscillator

Applications

Crystal oscillators

Figure 38.137 shows a simple crystal oscillator using one half of an LT1720. The 2k-620Ω resistor pair set a bias point at the comparator's noninverting input. The 2k-1.8k-0.1μF path sets the inverting input node at an appropriate DC average level based on the output. The crystal's path provides resonant positive feedback and stable oscillation occurs. Although the LT1720 will give the correct logic output when one input is outside the common mode range, additional delays may occur when it is so operated, opening the possibility of spurious operating modes. Therefore, the DC bias voltages at the inputs are set near the center of the LT1720's common mode range and the 220Ω resistor attenuates the feedback to the noninverting input. The circuit will operate with any AT-cut crystal from 1MHz to 10MHz over a 2.7V to 6V supply range.

The output duty cycle for the circuit of Figure 38.137 is roughly 50% but it is affected by resistor tolerances and, to a lesser extent, by comparator offsets and timings.

Timing skews

For a number of reasons, the LT1720 is an excellent choice for applications requiring differential timing skew. The two comparators in a single package are inherently well matched, with just 300ps Δt_{PD} typical. Monolithic construction keeps the delays well matched vs supply voltage and temperature. Crosstalk between the comparators, usually a disadvantage in monolithic duals, has minimal effect on the LT1720 timing due to the internal hysteresis.

The circuits of Figure 38.138 show basic building blocks for differential timing skews. The 2.5k resistance interacts with the 2pF typical input capacitance to create at least ±4ns delay, controlled by the potentiometer setting. A differential and a single-ended version are shown. In the differential configuration, the output edges can be smoothly scrolled through $\Delta t = 0$ with negligible interaction.

Fast waveform sampler

Figure 38.139 uses a diode-bridge-type switch for clean, fast waveform sampling. The diode bridge, because of its inherent symmetry, provides lower AC errors than other semiconductor-based switching technologies. This circuit features 20dB of gain, 10MHz full power bandwidth and 100μV/°C baseline uncertainty. Switching delay is less than 15ns and the minimum sampling window width for full power response is 30ns.

The input waveform is presented to the diode bridge switch, the output of which feeds the LT1227 wideband amplifier. The LT1720 comparators, triggered by the sample command, generate phase-opposed outputs. These signals

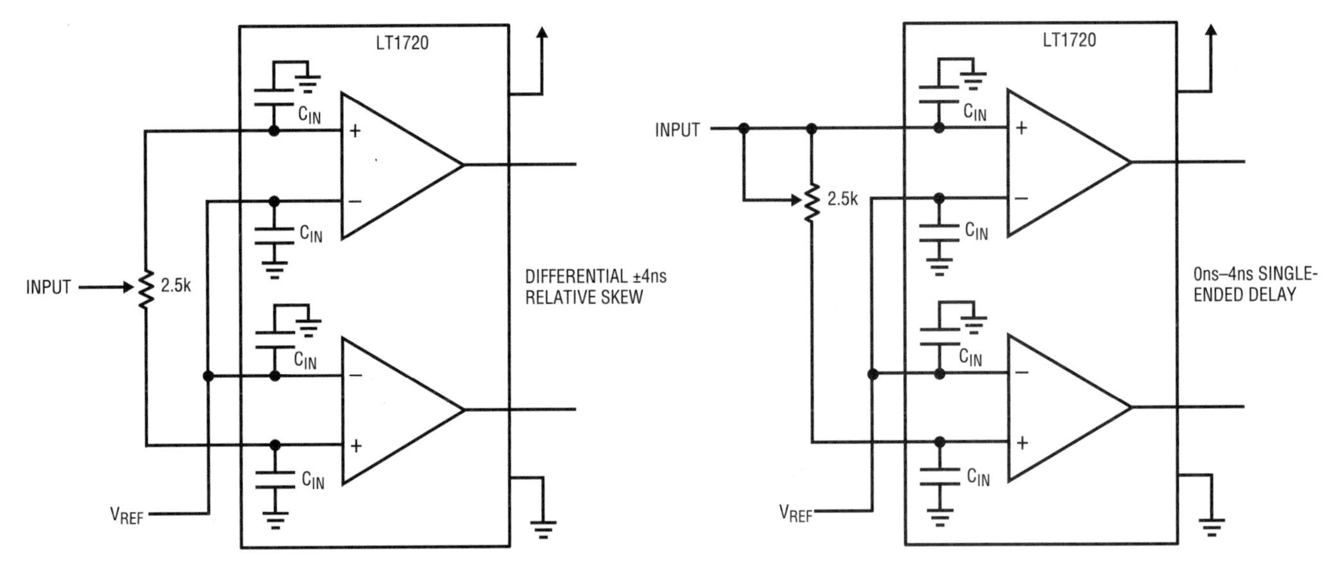

Figure 38.138 • Timing-Skew Generation is Easy with the LT1720

are level shifted by the transistors, providing complementary bipolar drive to switch the bridge. A skew compensation trim ensures bridge-drive signal simultaneity within 1ns. The AC balance corrects for parasitic capacitive bridge imbalances. A DC balance adjustment trims bridge offset.

The trim sequence involves grounding the input via 50Ω and applying a 100kHz sample command. The DC balance is adjusted for minimal bridge ON vs OFF variation at the output. The skew compensation and AC balance adjustments are then optimized for minimum AC disturbance in the output. Finally, unground the input and the circuit is ready for use.

Coincidence detector

High speed comparators are especially suited for interfacing pulse-output transducers, such as particle detectors, to logic circuitry. The matched delays of a monolithic dual are well suited for those cases where the coincidence of two pulses needs to be detected. The circuit of Figure 38.140 is a coincidence detector that uses an LT1720 and discrete components as a fast AND gate.

The reference level is set to 1V an arbitrary threshold. Only when both input signals exceed this will a coincidence be detected. The Schottky diodes from the comparator outputs to the base of the MRF-501 form the AND gate, while the other two Schottkys provide for fast turn-off. A logic AND gate could instead be used, but would add considerably more delay than the 300psec contributed by this discrete stage.

This circuit can detect coincident pulses as narrow as 2.5ns. For narrower pulses, the output will degrade gracefully, responding, but with narrow pulses that don't rise all the way to high before starting to fall. The decision delay is 4.5ns with input signals 50mV or more above the reference level. This circuit creates a TTL compatible output but it can typically drive CMOS as well.

Pulse stretcher

For detecting short pulses from a single sensor, a pulse stretcher is often required. The circuit of Figure 38.141 acts as a one-shot, stretching the width of an incoming pulse to a consistent 100ns. Unlike a logic one-shot, this LT1720-based circuit requires only 100pV-s of stimulus to trigger.

The circuit works as follows: Comparator C1 functions as a threshold detector, whereas comparator C2 is configured as a one-shot. The first comparator is prebiased with a threshold of 8mV to overcome comparator and system offsets and establish a low output in the absence of an input signal. An input pulse sends the output of C1 high, which in turn latches C2's output high. The output of C2 is fed back to the input of the first comparator, causing regeneration and latching both outputs high.

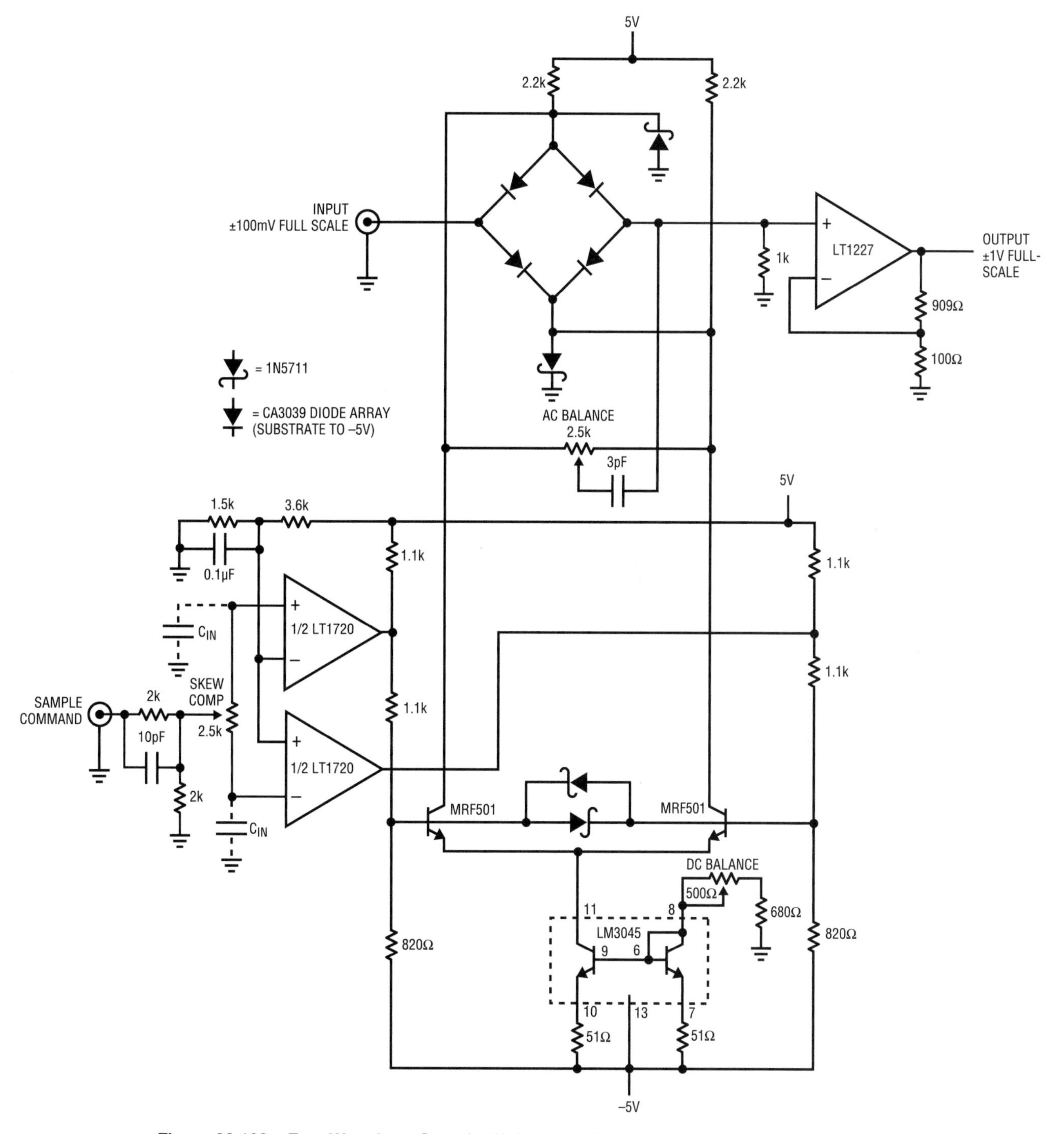

Figure 38.139 • Fast Waveform Sampler Using the LT1720 for Timing-Skew Compensation

Timing capacitor C now begins charging through R and, at the end of 100ns, C2 resets low. The output of C1 also goes low, latching both outputs low. A new pulse at the input of C1 can now restart the process. Timing capacitor C can be increased without limit for longer output pulses. This circuit has an ultimate sensitivity of better than 14mV with 5ns-10ns input pulses. It can even detect an avalanche generated test pulse of just 1ns duration with sensitivity better than 100mV[2] It can detect short events better than the coincidence detector above because the one-shot is configured to catch just 100mV of upward movement from C1's V_{OL}, whereas the coincidence detector's 2.5ns specification is based on a full, legitimate logic high.

Note 2: See Linear Technology Application Note 47, Appendix B. This circuit can detect the output of the pulse generator described after 40dB of attenuation.

Conclusion

The new LT1720 dual 4.5ns single-supply comparators feature high speeds and low power consumption. They are versatile and easy-to-use building blocks for a wide variety of system design challenges.

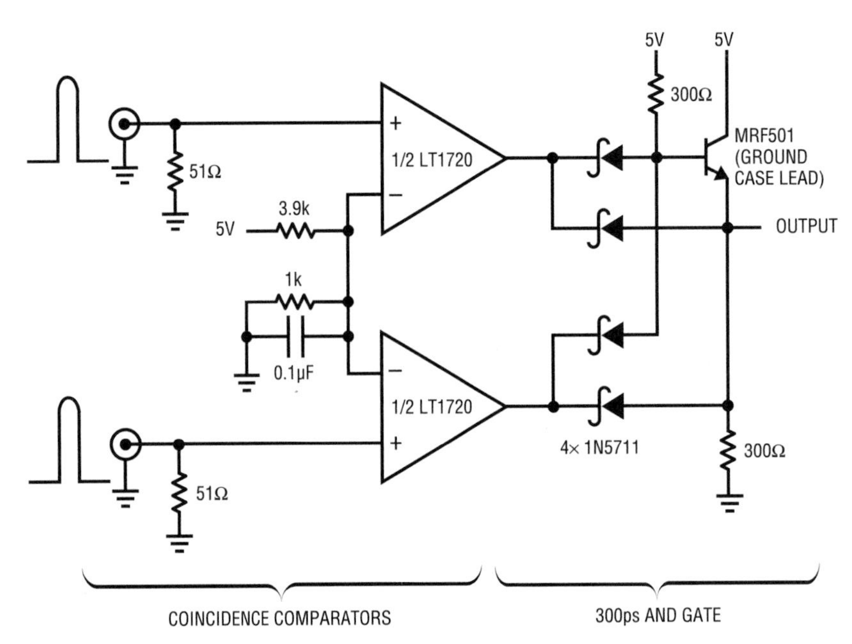

Figure 38.140 • A 2.5ns Coincidence Detector

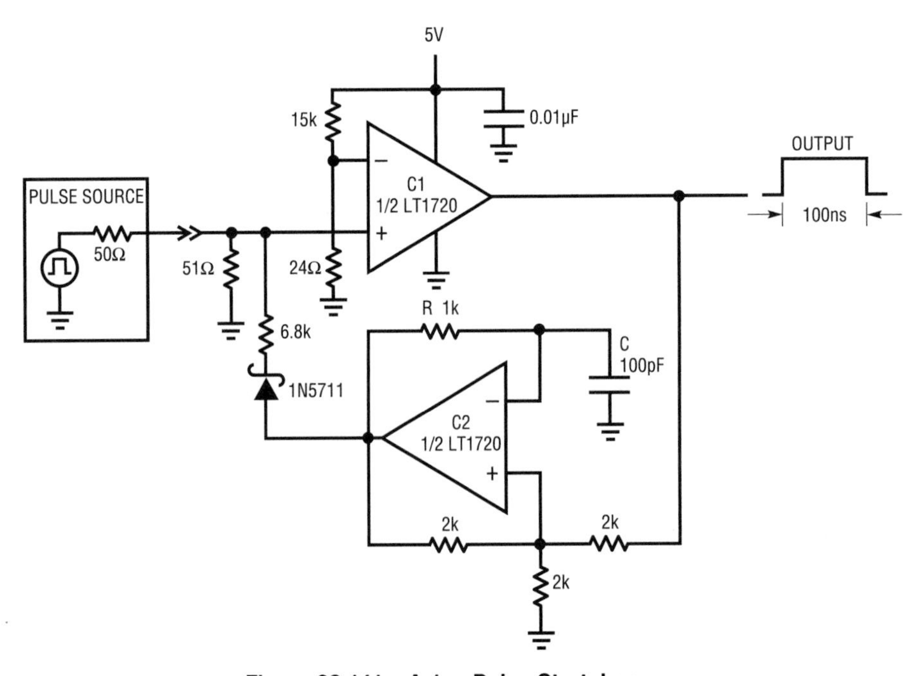

Figure 38.141 • A 1ns Pulse Stretcher

Instrumentation circuits

LTC1441-based micropower voltage-to-frequency converter

by Jim Williams

Figure 38.142 is a voltage-to-frequency converter: A 0V-5V input produces a 0Hz–10kHz output, with a linearity of 0.02%. Gain drift is 60ppm/°C. Maximum current consumption is only 26μA, 100 times lower than currently available units.

To understand the circuit's operation, assume that C1's negative input is slightly below its positive input (C2's output is low). The input voltage causes a positive-going ramp at C1's input (trace A, Figure 38.143). C1's output is high, allowing current flow from Q1's emitter, through C1's output stage to the 100pF capacitor. The 2.2μF capacitor provides high frequency bypass, maintaining low impedance at Q1's emitter Diode connected Q6 provides a path to ground. The voltage to which the 100pF unit charges is a function of Q1's emitter potential and Q6's drop. C1's CMOS output, purely ohmic, contributes no voltage error. When the ramp at C1's negative input goes high enough, C1's output goes low (trace B) and the inverter switches high (trace C). This action pulls current from C1's negative input capacitor via the Q5 route (trace D).

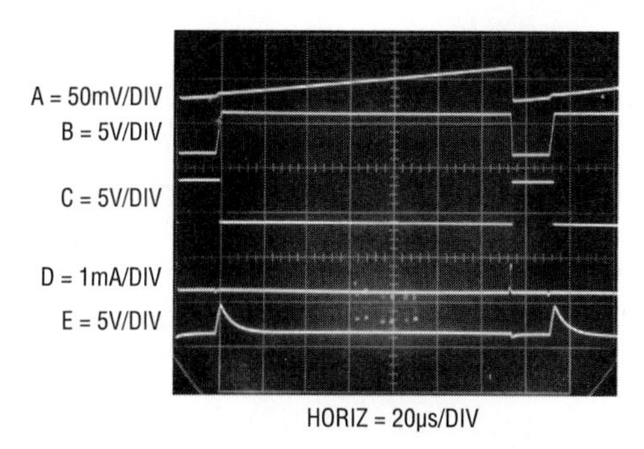

Figure 38.143 • Waveforms for the Micropower V/F Converter: Charge-Based Feedback Provides Precision Operation with Extremely Low Power Consumption.

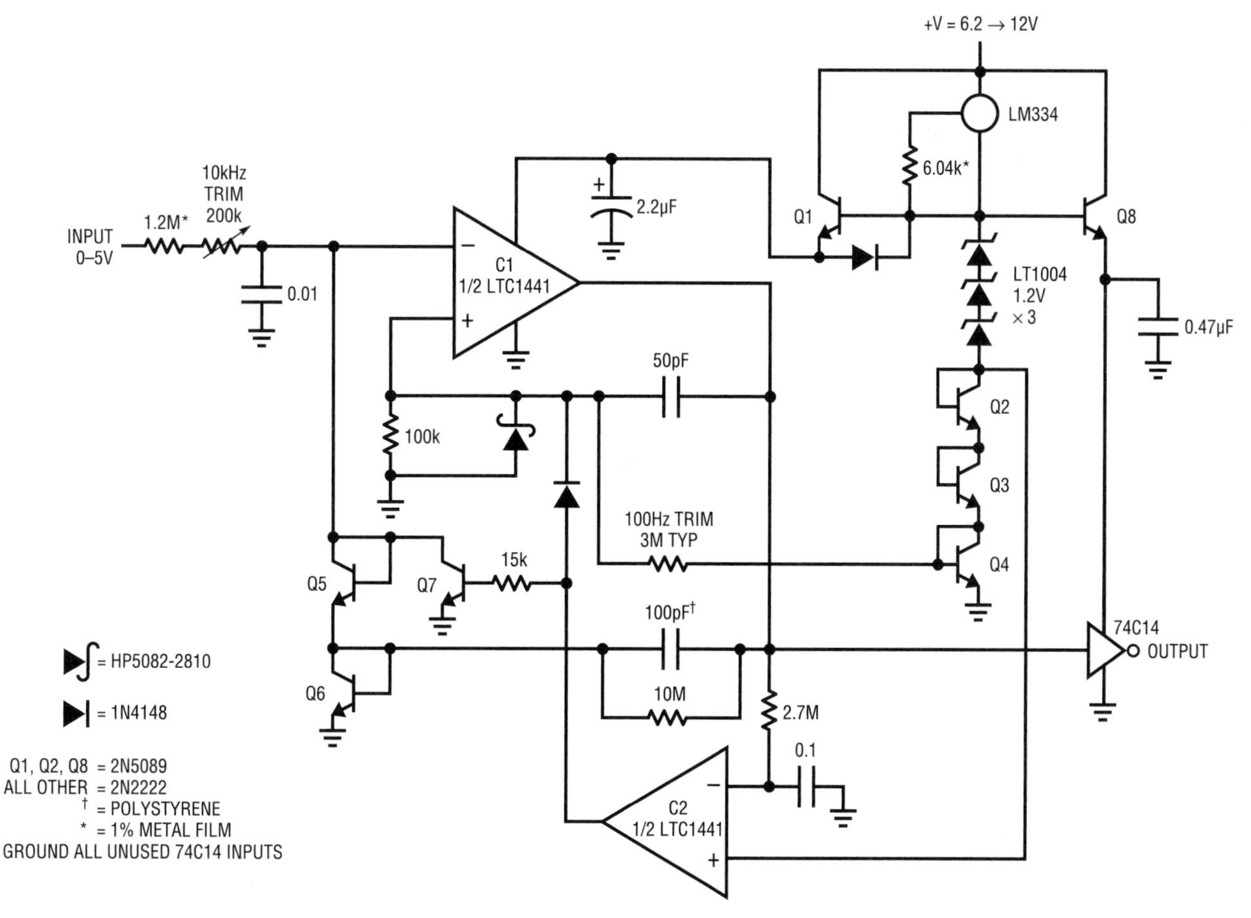

Figure 38.142 • 0.02% V/F Converter Requires only 26μA Supply Current

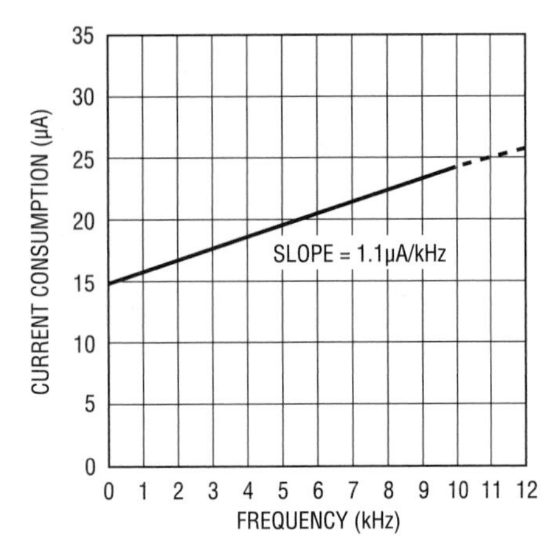

Figure 38.144 • Current Consumption vs Frequency for the V/F Converter: Charge/Discharge Cycles Account for 1.1µA/kHz Current Drain Increase

This current removal resets C1's negative input ramp to a potential slightly below ground. The 50pF capacitor furnishes AC positive feedback (C1's positive input is trace E) ensuring that C1's output remains negative long enough for a complete discharge of the 100pF capacitor. The Schottky diode prevents C1's input from being driven outside its negative common mode limit. When the 50pF unit's feedback decays, C1 again switches high and the entire cycle repeats. The oscillation frequency depends directly on the input-voltage-derived current.

Q1's emitter voltage must be carefully controlled to get low drift. Q3 and Q4 temperature compensate Q5 and Q6 while Q2 compensates Q1's V_{BE}. The three LT1004s are the actual voltage reference and the LM334 current source provides 12µA bias to the stack. The current drive provides excellent supply immunity (better than 40ppm/V) and also aids circuit temperature coefficient. It does this by using the LM334's 0.3%/°C tempco to slightly temperature modulate the voltage drop in the Q2-Q4 trio. This correction's sign and magnitude directly oppose the −120ppm/°C 100pF polystyrene capacitor's drift, aiding overall circuit stability. Q8's isolated drive to the CMOS inverter prevents output loading from influencing Q1's operating point. This makes circuit accuracy independent of loading.

The Q1 emitter-follower delivers charge to the 100pF capacitor efficiently. Both base and collector current end up in the capacitor. The 100pF capacitor, as small as accuracy permits, draws only small transient currents during its charge and discharge cycles. The 50pF-100k positive feedback combination draws insignificantly small switching currents. Figure 38.144, a plot of supply current versus operating frequency, reflects the low power design. At zero frequency, comparator quiescent current and the 12µA reference stack bias account for all current drain. There are no other paths for loss. As frequency scales up, the 100pF capacitor's charge-discharge cycle introduces the 1.1µA/kHz increase shown. A smaller value capacitor would cut power, but effects of stray capacitance and charge imbalance would introduce accuracy errors.

Circuit start-up or overdrive can cause the circuit's AC-coupled feedback to latch. If this occurs, C1's output goes low; C2, detecting this via the 2.7M-0.1µF lag, goes high. This lifts C1's positive input and grounds the negative input with Q7, initiating normal circuit action.

To calibrate this circuit, apply 50mV and select the indicated resistor at C1's positive input for a 100Hz output. Complete the calibration by applying 5V and trimming the input potentiometer for a 10kHz output.

Bridge measures small capacitance in presence of large strays

by Jeff Witt

Capacitance sensors measure a wide variety of physical quantities, such as position, acceleration, pressure and fluid level. The capacitance changes are often much smaller than stray capacitances, especially if the sensor is remotely placed. I needed to make measurements with a 50pF cryogenic fluid level detector, with only 2pF full-scale change, hooked to several hundred pF of varying cable capacitance. This required a circuit with high stability, sensitivity and noise rejection, but one insensitive to stray capacitance caused by cables and shielding. I also wanted battery operation and analog output for easy interfacing to other instruments. Two traditional circuit types have drawbacks: integrators are sensitive to noise at the comparator and voltage-to-frequency converters typically measure stray as well as sensor capacitance. The capacitance bridge presented here measures small transducer capacitance changes, yet rejects noise and cable capacitance.

The bridge, shown in Figure 38.145, is designed around the LTC1043 switched-capacitor building block. The circuit compares a capacitor, C_X, of unknown value, with a reference capacitor C_{REF}. The LTC1043, programmed with C1 to switch at 500Hz, applies a square wave of amplitude V_{REF} to node A, and a square wave of amplitude V_{OUT} and opposite phase to node B. When the bridge is balanced, the AC voltage at node C is zero, and

$$V_{OUT} = V_{REF}\frac{C_X}{C_{REF}}$$

Balance is achieved by integrating the current from node C using an op amp (LT1413) and a third switch on the LTC1043 for synchronous detection. With $C_{REF} = 500pF$ and $V_{REF} = 2.5V$ this circuit has a gain of 5mV/pF and when measured with a DMM achieves a resolution of 10fF for a dynamic range of 100dB. It also rejects stray capacitance (shown as ghosts in Figure 38.145) by 100dB. If this rejection is not important, the switching frequency f can be increased to extend the circuit's bandwidth, which is

$$BW = f\frac{C_{REF}}{C_{OUT}}$$

C_{OUT} should be larger than C_{REF}.

The circuit operates from a single 5V supply and consumes 800μA. If the capacitances at nodes A and C are kept below 500pF the LT1078 micropower dual op amp may be used in place of the LT1413, reducing supply current to just 160μA.

If the relative capacitance change is small, the circuit can be modified for higher resolution, as shown in Figure 38.146. A JFET input op amp (LT1462) amplifies the signal before demodulation for good noise performance, and the output of the integrator is attenuated by R1 and R2 to increase the sensitivity of the circuit. If $\Delta C_X << C_X$, and $C_{REF} \approx C_X$, then

$$V_{OUT} - V_{REF} \approx V_{REF}\frac{\Delta C_X(R1 + R2)}{C_{REF}R2}$$

With $C_{REF} = 50pF$ the circuit has a gain of 5V/pF and can resolve 2fF Supply current is 1mA. The synchronous detection makes this circuit insensitive to external noise sources and in this respect shielding is not terribly important. However; to achieve high resolution and stability, care should be taken to shield the capacitors being measured. I used this circuit for the fluid level detector mentioned above, putting a small trim cap in parallel with C_{REF} to adjust offset and trimming R2 for proper gain.

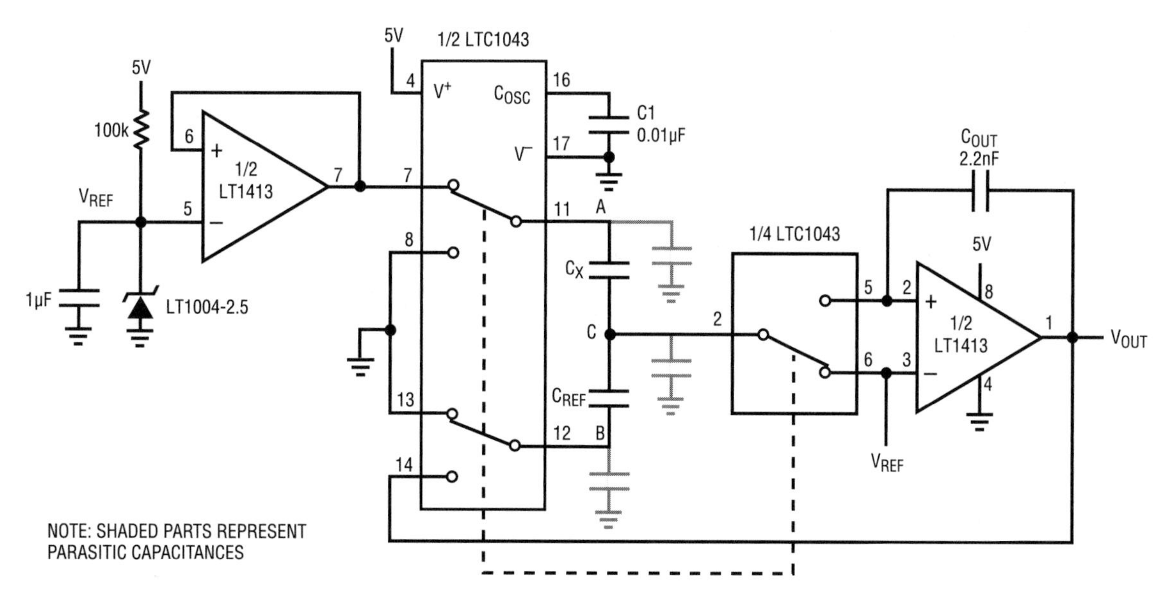

Figure 38.145 • A Simple, High Performance Capacitance Bridge

Bridge circuits are particularly suitable for differential measurements. When C_X and C_{REF} are replaced with two sensing capacitors, these circuits measure differential capacitance changes, but reject common mode changes. CMRR for the circuit in Figure 38.146 exceeds 70dB. In this case, however, the output is linear only for small relative capacitance changes.

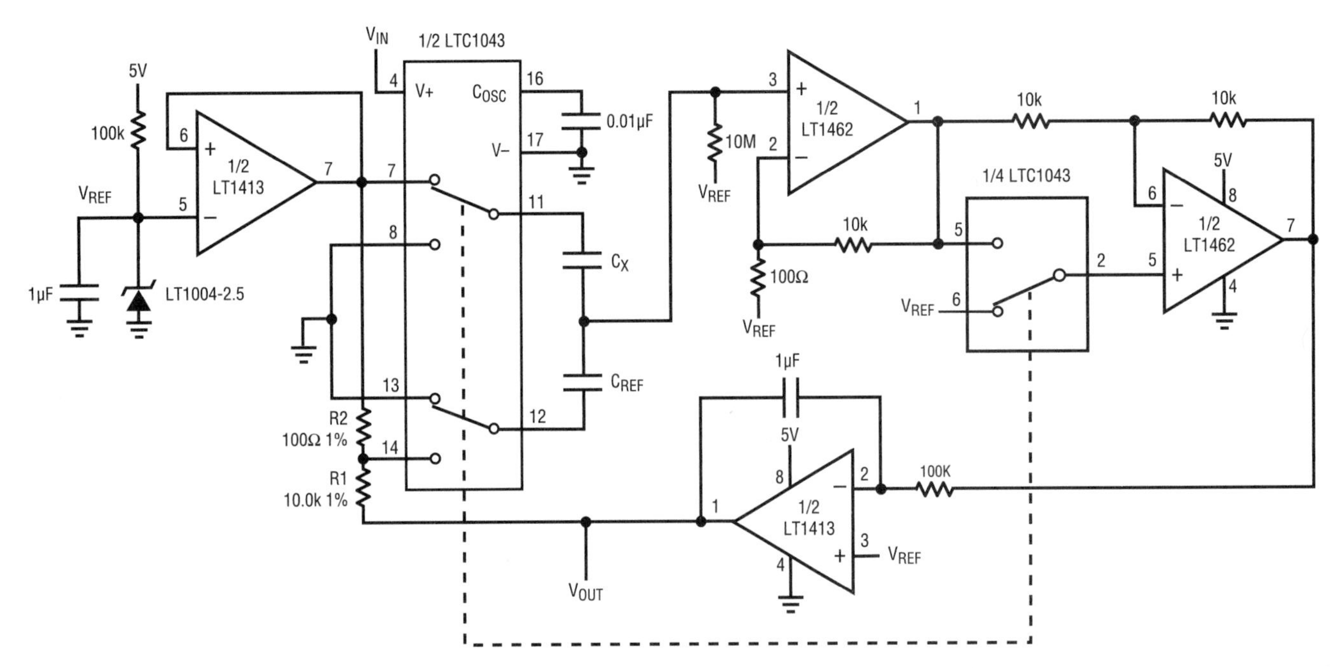

Figure 38.146 • A Bridge with Increased Sensitivity and Noise Performance

Water tank pressure sensing, a fluid solution

by Richard Markell

Introduction

Liquid sensors require a media compatible, solid state pressure sensor. The pressure range of the sensor is dependent on the height of the column or tank of fluid that must be sensed. This article describes the use of the E G & G IC Sensors Model 90 stainless steel diaphragm, 0 to 15psig sensor used to sense water height in a tank or column.

Because large chemical or water tanks are typically located outside in "tank farms," it is insufficient to provide only an analog interface to a digitization system for level sensing. This is because the very long wires required to interconnect the system cause IR drops, noise and other corruption of the analog signal. The solution to this problem is to implement a system that converts the analog to digital signals at the sensor: In this application, we implement a "liquid height to frequency converter."

Circuit description

Figure 38.147 shows the analog front-end of the system, which includes the LT1121 linear regulator for powering the system. The LT1121 is a micropower, low dropout linear regulator with shutdown. For micropower applications of this or other circuits, the ability to shut down the entire system via a single power supply pin allows the system to operate only when taking data (perhaps every hour), conserving power and improving battery life.

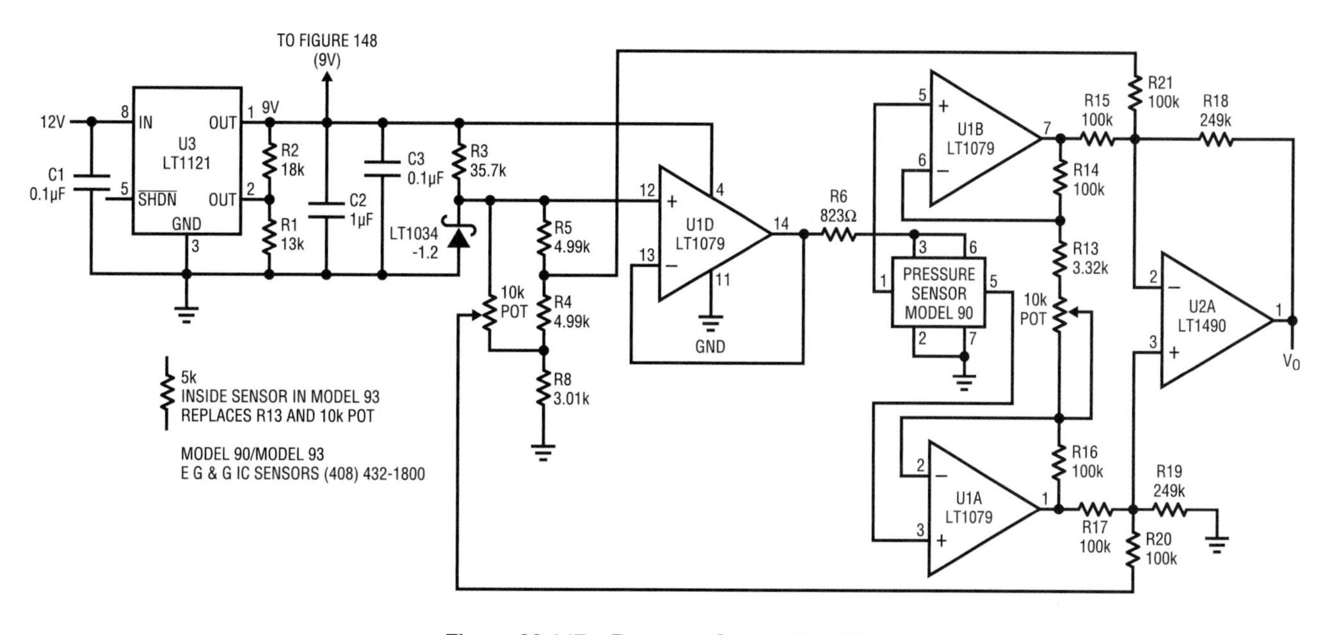

Figure 38.147 • Pressure-Sensor Amplifier

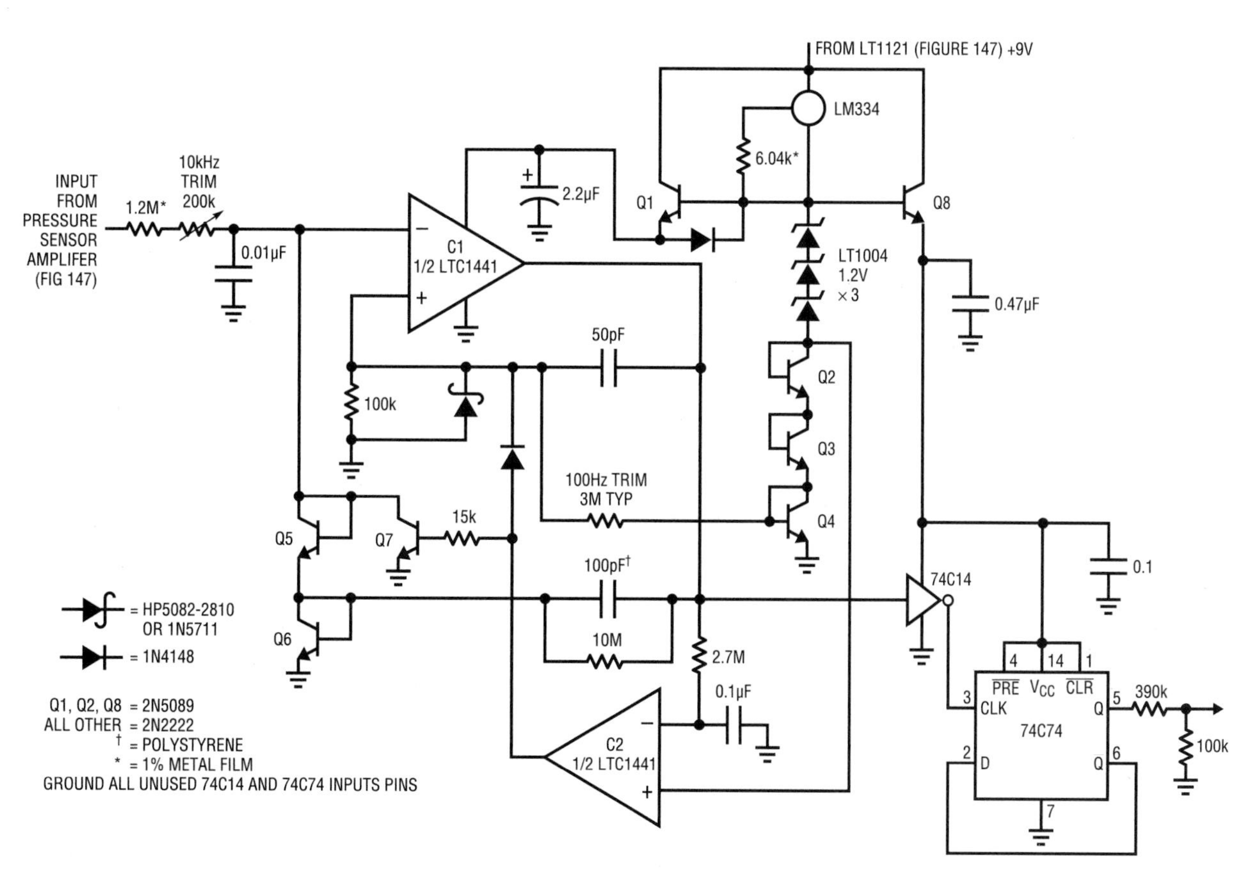

Figure 38.148 • This 0.02% V/F Converter Requires only 26µA Supply Current

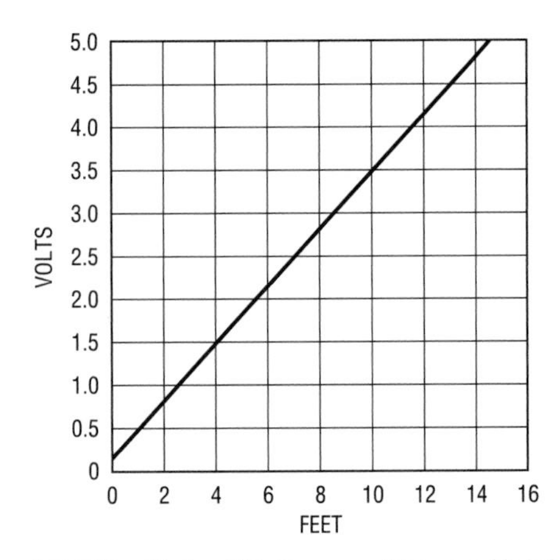

Figure 38.149 • Output Voltage vs Column Height

In Figure 38.147, U3, the LT1121, converts 12V to 9V to power the system. The 12V may be obtained from a wall cube or batteries.

The LT1034, a 1.2V reference, is used with U1D, 1/4 of an LT1079 quad low power op amp, to provide a 1.5mA current source to the pressure sensor The reference voltage is also divided down by R5, R8, R4 and the 10k potentiometer and used to offset the output amplifier; U2A, so that the signals are not too close to the supply rails.

Op amps U1A and U1B (each 1/4 of an LT1079) amplify the bridge pressure sensor's output and provide a differential signal to U2A (an LT1490). Note that U2A must be a rail-to-rail op amp. The system's analog output is taken from U2A's output.

Figure 38.149 plots the output voltage for the sensor system's analog front end versus the height of the water column that impinges on the pressure transducer Note that the pressure change is independent of diameter of the water column, so that a tank of liquid would produce the same resulting output voltage. Figure 38.150 is a photograph of our test setup.

The remainder of the circuitry, shown in Figure 38.148, allows transmission of analog data over long distances. The circuit was designed by Jim Williams. The circuit takes a DC input from 0V to 5V and converts it to a frequency. For the pressure circuit in Figure 38.147, this translates to approximately 0Hz to 5kHz.

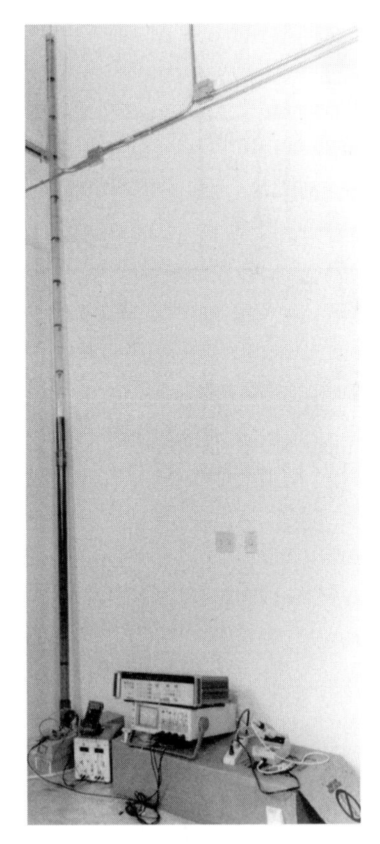

Figure 38.150 • Test Setup for Water-Column Sensor

The voltage-to-frequency converter shown in Figure 38.148 has very low power consumption (26μA), 0.02% linearity, 60ppm/°C drift and 40ppm/V power supply rejection.

In operation, C1 switches a charge pump, comprising Q5, Q6 and the 100pF capacitor, to maintain its negative input at 0V The LT1004s and associated components form a temperature-compensated reference for the charge pump.

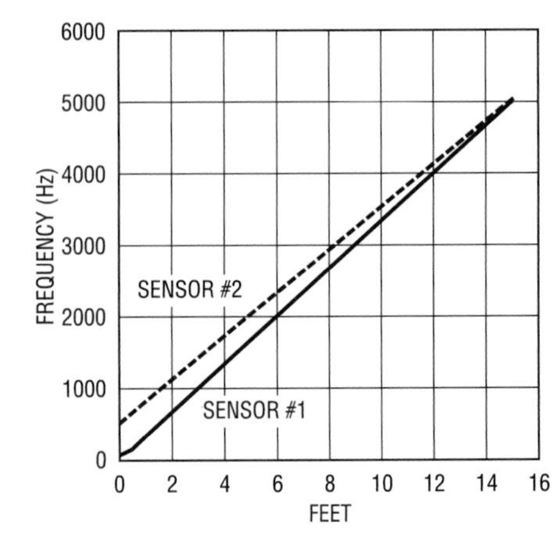

Figure 38.151 • Output Frequency vs Column Height for Two Model 90 Sensors

The 100pF capacitor charges to a fixed voltage; hence, the repetition rate is the circuit's only degree of freedom to maintain feedback. Comparator C1 pumps uniform packets of charge to its negative input at a repetition rate precisely proportional to the input-voltage-derived current. This action ensures that circuit output frequency is determined strictly and solely by the input voltage.

Figure 38.151 shows the output frequency versus column height for two different Model 90 transducers. Note the straight lines, which are representative of excellent linearity.

Conclusion

A cost effective system is shown here consisting of a fluid pressure sensor, IC Sensors Model 90. This sensor's output is fed to signal processing electronics that convert the low level DC output of the bridge-based pressure sensor to a frequency in the audio range depending on the height of the fluid column impinging on the pressure transducer.

05μV/°C chopped amplifier requires only 5μA supply current

by Jim Williams

Figure 38.152 shows a chopped amplifier that requires only 5.5μA supply current. Offset Voltage is 5μV with 0.05μV/°C drift. A gain exceeding 10^8 affords high accuracy, even at large closed-loop gains.

The micropower comparators (C1A and C1B) form a biphase 5Hz clock. The clock drives the input-related switches, causing an amplitude-modulated version of the DC input to appear at A1A's input. AC-coupled A1A takes a gain of 1000, presenting its output to a switched demodulator similar to the aforementioned modulator.

The demodulator output, a reconstructed, DC-amplified version of the circuit's input, is fed to A1B, a DC gain stage. A1B's output is fed back, via gain setting resistors, to the input modulator, closing a feedback loop around the entire amplifier. The configuration's DC gain is set by the feedback resistor's ratio, in this case 1000.

The circuit's internal AC coupling prevents A1's DC characteristics from influencing overall DC performance,

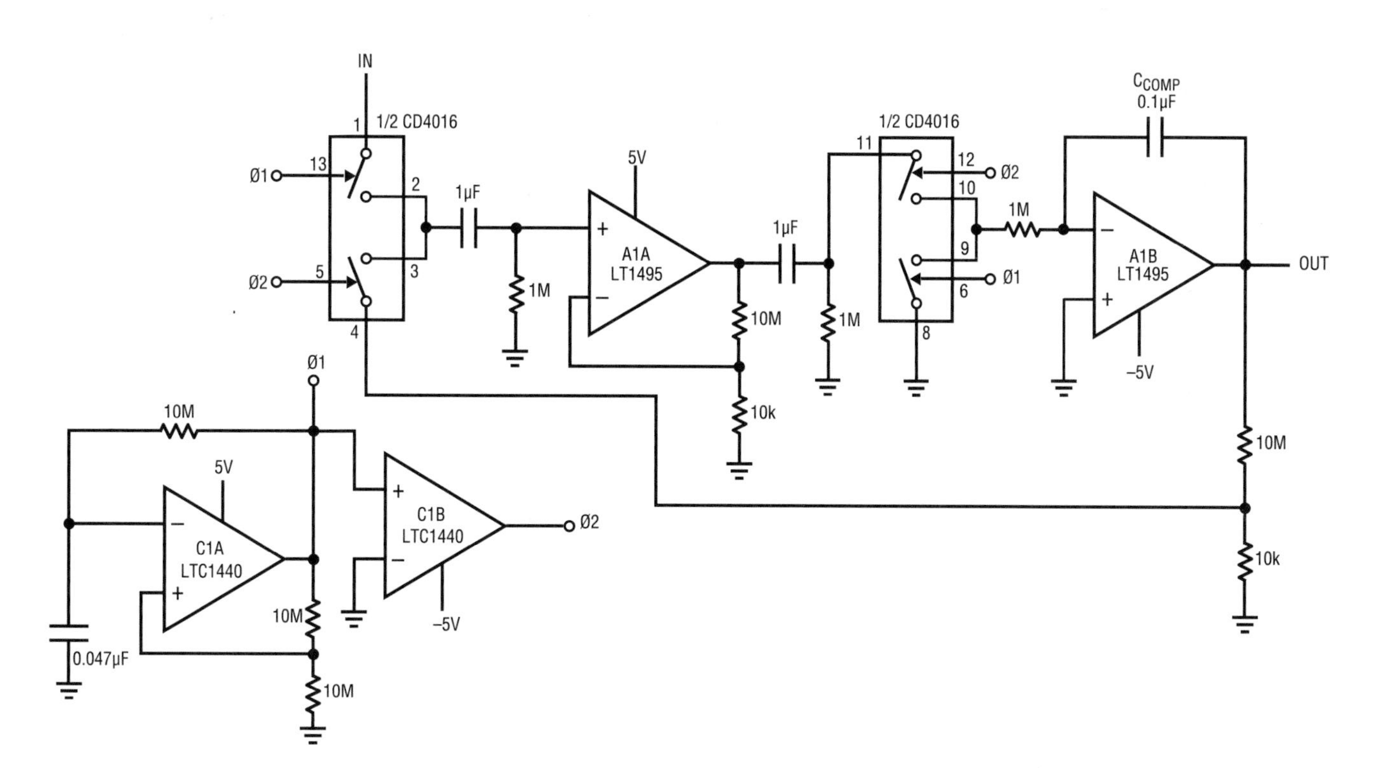

Figure 38.152 • 0.05μV/°C Chopped Amplifier Requires only 5μA Supply Current

accounting for the extremely low offset uncertainty noted. The high open-loop gain permits 10ppm gain accuracy at a closed-loop gain of 1000.

The desired micropower operation and A1's bandwidth dictate the 5Hz clock rate. As such, the resultant overall bandwidth is *low*. Full-power bandwidth is 0.05Hz with a slew rate of about 1V/s. Clock-related noise, about 5μV can be reduced by increasing C_{COMP}, with commensurate bandwidth reduction.

4.5ns dual-comparator-based crystal oscillator has 50% duty cycle and complementary outputs

by Joseph Petrofsky and Jim Williams

Figure 38.153's circuit uses the LT1720 dual comparator in a 50% duty cycle crystal oscillator. Output frequencies of up to 10MHz are practical.

The circuit of Figure 38.153 creates a pair of complementary outputs with a forced 50% duty cycle. Crystals are narrowband elements, so the feedback to the noninverting input is a filtered analog version of the square wave output. Changing the noninverting reference level can therefore vary the duty cycle. C1 operates as in the previous example, where the 2k-600Ω resistor pair sets a bias point at the comparator's noninverting input. The 2k-1.8k-0.1μF path sets the inverting input at the node at an appropriate DC-average level based on the output.

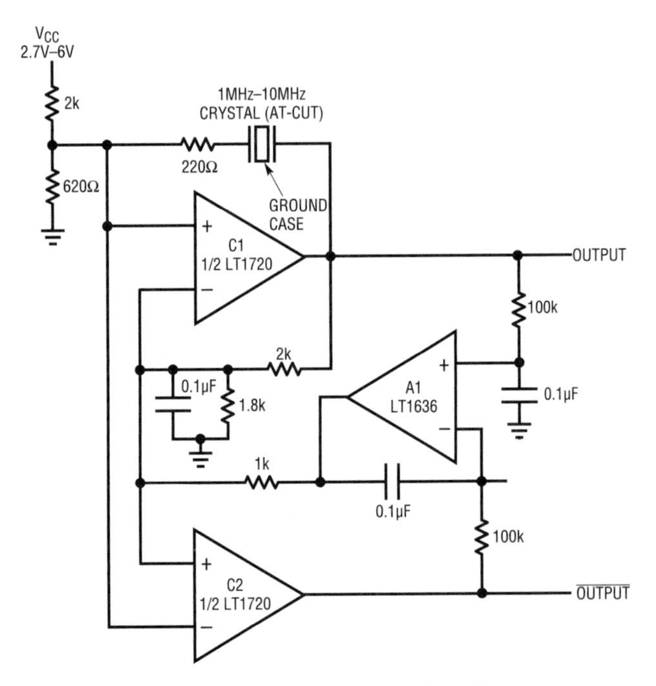

Figure 38.153 • Crystal Oscillator has Complementary Outputs and 50% Duty Cycle. A1's Feedback Maintains Output Duty Cycle Despite Supply Variations

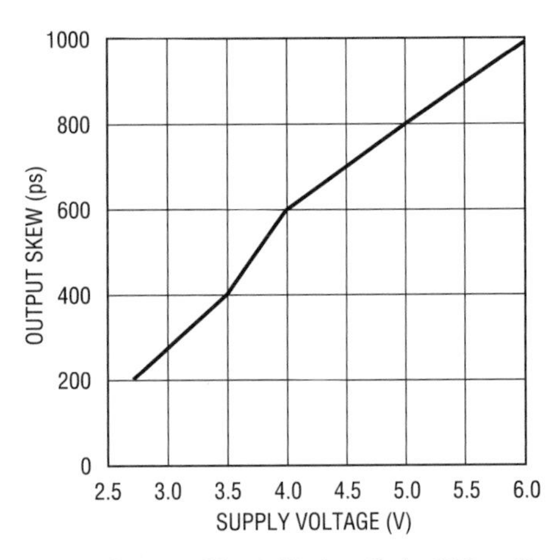

Figure 38.154 • Output Skew Varies Only 800ps Over a 2.7V–6V Supply Excursion

The crystal's path provides resonant positive feedback, and stable oscillation occurs. The DC bias voltages at the inputs are set near the center of the LT1720's common mode range and the 220Ω resistor attenuates the feedback to the noninverting input. C2 creates a complementary output by comparing the same two nodes with the opposite input polarity. A1 compares band-limited versions of the outputs and biases C1's negative input. C1's only degree of freedom to respond is variation of pulse width; hence, the outputs are forced to 50% duty cycle. The circuit operates from 2.7V to 6V and the skew between the edges of the of the two outputs is as shown in Figure 38.154. There is a slight duty-cycle dependence on comparator loading, so equal capacitive and resistive loading should be used in critical applications. This circuit works well because of the two matched delays and rail-to-rail-style outputs of the LT1720.

LTC1531 isolated comparator

by Wayne Shumaker

Introduction

The LTC1531 is an isolated, self-powered comparator that receives power and communicates through internal isolation capacitors. The internal isolation capacitors provide $3000V_{RMS}$ of isolation between the comparator and its output. This allows the part to be used in applications that require high voltage isolated sensing without the need to provide an isolated power source. The isolated side provides a 2.5V pulsed reference output that can deliver 5mA for 100μs using the power stored on the isolated external capacitor. A 4-input, dual-differential comparator samples at the end of the reference pulse and transmits the result back to the nonisolated side. The nonisolated, powered side latches the result of the comparator and provides a zero-cross comparator output for triggering a triac.

Applications

The LTC1531 can be used to isolate sensors such as in the isolated thermistor temperature controller in Figure 38.155. In this circuit, a comparison is made between the voltages across a thermistor and a resistor that is driven

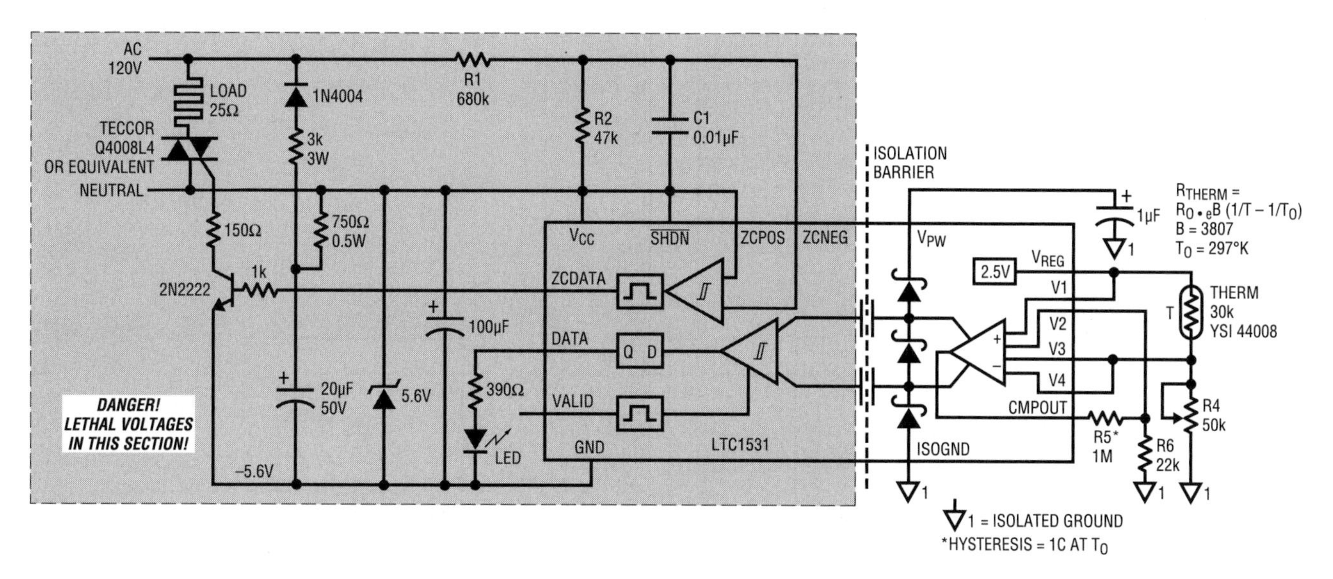

Figure 38.155 • Isolated Thermistor Temperature Controller

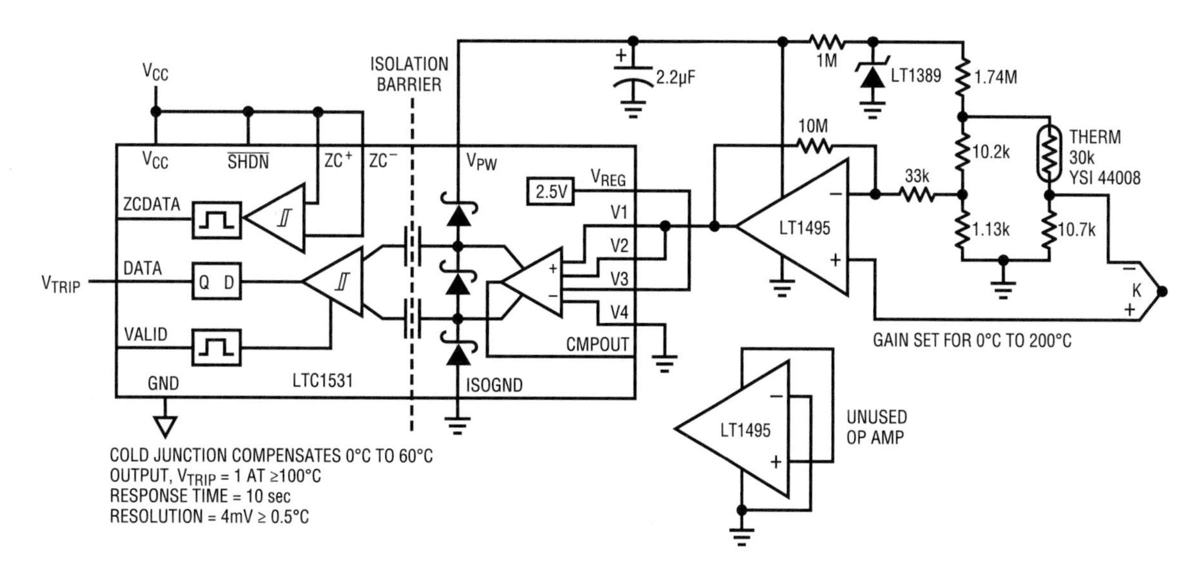

Figure 38.156 • Overtemperature Detect

by the 2.5V V_{REG} output. As the thermistor resistance rises with temperature, the voltage across the thermistor increases. When it exceeds the voltage across R4, the comparator output becomes zero and the triac control to the heater is turned off. Hysteresis can be added in the temperature control by using CMPOUT and R5. A 10° phase-shifted AC line signal is supplied through R1, R2 and C1 to the zero-cross comparator for firing the triac.

In the overtemperature detect application in Figure 38.156, an isolated thermocouple is cold junction compensated with the micropower LT1389 reference and the Yellow Springs thermistor. The micropower LT1495 op amp provides gain to give an overall 0°C–200°C temperature range, adjustable by changing the 10M feedback resistor. The isolated comparator is connected to compare at 1.25V or the center of the temperature range. In this case, V_{TRIP} goes high when the temperature exceeds 100°C.

The LTC1531 can use the high impedance nature of CMPOUT as a duty-cycle modulator, as in the isolated voltage sense application in Figure 38.157. The duty-cycle output of the comparator is smoothed with the LT1490 rail-to-rail op amp to reproduce the voltage at V_{IN}. The output time constant, R2 • C2, should approximately equal the input time constant, 35 • R1 • C1. The factor of 35 results from CMPOUT being on for only 100μs at an average sample rate of 300Hz.

Conclusion

The LTC1531 is a versatile part for sensing signals that require large isolation voltages. The ability of the LTC1531 to supply power through the isolation barrier simplifies applications; it can be combined with other micropower circuits in a variety of isolated signal conditioning and sensing applications.

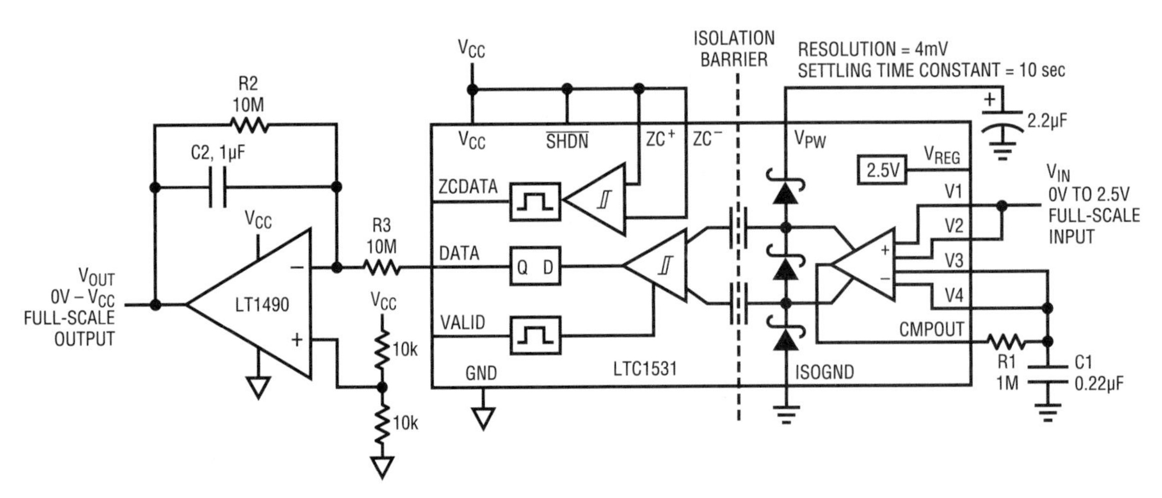

Figure 38.157 • Isolated Voltage Detect

Filters

The LTC1560-1: a 1MHz/500kHz continuous-time, low noise, elliptic lowpass filter

by Nello Sevastopoulos

Introduction

The LTC1560-1 is a high frequency, continuous-time, low noise filter in an SO-8 package. It is a single-ended input, single-ended output, 5th order elliptic lowpass filter with a pin-selectable cutoff frequency (f_C) of 1MHz or 500kHz.

The LTC1560-1 delivers accurate fixed cutoff frequencies of 500kHz and 1MHz without the need for internal or external clocks.

Applications and experimental results

The LTC1560-1 can be used as part of a more complete frequency-shaping system. Two representative examples follow.

Highpass-lowpass filter

As a typical application in communication systems, where there is a need to reject DC and some low frequency signals, a 2nd order RC highpass network can be inserted in front of the LTC1560-1 to obtain a highpass-lowpass response. Figures 38.158 and 38.159 depict the network and its measured frequency response, respectively. Notice that the second resistor in the highpass filter is the input resistance of the LTC1560-1, which is about 8.1k.

Delay-equalized elliptic filter

Although elliptic filters offer high Q and a sharp transition band, they lack a constant group delay in the passband, which implies more ringing in the time-domain step response. In order to minimize the delay ripple in the passband of the LTC1560-1, an allpass filter (delay equalizer) is cascaded with the LTC1560-1, as shown in Figure 38.160. Figures 38.161 and 38.162 illustrate the eye diagrams before and after the equalization, respectively.

An eye diagram is a qualitative representation of the time- domain response of a digital communication system. It shows how susceptible the system is to *intersymbol interference* (ISI). *Intersymbol interference* is caused by erroneous decisions in the receiver due to pulse overlapping and decaying oscillations of a previous symbol. A pseudorandom 2-level sequence has been used as the input of the LTC1560-1 to generate these eye diagrams. The larger eye opening in Figure 38.162 is an indication of the equalization effect that leads to reduced ISI. Note that in Figure 38.160, the equalizer section has a gain of 2 for driving and back-terminating 50Ω cable and load. For a simple unterminated gain-of-1 equalizer, the 40.2k resistor changes to 20k and the 49.9Ω resistor is removed from the circuit. The 22pF capacitors are 1% or 2% dipped silver mica or COG ceramic.

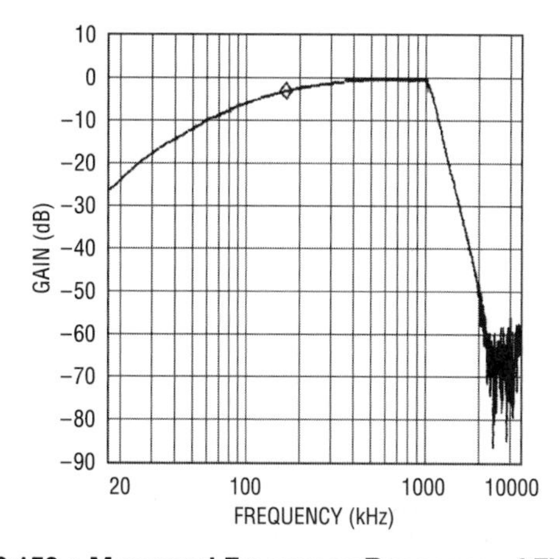

Figure 38.159 • Measured Frequency Response of Figure 38.158's Circuit

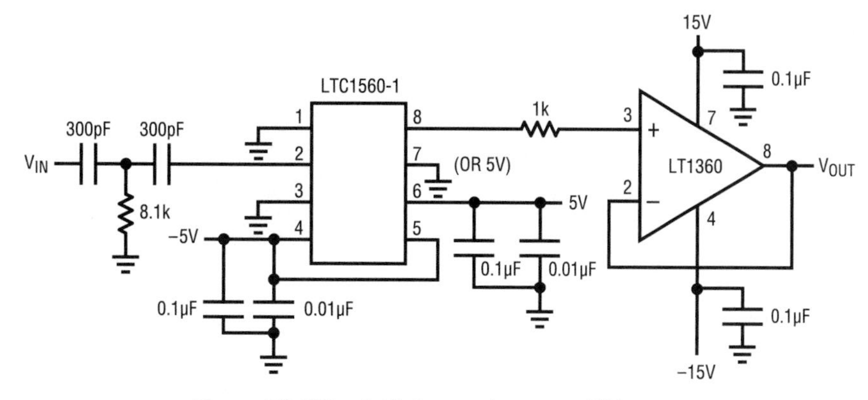

Figure 38.158 • A Highpass-Lowpass Filter

Conclusion

The LTC1560-1 is a 5th order elliptic lowpass filter that features a 10-bit gain linearity at signal ranges up to 1MHz. Being small and user friendly, the LTC1560-1 is suitable for any compact design. It is a monolithic replacement for larger, more expensive and less accurate solutions in communications, data acquisitions, medical instrumentation and other applications.

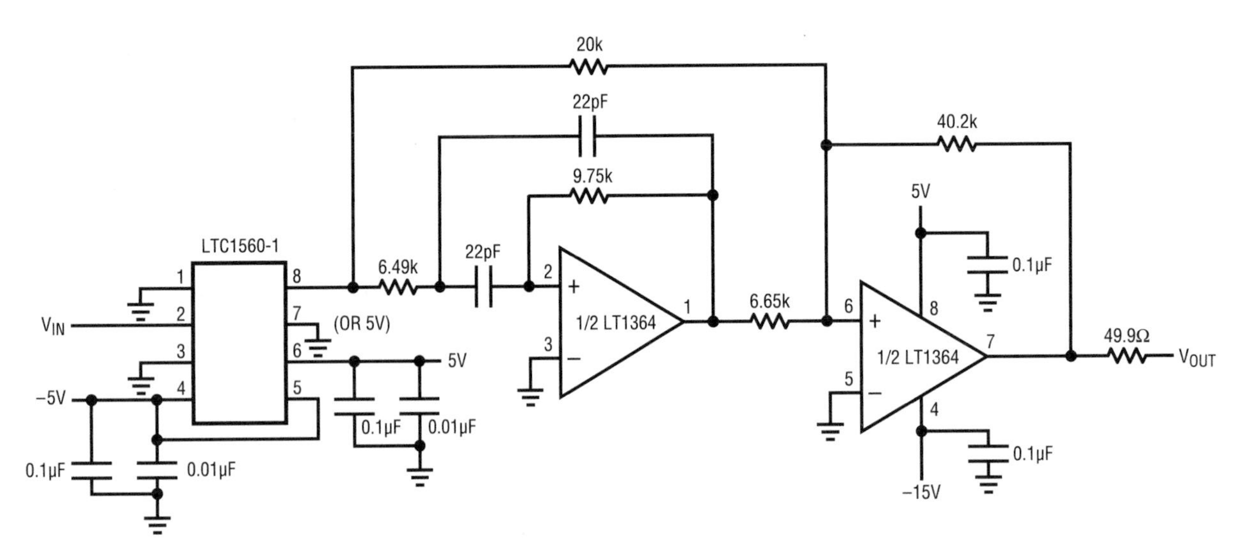

Figure 38.160 • Augmenting the LTC1560-1 for Improved Delay Flatness

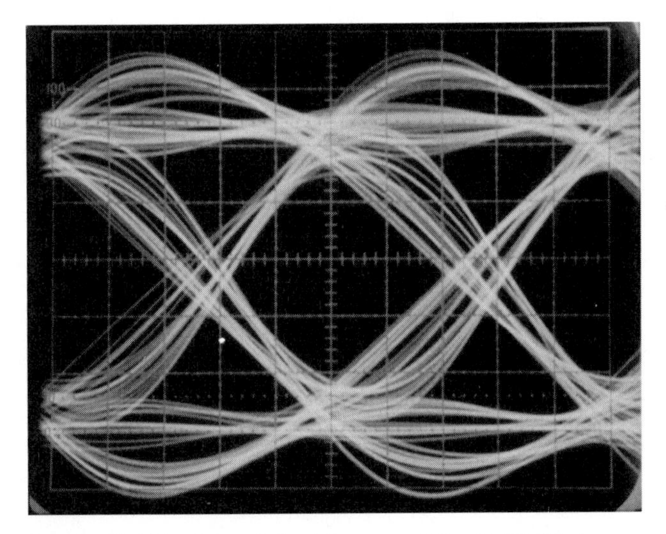

Figure 38.161 • 2-level Eye Diagram of the LTC1560-1 Before Equalization

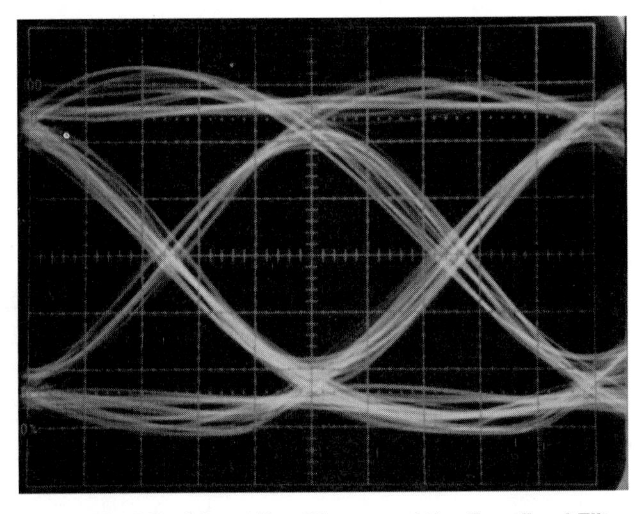

Figure 38.162 • 2-level Eye Diagram of the Equalized Filter

The LTC1067 and LTC1067-50: universal 4th order low noise, rail-to-rail switched capacitor filters

by Doug La Porte

LTC1067 and LTC1067-50 overview

The LTC1067 and the LTC1067-50 are universal, 4th order switched capacitor filters with rail-to-rail operation. Each part contains two identical, high accuracy, very wide dynamic-range 2nd order filter building blocks. Each building block, together with three to five resistors, provides 2nd order filter transfer functions, including lowpass, bandpass, highpass, notch and allpass. These parts can be used to easily design 4th order or dual 2nd order filters.

Linear Technology's FilterCAD™ for Windows filter design software fully supports designs with these parts.

The center frequency of each 2nd order section is tuned by an external clock. The LTC1067 has a 100:1 clock-to-center frequency ratio. The LTC1067-50's clock-to-center frequency ratio is 50:1.

Some LTC1067 and LTC1067-50 applications

High dynamic-range Butterworth lowpass filter with built-in track-and-hold challenges discrete designs

Figure 38.163 shows an LTC1067 configured as a 5kHz Butterworth lowpass filter. This circuit runs on a 3.3V power supply and uses an external logic gate to stop the clock for track-and-hold operation. The transfer function for this circuit, shown in Figure 38.164, is the classical Butterworth response. This circuit can be used with either the LTC1067 or the LTC1067-50. The broad-band noise for the LTC1067 circuit is $45\mu V_{RMS}$ and the DC offset is typically less than 10mV For the LTC1067-50, the broad-band noise is $55\mu V_{RMS}$ and the DC offset is typically less than 15mV.

This circuit has tremendous dynamic range, even on low supply voltages. Figure 38.165 shows a plot of the LTC1067's signal-to-noise plus total harmonic distortion (SINAD) vs input signal level for a 1kHz input at three different power supply voltages. SINAD is limited for small signals by the noise floor of the LTC1067, for medium signals by the part's linearity and for large signals by the output signal swing. The part's low noise input stage and excellent linearity allow the SINAD to exceed 80dB for signals as small as $700mV_{P-P}$, while the rail-to-rail output stage maintains this level for input signals approaching the supply rails. Previous parts could not attain this high

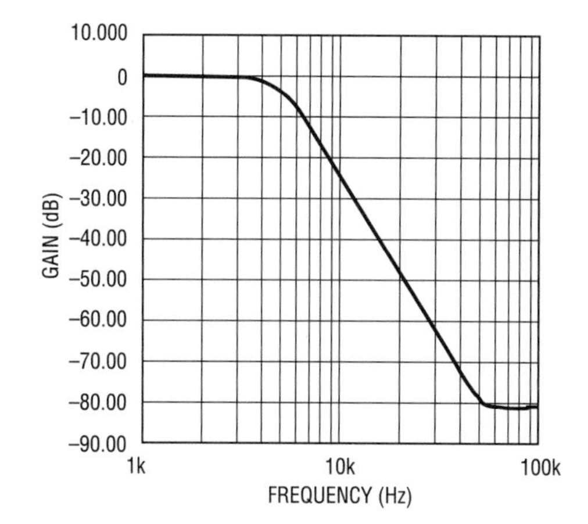

Figure 38.164 • Transfer Function of the LTC1067 5kHz Butterworth LPF

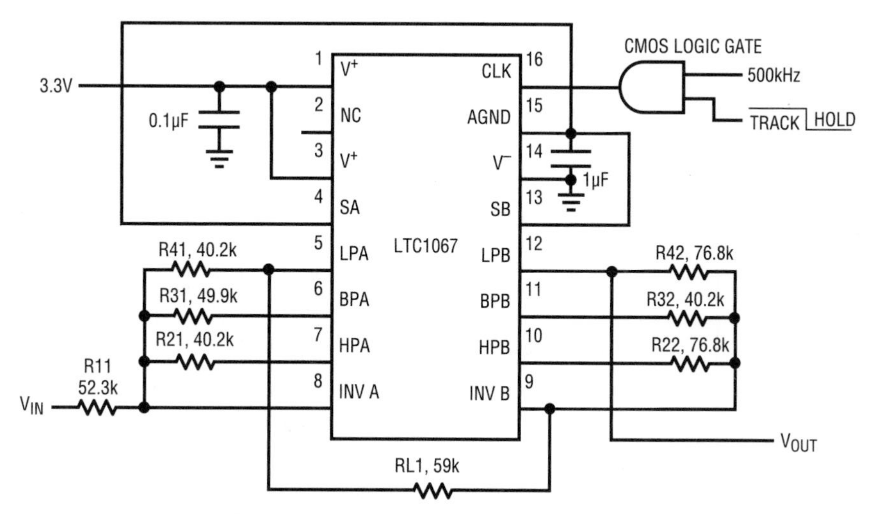

Figure 38.163 • High Dynamic-Range Butterworth LPF with Track-and-Hold Control

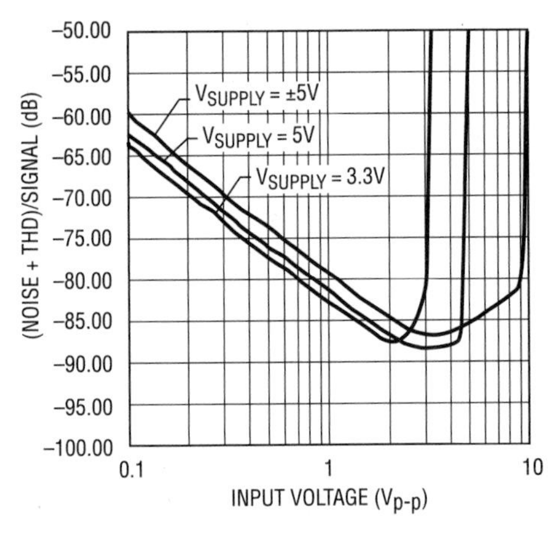

Figure 38.165 • Dynamic Range of LTC1067 Butterworth LPF

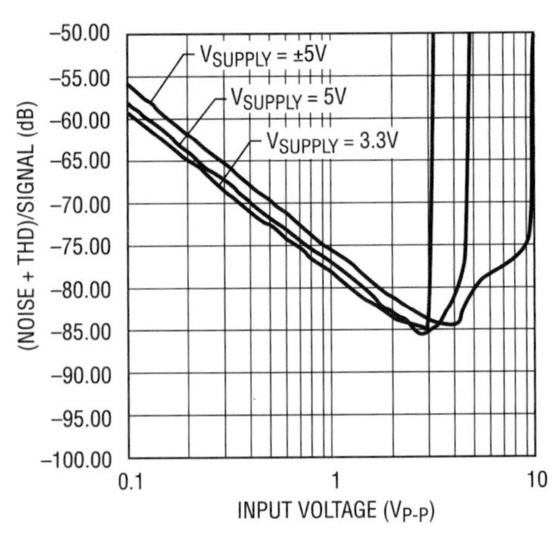

Figure 38.166 • Dynamic Range of LTC1067-50 Butterworth LPF

dynamic range due to higher input noise levels, poor linearity and limited output-stage signal swing. The low noise and rail-to-rail output swing are especially crucial on the lower 3.3V power supply, where every bit of detectable signal range is precious. Figure 38.166 shows the same plot for the LTC1067-50 circuit. The dynamic range is not quite equal to that of the LTC1067, but is still very good. Recall that, for the same clock frequency, the LTC1067-50 based filter has double the bandwidth and half the supply current of the LTC1067.

The LTC1067 and LTC1067-50 also perform a track-and-hold function. Stopping the clock holds the output of the filter at its last value. The LTC1067 is the best performing part in this area. The LTC1067's hold step is less than $-100\mu V$ and the droop rate is less than $-50\mu V/ms$ over the full temperature range. These numbers compare very favorably with dedicated track-and-hold amplifiers. When the clock is restarted, the filter resumes normal operation within ten clock cycles and the output will then correctly reflect the input as soon as the filter's mathematical response allows.

Elliptic lowpass filter

The LTC1067 family is capable of much more challenging filters. Figure 38.167 shows the schematic for a 25kHz elliptic lowpass filter using the LT1067-50 operating on a 5V supply. Maximum attenuation one octave from the −3dB corner is the design goal for this filter. Figure 38.168 shows the frequency response of the filter with the −3dB cutoff at 25kHz and −48dB of attenuation at 50kHz. The broad-band noise of the filter is $85\mu V_{RMS}$ and the DC offset is less than 15mV typically.

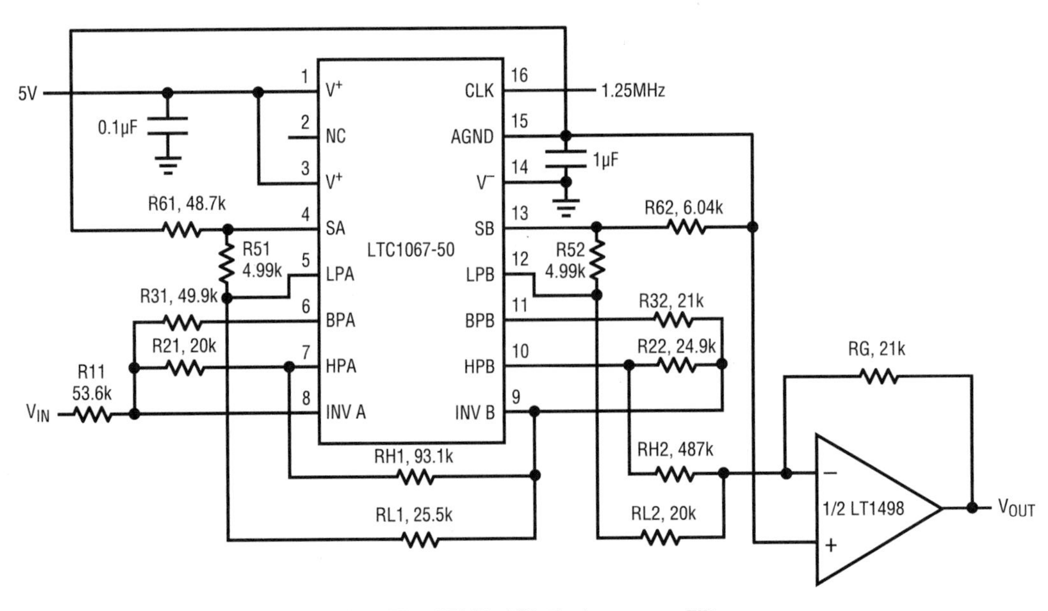

Figure 38.167 • 25kHz Elliptic Lowpass Filter

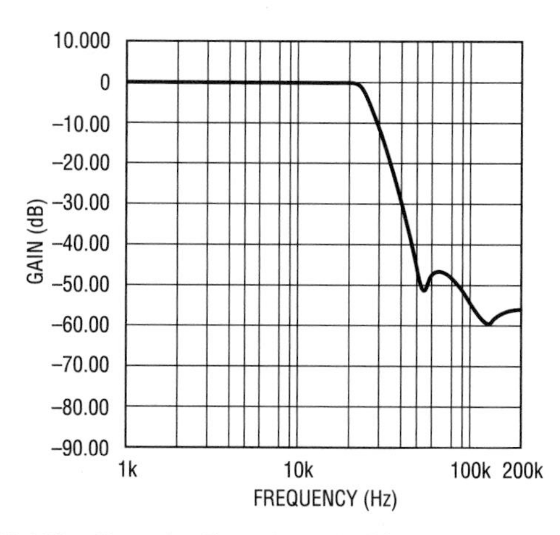

Figure 38.168 • Transfer Function of LTC1067-50 25kHz LPF

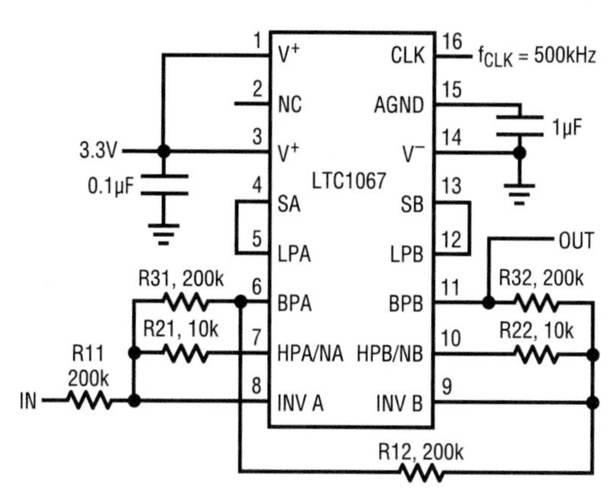

Figure 38.169 • Low Noise, Low Voltage Narrow BPF

Although Figure 38.167 shows the filter powered by a single 5V supply, 3.3V or ±5V supply operation is also supported. The maximum cutoff frequency is 15kHz for the 3.3V supply and 35kHz for the ±5V supply. The same design and schematic used with an LTC1067 will achieve a somewhat lower noise, lower DC-offset filter. With the LTC1067, the broad-band noise is $70\mu V_{RMS}$ and the DC offset is typically less than 10mV The maximum operating frequencies for the LTC1067 are one half of those for the LTC1067-50.

Narrow-band bandpass filter design extracts small signals buried in noise

Narrow-band bandpass filters are difficult to design but are easily achievable with these parts. Most applications for these filters involve extracting a low level signal from a noisy environment. The noise may be the standard broadband, Gaussian-type noise or it may consist of multiple interfering signals. For example, the signal may be a low level tone or a narrow-bandwidth modulated signal, in a voice-band system. The presence of the tone must be detected even while the large voice signals are present. A narrow-band bandpass filter will allow the tone to be separated and detected even in this hostile environment. Numerous systems also require a narrow bandpass filter to be swept across a band looking for the tones. Switched capacitor filters allow the filter to be swept by simply changing the clock frequency.

To achieve success in designing narrow-band bandpass filters, you must start with precision components. In an LC or RC design, you would have to start with 0.1% resistors, 1% inductors and 1% capacitors to have any hope of finishing with a successful, repeatable design in production. A competing solution, a digital filter implementation, also requires precision components. The full input signal (signal, noise and out-of-band interference) must be correctly digitized and then processed with a DSP device to finally determine the tone's presence. If an out-of-band interfering signal is 20dB greater than the desired tone, the ADC must have an extra 20dB of dynamic range above the signal's requirement. To pull a small-signal tone from a large signal interferer, you may need a 16-bit ADC to digitize the signal just to get 12-bit resolution of the tone after processing. The added cost, power, board space and development time make this approach unattractive.

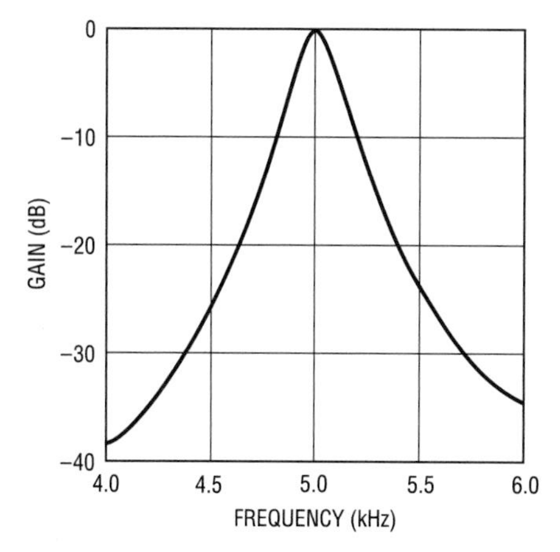

Figure 38.170 • Frequency Response of Narrow BPF

A precision switched capacitor filter provides a simple, small, low power, repeatable, inexpensive solution. The older MF-10-type parts do not have the necessary f_O accuracy to achieve a reliable, repeatable design. Figure 38.169 shows the schematic of a narrow-band bandpass filter centered at 5kHz. The design uses two identical cascaded sections, each with a Q of 20. Multiply the individual Q of each section by 1.554 to calculate the total Q of a filter with two identical f_O, identical Q sections. This filter has a total Q of 31. For tunable filter applications, simply lowering the clock frequency lowers the center frequency of the filter. Figure 38.170 shows the frequency response of this filter. The broad-band noise of this filter is only $90\mu V_{RMS}$. Highly selective bandpass filters are possible due to the LTC1067's excellent f_0 accuracy.

Higher Q, narrower bandwidth filters are achievable with 0.1% resistors or matched resistor networks. An LTC1067 mask-programmed part is ideal for these ultranarrow filters. The well matched, on-chip resistors, coupled with specified test conditions, yield a fully functioning filter module, in an SO-8 package, without any of the hassles or cost of procuring precision resistors or resistor networks.

Narrow-band notch filter design reaches 80dB notch depth

Narrow-band notch filters are especially challenging designs. The requirement for most notch filters is to remove a particular tone and not affect any of the remaining signal bandwidth. This requires an infinitesimally narrow filter that can only be approximated by a reasonably narrow bandwidth. These types of filters, like the narrow-band bandpass discussed above, require precision f_O accuracy. Figure 38.171 shows the schematic of this type of filter. This filter is a 1.02kHz notch filter that is often used in telecommunication test systems.

One of the challenges of designing a switched capacitor notch filter i nvolves the broad-band nature of a notch filter. The broad-band noise can be aliased down into the band of interest. Optimal high performance notch filters should employ some form of noise-band limiting. To accomplish the noise-band limiting, the design in Figure 38.171 places capacitors in parallel with the R2 resistors of each 2nd order section. This forms a pole, set at $f_P = 1/(2 \bullet \pi \bullet R2 \bullet C2)$, that will limit the bandwidth. This pole frequency must be low enough to have a band-limiting effect but must not be so low as to affect the notch filter's response. The pole should be greater than thirty times the notch frequency and less than seventy-five times the notch frequency for the best results. Figure 38.172 shows the frequency response of the filter. Note that the notch depth is greater than −80dB. Without the use of the C21 and C22, the notch depth is only about −35dB.

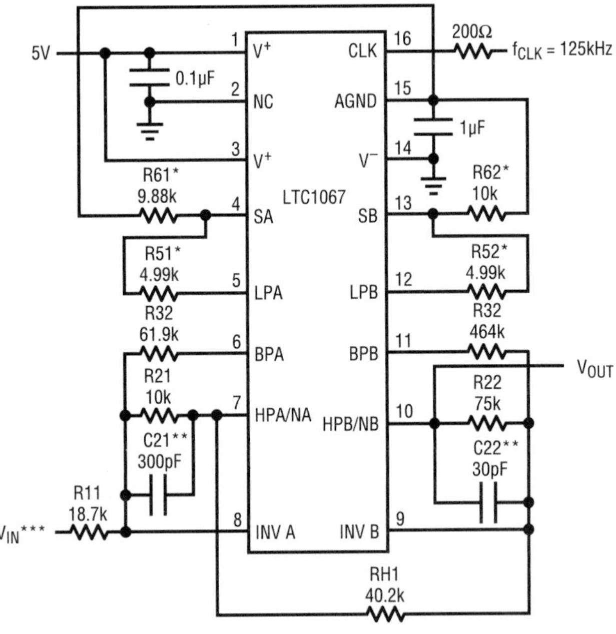

Figure 38.171 • Narrow-Band Notch Filter

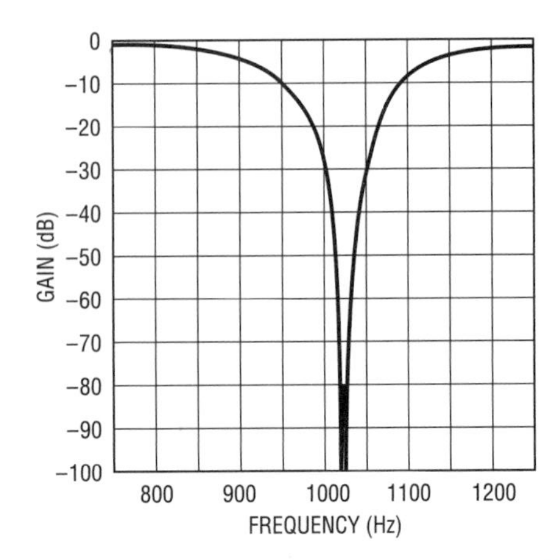

Figure 38.172 • Measured Frequency Response of Figure 38.171's Narrow-Band Notch Filter

Universal continuous-time filter challenges discrete designs

by Max Hauser

The LTC1562 is the first in a new family of tunable, DC-accurate, continuous-time filter products featuring very low noise and distortion. It contains four independent 2nd order, 3-terminal filter blocks that are resistor programmable for lowpass or bandpass functions up to 150kHz, and has a complete PC board footprint smaller than a dime. Moreover, the part can deliver arbitrary continuous-time pole-zero responses, including highpass, notch and elliptic, if one or more programming resistors are replaced with capacitors. The center frequency (f_0) of the LTC1562 is internally trimmed, with an absolute accuracy of 0.5%, and can be adjusted independently in each 2nd order section from 10kHz to 150kHz by an external resistor. Other features include:

- Rail-to-rail inputs and outputs
- Wideband signal-to-noise ratio (SNR) of 103dB
- Total harmonic distortion (THD) of −96dB at 20kHz, −80dB at 100kHz
- Built-in multiple-input summing and gain features; capable of 118dB dynamic range
- Single- or dual-supply operation, 4.75V to 10.5V total
- "Zero-power" shutdown mode under logic control
- No clocks, PLLs, DSP or tuning cycles required

The LTC1562 provides eight poles of programmable continuous-time filtering in a total surface mount board area (including the programming resistors) of 0.24 square inches (155 mm^2)—smaller than a U.S. 10-cent coin. This filter can also replace op amp-R-C active filter circuits and LC filters in applications requiring compactness, flexibility, high dynamic range or fewer precision components.

Each of the four 3-terminal Operational Filter™ building blocks in an LTC1562 has a virtual ground input, INV and two outputs, V1 and V2. These are described in detail in the LTC1562 data sheet.

Dual 4th order 100kHz Butterworth lowpass filter

The practical circuit in Figure 38.173 is a dual lowpass filter with a Butterworth (maximally-flat-passband) frequency response. Each half gives a DC-accurate, unity-passband- gain lowpass response with rail-to-rail input and output. With a 10V total power supply, the measured output noise for one filter is $36\mu V_{RMS}$ in a 200kHz bandwidth, and the large-signal output SNR is 100dB. Measured THD at $1V_{RMS}$ input is −83.5dB at 50kHz and −80dB at 100kHz. Figure 38.174 shows the frequency response of one filter.

8th order 30kHz Chebyshev highpass filter

Figure 38.175 shows a straightforward use of the highpass configuration. Each of the four cascaded 2nd order sections has an external capacitor in the input path. The resistors in Figure 38.175 set the f_0 and Q values of the four sections to realize a Chebyshev (equiripple-passband) response with 0.05dB ripple and a 30kHz highpass corner. Figure 38.176 shows the frequency response. Total output noise for this circuit is $40\mu V_{RMS}$.

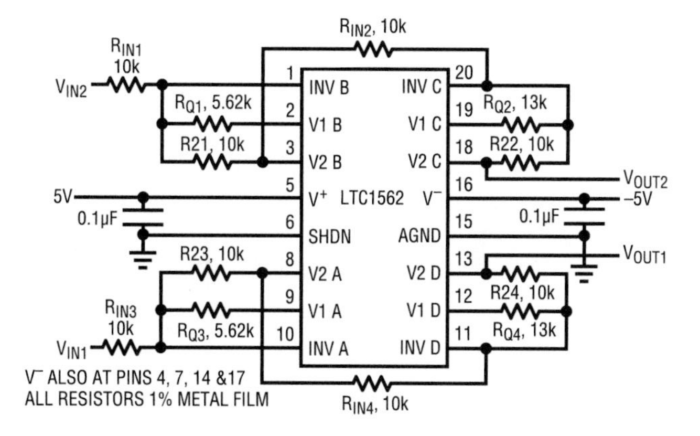

Figure 38.173 • Dual, Matched 4th Order 100kHz Butterworth Lowpass Filter

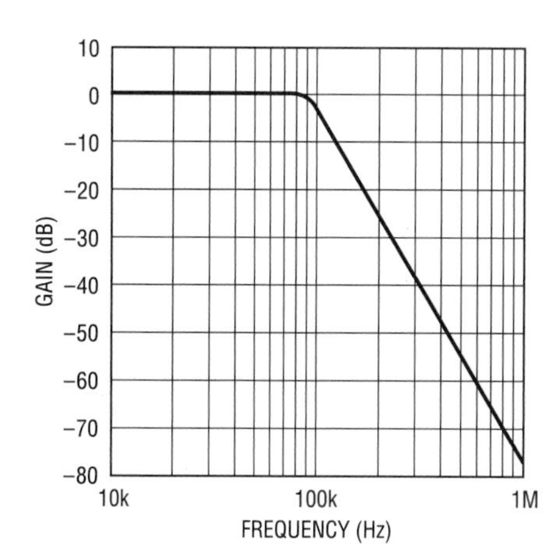

Figure 38.174 • Frequency Response of Figure 38.173's Circuit

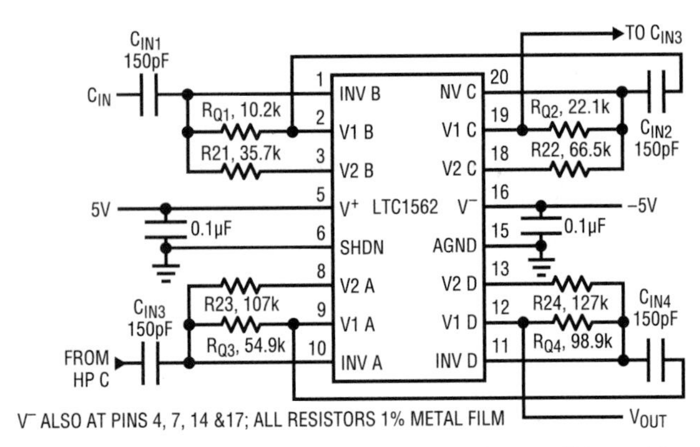

Figure 38.175 • 8th Order Chebyshev Highpass Filter with 0.05dB Ripple (f_{CUTOFF} = 30kHz)

50kHz, 100dB elliptic lowpass filter

Figure 38.177 illustrates how sharp-cutoff filtering can exploit the Operational Filter capabilities of the LTC1562. In this design, external capacitors are added and the virtual-ground inputs of the LTC1562 sum parallel paths to obtain three notches in the stopband of a lowpass filter, as plotted in Figure 38.178. This response falls 100dB in a little more than one octave; the total output noise is $60\mu V_{RMS}$ with the rail-to-rail output for a peak SNR of 95dB from ±5V supplies.

Quadruple 3rd order 100kHz Butterworth lowpass filter

Another example of the flexibility of the virtual-ground inputs is the ability to add an extra, independent real pole with an R-C-R "T" network. In Figure 38.179, a 10k input resistor has been split into two parts and the parallel combination of the two forms a 100kHz real pole with the 680pF external capacitor. Four such 3rd order Butterworth lowpass filters can be built from one LTC1562.

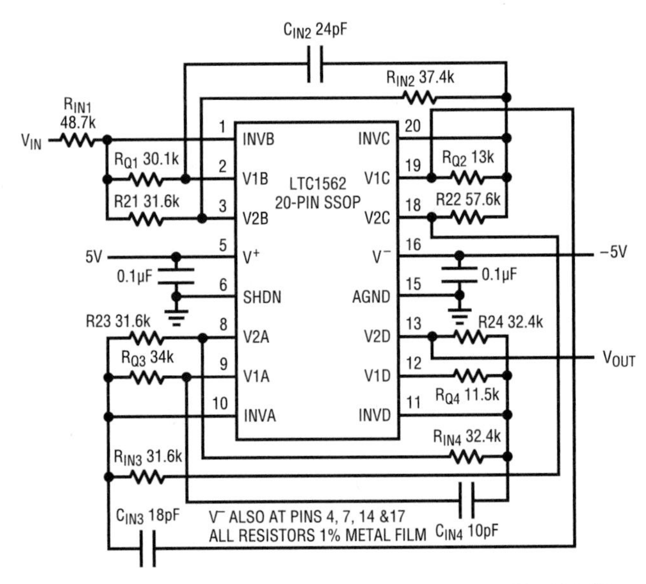

Figure 38.177 • 50kHz Elliptic Lowpass Filter with 100dB Stopband Rejection

The same technique can add additional real poles to other filter configurations as well, for example, augmenting Figure 38.173's circuit to obtain a dual 5th order filter from a single LTC1562.

Conclusion

The LTC1562 is the first truly compact universal active filter, yet it offers instrumentation-grade performance rivaling much larger discrete-component designs. It serves applications in the 10kHz–150kHz range with an SNR as high as 100dB or more (16+ equivalent bits). The LTC1562 is ideal for modems and other communications systems and for DSP antialiasing or reconstruction filtering.

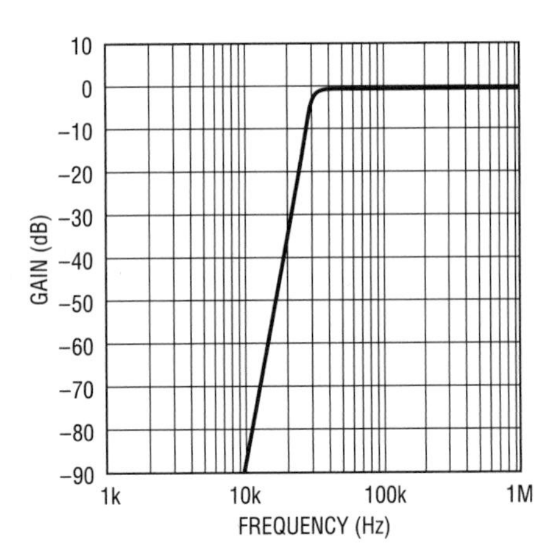

Figure 38.176 • Frequency Response of Figure 38.175's Circuit

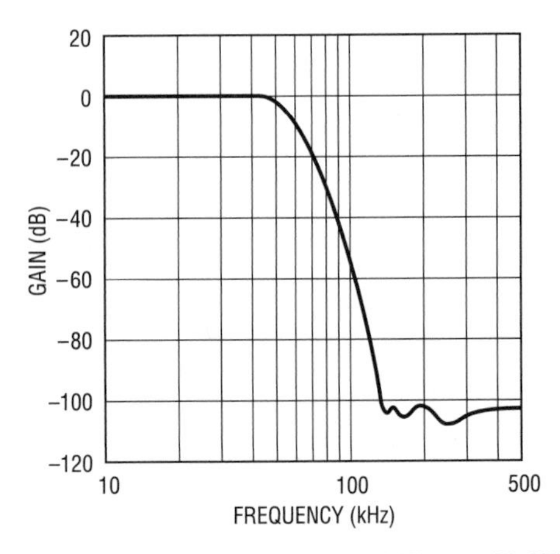

Figure 38.178 • Frequency Response of Figure 38.177's Circuit

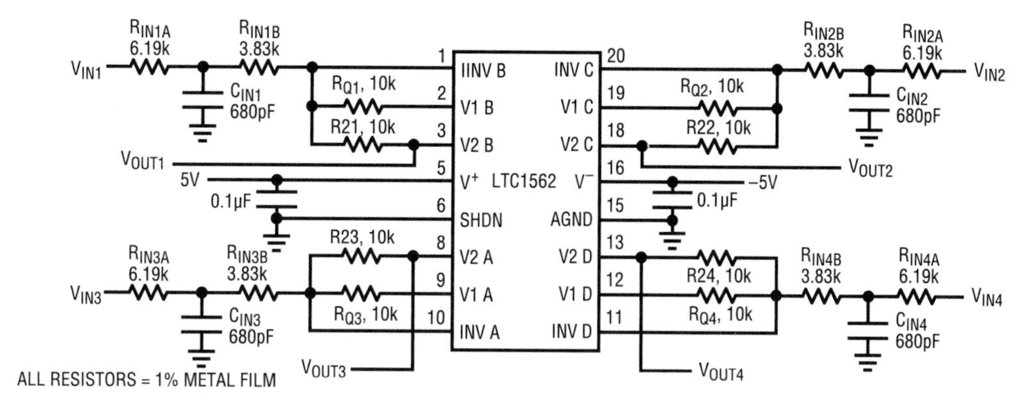

Figure 38.179 • Quad 3-Pole 100kHz Butterworth Lowpass Filter

High clock-to-center frequency ratio LTC1068-200 extends capabilities of switched capacitor highpass filter

by Frank Cox

The circuit in Figure 38.180 is a 1kHz 8th order Butterworth highpass filter built with the LTC1068-200, a switched capacitor filter (SCF) building block. In the past, commercially available switched capacitor filters have had limited use as highpass filters because of their sampled-data nature. Sampled-data systems generate spurious frequencies when the sampling clock of the filter and the input signal mix. These spurious frequencies can include sums and differences of the clock and the input, in addition to sums and differences of their harmonics. The input of the filter must be band limited to remove frequencies that will mix with the clock and end up in the passband of the filter. Unfortunately, the passband of a highpass filter extends upward in frequency by its very nature. If you have to band limit the input signal too much you will also limit the passband of the filter, and hence its usefulness.

What makes this filter different is the 200:1 clock-to-center frequency ratio (CCFR) and the internal sampling scheme of the LTC1068-200. Figure 38.181a shows the amplitude response of the filter plotted against frequency

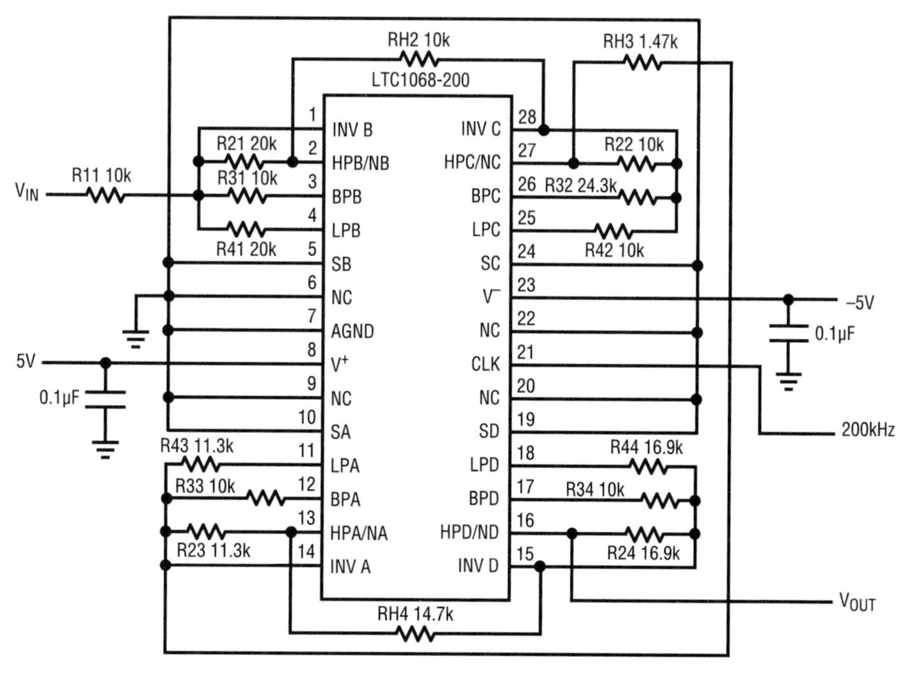

Figure 38.180 • LTC1068-200 1kHz 8th Order Butterworth Highpass Filter

from 100Hz to 10kHz. For comparison, Figure 38.181b shows the same filter built with an LTC1068-25. This is a 25:1 CCFR part. The 200:1 CCFR filter delivers almost 30dB more ultimate attenuation in the stopband. A standard amplitude vs frequency plot of a highpass filter can be misleading because it masks some of the aforementioned spurious signals introduced into the passband.

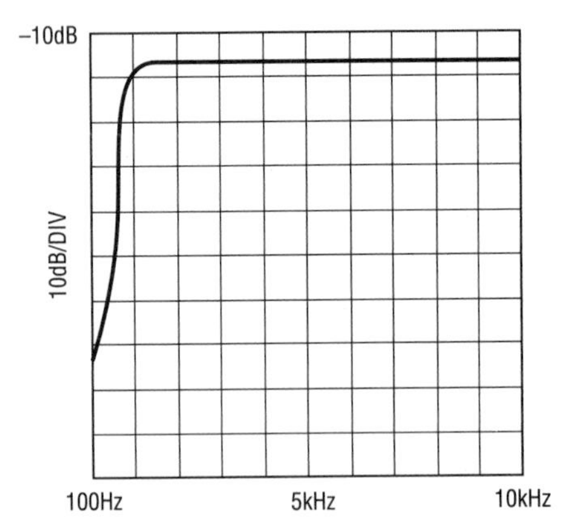

Figure 38.181a • Amplitude vs Frequency Response of Figure 38.180's Circuit

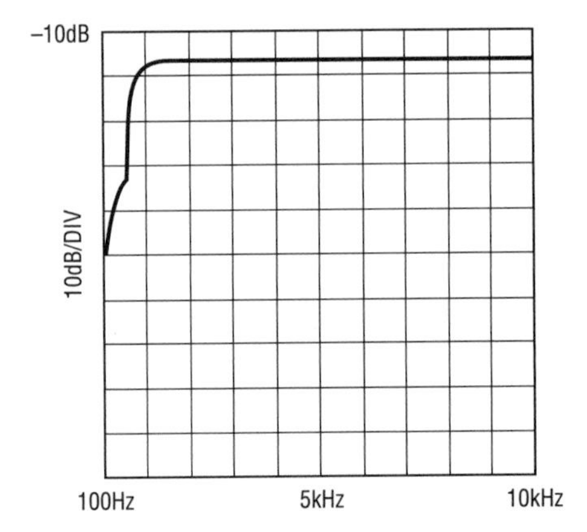

Figure 38.181b • Amplitude vs Frequency Response of Comparable Filter Using the LTC1068-25

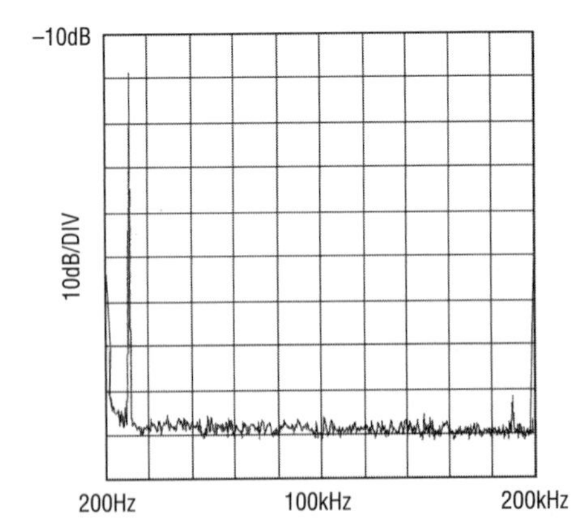

Figure 38.182a • Spectrum Plot of Figure 38.180's Circuit with a Single 10kHz Input

Figure 38.182a is a spectrum plot of the 200:1 filter with a single 10kHz tone on the input. This plot shows that the spurious free dynamic range (SFDR) of the LTC1068 highpass filter is in excess of 70dB. In fact, the filter has a 70dB SFDR for all input signals up to 100kHz. In a 200kHz sampled-data system, you would normally need to band limit the input below 100kHz, the Nyquist frequency. Because the LTC1068 uses double sampling techniques, its useful input frequency range extends to the Nyquist frequency and even above, albeit with some care. Figure 38.182b shows the LTC1068-200 highpass filter with an input frequency of 150kHz. There is a spurious signal at 50kHz, but even though there is no input filtering, the SFDR is still 60dB. For input signals from 100kHz to 150kHz, the filter demonstrates an SFDR of at least 60dB. The SFDR plot of the same filter built with the LTC1068-25 is shown in Figure 38.183. Note that the lower CCFR (25:1) part still manages a respectable 55dB SFDR with a 10kHz input. The LTC1068-25 is used primarily for band-limited applications, such as lowpass and bandpass filters.

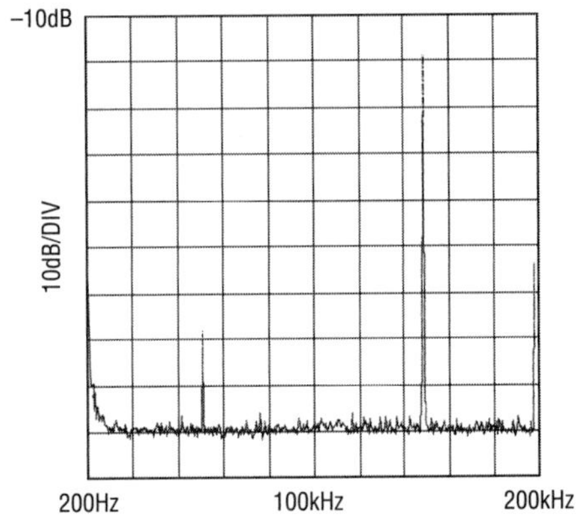

Figure 38.182b • Spectrum Plot of Figure 38.180's Circuit with a Single 150kHz Input

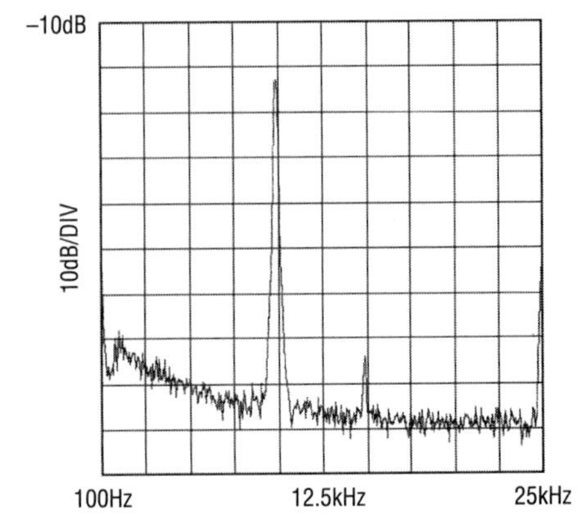

Figure 38.183 • Spectrum Plot of a Comparable Filter Using the LTC1068-25 with a Single 10kHz Input Shows a Respectable 55dB SFDR

Note: The filters for this article were designed using Linear Technology's FilterCAD™ (version 2.0) for Windows®. This program made the design and optimization of these filters fast and easy.

Clock-tunable, high accuracy, quad 2nd order, analog filter building blocks

by Philip Karantzalis

Introduction

The LTC1068 product family consists of four monolithic, clock-tunable filter building blocks. Each product contains four matched, low noise, high accuracy 2nd order switched capacitor filter sections. An external clock tunes the center frequency of each 2nd order filter section. The LTC1068 products differ only in their clock-to-center frequency ratio. The clock-to-center frequency ratio is set to 200:1 (LTC1068-200), 100:1 (LTC1068), 50:1 (LTC1068-50) or 25:1 (LTC1068-25). External resistors can modify the clock-to-center frequency ratio. Designing filters with an LTC1068 product is fully supported by the FilterCAD 2.0 design software for Windows. The internal sampling rate of all the LTC1068 devices is twice the clock frequency. This allows the frequency of input signals to approach twice the clock frequency before aliasing occurs. Maximum clock frequency for LTC1068-200, LTC1068 and LTC1068-25 is 6MHz with ±5V supplies; that for the LTC1068-50 is 2MHz with a single 5V supply. For low power filter applications, the LTC1068-50 power supply current is 4.5mA with a single 5V supply and 2.5mA with a single 3V supply. The LTC1068 products are available in a 28-pin SSOP surface mount package. The LTC1068 (the 100:1 part) is also available in a 24-pin DIP package.

LTC1068-200 ultralow frequency linear-phase lowpass filter

Figure 38.184 shows an LTC1068-200 linear-phase 1Hz low-pass filter schematic and Figure 38.185 shows its gain and group delay responses. The clock frequency of this filter is 400 times the −3dB frequency (f_{-3dB} or f_{CUTOFF}). The large clock-to-f_{CUTOFF} frequency ratio of this filter is useful in ultralow frequency filter applications when

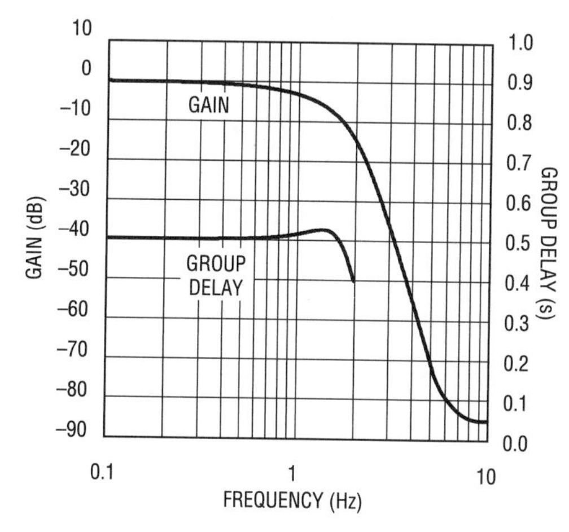

Figure 38.185 • Gain and Group Delay Response of Figure 38.184's Circuit

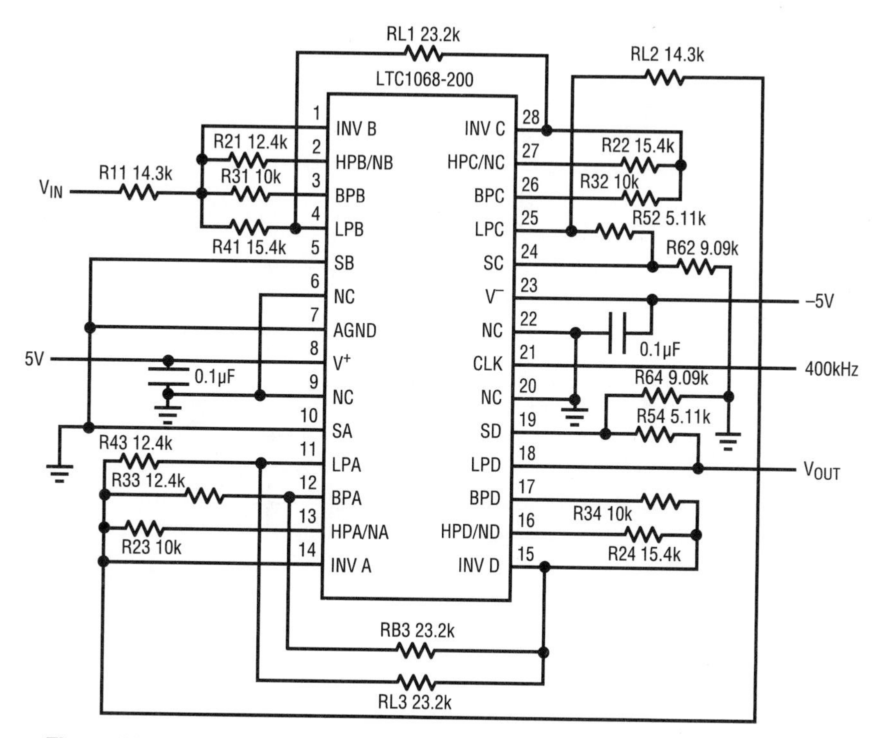

Figure 38.184 • Linear-Phase Lowpass Filter: f_{-3dB} = 1Hz = f_{CLK}/400

minimizing aliasing errors could be an important consideration. For example, the 1Hz lowpass filter shown in Figure 38.184 requires a 400Hz clock frequency. For this filter, the input frequencies that can generate aliasing errors are in a band from 795Hz to 805Hz (2 • f_{CLK} ±5 • f−3dB). For most very low frequency signal-processing applications, the signal spectrum is less than 100Hz. Therefore, Figure 38.184's filter will process very low frequency signals without significant aliasing errors, since its clock frequency is 400Hz and the aliasing inputs are in a small band around 800Hz.

LTC1068-50 single 3.3V low power linear-phase lowpass filter

Figure 38.186 is a schematic of an LTC1068-50-based, single 3.3V low power; lowpass filter with linear phase. The clock- to-f_{CUTOFF} ratio is 50 to 1 (f_{CUTOFF} is the −3dB frequency). Figure 38.187 shows the gain and group delay response. The flat group delay response in the filter's passband implies a linear phase. A linear-phase filter has a transient response with very small overshoot that settles very rapidly. A linear-phase lowpass filter is useful for processing communication signals with minimum intersymbol interference in digital communications transmitters or receivers. The maximum clock frequency for this filter is 1MHz with a single 3.3V supply and 2MHz with a single 5V supply. Typical power supply current is 3mA with a single 3.3V supply and 4.5mA with a single 5V supply.

LTC1068-25 selective bandpass filter is clock tunable to 80kHz

Figure 38.188 shows a 70kHz bandpass filter based on the LTC1068-25 operating with dual 5V power supplies. The clock-to-center frequency ratio is 25 to 1. Figure 38.189 shows the gain response of Figure 38.188's bandpass filter. The passband of this filter extends from 0.95 • f_{CENTER} to

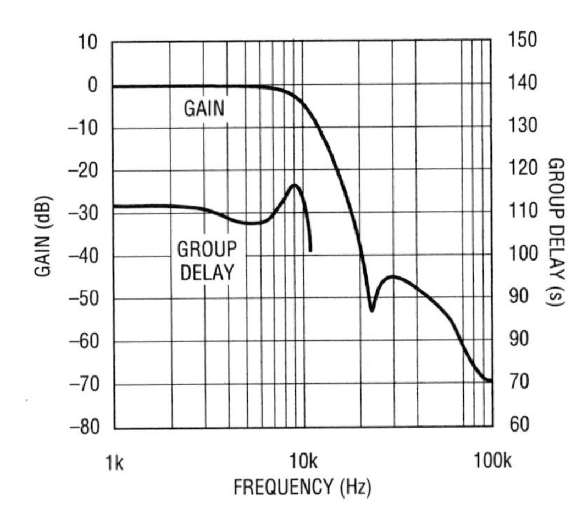

Figure 38.187 • Gain and Group Delay Response of Figure 38.186's Filter

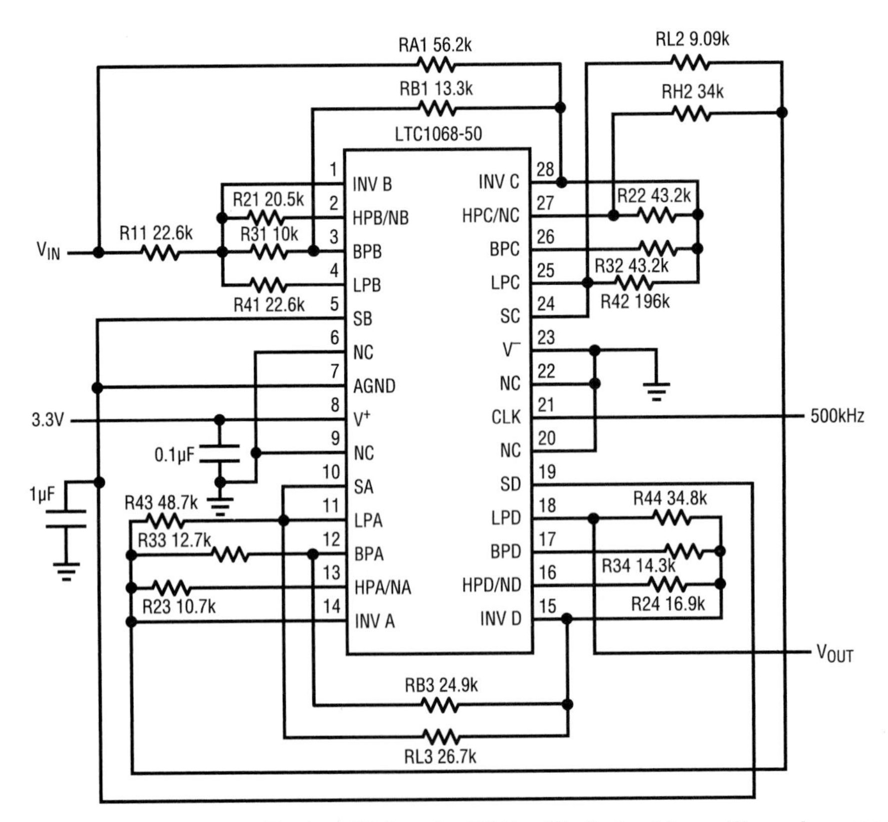

Figure 38.186 • Low Power, Single 3.3V Supply, 10kHz, 8th Order, Linear-Phase Lowpass Filter

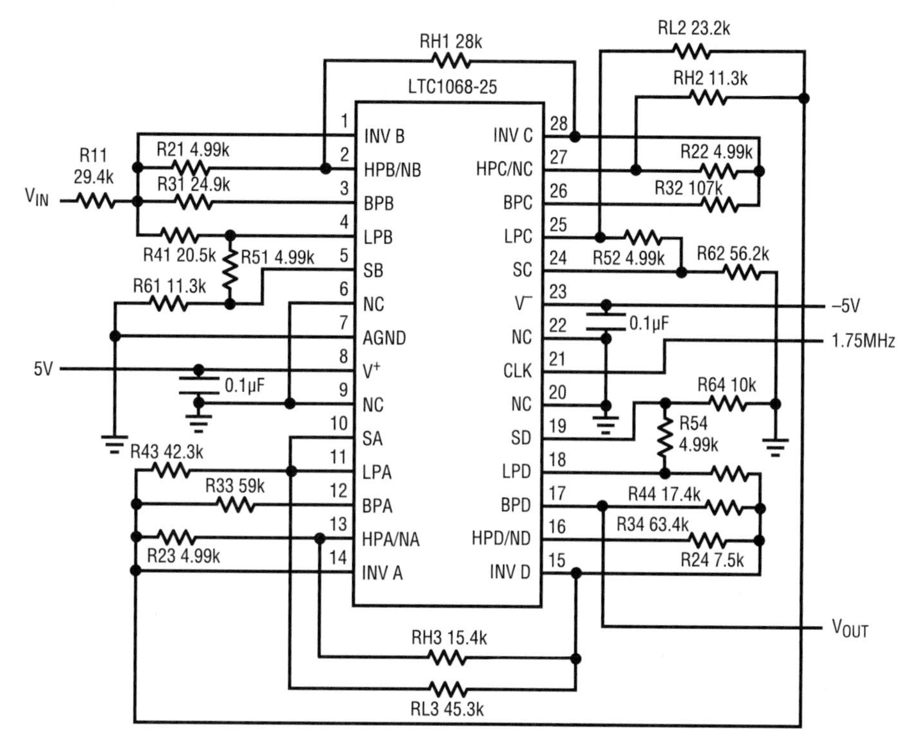

Figure 38.188 • 70kHz, 8th order, Bandpass Filter

1.05 • f_{CENTER}. The stopband attenuation is greater than 40dB at 0.8 • f_{CENTER} and 1.15 • f_{CENTER}. The center frequency can be clock tuned to 80kHz with dual 5V supplies and to 40kHz with a single 5V supply. With FilterCAD, the LTC1068-25 can be used to realize bandpass filters less selective than that shown in Figure 38.188, which can be clock tuned up to 160kHz with dual 5V supplies.

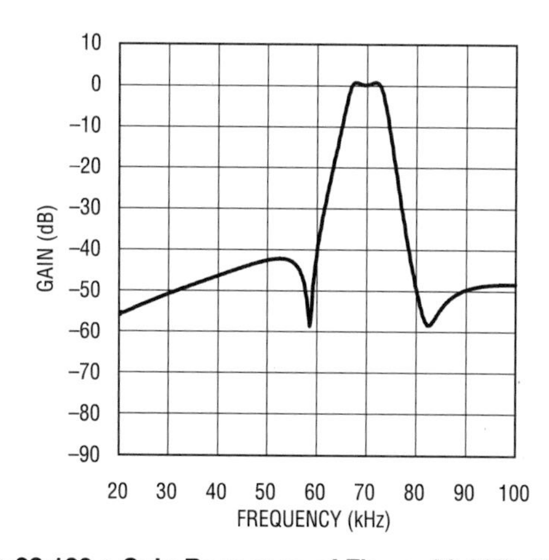

Figure 38.189 • Gain Response of Figure 38.188's Filter

LTC1068 square-wave-to-quadrature oscillator filter

Figure 38.190 shows the schematic of a LTC1068 based filter that is specifically designed to produce a low harmonic distortion sine and cosine oscillator from a CMOS-level square wave input. The reference sine wave output of Figure 38.190's circuit is on pin 15 (BPD on the 24-pin LTC1068 package) and the cosine output is on pin 16 (LPD on the 24-pin LTC1068 package). The output frequency of this quadrature oscillator is the filter's clock frequency divided by 128. The output of a CMOS CD4520 divide-by-128 counter is coupled with a 0.47μF capacitor to the input to the LTC1068 filter operating with dual 5V power supplies. The filter's clock frequency is the input to the CD4520 counter. The LTC1068 filter is designed to pass the fundamental frequency component of a square wave and attenuate any harmonic components higher than the fundamental. An ideal square wave (50% duty cycle) will have only odd harmonics (3rd, 5th, 7th and so on), whereas a typical practical square wave has a duty cycle less or more than 50% and will also have even harmonics (2nd, 4th, 6th and so on). The filter of Figure 38.190 has a stopband notch at the 2nd and 3rd harmonics for a square wave input with a frequency equal to the filter's clock frequency divided by 128. The filter's sine wave output (pin 15) is 1V_{RMS} for a ±2.5V square wave input and has less than 0.025% THD (total harmonic distortion) for input frequencies up to 16kHz and less than 0.1% THD for frequencies up to 20kHz. The cosine output (on pin 16, referenced to pin 15's sine wave output) is 1.25V_{RMS} for a ±2.5V square wave input and has less than 0.07% THD for frequencies up to 20kHz.

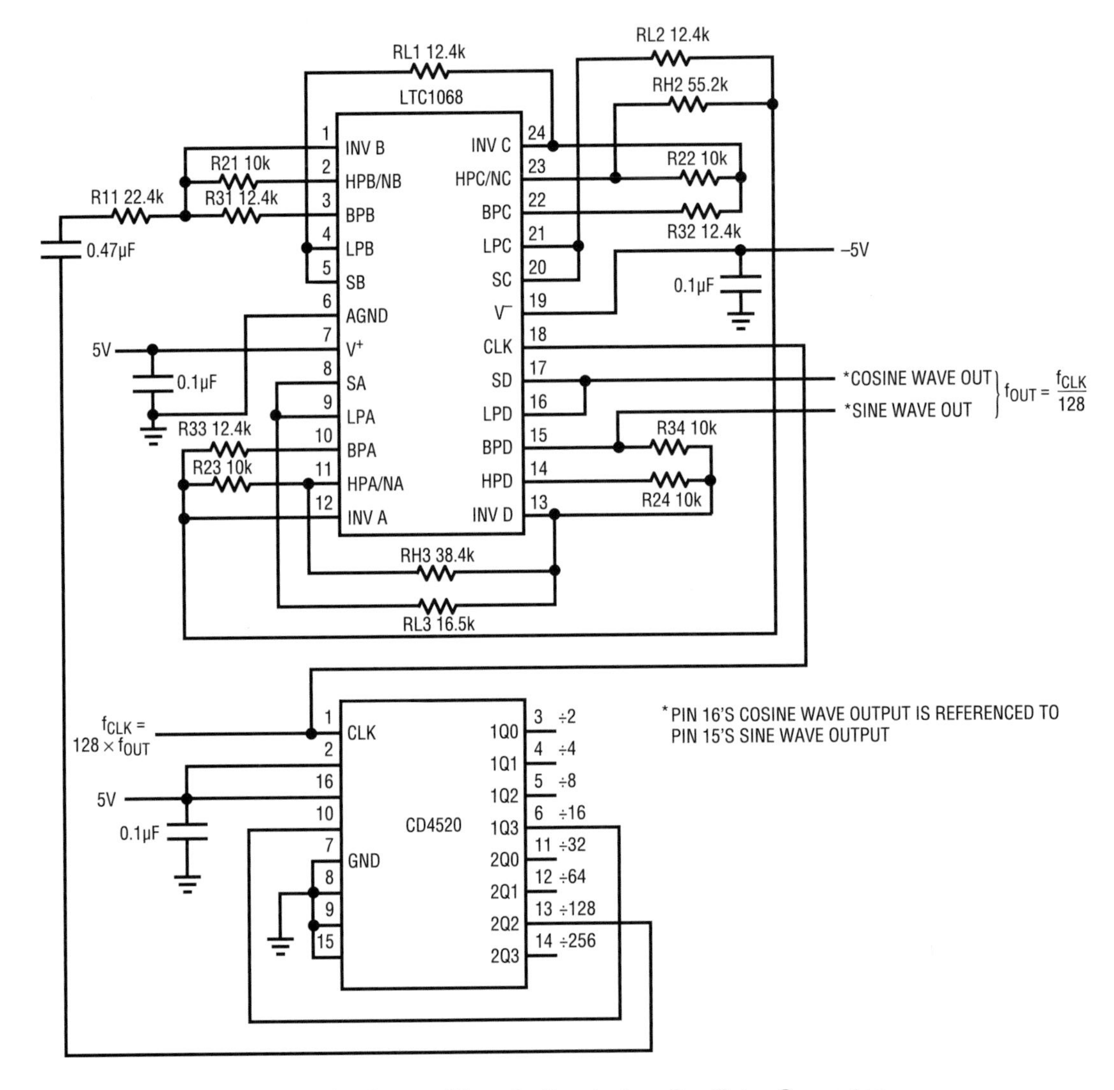

Figure 38.190 • Square-Wave-to-Quadrature Oscillator Converter

The 20kHz frequency limit is due to the CD4520; with a 74HC type divide-by-128 counter, sine and cosine waves up to 40kHz can be generated with the LTC1068-based filter of Figure 38.190.

Miscellaneous

Biased detector yields high sensitivity with ultralow power consumption

by Mitchell Lee

RF ID tags, circuits that detect a "wake-up" call and return a burst of data, must operate on very low quiescent current for weeks or months, yet have enough battery power in reserve to answer an incoming call. For smallest size, most operate in the ultrahigh frequency range, where the design of a micropower receiver circuit is problematic. Familiar techniques, such as direct conversion, super regeneration or superheterodyne, consume far too much supply current for long battery life. A better method involves a technique borrowed from simple field-strength meters: a tuned circuit and a diode detector.

Figure 38.191 shows the complete circuit, which was tested at 470MHz. It contains a couple of improvements over the standard L/C-with-whip field-strength meter: Tuned circuits aren't easily constructed or controlled at UHF so a transmission line is used to match the detector diode (1N5711) to a 6" whip antenna. The 0.22-wavelength section presents an efficient, low impedance match to the base of the quarter-wave whip, but transforms the received energy to a relatively high voltage at the diode for good sensitivity.

Biasing the detector diode improves the sensitivity by an additional 10dB. The forward threshold is reduced to essentially zero, so a very small voltage can generate a meaningful output change. The detector diode's bias point is monitored by an LTC1440 ultralow power comparator, and by a second diode, which serves as a reference.

When a signal at the resonant frequency of the antenna is received, Schottky diode D1 rectifies the incoming carrier and creates a negative-going DC bias shift at the noninverting input of the comparator. Note that the bias shift is sensed at the base of the antenna where the impedance is low, rather than at the Schottky where the impedance is high. This introduces less disturbance into the tuned antenna and transmission-line system. The falling edge of the comparator triggers a one-shot, which temporarily enables answer-back and other pulsed functions.

Total current consumption is approximately 5μA. Monolithic one-shots draw significant load current, but the venerable '4047 is about the best in this respect. Alternatively, a discrete one-shot constructed from a quad NAND gate draws negligible power.

Sensitivity is excellent. The finished circuit can detect 200mW radiated from a reference dipole at 100'. Range, of course, depends on operating frequency, antenna orientation and surrounding obstacles; in the clear, a more reasonable distance, such as 10', can be covered at 470MHz with only a few milliwatts.

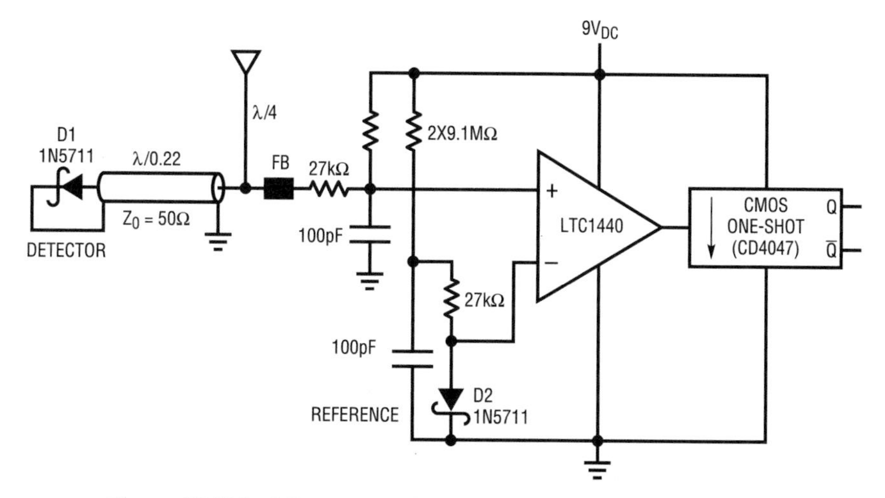

Figure 38.191 • Micropower Field Detector for Use at 470MHz

All selectivity is provided by the antenna itself. Add a quarter-wave stub (shorted with a capacitor) to the base of the antenna for better selectivity and improved rejection of low frequency signals.

Zero-bias detector yields high sensitivity with nanopower consumption

by Mitchell Lee

RF ID tags, circuits that detect a "wake-up" call and return a burst of data, must operate on very low quiescent current for months or years, yet have enough battery power in reserve to answer an incoming call. For smallest size, most operate in the ultrahigh frequency range, where the design of a micropower receiver circuit is problematic. Familiar techniques, such as direct conversion, super regeneration or superhetrodyne, consume far too much supply current for long battery life. A better method involves a technique borrowed from simple field-strength meters: a tuned circuit and a diode detector.

Figure 38.192 shows the complete circuit, which was tested for proof-of-concept at 445MHz. This circuit contains a couple of improvements over the standard L/C-with- whip field-strength meter. Tuned circuits aren't easily constructed or controlled at UHF so a transmission line is used to match the detector diode (1N5712) to a quarter- wave whip antenna. The 0.23λ, transmission-line section transforms the 1pF (350Ω) diode junction capacitance to a virtual short at the base of the antenna. At the same time, it converts the received antenna current to a voltage loop at the diode, giving excellent sensitivity.

Biasing the detector diode can improve sensitivity,[3] but only when the diode is loaded by an external DC resistance. Careful curve-tracer examination of the 1N5712 at the origin reveals that it follows the ideal diode equation, with scales of millivolts and nanoamperes. To use a zero-bias diode at the origin, the external comparator circuitry must not load the rectified output.

The LTC1540 nanopower comparator and reference is a good choice for this application because it not only presents no load to the diode, but also draws only 300nA from the battery. This represents a 10-times improvement in battery life over biased detector schemes.[4] The input is CMOS, and input bias current consists of leakage in a small ESD-protection cell connected between the input and ground. The input leakage measures in the picoampere range, whereas the 1N5712 leaks hundreds of picoamperes. Any rectified output from the diode is loaded by the diode itself, not by the LTC1540, and the sensitivity can match that of a loaded, biased detector.

The rectified output is monitored by the LTC1540 comparator. The LTC1540's internal reference is used to set up a threshold of about 18mV at the inverting input. A rising edge at the comparator output triggers a one-shot, which temporarily enables answer-back and any other pulsed functions.

Note 3: Eccles, W.H. Wireless Telegraphy and Telephony, Second Edition. Ben Brothers Limited, London, 1918, page 272.

Note 4: Lee, Mitchell. "Biased Detector Yields High Sensitivity With Ultralow Power Consumption." Page 110 of this application note.

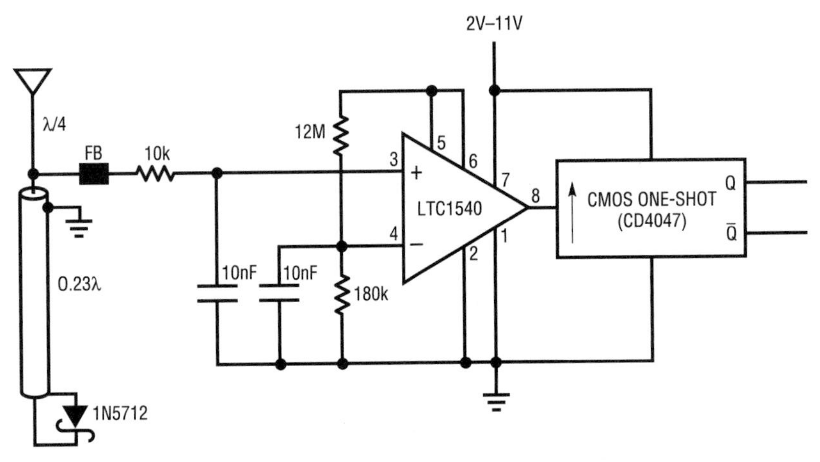

Figure 38.192 • Nanopower Field Detector

Total supply current is 400nA, consuming just 7mAH battery life over a period of five years. Monolithic one-shots draw significant load current, but the '4047 is about the best in this respect. A one-shot constructed from discrete NAND gates draws negligible power.

Sensitivity is excellent, and the circuit can detect about 200mW from a reference dipole at 100 feet. Range, of course, depends on operating frequency, antenna orientation and surrounding obstacles. Sensitivity is independent of supply voltage; this receiver will work just as well with a 9V battery as with a single lithium cell.

The length of the transmission line does not scale with frequency. Owing to a decrease in diode reactance, the electrical length will shorten as frequency increases. Adjust the line length for minimum feed-point impedance at the operating frequency. If an impedance analyzer is used to measure the line, a 1pF capacitor can be substituted for the diode to avoid large signal effects in the diode itself. Consult the manufacturer's data sheet for accurate characterization of diode impedance at the frequency of interest.

Transparent class-D amplifiers featuring the LT1336

by Dale Eagar

Introduction

Efficiency in the field of power conversion is like transparency in the field of light transmission. It is no wonder, then, that Class-D amplifiers are often called transparent, since they have no significant power losses. In contrast to class-D amplifiers' nearly lossless switching, class-A through class-C amplifiers are throttling devices that waste significant energy. Amplifiers of the "lower classes" (A-C) are modeled as rheostats (variable resistors), whereas class-D amplifiers are modeled as variacs (variable transformers). The ideal resistor dissipates power; whereas the ideal transformer does not. Like transformers (variacs), many class-D amplifiers can transfer energy in both directions—input to output and output to input.

Class-D amplifiers also have a way of ignoring reactive loads that can be uncanny. A class-D amplifier operating with an AC output will draw very little additional input power when a sizable capacitive or inductive load is placed at its output. This is because the reactive load has AC voltage across it and AC current flowing through it, but the phase angle of the voltage and current is such that no real power is dissipated. The class-D amplifier ends up shuttling power back and forth between its input and its output, doing both with minimal loss. An ideal class-D amplifier can be thought of as having no place to dissipate power, since all of its components are lossless; that is, it contains no resistors.

The electric heater—a simple class-D amplifier

Class-D amplifiers can be simple or complex, depending on what is required by the application. A simple class-D amplifier is the thermostatic switch in an electric heater. The thermostat controls the heater by turning it on or off. The switch is essentially lossless, dissipating practically no power. This class-D amplifier is remarkably efficient, since even the energy lost in the switch, power cord and house wiring contributes to the desired result. The duty factor, and hence the average amount of power delivered to the heater, can assume an infinite number of values. This is true even though a constant amount of heat is delivered when the heater is on.

Quadrants of energy transfer

Class-D amplifiers have a property that requires new terminology, a property that generally isn't considered in lower-class amplifiers. This property, quadrants of energy transfer, describes the output characteristics of the class-D amplifier. The output characteristics are plotted on a imaginary X-Y plot (I've yet to see someone actually do one on paper), one axis representing output voltage and the other axis representing output current, with the intersection of the axes representing zero volts and zero amps.

A simple switcher that can only provide a positive output current into a positive output voltage can be described as a 1-quadrant device. This 1-quadrant device could be a computer power supply, a battery charger or any supply that delivers a positive voltage into a device that can only consume power.

The 2-quadrant converter can be one of two different things: 1) A positive output voltage that can both source and sink current, or 2) A positive current that can comply both positive and negative output voltage. Finally, the 4-quadrant converter can both source and sink current into both positive and negitive output voltages.

1-Quadrant class-D converter

To illustrate the 1-quadrant class-D amplifier; we will focus on the boost mode converter detailed in Figure 38.193 This circuit removes power from the source (12V automotive battery) and delivers it to the load (some as-yet-unknown 55V device) This circuit is classified as "1 quadrant" because it can only regulate output voltage in one polarity (positive) and it can output current in only one polarity (positive).

Introducing the LT1336 half-bridge driver

Taking a side step from our main discussion, we will introduce a component, the half-bridge power amplifier. Figure 38.194 details the LT1336 driving power MOSFETs and shows the symbolic representation of this subcircuit that will appear in subsequent figures. Table 38.7 shows the logical states of this half-bridge power driver.

4-Quadrant class-D amplifier

Class-D amplifiers are commonly used in subwoofer drivers. This is because subwoofers require a great deal of power. A class AB amplifier driving a subwoofer will put

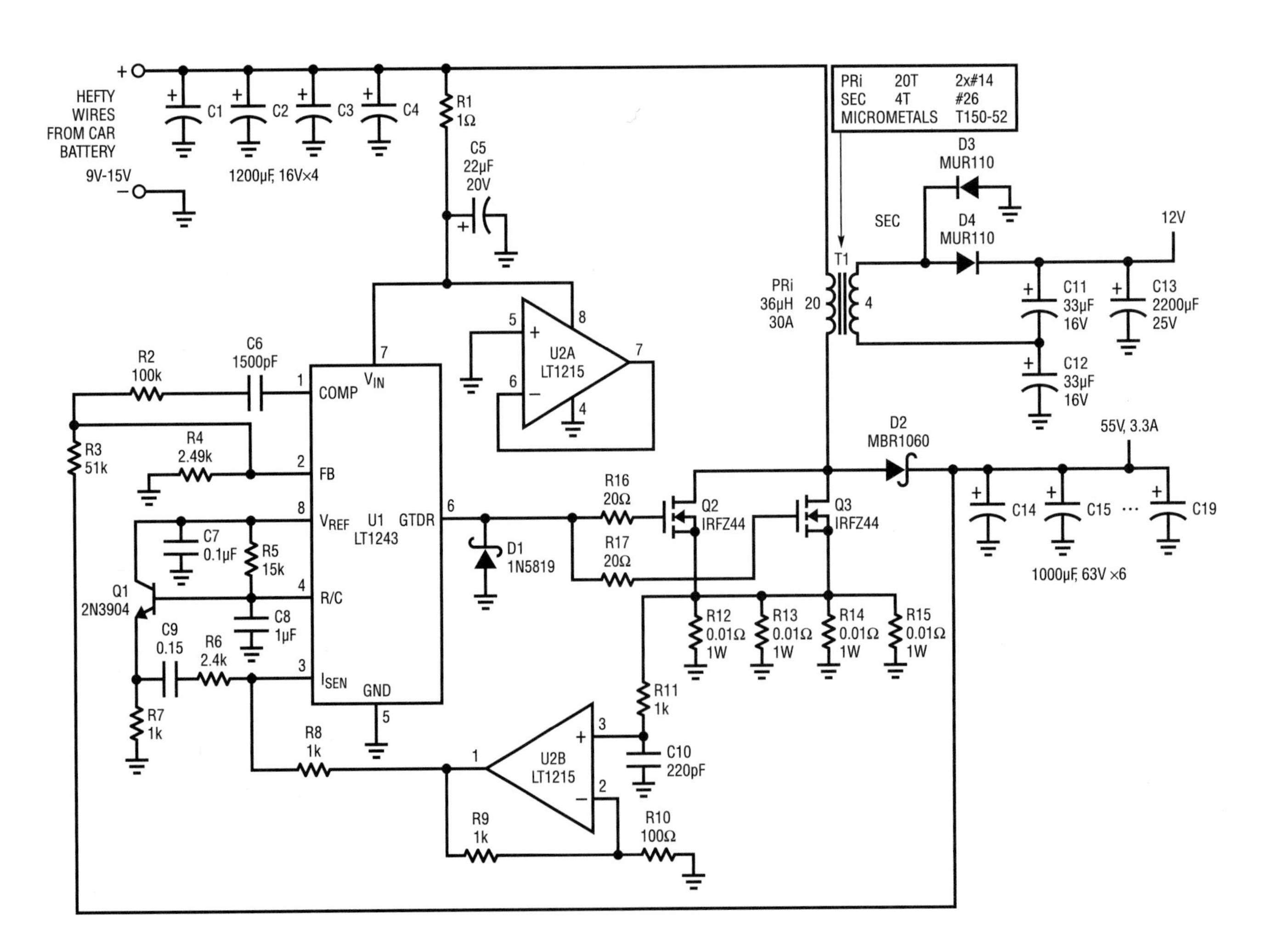

Figure 38.193 • 200W, 12V to 55V Front End for Automotive Applications

about half of its input power into its heat sink. Driving the same subwoofer at the same volume with the same music, a class-D amplifier will put about five percent of its input power into the heat sink. The difference is ten to one on the heatsink size and two to one on the input power supply. Figure 38.195 is the 200W class-D subwoofer driver. This circuit uses the 200W front end developed in Figure 38.193 as its power source. The circuit in Figure 38.195 performs as follows: U1a, R1–R4 and C7 implement a 75kHz pseudosawtooth oscillator. U1d is the input amplifier/ filter, with a gain of 6.1 and 200Hz Butterworth lowpass response. U1b and U1c are comparators that compare the sawtooth and the amplified/filtered input signal to form two complimentary, pulse-width modulated square waves. X1 and X2 are two half-bridge power drivers and M1 is the subwoofer driver.

One of the properties of Class-D, 4-quadrant amplification is the ability to transfer power both to and from the load. In our subwoofer driver, this happens when the driver reaches the end of any given excursion and the combination of the driver spring and the acoustic spring drive the cone back to center. During this time, energy is transferred from the driver back to the input of the class-D amplifier stage. In the case shown in Figure 38.195, the energy ends up on the 55V bus, where the bus voltage climbs during these periods of "negative energy delivered to the load." Fortunately, C14–C19 of Figure 38.193 can store this energy; otherwise the 55V bus would subject to excessive voltage until someplace was found for the energy to go.

Table 38.7 Half-Bridge Power Driver Truth Table

In Top	In Bottom	Output
L	L	Floating
L	H	Ground
H	L	55V
H	H	Floating

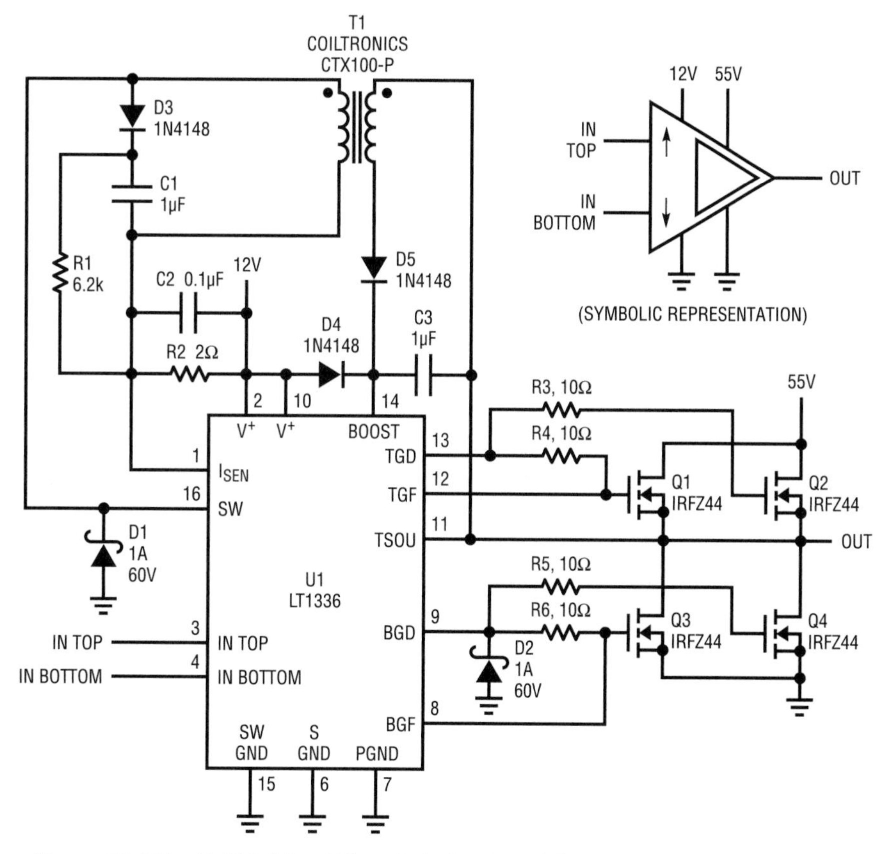

Figure 38.194 • Half-Bridge Driver Subcircuit and Symbolic Representation

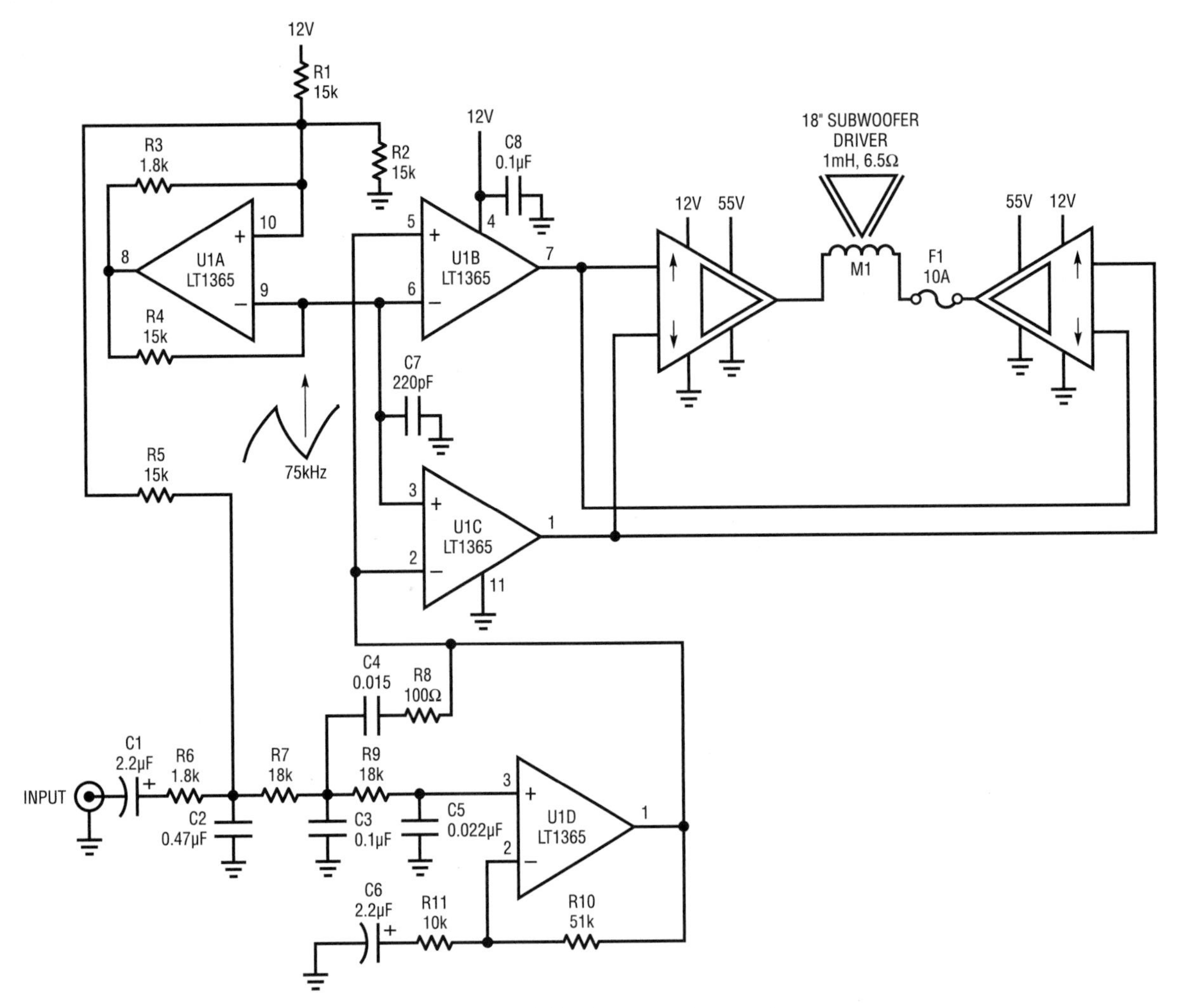

Figure 38.195 • 200W-Powered Subwoofer

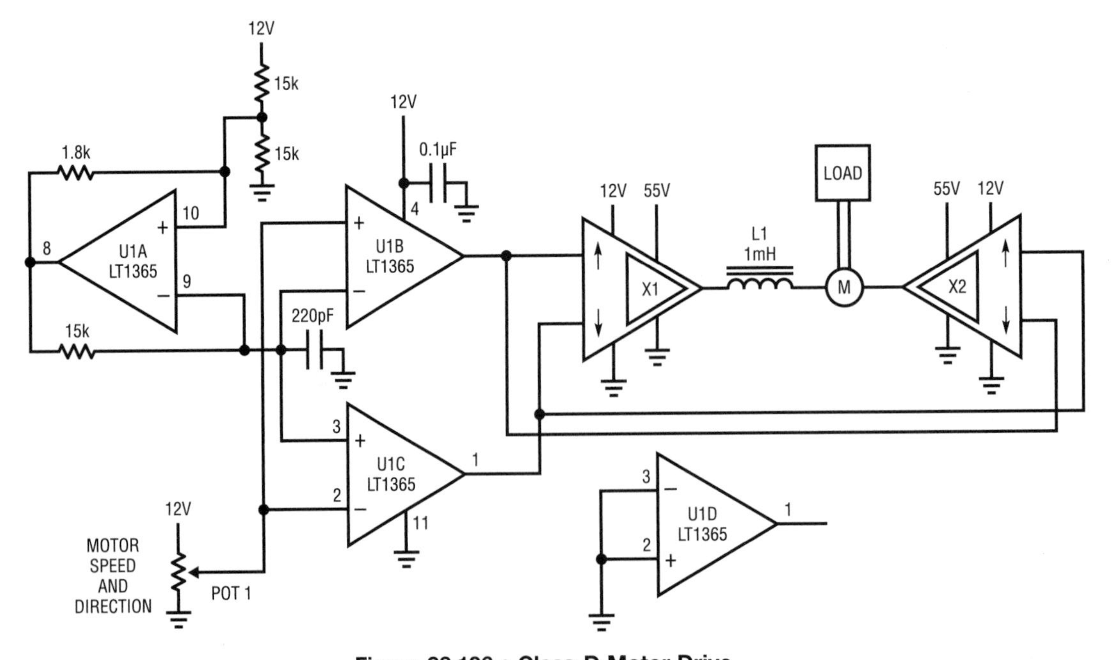

Figure 38.196 • Class-D Motor Drive

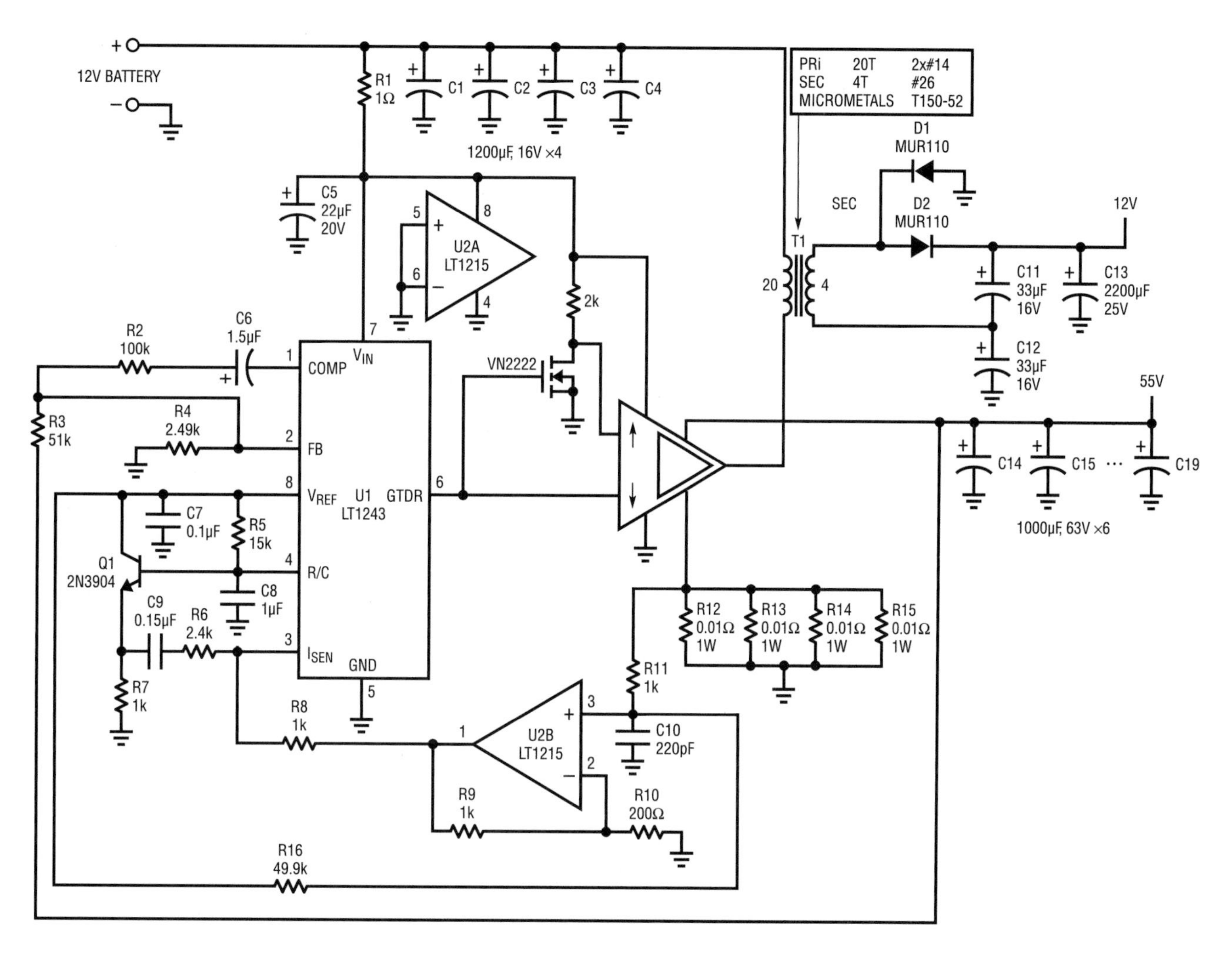

Figure 38.197 • 200W, 2-Quadrant Front End for Automotive Applications

Class-D for motor drives

Substituting a motor and an inductor for the subwoofer in Figure 38.195 and simplifying the control, we arrive at the circuit shown in Figure 38.196. Connecting this circuit to the front end shown in Figure 38.193 and then getting the motor up to speed is no problem, but when one wants to slow the motor down by turning pot 1 back toward its center, disaster strikes. Rotational energy stored in the inertia of the motor is converted back into electrical energy by the motor and is presented to the output of the class-D amplifier. L1, X1 and X2 do their job by transferring the energy back into the 55V bus. The energy goes into C14-C19 of Figure 38.193, charging them to some voltage significantly above 55V and something breaks. The problem here is that the circuit in Figure 38.193 is only a 1-quadrant class-D amplifier.

Managing the negative energy flow

Sound like a course in management? The negative energy transferred through the class-D amplifier needs a home. One simple home is a 62V power Zener diode strapped across the 55V bus and bolted to a massive heat sink. One could easily imagine the heat sink as the brake shoes heating up as the electric vehicle winds down the mountain road. Another place to put the energy is back into the 12V battery. This will require upgrading the 12V to 55V front-end power converter from 1 quadrant to 2 quadrants.

The 2-quadrant class-D converter

Converting Figure 38.193 to two quadrants involves replacing D2 with a switch and activating the switch out of phase with the switch formed by Q2 and Q3. The half-bridge power driver shown in Figure 38.195 is just such a switch. Refer to Figure 38.197. The I_{SENSE} signal (U1, pin 3) needs to be offset to accommodate negative current (add R16, Figure 38.197) The I_{SENSE} signal needs to be scaled for twice the range (−30A to 30A rather than 0A to 30A); this is done by changing R10.

Now we are happily winding down the mountain road, watching the scenery unfold before us. We are happy in knowing that we are recycling the energy released from the descent by charging our batteries, while watching the mountain bikers burn their descent energy off in brake linings. Once again technology wins over sweat and brawn.

A trip over the great divide

Climbing the great divide in an electric vehicle requires some planning. Stops to recharge are necessary. Once on top, the whole scheme changes: descending the hill, charging our battery, all goes well until the battery is fully charged; then we have to stop. Further descent would overcharge our battery, boiling out the electrolyte. Not only would this ruin our battery, in the end we would have no place to put the energy and our class-D amplifier would find some way to fail. We need to stop and drain off some charge, trade batteries with someone climbing the other side or put a power Zener on our battery. Figure 38.198 details the active Zener circuit. Using the reference in U1 of Figure 38.197 and the unused half of U2 we are able to make a hysteretic clamp that puts all of the heat into a resistor, R5. This circuit will save the battery from destruction and drop our level of smugness back to that of the mountain bikers.

Conclusion

Class-D has been around for a long time: the venerable electric heater with its bang-bang controller is a remarkably efficient and reliable class-D amplifier. Class-D drives have been used for decades in golf carts, fork lifts, cranes and

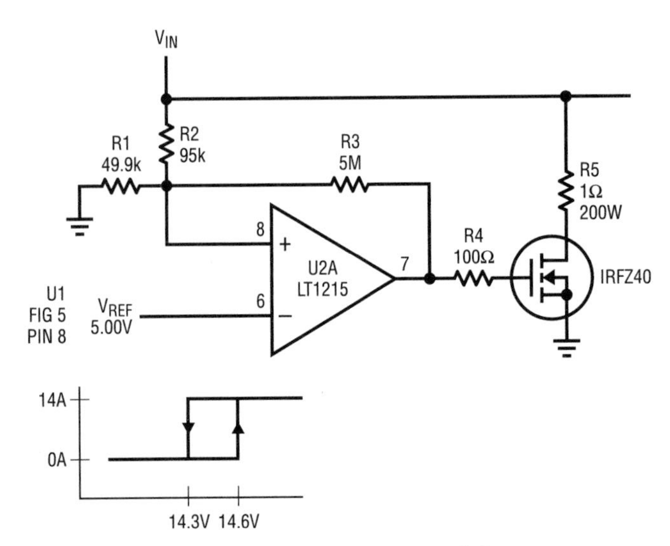

Figure 38.198 • Wolf Creek Pass Adapter

industry. The advent of the half-bridge driver greatly simplifies the Class-D Amplifier. Here at Linear Technology we have a family of half/full bridge MOSFET drivers. For further information, contact us at the factory or refer to the LT1158, LT1160, LT1162 or LT1336 data sheets.

Single-supply random code generator

by Richard Markell

Presented here is a truly random code generator that operates from a single supply. The circuit allows operation from a single 5V supply with a minimum of adjustments.

The circuit produces random ones and zeroes by comparing a stream of random noise generated in a Zener diode to a reference voltage level. If the threshold is correctly set and the time period is long enough, the noise will consist of a random but equal number of samples above and below threshold.

That fuzz is noise

The circuit shown in Figure 38.199 is the random noise generator. Optimum noise performance is obtained from a 1N753A Zener diode, which has a 6.2 volt Zener "knee." The diode is used to generate random noise. We have found that optimum noise output for this diode occurs at the "knee" of the I-V curve, where the Zener just starts to limit voltage to 6.2 volts.

Operating a 6.2V Zener from a 5V supply required some thought. Obviously, some type of voltage boosting scheme was needed to provide the diode with the 8V or more that it requires in this circuit. U1, an LTC1340 low noise, voltage-boosted varactor driver, provides 9.2V at 20μA from an input of 5V This Zener current is the optimal for noise output from the diode (at 20μA the output is about 20m$V_{P\text{-}P}$).

The 1M and 249k resistors bias the input to operational amplifier U2 to 1.25V to match the input common mode range of comparator U3. The 1μF capacitor provides an AC path for the noise. Note: be careful where you place any additional capacitors in this part of the circuit or the noise may be unintentionally rolled off. This is one circuit where noise is desirable.

U2 is an LT1215 23MHz, 50V/μs, dual operational amplifier that can operate from a single supply. It is used as a wideband, gain-of-eleven amplifier to amplify the noise from the Zener diode; the second op amp in U2 is unused. U3, an LT1116 high speed, ground-sensing comparator, receives the noise at its positive input. A threshold is set at the negative comparator input and the output is adjusted via the 2k potentiometer for an equal number of ones and zeroes. The 5k resistor and the 10μF capacitor provide limited hysteresis so that the adjustment of the potentiometer is not as critical. Latch U4, a 74HC373, ensures that the output remains latched throughout one clock period. The circuit's output is taken from U4's Q0 output.

Some thoughts on automatic threshold adjustment

Several circuit designers have asked about threshold adjustment without manual knobs or potentiometers. One way to implement this would be to have the microprocessor count the number of ones and zeroes over a given time period and adjust the threshold (perhaps via a digital pot) to produce the required density of ones.

A more "analog" method of adjusting threshold might be to implement an integrator with reset. This circuit integrates the number of ones and zeroes over time to produce a zero result for an adjustment that produces equal numbers of ones and zeroes. Again, a digital pot could be used to adjust threshold, with the threshold being decreased for the case of "not enough ones" and increased for the case of "too many ones."

After many more conversations with the "cyber illuminati," the circuit in Figure 38.200 was devised. This circuit can be used to replace the pot shown in the dashed box in Figure 38.199. In operation, an LT1004-2.5 is used as a reference at the front end of a precision voltage divider string. A series of voltages is generated along the divider string and a jumper is used to connect this voltage to a buffer and then to the negative input of the LT1116 comparator. As was the case with the 2k pot, the voltage at pin 2 (the negative input of the comparator) sets the threshold for the comparator. The selection of voltage taps on the resistor string is arbitrary; they were selected to allow a good adjustment range (defined as allowing jumper adjustment to 50% ones and 50% zeros) for a sample of ten 1N753A Zener diodes used to produce noise. The jumper could (and probably should) be replaced with analog switches controlled by a microprocessor in medium- to high-volume applications.

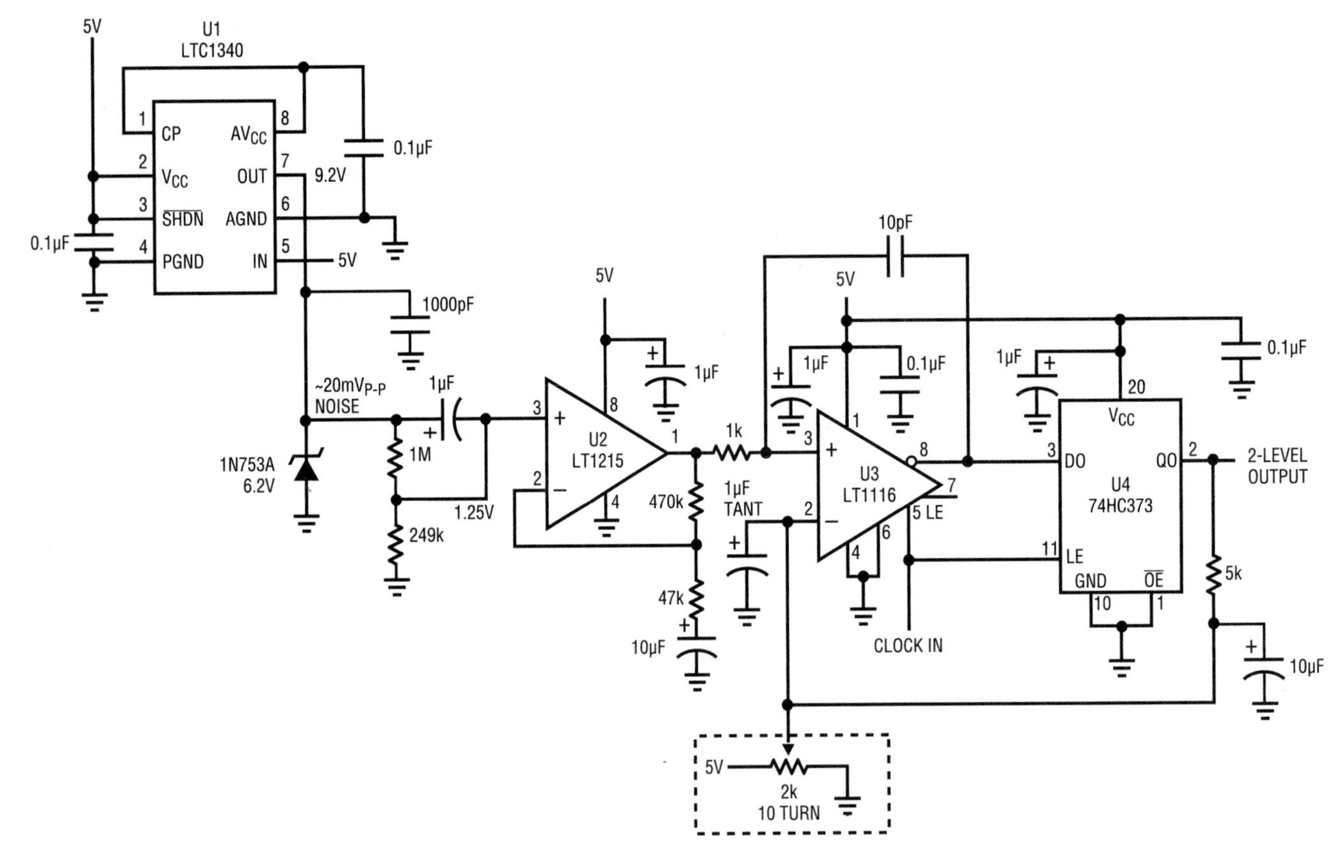

Figure 38.199 • Single-Supply Random Code Generator

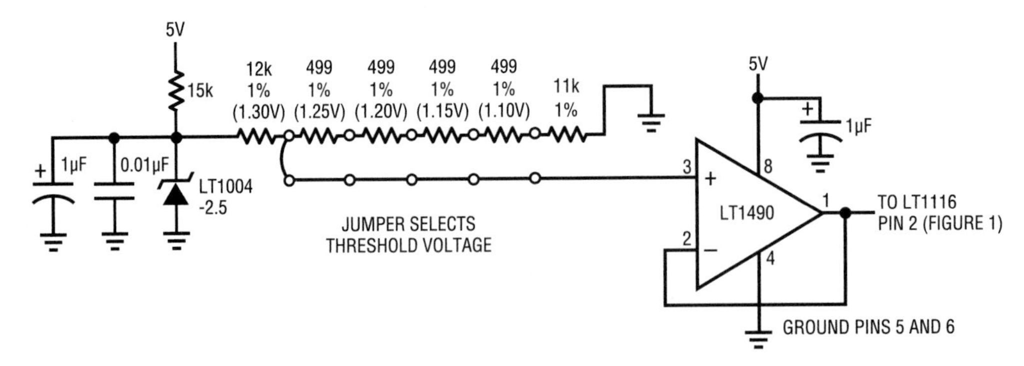

Figure 38.200 • Jumper Selects Threshold for Figure 38.199's Circuit

Appendix A
Component vendor contacts

The tables on this and the following pages list contact information for vendors of non-LTC parts used in the application circuits in this publication. In some cases, components from other vendors may also be suitable. For information on component selection, consult the text of the respective articles and the appropriate LTC data sheets.

Capacitors

Vendor	Product	Phone	URL
AVX	Chip Capacitors	(843) 946-0362	www.avxcorp.com/products/capacitors
AVX	Tantalum Capacitors	(207) 282-5111	
Electronic Concepts	400V Film Capacitors	(908) 542-7880	www.eci-capacitors.com
Kemet	Tantalum Capacitors	(408) 986-0424	www.kemet.com
Marcon	High C/V Capacitors	(847) 696-2000	www.chemi-con.com/main/company/marcon.html
Murata Electronics	Capacitors	(770) 436-1300	www.iijnet.or.jp/murata/products/english
Nichicon	Electrolytic Capacitors	(847) 843-7500	www.nichicon-us.com
Panasonic	Poly Capacitors	(714) 373-7334	www.panasonic.com/industrial_oem/electronic_components/electronic_components_capacitors_home.htm
Sanyo	Oscon Capacitors	(619) 661-6835	www.sanyovideo.com
Sprague	Capacitors	(207) 324-4140	www.comsprague.com
Taiyo Yuden	Chip Capacitors	(408) 573-4150	www.t-yuden.com
Tokin	Capacitors	(408) 432-8020	www.tokin.com
United Chemicon	Electrolytic Capacitors	(847) 696-2000	www.chemi-con.com/main
Vitramon	Ceramic Chip Capacitors	(203) 268-6261	www.vishay.com
Wima	Paper/Film Capacitors	(914) 347-2474	www.wimausa.com

Diodes

Vendor	Product	Phone Number	URL
Agilent (formerly Hewlett Packard)	IR LEDs	(800) 235-0312	www.semiconductor.agilent.com/ir
Central Semiconductor	Small Signal Discretes	(516) 435-1110	www.centralsemi.com
Chicago Miniature Lamp	LEDs	(201) 489-8989	www.sli-lighting.com/cml
Data Display Products	LEDs	(800) 421-6815	www.ddp-leds.com
Fuji	Schottky Diodes	(201) 712-0555	www.fujielectric/co/jp/eng/index-e.html
General Semiconductor	Diodes	(516) 847-3000	www.gensemi.com
Motorola*	Discretes	(800) 441-2447	www.mot-sps.com/products/index.html
ON Semiconductor*	Discretes	(602) 244-6600	www.onsemi.com/home
Panasonic	LEDs	(201) 348-5217	http://www.panasonic.com/industrial_oem/semiconductors/semiconductor_home.htm
Temic	IR Photo Diodes	(408) 970-5700	www.temic.com
Vishay	Zener/Small Signal Diodes	(408) 241-4588	www.vishay.com
Zetex	Small Signal Discretes	(516) 543-7100	www.zetex.com

*Discretes formerly manufactured by Motorola are now manufactured by ON Semiconductor. Part numbers have not been chanaged as of January 2000

Inductors and Transformers

Vendor	Product	Phone Number	URL
API Delevan	Inductors	(716) 652-3600	www.delevan.com
BH Electronics	Inductors	(612) 894-9590	www.bhelectronics.com
BI Technologies	Transformers	(714) 447-2656	www.bitechnologies.com
Coilcraft	Inductors	(847) 639-6400	www.coilcraft.com
Cooper	Inductors/Transformers	(561) 752-5000	www.coiltronics.com
Dale	Inductors/Transformers	(605) 665-1627	www.vishay.com/fp/fp.html#inductors
Gowanda	Inductors	(716) 532-2234	www.gowanda.com
Midcom	Inductors/Transformers	(605) 886-4385/ (800) 643-2661	www.midcom-inc.com
Murata Electronics	Inductors	(814) 237-1431	www.murata.com
Panasonic	Inductors/Transformers	(714) 373-7334	www.panasonic.com/industrial_oem/electronic_components/ electronic_components_inductors_coils_and_transformers.htm
Philips	Inductors	(914) 246-2811	www.acm.components.philips.com
Philips	Planar Inductors	(914) 247-2036	www.acm.components.philips.com
Pulse	Inductors	(619) 674-8100	www.pulseeng.com
Sumida	Inductors	(847) 956-0667	www.japanlink.com/sumida
Tokin	Inductors	(408) 432-8020	www.tokin.com

Logic

Vendor	Product	Phone Number	URL
Fairchild	Logic	(207) 775-4502	www.fairchildsemi.com
Intersil (formerly Harris)	Logic	(800) 442-7747	www.intersil.com
*Motorola	Logic	(800) 441-2447	www.mot-sps.com/products/index.html
*ON Semiconductor	Logic	(602) 244-6600	www.onsemi.com/home
Toshiba	Logic Single Gate Logic	(949) 455-2000/ (714) 455-2000	www.toshiba.com/taec

*Logic Devices formerly manufactured by by Motorola are now manufactured by ON Semiconductor; there have been no changes in part numbers as of January 2000

Resistors

Vendor	Product	Phone Number	URL
Allen Bradley	Carbon Resistors	(800) 592-4888	www.ab.com
AVX	Chip Resistors	(843) 946-0524	www.avxcorp.com/products/resistors/chiprstr.htm
BI Technologies	Resistors/Resistor Networks	(714) 447-2345	www.bitechnologies.com
Bourns	Potentiometers, SIPs	(801) 750-7253	www.bourns.com
Dale	Sense Resistors	(605) 665-9301	www.vishayfoil.com or www.vishay.com
IRC	Sense Resistors	(361) 992-7900	www.irctt.com
RG Allen	Metal Oxide Resistors	(818) 765-8300	www.rgaco.com
TAD	Chip Resistors	(800) 508-1521	www.tadcom.com
Taiyo Yuden	Chip Resistors	(408) 573-4150	www.t-yuden.com
Thin Film Technology	Thin Film Chip Resistors	(507) 625-8445	www.thin-film.com
Tocos	SMD Potentiometers	(847) 884-6664	www.tocos.com

Transistors

Vendor	Product	Phone Number	URL
Central Semiconductor	Small Signal Discretes	(516) 435-1110	www.centralsemi.com
Fairchild	MOSFETs	(408) 822-2126	www.fairchildsemi.com
IR	MOSFETs	(310) 322-3331	www.irf.com
Motorola*	Discretes	(800) 441-2447	www.mot-sps.com/products/index.html
ON Semiconductor*	Discretes	(602) 244-6600	www.onsemi.com/home
Philips	Discretes	(401) 767-4427	www-us.semiconductors.philips.com
Siliconix	MOSFETs	(800) 554-5565	www.siliconix.com
Zetex	Small Signal Discretes	(631) 543-7100	www.zetex.com

*Discretes formerly manufacured by Motorola are now manufactured by ON Semiconductor; There are no changes in part numbers as of January 2000.

Miscellaneous

Vendor	Product	Phone Number	URL
Aavid	Heat Sinks	(714) 556-2665	www.aavid.com
Epson	Crystals	(310) 787-6300	www.eea.epson.com
Infineon (formerly Siemens Semiconductor)	Optoelectronics	(108) 257-7910	www.infineon.com/us/opto/content.htm
Magnetics, Inc.	Toroid Cores, etc.	(800) 245-3984	www.mag-inc.com
MF Electronics	Crystal Oscillators	(914) 576-6570	www.mfelec.com
Murata Electronics	RF Devices	(770) 433-5789	www.murata.com
QT Optoelectronics	RF Switches	(408) 720-1440	www.qtopto.com
Raychem	Fuses	(800) 227-4856	www.raychem.com
RF Micro Devices	RF Semiconductors	(336) 664-1233	www.rfmd.com
RTI/Ketema	Surge Suppressors	(714) 630-0081	www.rtie.rti-corp.com
Schurter	Fuses and Holders	(707) 778-6311	www.schurterinc.com
Thermalloy	Heat Sinks	(972) 243-4321	www.thermalloy.com
Toko	RF Products	(847) 699-3430	www.tokoam.com

Linear Technology Corporation

Product	Phone Number	URL
High Performance Analog ICs	(408) 432-1900	www.linear.com

Signal sources, conditioners and power circuitry

39

Jim Williams

Introduction

Occasionally, we are tasked with designing circuitry for a specific purpose. The request may have customer origins or it may be an in-house requirement. Alternately, a circuit may be developed because its possibility is simply too attractive to ignore[1]. Over time, these circuits accumulate, encompassing a wide and useful body of proven capabilities. They also represent substantial effort. These considerations make publication an almost obligatory proposition and, as such, a group of circuits is presented here. This is not the first time we have displayed such wares and, given the encouraging reader response, it will not be the last[2]. Eighteen circuits are included in this latest effort, roughly arranged in the categories given in this publication's title. They appear at the next paragraph.

Voltage controlled current source—ground referred input and output

A voltage controlled current source with ground referred input and output is difficult to achieve. Executions exist, but are often cumbersome, involving numerous components. Figure 39.1's conceptual design utilizes a differential amplifier featuring differential, uncommitted feedback inputs. The independent feedback inputs permit the differential signal inputs to operate anywhere inside their common mode range, unencumbered by feedback interaction. Similarly, the differential feedback ports may sense referred to any point within their common mode range. In both cases, common mode range extends from V^- to within 2V of the positive rail. Output swing extends to both rails.

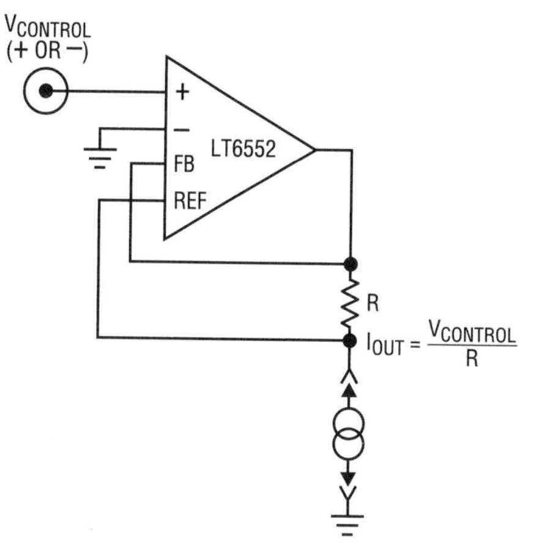

Figure 39.1 • Conceptual Ground Referred Voltage Controlled Bipolar Current Source Utilizes Differential Amplifier's Separate Feedback Inputs. Compliance Limits are Imposed by Supply Voltage, Output Current Capacity and Input Common Mode Range

The freedoms described above invite Figure 39.1's configuration. The amplifier is biased by a control voltage input, which feedback action impresses across the resistor. Scaling is set by the equation given, which will be recognized as a dressed version of Ohm's Law. Note that this circuit will produce current outputs of either polarity, as dictated by the control input. Compliance limits are imposed by power supply voltage, output current capacity and input common mode range.

Figure 39.2 puts Figure 39.1's thesis to work. The test circuit (figure left) produces control signals to exercise the current source (figure right), which drives a capacitive load. Figure 39.3's waveforms describe circuit activity. Trace A is the clock, trace B A1's control input and trace C is capacitor voltage. The test circuit presents alternating polarity

Note 1: "When you see something technically sweet, you do it" (Robert J. Oppenheimer).

Note 2: Previous efforts of this ilk include AN45, AN52, AN61, AN66, AN67 and AN75. See References 14 to 19.

Analog Circuit and System Design: Immersion in the Black Art of Analog Design. http://dx.doi.org/10.1016/B978-0-12-397888-2.00039-0

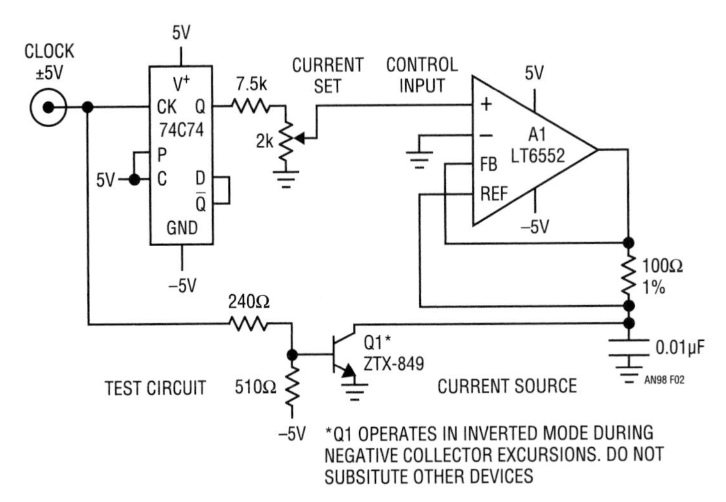

Figure 39.2 • Practical Version of Figure 39.1 Sources Bipolar Current to Capacitive Load. Test Circuit Provides Bipolar Control Input and Resets Capacitor. Result is Alternating, Opposed Polarity Ramps Across Capacitor

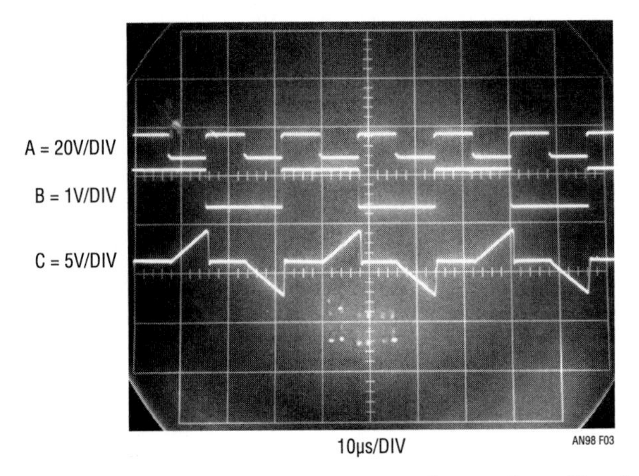

Figure 39.3 • Voltage Controlled Current Source Test Circuit Waveforms Include Clock (Trace A), Control Input (Trace B) and Capacitor Voltage (Trace C). Bipolar Control Input Voltage Results in Complementary Capacitor Ramps

control inputs (trace B) after each Q1 directed capacitor reset to zero (trace C). The result, alternating, equal amplitude, opposed polarity linear capacitor ramps, clearly demonstrates the current sources capabilities.

Stabilized oscillator for network telephone identification

Some telephone networks require an amplitude and frequency stabilized 100Hz carrier to indicate the status of any phone in the network. Figure 39.4, operating from a single 5V supply, provides this function using only two dual op amps and attendant discrete components. A1, a conventional multivibrator, operates at 100Hz. Its square and triangle outputs appear in Figure 39.5, traces A and B, respectively. The 100Hz triangle, heavily filtered by A2's 16Hz RC input pair, appears as a sine wave at A2's amplified output (trace C). A2's output, in turn, is applied to A3, configured as a half wave rectifier. A3's input attenuation keeps the sine wave's negative excursions within the amplifier's input range ($V_{CM(LIMIT)} = -0.3V$). Single rail powered A3's output can't track the sine wave's negative portion; it simply saturates within millivolts of ground, producing trace D's half-wave rectified output. This output, representing A2's amplitude, is compared to a DC reference by band-limited A4-Q1. Q1's collector biases A1's power pin, closing an amplitude stabilization loop which

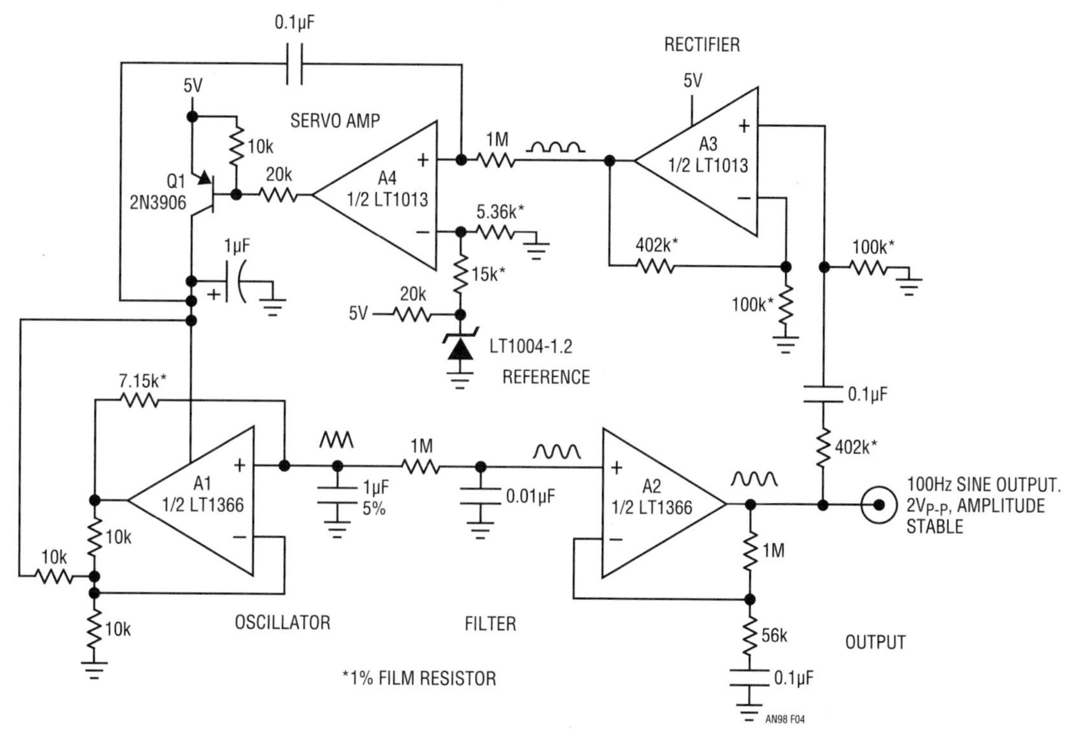

Figure 39.4 • Amplitude/Frequency Stabilized Sine Wave Oscillator, Developed for Network Telephone Identification, Suits General Purpose Use. A1's Filtered Triangle Output Produces a $2V_{P-P}$ Sinewave at A2. A3's Rectified Output is Balanced Against Reference at A4. Q1 Closes Regulation Loop by Modulating A1's Power Pin

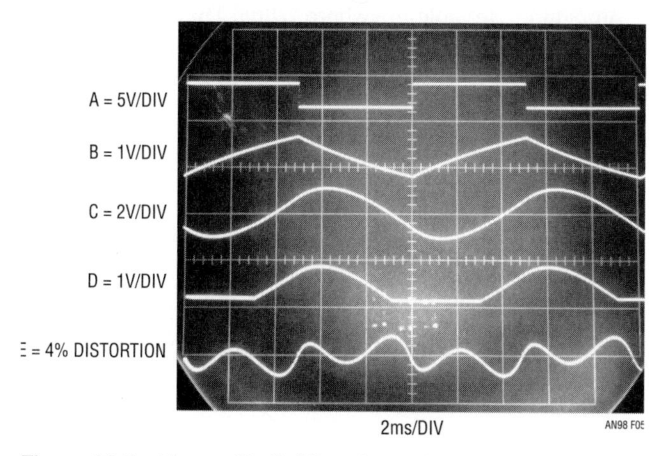

Figure 39.5 • Figure 39.4's Waveforms Include A1's Square (Trace A) and Triangle (Trace B) Outputs, A2's Sinewave (Trace C), A3's Rectified Output (Trace D) and Distortion Residue (Trace E). 1M-0.01μF Filter at A2 Permits 4% Distortion, Despite Triangle Wave Infidelity

regulates the circuit's sine wave output. Sine wave distortion, appearing in trace E, is only 4% despite the originating triangle waves infidelity. Other specifications include less than 0.15% amplitude variation for supply shifts of 3.4V to 36V, frequency stability inside 0.01% over the same supply range and initial frequency accuracy of 6%.

Micro-mirror display pulse generator

Some "micro-mirror" displays require high voltage pulses for biasing. Pulse amplitude must be adjustable anywhere within a 0V to −50V window, with pulse top and bottom amplitude independently settable. Additionally, rise and fall times must be within 150ns into the 1500pF micro-mirror load, with absolutely no overshoot permissible. The input pulse is supplied from 5V powered positive going logic. These requirements dictate a very carefully considered level shifter.

Figure 39.6's circuit meets display requirements. The input pulse is applied to both sections of an LTC®1693 noninverting driver. The LTC1693 output reproduces the input pulse at a much lower source impedance. The LTC1693 output, referenced to the negative rail by the RC-diode combination, drives level shifter Q1. Q1, utilizing Baker clamping and base speed-up capacitance, provides wideband voltage gain with pulse amplitude set by collector and emitter supply potentials. Q1's collector capacitance is isolated by Q2-Q3. These transistors, in turn, drive output stage Q4-Q5 via a resistor. This resistor combines with Q4-Q5 input capacitance to control edge times and overshoot. Its value, nominally 200Ω, will vary somewhat with layout and should be selected for best output waveform purity. Q4 and Q5, high current types, drive the capacitive load.

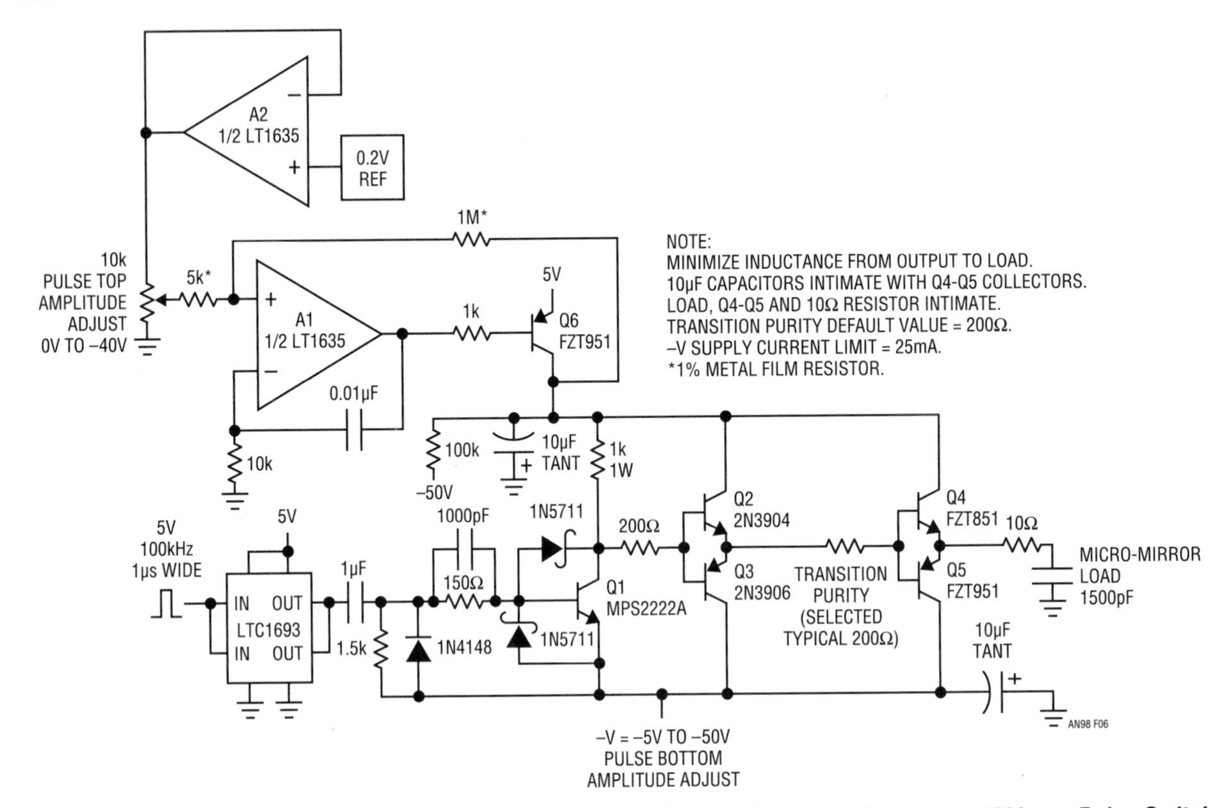

Figure 39.6 • High Voltage, Wideband Level Shift for Micro-Mirror Biasing Precludes Overshoot. 5V Input Pulse Switches Q1 Voltage Gain Stage via LTC1693 Driver. Q2-Q3 Isolate Q1's Collector, Bias Q4-Q5 Output. A1-Q6 Regulate Pulse Top Amplitude; –V Potential Sets Pulse Bottom Voltage. Output Pulse Amplitude, Settable Anywhere Within These Limits, Has No Overshoot

The 5-transistor stage swings to potentials established by Q1's emitter and collector rails[3]. Emitter rail voltage, hence "pulse bottom" amplitude, is set by the DC potential of its power supply, variable between −5V and −50V. The collector rail is controlled by A1, operating in the Wu configuration[4]. A1, containing an amplifier and a 0.2V reference, drives Q6 to regulate the collector rail anywhere between zero and −40V in accordance with the 10k potentiometer's setting. The settability of both power rails, combined with the transistor stages wide operating region, permits pulse amplitude control over the desired range.

Figure 39.7 shows the level shift output (trace B) responding to an input pulse (trace A) with amplitude limits adjusted for zero and −50V. The high voltage output transitions, occurring within 100ns, are exceptionally pure.

A = 5V/DIV
B = 10V/DIV
100ns/DIV
AN98 F07

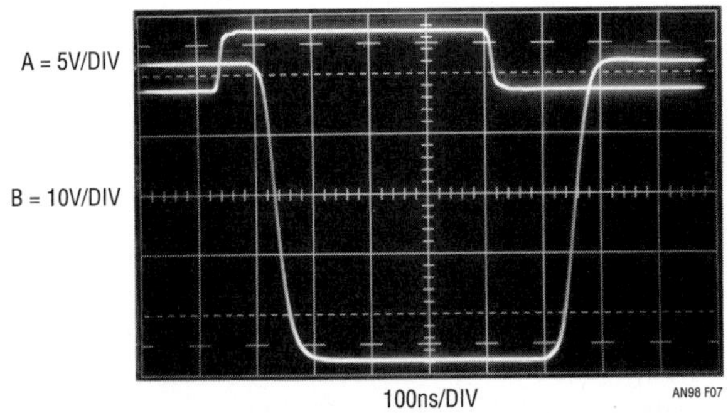

Figure 39.7 • Level Shift Responds (Trace B) to Input Pulse (Trace A) with Amplitude Limits Adjusted for Zero and –50V. Fast, High Voltage Transitions are Exceptionally Pure

Note 3: Transistor data sheet aficionados may notice that the −50V potential exceeds Q1, Q2, Q3 V_{CEO} specifications. The transistors operate under V_{CER} conditions, where breakdown is considerably higher.
Note 4: The collector rail regulation scheme was suggested by Albert Wu of Linear Technology Corporation.

Simple rise time and frequency reference

A frequent requirement in wideband circuit work is a rise time/frequency reference. The LTC6905 oscillator provides a simple way to realize this. This device, programmable by pin strapping and a single resistor, achieves outputs over a continuous 17MHz to 170MHz range with accuracy inside 1%. Additionally, output stage transitions are typically within 500ps.

Figure 39.8's circuit is delightfully simple. The LTC6905 is set for 100MHz output by the pin strapping and resistor value shown. The 953Ω resistor isolates the IC's output from the 50Ω oscilloscope input and any parasitic capacitance, promoting the fastest possible transitions. Figure 39.9 shows circuit output in a 1GHz real-time bandwidth (t_{RISE} = 350ps). The 100MHz square wave displays sub-nanosecond transitions. Determining transition rise and fall times requires a faster oscilloscope[5]. Figures 39.10 and 39.11, measured in a 3.9GHz sampled bandpass, record a 400ps rise time (Figure 39.10) and a 320ps fall time (Figure 39.11).

Note 5: See Appendix A, "How Much Bandwidth is Enough?" and Appendix B, "Connections, Cables, Adapters, Attenuators, Probes and Picoseconds."

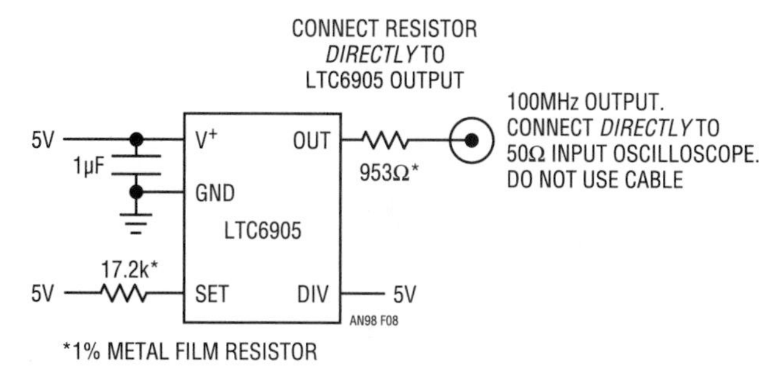

Figure 39.8 • LTC6905 Oscillator Configured for Sub-Nanosecond Transitions and 100MHz Output is Rise Time/Frequency Reference

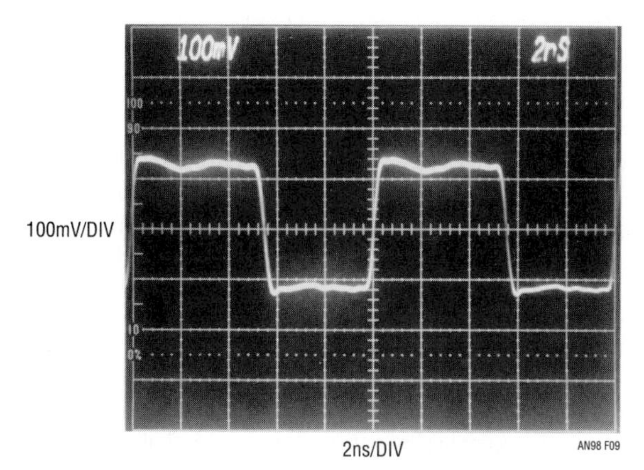

Figure 39.9 • 100MHz Output Viewed in 1GHz Real-Time Bandwidth Displays Sub-Nanosecond Transitions

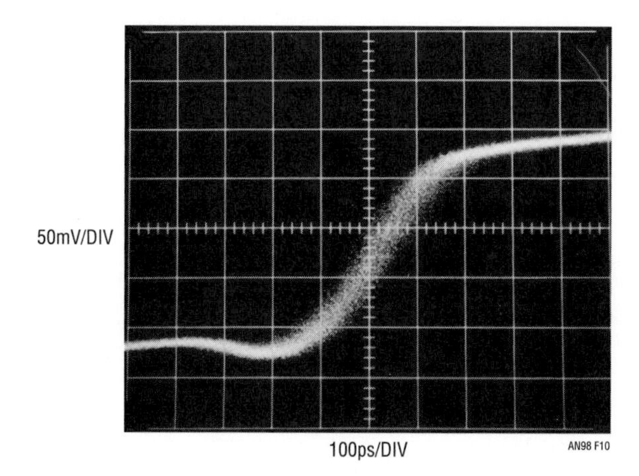

Figure 39.10 • Transition Rise Time Measures 400ps in 3.9GHz (t_{RISE} = 90ps) Sampled Bandpass. Trace Granularity Derives from Sampling Oscilloscope Operation

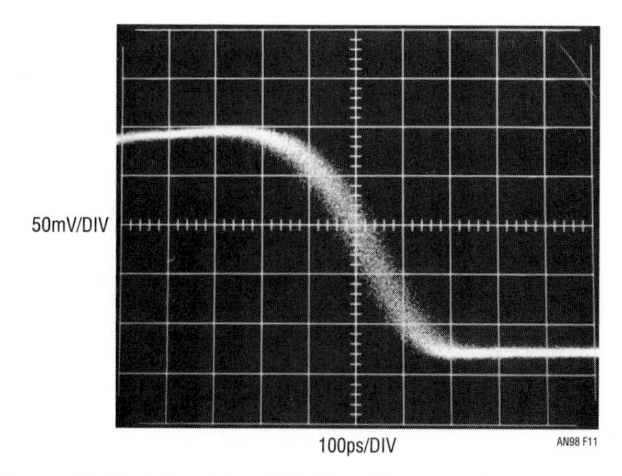

Figure 39.11 • Transition Fall Time Measures 320ps in 3.9GHz (t_{RISE} = 90ps) Sampled Bandpass. Trace Granularity Derives from Sampling Oscilloscope Operation

850 picosecond rise time pulse generator with <1% pulse top aberrations

Impulse response and rise time testing often require a fast rise time source with a high degree of pulse purity. These parameters are difficult to simultaneously achieve, particularly at sub-nanosecond speeds. Figure 39.12's circuit, derived from oscilloscope calibrators, meets these criteria, delivering an 850ps output with less than 1% pulse top aberrations.

Oscillator 01 delivers a 10MHz square wave to current mode switch Q2-Q3. Note that 01 is powered between ground and −5V to meet transistor biasing requirements. Q1 provides current drive to Q2-Q3. When 01 biases Q2, Q3 goes off. Q3's collector rises rapidly to a potential determined by Q1's collector current, D1, the resistors at the circuits output and the 50Ω termination. When 01 goes low, Q2 turns off, Q3 comes on and the output settles to zero. D2 prevents Q3 from saturating.

The circuit's positive output transition is extremely fast and singularly clean. Figure 39.13, viewed in a 1GHz real-time bandwidth, shows 850ps rise time with exceptionally pure pre- and post-transition characteristics[6]. Figure 39.14 details pulse top settling. The photo shows the pulse-top region immediately following the positive

Note 6: The measured 850ps rise time, influenced by the monitoring 1GHz oscilloscopes 350ps rise time, is almost certainly pessimistic. A root-sum-square correction applied to the measurement indicates a 775ps rise time. See Appendix A for detailed discussion.

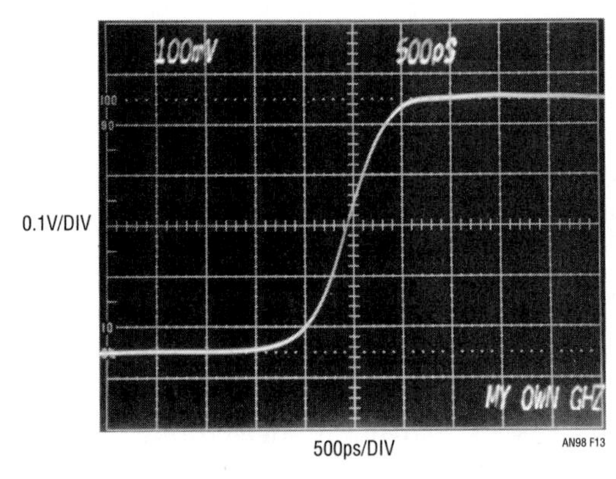

Figure 39.13 • Figure 39.12's Displayed 850ps Transition Time is Free of Discontinuities when Viewed in a Real-Time 1GHz (t_{RISE} = 350ps) Bandwidth. Root-Sum-Square Correction Applied to Measurement Indicates 775ps Rise Time

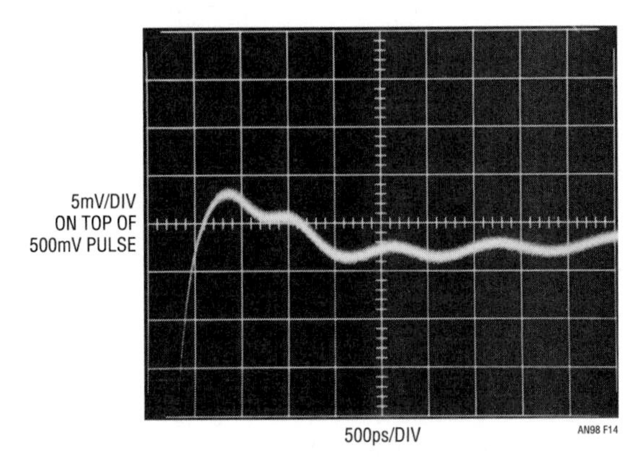

Figure 39.14 • Pulse Top Aberrations Remain Inside 4mV Within 400ps of Transition Completion. 1GHz Ring-Off is Probably Due to Breadboard Limitations. Trace Granularity Derives from Sampling Oscilloscope Operation

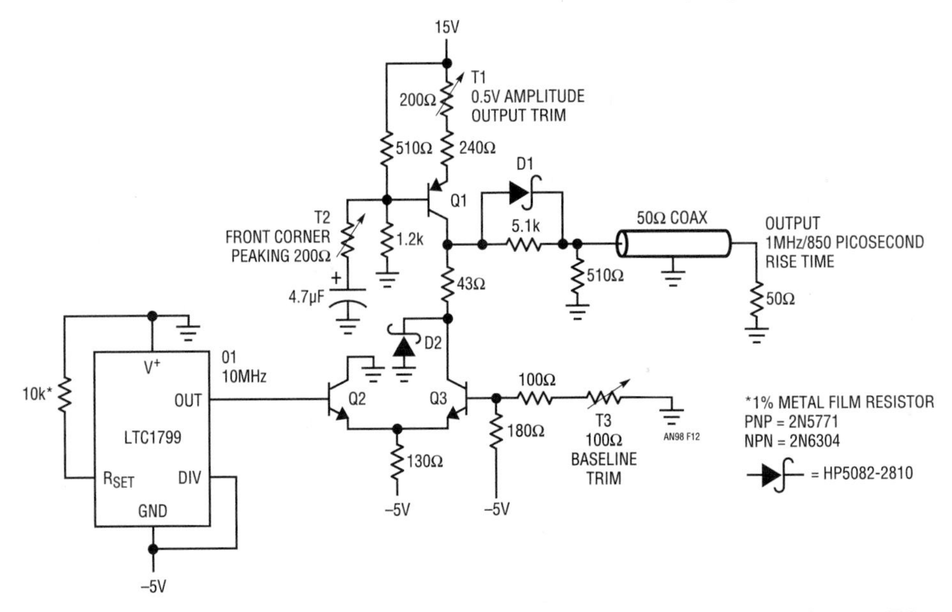

Figure 39.12 • Oscillator 01 Drives Q2-Q3 Current Mode Switch, Producing 850ps Rise Time Output. Trims Facilitate Clean Transition with <1% Pulse Top Aberrations

500mV transition. Settling occurs within 400ps of the edge's completion, with all undesired activity within ±4mV. The 1mV, 1GHz ring-off is probably due to breadboard construction limitations, and could be eliminated with stripline layout techniques.

This level of performance requires trimming. The oscilloscope used should have at least 1GHz of bandwidth. T2 and T3 are adjusted for best pulse presentation while T1 sets 500mV output amplitude across the 50Ω termination. The trims are somewhat interactive, although not unduly so, converging quickly to give the results noted.

20 picosecond rise time pulse generator

Figure 39.15, another fast rise time pulse generator, switches a high grade, commercially produced tunnel diode mount to produce a 20ps rise time pulse. 01's clocking (trace A, Figure 39.16) causes Q1's collector (trace B) to switch the capacitively loaded Q2-Q3 current source. The resultant repetitive ramp at Q3's collector (trace C), buffered by Q4, biases the tunnel diode mount via the output resistors. The tunnel diode driven output (trace D) follows the ramp until abruptly rising (trace D, just prior to 4th vertical division). This departure is caused by tunnel diode triggering. The edge associated with this triggering is extremely steep, with a specified rise time of 20ps and clean settling. Figure 39.17 examines this edge within the limitations of a 3.9GHz (t_{RISE} = 90ps) sampling oscilloscope. The trace shows the tunnel diode's switching,

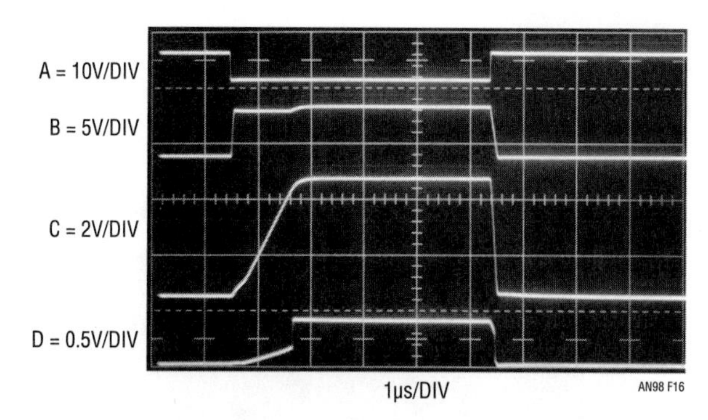

Figure 39.16 • 01 (Trace A) Clocks Q1's Collector (Trace B), Switching Capacitively Loaded Q2-Q3 Current Source. Resultant Repetitive Ramp at Q3's Collector (Trace C), Buffered by Q4, Biases Tunnel Diode via Output Resistors. Tunnel Diode Output (Trace D) Follows Ramp Until Abruptly Triggering

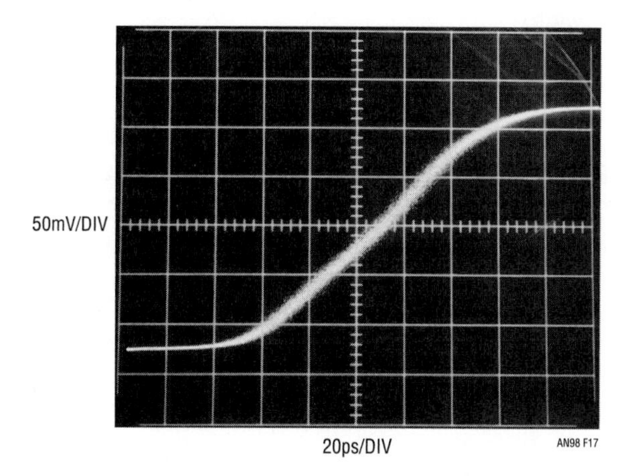

Figure 39.17 • Figure 39.15's 20ps Edge Drives a 3.9GHz Sampling 'Scope to its 90ps Rise Time Limit. Trace Granularity is Characteristic of Sampling Oscilloscope Display

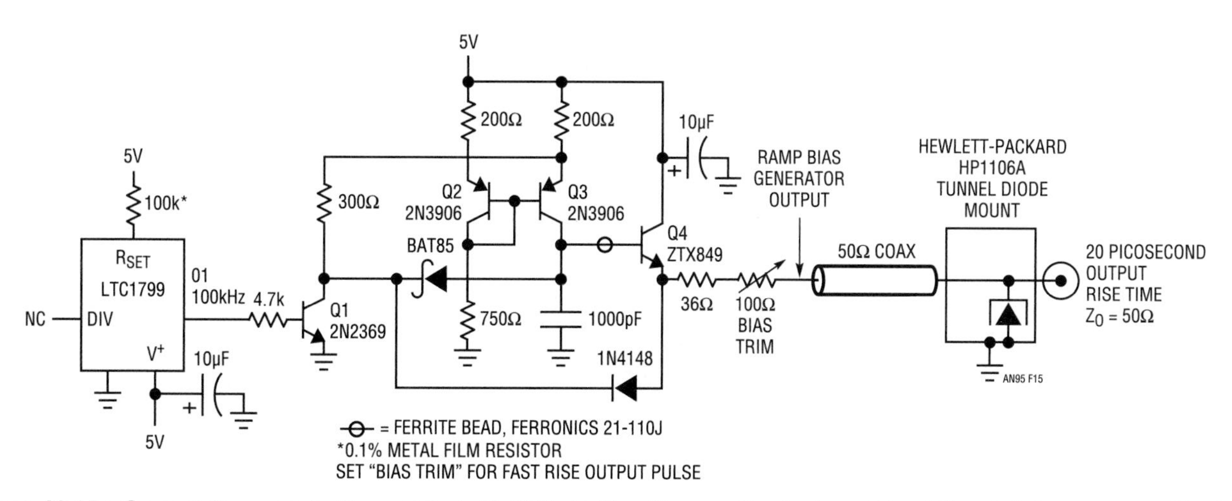

Figure 39.15 • Current Ramps Into Tunnel Diode Until Switching Occurs, Producing a 20ps Edge. Q1, Squarewave Clocked from 01, Switches Q2-Q3 Capacitively Loaded Current Source, Producing Repetitive Ramps at Q4. Ascending Current Through Output Resistors Triggers Tunnel Diode

driving the oscilloscope to its 90ps rise time limit[7]. Figure 39.18, slowing sweep speed to 100ps/divison, shows pulse top settling (in a 3.9GHz bandwidth) within 4% inside 100ps[8].

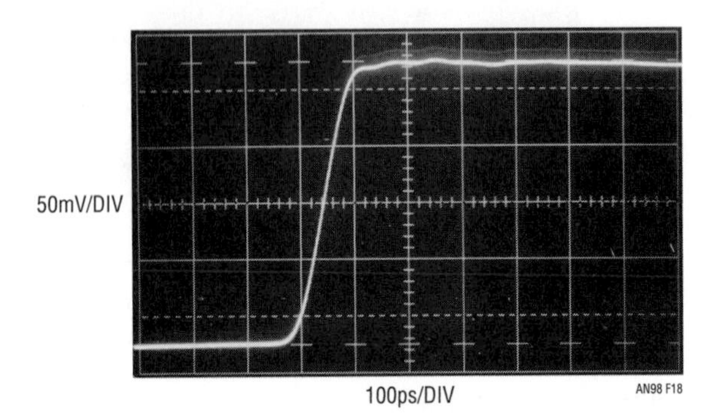

Figure 39.18 • Reducing Sweep Speed Shows 4% Pulse Top Flatness Within Oscilloscope's 3.9GHz (t_{RISE} = 90ps) Bandwidth

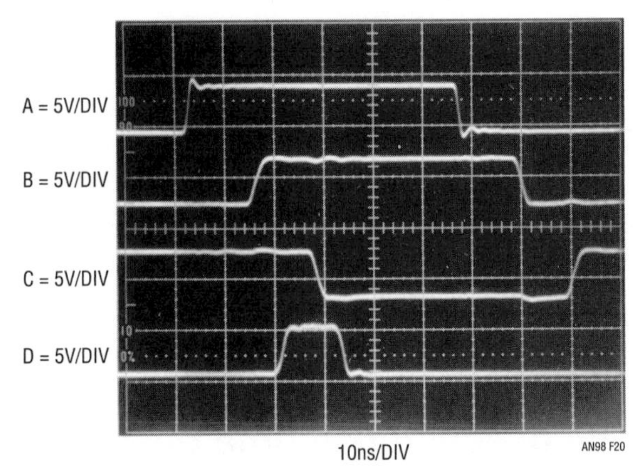

Figure 39.20 • Pulse Generator Waveforms, Viewed in 400MHz Real-Time Bandwidth, Include Input (Trace A), C3 (Trace B) Fixed and C2 (Trace C) Variable Outputs. Circuit Output Pulse is Trace D. RC Network's Differential Delay Manifests as C2-C3 Positive Overlap. G1 Extracts This Interval, Presents Output

Nanosecond pulse width generator

The previous three circuits were optimized for fast rise time. It is sometimes desirable to produce extremely short width pulses in response to an input trigger. Such a predictable, programmable short time interval generator has broad use in fast pulse circuitry, particularly in sampling applications[9]. Figure 39.19, built around a quad high speed comparator and a fast gate, has a settable 0ns to 10ns output width with 520ps, 5V transitions. Pulse width varies less than 100ps with 5V supply variations of ±5%. Minimum input trigger width is 30ns and input-output delay is 18ns[10].

The input pulse (Figure 39.20, trace A) is inverted by C1, which also isolates the 50Ω termination. C1's output drives fixed and variable RC networks. The networks charge time difference, and hence delay, is primarily determined

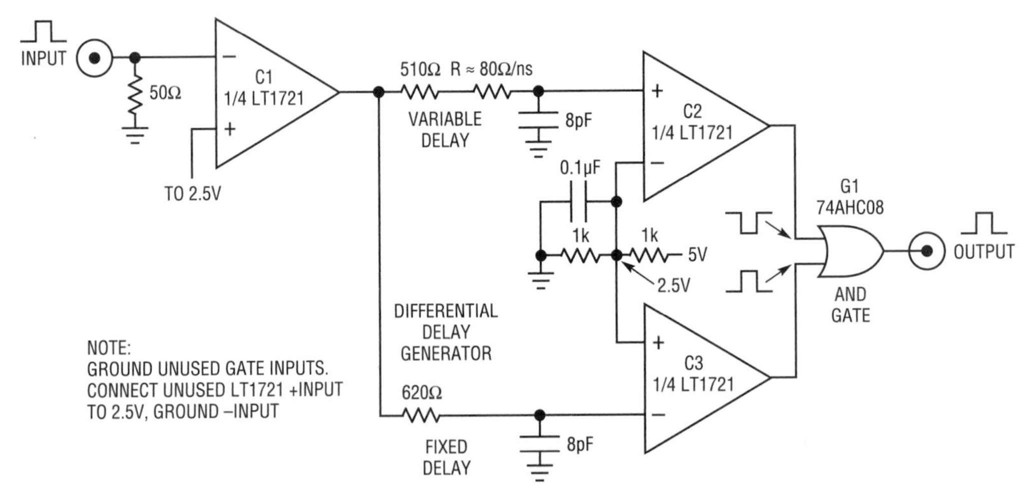

Figure 39.19 • Pulse Generator Has 0ns to 10ns Width, 520ps Transitions. C1 Unloads Termination, Drives Differential Delay Network. C2-C3 Complementary Outputs Represent Delay Difference as Edge Timing Skew. G1, High During C2-C3 Positive Overlap, Presents Circuit Output

Note 7: Sorry, but 3.9GHz is the fastest 'scope in my house. See Appendix A for relevant comment.

Note 8: The HP1106 is no longer produced, although available on the secondary market. The TD1107, currently manufactured by Picosecond Pulse Labs, is an equivalent unit, although we have no experience with it.

Note 9: Pedestrian laboratory argot for interval generator is "one-shot."

Note 10: This circuit is a considerably improved extension of earlier work. See References 4 and 5.

by programming resistor R, at a scale factor ≈80Ω/ns. C2 and C3, arranged as complementary output level detectors, represent the network's delay difference as edge time skew. Trace B is C3's ("fixed") output and trace C is C2's ("variable") output. Gate G1's output (trace D), high during C2-C3 positive overlap, presents the circuit's output pulse. Figure 39.21 shows a 5V, 5ns width (measured at 50% amplitude) output pulse with R = 390Ω. The pulse is clean, with well defined transitions. Post-transition aberrations, within 8%, derive from G1's bond wire inductance and an imperfect coaxial probing path. Figure 39.22 shows the narrowest full amplitude (5V) pulse obtainable. Width measures 1ns at the 50% amplitude point and 1.7ns at the base in a 3.9GHz bandwidth. Shorter widths are obtainable if partial amplitude pulses are acceptable. Figure 39.23 shows a 3.3V, 700ps width (50%) with a 1.25ns base. G1's rise time limits minimum achievable pulse width. Figure 39.24, taken in a 3.9GHz sampled bandpass, measures 520ps rise time. Fall time is similar.

Single rail powered amplifier with true zero volt output swing

Many single supply powered applications require amplifier output swings within millivolt or even sub-millivolt levels of ground. Amplifier output saturation limitations normally preclude such operation. Figure 39.25's power supply bootstrapping scheme achieves the desired characteristics with minimal component addition[11].

A1, a chopper stabilized amplifier, has a clock output. This output switches Q1, providing drive to the diode-capacitor charge pump. The charge pump output feeds

Note 11: See Reference 8, Appendix D.

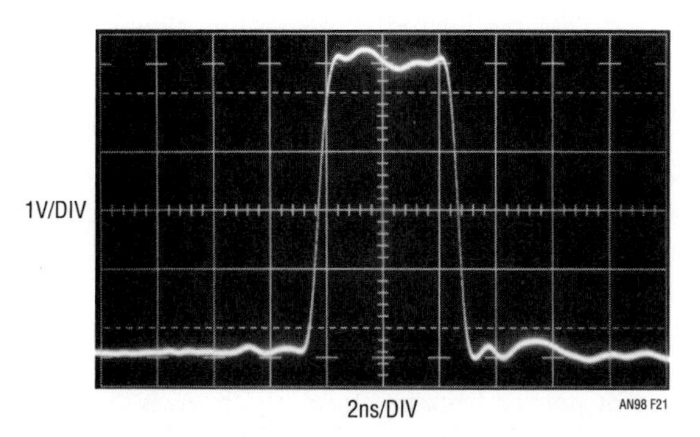

Figure 39.21 • 5ns Wide Output with R = 390Ω is Clean, with Well Defined Transitions. Post-Transition Aberrations, Within 8%, Derive from G1 Bond Wire Inductance and Imperfect Coaxial Probe

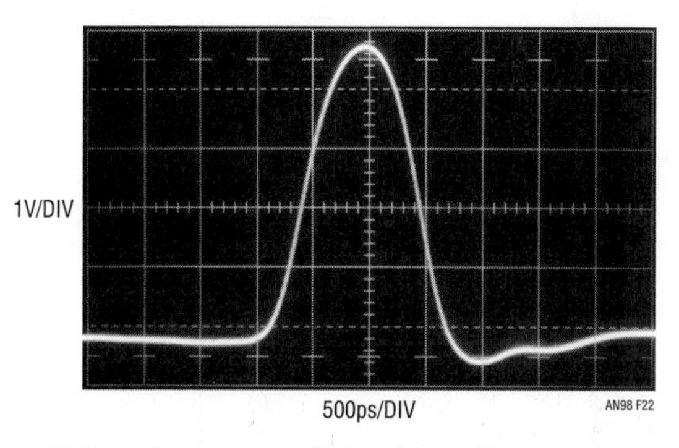

Figure 39.22 • Narrowest Full Amplitude Pulse Width is 1ns; Base Width Measures 1.7ns. Measurement Bandwidth is 3.9GHz

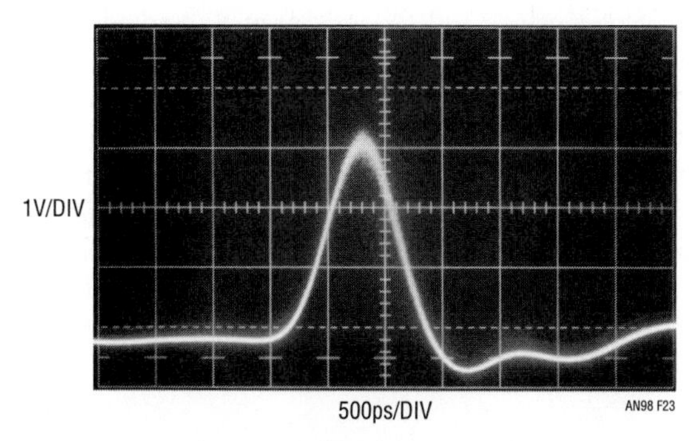

Figure 39.23 • Partial Amplitude Pulse, 3.3V High, Measures 700ps Width with 1.25ns Base. Trace Granularity is Artifact of 3.9GHz Sampling Oscilloscope Operation

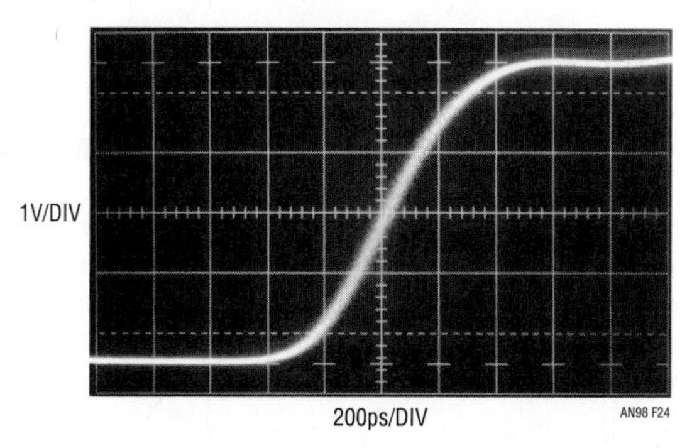

Figure 39.24 • Transition Detail in 3.9GHz Bandpass (t_{RISE} = 90ps) Shows 520ps Rise Time. Fall Time is Similar. Trace Granularity Derives from Sampling Oscilloscope Operation

A1's V⁻terminal, pulling it below zero, permitting output swing to (and below) ground. If desired, the negative output excursion can be limited by either clamp option shown.

Reliable start-up of this bootstrapped power supply scheme is a valid concern, warranting investigation. In Figure 39.26, the amplifier's V^- pin (trace C) initially rises at supply turn-on (trace A) but heads negative when amplifier clocking (trace B) commences at about midscreen.

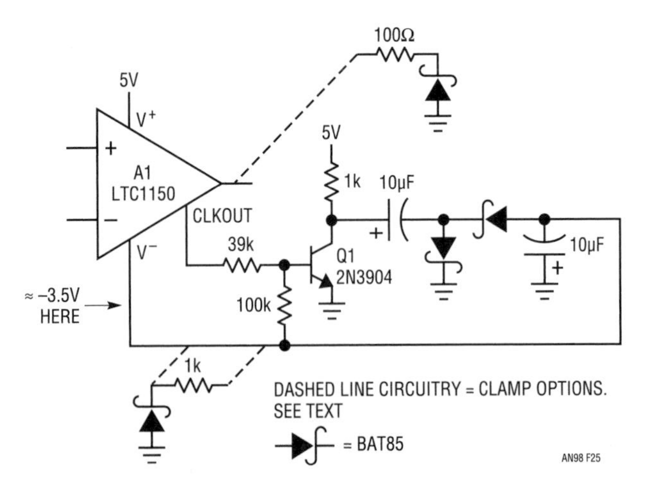

Figure 39.25 • Single Rail Powered Amplifier Has True Zero Volt Output Swing. A1's Clock Output Switches Q1, Driving Diode-Capacitor Charge Pump. A1's V– Pin Assumes Negative Voltage, Permitting Zero (and Below) Volt Output Swing

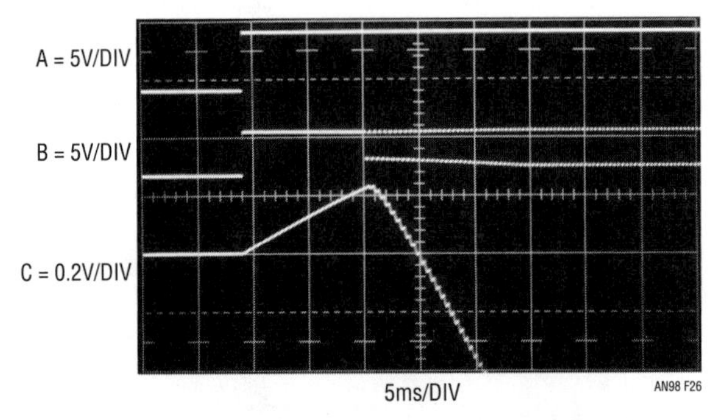

Figure 39.26 • Amplifier Bootstrapped Supply Start-Up. Amplifier V– Pin (Trace C) Initially Rises Positive at 5V Supply (Trace A) Turn-On. When Amplifier Internal Clock Starts (Trace B, 5th Vertical Division), Charge Pump Activates, Pulling V⁻ Pin Negative

The circuit provides a simple way to obtain output swing to zero volts, permitting a true "live at zero" output.

Milliohmmeter

Resistance measurement of contacts, PC traces and vias requires a low resistance ohmmeter. Figure 39.27's 9V battery-powered design has a 1Ω full-scale range, with resolution down to 1mΩ. It produces a 0V to 1V output for a 0Ω to 1Ω resistance at its 4-terminal Kelvin sensed input with 0.1% accuracy over a 5.25V to 9.5V power supply range. An AC carrier modulation scheme is employed to reject noise and error inducing DC offsets due to parasitic thermocouples (Seebeck effect)[12].

A1 and associated components form a 10mA current source that is alternately steered between Rx, the unknown resistance, and ground by LTC6943 switch pins 10, 11 and 12. The LTC6943's control pin (Pin 14) is clocked at ≈45Hz from the CD4024 divider output. This action causes a carrier modulated 10mA current flow through Rx. Rx's value determines the resultant AC voltage across it. This AC signal is capacitively coupled to LTC6943 switch pins 1, 4 and 5, driven synchronously with the current source modulation. These pins switching forms a synchronous rectifier, demodulating the AC signal back to DC across A2's input capacitor. A2 amplifies this DC potential at a gain of 1mV per milliohm, or 1V full scale. Note that single-rail powered A2's output can swing to true "zero" because it utilizes a variant of the supply boostrapping scheme presented back in Figure 39.25. A2's clock output drives Q2, which pulses the CD4024 divider. One divider output switches the LTC6943 modulator-demodulator while another output drives the bootstrapped charge pump to supply A2's V^- pin with about −7V.

Diode clamps prevent accidental overvoltage at the probe inputs without introducing loading error to the 10mV maximum Rx carrier waveform. Circuit calibration involves placing a 1Ω, 0.1% resistor at Rx and adjusting the 200Ω trimmer for $1.000V_{OUT}$. The synchronously demodulated AC carrier technique displays the inherent narrow band noise rejection characteristics of "lock-in" type measurements. Figure 39.28 shows a normal waveform across Rx for Rx = 1Ω. The 10mV signal is clean, and circuit output reads 1.000V. In Figure 39.29 noise is deliberately injected into the Rx probes, burying the carrier in a 6× noise-to-signal ratio. Despite this, circuit output remains at 1.000V.

Note 12: This circuit's operation is derived from the Hewlett-Packard HP-4328A. See Reference 7.

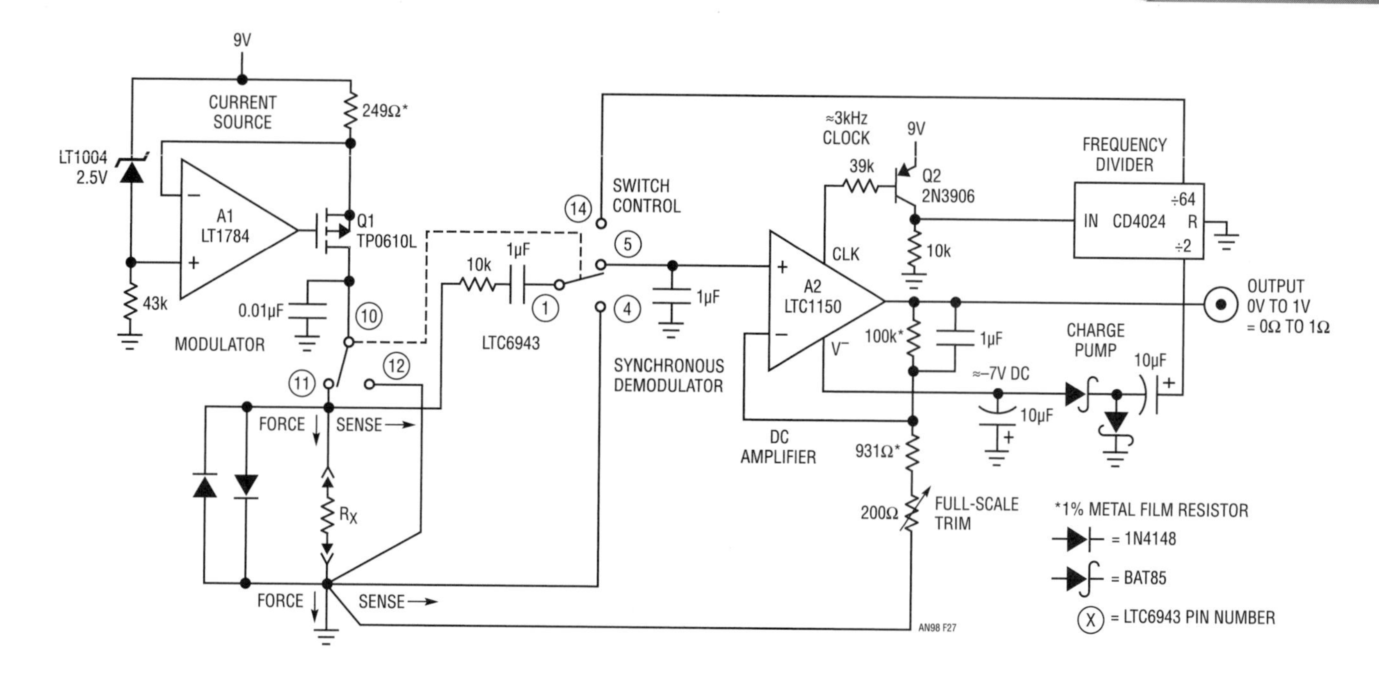

Figure 39.27 • 1Ω Full-Scale Ohmmeter Accurately Resolves 0.001Ω for PC Board Trace/Via Resistance Measurement. Carrier Modulation of Unknown Resistance Permits Narrowband Synchronous Demodulation, Rejecting Noise and Parasitic DC Offsets. Kelvin Sensing at Rx Prevents Test Lead Induced Errors

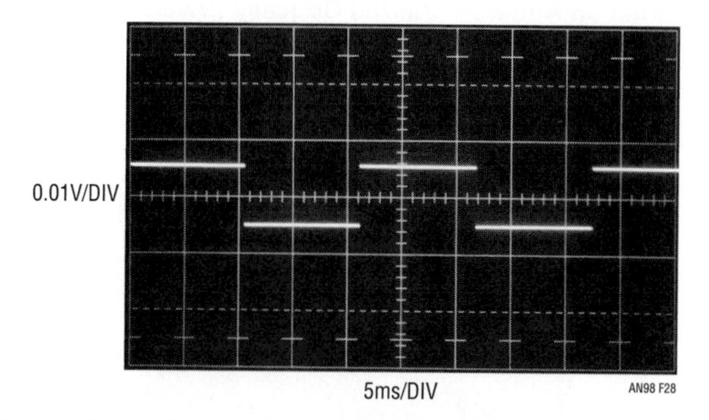

Figure 39.28 • Normal Waveform at Rx with Rx = 1.000Ω. Circuit Output Correctly Reads 1.000V

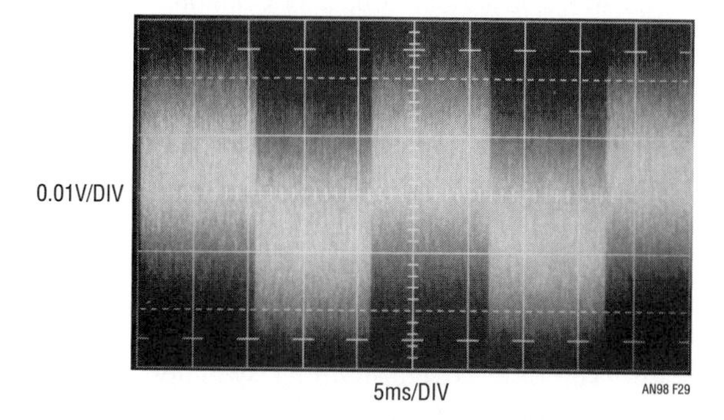

Figure 39.29 • Rx Waveform with Rx = 1.000Ω and Noise Added. Circuit Output Remains 1.000V Despite 6× Noise-to-Signal Ratio

0.02% accurate instrumentation amplifier with $125V_{CM}$ and 120dB CMRR

Figure 39.30's circuit may be used when high accuracy differential input measurement is required[13]. It is particularly suited to transducer signal conditioning where high common mode voltage may occur. The circuit has the low offset and drift of chopper stabilized A1, but also incorporates a novel optically coupled, switched capacitor input stage to achieve specifications unavailable in conventional designs. DC common mode rejection exceeds 120dB over a ±125V input range and gain accuracy and stability are set by A1. Error from all sources is inside 0.02%. The design's high common mode voltage capability allows it to reliably extract small signals while withstanding transient and fault conditions often encountered in industrial environments.

This scheme measures input difference voltage by switching (S1A, S1B) a capacitor across the input ("ACQUIRE"). After a time the capacitor charges to the voltage across the input. S1A and S1B open and S2A and S2B close ("READ"). This grounds one capacitor plate and

Note 13: Sharp-eyed devotees of LTC publications will recognize this as a mildly modified variant of Reference 8 (pp. 10-11) and Reference 13 (pp. 1-2).

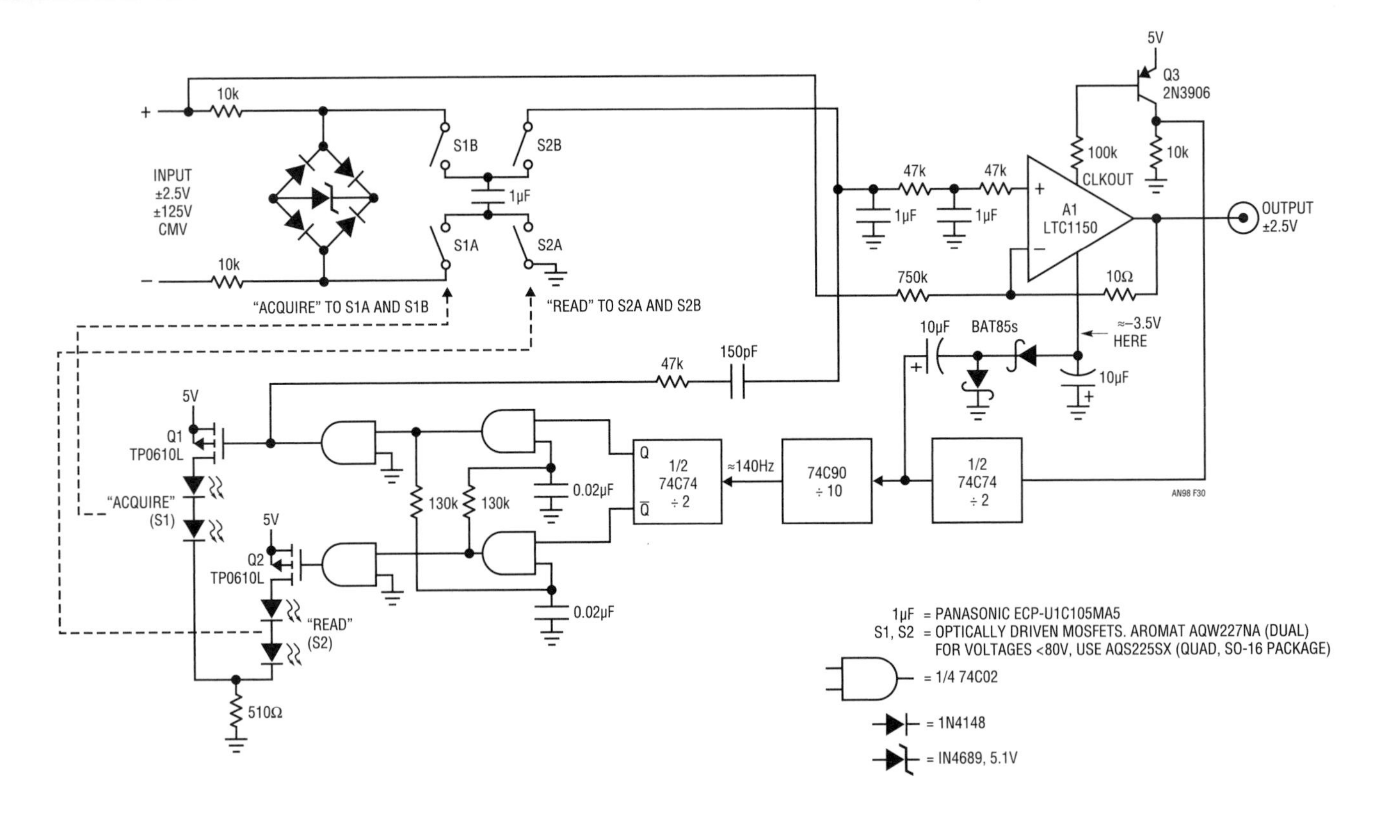

Figure 39.30 • 0.02% Accurate, 125V Common Mode Range Instrumentation Amplifier Utilizes Optically Driven FETs and Flying Capacitor. Logic Driven Q1-Q2 Provides Nonoverlapping Clocking to S1-S2 LEDs. Clock Derives from A1's Internal Oscillator

the capacitor discharges into the grounded 1µF unit at S2B. This switching cycle is continuously repeated, resulting in A1's ground referred positive input assuming the input difference voltage. The common mode voltage is rejected by the optical switching of the ungrounded 1µF capacitor. The LED driven MOSFET switches specified do not have junction potentials and the optical drive contributes no charge injection error. A nonoverlapping clock prevents simultaneous conduction in S1 and S2, which would result in charge loss, causing errors and possible circuit damage. The 5.1V zener prevents switched capacitor failure if the inputs are subjected to differential overvoltage.

A1, a chopper stabilized amplifier, has a clock output. This clock, level shifted and buffered by Q3, drives a logic divider chain. The first flip-flop activates a charge pump, pulling A1's V^- pin negative, permitting amplifier swing to (and below) zero volts[14]. The divider chain terminates into a logic network. This network provides phase opposed charging of the 0.02µF capacitors (Traces A and B, Figure 39.31). The gating associated with these capacitors is arranged so the logic provides nonoverlapping, complementary biasing to Q1 and Q2. These transistors supply this nonoverlapping drive to the S1 and S2 actuating LEDs (Traces C and D). The extremely small parasitic error terms in the LED driven MOSFET switches results

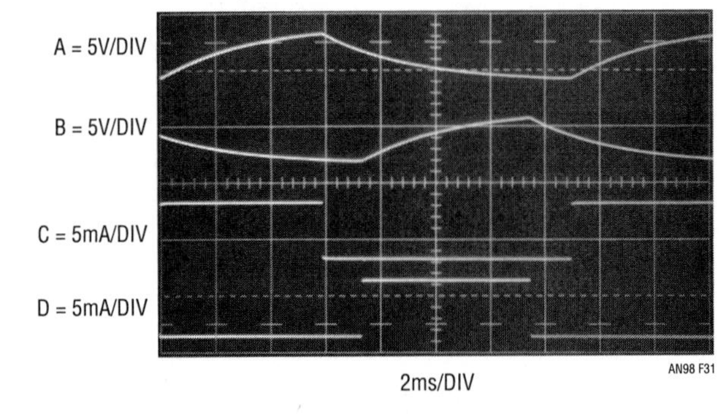

Figure 39.31 • Clocked, Cross Coupled Capacitors (Traces A and B) in 74C02 Based Network Result in Nonoverlapping Drive (Traces C and D) to S1-S2 Actuation LEDs

Note 14: This arrangement will be recognized from Figures 39.25 and 39.27. See also Reference 8, Appendix D.

in nearly theoretical circuit performance. However, residual error (≈0.1%) is caused by S1A's high voltage switching pumping S2B's 3pF to 4pF junction capacitance. This results in a slight quantity of unwanted charge being transferred to the 1μF capacitor at S2B. The amount of charge transferred varies with the input common mode voltage and, to a lesser extent, the varactor-like response of S2B's off-state capacitance. These terms are partially cancelled by DC feedforward to A1's negative input and AC feedforward from Q1's gate to S2B. The corrections compensate error by a factor of five, resulting in 0.02% accuracy.

Optical switch failure could expose A1 to high voltage, destroying it and possibly presenting destructive voltages to the 5V rail. This most unwelcome state of affairs is prevented by the 47k resistors in A1's positive input.

Wideband, low feedthrough, low level switch

Rapid switching of wideband, low level signals is complicated by switch control artifacts corrupting the signal channel. FET-based designs suffer large charge injection-based errors, often orders of magnitude larger than the signal of interest. The classic diode bridge switch has much lower error, but requires substantial support circuitry and careful trimming[15]. Figure 39.32's circuit takes a different approach to synthesize a switch with minimal control channel feedthrough. This design switches signals over a ±30mV range with peak control channel feedthrough of millivolts and settling times inside 40ns. This capability, developed for amplifier and data converter settling time measurement, has broad implication in instrumentation and sampling circuitry.

The circuit approximates switch action by varying the transconductance of an amplifier, the maximum gain of which is unity. At low transconductance, amplifier gain is nearly zero, and essentially no signal is passed. At maximum transconductance, signal passes at unity gain. The amplifier and its transconductance control channel are very wideband, permitting them to faithfully track rapid variations in transconductance setting. This characteristic means the amplifier is never out of control, affording clean response and rapid settling to the "switched" input's value.

A1A, one section of an LT®1228, is the wideband transconductance amplifier. Its voltage gain is determined by its output resistor load and the current magnitude into its "I_{SET}" terminal. A1B, the second LT1228 section, unloads A1A's output. As shown it provides a gain of two, but when driving a back-terminated 50Ω cable, its effective gain is unity at the cable's receiving end. Current source Q1, controlled by the "switch control input, " sets A1A's transconductance, and hence gain. With Q1 gated off (control input at zero), the 10MΩ resistor supplies about 1.5μA into A1A's I_{SET} pin, resulting in a voltage gain of nearly zero, blocking the input signal. When the switch control

Note 15: See References 20 and 21 for practical examples of diode bridge switches.

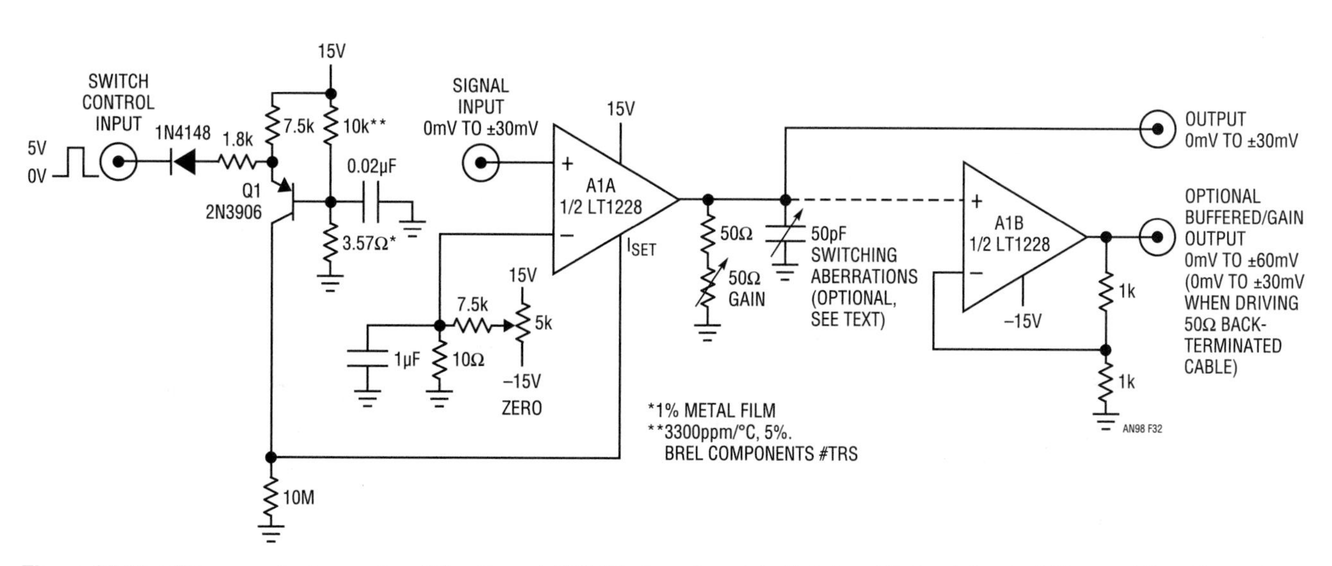

Figure 39.32 • Transconductance Amplifier Based 100MHz Low Level Switch has Minimal Control Channel Feedthrough. A1A's Unity-Gain Output is Cleanly Switched by Logic Controlled Q1's Transconductance Bias. Optional A1B Provides Buffering and Signal Path Gain

input goes high, Q1 turns on, sourcing ≈1.5mA into the I_{SET} pin. This 1000:1 set current change forces maximum transconductance, causing the amplifier to assume unity gain and pass the input signal. Trims for zero and gain ensure accurate input signal replication at the circuit's output. The optional 50pF variable capacitor can be used to damp residual settling transients. The specified 10k resistor at Q1 has a 3300ppm/°C temperature coefficient, compensating A1A's complementary transconductance tempco to minimize gain drift.

Figure 39.33 shows circuit response for a switched 10mV DC input and $C_{ABERRATION}$ = 35pF. When the control input (trace A) is low, no output (trace B) occurs. When the control input goes high, the output reproduces the input with "switch" feedthrough settling in about 20ns. Note that turn-off feedthrough is undetectable, due to the 1000× transconductance reduction and attendant 25× bandwidth drop. Figure 39.34 speeds the sweep up to 10ns/division to examine settling detail. The output (trace B) settles inside 1mV 40ns after the switch control (trace A) goes high. Peak feedthrough excursion, damped by $C_{ABERRATION}$, is only 5mV. Figure 39.35 was taken under identical conditions, except that $C_{ABERRATION}$ = 0pF. Feedthrough increases to ≈20mV, although settling time to 1mV remains at 40ns. Figure 39.36, using double exposure technique, compares signal channel rise times for

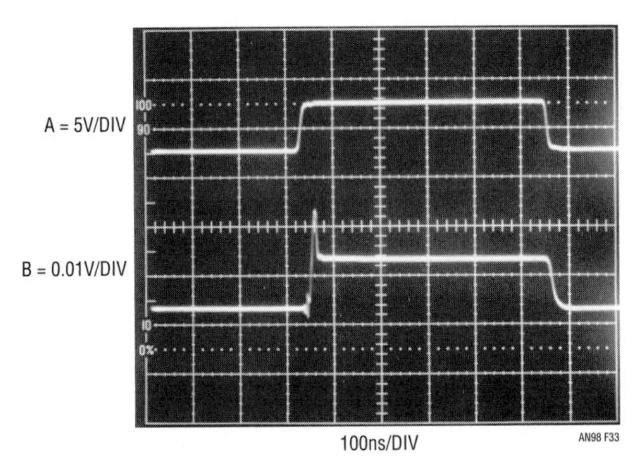

Figure 39.33 • Control Input (Trace A) Dictates Switch Output's (Trace B) Representation of 0.01V DC Input. Control Channel Feedthrough, Evident at Switch Turn-On, Settles in 20ns. Turn-Off Feedthrough is Undetectable Due to Decreased Signal Channel Transconductance and Bandwidth. $C_{ABERRATION}$ ≈ 35pF for This Test

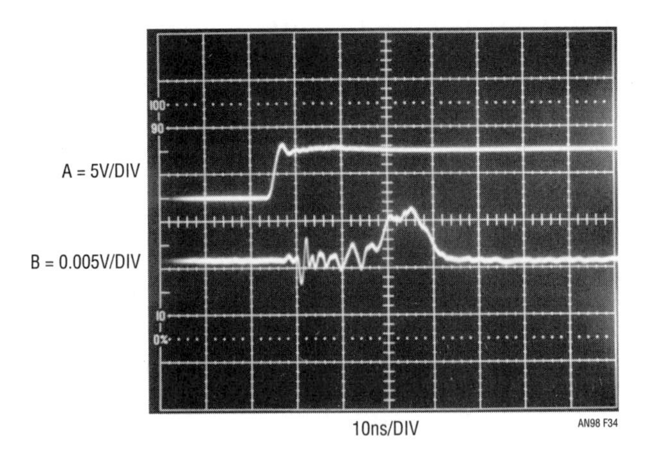

Figure 39.34 • High Speed Delay and Feedthrough for 0V Signal Input. Output (Trace B) Peaks Only 0.005V Before Settling Inside 0.001V 40ns After Switch Control Command (Trace A). $C_{ABERRATION}$ ≈ 35pF for This Test

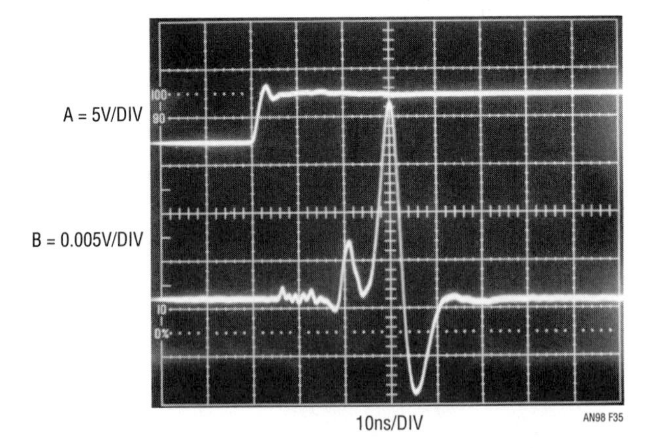

Figure 39.35 • Identical Conditions as Figure 39.34 Except $C_{ABERRATION}$ = 0pF. Feedthrough Related Peaking Increases to ≈0.02V; 0.001V Settling Time Remains at 40ns

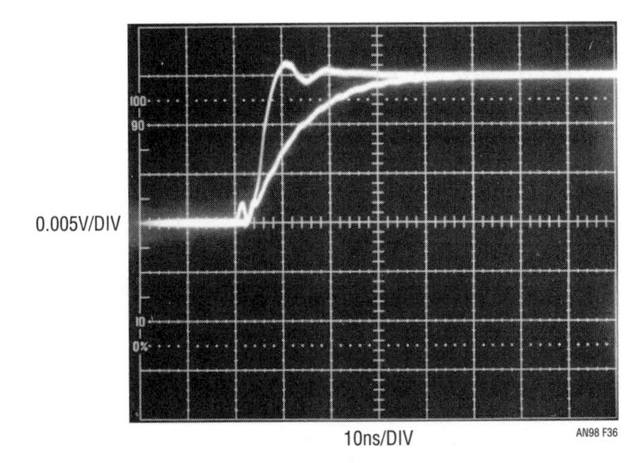

Figure 39.36 • Signal Channel Rise Time for $C_{ABERRATION}$ = 0pF (Leftmost Trace) and ≈35pF (Rightmost Trace) Record 3.5ns and 25ns, Respectively. Switch Control Input High for this Measurement. Photograph Utilizes Double Exposure Technique

$C_{ABERRATION}$ = 0pF (leftmost trace) and ≈35pF (rightmost trace) with the control channel tied high. The larger $C_{ABERRATION}$ value, while minimizing feedthrough amplitude (see Figure 39.34), increases rise time by 7× versus $C_{ABERRATION}$ = 0pF.

To calibrate this circuit, ground the signal input and tie the control input to 5V. Set the "zero" trim for a zero volt output within 500μV. Next, put 30mV into the signal input and adjust the gain trim for exactly 60mV at A1B's unterminated output. Finally, if $C_{ABERRATION}$ is used, adjust it for minimum feedthrough amplitude with the signal input grounded and the control input fed with a 1MHz square wave.

5V powered, 0.0015% linearity, quartz-stabilized V→F converter

Almost all precision voltage-to-frequency converters (V→F) utilize charge pump based feedback for stability. These schemes rely on a capacitor for stability. A great deal of effort towards this approach has resulted in high performance V→F converters (see Reference 31). Obtaining temperature coefficients below 100ppm/°C requires careful attention to compensating the capacitor's drift with temperature. Although this can be done, it complicates the design. Similarly, capacitor dielectric absorption causes errors, limiting linearity to typically 0.01%.

Figure 39.37's 5V powered design, derived from Reference 31's ±15V fed circuit, reduces gain TC to 8ppm/°C and achieves 15ppm linearity by replacing the capacitor with a quartz-stabilized clock.

In charge pump feedback-based circuits, the feedback is based on Q = CV. In a quartz-stabilized circuit, the feedback is based on Q = IT, where I is a stable current source and T is an interval of time derived from a clock. No capacitor is involved.

Figure 39.38 details Figure 39.37's waveforms of operation. A positive input voltage causes A1 to integrate in the negative direction (trace A, Figure 39.38). The flip-flop's Q1 output (trace B) changes state at the first positive-going clock edge (trace C) after A1's output has crossed the D input's switching threshold. C1 provides the quartz-stabilized clock. The flip-flop's Q1 output controls the gating of a precision current

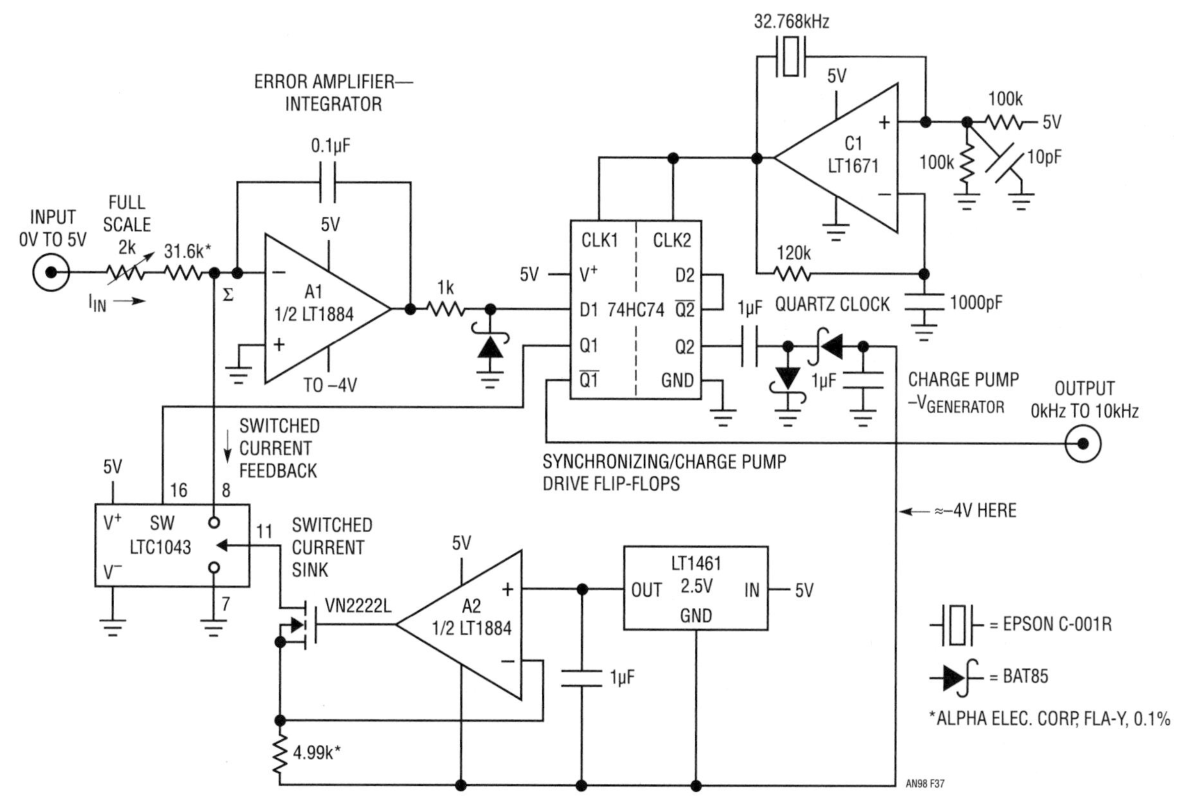

Figure 39.37 • 5V Powered, Quartz-Stabilized 10kHz V→F Converter has 0.0015% Linearity and 8ppm/°C Temperature Coefficient. A1 Servo Controls A2 FET Switched Current Sink Via Clock Synchronized Flip-Flop to Maintain Zero Volt Summing Junction (Σ). Loop Repetition Frequency Directly Conforms to Input Voltage

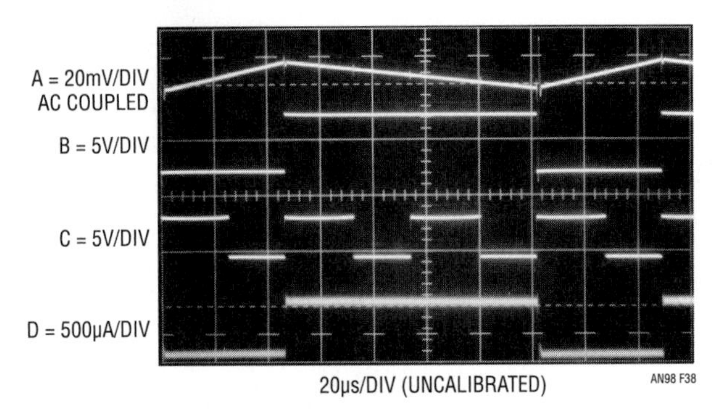

Figure 39.38 • Quartz-Stabilized V→F Converter Waveforms Include A1 Output (Trace A), Flip-Flop Q1 Output (Trace B), Clock (Trace C) and Switched Current Feedback (Trace D). Current Removal (Trace D) from Summing Junction Commences When Clock Goes High with Q1 Low

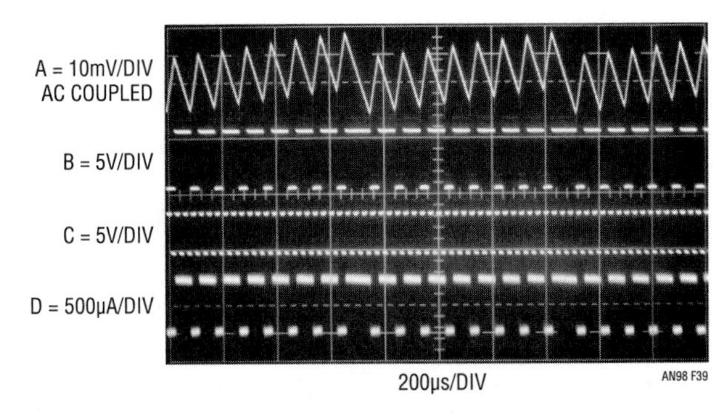

Figure 39.39 • Same Trace Assignments as Figure 39.38. Reduced Oscilloscope Sweep Speed Shows Effect of Timing Uncertainty Between Loop and Clock. Loop Pulse Position is Occasionally Irregular, But Frequency is Constant Over Practical Measurement Intervals

sink composed of A2, the LT1461 voltage reference, a FET and the LTC1043 switch. A negative bias supply, derived from the flip-flop's Q2 output driving a charge pump, furnishes the sink current. When A1 is integrating negatively, Q1's output is high and the LTC1043 directs the current sink's output to ground via Pins 11 and 7. When A1's output crosses the D input's switching threshold, Q1 goes low at the first positive clock edge. LTC1043 Pins 11 and 8 close and a precise, quickly rising current flows out of A1's summing point (trace D).

This current, scaled to be greater than the maximum signal-derived input current, causes A1's output to reverse direction. At the first positive clock pulse after A1's output crosses the D input's trip point, switching again occurs and the entire process repeats. The repetition frequency depends on the input-derived current, hence the frequency of oscillation is directly related to the input voltage. The circuit's output is taken from the flip-flop's $\overline{Q}$ 1 output. Because this circuit replaces a capacitor with a quartz-locked clock, temperature drift is low, typically inside 8ppm/°C. The quartz crystal contributes about 0.5ppm/°C, with most drift contributed by the current source components, the input resistor and switching time variations.

Short term frequency jitter occurs because of the uncertain timing relationship between loop frequency and clock phase. This is normally not a problem because the circuit's output is usually read over many cycles, e.g., 0.1 to 1 second. Figure 39.39 shows the effects of the timing uncertainty. Reduced sweep speed allows viewing of phase uncertainty induced modulation of A1's output ramp (trace A). Note pulse position (traces B and D) irregularity during A1's major excursions. This behavior causes short term pulse displacement, but output frequency is constant over practical measurement intervals.

Circuit linearity is inside 0.0015% (0.15Hz), gain temperature coefficient is 8ppm/°C (0.08Hz/°C) and power supply rejection better than 100ppm (1Hz) over a 4V to 6V range. The LT1884's low input bias and drift reduce zero point originated errors to insignificant levels. To trim this circuit, apply 5.0000V in and adjust the 2kΩ potentiometer for 10.000kHz output.

Basic flashlamp illumination circuit for cellular telephones/cameras

BEFORE PROCEEDING ANY FURTHER, THE READER IS WARNED THAT CAUTION MUST BE USED IN THE CONSTRUCTION, TESTING AND USE OF THIS CIRCUIT. HIGH VOLTAGE, LETHAL POTENTIALS ARE PRESENT IN THIS CIRCUIT. EXTREME CAUTION MUST BE USED IN WORKING WITH, AND MAKING CONNECTIONS TO, THIS CIRCUIT. REPEAT: THIS CIRCUIT CONTAINS DANGEROUS, HIGH VOLTAGE POTENTIALS. USE CAUTION.

Next generation cellular telephones will include high quality photographic capability. Flashlamp-based lighting is crucial for good photographic performance. A previous full-length Linear Technology publication detailed flash illumination issues and presented flash circuitry equipped with "red-eye" reduction capability.[16,17] Some applications do not require this feature; deleting it results in an extremely simple and compact flashlamp solution.

Figure 39.40's circuit consists of a power converter, flash-lamp, storage capacitor and an SCR-based trigger. In operation the LT3468-1 charges C1 to a regulated 300V at about 80% efficiency. A "trigger" input turns the SCR on, depositing C2's charge into T2, producing a high voltage trigger event at the flashlamp. This causes the lamp to conduct high current from C1, resulting in an intense flash of light. LT3468-1 associated waveforms, appearing in Figure 39.41, include trace A, the "charge input," going high. This initiates T1 switching, causing C1 to ramp up (trace B). When C1 arrives at the regulation point, switching ceases and the resistively pulled-up "DONE" line drops low (trace C), indicating C1's charged state. The "TRIGGER" command (trace D), resulting in C1's discharge via the lamp, may occur any time (in this case ≈600ms) after "DONE" goes low. Normally, regulation feedback would be provided by resistively dividing down the output voltage. This approach is not acceptable because it would require excessive switch cycling to offset the feedback resistor's constant power drain. While this action would maintain regulation, it would also drain excessive power from the primary source, presumably a battery. Regulation is instead obtained by monitoring T1's flyback pulse characteristic, which reflects T1's secondary amplitude. The output voltage is set by T1's turns ratio. This feature permits tight capacitor voltage regulation, necessary to ensure consistent flash intensity without exceeding lamp energy or capacitor voltage ratings. Also, flashlamp energy is conveniently determined by the capacitor value without any other circuit dependencies.

Figure 39.42 shows high speed detail of the high voltage trigger pulse (trace A), the flashlamp current (trace B) and the light output (trace C). Some amount of time is required for the lamp to ionize and

Note 16: See References 9 and 10.

Note 17: "Red-eye" in a photograph is caused by the human retina reflecting the light flash with a distinct red color. It is eliminated by causing the eye's iris to constrict in response to a low intensity flash immediately preceding the main flash.

DANGER! Lethal Potentials Present — Use Caution

C1: RUBYCON 330FW13AK6325
D1: TOSHIBA DUAL DIODE 1SS306, CONNECT DIODES IN SERIES
D2: PANASONIC MA2Z720
SCR: TOSHIBA S6A37
T1: TDK LDT565630T-002
T2: TOKYO COIL-BO-02
FLASHLAMP: PERKIN ELMER BGDC0007PKI5700

Figure 39.40 • Complete Flashlamp Circuit Includes Capacitor Charging Components, Flash Capacitor C1, Trigger (R1, C2, T2, SCR) and Flashlamp. TRIGGER Command Biases SCR, Ionizing Lamp via T2. Resultant C1 Discharge Through Lamp Produces Light

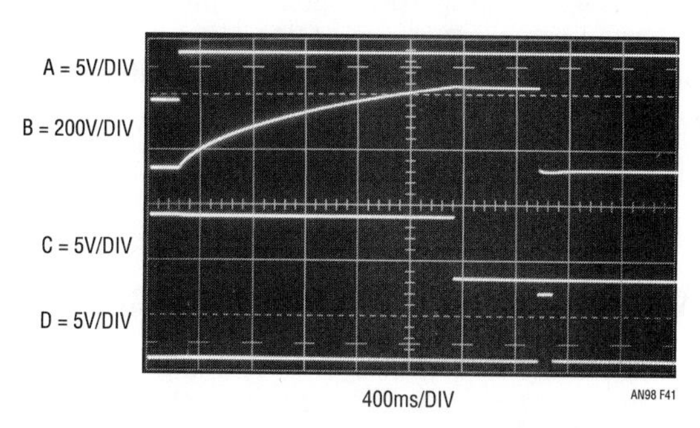

Figure 39.41 • Capacitor Charging Waveforms Include Charge Input (Trace A), C1 (Trace B), $\overline{\text{DONE}}$ Output (Trace C) and TRIGGER Input (Trace D). C1's Charge Time depends Upon Its Value and Charge Circuit Output Impedance. TRIGGER Input, Widened for Figure Clarity, May Occur any Time After $\overline{\text{DONE}}$ Goes Low

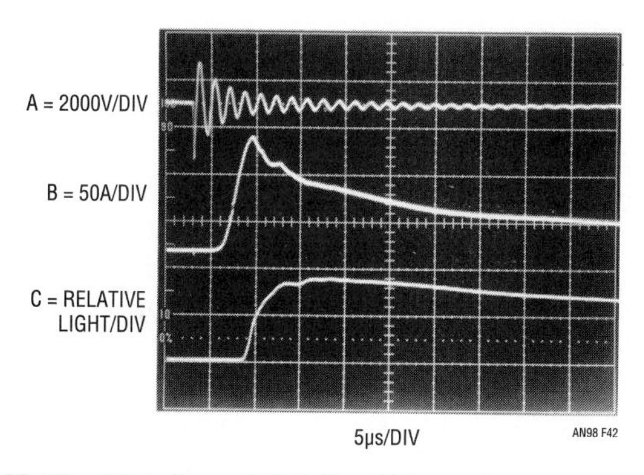

Figure 39.42 • High Speed Detail of Trigger Pulse (Trace A), Resultant Flashlamp Current (Trace B) and Relative Light Output (Trace C). Current Exceeds 100A After Trigger Pulse Ionizes Lamp

begin conduction after triggering. Here, 3μs after the $4kV_{P-P}$ trigger pulse, flashlamp current begins its ascent to over 100A. The current rises smoothly in 3.5μs to a well defined peak before beginning its descent. The resultant light produced rises more slowly, peaking in about 7μs before decaying. Slowing the oscilloscope sweep permits capturing the entire current and light events. Figure 39.43 shows that light output (trace B) follows lamp current (trace A) profile, although current peaking is more abrupt. Total event duration is ≈200μs with most energy expended in the first 100μs.

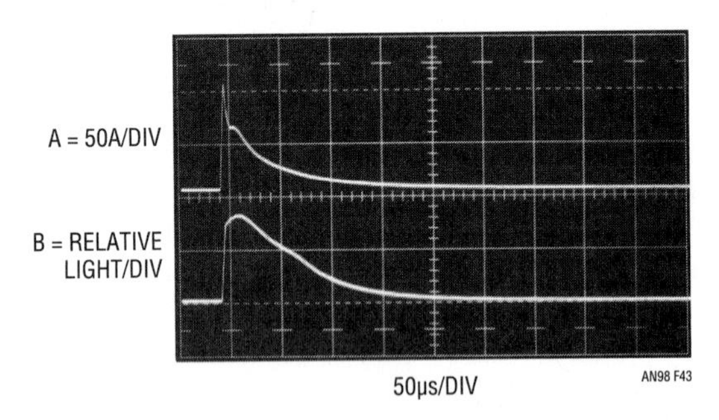

Figure 39.43 • Photograph Captures Entire Current (Trace A) and Light (Trace B) Events. Light Output Follows Current Profile Although Peaking is Less Defined. Waveform Leading Edges Enhanced for Figure Clarity

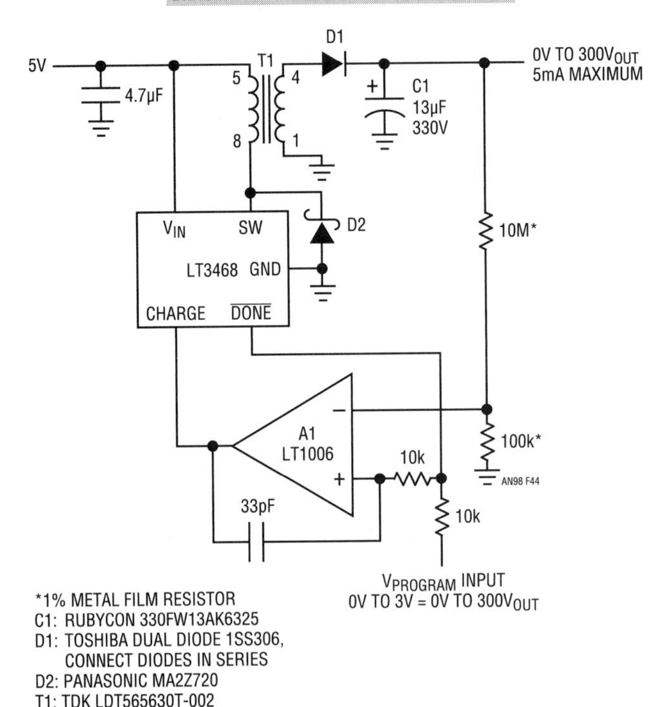

Figure 39.44 • A Voltage Programmable 0V to 300V Output Regulator. A1 Controls Regulator Output by Duty Cycle Modulating LT3468/T1 DC/DC Converter Power Delivery

0V to 300V output DC/DC converter

BEFORE PROCEEDING ANY FURTHER, THE READER IS WARNED THAT CAUTION MUST BE USED IN THE CONSTRUCTION, TESTING AND USE OF THIS CIRCUIT. HIGH VOLTAGE, LETHAL POTENTIALS ARE PRESENT IN THIS CIRCUIT. EXTREME CAUTION MUST BE USED IN WORKING WITH, AND MAKING CONNECTIONS TO, THIS CIRCUIT. REPEAT: THIS CIRCUIT CONTAINS DANGEROUS, HIGH VOLTAGE POTENTIALS. USE CAUTION.

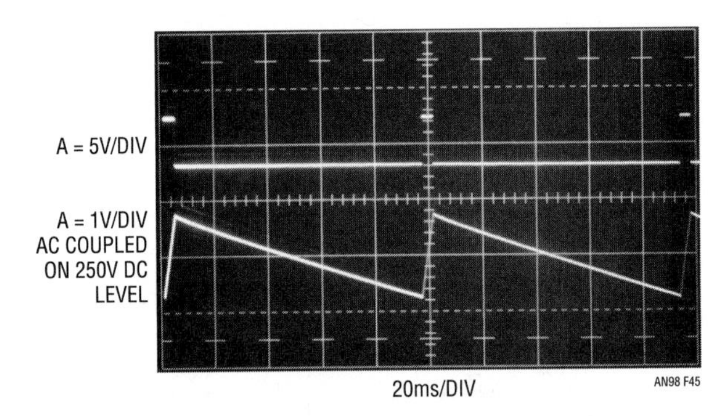

Figure 39.45 • Details of Figure 39.44's Duty Cycle Modulated Operation. High Voltage Output (Trace B) Ramps Down Until A1 (Trace A) Goes High, Enabling LT3468/T1 to Restore Output. Loop Repetition Rate Varies with Input Voltage, Output Set Point and Load

Figure 39.44 shows the LT3468 photoflash capacitor charger, described in the previous application, used as a general purpose, high voltage DC/DC converter. Normally, the LT3468 regulates its output at 300V by sensing T1's flyback pulse characteristic. This circuit forces the LT3468 to regulate at lower voltages by truncating its charge cycle before the output reaches 300V. A1 compares a resistively divided down portion of the output with the program input voltage. When the program input voltage (A1 + input) is exceeded by the output derived potential (A1 – input) A1's output goes low, shutting down the LT3468. The feedback capacitor provides AC hysteresis, sharpening A1's output to prevent chattering at the trip point. The LT3468 remains shut down until the output voltage drops low enough to trip A1's output high, turning it back on. In this way, A1 duty cycle modulates the LT3468, causing the output voltage to stabilize at a point determined by the program input. Figure 39.45 shows a 250V DC output (trace B) decaying down about 2V until A1 (trace A) goes high, enabling the LT3468 and restoring the loop. This simple circuit works well, regulating over a programmable 0V to 300V range, although its inherent hysteretic operation mandates the 2V output ripple noted. Loop repetition rate varies with input voltage, output set point and load but the ripple is always present. The next circuit essentially eliminates the ripple at the cost of increased complexity.

Low ripple and noise 0V to 300V output DC/DC converter

BEFORE PROCEEDING ANY FURTHER, THE READER IS WARNED THAT CAUTION MUST BE USED IN THE CONSTRUCTION, TESTING AND USE OF THIS CIRCUIT. HIGH VOLTAGE, LETHAL POTENTIALS ARE PRESENT IN THIS CIRCUIT. EXTREME CAUTION MUST BE USED IN WORKING WITH, AND MAKING CONNECTIONS TO, THIS CIRCUIT. REPEAT: THIS CIRCUIT CONTAINS DANGEROUS, HIGH VOLTAGE POTENTIALS. USE CAUTION.

Figure 39.46 uses a post-regulator to reduce Figure 39.44's output ripple and noise to only 2mV. A1 and the LT3468 are identical to the previous circuit, except for the 15V zener diode in series with the 10M-100k feedback divider. This component causes C1's voltage, and hence Q1's collector, to regulate 15V above the $V_{PROGRAM}$ inputs dictated point. The $V_{PROGRAM}$ input is also routed to the A2-Q2-Q1 linear post-regulator. A2's 10M-100k feedback divider does not include a zener, so the post-regulator follows the $V_{PROGRAM}$ input with no offset. This arrangement forces 15V across Q1 at all output voltages. This Figure is high enough to eliminate undesirable ripple and noise from the output while keeping Q1 dissipation low.

Q3 and Q4 form a current limit, protecting Q1 from overload. Excessive current through the 50Ω shunt turns Q3 on. Q3 drives Q4, shutting down the LT3468. Simultaneously a portion of Q3's collector current turns Q2 on hard, shutting off Q1. This loop dominates the normal regulation feedback, protecting the circuit until the

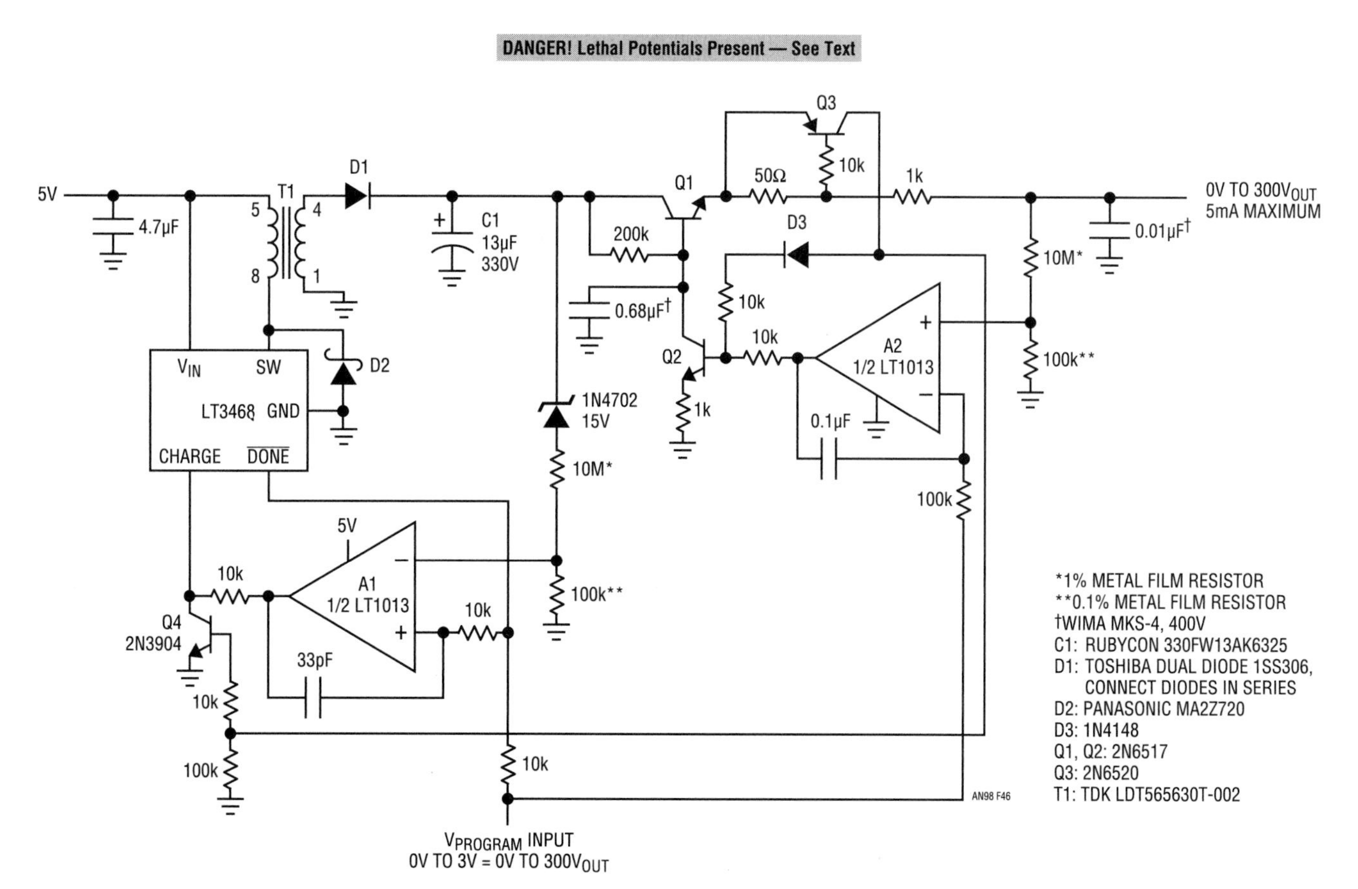

Figure 39.46 • Post-Regulation Reduces Figure 39.44's 2V Output Ripple to 2mV. LT3468-Based DC/DC Converter, Similar to Figure 39.44, Delivers High Voltage to Q1 Collector. A2, Q1, Q2 Form Tracking, High Voltage Linear Regulator. Zener Sets Q1 V_{CE} = 15V, Ensuring Tracking with Minimal Dissipation. Q3-Q4 Limit Short-Circuit Output Current

overload is removed. Figure 39.47 shows just how effective the post regulator is. When A1 (trace A) goes high, Q1's collector (trace B) ramps up in response (note LT3468 switching artifacts on ramps upward slope). When the A1-LT3468 loop is satisfied, A1 goes low and Q1's collector ramps down. The circuits output post-regulator (trace C), however, rejects the ripple, showing only 2mV of noise. Slight trace blurring derives from A1-LT3468 loop jitter.

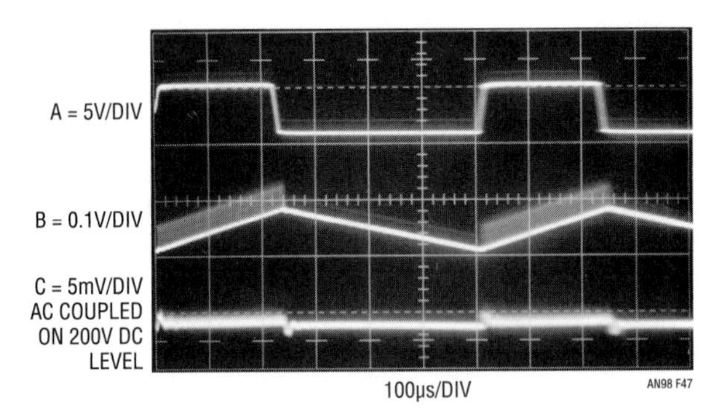

Figure 39.47 • Low Ripple Output (Trace C) is Apparent in Post-Regulator's Operation. Traces A and B are A1 Output and Q1's Collector, Respectively. Trace Blurring, Right of Photo Center, Derives from Loop Jitter

5V to 200V converter for APD bias

BEFORE PROCEEDING ANY FURTHER, THE READER IS WARNED THAT CAUTION MUST BE USED IN THE CONSTRUCTION, TESTING AND USE OF THIS CIRCUIT. HIGH VOLTAGE, LETHAL POTENTIALS ARE PRESENT IN THIS CIRCUIT. EXTREME CAUTION MUST BE USED IN WORKING WITH, AND MAKING CONNECTIONS TO, THIS CIRCUIT. REPEAT: THIS CIRCUIT CONTAINS DANGEROUS, HIGH VOLTAGE POTENTIALS. USE CAUTION.

Avalanche photodiodes (APD) require high voltage bias. Figure 39.48's design provides 200V from a 5V input. The circuit is a basic inductor flyback boost regulator with a major important deviation. Q1, a high voltage device, has been interposed between the LT1172 switching regulator and the inductor. This permits the regulator to control Q1's high voltage switching without undergoing high voltage stress. Q1, operating as a "cascode" with the LT1172's internal switch, withstands L1's high voltage flyback events[18]. Diodes associated with Q1's source

Note 18: See References 8 (page 8), 11 (Appendix D) and 22.

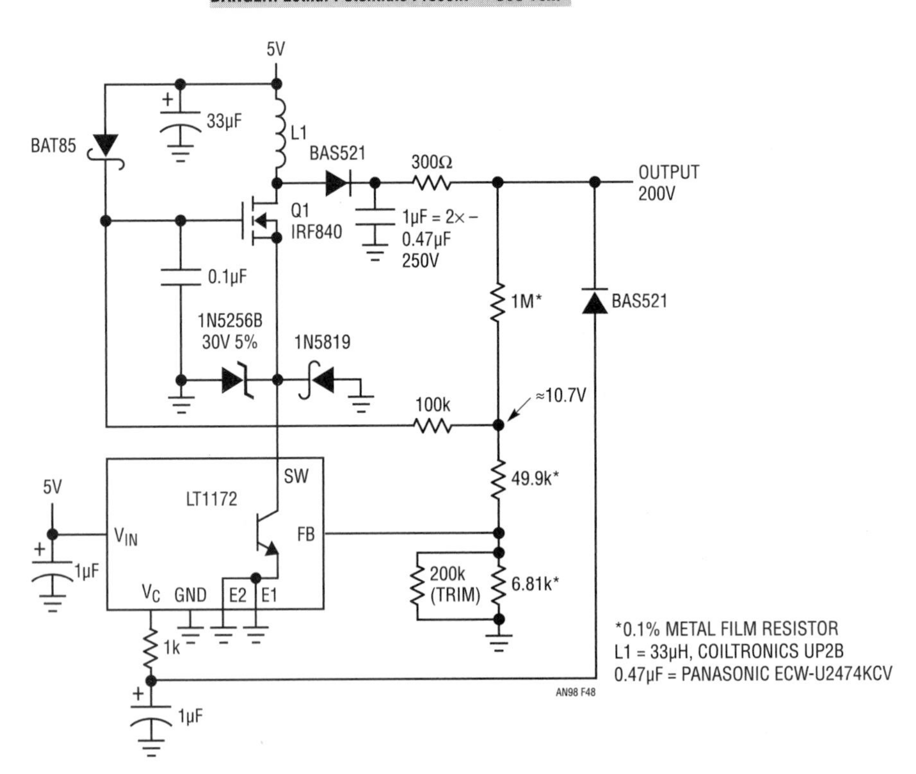

Figure 39.48 • 5V to 200V Output Converter for APD Bias. Cascoded Q1 Switches High Voltage, Allowing Low Voltage Regulator to Control Output. Diode Clamps Protect Regulator from Transient Events; 100k Path Bootstraps Q1's Gate Drive from Output. Output Connected 300Ω-Diode Combination Provides Short-Circuit Protection

terminal clamp L1 orginated spikes arriving via Q1's junction capacitance. The high voltage is rectified and filtered, forming the circuit's output. Feedback to the regulator stabilizes the loop and the RC at the V_C pin provides frequency compensation. The 100k path from the output divider bootstraps Q1's gate drive to about 10V, ensuring saturation. The output connected 300Ω-diode combination provides short-circuit protection by shutting down the LT1172 if the output is accidentally grounded. The 200k trim resistor sets the 200V output ±2% while using standard values in the feedback divider.

Figure 39.49 shows operating waveforms. Traces A and C are LT1172 switch current and voltage, respectively. Q1's drain is trace B. Current ramp termination results in a high voltage flyback event at Q1's drain. A safety attenuated version of the flyback appears at the LT1172 switch. The sinosoidal signature, due to inductor ring-off between conduction cycles, is harmless.

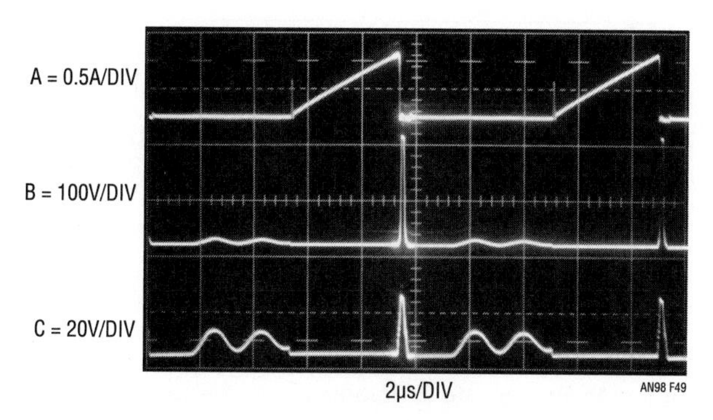

Figure 39.49 • Waveforms for 5V to 200V Converter Include LT1172 Switch Current and Voltage (Traces A and C, Respectively) and Q1's Drain Voltage (Trace B). Current Ramp Termination Results in High Voltage Flyback Event at Q1 Drain. Safely Attenuated Version Appears at LT1172 Switch. Sinosoidal Signature, Due to Inductor Ring-Off Between Current Conduction Cycles, is Harmless. All Traces Intensified Near Center Screen for Photographic Clarity

Wide range, high power, high voltage regulator

BEFORE PROCEEDING ANY FURTHER, THE READER IS WARNED THAT CAUTION MUST BE USED IN THE CONSTRUCTION, TESTING AND USE OF THIS CIRCUIT. HIGH VOLTAGE, LETHAL POTENTIALS ARE PRESENT IN THIS CIRCUIT. EXTREME CAUTION MUST BE USED IN WORKING WITH, AND MAKING CONNECTIONS TO, THIS CIRCUIT. REPEAT: THIS CIRCUIT CONTAINS DANGEROUS, HIGH VOLTAGE POTENTIALS. USE CAUTION.

Figure 39.50 is an example of a monolithic switching regulator making a complex function practical. This regulator provides outputs from millivolts to 500V at 100W with 80% efficiency[19]. A1 compares a variable reference voltage with a resistively scaled version of the circuits output and biases the LT1074 switching regulator configuration. The switcher's DC output drives a DC/DC converter comprised of L1, Q1 and Q2. Q1 and Q2 receive out-of-phase square wave drive from the 74C74 ÷ 4 flip-flop stage and the LTC1693 FET drivers. The flip-flop is clocked from the LT1074 V_{SW} output via the Q3 level shifter. The LT3010 provides 12V power for A1, the 74C74 and the LTC1693. A1 biases the LT1074 regulator to produce the DC input at the DC/DC converter required to balance the loop. The converter has a voltage gain of about 20, resulting in high voltage output. This output is resistively divided down, closing the loop at A1's negative input. Frequency compensation for this loop must accommodate the significant phase errors generated by the LT1074 configuration, the DC/DC converter and the output LC filter. The 0.47μF roll-off term at A1 and the 100Ω-0.15μF RC lead network provide the compensation, which is stable for all loads.

Note 19: This circuit is an updated version of Reference 12.

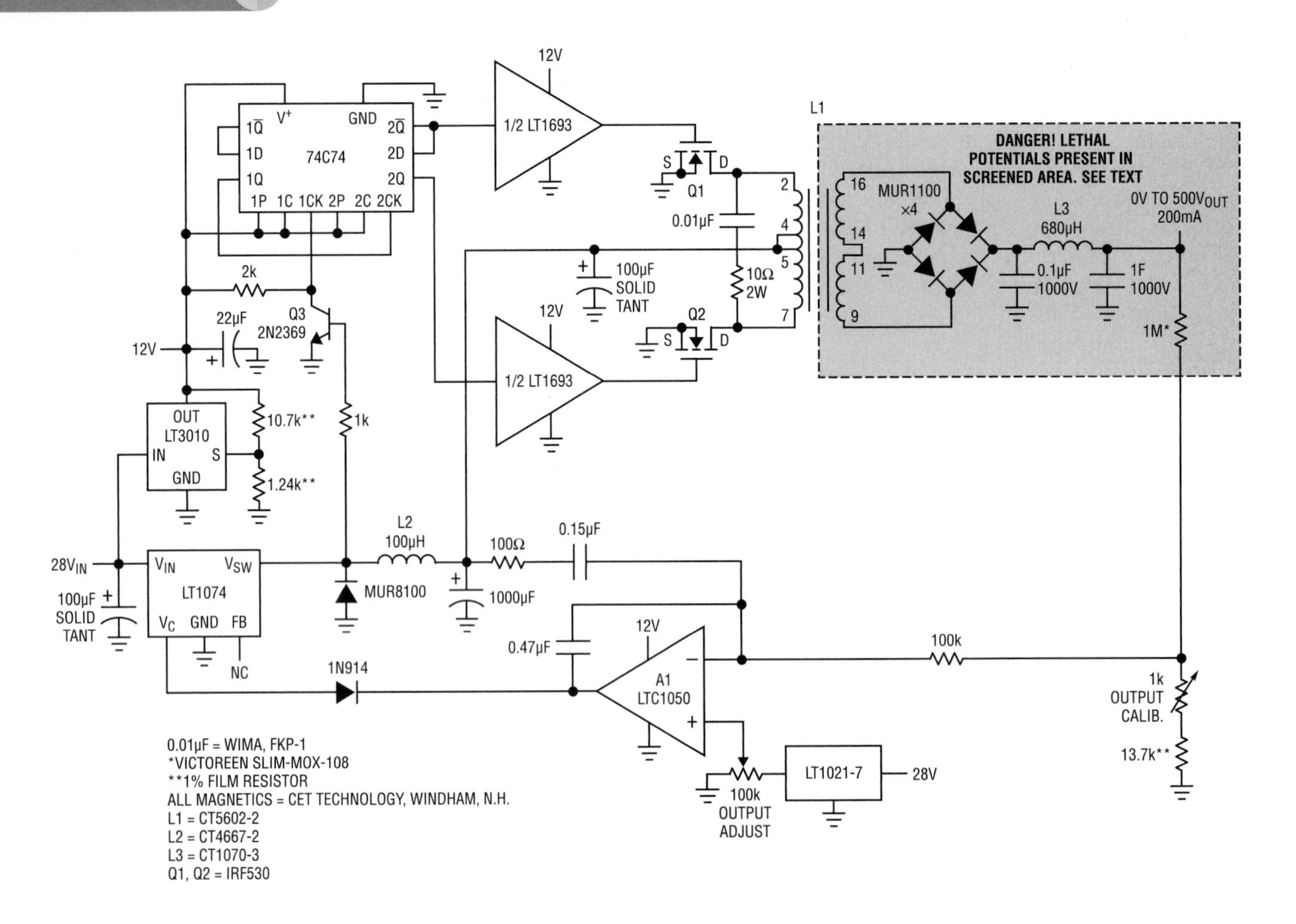

Figure 39.50 • LT1074 Permits High Voltage Output Over 100dB Range with Power and Efficiency. DANGER! Lethal Potentials Present—See Text

Figure 39.51 gives circuit waveforms at 500V output into a 100W load. Trace A is the LT1074 V_{SW} pin while trace B is its current. Traces C and D are Q1 and Q2's drain waveforms. The disturbance at the leading edges is due to cross-current conduction, which lasts about 300ns—a small percentage of the cycle. Transistor currents during this interval remain within reasonable values, and no overstress or dissipation problems occur. This effect could be eliminated with non-overlapping drive to Q1 and Q2[20], although there would be no reliability or significant efficiency gain.

All waveforms are synchronous because the flip-flop drive stage is clocked from the LT1074 V_{SW} output. The LT1074's maximum 95% duty cycle means that the Q1-Q2 switches can never see destructive DC drive. The only condition allowing DC drive occurs when the LT1074 is at zero duty cycle. This case is clearly nondestructive, because L1 receives no power.

Figure 39.52 shows the same circuit points as Figure 39.51 but at only 5mV output. Here, the loop restricts drive to the DC/DC converter to small levels. Q1 and Q2 chop just 60mV into L1. At this level L1's output diode drops look large, but loop action forces the desired 0.005V output.

The LT1074's switched mode drive to L1 maintains high efficiency at high power, despite the circuits wide output range[21].

Note 20: See Reference 24 for an example of this technique.

Note 21: A circuit related to the one presented here appears in Reference 13. Its linear drive to the step-up DC/DC converter forces dissipation, limiting output power to about 10W.

Figure 39.53 shows output noise at 500V into a 100W load. Q1-Q2 chopping artifacts are clearly visible, although limited to about 50mV. The coherent noise characteristic is traceable to the synchronous clocking of Q1 and Q2 by the LT1074.

A 50V to 500V step command into a 100W load produces the response of Figure 39.54. Loop response on both edges is clean, with the falling edge slightly underdamped. This slew asymmetry is typical of switching configurations, because the load and output capacitor determine negative slew rate. The wide range of possible loads mandates a compromise when setting frequency compensation. The falling edge could be made critically or even over damped, but the response time for other conditions would suffer. The compensation used seems a reasonable compromise.

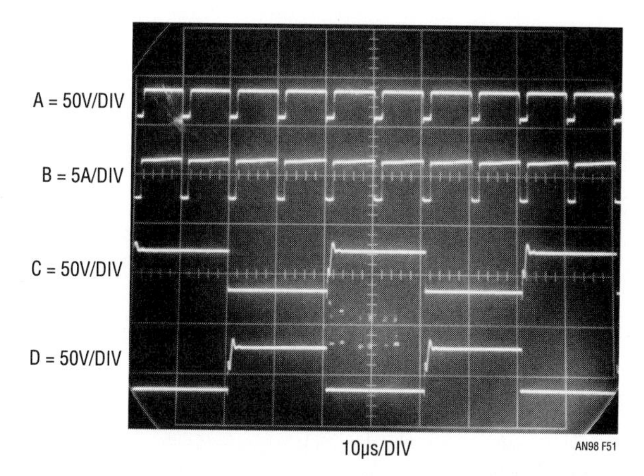

Figure 39.51 • Figure 39.50's Operating Waveforms at 500V Output Into a 100W Load

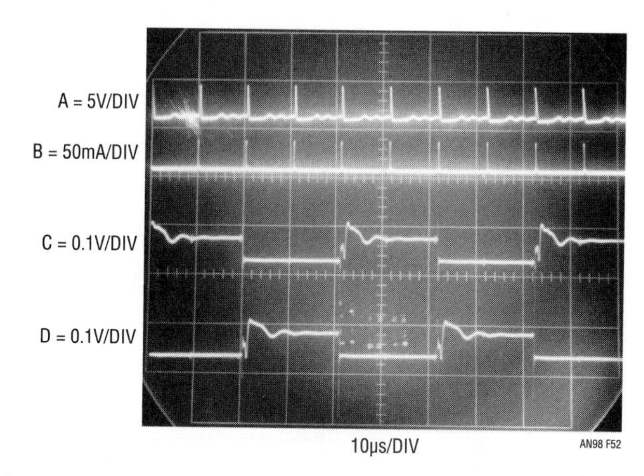

Figure 39.52 • Operating Waveforms at 0.005V Output

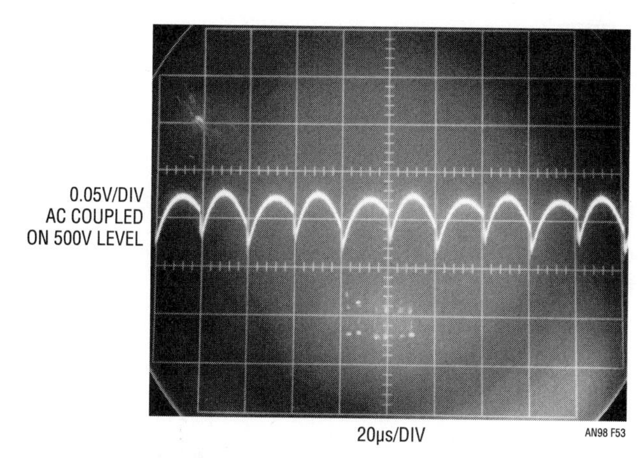

Figure 39.53 • Output Noise at 500V into a 100W Load. Residue is Composed of Q1-Q2 Chopping Artifacts. DANGER! Lethal Potentials Present—See Text

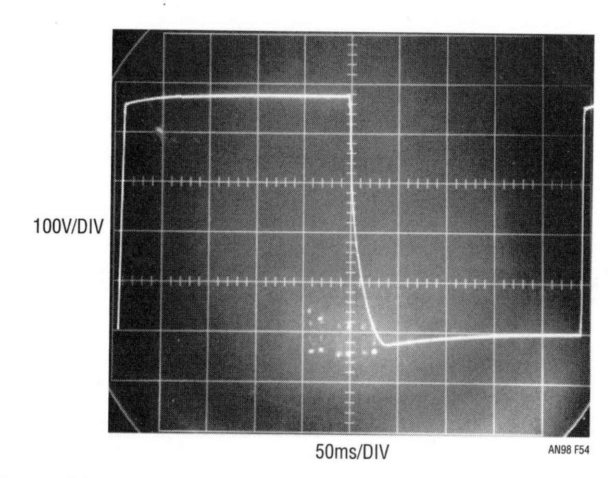

Figure 39.54 • 500V Step Response with 100W Load (Photo Retouched for Clarity). DANGER! Lethal Potentials Present—See Text

5V to 3.3V, 15A paralleled linear regulator

Figure 39.55 is another high power supply; unlike the previous example, it is a linear regulator. Two 7.5A regulators are paralleled in a "master-slave" arrangement. The "master" regulator is wired to produce a 3.3V output in the conventional manner. The 124Ω feedback resistor senses at the 0.001Ω shunt located directly before the circuit output. The "slave" regulator, also wired for a nominal 3.3V output, sources the circuit output in identical fashion. A1, sensing the regulators difference voltage, adjusts the "slave" regulator to equal the master's output voltage. This allows the regulators to equally share the load current. The 0.001Ω shunts cause negligible regulation loss, but provide adequate signal for A1.

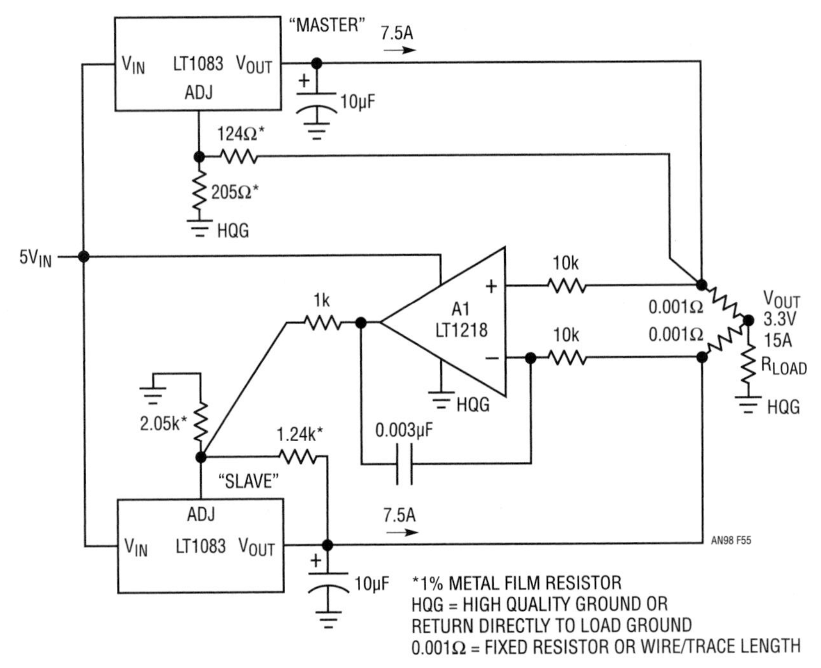

Figure 39.55 • Paralleled Regulators Share Load Current. Amplifier Senses Differential Regulator Voltage; Biases "Slave" to Equalize Output Currents. Remote Sensing Negates Lead Wire Voltage Drops

References

1. LTC6905 Data Sheet, Linear Technology Corporation
2. Tektronix, Inc., "Calibrator," Type 485 Oscilloscope Service and Instruction Manual, 1973, p. 3-15
3. Hewlett-Packard Company, HP1106A/1108A Tunnel Diode Mount, Hewlett-Packard Test and Measurement Catalog, 1970, p. 513
4. Williams, Jim, "A Seven-Nanosecond Comparator for Single Supply Operation," Linear Technology Corporation, Application Note 72, May 1998, p. 32
5. Williams, Jim, "High Speed Comparator Techniques," Linear Technology Corporation, Application Note 13, April 1985, p. 17-18
6. Balasubramaniam, S., "Advanced High Speed CMOS (AHC) Logic Family, " "Ground Bounce Measurement," Texas Instruments, Inc., Publication SCAA034A, 1997
7. Hewlett-Packard Company, HP4328A Milliohmmeter Operating and Service Manual, 1967
8. Williams, Jim, "Bias Voltage and Current Sense Circuits for Avalanche Photodiodes," Linear Technology Corporation, Application Note 92, November 2002, p 8, 11, 30
9. Williams, Jim and Wu, Albert, "Simple Circuitry for Cellular Telephone/Camera Flash Illumination," Linear Technology Corporation, Application Note 95, March 2004
10. Williams, Jim, "Basic Flashlamp Illumination Circuitry for Cellular Telephones/Cameras," Linear Technology Corporation, Design Note 345, September 2004
11. Williams, Jim, "Switching Regulators for Poets," Appendix D, Linear Technology Corporation, Application Note 25, September 1987
12. Williams, Jim, "Step Down Switching Regulators," Linear Technology Corporation, Application Note 35, August 1989, p. 11-13
13. Williams, Jim, "Applications of New Precision Op Amps," Linear Technology Corporation, Application Note 6, January 1985, p. 1–2, 6-7
14. Williams, Jim, "Measurement and Control Circuit Collection," Linear Technology Corporation, Application Note 45, June 1991
15. Markell, R. Editor, "Linear Technology Magazine Circuit Collection, Volume 1," Linear Technology Corporation, Application Note 52, January 1993
16. Williams, Jim, "Practical Circuitry for Measurement and Control Problems," Linear Technology Corporation, Application Note 61, August 1994
17. Markell, R. Editor, " Linear Technology Magazine Circuit Collection, Volume II," Linear Technology Corporation, Application Note 66, August 1996
18. Markell, R. Editor, " Linear Technology Magazine Circuit Collection, Volume III," Linear Technology Corporation, Application Note 67, September 1996
19. Williams, Jim, "Circuitry for Signal Conditioning and Power Conversion," Linear Technology Corporation, Application Note 75, March 1999
20. Williams, Jim, "Component and Measurement Advances Ensure 16-Bit DAC Settling Time," Linear Technology Corporation, Application Note 74, July 1998
21. Williams, Jim, "30 Nanosecond Settling Time Measurement for a Precision Wideband Amplifier," Linear Technology Corporation, Application Note 79, September 1999
22. Hickman, R. W. and Hunt, F. V., "On Electronic Voltage Stabilizers," "Cascode," Review of Scientific Instruments, January 1939, p. 6-21, 16
23. Seebeck, Thomas Dr., "Magnetische Polarisation der Metalle und Erze durch Temperatur-Differenz," Abhaandlungen der Preussischen Akademic der Wissenschaften, 1822-1823, p. 265-373
24. Williams, J. and Huffman, B., "Some Thoughts on DC-DC Converters," Linear Technology Corporation, Application Note 29, October 1988
25. Meade, M. L., "Lock-In Amplifiers and Applications, " London, P. Peregrinus, Ltd
26. Williams, J., "Designs for High Performance Voltage-to-Frequency Converters," Linear Technology Corporation, Application Note 14, March 1986

Appendix A
How much bandwidth is enough?

Accurate wideband oscilloscope measurements require bandwidth. A good question is just how much is needed. A classic guideline is that "end-to-end" measurement system rise time is equal to the root-sum-square of the system's individual component's rise times. The simplest case is two components; a signal source and an oscilloscope.

Figure A1's plot of $\sqrt{\text{signal}^2 + \text{oscilloscope}^2}$ rise time versus error is illuminating. The figure plots signal-to-oscilloscope rise time ratio versus observed rise time (rise time is bandwidth restated in the time domain, where:

$$\text{Rise Time (nanoseconds)} = \frac{350}{\text{Bandwidth(MHz)}}).$$

The curve shows that an oscilloscope 3 to 4 times faster than the input signal rise time is required for measurement accuracy inside about 5%. This is why trying to measure a 1ns rise time pulse with a 350MHz oscilloscope (t_{RISE} = 1ns) leads to erroneous conclusions. The curve indicates a monstrous 41% error. Note that this curve does not include the effects of passive probes or cables connecting the signal to the oscilloscope. Probes do not necessarily follow root-sum-square law and must be carefully chosen and applied for a given measurement. For details, See Appendix B. Figure A2, included for reference, gives 10 cardinal points of rise time/bandwidth equivalency between 1MHz and 5GHz.

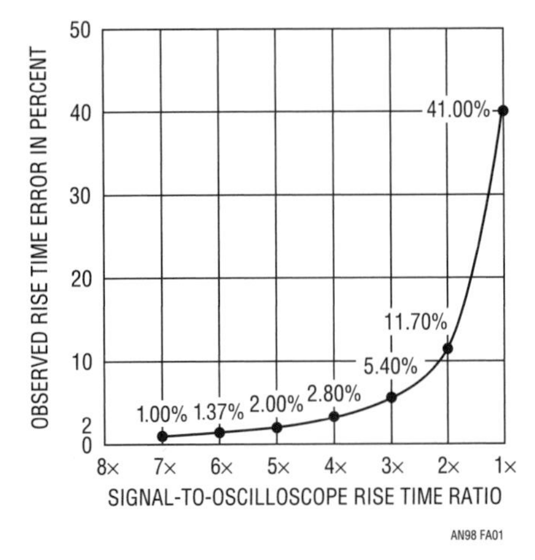

Figure A1 • Oscilloscope Rise Time Effect on Rise Time Measurement Accuracy. Measurement Error Rises Rapidly as Signal-to-Oscilloscope Rise Time Ratio Approaches Unity. Data, Based on Root-Sum-Square Relationship, Does Not Include Probe, Which Does Not Follow Root-Sum-Square Law

RISE TIME	BANDWIDTH
70ps	5GHz
350ps	1GHz
700ps	500MHz
1ns	350MHz
2.33ns	150MHz
3.5ns	100MHz
7ns	50MHz
35ns	10MHz
70ns	5MHz
350ns	1MHz

Figure A2 • Some Cardinal Points of Rise Time/Bandwidth Equivalency. Data is Based on Rise Time/Bandwidth Formula in Text

Appendix B
Connections, cables, adapters, attenuators, probes and picoseconds

Subnanosecond rise time signal paths must be considered as transmission lines. Connections, cables, adapters, attenuators and probes represent discontinuities in this transmission line, deleteriously effecting its ability to faithfully transmit desired signal. The degree of signal corruption contributed by a given element varies with its deviation from the transmission lines nominal impedance. The practical result of such introduced aberrations is degradation of pulse rise time, fidelity, or both. Accordingly, introduction of elements or connections to the signal path should be minimized and necessary connections and elements must be high grade components. Any form of connector, cable, attenuator or probe must be fully specified for high frequency use. Familiar BNC hardware becomes lossy at rise times much faster than 350ps. SMA components are preferred for the rise times described in the text. Additionally, cable should be 50Ω "hard line" o r, at least, teflon-based coaxial cable fully specified for high frequency operation. Optimal connection practice eliminates any cable by coupling the signal output *directly* to the measurement input.

Mixing signal path hardware types via adapters (e.g. BNC/ SMA) should be avoided. Adapters introduce significant parasitics, resulting in reflections, rise time degradation, resonances and other degrading behavior. Similarly, oscilloscope connections should be made directly to the instrument's 50Ω inputs, avoiding probes. If probes must be used, their introduction to the signal path mandates attention to their connection mechanism and high frequency compensation. Passive "Z_0" types, commercially available in 500Ω (10×) and 5kΩ(100×) impedances, have input capacitance below 1pf. Any such probe must be carefully frequency compensated before use or misrepresented measurement will result. Inserting the probe into the signal path necessitates some form of signal pick-off which nominally does not influence signal transmission. In practice, some amount of disturbance must be tolerated and its effect on measurement results evaluated. High quality signal pick-offs always specify insertion loss, corruption factors and probe output scale factor.

The preceding emphasizes vigilance in designing and maintaining a signal path. Skepticism, tempered by enlightenment, is a useful tool when constructing a signal path and no amount of hope is as effective as preparation and directed experimentation.

Current sense circuit collection

Making sense of current

40

Tim Regan Jon Munson Greg Zimmer Michael Stokowski

Introduction

Sensing and/or controlling current flow is a fundamental requirement in many electronics systems, and the techniques to do so are as diverse as the applications themselves. This Application Note compiles solutions to current sensing problems and organizes the solutions by general application type. These circuits have been culled from a variety of Linear Technology documents.

Circuits organized by general application

Each chapter collects together applications that tend to solve a similar general problem, such as high side current sensing, or negative supply sensing. The chapters are titled accordingly. In this way, the reader has access to many possible solutions to a particular problem in one place.

It is unlikely that any particular circuit shown will exactly meet the requirements for a specific design, but the suggestion of many circuit techniques and devices should prove useful. Specific circuits may appear in several chapters if they have broad application.

Analog Circuit and System Design: Immersion in the Black Art of Analog Design. http://dx.doi.org/10.1016/B978-0-12-397888-2.00040-7

Current sense basics

This chapter introduces the basic techniques used for sensing current. It serves also as a definition of common terms. Each technique has advantages and disadvantages and these are described. The types of amplifiers used to implement the circuits are provided.

Low side current sensing (Figure 40.1)

Current sensed in the ground return path of the power connection to the monitored load. Current generally flows in just one direction (unidirectional). Any switching is performed on the load-side of monitor.

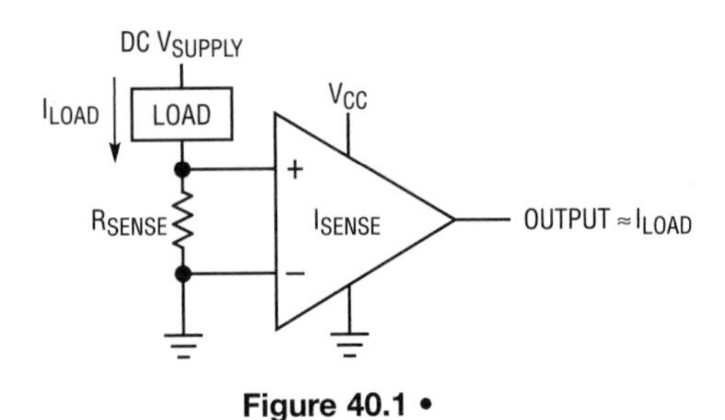

Figure 40.1 •

Low side advantages

- Low input common mode voltage
- Ground referenced output voltage
- Easy single-supply design

Low side disadvantages

- Load lifted from direct ground connection
- Load activated by accidental short at ground end load switch
- High load current caused by short is not detected

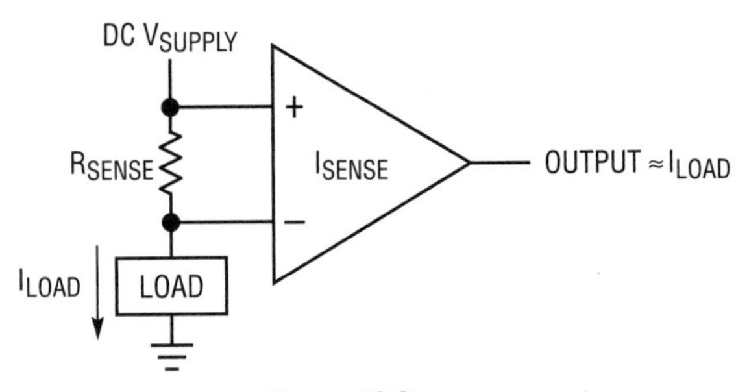

Figure 40.2 •

High side current sensing (Figure 40.2)

Current sensed in the supply path of the power connection to the monitored load. Current generally flows in just one direction (unidirectional). Any switching is performed on the load-side of monitor.

High side advantages

- Load is grounded
- Load not activated by accidental short at power connection
- High load current caused by short is detected

High side disadvantages

- High input common mode voltages (often very high)
- Output needs to be level shifted down to system operating voltage levels

Full-range (high and low side) current sensing (Figure 40.3)

Bidirectional current sensed in a bridge driven load, or unidirectional high side connection with a supply side switch.

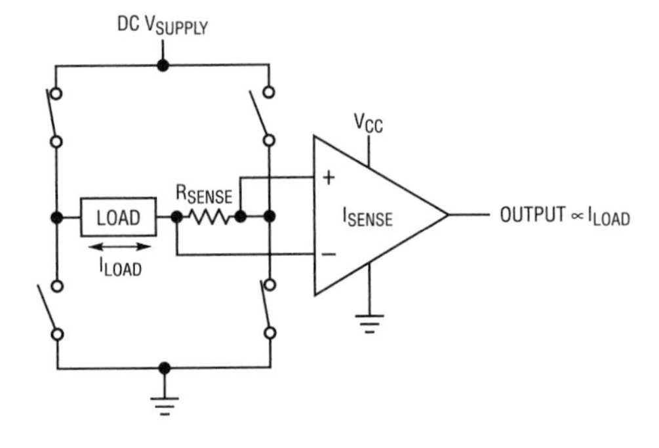

Figure 40.3 •

Full-range advantages

- Only one current sense resistor needed for bidirectional sensing
- Convenient sensing of load current on/off profiles for inductive loads

Full-range disadvantages

- Wide input common mode voltage swings
- Common mode rejection may limit high frequency accuracy in PWM applications

High side

This chapter discusses solutions for high side current sensing. With these circuits the total current supplied to a load is monitored in the positive power supply line.

LT6100 load current monitor (Figure 40.4)

This is the basic LT6100 circuit configuration. The internal circuitry, including an output buffer, typically operates from a low voltage supply, such as the 3V shown. The monitored supply can range anywhere from V_{CC} + 1.4V up to 48V. The A2 and A4 pins can be strapped various ways to provide a wide range of internally fixed gains. The input leads become very Hi-Z when V_{cc} is powered down, so as not to drain batteries for example. Access to an internal signal node (Pin 3) provides an option to include a filtering function with one added capacitor. Small-signal range is limited by V_{OL} in single-supply operation.

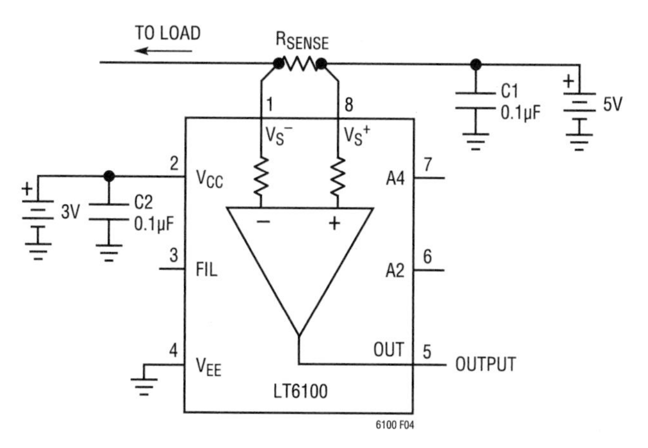

Figure 40.4 •

"Classic" positive supply rail current sense (Figure 40.5)

This circuit uses generic devices to assemble a function similar to an LTC6101. A rail-to-rail input type op amp is required since input voltages are right at the upper rail. The circuit shown here is capable of monitoring up to 44V applications. Besides the complication of extra parts, the V_{OS} performance of op amps at the supply is generally not factory trimmed, thus less accurate than other solutions. The finite current gain of the bipolar transistor is a small source of gain error.

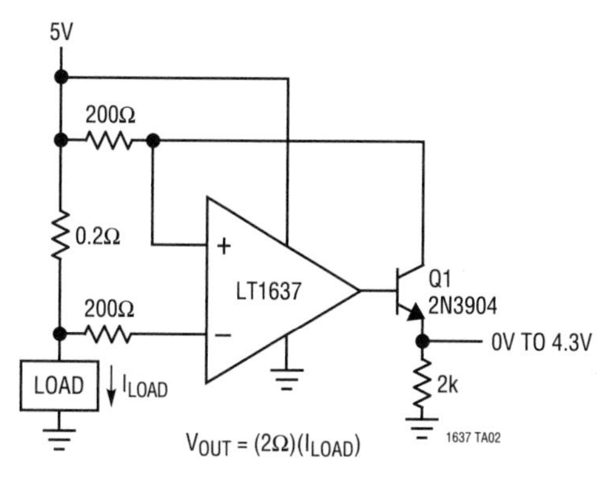

Figure 40.5 •

Over-the-Top current sense (Figure 40.6)

This circuit is a variation on the "classic" high side circuit, but takes advantage of Over-the-Top input capability to separately supply the IC from a low voltage rail. This provides a measure of fault protection to downstream circuitry by virtue of the limited output swing set by the low voltage supply. The disadvantage is V_{OS} in the Over-the-Top mode is generally inferior to other modes, thus less accurate. The finite current gain of the bipolar transistor is a source of small gain error.

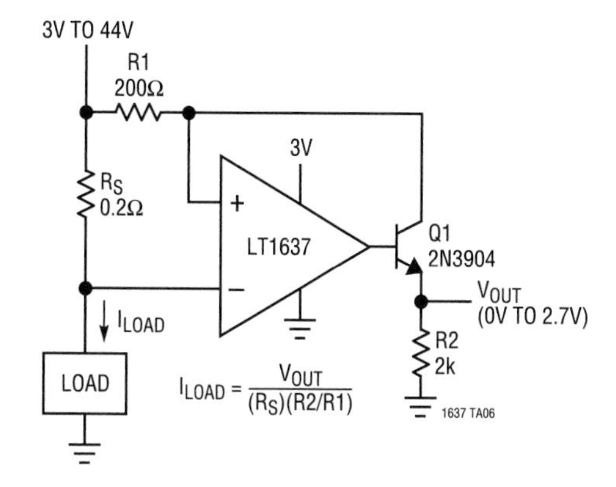

Figure 40.6 •

Self-powered high side current sense (Figure 40.7)

This circuit takes advantage of the microampere supply current and rail-to-rail input of the LT1494. The circuit is simple because the supply draw is essentially equal to the load current developed through R_A. This supply current is simply passed through R_B to form an output voltage that is appropriately amplified.

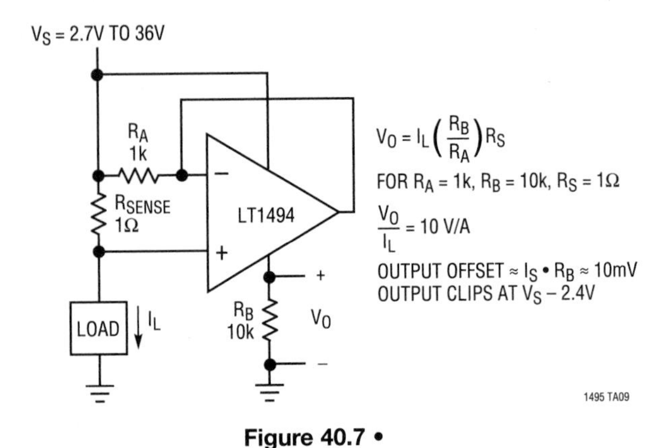

Figure 40.7 •

High side current sense and fuse monitor (Figure 40.8)

The LT6100 can be used as a combination current sensor and fuse monitor. This part includes on-chip output buffering and was designed to operate with the low supply voltage (≥2.7V), typical of vehicle data acquisition systems, while the sense inputs monitor signals at the higher battery bus potential. The LT6100 inputs are tolerant of large input differentials, thus allowing the blown-fuse operating condition (this would be detected by an output full-scale indication). The LT6100 can also be powered down while maintaining high impedance sense inputs, drawing less than 1μA max from the battery bus.

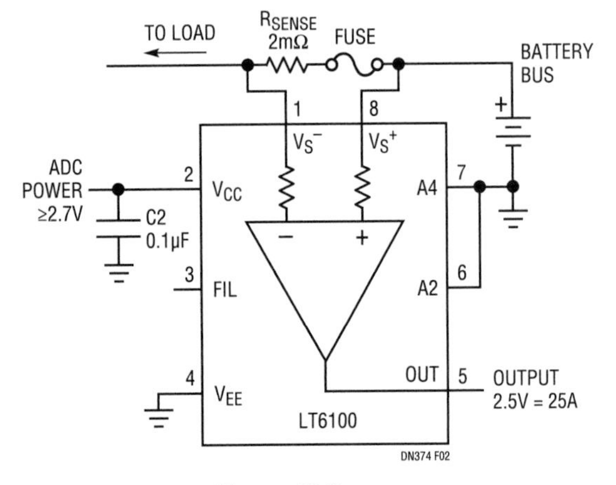

Figure 40.8 •

Precision high side power supply current sense (Figure 40.9)

This is a low voltage, ultrahigh precision monitor featuring a zero-drift instrumentation amplifier (IA) that provides rail-to-rail inputs and outputs. Voltage gain is set by the feedback resistors. Accuracy of this circuit is set by the quality of resistors selected by the user, small-signal range is limited by V_{OL} in single-supply operation. The voltage rating of this part restricts this solution to applications of <5.5V. This IA is sampled, so the output is discontinuous with input changes, thus only suited to very low frequency measurements.

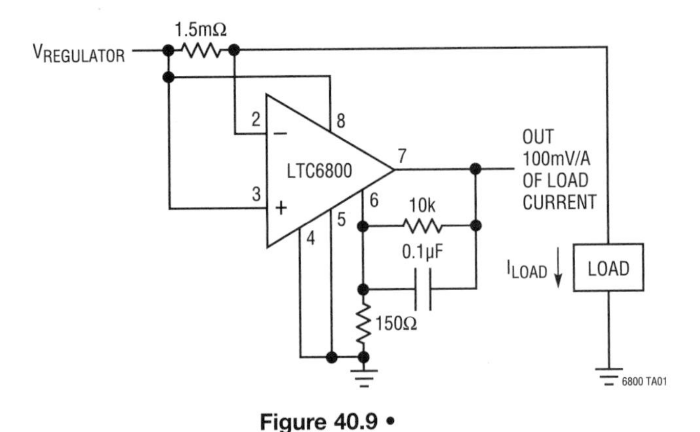

Figure 40.9 •

Positive supply rail current sense (Figure 40.10)

This is a configuration similar to an LT6100 implemented with generic components. A rail-to-rail or Over-the-Top input op amp type is required (for the first section). The first section is a variation on the classic high side where the P-MOSFET provides an accurate output current into R2 (compared to a BJT). The second section is a buffer to allow driving ADC ports, etc., and could be configured with gain if needed. As shown, this circuit can handle up to 36V operation. Small-signal range is limited by V_{OL} in single-supply operation.

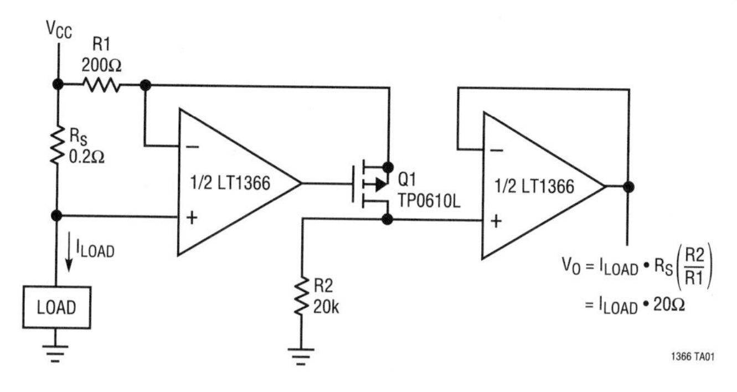

Figure 40.10 •

Precision current sensing in supply rails (Figure 40.11)

This is the same sampling architecture as used in the front end of the LTC2053 and LTC6800, but sans op amp gain stage. This particular switch can handle up to 18V so the ultrahigh precision concept can be utilized at higher voltages than the fully integrated ICs mentioned. This circuit simply commutates charge from the flying sense capacitor to the ground-referenced output capacitor so that under DC input conditions the single-ended output voltage is exactly the same as the differential across the sense resistor. A high precision buffer amplifier would typically follow this circuit (such as an LTC2054). The commutation rate is user set by the capacitor connected to Pin 14. For negative supply monitoring, Pin 15 would be tied to the negative rail rather than ground.

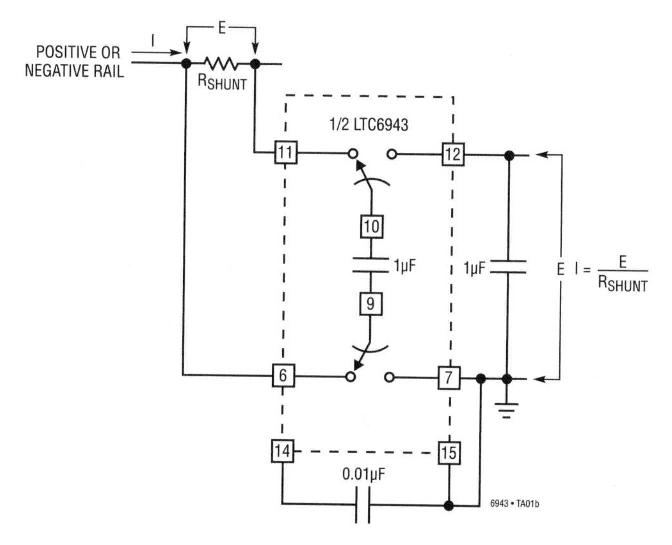

Figure 40.11 •

Measuring bias current into an avalanche photo diode (APD) using an instrumentation amplifier (Figures 40.12a and 40.12b)

The upper circuit (a) uses an instrumentation amplifier (IA) powered by a separate rail (>1V above V_{IN}) to measure across the 1kΩ current shunt. The lower figure (b) is similar but derives its power supply from the APD bias line. The limitation of these circuits is the 35V maximum APD voltage, whereas some APDs may require 90V or more. In the single-supply configuration shown, there is also a dynamic range limitation due to V_{OL} to consider. The advantage of this approach is the high accuracy that is available in an IA.

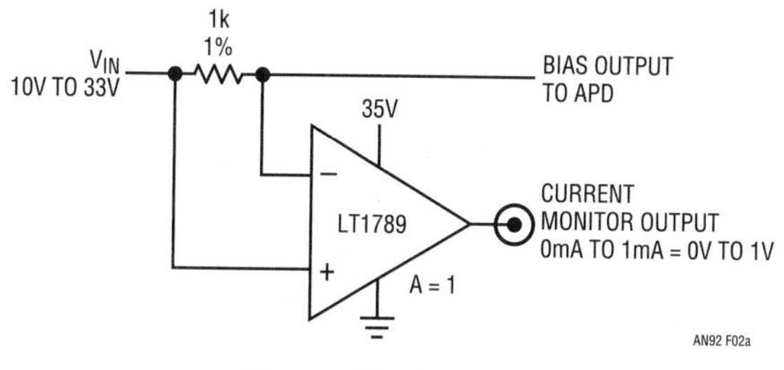

Figure 40.12a •

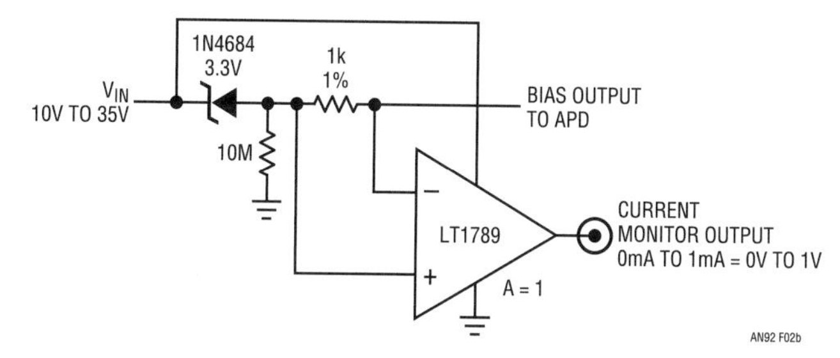

Figure 40.12b •

Simple 500V current monitor (Figure 40.13)

Adding two external MOSFETs to hold off the voltage allows the LTC6101 to connect to very high potentials and monitor the current flow. The output current from the LTC6101, which is proportional to the sensed input voltage, flows through M1 to create a ground referenced output voltage.

Bidirectional battery-current monitor (Figure 40.14)

This circuit provides the capability of monitoring current in either direction through the sense resistor. To allow negative outputs to represent charging current, V_{EE} is connected to a small negative supply. In single-supply operation (V_{EE} at ground), the output range may be offset upwards by applying a positive reference level to V_{BIAS} (1.25V for example). C3 may be used to form a filter in conjunction with the output resistance (R_{OUT}) of the part. This solution offers excellent precision (very low V_{OS}) and a fixed nominal gain of 8.

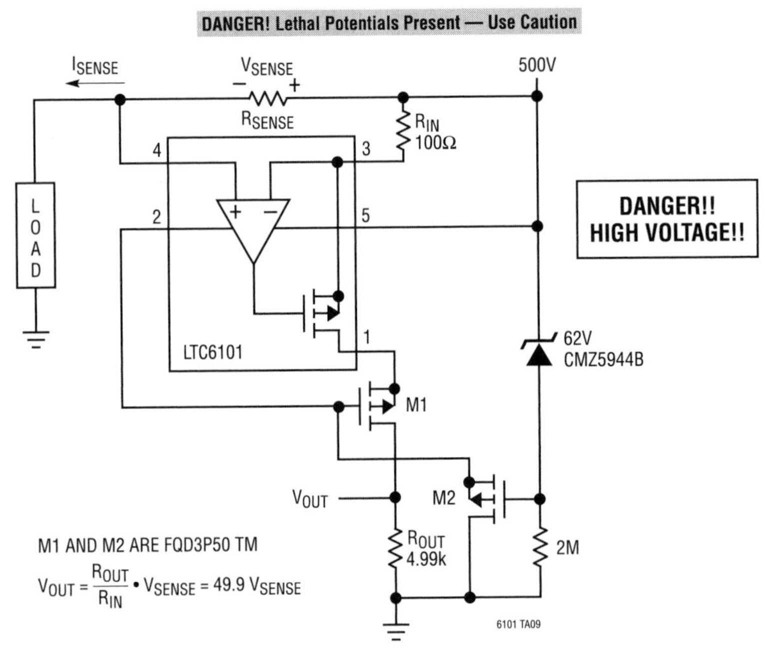

Figure 40.13 •

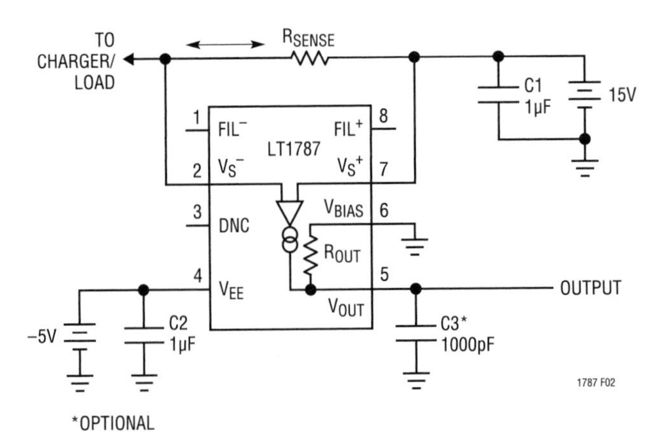

Figure 40.14 •

LTC6101 supply current included as load in measurement (Figure 40.15)

This is the basic LTC6101 high side sensing supply-monitor configuration, where the supply current drawn by the IC is included in the readout signal. This configuration is useful when the IC current may not be negligible in terms of overall current draw, such as in low power battery-powered applications. R_{SENSE} should be selected to limit voltage drop to <500mV for best linearity. If it is desirable not to include the IC current in the readout, as in load monitoring, Pin 5 may be connected directly to V^+ instead of the load. Gain accuracy of this circuit is limited only by the precision of the resistors selected by the user.

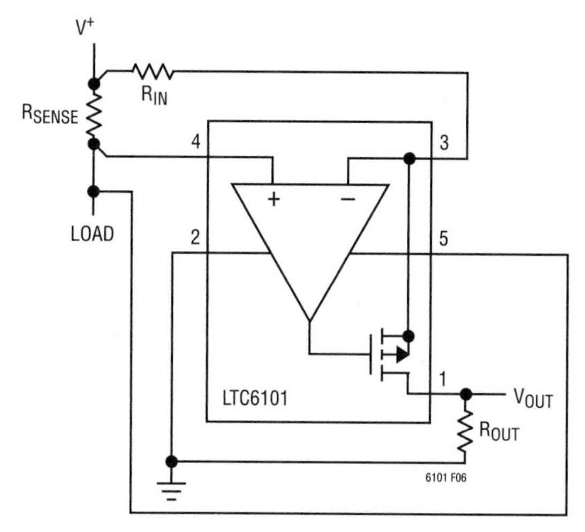

Figure 40.15 •

Simple high side current sense using the LTC6101 (Figure 40.16)

This is a basic high side current monitor using the LTC6101. The selection of R_{IN} and R_{OUT} establishes the desired gain of this circuit, powered directly from the battery bus. The current output of the LTC6101 allows it to be located remotely to R_{OUT}. Thus, the amplifier can be placed directly at the shunt, while R_{OUT} is placed near the monitoring electronics without ground drop errors. This circuit has a fast 1μs response time that makes it ideal for providing MOSFET load switch protection. The switch element may be the high side type connected between the sense resistor and the load, a low side type between the load and ground or an H-bridge. The circuit is programmable to produce up to 1mA of full-scale output current into R_{OUT}, yet draws a mere 250μA supply current when the load is off.

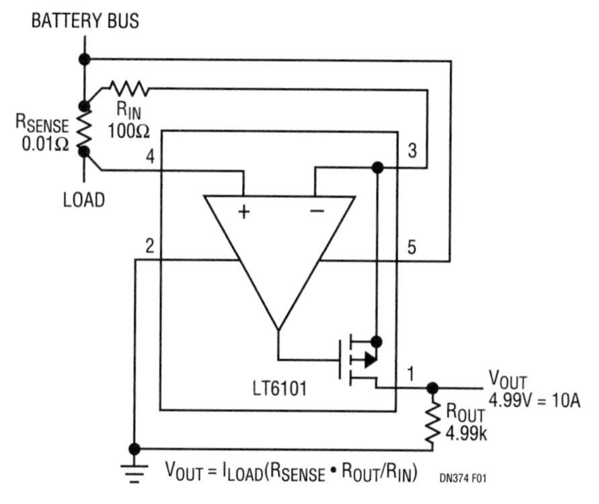

Figure 40.16 •

High side transimpedance amplifier (Figure 40.17)

Current through a photodiode with a large reverse bias potential is converted to a ground referenced output voltage directly through an LTC6101. The supply rail can be as high as 70V Gain of the I to V conversion, the transimpedance, is set by the selection of resistor R_L.

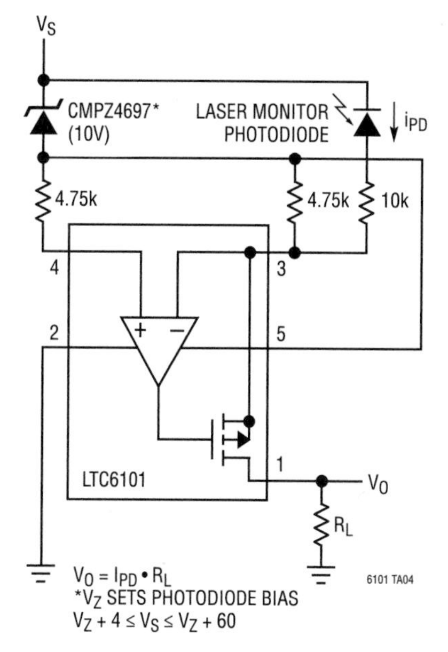

Figure 40.17 •

Intelligent high side switch (Figure 40.18)

The LT1910 is a dedicated high side MOSFET driver with built in protection features. It provides the gate drive for a power switch from standard logic voltage levels. It provides shorted load protection by monitoring the current flow to through the switch. Adding an LTC6101 to the same circuit, sharing the same current sense resistor, provides a linear voltage signal proportional to the load current for additional intelligent control.

48V supply current monitor with isolated output and 105V survivability (Figure 40.19)

The HV version of the LTC6101 can operate with a total supply voltage of 105V. Current flow in high supply voltage rails can be monitored directly or in an isolated fashion as shown in this circuit. The gain of the circuit and the level of output current from the LTC6101 depends on the particular opto-isolator used.

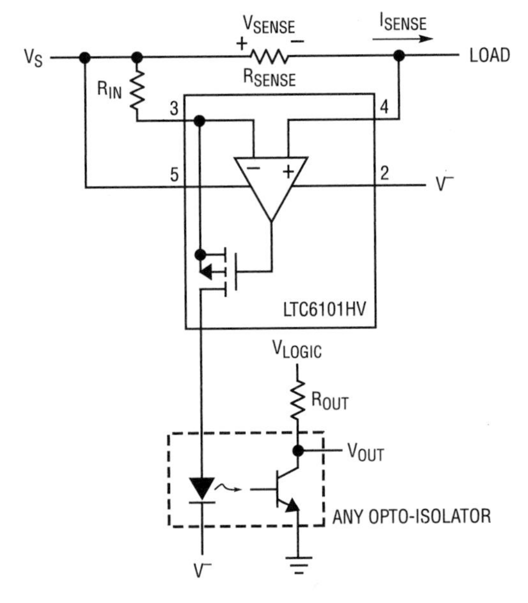

Figure 40.19 •

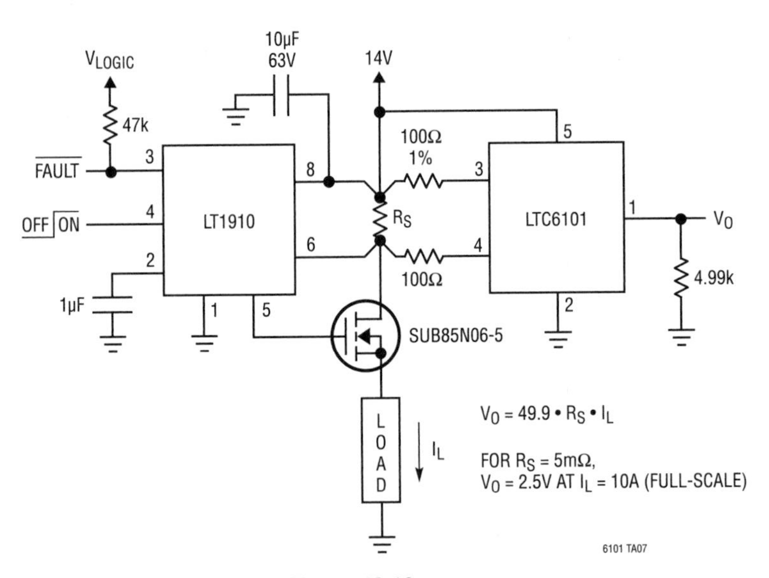

Figure 40.18 •

Precision, wide dynamic range high side current sensing (Figure 40.20)

The LTC6102 offers exceptionally high precision ($V_{os} < 10\mu V$) so that a low value sense resistor may be used. This reduces dissipation in the circuit and allows wider variations in current to be accurately measured. In this circuit, the components are scaled for a 10A measuring range, with the offset error corresponding to less than 10mA. This is effectively better than 10-bit dynamic range with dissipation under 100mW.

Sensed current includes monitor circuit supply current (Figure 40.21)

To sense all current drawn from a battery power source which is also powering the sensing circuitry requires the proper connection of the supply pin. Connecting the supply pin to the load side of the sense resistor adds the supply current to the load current. The sense amplifier operates properly with the inputs equal to the device V+ supply.

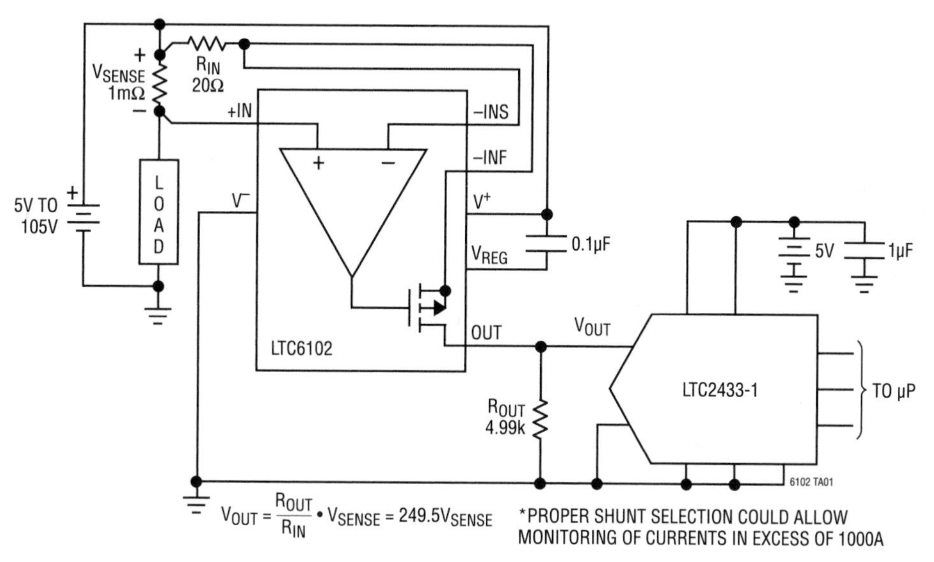

Figure 40.20 •

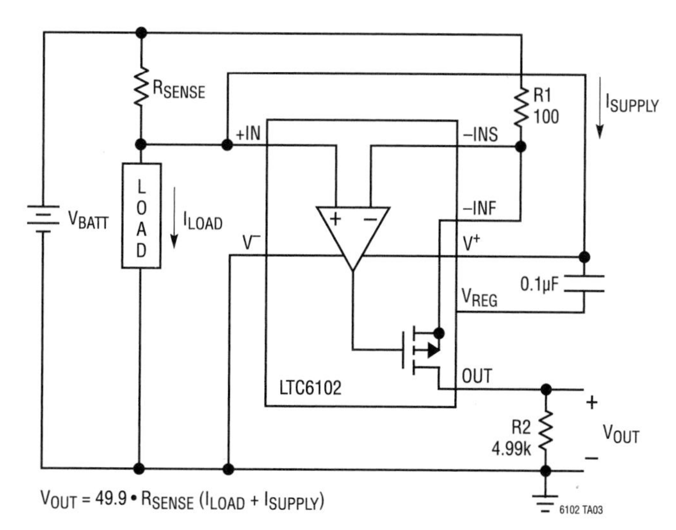

Figure 40.21 •

Wide voltage range current sensing (Figure 40.22)

The LT6105 has a supply voltage that is independent from the potential at the current sense inputs. The input voltage can extend below ground or exceed the sense amplifier supply voltage. While the sensed current must flow in just one direction, it can be sensed above the load, high side, or below the load, low side. Gain is programmed through resistor scaling and is set to 50 in the circuit shown.

Smooth current monitor output signal by simple filtering (Figure 40.23)

The output impedance of the LT6105 amplifier is defined by the value of the gain setting output resistor. Bypassing this resistor with a single capacitor provides first order filtering to smooth noisy current signals and spikes.

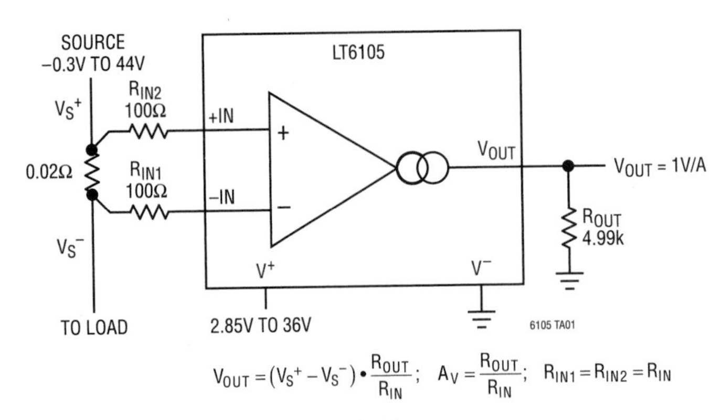

$$V_{OUT} = (V_S^+ - V_S^-) \bullet \frac{R_{OUT}}{R_{IN}}; \quad A_V = \frac{R_{OUT}}{R_{IN}}; \quad R_{IN1} = R_{IN2} = R_{IN}$$

Figure 40.22 •

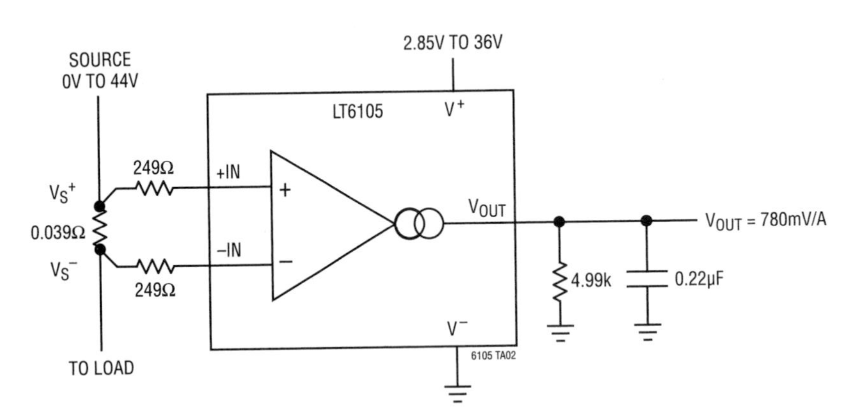

Figure 40.23 •

Power on reset pulse using a TimerBlox device (Figure 40.24)

When power is first applied to a system the load current may require some time to rise to the normal operating level. This can trigger and latch the LT6109 comparator monitoring undercurrent conditions. After a known startup time delay interval, R7 and C1 create a falling edge to trigger an LTC6993-3 one-shot programmed for 10μs. This pulse unlatches the comparators. R8 and Q2 will discharge C1 on loss of the supply to ensure that a full delay interval occurs when power returns.

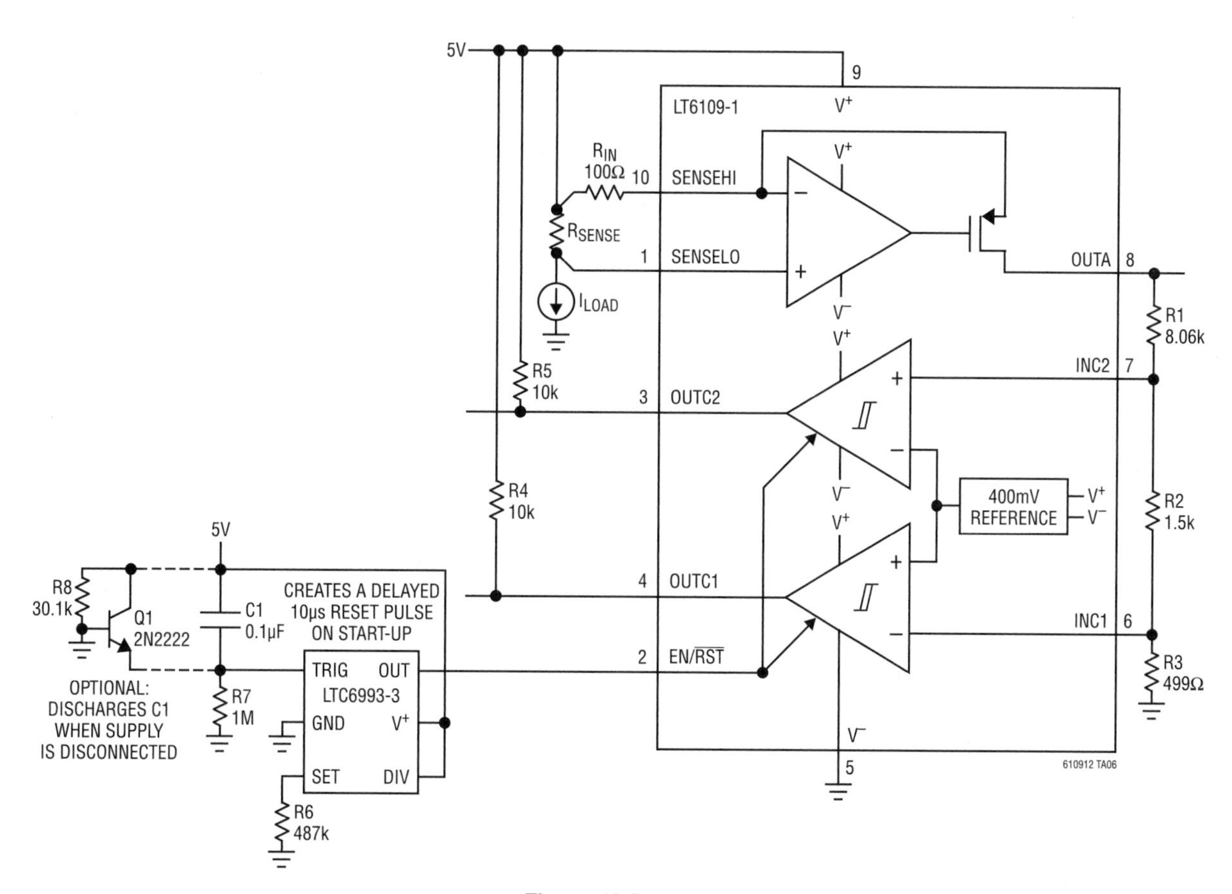

Figure 40.24 •

Accurate delayed power on reset pulse using TimerBlox devices (Figure 40.25)

When power is first applied to a system the load current may require some time to rise to the normal operating level. This can trigger and latch the LT6109 comparator monitoring undercurrent conditions. In this circuit an LTC6994-1 delay timer is used to set an interval longer than the known time for the load current to settle (1 second in the example) then triggers an LTC6993-3 one-shot programmed for 10μs. This pulse unlatches the comparators. The power-on delay time is resistor programmable over a wide range.

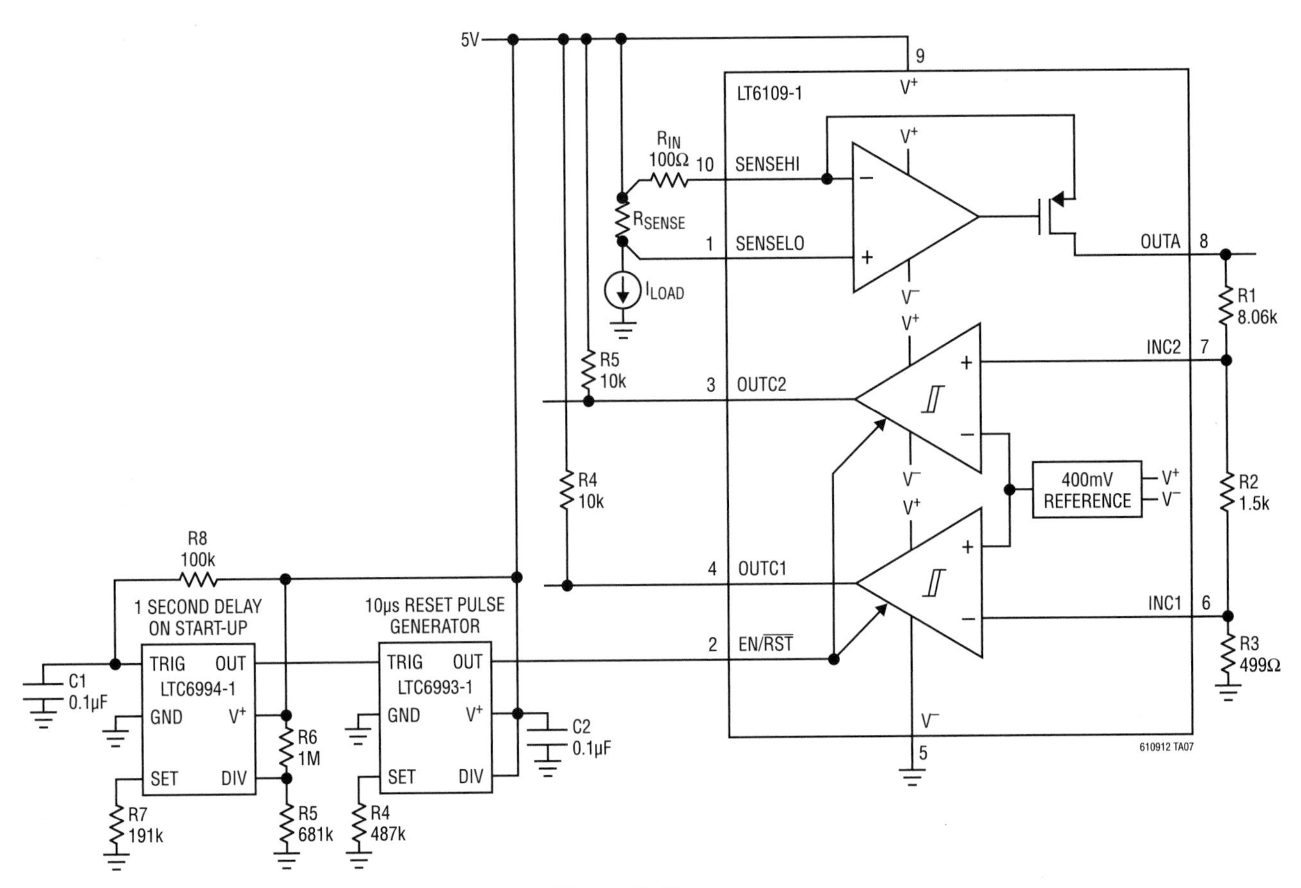

Figure 40.25 •

Low side

This chapter discusses solutions for low side current sensing. With these circuits the current flowing in the ground return or negative power supply line is monitored.

"Classic" high precision low side current sense (Figure 40.26)

This configuration is basically a standard noninverting amplifier. The op amp used must support common mode operation at the lower rail and the use of a zero-drift type (as shown) provides excellent precision. The output of this circuit is referenced to the lower Kelvin contact, which could be ground in a single-supply application. Small-signal range is limited by V_{OL} for single-supply designs. Scaling accuracy is set by the quality of the user-selected resistors.

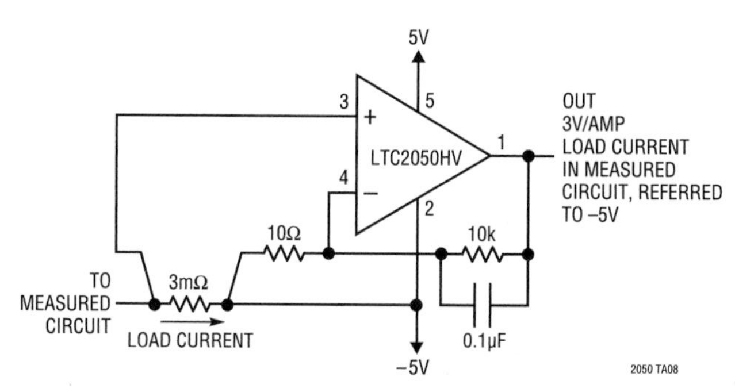

Figure 40.26 •

Precision current sensing in supply rails (Figure 40.27)

This is the same sampling architecture as used in the front end of the LTC2053 and LTC6800, but sans op amp gain stage. This particular switch can handle up to 18V so the ultrahigh precision concept can be utilized at higher voltages than the fully integrated ICs mentioned. This circuit simply commutates charge from the flying sense capacitor to the ground-referenced output capacitor so that under DC input conditions the single-ended output voltage is exactly the same as the differential across the sense resistor. A high precision buffer amplifier would typically follow this circuit (such as an LTC2054). The commutation rate is user-set by the capacitor connected to Pin 14. For negative supply monitoring, Pin 15 would be tied to the negative rail rather than ground.

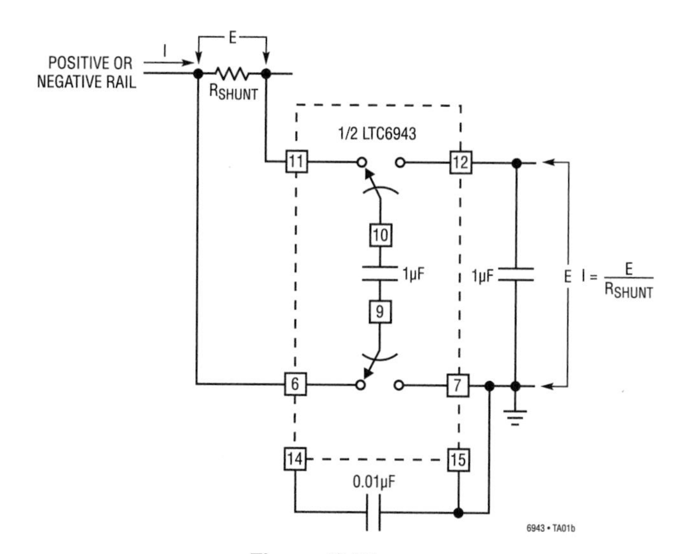

Figure 40.27 •

–48V Hot Swap controller (Figure 40.28)

This load protecting circuit employs low side current sensing. The N-MOSFET is controlled to soft-start the load (current ramping) or to disconnect the load in the event of supply or load faults. An internal shunt regulator establishes a local operating voltage.

–48V low side precision current sense (Figure 40.29)

The first stage amplifier is basically a complementary form of the "classic" high side current sense, designed to operate with telecom negative supply voltage. The Zener forms an inexpensive "floating" shunt-regulated supply for the first op amp. The N-MOSFET drain delivers a metered current into the virtual ground of the second stage, configured as a transimpedance amplifier (TIA). The second op amp is powered from a positive supply and furnishes a positive output voltage for increasing load current. A dual op amp cannot be used for this implementation due to the different supply voltages for each stage. This circuit is exceptionally precise due to the use of zero-drift op amps. The scaling accuracy is established by the quality of the user-selected resistors. Small-signal range is limited by V_{OL} in single-supply operation of the second stage.

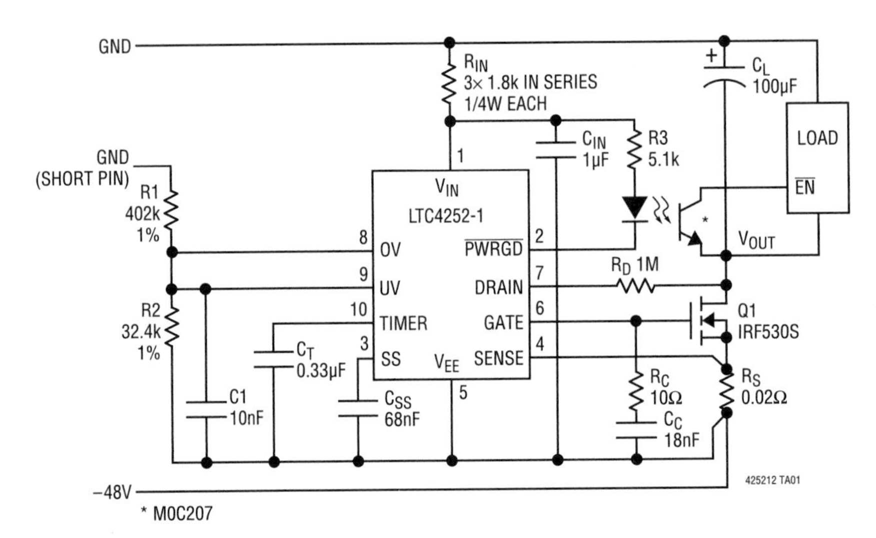

Figure 40.28 •

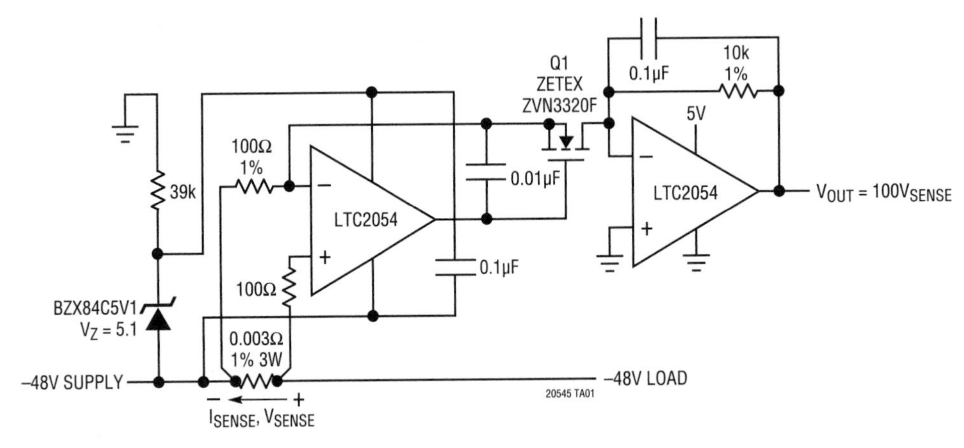

Figure 40.29 •

Fast compact −48V current sense (Figure 40.30)

This amplifier configuration is essentially the complementary implementation to the classic high side configuration. The op amp used must support common mode operation at its lower rail. A "floating" shunt-regulated local supply is provided by the Zener diode, and the transistor provides metered current to an output load resistance (1kΩ in this circuit). In this circuit, the output voltage is referenced to a positive potential and moves downward when representing increasing −48V loading. Scaling accuracy is set by the quality of resistors used and the performance of the NPN transistor.

−48V current monitor (Figures 40.31a and 40.31b)

In this circuit an economical ADC is used to acquire the sense resistor voltage drop directly. The converter is powered from a "floating" high accuracy shunt-regulated supply and is configured to perform continuous conversions. The ADC digital output drives an opto-isolator, level-shifting the serial data stream to ground. For wider supply voltage applications, the 13k biasing resistor may be replaced with an active 4mA current source such as shown in Figure 40.31b. For complete dielectric isolation and/ or higher efficiency operation, the ADC may be powered from a small transformer circuit as shown in Figure 40.31b.

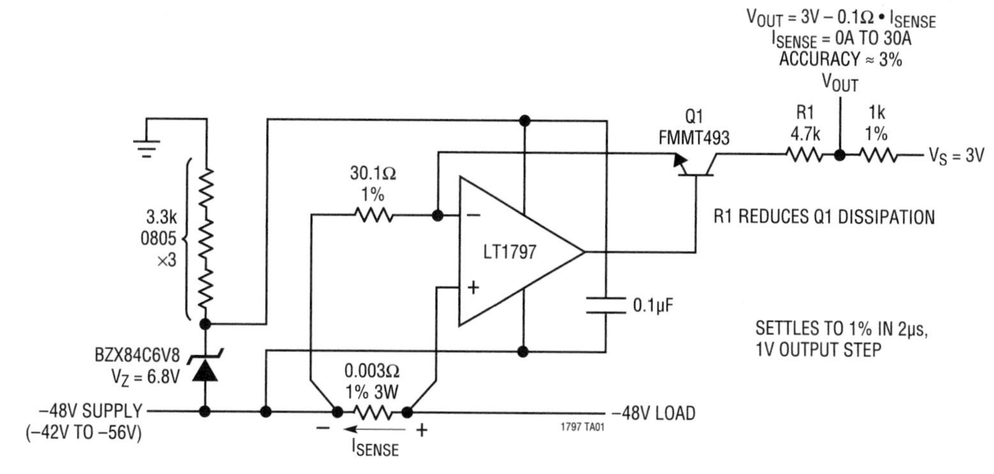

Figure 40.30 •

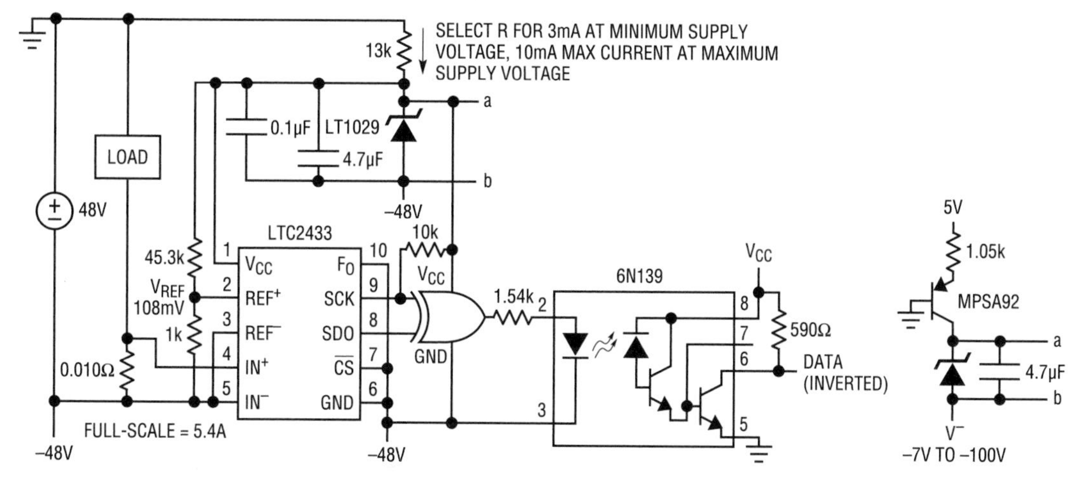

Figure 40.31a •

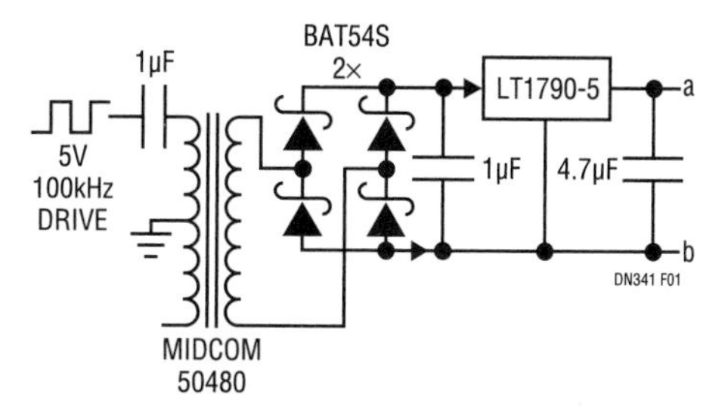

Figure 40.31b •

–48V Hot Swap controller (Figure 40.32)

This load protecting circuit employs low side current sensing. The N-MOSFET is controlled to soft-start the load (current ramping) or to disconnect the load in the event of supply or load faults. An internal shunt regulator establishes a local operating voltage.

Simple telecom power supply fuse monitor (Figure 40.33)

The LTC1921 provides an all-in-one telecom fuse and supply voltage monitoring function. Three opto-isolated status flags are generated that indicate the condition of the supplies and the fuses.

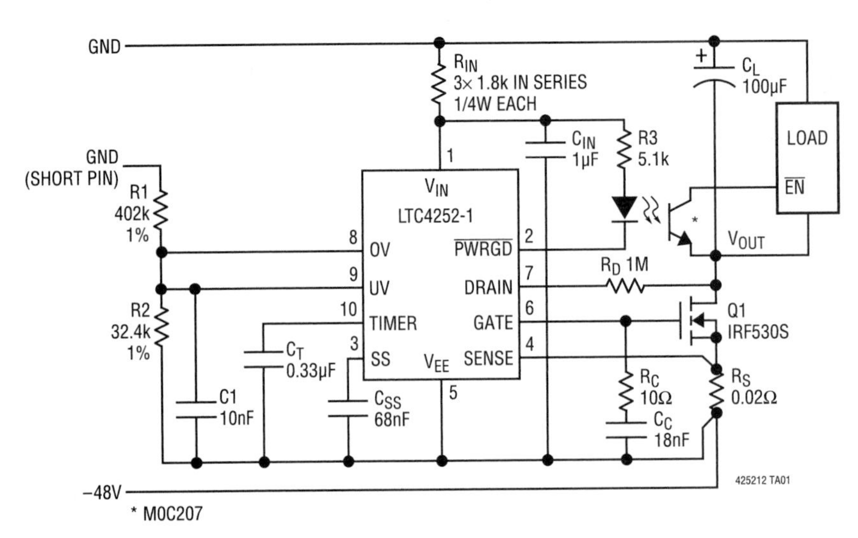

Figure 40.32 •

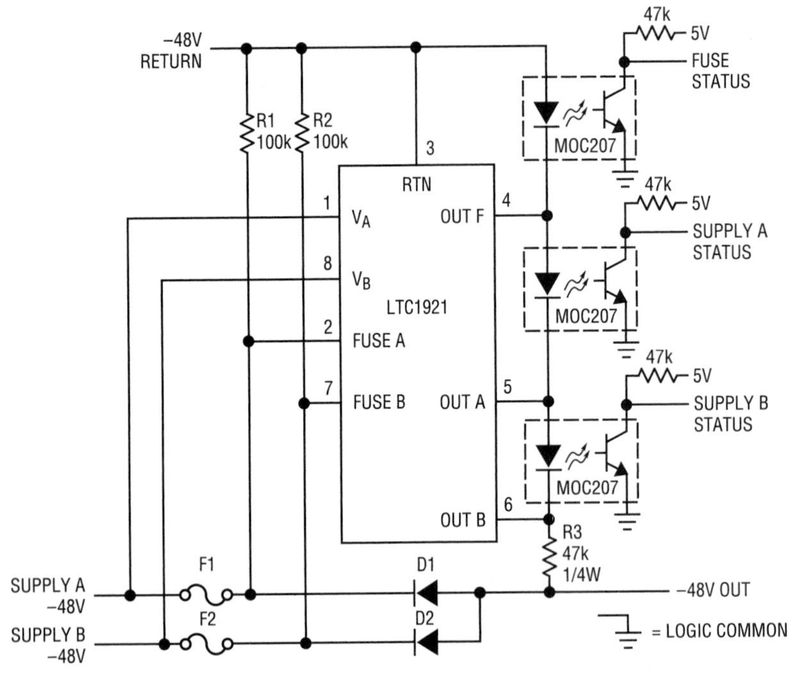

V_A	V_B	SUPPLY A STATUS	SUPPLY B STATUS
OK	OK	0	0
OK	UV OR OV	0	1
UV OR OV	OK	1	0
UV OR OV	UV OR OV	1	1

OK: WITHIN SPECIFICATION
OV: OVERVOLTAGE
UV: UNDERVOLTAGE

$V_{FUSE\,A}$	$V_{FUSE\,B}$	FUSE STATUS
$= V_A$	$= V_B$	0
$= V_A$	$\neq V_B$	1
$\neq V_A$	$= V_B$	1
$\neq V_A$	$\neq V_B$	1*

0: LED/PHOTODIODE ON
1: LED/PHOTODIODE OFF
*IF BOTH FUSES (F1 AND F2) ARE OPEN, ALL STATUS OUTPUTS WILL BE HIGH SINCE R3 WILL NOT BE POWERED

Figure 40.33 •

Negative voltage

This chapter discusses solutions for negative voltage current sensing.

Telecom supply current monitor (Figure 40.34)

The LT1990 is a wide common mode range difference amplifier used here to amplify the sense resistor drop by ten. To provide the desired input range when using a single 5V supply, the reference potential is set to approximately 4V by the LT6650. The output signal moves downward from the reference potential in this connection so that a large output swing can be accommodated.

−48V Hot Swap controller (Figure 40.35)

This load protecting circuit employs low side current sensing. The N-MOSFET is controlled to soft-start the load (current ramping) or to disconnect the load in the event of supply or load faults. An internal shunt regulator establishes a local operating voltage.

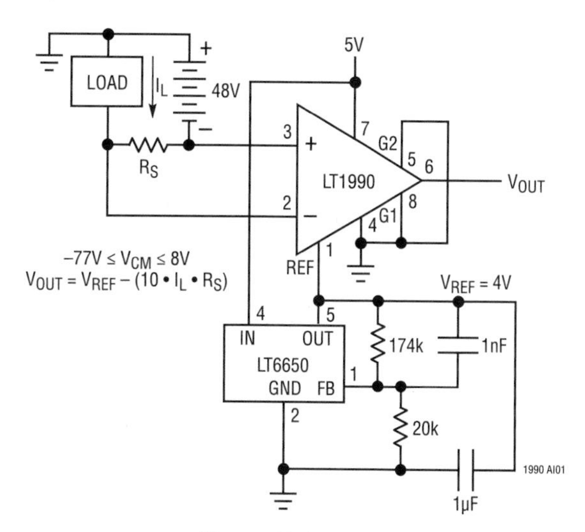

Figure 40.34 •

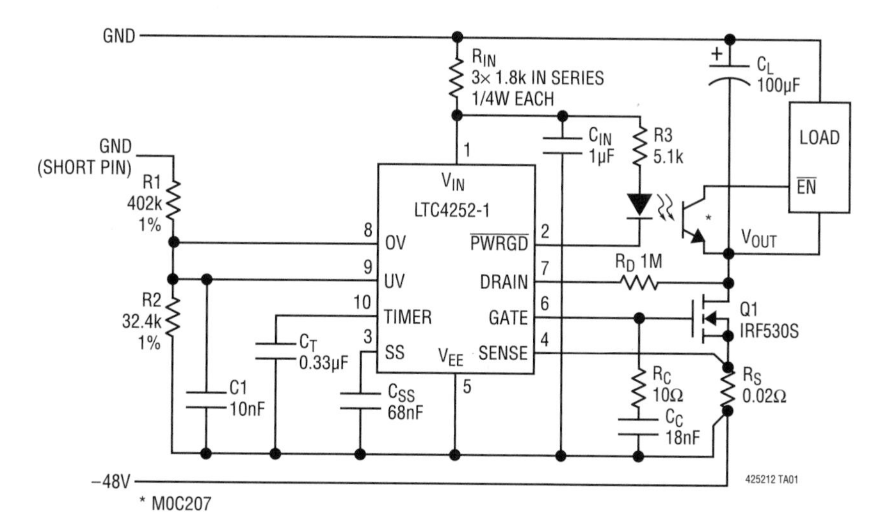

Figure 40.35 •

−48V low side precision current sense (Figure 40.36)

The first stage amplifier is basically a complementary form of the "classic" high side current sense, designed to operate with telecom negative supply voltage. The Zener forms an inexpensive "floating" shunt-regulated supply for the first op amp. The N-MOSFET drain delivers a metered current into the virtual ground of the second stage, configured as a transimpedance amplifier (TIA). The second op amp is powered from a positive supply and furnishes a positive output voltage for increasing load current. A dual op amp cannot be used for this implementation due to the different supply voltages for each stage. This circuit is exceptionally precise due to the use of zero-drift op amps. The scaling accuracy is established by the quality of the user-selected resistors. Small-signal range is limited by V_{OL} in single-supply operation of the second stage.

Fast compact −48V current sense (Figure 40.37)

This amplifier configuration is essentially the complementary implementation to the classic high side configuration. The op amp used must support common mode operation at its lower rail. A "floating" shunt-regulated local supply is provided by the Zener diode, and the transistor provides metered current to an output load resistance (1kΩ in this circuit). In this circuit, the output voltage is referenced to a positive potential and moves downward when representing increasing −48V loading. Scaling accuracy is set by the quality of resistors used and the performance of the NPN transistor.

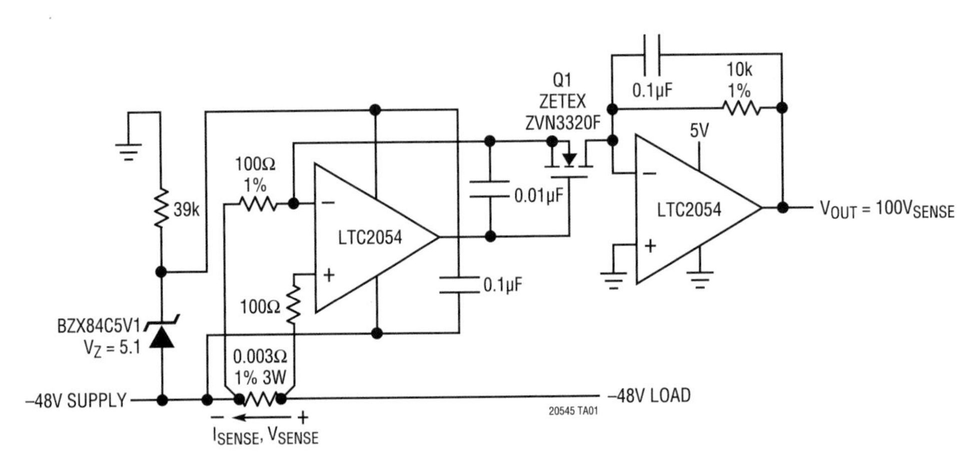

Figure 40.36 •

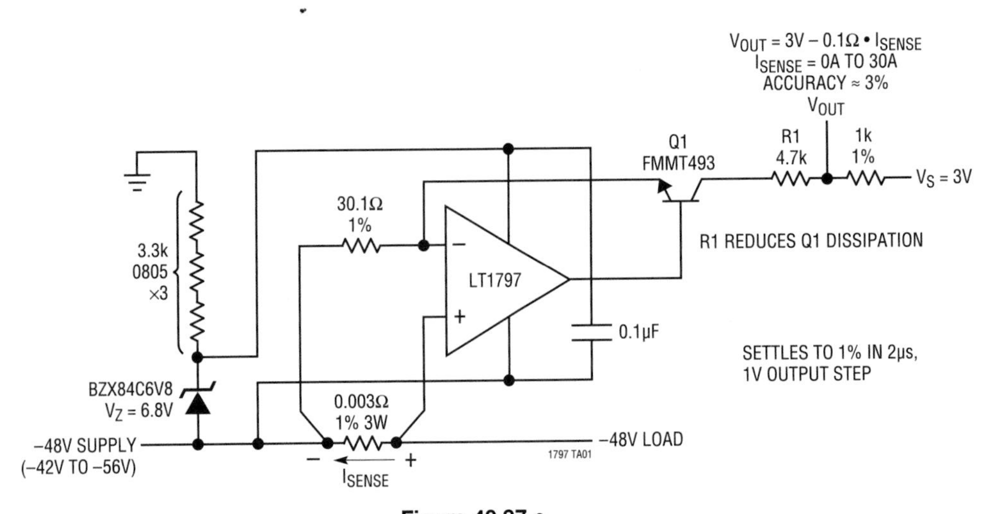

Figure 40.37 •

−48V current monitor (Figures 40.38a and 40.38b)

In this circuit an economical ADC is used to acquire the sense resistor voltage drop directly. The converter is powered from a "floating" high accuracy shunt-regulated supply and is configured to perform continuous conversions. The ADC digital output drives an opto-isolator, level-shifting the serial data stream to ground. For wider supply voltage applications, the 13k biasing resistor may be replaced with an active 4mA current source such as shown to the right. For complete dielectric isolation and/or higher efficiency operation, the ADC may be powered from a small transformer circuit as shown in Figure 40.38b.

Simple telecom power supply fuse monitor (Figure 40.39)

The LTC1921 provides an all-in-one telecom fuse and supply voltage monitoring function. Three opto-isolated status flags are generated that indicate the condition of the supplies and the fuses.

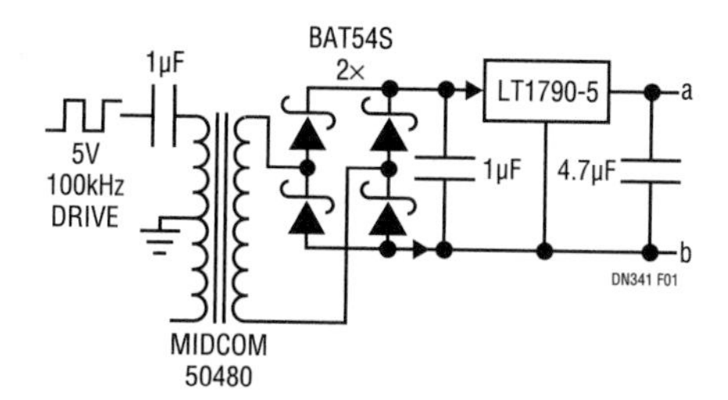

Figure 40.38b •

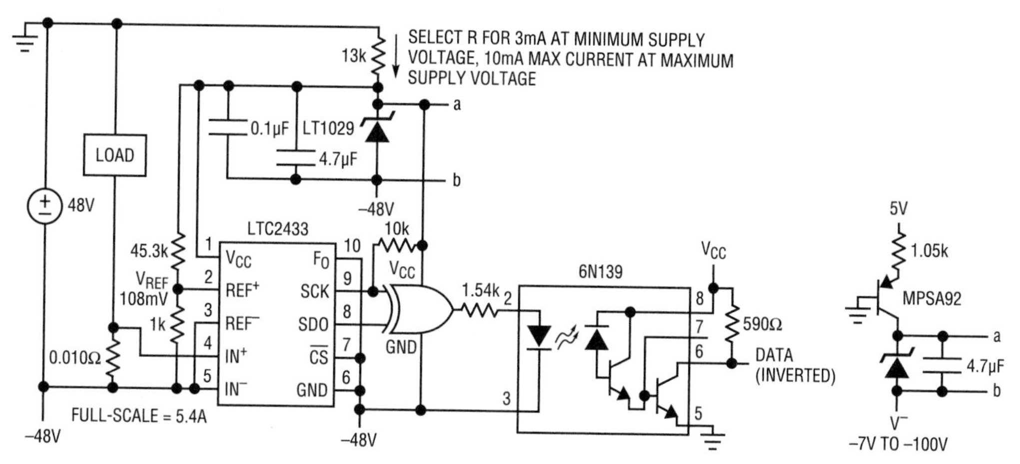

Figure 40.38a •

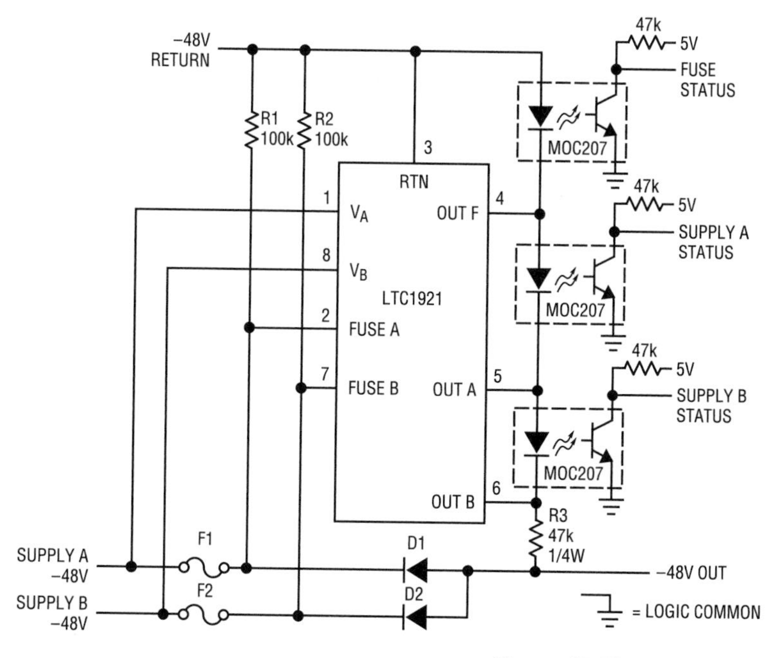

V_A	V_B	SUPPLY A STATUS	SUPPLY B STATUS
OK	OK	0	0
OK	UV OR OV	0	1
UV OR OV	OK	1	0
UV OR OV	UV OR OV	1	1

OK: WITHIN SPECIFICATION
OV: OVERVOLTAGE
UV: UNDERVOLTAGE

$V_{FUSE\ A}$	$V_{FUSE\ B}$	FUSE STATUS
$= V_A$	$= V_B$	0
$= V_A$	$\neq V_B$	1
$\neq V_A$	$= V_B$	1
$\neq V_A$	$\neq V_B$	1*

0: LED/PHOTODIODE ON
1: LED/PHOTODIODE OFF
*IF BOTH FUSES (F1 AND F2) ARE OPEN, ALL STATUS OUTPUTS WILL BE HIGH SINCE R3 WILL NOT BE POWERED

Figure 40.39 •

Monitor current in positive or negative supply lines (Figure 40.40)

Using a negative supply voltage to power the LT6105 creates a circuit that can be used to monitor the supply current in a positive or negative supply line by only changing the input connections. In both configurations the output is a ground referred positive voltage. The negative supply to the LT6105 must be at least as negative as the supply line it is monitoring.

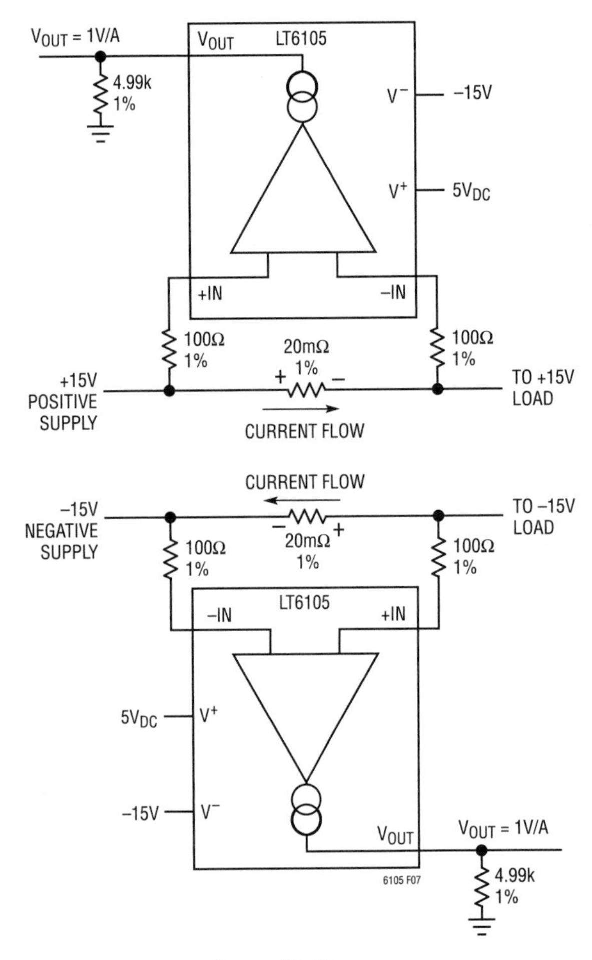

Figure 40.40 •

Unidirectional

Unidirectional current sensing monitors the current flowing only in one direction through a sense resistor.

Unidirectional output into A/D with fixed supply at V_S^+ (Figure 40.41)

Here the LT1787 is operating with the LTC1286 A/D converter. The −IN pin of the A/D converter is biased at 1V by the resistor divider R1 and R2. This voltage increases as sense current increases, with the amplified sense voltage appearing between the A/D converters −IN and +IN terminals. The LTC1286 converter uses sequential sampling of its −IN and +IN inputs. Accuracy is degraded if the inputs move between sampling intervals. A filter capacitor from FIL^+ to FIL^- as well as a filter capacitor from V_{BIAS} to V_{OUT} may be necessary if the sensed current changes more than 1LSB within a conversion cycle.

Unidirectional current sensing mode (Figures 40.42a and 40.42b)

This is just about the simplest connection in which the LT1787 may be used. The V_{BIAS} pin is connected to ground, and the V_{OUT} pin swings positive with increasing sense current. The output can swing as low as 30mV. Accuracy is sacrificed at small output levels, but this is not a limitation in protection circuit applications or where sensed currents do not vary greatly. Increased low level accuracy can be obtained by level shifting V_{BIAS} above ground. The level shifting may be done with resistor dividers, voltage references or a simple diode. Accuracy is ensured if the output signal is sensed differentially between V_{BIAS} and V_{OUT}.

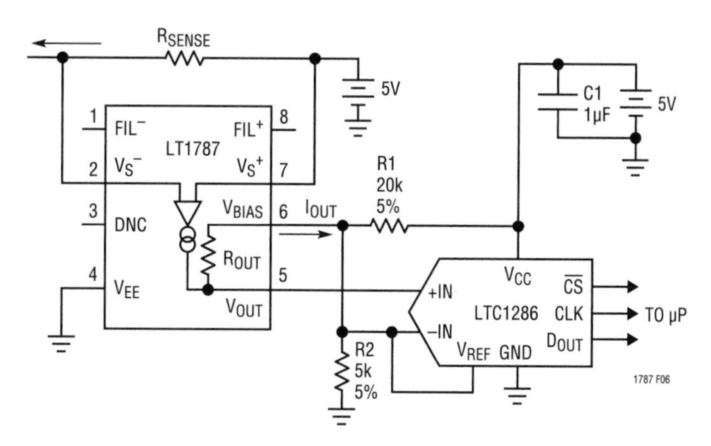

Figure 40.41 •

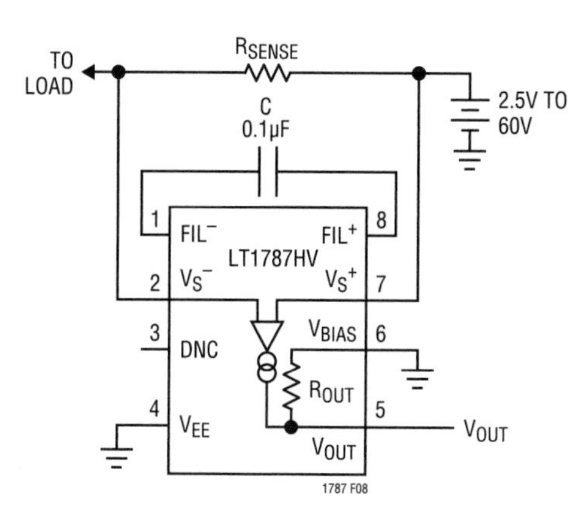

Figure 40.42a •

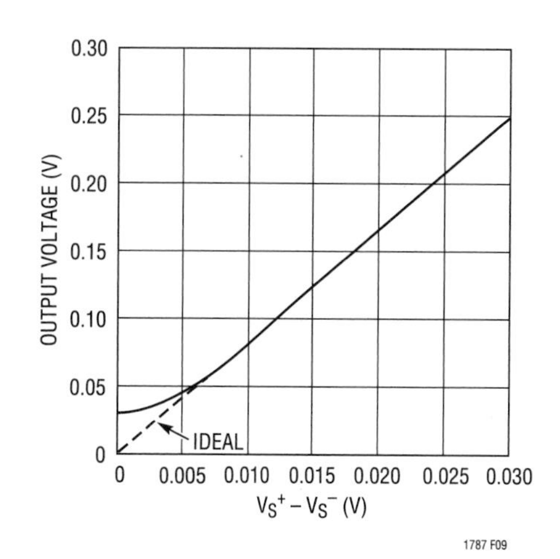

Figure 40.42b •

16-bit resolution unidirectional output into LTC2433 ADC (Figure 40.43)

The LTC2433-1 can accurately digitize signal with source impedances up to 5kΩ. This LTC6101 current sense circuit uses a 4.99kΩ output resistance to meet this requirement, thus no additional buffering is necessary.

Intelligent high side switch (Figure 40.44)

The LT1910 is a dedicated high side MOSFET driver with built in protection features. It provides the gate drive for a power switch from standard logic voltage levels. It provides shorted load protection by monitoring the current flow to through the switch. Adding an LTC6101 to the same circuit, sharing the same current sense resistor, provides a linear voltage signal proportional to the load current for additional intelligent control.

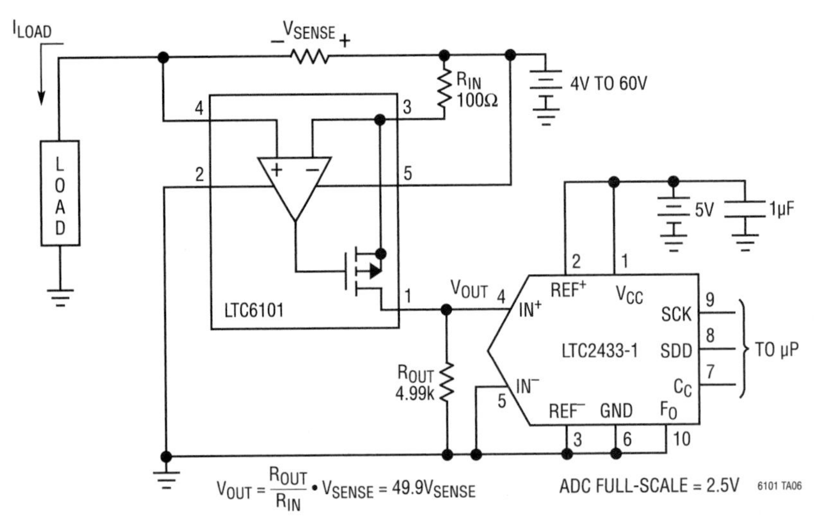

Figure 40.43 •

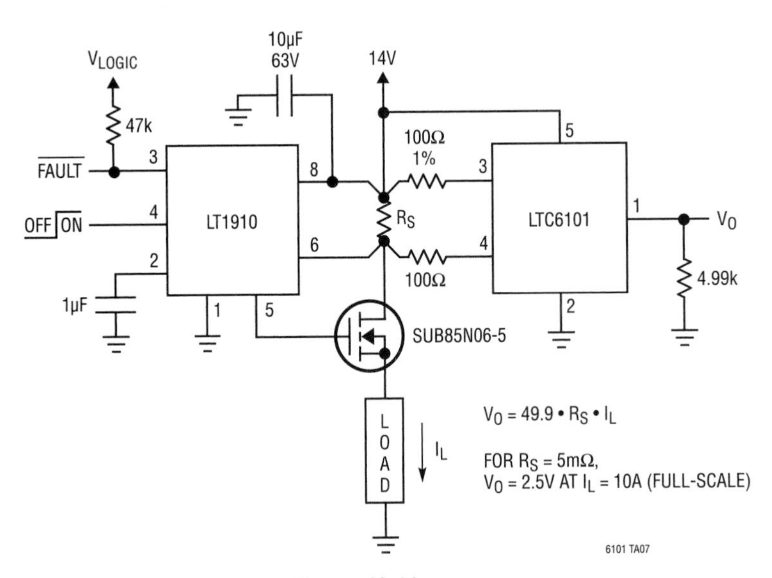

Figure 40.44 •

48V supply current monitor with isolated output and 105V survivability (Figure 40.45)

The HV version of the LTC6101 can operate with a total supply voltage of 105V. Current flow in high supply voltage rails can be monitored directly or in an isolated fashion as shown in this circuit. The gain of the circuit and the level of output current from the LTC6101 depends on the particular opto-isolator used.

12-bit resolution unidirectional output into LTC1286 ADC (Figure 40.46)

While the LT1787 is able to provide a bidirectional output, in this application the economical LTC1286 is used to digitize a unidirectional measurement. The LT1787 has a nominal gain of eight, providing a 1.25V full-scale output at approximately 100A of load current.

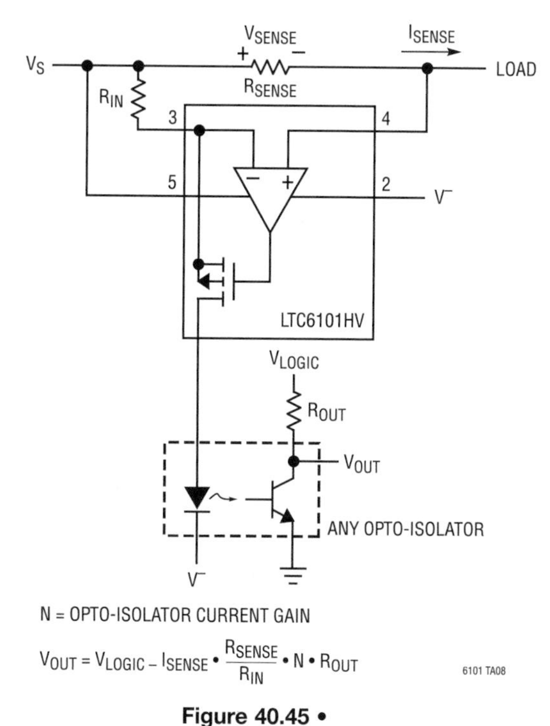

Figure 40.45 •

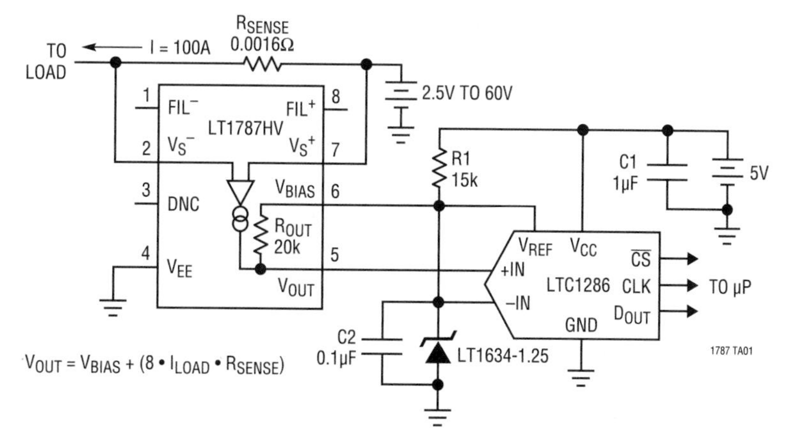

Figure 40.46 •

Bidirectional

Bidirectional current sensing monitors current flow in both directions through a sense resistor.

Bidirectional current sensing with single-ended output (Figure 40.47)

Two LTC6101's are used to monitor the current in a load in either direction. Using a separate rail-to-rail op amp to combine the two outputs provides a single ended output. With zero current flowing the output sits at the reference potential, one-half the supply voltage for maximum output swing or 2.5V as shown. With power supplied to the load through connection A the output will move positive between 2.5V and V_{CC}. With connection B the output moves down between 2.5V and 0V.

Practical H-bridge current monitor offers fault detection and bidirectional load information (Figure 40.48)

This circuit implements a differential load measurement for an ADC using twin unidirectional sense measurements. Each LTC6101 performs high side sensing that rapidly responds to fault conditions, including load shorts and MOSFET failures. Hardware local to the switch module (not shown in the diagram) can provide the protection logic and furnish a status flag to the control system. The two LTC6101 outputs taken differentially produce a bidirectional load measurement for the control servo. The ground-referenced signals are compatible with most ΔΣADCs. The ΔΣADC circuit also provides a "free" integration function that removes PWM content from the measurement. This scheme also eliminates the need for analog-to-digital conversions at the rate needed to support switch protection, thus reducing cost and complexity.

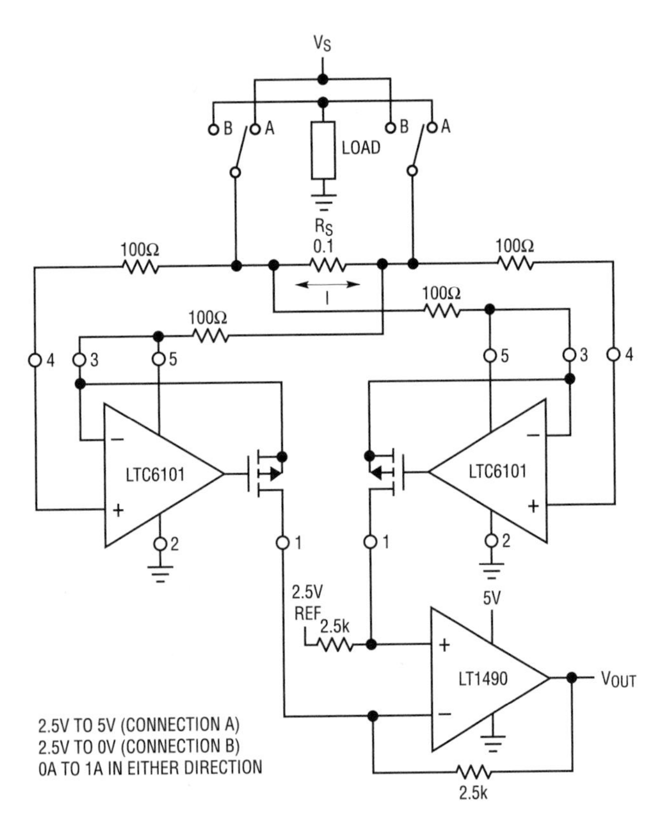

Figure 40.47 •

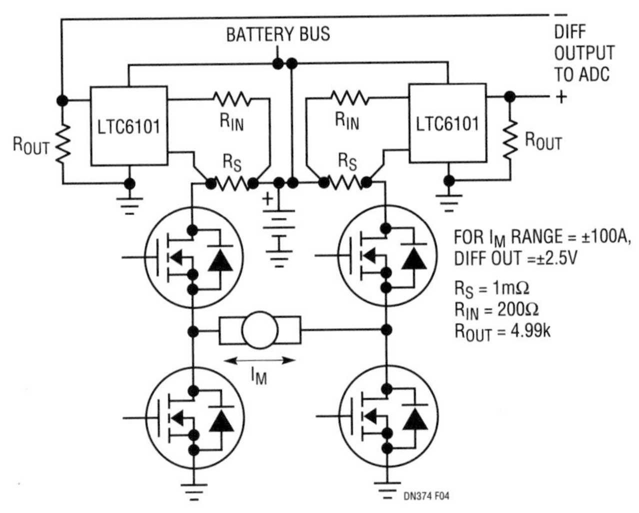

Figure 40.48 •

Conventional H-bridge current monitor (Figure 40.49)

Many of the newer electric drive functions, such as steering assist, are bidirectional in nature. These functions are generally driven by H-bridge MOSFET arrays using pulse-width modulation (PWM) methods to vary the commanded torque. In these systems, there are two main purposes for current monitoring. One is to monitor the current in the load, to track its performance against the desired command (i.e., closed-loop servo law), and another is for fault detection and protection features.

A common monitoring approach in these systems is to amplify the voltage on a "flying" sense resistor, as shown. Unfortunately, several potentially hazardous fault scenarios go undetected, such as a simple short to ground at a motor terminal. Another complication is the noise introduced by the PWM activity. While the PWM noise may be filtered for purposes of the servo law, information useful for protection becomes obscured. The best solution is to simply provide two circuits that individually protect each half-bridge and report the bidirectional load current. In some cases, a smart MOSFET bridge driver may already include sense resistors and offer the protection features needed. In these situations, the best solution is the one that derives the load information with the least additional circuitry.

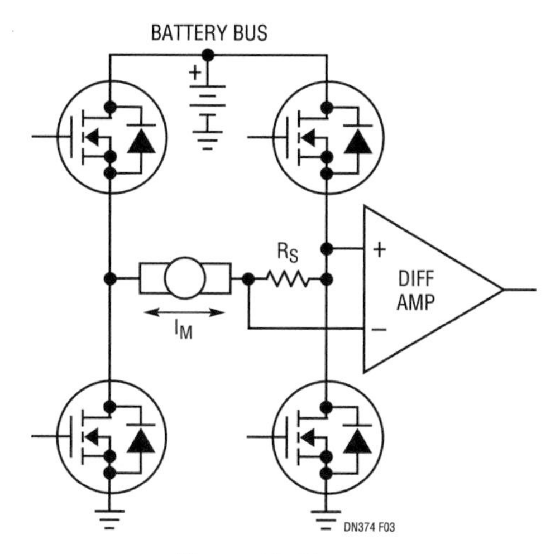

Figure 40.49 •

Single-supply 2.5V bidirectional operation with external voltage reference and I/V converter (Figure 40.50)

The LT1787's output is buffered by an LT1495 rail-to-rail op amp configured as an I/V converter. This configuration is ideal for monitoring very low voltage supplies. The LT1787's V_{OUT} pin is held equal to the reference voltage appearing at the op amp's noninverting input. This allows one to monitor supply voltages as low as 2.5V. The op amp's output may swing from ground to its positive supply voltage. The low impedance output of the op amp may drive following circuitry more effectively than the high output impedance of the LT1787. The I/V converter configuration also works well with split supply voltages.

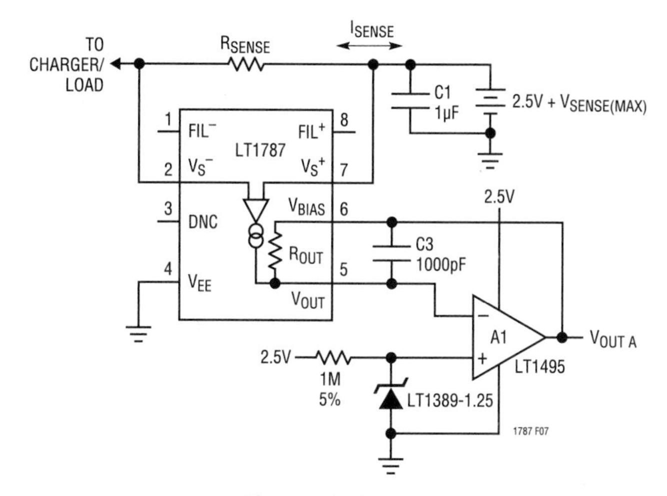

Figure 40.50 •

Battery current monitor (Figure 40.51)

One LT1495 dual op amp package can be used to establish separate charge and discharge current monitoring outputs. The LT1495 features Over-the-Top operation allowing the battery potential to be as high as 36V with only a 5V amplifier supply voltage.

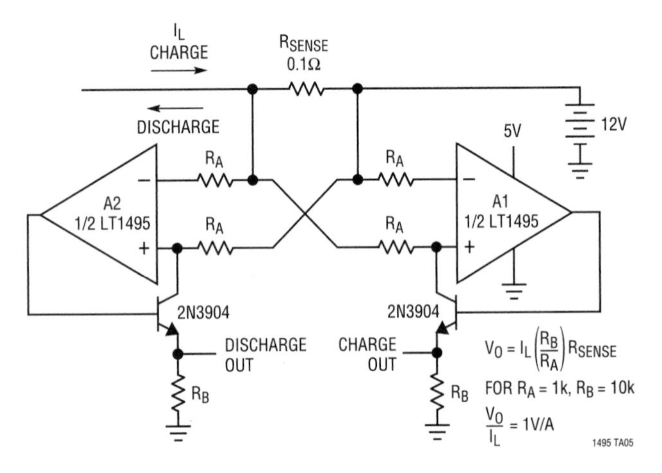

Figure 40.51 •

Fast current sense with alarm (Figure 40.52)

The LT1995 is shown as a simple unity gain difference amplifier. When biased with split supplies the input current can flow in either direction providing an output voltage of 100mV per Amp from the voltage across the 100mΩ sense resistor. With 32MHz of bandwidth and 1000V/μs slew rate the response of this sense amplifier is fast. Adding a simple comparator with a built in reference voltage circuit such as the LT6700-3 can be used to generate an overcurrent flag. With the 400mV reference the flag occurs at 4A.

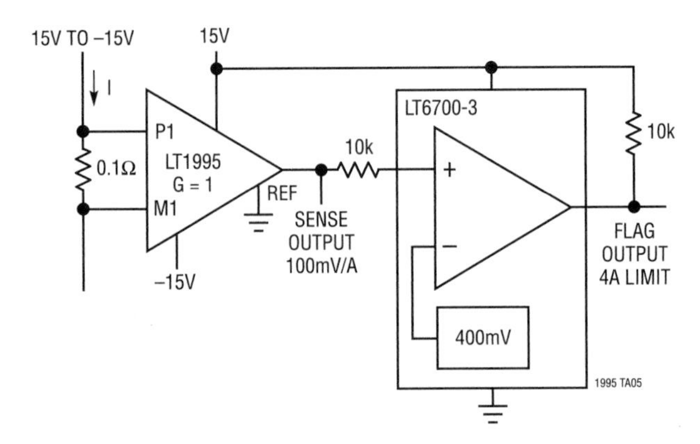

Figure 40.52 •

Bidirectional current sense with separate charge/discharge output (Figure 40.53)

In this circuit the outputs are enabled by the direction of current flow. The battery current when either charging or discharging enables only one of the outputs. For example when charging, the V_{OUT} D signal goes low since the output MOSFET of that LTC6101 turns completely off while the other LT6101, V_{OUT} C, ramps from low to high in proportion to the charging current. The active output reverses when the charger is removed and the battery discharges into the load.

Bidirectional absolute value current sense (Figure 40.54)

The high impedance current source outputs of two LTC6101's can be directly tied together. In this circuit the voltage at V_{OUT} continuously represents the absolute value of the magnitude of the current into or out of the battery. The direction or polarity of the current flow is not discriminated.

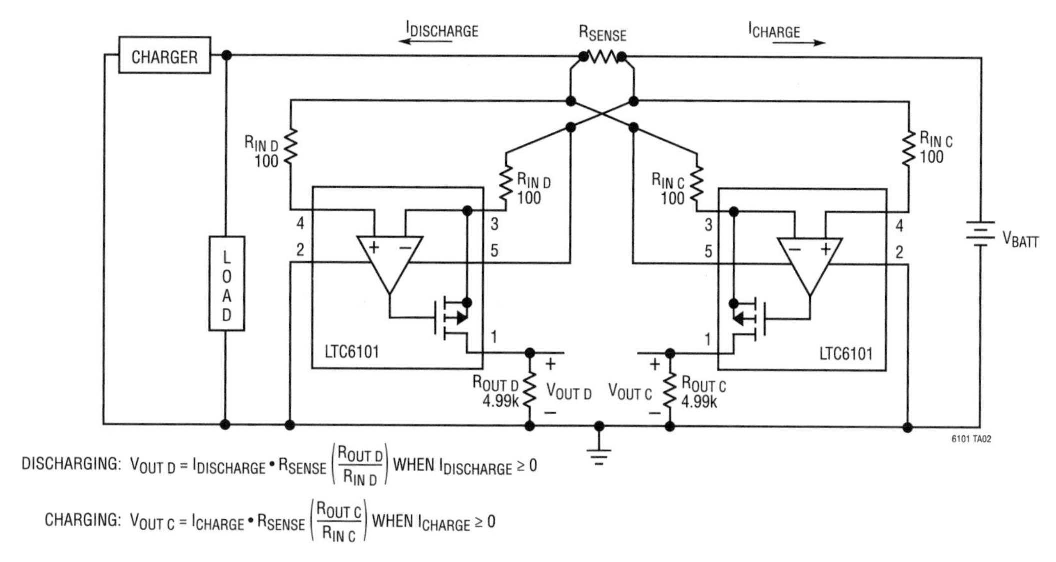

Figure 40.53 •

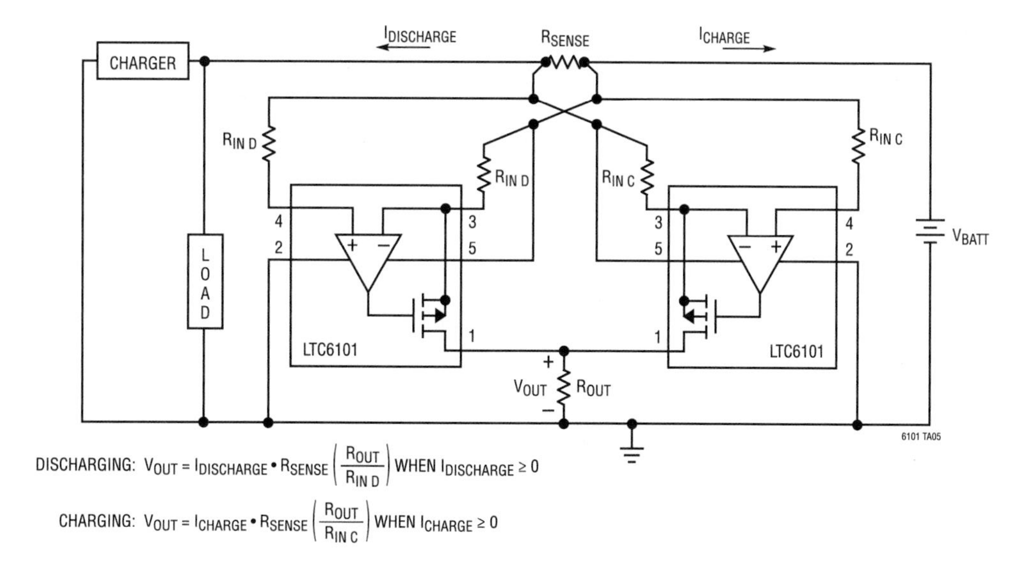

Figure 40.54 •

Full-bridge load current monitor (Figure 40.55)

The LT1990 is a difference amplifier that features a very wide common mode input voltage range that can far exceed its own supply voltage. This is an advantage to reject transient voltages when used to monitor the current in a full-bridge driven inductive load such as a motor. The LT6650 provides a voltage reference of 1.5V to bias up the output away from ground. The output will move above or below 1.5V as a function of which direction the current in the load is flowing. As shown, the amplifier provides a gain of 10 to the voltage developed across resistor R_S.

Low power, bidirectional 60V precision high side current sense (Figure 40.56)

Using a very precise zero-drift amplifier as a pre-amp allows for the use of a very small sense resistor in a high voltage supply line. A floating power supply regulates the voltage across the pre-amplifier on any voltage rail up to the 60V limit of the LT1787HV circuit. Overall gain of this circuit is 1000. A 1mA change in current in either direction through the 10mΩ sense resistor will produce a 10mV change in the output voltage.

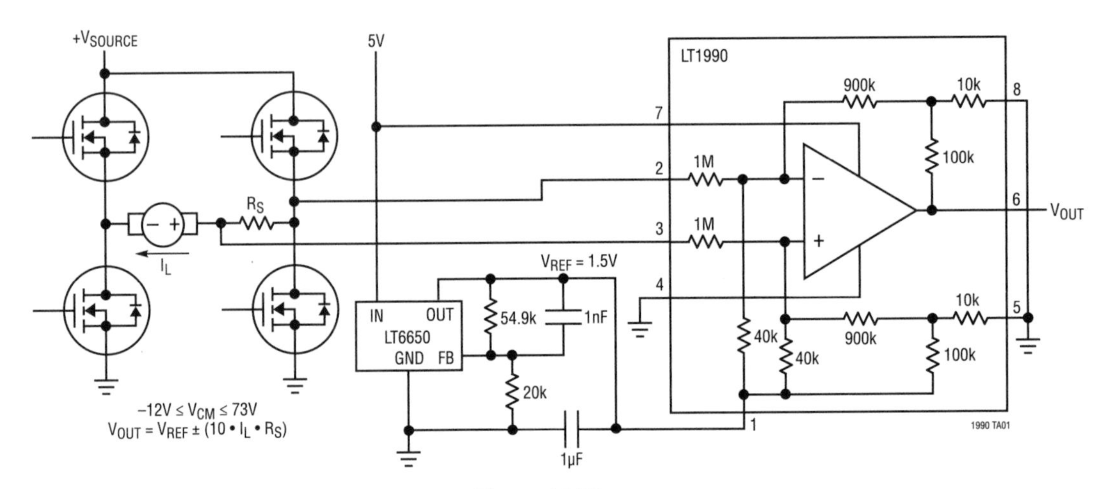

Figure 40.55 •

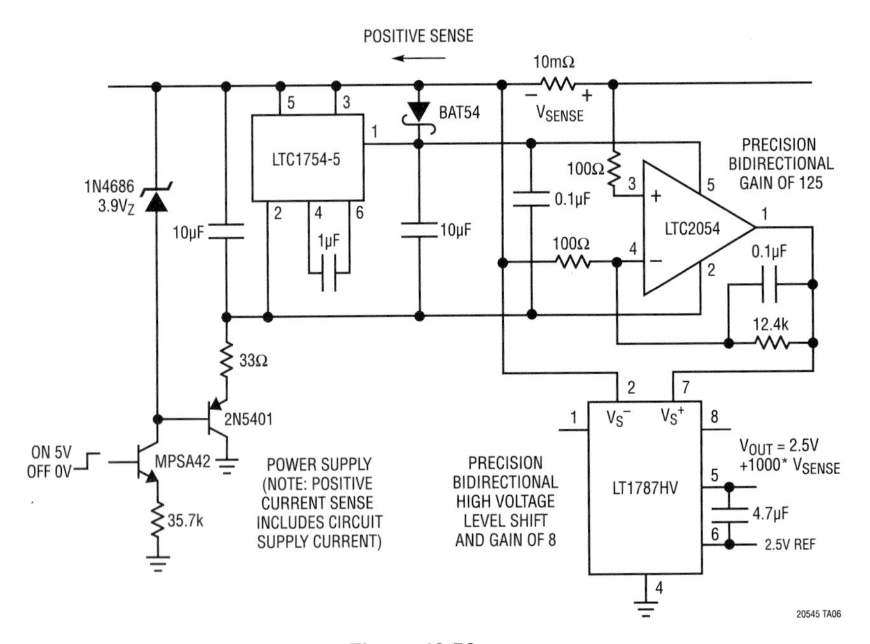

Figure 40.56 •

Split or single supply operation, bidirectional output into A/D (Figure 40.57)

In this circuit, split supply operation is used on both the LT1787 and LT1404 to provide a symmetric bidirectional measurement. In the single-supply case, where the LT1787 Pin 6 is driven by V_{REF}, the bidirectional measurement range is slightly asymmetric due to V_{REF} being somewhat greater than midspan of the ADC input range.

Bidirectional precision current sensing (Figure 40.58)

This circuit uses two LTC6102 devices, one for each direction of current flow through a single sense resistance. While each output only provides a result in one particular direction of current, taking the two output signals differentially provides a bipolar signal to other circuitry such as an ADC. Since each circuit has its own gain resistors, bilinear scaling is possible (different scaling depending on direction).

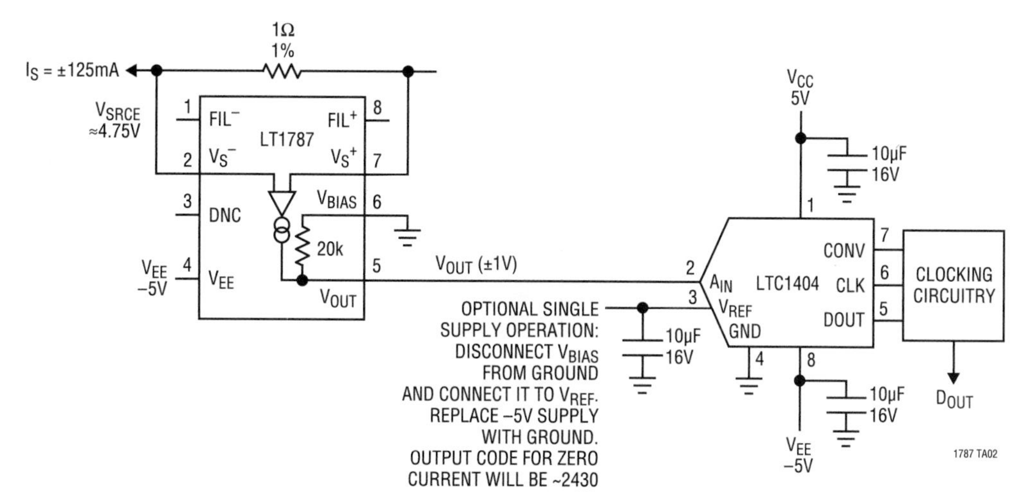

Figure 40.57 •

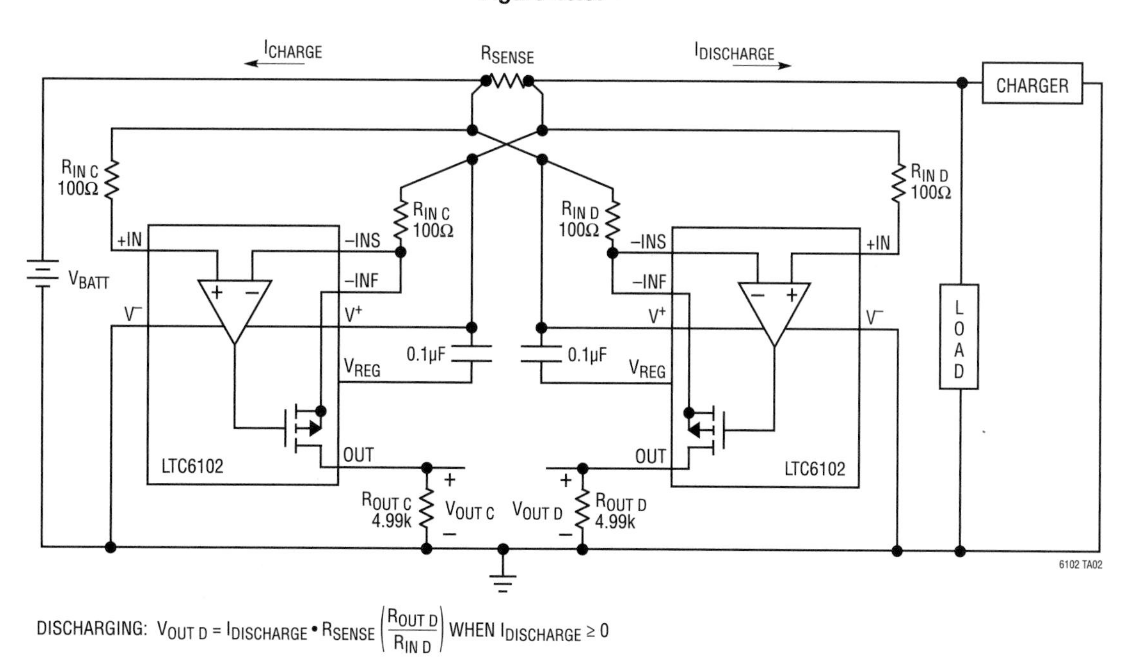

$$\text{DISCHARGING: } V_{OUT\,D} = I_{DISCHARGE} \cdot R_{SENSE}\left(\frac{R_{OUT\,D}}{R_{IN\,D}}\right) \text{ WHEN } I_{DISCHARGE} \geq 0$$

$$\text{CHARGING: } V_{OUT\,C} = I_{CHARGE} \cdot R_{SENSE}\left(\frac{R_{OUT\,C}}{R_{IN\,C}}\right) \text{ WHEN } I_{CHARGE} \geq 0$$

Figure 40.58 •

Differential output bidirectional 10A current sense (Figure 40.59)

The LTC6103 has dual sense amplifiers and each measures current in one direction through a single sense resistance. The outputs can be taken together as a differential output to subsequent circuitry such as an ADC. Values shown are for 10A maximum measurement.

Absolute value output bidirectional current sensing (Figure 40.60)

Connecting an LTC6103 so that the outputs each represent opposite current flow through a shared sense resistance, but with the outputs driving a common load, results in a positive only output function while sensing bidirectionally.

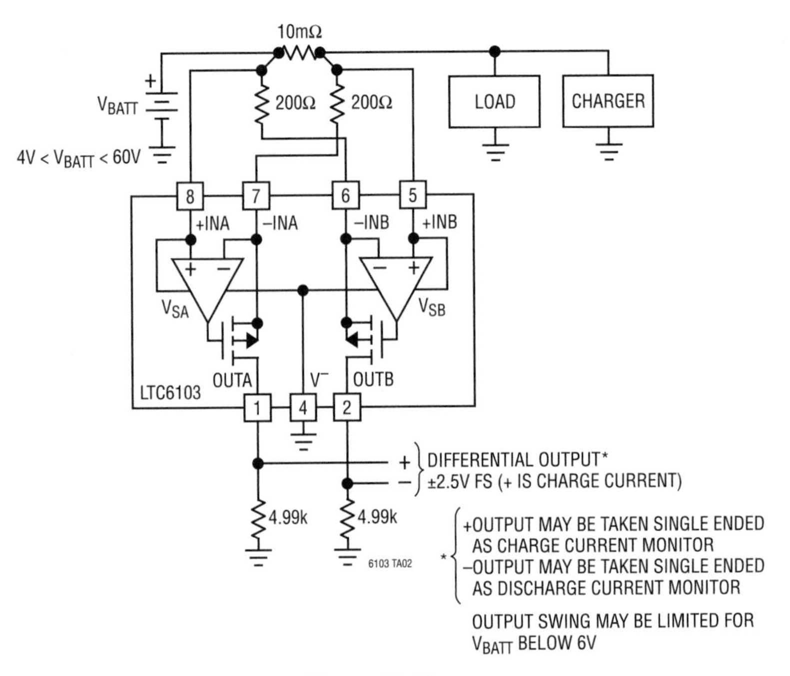

Figure 40.59 •

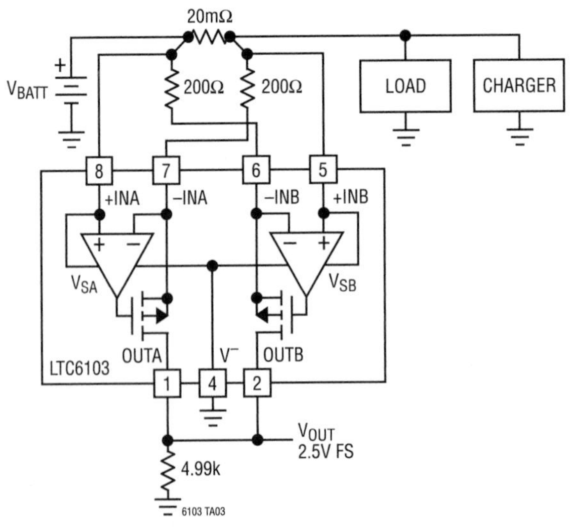

Figure 40.60 •

AC

Sensing current in AC power lines is quite tricky in the sense that both the current and voltage are continuously changing polarity. Transformer coupling of signals to drive ground referenced circuitry is often a good approach.

Single-supply RMS current measurement (Figure 40.61)

The LT1966 is a true RMS-to-DC converter that takes a single-ended or differential input signal with rail-to-rail range. The output of a PCB mounted current sense transformer can be connected directly to the converter. Up to 75A of AC current is measurable without breaking the signal path from a power source to a load. The accurate operating range of the circuit is determined by the selection of the transformer termination resistor. All of the math is built in to the LTC1966 to provide a DC output voltage that is proportional to the true RMS value of the current. This is valuable in determining the power/energy consumption of AC-powered appliances.

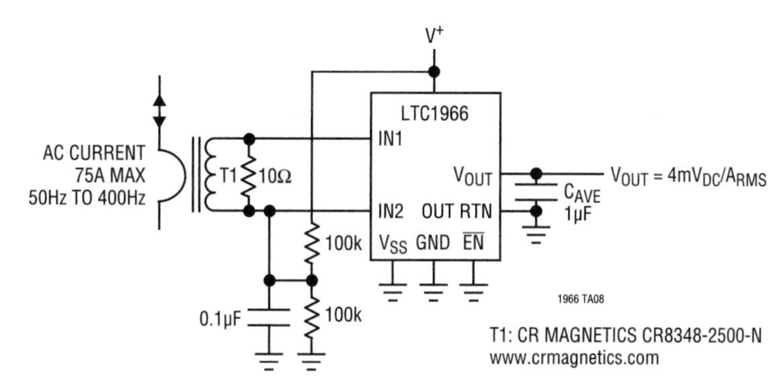

Figure 40.61 •

DC

DC current sensing is for measuring current flow that is changing at a very slow rate.

Micro-hotplate voltage and current monitor (Figure 40.62)

Materials science research examines the properties and interactions of materials at various temperatures. Some of the more interesting properties can be excited with localized nano-technology heaters and detected using the presence of interactive thin films.

While the exact methods of detection are highly complex and relatively proprietary, the method of creating localized heat is as old as the light bulb. Shown is the schematic of the heater elements of a Micro-hotplate from Boston Microsystems (www.bostonmicrosystems.com). The physical dimensions of the elements are tens of microns. They are micromachined out of SiC and heated with simple DC electrical power, being able to reach 1000°C without damage.

The power introduced to the elements, and thereby their temperature, is ascertained from the voltage-current product with the LT6100 measuring the current and the LT1991 measuring the voltage. The LT6100 senses the current by measuring the voltage across the 10Ω resistor, applies a gain of 50, and provides a ground referenced output. The I to V gain is therefore 500mV/mA, which makes sense given the 10mA full-scale heater current and the 5V output swing of the LT6100. The LT1991's task is the opposite, applying precision attenuation instead of gain. The full-scale voltage of the heater is a total of 40V (±20), beyond which the life of the heater may be reduced

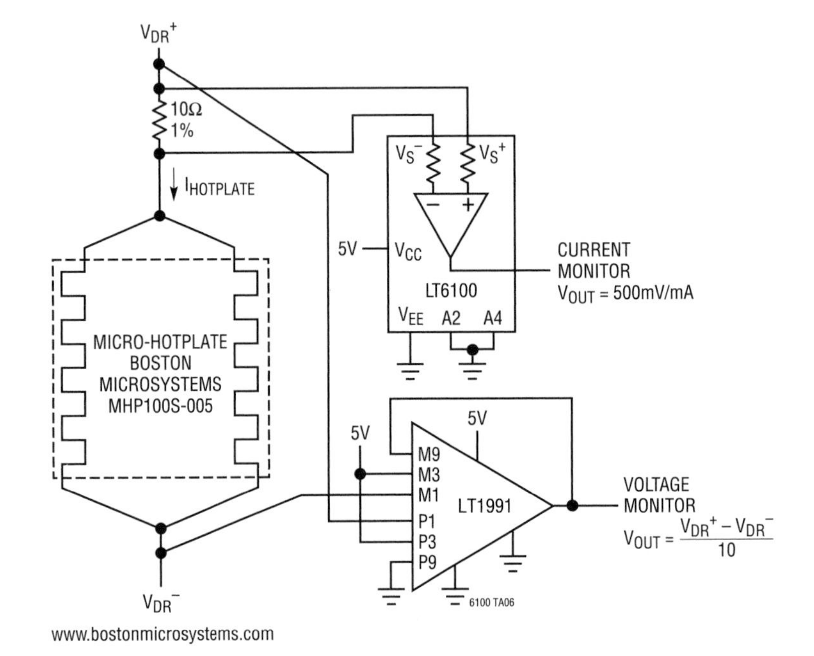

Figure 40.62 •

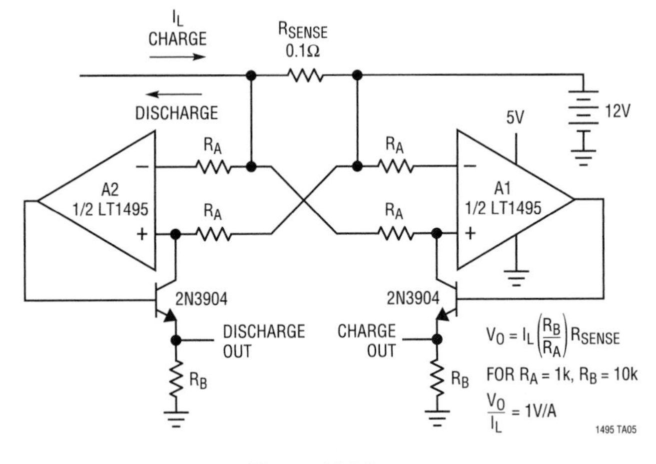

Figure 40.63 •

in some atmospheres. The LT1991 is set up for an attenuation factor of 10, so that the 40V full-scale differential drive becomes 4V ground referenced at the LT1991 output. In both cases, the voltages are easily read by 0V–5V PC I/O cards and the system readily software controlled.

Battery current monitor (Figure 40.63)

One LT1495 dual op amp package can be used to establish separate charge and discharge current monitoring outputs. The LT1495 features Over-the-Top operation allowing the battery potential to be as high as 36V with only a 5V amplifier supply voltage.

Bidirectional battery-current monitor (Figure 40.64)

This circuit provides the capability of monitoring current in either direction through the sense resistor. To allow negative outputs to represent charging current, V_{EE} is connected to a small negative supply. In single-supply operation (V_{EE} at ground), the output range may be offset upwards by applying a positive reference level to V_{BIAS} (1.25V for example). C3 may be used to form a filter in conjunction with the output resistance (R_{OUT}) of the part. This solution offers excellent precision (very low V_{OS}) and a fixed nominal gain of 8.

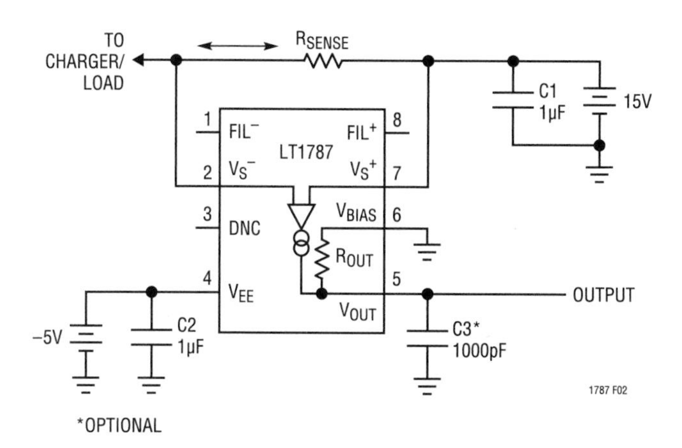

Figure 40.64 •

"Classic" positive supply rail current sense (Figure 40.65)

This circuit uses generic devices to assemble a function similar to an LTC6101. A rail-to-rail input type op amp is required since input voltages are right at the upper rail. The circuit shown here is capable of monitoring up to 44V applications. Besides the complication of extra parts, the V_{OS} performance of op amps at the supply is generally not factory trimmed, thus less accurate than other solutions.

The finite current gain of the bipolar transistor is a small source of gain error.

High side current sense and fuse monitor (Figure 40.66)

The LT6100 can be used as a combination current sensor and fuse monitor. This part includes on-chip output buffering and was designed to operate with the low supply voltage (≥2.7V), typical of vehicle data acquisition systems, while the sense inputs monitor signals at the higher battery bus potential. The LT6100 inputs are tolerant of large input differentials, thus allowing the blown-fuse operating condition (this would be detected by an output full-scale indication). The LT6100 can also be powered down while maintaining high impedance sense inputs, drawing less than 1μA max from the battery bus.

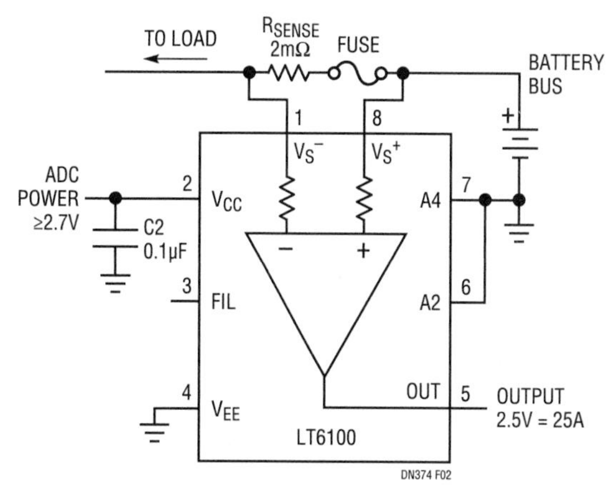

Figure 40.66 •

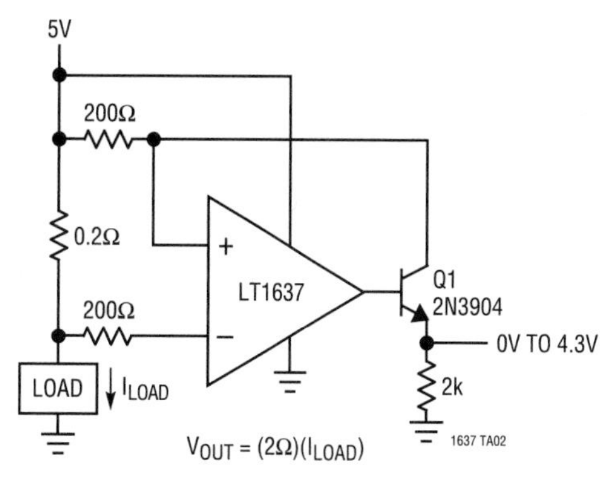

Figure 40.65 •

Gain of 50 current sense (Figure 40.67)

The LT6100 is configured for a gain of 50 by grounding both A2 and A4. This is one of the simplest current sensing amplifier circuits where only a sense resistor is required.

Dual LTC6101s allow high-low current ranging (Figure 40.68)

Using two current sense amplifiers with two values of sense resistors is an easy method of sensing current over a wide range. In this circuit the sensitivity and resolution of measurement is 10 times greater with low currents, less than 1.2A, than with higher currents. A comparator detects higher current flow, up to 10A, and switches sensing over to the high current circuitry.

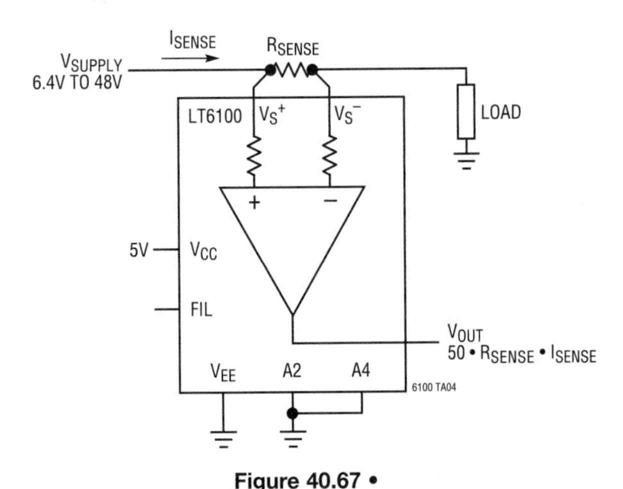

Figure 40.67 •

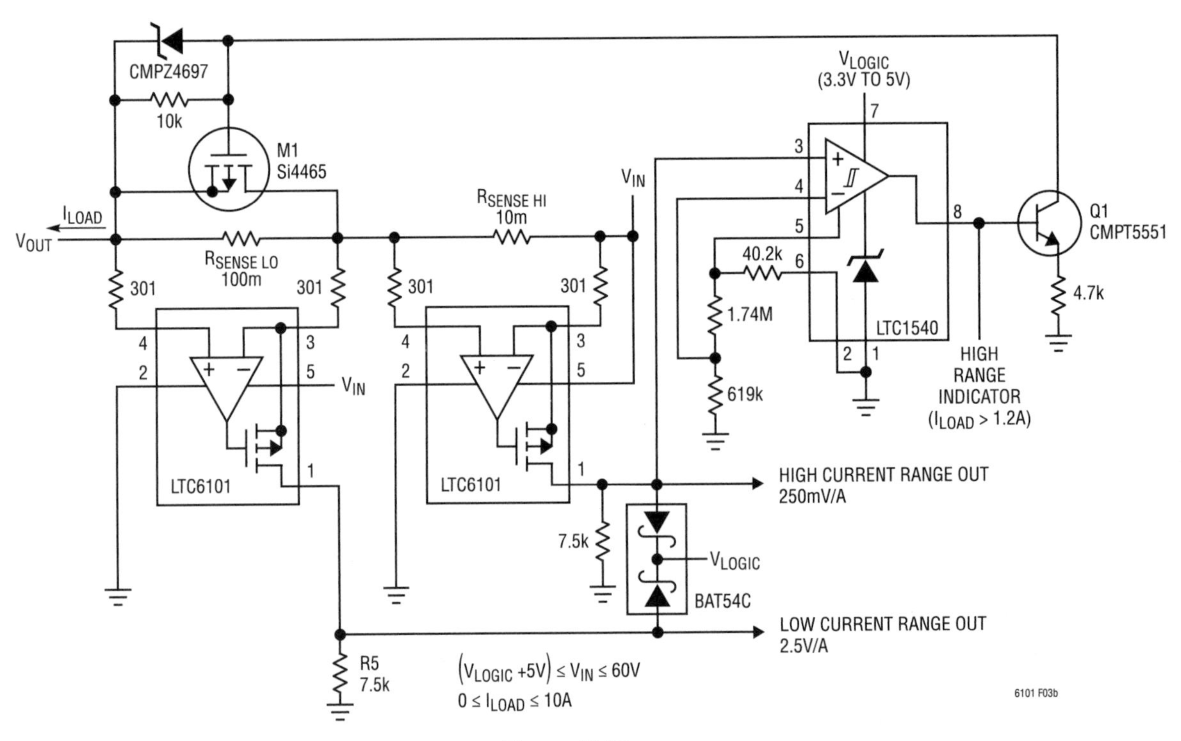

Figure 40.68 •

Two terminal current regulator (Figure 40.69)

The LT1635 combines an op amp with a 200mV reference. Scaling this reference voltage to a potential across resistor R3 forces a controlled amount of current to flow from the +terminal to the −terminal. Power is taken from the loop.

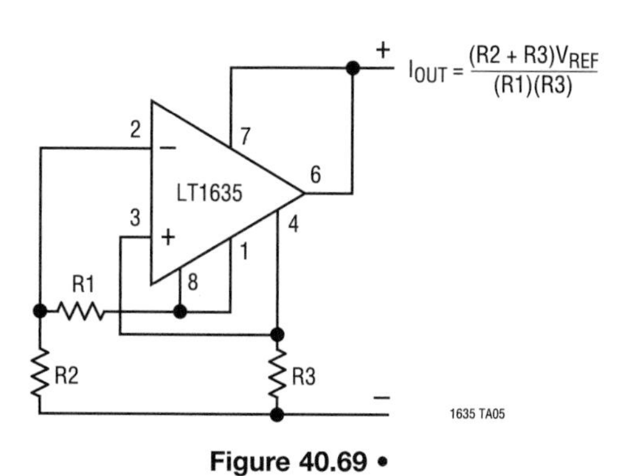

Figure 40.69 •

High side power supply current sense (Figure 40.70)

The low offset error of the LTC6800 allows for unusually low sense resistance while retaining accuracy.

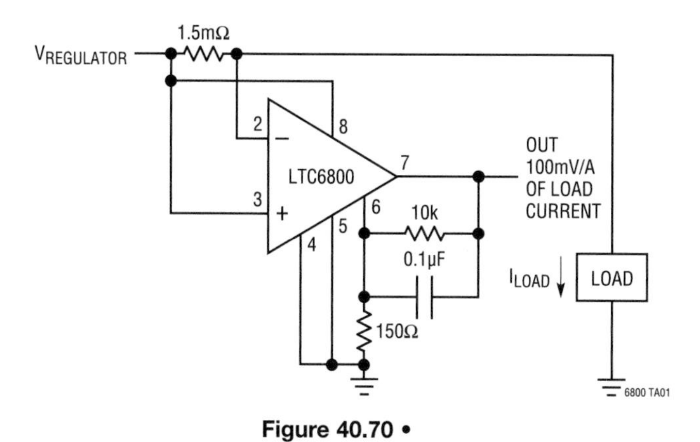

Figure 40.70 •

0nA to 200nA current meter (Figure 40.71)

A floating amplifier circuit converts a full-scale 200nA flowing in the direction indicated at the inputs to 2V at the output of the LT1495. This voltage is converted to a current to drive a 200μA meter movement. By floating the power to the circuit with batteries, any voltage potential at the inputs are handled. The LT1495 is a micropower op amp so the quiescent current drain from the batteries is very low and thus no on/off switch is required.

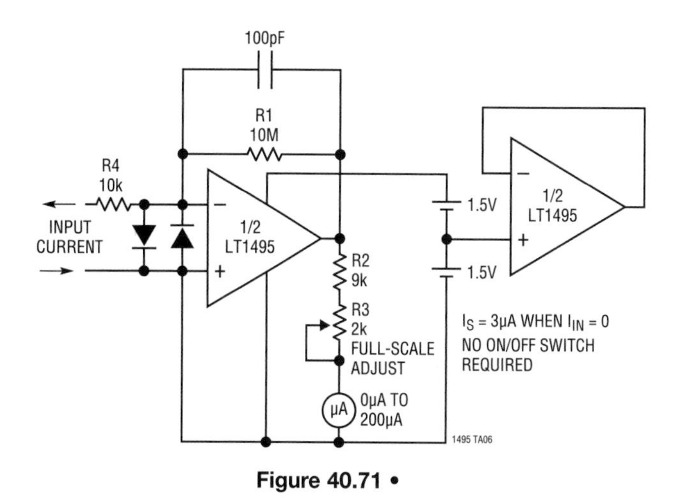

Figure 40.71 •

Over-the-Top current sense (Figure 40.72)

This circuit is a variation on the "classic" high side circuit, but takes advantage of Over-the-Top input capability to separately supply the IC from a low voltage rail. This provides a measure of fault protection to downstream circuitry by virtue of the limited output swing set by the low voltage supply. The disadvantage is V_{OS} in the Over-the-Top mode is generally inferior to other modes, thus less accurate. The finite current gain of the bipolar transistor is a source of small gain error.

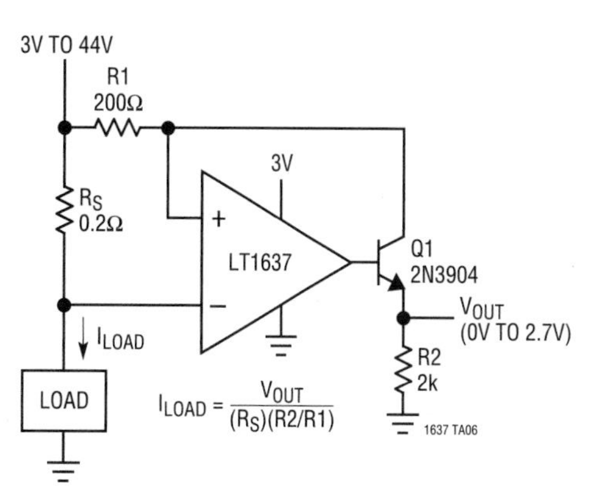

Figure 40.72 •

Conventional H-bridge current monitor (Figure 40.73)

Many of the newer electric drive functions, such as steering assist, are bidirectional in nature. These functions are generally driven by H-bridge MOSFET arrays using pulse- width modulation (PWM) methods to vary the commanded torque. In these systems, there are two main purposes for current monitoring. One is to monitor the current in the load, to track its performance against the desired command (i.e., closed-loop servo law), and another is for fault detection and protection features.

A common monitoring approach in these systems is to amplify the voltage on a "flying" sense resistor; as shown. Unfortunately, several potentially hazardous fault scenarios go undetected, such as a simple short to ground at a motor terminal. Another complication is the noise introduced by the PWM activity. While the PWM noise may be filtered for purposes of the servo law, information useful for protection becomes obscured. The best solution is to simply provide two circuits that individually protect each half-bridge and report the bidirectional load current. In some cases, a smart MOSFET bridge driver may already include sense resistors and offer the protection features needed. In these situations, the best solution is the one that derives the load information with the least additional circuitry.

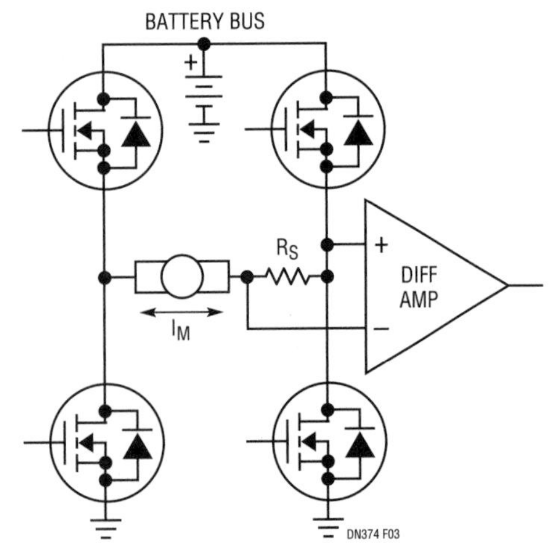

Figure 40.73 •

Single-supply 2.5V bidirectional operation with external voltage reference and I/V converter (Figure 40.74)

The LT1787's output is buffered by an LT1495 rail-to-rail op amp configured as an I/V converter. This configuration is ideal for monitoring very low voltage supplies. The LT1787's V_{OUT} pin is held equal to the reference voltage appearing at the op amp's non-inverting input. This allows one to monitor supply voltages as low as 2.5V. The op amp's output may swing from ground to its positive supply voltage. The low impedance output of the op amp may drive following circuitry more effectively than the high output impedance of the LT1787. The I/V converter configuration also works well with split supply voltages.

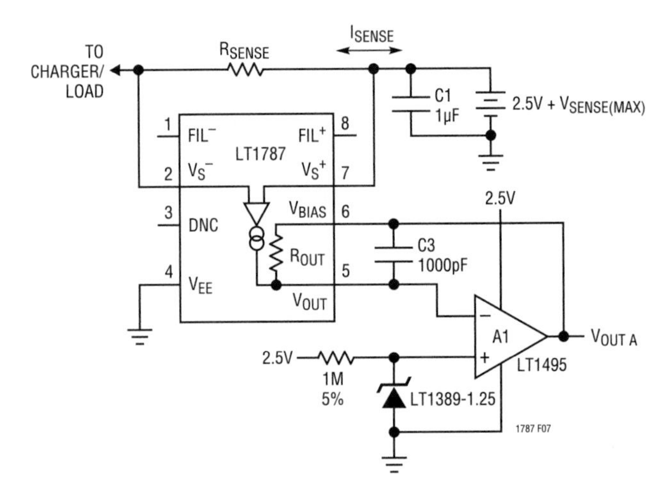

Figure 40.74 •

Battery current monitor (Figure 40.75)

One LT1495 dual op amp package can be used to establish separate charge and discharge current monitoring outputs. The LT1495 features Over-the-Top operation allowing the battery potential to be as high as 36V with only a 5V amplifier supply voltage.

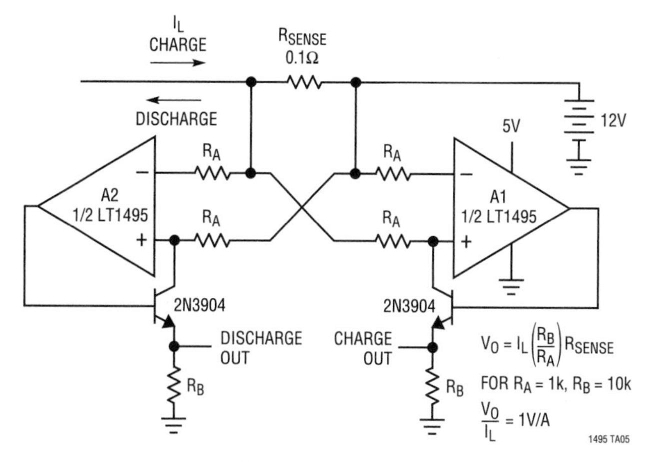

Figure 40.75 •

Fast current sense with alarm (Figure 40.76)

The LT1995 is shown as a simple unity-gain difference amplifier. When biased with split supplies the input current can flow in either direction providing an output voltage of 100mV per Amp from the voltage across the 100mΩ sense resistor With 32MHz of bandwidth and 1000V/μs slew rate the response of this sense amplifier is fast. Adding a simple comparator with a built in reference voltage circuit such as the LT6700-3 can be used to generate an over current flag. With the 400mV reference the flag occurs at 4A.

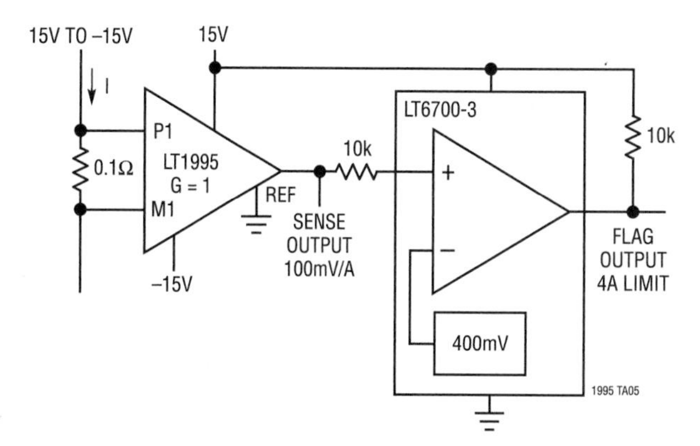

Figure 40.76 •

Positive supply rail current sense (Figure 40.77)

This is a configuration similar to an LT6100 implemented with generic components. A rail-to-rail or Over-the-Top input op amp type is required (for the first section). The first section is a variation on the classic high side where the P-MOSFET provides an accurate output current into R2 (compared to a BJT). The second section is a buffer to allow driving ADC ports, etc., and could be configured with gain if needed. As shown, this circuit can handle up to 36V operation. Small-signal range is limited by V_{OL} in single-supply operation.

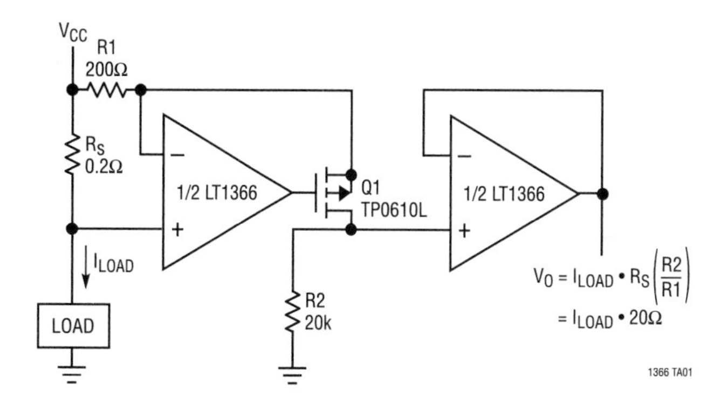

Figure 40.77 •

LT6100 load current monitor (Figure 40.78)

This is the basic LT6100 circuit configuration. The internal circuitry, including an output buffer, typically operates from a low voltage supply, such as the 3V shown. The monitored supply can range anywhere from V_{CC} + 1.4V up to 48V. The A2 and A4 pins can be strapped various ways to provide a wide range of internally fixed gains. The input leads become very Hi-Z when V_{CC} is powered down, so as not to drain batteries for example. Access to an internal signal node (Pin 3) provides an option to include a filtering function with one added capacitor. Small-signal range is limited by V_{OL} in single-supply operation.

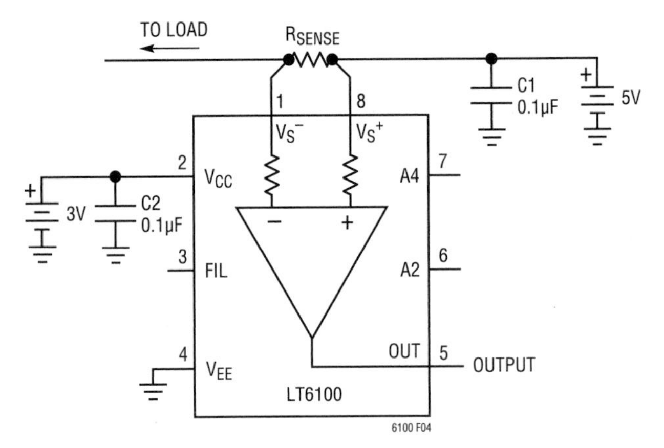

Figure 40.78 •

1A voltage-controlled current sink (Figure 40.79)

This is a simple controlled current sink, where the op amp drives the N-MOSFET gate to develop a match between the 1Ω sense resistor drop and the V_{IN} current command. Since the common mode voltage seen by the op amp is near ground potential, a "single-supply" or rail-to-rail type is required in this application.

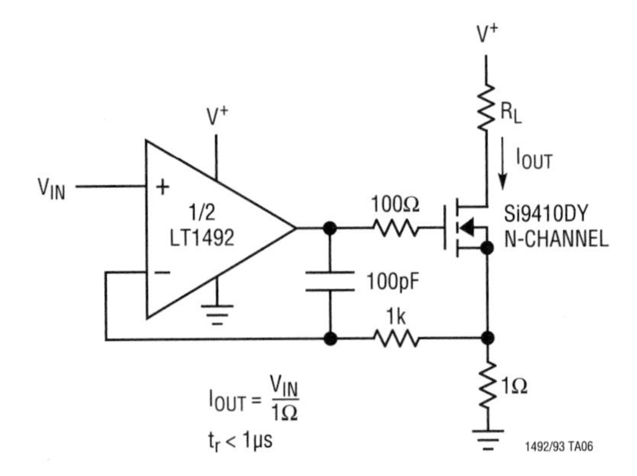

Figure 40.79 •

LTC6101 supply current included as load in measurement (Figure 40.80)

This is the basic LTC6101 high side sensing supply-monitor configuration, where the supply current drawn by the IC is included in the readout signal. This configuration is useful when the IC current may not be negligible in terms of overall current draw, such as in low power battery-powered applications. R_{SENSE} should be selected to limit voltage drop to <500mV for best linearity. If it is desirable not to include the IC current in the readout, as in load monitoring, Pin 5 may be connected directly to V^+ instead of the load. Gain accuracy of this circuit is limited only by the precision of the resistors selected by the user.

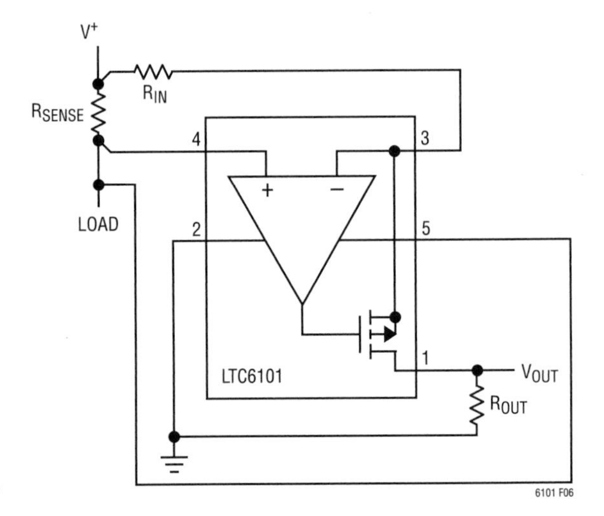

Figure 40.80 •

V^+ powered separately from load supply (Figure 40.81)

The inputs of the LTC6101 can function from 1.4V above the device positive supply to 48V DC. In this circuit the current flow in the high voltage rail is directly translated to a 0V to 3V range.

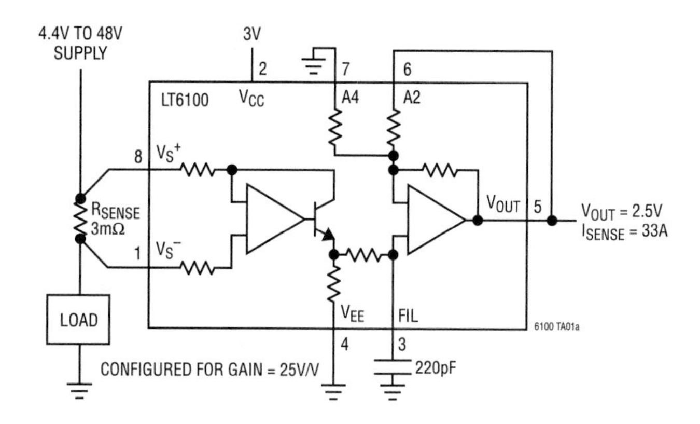

Figure 40.81 •

Simple high side current sense using the LTC6101 (Figure 40.82)

This is a basic high side current monitor using the LTC6101. The selection of R_{IN} and R_{OUT} establishes the desired gain of this circuit, powered directly from the battery bus. The current output of the LTC6101 allows it to be located remotely to R_{OUT}. Thus, the amplifier can be placed directly at the shunt, while R_{OUT} is placed near the monitoring electronics without ground drop errors. This circuit has a fast 1μs response time that makes it ideal for providing MOSFET load switch protection. The switch element may be the high side type connected between the sense resistor and the load, a low side type between the load and ground or an H-bridge. The circuit is programmable to produce up to 1mA of full-scale output current into R_{OUT}, yet draws a mere 250μA supply current when the load is off.

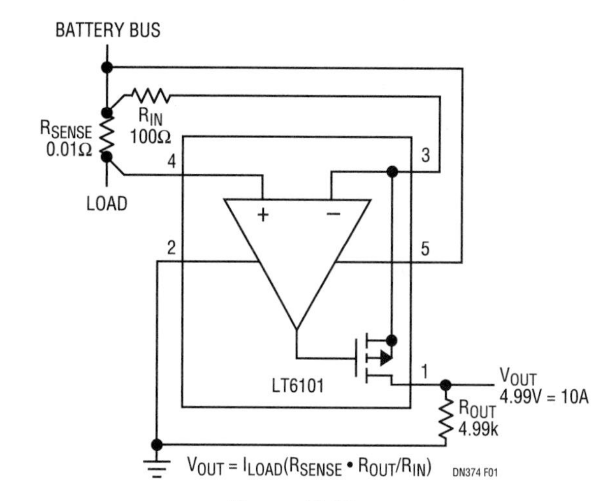

Figure 40.82 •

"Classic" high precision low side current sense (Figure 40.83)

This configuration is basically a standard noninverting amplifier. The op amp used must support common mode operation at the lower rail and the use of a zero-drift type (as shown) provides excellent precision. The output of this circuit is referenced to the lower Kelvin contact, which could be ground in a single-supply application. Small-signal range is limited by V_{OL} for single-supply designs. Scaling accuracy is set by the quality of the user-selected resistors.

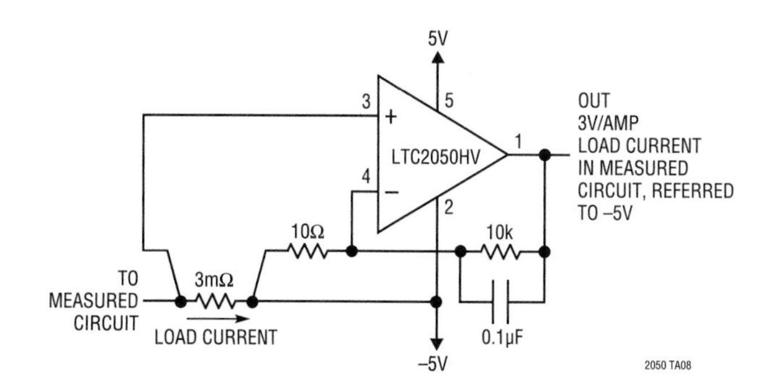

Figure 40.83 •

Level shifting

Quite often it is required to sense current flow in a supply rail that is a much higher voltage potential than the supply voltage for the system electronics. Current sense circuits with high voltage capability are useful to translate information to lower voltage signals for processing.

Over-the-Top current sense (Figure 40.84)

This circuit is a variation on the "classic" high side circuit, but takes advantage of Over-the-Top input capability to separately supply the IC from a low voltage rail. This provides a measure of fault protection to downstream circuitry by virtue of the limited output swing set by the low voltage supply. The disadvantage is V_{OS} in the Over-the-Top mode is generally inferior to other modes, thus less accurate. The finite current gain of the bipolar transistor is a source of small gain error.

V$^+$ powered separately from load supply (Figure 40.85)

The inputs of the LTC6101 can function from 1.4V above the device positive supply to 48V DC. In this circuit the current flow in the high voltage rail is directly translated to a 0V to 3V range.

Voltage translator (Figure 40.86)

This is a convenient usage of the LTC6101 current sense amplifier as a high voltage level translator. Differential voltage signals riding on top of a high common mode voltage (up to 105V with the LTC6101HV) get converted to a current, through R_{IN}, and then scaled down to a ground referenced voltage across R_{OUT}.

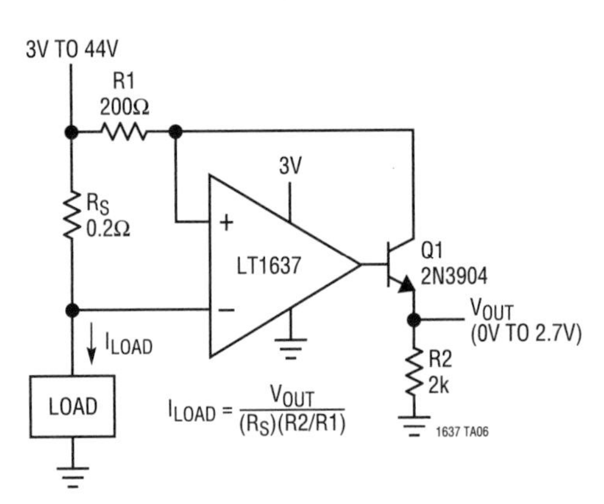

Figure 40.84 •

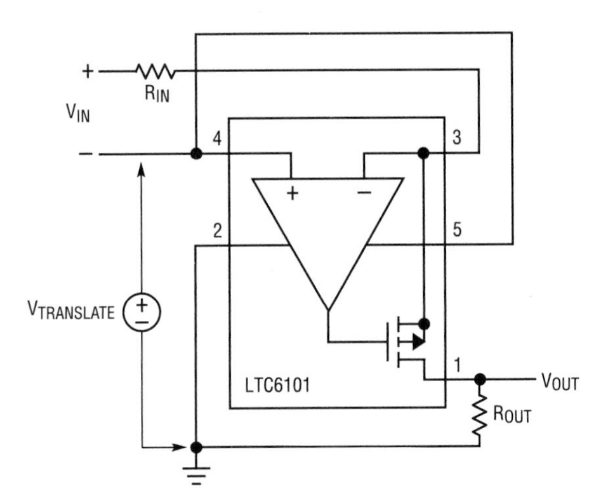

Figure 40.86 •

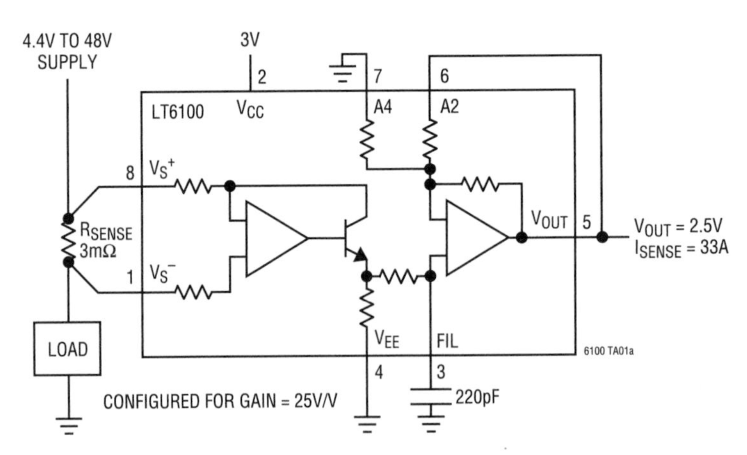

Figure 40.85 •

Low power, bidirectional 60V precision high side current sense (Figure 40.87)

Using a very precise zero-drift amplifier as a pre-amp allows for the use of a very small sense resistor in a high voltage supply line. A floating power supply regulates the voltage across the pre-amplifier on any voltage rail up to the 60V limit of the LT1787HV circuit. Overall gain of this circuit is 1000. A 1mA change in current in either direction through the 10mΩ sense resistor will produce a 10mV change in the output voltage.

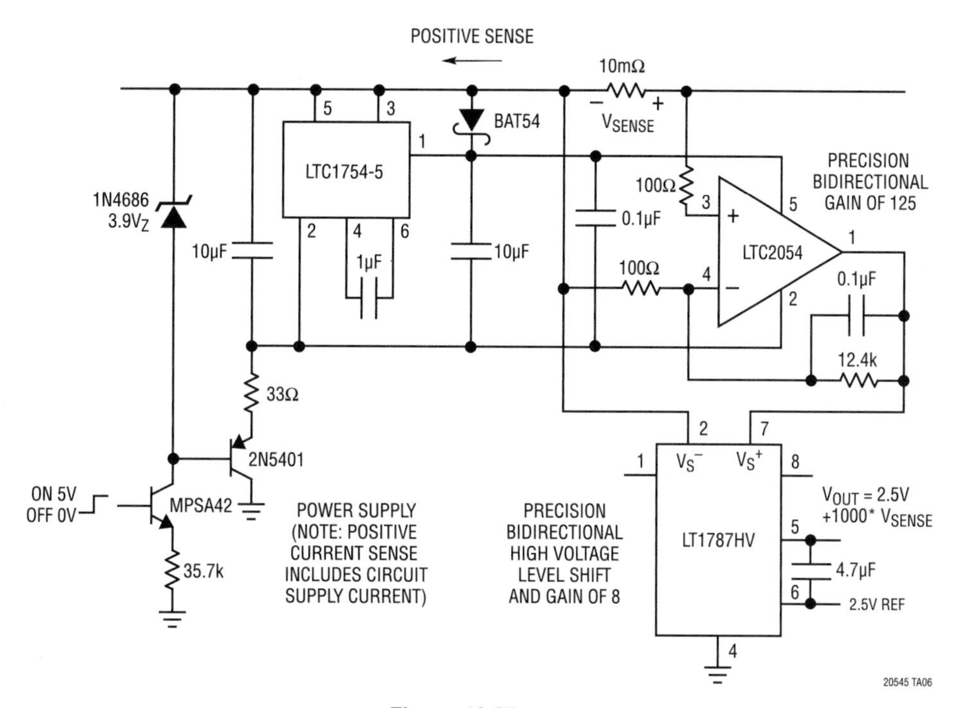

Figure 40.87 •

High voltage

Monitoring current flow in a high voltage line often requires floating the supply of the measuring circuits up near the high voltage potentials. Level shifting and isolation components are then often used to develop a lower output voltage indication.

Over-the-Top current sense (Figure 40.88)

This circuit is a variation on the "classic" high side circuit, but takes advantage of Over-the-Top input capability to separately supply the IC from a low voltage rail. This provides a measure of fault protection to downstream circuitry by virtue of the limited output swing set by the low voltage supply. The disadvantage is V_{OS} in the Over-the-Top mode is generally inferior to other modes, thus less accurate. The finite current gain of the bipolar transistor is a source of small gain error.

Measuring bias current into an avalanche photo diode (APD) using an instrumentation amplifier (Figures 40.89a and 40.89b)

The upper circuit (a) uses an instrumentation amplifier (IA) powered by a separate rail (>1V above V_{IN}) to measure across the 1kΩ current shunt. The lower figure (b) is similar but derives its power supply from the APD bias line. The limitation of these circuits is the 35V maximum APD voltage, whereas some APDs may require 90V or more. In the single-supply configuration shown, there is also a dynamic range limitation due to V_{OL} to consider. The advantage of this approach is the high accuracy that is available in an IA.

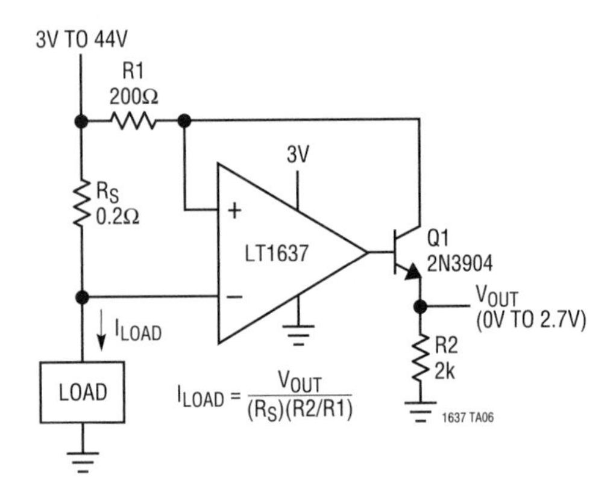

Figure 40.88 •

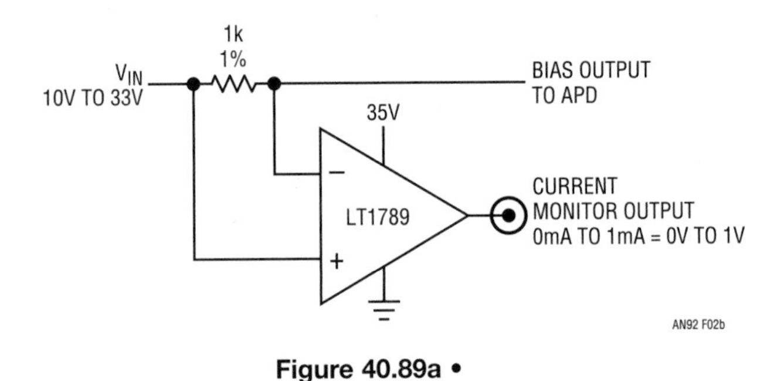

Figure 40.89a •

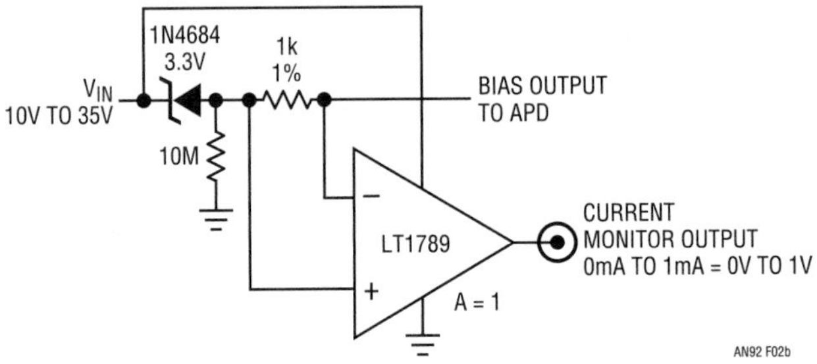

Figure 40.89b •

Simple 500V current monitor (Figure 40.90)

Adding two external MOSFETs to hold off the voltage allows the LTC6101 to connect to very high potentials and monitor the current flow. The output current from the LTC6101, which is proportional to the sensed input voltage, flows through M1 to create a ground referenced output voltage.

48V supply current monitor with isolated output and 105V survivability (Figure 40.91)

The HV version of the LTC6101 can operate with a total supply voltage of 105V. Current flow in high supply voltage rails can be monitored directly or in an isolated fashion as shown in this circuit. The gain of the circuit and the level of output current from the LTC6101 depends on the particular opto-isolator used.

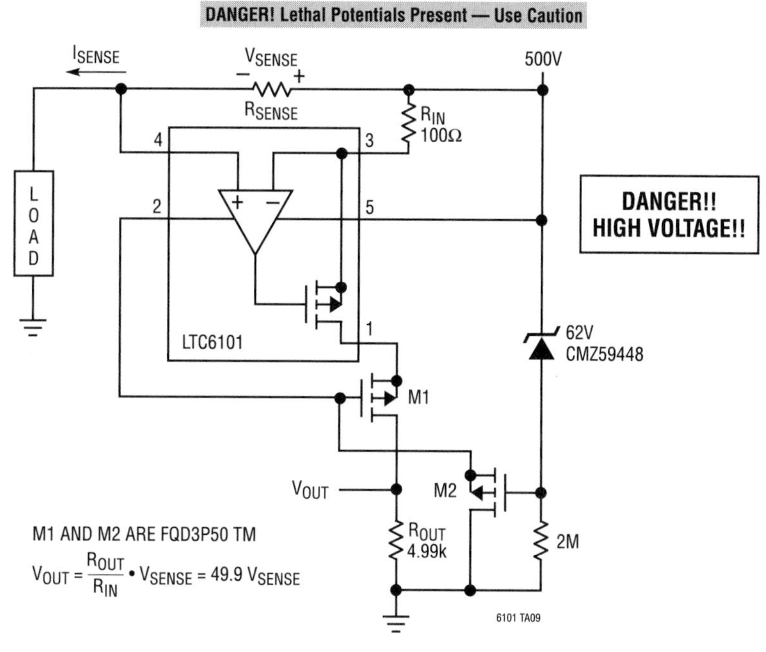

Figure 40.90 •

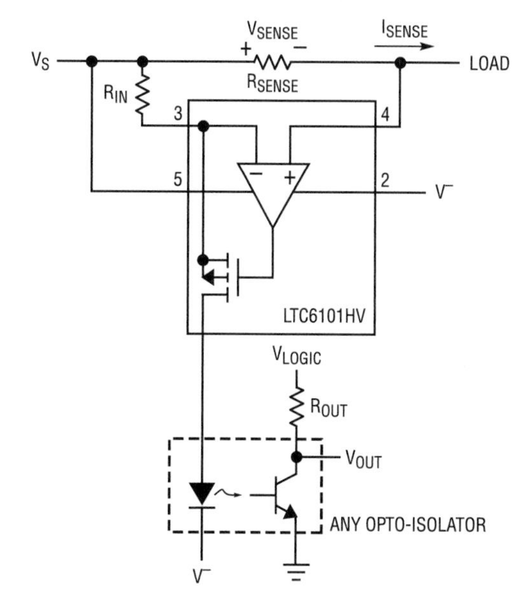

Figure 40.91 •

Low power, bidirectional 60V precision high side current sense (Figure 40.92)

Using a very precise zero-drift amplifier as a pre-amp allows for the use of a very small sense resistor in a high voltage supply line. A floating power supply regulates the voltage across the pre-amplifier on any voltage rail up to the 60V limit of the LT1787HV circuit. Overall gain of this circuit is 1000. A 1mA change in current in either direction through the 10mΩ sense resistor will produce a 10mV change in the output voltage.

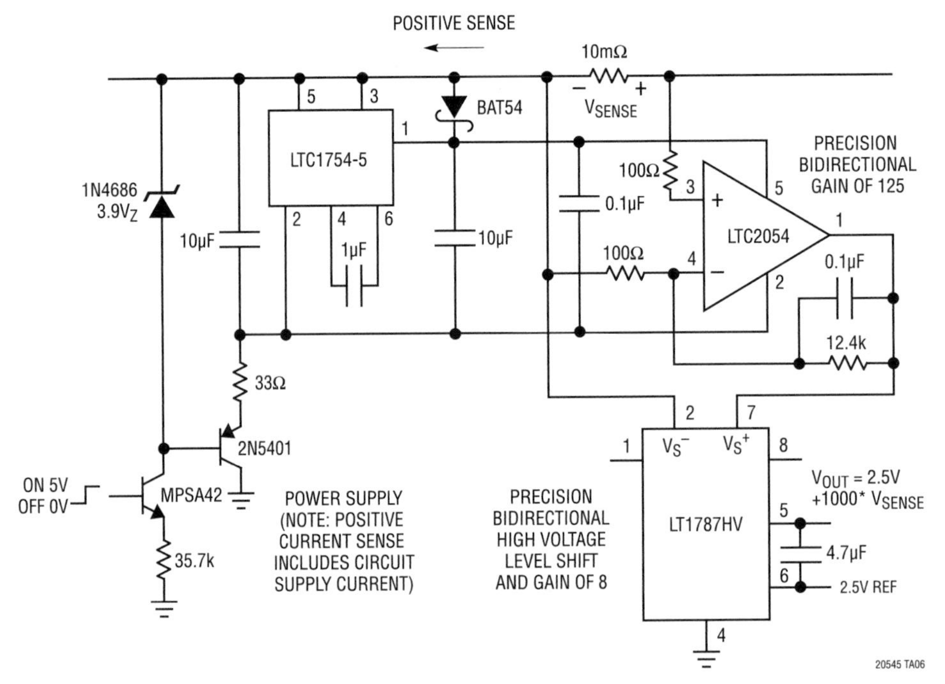

Figure 40.92 •

High voltage current and temperature monitoring (Figure 40.93)

Combining an LTC2990 ADC converter with a high voltage LTC6102HV current sense amplifier allows the measurement of very high voltage rails, up to 104V, and very high current loads. The current sense amplifier outputs a ground referenced voltage proportional to the load current and is measured as a single ended input by the ADC. A divided down representation of the supply voltage is a second input. An external NPN transistor serves as a remote temperature sensor.

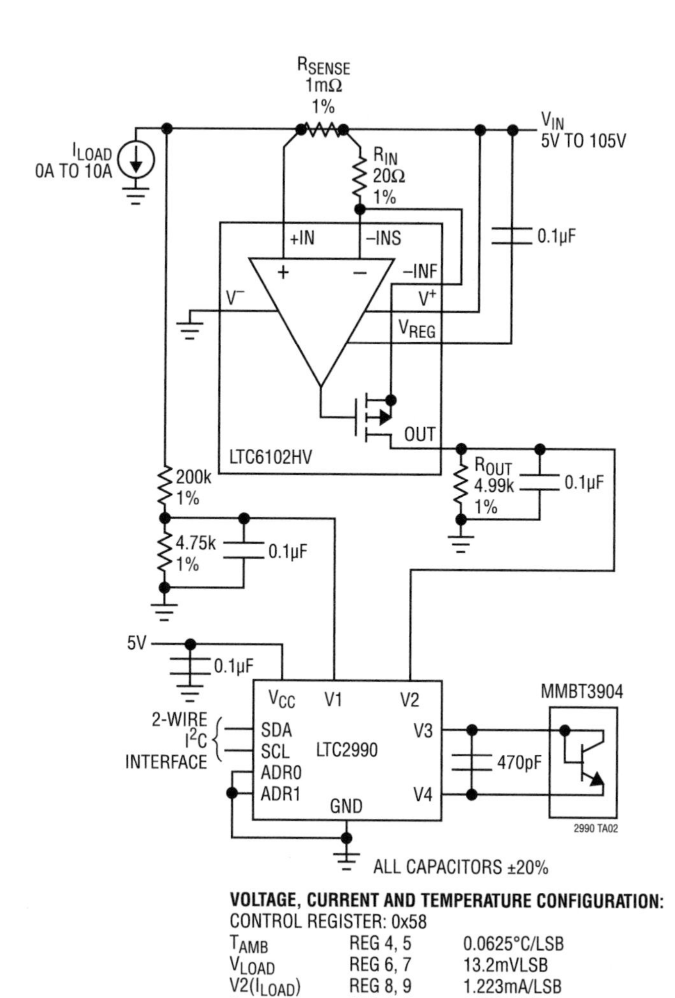

Figure 40.93 •

Low voltage

Single-supply 2.5V bidirectional operation with external voltage reference and I/V converter (Figure 40.94)

The LT1787's output is buffered by an LT1495 rail-to-rail op amp configured as an I/V converter. This configuration is ideal for monitoring very low voltage supplies. The LT1787's V_{OUT} pin is held equal to the reference voltage appearing at the op amp's noninverting input. This allows one to monitor supply voltages as low as 2.5V. The op amp's output may swing from ground to its positive supply voltage. The low impedance output of the op amp may drive following circuitry more effectively than the high output impedance of the LT1787. The I/V converter configuration also works well with split supply voltages.

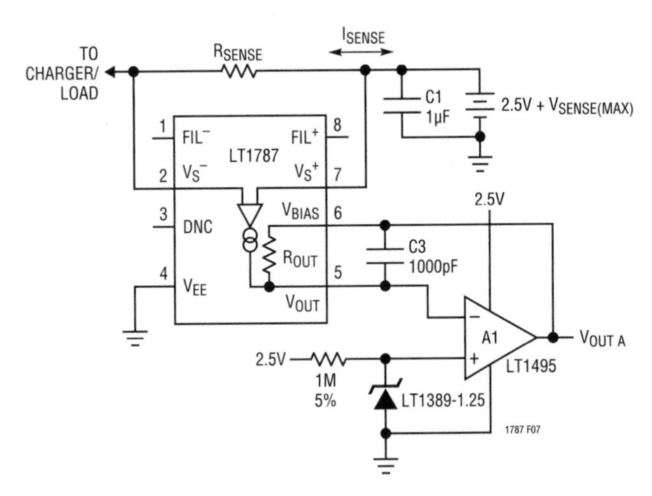

Figure 40.94 •

1.25V electronic circuit breaker (Figure 40.95)

The LTC4213 provides protection and automatic circuit breaker action by sensing drain-to-source voltage drop across the N-MOSFET The sense inputs have a rail-to-rail common mode range, so the circuit breaker can protect bus voltages from 0V up to 6V Logic signals flag a trip condition (with the READY output signal) and reinitialize the breaker (using the ON input). The ON input may also be used as a command in a "smart switch" application.

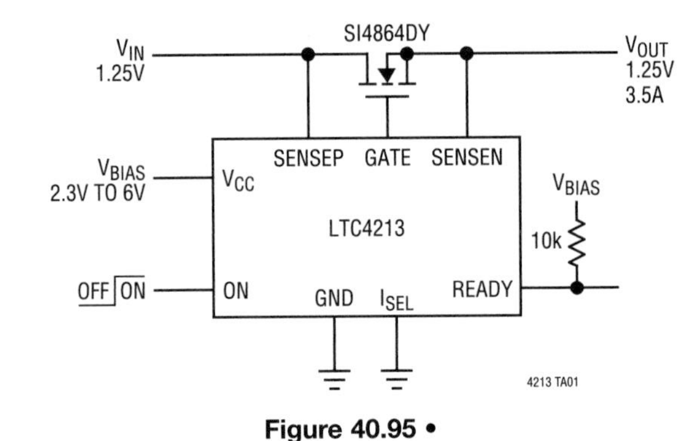

Figure 40.95 •

High current (100mA to Amps)

Sensing high currents accurately requires excellent control of the sensing resistance, which is typically a very small value to minimize losses, and the dynamic range of the measurement circuitry

Kelvin input connection preserves accuracy despite large load currents (Figure 40.96)

Kelvin connection of the −IN and +IN inputs to the sense resistor should be used in all but the lowest power applications. Solder connections and PC board interconnections that carry high current can cause significant error in measurement due to their relatively large resistances. By isolating the sense traces from the high current paths, this error can be reduced by orders of magnitude. A sense resistor with integrated Kelvin sense terminals will give the best results.

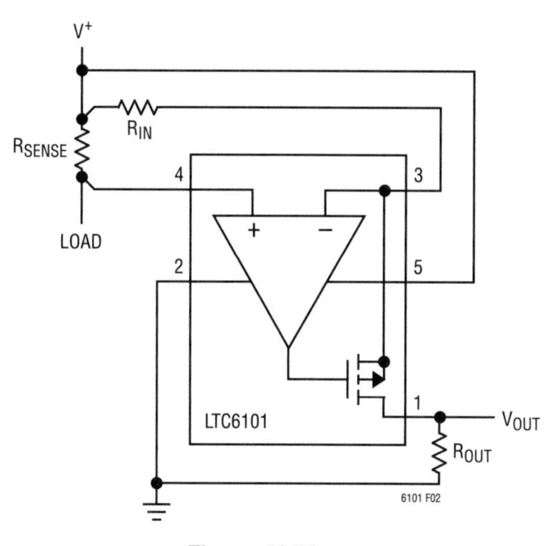

Figure 40.96 •

Shunt diode limits maximum input voltage to allow better low input resolution without over-ranging the LTC6101 (Figure 40.97)

If low sense currents must be resolved accurately in a system that has very wide dynamic range, more gain can be taken in the sense amplifier by using a smaller value for resistor R_{IN}. This can result in an operating current greater than the max current spec allowed unless the max current is limited in another way, such as with a Schottky diode across R_{SENSE}. This will reduce the high current measurement accuracy by limiting the result, while increasing the low current measurement resolution. This approach can be helpful in cases where an occasional large burst of current may be ignored.

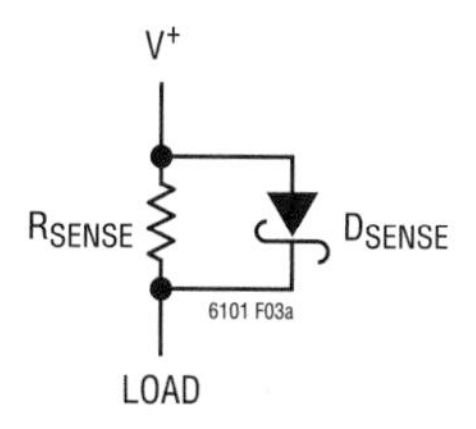

Figure 40.97 •

Kelvin sensing (Figure 40.98)

In any high current, >1A, application, Kelvin contacts to the sense resistor are important to maintain accuracy. This simple illustration from a battery charger application shows two voltage-sensing traces added to the pads of the current sense resistor. If the voltage is sensed with high impedance amplifier inputs, no IxR voltage drop errors are developed.

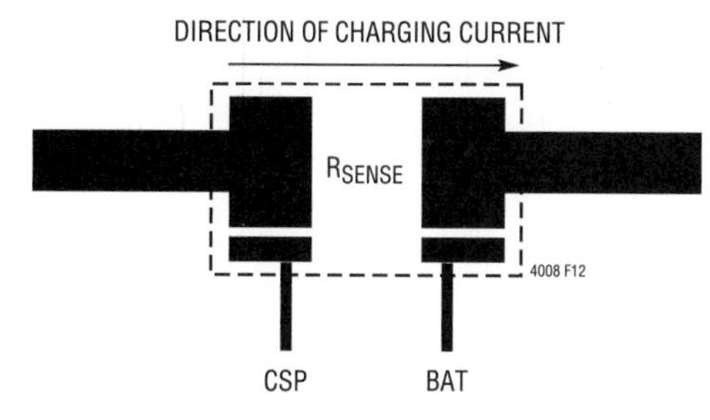

Figure 40.98 •

0A to 33A high side current monitor with filtering (Figure 40.99)

High current sensing on a high voltage supply rail is easily accomplished with the LT6100. The sense amplifier is biased from a low 3V supply and pin strapped to a gain of 25V/V to output a 2.5V full-scale reading of the current flow. A capacitor at the FIL pin to ground will filter out noise of the system (220pF produces a 12kHz lowpass corner frequency).

Single supply RMS current measurement (Figure 40.100)

The LT1966 is a true RMS-to-DC converter that takes a single-ended or differential input signal with rail-to-rail range. The output of a PCB mounted current sense transformer can be connected directly to the converter. Up to 75A of AC current is measurable without breaking the signal path from a power source to a load. The accurate operating range of the circuit is determined by the selection of the transformer termination resistor. All of the math is built in to the LTC1966 to provide a DC output voltage that is proportional to the true RMS value of the current. This is valuable in determining the power/energy consumption of AC-powered appliances.

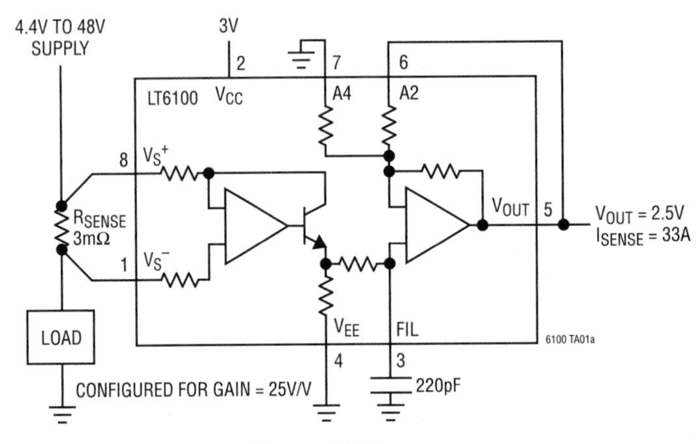

Figure 40.99 •

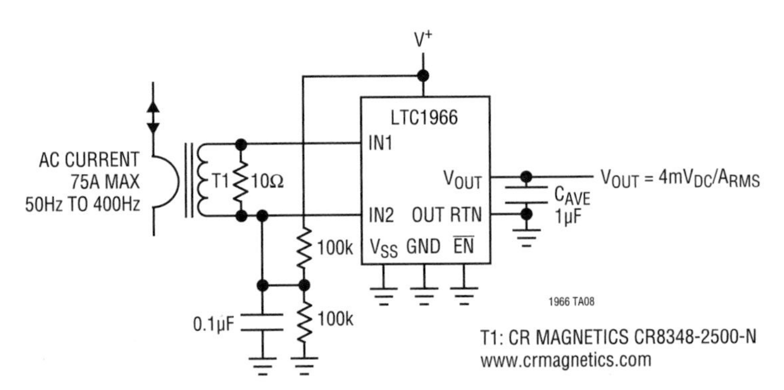

Figure 40.100 •

Dual LTC6101s allow high-low current ranging (Figure 40.101)

Using two current sense amplifiers with two values of sense resistors is an easy method of sensing current over a wide range. In this circuit the sensitivity and resolution of measurement is 10 times greater with low currents, less than 1.2A, than with higher currents. A comparator detects higher current flow, up to 10A, and switches sensing over to the high current circuitry.

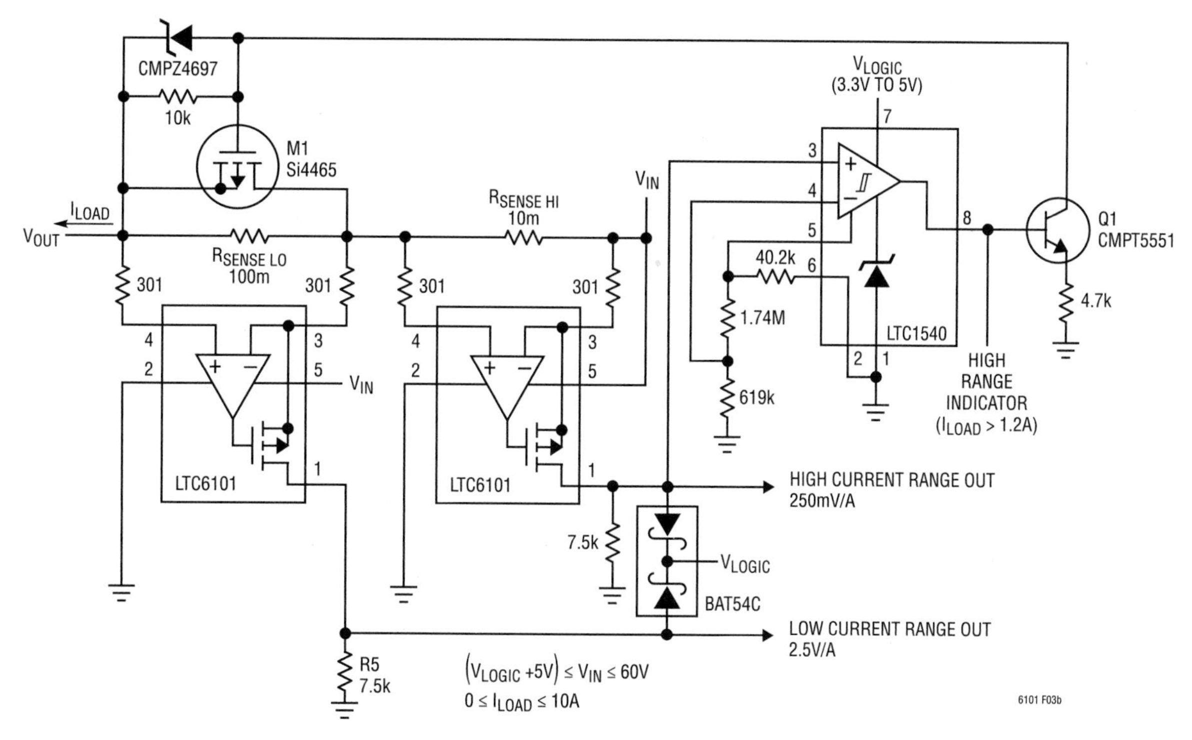

Figure 40.101 •

LDO load balancing (Figure 40.102)

As system design enhancements are made there is often the need to supply more current to a load than originally expected. A simple way to modify power amplifiers or voltage regulators, as shown here, is to parallel devices. When paralleling devices it is desired that each device shares the total load current equally. In this circuit two adjustable "slave" regulator output voltages are sensed and servo'ed to match the master regulator output voltage. The precise low offset voltage of the LTC6078 dual op amp (10μV) balances the load current provided by each regulator to within 1mA. This is achieved using a very small 10mΩ current sense resistor in series with each output. This sense resistor can be implemented with PCB copper traces or thin gauge wire.

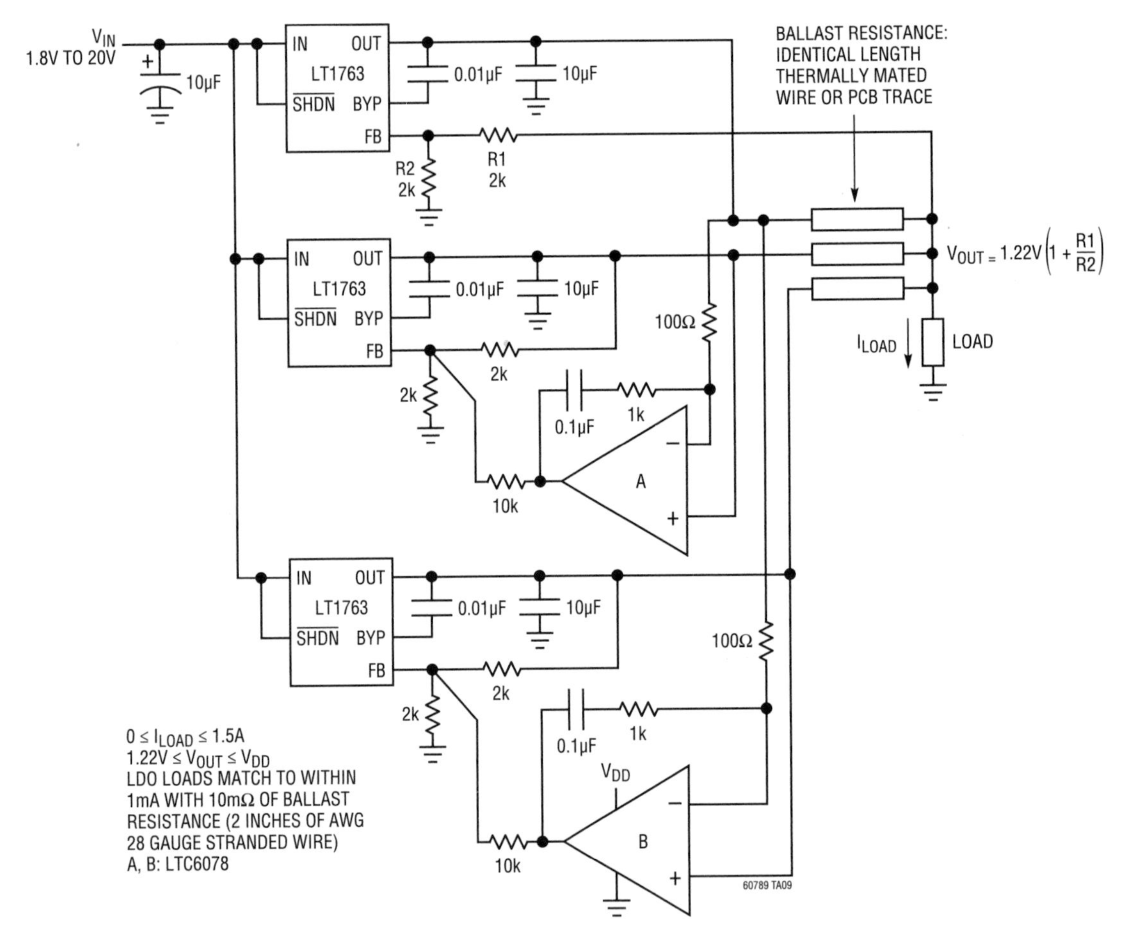

Figure 40.102 •

Sensing output current (Figure 40.103)

The LT1970 is a 500mA power amplifier with voltage programmable output current limit. Separate DC voltage inputs and an output current sensing resistor control the maximum sourcing and sinking current values. These control voltages could be provided by a D-to-A converter in a microprocessor controlled system. For closed loop control of the current to a load an LT1787 can monitor the output current. The LT1880 op amp provides scaling and level shifting of the voltage applied to an A-to-D converter for a 5mV/mA feedback signal.

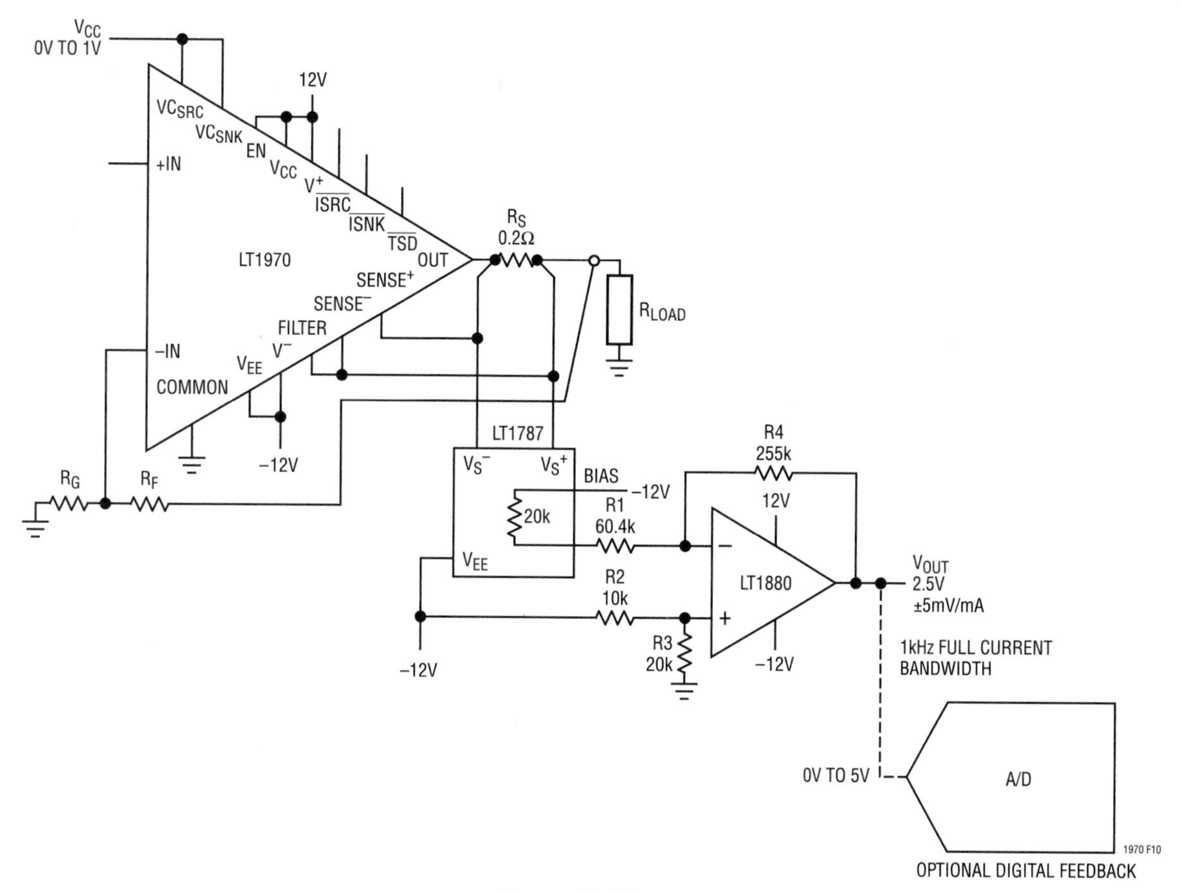

Figure 40.103 •

Using printed circuit sense resistance (Figure 40.104)

The outstanding LTC6102 precision allows the use of sense resistances fabricated with conventional printed circuit techniques. For "one ounce" copperclad, the trace resistance is approximately (L/W). 0.0005Ω and can carry about 4A per mm of trace width. The example below shows a practical 5A monitoring solution with both L and W set to 2.5mm. The resistance is subject to about +0.4%/°C temperature change and the geometric tolerances of the fabrication process, so this will not generally be for high accuracy work, but can be useful in various low cost protection and status monitoring functions.

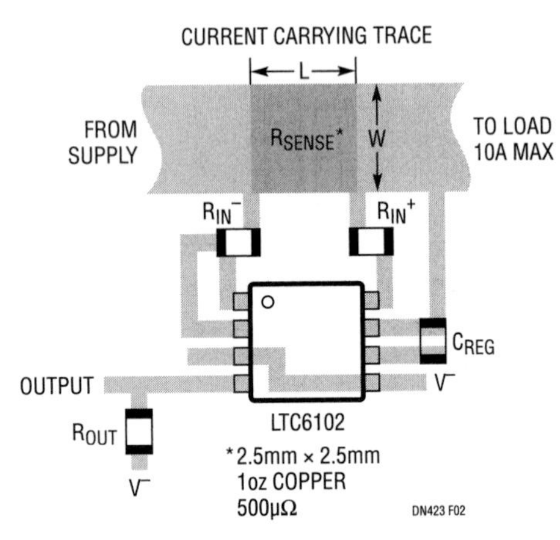

Figure 40.104 •

High voltage, 5A high side current sensing in small package (Figure 40.105)

The LT6106 is packaged in a small SOT-23 package but still operates over a wide supply range of 3V to 44V. Just two resistors set the gain (10 in circuit shown) and the output is a voltage referred to ground.

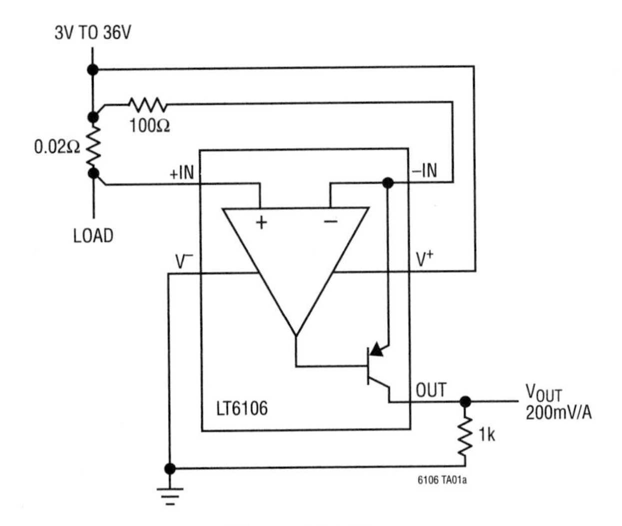

Figure 40.105 •

Low current (picoamps to milliamps)

For low current applications the easiest way to sense current is to use a large sense resistor. This however causes larger voltage drops in the line being sensed which may not be acceptable. Using a smaller sense resistor and taking gain in the sense amplifier stage is often a better approach. Low current implies high source impedance measurements which are subject approach. Low current implies high source impedance measurements which are subject to noise pickup and often require filtering of some sort.

Filtered gain of 20 current sense (Figure 40.106)

The LT6100 has pin strap connections to establish a variety of accurate gain settings without using external components. For this circuit grounding A2 and leaving A4 open set a gain of 20. Adding one external capacitor to the FIL pin creates a lowpass filter in the signal path. A capacitor of 1000pF as shown sets a filter corner frequency of 2.6KHz.

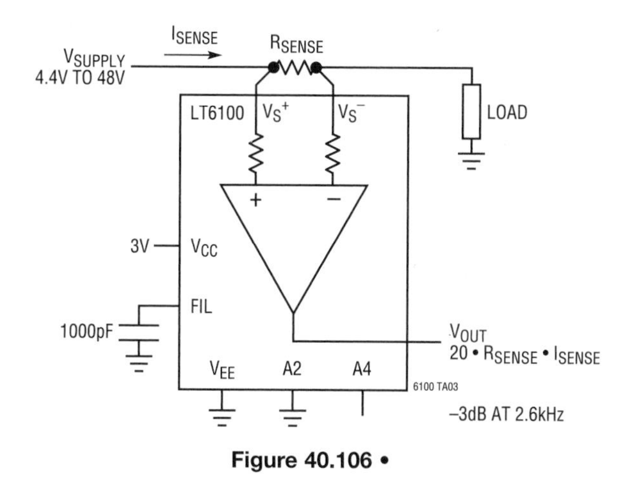

Figure 40.106 •

Gain of 50 current sense (Figure 40.107)

The LT6100 is configured for a gain of 50 by grounding both A2 and A4. This is one of the simplest current sensing amplifier circuits where only a sense resistor is required.

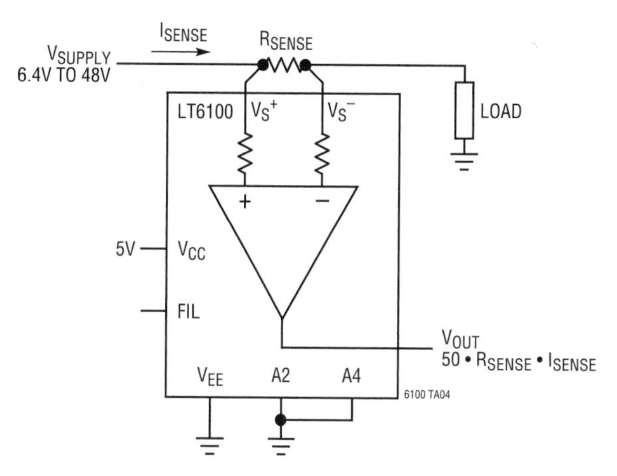

Figure 40.107 •

0nA to 200nA current meter (Figure 40.108)

A floating amplifier circuit converts a full-scale 200nA flowing in the direction indicated at the inputs to 2V at the output of the LT1495. This voltage is converted to a current to drive a 200μA meter movement. By floating the power to the circuit with batteries, any voltage potential at the inputs are handled. The LT1495 is a micropower op amp so the quiescent current drain from the batteries is very low and thus no on/off switch is required.

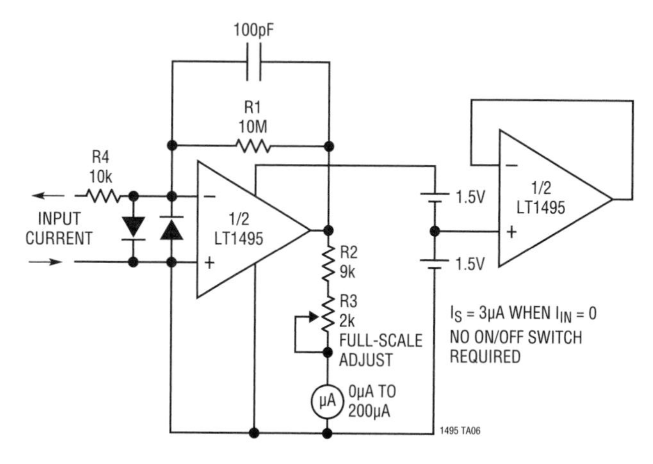

Figure 40.108 •

Lock–in amplifier technique permits 1% accurate APD current measurement over 100nA to 1mA range (Figure 40.109)

Avalanche Photodiodes, APDs, require a small amount of current from a high voltage supply. The current into the diode is an indication of optical signal strength and must be monitored very accurately. It is desirable to power all of the support circuitry from a single 5V supply.

This circuit utilizes AC carrier modulation techniques to meet APD current monitor requirements. It features 0.4% accuracy over the sensed current range, runs from a 5V supply and has the high noise rejection characteristics of carrier based "lock in" measurements.

The LTC1043 switch array is clocked by its internal oscillator. Oscillator frequency, set by the capacitor at Pin 16, is about 150Hz. S1 clocking biases Q1 via level shifter Q2. Q1 chops the DC voltage across the 1k current shunt, modulating it into a differential square wave signal which feeds A1 through 0.2μF AC coupling capacitors. A1's single-ended output biases demodulator S2, which presents a DC output to buffer amplifier A2. A2's output is the circuit output.

Switch S3 clocks a negative output charge pump which supplies the amplifier's V− pins, permitting output swing to (and below) zero volts. The 100k resistors at Q1 minimize its on-resistance error contribution and prevent destructive potentials from reaching A1 (and the 5V rail) if either 0.2μF capacitor fails. A2's gain of 1.1 corrects for the slight attenuation introduced by A1's input resistors. In practice, it may be desirable to derive the APD bias voltage regulator's feedback signal from the indicated point, eliminating the 1kΩ shunt resistor's voltage drop. Verifying accuracy involves loading the APD bias line with 100nA to 1mA and noting output agreement.

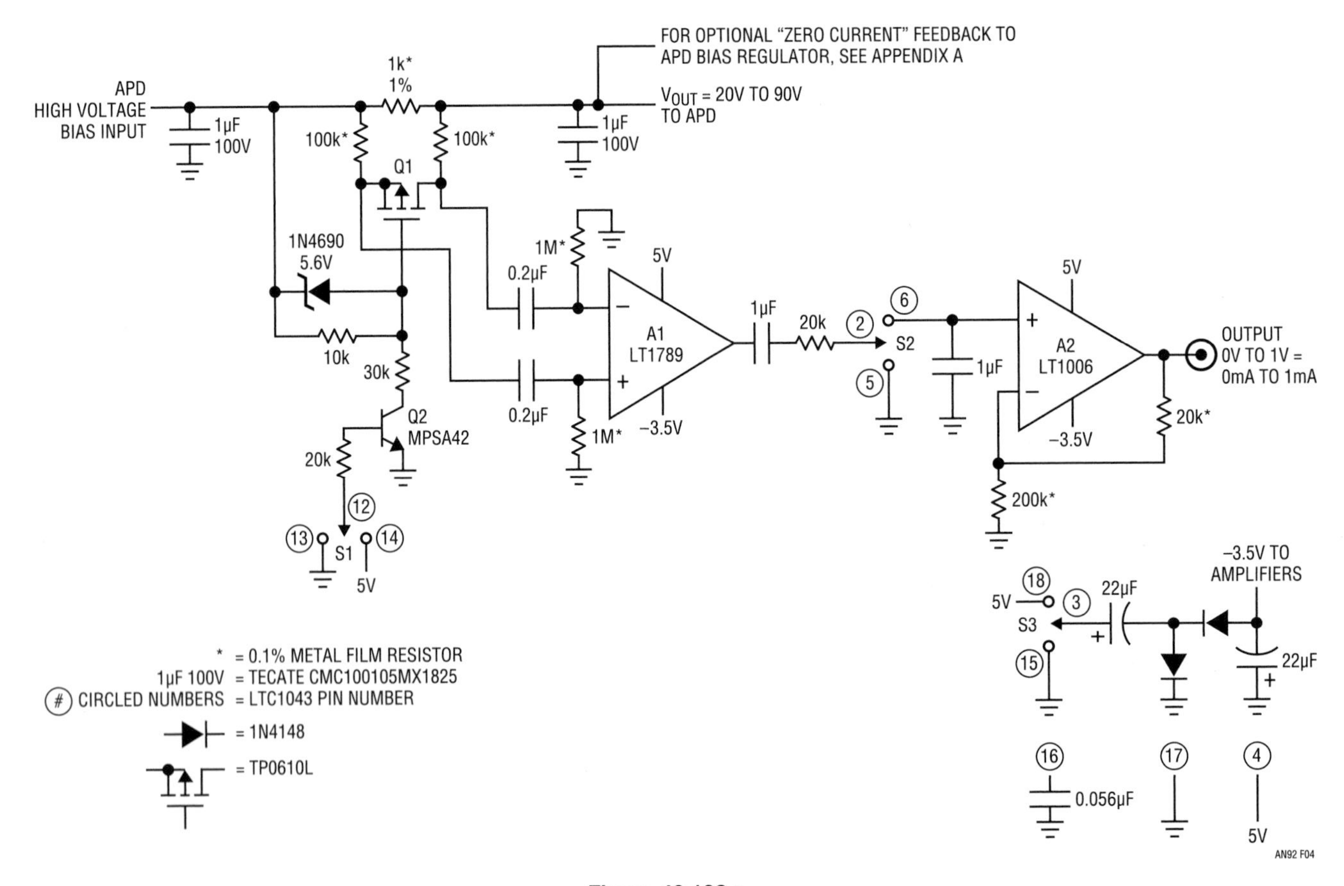

Figure 40.109 •

DC-coupled APD current monitor (Figure 40.110)

Avalanche Photodiodes, APDs, require a small amount of current from a high voltage supply. The current into the diode is an indication of optical signal strength and must be monitored very accurately. It is desirable to power all of the support circuitry from a single 5V supply.

This circuit's DC-coupled current monitor eliminates the previous circuit's trim but pulls more current from the APD bias supply. A1 floats, powered by the APD bias rail. The 15V Zener diode and current source Q2 ensure A1 never is exposed to destructive voltages. The 1kΩ current shunt's voltage drop sets A1's positive input potential. A1 balances its inputs by feedback controlling its negative input via Q1. As such, Q1's source voltage equals A1's positive input voltage and its drain current sets the voltage across its source resistor. Q1's drain current produces a voltage drop across the ground referred 1kΩ resistor identical to the drop across the 1kΩ current shunt and, hence, APD current. This relationship holds across the 20V to 90V APD bias voltage range. The 5.6V zener assures A1's inputs are always within their common mode operating range and the 10MΩ resistor maintains adequate Zener current when APD current is at very low levels.

Two output options are shown. A2, a chopper stabilized amplifier, provides an analog output. Its output is able to swing to (and below) zero because its V^- pin is supplied with a negative voltage. This potential is generated by using A2's internal clock to activate a charge pump which, in turn, biases A2's V^- Pin 3. A second output option substitutes an A-to-D converter, providing a serial format digital output. No V^- supply is required, as the LTC2400 A-to-D will convert inputs to (and slightly below) zero volts.

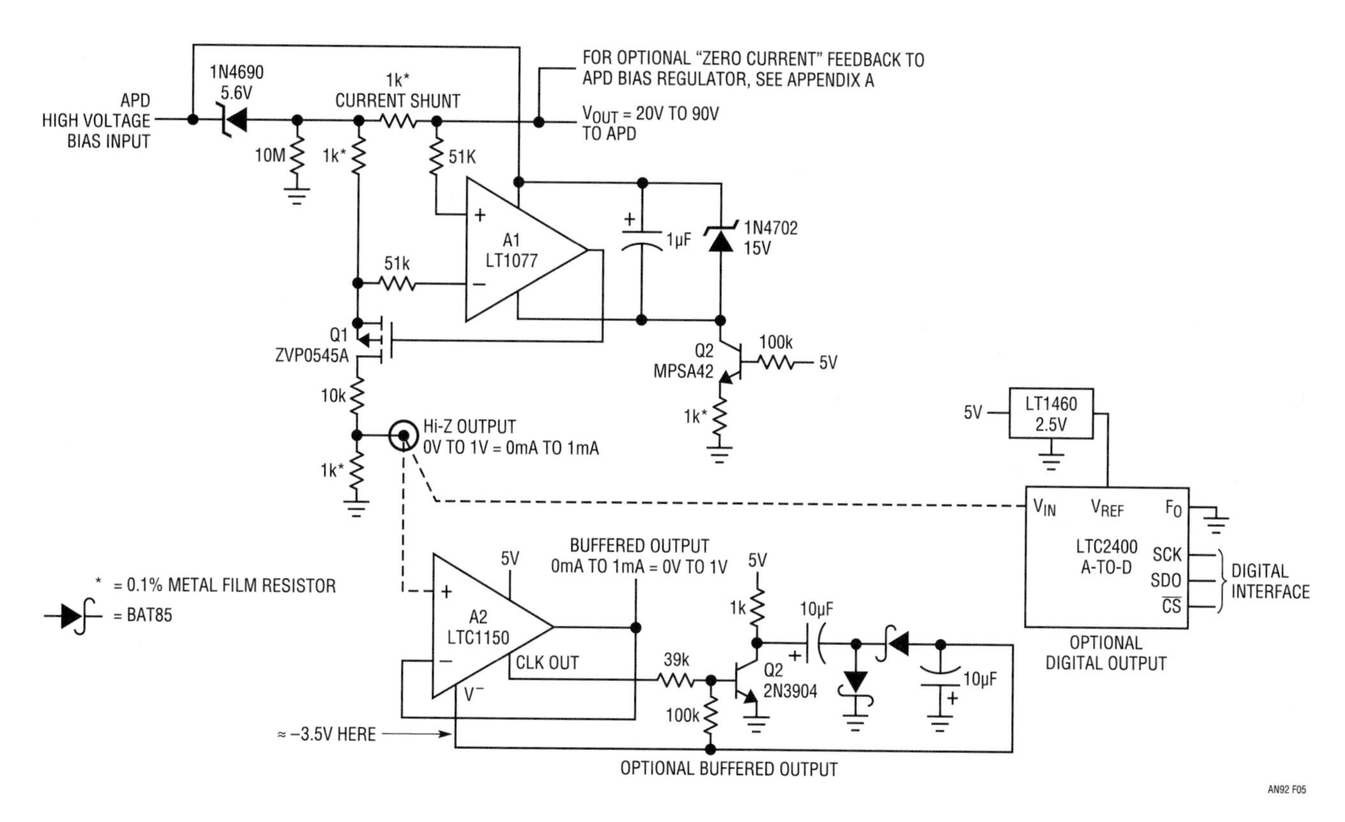

Figure 40.110 •

Six decade (10nA to 10mA) current log amplifier (Figure 40.111)

Using precision quad amplifiers like the LTC6079, (10μV offset and <1pA bias current) allow for very wide range current sensing. In this circuit a six decade range of current pulled from the circuit input terminal is converted to an output voltage in logarithmic fashion increasing 150mV for every decade of current change.

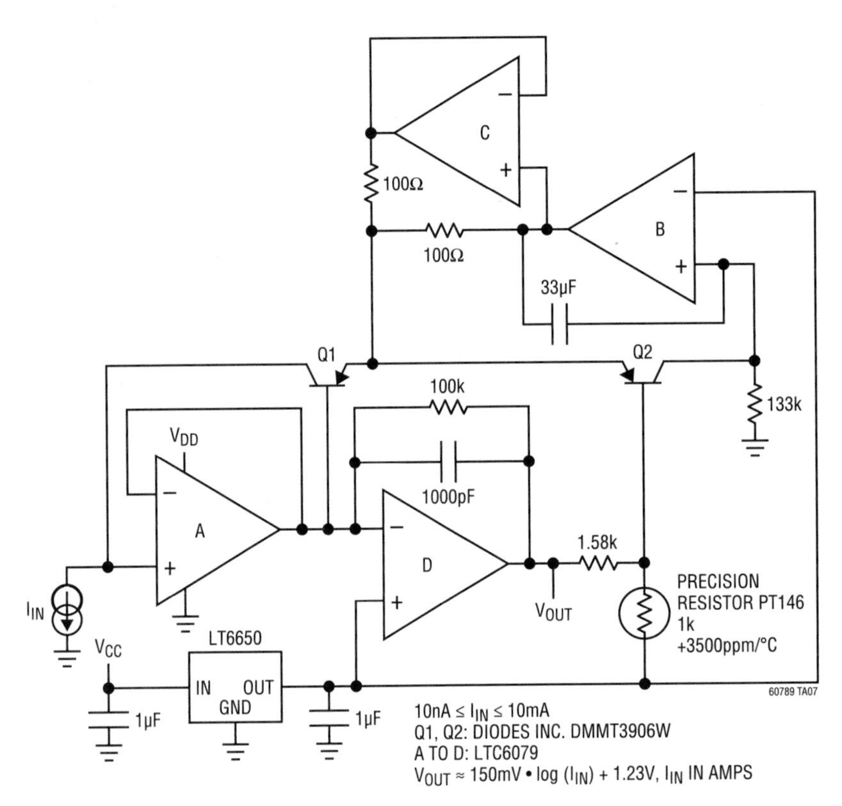

Figure 40.111 •

Motors and inductive loads

The largest challenge in measuring current through inductive circuits is the transients of voltage that often occur. Current flow can remain continuous in one direction while the voltage across the sense terminals reverses in polarity.

Electronic circuit breaker (Figure 40.112)

The LTC1153 is an electronic circuit breaker. Sensed current to a load opens the breaker when 100mV is developed between the supply input, V_S, and the drain sense pin, DS. To avoid transient, or nuisance trips of the break components RD and CD delay the action for 1ms. A thermistor can also be used to bias the shutdown input to monitor heat generated in the load and remove power should the temperature exceed 70°C in this example. A feature of the LTC1153 is timed automatic reset which will try to reconnect the load after 200ms using the 0.22μF timer capacitor shown.

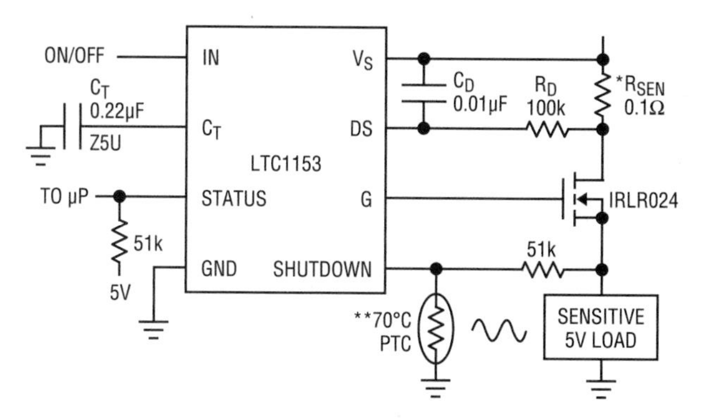

Figure 40.112 •

Conventional H-bridge current monitor (Figure 40.113)

Many of the newer electric drive functions, such as steering assist, are bidirectional in nature. These functions are generally driven by H-bridge MOSFET arrays using pulse-width modulation (PWM) methods to vary the commanded torque. In these systems, there are two main purposes for current monitoring. One is to monitor the current in the load, to track its performance against the desired command (i.e., closed-loop servo law), and another is for fault detection and protection features.

A common monitoring approach in these systems is to amplify the voltage on a "flying" sense resistor, as shown. Unfortunately, several potentially hazardous fault scenarios go undetected, such as a simple short to ground at a motor terminal. Another complication is the noise introduced by the PWM activity. While the PWM noise may be filtered for purposes of the servo law, information useful for protection becomes obscured. The best solution is to simply provide two circuits that individually protect each half-bridge and report the bidirectional load current. In some cases, a smart MOSFET bridge driver may already include sense resistors and offer the protection features needed. In these situations, the best solution is the one that derives the load information with the least additional circuitry.

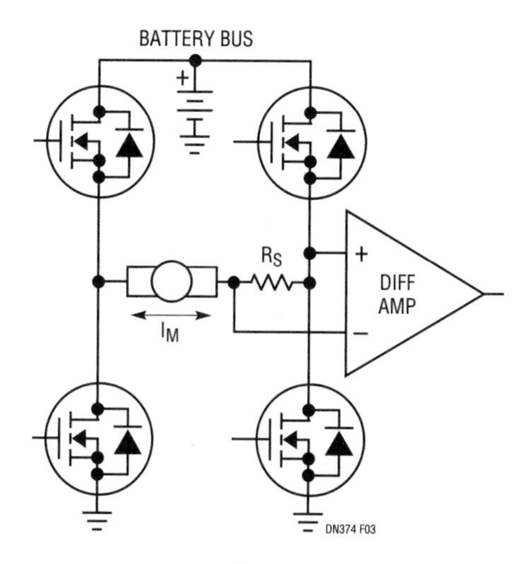

Figure 40.113 •

Motor speed control (Figure 40.114)

This uses an LT1970 power amplifier as a linear driver of a DC motor with speed control. The ability to source and sink the same amount of output current provides for bidirectional rotation of the motor: Speed control is managed by sensing the output of a tachometer built on to the motor. A typical feedback signal of 3V/1000rpm is compared with the desired speed-set input voltage. Because the LT1970 is unity-gain stable, it can be configured as an integrator to force whatever voltage across the motor as necessary to match the feedback speed signal with the set input signal. Additionally, the current limit of the amplifier can be adjusted to control the torque and stall current of the motor.

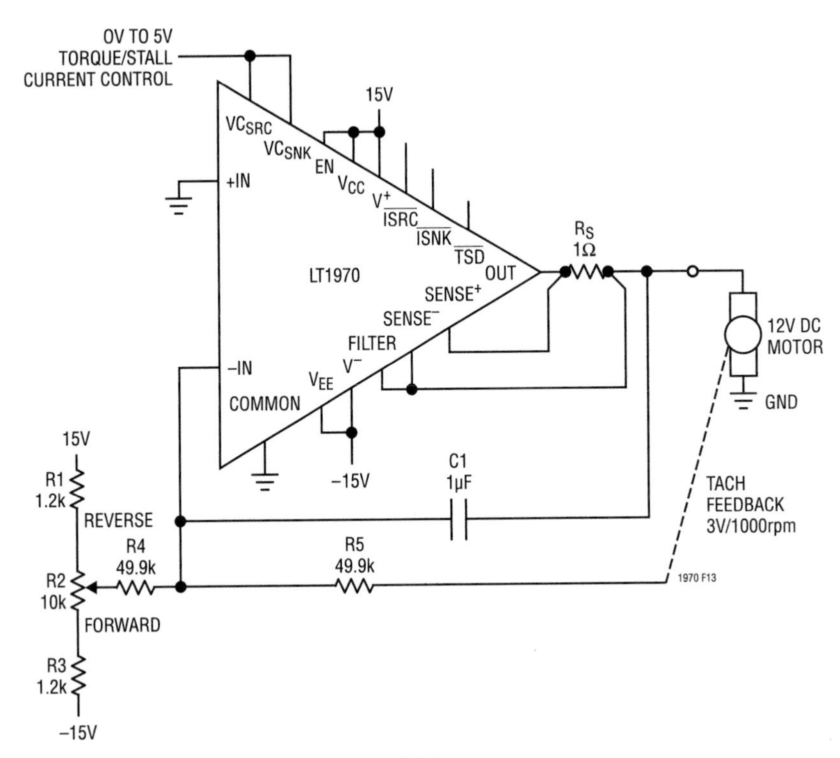

Figure 40.114 •

Practical H-bridge current monitor offers fault detection and bidirectional load information (Figure 40.115)

This circuit implements a differential load measurement for an ADC using twin unidirectional sense measurements. Each LTC6101 performs high side sensing that rapidly responds to fault conditions, including load shorts and MOSFET failures. Hardware local to the switch module (not shown in the diagram) can provide the protection logic and furnish a status flag to the control system. The two LTC6101 outputs taken differentially produce a bidirectional load measurement for the control servo. The ground-referenced signals are compatible with most ΔΣADCs. The ΔΣADC circuit also provides a "free" integration function that removes PWM content from the measurement. This scheme also eliminates the need for analog-to-digital conversions at the rate needed to support switch protection, thus reducing cost and complexity.

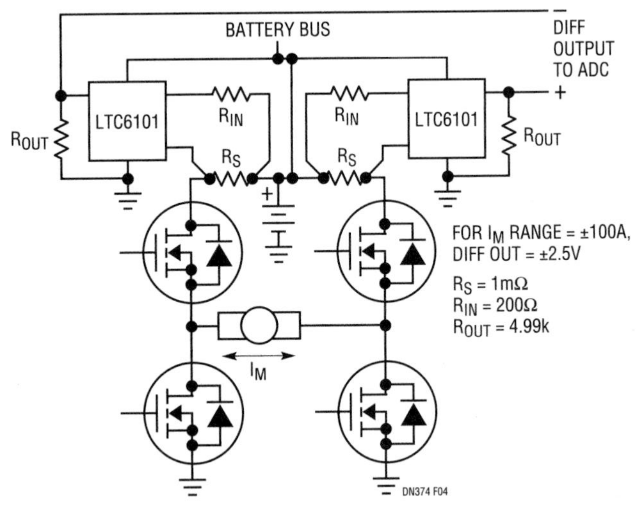

Figure 40.115 •

Lamp driver (Figure 40.116)

The inrush current created by a lamp during turn-on can be 10 to 20 times greater than the rated operating current. This circuit shifts the trip threshold of an LTC1153 electronic circuit breaker up by a factor of 11:1 (to 30A) for 100ms while the bulb is turned on. The trip threshold then drops down to 2.7A after the inrush current has subsided.

Intelligent high side switch (Figure 40.117)

The LT1910 is a dedicated high side MOSFET driver with built in protection features. It provides the gate drive for a power switch from standard logic voltage levels. It provides shorted load protection by monitoring the current flow to through the switch. Adding an LTC6101 to the same circuit, sharing the same current sense resistor, provides a linear voltage signal proportional to the load current for additional intelligent control.

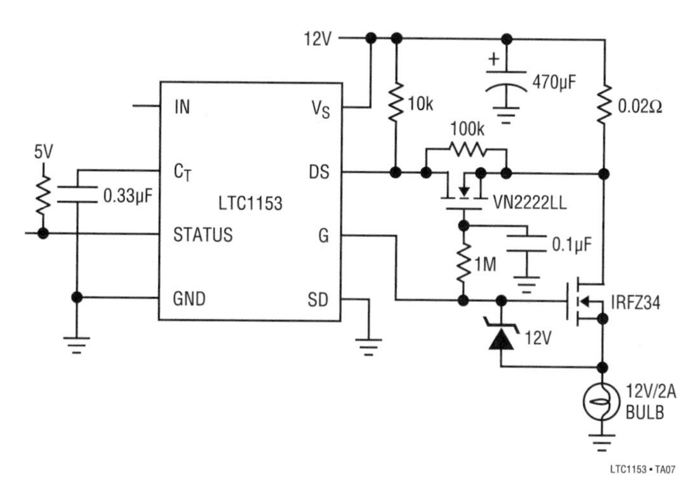

Figure 40.116 •

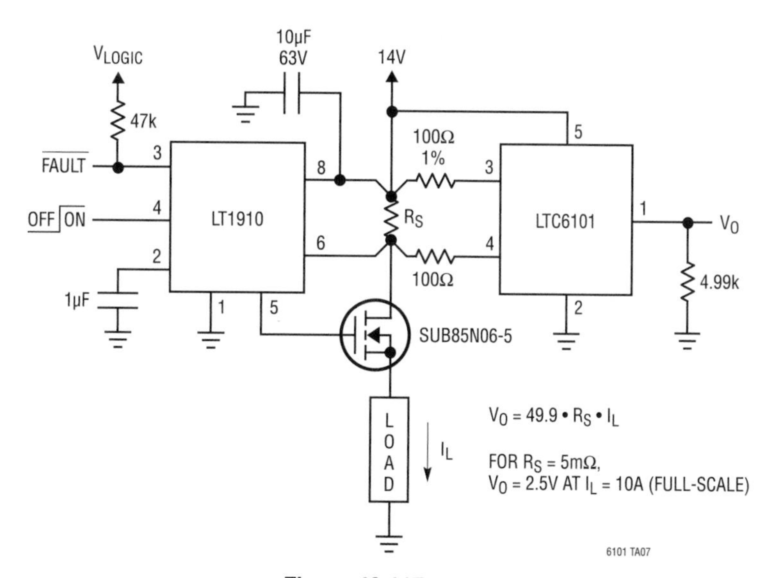

Figure 40.117 •

Relay driver (Figure 40.118)

This circuit provides reliable control of a relay by using an electronic circuit breaker circuit with two-level overcurrent protection. Current flow is sensed through two separate resistors, one for the current into the relay coil and the other for the current through the relay contacts. When 100mV is developed between the V_S supply pin and the drain sense pin, DS, the N-channel MOSFET is turned off opening the contacts. As shown, the relay coil current is limited to 350mA and the contact current to 5A.

Full-bridge load current monitor (Figure 40.119)

The LT1990 is a difference amplifier that features a very wide common mode input voltage range that can far exceed its own supply voltage. This is an advantage to reject transient voltages when used to monitor the current in a full-bridge driven inductive load such as a motor. The LT6650 provides a voltage reference of 1.5V to bias up the output away from ground. The output will move above or below 1.5V as a function of which direction the current in the load is flowing. As shown, the amplifier provides a gain of 10 to the voltage developed across resistor R_S.

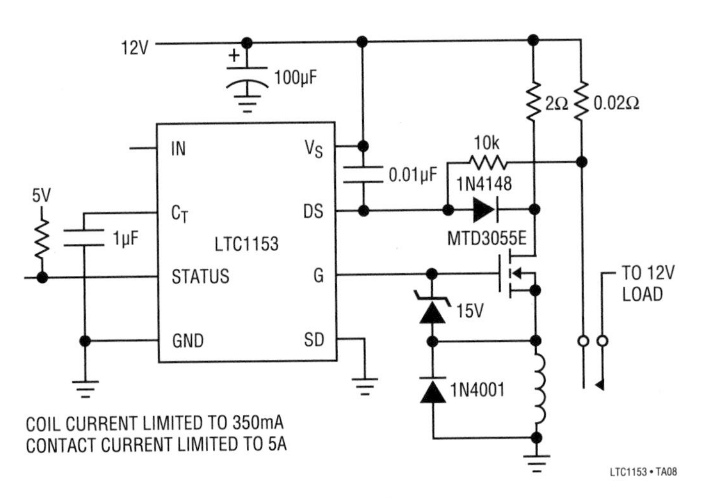

Figure 40.118 •

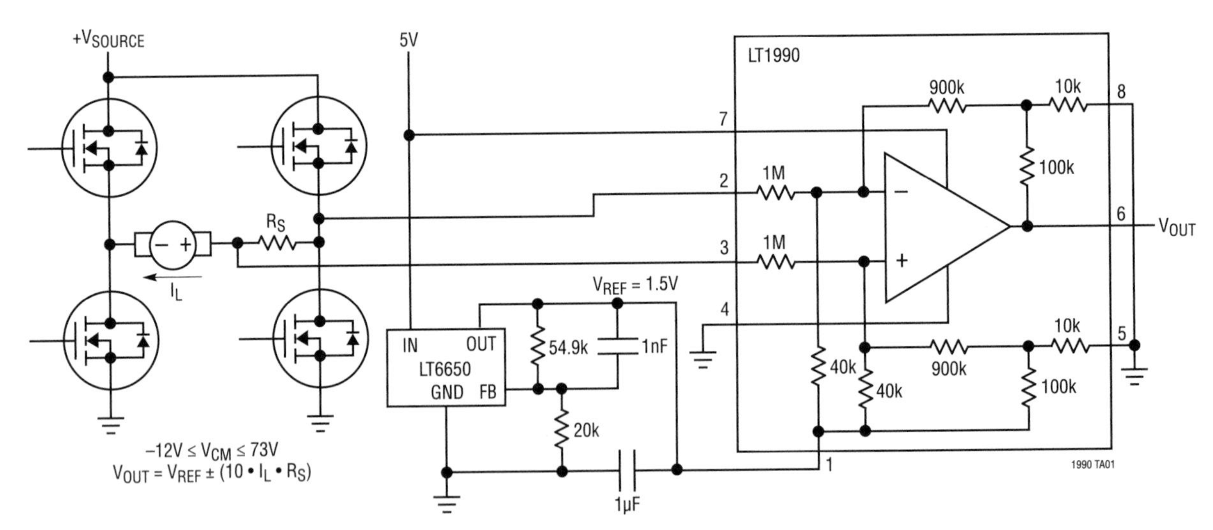

Figure 40.119 •

Bidirectional current sensing in H-bridge drivers (Figure 40.120)

Each channel of an LTC6103 provides measurement of the supply current into a half-bridge driver section. Since only one of the half-bridge sections will be conducting current in the measurable direction at any given time, only one output at a time will have a signal. Taken differentially, the two outputs form a bidirectional measurement for subsequent circuitry, such as an ADC. In this configuration, any load fault to ground will also be detected so that bridge protection can be implemented. This arrangement avoids the high frequency common mode rejection problem that can cause problems in "flying" sense resistor circuits.

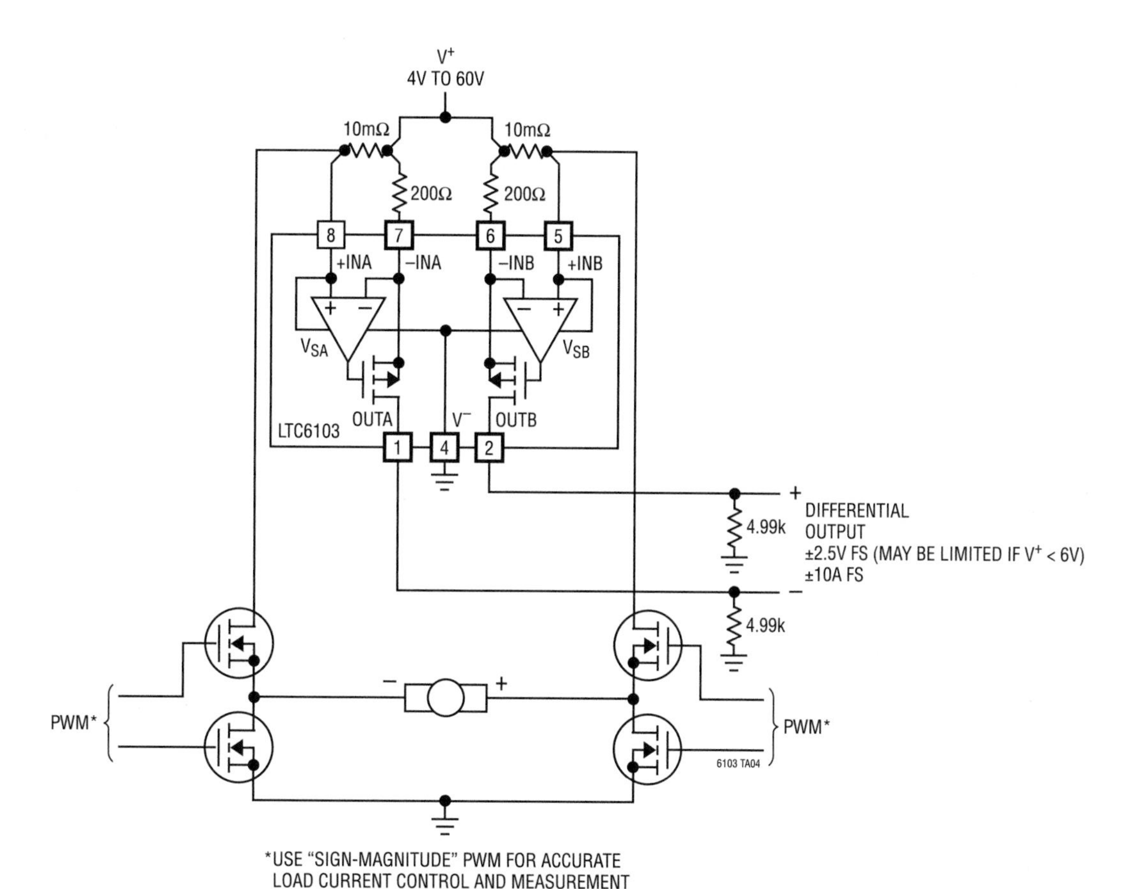

Figure 40.120 •

Single output provides 10A H-bridge current and direction (Figure 40.121)

The output voltage of the LTC6104 will be above or below the external 2.5V reference potential depending on which side of the H-bridge is conducting current. Monitoring the current in the bridge supply lines eliminates fast voltage changes at the inputs to the sense amplifiers.

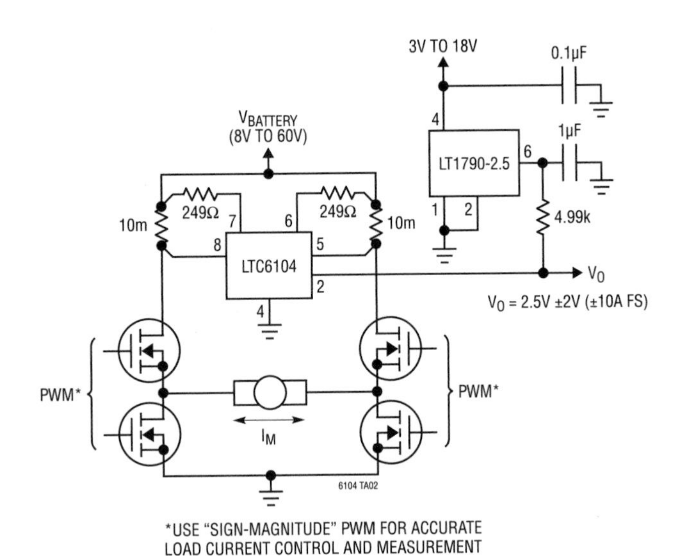

Figure 40.121 •

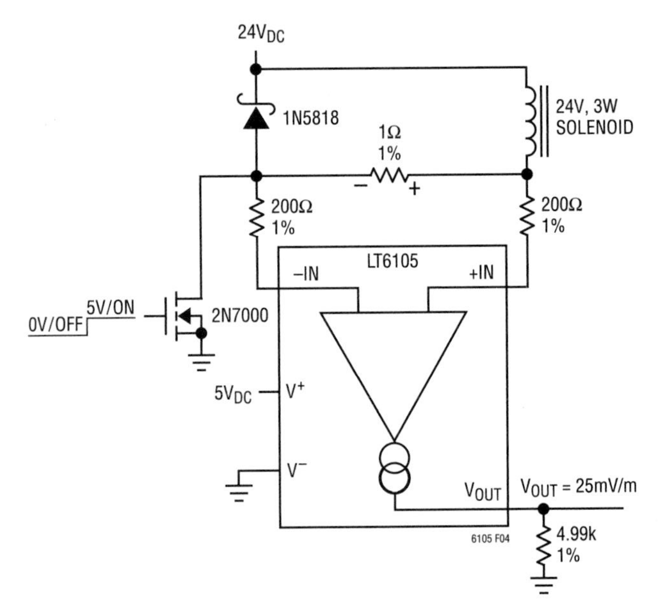

Figure 40.122 •

Monitor solenoid current on the low side (Figure 40.122)

Driving an inductive load such as a solenoid creates large transients of common mode voltage at the inputs to a current sense amplifier. When de-energized the voltage across the solenoid reverses (also called the freewheel state) and tries to go above its power supply voltage but is clamped by the freewheel diode. The LT6105 senses the solenoid current continuously over an input voltage range of 0V to one diode drop above the 24V supply.

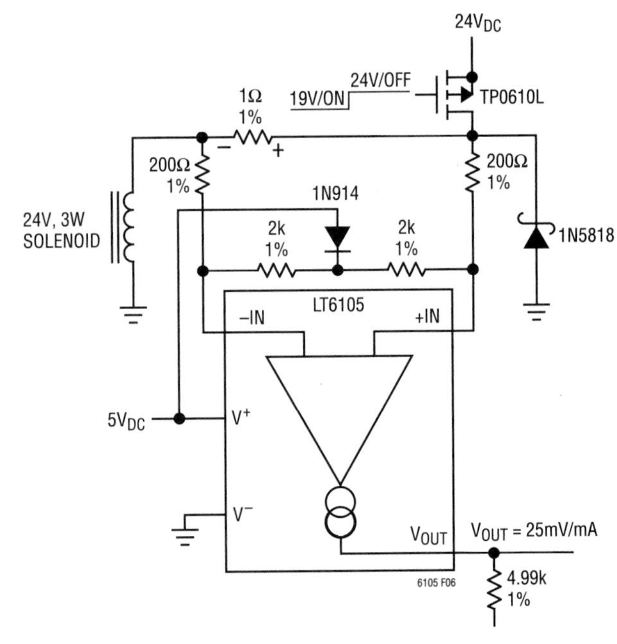

Figure 40.123 •

Monitor solenoid current on the high side (Figure 40.123)

Driving an inductive load such as a solenoid creates large transients of common mode voltage at the inputs to a current sense amplifier. When de-energized the voltage across the solenoid reverses (also called the freewheel state) and tries to go below ground but is clamped by the freewheel diode. The LT6105 senses the solenoid current continuously with pull-up resistors keeping the inputs within the most accurate input voltage range.

Monitor H-bridge motor current directly (Figures 40.124a and 40.124b)

The LT1999 is a differential input amplifier with a very wide, −5V to 80V, input common mode voltage range. With an AC CMRR greater than 80dB at 100kHz allows the direct measurement of the bidirectional current in an H-bridge driven load. The large and fast common mode input voltage swings are rejected at the output. The amplifier gain is fixed at 10, 20 or 50 requiring only a current sense resistor and supply bypass capacitors external to the amplifier.

Large input voltage range for fused solenoid current monitoring (Figure 40.125)

The LT1999 has series resistors at each input. This allows the input to be overdriven in voltage without damaging the amplifier. The amplifier will monitor the current through the positive and negative voltage swings of a solenoid driver. The large differential input with a blown protective fuse will force the output high and not damage the LT1999.

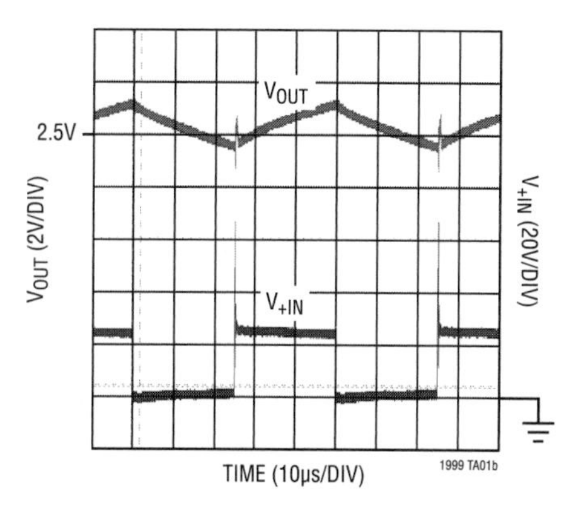

Figure 40.124b •

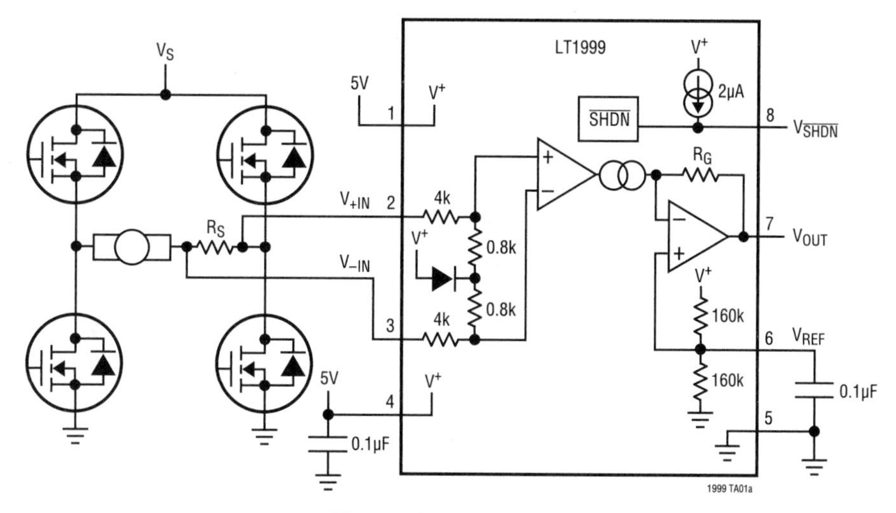

Figure 40.124a •

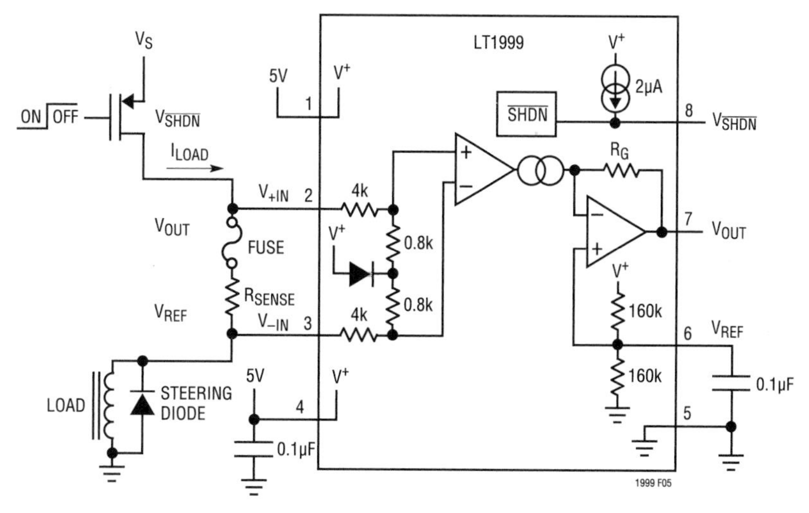

Figure 40.125 •

Monitor both the ON current and the freewheeling current through a high side driven solenoid (Figure 40.126)

Placing the current sense resistor inside the loop created by a grounded solenoid and the freewheeling clamp diode allows for continuous monitoring of the solenoid current while being energized or switched OFF. The LT1999 operates accurately with an input common mode voltage down to −5V below ground.

Monitor both the ON current and the freewheeling current in a low side driven solenoid (Figure 40.127)

Placing the current sense resistor inside the loop created by a grounded solenoid and the freewheeling clamp diode allows for continuous monitoring of the solenoid current while being energized or switched OFF. The LT1999 operates accurately with an input common mode voltage up to 80V. In this circuit the input is clamped at one diode above the solenoid supply voltage.

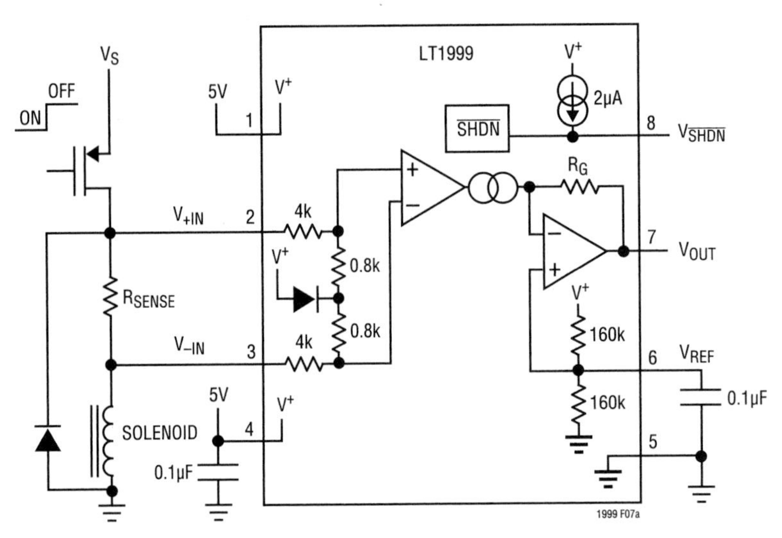

Figure 40.126 •

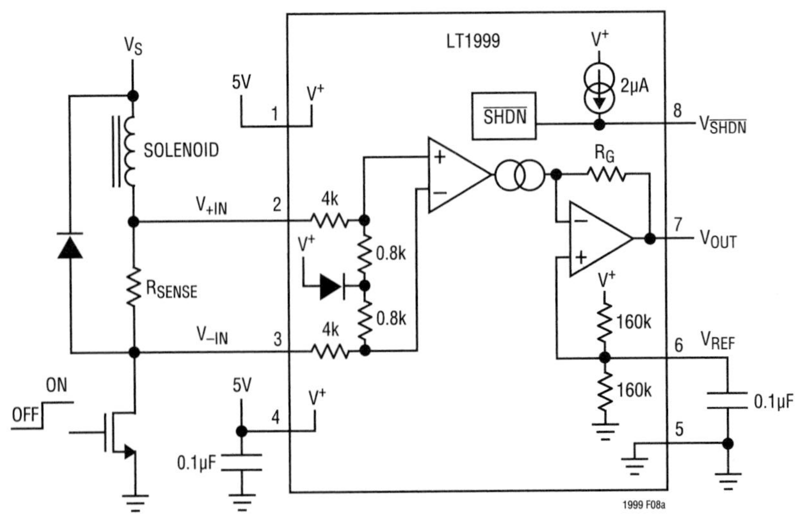

Figure 40.127 •

Fixed gain DC motor current monitor (Figure 40.128)

With no critical external components the LT1999 can be connected directly across a sense resistor in series with an H-bridge driven motor. The amplifier output voltage is referenced to one-half supply so the direction of motor rotation is indicated by the output being above or below the DC output voltage when stopped.

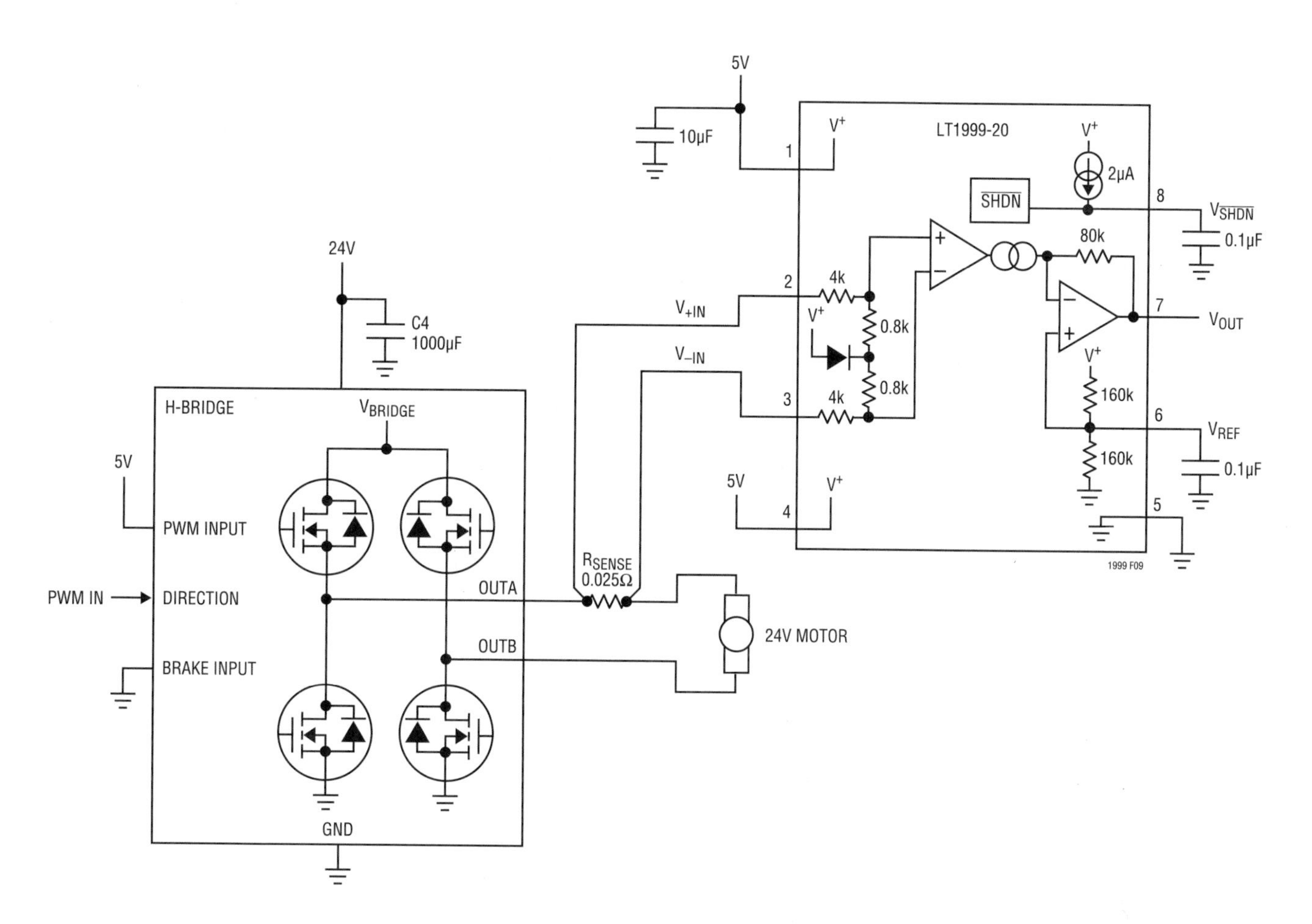

Figure 40.128 •

Simple DC motor torque control (Figure 40.129)

The torque of a spinning motor is directly proportional to the current through it. In this circuit the motor current is monitored and compared to a DC set point voltage. The motor current is sensed by an LT6108-1 and forced to match the set point current value through an amplifier and a PWM motor drive circuit. The LTC6992-1 produces a PWM signal from 0% to 100% duty cycle for a 0V to 1V change at the MOD input pin.

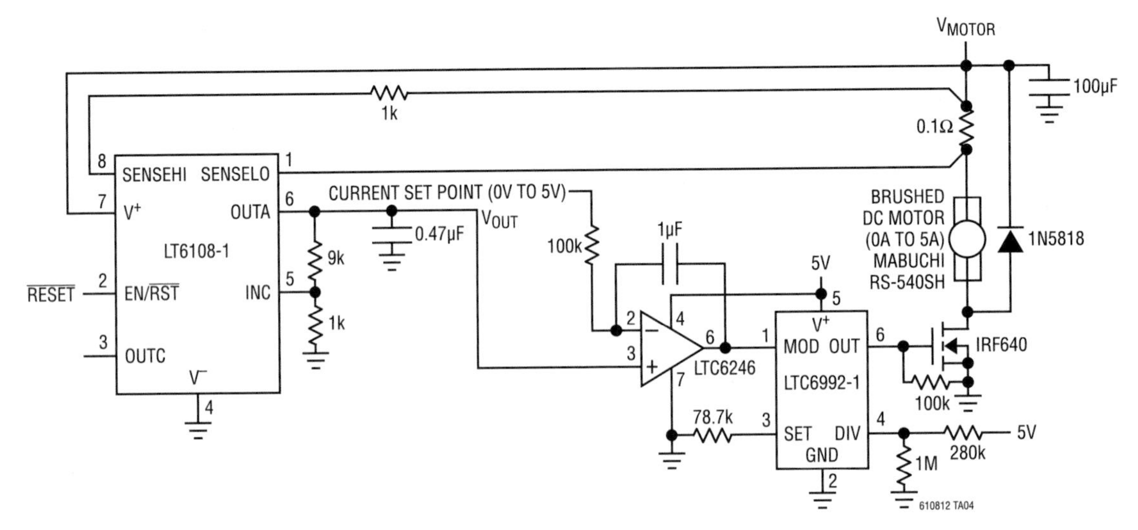

Figure 40.129 •

Small motor protection and control (Figure 40.130)

DC motor operating current and temperature can be digitized and sent to a controller which can then adjust the applied control voltage. Stalled rotor or excessive loading on the motor can be sensed.

Large motor protection and control (Figure 40.131)

For high voltage/current motors, simple resistor dividers can scale the signals applied to an LTC2990 14-bit converter. Proportional DC motor operating current and temperature can be digitized and sent to a controller which can then adjust the applied control voltage. Stalled rotor or excessive loading on the motor can be sensed.

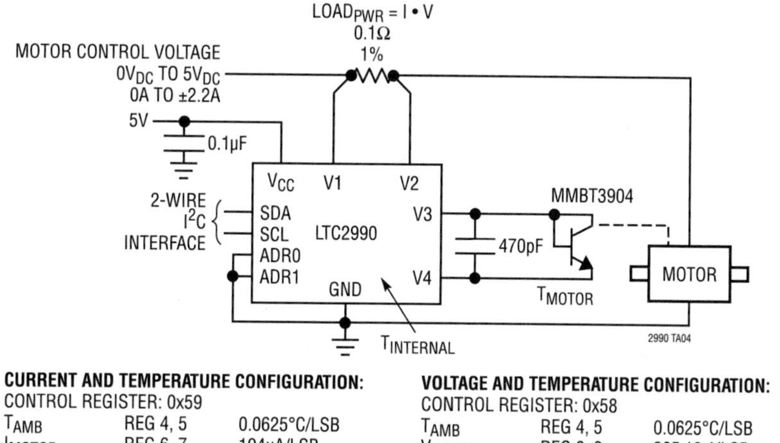

CURRENT AND TEMPERATURE CONFIGURATION:
CONTROL REGISTER: 0x59

T_{AMB}	REG 4, 5	0.0625°C/LSB
I_{MOTOR}	REG 6, 7	194μA/LSB
T_{MOTOR}	REG A, B	0.0625°C/LSB
V_{CC}	REG E, F	2.5V + 305.18μV/LSB

VOLTAGE AND TEMPERATURE CONFIGURATION:
CONTROL REGISTER: 0x58

T_{AMB}	REG 4, 5	0.0625°C/LSB
V_{MOTOR}	REG 8, 9	305.18μVLSB
T_{MOTOR}	REG A, B	0.0625°C/LSB
V_{CC}	REG E, F	2.5V + 305.18μV/LSB

Figure 40.130 •

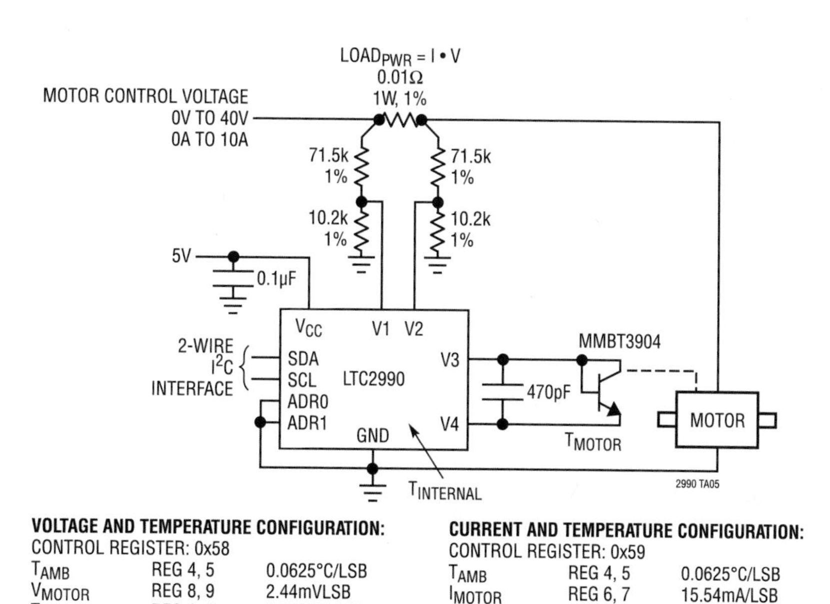

VOLTAGE AND TEMPERATURE CONFIGURATION:
CONTROL REGISTER: 0x58

T_{AMB}	REG 4, 5	0.0625°C/LSB
V_{MOTOR}	REG 8, 9	2.44mVLSB
T_{MOTOR}	REG A, B	0.0625°C/LSB
V_{CC}	REG E, F	2.5V + 305.18μV/LSB

CURRENT AND TEMPERATURE CONFIGURATION:
CONTROL REGISTER: 0x59

T_{AMB}	REG 4, 5	0.0625°C/LSB
I_{MOTOR}	REG 6, 7	15.54mA/LSB
T_{MOTOR}	REG A, B	0.0625°C/LSB
V_{CC}	REG E, F	2.5V + 305.18μV/LSB

Figure 40.131 •

Batteries

The science of battery chemistries and the charging and discharging characteristics is a book of its own. This chapter is intended to provide a few examples of monitoring current flow into and out of batteries of any chemistry.

Input remains Hi-Z when LT6100 is powered down (Figure 40.132)

This is the typical configuration for an LT6100, monitoring the load current of a battery. The circuit is powered from a low voltage supply rail rather than the battery being monitored. A unique benefit of this configuration is that when the LT6100 is powered down, its battery sense inputs remain high impedance, drawing less than 1μA of current. This is due to an implementation of Linear Technology's Over-The-Top input technique at its front end.

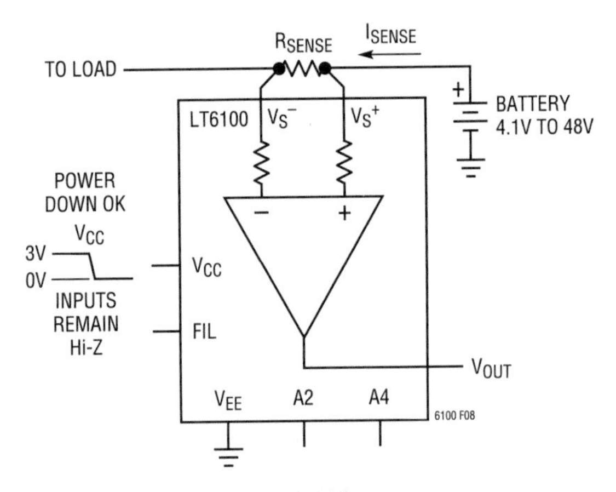

Figure 40.132 •

Charge/discharge current monitor on single supply with shifted V_{BIAS} (Figure 40.133)

Here the LT1787 is used in a single-supply mode with the V_{BIAS} pin shifted positive using an external LT1634 voltage reference. The V_{OUT} output signal can swing above and below V_{BIAS} to allow monitoring of positive or negative currents through the sense resistor. The choice of reference voltage is not critical except for the precaution that adequate headroom must be provided for V_{OUT} to swing without saturating the internal circuitry. The component values shown allow operation with V_S supplies as low as 3.1V.

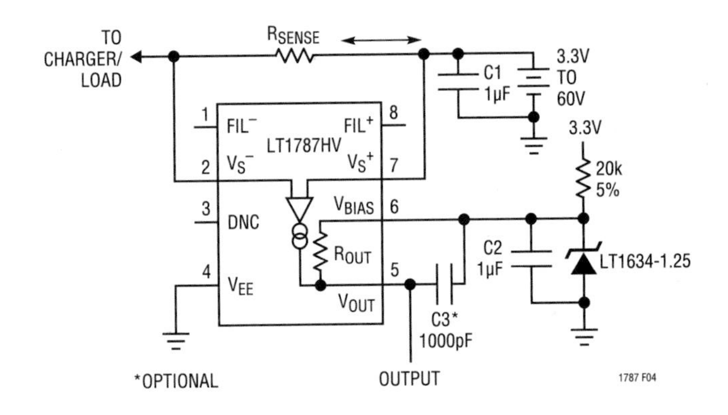

Figure 40.133 •

Battery current monitor (Figure 40.134)

One LT1495 dual op amp package can be used to establish separate charge and discharge current monitoring outputs. The LT1495 features Over-the-Top operation allowing the battery potential to be as high as 36V with only a 5V amplifier supply voltage.

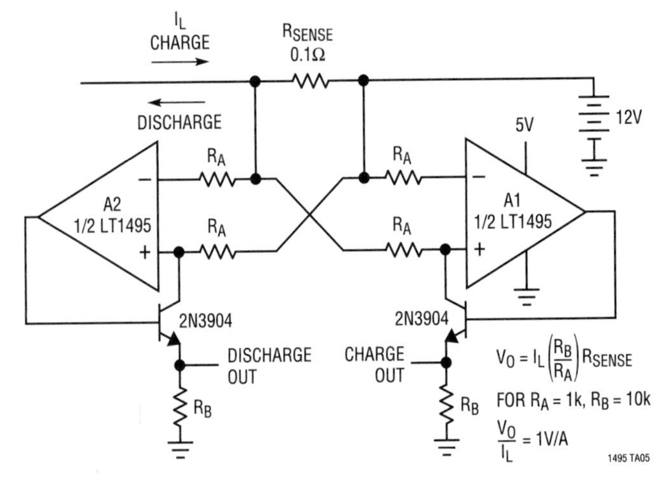

Figure 40.134 •

Input current sensing application (Figure 40.135)

The LT1620 is coupled with an LT1513 SEPIC battery charger IC to create an input over current protected charger circuit. The programming voltage ($V_{CC} - V_{PROG}$) is set to 1.0V through a resistor divider (R_{P1} and R_{P2}) from the 5V input supply to ground. In this configuration, if the input current drawn by the battery charger combined with the system load requirements exceeds a current limit threshold of 3A, the battery charger current will be reduced by the LT1620 such that the total input supply current is limited to 3A.

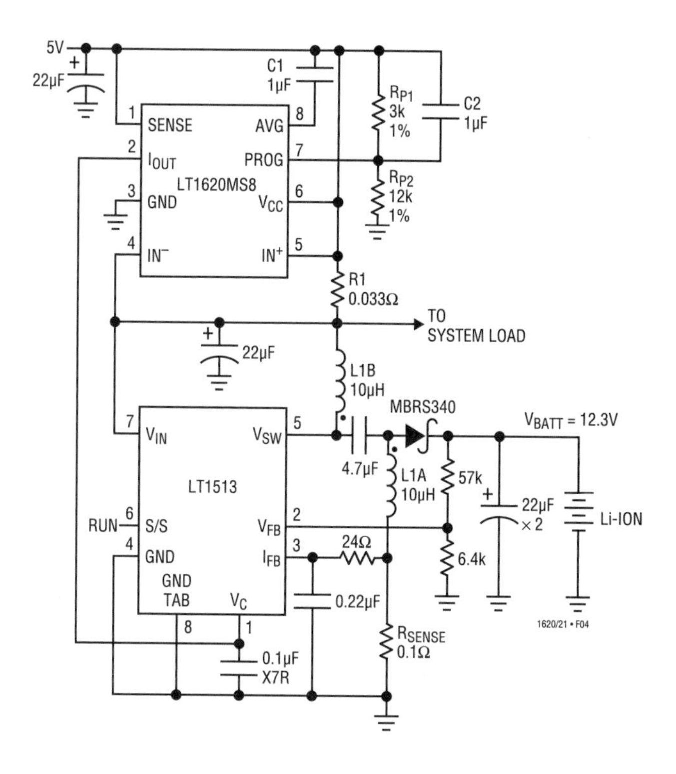

Figure 40.135 •

Coulomb counter (Figure 40.136)

The LTC4150 is a micropower high side sense circuit that includes a V/F function. Voltage across the sense resistor is cyclically integrated and reset to provide digital transitions that represent charge flow to or from the battery. A polarity bit indicates the direction of the current. Supply potential for the LTC4150 is 2.7V to 8.5V. In the free-running mode (as shown, with CLR and INT connected together) the pulses are approximately 1ps wide and around 1Hz full-scale.

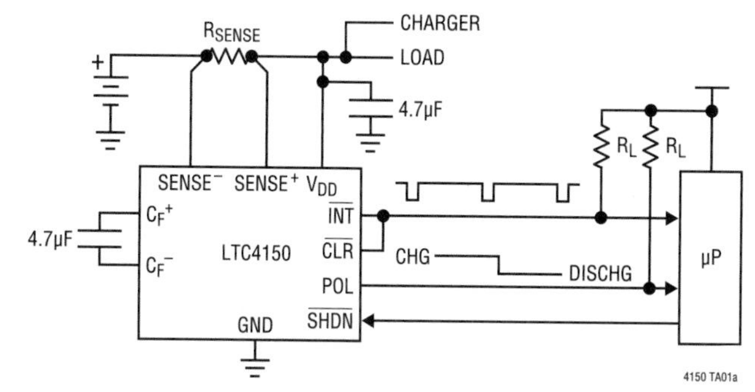

Figure 40.136 •

Li-Ion gas gauge (Figure 40.137)

This is the same as the coulomb counter circuit, except that the microprocessor clears the integration cycle complete condition with software, so that a relatively slow polling routine may be used.

NiMH charger (Figure 40.138)

The LTC4008 is a complete NiMH battery pack controller. It provides automatic switchover to battery power when the external DC power source is removed. When power is connected the battery pack is always kept charged and ready for duty.

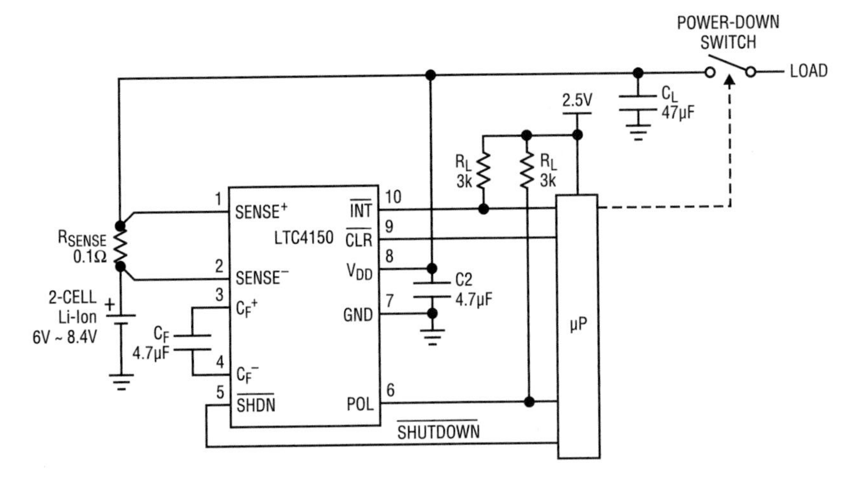

Figure 40.137 •

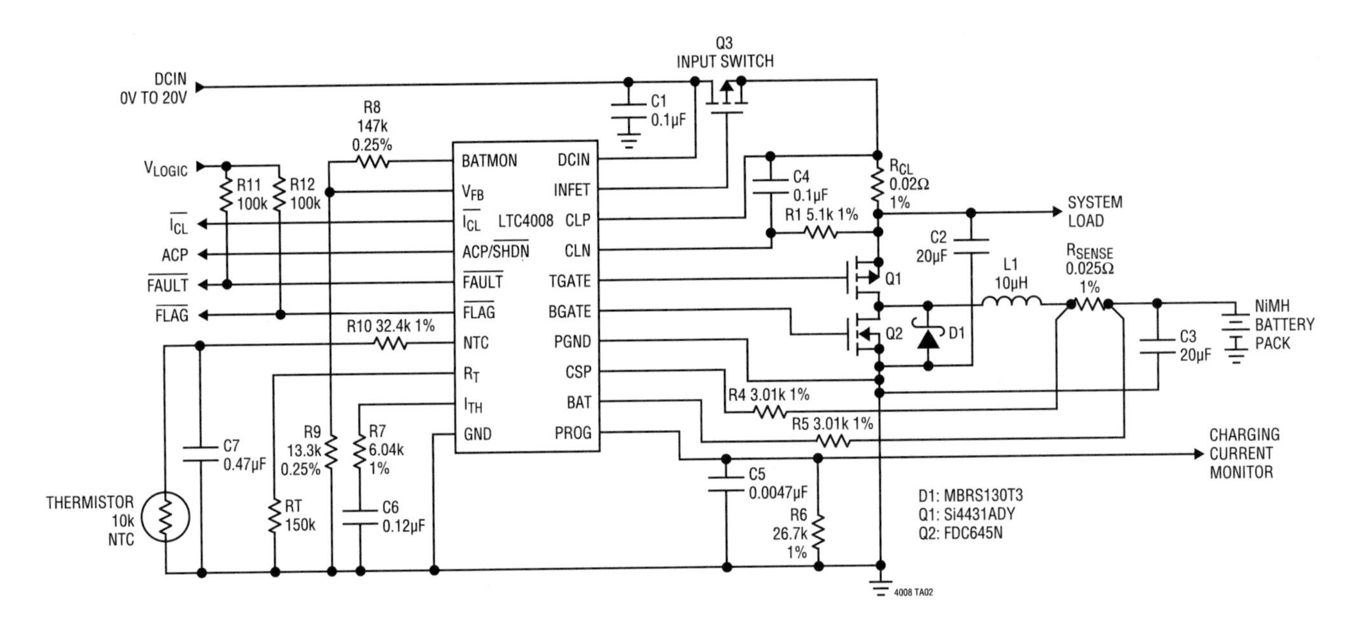

Figure 40.138 •

Single cell Li-Ion charger (Figure 40.139)

Controlling the current flow in lithium-ion battery chargers is essential for safety and extending useful battery life. Intelligent battery charger ICs can be used in fairly simple circuits to monitor and control current, voltage and even battery pack temperature for fast and safe charging.

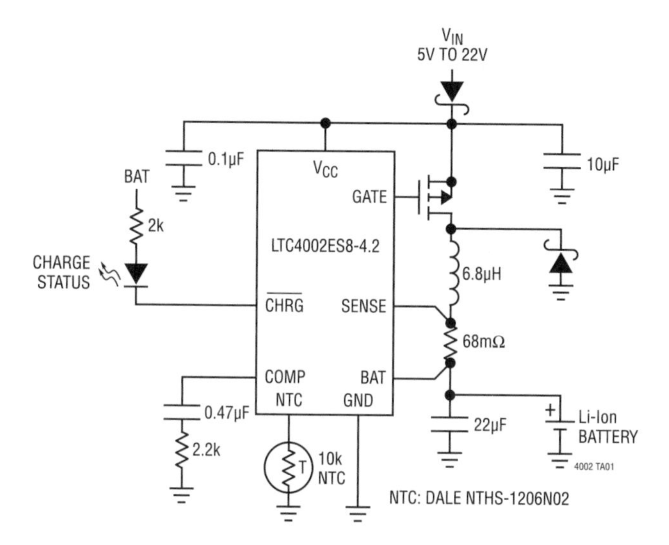

Figure 40.139 •

Li-Ion charger (Figure 40.140)

Just a few external components are required for this single Li-Ion cell charger. Power for the charger can come from a wall adapter or a computer's USB port.

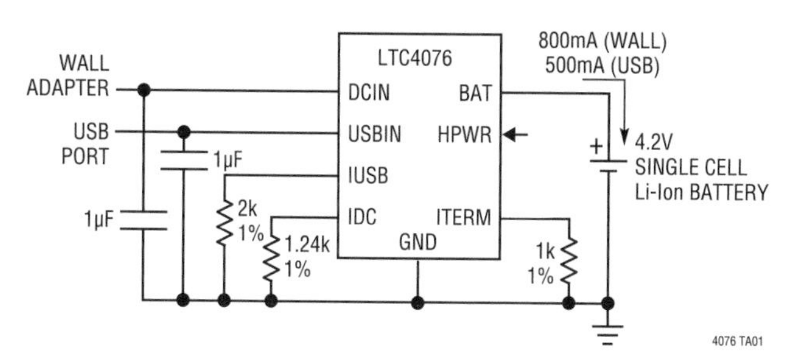

Figure 40.140 •

Battery monitor (Figure 40.141)

Op amp sections A and B form classical high side sense circuits in conjunction with Q1 and Q2 respectively. Each section handles a different polarity of battery current flow and delivers metered current to load resistor R_G. Section C operates as a comparator to provide a logic signal indicating whether the current is a charge or discharge flow. S1 sets the section D buffer op-amp gain to +1 or +10. Rail-to-rail op amps are required in this circuit, such as the LT1491 quad in the example.

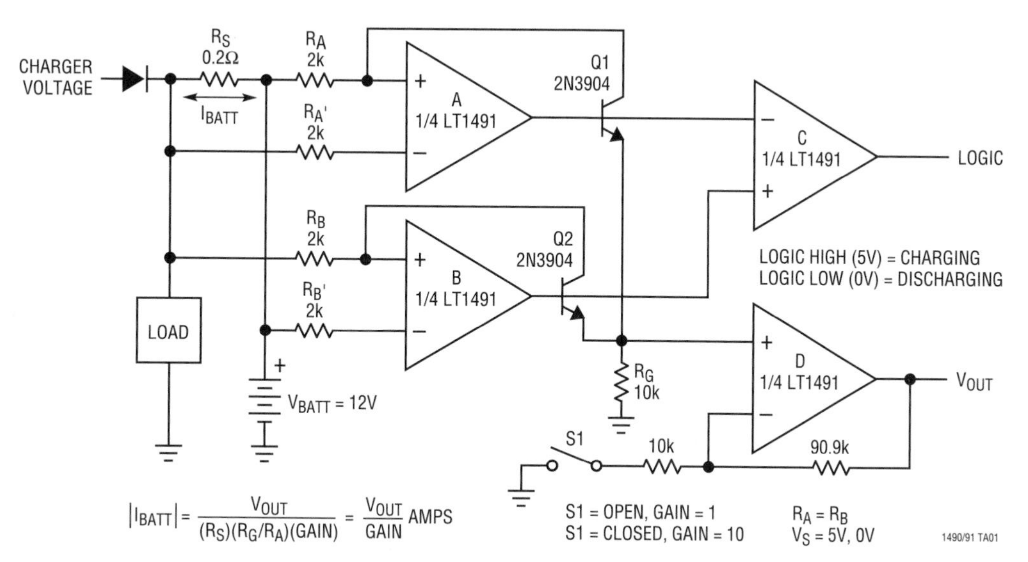

Figure 40.141 •

Monitor charge and discharge currents at one output (Figure 40.142)

Current from a battery to a load or from a charger to the battery can be monitored using a single sense resistor and the LTC6104. Discharging load current will source a current at the output pin in proportion to the voltage across the sense resistor. Charging current into the battery will sink a current at the output pin. The output voltage above or below the voltage V_{REF} will indicate charging or discharging of the battery.

Battery stack monitoring (Figure 40.143)

The comparators used in the LT6109 can be used separately. In this battery stack monitoring circuit a low on either comparator output will disconnect the load from the battery. One comparator watches for an overcurrent condition (800mA) and the other for a low voltage condition (30V). These threshold values are fully programmable using resistor divider networks.

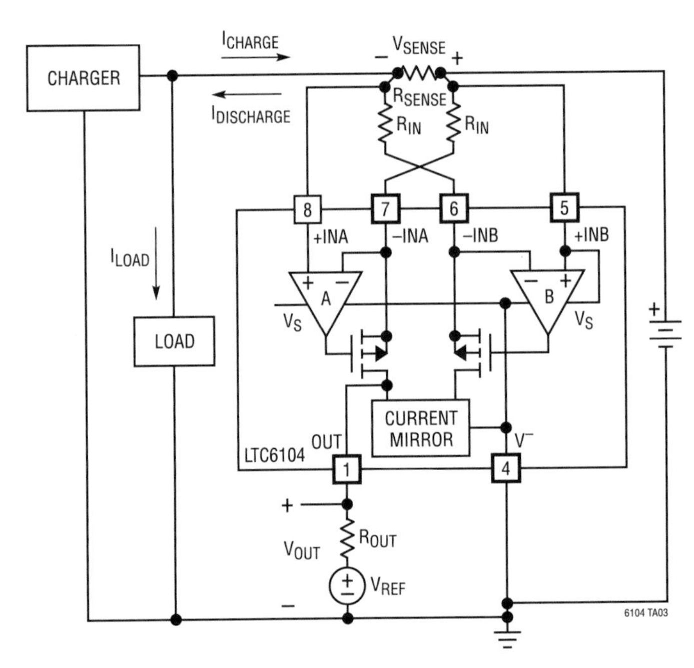

Figure 40.142 •

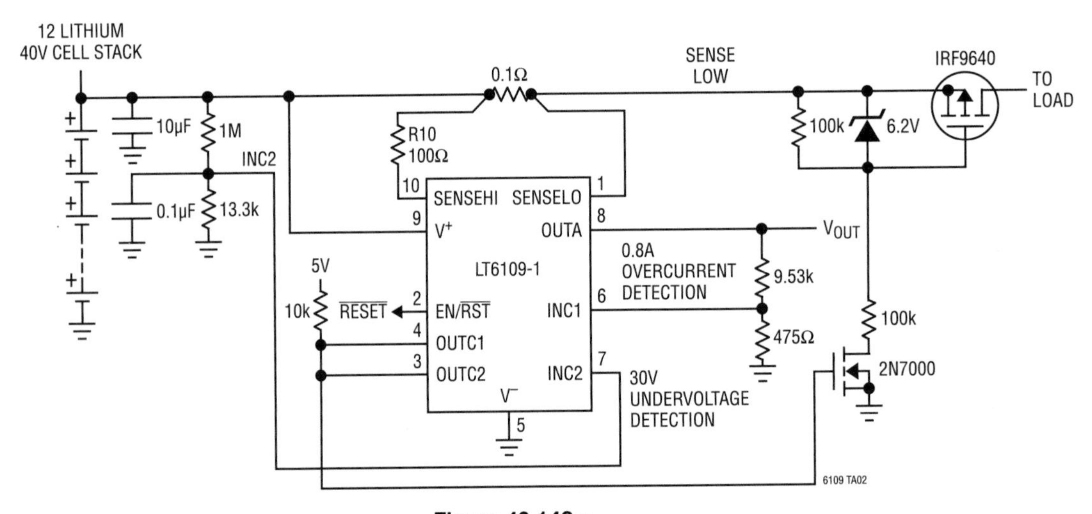

Figure 40.143 •

Coulomb counting battery gas gauge (Figure 40.144)

The LTC4150 converts the voltage across a sense resistor to a microprocessor interrupt pulse train. The time between each interrupt pulse is directly proportional to the current flowing through the sense resistor and therefore the number of coulombs travelling to or from the battery power source. A polarity output indicates the direction of current flow. By counting interrupt pulses with the polarity adding or subtracting from the running total, an indication of the total change in charge on a battery is determined. This acts as a battery gas gauge to indicate where the battery charge is between full or empty.

High voltage battery coulomb counting (Figure 40.145)

When coulomb counting, after each interrupt interval the internal counter needs to be cleared for the next time interval. This can be accomplished by the μP or the LTC4150 can clear itself. In this circuit the IC is powered from a battery supply which is at a higher voltage than the interrupt counting μP supply.

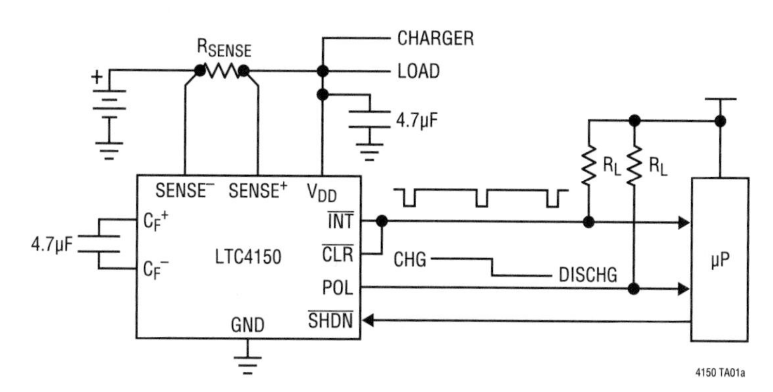

Figure 40.144 •

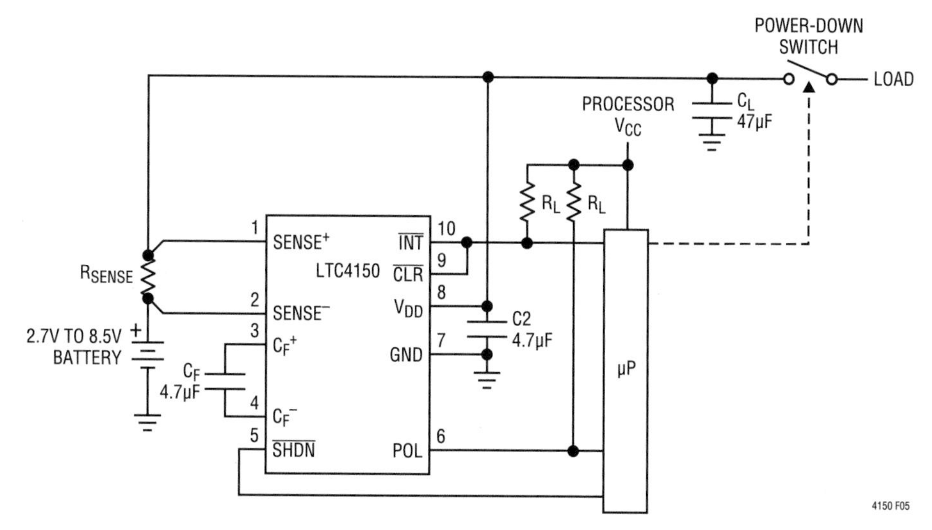

Figure 40.145 •

Low voltage battery coulomb counting (Figure 40.146)

When coulomb counting, after each interrupt interval the internal counter needs to be cleared for the next time interval. This can be accomplished by the μP or the LTC4150 can clear itself. In this circuit the IC is powered from a battery supply which is at a lower voltage than the interrupt counting μP supply. The CLR signal must be attenuated because the INT pin is pulled to a higher voltage.

Single cell lithium-ion battery coulomb counter (Figure 40.147)

This is a circuit which will keep track of the total change in charge of a single cell Li-Ion battery power source. The maximum battery current is assumed to be 500mA due to the 50mV full-scale sense voltage requirement of the LTC4150. The μP supply is greater than the battery supply.

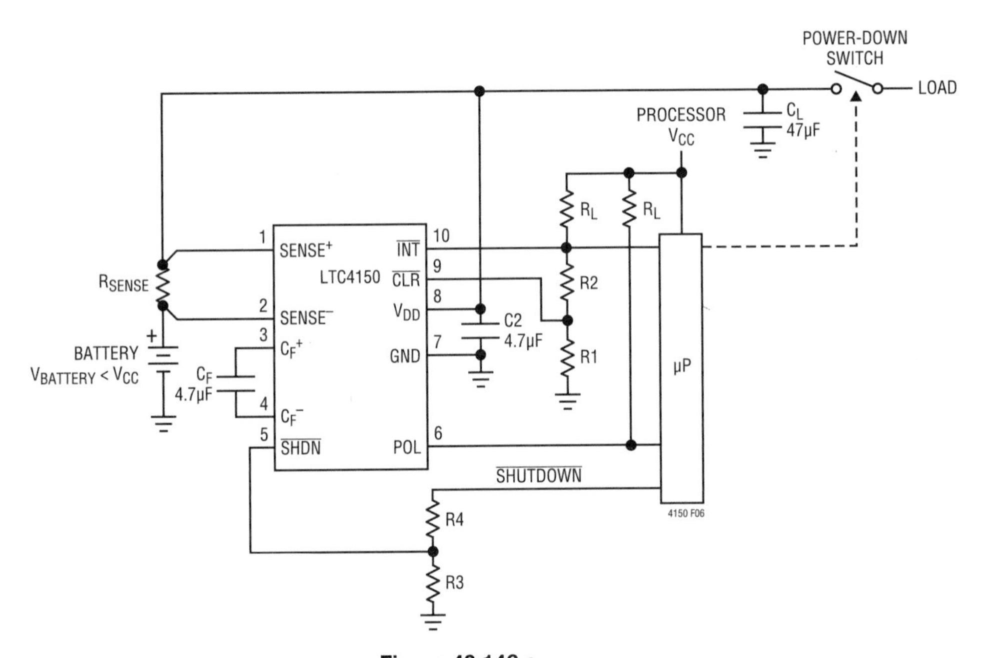

Figure 40.146 •

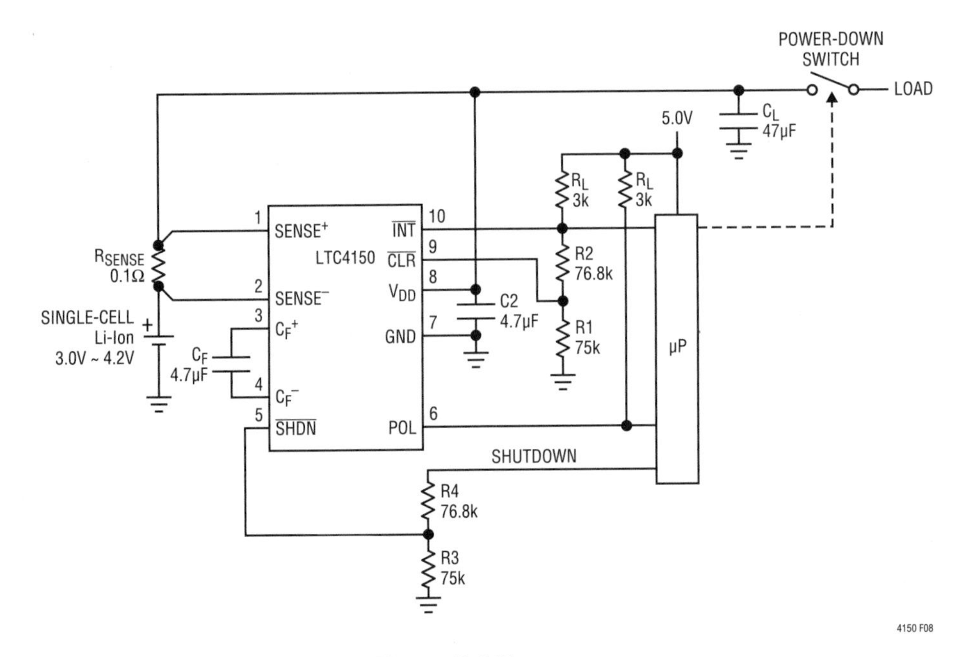

Figure 40.147 •

Complete single cell battery protection (Figure 40.148)

Voltage, current and battery temperature can all be monitored by a single LTC2990 ADC to 14-bit resolution. Each of these parameters can detect an excessive condition and signal the termination or initiation of cell charging. The ADC can be continually reconfigured for single-ended or differential measurements to produce the required information.

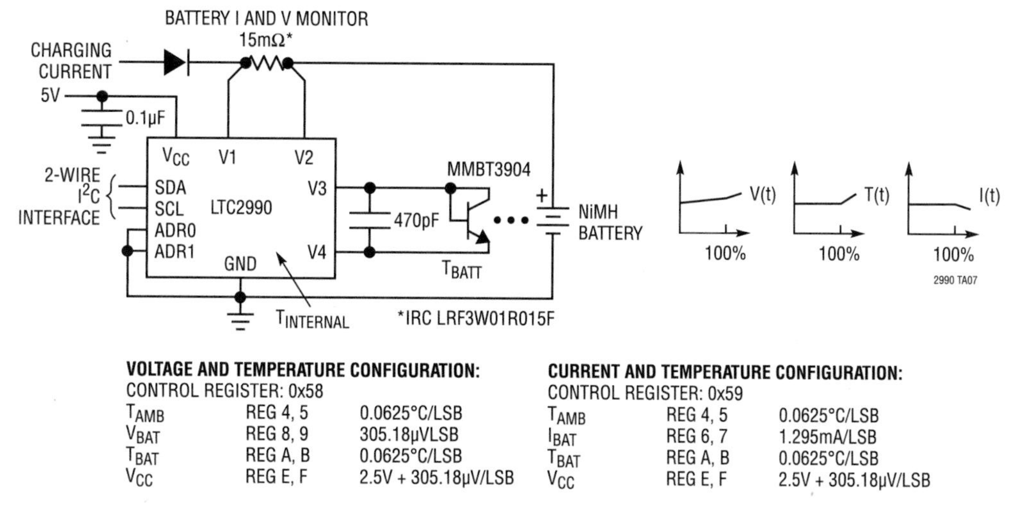

Figure 40.148 •

High speed

Current monitoring is not normally a particularly high speed requirement unless excessive current flow is caused by a fault of some sort. The use of fast amplifiers in conventional current sense circuits is usually sufficient to obtain the response time desired.

Fast compact −48V current sense (Figure 40.149)

This amplifier configuration is essentially the complementary implementation to the classic high side configuration. The op amp used must support common mode operation at its lower rail. A "floating" shunt-regulated local supply is provided by the Zener diode, and the transistor provides metered current to an output load resistance (1kΩ in this circuit). In this circuit, the output voltage is referenced to a positive potential and moves downward when representing increasing −48V loading. Scaling accuracy is set by the quality of resistors used and the performance of the NPN transistor.

Conventional H-bridge current monitor (Figure 40.150)

Many of the newer electric drive functions, such as steering assist, are bidirectional in nature. These functions are generally driven by H-bridge MOSFET arrays using pulse-width modulation (PWM) methods to vary the commanded torque. In these systems, there are two main purposes for current monitoring. One is to monitor the current in the load, to track its performance against the desired command (i.e., closed-loop servo law), and another is for fault detection and protection features.

A common monitoring approach in these systems is to amplify the voltage on a "flying" sense resistor, as shown. Unfortunately, several potentially hazardous fault scenarios go undetected, such as a simple short to ground at a motor terminal. Another complication is the noise introduced by the PWM activity. While the PWM noise may be filtered for purposes of the servo law, information useful for protection becomes obscured. The best solution is to simply provide two circuits that individually protect each

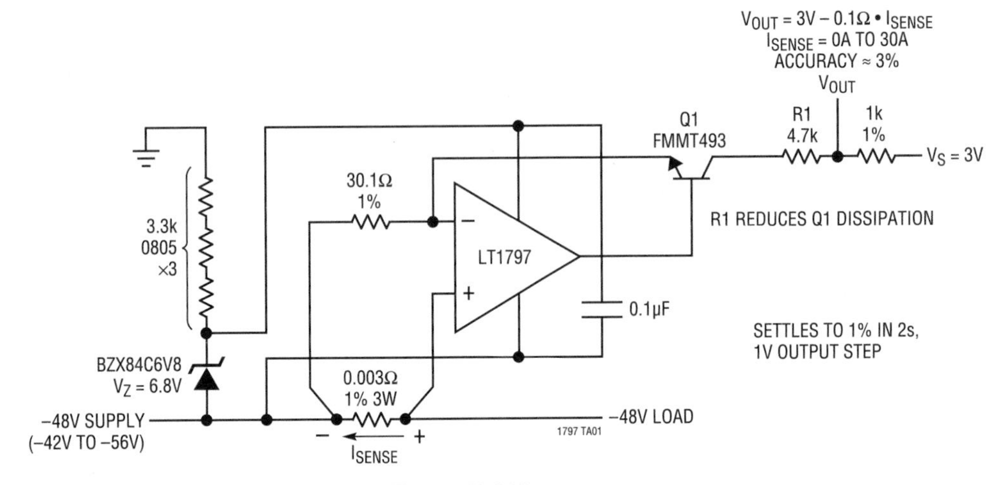

Figure 40.149 •

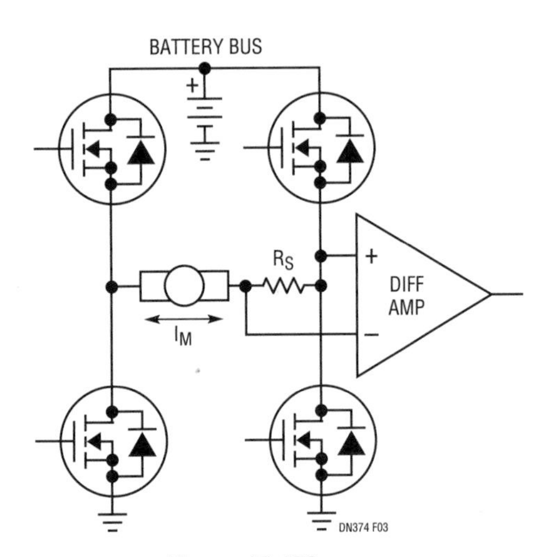

Figure 40.150 •

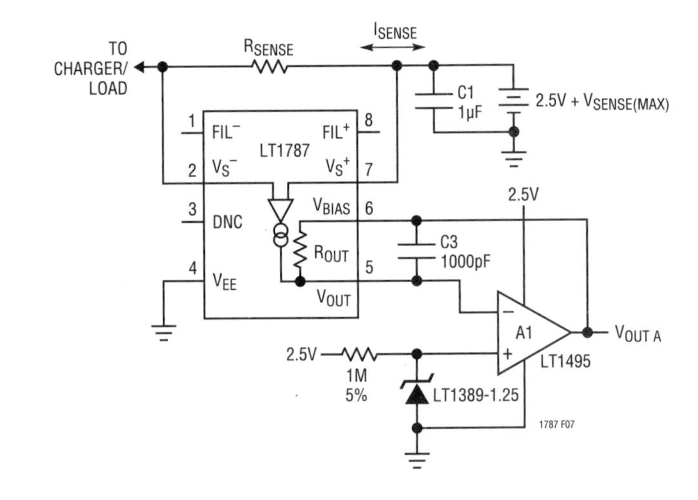

Figure 40.151 •

half-bridge and report the bidirectional load current. In some cases, a smart MOSFET bridge driver may already include sense resistors and offer the protection features needed. In these situations, the best solution is the one that derives the load information with the least additional circuitry.

Single-supply 2.5V bidirectional operation with external voltage reference and I/V converter (Figure 40.151)

The LT1787's output is buffered by an LT1495 rail-to-rail op amp configured as an I/V converter. This configuration is ideal for monitoring very low voltage supplies. The LT1787's V_{OUT} pin is held equal to the reference voltage appearing at the op amp's non-inverting input. This allows one to monitor supply voltages as low as 2.5V. The op amp's output may swing from ground to its positive supply voltage. The low impedance output of the op amp may drive following circuitry more effectively than the high output impedance of the LT1787. The I/V converter configuration also works well with split supply voltages.

Battery current monitor (Figure 40.152)

One LT1495 dual op amp package can be used to establish separate charge and discharge current monitoring outputs. The LT1495 features Over-the-Top operation allowing the battery potential to be as high as 36V with only a 5V amplifier supply voltage.

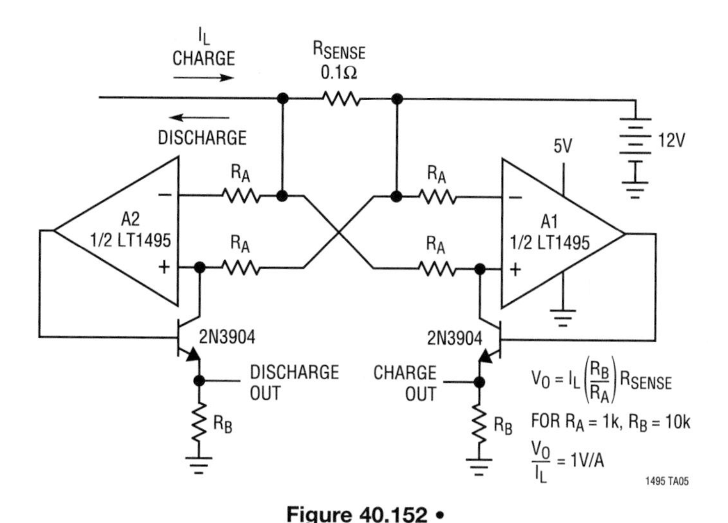

Figure 40.152 •

Fast current sense with alarm (Figure 40.153)

The LT1995 is shown as a simple unity gain difference amplifier. When biased with split supplies the input current can flow in either direction providing an output voltage of 100mV/A from the voltage across the 100mΩ sense resistor. With 32MHz of bandwidth and 1000V/μs slew rate the response of this sense amplifier is fast. Adding a simple comparator with a built in reference voltage circuit such as the LT6700-3 can be used to generate an overcurrent flag. With the 400mV reference the flag occurs at 4A.

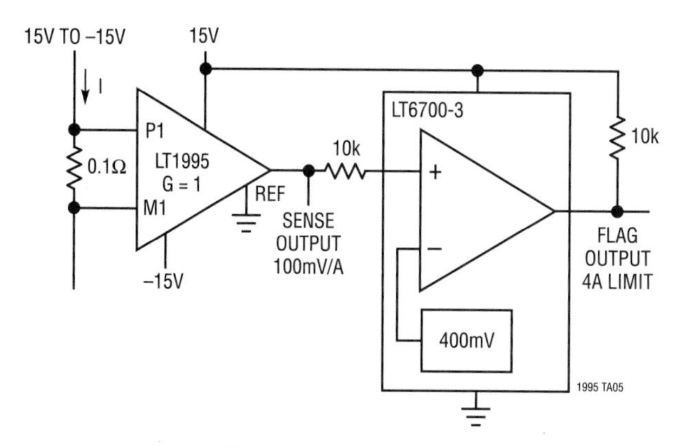

Figure 40.153 •

Fast differential current source (Figure 40.154)

This is a variation on the Howland configuration, where load current actually passes through a feedback resistor as an implicit sense resistance. Since the effective sense resistance is relatively large, this topology is appropriate for producing small controlled currents.

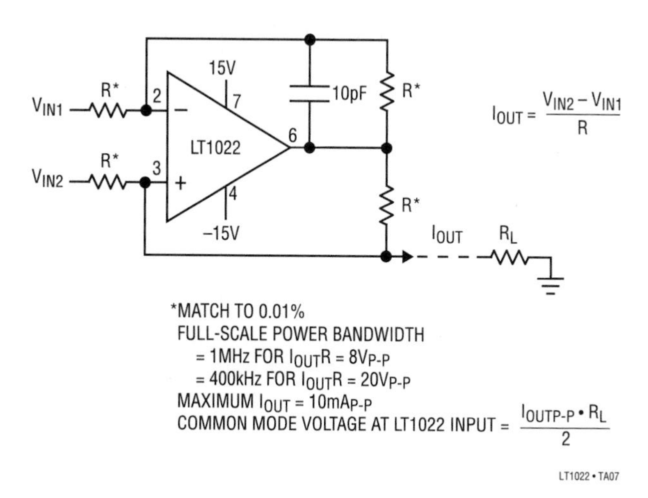

Figure 40.154 •

Fault sensing

The lack of current flow or the dramatic increase of current flow very often indicates a system fault. In these circuits it is important to not only detect the condition, but also ensure the safe operation of the detection circuitry itself. System faults can be destructive in many unpredictable ways.

High side current sense and fuse monitor (Figure 40.155)

The LT6100 can be used as a combination current sensor and fuse monitor. This part includes on-chip output buffering and was designed to operate with the low supply voltage (≥2.7V), typical of vehicle data acquisition systems, while the sense inputs monitor signals at the higher battery bus potential. The LT6100 inputs are tolerant of large input differentials, thus allowing the blown-fuse operating condition (this would be detected by an output full-scale indication). The LT6100 can also be powered down while maintaining high impedance sense inputs, drawing less than 1μA max from the battery bus.

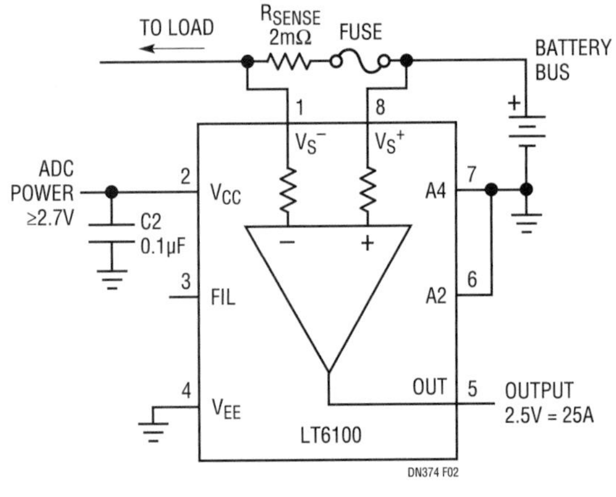

Figure 40.155 •

Schottky prevents damage during supply reversal (Figure 40.156)

The LTC6101 is not protected internally from external reversal of supply polarity. To prevent damage that may occur during this condition, a Schottky diode should be added in series with V^-. This will limit the reverse current through the LTC6101. Note that this diode will limit the low voltage performance of the LTC6101 by effectively reducing the supply voltage to the part by V_D.

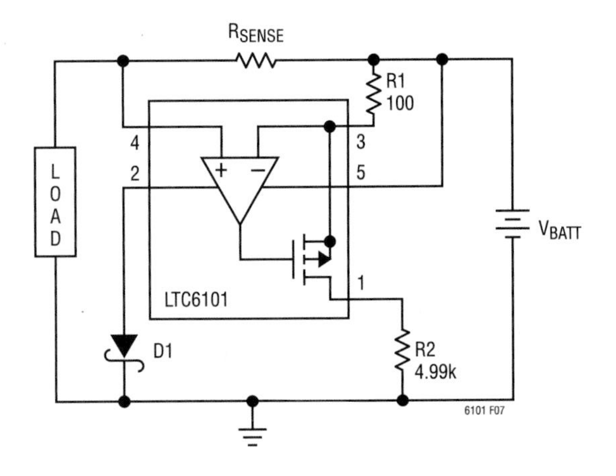

Figure 40.156 •

Additional resistor R3 protects output during supply reversal (Figure 40.157)

If the output of the LTC6101 is wired to an independently powered device that will effectively short the output to another rail or ground (such as through an ESD protection clamp) during a reverse supply condition, the LTC6101's output should be connected through a resistor or Schottky diode to prevent excessive fault current.

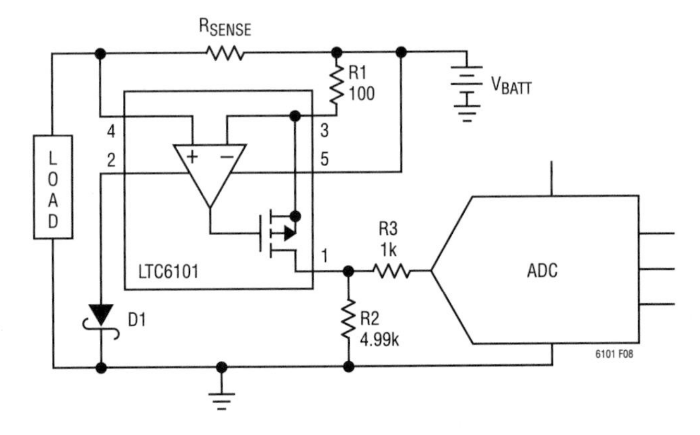

Figure 40.157 •

Electronic circuit breaker (Figure 40.158)

The LT1620l current sense amplifier is used to detect an overcurrent condition and shut off a P-MOSFET load switch. A fault flag is produced in the overcurrent condition and a self-reset sequence is initiated.

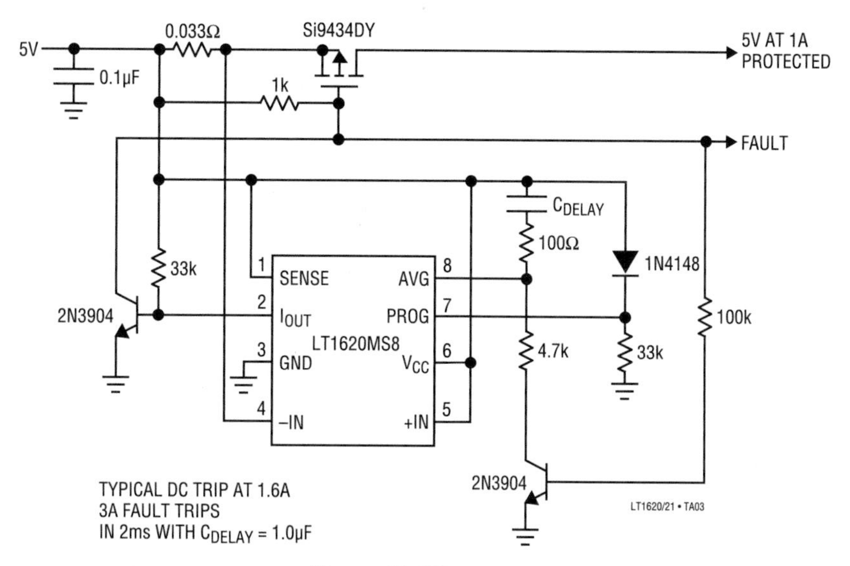

Figure 40.158 •

Electronic circuit breaker (Figure 40.159)

The LTC1153 is an electronic circuit breaker. Sensed current to a load opens the breaker when 100mV is developed between the supply input, V_S, and the drain sense pin, DS. To avoid transient, or nuisance trips of the break components RD and CD delay the action for 1ms. A thermistor can also be used to bias the shutdown input to monitor heat generated in the load and remove power should the temperature exceed 70°C in this example. A feature of the LTC1153 is timed automatic reset which will try to reconnect the load after 200ms using the 0.22μF timer capacitor shown.

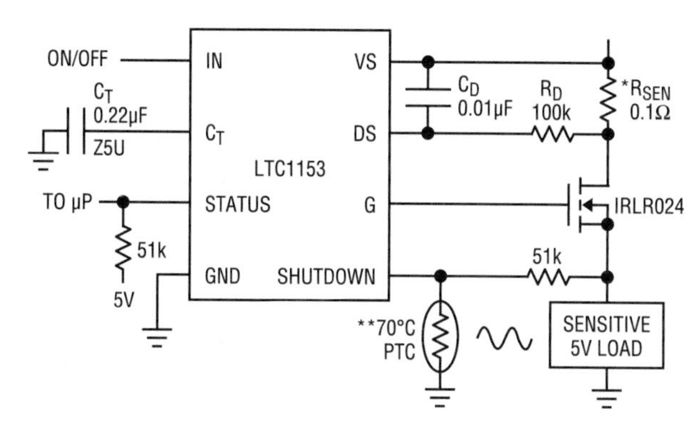

Figure 40.159 •

1.25V electronic circuit breaker (Figure 40.160)

The LTC4213 provides protection and automatic circuit breaker action by sensing drain-to-source voltage drop across the N-MOSFET The sense inputs have a rail-to-rail common mode range, so the circuit breaker can protect bus voltages from 0V up to 6V Logic signals flag a trip condition (with the READY output signal) and reinitialize the breaker (using the ON input). The ON input may also be used as a command in a "smart switch" application.

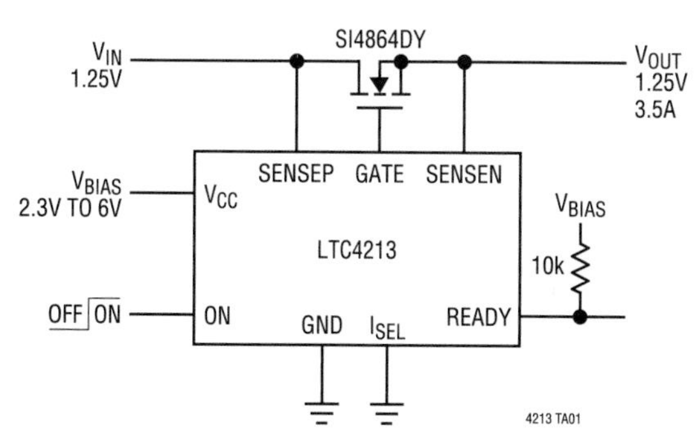

Figure 40.160 •

Lamp outage detector (Figure 40.161)

In this circuit, the lamp is monitored in both the on and off condition for continuity. In the off condition, the filament pull-down action creates a small test current in the 5kΩ that is detected to indicate a good lamp. If the lamp is open, the 100kΩ pull-up, or the relay contact, provides the op amp bias current through the 5kΩ, that is opposite in polarity. When the lamp is powered and filament current is flowing, the drop in the 0.05Ω sense resistor will exceed that of the 5kΩ and a lamp-good detection will still occur. This circuit requires particular Over-the-Top input characteristics for the op amp, so part substitutions are discouraged (however, this same circuit also works properly with an LT1716 comparator, also an Over-the-Top part).

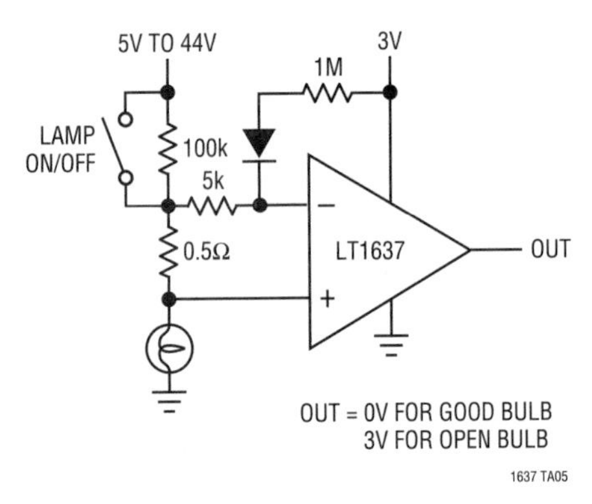

Figure 40.161 •

Simple telecom power supply fuse monitor (Figure 40.162)

The LTC1921 provides an all-in-one telecom fuse and supply-voltage monitoring function. Three opto-isolated status flags are generated that indicate the condition of the supplies and the fuses.

Conventional H-bridge current monitor (Figure 40.163)

Many of the newer electric drive functions, such as steering assist, are bidirectional in nature. These functions are generally driven by H-bridge MOSFET arrays using pulse-width modulation (PWM) methods to vary the commanded torque. In these systems, there are two main purposes for current monitoring. One is to monitor the current in the load, to track its performance against the desired command (i.e., closed-loop servo law), and another is for fault detection and protection features.

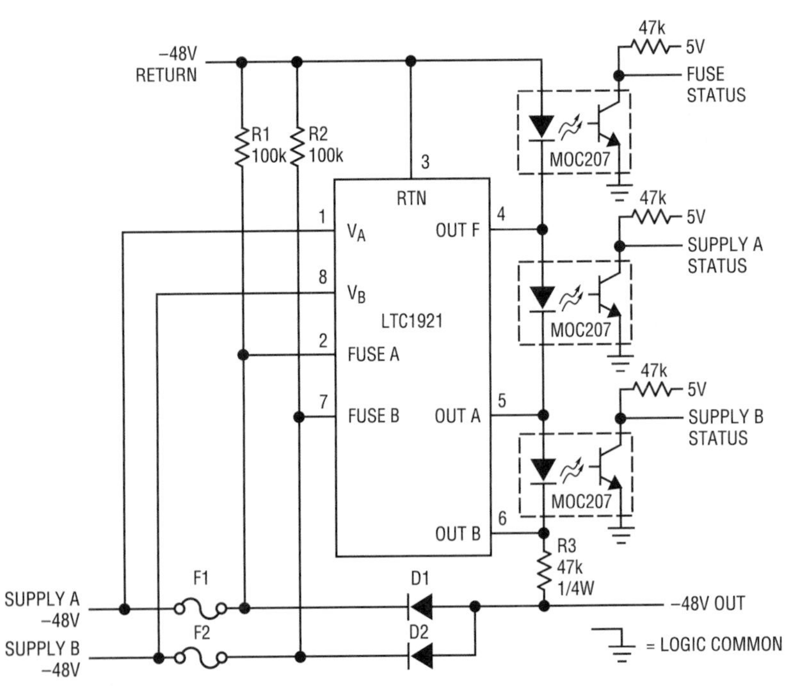

V_A	V_B	SUPPLY A STATUS	SUPPLY B STATUS
OK	OK	0	0
OK	UV OR OV	0	1
UV OR OV	OK	1	0
UV OR OV	UV OR OV	1	1

OK: WITHIN SPECIFICATION
OV: OVERVOLTAGE
UV: UNDERVOLTAGE

$V_{FUSE\ A}$	$V_{FUSE\ B}$	FUSE STATUS
$= V_A$	$= V_B$	0
$= V_A$	$\neq V_B$	1
$\neq V_A$	$= V_B$	1
$\neq V_A$	$\neq V_B$	1*

0: LED/PHOTODIODE ON
1: LED/PHOTODIODE OFF
*IF BOTH FUSES (F1 AND F2) ARE OPEN, ALL STATUS OUTPUTS WILL BE HIGH SINCE R3 WILL NOT BE POWERED

Figure 40.162 •

A common monitoring approach in these systems is to amplify the voltage on a "flying" sense resistor, as shown. Unfortunately, several potentially hazardous fault scenarios go undetected, such as a simple short to ground at a motor terminal. Another complication is the noise introduced by the PWM activity. While the PWM noise may be filtered for purposes of the servo law, information useful for protection becomes obscured. The best solution is to simply provide two circuits that individually protect each half-bridge and report the bidirectional load current. In some cases, a smart MOSFET bridge driver may already include sense resistors and offer the protection features needed. In these situations, the best solution is the one that derives the load information with the least additional circuitry.

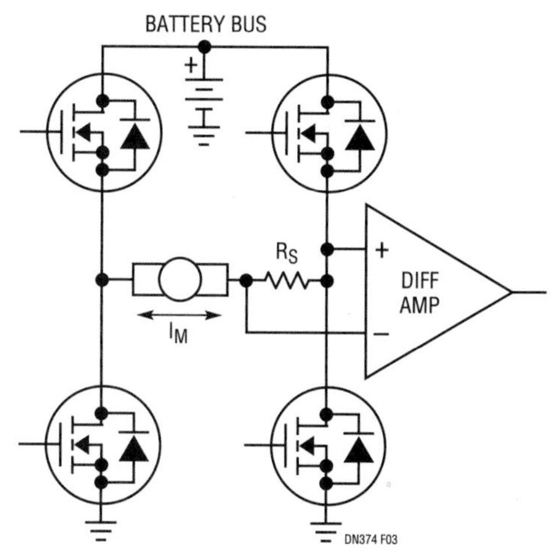

Figure 40.163 •

Single-supply 2.5V bidirectional operation with external voltage reference and I/V converter (Figure 40.164)

The LT1787's output is buffered by an LT1495 rail-to-rail op amp configured as an I/V converter. This configuration is ideal for monitoring very low voltage supplies. The LT1787's V_{OUT} pin is held equal to the reference voltage appearing at the op amp's non-inverting input. This allows one to monitor supply voltages as low as 2.5V. The op amp's output may swing from ground to its positive supply voltage. The low impedance output of the op amp may drive following circuitry more effectively than the high output impedance of the LT1787. The I/V converter configuration also works well with split supply voltages.

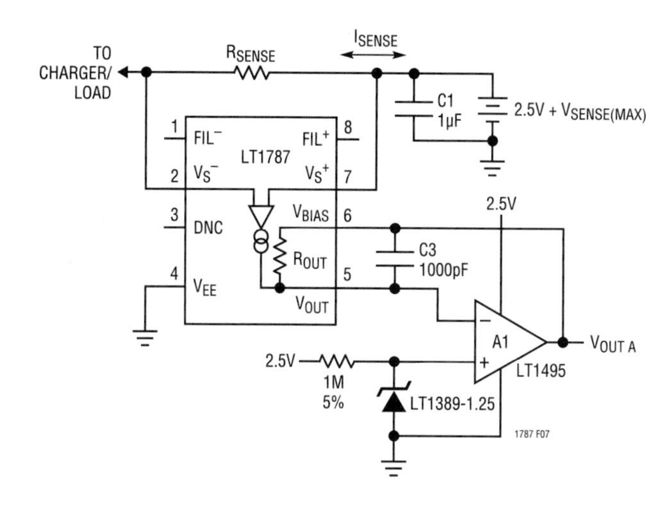

Figure 40.164 •

Battery current monitor (Figure 40.165)

One LT1495 dual op amp package can be used to establish separate charge and discharge current monitoring outputs. The LT1495 features Over-the-Top operation allowing the battery potential to be as high as 36V with only a 5V amplifier supply voltage.

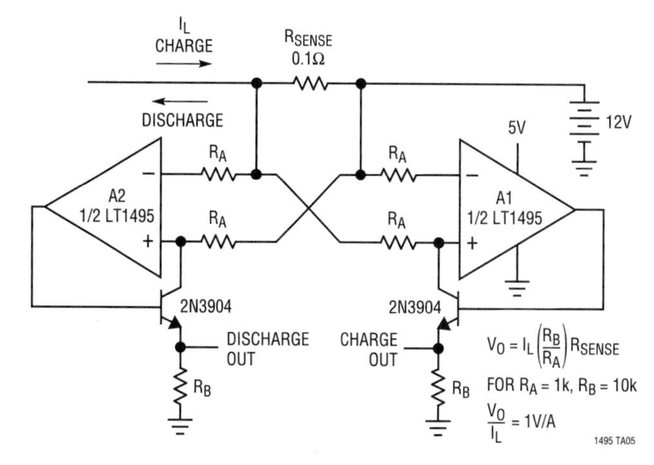

Figure 40.165 •

Fast current sense with alarm (Figure 40.166)

The LT1995 is shown as a simple unity gain difference amplifier. When biased with split supplies the input current can flow in either direction providing an output voltage of 100mV/A from the voltage across the 100mΩ sense resistor. With 32MHz of bandwidth and 1000V/μs slew rate the response of this sense amplifier is fast. Adding a simple comparator with a built in reference voltage circuit such as the LT6700-3 can be used to generate an overcurrent flag. With the 400mV reference the flag occurs at 4A.

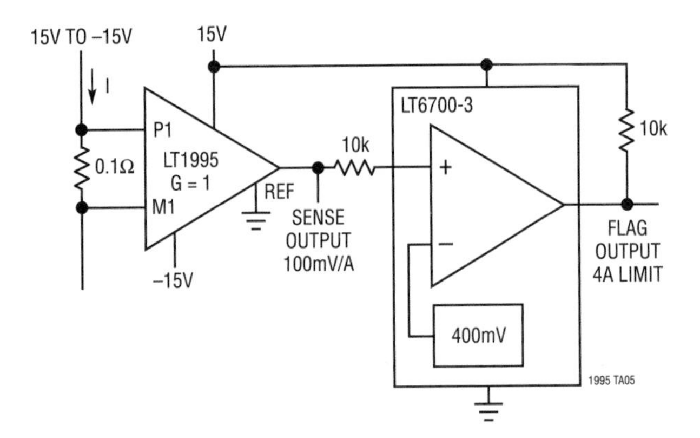

Figure 40.166 •

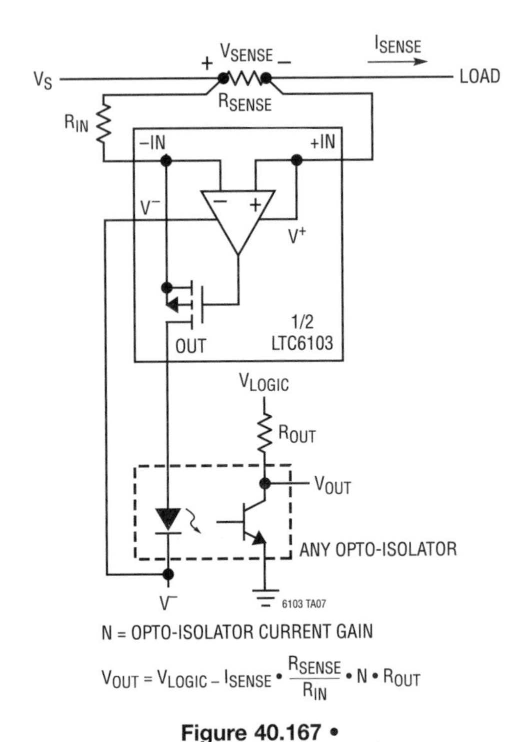

Figure 40.167 •

Monitor current in an isolated supply line (Figure 40.167)

Using the current sense amplifier output current to directly modulate the current in a photo diode is a simple method to monitor an isolated 48V industrial/telecom power supply. Current faults can be signaled to nonisolated monitoring circuitry.

Monitoring a fuse protected circuit (Figure 40.168)

Current sensing a supply line that has a fuse for overcurrent protection requires a current sense amplifier with a wide differential input voltage rating. Should the fuse blow open the full load supply voltage appears across the inputs to the sense amplifier. The LT6105 can work with input voltage differentials up to 44V. The LT6105 output slews at 2V/μs so can respond quickly to fast current changes. When the fuse opens the LT6105 output goes high and stays there.

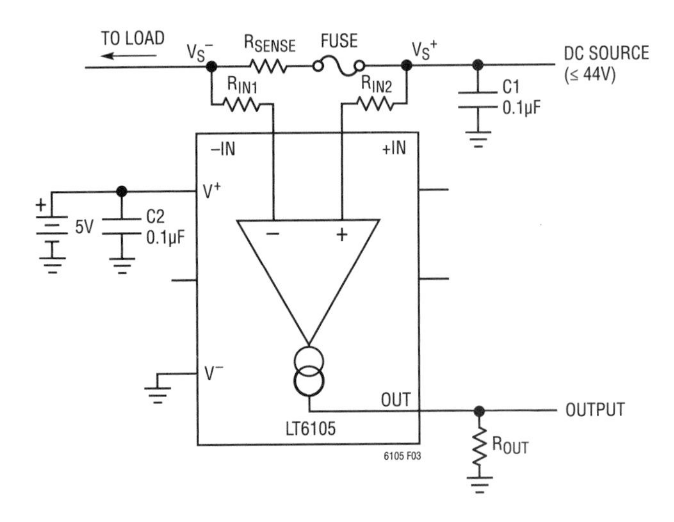

Figure 40.168 •

Circuit fault protection with early warning and latching load disconnect (Figure 40.169)

With a precision current sense amplifier driving two built in comparators, LT6109-2 can provide current overload protection to a load circuit. The internal comparators have a fixed 400mV reference. The current sense output is resistor divided down so that one comparator will trip at an early warning level and the second at a danger level of current to the load (100mA and 250mA in this example). The comparator outputs latch when tripped so they can be used as a circuit breaker to disconnect and protect the load until a reset pulse is provided.

Use comparator output to initialize interrupt routines (Figure 40.170)

The comparator outputs can connect directly to I/O or interrupt inputs to any microcontroller. A low level at OUTC2 can indicate an undercurrent condition while a low level at OUTC1 indicates an overcurrent condition. These interrupts force service routines in the microcontroller.

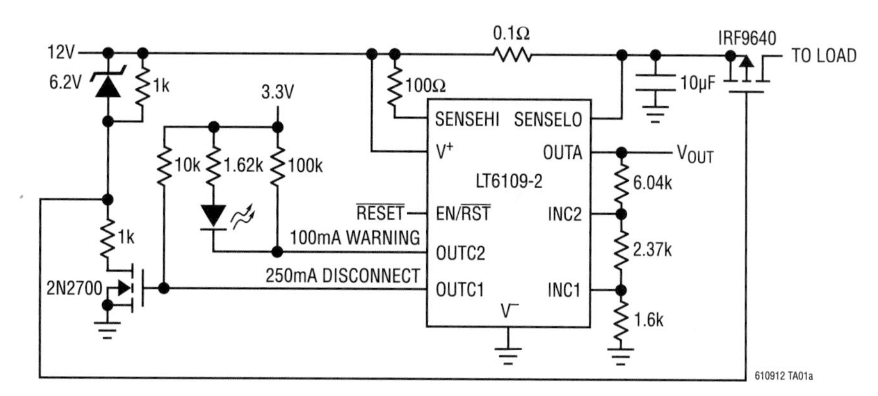

Figure 40.169 •

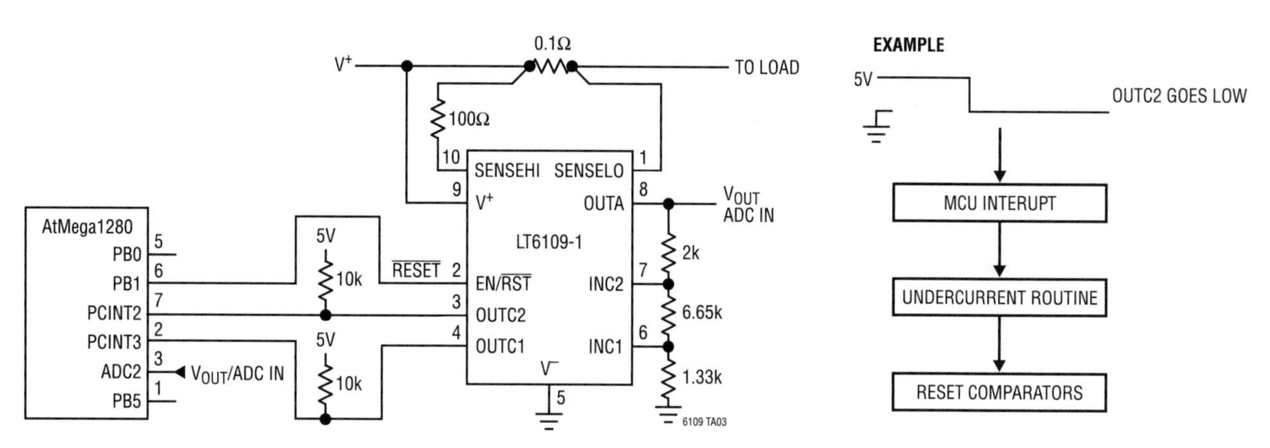

Figure 40.170 •

Current sense with overcurrent latch and power-on reset with loss of supply (Figure 40.171)

The LT6801-2 has a normal nonlatching comparator built in. An external logic gate configured in a positive feedback arrangement can create a latching output when an overcurrent condition is sensed. The same logic gate can also generate an active low power-on reset signal.

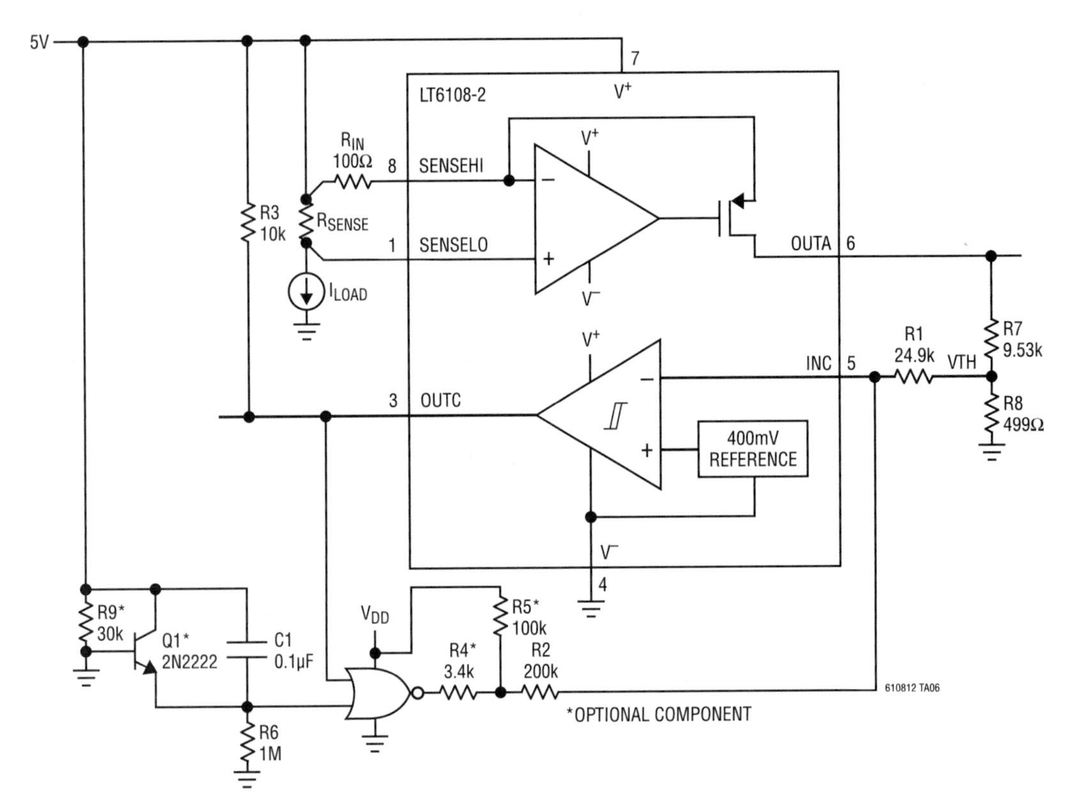

Figure 40.171 •

Digitizing

In many systems the analog voltage quantity indicating current flow must be input to a system controller. In this chapter several examples of the direct interface of a current sense amplifier to an A to D converter are shown.

Sensing output current (Figure 40.172)

The LT1970 is a 500mA power amplifier with voltage programmable output current limit. Separate DC voltage inputs and an output current sensing resistor control the maximum sourcing and sinking current values. These control voltages could be provided by a D-to-A converter in a microprocessor controlled system. For closed loop control of the current to a load an LT1787 can monitor the output current. The LT1880 op amp provides scaling and level shifting of the voltage applied to an A-to-D converter for a 5mV/mA feedback signal.

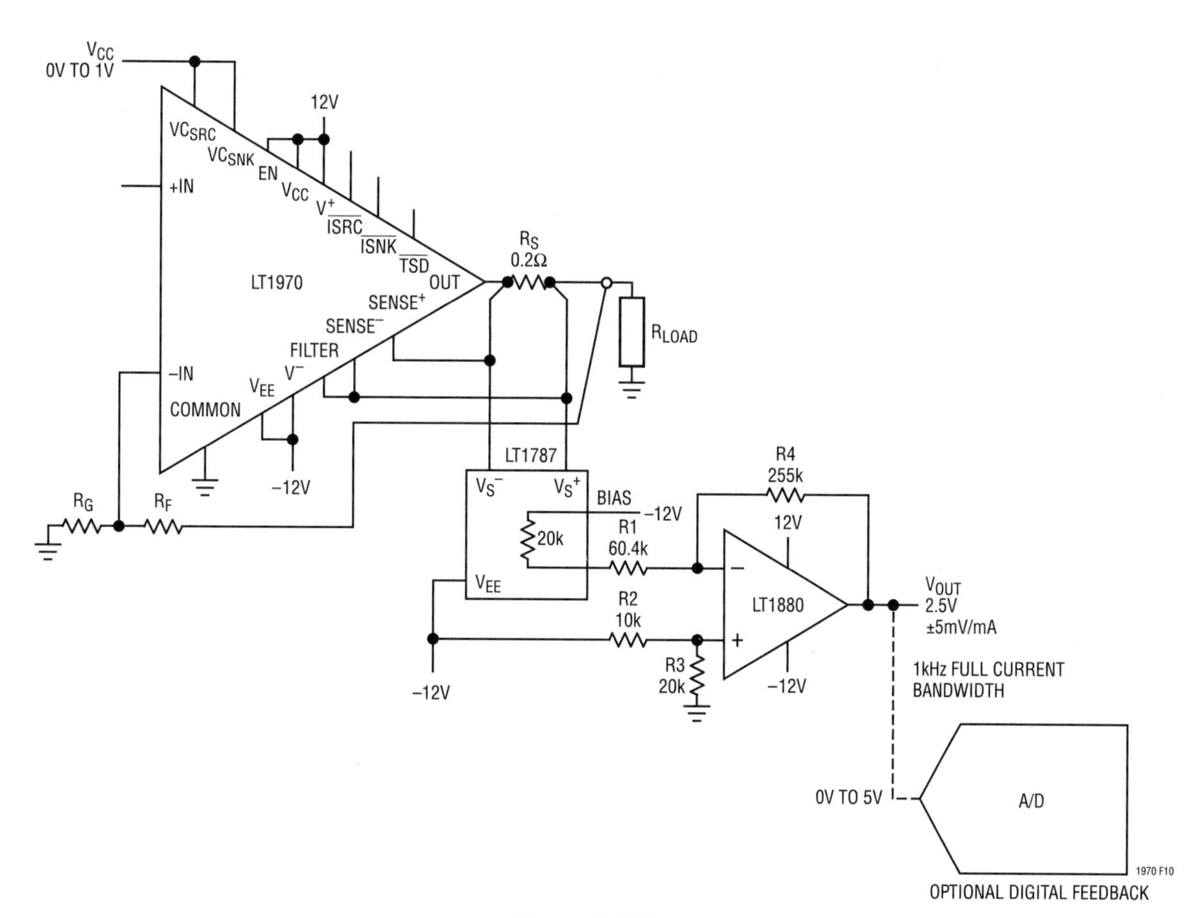

Figure 40.172 •

Split or single-supply operation, bidirectional output into A/D (Figure 40.173)

In this circuit, split supply operation is used on both the LT1787 and LT1404 to provide a symmetric bidirectional measurement. In the single-supply case, where the LT1787 Pin 6 is driven by V_{REF}, the bidirectional measurement range is slightly asymmetric due to V_{REF} being somewhat greater than midspan of the ADC input range.

16-bit resolution unidirectional output into LTC2433 ADC (Figure 40.174)

The LTC2433-1 can accurately digitize signal with source impedances up to 5kΩ. This LTC6101 current sense circuit uses a 4.99kΩ output resistance to meet this requirement, thus no additional buffering is necessary.

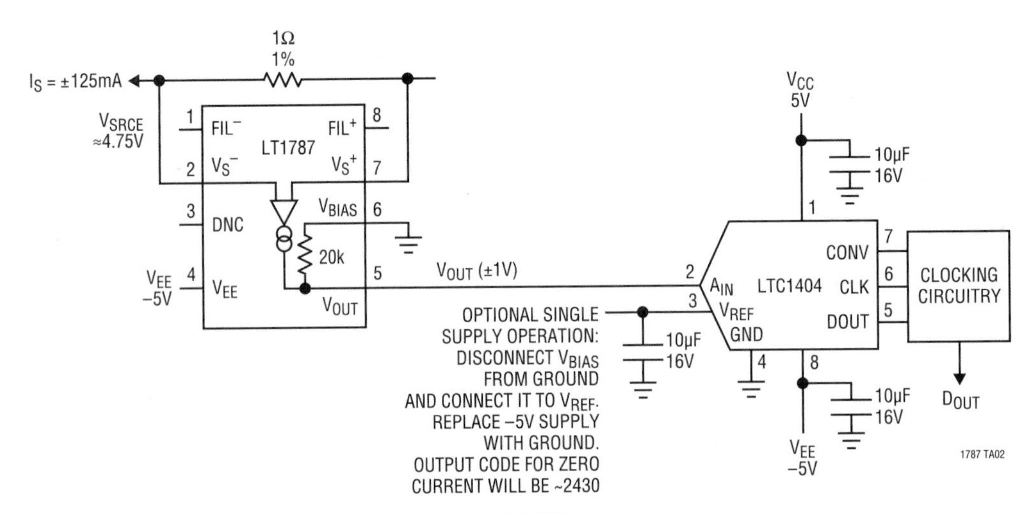

Figure 40.173 •

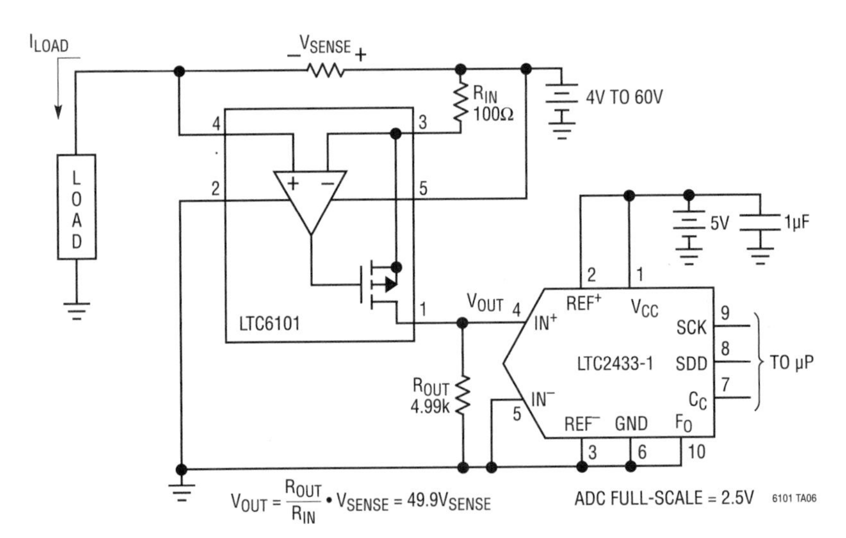

Figure 40.174 •

12-bit resolution unidirectional output into LTC1286 ADC (Figure 40.175)

While the LT1787 is able to provide a bidirectional output, in this application the economical LTC1286 is used to digitize a unidirectional measurement. The LT1787 has a nominal gain of eight, providing a 1.25V full-scale output at approximately 100A of load current.

Directly digitize current with 16-bit resolution (Figure 40.176)

The low offset precision of the LTC6102 permits direct digitization of a high side sensed current. The LTC2433 is a 16-bit delta sigma converter with a 2.5V full-scale range. A resolution of 16 bits has an LSB value of only 40μV In this circuit the sense voltage is amplified by a factor of 50. This translates to a sensed voltage resolution of only 0.8μV per count. The LTC6102 DC offset typically contributes only four LSB's of uncertainty.

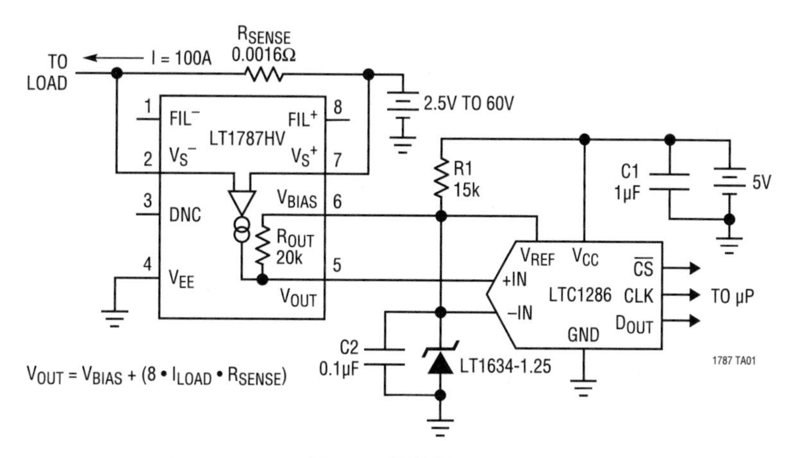

Figure 40.175 •

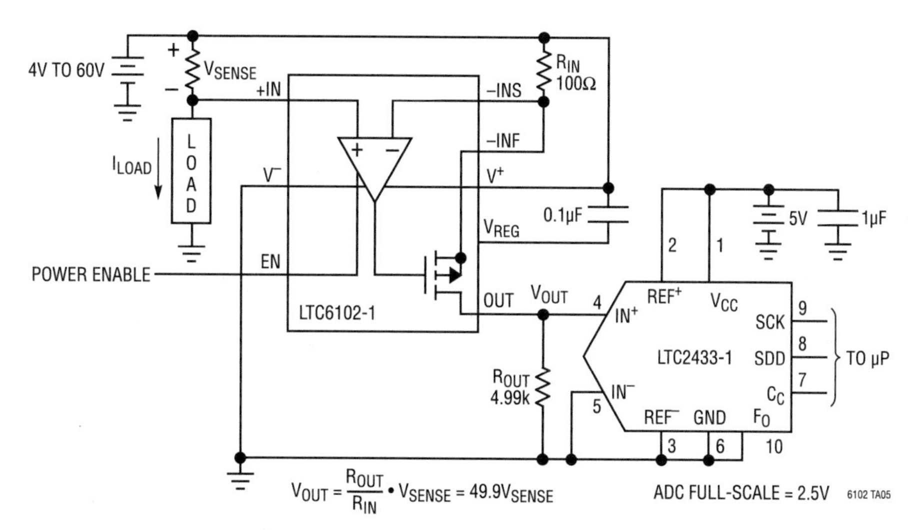

Figure 40.176 •

Directly digitizing two independent currents (Figure 40.177)

With two independent current sense amplifiers in the LTC6103, two currents from different sources can be simultaneously digitized by a 2-channel 16-bit ADC such as the LTC2436-1. While shown to have the same gain on each channel, it is not necessary to do so. Two different current ranges can be gain scaled to match the same full-scale range for each ADC channel.

Digitize a bidirectional current using a single-sense amplifier and ADC (Figure 40.178)

The dual LTC6104 can be connected in a fashion to source or sink current at its output depending on the direction of current flow through the sense resistor. Biasing the amplifier output resistor and the V_{REF} input of the ADC to an external 2.5V LT1004 voltage reference allows a 2.5V full-scale input voltage to the ADC for current flowing in either direction.

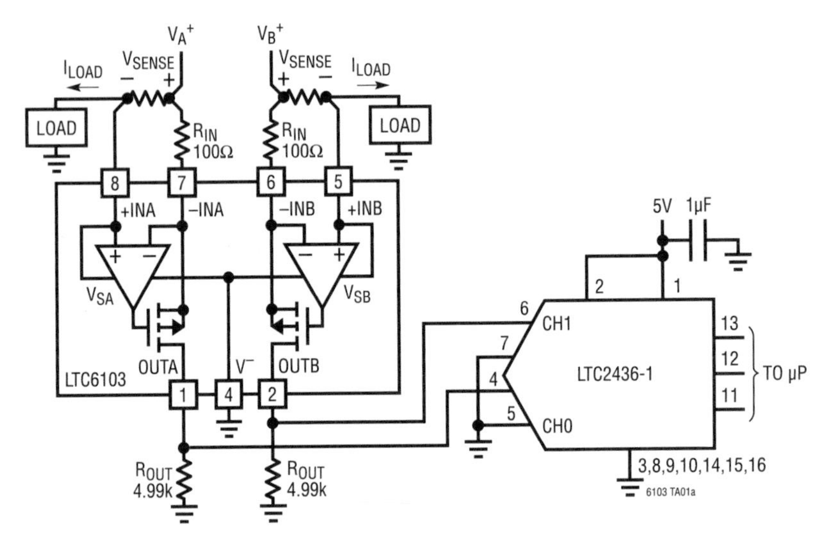

Figure 40.177 •

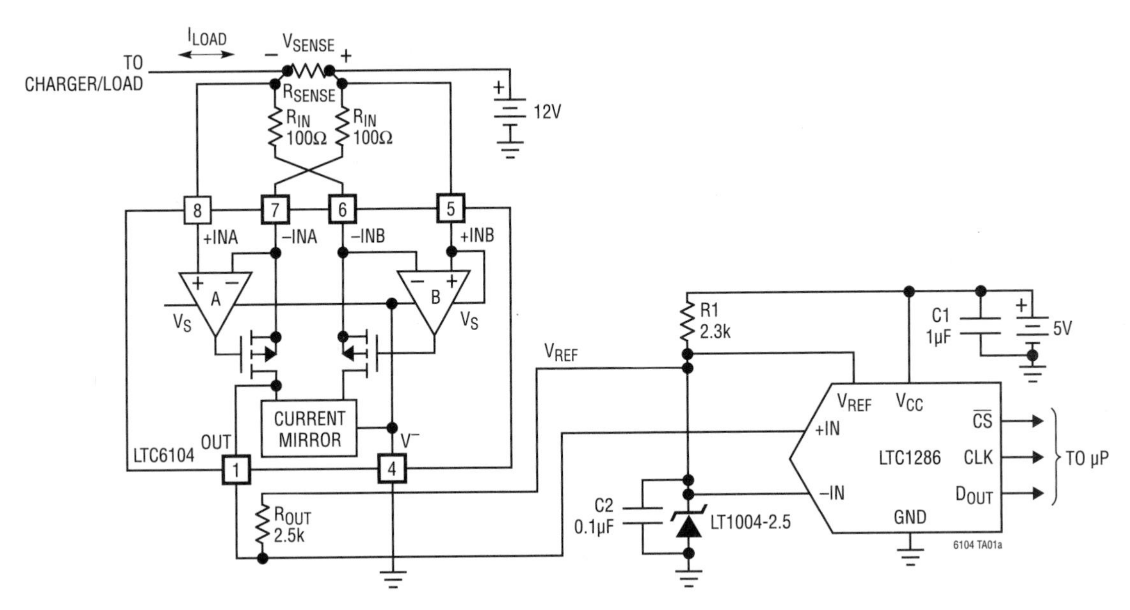

Figure 40.178 •

Digitizing charging and loading current in a battery monitor (Figure 40.179)

A 16-bit digital output battery current monitor can be implemented with just a single sense resistor, an LT1999 and an LTC2344 delta sigma ADC. With a fixed gain of ten and DC biased output the digital code indicates the instantaneous loading or charging current (up to 10A) of a system battery power source.

Complete digital current monitoring (Figure 40.180)

An LTC2470 16-bit delta sigma A-to-D converter can directly digitize the output of the LT6109 representing a circuit load current. At the same time the comparator outputs connect to MCU interrupt inputs to immediately signal programmable threshold over and undercurrent conditions.

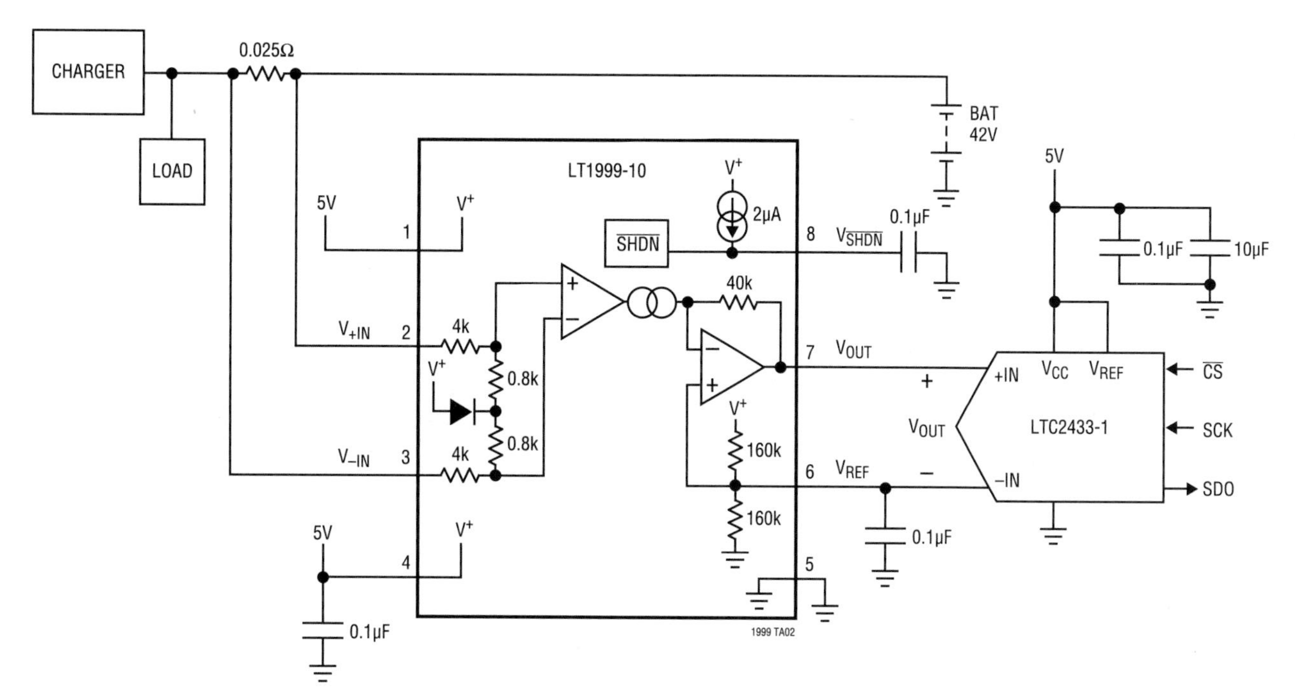

Figure 40.179 •

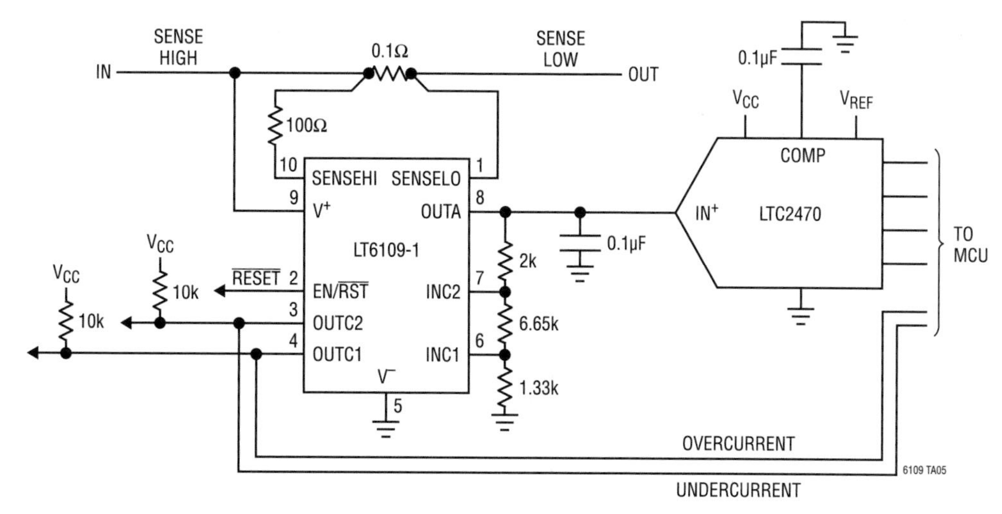

Figure 40.180 •

Ampere-hour gauge (Figure 40.181)

With specific scaling of the current sense resistor, the LTC4150 can be set to output exactly 10,000 interrupt pulses for one Amp-hr of charge drawn from a battery source. With such a base-10 round number of pulses a series of decade counters can be used to create a visual 5-digit display. This schematic is just the concept. The polarity output can be used to direct the interrupt pulses to either the count-up or count-down clock input to display total net charge.

Power sensing with built-in A-to-D converter (Figure 40.182)

The LTC4151 contains a dedicated current sense input channel to a 3-channel 12-bit delta-sigma ADC. The ADC directly and sequentially measures the supply voltage (102V full-scale), supply current (82mV full-scale) and a separate analog input channel (2V full-scale). The 12-bit resolution data for each measurement is output through an I²C link.

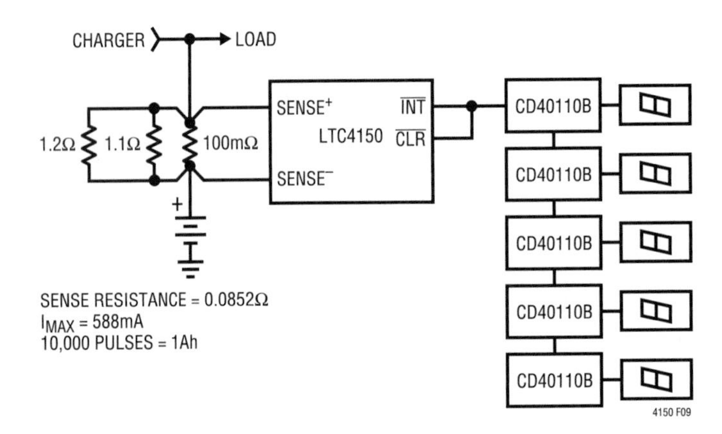

Figure 40.181 •

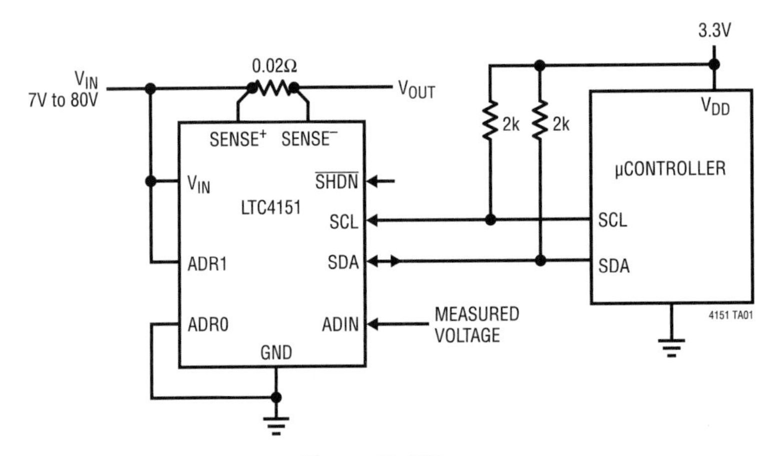

Figure 40.182 •

Isolated power measurement (Figure 40.183)

With separate data input and output pins, it is a simple matter to fully isolate the LTC4151-1/LTC4151-2 from a controller system. The supply voltage and operating current of the isolated system is digitized and conveyed through three opto-isolators.

Fast data rate isolated power measurement (Figure 40.184)

With separate data input and output pins, it is a simple matter to fully isolate the LTC4151-1/LTC4151-2 from a controller system. The supply voltage and operating current of the isolated system is digitized and conveyed through three high speed opto-isolators.

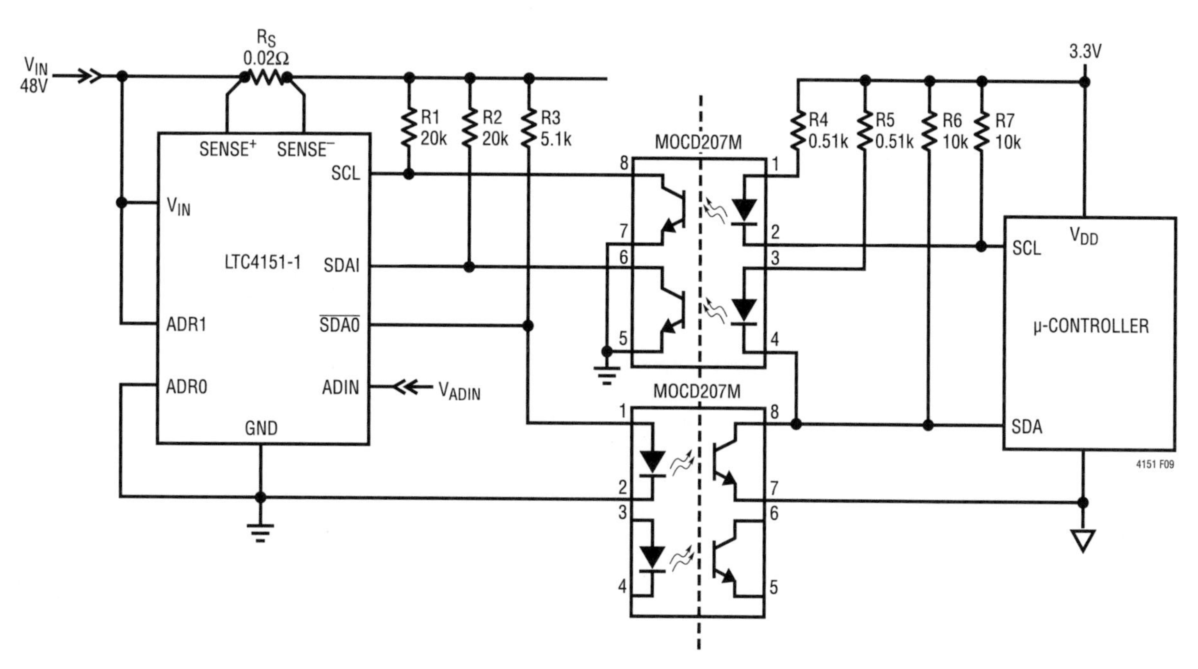

Figure 40.183 •

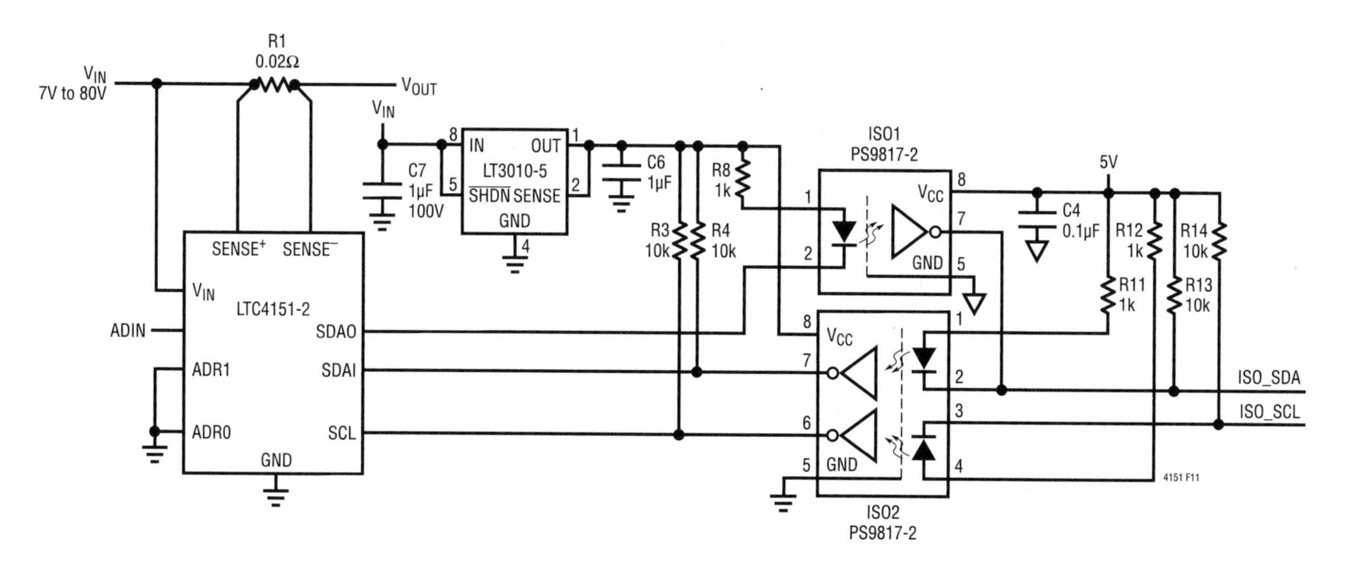

Figure 40.184 •

Adding temperature measurement to supply power measurement (Figure 40.185)

One use for the spare analog input of the LTC4151 could be to measure temperature. This can be done by using a thermistor to create a DC voltage proportional to temperature. The DC bias potential for the temperature network is the system power supply which is also measured, Temperature is derived from both measurements. In addition the system load current is also measured.

Current, voltage and fuse monitoring (Figure 40.186)

Systems with redundant back-up power often have fuse protection on the supply output. The LTC4151, with some diodes and resistors can measure the total load current, supply voltage and detect the integrity of the supply fuses. The voltage on the spare analog input channel determines the state of the fuses.

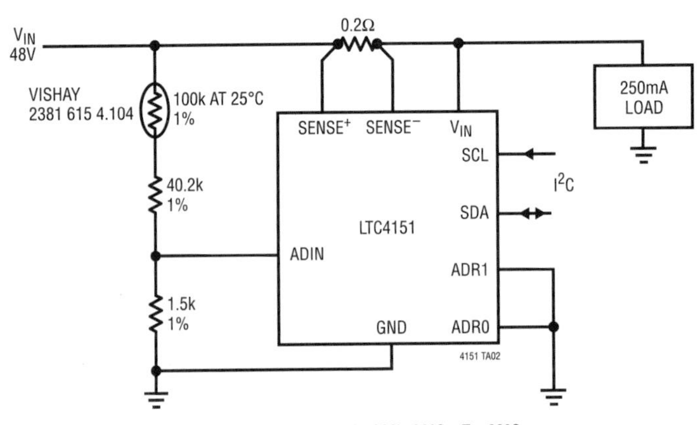

T(°C) = 58.82 • (N_{ADIN}/N_{VIN} − 0.1066), 20°C < T < 60°C.
N_{ADIN} AND N_{VIN} ARE DIGITAL CODES MEASURED BY THE ADC AT THE ADIN AND V_{IN} PINS, RESPECTIVELY

Figure 40.185 •

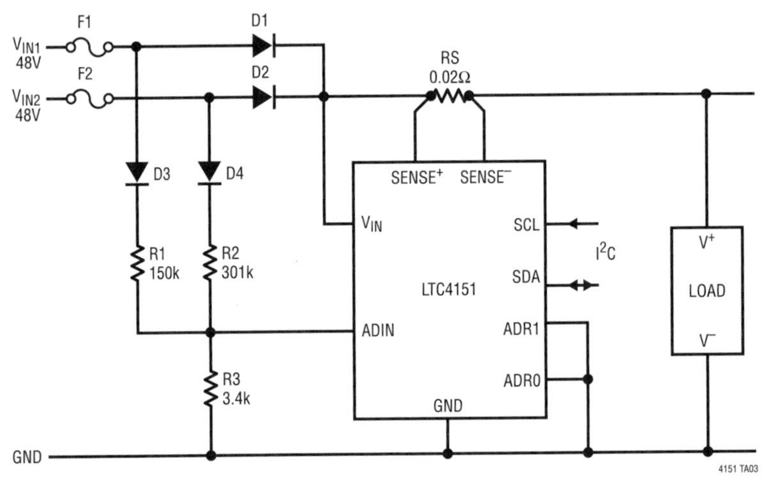

CONDITION	RESULT
$N_{ADIN} \geq 1.375 \bullet N_{VIN}$	Normal Operation
$0.835 \bullet N_{VIN} \leq N_{ADIN} < 1.375 \bullet N_{VIN}$	F2 is Open
$0.285 \bullet N_{VIN} \leq N_{ADIN} < 0.835 \bullet N_{VIN}$	F1 is Open
(Not Responding)	Both F1 and F2 are Open

V_{IN1} AND V_{IN2} ARE WITHIN 20% APART. N_{ADIN} AND N_{VIN} ARE DIGITAL CODES MEASURED BY THE ADC AT THE ADIN AND V_{IN} PINS, RESPECTIVELY.

Figure 40.186 •

Automotive socket power monitoring (Figure 40.187)

The wide operating voltage range is adequate to permit the transients seen in automotive applications. The power consumption of anything plugged into an auto power socket can be directly digitized.

Power over Ethernet, PoE, monitoring (Figure 40.188)

The power drawn by devices connected to an isolated telecom power supply can be continually monitored to ensure that they comply with their power class rating. A voltage proportional to the powered device rating is digitized by the spare analog input of the LTC4151-1.

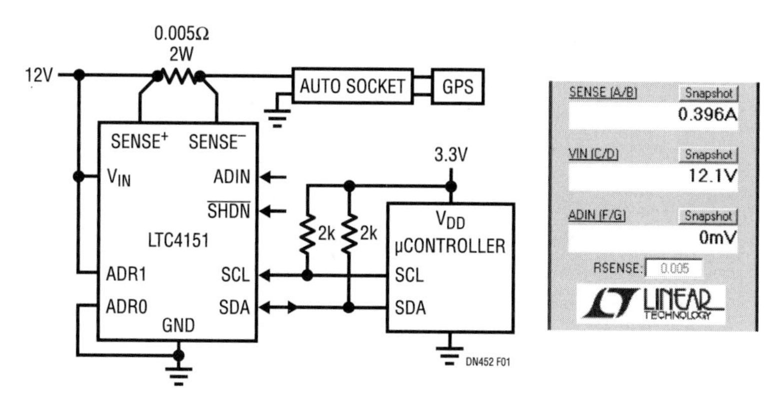

Figure 40.187 •

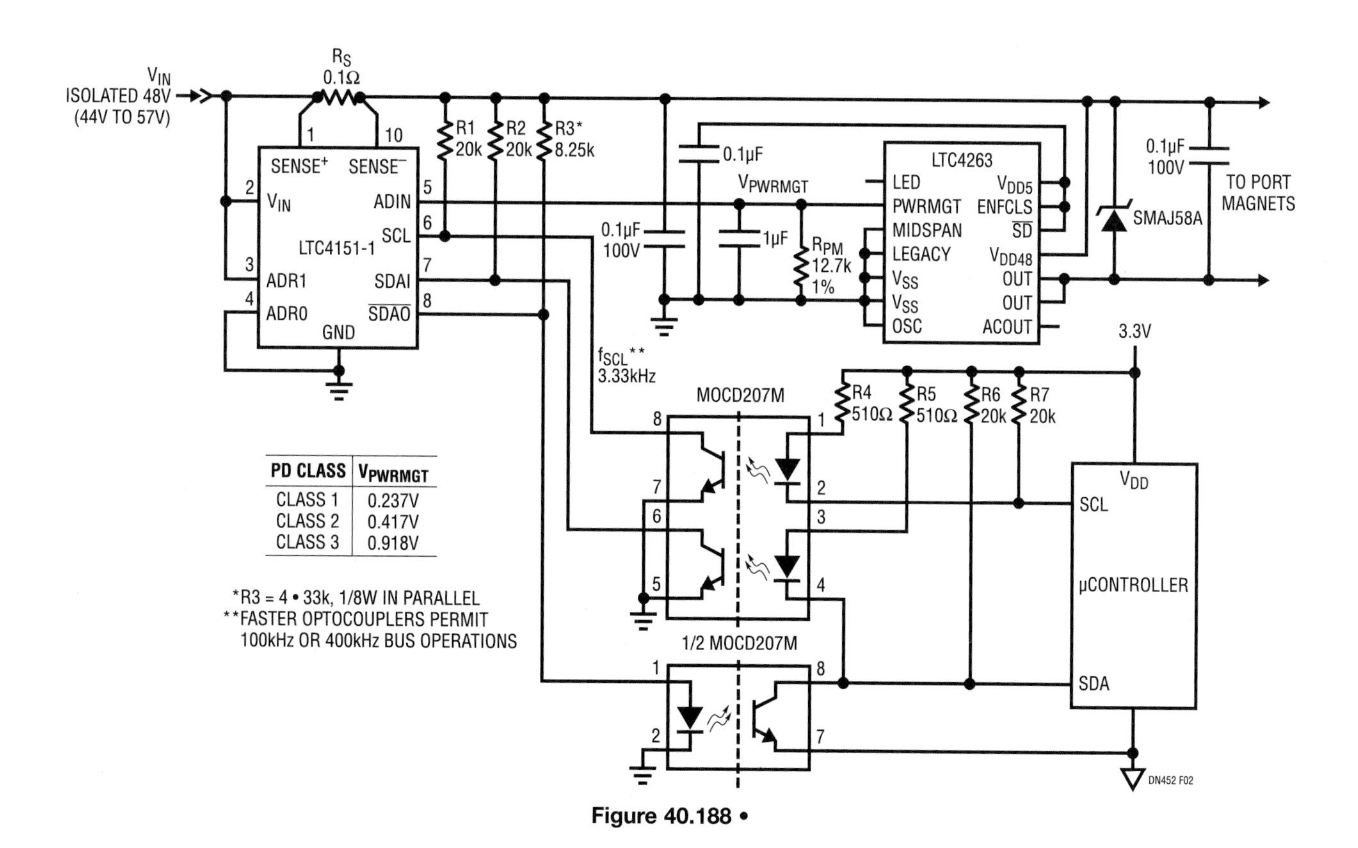

PD CLASS	V_{PWRMGT}
CLASS 1	0.237V
CLASS 2	0.417V
CLASS 3	0.918V

Figure 40.188 •

Monitor current, voltage and temperature (Figure 40.189)

The LTC2990 is a 4-channel, 14-bit ADC fully configurable through an I^2C interface to measure single-ended, differential voltages and determine temperature from internal or external diode sensors. For high side current measurements two of the inputs are configured for differential input to measure the voltage across a sense resistor. The maximum differential input voltage is limited to ±300mV. Other channels can measure voltage and temperature for a complete system power monitor.

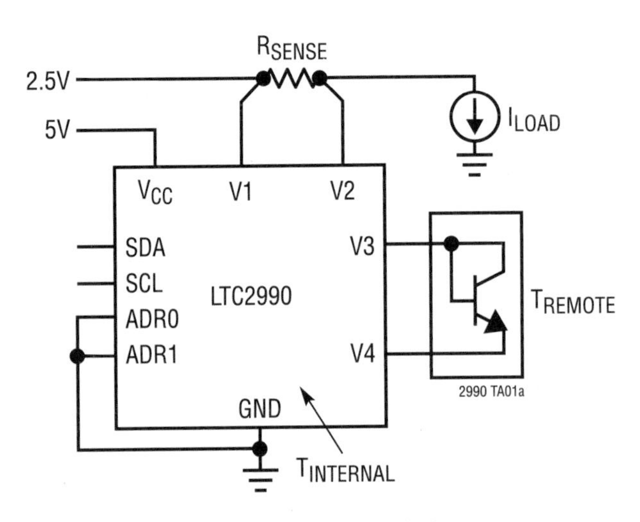

Figure 40.189 •

Current control

This chapter collects a variety of techniques useful in generating controlled levels of current in circuits.

800mA/1A white LED current regulator (Figure 40.190)

The LT6100 is configured for a gain of either 40V/V or 50V/V depending on whether the switch between A2 and

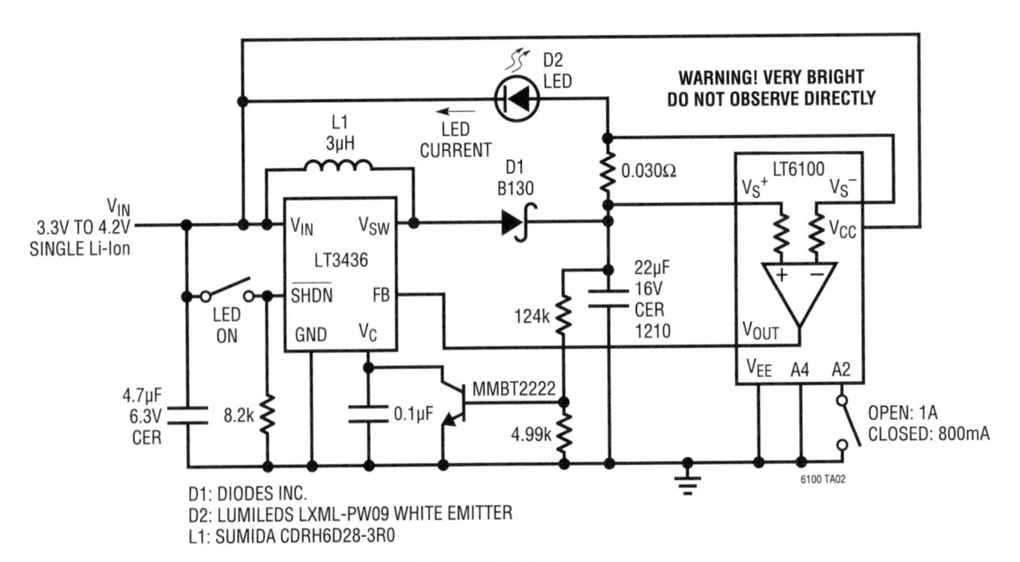

Figure 40.190 •

V_{EE} is closed or not. When the switch is open (LT6100 gain of 40V/V), 1A is delivered to the LED. When the switch is closed (LT6100 gain of 50V/V), 800mA is delivered. The LT3436 is a boost switching regulator which governs the voltage/current supplied to the LED. The switch "LED ON" connected to the SHDN pin allows for external control of the ON/OFF state of the LED.

Bidirectional current source (Figure 40.191)

The LT1990 is a differential amplifier with integrated precision resistors. The circuit shown is the classic Howland current source, implemented by simply adding a sense resistor.

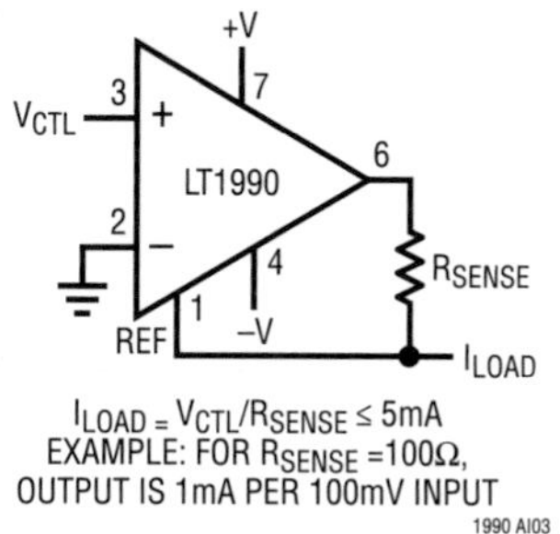

Figure 40.191 •

2-terminal current regulator (Figure 40.192)

The LT1635 combines an op amp with a 200mV reference. Scaling this reference voltage to a potential across resistor R3 forces a controlled amount of current to flow from the +terminal to the −terminal. Power is taken from the loop.

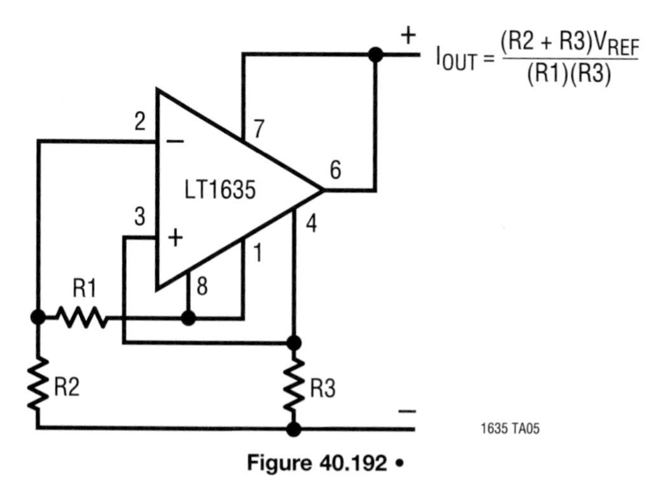

Figure 40.192 •

Variable current source (Figure 40.193)

A basic high side current source is implemented at the output, while an input translation amplifier section provides

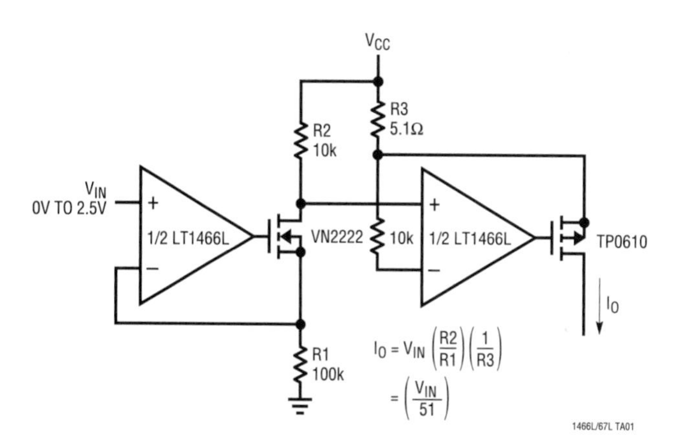

Figure 40.193 •

for flexible input scaling. A rail-to-rail input capability is required to have both amplifiers in one package, since the input stage has common mode near ground and the second section operates near V_{CC}.

Precision voltage controlled current source with ground referred input and output (Figure 40.194)

The LTC6943 is used to accurately sample the voltage across the 1kΩ sense resistor and translate it to a ground reference by charge balancing in the 1μF capacitors. The LTC2050 integrates the difference between the sense voltage and the input command voltage to drive the proper current into load.

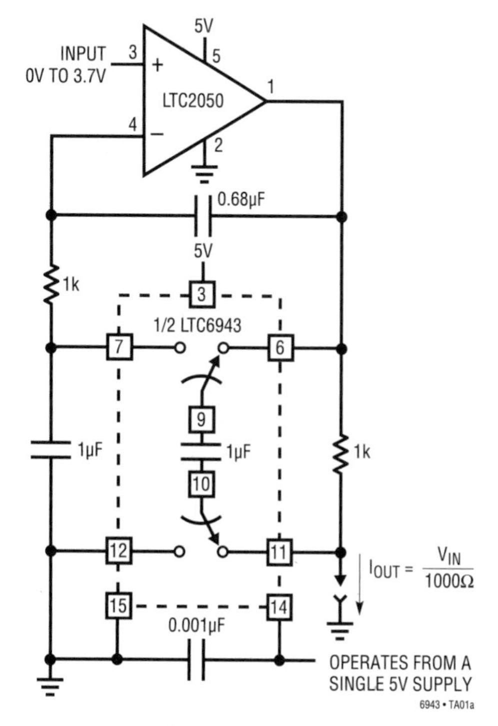

Figure 40.194 •

Precision voltage controlled current source (Figure 40.195)

The ultra-precise LTC2053 instrumentation amplifier is configured to servo the voltage drop on sense resistor R to match the command V_C. The LTC2053 output capability limits this basic configuration to low current applications.

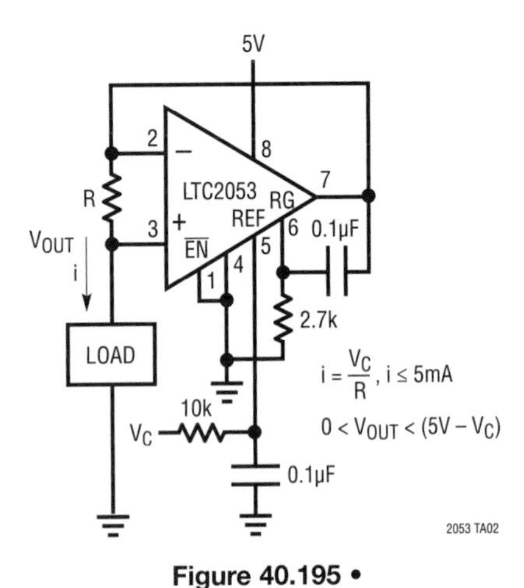

Figure 40.195 •

Switchable precision current source (Figure 40.196)

This is a simple current-source configuration where the op amp servos to establish a match between the drop on the sense resistor and that of the 1.2V reference. This particular op amp includes a shutdown feature so the current source function can be switched off with a logic command. The 2kΩ pull-up resistor assures the output MOSFET is off when the op amp is in shutdown mode.

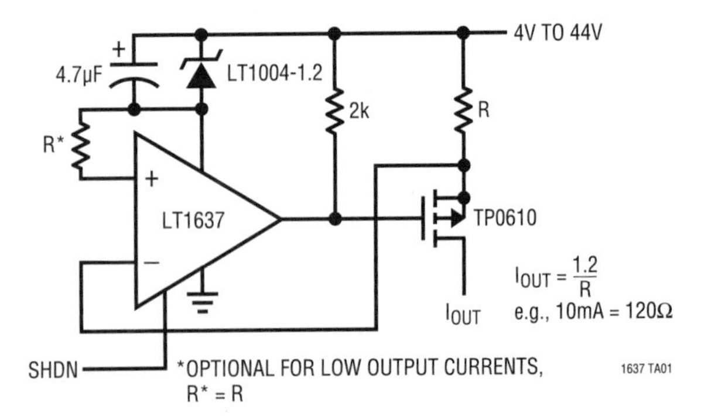

Figure 40.196 •

Boosted bidirectional controlled current source (Figure 40.197)

This is a classical Howland bidirectional current source implemented with an LT1990 integrated difference amplifier. The op amp circuit servos to match the R_{SENSE} voltage drop to the input command V_{CTL}. When the load current exceeds about 0.7mA in either direction, one of the boost transistors will start conducting to provide the additional commanded current.

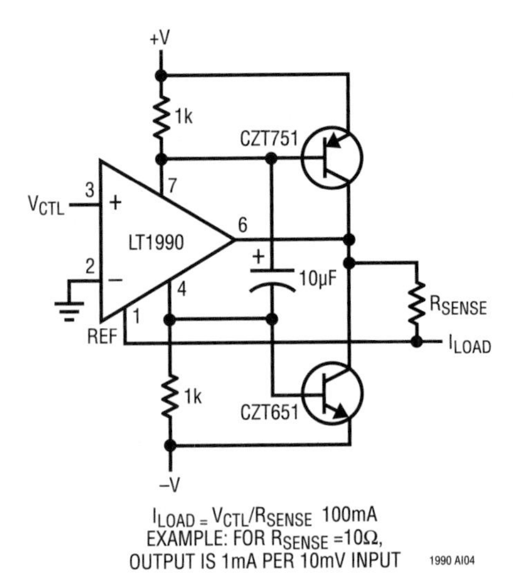

Figure 40.197 •

0A to 2A current source (Figure 40.198)

The LT1995 amplifies the sense resistor drop by 5V/V and subtracts that from V_{IN}, providing an error signal to an LT1880 integrator. The integrated error drives the P-MOSFET as required to deliver the commanded current.

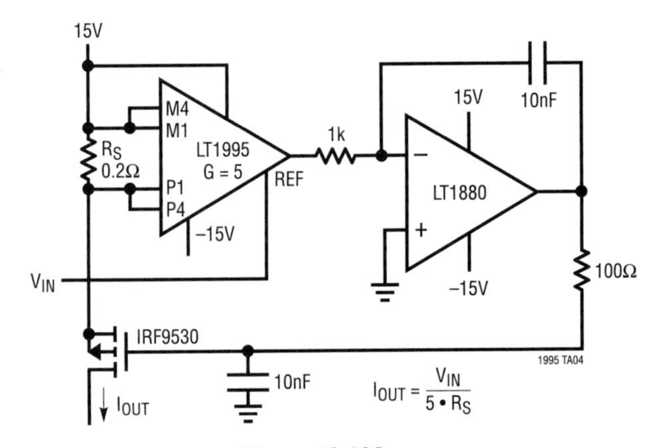

Figure 40.198 •

Fast differential current source (Figure 40.199)

This is a variation on the Howland configuration, where load current actually passes through a feedback resistor as an implicit sense resistance. Since the effective sense resistance is relatively large, this topology is appropriate for producing small controlled currents.

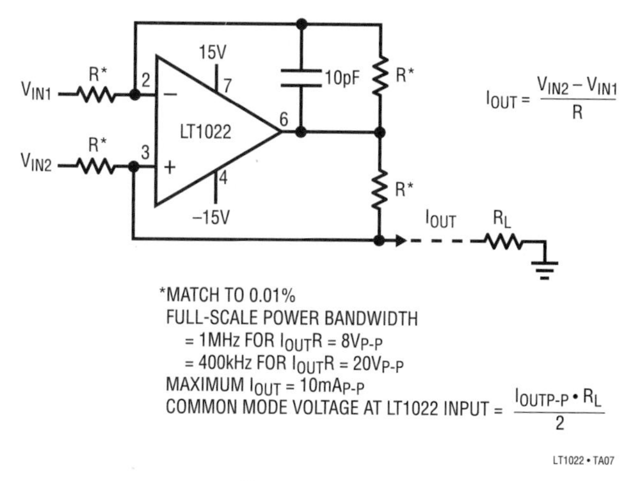

Figure 40.199 •

1A voltage-controlled current sink (Figure 40.200)

This is a simple controlled current sink, where the op amp drives the N-MOSFET gate to develop a match between the 1Ω sense resistor drop and the V_{IN} current command. Since the common mode voltage seen by the op amp is near ground potential, a "single supply" or rail-to-rail type is required in this application.

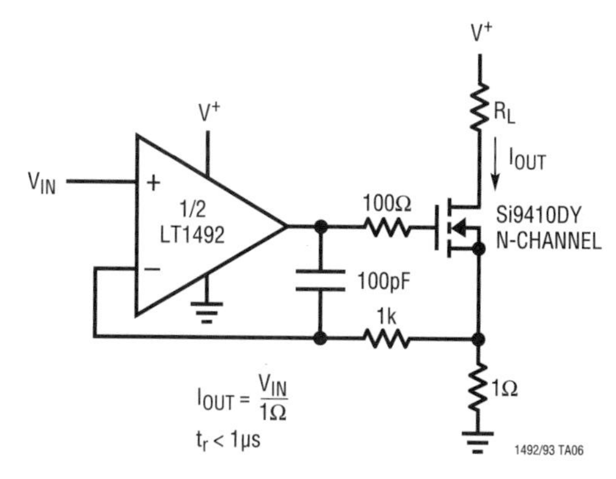

Figure 40.200 •

Voltage controlled current source (Figure 40.201)

Adding a current sense amplifier in the feedback loop of an adjustable low dropout voltage regulator creates a simple voltage controlled current source. The range of output current sourced by the circuit is set only by the current capability of the voltage regulator. The current sense amplifier senses the output current and feeds back a current to the summing junction of the regulator's error amplifier. The regulator will then source whatever current is necessary to maintain the internal reference voltage at the summing junction. For the circuit shown a 0V to 5V control input produces 500mA to 0mA of output current.

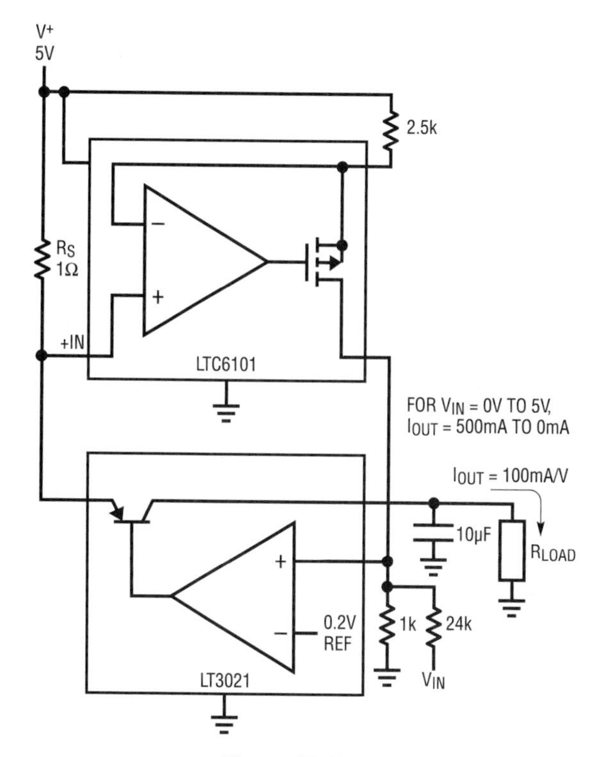

Figure 40.201 •

Adjustable high side current source (Figure 40.202)

The wide-compliance current source shown takes advantage of the LT1366's ability to measure small signals near the positive supply rail. The LT1366 adjusts Q1's gate voltage to force the voltage across the sense resistor (R_{SENSE}) to equal the voltage between V_{DC} and the potentiometer's wiper. A rail-to-rail op amp is needed because the voltage across the sense resistor is nearly the same as V_{DC}. Q2 acts as a constant current sink to minimize error in the reference voltage when the supply voltage varies. At low input voltage, circuit operation is limited by the Q1 gate drive requirement. At high input voltage, circuit operation is limited by the LT1366's absolute maximum ratings.

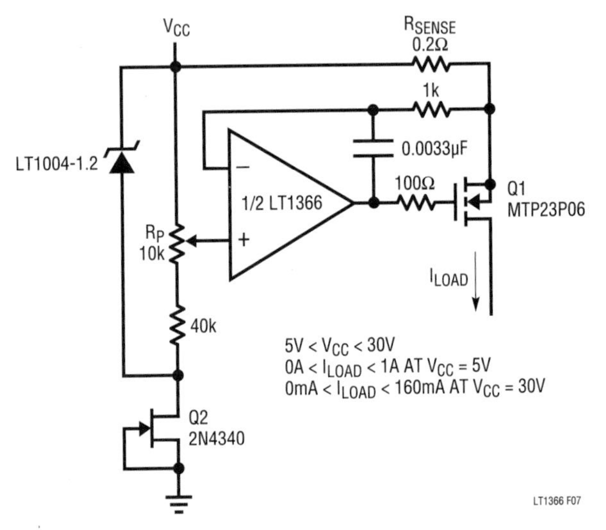

Figure 40.202 •

Programmable constant current source (Figure 40.203)

The current output can be controlled by a variable resistor (R_{PROG}) connected from the PROG pin to ground on the LT1620. The LT1121 is a low dropout regulator that keeps the voltage constant for the LT1620. Applying a shutdown command to the LT1121 powers down the LT1620 and eliminates the base drive to the current regulation pass transistor, thereby turning off I_{OUT}.

Snap back current limiting (Figure 40.204)

The LT1970 provides current detection and limiting features built-in. In this circuit, the logic flags that are produced in a current limiting event are connected in a feedback arrangement that in turn reduces the current limit command to a lower level. When the load condition permits the current to drop below the limiting level, then the flags clear and full current drive capability is restored automatically.

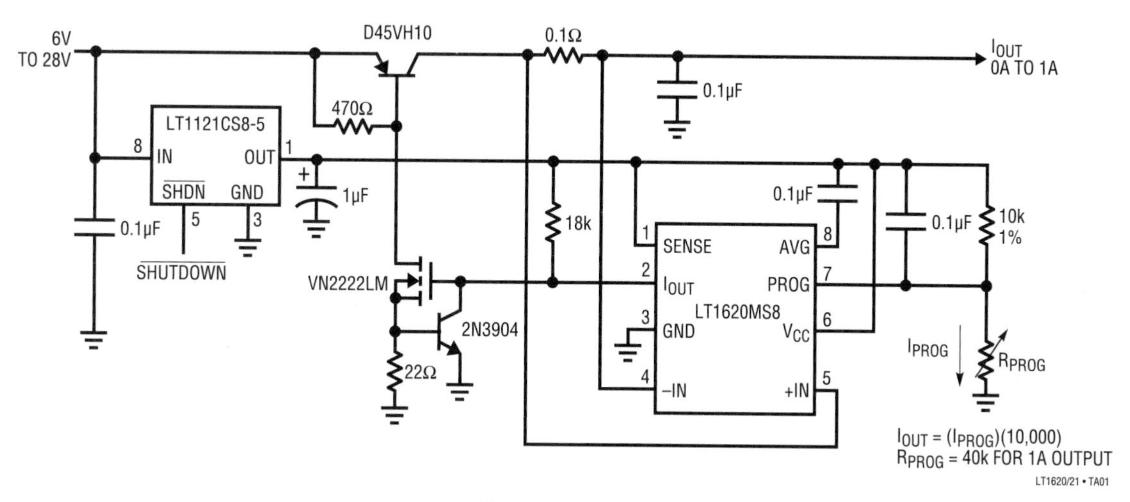

Figure 40.203 •

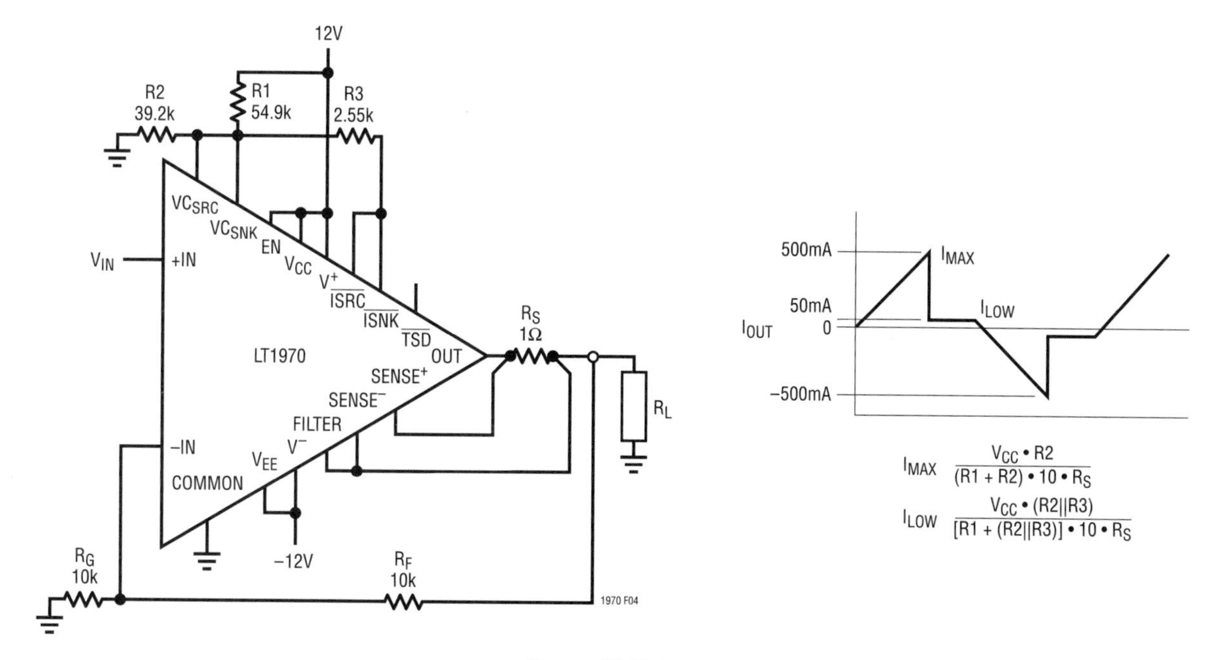

$$I_{MAX} = \frac{V_{CC} \cdot R2}{(R1 + R2) \cdot 10 \cdot R_S}$$

$$I_{LOW} = \frac{V_{CC} \cdot (R2||R3)}{[R1 + (R2||R3)] \cdot 10 \cdot R_S}$$

Figure 40.204 •

Precision

Offset voltage and bias current are the primary sources of error in current sensing applications. To maintain precision operation the use of zero drift amplifier virtually eliminates the offset error terms.

Precision high side power supply current sense (Figure 40.205)

This is a low voltage, ultra high precision monitor featuring a zero drift instrumentation amplifier (IA) that provides rail-to-rail inputs and outputs. Voltage gain is set by the feedback resistors. Accuracy of this circuit is set by the quality of resistors selected by the user, small-signal range is limited by V_{OL} in single-supply operation. The voltage rating of this part restricts this solution to applications of <5.5V This IA is sampled, so the output is discontinuous with input changes, thus only suited to very low frequency measurements.

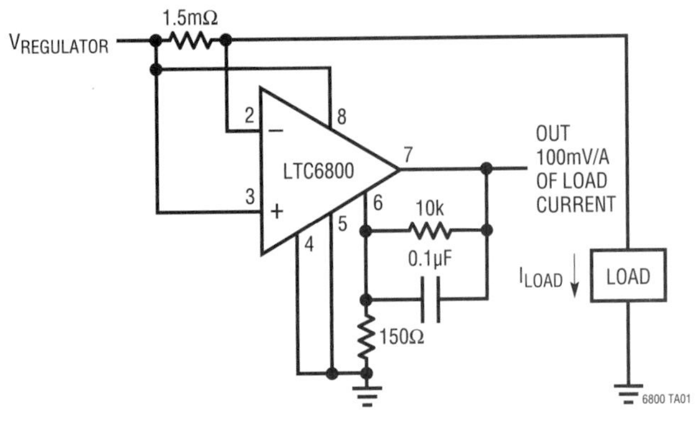

Figure 40.205 •

High side power supply current sense (Figure 40.206)

The low offset error of the LTC6800 allows for unusually low sense resistance while retaining accuracy.

Second input R minimizes error due to input bias current (Figure 40.207)

The second input resistor decreases input error due caused by the input bias current. For smaller values of R_{IN} this may not be a significant consideration.

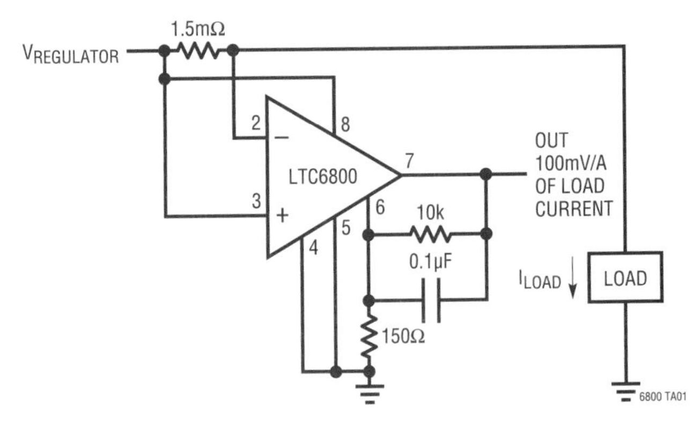

Figure 40.206 •

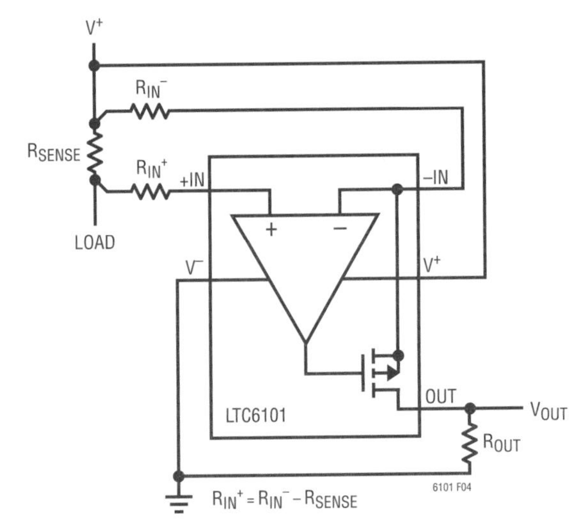

Figure 40.207 •

Remote current sensing with minimal wiring (Figure 40.208)

Since the LTC6102 (and others) has a current output that is ordinarily converted back to a voltage with a local load resistance, additional wire resistance and ground offsets don't directly affect the part behavior. Consequently, if the load resistance is placed at the far end of a wire, the voltage developed at the destination will be correct with respect to the destination ground potential.

Use Kelvin connections to maintain high current accuracy (Figure 40.209)

Significant errors are caused by high currents flowing through PCB traces in series with the connections to the sense amplifier. Using a sense resistor with integrated V_{IN} sense terminals provides the sense amplifier with only the voltage across the sense resistor. Using the LTC6104 maintains precision for currents flowing in both directions, ideal for battery charging applications.

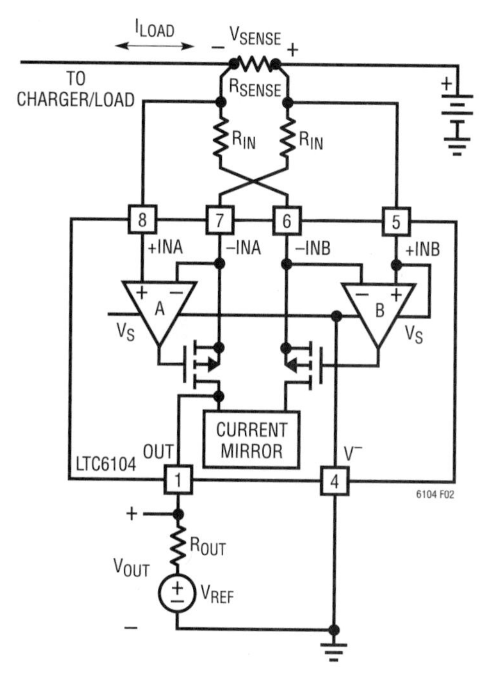

Figure 40.209 •

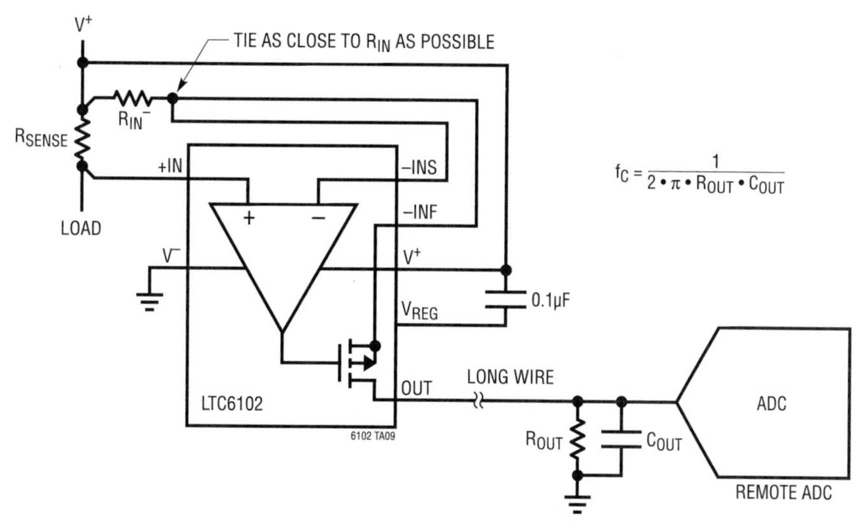

$$f_C = \frac{1}{2 \bullet \pi \bullet R_{OUT} \bullet C_{OUT}}$$

Figure 40.208 •

Crystal/reference oven controller (Figure 40.210)

High precision instrumentation often use small ovens to establish constant operation temperature for critical oscillators and reference voltages. Monitoring the power (current and voltage) to the heater as well as the temperature is required in a closed-loop control system.

Power intensive circuit board monitoring (Figure 40.211)

Many systems contain densely populated circuit boards using high power dissipation devices such as FPGAs. 8-channel, 14-bit ADC LTC2991 can be used to monitor device power consumption with voltage and current measurements as well as temperatures at several points on the board and even inside devices which provide die temp monitoring. A PWM circuit is also built-in to provide closed-loop control of PCB operating temperature.

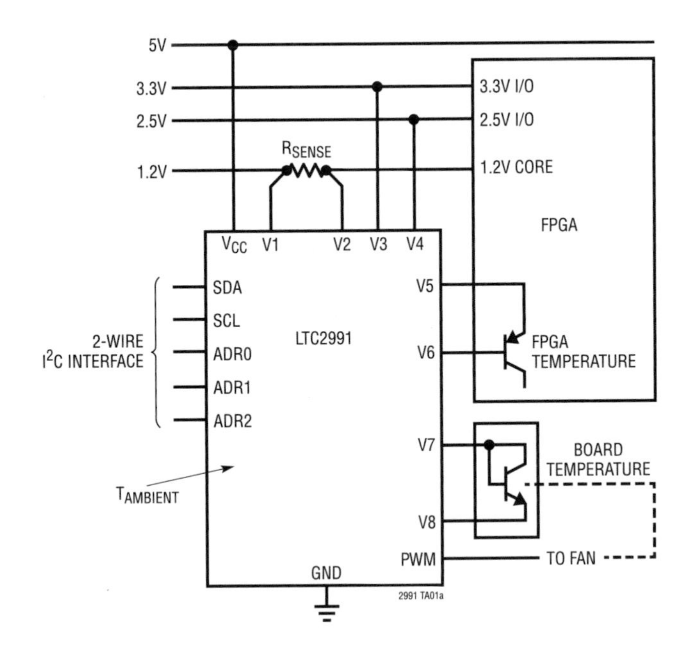

Figure 40.211 •

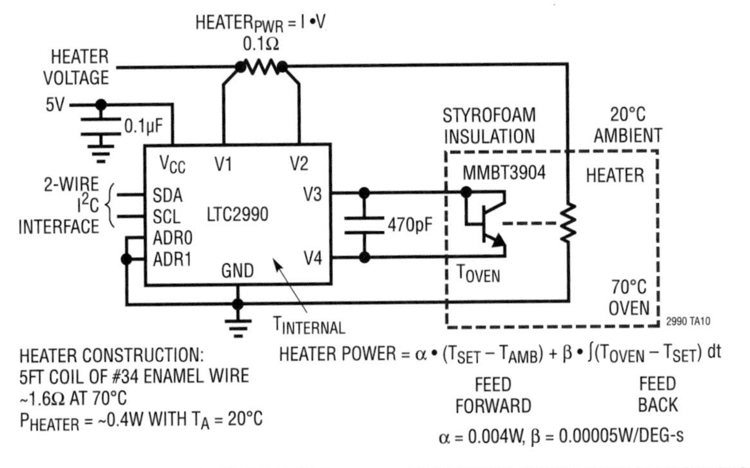

VOLTAGE AND TEMPERATURE CONFIGURATION:
CONTROL REGISTER: 0x58

T_{AMB}	REG 4, 5	0.0625°C/LSB
V1, V2	REG 8, 9	305.18μVLSB
T_{OVEN}	REG A, B	0.0625°C/LSB
V_{CC}	REG E, F	2.5V + 305.18μV/LSB

CURRENT AND TEMPERATURE CONFIGURATION:
CONTROL REGISTER: 0x59

T_{AMB}	REG 4, 5	0.0625°C/LSB
I_{HEATER}	REG 6, 7	269μVLSB
T_{HEATER}	REG A, B	0.0625°C/LSB
V_{CC}	REG E, F	2.5V + 305.18μV/LSB

Figure 40.210 •

Crystal/reference oven controller (Figure 40.212)

High precision instrumentation often use small ovens to establish constant operation temperature for critical oscillators and reference voltages. Monitoring the power (current and voltage) to the heater as well as the temperature is required in a closed-loop control system. The LTC2991 includes a PWM output which can provide closed-loop control of the heater.

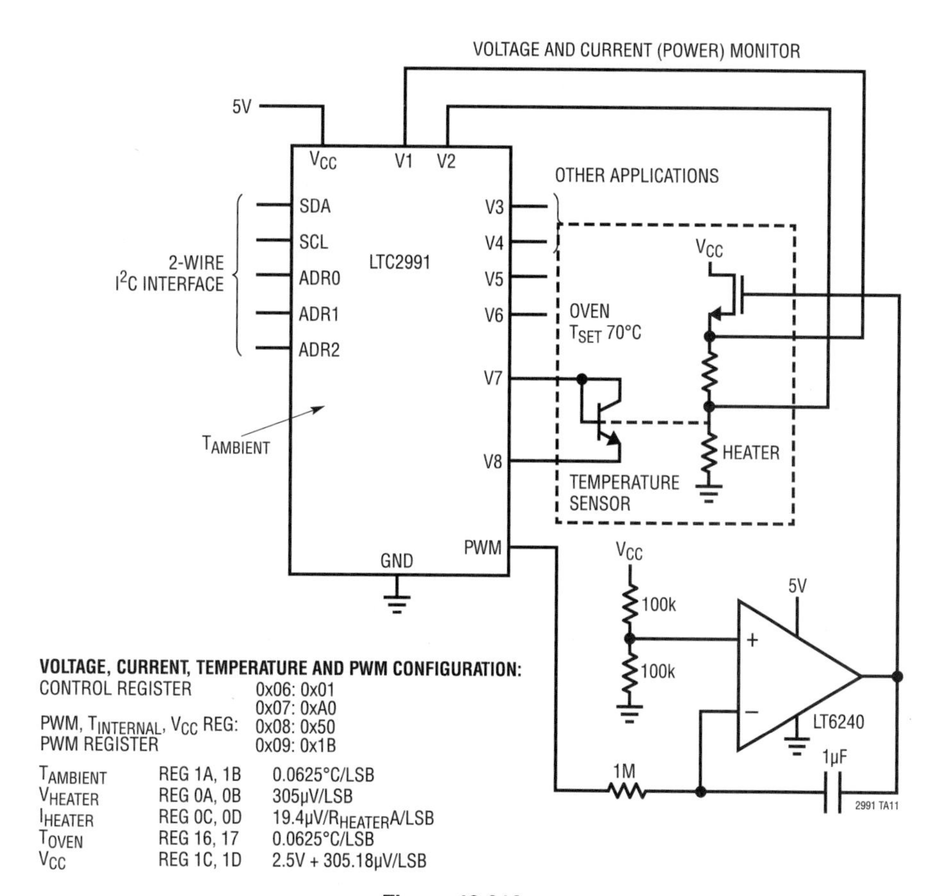

Figure 40.212 •

Wide range

To measure current over a wide range of values requires gain changing in the current sense amplifier. This allows the use of a single value of sense resistor: The alternative approach is to switch values of sense resistor. Both approaches are viable for wide range current sensing.

Dual LTC6101s allow high-low current ranging (Figure 40.213)

Using two current sense amplifiers with two values of sense resistors is an easy method of sensing current over a wide range. In this circuit the sensitivity and resolution of measurement is 10 times greater with low currents, less than 1.2A, than with higher currents. A comparator detects higher current flow, up to 10A, and switches sensing over to the high current circuitry.

Adjust gain dynamically for enhanced range (Figure 40.214)

Instead of having fixed gains of 10, 12.5, 20, 25, 40, and 50, this circuit allows selecting between two gain settings. An N-MOSFET switch is placed between the two gain-setting terminals (A2, A4) and ground to provide selection of gain = 10 or gain = 50, depending on the state of the gate drive. This provides a wider current measurement range than otherwise possible with just a single sense resistor.

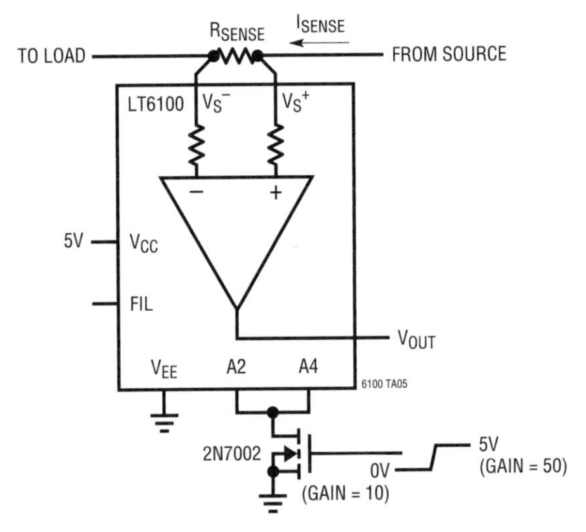

Figure 40.214 •

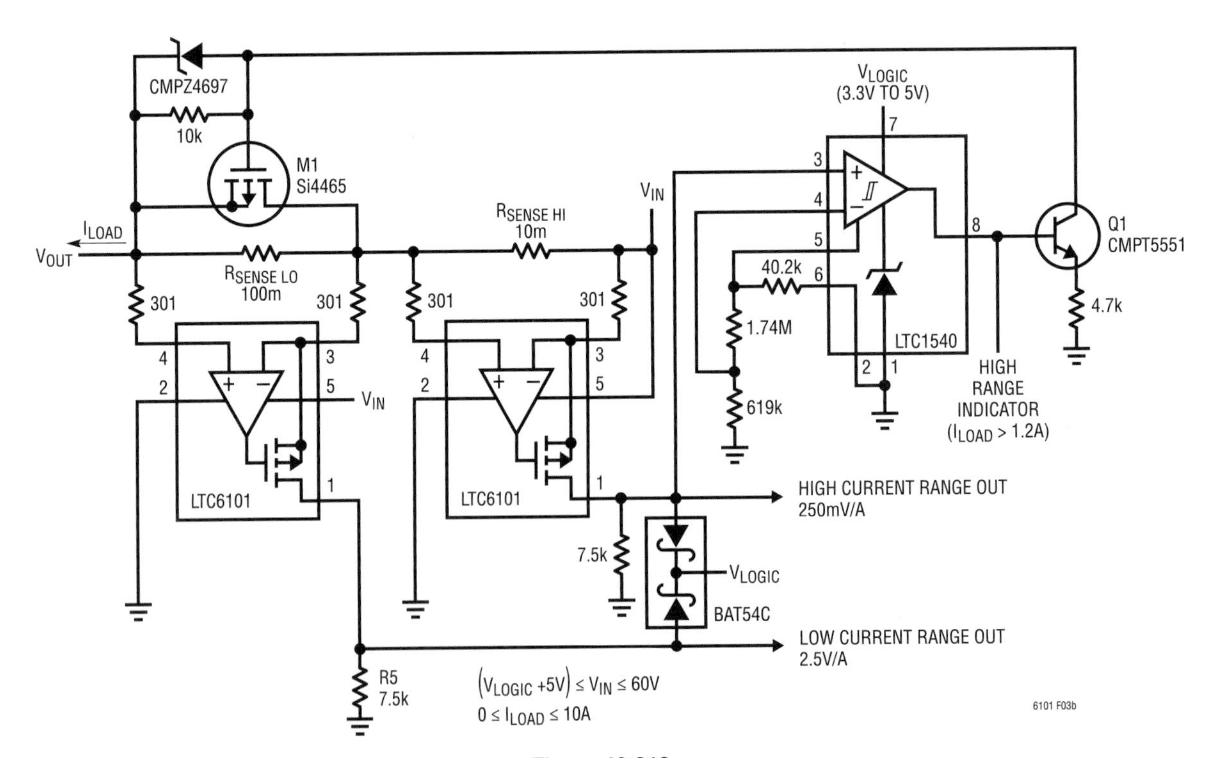

Figure 40.213 •

0 to 10A sensing over two ranges (Figure 40.215)

Using two sense amplifiers a wide current range can be broken up into a high and low range for better accuracy at lower currents. Two different value sense resistors can be used in series with each monitored by one side of the LTC6103. The low current range, less than 1.2A in this example, uses a larger sense resistor value to develop a larger sense voltage. Current exceeding this range will create a large sense voltage, which may exceed the input differential voltage rating of a single sense amplifier. A comparator senses the high current range and shorts out the larger sense resistor. Now only the high range sense amplifier outputs a voltage.

Dual sense amplifier can have different sense resistors and gain (Figure 40.216)

The LTC6104 has a single output which both sources and sinks current from the two independent sense amplifiers. Different shunt sense resistors can monitor different current ranges, yet be scaled through gain settings to provide the same range of output current in each direction. This is ideal for battery charging application where the charging current has a much smaller range than the battery load current.

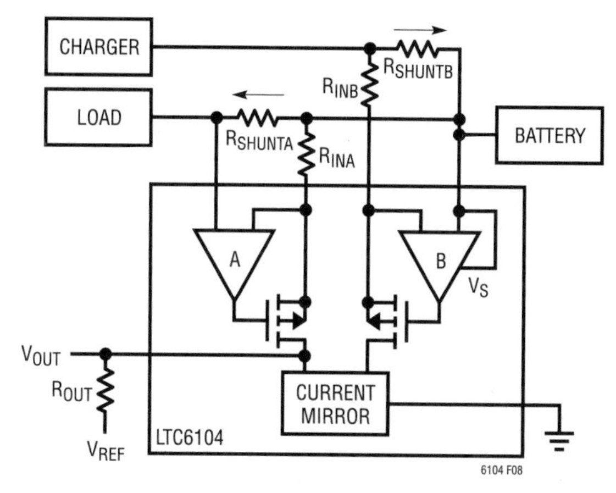

Figure 40.216 •

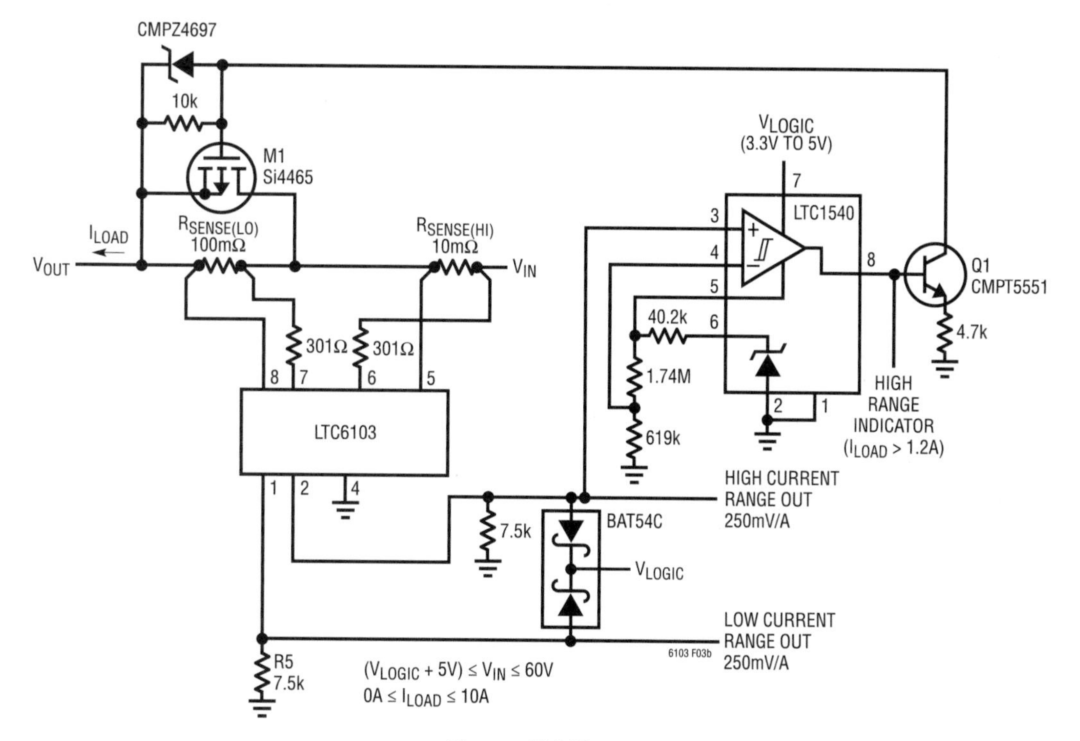

Figure 40.215 •

Power conversion, measurement and pulse circuits

41

Tales from the laboratory notebook

Jim Williams

Introduction

This ink marks LTC's eighth circuit collection publication.[1] We are continually surprised, to the point of near mystification, by these circuit amalgams seemingly limitless appeal. Reader requests ascend rapidly upon publication, remaining high for years, even decades. All LTC circuit collections, despite diverse content, share this popularity, although just why remains an open question. Why is it? Perhaps the form; compact, complete, succinct and insular. Perhaps the freedom of selection without commitment, akin to window shopping. Or, perhaps, simply the pleasure of new recruits for the circuit aficionados intellectual palate. Locally based electrosociolgists, spinning elegantly contrived theories, offer explanation, but no convincing evidence is at hand. What is certain is that readers are attracted to these compendiums and that calls us to attention. As such, in accordance with our mission to serve customer preferences, this latest collection is presented. Enjoy.

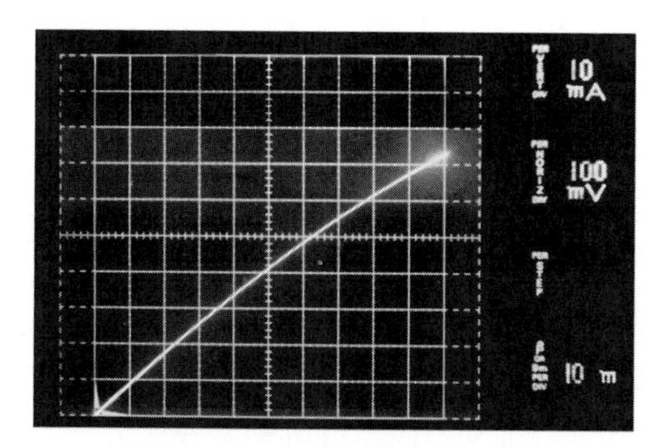

Figure 41.1 • Zero Volt. Biased JFET I-V Curve Shows 10mA Conduction at 100mV, Rising Above 40mA at 500mV. Characteristic Permits DC/DC Converter Powered From 300mV Supply

Note 1: Previous efforts include References 4,6,7 and 23–26.

JFET-based DC/DC converter powered from 300mV supply

A JFET's self-biasing characteristic can be utilized to construct a DC/DC converter powered from as little as 300mV Solar cells, thermopiles and single-stage fuel cells, all with outputs below 600mV are typical power sources for such a converter.

Figure 41.1, an N-channel JFET I-V plot, shows drain-source conduction under zero bias (gate and source tied together) conditions. This property can be exploited to produce a self-starting DC/DC converter that runs from 0.3V to 1.6V inputs.

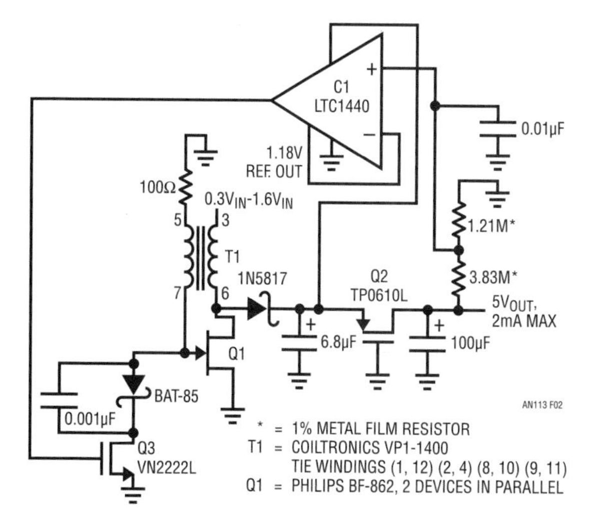

Figure 41.2 • JFET-Based DC/DC Converter Runs From 300 Millivolt Input. Q1-T1 Oscillator Output Is Rectified and Filtered. Load Is Isolated Until Q2 Source Reaches ≈2V, Aiding Start-Up. Comparator and Q3 Close Loop Around Oscillator, Controlling Q1's On-Time to Stabilize 5V Output

Analog Circuit and System Design: Immersion in the Black Art of Analog Design. http://dx.doi.org/10.1016/B978-0-12-397888-2.00041-9

Figure 41.2 shows the circuit. Q1 and T1 form an oscillator with T1's secondary providing regenerative feedback to Q1's gate. When power is applied, Q1's gate is at zero volts and its drain conducts current via T1's primary. T1's phase inverting secondary responds by going negative at Q1's gate, turning it off. T1's primary current ceases, its secondary collapses and oscillation commences. T1's primary action causes positive going "flyback" events at Q1's drain, which are rectified and filtered. Q2's $\approx$ 2V turn-on potential isolates the load, aiding start-up. When Q2 turns on, circuit output heads towards 5V. C1, powered from Q2's source, enforces output regulation by comparing a portion of the output with its internal voltage reference. C1's switched output controls Q1's on-time via Q3, forming a control loop.

Waveforms for the circuit include the AC coupled output (Figure 41.3, trace A), C1's output (trace B) and Q1's drain flyback events (trace C). When the output drops below 5V, C1 goes low, turning on Q1. Q1's resultant flyback events continue until the 5V output is restored. This pattern repeats, maintaining the output.

The 5V output can supply up to 2mA, sufficient to power circuitry or furnish bias to a higher power switching regulator when larger current is required. The circuit will start into loads of 300μA at 300mV input; 2mA loading requires a 475mV supply. Figure 41.4 plots minimum input voltage versus output current over a range of loads.

Q3's shunt control of Q1 is simple and effective, but results in 25mA quiescent current drain. Figure 41.5's modifications reduce this figure to 1mA by series switching T1's secondary. Here, Q3 switches series connected Q4, more efficiently controlling Q1's gate drive. Negative turn-off bias for Q4 and Q1 is bootstrapped from T1's secondary; the 6.8V zener holds off bias supply loading during

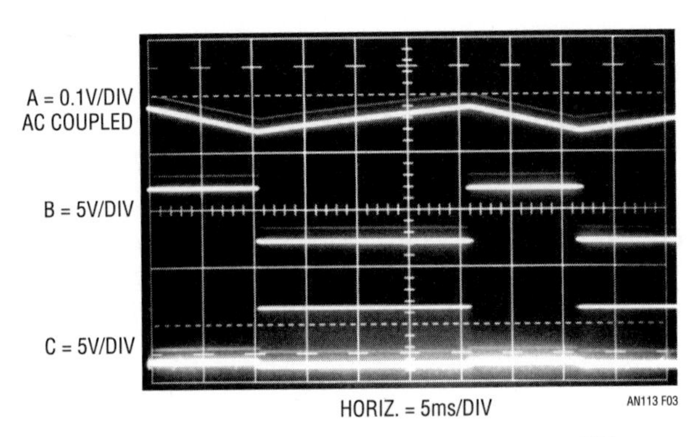

Figure 41.3 • JFET-based Converter Waveforms. When Supply Output (Trace A) Decays, C1 (Trace B) Switches, Allowing Q1 to Oscillate. Resultant Flyback Events at Q1 Drain (Trace C) Restore Supply Output

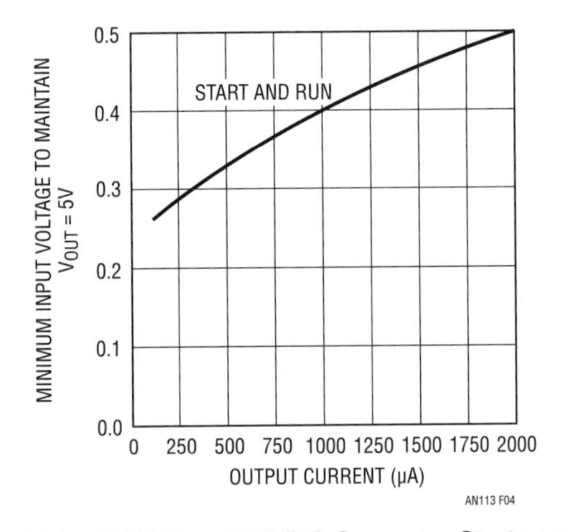

Figure 41.4 • JFET-Based DC/DC Converter Starts and Runs into 100µA Load at V_{IN} = 275mV. Regulation to 2mA Is Possible, Although Required V_{IN} Rises to 500mV

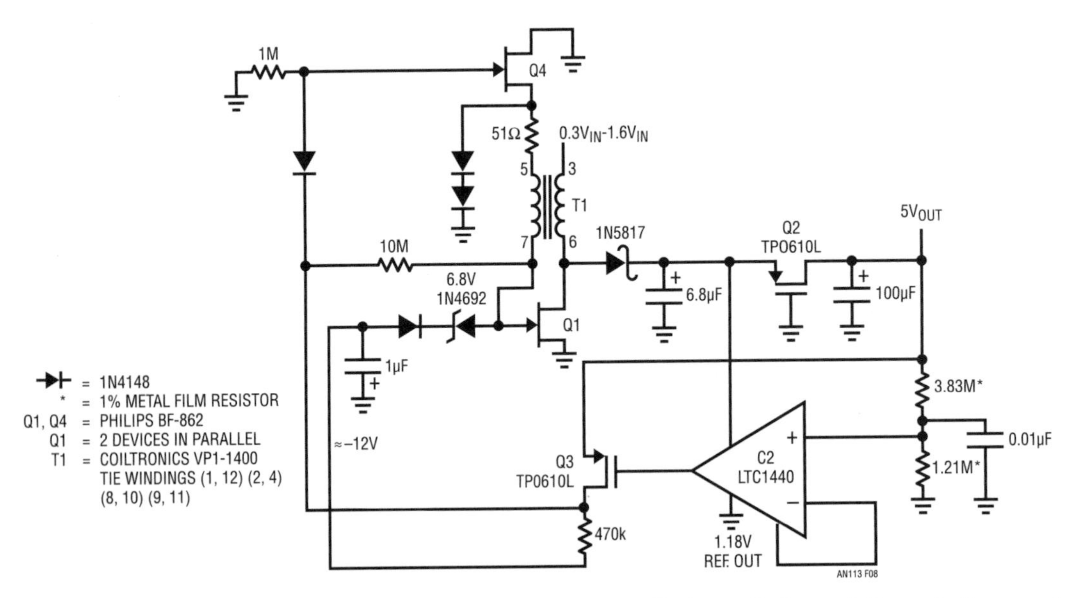

Figure 41.5 • Adding Q3, Q4 and Bootstrapped Negative Bias Generator Reduces Quiescent Current. Comparator Directed Q3 Switches Q4, More Efficiently Controlling Q1's Gate Drive. Q2 and Zener Diode Isolate All Loading During Q1 Start-Up

initial power application, aiding start-up. Figure 41.6's plot of minimum input voltage versus output current shows minimal penalty (versus Figure 41.4's data) imposed by the added quiescent current control circuitry.

Bipolar transistor-based 550mV input DC/DC converter

Bipolar transistors may be used to obtain higher output currents, although their V_{BE} drop raises input supply requirements to 550mV. Figure 41.7's curve tracer plot shows base-emitter conduction just beginning at 450mV (25°C) with substantial current flow beyond 500mV Figure 41.8's circuit operates similarly to FET-based Figure 41.2, although the bipolar transistor's normally off characteristic allows more efficient operation. Figure 41.9's operating waveforms are similar to Figure 41.3, except the comparator's output state is reversed to accommodate the bipolar transistor. Figure 41.10's start-run curves show 6mA output current at 550mV input—3 times the FET circuit's capacity. The "run" curve indicates that, once started, the

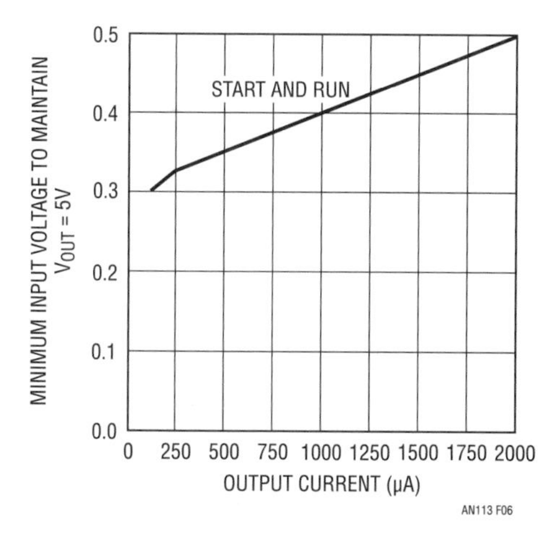

Figure 41.6 • Start/Run Curve for Low Quiescent Current JFET-Based DC/DC Converter. Quiescent Current Control Circuitry Slightly Increases Input Voltage Required to Support Load

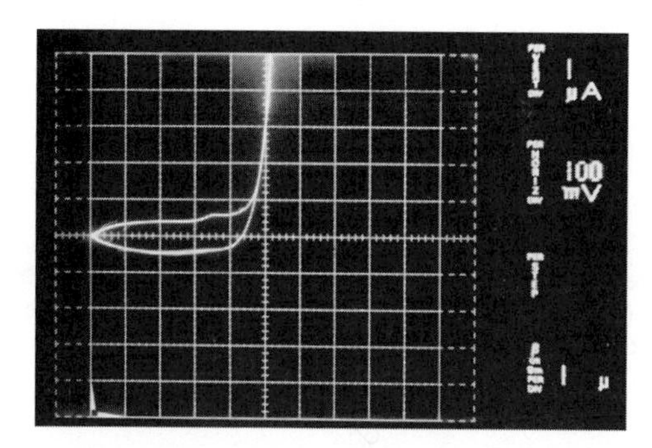

Figure 41.7 • Bipolar Transistor Base Emitter Junction I-V Curve Shows Conduction Beginning at 450 Millivolts (25°C). Characteristic Forms Basis of DC/DC Converter Powered From 550 Millivolts

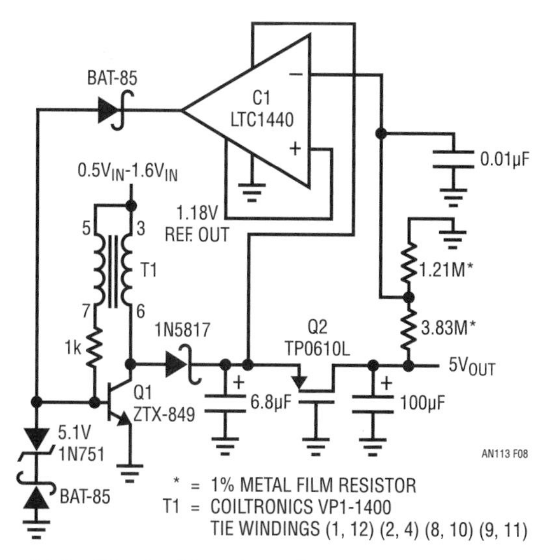

Figure 41.8 • Bipolar Transistor-Based DC/DC Converter Runs From 500 Millivolt (25°C) Input. Q1-T1 Oscillator Output is Rectified and Filtered. Load is Isolated Until Q2's Source Reaches ≈2V, Aiding Start-Up. Comparator Closes Loop Around Oscillator, Controlling Q1's On-Time to Stabilize 5V Output

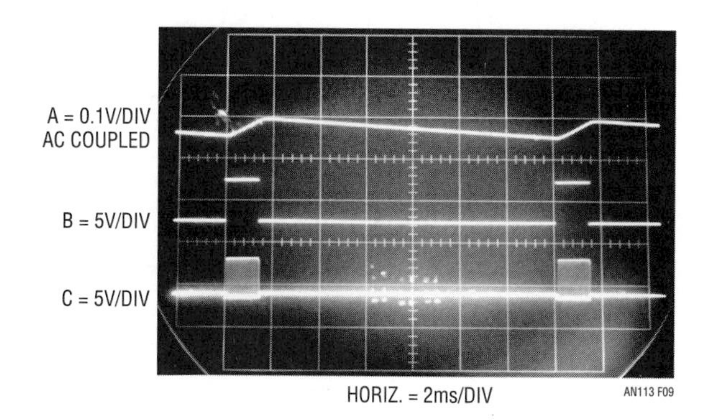

Figure 41.9 • Converter Waveforms. When Output (Trace A) Decays, C1 (Trace B) Switches, Allowing Q1 to Oscillate. Resultant Flyback Events at Q1 Collector (Trace C) Restore Output

circuit will run at input voltages as low as 300mV depending on loading.

When considering these circuits' extremely low input voltages and output power limits it is worth noting that the transformer specified is a standard product. A transformer specifically optimized for these applications would likely enhance performance.

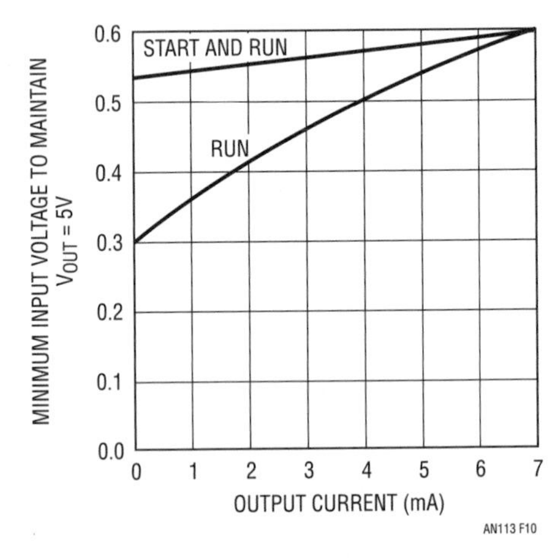

Figure 41.10 • Bipolar Transistor-Based DC/DC Converter Requires ≈550mV (25°C) Input to Start into 0mA to 6mA Loading. Once Running, Converter Maintains Regulation Down to 300mV Inputs for 100µA Load

5V to 200V converter for APD bias

BEFORE PROCEEDING ANY FURTHER, THE READER IS WARNED THAT CAUTION MUST BE USED IN THE CONSTRUCTION, TESTING AND USE OF THIS CIRCUIT: HIGH VOLTAGE, LETHAL POTENTIALS ARE PRESENT IN THIS CIRCUIT: EXTREME CAUTION MUST BE USED IN WORKING WITH, AND MAKING CONNECTIONS TO, THIS CIRCUIT. REPEAT: THIS CIRCUIT CONTAINS DANGEROUS, HIGH VOLTAGE POTENTIALS. USE CAUTION.

Avalanche photodiodes (APD) require high voltage bias. Figure 41.11's design provides 200V from a 5V input. The circuit is a basic inductor flyback boost regulator with a major important deviation. Q1, a high voltage device, has been interposed between the LT1172 switching regulator and the inductor. This permits the regulator to control Q1's high voltage switching without undergoing high voltage stress. Q1, operating as a "cascode" with the LT1172's internal switch, withstands L1's high voltage flyback events.[2] Diodes associated with Q1's source terminal clamp L1 originated spikes arriving via Q1's

Note 2: See References 1 (page 8), 2 (Appendix D), and 3.

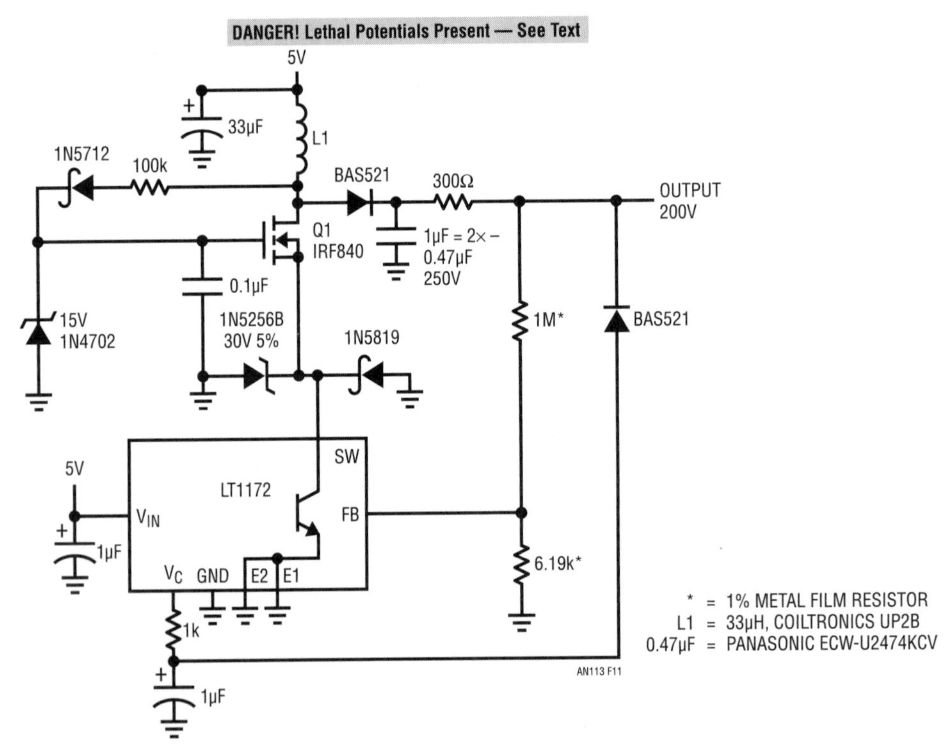

Figure 41.11 • 5V to 200V Output Converter for APD Bias. Cascoded Q1 Switches High Voltage, Allowing Low Voltage Regulator to Control Output. Diode Clamps Protect Regulator from Transient Events; 100k Path Bootstraps Q1's Gate Drive from L1s Flyback Events. Output Connected 300Ω-Diode Combination Provides Short-Circuit Protection

junction capacitance. The high voltage is rectified and filtered, forming the circuit's output. Feedback to the regulator stabilizes the loop and the RC at the Vc pin provides frequency compensation. The 100k path from L1 bootstraps Q1's gate drive to about 10V ensuring saturation.[3] The output connected 300Ω-diode combination provides short-circuit protection by shutting down the LT1172 if the output is accidentally grounded.

Figure 41.12 shows operating waveforms. Traces A and C are LT1172 switch current and voltage, respectively. Q1's drain is trace B. Current ramp termination results in a high voltage flyback event at Q1's drain. A safely attenuated version of the flyback appears at the LT1172 switch. The sinosoidal signature, due to inductor ring-off between conduction cycles, is harmless.

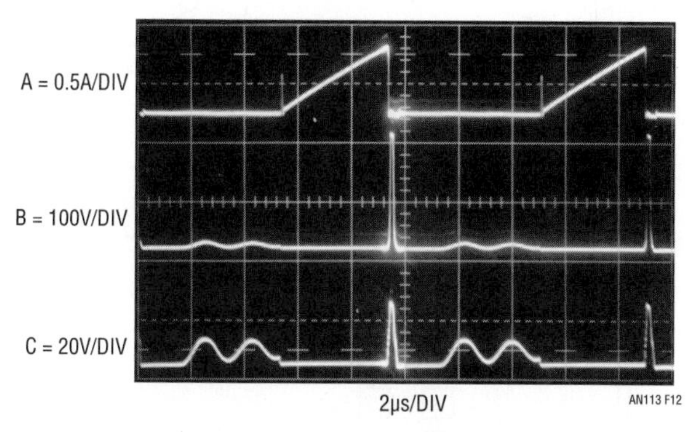

Figure 41.12 • Waveforms for 5V to 200V Converter Include LT1172 Switch Current and Voltage (Traces A and C, Respectively) and Q1's Drain Voltage (Trace B). Current Ramp Termination Results in High Voltage Flyback Event at Q1 Drain. Safely Attenuated Version Appears at LT1172 Switch. Sinosoidal Signature, Due to Inductor Ring-Off Between Current Conduction Cycles, is Harmless. All Traces Intensified Near Center Screen for Photographic Clarity

Battery internal resistance meter

It is often desirable to determine a battery's internal resistance to evaluate its condition or suitability for an application. Accurate battery resistance determination is complicated by inherent capacitive terms which corrupt results taken with AC-based milliohmmeters operating in the kHz range. Figure 41.13, a very simplistic battery model, shows a resistive divider with a partial shunt capacitive term. This capacitive term introduces error in AC-based measurement. Additionally, the battery's unloaded internal resistance may significantly differ from its loaded value. As such, a realistic determination of internal resistance must be made under loaded conditions at or near DC.

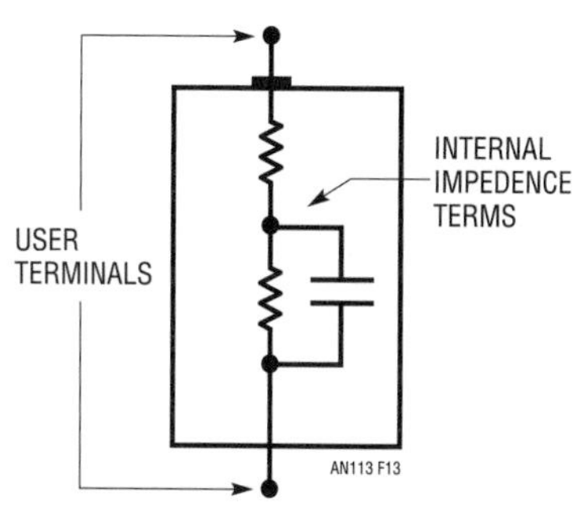

Figure 41.13 • Simplistic Model Shows Battery Impedance Terms Including Resistive and Capacitive Elements. Capacitive Component Corrupts AC-Based Measurement Attempts to Determine Internal DC Resistance. More Realistic Results Occur if Battery Voltage Drop Is Measured Under Known Load

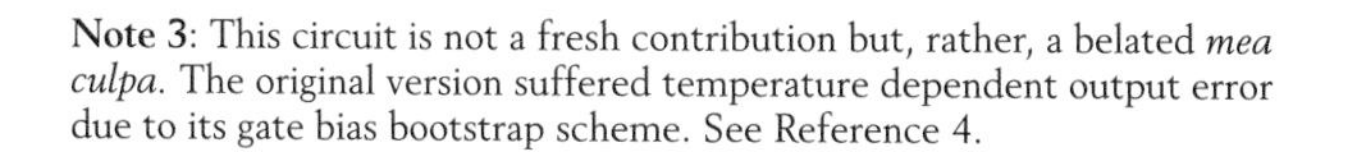
Note 3: This circuit is not a fresh contribution but, rather, a belated *mea culpa*. The original version suffered temperature dependent output error due to its gate bias bootstrap scheme. See Reference 4.

Figure 41.14's circuit meets these requirements, permitting accurate internal resistance determination of batteries up to 13V over a range of 0.001Ω to 1.000Ω. A1, Q1 and associated components form a closed loop current sink which loads the battery via Q1's drain. The 1N5821 provides reverse battery protection. The voltage across the 0.1Ω resistor, and hence the battery load, is determined by A1's "+" input voltage. This potential is alternately switched, via S1, between 0.110V and 0.010V derived from the 2.5V reference driven resistor string. S1's 0.5Hz square wave switching drive comes from the CD4040 frequency divider. The result of this action is a 100mA biased 1A 0.5Hz square wave load applied to the battery. The battery's internal resistance causes a 0.5Hz amplitude modulated square wave to appear at the Kelvin-sensed S2-S3-A2 synchronous demodulator. The demodulator DC output is buffered by chopper stabilized A2 which provides the circuit output. A2's internal 1kHz clock, level shifted by Q2, drives the CD4040 frequency divider. One divider output supplies the 0.5Hz square wave; a second 500Hz output activates a charge pump, providing a -7V potential to A2. This arrangement allows A2 output swing to zero volts.

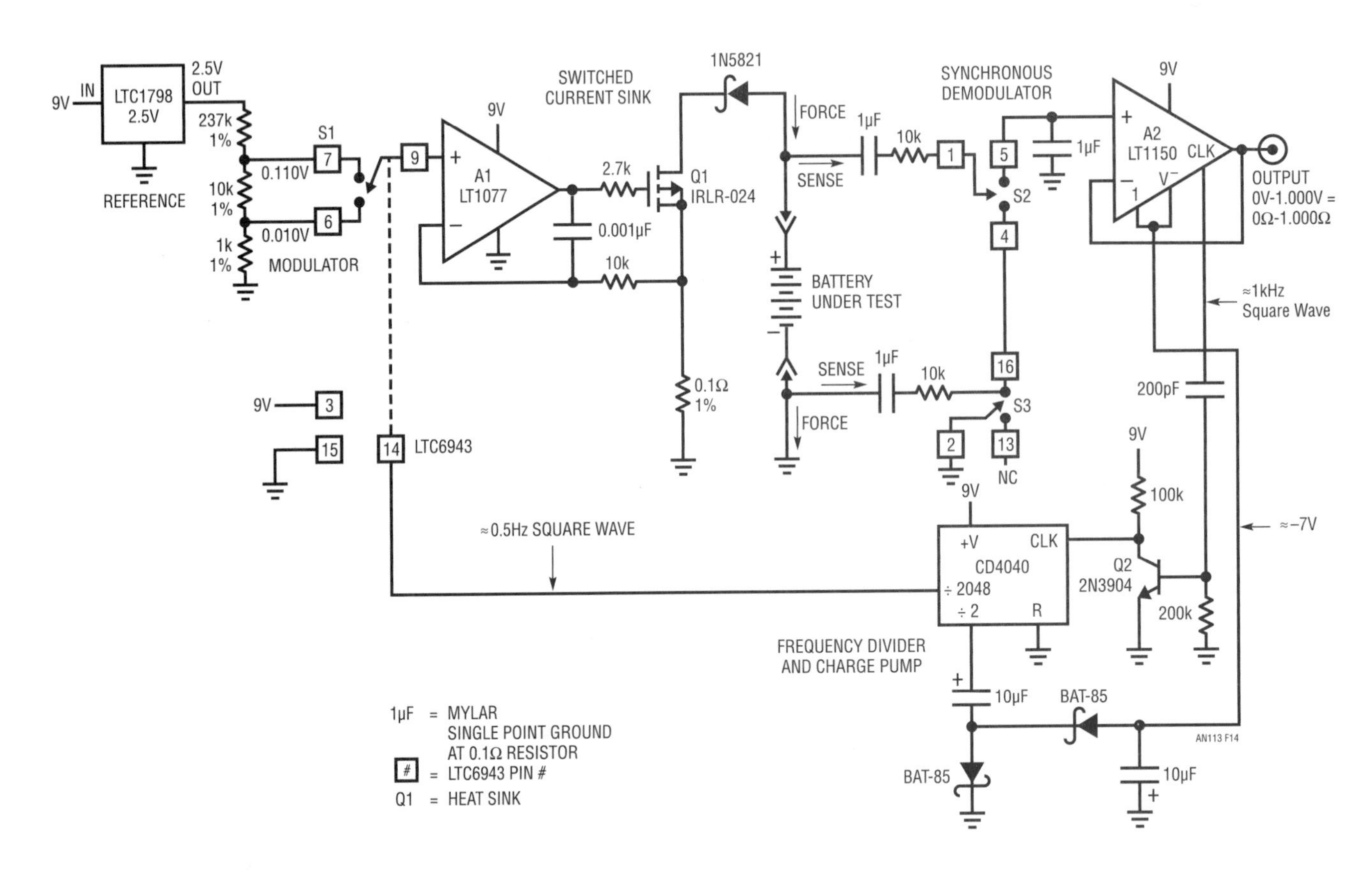

Figure 41.14 • Battery Internal Resistance Is Determined by Repetitively Stepping Calibrated Discharge Current and Reading Resultant Voltage Drop. S1-Based Modulator, Clocked From Frequency Divider, Combines with A1-Q1 Switched Current Sink to Generate Stepped, 1 Ampere Battery Discharge Cycles. S2-S3-A2 Synchronous Demodulator Extracts Modulated Voltage Drop Information, Provides DC Output Calibrated in Ohms

The circuit pulls 230μA from its 9V battery power supply, permitting about 3000 hours battery life. Other specifications include operation down to 4V with less than 1mV (0.001 Ω) output variation, 3% accuracy and battery- under-test range of 0.9V to 13V. Finally, note that battery discharge current and repetition rate are easily varied from the values given, permitting observation of battery resistance under a variety of conditions.

Floating output, variable potential battery simulator

Battery stack voltage monitor development (Reference 5) is aided by a floating, variable potential battery simulator. This capability permits monitor accuracy verification over a wide range of battery voltage. The floating battery simulator is substituted for a cell in the stack and any desired voltage directly dialed out. Figure 41.15's circuit is simply a battery powered follower (A1) with current boosted (A2) output. The LT1021 reference and high resolution potentiometric divider specified permits accurate output setting within 1mV. The composite amplifier unloads the divider and drives a 680μF capacitor to approximate a battery. Diodes preclude reverse biasing the output capacitor during supply sequencing and the 1μF-150k combination provides stable loop compensation. Figure 41.16 depicts loop response to an input step; no overshoot or untoward dynamics occur despite A2's huge capacitive load. The battery monitor determines battery voltage by injecting current into the battery and measuring resultant clamp

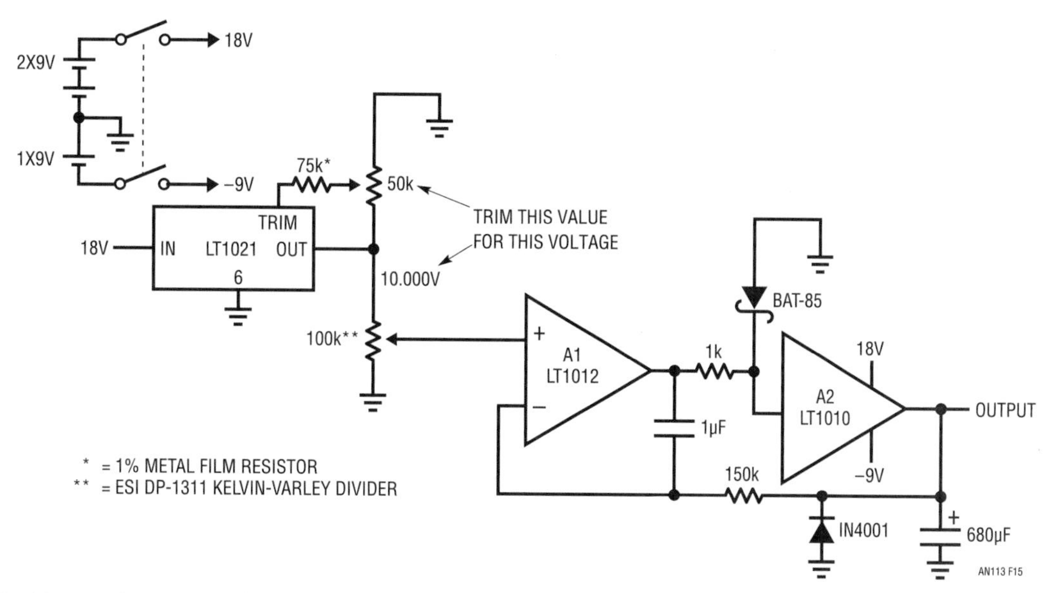

Figure 41.15 • Battery Simulator Has Floating Output Settable within 1mV. A1 Unloads Kelvin-Varley Divider; A2 Buffers Capacitive Load

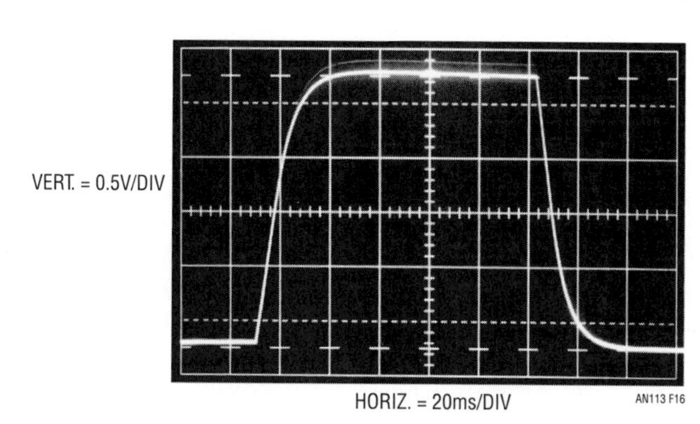

Figure 41.16 • 150k-1μF Compensation Network Provides Clean Response Despite 680μF Output Capacitor

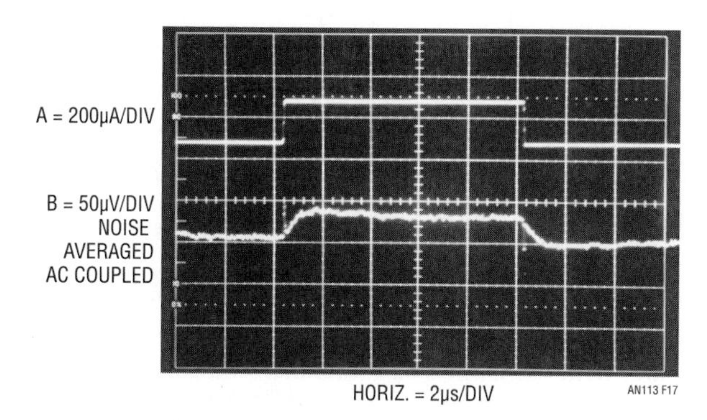

Figure 41.17 • Battery Simulator Output (Trace B) Responds to Trace A's Monitor Current Pulse. Closed Loop Control and 680μF Capacitor Maintain Simulator Output within 30μV. Noise Averaged, 50μV/Division Sensitivity Is Required to Resolve Response

voltage (again, see Reference 5). Figure 41.17 shows battery simulator response (trace B) to trace A's monitor current pulse into the output. Closed loop control and the 680μF capacitor limit simulator output excursion within 30μV. This error is so small that noise averaging techniques and a high gain oscilloscope preamplifier are required to resolve it.[4]

40nV$_{p\text{-}p}$ noise, 0.05µV/°C drift, chopped FET amplifier

Figure 41.18's circuit combines the LTC6241's rail-to-rail performance with a pair of extremely low noise JFETs configured in a chopper-based carrier modulation scheme to achieve extraordinarily low noise and DC drift. This circuit's performance suits demanding transducer signal conditioning situations such as high resolution scales and magnetic search coils.

The LTC1799's output is divided down to form a 2-phase 925Hz square wave clock. This frequency, harmonically unrelated to 60Hz, provides excellent immunity to harmonic beating or mixing effects which could cause instabilities. S1 and S2 receive complementary drive, causing the FET-A1 based stage to see a chopped version of the input voltage. A1's square wave output is synchronously demodulated by S3 and S4. Because these switches are synchronously driven with the input chopper; proper amplitude and polarity information is presented to DC output amplifier A2. This stage integrates the square wave into a DC voltage, providing the output. The output is divided down (R2 and R1) and fed back to the input chopper where it serves as a zero signal reference. Gain, in this case 1000, is set by the R1-R2 ratio. The AC coupled input stage's DC errors do not affect overall circuit characteristics, resulting in the extremely low offset and drift noted.

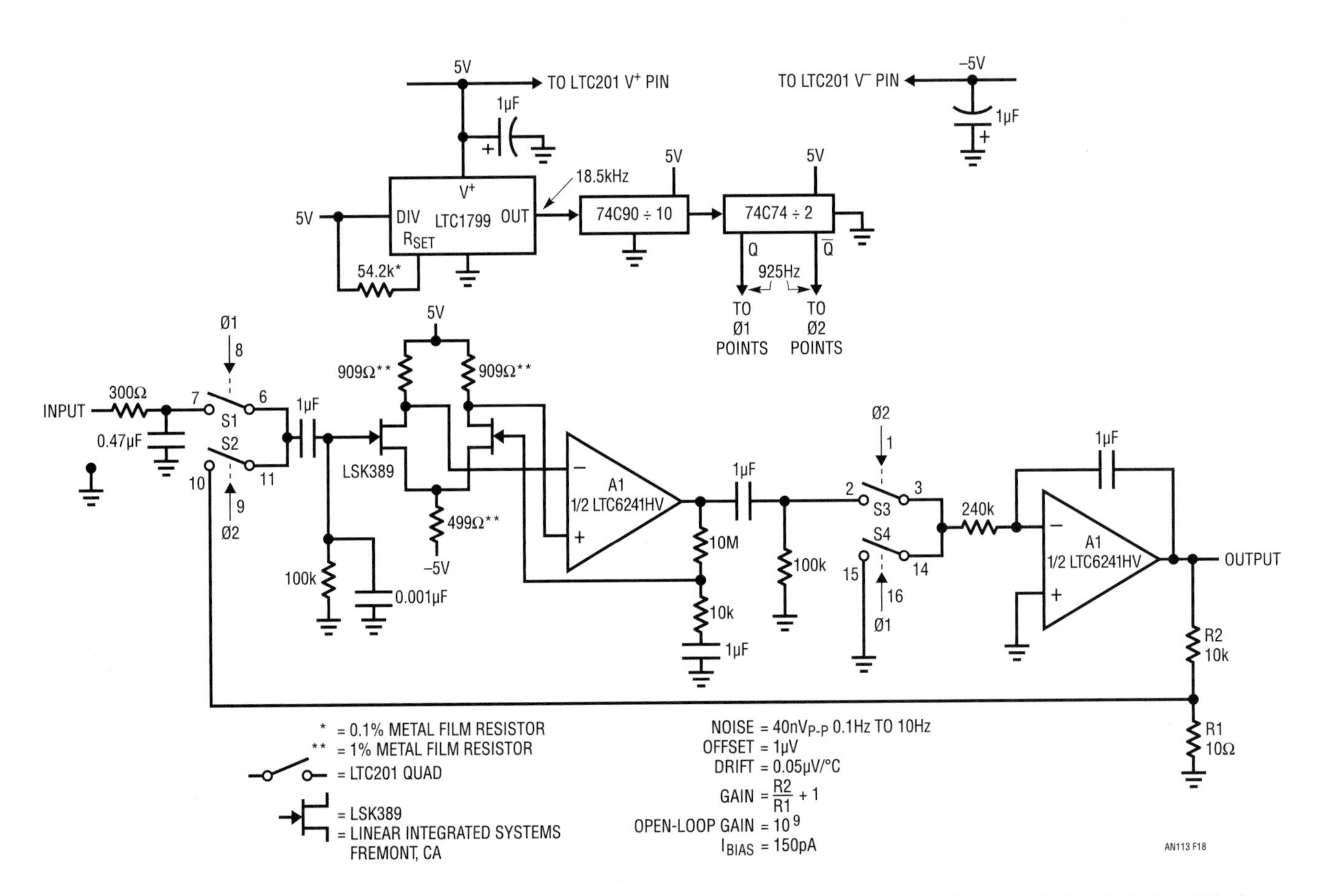

Figure 41.18 • Chopped FET Amplifier Has 40nV$_{P\text{-}P}$ Noise and 0.05µV/°C Drift. DC Input Is Carrier Modulated, Amplified by A1, Demodulated to DC and Fed Back From A2. 925Hz Carrier Clock Prevents Interaction with 60Hz Line Originated Components

Note 4: This may be viewed as a historic event in some thinly populated circles. Figure 41.17 marks the author's first published use of a digital oscilloscope (Tektronix 7603/7D20), updating him to the 1980s.

Figure 41.19, measured over a 50 second interval, shows $40nV_{P-P}$ noise in a 0.1Hz to 10Hz bandwidth. This is spectacularly low noise for a JFET-based design and is directly attributable to input pair area and current density.

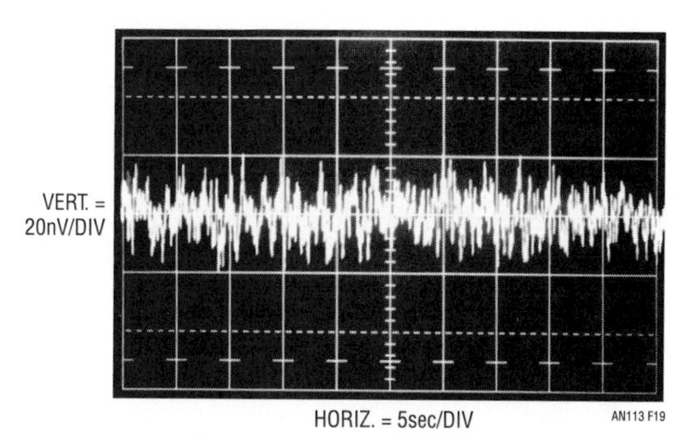

Figure 41.19 • Amplifier 0.1Hz to 10Hz Noise Measures $40nV_{P-P}$ in 50 Second Sample Period

Wideband, chopper stabilized FET amplifier

The previous circuit's bandwidth is limited because the chopping occurs within the signal path. Figure 41.20's circuit circumvents this restriction by placing the stabilizing element in parallel with the signal path. This maintains DC performance although noise triples to 125nV in a 0.1Hz to 10Hz bandpass.

FET pair Q1 differentially feeds A2 to form a simple low noise op amp. Feedback, provided by R1 and R2, sets closed loop gain (in this case 1000) in the usual fashion. Although Q1 has extraordinarily low noise characteristics, its offset and drift are relatively high. A1, a chopper stabilized amplifier, corrects these deficiencies. It does this by measuring the difference between the amplifier's

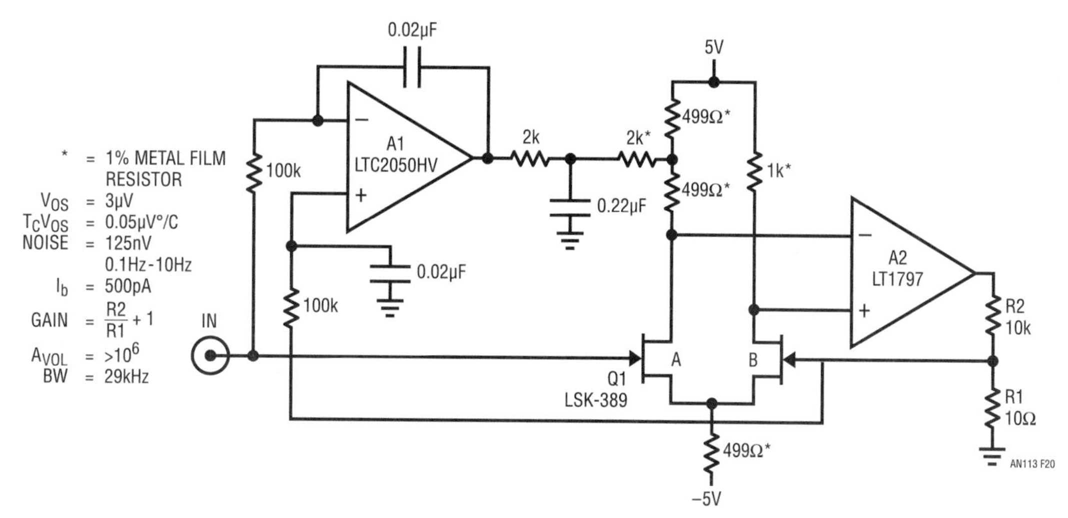

Figure 41.20 • Placing Stabilizing Amplifier Outside Signal Path Permits Bandwidth Increase over Previous Circuit. Noise Triples to 125nV in 0.1Hz to 10Hz Bandpass

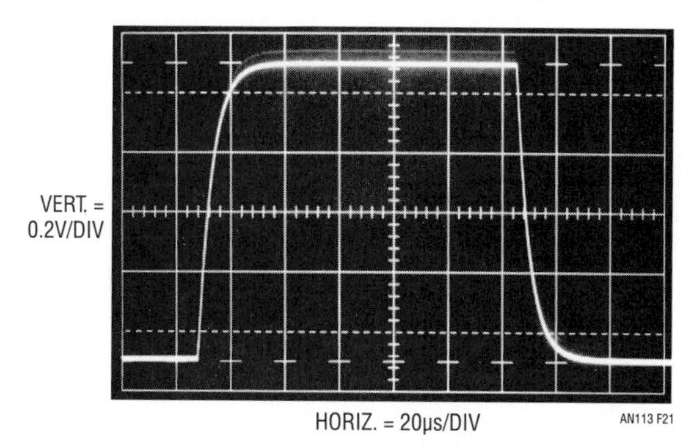

Figure 41.21 • Figure 41.20 Responds to a 1mV Input. 12µs Rise Time Indicates 29kHz Bandwidth at A = 1000

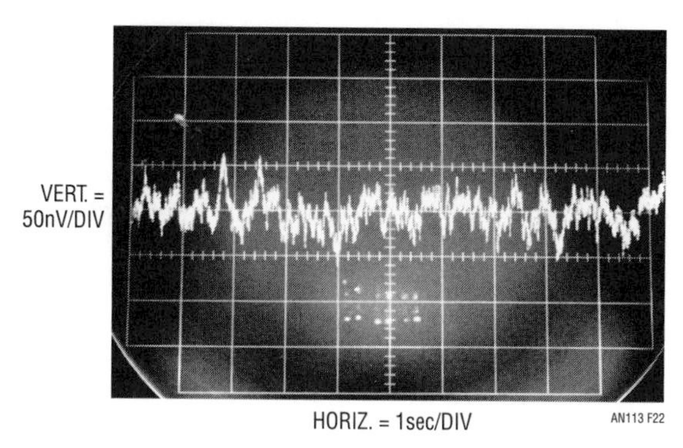

Figure 41.22 • Chopper Stabilized FET Pair Noise Measures 125nV in 0.1Hz to 10Hz Bandpass

inputs and adjusting Q1A's channel current to minimize the difference. Q1's drain values ensure that A1 will be able to capture the offset. A1 supplies whatever current is required to Q1A's channel to force offset within 5μV Additionally, A1's low bias current does not appreciably add to the overall 500pA amplifier bias current. As shown, the amplifier is set up for a noninverting gain of 1000, although other gains and inverting operation are possible.

Placing the offset correction in parallel with the signal path permits high bandwidth. Figure 41.21 shows response to a 1mV input. The 12μs risetime indicates 29kHz bandwidth at A = 1000.

Figure 41.22's photo measures noise in a 0.1Hz to 10Hz bandwidth. The performance obtained is almost 6 times better than any monolithic chopper stabilized amplifier, while retaining low offset and drift.

Submicroampere RMS current measurement for quartz crystals

Quartz crystal RMS operating current is critical to long term stability, temperature coefficient and reliability. Accurate determination of RMS crystal current, especially in micropower types, is complicated by the necessity to minimize introduced parasitics, particularly capacitance, which corrupt crystal operation. Figure 41.23's high gain, low noise amplifier combines with a commercially available

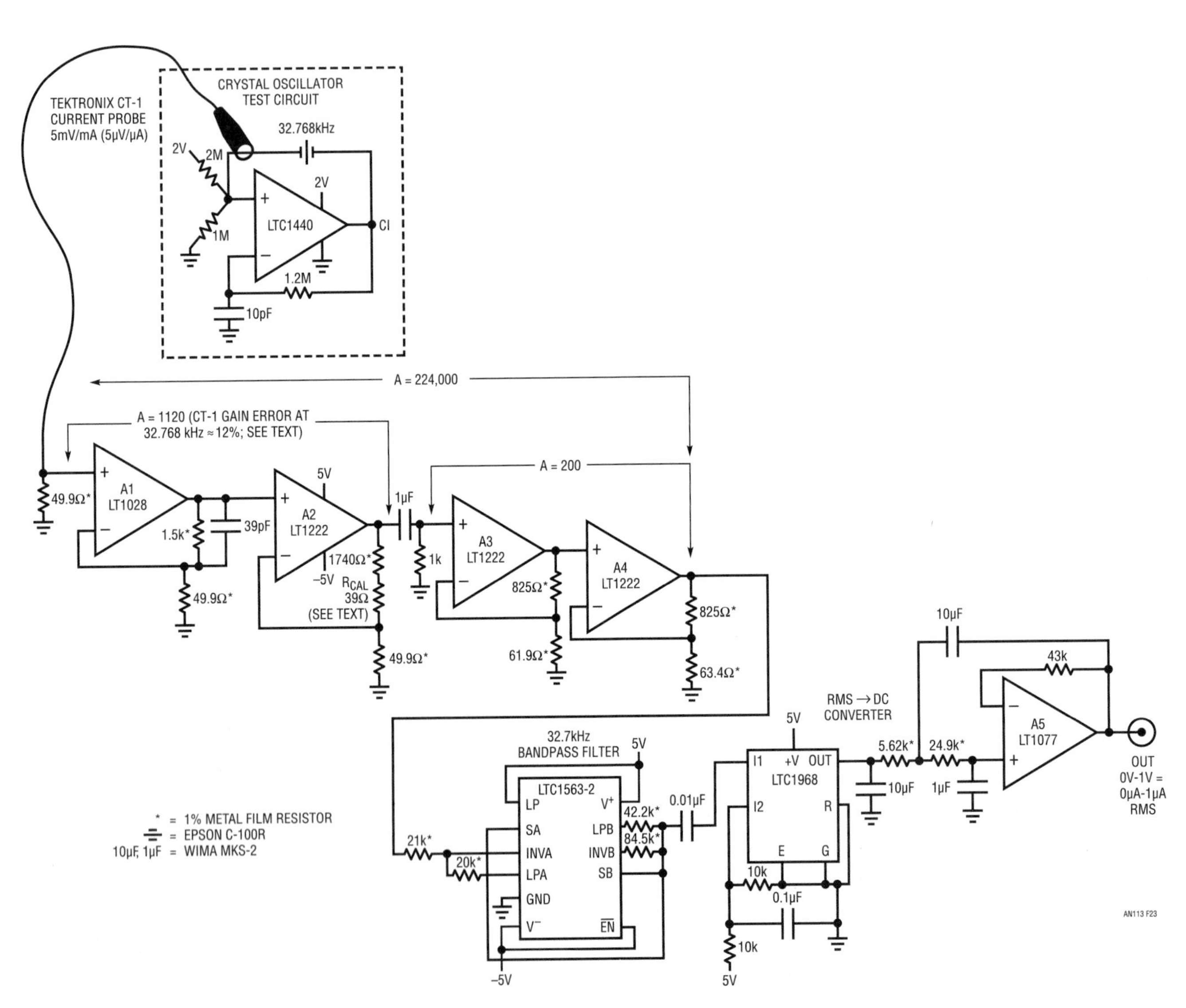

Figure 41.23 • A1 to A4 Furnish Gain of >200,000 to Current Probe, Permitting Submicroamp Crystal Current Measurement. LTC1563-2 Bandpass Filter Smooths Residual Noise While Providing Unity Gain at 32.768kHz. LTC1968 Supplies RMS Calibrated Output

closed core current probe to permit the measurement. An RMS-to-DC converter supplies the RMS value. The quartz crystal test circuit shown in dashed lines exemplifies a typical measurement situation. The Tektronix CT-1 current probe monitors crystal current while introducing minimal parasitic loading. The probe's 50Ω terminated output feeds A1. A1 and A2 take a closed loop gain of 1120; excess gain over a nominal gain of 1000 corrects for the CT-1's 12% low frequency gain error at 32.768kHz.[5] A3 and A4 contribute a gain of 200, resulting in total amplifier gain of 224,000. This figure results in a 1V/μA scale factor at A4 referred to the gain corrected CT-1's output. A4's LTC1563-2 bandpass filtered output feeds an LTC1968-A5 based RMS→DC converter which provides the circuits output. The signal processing path constitutes an extremely narrow band amplifier tuned to the crystal's frequency. Figure 41.24 shows typical circuit waveforms.

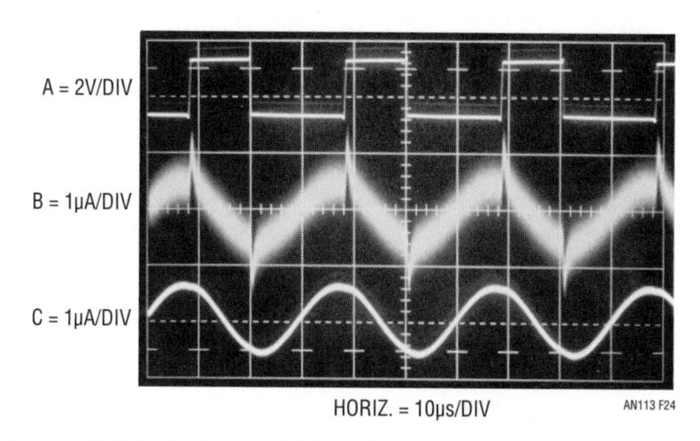

Figure 41.24 • C1's 32.768kHz Output (Trace A) and Crystal Current Monitored at A4 Output (Trace B). RMS Converter Input Is Trace C. Peaks in Trace B's Unfiltered Waveform Derive From Inherent and Parasitic Paths Shunting Crystal

Crystal drive, taken at C1's output (trace A), causes a 530nA RMS crystal current which is represented at A4's output (trace B) and the RMS→DC converter input (trace C). Peaking visible in trace B's unfiltered presentation derives from parasitic paths shunting the crystal.

Typical circuit accuracy is 5%. Uncertainty terms include the transformer's tolerances, its ≈1.5pF loading and resistor/RMS→DC converter error. Calibrating the circuit reduces error to less than 1%. Calibration involves driving the transformer with 1μA at 32.7kHz. This is facilitated by biasing a 100k, 0.1% resistor with an oscillator set at 0.100V output. The output voltage should be verified with an RMS voltmeter having appropriate accuracy (see Reference 8's Appendix B). Figure 41.23 is calibrated by padding A2's gain with a small resistive correction, typically 39Ω.

Direct reading quartz crystal-based remote thermometer

Although quartz crystals have been used as temperature sensors (see References 7 and 10) there has been almost no adaptation of this technology. The primary impediment has been lack of standard product quartz crystal temperature sensors. The advantages of quartz-based temperature sensing include nearly purely digital signal conditioning, good stability and a direct, noise immune digital output ideally suited to remote sensing.

Figure 41.25 utilizes an economical, commercially available (Reference 9) quartz temperature sensor in a direct reading thermometer scheme suited to remote data collection. An LTC485 transceiver; set up in transmit mode, forms a quartz-based, Pierce class oscillator. The transceiver's differential line driving outputs provide frequency coded temperature data to a 1000 foot cable run. A second RS485 transceiver differentially receives the data, presenting a single ended output to the PIC-16F73 processor. The processor converts the frequency coded temperature data to its °C equivalent, which appears on the display. Figure 41.26 is a software listing of the processor's program.[6] Accuracy over a sensed −40°C to 85°C range is about 2%.

Note 5: The validity of this gain error correction at one sinusoidal frequency -32.768kHz—was investigated with a 7-sample group of Tektronix CT-1s. Device outputs were collectively within 0.5% of 12% down for a 1μA 32.768kHz sinusoidal input current. Although this tends to support the measurement scheme, it is worth noting that these results are as measured. Tektronix does not guarantee performance below the specified -3dB 25kHz low frequency roll-off.

Note 6: Mark Thoren of LTC designed the processor-based circuitry and authored Figure 41.26's software.

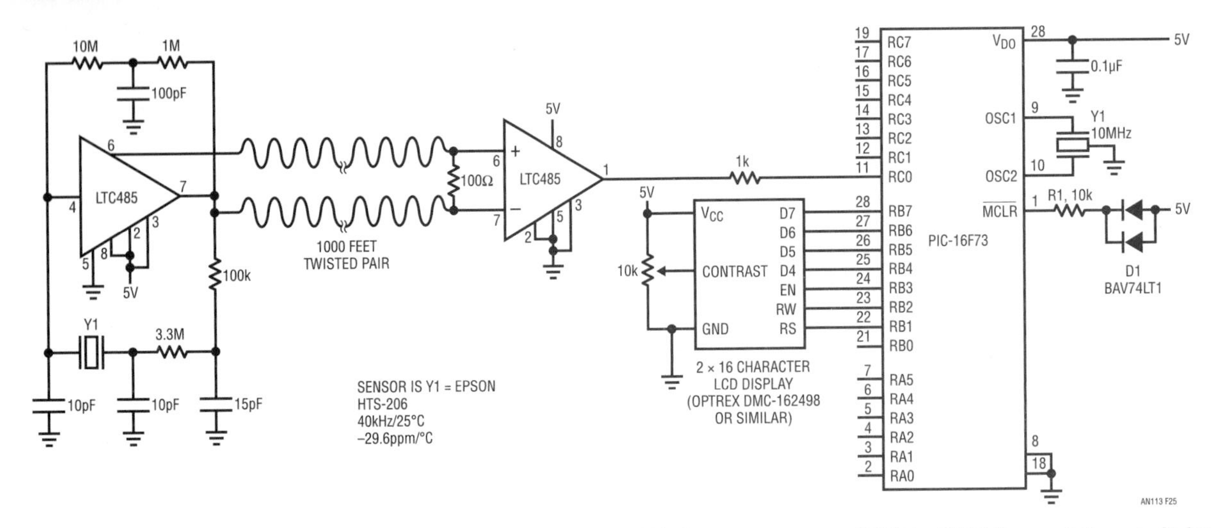

Figure 41.25 • Quartz Crystal-Based Remote Thermometer Has 2% Accuracy over −40°C to 85°C Sensed Range, Drives 1000ft Cable. RS-485 Transceiver Oscillates at Y1 Quartz Sensor Determined Frequency and Drives Cable. Second Transceiver Receives Data and Feeds Processor. Display Reads Directly in °C

```
/*
Thermometer based on Epson HT206 temperature sensing crystal.
Output is to a standard Epson HD447980 based alphanumeric
LCD display. LCD driver functions are part of compiler library

Written for CSS compiler version 3.182

*/

#include <16F73.h>
#device adc=8
#fuses NOWDT, HS, PUT, NOPROTECT, NOBROWNOUT
#use delay (clock=10000000) // Tell compiler how fast we're going
#use rs232(baud=9600, parity=N, xmit=PIN_C6, rcv=PIN_C7, bits=8)
#include "lcd.c" // LCD driver functions

void main( )
{
    int16 temp;
    unsigned int16 f;
    setup_adc_ports(NO_ANALOGS);
    setup_adc(ADC_OFF);
    setup_spi (FALSE);
    setup_counters (RTCC_INTERNAL, RTCC_DIV_1);
    setup_timer_1 (T1_EXTERNAL|T1_DIV_BV_1);
    setup_timer_2 (T2_DISABLED, 0, 1);
    lcd_init( ); // Initialize LCD

while(1)
    (
    set_timer1(0);                              // Reset counter
    setup_timer_1(T1_EXTERNAL|T1_DIV_BY_1);     // Turn on counter
    delay_ms(845);                              // 0.845412 is the magic time to count
    delay_us(412);                              // -it gives 1 less plus per degree C
    setup_timer_1(T1_DISABLED);                 // turn off counter
    f = get_timer1();                           // Read result
    temp = 33770 - f + 25;                      // Convert to temperature

    //**  At this point 'f' is the temperature in degrees C          **//
    //**  For this experiment, dump to standard HD44780 type LCD display **//

    lcd_putc('\f');                             // Clear screen
    lcd_gotoxy(1, 1);                           // Goto home position
    printf(lcd_putc, "%ld", temp);              // And display result
    }
}
```

Figure 41.26 • Software Listing for PIC Processor Program. Code Converts Frequency to °C Equivalent, Drives Display

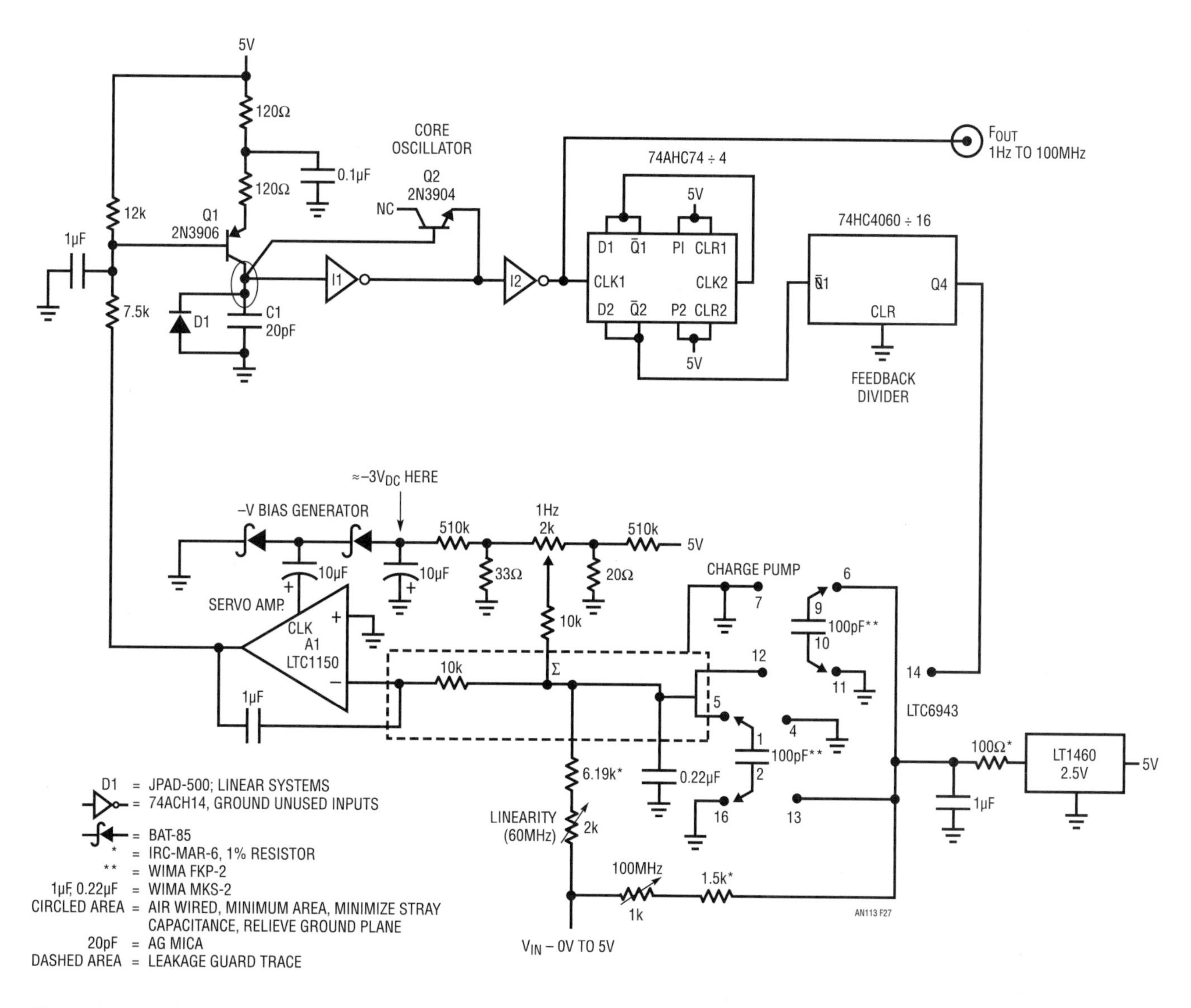

Figure 41.27 • 1Hz to 100MHz V→F Converter Has 160dB Dynamic Range, Runs From 5V Supply. Input Biased Servo Amplifier Controls Core Oscillator, Stabilizing Circuits Operating Point. Wide Range Operation Derives From Core Oscillator Characteristics, Divider/Charge Pump-Based Feedback and A1's Low DC Input Errors

1Hz–100MHz V→F converter

Figure 41.27's circuit achieves a wider dynamic range and higher output frequency than any commercially available voltage to frequency (V→F) converter. Its 100MHz full scale output (10% overrange to 110MHz is provided) is at least ten times faster than available units. The circuit's 160dB dynamic range (8 decades) allows continuous operation down to 1Hz. Additional specifications include 0.1% linearity, 250ppm/°C gain temperature coefficient, 1Hz/°C zero shift, 0.1% frequency shift for $V_{SUPPLY} = 5V \pm 10\%$ and a 0V to 5V input range. A single 5V supply powers the circuit.[7]

A1, a chopper stabilized amplifier, servo biases a crude but wide range core oscillator in Figure 41.27. The core oscillator drives a charge pump via digital dividers. The averaged difference between the charge pump's output and the circuit's input appears at a summing node (Σ) and biases A1, closing a control loop around the wide range core oscillator. The circuit's extraordinary dynamic range and high speed derive from core oscillator characteristics, divider/charge pump-based feedback and A1's low DC input errors. A1 and the LTC6943-based charge pump stabilize circuit operating point, contributing high linearity and low drift. A1's low offset drift allows the circuit's 50nV/Hz gain slope, permitting operation down to 1Hz at 25°C.

The positive input voltage causes A1 to swing negatively, biasing Q1. Q1's resultant collector current ramps C1 (trace A, Figure 41.28) until Schmitt trigger inverter I1's output (trace B) goes low, discharging C1 via Q2. C1's discharge resets I1's output to its high state, Q2 goes off and the ramp-and-reset action continues. D1's leakage dominates all parasitic currents in the core oscillator, ensuring operation down to 1Hz. The ÷64 divider chain's output clocks the LTC6943-based charge pump. The charge pump's two sections operate out-of-phase, resulting in charge transfer at each clock transition. Charge pump stability is primarily determined by the LT1460 2.5V reference, the switches low charge injection and the 100pF capacitors. The 0.22μF capacitor averages the pumping action to DC. The averaged difference between the input derived current and the charge pump feedback signal is amplified by A1, which biases Q1 to control circuit operating point. Core oscillator nonlinearity and drift are compensated by A1's servo action, resulting in the high linearity and low drift previously noted. A1's 1μF capacitor provides stable loop compensation. Figure 41.29 shows loop response (trace B) to an input step (trace A) is well controlled.

Some special techniques enable this circuit to achieve its specifications. D1's leakage current dominates all parasitic currents at I1's input; hence Q1 must always source current to sustain oscillation, assuring operation down to 1Hz. The 100MHz full-scale frequency sets stringent restrictions on core oscillator cycle time. Only 10ns

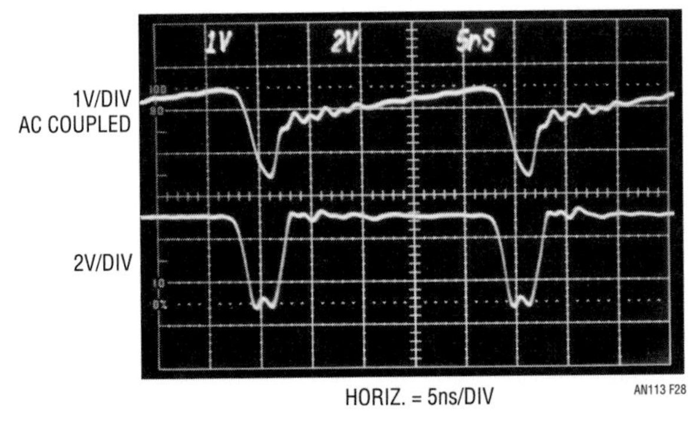

Figure 41.28 • V→F Operation at 40MHz. Core Oscillator Waveforms Viewed in 670MHz Real Time Bandwidth Include Q1 Collector (Trace A) and Q2 Emitter (Trace B). Ramp-and-Reset Operating Characteristic Is Apparent; Reset Duration of 6ns Permits 100MHz Repetition Rate

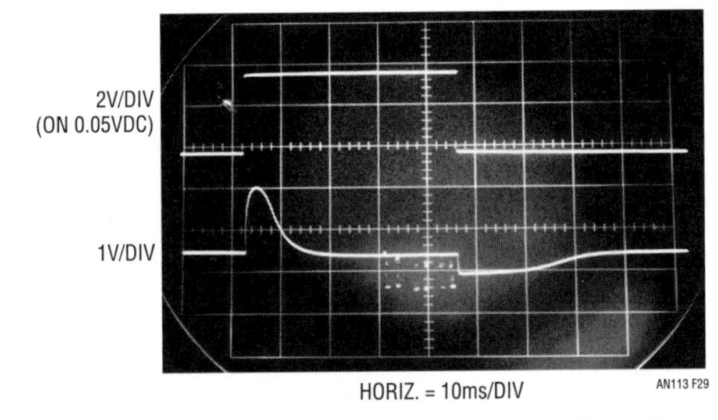

Figure 41.29 • Response (Trace B) to an Input Step (Trace A) Shows 30ms Settling Time at Summing Junction (Σ). A1's 1μF Capacitor Shapes Response, Stabilizing Feedback Loop. Clamped Response on Negative Going Input Step is Due to Summing Junction Limiting

Note 7: Reference 12 (1986) contains a circuit with comparable specifications although considerably more complex than the one presented here. The advent of high speed CMOS logic permitted replacing the earlier designs ECL elements, facilitating a dramatic decrease in complexity. Comparing the designs permits viewing the impact a technology shift can have in realizing a circuit function. In this case, the effect is pervasive, directly or indirectly influencing nearly every aspect of circuit operation. While circuit architecture is consistent, this incarnation is substantially and favorably altered.

is available for a complete ramp-and-reset sequence. The ultimate speed limitation is the reset interval. Figure 41.28, trace B, shows a 6ns interval, comfortably within the 10ns limit.

A scaled resistive path from the input to the charge pump corrects small nonlinearities due to residual charge injection. This input derived correction is effective because the charge injections effect varies directly with input determined frequency.

Prototype and small lot construction may proceed using the schematic and its notes, but component selection should be considered for volume production. Figure 41.30 lists applicable components and their selection targets.

To calibrate this circuit apply 5.000V and trim the 100MHz adjustment for a 100.0MHz output. Next, ground the input and adjust the 1Hz trim for 1Hz output. Allow for long settling time, as charge pump update rate at this frequency is once every 32 seconds. Note that this trim accommodates either offset polarity because of the -V bias derived from A1's clock output. Finally, set the 60MHz adjustment for 60.0MHz with $3.000V_{IN}$. Repeat these adjustments until all three points are fixed.

Delayed pulse generator with variable time phase, low jitter trigger output

Fast circuitry often requires a pulse generator that also supplies a variable time phase trigger output. It is desirable that the main output pulse occurrence be continuously settable from before to after the trigger output with low time jitter. Figure 41.31's circuit produces a 360ps rise-time output pulse with trigger output time phase variable from −30ns to 100ns. Jitter is 40ps.

Q1 and Q2 form a current source that charges the 1000pF capacitor. When the LTC1799 clock is high (trace A, Figure 41.32) both Q3 and Q4 are on. The current source is off and A2's output (trace B) is at ground. C1's latch input prevents it from responding and its output remains high. When the clock goes low, C1's latch input is disabled and its output drops low. The Q3 and Q4 collectors lift and Q2 comes on, delivering constant current to the 1000pF capacitor. The resulting linear ramp at A2 (trace B) is applied to bounded current summing amplifiers A3 and A4. Both amplifiers compare ramp induced current with fixed, opposite polarity currents derived from A1-Q6. A1-Q6, in turn, is referred to the +5 supply rail which also sets Q1-Q2 current and hence, ramp slope. This ratiometric connection promotes supply rejection. When A4 and A3 (traces C and F, respectively) come out of diode bound and cross zero, comparators C2 and C1 (traces D and G, respectively) are heavily overdriven and switch rapidly. C2's output path includes components which form trace E's trigger output pulse. C1 triggers output pulse generator Q5, operating in avalanche mode (trace H).[8]

The "delay adjust" control sets the ramp amplitude that A3-C1 switches the main output at, providing the desired variable time phase with respect to the A4-C2 controlled trigger output. Time jitter between C1 and C2 outputs is minimized because A3 and A4 effectively multiply ramp transition rate as their outputs enter the active region, provide gain and cross zero.

COMPONENT	SELECTION PARAMETER (25°C)	TYPICAL YIELD (%)
Q1	I_{CER} < 20pA at 3V	90
Q2	I_{EBO} < 20pA at 3V	90
D1	I_{REV} < 500pA, > 75pA at 3V	80
I1	I_{IN} < 25pA	80
A1	I_B < 5pA at V_{SUPPLY} = 5V	90
74ACH74	Operate with 3.6nS Wide (50% Point) Input Pulse	80

Figure 41.30 • Selection Criteria for Components Ensure V→F Performance. First Five Entries Enhance Operation Below 100Hz. Last Entry Assures Reliable Feedback Divider Operation

Note 8: Avalanche mode pulse generation is a subtle, arcane technique requiring extensive discussion. The text's cavalier treatment is deliberately brief in order to maintain focus on this circuit's low timing jitter characteristics. More studious coverage can be found in References 13–22.

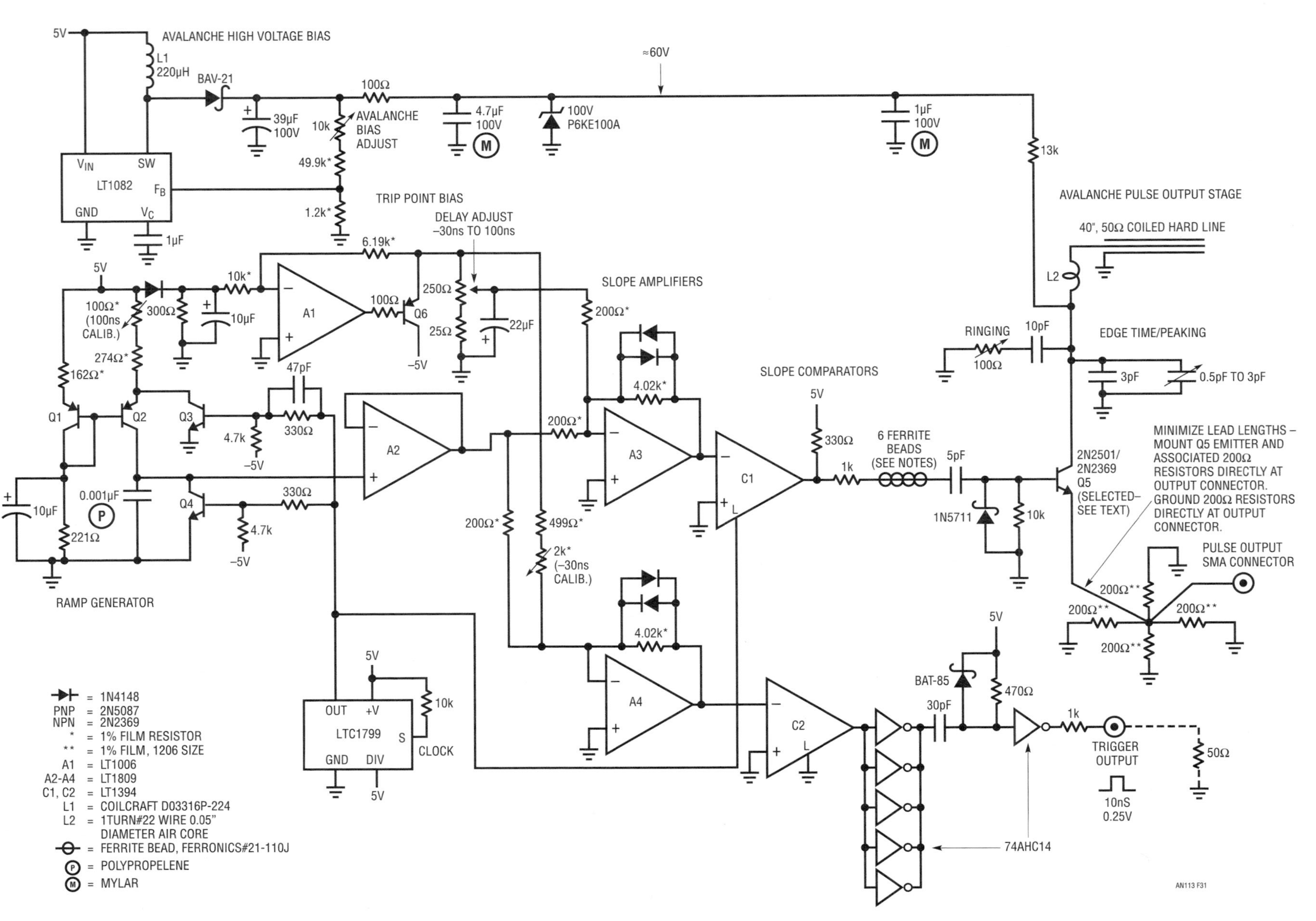

Figure 41.31 • Pulse Generator Output Time Phase Varies—30ns to 100ns with Respect to Trigger Output; Jitter Is 40ps. Clocked Ramp at A2 Produces Variable (A3-C1) and Fixed (A4-C2) Delays Driving Pulse and Trigger Outputs, Respectively. A3-A4 Provide Gain to Comparators at Trip Point, Minimizing Time Jitter Between Outputs

The A3-A4 amplifier gain is the key to low jitter between C1 and C2's switching times. The amplifiers augment the comparator's relatively low gain, promoting decisive switching despite the ramp input. Figure 41.33 shows A4 (trace A)-C2 (trace B) response to the ramp crossing the trip point. As the ramp nears the trip point A4 comes out of bound, providing an amplified version of the ramp's transition rate to C2. C2 responds by switching decisively 6ns after A4 crosses zero volts at center screen. A3-C1 waveforms are identical. Figure 41.34, Q5's pulse output, taken with the oscilloscope synchronized to the trigger output, shows 40ps jitter in a 3.9GHz sampled bandpass.

Circuit calibration is accomplished by first adjusting the "-30ns cal" so the main pulse output occurs 30ns before the trigger output with the "Delay Adjust" set to minimum. Next, with the "Delay Adjust" set to maximum, trim the "100ns cal" so the main pulse output occurs 100ns after the trigger output. Slight interaction between the 30ns and 100ns trims may require repeating their adjustments until both points are calibrated. As mentioned, the avalanche output stage is illustrative only and not detailed in this discussion. Its optimization and calibration are covered in Reference 13.

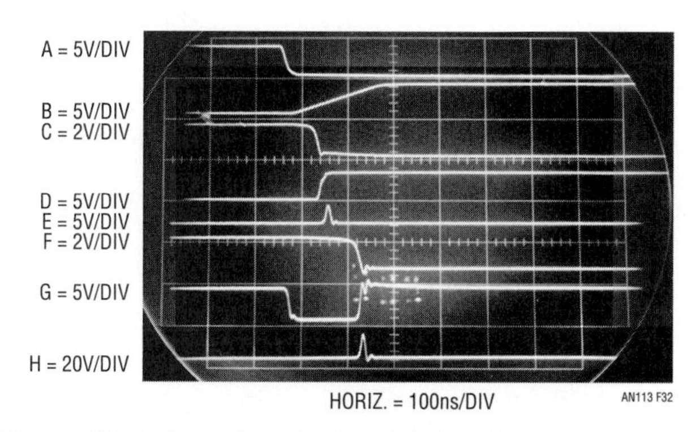

Figure 41.32 • Low Jitter Delayed Pulser Waveforms Include Clock (Trace A), A2 Ramp (B), A4 (C), C2 (D), Trigger Output (E), A3 (F), C1 (G) and Delayed Output Pulse (H). Trigger-to-Output Pulse Delay Is Continuously Variable From −30ns to 100ns

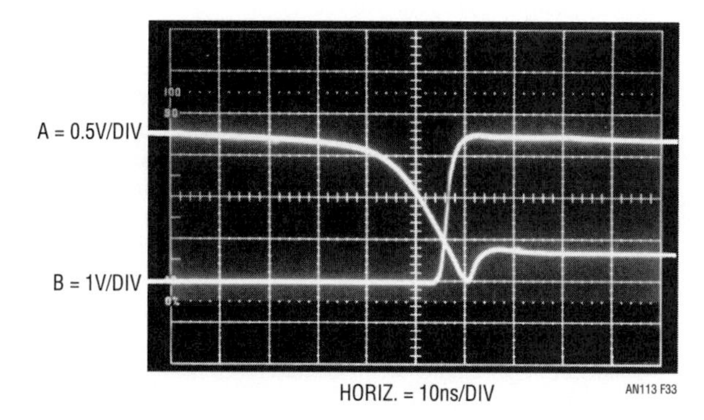

Figure 41.33 • A4 (Trace A) - C2 (Trace B) Response to A2's Ramp Crossing Trip Point. C2 Goes High 6ns after A4 Crosses Zero (Center Screen). A3-C1 Waveforms Are Identical

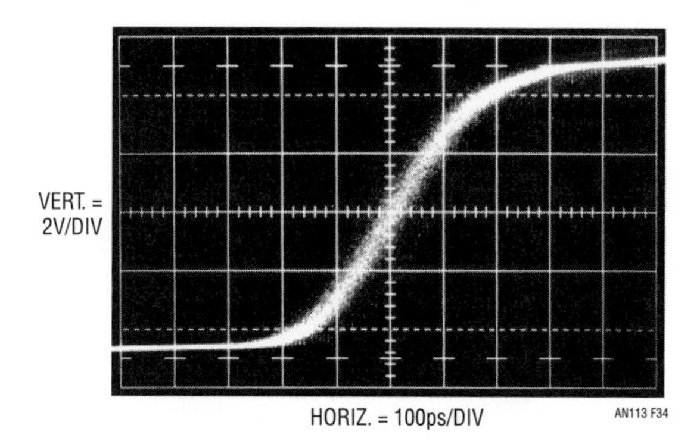

Figure 41.34 • Main Pulse Output Synchronized to Trigger Output Shows 40ps Jitter in 3.9GHz. Sampled Bandpass

References

1. Williams, Jim, "Bias Voltage and Current Sense Circuits for Avalanche Photodiodes," Linear Technology Corporation, Application Note 92, November 2002, p. 8
2. Williams, Jim, "Switching Regulators for Poets," Appendix D, Linear Technology Corporation, Application Note 25, September 1987
3. Hickman, R. W. and Hunt, F V, "On Electronic Voltage Stabilizers," "Cascode," Review of Scientific Instruments, January 1939, p. 6-21, 16
4. Williams, Jim, "Signal Sources, Conditioners and Power Circuitry," Linear Technology Corporation, Application Note 98, November 2004, p. 20-21
5. Williams, Jim, and Thoren, Mark, "Developments in Battery Stack Voltage Measurement," Application Note 112, Linear Technology Corporation, March 2007
6. Williams, Jim, "Measurement and Control Circuit Collection," Linear Technology Corporation, Application Note 45, June 1991. p. 1-3
7. Williams, Jim, "Practical Circuitry for Measurement and Control Problems," Linear Technology Corporation, Application Note 61, August 1994. p. 13-15
8. Williams, Jim, "Instrumentation Circuitry Using RMS- to-DC Converters," Linear Technology Corporation, Application Note 106, February 2007. p. 8-9
9. Seiko Epson Corp. Crystal Catalog. Models HTS-206 and C-100R See also, p. 10-11, "Drive Level."
10. Benjaminson, Albert, "The Linear Quartz Thermometer—A New Tool for Measuring Absolute and Difference Temperatures," Hewlett-Packard Journal, March 1965
11. Williams, Jim, "Applications Considerations and Circuits for a New Chopper Stabilized Op Amp," 1Hz-30MHz V→F Linear Technology Corporation, Application Note 9, p. 14-15
12. Williams, Jim, "Designs for High Performance Voltage-to-Frequency Converters," "1Hz-100MHz V→F Converter," p. 1-3, Linear Technology Corporation, Application Note 14, March 1986
13. Williams, Jim, "Slew Rate Verification for Wideband Amplifiers," Linear Technology Corporation, Application Note 94, May 2003
14. Braatz, Dennis, "Avalanche Pulse Generators," Private Communication, Tektronix, Inc. 2003
15. Tektronix, Inc., Type 111 Pretrigger Pulse Generator Operating and Service Manual, Tektronix, Inc. 1960
16. Hass, Isy, "Millimicrosecond Avalanche Switching Circuit Utilizing Double-Diffused Silicon Transistors," Fairchild Semiconductor, Application Note 8/2, December 1961
17. Beeson, R. H., Haas, I., Grinich, V H., "Thermal Response of Transistors in Avalanche Mode," Fairchild Semiconductor, Technical Paper 6, October 1959
18. G. B. B. Chaplin, "A Method of Designing Transistor Avalanche Circuits with Applications to a Sensitive Transistor Oscilloscope," paper presented at the 1958 IRE-AIEE Solid State Circuits Conference, Philadelphia, PA., February 1958
19. Motorola, Inc., "Avalanche Mode Switching," Chapter 9, p. 285-304. Motorola Transistor Handbook, 1963
20. Williams, Jim, "A Seven-Nanosecond Comparator for Single Supply Operation," "Programmable Subnanosecond Delayed Pulse Generator," p. 32-34, Linear Technology Corporation, Application Note 72, May 1998
21. D. J. Hamilton, F H. Shaver, P G. Griffith, "Avalanche Transistor Circuits for Generating Rectangular Pulses," Electronic Engineering, December 1962
22. R. B. Seeds, "Triggering of Avalanche Transistor Pulse Circuits," Technical Report No. 1653-1, August 5, 1960, Solid-State Electronics Laboratory, Stanford Electronics Laboratories, Stanford University, Stanford, California
23. Markell, R. Editor, "Linear Technology Magazine Circuit Collection, Volume I," Linear Technology Corporation, Application Note 52, January 1993
24. Markell, R. Editor, "Linear Technology Magazine Circuit Collection, Volume II," Linear Technology Corporation, Application Note 66, August 1996
25. Markell, R. Editor, "Linear Technology Magazine Circuit Collection, Volume III," Linear Technology Corporation, Application Note 67, September 1996
26. Williams, Jim, "Circuitry for Signal Conditioning and Power Conversion," Linear Technology Corporation, Application Note 75, March 1999
27. Williams, Jim, "Instrumentation Applications for a Monolithic Oscillator," Linear Technology Corporation, Application Note 93, February 2003. "Chopped Amplifiers," p. 9-10

Index

C

D

E

F

G

H

I

J

K

L

M

N

O

P

Q

R

S

T

U

V

W

Z